Program of the

# Thirteenth Annual Conference of the Cognitive Science Society

7-10 August 1991
Chicago, Illinois

LEA
1991
LAWRENCE ERLBAUM ASSOCIATES, PUBLISHERS
Hillsdale, New Jersey Hove and London

Published and distributed by

Lawrence Erlbaum Associates, Inc.
365 Broadway
Hillsdale, New Jersey 07642

ISBN 0-8058-1138-9

ISSN 1047-1316

Printed in the United States of America

## THE THIRTEENTH ANNUAL CONFERENCE OF THE COGNITIVE SCIENCE SOCIETY

**Kristian J. Hammond**

University of Chicago
Artificial Intelligence Laboratory
1100 E. 58th Street
Chicago, Illinois 60637

Conference Co-Chair

**Dedre Gentner**

Northwestern University
Department of Psychology
2029 Sheridan Road
Evanston, Illinois 60208

Conference Co-Chair

### STEERING COMMITTEE



### OFFICERS OF THE COGNITIVE SCIENCE SOCIETY



This conference is supported in part by funds from the University of Chicago Division of the Social Sciences, the University of Chicago Computer Science Department, and the Hugnal Lecturer's Endowment.

## Table of Contents

The Thirteenth Annual Conference of the
Cognitve Science Society
The University of Chicago
Chicago, Illinois 60637
August 7th–10th, 1991

# Paper Presentations

## Discourse and Text



## Imagery



## Neuroscience Models of Language



## Planning and Action



## Reasoning and Mental Models

## Case Representation and Adoptation



## Memory for Objects



## Regularities and Estimation



## Word and Concept Learning



## Category Formation and Similarity

## Philosophical Perspectives



## Reminding and Case Retrieval



## Attention and Learning



## Computer Interfaces



## Decision Making



## Discovery Learning



## Heuristics in Reasoning

## Poster Presentations

# REVIEWERS

Yigal Arens
Terry Au
Bernard Baars
Norman I. Badler
John A. Barnden
William Bechtel
Douglas A. Behrend
Irving Biederman
Dorrit O. Billman
John B. Black
Daniel Bobrow
William F. Brewer
Mark Burstein
William Clancey
Catherine Clement
Gregg Collins
Paul Cooper
Gary Cottrell
Stephanie Doane
Kevin Dunbar
Jeff Elman
Brian Falkenhainer
Martha J. Farah
Jeremiah Faries
Sylvia Farnham-diggory
Paul J. Feltovich
R. James Firby
Ken Forbus
Susan M. Garnsey
Dedre Gentner
Morton A. Gernsbacher
Arthur C. Graesser
Jordan Grafman
Richard Granger
James G. Greeno
Kristian J. Hammond
Reid Hastie
James A. Hendler
Geoffrey E. Hinton
Douglas R. Hofstadter
Keith J. Holyoak
Eduard H. Hovy
William J. Hoyer
Edwin L. Hutchins
Alex Kass
Susan Kemper
Josh Klayman
Phyllis A. Koton
Pat Langley
David Leake
Jill Lehman
Wendy G. Lehnert
Clayton Lewis
Steven Lytinen
Colin M. Macleod
Barbara Malt
James Martin
Charles Martin
Richard E. Mayer
Jim Mccawley
Ron Mcclamrock
James L. Mcclelland
Doug Medin
Melanie Mitchell
Michael C. Mozer
Paul Munro
Gregory Murphy
Ulric Neisser
Nancy J. Nersessian
Denis Newman
Laura R. Novick
Howard Nussbaum
Gregg C. Oden
Stellan Ohlsson
Gary Olson
Judith Oresanno
Christopher Owens
Vimla L. Patel
Michael J. Pazzani
Jordan Pollack
Michael I. Posner
Ashwin Ram
William Rapaport
Arthur Reber
Brian J. Reiser
Lance J. Rips
Scott Robertson
Brian Ross
Jerry Sadok
Jeffrey C. Schlimmer
Walter Schneider
Alberto Segre
Colleen M. Seifert
Stuart C. Shapiro
Marilyn Shatz
Wei-min Shen
Jeffrey C. Shrager
Thomas Shultz
Steven Small
Paul Smolensky
Elliot M. Soloway
Michael Swain
Copy Of Michael Swain
David A. Swinney
Katia Sycara
David Touretzky
Thomas R. Trabasso
Elise Turner
Kurt Van Lehn
Yorick Wilks
Ed Wisniewski
Ilan Yaniv
Frank Yates

# Paper Presentations

# Empirical analysis of a discourse model for natural language interfaces

**Nils Dahlbäck**
Department of Computer and Information Science
Linköping University
S-581 83 Linköping, Sweden
nda@ida.liu.se

## Abstract

**A structural model of discourse for natural language interaction developed for the LINLIN-system is evaluated using the Wizard-of-Oz metod. 21 dialogues were collected using five different background systems, making it possible to vary the type and number of tasks possible to perform by the users. The results indicate that the structural complexity of the discourse in man-machine dialogues is simpler than most human dialogues, at least for information retrieval and some types of ordering systems, suggesting that computationally simpler discourse models can be used in these domains.**

## Introduction

The aim of the LINLIN-project is to develop a natural language interface for Swedish. One important characteristic of the project is that it is based on a sublanguage approach (Grishman & Kittredge, 1987). The system is viewed as a shell that needs to be customized to different applications. The general parts are the different processing modules, whereas the static knowledge bases used by the processing modules need to be adapted to the different applications. The purpose of the present paper is to describe the empirical foundations of the project's approach to discourse structure. The linguistic and computational aspects of the project have been described elsewhere (Ahrenberg, 1989, Ahrenberg, Jönsson & Dahlbäck 1990, Jönsson 1991). Other aspects of the empirical work are presented in Jönsson & Dahlbäck (1988), Dahlbäck & Jönsson (1989).

The LINLIN-approach to discourse differs in some important respects from the 'standard model' of discourse within computational linguistics (Grosz&Sidner 1986). Our intention is not to create a general discourse theory. It is only concerned with man-machine interaction in natural language; there is no claim that it is applicable to all types of discourse, or even all types of dialogue. In a sense, computational efficiency within a limited domain is preferred to generality with unknown properties. It is thus a computational theory in the sense that its focus is on interaction with computers, but also in the sense that it is implemented (albeit not as yet integrated with the other modules in the system) (Jönsson, 1991).

## The Dialogue Model

The dialogue model of the LINLIN-system (called LINDA, for LINköping DiAlogue) is an abstract characterization of the dialogue structure in NLI-dialogues. In its pure form, the model is only about the transfer of content between interlocutors in so-called 'transactional dialogues' (Brown & Yule, 1983, p 1). I will first describe this pure model, and then return to the necessary amendments. The description below is based on Ahrenberg, Jönsson & Dahlbäck (1990).

The dialogue objects are divided into three main classes on the basis of structural complexity. There is one class corresponding to the size of the dialogue, one corresponding to the size of a discourse segment, and a third corresponding to the size of a single speech act or dialogue move. The dialogue is thus structured in terms of discourse segments, and discourse segments in terms of moves and embedded segments. Utterances are not analyzed as dialogue objects, but as linguistic objects which function as vehicles of one or more moves.

The basic discourse segment unit is an initiative-response-unit (*IR-unit*). i.e. an initiative (which opens a segment) and a response (which closes the segment). Initiatives are basically directives in standard speech-act terminology. An initiative can be seen as an act where the speaker introduces a goal and the expectation that it will be satisfied. A response is an act where the goal is satisfied, i.e. the value is given, or the expectation is eliminated by other means. The prototypical case of this is where a question is followed by a content answer or saying that the speaker does not know the answer (elimination of expectation by other means). The response closing the segment need not be adjacent to the initiative, since interruptions and clarification sub-dialogues may intervene, but it must follow at some point. The IR-units are

---

This work was supported by the Swedish Council for Research in the Humanities and Social Sciences (HSFR).

then sub-classified into different types, such as Q/A (question-answer) *do you have any books on AI? / No*, Q/AS (question-assertion) *who is the author of Winnie-the-Pooh? / PUB has no information about that book*, D/ACK (directive-acknowledgement) *Luxor Symphony ordered / OK*, etc.[1]

The basic categories in the dialogues are turn and move. A turn is a continuous period of the dialogue where one participant "speaks". A turn is co-extensive with an utterance by user or system, since mid-turn interruptions cannot occur. Each turn consists of one or more moves. A move consists of one (speech) act.

The entire dialogue is thus structured by the recursive use of IR-units, which give the dialogue a tree structure. In contradistinction to some other models intended for computational use (e.g. Wachtel, 1986), there is no intermediate topic-based descriptive level. Instead a number of topical parameters are distinguished, e.g. CurrentObject, CurrentAttribute, and CurrentSet, which may vary independently of each other. The managemet of topic and topic structure is thus more elaborate than in the Grosz and Sidner (1986) model. A basic distinction is made between different topic domains, e.g. task-related (T) and discourse related (D) moves and IR-units. Other topic domains are also used, but the types and numbers vary between different dialogue domains.

It should be noted that Levinson (1981) has convincingly argued against the possibility of basing general structural models of dialogues on speech act categories and an adjacency pairs structure, i.e. that there can be no syntax of conversational structure. Two important parts of the argument is that there are no means of determining what counts as what in a dialogue independently of the situation in which it takes place, and that different speech-acts within an utterance need not be separated into different sub-parts, but that the same part of an utterance can perform more than one such act. A rather strong assumption of the LINDA model is that this need not be true for man-computer dialogues.

## Empirical analysis

The LINDA-model has some obvious computational advantages, and some obvious deficiencies as a general computational model of discourse. The question, then, is to what extent the LINDA model is sufficient for this application domain in its entirety, or to some subsets thereof. Since we lack an adequate sub-language description of this domain[2], this question can only be answered by an empirical analysis, in the present case we have used so-called Wizard-of-Oz experiments. (For a review, see Jönsson & Dahlbäck (1988), and Fraser and Gilbert (in press))

The basic idea is simple. We have our subjects sit down at a terminal to solve some problem which requires them to use the computer. But instead of interacting with the system through our "newly developed natural language interface", we have one person sitting at the other end of the line, simulating such a device.

The idea might be simple. But to actually run such an experiment is not quite that simple, since you want the output from the simulated 'system' to be everything that the person simulating it is not — providing quick, consistent answers without spelling mistakes and other signs of human fallibility. To achive this, we use a simulation environment with pre-stored answers and other help systems. We consider the quality of the simulation environment a crucial factor behind the ecological validity of the dialogues collected and the concomitant applicability of conclusions obtained from analysing them. It can also be noted that researchers not having such a method at their disposal often comment on the need for it when discussing their work (e.g. Mc Kevitt, 1990). None of our subjects have realized that the 'system' was not a system.

### Background systems

To circumvent the risk of drawing general conclusions that in fact are only a reflection of the specific experimental setting used, we have used five different background systems. We have varied not only the content domain, but also the 'intelligence' of the systems, and the number and types of tasks possible to perform by the user.

We have used two database systems. *PUB* is a library DB in use at our department. *C-line* is a simulated DB containing information about the computer science curriculum at Linköping University.

In the *HiFi*-system the user can order HiFi-equipment after having queried a (simulated) DB containing information about the available equipment. The *Travel* system simulates an automated travel agency offering charter holidays to Greek islands. These systems differ from the two above in two respects; the system is more 'cognitively' advanced, and there are more actions that can be performed by the user, i.e. not only asking for information but also order something.

The *Wine* system is a simulated advisory system, capable of suggesting suitable wines for different dishes, if necessary within a specific price range.

The total number of dialogues is 21; PUB: 4, HiFi: 5, C-line: 5, Wine: 4, Travel: 3. The total number of utterances is 1055.[3] This gives us an average of 50 utterances/dialogue. The longest are in the travel

[1] For at least some spoken dialogues, the minimal segment seems to consist of three units; (I)nitiative, (R)esponse, (C)onfirmation, cf Stubbs, 1983, Bilange, 1991.

[2] For a theoretical analysis of the differences between human dialogues and NLI dialogues, see Jönsson&Dahlbäck (1988), Dahlbäck (1991). An experimental investigation of this is currently in progress at our lab.

[3] Apart from the dialoges analyzed here, we have collected more than 60 others, using four other real or simu-

domain, where the average dialogue is 92 utterances long, and the shortest are the PUB dialogues with an average of 25 utterances. (The dialogues are available on file or in print on request. All are in Swedish.)

### Analysis criteria

Our analysis is based on the same IR categories that are used in the LINLIN system. One important reason for this is that the categories are domain independent. We can therefore compare dialogues from different domains. Another advantage is that the categories are (fairly) simple to define and identify, making it possible to code the dialogues with sufficient inter-rater reliability (97.1 %). The computational advantage of not having to infer the invisible intentions behind an utterance in order to structure the dialogue is thus parallelled by a similar advantage from an analysis point of view.

We use only two basic types of moves, initiatives (I) and responses (R). The definiton of the categories is based on local information only. If the move is seen as introducing a goal it is scored as an initiative, if it is a goal-satisfying move, it is scored as a response. In the analysis of the dialogues we do not take into consideration the fact that in the LINDA model only segments of the I-R type are allowed, but not for instance I-R-R. It then becomes an empirical question, which we will return to, as to what extent the dialogues follow the structural constraints postulated by the model, e.g. its tree structure.

In most cases the I and R categories of the LINDA model can be used in a fairly straightforward way. There are however some cases requiring further consideration. One case is where the system cannot answer the question and replies with phrases such as "Don't understand", or "What do you mean by X?". We code the first case, where the utterance concerns the entire phrase, as a response, and the latter as an initiative. Thus we only identify subdialogues of clarification if part of the previous input is understood by the system. When the system replies that it cannot provide an answer because the parameter or value can not be found in the background system,we code this as a response, and not an initiative for clarification, even if the system supplies information about what value or field was missing.

To illustrate these definitions, consider the dialogue below (PUB1:17-20). Note that the second move spans two sentences. The dialogue is scored I, I, R, R. The reason for not scoring the first part of the second turn as a response, is that it does not in a strict sense answer the question about *where* these books can be found. Nor is it a response giving reasons for why an answer cannot be provided, but rather a question about how the answer should be presented to the user, or if any further processing should be done before presenting it.

---

lated background systems. Dahlbäck (1991) describes some of these in more detail.

```
U> Where can I find books about Artificial
Intelligence?
S> There are five books on Artificial
 Intelligence. Do you want me
 to find out where all are?
U> Yes
S> (LIST OF BOOKS)
```

Discourse management moves such as "Can I help you with anything more?" are all scored as initiatives. We subcategorize them as DO (discourse opening), DC (discourse continuation), and DE (discourse ending), to make it possible to exclude them from some of the analysis presented below. (Responses to these kind of initiatives are optional in the model). Only one type of move cannot be coded as initiative or response, namely acknowledgements of input (AI) where the system replies "PUB searching" or something similar.

## Results

The results of the analysis will be presented in two main parts. This section gives an overview of the discourse in the different domains, and at the same time discusses the applicability of the LINDA model to them. The next section raises some general issues emerging from the analysis.

### Database dialogues

The PUB dialogues can be characterized as a sequence of isolated question-answer sequences with no global structure. A large number of the utterances do not, however, follow the simple pattern of I followed by R. The reason for this is that the system, when asked about e.g. books about a specific topic, often does not print all items that have been found, but instead tells how many books have been found and asks if the user wants to see all of them, as in the example above.

These dialogues all fit the LINDA model very well. In fact, they all consist of sequences of only three types; (I R), (I (I R) R), or (I (I R) (I R) R), with only one instance of the last type. Almost all of the embedded segments concern the presentation of the information found by the system. The sub-dialogues never go more than one level down, the longest being six moves long.

The C-line dialogues, also database dialogues, have a clear and simple IR structure. Apart from the ending of the dialogues, where the system does not always send some kind of good-bye response when the user declares he has finished, there are only three cases where the dialogues deviate from being a sequence of pure IR-units with no embeddings. One occurs because of a mistake by the Wizard (or malfunctioning of the system). The other two occur because of requests of clarification from the system. In one of these user supplies the requested information, in the other the user rephrases the entire question, and thus aborts the current IR-segment and creates a new one. The sub-dialogues are never more than two moves long, and their maximum depth is consequently one level down.

**Summary of the database dialogues** Both sets of database dialogues show a simple and clear discourse structure, characterized by only local and no global coherence. Or, if you so wish, there is one global structure, but no intermediate structures, something which was also found in Grosz' (1977) analysis of her DB-dialogues. The sub-dialogues are of a simple character, being concerned either with requests for clarification or questions of how the result should be presented on the screen. These sub-dialogues are the only cases where the system takes over the initiative; for the major part the dialogues are user driven.

## Ordering dialogues

When we come to the dialogues where the user can place an order, HiFi and Travel, the situation is somewhat more complicated. The main reason for this is that the user and the system in these cases have more tasks to perform; seek information, order, and tell where to send the ordered goods. As a consequence there are more kinds of topics to consider.

The HiFi dialogues all have a simple discourse structure that can easily be accommodated to the structural constraints of the LINDA model. In most dialogues the user discusses the available equipment of one kind, e.g. receivers, and thereafter orders one of these before continuing to the next kind of equipment. In all cases (but one) the move where the user orders something is immediately followed by an explicit confirmation of the order, either by telling which item that has been ordered, or simply by a confirmatory OK. The vast majority of dialogue segments consist of simple IR-units with no embeddings. There are in fact only two embeddings in the entire HiFi corpus, both caused by requests for clarification, making the sub-dialogues in HiFi as simple as in the previously discussed systems. Apart from the initiatives sub-classified DO, DC, and DE, there are only three moves which do not fit into the structural constraints of the LINDA model, in a total of more than 300 utterances and even more moves.

There is in fact one exception to the discourse pattern described here, namely HiFi1. The strategy of this user differs from the others as she uses a kind of "zooming in" strategy. Instead of choosing one item at the time, she asks initially for a set of models suitable for a small room. This set turns out to be too expensive, and the rest of the dialogue concerns which items to change to get an optimal set given the stipulated cost constraints. When she is satisfied, she declares this explicitly "this is what I want to order" (HiFi1:27), followed by the system's confirmation of the order, after which the sytem takes the initiative to obtain information about the delivery in the usual way.

This stategy is also seen in all the Travel dialogues. They contain a number of utterances where the user niether seeks information, nor orders something, but instead declares some intents or preferences. These moves have an important communicative function, since they constrain the set of alternatives to discuss further, but also the set of possible referents for referring expressions. After having declared the intent to go to Crete, a question like "are there any cheaper hotels?" is actually about hotels on Crete, not about hotels in Greece. Because of this gradual 'zooming in' on the order, there are no moves in the dialogue where the system explicily acknowledges what has been decided upon or ordered so far.

The way these moves are categorized and analyzed obviously affects the structural analysis of these dialogues. It is in fact possible to make the three dialogues all fit the structural constraints of the LINDA model. The way this can be done is to be somewhat 'liberal' in the assignement of responses to the initiatives where the user declares his intent. If we accept such an analysis strategy, there is actually only one case of an incomplete IR-unit in the entire corpus of Travel dialogues comprising some 300 moves, and this occurs because of a simulation mistake by the Wizard.

The problem with this analysis is that it fails to capture that which makes these dialogues differ from the database and HiFi dialogues (except for HiFi1). The problem of how to handle the confirmations of the orders or the bookings is from a system development point of view perhaps no large obstacle, since it concerns what the system does, and therefore its interpretation seems not to be all that critical. From an analysis point of view, the problem is that we seem to touch the limits of the model used.

Because of the difficulties with the structural analysis, it is somewhat difficult to assess the length and depth of the sub-dialogues here. If we restrict ourselves to the clear-cut cases, we find here such insertion sequences with a maximum length of 8 utterances and depth of three levels.

**Summary of the ordering dialogues** Perhaps the most interesting result emerging from the analysis of these dialogues is that, despite being conducted in a more complex situation than the pure database dialogues, they have a surprisingly simple discourse structure. The number of sub-dialogues are few and they are all rather short. There is one case going three levels down, but most of them are only one level deep.

The distribution of initiatives differs between the two dialogue domains. The HiFi dialogues are almost exclusively user-driven, whereas the opposite is more accurate for the Travel dialogues, though not to the same high extent. The important consequence of this is that there does not seem to be any simple connection between who has the initiative and the type of discourse strategy, since we find the 'zoom' strategy both in user-driven and system-driven dialogues.

## Advisory systems

The Wine system is the only advisory system in our corpus of dialogues. This means that we have some difficulties in discerning how much of the patterns found in the present corpus are reflections of the class of system it represents and how much it reflects the specific domain used.

These dialogues are by far the most difficult to analyze using our IR dialogue model. There are many reasons for this, on many different levels. Since the domain for these dialogues is not as clearly circumscribed as for the other systems under study, both as regards the competence of the system and as regards the mutual knowledge of the interlocutors, the users sometimes find it difficult to understand the limits of the linguistic and conceptual competence of the system.

Another important reason for the difficulties we encounter when analyzing the dialogues is that the dialogues have a mixed initiative structure. We saw previously that mixed initiative dialogues are not necessarily difficult to analyze using our model. The structures we find in the Wine dialogues are, however, of a different kind. They are not sub-dialogues of clarification, but rather of development of the topic discussed. This makes some utterances difficult to classify as consisting of separate moves consisting of pure initiatives or responses. There are 17 instances of the type of cases discussed here in the Wine dialogues. In other words they are quite frequent, covering approximately 40 % of the entire Wine corpus.

There is also another kind of move occurring in these dialogues that makes their structure deviate from some simple and 'pure' IR-pattern. This is when the system solicits information not requested by the user, and which shapes the continuation of the dialogue without being initiatives that makes the partner issue a response. These cases are the only ones in our dialogues where our simplified classificatory scheme is insufficient. There are five such utterances in the Wine dialogues. The surprising thing is not that we find these kind of sequences in our corpus, but rather that they are so few, and that they only occur in the Wine dialogues. In ordinary human discourse they are probably the rule rather than the exception

Using the criteria described above for the structural analysis, we find in this system sub-dialogues of a complexity similar to the one found in the Travel dialogues. The maximum length being 8 moves and the maximum depth three levels.

## Evaluating the LINDA model

In its most simple form the dialogue structure is a sequence of IR-segments. It is one remarkable result of our analysis that such large parts of our dialogues actually fit this simple structure. One way of expressing this is to compute how much of the different dialogues that fit this pattern, using some simple complexity index. To do this we first exclude all initiatives that are coded DI, DC, or DE, and also all IA-moves, as the structure of the dialogues at these points is more a reflection of the design of the simulated system than anything else. We then count the number of times that the dialogue adheres to the prototypical adjacency-pair sequence of initiative followed by a response, divided by the number of pairs of moves(i.e. moves divided by two).

This complexity-index, which shows the extent to which the dialogues conform to the basic adjacency-pair sequence of an initiative followed be a response, gives the following figures: PUB 75.0; C-line 95.7; HiFi 98.0; Travel 87.5; Wine 78.1.

We can see that 75% or more of the dialogues consist of sequences of simple IR-units, a surprisingly high figure. This does not mean that the dialogues consist of a sequence of isolated questions and answers, as there is frequent use of anaphoric expressions. In fact 49 % of the *initiatives* contain some kind of anaphoric expression (Dahlbäck & Jönsson, 1989).[4] What the figures show is rather that in spite of being clear cases of connected discourse, these dialogues have a much simpler structural complextity than most other genres. It thus seems as if most man-machine dialogues in natural language, even when no restrictions on the users' way of expressing themselves, lack most of the complexity found in other types of discourse. Our corpus is admittedly of a limited size, but it covers some of the most typical possible applications for NLI technology, and is not tied to one particular topic domain. This gives us some confidence in believing that the results have some generalizability.

75 % is not 100 %, and a system that fails in one case out of four is obviously not sufficient. But the model is not as simple as this; recursive embedding of IR units is one of its central features. The low index figure for the PUB dialogues is due to the large frequency of simple clarification sub-dialogues. The index is thus misleading for this group, which means that for three of the dialogue systems the LINDA model can probably be used here without amendments.

The fit to the model, even when insertion sequences are taken into consideration, is not perfect even in these three sets of dialogues, but the causes for this are all rather simple. First, we have the opening and closings of the dialogues, but here we believe that careful design of these aspects of the dialogue will eliminate most of those cases. The other case of deviation from the pattern of strict and embedded IR-sequences is when the system initiates a clarification sub-dialogue which leads the user to abort this sequence and either rephrasing the question or even bringing up a new issue, leading to a ....R)(I (I ))(I... sequence. What is interesting about these cases is that they seem to be more frequent in man-computer dialogues than in dialogues between people, the reason obviously being that there is no need to be polite to a computer!

If we expand the LINDA model to handle these cases

[4]Work is currently in progress on the relationship between the IR-structure and the use of anaphoric expressions in these dialogues.

too, we claim that our results support the notion that such a model can be used in an NLI for database applications.

When we come to the class of ordering system, the issue gets a bit more complicated. The results from the HIFi system seem to indicate that the model can also accommodate these kinds of dialogues, but when we come to the Travel system we encountered some difficulties. The sample is too limited to make it possible to draw any definite conclusions on this issue, but as a working hypothesis we suggest that it is possible to allow for more than one task to be performed by the system, given that it is clear to which task each move belongs. From a system development point of view, there is another possible solution, namely to physically separate the two tasks into two windows on the screen. (These isssues are discussed further in Dahlbäck 1991)

So far we have not found any cases that cannot be handled by a slightly modified LINDA model. But in the advisory Wine dialogues we seem to touch the borders of the model, and perhaps even cross them. There are a number of interrelated issues here. Space limitations prohibit a detailed discussion of them here, but not mentioning them would make us paint too bright a picture of the model discussed.

Structurally, we found some cases of IRR-sequences and other deviations from the basic IR-structure. Expanding the model to allow for this kind of structure is perhaps possible to do and still preserve its basic assumptions. What is more troublesome is that we found some cases where it was difficult to classify an utterance as a clear case of an initiative or a response, it was rather a little of both. This is of course something quite natural in most dialogues, but it creates problems for the LINDA model. The difficulties above can be seen as illustrating the limitations of the LINDA model in its pure form. It was never meant to be a general discourse model, so the fact that it cannot handle some discourse types is not critical. But the interesting point is then in a sense not that cases of these kinds are found in our corpus, but rather that they are so few and confined to one type of dialogue.

## Summing up

The most important result of the present study is the demonstration that man-machine dialogues in natural language do not show the structural complexity often found in human dialogues. A caveat is that a sublanguage approach to this domain suggests that the successful management of connected discorse in natural language interfaces need not be that far off, at least for some of the most important application domains.

## Acknowledgements

The discourse manager was developed by Lars Ahrenberg and Arne Jönsson, and they also contributed to the conceptual and empirical analysis presented here.

## References

Ahrenberg, L. 1989. On the integration of linguistic knowledge and world knowledge in natural language understanding. In Dahl&K. Fraurud (eds) *Papers from the First Nordic Conference on Text Comprehension in Man and Machine.* Institute of Linguistics, Stockholm University.

Ahrenberg, L. Jönsson, A. & Dahlbäck, N. 1990 Discourse Representation and Discourse Management for a Natural Language System. In Dahl&K. Fraurud (eds) *Papers from the Second Nordic Conference on Text Comprehension in Man and Machine.* Institute of Linguistics, Stockholm University.

Bilange, E. 1991 A Task Independent Dialogue Model. *Proceedings of the 5th Conference of the European Chapter of the Association for Computational Linguistics (E-ACL '89)*, Berlin, Germany.

Brown, G. & Yule, G. 1983 *Discourse Analysis.* Cambridge: Cambridge University Press.

Dahlbäck, N. 1991 *Representations of Discourse; Cognitive and Computational aspects.* Forthcoming.

Dahlbäck, N. & Jönsson, A. 1989 Empirical Studies of Discourse Representation for Natural Language Interfaces. *Proceedings of the 4th Conference of the European Chapter of the Association for Computational Linguistics (E-ACL '89)* Manchester, England.

Fraser, N. & Gilbert, N. (in press) Simulating Speech Systems. *Computer Speech and Language*

Grishman, R & Kittredge, R. (eds) 1987 *Analyzing Language in Restricted Domains* Hillsdale, N.J.: Erlbaum.

Grosz, B. & Sidner, C. 1986 Attention, Intention, and the Structure of Discourse. *Computational Linguistics* 12(3): 175-204.

Jönsson, A. 1991 A Dialogue Manager Using Intiative-Response Units and Distibuted Control. *Proceedings of the 5th Conference of the European Chapter of the Association for Computational Linguistics (E-ACL '89)*, Berlin, Germany.

Jönsson, A & Dahlbäck, N. 1988 Talking to a computer is not like talking to your best friend. *Proc. of the 1st Scandinavian Conference on Artificial Intelligence (SCAI' 88)* Tromsö, Norway.

Levinson, S. 1981 Some pre-observations on the modeling of dialogue. *Discourse Processes* 4, 93-116.

Mc Kevitt, P. 1990 Data acquisiton for natural language interfaces. Memoranda in Computer and Cognitive Science MCCS-90-178, New Mexico State University.

Stubbs, M. 1983. *Discourse Analysis.* Oxford: Blackwell.

Wachtel, T. 1986 Pragmatic sensitivity in NL interfaces and the structure of conversation. In *Proc. of the 11th International conference of Computational Linguistics* University of Bonn.

# A Distributed Representation and Model for Story Comprehension and Recall

**Richard M. Golden**

School of Human Development
University of Texas at Dallas
GR41, Box 830688
Richardson, Texas, 75083–0688
golden@utdallas.edu

**David E. Rumelhart**

Stanford University
Psychology Department
Stanford, California 94305
der@psych.stanford.edu

**Abstract**

An optimal control theory of story comprehension and recall is proposed within the framework of a "situation" state space. A point in situation state space is specified by a collection of propositions each of which can have the values of either "present" or "absent". Story comprehension is viewed as finding a temporally-ordered sequence of situations or "trajectory" which is consistent with story-imposed constraints. Story recall is viewed as finding a trajectory consistent with episodic memory constraints. A multi-state probabilistic (MSP) machine representational scheme is then introduced for compactly and formally assigning a "degree of belief" (i.e., a probability value) to each trajectory in the state space. A connectionist model is also introduced which searches for trajectories which are highly probable with respect to a set of constraints and an MSP machine representation. Like human subjects, the model (i) recalls propositions with greater causal connectivity as retention interval is increased, and (ii) demonstrates how misordered propositions tend to "drift" more towards their canonical position in a text as retention interval is increased.

## General Theory

We have currently been approaching the problem of story comprehension and recall within the framework of a special high dimensional state space which is called the "situation" state space. A point in situation state space consists of a collection of $d$ facts about the world, each of which can be classified as being either "present" or "absent." The reader's current mental state is therefore modelled as a single point in a $d$-dimensional situation state space. At some later point in time, the reader's mental state would be modelled as another point in the same $d$-dimensional situation state space. It will be convenient, therefore, to order these points according to a time index. Thus, the "unfolding" or "evolution" of the reader's mental state as a function of time, may be represented as an ordered sequence of points or *trajectory* in situation state space.

Some trajectories in situation state space will be more likely to occur than others. For example, a trajectory in a four-dimensional situation state space might be specified by the sequence of three four–dimensional situation state vectors depicted in Table 1. World knowledge is represented by a belief function which assigns a specific "degree of belief" to different trajectories in the situation state space. In particular, trajectories which are more likely are assigned values close to one, while trajectories which are less likely are assigned values close to zero.

A story is a set of constraints upon the class of possible trajectories in situation state space. The features possessing the value "present" along the trajectory are fixed in value. The remaining features possessing the value "absent" may have their values modified.

The problem of story comprehension is now formally defined as finding a trajectory in situation state space which is *most probable* with respect to some probabilistic belief function (for some closely related ideas see Rumelhart, 1977). We suggest that readers may find highly probable trajectories by "running" their probabilistic mental models forwards and backwards in time. The resulting trajectory is then used to update the parameters of the belief function via some learning process, and then a reconstructed story trajectory is recalled from memory by using another partially specified trajectory (perhaps the title of the story) as a retrieval cue, and "rerunning" the probabilistic mental model. Note that once a trajectory has been constructed following the comprehension process or reconstructed following the recall process, the resulting representation may be used by other systems to answer questions about the story, summarize the story, or recall the original story.

This research was supported in part by National Institute of Health post-doctoral fellowship HD06943 to the first author while he was at Stanford University, and the Program in Cognition and Neuroscience at the University of Texas at Dallas where the first author is currently located.

| Story Feature | t=1 | t=2 | t=3 |
|---|---|---|---|
| Hear (M, IC Music) | 1 | 1 | 0 |
| Desire (M,Eat(M,IC) | 0 | 1 | 0 |
| Eat (M, IC) | 0 | 0 | 1 |
| Sleep (M) | 0 | 0 | 0 |

*Table 1. A situation state space trajectory. M=Mary. IC=Ice-Cream. 1 = Feature Present. 0=Feature Absent.*

## Representation and Model

### Multi-state probabilistic machines.

Gabrielian and his colleagues (for a review see Gabrielian and Iyer, 1991) have been developing the theoretical foundations of a new class for machines known as *hierarchical multi-state machines* for specifying the performance requirements of complex high-dimensional systems. We have found that a probabilistic extension of the multi-state machines studied by Gabrielian and his colleagues provides a useful way of implicitly, yet precisely, representing complex belief functions that assign degrees of belief to specific trajectories in situation state space. We will call this special notation for representing complex probabilistic knowledge structures a MSP (multi-state probabilistic) machine.

A MSP machine consists of three distinct types of entities: local states, transitions, and controls. The *local states* of the MSP machine correspond to the set of *d* features required to specify a point in a *d*-dimensional situation state space. The set of all states of the MSP machine is sometimes referred to as a *global state* of the machine. A *marking* of the MSP machine indicates the values of local states of the machine at an instant in time. Thus, a marking of the machine identifies a point in the d-dimensional situation state space. The *transitions* and the *controls* of the MSP machine specify how the global state of the machine evolves from one instant of time to the next.

Table 2 and Figure 1 show a more complicated example where the multi-state probabilistic machine notation is used to represent the causal knowledge structure underlying an actual story. The story was parsed into propositions following Trabasso, Secco, and Van Den Broek (1984). All probabilistic transitions are designed to "fire" with probability 0.9. Close inspection of Figure 3 reveals a very compact notation for specifying complex causal relations over time. Like the notation of Trabasso and his colleagues, the links in this representation are derived from a combination of intuitive considerations supported by counterfactual arguments. This notation has the advantage, however, of permitting multiple local states to be simultaneously active, and permitting multiple local transitions to simultaneously fire. We also exploit a Markov random field framework to formally link the firing of the "local probabilistic" transitions, with the global subjective probability function which assigns a probability to each trajectory in the system. Thus, once the probabilistic causal chain representation has been constructed, it is possible to implicitly assign a degree of belief to all possible trajectories in situation state space through the use of interpretable local probability distributions.

### The parallel distributed processing model.

The theory of story comprehension and recall we have been describing is based on a two-step process. First, a highly probable (believable) trajectory is computed by the reader which is consistent with the constraints of the story. This story trajectory is the reader's mental model of the story. The reader then learns the constructed mental model. During the recall process, the reader is given the initial portion of the trajectory (story) as a retrieval cue, and the remainder of the story is retrieved.

**Network Architecture.** We have devised a recurrent parallel distributed processing model which searches for highly probable trajectories where the probability of a trajectory is formally defined with respect to a MSP representation consisting of *d* local states or propositions.. Figure 1 shows a convenient way of thinking about this system. Instead of trying to visualize the complex temporal dynamics of a *d*-unit system, a system of *dM* units is considered where the unit in the *i*th row and *j*th column of the array corresponds to the value (absent or present) of the *i*th feature at time *j*. Thus, the each column of the array represents a point in the *d*-dimensional situation state space, and the *M* columns of the array correspond to an ordered sequence of *M* points in situation state space.

| Feature # | Sentence Fragment associated with Story Feature |
|---|---|
| 1. | Once there was a little boy |
| 2. | who lived in a hot country. |
| 3. | One day his mother told him to take some cake to his grandmother. |
| 4. | She warned him to hold it carefully |
| 5. | so it wouldn't break into crumbs. |
| 6. | The little boy put the cake in a leaf under his arm |
| 7. | and carried it to his grandmother's. |
| 8. | When he got there |
| 9. | the cake had crumbled into tiny pieces. |
| 10. | His grandmother told him he was a silly boy |
| 11. | and that he should have carried the cake on top of his head |
| 12. | so it wouldn't break. |
| 13. | Then she gave him a pat of butter to take back to his mother. |
| 14. | The little boy wanted to be very careful with the butter |
| 15. | so he put it on top of his head |
| 16. | and carried it home. |
| 17. | The sun was shining hard |
| 18. | and when he got home |
| 19. | the butter had all melted. |
| 20. | His mother told him that he was a silly boy |
| 21. | and that he should have put the butter in a leaf |
| 22. | so that it would have gotten home safe and sound. |

*Table 2. Situation State Space Representation of Epaminondas Story*

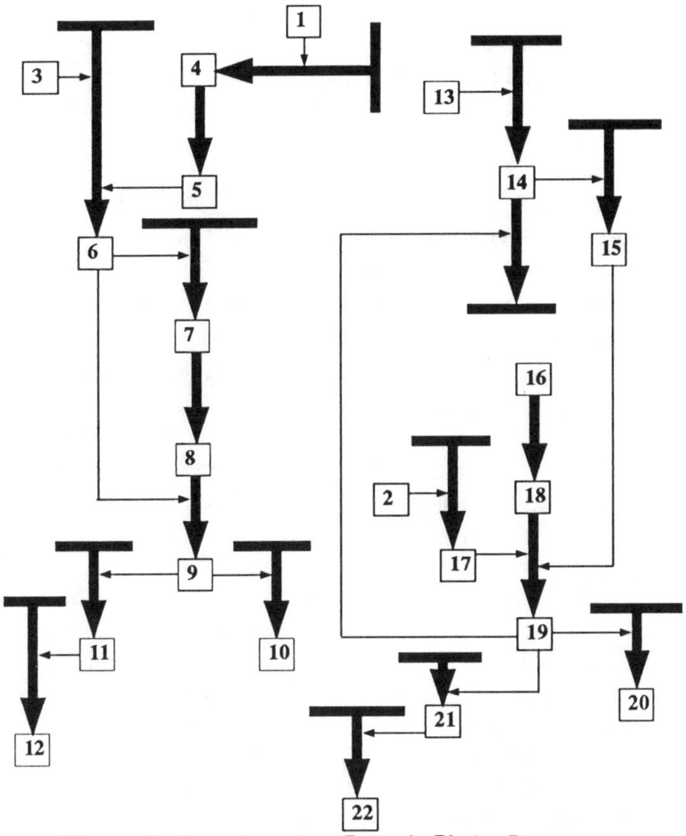


*Figure 1. Epaminondas Causal Chain Representation. The numbered states refer to propositions in Table 2.*

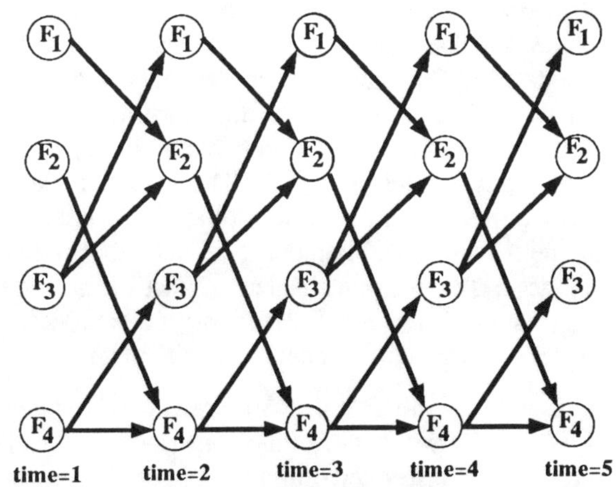


*Figure 2 An "expanded" version of a recurrent PDP network. The recurrent network may be visualized as a matrix of units where the pattern of connectivity from one time instant to the next is identical. In this example, a four unit recurrent network has been rewritten as a sequential 20 unit network which can reconstruct state space trajectories of length five or less.*

Notice that the connectivity pattern between units at adjacent time intervals is identical because this two-dimensional array is intended to model a two-layer recurrent network. The network shares important similarities with recurrent back-propagation (Rumelhart, Hinton, and Williams, 1986) and the brain-state-in-a-box model (Anderson et al., 1977; Golden, 1986). Golden (submitted) has found a discrete-time Liapunov function for this algorithm which explicitly states when all trajectories will converge to the set of system equilibrium points.

**Comprehension Process.** The comprehension process is modelled by beginning with a story which has already been parsed into a situation state space trajectory as in Table 1. If situation feature $i$ at time $j$ is equal to one (indicating feature $i$ at time $j$ is present), then the activation value of the unit in the $i$th row and $j$th column of the array in Figure 4 is clamped to the value of one during the comprehension process. The term *clamped* refers to the case where the activation of a unit is not permitted to change. The unclamped unit representing situation feature $i$ at time $j$ then computes a weighted sum of the activations of the units at time $j-1$ and time $j+1$, and uses this weighted sum to increment or decrement the unclamped unit's activation value. The unclamped unit's activation value is also decremented by an amount inversely proportional to $\rho$ so that the system searches the high-dimensional trajectory space in a region near the original story trajectory. Finally, the activations of all units in the system are constrained to lie in the range of zero to one. When the activations of the unclamped units are updated in this manner, then the network may be formally viewed as seeking a trajectory in situation state space which is highly probable with respect to a probability (belief) function. The system updates the activations of the units for some pre-determined maximum number of iterations, or until the system stabilizes. The resulting pattern of activation over the $dM$ units is the constructed story trajectory, and represents the system's "understanding" of the story.

**Learning and Recall Processes.** It is assumed that people learn the story trajectory (mental model) which they constructed during the comprehension process. Rather than attempt to model the details of this learning process, we are currently content to model the end results of that process. In particular, it is assumed that when people are asked to recall a story from memory, an episodic memory trace of the story trajectory is available to guide the recall process. In the model, the strength of this episodic memory trace is the $\rho$ parameter. Large values of $\rho$ correspond to long retention intervals so the strength of the episodic memory trace of the story trajectory is strong. Small values of $\rho$ correspond to short retention intervals so the strength of the episodic memory trace of the story trajectory is weak.

The recall process is then similar to the comprehension process, except that only the first few features of the story in the first three time steps of the trajectory are clamped as a retrieval cue. The network must then reconstruct the remainder of the trajectory. Although this is a very high dimensional parameter estimation problem, the introduction of the learning constraint helps by restricting the search to trajectories close in an Euclidean distance sense to the original story trajectory. In particular, each unit at time $t$ uses a weighted sum of the activations of the units at times $t-1$ and $t+1$ to update its activation value, but each unit at time $t$ also has its activation value modified by the episodic memory trace of the story weighted by the $\rho$ parameter.

The reconstructed recall trajectory is then used to generate the model's free recall responses. In particular, the "most active" unit in each situation is assumed to be the model's summary of that situation. Moreover, since the situation vectors are ordered in the recall trajectory, the responses of the model will be ordered as well. If the maximally active unit's value was less than 0.5, the model does not recall a situation feature summarizing that situation. If several units in a situation are maximally active, then exactly one of the units is randomly chosen to summarize the situation.

## Simulation Experiments

Human memory for simple causally coherent stories is characterized by at least four fundamental phenomena (Trabasso, Secco, & Van Den Broek, 1984; Van Den Broek & Trabasso, 1985). First, propositions with more causal connections are more likely to be on the main causal chain of the story and are more likely to be recalled. Second, as the retention interval between reading and recalling the story is increased, the percentage of propositions recalled decreases but those propositions will be more likely to lie on the main causal chain of the story. And third, more propositions will be recalled from stories which are more causally coherent (i.e., stories possessing more propositions on the main causal chain of the story). In this first set of simulations, we were interested in whether the proposed model would exhibit some of these qualitative phenomena.The average *causal connectivity* of a group of propositions was computed by counting the number of forward and backward connections in the MSP representation each statement in the group possessed, and then averaging over the group of propositions.

Each of the four stories were individually "comprehended" by the model, "learned" by the model, and "recalled" by the model as previously described. The $\rho$ parameter during the recall process was varied, and took on the values: $\rho = 0.1$, $\rho = 0.2$, and $\rho = 0.3$ corresponding to short, medium, and long retention intervals respectively.

### Effects of causal connectivity.

Figure 3 shows how variation of the retention interval parameter of the model affects which propositions are recalled by the model from memory. The dependent measure was *relative causal connectivity* which was defined as the average number of causal connections per proposition in a story subtracted from the average number of causal connections per proposition in the set of propositions recalled by the subjects. Like human subjects (e.g., Trabasso et al., 1984; Van Den Broek & Trabasso, 1985), propositions with greater causal connectivity (greater causal cohesiveness) are more likely to be recalled as retention interval increases. We would expect to see these effects in the model since the algorithm is minimizing an energy function which becomes more closely related to the relative causal connectivity as the retention interval parameter $\rho$ is increased (Golden, submitted).

An analysis of variance of the data using a story by retention interval design where the story factor was treated as random, however, did not support the hypothesis that statements with greater causal connectivity are more likely to be recalled as retention interval increases although the effect was marginally significant ($F(2,6) = 4.2$, $p < 0.10$). On the other hand, the average relative causal connectivity was significantly greater than zero ($t(11) = 2.4$, $p < 0.05$), indicating statements with greater causal connectivity are more likely to be recalled (e.g., Trabasso et al., 1984; Trabasso and Van Den Broek, 1986) was replicated by the model.

### Migration of misordered propositions.

Another important aspect of human memory for stories with a strong causal structure is that if propositions in such stories are "displaced," they will "migrate" back towards their canonical positions in the text (Bischofshausen, 1985; Bower, Black, and Turner, 1979; Mandler, 1978). Moreover, the magnitude of these effects tend to increase as retention interval is increased (Bischofshausen, 1985). We would expect to see these effects in the model as well since as the retention interval factor is decreased, the local probability of predicting a feature at the next instant of time becomes less dependent upon the episodic memory trace of the story formed during the comprehension process.

To model these phenomena, the simulations described above were repeated using exactly the same four causal knowledge structures (i.e., sets of connections among the units), but using stories which were slight distortions of the original four stories. For example, a "distorted" story was generated from the original *Fox and Bear* story by switching statement #8 with statement #21. Three distorted stories for each of the original four stories were generated in this manner, and the recall of the model for each of the twelve stories was recorded. Because the retention interval factor was varied in this experiment as well, thirty six independent simulation runs of the model were done. The number of steps along the recalled trajectory a misordered statement "drifted" towards its canonical position was used as a dependent measure.The results of these experiments are shown in Figure 4.

An analysis of variance of the data using a story by retention interval design where the story factor was treated as random supported the hypothesis that as retention interval increased, the magnitude of the drift increased as well ($F(2,6) = 15.6$, $p < 0.01$). Also the drift measure was significantly greater than zero ($t(11) = 6.1$, $p < 0.01$) indicating that misordered statements did indeed "drift" towards their canonical positions in the text rather than in the opposite direction.

## References

Anderson, J. A., Silverstein, J. W., Ritz, S. A., and Jones, R. S. 1977. Distinctive features, categorical perception, and probability learning: Some applications of a neural model. *Psychological Review*, 84:413–451.

Bischofshausen, S. 1985. Developmental differences in schema dependency for temporally ordered story events. *Journal of Psycholinguistic Research*, 14:543–556.

Bower, G. H., Black, J. B., & Turner, T. J. 1979. Scripts in memory for text. *Cognitive Psychology*, 11: 177–220.

Gabrielian, A. and Iyer, R. 1991. Integrating automata and temporal logic: A framework for specification of real-time systems and software. In C.M.I. Rattray and R. G. Clark, editors, *The Unified Computation Laboratory, Institute of Mathematics and its Applications*. Oxford University Press, Stirling, Scotland. Forthcoming.

Golden, R. M. Stability and optimization analyses of the generalized brain-state-in-a-box neural network model. Submitted.

Golden, R. M. 1986. The brain-state-in-a-box neural model is a gradient descent algorithm. *Journal of Mathematical Psychology*, 30: 73–80.

Mandler, J. 1978. A code in the node: The use of a story schema in retrieval. *Discourse Processes*,1: 14–35.

Rumelhart, D. E.1977. Understanding and summarizing brief stories. In D. LaBerge and S. J. Samuels, editors, *Basic processes in reading: Perception and comprehension*, Hillsdale: NJ.

Rumelhart, D. E., Hinton, G., and Williams, R. 1986. Learning internal representations by error propagation. In D. E. Rumelhart and J. L. McClelland, editors, *Parallel distributed processing. Volume 1: Foundations*, pages 318–362. MIT Press, Cambridge, MA.

Trabasso, T., Secco, T., and Van Den Broek, P. 1984. Causal cohesion and story coherence. In H. Mandl, N. L. Stein, and T. Trabasso, editors, *Learning and comprehension of text*, pages 83–111. Lawrence Erlbaum, Hillsdale, NJ.

Trabasso, T., and Van Den Broek, P. 1985. Causal thinking and the representation of narrative events. *Journal of Memory and Language*, 24: 612–630.


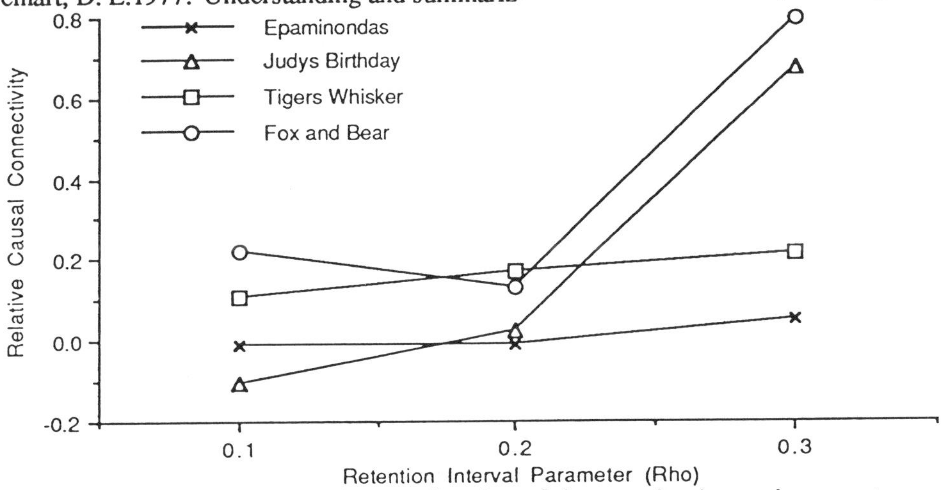


*Figure 3 Relative causal connectivity as a function of the retention interval parameter ρ.*

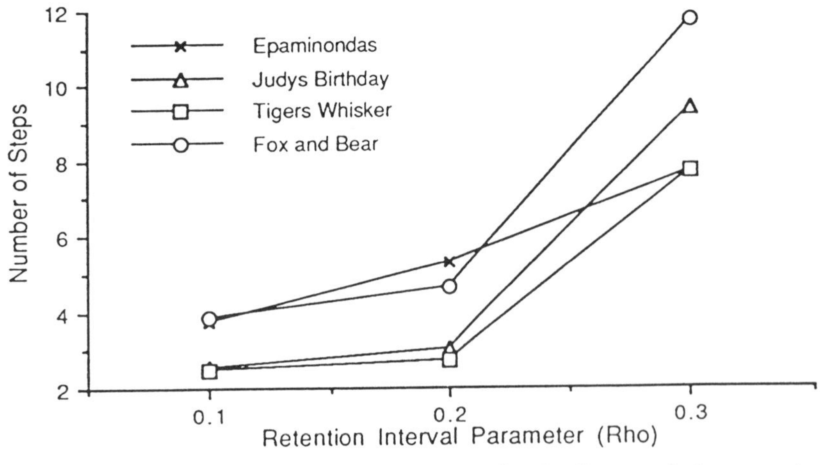


*Figure 4 Number of steps a misordered statement "drifted" towards its canonical position as a function of the retention interval parameter ρ.*

# TESTS OF SOME MECHANISMS THAT TRIGGER QUESTIONS

**Arthur C. Graesser, Cathy L. McMahen, and Brenda K. Johnson**

Department of Psychology
Memphis State University
Memphis, TN 39152
bitnet: graesserac@memstvx1

## Abstract

We have identified mechanisms that generate questions when individuals solve problems, comprehend text, and engage in conversations. Some of these mechanisms have been discussed in previous research in cognitive science and discourse processing, whereas other mechanisms were discovered when we analyzed videotapes of student-tutor interactions. The present study tested whether anomalous information causes an increase in questions when individuals solve mathematics problems and comprehend stories. College students were instructed to generate questions while they were solving problems (i.e., algebra and statistics) or while they were comprehending stories (e.g., fables and parables). There were several different versions of each problem or story: (1) complete original, (2) deletion of critical information, (3) addition of contradictory information, and (4) addition of irrelevant information. The deletion versions elicited the most questions whereas the original versions elicited the fewest questions; the addition versions were in-between. The validity of some of the question generation mechanisms is supported by the fact that these transformations of content caused an increase in questions.

Question generation has had a controversial status in cognitive science and education. At one extreme, there are models of cognition which postulate that question generation plays a critical role in comprehension, curiosity, and learning (Collins, 1988; Collins, Brown, & Larkin, 1980; Olson, Duffy, & Mack, 1985; Palinscar & Brown, 1984; Ram, 1990; Schank, 1986; Sternberg, 1987). According to Schank's SWALE model, for example, learning occurs when an individual observes an anomalous event and generates questions that lead to an explanation of the event (Schank, 1986).

A critical part of learning consists of generating illuminating questions and of transforming questions to become more illuminating. Collins (1988) has discussed how learning may be enhanced by having tutors ask students questions that expose the students' misconceptions during the course of answering the questions. The learning environment should also encourage students to ask questions and to model effective question asking (hopefully displayed by teachers). In the best of worlds, the learner would be an active, inquisitive, creative individual who energetically asks questions and insists on good answers.

The other end of the continuum provides a more pessimistic picture. It is well documented that student-generated questions are both infrequent and unsophisticated (Dillon, 1988; Flammer, 1981; Kerry, 1987). On the average, only one question is asked by a student per hour in a classroom setting. These questions are normally shallow questions that address the content and interpretation of explicit material rather than high-level questions that involve inferences, application, synthesis, and evaluation. A learner needs to master a significant amount of the material before the learner formulates good questions (Miyake & Norman, 1979). There are costs to posing questions in a classroom setting. The student is humiliated for revealing his or her lack of knowledge. The student imposes on the teacher, who rarely wants to be bothered by students' questions. Teachers frequently have trouble understanding the students' questions when the student has low domain-specific knowledge (Coombs & Alty, 1980), so the teacher ends up dismissing the question or answering the wrong question.

Unfortunately, teachers are normally poor role models for generating good questions. Less than 4% of the teacher-generated questions are higher-level demanding questions, i.e., those that involve

---

This research was funded by grants awarded to the first author by the Office of Naval Research (N00014-88-K-0110 and N00014-90-J-1492).

inferences, the application of an idea to a new domain of knowledge, or the synthesis of a new idea from multiple information sources (Kerry, 1987). Teachers are rarely sophisticated Socratic tutors who ask carefully planned sequences of thought-provoking questions, thereby exposing the students' misconceptions and contradictions.

The nature of question generation mechanisms in humans has profound implications for expert systems, decision support systems, and intelligent tutoring systems (Buchanan & Shortliffe, 1984; Sleeman & Brown, 1982; Wenger, 1987). The users of these sophisticated computer systems may need to radically alter their styles of learning and thinking to the extent that the users are expected to be active, creative question askers. An inept question asking or question answering facility on a computer will rapidly kill any incentive for a user to ask the computer questions.

Given the above theoretical and practical issues surrounding question generation, cognitive scientists need to understand the mechanisms that trigger questions when individuals genuinely seek information. To the extent that we understand these question generation mechanisms, we can design programs that tap into these mechanisms and maximize the generation of good questions. Learning, creativity, and problem solving will be facilitated to the extent that good questions are asked by students and computer users.

## Mechanisms that Generate Questions

Graesser, Person, and Huber (in press) identified four groups of mechanisms that generate questions. Some of these mechanisms exist in computer models of question generation, problem solving, and reasoning, such as AQUA (Ram, 1990; Reisbeck, 1988), SWALE (Schank, 1986), SOAR (Laird, Rosenbloom, & Newell, 1987), and SDDS (Klahr & Dunbar, 1988). Other mechanisms exist in theories of conversation (Allen, 1983; Clark & Schaefer, 1989), education (Collins, 1988; Sleeman & Brown, 1982), and question generation per se (Graesser, Lang, & Horgan, 1988; van der Meij, 1987). In addition to these theoretical contributions, Graesser et al. (in press) discovered question generation mechanisms when they analyzed transcripts of tutoring sessions on a variety of topics (e.g., research methods, mathematics, computer systems, climate, agriculture). The four groups of question generation mechanisms are summarized below.

**(1) Correction of knowledge deficits.** When individuals discover that their knowledge base is either incomplete or in error, they attempt to correct such knowledge deficits by asking questions from information sources that/who have useful, relevant information. Five question generation mechanisms are associated with this group:

(a) An obstacle in planning or problem solving. The question elicits information that removes or circumvents the obstacle.
(b) A decision among alternatives that are equally attractive. The question elicits information that breaks the tie.
(c) Glitch in an explanation of an event. The question elicits information that explains why an anomalous event occurred in a situation.
(d) Gap in knowledge that is needed for comprehension. The question elicits information that fills a gap in the questioner's knowledge base when he or she attempts to comprehend a written message, event, or conversation.
(e) Contradiction. The question elicits information that resolves the contradiction.

We were particularly interested in this first group of question generation mechanisms in the present study.

**(2) Monitoring common ground.** Successful communication requires that the speech participants keep track of what information is in the "common ground", i.e., knowledge that the speech participants believe they share. Speech participants frequently ask questions in order to update, verify, and monitor the knowledge in the common ground. Graesser et al. (in press) identified several question generation mechanisms which perform these functions: estimating or establishing common ground (e.g., "Do you know about X?"), confirming whether some belief is true ("Is X really true?"), and gauging whether the questioner or answerer comprehends something ("Do you understand what I'm saying?", "Am I understanding what you are saying?").

**(3) Social coordination of action.** Questions are frequently generated to coordinate actions of speech participants, to manipulate people, or to get permission to do something. Graesser et al. identified the following mechanisms in this group: indirect requests (e.g., "Would you do X?"), indirect advice ("Why don't you do X?"), asking permission ("Can I do X?"), offers ("Would you like me to do X for you?"), and negotiations ("If I do X, would you do Y?").

**(4) Control of conversation and attention.** Questions are frequently asked to monitor the flow of conversation among speech participants. This includes getting a speaker's attention, maintaining attention on a speaker, and changing speakers. Graesser at al. identified the following mechanisms in this group: greetings, replies to summons, changing speakers, maintaining focus on an agent's actions, rhetorical questions, and gripes.

The purpose of the present study was to test some of the question generation mechanisms in the first group, i.e., correction of knowledge deficits. In two experiments, college students were instructed to generate questions while they solved mathematics problems or while they comprehended stories. There were alternative versions of each problem or story. One version was the original complete problem or story. The other versions were transformations of the original version that were designed to elicit additional questions. A critical piece of information was removed in deletion versions. Addition versions contained an extra piece of information that introduced either a contradiction or an irrelevancy. If the question generation mechanisms are valid, then the transformed versions should yield more questions than the original versions. Moreover, a subset of the questions should address the particular transformations.

## Question Generation in the Context of Mathematics Problems.

**Methods.** Twelve original mathematics problems were selected from textbooks. Six of the problems were word problems from a basic algebra text at the seventh grade level; the other six were statistics problems from a college-level statistics textbook. In addition to the original version of each problem, we prepared a deletion version and an addition version that contained irrelevant information. An example of these three versions is presented below.

ORIGINAL. A bottle of Harvest Time Apple Juice contains 64 ounces and costs 99 cents. Farm Fresh Juice is available in bottles that contain 1 gallon for $1.88 each. Which is the better buy?

DELETION. A bottle of Harvest Time Apple Juice costs 99 cents. Farm Fresh Juice is available in bottles that contain 1 gallon for $1.88 each. Which is the better buy?

ADDITION. A bottle of Harvest Time Apple Juice contains 64 ounces and costs 99 cents. There are coupons which give the customer a 10% savings on all items. Farm Fresh Juice is available in bottles that contain 1 gal for $1.88 each. Which is the better buy?

The subjects were 28 college students who were enrolled in a Research Methods course at Memphis State University. Each subject received 4 problems in each of the 3 versions, yielding 12 problems altogether. The assignment of problem versions to subjects was counterbalanced so that approximately the same number of subjects received each version of a particular problem. Each subject received a 12-page booklet, with one problem per page.

The subjects were instructed to work on each problem in three phases, which were timed by the experimenter. The subjects read the problem during phase 1 for 30 seconds. The subjects wrote down questions that came to mind about the problem during phase 2 (90 seconds). The subjects solved the problem and generated additional questions during phase 3 (150 seconds). The subjects were given 30 seconds of rest before starting on the subsequent problem. Therefore, 5 minutes was allocated to each of the 12 problems.

**Results.** We scored the number of questions that each subject generated for each problem. An expression counted as a question if it was in an interrogative form, as opposed to an assertion or other type of speech act. Questions can be identified by trained judges with a high degree of reliability (Graesser et al., in press). The mean number of questions per problem significantly differed among versions, with means of 1.81, 2.49, and 2.09 in the original, deletion, and addition versions, respectively, $F(2, 54) = 5.61$, $p < .05$. Planned comparisons revealed the following differences: deletion > addition = original. Therefore, the deletion transformations had a more robust impact on question generation than did the addition transformations.

In a follow up analysis, we scored those questions that addressed the transformations in the deletion and addition versions. For example, a question that would address the deletion version would be "How big is the bottle of the Harvest Time apple juice?"; a question that would address the addition version would be "Why is there this information about coupons when it makes no difference as to the true value?" The mean number of transformation-relevant questions per problem was significantly higher in the deletion versions than the addition versions, .81 versus .41, respectively, $t(27)= 4.32$, $p < .05$. Both the .81 and the .41 significantly differed from 0 (i.e., the baseline level of the original version). Therefore, both types of transformations caused a significant increase in the number of transformation-relevant questions.

## Question Generation in the Context of Stories.

**Methods.** The original stories were 5 parables and 5 fables selected from anthologies of classical or famous writers. The stories had a short length of approximately 100 words. An example story is presented below.

> A hunter caught a wild rooster and brought it to the city for sale. A merchant asked: "What kind of bird is this? " "A phoenix," cheated the hunter. "That is a lucky symbol, which I see for the first time. Do you want to sell it to me?" "Sure, if you pay a high price." The merchant gave him 20 tael because he wanted to bring it to the king. But on the way the rooster died. He cried loudly, because he could not bring it to the king. People heard of this. Later, even the king knew of it. The king then gave him 200 gold pieces for his kindness, although he did not know it had only been a rooster.

There were five versions of each story as specified below.

ORIGINAL. This was the original, complete story.

DELETION. A critical piece of information was deleted; the point or moral of the story could not be grasped without this information. In the context of the example story, we deleted this sentence: "But on the way the rooster died."

ADDITION-CONTRADICTION. Information was inserted that contradicted a major idea in the original story. For example, we added this sentence: "The merchant had the bird shipped to his daughter."

ADDITION-SUBTLE-IRRELEVANCY. The additional information was an episode that could occur in the story, but was not relevant to the main plot and point of the story. For example, the story in this condition reads: "The merchant locked all the doors properly. But on the way the rooster died."

ADDITION-SALIENT-IRRELEVANCY. The additional information was an episode that would probably not occur in the story, and was not relevant to the main plot and point of the story. For example, the following statement was added to the text: "The teacher showed him how to do the equations."

The subjects were 24 students enrolled in an Introductory Psychology class at Memphis State University. Each subject completed a 10-page booklet, with one page assigned to each story. Each subject had one fable assigned to each of the 5 versions and one parable assigned to each of the 5 versions. The materials were counterbalanced so that each story had approximately the same number of subjects assigned to each of the 5 versions. The subjects' task was to read each story and to generate questions that came to mind while reading the stories.

**Results.** The mean number of questions per story was 3.13, 3.63, 3.21, 3.65, and 3.54 in the original, deletion, addition-contradiction, addition-subtle-irrelevancy, and addition-salient-irrelevancy conditions, respectively. The mean of the four transformation conditions was significantly higher than the mean of the original condition, $F(1, 23)= 4.67$, $p < .05$; the four transformation conditions did not significantly differ from one another.

We computed the mean number of questions that were relevant to the transformations of the story content. The four transformation conditions significantly differed on this measure, with means of .77, .47, .38, and .67 in the deletion, addition-contradiction, addition-subtle-irrelevancy, and addition-salient-irrelevancy conditions, respectively. All of these scores significantly differed from a zero baseline (which would correspond to the original version that had no transformations). Therefore, the transformations in content caused an increase in generated questions and a subset of the questions addressed the transformations. Our data suggest that the deletions and additions of salient irrelevancies were more obvious to the students than were the contradictions and additions of subtle irrelevancies.

## Conclusions

We found that the transformed versions of the mathematics problems and stories elicited more questions than did the original versions. Moreover, in the transformed versions, a subset of the questions addressed the transformations. These findings are consistent with the models of question generation that were discussed earlier. For example, Schank's SWALE model predicts that anomalous events trigger questions (Schank, 1986). The contradictions and irrelevant statements that were added in our transformed versions would be construed as anomalous information. Our findings support the claim that some of the question generation mechanisms identified by Graesser at al. (in press) are causally related to the volume and qualitative characteristics of generated questions. In particular, support was found for those question generation mechanisms that involve "corrections of knowledge deficits."

Graesser at al. (in press) identified five question generation mechanisms associated with corrections of knowledge deficits. Four of these five mechanisms were manipulated by the transformations in this study. The deletion transformations involved both (a) an obstacle in a plan or problem solving and (b) a gap in knowledge that is needed for comprehension. Clearly, the students faced an insurmountable obstacle when a problem omitted information that is critical for solving the problem (or alternatively, comprehending the point of a story). The transformation that added a contradiction obviously involved Graesser et al.'s contradiction mechanism. The transformations that added irrelevancies involved either (a) a gap in knowledge that is needed for comprehension or (b) a glitch in an explanation of an event. There is a pragmatic assumption that all information is relevant and important when short stories and problems are constructed. Students presumably would not believe that they completely comprehended a story or problem unless they had some explanation of an irrelevancy.

According to the results of the present study, the above question generation mechanisms are causally related to the volume and qualitative questions that students generate. Person (1990) has also reported that these question generation mechanisms naturally occur when students generate questions in a tutoring session on research methods (e.g., statistics, measuring variables, testing hypotheses, factorial designs). Nearly half of the student-generated questions in Person's videotaped tutoring sessions were in the major category of "corrections in knowledge deficits." Therefore, there is support for these question generation mechanisms both in controlled laboratory environments and in naturalistic environments.

## References

Allen, J. 1983. Recognizing Intentions from Natural Language Utterances. In M. Brady, and R.C. Berwick (Eds.). *Computational Models of Discourse.* Cambridge, MA: MIT Press.

Buchanan, B.G., and Shortliffe, E.H. eds. 1984. *Rule-Based Expert Systems.* Menlo Park, CA: Addison-Wesley.

Clark, H.H., and Schaefer, E.F. 1989. Contributing to Discourse. *Cognitive Science* 13:259-294.

Collins, A. 1988. Different Goals of Inquiry Teaching. *Questioning Exchange* 2: 39-46.

Collins, A., Brown, J.S., and Larkin, K.M.. 1980. Inference in Text Understanding. In R.J. Spiro, B.C. Bruce, and W.F. Brewer (Eds.). *Theoretical Issues in Reading Comprehension* (pp.385-407). Hillsdale NJ: Erlbaum.

Coombs, M.H., and Alty, J.L. `1980. Face-to-Face Guidance of University Computer Users-II. Characterizing Advisory Interactions. *International Journal of Man-Machine Studies* 12:407-429.

Dillon, J.T. 1988. *Questioning and Teaching: A Manual of Practice.* New York: Teachers College Press.

Flammer, A. 1981. Towards a Theory of Question Asking. *Psychological Research* 43: 407-420.

Graesser, A.C., Lang, K.L., and Horgan, D. 1988. A Taxonomy for Question Generation. *Questioning Exchange* 2(1): 3-15.

Graesser, A.C., Person N.K., and Huber J.D. (in press). Mechanisms That Generate Questions. In T. Lauer, E. Peacock, and A.C. Graesser (Eds.). *Questions and Information Systems.* Hillsdale, HJ: Erlbaum.

Kerry, T. 1987. Classroom Questions in England. *Questioning Exchange* 1(1):32-33.

Klahr, D and Dunbar, K. 1988. Dual Search Space During Scientific Reasoning. *Cognitive Science* 12(1):1-48.

Laird, J., Rosenbloom, P. S., and Newell, A. 1987. SOAR: An Architecture for General Intelligence.

*Artificial Intelligence* 33(1): 1-64.

Miyaki, N. and Norman, D.A. 1979. To Ask a Question, One Must Know Enough to Know What is Not Known. *Journal of Verbal Learning and Verbal Behavior* 18:357-364.

Olson, G.M., Duffy, S.A., and Mack, R.L. 1985. Question Asking as a Component of Text Comprehension. In A.C. Graesser and J.B. Black (Eds.). *The Psychology of Questions.* (pp. 219-226). Hillsdale NJ: Erlbaum.

Palinscar, A.S., and Brown, A.L. 1984. Reciprocal Teaching of Comprehension-Fostering and Comprehension-Monitoring Activities. *Cognition and Instruction* 1(2):117-175.

Person, N.K. 1990. The Documentation of Questioning Mechanisms and Types of Questions in Tutoring Protocols. Unpublished master's thesis, Memphis State University, Memphis, TN.

Ram, Ashwin. 1990. Decision Models: A Theory of Volitional Explanation. In Proceedings of the Twelfth Annual Conference of the Cognitive Science Society, 198-205. Hillsdale NJ: Erlbaum.

Reisbeck, C.K. 1988. Are Questions Just Function Calls? *Questioning Exchange* 2 (1):17-24.

Schank, R.C. 1986. *Explaining Problems: Understanding Mechanically and Creatively.* Hillsdale, NJ: Lawrence Earlbaum Associates.

Sleeman, D.H. and Brown, J.S. eds. 1982. *Intelligent Tutoring Systems.* New York: Academic Press.

Sternberg, R.J. 1987. Questioning and Intelligence. *Questioning and Exchange* 1:11-13.

van der Meij, H. 1987. Assumptions of Information-Seeking Questions. *Questioning Exchange* 1(2):111-117.

Wenger, E. 1987. *Artificial Intelligence and Tutoring Systems.* Los Altos, CA: Morgan Kaufman.

# In the Eye of the Beholder: The Coherence of Nonstandard Discourse

**Richard M. Roberts and Roger J. Kreuz**

Department of Psychology
Memphis State University
Memphis, TN 38152
kreuzrj@memstvx1.bitnet

**Abstract**

Researchers investigating discourse coherence typically examine the various mechanisms that bring about coherence. This body of research has acknowledged that the specific coherence relations which unite the individual discourse units work as a result of an assumption about the coherence of discourse in general. The standard approach to coherence investigation has been to analyze conventional texts and conversations in which both coherence relations and the assumption of coherence are present. By limiting themselves to the analysis of standard discourse, researchers have ignored nonstandard sources, which can provide insight into the necessity and sufficiency of these mechanisms. This paper provides several examples of nonstandard discourse. From these examples, we conclude that an assumption of coherence is the only necessary and sufficient mechanism required for judgments of coherence.

Although coherence is crucial for the perception of well-formed discourse (Dahlgren 1988, 1989; Green 1989; Hobbs 1978, 1979, 1985; Kintsch & van Dijk 1978; Mann & Thompson 1986), definitions of coherence are often circular. If coherence is defined as text which "hangs together," no real insight is gained. Definitions which merely replace the idea of coherence with other imprecise terms like "unified" or "related" are not useful.

Coherence researchers usually begin with discourse which is intuitively coherent and then describe the mechanisms that are present (e.g., Dahlgren 1988, 1989; Hobbs 1978, 1979, 1985; Mann & Thompson 1986). In other words, whenever discourse strikes one as coherent, coherence mechanisms are assumed to be present and derivable. Because noncoherent discourse should not contain these mechanisms, it has been dismissed as only the logical opposite of coherent discourse. As a result, it has been ignored by coherence researchers.

The intuitively coherent discourse used in coherence analysis is usually taken from conversations involving adult speakers (Craig & Tracy 1983), or from well-known publications such as *The Wall Street Journal* (e.g., Dahlgren 1988, 1989). From such analyses, researchers have established a variety of coherence mechanisms. For our purposes, these mechanisms can be divided into two general categories: structural relations, and the assumption of coherence.

Partial support for the preparation of this article was provided by a Center of Excellence grant from the State of Tennessee to the Department of Psychology at Memphis State University.

## Coherence Mechanisms

*Structural relations* refer to the ways in which words, phrases, sentences, and larger units keep the discourse unified and connected. These are relations which speakers and writers build into their discourse so that readers and listeners can unite potentially disjointed segments into a coherent whole. Structural relations may be explicitly marked in the syntax, or inferred. Examples of *explicit* relations include argument overlap (Kintsch & van Dijk 1978), and some of the "lexicogrammatical" devices of Halliday and Hasan (1976). Relations which *implicitly* connect discourse occur when inferences are made about the presented information. In general, these implicit relations require participants to make inferences based on world knowledge. For example, participants make inferences based on an understanding of causality or logical reasoning. Because structural relations must be recognized as such in order for a discourse to be judged coherent, there are probably only a finite number of them (see Dahlgren 1989; Hobbs 1978, 1979, 1985; Mann & Thompson 1986).

The *assumption of coherence* is part of a participant's general understanding about the way discourse is supposed to work—even when discourse execution is technically flawed. The assumption of coherence is maintained through the common desire of the discourse participants (see Grice 1975; Daly, Weber, Vangelisti, Maxwell, & Neel 1989; Mann & Thompson 1986). By assuming that a discourse is intended to be coherent, participants work to unite potentially disjointed pieces of it (see de Beaugrande 1987). Moreover, participants actively work to achieve coherence for all presented information. Vuchinich (1977), for example, reported increased response latencies after unrelated information was presented during conversations. Presumably, response time increased because the respondent needed more time to try and incorporate the irrelevant utterances into the ongoing conversation.

Because the structural relations and the assumption of coherence normally work together in coherent discourse, researchers usually do not consider separating them. As a result, it is difficult to assess the relative importance of these two mechanisms. Intuitive judgments of coherence may not necessarily be tied to both kinds of mechanisms.

There are actually *four* possible combinations between intuitive judgments of coherence and the presence or absence of coherence mechanisms. The two most obvious combinations were previously mentioned: (1) coherence mechanisms are present in the discourse and the discourse is judged as coherent; and (2) no coherence mechanisms are present in the discourse and the discourse is judged as noncoherent. In addition, there are two other possible combinations which have generally been overlooked. Specifically, it is possible that (3) a discourse is judged as coherent even though no coherence mechanisms are present; and (4) a discourse is judged as noncoherent when there are coherence mechanisms present.

It is the examination of the latter two combinations which also should be of interest to investigators of coherence. This paper will demonstrate the utility of examining discourse coherence by using discourse which could be judged as either coherent or noncoherent. In general, discourse which contains various coherence mechanisms, but which is still considered noncoherent, should establish the insufficiency of these mechanisms. Similarly, discourse that lacks various coherence mechanisms, but which is considered coherent, can be used to establish the non-necessity of these mechanisms.

## Coherence and Nonstandard Discourse

Discourse which may or may not be judged as coherent is found in nonstandard sources. The ambiguity of coherence judgments arises because the coherence of nonstandard passages may only exist for a limited number of readers or speakers. Therefore, nonstandard discourse is not generally studied in discourse processing, artificial intelligence, or computational linguistics. Passages in which coherence is less well defined could serve as powerful tests of the necessity and sufficiency of the two coherence mechanisms.

Nonstandard discourses which are generally not studied for discourse coherence include poetry, discourse from linguistically deviant populations (e.g., schizophrenics and aphasics), and literary texts from purposefully nonstandard authors (e.g., the experimental writing of James Joyce, Gertrude Stein, or Eugene Ionesco).

The following excerpt from "Tender Buttons" by Gertrude Stein is an example of a nonstandard discourse that some readers would consider coherent and others would not (cited in DeKoven 1983, p. 80):

> A cushion has that cover. Supposing you do not like to change, supposing it is very clean that there is no change in appearance, supposing that there is regularity and a costume is that any of the worse than an oyster and an exchange. Come to season that is there any extreme use in feather and cotton. Is there not much more joy in a table and more chairs and very likely roundness and a place to put them.

If this discourse is coherent, it is not immediately obvious. Such nonstandard examples should be useful in determining how judgments of coherence are made. Comparing the presence or absence of coherence mechanisms in nonstandard examples provides a more complete understanding of coherence in standard examples.

The following table shows the relation between the presence or absence of coherence mechanisms and types of discourse. For nonstandard discourse, one mechanism is present while the other is absent.

| | | Assumption of Coherence | |
|---|---|---|---|
| | | Present | Absent |
| Structural Relations | Present | Standard texts and conversations | Schizophrenic discourse |
| | Absent | Experimental writing (Stein) Theater of the absurd (Ionesco) Nonstandard poetry (cummings) | Gibberish |

### Schizophrenic Speech

The psychotic disorder known as schizophrenia is characterized by noncoherent speech (American Psychiatric Association 1987). This is true even though schizophrenics can produce discourse which uses explicit and implicit structural coherence relations. Therefore, these structural coherence relations must be *insufficient* for the realization of coherence.

There is no overall coherence in the following letter, written by a schizophrenic patient. However, explicit structural relations are present, and appropriate inferences can be drawn (Bleuler 1950, p. 17):

> I am writing on paper. The pen which I am using is from a factory called 'Perry & Co.' This factory is in England. I assume this. Behind the name of Perry Co. the city of London is inscribed; but not the city. The city of London is in England. I know this from my school days. Then, I always liked geography. My last teacher in the subject was August A. He was a man with black eyes. I also like black eyes. There are also blue and grey eyes and other sorts, too. I have heard it said that snakes have green eyes. All people have eyes. There are some, too, who are blind. These blind people are led about by a boy. It must be very terrible not to be able to see. There are people who can't see, and in addition, can't hear. I know some who hear too much. One can hear too much. There are many sick people in Burgholzli (sic); they are called patients.

In this example, arguments found in each sentence are linked to arguments found in following sentences. For example, the word "eyes" appears in five consecutive sentences. Although the reader may be able to understand the meaning of the individual sentences, there does not appear to be any overall unity or global coherence. As Hobbs (1982) has pointed out, "coherence isn't mere comprehensibility" (p. 225).

These explicit relations may actually contribute to the incoherence of the passage: The writer spells out inferences which should be obvious (e.g., sick people are called patients). The reader also must make inferences, based on world knowledge, that are *not* explicitly stated. For example, the reader must infer that geography is a subject taught in school. This letter is a good example of how explicit and implicit coherence relations alone are insufficient to establish coherence.

Structural relations are also *unnecessary* for judgments of coherence to be made. The ability of participants to generate inferences demonstrates that explicit coherence relations can be eliminated without losing coherence. These inferences, however, are also unnecessary for coherence. This conclusion can be supported by looking at another type of nonstandard discourse, in this case from the theater of the absurd.

### Theater of the Absurd

According to Altenbernd and Lewis (1969), the theater of the absurd is based in nihilism: Language is "employed illogically—fragmented, made into mechanical nonsense, kept *non sequitur*, kept banal and worn out. Once the sense of association and congruity necessary to keep language communicative has been removed, the attempt at communication itself becomes an absurdity" (p. 517).

Consider the following example taken from Ionesco's play *The Chairs* (in Altenbernd & Lewis 1969, p. 519):

> Old Woman: Come on now, imitate the month of February.
>
> Old Man: I don't like the months of the year.
>
> Old Woman: Those are the only ones we have, up till now. Come on, just to please me. . .
>
> Old Man: All right, here's the month of February. (*He scratches his head like Stan Laurel.*)

In this example, explicit coherence relations are present since the arguments in each line of the dialog overlap. However, new inferences about the months of the year (e.g., February looks like Stan Laurel scratching his head) must be made: The usual inferences (e.g., February is a cold month) no longer apply. The plays of the theater of the absurd use the audience's expectations of reality in order to violate them.

Such plays are understood coherently because the audience maintains the general communicative assumption of coherence. Theater of the absurd is an example of how normal inferencing mechanisms can be rejected without a loss of coherence, as long as the audience believes that the discourse is intended coherently.

### Nonstandard Poetry

**The poetry of e e cummings**. It is possible for a discourse to be considered coherent when both the explicit *and* implicit structural relations are missing or disordered. Such is the case in some forms of poetry. For example, the e e cummings poem "a like a" is nonstandard both linguistically and typographically (in Fairley 1980, p. 243):

a like a
grey
rock wanderin

g
through
pasture
wom

an creature whom
than
earth hers

elf
could
silent more no
be

Because normal word order is not used, the syntax and semantics on which coherence relations are based cannot account for the coherence of the poem. It can be argued, however, that the poem describes an image very effectively. A standard rendering of the poetic image would be much different, and arguably less powerful: "A woman creature is like a grey rock, she (or it) is wandering through pasture. . . earth herself could not be more silent than her" (Fairley 1980, p. 244).

The coherence of this poem results from the abandonment of standard, prescribed, coherence relations. Fairley (1980) points out that it is precisely the deviations from normal structure which make the poem coherent. Although words which explicitly signal structural relations are present (e.g., "whom" is a referent for "wom an creature"), the poem would remain disjointed and fragmented unless the reader worked to relate the pieces into a unified representation.

The following stanza from another cummings poem succinctly sums up this argument (cummings 1968, p. 290):

since feeling is first
who pays any attention
to the syntax of things
will never wholly kiss you;

In much of cummings' poetry, syntax and semantics are used bizarrely. The reader cannot resort to the conventional mechanics of language, and cummings encourages us not to try.

**The poetry of André Breton**. Another example of nonstandard poetry may be found in the surreal writings of André Breton (cited in Scholes, Comley, & Ulmer 1988, p. 66):

. . . there go the fuses blown again
Here's the squid with his elbows on the window sill
And wondering where to unfold his sparkling sewer grill
Is the clown of the eclipse in his white outfit
Eyes in his pocket. . .

Unlike cummings' poem, which is an unconventional rendering of an image, Breton presents a conventional rendering of a "non-image." If the passage is intended metaphorically, then the reader must actively work to decode it. The reader is willing to make this effort because he or she believes that there is a coherent message to be derived.

## Nonstandard Discourse and Mental Models

These examples demonstrate that readers of nonstandard plays and poetry are able to construct coherent representations without recourse to conventional structural relations. According to Johnson-Laird (1983), the coherence of a discourse is ultimately dependent upon the construction of a mental model or representation for that discourse. In other words, coherent discourse is different from a random collection of sentences because a mental model is constructed.

The establishment of a mental model should not imply that only one mental model is available for each coherent discourse. Discourses simultaneously may be represented by competing mental models, and each one may be coherent or not coherent in varying degrees. For example, the novel *Gulliver's Travels* by Swift (1726/1967) could be simultaneously represented as a series of fantastic adventures or as a critique of English society. The construction of a mental model of the story at one level does not preclude a coherent representation of the story at another level. Readers who do not recognize the satire can nonetheless construct a coherent mental model of the story.

A definition of coherence based upon a mental model does not require explicit structural relations, implicit structural relations, or knowledge about the world. The essential requirement for coherence is the willingness of participants to construct such a model. Nonstandard discourse, such as poetry, is a good example of how mental models can be built without any structural coherence relations. In addition, the language of schizophrenics demonstrates that these relations alone do not guarantee coherence.

## Conclusions

Normally, the structural relations and the assumption of coherence work in tandem to make coherence obvious and robust. Writers and speakers try to build coherence into their texts and utterances by using various structural coherence relations. Indeed, some writers and speakers may be better communicators because they are better at signalling how their message coheres (see Britton 1990). Similarly, some readers and listeners may be better comprehenders because they are better at deciphering these explicit and implicit devices.

Although structural coherence relations would not be interpretable without the assumption of coherence, it is not easy to continue assuming coherence when linguistic structure seemingly belies it. Furthermore, with the exception of schizophrenic speech, it is difficult to imagine a discourse in which the structural coherence relations are present, and yet the discourse is not considered coherent. This must be the case, however, if the assumption of coherence is abandoned. Even the most compelling arguments would be unrecognizable.

The popular arts have used the reliance upon the assumption of coherence for humorous effect. In the movie *Big* and the novel *Being There* (Kosinski 1970) the main characters are not linguistically competent adults: They are "children" thrust into an adult world, in which their linguistic competence is assumed. Consequently, their bizarre linguistic productions are not perceived as such: Their listeners put in the necessary work to make these productions coherent. As a result, their discourse is considered remarkably insightful, when in fact the insight comes from the listener, and not from the child-like speaker.

By assuming that other participants are trying to be coherent, participants work to make the discourse coherent for themselves. Coherence structures facilitate this process for participants; however, the failure to use coherence structures does not mean that coherence fails. Examples from nonstandard discourse have demonstrated that these structures are neither necessary nor sufficient for coherence. The only necessary component a discourse requires to be judged as coherent is the prior assumption that it *will be* coherent. Therefore, the assumption of coherence alone is sufficient to establish coherence.

It has been shown that the necessity and sufficiency of coherence mechanisms can be understood through the investigation of nonstandard discourse. Typically, researchers who study language examine linguistic samples which are selected for their normalcy. Investigations of coherence which are limited to ordinary discourse are unnecessarily restrictive. Because coherence is in the eye of the beholder, participants will manage to make even the most bizarre linguistic productions cohere.

## Acknowledgements

The authors would like to thank Bob Cohen, Leo Connolly, Art Graesser, and Marie Sell for their comments on earlier drafts of this paper.

## References

Altenbernd, L. and Lewis, L. L. 1969. *Introduction to literature: Plays*. Toronto: The MacMillan Company.

American Psychiatric Association. 1987. *Diagnostic and statistical manual of mental disorders* (3rd ed. revised). Washington, DC: Author.

Bleuler, E. 1950. *Dementia praecox or the group of schizophrenias*. New York: International Universities Press.

Britton, B. K. 1990. Kintsch's computational model of coherence processing: Diagnosing inferences in naturally-occurring text, and repairing text accordingly to increase learning. In Proceedings of the Twelfth Annual Conference of the Cognitive Science Society, 1056. Hillsdale, NJ: Erlbaum.

Craig, R. T., and Tracy, K. eds. 1983. *Conversational coherence: form, structure, and strategy*. Beverly Hills, CA: Sage.

cummings, e e 1968. *Complete poems 1913-1962*. New York: Harcourt Brace Jovanovich.

Dahlgren, K. 1988. *Naive semantics for natural language understanding*. Boston: Kluwer Academic Press.

Dahlgren, K. 1989. Coherence relation assignment. In Proceedings of the Eleventh Annual Conference of the Cognitive Science Society, 588-596. Hillsdale, NJ: Erlbaum.

Daly, J. A., Weber, D. J., Vangelisti, A. L., Maxwell, M., and Neel, H. 1989. Concurrent cognitions during conversations: Protocol analysis as a means of exploring conversations. *Discourse Processes* 12:227-244.

de Beaugrande, R. 1987. Schemas for literary communication. In L. Halász ed. *Literary discourse: Aspects of cognitive and social psychological approaches*, 49-99. Berlin: de Gruyter.

DeKoven, M. 1983. A different language: Gertrude Stein's experimental writing. Madison, WI: The University of Wisconsin Press.

Fairley, I. R. 1980. Syntactic deviation and cohesion. In M. K. L. Ching, M. C. Haley, and R. F. Lunsford eds. *Linguistic perspectives on literature*, 242-256. London: Routledge and Kegan Paul.

Green, G. M. 1989. *Pragmatics and natural language understanding*. Hillsdale, NJ: Erlbaum.

Grice, H. P. 1975. Logic and conversation. In P. Cole and J. L. Morgan, eds., *Syntax and semantics 3: speech acts*, 41-58. New York: Academic Press.

Halliday, M. A. K., and Hasan, R. 1976. *Cohesion in English.* London: Longmans.

Hobbs, J. R. 1978. *Why is discourse coherent?* Technical Note 176. Menlo Park, CA: SRI International.

Hobbs, J. R. 1979. Coherence and coreference. *Cognitive Science* 3:67-90.

Hobbs, J. R. 1982. Towards an understanding of coherence in discourse. In W. G. Lehnert and M. H. Ringle eds. *Strategies for natural language processing*, 223-243. Hillsdale, NJ: Erlbaum.

Hobbs, J. R. 1985. On the coherence and structure of discourse. Report No. CLSI-85-37. Stanford, CA: Stanford University, Center for the Study of Language and Information.

Johnson-Laird, P. N. 1983. *Mental models.* Cambridge, MA: Harvard University Press.

Kintsch, W., and van Dijk, T. A. 1978. Toward a model of text comprehension and production. *Psychological Review* 85:363-394.

Kosinski, J. 1970. *Being there.* New York: Bantam.

Mann, W. C., and Thompson, S. A. 1986. Relational propositions in discourse. *Discourse Processes* 9:57-90.

Scholes, R., Comley, N. R., and Ulmer, G. L. 1988. *Textbook: An introduction to literary language.* New York: St. Martin's Press.

Swift, J. 1967. *Gulliver's Travels.* (P. Dixon and J. Chalker, eds.). Baltimore, MD: Penguin Books. (Original work published 1726)

Vuchinich, S. 1977. Elements of cohesion between turns in ordinary conversation. *Semiotica* 20:229-257.

# The Story Gestalt
# A Model of Knowledge Intensive Processes in Text Comprehension

**Mark F. St. John**

Department of Cognitive Science
University of California, San Diego
La Jolla, CA 92093-0515 USA
stjohn@cogsci.ucsd.edu

### Abstract

How are knowledge intensive text comprehension processes computed? Specifically, how are 1) explicit propositions remembered correctly, 2) pronouns resolved, 3) coherence and prediction inferences drawn, 4) on-going interpretations revised as more information becomes available, and 5) how is information learned in specific contexts generalized to novel texts? The Story Gestalt model, which uses a constraint satisfaction process to compute these processes, is successful because each of the above processes can be seen as examples of the same process of constraint satisfaction, constraints can have strengths to represent the degrees of correlation among information, and the independence of constraints provides insight into generalization. In the model, propositions describing a simple event, such as going to the beach or a restaurant, are sequentially presented to a recurrent PDP network. The model is trained to process the texts by requiring it to answer questions about the texts. Each question is the bare predicate from a proposition in the text or a proposition that is inferrable from the text. The model answers the question by completing the proposition to which the predicate belongs. The model accomplishes the five processing tasks listed above and provides insight into how a constraint satisfaction model can compute knowledge intensive processes in text comprehension.

How do readers comprehend text? How do they build a representation of the event to which the text refers? How do they infer important causal information that the text does not explicitly mention, and how is information from different past experiences combined and modified to fit a new current context? For example, how do readers comprehend, "Jolene raced down the track ahead of her rivals. At the finish line, the judge handed her the trophy," and infer that Jolene won the race? In general, how are knowledge intensive text comprehension processes computed?

In Charniak's (1983) model of text comprehension, a sentence is processed into a proposition, then a semantic network is traversed to find a path connecting the proposition to previous propositions. Concepts along the connecting path constitute inferences drawn to make a coherent interpretation. Alternately, as in Schank's (1981) model, a detailed script is retrieved from memory and the text is mapped onto it.

These models are problematic in several ways. First, in both of these approaches, processing the explicit text, resolving pronouns, and drawing inferences work separately. Second, information is represented in an all-or-none fashion that does not reflect the gradedness of information in the real world. For example, the likelihood of a causal relationship is not represented. Third, revision is difficult because changes to an all-or-none representation are necessarily drastic. Consequently, these systems are conservative and rarely make predictions or other risky inferences that might require revision.

A more workable approach to comprehension processes is offered by the idea of *weak constraint satisfaction* (Rumelhart, Smolensky, McClelland, & Hinton, 1986). The explicit text places constraints on an interpretation, and the constraint satisfaction process computes the interpretation that best satisfies these constraints. According to this approach, representing the explicit text, resolving pronouns, and drawing inferences are part and parcel of the same process of constraint satisfaction. Second, because constraints can be graded, degrees of likelihood and strengths of relationships can be represented and used. Third, the gradedness of the representation makes revision easier and prediction inferences less risky.

Constraint satisfaction has only recently been applied to higher levels of text comprehension (Allen, 1987; Dolan & Dyer, 1989; Kintsch, 1988; Miikkulainen & Dyer, 1989). The most comprehensive, both in processing and learning, is the Miikkulainen & Dyer model. One portion of their model combines propositions into a schema representation. The model represents the propositions as vectors of activation across an input layer, sequentially integrates these propositions into a schema representation, and learns the mapping from propositions to schemas through practice. The model is able to perform a number of important knowledge intensive comprehension processes.

One shortcoming of the Miikkulainen & Dyer model, however, is that it does not learn its schema representation through training. One advantage to learning is that a model would not be restricted to any *a priori* schema representation defined by a programmer. A second advantage is that the burden of developing a schema representation falls to the model and its training environment, where it ultimately belongs.

The goals of this paper are to describe a constraint satisfaction model of text comprehension similar to the

Miikkulainen & Dyer model, and to tie its processing more closely to a greater set of psychological results concerning 1) representing the explicit text, 2) resolving pronouns, 3) drawing coherence and prediction inferences, 4) revising on-going interpretations as new information becomes available, and 5) sharing knowledge learned in different contexts. Additionally, the model is designed to learn its own schema representation in a hidden layer. This larger learning component in the model should enhance our understanding of what is required from a corpus of experiences, and from the training task posed to the model, to achieve good performance.

## Comprehension Tasks

**Representing Multiple Propositions.** A basic requirement for text comprehension is the ability to represent more than one proposition at a time without becoming confused about who did what. In some models, however, these role assignments are lost when causal inferences are drawn, and expensive additional processing is required to recover them (Charniak, 1983).

**Resolving Pronouns.** Pronouns create substantial ambiguity in text. Language cues, such as focus and the gender and number of a pronoun provide some constraint on an interpretation (Corbett & Chang, 1983; Carpenter & Just, 1977). Additionally, the situation itself can often provide some constraint (Hirsh & Brill, 1980). The comprehension process should be able to use every available constraint to help compute the referent. Finally, the referent for a pronoun may remain ambiguous for some time. The model should be able to tolerate this ambiguity and then use information as it becomes available to resolve the pronoun.

**Inferring propositions.** Inferred propositions can be divided into two categories. Coherence inferences are drawn to explain or justify the text, and prediction inferences are drawn to predict additional information, such as future actions, that fit the context.

Empirical evidence suggests that both coherence and prediction inferences are drawn, but that prediction inferences are only activated weakly according to the amount of support provided by the text (Graesser, 1981; McKoon & Ratcliff, 1986; but see Potts, Keenan, & Golding, 1988). Coherence inferences, on the other hand, are fully activated because of their stronger support from the text. Models of text comprehension should consequently incorporate a mechanism that uses the degree of support provided by the text to determine the activation level of inferences.

**Revising an on-going interpretation.** Readers revise their interpretations when new information makes their initial interpretations unlikely (Rumelhart, 1981). The idea is that readers immediately updating their interpretations as each new piece of information is processed (Carpenter & Just, 1977), and that carefully written "garden path" texts can precipitate more dramatic revisions.

**Knowledge sharing and generalization.** A critical question for any system that learns is how it fares on novel examples. Specifically in text comprehension, we can ask whether knowledge about people and events learned in specific situations remains tied to those situations, or whether that knowledge can be shared among all situations? Will the model be able to represent and remember novel texts composed of familiar pieces, or will it forget or even regularize novel texts to look more like familiar texts? Bartlett (1932), for example, presented English college students with highly unusual Native-American folk tales. The students forgot bizarre events, reorganized sequences of actions, and invented information that fit their interpretations. On the other hand, unusual stories often spur our memories. It is not clear when people represent new texts correctly and when they regularize. A good possibility is that forgetting and regularization increase with the unusualness of the text since unusual stories depart more from the reader's repertoire of schemas. The comprehension model should demonstrate this effect.

## Simulation Method

The model's task is to take a text as input and understand the text so that the model can answer questions. A text may leave out information or contain pronouns which the model must resolve to answer questions correctly. The model learns to comprehend texts through experience comprehending example texts. Once trained, the model's performance on each task can be evaluated by using example texts from the corpus or by using novel texts.

**Corpus.** The corpus consists of a large number of texts that describe events in six different contexts: going to the beach, restaurant, bar, park, airport, or race. Each text is represented in the input as a sequence of propositions that describe actions or attributes of a character or of the situation.

A proposition is represented as a set of thematic roles: agent, predicate, patient or theme, recipient or destination, location, manner, and attribute. For example, the proposition that conveys the fact that *the judge gave the trophy to Jolene*, would be (agent=judge, predicate=gave, patient=trophy, recipient=Jolene).

A proposition is represented in the network as a vector of activation values. The concepts are represented locally. There are 19 units to represent each of the 19 agents, 34 different units to represent each of the 34 patients, and so on. It is important to understand that the localist representation pertains only to the input and output layers. The internal hidden layers are free to develop distributed representations.

The corpus of texts is generated from a set of six scripts, one for each context. Propositions in each script are chosen probabilistically to be included in a particular text.

Across the six scripts, the total number of different events is 28,480. For example, in the restaurant script, there are 20 different sequences of events involving two of ten possible characters and one of four possible vehicles for a total of 7200 restaurant events (20 * 10 * 9 * 4). The number of possible input texts is much higher because of pronouns and missing propositions.

**Architecture and Training Regime.** The architecture is similar to that of St. John & McClelland's (1990) sentence processing model and Miikkulainen & Dyer's (1989) text processor. The model processes propositions one at a time to iteratively build and refine a representation of the whole text. The network is trained to perform this task by receiving feedback on comprehension questions. The questions provide a predicate from a proposition in the event and the network is required to complete the whole proposition.

The network can be divided into two parts according to their functions. Part A of the network (see Figure 1) sequentially processes propositions to refine its interpretation of the text. The input consists of the *current proposition* and a copy of the activations of the units in the story gestalt layer on the previous cycle. At the beginning of each text, the activations in the *previous story gestalt* layer are 0.0. Activation feeds forward to produce a new *story gestalt* that represents what is known about the text interpretation at that point in processing. On each new cycle, the next proposition in the text replaces the last, and the *story gestalt* activations are copied to the *previous story gestalt* layer.

Part B of the network performs the question answering computation. A question is created by removing everything except the predicate from a proposition. Activation from the *question* and the *story gestalt* feed forward to the *complete proposition* layer.

There are 136 units in the *current proposition* layer, 100 units in the *story gestalt* and *previous story gestalt* layers, 100 units in the *intermediate* layers, 34 units in the *question* layer, and 136 units in the *complete proposition* layer.

After a proposition is presented, the model is trained on the part of a text that has been presented in the input up to that point. After the first proposition, the model is asked only about the first proposition. After the second proposition is presented, the model is asked about both the first and second propositions. Propositions that are missing from the input are only asked about after they have been skipped in the input. For example, if the third proposition were missing, the model would only be asked about it after the fourth proposition were presented.

Before training on the corpus began, the model was trained on single, random propositions. The propositions were generated by randomly activating one unit in each thematic role, and the model was required to reproduce that proposition in the output. The purpose of this training was to help the model learn to remember the propositions actually presented by learning to map single propositions from the input to the output. Once this mapping was been learned, the model was trained on whole texts. So that the model would not immediately lose the mapping it learned in the pretraining, the random, single proposition training continued on 20% of the training trials.

## Simulation Results

The model was pretrained on the single propositions until the total sum squared error across the output fell below .05. This training itself required 145,000 training trials with a learning rate of .001. The model was then trained on texts for 160,000 trials with more single propositions mixed in between trials. The learning rate was .0001, and the momentum was .9. Finally, 20,000 more texts were trained using a momentum of 0. The model's performance was then evaluated.

**Pronoun resolution and inference.** The model's performance here is described by example. A text about visiting a restaurant is presented to the model

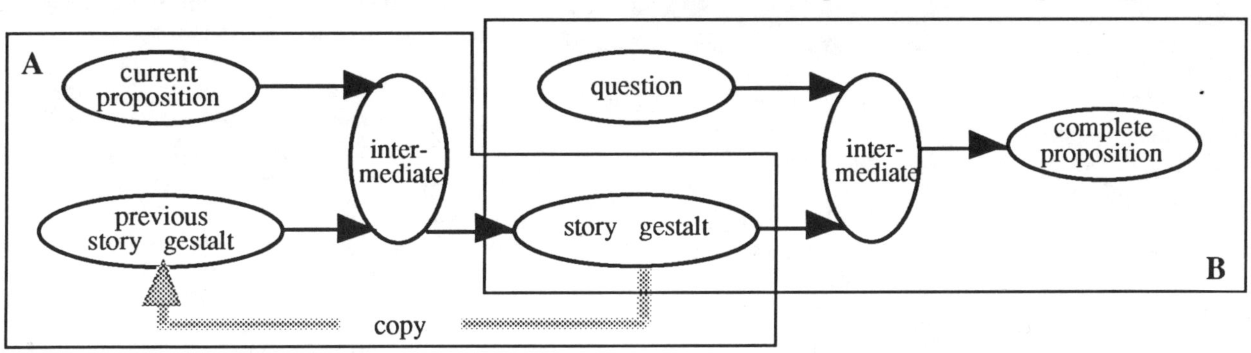


**Figure 1. The architecture of the Story Gestalt network.** The boxes highlight the functional parts: Area A process the propositions into the story gestalt. Area B processes the story gestalt into the output propositions

that contains a pronoun and that skips several propositions.

The restaurant script contains a number of regularities, that if learned, can be used to help perform these tasks. One such regularity is that the character who orders also pays the bill and receives back his credit-card. Given which character performed any of these tasks, the model can infer who performed the other two. As the example below shows, given that Clement paid the bill, the model is able to infer that he also ordered, and that he was returned his credit-card. In effect, the model has assigned Clement to the "payer" script role and has instantiated the character in each relevant proposition to be Clement.

**Input Text**

Albert and Clement, decided to go, restaurant
restaurant, quality, expensive
Clement, paid, bill, restaurant
He, tipped, waiter, restaurant, small

**Some Missing but Inferred Propositions**

restaurant, distance, far
Clement, ordered, cheap wine
waiter, returned, Clement, credit-card

Another regularity is that, in the corpus, expensive restaurants are always far away and cheap restaurants are always nearby. If the model has learned this regularity, it can use the explicit information that the restaurant is expensive to infer that it is far away. As the example shows, the model draws this conditional inference.

Finally, the model can draw inferences based on a previous experience with particular characters. The model uses previous experience with Clement to infer that he will order cheap wine.

Note that the model draws both coherence and prediction inferences. Both inference types are drawn by the same mechanism. Constraints from information explicit in the stories activates correlated information regardless of whether that information involves past or future events. The only limit on drawing inferences is the degree of correlation: as the correlations diminish, so do the activation of inferences.

**Revision.** Next, the model's ability to revise it's on-going interpretation of a text was tested (See Table 1.). In the bar text, the second and third propositions predict the fourth. Before either proposition is presented, rubbed lipstick and rubbed cheek are equally likely: 50%. The second proposition predicts the fourth proposition probabilistically. A polite pass predicts rubbed lipstick 70% and rubbed cheek 30%, while an obnoxious pass predicts rubbed lipstick 30% and rubbed cheek 70%. The third proposition predicts the fourth proposition absolutely: kissed predicts rubbed lipstick 100% and slapped predicts rubbed cheek 100%. These probabilities derive from the frequency of training of the various scenarios.

Table 1 shows the activations of lipstick versus cheek in the fourth proposition averaged across three test cases per cell. The model revises its activations as evidence accumulates roughly in accord with the probabilities.

The contradictory cases (polite-slapped and obnoxious-kissed) show that the model assigns greater weight to the more reliable third proposition and reverses its predictions accordingly. The conflict in the evidence, however, is visible in the moderate activation values.

**Table 1**
**Successive Revision in the Bar Script**

Andrew decided to go to a bar.
He made a polite/obnoxious pass at Roxanne.
Roxanne kissed/slapped him.
Andrew rubbed lipstick/cheek.

| **Propositions** | | | | |
|---|---|---|---|---|
| 2nd | none | polite | polite | polite |
| 3rd | none | none | kissed | slapped |
| **Activations** | | | | |
| lipstick | .7 (.5)* | .6 (.7) | .7 (.9) | .2 (.0) |
| cheek | .2 (.5) | .2 (.3) | .1 (.0) | .5 (.9) |

| **Propositions** | | | | |
|---|---|---|---|---|
| 2nd | none | obnoxious | obnoxious | obnoxious |
| 3rd | none | none | slapped | kissed |
| **Activations** | | | | |
| lipstick | .7 (.5) | .3 (.3) | .1 (.0) | .6 (.9) |
| cheek | .2 (.5) | .5 (.7) | .7 (.9) | .2 (.0) |

*Probabilities are in parentheses.

**Novel texts.** Finally, how might novel stories be understood? Under the constraint satisfaction view, the comprehension of novel stories simply consists of combining novel sets of constraints to produce an interpretation of a novel text. The constraints on a text interpretation can be divided into two categories: direct constraints and associative constraints. Direct constraints represent the correlations of information with itself, and they are responsible for the recall of the information explicitly mentioned in a text. Associative constraints represent the correlations between pieces of information, and they are responsible for inferences and pronoun resolution.

For novel stories, direct constraints from each part of the text are used to activate parts of the story gestalt. The difficulty for novel stories is that each explicit proposition, through its associative constraints, actually constrains many parts of the text in context specific ways. When propositions are placed in new combinations, many of these correlated parts of the text interpretation will contradict each other. The difficulty, then, is keeping the associative constraints, that activate the correlated parts, from overriding the direct constraints.

These ideas were tested in a new corpus pared down to the essential details. First, a corpus was designed to demonstrate the ability of strong associative constraints to override information in novel stories. This corpus contained only two stories: one about an expensive restaurant and one about a cheap restaurant. The stories were four propositions long. Each proposition contained a role that could take on one of two values. The cheap restaurant had the values cheap, near, change, and small. The expensive restaurant had the values expensive, far, credit-card, and large.

**Organization of the New Restaurant Stories**
restaurant, quality, {cheap, expensive}
restaurant, distance, {near, far}
Andrew, paid, bill, restaurant, {change, credit-card}
Andrew, tipped, restaurant, {small, large}

Once trained, the model was tested with stories containing combinations of role values the model had not seen: e.g. an expensive restaurant where Andrew pays with change. The model strongly overrides change and activates credit-card. The associative constraints do not allow the irregular role value to be activated by the much weaker direct constraint from change.

A fresh model was then trained on an expanded corpus. The expanded corpus again contained the two restaurant stories. In addition, it contained stories in four other locations. Each of these stories again consisted of the same four propositions, except that the context was changed to the new location. In each of these four new locations, all 16 of the possible combinations of role values ($2^4$) were trained. With this new corpus, the model cannot rely on associative constraints to predict one part of a text from the others because there are no correlations between role values. Instead, the model must learn strong direct constraints to be able to comprehend these stories. The hope is that the stronger direct constraints will override the associative constraints in the restaurant context, and allow the novel restaurant stories to be recalled accurately.

Once trained, the model was again tested with the expensive-restaurant-change text. The model now can represent the text despite never having seen that combination of role values before, and despite the strong regularities in the restaurant context fighting against that combination. The direct constraints win, but the activation of change is reduced due to the competition from the associative constraints. Interestingly, if the proposition about paying is missing, the associative constraints are applied, and the model infers credit-card, the regular value.

What the model has learned from this corpus, then, is something like a variable slot that has a default value. The model will recall the actual input, but if the input is missing, it will infer the default. However, the default is felt even when a value is provided in the input. These "graded slots" occur where constraints compete according to their strengths.

Of course sharing knowledge runs in both directions. While the restaurant context gains the use of the direct constraints, the other contexts gain the constraints among role values. When a value is missing in one of these contexts, the script-appropriate default value from the restaurant context is inferred. So, in an expensive beauty salon where the tip was large, the model will infer that a credit-card was used.

## Conclusions: Limitations and Abilities

The model successfully performs a number of knowledge intensive comprehension processes, but several of the model's specific features are problematic. One problem is that the propositional representation is clumsy and will not scale. First, each text can use a particular predicate only once because the predicates must be used as unambiguous questions for training. Second, the seven thematic roles in the propositions are both arbitrary and confining. Ideally, the model would learn its own representation for propositions just as it learns its own representation for whole texts.

Another problem is that learning is very slow due, in part, to the depth and size of the network. The error must be passed back through several layers of weights, and each layer contains a large number of weights, so credit assignment is difficult.

A larger issue here is that the model has no intrinsic understanding about what can and cannot occur together in a text. It does not intrinsically know that you cannot be at a beach and an airport at the same time. The model can acquire this knowledge only from the fact that it never occurs in its training corpus.

In spite of these difficulties, the model performs well on the comprehension tasks and demonstrates the utility of a constraint satisfaction approach for text comprehension generally, and for modeling psychological processes specifically. Weak constraint satisfaction brings several new ideas to our understanding of text comprehension. The most basic idea is that the interpretation of the text is supported by constraints from the text rather than build from it by appending successive propositions to some structure. The constraint view carries with it the idea that resolving pronouns, drawing inferences, and representing the explicit text result from the same process: the constraining of an interpretation. Each constraint supports or inhibits features of the story gestalt: it constrains what the gestalt will represent. The gestalt that results from applying all of the constraints is the one that satisfies these constraints the best. Inferences and pronoun resolution occur when one part of the text provides constraints on other parts that were missing from the text. This process is more sophisticated than simple pattern completion since it is capable of binding characters to their roles in a script.

A second idea is that these constraints can be graded, and their support of an interpretation can be graded. Pieces of text often may be only partially reliable as constraints on an interpretation. A polite or obnoxious pass, for instance, is only partially reliable at predicting the outcome in the bar texts. Allowing constraints to take on strength values according to their reliability, therefore, allows the model to utilize even partially reliable information.

The constraint view also carries with it the idea that each constraint makes its own contribution to the interpretation. Each new constraint adds its own support to the evolving interpretation. This independence of constraints allows the model to update its interpretation as each new constraint is processed.

The idea of independent constraints also shows how novel stories can be understood. The direct constraints from each part of the input add their own support and influence to the interpretation. The associative constraints from each part add extra information, but they may support conflicting information. Misremembering a text to better fit prior experience can be one result of conflicting associative constraints, as Bartlett's (1932) study demonstrates.

The question as to whether a system should remember or misremember depends, in the model, on the relative strengths of the direct and associative constraints. Stronger associative constraints produce regularizations. The corpus of training examples has a strong influence on the constraints that are learned. A wide range of examples is important for weakening associative constraints and strengthening direct constraints, as the simulation experiments with the restaurant stories showed.

Finally, the graded strengths of constraints allows them to be learned incrementally with experience. The model does not have to decide all-or-none whether a constraint is valuable. It simply assigns a strength to that constraint commensurate with its reliability for comprehension.

To sum up, weak constraint satisfaction is a useful approach to knowledge intensive processes in text comprehension. Some of the difficulties found in other models of text comprehension are ameliorated by the qualities of this approach. There is much to text comprehension that the model does not yet do, and the model faces some difficulties in learning speed and architecture, but the fundamental characteristics of the approach are successful and promising.

## References

Allen, R. B., (1987). Several studies on natural language and back-propagation. *Proceedings of the International Conference on Neural Networks, vol. 2*, p. 335-341, June 21-24, 1987, San Diego, CA.

Bartlett, F. C. (1932). *Remembering: An experimental and social study.* Cambridge: Cambridge University Press.

Carpenter, P. A. & Just, M. A. (1977). Reading comprehension as the eyes see it. In M. A. Just & P. A. Carpenter (Eds.), *Cognitive processes in comprehension.* Hillsdale, NJ: Erlbaum.

Charniak, E. (1983). Passing markers: A theory of contextual influence in language comprehension. *Cognitive Science*, 7, 171-190.

Corbett, A. T., & Chang, F. R. (1983). Pronoun disambiguation: Accessing potential antecedents. *Memory and Cognition, 11*, 283-294.

Dolan, C. P. & Dyer, M. G. (1989). Parallel retrieval and application of conceptual knowledge. In D. Touretzky, G. Hinton, & T. Sejnowski (Eds.), *Proceedings of the 1988 Connectionist Models Summer School.* San Mateo, CA: Morgan Kaufmann.

Graesser, A. C. (1981). *Prose comprehension beyond the word.* New York, NY: Springer-Verlag.

Hirst, W. & Brill, G. A. (1980). Contextual aspects of pronoun assignment. *Journal of Verbal Learning and Verbal Behavior, 19*, 168-175.

Kintsch, W. (1988). The role of knowledge in discourse comprehension: A construction-integration model. *Psychological Review*, 95, 163-182.

McKoon, G. & Ratcliff, R. (1986). Inferences about predictable events. *Journal of Experimental Psychology: Learning, Memory, and Cognition*, 12, 82-91.

Miikkulainen, R. & Dyer, M. G. (1989b). A modular neural network architecture for sequential paraphrasing of script-based stories. In the *Proceedings of the International Joint Conference on Neural Networks.* Piscataway, NJ: IEEE

Potts, G. R., Keenan, J. M., Golding, J. M. (1988). Assessing the occurrence of elaborative inferences: Lexical decision versus naming. *Journal of Memory and Language*, 27, 399-415.

Rumelhart, D. E. (1981). Understanding understanding. In H. W. Dechert and M. Raupach (Eds.), *Psycholinguistic models of production.* Norwood, NJ: Ablex Publishing.

Rumelhart, D. E., Smolensky, P., McClelland, J. L., & Hinton, G. E. (1986). Schemata and sequential thought processes in PDP models. In D. E. Rumelhart, J. L. McClelland, and the PDP Research Group (Eds.), *Parallel distributed processing: Explorations in the microstructure of cognition, Volume 2.* Cambridge, MA.: MIT Press.

Schank, R. C. (1981). Language and memory. In D. A. Norman (Ed.) *Perspectives on cognitive science.* Norwood, NJ: Ablex Publishing.

St. John, M. F. & McClelland, J. L. (1990). Learning and applying contextual constraints in sentence comprehension. *Artificial Intelligence, 46*, 217-257.

# Encoding Images into Constraint Expressions

**Kazuo Hiraki, John Gennari, Yoshinobu Yamamoto, Yuichiro Anzai**
Department of Computer Science, Keio University
3-14-1, Hiyoshi Kohoku-ku Yokohama 223 Japan
E-Mail: {hiraki,gennari,yoshinov, anzai}@aa.cs.keio.ac.jp

**ABSTRACT**

This paper presents a method, **generalization to interval**, that can encode images into symbolic expressions. This method generalizes over instances of spatial patterns, and outputs a *constraint program* that can be used declaratively as a learned concept about spatial patterns, and procedurally as a method for reasoning about spatial relations. Thus our method transforms numeric spatial patterns to symbolic declarative/procedural representations. We have implemented **generalization to interval** with ACORN,[1] a system that acquires knowledge about spatial relations by observing 2-D raster images. We have applied this system to some layout problems to demonstrate the ability of the system and the flexibility of constraint programs for knowledge representation.

## 1. Introduction

Representation of spatial knowledge is an important task for intelligent agents. This task can arise in many domains: visual scene understanding, problem solving, robot navigation and so on. One important aspect of spatial knowledge is the use of a set of symbolic predicates that define spatial relations among objects in a scene. Classic system such as STRIPS [Fikes 71] or GPS [Newell 63] use symbolic predicates for representing spatial knowledge. For example, in the blocks world domain, such systems usually use primitive predicates such as: **on(block1, table)**, **right-of(block1,block2)**, and **top-of(block3)**. These primitives are an abstraction of the actual scene – they give only approximate information about the location of objects – but they are important abstractions for reasoning about objects in the environment.

However, in order to apply such system to real-world domains, one would need a perceptual system to provide the appropriate symbolic primitives for reasoning (see Figure 1a). Winston's ARCH system [Winston 75] and Connell's system [Connell 87] are designed in this fashion: a vision system translated images into a set of symbolic facts, which were then used by a concept learning system. This approach leads to some difficult questions for the perceptual system: the number of primitives is not known, nor methods for mapping continuous image information into symbolic primitives. Ideally, a perceptual system should be general-purpose, and not restricted to a particular set of symbolic predicates.

Figure 2 demonstrates one problem with using *a priori* primitives for concept learning. Suppose that a vision system has six primitives to represent the distance between two objects: **very very near, very near, near, far, very far, very very far**. Given two scenes of two objects **A** and **B**,their distances are **11** and **19**, as examples for target concept. The vision system encodes these values to predicates: {**very-near(A,B), near(A,B)**} and { **far(A,B)** }. Some learning systems simply apply a kind of *dropping condition rule* to these predicates. This would mean that the system could not describe both examples with a single concept (all conditions would be dropped), even if this may be an appropriate decision.

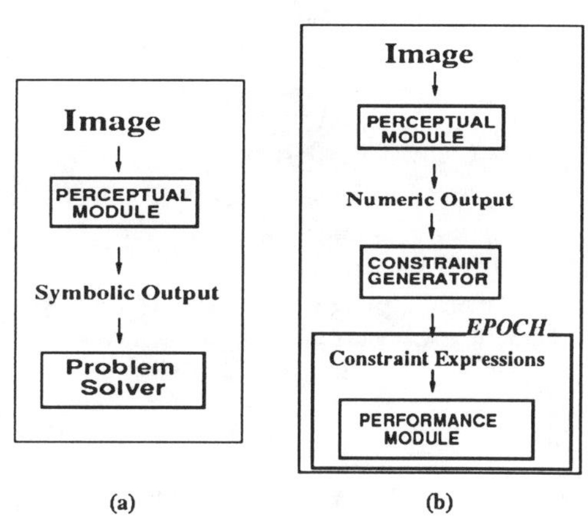


Figure 1: Datastream for classic problem solver (a) and ACORN (b).

[1] ACquisition Of Relations from Numeric data

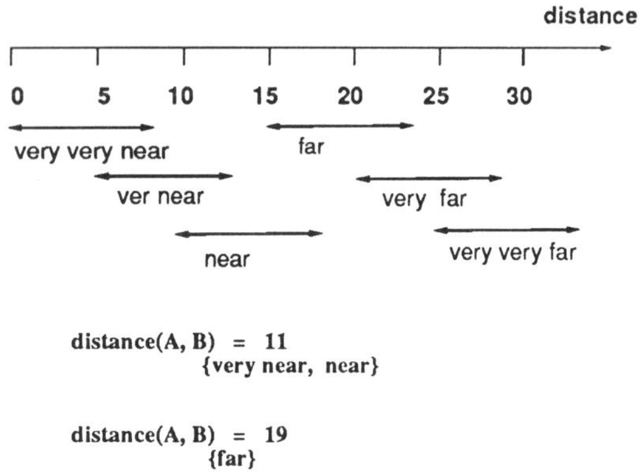


Figure 2: An inappropriateness of using *a priori* primitives

In contrast with defining *a priori* primitives, we assume all symbolic expressions are grounded in the environment. Primitive expressions should be learned from spatial patterns. According to this assumption, we have developed ACORN, a system that addresses the task of learning such expressions directly from visual scenes. In particular, our learning task can be described as follows:

Given: A set of positive training instances, where instances are 2-dimensional raster images of a scene of labeled objects.
Find: Constraint expressions among given images.

There are two important features of this task. First, it requires a translation from continuous, numeric information to more approximate symbolic knowledge.[2] The learned expressions should be invariant under translation, and should hold true over a variety of minor perturbations to the numeric coordinate input. Second, the learned knowledge should be useful for a variety of performance tasks. It should be possible to recognize examples, to generate scenes that satisfy a set of constraints, to make inferences from a set of relations, or to detect redundancy and inconsistencies in a set of relations. To this end, ACORN uses *constraint programming* [Sussman 80][Leler 88] to represent its acquired knowledge.

In this paper, we focus on layout problems to demonstrate ACORN's performance. The layout task has been studied in some fields ([Haar 82] [Yamada 90]). However, they lack the generality of ACORN's learning mechanism.

[2] Note that this is not the first system to learn symbolic information from numeric data. Aha's IBL system [Aha 91], Quinlan's C4 system [Quinlan 87] and the CART system [Breiman 84] are classifiers that work with numeric data. However, these have not been applied to spatial relations, and their learned knowledge structures have only been used for classification.

ACORN is a general-purpose system for learning symbolic relational predicates directly from numeric, image information. The next section describes our system more completely, focusing on knowledge representation and a description of the learning method. Section 3 describes the performance system and our experimentation thus far, and we conclude the paper with a discussion of future work with ACORN.

## 2. Description of ACORN

ACORN consists of three modules: the *perceptual module, the constraint generator* (or learning module) and the *performance module*. Figure 1b shows the data stream between these modules. The perceptual module generates initial, numeric scene descriptions directly from raster information. From these descriptions, the constraint generator learns a set of *constraint expressions* that describe the spatial relations between objects in the scene. Finally, the performance module can use these constraints to make predictions about a scene, make inferences with the learned relations, or generate scenes given a set of objects and relationships among those objects.

### Perceptual Module

Although ACORN uses numeric input, it does not work with images directly and requires an intermediate representation of objects. Currently, we have limited our domains to two-dimensional images of rectangles.[3] For these images, the perceptual module's task is to scan a scene of dark rectangles on light paper, identify the rectangles and output a numeric representation and an arbitrary label for each rectangle. Although nontrivial, this is well within the ability of current vision systems.

In more detail, the numeric representation of each rectangle includes:

- Coordinates of each vertex of the rectangle on the global x-y coordinate system.
- Coordinates of the center of the rectangle on the global x-y coordinate system.
- Object-centered coordinate axes (major axis and minor axis) for the rectangle.
- Height of rectangle.
- Width of rectangle.

Note that only the first item need be observed from the raster image; once the four vertices are known, all other representational data can be computed. There is nothing special-purpose about this representation; indeed, we expect that the system could use other types of representations for other shapes.

[3] In Section 4, we discuss the possibility of extending ACORN to work with a richer variety of shapes.

## Constraint Expression

ACORN represents learned knowledge with *constraint logic program*[Jaffar 87]. In general, constraint logic program consists of the set of *extended horn clauses* in the form:

$$Relation(\langle args\rangle) \leftarrow P_1 \wedge P_2 \wedge \ldots \wedge P_n \wedge C_1 \wedge C_2 \wedge \ldots \wedge C_m$$

where $P_i$ is a *term* (as in Prolog) and $C_j$ is an algebraic constraint expression such as an equation or inequality.

One of the most remarkable features of constraint-based language is that the user does not have to write a procedure explicitly in order to solve constraints. The user has only to give *declarative* relations among variables. Therefore constraints can be used to make a variety of inferences. For example, the constraint

$$convert(C, F) \leftarrow (F = \frac{5C}{9} + 32)$$

can be used to make predictions either about $F$ given $C$ or $C$ given $F$, without requiring separate rules. In general, constraint expressions make use of a powerful *constraint solver* that is used to make inferences or solve problems. This approach lets declarative knowledge be used procedurally, allowing ACORN to reason about its learned spatial relations in a variety of tasks.

One of the advantages of using a constraint logic programming language is that semantics can be formulated more easily than other constraint-based languages[Sussman 80][Leler 88][Borning 81], using logic programming semantics (see [Jaffar 87] for further discussion). Constraint logic programming has proved valuable in a number of domains, such as option trading[Lassez 87] and scheduling [Dincbas 88].

ACORN uses EPOCH, a constraint logic programming system developed in our laboratory.[4] EPOCH's constraint solver uses the *simplex algorithm* to solve linear programming problems. EPOCH is similar to CLP($\Re$) [Lassez 87] in its ability as a constraint solver. However, EPOCH uses two types of constraints: *necessary constraints* that must be satisfied and *preferred constraints* that should be, but do not have to be satisfied. Preferred constraints are used to choose an optimal solution when there is more than one possible answer (see [Hiraki, in press] for more details).

## Learning Method

As described in the previous section, learned knowledge is represented with constraint expressions. ACORN uses binary constraints – it looks at information about a pair of rectangles. Note that these can be used to model n-ary spatial relations, simply by learning about all $C_2^n$ pairs of objects in the scene. Currently, the learning module of ACORN takes the information from the perceptual module and generates constraints based on the following *relational attributes*:

[4] Both EPOCH and ACORN are implemented in Quintus Prolog.

- distance ($\gamma$) between corresponding vertices of the objects ($0 < \gamma$).
- direction ($\theta_{ij}$) from $vertex_i$ to $vertex_j$ ($0 < \theta_{ij} \leq \pi$).

Certainly, other relational attributes could be used (angle between major axes, relative height or width, etc.), but thus far, these attributes have proved sufficient.

As the system observes positive instances of the relation, it builds both necessary and preferred constraints. Necessary constraints represent the necessary conditions that any instance of the relation must satisfy. They are expressed by a set of ranges defined by algebraic inequalities. For example, the necessary constraints for "point a is right-of point b" is expressed by the fan-shape region illustrated in Figure 3a. Because EPOCH's constraint solver only works with linear problems, we approximate this region with four linear inequalities.[5] EPOCH incorporates these inequalities in a constraint program that represents the fan-shaped region; Figure 3b shows the program in Prolog form. The four inequalities (the necessary constraints) correspond to four lines delimiting the region used to define "right-of". The satisfying region shrinks as the number of constraints increases (e.g. by adding "point b is right-of point c"). If the region becomes null, the constraints cannot be satisfied.

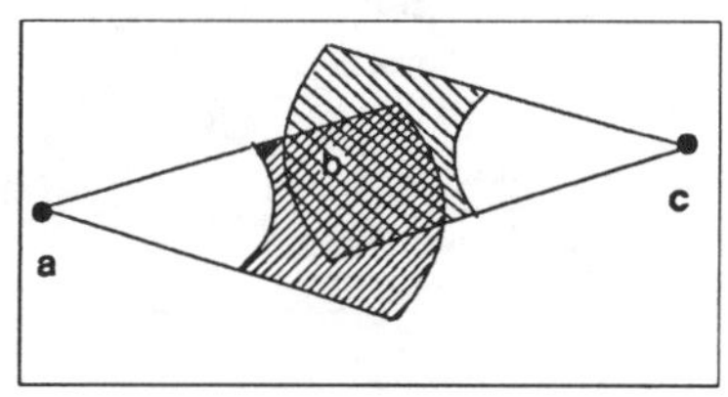


(a)

```
constraint( a([Xa,Ya]),b([Xb,Yb]) ) :-
        X = (Xb - Xa), Y = (Yb - Ya),
        Y =< 20.0*X - 100.0,
        Y >= -0.6*X, Y =< 0.4*X,
        Y >= 20.0*X - 200.0,
        pref:(X = 7.5), pref:(Y = 0.0).
```

(b)

Figure 3: (a) A region defined by some constraints. (b) A simple constraint program for the region.

On the other hand, preferred constraints represent a typical member of the learned spatial relation. In Figure 3b, *pref:* indicates a preferred constraint. ACORN uses preferred constraints for finding the optimal location of an object given a region that satisfies the necessary constraints of the learned relation. Preferred constraints are created by computing averages over the set of training instances.

[5] There are constraint solvers that can deal with nonlinear forms, but these are more expensive [Sakai 89].

In order to learn the necessary constraints for a spatial relation, ACORN uses a simple learning rule as it observes training instances. This rule is an application to constraint programming of Michalski's "closing interval generalization" heuristic [Michalski 83]; we will refer to this as **generalization to interval:**

> For each relational attribute, given numeric values, $x_1$ and $x_2$, from two positive training instances, assume that all numeric values between $x_1$ and $x_2$ are also examples of positive instances.

This can be used directly with linear attributes such as distance, but for cyclical attributes, such as direction defined as an angle, two positive instances produce a pair of intervals. In this case ACORN adopts the heuristic of generalizing over the smaller interval.

Table 1 presents the incremental algorithm for a single linear attribute in more detail. Given the old constraint for an attribute, $C_j$, and the value of the attribute for the new positive instance, $v_j$, this algorithm returns a new constraint, $C'_j$. A constraint, $C$, is defined by the bounding interval $[\alpha, \beta]$.

Table 1: ACORN's learning algorithm for a linear relational attribute.

**generalization-to-interval**($C_j, v_j$)
  **If** $v_j$ satisfies the constraint $C_j$
    **Then** return $C'_j := C_j$
  **Else If** $v_j \leq \alpha$ **Then** return $C'_j := [v_j, \beta]$
  Else return $C'_j := [\alpha, v_j]$

This learning method is only an initial heuristic; it is a simple method, and has a number of drawbacks. One problem occurs with negative training instances: it is unclear how to modify constraints when the negative instance includes values between $\alpha$ and $\beta$. Additionally, $\alpha$ and $\beta$ are hard thresholds: after learning, a test instance with values $\alpha - \varepsilon$, or $\beta + \varepsilon$ does not satisfy the constraint. Especially in noisy domains, one would prefer a more flexible approach. Despite its limitations, we have found that even this simple form of learning can be very useful for recognizing and reasoning with spatial relations.

## 3. Performance of ACORN

As we described earlier, the advantage of using constraint expressions is that the knowledge can be used in a variety of performance tasks without requiring special purpose representations or rules. If knowledge about spatial relations is represented as a set of constraint expressions, the performance system should be able to recognize examples of a given relation, generate scenes that satisfy a set of relations, make inference from a set relations, or detect redundancy and inconsistencies in a set of relations.

(a)



Figure 4: Face domain input for ACORN: (a) Images (b) Partial list of training relations.

In this section, we describe ACORN's ability with two performance tasks. We begin with the layout task: generating scenes from a set of relational specifications. Next, we consider a variation on this task: completing partial scenes. In this task, the system learns a set of constraints that represent the spatial relations among a set of objects in a scene. Then, the system can use these constraints to predict the location of missing objects.

### The Layout Task

For the layout problem, we use the simple task of arranging a set of six rectangles into a 'face'. Figure 4a shows the training instances given to ACORN. In order to learn relations, the system needs the relation names, as well as labeled scenes. Thus, the rectangles in these figures are labeled *eye1*, *eye2*, *eyebrow1*, *eyebrow2*, *nose*, *mouth* and there are two relations given: **above(X,Y)** and **right-of(X,Y)**. Figure 4b gives a partial list of the relations that were associated with each scene. ACORN uses these as positive training instances for building the necessary constraints that define the spatial relation. In particular, it builds a *constraint program* for each relation as follows:[6]

$$\begin{aligned}&constraint(Relation\text{-}name,\\ &\quad [Vertex_1^1, Vertex_2^1, Vertex_3^1, Vertex_4^1],\\ &\quad [Vertex_1^2, Vertex_2^2, Vertex_3^2, Vertex_4^2]) \leftarrow\\ &necessary\_constr_1, \cdots, necessary\_constr_n,\\ &preferred\_constr_1, \cdots, preferred\_constr_m.\end{aligned}$$

Note that the two arguments to this relation are represented as lists of vertices.

After building constraint programs for the relations **above** and **right-of**, ACORN can construct any arrangement of rectangles given a specification in terms of the two acquired relations. For example, given the specification:

**{right-of(eye2,eye1),**
**above(eyebrow1,eye1),**

[6] We use standard Prolog notation here.

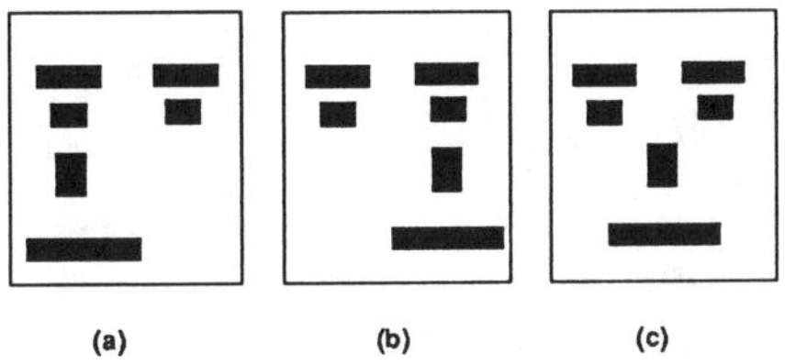


Figure 5: Faces generated from specifications.

**above(eyebrow2,eye2),**
**above(eye1,nose), above(nose,mouth)}** ,
ACORN generates the scene in Figure 5a, while
**{right-of(eye2,eye1),**
**above(eyebrow1,eye1),**
**above(eyebrow2,eye2),**
**above(eye2,nose), above(nose,mouth)}** ,
creates the scene in Figure 5b. Note that the nose and mouth are 'off-center' in these scenes because the preferred constraint for **above** is straight above. If we add the constraints **right-of(eye1,nose), right-of(nose,eye2)** to the specification, we get the 'corrected' scene as in Figure 5c.

## Completing Partial Scenes

In addition to binary relations, ACORN can learn more complex n-ary spatial relations. As described in Section 2, the system simply models $n$-ary relations by building constraints for every pair of arguments. Although this means that there is a computational limit to the number of arguments in a relation (there are $O(n^2)$ constraints needed for an $n$-ary relation), we imagine that most useful relations do not have too many arguments. For example, the faces from the previous section can be used to learn a six-argument spatial relation: **face(eye1,eye2,eyebrow1,eyebrow2,mouth,nose)**.

Since there are six components of each face, there are 15 constraints that make up the constraint program for learning the **face** relation. Each constraint must include the component labels, but is otherwise similar to that presented in the previous section:

$$\begin{array}{l} constraint( \\ obj\text{-}name1([Vertex_1^1, Vertex_2^1, Vertex_3^1, Vertex_4^1]), \\ obj\text{-}name2([Vertex_1^2, Vertex_2^2, Vertex_3^2, Vertex_4^2])) \\ \leftarrow \\ \quad necessary\_constr_1, \cdots, necessary\_constr_n, \\ \quad preferred\_constr_1, \cdots, preferred\_constr_m. \end{array}$$

(The relation name can be omitted since the system is learning only one relation.)

After acquiring a definition for this relation, it can be used in a number of ways. First, scenes can be 'recognized' as satisfying the constraints that define a face. Alternatively, the constraints can be used to predict the location of missing objects in a 'face' scene. That is, given a partial face, and knowledge about the **face** relation, ACORN can complete the scene. Figure 6 shows a pair of incomplete 'faces', and the resulting scenes as completed by ACORN. This task is similar to the layout problem: the system uses the preferred constraints to generate the location of the missing object.

Figure 6: Two incomplete faces, and their completion by ACORN.

In general, ACORN can predict the location of any number of missing components. As more and more components are missing, the necessary constraints become looser and looser. If all components are missing, then the system uses only the preferred constraints, and this becomes a form of the layout task.

This is a simple demonstration of the generality of ACORN's acquired knowledge, and the versatility of constraint expressions. Once the system has learned a few key spatial relations, it can use this knowledge in a variety of domains and performance tasks.

## 4. Discussion and Future Work

As described in the above section, ACORN can perform lots of tasks with acquired constraint programs. However, ACORN includes a number of limitations that should be addressed in the future. In this section, we discuss the limitations of ACORN's learning method and indicate the directions in which the research is proceeding.

One task that seems important is to improve the learning algorithm. The system should be able to learn from negative instances and build constraints that define disjunctive spatial relations. For examples, the spatial relation **next-to(X,Y)** has to be represented by disjunctive constraint expressions. Hiraki and Anzai [Hiraki 90] describe an initial version of ACORN that was able to learn from negative instances, but only with user input, and objects that were represented as points.

A second area for improvement is the range of input images. ACORN should be able to work with triangles, circles or any concave polygon, rather than only simple rectangles. One possibility is to use 2-D generalized cylinders; these provide a consistent method for representing a variety of shapes. The potential difficulty of this approach is that it greatly increases the number of relational attributes between each pair of objects, and thus the complexity of the constraint programs. For this problem, we should consider methods that can find the importance of relational attributes of a spatial relation and then reduce the number of given attributes.

Ultimately, knowledge about spatial relations should be used by higher-level systems such as problem solvers, natural language system, and navigation systems. In the future, we hope to connect ACORN with such a system. For example, Yamada's SPRINT[Yamada 90] system reconstructs spatial configurations from a given natural language description, can be a good application for ACORN. SPRINT uses hard-coded constraint expressions corresponding to special words that have spatial information. ACORN may be used to learn the meaning of such words and any new spatial concepts that may be used. In general, our system may be useful as an initial phase for any of the higher-level uses of spatial knowledge: route knowledge, spatial maps, and path planning.

We believe that ACORN addresses a critical task for an intelligent learning system: acquiring symbolic relational knowledge from numeric image data. Certainly, our current system includes many limitations, but we hope that it is a good start toward a general-purpose learning system for spatial relations.

## Acknowledgment

This research was supported in part by a grant from IBM Japan. The authors would also like to acknowledge Pat Langley for his comments during the initial stages of this paper.

## References

[Aha 91] Aha, D.W., Kibler, D. and Albert, M.K.: Instanced-based learning algorithms, *Machine Learning*, **6**, 37-66, 1991.

[Borning 81] Borning,A.: The programming language aspects of ThingLab: A constraint-oriented simulation laboratory. *ACM Transactions on Programming Languages and Systems*, **3**, 353-387, 1981.

[Breiman 84] Breiman, L., Friedman, J. H., Olshen, R. A. and Stone, C. J.:*Classification and Regression Trees*, Wadsworth International Group, 1984.

[Connell 87] Connell,J. H. and Brady,M.: Generating and generalizing models of visual objects, *Artificial Intelligence*, **31**, 159-183, 1987.

[Dincbas 88] Dincbas, M., Van Hentenryck, P., Simmonis, H., Aggoum, A., Graf, T. and Berthier, F.: The constraint logic programming language CHIP, *Proc. of the International Conference on Fifth Generation Computer Systems*, 693-702, 1988.

[Fikes 71] Fikes, R. E. and Nilson, N. J. : STRIPS: A new approach to the application of theorem proving to problem solving, *Artificial Intelligence*, **2**, 1971.

[Haar 82] Haar, R. L.: Sketching: Estimating object position from relational descriptions, *Computer Graphics and Image Processing*, **19**, 227-247, 1982.

[Hiraki 90] Hiraki, K. and Anzai, Y.: Knowledge acquisition for spatial knowledge-based system, *Proc. of 5th Knowledge Acquisition for Knowledge-Based Systems Workshop*, 1990.

[Hiraki, in press] Hiraki, K., Nishizawa, T. and Anzai, Y.: Acquisition and use of constraint expressions for spatial knowledge, (in Japanese) *Computer Software* (Journal of the Japan Society for Software Science and Technology), *in press.*

[Jaffar 87] Jaffar,J. and Lassez,J.: From unification to constraints, In Furukawa,K., Tanaka,H. and Fujisaki,T. (eds.), *Logic Programming '87, Lecture Notes in Computer Science*, Springer, 1987.

[Lassez 87] Lassez, C.: Constraint logic programming and option trading, *IBM Technical Report*, 1987.

[Leler 88] Leler, W.: *Constraint Programming Languages: Their Specification and Generation*, Addison-Wesley, 1988.

[Michalski 83] Michalski, R.: A theory and methodology of inductive learning, In Michalski, R., Carbonell, J. and Mitchell, T. (eds.), *Machine Leaning: An artificial intelligence approach.* Los Altos, CA: Morgan Kaufmann, 1983.

[Newell 63] Newell, A. and Simon, H. A.: GPS, A program that simulates human thought, In Feigenbaum, E.A and Feldman, J. (eds.), *Computers and Thought*, McGraw-Hill, New York, 1963.

[Quinlan 87] Quinlan, J. R.: Simplying decision trees, *Int. Journal Man-Machine Studies*, **27**, 221-234, 1987.

[Sakai 89] Sakai, K., and Aiba, A.: CAL: A theoretical background of constraint logic programming and its applications, *J. Symbolic Computation*, **8**, 589-603, 1989.

[Sussman 80] Sussman, G. J. and Steel, G. L.: CONSTRAINTS - A language for expressing almost-hierarchical descriptions, *Artificial Intelligence*, **14**, 1-39, 1980.

[Winston 75] Winston, P.H.: Learning structural descriptions from examples, In Winston, P.H. (ed.), *The Psychology of Computer Vision*, McGraw-Hill, 1975.

[Yamada 90] Yamada, A., Amitani, K., Hoshino, T., Nishida, T., Doshita, S.,: The analysis of the spatial description in natural language and the reconstruction of the scene, (in Japanese) *Journal of Information Processing Society of Japan*, **31**, No.5, 660-672, 1990.

# Imagery and Categories: The Indeterminacy Problem

**Thomas R. Ioerger***
Beckman Institute
University of Illinois
405 N. Mathews Ave.
Urbana, IL 61801
ioerger@cs.uiuc.edu

## Abstract

One of the classical problems faced by theories of mental imagery is the Indeterminacy Problem: a certain level of detail seems to be required to construct an image from a generating description, but such detail might not be available from abstract, categorical descriptions. If we commit to unjustified details and incorporate them into an image, subsequent queries of the image might indiscriminantly report not only information implied by the description but also information that was arbitrarily fixed. The Indeterminacy Problem is studied in a simplified domain, and a computational model is proposed in which images can be incrementally adjusted to satisfy a set of inter-constraining assertions as well as possible. In this model, queries can discriminate between those details in an image which are necessary (implied by the generating description) and those which are incidental (consistent but arbitrarily fixed). The computational model exploits the graded prototypicality of the categorical relations in the simplified domain, and suggests the importance of a grounded language for reasoning with categories.

## Introduction

Although the psychological validity of mental imagery has long been debated, images have often served usefully for modeling knowledge in computational systems (Funt 1980; Waltz & Boggess 1979). One of the arguments against imaginal representations and for propositional ones has been that images seem to possess a certain level of detail in representation that cannot always be provided by an abstract description being used to generate the image (Pylyshyn 1973). For example, in imagining a tiger, people often report that the tiger has stripes but they cannot say how many (Kosslyn 1980). Yet if an image of a striped tiger had been generated, it is argued, surely its stripes could have be counted. If default information had been used to fill in the required details, the image would have been unjustly over-committed; the necessary information in the image would have been indistinguishable from those details which were consistent but arbitrary. Pinker (1984) claims that this problem "would speak against a totally factored structural description," suggesting that imaginal representations are inherently incapable of handling such indeterminacy. We call this the Indeterminacy Problem.

The Indeterminacy Problem has been acknowledged in at least one computational system that uses images for reasoning (Waltz & Boggess 1979). Boggess's computer program takes a sequence of (restricted) natural language sentences as input, constructs an internal image of the spatial relationships implied by the prepositional phrases, and answers questions about these spatial relationships by examining the image. Nouns like "table," "box," and "goldfish bowl" are drawn as right parallelpipeds with default height, width, and depth in a three-dimensional bitmap representations of space.[1] Prepositions like "on" and "in" are realized as procedures for placing and locating objects in the image relative to other objects.

In Boggess's program, the meaning of a preposition can be sensitive to the kinds of objects being related, so a shadow can be "on" a wall in a different way than a book can be "on" a table. Nevertheless, for any instantiated proposition, the program has default knowledge for drawing and evaluating images. For example, to draw a book on a table, a specific spot which "tends to a particular corner" of the table is used. This of course simplifies the task of finding the book on the table later, but the use of such default information can lead to some embarrasing conclusions. Consider Waltz and Boggess's own example: the program is given "The shelf is on the wall" and "The fly is on the wall" as input. Using default knowledge, the program constructs an acceptable model in which the fly is below the shelf. When the program is queried with "Is the fly below

*This material is based upon work supported under a National Science Foundation Graduate Fellowship and ONR grant N00014-88K0124.

[1]Bitmaps are quantized representations of space and its occupation by objects. In this paper, a Cartesian coordinate system will be assumed.

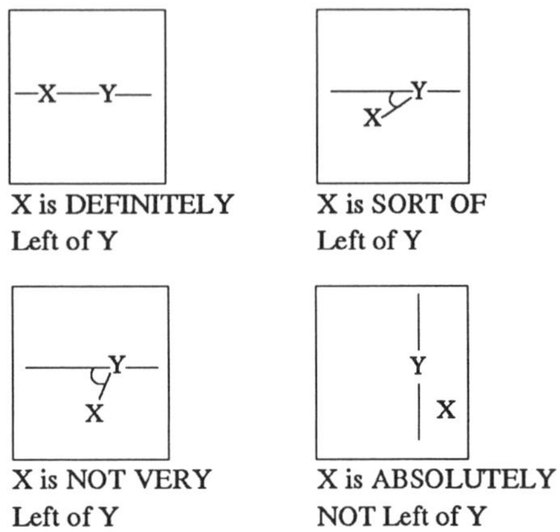


Figure 1: The categorical preposition Left has many meanings, which are distinguished by the use of various hedges in natural language.

the shelf?" it responds affirmatively. Yet the program was not told this directly, nor was it implied by the input. The information was derived as an artifact of using defaults to construct the image; this is a direct demonstration of the Indeterminacy Problem.

## A Simplified Domain

To study the Indeterminacy Problem, we define a simplified domain in which categorical relations are used to generate and query images. The domain consists of three structureless points, labelled X, Y, and Z, initially assumed to exist in some arbitrary spatial arrangement on a plane. We decompose the use of images into two processes that interact. There is a client process which can make a sequence of (possibly interleaved) assertions and queries to an image server in a propositional language. The language consists of three constants–X, Y, and Z–and four two-place predicates–Left, Right, Above, and Below. The image server must update the image when given assertions, and must derive the answers to queries by examining the image (as opposed to proving theorems with the assertions).

The categorical nature of the four relations can be explored by first making some intuitive observations on how people might hedge descriptions in this simplified domain (see Figure 1). There is a technical sense of the Left relation between two points, say X and Y: X is "definitely" left of Y if X is on the left half-plane with respect to Y and the line XY is horizontal. However, if the absolute value of the angle between the line XY and the horizontal is small but non-zero, we might say that X is "sort of" left of Y. As the line XY approaches verticality, we might say that X is "not very" left of Y. And finally, if X is anywhere on the right half-plane with respect to Y, then X is "absolutely not" left of Y. The other relations can be defined symmetrically. Note that these definitions do not depend on the distance between X and Y and are invariant under translation. Also notice the internal symmetry in each relation; for example, given any image, X is just as much to the left of Y as the image in which X is flipped across the horizontal passing through Y.

Lakoff (1973) has shown that the use of hedges with propositions constructed from a particular predicate strongly suggests that the propositions should be allowed to take on not only True and False as truth values, but also intermediate degrees of truth. If we follow the conventions of Fuzzy Logic (Zadeh 1965) and map truth values into the unit interval, [0,1], where 0 represents falsity and 1 represents truth in their classical senses, the intermediate values can be interpreted as prototyicality ratings–estimates of the closeness of some state of the world (in this case, modeled by an image) to the "central meaning" of the categorical predicate (Smith & Medin 1984). Given these two representation languages–images and fuzzy-truth-valued propositions–we must specify how they are related so that asserted propositions can be used to update images and so images can be used to answer queries. Specifically, for any instantiated proposition, such as '(Left X Y), we must describe how images are related to truth values.

The examples of hedging in Figure 1 suggest that the prototypicality rating of an image for the the proposition '(Left X Y) is a function of only the absolute value of the angle between the line XY and the horizontal (based on independence from distance and on internal symmetry).[2] Prototypicality is highest when the angle is zero, decreases monotonically as the absolute value of the angle increases to 90 degrees, and is essentially zero when X is in the right half-plane with respect to Y (although there are still slight differences in distances from the prototypical Left relationship). Figure 2 describes a method for computing the truth value of an image for the proposition '(Left X Y) using a function LEFT that has these properties. The other relations can be computed similarly to the Left relation by first rotating the image. The negation of a proposition with a fuzzy truth value q is defined to be the same proposition with the truth value 1-q (Zadeh 1965).

This method for computing prototypicality only models the "perception" of images in terms of fuzzy-truth-valued propositions. The inverse relation is one-to-many because, for any instantiated proposition and

[2] We assume that the hedges in Figure 1 roughly reflect prototypicality. That is, for any proposition P, an example of "definitely" P is more prototypical than an example of "sort of" P, which is more prototypical than an example of "not very" P, which is more prototypical than an example of "absolutely not" P. See Lakoff (1973) for a formal discussion of the relationships of hedges to prototypicality based on possibility distributions.

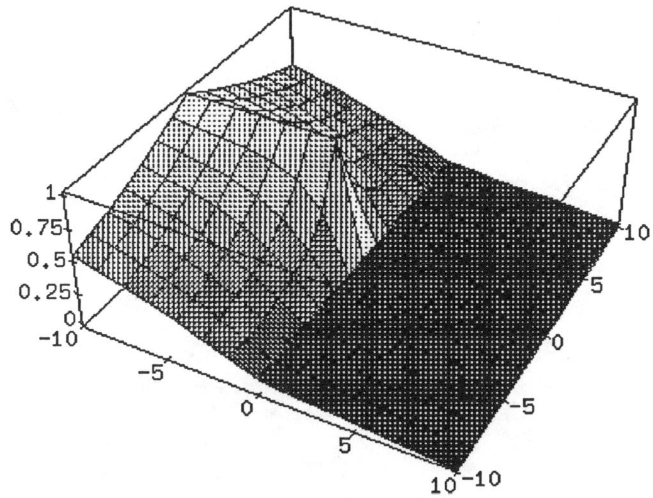


Figure 2: To compute the truth value of '(Left X Y) for a particular image, translate both X and Y so that Y is at the center of the image. Then apply a function LEFT, which might look like this plot but must be investigated empirically, to the translated location of X. The number that is returned represents the prototypicality of the image as an example of '(Left X Y). The function LEFT decreases monotonically as the absolute value of the angle between the line XY and the horizontal increases from 0 to 90 degrees. There is a slight gradient in LEFT when X is in the right half-plane with respect to Y because, eventhough X is "definitely not" left of Y for any of these positions, they are not equivalent; some relative positions of X are closer to the prototypical Left relationship with Y than others.

some truth value, there are many equivalent images which differ in distance between the arguments or which are symmetric. Also note that the "meaning" of a proposition becomes a possibility function among images such that the distribution reflects prototypicality. This property will later be exploited to generate default information when updating images with asserted propositions.

In this simplified domain, queries should return one of four values:

- YES means the queried proposition was perviously asserted verbatim, or the image is sufficiently constrianed by the previous assertions that the queried proposition must be true (have an acceptably high truth value). For example, after asserting '(Above X Y) and '(Above Y Z), the query '(Below Z X) should return YES.
- NO means the previous assertions sufficiently constrain the image so that the negation of the queried proposition must be true. For example, after asserting '(Above X Y), the query '(Above Y X) should should return NO.
- UNSPECIFIED indicates that the assertions have not constrained the model enough to favor either the queried proposition or its negation; both are consistent with the previous assertions. For example, initially X, Y, and Z are in some arbitrary arrangement. Since no assertions have been made, any query should return UNSPECIFIED.
- CONTRADICTION should be returned if the previous assertions contained two contradictory propositions, like '(Above X Y) and '(Below X Y), or were similarly overconstrained to make any sense.

## A Computational Model

We propose two procedures as a computational model of how the image server can process a sequence of assertions and queries and still avoid the Indeterminacy Problem. Consider the case when the image server is given an assertion that is categorical and has some choice as to how to update the image to satisfy the assertion. As a working example, suppose we first assert '(Above X Y). As long as X is on the upper half-plane with respect to Y, the image would satisfy the assertion somewhat. But, since there are no other constraints, the best image is one in which X is directly above Y. Even if we choose this spatial arrangement we must arbitrarily choose a distance between X and Y to construct the image. To see how these choices can be defeated with additional constraints, suppose we now assert '(Right X Y). As long as X is on the half-plane to the right of Y, the image would satisfy this new proposition somewhat, and the way to satisfy the proposition in the most prototypical sense would be to have X directly to the right of Y. But these constraints interact; intuitively, the most appropriate spatial arrangement in which X is both above and to the right of Y has X on a 45 degree angle above the horizontal from Y. Thus we have to modify some of the details that we arbitrarily chose for modeling the first assertion.

Figure 3 shows the Adjustment Procedure, which can be used by the image server to handle a sequence of assertions properly. The Adjustment Procedure assumes the image server is maintaining a record of the previous assertions and has a current image which models the previous assertions as well as possible. To determine how well an image models a set of propositions, we take the minimum[3] prototypicality rating of the image for any previously asserted proposition; this is called the evaluation of the image with respect to the set of propositions. When an assertion is made to the image server, it calls the Adjustment Procedure, passing to it the set of previous assertions, the new assertion, and the current image. The image server resets the current image to the one which is returned by

[3]This is a standard aggregator function for conjunctions in Fuzzy Logic (see Zadeh 1965).

1. Generate a set $\Phi$ of adjusted images.
2. If no image in $\Phi$ has a better evaluation with respect to $\Gamma \cup \{P\}$ than I, then return I.
3. Else reset I to some image J in $\Phi$ such that no other image in $\Phi$ has a better evaluation with respect to $\Gamma \cup \{P\}$, and goto step 1.

Figure 3: The Adjustment Procedure takes as input a set of propositions $\Gamma$, a new proposition P, and a starting image I, which is expected to be the best model of $\Gamma$ that could be found. The procedure returns an updated image that models the propositions $\Gamma \cup \{P\}$ as well as possible.

the Adjustment Procedure, and it also adds the new assertion to the set of previous assertions.

Rather than search all possible spatial arrangements for an image for which no other image has a better evaluation, it is a central proposal of the Adjustment Procedure that the search proceed in a directed and incremental way. The only images that are considered are the ones constructed by adjusting the position of the first argument of any least-satisfied assertion (with respect to the current image) a small distance in any direction,[4] and the ones formed by similarly adjusting the position of the second argument of such an assertion. If none of these evaluates better with respect to the set of propositions (including the most recent assertion), then we are done so we return the current image. Otherwise we reset the current image to one of the adjusted images such that none of the other adjusted images has a better evaluation, and then we iterate. This search procedure is called hill-climbing (Winston 1984). It is guaranteed to halt, possibly at a locally but not globally best image, if the representation of images admits only a finite number of states, which is true for bitmaps.

It should be clear that the Adjustment Procedure will find an appropriate spatial arrangement for our working example. Given an image that satifies the first assertion '(Above X Y) prototypically, the prototypicality of the second assertion '(Right X Y) will be nearly 0. According the the Adjustment Procedure, we construct and evaluate the images in which X or Y are adjusted independently, and we discover that moving X to the right or Y to the left increases the evaluation of the image. This process iterates until X is on a 45 degree angle above the horizontal with respect to Y; any adjustments of this image would have equivalent evaluations (if X were moved directly toward or away from Y) or the prototypicality of one of the asserted propositions would drop, decreasing the overall evaluation.

[4]With bitmap representations, space is already quantized (at a hopefully adequate resolution) so the finite number of possible adjustments to the position of an object is explicit.

1. Call the Adjustment Procedure with $\Gamma$, P, and I. Let A be the evaluation of the returned image with respect to $\Gamma \cup \{P\}$.
2. Call the Adjustment Procedure with $\Gamma$, the negation of P, and I. Let B be the evaluation of the returned image with respect to $\Gamma \cup$ {the negation of P}.
3. If $A \geq C$ and $B < C$, then return YES.
4. If $A < C$ and $B \geq C$, then return NO.
5. If $A \geq C$ and $B \geq C$, then return UNSPECIFIED.
6. IF $A < C$ and $B < C$, then return CONTRADICTION.

Figure 4: The Decision Procedure takes as input a set of propositions $\Gamma$, a queried proposition P, and an image I, which is expected to be the best model of $\Gamma$ that could be found. The Decision Procedure returns one of four responses. A represents how consistent P is with $\Gamma$ since it is the evaluation of the best model of $\Gamma \cup \{P\}$ that could be found by the Adjustment Procedure. Similarly, B represents how consistent the negation of P is with $\Gamma$. Since these consistency values come from evaluations of images, which are minima of prototypicality ratings, they fall into [0,1]. We differentiate high consistency from low consistency in this range with a cutoff C, which should be determined empirically.

Figure 4 shows the Decision Procedure, which uses the Adjustment Procedure to avoid the Indeterminacy Problem when answering queries about an image. To answer a query, we must first check to see how the queried proposition interacts with the previous assertions. So we call the Adjustment Procedure to find a best adjusted image if we were to assert the proposition, and we evaluate that image. If it has a high evaluation, then the Adjustment Procedure was able to find a spatial arrangement in which all of the asserted propositions plus the queried one could be satisfied fairly well. This alone does not imply that the previous assertions necessitate the queried one, however; it might spuriously be consistent or just not interact with the previous assertions. Thus we have to call the Adjustment Procedure to see what the effect would be if we had asserted the negation of the queried proposition, and we evaluate the image that is returned this time as well.

The two evaluations are enough to make a decision. To respond affirmatively to the query, it must be the case that the queried proposition is consistent with the previous assertions, but its negation over-constrains the image, producing a low evaluation for the best model that could be found if the negated proposition had been asserted. The analysis is symmetrical for the negative response. If both the queried proposition and its negation are consistent with the previous assertions, then the previous assertions do not constrain the answer, and, regardless of the actual details of the image maintained by the server, the response is UNSPECIFIED. Finally, if neither the queried proposition nor its negation are consistent with the previous assertions, the previous assertions must be contradictory (over-constrained) since for any not-over-constrained model we expect at least one of the propostions to be true.

This computational model has been implemented for the simplified domain in ISR, a computer program written in Common LISP. An image is represented with a 10 × 10 bitmap, which is a list of ten lists of ten symbols. ISR avoids the Indeterminacy Problem by using the Adjustment Procedure to update images when assertions are made, and by using the Decision Procedure to answer queries about the current image. As an example, ISR begins with an arbitrary spatial arrangement of the points X, Y, and Z; all queries respond with UNSPECIFIED. If we assert '(Left X Y) and then '(Left Y Z), ISR can tell that us '(Left X Z) and '(Right Z X) are true but '(Right X Y) is false. If we query '(Above X Y), ISR responds UNSPECIFIED because X could be either above or below Y and still satisfy the constraints of the previous assertions. ISR can also handle interacting assertions and can find an appropriate image (with X on a 45 degree angle above and to the right of Y) in our working example, after '(Above X Y) and '(Right X Y) have been asserted.

## Discussion

Kosslyn (1980) has developed a detailed computational model of imagery based on an imaginal representation called a visual buffer, which is essentially a two-dimensional bitmap. However, the Indeterminacy Problem is not a difficulty in this model because of the restricted range of tasks it was designed to explain. The phenomenon Kosslyn was trying to model is how, given a question about the appearance of some object, people seem to retrieve an image and examine it to find the answer. For example, it was hypothesized that, to compute how far apart the front and rear wheels of a car are, one simply scans the distance in the visual buffer between the tires in a default image of a car. Thus, default information was expected to be used to answer queries, and no subsequent assertions were made which could interact with and defeat such details in an image. When addressing the common introspection that images can somehow be sketchy and abstract, Kosslyn claims that this effect could be achieved in visual buffers by simply leaving out details, but it remains unclear how this would work.

From a general computational perspective, we might want to use images as mental models (Johnson-Laird 1983) to reason about knowledge in other formats. For example, images can form an oracle for inferring things from an alternate description because their fixed level of detail in representation makes implications explicit. The source of the Indeterminacy Problem lies in translating between the language in which the knowledge is described and the "language" of images. For propositional representations, descriptions might be indeterminate because they simply do not mention some aspect of the world, or because they are categorical in the sense that they can only give a possibility distribution over some states rather than implying which are necessarily true and which are necessarily false. This abstraction away from insignificant details is the power of propositional representations, but modeling knowledge in this format with images is not straightforward.

Our solution to using detailed images to model indeterminate propositional descriptions is to incrementally adjust an image to satisfy as many of the previously asserted constraints as possible, making and modifying consistent but not necessary choices as needed. To distinguish the necessitated from the just-consistent information in an image when we query it, we consider both the consistency of the queried proposition and of its negation with the previous assertions, again relying on adjustment to find the most satisfactory models. This computational model discredits a classical argument against imagery by showing how detailed models of indeterminate descriptions can be maintained and used without committing to consistent but unnecessary details required by the representation. It has been acknowledged that statements about the limitations of imaginal representations often fail to address the variety of possible processes that could operate on them (Anderson 1978; Johnson-Laird 1983). Our computational model suggests an interesting process that surprisingly would provide imaginal representations with the potential to handle indeterminacy.

If we view the image server from the perspective of the propositional client, it appears to be doing default reasoning with categories. This capability can be explained by the grounded nature (Lakoff 1987) of the "language" of images; this language is so close to perception that it can distinctly represent all the relevant states of the world, and the interactions among categories can be explicity described at this level. Our computational model demonstrates how a grounded language can be used to model categorical knowledge in such a way that 1) defaults can be assumed at one time and retracted or modified at a later time, and 2) implications of the generating description found by examining a detailed model can be distinguished from information in the model that is consistent but not necessary. Thus, if knowledge can be translated into a grounded language, default reasoning with categories reduces to a Constraint Satisfaction Problem (Mackworth 1990).

The domain in which we developed our computational model was greatly simplified; we should consider the possibility of extending the model to handle more interesting categories used in imagery (Herskowitz 1986). Even for Left, Right, Above, and Below, there are complications we have ignored, such as the fact that the sense of these predicates can be influenced by their arguments or by context. One subtle aspect of determining the meaning of spatial prepositional phrases, for example, is choosing a frame of reference (Clark 1973). As another example, the spatial preposition "over" has been shown to have nearly one hundred qualitatively different senses, as opposed to quantitative differences in distance or angle (Brugman 1983). Hopefully, these can be disambiguated prior to assertion to keep the prototypicality functions amenable to hill-climbing. Even still, it is possible that hill-climbing cannot find the globally best adjusted image. Maybe we can call on prototypes to help us "jump" out of local maxima. Or perhaps we can detect the need to change representations, like switching to higher resolution, three dimensions, or polar coordinates.

In the simplified domain, objects are structureless points, which we can easily locate. Imaging extended objects introduces a number of interesting complications. First, the objects themselves will be described by possibly categorical predicates. For example, our default image for a tiger could have exactly nineteen stripes, four legs, and a tail three feet long. Similarly, size, posture, and perspective (for projecting onto two dimensions) would have to be determined. Simply finding objects and features, which might have been adjusted, could be an extremely difficult recognition task. However, Kosslyn's computational model gives a simplified method for scanning for objects and features, and perhaps we could exploit the fact that, in images we construct, we have knowledge of what things are where. Additionally, we need to know a very limited number of relevant adjustments for each categorical predicate so the Adjustment Procedure will not degenerate into a full search.

## Conclusion

We have presented a computational model of how detailed images can be used to model spatially indeterminate information. The central thesis of the model is that images should be incrementally adjusted to satisfy interacting constraints in the form of previously asserted categorical propositions. By considering the consistency not only of a queried proposition but also its negation with the set of previous assertions, the necessary information in the image can be distinguished from the details which were required for representation but not implied by the description. This computational model specifically relies on graded prototypicality functions which guide the search for the best model of a set of categorical propositions. The use of a grounded language like images for modeling knowledge reduces reasoning with categories to a Constraint Satisfaction Problem.


## Acknowledgements

The author thanks Sheldon Nicholl, David C. Wilkins, and the Knowledge-Based Systems Group at the University of Illinois for helpful comments. Additional comments were kindly provided by Gordan Logan, Lois Boggess, and Anthony Maida.


## References


Anderson, J. 1978. Arguments Concerning Representations for Mental Imagery. *Psychological Review* 85(4):249-277.

Brugman, C. M. 1983. *The Story of Over.* Bloomington, IN: Indiana University Linguistics Club.

Clark, H. H. 1973. Time, Space, Semantics, and the Child. In Moore, T. E. ed. *Cognitive Development and the Acquisition of Language.* New York: Academic Press.

Funt, B. V. 1980. Problem-Solving with Diagrammatic Representations. *Artificial Intelligence* 13:203-230.

Herskowitz, A. 1986. *Language and Spatial Cognition.* New York: Cambridge University Press.

Johnson-Laird, P. N. 1983. *Mental Models.* Cambridge, MA: Harvard University Press.

Kosslyn, S. M. 1980. *Image and Mind.* Cambridge, MA: Harvard University Press.

Lakoff, G. 1973. Hedges: A Study in Meaning Criteria and the Logic of Fuzzy Concepts. *Journal of Philosophical Logic* 2:458-508.

Lakoff, G. 1987. *Women, Fire, and Dangerous Things.* Chicago: University of Chicago Press.

Mackworth, A. K. 1990. Constraint Satisfaction. In Shapiro, S. C. ed. *Encyclopedia of Artificial Intelligence.* 2nd edition. New York: John Wiley & Sons.

Pinker, S. 1984. Visual Cognition: An Introduction. *Cognition* 18:1-63.

Pylyshyn, Z. W. 1973. What the Mind's Eye Tells the Mind's Brain: A Critique of Mental Imagery. *Psychological Bulletin* 80:1-24.

Smith, E. and Medin, D. 1981. *Categories and Concepts.* Cambridge, MA: Harvard University Press.

Waltz, D. L. and Boggess, L. 1979. Visual Analog Representations for Natural Language Understanding. In *Proceedings of the Sixth International Joint Conference on Artificial Intelligence* 926-934.

Winston, P. H. 1984. *Artificial Intelligence.* Reading, MA: Addison-Wesley.

Zadeh, L. 1965. Fuzzy Sets. *Information and Control* 8:338-353.

# Telling where one is heading and where things move independently

**Niels da Vitoria Lobo and John K. Tsotsos***
Dept. of Computer Science, University of Toronto, Toronto, Canada M5S 1A4.
*e-mail: niels@vis.toronto.edu, tsotsos@vis.toronto.edu*

### Abstract

We summarize our recent novel approach to computing the Focus of Expansion for an observer moving with unrestricted motion in a scene with objects of unrestricted shape. This method also detects points not moving rigidly with the scene. The approach, using collinear image points, is based on an exact method for cancelling effects of the observer's rotation from optic flow. The computational results are being presented elsewhere (da Vitoria Lobo & Tsotsos 1991). Here, we argue that this algorithm is biologically plausible.

## Introduction

Many years ago researchers (Helmholtz 1925, Gibson 1957) hypothesized that the 3-D motion and the shape of the environment are perceivable from the projected motion arising out of the relative motion between a monocular observer and the scene. In this paper, we summarize our recent computational solution to this problem and discuss its biological plausibility.

The paradigm we work within assumes that an approximation to instantaneous image velocity (also termed *image flow* or *optic flow*) can be measured, and some progress has been made towards achieving such measurements (see Watson & Ahumada 1985, Heeger 1988, Fleet 1990). The subsequent step, that of computing the 3-D motion parameters and shape information from the instantaneous image velocity, has received ample attention from researchers. However, in addition to the fact that every 3-D algorithm proposed so far is not robust to noise in the input instantaneous velocity, the algorithms typically suffer from other important problems that disqualify them from biological candidacy. Algorithms that permit general rigid motion and unrestricted shapes have to search in spaces that have at least three dimensions (Bruss & Horn 1983), and the non-linear numerical methods used are very sensitive to the initial guess. Others assume some restricted form of motion (eg., no rotation in a certain dimension; see Barron 1988), or restrict the allowable shapes (eg., planarity), in order to get closed-form solutions for the unknowns (Waxman & Wohn 1987), or assume that some parameters are known and solve for the others (Ballard & Kimball 1983, Matthies *et al.* 1989) — all of them too restrictive for biological plausibility.

The basis of our approach is a technique for combining collinear image points which allows rotation to be cancelled in an exact manner. Along any straight line in the image, the rotational contribution to the image velocity component orthogonal to the line varies linearly with length, so that taking approximations to the second derivative of this component of velocity cancels out rotation. Thus despite the unrestricted motion and unrestricted shape involved in the problem, the motion parameters can be unlocked by a search for the correct direction of translation, which is a mere 2-dimensional search and incurs a far lower computational cost than other algorithms that search in higher dimensions. The algorithm, termed the *FOE Algorithm*, uses an operator that simultaneously cancels out rotation exactly and samples the translation contribution to find the direction of translation. Earlier, da Vitoria Lobo & Tsotsos (1990) showed that for three non-collinear image points, the pair-wise relative depths of the three scene points are dependent only on the unknown 3-D direction of translation and the known image velocities and image positions (i.e., that, in principle, knowledge of rotation is irrelevant to the calculation of shape from motion). For other work that cancels rotation see Prazdny (1983), Nelson & Aloimonos (1988) and Heeger & Jepson (1990). Our approach also straightforwardly detects points that do not move in a manner consistent with the assumption of a rigid scene. In this paper we argue that due to its simplicity, low computational cost, inherent parallelism, and robustness to noise, this method is biologically plausible, and we explore its consequences for research in biological systems. We first review our *FOE algorithm* and its extension to detecting independent motion. Then we discuss the importance of the FOE and present the arguments for biological plausibility, along with predictions that result from the model.

## Locating the FOE and independent motion

In this section, we summarize work appearing in da Vitoria Lobo & Tsotsos (1991).

### Image velocity and scene parameters

When the relative motion between the viewer and a scene point (at depth $Z$) is described by an instantaneous translation $(U, V, W)$, and an instantaneous rotation $(A, B, C)$, the projected velocity in the image plane at position $(x, y)$ is $(u, v)$, (Longuet-Higgins & Prazdny 1980), where

$$\begin{aligned} u &= \tfrac{-U+xW}{Z} - Axy + B(x^2+1) - Cy, \\ v &= \tfrac{-V+yW}{Z} - A(y^2+1) + Bxy + Cx. \end{aligned} \qquad (1)$$

### The Focus of Expansion

We define the Focus of Expansion (FOE) to be the position of intersection of the imaging surface[1] and the direction of

*John K. Tsotsos is the Canadian Pacific Fellow of the Canadian Institute for Advanced Research. This work was supported by the Natural Sciences and Engineering Research Council and the Information Technology Research Center, a Province of Ontario Center of Excellence. The range data came from the Range Image Database of NRC Canada.

[1] This surface could be planar, hemi-spherical, etc.

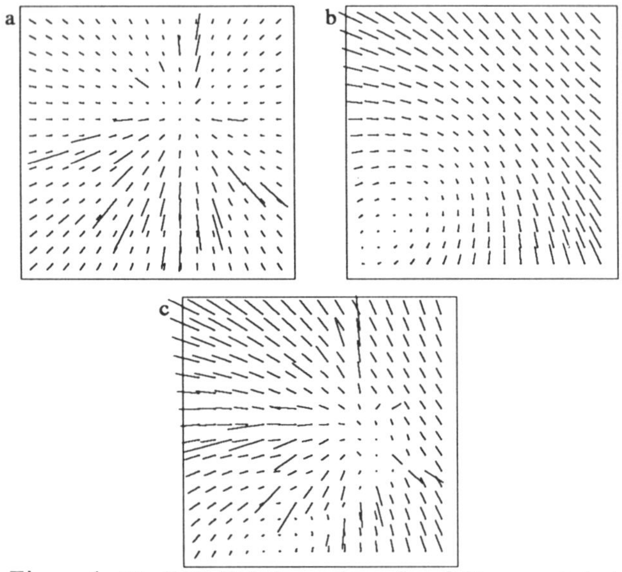

Figure 1: The FOE: This figure shows flow fields generated when observer a) only translates, b) only rotates, and c) moves with both rotation and translation, the most typical situation. In the second case there is no FOE because there is no observer translation, while in the first and third cases the FOE is in the same position for both cases, just above and to the right of the image center. That is, our definition of FOE renders its position in the image completely independent of observer rotation.

instantaneous observer translation. This[2] defines the FOE to be independent of the viewer's instantaneous rotation. Figure 1 illustrates how our definition of FOE makes its position invariant to rotation.

Several authors have researched the computation of the FOE either by restricting the allowable motion to translation or by approximately cancelling rotation(Jain 1982, Reiger & Lawton 1985). In work related to ours, Weinshall (1990) finds the FOE but needs to find elliptical surface patches beforehand. The contribution of the work in da Vitoria Lobo & Tsotsos (1991) is that we find the FOE in an exact manner, even when the motion includes general rotation, with no restriction on surface shape.

## FOE From Collinear Measurements

We refer to three image points as a *triplet*. The unit computation of our algorithm is a collinear triplet computation. The computation is a generalization of an approximation to the second derivative of the velocity component that is normal to the line joining the three collinear image points, the "derivative" being taken in the direction of the line. For points subscripted by $i$, $(i = 1, 2, 3)$, let the image coordinates of the points be $(x_i, y_i)$, their image velocities be $(u_i, v_i)$, and the depths to their counterparts in the scene be $Z_i$.

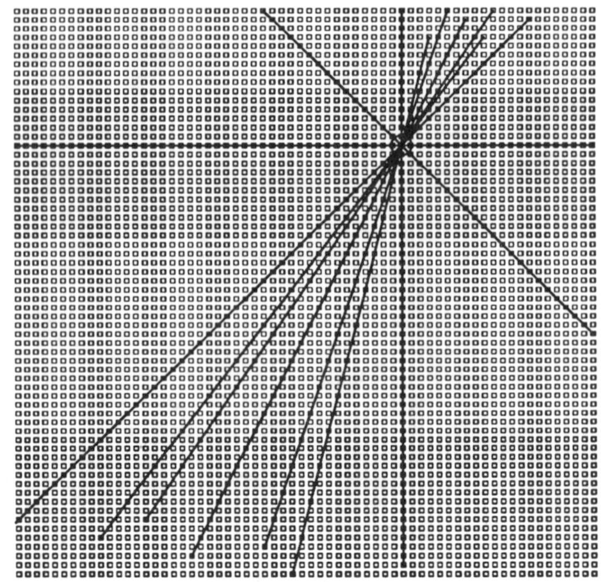

Figure 2: The FOE Operator: This shows how an operator is made up of intersecting lines of points in the image. On a regular 256x256 square optic array, at a single point we can have about 60-70 intersecting lines each of which passes through at least several image points. Here, only 9 lines are shown. The image is a dense grid of points, shown as hollow squares, each with a flow estimate associated with it. Some of the points used by each line have been blackened to identify them. Along each line a *Line Sum* is calculated (see Fig 3). The Line Sums are added to give the response of the operator at position (x,y) which is where the lines intersect.

The computation is the weighted[3] sum[4],

$$Sum \stackrel{\text{def}}{=} \begin{array}{l}(-\sin\theta)(nu_1 - (m+n)u_2 + mu_3) + \\ (\cos\theta)(nv_1 - (m+n)v_2 + mv_3)\end{array} \qquad (2)$$

where $\theta$ is the angle that the line through the three image points makes with the image x-axis, going from the x-axis to the line in a counterclockwise manner, and where $m$ is the distance between the first and second image points and $n$ the distance between the second and third. By substitution of $(u_i, v_i)$, this generalized *Sum* can be verified to be

$$Sum = \begin{pmatrix} U\sin\theta - V\cos\theta \ + \\ Wy_3\cos\theta - Wx_3\sin\theta \end{pmatrix}\left(\frac{n}{Z_1} - \frac{(n+m)}{Z_2} + \frac{m}{Z_3}\right)$$

i.e., *Sum* = product of translation factor and depth factor.

*Sum* is zero either if the scene points are collinear, or the translation factor is zero, or if the line passing through the three image points also passes through the FOE. Thus, in general, for a scene containing sufficient depth variation, if we compute many such collinear triplet sums across the complete image, the FOE will be in the position of intersection

[2]This definition differs from that of Regan & Beverley (1982) and Cutting (1986) who define the FOE as the intersection of the image and the direction of motion, thus making it dependent on rotation.

[3]When the collinear points are equi-spaced, the weights are (1,–2,1) which is an approximation to the second derivative (Rektorys 1969).

[4]We use the symbol $\stackrel{\text{def}}{=}$ to define a calculation; this is to distinguish it from an expression of equality.

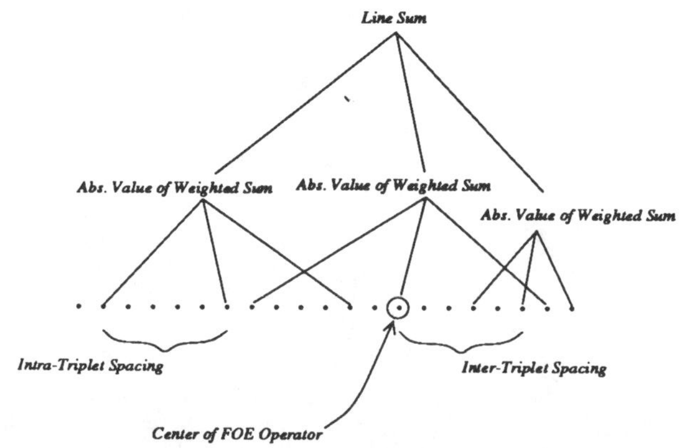


Figure 3: This shows how points on a line were processed. Flow estimates at points are grouped into triplets and summed according to the weighted sum of equation (2). Then the absolute values of these weighted sums are summed to give the *Line Sum* for the line.

of many triplet lines for which, in each case, the triplet sum is zero. Next, we describe an operator and an algorithm to locate the FOE.

## The FOE Operator

Consider an operator (called the *FOE operator*) with many lines passing through its center (centered at some $(x,y)$ position in the image) in many directions in the image (see figure 2), such that each line passes through several image points. Along the line, overlapping triplets of image velocity are used and summed (see figure 3).

The absolute values of all triplet sums (as defined by equation 2) along a line are summed to give a *LineSum*, and then the LineSums are summed to get a response at the center. For a rigid scene, there are three reasons why the response could be zero. First, each of the triplets summed by this operator could be a collinear triplet in the scene; however, since each line contains overlapping triplets, for the zero response to be caused by collinear scene points, the whole scene would have to be a single plane, which is quite rare. Second, there could be no translation. Third, the true FOE could be at the position of the center of the operator. Fortunately, if either of the first two cases were to hold, the operator would respond with zero everywhere; whereas, in the third case, assuming the first two do not hold, there will be a unique zero. So, the first two cases can be detected easily and eliminated. Thus, we can sweep this operator across the whole flow field to obtain a response map, detecting where it gives a zero response surrounded by sufficiently non-zero responses.

## Testing the FOE computation

Here tests of the FOE algorithm are described. The range image shown in Fig 4 was used to generate the synthetic optic flow field shown in Fig 5. This flow was used as input to our FOE algorithm. For the implementation of the FOE algorithm, an FOE operator was centered at each element in the optic array. Each FOE operator consisted of 24 lines passing through it (see Fig 2). Along each line we centered

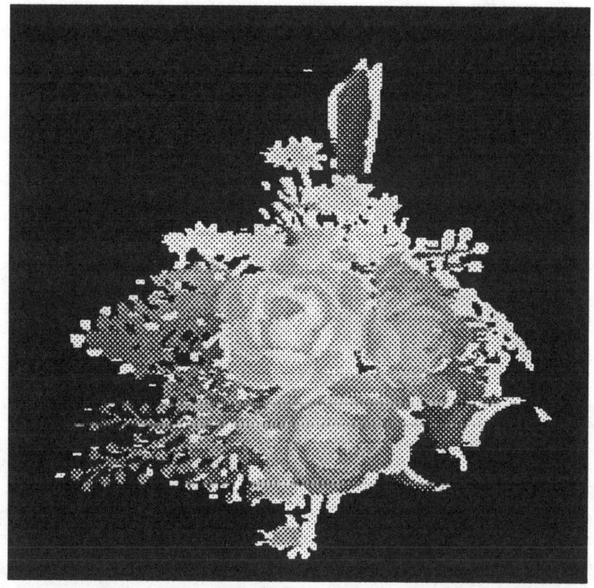

Figure 4: Range data that was used to generate the synthetic flow field used in these experiments. Observer moves with some instantaneous 3-D rotation and translation parameters with respect to these depth points and the flow equations are used to give an image velocity vector at each grid point on the optic array.

a triplet at each point on that line. Within each triplet, the intra-triplet spacing was such that three alternate points on the line were used.

Fig 6 shows the response map for the operator as a function of $(x,y)$. Here, response maps are shown with brightness proportional to the *Log* (response). The darkest point, the global minimum, corresponds to the computed FOE and it is not surprising that it is exactly correct. Even with noise (as high as 8% on average) in the flow, the FOE is found easily. With noise, the pit of the minumum broadens out but is still pronounced (somewhat like Fig 8.)

## Functioning in a non-rigid scene

Objects in a moving observer's view will often move independently. Our computation remains competent in these situations. Fig 7 depicts the flow field generated by combining the original rigid flow field used in the previous section with the flow for an independently moving rectangular patch in the upper middle of the image. The frontoparallel patch translates upward and to the right, with an image flow magnitude comparable to those of the flower petals. The FOE algorithm was tested on this combined flow field. The output direction of translation is still correct (filter response map in Fig 8.) Tests indicate that even with larger patches the FOE computation is accurate, demonstrating robustness to patches of non-rigidity in the scene.

## Finding points that thwart rigidity

A biological agent needs the ability to detect parts of the

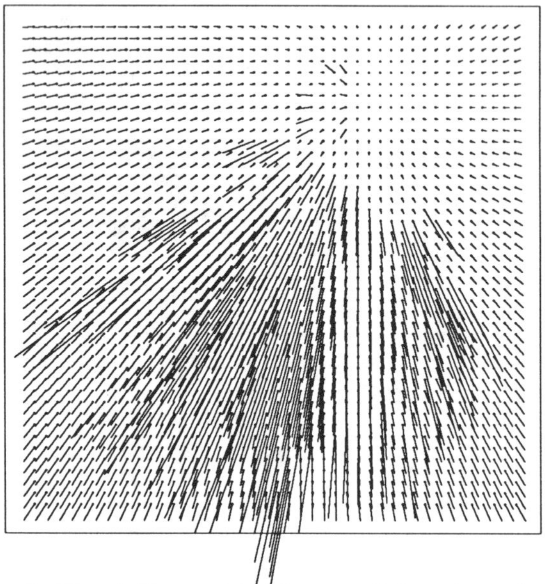

Figure 5: Flow field generated using some 3-D motion and range data of Fig 4. Field shown sub-sampled.

Figure 6: Response map for noise-free flow. In these maps, we show brightness proportional to *Log*(response). For this input, the global minimum is exactly at the true FOE.

scene where the overall-rigidity assumption does not hold. A point may be measured as moving differently from the rest of the scene either because it may be a noisy measurement – in which case any subsequent use of this erroneous flow value such as in the shape reconstruction step, should be approached with caution – or because it could legitimately belong to an independently moving part of the scene. If this is the case, this moving part will probably need additional attention and possibly some special purpose processing, such

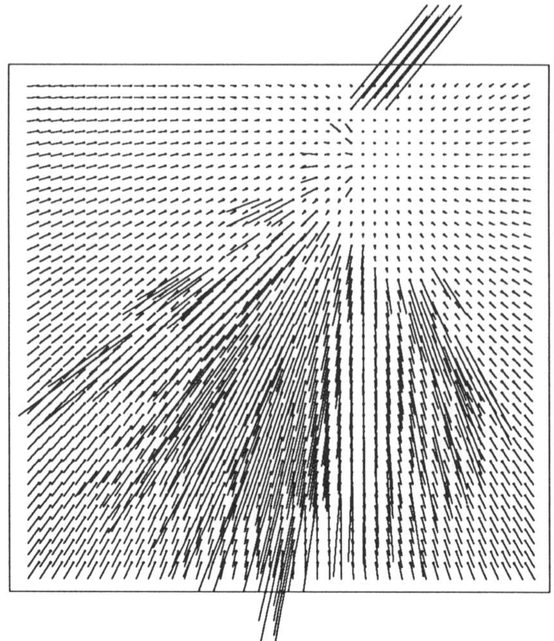

Figure 7: Flow field combining the original rigid flow field and an independently moving patch in upper right part of image. The frontoparallel rectangular patch translates upward, to the right, with an image flow magnitude that is comparable to those of the flower petals.

as that which will segment it from its background.

Our algorithm can easily signal that certain points are moving inconsistently relative to the rest of the scene. We achieve this by using the operator that gave the minimum response indicating that the FOE is at its center. We traverse each of its various lines searching for triplets that don't give near-zero sums. Such triplets correspond to points such that at least one of the three points moves inconsistently with the rest of the scene[5]. Fig 9 shows the region around the patch in the flow field of Fig 7 being marked by the program as sets of inconsistent triplets. Note that because the patch itself is moving in a planar fashion, the triplet sum, when all three points are inside the patch, is zero. So, at an independently moving planar region, only the boundary areas of the region will be detected (this depends on the intra-triplet spacing used). For an independently moving non-planar region, all the triplets, in any way overlapping the region's points, will be detected.

## Importance of the FOE

As mentioned earlier, we have already shown in principle that relative depth is a function of the direction of translation and is independent of rotation. In other work, we are proposing that for the noisy cases shape information be recovered in stages, involving qualitative and quantitative recovery based on knowledge of the FOE. The qualitative recovery work has already appeared in Weinshall (1990), and in da Vitoria Lobo & Tsotsos (1990).

[5] We have already stated earlier in this section that if such a collinear triplet were moving with the same rigid parameters as the scene, the triplet *Sum* must be zero.

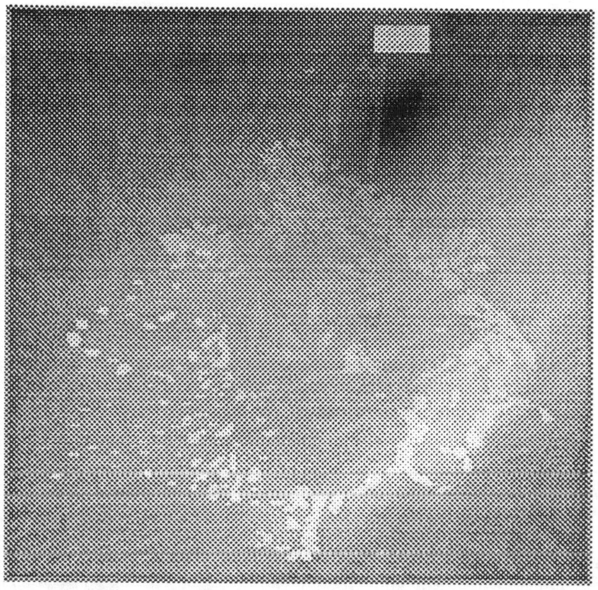

Figure 8: Response map for the flow associated with the nonrigid scene in which a patch moves independently. The global minimum is in the same position, indicating that the FOE computation is robust to non-rigidity in the scene.

In addition to its use in computing shape, the FOE tells where one is heading. This could be useful for navigation tasks. Also, since our computation is a monocular one, a binocular system could compute an FOE for each eye seperately, and this could be used to obtain information about the relative poses of the two eyes.

## Relation to biological systems

We suggest that the FOE Algorithm summarized above is biologically plausible because our simple calculations can easily be implemented by the visual cortex. Consider Fig 10. On the left we show one of the motion pathways from Maunsell & Newsome (1987). On the right we show the stages we use to compute our FOE operator response map, suggesting points in the cortical pathway at which our computations could be implemented. Cells in area V1 could compute normal velocities for moving intensity structure, and then connect to area MST either directly or via area MT. (Maunsell & Van Essen 1983, Ungerleider & Desimone 1986).

Tanaka *et al.* (1989) have described cells in macaque dorsal MST that appear to be responsive to patterns composed of points moving outwardly or inwardly along radial lines. These cells have been termed "changing size" or "expansion/contraction" cells. A family of such cells, similar to each other, could be computing in parallel an inverse of our response map, with each cell responding strongly when the FOE is at its center. That is, each cell would be encoding a different direction of translation. If such a cell were computing a sum of our *LineSums*, then a cell that responds to a strictly expanding/contracting radial flow should continue its response even when a rotational field of the kind shown in Fig 1b is added to the expanding/contracting pattern. If this were to be found, this would be very strong evidence for a computation similar to our algorithm.

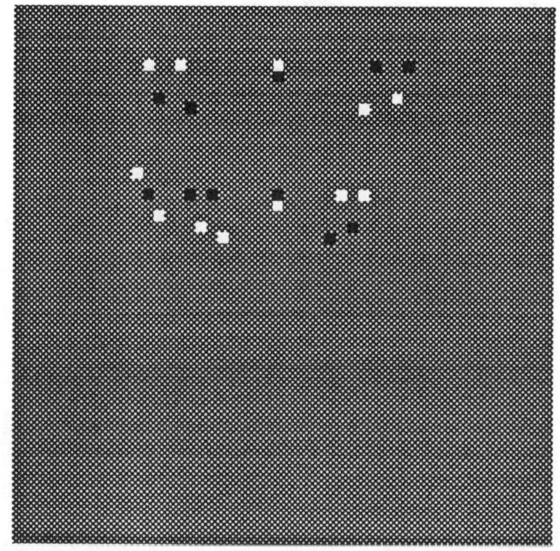

Figure 9: An enlarged portion of the output from the program that detects triplets containing inconsistently moving points, when it was run on the flow of Fig 7. The detected points correspond to areas around the independently moving patch. Because the patch itself is moving in a planar fashion, the triplet sum, when all three points are inside the patch, is zero. So, these internal points are not marked.

To compute our FOE operator response at nearby positions, neighboring MST cells would receive input from overlapping triplet *Sums*. Hence it would be reasonable to expect that these *Sums* are not being re-computed each time, but rather that some intermediate cells compute something akin to triplet *Sums*. There are cells in area MT that are known to respond to patterns of activity in which the center differs from its surround. This would be an appropriate substrate in which our triplet *Sums* could be computed. These could appear in the form of elongated cells computing an approximation to the second derivative of the normal velocities, the derivative taken along the long axis. An analogue of such receptive fields has been proposed by Dobbins *et al.* (1987) for curvature detection. To reduce connectivity to the MST cell computing the response for the map, the flow field may be sampled quite sparsely, and we need to study the computational effects of such sparse sampling.

If the response map is being computed in parallel using a family of MST cells, then some subsequent mechanism would need to find the global minimum. This could possibly be accomplished using computations at multiple scales across the map, so that a coarse sampling of the map would indicate the ballpark of the minimum, while further finer grain spatial sampling would give better resolution of the position of the FOE. A framework involving attention would be suitable for achieving this (Tsotsos 1990).

Detecting independently moving points requires top-down activation of the particular MST cell positioned at the FOE. We hypothesize that such feedback exists under attentional control, but its exact nature would need to be discovered.

Finally, there are connections from area MST to areas 7a

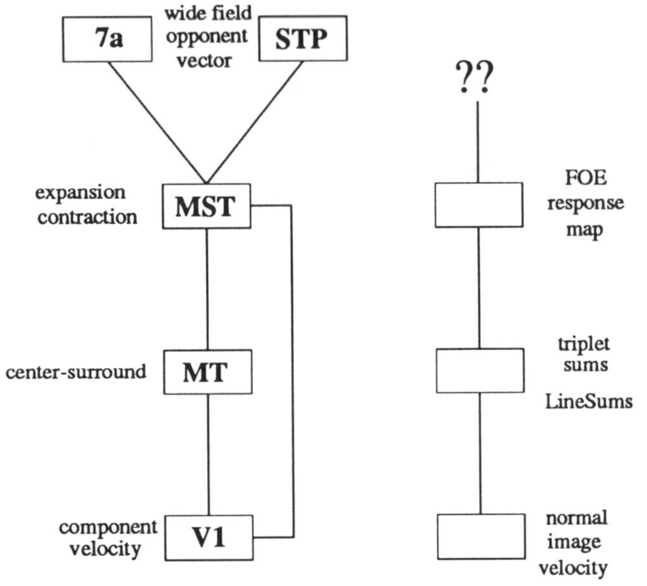


Figure 10: Is the FOE algorithm biologically plausible? On the left, an abstraction of one of the motion pathways described in Maunsell & Newsome (1987). On the right, stages leading up to the computation of our FOE operator response map.

and STP where wide field, "opponent vector" organizations have been found. These cells respond to patterns of radial flow to and from the fixation point (Maunsell & Van Essen 1987). We are studying the possible roles these would play in our framework.

## Conclusion

We have summarized a recent novel approach to computing the direction of the translation component of egomotion, and detecting points not moving rigidly with the scene, in the view of an observer moving with unrestricted motion. The detailed computational work appears elsewhere, but here we argued that this approach is biologically plausible and sketched some of its consequences.

## References


Ballard D.H. and Kimball O.A. (1983), "Rigid body motion from depth and optical flow," *CVGIP*, **22**, 95-115.

Barron J. (1988), "Determination of egomotion and environmental layout from noisy time-varying image velocity information in monocular image sequences," *Ph.D. Thesis*, Dept. of Computer Science, Univ. of Toronto. Available as RBCV-TR-88-24.

Bruss A.R. and Horn B.K.P. (1983), "Passive navigation," *CVGIP*, **21**, 3-20.

Cutting J. (1986), **Perception with an Eye for Motion**, MIT Press.

da Vitoria Lobo N. and Tsotsos J.K. (1990), "Extracting qualitative shape from image motion: applications to stereo-pairs", *Proc. AAAI Workshop on Qualitative Vision*, pp. 36-40, Boston, July.

da Vitoria Lobo N. and Tsotsos J.K. (1991), "Using collinear points to compute egomotion and detect nonrigidity," *in press, Proc. of Computer Vision and Pattern Recognition*, Hawaii, June 1991.

Dobbins A., Zucker S.W., and Cynader M.S. (1987), "Endstopping in the visual cortex as a substrate for calculating curvature," *Nature*, **329**, 438-441.

Fleet D. (1990), "The Measurement of Image Velocity", *Ph.D. Thesis*, Dept. of Computer Science, Univ. of Toronto.

Gibson J.J. (1957), "Optical motions and transformations as stimuli for visual perception", *Psychological Review*, **64**, No 5, 288-295.

Heeger D. (1988), "Optical flow using spatiotemporal filters", *IJCV*, **1**, 279-302.

Heeger D. and Jepson A. (1990), "Simple method for computing 3-D motion and depth," *Proc. ICCV*, Osaka, Japan, 96-100.

Helmholtz H.(1925), **Treatise on Physiological Optics**, Optical Society of America.

Jain (1982), "An approach for the direct computation of the focus of expansion," *Proc. PRIP-82*, 262-268.

Longuet-Higgins H.C. and Prazdny K. (1980), "The interpretation of a moving retinal image," *Proc. Roy. Soc. Lond.*, **B 208**, 385-397.

Matthies L., Szeliski R., Kanade T. (1989), "Kalman filter-based algorithms for estimating depth from image sequences," *IJCV*, **3**, 181-208.

Maunsell J.H.R., and Newsome W.T. (1987), "Visual processing in monkey extrastriate cortex," *Ann. Rev. Neurosci.*, **10**, 363-401.

Maunsell J.H.R., and Van Essen D.C. (1983), "The connections of the middle temporal visual area (MT) and their relationship to a cortical hierarchy in the macaque monkey," *Jour. of Neurosci.*, **3**, 2563-2586.

Nelson R. and Aloimonos J. (1988), "Using flow field divergence for obstacle avoidance: towards qualitative vision," *Proc. ICCV* Florida, 188-196.

Prazdny K. (1983), "On the information in optical flows", *CVGIP*, **22**(2), 239-259.

Regan D.M. and Beverley (1982), "How do we avoid confounding the direction we are looking and the direction we are moving?" *Science*, **215**, 194-196.

ReigerJ.H. and Lawton D.T. (1985), "Processing differential image motion", *J. Optical Society of America*, **A2**, 354-359.

Rektorys K. (1969), **Survey of Applicable Math**, MIT Press.

Tanaka K., Fukada Y., Saito H. (1989), "Underlying mechanisms of the response specificity of expansion/contraction and rotation cells in the dorsal part of the Medial Superior Temporal area of the macaque monkey," *Journ. of Neurophysiology*, **62**, No. 3, 642-656.

Tsotsos J.K. (1990), "Analyzing Vision at the Complexity Level," *Behavioral and Brain Sciences*,.

Ungerleider L.G., and Desimone R. (1986), "Cortical connections of area MT in the macaque", *J. Comp. Neurol.*, **248**, 190-222.

Watson A. and Ahumada J. (1985) "Model of human visual motion sensing," *J. Optical Society of America*, **A2**, 322-342.

Waxman A.M. and Wohn K. (1987), "Image flow theory: a framework for 3-D inference from time-varying imagery," Chapter Three in **Advances in Computer Vision**, ed. C. Brown, (Erlbaum Publishers).

Weinshall D. (1990), "Direct computation of qualitative 3-D shape and motion invariants," *ICCV*, Osaka, Japan, 230-237.

# A Knowledge Representation Scheme for Computational Imagery

Dimitri Papadias Janice Glasgow *
Department of Computing and Information Science
Queen's University, Kingston, Canada
janice@qucis.queensu.ca

## Abstract

After many years of neglect, the topic of mental imagery has recently emerged as an active area of debate. One aspect of this ongoing debate is whether an image is represented as a *description* or a *depiction* of its components. This paper is not so concerned with how mental images are stored, as with what machine representations will provide a basis for imagery as a problem solving paradigm in artificial intelligence. In fact, we argue that a knowledge representation scheme that combines the ability to reason about both descriptions and depictions of images best facilitates the efficient implementation of the processes involved in imagery.

## 1 Introduction

Numerous experimental and neuropsychological studies have been carried out and several, often conflicting, models of mental imagery have been proposed. This paper does not present another computational model, but instead treats imagery as a problem solving paradigm in artificial intelligence (AI). Our thesis is that there exists a concept of *computational imagery*, which has potential applications to problems whose solutions by humans involve the use of mental imagery. In order to define the concept of computational imagery, we require a knowledge representation scheme. Although psychological theories are used as a guide to the properties of mental imagery that our scheme should preserve, whenever possible we attempt to overcome the limitations of the human information processing system. Thus, our primary concerns are efficiency, expressive power and inferential adequacy.

The knowledge representation scheme for computational imagery includes three representations - the deep representation, the symbolic array representation and the surface representation - each appropriate for a different kind of processing. While the deep representation is used as a permanent store for information, the symbolic array and surface representations are only constructed when retrieving spatial information not explicitly stored in the deep representation.

## 2 Mental Imagery

Although no one seems to deny the existence of the phenomenon called imagery, there has been an ongoing debate about the structure and the function of imagery in human cognition. The imagery debate is primarily concerned with whether images are represented as *descriptions* or *depictions*. It has been suggested that descriptive representations contain symbolic, interpreted information, whereas depictive representations contain geometric, uninterpreted information (Finke, Pinker & Farah 1989). The debate can also be expressed as whether images are epiphenomenal or not, where epiphenomena are effects that do not play any causal role in the brain's information processing (Block 1981).

This paper does not attempt to debate the issues involved in mental imagery, but to describe effective computational techniques for storing and manipulating imagery-related representations. To achieve this goal, though, we believe it is necessary to better understand the processes involved in mental imagery.

Results in the area of cognitive psychology have suggested several properties of mental imagery that the proposed knowledge representation scheme should attempt to capture. Finke summarizes these properties by proposing five "unifying principles" of mental imagery: the principles of implicit encoding, perceptual equivalence, spatial equivalence, transformational equivalence and structural equivalence (Finke 1989).

While individual experiments attempt to explain a limited number of imagery phenomena, theories are

*Support for this research was provided by the Institute for Robotics and Intelligent Systems Federal Center of Excellence of Canada and the Natural Science and Engineering Research Council of Canada.

more general and attempt to explain a large number of such phenomena. Pylyshyn, a forceful proponent of the descriptive view, argues that mental imagery simply consists of the use of general thought processes to simulate perceptual events, based on tacit knowledge of how these events happened (Pylyshyn 1981). He disputes the idea that mental images are stored in a raw uninterpreted form resembling mental photographs and argues for an abstract format of representation called propositional code. Kosslyn's model of mental imagery (Kosslyn 1980) is a depictive theory, which claims that images are quasi-pictorial. According to Kosslyn's model, mental images are working memory, surface representations generated from long-term memory, deep representations. A set of procedures serves as an interface between the surface representations and the underlying data structures, which may be decidedly non-pictorial in form.

Interesting findings, provided by the study of patients with visual impairments, have been interpreted by some researchers, for example (Kosslyn 1987), as suggestive of two distinct components of mental imagery: the spatial and the visual. The *spatial* component of imagery preserves information about the relative positions of the meaningful pars of an image. The *visual* component preserves information about *how* (size, shape, etc.) an image looks.

We do not propose a computational model for imagery. Rather, we present a knowledge representation scheme that attempts to capture the functionality of mental imagery. Although the scheme is similar to Kosslyn's in some ways, it extends his approach by defining image representations that are *3D* and hierarchical. Our approach also considers two working memory representations, corresponding to the visual and spatial components of imagery.

# 3 Knowledge Representation

AI programs rely on the ability to store domain descriptions and formally manipulate these descriptions to derive new knowledge about the given domain. Traditional approaches to knowledge representation include logic representations, which denote the objects and relations in the world in terms of logic rules, and structural representations, which denote concepts and relations in terms of hierarchies. In addition to general knowledge representation schemes, there exist specialized schemes concerned with the representation of shape, volume and other spatial qualities of images. These include discrimination trees (McDermott & Davis 1984), quadtrees (Samet 1980) and simple primitive volumes (Biederman 1985). A major contribution in representational formalisms for visual images is the progression of primal sketch, *2-1/2D* sketch and *3D* sketch (Marr & Nishihara 1978).

Kosslyn, in a discussion of the implementation of his computational model for mental imagery, expresses frustration with existing computational tools as well as the need for a metadescription for theories of imagery:

> *There is a major problem with this approach however; the program will not actually run without numerous "kluges", numerous ad hoc manipulations required by the realities of working with a digital computer and a programming language like ALGOL or LISP... The ideal would be a precise, explicit language in which to specify the theory and how it maps into the program.* (Kosslyn 1980)

A formal theory of arrays provides such a metalanguage. Array theory is the mathematics of nested, rectangularly-arranged collections of data objects (More 1979)(Jenkins & Glasgow 1989). Similar to set theory, array theory is concerned with the concepts of nesting, aggregation and membership. Array theory is also concerned with the relative spatial positions of data objects in a collection.

Nial, an interactive programming language that combines concepts from LISP and APL, constitutes an implementation of array theory (Jenkins, Glasgow & McCrosky 1986). Thus, the language of array theory provides for a specification of the representations and functions for computational imagery as well as a mapping into Nial programs.

## 3.1 Deep Representation

Research in vision and imagery has focused almost exclusively on the format of the on-line conscious representations, to the exclusion of that entailed in long-term storage. In our view, the deep representation falls more within the limits of research in long-term memory than imagery. Thus, we base its implementation on the hierarchical network model of semantic memory (Collins & Quillian 1969). This network model is suitable for storing images since it can be used to denote the decomposition and classification hierarchies involved in imagery, as well as the features of concepts and instances of images.

There are two kinds of hierarchies in the network representation: the AKO (a kind of) and the PARTS. The AKO hierarchy provides property inheritance: an

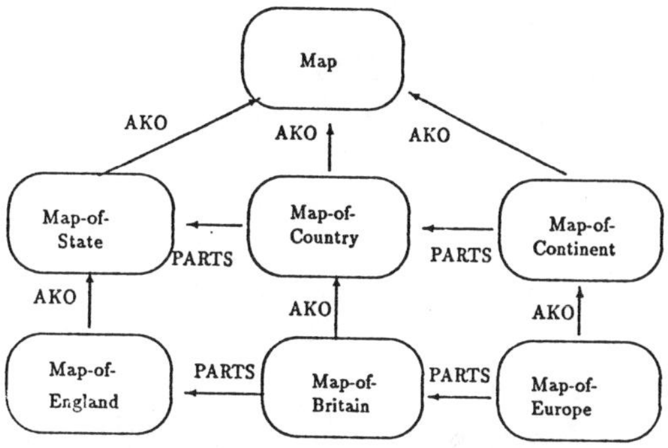


*a) Semantic network representation*

| FRAMENAME | Map-of-Europe |
|---|---|
| AKO | Map-of-Continent |
| PARTS | Sweden (0 4)<br>Britain (1 0)<br>... |
| POPULATION | 'find-population' |
| ... | ... |

*b) Frame representation*

Figure 1: Deep representation

image can inherit features from more generic images. The PARTS hierarchy is used to denote the structural decomposition of images into their meaningful components.

For the implementation of the deep representation in our scheme we use a frame language, in which each image frame contains (or inherits) all the salient information about the image. This information includes propositional knowledge, in addition to encodings that permit the reconstruction of the working-memory representations.

A frame corresponding to the image of a map of Europe and part of the semantic network for a geographic map domain is illustrated in Figure 1. Each node in the network corresponds to an individual frame and the links describe the structural and conceptual relationships between frames.

The frame implementation of the deep representation has several attractive properties. First it provides a natural way to represent knowledge since the hierarchical structure of an image is captured by the network structure. It also incorporates the concept of semantic networks in a computational implementation that provides features, such as multiple hierarchies and procedural attachment (ability to attach procedures to slot values). The non-monotonic feature of defaults in the frame structure corresponds to the cognitive ability to make conjectures and infer information when the existent knowledge is incomplete.

Despite its attractive properties, the deep representation is not the most suitable representation for all computations. Certain tasks, such as determining the spatial relations between image components, surface pattern matching and image scanning, require information that is not explicitly stored in long-term memory. Thus, the construction of representations corresponding to the visual and spatial components of imagery are required to facilitate the efficiency of the scheme.

## 3.2 Working-Memory Representations

Unlike the deep representation, which is used for the permanent storage of information, the working-memory representations of an image exist only while the image is active (when visual or spatial information processing takes place).

### 3.2.1 Surface Representation

The surface representation corresponds to the visual component of imagery, and it can either be reconstructed from the underlying deep representation or generated from low level vision processes. Similar to Kosslyn's skeletal image (Kosslyn 1980), the surface representation is depictive and incorporates geometric information, including shape and volume. Unlike these representations, we assume that the surface representation can be *3D*. In such cases, we assume an object-centered description of an image.

For the current implementation of the surface representation we use occupancy arrays. An occupancy array consists of cells, each mapped to a local region of the visual field and representing information such as lightness, color, texture, edge orientation or depth about this region. Objects are depicted in the arrays by patterns of filled cells isomorphic in shape to the objects. Figure 2 illustrates a *2D* occupancy array corresponding to a map of Italy. Even at low resolution, this surface representation contains information about the shape and size of the country.

### 3.2.2 Symbolic Array Representation

A primary characteristic of a good formalism for knowledge representation is that it makes relevant properties explicit. While an occupancy array provides a representation for the visual component of imagery, it is basically uninterpreted. For the spatial component of imagery we

|  |  | 1 | 1 | 1 | 1 |  |  |  |  |
|---|---|---|---|---|---|---|---|---|---|
|  | 1 | 1 | 1 | 1 | 1 | 1 |  |  |  |
| 1 | 1 | 1 | 1 | 1 | 1 | 1 |  |  |  |
|  |  |  | 1 | 1 | 1 |  |  |  |  |
|  |  |  | 1 | 1 | 1 |  |  |  |  |
|  |  |  |  | 1 | 1 | 1 |  |  |  |
|  |  |  |  | 1 | 1 | 1 |  |  |  |
|  |  |  |  |  | 1 | 1 | 1 |  |  |
|  |  |  |  |  | 1 | 1 | 1 |  |  |
|  |  |  |  |  |  | 1 | 1 |  |  |
|  |  |  |  |  | 1 | 1 | 1 |  |  |
|  |  |  |  | 1 | 1 |  |  | 1 |  |
|  |  |  | 1 | 1 | 1 |  |  |  | 1 |

Figure 2: Occupancy array

|  |  |  |  | Sweden |
|---|---|---|---|---|
| Britain |  |  | Denmark |  |
|  |  | Holland | Germany | Germany |
|  |  | Belgium |  |  |
|  | France | France | Italy | Yugoslavia |
| Portugal | Spain |  | Italy |  |

Figure 3: Symbolic array representation

|  |  |  |  | Sweden |
|---|---|---|---|---|
| Scotland<br>Wales | England |  |  | Denmark |  |
|  |  | Holland | Germany | Germany |
|  |  | Belgium |  |  |
|  | France | France | Italy | Yugoslavia |
| Portugal | Spain |  | Italy |  |

Figure 4: Embedded array representation

are best served by a descriptive representation that denotes the spatial relations between meaningful parts of an image. Thus we use a multidimensional symbolic array to depict the spatial structure of an image, where elements of the array denote the meaningful components of an image. The symbolic array captures the spatial relationships of image components, but not necessarily relative sizes or distances. The interpretation of the spatial relations in a symbolic array is dependent on the given domain. If, for example, we use the scheme to represent mental geographic maps, interpretations could include predicates such as *north, east, south* and *west*; if the array is used to represent the image of a room, then the interpretation would involve predicates such as *above, behind, left-of, beside,* etc.

Symbolic arrays can be generated using the frame's procedural attachment facility. The information to define this representation can either be stored explicitly, as in the example of Figure 1 where the PARTS slot contains the indices for the subimages, or derived from a geometric description of an image. Figure 3 illustrates the symbolic array corresponding to the image of a map of Europe. Note that some parts occupy more than one element in an array (e.g., Italy, France). This is necessary to capture all the spatial relationships of the image components. Also note that there is no attempt to capture size or distance information in this representation.

According to Pylyshyn, images are not raw uninterpreted mental pictures, but are organized into meaningful parts, which are remembered in terms of their spatial relations. Furthermore, we can access the meaningful parts; we are able to focus attention on a specific feature of an image (Pylyshyn 1973). Nested symbolic arrays capture this property of mental imagery by representing images at various levels of abstraction (as suggested by the structural hierarchy of the deep representation). For instance, focusing attention on Britain in the array of Figure 3 would result in a new array in which the symbol for Britain is replaced by the symbolic array specifying its parts (see Figure 4). This subimage is generated from the frame representation for the image of Britain.

Similar to the surface representation, a symbolic array can be *2D* or *3D* depending on the application. The symbolic representation can be thought of as descriptive since it can be expressed as a propositional representation, where the predicates are spatial relationships and the arguments are concrete, imaginable, objects. Although information in the symbolic representation can be expressed as propositions, the two representations are not computationally equivalent. The spatial structure of images possesses properties not found in deductive propositional representations. These properties help avoid the combinatorial explosion of correct but trivial inferences that must be explicitly represented in a propositional system. Lindsay argues that spatial image representations (symbolic representations in our case) support non-deductive inference by a constraint satisfaction mechanism built into the processes that construct and access them (Lindsay 1988). Consider, for example, the image of the map of Europe. Determining the countries north of Germany can be achieved by mentally scanning a small portion of the mental image of a map. Similarly, in the corresponding symbolic array we only need to search the subarray above Germany.

Another advantage of symbolic arrays, with respect to propositional representations, concerns the frame problem. Any cognitive system, natural or artificial, should be able to deal with a dynamic environment in which a change in a single item of knowledge might have widespread effects on many other items. Suppose that we change the position of a country in our map of Europe. Using propositions, we might need to change several relations involving the given country, and perhaps relations derived from the initial propositions. Using a symbolic array to store the map, we need only delete the country from its previous position and insert it in the new one. Since spatial relationships are determined functionally (not logically inferred) from image representations, we eliminate the problem of nonmonotonicity in domains involving spatial reasoning.

### 3.3 Primitive Functions for Imagery

Approaches to knowledge representation are distinguished by the the operations performed on the representations. Thus, the effectiveness of our scheme can be partially measured by how well it facilitates the implementation of imagery related processes. Several primitive functions for imagery have been specified in array theory and implemented in Nial. These include functions for constructing images, transforming images and accessing images.

Images can be constructed in unique and creative ways by defining a symbolic array whose components correspond to existing frames in the deep representation. New images can also be created through low level perception techniques that transform an uninterpreted surface representation into a symbolic array. Operations that transform the symbolic and surface representations through rotation and translation have been implemented. As well, a focus function, which transforms a symbolic array by replacing a symbol with its corresponding symbolic array, has been defined. Operations that extract propositional information, relative to a particular domain, have also been implemented. For a more complete and detailed list of functions for computational imagery see (Glasgow 1990).

## 4 Conclusions

This paper treats imagery as a problem solving paradigm in AI and proposes a knowledge representation scheme for computational imagery. Aside from related research in vision, the AI community has given little attention to the topic of imagery. Thus we rely on the relevant theories of cognitive psychology to provide initial guidance for our research. Although psychological plausibility is one of its goals, the scheme was not designed to be a model of human information processing but a tool for practical applications.

The knowledge representation scheme for computational imagery includes three image representations, each appropriate for a different kind of processing: an image is stored in long-term memory as a deep representation that contains relevant descriptive information about the image; the surface representation contains geometric information about the image components; the symbolic array representation is a spatially indexed description of the meaningful parts of the image. A set of primitive functions, which correspond to the fundamental processes involved in mental imagery, have been designed and implemented.

Since the representation scheme is not intended to be a psychological model of mental imagery, we do not suggest that in human working-memory two "mind's eyes" exist that process symbolic and surface representations identical to the ones that we have implemented. What we do claim is that the scheme is informationally equivalent to the representations involved in mental imagery, where two representations are said to be informationally equivalent if all the information in one representation is also inferable from the other and vice versa (Larkin & Simon 1987).

Like Kosslyn's model (Kosslyn 1980), the scheme is based on an array theory of imagery. According to such theories, mental images are array-like surface representations generated through perception, or from long-term, deep representations. Several types of image arrays have previously been suggested. Kosslyn's model uses *2D* arrays for the implementation of the surface representation. Pinker has suggested *2D* arrays combined with information about the *3D* shapes of objects in long-term memory files from which array patterns are generated (Pinker 1980). A recent model based on an array theory of imagery describes how visual information can be represented within the computational framework of discrete symbolic representations in such a way that both mental images and symbolic thought processes can be explained (Chandrasekaran & Narayanan 1990).

Mental imagery can provide insights that contribute

to effective problem solving techniques. A goal of our research is to develop knowledge-based systems that integrate computational imagery with other AI problem solving paradigms. One such system, which is currently under development, is an application to the problem of molecular scene analysis (Glasgow, Fortier and Allen 1991). Other potential applications include vision and tactile perception, medical imaging, motion planning and game playing.

### References


Block, N. 1981. *Imagery.* MIT Press.

Chandrasekaran, B. & Narayanan, N. H. 1990. Integrating imagery and visual representations. *Proceedings of the 12th Annual Conference of the Cognitive Science Society.* Lawrence Erlbaum Associates.

Collins, A.M. & Quillian, M.P. 1969. Retrieval time from semantic memory. *Journal of Verbal Learning and Verbal Behavior,* 8:240-247.

Farah, M.J. 1988. Is visual imagery really visual? Overlooked evidence from neuropsychology. *Psychological Review,* 95:307-317.

Finke, R.A. 1989. *Principles of Mental Imagery.* MIT press.

Finke, R.A., Pinker, S. & Farah, M., J. 1989. Reinterpreting visual patterns in mental imagery. *Cognitive Science,* 13:51-78.

Glasgow, J.I. 1990. Artificial intelligence and imagery. *Proceedings of the Second IEEE Conference on Tools for AI.*

Glasgow, J.I., Fortier S. & Allen F.H. 1991. Crystal and molecular structure determination through imagery. *Proceedings of the Seventh IEEE Conference on Artificial Intelligence Applications,* 98-105.

Jenkins, M.A. & Glasgow, J.I. 1989. A logical basis for nested array data structures. *Computer Languages,* 14:35-49.

Jenkins, M.A., Glasgow, J.I. & McCrosky, C. 1986. Programming styles in Nial. *IEEE Software,* 86, 46-55.

Jenkins, M.A., Glasgow, J.I., Blevis, E., Chau, R., Hache, E. & Lawson, E. 1988. The Nial AI toolkit. *Proceedings of Avignon '88 Eighth International Workshop on Expert Systems and their Applications.*

Kosslyn, S.M. 1980. *Image and Mind.* Harvard University Press.

Kosslyn, S.M. 1987. Seeing and imagining in the cerebral hemispheres: a computational approach. *Psychological Review,* 94, 148-175.

Kubovy, M. & Podgorny, P. 1981. Does pattern matching require the normalization of size and orientation? *Perception and Psychophysics,* 30, 24-28.

Larkin, J. & Simon, H.A. 1987. Why a diagram is (sometimes) worth ten thousand words. *Cognitive Science,* 10:65-99.

Lindsay, R.K. 1988. Images and inference. *Cognition,* 29:229-250.

Marr, D. & Nishihara, H.K. 1978. Representation and recognition of the spatial organization of three dimensional shapes. *Proceedings of the Royal Society,* B200:269-294.

McDermott, D. V. & Davis, E. 1984. Planning routes through uncertain territory. *Artificial Intelligence,* 22:107-156.

More, T. 1979. The nested rectangular array as a model of data. From proc. APL79, *APL Quote Quad,* 9.

Paivio, A. 1975. Perceptual comparisons through the mind's eye. *Memory and Cognition,* 3:635-647.

Pinker, S. 1980. Mental imagery and the third dimension. *Journal of Experimental Psychology: General,* 109, 354-371.

Pinker, S. 1988. A computational theory of the mental imagery medium. In Denis, Engelkamp & Richardson (Eds), *Cognitive and Neuropsychological Approaches to Mental Imagery.* Martinus Nijhorff Publishers.

Pylyshyn, Z.W. 1973. What the mind's eye tells the mind's brain: a critique of mental imagery. *Psychological Bulletin,* 80:1-24.

Pylyshyn, Z.W. 1981. The imagery debate: Analogue media versus tacit knowledge. *Psychological Review,* 88:16-45.

Samet, H. 1980. Region Representation: Quadtrees from boundary codes. *Comm. of the ACM,* 23: 163-170.

# Can Images Be Rotated and Inspected? A Test of the Pictorial Medium Theory.

**Peter Slezak**

Centre for Cognitive Science
University of New South Wales
P.O. Box 1, Kensington 2033 NSW., Australia
cogsci@spinifex.eecs.unsw.oz.au

**Abstract**

Since the "equivalence" of imagery and perception has been one of the central tenets of the pictorial theory, the negative results of Chambers and Reisberg (1985) on an image reinterpretation task may be seen as posing a fundamental challenge for the pictorial account. Finke, Pinker and Farah's (1989) claimed refutation of these negative results may be questioned on a number of methodological grounds. In addition to examining these issues, we report results of an experiment which tests what is seemingly another direct prediction of pictorial theories. Our investigation employs newly devised imagery tasks whose success *depends* on being able to "rotate", "inspect" and reinterpret images. Our negative results add further weight to a tacit knowledge account of images as intrinsically interpreted, abstract symbols.

## New Focus for Imagery Debate

After twenty years of the recent controversy, the 'imagery debate' is widely regarded as having become stalled, with no new ideas on how to break the impasse by bringing some decisive evidence to bear. It is in this regard that the possibility of reinterpreting visual patterns in mental imagery has recently emerged as new focus for the long-standing controversy. The question of whether, and under what conditions, novel information may be discovered from images may shed new light on the debate, since it provides a new means for testing the claimed parallel between imagery and perception and the properties of the conjectured pictorial medium.

## "Equivalence" of Imagery and Perception

Specifically, the possibility of reinterpreting an image follows as a direct implication of the pictorial theory which posits an "equivalence" between imagery and perception. On the pictorial view, a mental image is conceived to be a "surrogate percept, allowing people to detect some pattern or property in a remembered scene that they did not encode explicitly when they saw the scene initially" (Pinker and Finke 1980, p. 246). It is in this sense that the uninterpreted images in a spatial medium are themselves "functionally equivalent to physical objects or events" (Finke, 1980 p. 113), and cause the same mechanisms to be activated as in actual visual perception itself (ibid, p. 130). Kosslyn (1987, p. 149) explains, one purpose of imagery involves "recognition processes" to discover information which is not stored explicitly in memory and thus we "look" at our images in a way which is analogous to the way we look at external objects in order to inspect them.

## Divergent Predictions

By assimilating imagery so closely with vision, indeed by claiming their "equivalence", pictorialism is committed to predicting closely similar "perceptual" phenomena in imagery to those found in perception itself. It is this deep commitment to the perceptual character of imagery which is the source of its vulnerability to such asymmetries as those of Chambers and Reisberg (1985) and our own results.

Thus, Kosslyn suggests that "image interpretation is at the heart of the role of imagery in cognition (if one cannot inspect imaged patterns, they are useless)" (1988, p. 249). In line with the remarks we have seen by Pinker and Finke, Kosslyn too suggests, "The recognition mechanisms [of vision] can be used in imagery as a way of accessing stored information" (1988, p. 264) and his model entails that "images depict visual information, and that this information is interpreted by some of the same sorts of classificatory procedures used in classifying sensory input during vision" (1980, p. 32). Kosslyn explains further that "the purposes of imagery, in large part, parallel those of vision" and "one may 'recognize' parts and properties of imaged objects that had not been previously considered" through the "use of recognition processes" (1987, p. 149). Thus, the significance of reinterpreting images as a crucial test of the pictorial theory is evident in Kosslyn's explicit predictions:

> The image is formed by forcing a change of state in the visual buffer in the attended region, *which can then be reprocessed as if it were perceptual input (e.g., the shape could be recategorized)*, thereby accomplishing the purposes of imagery that parallel those of perception. (1987, p. 155; emphasis added)

By contrast with these direct implications of the quasi-perceptual, pictorial-medium theory, the 'tacit knowledge' account would predict that the re-interpretation of images is difficult because it construes the mental representations as highly abstract entities which are the output of 'higher' cognitive processes; these are encodings of conceptualizations or beliefs and, in this sense, already meaningful and not requiring interpretation, - nor susceptible of easy *re*-interpretation (Pylyshyn 1973, 1978).

### New Paradigm: Non-Chronometric Tests

At least part of the reason for the persistence of the imagery debate has been the fact that the dispute has involved alternative explanations of the *same* chronometric evidence. Accordingly, experiments are needed on which the contending accounts deliver *different* predictions. In particular, more directly illuminating would be data concerning a task whose success is directly dependent upon the operation of the postulated process. Julesz's (1971) random-dot stereograms served in this way as an unambiguous and unfakeable test for stereopsis. Divergent predictions of the contending theories on an 'all-or-none' task is a significant feature of the experiment which we report here. Our own evidence concerns a perceptual organisation task which provides unequivocal criteria of the successful rotation, inspection and re-interpretation of images using "recognition processes" and "shape classification" procedures. Despite the demonstrated ease of our task under *perceptual conditions,* naive subjects have been uniformly unable to succeed in the task under imagery conditions as would be predicted on the pictorial theory.

## Chambers & Reisberg Negative Results

This question of reinterpreting visual images had been brought into sharp relief with the work of Chambers and Reisberg (1985) who found that subjects were uniformly unable to reverse their mental images of the familiar ambiguous figures such as the duck/rabbit and Necker cube. Chambers and Reisberg see their results as supporting the "philosophical" arguments for taking imagery to be more conceptual and cognitive, in the sense that they are intrinsically interpreted symbols which do not need, and do not easily permit, further interpretation. Since the close parallel, indeed "equivalence" of imagery with the mechanisms of perception has been one of the central tenets of the pictorialist theory (Podgorny and Shepard 1978, Finke 1980, Farah 1988), the results of Chambers and Reisberg are surprising and may be seen as posing a fundamental challenge for the pictorial theory.

Of course, the negative results of Chambers and Reisberg provide experimental support for the specific claims made over a decade ago by Pylyshyn (1973, 1978) in his critiques of pictorial theories. Despite these earlier skeptical claims, however, in the intervening period there has been experimental evidence of just such abilities of people to detect novel properties in imagined scenes. Pinker and Finke (1980) report subjects' ability to "see" novel properties which "can be 'read off' the display" and should emerge from images after mental rotation. Finke and Slayton (1988) have extended this work, providing further evidence "that people are capable of making unexpected discoveries in imagery" and that novel patterns can "emerge" from within imaged patterns.

## Response by Finke, Pinker and Farah.

Most recently, Finke, Pinker and Farah (1989) have sought to reinforce these claims with new experiments which also purport to show that subjects can inspect and reinterpret their images by "applying shape classification procedures to the information in imagery" (1989, p. 51). This latter work is of particular interest because it has been specifically designed to counter the skeptical conclusions warranted by the negative results of Chambers and Reisberg and sets the scene for our own investigation. Thus, it is in the light of this clash of experimental results and theoretical claims that our own experiment is to be understood: Our new experiment avoids the specific objections to Chambers and Reisberg results while also addressing potential problems of their own procedures. In this way, by falsifying central entailments of the pictorial medium theory in a different manner, our results can be seen as illuminating the precise conditions under which such seemingly contradictory results can be obtained.

### Methodology of Falsification

Discounting "philosophical" considerations, Finke et al. claim to "refute" Chambers and Reisberg by showing them to have been mistaken on the "strictly empirical" claim that mental images are not subject to reconstrual. Construing the question at stake in this way as a "strictly empirical" one, Finke et al. assert "our experiments show that the answer to it is that such reinterpretation is possible" (1989, p. 74). But this notion of refutation uncharitably construes the Chambers and Reisberg case in its most implausible and most uninteresting form. The logic of their claim is not merely that images cannot be interpreted in some unqualified and completely general sense, and consequently refuted by finding a contrary case. The Chambers and Reisberg investigation derives its significance from showing the inability of image reconstrual precisely *under conditions in which we would have expected it according to the quasi-perceptual account.* By demonstrating the possibility or reinterpretation under entirely different conditions, Finke et al. have merely avoided the issue. It is not the "strictly empirical question" of whether or not images can be reconstrued *simpliciter*. To be sure, *this* is an empirical question in an unproblematic sense, but it is not the question raised by the investigation of Chambers and Reisberg. Undeniably, *"given suitable conditions"*, as Finke et al. say, people *are* able to reconstrue their images. But clearly, it only begs the deeper questions raised by

Chambers and Reisberg, - questions which are, moreover, inescapably the "philosophical", that is, *theoretical*, ones Finke et al., have wished to dismiss (See Slezak 1990). The apparent conflict among the data requires careful analysis which will clarify the precise conditions of discrepant results.

## Interpretation and Nothing But

In their attempt to deflect the consequences of the negative findings of Chambers and Reisberg, Finke et al. claim that they "refute one kind of explanation for this difficulty: [namely] that visual images *do not contain information about the geometry of a shape* necessary for reinterpreting it..." (1989, p. 51; emphasis added). That is, Finke et al. devote much of their effort to refuting what they refer to as "the strong position that images are nothing but interpretations" and consistently attribute to Chambers and Reisberg the conclusion that "images *do not contain* uninterpreted information" (1989, p. 54). They take Chambers and Reisberg to claim "that images lack 'uninterpreted' information pertaining to the geometry of an object" (ibid, p. 54). This is misleading because Chambers and Reisberg do not need to go so far as to deny the presence of any purely geometric information in images. They only assert the intrinsic association of such information with a semantic interpretation and that this information is abstractly encoded, inherently interpreted and, therefore, not readily dissociated from the old interpretation and re-associated with a new one by visual mechanisms. That is, the claim of Chambers and Reisberg is only that images are "created as symbols of something and hence *need no interpretive process*" (1985, p. 317).

## Interpretation as Semantic Content

It is essential to notice that the notion of interpretation at issue for Chambers and Reisberg is explicitly 'semantic' in the sense of taking the symbol (the image) to have a referent or meaning. Though they dismiss such "philosophical" considerations as irrelevant, this point is crucial in appreciating both the failure of the critique by Finke et al. and also the manner in which their own experimental design is problematic. The relevant notion of interpretation certainly does not preclude the encoding of geometric information as Finke et al. have claimed, and is a semantic one, being referential or 'intentional' in Brentano's sense that every mental phenomenon has "reference to a content" or "direction toward an object" (1985, p. 318). The interpretation of an image in this conception consists in the fact that images are *about* some object or scene

Consistent with having neglected this "philosophical" point in their critique, Finke et al. also design their experiment in a way which minimises the crucial relevance of the intentionality of symbols. The departure of the experiments of Finke et al. in this regard from the paradigm, serves to weaken their relevance for those of Chambers and Reisberg. Finke et al. want to show that "people *can* assign novel interpretations to ambiguous images" (1989, p. 51) but the "stimulus" figures or patterns employed were all of a *purely* geometric character including letters and numbers. That is, there is, at best, only a very attenuated sense in which one can speak of 'interpretation' and 're-interpretation' in regard to these patterns at all, and certainly there is nothing which corresponds to the full-bodied meanings or semantic content of the ambiguous duck/rabbit, for example. This crucial difference in the stimuli used by Finke et al., is such as to weaken the force of their claimed refutation of the Chambers and Reisberg results. It is significant, therefore, that in our own experiment, when the stimuli have a semantic interpretation in the fullest sense, once again the subjects were unable to accomplish the reconstruals which are easy and instantaneous in perception.

Finke et al. anticipate a possible objection that their stimuli had no initial construal of their stimuli to be switched away from. They answer by claiming that all of their "stimuli" had at least two interpretations, notwithstanding "The fact that one of the two interpretations was invariably characterised by a complex articulated description rather than by a single word" (1989, p.69). However, the issue is not the complexity or otherwise of *descriptions* but rather the nature of a *perceptual gestalt*. In particular, it is not the one-word *labels* which might attach to the ambiguous figures such as 'duck' or 'rabbit' which is their salient property, but rather the fact that the figure permits a *visual* identification and interpretation *as* something in the first place - that is, other than the geometrical figure itself. It is the claimed "equivalence" with the interpretive mechanisms of visual perception which is at stake here, and it is precisely this possibility of semantic interpretation which is absent from the geometrical figures in Finke et al.'s experiment.

## "Suitable Conditions"

Finke et al. claim that their experimental results "show that the kind of object a mental image corresponds to need not be assigned during an act of perception" (1989, p. 67) in a way which might prevent its easy re-interpretation, as would be expected on Pylyshyn's view. However, it must be noticed that, in a further significant departure from the experimental paradigm of Chambers and Reisberg, the "stimuli" and "patterns" used in their experiments were not visually presented perceptual objects at all. Finke et al. use *verbal descriptions* of certain patterns. As is well known, seeing the film and reading the book are radically different in their consequences for memory encoding and in the constraints placed on imagination. As a matter of experimental procedure, Finke et al. rightly require that any *new* interpretation of a pattern should not have been the result of "having encoded that interpretation while the stimulus was *actually visible*" (1989, p.55; emphasis added), but they manage to avoid this problem of perceptual confounding by avoiding relevant perceptual encoding altogether. In the absence of explicit theoretical justification, this further departure from the experimental conditions which have been central to the dispute obscures the relevance of their results. A particular virtue of our own experiment is its reverting to visual stimuli, while avoiding the confounding by perceptual reconstrual in other ways.

### Episodic and Semantic Images

The relevant distinctions here correspond closely with Tulving's (1972) distinction between 'episodic' and 'semantic' memory. Tulving's 'episodic' memory stores information about dated episodes or events such as specific perceptual experiences coded in terms of perceptible properties or attributes of the events and their autobiographical reference. Of particular relevance in connection with the debates over re-interpretation of visual images is the supposed difference between episodic and semantic memory regarding the retrieval of information. Tulving writes:

> The episodic memory system does not include the capabilities of inferential reasoning or generalization. Inferential reasoning, generalization, application of rules and formulas, and use of algorithms, on the other hand, represent important methods of utilization of information stored in semantic memory. By relying on his semantic memory, it is literally quite possible for a person to know something he did not learn. (1972, p. 390)

## The Experiment

### Predictions of the Pictorial Theory

The experiment[1] is concerned to test the 'quasi-perceptual' claims of the pictorial theory according to which "the purposes of imagery, in large part, parallel those of vision" in "the use of recognition processes" through which Kosslyn claims "one may 'recognize' parts and properties of imaged objects that had not been previously considered" (1987, p. 149) The abilities being tested are exactly those which are claimed to have been demonstrated by Pinker and Finke (1980) who reported subjects' ability to "see" novel properties which could emerge from images only after they had been mentally rotated where new properties of images "can be 'read off' the display" (1980, p. 262). The ability to "re-parse" an image by using perceptual shape classification procedures has been claimed by Finke et al. (1989), just as Finke and Slayton (1988) claim to have shown "that people are capable of making unexpected discoveries in imagery".

### Non-Chronometric Test

The new stimulus materials have been designed to have two distinct interpretations which are highly orientation specific. Thus, the figures are recognizable as a certain object in one orientation, but are interpretable as an entirely different object when rotated by 90 degrees. These stimuli are variants of the stimuli used by Rock (1973), and are considerably improved in their recognizability. In this respect, the shapes have the important feature that the alternative interpretations are readily obtained by rotation under *perceptual* conditions.

Figure 1.

It is important that the task of reinterpretation can be readily accomplished in this way during perception because this makes the conditions for reconstrual under imagery conditions as favourable as possible. Thus, for example, when subjects are shown figure 1 in one orientation, it is immediately recognized as the duckling; then upon rotating the figure by 90 degrees, subjects immediately notice (with frequent expressions of surprise and delight) the alternative interpretation, the rabbit.

Figure 2.

The direct expectation of the pictorial medium theory is that the same effect should be obtainable under imagery conditions. That is, if subjects are shown the figures in only one orientation, it would be expected that they could *rotate their image* and *discover* the alternative construal by inspection *from their rotated image*. Of course, the tacit knowledge alternative account takes images to be abstract, intrinsically interpreted conceptualizations and would predict that such reinterpretation would be difficult or impossible for subjects to perform in this way on their rotated images.

[1]Independently, Reisberg and Chambers (1990) have conducted methodologically similar experiments, but they are not concerned with the "format" question as such, and, in fact, appear to endorse Kosslyn's pictorial medium view as "fully persuasive".

## Method

**Subjects.** The subjects were 15 students and staff at the University of New South Wales who participated on a voluntary basis.

**Materials and Viewing Arrangement.** The stimulus figures were presented in the form of black and white computer displays to subjects who were seated at a normal viewing distance of approximately 40 cm from a high-resolution Apple Macintosh screen. The display was under the control of the experimenter.

**Procedure.** Without marking any distinction among the stimulus figures, subjects were first shown several distractors in the form of sillhouettes of easily recognizable animals such as an elephant, ostrich and marlin, none of which were orientation-dependent shapes having any alternative interpretation. This was to set subjects' expectations for the subsequent stimuli which were, in fact, orientation-dependent figures. For each presented figure, subjects were asked if they could recognize it and to name it. Subjects were then shown an orientation-dependent figure and were asked to remember it in order to answer a question afterwards. After 10 seconds the stimulus figure was then removed from view and the subject was asked to imagine rotating it by 90 degrees in a clockwise direction. When the subject confirmed that the figure was being imagined in this rotated position, he/she was asked if it could be interpreted as anything else from this viewpoint. This procedure was repeated with each of the orientation-dependent shapes in turn.

As Finke et al. rightly emphasize, an important methodological problem arises from the danger that subjects might encode both interpretations during the original perceptual exposure since in this case they would not be relying on imagery processes to discover the alternative interpretation. In particular, it is evident that the problem may arise when subjects are set the imaginal rotation task more than once, since after the first one they are no longer naive concerning the possibility of a second construal of the figure. On any subsequent test, they may unavoidably seek a second interpretation during the initial perceptual exposure. This possibility can be minimised by restricting exposure time as Chambers and Reisberg have done to a duration which is long enough to establish the image and yet not long enough to seek alternative interpretations. However, this is a risky procedure since subjects who are aware of a possible reconstrual are frequently able to notice both interpretations almost immediately and simultaneously.

**Results.** At first glance, the experimental results appear somewhat equivocal on the question of reinterpretation since subjects were generally able to reconstrue in imagination about *one third* (35%) of the figures they were presented. Even on these data it is clear that reinterpretation of the rotated image was difficult to perform, even if not always impossible. However, these results across multiple presentations take on a greater significance when the *order of presentation* is taken into account: It is most significant that no subject was able to reconstrue the *first* stimulus presented, which is, of course, the only one for which they did not know in advance that there might be an alternative interpretation. This striking relevance of stimulus order confirms the anticipated effects due to loss of naivete concerning the task. Subsequent stimuli were, nevertheless, worth testing and it is significant that even in these cases the difficulty of reconstrual is apparent.

The interpretation of these preliminary results will be clarified in subsequent testing which can eliminate possible confounding due to two factors. The ordering effect in which subjects showed an improvement from initial failure could be due to practice in the task rather than perceptual confounding. This will be controlled in follow-up experiments by giving subjects prior practice with image rotation using Cooper's (1975) random polygons. The uniform observed difficulty with the first naive exposure to an orientation-dependent shape will then be less likely due to the inherent difficulty of the task itself. Present data showing slight improvement following the first stimulus could also have been due to the fact that the shapes were not presented in random order. Given the considerable difficulty in devising suitable figures, they are not equally good in their orientation specificity. Stimuli were presented in order of decreasing suitability and this could account for the slightly improved success rate across subjects. In repeating the experiment, the shapes can be independently rated for their recognizability and orientation-specificity, and the order of presentation can be randomized.

Even if we disregard ordering and other effects such as practice which might have made the task easier, the mean success rate overall was still only 35 per cent. Typical of the predicted difficulty of reconstrual was the reaction of subjects when pressed to reinterpret their rotated image of the duckling: just as we would expect on a tacit knowledge account according to which the image is intrinsically bound to its interpretation, many subjects would volunteer the response that it is a "duckling on its back"!

**Discussion and Conclusion.** We have noted that the significance of our negative results derives from the fact that, unlike chronometric data, our own experiment permits a *direct* inference about the rotation and inspection of imagery. Of additional importance is the fact that the task is readily performed under perceptual conditions, thereby entitling us to expect it in imagery as well according to the pictorial account. Further favouring reconstrual of images is the fact that our figures are considerably simpler than the representations of blocks stacked in three dimensions employed by Shepard (1971) for which the mental rotation has been claimed, and our own shapes are geometrically no more complex that those of Cooper (1975) for which complexity was specifically found *not* to be a factor in the claimed ease of rotation.

Rock, Wheeler and Tudor (1989) have recently obtained negative results which also challenge the claims of mental rotation and scanning in a pictorial medium. Although supporting his results and conclusions, our experiment has the advantage of utilizing *recognizable* shapes whose identification is the unambiguous criterion of success in the imagery task. As Rock et al. have acknowledged, their amorphous-shaped wire stimuli are unusual and not representative of objects typically encountered. Although this factor is discounted by Rock et al., the issue does not arise as a potential objection with our stimuli.

As distinct from reaction-time data, when tested on a task which actually *requires* these abilities for success we have found subjects to perform very poorly or not at all. Although these results are contrary to the explicit predictions and direct implications of the pictorial theory, they are precisely as would be expected on a tacit knowledge account. It must be admitted that the tacit knowledge theory is little more than a sketch or hint of the *kind* of theory required rather than a worked out alternative to the quasi-perceptual account. Nevertheless, filling out this sketch will draw on theories of higher cognitive processes in long term memory which have independent theoretical justification.

Contrary to some views (Anderson 1978), we have shown that there *are* significant empirical differences between these two theories of imagery, and their markedly divergent predictions may be readily tested. As a follow-up, we have devised experiments which reconstruct other perceptual phenomena in imagery. Preliminary investigations suggest that *figure-ground separation* is exceedingly difficult or impossible to perform in imagery, although once again the parallel would be expected on the pictorial account. Likewise, the Kanizsa *illusory contours* appear not to be imposed under imagery conditions. We are also repeating an earlier image inspection experiment of Binet's using an array of letters which has not received discussion in the recent debate, but which has the same negative implications regarding the claimed ability to "inspect" images. A pattern of such failures on diverse perceptual phenomena would leave only *ad hoc* ways of avoiding their significance for the pictorial theory of imagery.

## References


Anderson, J.R. 1978. Arguments Concerning Representations for Mental Imagery. *Psychological Review* 85:249-277.

Chambers, D. and Reisberg, D. 1985. Can Mental Images Be Ambiguous? *Journal of Experimental Psychology: Human Perception and Performance* 11:317-28.

Cooper, L.A. 1975. Mental Rotation of Random Two-Dimensional Shapes. *Cognitive Psychology* 7:20-43.

Farah, M.J. 1988. Is Visual Imagery Really Visual? Overlooked Evidence From Neuropsychology. *Psychological Review* 95:307-317.

Finke, R.A. 1980. Levels of Equivalence in Imagery and Perception. *Psychological Review* 87:113-132.

Finke, R.A., Pinker, S. and Farah, M.J. 1989. Reinterpreting Visual Patterns in Mental Imagery. *Cognitive Science* 13:51-78.

Finke, R.A., & Slayton, K. 1988. Explorations of Creative Visual Synthesis in Mental Imagery. *Memory and Cognition*. 16:252-257.

Julesz, B. 1971. *Foundations of Cyclopean Perception*. Chicago: University of Chicago Press.

Kosslyn, S.M. 1980. *Image and Mind*. Cambridge, Mass.: Harvard University Press.

Kosslyn, S.M. 1981. The Medium and the Message in Mental Imagery: A Theory. *Psychological Review* 88:46-66.

Kosslyn, S.M. 1987. Seeing and Imagining in the Cerebral Hemispheres: A Computational Approach. *Psychological Review* 94:148-175.

Kosslyn, S.M., Pinker, S., Smith, G.E., and Schwartz, S.P. 1979. On the Demystification of Mental Imagery. *The Behavioral and Brain Sciences* 2:535-581.

Kosslyn, S.M. 1988. Imagery in Learning. In Michael S. Gazzaniga ed. *Perspectives in Memory Research*, Cambridge, Mass.: Bradford/MIT Press.

Pinker, S. and Finke, R. 1980. Emergent two-Dimensional Patterns in Images Rotated in Depth, *Journal of Experimental Psychology: Human Perception and Performance* 6(2):244-264.

Podgorny, P., and Shepard, R.N. 1978. Functional Representations Common to Visual Perception and Imagination. *Journal of Experimental Psychology: Human Perception and Performance* 4:21-35.

Pylyshyn, Z. 1973. What the Mind's Eye Tells the Mind's Brain: A Critique of Mental Imagery. *Psychological Bulletin*. 80(1):1-24.

Pylyshyn, Z. 1978. Imagery and artificial intelligence. In *Minnesota Studies in the Philosophy of Science*, C. Wade Savage ed. Vol. IX. Minnesota: University of Minnesota Press.

Reisberg, D. and Chambers, D. 1990. Neither Pictures Nor Propositions: What Can We Learn from a Mental Image? *Canadian Journal of Psychology*, Forthcoming.

Rock, I., Wheeler, D. and Tudor, L. 1989. Can we Imagine How Objects Look From Other Viewpoints? *Cognitive Psychology*. 21(2):185-210.

Shepard, R.N., and Metzler, J. 1971. Mental rotation of three-dimensional objects. *Science* 171:701-703.

Slezak, P., 1990. Re-Interpreting Images. *Analysis* 50(4):231-243.

# A Connectionist Model of Alphabetic Spelling Development and Developmental and Acquired Dysgraphia

**Richard P. W. Loosemore, Gordon D. A. Brown and Frances L. Watson**

Cognitive Neurocomputation Unit
Department of Psychology
University of Wales, Bangor
Gwynedd, LL57 2DG
United Kingdom
Earn/BITNET: PSSØ3Ø@uk.ac.bangor.vaxa

**Abstract**

In this paper we describe a connectionist model of the development of alphabetic spelling. The model learns to spell regular words more quickly than words with irregular spellings. When the computational resources available to the model are restricted, the model learns more slowly and, analogously to developmental dyslexics, fails to learn some of the irregular items in its vocabulary. Experimental evidence is reported, which shows that both normal and dyslexic children of various ages have difficulty with particular word types that are similar to the problems experienced by the model on the same words. Finally, the model is "lesioned," and its performance is then similar to that of "surface dysgraphics." The good fit between model and data is taken as evidence that, throughout much of the relevant developmental period, the task facing children can be usefully viewed as a statistical one.

## Introduction

Our purpose in the present paper is to apply the connectionist modelling approach, which has provided useful unifying models of spelling-to-sound translation in reading (e.g. Seidenberg & McClelland, 1989a), to the case of spelling development. Here, the computational task faced by children is one of mapping phonology onto orthography. This mapping is more "irregular" than the mapping from spelling to sound investigated by Seidenberg & McClelland, and one of our aims is to determine whether a connectionist system could learn this more demanding set of associations.

The spelling of many English words cannot reliably be derived from phonological information alone, because of words such as SOAP (cf. ROPE, COPE, HOPE), and pairs or triples of words that are pronounced identically but spelled differently (e.g. BARE/BEAR and HAIR/HARE).

The research reported in this paper was supported by grants to the second author from the Medical Research Council (U.K.), the Economic and Social Research Council (U.K.) and the Leverhulme Trust.

Pronunciation information nevertheless partially constrains the written forms of words, and there is abundant psychological evidence, from the study of adult and child spelling errors, that pronunciation information is used in spelling. The spelling of non-words is also generally assumed to implicate mechanisms for translating sound into spelling, as are the characteristic patterns of breakdown exhibited by patients suffering from various forms of head injury (see Shallice, 1988, for a review).

These considerations have lead psychologists to produce models of spelling in which there are two separate routines available to translate sound into spelling. Early spelling is assumed to occur by learning the visual forms of a small vocabulary. Subsequent development of "alphabetic spelling" occurs with the realisation that there are useful regularities in the sound-to-spelling mapping system. When a child has grasped this principle, the ability to spell non-words will emerge, along with an advantage in spelling words which conform to the sound-to-spelling "rules" of English. For the reasons discussed above, however, rules alone are not sufficient. A second "lexical" routine is required to enable the correct spelling of irregular words, and to distinguish between non-homographic homophones such as THEIR and THERE. It is assumed that skilled spellers have both strategies available to them. This "dual-route" approach has provided a successful framework for describing spelling development and classifying developmental and acquired spelling disorders (Ellis, 1984; Frith, 1985; Ellis & Young, 1988).

A consequence of the "dual-route" approach is that word classifications have tended to label items as either "regular" or "irregular," according to whether or not they conform to the sound-to-spelling translation rules of English. However, a connectionist model should be able to exploit regularities at various levels in the mapping from sound to spelling without making use of explicit rules. In the case of reading, the connectionist approach has lead to a different classification of words in terms of the spelling-to-sound "friends" and "enemies" that they possess (Brown, 1987; Jared, McRae & Seidenberg, 1990). According to this approach, only words with enemies would be counted as irregular, but regular words can differ according to the number of friends they have. Table 1 illustrates this classification as applied to the case of spelling, and shows that "number of friends" and "number of enemies" reflect different dimensions (here, as elsewhere, we use the "rime" of a word to classify it: there is ample evidence that this is the relevant psychological dimension: Treiman & Chafetz, 1987).

| Word | Friends | Enemies |
|---|---|---|
| SOAP | none | HOPE, COPE etc |
| LOCK | DOCK, ROCK etc | none |
| BULB | none | none |

Table 1: Classification of Word Types

Thus friends and enemies effects can be assessed independently: a difference between "bulb" type words and "soap" type words would reflect an independent effect of number of sound-to-spelling enemies, and a difference between "bulb" type words and "lock" type words would reflect an independent effect of number of sound-to-spelling friends. In the case of single word reading it now appears that there are independent friends and enemies effects on reading time (Brown, 1987, Jared, McRae & Seidenberg, 1990). Many experimental investigations of spelling have, however, not distinguished between sound-to-spelling friends and enemies, and empirically it is not clear which is the relevant factor, although several recent experiments on spelling have found influences of factors other than simple regularity on spelling performance (e.g. Barry & Seymour, 1988). In this paper we report an experimental investigation of this question which focusses on the three word types illustrated in Table 1. First, however, we describe a connectionist model that we have implemented to investigate learning of the English sound-to-spelling mapping system.

## A Connectionist Model of Spelling

The model is a feedforward backpropagation net with 50 input, 30 hidden and 50 output units. The training set consisted of 250 words. These included all 63 words used in the experiment discussed below. The remaining words were friends and enemies of words in the group of 63. Input patterns encoded the phonological forms of words, and output patterns represented the corresponding orthographic forms. These patterns were created using a scheme similar to that employed by Seidenberg & McClelland (1989a): words were treated as sets of "triples," (e.g. **SOAP** = **_SO** + **SOA** + **OAP** + **AP_** ; */swp/* = *_sw* + *swp* + *wp_* ). Patterns were created independently for all the triples that occurred in the training set. Each triple pattern contained four ON bits and 46 OFF bits: these were chosen quasi-randomly in such a way that similar triples had similar patterns. The pattern for a word was made by superposing the patterns for its triples. The learning rate was set at 0.01, and the momentum parameter was 0.9. The model was given 2000 epochs of learning with the training set.

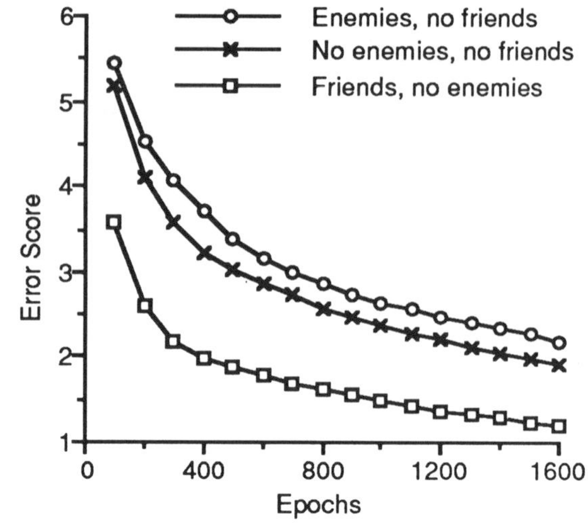


Figure 1: Error Score for 30-Hidden-Unit Model

The vocabulary of the model contained 21 matched word triplets. The words were selected with a view to their later use in the experimental investigation, and corresponded to the three word types in Table 1. Words within each triplet varied in their sound-to-spelling correspondences but were matched as closely as possible on word frequency, positional bigram frequency, and word length. No word in the sample was homophonic with any other English word.

In addition to examining the development of spelling in normal children, we wanted to assess the possibility that the spelling problems experienced by developmental dyslexics could be characterised in terms of reduced computational resources being devoted to the learning process. Similar claims have been made in the case of reading: Seidenberg & McClelland (1989b) found that a version of their spelling-to-sound translation model, which was given fewer hidden units over which to represent its regularities, produced a pattern of performance that was

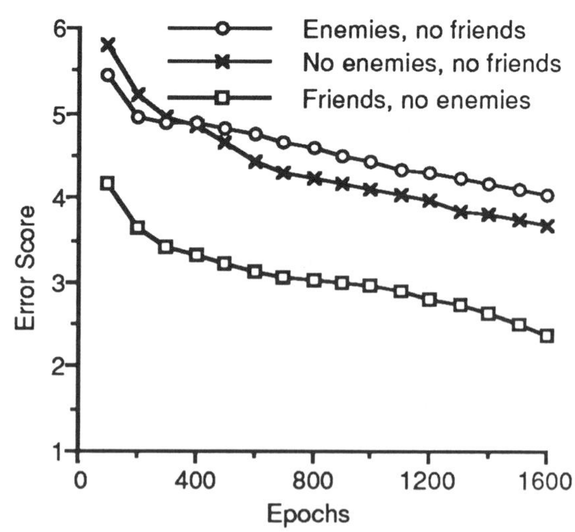


Figure 2: Error Score for 20-Hidden-Unit Model

analogous to that of developmentally dyslexic or delayed readers. In our simulations we adopted a similar approach. The "normal" model was given 30 hidden units, while the "developmentally dyslexic" model was provided with only 20 hidden units during the learning process.

## Results of Simulations

The behaviour of the model during learning is presented in Figures 1 through 4. Figure 1 shows the mean summed squared error for each of the three word types during learning in the "normal" model, and Figure 2 shows the same measure for the "developmentally dyslexic" model. In both cases the same ordering of word type difficulty is seen throughout, but the dyslexic model learns more slowly and has higher error scores at any given point in learning.

Of more direct relevance is the actual number of items that the network was able to spell correctly at any point. In order to assess whether a word was spelled correctly, the following procedure was carried out. First, 30 "competitor" spellings were generated for each of the 63 words by randomly changing one of the letters in the correct, "target" spelling. For the words with enemies, one competitor was always the regularised spelling of the word. Thus, the competitors for SOAP included SOPE, SSAP, SOOP, SOAB, TOAP and so on. At each stage of learning, a summed squared error was calculated for all competitors of each word, as well as for the target spelling. A word was classified as correctly spelled, and counted as such in Figures 3 and 4, when the error score for the correct spelling was lower than the error score for *all* of the competitor spellings. Figure 3 shows the *number of words correct* (out of the 21 words of each type) as a function of training for the normal (30 hidden unit) model, while Figure 4 shows the same for the restricted version.

Inspection of Figure 3 reveals that by 2000 epochs of learning the normal network was producing the correct spelling of every word in its vocabulary. At all earlier stages in learning, the model produced the correct spelling of more of the words with friends and no enemies than words with neither friends nor enemies, and even fewer of the words with enemies and no friends.

Comparison of Figures 1 and 3 with Figures 2 and 4 shows that the dyslexic model exhibits a qualitatively similar pattern of performance to the non-dyslexic model, but that performance lags behind the non-dyslexic model at all stages. By the end of 2000 epochs of learning, the dyslexic model correctly spells 20 out of the 21 words with friends only, 17/21 of the words with neither friends nor enemies, and also 17/21 of the words with enemies and no friends. This pattern of selective difficulty with exceptional items is similar to the pattern observed in developmentally dyslexic children.

It is clear that the non-dyslexic model was successful in learning the relevant sound-to-spelling mapping characteristics within its limited vocabulary. Furthermore, the performance of the model leads to a clear prediction about the ordering of word type difficulty during learning.

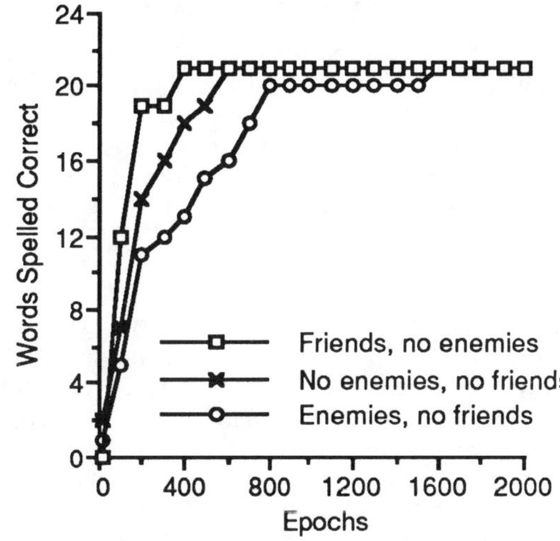


Figure 3: Number of Correct Spellings ("Normal" Model)

The results of the two simulations suggest that the difference between normal and dyslexic spelling development can be well characterised in terms of the amount of computational resources devoted to the task. When insufficient resources are allocated to learning the relevant sound-to-spelling associations, the result is that a lower overall level of performance is achieved at any given stage in learning, but the ordering of the different word types in terms of accuracy is the same. This is consistent with a "delay" rather than a "deviance" picture of dyslexic spelling: the model predicts that when dyslexic and non-dyslexic subjects are matched for overall level of performance, they should show the same qualitative pattern of errors on the three word types. By contrast, "deviance" models assume that qualitatively different processing is involved in developmental dyslexia.

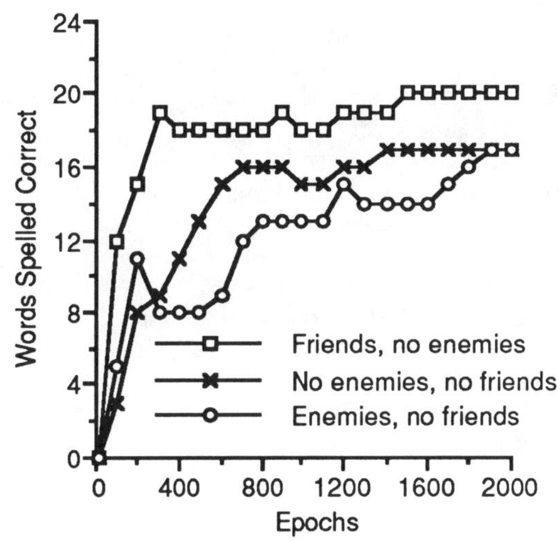


Figure 4: Number of Correct Spellings ("Dyslexic" Model)

## Experimental Studies

We now report the results of experiments carried out on both normal and dyslexic children to test the predictions of the models reported above. We describe the spelling performance of four different age-groups of normal children, and a separate comparison between dyslexic and spelling-level matched non-dyslexic subjects.

### Subjects

132 non-dyslexic children aged between 91 and 143 months were tested. Data were also collected from 12 "junior" dyslexics and 13 "senior" dyslexics from a specialist school. One of the senior subjects (with the lowest total spelling correct score) was excluded from the analysis to make equal sized groups. All the dyslexic subjects tested in this study had been formally diagnosed as having specific learning difficulties by an independent examiner and were attending a special school for dyslexic boys. Additional tests showed that the junior dyslexics had a reading age 30 months behind their chronological age, and the senior dyslexics were 37 months delayed. Reading ages were assessed by the British Ability Scales.

### Materials and Procedure

The stimulus materials for the experiment were the 21 word triplets used in the models. For the spelling test, each stimulus word was presented in a short sentence that used the word in a meaningful context but did not define its meaning. A separate comprehension test was conducted and error rates were examined only for words that were known to individual subjects.

### Analysis and Results

The 132 non-dyslexic subjects were divided into four groups of 33 subjects on the basis of spelling ability. To compare normal and dyslexic spellers, control subjects were selected from the sample of normal spellers. They were matched with individual dyslexics on the basis of total correct spelling scores. The dyslexics were on average around two years older than the control subjects spelling at the same level.

The results were examined in terms of the proportion of errors made to *known* words only. Two triplets were withdrawn before analysis because insufficient subjects knew the meanings of one of the words in the triplets. Spelling scores were expressed as the number of correctly spelt words as a proportion of the items known, and subjected to angular transformation prior to analysis. Figure 5 presents the (untransformed) error proportions for the four different age groups of control subjects, and Figure 6 gives the proportions for the junior and senior dyslexics and controls.

For normal subjects (data in Figure 5), ANOVA revealed main effects of age-group and word-type, and a significant group X word type interaction reflecting a ceiling effect in

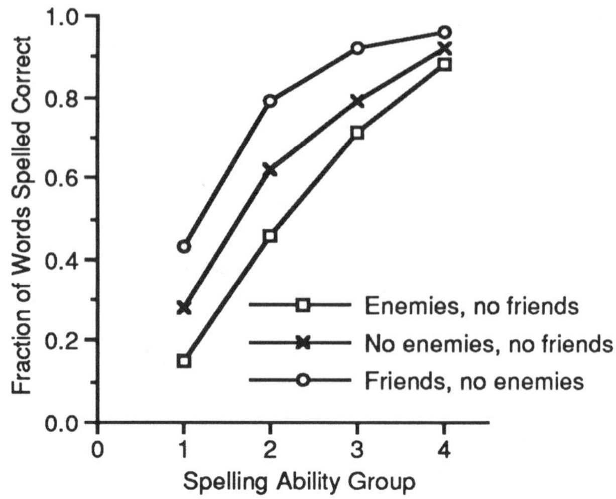


Figure 5: Fraction of Correct Spellings (Normal Subjects)

the more able groups. Post-hoc analyses of the error rates revealed independent effects of both sound-to-spelling friends and sound-to-spelling enemies within the main effect of word type. ("Friends effects" were assessed by comparing error rates to words with many friends and no enemies to words with neither friends nor enemies; "enemies effects" were assessed by comparing error rates for words with neither friends nor enemies to words with enemies but no friends).

In the error analyses comparing dyslexic and control subjects, there was no significant difference between control and dyslexic subjects (as expected, given that groups were matched on total spelling score) and no significant interactions. There were, however, main effects of both ability group and word type. Post-hoc comparisons of the word type means indicated that there were independent sound-to-spelling friends and enemies effects.

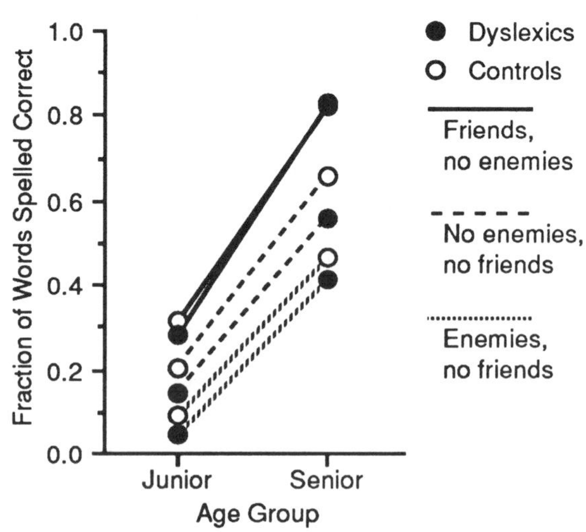


Figure 6: Fraction of Correct Spellings (Dyslexic Subjects)

The results may be summarised as follows. Both dyslexic and control subjects showed independent effects of sound-to-spelling friends and sound-to-spelling enemies on error rates. Furthermore, dyslexic subjects showed the same pattern of effects as younger non-dyslexic subjects matched on overall spelling scores.

These results are in accord with the predictions made by the model. The ordering of difficulty of the three different word types, in terms of the proportion of errors made at different times during development, is exactly the same for the connectionist model and for our human subjects.

Furthermore, despite the fact that we have used a more fine-grained classification of word types than has previously been used to compare normal and dyslexic spelling performance, we find that the dyslexics perform similarly to younger control subjects spelling at the same overall level. This is consistent with the behaviour of our "developmentally dyslexic" model.

## Lesioning the Model

We wish to argue that the development of normal spelling, and the delayed development in young dyslexics, can be characterised in terms of the computational demands of mastering a statistical mapping system. A natural question, then, is whether the performance of head-injured patients can be accounted for in terms of a loss of the computational resources available to achieve the requisite mapping (cf. Patterson, Seidenberg & McClelland, 1988).

The relevant phenomena may be summarised as follows. Some patients, generally known as "phonological dysgraphics," lose the ability to spell non-words while the ability to spell real words (whether regular or irregular) is relatively well preserved (e.g. Shallice, 1981). In terms of dual-route models of spelling, this is taken as evidence for loss of the non-lexical sound-to-spelling translation pathway. So-called "deep dysgraphics" exhibit similar problems but also produce semantically-related errors. The complementary syndrome, variously known as "surface dysgraphia" (Ellis, 1984), "lexical dysgraphia" (Beauvois & Derouesné, 1981) or "phonological spelling" (Hatfield & Patterson, 1983) appears to involve a relative sparing of the sound-to-spelling translation routine along with impairment of the lexical spelling routine. These patients therefore have particular difficulty in spelling words with exceptional sound-spelling correspondences. The picture is of course more complex than the simple one presented above (see Ellis & Young, 1988, for a review), and patients vary in the extent of dissociation which they exhibit. For present purposes, however, we are concerned to examine the possibility that our connectionist model of spelling could become "surface dysgraphic" and selectively lose the ability to spell irregular words correctly.

We simulated acquired spelling disorders in the model by setting 2%, 5%, and 10% of the weights in either the input or the output layer to zero. The performance of the model was then examined to see whether it produced the correct spellings of the words in its vocabulary. A total of 10 simulations were carried out at each level of damage.

### Results of Lesioning the Model

The results of zeroing 2%, 5% or 10% of the connections in the input or output layers are shown in Figure 7. Performance on words with enemies (i.e. words with exceptional sound-spelling patterns) does indeed drop off more quickly with increasing damage. Thus with 10% damage to the connections between input and hidden

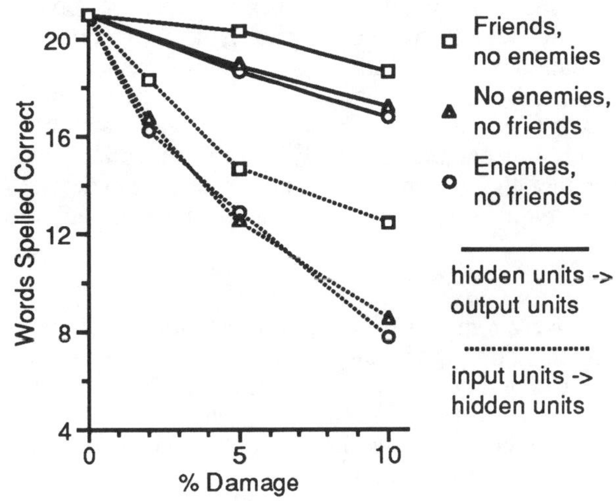


Figure 7: Performance of the Model After Lesioning

layers, the model still correctly spells 13 out of the 21 friends-only words, but only 8 of the 21 words with enemies.

How does this compare with the pattern of performance produced by surface dysgraphic patients? Some patients show a sharper dissociation than our model: Shallice (1988) summarises the results of several studies on the relevant patients. Most patients show very good performance on non-words and words with low ambiguity (i.e. regular items), performing at over 90% on these categories. Performance on words with medium levels of ambiguity ranged from about 70% to 100%, and performance on highly ambiguous items varied from 40% to 90%. Thus only a very few patients show a sharp dissociation, although all perform worse on the ambiguous items.

The average performance of our model at low levels of damage was similar to the relatively unimpaired patients, but Figure 7 shows that a 40% level of performance on our exception words (those with enemies) was accompanied by just over 60% correct performance on words with friends only. This appears to contrast with the sharper dissociations exhibited by some of the patients, but two further factors need to be taken into consideration. First, Figure 7 represents the average of 10 lesioning simulations. There was considerable variability in the simulations, as in the patient populations, and some of our individual runs showed more extreme dissociations. In several simulations no friends-only words were lost but 5 exception words were lost. In one run, after just 2% damage, 20/21 of the regular words (with friends) were spelled correctly but only 14/21 of the exceptional items. A further point is that the vocabulary of our model was relatively small, and so it included only a few enemies for our exception words. The patient data shows that much less severe impairments are found with words of only medium level ambiguity, and this perhaps provides the best characterisation of the words in our model, given its limited vocabulary. With this assumption, the model's performance provides a good fit to the empirical data. Furthermore, the inclusion in our simulation of words with neither friends nor enemies leads to a prediction (untested to our knowledge): surface dysgraphics will have difficulty with words with neither friends nor enemies as well as with "exception" words, even though the words with neither friends nor enemies are not ambiguous in their sound-to-spelling correspondence.

## Conclusions

We have attempted to use the connectionist approach to provide a computationally explicit characterisation of one aspect of human spelling development. We should note that we see the model as complementing rather than replacing the descriptive-level insights from developmental and cognitive psychology.

The model is successful in that its performance closely mirrors that of human subjects under a wide range of circumstances. We conclude that much of the process of learning to spell is usefully viewed as a statistical one. Observed performance in both developmental and acquired disorders can be characterised as being due to a lack of, or loss of, computational resources over which to represent the statistical regularities and sub-regularities implicit in the English sound-to-spelling mapping system. This abstract, high-level characterisation is, of course, far removed from any strictly neurobiological interpretation of the connectionist approach. However some researchers, (e.g. Galaburda, 1989), have pointed to the parallels between connectionist characterisations of developmental reading disorders in terms of provision of computational resources, and the atypical symmetry distribution of the *planum temporale* found in dyslexic brains.

Our model is clearly limited in a number of respects. It operates with a small and constant vocabulary, and contains no mechanism for modelling early "logographic" spelling or strategic change. Rather, it is concerned only with the development of *alphabetic* spelling, and so does not address the issue of developmental stages and strategies. Furthermore, it only accounts for one type of acquired spelling disorder: extensions to the model would be necessary to account for complementary syndromes such as phonological dysgraphia, and also to enable the model to distinguish non-homographic homophones such as THEIR and THERE. A final comment concerns the nature of the "triple" representation used on both input and output units. As Prince & Pinker (1988) have shown, such representations are not adequate for representing the whole of English vocabulary, although they suffice for the 250 word vocabulary we have modelled to date.

We believe that these limitations do not compromise our main conclusion, however. This conclusion is that much psychological data concerning normal spelling development, developmental dyslexia, and acquired spelling disorders can be well described in terms of the degree to which the various subject populations have mastered the statistical properties of the English sound-spelling mapping system.

## References


Barry, C., and Seymour, P. H. K. 1988. Lexical Priming and Sound-to-Spelling Contingency Effects in Nonword Spelling. *Quarterly Journal of Experimental Psychology* 40(A):5-40.

Beauvois, M-F, and Derouesné, J. 1981. Lexical or Orthographic Dysgraphia. *Brain* 104:21-50.

Brown, G. D. A. 1987. Resolving Inconsistency: A Computational Model of Word Naming. *Journal of Memory and Language* 23(1):1-23.

Ellis, A. W. 1984. *Reading, Writing and Dyslexia.* London: LEA.

Ellis, A. W., and YOUNG, A. W. 1988. *Human Cognitive Neuropsychology.* London: LEA.

Frith, U. 1985. Beneath the Surface of Developmental Dyslexia. In K. E. Patterson, J. C. Marshall and M. Coltheart, eds. *Surface Dyslexia.* Hillsdale, NJ: LEA.

Galaburda, A. M. ed. 1989. *From Reading to Neurons.* Cambridge, Mass.: MIT Press.

Hatfield, F. M., and Patterson, K. E. 1983. Phonological Spelling. *Quarterly Journal of Experimental Psychology* 35(A):451-468.

Jared, D., McRae, K., and Seidenberg, M. S. 1990. The Basis of Consistency Effects in Word Naming. *Journal of Memory and Language* 29:687-715.

Patterson, K. E., Seidenberg, M. S., and McClelland, J. L. 1988. Connections and Disconnections: Acquired Dyslexia in a Computational Model of Reading. In R. G. M. Morris, ed. *Parallel Distributed Processing: Implications for Psychology and Neurobiology.* Oxford: OUP.

Prince, A., and Pinker, S. 1988. Wickelphone Ambiguity. *Cognition* 30:189-190.

Seidenberg, M. S., and McClelland, J. L. 1989a. A Distributed, Developmental Model of Word Recognition and Naming. *Psychological Review* 96:523-568.

Seidenberg, M. S., and McClelland, J. L. 1989b. Visual Word Recognition and Pronunciation: A Computational Model of Acquired Skilled Performance and Dyslexia. In A. Galaburda, ed. *From Reading to Neurons.* Cambridge, Mass.: MIT Press.

Shallice, T. 1981. Phonological Agraphia and the Lexical Route in Writing. *Brain* 104:413-429.

Shallice, T. 1988. *From Neuropsychology to Mental Structure.* Cambridge: CUP.

Treiman, R., and Chafetz, J. 1987. Are There Onset- and Rime-Like Units in Written Words? In M. Coltheart ed. *Attention and Performance XII.* London: LEA.

# Generating Expressions Referring to Eventualities

**Jon Oberlander**[1] and **Robert Dale**[2]
Centre for Cognitive Science
University of Edinburgh
2 Buccleuch Place
Edinburgh EH8 9LW Scotland

## Abstract

We note (a) the well-rehearsed linguistic observation that eventualities can be referred to by using either noun phrases or sentences, and (b) the seductive ontological parallels drawn by Bach [1986] between eventualities and individuals. We show how the mechanisms for knowledge representation and referring expression generation in an existing natural language generation system [Dale 1988, 1989] can be easily extended to combine these two insights in the generation of a wide variety of forms of reference to eventualities.

## Referring to Eventualities

Most domains consist both of things and events. In order to communicate effectively, it is necessary to be able to generate appropriate references to ordinary physical objects, like chairs, cats and carrots, and to **eventualities**, like being in love, chopping carrots and having two hours of negotiations. A natural language generation system will thus have to encode information about both things and eventualities in its underlying knowledge base (KB) (cf. [Kowalski and Sergot 1986; Moens and Steedman 1987]). Now, consider the following examples:

(1) a. John and Bill discussed the paper.
    b. John and Bill had a discussion about the paper.

(2) a. Caesar died slowly.
    b. Caesar's slow death upset Mary.

The data indicate that very similar information about eventualities can be conveyed via either sentences or noun phrases. In the context of theoretical linguistics, Chomsky [1970] observed that to avoid problems arising from syntactic parallelism, verbs and their derived nominals could be made to share lexical entries.[3] In the context of NL understanding, Dahl *et al* [1987] urge the desirability of returning the same semantic representation when interpreting clauses and their nominalizations. In a generation system, we want to be able to generate both syntactic forms. It is convenient, for instance, to be able to refer anaphorically to an eventuality (introduced by a clause) using a pronoun, as in this example from Schuster [1988:602]:

(3) I want to move a block of text as a unit. How do I do it?

On this basis, we may impose the following condition on a generation system. Desideratum 1: we must select a sufficiently neutral KB format, and provide sufficiently strong mapping rules from KB entities to surface strings to allow the generation of both forms. We urge below that this in turn leads us towards another relevant desideratum.

## Eventualities and the Mass–Count Distinction

The space of possible eventualities possesses considerable structure, and this structure has been taxonomised in various ways. Following Vendler [1967], much consideration has been given to the 'aspectual types' of utterances of English sentences (cf. [Mourelatos 1978; Hinrichs 1986; Dowty 1986; Moens and Steedman 1987]). Bach [1986] takes the space to include **states** and **non-states**; in turn, states consist of **dynamic** and **static** states, while non-states consist of **processes** and **events**. Events are then either **protracted** or **momentaneous**; momentaneous events are either **happenings** or **culminations**.

[1]Also of the Human Communication Research Centre at the University of Edinburgh. HCRC is supported by the UK Economic and Social Research Council. The work was also supported in part by the UK Science and Engineering Research Council through project grant GR/G22077. Email: J.Oberlander@uk.ac.edinburgh

[2]Also of the Department of Artificial Intelligence at the University of Edinburgh. Email: R.Dale@uk.ac.edinburgh

[3]The issues involved in nominalization are broad; some provisos are therefore in order. We here consider only NPs which explicitly refer to eventualities; thus we do not discuss nominals with 'hidden events' in their analyses (cf. [Pustejovsky and Anick 1988]). We do not here consider the relation between tense and aspectual systems and NPs referring to eventualities. Further, although we pursue a relationship between nominal reference and eventualities, there is no direct relation to Partee [1984], where a DRT treatment is given of the thesis that tense is an expression allowing anaphoric reference to *times*.

Putting states to one side, we wish here to focus on Bach's observation of the seductive parallels between the mass–count distinction in nominal systems and the aspectual classification of verbal expressions.[4] Bach exploited Link's [1983] treatment of nominal systems in the context of verbal expressions. In particular, he expounded 'this proportion: events: processes :: things: stuff' [1986:5]. The algebra of events and processes derived has a number of appealing features, and explains why 'temporal mass' indicators, like *twice last night*, occur with some VPs but not others, as well as indicating under what circumstances we can switch from a count-expression to a mass-expression, and *vice versa*.

However, Bach does not discuss a class of expressions which share some of the features of both sides of the analogy, and these are precisely the NPs which refer to eventualities we adverted to above. Further examples include: *war*, *concert*, *discussion* and *the destruction of Carthage by the Romans*. We would argue that just as Bach's interpretation of the mass–count distinction applies to verb phrase reference to eventualities, so too it applies to nominal reference to eventualities; and that the distinction should again be understood in terms of a process–event division. For example:

(4) I had a discussion with Fred.

(5) I had two discussions with Fred.

(6) I had a lot of discussion with Fred.

(7) I had two hours of discussion with Fred.

Here, (4) and (5) represent the use of *discussion* as a count noun, appearing with an indefinite and a number expression respectively; (6) and (7) represent its use as a mass noun, appearing with the quantifier *a lot of* and a measure expression respectively. Lewis's [1983] 'Universal Grinder', which converts a count expression into a mass one, seems to apply rather easily to eventualities, and, following Bach, we may say that we have moved from the discrete event of discussion into the process stuff that composes it. Since grinding and packaging seem to be zero-morphological operations, leaving no trace of their operation in lexical realisations, it is perhaps unsurprising that English should contain many eventuality-denoting words that appear both as count nouns and mass nouns; *killing*, for example, can co-occur both with mass-measures and count-measures.

We therefore wish to place a new condition to our generation system. Desideratum 2: if our generation system is to produce reference to eventualities, it must be able to generate both mass and count eventuality expressions. Given the two desiderata, we want to be able to generate sentential and noun phrase references to eventualities, in the latter case allowing singular, mass or plural expressions. We demonstrate below that this is possible by relatively simple augmentation of the ontological structures and generation mechanisms in an existing natural language generation system [Dale 1988, 1989].

## The Representation of Entities

Dale's [1988, 1989] EPICURE system contains knowledge base structures appropriate for generating reference to count, mass and plural reference to things. The system makes use of a notion of a generalized physical object or **physobj**. This permits a consistent representation of entities irrespective of whether they are viewed as individuals, masses or sets, by representing each as a **knowledge base entity** (KBE) with an appropriate **structure** attribute. To construct a referring expression corresponding to a KBE, we first build a **deep semantic structure** which specifies the semantic content of the noun phrase to be generated. We call this the *recoverable* semantic content, since it consists of just that information the hearer should be able to derive from the corresponding utterance, even if that information is not stated explicitly: in particular, elided elements and instances of *one*-anaphora are represented in the deep semantic structure by their more semantically complete counterparts. From the deep semantic structure, a **surface semantic structure** is then constructed. Unlike the deep semantic structure, this closely matches the syntactic structure of the resulting noun phrase, and is suitable for passing directly to a PATR-like unification grammar. It is at the level of surface semantic structure that processes such as elision and *one*-anaphora take place. The transitions from KBE to deep semantic structure (the issue of content determination) and from deep semantic structure to surface semantic structure (the choice of anaphoric strategies and broad linguistic choice) are performed by means of independent sets of **mapping rules** which rewrite the appropriate structures from one level into those of the other.[5]

Dale's [1988] system encoded only rudimentary information about eventualities, although still in the same general form as the encoding of physobjs. Eventualities were treated simply as operators which change the properties of objects. Hence (as appropriate to the sublanguage chosen for the system), eventualities appeared as (imperative) sentences, such as *peel and slice the onions* or *simmer the soup*. Effectively, they were treated, in accordance with situation-calculus based planning, as transitions from global state to global state. However, by representing eventualities as the (near) equals of physical objects, we can satisfy both of our desiderata. Below, we extend the parallels between the two types of entities.

[4] In a fuller analysis, we would argue that states are to be represented via objects possessing temporally-constrained properties.

[5] There is insufficient space in the present paper to provide examples of these two levels of representation: see Dale [1988, 1989] for examples.

## Common Features of Physobjs and Eventualities

The domain in which the system operates consists of a finite set of entities. Each entity is represented by a distinct symbolic constant called its **index.**[6] An entity is either a **physobj** or an **eventuality**, where these are defined as follows: a **physobj** is any (not necessarily contiguous) collection of contiguous regions of space occupied by matter; a **eventuality** is any (not necessarily contiguous) collection of contiguous regions of time occupied by process stuff. Physobjs have spatial parts; if a physobj can be decomposed into parts, those parts will be physobjs. Eventualities have temporal parts; if an eventuality can be decomposed into parts, those parts will be eventualities. Every entity has, in addition to its index and type, a **specification**. The specification of an entity provides all the information known to the system about that entity.

An entity may have as part of its specification a **location**. In a physobj, this will typically be its spatial location; in an eventuality, this will typically be its temporal location, given via begin and end points (which may coincide). In both cases, the system need not know any value for the location. However, every entity must have, as part of its specification, a **substance**. In the case of physobjs, the substance is the kind of matter from which the object is made; in the case of eventualities, the substance is the kind of process stuff that makes up the eventuality. There is a finite but extensible set of substances and process stuffs represented within the system by means of symbolic constants, which are organised into a taxonomic graph structure.

As a further part of its specification, every entity has a **structure**. This corresponds to the way in which the entity is perceived. Whether the entity is a physobj or an eventuality, its structure is either **individual, set** or **mass**. An eventuality is treated on a par with a physobj. Whether it is referred to as a process or an event is determined by whether its structure is mass or individual. Any entity may have any number of additional properties specified as part of its specification, where those additional properties are drawn from a finite but extensible set of properties. These properties are binary valued features with + and − being the possible values. For example, for either *Caesar died slowly* or *Caesar's slow death,* we would expect the KB specification to include [slow = +] as part.

If an entity has structure **individual**, then it also has a **packaging** as part of its specification. A packaging is a tuple consisting of a **shape** and a **size**. The possible values of **shape** and **size** are each drawn from two finite but extensible sets: one in each case relevant to physobjs, the other to eventualities. If an entity is a **mass**, it may or may not have a **quantity**. If the

$$
\left[\begin{array}{ll}
\textit{index} = x_0 & \\
\textit{stage} = s_0 & \\
\textit{spec} = & \left[\begin{array}{ll}
\textit{structure} = \textit{set} & \\
\textit{quantity} = & \left[\begin{array}{l} \textit{unit} = \textit{kilogram} \\ \textit{number} = 2 \end{array}\right] \\
\textit{elt} = & \left[\begin{array}{ll}
\textit{structure} = \textit{individual} & \\
\textit{substance} = \textit{onion} & \\
\textit{pack} = & \left[\begin{array}{l} \textit{shape} = \textit{onion} \\ \textit{size} = \textit{regular} \end{array}\right]
\end{array}\right]
\end{array}\right]
\end{array}\right]
$$

Figure 1: The knowledge base entity corresponding to *two kilograms of onions*

$$
\left[\begin{array}{ll}
\textit{index} = x_0 & \\
\textit{stage} = s_1 & \\
\textit{spec} = & \left[\begin{array}{ll}
\textit{structure} = \textit{mass} & \\
\textit{quantity} = & \left[\begin{array}{l} \textit{unit} = \textit{kilogram} \\ \textit{number} = 2 \end{array}\right] \\
\textit{substance} = \textit{onion} & \\
\textit{chopped} = + &
\end{array}\right]
\end{array}\right]
$$

Figure 2: The knowledge base entity corresponding to *two kilograms of chopped onion*

entity is a **set**, it may have either a **cardinality**, a **quantity**, an explicit list of **constituents**, or none of these (in which case it can only be described by a bare plural).

Cardinality is specified as a numerical value, or a **range** of numerical values where the range consists of a **lowerlimit** and an **upperlimit**. The quantity of an entity is specified as a tuple consisting of some **unit** of of measurement, and a number; or a range consisting of two such tuples, one for the lower limit and one for the upper. The possible values for the **unit** feature are drawn from finite but extensible sets appropriate to physobjs and eventualities. If an entity whose structure is **set** has the feature **constituents,** the value of this feature is a list of symbolic constants which are the indices of other entities. If an entity whose structure is **set** does not have an explicit list of constituents, then it will have an **elements** feature, which provides a **specification** which is true of all the elements of the entity.

## Special Features of Physobjs

Since a physobj can change its properties through time, we let properties hold true of physobjs at **stages.**[7]

[6] We will not discuss here the obvious benefits of this generalisation for anaphoric reference to eventualities.

[7] This was termed 'state' in Dale [1988, 1989].

The invariant properties (**substance** and **quantity**) hold true of an object at all stages; all other properties of objects may change. In general, any change in a physobj's properties is mediated by the occurrence of an eventuality. Figures 1 and 2 show two different KBEs, which might be realized by the noun phrases *two kilograms of onions* and *two kilograms of chopped onion* respectively; these correspond to the same physobj at different stages.

### Special Features of Eventualities

As stated above, eventualities have temporal parts, while physobjs have spatial parts. Just as we don't always know the spatial location of a physobj, we don't always know the temporal location of an eventuality. But it is an option: for some eventualities the **location** will take as value a **begin** time and/or an **end** time. Then, an eventuality will have, as part of its specification, a finite but extensible set of **participants.** These are collected into two substructures, corresponding to the **in** participants and the **out** participants: the **in** participants are those participants which exist in the **begin** location of the eventuality, and the **out** participants are those participants which exist in the **end** location of the eventuality. This allows us to model eventualities which result in the destruction of existing entities or the creation of new ones.[8] The participants in each set will include those entities known by the system to fill various participant roles within the eventuality, such as **agent**, **patient** or **theme**. Each role has as its value a symbolic constant that corresponds to an entity in the domain. Note that we can capture the notion of causation in this way: an eventuality may contain other eventualities amongst its participants.

## Generating Reference to Eventualities

### Sentences vs NPs

Consider the KBE in Figure 3, where $c$ is the index of the KBE corresponding to the individual named *Caesar*; this represents the event of Caesar dying. The choice of whether this is to be realized as a sentence or as a noun phrase is decided by the mapping rules that determine the content of the deep semantic structures underlying a text. If a sentence realisation is chosen, then, given an appropriate speech time, the sentence will be something like

(8) Caesar died slowly.

Now consider the KBE in Figure 4, where $m$ is the index of the KBE corresponding to the individual named *Mary*, and the agent of the event is the eventuality shown in Figure 3; this represents the event of Mary being upset by Caesar's death. This allows us various possible realizations:

(9) Caesar died slowly. It upset Mary.

[8] Noun phrase reference to eventualities will not in general require this; but *Caesar's death* may well do.

$$\left[\begin{array}{l} \mathit{index} = e_0 \\ \mathit{location} = \left[\begin{array}{l} \mathit{begin} = t_1 \\ \mathit{end} = t_2 \end{array}\right] \\ \mathit{spec} = \left[\begin{array}{l} \mathit{structure} = \mathit{individual} \\ \mathit{substance} = \mathit{dying} \\ \mathit{particips} = \left[\mathit{in} = \left[\mathit{agent} = c\right]\right] \\ \mathit{slow} = + \end{array}\right] \end{array}\right]$$

Figure 3: The KBE corresponding to *Caesar dies slowly*

$$\left[\begin{array}{l} \mathit{index} = e_1 \\ \mathit{location} = \left[\begin{array}{l} \mathit{begin} = t_6 \\ \mathit{end} = t_7 \end{array}\right] \\ \mathit{spec} = \left[\begin{array}{l} \mathit{structure} = \mathit{individual} \\ \mathit{substance} = \mathit{upsetting} \\ \mathit{particips} = \left[\mathit{in} = \left[\begin{array}{l} \mathit{agent} = e_0 \\ \mathit{affected} = m \end{array}\right]\right] \end{array}\right] \end{array}\right]$$

Figure 4: Mary's being upset by Caesar's death

(10) The fact that Caesar died slowly upset Mary.

(11) Caesar's slow death upset Mary.

(12) The slow death of Caesar upset Mary.

(13) Mary was upset by Caesar's slow death.

Here, the alternation between (9) and (10) is determined by the specific mapping rules chosen to construct deep semantic structures from the underlying KBEs (a question of 'chunking' and content determination);[9] and the alternation between (11), (12) and (13) is determined by the specific mapping rules chosen to construct surface semantic structures from the corresponding deep semantic structures (a question of the choice of gross syntactic structure and clause-internal information structure).

This characterisation misses a crucial element, of course: *how* do we choose between these different possible realisations? In particular, it is not yet clear what will lead a speaker to choose an NP rather than an S as the resource for realising a particular eventuality specification. Two kinds of considerations might be brought to bear. First, where the available information is less than complete, syntactic constraints may rule out certain realisations. Facts about obligatory arguments roles are particularly relevant here. For example, Rappaport [1983:137] observes that 'Bach's generalization

[9] An alternative approach to this question, which partitions the task of language generation slightly differently, is described in Meteer [1990].

$$\begin{bmatrix} index = e_2 & \\ spec = & \begin{bmatrix} structure = individual \\ substance = discussing \\ particips = \begin{bmatrix} in = [agent = \langle s, f \rangle] \end{bmatrix} \end{bmatrix} \end{bmatrix}$$

Figure 5: I had a discussion with Fred

$$\begin{bmatrix} index = e_3 & \\ spec = & \begin{bmatrix} structure = set & \\ cardinality = 2 & \\ elt = & \begin{bmatrix} structure = individual \\ substance = discussing \\ particips = \begin{bmatrix} in = [agent = \langle s, f \rangle] \end{bmatrix} \end{bmatrix} \end{bmatrix} \end{bmatrix}$$

Figure 6: I had two discussions with Fred

$$\begin{bmatrix} index = e_4 & \\ spec = & \begin{bmatrix} structure = mass \\ substance = discussing \\ quantity = extended\_amount \\ particips = \begin{bmatrix} in = [agent = \langle s, f \rangle] \end{bmatrix} \end{bmatrix} \end{bmatrix}$$

Figure 7: I had a lot of discussion with Fred

$$\begin{bmatrix} index = e_5 & \\ spec = & \begin{bmatrix} structure = mass & \\ quantity = & \begin{bmatrix} unit = hour \\ number = 2 \end{bmatrix} \\ substance = discussing & \\ particips = \begin{bmatrix} in = [agent = \langle s, f \rangle] \end{bmatrix} & \end{bmatrix} \end{bmatrix}$$

Figure 8: I had two hours of discussion with Fred

[Bresnan 1982:418], that verbs of obligatory object control may never be detransitivized, does not hold for the corresponding nominals'. In the case of interpretation, the PUNDIT system [Dahl *et al* 1987:137] analyses both clauses and nominalizations; analysis fails in the case of a clause with an unfilled obligatory role (such as *John discussed with Bill*), but succeeds with the nominalization (*John and Bill's discussion*). The moral for generation should be clear: where the system lacks critical information about participants in an eventuality, an NP should be generated in preference to a clause.

The second kind of consideration concerns issues that might be regarded as rather more stylistic in nature, although we would prefer to follow Scott and Souza [1990:47] in viewing these constraints as being cognitive rather than aesthetic. In particular, several of Scott and Souza's heuristics for generating effective texts in the framework of Rhetorical Structure Theory (RST) may help here. For example, they suggest that a generator (a) should attempt to make a single sentence out of every rhetorical relation [1990:54], and (b) should prefer syntactically simple embeddings over more complex ones [1990:60]. Taken together, in appropriate rhetorical circumstances these heuristics support the use of an NP reference to an eventuality embedded within a simple S. This would be preferred to an S reference to that same eventuality, either embedded within a complex S or within a multi-sentential structure; on these grounds, (11) is preferable to (9).

One might wish to extend such a cognitively-oriented account to claim that, in general, an NP is to be preferred to an S on the grounds of efficiency, since NPs offer rather convenient reduced forms, such as pronouns, for referring to previously mentioned entities. But such a claim cannot be supported: Ss are equally flexible, since they allow VP ellipsis. It should also be borne in mind that the use of more concise forms carries a price. Just as *an antique lover* is ambiguous as to whether we mean a lover of antiques or a very old lover, using an NP in preference to an S may introduce an ambiguity in interpretation: *Sue Ellen's betrayal* could refer to Sue Ellen betraying J. R. or to J. R. betraying Sue Ellen.

Considerations of efficiency are more likely to choose between NPs than to choose between an NP and an S; it is to the choice between NPs that we now turn.

## Mass vs Count Descriptions

Consider examples (4)–(7), repeated below as (14)–(17):

(14) I had a discussion with Fred.

(15) I had two discussions with Fred.

(16) I had a lot of discussion with Fred.

(17) I had two hours of discussion with Fred.

The KBEs underlying these eventualities are shown in Figures 5–8 (with $s$ corresponding to the speaker and $f$ corresponding to *Fred*).

In each case, the structure of the underlying KBE determines, as in the case of descriptions of objects, whether a mass or count noun phase is to be used in the realization. Note, however, that we have four distinct KBEs here; perhaps at least some of these ought to be represented by the same underlying object. Arguably, one and the same eventuality underlies (14) and (17) above; but we have represented these by distinct KBEs. Recall that our KBEs correspond to the way in which entities are *perceived*; this same issue arises, although less obviously so, in the case of physobjs, since we can often describe one and the same object, quite apart from any physical transformations that might be applied to it, by means of count or mass descriptions (again, cf. Lewis's [1983] 'Universal Grinder' and 'Universal Packager'). Ultimately, we require some means of tying different KBEs together as different **perspectives** on the same entity. We are then faced with a question analogous to that raised in the previous section: how we choose between the different perspectives? There are clear differences amongst the candidates in terms of informational content and specificity, and so the criteria would appear to be more a matter of what to say rather than how to say it. Such issues are beyond the scope of the present paper, but must ultimately be addressed, since strategic decisions have an impact on subsequent tactical decisions; indeed, in some models of the generation process, the two mutually constrain each other.

## Conclusions and Future Directions

Generation of appropriate reference to eventualities in discourse will involve a number of interacting factors. Whether we refer to an eventuality with a mass or count noun (in NP reference), and the aspectual class of a reference (in sentential reference), is currently a matter decided within the KB, where the entity's structure is encoded as **individual, set** or **mass**. We have suggested here that it is a relatively simple matter to *permit* the generation of either sentences or NPs referring to eventualities. This leaves open the question of 'chunking' information: issues that have been touched upon in the use of RST relations in generation are relevant here (cf. [Hovy 1990; Scott and Souza 1990]).

The major area for future study is the generation of appropriate reference to eventualities in connected discourse. The existing generation system is geared toward the production of reference to objects in multi-sentence texts; hence we would argue that the extensions to it described here should allow a natural path into the area. A current limitation of our approach is that the nominals are de-coupled from the generation of tense and aspectual class; nonetheless, the KB entities contain the appropriate attributes, and we hope to remedy this limitation.

## Acknowledgments

We thank Judy Delin, Ewan Klein, Alex Lascarides, and members of the Centre for Cognitive Science's Discourse Workshop for helpful discussions.

## References

Bach, E [1986] The algebra of events. *Linguistics and Philosophy*, **9**, 5–16.

Bresnan, J [1982] Control and complementation. *Linguistic Inquiry*, **13**, 343–434.

Chomsky, N [1970] Remarks on nominalization. In R Jacobs and P Rosenbaum (eds) *Readings in English transformational grammar*, pp184–221. Waltham, Mass.: Ginn and Co.

Dahl, D, Palmer, M and Passonneau, R [1987] Nominalizations in PUNDIT. In *Proceedings of the 25th Annual Meeting of the Association for Computational Linguistics*, Stanford, Ca., July, 1987, pp131–139.

Dale, R [1988] Generating Referring Expressions in a Domain of Objects and Processes. PhD Thesis, Centre for Cognitive Science, University of Edinburgh.

Dale, R [1989] Cooking up Referring Expressions. In *Proceedings of the 27th Annual Meeting of the Association for Computational Linguistics*, Vancouver BC, pp68–75.

Dowty, D [1986] The effects of aspectual class on the temporal structure of discourse: semantics or pragmatics? *Linguistics and Philosophy*, **9**, 37–61.

Hinrichs, E [1986] Temporal anaphora in discourses of English. *Linguistics and Philosophy*, **9**, 63–82.

Hovy, E H [1990] Unresolved Issues in Paragraph Planning. In R Dale, C Mellish and M Zock (eds) *Current Research in Natural Language Generation*, pp17–45. London: Academic Press.

Kowalski, R and Sergot, M [1986] A logic-based calculus of events. *New Generation Computing*, **4**, 67–95.

Lewis, D [1983] Scorekeeping in a Language Game. Chapter 13 in *Philosophical Papers*, Volume 1. Oxford: Oxford University Press.

Link, G [1983] The logical analysis of plurals and mass terms: a lattice-theoretical approach. In R Bauerle, C Schwarze and A von Stechow (eds) *Meaning, Use and Interpretation of Language*. Berlin: de Gruyter.

Meteer, M [1990] The "Generation Gap": The Problem of Expressibility in Text Planning. COINS Technical Report 90–04, Computer and Information Science Department, University of Massachusetts, February 1990.

Moens, M and Steedman, M [1987] Temporal Ontology in Natural Language. In *Proceedings of the 25th Annual Meeting of the Association for Computational Linguistics*, Stanford, Ca., 6–9 July, 1987, pp1–7.

Mourelatos, A P D [1978] Events, Processes and States. *Linguistics and Philosophy*, **2**, 415–434.

Partee, B H [1984] Nominal and temporal anaphora. *Linguistics and Philosophy*, **7**, 243–286.

Pustejovsky, J and Anick, P G [1988] On the semantic interpretation of nominals. In *Proceedings of COLING–88*, Budapest, Hungary, 22–27 August, 1988, pp518–523.

Rappaport, M [1983] On the Nature of Derived Nominals. In L Levin, M Rappaport and A Zaenen (eds) *Papers in Lexical Functional Grammar*. Bloomington, Indiana: Indiana University Linguistics Club.

Schuster, E [1988] Anaphoric reference to events and actions: a representation and its advantages. In *Proceedings of COLING–88*, Budapest, Hungary, 22–27 August, 1988, pp602–607.

Scott, D R and Souza, C S [1990] Getting the Message Across in RST-based Text Generation. In R Dale, C Mellish and M Zock (eds) *Current Research in Natural Language Generation*. London: Academic Press.

Vendler, Z [1967] Verbs and times. Chapter 4 in *Linguistics in Philosophy*. Ithaca, NY: Cornell University Press.

Vendler, Z [1968] *Adjectives and Nominalization*. The Hague: Mouton.

# Effects of Word Abstractness in a Connectionist Model of Deep Dyslexia*

**David C. Plaut**
School of Computer Science
Carnegie Mellon University
Pittsburgh, PA 15213
dcp@cs.cmu.edu

**Tim Shallice**
Department of Psychology
University College
London, England WC1E 6BT
ucjtsts@ucl.ac.uk

## Abstract

Deep dyslexics are patients with neurological damage who exhibit a variety of symptoms in oral reading, including semantic, visual and morphological effects in their errors, a part-of-speech effect, and better performance on concrete than abstract words. Extending work by Hinton & Shallice (1991), we develop a recurrent connectionist network that pronounces both concrete and abstract words via their semantics, defined so that abstract words have fewer semantic features. The behavior of this network under a variety of "lesions" reproduces the main effects of abstractness on deep dyslexic reading: better correct performance for concrete words, a tendency for error responses to be more concrete than stimuli, and a higher proportion of visual errors in response to abstract words. Surprisingly, severe damage within the semantic system yields better performance on *abstract* words, reminiscent of CAV, the single, enigmatic patient with "concrete word dyslexia."

## Introduction

Extensive work within cognitive neuropsychology suggests that there are (at least) two separable processing routes for pronouncing a written word: a "semantic" route that recognizes the word and accesses its pronunciation from its meaning, and a "phonological" route that obtains the pronunciation based on spelling-to-sound correspondences. Strong evidence for the separability of these routes comes from the existence of two complementary sets of neurological patients (Marshall & Newcombe, 1973). "Surface" or "semantic" dyslexics (Patterson et al., 1985) have an impaired semantic route and so must rely primarily on spelling-to-sound correspondences in oral reading. In contrast, "deep" dyslexics (Coltheart et al., 1980) have lost the ability to derive phonology directly from print and so can pronounce words only via their meaning. These latter patients shows a number of characteristics. Five types of errors typically occur: semantic (e.g. CHEER ⇒ "laugh"), visual (SWORD ⇒ "words"), morphological (GROWN ⇒ "growing", also called "derivational"), visual-and-semantic (PAPER ⇒ "page") and visual-then-semantic (SYMPATHY ⇒ "orchestra"). In addition, there is a pronounced part-of-speech effect, with the ordering of correct performance being roughly: nouns > adjectives > verbs > function words. Furthermore, concrete, highly imageable words are read much better than abstract, less imageable words. Finally, pronounceable nonwords cannot be read. Of these effects, the morphological errors and part-of-speech effects may well be secondary to other characteristics (see Funnell, 1987), but any account of the disorder needs to explain all the other apparently independent symptoms.

In the conclusion of their review article, "Deep Dyslexia since 1980," Coltheart, Patterson & Marshall (1987) argue that deep dyslexia presents cognitive neuropsychology with a major challenge. They raise two main issues specific to the domain of reading. First, they argue that standard "box-and-arrow" information-processing accounts of deep dyslexia (e.g. Morton & Patterson, 1980) provide no explanation for why such a variety of symptoms should virtually always occur in patients who make semantic errors. Second, they point out that the standard explanations for semantic errors and for effects of abstractness involve *different* impairments along the semantic route.

> The loss of semantic information for abstract words that explained visual errors in oral reading cannot readily explain semantic errors in oral reading, since semantic errors typically occur on moderately concrete words.... The deficit in the semantic routine that gives a pretty account of semantic errors is, rather, an abnormal sloppiness in the procedure of addressing a phonological output code from a set of semantic features. .... Must we now postulate several different semantic-routine impairments in deep dyslexia, and if so, why do we not observe patients who have one but not the other: in particular, patients who make semantic errors but do not have difficulty with abstract words? [Coltheart et al., 1987, pp. 421-422]

Hinton & Shallice (1991, hereafter H&S) put forward a connectionist account that addresses the first issue—why semantic, visual and mixed visual-and-semantic errors co-occur. Based on previous work by Hinton & Sejnowski (1986) with Boltzmann Machines, they trained a recurrent back-propagation network to map from the orthography of

*Most of the research presented in this paper was carried out while the authors were visiting researchers in the Departments of Psychology and Computer Science at the University of Toronto under the generous support and guidance of Geoff Hinton, whom we believe deserves to be a co-author of this paper but would not be persuaded to be included. We also wish to thank Marlene Behrmann and Angela Hickman for their help. All of the simulations described in this paper were run on a Silicon Graphics Iris-4D/240S using the Xerion simulator developed by Tony Plate. This research was supported by grant 87-2-36 from the Alfred P. Sloan Foundation.

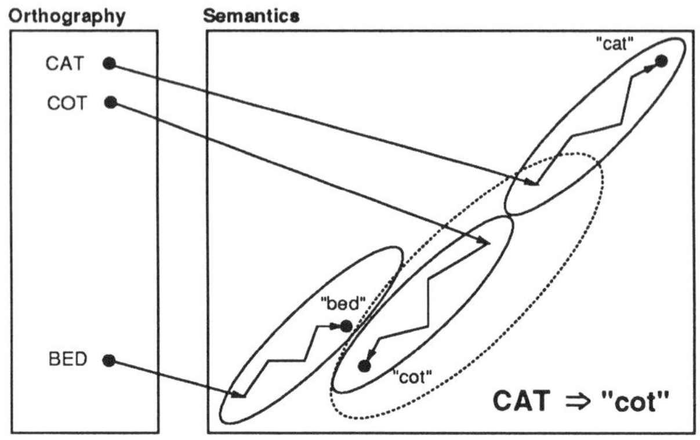


Figure 1: How damage to semantic attractors can cause visual errors. The solid ovals depict the normal basins of attraction; the dotted one depicts a basin after semantic damage.

40 three- or four-letter words to a simplified representation of their semantics, described in terms of 68 predetermined semantic features. They then systematically lesioned the network, by removing proportions of units or connections, or by adding noise to the weights, and found that the damaged network occasionally settled into a pattern of semantic activity that satisfied the response criteria for a word other than the one presented. These errors were more often semantically and/or visually similar to presented stimuli than would be expected by chance. While the network showed a greater tendency to produce visual errors with damage near the input layer and semantic errors with damage near the output layer, both types of error occurred for almost all sites of damage.

H&S explain the co-occurrence of visual and semantic errors in terms of the effects of damage in a network that builds attractors in mapping between two arbitrarily related domains (see Figure 1). The network can generate completely different meanings from visually similar words (e.g. CAT and COT) by constructing large basins of attraction around each familiar meaning, such that any initial semantic pattern within the basin will move to that meaning. Visually similar words are free to generate similar initial semantic patterns as long as they each fall somewhere within the appropriate basin of attraction. Damage within the semantic system distorts these basins, occasionally causing the normal initial semantic pattern of a word to be "captured" within the basin of a visually similar word. Essentially, the layout of attractor basins must be sensitive to both visual and semantic similarity, and so these metrics are reflected in the types of errors that occur as a result of damage.

The nature of attractors in the H&S model addresses the first challenge raised by Coltheart, Patterson & Marshall—why various error types virtually always co-occur in deep dyslexia.The current research investigates whether essentially the same account can be extended to successfully address the second difficulty they raise, relating to the abstractness of the stimuli. The next section describes in more detail the effects of abstractness on the reading behavior of deep dyslexics. Following this, we define a semantic representation capable of describing both concrete and abstract words, and develop a network that maps from orthography to phonology via semantics. We then subject the network to a variety of lesions and compare its impaired performance with that of deep dyslexics.

## Effects of abstractness in deep dyslexia

The effect of the abstractness of the stimulus on deep dyslexic reading has been investigated in a number of ways. The most basic is its effect on the probability that a word will be read correctly. Coltheart et al. (1987) claim that all patients who make semantic errors find concrete words easier to read than abstract ones. In many patients a very large difference is observed: 73% vs. 14% for KF (Shallice & Warrington, 1980), 67% vs. 13% for PW and 70% vs. 10% for DE (Patterson & Marcel, 1977).

A more subtle effect is the way that the concreteness of a word can affect the probability of the occurrence of visual errors. In the three patients in which the relative concreteness of the stimulus and response for visual errors has been investigated, the response was consistently more concrete for two of them (88% for GR (Barry & Richardson, 1988); 73% for KF) and showed a similar trend in the third ($p < .06$ for PS (Shallice & Coughlan, 1980)). The same effect is also apparent in the corpora of visual errors made by PW and DE (see Coltheart et al., 1980, Appendix 2) and by FM (Gordon et al., 1987)). In addition, stimuli with a low concreteness rating produce more visual errors than stimuli with high concreteness in the two patients in which it has been examined (KF and PS). Also, stimuli producing visual errors were less concrete than those producing semantic errors for PD (Coltheart, 1980) and FM (but not for GR). Thus a semantic variable—concreteness—clearly influences the nature of *visual* errors.

There is a single known exception to the advantage for concrete words shown by deep dyslexics: patient CAV with "concrete word dyslexia" (Warrington, 1981). CAV failed to read words like MILK and TREE but succeeded on highly abstract words such as APPLAUSE, EVIDENCE, and INFERIOR. Overall, abstract words were more likely to be correctly read than concrete (55% vs. 36%). In complementary fashion, 63% of his visual error responses were more abstract than the stimulus. However, the incidence of visual errors was approximately equal for words above and below the median in concreteness. While CAV made no more semantic errors than might be expected by chance, he appeared to be relying at least in part on the semantic route because his performance improved when given a word's semantic category. CAV is clearly a very unusual patient, but any account of the relation between visual errors and concreteness can hardly ignore him.

## A semantic representation for concrete and abstract words

The type of semantic feature representation used by H&S is quite similar to that frequently employed in psychological theorizing on semantic memory (e.g. Smith & Medin, 1981). More complex, frame-like representations can be implemented using this approach if units can represent a conjunction of a role and a property of what fills it (Hinton, 1981).

| TART | TACT | GRIN | GAIN | FLAN | PLAN | REED | NEED |
|---|---|---|---|---|---|---|---|
| TENT | RENT | LOCK | LACK | HIND | HINT | LOON | LOAN |
| FACE | FACT | ROPE | ROLE | WAVE | WAGE | CASE | EASE |
| DEER | DEED | HARE | HIRE | FLEA | PLEA | FLAG | FLAW |
| COAT | COST | LASS | LOSS | STAR | STAY | POST | PAST |

Figure 2: The 40 words used in the simulation.

More critically for the present purpose, there is a natural extension to the problem of the effects of imageability. Jones (1985) has argued that words vary greatly in the ease with which predicates about them can be generated, and that this measure reflects a psychologically important property of semantic representation. For example, more predicates can be generated for basic-level words than for subordinate or superordinate words (Rosch et al., 1976). Jones showed that there is a very high correlation (0.88) between a measure of ease-of-predication and imageability, and that the relative difficulty of parts-of-speech in deep dyslexia maps perfectly onto their ordered mean ease-of-predication scores. He argued that the effects of both imageability and part-of-speech in deep dyslexia can be accounted for by assuming that the semantic route is sensitive to ease-of-predication. Within the present framework, the natural way to realize this distinction is by representing the semantics of concrete and abstract words in terms of different numbers of features.

To examine the effect of concreteness on visual errors, a set of 20 abstract and 20 concrete words were chosen such that each pair of words differed by a single letter (see Figure 2). We represented the semantics of each of these words in terms of 98 semantic features.[1] Sixty-seven of these are taken from the H&S semantic features for concrete words (e.g. *main-shape-3d, found-woods, living*). The 31 additional features (e.g. *has-duration, relates-location, quality-difficulty*) are required to make distinctions among abstract words, but occasionally apply to concrete words as well. Overall, concrete and abstract words differ systematically in their semantic representations: concrete words have an average of 18.2 features while abstract words have an average of only 4.7 features. We do not claim that this type of representation adequately captures the richness and subtlety of the true meanings of any of these words. Rather, we claim that it captures important qualitative distinctions about the relationships *between* word meanings—namely, that similar words (e.g. LACK and LOSS) have similar representations, and that there is a systematic difference between the semantics of concrete and abstract words reflecting their relative ease of predication.

## Mapping from orthography to phonology via semantics

A network that maps from orthography to phonology via semantics was developed incrementally. An "input" network, analogous to the H&S model, was trained to map from orthography to semantics. A similarly structured "output" network

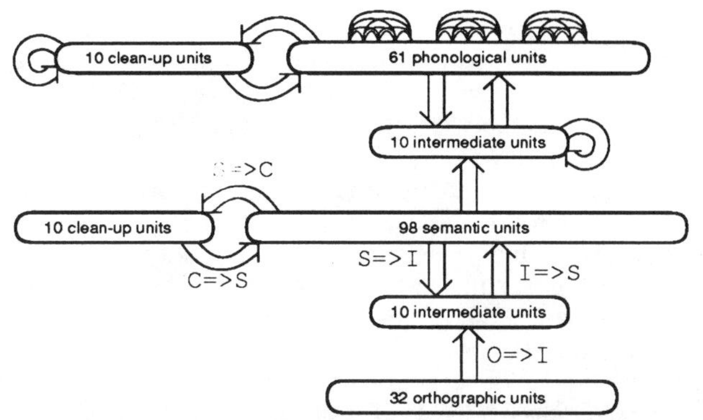


Figure 3: The network for mapping from orthography to phonology via semantics. Arrows represent unidirectional sets of connections between groups of units—sets that will be lesioned are labeled by the initials of the source and destination unit groups (e.g. O⇒I for orthographic-to-intermediate connections).

was trained separately to map from semantics to phonology. These two networks were then combined into the complete network, shown in Figure 3.

The task of the input network is to generate the semantics of each word from its orthography. Orthography is represented in terms of 4 groups of 8 features, with a separate group for each letter in a word. The set of features was designed to ensure that visually similar letters (e.g. E and F) have similar representations, while keeping the number of features to a minimum. The architecture of the input network is shown in the bottom half of Figure 3. It was trained with an iterative version of the back-propagation learning procedure, known as "back-propagation through time" (Rumelhart et al., 1986), to activate the appropriate semantic units for a word when presented with the word's orthography corrupted by independent gaussian noise with mean 0.0 and standard deviation $\sigma = 0.1$. The purpose of training on noisy input is to encourage the development of strong semantic attractors by enforcing a particular kind of generalization: inputs that are *near* known patterns must give identical responses. The network was required to activate each semantic unit to within 0.1 of its correct state over the last 3 of 8 iterations to ensure it had developed stable, accurate fixedpoints for each word. The network satisfied these criteria reliably after 4700 sweeps through the training set.

Some procedure is needed for converting the pattern of semantic activity produced by the input network into an explicit response. H&S use an *external* procedure, comparing the semantic activity produced by the network with the correct semantics of all known words, selecting the closest-matching word as long as the match is sufficiently good (the *proximity* criterion) and sufficiently better than any other match (the *gap* criterion). However, H&S provide no evidence that these criteria adequately approximate the input requirements of a network that can generate actual phonological output. An even more severe problem is that the criteria are based on the semantic representations themselves. Any differences found in performance on concrete and abstract words might simply be due to an inherent bias in the response criteria. However,

[1]See Plaut & Shallice (1991) for precise details of the orthographic, semantic, and phonological representations of words, as well as for how the network architecture and training procedure were motivated.

by developing an output network that pronounces concrete and abstract words equally well under normal operation, any systematic differences observed under damage must be due to properties of the network itself and not some external interpretation procedure.

Phonology is represented in terms of 7 "slots," each consisting of a group of position-specific, mutually-exclusive phoneme units (including one for the "null" phoneme). There are three slots for the initial (onset) consonant cluster, one slot for the vowel, and three slots for the final (coda) consonant cluster. The task of the output network is to generate the phonological representation of each word from its semantic representation. The architecture of this network is shown in the top half of Figure 3. In addition to the major sets of connections, phoneme units in the same consonant (or vowel) cluster are fully interconnected. This connectivity allows units within a slot to develop a "winner-take-all" strategy while still cooperating with units in other slots within the same cluster. The clean-up units provide for coordination and competition between clusters.

The output network was trained in a way that maximizes the strength of the attractors it develops—no attempt was made to simulate the development or mode of operation of the human speech production system. Specifically, the "direct" pathway (from semantics to phonology) was trained to produce the correct phonemes of each word during the last 2 of 5 iterations when presented with its semantics corrupted by noise ($\sigma = 0.1$). After about 3000 sweeps through the training set, the activity of each phoneme unit was accurate to within 0.2 of its correct value for each word. At this point, intra-phoneme connections and the clean-up pathway were added and the amount of noise was increased to 0.2. In this way the clean-up pathway learned to compensate for the limitations of the direct pathway when pressed by severely corrupted input. The network was trained to produce the correct phonemes over the last 3 of 8 iterations to within 0.1 of their correct values. The amount of noise prevented the network from achieving this criterion consistently, and after 18,000 training sweeps performance had ceased to improve. However, the network easily satisfied the criterion for every word given uncorrupted input.

Finally, the output network was attached to the input network and given about 100 sweeps of additional training with the weights of the input network held fixed. This ensured that the output network could generate the correct pronunciation of each word over the last 3 of 14 iterations with semantics generated by the input network rather than being clamped.

## The effects of lesions

After training, the complete network successfully derives the semantics and phonology of each word when presented with its orthography. We model the neurological damage of deep dyslexic patients by removing a proportion of the connections between groups of units in the network. As in patients, this damage impairs the ability of the network to derive the correct pronunciations of words. In order to directly compare the behavior of the damaged network with the reading responses of patients, we used the following procedure to interpret the

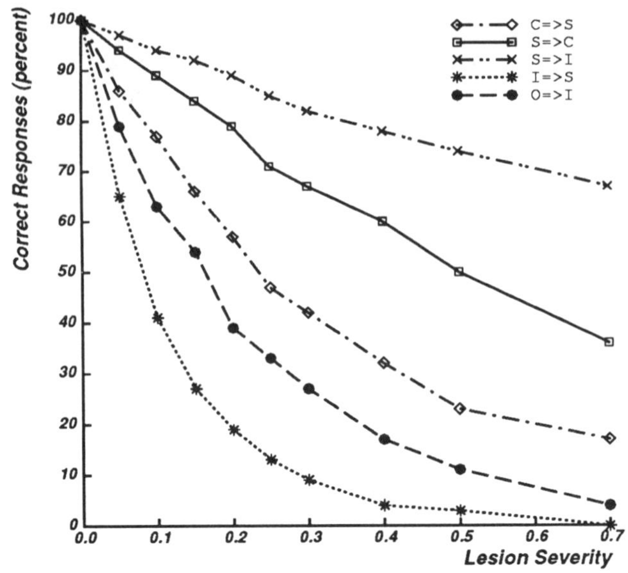


Figure 4: Correct response rates as a function of severity of lesions to the 5 main sets of connections in the input network.

corrupted output of the network as an oral response. Given the pattern of activity over phoneme units produced by the stimulus, we determined the most likely binary output vector for each slot, interpreting unit states as independent probabilities. If each of these vectors had exactly one phoneme active and probability greater than 0.6, the set of active phonemes constituted the response of the network (which might be correct or an error). Otherwise, the network was considered to have made an omission—in fact, patients frequently produce no response to a word, or respond, "I don't know." This procedure is closely related to the maximum-likelihood interpretation of the cross-entropy error function that was used to train the network (Hinton, 1989). In contrast to the response criteria that H&S applied to semantics, it does not rely on any knowledge of what the network has been trained on—it only considers the *form* of the output representation. In particular, it cannot distinguish concrete from abstract words.

Each of the 5 main sets of connections in the input network was subjected to "lesions" of a wide range of severity, in which a proportion of the connections were chosen at random and removed. Fifty instances of each location and severity of lesion were carried out, and correct, omission, and error responses were accumulated. Figure 4 shows the overall correct performance of the network as a function of lesion severity. Considering correct responses to concrete and abstract words separately, there is a significant advantage for concrete words (55% correct) over abstract words (49% correct, $F(1, 2205) = 489.3, p < .001$). The relative difference in correct performance between these two sets is shown in Figure 5. Two main results are apparent from the figure. The first is that the advantage for concrete over abstract words arises almost entirely from lesions to the direct pathway, where the majority (88%) of errors are produced. The second, unexpected result is that severe lesions of the clean-up pathway produce the reverse advantage—abstract words are responded to more

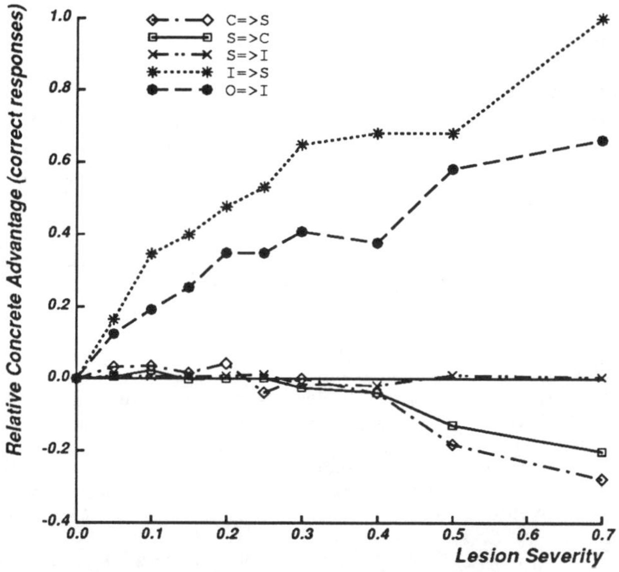


Figure 5: The relative difference in correct performance between concrete and abstract words, $(C - A)/(C + A)$, where $C$ and $A$ are the number of correct responses to concrete and abstract words, respectively. Positive values reflect a concrete advantage.

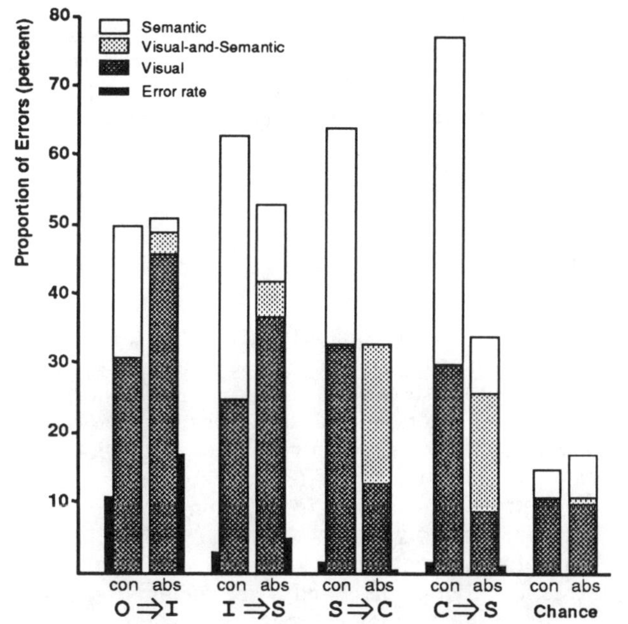


Figure 6: Overall proportions of error types for concrete (con) and abstract (abs) words for each lesion location (except S⇒I lesions which produce so few errors). Remaining errors (summing to 100%) are unrelated to the stimulus.

accurately than concrete words ($F(1,49) > 22$, $p < .001$ for each of S⇒C(0.5,0.7) and C⇒S(0.5,0.7)). This result is consistent with what is known about the concrete word dyslexic, CAV (Warrington, 1981). His reading disorder was quite severe initially, and he also showed an advantage for abstract words in picture-word matching tasks and with auditory presentation, suggesting modality-independent impairment at the level of the semantic system.

Error responses were categorized in terms of their visual and semantic similarity to the stimulus. Words were considered visually similar if they overlapped in two or more letters, and semantically similar if their semantic representations overlapped by at least 84% for concrete words and 95% for abstract words.[2] Figure 6 shows, for concrete and abstract words separately, the overall error rates and the proportions of error types for each lesion location (and for "chance" error responses chosen randomly from the word set). Overall, the network replicates (on a different word set) the H&S finding of mixtures of error types for lesions throughout the network, including purely visual errors for lesions entirely within the semantic clean-up system. Interestingly, a number of the unclassified errors are actually of the visual-then-semantic type found in deep dyslexia (e.g. PLAN ⇒ (flan) ⇒ "tart"). When this type of error occurs, the semantic activity tends to match the intermediate word only moderately well.

A comparison of error types for concrete and abstract words revealed that the proportion of errors which are visual is higher for abstract words ($F(1,2205) = 92.24$, $p < .001$), while the proportion of errors which are semantic is higher for concrete words ($F(1,2205) = 228.8$, $p < .001$). This effect is most clearly shown in Figure 6 for lesions of the direct pathway (which produce the majority of errors). As a measure of the "abstractness" of the errors produced by a lesion, we used the number of errors to abstract words minus the number of errors to concrete words. Applying this measure to visual and semantic errors separately revealed that visual errors are more abstract (mean 0.1644) than semantic errors (mean -0.1458, $F(1,2205) = 249.6$, $p < .001$). Finally, for each pair of visually similar words of contrasting types (e.g. TART and TACT), we compared how often each word produced the other as an error. Overall, abstract words are more likely to produce the paired visually similar concrete word as an error than *vice versa* (Wilcoxon signed-ranks test, $n = 900$, $Z = 3.68$, $p < .001$). However, severe lesions of the clean-up pathway produce the opposite effect ($n = 80$, $Z = 1.98$, $p < .025$).

These effects can be understood in the following way. As abstract words have fewer semantic features, they are less effective than concrete words at engaging the semantic clean-up mechanism and must rely more heavily on the direct pathway. Concrete words are read better under lesions to this pathway because of the stronger semantic clean-up they receive. Abstract words are more likely to produce visual errors as the influence of visual similarity is strongest in the direct pathway. Slight or moderate damage to the clean-up pathway impairs what little support abstract words receive from this system, but also impairs concrete words, producing no relative differ-

[2] The definition of semantic similarity is more complicated because of the systematic differences between concrete and abstract semantics and because the semantic representations are not organized into categories as in the H&S simulations. Note that two typical unrelated words have roughly 67% overlap if both are concrete and 91% if both are abstract.

ence. Under severe damage to this pathway, the processing of most concrete words is impaired but many abstract words can be read solely by the direct pathway, producing an advantage of abstract over concrete words in correct performance.

## Discussion

> No architecture which anyone has proposed contains anything remotely approaching a component damage to which would produce all the nine symptoms we are considering. [Coltheart et al., 1987, p. 417]

The symptoms of deep dyslexia that Coltheart, Patterson & Marshall are referring to are: (1) semantic errors, (2) visual errors, (3) function word substitutions, (4) morphological/derivational errors, (5) advantage for concrete over abstract words, (6) advantage for content over function words, (7) inability to read non-words, (8) inability to access phonology from print for words or in non-naming tasks, and (9) impaired writing and spelling. This last symptom is beyond the scope of the present simulation, which is concerned solely with reading. Symptoms 7 and 8 are commonly attributed to the loss of a phonological route employing spelling-to-sound correspondences. The current research demonstrates how symptoms 1, 2, and 5 can arise from unitary lesions to a network trained to map from orthography to phonology via semantics, in which abstract words have far fewer semantic features. From the arguments of Jones (1985), Funnell (1987) and others, it seems likely that the morphological/derivational errors reduce to a special case of mixed visual-and-semantic errors, and part-of-speech effects can be accounted for in terms of the ease-of-predication variable on which our simulation was based. Thus the effects we have demonstrated may also account for the remaining symptoms (3, 4, 6). Most critically, the present simulation provides an explanation for the puzzling cross-domain interactions that occur between the abstractness of stimuli/responses and the occurrence of visual errors. The explanation has some similarities to those previously offered for the interaction (e.g. Morton & Patterson, 1980; Shallice & Warrington, 1980) but these were essentially *ad hoc* verbal extrapolations from cascade notions unrelated to other aspects of the syndrome, without even a principled account of the abstract/concrete difference. The present account is supported by a simulation, is linked to explanations of other aspects of the syndrome, and offers the possibility of also addressing concrete word dyslexia.

Are the differences obtained for performance on abstract and concrete words simply due to differences we built in? The answer is yes, but not *simply*. The way the contrast was realized—in terms of a difference in number of features—was independently motivated, and the effects obtained were complex and not *transparent* from the characteristics of the semantic representation. In addition, we ensured that normal performance on the two types of words was equivalent, so that the contrast is reflected only in the performance under damage, over which we had no direct control. Taken together, the replication of the diverse set of symptoms of deep dyslexia through unitary lesions of a network that pronounces words via their meanings strongly suggests that the computational principles underlying the network's behavior may shed light on normal and impaired reading mechanisms in humans.

## References


Barry, C. & Richardson, J. T. E. (1988). Accounts of oral reading in deep dyslexia. In Whitaker, H. A., editor, *Phonological Processing and Brain Mechanisms*. Springer-Verlag, New York.

Coltheart, M. (1980). Deep dyslexia: A right-hemisphere hypothesis. In Coltheart, M., Patterson, K. E., & Marshall, J. C., editors, *Deep Dyslexia*, chapter 16, pages 326–380. Routledge, London.

Coltheart, M., Patterson, K., & Marshall, J. C. (1987). Deep dyslexia since 1980. In Coltheart, M., Patterson, K. E., & Marshall, J. C., editors, *Deep Dyslexia*, chapter 18, pages 407–451. Routledge, London.

Coltheart, M., Patterson, K. E., & Marshall, J. C. (1980). *Deep Dyslexia*. Routledge, London.

Funnell, E. (1987). Morphological errors in acquired dyslexia: A case of mistaken identity. *Quarterly Journal of Experimental Psychology*, 39A:497–539.

Gordon, B., Goodman-Schulman, R., & Caramazza, A. (1987). Separating the stages of reading errors. Technical Report 28, Cognitive Neuropsychology Laboratory, Johns Hopkins University, Baltimore, MD.

Hinton, G. E. (1981). Implementing semantic networks in parallel hardware. In Hinton, G. E. & Anderson, J. A., editors, *Parallel Models of Associative Memory*, pages 161–188. Lawrence Erlbaum Associates, Hillsdale, NJ.

Hinton, G. E. (1989). Connectionist learning procedures. *Artificial Intelligence*, 40:185–234.

Hinton, G. E. & Sejnowski, T. J. (1986). Learning and relearning in Boltzmann Machines. In Rumelhart, D. E., McClelland, J. L., & the PDP research group, editors, *Parallel Distributed Processing: Explorations in the Microstructure of Cognition. Volume 1: Foundations*, pages 282–317. MIT Press, Cambridge, MA.

Hinton, G. E. & Shallice, T. (1991). Lesioning an attractor network: Investigations of acquired dyslexia. *Psychological Review*, 98(1):74–95.

Jones, G. V. (1985). Deep dyslexia, imageability, and ease of predication. *Brain and Language*, 24:1–19.

Marshall, J. C. & Newcombe, F. (1973). Patterns of paralexia: A psycholinguistic approach. *Journal of Psycholinguistic Research*, 2:175–199.

Morton, J. & Patterson, K. (1980). A new attempt at an interpretation, or, and attempt at a new interpretation. In Coltheart, M., Patterson, K. E., & Marshall, J. C., editors, *Deep Dyslexia*, chapter 4, pages 91–118. Routledge, London.

Patterson, K. E., Coltheart, M., & Marshall, J. C. (1985). *Surface Dyslexia*. Lawrence Erlbaum Associates, Hillsdale, NJ.

Patterson, K. E. & Marcel, A. J. (1977). Aphasia, dyslexia and the phonological coding of written words. *Quarterly Journal of Experimental Psychology*, 29:307–318.

Plaut, D. C. & Shallice, T. (1991). Deep dyslexia: A case study of connectionist neuropsychology. *Cognitive Neuropsychology*. Submitted.

Rosch, E., Mervis, C., Gray, W., Johnson, D., & Boyes-Braem, P. (1976). Basic objects in natural categories. *Cognitive Psychology*, 8:382–439.

Rumelhart, D. E., Hinton, G. E., & Williams, R. J. (1986). Learning representations by back-propagating errors. *Nature*, 323(9):533–536.

Shallice, T. & Coughlan, A. K. (1980). Modality specific word comprehension deficits in deep dyslexia. *Journal of Neurology, Neurosurgery and Psychiatry*, 43:866–872.

Shallice, T. & Warrington, E. K. (1980). Single and multiple component central dyslexic syndromes. In Coltheart, M., Patterson, K. E., & Marshall, J. C., editors, *Deep Dyslexia*. Routledge, London.

Smith, E. E. & Medin, D. L. (1981). *Categories and concepts*. Harvard University Press, Cambridge, MA.

Warrington, E. K. (1981). Concrete word dyslexia. *British Journal of Psychology*, 72:175–196.

# Language And The Primate Brain

**Martin I. Sereno**

Cognitive Science D-015
University of California, San Diego
La Jolla, CA 92093
sereno@cogsci.ucsd.edu

**Abstract**

New data on the large number of modality-specific areas in the post-central cortex of several non-human primates, and recent anatomical and functional studies of the human brain suggest that very little of the cortex consists of poly-modal 'association' areas. These observations are used to reinterpret psychological and neuropsychological data on language comprehension in normal and brain-damaged humans. I argue that language comprehension in sighted people might best be thought of as a kind of code-directed scene comprehension that draws heavily upon specifically visual, and probably largely prelinguistic processing constraints. The key processes of word-recognition and the assembly of visual word meaning patterns into interacting chains, however, may be mediated in part by species-specific activity patterns in secondary auditory cortex similar to those generated by uninterpreted speech-sound sequences.

One obvious reason to study non-human primate brains is that they resemble the human primate brain in many ways. Yet humans exhibit behaviors--especially the comprehension of linguistic discourse--that are qualitatively very different from behaviors of primates and other animals. Because of this, some have concluded that animal brains may be poor models for the human brain. There are presently quite substantial rifts between psychological, neuropsychological, and neurobiological approaches to language. Recent developments in studying human and animal brains, however, provide a strong impetus to re-open discourse among these disciplines.

The neocortex of all mammals is now known to consist primarily of a mosaic of visual, auditory, somatosensory, motor, and limbic areas. Primitive mammals have a small number of areas in each of these modalities, while carnivores and primates have many. In monkeys, for example, a mosaic of 25 visual areas occupies more than half of the entire neocortex (Merzenich & Kaas, 1980; Sereno, 1988; Felleman & Van Essen, 1991; Sereno and Allman, 1991). The traditional site for higher-level functions--"polymodal association cortex"--has been reduced to a few diminutive strips in between large expanses of unimodal visual, auditory, and somatosensory areas. The potential significance of this reparcellation of cortex for the study of language and the brain has hardly been explored. The aim of this paper is to re-introduce a thoroughly comparative perspective into the evolutionary acquisition of the capacity for language, but one that does not back away from the obvious cognitive differences between humans and other animals. The anatomical and physiological organization of cortical areas in primates, including recent work on human cortex, is reviewed first. The implications of this work for theories of human language comprehension are then explored.

## Cortical Sensory Areas in Primates

**Definition of a Visual Area.** Cortical sensory areas are best defined by multiple converging criteria (Van Essen, 1985; Sereno and Allman, 1991). I begin here with visual areas, since they constitute the largest of the primary subdivisions of the cortex. Criteria for the definition of a visual area presently include architectonic features (e.g., degree of myelination, cell size, cell morphology, and cell packing density in cortical layers, histochemical features), connection patterns (e.g., input and output areas, laminar origins and targets of connections), visuotopic organization (e.g., mirror-image or non-mirror-image map of hemifield, bounding areas, pattern of map discontinuities, degree of retinotopy), and physiological properties (e.g., excitatory receptive field size, direction selectivity, attention-related modulation). Areas differ in the degree to which these criteria have been explored. V1 (primary visual cortex) and MT (middle temporal area) are distinct, well-studied areas in primates that are convergently identified by many of these criteria. Other areas--e.g., in inferotemporal cortex--are less well studied. There is no evidence to suggest that they are any less distinct.

**Visual Areas in Prosimians and Monkeys.** The first primates were probably nocturnal, judging from the large size of their orbits. The primates living today most closely related to these early primates are also nocturnal or crepuscular. The bush baby or galago, the only prosimian primate studied in detail (Allman and McGuinness, 1983), has on the order of 16 visual areas. Almost all visual areas in galagos exhibit a substantial degree of retinotopic organization, including areas in the inferotemporal cortex. In these studies, the entire extent of visual cortex was physiologically mapped in detail for the first time. In a passive animal, visual areas only respond to visual stimuli, auditory areas only to auditory stimuli, and somatosensory areas only to somatosensory stimuli. Visual cortical areas border almost directly upon somatosensory areas (dorsally) and auditory areas (ventrally). The transitional strip between, for example, auditory and visual areas (in which neurons have both a visual and an auditory receptive field) is less than one mm wide.

Monkeys (anthropoids) are thought to have diverged from the ancestors of galagos at least 40 million years ago. All but one of the anthropoids are diurnal (day-living), suggesting strongly that day-living habits evolved early in the monkey lineage. The one nocturnal monkey, the New World owl monkey, lacks a tapetum, suggesting that its ancestors had diurnal habits. The organization of visual cortex has been studied in detail in two different monkeys--the owl monkey and the macaque monkey. Figure 1B shows a flattened summary map of visual areas in the owl monkey (Weller and Kaas, 1987; Sereno and Allman, 1991). As in galagos, V1 is the largest area, followed by V2. There appear to be at least three somewhat separate 'streams' of information passing through V1 and V2--the magnocellular, parvocellular interblob, and parvocellular blob streams (named after their relay structures in the dorsal lateral geniculate nucleus and area V1)--that remain somewhat separated as one moves on to higher areas (Livingstone and Hubel, 1984; DeYoe and Van Essen, 1988). These pathways process different aspects of the visual signal in parallel--roughly, motion, location, and depth in the magnocellular pathway, and color, shape, and shading in the parvocellular pathways. The pathways pass through layer 4B, layer 2-3 interblobs, and layer 2-3 blobs in V1, and the thick stripes, interstripes, and thin stripes in V2, respectively. There is a broad subdivision of the more rostral visual areas into parietal (e.g., TP, ST--receiving primarily magnocellular stream input) and inferotemporal (e.g., ITcd, ITr--receiving primarily parvocellular interblob and

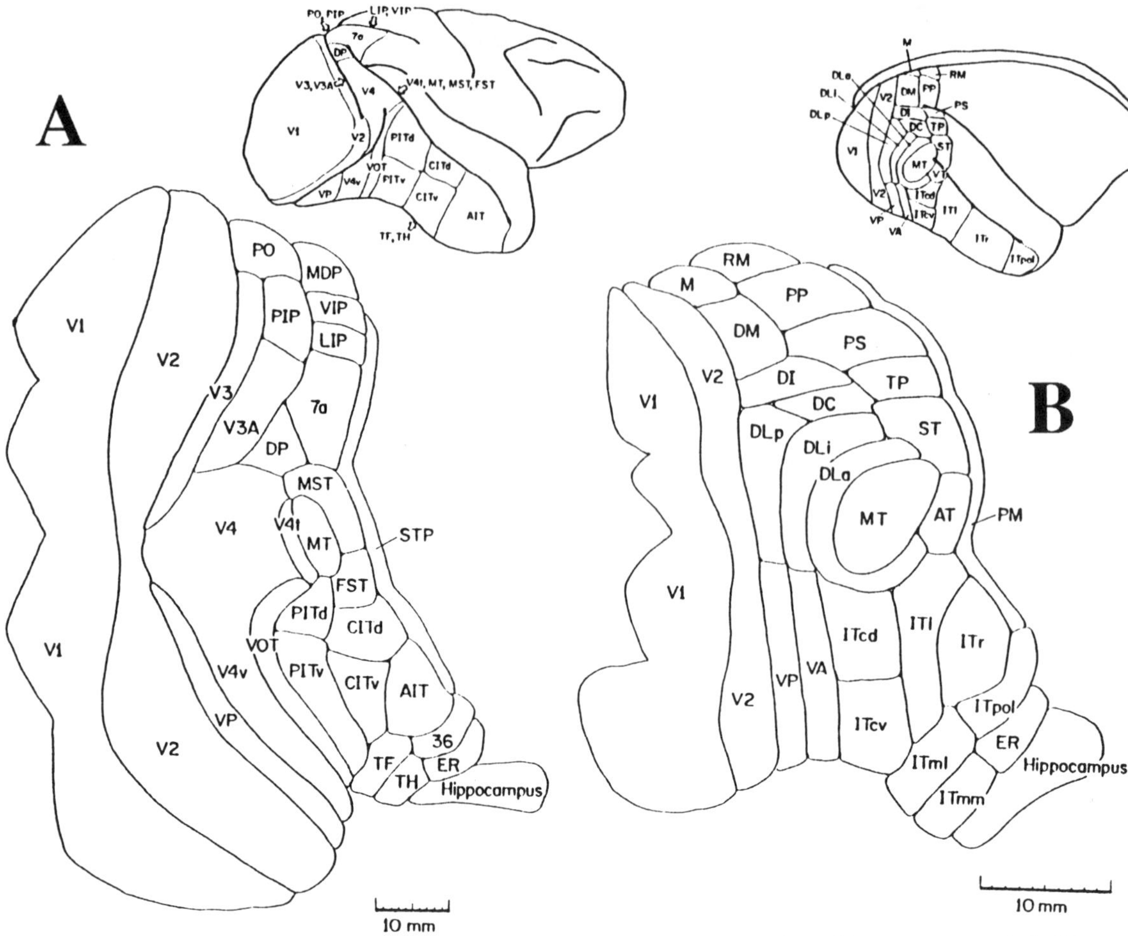


**Figure 1.** Cortical visual areas in the macaque monkey (A) and the owl monkey (B). A cut was made in V1 medially to allow the cortex to lie flat. The insets illustrate the location these areas in occipital, parietal, and temporal cortex (after Sereno and Allman, 1991). All areas shown are visual except for area PM (owl monkey) and area STP (macaque), which border on somatosensory and auditory cortices (not shown).

parvocellular blob input). Retinotopy is lost in the most anterior members of these two streams. One can define a hierarchy of visual areas based on the laminar targets of corticocortical projections; feedforward projections synapse mainly in layer 4 of the target area, while feedback projections avoid layer 4 (Rockland and Pandya, 1979; Felleman and Van Essen, 1991). The border between different modalities appears to be as sharp as in galagos; detailed mapping experiments at the anterior border of visual cortex reveal that the transitional strip between visual and somatosensory areas in parietal cortex as well as the strip between visual and auditory areas in temporal cortex is less than one millimeter wide (Sereno and Allman, 1991).

Figure 1A shows a similar summary map for the macaque monkey (an Old World monkey) (based on Van Essen, 1985; Desimone and Ungerleider, 1986; Felleman et al., 1986; 1987; and personal communication; Colby et al., 1988). Although many of the areal names are not the same, and though the relative sizes of similar areas differ, the overall configuration of the map, the retinotopic and functional organization of individual areas, and the interareal connection pattern is remarkably similar to our results in the owl monkey. New and Old World monkeys diverged over 30 million years ago. The main difference between the maps is the reduced size of the areas between V2 and MT in owl monkeys, the shape of V3 (owl monkey DM, its probably homologue, is much less elongated than the macaque area), and the somewhat larger size of several inferotemporal areas. Most of these differences reflect the reduced emphasis on the center of gaze in the retina of the secondarily nocturnal owl monkey. An important point is that there does not appear to be any substantial increase in the area of overlap between modalities. The zone in the dorsal bank of the superior temporal sulcus that responds to more than one modality is several millimeters wide (Seltzer and Pandya, 1989); this is in line with the greater overall area of the primary cortical areas in the macaque compared to the owl monkey.

**Auditory and Somatosensory Areas in Monkeys.** Auditory and somatosensory areas have been studied in parallel with visual areas. The main differences are the basis for topography (tonotopy and somatotopy vs. retinotopy), the one-dimensional nature of tonotopy (in contrast to two-dimensional retinotopy and somatotopy), the smaller overall size of auditory and somatosensory cortex, and the greater diversity of types of information collected by somatosensory receptor types (light touch, pain and temperature, muscle length changes, force on tendons, joint position). In both New and Old World monkeys, there are about 9 auditory cortical areas (Merzenich and Brugge, 1973; Pandya and Yeterian, 1985) and about 9 somatosensory cortical areas (Merzenich et al., 1978; Burton, 1986; Cusick et al, 1989). As in visual cortex, one can define a hierarchy of areas based on the laminar targets of between-area projections, and, as in vision, there is a successive loss of receptotopy as one progresses to higher levels in the two systems. Most of the somatosensory maps are based on responses to cutaneous stimulation (it is difficult to stimulate muscle and tendon receptors without also stimulating the skin).

These maps (and data from other species) suggest that the parcellation of most of the cortex has not changed radically during the evolution of the primate order. Notably, there does not seem to be any significant increase in regions where modalities overlap; rather, modality-specific areas have increased in size, and quite moderately in number; the number of cortical areas has probably not changed in New and Old World monkeys, which have evolved independently for over 30 million years.

**Visual Areas in Apes and Humans.** The organization of the cortex in a variety of mammals including humans was studied extensively by Brodmann and others at the beginning of the century using stains for cell bodies and myelin (Brodmann, 1909). Since then, anatomical and physiological studies have revised many of Brodmann's conclusions with respect to non-human primate brains (e.g., Brodmann's area 18 in Old World monkeys is twice as wide as it should have been; Brodmann's area 19 actually contains many distinct cortical areas). But it is only very recently that human cortex has been approached from a modern perspective. Preliminary results suggest that human visual cortical areas are organized quite similarly to those of other primates.

The human visual area whose borders are best known is V1--by far the most distinct visual area on architectonic grounds. Fixed- tissue injections of membrane-intercalating dyes suggest that local circuit connections within, and long range connections between, human areas V1 and V2 are very similar to those of other primates (Burkhalter and Bernardo, 1989). There is a densely myelinated, ellipsoidal area in a dorsolateral occipital sulcus that may correspond to human visual area MT, an area found in all primates (Sereno et al., 1988; Sereno and Allman, 1991) (see Figure 2). Studies using PET to monitor blood flow and a stimulus designed to selectively activate MT (based on animal studies) have uncovered an active locus near the densely myelinated region (Miezin et al., 1987). Now clearly, there is a great deal of 'additional' non-primary cortex in humans. Despite the fact that monkeys, apes and humans all have about the same number of cells in the retina, the dorsal lateral geniculate nucleus, and in V1 (Frahm et al., 1984; Tolhurst and Ling, 1988), V1 comes to occupy a smaller and smaller proportion of the total neocortex--about 10-12% or the neocortex in monkeys, about 6% of the neocortex in apes, but only about 2.5% of the total neocortex in humans. The preliminary studies cited above suggest a new answer to the problem of this 'extra' cortex in humans--it may be occupied mostly by larger versions of areas already familiar from work in monkeys (as opposed, for example, to an evolutionarily unprecedented 'language organ'). V2 in humans, for example, is much wider than would be expected when normalized with respect to the area of V1. Similarly, there is much more area between V1 and the putative human MT than would be expected (this region is mostly occupied by area V4 in Old World monkeys). Finally, the area of the putative human MT is about 3 to 4 times as big as would be predicted on a macaque model. If the other 25 or so extrastriate areas in human visual cortex increased in size (relative to V1) as much as this preliminary data suggests that V2, V4, and MT have, we could almost completely account for the 'extra' non-primary cortex in humans relative to monkeys. These observations, combined with the lack of any trend toward increased polymodal cortex in neocortical evolution, suggest a radical revision of current neuropsychological theories of human cognitive processing.

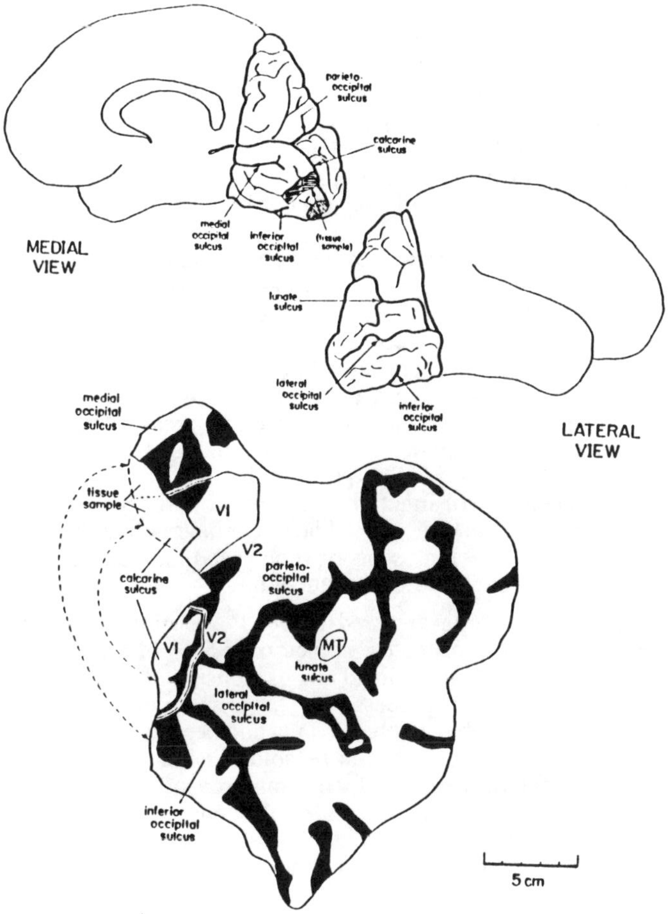


**Figure 2.** Cortical visual areas in the human (preliminary). A left occipital lobe (reversed here to aid comparison with previous figures) was physically flattened, sectioned, and stained for myelin. The exposed crowns of the gyri are colored black. A cut was made in V1 medially to allow the cortex to lie flat. The insets illustrate the location these areas in the intact brain (after Sereno and Allman, 1991). Note that the scale is now in centimeters.

## Language Processing in the Primate Brain

**Modularity and Levels of Explanation.** The question of what language processing looks like in the brain is a contentious one, especially given the preliminary state of our current knowledge in this area. A certain tradition in cognitive science and neuropsychology seems to have taken as its goal, the isolation of higher levels of explanation from their lower level implementation. Such a so-called 'functional' approach is quite curious from a biological perspective. Surely, biologists are interested in function (e.g., the heart serves as a pump for blood). But the goal there is to try to explain how it is that the structure of the heart gives rise to its function--not to ignore that structure and build an independently motivated theory in a different language (a language of 'heart'?!). The fact that the same program can run on somewhat differently designed von Neumann machines (e.g., Fodor and Pylyshyn, 1988) seems an insufficient reason to abandon a biological and evolutionary approach to the functional organization of the human brain.

This tendency to ignore the structure of the brain is quite unfortunate in light of the recent progress made in primate neurobiology. Most current texts of physiological psychology, neuropsychology, and cognitive neuroscience (e.g., Caplan, 1987; Ellis and Young, 1988) still implicitly employ a model of the organization and evolution of the cortex that dates to the associationists of the late nineteenth century. In this way of thinking, 'primitive' mammals like rats start out with primary visual, auditory, and somatosensory areas almost touching. Next up the rung of an essentially pre-evolutionary *scala natura* come animals like cats, which have a small amount of 'uncommitted' space in between. Finally, at the top, are primates and especially humans, where we find a great deal of uncommitted 'association' cortex, properly situated to integrate and associate the modality-specific information presented to it by visual, auditory and somatosensory cortices (see e.g., Fodor, 1983; Ellis and Young, 1988, on the 'semantic system' postulated in most models of word processing; Damasio, 1989).

Fine-grained mapping experiments in hedgehogs, rodents, cats, and primates, during the past decade have shown this picture of the evolution of the cortex to be incorrect. Cats and primates do have more cortex in between the primary sensory areas; but that cortex consists not of poly-modal association areas, but rather larger and more numerous modality-specific (i.e., visual, auditory, and somatosensory) areas. The studies discussed above provide no indication that humans are any different in this regard. The problem is, then, in the spirit of biological studies of functional organization, to try to describe how the basic anatomical modules of primate cortex--namely visual, auditory, somatosensory, motor, and limbic areas--support a new, peculiarly human function.

**Language as Code-directed Scene Perception.** Vision is very important to primates; in fact, over 50% of the cortex in primates, probably including humans, consists of areas devoted to specifically visual processing. This is not to deny that information about an object perceived via another modality--say the somatosensory system--might be able to enter visual areas in the form of a visual copy of the somatosensory areas' activity pattern (see e.g., experiments by Haenny et al. (1988) in macaque visual area V4 using a somatosensory-visual matching task). But it does suggest that we carefully distinguish a visual copy of a somatosensory stimulus (in a visual area with a visual map) from a somatosensory copy of a visual stimulus (in a somatosensory area with a somatosensory map).

Some linguists have independently suggested that visual representations may be very important in the semantics of natural language (Jackendoff, 1987; Fauconnier, 1985; Lakoff, 1987; Langacker, 1987). An idea common to several different approaches is that more concrete visual meanings may have been extended by analogical processes to deal with more abstract objects and relations. The present proposal goes further in suggesting a particularly direct relationship between the mechanisms of scene and discourse comprehension.

The integration of successive glances in the comprehension of a visual scene requires a kind of serial assembly operation similar in some respects to the integration of word meanings in discourse comprehension. Primates (but also many other animals) make long series of fixations at the rate of several new views per second during scene comprehension. Each fixation brings the retina to a new part of the visual scene and generates a burst of activity in V1, which largely replaces the burst caused by the previous fixation. Higher visual areas with less precise retinotopy somehow integrate information from these disconnected activity sequences to generate an internal representation of the location and identity of the relevant objects in the current scene (e.g., predators, food items, particular conspecifics, escape routes, suitable sleeping trees, etc.) that can serve as a basis for action. Many aspects of this process are redolent of linguistic integration--e.g., the underspecified, context-free information in an isolated glance is sharpened and focused by context (cf. polysemy); information from temporally distant glances must be tied together (cf. anaphora). None of this implies that scene representations (or their presumed linguistic fellows) need look anything like pictures; the patterns in question would be distributed across many areas, some of which show little retinotopy.

One main difference between scene and discourse comprehension is, of course, that scene comprehension is tied closely to the current scene. Discourse comprehension might best be thought of as a kind of fictive visual scene comprehension directed, in the case of spoken language comprehension, by sequences of phoneme representations in secondary auditory cortex. The advantage of linguistic discourse comprehension is that we are no longer tied to the current scene. However, once the appropriate visual word meaning patterns have been called up and bound together, the nature and interactions of the composite pattern may be conditioned mainly by the prelinguistic rules of interaction of scene representations in primate visual areas networks. In this sense, a large part of what has been called linguistic syntax and semantics might not be modular with respect to the neurobiology of vision.

There is in fact substantial evidence that visual areas in humans are involved in specifically linguistic functions. There is a kind of aphasia confusingly called 'transcortical sensory' aphasia (i.e., 'across-from-the-language-cortex' aphasia!) that is generated by a lesion in left human inferotemporal cortex (Rubens and Kertesz, 1983). Many of these lesions are so posterior and ventral that they are associated with overt visual field defects. Transcortical sensory aphasics have poor, "Wernicke's-like" comprehension, yet paradoxically (at least in the context of traditional models of language comprehension), can repeat words effortlessly. Far from being 'across from the language cortex', the visual areas in posterior inferotemporal cortex damaged in these patients may be the primary site of semantic processing in sighted humans. Transcortical sensory aphasics recover more quickly than patients with more dorsal lesions; this may only be an indication that the functions performed by visual cortex in language comprehension are less lateralized than those performed by auditory cortex. This is consistent with what we know about primate visual areas; permanent deficits in visual pattern recognition in monkeys require bilateral inferotemporal cortex lesions (Gross, 1973). There is no need to assume that all the cortical areas involved in language comprehension are equally lateralized; for example, the functions performed by the superior temporal gyrus (see below) may be more lateralized than the functions performed by the inferotemporal cortex.

Psycholinguistic experiments using pictures inserted into sentences and picture-word priming (e.g., Potter et al., 1986) suggest that it is surprisingly easy for visually represented concepts to be integrated into ongoing linguistic discourse comprehension. This may be another indicator of the closeness of visual category representations to linguistic meanings.

**Some PET Experiments.** Recently, it was suggested on the basis of PET experiments that semantic processing may be localized instead in the frontal lobe, just in front of "Broca's area (Petersen et al., 1988; Posner et al., 1988). In the key experiment, subjects performed two tasks--1) repeating visually presented nouns, and 2) generating "uses" (related verbs) upon viewing an otherwise comparable series nouns. Upon subtracting these two conditions, an activated locus was uncovered in frontal cortex, just anterior to the representation of face, tongue, and throat muscles in primary motor cortex. Given the ease with which preparation for movement elicits strong activation in premotor areas (see e.g., Roland et al., 1980), however, it seems likely that the activity uncovered in this experiment actually represents the different motor programming demands of the two tasks. In the first case, a motor pattern is called up directly via overlearned connections between visual word shape and articulatory movements. In the second case, by contrast, the subject must make a new motor plan to say a word that is different from that which was viewed. In fact, the subject must also suppress an output that would normally be generated by looking at the first word (in the context of reading words aloud). Frontal cortex lesions in monkeys and man are known to especially impair the ability to make delayed responses. Given that posterior inferotemporal cortex has rarely if ever been selectively activated in a blood flow experiment, and that the PET technique has limited resolution, the activation underlying semantic processing may not yet have been seen. A posterior locus for semantics is more in line with the observation made long ago (and not overturned by more recent studies) that patients with large posterior lesions are generally much more impaired in extracting meaning from linguistic discourse--and surely seem to have a much more severe derangement of thought processes--than patients with large anterior lesions.

**What's in Wernicke's Area?** Wernicke's area has occupied several different gyri over the years. Sometimes it is placed on the angular gyrus; sometimes it sits more anteriorly on the superior temporal gyrus; and often it sneaks across the superior temporal sulcus (the boundary between auditory cortical areas dorsally and visual cortical areas ventrally in primates) to sit partly in inferotemporal cortex. The left-right asymmetry originally demonstrated by Geschwind and Levitsky (1968) was in yet a different place--on the planum temporale (not even clearly visible in a lateral view). Several architectonic studies (Braak, 1978; Galaburda and Sanides, 1980) have identified a distinct area that shows a considerable left-right asymmetry (Braak's temporal magnopyramidal zone; Galaburda and Sanides' area Tpt) confined entirely to the posterior part of the lateral superior temporal gyrus. By comparison with other primates, this area is very likely to be a unimodal, secondary or tertiary auditory cortical area. Merzenich and Brugge (1973) recorded diffuse auditory responses from a geographically similar area in macaques.

If Wernicke's area proper (e.g., of Braak) is in fact a secondary or tertiary auditory area, we are left with something of a conundrum. Why should a lesion in an auditory area cause deficits in the assembly of the meaningful units of language? The deficits exhibited by many patients with a lesion in this area seem to extend beyond mere problems with auditory representations of words--their thoughts seem disarranged; often they are unable to manipulate even words with concrete visual meanings. The traditional conclusion has thus been that Wernicke's area must be an evolutionarily new 'language organ' not tied to one modality. A new interpretation more in line with the animal literature, is that the internal representations of speech sound sequences that a primate neurobiologist would expect to find in Wernicke's area proper must have some other function besides merely serving as internal copies of the speech stream; these uninterpreted speech sound representations must also be involved in word recognition and assembly of (primarily visual) meaning into coherent discourse structures. By this account, what distinguishes humans is the ability to use a sequence of symbol patterns from another modality to cause the assembly of meaning patterns in tertiary visual cortex. But the product of that assembly may be very similar to patterns assembled from direct visual inputs arriving via V1 during scene comprehension. The implication is that the trick of language was not to have invented the basic meaningful units but to have found a way of making standardized connections between them (see Sereno, 1986; 1991a; 1991b). In monkeys, the superior temporal sulcus forms, as noted, the border between auditory and visual cortices. Since clinically defined Wernicke's-like aphasics often have lesions that extend into the inferotemporal region on the middle and inferior temporal gyri, a typical 'Wernicke's aphasia' may require damage to both the auditory cortex meaning assemblers and the visual cortex meanings they assemble.

**New Routes Between Modalities.** In monkeys, one pathway responsible for cross-modal matching performance has been well-defined. Performance on somatosensory-visual matching tasks is catastrophically impaired by lesions to the basolateral amygdala (Murray and Mishkin, 1985). This part of the amygdala receives projections from secondary and tertiary visual, somatosensory, and auditory areas, and projects back to them. There is also a small polymodal strip on part of the upper bank of the superior temporal sulcus (e.g., Seltzer and Pandya, 1989). But this strip cannot by itself support cross-modal matching in monkeys.

The situation in humans must be somewhat different, at least with regard to the relative importance of the amygdala in one particular kind of cross-modal mapping that characterizes human language--the mapping between speech sounds and visual word meanings. The patient H.M. who had his amygdala removed bilaterally is quite unimpaired in recognizing visual objects named for him (or in naming visual objects himself). This suggests that humans must have a more robust connection between areas on either side of the superior temporal sulcus than monkeys do. Cross-modal matching experiments of the kind that amygdala-lesioned monkeys fail to perform have not yet been tried with H.M., and so the cross-modal pathway through the amygdala could very well still be important for some tasks in humans.

## Conclusion

Language is recently derived; based on the evidence of stone tools and other more spectacular artifacts like cave paintings, it seems likely that peculiarly human cognition and presumably language use originated rather suddenly less than 50,000 to 100,000 years ago. In view of our knowledge of the strong similarities between the brains of various non-human primates, it seems unlikely that the cortex could have been completely reorganized in so short a time. Surely, there is no positive evidence for such a major reorganization. Recent evidence instead suggests that human and non-human primate brains are organized quite similarly. We need more attempts to explain the large qualitative differences between animal cognition and human language-based cognition as the result of relatively minor modifications and re-use of pre-existing primate neural circuitry.

This paper suggests that it might be profitable to view language comprehension in sighted people as a kind of code-directed scene comprehension taking place primarily in unimodal visual areas in posterior inferotemporal cortex. A second suggestion is that internal representation of speech sound chains in secondary auditory cortical areas (Wernicke's area proper) may have other functions besides merely serving as internal copies of the speech code chain; they may be intimately involved in word recognition and the binding together of visual cortex meaning patterns. Code-directed pattern binding is clearly a specifically human faculty; but many of the constraints on the resulting bound-together patterns may reflect prelinguistic (non-modular) constraints on interactions between activity patterns in tertiary visual areas. Studies of the connections of superior temporal sulcus region in humans--just now becoming possible--may throw more light on the presently obscure neural substrate of language and human thought.

## References


Allman, J.M. and E. McGuinness (1983) The organization of visual areas in a strepsirrhine primate, *Galago senegalensis. Society for Neuroscience, Abstracts* **9**:957.

Braak, H. (1978) On magnopyramidal temporal fields in the human brain--probable morphological counterparts of Wernicke's sensory speech region. *Anat. Embryol.* **154**:141-169.

Brodmann, K. (1909) *Vergleichende Lokalisationslehre der Grosshirnrinde.* Leipzig: J.A. Barth.

Burkhalter, A. and K.L. Bernardo (1989) Organization of corticocortical connections in human visual cortex. *Proc. Nat. Acad. Sci.* **86**:1071-1075.

Burton, H. (1986) Second somatosensory cortex and related areas. In E.G. Jones and A. Peters (eds.), *Cerebral Cortex,* Volume 5. Plenum Press, pp. 31-98.

Caplan, D. (1987) *Neurolinguistics and linguistic aphasiology.* Cambridge University Press.

Cusick, C.G., J.T. Wall, Felleman, D.J., and J.H. Kaas (1989) Somatotopic organization of the lateral sulcus of owl monkeys: area 3b, S-II, and a ventral somatosensory area. *Jour. Comp. Neurol.* **282**:169-190.

Damasio, A.R. (1989) The brain binds entities and events by multiregional activation from convergence zones. *Neural Computation* **1**:123-132.

Desimone, R. and L.G. Ungerleider (1986) Multiple visual areas in the caudal superior temporal sulcus of the macaque. *Jour. Comp. Neurol.* **248**:164-189.

DeYoe, E.A. and D.C. Van Essen (1988) Concurrent processing streams in monkey visual cortex. *Trends Neurosci.* **11**:219-226.

Fauconnier, G. (1985) *Mental Spaces.* MIT Press.

Felleman, D.J., J.J. Knierim, and D.C. Van Essen (1986) Multiple topographic and non-topographic subdivisions of the temporal lobe revealed by connections of area V4 in macaques. *Soc. Neurosci., Abstr.* **12**:1182.

Felleman, D.J., A. Burkhalter, and D.C. Van Essen (1986) Visual area PIP: an extrastriate cortical area in the posterior intraparietal sulcus of the macaque monkey. *Soc. Neurosci., Abstr.* **13**:626.

Felleman, D.J. and D.C. Van Essen (1990) Distributed hierarchical processing in primate visual cortex. *Cerebral Cortex.*

Fodor, J.A. (1983) *The Modularity of Mind.* MIT Press.

Fodor, J.A. and Z.W. Pylyshyn (1988) Connectionism and cognitive architecture: a critical analysis. *Cognition* **28**:3-71.

Frahm, H.D., H. Stephan, and G. Baron (1984) Comparison of brain structure volumes in Insectivora and Primates. V. area striata (AS). *Journal fur Hirnforschung* **25**:537-557.

Galaburda, A.M. and F. Sanides (1980) Cytoarchitectonic organization of the human auditory cortex. *Jour. Comp. Neurol.* **190**:597-610.

Garfield, J.L. (ed.) (1987) *Modularity in Knowledge Representation and Natural-Language Understanding.* MIT Press.

Gross, C.G. (1973) Inferotemporal cortex and vision. *Progress in Psychobiology and Physiological Psych.* **5**:77-123.

Haenny, P.E., J.H. Maunsell, and P.H. Schiller (1988) State dependent activity in monkey visual cortex. II. retinal and extraretinal factors in V4. *Exp. Brain Res.* **69**:245-259.

Jackendoff, R. (1987) *Consciousness and the Computational Mind.* MIT Press.

Lakoff, G. (1987) *Women, Fire, and Dangerous Things.* University of Chicago Press.

Langacker, R. (1987) *Foundations of Cognitive Grammar.* Stanford University Press.

Livingstone, M.S. and D.H. Hubel (1984) Anatomy and physiology of a color system in the primate visual cortex. *Jour. Neurosci.* **4**:309-356.

Merzenich, M.M. and J.F. Brugge (1973) Representation of the cochlear partition on the superior temporal plane of the macaque monkey. *Brain Res.* **50**:275-296.

Merzenich, M.M., J.H. Kaas, M. Sur, and C.-S. Lin (1978) Double representation of the body surface within cytoarchitectonic areas 3b and 1 in "S-I" in the owl monkey. *Jour. Comp. Neurol.* **181**:41-74.

Merzenich, M.M. and J.H. Kaas (1980) Principles of organization of sensory-perceptual systems in mammals. *Progress in Psychobiology and Physiological Psychology* **9**:1-42.

Miezin, F.M., P.T. Fox, M.E. Raichle, and J.M. Allman (1987) Localized responses to low contrast moving random dot patterns in human visual cortex monitored with positron emission tomography. *Soc. Neurosci., Abstr.* **13**:631.

Murray, E.A. and M. Mishkin (1985) Amygdalectomy impairs crossmodal association in monkeys. *Science* **28**:604-606.

Pandya, D.N. and E.H. Yeterian (1985) Architecture and connections of cortical association areas. In A. Peters and E.G. Jones (eds.), *Cerebral Cortex,* Volume 4. Plenum Press, pp. 3-61.

Petersen, S.E., P.T. Fox, M.I. Posner, M. Mintun, and M.E. Raichle (1988) Positron emission tomographic studies of the cortical anatomy of single-word processing. *Nature* **331**:585-589.

Posner, M.I., S.E. Petersen, P.T. Fox, and M.E. Raichle (1988) Localization of cognitive operations in the human brain. *Science* **240**:1627-1631.

Potter, M.C., J.F. Kroll, B. Yachzel, E. Carpenter, and J. Sherman (1986) Pictures in sentences: understanding without words. *Jour. Exp. Psych.: Gen.* **115**:281-294.

Rockland, K.S. and D.N. Pandya (1979) Laminar origins and terminations of cortical connections of the occipital lobe in the rhesus monkey. *Brain Research* **179**:3-20.

Roland, P.E., B. Larson, N.A. Lassen, and E. Skinhoj (1980) Supplementary motor area and other cortical areas in organization of voluntary movements in man. *Jour.Neurophysiol.* **43**:118-136.

Rubens, A.B. and A. Kertesz (1983) The localization of lesions in transcortical aphasias. In A. Kertesz (ed.), *Localization in Neuropsychology.* Academic Press, pp. 245- 268.

Seltzer, B, and D.N. Pandya (1989) Intrinsic connections and architectonics of the superior temporal sulcus in the rhesus monkey. *Jour. Comp. Neurol.* **290**:451-471.

Sereno, M.I. (1986) A program for the neurobiology of mind. *Inquiry* **29**:217-240.

Sereno, M.I. (1988) The visual system. In I.W.v. Seelen, U.M. Leinhos, and G. Shaw (eds.), *Organization of Neural Networks.* Weinheim: VCH Verlagsgesellschaft, pp. 167-184.

Sereno, M.I., C.T. McDonald, and J.M. Allman (1988) Myeloarchitecture of flat-mounted human occipital lobe: Possible location of visual area MT. *Soc. Neurosci., Abstr.* **14**:1123.

Sereno, M.I. and J.M. Allman (1991) Cortical visual areas in mammals. In A. Leventhal (ed.), *Neural Basis of Visual Function.* London: Macmillan, pp. 160-172.

Sereno, M.I. (1991a) Four analogies between biological and cultural/linguistic evolution. *Jour. Theoret. Biol.* (in press).

Sereno, M.I. (1991b) *DNA and Language.* (to be published by MIT Press).

Tolhurst, D.J. and L. Ling (1988) Magnification factor and the organization of the human striate cortex. *Human Neurobiol.* **6**:247-254.

Van Essen (1985) Functional organization of primate visual cortex. In E.G. Jones and A. Peters (eds.), *Cerebral Cortex,* Volume 3. Plenum Press, pp. 259-329.

Weller, R.E. and J.H. Kaas (1987) Subdivisions and connections of inferior temporal cortex in owl monkeys. *Jour. Comp. Neurol.* **256**:137-172.

# Focal and Diffuse Lesions of Cognitive Models

**Steven L. Small**

Cognitive Modelling Laboratory
Department of Neurology
University of Pittsburgh

## Abstract

With the recent ability to construct fault tolerant computer models using connectionist approaches, researchers are now able to investigate the effects of damage to these models. This has great appeal for cognitive science as it provides a further way to verify or falsify a computer model. Existing studies employ a concept of network "lesioning" that fails to have explanatory adequacy for neurobiology. While using anatomically plausible architectures for cognitive models, they nonetheless use biologically implausible methods for simulating neurological damage to these networks. This paper examines the different objects of computational networks and their analogical neurobiological counterparts, and suggests a taxonomy of connectionist network lesion methods. Finally, an existing visual system model is used as a testbed to study the differential effects of focal and diffuse lesions. The experiments with focal damage versus diffuse damage suggest that while the effects of focal brain injury may be due to the particular computations performed in some brain area, the effects of diffuse brain injury or degeneration may cause cognitive deficits because of the inherent nature of the brain as a distributed computational device, and not through differential local effects.

## Part I: Introduction

One feature of connectionist networks that makes them particularly interesting for cognitive modelling is their computational relationship to the human brain. This relationship, expressed to a greater or lesser degree in different types of network architectures and domains of modelling (i.e, cognitive and/or neurobiological), contributes a new class of constraints to the validation of such models (Sejnowski and Churchland, 1988). One valuable new constraint is that of lesionability. Instead of evaluating a model solely on its performance with respect to the normal processing behavior of the object being modelled (e.g., input/output behavior, intermediate representations), a model may now be subject to computational disruption and be expected to produce behaviors that are analogous to known abnormal processing behaviors as well.

The analogy between the computational structures of a cognitive model and the neurobiological structures of nature can be of variable strength, due in part to (a) the nature of the computational architectures themselves; and (b) the specification of the analogical relationships by the modeller. This holds equally for the lesions inflicted on computational networks in the name of neurological or cognitive impairment. In this paper, we explore the ways in which a connectionist network can be lesioned, and discuss their plausibility in terms of both basic neurobiology and clinical neurology.

Address: Cognitive Modelling Laboratory, Department of Neurology, University of Pittsburgh, 325 Scaife Hall, Pittsburgh, PA 15261. Phone: (412) 648-9200. Network: sls@dsl.pitt.edu.

### Neurological Lesions

Acquired damage to the human brain leads to dysfunctional performance in a variety of modalities, depending on both the quality of the damage and its quantity. The damaged area of the brain is considered the "lesion" (Damasio and Damasio, 1989) and neurologists focus on using various intellectual and radiographic techniques to characterize the location, size, and cause of a particular lesion. In the case of visual or linguistic disorders, neuropsychologists employ specialized examination techniques to make these inductions (Boller and Grafman, 1990).

While it has always been questioned to what extent different functions localize to different parts of the brain, the currently prevailing view is that there is in fact a tremendous localization of function (Galaburda, et al., 1978). Such a view is necessary for the inductive reasoning steps described above to be meaningful. Although localization of function definitely exists to some extent (e.g., primary motor area), there is ample reason to ask two things: (1) What functions are the ones that localize? (2) Over how large an area do functions localize, given the probable distributed representations of the neural implementations of these functions? A corollary to question (2) is how minutely can we attempt to localize particular functions?

Many causes exist for brain lesions. Examples include strokes, tumors, intoxications, and degenerative diseases. A useful way to classify them is to divide them into two main categories: Focal lesions represent damage to well circumscribed regions of brain substance; while the extent of such lesions cannot always be perfectly demarcated, and frequently changes over time, there is nonetheless a focus of impaired nervous system functioning. Strokes constitute the most prevalent cause of focal brain lesions. Diffuse lesions involve damage to a large number of discrete neural elements over a widespread area of the brain, involving one or more particular classes of neurons or neuronal elements. Alzheimer's Disease is the most prevalent cause of diffuse brain lesions.

| Study | Function | Damage | Nature of Model Disruption(s) |
|---|---|---|---|
| (a) | Dyslexia | Focal | (1) Proportion of weights set to zero<br>(2) Random number added to all weights<br>(3) Removal of some hidden units |
| (b) | Spatial neglect | Focal | Unspecified "damage" to "connections" |
| (c) | Schizophrenia | Diffuse | Decrease gain in particular subnetwork |
| (d) | Aphasia | Focal | Unspecified "noise" to "connections" |

Table 1: Existing Computer Model Lesioning Experiments

**Computational Model Lesions**

With the recent ability to construct fault tolerant computer models using connectionist approaches, researchers are now able to investigate the effects of damage to these models. Several investigators have taken this approach to the study of cognition, and have lesioned particular cognitive models, and then attempted to compare the resulting model performance with that of people who have suffered brain lesions. This method has great appeal for cognitive science as it provides a further way to support or falsify a computer model.

This method has been attempted in several areas, including (a) acquired dyslexia (alexia) (Hinton and Shallice, 1989); (b) "neglect dyslexia" (spatial neglect in reading) (Mozer and Behrmann, 1989); (c) schizophrenia (Cohen and Servan-Schreiber, 1989); and (d) aphasia (dysphasia) (Miikkulainen, 1990). In each case, an interesting cognitive model was disrupted to produce one or more behaviors that resemble human information processing under some condition of neurological damage. Table 1 shows for each model whether the human condition reflects diffuse or focal damage and how the model was disrupted to simulate that condition.

Note that these studies employ a concept of network "lesioning" that fails to have explanatory adequacy (Chomsky, 1965) in the neurobiological sense. Whereas these researchers go to great trouble to build anatomically plausible architectures for cognitive models, they nonetheless use biologically implausible methods for simulating neurological damage to these networks. For example, while aphasia is typically the result of focal neurological damage (e.g., a stroke), Miikkulainen (1990) adds "noise" to the connections of his network, which is a diffuse strategy. Spatial neglect also arises most typically from focal brain damage, yet Mozer and Behrmann (1989) nonspecifically "damage" some connections. In applying their notion of "gain" in a diffuse manner to a subnetwork of their model, Cohen and Servan-Schreiber (1989) do in fact meet the explanatory criteria suggested here, as they are using diffuse lesions to account for impaired problem-solving behavior in patients with diffusely damaged brains (i.e., schizophrenia).

The study of Hinton and Shallice (1989) investigates the effects of three different methods of network lesioning. They applied each method to the different layers of their model to observe the effects, and explored empirically the relationships between the reading behaviors of differently impaired networks and dyslexic patients. While acquired dyslexia typically results from focal brain insults, the study explored diffuse model lesions (i.e., resetting a proportion of weights, adding a random number to all weights) in addition to a method of focal lesioning (i.e., removing some hidden units, which is only focal if these units are physically adjacent to each other).

## Part II: A Taxonomy of Lesions

Connectionist models of cognitive processing typically incorporate architectures that limit specifically the types of lesions one might consider for computational experiments of dysfunctional cognitive performance. Table 2 lists some neurobiological concepts and their analogues in connectionist models. It does not matter that a single unit in a parallel distributed (PDP) cognitive model does not represent a single neuron in the brain (Sejnowski, et al., 1988). Biological nervous systems have motivated massively parallel approaches (Feldman, 1989) and in this paper, the

| CNS Concept | Model Analogue | Nature | Description |
|---|---|---|---|
| Neuron | Unit | Abstraction | Associated values and functions |
| Synaptic strength | Connection weight | Value | Real number |
| Axon firing rate | Unit potential | Value | Real number |
| Synapse | Unit input | Value | Weighted unit potential |
| Inhibition | Negative weight | Value | Negative real number |
| Excitation | Positive weight | Value | Positive real number |
| Depolarization | Potential function | Function | Adjusted sum of inputs |
| Threshold | Bias | Value | Real number |

Table 2: Computer Model Correlates of Neurobiological Concepts

analogy between neurobiological and computational networks will be analyzed in terms of lesions that can be produced.

For each biological entity shown in the Table, a variety of neurological disruptions naturally occur in human illness. Furthermore, in vitro and in vivo studies of the neuroanatomy, neurophysiology, and neuropharmacology of these entities have led to some valuable information.

| Structures to Lesion |
|---|
| Processing Units |
| ConnectionWeights |
| Activation Functions |

Table 3: Objects to Lesion

In focal neurological disease, whole collections of neurons, supporting structures, and connections are lost in one brain area. By contrast, in diffuse disease, one particular aspect of (one type of) neuronal functioning might be lost throughout the entire brain.

### Lesion Objects

Lesions to network models can occur either focally or diffusely to the three general structures listed in Table 3. Each network object has a neurobiological correlate, and different lesions to these objects have neurological analogues.

The specific subpopulation for focal lesions depends on the architecture of the particular network representations in the model. One method of classification is by location in the overall network. Table 4 shows that the concept of location has several possible interpretations, all of which are interesting and relevant.

| Locations |
|---|
| Functional Location in Cognitive System |
| Functional Location in Computational Processing |
| Spatial Location |

Table 4: Object Selection by Location

The input and output layers of a network are locations from the vantage point of computational function. The subpart of the input layer that represents the phonological input lexicon for a model of lexical access is a location in the cognitive domain. The bottom left quadrant of a two dimensional drawing of a network is a topographically specified location. Note that this latter specification, which sounds more arbitrary and untheoretical, may in fact be the most accurate way of (focally) lesioning a cognitive (neuroscientific) model. This is a consequence of the vascular organization of the brain, which leads to brain lesions that follow a spatial pattern, rather than a functional one.

| Values |
|---|
| Range of Absolute Values |
| Range of Signed Values |
| All Values of a Particular Sign |

Table 5: Object Selection by Value

Dynamic aspects of computational networks include the values associated with the fixed structures over time. Table 5 lists several specific categories of values that can be used as selection criteria for choosing structures to lesion. Note that both ranges of values and their signs alone are useful.

| Network Lesions |
|---|
| Ablation (Deletion) |
| Attenuation |
| Augmentation |
| Resetting |
| Addition of Noise |

Table 6: Lesion Methods

The neurobiological plausibility of the different selections by value depends to some extent on the specific intended analogy between the lesioned structures and the brain or cognitive system. However, the existence of anatomical differences between (certain) inhibitory and excitatory synapses in the brain (Shepherd and Koch, 1990) suggests that a strategy based on signed values (or signs alone) may have validity for different classes of structures.

### Lesion Methods

A network model can be computationally lesioned by a number of different operations. Table 6 lists a few possibilities, involving (a) removal of an object; (b) increasing or decreasing its value by a fixed percentage or by adding random noise; and (c) setting the value of an object to some previous value.

### Units

Insofar as the computational units are analogous to neurons in the brain, they represent a well founded candidate population for lesioning experiments. Since a typical computational unit represents a collection of values and functions, the designation of a lesion must be more specific, e.g., unit output value, as shown in Table 7.

A straightforward disruption of the functioning of an individual unit is to delete it from the network. This is achieved by removing all of its connections, such that no further network processing includes computations originating there. Less drastic manipulations include increasing or decreasing its output by a certain percentage or a fixed amount or adding noise (random values within some range) to the output.

Unit deletion has a neurobiological correlate in destructive brain damage (e.g., stroke). The biological correlate of unit potential is the axonal firing rate. Augmentation and attenuation of this value are analogous to increases and decreases in firing rate, which could be the result of changes in (a) neurotransmitter (or neuromodulator) concentration; (b) excitatory inputs; (c) inhibitory inputs; or (d) axonal conduction. Certain conditions lead to changes in neurotransmitter concentrations in particular brain areas, e.g., dopamine concentration in the striatum in Parkinson's disease. An analogy to random noise in affecting axonal firing rate might be brain intoxications of various kinds, either by external toxins (e.g., drugs) or internal ones (e.g., metabolic derangements).

| Unit Lesion Sites |
|---|
| Potential (Output Value) |
| Input Values |
| Activation Functions |

Table 7: Aspects of Units

The biological correlate of a unit input value is the synapse. Since the synapses of a neuron are spatially distributed along a variety of dendrites, it makes more sense to talk about subpopulations of units than either individual synapses or all synapses. Relevant subpopulations include the synapses on a particular dendrite or ones that utilize a particular neurotransmitter. Attenuation or augmentation of the analogous values in the model corresponds to the under- or over-sensitivity of particular synaptic transmission (e.g., post-synaptic receptor density), and could be caused by local neurotransmitter changes. Toxins could produce effects analogous to the addition of noise.

### Activation Functions

Activation functions take unit inputs and produce an output. The neurobiological correlate of the activation function is membrane depolarization, and many different factors bear on the ability of the neuronal membrane to depolarize. While the absolute number of active synapses plays a role, perhaps more important are their spatial distribution and chemical characteristics. This was noted above, and lesions to subpopulations of unit inputs constitute reasonable analogies to neurological damage to neurotransmitter systems or dendritic locales.

Table 8 lists four aspects of activation functions that are subject to lesioning. Each of these aspects corresponds to a stage in the application of the function, and affects the ultimate behavior of the unit. The unit first computes the (weighted) sum of all inputs, corresponding roughly to the combined membrane electrical effects of the chemistry of the synapses. This number is then manipulated arithmetically to produce another value, which corresponds to the total membrane depolarization. In connectionist networks, the summed values are combined by a "squashing function" (Rumelhart and McClelland, 1986), and the precise sigmoidal curve obtained depends on a value called the "gain", the adjustment of which is the subject of the interesting lesioning experiments of Cohen and Servan-Shreiber (1989) discussed earlier. Lastly, action potential propagation depends on this final value exceeding a threshold, which can also be varied. The presence of different physiological properties of neurons and responses to neuromodulators (Shepherd, 1990), e.g., peptides, provide ample neurobiological correlation of these mathematical manipulations.

| Activation Function Lesion Sites |
|---|
| Unit Input Sum |
| Adjustment of Input Sum |
| Gain |
| Threshold (Bias) |

Table 8: Activation Functions

### Weights

The connection strengths or weights of an artificial neural network correspond to synaptic strengths of connections. Negative weights are analogous to inhibitory synapses, positive weights to excitatory synapses. Weight lesions thus correspond to disruptions in the role of particular synapses in effecting the action potential.

## Part III: Experimental Results

We have recently been investigating network lesions in a general manner by incorporating lesioning into a connectionist simulator (called DYSNET) and using it to study various models in cognitive neuroscience. Lesion specification in DYSNET involves selection of one feature from each column of Table 9, and then declaration of the appropriate parameter(s). Particularly useful selection criteria include specific network partitions (subnetworks), which may be lesioned independently, and value ranges, either absolute (e.g., all weights with absolute value less than 2.0), signed (e.g., all values between -1.0 and 1.5), or signs themselves (e.g., all inhibitory weights).

### Motivation

One computational experiment was motivated by an interesting difference in visual system performance under different conditions of damage. Focal damage to the parieto-occipital junction can lead to a syndrome of visuospatial

| Object Type | Functional Location | Range | Lesion Type |
|---|---|---|---|
| Weight | Layer | Absolute Value | Deletion |
| Unit | Partition | Signed Value | Attenuation |
| | | Sign | Reset |
| | | | Noise |

Table 9: Parameters Characterizing Computational Lesions

disruption known as Balint's Syndrome (Balint, 1909). Patients with this syndrome are unable visually to guide their hands to a particular location in space in order to grasp an object. They are able to tell what object was shown to them, but seem to have difficulty in using visuospatial knowledge. Focal damage to the posterior temporal lobes can lead to object recognition problems, without visuospatial difficulties.

Patients with diffuse neurological damage from Alzheimer's Disease have problems with both object recognition and spatial orientation. However, problems with object recognition both precede the development of visuospatial dysfunction and are more serious than the spatial problems (Mendez, et al., 1990).

### Two Pathways

In order to model this difference, a number of lesion experiments were conducted using the visual system model of Rueckl and his colleagues (Rueckl, et al., 1989). Their network classifies two-dimensional visual images into two categories, (1) what object was shown, and (2) where in the visual image the object appeared. The empirical studies of Mishkin and his colleagues (1983) on macaque visual processing constrained the architecture of the connectionist model and led to computational hypotheses. When required to perform the dual task of visual object recognition and spatial localization, the macaque uses two separate visual systems to perform the two tasks, a temporal "what" system and a parietal "where" system (Desimone, et al., 1985; Mishkin, et al., 1983). Rueckl, Cave, and Kosslyn (1989) showed that a computational neural network learned the two tasks much faster if the network were subdivided into two parallel networks, one to perform the object recognition and the other to perform the spatial localization.

### Hypotheses

The conclusion of these researchers is that the learning of an object recognition and spatial localization task is easier with separate "what" and "where" networks than with a single integrated network. The current investigation of lesioning aims to build upon this research, and to test further its neurobiological plausibility by comparing its functioning when lesioned to several general features of impaired human functioning.

As noted, Balint's Syndrome (Balint, 1909) involves significant problems in visuospatial analysis (Newcombe and Ratcliff, 1989). Brain lesions that cause such problems are in the junction of the occipital and parietal lobes, and are typically the result of focal damage such as stroke. Patients with Alzheimer Disease (AD) can also get Balint's syndrome from their diffuse degenerative disease, but early deficits in AD involve object recognition and not visuospatial tasks. In fact, the appearance of Balint's syndrome in AD is accompanied by very impaired object recognition in all cases (Mendez, et al., 1990).

| NUMBER OF EXPERIMENTS | Lesion Method | Attenuate | Delete | Noise | TOTAL |
|---|---|---|---|---|---|
| Weight Selection Method | All | 5 | 7 | 5 | 17 |
| | Unselected | 2 | 2 | 1 | 5 |
| | Layer | 1 | 1 | 2 | 4 |
| | Partition | 0 | 0 | 0 | 2 |
| | Value | 0 | 2 | 0 | 2 |
| | Sign | 2 | 2 | 0 | 4 |
| Unit Selection Method | All | 19 | 6 | 1 | 26 |
| | Unselected | 4 | 1 | 0 | 5 |
| | Layer | 9 | 3 | 1 | 13 |
| | Partition | 6 | 2 | 0 | 8 |
| Gain | All | 2 | 0 | 0 | 2 |

Table 10: Lesion Experiments Conducted

This information, combined with the modelling results described above, has led to two hypotheses regarding lesions to the Rueckl network:

*Focal Lesion Hypothesis*: Focal lesions to an object recognition and spatial localization task can disproportionately affect either spatial localization or object recognition, depending on the site of damage.

*Diffuse Lesion Hypothesis*: Diffuse lesions to an object recognition and spatial localization task initially affect object recognition, but with sufficient damage, can disrupt spatial localization as well.

Multiple computer model lesioning experiments were conducted to test these hypotheses.

### Lesioning the Rueckl Model

For this project, the (non-recurrent) feed forward connectionist network described in (Rueckl, et al., 1989) was reimplemented using the DYSNET simulator. Specific choices regarding potential functions, learning parameters, error measure, and weight updating function may differ from those in the original model (available on request). The "what" and "where" components of the network were implemented as distinct partitions of the network that share common input nodes.

Many different lesions were introduced into the model, and a simple analysis of the resulting behavior was recorded. This included (a) the number of spatial location errors (false positive and false negative); (b) the number of object identification errors; (c) the sum squared error of the "where" subnetwork; (d) the sum squared error of the "what" subnetwork; and (e) the sum squared error of the entire network.

The different lesions performed on the network are summarized in Table 10. This Table lists the number of specific experiments conducted using each selection method (e.g., selection by location or value) and specific network

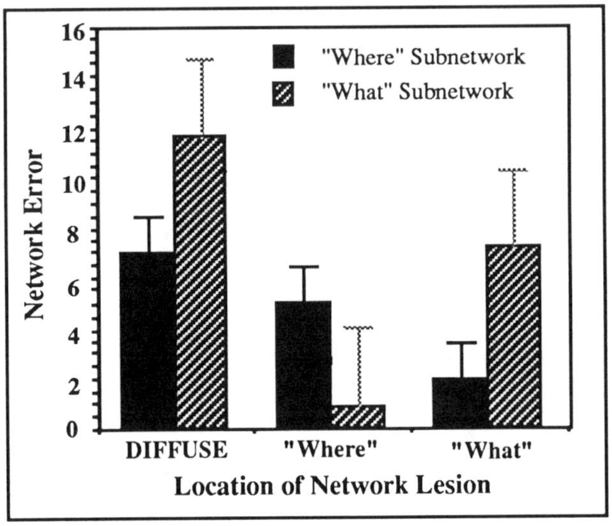


Figure 1: Lesions to Visual System Model

lesioning type (e.g., attenuation of value).

The pooled results of the experiments with focal damage ("object selection by partition" in Table 10) show that focal damage to the "parietal" (or "where") part of the network produces computational problems with the spatial orientation task and that focal damage to the "temporal" (or "what") part of the network produces problems with the object recognition part of the network. In the diffuse lesioning experiments, the pooled data show that while errors occurred in both the object recognition and spatial discrimination subtasks, the object recognition task was disproportionately affected. The results from all experiments are shown graphically in Figure 1. The sum squared error of each subnetwork across all focal and diffuse experiments was averaged and this value is shown in the graph, along with the standard error.

### Discussion

The connectionist model study suggests that the neuropsychological deficits of Alzheimer's Disease and other central nervous system disruptions caused by diffuse (rather than focal) damage may have a computational basis. In other words, the particular pattern of cognitive disruption may not be due to a special predilection of the disease process for one or another part of the brain. Rather, diffuse brain injury or degeneration may cause these cognitive deficits because of the inherent nature of the brain as a distributed computational device, with patterns of connectivity that implement specific processing tasks.

## Acknowledgements

Many thanks to Audrey Holland for many stimulating discussions leading to some of the work presented here. Thanks to the members of Pat Carpenter's and Audrey Holland's aphasia seminar for provocative conversations. Special thanks to Gloria Hoffman and Mark Fitzsimmons for discussions, support, and manuscript suggestions.

## References

Balint, R. 1909. Seelenlähmung des 'Schauens', Optische Ataxie, raümliche Störung der Aufmerksamkeit. *Monatsschr Psychiatr Neurol* 25:51-81.

Boller, F.; and Grafman, J. ed. 1990. *Handbook of Neuropsychology*. Amsterdam: Elsevier Science Publishers.

Chomsky, N. 1965. Aspects of the Theory of Syntax. Cambridge, Massachusetts: The MIT Press.

Damasio, H.; and Damasio, A. R. 1989. Lesion Analysis in Neuropsychology. New York: Oxford University Press.

Desimone, R.; Schein, S. J.; Moran, J.; and Ungerleider, L. G. 1985. Contour, Color, and Shape Analysis Beyond the Striate Cortex. *Vision Res* 25:441-452.

Feldman, J. A. 1989. Neural Representation and Neural Computation. In *Neural Connections, Mental Computation*, Nadel, L.; Cooper, L. A.; Culicover, P.; and Harnish, R. M. (ed.), Cambridge, Massachusetts: The MIT Press.

Galaburda, A. M.; LeMay, M.; Kemper, T. L.; and Geschwind, N. 1978. Right-Left Asymmetries in the Brain: Structural Differences between the Hemispheres may Underlie Cerebral Dominance. *Science* 199:852-856.

Mendez, M. F.; Mendez, M. A.; Martin, R.; and Smyth, K. A., Whitehouse, P.J. 1990. Complex Visual Disturbances in Alzheimer's Disease. *Neurology* 40:439-443.

Miikkulainen, R. 1990. A Distributed Feature Map Model of the Lexicon. In *Twelfth Annual Meeting of the Cognitive Science Society*, Boston, Massachusetts.

Mishkin, M.; Ungerleider, L. G.; and Macko, K., A. 1983. Object Vision and Spatial Vision: Two Cortical Pathways. *TINS* 6:414-417.

Newcombe, F.; and Ratcliff, G. 1989. Disorders of Visuospatial Analysis. In *Handbook of Neuropsychology*, Boller, F.; and Grafman, J. (ed.), Amsterdam: Elsevier Science Publishers.

Rueckl, J. G.; Cave, K. R.; and Kosslyn, S. M. 1989. Why are "What" and "Where" Processed by Separate Cortical Visual Systems? A Computational Investigation. *J Cog Neurosci* 1:171-186.

Rumelhart, D. E.; and McClelland, J. L. 1986. Parallel Distributed Processing: Explorations in the Microstructure of Cognition: Volume 1: Foundations. Cambridge, Massachusetts: The MIT Press.

Sejnowski, T.; Koch, C.; and Churchland, P. 1988. Computational Neuroscience. *Science* 241:1299-1306.

Sejnowski, T. J.; and Churchland, P. S. 1988. Brain and Cognition. In *Foundations of Cognitive Science*, Posner, M. I. (ed.), Cambridge, Massachusetts: MIT Press.

Shepherd, G. ed. 1990. *The Synaptic Organization of the Brain*. New York: Oxford University Press.

Shepherd, G. M.; and Koch, C. 1990. Introduction to Synaptic Circuits. In *Synaptic Organization of the Brain*, Shepard, G. M. (ed.), New York: Oxford University Press.

# Incorporating Resource Analyses into an Action System

**Matthew Brand**
Northwestern University
The Institute for the Learning Sciences
Evanston, Illinois
brand@ils.nwu.edu

## Abstract

OOPS is a reactive planner which integrates sensory perception and action selection. A principal feature of the OOPS architecture is the use and discovery of cheap, diagnostic features to indicate opportunities, which are then verified in more expensive computations. This diagnostic relationship is established analytically and refined using tools from decision theory. The use and refinement of diagnostic features depends upon assumptions of conditional independence. However, in the case of a multiplanning agent (one which simultaneously pursues several goals), while conditional independence holds true for features vis-a-vis individual opportunities, plans may interact in the world, and conditional independence may not hold. In this paper we discuss how the knowledge needed to avoid detrimental action interactions can be incorporated into OOPS's inexpensive diagnostic computations, with benefits for robustness, performance, and learning.

## Realtime action and multiplanning

The Optimizing Opportunistic Planning System (OOPS) is an action architecture designed to propel a purposeful agent through an environment that is rich with opportunities and threats. It is designed for realtime perception and multiplanning: at any given time, OOPS is juggling several ongoing plans, each in the service of a different active goal, and all selected by an integrated perception/decision system.

Multiplanning poses challenges to any planner, synthetic or reactive, that go beyond the normal difficulties of protecting satisfied subgoals. Interactions between concurrently executing plans can demolish any plan's hope for success. This is a particular concern for memory-based and reactive planners, which generally substitute pre-specified plans with protections already worked out for the machinery which ensures protections. Such planners are ill-equipped to deal with resource conflicts between unrelated plans. This is a particular concern in our research, since the design of OOPS emphasizes the production of behavior from perceptual rather than inferential mechanisms.

One of the central thrusts of the design of OOPS is finding and relying upon very cheap perceptual features that can trigger actions with high potential utility[1]. This reflects a research assumption that an agent may be successful in its world with only a limited ability to consider complex interactions between states and actions. In order for this to be true, there must exist a wide range of cheaply computed and conditionally independent features, which can account for the greater part of the agent's useful behaviors. In the following discussion, we discuss the strengths and apparent limitations of this hypothesis, and how to deal with the fact that multiplanning violates the conditional independence assumption, without departing significantly from our perception-intensive framework.

### OOPS' view of the world

The world that OOPS is being designed for is not unlike the world of an animal: changing, perceptually rich and noisy, requiring constant monitoring and rapid action. As previous work in reactive planning and action architectures [Brooks, Connell, & Ning, 88, Maes & Brooks 90, Firby 89, Hammond Converse & Martin 90, Agre & Chapman 87] has shown, we have found that being tightly coupled with such a demanding environment shifts the emphasis in planning from correctness and completeness to concerns about ecological adequacy, flexibility, and reaction speed. Realtime planning means that quickly identifying usable and desirable actions is preferable to synthesizing and verifying plans, and that the commitment of resources to potential actions—including computational attention—must reflect those actions' applicability and utility. There is a premium on the ability to rapidly identify and execute, as well as suspend, resume, or abandon useful plans in response to the face of novel circumstances.

[1]For similar views on the significance of low-level features in representation, see [Forbus & Gentner 90]

### Integral perception and action

The tight coupling between an action agent and its world should be reflected by a tight coupling between sensory and motor processing. OOPS meshes the ethological model of drive hierarchies [Tinbergen 51, Baerends 76, Dawkins 76] with connectionist-like perception networks [Rumelhart &McClelland 86], in order to achieve a smooth integration of perception and action selection. The architecture embodies the idea of *computational economy*: an expected value can be calculated for every prediction of a feature or action opportunity, and then used to determine which features and proposed actions should receive computational attention. This yields a uniform and principled procedure for allocating resources to promising perceptual and action hypotheses.

### OOPS epistemics: diagnostic features

The OOPS network includes a hierarchy of "quick and dirty" diagnostic features and more expensive, precise features. A cheap feature, when signalled, can be validated by a more expensive feature [Birnbaum 86, Sussman 75]. Features can be more expensive in their CPU requirements, the larger array of subfeatures they depend on, or in that they take longer to detect in the world and thus have less predictive value. In OOPS the main thrust of learning is to validate existing feature relationships, and to find ever cheaper features that detect situations of utility to the agent. As this proceeds, attention focusing and its motor equivalent, action selection, become increasingly efficient and accurate.

Action selection is a two-tiered process. First, values are obtained for all features and drives,[2] then used to identify the most highly applicable and desired actions. This computation is very cheap; the update cycle is constrained to yield a result in time linear in the size of the OOPS network. In the second phase, proposed actions are verified according to more elaborate specifications of conditions of applicability, and then effectors are allocated to selected actions. The specifics of attention focusing, representation, and information flow are discussed in [Brand & Birnbaum 90].

**Conditional independence** The use of cheap diagnostic features to rapidly winnow the range of action down to a few promising choices is a response to the need for a agent to be "on the ball," able to act fast on new information. In order for diagnostic features to be worth their costs, assumptions of conditional independence must hold: features must correctly predict opportunities in the *vast majority* of common situations, without requiring heaving computation. This means that the feature should be computable with little regard to context, yet still be applicable in many contexts. For example, innate releasing mechanisms in animals such as the red breast of the courting frigate bird are good conditionally independent features, because they are easily detected, yet have the same meaning in almost all contexts [Eibl-Eibesfeldt 75]. As will be discussed below, assumptions of conditional independence are even more important to learning.

**Conditional independence and learning** For learning, a third phase follows in which the results of the two phases are compared, and the resultant information is incorporated into estimates of the predictive value of features using signal-detection theory [Green & Swets 66]. The diagnosis of a feature with regard to any other feature or action is classified into True Hit, False Alarm, Correct Rejection, or Missed Opportunity, and this is used to derive the expected value of the feature [Brand 91]. This information is then incorporated into connection strengths between them.

**Resource conflicts** The rate and success of learning is partly determined by the degree that features actually are conditionally independent. A problem in the operation of the OOPS simulator is that though its first pass correctly chooses a set of applicable actions, many of these actions fail to execute after inspection because of conflicts over effectors and external resources. That is, with regard to resource management, these actions are *not* conditionally independent. At first blush, this merely means a waste of computation; it would be preferable if the first pass proposed fewer actions with likely conflicts. However, the problem is more serious in that this generates unwanted false alarms: Proposed actions fail even though their conditions of applicability are satisfied. Unwanted false alarms retard learning the relationships of predictive features, because the learning algorithm construes false alarms as incorrect diagnoses of opportunities, *independent* of other plans that may be executing. Such failures cannot simply be discounted or re-interpreted as semi-successes. Discounting means losing information important for learning. Re-interpreting means trying to represent in a connection a relation that does not hold between the feature and action it connects: that under unknown conditions some other proposed action may usurp effectors needed to execute the action.

## Multiplanning

Inside the agent, the primary challenge of multiplanning is the selection, execution, and retirement of plans. In OOPS this is achieved through a simple extension of the attention focusing system, which enforces an economy of internal resources, notably

[2] In actuality, one of the principal advantages of the notion of *computational economy* is that only selected features are computed, according to their expected utility and the availability of CPU resources. Nonetheless, values are *available* for all features, and are accurate in proportion to the agent's interest in each feature.

computation and memory.

Outside the agent, the primary challenges arise from the need to coordinate actions: the planner must avoid conflicts over resources. Depending on the number of effectors an agent has under independent motor control, it may be pursuing several plans simultaneously, interweaving plans, or using different plans to control separate effectors in the service of the same goal.

Where there is a concurrency of action, interactions between plans under execution become a serious concern. One may walk and chew gum successfully, but whistling and chewing gum is an uncertain proposition. This is a simple conflict over an effector. More subtle conflicts arise over resources frequently manipulated by separate effectors: A musician can certainly sing and play at the same time, but doing vocals over a solo is a bad idea because one may drown out the other, or both may be too complicated to listen to simultaneously. There is limited space in the dynamical and informational bandwidth of a listener's hearing. This must be learned by musicians just as toddlers learn that banging pots obscures the content of their screams for attention.

Interactions are an equally difficult problem for classical planners. Consider a case-based classical planner which must pursue multiple plans. Using a memory store of useful plans instead of synthesizing new ones becomes less and less advantageous, since the overhead of checking interactions between known plans is just as bad as the cost of doing such checking when synthesizing new plans. Nor will simple solutions help. As pointed out above, it is not satisfactory to simply block actions that require already allocated resources. It is entirely reasonable to interrupt a walk to kick a soccer ball back to the game it rolled out of, but somewhat less reasonable to do so when jogging, and not very wise to do so when being chased. The process of resource allocation must account for varying priorities.

The fact that plans have different resource usages over their execution course further complicates the matter: One cannot lock up every effector that an action will require over its execution life right from the outset. Some actions have very long execution courses, and will run concurrently with, or be punctuated by, more staccato actions. The effectors a plan uses near the end of execution shouldn't be reserved from the beginning. For example, running deep to catch a football should not preclude using one's hands to repel opposing players on the way. Nor should resources be completely available to be usurped by an unrelated plan: while in the workshop gluing wood, one does not want to "notice" an opportunity to fix a table using nearby clamps, only to find that none are left to take the place of one's hand holding the wood together.

### Recognizing Resource Conflicts

The scope of this problem has had a substantial impact on our assumptions about what makes for an adequate epistemology of the dynamics of an agent's world. It is prohibitively expensive to inferentially predict the consequences of action interactions, yet there is a strong motivation to incorporate some sort of knowledge about conditional dependence into the our planner.

Noting that the greater range of human activity does not require provably correct planning—people routinely make planning errors and action slips[3] [Norman 88] and survive—one plausible hypothesis is that much of the knowledge about conditional dependencies is brought to bear on planning in a manner more associative, and less accurate, than truth-preserving inferential reasoning. That is, up to a point, a planner benefits from from knowledge about how its actions lock up and consume resources, but that point does not extend as far as costly inferential reasoning in all cases. Thus we have focused on moving such knowledge out of the expensive action-validation process and pushing it down into the computation-efficient process of action-proposal.

## Looking to ethology

> *Animals avoid the potential problem of ambivalence; stimuli appropriate for more than one behavior may be present but the animal usually can only perform one behavior at a time.*[Toates 80]

What makes an animal so good at picking a workable behavior for each situation? The knowledge-poor answer is that things are stacked so that all possibilities can be considered in parallel. The knowledge-rich answer is that things are arranged so that the most appropriate possibilities are considered first. Cognitive scientists might attribute this to *priming*, AI researchers might say *indexing*. Ethologists tend to talk about *suppression*: Given the kind of situation the animal is in, a whole range of behaviors will be suppressed. For simple animals, there are perhaps 10-50 kinds of situations: feeding, predation, fight, flight, courtship, mating, gestation, sleep, birthing, patrolling, etc. Each of these has characteristic resource needs, and behaviors with inimical resource-consumption patterns are suppressed.

### Avoiding resource conflicts

Consider a very simple example from the low end of the food chain: The gastropod *Pleurobranchaea* is a voracious eater. It will routinely abandon mating if food appears nearby. In its evolutionary history, energy intake probably was its greatest challenge. When laying eggs, however, it must suppress feeding

[3] Planning errors and action slips correspond to failures to preserve external and internal state, respectively.

to protect its own progeny. In its very simple world, the short-term resource strategy of energy intake (eat everything within range) must be subordinated to the long-term resource strategy of propagating its genes (stabilize the environment until the eggs hatch).

To take another example, mobility is a resource that many animals will temporarily sacrifice for extra manipulative capacity (e.g. woodpeckers wedge nuts into corners so that they can peck them, beavers wedge themselves against branches to gain leverage in fitting together new dam parts). When frightened, woodpeckers will avoid places in their environment that have been useful for wedging, but they will go near other places that are significant for them. Presumably, fear-related behaviors suppress behaviors that relate to reduction of mobility.

The key observation here is that suppression inhibits behaviors with incompatible resource usage patterns. Usually, these rival behaviors fail to become sufficiently active to claim effectors. Mapping this onto the two-tiered operation of OOPS, it is likely that suppression operates in the cheaper computations which *find* applicable actions, rather than the more expensive *validation* of proposed actions. In this view, suppression operates in the realm of memory more than it does in realm of reasoning. Although this provides only a weak protection from detrimental interactions, it may suffice to account for enough successes to make an agent environmentally viable.

Finally, in business planning, time-sensitive behaviors should suppress the consideration of behaviors that treat time as a limitless or free resource. An example lies in the reaction of liquor retailers to the alcohol sales tax imposed at the beginning of 1991. An owner of a large chain of liquor stores in Milwaukee was interviewed a few months prior to the date. He was stocking up in September for a sale in December. The reporter asked him why not have a long, huge sale to beat the deadline. The retailer answered that sales take an unpredictable amount of time to unload inventory; it was better to rent a warehouse, stock it up with pre-tax beer, and sell it off when consumer awareness of the tax was at its highest. In this case, a sale with specific time costs is preferable to one with indeterminate time costs, because the latter may not satisfy its goal before the deadline. The latter plan will also fail because it ties up many other resources, including store space, for its duration. Although the retailer's explanation may refer to more sophisticated reasoning about resource tradeoffs, suppression of inimical resource-strategies is *sufficient to produce the preference.*

## Suppressing resource competitors

Most behaviors can be characterized in terms of the resources that they conserve and the resources that they expend. For example, hoarding conserves the potential energy of food at the expense of the time spent scavenging, retrieving, and hiding. Once these characterizations are available, mutually inhibitory relations should hold between behaviors with incompatible resource characterizations. This will reduce the likelihood of conflict, and thus failed actions.

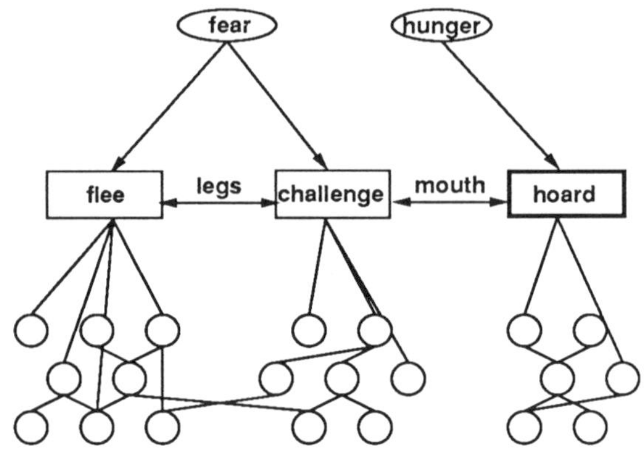


Figure 1: A schematic of a subnetwork from an OOPS simulation of an animal that must keep up both its territory and its food supply. The circles represent feature nodes; flee, challenge, and hoard are action nodes; mouth and legs represent effectors. Fear and hunger are drive nodes that boost the chances of their servant actions threshholding.

Resource characterizations can in many cases be deduced from the effectors that are involved. In the current set of OOPS simulations, where the domain is animal behavior, all the needed resource characterizations thus far have been available this way. Each effector is good for manipulating a particular kind of resource. A plan's resource consumption can be derived from the spans of time that it monopolizes various effectors. For example, consider the plan carry-back-to-nest. This monopolizes the jaw for the duration of the trip, and thus blocks access to, and use of, resources associated with the mouth: dexterity, weaponry, food. In simple animals, where all effectors are just bodily parts, one can produce a resource analysis purely by examining all the behaviors that compete for the same effectors. In many human domains, there are resources that are not explicitly manipulated by effectors, thus the analyses are more complicated.

Consider a dog carrying a tasty bone. Another dog approaches. Normally the first dog would approach to inspect and challenge, out of territoriality. In this case, it makes more sense to run away with the bone. Why? What is it that causes the flee action to threshold before the challenge action? We can take figure 1 to be a simplified model of the drives and behaviors involved. Normally, when fear is active and a foreign dog is perceived, challenge would threshhold higher than flee. It would be validated first and gains control of the legs. Flee would also be validated, but then fail to claim the legs, and thus signal a false alarm. But in this case, challenge is being suppressed by hoard, because the two actions tend to have resource conflicts. It fails to threshhold. Flee threshholds and the the system chooses to run, without the additional overhead of considering challenge. In this example, the conditionality expressed in the lateral inhibition between actions suffices to bias the network in favor of flight.

### Generalized resource analyses

In animals, effectors are synonymous with body parts, and it is easy to propagate information about the full catalog of effectors back into the network, where it takes the form of suppression links between actions. In modelling more abstract human tasks, the use of tools expands the range of potential effectors. In these cases, the network must be updated as new tools become available. However, this is probably a low overhead operation, compared to savings in computation gained from such additional knowledge.

```
for each behavior B
    for each effector E used by behavior B
        for each resource R consumed by effector E
            for each effector E' indexed under resource R
                for each behavior B' which uses effector E'
                    if R is a { monopolizable, exhaustable,
                                core, time } resource
                        increase inhibition between B and B'
                    if R is a { replenishable,
                                self-replenishing, sharable,
                                renewable, } resource
                        decrease inhibition between B and B'
```

Figure 2: A simple heuristic for establishing a priori resource conflicts when constructing an action network. This algorithm depends on an abstract characterization of resource depletion and replenishment. *Renewable*, *replenishable*, and *self-replenishing* refer to resources that can be restored instantaneously, willfully, and automatically, respectively. An example of each is river water, food stores, and food crops in the management of a town. A preliminary typology of resources appears in [Brand 90].

For a model of an agent in the financial industry, effectors are what economists call financial "instruments." Options, contracts, holds, voting blocks, etc., are all instruments. An interesting aspect of instruments is that they are largely understood by those who use them as devices for exchanging resources. In any newspaper discussion of devices and their use, they are discussed almost exclusively in terms of the resource trade-offs that they embody. Options preserve opportunities at the loss of cash or credit. Voting blocks are nonliquid assets that reduce opponents' control of other companies.

Consider two corporations in a takeover battle, predator and prey. Two of the many plans for fighting a takeover are (1) raise one's own stock price through buy-backs, and (2) achieve a voting bloc of the predator's stock, and lead a shareholder vote to challenge the takeover internally. The instruments used here are voting blocks and options, particularly options on one's own stock and on that of potential predators.

Both of these instruments consume liquidity. In fact, to be effective, both devices consume indeterminate amounts of liquidity, since it is not clear in advance how much one's own share price must be driven up to drive off a suitor, nor how many voting shares are necessary to lead a shareholder revolt against the suitor's takeover plans. Even if sufficient liquidity is available for exercising both options at the outset, since both plans may require more cash or credit, it would be foolish to pursue both tracks.

An analysis of either instrument shows that they both require large amounts of liquidity. This information can be propagated back to both plans, where the *lack* of a specification of how much liquidity is needed shows that both plans require indeterminate amounts of money. Thus labelled, an inhibitory connection is added so that in the future, they are less likely to undermine each other through concurrent execution. Note that this is not a conflict over instruments, but the underlying resources that they consume.

Although both plans consume a resource of general universality—financial liquidity—the same treatment applies to more specific resources. Alternative plans for taking legal action should have the same mutually inhibitory relationship because of possible conflicts over the resources of the legal department, money, and intra-office information flow.

## Summary

OOPS is an action architecture designed with an emphasis on the perception of opportunities. It reflects the belief that inferential mechanisms are not efficient enough to produce realtime behavior over a course of situations routinely encountered in a robot's world. The system attempts where ever possible to substitute inexpensive computations for expensive computations, without loss of utility. Cheap diagnostic features yield probablistic indications of opportunities, and then more expensive mechanisms are invoked to verify the opportunities and select among proposed actions. Decision-theoretic methods are used to learn the probablistic value of each feature's diagnosis; the results are used to allocate computation to hypotheses about opportunities in the world. Both the utility and the learnability of these features depends upon assumptions of conditional independence. Conditional relations between features are only handled after the winning hypotheses have thresholded. However, the conditional independence of hypotheses does not necessarily hold; frequently, viable plans are considered that have conflicting requirements. They may need the same effectors, or need different effectors that usurp the same resources. Anticipating and handling such conflicts is far to expensive to do inferentially. Rather than leave it up to expensive action validation to protect against such conflicts, it is much more economical to "push down" this knowledge into the cheap pre-thresholding computation, where inhibitory connections between action hypotheses bias perception against seeing conflicting opportunities. This insulates the multiplanner against the twin problems of ambivalence and overextension.

## Acknowledgments

The ideas presented here were developed in collaboration with Larry Birnbaum. Thanks to Bruce Krulwich, Gregg Collins, Louise Pryor, and Michael Freed for many useful discussions. This work was supported in part by the Defense Advanced Research Projects Agency, monitored by the Air Force Office of Scientific Research under contract F49620-88-C-0058. The author is supported by a National Science Foundation Graduate Fellowship. The Institute for the Learning Sciences was established in 1989 with the support of Andersen Consulting, part of The Arthur Andersen Worldwide Organization. The Institute receives additional support from Ameritech, an Institute Partner, and from IBM.

## References

[Agre & Chapman 87] Philip E. Agre and David Chapman. Pengi: An Implementation of a Theory of Activity. In *Proceedings of AAAI-87*, 1987.

[Baerends 76] G. P. Baerends. The functional organization of behavior. *Animal Behavior.* **24**:726-735. 1976.

[Birnbaum 86] Lawrence Birnbaum. *Integrated Processing in Planning and Understanding.* PhD Thesis, Yale University Computer Science Department, 1986. Available as Report RR-489.

[Brand 90] Matthew Brand. A Typology of Planning. In R. Schank, *et. al.*, A Content Theory of Memory Indexing. Institute for the Learning Sciences Tech Report #2. 1990.

[Brand 91] Matthew Brand. Decision-Theoretic Learning in an Action System. To appear in the *Proceedings of the 1991 Machine Learning Workshop.*

[Brand & Birnbaum 90] Matthew Brand and Lawrence Birnbaum. Noticing opportunities in a rich environment. In *Proceedings of the 12th Annual Conference of The Cognitive Science Society*, 1990.

[Brooks, Connell, & Ning, 88] Rodney Brooks, Jonathon Connell, and Herbert Peter Ning. A second generation mobile robot, AI Memo 1016, MIT Artificial Intelligence Laboratory, 1988.

[Eibl-Eibesfeldt 75] Irenaus Eibl-Eibesfeldt. *Ethology: The Biology of Behavior.* New York: Holt, Rinehart and Winston, Inc., 1975.

[Dawkins 76] Richard Dawkins. Hierarchical organization: a candidate principle for ethology. In P.P.G. Bateson and R.A. Hinde *(eds.) Growing Points in Ethology*, pp. 7-54. Cambridge University Press, 1976.

[Firby 89] R. James Firby. *Adaptive Execution in Complex Dynamic Worlds.* PhD Thesis, Yale University Computer Science Department, 1989. Available as Report RR-672.

[Forbus & Gentner 90] Kenneth Forbus and Dedre Gentner. Structural Evaluation of Analogies: What Counts? In *Proceedings of the 11th Annual Conference of The Cognitive Science Society*, 1989.

[Green & Swets 66] David M. Green and John A. Swets. *Signal detection theory and psychophysics.* New York: J. Wiley & Sons. 1966.

[Hammond Converse & Martin 90] Kristian Hammond, Tim Converse, and Charles Martin. Integrating planning and acting in a case-based framework. *Proceedings of the Eighth National Conference on Artificial Intelligence,* Boston, MA, pp. 292-297. 1990.

[Maes & Brooks 90] Pattie Maes. Learning to Coordinate Behaviors. In *Proceedings, AAAI-89.* 1990.

[Norman 88] Donald A. Norman. *The Psychology of Everday Things*, New York: Basic Books, Inc., 1988.

[Rumelhart &McClelland 86] David Rumelhart and James McClelland *(eds.). Parallel Distributed Processing.* Cambridge, MA: Bradford Books. 1986.

[Sussman 75] Gerald J. Sussman. *A Computer Model of Skill Acquisition.* American Elsevier, New York, 1975.

[Tinbergen 51] Niko Tinbergen. *The Study of Instinct.* Oxford University Press. Oxford. 1951.

[Toates 80] Frederick M. Toates. *Animal Behavior—A Systems Approach* New York: John Wiley & Sons, 1980.

# Opportunistic memory and visual search

**Timothy M. Converse** and **Kristian J. Hammond**
Department of Computer Science
The University of Chicago
1100 East 58th Street
Chicago, IL 60637
converse@cs.uchicago.edu

## Abstract

In earlier work, we proposed a memory model that would facilitate the detection of opportunities to satisfy suspended goals. In this *opportunistic memory* model, suspended goals are indexed under feature sets that are predictive of the presence of the opportunity, and which are likely to be encountered in the normal course of future activity. The functional benefit of such encoding depends crucially on the particular vocabulary of features used, the costs of their detection, and the overlap of features relevant to the pursuit of different goals. In this paper we investigate the feature vocabulary implied by recent work on visual search [Treisman, 1985, Tsotsos, 1990], and its use in indexing goals suspended due to the lack of a particular object.

## Opportunism

In daily life, people pursue a large number of different goals, and display great flexibility in shifting attention from one goal to another. These shifts in focus are often in response to new information that could not have been predicted in advance. We all capitalize on chance encounters with people we have been intending to talk to, buy things because we happen to see them displayed in a store, realize that we can perform one errand while "on the way" to perform another, and so on.

There is good reason to believe that a better understanding of this sort of opportunistic behavior is important for designing artificial agents as well. *Any* agent that must pursue multiple goals in an unpredictable world faces a problem of focus — which goals should be acted on now? If there are too many goals to plan for completely, or if not all of them are available when action must begin, then there must be some heuristic selection of a subset of goals that should be actively pursued. On the other hand, the agent needs to be ready to reconsider these choices in the face of unanticipated opportunities.

The central problem for accounts of opportunistic behavior is how to reconcile the background status of a goal that is not under active consideration with the fact that an opportunity to satisfy it can be recognized at all. Recognition requires perceptual and inferential work. If recognizing an opportunity to satisfy a background goal assumes as much work as actively seeking to satisfy it does, then the functional benefit of focusing on a small set of goals is lost.

## Opportunistic Memory

In earlier work [Hammond, 1989b, Hammond *et al.*, 1988] we proposed a memory architecture that supports remindings of background goals at appropriate times. The idea of *opportunistic memory* relies on the fact that normal work toward the satisfaction of an active goal demands perceptual and inferential effort. In contrast to accounts of opportunistic behavior where goals are imbued with autonomous processing power [Birnbaum and Collins, 1984], the opportunistic memory model relies on encoding-time work to appropriately index passive notations in the same memory used in execution, so that suspended goals can be cheaply reawakened in the course of execution. Once "awakened", a suspended goal can be more thoroughly considered to see if it is plausible to pursue now. The potential benefit is that given proper encoding-time work, opportunity recognition can exploit the effort already being expended on goals that are currently active.

In brief, the opportunistic memory "algorithm" is as follows:

- Goals that cannot be immediately satisfied or integrated into a currently active plan are considered blocked, and are *suspended* [Schank, 1982].
- Suspended goals are associated with features in memory that, if encountered in the world, would indicate that the goal may now be satisfiable.
- The same memory structures are used to parse the world and make routine execution-time decisions. When the features that a suspended goal has been

associated with are activated in the course of this, the goal is brought back under consideration.

The objective of this encoding is that the agent will have goals "resubmitted" to attention by memory at exactly those times when they can be acted on. This will be successful to the extent that

1. the agent can characterize the ways in which the world would have to change to make the goal satisfiable, and
2. the agent can associate the goal with a set of easily computed operational features that strongly predict that such a change has taken place.

We now treat the first of these briefly, and then spend the remainder of the paper on the second.

### Reasons for goal blockage

The simplest way that a goal can become blocked is if a standard plan associated with it is presently unusable for some reason. The possibility of encountering such a situation is an inevitable consequence of a case-based planning framework [Hammond, 1989a], where the attempt is made to reuse plans in situations that may not be exactly the ones for which they were designed.

Among the things which can block the use of a standard plan are:

- Lack of time.

  Due to other pressing goals and activities, there may not be time at the moment to run the standard plan. For example, the goal to run an errand to purchase something may have to be suspended when running late for an appointment.

- Lack of tool or resource.

  The plan depends on something we don't possess at the moment. For example, our plan to drink wine with our picnic lunch might be blocked by lack of a corkscrew.

- Lack of knowledge.

  In this case there is no state in the world that is blocking our efforts, but we lack the knowledge to properly exploit it. For example, the plan to get downtown via the infrequent commuter train might be blocked by lack of knowledge of the schedule.

- Lack of proximity.

  Being in a particular location is a common sort of prerequisite for a host of mundane tasks. Remembering that your plants need watering does not help you when you are away from home.

The features that should cause a goal to become reactivated don't have to be limited to those produced by a characterization of what originally blocked its satisfaction — it might be reasonable to index a goal under features that indicate unusual preconditions of a standard plan, on the theory that it is worth checking the other preconditions at that point. But the realization that a blocking condition is no longer true should be a strong predictor that the goal is now satisfiable.

Some of the categories above are abstract, and the corresponding opportunities are probably difficult to recognize (for example "having time"). From here on we will focus on opportunity recognition that corresponds simply to visually recognizing a particular object.

### Functional requirements on indexing features

Clearly even with a good characterization of what would constitute an opportunity, the benefit of the encoding depends strongly on the particular vocabulary of features and the costs of detecting them. If a suspended goal is to be indexed under a given set of features, then

1. The feature set should be cheap to compute relative to the cost of completely verifying the existence of the opportunity.
2. As much as possible, the features should be ones that are likely to be checked anyway in pursuing other courses of action. (This consideration trades off with the previous one.)
3. The ratio of false negatives (missed opportunities) to false positives (inappropriate activations of suspended goals) should be in line with the urgency of the goal and the cost of verification.

We now discuss these requirements in the context of recent experimental results on visual search.

## Visual search

It is extremely difficult to design experiments in cognitive and perceptual psychology that balance the conflicting demands of experimental control and relevance to real-world activity. Visual search is an attractive paradigm in part because, although people rarely face tasks exactly like the experimental task, it seems plausible that the faculty being tested is a building block (and possibly a bottleneck) in larger-scale visual activity.

In visual search experiments, subjects are given the task of quickly detecting a target image amid a field of distractors, and response time is measured. Treisman's interpretation of such experiments is to suggest that there is a small set of attributes of the visual field which are computed preattentively and in parallel [Treisman, 1985]. When the search task involves simply identifying that such a "pop-out" property is present, response times tend to be independent of the number of distractors. When identification of the target requires verifying a more complex property, or a conjunction of pop-out properties, response times vary linearly with the number of distractors, suggesting that serial attention is required in the verification. A small number of properties pop out, among them color, brightness, closure, curvature and tilt.

Tsotsos [Tsotsos, 1990] has argued for the intrinsic difficulty of *unbounded* visual search, where success depends on satisfying a function that, rather than being a straightforward match to a target, may depend on arbitrary interrelations of portions of the test image. Examples are tasks such as identifying the "odd man out" of a field of distractors, where the properties that will be unique are not identified in advance. Tsotsos proves an abstraction of unbounded visual search to be NP-hard, and uses this result to motivate a particular attentional architecture.

One implication of these results is, of course, that it is very unlikely that we can instantly identify all the objects in a complex visual scene. Ballard [Ballard, 1990] suggests that in general, the problem of simultaneously relating many different models to many visual locations may be too hard, and argues for the separation of *location* and *identification* algorithms, where location algorithms attempt to find a known object in the visual field, and identification algorithms do object recognition in situations where the location is not the issue. (In this he makes a functional argument for a specialization that is believed to exist in biological systems [Maunsell and Newsome, 1987, Mishkin, 1982, Mishkin *et al.*, 1983].) When incorporated in task-dependent behaviors, location and identification algorithms can use methods that are tailored to the outstanding attributes that are likely to be present in the context of the task, further avoiding visual reconstruction in its full generality. See *e.g.* [Swain, 1990] for real-time algorithms for identification or location of objects by their colors.

## Visual search and opportunism

One way in which visual search experiments idealize real-world search is that, regardless of how the object of search is specified, finding the object is the only goal. It seems likely that in human activity, while there may often be a particular object being sought or identified, there are a number of other things that it is worthwhile to notice if they are seen. Visual cues that indicate opportunities to satisfy suspended goals may be among them.

### An example

To motivate the rest of the discussion, we will tell a story:

> When I was in college, my black leather jacket was stolen. For some weeks after that, whenever I saw anyone wearing anything black and shiny, it "caught my eye", and I checked to see if it was my jacket. This happened even though, to my knowledge, I wasn't thinking about the jacket beforehand.

Depending on the importance of the goal of recovering the jacket, this is a functional response. The urgency and the chances of seeing the jacket at any one time don't warrant actively searching for it all the time. Still, if by chance it is encountered, it might be worth having a word with the current wearer.

### Choosing visual indices

How should the opportunity represented by seeing the jacket be indexed in memory?

Our aim here is to have as much as possible of the work needed for opportunity recognition be done in the normal course of activity, with our suspended goal remaining a passive annotation. At the risk of drastic oversimplification, we can use the results cited above to sketch a hierarchy of effort that might be expended in recognizing an opportunity associated with a particular object.

1. Recognition of a particular pop-out property in the visual scene. For example, simply detecting an object of a certain color. Also, it may be possible to preattentively detect the conjunctive presence of more than one pop-out property, without necessarily verifying that they are in the same location or due to the same object.
2. Verification of a conjunct of pop-out properties in the same location, or any other process that demands serial visual attention, whether overt or covert.
3. True object recognition, i.e. classifying a seen object as a jacket or a fork or a car.
4. Inference based on object recognition. Here we move into processing that may or may not be completely non-visual, but that has the recognition of a particular object as a prerequisite. Note that much of the work of actually evaluating an opportunity represented by seeing an object falls into this category, since even if the object seen is exactly what was being sought at encoding time, circumstances may have changed so that the goal is not satisfiable for some other reason.

The point of this hierarchy is that, while the processes occurring at a given level may be very complex (and poorly understood), it seems reasonable to assume that the higher-numbered levels are dependent on the lower ones, and work at all levels will have to be done at some point to detect, verify, and evaluate opportunities represented by seeing particular objects. If a description that falls at one of these levels is associated with a suspended goal, we assume that activation of those features will be enough to activate the goal, which in turn will lead to the direct suggestion of the rest of the work necessary in its verification. So the question in determining indexing features then becomes: what is the lowest level at which a given goal should be indexed?

As we said above, a good recommendation for a potential index is that it will be computed anyway. Almost by definition, pop-out properties qualify (although detecting a particular one may well require some non-spatial attention, e.g. to a particular feature map). There are two reasons why pop-out properties may not provide good indices in particular cases. The first is simply that many of these properties are quite low-level and even in conjunction are not expressive enough to characterize interesting opportunities. The most promising ones seem to be motion, brightness, and especially color – but if the relevant type of opportunity is seeing a restaurant, then they do not suffice. The second reason is the possibility of an unacceptably high rate of false alarms.

In our example story, the suspended goal of recovering the jacket might simply have been indexed in association with the color black.[1] Whenever that color appeared in the visual field, the goal would be reactivated, and further verification that in fact the color was not due to the missing jacket would have to be performed, probably resulting at least in redirection of visual attention. One way to interpret the story is that in fact the goal was associated with the conjunction of a color and a texture, and that even that requirement gave a sufficiently high false alarm rate that it caused the experience to be memorable.

Indexing an opportunity at the level of full object recognition may offer a better characterization and result in fewer false alarms, but it may also then restrict discovery to things that are fully attended to and identified. This may be acceptable when the cost of completely missing the opportunity is low. It may also be acceptable when enough is known about the structure of future activities and tasks to predict that the given object *will* be attended to at some point [Hammond, 1990]. For example, it may be reasonable to index the goal of responding to a particular letter lost on a messy desk by actually seeing and identifying the letter, if you know that you will be cleaning the desk at some point in the future.

No strong theory of how to determine the appropriate level of indexing is offered here – in part it may just have to be a process of adding features to conjuncts or moving up the hierarchy given above in response to a high false alarm rate. (For a decision-theoretic approach to tuning opportunity recognition, see [Brand, 1991].) The one thing that seems clear is that if the feature set under which a goal is indexed needs to be refined, there is a functional benefit to starting on the overly inclusive side. It is easier to notice inappropriate remindings than it is to notice the lack of appropriate ones.

More realistic theories of attention and incremental object recognition doubtless allow for better-defined intermediate "hooks" on which annotations about opportunities can be hung. But to a great extent the lower-level visual vocabulary seems likely to be fixed, and choices of appropriate indexing level need to be made from the options provided by it; the visual system cannot be dynamically reconfigured to detect leather jackets preattentively.

[1]Here we make the simplifying assumption that the properties are binary—of course, the actual responses of "detectors" for the features are likely to be complicated and graded. The point here is that the goal is associated with some visual characterization of the object that can be detected preattentively.

# Discussion

## Relevance to artificial systems

All the discussion so far has been grounded in the particular feature vocabularies resulting from experiments on human vision. How do these ideas transfer to constructing systems that may operate under different perceptual constraints?

There seem to be strong design advantages to giving autonomous agents highly optimized peripheral perceptual systems, which compute a wide (but fixed) variety of low-level features all the time, very efficiently. More recently, there has been growing interest in the vision community in task-directed short-term control of perception, which actively directs the orientation of the perceptors, and uses attention to select a very small task-relevant subset of the provided information. At the same time, there is increased interest in the planning-and-action community in the realities and costs of perception

At its most abstract, our argument here is for creating indices for suspended goals on the fly, where the indices are composed of primitive perceptual features (whatever they happen to be) or features that may be derived due to attention to other goals. These indexed goals become activated when all their features are seen, but until then do not compete for attentional resources.

## A closing story

There are types of opportunistic behavior which are not covered in the framework we've been discussing, one of which is exemplified by the following story:

> While on vacation at a lakeside cabin, the frisbee I was throwing became lodged in a tree. I moved down to the nearby shoreline to look for a small rock or piece of wood I could throw at the frisbee to dislodge it. As I looked, I saw a long paddle from a paddle-board, which I realized I could use to poke the frisbee free.

In effect, the opportunism here is in the recognition that the means to pursue an alternate plan is at hand. In part this is outside the framework we've presented because our focus has been on what goals should be attended to at all, rather than on suggestion of plans. The type of recognition that needs to be done is also subtly different, however. The recognition depends not only on object identification (as a paddle), but on further inference about the uses to which the object could be put (its suitability as a reach-extender), and as such falls into the highest category in the hierarchy we sketched above.

The only connection between the object and the current goal is through the plan that is eventually suggested. For some suggestions how of this integration of top-down and bottom-up inference might be implemented, see [Hammond *et al.*, 1990].

## Acknowledgements

We'd like to thank Jim Firby, Patricia Goldweic, Mike Swain, Ron McClamrock, and Steve Whitehead for comments and stimulating conversations; they are not responsible for our errors and oversimplifications. We also thank anonymous reviewers for their comments.

## References

[Ballard, 1990] Dana Ballard. Animate vision. *Artificial Intelligence*, 48(1), 1990.

[Birnbaum and Collins, 1984]
L. Birnbaum and G. Collins. Opportunistic planning and Freudian slips. In *Proceedings of the Sixth Annual Conference of the Cognitive Science Society*, Boulder, CO, 1984.

[Brand, 1991] Matthew Brand. Decision-theoretic learning in an action system. Paper submitted to the ML91 Workshop on Learning Reaction Strategies, 1991.

[Hammond *et al.*, 1988] Kristian J. Hammond, Tim Converse, and Mitchell Marks. Learning from opportunities: Storing and reusing execution-time optimizations. In *The Proceedings of the Seventh National Conference on Artificial Intelligence*, pages 536–40. AAAI, 1988.

[Hammond *et al.*, 1990] Kristian Hammond, Timothy Converse, and Charles Martin. Integrating planning and acting in a case-based framework. In *The Proceedings of the 1990 National Conference of Artificial Intelligence*, August 1990.

[Hammond, 1989a] Kristian Hammond. *Case-Based Planning: Viewing Planning as a Memory Task*, volume 1 of *Perspectives in Artificial Intelligence.* Academic Press, San Diego, CA, 1989.

[Hammond, 1989b] Kristian Hammond. Opportunistic memory. In *Proceedings of the Eleventh International Joint Conference on Artificial Intelligence.* IJCAI, 1989.

[Hammond, 1990] Kris Hammond. Learning and enforcement: Stabilizing environments to facilitate activity. In *The Proceedings of the Seventh International Conference on Machine Learning*, July 1990.

[Maunsell and Newsome, 1987] J.H.R. Maunsell and W.T. Newsome. Visual processing in monkey extrastriate cortex. *Annual Review of Neuroscience*, 10, 1987.

[Mishkin *et al.*, 1983] M. Mishkin, L.G. Ungerleider, and K.A. Macko. Object vision and spatial vision: two cortical pathways. *Trends Neurosci.*, 6, 1983.

[Mishkin, 1982] M. Mishkin. A memory system in the monkey. *Philos. Trans. Royal Society London*, 298, 1982.

[Schank, 1982] R.C. Schank. *Dynamic Memory: A Theory of Reminding and Learning in Computers and People.* Cambridge University Press, 1982.

[Swain, 1990] Michael J. Swain. *Color Indexing.* PhD thesis, University of Rochester, November 1990. Technical Report 360.

[Treisman, 1985] Anne Treisman. Preattentive processing in vision. *Computer Vision, Graphics, and Image Processing*, 31, 1985.

[Tsotsos, 1990] John K. Tsotsos. Analyzing vision at the complexity level. *Behavioral and Brain Sciences*, September 1990.

# Determining What to Learn in a Multi-Component Planning System*

**Bruce Krulwich**
Northwestern University
The Institute for the Learning Sciences
1890 Maple Avenue
Evanston, Illinois 60201

## Abstract

An intelligent agent which is involved in a variety of cognitive tasks must be able to learn new methods for performing each of them. We discuss how this can be achieved by a system composed of sets of rules for each task. To learn a new rule, the system first isolates the rule set which should be augmented, and then invokes an explanation-based learning mechanism to construct the new rule. This raises the question of how appropriate *target concepts* for explanation can be determined for each task. We discuss the solution to this problem employed in the CASTLE system, which retrieves target concepts in the form of performance specifications of its components, and demonstrate the system learning rules for several different tasks using this uniform mechanism.

## Cognitive tasks and components

In the course of determining its behavior, an intelligent agent must activate goals, notice opportunities, and devise and select among plans of action that will accomplish its goals. In competitive planning domains, including such games as chess (in which our system operates), an agent must additionally notice threats posed by other agents and develop plans to respond to them. Such an agent should be able to learn new methods for accomplishing each of these cognitive tasks.

The CASTLE system[1] engages in these tasks while playing chess [Birnbaum *et al.,* 1990; Collins *et al.,* 1991]. In the course of its decision-making, CASTLE generates explicit *expectations* about future actions and abilities, and learns from failures of these expectations. A variety of decision-making methods are represented explicitly, and an explicitly represented model of these procedures is used to diagnose expectation failures.

CASTLE is broken up into a number of *components*, which reflect a functional decomposition of the decision-making process [Collins *et al.,* 1991]. Each component is dedicated to a particular cognitive task, and is implemented as a set of rules which provide different methods for performing the task. This decomposition of the agent into distinct components enables a uniform learning process to learn methods for any cognitive task in which the system is engaged.

For example, one such component has the task of noticing threats and opportunities as they become available. Rather than recomputing these at each turn, CASTLE maintains a set of active threats and opportunities which is updated over time. To accomplish this incremental threat detection, the system uses a *detection focusing* component, which consists of *focus rules* that specify the areas in which new threats may have been enabled. Then, a separate *threat detection* component, consisting of rules for noticing specific types of threats, detects the threats that have been enabled. A sample focus rule is shown in figure 1, which embodies the system's knowledge that the most recently moved piece, in its new location, may be a source of new threats. Another focus rule, not shown, specifies that the move recently moved piece can also be a target of newly enabled attacks. Using focus rules such as these, the actual threat detector rules will only be invoked on areas of the board which can possibly contain new threats.

A second system component has the task of generating responses to enemy threats. This *counterplanning*

```
(def-brule focus-new-source
  (focus focus-moved-piece ?player
       (move ?player ?move-type ?piece ?loc1 ?loc2)
       (world-at-time ?time))
 <=
  (move-to-make (move ?player ?prev-move-type
                      ?piece ?old-loc ?loc1)
              ?player ?goal (1- ?time)) )
```

Figure 1: Focusing on new moves by a moved piece

*This work was supported in part by the Office of Naval Research under contract N00014-89-J-3217, and by the Defense Advanced Research Projects Agency, monitored by the Air Force Office of Scientific Research under contract F49620-88-C-0058. The Institute for the Learning Sciences was established in 1989 with the support of Andersen Consulting, part of The Arthur Andersen Worldwide Organization. The Institute receives additional support from Ameritech, an Institute Partner, and from IBM.

[1] CASTLE stands for *C*oncocting *A*bstract *S*trategies *T*hrough *L*earning from *E*xpectation-failures.

```
(def-brule counterplan-1
  (counterplan ?player
     (goal-capture ?piece ?loc
        (move opponent (capture ?piece)
                       ?opp-piece ?opp-loc ?loc))
     ?time ?the-response)
<=
 (and (move-legal
        (move ?player move ?piece ?loc ?new-loc))
   (no (and (at-loc ?opponent ?other-opp-piece
                    ?other-opp-loc ?time)
            (move-legal
             (move ?opponent (capture ?piece)
                   ?other-opp-piece ?other-opp-loc
                   ?new-loc))))
   (is-seq ?the-response
      (move ?player move ?piece ?loc ?new-loc))))
```

Figure 2: Reacting to a threat by running away

*component* is invoked whenever an enemy threat is noticed, and generates moves that the system can make in response. It is additionally invoked whenever the system has to reason about responses to an action, as we will see later. One such counterplanning rule is shown in figure 2, which says that a possible response to an enemy threat is to move the threatened piece to a safe location.

These two components, and others like them, span the spectrum of cognitive tasks, including monitoring the environment, reacting to external actions, planning computer actions, and selecting among possible actions. Each component consists of a set of rules which collectively achieve the desired task. For each of these components, learning new methods for achieving their tasks is accomplished by constructing new rules for their rule sets.

## Counterplanning by interposition

Consider one way in which CASTLE can learn a new rule for the *counterplanning* component. In the chess situation shown in figure 3(a), the computer has two options — it can capture the opponent's knight or it can construct a plan to take the opponent's bishop. One plan to capture the opponent's bishop would be to move the queen two squares to the right and capture the bishop on the subsequent turn. In deciding whether this is a viable plan, the computer considers whether the opponent would have a response to its move that would disable the attack before it could be carried out. In other words, the computer checks whether there is a possible counterplan for the opponent [Krulwich, 1991].

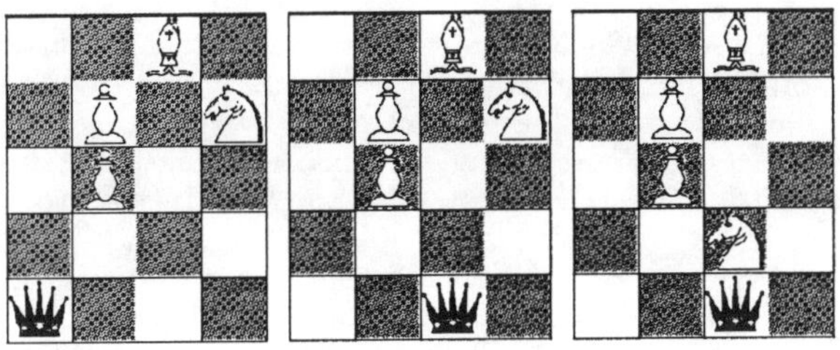

(a) (b) (c)

Figure 3: *Interposition:* Computer (black) to move

It is clear in figure 3(b) that the opponent would in fact have a possible counterplan, namely moving its knight to block the attack. Suppose, however, that the system had an incomplete collection of counterplanning rules that included rules for moving a piece to a safe location (the rule shown in figure 2) and capturing the attacking piece, but did not include a rule for interposing a piece to block a threat. Under these circumstances, when considering the viability of its plan, the system would conclude that the opponent will not have a response to the threat. Given this assumption, the system would consider the plan to take the bishop foolproof, and would proceed to carry it out.

Of course, this plan will fail, because the opponent will block the attack by interposing its knight, as we see in figure 3(c). Because the plan failure came about due to the system's lack of counterplanning knowledge, our system should respond to the failure by constructing a rule for counterplanning against an attack by interposing a piece.

## Learning from the component failure

To learn from this plan failure, the system must first determine which component is at fault. The system's diagnosis engine has this task of identifying the faulty component. CASTLE accomplishes this using a *model-based reasoning* approach [Davis, 1984; deKleer and Williams, 1987; Simmons, 1988]. Currently, our model is limited to the case in which only one component is at fault. Failures of more than one component could be handled by repairing the two components independently, if the failure reflects separable faults in the two components. Alternatively, the higher-level component which made use of the two faulty components may need to be repaired, to account for unanticipated interactions between the two components. In this paper we are limiting our discussion to the case of single-component faults.

CASTLE performs its diagnosis using explicit *justification structures* [deKleer *et al.*, 1977; Doyle, 1979], which record how the planner's expectations are inferred from the rules that constitute its decision-making mechanisms, in conjunction with the policies and underlying assumptions which it has adopted. Diagnosing the failure then involves "backing up" through the justification structures, recursively explaining the failure in terms of faulty rule antecedents [Smith, Winston, Mitchell, and Buchanan, 1985; Simmons, 1988; Birnbaum, Collins, and Krulwich, 1989]. In the cases we are considering, this procedure will "bottom out" by faulting a completeness assumption for a component. In other words, many of the underlying assumptions of the system will refer to the completeness of its component rule sets, and faulting one of these assumptions is a signal that a rule set must be augmented. There are, of course, other types of assumptions that could be faulted. CASTLE examines the form of the faulted assumption to determine which type of repair is necessary. In our example, the diagnosis will indicate that the failure was due to the lack of a counterplanning rule to predict the opponent's response. The structure of the faulted assumption will direct CASTLE to repair the failure by constructing a new counterplanning rule.

The task now at hand is for the system to analyze the failure and encapsulate the relevant information into a new rule for counterplanning by interposition. One common approach to learning from failures is to equip the system with *critics,* which map directly from failure descriptions to repairs [Sussman, 1975]. CASTLE's more general approach is to use explanation-based learning ("EBL") [Mitchell, Keller, and Kedar-Cabelli, 1986; DeJong and Mooney, 1986]. In its general form, EBL constructs new rules for category membership by explaining why a particular instance is a member of a category, and then determining the most general preconditions under which the explanation will still be correct. To accomplish this, EBL is given a *target concept,* which is a description of the category being learned. As has been widely discussed, target concepts must relate to the purpose to which the acquired category membership rules will be put. In previous work, however, this observation has been applied in a framework in which the purpose of the rules to be learned has been fixed, such as recognizing objects for use in plans [Kedar-Cabelli, 1987] or speeding up a search process [Keller, 1987; Minton, 1988]. We would like to extend the notion of EBL target concept formulation for an intelligent agent which is involved in many cognitive tasks.

## Constructing a counterplanning rule

Once the system has determined that the plan failure is due to its set of counterplanning rules being incomplete, it knows that it needs to construct a new rule to perform counterplanning, i.e., a rule whose consequent matches the generic form of counterplanner queries. The antecedent of this new rule should encode the type of counterplan that the opponent used in the situation that caused the failure. More specifically, the antecedent should be a generalized description of the counterplan (i.e., the move) that the opponent used. To construct this rule, CASTLE uses explanation-based learning to generalize a description of the move that the opponent made. The EBL target concept should thus be an expression which implies the assertion that the counterplanner should have generated.

CASTLE has expressions which fit these requirements, in the form of *component performance specifications.* These specifications are part of a larger planner self-model [Collins *et al.,* 1991], which is used for planning, expectation formulation, failure diagnosis, and, as described here, fault repair. In general, performance specifications describe the correct behavior of each component, in a way that can be recognized after the fact in a situation in which the component should have produced a particular behavior.[2] In other words, when the system has reason to believe that a component should have been executed in a given situation, based on information from the diagnosis module, this specification will enable the system to conclude that the component should have behaved in a particular way, and to construct an explanation of why it should have done so.[3] An explanation (proof) of the correctness of the specification as applied to a particular situation can then be generalized using EBL, and the new rule can be constructed.

For example, the system's performance specification for the *counterplanning* component says, simply, that a correct counterplan to an attack is a one-move sequence consisting of a move that disabled the attack. Since this specification is only used to explain why a particular move is a correct counterplan, MOVE-DISABLED-MOVE need not be implemented in a way which can be evaluated before the fact. In other words, even without a specific rule for disabling an attack by interposing a piece, the system has the ability to observe the game board sequence after the fact and use its domain physics knowledge to realize how the attack was blocked.

In our example from figure 3, after the opponent has responded to the computer's attack by interposing its knight, the diagnosis module determines that the counterplanning component must be augmented. To construct a new counterplanning rule, the system uses its counterplanner specification as a target concept, and invokes a deductive explanation mechanism to construct an explanation of the opponent's counterplan. This explanation, shown in figure 4, says roughly that *the opponent counterplanned against the queen-bishop attack,* because *the attack was disabled by the opponent's move,* because *the line of attack was no*

[2]CASTLE's component performance specifications serve the purpose of PRODIGY's *target concept specifications* [Minton, 1988], which specify EBL target concepts for each type of search-control rule. In PRODIGY terms, CASTLE is (a) generalizing from search-control rules to functional agent components, (b) embedding the specifications in a general planner self-model, and (c) replacing *example recognizers* with expectation-failure diagnosis.

[3]CASTLE's component performance specifications, along with the desired component result which is provided by the diagnosis module, can be viewed as a "performance improvement objective" as discussed in Chapter 5 of [Keller, 1987] and in [Keller, 1989]. The leverage gained from being *failure-driven* enables CASTLE to generate these performance objectives automatically as a consequence of failure diagnosis.

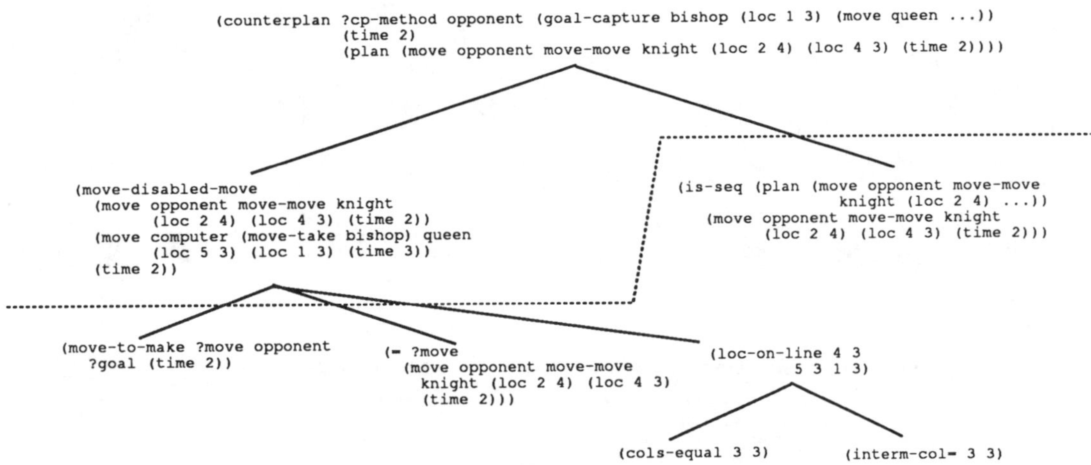


Figure 4: Explanation of desired counterplanner performance

*longer clear,* because *the opponent's knight was moved into the line of attack.*

Explanation-based learning is then invoked, which generalizes the details of the explanation which are not relevant to its correctness. For example, it is irrelevant what type of piece is interposed, because all links in the explanation which relate to the interposed piece will remain valid for any type of piece. The system then collects the most general operational expressions in the proof tree (operational assertions are shown below the dotted line). These expressions are then "backed up" in time so as to be predictive instead of explanatory. An example of a leaf which must be backed up is the MOVE-TO-MAKE assertion about the opponent's actual move, which took place after the counterplanner was invoked. CASTLE accomplishes this transformation using a set of rules for backing-up propositions in time.[4]

These generalized and adjusted proof-tree leaves will then form the antecedent of a new rule. Since the leaves imply the target concept, and the target concept implies the correct behavior of the counterplanner, the leaves are sure to specify a correct counterplanning method. The new counterplanning rule which the system constructs says, roughly,

TO COMPUTE: A reactive counterplan to a capture

DETERMINE: A location on the line of attack that another piece can move to

When this rule is added to the set of counterplanning rules, the system will correctly predict the opponent's response in a situation where interposition is viable, such as the situation of figure 3(b). Furthermore, because the same set of rules are used for actual counterplanning against the opponent's threats, the system will be able to respond to threats in a new way [Krulwich, 1991].

Tangentially, it should be noted that the explanation from which the new rule was generated will form a *justification* for the rule's correctness [Mitchell, Keller, and Kedar-Cabelli, 1986]. This justification structure could then be used for maintaining correctness when underlying assumptions change [deKleer *et al.*, 1977; Doyle, 1979; McDermott, 1989]. More importantly, CASTLE can use these justifications for constructed rules to facilitate further failure diagnosis. This will enable the system to further refine its learned rules, in the event that a constructed rule fails in the future.

## Learning threat detection

The main point of our model of target concept retrieval is, of course, that it enables a system to learn new rules for a variety of cognitive tasks. Consider, for example, how CASTLE can learn a new rule for the *detection focusing* component which we discussed earlier. As we saw, this component determines the areas of the board in which threats may have been recently enabled.

Suppose CASTLE had an incomplete set of detection focusing rules, in particular that it knew about threats being enabled either to or from the new location of a piece that has just been moved (as we discussed earlier), but did not know that threats could be enabled through *discovered attacks,* namely, by moving a third piece out of the line of attack. If this were the case, there could be threats enabled through discovered attacks which CASTLE would not be aware of. In the situation shown in figure 5(a), the opponent advances its pawn and thereby enables an attack by its bishop on the computer's rook. When the system updates its set of active threats and opportunities, its threat focusing

[4]These rules are merely an approximation of the ideal behavior: A system should examine the *implicit* assumptions that underly an assertion's being true (e.g., that the legality of a move underlies its being made), and select from them the appropriate predictive tests.

(a) (b) (c) (d)

Figure 5: *Discovered attacks* example: Opponent (white) to move

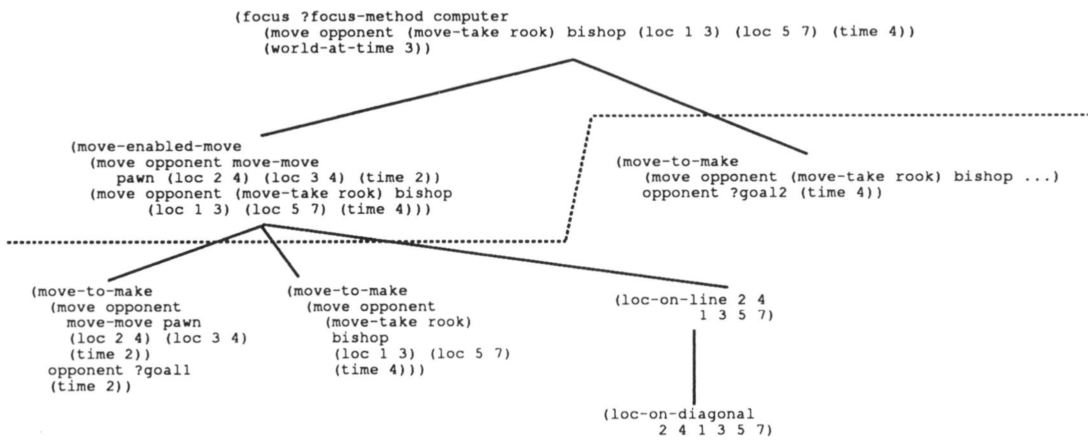


Figure 6: Explanation of desired detection focusing

rules will enable it to detect its own ability to attack the opponent's pawn, but it will not detect the threat to its rook. Because of this, when faced with the situation in figure 5(b), the computer will capture the opponent's pawn instead of rescuing its own rook, and it will expect that the opponent's response will be to execute the attack which it believes to the only one available, namely to capture the computer's pawn. Then, in the situation shown in figure 5(c), when the opponent captures the computer's rook, the system has the task of diagnosing and learning from its failure to detect the threat which the opponent executed.

As in the *interposition* example discussed earlier, the system invokes its diagnosis engine to determine the component which is at fault. In this case it will conclude that the failure resulted from having an incomplete set of detection focusing rules, and it will try to learn from the failure by constructing a new one. To accomplish this, the system retrieves a specification of the detection focusing component, which says that focus rules will indicate any moves which have been enabled. To learn from the failure, CASTLE uses this specification as a target concept to construct an explanation of what the focus rules should have computed. This explanation, which is shown in figure 6, says roughly that *the focus rules should have focussed on the bishop-rook attack,* because *a new threat was enabled,* because *the line of attack was opened up,* because *the opponent's pawn was moved out of the line of attack.*

As in the previous example, explanation-based learning is used to generalize this explanation and collect the operational leaves of the proof tree. CASTLE then constructs a rule from the proof tree leaves which will correctly focus on discovered attacks. This rule says roughly:

TO COMPUTE: Moves that may have been enabled

DETERMINE: Moves through the square
vacated by the most recent move

While the two rules that we have seen learned are very similar in terms of their structure and the computation which they perform, they fulfill very different purposes in the CASTLE system, and more generally they relate to very different cognitive tasks. Using component performance specifications, CASTLE can learn rules to focus perceptual detection by means of the same mechanism that it uses to learn counterplanning rules.

## Conclusions

We have given a framework for determining appropriate explanation-based learning target concepts from explicit performance specifications of agent components. This framework enables explanation-based learning to be applied to the spectrum of cognitive tasks which are performed by intelligent agents. We have demonstrated this within the CASTLE system, which learns new rules for such tasks as counterplanning and threat detection focusing.

Applying this technique to learning rules for other cognitive tasks involves extending CASTLE's model in two ways. First, the explicit decision-making model must be extended to include the cognitive task in question. Secondly, a performance specification must be given for any new components. We are in the process of extending the model to include components for goal-regression planning, execution scheduling, and case-based planning. When these and other tasks are expressed in CASTLE's decision-making vocabulary, and performance specifications are given for them, CASTLE will be able to learn new methods for them in response to failures. Future work will determine how well the technique applies to other cognitive tasks, as well as to other types of intelligent agents.

**Acknowledgements:** The ideas presented here were developed in collaboration with Larry Birnbaum, Gregg Collins, and Mike Freed. Additional thanks go to Matt Brand, Eric Jones, and Louise Pryor for comments on this paper and many discussions on the research presented.

## References

Birnbaum, L., Collins, G., Freed, M., and Krulwich, B. 1990. Model-based diagnosis of planning failures. *Proceedings of the Eighth National Conference on Artificial Intelligence,* Boston, MA, pp. 318-323.

Birnbaum, L., Collins, G., and Krulwich, B. 1989. Issues in the justification-based diagnosis of planning failures. *Proceedings of the Sixth International Workshop on Machine Learning,* Ithaca, NY, pp. 194-196.

Collins, G., Birnbaum, L., and Krulwich, B. 1989. An adaptive model of decision-making in planning. *Proceedings of the Eleventh IJCAI,* Detroit, MI, pp. 511-516.

Collins, G., Birnbaum, L., Krulwich, B., and Freed, M. 1991. Plan debugging in an intentional system. To appear in *Proceedings of the Twelfth International Joint Conference on Artificial Intelligence,* Sydney, Australia.

Davis, R. 1984. Diagnostic reasoning based on structure and behavior. *Artificial Intelligence,* Vol. 24, pp. 347-410.

DeJong, G., and Mooney, R. 1986. Explanation-based learning: An alternative view. *Machine Learning,* vol. 1, pp. 145-176.

deKleer, J., Doyle, J., Steele, G., and Sussman, G. 1977. AMORD: Explicit control of reasoning. *Proceedings of the ACM Symposium on Artificial Intelligence and Programming Languages,* Rochester, NY, pp. 116-125.

deKleer, J., and Willams, B.C. 1987. Diagnosing multiple faults. *Artificial Intelligence* **32**, pp. 97-130.

Doyle, J. 1979. A truth maintenance system. *Artificial Intelligence,* vol. 12, pp. 231-272.

Kedar-Cabelli, S. 1987. Formulating concepts according to purpose. *Proceedings of the 1987 AAAI Conference,* Seattle, WA, pp. 477-481.

Keller, R. 1987. *The role of explicit contextual knowledge in learning concepts to improve performance.* Technical report ML-TR-7, Rutgers University Department of Computer Science, New Brunswick, NJ.

Keller, R. 1989. Compiling learning vocabulary from a performance system description. *Proceedings of the Sixth International Workshop on Machine Learning,* Ithaca, NY, pp. 492-495

Krulwich, B. 1991. Learning from deliberated reactivity. To appear in *Proceedings of the Eighth International Workshop on Machine Learning,* Evanston, IL.

Krulwich, B., Collins, G., and Birnbaum, L. 1989. Improving decision-making on the basis of experience. *Proceedings of the Sixth International Workshop on Machine Learning,* Ithaca, NY, pp. 55-57.

Krulwich, B., Collins, G., and Birnbaum, L. 1990. Cross-Domain Transfer of Planning Strategies: Alternative Approaches. *Proceedings of the Twelfth Annual Conference of the Cognitive Science Society,* Cambridge, MA, pp. 954-961.

McDermott, D. 1989. *A general framework for reason maintenance.* Technical report YALEU/CSD/RR-691, Yale University Department of Computer Science, New Haven, CT.

Minton, S. 1988. *Learning Effective Search Control Knowledge: An Explanation-Based Approach.* Technical report CMU-CS-88-133, Carnegie Mellon University School of Computer Science, Pittsburgh, PA.

Mitchell, T., Keller, R., and Kedar-Cabelli, S. 1986. Explanation-based generalization: A unifying view. *Machine Learning,* vol. 1, pp. 47-80.

Simmons, R. 1988. A theory of debugging plans and interpretations. *Proceedings of the 1988 AAAI Conference,* St. Paul, MN, pp. 94-99.

Smith, R., Winston, H., Mitchell, T., and Buchanan, B. 1985. Representation and use of explicit justifications for knowledge base refinement. *Proceedings of the Ninth IJCAI,* Los Angeles, CA, pp. 673-680.

Sussman, G. 1975. *A Computer Model of Skill Acquisition.* American Elsevier, New York.

# Adaptive Action Selection

**Pattie Maes**
MIT Media-lab (E15-495)
20 Ames Street
Cambridge, MA 02139
pattie@media.mit.edu

## Abstract

In earlier papers we presented a distributed model of action selection in an autonomous intelligent agent (Maes, 1989a, 1989b, 1991a, 1991b). An interesting feature of this algorithm is that it provides a handful of parameters that can be used to tune the action selection behavior of the algorithm. They make it possible, for example, to trade off goal-orientedness for data-orientedness, speed for quality, bias (inertia) for adaptivity, and so on. In this paper we report on an experiment we did in automating the tuning and run-time adaptation of these parameters. The same action selection model is used on a meta-level to select actions that alter the values of the parameters, so as to achieve the action selection behavior that is appropriate for the environment and task at hand.

## Introduction

An important problem to be addressed when modeling an intelligent agent is the problem of *action selection*: what mechanism can an agent use to determine what to do next, given that it has a repertoire of actions it can engage in, wants to achieve certain goals (or has certain motivations) and is facing a particular situation. Computational models of action selection are important both for producing actions in an artificial agent (e.g. an autonomous mobile robot) as well as for understanding the action selection behavior of biological agents (human and animal). In previous papers we argued that none of the models of action selection that have been presented in the literature on AI present a satisfactory solution to the problem (Maes, 1991a, 1991b). Two classes of approaches can be identified: traditional *planners* and *reactive systems*. The former are not satisfactory because they do not produce fast (re)action, are completely goal-driven and not adaptive. The problem with reactive systems is that they are completely sensor-data driven, do not have a notion of goals, cannot anticipate or predict the outcome of actions and are hard to design.

In (Maes, 1989a, 1989b, 1991a, 1991b) we presented an action selection algorithm which tries to combine the best of previous models. This algorithm produces fast action in a tight connection with the environment, while providing a notion of goals and run-time arbitration. Just like traditional planners, our algorithm uses an explicit description of the conditions and expected effects of actions (cf. STRIPS operators). But instead of implementing action selection as a deliberative, sequential reasoning process, we use this information to construct a network, in which action selection is an emergent property of an activation/inhibition dynamics among the nodes (operators/actions). The algorithm provides a distributed solution to the problem of action selection through the use of a connectionist computational model on a symbolic representation. The results of experiments prove that our algorithm does arbitrate among actions, and implements a form of prediction, planning and anticipation.

One of the interesting properties of the algorithm is that its behavior can be tuned by hand according to the requirements of a particular application. The algorithm has five global, numerical parameters which make it possible to make the action selection more goal-driven versus more data-driven, faster versus more cautious, biased towards the recent history versus adaptive, and more or less sensitive to goal and action conflicts. This is useful because different tasks and applications require different action selection behavior. For example, if an agent has very unreliable sensors, we want it to bias its action selection towards the past history of action selection. Or if the environment is very dynamic, we want the action selection behavior to be very "reactive" and data-driven.

Until recently, we set the values of these parameters by hand by means of a generate and test iterative process. This solution is far from ideal, first of all, because this process turned out to be very difficult in certain cases, and second, because the parameters did not adapt to changes in the task or environment at run time (unless the programmer would go through the same process again). In this paper we describe an extension of the algorithm which allows the system to autonomously adapt the parameters to the characteristics of the task and environment. This functionality is achieved by using the same action selection algorithm on a "meta-level". A meta-network was constructed which consists of actions that observe the behavior of the object-level network and control its parameters so as to achieve some preset goals (in terms of performance).

## The Basic Algorithm

This section summarizes the basic action selection model and algorithm, without the meta-level control. More elaborate descriptions of the algorithm and its results can be found in (Maes, 1989b, 1991a, 1991c). An intelligent agent is viewed as a collection of *competence modules*. Competence modules are like the (mindless) agents populating the Society of Mind (Minsky, 1986): they are autonomous modules that are expert in achieving a particular competence, such as grasping a cup. Every competence has an associated *activation level*, which is a real number. A competence module further also has a set of *conditions* which have to be observed in order for the competence module to be *executable*. And finally, a competence module has an *add-list* and *delete-list* which describe the expected effects on the state of the world that the module has. An executable competence module whose activation level surpasses a preset threshold can be selected (become active). This means that a set of processes start running which try to "realize" the competence (certain real world actions are performed at that moment). Consider as an example the pick-up-board competence module described below.

```
(defmodule PICK-UP-BOARD
  :condition-list '(board-within-reach hand-empty)
  :add-list '(board-in-hand)
  :delete-list '(board-within-reach hand-empty)
  :processes <a set of processes that implement
             the action of picking up the board>)
```

An agent also has a set of *goals* (or motivations), which have an associated number representing the strength of that goal at a particular moment in time. For example, the agent can have the goal of sanding the board with associated strength 40 and the goal of spray painting itself with strength 90. An agent also has a set of *protected goals*, which are goals that are achieved and that should remain achieved. E.g. once the board has been sanded, this becomes a protected goal. The set of goals and their associated strengths vary over time as the agent takes actions. Finally an agent has a set of *perceptual conditions* (or sensor data) that it observes at a particular moment in its environment

The different competence modules of an agent are linked in a network through "predecessor", "successor" and "conflicter" links. In this paper we assume that these links are known. The idea is that these links among competence modules are either innate (i.e. programmed) or learned through experience (Maes & Brooks, 1990). There is a predecessor link from competence module A to competence module B (competence module A has competence module B as predecessor) if competence module B (through the actions it takes), makes certain conditions of competence module A come true (the intersection of B's add-list with A's condition-list is non empty). For example, 'pick-up-board' may make the condition of 'sand-board' that the board has to be in the hand of the agent, become true. Second, there is a matching successor link in the opposite direction for every predecessor link. Finally there is a conflicter link from competence module A to competence module B (competence module B is said to conflict with competence module A) if competence module B makes a certain condition of competence module A undone (the intersection of B's delete-list with A's condition-list is non empty). For example, 'put-down-board' may make the condition of 'sand-board', that board has to be in the hand, untrue.

The intuitive idea behind the action selection mechanism is that competence modules use the links of the network to activate and inhibit each other (respectively increase and decrease each other's activation level), so that after some time the activation energy accumulates in the competence module that represents the "best" choice, given the current situation and motivational state of the agent. Once the activation level of a competence module reaches a certain threshold, it may be selected, and its processes start operating. The pattern of spreading activation among competence modules, as well as the input of new activation energy into the network is determined by the current situation and the motivational state of the agent. The input of new activation energy into the network of competence modules is defined as follows:

- *Activation by the Current Situation.* The currently observed perceptual conditions (sensor data) increase the activation level of the competence modules that have those conditions. For example, if the agent observes a board within its reach, then the activation level of the 'pick up board' competence module is increased.
- *Activation by the Goals.* The goals of the agent increase the activation level of the competence modules that might achieve them (have the goal in their add-list) with an amount which is relative to the strength of the goal. For example, the activation level of 'sand-board' is increased with some amount proportional to the strength of the goal 'board-sanded'.
- *Inhibition by the Protected Goals.* Finally the protected goals of the agent decrease the activation level of the competence modules that might undo them (have the protected goal in their delete-list) with an amount that is relative to the strength of the goal.

These processes are continuous: there is a continual flow of activation energy towards the competence modules whose condition list partially matches the current situation and towards the modules whose add-list matches the goals. The modules that would undo protected goals continuously weakened. If the situation or the motivational state changes (e.g. the perceived conditions change or the strengths of the goals change), the input of activation energy automatically flows to other competence modules. Besides the impact on activation levels from the current situation and goals, competence modules also activate and inhibit each other. Competence modules spread activation along their links as follows:

- *Activation of Successors.* An executable competence module spreads activation forward. It increases (by a fraction of its own activation level) the activation level of its successors. Intuitively, we want these successor competence modules to become more activated because they are "almost executable", since more of their conditions will be fulfilled after the competence module has become active. For example, if the 'pick up board' competence module is executable (i.e. a board is withing reach of the agent and it's hand is empty), this will increase the activation level of its successor, the 'sand-board' competence module.
- *Activation of Predecessors.* A competence module that is not executable spreads activation backward. It increases (by a fraction of its own activation level) the activation level of its predecessors. Intuitively, a non-executable competence module spreads to the competence modules that can fulfill its conditions that are not yet true, so that the competence module itself may become executable afterwards. For example, if 'sand-board' is not executable because the agent is not holding the board, it will spread activation energy to 'pick up board' to encourage this competence module to become active (so that afterwards 'sand-board' can become active).
- *Inhibition of Conflicters.* Every competence module (executable or not) decreases (by a fraction of its own activation level) the activation level of its conflictors. Intuitively, every competence module will try to prevent a competence module that undoes one of its true conditions from becoming active. For example, the 'sand-board' competence module will decrease the activation energy of its conflicter 'put-board-down', because the latter would undo the condition that the board has to be in the hand.

The global algorithm performs a loop, in which at every timestep the following computation takes place over all of the competence modules:

1. The impact of the current situation and goals on the activation level of a competence module is computed.
2. The way the competence module activates and inhibits related competence modules through its successor links, predecessor links and conflicter links is computed.
3. A decay function ensures that the overall activation level remains within some boundaries.
4. The competence module that fulfills the following three conditions becomes active: (i) it has to be executable, (ii) its level of activation has to surpass the threshold and (iii) it must have a higher activation level than all other competence modules that fulfill conditions (i) and (ii). When two competence modules fulfill these conditions (i.e., they are equally strong), one of them is chosen randomly. Once a competence module has been activated (it actually performs its actions at that moment), its activation level is reset to 0 (but it may quickly increase again, since the spreading activation process goes on continuously).

We have evaluated the algorithm empirically by performing a wide series of experiments using several example applications. The networks cannot be said to show a 'jump-first think-never' behavior. They do exhibit planning or "reasoning" capabilities. The effects of a sequence of actions are considered before actually embarking on its execution. If a sequence of modules exists that transforms the current situation into the goal state, then this sequence becomes highly activated through the cumulative effect of the forward spreading (starting from the current state) and the backward spreading (starting from the goals). If this sequence potentially implies negative effects, it is weakened by the inhibition rules.

More specifically, goal-relevance of the selected modules is obtained through the input from the goals and the backward spreading of activation. Situation relevance and opportunistic behavior are achieved through the input of the sensor data and the forward spreading of activation. Conflicting and interacting goals are taken into account through inhibition by the protected goals and inhibition among conflicting modules. Further, local maxima in the action selection are avoided, provided that the spreading of activation can go on long enough (the threshold is high enough), so that the network can evolve towards the optimal activity pattern. And finally, the algorithm automatically biases towards ongoing "plans", because these are favored by the remains of the past spreading activation patterns.

## The Need for Adaptive Action Selection

In the recent past, researchers have been arguing about different action selection models, trying to convince their peers that one model is more appropriate than another (AI-magazine, 1990). For example, advocates of the traditional planning model argue that reactive systems are not very "thoughtful" because they do not model and reason about the consequences of actions. They argue that such systems would fail in "critical" environments such as a nuclear power plant. On the other hand, advocates of the reactive systems model claim that traditional planners are slow, brittle and not flexible and will fail in environments that are very dynamic. We believe that the desired characteristics for an action selection model differ from application to application. The degree to which an environment or task is "critical' (the cost of non-optimal actions) determines how much prediction is desired. The degree to which an environment is "dynamic" (changes rapidly) determines how reactive or fast the action selection has to be. The degree to which an environment is predictable, determines the usefulness of internal models, and so on.

Therefore any action selection model that aims to be "generally useful" should be tunable to the characteristics of an environment and task. This is possible for the action selection algorithm that we described above. Its behavior can be tuned to the characteristics of a particular application by a handful of global parameters (Maes, 1989b, 1991a):

- *Threshold* $\theta$. This parameter determines how much activation energy has to be accumulated by a module before it can be selected. By varying $\theta$ one can tradeoff *speed* for *thoughtfulness*. The higher $\theta$, the longer the activation spreading goes on before a module has accumulated enough activation energy. As such, it allows the network to look ahead further, thereby avoiding local maxima (in time) of activation levels.
- *Goal-orientedness* $\gamma$. This parameter determines the strength of the backward spreading of activation energy. It determines (i) how much activation energy is put into the network by the goals (as opposed to the sensor data) and (ii) what the (global) weight is of the predecessor links. By varying $\gamma$ the action selection behavior can be made more or less goal-driven.
- *Data-orientedness* $\phi$. This parameter determines the strength of the forward spreading of activation energy. It determines (i) how much activation energy is put into the network by the sensor data (as opposed to the goals) and (ii) what the (global) weight is of the successor links. By varying $\phi$ the action selection behavior can be made more or less data-driven.
- *Goal-conflict sensitivity* $\delta$. This parameter determines the strength of the inhibitory spreading of activation. It determines (i) how much activation energy is taken away from the network by a protected goal (as opposed inserted by a goals) and (ii) what the (global) weight is of the conflictor links. By varying $\delta$ the action selection behavior can be made more or less sensitive to goal conflicts. For example, if $\delta$ is larger than $\gamma$, the system cares more about avoiding goal conflicts than about achieving goals.
- *Bias towards past history* $\pi$. The parameter $\pi$ determines what the mean level of activation is that the activation levels are reduced to (decayed) at every timestep (activation levels are not reinitialized at every timestep). It determines in how far the 'history' of spreading activation plays a role in the action selection. This parameter can be used to tradeoff adaptivity (quick response to changes in the sensor data or goals) for bias towards the ongoing plan (inertia).

In the original algorithm presented in the previous section, these parameters have to be set by hand. There are several problems with this approach. One problem is that it is difficult to come up with the "optimal" parameter settings. It is hard to understand the characteristics of the environment and task well enough to decide what the optimal parameter values are. A second problem is that the parameters are not independent from one another, so changing the value of one also affects the others. Typically, the values for the parameter were chosen by running an experiment, noticing whether something goes wrong, tuning the parameters and starting anew. This process was iterated until the parameter values were stable. A third problem with this solution is that the parameters did not adapt to changes in the task and environment. E.g. if something about the network or the environment changed at run-time, the parameter values were not adapted to the new situation.

Therefore we decided to automate the continuous run-time adaptation of these parameters. The same agent selection algorithm is used in a "meta-level network", which runs in parallel with the network we had before. The nodes of this network do not take actions in the environment, but instead alter the parameter values for the first network so as to adapt the action selection behavior of the first network to the characteristics of the particular task and environment at hand.

The meta-level network also proved to be useful for dealing with a related problem. Occasionally we observe problems of a "non-local" nature in the action selection networks: a network can get stuck in a deadlock (the activation levels of modules are not changing anymore and none of the modules has enough energy to exceed the threshold), or in a loop (the network keeps activating the same set of modules without making progress). Noticing these problems requires a more global perspective: it is not possible for a single module to decide that the network is stuck in a deadlock or a loop. However, this proved to be easily diagnosed by the meta-level network.

## The Meta-Network

The idea of the meta-net is related to other research in the Artificial Intelligence and Cognitive Science literature. Many of the AI systems that have been built have a *meta-level architecture*. They consist of two distinct levels, where one level, called the *object-level*, solves problems about and acts upon an external problem domain, while the *meta-level* solves problems about and acts upon the object-level problem solver. For an overview see (Maes & Nardi, 1988).

In our model, the meta-level control is implemented using the same action selection algorithm in a *meta-network*. The nodes (or competence modules) of this network are experts in diagnosing and curing typical problems with action selection networks. The meta network contains nine competence modules. Four of those increase and decrease the threshold and mean activation level. Three more increase the goal orientedness, data orientedness and goal conflict sensitivity. There are no meta modules that decrease these three parameters. This is the case because whenever one parameter is increased, the other parameters are decreased a little (i.e. the total sum of all the parameters is constant). We only experienced the need for increasing the sensitivity to sensor data, goals and goal conflicts. It does not seem to be necessary ever to decrease their sensitivity, except to change their respective ratios (make the action selection more goal oriented than data oriented, or more goal conflict sensitive than goal oriented, and so on). Finally there are two more meta-nodes that deal with "global" problems with the object level network: one node decreases the threshold when the object level is in a deadlock, and the other one restores the threshold to its original value afterwards. The following is a description of the different meta-nodes:

- *Make-faster* decreases the threshold $\theta$ when activated. It is triggered when the object-level network is too slow. In particular, when the average time it takes for the object network to select an action is longer than the average time it takes for the environment to change spontaneously. This is what the definition of this meta-node looks like:

```
(defmodule MAKE-FASTER
  :condition-list '(slow)
  :add-list '(appropriate-speed)
  :delete-list '(slow)
  :processes '((decrease (network-threshold
                   *object-network*))))
```

- *Make-more-thoughtful* increases the threshold $\theta$ when activated. It is triggered when the object-level is too fast, i.e. when the average time needed to select an action is shorter than the average width of the object-network (the average length of a path from executable nodes to the goal nodes). Things are slightly more complicated in that this module also takes into account how unpredictable the environment is. In a very unpredictable (dynamic) environment the threshold has to be set higher for the network to be thoughtful than in a predictable environment (one for which the object level network reliably models how it changes).
- *Make-more-biased* increases the mean-activation-level $\pi$ when activated. It is triggered when the object-network is not biased enough towards the past history. This is true in two cases: (i) when the object-level "jumps" from working on one goal (selecting an action in one goal tree) to working on another goal (activating an action in a disjunct goal tree) without any good reason to do so (without the situation representing a unique opportunity, i.e. the two paths are equally long), and (ii) when the sensor data seem to be very noisy and unstable (something is true all the time and then false for just one timestep, or vice versa, something is false all the time and then true for just one timestep).
- *Make-more-adaptive* decreases the mean activation level $\pi$ when activated. It is triggered when the object-level is not adaptive enough. This is the case if the object level does not react enough to important changes in the goals or the sensor data (e.g. a sudden opportunity to achieve a different goal than the one it is working on easily).
- *Make-more-goal-oriented* increases the goal-orientedness $\gamma$ when activated. It is triggered when the object-level is not goal oriented enough. In particular, this is the case when, after selecting an object-level action, the total length of the tree (branching paths) from the goals to executable modules has increased.
- *Make-more-data-oriented* increases the data-orientedness $\phi$ when activated. It is triggered when the object-level is not data-oriented enough, i.e. when the action selection does not respond to changes in the environment fast enough, even thought the mean activation level (bias towards past history) $\pi$ is low.
- *Make-more-sensitive-to-goal-conflicts* increases the goal-conflict-sensitivity $\delta$ when activated. It is triggered when the object-level is not enough sensitive to goal conflicts. This is the case when the object level is no longer able to satisfy a particular (sub-)goal (a particular proposition can theoretically not be made true anymore). This is the case when the network contains a lot of conflictor links and negative (protected) goals for which no "reversible (restoring) actions" exist.
- *Deal-with-deadlock* decreases the threshold $\theta$ when activated. It is triggered when the object-level is in a deadlock situation. This is the case when the activation levels of the object level modules change less then a small value $\epsilon$ and none of the modules has accumulated enough activation energy to exceed the threshold. Deal-with-deadlock is activated repeatedly until the threshold is low enough so that some executable module has enough activation energy.
- *Restore-threshold-after-deadlock* increases the threshold $\theta$ again after the deadlock problem has been dealt with (it restores it to its previous value). It is activated whenever deal-with-deadlock has been activated, except if deal-with-deadlock is being activated very often (in half or more of the cases that an object level module gets selected). In the latter case, the effect of deal-with-deadlock (the decrease

of the threshold) is permanent.

The meta-level network does not have any goals. It is completely data driven. Whenever the conditions for a meta-module are fulfilled (the meta-module is executable) and the meta-module has accumulated enough activation energy (from the sensor data) to exceed the meta-net threshold, the module is selected. The latter condition means that the meta-node's triggering conditions should be observed for a sufficiently long time, so that they can increase the node's activation level to a point where it exceeds the threshold. This implements the desirable feature that meta-nodes look at the average recent behavior of the object network (where the relevance of observations is weighed over time).

## Discussion

The meta-level network was tested by running several experiments with very different applications. Some applications required very fast action selection, others required very thoughtful action selection, yet others emphasized goal conflict sensitivity and so on. We evaluated the system by adopting random values for the object-level parameters and then running the two networks (in parallel) until the meta-network did not select any actions anymore. Gradually the object level network displays a better and better action selection behavior, until it does not make any "wrong selections" anymore according to the meta-net's standards.

These experiments proved that the meta-net provides an effective way of setting and adapting the object-level parameters. Each time, the meta-net was able to fairly quickly settle on a selection of parameters that was satisfactory (in that no meta-level modules got triggered after a while). The meta-net does not necessarily convergence to the same parameter values for two experiments with the same application. This is the case because the same action selection behavior can be produced by "similar" parameter values (values that are within a certain interval).

The question can be raised whether there is a need for a meta-meta-net (or an infinite tower of meta-levels). In the current implementation the parameter values for the meta-network are set by hand. This is an acceptable solution because (i) the meta-net is very simple, and because of that the parameters values do not make much of a difference in its behavior, (ii) the meta-net always solves the same problem (operates in the same domain), namely that of setting the parameters of other networks, as such there is less of a need for it to adapt its action selection behavior to its environment (the environment is always the same). The only meta-level parameter which we found to be important is the threshold (and how it compares to the data-orientedness). It determines how much "evidence" has to be built up for a particular problem before the meta-net takes action and changes certain object-level parameters. It would be nice if the threshold would increase as the system proceeds, so as to implement some notion of "temperature" as in simulated annealing. This way the system would make changes to the object-level parameters more easily in the beginning and be more reluctant to change things as the object-level action selection behavior improves.

In conclusion, this experiment proves that the meta-net is a convenient method for adapting the parameters of our action selection algorithm to the characteristics of the task and environment. In addition, it proves that meta-level control does not necessarily imply a rigid, hierarchical control structure, but instead can be implemented in a distributed way. The meta-net runs in parallel with the object-net and as such does not make the latter less adaptive or reactive. Finally, more experiments will have to be performed to determine whether the current set of meta-nodes is sufficient for dealing with a wide range of tasks and environments characteristics. To this end, we plan to use "TileWorld" (Pollack & Ringuette, 1990), which is a highly parametrized environment simulator, enabling to control the characteristics of an environment.

## References


AI magazine. Special issue on Universal Planning, Vol. 10 (4), 1989.

Maes, P. & Nardi, D. (editors). Meta-level Architectures and Reflection. North-Holland, 1988.

Maes, P. The Dynamics of Action Selection. Proceedings of the IJCAI-89 conference, Detroit, 1989a.

Maes, P. How To Do the Right Thing. Connection Science Journal. 1(3), 1989b.

Maes, P. Situated Agents can Have Goals. In: Designing Autonomous Agents: Theory and Practice from Biology to engineering and Back. P. Maes (editor). Special Issue of the Journal Robotics and Autonomous Systems. 6(1&2), 1990. Also MIT-Bradford Press book, 1991a.

Maes, P. & Brooks, R.A. Learning to Coordinate Behaviors. Proceedings of AAAI-90, 1990.

Maes, P. A Bottom-Up Mechanism for Action Selection in an Artificial Creature. In: Adaptive Behavior, from Animals to Animats. Edited by Stewart Wilson and Jean-Arcady Meijer. MIT-Press, 1991b.

Maes, P. The Agent Network Architecture (ANA). Proceedings of the AAAI Spring Symposium on Integrated Intelligent Architectures. AAAI-Press, 1991c.

Minsky, M. The Society of the Mind. Simon and Schuster, New York, New York, 1986.

Pollack, M. & Ringuette, M. Introducing the TileWorld: Experimentally Evaluating Agent Architectures. Proceedings of AAAI-90, 1990.

# Memory for Incomplete Tasks: A Re-examination of the Zeigarnik Effect*

**Colleen M. Seifert**
Department of Psychology
University of Michigan
330 Packard Road
Ann Arbor, MI 48104
(313) 763-0210
seifert@um.cc.umich.edu

**Andrea L. Patalano**
Department of Psychology
University of Michigan
330 Packard Road
Ann Arbor, MI 48104
(313) 763-5710
patalano@ub.cc.umich.edu

**Abstract**

An important feature of human memory is the ability to retrieve previously unsolved problems, particularly when circumstances are more favorable to their solution. Zeigarnik (1927) has been widely cited for the finding that interrupted tasks are better remembered than completed ones; however, frequent replications and non-replications have been explained in terms of social psychological variables (Prentice, 1944). The present study examines differences in memory for tasks based on completion status by appealing to cognitive variables such as the nature of interruption, time spent during processing, and set size. In one experiment using word problems, subjects were interrupted on half of the problems after a short interval of active problem solving, and completed tasks were in fact better remembered than interrupted ones. However, less processing time was necessarily spent on problems that were interrupted. A second experiment held time constant, allowing subjects to abandon tasks they could not complete. In this experiment, the opposite result occurred, replicating Zeigarnik and showing better access to unsolved problems in free recall. However, enhanced memorability in this study may have resulted from a subject-generated impasse in problem solving rather than "interruption" per se. This successful replication also included set size differences in favor of incomplete problems. Under these conditions, the status of completion can serve as a useful index to past problem situations. These experiments are successful in identifying cognitive variables that explain when one can suspend effort on a failed problem, and recall it at a later time.

## Introduction

In the world, we are constantly presented with problems to solve. Learning to operate a new appliance, balancing a checkbook, or doing a crossword puzzle, for example, are just a few of the problems we typically encounter. Sometimes, we are able to solve a problem at the moment that it presents itself; one might have a new appliance operating soon after taking it home from the store. At other times, due to interruption, lack of needed resources, or failed attempts, we are forced to give up work on a problem before we have successfully arrived at a solution. It is easy, for instance, to imagine getting stuck on a crossword puzzle problem and eventually giving up in favor of pursuing other activities. Obviously, it would be to great advantage to be able to retrieve previously unsolved problems in order to retry tasks. There is some suggestion that human cognition is able to take advantage of improved circumstances in order to reattempt previously failed task goals. In order to do so, however, incomplete problems must be stored and retrieved from memory. What, if any, are the differences in the way in which we encode and remember completed versus interrupted problems? Is there a special status in memory for problems we meant to solve, versus those we have already solved?

Widely-cited results of a classic experiment by Zeigarnik (1927) claim that interrupted problems do indeed hold some special status in memory. In Zeigarnik's experiment, subjects were given approximately 20 tasks to perform. These tasks included mental problems such as arithmetic and puzzles, as well as manual skills including constructing cardboard boxes and creating clay figures. In the course of half of these tasks, subjects were interrupted before finishing the task and were forced to put it aside. The interruption came "when the subject looked most engrossed in his work." This was reported to have occurred when the subject discovered how the problem was to be done but had not yet envisioned the ultimate result. Subjects were allowed to complete the other half of the tasks.

After performing all of the tasks, though not always to completion, the subjects were asked to report all of the

*This research was supported by the Office of Naval Research under Contracts N00014-88-0295 and N0014-91-1128. Gretchen Dettloff and Michelle Berris provided valuable assistance with the first experiment.

problems using a free recall method. Zeigarnik found that unfinished tasks were 90% more likely to be recalled than finished ones. Additionally, in both first and second recall positions, unfinished tasks were mentioned three times as often as completed tasks. Zeigarnik concluded that a significant memory advantage exists for interrupted tasks as compared with completed ones.

While the theory that unsolved problems hold some special status in memory is an appealing one, the results of Zeigarnik's experiment appear somewhat counterintuitive. Any memorial advantage in the Zeigarnik experiment should lie with completed tasks, since a subject logically must spend more time on average on completed tasks. However, though less processing time is presumably spent on interrupted problems, they are recalled more frequently in Zeigarnik's experiment. Zeigarnik accounted for this effect in terms of motivational factors, suggesting that when a subject sets out to perform the operations required by one of the tasks, there develops a "quasi-need" within the subject for the completion of the task. This is like the occurrence of a tension system, where completing the task means resolving the tension system or discharging the quasi-need. Thus, claims Zeigarnik, the memorial advantage enjoyed by interrupted tasks must be due to the continuation of that quasi-need, which motivates retrieval of unsatisfied tasks.

Additional social, motivational, and personality factors have since been suggested to account for results of variations and modifications of the original Zeigarnik experiment. Bogoslavsky and Guthrie (1941) suggested that tension present during the solving of a problem increases the problem's memorability. This hypothesis accounts for their findings that subjects best remembered tasks that followed interrupted tasks, regardless of whether the "follow-up" tasks were themselves interrupted or completed. Other studies discovered contexts where the Zeigarnik effect did not occur. Rosenzweig (1943) hypothesized a form of repression to account for the non-replication of Zeigarnik's results. In this study, in which subjects were told that the tasks comprised an intelligence test, they remembered more completed than interrupted tasks. Others have suggested stress-related factors (Glixman, 1949), individual differences (Alper, 1946), and subject fatigue (Zeigarnik, 1927) to account for discrepancies between their results and the original findings of Zeigarnik. The use of theories based on social, motivational, and personality-related variables to account for memory differences, has met with limited success. Such theories have been unable to explain numerous seemingly contradictory findings (see Prentice, 1944).

A greater degree of success might be met in trying to account for Zeigarnik's original results and some subsequent manipulations in terms of a cognitive model of problem solving. By re-examining the Zeigarnik effect in terms of modern theories of problem representations, goals, and context effects, perhaps we can explain the circumstances under which the Zeigarnik effect will occur, and how it may function within a broader memory and problem solving cognitive architecture. In this article, we examine the Zeigarnik effect, and explore factors including the nature of the interruption, the processing time spent on problems, and the context of set size of the incomplete problems. By examining these cognitive factors, we attempt to account for both the original effect and the variety of studies that at times failed to replicate it.

## Experiment 1

In this first experiment, we attempted to match Zeigarnik's methods (1927) as closely as possible, and to replicate the effect of better memory for incomplete problems. However, one necessary change was to employ only problem solving tasks, rather than including manual and artistic tasks as in Zeigarnik. In order to look at the role of cognitive factors in predicting and explaining problem memorability, the following experiment will attempt to replicate Zeigarnik's original results using exclusively cognitive problem solving tasks. Using word problems, we manipulated task interruption versus completion on each problem. Our goal was to determine whether the counterintuitive effect Zeigarnik observed can hold as a memory phenomenon under controlled laboratory conditions.

### Method

**Subjects.** The subjects were 39 undergraduate students (25 female and 14 male) from the University of Michigan. The subjects received credit from an introductory psychology course.

**Materials.** Twenty word problems, including mathematical, logical, and insight reasoning (from Mosler, 1977), were used in this study. All of the problems were pretested on a separate group of subjects and were selected for successful completion rate, each requiring between 15 seconds and four minutes for solution. Each problem was presented on a separate half-sheet of paper with space below the problem to write a solution. Each problem was presented with a short title such as "The Bridge". Following each problem was a rating scale. For each problem, subjects were asked to rate how confident they were that their answer was correct. A rating of 1 meant "certain it is incorrect" and a score of 7 meant "certain it is correct," and subjects were told to use intermediate values when appropriate.

**Design and Procedure.** Each subject was tested individually in a one hour session. The subjects were given the following instructions:

> You will be presented with a series of problems. You will work on the problems one at a time. Please work as quickly and accurately as you can. Show your work. Don't guess at a problem; try to solve each one and indicate your best answer. Once you have completed a problem I will give you another one to

work on. Each of these problems is randomly paired with a time interval of varying length. I will be stopping you according to these times. Don't worry if you do not get to finish a problem. I will administer another one and you can proceed as instructed. Do you have any questions?

Following these instructions, the subjects were given two practice problems. The first was simple, and every subject completed it between 30 and 210 seconds. The second was very difficult, and every subject was successfully interrupted by the experimenter saying, "Please stop now" between 15 and 60 seconds into the problem. The experimenter followed this practice set procedure with each of the 20 test problems. The test problems were presented in a single random order for all subjects. Each subject was interrupted on half of the problems and allowed to complete the other half. The order of interruption within the problems was counterbalanced by subject. If a problem was to be completed, the subject was allowed enough time to finish and to indicate so (maximum time to completion was 4 minutes). If a problem was scheduled to be interrupted, the experimenter attempted to interrupt the subject when she was "most engrossed in the problem" (Zeigarnik, 1927) (after she had read the problem all the way through, but before she had written a complete answer). Minimum time to interruption was 15 seconds on each problem, and between 15 and 30 seconds for most trials in order to ensure subjects would be stopped before solution. When a subject finished a problem, the subject gave a confidence rating (no ratings were given for unfinished problems). After a subject was interrupted, or a problem was finished and rated, the experimenter removed the problem and administered another. This cycle repeated until all of the problems had been presented.

Immediately after exhausting the problem set, the subject was given a free recall test. Subjects were asked to recall all of the problems that they could remember. They were asked to write only enough to uniquely identify the problem they had in mind. Following Zeigarnik's procedure, the point in free recall was recorded where subjects seemed to exhaust an initial recall spurt.

## Results

All of the problem solving answers were scored as "completed" or "not completed" by an independent rater based on the information written by the subjects. All trials in which the planned interruption or completion was not successful were eliminated from the analysis. This constituted 11 out of 680 attempts or 1.6% of the data. Ten out of these eleven discarded trials were trials that should have been interrupted, but were actually completed by the subject.

The free recall was scored by counting a problem as remembered if the written protocol uniquely identified one of the 20 test problems. The range of recall proportions for interrupted tasks was 0 to 0.9 with a mean of .39. The recall proportions for the completed tasks ranged from 0.1 to 0.9 with a mean of .54. This difference was significant, $t(38) = 4.368$, $p=0.00$, with completed tasks recalled more frequently than incomplete problems. The same result obtains when comparing the free recall only up to the point of first pause, as in Zeigarnik's (1927) analysis. For the completed problems, there were no differences in recall whether or not the solutions given were in fact correct (mean proportion of correct problems recalled was .60, of incorrect, .52, $t < 1$).

Another indicator of completion status may be the subject's level of confidence in her answer. Therefore, an analysis was performed using the subject's confidence rating as an independent variable. Due to the concentration of ratings at the ends of the scale (perhaps because, with subjects deciding when to stop working on a problem, they may have had a clear idea of when their answers were correct or incorrect), the seven-point confidence scale was broken down into three parts. There was a main effect of confidence rating, where subjects remembered the problems they answered confidently (above four on the seven point scale) with a mean of .56 recalled, better than those for which they lacked confidence (rated less than four on the seven point scale), with a mean recall of .27, $t(38)=4.349$, $p=0.000$. The completed-confident problems were also recalled better than those rated at the midpoint (for problems rated 4 on the seven point scale, mean number recalled = .32), $t(38)=3.434$, $p=.001$. These completed-confident problems were also recalled significantly more often than the interrupted problems ($t(38) = 3.6$, $p = .001$) However, there were no differences in recall proportions for interrupted, completed-unconfident and completed-midpoint problems.

While correctness had no effect on the finding that completed tasks are recalled better than interrupted tasks, the level of confidence the subject has regarding the completion of the tasks does correlate with recall. There is difference in recall for completed-confident and completed-unconfident categories, with the "confident" rated problems more frequently recalled. Confidence may be considered a reflection of how "complete" a subject felt their answer was, so that problems with lower confidence ratings are recalled as frequently as interrupted problems. Because subjects themselves terminated their problem solving efforts, low confidence ratings may indicate having abandoned solution attempts even though incomplete. Higher confidence ratings may indicate subjects' feelings of satisfactory completion for the problems they were allowed to finish.

The above analyses show that free recall memory for completed tasks is better than memory for interrupted tasks. However, this is not surprising given that subjects spent substantially more time, both when solving correctly and solving incorrectly, on the completed problems than on the incomplete problems. This difference is logically required if one wants to manipulate which problems are completed, and keep the set size of the two conditions equal. In Zeigarnik's description of her methodology, there is no information on how long subjects were allowed to work on interrupted compared to completed problems.

It is, of course, possible to correct for the confound of differing times spent on the problem if one is willing to forego control over which problems are interrupted. This can be accomplished by allowing subjects a specified amount of time to work on each problem. One can then compare the memorability of problems completed in the short time interval with those problems left incomplete. In this procedure, subjects are determining which condition a given problem falls into, by either completing or getting stuck on the problem within the time interval. Therefore, the nature of the interruption in this procedure may be more likely to be based on a failure in problem solving attempt, rather than on interruption of ongoing problem processes. In the following experiment, the time interval was kept fairly short (1 minute) in order to ensure that no subject would successfully solve all problems. In addition, because some problems might be quickly solved, subjects were told to continue to check their answers when completed for the full amount of time available for each problem.

## Experiment 2

The following experiment uses this timed procedure to look at the effect of holding time constant across completed and incomplete problems. If the greater amount of time spent on completed problems had been preventing an emergence of the Zeigarnik effect in Experiment 1, we expect that the effect will now emerge under equal processing time conditions.

### Method

**Subjects.** Sixty-nine undergraduates at the University of Michigan (34 female and 35 male) participated in the experiment. The subjects received credit towards an introductory psychology course.

**Materials.** Thirty word problems, requiring mathematical, logical, and spatial reasoning skills, were used in this study. Problems were drawn from numerous published collections (e.g. Friedland, 1970; Müller, 1989; Morris, 1988). All of the problems were pretested on a separate group of subjects, and were found to require approximately thirty seconds to two minutes for solution. Each problem was presented to subjects on a separate sheet of paper with space below the problem for work and solution to be recorded. The problems were contained in workbooks. Each workbook consisted of a problem solving instruction sheet, thirty word problems, and a subsequent recall task.The problems were presented in a different random order to each subject.

**Design and Procedure.** Subjects were tested in groups of 10 to 20 in one-hour sessions. At the outset of a session, subjects were told that they would be presented with a series of word problems, one on each of the subsequent pages of the workbook. They were to work consistently and diligently throughout the experiment, making every attempt to solve each problem. All work was to be recorded in the space provided underneath each problem. Subjects were told that they would be given exactly one minute in which to work on each problem. To assist in determining when an answer was completed, subjects were asked to circle their answer when finished, and to spend the remainder of the minute checking their work. They were instructed not to circle any part of their work unless they truly believed they had arrived at an adequate solution to the problem posed.

In accordance with the instructions, subjects were given one minute in which to work on each problem. At the end of each minute, they were reminded to circle their solution if they had arrived at one and to go on to the next problem. This procedure continued until all thirty problems had been exhausted.

Immediately following the last problem, subjects were administered a free recall test. Test instructions were as follows:

> Your next task will be to recall as many problems as you can from the first part of the experiment. When you are told to do so, please jot down problems in the space provided below in the order in which they come to your mind. Write only enough information so that someone else could recognize which of the earlier problems you are referring to. You will have 3 minutes in which to recall the problems.

After three minutes of free recall, subjects were told to stop work on the task.

### Results

All of the problem solving answers were scored as "completed" or "incomplete" by the experimenter based on the solution information written and circled by the subjects. Additionally, completed problems were scored as either "correct" or "incorrect" in their solutions. On average, 20 out of the 30 problems were completed by each subject. The remaining 10 problems, a proportion of .33 problems, were left incomplete. Of the completed problems, a mean of 11 were answered correctly, while 8.9 were answered incorrectly.

In free recall responses, subjects recalled a mean of .45 incomplete problems, with proportions ranging from 0 to 1. The recall proportions for completed problems, also ranging from 0 to 1, had a mean of .33. This difference was significant, $t(68)=3.66$, $p=0.000$, indicating that incomplete problems were recalled more frequently than completed ones, and replicating the Zeigarnik effect. A significant difference was found between recall proportions for completed, correctly-answered problems (.37) and those for completed, incorrectly-answered problems (.28), $t(68)=2.59$, $p=.012$, indicating that subjects best recalled the correctly-answered problems. This difference in memorability for correctly answered problems can be contrasted with the results of Experiment 1 in which no difference was found. Finally, memorability of completed, correctly-answered problems and of completed-incorrectly answered problems (the two subcategories of completed problems) can be individually compared to incomplete problems. A significant difference is found between incomplete (.45) versus correctly-answered (.37) problems ($t(68)=2.48$, $p=0.016$), as well as between incomplete (.45) versus incorrectly answered (.28) problems ($t(68)=4.61$, $p=.000$). Thus, incomplete problems are better remembered than either category of completed problems.

From this second experiment, we conclude that the Zeigarnik effect (better memory for incomplete problems) does not depend on a time difference in favor of incomplete problems. At most, subjects who completed the problems and checked their answers for the remaining time spent only marginally less time on the problems than on the ones where they were not able to provide an answer within the one minute interval. With time as nearly equivalent as possible, we still observed enhanced memorability for incomplete problems.

## Discussion

There were important differences in the two experiments presented that appear to be critical factors in the occurrence of the Zeigarnik effect. When comparing the replication of Zeigarnik in Experiment 2 to the non-replication of Experiment 1, the large time difference in favor of completed problems in Experiment 1 may account for the finding of better memory for completed problems. In Experiment 2, where time on the problem was equivalent, a memory advantage occurred for incomplete problems that cannot be based on time. Therefore, a possible explanation for Zeigarnik's result, that of unintentionally allowing subjects to spend more time on incomplete problems thereby producing better recall, can be ruled out. Substantial differences in time spent on processing, however, as in Experiment 1, may wash out any existing memorial advantage for incomplete problems.

An additional factor is the nature of interruption in the tasks. In Experiment 1, subjects were stopped in their processing before it could reach completion. From pilot sessions, we learned that subjects must be stopped early in their work on a problem in order to ensure that a particular problem be successfully interrupted. In contrast, in Experiment 2, subjects themselves determined which problems were interrupted through failure to complete the problem. While they may sometimes have simply run out of processing time, there was more frequent opportunity to become "stuck" on a problem, to reach an impasse from which no further processing direction was apparent.

This "stuck" state may be of more interest than simple interruption in terms of comparison to real world problem solving. It is impossible to tell, however, from Zeigarnik's reports, whether her subjects were actually of the interrupted (knowing how to proceed if more time is given) or of the "stuck" (at a problem solving impasse) state in her incomplete conditions. Given the reported results, we might expect that if her subjects were allowed to continue to the point of their own "stuck" state, better memorability (and more time on the problem) may have resulted. If instead, interruption was utilized in other studies, the timing differences alone might account for better memory (and more time) on the completed problems, resulting in the non-replication of the effect.

A final factor that must be considered is the set size of the completed vs. incomplete problems. In Experiment 1, we carefully controlled the two sets to be equivalent by manipulating which problems fell into the two conditions. Even so, a memory advantage for completed problems resulted (again, with longer times on completed problems). In Experiment 2, however, we allowed subjects to determine how many problems fell into completed vs. incomplete categories. The resulting ratio of 2/3 completed to 1/3 incomplete characterized the situation where better memory for incomplete tasks was replicated. Possibily, under these conditions, completion status acts as a recall cue, increasing the likelihood that particular problems from a smaller set will be recalled relative to problems from a larger set. Further experiments are under way attempting to manipulate set size. Results obtained thusfar suggest that when time is held constant and set sizes are equal, no recall advantage is found for either completed or incomplete problems. Thus, set size appears to be a critical factor in determining when the Zeigarnik effect will occur.

That set size equivalence results in elimination of the Zeigarnik effect should not, however, imply that the Zeigarnik effect is irrelevant or unimportant to real-world problem solving. For example, in a set of tasks to be accomplished in a list of errands, most may be completed successfully, leaving as pending problems only a subset of tasks. So, if the completion status alone is working as a cue to facilitate retrieval of past problems, even if only when that set is smaller than the completed, the status of intended tasks has been successfully shown to be an important factor in memory. If information about intention to complete, or failure in completion, can be used as a general memory cue, then task status can be successfully used in retrieval of target items from memory.

## Conclusion

We have argued from the evidence of two experiments that the effect first identified by Zeigarnik does in fact have demonstrated replicability as a memory phenomenon. Free recall access to incomplete problems may in fact be better than for completed problems under certain circumstances. In particular, we point to the need to control the amount of time spent in processing, the nature of the interruption, and the relative set size of the two problem conditions. We believe these factors explain the difficulties some studies may have had in replicating the effect, and will serve as a methodological guideline for the study of task interruption. This cognitive explanation is sufficient and more straightforward than appealing to more complex explanations involving individual differences, threat of evaluation, or states of "tension" resulting from blocked tasks (Prentice, 1944).

However, the results we describe have an even greater importance in work on memory for tasks. This is apparent by examining the implications of the finding. Why might memory be designed to note and take advantage of task status in retrieving past problems? What purpose would this memory advantage serve? Obviously, keeping track of the status of problem attempts and being able to use that information in retrieving those past failures would be very useful in the later solution of the interrupted problems. If a goal to solve a problem is not satisfied, information about that failure can be used to preserve and encode the problem in a way that might facilitate its later retrieval. Consequently, failed problems may be more likely to be recalled, and to be pursued for a second time. Such a memory enhancement would assist in bringing to mind past failures so that solution can be reattempted at a later time, perhaps when circumstances will favor success. This cognitive ability is critical for the optimal satisfaction of goals given that many tasks are being pursued. Rather than persist at a difficult problem, effort can be suspended, and work on the goal can be resumed at a later time. The memory effect we have been discussing is an important factor in such an ability.

## References


Alper, T. (1946). Memory for completed and incompleted tasks as a function of personality: An analysis of group data. *Journal of Abnormal Social Psychology*, 41, 403-421.

Boguslavsky, G. & Guthrie, E. (1941). The recall of completed and interrupted activities: An investigation of Zeigarnik's experiment. *Psychological Bulletin*, 33, 575-576.

Friedland, A. (1970). *Puzzles in math and logic*. New York: Dover Publications.

Glixman, A. (1949). Recall of completed and incompleted activities under varying degrees of stress. *Journal of Experimental Psychology*, 39, 281-295.

Morris, S. (1988). *Omni games*. New York: Omni Publications.

Mosler, G. (1977). *The puzzle school*. New York: Abelard-Schuman.

Müller, R. (1989). *The great book of math teasers*. New York: Sterling Publications.

Prentice, W. (1944). The interruption of tasks. *Psychological Review*, 51, 329-340.

Rosenzweig, S. (1943). An experimental study "of repression" with special reference to need-persistence and ego-defense reactions to frustration. *Journal of Experimental Psychology*, 32, 64-74.

Zeigarnik, B. (1927). Das Behalten erledigter und unerledigter Handlungen. *Psychologische Forschungen*, 9, 1-85.

# Characterizing, Rationalizing, and Reifying Mental Models of Recursion

**Shawkat H. Bhuiyan Jim E. Greer Gordon I. McCalla**

ARIES Laboratory
Department of Computational Science
University of Saskatchewan
Saskatoon, SK, Canada S7N 0W0
aries@cs.usask.ca

## Abstract

Mental models reflect people's knowledge about entities and systems around them. Therefore, knowing and understanding mental models can help in exploring cognitive issues in instruction including why a student takes a certain approach or applies a particular strategy to solve a problem, why a student makes mistakes, and why and how misconceptions are developed. Four different mental models of recursion, used for synthesizing solutions to recursive programming problems, have been identified through students' protocols. Each model has been characterized in a way consistent with the students' protocols. Various problem solving behaviours are rationalized in terms of the models. Suggestions are made as to how the mental models develop and evolve in the course of learning. We also present a learning environment in which these mental models are reified and we show how mental models can be incorporated into an intelligent tutoring system.

## Introduction

A *mental model* is a coherent collection of knowledge held by a person about some aspect, entity or concept of the world. People use mental models to interpret the world, and therefore, aspects of human behaviour can be explained through these models (Gentner and Stevens 1983). Since mental models reflect people's knowledge about entities around them, knowing and understanding mental models can help in understanding cognitive issues in instruction such as *why* a student takes a certain approach or applies a particular strategy to solve a problem, *why* a student makes mistakes, and *why* and *how* misconceptions are developed. By understanding the mental models of a student, a tutor or mentor can design appropriate individualized pedagogical guidance.

Most mental model research has been conducted in domains involving physical devices with well-structured physical principles. Working on the Student Computing ENvironmenT (SCENT) LISP programming advisor (McCalla *et al* 1988), an ongoing intelligent tutoring project at the ARIES Laboratory, we became interested in understanding mental models of abstract concepts, especially recursion in computer programming. Recursion is a very interesting domain to study because many students claim to experience a significant cognitive change (often describing this as a radical re-organization of knowledge) as they gain understanding of recursive concepts.

In this regard, an empirical study was carried out to study mental models of recursion with the following goals in mind: 1) What are mental models of recursion in the context of computer programming? 2) How do these models develop and how are they modified? 3) How does migration from one model to another model take place? 4) How do different models relate to one another? 5) When multiple models exist, are they all applicable or are they situation dependent? 6) What are the transformation operators or mapping functions among different mental models? 7) Are changes in mental models evolutionary or revolutionary in nature?

The protocol analysis and preliminary findings of our empirical study were discussed in (Bhuiyan, Greer & McCalla 1989). This paper takes the work further by characterizing mental models of recursion, rationalizing students' problem solving in light of these models, and providing a system for reifying these mental models.

## Background

Various researchers have studied students' conceptions of recursion in the context of computer programming. Their findings can be summarized as follows: 1) Novices at early stages of learning about recursive programming possess a *loop model;* that is, they view recursion as some sort of iterative process (Anzai and Uesato 1982; Kahney 1982; Kurland and Pea 1985; Kessler & Anderson 1986). 2) A *syntactic model* of recursion has been suggested by Pirolli (1985) as an ideal novice model of recursion, yet with a *syntactic model* a student cannot explain the behaviour of a recursive procedure (Kahney 1982). 3) First learning iteration may help students learning recursion, but not vice versa (Anzai and Uesato 1982; Kessler and Anderson 1986). 4) Experts' mental models of recursion are different from those of novices (Kahney 1982). 5) There are no significant difference in students' performance with recursive tasks when they were taught with theoretical *vs* syntactic methods, although there is some indication that transfer is enhanced through a theoretical approach (Greer 1987).

Although the literature hints at several different mental models of recursion, no comprehensive work has yet been done in this domain. Therefore, we believe it is important to investigate mental models of recursion more carefully, especially how the models are used in synthesizing recursive programs, and hence we carried out the empirical study. The intent of the study was not to statistically validate mental models of recursion, but rather to explore the goals mentioned above.

## Mental Models of Recursion

Our empirical study involved monitoring the problem solving of six non-major computer science students on a weekly basis to analyse their knowledge of recursion as they learned the concept in a programming course. Four different mental models of recursion: *loop model, syntactic model, analytic model* and *analysis/synthesis (A/S) model* have been identified through analysis of the students' verbal protocols. These four mental models of recursion were used by various students at varying times as they synthesized solutions for recursive problems. Earlier papers (Bhuiyan et al. 1989; Greer 1987) provide evidence justifying these mental models. A number of analysis models were also used by the students for analyzing and tracing recursive programs, but these are not discussed here. Here an attempt is made to characterize students' problem-solving behaviour consistent with each model.

### Loop Model

The *loop model* is a flawed model of recursion that many students develop early in their study of recursive programming (Anzai and Uesato 1982; Kahney 1982; Kurland and Pea 1985; Kessler and Anderson 1986). This model comes into existence when the student tries to understand and explain recursion in terms of prior knowledge about iterative programming.

Typically, the beginner has some basic declarative knowledge about recursion either from a text book or from a teacher. This knowledge could include facts such as: a recursive function calls itself, a recursive function has a base case and a recursive case, and so forth. The student may not have acquired any means to apply this declarative knowledge, i.e, he/she may not have the procedural counterpart of the knowledge. On the other hand, his/her knowledge about iteration typically consists of both declarative and procedural knowledge. The student's knowledge about iteration dominates his/her knowledge about recursion, and it is likely that the student tries to understand recursion by drawing analogies to loop or iterative structures.

Usually students develop informal solution plans before writing programs. A student with a *loop model* is likely to develop an iterative solution plan. If he/she attempts to translate this to a recursive algorithm, there is a good chance that the algorithm will be flawed, incorporating such features as declaration of loop index variables, initialization of loop index variables, update (increment/decrement) of these variables, and a termination test. Manifestation of these features depends upon the initial selection of the loop structure that the student had in mind while developing the algorithm. The student's program usually does not contain an explicit loop, because the student knows that recursions are different from iterations, but instead substitutes the intended loop structure with a recursive call.

Protocol analysis revealed several issues related to students' use of the *loop model.* Learning iteration may assist later learning of recursion, as Kessler and Anderson (1986) and Uesato and Anzai (1982) claim, but we observed that the *loop model* did not help students in understanding recursion; rather students became confused as they tried to apply the *loop model* to synthesizing recursive solutions. By the end of the second week of our investigation, all students who were initially using the loop model seemed to have moved on to more suitable mental models. Greer (1987) made a similar observation in his experiment.

### Syntactic Model

Students acquire the *syntactic model* by abstracting structural features of recursion when they have little conceptualization of recursion as a problem-solving technique, but have considerable declarative knowledge about recursion. At this stage, their declarative knowledge focuses on two structural features of recursive functions: a base case and a recursive case; where a process to solve a problem is repeatedly called in the recursive case, and the base case stops the process. When a student transforms such knowledge of recursion into problem solving for simple problems, he/she usually abstracts these two features into a *recursion template*, a schema with slots for base case and recursive case as shown in Figure 1. Again, each case has a condition part and an action part.

```
(defun <function-name> (<arg1> ... <arg n>)
   (cond
      (<condition-1>  <action-1>)   ;; base case
      (<condition-2>  <action-2>)   ;; base / recursive case
      •
      •
      •
      (<condition-k>  <action-k>)))  ;; recursive case
```

**Figure 1:** *Syntactic model* template

To synthesize a solution to a recursive programming problem using the *syntactic model*, first a template is selected. The selected template can be a 'basic template' like the one in Figure 1 or a template from an earlier problem analogous to the current problem. After selecting the template the student fills in the condition and the action parts in the case slots with programming language specific code chunks. In general, a student makes several attempts to fill in recursive case slots when partial results are to be passed back. The protocols in Figure 2 illustrate the use of the *syntactic model.*

A serious shortcoming of the *syntactic model* is that although a student is aware of the conditions and actions in a template, he/she may not know *how* to derive them. This frequently results in nearly random slot-filling behaviour by novice programmers. Moreover, while synthesizing programs using this model, novices, generally, think of the solutions at the programming language code level. This may be due to the fact that they are simply concerned with filling in the template slots.

(a) Student K is looking for an arbitrary recursion template
K: .. I don't know. I have to ... I can't remember the "format" that goes here. You showed me a couple of minutes ago.
Tutor: Okay. Why do you have to see a format?
K: Well, the base case and the recursive case - how they are set up, I forgot.

(b) An example of template filling behaviour
K: So, we need a base case. The base case would be: if the list is empty then return 0, else .. (over a minute) .. Else take rest of L. ... return Count + 1.
....This is not how it's done

**Figure 2:** Protocol demonstrating the *syntactic model*

Although the *syntactic model* is better than the *loop model* it does not constitute a complete understanding of recursion. Nevertheless, we observed that novice students frequently used the *syntactic model* in synthesizing recursive solutions. This corresponds to Escott's (1988) research, where she found 80% of students' programs were structural analogies of earlier programs. The LISP Tutor (Anderson and Reiser 1986) encourages students to use recursion templates which they fill in to arrive at a final program. We have found that for solutions to routine problems, the *syntactic model* is sufficient, although its usefulness diminishes when students are faced with novel or difficult problems. Perhaps for this reason the LISP Tutor augments the problem solving by guiding students through a planning dialog to formulate solutions.

## Analytic Model

Although a student may solve simple recursive problems employing the *syntactic model,* complex problems require a deeper understanding in order to map the problem to a recursive construct. This "deep understanding" involves the ability to analyze input-output behaviour of a problem and to determine input conditions and corresponding output actions together with associated transformations. Such analytic cognitive ability to synthesize a recursive solution gives rise to the *analytic model.*

With the *analytic model*, a student does not view recursion simply as a physical construct like the syntactic template; rather he/she views recursion as a problem-solving technique. Therefore, the significance of this model is to analyse the input-output behaviour of the given problem.

Solution synthesis using the *analytic model* has the following three steps: 1) Determine cases: Analyze the various input cases for the given problem. Determine the input cases and the corresponding output strategies. In this step, the student specifies his/her intentions, an informal solution specification, for the problem. 2) Translate input cases to input conditions and translate output strategies to output actions: In this step, the student plans a high-level solution, which can be translated to any programming language. Students in the empirical study were found to use natural language phrases to outline the solution plan. 3) Translate input conditions and output actions into programming language code.

The *analytic model* provides two levels of intermediate representations between the problem statement and the solution (step 1 and step 2 above). The step 2 intermediate representation is particularly interesting because it supports a pseudo-language; a mixture of natural language and programming language primitives, which students use to elaborate the input conditions and the corresponding output actions for the problem. To describe a solution in natural language, a student needs to attain a deeper level of cognition than for the *syntactic model.* The protocol in Figure 3 shows a student's solution plan for the *LIST-B* problem, to *return a list of all top level B's in a list*:

S:....So, these are lists, then? O.K. If Null L then Return the empty list. If First of the list is a B then return B, and work (on) the rest. Else Gather ..

The solution plan consisted of the following three cases:
Case 1: If NULL L then return an empty list.
Case 2: If the first of the list is a B then return B, and work on the rest of the input list.
Case 3: Else (for none of the above conditions), Gather

**Figure 3:** Protocol demonstrating the *analytic model*

In Figure 3, student S was not working directly with base case(s) or recursive case(s) as in the *syntactic model.* Moreover, she was also not concerned with transforming the solution plan into program code. For example, in Case 2, she did not articulate the exact code for the condition part or for the action part. Interestingly, she made no mention of constructing an output list of B's, which the problem asked for, but she wanted to process ("work on") the rest of the input list. This suggests her intention (or goal) was implicit from the problem statement.

The main difference between the *analytic model* and the *syntactic model* is that with the *syntactic model* a student writes program code directly from the problem statement, whereas with the *analytic model* the student first determines a solution plan from the problem statement using an intermediate language and then transforms the plan into program code.

## Analysis/Synthesis Model

The *analysis/synthesis (A/S) model* is the most powerful model of recursion encompassing both structural and functional properties. Expert programmers seem to possess this model. Only one student in our study began to acquire this model and attempted to use it in the solution of a challenging problem in the last week of our study. We have uncovered little direct evidence for this model, although other work (cf. Greer 1987) supports the existence of the *A/S model.* We include a brief description of the *A/S model* in this paper for the sake of completeness.

The *A/S model* gives the ability to reduce a given problem into smaller ones and to synthesize the corresponding solutions into a global solution for the problem. The usual steps of solving a problem using the *A/S model* are as follows: 1) Determine the smallest pieces that can be solved instantaneously. 2) Break the problem into subproblems where solution to a subproblem is the solution to a smallest piece plus solution to a smaller subproblem. 3) Determine stopping case(s).

Even expert programmers do not apply the *A/S model* every time they solve a problem recursively. If the problem is simple then there is a good chance that a similar problem has been previously solved, thus providing a template or a chunk to embody the solution. In other words, for familiar problems, most programmers seem to employ a *syntactic model.*

## Evolution of Mental Models

Students acquire (or develop) mental models of recursion during instruction and problem solving. Mental models seem to evolve in a sequential fashion, roughly in the sequence of *loop model,* then *syntactic model, analytic model* and finally the *A/S model.*

Over the five weeks of our study students were asked (among other things) to solve nine problems with recursive solutions. From their protocols, the mental model that they were trying to use for formulating each problem solution was identified. The problems generally increased in difficulty through the first four weeks and a difficult, challenging problem was given in the final week. Table 1 shows the mental models that the six students used while attempting to formulate a solution to the problems. In a number of cases one mental model proved insufficient for the student, and a second attempt using a different mental model was observed. Table 1 demonstrates that in most cases students evolved more sophisticated mental models over time.

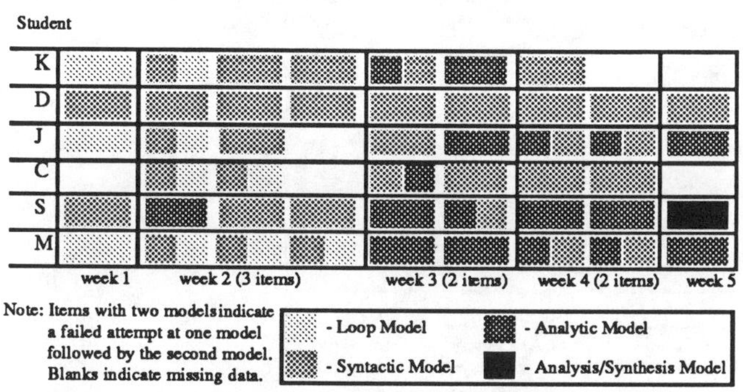


**Table 1:** Evolution of Mental Models of Recursion

Our empirical study supports the hypotheses that students construct particular mental models, that the models are enhanced by new material that they learn from class lectures or from other sources, that the models become inappropriate for certain problem solving situations, and finally that new models evolve. It seems that mental models, once formed, are persistent. Students frequently return to more established mental models when a first attempt to solve a problem with a new model fails. It also seems that some students will attempt to apply a well established mental model, find that it is not sufficient to conceptualize the solution to the problem, and only then will they apply a new mental model. Students seem to adopt more sophisticated mental models of recursion as they meet with more difficult recursive problems.

## A Learning Environment to Support Mental Models

We have determined that mental models play an important role in students' understanding of recursive programming. It follows that mental models can be utilized in better understanding students' intermediate problem solving behaviour and also can help in promoting students' learning of recursion. In essence, mental models, if properly used in intelligent tutoring systems, will increase the diagnostic precision and the appropriateness of pedagogical guidance offered by the system. To this end, we have constructed a prototype learning environment (named PETAL) in which students may choose from a variety of **P**rogramming **E**nvironment **T**ools (PETs) which correspond to various mental models of recursion.

At the present time PETAL is being used to further explore mental models of recursion by observing student problem-solving behaviour when they are constrained by an environment that supports a particular mental model. PETAL does not simulate the execution of explicit runnable mental models; rather it uses what we believe about the existence of these models in the learner's mind to provide scaffolding to support the learner's use of the models. It is important to realize that it is the learner who "runs" the mental model on his/her mind with PETAL merely providing support and constraint. PETAL directly supports specific models of recursion by providing an intelligent computer-based environment in which students formulate solutions to programming problems. The student first selects a problem to work on and then specifies the parameters for the function to be synthesized. Next the student selects among the available PETs to choose an environment corresponding to a mental model in which to solve the problem. Students may switch from one PET to another as they attempt to solve a problem. Since the *loop model* is not a viable model of recursion, there is no corresponding PET. In our prototype implementation of PETAL, only the *syntactic* and the *analytic* PETs have been built. Twenty-three programming tasks have been included so far.

The *syntactic* PET supports solution synthesis from the viewpoint of the *syntactic model.* The student using this PET must construct a recursive template of base case(s) and recursive case(s). Next he/she must fill in the case slots with problem specific code chunks. In order to assist the student, the *syntactic* PET provides a menu of available code chunks specific to the selected problem. The code chunks for a particular problem are created in advance by a domain expert. Code chunks may contain distractors, which can be useful in diagnosing and clarifying potential misconceptions which the student might possess. Once the code chunks have been filled into the template, corresponding LISP code is automatically generated by the PET.

The *analytic* PET corresponds to the *analytic model* for solution synthesis. With the *analytic* PET, the student views recursion as a problem solving technique based on input/output (I/O) analysis of the problem whereupon these I/O behaviours are mapped to properties of recursion: base case(s) and recursive case(s). Unlike the *syntactic model*, the *analytic model* provides cognitive support to the student to derive the cases through I/O analysis. Program synthesis using the *analytic* PET consists of three stages. **Intention stage**: Determine input cases and corresponding output strategies for the given problem. **Plan stage:** Determine the solution plan by deriving input conditions and output actions corresponding to the input cases and the output strategies respectively. **Code translation stage:** Translate the solution plan into program code.

Figure 4 shows the *analytic* PET initialized with the *count-atoms* problem (counting the number of top level atoms in a list). The *analytic* PET manifests these three stages of solution synthesis. First, at the **intention stage**, the student determines the possible input cases and the corresponding output strategies for the problem. Lists of such choices of the input cases and the output strategies are displayed in Input Case Choices and Output Strategy Choices tables (in the upper left part of the screen display). As input case choices and output strategies are selected, each is displayed in the I-Cases and O-strategies tables (in the lower left part of the screen display). Together the cases and the strategies constitute an informal and coarse-grained solution to the problem. In a broad sense, the student's intentions are captured in this informal solution.

During the **plan stage**, the student develops a solution plan consisting of input conditions and output actions, derived from the input cases and the output strategies respectively. An input condition and its corresponding output action together make a subplan, where the condition is the context of the subplan and the action is the goal of the subplan. The *analytic* PET provides natural language phrases to elaborate the solution plan. These phrases are problem specific and are supported by empirical evidence from our earlier study of problem-solving protocols.

After developing the solution plan, the **code translation stage** begins. The translation is not fully automatic since precise code cannot always be inferred from the student's natural language plans. When the student presses the **Edit Program** button, a LISP Editor appears which shows all the input-output analysis (as commentary) together with a partially filled code template for the solution. It is left to the student to flesh out the program code derived from the solution plan.

The example above, shows how the *analytic* PET provides students support for synthesizing solutions to recursive problems consistent with the *analytic model* of recursion. PETAL currently acts as a stand-alone programming environment, and not as a complete instructional system like the Bridge System (Bonar & Cunningham, 1988) or the LISP Tutor (Anderson & Reiser 1986). PETAL is designed to ease students' transitions through increasingly sophisticated mental models of recursion. Even without knowledgeable feedback to the student, PETAL is proving to be a useful programming environment, since students explore solutions to problems in a rich conceptual environment and can produce LISP code with little effort. After pilot-testing PETAL with a small number of students we are now beginning to consider its potential as an interface to an intelligent tutoring system (specifically the SCENT-3

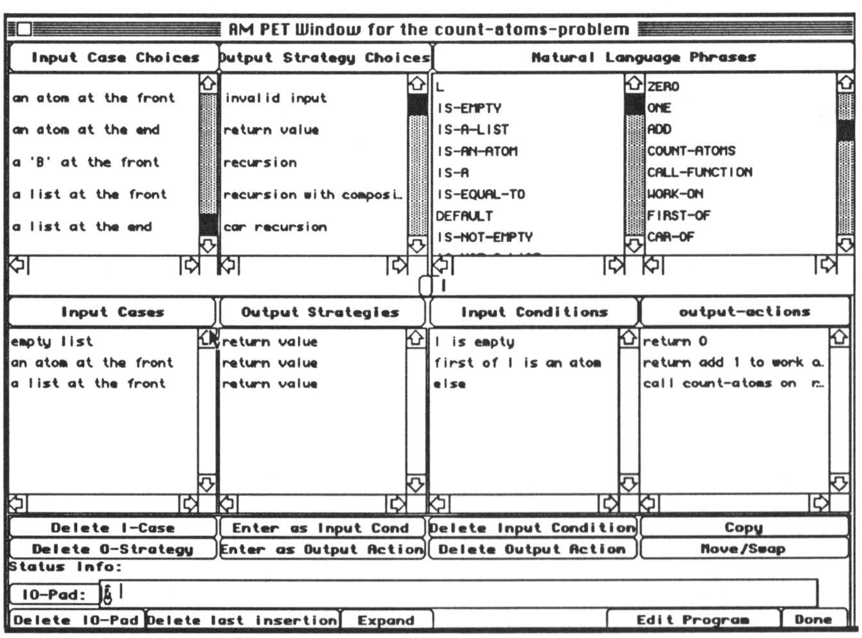


**Figure 4:** *Analytic* PET for the count-atoms problem

system). Interacting with PETAL, students make explicit the general strategies and specific plans they are considering. This knowledge about strategies and plans is extremely useful and difficult to infer in a standard intelligent tutoring system. The additional knowledge available to a tutoring system that uses a mental model-based interface is considerable. At the same time, the added benefit of knowledgeable feedback to the student using PETAL should prove to be substantial. Although much more research is needed, PETAL promises to be a viable front-end for the SCENT intelligent tutoring system.

## Conclusion

Intelligent tutoring systems will achieve their fullest potential when they can support students' learning at the deep "mental modelling" levels, rather than merely support their understanding at a surface level. We have examined how to create such support tools for the domain of programming, in particular for recursive programming concepts. A prerequisite to the success of our endeavour has been to carry out an empirical study delineating the space of mental models actually used by students. Then support tools for each identified mental model have been defined and are currently being tested on students. Student use of these tools is feeding back into our understanding of the mental models themselves. The symbiotic relationship between empirical studies and the system that grows out of these studies is an important lesson of this research.

This research also contributes to the burgeoning body of knowledge about mental models. In contrast to the usual mental models of physical systems, we investigate mental models in an abstract domain. Recursion in computer programming has two features: first, it is a problem solving technique used in synthesizing programs, and second, it is a process or control structure that controls the execution order of the program. Using the synthesis models of recursion, discussed in this paper, students formulate recursive solutions. Our experiments have shown that students do indeed employ mental models in recursive problem solving, that they adopt more sophisticated mental models as the need arises, and that they can switch among mental models as appropriate.

Moreover, our work on PETAL adds to the repertoire of tools to support deep conceptualization by students as they learn. Our PETs are similar to so-called intermediate representations employed in various intelligent tutoring systems such as the Bridge Tutor (Bonar & Cunningham 1988), GIL (Reiser et al 1990), and Angle (Koedinger & Anderson 1990). However, we provide several different representations, rather than just one as in these other systems. These multiple representations are proving to be valuable pedagogically, and are being used to help us refine and clarify our understanding of the structure and use of mental models. They also provide us with a precise lens through which to view the genetic relationships students evolve among several different mental models.

Our future work will continue to refine and elaborate these mental models. We will explore how students actually use the mental models, especially how use of the models is altered as student knowledge evolves. Finally, we will continue to investigate how to build effective support tools for the mental model level of learning.

### Acknowledgements

We would like to thank NSERC and the University of Saskatchewan for their financial support.

### References

Anderson, J.R. and Reiser, B. 1986. The LISP Tutor, *Byte*, April, 159-175.

Anzai, Y. and Uesato, Y. 1982. Learning recursive procedures by middle school children. *The 4th Cognitive Science Conference*, 100-102, Ann Arbor, MI, USA.

Bhuiyan, S.H., Greer, J.E. and McCalla, G.I. 1989. Mental models of recursion and their use in SCENT. In Ramani et al. (Eds.), *Knowledge-based Computer Systems*, 135-144. Bombay, India.

Bonar, J.G. and Cunningham, R. 1988. Intelligent tutoring with intermediate representations. In *Proceedings of ITS'88*, 25-32. Montreal, Canada..

Escott, J. 1988. *Problem Solving by Analogy in Novice Programming*. ARIES LAB Research Report 88-3. Dept of Comp. Science, University of Saskatchewan, Canada.

Gentner, D. and Stevens, A. Eds. 1983. *Mental Models*. Hillside, NJ: Lawrence Erlbaum.

Greer, J.E. 1987. *Empirical Comparison of Techniques for Teaching Recursion in Introductory Computer Science*. Ph.D. Thesis. The University of Texas at Austin.

Kahney, H. 1982. *An In-depth Study of the Cognitive Behaviour of Novice Programmers*. Tech. Report No. 5. Milton-Keynes, England: The Open University.

Kessler, C. and Anderson, J. 1986. Learning flow of control: Iterative and recursive procedures. *Human-Computer Interaction*, **2**, 135-166. Hillside, NJ: Lawrence Erlbaum.

Koedinger, K.R. and Anderson, J. 1990. Theoretical and empirical motivations for design of ANGLE. *AAAI Spring Symposium* , Stanford University.

Kurland, D.M. and Pea, R.D. 1985. Children's mental models of recursive LOGO programming. *Journal of Educational Computing Research*, **1**(2):235-243.

McCalla, G.I., Greer, J.E. et al. 1988. Intelligent advising in problem solving domains: The SCENT-3 architecture. *Proceedings of ITS'88*, 124-131. Montreal, Canada.

Norman, D. 1983. Some observations on mental models. In D. Gentner and A. Stevens (Eds). *Mental Models*. Hillside, NJ: Lawrence Erlbaum.

Pirolli, P. 1988. A Cognitive model of computer tutor for programming recursion. *Human-Computer Interaction*, **2**, 329-355. Hillside, NJ: Lawrence Erlbaum.

Reiser, B.J., Ranney, M. Lovett, M.C. and Kimberg, D.Y. 1990. Facilitating students' reasoning with causal explanations and visual representations. *4th Conference on AI and Education*, 228-235. Amsterdam, Netherlands.

# Extending a Model of Human Plausible Reasoning

**Mark H. Burstein**
**Allan M. Collins***
**Bolt Beranek and Newman Inc.**
**10 Moulton St.**
**Cambridge, MA 02138**
**EMAIL: BURSTEIN@BBN.COM**
*** also the Institute for the Learning Sciences**

## Abstract

When one looks at transcripts of people answering questions or carrying on dialogues about everyday matters, their comments are filled with plausible inferences -- inferences that are not certain, but that make sense. Often, in forming these inferences, generalizations are made that are equally uncertain, but are nevertheless useful as a guide to their reasoning. This paper describes some extensions to our earlier description of a core theory of plausible reasoning (Collins and Michalski, 1989), based in large part on a recent protocol study. The primary focus is on the inductive inference patterns people use to form *plausible generalizations*, weakly held beliefs based on few examples. We also show how the model was extended to deal with plausible inferences involving continuous quantities and inequalities.

## Introduction

In a previous paper (Collins & Michalski, 1989) we presented a core theory of human plausible reasoning based on analysis of a large set of people's answers to everyday questions about the world (Collins 1978). The theory was characterized in terms of Michalski's (1987) variable-valued logic notation. This paper briefly describes several kinds of revisions to the formalism developed in that earlier paper, based on the results of a more recent protocol study.

There are four types of expressions in the core theory of Collins and Michalski (1989) that are shown in Table 1. The first are simple statements consisting of a *descriptor* **d** (e.g. means-of-locomotion) applied to an *argument* **a** (e.g. birds) and realized by a *referent* **r** (e.g. flying). The brackets and dots around the referent indicate that there may be other means of locomotion for birds, such as walking. The second kind of expression involves one of four relations: **GEN** for generalization, **SPEC** for specialization, **SIM** for similarity, and **DIS** for dissimilarity. Each relational statement specifies a context (CX) where the first variable is the domain over which typicality or similarity are computed, and the second variable is the descriptor(s) with respect to which typicality or similarity are computed. The last two examples of relational statements represent the fact that ducks and geese are similar in their habitats, but dissimilar in neck length.

The other two types of expressions in Table 1 are called mutual implications and mutual dependencies.A mutual implication specifies how one statement is related to another statement. The example states that warm temperature and heavy rainfall imply rice growing, and vice versa. A mutual dependency relates two *terms* (e.g. latitude(place) and temperature(place)). Our model of dependencies and their use in plausible inferences is much like Russell's(1986) notion of determinations. The example represents the belief that the average temperature of a place is inversely related to its latitude.

Each type of statement carries with it various kinds of certainty information that affects inference conclusion certainty :

- **Certainty** (γ) annotates all state- ments. It indicates overall level of belief in a statement.
- ***Typicality*** (τ) annotates GEN and SPEC statements. The more typical England is of Europe, or Surrey is of England, with respect to climate (or any variable that affects flower growing), the more certain the inference.
- ***Similarity*** (σ) annotates the SIM and DIS statements. Hence, the more similar Holland is to England, and the less similar Java is to England, with respect to climate, the more certain the inference.
- ***Conditional Likelihood*** (α,β) modifies dependencies and implications. For example, α indicates how much latitude depends on temperature, and β how much temperature depends on latitude.

  ***Frequency*** (ϕ) reflects the all/some distinction in logic, but as a continuous variable. I.e., the frequency of daffodils and roses in different parts of England. The more frequent daffodils and roses are in England, the more likely they are found in Europe, Surrey, Holland, or even Java.

**Table 1:** Different Types of Expressions in the Core Theory

| | Form | Examples |
|---|---|---|
| **Statements (S):** | d(a) = r | means-of-locomotion(birds)={flying...} |
| **Relational Statements (R)** | $a_1$ REL $a_2$ in CX (A,d) | |
| | $a_1$ GEN $a_2$ in CX (A,d) | bird GEN robin in CX (birds, all characteristics) |
| | $a_1$ SPEC $a_2$ in CX (A,d) | chicken SPEC fowl in CX (birds, biological chars.) |
| | $a_1$ SIM $a_2$ in CX (A,d) | duck SIM goose in CX (birds, habitat) |
| | $a_1$ DIS $a_2$ in CX (A,d) | duck DIS goose in CX (birds, neck length) |
| **Mutual Implications (I)** | $d_1(a) = r_1$ <==> $d_2(a) = r_2$ | temperature(place)=warm & rainfall(place)=heavy <==> grain(place)=rice |
| **Mutual Dependencies (D)** | $d_1(a)$ <--±--> $d_2(a)$ | average temperature (place) <--⁻--> latitude (place) |

- ***Dominance*** ($\partial$) applies to GEN and SPEC inferences and reflects the degree the subset comprises a large part of the set. For example, since Surrey in only a small part of England, growing daffodils and roses is less certain than for Southern England as a whole.
- ***Multiplicity of the argument*** ($\mu_a$) reflects the degree to which more than one country (the superordinate of the argument) has daffodils and roses. Since many countries presumably have daffodils and roses, $\mu_a$ is high and the argument-based inferences are more certain (except for the DIS inference).
- ***Multiplicity of the referent*** ($\mu_r$) reflects the degree to which England has more types of flowers (the superordinate of the referent) than daffodils and roses. Since countries usually have many different types of flowers, $\mu_r$ is high and the referent-based inferences more certain (except for the DIS inference).

A *similarity transform* (a basic inference type) would make use of these as follows:

flower-type(England) = roses
: $\gamma$ = high, $\mu_a$ = high, $\mu_b$ = high, $\phi$ = high
England SIM Holland in CX (place, climate)
: $\gamma$ = high, $\sigma$ = high
climate(place) <---> flower-type(place)
: $\gamma$ = high, $\alpha$ = high, $\beta$ = moderate
England, Holland SPEC place
: $\gamma$ = high, $\tau$ = high, $\delta$ = low

---

flower-type(Holland) = roses: : $\gamma$ = high

Table 2 shows a pattern of eight statement transforms from the core theory (Collins & Michalski, 1989). Given a person believes that the flowers of England include daffodils and roses, the first four transforms all vary the argument, England. Given no other information, it is a plausible inference that daffodils and roses are flowers of Europe in general (a *generalization transform*); that Surrey, which is a small county in England, has daffodils and roses (a *specialization transform*); that Holland, which is similar to England in its climate, has daffodils and roses (a *similarity transform*); and that Java, which is quite dissimilar to England in climate, does not have daffodils and roscs (a *dissimilarity transform*). In general, when an inference involves a dependancy, relationships between terms (GEN, SPEC, SIM, DIF) these relationships are specified within a context, CX, that relates to the dependancy governing the inference. The overall certainty of the conclusion will vary depending on the similarity/difference between the related terms *in that context*, not simply with respect to overall similarity (which is written as "all characteristics").

The other four transforms vary the referent, daffodils and roses. If you believe that daffodils and roses are flowers of England, it is plausible that most temperate flowers grow there (a *generalization transform*), that yellow roses grow there (a *specialization transform*), that peonies grow there (a *similarity transform*), and that bougainvillea, a tropical plant, does not grow there (a *dissimiliarity transform*). These eight transforms were one of four classes of inferences in the core theory.

On the right side of the table, is a matrix showing how and in which direction each certainty parameter affects each of these transforms. Thus, typicality ($\tau$), which relates terms to their specializations, affects GEN and SPEC transforms, while similarity ($\sigma$) affects SIM and DIS transforms. The certainty of each of these inferences is affected by the certainty ($\gamma$) of the person's belief in each of the premises in the inference. For example, the more certain the person is that England produces daffodils and roses, and that flowers depend on climate, the more certain the inference. These parameters are described in more detail in the Collins & Michalski (1989).

There are a number of other inference types over these forms of statements that are described in detail in Collins and Michalski (1989), and a computer implementation of the core theory was described in Baker et al (1987).

**Table 2:** Eight Transforms on the Statement "flower-type(England)={daffodils, roses...}"

| | **Parameter:** | $\tau$ | $\sigma$ | $\alpha$ | $\beta$ | $\delta$ | $\mu_a$ | $\mu_r$ |
|---|---|---|---|---|---|---|---|---|
| **Argument-based Transforms** | | | | | | | | |
| (1) **GEN** | flower-type(Europe)={daffodils, roses...} | + | 0 | + | + | + | + | 0 |
| (2) **SPEC** | flower-type(Surrey)={daffodils, roses...} | + | 0 | + | + | + | 0 | 0 |
| (3) **SIM** | flower-type(Holland)={daffodils, roses...} | 0 | + | + | + | 0 | + | 0 |
| (4) **DIS** | flower-type(Java)={daffodils, roses...} | 0 | - | + | - | 0 | - | 0 |
| **Referent-based Transforms** | | | | | | | | |
| (5) **GEN** | flower-type(England)={temperate flowers...} | + | 0 | + | + | + | 0 | + |
| (6) **SPEC** | flower-type(England)={yellow roses...} | + | 0 | + | + | + | 0 | 0 |
| (7) **SIM** | flower-type(England)={peonies...} | 0 | + | + | + | 0 | 0 | + |
| (8) **DIS** | flower-type(England)={bougainvillea...} | 0 | - | + | - | 0 | 0 | - |

+ means higher values of parameter increase the certainty of the inference, and
- means higher values of parameter decrease the certainty of the inference.

## A Protocol Experiment

The extensions to the core theory of Collins and Michalski (1989) described in this paper are based on a new set of protocols collected as subjects tried to reason about geographical attributes. This experiment provided lengthy examples of plausible reasoning from given to unknown information. In this experiment, designed and conducted by Michell Baker, we gave human subjects a partially-specified matrix of geographical variables crossed by countries, shown in Table 3, that we had developed for the computer simulation. We then interviewed five different scientists (who were not geographers) as they attempted to fill in the missing cells in the matrix. The resulting protocols forced us to consider how people reason about quantities using "less than" and "greater than", and how they generalize to form new knowledge.

Subjects were shown the entire matrix and asked to fill in the missing cells, in whatever order they chose. They were asked to verbalize their reasoning as they tried to fill in each cell, and were prompted to expand on that reasoning anytime the reasoning was unclear. The sessions varied in length from one half hour to one and one half hours for different subjects. Each session was recorded on audio tape and transcribed.

While the tasks for subjects was less natural than the teaching dialogue and question answering tasks used earlier, we think much of the flavor of natural plausible reasoning was captured by the task. There are two reasons we believe this is so: First the same kinds of inference forms that we have specified in earlier papers (Collins 1978; Collins and Michalski, 1989) were predominant in the reasoning on this task. Second, the subjects talked quite naturally and at length about why they made their guesses, so there does not seem to be anything artificial about the discourse they produced. The task did have two aspects that made it different from earlier protocols we had collected. Because the subjects had the matrix in front of them, they could consider more cases and variables than the one or two they usually do. Second, because subjects were asked a number of questions with respect to the same set of variables, they accumulated more knowledge about the same topic area in this context. So these protocols had more a flavor of scientific investigation to form generalizations than earlier protocols.

In analyzing the transcripts of these sessions, we sought to identify and formalize as many of the plausible inferences as we could find, each time considering whether the formal theory, as described in Collins and Michalski (1989) accounted for the observed behavior. Where there were discrepancies, we considered how the theory was inadequate, and considered possible extensions and their ramifications.

One issue that arose was found in the protocol of Subject 1, who tried to use the value for the amount of available fresh water supply in Italy to reason backward to infer the value for precipitation in Italy. In the matrix, there were two variables that directly affected water supply. The principal one was a qualitative value (light, moderate, abundant) for the average amount of precipitation of the country. The second was a variable indicating whether there were any rivers in the country or not (yes or no). For Italy, the water supply was listed as being moderate, and the column labeled HAS-RIVER? had the value YES.

**S1:** Let's go back and do Italy first then... What the mountains tell you is that increases the precipitation. And the Mediterranean climate tells you that it doesn't typically have a lot; Mediteraneans tend to be fairly dry climates. So my guess about Italy is that it probably... but the fresh water supply also implies... well it could get its fresh water all from the rivers, so the moderate fresh water supply... because with Egypt had moderate and that other one I inferred was moderate. My inclination would be to say that implies that there is not a lot of rainfall, okay. But the mountains imply that there is rainfall, okay. So that leads me to... I'm not sure what variables I have for rainfall, very light and light, so I'd go for light.

**Table 3:** Experimental Matrix of Geographic Data

| 12 x 9 | Climate | Water Supply | Grain Grown | Has River? | Precip-itation | Season Description | Soil Type | Temperature Range | Terrain |
|---|---|---|---|---|---|---|---|---|---|
| Afghanistan | ? | ? | NONE | ? | ? | ? | Brown Grey | Hot Very Hot | Mountains |
| Angola | ? | Moderate Abundant | Corn | YES | Abundant | Summer Rain | Dark Brown Grey | Hot | ? |
| Egypt | Dry Climate | Moderate (Irrigated) | Wheat | YES | Very Light | ? | Grey | Very Hot | Plains |
| Florida | Subtropical Humid Trop. | ? | Corn | ? | Moderate Abundant | Mild Winter Long Summer Even Rain | ? | ? | Lowlands Plains |
| Iran | Semi-Arid Mediteranean | ? | ? | NO | Light | Winter Rain | Grey | ? | ? |
| Italy | Mediteranean | Moderate | ? | YES | ? | Mild Winter Hot Summer Winter Rain | Complex Red-Yellow | Mild Hot | Mountains Plains |
| Java | Humid Tropics | ? | Rice Corn | NO | Abundant Very Wet | No Winter Even Rainfall | ? | Hot | Mountains Lowlands |
| Louisiana | Subtropical | Abundant | ? | YES | ? | Mild Winter Long Summer Even Rainfall | Red-Yellow Black | ? | Lowlands Plains |
| Peru | Highland Arid | Moderate (Irrigated) | Corn Rice | ? | Very Light Light | Summer Rain | Complex | ? | Mountains |
| Saskat-chewan | Dry Climate | ? | Wheat Oats,Rye | YES | Light | Winter Rain | Dark Brown Brown Complex | Cool Mild | Plateau |
| Upper Volta | ? | Abundant | Rice Millet | YES | Very Wet | ? | ? | Hot Very Hot | Lowlands Plains |
| West Indies | Humid Tropics | Abundant | Rice Corn | NO | Abundant Very Wet | No Winter Even Rainfall | Red-Yellow | Hot | ? |

In the first part of this response, the subject focused on the evidence that the Italian climate was Mediterranean, and the fact that there were mountains. The Mediterranean climate led the subject to infer that Italy had limited precipitation, while the presence of mountains indicated that there would be more rain than other, similar, lowland areas with the same general climate. In the second half of the response, the subject based his inferences on the evidence that the fresh water supply given for Italy was moderate, and the fact that there were rivers. There are two kinds of uncertain information here. One is the question of how much of a contribution each of those factors can make to overall water supply, a question for which the subject presumably had little direct knowledge. The other problem is the lack of information of even a qualitative kind as to the amount of water available from rivers. All the matrix supplied was the fact that there were at least *some* rivers.

It appears from the pattern of this subject's protocols, and from subsequent questioning of the subject, that he normally treated water supply as if it was directly dependent on precipitation, independently of the presence of rivers. Other subjects treated the two as more "additive". In general, either precipitation or rivers could account for the water supply of a place, and this subject generally assumed water supply was directly correlated with precipitation, unless there was evidence to the contrary. In making a comparison to Egypt, the subject reasoned that Egypt had moderate water supply even though the precipitation there was very light, because there was a river. Thus the analogy to Egypt supported the conclusion that the precipitation was **very light**. By combining evidence from three sources: the analogy to Egypt, the presence of mountains, and the Mediterranean climate, the subject concluded that the precipitation was probably **light.**

## Refining the Core Theory

Several issues were raised in this protocol that were not specifically covered by the core theory. The first was the reasoning about inequalities on ordinal referents. Another, related issue was the use of reasoning about deviation from a default value.The logic of this inference was: Whatever is determined to be the normal value of rainfall in a place based on other variables, mountains tend to make the rainfall higher. So if Mediterranean climates have light rainfall, the mountains would make the rainfall greater than light. A third problem for the core theory occurred when the "counter evidence" to this was considered. When the subject saw that Italy had rivers, and that accounted for Italy's moderate water supply (by analogy to Egypt), then that decreased the certainty of the inference that Italy's moderate water supply implied moderate precipitation. This inference type has been called a **Functional alternative** meta-inference in Collins (1978), and is quite common (Pearl, 1987). The pattern occurs when there are several variables that independently can influence a dependent variable. The conclusion is not that the original inference was wrong, but that the evidence used to make the inference could be accounted for by other means.

The next portion of the protocol of Subject 1 illustrates a new component of the theory, the formation of a new implication from the water supply variables for Italy and Egypt, and how that knowledge is used to guide his inferences about Louisiana. In the matrix, Louisiana was given as having a subtropical climate, abundant water supply, rivers, and a terrain of lowlands and plains. This protocol reveals that sometime between the earlier protocol, where he reasoned about Egypt and Italy, and this one, he made a generalization that what was true of Egypt and Italy was true of all places.

**S1**: Louisiana ... Precipitation, what is the precipitation? So the places with just a river and very little rainfall were moderate in their fresh water supply, and this is abundant. Now, unfortunately that is a case where I really know that Louisiana has a lot of rainfall. But that would be the nature of my inference, that it at least has a moderate precipitation ... from the fresh water supply.

The generalization from Egypt and Italy is formalized as follows:

Has-river(place) & Precipitation(place) <--->
    Water-supply(place)
Has-river(Egypt) = yes
Precipitation(Egypt) = very light
Water-supply(Egypt) = moderate
Has-river(Italy) = yes
Precipitation(Italy) = light
Water-supply(Italy) = moderate
―――――――――――――――――――
Has-river(place) = yes & Precipitation(place) = light
    <==> Water-supply(place) = moderate

This generalization is one of the new rules added to the theory, described in detail in (Burstein, Collins and Baker, 1991). It is called **Refining a dependency to form an implication** (Table 10, Rule 3). Generalizations like above inference, where an existing dependency is combined with specific examples to form an intermediate statement (the implication), are essentially the analogs in our plausible reasoning theory of the *explanation-based generalizations* described by DeJong (1981) and Mitchell (1983), except that the initial knowledge is in the form of dependencies, not full causal or inferential rules. All of these generalization mechanisms hinge on the combination of general causal or explanatory background knowledge with a new specific case or cases to form a new, potentially more useful general rule. We call the class of such generalizations *refinements*.

In this second protocol, the subject reasons that places with rivers and a little rainfall have moderate water supply, so places with abundant water supply must get more rain. Louisiana's abundant water supply, being greater than both Egypt and Italy's, means that it should have greater precipitation as well. The conclusion was that if Louisiana had an abundant water supply, then it must have "at least moderate" precipitation. In the formalization of this inference, we have described the pattern as predicting simply that it should be "greater than light", light being the corresponding value in the implication. We take these as equivalent with respect to a {low, medium, high} scale. (For precipitation, the term **light** corresponds to the more neutral referent **low**, and **abundant** is the same as **high**.)

Has-river(place) & Precipitation(place) <--+-->
    Water-supply(place)
Has-river(place) = yes & Precipitation(place) = light
    <==> Water-supply(place) = moderate
Has-river(Louisiana) = yes
Water-supply(Louisiana) = abundant (or > moderate)
―――――――――――――――――――
Precipitation(Louisiana) > light (or $\geq$ moderate)

This is an example of a class of inferences that was not explicitly dealt with previously. The issue is reasoning with inequalities on continuous or ordered variables, in conjunction with dependencies between those types of variables. These inferences all depend on the presence of a *specified* dependency: specified dependencies are those labeled with a + or - to indicate that an increase/decrease in the values for a term on one side has a corresponding positive or negative effect on the other. However, by extending the formalism to include explicit orderings on referents and allowing all inferences that contain statements of the form d(a) = r to also allow the = to be replaced by any of <, >, $\leq$, and $\geq$, we were able to rewrite many of the original rules to cover this kind of inference (Burstein, Collins and Baker, 1991). For example, in a SIM-based argument transform, we would rewrite the rule as:

$d(a_1) \sim r$    where $\sim$ is one of =, < , >, $\leq$, or $\geq$.
$a_2$ SIM $a_1$ in CX (A, D)
D(A) <-----> d(A)
$a_1, a_2$ SPEC A
―――――――――――――――――――
$d(a_2) \sim r$

This reformulation to introduce inequalities also yields some new derivation, transformation and generalization rules. All of the new inferences involve specified dependencies (dependencies of the form $d_1(A)$<--+-->$d_2(A)$ or $d_1(A)$<--⁻-->$d_2(A)$), like the inference about precipitation above. The new rules for *creating* inequalities with SIM-based transforms on dependencies and implications are much like the SIM-based referent transform rules described in Collins and Michalski (1989). In the new rules for generating inequalities from directed dependencies, the two arguments $a_1$ and $a_2$ are related by an inequality instead of by similarity (as by $d_1(a_1) < d_1(a_2)$). For example, the relationship between latitude and temperature range might be inferred as follows:

latitude(Alaska) = high : $\gamma$ = hi, $\mu_r$ = lo, $\mu_a$ = hi
temp-range(Alaska) = cold: $\gamma$ = hi, $\mu_r$ = lo, $\mu_a$ = hi
latitude(Equador) = low: $\gamma$ = hi, $\mu_r$ = lo, $\mu_a$ = hi
temp-range(Equador) = hot: $\gamma$ = hi, $\mu_r$ = lo, $\mu_a$ = hi
Equador, Alaska SPEC place: $\gamma_i$ = hi, $\tau_i$ = lo, $\delta_i$ = lo
hot > cold: $\gamma$ = high
low < high: $\gamma$ = high

---

latitude(place) <--$^{-}$--> temp-range(place)
: $\gamma$ = low, $\alpha$ = moderate, $\beta$ = low

## Open Issues

The revisions to the core theory of human plausible inference developed by Collins and Michalski (1989), as described here and in (Burstein, Collins and Baker, 1991), were part of our continuing attempt to formalize the plausible inferences observed in people's answers to questions for which they do not have ready answers. The recent work was motivated in large part by a protocol experiment that was designed to bring forth more clearly people's uses of multiple sources of evidence in forming plausible conclusions. We have shown how this experimental evidence supported the introduction of a mechanism for reasoning about inequalities, and clarified the need for an integrated theory of generalization to accompany the core plausible reasoning theory.

There are a number of other issues apparent to us in the experiment and earlier protocols that we have not yet addressed. We think they are amenable to the kind of analysis we have been using, but the solutions were not as apparent. We list some of them here:

- Reasoning with complex dependencies and implications. Subjects frequently reasoned backwards over dependencies where a number of variables (e.g. precipitation and rivers) affected a particular variable (e.g. water supply). Sometimes these were treated as disjunctive variables, sometimes as conjunctive (neither alone was predictive), and sometimes as if their effects were additive. This may require several alternative mechanisms be introduced into the model, based on the form of the causal knowledge used.

- The tradeoff between range and certainty. Subjects appear to trade off certainty about a belief against the range of the referent. For example, one might be very certain that the average rainfall in Louisiana is "at least moderate," somewhat less certain that it is "heavy," still less certain that it is between 40 and 60 inches a year.

- Merging of qualitative and quantitative reasoning. Sometimes subjects bring in various quantitative relationships to guide their qualitative reasoning (e.g. the Amazon jungle averages 85° or 1 mile in altitude affects temperature as much as 800 miles in latitude).

- The extent parameter. Collins (1978) identified a parameter he called "extent" which was particularly prevalent in temporal and spatial inferences. It is necessary because people have a notion of how far rainstorms vs. parades vs. continents extend in space, and how long they extend in time. This notion is central to people's reasoning about space and time, and features of categories.

## References

Baker, M., Burstein, M. H., and Collins, A. 1987. Implementing a model of human plausible reasoning. In *Proceedings of the Tenth International Joint Conference of Artificial Intelligence, 1*, 185-188. Morgan Kaufmann.

Burstein, M. H., Collins, A. and Baker, M., 1991. Plausible Generalization: Extending a Model of Human Plausible Reasoning. *Journal of the Learning Sciences.* (Forthcoming.)

Collins, A. 1978. Fragments of a theory of human plausible reasoning. In D. Waltz ed., *Theoretical Issues in Natural Language Processing II* 194-201. Urbana, IL: University of Illinois.

Collins, A. and Michalski, R. S. 1989. The logic of plausible reasoning: A core theory. *Cognitive Science*, 13, 1-49.

DeJong, G. and Mooney, R. 1986. Explanation-Based Learning: An Alternative View. *Machine Learning* 1:145-176.

Mitchell, T. M., Keller, R. M. and Kedar-Cabelli, S. T. 1986. Explanation-Based Generalization: A Unifying View. *Machine Learning* 1:1.

Pearl, J. 1987. Embracing causality in formal reasoning. In *Proceedings of the Sixth National Conference on Artificial Intelligence* 369-373. Morgan Kaufmann.

Russell, S. J. 1986. *Analogical and Inductive Inference.* Ph.D. thesis, Dept. of Computer Science, Stanford University.

## Acknowledgements

We would like to thank Michelle Baker for her contributions to this work and her comments on an earlier draft of this paper.

This research was supported by the Army Research Institute under Contract No. MDA 903-85-C-0411.

# Learning Strategic Concepts from Experience: A Seven-Stage Process

Michael Freed
Northwestern University
The Institute for the Learning Sciences
Evanston, Illinois 60201

## Abstract

One way novices improve their skill is by learning not to repeat mistakes. Often this requires learning entirely new concepts which must be operationalized for use in plans. We model this learning process in seven stages, starting with the generation of expectations which, when proven faulty, invoke mechanisms to modify decision-making mechanisms in order to prevent the failure from occurring again. This process is demonstrated in the context of our testbed system which learns new rules for detecting threats, formulating counterplans and other cognitive tasks. It is shown how this process may be used to learn the concept of *immobility* as it occurs in the domain of chess.

## Novice Acquisition of Strategic Concepts

Some of the concepts needed to successfully practice a skill will be unknown to a novice when beginning to learn the skill. A novice basketball player, for instance, will be made aware of the concept of *scoring a basket* in the course of learning the rules of the game. However, the concept of a *fake*, in which a player pretends to go one way but actually goes another, may not be evident unless the player is told about it or experiences it first hand[1]. Novices advance their level of skill by discovering such concepts and using them in plans.

One way human novices improve their abilities is by learning not to repeat mistakes. A novice basketball player with the job of preventing an opponent with the ball from moving past him might believe that his opponent's leftward glance indicates imminent travel in that direction. In anticipation, the novice player may go left to block his opponent's motion. However, the opponent may look left and then go right, employing a fake to get past. The novice's mistake stems from an otherwise plausible assumption: namely that the direction of an opponent's gaze is predictive of the opponent's direction of motion. This assumption makes the novice vulnerable to a fake.

### An approach to learning from failures

We have been exploring an approach to these problems in a system that learns to improve its chess playing abilities (see, e.g., Collins, Birnbaum and Krulwich, 1989; Birnbaum, Collins, Freed and Krulwich, 1990). The system learns when expectations about events in a chess game prove faulty. For example, the system expects to be able to block all capture-threats to pieces. If the opponent captures a piece, the expectation fails. Expectations are justified by a set of assumptions about what can and cannot happen in the domain and about what the system will do to insure that expectations succeed. Often, an expectation will fail because an opponent employs some tactic which the system was unaware of and which was therefore not accounted for by the assumptions underlying the expectation.

An expectation's justification makes assumptions not only about the 'physics' of the domain but also about the decision-making process employed by the planner (Birnbaum, Collins, Freed, Krulwich, 1990). For instance, the expectation that all threats against pieces will be blocked is justified in part by the assumption that all threats will be detected. Decision-making in our system is performed by a set of components, each designed to perform some domain-independent task such as detecting threats or formulating counterplans. Components are implemented as sets of rules which provide different methods for achieving the component's purpose. Learning to improve performance involves adding to or altering the set of rules in one or more of these components.

## Case Study: The Pin

In the course of learning chess, novice players are almost invariably victimized by *the pin* which is a configuration of pieces achieved by positioning one piece

[1]Concepts may also be imported from other domains (see e.g. Collins and Birnbaum, 1988). The fake, for instance, may be learned from experience with almost any sport.

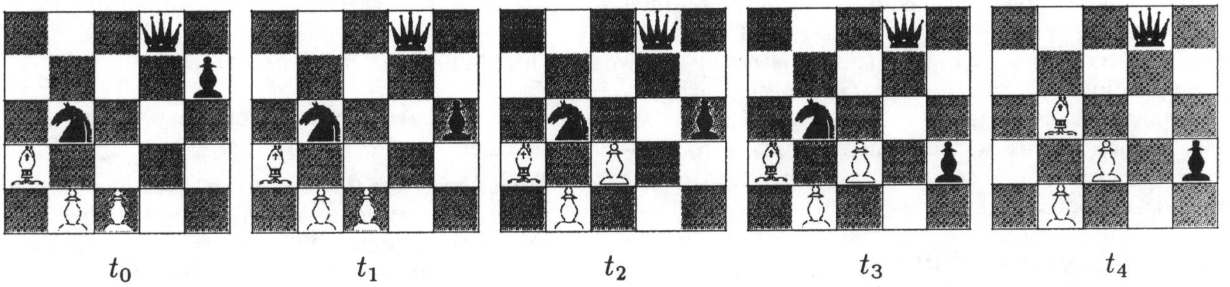

Figure 1: The pin: computer (black) to move at time $t_0$

on a line of attack that includes two opposing pieces so that the closer of the two pieces cannot be moved without exposing the other piece to the threat of imminent capture. In figure 1 at $t_0$, white's bishop is pinning black's knight; moving the knight would allow black's queen to be captured. In effect, the knight is immobilized — not because the rules of the game prevent movement, but because the consequent loss of the queen is undesirable. The opponent can exploit this situation in several ways; the simplest is to attack the immobilized knight. The player is then left in the position of choosing between leaving the pinned knight in place and allowing it to be captured, or moving it away, thereby permitting capture of the queen.

Not knowing about the pin tactic or how a pin might be exploited by the opponent, the computer might decide to advance its pawn at $t_1$ reasoning that there are no particularly dangerous threats or appealing opportunites and that advancing the pawn might pay off in the future. In coming to this conclusion the system looks at several moves including using the knight to capture the rightmost pawn. This move is rejected because it would expose the queen to attack from the opponent's bishop. In rejecting the pawn capture, the computer has noted the role of its knight in protecting the queen and has effectively committed itself to keeping the knight immobile. However, it does not yet construe this situation as a threat or have even implicit knowledge of the tactical concept of a pin. The opponent is in a position to exploit this situation by advancing its pawn to attack the knight. It does so after $t_1$, leaving the system in a *double bind* — i.e. a situation in which all plans to prevent one loss are mutually exclusive with plans to prevent the other loss. Since the queen is more valuable the computer decides to sacrifice the knight.

## Learning a New Strategic Concept

In our model of the learning process, a strategic concept is discovered and operationalized following the failure of an expectation (see, e.g. Birnbaum, Collins, Krulwich 1989; Birnbaum, Collins, Freed, Krulwich, 1990). This process can be decomposed into a series of steps as summarized in figure 2, each of which will be described following an abbreviated account of how this process is used to learn about pins.

Before encountering the scenario of figure 1, the system prepares to learn by deciding what to expect and how often to check whether its expectations are true. In particular, it decides to have the expecation that all capture threats against its pieces will be blocked and to monitor this expectation by checking it after every opponent move. When the opponent takes the computer's knight after $t_3$, the computer's expectation monitor signals a failure immediately afterward at $t_4$. The signalled failure invokes diagnostic machinery which examines the computer's last few moves in search of an alternative move which would have avoided the loss of the knight (or anything worse). It discovers that advancing its pawn at $t_0$ was a mistake; if it had moved its queen out of the bishops's path at $t_0$, no pieces would have been captured.

Having identified a better action, the system uses its explicit model of the decision-making process to determine which component must be modified to allow the

1. Generate expectation
2. Note an expectation-failure
3. Determine what action or inaction caused the failure
4. Determine what bad underlying decision(s) led to the faulty action
5. Generate one or more candidate modifications for preventing recurrence of the failure
6. Select the best modification
7. Inductively merge with similar modifications

Figure 2: Seven-stage failure-driven learning process

planner to make the better move in similar circumstances in the future. Put another way, the diagnostic machinery determines which component of the planner is to blame for making the wrong move. The underlying cause is found to be the threat-detection component's failure to construe a certain configuration of pieces (the pin) as a threat. To prevent this failure in the future, a rule which identifies such a configuration as a threat is generated and then added to the threat-detection component[2].

## Generating expectations

Expectations are explicitly represented beliefs about what will or will not occur. When an expectation fails, i.e. when the belief is demonstrably in error, the system invokes learning machinery to prevent recurrence of the failure[3]. For novices, it is worth asking how the initial set of expectations is arrived at. One way to learn which expectations to monitor is to be told. Since all initial expectations are entered by hand, our system learns what to monitor in essentially this way. Another way is to generate them automatically (Doyle, Atkinson and Doshi, 1986). Automatic generation of expectations should accompany the generation of new goals[4]. For example, in learning to detect and avoid pins, the system should also learn to *expect* to avoid being pinned. The new expectation provides a basis for future learning.

## Notice an expectation failure

Having generated a set of expectations, the system must devote resources to monitoring these expectations and detecting when they fail. If the system has a lot of expectations and checks them frequently, looking for failures can be very costly. Therefore decisions must be made about which expectations to monitor and when.

Many factors constrain the set of expectations which can profitably be monitored at any particular time. Clearly irrelevant expectations such as those used exclusively in one domain — say checkers — can not be profitably monitored while engaged in another domain such as chess. Even within a domain, the best time to monitor an expecation will vary. For instance, the expectation in our chess-playing planner that all capture-threats will be blocked is monitored by checking it after each opponent move; an expectation that a plan will prove workable should be monitored during the plan evaluation process.

## Determining what action or inaction caused the failure

When an expectation-failure is detected, the learning machinery is invoked to diagnose the problem and repair the system. The first step in diagnosis is determining which executed action led to the failure. Locating this action is a matter of searching backward from the time of the failure for a time when the system had an alternative which, if taken, would have prevented the failure and would not have led to an equally bad or worse failure (cf. Minton, 1985).

Let us consider the pin example once again. At $t_4$ the expectation failure is noticed and the search for a faulty action begins. The search proceeds by examining the system's opportunites to act in reverse temporal order, starting with the time of the failure. The most recent opportunity to act before $t_4$ occurred at $t_2$, when the planner advanced its pawn. At this point, the system could have saved the knight and prevented the failure by moving it away from the attacking bishop and pawn, but this move would have been a poor choice since it would have exposed the queen and led to the even worse failure. At $t_2$, the computer was in a *double bind* — nothing could have prevented the loss of a piece.

The search continues until an opportunity to prevent the failure is found in which no double bind would negate the value of preventive action. At $t_0$, the system could have prevented the failure by moving its queen 2 spaces to the left, thereby defusing the pin and maintaining the guard on its knight. This would have prevented *any* expectation failure from occurring; therefore the action taken at this time is blamed for the failure.

This method of locating failure-producing actions is adequate for the simple turn-taking games we are studying. Similar methods have been proposed by Minton (1985) for learning about the *fork* and by Utgoff (1983) for constraint back-propogation.

## Determining which component caused the failure

Every action in our system is the product of decisions made by components of the system's planner. One component is responsible for detecting threats and opportunities to achieve the system's goals. Another is responsible for formulating plans, another for comparing candidate plans to determine which is best. Our current formulation of the planning process includes 22 classes of decision and therefore 22 distinct planning components (Freed, 1991). When a faulty action is made, one or more of these components is responsible for the failure and should be modified to prevent

[2] Actually, many components besides the threat-detection component should be modified as a result of learning about pins. For example the counterplanning component will be modified so that detected pin-threats can be avoided. Other modifications cause pin-avoidance plans to be ranked against other plans, pinning plans to be employed tactically against opponents and expectations that the computer will not get pinned to be generated. The need to separate planning knowledge among components in this way is discussed in (Collins, Birnbaum, Krulwich, Freed, 1991)

[3] Alternately, the system may learn that an expectation is unrealistic in some respect and modify it (Krulwich, Birnbaum, Collins, 1988).

[4] Expectations are generated by inferentially determining which observable events indicate the success or failure of achievable goals.

$$\begin{aligned}
&\text{has-legal-path}(p_1\ \text{loc}(p_1)\ \text{loc}(p_2)\ path_A)\\
\wedge\ &\text{has-legal-path}(p_1\ \text{loc}(p_1)\ \text{loc}(p_3)\ path_B)\\
\wedge\ &\text{has-legal-path}(p_4\ \text{loc}(p_4)\ sq\ path_C)\\
\wedge\ &\text{has-legal-path}(p_4\ sq\ \text{loc}(p_2)\ path_D)\\
\wedge\ &\text{blocks-path}(p_2\ path_B)\\
\wedge\ &\neg\ \text{blocked}(path_A)\\
\wedge\ &\neg\ \text{blocked}(path_C)\\
\wedge\ &\neg\ \text{blocked}(path_D)\\
\wedge\ &\text{I-own}(p_2)\\
\wedge\ &\text{I-own}(p_3)\\
\wedge\ &\neg\ \text{I-own}(p_1)\\
\wedge\ &\neg\ \text{I-own}(p_4)\\
&\longrightarrow \text{isa-threat}(<\ldots>)
\end{aligned}$$

Figure 3: Rule for detecting pin-threats against $p_2$

recurrence of the the faulty behavior. See (Birnbaum, Collins, Freed, Krulwich, 1990) for a description of the diagnostic process used by our system to determine which component is responsible for a faulty action.

In some cases, no single component can be held solely accountable for the failure. Diagnosing the failure in these cases means locating a (preferably minimal) set of components which may be modified to prevent recurrence of the failure. As the system was unaware of the pin tactic, many components require modification in order to defend against it and employ it against other agents. The component responsible for threat detection should be modified to detect the pin; the component for counerplanning must be able to generate a response to a pin-threat if one is has been detected. Learning and operationalizing knowledge of the pin entails modifying each component. For simplicity we will only consider modifications to the threat-detection component.

## Generating candidate modifications

Once the cause of the failure has been traced to a faulty component, an explanation-based learning algorithm (cf. Mitchell, Keller and Kedar-Cabelli, 1986; DeJong and Mooney, 1986) is used to construct a modification to the component which will prevent recurrence of the failure. Modifying the threat-detection component to detect pins entails constructing a rule to signal a threat when the opponent is in a position to pin one of the computer's pieces.

Figure 3 shows a threat-detection rule which, if added to the rule set employed by the threat-detection component, will prevent recurrence of the piece-loss seen in the example. It should be noted that the rule detects a more specific class of threats than would be expected given a full understanding of the danger of being pinned; only pins which the opponent can exploit by attacking the pinned piece are detected. The process used by our system to produce component modifications is described by Krulwich (1991).

As it stands, the system learns to detect an overly specific class of threats because its single experience being pinned could not illustrate other ways being pinned could be costly. We will discuss how the system might learn the more general rule below.

## Choosing the best modification

The learning element will frequently be able to generate several modifications that vary in effectiveness and cost. To see why this would be the case, consider the following informal example: a person breaks a traffic law — say by making a U-turn in an intersection — and consequently collides with another car. Attempting to prevent future collisions, the driver could decide: to be more careful when breaking that law in the future, not to break that law at all, not to break traffic laws in general or simply not to drive because it is too dangerous. In deciding which lesson to learn, the driver must reason about the tradeoff between effectiveness (at reducing the likelihood of future accidents) and cost (inconvenience).

In deciding which rule to learn, the system should consider such factors as generality and computational expense. The more general rules will typically be more expensive to use. For example, the threat-detection rule of figure 3 is more general and more expensive than a similar rule which only considers pins arising from bishops. Such a rule would prevent a recurrence of the exact failure seen in the example (in which the pinning piece was a bishop), but fail to detect pins made by rooks and queens. The more specific rule is cheaper to use because it only has to be checked on bishops.

Extensive knowledge of the domain may be used to predict what level of generality optimizes the tradeoff between effectiveness and computational expense. For the novice, such knowledge will normally be unavailable; however, conservative guesses about this level can be adjusted in light of new experience by inductively generalizing similar modifications (cf. Pazzani 1989).

## Merging similar modifications

The rule for detecting pins illustrated in figure 3 detects too narrow a range of threats. In effect, it assumes that the only pins worth responding to are those which the opponent can exploit by attacking the pinned piece. However, the mobility-limit produced by the pin can have other costly consequences. For example, the computer will be unable to move a pinned piece to capture an opponent's piece without exposing itself to attack. Essentially any chess goal which can be served by moving a piece is compromised if the piece is pinned.

Although the necessary mechanisms are not yet implemented in our system, I will discuss how the more general rule may be produced after a second experience of being pinned. Briefly, this second experience might take the following form:

One of the system's general expectations is that it will always be able to capture unguarded pieces. If the attack piece is pinned, the attack would have to be aborted, causing the expectation to fail. A learning episode, similar to the initial pin example, would then occur. The new threat-detection rule would be very similar to the initial pin rule, and could then be combined with it to yield a new rule, detecting all pins.

The crucial elements of this step are methods for identifying similar rules and for merging them (cf. Minton, 1988). For each of these steps we assume a concept called *reasons-to-move* for which *avoid-capture* and *capture-opponent-piece* are specializations. Associated with the latter concepts are abstract descriptions of the chessboard positions which enable them, represented in the same language as the rules employed by the detection component.

To notice the similarity of the two detection rules, the system must first label a subset of the propositions in the first rule (figure 3) as equivalent to the enabling conditions of *avoid-capture* and similarly label a portion of the second rule as equivalent to the enabling conditions for *capture-opponent-piece.* Treating these labels and the unlabelled propostions in each rule as features, a simple inductive generalization algorithm (e.g. Winston 1970) can be used to produce a new rule which includes only the common features of the first two. The new rule detects all pins.

## Conclusion

We have presented a seven-stage process for discovering strategic concepts from expectation failures and constructing methods for utilizing this knowledge. We have demonstrated this within the context of our testbed system which learns new rules for detecting threats, counterplanning and other cognitive tasks. Future work will expand the more speculative stages of this process and elaborate the functional decomposition of planning tasks employed by the system.

## Acknolwedgements

The ideas presented here were developed in collaboration with Gregg Collins, Larry Birnbaum and Bruce Krulwich. Additional thanks go to Matt Brand, Robin Burke, Eric Domeshek, Bill Ferguson and Louise Pryor for comments on this paper and discussions on the research presented. This work was supported in part by the Office of Naval Research under contract N00014-89-J-3217, and by the Defense Advanced Research Projects Agency, monitored by the Air Force of Scientific Research under contract F49620-88-C-0058. The Institute for the Learning Sciences was established in 1989 with the support of Andersen Consulting, part of the Arthur Andersen Worldwide Organization.

## References

Birnbaum, L., and Collins, G. 1988. The transfer of experience across planning domains throught the acquisition of abstract strategies. *Proceedings of the 1988 Case-Based Reasoning Workshop*

Birnbaum, L., Collins, G., and Krulwich, B. 1989. Issues in the justification-based diagnosis of planning failures. *Proceedings of the Sixth International Workshop on Machine Learning,* Ithaca, NY, pp. 194-196.

Birnbaum, L., Collins, G., Freed, M., and Krulwich, B. 1990. Model-based diagnosis of planning failures. *Proceedings of the Eighth National Conference on Artificial Intelligence,* Boston, MA, pp. 318-323.

Collins, G., Birnbaum, L., and Krulwich, B. 1989. An adaptive model of decision-making in planning. *Proceedings of the Eleventh IJCAI,* Detroit, MI, pp. 511-516.

Collins, G, Birnbaum, L., Krulwich, B., and Freed, M. 1991. Plan debugging in an intentional system. Accepted for publication in the *Proceedings of Thirteenth IJCAI.*

DeJong, G., and Mooney, R. 1986. Explanation-based learning: An alternative view. *Machine Learning,* vol. 1, pp. 145-176.

Doyle, R., Atkinson, D., and Doshi, R. 1986. Generating perception requests and expectations to verify the execution of plans. *Proceedings of the Fifth National Conference on Artificial Intelligence,* pp. 81-88.

Freed, M. 1991. A Taxonomy of Planning and Learning Decisions. Unpublished manuscript.

Krulwich, B. 1991. Determining appropriate EBL target concepts in a multi-component planning system. Accepted for publication in the *Proceedings of the Thirteenth Annual Conference of the Cognitive Science Society.*

Minton, S. 1985. *A Game-Playing Program that Learns by Analyzing Examples* Technical report CMU-CS-85-130, Carnegie Mellon University School of Computer Science, Pittsburgh, PA.

Minton, S. 1988. *Learning effective search control knowledge: an explanation-based approach.* Technical Report CMU-CS-88-133, Carnegie Mellon University School of Computer Science, Pittsburgh, PA.

Mitchell, T., Keller, R., and Kedar-Cabelli, S. 1986. Explanation-based generalization: A unifying view. *Machine Learning,* vol. 1, pp. 47-80.

Utgoff, P. 1983. Adjusting bias in concept learning. *Proceedings International Machine Learning Workshop.*

Winston, P.H. 1970. Learning structural descriptions form examples. Reps No. TR-231, AI Laboratory, Massachusetts Institute of Technology. (Reprinted in P.H. Winston (Ed.). 1975. *The psychology of computer vision.* New York: McGraw-Hill, 157-209)

# Modeling the Self-explanation Effect with Cascade 3

**Kurt VanLehn, Randolph M. Jones, & M. T. H. Chi**

Learning Research and Development Center
University of Pittsburgh
Pittsburgh, PA 15260
VanLehn@cs.pitt.edu, Jones@cs.pitt.edu, Chi@vms.cis.pitt.edu

**Abstract**

**Several investigations have found that students learn more when they explain examples to themselves while studying them. Moreover, they refer less often to the examples while solving problems, and they read less of the example each time they refer to it. These findings, collectively called the self-explanation effect, have been reproduced by our cognitive simulation program, Cascade. Cascade has two kinds of learning. It learns new rules of physics (the task domain used in the human data modeled) by resolving impasses with reasoning based on overly-general, non-domain knowledge. It acquires procedural competence by storing its derivations of problem solutions and using them as analogs to guide its search for solutions to novel problems. This paper discusses several runs of Cascade wherein the strategies for explaining examples is varied and the initial domain knowledge is held constant. These computational experiments demonstrate the computational sufficiency of a strategy-based account for the self-explanation effect.**

## Introduction

The long-term goal of the Cascade project is to develop a model of knowledge acquisition in scientific task domains. The short-term goal is to model the self-explanation effect. Cascade has been tested using data from a study by Chi, Bassok, Lewis, Reimann and Glaser (1989) who studied students learning classical particle dynamics, the first topic in a typical first-year college physics course. The 8 subjects studied the first three chapters of college textbook, then read the prose part of a chapter on Newton's laws. They took a test on their understanding of the chapter, then studied 3 worked examples and solved 25 problems. Protocols were taken as they studied the examples and solved the problems. On the basis of the scores on problem solving, the 8 subjects were divided two groups. The 4 students with the highest scores were called the Good solvers; the others were called Poor solvers. Since the students in both groups scored the same on pre-tests, the Good solvers seem to have learned more during the experiment (but see discussion below). Using protocol analysis, Chi et al. attempted to find out how the Good solvers managed to learn more than the Poor solvers from the same material. They found five differences:

1. The Good solvers uttered more self-explanations as they studied examples, whereas the Poor solvers' comments were mostly paraphrases of the examples' statements.
2. All students commented frequently on whether they understood what they had just read. The Good solvers tended to say that they did not understand what they had just read, whereas the Poor solvers tended to say that they did understand. Since the Poor solvers scores show that they understood less than the Good solvers, their self-monitoring was less accurate than the Good solvers'.
3. During problem solving, the Poor solvers tended to refer back to the examples more often than the Good solvers.
4. When the Good solvers referred to the examples, they read fewer lines then the Poor solvers. The Poor solvers tended to start at the beginning of the example and read until they found a useful line, whereas the Good solvers started reading in the middle of the example and read only one line.

The first two findings have also been observed in similar studies of students learning Lisp (Pirolli & Bielaczyc, 1989) and electrodynamics (Fergusson-Hessler & de Jong, 1990). This cluster of findings is called the self-explanation effect.

## Hypothesized sources of the self-explanation effect

Cascade is based on the hypothesis that the self-explanation effect is caused by knowledge-level learning that occurs as the students explain examples and solve problems. This section introduces the hypothesis by first examining competing hypotheses.

A plausible hypothesis is that the two groups of students had accumulated different knowledge of physics just prior to studying the examples. This difference might be due to prior exposure to physics or to reading the text of the chapter more carefully. The ones who had more prior knowledge solved more problems correctly, and thus were classified as Good solvers. Under this prior-knowledge hypothesis, all subjects try to explain the text and the example lines, but those who already know a lot are better able to explain the example and so produce more self-explanations (finding 1 in the list above). Moreover, because they produce more derivations during example processing, they use fewer (finding 3) and more economical (finding 4) references during analogical problem solving. Thus, the prior-knowledge hypothesis is consistent with 3 of the 4 findings. There are, however, three sets of evidence against the prior knowledge hypothesis.

1. After reading the text of the target chapter, the students in the Chi et al. study took a test on their knowledge of Newton's laws. The mean scores of the Good and Poor students on this test were exactly the same (Chi et al., 1989). This suggest that

both groups of students had roughly the same prior knowledge.

2. Chi and VanLehn (1991) conducted a fine-grained analysis of all 258 self-explanations in the protocols, reducing each to a set of propositions. For each proposition, they attempted to determine whether it was inferred (a) from the example line, (b) from common sense knowledge, (c) from knowledge acquired from previous example lines, or (d) from the text. The first three categories are considered non-text sources. Of the propositions whose source could be determined, 68.5% were inferred from non-text sources. More importantly for the present argument, this proportion was the same for both Good and Poor students. If the Good students had more prior knowledge, more of their propositions would be encoded as coming from the text. Thus, this result is inconsistent with the prior knowledge hypothesis.
3. The prior-knowledge hypothesis predicts that Poor subjects would utter more negative self-monitoring statements because they more often fail to explain a line. In fact, they utter fewer negative self-monitoring statements.

Although it is unlikely that all 9 students had exactly the same prior knowledge, the above difficulties indicate that variations in prior knowledge cannot be the sole source of the self-explanation effect. There must be some kind of learning going on.

Because the subjects are explaining examples, a plausible type of learning is EBL (Mitchell, Keller & Smadar-Cabelli, 1986). EBL is symbol-level learning (Dietterich, 1986), in that all the knowledge is assumed to be present in some form before the learning begins. Learning consists of making the knowledge more efficiently usable. EBL, knowledge compilation (Anderson, 1983) and chunking (Newell, 1990) are all instances of symbol-level learning. However, the hypothesis that self-explanation is caused by symbol-level learning has two difficulties.

1. When the subjects took an untimed test on the content of the text, their mean score was only 5.5 out of a possible 12. Moreover, after studying the examples and solving the problems, the Good student's score increased to 8.5. This suggests that students did not learn much physics from studying the text, and that some more physics was learned by studying the examples and working the problems.
2. The text does not contain all the information needed by the subjects to explain the examples or solve the problems. To quantify this inadequacy, an extensive task analysis and simulation conducted with the aid of Bernadette Kowalski and William Ball. Starting with the task analyses of Bundy, Byrd, Luger, Mellish and Palmer (1979) and Larkin (1983), we developed a set of rules and a representation of physics problems that was simple and yet sufficient for solving all but 2 of the 25 problems in the Chi study. (Solving the 2 problems would require a type of mathematical reasoning that we did not bother to implement). During this time, extensive, albeit informal, analyses of the Chi protocols were conducted in an effort to align the proposed knowledge representations with the subject's comments. The resulting target knowledge base contained 62 physics rules. Next, two people who were not involved in the development of the target knowledge base were asked to judge each rule and determine whether it was mentioned anywhere in the textbook prior to the examples. There was 95% agreement between the judges. Disagreements were settled by a third judge. Of the 62 rules in the target knowledge base, 29 (47%) were judged to be present in the text prior to studying the examples. Thus, more than half the knowledge required for explaining the examples and solving the problems is not presented in the text, and presumably is not known by the subjects prior to explaining the examples and solving the problems.

These results suggest that the major prerequisite of symbol-level learning is not met, for the students did not seem to have complete knowledge before the example studying and problem solving began. Thus, some kind of knowledge-level learning must be going on during the explanation of examples and the solving of problems.

Because the examples contain more information than the problems, a plausible hypothesis is that all knowledge-level learning occurs during the explanation of examples. Using the 33 rules that did not occur in the text, we estimated that only 11 of the rules were used during the examples. The other 22 were first used during the problems. This suggests that two-thirds of the rules are acquired during problem solving. Thus, it appears that some kind of knowledge-level learning is going on during both example explaining and problem solving. This is the hypothesis upon which Cascade is based.

The hardest technical challenge is to find a knowledge-level learning method that can learn correct rules during problem solving. Learning during problem solving is harder than learning during example explaining, because the examples provide partial description of their solution paths which allows the program to do less guessing than it must do when learning during problem solving. Because students often refer to examples as they solve problems, we assume that they are using the example's solutions to constrain their generation of the problems' solutions, and this in turn facilitates correct learning during problem solving. We hypothesize that knowledge-level learning takes place in the context of analogical problem solving and example explaining.

Since the Poor students learn fewer rules during analogical problem solving than the Good students, and the Poor students generated fewer self-explanations during example studying, we hypothesize that there is something about explaining the examples that causes analogical problem solving to be more effective. It generates solution paths that are more often correct and this in turn establishes a better context for knowledge-level learning. We built Cascade order to work out the interactions between example explaining, analogical problem solving and learning, and thus provide a test of the computational and empirical sufficiency of our hypotheses.

## The Cascade model

Cascade models two basic activities: explaining examples and solving problems. Knowledge-level learning goes on during both. Because the type of physics problems used in Chi et al.'s study involve only monotonic reasoning in a single state, Cascade uses a rule-based, backchaining theorem prover (similar to Prolog) to implement both activities. A physics example is presented to Cascade as a set of facts representing the givens of the problem and a list of propositions representing the force diagram and the lines of the problem's solution. Cascade explains each proposition by proving that it follows from the givens and the preceding propositions. To solve a problem, Cascade is presented with facts representing the problem's givens and is asked to prove a proposition involving the quantities the problem seeks. For instance, the translation of, "What is the tension of string A?" is "Prove the proposition value(tension(stringA), X)." In the process of proving the proposition, Cascade derives a value for the variable X, thus solving the problem. Although this model of problem solving and example explaining is clearly too simple to cover all task domains, it suffices for physics and other task domains dominated by monotonic reasoning.

Cascade includes two kinds of analogical problem solving. One kind of analogy is used when Cascade has multiple rules that can be applied to achieve a goal and it does not know which one to choose. To get advice, it refers to its derivation of the examples' lines, which are stored as the examples are explained. (A derivation is represented by a set of pairs, each containing a goal from the proof of a line and the rule used to achieve that goal.) This type of analogy begins by retrieving an example (retrieval is currently not modeled in a psychologically plausible way), establishing a mapping between the example's givens and the current problem's givens, then using the mapping to see if the example's derivation has a goal that is equivalent to the goal that it is currently worried about. If it finds an equivalent goal, the rule that achieves that goal is chosen for attempting to prove the worrisome goal. This type of analogy is called analogical search control, because it uses the example as a source of advice on which of several alternative to try first. For instance, a student might say, "I cannot tell whether I should project this onto the x-axis or the y-axis. At an analogous point in the example, they projected onto the x-axis, so I'll try that too." Analogical search control is also used in the Eureka (Jones, 1989).

The second type of analogy is used when Cascade cannot find a rule that will apply to the current goal. Here, it tries to find a line in an old example that it can convert into an appropriate rule. It begins just like analogical search control by retrieving an example and forming a mapping between the givens of the example and the problem. Next it now looks for a line in the example's solution that mentions the current goal (or rather, a goal equivalent to the current goal under the mapping). Most lines are equations, so it is simple to convert a line to a temporary rule which can then be used to try to achieve the goal. For instance, a student might say, "I need some way to get the tension of string A. The example has a line saying that string 1's tension is mg sin 30. Those two strings are analogous, and 30 degrees is analogous to 45 degrees in this problem, so I bet that the tension of string A is mg sin 45 degrees." This type of analogy is called transformational analogy, after a similar method explored by Carbonell (1986). As Carbonell discovered, transformational analogies often yield wrong answers.

Cascade's mechanism for learning at the knowledge level is called explanation-based learning of correctness or EBLC (VanLehn, Ball & Kowalski, 1990). The basic idea is divide knowledge into domain knowledge and non-domain knowledge. Domain knowledge contains rules that are believed to be correct and appropriate for the task domain. Non-domain knowledge contains rules that are believed to be incorrect or relevant only to other task domains. The most important non-domain rules for learning are overly general rules. The basic process of EBLC is to use overly general rules whenever domain rules fail, then to save the particular usage of that rule if its use turns out to be correct. EBLC begins when Cascade reaches a knowledge-level impasse. A knowledge-level impasse occurs when there is no domain rule for achieving a goal and there is no successful alternative solution path that uses only domain rules (VanLehn & Jones, in press, describe how such impasses are detected). To resolve a knowledge-level impasse, Cascade tries to use non-domain rules, such as, "If an object has a part, then the property values of the part and the whole are the same." If the use of such non-domain rules ultimately leads to a successful explanation of an example line or a successful solution to a problem, then Cascade forms a new domain rule that is a specialization of the overly general one. The specialization is chosen so that it is also a generalization of the particular usage. For instance, on one problem Cascade could not determine the pressure in a part of a container even though it knew the pressure in the whole. Since there was no alternative solution to the problem, Cascade was at a knowledge-level impasse. It used the overly-general rule just mentioned, which ultimately led to a solution of the problem. Cascade then formed a new domain rule, "If a container has a part then the pressure in the part is equal to the pressure in the whole."

From a machine learning point of view, Cascade does both knowledge-level learning (via EBLC) and symbol-level learning (via analogical search control). Analogical search control is a form of learning, because each time an example is explained or a problem is solved, the system gains another derivation that can be used by analogical search control. However, analogical search control does not change the set of problems solvable with infinite resources. It only changes the efficiency of the search. Thus, it is a symbol-level learning mechanism.

Cascade's learning is similar to those proposed by existing theories of skill acquisition. We believe that analogical search control can eventually provide an account for the practice effects usually explained by chunking (Newell, 1990), knowledge compilation (Anderson, 1983), and other also symbol-level learning mechanisms. EBLC is similar to proposals by Schank (1986), Lewis (1988), Anderson (1990) and others, which also acquire new knowledge by specializing existing, overly general knowledge. Although all these models of skill acquisition are similar in spirit, they differ in significant ways. For more on the Cascade system and a detailed comparison with its predecessors, see VanLehn and Jones (in press).

## Modeling the self-explanation effect with Cascade

Given the learning mechanism of Cascade, a simple hypothesis for explaining the difference between Good and Poor solvers is that Good solvers chose to explain more example lines than Poor solvers. To test this, several simulation runs were made, varying the number of example lines explained and turning on and off various learning mechanisms. All these simulations began with the same initial knowledge state. The initial domain knowledge consisted of the 27 rules that three judges found to be present in the text (see above). The rest of the initial knowledge base consists of 45 non-domain rules, of which 28 represented common sense physics (e.g., a taught rope tied to a object pulls on it) and 17 represented over-generalizations, such as "If there is a push or a pull on an object at a certain angle, then there is a force on the object at the same angle." See VanLehn, Jones and Chi (in press) for a list of the 45 non-domain rules.

### The simulation runs

Run 1 was intended to simulate a very good student who explains every line of every example. Cascade first explained the 3 examples in the study, then it solved the 23 problems. (The 2 problems that are not solvable by the target knowledge were excluded.) It was able to correctly solve all the problems. It acquired 22 rules: 7 while explaining examples and 15 while solving problems. The new rules are correct physics knowledge, allowing for the simplicity of the knowledge representation. Moreover, they seem to have the right degree of generality in that none were applied incorrectly and none were inapplicable when they should have been. However, some of the rules dealt with situations that only occurred once in this problem set, so they were never used after their acquisition.

Run 2 was intended to simulate a very poor student who explains none of the example lines. To simulate a student who merely reads an example without explaining it, the lines from the 3 examples were placed in Cascade's memory without explaining them. Thus, there was no opportunity for EBLC to learn new rules nor were any derivations left behind to act as search control for later problem solving. Cascade was given the same 23 problems given to it in run 1. It correctly solved 9 problems. Apparently these problems require only knowledge that Cascade had been given initially. As it solved these problems, Cascade learned 3 correct rules via EBLC. On 6 other problems, Cascade found an incorrect solution. EBLC did not occur on these problems. On the remaining 8 problems, Cascade failed to find any solution or its search went on for so long that it was cut it off after 20 minutes. Although EBLC was used extensively on these problems problems, the rules produced were always incorrect. On the assumption that a poor student would not believe a rule unless it led to a correct solution to a problem, rules acquired during failed solution attempts were deleted.

Run 3 was intended to separate the benefits of EBLC from the benefits of analogy. Cascade studied the examples as in run 1, learning the same 7 rules as on run 1. During problem solving, both analogical search control and transformational analogy were disabled. As would be expected of a symbol-level learner whose learning was turned off, Cascade was slower on run 3 than on run 1 (249 seconds per correctly solved problem vs. 154 seconds for run 1), and it answered only 19 of 23 problems correctly. More importantly, a large interaction was found with EBLC. When analogy is not used during problem solving, EBLC learned 10 rules, only 6 of which were correct. Moreover, three of the 6 were the same three that it learned on run 2. Thus, of the 15 rules learned during problem solving on run 1, 3 can be learned without benefit of the rules learned during example studying, 3 others require the example studying rules but can be learned without analogy, and the remaining 9 require both analogy and the example-studying rules. This finding makes sense. Analogical search control and, to a lesser extent, transformational analogy influence the exact location of impasses, which in turn determine the rules learned by EBLC. Their influence is strong enough that analogy is necessary for EBLC to learn 9 of the 15 rules (60%) acquired during run 1's problem solving.

In order to determine whether this effect is due to transformational analogy or analogical search control, a fourth run was conducted that was similar to run 3 except that only analogical search control was disabled. Cascade still used transformational analogy. This allowed it to get two more problems correct, raising its score to 21 of 23 problems. More importantly, EBLC acquired the same 6 correct rules as on run 3. The fact that no further correct rules were acquired implies that it is analogical search control and not transformational analogy that helped EBLC during run 1. Thus, it appears that analogical search control (or some other kind of search control) is necessary during problem solving if EBLC is to learn successfully.

## Explaining the self-explanation correlations

Cascade should be able to explain the four differences observed by Chi et al. (1989) between Good and Poor solvers. Assuming that the number of self-explanatory utterances is directly proportional to the number of lines explained during example studying, the job facing Cascade is to explain why explaining more lines causes better scores on quantitative post-tests (finding 1), more accurate self-monitoring (finding 2) and more frequent (finding 3) and more economical reference to the examples (finding 4).

The contrast between runs 1 and 2 indicates that Cascade can reproduce the positive correlation between the number of example lines explained and the number of problems solved correctly. On run 1, it explained all the example lines and got all 23 problems correct; on run 2, it explained none of the example lines and got 9 of the problems correct. Knowing the operation of Cascade, it is clear that having it explain an intermediate number of lines would cause it to correctly answer an intermediate number of problems. So the two extreme points (runs 1 and 2) plus Cascade's deterministic design are sufficient to demonstrate the main finding of the self-explanation effect.

Several mechanisms contributed to this result, and each will be examined in turn. First, when more lines are explained, Cascade is more likely to stumble across a gap in its domain knowledge. Such missing knowledge causes impasses, which lead to EBLC and the acquisition of new rules during example explaining. Of the 19 rules that were learned during run 1 and not run 2, 7 (37%) were learned while explaining examples. As the domain knowledge becomes more complete, performance on problem solving rises. Thus, the more self-explanation, the more EBLC during example studying, and hence the more improvement in problem solving.

The acquisition of rules during example studying helps produce contexts during problem solving that allow EBLC to learn more rules during problem solving even without the aid of analogical search control. Of the 19 rules, run 3 shows that 3 (16%) were acquired in this fashion. These new rules also contributed to the improvement in problem solving.

Analogical search control contributes to the correlation both directly and indirectly. When more lines are explained, more derivations available for analogical search control. Because analogical search control prevents Cascade from going down some dead ends, it directly helps raise the score during problem solving (compare runs 1 and 4). There is an indirect effect as well. Analogical search control causes impasses to occur at places where knowledge is truly missing, rather than at local dead ends in the search space, so EBLC is more often applied to appropriate impasses, and thus more often generates correct domain rules. The remaining 9 of the 19 rules (47%) require analogical search control for their acquisition.

Cascade provides a simple explanation of the correlation between the amount of self-explanation and the accuracy of self-monitoring statements, assuming that negative self-monitoring statements (e.g., "I don't understand that") correspond to impasses, and that positive self-monitoring statements (e.g., "Ok, got that.") occur with some probability during any non-impasse situation. When more example lines are explained, there are more impasses, and hence the proportion of negative self-monitoring statements will be higher. In the extreme case of run 2, where no example lines are explained, all the self-monitoring statements during example processing would be positive, which is not far off from Chi et al.'s observation that 85% of the Poor solver's self-monitoring statements were positive.

The third finding involves the frequency of analogical references during problem solving. Chi et al. observed that during problem solving, the Good solvers make fewer references to the examples than the Poor solvers (2.7 references per problem vs. 6.7). These were mostly physical references, wherein the solver turned to the example and reread part of it. Cascade does not distinguish memory references from physical references. However, it does have two different kinds of analogical references. Analogical search control searches for a sought quantity in the derivation of a solution, while transformational analogy reads consecutive lines in an example looking for one that contains the sought quantity. Suppose we assume that all of the transformational analogy references are physical and that few, say P, of the references due to analogical search control are physical. On the Good solver run, Cascade made 551 references for analogical search control and 40 for transformational analogy. Using the assumption, this would yield $551 \times P + 40$ physical references. On the Poor solver run, Cascade could not use analogical search control because no derivations were available from explaining examples. However, it made 91 references for transformational analogy. If $P < .092$, then $551 \times P + 40 < 91$ and Cascade correctly predicts that the Good solvers make fewer physical references than Poor solvers.

Chi et al. observed that the Good solvers read fewer lines when they referred to examples than the Poor solvers (1.6 lines per reference vs. 13.0 lines per reference). Cascade can model this effect, although an assumption is again needed about the percentage of analogical search control references that are physical. Suppose we assume as before that P of the analogical search control references are physics, and furthermore, assume that a physical reference by analogical search control reads only one line. On the Good solver run, Cascade read 340 lines during transformational analogy and $551 \times P$ lines during analogical search control, for a total of $(551 \times P + 340)/(551 \times P + 40)$ lines per reference. On the Poor solver run, Cascade read 642 lines during its 91 transformational analogy references, for $642/91 = 7.1$ lines per reference. If $P > .017$, then $(551 \times P + 340)/(551 \times P + 40) > 7.1$ and Cascade correctly predicts that the Good solvers read fewer lines per references than the Poor solvers.

Notice that the lower bound (.017) on P does not have to be beneath the upper bound (.092). If P had to be above, say .1 in order to get the lines-per-reference finding correct and below .05 in order to get the reference frequency finding correct, then Cascade could not model both these findings. Thus, these findings jointly have the power to test Cascade, and yet it passed their test.

## Discussion

The major technical hurdle in developing Cascade was finding a way to constrain problem solving so that knowledge-level learning could operate correctly during it. This was achieved by adding analogical search control, a form of symbol-level learning. There was no way to tell in advance of running Cascade whether analogical search control was sufficient. Fortunately, it was, and Cascade was able to learn all 15 rules that it needed to learn.

As a model of the self-explanation effect, Cascade is both qualitatively and quantitatively adequate. It exhibits the same qualitative behavior as subjects: It can self-explain examples as well as "paraphrase" them. It can solve problems with and without referring to examples, and its analogical references can both dive into the middle of the example to pick out a single fact (analogical search control) or read the example from the beginning searching for a useful equation (transformational analogy). Cascade's performance is quantitatively similar to the Good and Poor solvers as its reproduction of the 4 self-explanation findings shows.

Cascade is based on the assumption that the self-explanation effect is due solely to a difference in example studying habits rather than a difference is prior knowledge. This is probably too extreme. We plan to explore the tradeoff by fitting protocols of each individual subject. Cascade will be forced to explain exactly the lines that the subject explains. When given problems to solve, Cascade should reach impasses in the same places that the subject does. However, the subject will probably display more impasses than Cascade, thus indicating gaps in the prior knowledge. Thus, we should be able to tell exactly how much variation in performance is due to prior knowledge and how much is due to learning strategies.

As a general model of cognitive skill acquisition, Cascade shows promise but needs considerable work. In order to be a more complete account of the phenomena at hand, it needs a model of analogical retrieval and of the difference between physical and mental references to the examples. We believe the existing mechanisms can also handle some well-known phenomena of skill acquisition, such as practice and transfer effects, but this needs to be demonstrated. The major limitation on the generality of Cascade 3 is its use of monotonic single-state reasoning. With the help of Rolf Ploetzner, we are currently incorporating a version of the situation calculus which will greatly enhance the types of reasoning Cascade can model, and thus the number of task domain that it can model. We are encouraged to extend Cascade to become a more complete, more general model of learning by its similarity to other theories of cognitive skill acquisition (e.g., Anderson, 1990; Schank, 1986). It is considerably simpler than those theories and probably more thoroughly implemented and tested. We hope that its simplicity and empirical adequacy remain intact as it is extended.

## Acknowledgements

This research was supported by the Cognitive Sciences division of ONR (N00014-88-K-0086) and the Information Sciences division of ONR (N00014-86-K-0678).

## References

Anderson, J.R. (1983) *The Architecture of Cognition*, Cambridge, MA: Harvard.

Anderson, J.R. (1990) *Adaptive Control of Thought.* Hillsdale, NJ: Erlbaum.

Bundy, A., Byrd, L., Luger, G., Mellish, C. & Palmer, M. (1979) Solving mechanics problems using meta-level inference. In *Proceedings of the Sixth IJCAI.* Los Altos, CA: Morgan-Kaufman.

Carbonell, J. (1986). Derivational analogy: A theory of reconstructive problem solving and expertise acquisition. In R.S. Michalski, J.G. Carbonell & T.M. Mitchell (Eds.) *Machine Learning, An AI Approach: Vol. 2.* Los Altos, CA: Morgan-Kaufman.

Chi, M.T.H., Bassok, M., Lewis, M.W., Reimann, P. & Glaser, R. (1989). Self-explanations: How students study and use examples in learning to solve problems. *Cognitive Science*, 13, 145-182.

Chi, M.T. H. & VanLehn, K. (1991) The content of physics self-explanations. *Journal of the Learning Sciences*, 1(1), 69-106.

Dietterich, T.G. (1986) Learning at the knowledge level. *Machine Learning*, 1, 287-316.

Ferguson-Hessler, M.G.M. & de Jong, T. (1990). Studying physics texts: Differences in study processes between good and poor solvers. *Cognition and Instruction*, 7, 41-54.

Jones, R.M. (1989) *A model of retrieval in problem solving.* PhD Thesis, Information and Computer Science, University of California at Irvine.

Larkin, J. (1983) The role of problem representation in physics. In D. Gentner & A. Collins (Eds.) *Mental Models*, Hillsdale, NJ: Lawrence Erlbaum Associates.

Lewis, C. (1988) Why and how to learn why: Analysis-based generalization of procedures. *Cognitive Science*, 12, 211-256.

Mitchell,T.M., Keller, R.M. & Kedar-Cabelli, S.T. (1986) Explanation-based generalization. *Machine Learning*, 1, 47-80.

Newell, A. (1990) *Unified Theories of Cognition*, Cambridge, MA: Harvard.

Pirolli, P. & Bielaczyc, K. (1989). Empirical analyses of self-explanation and transfer in learning to program. In *Proceedings of the 11th Annual Conference of the Cognitive Science Society*, Hillsdale, NJ: Erlbaum.

Schank, R.C. (1986) *Explanation Patterns: Understanding Mechanically and Creatively*, Hillsdale, NJ: Erlbaum.

VanLehn, K., Ball, W. & Kowalski, B. (1990). Explanation-based learning of correctness: Towards a model of the self-explanation effect. In *Proceedings of the 12th Annual Conference of the Cognitive Science Society. Hillsdale, NJ: Erlbaum.*

VanLehn, K., Jones, R.M. & Chi, M.T.H. (in press). A model of the self-explanation effect. *Journal of the Learning Sciences*

VanLehn, K. & Jones, R.M. (in press). Integration of analogical search control and explanation- based learning of correctness. In S. Minton & P. Langley (Eds.) *Planning, Scheduling and Learning.* Los Altos, CA: Morgan Kaufman.

# A Model-Based Approach to Case Adaptation

**Ashok K. Goel***

College of Computing
Georgia Institute of Technology
Atlanta, Georgia 30332

**Abstract**

In case-based reasoning, a given problem is solved by adapting the solutions to similar problems encountered in the past. A major task in case-based problem solving is to generate modifications that are useful for adapting a previous solution to solve the present problem. We describe a model-based method that uses qualitative models of cases for generating useful modifications. The qualitative model of a case expresses a problem-solver's comprehension of how the solution satisfies the constraints of the problem. We illustrate the model-based method in the context of case-based design of physical devices. A designer's understanding of how the structure of a previously encountered design produces its functions is expressed in the form of a function-structure model. The functional differences between a given problem and a specific case are mapped into structural modifications by a family of modification-generation plans. Each plan is applicable to a specific type of functional difference, and uses the function-structure model to identify the specific components that need to be modified. We discuss the evaluation of this model-based method in an experimental case-based system called KRITIK.

## Case Adaptation in Case-Based Reasoning

Much of real-world problem solving appears to be case-based. Cognitive agents often seem to solve new problems by retrieving and adapting solutions to similar problems that they have encountered in the past. These observations have recently led to the development of several computational models of case-based reasoning [Ashley and Rissland 1988] [Hammond 1989] [Kolodner and Simpson 1990] [Riesbeck and Schank 1989]. These computational models posit different methods for adapting previous cases for solving new problems. For example, *heuristic search* [Stallman and Sussman 1977], *heuristic association* [Hammond 1989], and *derivational analogy* [Carbonell 1986]. The case-based method itself has been recursively used to adapt cases [Kolodner and Simpson 1990].

Methods for case adaptation differ in the types of knowledge, inference, and control they use. The method of heuristic-association, for example, uses knowledge of situation-specific associations that directly map differences between the specifications of the new and the known problems into modifications to the solution of the known problem. In addition, case-adaptation methods differ in the types of knowledge stored in the cases. In the method of derivational analogy, for example, a case contains a specification of a problem, a specification of a solution to the problem, and a specification of the problem-solving process by which the cognitive agent arrived at the solution.

In this paper, we describe another computational model for case-based reasoning in which *qualitative models of cases* are used to generate case modifications.

## A Model-Based Method for Case Adaptation

The core idea in the model-based approach to case adaptation is that in addition to their knowledge of solutions to problems encountered previously, cognitive agents often also *comprehend* how a solution actually satisfies the constraints of the problem. This comprehension can be expressed in the form of a qualitative model for the case and used for adapting the case in solving a new problem. A designer, for example, may not only know the functions and the structure of a previously encountered artifact, but may also comprehend how the structure of the artifact produces its functions. If and when this type of knowledge is available in memory, it can be accessed and used in adapting the structure of the known artifact to design new ones.

The main issues in this model-based approach to case adaptation, then, are (i) how to represent the qualitative models of cases, (ii) how to index the models in memory, (iii) how to access them when needed, and (iv) how to use them in case adaptation. The KRITIK project investigates these issues in the context of solving design

*Some of this work was conducted when the author was with the Laboratory for Artificial Intelligence Research at the Ohio State University. There this work was supported by the Defense Advanced Research Projects Agency, contract F-49620-89-C-0110, and the Air Force Office of Scientific Research, grant 89-0250. At Georgia Institute of Technology, this work has been partially supported by the Defense Advanced Research Projects Agency, contract F-49620-88-C-0058.

problems in the domain of physical devices [Goel 1989]. KRITIK is a fully operational system that uses the case-based method for designing physical devices, and the model-based method for adapting design cases. It thus integrates model-based reasoning with case-based reasoning.

In KRITIK, the case-adaptation task is to map the differences between the function desired of a device and the function delivered by the retrieved design case into candidate modifications to the structure of the known design (*functional differences → structural modifications*). KRITIK uses knowledge of how the structure of the known device achieves its functions (*structure → function*) for generating useful candidate modifications. This knowledge is expressed as a qualitative *function-structure model* that specifies how the internal causal mechanisms of the known device compose the functional abstractions of its structural components into the functions of the device as a whole.

A design case in KRITIK contains three types of knowledge: (i) a specification of the functions delivered by a previously encountered design, (ii) a specification of the structure of the stored design, and (iii) a pointer to the qualitative function-structure model of the design. The cases are indexed by the functions delivered by the stored designs, and themselves act as indices to the function-structure-models of the designs. The knowledge content of a given function-structure model is specific to a particular design case but the representation language is useful for specifying the models of a large class of physical devices.

KRITIK uses *a family of modification-generation plans* for mapping the differences between the function desired of and delivered by the retrived design case into candidate modifications to the structure of the known design. These plans are indexed by the *types of functional differences* to which they are applicable. Each plan knows of the *types of structure modifications* that can help to reduce a specific functional difference but does not know what components in the specific design case need to modified. The modification-generation plans access the case-specific qualitative model, and use it to identify the specific components that need to be modified.

## An Illustrative Example from KRITIK

Let us consider, as a simple illustrative example from KRITIK, the task of designing a device to cool high-acidity Sulfuric Acid by the case-based method. Let us assume that a number of previously encountered designs are available in memory, and that the case-retrieval task results in retrieving the design of a Nitric Acid Cooler (NAC) that cools low-acidity Nitric Acid. The design of NAC is shown in Figure 1.

The desired function and the delivered function in this example differ in (i) the substance to be cooled (Sulfuric Acid instead of Nitric Acid), and (ii) a property of the substance (high-acidity instead of low-acidity). The task of case adaptation is to map these functional differ-

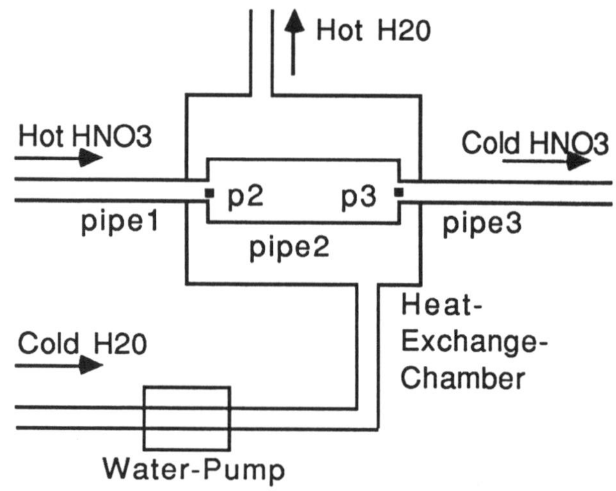


Figure 1: The Nitric Acid Cooler

ences into modifications to the structure of NAC such that the modified design can deliver the desired function of cooling high-acidity Sulfuric Acid. The task of modification generation is to propose candidate modifications that can help to deliver the desired function.

## Model-Based Case Adaptation: Knowledge, Inference, and Control

The generation of useful candidate structural modifications is in general computationally complex because (i) the differences between the function desired of and delivered by the retrieved design case can be large and many, (ii) the needed structural modifications can be non-local and many, (iii) there may be no simple correspondence between the functional differences and the structural modifications, and (iv) the structural modifications can interact with one another and with the components in the structure of the known design. Therefore, solving the modification-generation task computationally efficiently and effectively in general requires knowledge that can help to constrain and focus the process of generating useful modifications.

KRITIK uses three types of knowledge for adapting design cases:

*1. Typologies of Functional Differences and Structural Modifications:* The typology of functional differences classifies differences between the functions desired of and delivered by a design case; the typology of structural modifications similarly classifies modifications to the structure of a design.

*2. Case-Specific Function-Structure Models:* These models describe how the solution in a design case (the design structure) satisfies the constraints of the design (the functions of the design).

*3. A Family of Modification-Generation Plans:* These plans specify compiled sequences of abstract operations for mapping functional differences into candidate structure modifications.

Given a specific difference between the function desired of (e.g., to cool high-acidity Sulfuric Acid) and delivered by (e.g., to cool low-acidity Nitric Acid) a design case, KRITIK sets up four subtasks of the modification-generation task:

*(i) Plan Selection:* First, the functional difference is used as a probe into the functionally-indexed memory of modification-generation plans to select the applicable plan.

*(ii) Plan Instantiation:* Next, the retrieved plan is instantiated in the context of the specific design case to be adapted.

*(iii) Model Retrieval:* Then, the instantiated plan uses the case as a pointer to retrieve the function-structure model for the design case.

*(iv) Plan Execution:* Finally, the plan uses the case-specific function-structure model to generate candidate structure modifications that can help to achieve the desired function.

## A Function-Structure Model

A case-specific function-structure model in KRITIK is an instantiation of a more general *component-substance model* of the functioning of a large class of physical devices. The component-substance model extends and generalizes the component and substance ontology used earlier in the *consolidation method* for deriving the behaviors of a device from the behavioral interactions between its structural components [Bylander and Chandrasekaran 1985]. Knowledge of a case-specific function-structure model in KRITIK is represented and organized in a *behavioral representation language.* The behavioral representation language extends and generalizes the *functional representation scheme* for representing knowledge of the functioning of a device [Sembugamoorthy and Chandrasekaran 1986].

The function-structure model of a given design case in KRITIK explicitly represents the structure, the functions, and the internal causal behaviors of the design. The behaviors compose the structural and behavioral interactions between the structural components into the functions of the device as a whole.

**Structure:** The structure of a device is viewed as constituted of components (e.g., battery, pipe), substances (e.g., water, electrical charge), and structural relations between them (e.g., connection, containment). The substances can be abstract, e.g., heat. The components and substances can have behavioral interactions. For example, substances can flow from one component to another if there is a certain type of structural relation between the two components, viz., the connection relation. The model borrows a typology of behavioral

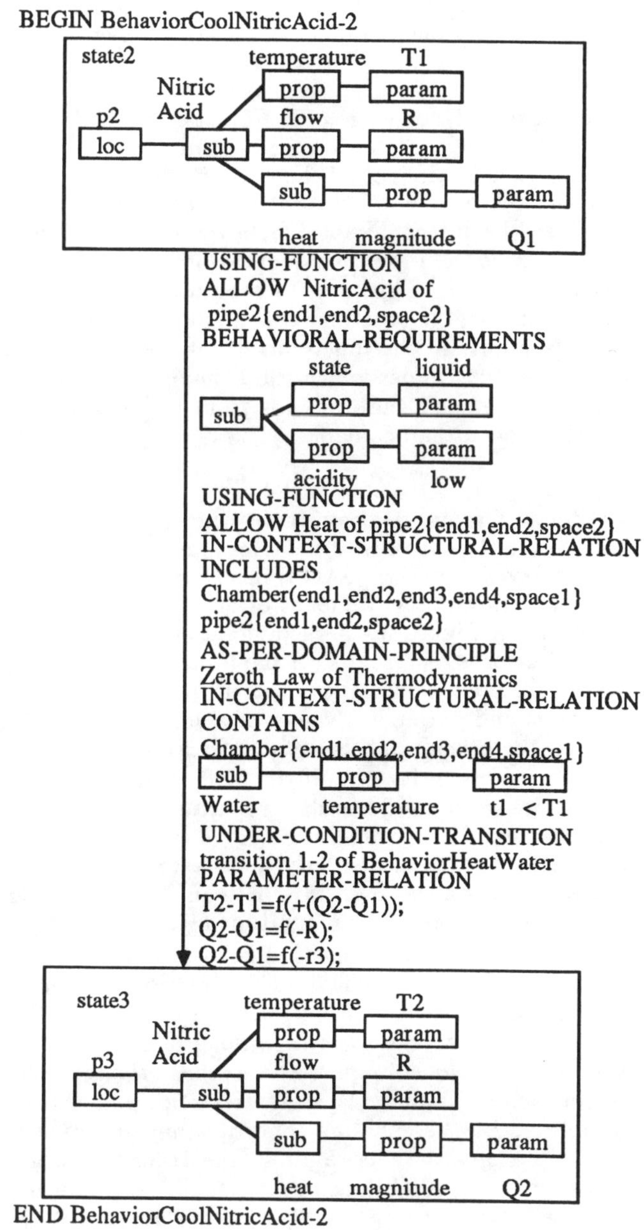


Figure 2: Behavior CoolNitricAcid-2

interactions from the consolidation method. For example, a battery *pumps* electrical charge and a pipe *allows* substances with certain properties. Knowledge of the structure of the device is organized in a structure-substructure hierarchy. A substructure is represented as a schema that specifies its functional abstractions, structural relations, modalities, parameters, etc.

**Functions:** Knowledge of the functions of a device is also represented in the form of schemas. The schema for function specifies the behavioral state it takes as input and the behavioral state it gives as output. It also specifies the internal causal behaviors responsible for transforming the input behavioral state into the output behavioral state. Thus, like in the functional representation scheme, the functions act as indices to the behaviors. In addition, the function schema specifies the device conditions under which the behavior accomplishes the function, and the stimulus from the environment which triggers the behavior.

**Internal Causal Behaviors:** Knowledge of the internal causal behaviors of a device is represented as directed acyclic graphs (DAGS) of behavioral states and state transitions. A fragment of one behavior of NAC is shown in Figure 2. A behavioral state in an internal behavior is represented in the form of schemas. The state schema specifies the location, property, and parameters and parameter values of a substance. For example, the schema labeled *state*2 in Figure 2 specifies that (some quantity of) Nitric Acid is at point $p2$ in the device space, and has the properties of temperature and flow rate with values $T1$ and $R$ respectively. The state schema may also specify the substances contained within a substance, e.g., in *state*2 in Figure 2, Nitric Acid contains heat with magnitude $Q1$.

A behavioral-state transition in an internal behavior is annotated by the behavioral interactions that cause the transition to occur, e.g., the transition *state*2 $\Rightarrow$ *state*3 in Figure 2 is caused (among other causes) by the function *allow* of *pipe*2, where *pipe*2 *allows* the flow of low-acidity liquids. A state transition may also be annotated by the enabling conditions under which the behavioral interactions result in the transition. Some of the enabling conditions may pertain to structural relations between components, e.g., the structural relation that the *heat-exchange-chamber includes pipe*2 in transition *state*2 $\Rightarrow$ *state*3 (see Figure 1). Also, the enabling conditions in one behavior may act as indices to another other behavior, e.g., the enabling condition *under − state − transition* in transition *state*2 $\Rightarrow$ *state*3 refers to another behavior of NAC (BehaviorHeatWater, not shown here). In addition, a transition may be annotated by knowledge of deeper domain principles and qualitative equations as indicated in Figure 2.

## Functional Differences and Structure Modifications

The component-substance model provides a vocabulary for expressing (i) certain types functional differences between design cases, and (ii) certain types of modifications to the structure of a design. The typology of functional differences includes the categories of substance difference, substance property difference, substance location difference, component difference, component modality difference, and component parameter difference. The typology of structure modifications includes the categories of substance substitution (including substance generalization and specialization), component modification (including component replacement, component modality change, and component parameter adjustment), relation modification (such as series-to-parallel and parallel-to-series conversion), substructure deletion (such as component deletion), and substructure insertion (such as substructure replication).

Given a function desired of a design, (e.g., to cool high-acidity Sulfuric Acid) and the function delivered by a specific design case (e.g., to cool low-acidity Nitric Acid), KRITIK classifies the differences between the two functions according to its typology of functional differences. If the desired and the delivered functions differ in more than feature, then it heuristically ranks the differences in order of the difficulty of reducing them. In the NAC example, for instance, the desired function and the delivered function differ in two features: *substance*1 $\rightarrow$ *substance*2 (Nitric Acid $\rightarrow$ Sulfuric Acid), and *property*1 $\rightarrow$ *property*2 (low-acidity $\rightarrow$ high-acidity). Since, in the domain of physical devices that can be modeled in terms of flow of substances between components, reducing the difference *substance*1 $\rightarrow$ *substance*2 is in general less difficult than reducing *property*1 $\rightarrow$ *property*2, KRITIK reduces the latter before the former.

## A Family of Modification-Generation Plans

The types of knowledge, inference, and control, needed for reducing different types of functional differences are in general different. This implies a *family* of plans for generating structural modifications, where each plan in this family is applicable to a specific type of functional difference. For example, the *substance-property-difference plan* is useful for reducing the difference in the property of a substance, e.g., low-acidity $\rightarrow$ high-acidity. KRITIK uses modification-generation plans for several different types of functional differences. These plans are indexed by the type of functional differences they can help to reduce. The substance-property-difference plan, for example, is indexed by the functional difference *property*1 $\rightarrow$ *property*2. Each plan knows of the types of structure modifications that can help to reduce a specific functional difference. The substance-property-difference plan, for example, knows that a difference in the property of a substance can potentially be reduced by the structural modifications of (i) component parameter adjustment, (ii) component modality change, and (iii) component replacement.

The modification-generation plans, however, do not know what components in the given design need to modified. The substance-property-difference plan, for example, does not know which components in NAC's design need to be modified. This is determined by the component-substance model for NAC. First, the substance-property-difference plan uses the NAC case as a pointer to retrieve the component-substance model for NAC (recall that a case contains a pointer to its qualitative model). Next, since the substance property difference *low-acidity* → *high-acidity* occurs in the function of cooling low-acidity Nitric Acid, the substance-property-difference plan uses this function to access the internal causal behavior responsible for it, a fragment of which is shown in Figure 2 (recall that in the component-substance model, functions act as indices to the behaviors responsible for them). Then, the substance-property-difference plan traces through the retrieved behavior, checking each state transition in it to determine whether reducing the substance property difference *low-acidity* → *high-acidity* requires any component in the transition to be modified (or replaced). If so, it generates the corresponding structure modification. For example, when the plan arrives at the transition *state*2 ⇒ *state*3 shown in Figure 2, it finds that *pipe*2 allows the flow of only low-acidity substances. It therefore generates the structure modifications of (i) component parameter adjustment (in case *pipe*2 can allow the flow of high-acidity substances in a different parameter setting), (ii) component modality change (in case *pipe*2 can allow the flow of high-acidity substances in a different mode of operation), and (iii) component replacement (in case the first two modifications are not possible and *pipe*2 has to be replaced with some $new - pipe2$ which can allow the flow of high-acidity substances). The generated modifications can now be evaluated but that is different task altogether (see Goel [1989] for how KRITIK uses the model-based approach to solve it).

In this way, the substance-property-difference plan uses the component-substance model for NAC to generate structural modifications that can help to reduce the functional difference *low-acidity* → *high-acidity*. The substance-difference plan corresponding to the functional difference of *substance*1 → *substance*2 (Nitric Acid → Sulfuric Acid) similarly results in the generation of the structure modification of substance substitution $NitricAcid \rightarrow SulfuricAcid$.

## Evaluation of the Model-Based Method

The KRITIK system evaluates model-based method for case adaptation in two different domains: heat exchangers such as NAC, and electrical circuits such as the electrical circuit in a flashlight. This insures that the method is not specific to any narrow domain. The design tasks KRITIK solves range from simple "naive" design tasks in the domain of electrical circuits to complex "expert" tasks in the domain of heat exchangers. We are currently extending the KRITIK system to two new domains: electromagnetic devices such as buzzers on house doors, and rotational devices such as reaction wheels.

KRITIK shows that the model-based method is quite effective for adapting design cases because the case-specific component-substance model explicitly represents the behavioral states of the known device, the functional role played by each structural component in the state transitions, the enabling conditions for the transitions, and so on. It also shows that the model-based method is quite efficient because each modification-generation plan needs to search only a small portion of the design. For example, the substance-property-difference plan searches only the internal causal behavior responsible for the function of cooling low-acidity Nitric Acid (ignoring all structural components that do not play any functional role in this behavior). The organization of the component-substance model thus constrains and focuses the search for generating structure modifications that can help to reduce a given functional difference.

Of course, the model-based method, like any other method, is applicable only when the types of knowledge it uses are available in the domain of interest. Our experience with KRITIK suggests that case-specific component-substance models are readily available in the domain of physical devices. Whether the model-based method is useful outside physical domains is an empirical issue that KRITIK does not answer. In any event, the domain of physical devices is very large and important; it includes, for example, all engineering devices.

## Related Research

Goel and Chandrasekaran [1989] have earlier proposed the use of function-structure models for generating modifications to the designs of physical devices for achieving new device functionalities. The KRITIK system transforms their proposal into a computational model for case-based problem solving.

Simmons and Davis [1987] have used causal domain models for debugging plans but only for testing modifications to a plan, not for generating the modifications. Koton [1988] has used causal domain models for comprehending diagnostic problems in internal medicine and retrieving appropriate diagnostic cases from memory. Sycara and Navinchandra [1989] similarly have proposed the use of causal domain models for elaborating engineering design problems and retrieving appropriate cases from memory. The KRITIK system uses the model-based approach for three subtasks in case-based reasoning: retrieval of relevant cases from memory, generation of modifications to the retrieved case, and evaluation of the generated modifications. In this paper, we have described only its use of the model-based method for generation of case modifications.

In addition to the differences in tasks and domains, our model-based approach to case-based reasoning differs from that of Simmons and Davis, Koton, and Sycara and Navinchandra, in the representation and organization of the qualitative model it uses. Below, we describe these differences by comparing our model-based method to Sycara and Navinchandra's. First, while their methods use *causal models*, our method uses deeper *function-structure models.* The states and the state transitions in their models are grounded neither in the function nor in the structure of the system. In contrast, the component-substance model explicitly relates the internal causal behaviors to both the function and the structure of a device, and thus constrains them both from the top and the bottom. Second, the states and the state transitions in their models are represented as character strings. In contrast, the behavioral representation language provides a specific vocabulary for expressing the semantics of behavioral states and state transitions. Third, their methods do not provide any scheme for organizing knowledge of the internal causal behaviors of a system. In contrast, the behavioral representation language provides a specific vocabulary for organizing knowledge of the behaviors around the functions they achieve. Fourth, their models do not provide any typology of structural or behavioral interactions between the components of a system, or of the functional differences and structural differences between cases. In contrast, the component-substance model explicitly provides these typologies. Fifth, as a consequence of above, their model-based methods can handle only very simple forms of causal reasoning. In contrast, our model-based method involves teleological reasoning about the functions of physical devices and topographic reasoning about their structures, in addition to causal reasoning.

As mentioned earlier, the component-substance model and the behavioral representation language integrate and generalize two earlier device representations: the functional representation scheme and the consolidation method. The model and the language are complementary to the commonsense algorithms of Rieger and Grinberg [1978]. The representation of behavioral states and state transitions in our scheme is similar to their representations. The typology of structural, causal, and behavioral relations in our scheme and the types of inferences this scheme enables, complement those used in the commonsense algorithms.

Finally, the model-based method for case adaptation is complementary to Carbonell's [1986] method of derivational analogy. The method of derivational analogy uses knowledge of the problem-solving process by which the agent arrived at the solution to a previous problem. The model-based based, in contrast, uses knowledge of how the solution of the previous problem satisfies its constraints. The interaction between the two approaches is an open research issue.

**Acknowledgements** This work has benefited from many discussions with B. Chandrasekaran. S. Prabhakar contributed to the programming of the KRITIK system.

## References

[Ashley and Rissland 1988] K. Ashley and E. Rissland. A Case-Based Approach to Modeling Legal Expertise. *IEEE Expert*, Summer, 1988.

[Bylander and Chandrasekaran 1985] T. Bylander and B. Chandrasekaran. Understanding Behavior Using Consolidation. *Proc. Ninth International Joint Conference on Artificial Intelligence*, pp. 450-454, 1985, Los Altos, CA: Morgan Kaufmann.

[Carbonell 1986] J. Carbonell. Derivational Analogy: A Theory of Reconstructive Problem Solving and Expertise Acquisition. *Machine Learning: An Artificial Intelligence Approach, Volume II*, R. Michalski, J. Carbonell and T. Mitchell (editors). San Mateo, CA: Morgan Kauffman, 1986.

[Goel 1989] A. Goel. Integration of Case-Based Reasoning and Model-Based Reasoning for Adaptive Design Problem Solving. Ph.D. Dissertation, Dept. of Computer and Information Science, The Ohio State University, Fall 1989.

[Goel and Chandrasekaran 1989] A. Goel and B. Chandrasekaran. Use of Device Models in Design Adaptation. *Proc. of the Second DARPA Case-Based Reasoning Workshop*, pp. 100-109, 1989, Los Altos, CA: Morgan Kaufmann.

[Hammond 1989] K. Hammond. *Case-Based Planning: Viewing Planning as a Memory Task*, 1989, Boston, MA: Academic Press.

[Kolodner and Simpson 1990] J. Kolodner and R. Simpson. The MEDIATOR: Analysis of an Early Case-Based Reasoner. *Cognitive Science*, 13, 1990.

[Koton 1988] P. Koton. Integrating Case-Based and Causal Reasoning. *Proc. Tenth Conference of the Cognitive Science Society*, pp. 167, 1988, Hillsdale, NJ: Lawrence Erlbaum.

[Rieger and Grinberg 1978] C. Rieger and M. Grinberg. A System for Cause-Effect Representation and Simulation for Computer-Aided Design. *Artificial Intelligence and Pattern Recognition in Computer-Aided Design*, J. Latombe (editor), pp. 299-334. Amsterdam, Netherlands: North Holland, 1978.

[Riesbeck and Schank 1989] C. Riesbeck and R. Schank. *Inside Case-Based Reasoning.* Hillsdale, NJ: Erlbaum, 1989.

[Sembugamoorthy and Chandrasekaran 1986] V. Sembugamoorthy and B. Chandrasekaran. Functional Representation of Devices and Compilation of Diagnostic Problem Solving Systems. In *Experience, Memory, and Reasoning*, J. Kolodner and C. Riesbeck (editors), pp. 47-73, Hillsdale, NJ: Erlbaum, 1986.

[Simmons and Davis 1987] R. Simmons and R. Davis. Generate, Test and Debug: Combining Associational Rules and Causal Models. *Proc. Tenth International Joint Conference on Artificial Intelligence*, 1987, Los Altos, CA, Morgan Kaufmann.

[Stallman and Sussman 1977] R. Stallman and G. Sussman. Forward Reasoning and Dependency-Directed Backtracking in a System for Computer-Aided Circuit Analysis. *Artificial Intelligence*, 9:135-196, 1977.

[Sycara and Navinchandra 1989] K. Sycara and D. Navinchandra. A Process Model of Case-Based Design. *Proc. Eleventh Cognitive Science Society Conference*, 1989, Hillsdale, NJ: Lawrence Erlbaum.

# Towards a content model of strategic explanation.*

**Kristian J. Hammond**
Department of Computer Science
The University of Chicago
1100 East 58th Street
Chicago, IL 60637

**Colleen M. Seifert**
Department of Psychology
University of Michigan
330 Packard Road
Ann Arbor, MI 48104

## Abstract

Over the past few years there has been a growing interest in the notion of using causal explanations in both learning ([Dejong and Mooney, 1986] [Mitchell *et al.*, 1986]) and planning ([Hammond, 1989], [Hammond, 1987] and [Simmons and Davis, 1987]). The study of how complex causal explanations can be used in learning has turned into something of a cottage industry in AI; however, little attention has paid to how explanations may be constructed. In this paper, we will examine some of the current proposals concerning the process of explanation, augment them with a few ideas of our own, and suggest a new, more strategic level of knowledge about explanations that can be used to guide the explanation process. In particular, we are interested in the problems involved with integrating rule-based methods of explanation construction with memory-based approaches.

## Explanation: Use vs. Construction

In recent years, explanation has been rediscovered as a theoretical issue in Cognitive Science. The stress, however, in the study of explanation has been aimed at the *use* of explanations, rather than their construction. Little work has been done on the issue of explanation as a constructive task. Our goal is to develop a model of the task that is a content theory of the *process* of explanation. In particular, we are interested in a theory that includes a strategic level of knowledge about explanations that can be used to guide the construction as well as the use of explanations.

The need for such a theory is straightforward: Explanations are being billed as the answer to the problem of credit assignment in learning ([Dejong and Mooney, 1986] and [Mitchell *et al.*, 1986]) as well as a credible approach to avoiding extensive search in planning ([Hammond, 1989], [Hammond, 1987] and [Simmons and Davis, 1987]). For these claims to be anything more than the exchange of one set of intractable problems for another, we need a theory that provides us with a method for constructing as well as for using causal explanations. As a first step towards this, we will examine the different proposals that have been made concerning explanation as a task and attempt to organize the operations required by each in into a single knowledge base. We will further propose a three-leveled organization of that knowledge base into **Task**, **Method** and **Domain** level knowledge.

## Explanation in Learning: Backward Chaining

The most recent impact that causal explanation has had in Artificial Intelligence is as a mechanism for credit assignment in various theories of learning. Initially proposed as a method for learning macro operators in STRIPS ([Fikes *et al.*, 1972]) it was rediscovered by both DeJong and Mitchell and implemented in EGGs and EBG respectively. Although both researchers suggest an interesting use for explanations, neither has gone on to suggest an advancement over existing methods for the construction of the explanations that they use.

DeJong, because he has been concerned with domains involving planful action, has relied on a PAM-like ([Wilensky, 1978]) process of explanation that performs goal regression; that is, backward chaining from a perceived goal through the steps that are input to the system. The back chaining is done using a set of causal rules that include provide links between actions, states and intentions. As in PAM, an action is explained when it is linked to a plan, a plan is explained when it is linked to a goal and a goal is explained when it is either linked to an existing theme or has been provided to the system as part of its input.

Unfortunately, also like PAM, this process is non-deterministic and therefore potentially exponential. DeJong handles this problem by introducing an arbitrary length limit to the chains that the explanation

*This work was supported in part by the Defense Advanced Research Projects Agency, monitored by the Air Force Office of Scientific Research under contract F49620-88-C-0058, and the Office of Naval Research under contract N0014-85-K-010.

process is allowed to construct. As a result, any set of inputs that include gaps larger than this limit simply cannot be explained. This limit is, in some sense, essential to the overall argument that DeJong proposes for the use of explanations in learning. Without it, all inputs would fall into the class of "explainable" cases, and there would be no reason to build new scripts that are the result of the learning process. With it, the only inputs that can be explained are those which are either sufficiently detailed to allow the explanation process to link the steps, or those that are examples of previously constructed scripts.

EBL is not the only theory with this drawback; Mitchell's EBG, though concerned with more structural explanations and categories, also faces this problem. The fact that Mitchell has tended to use theorem provers as the basis for the explanation process tends to make this fact even more apparent because the techniques he uses are known to be worst-case exponential. While these techniques are known to be potentially exponential, their extensive use has minimized the problematic status of these solutions to the explanation problem.

## Explanation Patterns: Reminding and Transformation

Outside of the realm of EBL, some work has been done in the area of explanation generation. Schank has suggested a memory intensive approach that uses structures called Explanations Patterns ([Schank, 1986]) that record existing explanations in a form that allows them to be reused analogically. Schank's approach combines a retrieval stage in which the features of an unexplained event are used to find an explanation pattern (or XP) in memory, followed by a transfer or "tweaking stage" in which the existing explanation is transformed to fit the current case. This approach differs from backward chaining because it involves search of a space of transformations rather than a space of operators. Given the nature of most domains (non-homogeneous with clusters of viable explanations), this approach is a potential winner in terms of efficiency of search.

It appears that the techniques that Schank suggests may be directly adaptable to the problem of building the causal *dependency structures* that the EBL methods require. A much clearer case for the reuse of packaged explanations in the construction of dependency structures has been provided by Simmons' GTD, an explanatory system that uses associational rules to generate approximate explanations and detailed causal rules to debug them ([Simmons and Davis, 1987]). While different in form, this approach depends on the same principle as Schank's XP's. That is, viable explanations for events in a domain tend to be found in clusters, making the approach of "approximate and debug" an effective one.

Both of these approaches rest on the overall coherence of most domains. They construct approximate solutions - Schank through recall of existing patterns and Simmons through the use of associational rules - that are then modified to fit the facts that require explanation. The interesting aspect of both is that they are alternatives to simple backward chaining, and provide a different means by which the space of possible explanations can be searched. Specifically, the space of explanations is searched via transformational rules that allow the explanation process to change the structure of an explanation without any backtracking.

## Understanding as Explanation: Script Application

There is a third view of explanation that is, in some sense, the simpliest theory of how explanations are constructed. This is the view of *understanding* as an explanation process, in which new events are explained by virtue of their identification within a script. The result of this view is the notion that script application is an explanatory process in which attention is focused on relevant aspects of the situation by the script or frame ([Minsky, 1975], [Charniak, 1977], [Schank and Abelson, 1977], [Wilensky, 1978] and [DeJong, 1979]).

The basic process of script (or frame) application has two parts: script selection and script application. Script selection requires the use of *semantic indexing* of scripts in a knowledge base that is traversed in the initial phases of understanding. Once a script is found, it is applied (or instantiated) by having empty fields in the script filled in with information gleaned from a piece of text or set of conceptual structures.

Like the process of explanation transformation suggested by Schank and Simmons, this process rests of the existence of structures in memory that approximate the final form of the explanation. Like backward chaining, however, it is essentially non-deterministic in both selection and instantiation of the scripts, opening the door to the possibility of exponential search.

## Rules, Memories and Scripts

Each of these three approaches to explanation makes use of a different knowledge base; specifically, domain level inference rules for backward chaining, explicit memories for reminding/transformation and scripts for selection/instantiation. But what is the relationship between these approaches? Is there a way to combine them into a single model of explanation?

An obvious approach would be to characterize scripts and episodic information as macro versions (combinations) of the inference rules. This approach is immediately problematic: The basic algorithm of transformational (memory-based) explanation is very different than that of backward chaining, in that the internal structure of the reminding is altered using domain neutral transformation rules ([Kass *et al.*, 1986]). Likewise, the process of script selection and instantiation makes use of control heuristics that have little to do with the search control rules of backward chaining. Since the methods are incompatible, simply combining the rule sets is unworkable.

The opposite approach, that of characterizing backward chaining as a degenerate form of reminding plus transformation, seems to be equally problematic. There are dynamics and technology involved in backward chaining that cannot currently be reproduced within the confines of the reminding/transformation approach.

The only alternative, then, is to view these three approaches to explanation as exactly that, three separate and distinct methods for constructing explanations. This implies, however, that we must now think in terms of how they interact, and how we can integrate them into a single theory of explanation. In effect, it implies the need for a theory of how to control the movement between these approaches in constructing viable explanations. In some ways, this theory will be analogous to content theories of search control; however, the basic operations that make up the right-hand-sides of the control rules will differ in that they will refer to the components of reminding – transformation, selection and instantiation – as well as rule application. The intention here is to provide for explanation what Stefik did for planning ([Stefik, 1981a]); that is, suggest a content theory of the processes involved in the construction of explanations.

## Step One: Representation

Our basic notion of explanation comes out of the literature on plan understanding ([Wilensky, 1978] and [Charniak, 1983]) and DeJong's work in explanation-based learning ([Dejong and Mooney, 1986]). In this work, an explanation is a set of dependencies in which states are supported by action/rule pairs, actions are supported by plan/rule pairs, plans are supported by goal/rule pairs and goals are supported by the theme/rule pairs. We will ignore for a moment the support structure of the rules themselves.

While, each support requires both a rule and a fact, either can be an assumption required to construct the explanation. This form is exactly what is produced through backward chaining, and is stored in memory for use in the construction of new explanations. Scripts take a slightly different form in that they organize multiple *part-of* relationships into a single structure that has a standard goal associated with it. The final form of the type of explanation associated with scripts, however, still consists of these dependency structures.

Although there is a single final form for each of these methods, each uses a different sort of knowledge base. Explanation via chaining requires a base of inference rules that allows the incremental addition of individual links to the final dependency chain. Explanation via reminding plus transformation requires a knowledge base of known explanations and a set of transformation rules that allows these explanations to be modified. And script based explanation requires a knowledge base of abstracted scripts that can be refined through the use of slot filling mechanisms. These different kinds of facts about a domain make up what we call the **Domain Level** of knowledge used in explanation.

## Step Two: The Task

The task of explanation begins with the notion of an anomaly ([Schank, 1986]) and ends when that anomaly supported by a dependency structure that serves to link it to known or assumed facts. As we have said, this can be done through the use of any one of the three methods we have discussed. But each of these methods brings its own characteristic method and dynamic to the problem.

Backward chaining involves the use of domain rules to traverse a virtual AND/OR tree, using control rules to guide the search. Reminding/transformation requires the retrieval of explanations from memory and the subsequent transformation of those explanations. This is supported by rules that select the appropriate features from the initial problem for use in indexing, and rules that control the application of the transformation rules themselves. Script application also depends on the use of indexing techniques to find the appropriate script in a library of possibilities.

## Step Three: The Actions

In this section, the central issue of the set of actions that can be applied during the construction of a explanation is addressed. What we examine here are the "right hand sides" of the control rules that guide the construction of explanations. Some of these actions, such as rule or script selection, are already part of the standard repertoire of understanding and explanation. Others, such the feature selection and *recasting of representation*, are more recent additions to this list. For presentation purposes, the more familiar of these actions are considered first, followed by more recent additions. However, the unifying notion is that these actions may be taken at any time during the construction of an explanation. While some are logically dependent on others (e.g., you have to select a set of features to use in indexing before looking for an explanation in memory), the idea is that these actions are on the right hand side of rules that theoretically can be fired at any time during the explanation process. In particular, we want to stress that the selection of an explanation method (chaining, script application or reminding/transformation) is not fixed: An explainer can (and almost always will) move between the different methods.

There really are only two actions (aside from backing up) that can be taken in constructing an explanation out of a base of domain rules; namely, rule selection and rule application. While they are not the most important rules in the library we are constructing, search control rules are the most familiar. These are rules that guide the selection of the individual domain rules that are applied to explain an anomaly when a disjunct of possibilities is presented. For example, in trying to explain why Jack is out of orange juice, we could have attempted to find an explanation that involved some use of orange juice other than as an ingestible liquid. But this line of reasoning is cut short by control rules that give preference to the explanation that includes the "standard" use. But chaining is only a small part of the overall explanation process and is really only attempted after many other steps are taken. In fact, there are many steps involved with constructing an explanation that come well before the selection of a backward chaining rule or even the decision to apply backward chaining as a method in the construction of the explanation.

Before building an explanation, there must be something to explain; therefore, the first step is the selection of the anomaly or set of anomalies to explain. What is needed at this stage is a set of rules for deciding which anomalies are even candidates for inclusion in a single explanation. These rules are needed to initially focus the explainer's attention on the events and states that it should attempt to connect. In an earlier paper ([Hammond and Hurwitz, 1988]), we suggested a set of heuristics based on temporal, physical and systemic proximity that guided the attempts at joining separate anomalies into a single coherent explanation. These rules are particularly important when the explainer does not have a complete set of domain rules, and thus must make assumptions about the connectivity (or lack of connectivity) of a set of input features.

These rules allow a system to propose causal relationships between anomalous features that were proximate from one of many possible points of view. Schank has also proposed this type of heuristic ([Schank, 1986]); in particular, attempting to coordinate anomalies that occur at the same time as well as those that are of the same type. Likewise, Falkenhainer ([Falkenhainer, 1988]) has proposed a similar heuristic that looks for particular differences in the input (as compared to a standard model) to try to connect.

The important point here is that these heuristics focus the explainer's attention of a select sub-set of features that will be included in the explanation itself. Of course, these are heuristics, and, as such, are open to error. But it is important to understand that, like the search control rules of backward chaining, they are essential to controlling the movement through the space of possible explanations that can be constructed for any one set of inputs.

Once a set of anomalies are selected, the next step is to decide on the basic strategy that should be taken in trying to join them into a single explanation. This is just the decision as to whether to try to find a single cause for all of the anomalies, to explain them in terms of a flow of cause and effect or to find an explanation that includes multiple actors and goals that tie the anomalies together. For example, in trying to explain the correlation between a rise in ice cream sales and a rise in the occurrence of drownings, one important step is to see the two facts as the dual effects of a single cause; namely, the onset of summer. An attempt to explain these events by creating a causal chain leading from one to the other ("eating ice cream gives you stomach cramps when you try to swim" or "people who watch drowning victims being taken away tend to eat a lot of ice cream") leads the explainer away from a more reasonable explanation.

Examples of useful heuristics include attempting to build chains forward from earlier actions, building out from intentional actions to states, and attempting to construct the typical **theme->goal->plan->action** chains that are often used to explain human behavior ([Schank and Abelson, 1977]). Like the rules determining the selection of the anomalies, these heuristics are a large part of what controls the search for the explanation.

Next, somewhat dependent on the selected strategy, comes the selection of actual method to use in the construction of the explanation (or part of an explanation). This involves deciding on which of chaining, script application or reminding/transformation to apply to the problem. As we mentioned earlier, this decision is not a fixed one that will remain unchanged throughout the explanation process. Existing scripts or explanations often have to be extended using individual causal rules. Likewise, a partial explanation constructed using backward chaining will uncover a new set of features that may be used to find a script or existing explanation.

This flexibility in method application is an important point in this model: We are proposing an integrated approach, not advocating one method over others. It is clear that each of these methods has its advantages, and that each has its problems. Script and memory-based methods provide a good means for searching very regular spaces of explanations, and chaining gives us the ability to do a formal analysis of the search. However the complexity of script and memory-based approaches when applied to less regular domains has yet to be solved, while simple chaining methods do not allow for the possibility of reuse of already formed explanations. So, it is clear that the best approach is one that attempts to integrate these into a single methodology for constructing and saving explanations. It may in fact be the case that, as a model of human cognition, a variety of such approaches are necessary to replicate the flexibility of the explanation process. Depending upon how these separate processes are defined, there may be functional justifications for their existence within an arsenal of methods to be applied to problems. Finally, some psychological evidence suggests that problem solving ability, and creativity, may be linked to the individual's ability to flexibly move between approaches to a problem.

Along with explanation retrieval and transformation, there is another sort of action that needs to be taken by memory based explanation systems that is often ignored: This is the *recasting* of initial representations into forms that might provide better feature sets for use in retrieval of existing explanations. Intuitively, this is what we do when we "see something from a different point of view" or "in a different light". In order for this to make sense, however, it must be driven by a set of heuristics that makes these changes for a reason.

Kass, Leake and Owens have done some tentative work in this area in the SWALE project ([Kass *et al.*, 1986]). In this work, they looked at the issue of transforming slots fillers of partially filled structures in order to generate more indices for search into memory. For example, viewing a race horse as an athlete in order to find explanations related to athletes that might be applied to horses. Most of this work centered on the notion of moving around in a semantic net however and was not concerned with either the notion of generating characterizations of an overall situation or in using aspects of the macro structure of the representation to generate new features. An example of the first would include representational changes such as seeing a take-over of space in an office complex as imperialism ([Schank, 1986]) or seeing the decision to go to sleep or keep working as a resource allocation problem. The point in both of these is that there is a representation that captures aspects of the situation that are not captured by a simple listing of the actions involved.

Recasting, as we are defining it, involves looking at event and goal configurations at a more abstract level. For example, in the following story:

> Bob was sitting in a hallway when a woman came out and took a drink from the near by water fountain. A few moments after she left, she returned for another drink. A few moments after this, she came back for a third drink. As she was drinking, a man came up behind her and held a knife above her, poised to strike. She turned, screamed, and then the two of them laughed and walked away together.

As with the single bug assumption, a strategy selection decision is made to try to find a single goal that both drinking water and being attacked might satisfy. How to find such an explanation? The flat representation of individual actions can be recast as the *repetition* of a single event – leading to a memory search that attempts to explain why events are repeated – followed by the threat of attack and laughter. This leads to an attempt to explain the story in terms of the possible reasons for multiple versions of the same action.

In protocol studies [Seifert, 1989] subjects seem to perform this particular recasting of representation whether or not they go on to find an explanation for the story. In particular, subjects often propose the notion of rehearsal to explain the repeated events, either as part of a play or movie, or for a real attack. In either case, the structural form of the actions is used to recognize a feature which can then be used to either find an existing explanation or help in the construction of a new one. The point is that there is a basic piece of vocabulary being used to store information in memory (and as part of backward chaining rules) that must be derived from even the deepest representations of events.

A second piece of recasting appears to be critical in the development of indices that will successfully retrieve a particularly apt explanation from memory. In representing the attack, the outcome (shared laughter) is not the expected one, and that it therefore did not achieve the desired result. Therefore, the notion of a different intention or action may be considered through a variety of strategies: the outcome that did occur can be considered the intended one, and the actions recast to fit; or the outcome can be considered a side effect of the actions rather than the intended outcome. In both cases, if the attempted attack is recast in terms of its actual outcome, the scream is viewed as a successful outcome of an attempt to scare the woman. The addition to the representation through recasting – that the attack "scared" the woman – facilitates the connection of the repeated actions with the attack. Attempting to incorporate "scaring someone" and "someone drinking water" as plans in service of the same goal are much better indices to retrieve from memory the goal of curing the hiccups, an explanation satisfying all of the information in the problem.

This process of recasting representation, although presented here as an aspect of reminding/transformation, actually belongs at the level of anomaly, strategy and method selection. This is because the recasting of representation has a global effect on the course of the explanation in much the same way that it does in problem solving. That is, it allows the introduction of new features that themselves present alternative paths for chaining and/or script and explanation retrieval. These four basic actions make up what we call the **Task Level** knowledge in explanation. That is, a level of knowledge about how to proceed in building an explanation in general. The other actions are part of what we call the **Method Level** in that they are the subparts of specific approaches. These division play the same role in explanation as the divisions suggested by both Stefik ([Stefik, 1981b]) and Hayes-Roth&Hayes-Roth ([Hayes-Roth and Hayes-Roth, 1979]) play in planning. They allow different types of knowledge to have access to the process control during the entire process, rather than just at a single stage.

The final set of actions in the method level are those associated with script application. The two most relevant are script selection and slot instantiation. We hold with the idea that script selection is an issue open to guidance by both semantics and pragmatics ([Schank and Birnbaum, 1980]) and that slot instantiation is best guided by the semantic constraints provided by the script itself ([DeJong, 1979]). A full treatment of either of these issues is well beyond the scope of this paper, but it is important to note that these are the two most relevant actions in script application, even though each is further decomposable into subactions.

The actions involved with explanation construction fall into two main groups: **Task Level** action and **Method Level** actions. Task level actions relate to global issues of representation and strategy. Method level actions are the actions used in the control of the actual building of the explanations within a particular method being used. An important point is that method level actions tend to be used only within the confines of particular methods while task level actions are always available to guide the construction of the explanation. Knowledge of rules, individual scripts and memories of existing explanations is stored on its own **Domain Level**.

## Conclusions and the next step

Our goal in this paper was to outline the process of explanation as a task. We have tried to accomplish this in three basic steps. First, we have posited a general representation for explanations that can be (and in many cases is) used for work in both EBL and plan debugging. Second, we have tried to at least partially explicate the different process models that have been proposed. And third, we have suggested an integration of those proposals through the decomposition of the primitive actions that fit into a three level organization of **Task**, **Method** and **Domain** level knowledge. This decomposition is the first step in a true integration of the different methods into a single theory of explanation construction. We suggest that the final form of this integration will be similar to the distributed systems suggested by Stefik, and Hayes-Roth and Hayes-Roth, for use in planning.

## References

E. Charniak. Ms. malaprop: A language comprehension program. In *The Proceedings of the Fifth International Joint Conference on Artificial Intelligence*, Cambridge, Mass., August 1977. IJCAI.

E. Charniak. Passing markers: A theory of contextual influence in language comprehension. *Cognitive Science*, 7:171–190, 1983.

G. Dejong and R. Mooney. Explanation-based learning: An alternative view. *Machine Learning*, 1(2):145–176, 1986.

G. DeJong. *Skimming Stories in Real Time: An Experiment in Integrated Understanding.* PhD thesis, Yale University, May 1979.

B. Falkenhainer. The utility of difference-based reasoning. In *The Proceedings of the Seventh Annual Conference on Artificial Intelligence*, St. Paul, Minnesota, August 1988. AAAI.

R.E. Fikes, P.E. Hart, and N.J. Nilsson. Learning and executing generalized robot plans. *Artificial Intelligence*, 3:251–288, 1972.

K. Hammond. Learning and re-using explanations. In *Proceedings of the Fourth International Conference on Machine Learning*, Irvine, CA, June 1987. ML.

K. Hammond. *Case-Based Planning: Viewing Planning as a Memory Task.* Academic Press, 1989.

B. Hayes-Roth and F. Hayes-Roth. A cognitive model of planning. *Cognitive Science*, 3(4), 1979.

A. Kass, D. Leake, and C. Owens. Programming the theory: Swale, a program that explains. In R. Schank, editor, *Explanation Patterns: Understanding Mechanically and Creatively.* Erlbaum, 1986.

M. Minsky. A framework for representing knowledge. In P. Winston, editor, *The Psychology of Computer Vision*, chapter 6, pages 211–277. McGraw-Hill, New York, 1975.

T.M. Mitchell, R.M. Keller, and S.T. Kedar-Cabelli. Explanation-based generalization: A unifying view. *Machine Learning*, 1(1):47–80, 1986.

R. Schank and R. Abelson. *Scripts, Plans, Goals and Understanding.* Lawrence Erlbaum Associates, Hillsdale, New Jersey, 1977.

M. Schank, R. Lebowitz and L. Birnbaum. An integrated understander. *American Journal of Computational Linguistics*, 6(1):13–30, 1980.

R. Schank. *Explanation Patterns: Understanding Mechanically and Creatively.* Lawrence Erlbaum Associates, 1986.

C. M. Seifert. Inference in problem solving. *Unpublished manuscript*, 1989.

R.F. Simmons and R. Davis. Generate, test, and debug: combining associational rules and causal models. In *The Proceedings of the Tenth International Joint Conference on Artificial Intelligence*, Milan, Italy, 1987. Morgan Kaufman.

M.J. Stefik. Planning and meta-planning. *Artificial Intelligence*, 16:141–169, 1981.

M.J. Stefik. Planning with constraints. *Artificial Intelligence*, 16:141–169, 1981.

R. Wilensky. *Understanding Goal-Based Stories.* PhD thesis, Yale University, 1978. Research Report #140.

# Adapting Abstract Knowledge*

**Eric K. Jones**
The Institute for the Learning Sciences
Northwestern University
Evanston, IL 60201
jones@ils.nwu.edu

## Abstract

For a case-based reasoner to use its knowledge flexibly, it must be equipped with a powerful case adapter. A case-based reasoner can only cope with variation in the form of the problems it is given to the extent that its cases in memory can be efficiently adapted to fit a wide range of new situations.

In this paper, we address the task of adapting abstract knowledge about planning to fit specific planning situations. First we show that adapting abstract cases requires reconciling incommensurate representations of planning situations. Next, we describe a representation system, a memory organization, and an adaptation process tailored to this requirement. Our approach is implemented in BRAINSTORMER, a planner that takes abstract advice.

## Introduction

Most knowledge-based systems are unable to use their knowledge flexibly: they can only solve problems that are stated in exactly the right way. This is one aspect of the *brittleness* problem [Feigenbaum *et al.*, 1971], which limits the scope of many existing artificial intelligence systems both as practical tools and as cognitive models.

Even if a system contains the knowledge it needs to solve a problem, it may not be able to bring this knowledge to bear. At one extreme are systems that use radically incomplete inference procedures such as schema or script application [Cullingford, 1977], which although efficient, can only solve a limited class of problems. At the other extreme are systems that countenance multistep inference chaining [Rieger, 1976; Wilensky, 1978]. In principle, these systems are capable of solving a large number of problems, but in practice they make so many useless inferences that the number of problems they can actually solve under reasonable resource constraints remains small.

Case-based reasoning has been proposed as a promising middle ground [Hammond, 1986; Kolodner *et al.*, 1985]. Like schema appliers, case-based reasoners trade completeness for efficiency. Unlike schema appliers, however, case-based reasoners come with an adaptation component for resolving mismatches between past cases and current problematic situations. If the adapter can efficiently resolve an interesting class of commonly-occurring mismatches, then the case-based reasoner will be able to solve a wider range of problems than a schema applier. The flexibility of a case-based reasoner thus depends heavily on the power of its adapter.

In this paper, we consider a particular adaptation task: adapting abstract knowledge about planning to fit specific situations. A real-world activity in which this task arises is taking advice. People often communicate advice about planning in terms of high-level culturally-shared models of the planning process [White, 1987]. The vocabulary of these models may be very different from the vocabulary of the representations the person actually uses to solve the problem. In other words, the initial form of the advice is not necessarily operational for problem solving. The person taking the advice is therefore faced with the task of adapting abstract knowledge to fit her current problem solving needs.

We have built the BRAINSTORMER system to investigate the problem of adapting abstract knowledge [Jones, 1990; Jones, 1991]. BRAINSTORMER is a planner that takes advice about problems in the domain of terrorist crisis management. Whenever it encounters a difficulty, it asks for advice. A user then presents advice in the form of a proverb or aphorism, represented in a high-level vocabulary of culturally-shared planning concepts. BRAINSTORMER adapts the advice it receives by converting it into specific information in the operational vocabulary of the planner, which the planner can then use to resolve its difficulty and continue planning.

We have focused on proverbs for two reasons. First, proverbs are a well-defined class of abstract cases that encode useful culturally-shared knowledge about the idiosyncrasies of human planning and social interaction [Owens, 1988; Schank, 1986; White, 1987]. *The grass is always greener on the other side of the fence*, for example, expresses a peculiarity of the human planning process: when comparing options, people tend to be biased in favor of the unfamiliar or the unpossessed.

Second, representing a large number of proverbs has proved an effective strategy for developing and testing our representational vocabulary for advice. *The grass is always greener on the other side of the fence*, for example, not only encodes an important truth about human planning, but also points to the need for a conceptual

*This research was supported in part by DARPA contract F49620-88-C-0058 and AFOSR contract 89-0493. The Institute for the Learning Sciences was established in 1989 with the support of Andersen Consulting, part of The Arthur Andersen Worldwide Organization. Additional funding is provided by Ameritech, an Institute Partner, and by IBM.

vocabulary sufficient to encode the idea of *choosing between options in terms of estimates of their utility.*

BRAINSTORMER is capable of accepting a range of advice expressed in its high-level, culturally-shared vocabulary of planning concepts: proverbs are just one useful class of expressions representable in this vocabulary. Thus our approach to abstract case adaptation not only contributes to the field of case-based reasoning, but also addresses the more general problem of flexible understanding of arbitrary abstract advice.

## The Problem

Taking abstract advice imposes particular inferential requirements that constrain the design of an adapter. BRAINSTORMER's descriptions of the problematic situations it faces are typically encoded in a very different vocabulary than the advice it receives; therefore, its adapter has to be able to infer the connection between the two. For example, one of BRAINSTORMER's goals is preventing terrorism, which it pursues by trying to analyze and counterplan against particular past terrorist attacks. There are many ways that a piece of abstract advice might characterize a terrorist attack, including *an act of frustration, an act of defiance, an act of revenge, an attack on civilians, a goal conflict, an illegal action, an immoral action, part of a propaganda campaign*, and *a political statement*, among others.

Conversely, there are many ways in which any of the above abstractions could be plausibly inferred from different specific planning situations as a planner might initially represent them. For example, a goal conflict can be plausibly inferred by observing an action on the part of one person that violates a goal of someone else, or from the knowledge that two different people hold incompatible goals, or by noticing that someone is angry.

It follows that adapting abstract knowledge requires being able to reconcile descriptions of given objects expressed in incommensurate vocabularies. This is essentially a problem of bidirectional search. As the above examples indicate, this problem is potentially serious, because the branching factor is high in both directions: on the one hand, there are many possible abstractions of a given concrete situation; on the other, there are many ways to plausibly recognize an instance of any particular abstract concept in a concrete situation.

In the remainder of this paper, we present an approach to adaptation sufficient to solve this problem. First, we give an example showing how the need to reconcile representations in incommensurate vocabularies arises in BRAINSTORMER. Second, we show that our underlying representation system is sufficiently powerful to represent the results of this process. As we will see shortly, this is a nontrivial requirement. Third, we sketch a memory organization that allows alternate encodings of a given situation in different vocabularies to be efficiently related. Finally, we specify an algorithm for this task, which works by redescribing one representation in the vocabulary of the other.

## The Need for Redescription: An Example

A typical interaction between BRAINSTORMER and a user is as follows. BRAINSTORMER issues a query for an explanation of a particular past terrorist attack, as part of trying to come up with plans for preventing terrorism. The user responds with advice in the form of a representation of a proverb; let us suppose that the proverb is *no trouble but a priest is at the bottom of it*, which BRAINSTORMER represents as a causal connection between religious leaders and goal conflicts. Figures 1 and 2 depict the query and the initial representation of the advice.[1] It is the task of BRAINSTORMER's adapter to turn the advice into an answer to the query.

```
?explanation
   explained terrorist-attack
               actor group
                       typical-elt terrorist
```

Figure 1: A query from the planner.

```
cause
   cause  action
             actor religious-figure
   caused goal-conflict
```

Figure 2: Initial advice representation.

The adapter begins by transforming the advice into a form that could conceivably match the system's query. As the query asks for an explanation, the adapter converts the initial causal relation into an explanation structure based on that causality, as shown in figure 3.

The need for redescription arises in the next stage of adaptation, when the system tries to match or unify this newly created explanation structure with the query. No difficulty is encountered until an attempt is made to match the `terrorist-attack` to the `goal-conflict`. At that point, the matcher complains that these are instances of incommensurate concepts and starts trying to redescribe the terrorist attack in terms of goal conflicts.

```
causal-explanation
  explained goal-conflict =gc
  causals   (cause
               cause  action
                         actor religious-figure
               caused =gc)
```

Figure 3: The advice transformed into an explanation.

[1]Brainstormer uses a frame-based representation system with a slot-filler notation:

```
<frame>
    <slot1> <filler1>
    <slot2> <filler2>
    ...
```

Answers to queries are bindings to the free variables in the query, which are prefixed by "?"s. Variable binding constraints are expressed using the notation `=<var>`.

## Redescription, Vagueness, and Dynamic Concept Formation

Before we can talk about how to efficiently carry out this redescription, we have to make clear what we want the result to look like. What did the user really mean in using the proverb ***no trouble but a priest is at the bottom of it*** to explain a terrorist attack? Presumably, that the religious figure caused the terrorist attack as part of causing a goal conflict. Notice that the intended meaning is *vague*, in that it doesn't specify exactly how the religious figure was involved. The output of the adapter should be a representation of this information at the right level of abstraction. What should this look like? We start with three inadequate answers to this question that point the way to a more acceptable one.

In BRAINSTORMER's ontology, a goal conflict exists if two different agents hold incompatible goals, the pursuit of one of which negatively impacts the other. Here we can safely assume that the terrorists had the goal to carry out a terrorist attack and that carrying out the attack violated goals of other agents: goals of people injured or killed in the attack, for example, and also BRAINSTORMER's goal to prevent terrorism. Therefore, various goal conflicts obtain as a consequence of carrying out the terrorist attack.

A first attempt to represent the result of adaptation might involve a four-step causal chain: a religious figure incited the terrorists to carry out the terrorist attack, causing the attack to occur, which violates a goal of someone else, thereby producing a goal conflict. Unfortunately, this representation is more specific than the user intends by the advice, as it asserts that the religious figure actually caused the terrorists to have the goal to carry out the attack. What the user really intends is vaguer. In particular, the religious figure may have been causally implicated in producing the terrorist attack in other ways as well: by facilitating its execution, for example, or by helping to ensure that its results would be effective.

A second possibility is to represent the advice using a one-step causal chain: the religious figure caused a goal conflict involving a terrorist attack. This representation, however, is too general. The output of adaptation is supposed to explain the terrorist attack, not the goal conflict. The proposed representation would cover a situation in which a religious figure caused someone to have the goal to prevent the terrorist attack; while this may well cause a goal conflict to exist once the terrorist attack occurs, it is hardly an explanation for the terrorist attack.

A third possibility is a disjunctive representation: list all of the ways that the religious figure could have helped the goal conflict come about that were also causally implicated in producing the terrorist attack, and represent the result of adaptation as the assertion that the religious figure interfered in one of these ways. While this approach adequately captures the meaning of the advice, it is still somewhat unsatisfying. Representing a vague piece of advice by cashing out all the ways it can be made more specific, while epistemologically adequate, is hardly concise or easy to reason with.

A truly satisfactory approach should allow us to represent the intended vague meaning of the advice non-disjunctively, by creating a first-class object to represent "the thing that the religious figure caused, leading to the terrorist attack and the goal conflict." That way, the advice can be represented and reasoned about directly, without having to enumerate the various ways it could be specialized.

But what in fact does the advice assert that the religious figure caused? The answer, we believe, is an instance of a brand new concept: an instance of the class of actions which, like the terrorist attack, are performed intentionally and violate a goal of someone else, thereby signaling a goal conflict. In other words, we suggest that the advice asserts that a religious figure caused the terrorist attack ***under a new description*** in terms of a set of recognition conditions or features for plausibly inferring a goal conflict.

More generally, an important part of using knowledge flexibly is being able to create and reason with novel concepts. Even if BRAINSTORMER has never before been asked to describe a terrorist attack in terms of goal conflict, it should be able to do so if asked. This imposes new demands on our representation language. Many knowledge representation systems are incapable of representing new concepts on the fly. Usually, all possible concepts are hard-wired into a type hierarchy in advance, and the system can only categorize inputs as being instances of these fixed concepts.

BRAINSTORMER, in contrast, can dynamically extend its base set of concepts by the mechanism of $\lambda$-abstraction [Sowa, 1984]: $\lambda(a)$(*representations mentioning a*) defines a new concept, whose instances are all of the $a$'s that satisfy the conditions in the body of the lambda. In BRAINSTORMER, we represent $\lambda$-abstractions using the notation of *views*, which encode relationships between incommensurate representations of a single situation. Views are instances of $\lambda$-abstractions that permit one concept to be redescribed in terms of recognition conditions for another.[2]

Figure 4 shows our preferred representation of the advice after it has been adapted to fit the planner's query. The **`part-whole-view`** frame in the figure encodes a redescription of the **`terrorist-attack`** as an instance of the dynamically-constructed abstract concept $\lambda(a)$(*a is an intentional action of some agent that violates a goal of someone else*). This new concept was

[2]Actually, there are two kinds of views: *part-whole views* and *whole-whole views*. Only part-whole views create new concepts dynamically. Whole-whole views, in contrast, reference only preexisting concepts. For example, a terrorist attack can also be redescribed in terms of the preexisting concept of illegal action using a whole-whole view. Whole-whole views are similar in spirit to the views in Jacob's ACE system [Jacobs, 1987].

```
causal-explanation
  explained --
   part-whole-view =view
     source      terrorist-attack =attack
                   actor group =terrorists
                           typical-elt terrorist
     recog-conds (cause
                    cause achieve-goal =g1
                            actor =terrorists
                            state =attack
                    caused =attack
                  violates-goal
                    state =attack
                    goal prevent-goal =g2
                          actor brainstormer
                          state terrorist-attack)
     target      goal-conflict
                   actor1 =terrorists
                   actor2 brainstormer
                   goal1  =g1
                   goal2  =g2
     concept     (gc-schema2 lhs3)
  causals (cause
            cause  action
                     actor religious-figure
            caused =view)
```

Figure 4: The trace of redescription.

built out of a configuration of plausible recognition conditions for goal conflicts (determined, as we will see in the next section, by the schema **gc-schema2**). The filler of the `recog-conds` slot shows how this concept is instantiated by the terrorist attack: the terrorists had the goal to carry out the terrorist attack, and the attack violates BRAINSTORMER's goal to prevent terrorism. The **source** slot stores the **terrorist-attack** prior to redescription, while the **target** slot holds the **goal-conflict** that follows from the instantiated recognition conditions. The latter is the original `goal-conflict` of the advice, fleshed out to include the information supplied by the terrorist attack.

The **part-whole-view** in our example reifies "the thing the religious figure caused" as an instance of a new class of action, so that the system can reason about it without making any definite commitment as to how the religious figure caused it. This representation of the advice compactly encodes an answer to the planner's original query, which BRAINSTORMER can use to continue planning. As we saw above, the only way that BRAINSTORMER could represent this somewhat vague information without the benefit of views is in terms of a clumsy disjunction of all the ways the advice might be more specific. Of course, the system might later wish to postulate one or more of these specializations itself; if it does, the information necessary to do so is stored in the `recog-conds` slot of the view.

## Memory Organization for Redescription

Now that we have specified the results we want from adaptation, we demonstrate a memory organization that allows results like this to be efficiently computed. The basic task to be addressed is redescribing specific planning situations in terms of abstract concepts present in advice. As we have seen, this involves bidirectional search, and unfortunately, the branching factor is high in both directions. Therefore, neither chaining backwards from the advice nor forwards from the planning situation looks attractive.

```
gc-schema2
  lhs1 goal =goal1               ; Someone has a goal
         actor agent =actor1     ; that an action occur
         state action =act
  lhs2 cause =cs                 ; which causes the
         cause  =goal1           ; action to occur.
         caused =act
  lhs3 action =act
         actor agent
  lhs4 violates-goal =v          ; The action violates a
         state =act              ; goal of someone else.
         goal  =goal2
  lhs5 goal =goal2
         actor agent =actor2
  rhs  goal-conflict
         actor1 =actor1
         actor2 =actor2
         goal1  =goal1
         goal2  =goal2
  abductively-infer (=goal1 =cs)
```

Figure 5: A viewing schema for goal conflicts.

One way to achieve greater efficiency would be to make the abstract concept work together with the specific planning situation to guide inference. This suggests the following general strategy for organizing memory to support redescription inference: create schemas for resolving differences in vocabulary and index them in terms of the endpoints of the inference chains they help to construct. Search can then take place in a much smaller space of ways to instantiate these schemas.

We have followed this approach in BRAINSTORMER. Associated with each of BRAINSTORMER's hard-wired abstract concepts such as goal conflict are **viewing schemas**, which are special rules for inferring that abstract concept. The antecedents of these rules are plausible recognition conditions for instances of the abstract concept; their consequents specify how the abstract concept should be instantiated. A viewing schema that can relate the terrorist attack to the goal conflict is shown in figure 5. The **lhs** slots are the schema's antecedents, the **rhs** slot its consequent. The **abductively-infer** slot is described below.

Viewing schemas are indexed in memory in multiple ways, under every conjunction of the form ($concept, recog_i$), where *concept* is the concept that the schema infers, and the $recog_i$ are components of the schema's recognition conditions whose presence is predictive of the schema's applicability. The schema shown in figure 5, for example, is indexed in terms of (**goal-conflict**,**action**), (**goal-conflict**,**goal**), and (**goal-conflict**,**violates-goal**).

## The Redescription Inference Process

BRAINSTORMER's memory of viewing schemas helps it to adapt abstract advice it receives to fit queries from the planner: whenever the system tries to match pairs of incommensurate concepts, it uses viewing schemas to efficiently resolve its difficulty. For example, we saw BRAINSTORMER attempt to match a `goal-conflict` in advice it is handed to a `terrorist-attack` in a query. These two concepts are incommensurate, so BRAINSTORMER retrieves all viewing schemas indexed in terms of them, or generalizations thereof.

The system's predefined concepts are arranged in a type hierarchy; `action` dominates `terrorist-attack` in the hierarchy.[3] Therefore, the system retrieves `gc-schema2`, as shown above, using (`goal-conflict`, `terrorist-attack`) as an index. Next, the system matches the `goal-conflict` in the advice to the consequent of the viewing schema and matches the `terrorist-attack` to its `lhs3` antecedent, then attempts to satisfy the schema's remaining antecedents.

BRAINSTORMER tries to match all of these antecedents to existing memory representations, but is prepared to abductively infer some of them if necessary. First, the system attempts to retrieve items from memory that satisfy the antecedents of the viewing schema. In this example, it retrieves its own goal to prevent terrorist attacks. Next, the system tries one-step backward chaining to satisfy the remaining antecedents. Limiting the system to one-step backward inference is somewhat arbitrary, but it has proved sufficient for our needs, while putting a hard limit on the adapter's computation. Finally, if some antecedents are still unsatisfied, then the system uses the `abductively-infer` slot of the viewing schema to determine if it is reasonable to abductively hypothesize them: abductive inference can proceed if the remaining antecedents are a subset of the filler of this slot. (See figure 5.)

If all antecedents of the viewing schema are now satisfied, the resulting variable bindings are then used to instantiate the schema's consequent, which is already bound to a component of the advice (here, the `goal-conflict`). Finally, a `part-whole-view` frame is built from this instantiated schema, as shown earlier in figure 4, and redescription inference is complete.

Matching antecedents of a viewing schema can recursively trigger further redescription inference. In the current example, the `lhs3` recognition condition of the viewing schema requires that the actor of the `action` be of type `agent`. The actor of the `terrorist-attack`, however, is a *group* of terrorists, and groups are not agents. BRAINSTORMER is, however, capable of redescribing a group of agents as a composite agent by retrieving and applying an appropriate viewing schema.

The possibility of recursive redescription inference means that BRAINSTORMER's matching algorithm is actually a kind of *means-ends analysis* process [Fikes and Nilsson, 1971], where the differences to be resolved are incommensurate type labels on concepts being matched, and the operators for resolving differences are viewing schemas. BRAINSTORMER can thus engage in a nontrivial amount of search in attempting to match two incommensurate items. It is important to note, however, that this means-ends analysis approach is considerably more efficient than any obvious forward or backward chaining alternative, because search is much more tightly focused. Inference is only attempted if an important recognition condition of a concept is already known, and the system already wants to relate this condition to the concept. Figure 6 sketches BRAINSTORMER's algorithm for matching advice to queries.

```
(MATCH adv qry):
  IF the type labels of adv and qry match
  THEN recursively MATCH corresponding
       slot fillers of adv and qry
  ELSE (REDESCRIBE adv qry).

(REDESCRIBE adv qry):
  1. Retrieve all viewing schemas indexed under
     (adv',qry'), where adv' and qry' are general-
     izations of adv and qry in the type hierarchy
  2. Until successful, or no more viewing
     schemas, pick a viewing schema and
     a. MATCH adv to the schema's consequent;
        MATCH qry to the appropriate antecedent.
     b. Attempt to satisfy the other antecedents:
          i.   Try memory retrieval.
          ii.  Try one-step backward chaining.
          iii. Try abductive inference.
     c. IF all antecedents are satisfied
        THEN schema application is successful.
          Construct an appropriate view frame
          as the binding associated with adv.
```

Figure 6: Matching and the redescription process.

## Discussion

BRAINSTORMER's task is adapting abstract cases to fit planning specific situations; we have seen that this requires being able to dynamically categorize these situations as instances of abstract concepts. Existing representation systems are generally inadequate to this task. Like many frame-based representation systems, BRAINSTORMER starts with a fixed set of base concepts arranged in a taxonomic hierarchy that supports property inheritance. The system's representations of planning situations and advice are initially encoded as instances of these concepts. In contrast to many other systems, however, the redescription process allows BRAINSTORMER to flexibly and dynamically redescribe instances in terms of other concepts, thereby categorizing them in new ways. BRAINSTORMER can categorize instances both in terms of other base concepts and in terms of new concepts that it dynamically constructs from recognition conditions for other concepts.

BRAINSTORMER's redescription mechanism relates to three areas of past work in cognitive science and ar-

[3] Two type labels are commensurate if and only if they can be related in the system's type hierarchy.

tificial intelligence. First are investigations into the idea of viewing or redescription inference, starting with MERLIN [Moore and Newell, 1973] and KRL [Bobrow and Winograd, 1977] in the 1970's. More recently, Jacobs has resurrected these ideas in the context of natural language generation [Jacobs, 1987]. His ACE system redescribes concepts with no associated generation method in terms of other concepts that have one. ACE cannot create new concepts on the fly, but it would be pointless to do so in any case, as the newly created concepts would lack generation methods. Redescription in BRAINSTORMER, in contrast, is driven by the need to relate abstract advice to concrete situations. We have seen that BRAINSTORMER needs to introduce new concepts to adequately express the intended meaning of abstract advice as it applies to a specific situation.

Our research also relates to past work on analogical reasoning. For example, recategorizing a terrorist attack in terms of recognition conditions for a goal conflict can be thought of as answering the question "in what way is a terrorist attack like a goal conflict?" Redescription maps the terrorist attack into a target concept that is simultaneously (1) an instance of a covering class of intentional actions that violate goals of others and (2) a set of conditions sufficient for inferring a goal conflict. This is a kind of analogical reasoning. Unlike many existing systems, however (e.g., Gentner's structure mapping engine [Falkenhainer *et al.*, 1986]), BRAINSTORMER does not depend on preexisting structural or predicate correspondences between base and target domains. All that is required is that the source can be extended to a set of plausible recognition conditions for the target.

Finally, our work extends earlier research on case adaptation in the field of case-based reasoning. In particular, Kass describes his research on ABE [Kass, 1989] as extending script/frame theory to handle a wider range of input situations; similarly, we are interested in flexibly relating past cases to new situations. BRAINSTORMER's cases, however, are much more abstract than ABE's. Consequently, a principal concern of BRAINSTORMER is making abstract descriptions more specific; in contrast, ABE spends most of its time trying to replace components of past explanations that do not apply in a new situation with different components that do. Therefore, while dynamic redescription is of central importance in BRAINSTORMER, it does not exist in ABE, and if it did, it would probably play a peripheral role.

## Conclusion

In this paper, we have focused on the problem of adapting abstract knowledge about planning to fit specific planning situations. We have shown this to entail redescribing aspects of particular planning situations in terms of a dynamically-constructed abstract concepts. We have presented a memory organization and a redescription process that efficiently accomplishes this task. Our redescription process is implemented as part of the BRAINSTORMER system, a planner that takes abstract advice in the context of ongoing problem solving.

**Acknowledgements.** Thanks to Eric Domeshek, Andy Fano, Bruce Krulwich, and David Pautler for comments on drafts of this paper.

### References

[Bobrow and Winograd, 1977] D. G. Bobrow and T. Winograd. An overview of KRL, a knowledge representation language. *Cognitive Science*, 1(1):3–46.

[Cullingford, 1977] R. Cullingford. *Script Application: Computer Understanding of Newspaper Stories.* PhD diss., Yale Univ.

[Falkenhainer *et al.*, 1986] B. Falkenhainer, D. K. Forbus, and D. Gentner. The structure-mapping engine. In *Proceedings AAAI-86*, 272–277, Philadelphia, PA. AAAI.

[Feigenbaum *et al.*, 1971] E. A. Feigenbaum, B. Buchanan, and J. Lederberg. On generality and problem solving: a case study using the DENDRAL program. In B. Meltzer and D. Michie, editors, *Machine Intelligence 6*, 165–190. Edinburgh University Press, Edinburgh.

[Fikes and Nilsson, 1971] R. E. Fikes and N. J. Nilsson. STRIPS: A new approach to the application of theorem proving to problem solving. *Artificial Intelligence*, 2:189–208.

[Hammond, 1986] K. J. Hammond. *Case-based Planning: An Integrated Theory of Planning, Learning and Memory.* PhD diss., Yale Univ.

[Jacobs, 1987] P. S. Jacobs. Knowledge-intensive natural language generation. *Artificial Intelligence*, 33:325–378.

[Jones, 1990] E. K. Jones. BRAINSTORMER: A model of advice taking. In *Proceedings of the Twelfth Annual Conference of the Cognitive Science Society*, 269–276, Cambridge, MA. Cognitive Science Society.

[Jones, 1991] E. K. Jones. Knowledge refinement using a high-level, non-technical vocabulary. In *Machine Learning: Proceedings of the Eighth International Workshop*, Chicago, IL. Morgan Kaufmann.

[Kass, 1989] A. Kass. Adaptation-based explanation: Extending script/frame theory to handle novel input. In *Proceedings of the Eleventh International Joint Conference on Artificial Intelligence*, 143–147, Detroit, MI. IJCAI.

[Kolodner *et al.*, 1985] J. L. Kolodner, R. L. Simpson, and K. Sycara-Cyranski. A process model of case-based reasoning in problem-solving. In *Proceedings of the Ninth International Joint Conference on Artificial Intelligence*, Los Angeles, CA. IJCAI.

[Moore and Newell, 1973] J. Moore and A. Newell. How can MERLIN understand? In L. W. Gregg, editor, *Knowledge and Cognition.* Lawrence Erlbaum, NJ.

[Owens, 1988] C. Owens. Domain-independent prototype cases for planning. In *Proceedings of a Workshop on Case-Based Reasoning*, 302–311, Clearwater, FL. Defense Advanced Research Projects Agency, Morgan Kaufmann.

[Rieger, 1976] C. Rieger. An organization of knowledge for problem solving and language comprehension. *Artificial Intelligence*, 7:89–127.

[Schank, 1986] R. C. Schank. *Explanation Patterns: Understanding Mechanically and Creatively.* Lawrence Erlbaum Associates, Hillsdale, NJ.

[Sowa, 1984] J. F. Sowa. *Conceptual Structures: Information Processing in Mind and Machines.* Addison-Wesley, Reading, MA.

[White, 1987] G. M. White. Proverbs and cultural models. In Dorothy Holland and N. Quinn, editors, *Cultural Models in Language and Thought*, 151–172. Cambridge University Press, New York.

[Wilensky, 1978] R. Wilensky. *Understanding Goal-Based Stories.* PhD diss., Yale Univ.

# Adaptation Strategies for Case-Based Plan Recognition*

Alex Kass
The Institute for the Learning Sciences
Northwestern University
1890 Maple Avenue, Evanston, IL 60201

## Abstract

Figuring out what plan another agent might be executing is an important and difficult type of explanation problem which involves a special type of knowledge about plans and goals. Elsewhere, we have discussed an approach to the general explanation problem that involves adapting stored explanations to new situations. In this paper, we review that discussion very briefly and then examine the special issues that arise when applying that approach to explaining intentional actions, focusing on adaptation strategies that are specifically relevant to plan recognition.

## Introduction

Making sense of other agents' actions can be hard. Reconstructing the chain of reasoning that led someone to believe that an action might service its goals is complex for several reasons: (1) the other agent's goals may be obscure; (2) the set of goals that a given action might conceivably serve can be large; (3) the chain of reasoning needed to connect actions to goals can be long and complex; and (4), the other agent's knowledge may be incomplete or even incorrect.

Plan-recognition is a type of explanation problem in which the explanations are cast in terms of goals and intentional actions. The principal problem in designing any explanation-building system is how to bring the system's causal knowledge to bear on a problem so that it can efficiently infer an unseen cause of observed events. In [Kass *et al.*, 1986, Kass, 1990a, Kass, 1990b] we have discussed an approach to the explanation problem that involves adapting stored explanations to new situations. In this paper we review that discussion and then examine the special issues that arise when applying that approach to explaining intentional actions, focussing on adaptation strategies that are specifically relevant to plan recognition.

## Previous approaches

Approaches to plan recognition have essentially fallen into two categories. The first approach was to chain inferences together on the fly, in response to each new problem. Examples of this approach include [Rieger, 1975] and [Wilensky, 1978]. Researchers who take this approach are led to concentrate on methods of controlling the combinatorial explosion inherent in the inference-chaining process. Parallel methods of rule activation, bi-directional search, and application of either domain-dependent or abstract, thematic search-control knowledge are among their important methods.

There is another school of research that seeks to view the entire explanation process as an indexing problem. These researchers postulate the existence of ready-made knowledge schemas, such as scripts or frames ([Schank and Abelson, 1977], [Cullingford, 1978], [Minsky, 1975], [Charniak, 1977]) that will provide all of the proper inferences, if only the right structure can be found in memory. This is the "retrieve and apply" school of explanation construction. Its concerns revolve around how these knowledge structures are represented, and especially how they are organized in memory so that the right one can be found at the right time.

Both approaches have crippling limitations. The script-application approach originally arose out of the realization that the "chaining + control" method was ignoring issues of memory. Rule-chaining systems treat each experience as completely novel, dooming them to perform expensive analyses every time they recognized a new instance of a plan. However, the script-application approach can only be employed when memory contains a script that *precisely* matches the plan to be recognized. In order for the matching process itself to be relatively inference-free, it is important that the expectations represented in a script be at a very specific level. This means that a "retrieve and apply" system is stymied when confronted with situations that stray even slightly from its specific expectations. In short, "chaining + control" is a weak approach because it knows nothing of stereotypes, and "retrieve and apply" is limited because it knows *only* its stereotypes. What's needed is a system that knows stereotypes, but also knows how to go beyond them when necessary.

## Adaptation-based Explanation

It is possible to build a system that enjoys both the efficiency benefits of employing stored knowledge structures *and* the flexibility to handle novel input by extending script/frame theory in the following ways:

*This work was supported in part by the Defense Advanced Research Projects Agency and monitored by the Air Force Office for Scientific Research under contract number F49620-88-C-0058. The Institute for the Learning Sciences was established in 1989 with the support of Andersen Consulting, part of The Arthur Andersen Worldwide Organization.

- Augmenting the knowledge structures stored in memory, so that in addition to representing specific expectations to be matched, each structure also includes a representation of the causal inferences that underlie those specific expectations.
- Building into the system strategies that make use of this causal annotation to *adapt* the structures it retrieves when they don't precisely match the plan being recognized.

In this augmented theory, scripts evolve into variablized explanations (called Explanation Patterns, or XPs for short), and "retrieve and apply" evolves into "retrieve, evaluate, adapt, apply". This is essentially an application of the case-based methodology applied to constructing explanations. (For a discussion of XPs see [Schank, 1986], [Kass, 1986], and [Leake and Owens, 1986]. For examples of some previous work relevant to case-based reasoning see, [Kolodner *et al.*, 1985], [Simpson, 1985], [Hammond, 1986], and [Sycara, 1987]). The system uses its pre-stored stereotypes as a starting point, but when it determines that its stereotyped explanations don't quite fit a new situtION, it can learn a new variation on one of those explanations, which, in addition to enabling it to understand the current input, can be stored in memory to help process future inputs.

ABE (Adaptation-Based Explainer) [Kass, 1990b] is an implementation of this theory that can develop explanations of unexpected deaths and disasters by adapting XPs stored in its memory to new contexts. For instance, ABE can adapt an explanation of a human dying from the exertion of recreational jogging to explain the death of a famous racehorse, or it can explain death of a star athlete by adapting an explanation of why rock stars die of drug overdoses.

ABE employs a library of XPs containing the explanations that the system has generated to solve past problems, as well as a set of patterns that is hand-coded into it. ABE has a number of built-in XPs about death, some volitional, others non-volitional. For instance, ABE knows explanations of death involving drug overdoses, overexertion, and terrorism. In addition to its XP library, ABE contains a large knowledge-base of specific facts and causal rules, and a semantic hierarchy for objects and actions.

The core of ABE is its library of tweaking strategies. Each strategy encodes an algorithm for making a small change to an XP in response to a specific mismatch between the XP and a particular anomaly. There are three main categories of tweaking strategies: Those which replace slot fillers with causal equivalents, those which generalize over-specific explanations, and those which elaborate incomplete explanations. Tweaks are more general than domain-specific heuristics, but they are quite knowledge-based. Each tweak knows how to search the system's semantic memory for domain-specific knowledge, and to use that knowledge to produce a slightly altered variation of a given XP. Thus, tweaking strategies relevant to plan recognition do not encode knowledge that is specific to any particular planning domain (that knowledge is encoded in the knowledge base). However, they do encode general knowledge about planning, about what can make a plan-based explanation more plausible, and about how plan-related knowledge can be found in the knowledge base. Since explanations involving intentional actions involve special types of beliefs and rules, it is useful for a system to employ special memory-search and inference techniques when modifying these explanations. The remainder of this paper will be devoted to illustrating these techniques through a discussion of three such strategies.

## Three plan recognition strategies

Explanations that involve volitional actions only make sense if the agents involved could reasonably be imagined to have performed each of the actions that the explanation hypothesizes they performed. In other words, evaluating whether such an explanation makes sense involves simulating the various agents' decision-making processes. This requires some specialized knowledge about motivation and decision-making. The simple utilitarian model of decision-making which underlies all three of the strategies described here is that an agent weighs the positive effects of a prospective action against the negative effects, and then decides whether to perform the action based on whether the positive outweighs the negative. These weighting functions are very subjective, of course, so that in order to accept the hypothesis that a given agent might have carried out a particular plan, an evaluator must attempt to understand the situation from the other agent's point of view. The system must simulate the agent's ability to project the effects of the action, then it must decide whether the action would be worth performing from the agent's point of view by summing the projected positive and projected negative effects and seeing whether the grand total is positive or negative. When the system's simulation of the agent predicts that the agent would give a negative score to an action which explanation claims the agent performed, the explanation will not be acceptable. Building an acceptable variant on such an explanation involves accounting for the discrepancy.

Let's consider this in terms of an example story, and how explanations can be adapted to provide an explanation of the story.

- **The Suicide Bomber Story**

  A teenage girl exploded a car bomb at a joint post of Israeli troops and pro-Israeli militiamen in southern Lebanon. The bomber and a number of Israeli soldiers were killed by the blast.[1]

ABE knows a standard TERRORIST BOMBING XP, which essentially states that people who feel oppressed will sometimes explode bombs in the vicinity of those

[1]See [Ram, 1989] for an in-depth discussion of many issues related to understanding this story.

who they consider the oppressor. Applying that XP to this story yields a plausible reason why the bomber killed her victims. But that explanation has a major inadequacy when applied to this particular story; the fact that the bomber did something which resulted in her own death, which is a central part of the anomaly raised by this is not addressed by the XP. In other words, the standard TERRORIST BOMBING XP is inadequate because the negative side effects of the action seem to outweigh the benefits.

The TERRORIST BOMBING XP is clearly a reasonable starting point for understanding this story even though it does not explain the entire anomaly. It would be a mistake to ignore the XP entirely. But in order to produce an explanation which adequately covers this story the TERRORIST BOMBING XP must be amended with sub-explanations that explain why the bomber might have decided that objective(s) achieved outweighed the negative side-effects.

ABE knows several strategies for elaborating its explanations in cases like this. Each strategy encodes a method of producing a reason that the system's simulation of the agent's plan evaluator might have been inadequate. Each can be conceptualized as an algorithm for asking — and answering — a specific question about the explanation. Such strategies do inference chaining, which can potentially be expensive. However, since they only add small sections to otherwise preconstructed explanations, the expense is much less than chaining together entire explanations from scratch.

In the sections that follow, we describe three plan-recognition-related adaptation strategies in terms of the general question that each strategy asks, examples of how that question is instantiated to build variations on the TERRORIST BOMBING XP that account for the SUICIDE BOMBER STORY, and some of the details of the memory-search and inference algorithms the strategy uses to answer that question.

## Elaborate motivation : Downplay negative

**Underlying question:** TWEAK 1 involves asking, **"Why would the negative effects of the action be particularly unimportant to the agent?"**

**Example:** TWEAK 1 searches for rules that indicate reasons why one of the bad side-effects of a problematic action might have been given less weight by the agent than the evaluator would have expected. In the suicide-bombing example, dying might have been less important to the suicide bomber if she were convinced that she knew she would die soon anyway, or if she were depressed enough to have decided that it wasn't particularly important to go on living.

**Discussion:** The system's knowledge base contain a set of inference rules called value-system adjustment rules. These rules indicate ways in which the desirability of certain states can become raised or lowered for a particular individual. The subcategory of value-system adjustment rules used by this strategy are those concerned with reasons that the importance of a preservation goal can become reduced. They are called *goal-state value decreasers*, and have the following form:

- **General form of a goal-state value decreaser:** Preserving goal $G$, will decrease in importance for agent $A$ if fact $F$ is true of $A$.

  Examples of goal-state value-decreaser rules:

1. The goal of living in a neighborhood with a good school system will become less important to $A$ when $A$'s children get beyond school age.
2. The goal of staying alive may become less important to $A$ if $A$ has a chronic, painful illness.
3. The goal of staying alive may become less important to $A$ if $A$ is deeply depressed.

Goal-state value decreasers are used by TWEAK 1 to build a causal chain linking a state that the agent is in, to a reason why he might not have attached much importance to the projected negative effect. For each of the projected negative side effects of the problematic action, the system collects the value-decreasers relevant to that event-type. It then tries using those rules to construct a causal chain linking a fact about the agent to a reason he might value the negative side effect less than the evaluator originally predicted.

When trying to process the suicide bombing story, the system would retrieve rules 2 and 3, and attempt to infer that the agent was depressed or terminally ill.

**Inference and memory search:** There are two components to the task of building the causal chain mentioned above. The first is to retrieve relevant value-decreaser rules, and the second is to build the inference chain necessary to satisfy the antecedent portion of the rules. Recall that each of the rules has three slots, $G$, $A$, and $F$, corresponding to the *goal* whose value is decreased by the rule, the *agent* for whom it is decreased, and the *fact* which must be true about the agent in order to put the rule into effect. Retrieving rules which may be applicable involves finding all the rules whose $G$ slot can be filled by any of the event types that any of the negative side effects of the action fall into.

The steps followed in the attempt to build this chain are as follows:

1. The first step in the process is to collect all the negative side-effects of the action.
2. These events are used to retrieve goal-state value-decreasing rules. Event categories all have links to rules which call for events of that category to fill their $G$ slots. (For example, the event category, M-DIE has links pointing to rules like 2 and 3 above).
3. If any appropriate rules are found, the system then checks to see if the fact which fills the $F$ slot is known to be true of the agent in question. This means checking to see if the knowledge-base explicitly states that this fact is true of the agent himself, or of a category which the agent is in. If no such fact is retrieved, a limited amount of bi-directional inference

is performed to try to infer that $F$ is true of $A$. The system forward chains from the facts which are explicitly stated about $A$, and backward chains from rules which would imply $F$, attempting to find an intersection between the two chains. The amount of chaining is an adjustable parameter set by the system at large.

### Elaborate motivation : Amplify positive

**Underlying question:** TWEAK 2 involves asking, **"Why would the positive effects of the anomalous action be particularly important to the agent involved?"**

**Example:** TWEAK 2 is like TWEAK 1 in that it also looks for value-system adjustment rules which might be relevant to the story at hand, but TWEAK 2 looks for a different kind of rule. While TWEAK 1 looks for a rule indicating why a bad side could have been less important than normally expected, TWEAK 2 assumes that the negative side effects have their normal value, and instead searches for a reason to believe that one of the positive effects might have been *more* important than expected.

For example, the TERRORIST BOMBING XP could be made suitable to the suicide bombing story if it were elaborated to include a sub-explanation that showed why the death of the Israeli soldiers was especially important to the bomber; so important that it outweighed the negative effect of the bomber's own death. For instance, since the system knows that the importance of killing an enemy will increase for an agent if the enemy kills a member of the agent's family or one of his close friends, and it knows that many Palestinians have had family and friends killed by Israeli soldiers, it could hypothesize that one of the soldiers might have killed a friend or family member of the bomber.

**Discussion:** It is important to recognize that the main focus of this strategy is *not* to identify a positive effect of the action which the evaluator didn't originally project (ABE has another strategy — not described in this paper — which does that). The main idea behind TWEAK 2 is to isolate one of the positive effects which the original explanation projects, and to propose a previously-overlooked reason why the agent involved might have placed more importance than expected on that positive effect; in other words, this strategy works by proposing that an agent employed a different value system than expected, rather than proposing that he had different expectations.

The rules used to suggest an increase in how much importance an agent will place on a certain outcome are called *goal-state value increasers*. They have the following form:

- **General form of a goal-state value increasers**: Achieving goal $G$ will increase in importance for agent $A$ if fact $F$ is true of $A$.

Some examples of value-system modification rules of this type are:

1. The goal of acquiring alcohol will increase in importance for $A$ if $A$ is an alcoholic.
2. The goal of living in a safe neighborhood will increase in importance for $A$ if $A$ has children.
3. The goal of killing a military enemy will increase in importance for $A$ if the enemy directly harms someone close to $A$.

TWEAK 2 uses goal-state value increasers in a manner very similar to the way in which TWEAK 1 uses the value decreasers; the rules guide the strategy in its search for a causal connection between the problematic action and a state which the agent might have had an exceptionally strong desire to achieve. TWEAK 2 examines the states which typically result from the action and then searches for a goal-state value increaser which admits one of those states in its $G$ slot and whose $F$ slot matches a fact true about the agent.

It is important to emphasize that the system does *not* attempt to foresee every conceivable effect of the action. This would be too hard (the combinatorial explosion which results from considering the effects, and then the effects of the effects, and so on would quickly overwhelm the system); it would also be inappropriate because the idea is to simulate the agent's thinking, and the agent cannot be expected to have thought through every possible implication of his actions either. The set searched is a smaller one: those indexed as **stereotypical** effects of the action, and those which are mentioned as effects of the action by either the XP or the current story.

With regard to the suicide bomber example, the system would examine the effects of the bombing, principal among these being the deaths it caused. It would then retrieve the third of the above rules. Next it would try to determine if the soldiers might have killed anyone close to the bomber. Since it doesn't know anything directly about the bomber herself, it would examine what it knew about categories the bomber is in. Since the bomber is a Palestinian, and the system knows that many Palestinians have had friends and family killed by Israeli soldiers, it would construct the hypothesis that perhaps the bomber had had someone close to her killed by the Israeli soldiers, and that because of this, the value of killing the soldiers had become elevated above the counter-balancing desire to keep herself alive.

**Inference and memory search:** The following outlines how these rules can be put to use by the adapter to build the sub-explanations needed. Most of the inference algorithm is quite analogous to the one used by TWEAK 1; the main differences are that TWEAK 2 uses positive effects and value increasers where TWEAK 1 uses negative effects and value decreasers. There is one significant addition to the algorithm; if the explicitly-mentioned positive effects don't index any useful value-increasers, the strategy adds an extra step, in which inference is performed to determine implications of the positive effects which are explicitly mentioned by the XP, in order to see if the implications link to any goal-state value increasers.

1. The process begins by collecting the events which the original XP mentions as causal consequences of the problematic action. These events serve as the first candidates in the search for an event which may have had special importance to the agent.
2. The system then uses these candidate events collected as indexes into the library of goal-state value increasing rules, to see if there any rules in the knowledge base which call for the collected event to fill the $G$ slot.
3. If any appropriate rules are found, the system then checks to see if the fact which fills the $F$ slot is true of the particular agent involved in this situation. This involves checking if the knowledge-base explicitly states that this fact is true of the agent himself, or of a category which the agent is in. If that doesn't pan out, a limited amount of forward chaining from the facts which are true of the categories the agent is in is done to see if $F$ can be inferred to be true about the agent (the amount of forward chaining is one of the adjustable parameters set by the system at large).
4. If any of the events collected in step 1 index a rule which can be connected up to a category which the agent is a member of, then a new variation on the XP is built, incorporating this new causal chain. If no such chain can be built then a step of forward chaining from the events mentioned in the XP is performed, to form a new set of events to use as candidates, and the process goes back to step 2 to look for relevant rules again. In other words, if the immediate effects of the action don't yield any results, then the effects of those effects are tried. The loop continues until the forward-chaining limit is met, or until a reasonable chain is found.

## Elaborate motivation : Identify gap

**Underlying question:** TWEAK 3 involves asking, **"Which negative effect of his decision might the agent not have known about, and why?"**

**Example:** This strategy can be applied to suicide bombing example to give a different sort of answer than that given by the previous two strategies. Instead of looking for reasons why she would not have cared about dying (as TWEAK 1 does) or reasons why she would have been especially anxious to kill the Israeli soldiers (as TWEAK 2 does), TWEAK 3 would elaborate the TERRORIST BOMBING XP by identifying possible reasons that the bomber would have failed to predict one of the negative effects of the action, and therefore wouldn't have brought its negative value into the equation at all. For instance, by using this strategy the system could hypothesize that perhaps the bomber did not predict that she would die because she didn't realize that she would be inside the car at the moment of the explosion. Perhaps the people who convinced her to perform the bombing had not told her that the car would explode while she was inside.

**Discussion:** TWEAK 3 puts forth the hypothesis that an agent performed an anomalous action because he failed to project some negative effect of his actions. It attempts to explain why this would be by identifying which of the pieces of knowledge needed to infer the negative effect the agent might have been missing

There are two phases to this strategy. So the simpler part of the process is identifying a knowledge gap which might have been involved. The harder part of the process is to produce an explanation of *why* the knowledge gap might have existed. It isn't necessary to do this to produce a coherent explanation, but if such a sub-explanation is included it will make the overall explanation more convincing.

**Inference and memory search:** The first phase involves identifying a fact, $F$, such that if the agent hadn't known $F$, it would have caused him not to predict one of the negative effects of the action. This works as follows:

- First, collect up each expected negative side-effect of the action, along with the causal chain which allows one to infer that the negative side-effect.
- For each negative side-effect, $F1$, there is a set of other beliefs which comprise the inference chain which leads to the belief in $F1$. Each belief, $F2$ which appears on the inference chain leading to $F1$ is a candidate for a hypothesis of the form, "The agent did not anticipate $F1$ because he did not know that $F2$ was true."

The second phase involves building a sub-explanation to explain why it would be the case that "The agent did not know that $F2$ was true." This involves explaining why the inference leading to $F2$ would not take place when the agent projected the results of his action.

Building sub-explanations must be handled by another strategy, the general sub-explanation builder (not described in this paper, see [Kass, 1990b]). Some of the kinds of explanations relevant to why an agent might not believe a particular fact are as follows:

- Is there a plausible contradictory belief that the agent might have held? For example, the bomber could have believed that the car wouldn't blow up while he was inside because he believed that there would be a time delay between when the car stopped and when it exploded.
- Could the agent have lacked the time necessary to infer the relevant belief. Perhaps in the heat of the moment the agent had failed to think about the fact that her plan implied that she would be inside the car at time of the explosion.
- Could the agent have lacked the inference rule necessary to infer $F2$? Could the suicide bomber have been missing the rule that proximity to the bomb is

what determines how much damage someone is likely to suffer, and therefore failed to realize that someone inside the car would be killed by the explosion?

A prospective knowledge gap is not considered a viable candidate if ignorance of the belief would have caused the agent to fail to predict the main positive effect of his action. For instance, it is not reasonable to hypothesize that the bomber did not know that explosives would cause damage — even though this knowledge gap would account for the bomber's failure to predict her own death — because this would also have caused the bomber to fail to predict that her enemy would be killed. On the other hand, it would be reasonable to entertain the hypothesis that the bomber did not know she would be inside the car when the bomb went off since this would break the causal chain which leads to one of the negative side effects without breaking the chain leading to the main desired effect.

## Conclusion

A case-based explainer depends on the availability of adaptation strategies that know enough about the structure of the knowledge base to efficiently perform the memory search and inference necessary to build variations on stored explanations in a variety of domains. Because plan recognition calls for explanations containing the special classes of inference rules and memory links that are relevant to understanding intentional actions, it also calls for a specialized set of adaptation strategies which know how to employ those classes of rules and links.

In this paper we have attempted to contribute to the understanding of case-based plan recognition by presenting three such strategies. While the three strategies described here represent only a fraction of what is necessary in order to account for the full range of failures that can occur when applying old explanations to new situations ([Kass, 1990b] describes 21 strategies; there are probably more waiting to be discovered) they illustrate the kinds of knowledge that the knowledge base must contain, and the kinds of search and inference techniques that can be used to apply such knowledge in building variations on stored explanations when attempting to understand novel plans.

## References


[Charniak, 1977] E. Charniak. Ms. malaprop: A language comprehension program. In *Proceedings of the Fifth International Joint Conference on Artificial Intelligence*, Cambridge, Mass., August 1977. IJCAI.

[Cullingford, 1978] R. Cullingford. *Script Application: Computer Understanding of Newspaper Stories*. PhD thesis, Yale University, 1978. Technical Report 116.

[Hammond, 1986] K.J. Hammond. *Case-based Planning: An Integrated Theory of Planning, Learning and Memory*. PhD thesis, Yale University, 1986. Technical Report 488.

[Kass *et al.*, 1986] A. M. Kass, D. B. Leake, and C. C. Owens. Swale: A program that explains. In *Explanation Patterns: Understanding Mechanically and Creatively*, pages 232–254. Lawrence Erlbaum Associates, Hillsdale, NJ, 1986.

[Kass, 1986] A. Kass. Modifying explanations to understand stories. In *Proceedings of the Eighth Annual Conference of the Cognitive Science Society*, Amherst, MA, August 1986. Cognitive Science Society.

[Kass, 1990a] Alex Kass. Adaptation-based explanation: Extending script/frame theory to handle novel input. In *Proceedings of the Eleventh International Joint Conference on Artificial Intelligence*, pages 143–147, Detroit, MI, August 1990. IJCAI.

[Kass, 1990b] A.M. Kass. *Developing Creative Hypotheses by Adapting Explanations*. PhD thesis, Yale University, 1990. Also available as ILS Tech Report No. 6, Northwestern University.

[Kolodner *et al.*, 1985] J. Kolodner, R. Simpson, and K. Sycara. A process model of case-based reasoning in problem solving. In A. Joshi, editor, *Proceedings of the Ninth International Joint Conference on Artificial Intelligence*, pages 284–290, Los Angeles, CA, August 1985. IJCAI.

[Leake and Owens, 1986] D. Leake and C. Owens. Organizing memory for explanation. In *Proceedings of the Eighth Annual Conference of the Cognitive Science Society*, Amherst, MA, August 1986. Cognitive Science Society.

[Minsky, 1975] M. Minsky. A framework for representing knowledge. In P. Winston, editor, *The Psychology of Computer Vision*, chapter 6, pages 211–277. McGraw-Hill, New York, 1975.

[Ram, 1989] Ashwin Ram. *Question-driven understanding: An integrated theory of story understanding, memory and learning*. PhD thesis, Yale University, New Haven, CT, 1989. In preparation.

[Rieger, 1975] C. Rieger. Conceptual memory and inference. In *Conceptual Information Processing*. North-Holland, Amsterdam, 1975.

[Schank and Abelson, 1977] R.C. Schank and R. Abelson. *Scripts, Plans, Goals and Understanding*. Lawrence Erlbaum Associates, Hillsdale, New Jersey, 1977.

[Schank, 1986] R.C. Schank. *Explanation Patterns: Understanding Mechanically and Creatively*. Lawrence Erlbaum Associates, Hillsdale, NJ, 1986.

[Simpson, 1985] R.L. Simpson. *A Computer Model of Case-based Reasoning in Problem-solving: An Investigation in the Domain of Dispute Mediation*. PhD thesis, School of Information and Computer Science, Georgia Institute of Technology, 1985.

[Sycara, 1987] E. P. Sycara. *Resolving Adversarial Conflicts: An Approach Integrating Case-based and Analytic Methods*. PhD thesis, School of Information and Computer Science, Georgia Institute of Technology, 1987.

[Wilensky, 1978] R. Wilensky. *Understanding Goal-Based Stories*. PhD thesis, Yale University, 1978. Technical Report 140.

# A Functional Taxonomy of Abstract Plan Failures

**Christopher Owens**
The University of Chicago, Department of Computer Science
1100 East 58th Street, Chicago, IL 60637
E-mail: owens@gargoyle.uchicago.edu

## Abstract

To reason about a plan failure requires an appropriate explanation of the failure. Such explanations, as has been argued elsewhere, can be generated by instantiating and adapting abstract knowledge structures. But the two fundamental goals in reasoning about failure, *recovering* from the failure and *repairing* the knowledge that led to the failure, require qualitatively different types of explanation. This paper presents an abstract taxonomization of failures, oriented toward the latter problem. Each abstract failure type in the taxonomy is tied to heuristics for recognizing occurrences of failures of that type and for identifying the knowledge that can be repaired to avoid future occurrences of the same failure.

## Failures and explanations

Autonomous agents operating in complex environments will inevitably experience failures. A robot vehicle, for example, may find an expected fuel cache missing, run low on fuel, and therefore be unable to continue operation.

A standard approach to handling failure (see, for example, [Sussman, 75], [Hayes-Roth, 83], [Hammond, 89], [Birnbaum and Collins, 88]) dictates that an agent experiencing a failure must both **recover** from the immediate failure, and **repair** its knowledge so as to avoid similar situations in the future. Recovery and repair are predicated upon building a good causal **explanation** of the failure.

For the purposes of reasoning about failure, an explanation is a pattern of inference that causally connects the anomalous experience to previously known facts. A good explanation is one that correctly assigns blame for the failure, and that assigns blame to some condition under the planner's control, therby leading to an appropriate recovery strategy. If the autonomous vehicle, for example, explains its predicament as being due to its own poor navigation and inability to find the cache, the appropriate recovery may involve gathering more position information. If the explanation posits that the fuel cache is actually missing, for example having been stolen, then the appropriate recovery involves obtaining fuel from some other source (turning back, calling for assistance).

Similarly, an appropriate explanation can lead to an appropriate repair. The robot, for example, could explain its current predicament in terms of a prior decision not to take on extra fuel, which was made with the expectation that the fuel cache would be present and easy to find. The explanation implicates the agent's reasoning about the persistence of fuel caches ("I should have known someone might have stolen the fuel."), its predictions of its own navigation abilities ("I should have realized it would be hard to find the fuel."), and its reasoning about resource reserves ("I should have taken on a full tank when it was available.").

### Memory-based explanation

An obvious way to build an explanation is to chain through a space of explanatory inference rules so as to connect the anomalous experience to known facts. Work on memory-based explanation (e.g. [Schank, 86],[Kass, 90], [Leake, 90], [Owens, 90a]) has argued that a better way to build explanations is to retrieve, instantiate, and adapt knowledge structures that characterize familiar, recurring patterns of causality. These knowledge structures are referred to as **explanation patterns**, or XPs. One advantage of explanation by adaptation and instantiation is that, given an incomplete and imperfect domain theory, explanations derived by incremental modification of known valid causal patterns are more likely to be useful than explanations built by chaining through the space of primitive rules. Another advantage is that storing successful explanations for re-use provides a natural learning mechanism.[1]

### Memory, recovery, and repair

In [Owens, 90a], and [Owens, 90b] I present an additional argument in favor of memory-based explanation. I argue that abstract explanatory knowledge structures known as **plan failure explanation patterns**, or

[1]These arguments are made in more detail in [Kass and Owens, 88].

PFXPs, can play a direct role in both recovery and repair. For each known category of failures, the corresponding PFXP provides a place in memory to store repair and recovery strategies appropriate to that class of failure. Instead of dynamically calculating a repair or recovery strategy from the explanation, a pre-existing strategy, stored in memory with the causal pattern, can be instantiated and used. For example, if a PFXP stores the causal pattern describing situations in which some needed object is not found where expected, it can package the recovery strategies:

- Gather more information about your current position.
- Gather more information about the location of the needed object.
- Find an alternate way to obtain the object.

A key question associated with such explanatory patterns is *indexing*: how should structures be labeled, and what features should the system use to describe the world. A step towards the development of labeling features is to taxonomize the types of explanations the system will be required to make. The appropriate basis for such a taxonomy is functional: the system should discriminate between different abstract patterns of causality occurring in the world, to the degree that different patterns of causality suggest different recovery and repair strategies.

## Taxonomizing for repair

Recovery depends upon the concrete particulars of the situation. Failures involving a resource shortfall suggest one class of recovery strategies; failures involving being in an inappropriate location dictate another; failures involving lacking some piece of information, a third. The appropriate explanation for selecting a recovery strategy is one that attributes blame to some condition in the world that is under the planner's control.

Repair, on the other hand, requires an explanation that ties the current failure to some inappropriate decision made by the planner, and which in turn ties that inappropriate decision to either a missing piece of knowledge or an inappropriate control strategy. Once the appropriate missing knowledge or inappropriate control strategy has been identified, the planner can learn by acquiring the knowledge or modifying the strategy.

The remainder of this paper outlines a taxonomy of abstract descriptions of plan failures. These abstract descriptions are presented at a domain-independent, but task-specific level of abstraction, and are oriented toward repair rather than recovery. This particular taxonomization breaks failures into **goal-related**, **task-related**, **resource-related**, and **time-related** failures. Each failure type is briefly described, along with strategies for recognizing and repairing failures of that type.

## Goal-related failures

Running out of fuel is on the surface a resource management problem, but not necessarily. It can also be the presenting symptom for goal-related problems. The robot, for example, may have failed to generate the goal of preserving its fuel supply, or it may have inappropriately let another goal take precedence. Agents mismanage goals by neglecting goals, by pursuing irrelevant goals, or by inappropriately mediating among conflicting goals. Many of the goal relationships described in [Schank and Abelson, 77] and [Wilensky, 83] can be useful in building goal-related explanations of failures, and it is in the context of that work that the the goal-related failure explanations below are described.

### Neglected goals

The simplest goal-related explanation for the out-of-fuel situation is that the agent did not have the goal of preserving its fuel supply. This could reflect such underlying flaws in the agent's knowledge as:

- Having no knowledge about the consequences of having no fuel.
- Failing to infer a threat to the goal of preserving fuel.

**Recognizing neglected goal failures** Recognizing a failure resulting from a totally absent goal is difficult. Since the vehicle does not have the goal of preserving its fuel supply, it will never notice nor reason about threats to that goal. A person who doesn't know that automobiles require gasoline, will be at a loss to explain why his car suddenly stops working. Only once an explanation is obtained — either by consulting an external knowledge source or, less likely, by reasoning from known principles, can the system build the causal connection between having an adequate fuel supply and being able to perform operations. At that point it can recognize that preserving a fuel supply is goal that should have been present.

If the agent has the goal of preserving its fuel supply but does not correctly infer a threat to that goal, the failure is easier to detect simply because the agent already has a goal structure in place against which to explain the failure: "My goal of preserving my fuel supply has been violated, and I have no record of having reasoned about any threats to that goal." Detecting this class of neglected goal failures requires maintaining a trace of reasoning about threats to goals, indexed by the goal that was threatened.

**Repairing plans that neglect goals** If a failure results from a system totally lacking a certain goal, then the obvious repair is to add that goal. But goals, in their detailed, specific instantiations, are not static objects in a system's memory. If one wanted to add, as a static entity, the goal to preserve one's fuel supply when operating in uncertain territory, it is not clear where in memory this static entity should reside or how it should be activated at the appropriate time.

If, on the other hand, a system explains its failure in terms of failing to notice a threat to a goal, the repair involves building a detailed causal explanation of the failure, and using that explanation to learn features that are predictive of the threat.

### Inappropriate goal priorities

More common goal-related failures involve assigning inappropriate priorities to conflicting goals. Many goal prioritization failures result from inappropriately resolving **goal conflict** or other goal interrelationships.

For example, the robot vehicle might have placed more emphasis on completing its mission quickly than on preserving its fuel supply, and therefore decided not to take on additional fuel from a known supply and chance the availability of the fuel cache.

### Detecting goal prioritization errors

Just as the system must keep a trace, indexed by the involved goals, of all reasoning about inferred threats to goals, it must also keep a trace of all goal conflict resolutions if it is to detect goal prioritization errors. If a plan fails and the initial explanation relates the failure to a violated goal, and the system has a record of having mediated a conflict between that goal and some other goal, then the goal mediation is suspect.

**Repairing goal prioritization errors** Explicit reasoning about conflicting goals is often expressed in terms of **costs** — paying too high a price to obtain some goal, or wrongly failing to undertake a plan when because the cost was (erroneously) considered too high.

But characterizing an error in terms of, for example, *being too conservative about costs*, does not repair the incorrect knowledge that caused the error. If the priorities of two goals are inappropriate (e.g. if the vehicle inappropriately rates the time necessary to take on extra fuel as more costly than the consequences of running out of fuel far from home, then the repair is to change the priorities of those goals. But one cannot simply find the "preserve fuel" goal and increase its weight, because there is no preserve fuel goal except when the goal is dynamically spawned by the interaction of knowledge about fuel supply and the knowledge about the likely fuel utilization of the current mission. If a failure results from goal conflicts being resolved inappropriately, than the repair must look at and modify the source of the goals.

## Task-related failures

The second major category of planning failures captures the errors that arise from choosing an inappropriate action in service of some goal, from pursuing an otherwise appropriate course of action either too vigorously or not vigorously enough, from wrongly combining two actions into a single plan, and from splitting a task into subtasks inappropriately.

### Task selection

Task selection errors involve choosing an action that is inappropriate either because it fails to achieve its goal or because, despite achieving the goal, it has an unacceptable side effect. A task-selection explanation of the out-of-fuel situation would be that searching for a fuel cache was an inappropriate task in service of the goal of obtaining fuel.

Although the action turned out to be inappropriate, the agent had good reason to expect the approach to work: searching for a cache is a known plan for obtaining fuel. It just didn't work here — either because another agent took the fuel or because the location of the fuel cache was not as well known as had been believed.

**Recognizing task selection failures** Whenever an action fails to achieve its intended goal, by definition the action was an inappropriate choice given the state of the world prior to undertaking that action.[2] The question, from the point of view of repair, is whether the choice was inappropriate given the **known** or **inferrable** state of the world. Certain conditions can strengthen the planner's belief that the choice of actions was faulty, the most significant being the planner's knowledge that it made a difficult choice. Just as with the traces of goal mediation described above, the agent can keep a trace of all decisions involving substitution of one action for another, indexed by the action that was the final result of the choice. For example, if the standard plans for obtaining fuel are to take on extra at a known supply point or to rely on a cache, and the system made a close call in choosing between them, that choice becomes suspect in building a repair-oriented explanation of the failure. Keeping track of selections that were close calls can be integrated into the framework for saving the justifications for decisions described in [Carbonell, 83] and [Carbonell and Veloso, 88].

**Repairing task selection errors** Merely recognizing that the planner chose an inappropriate action is not a functionally useful explanation because it does not suggest a repair to the planner's knowledge. Any of several causes might underlie the inappropriate choice of actions:

1. It was the only action known to satisfy that goal.
2. It was one of several actions known to satisfy that goal, but it was the only one without known unacceptable side-effects.
3. It was one of several apparently equally appropriate actions known to satisfy this goal, and the planner chose arbitrarily.
4. There was no action known to satisfy this goal, but the planner selected this action because it is known to satisfy a similar goal.

The repair consists of augmenting or refining those specifications. The planner can, for example, repair its

[2] Barring, of course, the explanation that the failure resulted from some unforseeable fluke beyond the control of the agent.

knowledge of the domain by adding new side-effects or constraints to the description of a particular action, such as "Searching for a cache is unreliable if the territory is known to be used heavily by other agents, or if the description of the location of the cache is known to be approximate."

### Task scaling

A different class of task-related failures results from pursuing a course of action too vigorously or from not pursuing it vigorously enough. For a robot vehicle, a clear example of performing an action too intensely is traveling too quickly. The actual failure can take the form either of a bad side effect of the action that would not be a problem at a reduced degree of intensity (running off the road) or from the waste of resources that is entailed.

Failing from insufficient intensity is a more obvious type of failure. The critical question in this latter class of situation is **why** the agent failed to perform the action with sufficient intensity — whether the required level of intensity was beyond the agent's capabilities, whether the agent did not assess the required level correctly, or whether the agent was overly conservative about a potential side effect.

**Recognizing task scaling failures** There are two different kinds of situations in which task scaling failures occur. In the first, the system made an explicit decision about the degree of intensity with which to pursue the task (such as deciding to travel faster than the "normal"[3] speed in order to accomplish some mission more quickly). In the second, the original "normal" level of intensity with which the action was performed was inappropriate — for example the normal rate of travel for a given road is inappropriate if it is snowing.

In the former case, with an explicit decision to modify the level of intensity, that decision can be remembered and used as a starting point for explanations of failures involving that task. In the latter case, failures are difficult to notice. "Overload" failures resulting from increasing the intensity of a particular action (e.g. running off the road from excessive speed) tend to have distinctive presenting symptoms, but failures that result from not performing an action intensely enough do not have their own set of presenting symptoms.

**Repairing task scaling errors** Obviously the repair to a plan causing a task scaling failure is to change the intensity with which the task is pursued. But, as was the case with goal prioritization errors, that dodges the issue of knowing **when** the change should be applied. If the vehicle runs off the road, does that mean that the vehicle should travel more slowly in general? Probably not.

A noteworthy aspect of the failures resulting from performing an action too intensely is that they result from taking a good plan (a travel plan), and applying a good plan transformation to it (get the job done faster by traveling faster.) Despite the desirable plan transformation strategy, the resulting, transformed plan, is inappropriate. The problem is analogous to the problem with task selection: either that the applicability conditions for that plan transformation were not well enough defined, or that the description of the current situation was not rich enough. The repair is to learn features that make sharper discriminations between situations in which the transformation is applicable and those in which it is not.

### Task combination and splitting

The final class of task-related planning failures involves splitting and combining tasks. Several transformations that are normally desirable can have bad results if applied indiscriminately. These are:

- Make maximum use of your time by working on more than one task at a time. When this fails, symptoms include missing deadlines and failing to complete subtasks because one lacks the capacity. The failure comes from the system's inability to predict the interaction of the two tasks; the repair is to enrich the descriptions of each of the tasks involved so that bad interactions can be more easily predicted.
- Subdivide a task and work on part of it at a time such that the entire task is done more slowly. Not all tasks can be subdivided because some tasks require continuity. One cannot, for example, strip the old paint off a house one season and paint it the next, because the intervening weather will destroy the exposed surfaces. Again, this failure comes from an inadequate representation of the interaction of the two tasks.

## Resource-related failures

The third major category of planning failures reflects to errors in resource management. Once a planner has identified a worthwhile goal and chosen a plan to satisfy this goal, errors arise in choosing the resources and the scale of the resources that will be used in service of this plan.

### Resource selection

There are two fundamental resource selection failures; they are analogous to the task selection failures described above. The first, **inappropriate resource failure**, occurs when an agent uses an inappropriate resource in service of an otherwise acceptable plan. The failure results from inadequate representations of the resources involved and inadequate specifications of the relationship of the resources to the task. If its knowledge about fuel and vehicles were underspecified, the robot vehicle, unable to obtain its usual fuel (diesel fuel), might inappropriately apply the transformation **substitute similar resource** and generate a plan to fill its tanks with some other kind of fuel (gasoline), or,

[3] A great deal has been swept under the representational rug here.

if its knowledge were even less rich, with some other liquid that is put in tanks in other situations (water). The problem is that the system's representations of its fuel needs, of gasoline, and of diesel fuel were not sufficiently rich.

The second, **resource substitution conservatism** occurs when an agent fails to recognize that one resource is an acceptable substitute for another. For example, the vehicle might have avoided filling its tanks with an acceptable substitute (winterized diesel fuel) if its normal fuel (straight diesel fuel) were unavailable.

**Recognizing resource selection failures** Resource selection failures are somewhat difficult to separate from task selection failures. If an agent chooses a particular task and a resource to implement that task and the task fails, it is difficult to say whether the task or the resource was inappropriate.

One heuristic solution is to look at which choice (action or resource) was more questionable at the time it was made (e.g. which involved a substitution or second choice, or which was decided arbitrarily). If the choice of tasks was less problematic than the choice of resources, then any subsequent plan failure can try to implicate the resource choice as a possible explanation. As with task selection, resource selection failures involving inappropriate conservatism are difficult to detect.

**Repairing resource selection errors** In a manner analogous to task selection errors, the repair of a resource selection error depends upon an explanation of why the system made that selection:

- If the resource was the only possible choice (if no substitute was available) there is no repair.
- If the resource was chosen arbitrarily from among several apparently equal possibilities, then the refine the description of the resource. For example, if gasoline and diesel fuel are both described as fuel for running engines, and if the system failed because it chose gasoline to fill the tanks of a diesel vehicle, then the selection criteria for fueling need to be modified to reflect vehicle engine type as a discriminator.
- If the resource was chosen as a substitute for another unavailable resource, then the substitution procedure needs to be modified.

### Resource scaling

Resource shortfalls provide a a good example of how explaining toward repair differs from explaining toward recovery. If a vehicle is low on fuel, the recovery strategies involve either finding more fuel or altering the operation so as to reduce the rate of fuel consumption. The repair strategies, however, require a different kind of explanation:

- The agent might have mis-estimated the amount of the resource necessary to complete the task, either by mis-estimating the scale of the task or by mis-estimating the rate at which the task consumes the resource.
- The agent might have mis-estimated the available stock of resource before commencing the task.
- The agent might have started the task without enough of the resource on hand, under the mistaken assumption that more would be available.

**Recognizing resource scaling failures** A resource shortfall is easy to detect. When a system goes to use a particular resource and finds it unavailable, the current plan will block. This condition will be the symptom by which the failure is made known to the planner. Unfortunately, as seen above, a resource shortfall can often be the symptom for other failures as well.

A resource surplus is more difficult to detect because there is no reason for a system to check its supply of a resource that it is not currently using. A resource surplus can be detected if it is explicitly searched for (when finishing a trip, check to see if there is excessive reserve fuel) but it does not present itself as a failure as does a resource shortfall.

**Repairing resource scaling errors** Repairing resource scaling errors is very much like repairing task scaling errors: the obvious repair is to apply more or less of this resource to the task next time. Analogously to task scaling repairs, there is a problem with situating the repair at the appropriate level of generality. If, for example, the vehicle finds that it did not take on enough fuel to complete the full range of planned activities, it should not always take on more fuel in the future. The system must explain why it predicted the required amount of the resource incorrectly, and it must modify the erroneous knowledge that led to that prediction.

## Time-related failures

The fourth major category of plan failures relates broadly to to time. Along with scaling errors involving decisions about how much time to spend on a particular task (which look very much like resource scaling errors), time-related errors include errors resulting from decisions about task scheduling and task ordering.

### Task ordering

"Goal Clobbers Brother Goal" and related characterizations in [Sussman, 75] and [Sacerdoti, 77] are the prototypical **task ordering** failures. Two specializations of task ordering failures are *failure to establish precondition* and *prematurely establishing precondition*. They describe the potential results of decisions about when to schedule the task that will establish a precondition for some larger goal. The system can establish this precondition as soon as possible, which guarantees that the precondition will be enabled in time to satisfy the larger goal but which risks wasting resources — establishing that precondition for nothing if the larger goal is deferred. On the other hand, the system can defer enabling a precondition until it becomes necessary, which avoids the risk of wasting the effort, but which risks the precondition not being enabled in time to meet the need..

**Recognizing task ordering failures** Failure to establish a precondition is its own presenting symptom. Consider, for example, arriving at a door and being unable to open it because one lacks a key. This failure can be a symptom of an improper task ordering, e.g. "I should have stopped by the hardware store to pick up the keys that were being copied *before* going home for lunch rather than on the way back to work." When the system is blocked because some precondition has not been met, it can check to see if it ever explicitly planned to achieve that precondition. If so, then the failure resulted from having scheduled that plan inappropriately. If, on the other hand, the system never planned to achieve the precondition, the problem is not one of task ordering but rather of a neglected goal described above.

**Repairing task ordering errors** As with other failure types, the correct repair to a task ordering error depends upon how the tasks came to be misordered. For example, if the tasks are mis-scheduled because some transformation put them out of order (see, e.g. [Collins, 87] for a taxonomy of task reordering transformations), then the transformation was insensitive to the dependence of one of the tasks on the other. As with task selection and resource selection errors, the problem is either that the descriptions of the tasks being reordered is underspecified, or that the applicability criteria for the re-ordering transformation is underspecified. The repair is to add discriminating features to either of these descriptions.

**Other scheduling errors** Other types of scheduling errors include making plans contingent upon an event whose occurrence has not been ascertained (counting on the availability of the fuel cache), for which the repair is to patch the knowledge used to predict the likelihood of the event.

## Conclusions

The purpose behind taxonomizing plan failures is that certain abstractly defined failures tend to recur. If a system possesses a canonical set of abstract descriptions of plan failures, it can use those descriptions to recognize occurrences of familiar failures, and can use the recognition as a basis for selecting appropriate recovery and repair strategies. Accordingly, the appropriate basis for taxonomizing plan failures is functional: it must be informed by a model of recovery and repair.

Examples of planning failures are important data that can serve two purposes. Using recovery as a functional basis for taxonomization can drive the development of domain-level knowledge about the relationship between an agent, its actions, and the environment. Using repair as a functional basis, on the other hand, can drive the development of a domain-independent model of the agent's interaction with its own knowledge.

## References


[Birnbaum and Collins, 88] L. Birnbaum and G. Collins. The transfer of experience across planning domains through the acquisition of abstract strategies. *Proceedings of a workshop on Case-based Reasoning*, Palo Alto, 1988. Morgan Kauffmann.

[Carbonell and Veloso, 88] J. Carbonell and M. Veloso. Integrating derivational analogy into a general problem solving architecture. *Proceedings of a Workshop on Case-Based Reasoning*, Palo Alto, 1988. Morgan Kaufmann, Inc.

[Carbonell, 83] J. Carbonell. Derivational analogy and its role in problem solving. In *aaai83*, Washington, DC, August 1983. AAAI.

[Collins, 87] G. Collins. *Plan creation: using strategies as blueprints*. PhD thesis, Yale University, 1987. Tech Report 599.

[Hammond, 89] K. Hammond. *Case-Based Planning: Viewing Planning as a Memory Task*, volume 1 of *Perspectives in Artificial Intelligence*. Academic Press, San Diego, CA, 1989.

[Hayes-Roth, 83] F. Hayes-Roth. Using proofs and refutations to learn from experience. In R. Michalski, J. Carbonell, and T. Mitchell, editors, *Machine Learning: An Artificial Intelligence Approach*, pages 221–240. Tioga, Palo Alto, 1983.

[Kass and Owens, 88] A. Kass and C. Owens. Learning new explanations by incremental adaptation. In *Proceedings of the 1988 AAAI Spring Symposium on Explanation-Based Learning*. AAAI, 1988.

[Kass, 90] A. Kass. Developing creative hypothesis by adapting explanations. Technical Report 6, Institute for the Learning Sciences, Northwestern University, 1990.

[Leake, 90] D. Leake. *Evaluating Explanations*. PhD thesis, Yale University, 1990. Technical Report 769.

[Owens, 90a] C. Owens. *Indexing and Retrieving Abstract Planning Knowledge*. PhD thesis, Yale University, Department of Computer Science, 1990. In preparation.

[Owens, 90b] C. Owens. Representing abstract plan failures. In *Proceedings of the Twelfth Annual Conference of the Cognitive Science Society*. CSS [?], 1990.

[Sacerdoti, 77] E. Sacerdoti. Language access to distributed data with error recovery. In *Proceedings of the Fifth International Joint Conference on Artificial Intelligence*, Cambridge, Mass., August 1977. IJCAI.

[Schank and Abelson, 77] R. C. Schank and R. Abelson. *Scripts, Plans, Goals and Understanding*. Lawrence Erlbaum Associates, Hillsdale, New Jersey, 1977.

[Schank, 86] R. Schank. *Explanation Patterns: Understanding Mechanically and Creatively*. Lawrence Erlbaum Associates, Hillsdale, NJ, 1986.

[Sussman, 75] G. Sussman. *A computer model of skill acquisition*, volume 1 of *Artificial Intelligence Series*. American Elsevier, New York, 1975.

[Wilensky, 83] R. Wilensky. *Planning and Understanding: A Computational Approach to Human Reasoning*. Addison-Wesley Publishing Company, Reading, Massachusetts, Reading, Mass, 1983.

# Using Semi-Distributed Representations to Overcome Catastrophic Forgetting in Connectionist Networks

Robert M. French
Center for Research on Concepts and Cognition
Indiana University
510 North Fess
Bloomington, Indiana 47408
e-mail: french@cogsci.indiana.edu

## Abstract

In connectionist networks, newly-learned information can completely destroy previously-learned information unless the network is continually retrained on the old information. This behavior, known as catastrophic forgetting, is unacceptable both for practical purposes and as a model of mind. This paper advances the claim that catastrophic forgetting is a direct consequence of the overlap of the system's distributed representations and can be reduced by reducing this overlap. A simple algorithm is presented that allows a standard feedforward backpropagation network to develop semi-distributed representations, thereby significantly reducing the problem of catastrophic forgetting.

## Introduction

Catastrophic forgetting is the inability of a neural network to retain old information in the presence of new. New information destroys old unless the old information is continually relearned by the net. McCloskey & Cohen [1990] and Ratcliff [1989] have demonstrated that this is a serious problem with connectionist networks. A related problem is that connectionist networks are not sensitive to overtraining. A network trained 1000 times to associate a pattern A with a pattern A' will forget that fact just as quickly as would a network trained on that association for 100 cycles. Clearly, this behavior is unacceptable as a model of mind, as well as from a purely practical standpoint. Once a network has thoroughly learned a set of patterns, it should be able to learn a completely new set and still be able to recall the first set with relative ease. In this paper I will suggest that catastrophic forgetting arises because of the overlap of distributed representations and I will present an algorithm that will allow a standard feedforward backpropagation (FFBP) network to overcome to a significant extent the problems of catastrophic forgetting and insensitivity to overtraining.

## Catastrophic forgetting and the overlap of representations

I suggest the following relation between catastrophic forgetting and representations in a distributed system:

> Catastrophic forgetting is a direct consequence of the overlap of distributed representations and can be reduced by reducing this overlap.

Very local representations will not exhibit catastrophic forgetting because there is little interaction among representations. Consider the extreme example of a look-up table where there is no overlap at all among representations. There is no catastrophic forgetting; new information can be added without interfering at all with old information. However, because of its completely local representations, a look-up table lacks the all-important ability to generalize.

At the other extreme are fully distributed networks where there is considerable interaction among representations. This interaction is responsible for the networks' generalization ability. On the other hand, these networks are severely affected by catastrophic forgetting.

The moral of the story is that you can't have it both ways. A system that develops highly distributed representations will be able to generalize but will suffer from castastrophic forgetting; conversely, a system that develops very local representations will not suffer from catastrophic forgetting, but will lose some of its ability to generalize. The challenge is to develop systems capable of producing semi-distributed representations that are local enough to overcome catastrophic forgetting yet that are sufficiently distributed to nonetheless allow generalization.

In what follows, I will examine two distributed systems that do not suffer from catastrophic forgetting. Both of these systems work because their representations are not fully distributed over the entire memory, but rather are semi-distributed and hence exhibit limited representation overlap, at least prior to memory saturation. Finally, I will present a simple method that allows standard layered feedforward backpropagation networks to develop semi-distributed representations in the hidden layer. Not only does this method appear to dramatically reduce catastrophic forgetting but it also allows the system's representations to partially reflect the degree to which a particular pattern has been learned. Even after a particular pattern has been learned, overlearning continues to modify connection weights in such a way that unlearning of the pattern will be made more difficult.

## Two examples of semi-distributed representations

I will briefly examine two systems that produce semi-distributed representations. In both systems, assuming that they are not saturated, there is little overlap of the representations produced. For this reason, they exhibit little catastrophic forgetting.

### Sparse Distributed Memory

Sparse Distributed Memory (hereafter, SDM [Kanerva 1988]) is an auto-associative, content-addressable memory typically consisting of one million 1000-bit "memory locations". The memory is called "sparse" because it uses only one million locations out of a possible $2^{1000}$ (i.e., $10^6$ of approximately $10^{300}$ possible locations). At each of these locations there is a vector of 1000 integers, called "counters". New data are represented in the system as follows: If we wish to write a particular 1000-bit string to this memory, we select all memory locations that are within a Hamming distance of 450 bits of the write address. This gives us approximately 1000 locations (i.e. 0.1% of all of the entire address space). Wherever there is a 1 in the bit-string to be written to memory, we increment the corresponding counter in each of the vectors at the 1000 memory locations; wherever there is a 0, we decrement the corresponding counter. This is clearly a semi-distributed representation of the input data: storage of the bit-string is distributed over 1000 different memory locations but these 1000 memory locations account for a mere 0.1% of the total available memory.

This system can easily store new information without interfering with previously stored information as long as the representations do not overlap too much. As soon as the memory starts to become saturated (at somewhat less than 100,000 words written to memory), there is interference among representations, and learning new information begins to interfere with the old. In this case, not only is there forgetting of the old information but the new information cannot be stored either.

### ALCOVE

ALCOVE [Kruschke 1990] is a computer memory model based on Nosofsky's exemplar memory model [Nosofsky 1984]. This model does not suffer from the phenomenon of catastrophic forgetting noted by Ratcliff and McCloskey & Cohen. As we will see, ALCOVE, like SDM, uses semi-distributed representations.

ALCOVE is a three-layer feed-forward network in which the activation of a node in the hidden layer is inversely exponentially proportional to the distance between the hidden node position and the input stimulus position. The hidden layer can be regarded as a "covering" of the input layer. The inverse exponential activation function has the effect of producing a localized receptive field around each hidden node, causing it to respond only to a limited part of the input field. This kind of localization does not exist in standard FFBP networks. This system therefore represents its inputs in a semi-distributed manner, with only a few hidden nodes taking part in the representation of a given input.

The architecture of ALCOVE is such that the representation of new inputs, especially of new inputs that are not close to already-learned patterns, will not overlap significantly with the old representations. This means that the set of weights that produced the old representations will remain largely unaffected by new input.

As in SDM, the representations in ALCOVE are also somewhat distributed, conferring on the system its ability to generalize. When the width of the receptive fields at each node is increased, thereby making each representation more distributed and causing greater overlap among representations, the amount of interference among representations increases.

## Semi-distributed representations in FFBP networks

If catastrophic forgetting could be reduced, the order in which inputs are presented to the network would be less important. Training could be done either sequentially or concurrently. In other words, the artificial constraint of requiring training data to be presented to the network in an interleaved fashion could be relaxed. If, in addition, the representations also reflected the amount of training required to produce them, it might be possible to produce a system that would better model overlearning than standard FFBP networks. An initial attempt to reduce catastrophic forgetting with semi-distributed representations by differentially modifying the learning rates of the connections in the network was described in [French & Jones 1991]. While this technique gave promising results on very small networks, it failed to scale up to larger networks. The algorithm presented below, using a different technique, allows semi-distributed representations to evolve that significantly reduce catastrophic forgetting.

## Activation overlap and representational interference in FFBP networks

Catastrophic forgetting is closely related to the much-studied phenomenon of crosstalk. The discussion of crosstalk has traditionally involved the capacity of a network to store information [Willshaw

1981]: above a certain capacity, distributed networks can no longer store new information without destroying old. In standard backpropagation models, there is a much more serious problem. As things currently stand, FFBP networks will not work at all without artificially interleaved training sets. Even when the network is nowhere near its theoretical storage capacity, learning *a single new input* can completely disrupt all of the previously learned information. Catastrophic forgetting is crosstalk with a vengeance.

A feedforward backpropagation network represents its inputs as activation patterns of units in the hidden layer. The amount of interaction among representations will be measured by their degree of "activation overlap". The activation overlap of a number of representations in the hidden layer is defined as their average shared activation over all of the units in the hidden layer. For example, if there are four hidden units and the representation for one input is (0.2, 0.1, 0.9, 0.1) and for a second is (0.2, 0.0, 1.0, 0.2), we calculate activation overlap by summing the smaller of the two activations (the "shared" activation) of each unit and averaging over all of the units. Here the activation overlap would be $(0.2 + 0.0 + 0.9 + 0.1)/4 = 0.3$.

I suggest that the amount that two representations interfere with one another is directly proportional to their amount of activation overlap. For example, consider the two following activation patterns: (1, 0, 0, 0) and (0, 0, 1, 0). Their activation overlap is 0. Regardless of the weights of the connections between the hidden layer and the output layer, there will be no interference in the production of two separate output patterns. But as activation overlap increases, so does the level of interference.

Therefore, if we can find a way to coax the network to produce representations with as little activation overlap as possible, we should be able to significantly reduce catastrophic forgetting.

## Sharpening the activation of hidden units

A technique that I call "activation sharpening" will allow an FFBP system to gradually develop semi-distributed representations in the hidden layer. Activation sharpening consists of increasing the activation of some number of the most active hidden units by a small amount, slightly decreasing the activation of the other units in a similar fashion, and then changing the input-to-hidden layer weights to accommodate these changes. The new activation for nodes in the hidden layer is calculated as follows:

$$A_{new} = A_{old} + \alpha(1 - A_{old}) \quad \text{for the nodes to be sharpened;}$$

$$A_{new} = A_{old} - \alpha A_{old} \quad \text{for the other nodes;}$$

where $\alpha$ is the sharpening factor.

The idea behind this is the following. Nodes whose activation values are close to 1 will have a far more significant effect on the output, on average, than nodes with activations close to 0. If the system could evolve representations with a few highly activated nodes, rather than many nodes with average activation levels, this would reduce the average amount of activation overlap among representations. This should result in a decrease in catastrophic forgetting. In addition, because sharpening occurs gradually over the course of learning and continues even after a particular association has been learned, the representations developed will reflect the amount of training that it took to produce them.

Let us consider one-node sharpening. On each pass we find the most active node, increase its activation slightly and decrease the activations of the other nodes. To preserve these changes we then backpropagate the difference between the pre-sharpened activation and the sharpened activation to the weights between the input layer and the hidden layer. Here are the details of this activation sharpening algorithm for *k*-node sharpening:

- Perform a forward-activation pass from the input layer to the hidden layer. Record the activations in the hidden layer;
- "Sharpen" the activations of *k* nodes;
- Using the difference between the old activation and the sharpened activation on each node as "error", backpropagate this error to the input layer, modifying the weights between the input layer and the hidden layer appropriately;
- Do a full forward pass from the input layer to the output layer.
- Backpropagate as usual from the output layer to the input layer;
- Repeat.

## Alternative Implementations of the Algorithm

The experiments described below were run using the "backpropagation-and-a-half" algorithm described above. However, for one-node sharpening, lateral inhibitory links among the nodes of the hidden layer might achieve a similar effect. These inhibitory links would allow the node with the highest activation to damp the activation of the other less active nodes, thereby effectively "sharpening" the activation of the most active node with respect to the others. It is less clear how sharpening of two or more nodes could be achieved in a straightforward manner by means of lateral inhibitory connections.

## Experimental results

The experiments consisted of training (and overtraining) an 8–8–8 feedforward backpropagation network on a set of eleven associations. The learning rate was 0.2 and momentum 0.9. The network was then presented with a new association. After this new association had been learned, one of the associations from the first set was chosen and tested to see how well the system remembered it. On the first presentation of this previously learned association,

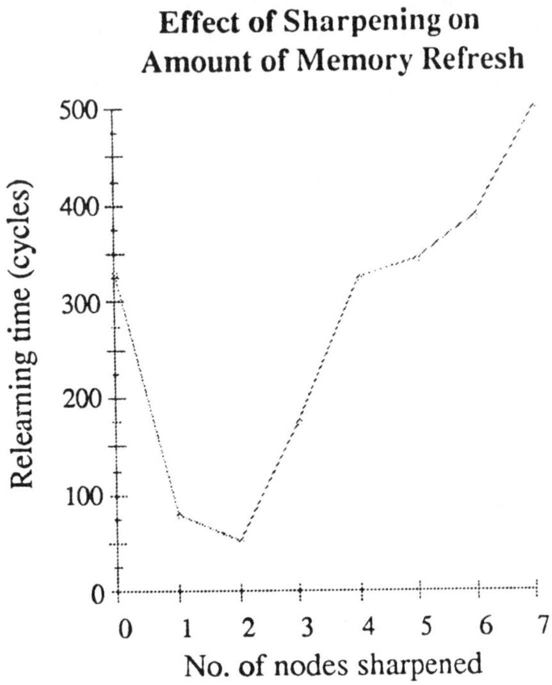


*Figure 1a*

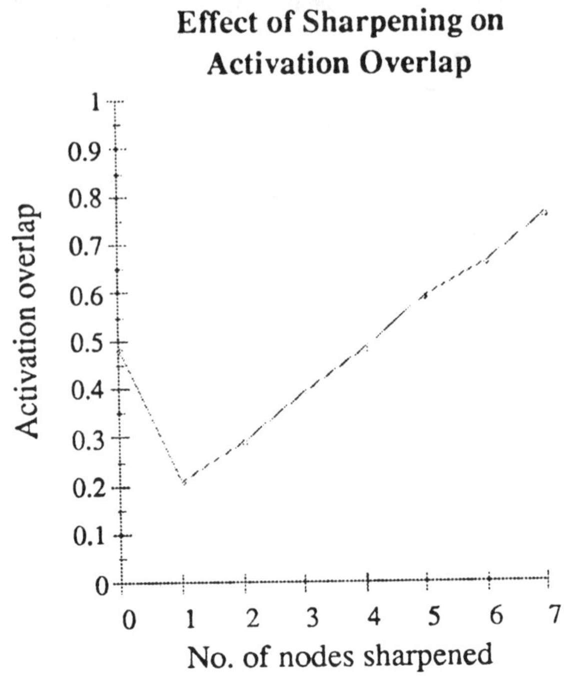


*Figure 1b*

the network invariably did very badly. The maximum error over all output nodes was almost always greater than 0.95 and the average error greater than 0.5. The amount of memory refresh required for a standard backpropagation network to relearn this association was recorded and compared to a network with one-node, two-node, three-node, etc. sharpening. In each case the sharpening factor was 0.2. The results are given in Figure 1a. (Note: 0-node sharpening is standard backpropagation.) It can be seen that one-node, two-node and three-node sharpening perform dramatically better than a standard FFBP network.

Over twenty separate runs, the standard FFBP network required an average of 330 cycles to relearn the previously-learned association. This figure dropped to 81 cycles for one-node sharpening and to 53 cycles for two-node sharpening. (Note: all runs were terminated at 500 cycles.) When the activations of three or more nodes were sharpened, the amount of relearning began to rise again. With three-node sharpening 175 cycles were required. With four-node (326 cycles) and five-node (346 cycles) sharpening, the modified system does no better than standard backpropagation. Above this, it does significantly worse. [Figure 1a]

The two graphs in Figures 1a and 1b suggest that amount of memory refresh required varies directly with the amount of activation overlap among representations. Figure 1b shows the amount of activation overlap of the eleven originally-learned inputs with various degrees of activation-sharpening. (As before "0 nodes sharpened" indicates standard backpropagation.) In general, the less activation overlap, the less the catastrophic forgetting as measured by the number of cycles required to relearn a previously-learned pattern.

In Figure 2 we can see the effect of this sharpening algorithm on the representations of one association. For each of twenty runs, the activation patterns on the hidden nodes at the end of the initial training period were recorded. The nodes in each of the twenty runs were sorted according to their activation levels and these figures were then averaged. As might be expected, for standard backpropagation the distribution of activations over the eight nodes was approximately uniform. This gives an activation profile from the most active nodes to least active nodes of approximately constant slope. However, the result of one-node sharpening is quite dramatic; one of the eight nodes was much more active than the other seven. The same phenomenon can be observed for the other experiments where two or more nodes where sharpened.

## Why does activation sharpening work?

Let us examine why activation sharpening reduces catastrophic forgetting. Consider two-node sharpening. As the system learns the first set of associations, it develops a set of sharpened representations in the hidden layer. A new association is then presented to the network. Activation sharpening immediately starts to coax the new representation into a sharpened form where two of the eight hidden nodes are highly active and six are not. Thus, very early on, the newly developing representation will have less chance of

**Effect of Sharpening on Hidden-Layer Activation Profiles**

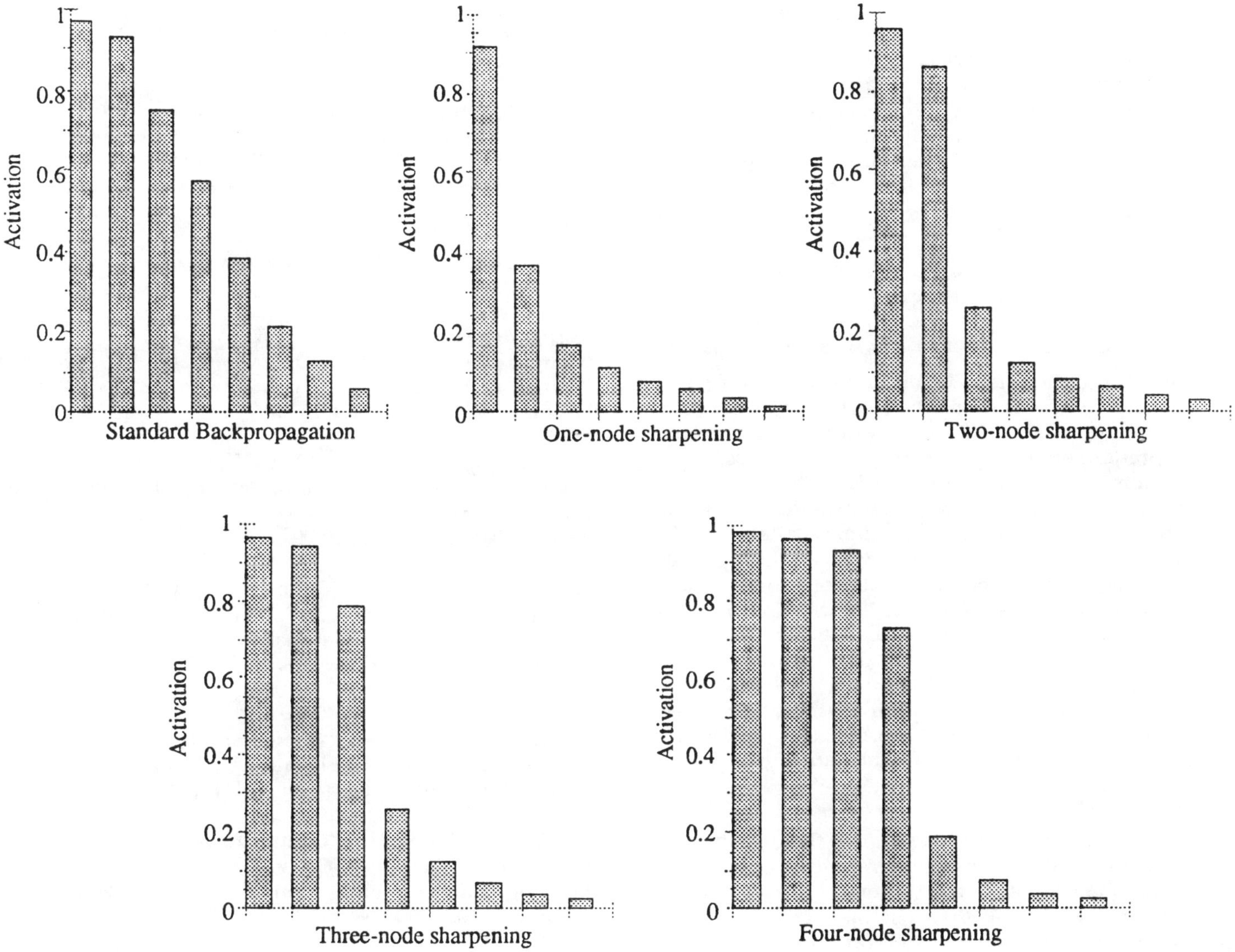


*Figure 2*

activation overlap with the already formed representations than in standard backpropagation, where the activation is spread out over all eight nodes.

Sharpened activation patterns interfere less with the weights of the network than unsharpened ones. The reason for this has to do with the way the backpropagation algorithm changes weights. When the activation of a node is near zero, the weight changes of the links associated with it are small. Thus, if a significant number of the nodes in the new representation have a very low activation, then the weights on the connections to and from that node will be modified much less, on average, than the weights associated with a highly active node. Therefore, the only representations significantly affected by the new representation will be those in which highly active nodes overlap. Consequently, if we reduce the probability of this overlap by activation sharpening, there will be a decrease in the amount of disruption of the old weights and catastrophic interference will be reduced.

The idea is to sharpen new activation patterns as quickly as possible, thereby decreasing their potential to interfere with already learned patterns. Keeping the learning rate low (≤ 0.2) with a relatively high sharpening factor (0.2) allows new activation patterns to become sharpened before they have a chance to do much damage to previously-learned weights. Preliminary experiments in fact indicate that as the learning rate is decreased with the sharpening factor held constant, catastrophic forgetting decreases.

It seems likely that semi-distributed

representations will cost the network some of its ability to generalize. Optimal generalization depends on as much information as possible taking part in mapping from the input space to the output space. Any mechanism tending to reduce the amount of information brought to bear on a new association would most likely reduce the quality of the mapping. In some sense, activation sharpening forces the input data through a representational bottleneck and this results in information being lost. The extent and severity of this loss and its effect on generalization is a subject of ongoing study.

## How many nodes should be sharpened?

This is an open question. For $n$ nodes in the hidden layer, the answer might be $k$ where $k$ is the smallest integer such that ${}_nC_k$ is greater than the number of inputs. In other words, a sufficient number of nodes should be sharpened to allow the existence of enough distinct sharpened representations to cover the input space. To minimize the activation overlap, the least such sufficient number of sharpened nodes should be chosen. If the number of input patterns to be learned is not known in advance, it might be reasonable to sharpen approximately log $n$ nodes. This estimate is based on work on crosstalk [Willshaw 1981]. This work indicates that in a distributed memory crosstalk can be avoided when the number of active units for each input pattern is proportional to the logarithm of the total number of units. It would seem reasonable to apply this result to the sharpening of hidden-unit activations.

## Conclusion

In this paper I have argued that catastrophic forgetting in distributed systems is a direct consequence of the amount of overlap of representations in that system. I have further suggested that the trade-off between catastrophic forgetting and generalization is inevitable. It is claimed that one way to maintain generalization capabilities while reducing catastrophic forgetting is to use semi-distributed representations. To this end, I presented a simple method to allow a feedforward backpropagation network to dynamically evolve its own semi-distributed representations.

## Acknowledgments

I would like to thank Mark Weaver for his invaluable assistance with the ideas and emphasis of this paper. I would also like to thank David Chalmers, Terry Jones, and the members of CRCC and SESAME for their many helpful comments.

## Bibliography

Feldman, J. A., [1988], "Connectionist Representation of Concepts", In *Connectionist Models and Their Implications*, Waltz, D. and Feldman, J. (eds.), 341–363.

French, R. M. and Jones, T. C., [1991], "Differential hardening of link weights: A simple method for decreasing catastrophic forgetting in neural networks", CRCC Technical Report 1991–50. [To appear in the *Proceedings of IJCNN-91*.]

Hetherington, P. A. and Seidenberg, M. S., [1989], "Is there 'catastrophic interference' in connectionist networks?", *Proceedings of the 11th Annual Conference of the Cognitive Science Society*, Hillsdale, NJ: Erlbaum, 26–33.

Kanerva, Pentti, [1988], *Sparse Distributed Memory*, Cambridge, MA: MIT Press.

Kortge, Chris A., [1990]. "Episodic Memory in Connectionist Networks", *Proceedings of the 12th Annual Conference of the Cognitive Science Society*, Hillsdale, NJ: Erlbaum, 764-771.

Kruschke, J. K., [1990], "ALCOVE: A exemplar-based connectionist model of category learning", Indiana University Cognitive Science Research Report 19, February 22, 1991.

McCloskey, M. and Cohen, N. J., [1989], "Catastrophic interference in connectionist networks: The sequential learning problem", *The Psychology of Learning and Motivation*, Vol. 24, 109-165.

Nosofsky, R. M., [1984], "Choice, similarity and the context theory of classification", *J. Exp. Psych. Learning, Memory and Cognition*, Vol. 10, 104-114.

Ratcliff, R., [1990], "Connections models of recognition memory: Constraints imposed by learning and forgetting function", *Psychological Review*, Vol. 97, 285-308.

Sloman, S. and Rumelhart, D., [1991], "Reducing interference in distributed memories through episodic gating", In A. Healy, S. Kosslyn, and R. Shiffrin (eds.), *Essays in Honor of W. K. Estes* (in press).

Weaver, M., [1990], "An active symbol connectionist model of concept learning" (unpublished manuscript).

Willshaw, D., [1981], "Holography, associative memory, and inductive generalization", In G.E. Hinton & J. A. Anderson (eds.), *Parallel models of associative memory*. Hillsdale, NJ: Erlbaum.

# The Role of Physical Properties in Understanding the Functionality of Objects

**Daniel S. Jordan**
Department of Psychology
Stanford University

**Jeff Shrager**
System Sciences Laboratory
Xerox Palo Alto Research Center

**Abstract**

We investigate the role of physical properties in determining how people select objects for use in physical activities. We propose a geometric model in which dimensions represent properties relevant to the goals of the activity and objects occur as points in this property space. An object's proximity to an ideal value on each property is additively combined across properties to produce a measure of the usefulness of the object for that activity. We report an experiment that shows that this ideal-point model successfully describes how people select an object for use in a physical activity by using physical properties as an intermediary factor. This model is derived from models of preference choice in which an individual selects objects that he or she prefers.

## Properties and Functionality

If you want to pound in a nail, but have no hammer, would you rather use a pillow, a metal paperweight, or a table? Although we typically think of using an object for its intended purpose, most objects can be used in different ways, depending upon the demands of the activity. The classic studies of functional fixedness (Duncker, 1945) show that people can decide to use objects for unusual purposes, but that the intended (or most frequent or recent) use of an object strongly interferes with the person's ability to recognize unusual uses.

When people list attributes of objects, they frequently include the functional uses of an object among its characteristics (e.g., "makes music" for a piano; Rosch & Mervis, 1975; Rosch, Mervis, Gray, Johnson, & Boyes-Braem, 1976; B. Tversky & Hemenway, 1984). Other listed attributes include physical properties (e.g., "is made of wood") and parts (e.g., "has keys"). Richards, Goldfarb, Richards & Hassen (1989) found that affecting an object's ability to perform its main function strongly affected the classification of that object. For example, few people still called an object "that is just like a shower cap but had big holes in it" a shower cap. Thus, how people categorize an object depends upon how physical properties relate to the object's functional use. In this paper we present and empirically validate a mathematical model that describes how physical properties influence how people select objects for use in physical activities.

### *A Strategy for Object Choice*

One strategy for deciding what to use to pound a nail is to first prioritize the physical properties required of any object to be used in the task. For example, you might identify the physical properties "hard" and "graspable" as being important aspects of the object. Next, order the set of available objects relative to an ideal value for each of these physical properties for the activity. Finally, combine this ordering information to select an object for use. Thus, in the absence of a hammer, you might prefer a metal paperweight over a pillow (which is not hard) or a table (which is not

---

This report is based on the first author's doctoral dissertation (Jordan, 1991). We are grateful to Gordon Bower and Misha Pavel for extensive support and advice in this work. Correspondence should be addressed to Daniel S. Jordan, Xerox PARC, 3333 Coyote Hill Rd., Palo Alto, CA, 94304 graspable).

To make our hypothesis more specific, we identify the following steps for selecting an object for use in an activity. First, physical properties that the goals of the activity require be true of the object are identified and weighted according to their importance for achieving the goals. Second, the object is rated on each physical property dimension. Independently, an optimal value, or "ideal point", is selected on each property dimension. The ideal point specifies the value that any object should have on the dimension to best achieve the goals of the activity. Finally, this information is combined to produce the "usefulness" of the object in the given activity.

The weightings of properties according to the goals of the activity are independent of specific objects, and hence can be used for any set of objects. Also, the ratings of an object on each of the property dimensions are independent of any activity, and hence can be used for any activity. However, because the placement of an ideal point on a property dimension is dependent on the activity, object ratings relative to the ideal point are dependent on the activity.

*A Preference Choice Model*

This process of determining the usefulness of objects in a physical activity is similar to that described by models of preferential choice in which an individual selects items that he or she prefers based on relevant feature dimensions. Comparing our model to the preference choice models, we replace the agent that does the selecting, i.e., the individual, with the physical activity. Thus, the physical activity "selects" the object that best satisfies its goals (i.e., that it "prefers") based on relevant physical properties. To emphasize the difference between the two scenarios, in a preferential choice situation, preferences are assumed to vary across individuals as well as across contexts, whereas in selecting an object in a physical activity, preferences are assumed to only vary across activity contexts. In particular, it is assumed that for each activity the usefulness of an object is objectively determined by the constraints of the physical universe. Differences between individuals in their assessment of an objects usefulness could arise from individual experience with the objects or in the activities.

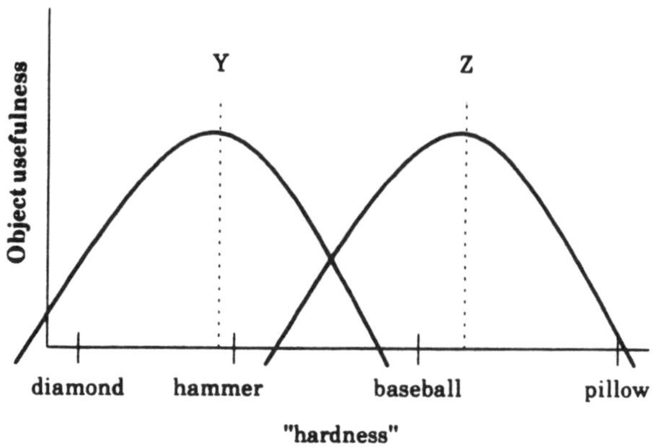


Figure 1: Illustration of the ideal-point model in one dimension. Objects "diamond" through "pillow" are ordered on the dimension of "hardness." Y and Z are ideal points for different activities. Usefulness increases upward.

The notion of an ideal point, relative to which preferences for items are determined, was first studied by Coombs (1950, 1964) with his unidimensional unfolding model. Coombs proposed that preference functions are single-peaked and symmetric with respect to items that lie on some ordered dimension. We follow Coombs' approach, and show in figure 1 an example of a function that specifies the usefulness of an object in an activity. Several objects are ordered along the single dimension of "hardness." The usefulness of an object is displayed in the dependent axis (increasing upward) and is assumed to peak, for a given activity, at some ideal point on the item dimension, such as Y or Z. For the activity with ideal point Y the usefulness ordering of objects is: hammer, diamond, baseball, pillow; whereas for the activity with ideal point Z the usefulness ordering of objects is: baseball, pillow, hammer, diamond.

The single peaked preference function of Coombs was extended to the multi-dimensional case by Bennett and Hay (1960) and Carroll (1972, 1980). The current model for object functionality follows the computational analysis presented by Carroll. As in Carroll's model, the current model represents the selection or choice process in a geometric model, where the property values of objects occur as points in a multi-dimensional space with physical properties as dimensions. We assume the standard Euclidean distance metric in this multi-dimensional property space. Then the distance $d(i,j)$ between a point $x$ and an ideal point $y$ in this space is given by:

$$d^2(i,j) = \sum_k^r w(i,k) \times (y(i,k) - x(j,k))^2 \quad (1)$$

where $d(i,j)$ is the distance of object $j$ from the ideal point for activity $i$, $y(i,k)$ is the coordinate of ideal point $y$ on dimension $k$ for activity $i$, $x(j,k)$ is the coordinate of item $j$ on dimension $k$, $w(i,k)$ is the importance that dimension k has in activity $i$, and $r$ is the number of dimensions. We further assume that the usefulness of object $i$ in activity $j$ is proportional to the square of this distance, $d^2(i,j)$, and is linear. Then:

$$u(i,j) = a(i) \times d^2(i,j) + b(i) \quad (2)$$

where $u(i,j)$ is the usefulness rating of object $j$ in activity $i$, and $a(i)$ and $b(i)$ are constants specific to activity $i$. The assumptions of a Euclidean metric in the property space and a linear relation between $u(i,j)$ and $d^2(i,j)$ implies that the usefulness function as shown in Figure 1 will be parabolic and centered at the ideal point $y$.

Equation 2 provides the computational form of a quantitative ideal-point model for determining the usefulness of objects in different activities. The ideal points $y(i,k)$, the weights $w(i,k)$, and the constants $a(i)$ and $b(i)$ are the unknown parameters to be estimated from the data.

## An Empirical Test

We have tested the ideal-point model by collecting data for the right hand side of Eq. 2 to calculate a predicted value for $u(i,j)$. We also collected data for the left hand side of Eq. 2 to provide observed values for $u(i,j)$ to which predicted values were compared and used to evaluate the fit of the model. To reduce the number of unknown parameters in the model, we assume the weights $w(i,k)$ as given; therefore, the total number of unknown parameters is $r+2$. Thus we obtained data for three different rating tasks: rating objects on properties $(x(j,k))$, rating objects in an activity context $(u(i,j))$, and rating the importance of properties in an activity context $(w(i,k))$.

> *Objects*: boomerang, bowl, cap, diamond, hammer, helmet, knife, needle, pillow, plate, rolling pin, screwdriver, sofa, table, tennis ball.
>
> *Properties*: hardness/softness, flexibility, thickness/thinness, smoothness/roughness, fragility, size, flatness/curvedness, graspability, weight.
>
> *Activities*: unscrew a 1/2-inch screw, anchor a helium-filled balloon that is attached to a string, flatten dough for baking cookies, slice a carrot, poke a hole in the side of a soda can, protect your head from harmful blows, pound a 1-inch nail into a wooden wall, sit down comfortably on an object, pry open a can of house paint.

Table 1: The objects, properties, and activities used in the experiment.

### *Method*

Fifteen commonly known objects were chosen so as to provide a variety of similar and dissimilar object pairs, as well as to have physical properties that are highly associated with the object's main function. In addition, 10 physical activities were chosen. All objects and activities were represented by line drawings. Finally, nine properties were obtained in a preliminary session where eight subjects were instructed to list those physical properties that the activity suggested should be true of objects. Table 1 gives the objects, properties, and activities. Using these stimuli, 29 subjects gave three different types of ratings: (1) object-on-property ratings, in which subjects rated objects on the given properties, (2) property-on-activity ratings, in which subjects rated how much the property mattered for an object that was to be used for the given activity, and (3) object-on-activity ratings, in which subjects rated the usefulness of objects in the different activities. Objects (other than these stimuli objects) were given as endpoint references for the object-on-property rating task in order to make the scale endpoints more determinate.

### *Results*

To test that subjects sufficiently discriminated in their ratings of the selected items (objects, properties) on their respective scales (property, activity), we calculated all pairwise T-tests for ratings between items on a scale. The percentage of item pairs whose ratings on a scale were significantly different at the $\alpha = 0.05$ level averaged 59% across properties for object-on-property ratings, 57% across activities for object-on-activity ratings, and 43% across activities for property-on-activity ratings. The averages above 50% for the object-on-property and object-on-activity ratings were taken to imply that subjects sufficiently discriminated between items on these scales, and that the data could be used in the model. (The only exception was the object-on-activity ratings for the activity "to protect your hand while holding a hot pan", which was eliminated from further analyses due to a small percentage of significant T-tests.) The below-50% average for the property-on-activity scales indicates that, at least in this rating task, properties do not vary much in their relative importance in a given activity, or equivalently that many property weights are about equal. For this reason we report results in which the $w(i,k)$ property weights in Eq. 2 are assumed to be equal.

We fitted the ideal point model given in Eq. 2 to the data for each of the activities. Observed data were used for the known parameters in the model, the $u(i,j)$ and $x(i,k)$. Averaged values (over the 29 subjects) were used for these parameters. The $w(i,k)$ or property-on-activity ratings were set equal to 1 (without loss of generality). We used an iterative weighted non-linear regression method to estimate the unknown parameters, the $y(i,k)$ or ideal points, and the constants $a(i)$ and $b(i)$. Although Carroll (1972, 1980) showed that Eq. 2 reduces to a standard linear regression equation from which the unknown parameters (the $y(i,k)$, $a(i)$ and $b(i)$) can be analytically determined from an exact least-squares solution, we could not use this approach because we wanted to restrict the ideal point estimates to the range of the object-on-property ($x(i,k)$) scales. Also, the reduced linear regression equation is too sensitive to collinearities between the property dimensions that do not exist in the original non-linear regression equation (Eq. 2). The non-linear regression procedure minimized the weighted sum of squared errors between the predicted and average observed $u(i,j)$ values with each error term in this sum divided by the variance (across subjects) of the observed $u(i,j)$ value. We used a chi-square test to determine the goodness-of-fit of the model to the data. The chi-square term was equal to the sum, over objects, of squared Z-scores, where Z-scores were calculated by dividing the difference between the predicted and averaged observed $u(i,j)$ values by the standard error of the mean of the observed $u(i,j)$ values.

We performed this analysis for each of the nine activities using from one to all nine properties in the ideal point model (i.e., in Eq. 2 the index $r$ ranged from 1 to 9). We used the average observed property-on-activity ratings ($w(i,k)$) to determine the order in which properties were added to the model for a given activity. Properties rated as more important for that activity were added to the model first. (Other methods may also be used to determine an ordering.) For each activity, Table 2 gives the estimated ideal point values for the minimum number of properties needed to achieve chi-square significance using the ideal point model in Eq. 2. For example, for the activity "to unscrew a screw," six properties were needed before the model achieved significance with the data. The average minimum number of properties needed to achieve significance was 6.22, which is greater than 1 ($t(8) = 6.56, p < .01$) and less than 9 ($t(8) = -3.48, p < .01$).

## Discussion

In all activities the ideal-point model fit the data, and in general more than one physical property was needed to determine an object's usefulness. In addition, the set of properties needed is determined by the context of the activity. For example, only one property, weight, is important for determining how useful an object is for anchoring a balloon; however, this property is not important for determining how useful an object is for unscrewing a screw, although six other properties are important.

The ideal point values given in Table 2 vary widely across activities, as expected, and are intuitively reasonable.[1] For example, the ideal point values for the property "thickness" imply that a thick object is most useful for protecting your head from harmful blows, but that a thin

[1] Only ideal point values for the property "hardness" appear to be counter-intuitive, a result for which we currently do not have a good explanation.

| Activity | Minimum Number of Dimensions ($\chi^2$) | *Flatness* | *Fragility* | *Grasp-ability* | *Hardness* | *Weight* | *Rigidity* | *Size* | *Smooth-ness* | *Thick-ness* |
|---|---|---|---|---|---|---|---|---|---|---|
| | | child's ball | light bulb | police pistol | human hair | kitchen stove | wine glass | auto-mobile | nail file | bedroom dresser |
| Anchor a helium bal-loon | 1 (3.234) | | | | | 2.3 | | | | |
| Flatten cookie dough | 7 (6.928) | 2.4 | 5.8 | 1.0 | 2.5 | 1.2 | 1.0 | | 9.0 | |
| Protect your head from blows | 8 (6.170) | 6.2 | 6.0 | 7.1 | 9.0 | 9.0 | 6.7 | 3.3 | | 1.2 |
| Pound a nail into a wall | 9 (7.047) | 6.2 | 7.5 | 4.6 | 1.6 | 3.4 | 1.0 | 9.0 | 3.7 | 6.3 |
| Pry open a paint can | 7 (7.411) | 7.4 | 9.0 | 1.7 | 1.9 | | 1.1 | 3.9 | | 6.8 |
| Poke a hole in a soda can | 5 (7.557) | | 2.1~ | 4.8~ | 8.3~ | | 9.0~ | | | 1.9~ |
| Unscrew a screw | 6 (11.119) | 7.5 | 8.9 | 2.5 | 1.7 | | 1.4 | | | 6.9 |
| Sit comfortably | 5 (4.157) | | 5.9~ | | 7.6~ | | 4.5~ | 9.0~ | 5.6~ | |
| Slice a carrot | 8 (9.940)[a] | 9.0 | 9.0 | 1.5 | 2.0 | 1.2 | 1.5 | 9.0 | | 7.4 |
| | | ironing board top | tanker anchor | rain barrel | diamond | maple leaf | bow (ribbon) | shirt button | glass bottle | razor blade |

1 ↑ physical property scale ↓ 9

Table 2: Ideal points occur on the rating scale from 1 to 9 shown at the right. For each property, the objects used as scale endpoints are shown at the top (for "1") and bottom (for "9") of each column.

a. For all reported chi-squares, the null hypothesis that the predicted and observed $u(i, j)$ values are equal could not be rejected at the $\alpha = .05$ level (i.e., $p > .05$), except for the last item, marked by "a", for which $p > .01$.

~ Values indicated by tilde (~) are anti-ideal points (see Carroll, 1980), i.e., values that are the *least* useful rather than the *most* useful. Comparing to Figure 1, the usefulness curve for an anti-ideal point is inverted and concave upward. Hence objects become *more* useful the farther they are from an anti-ideal point in either direction along the dimension.

object is most useful for prying the lid off a paint can. Our ideal point values suggest design strategies for "ideal objects" as well as an account of the affordance characteristics of objects in particular activities (Gibson, 1977; Norman, 1988). For example, Table 2 suggests that the ideal object for protecting your head from harmful blows should be light but rigid. In fact, manufacturers of helmets are constantly searching for new materials to satisfy these constraints; the problem being, of course, that lighter materials tend to be flexible.

This phenomenon can also be viewed from the standpoint of ad hoc categories (Barsalou, 1983). Selecting objects for an activity such as "to pound a nail" can be viewed as rating objects in the ad hoc category "things to use to pound a nail." From this standpoint, our research describes how a set of objects leads to a particular graded structure for particular activities. Also, this work can be viewed as a study of the effect of activtyi context on properties and concept definition. Previous results have shown that context affects the salience of properties (Barsalou, 1982; A. Tversky, 1977) as well as concept definition (Roth & Shoben, 1983).

In summary, the ideal-point model used to describe preferential choice data for individuals was successful in describing how people use physical properties to determine the usefulness of objects in different activities. Properties relevant to the goals of the activity are identified and an ideal value for each property is selected. An object's proximity to the ideal value on each property is additively combined across properties to produce a measure of the usefulness of the object for that activity. This characterization provides us with a better understanding of how people come to use objects in different activities, as well as how they perceive functionality in general.

## References

Barsalou, L.W. (1982). Context-independent and context-dependent information in concepts. *Memory & Cognition*, 10, 82-93.

Barsalou, L.W. (1983). Ad hoc categories. *Memory & Cognition*, 11, 211-227.

Bennett, J.F., & Hay, W.L. (1960). Multidimensional unfolding: Determining the dimensionality of ranked preference data. *Psychometrika*, 25, 27-43.

Carroll, J.D. (1972). Individual differences and multidimensional scaling. In R.N. Shepard, A.K. Romney, and S. Nerlove (Eds.), Multidimensional scaling: Theory and applications in the social sciences, Volume I: Theory. New York: Seminar Press, Inc., 105-155.

Carroll, J.D. (1980). Models and methods for multidimensional analysis of preferential choice (or other dominance) data. In E.D. Lantermann and H. Feger (Eds.), Similarity and choice. Bern: Hans Huber Publishers, 234-289.

Coombs, C.H. (1950). Psychological scaling without a unit of measurement. *Psychological Review*, 57, 148-158.

Coombs, C.H. (1964). A theory of data. New York: Wiley.

Duncker, K. (1945). On problem solving. *Psychological Monographs*, 58, No. 270.

Gibson, J.J (1977). The theory of affordances. In R.E. Shaw & J. Bransford (Eds.), *Perceiving, acting, and knowing.* Hillsdale, NJ: Lawrence Erlbaum Associates.

Jordan, D.S. (1991). *The role of physical properties in understanding the functionality of objects.* Unpublished doctoral dissertation, Stanford University.

Norman, D. (1988). The Design of Everyday Things. New York: Basic Books, Inc.

Richards, D.D., Goldfarb, J., Richards, A.L, & Hassen, P. (1989) The role of the functionality rule in the categorization of well-defined concepts. *Journal of Experimental Child Psychology*, 47, 97-115.

Rosch, E.H. & Mervis, C.B. (1975). Family resemblances: Studies in the internal structure of categories. *Cognitive Psychology*, 7, 573-605

Rosch, E.H., Mervis, C.B., Gray, W.D., Johnson, D.M. & Boyes-Braem, P. (1976). Basic objects in natural categories. *Cognitive Psychology.* 8, 382-439.

Roth, E.M. & Shoben, E.J. (1983). The effect of context on the structure of categories. *Cognitive Psychology.* 15, 346-378.

Tversky, A. (1977). Features of similarity. *Psychological Review*, 84, 327-352.

Tversky, B. & Hemenway, K. (1984). Objects, parts, and categories. *Journal of Experimental Psychology: General*, 113, 169-193.

# Blend Errors During Cued Recall

**Leigh Erik Nystrom** **James L. McClelland**

Department of Psychology
Carnegie Mellon University
Pittsburgh, PA 15213
ln0c+@andrew.cmu.edu
jlm+@andrew.cmu.edu

### Abstract

Connectionist models of memory account for recall behavior using processes which simultaneously access multiple memory traces and interactively construct the recalled information. This also allows the models to account for prototyping phenomena, but seems to predict retrieval of composite or "blended" information during ordinary recall. By contrast, models that simulate recall as a probabilistic selection of a single trace would not predict recall blend errors. To examine memory blending during recall, three experiments were performed; in each, subjects read sentences, some sharing words with one other sentence. They later recalled the sentences given partial-sentence cues. In all experiments subjects made blend errors, recalling one word from each of two similar sentences more often than one word from each of two dissimilar sentences, as predicted by multiple-trace models. The frequency of blend errors was relatively low, but a good account of this and other aspects of the results was provided by a multiple-trace model based on an Interactive Activation network as applied to memory retrieval in McClelland (1981).

## Introduction

Much of the current interest in connectionist models of memory stems from their ability to capture not only recall and recognition behavior, but also the prototyping and generalization found in concept formation experiments (Knapp & Anderson, 1984; McClelland, 1981; McClelland & Rumelhart, 1985). In all of these models, memory retrieval processes can lead to prototyping because they involve an activation and synthesis of multiple memory traces to produce a pattern of activation that may not necessarily correspond to a single trace as originally stored.

For example, McClelland (1981) outlined a connectionist model of memory wherein each memory trace consists of one "instance" unit representing the trace as a whole along with "property" units for each of that trace's properties (see Figure 1). Property units within a trace reinforce each others' activation through the centralized instance unit, while inhibiting the activation of competing property units from other traces. During retrieval, traces in memory become active to the extent that they include the properties given as a recall cue, and to the extent that their properties reinforce one another. The inhibition between the property and instance units of different traces also influences the final state of activation as traces compete for activation of shared properties. Properties that ultimately remain active constitute the recalled information, regardless of whether they all occurred in the same trace, although mutual reinforcement of same-trace properties often leads to the retrieval of a single trace anyway. It would seem that this sort of "multiple-trace" access model would predict abundant errors during everyday recall -- "blend" errors, consisting of mixtures of more than one trace.

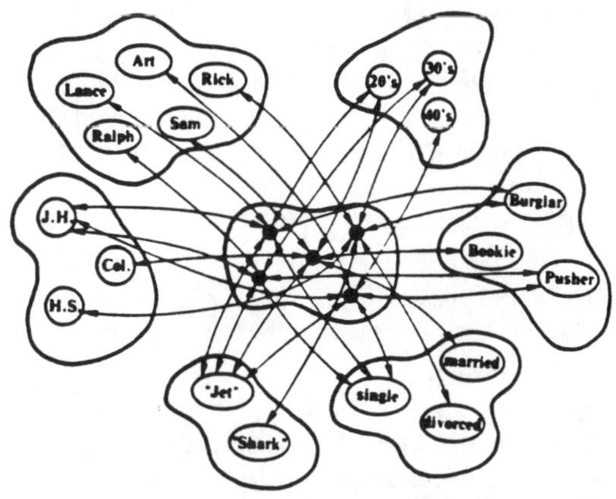


Figure 1: Multiple-trace model from McClelland (1981)

Not all theories of memory retrieval assume that recall involves synthesis of multiple memory traces. Some models view the recall process as a search through memory to find a single trace; most memory models before the 1980's had this character. Without the joint contributions of multiple traces, single-trace models do not automatically lead to prototyping or generalization -- and as a side effect, memory blends would be avoided. Shiffrin's SAM model of associative memory (Raaijmakers & Shiffrin, 1981; Gillund & Shiffrin, 1984) provides one example of a more recent single-trace model. In the SAM model, cued recall proceeds as a probabilistic selection of a single trace from the collection of all traces. The probability of selecting a given trace is a function of the baseline "strength" of the trace as well as the degree of match between the cue's properties and the trace's properties. Once selected, all properties of the trace are accessible, and no other traces will be accessed. Another single-trace model, Anderson's ACT theory (Anderson, 1976; 1983), represents related information as being subsumed under a single "trace" node; the goal of recall is the selection of one of the possible trace nodes, as a function of the activation of its subsumed

## Credits

This research was supported in part by an NSF Graduate Fellowship awarded to the first author, and NSF Grant BNS 88-12048 awarded to the second author.

nodes. As with Shiffrin's model, once a single trace is selected, all of its constituent information can be retrieved without interference from other traces.

So the question arises: Do blend errors actually exist? Or more specifically, do they occur as several memory traces combine at the time of recall? While memory blends have been studied in the context of eyewitness testimony (Loftus, 1977; Loftus, Miller, & Burns, 1978), the procedure used to induce blending in those and other experiments has left open the possibility that subjects integrated multiple traces at the time of encoding rather than retrieval. Both single- and multiple-trace models can easily account for blending in such circumstances -- any traces blended during encoding must necessarily be retrieved as blended information. Therefore, in order to distinguish single- from multiple-trace models, the focus of investigation was directed solely to blending of similar memory traces at the time of retrieval. The present experiments sought to find evidence for or against the existence of blend errors using sentence materials based on the classic studies of Anderson (1976), with methods designed to minimize the possibility that subjects might integrate sentences during encoding.

Imagine the situation wherein two sentences share three out of five content words, and the three shared words are used as a recall cue. This cue would match the two overlapping sentences equally well. If a subject were asked to recall the other two content words from one of the sentences, they could correctly answer with the two remaining words from either overlapping sentence. For example:

#1: *"The doctor gave the plumber the coat in the lobby."*
#2: *"The doctor gave the plumber the watch in the kitchen."*
cue: "The doctor gave the plumber the ____ in the ____."

Assuming that each sentence is encoded as a single trace, with its constituent words as properties, a multiple-trace model would predict difficulty for a subject's recall, because the ambiguous cue would equally activate the traces for both of the sentences and their constituent properties. As a consequence, subjects may be prone to making a "cross-over" intrusion error by responding *coat* + *kitchen*, or *watch* + *lobby*, mixing words from each overlapping sentence. By contrast, a single-trace model would predict that either of the two traces matching the ambiguous cue would be accessed in a probabilistic search with equal likelihood, yet the words from only one of these two traces would be retrieved. Therefore, a single-trace model would predict no blending of these highly similar memory traces despite the ambiguous cues.

In accordance with the logic of this example, the following experiments used a list of sentences in which half of the sentences shared several content words with one other sentence, while the other half served as completely dissimilar controls. Subjects studied the sentences individually under the pretense of a cover task, unaware that they would need to later recall any sentences. They therefore had no reason to continually rehearse sentences or actively integrate the traces of overlapping sentences, thereby allowing the experiment to plausibly investigate blending during retrieval alone. After subjects learned the list of sentences, a delay task ensured some degree of temporal separation between encoding and retrieval. Finally, subjects were given cues which consisted of the words corresponding to the positions shared by overlapping sentences. They were asked to recall the two missing target words from a single sentence studied earlier in the experiment.

In all three of the experiments, two general results were of primary interest, each following a hypothesis of multiple-trace models concerning recall blends. The first hypothesis was that subjects should blend similar memory traces, as evidenced by more cross-over intrusion errors between the similar overlapping sentences than between the dissimilar control sentences. If no difference was discovered, the null hypothesis of single-trace models would be supported. The second hypothesis was that subjects should be less likely with overlapping than with control sentences to correctly recall a complete sentence after having already accessed its trace. This result would be expected because while the overlapping sentence traces would continually influence each other during the independent retrieval of the trace properties, the control sentence traces would be immune from such post-access interference. This hypothesis was tested by examining the conditional probability that a subject would correctly recall both target words given that they had already correctly recalled one of the words. If no difference was found in this conditional probability as a function of sentence similarity, then single-trace models would be supported. Such models account for interference effects during the search for a trace, but once one is accessed, other traces are assumed to be of no further relevance.

## Experiment 1

Thirty-two experimental sentences were randomly created for every subject. Each sentence included five word-positions which were filled by random selection (without replacement) from a prearranged list of words of the appropriate semantic type. Care was taken to avoid words closely-related across semantic types, such as "barber" and "barbershop". Two sentences from each set of four were selected to become overlapping sentences, the other two left as matched control sentences. Three of the word-positions in each set were randomly chosen for the overlap; any three of the five were equally likely to be chosen for overlap. The three words in the overlap positions of one of the two overlapping sentences were then replaced with the corresponding words from the other sentence, thereby creating a pair of sentences differing by only two words. The order of presentation of the 32 sentences was randomized for each subject with the constraint that a minimum of 12 sentences intervened between any two matched overlapping or matched control sentences from a matched set of four.

The random generation of sentences created many unlikely combinations of sentence constituents, enabling the experiment to use a convincing cover story: subjects were told that the experiment was designed to explore how people read sentences that "sound strange," those in which the content words do not seem to belong together in real life. They therefore had no cause to believe that they would later need to remember anything about the sentences. Subjects read each sentence and made a judgement about the overall "plausibility" of the sentence, followed by judgements of how appropriate each of the five main words seemed, given the content of the rest of the sentence. First the sentence appeared along with a prompt to rate it as a whole, then each of the five words appeared sequentially underneath the sentence along with a prompt to rate it. The purpose of these ratings was both to make the cover story more convincing and to ensure that the subjects paid attention to every sentence and its primary words, forming memory traces through processing them at a "deep," semantic level.

After a 5-minute delay task, subjects were given recall cues containing all words from a previously-presented sentence except the two words in the non-overlapping word-positions (replaced by blanks of a fixed length). Presenting any one sentence from an overlapping pair in this manner therefore acted as a cue for either of the two paired sentences. Subjects were told that they may have seen two sentences that fit a cue equally well, but that in those cases, they would have to remember both of them, one at a time. However, they were never told that a particular test cue matched two sentences until they had first already recalled one sentence. It was stressed that in choosing each response, they should take care to recall two words from within the same sentence from the first phase. Therefore, with overlapping sentence cues, after the subject first recalled two words from one sentence, they were asked to recall two more words belonging to the other matching sentence. In this manner, all 16 of the overlapping sentences from the first phase were tested, along with 8 randomly-selected control sentences, all in a random order.[1]

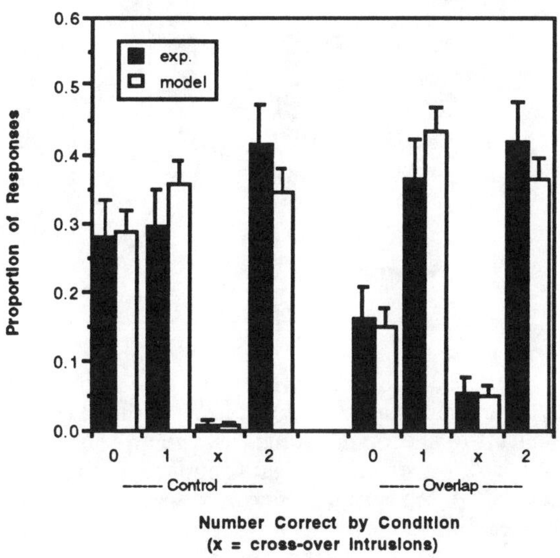


Figure 2: Experiment 1 Data

The overall recall performance of the 35 subjects can be seen in Figure 2. Subject data can be seen in the solid columns (error bars indicate 95% confidence intervals), illustrating the proportion of responses for both control and first overlap conditions as a function of the number of words correctly recalled. Results of the first hypothesis test supported a multiple-trace model: subjects made cross-over errors more often (using a sign test across subjects, $p < .01$) when tested in the overlap conditions (5.4% with the first, 2.1% with the second) than in the control condition (0.7%). The difference between the rates in the second overlap versus control conditions was not significant ($p > .05$) while the difference between the first and second overlap conditions was significant ($p < .05$). The greater frequency of intrusions between overlapping sentences was not simply a result of a greater tendency to recall words from overlapping sentences, because words from overlapping sentences intruded into recall of a matched control sentence only 1.4% of the time; words from control sentences intruded into recall of a matched overlapping sentence 2.7% of the time.

Results of the second hypothesis test also supported multiple-trace models: the conditional probability of recalling two words from the same sentence given the correct recall of one word was lower in the overlap condition ($p = .50$, first overlap; $p = .36$ second overlap) than with controls ($p = .58$). Using a Wilcoxon signed-ranks tests across subjects, the conditional probability in the control condition was significantly higher than in either the first overlap ($z = 1.81$, $p < .05$) or the second overlap conditions ($z = 4.11$, $p < .0001$); the conditional probability in the first overlap condition was significantly higher than in the second overlap condition ($z = 3.10$, $p < .001$).

## Experiment 2

It was difficult to know how to best interpret the data from the two overlap conditions in Experiment 1. The fact that the first sentence recalled in the overlap condition is likely to be the stronger of the two, while the second may be contaminated by the earlier recall of the first, makes quantitative comparison of the overlap and control conditions difficult. Experiment 2 was designed to generate clearer data, by using unambiguous overlapping-sentence test cues and thereby allowing only one correct answer in all cases. This was accomplished by reducing the number of content words shared among overlapping sentences from three to two, while continuing to use cues with only two blanks. For example:

#1: *"The doctor gave the plumber the coat in the lobby."*
#2: *"The doctor gave the lawyer the watch in the kitchen."*
cue: "The doctor gave the plumber the ____ in the ____."

Note that this cue only matches the first overlapping sentence completely, while offering a partial match to the second sentence. With this experiment, we expected to replicate the findings of Experiment 1, perhaps with less overall blending due to the lessened overlap between sentences.

Thirty-six sentences were randomly created for every subject, 24 overlapping and 12 control. One of the overlapping sentences in each overlapping pair was also used as a mate for a control sentence, since it did not share any content words with the control sentence. The order of presentation of the 36 sentences was randomized for each

[1]Post-test questionnaires indicated that most subjects were unaware of sentence overlapping during the first phase of the experiment. Only two subjects reported that they had recalled or integrated overlapping sentences at the time of encoding; excluding their data from the analyses had no significant effect. These results held across all three experiments.

subject by presenting 12 control sentences and 12 overlapping sentences (the "overlap-test" sentences, one from each pair of overlapping sentences) as the first 24 sentences, followed by the rest of the overlapping sentences (the "overlap-distractors"). This allowed the overlap-distractor sentences to provide retroactive inhibition, both for the overlapping-test sentences and the control sentences. The order was then randomized with the further constraint that a minimum of 12 sentences intervened between overlap-test or control sentences and their paired overlap-distractor sentence. The 24 test cues contained all but two words from an overlap-test or control sentence, and were presented in a completely random order.

Overall, the 38 subjects performed worse than those in Experiment 1 (see Figure 3.) Perhaps the lower overall level of performance in this experiment indicates that subjects had too difficult a time recalling the earlier of the two overlapping sentences due to the recency of the overlap-distractor sentence. In fact, fully 9% of subjects' answers with overlap-target cues consisted of recalling both words from its paired overlap-distractor sentence -- a sort of double cross-over intrusion -- despite the fact that the cue unambiguously matched the overlap-target sentence.

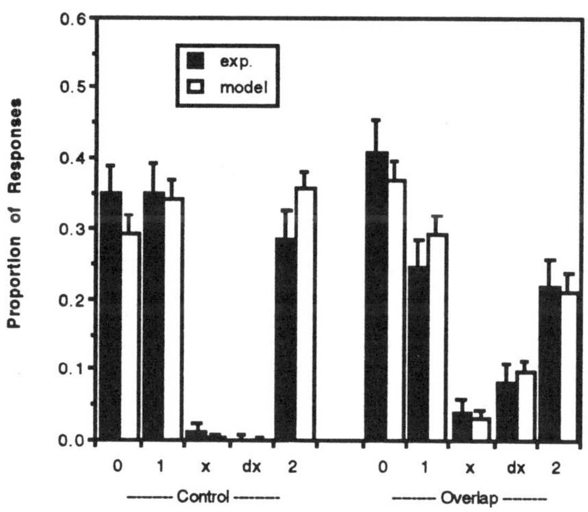


Figure 3: Experiment 2 Data

Results of the first hypothesis test showed that cross-over errors were more frequent between matched overlapping sentences (18 occurrences, or 4.0%) than between overlap-distractors and their matched controls (6 occurrences, 1.3%). As in Experiment 1, the difference between the cross-over error rates of the overlapping and the control sentences was significant (sign test, $p < .05$), as predicted by multiple-trace models.

The test of the second hypothesis supported single-trace models, however. The conditional recall probability was essentially the same in both conditions: .43 with overlapping-test sentences, and .44 with control sentences ($z = 1.07, p > .05$). This pattern of results would be expected if there were no post-access influences from similar memory traces.

## Experiment 3

One potential critique of Experiments 1 and 2 concerns the evaluation procedure used in the study phase of the experiments. Subjects had been asked to rate the plausibility of a whole sentence, and then of each individual content word; this latter requirement could conceivably have caused subjects to encode a separate individual memory trace for each content word as well as for the whole sentence. If this was the case, both multiple- and single-trace models might explain the conflicting pattern of results by proposing that subjects were occasionally able to exploit associations between traces for individual words along with more complete traces for the sentences containing them. Therefore, Experiment 3 was performed, using the same materials and procedure as Experiment 1, but with the elimination of subjects' ratings of individual words during the study phase.

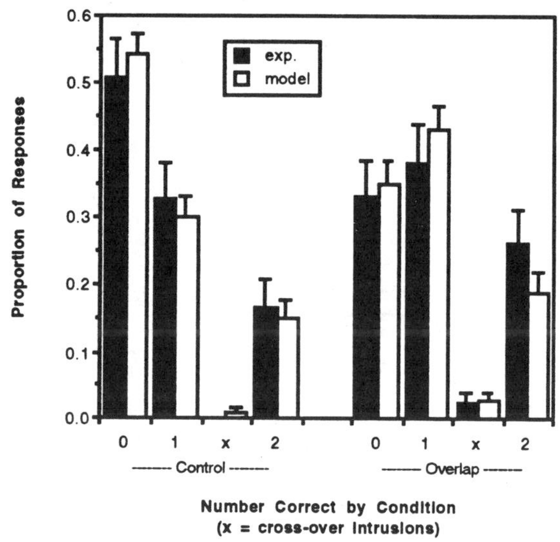


Figure 4: Experiment 3 Data

Overall, the 37 subjects performed worse than the subjects in the other experiments, no doubt because the new procedure led subjects to encode the stimuli less rigorously (see Figure 4). No cross-over errors were obtained between control sentences. Consequently, cross-over errors were significantly more frequent (sign test, $p < .01$) in the first overlap condition (2.4%) than in the control condition. They were also more frequent ($p < .05$) in the second overlap condition (1.7%) than in the control condition. There was no significant difference ($p > .05$) between the cross-over error rates in first and second overlap conditions. These differences again provided support for a multiple-trace model.

The conditional recall probabilities did not significantly differ ($z = -1.16, p > .05$) between the first overlap condition ($p = .35$) and the control condition ($p = .31$). On the other hand, the difference between the conditional probability in the control and second overlap ($p = .16$) conditions was significant ($z = 2.78, p < .005$), as was the difference between the first and second overlap conditions ($z = 2.93, p < .005$).

## Models

In all of the experiments, cross-over errors were more frequent between overlap sentences than controls, supporting the first hypothesis of multiple-trace models that similar sentence traces should blend during recall. On the other hand, the pattern of conditional recall probabilities, indicating the presence or absence of post-access interference between sentences, did not consistently support the second hypothesis of multiple-trace models. However, one important concept has been missing from the discussion of the models' predictions thus far: encoding failure. Subjects did not have perfect memory for every sentence in any of the experiments, considering the rate of 50% or more recall failure. In any model of memory, an item cannot be recalled if it was improperly encoded in the first place. Yet the models as discussed so far have simply assumed that subjects completely encoded all sentence information into well-integrated traces. Therefore, one needs to examine how both multiple- and single-trace models perform when a significant proportion of the properties in traces are either missing or too weak to be recalled.

How would a multiple-trace model deal with encoding failure? Because of the simultaneous access of all relevant traces in a multiple-trace model, whenever one of the two target words in an overlapping sentence had not been adequately encoded, the corresponding word from the other overlapping sentence trace should automatically be retrieved instead, as long as it had been properly encoded itself. However, multiple-trace models were already predicting blend errors when encoding failure was not considered; wouldn't they now predict entirely too many blend errors relative to the fairly low frequency obtained in the experiments?

In order to discover whether or not a multiple-trace model could exhibit behavior conforming to that of our subjects, we devised an interactive activation model based on the simple connectionist model discussed in McClelland (1981). This model was adapted for the current experiments by assuming that the learning of each sentence created a instance unit for the sentence trace as a whole, linked with bidirectional excitatory connections to five property units, one for each major content word in the sentence. All of the instance units together formed a "pool" of units, as did all of the property units corresponding to a given sentence position. All units within a pool were linked by bidirectional inhibitory connections (using negative weights), reflecting the fact that the units within them represent mutually-exclusive information that should not be recalled simultaneously.

Because the overlapping sentences shared words, the corresponding units in the property unit pools would also be shared. Simulation of recall proceeded by activating the property units representing the recall cue words. In the case of an overlapping sentence, activation from the cued units would be sent along excitatory connections to the instance units for both overlapping sentences. The two instance units would compete for activation, due to their inhibitory connections; they would also feed back activation to the cue words, as well as the target words. In the pools containing the target words, further competition would result from the activation of the mutually-inhibitory units. Eventually, the activation levels of all units will stop fluctuating, after the network has "settled", and the units with the highest activations can be chosen as the response. Note that if the cue had been words from a control sentence then the correct answer would be retrieved without competition, because activation would flow to only one word per pool.

Actually, as outlined above, the model would not be able to settle the competition between overlapping sentence words, because their inputs and resulting activations would always be equal. Based on this, the model led us to expect that our experiments would generate more blend errors than were actually obtained. However, there is a general problem with the interactive activation model that has recently been uncovered in other applications which appears to reflect a failure to take account of the important role of inherent variability in processing (Hinton & Sejnowski, 1986). Once this variability is introduced to the model, several difficulties are resolved (McClelland, 1991). In the present application, adding variability causes the model to tend to favor one of two complete sentence traces rather than a blend; the blend states represent less-optimal points in the "goodness" landscape of network states (c.f. Rumelhart et. al., 1986), and variability allows the network to escape such local minima. McClelland (1991) indicates that variability may be introduced in a variety of ways. In this case, we simply injected a small amount of normally distributed random noise into the input to each unit at each update.

The addition of intrinsic variability allowed the model to produce approximately correct blend rates, but tended to produce too large a difference in the probability of completely correct recall between the control and overlap conditions. (This is equivalent to over-predicting the difference in the conditional probability of recalling both words). The addition of the assumption of occasional encoding failures allowed the model to overcome this problem. To capture encoding failure, we assumed that each link between a property unit and its associated instance unit had a 20% chance of being absent entirely. In the experimental data, subjects' errors indicated that there were some associations between words from completely different sentences; these associations were captured by inserting randomly determined, non-negative weights ($M$ = 0.2, $SD$ = 0.25) between all word and instance units. The weights on connections within pools of word units were all set to -1.0, while those within the instance unit pool were stronger (-2.1), to counterbalance the activation coming from the three word units and ensure adequate competition between sentences.

The model networks for each experiment reflected the materials used in that experiment. For Experiments 1 and 3, the network therefore contained of a pool of 32 instance units, two pools of 32 target word units, and three pools of 24 word units (16 control words + 8 overlap words per pool). For Experiment 2, the network consisted of 36 instance units, two pools of 24 overlapping cue units (12 overlap + 12 control), one pool of 36 non-overlapping cue units, and two pools of 36 target units. Because the differing results from Experiments 1 and 3 were thought to have been a result of less rigorous encoding in Experiment 3, the only difference between the networks for Experiments 1 and 3 rested in increasing the model's encoding failure rate for Experiment 3 from 20% to 35%. To test recall, external activation was input to the three cue word units for each of the test sentences. The response of the computer subject was simply taken as the unit with the highest activation above a response threshold (.1) in each target word pool after 100 time cycles, enough time to allow the network to settle into an equilibrium and form a reasonable response hypothesis. A total of 100 computer subjects were run for each computer simulation; with each computer subject, a new random set of stimulus and random association weights was generated. The performance of each network was fit to the data as shown in Figures 2 through 4. In so doing, we ensured that the model not only had the correct overall recall rates, but also the appropriate performance with respect to both hypothesis tests – blend rates and post-access interference rates.

In the data fits illustrated in Figures 2 to 4 (in which error bars reflect 95% confidence intervals), the model's performance was not statistically different from that of our human subjects (for Exp.1, $\chi^2(7) = 7.0, p = .42$; for Exp.2, $\chi^2(9) = 15.9, p = .07$; for Exp. 3, $\chi^2(7) = 11.2, p = .13$). However, the fit comes close to failing significantly in two of the three cases. We were able to obtain even closer data fits by increasing the inhibition between instance units in Experiments 1 and 3 (to -2.5) and decreasing it for Experiment 2 (to -2.0). Thus, with the use of an additional free parameter one can achieve a nearly perfect correspondence between the performance of the model and experimental subjects (for Exp.1, $\chi^2(7) = 1.8, p = .97$; for Exp. 2, $\chi^2(9) = 2.5, p = .98$; for Exp.3, $\chi^2(7) = 3.0, p = .89$). It is not entirely clear why that particular parameter should differ between experiments, though it might reflect subjects' ability to strategically increase competitive inhibition between equivalently-cued overlapping traces in Experiments 1 and 3.

Given the success of our particular multiple-trace model, it is worth considering how a single-trace model would behave given similar assumptions about encoding failure. A single-trace model like Shiffrin's SAM model might also fit the data if multiple access attempts were allowed in cases where an initial trace selection resulted in incomplete recall. Successive attempts proceed just like the first, with the probability of selecting a given trace based on that trace's similarity to the cue. Since overlapping sentence traces equally match the cue, they have an equally good chance for selection. So if a first attempt accessed an overlapping sentence trace containing only one of the two target words, and a second attempt accessed the other overlapping sentence, the subject could make a cross-over error. It seems plausible that an account could be given for all aspects of the data based on such a possibility. However, this kind of extension of the single-trace model in some sense amounts to turning it into a multiple-trace account. The primary difference between this and our account is simply that the multiple traces are accessed in succession in one case and simultaneously in the other.

## Conclusion

In the end, what has been learned here about the nature of memory blending during recall? The present experiments and simulations have changed the way we approach multiple-trace models in general, as we have abandoned the original notion of unintended blending of complete traces in favor of the somewhat less dramatic notion of filling in missing parts of incompletely encoded traces.

What remains to be studied is how other instantiations of a multiple-trace model, such as the more superpositional matrix- or convolution-based models (for example, Knapp & Anderson, 1984, or Metcalfe, 1990), would manage the task of simulating the specific patterns of recall performance found in our experiments. It will be of interest to discover whether these types of models could produce as few blend errors as were obtained in our studies, given their tendency to blur the distinctions between traces. Conversely, it will be interesting to examine whether the prototyping and generalization properties of these other models can be captured with the model we have presented here. These complementary studies may take us some distance toward understanding what governs the human ability to generalize well and yet preserve relatively distinct access to particular prior events.

## References


Anderson, J. R. (1976). *Language, Memory, and Thought.* Hillsdale, NJ: Lawrence Erlbaum Associates.

Anderson, J. R. (1983). *The Architecture of Cognition.* Cambridge, MA: Harvard University Press.

Gillund, G., & Shiffrin, R. M. (1984). A retrieval model for both recognition and recall. *Psych. Review*, **91**, 1-65.

Hinton, G. E., & Sejnowski, T. J. (1986). Learning and relearning in Boltzmann machines. In Rumelhart, D. E. McClelland, J. L. & the PDP Research Group. *Parallel Distributed Processing*, Vol. 1. Cambridge, MA: MIT Press.

Knapp, A., & Anderson, J. A. (1984). A signal averaging model for concept formation. *JEP: Learning, Memory, and Cognition*, **10**, 616-637.

Loftus, E. F. (1977). Shifting human color memory. *Memory & Cognition*, **6**, 696-699.

Loftus, E. F., Miller, D. G., & Burns, H. J. (1978). Semantic integration of verbal information into a visual memory. *JEP: Human Learning & Memory*, **4**, 19-31.

McClelland, J. L. (1981). Retrieving general and specific information from stored knowledge of specifics. *Proceedings of the Third Annual Meeting of the Cognitive Science Society*, 170-172.

McClelland, J. L., & Rumelhart, D. E. (1985). Distributed memory and the representation of general and specific information. *JEP: General*, **114**, 159-188.

McClelland, J. L., & Rumelhart, D. E. (1988). *Explorations in parallel distributed processing.* Cambridge, MA: MIT Press/Bradford Books.

McClelland, J. L. (1991). Stochastic interactive processes and the effect of context on perception. *Cog. Psych.*, **23**, 1-44.

Metcalfe, J. (1990). Composite holographic associative recall model (CHARM) and blended memories in eyewitness testimony. *JEP: General*, **119**, 145-160.

Raaijmakers, J. G., & Shiffrin, R. M. (1981). Search of associative memory. *Psych. Review*, **88**, 93-134.

Rumelhart, D. E., Smolensky, P., McClelland, J. L., & Hinton, G. E. (1986). Schemata and sequential thought processes in PDP models. In McClelland, J. L., Rumelhart, D. E. & the PDP Research Group, *Parallel Distributed Processing*, Vol. 2. Cambridge, MA: MIT Press.

# Learning Object-Relative Spatial Concepts in the $L_0$ Project

**Terry Regier**
Computer Science Division
University of California at Berkeley
and
International Computer Science Institute
1947 Center Street, Berkeley, CA, 94704
(415) 642-4274 x 184
*regier@cogsci.Berkeley.EDU*

**Abstract**

This paper reports on the learning of spatial concepts in the $L_0$ project. The starting point is the identification of a visual primitive which appears to play a central role in the visually-based semantics for terms which express spatial relations between two objects. This primitive is simply the orientation of the imaginary ray connecting the two related objects where they are nearest each other. Given this, an important part of the learning consists of determining which other orientations this particular one should align with (e.g. it should align with upward vertical for "above"). These other orientations may be supplied by an object-centered coordinate frame, as in English "in front of" and Mixtec "čii", as well as by the upright coordinate frame. A central feature of the system design is the use of orientation-tuned Gaussian nodes which can learn their orientation and $\sigma$, and which perform the critical task of orientation comparison.

## Introduction

The $L_0$ project (Feldman et al. 1990; Weber & Stolcke 1990) concerns the computational task of acquiring natural language in the visually-based semantic domain of spatial relations between geometrical objects. The goal is to learn to determine, for any natural language, whether a scene description in that language is true of a particular scene. A significant part of this task is learning the perceptually grounded semantics for the individual spatial terms in the language. Thus, as a subtask, we would like to learn to associate scenes, containing several simple objects, with spatial terms describing the spatial relations in the scene. Languages differ widely in the perceptual features encoded in their spatial terms (Talmy 1983; Bowerman 1989), making this subtask a challenging one.

When learning a particular spatial concept, the system is supplied with a scene, and an indication of which object is the reference object (called the *landmark*, or LM) and which is the object located relative to the reference object (called the *trajector*, or TR). This is illustrated in Figure 1.

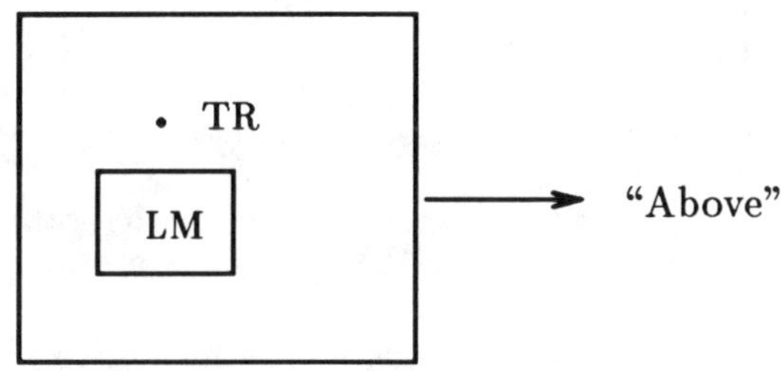


Figure 1: Learning to Associate Scenes with Spatial Terms

Earlier work on this subtask (Regier 1990; Regier 1991) used connectionist mechanisms to learn several basic spatial terms in English, and handled the problem of learning the semantics for these in the absence of explicit negative instances. It did not, however, cover *object-relative* terms, i.e. spatial terms which are sensitive to an inherent orientation that a LM may have, such as "in front of" in English. "In front of" makes reference to a coordinate frame centered in the LM itself: if the LM (a person, for example) has an inherent front, "in front of" is determined relative to that front. This is in contrast to terms like "above", which are unaffected by any inherent orientation of the LM.

This paper presents a way of viewing spatial concepts in which the semantics for object-relative terms and non-object-relative terms are learned in much the same manner. This is based on the identification of a useful visual primitive, and the specification of a mechanism for using it. The system presented here successfully learns the perceptually-based semantics of object-relative terms such as "in front of", as well as that of non-object-relative terms such as "above", "on", and the like.

## General Approach

A central assertion of this work is that one of the crucial primitives used in determining spatial relations is the orientation of the (imaginary) ray from the LM to the TR *where they are nearest to each other.* Consider Figure 2. Here we have small circles located relative

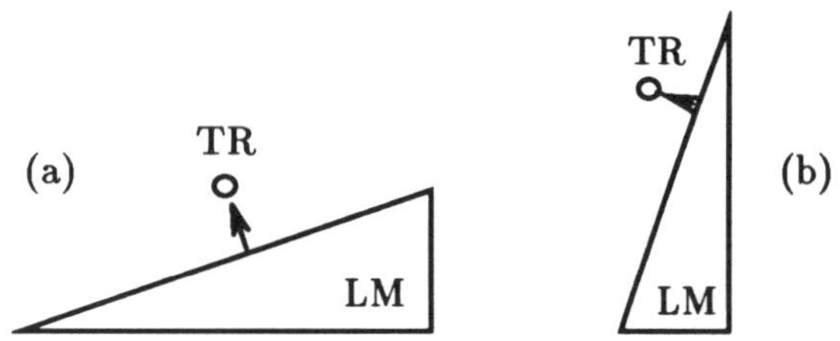


Figure 2: Good and Poor Instances of English "Above"

to two triangles, with the shortest possible connecting line between the LM and TR drawn in. Note that in both cases, the circle is above some part of the triangle. The assertion is that (a) is a better instance of "above" because the orientation of the ray is closer to upward vertical in this case than in (b). This has, at any rate, proven to be a productive working hypothesis. This orientation where the LM and TR are nearest is called the *TR orientation.*

Another potentially useful primitive, to be incorporated in the future, is the orientation of the ray connecting the centers of mass of the two objects (the *CoM orientation*). Figure 2 can actually be seen as supplying evidence that a combination of these two primitives might play a role in the determination of exactly when one object is "above" another. For if the small circle in (a) were moved down the slope of the triangle, it would eventually reach a point at which we would no longer feel comfortable calling the relation between the two objects "above"; however, the TR orientation would have remained unchanged. This can be explained by postulating that both the TR orientation and the CoM orientation must play a part in the determination of "above", and that the CoM orientation in this new case is too far from vertical to allow one to label the scene "above". The CoM orientation is not used in the system presented here; this is primarily to simplify the exposition.

Given orientation primitives of this sort, a large part of the task of learning the perceptually-based semantics for spatial terms is to determine which other orientations in the scene the TR (and/or CoM) orientation aligns with, and to what extent. As a very simple example, if the TR orientation aligns perfectly with upward vertical, we consider this an excellent instance of "above". If it deviates a little from upward vertical, this is a fair instance, and so on. Thus, what is required is an orientation comparison mechanism of some sort. Θ-nodes, to be described in detail below, fulfill this function.

There are in fact two distinct aspects to the process of learning orientations:

- Learning which of several available reference orientations the TR orientation should align with.
- Learning what the values of some of these reference orientations should be. For example, the system starts learning without knowledge of the fact that upward vertical is an important reference orientation. It must learn to tune itself so that upward vertical becomes one of the reference orientations available to the TR orientation for alignment.

These two aspects of the learning occur simultaneously in the system.

Some reference orientations are not learned, however. The orientations of the major and minor axes of the landmark are critical for learning object-relative terms, together with an indication of whether or not one end of each axis is inherently marked as the positive direction (e.g. the positive direction of the major axis for an erect human being is upwards, while the positive direction for the minor axis is toward the ventral side). The values of these orientations clearly change with each new landmark; thus, they are not learned.

Orientation alignment does not suffice to represent such important perceptual features as inclusion and contact. Another crucial primitive is a bitmap representation of the interior of the LM, i.e. a 2-D visual map such that all points that fall in the interior or on the boundary of LM are activated, and all others unactivated. Ideally, a fixed connectionist visual preprocessing stage will produce these primitives for the learning system, given the input image. Currently, they are computed in a non-connectionist fashion.

The system learns spatial concepts by combining evidence from these two forms of representation, orientation-based and bitmap-based. The detailed architecture will be presented below.

## The Mixtec Spatial System

As mentioned above, part of the goal of the $L_0$ project is to build a system that will be able to learn the spatial system of any natural language. For this reason, this paper focuses on the learning of spatial concepts from the Mixtec language, a system radically different from that of English. Both Mixtec and English concepts are learned by the system.

The Mexican Indian language Mixtec has a spatial system based on an extensive body-part metaphor (Brugman 1983), featuring a number of object-relative terms. In general, in Mixtec one would say "The [TR] is located (near/at) the [LM]'s [body-part-name]", even if the LM were not an animate object. Thus, the region above a tree would be described as "(near/at) the tree's head". One intriguing aspect of Mixtec is that it uses human body-part terms to describe relations relative to LMs which are vertically extended, and animal body-part terms for horizontally extended LM objects (since animals such as dogs and horses are generally seen on all fours, horizontally extended). There is a Mixtec term which is used for both humans and animals, however, and therefore, for both vertically and horizontally extended objects. This is the term *čii*, or "belly", which

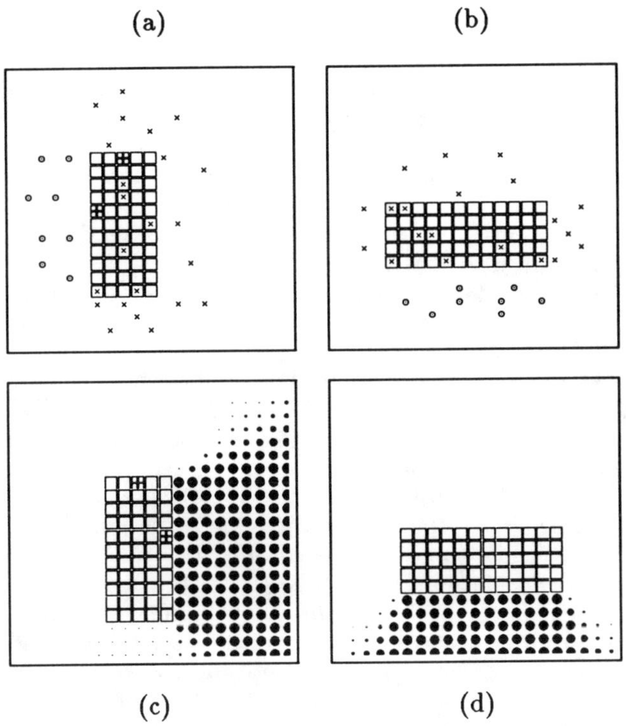

Figure 3: Some Training Data and Results for Mixtec "*čii*" (see text)

indicates the area in front of a vertically extended object (where human bellies are), and the region below a horizontally extended object (where animal bellies are, most of the time). Thus, this term is object-relative in the case of a vertically extended LM, and not object-relative in the horizontally extended case. I.e. the relevant reference orientations are supplied by the LM in the vertically extended case, and by the (learned) upright coordinate system otherwise.

## Training and Results

The TR is restricted to be a single point for the time being; current work is directed toward the more general case of an arbitrarily-shaped TR.

Figure 3 presents training data and results for Mixtec *čii* ("belly").

Figure 3 (a) contains some of the training data for Mixtec *čii* ("belly"). It shows an oriented upright LM (the large "+"'s on the LM mark the positive directions of the major and minor axes), and a number of point TRs: the circles indicate positive examples of *čii* with respect to this LM, and the small x-marks indicate negatives. Note that the positive examples are all located in the positive direction of the minor axis, or "in front of" the LM.

Figure 3 (b) also holds training data for *čii*. It shows an unoriented horizontally extended LM (no end of either axis is marked as positive), again with a number of point TRs indicating positive and negative examples. Recall that in the case of horizontally extended LMs, *čii* (or "belly") is associated with that region of space which is found where animal bellies usually are, viz the area below the LM. Thus, for horizontally extended LMs, it does not matter whether the axes of the object are marked for direction or not.

There is training data relative to other LMs as well (upright LMs pointing to the right, horizontal LMs with directed axes, such that the minor axis points downward, etc.), but considerations of space preclude presentation of all of these.

Figure 3 (c) illustrates the results of the learning. The size of the black circles indicates the appropriateness, as judged by the trained system, of using the term *čii* to describe the relation between a point TR at that location, and the LM shown. Note that the LM points to the right here, and that it is the region to the right that is considered to be *čii* the LM. In the case of an upright LM pointing to the left (like the LM presented with the training data in (a)), points to the left of the LM would be considered *čii* the LM.

Figure 3 (d) illustrates the results of the learning relative to an unoriented horizontally extended landmark. As desired, points under the LM are considered to be *čii* the LM. Again, space considerations rule out presentation of the results for all possible LMs, but the concept is successfully learned.

The system has learned the English concepts *in front of, in back of, above, below, left, right, in, out, off,* and *on*. It has also learned the following concepts from Chalcatongo Mixtec: *ini* ("spleen, gut", meaning inside), *šini* ("human head", above an upright LM), *čii* ("belly", already discussed), *haʔa* ("foot", at the base of an upright LM), *siki* ("animal back", above a horizontally extended LM), and *yata* ("human back", in front of an upright LM). Clearly, several of these are object-relative terms.

## System Design

One of the central mechanisms used in the system is one that determines to what extent a given orientation matches a reference orientation. A node mechanism is presented here which allows this sort of orientation comparison, and allows the node to learn to tune itself to embody an appropriate reference orientation and tolerance.

Once this node mechanism has been presented, the network as a whole will be presented and discussed.

### Θ-Nodes

All orientations in the system are represented in (sin,cos) pairs.[1] The structure of a single Θ-node, which learns to tune itself to a preferred orientation and tolerance using this representation, is presented in Figure 4. Note that $sin_\theta, cos_\theta$, and $\sigma_\theta$ are variables internal to

[1] This representation can be viewed as a minimalist version of coarse-coding.

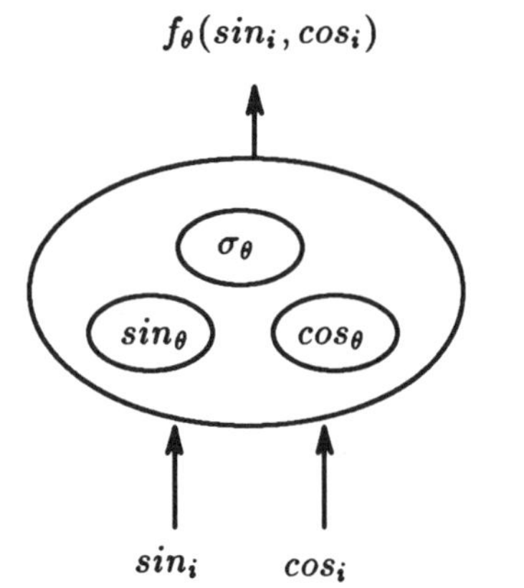


Figure 4: Internal Structure of a Single Θ-node

---

the Θ-node, defining its preferred orientation and tolerance. These internal values may be learned together with weights in the network as a whole, or may be imposed from the outside to order to "tune" the node.

The function of a Θ-node is simply a Gaussian (see (Moody & Darken 1988) for earlier work using Gaussian nodes in a somewhat different way in connectionist networks):

$$f_\theta(sin_i, cos_i) = \exp\left[-\frac{(sin_\theta - sin_i)^2 + (cos_\theta - cos_i)^2}{\sigma_\theta^2}\right] \tag{1}$$

The partial derivatives of $f_\theta$ with respect to each of the internal variables are:

$$\frac{\partial f_\theta}{\partial sin_\theta} = \frac{-2(sin_\theta - sin_i)}{\exp\left[\frac{((sin_\theta - sin_i)^2 + (cos_\theta - cos_i)^2)}{\sigma_\theta^2}\right]\sigma_\theta^2} \tag{2}$$

$$\frac{\partial f_\theta}{\partial cos_\theta} = \frac{-2(cos_\theta - cos_i)}{\exp\left[\frac{((sin_\theta - sin_i)^2 + (cos_\theta - cos_i)^2)}{\sigma_\theta^2}\right]\sigma_\theta^2} \tag{3}$$

$$\frac{\partial f_\theta}{\partial \sigma_\theta} = \frac{2((sin_\theta - sin_i)^2 + (cos_\theta - cos_i)^2)}{\exp\left[\frac{((sin_\theta - sin_i)^2 + (cos_\theta - cos_i)^2)}{\sigma_\theta^2}\right]\sigma_\theta^3} \tag{4}$$

From these, we can easily find $\frac{\partial E}{\partial sin_\theta}$, $\frac{\partial E}{\partial cos_\theta}$, and $\frac{\partial E}{\partial \sigma_\theta}$, which enables us to use back-propagation to train the internal variables of a given Θ-node, together with the weights of the network in which they are embedded.

Every Θ-node will learn its $\sigma_\theta$, and several, though not all, will learn their orientations as well $(sin_\theta, cos_\theta)$.

## Network Architecture

Figure 5 illustrates the architecture of the network used here. Note that all weights below the dotted line are frozen at 1.0, so learning occurs only above this line.

Recall that the system uses both orientation-based and bitmap-based primitives. We examine the halves of the network handling each of these in turn.

We first consider orientation-based processing. This is done by the left-hand half of the network, which receives input from the TR orientation and the major and minor axis orientations of the LM (here labeled "MAO" and "mAO", respectively). If the LM major axis has no inherent direction, then the two possibilities for that orientation are loaded into the MAO(1) and MAO(2) inputs (e.g. if it is a tall object with neither the top nor the bottom marked as the positive direction, the inputs are loaded with the (sin,cos) representations of 90 degrees and 270 degrees). If, on the other hand, the LM has an inherent direction for its major axis, *both* MAO(1) and MAO(2) are loaded with the representation for that direction. The minor axis orientation inputs (mAO(1) and mAO(2)) are handled analogously.

All hidden nodes marked Θ are Θ-nodes of the sort described above, and all receive their $(sin_i, cos_i)$ inputs (recall Figure 4) from the TR orientation input. Of these nodes, the middle three learn their $sin_\theta, cos_\theta$, and $\sigma_\theta$, while the others learn only $\sigma_\theta$, and have their $(sin_\theta, cos_\theta)$ internal variables set by the lines leading to them from the MAO and mAO inputs. Thus, the middle three nodes learn to tune themselves (generally to the upright vertical and to left and right), while the two on the left are always tuned to whatever the major axis orientation is, and the two on the right to whatever the minor axis orientation is, with ambiguously oriented LMs handled as described above.

Consider node H, above the Θ-node layer. It, like the other nodes not explicitly marked with a Θ, computes the usual sigmoid of its weighted and summed input. Notice that the two Θ-nodes associated with the major axis project to H, on links which are *constrained to be of the same weight*, denoted "r" (see (LeCun 1989) for details on this technique of weight-sharing). The two links from the MAO Θ-nodes to node J will also be constrained to be the same, though not necessarily the same as "r". Weights from the mAO Θ-nodes are similarly constrained.

Notice that under these constraints, node H will treat both MAO Θ-nodes identically, so that it can learn to respond to ambiguously directed LMs without worrying about which of the two possible orientations is represented by which of the two Θ-nodes. Note also that if the LM *does* have inherent orientation, then H can receive greater input from these nodes than it would in the case of an ambiguously oriented LM, since in the ambiguous case, at most one of the two Gaussian Θ-nodes will be responding strongly, while in an unambiguously directed case, both may be.

This played a role in the learning of Mixtec *čii*: if there was an inherent minor axis orientation, that direction was the correct one for *čii* as learned by the system. If there was not any inherent directionality to the minor axis however, the system responded to the coincidence with downward vertical of one of the two possibilities for the minor axis. Since a node such as H will receive greater input in the case of an unambigously oriented LM with a TR in that direction, but

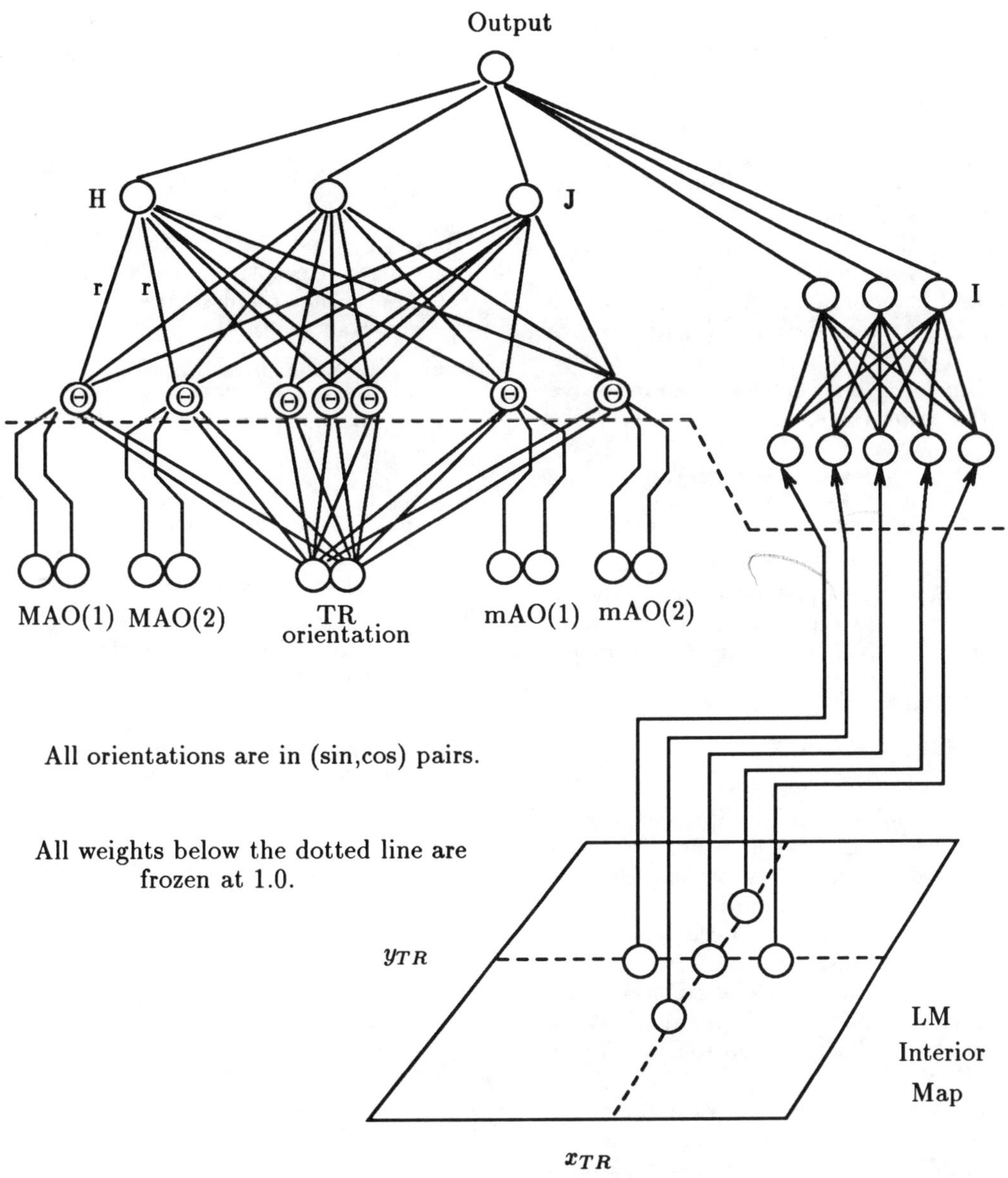


Figure 5: Network Architecture

will still respond somewhat in the case of an ambiguous LM with a TR in one of the two possible directions, this concept was learnable.

We now consider the processing of the bitmap-based LM representation, performed by the right-hand part of the network, which receives input from the LM interior map. Each of the three nodes in the cluster labeled "I" (for interior) has a receptive field of five pixels.

When a TR location is specified, the values of the five neighboring locations shown in the LM interior map, centered on the current TR location, are copied up to the five input nodes. The weights on the links between these five nodes and the three nodes labeled "I" in the layer above define the receptive fields learned. When the TR position changes, five new LM interior map pixels will be "viewed" by the receptive fields formed. This allows the system to detect the LM interior (or a border between interior and exterior) at a given point and to bring that to bear if that is a relevant semantic feature for the set of spatial terms being learned.

Note that the four outer links in this small receptive field are tied to the same value, so that this receptive field is radially symmetric. Thus, this half of the network handles strictly local features such as contact and inclusion, while the rest of the network handles directional features.

The orientation-based and bitmap-based representations are combined at the output level of the network, so that the learned semantics for a given spatial term may involve a mixture of evidence from the two representation types.

## Conclusions

A connectionist system has been presented which learns perceptually grounded semantics for both object-relative and non-object-relative spatial terms, from English and Mixtec. The system relies on the use of a particular visual primitive, the orientation of the imaginary ray connecting the two relevant objects where they are nearest to each other. This orientation is compared to various reference orientations, by Gaussian $\Theta$-nodes tuned for a particular orientation and tolerance. These $\Theta$-nodes learn their $\sigma$s, and some learn their reference orientations as well.

Immediate future work is directed toward extending the system to handle trajectors which are not simply a single point. In addition, moving trajectors, and a means for handling the resulting polysemy, are on the agenda.

## References


Bowerman, M. 1989. Learning a Semantic System: What Role do Cognitive Predispositions Play? In et al, M. L. R., ed., *The Teachability of Language*, 133–169. Paul H. Brookes, Baltimore.

Brugman, C. 1983. The Use of Body-Part Terms as Locatives in Chalcatongo Mixtec. in Report No. 4 of the Survey of California and other Indian Languages, pp. 235-90. University of California, Berkeley.

Feldman, J., Lakoff, G., Stolcke, A., and Weber, S. 1990. Miniature Language Acquisition: A Touchstone for Cognitive Science. Technical Report TR-90-009, International Computer Science Institute, Berkeley, CA. also in the Proceedings of the 12th Annual Conference of the Cognitive Science Society, pp. 686–693.

LeCun, Y. 1989. Generalization and Network Design Strategies. Technical Report CRG-TR-89-4, Department of Computer Science, University of Toronto.

Moody, J., and Darken, C. 1988. Learning with Localized Receptive Fields. In *Proceedings of the 1988 Connectionist Models Summer School*. Morgan Kaufmann.

Regier, T. 1990. Learning Spatial Terms Without Explicit Negative Evidence. Technical Report 57, International Computer Science Institute, Berkeley, California.

Regier, T. 1991. Learning Perceptually-Grounded Semantics in the $L_0$ Project. In *Proceedings of the 29th Annual Meeting of the Association for Computational Linguistics (to appear)*.

Talmy, L. 1983. How Language Structures Space. Technical Report 4, Institute of Cognitive Studies, University of California at Berkeley.

Weber, S. H., and Stolcke, A. 1990. $L_0$: A Testbed for Miniature Language Acquisition. Technical Report TR-90-010, International Computer Science Institute, Berkeley, CA.

# The Ontogeny of Units in Object Categories

**Philippe G. Schyns and Gregory L. Murphy**

Dept. of Cognitive and Linguistic Sciences
Brown University
Providence, RI, 02912
schyns@clouseau.cog.brown.edu

### Abstract

Theories of object recognition and categorization rely on a set of primitives to represent objects. The nature and the development of these primitives have been neglected in computational vision and in concept learning theories. We present a theory of part ontogeny in which not only perceptual, but also categorical constraints play a role. A two-phase experiment using categories of synthesized 3D objects (Martian rocks) was conducted to test the theory. The first phase tested the hypothesis that part identification is dependent on categorical context. The second phase tested whether the units extracted in the first phase played a conceptual role in learning a new category. In both phases, subjects interactively delineated the parts of the stimuli while learning the categories. The units subjects identified in the first phase were those that were predictive of the object's category. These units then influenced the perception of parts in the new categories of the second phase.

## Introduction

Most accounts of object recognition and categorization assume that in order to identify an object, one must first identify its components. Of course, not all components of an object must be identified before an object can be categorized, but many of them are probably recognized during the categorization process. For example, many accounts of categorization (see Smith & Medin, 1981) would claim that a person who sees a cat might identify a global shape, a size, body parts such as the head, tail and whiskers, a color, a texture, and so on. After these perceptual components of the object have been identified, they are compared with memory representations in order to categorize the object. Theories that embody this general approach include Rosch and Mervis's (1975) family resemblance model, Marr and Nishihara's (1978) generalized cones, and Biederman's (1987) Recognition By Components.

The topic of this paper is the nature of the components themselves. How are these components derived? Are the components independent entities, or does category learning influence them? In particular, we will be interested in what might informally be called the *parts* of an object. Tversky and Hemenway (1984) showed that the most useful categories (i.e., Rosch et al.'s 1976 basic categories) can be distinguished by parts whereas less useful categories cannot (see also Murphy, in press). Furthermore, parts have a privileged place in a number of accounts of object identification (Biederman, 1988; Biederman & Ju, 1988; Binford, 1981; Marr, 1982; Marr & Nishihara, 1978; Pentland, 1989).

Although parts may well be important in categorization, the question arises as to how people identify an object's parts and encode them in memory. The most common assumption concerning parts and other components of an object is that they are the result of perceptual structures. That is, our perceptual apparatus identifies the colors, size and parts of an object which then form the basis for conceptual tasks. Certainly, perceptual structures provide important constraints on what can be a part of an object. But important research in computationnal vision and psychophysics has shown that part extraction without constraints is essentially an ill-posed problem (Bertero et al., 1987; Biederman, 1987; Braunstein et al., 1989; Hoffman & Richards, 1985; Koenderink & van Doorn, 1982; Poggio et al., 1985). However, an additional constraint, which has received little attention, is the effect of category learning itself. The process of concept formation may alter the interpretation of objects such that new parts and combinations of parts are encoded.

To understand how this might occur, consider this thought experiment. Suppose you were a Martian with no experience of Earth objects. The first object you see is a cup which can be segmented into the analytical units *handle* and *container*. However, your Martian conceptual system does not possess these primitives. Supposing your low-level visual system were roughly like ours, you could describe the cup by mentioning at least that something seems to break the regularity of its surface: something sticks out. Suppose the second object you experience is a glass. This object is much smoother; nothing sticks out. From now on, you could use the primitive *Xd@#* (you certainly would not call it a handle), because it categorizes the two objects: one has a *Xd@#* and one doesn't. Before encountering the glass, you might have represented the cup as a whole, as a single unit. Thus, the category learning situation may select some parts and suppress others in representing objects.

This phenomenon illustrates what we call *the Homogeneity Principle*. This principle says that if a component or group of components occurs frequently throughout a category, then it will be treated as a single unit. In other words, the objects of the category provide a sufficient context to change a component's status from a potential part to an actual part--or unit--that is used to represent the object. We will call these analytical parts "conceptually instantiated units," or simply "units."

Units are the parts that are input into concept representation. Examples of units are *head*, *tail*, *handle* and so forth. Conceptually instantiated parts play a *conceptual role* by segmenting the external world into categories of objects. Once a unit is instantiated in the context of one particular category, it can then play a conceptual role in other contexts of categorization. To illustrate, if the Martian had extracted what we call a handle, it could use it in categorizing new objects that have a handle. The Homogeneity Principle also assumes that exposure to different contexts may result in the instantiation of different units. Therefore, this theory provides a framework to study the ontogeny of part/whole relationships. The representation of an object could eventually be decomposed into two units if the parts composing the object appear to play a conceptual role in another context; otherwise, the whole may be treated as one instantiated unit.

The study reported here examines the Homogeneity Principle through a two-part experiment using categories of novel objects. The first phase tested the hypothesis that part instantiation into units is dependent on categorical context. It also investigated a claim of *sufficiency*, that the context of one category is sufficient to instantiate the parts needed to represent the objects of the category. Components that were homogeneous throughout the category were expected to be instantiated as units, and other components were expected to be rejected, over the course of learning. The second phase introduced a new category, which overlapped somewhat with the previous one. This allowed a test of the conceptual role played by the units learned in the first phase. The overlapping units were expected to be parsed first, and thereby to bias the extraction of other components in the new category. A group with no prior experience with those units served as a control. It was expected to identify a different part structure (see below).

Studying the ontogeny of units presents certain methodological challenges. Using familiar objects is obviously inappropriate, because the theory concerns the effect of category learning on part analysis. However, most artificial categories used in previous experiment of this sort could not be used here because it was necessary to avoid familiar *units* as well as familiar objects. So, it would not do to teach subjects a new kind of furniture if they could already identify familiar units such as legs, seats, cushions, drawers, etc. One would not expect to counteract subjects' extensive prior knowledge of these parts. Therefore, the experiment used 3-dimensional spherical objects with various mountain-like projections. Each projection or group of projections could serve as a potential part. The experiment examined the degree to which subjects' perception of parts could be influenced by the context of one category of objects, and how the units acquired in this context would influence the extraction of new units in another category.

## Phase 1

Subjects were randomly assigned to one of two conditions in which they were exposed to exemplars of a particular category (A or B). Category A was a collection of objects on which a specific part (*a*) was always present. Category B was also a collection of objects, but they were characterized by another consistent part (*b*). We recorded the parts subjects considered important before, during and after category learning.

The Homogeneity Principle predicts that the parts subjects select after exposure to a category should be the ones that hold the category together. Therefore, subjects should pick part *a* or part *b* if they are in condition A or B, respectively. Also, the number of nontarget parts subjects pick should decrease as they gain more and more experience with the category. A corrolary of homogeneity is that subjects should not consider as relevant parts those that do not glue a category together. Thus, if subjects from group A were exposed to the target part of group B, they would not pick part *b* as a relevant part.

### Method

**Subjects.** Twelve Brown University undergraduates participated in the first phase of the experiment. Subjects were assigned randomly to condition A or B with the constraint that the number of subjects be equal in each condition.

**Stimuli.** "Martian rocks" were synthetized three-dimensional gray-level shaded objects displayed on the screen of a graphic mini-supercomputer (Stellar GS 1000). Each stimulus was generated on a Lisp Machine using the S-Geometry package. A Martian rock was created from a sphere by specifying eight equidistant regions along its equator. Each region covered a surface extending from the equator to the two poles of the sphere, and deformations were applied to each region. A deformation consisted of pulling out a facet of a region for a certain distance along the surface normal at that point. The deformations were deterministic for the target parts, while the seven remaining regions were randomly distorted. The top pictures of Figure 1 show the deformations that created unit *a* and unit *b*, and the bottom pictures present complete exemplars of categories A and B. Random Martian rocks were also created by randomly distorting *all* the regions. We will call the objects in the category being learned the M1 rocks.

**Procedure: Initial delinations.** Before experiencing objects of their category, subjects were asked to delineate

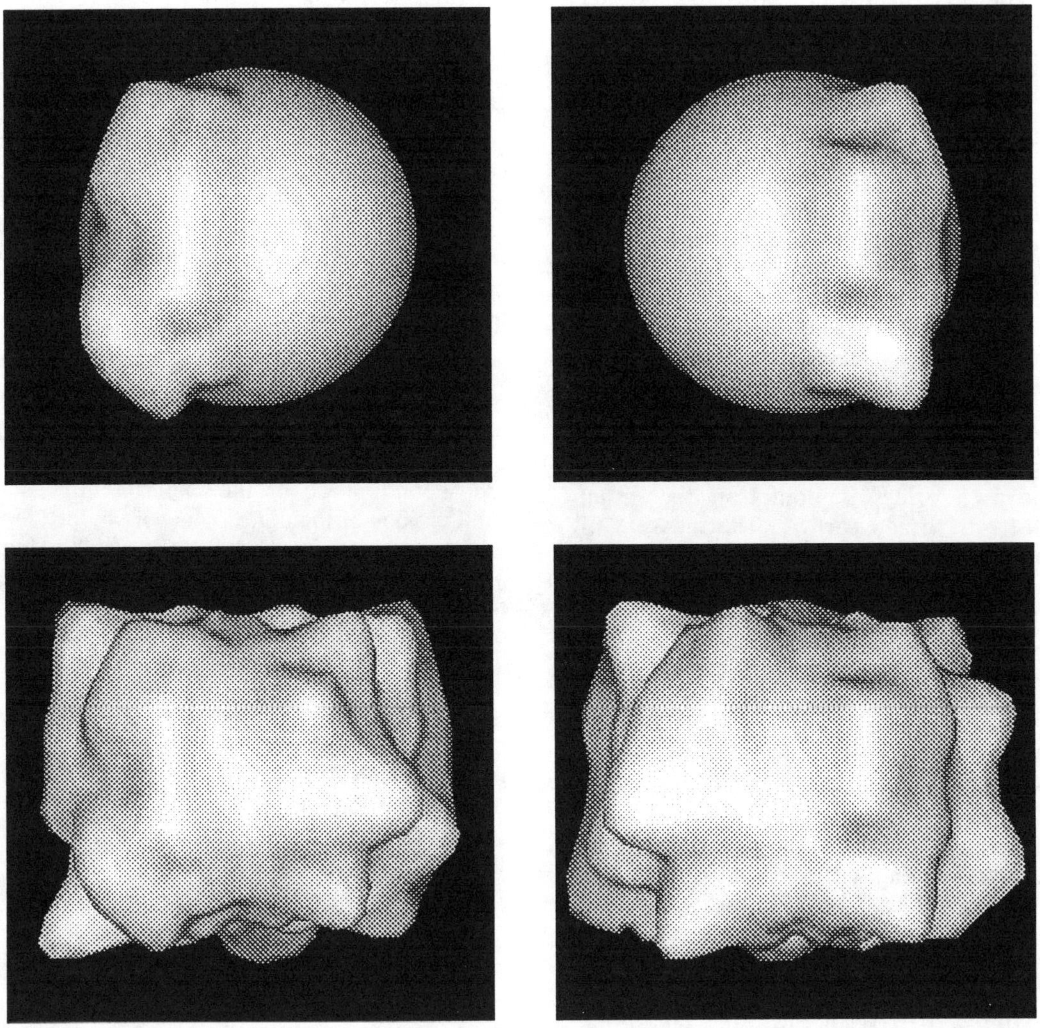

Figure 1: This figure illustrates the stimuli used in Phase 1. The top pictures show part a (left) and b (right) in isolation. The bottom pictures show examples of the complete objects.

the parts they thought were important on one M1 rock. Subjects were instructed to delineate at least one part per object, but no upper limit was specified. The purpose of this initial stage was to test for the salience of the target parts. Part delineation was done using a computer mouse. Using the left and right buttons of the mouse, subjects had to center the part they wanted to delineate with respect to a vertical bar that appeared on the screen. Then, subjects moved a cursor on the screen with the mouse in order to mark with a red line the part they had previously centered. That is, by clicking the mouse, subjects extended the line from its last vertex to the current location.

They continued this delineation until the entire part was indicated. Once one part was delineated, subjects could either center another part of the object or quit part delination for that object.

**Learning stage.** The learning stage was divided into 6 trials. On each trial, subjects first saw an M1 rock. This rock would rotate once clockwise and once counterclokwise in front of them. Subjects were told to study this rock to learn what M1 rocks were like in general. Then this stimulus was replaced with another rock on which subjects were instructed to delineate parts as described earlier. For five out of six delineation rocks, this item was an M1, and the other delineation rock was from the category of the other condition (category B for subjects in group A, and vice versa).

**Test stage.** In order to ensure that subjects learned the category, a categorization stage with 6 test rocks was conducted. Two of the test rocks were M1 rocks, and the other four were random rocks. One test rock at a time would rotate--once clockwise, once counterclockwise--on the screen and subjects had to decide, by pressing the "yes" or the "no" key on the keyboard, whether the rock was an M1. If subjects were correct on all six items, the first phase of the experiment was over. Otherwise, they repeated as many learning blocks as needed to perform the test stage without mistake. No feedback was given during the testing phase.

### Results

We constructed two grids of points in the exact configurations of part *a* and part *b*. The measure of subjects' accuracy in part identification was the number of points of the target parts contained within their delineations. Fifty percent or more of the target's surface had to be delineated in order to score the subject as having identified the target. To test the salience of the target parts, we collected the total number of part delineations in the initial delineation stage (before learning). A total of 44 parts were initially delineated; only one of them met the criterion for being a target part. Thus, the targets were not very salient units before learning.

To test the first aspect of the Homogeneity Principle, we considered the last learning session for all subjects (the one preceding a successful test stage). In this session, we computed the average number of points included in the delineation of the last target part (the M1 part), and compared it with the average number of points included in the delineation of the part from the other condition. Subjects delineated 93% of the target from the category they learned, but only 8% of the target from the other category on average. These means were significantly different $t(11) = 11.88, p < .001$.

The corrolary of homogeneity was tested by recording the number of nontarget delineations per object for all subjects. We expected the number of nontarget selections to decrease with learning. Thus, we divided each subject's learning stage into five blocks (since each subjects saw 5 M1 rocks for delineations per learning stage, the total number of M1 objects seen in learning was always a multiple of 5) and counted the number of nontargets delineated in each block. The number of nontargets decreased significantly from the first to fifth block ($F(4,44) = 18.63, p < .001$; means of 2.78, 2.09, 1.68, 0.70 and 0.21). A contrast showed that this decrease was linear ($F(1,44) = 73.59, p < .001$), and there was little remaining variance ($F(3,44) = .32$).

### Discussion

The results confirmed the two predictions of the Homogeneity Principle. The conjunction of perceptual constraints and the context of several objects forming a category provide a sufficient means to conceptually instantiate potential parts. At the end of learning, subjects picked the part that was redundant throughout the category. It is worth emphasizing that prior to the experiment, subjects didn't know the parts composing the objects. The corollary of homogeneity is that as subjects gain more and more experience with the context of one category, they should progressively reject infrequent potential parts as plausible candidates for a conceptual unit status. This prediction was corroborated with a linear decrease of the nontarget delineations.

If the unit instantiated in one particular context has a conceptual role, it should be used to describe objects of a new category. Once instantiated, these units have a status similar to *hand*, *wheel*, *handle* and so forth. That is, they can be used to characterize new objects. The purpose of the second phase of the experiment was to test both implications of the conceptual role by presenting subjects with a new category of objects in which a number of units are distributed throughout the category, with either one or none of them being already instantiated.

## Phase 2

Subjects from both groups of the first phase were exposed to a new group of objects, category C. In addition to groups A and B, a new group of subjects, group C, who did not participate in Phase 1, served in Phase 2. All objects of Phase 2 were characterized by part *a* and part *b* adjacent to one another.

### Method

**Stimuli.** New stimuli were created in which part *a* and part *b* were adjacent to one another and only 6 regions were random. Figure 2 shows different views of a Martian rock from this phase.

**Subjects.** 18 Brown undergraduates participated in the second phase of the experiment. 12 of them were the subjects of the previous phase and 6 subjects were new to the experiment.

**Procedure.** Phase 2 took place immediately following Phase 1 for groups A and B. As in the previous phase, subjects went through three main stages: an initial delineation stage, a learning stage and a testing stage. As before, subjects went through six trials in the learning phase, and they had to delineate 5 rocks of category C (the remaining rock was a random rock). In the test phase, 6 rocks were presented (2 of them with part *a* only, two with part *b*, and two with part *a and* part *b*), and subjects had to categorize them correctly to end the experiment. No feedback was given during the testing phase.

### Results

We recorded the number of times part *a* or part *b* was delineated in the only object tested before learning. The

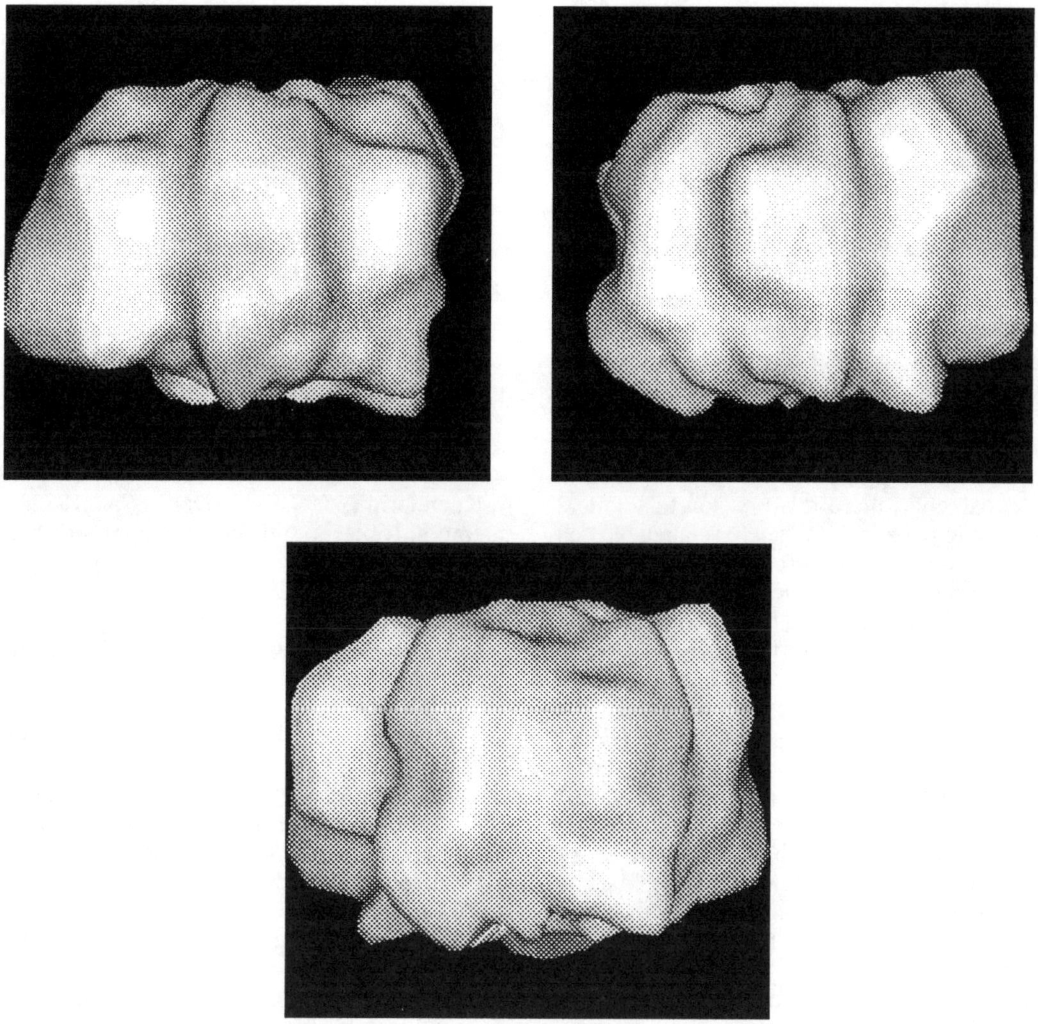

Figure 2: This figure illustrates the stimuli from Phase 2. Subjects in conditions A and B delineated the two parts presented in the top pictures. Subjects in condition C selected the part shown in the bottom picture as one unit. This part is the conjunction of the two parts shown above. During the experiment, subjects could rotate the object to center the part they wanted to delineate.

subjects of group A always delineated part *a*, and never *b*. The subjects of group B always delineated *b*, and never *a*. On the final learning trial, the subjects from groups A and B all delineated parts *a* and *b* as two distinct units, while all subjects from group C (the group with no prior experience with Martian rocks) delineated part *a and b* as a single unit. Typical delineations from groups A and B are presented on Figure 2 (the top pictures) and a typical delineation from group C is presented on Figure 2 (the bottom picture).

## Discussion

This study investigated the role of previous knowledge in parsing new objects. We can see that the units instantiated in the first phase were applied to categorize objects of category C. This shows that they played a conceptual role in the sense explained earlier. The bias created by units on part extraction is also illustrated in the second phase. Specifically, subjects with prior knowledge ended up instantiating two units--the one they had learned in the previous phase, either *a* or *b*, and the other part adjacent to the one they knew already. In contrast, subjects from condition C didn't segment *a* and *b* into two instantiated units. Instead, they considered *a* and *b* to be one single unit *c*. This is interesting for two reasons. First, *a* and *b* adjacent to one another can clearly be segmented as two units--as shown by the other subjects. This means that the two parts are separable on a perceptual basis. In fact, the parts could be extracted with a scheme along the lines of Hoffman and Richard's (1985) minima rule. Subjects in group C apparently encoded *c* as a perceptual unit which was not composed of the units *a* and *b*, which were not known as such to the subjects of condition C. This leads to the second point. The context of categorization may influence what is considered a unit in an *object representation*. The context enables the selection of a perceptual interpretation that maximizes the homogeneity of a category. Since parts *a* and *b* are always present on the object of the category, there is no reason to segment them into two separate parts, unless they have already been distinguished. This last result has far-reaching consequences for understanding the development of a conceptual system. Potential parts become instantiated units if the right categorical context offers them a functional or conceptual role. This provides, among other things, a theoretical framework within which the development of part/whole relationships can be understood, as well as the circumstances under which hierarchies occur in conceptual representations.

The theory of unit ontogeny presented in this paper may be considered an extension of the principles of concept learning. The arbitrary situation of the Martian rocks could be comparable to the kind of learning that takes place in infants when they start forging a conceptual system (Baldwin, 1989; Clark, 1973; Schyns, 1990). The conjunction of perceptual constraints with a context of categorization seems to be a promising framework to ground the development of a simple conceptual system on perceptual foundations. Clearly that will not account for all concepts, but these simple principles may account for the genesis of object concepts from "scratch."

## Acknowledgments

This work was supported in part by NIMH grant MH 41704. The authors would like to thank Heinrich Bülthoff and Shimon Edelman for useful help. Gregory Murphy is now at the University of Illinois.

## References

Baldwin, D. A. 1989. Priorities in children's expectations about object label reference: Form over color. *Child Development* 60:1291-1306.

Biederman, I. 1987. Recognition-by-components. A theory of image understanding, *Psychological Review* 94:115-147.

Biederman, I., & Ju, G. 1988. Surface vs. edge-based determinants of visual recognition. *Cognitive Psychology* 20:38-64.

Binford, T. E. 1981. Inferring surfaces from images. *Artificial Intelligence* 17:205-244.

Bertero, M., Poggio, T., & Torre, V. 1987. Regularization of ill-posed problems. AI memo. 924. AI Lab, MIT.

Braunstein, M. L., Hoffman, D. D., & Saidpour, A. 1989. Parts of visual objects: An experimental test of the minima rule. *Perception* 18:817-826.

Clark, E. V. 1973. What's in a word? On the child's acquisition of semantics in his first language. In T.E. Moore (ed.), *Cognitive development and the acquisition of language*. New York: Academic Press.

Hoffman, D. D., & Richards, W. 1985. Parts in recognition. *Cognition* 18:65-96.

Koenderink, J., & van Doorn, A. 1982. The shape of smooth objects and the way contours end. *Perception* 11:129-137.

Marr, D. 1982. *Vision*. San Francisco: Freeman.

Marr, D. & Nishihara, K. 1978. Representation and recognition of the spatial organization of three-dimensional shapes. *Proceedings of the Royal Society London B* 200:269-294.

Murphy, G. L. in press. Parts in object categories. *Memory & Cognition.*

Pentland, A. 1989. Part segmentation in object recognition. *Neural Computation* 1:82-91.

Poggio, T., Torre, V., & Koch, C. 1985. Computational vision and regularization theory. *Nature* 317:314-319.

Rosch, E., & Mervis, C. B. 1975. Family resemblances: Studies in the internal structure of cateogries. *Cognitive Psychology* 7:573-605.

Schyns, P. G. in press. Concept acquisition in a modular neural network architecture. *Cognitive Science.*

Smith, E. E., & Medin, D. L. 1981. *Categories and concepts*. Cambridge, MA: Harvard University Press.

Tversky, B., & Hemenway, K. 1984. Objects, parts, and categories. *Journal of Experimental Psychology: General* 113:169-193.

# Effects of Background Knowledge on Family Resemblance Sorting and Missing Features

**Woo-kyoung Ahn**

Department of Psychology
University of Michigan
Ann Arbor, MI 48104
Wookyoung_ahn@um.cc.umich.edu

**Abstract**

Despite people's strong bias to sort exemplars based on a single dimension, various situations where family resemblance (FR) categories tend to be created have been identified. In a previous study (Ahn 1990b), knowing prototypes or theories underlying categories led subjects to create FR categories. The current study investigates why existence of background knowledge encourages creation of FR categories. Comparison of results from two experiments indicates that there is no intrinsic tie between knowing theories or prototypes and FR structure. The role of background knowledge on FR sorting seems to lie in leading subjects to weight dimensions equally, in helping them to infer unavailable values in favor or FR sorting, and / or in relating surface dimensions in terms of a deeper feature.

## Introduction

The general consensus is that natural categories do not have defining features and members in the same category are related in terms of a family resemblance (FR) principle: exemplars in the same category are generally similar to each other, but exemplars in different categories are dissimilar to each other (Rosch & Mervis 1975; Smith & Medin 1981). Along with this principle, it has also been expected that when sorting unclassified exemplars, people achieve a compromise between maximizing within-category similarity and minimizing between-category similarity (Rosch 1975). Unlike Rosch's prediction, however, subjects in free sorting experiments show a strong bias to sort exemplars based on values on a single dimension and create categories with defining features (Ahn 1990a; Ahn & Medin 1989; Imai & Garner 1965 1968; Medin, Wattenmaker, & Hampson 1987).

When do we observe FR category construction? Previous research shows that FR categories can be created through two stages where the first stage involves uni-dimensional sorting of exemplars into a prescribed number of categories and the second stage involves assigning exceptional exemplars into initially created categories based on their overall similarities (Ahn 1990a; Ahn & Medin 1989). FR categories can also be created when people know underlying dimensions or theories of surface dimensions and perform uni-dimensional sorting based on the deeper dimensions (Ahn 1990b; Medin et al 1987).

The current study concerns why background knowledge helps people create FR categories. Four possible reasons are investigated: (a) There is some intrinsic link between prototypes or theories and FR structure, (b) the background knowledge changes weights of dimensions, (c) the background knowledge helps people infer unavailable information, and (d) a deeper dimension constructed by underlying theories can be serve as a basis as uni-dimensional sorting. This paper first briefly reviews previous studies on the effect of background knowledge in FR categories. Then two experiments testing the above four possibilities is presented.

## Previous Studies on FR Sorting

**Sorting in Knowledge-Poor Domains.** In previous category construction experiments, subjects have shown a strong tendency to create categories with defining features. To capture this bias, Ahn and Medin (1989) proposed a two-stage model of category construction; people first sort exemplars based on the most salient dimension and then in the second stage they assign remaining exemplars based on their overall similarity to the initially created categories.

Ahn and Medin argued that the reason why FR sorting was not observed in the previous experiments was that the examples had characteristic features that would appear in contrast categories. Therefore, if subjects carry out uni-dimensional sorting in the first stage based on these characteristic features, FR categories could not be created. Take an example from the sets used in Ahn and Medin (1989) shown in Figure 1 under Characteristic Set. (There are ten examples with four dimensions (D1, D2, D3, and D4) in each set. They are represented in terms of the example number (e.g., E1), and values (0, 1, or 2) for each dimension. The examples in the same column (e.g., E1, E2, E3, E4, and E5) belong to the same category if the set is grouped according to the FR principle. )

**Sufficient Set**

| | D1 | D2 | D3 | D4 | | D1 | D2 | D3 | D4 |
|---|---|---|---|---|---|---|---|---|---|
| E1 | 0 | 0 | 0 | 0 | E6 | 2 | 2 | 2 | 2 |
| E2 | 0 | 0 | 0 | 1 | E7 | 2 | 2 | 2 | 1 |
| E3 | 0 | 0 | 1 | 0 | E8 | 2 | 2 | 1 | 2 |
| E4 | 0 | 1 | 0 | 0 | E9 | 2 | 1 | 2 | 2 |
| E5 | 1 | 0 | 0 | 0 | E10 | 1 | 2 | 2 | 2 |

**Characteristic Set**

| | D1 | D2 | D3 | D4 | | D1 | D2 | D3 | D4 |
|---|---|---|---|---|---|---|---|---|---|
| E1 | 0 | 0 | 0 | 0 | E6 | 2 | 2 | 2 | 2 |
| E2 | 0 | 0 | 0 | 1 | E7 | 2 | 2 | 2 | 0 |
| E3 | 0 | 0 | 2 | 0 | E8 | 2 | 2 | 1 | 2 |
| E4 | 0 | 1 | 0 | 0 | E9 | 2 | 0 | 2 | 2 |
| E5 | 2 | 0 | 0 | 0 | E10 | 1 | 2 | 2 | 2 |

**Figure 1. Structure of Examples used in Ahn & Medin (1989)**

Suppose the task is to create two categories. Then regardless of which dimension is chosen as the initial defining feature, the resulting categories cannot be FR categories.

On the other hand, consider the Sufficient Set. The resulting FR categories from this set consisted only of sufficient features (i.e., 0's and 2's). Suppose the first dimension was chosen for the most salient dimension in the first stage. Then the model categorizes E1, E2, E3, and E4 into one category and E6, E7, E8, and E9 into another category. In the second stage, one of the remaining exemplars, E5, is grouped with E1, E2, E3, and E4, and E10 is grouped with E6, E7, E8, and D9, based on overall similarity. As a result, the final categories have an FR structure.

In Ahn and Medin's experiments, as predicted by the model, no subjects given the Characteristic Set created FR categories whereas more than half of those given the Sufficient Set created FR categories. This failure to obtain FR categories from the Characteristic Set was a very robust effect that appeared in various types of materials as long as subjects could not use any existing background knowledge. (See also Ahn, 1990a.)

**Sorting in Knowledge-Rich Domains.** Ahn (1990) extended these results using materials in which some hypothetical background theories could be developed. For example, in one set of stimuli, subjects received pictures of flowers and one group of subjects, called Theory Group, received a theory about how each class of flowers can attract a hypothetical group of birds or bees. In the other group (Prototype Group), subjects simply received prototypes of the resulting FR categories before they started sorting. The structure of the exemplars was the same as the Characteristic Set in Figure 1. Using these materials, the Theory and Prototype Groups mostly produced FR sorting. Providing knowledge or an underlying theory before sorting had the greatest influence on producing FR sorting. In contrast, when no background knowledge was provided (Control Group), few FR sortings were obtained.

It is not clear why prototypes and underlying theory helped people create FR categories. Ahn (1990) provided three suggestions. First, there might be some intrinsic link between prototypes or theories and FR sorting. If this is the case, then providing background knowledge should result in FR sortings regardless of types of materials or domains. The two experiments presented in this paper used materials from two different domains in order to test effectiveness of different types of background knowledge.

Second, the effect of background knowledge on FR sorting might derive from its role in helping people infer missing features. In Ahn (1990b), the 1's in Figure 1 did not have particular values and subjects were told that these values were not available for the exemplars. Many real world cases seem to have features that are unavailable to us. Then, background knowledge might help people infer these parts of instances in accordance with the values in prototypes or theories that are the most similar to the instances in other respects. To test this possibility, in the current experiments the non-characteristic value 1's were either present or absent. Given an instance, 0 0 0 1, for example, if the value 1 is absent and if background knowledge would help people assume it is 0, then it would be more likely for 0 0 0 1 to be grouped with 0 0 0 0 than when the value 1 takes an actual value.

Third, although both the Prototype Group and the Theory Group produced more FR sorting in the previous experiment, the processes involved in FR sorting might be different in the two groups. The subjects' protocols in the previous experiment indicate that the Prototype Group seemed to be able to get over the strong bias of uni-dimensional sorting and instead they seemed to assign equal weights on all four dimensions present. Still, they seemed to treat the dimensions independently. On the other hand, the Theory Group seemed to abstract an underlying dimension of each category (e.g., degree of attracting hypothetical birds or bees in the above example) and consider how values in a given exemplar as a whole satisfies this underlying dimension. The difference in strategies seems subtle but the results from the following two experiments will be discussed in terms of this conjecture.

## Experiment 1

### Method

**Procedure.** Subjects received a set of exemplars and were asked to sort them into two groups of any size. After they sorted the exemplars, they wrote down how and why they created the categories.

| Dimensions | Value 0 | Value 2 | Value 1 |
|---|---|---|---|
| 1. types of clothes | cotton | leather | silk |
| 2. types of religion | monotheism | polytheism | atheism |
| 3. types of leadership | hierarchical leaders | a single leader | three leaders |
| 4. types of funeral | buried the dead | cremated the dead | buried in river |

**Table 1. Dimensions and Values used in Experiment 1**

**Design and Material.** The design of the experiment was a 3 (types of background knowledge) X 2 (presence or absence of value 1) between-subject factorial design. The structure of exemplars are the same as the Characteristic Set in Figure 1. The four dimensions and the values used for 0, 2, and 1 are illustrated in Table 1. The value 1's were present only in Presence Condition. In Absence Condition, no value was present and subjects were told that these values were not available.

Within each Presence / Absence Condition, there were three groups depending on types of background knowledge they received before sorting. The types of background knowledge that the subjects received were either none (Control group), prototypes (Prototype Group), or underlying theories (Theory Group). The Control Group which did not receive any information was simply asked to sort ten cards in a way that seemed natural to them. The Prototype Group received eight non-prototype exemplars (i.e., eight exemplars except for 0 0 0 0 and 2 2 2 2 in Figure 1) to be sorted. They also received two prototypes of the potential FR categories before sorting and were told that these were the most typical exemplars of the two categories that they were to create. The Theory Group received only the eight non-prototype exemplars coupled with theories underlying each category they were to create. More specifically, they were told that one type of tribes were agricultural tribes and the other group were nomadic tribes. Then they were told that agricultural tribes wore cotton clothes obtained in their farms, were monotheists because they hardly had any chance to encounter other types of religion, had hierarchically organized leaders to control farmers living in their areas, and buried the dead near their farms. Subjects were also told that nomadic tribes wore leather clothes obtained from their hunting, were polytheists because they had contacted many types of religions while travelling, had a single leader to make flexible and quick decisions in their changing environments, and cremated the dead because they were always on the move.

There were 150 subjects randomly assigned to 6 groups with 25 subjects in each group.

**Results**

The results are summarized in Figure 2. Numbers indicate percentages of FR sorting within each group within each set. Other responses were uni-dimensional sorting except for one or two in each group.

**Comparison between Absence and Presence Condition.** Overall, the Absence Condition produced more FR sortings in all three groups. The chi-square test (or the Fisher's exact test where it is appropriate)

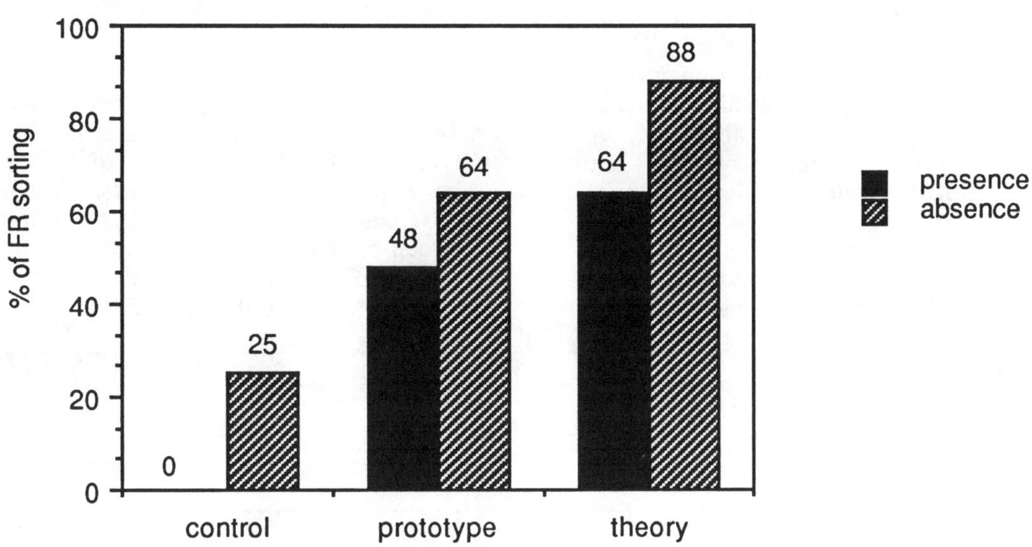


**Figure 2. Results of Experiment 1**

indicates a main effect of the Absence / Presence manipulation. Within each instructional group, the differences between the Absence and the Presence condition were significant for the Theory and the Control Groups ($p<0.05$) but it was not significant for the Prototype Group although the direction of the difference was consistent with other groups.

**Comparison among Groups within Each Condition.** Overall, in each condition, the Theory Group produced FR categories most frequently and the Prototype Group produced them next most frequently. Only a few FR sortings were obtained from the Control Group as in the previous studies (Ahn & Medin 1989). Within each condition (Presence / Absence), the pairwise comparisons among three groups were all significant ($p<.05$) except for the difference between the Prototype Group and the Theory Group in the Presence Condition.

## Experiment 2

### Method

The design and procedure of Experiment 2 were exactly the same as in Experiment 1. The materials used for Experiment 2 were descriptions of skin diseases instantiated for the Characteristic Set in Figure 1. The four dimensions and values are illustrated in Table 2. As in Experiment 1, value 1's were present only in the Presence Condition.

There was a slight change in instructions to make the sorting task more realistic. All subjects were told that each card had a description of one patient and that these patients suffered from inflammatory diseases of the skin and subcutaneous tissue characterized by tender red nodules. They were also told that their symptoms were very similar, but medical doctors had found that there were actually two different types of diseases with these symptoms and that it was important to distinguish these two skin diseases because they required different types of treatment. Then they were told about the four dimensions. The rest of the instructions for the Control Group and the Prototype Group were similar to those in Experiment 1.

For the Theory Group, subjects received the following additional instructions. They were told that Disease A was caused by Virus XB5 whereas Disease B involved symptoms associated with Ketasysm, which was a technical term for abnormal fat metabolism. Since Disease A was caused by an external virus in contact with the skin, the major infection usually, but not necessarily, appeared around exposed areas such as face, hands, and neck, it usually showed sudden onset because symptoms appeared as soon as a person contacted the virus, few family members of the patient had the same skin disease because the disease was produced by external agent, and the lesions were at first sharply limited to the site of contact and later spread to neighboring areas. On the other hand, since Disease B was produced due to Ketasysm, the major infection of Disease B usually appeared in hidden areas such as the abdomen and thighs where body fat was most concentrated, it usually showed gradual onset as Ketasysm progressed, about 50% of the patient's family members had history of the same skin disease due to genetic characteristics of Ketasysm, and the lesions were not sharply limited to a local area because body fat distribution was not sharply distinguished.

There were 120 subjects randomly assigned to 6 groups with 20 subjects in each group.

### Results

The results are summarized in Figure 3.

**Comparison between Absence and Presence Condition.** As in Experiment 1, the Absence Condition produced more FR sortings in all three groups. The chi-square test (or the Fisher's exact test where it is appropriate) indicates that the overall difference between the Absence and the Presence condition were marginally significant ($p=0.06$). Within each group with different background knowledge, only the difference between the Absence and the Presence Conditions of the Prototype Group was significantly different.

**Comparison among Control, Prototype, and Theory Groups.** There was a large main effect of type of background knowledge ($p<0.001$). Interestingly, however, the direction of the differences showed a pattern different from Experiment 1. Overall, the Prototype Group produced FR categories most frequently (80%) and the Theory Group produced next most frequently (72.5%). Again, only a few FR sortings were obtained

| Dimensions | Value 0 | Value 2 | Value 1 |
|---|---|---|---|
| 1. area of infection | face, hands, and neck | abdomen and thigh | hands and thigh |
| 2. onset | sudden | gradual | neither sudden nor gradual |
| 3. % of family members with the same disorder | no family members | 50 % of family members | 25% of family members |
| 4. boundaries of lesion | sharply limited | not sharply limited | somewhat moderately sharp boundaries |

**Table 2. Dimensions and Values used in Experiment 2**

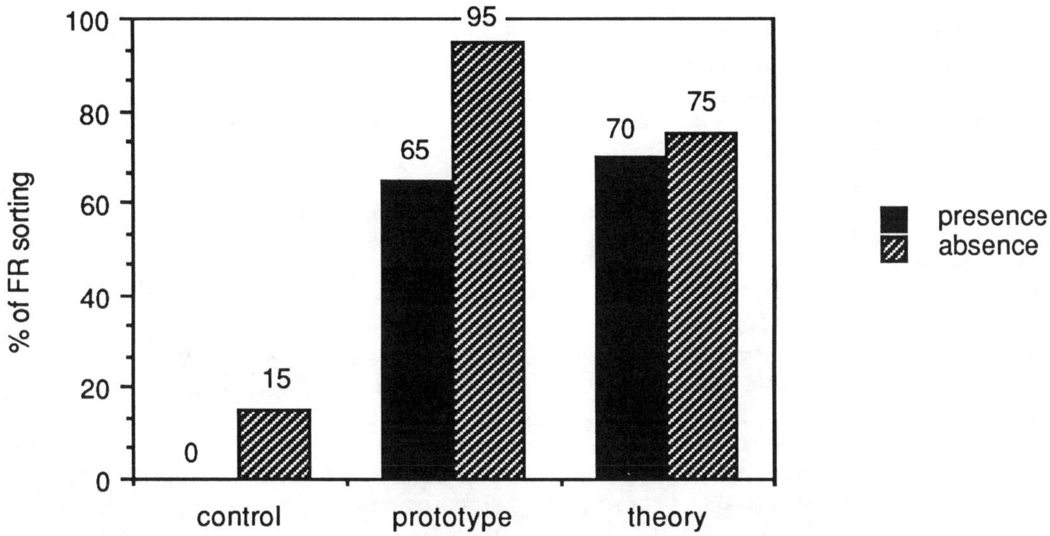


**Figure 3. Results of Experiment 2**

from the Control Group as before. Within the Absence Condition, the Prototype Group produced significantly more FR sortings (95%) than the Theory Group (75%), which in turn produced more FR sortings than the Control Group (15%). Within the Presence Condition, the Theory and the Prototype Groups produced more FR categories than the Control Group but no difference was found between the Theory and the Prototype Groups.

## General Discussion

### General Effect of Background Knowledge

The basic results of Experiments 1 and 2 replicated the previous findings in Ahn (1990). Knowing underlying theories or prototypes before sorting led to more FR category constructions.

### Differences Between Theory and Prototype

However, the Theory Group was not always more likely to produce FR sorting than the Prototype Group as shown in Experiment 2. This result suggests that knowing theories is not necessarily better than knowing prototypes in producing FR sortings. Instead, the advantage of knowing underlying theories in producing FR categories seems to depend on factors such as domains.

### Differences Between Theory and Prototype with Respect to Missing Features

Comparing the Absence and the Presence Conditions in Experiment 1, FR sortings were obtained more frequently when the value 1 was absent. In Experiment 2, however, this difference disappeared in the Theory Group whereas it still appeared in the Control and the Prototype Groups. These results suggest that there might be some differences between simply knowing surface features (Control and Prototype Groups) and knowing deeper features (Theory Group).

The reason why the Control and Prototype Groups produced more FR sorting in the Absence Condition seems to be the following. In these groups, subject might fill or infer unknown features with values that are consistent with correlational structure shown in other similar instances or prototypes. The other possibility is that if people follow the two-stage model, the absence of value 1 on the most salient dimension used in the first stage might force them to pay attention to the other dimensions, resulting in sorting based on all the present dimensions.

To investigate reasons why the Theory Group showed different patterns across the two experiments, the differences between the two experiments in materials and instructions were further analyzed. First, the types of features were different. In Experiment 1, the values were used as nominal values so that value 1's, when they were present, are not intermediate values between value 0's and 2's. In contrast, the values in Experiment 2 were continuous in that value 1's always were a middle or a mixture of the two extreme values on the same dimension.

How could the different types of dimensions lead to differences in the two Theory Groups? Perhaps the Theory Group is doing a special kind of uni-dimensional sorting where the defining dimension is the underlying dimension constructed by the theories. If this is the case, the continuous values used in Experiment 2 have more advantages over the nominal values in Experiment 1 because continuous values might be easier to be incorporated. Take an example of the tribe stimuli. If a person knows theories about why a tribe wearing cotton clothes tends to be monotheistic, it would be much more difficult to explain why a tribe wearing silk clothes (i.e., nominal value) tends to be monotheistic than to explain why a tribe sometimes wearing cotton clothes and sometimes leather clothes (i.e., continuous value) tends to be monotheistic. That is, for 'silk clothes', subjects had to develop a new explanation for the novel feature. This idea needs to be tested in more systematic ways in future studies.

The second difference between Experiments 1 and 2 is that in Experiment 2, the goal of sorting was specified (i.e., treatment of diseases). Also the theories in Experiment 2 were in terms of causes of diseases. Several investigations of lay people's belief in psychopathological disorders showed that people have a strong belief in the interaction between causes of disorders and their treatment. For example, if people believe schizophrenia is an innate disorder, they believe the appropriate treatment is hormonal and not behavioral (e.g. Furnham 1988). Due to this kind of existing belief, the Theory Group might have given more weight to those dimensions that seem to be a better indicator of causes no matter whether the value 1 was present or absent. This peculiarity of domains can also serve as an explanation for why the theories in Experiment 2 were not as effective as those in Experiment 1 in comparison to prototypes.

### Conclusion

All of the issues discussed above deserve further investigation. What can be concluded from the current studies seem to be the following. First, there seems to be no intrinsic tie between theories and family resemblance sorting. The advantage of theories over prototypes depends on domains, the goal of category construction, and the types of theories. Secondly, exemplars with missing features are more easily categorized in terms of the FR principle than exemplars with non-characteristic features. This result suggests that FR structure might be obtained because sometimes we do not have complete information about instances, and we infer these missing features in accordance with other similar features.

## Acknowledgement

I would like to thank Doug Medin and David Thau for their helpful comments on earlier drafts of the paper, Phil Webster, Judy Florian, and two anonymous reviewers for their comments on experimental materials, Rachael Jefferies, Leigh Elkins, Josh Rubinstein, Matt Kurbat, Jeannine Wells for collecting data. This work was supported by NSF grant BNS88-12913 given to Doug Medin.

## References

Ahn, W. 1990a. *A two-stage model of category construction.* Unpublished doctoral dissertation, University of Illinois, Urbana, IL.

Ahn, W. 1990b. Effects of background knowledge on family resemblance sorting, *Proceedings of the 12th Annual Conference of the Cognitive Science Society*, Boston, MA, 149-156.

Ahn, W., & Medin, D. L. 1989. A two-stage categorization model for family resemblance sorting, *Proceedings of the 11th Annual Conference of the Cognitive Science Society*, Ann Arbor, MI, 315-322.

Furnham, A. 1988. *Lay theories: Everyday explanations for problems in the social sciences.* Oxford: Pergamon.

Imai, S., & Garner, W. R. 1965. Discriminability and preference for attributes in free and constrained classification. *Journal of Experimental Psychology*, 69, 596-608.

Imai, S., & Garner, W. R. 1968. Structure in perceptual classification. *Psychonomic Monograph Supplements*, 2 (9, Whole No.2).

Medin, D. L., Wattenmaker, W. D., & Hampson, S. E. 1987. Family resemblance, concept cohesiveness, and category construction. *Cognitive Psychology*, *19*, 242-279.

Rosch, E. 1975. Universals and Cultural Specifics. In R. Brislin, S. Bochner, & W. Lonner (Eds.) *Cross-cultural perspectives on elarning.* New York: Halsted Press.

Rosch, E., & Mervis, C. B. 1975. Family resemblance: Studies in the internal structure of categories. *Cognitive Psychology*, *7*, 573-605.

Smith, E. E., & Medin, D. L. 1981. *Categories and concepts.* Cambridge, MA: Harvard University.

# Understanding and Improving Real-world Quantitative Estimation

**Norman R. Brown & Robert S. Siegler**

Department of Psychology
Carnegie Mellon University
Pittsburgh, PA 15213

nbrown@psy.cmu.edu
siegler@psy.cmu.edu

**Abstract**

One possible method for improving real-world quantitative estimation is to "seed the knowledge-base" with explicit quantitative facts. This method was employed in two population estimation experiments. In Experiment 1, subjects estimated the populations of 99 countries. They then studied the populations of 24 of these countries. Finally, they estimated the populations of all 99 countries a second time. As predicted, the post-learning estimates for the 75 "transfer" countries were much more accurate (48%) than the pre-learning estimates. However, the rank-order correlations between estimated population and true populations showed almost no improvement. These results suggested that there may be two analytically distinct components to estimation, a range component and a ranking component, and that an arbitrary set of quantitative facts is likely to affect the former but not the latter. The aim of Experiment 2 was to demonstrate that one can affect the ranking component by presenting subjects with a consistent set of population facts. In this experiment, one group of subjects was presented with facts that consistently confirmed their prior belief that European countries are quite large and Asian countries are quite small. Another group was presented with a set that consistently disconfirmed this view. As predicted, rank-order correlations between estimated and true populations were negatively affected by the bias-confirming facts and positively affected by the bias disconfirming facts.

The world is composed of innumerable discrete entities; each of these entities has many properties; some of these properties can be specified in quantitative terms. It is common for a person to known that a certain entity exists, to recognize that this entity must have a certain quantitative property, and still to have no knowledge of the value of that property. For example, the typical U.S. college student is likely to have heard of Malaysia, to be absolutely certain that some number of people live there, and to be very uncertain as to what that number might be. Nonetheless, when pressed, this student will produce an estimate that is at least more accurate than a random guess and often will express some confidence in the rough accuracy of the estimate.

How do people generate *real-world quantitative estimates* of this sort? The cognitive literature provides two distinct perspectives on the matter. One focuses on domain-specific knowledge. According to this view, the estimation process is driven by a cycle of retrieval and inference. This process first recovers a fact from long-term memory that is related to the "target" entity (e.g., Malaysia) and/or to the "target" property (e.g., population). If the fact is relevant, it triggers an inference that narrows the response range in some way. For example, a person attempting to estimate the population of Malaysia might recall that there are about 120 million Japanese. If this person also believes that Malaysia has fewer people than Japan, he or she can confidently infer that the population of Malaysia is less than 120 million. Generally, a single fact will not allow the response range to be narrowed enough to yield a precise estimate. As a result, the cycle of retrieval and inference may continue until a precise estimate has been reached, until all immediately relevant knowledge has been exhausted, or until some time or effort limit is exceeded (Collins, 1978; Collins & Michalski, 1989).

The second perspective on real-world estimation holds that people depend on a small number of general purpose heuristics (e.g., availability, representativeness) to estimate the quantitative properties of objects and events (e.g., Brown, Rips, & Shevell, 1985; Lichtenstein, Slovic, Fischhoff, Layman, & Combs, 1978). According to this view, a person attempting to estimate the population of Malaysia might realize that he or she knows relatively little about this country. This would imply, via the application of the availability heuristic (Tversky & Kahneman, 1973, 1974), that the population of Malaysia is relatively small. Alternatively, a person attempting to estimate the population of Norway might depend on its representativeness. This person might decide that Norway is typical of Scandinavian countries and then infer that the population of Norway is likely to be similar to the populations of other typical Scandinavian countries, such as Sweden.

Interestingly, the same estimation task often produces results consistent with both the knowledge-based and heuristics perspectives. For example, a number of studies

This research was supported in parted by a National Institute of Mental Health postdoctoral fellowship to Norman R. Brown, and in part by A.W. Mellon Foundation grant awarded to Robert S. Siegler.

have demonstrated that people consider well-known public and autobiographical events to have happened more recently than less well-known events of the same objective age (e.g., Brown et al., 1985; Wagenaar, 1986). These results indicate that availability plays a role in date estimation. There is also clear evidence that domain-specific knowledge plays a central role in this task. For example, Brown (1990) reported an experiment in which subjects were asked to think aloud as they estimated dates for a set of well-known public events. Analysis of these protocols revealed that 78% of the estimates were justified with reference to one or more domain-specific fact.

The existence of evidence within a single task that is consistent with both perspectives suggests that any satisfactory approach to estimation must account for the influences of both general heuristics and domain-specific knowledge. One way to reconcile these two approaches is to assume a knowledge-based architecture and to view a given estimate as a weighted blend of available relevant information. Thus, when estimating the population of Malaysia, one might recall the population of Japan and use it to truncate the response at 120 million. One might also use availability to classify Malaysia as a not-very-large country. Taken together these two sources of information suggest not only that the target population is smaller than 120 million, but that it probably is a good deal smaller.

Viewing estimates as a weighted blend of relevant information provides a plausible mechanism for reconciling knowledge-based and heuristic perspectives. However, when estimates are conceived in this way, one is confronted with the issue of how the estimation process weighs competing sources of information. Drawing on recent work carried out within the cue validity framework (e.g., Gigerenzer, Kleinbolting, & Hoffrage, in press; MacWhinney, 1987), we hypothesized that competing facts are weighed according to (a) their *predictive strength* or *validity*, and (b) their *specific relevance* to the task. In the current context, predictive strength reflects the ability of a fact or inference to correctly predict the value of the to-be-estimated quantity. In other words, we assumed that people believe some inferences to be more credible or more reliable than others and that they weigh the more credible inferences more heavily than the less credible inferences. We also believed that inferences based on specific quantitative facts (e.g., the population of Japan is 120 million) would be weighed more heavily than those that were not. The two experiments described below bear directly on the accuracy on this particular instantiation of the specificity assumption.

## Experiment 1

In an earlier experiment we found that (a) estimates of national populations tended to be very inaccurate in an absolute sense; the average population estimate was 3/4 of an order of magnitude away from the actual population. We also found that (b) better-known countries were considered to have larger populations than less well-known countries, other things being equal (Brown & Siegler, in preparation). We took the first finding to indicate that accurate knowledge of national populations is extremely uncommon, and the second to indicate that availability played a major role in this task.

The aim of the current experiment was to see if we could improve estimation performance and decrease the dependency on availability by introducing a set of relevant quantitative facts. In this experiment, subjects first estimated the populations of 99 countries. They then learned the actual populations of 24 of these countries. Finally, they were provided another set of estimates for all 99 countries. We expected that subjects would use the population facts they learned in two ways. First, these facts might serve as the basis for accurate generalizations about the geographical, social, economic and historical factors that result in populations of various sizes. Second, it seemed likely that the populations might be retrieved and used as quantitative reference points when subjects attempted to estimate the populations of the "transfer countries." These reference points would allow subjects either to truncate the range of possible responses or to select a numerically anchored region within the range that was likely to contain the target country's population.

Given these expectations, we predicted that "seeding the knowledge-base" would have the following effects on estimation performance. First, subjects should display a decreased dependence on availability. This follows from the specificity assumption made above; both the population facts and the newly derived demographic generalizations are more specifically relevant than availability judgments. Therefore, inferences based on this new information should be weighed more heavily than inferences based on availability. Second, we predicted that seeding the knowledge-based would lead to an improved rank-order correlation between true populations and estimated populations for the transfer countries. This would come about if subjects were able to use the seed set to correctly induce the factors that predict population size, or if they were able to select the most appropriate seed country or countries to serve as quantitative reference points. These same considerations led us to predict that mean estimates across the the transfer countries should be much more accurate in an absolute sense after the knowledge-base had been seeded.

## Method

Twenty-four Carnegie Mellon undergraduates participated in this experiment. Each of these subjects performed four tasks during a 1 hr experimental session: a knowledge rating task, an initial estimation task, a learning task, and a final estimation task. In all but the learning task, subjects were exposed once to each of 99 countries. These 99 countries represented all but one of the countries that had populations of at least 4 million in 1989 (*Information Please Almanac*, 1989). The one exception was the United States, whose population was given to subjects as an example before the first estimation task.

During the knowledge rating task, subjects were presented with the names of the 99 test countries, one at a time, on a computer controlled video display. They were instructed to evaluate their knowledge of each country on a 0-to-9 scale, with 0 indicating no knowledge of the country in question, 9 indicating a great deal of knowledge, and intermediate values indicating intermediate levels of knowledge. In this task, as in the others, subjects responded by typing their answers at the computer keyboard.

Following the knowledge rating task, subjects performed the initial estimation task. Again, the 99 test countries were presented, one at a time in a random order. Subjects were instructed to respond to each country with their best estimate of that country's current population.

Subjects were then presented with the four study-test blocks of the learning task. During each block, subjects were given the opportunity to study the actual population of each of 24 seed countries and were then tested on their knowledge of these populations. These countries were a subset of the full set of 99 countries. They were selected so that there were 6 countries in each cell of a 2 (country knowledge: High and Low) X 2 (estimation accuracy: High and Low) factorial design. Half of of the seed countries had received high knowledge ratings in a prior experiment and half received low ratings; half had received accurate estimates in the prior experiment and half had received inaccurate estimates. The high-knowledge, high-accuracy seed countries were: South Africa, Spain, Egypt, Italy, Great Britain, and West Germany. The the high-knowledge, low-accuracy seed countries were: Israel, Switzerland, Greece, Australia, Canada, and Vietnam. The low-knowledge, high-accuracy countries were: the Netherlands, Venezuela, Kenya, Romania, the Sudan, and Argentina. The low-knowledge, low-accuracy countries were: Bolivia, Zimbabwe, the Ivory Coast, Chile, Zaire, and Thailand.

The study-test blocks were divided into a study phase and a test phase. During the study phase, each seed country was presented with its population for 6 seconds. After subjects had studied all 24 study-test countries, they began the test phase. The task here was to respond to each country's name by typing its true population. When subjects could not recall a country's exact population, they were to respond with their closest approximation.

After completing the learning task, subjects began the second estimation task.The procedure followed during this task was identical to the one followed during the first estimation task.

## Results and Discussion

The results provided clear support for only one of the three predictions made above. As predicted, "seeding the knowledge-base" resulted in a large decrease in absolute error across 75 transfer countries (i.e., the 75 countries that were not presented during the learning task). Specifically, for each subject, we first computed the absolute difference between the estimated population and the true population for each transfer country. Then, we computed the median absolute difference (MAD), for each subject, over all 75 transfer countries. Average MAD for the transfer countries decreased 48%, from 20.9 million in the first estimation task, to 10.9 million in the second ($t(23)= -2.45$, $p < .05$).

In contrast to the large decrease in MAD, the seeding procedure had little effect on the correlation between estimated and true population. For the 75 transfer countries, the average (taken over subjects) rank-correlation between these two measures was .40 before the learning task, and .43 after the learning task ($t(23)=1.57$, $p > 1$).

Finally, the data completely failed to support the prediction that seeding the knowledge-base would decrease reliance on availability. The results relevant to this prediction come from a pair of regression analyses, one on the estimates before the learning task and one on the estimates after the learning task. In both cases, the dependent measure was the median estimated population for each of the transfer countries. The predictor variables were mean knowledge rating, true population, and true land area. The medians were computed over subjects. The three largest countries, China, India, and the Soviet Union, were excluded from these analyses because they unduly influenced the outcome of the regressions. The $R^2$ computed for the pre-learning estimates was .66, as was the $R^2$ for the post-learning estimates.

If subjects had depended less on availability after the learning task than before, then the knowledge variable should play a larger role in the analysis conducted on the the pre-learning estimates than in the analysis conducted on the post-learning estimates. Contrary to this prediction, the knowledge variable played a smaller role in the former than in the latter. In the pre-learning analysis, country knowledge accounted for 35% of the unique variance; in the post-learning analysis, it accounted for 42% ($p < .001$. in both cases). Neither actual population nor actual land area accounted for more than 4% of the unique variance in either analysis.

These results suggest that there may be two analytically distinct components to real-world quantitative estimation. One component is an absolute or *range* component, and the other is a relative or *ranking* component. In the population estimation task, the absolute component can be equated with the assumptions concerning the plausible range of national populations, and the relative component with the knowledge used to locate a country's population within the assumed range.

The seeding procedure primarily affected the absolute component. It makes sense that studying the populations of 24 countries would lead subjects to develop a better understanding of the response range. There is support for this claim. The subjects who decreased the size of their estimates to the greatest extent across the two estimation tasks were those who provided the largest overestimates in the pre-learning task. Similarly, the subjects who increased their estimates the most were those who provided the most extreme underestimates in the pre-learning task ($r$=.98).

At the outset of this study, we expected that the seeding procedure would affect both the relative and the absolute components. However, the two failed predictions described above indicate that this procedure had little or no effect on the relative component. We believe that there are two reasons for this outcome. First, it appears that high task specificity is not necessarily equivalent to high predictive strength. That is, quantitative reference points will not necessarily dominate performance when other credible source of information are available. Recalling the population of Japan may provide an upper bound for an estimate of Malaysia's population, but it may have no effect on a strongly held, availability-based belief that Malaysia has a fairly small population.

In addition to overestimating the potential usefulness of explicit reference points, we also seem to have overestimated the ability of subjects to act as "naive demographers." It appears that subjects were not able to use the population facts presented during the learning task to correctly update their beliefs about relative populations of different countries. In retrospect, this is not too surprising. Information provided by populations of the high-accuracy countries was consistent with subjects' prior beliefs, while the information provided by populations of the low-accuracy was inconsistent It seems unlikely that an ambiguous situation of this sort would compel subjects to reevaluate their beliefs.

## Experiment 2

The aim of Experiment 2 was to demonstrate that seeding the knowledge-base with sets of population facts that are consistent and informative can influance relative population estimates. In the current context, a consistent set of facts is one in which all of the seed countries from an identifiable geographical region have similar populations (e.g., all of the European populations are small), and an informative set of facts is one in which the generalizations implicit in the set always confirm or always disconfirm prior beliefs about populations.

We knew from prior experiments (Brown & Siegler, in preparation) that subjects tended to believe that small European countries have relatively large populations and that large Asian countries have relatively small populations. In the current experiment, we attempted to influence these beliefs, and hence the ranking of European and Asian countries, by seeding the knowledge-base with sets of facts that consistently confirmed or consistently disconfirmed them. The expectation was that exposure to disconfirming population facts (i.e., facts about small European countries and large Asian countries) would improve the rank-order correlation between estimated and true population, and that exposure to confirming population facts (i.e., facts about large European countries and small Asian countries) would decrease this correlation.

## Method

Three factors were varied: seed set, transfer region, and trial block. Seed set was a between subjects factor. It involvled the particular population facts that subjects learned. The 20 subjects in one group, the *bias-disconfirming* group, saw the populations of three small European countries (Switzerland, Sweden, the Netherlands) and three large Asian countries (Thailand, the Philippines, Vietnam). The 20 subjects in a second group, the *bias-confirming* group, saw the populations of three large European countries (Great Britain, Italy, West Germany) and three small Asian countries (Cambodia, Sri Lanka, Malaysia). Finally, 20 subjects were run in a control condition; these subjects were never exposed to the actual populations of any countries.

Trial block and transfer region were both within-subjects factors. Subjects in all three groups were required to estimate the populations of 36 countries once during each of 4 trial blocks. This set of countries consisted of the 12 seed countries and 24 transfer countries. Six transfer countries were selected from Asia (Burma, South Korea, Pakistan, Bangladesh, Japan, Indonesia), six from Europe (Norway, Denmark, Austria, Belgium, Greece, Portugal), six from Africa (Chad, Zimbabwe, the Ivory Coast, South Africa, Ethiopia, Nigeria) and six from Latin America (Honduras, Bolivia, Ecuador, Argentina, Mexico, Brazil). We refer to Asia and Europe as the *seeded* regions because both the confirming and disconfirming countries were drawn from them, and to Africa and Latin America as the *unseeded* regions, because subjects were not exposed to the populations of any of the countries from these regions.

The procedure followed on the first of the four trial blocks was identical to the procedure used during the Experiment 1 estimation tasks, except that subjects were only tested on the populations of 36 countries. As in the earlier experiment, subjects were instructed to enter a response that reflected their best estimate of each country's current population.

At the beginning of the second trial block, subjects in the bias-confirming and bias-disconfirming conditions. For the bias-disconfirming subjects, the computer display listed the names and populations of the smallest European seed country and the largest Asian seed country. Similarly, subjects in the bias-confirming condition saw the names and populations of the smallest Asian country and the largest European country. At the beginning of the third estimation block, two more country names and populations were added to the display. During this block, subjects in the experimental conditions were presented with information about the second largest of the small seed countries and the second smallest of the large seed countries. Finally, during the fourth block, the experimental subjects saw all six of the relevant seed set countries.

Subjects in the control condition simply provided four sets of estimates for the 36 stimulus countries and never saw the actual populations of any of these countries. Presentation order of the stimulus countries was randomized for each subject and each block.

## Predictions

The central prediction of this experiment concerned possible changes in rank-order correlations across blocks. Specifically, we predicted a Seed Set X Transfer Region X Trial Block interaction. We expected that subjects who received the disconfirming seeds might realize that some European countries have very small populations and that some Asian countries have quite large populations. This understanding should lead subjects to decrease estimates for the European transfer countries and increase them for the

Asian transfer countries. This in turn should lead to an increased rank-order correlation between estimated and true population for the countries in the seed regions.

In the bias-confirming condition, we expected that the seed facts would increase subjects' confidence in their prior biased beliefs. To the extent that these beliefs are strengthen by exposure to the confirming seed set, subjects should increase population estimates for the European countries and decrease them for the Asian counties, leading to an overall decrease in the rank-order correlation between estimated and true population.

In brief, for countries in the seeded regions, rank-order correlation between estimated and true population should increase over blocks in the bias-disconfirming condition and decrease over block in the bias-confirming condition. In contrast, we predicted no change in rank-order correlation for the countries in the unseeded regions, in either the bias-confirming or bias-disconfirming conditions. This is because neither seed set had clear implications for the accuracy of the beliefs that determine the relative size of African and Latin American countries. Finally, since the control subjects were given no information about national populations, we expected no change in the rank-orderings of the countries in either the seeded or unseeded regions.

We expected a large decrease in MAD for the countries from the seeded regions and for countries from the unseeded regions, in both the bias-confirming and bias-disconfirming conditions. This is because the seed facts should provide information that would allow subjects to evaluate and update their range assumptions. Since range assumptions play an important role in determining the absolute accuracy of all responses (at least when range assumptions can be off by several orders magnitude), the improvement in MAD should not be restricted to the countries drawn from the seeded regions. The MAD scores for the control subjects should not change over blocks.

## Results and Discussion

In order to test predictions concerning changes in the rank-ordering of countries, we first obtained eight rank-order correlations per subject. For each subject and each trial block, we computed one correlation between estimated and actual population for the 12 transfer countries drawn from the seeded regions, and one for the 12 transfer countries drawn from the unseeded regions. These correlations were then submitted to an ANOVA, which indicated that the interaction between seed set, transfer region and trial block was significant ($F(6,171)=5.21$, $p < .0001$).

As predicted, the rank-order correlation for the seeded regions increased when subjects were presented with bias-disconfirming seed populations and decreased when they were presented with bias-confirming seed populations (see Figure 1). Also as predicted, the rank-order correlations for the unseeded regions were unaffected by the presence of either confirming or disconfirming information. Finally, the control condition indicates subjects did not alter their ranking-ordering of countries in the absence of externally provided cues.

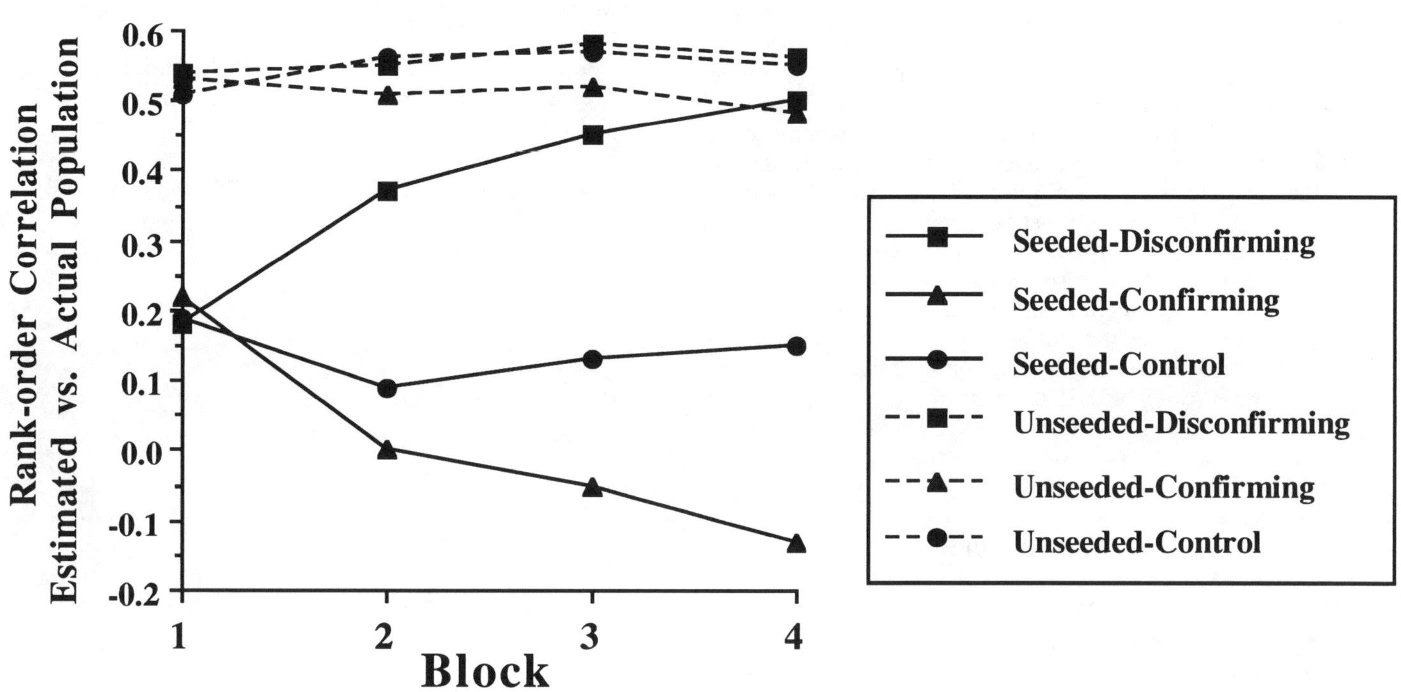


*Figure 1.* Mean rank-order correlations computed across countries in the seeded regions (Asia and Europe: solid lines) and the unseeded regions (Africa and Latin America: dashed lines), for subjects in the bias-disconfirming (squares), bias-confirming (triangles), and control conditions (circles). Experiment 2 data.

An ANOVA was also performed on the the accuracy data. In order to do this, we first computed the absolute difference between the estimated and actual population for each response. Next, for each subject and each block, we obtained medians of these absolute differences over the 12 seeded transfer countries and over the 12 unseeded transfer countries.

The most interesting result in this ANOVA was a significant Seed Set X Trial Block interaction ($F(6,171)=3.27$, $p < .01$). Across the four trial blocks, estimates in the control condition became slightly less accurate (MAD = 27.9 million in Block 1, MAD = 34.5 million in Block 4); estimates in the bias-confirming condition became somewhat more accurate (MAD = 35.6 million in Block 1, MAD = 24.0 million in Block 4); and estimates in the bias-disconfirming condition improved a great deal. (MAD = 34.5 million in Block 1, MAD = 18.5 million in Block 4). In contrast to the analysis performed on the rank-order correlations, the Seed Set X Transfer Region X Trial Block interaction was not signficant for the MAD measure ($F(6,171)=1.19$, $p > .1$). This is consitent with the prediction that the seeding proceedure would improve preformance in an absolute sense for both the seed and unseeded regions.

## Conclusions

There are three main points to take away from the research just described. First the process that generates real-world quantitative estimates often blends qualitative and quantitative information. It appears that this process does not necessarily grant a special status to information that is explicitly quantitative. Second, estimation can be seen as having two distinct components: an absolute or range component and a relative or ranking component. When the knowledge-base is seeded with an arbitrary or neutral set of quantitative facts, one can expect to improve performance in an absolute sense but not in a relative sense. Finally, it appears that seed facts must be carefully selected if one hopes to improve performance in both relative and absolute senses. Specifically, the set of facts presented to subjects must be consistent, informative, and valid. A consistent set of facts is one that allows for an obvious mapping between the items' qualitative and quantitative category; an informative set of facts is one that provides subjects with evidence that suggests changes in their beliefs; and a valid set of facts is one that leads to an increase of the predictive validity of the modified beliefs.

## References


Brown, N. R. (1990). The organization of public events in long-term memory. *Journal of Experimental Psychology: General, 119,* 297-314.

Brown, N. R., & Siegler, R. S. (in preparation). Quantitative estimation in knowledge-rich domains.

Brown, N. R., Rips, L. J., & Shevell, S. K. (1985). The subjective dates of natural events in very long-term emory. *Cognitive Psychology*, 17, 139-177.

Collins, A. M. (1978). Fragments of a theory of human plausible reasoning. In D. Waltz (Ed.), *Theoretical issues in natural language processing--2.* (pp. 194-201). Urbana, IL: University of llinois.

Collins, A. M., & Michalski, R. (1989). The logic of plausible reasoning: A core theory. *Cognitive Science, 13,* 1-49.

Gigerenzer, G., Kleinbolting, H., & Hoffrage, U. (in press). Confidence in one's knowledge: A Brunswikean view. *Psychological Review.*

*Information please almanac.* (1989). Boston: Houghton Mifflin.

Lichtenstein, S., Slovic, P., Fischhoff, B., Layman, M., & Combs, B. (1978). Judged frequency of lethal events.*Journal of Experimental Psychology: Human Learning and Memory, 4,* 551-578.

MacWhinney, B. (1987). The competition model. In B. MacWhinney (Ed.), *Mechanisms of language acquisition* (pp. 249-308). Hillsdale, NJ: Erlbaum.

Tversky, A., & Kahneman, D. (1973). Availability: A heuristic for judging frequency and probability. *Cognitive Psychology, 5,* 207-232.

Tversky, A., & Kahneman, D. (1974). Judgments under uncertainty: Heuristics and biases. *Science, 185,* 1124-1131.

Wagenaar, W. A. (1986). My memory: A study of autobiographical memory over six years. *Cognitive Psychology, 18,* 225-252.

# Implicit Detection of Event Interdependencies and a PDP Model of the Process

**Michael Kushner**
Department of Psychology
Brooklyn College of CUNY
Brooklyn, NY 11210
hmkushnr@bklyn.bitnet

**Axel Cleeremans**
Department of Psychology
Carnegie Mellon University
Pittsburgh, PA 15213
cleeremans@psy.cmu.edu

**Arthur Reber**
Department of Psychology
Brooklyn College of CUNY
Brooklyn, NY 11210
artreber@bklyn.bitnet

**Abstract**

We report on an experiment in which subjects were asked to predict the location of a stimulus based on observation of a series of five events. Unbeknownst to subjects, the location of the sixth event was determined by a double contingency between the second and fourth events in the sequence. This material is therefore highly complex, since the relevant events are embedded in a large number of irrelevant contexts. The results indicated that subjects improved their prediction performance over 10 sessions encompassing over 2400 trials of training, despite the fact that they remained completely unaware of the existence of the rule, and unable to verbalize their knowledge of the contingencies in the material. We propose a model of performance in this task, in the form of a PDP model of sequence processing. The model successfully accounts for performance and illustrates how knowledge about the temporal context may develop in a way that does not necessarily yield decomposable representations. Interestingly, the model also predicts that performance would be worse if subjects were required to predict successive events rather than simply observe them.

## Introduction

Implicit learning is the process whereby knowledge about complex, rule-governed stimulus environments is acquired without specific intentions to learn and largely independently of conscious knowledge about what was learned (Reber, 1989). This process has been explored in a wide variety of experimental contexts including artificial grammar learning (Reber 1967, 1989), patterned sequence learning (Lewicki, Hill, & Bizot, 1988; Nissen & Bullemer, 1987; Cleeremans & McClelland, in press), concept formation (Brooks, 1978), probability learning (Reber & Millward, 1968, 1971), and process control of simulated manufacturing plants (Berry & Broadbent, 1984). In all cases subjects learn to make decisions, classify novel stimuli, anticipate events, and solve problems that required knowledge of the regularities in the stimulus environment while showing little or no explicit, reportable knowledge about those regularities.

This research was supported by Grant BNS-89-07946 from the National Science Foundation and by a PSC-CUNY Grant from the City University of New York to Arthur S. Reber.

As has been argued elsewhere (Reber, 1989; Cleeremans & McClelland, in press), these experiments all have in common the property that the underlying knowledge base that subjects extract from their interactions with the stimulus environments can be captured by the notion of covariation. That is, subjects' behavior in all of these experiments appears to reflect a single process: the detection of covariations among events as they are instantiated in the stimulus display. The generality of this process is considerable and it has been observed across a wide range of stimulus materials (see Reber, 1989, for a review).

In this paper we explore this notion of the detection of covariation further by introducing a stimulus environment that is based on a complex array of events whose underlying structure is characterized by a remote, double- dependency rule that is far more complex than anything that has been studied to date. If implicit learning is as robust a process as some have suggested (Lewicki, 1986; Lewicki & Hill, 1989; Reber, 1989), then we ought to be able to observe this process emerging in situations where the associative links between events are complex and non-salient.

The procedure used in this study is a relatively simple prediction experiment in which subjects had to "guess" the successor of a sequence of similar events. Subjects were exposed to a series of five stimuli presented successively on a computer monitor and were asked to predict the location of the sixth stimulus. There were three possible locations, arranged as the vertices of a triangle, at which the stimuli could appear. The first five stimuli always appeared at random locations; the location of the sixth stimulus was determined on the basis of the relationship between the locations of the 2nd and 4th stimuli. The 1st, 3rd, and 5th stimuli were always irrelevant.

There are at least two reasons why this task may be quite hard. First, there are more irrelevant events than useful ones. Second, the rule that defines the location of the target stimulus is complex in that it involves a relationship between events rather than the particular events themselves. This results in each component of the rule being instantiated by different pairs of events, each of which may in turn be embedded in a large number of different irrelevant contexts. Nevertheless, we show that subjects do learn double dependencies of this kind, and that they do so independently of any explicit knowledge of the rules. We also show that once the pattern has been picked up subjects are capable of transferring their knowledge to a "shifted" rule despite the fact that they were unaware of the rule change.

Finally, we present a simulation of the human data using a Parallel Distributed Processing (PDP) model based on the simple recurrent network ("SRN") architecture first introduced by Elman (1990). Cleeremans and McClelland (in press) showed that this model could successfully account for implicit learning of sequential material in a choice reaction situation where the sequences were generated from an artificial grammar. This model therefore appears to be a natural candidate for modeling implicit learning processes in prediction tasks such as the one we describe in this paper. We show that the model is successful in accounting for some (but not all) aspects of performance in our situation. The model illustrates how successful prediction performance may emerge from representations that are not easily decomposable, and thus possibly hard to verbalize. The model also predicts that the particular training conditions used in this experiment are critical for successful performance. We conclude that the SRN model has the potential of giving a reasonable characterization of implicit learning in a variety of stimulus environments.

## A complex sequence prediction task

### Method

**Subjects.** Six Brooklyn College undergraduates participated in the experiment. They were paid $40, and received a bonus of one cent per correct prediction beyond chance level.

**Apparatus and display.** The experiment was run on an IBM microcomputer. The display consisted of three numbered boxes located at the vertices of an invisible inverted triangle, and measuring 6.5 cm X 7.5. cm each. A trial consisted of five successive events. Each event consisted of the appearance of a square stimulus (2.5 cm wide) in one of the boxes. The stimulus remained on screen for 250 msec. The inter-stimulus interval was 250 msec. During a series of 5 events, the stimulus moved from one box to another. After the fifth event had occurred, subjects were asked to predict in which box the stimulus would appear next. They entered their prediction by typing a number between 1 and 3 on the keyboard. The computer then displayed the correct response by presenting the stimulus in the correct box for 2000 msec. Subjects initiated the next trial by pressing the space bar.

**Design and stimulus generation.** We constructed 18 different random orders of the set of 243 ($3^5$) possible sequences of 5 stimuli. Each of these 18 sets of 243 sequences was then blocked in 3 groups of 81 trials, for a total of 54 blocks. Subjects were exposed to a total of 4374 trials over 6 days. There was a short pause between any two blocks. All subjects were exposed to the same 18 random orders. Intentionally vague instructions described the experiment as being about "prediction behavior".

The entire experiment was broken down in three phases (the existence of which subjects were naturally kept unaware). Phase I (the "training" phase) consisted of 2430 trials (i.e., 30 blocks of 81 trials). During this phase, the location of the sixth event could be predicted perfectly based on the relationship between the locations at which the second and fourth stimuli of the current trial had appeared. If these stimuli had appeared at the same screen location, then the sixth stimulus appeared in Box 1. If they had been in a clockwise relationship, the sixth stimulus appeared in Box 2. The sixth stimulus appeared in Box 3 if the second and fourth stimuli had been in a counter-clockwise relationship.

In Phase II (the "transfer" phase), the rule was surreptitiously modified by shifting the location of the sixth stimulus by one box for each component of the rule. For instance, sixth events which had appeared in Box 1 during Phase I now appeared in Box 2. Similarly, sixth events which should have appeared in Box 2 now appeared in Box 3, and those which should have appeared in Box 3 now appeared in Box 1. Subjects were asked to make 972 predictions (i.e., 12 blocks of 81 trials) in this phase.

During Phase III (the "random" phase), the rule was again modified: the location at which the sixth stimulus may appear was now simply determined at random. Subjects were again asked to make 972 predictions (i.e., 12 blocks of 81 trials) in this phase of the experiment.

Finally, we conducted extensive interviews with each subject immediately after completion of the experiment.

### Results

For each session of the experiment, we computed the average proportion of correct predictions about the location of the sixth stimulus. The results of this analysis are illustrated in Figure 1. Subjects become increasingly better at making accurate predictions over the first 10 sessions of training, and end up reaching about 45% correct responses in the 10th session. This is significantly above chance level (33%)[1], and clearly indicates that subjects have acquired knowledge about the relevant regularities embedded in the material. The second phase begins with a dramatic drop in performance (to chance level), but there is again evidence of learning over the next three sessions ($p < .01$). This suggests that subjects are able to transfer relatively easily from one stimulus-response set to another one. As expected, performance in the third, random, phase is low and fails to be significantly over chance level.

Despite this clear sensitivity to the complex regularities embedded in the material, none of the subjects exhibited explicit knowledge of the sequential structure when asked after the task. Post-experimental interviews revealed that subjects felt frustrated in their attempts to "learn the rule" that determined the location of the sixth stimulus. All subjects reported that they eventually abandoned the search for rules, and started predicting according to their "hunches" or according to "what felt right". Subjects were unable to specify which covariations were crucial in the sequence of five stimuli, not even in a general form such as "when the nth stimulus was in box X, the correct answer was usually

[1] All statistical tests reported in this section were conducted by using a normal approximation to the binomial distribution, at the .01 level.

Y". No subject reported awareness of the rule shifts between the three phases of the experiment.

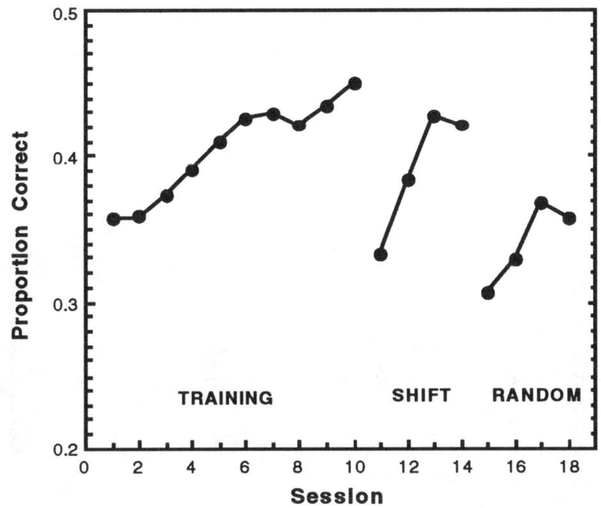


**Figure 1**: Mean proportion of correct predictions, over the 18 sessions of training, and for the three phases of the experiment ("Training", "Shift", & "Random").

Subjects' poor knowledge of the constraints embedded in the material was also confirmed by their performance on a ranking task, in which they were asked to rate each of the five stimuli in terms of their relevance in predicting the location of the sixth stimulus. The results failed to reveal sensitivity to the crucial events: On a scale of 1 (very important) to 5 (not important), the crucial events received average ranks of 3.5 (2nd event) and 2.67 (4th event), whereas the first, third and fifth events were ranked 3.33, 3.67 and 1.83 respectively. However, there was some evidence that particularly salient sequences which were reported by subjects also elicited very good predictions. For instance, sequences in which the first five stimuli had appeared at the same location always predicted Box 1 as the location of the sixth trial. (i.e., "11111 → 1", "22222 → 1", "33333 → 1"). Similarly, alternating sequences such as "12121" always predicted Box 1 as well. Subjects could correctly predict the successor of repeating sequences in about 61% of the cases of Phase I (49% for the alternating sequences) — considerably better than average. This result clearly indicates that, at least in some specific cases like this, subjects have become aware of some of the regularities embedded in the material. Subjects' successful prediction performance is far from being based only on these salient patterns, however: The average prediction score during Phase I only dropped by .0046 percentage points when the 3 possible "repeating" and 6 possible "single alternating" sequences were eliminated from the analysis. Clearly, subjects have become sensitive to contingencies about which they are unable to report.

## A simulation of human prediction performance

This study is a natural candidate for exploring how well a model of sequence processing first introduced by Elman (1990) may be used to simulate human prediction performance. Cleeremans & McClelland (in press) used the "Simple Recurrent Network", or "SRN", to model implicit learning processes in a choice reaction situation. The SRN (Figure 2) is a standard fully connected three-layers back-propagation network (see Rumelhart, Hinton & Williams, 1986), with the added property that the hidden unit layer is allowed to feed back on itself with a delay of one time step, so that the intermediate results of processing at time *t-1* can influence the intermediate results of processing at time *t*. In practice, the SRN is implemented by copying the pattern of activation on the hidden units onto a set of "context units", which feed back into the hidden layer along with the next input. All the forward-going connections in this architecture are modified by back-propagation. The recurrent connections from the hidden layer to the context layer implement a simple copy operation and are not subject to training.

As reported elsewhere (Cleeremans, Servan-Schreiber & McClelland, 1989, in press), we have explored the computational aspects of this architecture in considerable detail. Following Elman (1990), we have shown that an SRN trained to *predict* the successor of each element of a sequence presented one element at a time can learn to perform this "prediction task" perfectly on moderately complex material. For instance, the SRN can learn to predict optimally each element of a continuous sequence generated from small finite-state grammars such as those used by Reber (1989). After training, the network produces responses that closely approximate the optimal conditional probabilities of presentation of all possible successors of the sequence at each step. Note that the network is never presented with more than one element of the sequence at a time. Thus, it has to elaborate its own internal representations of as much temporal context as needed to achieve optimal predictions. Through training, the network progressively comes to discover which features of the previous sequence are relevant to the prediction task.

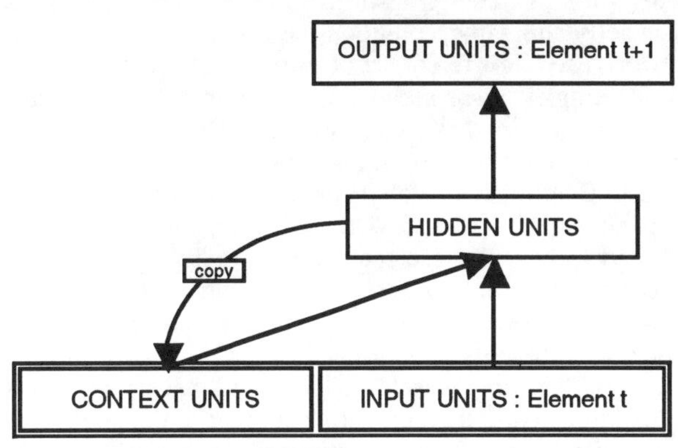


**Figure 2**: The simple recurrent network (SRN). Adapted from Cleeremans & McClelland (in press).

This architecture — as well as other connectionist architectures with which the SRN shares several basic features — appears to be a good candidate for modeling implicit learning phenomena. For instance, because all the knowledge of the system is stored in its connections, this knowledge may only be expressed through performance. Further, the back-propagation learning procedure implements the kind of elementary associative learning that seems characteristic of many implicit learning processes. However, there is also substantial evidence that knowledge acquired implicitly is very complex and structured (Reber, 1989) — not the kind of knowledge one thinks would emerge from associative learning processes. The work of Elman (1990), in which the SRN architecture was applied to language processing, has demonstrated that the representations developed by the network are highly structured and accurately reflect subtle contingencies, such as those entailed by pronominal reference in complex sentences. Thus, it appears that the SRN embodies two important aspects of implicit learning performance: elementary learning mechanisms that yield complex and structured knowledge. The SRN model shares these characteristics with many other connectionist models, but its specific architecture makes it particularly suitable for processing sequential material.

To model our experimental situation, we made the following assumptions: First, each of the three possible locations at which the stimulus may appear was represented by activating a single unit in either the input pool or the output pool. Second, we assumed that the activations of the output units represent the network's predictions about the location of the next stimulus. Third, since no predictions were required from subjects during presentation of the first five trials of a series, learning was turned off for those trials. The network was therefore merely storing successive events during presentation of the first four trials. When the fifth trial was presented as input to the network, learning was turned on, and the network was trained to activate the output unit corresponding to the location at which the sixth stimulus would appear. Finally, since trials (i.e. blocks of five events) were totally independent from each other, the context units were reset to zero at the beginning of each trial. Thus, the temporal context could influence processing within a block of five events, but it was prevented from carrying over to the next block.

**Procedure.** Three SRNs with 15 hidden units were trained in exactly the same conditions as human subjects. Each network used a different set of initial random weights, and the learning rate was set to 0.5. (the momentum parameter was not used in these simulations). On each of the 18 sessions of training (10 during Phase I, and 4 each during Phases II and III), each network was exposed to 243 sequences of five events. On each trial, the network was exposed to one of these sequences by activating the input unit corresponding to each event in turn. Note that the network was not trained at this point; it merely processed the sequence of events. When the fifth event was presented, however, learning was turned on, and the network was trained to activate the unit corresponding to the sixth event on its output layer. The error between its prediction and the actual sixth event (as specified by the rules appropriate for the current phase of training) was then computed and back-propagated to modify the weights. To evaluate the model's performance, we recorded the activation of the output units, and determined which unit was most active. A prediction response was considered correct if the activation of the unit corresponding to the actual sixth event was higher than the activation of the two other units. On the next trial, the context units were reset to zero, learning was turned off, and the network was presented with another sequence of five events.

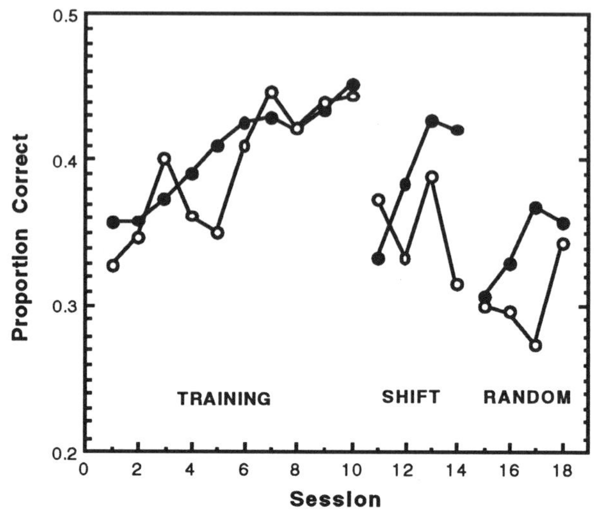


**Figure 3**: Mean proportion of correct predictions over the 18 sessions of training, and for the three phases of the experiment. Filled symbols represent human data; open symbols represent simulated data (epsilon = 0.5).

Figure 3 compares the human data with the average proportion of correct prediction responses produced by the networks, for each of the 18 sessions of training. It is clear that the model is learning the regularity in Phase I. There are other aspects of the data for which the correspondence with the simulations was far from perfect. In particular, the main discrepancy is located in Phase II: the model isn't nearly as good as the human subjects to readjust its performance to the new rule. This is interesting because it points to a shortcoming shared by most current simulation models of implicit phenomena: their relative inability to transfer to a different set of stimuli and responses which is structurally identical to the original, but instantiated by different tokens. We will return to this point in the discussion.

Figure 4a contrasts the model's performance in predicting the sixth event with its average performance in predicting events 2-5. Recall that the network was not trained to predict successive events during presentation of the first four events. However, it still produced responses, which may be used to compute a baseline "prediction" score against which to compare actual prediction responses of the sixth event. Figure 4a shows that this baseline curve remains flat and at about chance level throughout training, thereby confirming that the network is indeed acquiring information which is specific to predicting the sixth event.

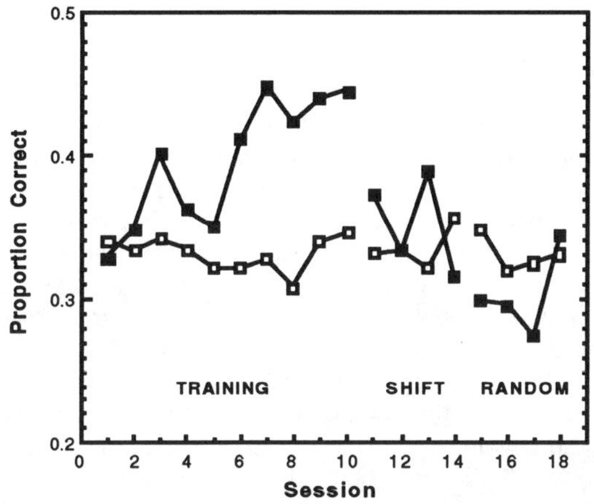


[a]

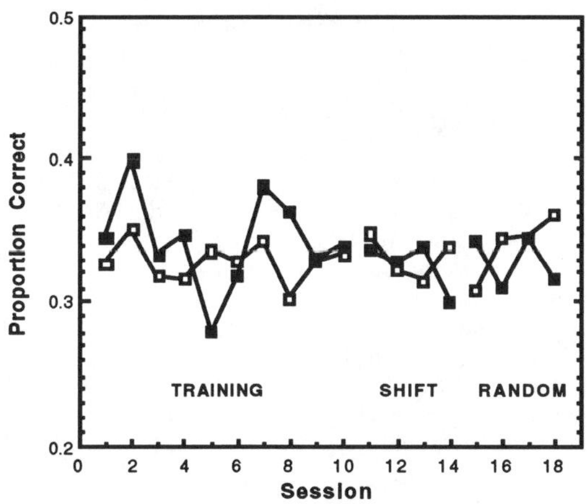


[b]

**Figure 4:** Proportion of correct prediction responses plotted separately for Events 2-5 (open symbols) and for Event 6 (filled symbols). [a]: Data from an SRN trained only on presentation of the 5th event. [b]: Data from an SRN trained on all events.

This analysis is most useful when comparing its results with those of the same analysis conducted on the responses of an SRN trained to predict the successor of *each* event. Figure 4b represents these data. Interestingly, the model's performance is much worse under these conditions than when it is trained to predict only the sixth event. Indeed, there is almost no difference between the curves corresponding to predictions of events 2-5 and to event 6: both curves remain flat and at about chance level throughout training.This is because the network is trained to predict random events (events 2 to 5) four times as much as it is trained to predict structured events (event 6). Because of the error minimization resulting from back-propagation, the responses of the network tend to represent the average probability of each output unit to be active *in the entire training set.* Since the probability of an output unit to be on in the entire training set is about 0.33, the responses of the SRN are essentially random when it is trained under these conditions. Thus, the SRN model makes the somewhat counter-intuitive prediction that subjects would fail to learn the contingencies embedded in the material if the task was such that they were required to guess the nature of *each* successive event.

## Discussion

Subjects were run on a complex task in which they were required to predict where the sixth event of a series would appear. Unbeknownst to them, the location of the sixth stimulus was determined based on the relationship between the second and fourth events of a sequence. Subjects' prediction performance improved with training, despite the fact that they remained unable to specify which events were relevant, or even that there was any kind of structure present in the material. Further, subjects could successfully transfer the knowledge acquired during training to a different, "shifted", rule.

The simulation work reported in this paper has demonstrated how the SRN model may be applied to this experimental situation. There are several interesting issues here worth commenting on:

First, the results show that the model is capable of learning the complex rule instantiated in the training material, and to do so at about the same rate as human subjects[2]. What are the mechanisms underlying this sensitivity? During the first five events of each trial, the model is merely storing information about each event, in the form of a time-varying pattern of activation over the hidden units. When the fifth trial is presented to the network, this pattern of activation now represents the entire sequence of five events. Different sequences will result in different such representations, even in the absence of any training . This is simply the result of the recurrent nature of the architecture (see Cleeremans, Servan-Schreiber & McClelland, in press, for a discussion of this point). It is on the basis of the differences and similarities between these internal representations that the network becomes capable of predicting the sixth event. To do so, however, it has to learn how to map clusters of internal representations corresponding to sequences resulting in the same prediction onto the output units corresponding to these predictions.

Another, more interesting point, is the observation that the representations developed by the network are completely opaque, in the sense that the information encoded by the network allows successful performance, but is not readily decomposable into critical features. Indeed, it would take sophisticated analysis methods (such as hierarchical clustering, see Cleeremans, Servan-Schreiber & McClelland, 1990) to uncover the regularities embedded in the internal representations developed by the network. This characteristic of the representational system of the SRN, and of PDP networks in general, provides a natural explanation for the fact that human subjects are unable to describe what

[2] Obviously, the learning rate is a free parameter in this model. We conducted a large number of simulations, each with different parameters. In some cases, the network failed to learn. In some other cases, performance was better than in the human data. The simulations discussed in this paper yielded the best fits with the human data.

elements of their knowledge are responsible for successful performance. Whether models are to be taken literally is arguable, particularly when it comes to their representational system, but this provocative interpretation of the SRN's behavior is quite compelling.

A further point is related to training. The results revealed that the network is unable to learn the task when trained on each event of a series. The corresponding experiment with human subjects remains to be done, but it is interesting to speculate on the reasons for this interference. A typical result in implicit learning experiments is that asking subjects to look for structure where there is none results in worse performance (see for instance Reber, 1976). In our situation, asking the network to predict successive events rather than simply observe them appears to have the same effects, and for the same reasons: the model tends to elaborate representations that fail to capture the relevant covariations, because the structure is deeply embedded in many irrelevant contexts.

One aspect of the SRN's performance in this task seems to suggest that it is not yet a complete model of the human data, however. Indeed, the lack of transfer to the new, shifted, rule system used in Phase II of the experiment may be incompatible with the idea that performance is based only on the kind of mechanisms instantiated by the SRN. Why is it hard for the SRN model to adjust its responses to the new rule? The answer lies in an examination of how the SRN model stores and processes knowledge about sequences. Basically, in the case of this particular task, the connections between the input units and the hidden units implement a mapping between sets of sequences of events and distinct internal representations. Optimally, each set of sequences which results in the same prediction about the sixth event should be associated to a unique code on the hidden units. More likely, there is a number of distinct clusters of internal representations, with each cluster grouping those internal representations which result in the same response. The connections between the hidden units and the output units, in turn, map these clusters of internal representations onto the responses. The rule shift introduced in the experiment is in effect a change in this latter mapping. Indeed, it does not result in any change in which events are important, or in the number of possible responses, etc. As a result, all the network really needs to do in order to produce the shifted responses is to adjust the connections from the hidden units to the output units. But, even though weight adjustments are known to be much faster in the last layer of connections than in the others, this process of remapping the internal representations appears to be quite a slow one: even though it would eventually learn the correct new mapping, it appears unable to do so within the limited amount of training available during Phase II.

This difficulty in dealing with "shifted" material exhibited by the SRN model is a very interesting shortcoming. Subjects do indeed seem to be able to readjust their responses to the new shifted rule rather quickly. Another, similarly intriguing transfer result has been repeatedly described in grammar learning experiments. For instance, Reber (1969) reported that transfer performance was significantly over chance with material generated from a grammar which had the same structural properties as the grammar used during training, but used a different set of letters. By contrast (but not surprisingly), transfer performance was much worse in a control condition in which the same set of letters but a different grammar were used. There appears to be no simple way of accounting for this kind of result. The basic problem is that successful models of implicit learning, including the SRN, base their performance on the processing of exemplars. If the representations that these models develop encode the relevant structural properties of the material, they are nevertheless expressed in terms of the exemplars themselves, and not as more abstract characterizations of these structural properties. The fact that subjects do transfer successfully in this situation suggests that some additional mechanisms may play an important role in implicit learning performance. There is no doubt that the current generation of simulation models of implicit learning phenomena will have to address this issue in the future.

## References


Berry, D.C. & Broadbent, D.E. (1984). On the relationship between task performance and associated verbalizable knowledge. *QJEP*, *36*, 209-231.

Brooks, L.R. (1978). Nonanalytic concept formation and memory for instances. In R. Rosch and B.B. Lloyd (Eds.), *Cognition and Categorization*, New York: Wiley.

Cleeremans, A. & McClelland, J.L. (in press). Learning the structure of event sequences. *JEP: General.*

Cleeremans, A., Servan-Schreiber, D. & McClelland, J.L. (1989). Finite state automata and simple recurrent networks. *Neural Computation*, *1*, 372-381.

Cleeremans A., Servan-Schreiber D., & McClelland J.L. (In press). Graded State Machines: The representation of temporal contingencies in simple recurrent networks, *Machine Learning.*

Elman, J.L. (1990). Finding structure in time. *Cognitive Science*, *14*, 179-211.

Lewicki, P. (1986). *Nonconscious Social Information Processing*. New York: Academic Press.

Lewicki, P., Hill, T., & E. Bizot, E. (1988). Acquisition of procedural knowledge about a pattern of stimuli that cannot be articulated. *Cognitive Psychology*, *20*, 24-37.

Lewicki, P. & Hill, T. (1989). On the status of nonconscious processes in human cognition: Comment on Reber. *JEP: General*, *118*, 239-241.

Nissen, M. J., & Bullemer, P. (1987). Attentional requirements of learning : Evidence from performance measures. *Cognitive Psychology*, *19*, 1-32.

Reber, A.S. (1967). Implicit learning of artificial grammars. *JVLVB*, *6*, 855-863.

Reber, A.S. (1976). Implicit learning of synthetic languages: The role of the instructional set. *JEP: Human Learning and Memory*, *2*, 88-94.

Reber, A.S. (1989). Implicit learning and tacit knowledge. *JEP: General*, *118*, 219-235.

Reber, A.S. & Millward, R.B. (1968). Event observation in probability learning. *Journal of Experimental Psychology*, *77*, 317-327.

Reber, A.S. & Millward, R.B. (1971). Event tracking in probability learning. *American Journal of Psychology*, *84*, 85-99.

Rumelhart, D.E., Hinton, G., & Willians, R.J. (1986). Learning internal representations by error propagation. In Rumelhart, D.E. and McClelland, J.L. (Eds.), *Parallel Distributed Processing, I : Foundations*. Cambridge, MA : MIT Press.

# Implicit understanding of functions in quantitative reasoning[1]

**Joyce L. Moore and James G. Greeno**

School of Education
Stanford University
Stanford, CA 94305
moore@leland.stanford.edu

## Abstract

We present a theoretical analysis of students' implicit understanding of the concepts of variables and functions, and present a cognitive model of this understanding based on the idea that reasoning involves a successful interaction between psychological agents and the things and other people in a situation. In the first part of the paper, we provide evidence that middle- and high-school students demonstrate implicit understanding of functional relations among quantities when they reason about a physical model of linear functions. Implicit understanding is knowledge of concepts or principles that enables and constrains performance, but is not articulate. In the second part of the paper, we describe several theoretical properties of our computational model: a.) activities are modelled as interactions between a person and a situation; b.) reasoning is modelled as a form of activity that produces new information; and c.) understanding is modelled as attunement to the constraints of conceptual activities.

In most information-processing models, the symbolic expressions of some programming language are assumed to correspond to cognitive representations that are constructed through perception and reasoning. We have developed a model based on the idea that reasoning involves a successful interaction between psychological agents and the things and other people in a situation, in which only some symbolic expressions are interpreted as cognitive representations. We interpret other symbolic expressions as statements that are assumed to be true about the situation, but we do not assume that these expressions correspond to representations constructed by the person whose reasoning the model simulates. We make this distinction in order to capture several important theoretical properties in our model. Firstly, activities are modelled as interactions between a person and a situation. Secondly, reasoning is modelled as a form of activity that produces new information. Finally, understanding is modelled as attunement to the constraints of conceptual activities.

The conceptual domain that we model involves the concepts of variables and functions. The first part of this paper summarizes an empirical investigation of middle- and high-school students' implicit understanding of functional relations in the context of a physical model. In the second part of the paper, we provide a theoretical analysis of students' understanding, and describe how our model functions within this theoretical framework.

[1]This research is supported by National Science Foundation Grants BNS-8718918 and MDR-9053605.

## Implicit Understanding

We use the term "implicit understanding" of a concept or principle to refer to reasoning in which the concept or principle plays a functional role, but the person does not give evidence of having an explicit representation of the concept or principle. That is, implicit understanding is knowledge of concepts or principles that enables and constrains performance, but is not articulate. The concept of implicit understanding has been used to account for successful reasoning in several domains. For instance, Gelman and Gallistel (1978) inferred from pre-school children's counting performance that they have implicit understanding of the principles of one-to-one correspondence, the order of numerals, and cardinality. Children's implicit understanding also has been studied in the areas of biology (Carey, 1985; Hatano & Inagaki, 1987), physical causality (Bullock, Gelman & Baillargeon, 1982), theory of mind (Wellman & Estes, 1986), ontological categories (Keil, 1979), and physical force and motion (diSessa, 1983; McCloskey, 1983).

This study is an attempt to demonstrate the same kind of understanding by middle- and high-school students of the concepts of variables and functions when they reason about a physical model of linear function. The particular physical model that we used is sketched in Figure 1, and is a variation of a device used by Piaget, Grize, Szeminska, & Bang (1977). The device consists of a board, about a yard long, with two grooves that run lengthwise. Beside each groove is a ruler, marked in inches, and inside each groove is a small metal block with a pointer on its top surface. At the end of each groove an axle with a handle is attached to a wheel so that the axle can be turned. Metal spools of various circumferences can be placed on the two axles, and each spool is attached by a string to a block. As a spool turns, the string winds around the spool and the block moves toward the spool. The two axles can be linked together so the spools turn simultaneously, or separated so that they turn independently.

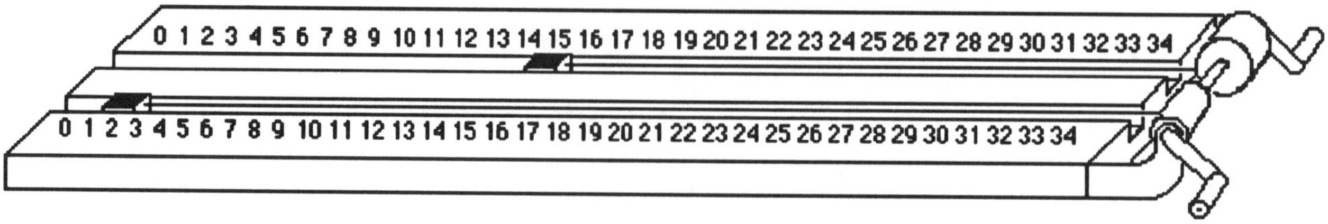


Figure 1. Physical model of linear functions.

In this physical model, the distance that a block moves is directly proportional to the number of turns of the spool, and the position of a block after a given number of turns can be found by adding the initial position to the distance travelled. Because of these properties, the winch device can be thought of as an embodiment of the linear function $y = mx + b$, where $y$ is the position of the block after a certain number of turns, $b$ is the block's starting position, $m$ is the circumference of the spool, and $x$ is the number of turns.

## Participants and Procedure

There were three groups of students: four pairs of seventh graders who had no prior algebra instruction, seven pairs of ninth graders who had one year of algebra instruction, and four pairs of 11th graders who had two years of algebra instruction. Sessions of approximately 45 minutes were conducted with pairs of students of the same sex. Students were asked a series of increasingly difficult problems by an interviewer.

## Results

Overall, students were quite successful on these tasks. Transcriptions of the interview sessions were used to infer whether students demonstrated implicit understanding of linear functions. Two examples are discussed below to provide a sense of the kind of interactions that we observed.

**Multiple-solution Problem.** Approximately 20 minutes into the interview, after an introduction to the winch, students were asked "How could you make both blocks get to 24 at the same time?", and then "Can you think of another way?". We then asked for solutions with unequal starting positions, with different spool sizes, and with different numbers of turns. Symbolically, this question required finding conditions such that $m_1x_1 + b_1$, the expression for one track of the winch, equals $m_2x_2 + b_2$, the expression for the second track of the winch, and both equal 24. The easiest solution is to have all the variables equal, which is the solution that every pair of students chose initially. If one of the variables is made unequal, such as making one spool size larger than the other, then another variable has to be made unequal in the opposite direction by an appropriate amount.

The operations that students performed to arrive at their answers were functionally equivalent to algebraic operations; however, the understanding they showed about variables and functions was implicit in their reasoning about quantities in the physical system, rather than being about the symbolic expressions of algebra. We analyzed students' solutions to infer features of the problem spaces in which they reasoned about the winch. Implicit understanding of a concept will be reflected in the problem-solving operators that are part of a person's problem space.

For this problem, we focussed on students' implicit understanding of variables. For example, one 7th-grade pair said:

S2: You could put a bigger one [spool] and put the other one [block] ahead. [...] 24, this one [a 6-spool] will be up there in 4 turns, this one [a 3-spool] in 8.
INT: Okay.
S2: So that's like 2 turns of this [red's spool] to 1 turn of this [blue's spool].

To say how the two blocks can get to 24 at the same time, given that one of the spools is larger than the other, students need to understand: (a) that a specified property functions as a variable, in this example, that spool size is a parameter of the system; (b) the way in which a variable influences the behavior of the system, in this case, that the block with the larger spool will move further on each turn; and (c) how the variables interact, for instance, that a variable other than spool size can be varied to compensate for the effect of the difference in spool sizes. All students were able to give answers in which variables compensated for one another.

**Quantitative Inference Problem.** The second example occurred approximately 35 minutes into the interview and had a situation with a 3-spool for the red block and a 6-spool for the blue block. The axles were linked and both blocks started at 0. The formulaic representation for this question is $y1 = 3x$ and $y2 = 6x$. The question had five parts, the first two are discussed briefly:

(a) When you turn the handle, will the red block ever get ahead? (When or why not?)
(b) How far ahead will the blue block be after 4 turns?

To answer questions such as these, students needed to make inferences based on functional relations among quantitative properties of the systems. To the extent that they reasoned successfully, we can infer that they had (at least) implicit understanding of those functions.

The first part of this question is about a general property of the two functions - one of the functions is monotonically greater than the other. An 11th grader said that the red block would never get ahead, and when asked "Why?" replied:

S1: Because right off from the start, after the first turn this one [6-spool] has a 3-inch lead, so it can never get ahead. Every turn it'll gain 3 inches.

The second part of this question, which called for a calculation, was "How far ahead will the blue block be after four turns?" or, in symbolic terms, evaluate $6x - 3x$ for $x = 4$. One of the 7th-grade pairs said:

S2: Let's see [measures intervals with fingers from 0 on blue side] 1, 2, 3, so 24. And then this one will be — what was that, 3? — so 3 times 4 is 12. So it'll be —

S1: 12 inches — double the amount.

S2: So, however many turns we turn together, this one [blue] will always be double the amount, so like if it was 30, then this one would be at 15.

INT: Ok, and how far ahead would the blue block be after 4 turns?

S1&2: 12.

This pair found the blue block's position by counting increments on the winch, and inferred the red block's position by multiplication. They then stated a general property of the two functions: that the position of the blue block will always be twice the position of the red. On this problem, all students recognized general properties of the relationship between the two functions, and were able to use the functional relations to calculate specific answers.

Overall, our results indicate that children have the capacity for reasoning about functional relations among quantities in the context of a physical model. We are currently exploring the hypothesis that this implicit understanding could be used as a foundation for learning about symbolic representations of algebra, and thus provide an environment for developing conceptual understanding of the mathematical symbols that denote variables and functions.

## Theoretical Analysis

In this section, we discuss the theoretical framework that we are using to develop a computational model of the reasoning described above. We discuss implicit understanding by developing the idea that reasoning involves a successful interaction between a person and a situation, which includes both objects and people (Brown, Collins & Duguid, 1989; Greeno, 1989; Jordan, 1987; Laboratory of Comparative Human Cognition, 1983; Lave, 1988; McCabe & Balzano, 1986; Shaw, Turvey, & Mace, 1982; Suchman, 1987). Within our analysis representations play a significant role in reasoning; however, we also assume some features of the environment can support cognitive activity without being mentally represented. Gibson's (1979) concepts of direct perception of affordances are used to describe this support. Our analysis extends Gibson's concept of affordances to include affordances for reasoning since we consider reasoning to be a form of activity that produces new information.

### Action Schemata

Figure 2 presents a taxonomy of some of the elements of our model. Each branch of the tree will be discussed in turn. A *state of affairs*, a term from situation theory (Barwise, 1989; Devlin, 1988), is represented by a true statement about a situation, such as "Emmy is eating yogurt". A state of affairs is a relation between objects that holds in the world. We characterize the space in which the model operates as a set of states of affairs, and will later distinguish between types of states of affairs and how they are used in reasoning.

An action schema is a description of an agent-situation interaction. This concept of a schema is different from that developed in cognitive science during the 1970's, where it was assumed to consist of a representation of abstract concepts which could be used to construct a representation of the current situation. In contrast, we assume that a schema is a pattern of interaction that a person has learned to participate in. This is consistent with Bartlett's use of the term: "'Schema' refers to an active organisation of past reactions, or of past experiences, which must always be supposed to be operating in any well-adapted organic response" (Bartlett, 1932, p. 201).

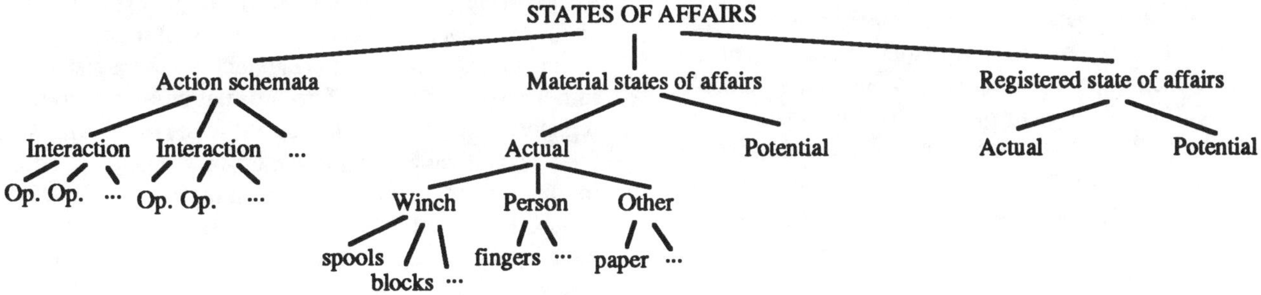


Figure 2. Taxonomy of model.

We have analyzed some of the activities that students engaged in with the winch into a network of interaction-operation units. The term *interaction* will be used to emphasize that activity is jointly constituted by an agent and the setting in which the activity occurs. In this view we follow other theorists, including Dewey and Bentley (1949), Lave (1988), McCabe and Balzano (1986), and Riegel and Meacham (1978). The term *operation* will be used to refer to a means or method of an interaction, following a distinction made by Leont'ev (1981) in an analysis of activity. The description of an interaction will focus on the function of activity, while the description of an operation will focus on the ways in which an activity is performed.

For example, imagine a situation in which the block starts at the ten-inch mark, a 4-spool is placed on the winch and the student is asked "Where will the block be after three turns?". An interaction that might occur in this situation could be described as "determining an end position after some number of turns from a starting position". We observed students use at least four different operations for this interaction: (a) turning the handle of the winch, and then attending to the resulting position of the block; (b) marking intervals of distance on the ruler to indicate where the block(s) would be after each turn of the handle; (c) making a table of values to indicate the numerical position of the block after each turn; and (d) multiplying the number of turns times the spool size, and adding that number to the starting position of the block.

The distinction between interactions and operations is relative to the level of description. That is, our analyses are nested, with units that are interactions at one level of description and operations at another. The description at any level distinguishes between an interaction (what is accomplished) and an operation (how it is accomplished). The distinction provides a convenient expression of relations between alternative methods for achieving a functional result.

What does it mean for an activity to be jointly constituted by an agent and the situation? The possibility of a person interacting in a certain way with a physical and/or conceptual system depends on the affordances provided by the situation and the abilities of the person. The term *affordances* (Gibson, 1979) refers to features of a situation that support certain activities - these are specified relative to an agent acting in that situation. *Abilities* are those activities that an agent can perform - these are specified relative to a situation. (We prefer the term *abilities* to the term *effectivities* that Shaw, Turvey, and Mace (1982) coined, but we believe that we mean the same thing.) For example, in order for a student to turn the handle of the winch, the handle must be "graspable" and "turnable", that is, it must be a certain size and shape, and must be connected to the rest of the device in a way that allows it to be rotated. However, a handle is only graspable and turnable by an agent with an appropriately shaped hand and the ability to turn something in a circular motion. An affordance and an ability are two sides of the same possibility: affordances for activities depend on the characteristics of an agent, and abilities for activities depend on characteristics of the situation.

### Reasoning as a Form of Activity

Figure 2 indicates the relationships between several kinds of states of affairs. Material states of affairs are those that hold in the physical world. Before distinguishing how different kinds of material states of affairs are used in reasoning, it is necessary to define some terms. A situation type is a parameterized state of affairs, that is, a partially specified state of affairs that applies in many situations. For instance, "Emmy is eating yogurt" is a state of affairs that may or may not hold in a given situation, while "someone is eating yogurt" is a situation type that holds in the class of situations in which a person is eating yogurt.

In any situation, various states of affairs hold. Because of affordances for activity, there are also *potential* states of affairs that hold. A state of affairs that might occur, such as "Emmy might eat yogurt" is a potential state of affairs. These depend on the affordances for an activity being available, as well as an agent with the appropriate abilities. Potential states of affairs designate ways in which an agent or something in the situation can contribute to an interaction.

A *constraint* is a relation between situation types such as "If someone is eating yogurt then that person might drop the spoon". It is this relation between situation types that supports inferences by a person who is attuned to the constraint. For example, if someone is eating yogurt with a spoon, then it can be inferred that the spoon might be dropped.

Actual states of affairs are those that are true about the material aspects of a situation. Potential states of affairs are generated from an actual state of affairs through a set of constraints. Figure 2 shows a further breakdown of kinds of actual states of affairs that we use in modelling interactions with the winch. The same distinctions could be made for the potential states of affairs.

The final distinction between states of affairs that we make is between material and registered states of affairs, indicated in the first level of Figure 2. We distinguish two aspects of a situation: what is happening - the material situation, and what is recognized about what is happening - the registered situation. Material states of affairs are relations in the world, while registered states of affairs are those states of affairs which have been attended to. Registering a state of affairs often involves saying it and communicating it in a conversational group. In general, it is the set of properties and relations that have symbolic representations. We distinguish registered states of affairs mainly to analyze activities that include inferences and communication. We assume that registration makes the information available for inferences; however, both material and registered states of affairs play a role in supporting certain kinds of activity. We are assuming, therefore, that the role of representation is to provide objects that function in inferences, and we use the notation of registered states of affairs to distinguish symbolic representations of information that are available for inferential reasoning.

We characterize reasoning within this framework as inference of new registered information using an inferential process that involves transformation of a representation. Reasoning about a situation can include generating potential states of affairs and registering those potential states of affairs. Representations of states of affairs afford reasoning activities that infer other states of affairs. In this view, material systems, registered states of affairs, and the person or persons participate interactively in the activity of reasoning.

## Understanding Based on Constraints

We hypothesize that activities, such as the interaction-operation units presented earlier, depend on constraints. Turning the handle to reach an end position works because certain relationships exist between states of affairs in the system, such as the handle being connected to the spool and a string being attached to both the spool and the block. Since activities can be conceptual as well as physical, "knowing a concept" in this framework is being attuned to the constraints of conceptual activities. This characterization of knowing a concept includes implicit understanding, as well as more explicit forms of knowing. A reasoner can represent a constraint and reason about it explicitly, as well as simply being able to use a constraint in an activity.

The concept of linear function includes several constraints:

1. *proportional quantities*: a change in one variable causes a proportional change in another variable;
2. *linearly-related measures*: each increment of a variable has the effect of adding a constant number to the previous numerical value of the function;
3. *multiplicative/additive properties*: the cumulative effect of $n$ increments of $k$ each is equal to $n \times k$; and a final quantity measure is equal to an initial measure added to $n \times k$.

In terms of the winch, "proportional quantities" means that each turn of the handle causes a constant change in the position of the block. "Linearly-related measures" means that for each turn of the handle a constant number is added to the numerical position of the block. The "multiplicative/additive" constraint means that the distance a block moves for some number of turns can be calculated by multiplying the spool size times the number of turns, and the final position of the block can be calculated by adding the starting position to the distance moved.

Since we are characterizing activities as being jointly constituted by affordances and abilities, the question arises as to what aspect of a situation is providing a certain constraint. We have analyzed the constraints for the interaction "determining the location of a block after some turns from a starting position". For the operation of turning the handle, all three constraints are embodied in the mechanism of the device. That is, the ending position of the block is determined by the physical constraints of the device. For the operation of measuring increments, the constraint of proportional quantities is provided through the abilities of the student, rather than through the affordances of the winch. The student must explicitly determine the size of the interval to be measured, and simulates the change in the position of the block for each turn. For the operation of creating a table, both the constraints of proportional quantities and linearly related measures are supplied by the abilities of the student. The student explicitly determines a numerical difference and adds that difference repeatedly to calculate where the block would be for each turn of the handle. The number in the last row of the table can then be read as the ending position of the block. For the operation of mental arithmetic, all three constraints have shifted to the abilities of the student. Proportional quantities and linearly related measures are captured as part of the calculations in the multiplicative/additive constraint.

## Conclusions

The analysis of this interaction indicates that alternative operations for an interaction differ in having constraints contributed by the situation's affordances or the agent's abilities. Across the four operations discussed above, the interaction is increasingly abstracted from the physical constraints of the device. Having more constraints of a concept embodied in a student's abilities, and reasoning more explicitly about constraints, constitute more understanding of the concept.

We hypothesize that an important aspect of understanding is a kind of correspondence between constraints of interactions. In the handle-turning interaction the proportional quantities constraint is met through the physical constraints of the device. In measuring increments the constraint is met through the size of the interval that the student has created. However, the process of measuring a series of intervals along the ruler only makes sense with respect to the constraints of the physical device. It is through the correspondence of the constraints of the two interactions that measuring increments gains its meaning.


## Acknowledgements

Other people who participated in this research are Jean-luc Guertner, Barbara Katzenberg, Jan Kerkhoven, Meg Korpi, and Richard Mander. The authors would also like to thank Richard Mander for extensive comments on an earlier draft of the paper.

## References

Bartlett, F. C. 1977. *Remembering: A study in experimental and social psychology*. Cambridge: Cambridge University Press. (Original published 1932).

Barwise, J. 1989. *The situation in logic*. Stanford, CA: Center for the Study of Language and Information, Stanford University.

Brown, J. S.; Collins, A.; and Duguid, P. 1989. Situated cognition and the culture of learning. *Educational Researcher, 18*(1).

Bullock, M.; Gelman, R.; and Baillargeon, R. 1982. The development of causal reasoning. In W. J. Friedman (Ed.), *The developmental psychology of time*. New York: Academic Press.

Carey, S. 1985. *Conceptual change in childhood*. Cambridge, MA: MIT Press/Bradford Books.

Devlin, K. 1988. *Logic and information. Volume 1: Situation theory*. (Draft). Stanford, CA: Center for the Study of Language and Information, Stanford University.

Dewey, J.; and Bentley, A. F. 1949. *Knowing and the known*. Boston, MA: Beacon Press.

diSessa, A. A. 1983. Phenomenology and the evolution of intuition. In D. Gentner & A. L. Stevens (Eds.) *Mental Models* (pp. 12-33). Hillsdale, NJ: Lawrence Erlbaum Associates.

Gelman, R.; and Gallistel, C. R. 1978. *The child's understanding of number*. Cambridge, MA: Harvard University Press.

Greeno, J. G. 1989. Situations, mental models, and generative knowledge. In D. Klahr & K. Kotovsky (Eds.), *Complex information processing: The impact of Herbert A. Simon* (pp. 285-318). Hillsdale: Lawrence Erlbaum Associates.

Hatano, G.; and Inagaki, K. 1987. Everyday and school biology: How do they interact? *The Quarterly Newsletter of the Laboratory of Comparative Human Cognition, 9*, 120-128.

Jordan, B. 1987. *Modes of teaching and learning: Questions raised by the training of traditional birth attendants* (Report No. IRL87-0004). Palo Alto, CA: Institute for Research on Learning.

Keil, F. C. 1979. *Semantic and conceptual development: An ontological perspective*. Cambridge MA: Harvard University Press.

Laboratory of Comparative Human Cognition. 1983. In W. Kessen (Ed.), *Handbook of Child Psychology. Volume 1: History, theory and methods* (pp. 295-356). New York, NY: John Wiley & Sons.

Lave, J. 1988. *Cognition in practice*. Cambridge: Cambridge University Press.

Leont'ev, A. N. 1981. The problem of activity in psychology. In J. V. Wertsch (Ed. & Trans.), *The concept of activity in Soviet psychology*. Armonk, N.Y.: M. E. Sharpe.

Gibson, J. J. 1979. *The ecological approach to visual perception*. Boston: Houghton Mifflin.

McCabe, V.; and Balzano, G. J. eds. 1986. *Event cognition: An ecological perspective*. Hillsdale, NJ: Lawrence Erlbaum Associates.

McCloskey, M. 1983. Intuitive physics. *Scientific American, 248,4*, 122-130.

Piaget, J.; Grize, J.; Szeminska, A.; and Bang, V. 1977. *The psychology and epistemology of functions*. (First published in French, 1968.)

Riegel, K. F.; and Meacham, J. A. 1978. Dialectic, transaction, and Piaget's theory. In L. A. Pervin & M. Lewis (Eds.), *Perspectives in Interactional Psychology*. N.Y.: Plenum Press.

Shaw, R.; Turvey, M. T.; and Mace, W. 1982. Ecological psychology: The consequence of a commitment to realism. In W. B. Weimer & D. S. Palermo (Eds.), *Cognition and the Symbolic Processes, Volume 2*. (pp. 159-226). Hillsdale, NJ: Lawrence Erlbaum Associates.

Suchman, L. A. 1987. *Plans and situated action*. Cambridge: Cambridge University Press.

Wellman, H. W.; and Estes, D. 1986. Early understanding of mental entities: A reexamination of childhood realism. *Child Development*.

# Does Memory Reflect Statistical Regularity in the Environment?

**Lael J. Schooler** **John R. Anderson**

**Department of Psychology**
**Carnegie Mellon University**
**Pittsburgh PA 15213**

## Abstract

Anderson and Milson (1989) derived optimal performance functions for memory based on assumptions about the goals of memory, the computational costs of achieving those goals, and the statistical structure of the environment. Based on these assumptions, and a good deal of Bayesian analysis, they accounted for a substantial number of empirical findings. Here we started with the same assumptions about the goals of memory, but instead of simulating the statistical structure of the environment, we analyzed it directly. It was found that the factors that govern memory performance also predict the probability with which words are spoken in children's linguistic environments. These factors include frequency, recency, and spacing between exposures. The ability of these factors to predict word use was analyzed in the context of four laboratory memory phenomena: 1) the power law of practice; 2) the power law of forgetting; 3) the interaction between study spacing and retention interval and 4) the combined effects of practice and retention. These factors predict information demand and lend strong support to Anderson and Milson's claim that memory behavior can be understood in terms of the statistical structure of the environment.

## Introduction

The environment constantly makes demands on memory. For instance, we may need to remember a particular friend's telephone number 23 times during the past semester and only once in the current semester, while we have had to recall another friend's number three times during the past semester and three times during the current semester. What do these two patterns predict about the probability of having to recall either friend's number today? Anderson & Milson (1989) proposed that the phenomena of human memory could be understood as memory's response to questions like this. They argued that human memory behaves as if it is performing a Bayesian inference to determine the probability that a given memory will be needed now. They contend that this probability is based on 1) the pattern of prior use of the memory (i.e., its history) and 2) the similarity of the current context to the previous contexts in which it had occurred.

Lacking any direct data about patterns of human memory use, Anderson and Milson adapted statistical models developed to describe patterns of use in other information retrieval systems, such as library borrowings (Burrelll, 1985) and computer file access (Stritter, 1977), to create a model of the demands that the environment might make on human memory. By applying Bayesian inference mechanisms to this model of the environment, they accounted for a number of memory phenomena including the effects of practice, retention interval, and study spacing on memory performance.

The model proposed by Anderson and Milson has basically two assumptions:

(1) Memories vary in their rate of use. Thus one of memory's goals is to estimate a particular memory's true rate of use, based on information about how often it has been needed in the past.

(2) Memories undergo fluctuations in their rate of use and some memories are more volatile than others. Thus, memory must infer changes in a particular memory's rate of use.

Assumption (1) helped to account for practice and forgetting effects. Memories which have been used often and recently would be identified as having high rates of use. Assumption (2) helped to account for the forgetting and spacing effects. Memories that have not been used recently would be identified as having undergone a loss in their rate of use and memories that occurred in massed patterns with long periods of disuse would be identified as undergoing rapid fluctuations in their rate of use. It should be emphasized that the Anderson and Milson model did more than just predict these effects; it generated the parametric form of the practice and forgetting functions and the exact interactions that occurred in spacing experiments.

There are, however, a number of problems with the Anderson and Milson analysis. First, one can question whether the statistics of library usage and file access, the basis of their environmental model, are representative of those faced by people. Second, the published analyses of these non-human systems only justify assumption 1. While Anderson and Milson argued that assumption 2 plausibly held, they could not provide hard evidence for it. The raw data from these sources is unavailable and this is needed to determine whether massing and spacing occurred. Third, the predictions were derived from an interaction between Bayesian inference procedures and a mathematical model of the environment. Replacing this model with actual statistics from the environment would help to remove an entire layer of assumptions from the theory.

These problems can be addressed by undertaking an analysis of the statistical properties of the environment; we need to get detailed records of informational demands being placed on human memory. We have analyzed three such sources of informational demand: speech to children, word usage in newspaper headlines, and electronic mail messages. For each of these sources we have conducted analyses that reveal strong similarities between a number of memory phenomena and statistical properties shared by each of these sources. To illustrate our methods, we report the results we have obtained from our analyses of speech to children. Each word a child hears is another demand to retrieve the meaning of that particular word. Hall & Tirre (1979) collected nearly 100 hours of preschool children's verbal interactions. These have been assembled and standardized in a larger database (MacWhinney & Snow, 1985). We analyzed roughly a quarter of Hall & Tirre's data, the transcripts of 9 white middle-class children. We excluded the children's own utterances from the analysis, because we are interested in the memory demands made by their environments and not on those demands resulting from production.

Building a case for each analysis requires first, characterizing the relevant results from the experimental literature; second, specifying the method for analyzing the environment; and third, comparing the results of the environmental analysis to its experimental counterpart. In particular, we will discuss four laboratory memory phenomena: 1) the power law of practice 2) the power law of forgetting 3) the interaction between study spacing and retention interval and 4) the effects of practice and retention.

## Power Law of Practice

Any memory experiment can be thought of as presenting some items to memory some number of times and then testing for recall at some later time. In the Anderson & Milson model, an experiment was encoded as a set of times that described the pattern of presentation and test. Presentation of information in the environment can be characterized similarly. A single exposure to a piece of information in the lab maps onto the occurrence of an item in the environment. The probability of recall or recognition in the lab maps onto the probability that an item in the environment will be used in the next time unit. This is what Anderson & Milson call "need probability".

Within this framework, the mappings between need probability and performance measures, such as probability of recall and recognition, is somewhat more complex then they might initially appear. One might expect that need probability and performance would simply be proportional to each other, but this is not so. Take for example the relationship between need probability and latency of recall. The model assumes that memories are considered in the order of their need probabilities. Thus, estimating the time it would take to retrieve a given memory requires an estimate of the number of memories with need probabilities greater than the given memory. Anderson and Milson argue that this implies that latency of recall should be a power function of need probability. They argue for a similar power function relating practice and probability of recall. Such power functions will preserve ordinal relationships between need probability and some parametric relationships.

Results from the empirical literature further constrain our expectations about the form of the relationship between frequency and need probability. The relevant experiments are those that examine the relationship between practice and memory performance. This relationship is often characterized as taking a power or exponential form. A power relationship, in its simplest form, means that performance, P, equals the amount of practice, L, raised to an exponent r, the learning rate:

$$P = k L^{r}$$

This can be conveniently transformed into a linear relationship by taking the natural logs of each side:

$$\log P = \log k + r \log L$$

In contrast, an exponential function takes the form:

$$P = k e^{bL}$$

Taking the natural log of this yields:

$$\ln P = \ln K + b L$$

These transformations allow the use of linear regression to determine whether power or exponential relationships best capture the relationship between practice and performance. The transformed power equation implies that if performance and practice are in a power relationship with each other, then when log performance is plotted against log practice the result should be a straight line. In contrast, the transformed exponential equation will be curved when plotted in log log, but straight when log performance is plotted against practice.

Using linear regression in combination with these transformations, we have analyzed the results of several learning experiments and in general the power relationship accounts for more variance than does the exponential form. These results are also consistent with Newel and Rosenbloom's (1981) detailed analyses of many other learning experiments. Based purely on these experimental results we can expect a power law relationship between word frequency, the environmental analog of practice, and the analog of performance, need probability.

We are not too concerned with whether the memory or environmental functions are truly power functions or just good approximations. Rather, the goal is to establish that the functional forms of statistical regularities in the environment are the same as the functional forms of human memory performance.

### Method

The analog of the practice curve amounts to calculating the conditional probability that a word will occur in an utterance given that it has occurred *n* times in the previous *k* utterances. We calculate this probability by looking at examples of words that occur *n* times in some period, say, the first 100 utterances and determining how often they occur in the 101'st utterance. Our analysis program first made a list of the words that occurred exactly once in the first 100 utterances. To store the partial results of the analysis the program created a variable, or bin, labelled *occurred-1-time*. Each time a word was found that occurred just once the *instances* field of the bin labeled *occurred-1-time* was incremented by 1. If the word occurred in the critical (101'st) utterance, the *uses* field of the *occurred-1-time* bin was incremented by 1. Dividing the number of uses by the number of instances, yields the need probability for items used exactly once. The analogous process was repeated to calculate the probabilities for words occurring exactly 2,3,4..100 times.

The focus of the analysis then shifted onto utterances 2 through 102, with utterance 102 becoming the critical utterance. Again, the program made a list of all the words occurring exactly once between utterance 2 and utterance 101, incrementing the appropriate fields of the *occurred-1-time* bin. As before, the process was repeated for those words that occurred exactly 2,3,4..100 times. The focus, or window, then covered utterances 3 through 103, with utterance 103 acting as the critical utterance. The process repeated until all utterances between utterance 101 and the last had played the role of the critical utterance.

### Results

The Hall corpus contained over 18500 utterances that were on average 7.88 words long. The size of the corpus resulted in many instances of infrequently occurring words; there were well over 1.6 million instances of words occurring 1 time in a 100 utterance window, and 525,000 words occurring 2 times. In figure 1 we have plotted the probability that a word will be used in the critical utterance against the number of times it has been mentioned in the previous 100 utterances. As the sample was insufficient to generate adequate numbers of high frequency instances, some cells corresponding to frequencies greater than 10 were collapsed together in blocks of five to achieve reasonable levels of accuracy. For example, one cell corresponds to words that were used between 11 and 15 times, another to words used between 16 and 20 times, etc.

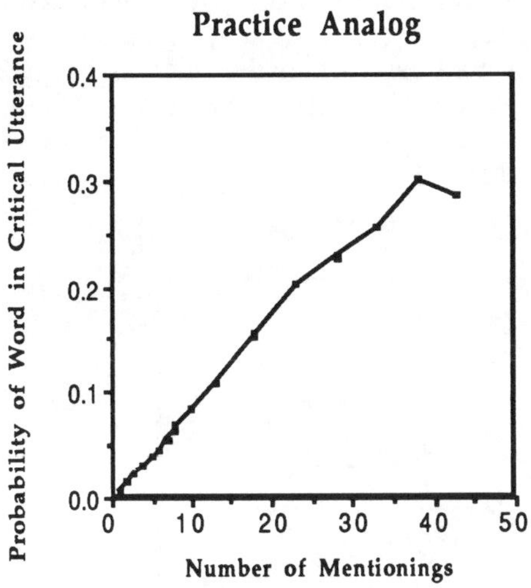


Figure 1. Need Probability (i.e., probability of a word occuring in the critical utterance) as a function of "practice" (i.e., frequency).

Consistent with the empirical and theoretical results the exponential ($R^2 = .78$) form faired poorly. However, there is nearly a perfect linear relationship between frequency and need probability, ($R^2 = .98$). A linear function is a special case of a power function, and so is consistent with the theory. It is worth noting that the Anderson & Milson model predicts exactly such a linear relationship between frequency and need probability.

We have identified three power relationships relating the Anderson & Milson model, the practice effect and children's linguistic environment.

1) Theoretically, Anderson & Milson predict that performance, P, should be related to need probability, N, by a power function: $P= a N^b$

2) Empirically, there is a power function relating practice, or frequency, L, to need probability in children's linguistic environments: $N= c L^d$

3) Empirically, there is a power function relating memory performance to practice: $P=f L^h$

Composing the power functions in (1) and (2) leads to a prediction of the power function in 3: $P= a(cL^d)^b = a c^d L^{db}$

In contrast, if we had found that need probability was in an exponential relationship with frequency in the environment, $N= ke^{mL}$, then this would imply that memory performance should be in an exponential relationship with practice: $P= a(ke^{mL})^b= a k^b e^{mbL}$

## Power Law of Forgetting

Another classic memory phenomenon is the retention function-the effect of the interval since the last exposure to a memory item on memory performance. Typically, this is studied by giving subjects a constant experience with one or more studies of an item, and then varying the retention interval. We fit a number of experiments that manipulated retention. Again it turns out that there is a power relationship between the independent measure of retention interval and a number of performance measures. The same form holds in time scales from seconds all the way up to years. This is again despite the frequent claim (e.g., Loftus 1975) that the forgetting function is exponential in nature.

As we have observed in the analysis of the practice function, a power function with respect to need probability implies a power function with respect to performance. Thus, the critical question is whether there is a power function in the environment. That is, if we look at the probability of an item being used as a function of how long it has been since it was last used, do we get a power function?

### Method

We are interested in calculating the conditional probability that a word will occur in the critical utterance given that it has not been mentioned in the last *k* utterances. Each time a word was found that had last occurred in the first utterance, the instances field of the bin labeled *last-seen-in-utterance-1* was incremented by 1. If the word occurred in the critical utterance, the *uses* field of the *last-seen-in-utterance-1* bin was incremented by 1. This process was repeated, in turn, for all items that had not been mentioned since the 2, 3,4...100 utterance. As in the practice analysis, the window shifted to cover utterances 2 through 102.

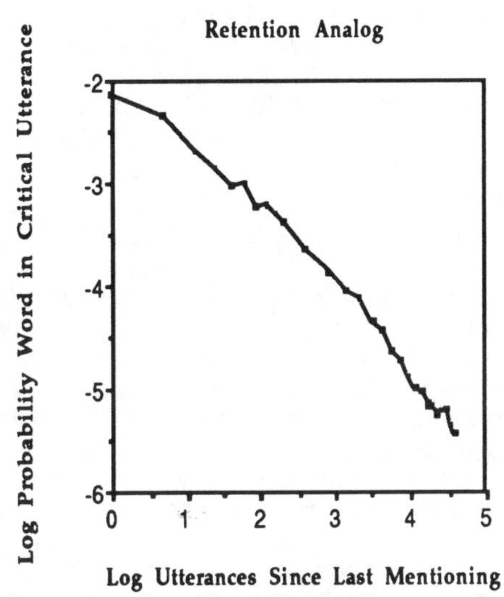


Figure 2. Log Need Probability as a function of log "retention" interval (i.e., the number of utterances since the word was last mentioned). Straight line indicates power function.

## Results

Figure 2 plots the power form of the data. Here the data is best captured by a power ($r^2$ =.98) relationship as compared to the exponential ($r^2$ = .88) and linear ($r^2$ =.55) forms. Note that here, unlike the practice function, the data is only fit by a power function and not by a linear function.

# Spacing Effect

Up to this point we have considered independently the contribution that retention and practice make on performance. Now we will look at how these interact in the spacing effect. The spacing effect results from manipulating the spacing between study presentations. Glenberg (1976) illustrated the nature of this interaction by orthogonally manipulating the lag between two study presentations and the lag between the second exposure and test (Figure 3). The size of the retention interval produces the largest effects; short test lags are always better than long ones, indicated by the separation between the various curves. It is apparent, however, that the lag between the two studies interacts with retention interval. If the test is going to be given after a long delay it is best to have a long lag between studies. In contrast, when testing occurs after short delays it is best to have short lags between studies. Apparently as the length of the retention interval increases better performance is obtained with longer study lags.

## Method

The parallel to the effect of study spacing is the probability that an item will occur in the critical utterance, given a particular combination of spacings between previous occurrences. To start, the program collected all the items used exactly twice in the first 100 utterances. The interval between the first occurrence and second occurrence corresponds to the "study" lag and the interval between the second occurrence and the critical (101 st) utterance the "test" lag. For example an item that first occurred in utterance 30 and then again in utterance 60, would have a study lag of 30 and a test lag of 41. As this is an example of a *study-30-test-41,* the instances field would be incremented by 1. In addition, if the word occurs in the critical utterance, the uses field would be incremented by 1. This process was repeated for each word that occurred twice. As in the previous analyses the focus next shifts to utterances 2 through 102.

## Results

Compared to the retention and practice analyses there were relatively few instances for each combination of study and test lag. For example, in the practice analysis there were over 1.6 million instances of items used exactly once, and in the retention analysis there were over 26 thousand instances of words that last occurred in the previous utterance. In contrast, there were only 607 instances of items with study and tests lags respectively of 1 and 5, the most common combination in the spacing analysis. To gain adequate statistical power we had to aggregate the data, collapsing it into a three by three matrix. This matrix consists of short (1-9 utterances), medium (10-30), and long (31-99) test and study lags.

The interaction between study and test lag that was apparent in the Glenberg experiment is evident in the child language data (Figure 4). The shorter the interval between the second occurence and the target utterance the higher the probability that the word will be mentioned. This corresponds to Glenberg's general result that the shorter tests lag lead to better performance. What is striking is the interaction between the spacing between the occurence of the words and the length of the interval between the second exposure and the critical utterance. When the interval between the second exposure and the critical utterance is long, the probability that the word will be included in the critical utterance increases with the spacing between occurences. In contrast, when the interval between the second exposure and the critical utterance is short, the probability that the word will be included in the critical utterance decreases with the spacing between occurences.

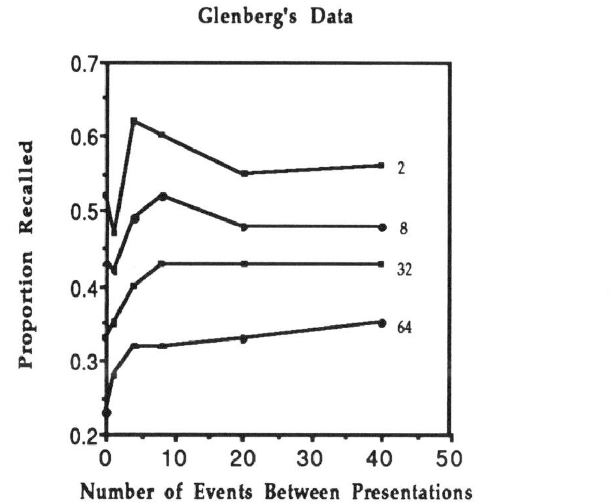


Figure 3. Proportion recalled as a function of the number of events between presentations and the number of events (2,8,32,64) between the second presentation and test.

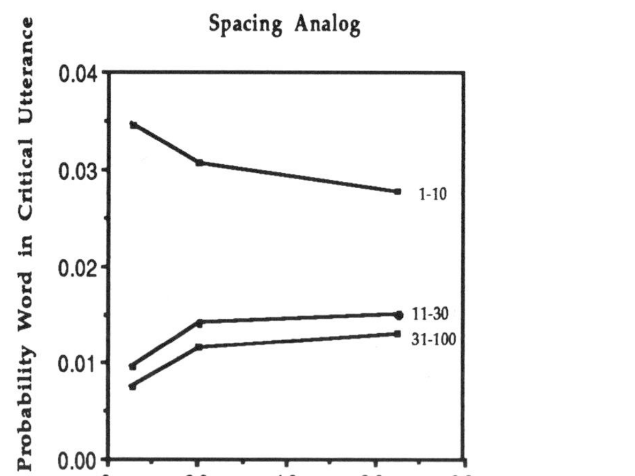


Figure 4. Need Prob. as a function of the num. of utterances between occurrences and the num. of utterances (1-10,11-30,31-100) between the 2nd occurrence and the critical utterance.

## Combinded Effects of Practice & Retention

Recently, there has been considerable interest in how different degrees of initial study affect the form of the retention function(Loftus, 1985). If a retention curve takes the form of a power relationship in standard coordinates, then there will be a linear relationship between log performance and log time. Further, we can interpret the slope of this line as a decay rate. Parallel slopes indicate that the retention functions associated with various levels of practice share the same decay rate.

A number of experiments bear directly on this question. For example, Krueger (1929) had subjects learn word lists to various degrees of over learning. Subjects were then asked to recall the lists after retention intervals that lasted between 1 and 28 days. More recently, Hellyer (1962) looked at similar issues but on a much shorter time scale. In this experiment subjects were presented with "syllables" composed of 3 consonants. These were presented from 1 to 8 times at a rate of 1 per second. Subjects were then required to recall the consonants after retention intervals ranging from between 3 and 27 seconds.

Subjects in the Hellyer experiment demonstrated nearly perfect recall at the three second retention interval. When measured in terms of the probability of recall, such near perfect performance approaches the limits of the measure (i.e., approaches 1). In contrast, a power function is unbounded above. Therefore a bounded performance measure, such as probability of recall, can obscure what could best be described in terms of a power relationship. This problem can be easily overcome by using an unbounded measure. Here, we used the odds of recall, another standard performance measure. If p is probability of recall, then $o = p/(1-p)$ will be the odds of recall.

The transformed results of both experiments are plotted in log log coordinates to test the degree to which they exhibit power law forgetting (Figures 5 & 6). In addition, we have estimated the decay parameters for the various retention curves. In these experiments the interaction between the degree of initial learning and the rate of forgetting is not statistically significant (Tables 1 & 2). Though perhaps a hint of an interaction can be seen in the Krueger data. In short, the effects of practice and retention appear to be essentially additive.

| Degree of Learning | Decay Parameter | 95% Confidence Interval | $R^2$ |
|---|---|---|---|
| 100% | -1.37 | (-1.88,-.87) | .96 |
| 150% | -1.02 | (-1.20,-.88) | .98 |
| 200% | -1.02 | (-1.22,-.82) | .98 |

Table 1. Decay parameter estimates (i.e., regression coefficients) for the various levels of over learning in Krueger (1929).

| Degree of Learning | Decay Parameter | 95% Confidence Interval | $R^2$ |
|---|---|---|---|
| 1 | -1.78 | (-2.72,-.85) | .99 |
| 2 | -1.70 | (-2.35,-1.05) | .98 |
| 4 | -1.33 | (-1.78,-.88) | .99 |
| 8 | -1.82 | (-2.66,-.98) | .98 |

Table 2. Decay parameter estimates (i.e., regression coefficients) for the various numbers of stimulus presentations in Hellyer (1962)

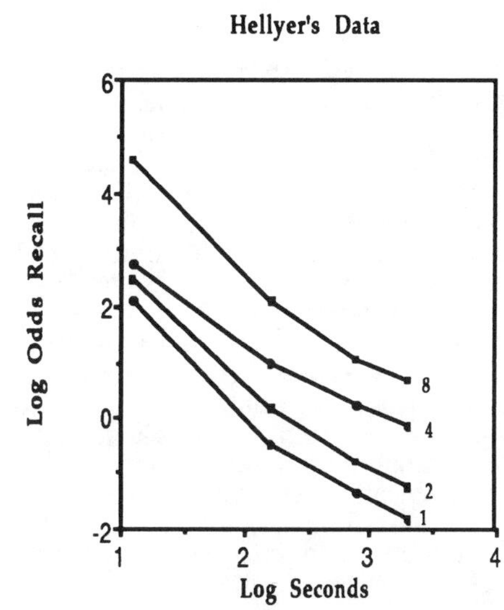


Figure 5. Retention curves at 4 practice levels.

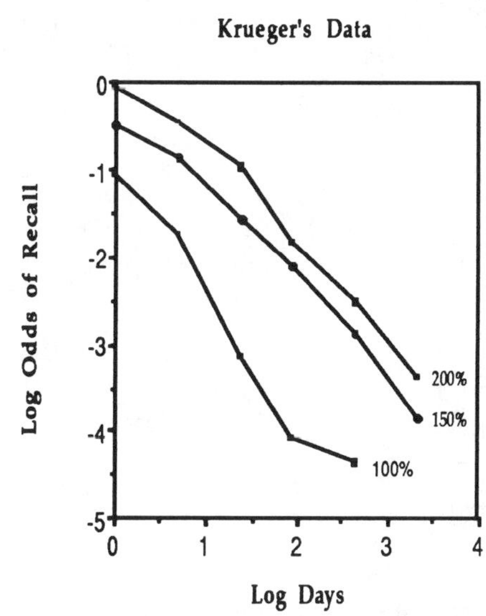


Figure 6. Retention curves at 3 practice levels.

## Method

This analysis involves calculating different retention functions for various levels of practice. The program gathered all the words that occurred exactly once in the first 100 utterances. From these it selected those words that last occurred in the first utterance. For each word in this list the *instances* field of the bin labeled *occurred-1-last-seen-1* was incremented by 1. In addition, if the word occurred in the critical (101'st) utterance, the *uses* field was incremented by 1. This process was repeated for all those words that had occurred 1 time and had not been seen since the 2, 3,4...100 utterance. The analogous process was repeated to calculate similar probabilities for words occurring exactly 2,3,4..100 times. As in all the analyses the window shifted to utterances 2 through 102 and so on.

## Results

As with the previous analyses, the data was aggregated to obtain reliable statistics. The data was collapsed into cells by the degree of practice and the length of the retention interval. Each cell had a range of 5 mentionings and 5 utterances. For example, one cell represented all instances of words that had been mentioned from 6 to 10 times and had not been mentioned for from 21 to 25 utterances, whereas another cell consisted of examples of words that had been mentioned from 16 to 20 times and were unmentioned for from 1 to 5 utterances. Each curve plots successively higher ranges of practice (Figure 7). The bottom curve represents words used from 1 to 5 times, the second from 6 to 10, etc. The plots use log retention for the abscissa and log odds of need probability as the ordinate. Certain combinations of practice and retention ranges were rare. For instance, there were only 5 instances of words that were used 21 to 25 times and had last been mentioned for between 46 and 50 utterances. So the graph contains only those points for which over 500 instances have been recorded.

Plotting the data in log log coordinates enables us to use regression to estimate the decay rates. These results indicate that the decay parameters associated with low frequency words are approximately the same as those associated with the high frequency words (Table 3). But, as with the Krueger results, there is a hint of an interaction between frequency and retention interval, with the lower frequency words associated with slightly faster decay rates.

| Degree of Learning | Decay Parameter | 95% Confidence Interval | R2 |
|---|---|---|---|
| 1-5 | -.63 | (-.67,-.59) | .98 |
| 6-10 | -.57 | (-.70,-.45) | .87 |
| 11-15 | -.39 | (-.49,-.28) | .92 |
| 16-20 | -.41 | (-.52,-.30) | .98 |
| 21-25 | -.46 | (-.67,-.25) | .98 |

Table 3. Decay parameter estimates (i.e., regression coefficients) for the various practice/retention curves.

## Conclusion

Anderson and Milson derived optimal, rational, behavior functions based on 1) assumptions about the goals of memory, 2) the computational costs of achieving those goals, and 3) the structure of the environment. Based on these three sets of assumptions, and a good deal of Bayesian analysis, they accounted for many empirical findings. Here we started with the same assumptions about the goals of memory, but instead of simulating the statistical structure of the environment based on plausible assumptions, we analyzed the patterns in the environment directly. In this analysis we found statistical analogs to some of the most robust memory phenomena. Not only did we find analogs to single variable functions, such as practice and retention curves, we found an analog to the interaction that underlies the spacing effect. That these analogs can be found in two distinct linguistic domains and one interpersonal one lends sound support to the framework that Anderson and Milson developed. In short, memory appears to be exquisitely tuned to regularities in the statistical structure of the environment.

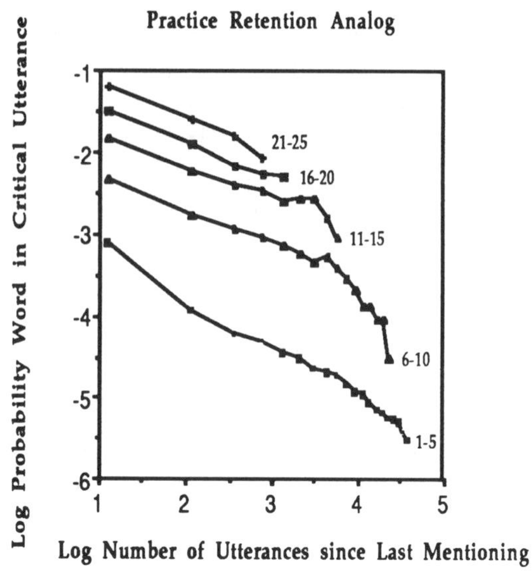


Figure 7. Retention curves at 4 frequency levels.

## References

Anderson, J.R. & Milson, R. (1989). Human memory: An adaptive perspective. *Psychological Review, 96 (4)*, 703-709.

Burrell, Q.L. (1985). A note on aging on a library circulation model. *Journal of Documentation, 41*, 100-115.

Glenberg, A. M. (1976). Monotonic and nonmonotonic lag effects in paired-associate and recognition memory paradigms. *JVLVB, 15*, 1-16.

Hall, W. S., & Tirre, W.C. (1979). *The communicative environment of young children: Social class, ethnic and situation differences.* University of Illinois, Center for the study of Reading.

Hellyer, S. (1962). Frequency of stimulus presentation and short-term decrement in recall. *JEP* 64, 650.

Krueger, G.R. (1985). The effects of overlearning on retention. *JEP, 12,71-78.*

Loftus, G.R. (1985). Evaluating forgetting curves. *JEP: LMC*, 11, 397-406.

MacWhinney, B., & Snow, C. (1990). The child language data exchange system: an update. *Journal of Child Language,* 17, 457-472.

Newell, A. & Rosenbloom, P. (1981). Mechanisms of skill acquisition and the law of practice. In J.R. Anderson (Ed.), *Cognitive skills and their acquisition* (1-55). Hillsdale, NJ: Lawrence Erlbaum Associates.

Stritter, E. P. (1977). *File Migration.* Unpublished doctoral dissertation, Stanford University, Stanford, CA.

# Shifting Novices' Mental Representations of Texts Toward Experts': Diagnosis and Repair of Novices' Mis- and Missing Conceptions

**Bruce K. Britton and Pamela Tidwell**

Institute for Behavioral Research
University of Georgia
Athens, GA 30602
(404) 542-3094; Email: IBR@UGA

**Abstract**[1]

The shape of the memory representation for a 1000 word text was measured for the text's author, 7 independent subject matter experts and 2 groups of novices (N = 83 Air Force recruits). To measure the shapes, we chose the 12 most important concepts from the text, and then collected proximity data on all possible pairs of them. Then we made maps of the mental representations from the proximities. In Experiment 1, results of empirical tests of text learning showed that the novices' mental representations after reading the Original Version of the text were correlated only +.1 with the author's or experts' representations. But for a Principled Revision the correlations were above +.5. In Experiment 2, the proximity data from Experiment 1 were used to diagnose specific misconceptions and missing conceptions in both the Original text and the Principled Revision. This revealed unsuspected cognitive misconceptions, as well as intrusions of affective and attitudinal factors into the novices' mental representations. These diagnoses were then used to revise both texts to repair the misconceptions and insert the missing conceptions. Results of empirical tests of these revisions (N = 160 Air Force Recruits) showed that novices' correlations with the author's and experts' representations were shifted close to ceiling (r = +.8 - + .9). These results show that novices' mental representations can be shifted to correspond with experts by using our methods to diagnose and repair mis- and missing conceptions.

The "shape" of the memory representations that are created by texts was investigated. By "shape" we mean the configuration of relations between the concepts in the text. The shape is measured by asking subjects to rate the relatedness of all possible pairs of important concepts from the text. This yields a set of proximity measures (i.e., distance measures), which are used to construct a graphic representation or map of the underlying mental structure. The map represents the shape of the reader's memory representation. We compare each reader's structure to the structure intended by the text's author and qualified experts in the subject matter. The result provides a quantitative measure of the goodness-of-fit of the readers' structure to the expert structure. Since this test measures cognitive structures, we will call it the Cognitive Structure Test.

## Experiment 1: Comparing the Mental Representations of Novices, Experts and the Text's Author

We measured the shape of mental representations for a text by collecting proximity data on the important concepts in a naturally occuring 1000 word text on Air Force history (Earhart, 1978).

### Method

The proximity data were collected as follows: (a) Two judges chose the 12 most important concepts from the text. (b) Then all possible pairs of the 12 concepts were constructed -- 66 in all. (c) Since these pairs represent all possible relations between pairs of important concepts in the text, they should capture all the relations between important concepts, and closely approximate a collection of all the important relations in the text. (d) Proximity data were

[1]This work was supported by the Learning Abilities Measurement Project at the Air Force Human Resources Laboratory under AFOSR grant 89-0515. We acknowledge indispensible assistance from Patrick Kyllonen and Abraham Tesser.

of Military Strategy, Military Advisers, and furthest away, Success. Those familiar with this period of history, or who have read the passage, (available from the authors) will recognize this representation.

Experiment 2: Diagnosing and Repairing Mis- and Missing Conceptions by Revising the Text

In this study we successfully used these proximity data to revise the Air Force Vietnam text, greatly improving the goodness-of-fit of novices' mental representations to the author's and experts' representation. For expository effectiveness, we report the results first, and then the revision method. Table 2 shows the results. The entries at the left replicate the results in the two rows of Table 1.

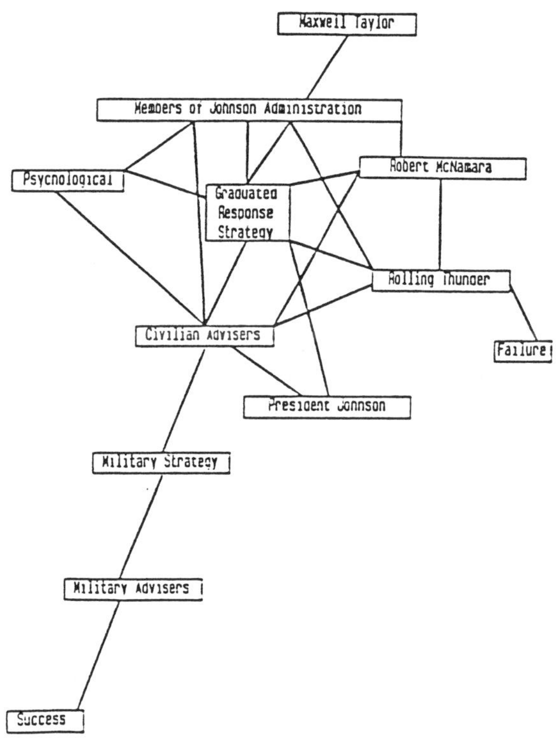


Figure 2 - Experts' Pathfinder Network

The revisions were done in 5 sentences (of 43 sentences in all) of the Original and Principled Revisions. The entries at the right of Table 2 show that these revisions had very large positive effects on the novices' correlations with the author's and experts' structure. These correlations were so close to the reliability of the Cognitive Structure Test that a ceiling effect is probably present. These results show that our techniques can be used successfully to shift novices' mental representations toward experts'.

How did we use the proximity data to decide which revisions to make?

To know what to revise in the text, we needed to know: (a) what novices knew correctly after they read the text; i.e., what correct conceptions they had; but more importantly (b) what they knew that wasn't so; i.e., what misconceptions they had after reading the text; and (c) what they didn't know after reading; i.e., what missing conceptions they had.

Table 2

Correlations of Four Recruit Groups with Author's and Experts' Structures on Cognitive Structure Test

| | Original Version | Principled Revision | Original Version with 5 Revision | Principled Version with 5 Revision |
|---|---|---|---|---|
| Author | .18 | .59* | .80* | .86* |
| Experts | .22 | .62* | .87* | .86* |

*$p < .05$, different from its control

To get this information, we further analyzed the cognitive structure data from Experiment 1. When we looked at the novices' network structure for the Original Version (Figure 3, built from the proximity data using Pathfinder (Schvaneveldt, 1990)) we were unable to interpret it plausibly. So we began to believe that the Original group might contain several subgroups, each with a somewhat different mental representation. If Figure 3 was those different representations superimposed on each other, the result would be a composite of different structures rather than a true representation of the central tendency of the Original group.

So, using factor analysis, we looked for subgroups in each condition who were correlated with each other. Then we took each subgroup's subjects, and looked at their structure. These structures were much more interpretable, and we derived our revisions from them for the present study.

Identifying subgroups by factor analysis. We proceeded as follows: (a) For the 46 subjects who read the Original Version, we took each subject's vector of 66 responses, and correlated it with the vector of each other subject who read the Original Version. This produced a large correlation matrix, showing which subjects were correlated with each other. (b) We then factor analyzed this matrix, using the individual subject as collected by having each subject rate the "relatedness" of each pair of concepts on a seven-point scale, ranging from "very closely related" to "very distantly related" (N=83 Air Force Recruits). This is the Cognitive Structure Test. (e) With proximity data for all pairs of entities, a map can be made of the configuration of relations between the entities, using any of a variety of algorithms; we chose network analysis (Schvaneneldt, 1990).

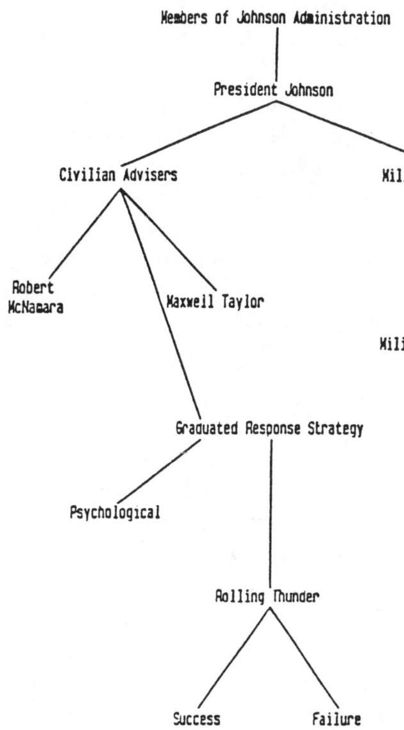


Figure 1 - 12 Terms Used in Test

These proximity data were collected from novices for two different versions of the text: The Original version and a Principled Revision (derived from the Original using the Kintsch and Van Dijk (1978) model, as described in Britton & Gulgoz, (in press) available from the first author). The same Cognitive Structure Test was also given to the text's author (a History Professor at the Air Force Academy), and to seven independent experts on the text's subject matter (an ambassador, three military historians, etc.)

The 12 Terms. To clarify the character of the test, the content of the passage is here described briefly with the terms used in the test arranged graphically in Figure 1 and underlined here. (Figure 1 was not shown to the subjects). The texts began by describing some members of the Johnson Administration, including President Johnson, who had civilian advisers, including Robert McNamara and Maxwell Taylor, as well as military advisers. The military advisers proposed the military strategy, which was to bomb North Vietnam very heavily. The civilian advisers proposed instead the graduated response strategy, which was to bomb North Vietnam a little and then pause to see if that had "broken their will"; if it hadn't, the bombing would be escalated gradually. Since the focus was on breaking the North Vietnamese will, this was described in the passage as a psychological strategy. Johnson chose the graduated response strategy, and the resulting operation was code-named Rolling Thunder. Success and failure could be attributed to the various persons, policies, actions, and consequences in the passage.

Results and Discussion

The recruits' 66 ratings of relatedness were correlated with those of the author and experts. Results showed that : (a) the author's and expert's representations were inter-correlated +.8, indicating they all had similar representations; (b) recruits who read the Original version correlated only +.10 with the author and experts (Table 1, column 1); and (c) recruits who read our Principled Revision correlated above +.5 with the author's and experts' representations, indicating that the Revision had improved the quality of their mental representation (Table 1, column 2), where "improved quality" means closer correspondence to the author's and experts' representation.

Table 1

Correlations of Two Recruit Groups with Author's and Experts' Structures on Cognitive Structure Test

(Author and Experts read Original Version)

| | Original Version | Principled Revision |
|---|---|---|
| Author | .08 | .52* |
| Expert | .10 | .55 |

*p < .05, different from Original

Figure 2 shows the network structure created from the average experts' proximity data by the Schvaneveldt Pathfinder program. It has the Graduated Response Strategy in the center, surrounded by a starlike pattern composed of: (a) the various persons involved in that strategy: President Johnson, Robert McNamara, Maxwell Taylor, Members of the Johnson Administration, and Civilian Advisers; (b) its main characteristic -- that it was Psychological; and (c) its consequence -- Rolling Thunder, which was a Failure. Toward the periphery of the representation are the notions the unit of analysis. The result was that we ended up with 3 subject factors (i.e., each factor was inhabited by a number of subjects). (The Scree test was used to select the number of factors; a standard varimax rotation was then used; subjects who loaded above +.40 on a factor were considered to be inhabitants of that factor.) (c) The subgroup of subjects who loaded above +.40 on any factor were then segregated from the rest of the subjects, into a factor of their own. (d) Then for each factor separately, we averaged the subgroup of subjects' responses for each item, producing an average vector for that factor, 66 items long, representing the cognitive structure characterizing that factor. (e) We also did steps a through d for the 37 subjects who had read the Principled Revision. (f) We now had the mental structure for each subject factor, for six factors in all, including three factors for each text.

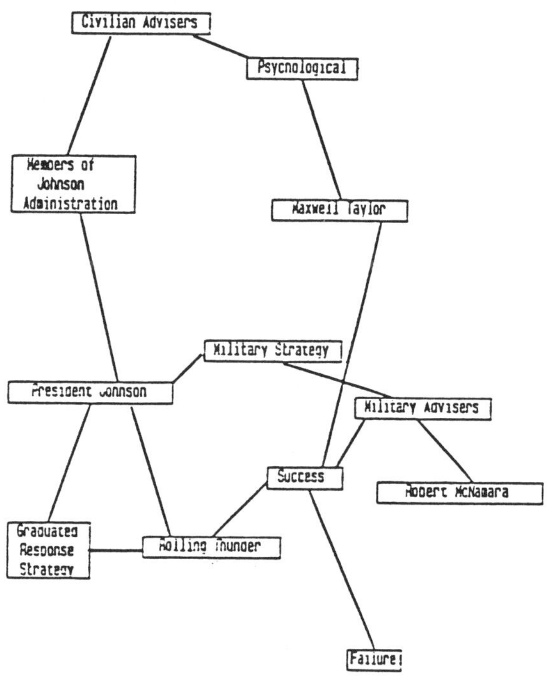


Figure 3 - Original Version Readers' Pathfinder Network

The next question was: What knowledge structures do these factors represent? The first answer was easy: surely some factors must be similar to the author's or experts' structure. To test this, we correlated the string of 66 numbers for each factor with the author's and experts' structure, yielding the results shown in Table 3.

Table 3

Correlations of Factor Scores in Groups with Author and Experts

| | Original Version | | | Principled Revision | | |
|---|---|---|---|---|---|---|
| | Factors | | | Factors | | |
| | 1 | 2 | 3 | 1 | 2 | 3 |
| | n=14 | n=12 | n=9 | n=12 | n=11 | n=6 |
| Author | -.10 | .17 | .47* | .77* | .35* | -.15 |
| Experts | -.10 | .21 | .50* | .77* | .41* | -.18 |

*p < .05, different from r = 0.

It was no surprise that the author's and experts' structure was found as Factor 1 --the largest factor-- in the Principled Revision. But we also found the experts' structure as Factor 3 in the Original Version, albeit as the factor with the smallest number of subjects of the three factors. Moreover, those two factors -- the two correlated with the author and experts -- were also highly correlated with each other (i.e., Original Factor 3 and Principled Factor 1 were correlated +.76 with each other) indicating that both of those factors had approximately the same pieces of the author's and experts' structure.

Identifying misconceptions and missing conceptions. What we needed to know in order properly to revise the text for the subjects in any factor was this: On which concepts did they deviate greatly from the experts' structure? Those concepts would represent misconceptions, or missing conceptions, and these we should specifically address in our revision, in order to repair those novices' faulty representations.

We identified each factor's mis- and missing conceptions by dividing each factor's string of 66 numbers into 12 equal parts. Each part represents the structure for one of the 12 concepts. Figure 4 shows this graphically for one concept on which the subjects in Factor 1 of the Principled Group were very similar to the experts (Graduated Response Strategy), and Figure 5 shows it for one concept on which Factor 3 of the Original Group was dissimilar to the experts (McNamara). (These figures have the 12 terms abbreviated along the side, and the rating scale that subjects used is shown along the bottom. Each plotted point represents the average relatedness rating between the term shown in the title of the figure and the corresponding term along the side. Missing in each figure is the relatedness rating between the term in the figure's title and itself.)

The first two panels in each figure show the pattern separately for the experts and the novices; the third panel shows the patterns superimposed to show the similarity and

Figure 4
Cognitive Structures for "Graduated Response Strategy" in Expert Group and Factor 1 of Principled Revision Group; Superimposition Shows Similar Conceptions

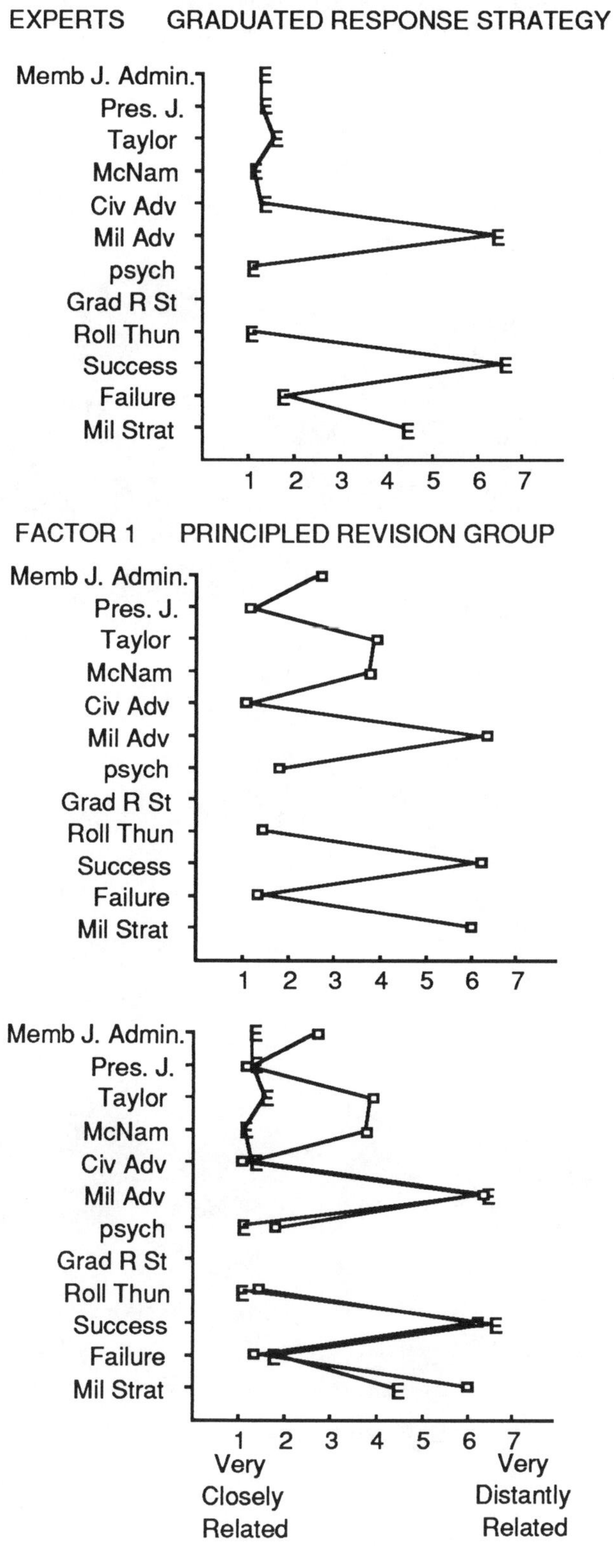


Figure 5
Cognitive Structures for "Robert McNamara" in Expert Group and Factor 3 of Original Group; Superimposition Shows Dissimilar Conceptions

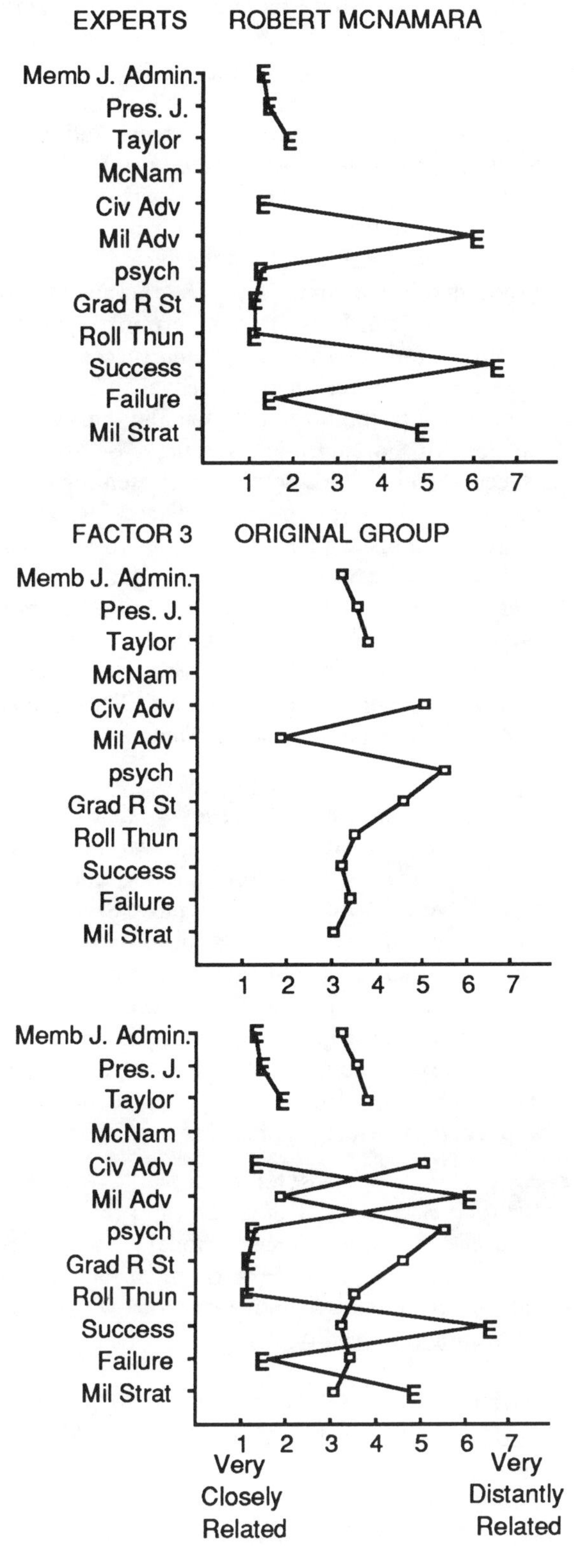

dissimilarity, respectively. The correlation between novices and experts (n=11 terms) for Graduated Response Strategy was +.87, for McNamara it was -.66. Obviously, these subjects have misconceptions about McNamara. Moreover, we can read off the figures what those misconceptions are.

The results of all 12 of these analyses showed that the novices' representations for McNamara and for Taylor were grossly discrepant from the experts'. So for our revisions we chose sentences mentioning those two terms.

Our analysis of all the factors in Table 3 proceeded in the same way. The recruits in two of the factors (one factor in each group) thought the whole operation was a resounding success, perhaps revealing an intrusion of affective and attitudinal factors due to these recruits being in their eleventh day of Air Force Basic Training. So the terms "success" and "failure" were also chosen for revision to help this group of novices with their representations. The final two factors indicated additional misconceptions for "military advisers" and "civilian advisers," and so the sentences involving those terms were also revised accordingly. As it turned out, all these revisions could be done in 5 of the 43 sentences in the text. The results of that study have been shown in Table 2.

## Conclusions

These studies show that novices' representations can be shifted toward experts' by repairing novices' mis- and missing conceptions. This was done by using the Cognitive Stucture Test to identify mis- and missing conceptions. A side benefit is that the problematic conceptions can be specified at a grain size that is the same as the grain size of appropriate revisions, i.e., at the grain size of text sentences that mention the concepts. This technology for improving instructional texts appears to be widely applicable.

## References

Britton, B. K., & Gulgoz, S. (in press). Using Kintsch's computational model to improve instructional text: Effects of repairing inference calls on recall and cognitive structures. Journal of Educational Psychology.

Earhart, R. C. (1978). History 202: Modern warfare and society. United States Air Force Academy.

Kintsch, W., & van Dijk, T. A. (1978). Toward a model of text comprehension and production. Psychological Review, 85, 363-394.

Schvaneveldt, R. W. (Ed.) (1990). Pathfinder associative networks: Studies in knowledge organization. Norwood, NJ: Ablex.

# Toward a Unified Theory of Lexical Error Recovery

**Kurt P. Eiselt**
College of Computing
Georgia Institute of Technology
Atlanta, GA 30332-0280
eiselt@cc.gatech.edu

**Jennifer K. Holbrook**
Department of Psychology
Albion College
Albion, MI 49224
jholbrook@albion.bitnet

### Abstract

The ambiguity inherent in natural language requires us to make many decisions about the meaning of what we hear or read. Yet most studies of natural language understanding have assumed that although language may be ambiguous, we always make the right choice when faced with a decision about ambiguity. Consequently, very little is said about how to recover from incorrect decisions. In this paper we look at two rare examples of investigations into recovering from erroneous decisions in resolving lexical ambiguity. After examining the corresponding theories, we find that what at first appear to be competing theories can in fact be resolved into a unified theory of lexical error recovery based upon a highly parallel architecture for language understanding.

## Why error recovery?

Researchers in natural language understanding have often assumed that although spoken or written text may be ambiguous in many ways, it is not misleading. In other words, it is assumed that when the understander resolves an ambiguity, the choice that is made is always the correct one. Consequently, the question of how an understander can correct its mistakes in interpretation is seldom asked. Yet to ignore the issue of error recovery in natural language understanding is to ignore one of the most vital aspects of the human language understander. As the following example, due to Lashley (1951), illustrates, spoken or written text is often misleading, and humans frequently appear to be able to recover from mistakes in semantic interpretation quite gracefully, without any conscious awareness that a mistake has been made:

**Text 1** Rapid righting with his uninjured hand saved from loss the contents of the capsized canoe.

People will often hear the second word as "writing" and realize their mistake only when they hear "the capsized canoe." This ability to recover from errors serves us well. Ambiguity in natural language allows us to economize in our communications by eliminating much of what could be said or written and relying on the listener or reader to supply the missing knowledge. By doing so, however, we also increase the potential for misunderstanding. The ability to recover from erroneous word sense decisions when resolving ambiguities is an important compensatory mechanism that allows useful communication to continue when the spoken or written text is misleading. Understanding the error recovery capability of the human language processor will aid in the development of more useful language understanding systems, and will lend insight into the architecture of the human language processor itself. In this paper we examine two models of lexical error recovery in text understanding, and show how the two models might be merged into a single unified model of lexical error recovery under a highly parallel architecture for language understanding.

## Error recovery without text reprocessing

As Text 1 above clearly indicates, it is not necessary to reprocess misleading input in order to recover from erroneous decisions. When Text 1 is presented aurally, the listener who hears "writing" instead of "righting" will be unable to reprocess earlier text when the mistake is later revealed after hearing "the capsized canoe." Because the memory of the verbatim text is available for only a short period of time (e.g., only until a clause boundary, according to Jarvella (1970)), some mechanism other than reprocessing must account for the listener's ability to recover from this mistake.

A theory of lexical access and disambiguation, called *conditional retention*, accounts for a human understander's ability to recover from an incorrect choice of word meaning without reprocessing the text (Granger, Holbrook, & Eiselt, 1984; Holbrook et al., 1988). According to this theory, all meanings of an ambiguous word are retrieved, the meaning most appropriate to the preceding context is chosen, and the other, less appropriate meanings are temporarily deactivated but retained. If later text contradicts the initially chosen meaning, the retained meanings are reconsidered in light of the updated context and a new meaning selected.

The combined deactivation and retention of the meanings not chosen accomplishes two goals for the language understander. First, it permits the processing at the lexical level to continue to make immediate decisions about the meanings of subsequent words in the context of a single, plausible interpretation of the preceding text instead of multiple interpretations of varying plausibility. If the unchosen meanings were not completely deactivated, there could be resulting confusion in making decisions about new word meanings. Second, it allows the retained decisions to be used by other processes in correcting wrong decisions made by the original process without reprocessing the original input text, at least for a short time. Tracking retained meanings allows the error recovery to be done without maintaining separate copies of all possible interpretations of the text processed so far, thus reducing both storage and processing overhead.

One argument against conditional retention is that several cross-modal lexical priming (CMLP) experiments have shown that, within 200 msec after a meaning of an ambiguous word has been selected, the alternate meanings are as inactive as unrelated concepts that are used for comparison—they are as inactive as if they had never been activated in the first place (Onifer & Swinney, 1981; Seidenberg et al., 1982; Swinney, 1979; Tanenhaus, Leiman, & Seidenberg, 1979). This model of lexical access and disambiguation is called *active suppression*. There are two problems with using this body of data to argue in favor of the active suppression model over conditional retention. The first is that most of these studies used texts that ended with the ambiguous word (cf. Onifer & Swinney, 1981).[1] For such texts, then, meaning selection will always occur at the end of the text when there is no chance of disconfirming information. Under this condition, the conditional retention theory also holds that unselected meanings will be forgotten. But these experiments do not test what happens when the ambiguous word is embedded in the text, with more text to follow the ambiguous word. Hudson and Tanenhaus (1984) found that meanings were available longer when preceded by neutral context and embedded within the text. Holbrook (1989) has found evidence that unselected meanings are retained when the ambiguous word is embedded in a longer text. Thus, the CMLP studies that best support active suppression alone did not include materials that would test for conditional retention under appropriate conditions. The second problem is that while the CMLP studies that support active suppression do in fact demonstrate that unselected word meanings are very quickly deactivated at the end of a text, these experiments do not thoroughly test whether the meanings have been completely forgotten. They use only one type of measure (reaction time) on two closely-related tasks (lexical decision and naming).

The experiments above do not address the question of how recently-deactivated word meanings might differ from those which have been inactive for a much longer duration. This difference might appear as different degrees of sensitivity to re-activation while processing additional input, or some other quality that does not correspond to the relative activation level and is therefore immeasurable in CMLP studies. Furthermore, lexical decision tasks and lexical naming tasks are good for acquiring data about subjects' response times, which in turn correlate to the degree of facilitation of the individual word meanings. However, the conditional retention theory suggests that retention may not show up as facilitation because retained meanings have been de-activated. This suggests the need for an experimental methodology other than the lexical decision or lexical naming task.

[1]Although Onifer and Swinney (1981) used texts that continued for several words after the ambiguous word, their materials were not as carefully controlled as has become common; specifically, meaning frequency, number of syllables of probes, and similar issues could have erased effects of meaning retention.

## New experimental evidence

We designed a study which uses a binary forced-choice task to test for conditional retention of unselected meanings across sentential boundaries. The forced-choice task offered the ability to detect indirectly the existence of conditional retention by studying the subjects' decisions instead of their response times. The use of texts with sentences following the ambiguous word allowed us to test whether suppression of unselected word meanings always occurs at the end of a sentence, or whether meanings can be retained when information with the potential to change the representation is at hand.

In this experiment, subjects were asked to read short texts of one to three sentences in length. Experimental texts were two sentences in length, and the others were filler texts. Filler texts were of varying length in order to ensure that subjects would not know whether there was more information to be added to the representation after their word choice. This provided the appropriate conditions under which retention is thought to occur. There were two types of experimental texts that are important to the current discussion:

- **Consistent bias surrounds ambiguous word:** The context preceding the ambiguity biases towards one meaning but does not preclude the other. The context following the ambiguity provides information requiring the original biased-for meaning.

  **Example:** Mary realized that she had examined the wrong bat. She took it back and got one that was aluminum.

- **Conflicting bias surrounds ambiguous word:** The context preceding the ambiguity biases towards one meaning but does not preclude the other. The context following the ambiguity provides information consistent with the early context but requiring the unselected meaning.

  **Example:** Mary realized that she had examined the wrong bat. She took it back and got one that was male.

The texts were presented on a computer monitor a few words at a time, with each group of words replacing the group before it. An information probe was displayed on the monitor at one of two points in the text: either between the ambiguous word and the disambiguating text or after the disambiguating text. The probe consisted of a pair of words, and the subject's task was to decide which of the two words was more related to the text. The choice was indicated by pressing one of two buttons, each corresponding to one of the two probe words. An example of materials presentation is shown below:

Mary realized
that she had examined
the wrong bat. She took
it back and got
CAVE    PITCH
one that was
male.

The first line would appear, centered on the monitor, for 640 msec. The second line would then replace the first line for

| consistent bias surrounds (n=18) | Cell 1 | | Cell 2 | | Cell 3 | |
|---|---|---|---|---|---|---|
| | correct word | incorrect word | correct word | unrelated word | incorrect word | unrelated word |
| CR | 75 | 25 | 100 | 0 | 100 | 0 |
| AS | 100 | 0 | 100 | 0 | 50 | 50 |
| data | 61 | 39 | 89 | 11 | 94 | 6 |
| **conflicting bias surrounds (n=27)** | **Cell 4** | | **Cell 5** | | **Cell 6** | |
| | correct word | incorrect word | correct word | unrelated word | incorrect word | unrelated word |
| CR | 25 | 75 | 100 | 0 | 100 | 0 |
| AS | 0 | 100 | 50 | 50 | 100 | 0 |
| data | 33 | 67 | 70 | 30 | 85 | 15 |

Table 1: Predictions and results at early probe point during forced choice task.

640 msec, and so on. When the two capitalized probe words appeared on the screen, the subject would press a designated key on the left side of the computer keyboard if he or she thought the word on the left was more appropriate to the text than the word on the right, or would press a designated key on the right side of the keyboard if the word on the right was thought to be more appropriate. Presentation of additional text did not continue until the subject pressed one of the two keys.

An early probe point and a late probe point were used for each text in a between-subjects design. The early probe point occurred some time after the ambiguous word but before the disambiguating information in the second sentence. The late probe point occurred at the end of the second sentence, after the disambiguating information had been presented. (For the sake of brevity, we will discuss only the results obtained at the early probe point in detail.) The probe word pairs were rotated between three types of words: a word related to the meaning of the ambiguous word that was correct at the end of the text, a word related to the incorrect meaning at the end of the text, and a word unrelated to either meaning or to the text as a whole. In the example above, the three words used were "PITCH," "CAVE," and "PAIR." "PITCH" is related to the baseball meaning of "bat," and is more appropriate at the early probe point.[2] "CAVE" is related to the animal meaning of "bat," and is more appropriate to the text at the late probe point. "PAIR" is unrelated to either meaning of the ambiguous word or the text.

In this experiment, the theories of conditional retention and active suppression are set up as opposing theories, so it is useful to compare the two theories' predictions of the outcomes for the different conditions of this experiment. The predictions of the subjects' responses at the early probe point, and the data gathered at this point are summarized in Table 1. This table refers to three types of word stimuli. **Correct** word stimuli are words related to the meaning of the ambiguous word that could be integrated with context at the conclusion of the text. **Incorrect** word stimuli are words related to the meaning of the ambiguous word that could not be integrated with context at the conclusion of the text. **Unrelated** word stimuli are words that were unrelated to either meaning of the ambiguous word at any point in the text. The tables give predicted choices as percentages of the total number of responses.

The first type of text reported in Table 1, called "consistent bias surrounds," was designed so that the context which occurs before the ambiguous word is encountered biases towards one meaning of the ambiguous word. Thus, the reader will have enough information from the text on which to base a decision, and will choose the meaning which is more related to the previous context. The context which follows the ambiguous words for these texts agrees with the context that precedes the ambiguous words, so the meaning choice that was made remains correct throughout the text.

Conditional retention (CR) predicts that the unselected meaning of the ambiguous word will be retained throughout the text, and that this will cause interference in the forced-choice task between the correct and incorrect probe words (Cell 1). The simplest model of this interference is that it will be reflected in 25% of the responses (the intermediate point between a prediction of no interference, which would be reflected in 0% of the choices, and complete interference, which would be reflected in 50% of the choices). However, when the correct word is paired with the unrelated word (Cell 2), there is no reason to select the unrelated word over the correct word because it is not being retained. Therefore, the correct

[2]The bias of the first sentence of each text was guaranteed with two groups of informants. The first group, naive judges, read and paraphrased each sentence. Those sentences which were paraphrased to reflect one meaning in 100% of the paraphrases were read by a second group of informants. These judges were informed that each sentence contained an ambiguous word. Each sentence read by this second group was followed by the two target words that were semantically related to the ambiguous word at the end of the sentence. The informants were asked to choose the word more related to the meaning of the sentence, or to indicate that both were equally related. A text was used in the experiment if the intended target word was chosen more than 80% of the time and was never judged as related to the other target word.

word should always be chosen in this condition. When the incorrect word is paired with an unrelated word (Cell 3), the incorrect word will always be chosen by virtue of its relationship to the retained meaning of the ambiguous word.

Active suppression (AS) makes a different set of predictions. When a meaning for the ambiguous word is selected, the unselected meaning will be actively suppressed. At the point of the forced-choice task, the suppressed meaning should have no effect on the word chosen in the task. Thus, in Cell 1, active suppression predicts that the correct word will be chosen 100% of the time. In Cell 2, the subject again will always choose the correct word. In Cell 3, there is no reason to suppose that the incorrect word is chosen with more probability than the unrelated word. The conditional retention theory is supported by the results for all three conditions, while the active suppression theory is supported only by the results of Cell 2.

The other type of text reported in Table 1, called "conflicting bias surrounds," was designed so that the context which occurs before the ambiguous word is encountered biases towards one meaning of the ambiguous word. During the meaning decision process for the ambiguous word, the meaning that is a better fit with the previous context will be chosen. The context which follows the ambiguous words for these texts disagrees with the context that precedes the ambiguous words, so the meaning that is contextually appropriate at the early probe point will be inappropriate at the end of the text. Correspondingly, the probe word that is correct at the end of the text in in "consistent bias" texts is incorrect at the end of "conflicting bias" texts, and the probe word that is incorrect in the former case is correct in the latter.

In Cell 4, conditional retention and active suppression both predict a significant effect of context type and target type. Although significance did not obtain, a binomial probability test supported the conditional retention theory. Conditional retention predicted this difference in Cell 5, but active suppression did not. In Cell 6, both theories predicted the finding. Conditional retention is again supported by all three conditions, whereas active suppression is only supported by one condition.

Overall, the weight of the evidence from this experiment clearly supports the conditional retention theory. The two theories predicted different results in four of the six different test conditions shown in Table 1: Cells 1, 3, 4, and 5. Analysis of the data showed that the conditional retention theory predicted the results better than the active suppression theory did in all four cells. An analysis of the data obtained with a late probe point also supported the conditional retention theory (Eiselt, 1989; Holbrook, 1989). Holbrook (1989) performed two additional related experiments which also support the conditional retention theory over active suppression. Still further support comes from another experiment which used a divided visual field methodology to find that the time course of the activation and suppression of two meanings of an ambiguous word followed significantly different paths in the different hemispheres of the brain (Burgess & Simpson, 1988). When the ambiguous word was presented only to the right visual field, and therefore to the left hemisphere of the brain, the more likely meaning was activated while the less likely meaning was suppressed, much like active suppression. When the ambiguous word was presented to the left visual field, however, the right hemisphere kept the less likely meaning active and suppressed the more likely meaning, indicating that a form of retention was taking place.

## Error recovery with text reprocessing

The evidence offered above strongly indicates that the human language processor does have a mechanism which would enable recovery from errors in lexical ambiguity resolution without text reprocessing. Yet common sense tells us that error recovery by backtracking and reprocessing is also a viable option. In fact, studies of human readers' eye fixations during the reading of misleading texts indicate that one method used by readers for recovery from lexical decision errors is simple reprocessing of the text. The following passage was used in an experiment by Carpenter and Daneman (1981, p. 137):

**Text 2** The young man turned his back on the rock concert stage and looked across the resort lake. Tomorrow was the annual one-day fishing contest and fishermen would invade the place. Some of the best bass guitarists in the country would come to this spot. The usual routine of the fishing resort would be disrupted by the festivities.

Subjects in this experiment were asked to read passages such as the one above while the duration and location of their eye fixations were automatically recorded. In the example above, most readers initially interpreted the word "bass" as a kind of fish because the preceding text is biased toward this interpretation. The interpretation is contradicted, however, by the next word, "guitarists," which forces a reinterpretation of "bass" as a low-frequency musical note. Carpenter and Daneman found that most readers' eyes fixated on "bass," then moved forward to and fixated on "guitarists," then *regressed* back to "bass," moved forward to "guitarists" again and continued reading the remainder of the passage.

The reprocessing heuristic is just one of several error recovery heuristics proposed by Carpenter and Daneman. Another heuristic involves making a larger-than-normal inference encompassing both the inconsistent concept with the preceding text. This is done, they say, if the contradiction is only "mildly semantically inconsistent" and does not involve a syntactic inconsistency (Carpenter & Daneman, 1981, p. 141). Still another error recovery heuristic is to continue reading the text with the expectation that later information will resolve the inconsistency.

In addition to permitting error recovery without reprocessing, Carpenter and Daneman's theory also demonstrates the need for retention of some sort. Their proposed heuristic of checking the previous text for words that caused processing difficulties, such as ambiguous words, does not specifically address how the reader knows which word or words to reread, but Carpenter and Daneman theorize that difficulties encountered during processing may leave a memory trace which makes finding the ambiguous word much easier. The Carpenter and Daneman model follows the premise that the activation levels of those concepts not selected for use in the interpretation either decay or are actively dampened to a base level, which would preclude the possibility that the memory trace is represented as activation, so this model strongly suggests a retention mechanism which is related to but distinct from the activation mechanism.

## Integrating the models

We have seen experimental evidence for two different theories of recovery from erroneous word sense choices, one relying on reprocessing of the text, and the other relying on conditional retention. If we believe the evidence, then both approaches to error recovery are used by the human language understander, although perhaps in different situations. This apparent dichotomy gives rise to the following question: how does the language understander know the difference between these two situations? That is, how does the reader know when to backtrack and when to use retention to recover from an erroneous lexical decision?

### Characterizing the differences

Consider again the two experiments discussed above. Target texts used by Carpenter and Daneman were multi-sentence passages containing an ambiguous word that had two different pronunciations and a different word sense associated with each pronunciation. The passages were constructed so that the initial context more strongly primed for one meaning of the ambiguous word. In half of the passages, information that was inconsistent with the primed meaning followed the homograph, while in the other half of the passages, the information following the ambiguous word was consistent with the primed meaning. A single passage was presented at once, thereby allowing the reader to backtrack as necessary. This construction is illustrated by Text 2, presented earlier. Of particular interest here is how closely the disambiguating information follows the ambiguity. In Text 2, the disambiguating information is the word immediately following the ambiguity. A survey of Carpenter and Daneman's target texts reveals that the disambiguating information was seldom separated from the ambiguity by more than two words, it was always in the same sentence as the ambiguity, and it was usually in the same clause as the ambiguity.

On the other hand, the target texts used in our experiment consisted of two sentences. The ambiguous word was always the last word of the first sentence, and the disambiguating information was contained in the second sentence. The text was presented to the subject in parts, so that only a few words of the entire text were available at any one time. In addition, the presentation was constructed so that there was at least one line of text presented on the monitor after the line containing the ambiguous word and before the line containing the disambiguating information. In other words, there was always at least a 640 msec delay between the reading of the ambiguity and the reading of the disambiguating information—long after active suppression of less appropriate word meanings should have taken place.

### Resolving the differences

The essential difference between the two experiments, at least for the purposes of this discussion, can be summarized as follows: Carpenter and Daneman's experiment establishes a situation conducive to error recovery by reprocessing, and subjects demonstrated reprocessing behavior. Our experiment prevented subjects from backtracking, and subjects exhibited behavior consistent with conditional retention. In either case, *the subjects appeared to make the best use of whatever information was available to them to interpret the target text correctly*.

As noted above, Carpenter and Daneman propose a suite of error recovery heuristics, of which reprocessing is just one example. We propose, however, that the two different error recovery techniques described herein, reprocessing and retention, can be explained by a single mechanism. Recall that Carpenter and Daneman suggest that processing difficulties, such as word sense ambiguities, leave a memory trace to the ambiguous word which is independent of the activation levels of the word senses themselves. This is supported by their observations of subjects' eye movements which indicate that subjects did not simply reread the text; instead, the eye movements indicate that the subjects searched selectively for the source of the ambiguity.

It is not entirely clear that this memory trace or pointer to the ambiguous word is different from the retained but unchosen meanings of that ambiguous word. However, because research indicates that memory for verbatim text is limited by clause boundaries (Jarvella, 1970), and Carpenter and Daneman's experiment tested only very small separations between the ambiguous word and the following disambiguating information, we can assume that the pointer to the ambiguous word has a relatively brief life span. In contrast, conditional retention effects persist across clause boundaries, so we assume that the pointer to the ambiguous word exists independently of the retained meanings (although this remains to be explored).

Following this assumption, we propose that a reader's error recovery process works as follows:

1. Upon reading an ambiguous word, the reader selects the context-appropriate meaning, deactivates but retains the unselected meanings, and retains a pointer to the source of the ambiguity.
2. If disambiguating information which conflicts with the chosen meaning follows the ambiguity, the reader attempts to reinterpret the text by first following the memory trace back to the source of the ambiguity and making a new choice of meaning based on the additional contextual information.
3. If the memory trace no longer exists, either because of a clause boundary or because the actual text is no longer available, the reader will then attempt to reinterpret the text in light of the additional context by re-evaluating the retained meanings, although the actual word which gives rise to those different meanings is no longer known. A computational model of such a mechanism, called ATLAST, has been implemented successfully using a marker passing (or spreading activation) architecture to retrieve and evaluate

the alternate meanings in parallel. This work is described in greater detail elsewhere (Eiselt, 1987; 1989).

4. The process as described so far is essentially automatic or unconscious. However, if the retained meanings are no longer available, perhaps because of time or working memory limitations, the reader will not be able to resolve the contradiction without devoting attentional cognitive resources to the problem (i.e., conscious and unselective rereading of the text).

We are currently in the process of revising ATLAST along these lines, in order to test the plausibility of our theory of lexical error recovery. ATLAST's parallel marker passing architecture is well suited to this task, making the revisions fairly straightforward. Adapting a more traditional serial language processing architecture to this task would be far more problematic.[3] In addition, we are designing an experiment to test the validity of this theory. Briefly, this experiment will involve using Carpenter and Daneman's target texts in the forced-choice paradigm we employed in our experiment described above. In this new paradigm, for example, the reader of a text like Text 2 would be unable to look back at the ambiguous word "bass" after reading "guitarists." If our theory is correct, the reader should still be able to arrive at a new correct interpretation of the text, even though the option of reprocessing the text has been eliminated, because of conditional retention.

## Conclusion

There has been very little study of lexical error recovery, and the two theories of lexical error recovery herein at first glance seem to be contradictory rather than complementary. We have demonstrated in this paper, however, that it is possible to combine these two theories into a simpler unified theory of lexical error recovery. This unified theory of lexical error recovery is consistent with a highly parallel model of human language understanding.

## Acknowledgments

This research was supported in part by the National Science Foundation under grant IST-85-12419.

## References

Burgess, C., & Simpson, G.B. 1988. Neuropsychology of lexical ambiguity resolution: The contribution of divided visual field studies. In S.L. Small, G.W. Cottrell, & M.K. Tanenhaus (Eds.), *Lexical ambiguity resolution: Perspectives from psycholinguistics, neuropsychology, and artificial intelligence* (pp. 411-430). San Mateo, CA: Morgan Kaufmann.

Carpenter, P.A., & Daneman, M. 1981. Lexical retrieval and error recovery in reading: A model based on eye fixations. *Journal of Verbal Learning and Verbal Behavior, 20*, 137-160.

Eiselt, K.P. 1987. A parallel-process model of on-line inference processing. *Proceedings of the Ninth International Joint Conference on Artificial Intelligence*, 863-869.

Eiselt, K.P. 1989. *Inference processing and error recovery in sentence understanding* (Technical Report 89-24). Doctoral dissertation. Irvine: University of California, Department of Information and Computer Science.

Eiselt, K.P., & Granger, R.H., Jr. Process independence and concurrency in a model of sentence understanding. In U. Hahn and G. Adriaens (Eds.), *Parallel natural language processing*. Norwood, NJ: Ablex Publishing. Forthcoming.

Granger, R.H., Holbrook, J.K., & Eiselt, K.P. 1984. Interaction effects between word-level and text-level inferences: On-line processing of ambiguous words in context. *Proceedings of the Sixth Annual Conference of the Cognitive Science Society*, 172-178.

Holbrook, J.K. 1989. *Studies of inference retention in lexical ambiguity resolution*. Unpublished doctoral dissertation. Irvine: University of California, School of Social Sciences.

Holbrook, J.K., Eiselt, K.P., Granger, R.H., Jr., & Matthei, E.H. 1988. (Almost) never letting go: Inference retention during text understanding. In S.L. Small, G.W. Cottrell, & M.K. Tanenhaus (Eds.), *Lexical ambiguity resolution: Perspectives from psycholinguistics, neuropsychology, and artificial intelligence* (pp. 383-409). San Mateo, CA: Morgan Kaufmann.

Hudson, S.B., & Tanenhaus, M.K. 1984. Ambiguity resolution in the absence of contextual bias. *Proceedings of the Sixth Annual Conference of the Cognitive Science Society*, 188-192.

Jarvella, R.J. 1970. Effects of syntax on running memory span for connected discourse. *Psychonomic Science, 19*, 235-236.

Lashley, K.S. 1951. The problem of serial order in behavior. In L.A. Jeffress (Ed.), *Cerebral mechanisms in behavior: The Hixon symposium* (pp. 112-136). New York: Hafner.

Onifer, W., & Swinney, D.A. 1981. Accessing lexical ambiguities during sentence comprehension: Effects of frequency of meaning and contextual bias. *Memory and Cognition, 9*, 225-236.

Seidenberg, M.S., Tanenhaus, M.K., Leiman, J.M., & Bienkowski, M. 1982. Automatic access of the meanings of ambiguous words in context: Some limitations of knowledge-based processing. *Cognitive Psychology, 14*, 489-537.

Swinney, D.A. 1979. Lexical access during sentence comprehension: (Re)consideration of context effects. *Journal of Verbal Learning and Verbal Behavior, 18*, 645-660.

Tanenhaus, M., Leiman, J., & Seidenberg, M. 1979. Evidence for multiple stages in processing of ambiguous words in syntactic contexts. *Journal of Verbal Learning and Verbal Behavior, 18*, 427-440.

[3] An in-depth comparison of serial and parallel architectures for natural language understanding is given by Eiselt and Granger (forthcoming).

# How Students Misunderstand Definitions: Some Evidence for Contextual Representations of Word Meanings

**Ute Fischer**
Vassar College
*Program in Cognitive Science*
Poughkeepsie, N.Y. 12601

**Abstract**

This study is concerned with the issue of whether word meanings are mentally represented in a decontextualized form, similar to dictionary definitions. If this assumption is correct then students should understand the meaning of an unfamiliar word when they read its definition. To test this hypothesis, German high school students were given unfamiliar English words and their monolingual English dictionary entries. Students used each target word in an English sentence, and then translated their sentences into German. The translations permitted the assessment of comprehension and the specification of its underlying components. The results indicate that students often did not understand the meaning of an unfamiliar word even though they did "understand" its definition. Information that specified in which contexts an entry word and its definition are synonymous promoted comprehension. Meaning representations are therefore best conceived of as contextual representations.

What do people "know" when they know the meaning of a word? One intuitively appealing answer calls upon dictionary definitions since dictionaries are believed to be the authority in matters of meaning. There you will find everything you need to know about the meanings of words. Conversely, this view implies that knowing the meaning of a word consists of knowing its definition. This position is certainly not new; rather it is the lexical twin of the classical account of concepts. Qualms about its validity are not new either. Critics have questioned the classical claim that concepts can be specified in terms of necessary and sufficient conditions (e.g., Rosch 1975). And what is criticized about concepts in general, also applies to lexical concepts. Other opponents have attacked the classical presupposition that there are primitive, unanalyzable lexical concepts that define more complex ones (Fodor et al. 1980). In this paper, I will take a somewhat different perspective. The focus of my argument will be on the assumption that definitions are models of people's mental lexicon, and I want to argue that this position is untenable.

When lexicographers define a word, they extract its meaning from a corpus of actual usages. The resulting definition, though derived from context, represents the meaning of the word in an abstract, context-independent form. Dictionary definitions presumably reflect what a word means, however, they do not state in which contexts you can use it. Knowing how to use a word appropriately, on the other hand, is exactly what we expect of a competent speaker of a language (Miller 1986). To illustrate this claim, consider the definition of the word *bachelor*. The definition tells you that *bachelor* means *unmarried man,* but it does not prevent you from producing non-idiomatic usages, such as *I planted a bachelor.* Of course, you wouldn't think of using *bachelor* in this context because you know that you can't use *man* as an argument of *to plant.* You could certainly claim that if you didn't know how to use *man*, you should look up its definition. But, as Fodor et al. (1980) have pointed out, "definitions have to stop somewhere," or, in terms of the issue at hand: definitions must provide context at some point. But do they? Dictionary definitions, instead, seem to trust the word knowledge that their readers bring to the task. If this evaluation correctly characterizes the state of affairs, then anyone who advocates definitions, i.e., decontextualized meaning representations, as a model of people's mental lexicon faces a serious dilemma.

The notion of decontextualized meaning representations has been widely adopted by psychologists interested in semantic development or vocabulary learning. In the developmental literature, for example, context-independent meanings are viewed as the end point of development (e.g., Keil & Carroll 1980; Keil & Batterman 1984). Learning the meanings of words is thus believed to result in acquiring their definitions. A similar argument is made by Sternberg who investigates the acquisition of word meanings from written context (Sternberg 1987; Sternberg & Powell 1983). Readers are assumed to formulate tentative definitions of unfamiliar words based on contextual information.

Psychologists, it seems, view learners as lay lexicographers, or vice versa, lexicographers are seen as having turned into a craft what people normally do. Whatever comparison one might favor, the same conclusions follow: Dictionary definitions reflect what people know when they know the meaning of a word, and they make easily accessible to learners what they would otherwise have to derive from numerous contexts. This line of reasoning also contributes, to a large extent, to the dominant role that dictionaries play in vocabulary instruction. But do dictionaries, in fact, provide students with the information they need in order to understand the meaning of a word?

Research on the use of dictionaries in vocabulary instruction has primarily focused on the efficiency of definitions in comparison to other teaching methods. It was found that students frequently failed to learn novel words from definitions (Johnson & Stratton 1966; Gipe 1979). The results, however, are inconclusive as to why students who had received definitions performed so poorly. Since students' vocabulary knowledge was assessed several days after the study period, comprehension and retention are confounded variables. It is thus unclear whether students could simply not remember the meaning of a target word, or whether its definition was inadequate for conveying its meaning in the first place. The fact that neither Johnson & Stratton (1966) nor Gipe (1979) addressed this issue suggests that they took the adequacy of definitions for granted.

Research by Miller & Gildea (1985), on the other hand, challenges the belief that students understand the meaning of a word when they read its definition. Ten-year-old American students were asked to look up unfamiliar words in a dictionary, and to compose a sentence using them. Half of the sentences that the students wrote were idiomatically unacceptable English sentences. This suggests that the students often did not achieve an adequate understanding of the unfamiliar words. Gross et al. (1989) found that many of the students' misuses of the novel words could be explained in terms of a substitution strategy. This strategy is thought to involve the following three steps: (1) Students take a familiar piece of the definition, (2) compose a sentence with it, and then (3) substitute the entry word for the familiar piece in their sentence. It is important to note that substitution is, in principle, an appropriate strategy for comprehending definitions. It is in line with lexicographical practice (Landau 1984), as well as the definition of synonymy (Lyons 1968). The students in the Miller & Gildea (1985) study frequently failed to apply the substitution strategy successfully because the definitions did not make sufficiently clear in which contexts an entry word and its definition were synonymous. What else could a student do when she reads that *erode* means *to eat out* but to assume that *Our family erodes a lot* is an appropriate usage of the word?

## Experiment

The research by Miller & Gildea (1985) suggests that students should be more likely to understand the meanings of unfamiliar words when their definitions include explicit contextual information. This hypothesis was tested in the present study by comparing the effectiveness of two British dictionaries. One, the *Collins Cobuild English Language Dictionary* (CCD), provides contextual information within definitions in addition to illustrative phrases, whereas the second dictionary, the *Oxford Advanced Learner's Dictionary of Contemporary English* (OALD) seldom does. The OALD is also inconsistent in including illustrative phrases in its entries. As an illustration of how the dictionaries typically define words, consider the following entries:

**CCD: accrue, accrues, accruing, accrued**
1. If money or interest accrues, it gradually increases in amount over a period of time. EG...*$100,000 plus accrued interest at 8%...tax benefits accruing to owner occupiers.*

**OALD: accrue ~ (to sb) (from sth)**
come as a natural growth or development. *If you keep your money in the Savings Bank, interest ~s. An ~ed interest is due, but not yet paid or received.*

As can be seen, the CCD but not the OALD definition, specifies which arguments (*money, interest*) the verb *accrue* can take. Students reading the CCD definition should therefore understand the meaning of *accrue*, and accordingly use it in appropriate contexts. Students who see the OALD definition, on the other hand, should encounter problems similar to those of Miller and Gildea's subjects. It was hypothesized that students who receive CCD-definitions will in general perform better than students who see OALD-definitions.

The second goal of the present study concerns the psychological reality of the substitution strategy. In the analysis by Gross et al. (1989), substitution was assumed to have occurred whenever it was possible to substitute part of the definition for the target word in a student's sentence to yield an idiomatically acceptable sentence. Even though substitution seems plausible, there was no independent evidence aside from the English sentences that the students had actually used this strategy. The present study attempts to test the hypothesis that students commonly employ a substitution strategy. In order to do so, German high school students who were learning English participated as subjects in a task similar to the one used by Miller & Gildea (1985). In addition to the English sentence production task, the students were also asked to translate their sentences into German. If substitution had occurred, then part of the German sentence should be a translation of the definition. German was thus used as a metalanguage to shed light on whether and how students understand definitions.

### Method and procedure

**Subjects:** One-hundred and thirty-four German high school students participated in the study. Subjects had seven or eight years of formal instruction in English. All subjects were native speakers of German, and were between 17 and 20 years old.

**Stimulus materials:** Sixty-three target words

(12 nouns, 18 verbs, 33 adjectives) were used in the study. Fifty-seven of the 63 words were selected from the Miller & Gildea (1985) study. The study employed entries from two dictionaries. The *Oxford Advanced Learner's Dictionary of Contemporary English* (OALD), and the *Collins Cobuild English Language Dictionary* (CCD). The 63 target words were assigned to three questionnaires. Words of the same part of speech were randomly distributed across the questionnaires, so that each included four nouns, six verbs, and eleven adjectives. Since there were both OALD and CCD entries for each word, six questionnaires were obtained: three OALD and three corresponding CCD questionnaires. Each page of a questionnaire stated a target word and its corresponding OALD or CCD entry. The order of the target words in the questionnaires was randomly varied. Every student in the study received one questionnaire.

**Procedure:** Students from eight different classes participated in the study. Each test session lasted 2 hours and consisted of three successive parts: (1) synonym test to assess familiarity with the target words; (2) dictionary test; (3) synonym re-test to evaluate vocabulary learning.[1] After the first synonym test, subjects received either an OALD or a CCD questionnaire. They were asked to compose an English sentence using a given target word and to complete the questionnaire in the given order. Since the students were only familiar with the OALD but not with the CCD, they were supplied with a hand-out explaining the abbreviations and symbols used in the CCD entries. After about an hour, subjects were interrupted. They were asked to return to their first English sentence, and to write the German translation for each sentence they had composed. Prior to this point, no mention had been made of translating the sentences.

## Coding of the data

**Coding of the English sentences:** Two native speakers of English independently rated whether a target word had been used in an idiomatically correct, questionable, or incorrect manner. Disagreements were resolved through discussion.

**Coding of the German translations:** A native speaker of German evaluated how well students' translations matched a standard; i.e., the German equivalent of a target as stated in bilingual dictionaries (Der kleine Muret-Sanders: Englisch-Deutsch and Collins German Dictionary: German-English, English-German). It was judged whether a translation was a *match* (translation was equivalent to the standard), a *near-match* (translation and standard were hypernyms, or one was a hyponym of the other), a *far-match*(meaning of standard entailed meaning of translation, or vice versa), or a *no-match* (translation was not semantically related to the standard). A monolingual German dictionary (Deutsches Universalwörterbuch) and a dictionary of synonyms (Duden sinn- und sachverwandter Wörter) were used to determine the adequacy of students' translations. The same rater also coded the strategies students seemed to have employed in the sentence production task (see below for a detailed description of the strategies). A second rater, also a native speaker of German, was asked to check the judgments of the first rater. Disagreements were resolved through discussion.

[1] Details about the synonym test can be found in Fischer (1990) and will not be described here since they are not relevant to the focus of the paper

## Results and Discussion

Since the results of the pre-test showed that most of the students were unfamiliar with the vast majority of the words, all data were included in the analysis. Overall 5628 responses, including no-responses, were obtained.[2]

**Adequacy of usage and comprehension of target words:** As can be seen in Tables 1 and 2, OALD-students performed similarly to students in the CCD-group.

| Adequacy | OALD | CCD |
|---|---|---|
| *Omissions* | 35.3 | 36.6 |
| *Incorrect* | 22.5 | 18.0 |
| *Questionable* | 3.9 | 2.9 |
| *Correct* | 38.3 | 42.6 |

**Table 1: Adequacy of usage of target words (in percent)**

| Adequacy | OALD | CCD |
|---|---|---|
| *Omissions* | 36.9 | 37.9 |
| *No-Match* | 26.4 | 22.9 |
| *Far-Match* | 3.7 | 5.3 |
| *Near-Match* | 5.4 | 5.9 |
| *Match* | 27.5 | 28.0 |

**Table 2: Adequacy of translations of target words (in percent)**

An Analysis of Variance was computed on the number of idiomatically correct English usages per subject in each dictionary group, and on the number of adequate translations per student in each group. Translations that were rated as *match, near-match,* and *far-match* counted as adequate translations. Dictionary type did neither influence word usage ($F(1,133) = 1.78$; ns) nor the

[2] No-responses were included in the analysis since students apparently omitted entries that they felt did not povide sufficient information.

adequacy with which students translated target words (F(1,133) = 0.59; ns). Overall the low percentages of adequate English usages and German translations indicate that students frequently did not understand the meanings of unfamiliar words when given their dictionary entries.

**Students' dictionary strategies:** Students' strategies were inferred from their translations of the target words and the dictionary entries they had seen. The occurrence of a substitution strategy was noted whenever the translation of a target word was either a literal or a non-literal translation of part of its dictionary entry. Based on the distinction between literal and non-literal, two classes of substitutions, literal substitutions and abstract substitutions, were observed. In addition to the literalness of a translation, the degree to which a translation took all or only part of the information that was stated in a definition into account was also coded. Complete string substitutions refer to translations (e.g., *verschmutzen*) in which students selected an appropriate part (*make dirty*) of the definition of a target word (*sully*). Substring substitutions are literal or non-literal translations of an inappropriately sized piece (e.g., *erlaubt* [= allowed]) of the definition (*[of materials] not allowing [water, etc.] to pass through)* of a target word (*impervious*). Altogether then there were four classes of substitutions: Complete string substitution, abstract complete string substitution, substring substitution, and abstract substring substitution.

Copying and modeling refer to responses in which students directly incorporated information from either definitions or illustrative phrases. Copying was coded when an English sentence involved part of the definition or an example. Besides utilizing information verbatim, students also modeled their sentences after an example or the definition. The strategies that were distinguished can account for 90 percent of the sentences that students in the OALD-group wrote, and for 87 percent of the sentences of students in the CCD-group.

As can be seen in Figure 1, students in both groups commonly adhered to a substitution strategy. More importantly, when disregarding the distinction between complete string and abstract complete string substitutions, it can be seen that students selected an appropriately sized piece of the definition 55 percent of the time. That is to say, about 55 percent of the time, did students choose an appropriate meaning equivalent to an entry word and had apparently understood the definitions.

Selecting an appropriate meaning equivalent from its definition, on the other hand, did not always lead to an adequate understanding of an entry word. It was found that about 40 percent of the incorrect English sentences of students in both dictionary groups could be accounted for by complete string and abstract complete string substitutions .

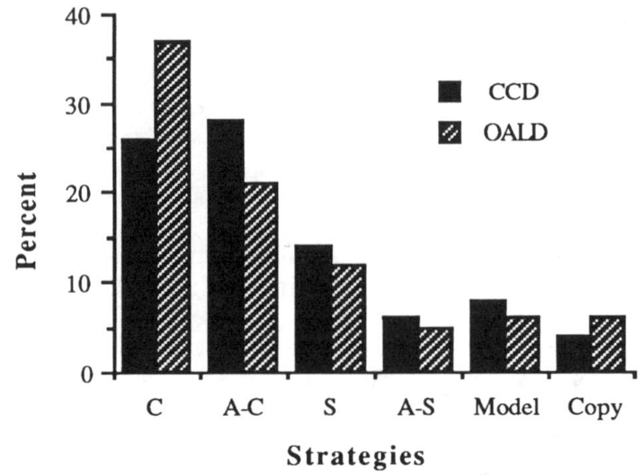


**Figure 1: Distribution of strategies for OALD- and CCD students**

**Explanation of Abbreviations:**

| | |
|---|---|
| C | Complete String Substitution |
| A-C | Abstract Complete String Substitution |
| S | Substring Substitution |
| A-S | Abstract Substring Substitution |
| Model | Modeling |
| Copy | Copying |

**Effect of contextual information:** The frequencies depicted for copying and modeling in Figure 1 refer to the number of instances in which students employed only these strategies and no additional substitutions. Overall, modeling was observed in 24 percent, and copying in 19 percent of the sentences that students in the OALD group wrote in response to 59 entries that specified contextual constraints in definitions and/or mentioned example phrases. In the CCD-group, modeling occurred in 43 percent and copying in 14 percent of the students' responses to all 63 dictionary entries. This indicates that students often did not consider contextual information even though it was available to them. Students apparently took contextual information into account when it was sufficiently constraining as to point to a unique lexical concept, and disregarded it otherwise as uninformative. In particular, students were more likely to make use of contextual information when it alluded to a context in which a specific German word frequently occurs. For instance, the CCD definition of *skim* mentions as context *remove cream, scum, etc.from a liquid: skim the fat from the milk*. This context readily maps onto one frequent usage of the German verb *abschöpfen: Rahm von der Milch abschöpfen*. Given the context, students could thus infer that *skim* means *abschöpfen*, and use it appropriately.

The previous example suggests that students performed better when they did consider contextual information. This indeed was found. A one-sample t-test was performed to test whether the difference between the number of correct responses based on contextual information and the number of correct responses without use of contextual information was greater than zero. For this analysis, a correct response refers to a response in which a target word was used correctly and its translation was a far-match, a near-match, or a match. For each student the number of correct responses without contextual usage was subtracted from the number of correct responses involving contextual information. It was found that OALD- and CCD-students gave more correct responses when they utilized contextual information than when they did not do so ($t_{OALD}$ (68) = 3.16, $p < .001$; $t_{CCD}$ (63) = 7.69, $p < .001$). The additional information that was provided in the context apparently helped students to understand what a given target word meant.

## Conclusion

The present research indicates that students who were learning English as a foreign language quite often did not understand the meaning of an unfamiliar English word even though they did "understand" its definition, i.e., did select an appropriate meaning equivalent. Furthermore, comprehension was better when students considered information that specified in which contexts an entry word and its definition were synonymous. These findings confirm earlier research by Miller & Gildea (1985) which showed that American fifth-graders frequently misunderstood dictionary definitions. Both studies suggest that learners need to know the contexts in which a word can be used in order to understand its meaning. Nonetheless one could argue that once you know the meaning of a word, you know its definition. Contextual information accordingly is assumed to be necessary only for learning. There are at least three arguments against this position. (1) Given that the subjects in the study were between 17 and 20 years old, it is safe to assume that they did not learn new concepts per se; rather they learned a new label, i.e., the English word, for an already familiar concept. That is to say, they merely had to identify to which concept a given definition referred (see Miller (1991) for a general discussion of this issue). The poor performance of the students therefore suggests that there was a mismatch between the definition of a word and students' mental representation of its meaning. (2) Research on the acquisition of word meanings from written context has shown that readers infer the meaning of a novel word by analogy to a familiar word that suits the context of the unfamiliar word (van Daalen-Kapteijns & Elshout-Mohr, 1981; McKeown, 1985; Fischer, 1990). Clearly, in order to do so people have to know in which contexts a word can be used. (3) Even though a definition does not explicitly specify in which contexts the entry word can be used, dictionary users are expected to know in which contexts the definition can be used. For example, to understand the OALD definition of *accrue* that was stated previously, readers have to know of which entities one can say that they *come as a natural growth.* Comprehending definitions thus presupposes mental representations of word meanings that are richer than the definitions themselves.

The claim that people know the definition of a word when they know its meaning could be refined to yield a weaker version: People know the definition of a word plus the contexts in which it is used. However, this assumption is also untenable. (1) Developmental psycholinguists (e.g., Litowitz, 1977; Wehren, de Lisi & Arnold, 1981; Watson, 1985) have pointed out that children achieve an appropriate understanding of the semantic and formal requirements of definitions relatively late in their language development. Wehren et al. (1981), for example, observed that even sixth-graders could not provide adequate definitions of such common words as *hat* and *clock*. On the other hand, no one would seriously want to claim that eleven-year-olds do not know the meanings of these words. They certainly know when it is appropriate to use them. (2) Definitions are metalinguistic statements about the meanings of words. They are derived from reflecting on how words are used. That is to say, you first have to know what a word means, i.e., how it is used, before you can define it. Definitional knowledge rests on contextual knowledge and it does not add anything new to it. Accordingly, when people know the meaning of a word, they need not know its definition plus its contextual usage. All there is to know is how the word is used in context.

The context of a word can be conceived of as a particular configuration of concepts that endow it with meaning. To illustrate this claim, consider the sentence *Only after his wife confronted him with the name of his mistress, did he confess his adulterous behavior.* The meaning of the sentence could be represented in the form of a network structure which specifies how the elements of the event are related, and what these relations imply. The meaning of a word, e.g., adulterous, thus results from its position in the network structure (see Berwick (1989) for a similar view).

Given that words become meaningful in context, meaning representations are therefore best thought of as representations of words in context. They are, as Miller & Charles (1991) propose, contextual and not decontextualized representations. The meaning of a word emerges from actual contextual usages. Its mental representation then reflects those contextual configurations in which it frequently occurs. Meaning representations could be conceived of as either schemata, i.e., generalized knowledge structures with default values, or alternatively as distributed representations. In the latter, PDP approach to meaning representations, various contextual usages of a word impose numerous constraints on its meaning. In adjusting the connection weights of its units, a PDP system would attempt to settle into a solution that maximally satisfies the various constraints. What is commonly referred to as the meaning of a word would accordingly correspond to a configuration of units with strong interconnections (see Rumelhart et al. (1986) for a discussion of schemata in PDP models). Common to both approaches is the notion that meaning representations are generalizations of contexts rather than abstractions from contexts. Meaning representations are thus not definitions but instead reflect the kinds of contexts in which a word is commonly used.

## References


Berwick, R.C. 1989. Learning word meanings from examples. In D. Waltz ed., *Semantic Structures: Advances in Natural Language Procecssing*. Hillsdale, N.J.: Erlbaum.

van Daalen-Kapteijns, M.M. & Elshout-Mohr, M. 1981. The acquisition of word meanings as a cognitive learning process. *Journal of Verbal Learning and Verbal Behavior* 20: 386-399.

Fischer, U. 1990. How students learn words from a dictionary and in context. Ph.D. diss., Dept. of Psychology, Princeton University.

Fodor, J. A., Garrett, M. F., Walker, E. C. T., & Parkes, C. H. 1980. Against definitions. *Cognition* 8: 263-367.

Gipe, J. 1979. Investigating techniques for teaching word meanings. *Reading Research Quarterly* 14: 624-645.

Gross, D., Kegl, J., Gildea, P. & Miller, G.A. 1989. A Coded Corpus and Commentary on Children's Dictionary Strategies. Technical Report 39, Cognitive Science Laboratory, Princeton University.

Hornby, A. S., et al. 1980. *Oxford Advanced Learner's* Dictionary of Current English, 3rd edition. London: Oxford University Press.

Johnson, D. M., & Stratton, R. P. 1966. Evaluations of five methods of teaching concepts. *Journal of Educational Psychology* 57: 48-53.

Keil, F. C., & Batterman, N. 1984. A characteristic-to-defining shift in the development of word meaning. *Journal of Verbal Learning and Verbal Behavior* 23: 221-236.

Keil, F. C., & Carroll, J. J. 1980. The child's conception of "tall": Implications for an alternative view of semantic development. *Papers and Reports on Child Language Development*19: 21-28.

Landau, S. I. 1984. *Dictionaries: The Art and Craft of Lexicography*. New York: Scribner.

Litowitz, B. 1977. Learning to make definitions. *Journal of Child Language* 4: 289-304.

Lyons, J. 1968. *Introduction to Theoretical Linguistics*. Cambridge: Cambridge University Press.

McKeown, M. G. 1985. The acquisition of word meanings from context by children of high and low ability. *Reading Research Quarterly* 20: 482-496.

Miller, G. A. 1986. Dictionaries in the mind. *Language and Cognitive Processes* 1: 171-185.

Miller, G. A. 1991. *The Science of Words*. New York: Scientific American Library.

Miller, G. A., & Charles, W. G. 1991. Contextual correlates of semantic similarity. *Language and Cognitive Processes* 6: 1-28.

Miller, G. A., & Gildea, P. M. 1985. How to misread a dictionary. In *AILA Bulletin*, Pisa.

Müller, W. ed. 1986[2]. *Duden: Sinn-und sachverwandte Wörter*. Mannheim: Dudenverlag.

Rosch, E., 1975. Cognitive representations of semantic categories. *Journal of Experimental Psychology: General* 104: 192-233.

Rumelhart, D. E., Smolensky, P., McClelland, J. L., & Hinton, G.E. 1986. Schemata and sequential thought processes in PDP models. In J. L. McClelland, D. E. Rumelhart & the PDP Research Group eds., *Parallel distributed processing, vol. 2*. Cambridge, Mass.: MIT Press, A Bradford Book.

Sinclair, J. ed. 1987. *Collins Cobuild English Language Dictionary*. London: Collins.

Sternberg, R. J. 1987. Most vocabulary is learned from context. In M. G. McKeown, & M. E. Curtis eds.. *The Nature of Vocabulary Acquisition*. Hillsdale, N.J.: Erlbaum.

Sternberg, R. J., & Powell, J. S. 1983. Comprehending verbal comprehension. *American Psychologist* 38: 878-893.

Terrell, P., et al. 1985. *Collins German Dictionary: German-English, English-German*. 5th reprint. London: Collins.

Watson, R. 1985. Towards a theory of definition. *Journal of Child Language* 12: 181-197.

Wehren, A., de Lisi, R., & Arnold, M. 1981. The development of noun definitions. *Journal of Child Language* 8: 165-175.

Willmann, H., Messinger, H., et al. 1988. *Langenscheidts Großwörterbuch der Englischen und Deutschen Sprache: Der kleine Muret-Sanders: Englisch-Deutsch*3. Auflage. Berlin: Langenscheidt.

# Learning Words: Computers and Kids*

**Peter M. Hastings**[†] **Steven L. Lytinen** **Robert K. Lindsay**

Artificial Intelligence Laboratory
University of Michigan
Ann Arbor, MI 48109

## Abstract

We present a computer-based model of acquisition of word meaning from context. The model uses semantic role assignments to search through a hierarchy of conceptual information for an appropriate meaning for an unknown word. The implementation of this approach has led to many surprising similarities with work in modelling human language acquisition. We describe the learning task and the model, then present an empirical test and discuss the relationships between this approach and the work in psycholinguistics.

## Introduction

This paper describes a computational model of acquisition of lexical items from context. The learning task is defined as follows: given a set of natural language sentences in which a previously unknown lexical item appears, infer the syntactic class and the meaning (or meanings) of the word. We assume that the vast majority of other words appearing in the set of sentences are already known.

Our approach has been implemented as part of a unification-based natural language processing system called LINK [Lytinen, 1990]. LINK's grammar rules are quite similar in form to those used in PATR-II [Shieber, 1986]. We have incorporated semantic information into LINK's grammar, along the lines of HPSG [Pollard and Sag, 1987]. The integration of syntactic and semantic knowledge into the same grammar formalism is key to our system's ability to infer information about unknown words.

We are using LINK in two prototype applications involving relatively narrow domains (i.e. the necessary domain knowledge can be described fairly completely), but the textual input is entered by a large number of users and is therefore subject to wide variations in the terminology used. Our system is able to infer the meanings of many unknown words in these applications. The examples in this paper will be taken from one of these applications. The texts in this application describe sequences of actions to be performed on an assembly line.

In this paper, we will provide a sketch of our word-learning approach. In particular, we will focus on the acquisition of word meanings. The reader is referred to [Lytinen and Roberts, 1989] for a more detailed discussion of syntactic learning in LINK. We also present the results of an empirical test, in which our approach was used to learn the meanings of 22 undefined verbs from a corpus of 100 inputs from one of our application domains.

Our approach to the word-learning task was not developed with the modeling of human behavior in mind. The constraints of the learning task, however, guided the implementation to a state that closely resembles theoretical and empirical linguistic explanations of language acquisition in humans. We will discuss these relationships after the presentation of the empirical test.

## The Learning Task

The LINK parser is often able to infer the syntactic category of an unknown word using grammatical constraints. Knowledge of the syntactic category allows LINK to make certain inferences about the semantic connections between the unknown word and other constituents of the sentence. This role-filling information is used in conjunction with a simple IS-A hierarchy in order to formulate hypotheses about the meaning of an undefined word. All of the semantic predicates defined in LINK's knowledge base are included in the hierarchy. Each concept definition includes a set of thematic roles or "slots" that can be (optionally or obligatorily) attached to the concept, as well as the type of concept which can fill each slot. The set of restrictions on fillers of slots for a concept must be at least as specific as the restrictions for its ancestors in the hierarchy (i.e. more general concepts). The ordering on generality of slot-filler constraints as well as other semantic information

*The research was funded in part by a grant from the Kellogg Foundation through the Presidential Initiatives Fund at the University of Michigan

[†]This author is also affiliated with EDS Center for Machine Intelligence, 2001 Commonwealth Boulevard, Suite 102, Ann Arbor, MI 48105

determines the structure of the semantic hierarchy.

Figure 1 presents a portion of the IS-A hierarchy for actions that is used in describing our assembly-line domain. Constraints on fillers of slots for actions are also represented in this figure. Slot-filling constraints on a concept are inherited from the concept's ancestors in the tree. For example, since GENERAL-FACTORY-ACTION requires an OBJECT that is a *FACTORY-OBJECT*, this restriction also implicitly holds for actions like *GET* and *INSPECT*. *RECORD-ACTION* is an example of a concept which makes a further restriction on a previously constrained slot. *RECORD*, the OBJECT of this action, must be a descendant of *FACTORY-OBJECT*.

LINK's concept hierarchy guides the process of learning word meanings. Initially, it is assumed that every concept in the hierarchy is a candidate hypothesis for the meaning of an unknown word. Example sentences can provide two types of restrictions on the set of candidate hypotheses. First, the unknown word may appear as the filler of a thematic role of another word, as in *Secure the flarge.* Because *flarge* is assigned as the direct object of *secure,* LINK's grammar suggests that it is the semantic OBJECT of *SECURE*. This condition places an upper bound on the generality of the word's meaning: *flarge* must be an AUTO-PART or one of its descendants in the hierarchy.

The second type of restriction that context may suggest is a filler for a thematic role of the unknown word, as in *Flarge the door.* In this case, LINK's unification grammar suggests that *DOOR* is the semantic OBJECT of *flarge.* Information about role-fillers of an unknown concept place a lower bound on the specificity of the concept: given that *DOOR* is the OBJECT, *flarge* can refer to concepts like *GENERAL-FACTORY-ACTION* and *ASSEMBLE*, but not to concepts like *FASTEN*, *REFILL*, or *TAPE-ACTION* (or any of its descendants) since a *DOOR* violates the restrictions that these concepts place on their OBJECTs.

Thus, two types of information are supplied by example sentences: information which provides a lower bound on the level in the hierarchy of the meaning of an unknown word, and information which provides an upper bound. This would suggest a least-commitment approach to learning, such as Mitchell's *candidate-elimination* algorithm [Mitchell, 1990]. Mitchell's algorithm used *version spaces* to represent the set of candidate hypotheses, and slowly narrowed the version space depending on the additional constraints provided by new examples. Unfortunately, in our word-learning task, often it is the case that particular kinds of words only appear in examples that provide one of the two types of restrictions. Nouns, which usually refer to things, almost always appear as role-fillers of actions or states; thus, examples only serve to limit the upper bound of the candidate hypotheses. Verbs, on the other hand, usually appear with role-fillers attached to them, and not as role-fillers themselves, since they refer to actions or states. Thus, examples only serve to place a lower bound on their candidate hypotheses. Thus, since examples only provide one of the two kinds of restrictions for many word classes, a least-commitment algorithm would not converge on a single hypothesis for the meaning of most unknown words.

Because of this, our algorithm is not a least-commitment algorithm. For nouns, we assume the most general candidate hypothesis is the correct one. Thus, the hypothesis for *Secure the flarge* is that *flarge* means *AUTO-PART*. In the case of verbs, the most specific candidate hypotheses are kept. From *flarge the nut,* then, *flarge* is assumed to mean *FASTEN* (since this concept requires a *NUT* as its OBJECT). A later example like *flarge the door* would eliminate the hypothesis *FASTEN*, since a *DOOR* cannot be its OBJECT. This would result in a generalization procedure which ascends the hierarchy and branches out until a concept (or concepts) whose constraints are satisfied by this set of slot-fillers (*DOOR* and *NUT*). The resulting set of hypotheses would then be *INSTALL*, *POSITION*, and *SECURE* since *DOOR* and *NUT* are both *AUTO-PART*s.

## Limitations of This Technique

Several artifacts of the learning mechanism limit the sort of word definitions that can be learned. The first is the assumption that the representation of the ontology is complete, i.e. that every concept which is part of the domain is *a priori* represented by some node in the semantic hierarchy. This clearly limits the range of concepts which can be learned.

In addition, this technique relies solely on one type of information, the semantic constraints of role-fillers. While this information is sufficient to differentiate between many of the word meanings, large classes of words exist that require additional information to distinguish the members of the class from one another.

As mentioned above, the learning algorithm can not handle ambiguous words. In such cases, an apparent contradiction is found between competing hypotheses, and an over-general concept is then chosen. Some sort of mechanism is needed to determine whether a more general concept or a disjunctive mapping is justified in specific situations.

Finally, the learning algorithm (as we have described it so far) often does not converge on a single hypothesis for the meaning of a word, especially in the case of verbs. To see why this is true, consider again the example *Flarge the nut.* Intuitively it seems that the best hypothesis for the meaning of *flarge* is *FASTEN*, since only nuts can be fastened, and *FASTEN* is the only action in the hierarchy which can be done to only nuts. However, many other hypotheses cannot be eliminated as possibilities: *flarge* might mean *INSTALL*, since according to our hierarchy nuts can be installed, too. Given the hierarchy as it stands, no examples

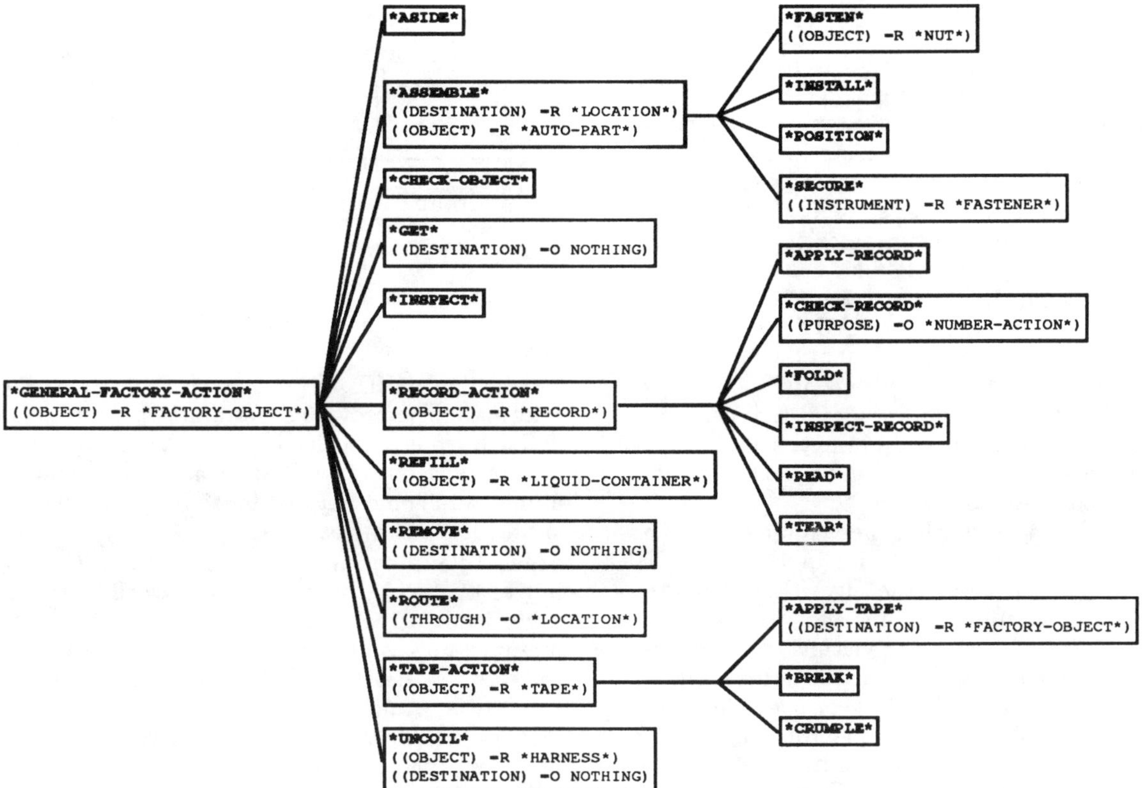


Figure 1: A portion of the action concept hierarchy

can be given which will eliminate all other candidate hypotheses (assuming *flarge* really does mean *FASTEN*), since nothing which meets the restrictions on the slots of *FASTEN* will violate any of the restrictions on the slots of these other candidates.

To remedy this problem, our algorithm ranks the list of candidate hypotheses according to how *tightly* each candidate's constraints on slots match with the actual slot fillers found in the examples. For the example *flarge the nut,* *FASTEN* is the highest-ranked candidate hypothesis for the meaning of *flarge,* since its restriction on the OBJECT slot exactly matches the OBJECT of *flarge* in the example sentence.

In addition, the final set of meaning hypotheses for an unknown word is checked at the end of the parse to see if all of the required slots for each hypothesis is present in the parse. Thus for an example like *flarge the door handle*, where *INSTALL*, *POSITION*, and *SECURE* would be hypotheses based on the filler of the OBJECT slot, *SECURE* would be eliminated from consideration because its required instrument slot is missing in the sentence.

## The Empirical Test

To evaluate the effectiveness of our approach, an empirical test was conducted. In the test, a set of 100 example inputs from one of our application corpora was chosen at random. The test corpus consisted of short descriptions of sequences of actions to be performed on an assembly line. This corpus was chosen for the test because we had already developed extensive sets of grammar rules and lexical entries for it.

We were particularly interested in evaluating our algorithm's performance on learning verbs, since they presented the largest challenge. For the test, we removed the definitions of all of the verbs that appeared in the 100 examples from LINK's lexicon. There were 22 verbs in this set of examples. We then ran the system on the 100 examples and inspected the definitions of the verbs to see whether the system had inferred their meanings correctly. Table 1 presents 2 typical verbs from the set of examples as well as the sentences in which they appeared.

Table 1: Typical verbs and example sentences

| | |
|---|---|
| secure | *Secure rr/dr hndl w/2 nuts* (secure right-rear door handle ...) |
| | *Secure hrns to rsb w/2 int clips* (Secure harness to right-side bolster with 2 int-clips) |
| get | *Get inspection record* |
| | *At bench, get manifest* |
| | *Get lock cylinder kit* |
| | *Get driver* |

A representative set of verbs along with their inferred meanings is presented in table 2. In table 3, the verbs are grouped according the quality of the result achieved. The 17 verbs in group 1 (77% of the total set) were learned to the maximum extent possible given the amount of knowledge that exists in the system. For 7 of these words, the correct meaning was the top-ranked hypothesis. For the others, the correct meaning is included among a set of hypotheses that are indistinguishable using only the role-filler constraints. For example, the actions *INSTALL* and *POSITION* are both defined as requiring an *AUTO-PART* for an object. Without additional information, there is no way to distinguish between these concepts. Thus, both of the concepts remain as hypotheses for the meanings of *install* and *position* at the end of the test run. Verbs of this type are counted as having been successfully learned in our test results.

Group 2 contains verbs that were ambiguous, i.e. that referred to two or more nodes in the semantic hierarchy. As stated above, the algorithm currently has no way of successfully handling such words.

The verbs in group 3 were the victims of shortcomings in the implementation. *Allow* always occurs with a sentential object, e.g. *Allow to load paper to printers.* This causes difficulty for the learning algorithm since it can only handle one word at a time (notice that *load* doesn't show up in the results). The word *preload* was only found in one sentence in this test set, so the hypothesis was overly specific.

The results of this test suggest that a large portion of the meaning of unknown words can be inferred automatically using only very basic conceptual information about the domain.

Table 2: Sample results of test run

| Verb | Ordered meaning hypotheses |
|---|---|
| check | *ASIDE* *CHECK-OBJECT* *GET* *INSPECT* *LOAD* *LUBRICATE* *OPEN* *PLACE* *REMOVE* *REPAIR* *RESTOCK* *ROUTE* *TOSS* |
| crumple | *BREAK* *CRUMPLE* |
| fasten | *FASTEN* |
| install | *INSTALL* *POSITION* |
| preload | *SECURE* |
| reach | *REACH* |
| uncoil | *UNCOIL* |

## Related Computer Models

Similar efforts at using machine learning techniques in lexical acquisition were reported in [Zernik, 1987]. Zernik described his approach as using a version space technique to learn phrasal lexicon rules. However, Zernik's system receives feedback from a teacher in the form of user-supplied "contexts" that explain what the input means. It is not clear if Zernik's approach can be adapted to a situation in which feedback is not available.

Table 3: Grouping of verbs in test results

| | |
|---|---|
| Group 1 | aside, break, crumple, fasten, fold, get, install, position, reach, remove, return, route, secure, step, toss, uncoil, walk |
| Group 2 | apply, check, place |
| Group 3 | allow, preload |

Selfridge's CHILD program [1986] used contextual information to provide constraints on definitions of undefined words in much the same way as our system does for nouns. However, CHILD learned from only one example, and could not further refine meanings based on subsequent examples.

Jacobs and Zernik [1988] describe the RINA system, in which a task very similar to our word-learning task is performed. RINA examines large corpora, extracting many examples of a given unknown word. Although they do not describe their algorithm in detail, it appears from examples discussed in the paper that word meaning acquisition in RINA is driven more heavily by discourse context than in LINK.

## Relationships to Developmental Psycholinguistics

Although this model was developed solely to allow efficient use of the limited information available to the natural language processing system, some of the challenges we faced in the development of the system bear a striking resemblance to issues brought up in the psycholinguistic literature. This suggests that these challenges are not unique to computational models but are inherent difficulties in language learning in general. Some of these issues are discussed below.

### The No-Negative-Evidence Problem

When children learn language, they must induce the structure of the language and the meanings of the words relying almost entirely on examples of utterances which are *within* the language. They don't have the benefit of negative evidence to help them in their learning task. This lack of discriminating information makes the learning process computationally very complex, yet children do learn language. The Subset Principle was described in [Berwick, 1985] as one way that children could reduce the complexity of the learning task. This principle suggests that children have a hierarchical mental representation of languages ordered on the specificity of the grammars. When learning syntax, children first hypothesize the most specific grammar that accounts for the input in order to avoid overgeneralization.

We are faced with a similar problem in our model of meaning acquisition. The lack of negative evidence about word meanings as well as the nature of the role-filler constraints provides a lower bound on the set of hypotheses, but no upper bound. Thus we are forced to choose the most specific hypotheses to be able to learn from a training set consisting of only positive examples.

Bowerman [1983] presents a model of how children deal with the no-negative-evidence problem in learning verb meanings. She describes a method in which children could use syntactic information to, in effect, subcategorize verbs according to aspects of their meanings (e.g. causation). Bowerman suggests that additional discriminatory information such as this can be used as pseudo-negative evidence in that children can make predictions about word usage from syntactic clues. The violation of their assumptions provides the negative evidence that makes the learning process less computationally overwhelming.

In our model, we try to find the most specific, falsifiable hypotheses. If a later example has a slot-filler that violates our original hypothesis, we choose one that can accommodate both the old and the new slot-fillers.

## Syntactic and Semantic Bootstrapping

Gleitman [1990] detailed a mechanism called "syntactic bootstrapping" that children might use to guide their search for meanings of verbs through the space of possible meanings that could be inferred from the immediate context. She suggested that children as young as 17 months have strong capabilities for recognizing syntactic distinctions and using them to constrain the meanings of verbs they are learning. For example, children who had no prior knowledge of the word *flex* were shown two videos, one of Big Bird and the Cookie Monster crossing and uncrossing their own arms, and another with one of them crossing the arms of the other. When the sentences *Big Bird is flexing with the Cookie Monster* and *Big Bird is flexing Cookie Monster* were broadcast through a speaker, the children showed a definite preference for the "syntactically congruent screen", i.e. the video that was showing the action that was being described, even though they had no semantic knowledge of the meaning of *flex*. Gleitman argued that without such a constraining mechanism, the task of word learning would be computationally infeasible. But while her approach relies solely on the syntactic structure of the sentence to yield semantic clues, our approach combines use of syntactic and semantic information (but no external context) to generate hypotheses.

In Shatz' [1987] description of a similar bootstrapping mechanism, she gives an example of a 4-year-old who said "I pricked my finger" after she had stuck herself with a needle, and then asked, "What does prick mean?" This suggests that children learning language can use their limited knowledge of the context in which a word is used to develop a partial hypothesis for the meaning of that word, just as our system incrementally refines inferred meanings over multiple examples.

## Later Language Acquisition

Although our approach presents many similar properties to some aspects of children's language acquisition, it cannot be seriously considered a model of the learning process of children because of the assumption that the system's domain knowledge is complete at the time of word meaning acquisition. In this sense, the model is more similar to human language acquisition that is done later in life. Two examples of this are Genie and second language learners.

As Curtiss explains [Curtiss, 1982], Genie, during her developmental years, was deprived of all of the linguistic input that children usually receive. She was also partially deprived of information about the world. She still had information about her own surroundings, however, and presumably the maturation of her cognitive abilities gave her a much more developed (though still quite limited) conceptual representation for the world than, say, a 2-year-old would have. But Genie didn't know the words that went with the concepts she knew. Because of this, Genie's task of learning language is very similar to the one that our model is faced with. Unfortunately, Genie's linguistic deprivation during her "sensitive years" appears to have rendered her syntactic ability permanently limited. Although Genie has done quite well in acquiring the meanings of words, there are still noticeable deficits. In the face of this computer model and the work on syntactic bootstrapping, it is easy to see why she would have difficulties in learning. A large part of the information that constrains the word-learning process is unavailable to her.

The learning of a second language is another case where a fully developed conceptual representation exists when word learning takes place. Unfortunately, the second language acquisition literature tends to concentrate on teaching methods and problems, and not on psychological or linguistic theories of the processes involved. One example of the former that leans toward the latter is Cornell's description [1985] of the difficulties of teaching second-language learners the meanings of phrasal verbs (verb-particle pairs). He cites many reasons for these difficulties, among them the subtle differences between meanings for these verbs, and varying syntactic constraints. Unfortunately, our model doesn't contain the answers to these problems either, since we're still trying to learn the gross differences in meanings of words in our limited domain. Cornell does give us motivation, however, stating, "Presumably what is needed is a computer intelligent enough to scan a corpus and recognize phrasal groupings and assign meanings to them."

As mentioned above, our model of language learning was not developed for the purpose of simulating language acquisition in humans. If the similarities found between our model and the psycholinguistic models are more than coincidence, however, then our model will provide a valuable testbed for the computational evaluation of language theories.

## Future Work

There are many ways in which our algorithm can be extended. First, the algorithm as it currently stands uses only information about semantic dependencies that the parser is able to identify between words in example sentences. It should be able take advantage of other information available from the examples, such as the syntactic constructions used with an unknown word, additional semantic contextual information, and so on. The use of such additional information would enhance the similarity between this approach and syntactic bootstrapping.

Second, the assumption that a word must map directly to a unique concept in the hierarchy is not a realistic one. Many words are ambiguous, and thus refer to two or more nodes in the hierarchy. Even an unambiguous word's meaning may not correspond exactly to an already existing node in the hierarchy. In fact the mutual exclusivity (contrast) assumption, described in [Markman, in press, Clark, 1989], suggests that children learning word meanings are biased against two words having the same meaning. Our system should be able to use a similar bias by entertaining disjunctive hypotheses for word meanings, and should also be able to consider "splitting" a node in the hierarchy (similar to the approach in [Winston, 1975]), so that a word can refer to a new subconcept. In addition, a mechanism could be added to the system to check for words that refer to particular concepts. If a concept already has a referent word, it can be skipped when looking for a meaning for an unknown word.

Finally, we will continue to examine the related issues found in the psycholinguistic literature and explore methods of incorporating these theoretical and experimental results into our computational model. Hopefully these relationships will allow us to make our model more efficient and more relevant to human learning.

## References


[Berwick, 1985] R. Berwick. *The Acquisition of Syntactic Knowledge.* MIT Press, 1985.

[Bowerman, 1983] M. Bowerman. How do children avoid constructing an overly general grammar in the absence of feedback about what is not a sentence? In *Proceedings of Research on Childrens Language Development*, volume 22, 1983.

[Clark, 1989] E. Clark. On the logic of constrast. *Journal of Child Language*, 15:317–335, 1989.

[Cornell, 1985] A. Cornell. Realistic goals in teaching and learning phrasal verbs. *International review of Applied Linguistics in Language Teaching*, 23(4):269–280, 1985.

[Curtiss, 1982] S. Curtiss. Developmental dissociations of language and cognition. In L. K. Obler & L. Menn, editor, *Exceptional language and linguistics*, pages 285–312. Academic Press, New York, 1982.

[Gleitman, 1990] L. Gleitman. The structural sources of verb meaning. *Language Acquisition*, I(1):3–55, 1990.

[Jacobs and Zernik, 1988] P. Jacobs and U. Zernik. Acquiring lexical knowledge from text: A case study. In *Proceedings of the Fourth National Conference on Artificial Intelligence*, pages 739–744, 1988.

[Lytinen and Roberts, 1989] Steven L. Lytinen and Susan N. Roberts. Unifying linguistic knowledge. AI Laboratory, Univ of Michigan, Ann Arbor, MI 48109, 1989.

[Lytinen, 1990] Steven L. Lytinen. Robust processing of terse text. In *Proceedings of the 1990 AAAI Symposium on Intelligent Text-based Systems*, pages 10–14, Stanford, CA, 1990.

[Markman, in press] E. Markman. The whole object, taxonomic, and mutual exclusivity assumptions as initial constraints on word meanings. In J. P. Byrnes & S. A. Gelman, editor, *Perspectives on language and cognition: Interrelations in development.* Cambridge University Press, New York, in press.

[Mitchell, 1990] T. Mitchell. Version spaces: A candidate elimination approach to rule learning. In *Proceedings of the Fifth International Joint Conference on Artificial Intelligence*, pages 305–309, 1990.

[Pollard and Sag, 1987] C. Pollard and I. Sag. *An Information-based Syntax and Semantics.* CLSI, Palo Alto, CA, 1987.

[Selfridge, 1986] Mallory Selfridge. A computer model of child language learning. *Artificial Intelligence*, 29:171–216, 1986.

[Shatz, 1987] M. Shatz. Bootstrapping operations in child language. *Children's Language*, 6:1–22, 1987.

[Shieber, 1986] S. Shieber. *An Introduction To Unification-Based Approaches To Grammar.* CSLI, Stanford, CA, 1986.

[Winston, 1975] Patrick Henry Winston. Learning structural descriptions from examples. In Patrick Henry Winston, editor, *The Psychology of Computer Vision.* McGraw-Hill Book Company, New York, NY, 1975.

[Zernik, 1987] Uri Zernik. How do machine language paradigms fare in language acquisition. In *Proceedings of the Fourth International Workshop on Machine Learning*, Los Altos, CA, 1987. Morgan Kaufmann.

# Ambiguity Resolution: Behavioral Evidence for a Delay

**Laurie A. Stowe**

Dept. of Linguistics
The Ohio State University
Columbus, OH 43210

## Abstract

This paper presents experimental evidence for a model of human language processing in which ambiguity resolution is delayed when there is a conflict between semantic contextual bias and the syntactically preferred interpretation. If there is no conflict, an immediate decision is made. Decision is not delayed indefinitely; the length of the delay is limited by available processing resources.

## Introduction

Are ambiguous sequences initially treated as if they were unambiguous? Alternatively are both or all of the possibilities considered until it is possible to make a choice? These questions continue to interest researchers concerned with language comprehension. In this paper I will propose that neither of these alternatives is correct. I will summarize recent evidence that ambiguities are recognized and decision may be temporarily delayed, waiting for evidence that can decide the issue. However, it appears (not surprisingly) that there are limits on the extent of delay that is possible. If no disambiguating evidence becomes available within the limits on delay, an arbitrary decision must be made. This model appears to extend to both syntactic ambiguity resolution, which will be the major focus of this paper, and lexical ambiguity resolution.

## Immediate Resolution Models

Many researchers have proposed that one analysis is initially pursued; the alternative initially pursued may be based on syntactic (e.g. Frazier & Rayner 1982) or on semantic criteria (e.g. Crain & Steedman 1985). The major type of evidence that has been adduced for initial treatment of syntactically ambiguous structures as if they were unambiguous is the "garden path" effect. In garden path sentences, a hearer or reader is "led up the garden path," that is the sentence appears initially to have one structure and then it turns out, to the surprise of the hearer, that the sentence in fact meant something else entirely. This effect suggests that at least under some circumstances people do **not** wait until they receive disambiguating evidence before selecting one of the alternatives. If they do so immediately, the initial unambiguous model is supported.

An example of a characteristic garden path structure is the reduced relative clause, shown in (1).

(1) a. The officers hidden in the bushes observed the practice attack.
b. The officers taught at the military academy were very demanding.

An easily interpreted example of such a structure appears in (1a). However, if the reduced relative contains a passive participle which is not formally differentiated from the past tense form, as in (1b), ambiguity results. It has long been recognized that reduced relatives containing such verbs are typically misanalyzed as main verb phrases, producing a garden path effect when the actual main verb phrase is read. This effect implies that the wrong structure has been calculated.

### Syntactic Resolution Procedures

An important question for any model of ambiguity resolution is what determines which structure is selected. For reduced relative clauses, the fact that people are surprised when they encounter the main verb phrase shows that they had assumed the reduced relative clause to be a main verb phrase. A specific proposal made by Frazier (1978) is that the main verb phrase analysis is initially pursued because it is the simplest, in terms of the number of nodes necessary to build a phrase structure representation: the Minimal Attachment Hypothesis.

Two other sentence types cited in support of the Minimal Attachment Hypothesis (Frazier & Rayner 1982; Rayner, Carlson & Frazier, 1983) are sentential complement sentences, as in (2) and prepositional phrase attachment ambiguities, as in (3).

Table 1: Mean Reading Times for Reduced Relative Clauses: Accumulating Presentation

| | RelVerb | Next | Next | MainVerb | Next |
|---|---|---|---|---|---|
| Animate Subject | | | | | |
| Ambiguous | 364 | 366 | 367 | 488 | 599 |
| Unambiguous | 371 | 377 | 385 | 475 | 495 |
| Inanimate Subject | | | | | |
| Ambiguous | 343 | 382 | 374 | 506 | 476 |
| Unambiguous | 358 | 378 | 394 | 493 | 511 |

(2) The reporter saw the woman who came in was not happy.

In (2), due to the absence of the grammatical marker **that**, it is not initially clear that the phrase following the verb **see** is the subject of a sentential complement rather than a direct object. Constructing the direct object structure does not require accessing as many phrase structure rules as the sentential complement structure, and therefore is simpler. Some people may feel that the verb phrase *was not happy* is unexpected (i.e., a garden path). Frazier and Rayner (1982) found that eye-tracking patterns suggested that people had to go back to and reprocess the ambiguous sequence only if it turned out that a sentential complement structure had to be assigned.

(3) The spy saw the cop with the binoculars.

As can be seen in (3), a prepositional phrase may frequently attach either to a verb phrase (e.g., *saw using the binoculars*) or to a direct object noun phrase (e.g., *the cop who had the binoculars*). Rayner, Carlson and Frazier (1983) argued that the lower (noun phrase modifier) attachment is not minimal, thus predicting that people tend to prefer the verb phrase attachment interpretation. When the meaning of the prepositional phrase forces the lower attachment, as *saw the cop with a revolver* (i.e., you can't see using a revolver), they found evidence for longer reading times and more regressions, apparently to "fix up" the initial error. Thus difficulties involving both of these types of structures can be predicted and explained under the Minimal Attachment Hypothesis.

### Semantic Resolution Procedures

However, before accepting that the Minimal Attachment Hypothesis is the correct account of the data, it is appropriate to consider if other types of information affect the structure initially assigned to the phrase, and if so how. The Minimal Attachment Hypothesis does not predict that semantic information has any initial effect on the structure that is built. Nevertheless, it is clear that the comparative interpretability of the two structures can have an impact on which is eventually constructed, as illustrated in (4).

(4) a. The officers taught at the military academy were very demanding.

b. The courses taught at the military academy were very demanding.

Moreover, Crain and Steedman (1985), Cupples and Stowe (1991), and Taraban and McClelland (1989), have argued that semantic information can have an early impact on how ambiguous reduced relative clauses and prepositional phrase attachment ambiguities are handled.

Cupples and Stowe (ms.) used the accumulating self-paced reading task to compare how subjects read sentences like those in (4). Subjects had to press a button to get each new word of the sentence, until the entire sentence was present on the screen. The time from the appearance of the word on the screen until the button was pressed is a measure of the time necessary to read the word in the sentential context. In (4a), the animate subject noun is equally plausible as either subject or object of the following verb; in (4b) the inanimate subject noun is anomalous as subject of the following verb, but plausible as its object. This semantic information, if available to the parser, can potentially decide between the active main clause reading and the passive reduced relative clause reading at the ambiguous verb itself, as soon as the ambiguity is encountered. We may call this the Immediate Semantic Decision Hypothesis. Certainly readers have the intuitive judgment that the semantic information makes the ambiguity easier to deal with.

Results for this experiment are summarized in Table 1. When the reading times for these sentences are compared with syntactically unambiguous controls (containing the phrase *that were* in the relative clause), a very large and significant increase in difficulty was found at the main verb phrase for (4a), as predicted by the Minimal Attachment Hypothesis. However, no significant difference was found at any point in the

sentence for (4b). Here, of course, semantic information is available to decide which structure is assigned to the relative clause. This evidence clearly indicates that semantic information can be used to assign the correct structure to the ambiguous sequence before reaching the syntactically disambiguating region.
A model in which semantic information is used immediately to decide immediately between the two possible structures, such as the Immediate Semantic Decision Hypothesis, is compatible with this data.

It might be suggested that this result is consistent with a version of the Minimal Attachment Hypothesis in which an unacceptable interpretation provides feedback to the parser and revision is started during the relative clause itself. However, this account suggests that the revision should cause increased reading times in the relative clause instead. No such difference was found, and to that extent the Immediate Semantic Decision Hypothesis is more compatible with the experimental results.

Taraban and McClelland (1989) examined the effect of sentential context preceding a prepositional phrase. They found that people consistently rated a nominal modifier completion as less expected than a main verb phrase completion for Frazier and Rayner's materials, like the sentence in (3). They constructed a set of materials for which the nominal attachment was rated more favorably instead, as in (5).

(5) a. The reporter exposed corruption in govern ment.
b. The reporter exposed corruption in the article.

In (5a), the expected low attachment reading is compatible with the ending (e.g. *corruption which is found in the government*), while in (5b), the final word is more compatible with verb phrase attachment (e.g., *exposed in the article*). According to the Minimal Attachment Hypothesis, the syntactic bias should take priority over contextual effects. Under the Immediate Semantic Decision Hypothesis, however, the context is used to make a decision between the alternatives as soon as the ambiguity is recognized; here, that point is the preposition.

Taraban and McClelland tested both sets of sentences using the self-paced reading paradigm. They found that there was no main effect of high vs. low attachment, providing no support for the hypothesis that Minimal Attachment is preferred. However, there was a significant interaction with sentential context; when the context predicted high attachment, verb phrase attachment was easier, but when the context predicted nominal attachment, attachment to the noun phrase was easier instead. These results support the Immediate Semantic Decision Hypothesis then, rather than Minimal Attachment.

## Delayed Resolution Procedures

The experiments that I have reviewed here suggest that the Minimal Attachment Hypothesis is in need of replacement or revision. At the very least, the model must be extended to include an early and almost cost-free revision procedure based on semantic information. When semantically biasing information is available, recovery from the incorrect analysis is not problematic

Do these experiments provide unequivocal evidence for the Immediate Semantic Decision Hypothesis? It would be difficult to argue that. As just pointed out, the Minimal Attachment Hypothesis with the above modifications can also account for the data. The Taraban and McClelland results provide the clearest evidence for the Immediate Semantic Decision model, since they show evidence of garden pathing **in both directions**, based on differences in context. However, in the Cupples and Stowe materials, animate subjects were explicitly selected that were rated as equally plausible when taken as the subject or object of the following verb. Thus, there is no basis for an Immediate Semantic Decision between the two alternatives and no garden path is predicted. This is clearly not an insuperable objection to the Immediate Semantic Decision Hypothesis. Crain and Steedman (1985) propose that discourse conditions can explain the clear preference for the main verb phrase interpretation. However, there is additional evidence that is worth considering before accepting this model.

Holmes, Stowe and Cupples (1989) looked at sentences like those in (6) using the self-paced reading task described above, with the exception that in this experiment only one word at a time appeared on the screen, although their positions were the same as if the entire sentence had been present. As discussed above, according to the Minimal Attachment Hypothesis, the direct object reading is preferred for ambiguities of this sort. This predicts that reading times during the verb phrase of the embedded clause should be longer than for versions of these sentences containing the disambiguating grammatical marker *that*.

(6) a. The reporter saw the woman was not very happy.
b. The student realized the answer was not clear.

The Immediate Semantic Decision Hypothesis, on the other hand, predicts that sentential context can be used to choose among the alternatives. In (6a), the verb *saw* has a primary meaning of physical perception, compatible with the direct object. In (6b), the verb *realize* has a primary meaning of comprehension of a concept, which is most compatible with a clausal complement. This information should be used to choose between the alternatives, according to the Immediate Semantic Decision Hypothesis. This in fact was the pattern that was found for the sentences in (6). However, in the Holmes, Stowe and Cupples experiment, the length of the ambiguous phrase was also varied (e.g. *the woman* vs. *the woman who cam in*). The differences in reading times between the ambiguous and control sentences for the complement verb phrase, which provides disambiguating information for the ambiguous sentences, are summarized in Table 2.

Table 2: Mean Difference between Ambiguous and Non-Ambiguous Sentences for Holmes, Stowe and Cupples Experiment 3

| | Verb | Next | Next |
|---|---|---|---|
| Object Preference | | | |
| Short NP | 97 | 49 | 8 |
| Long NP | 71 | 36 | 16 |
| Complement Preference | | | |
| Short NP | 1 | 25 | -17 |
| Long NP | 68 | 36 | -3 |

A positive difference means that the verb phrase causes increased parsing difficulties in the ambiguous sentences, in other words that people have garden pathed. For the short versions, the Immediate Semantic Decision Hypothesis appears to be correct. People only have difficulty in those cases where the semantic bias of the sentence should cause the choice of the wrong structure. However, for the long versions, it appears that the Minimal Attachment Hypothesis is correct. People have trouble with the more complex syntactic structure despite the semantic bias. Frazier and Rayner (1982) found a trend in the same direction. Clearly, the predictions of neither hypothesis are completely met.

How then are these results to be accounted for? The way that the results were described in the last paragraph perhaps provides a clue. Both the hypotheses appear to tell part of the story. That is, early in processing, semantic information has an impact that mimics an immediate decision in the correct direction. However, if the ambiguous region is long, it appears that semantic information loses its effect and the syntactic structure determines what happens. It hardly seems credible that semantic information is used to make an early decision and then a structurally based criterion causes a revision. However, the lack of a garden path effect at the verb phrase for the sentences containing a bias toward a complement completion may not be because the complement structure has bee **chosen** but because **no decision** has been made. That is, semantic information appears to **delay the decision**, rather than causing a decision to be made, for this ambiguity.

Clearly the Delayed Decision Model requires elaboration. First, if ambiguity is typically recognized and decision is delayed until disambiguating information is available, there is no reason to expect garden pathing for **any** of these sentences. The results do not support that prediction. In the case of the long ambiguous region, people typically do garden path. The most obvious explanation for that is that they cannot delay the decision any longer. That is, the hypothesis is that there is a **Limited Delayed Decision.**

There are two other points to notice about these results. First, it is clear that when the decision cannot be delayed any longer, the basis for the decision is syntactic structure rather than contextual bias, since Minimal Attachment effects are found for both context types. A second point is that the decision is delayed only for those sentences where the semantic bias is in opposition to the minimal attachment structure. Even short ambiguous regions for those sentences with a contextual bias toward the direct object structure show garden path effects.

Additional support for this model comes from further experiments reported by Cupples and Stowe (ms.). Sentences like those in the experiment discussed earlier were used, but the memory or processing load was altered in various ways. In one version, an introductory phrase was added to the beginning of the sentence, as illustrated in (7).

(7) a. According to the major, the officers taught at the military academy were very demand ing.

b. According to the major, the courses taught at the military academy were very demand ing.

This phrase must be integrated with the rest of the sentence and thus must be held in memory and accessed during comprehension of the rest of the clause. Another change in method was that the words did not accumulate on the screen; only one word was visible at a time, although each word appeared in the position that it would have if the entire sentence were visible. Presumably both of these alterations change the processing load associated with reading the sentence.

In another version, the sentences did not

Table 3: Mean Reading Times for Reduced Relative Clauses: Central Presentation

| | RelVerb | Next | Next | MainVerb | Next |
|---|---|---|---|---|---|
| Animate Subject | | | | | |
| Ambiguous | 416 | 401 | 396 | 497 | 492 |
| Unambiguous | 419 | 415 | 410 | 439 | 435 |
| Inanimate Subject | | | | | |
| Ambiguous | 428 | 421 | 402 | 485 | 459 |
| Unambiguous | 414 | 407 | 402 | 448 | 422 |

accumulate on the screen, but appeared only one word at a time. This means that subjects must rely on their memory of the preceding material to accomplish sentence comprehension. Table 3 contains data from this version of the experiment. It is obvious that people required additional reading time during the main verb phrase for both the animate and inanimate versions. This pattern of results suggests that people select the main verb phrase analysis despite the presence of semantic information that could guide the choice. This result was found for the version with an introductory phrase described above as well.

Thus, we again see a pattern where in the easier case, semantic information is used in a way that mimics a correct decision, but that structure appears to guide the decision under more difficult conditions. As with the direct object/complement clause ambiguity, these results can be accounted for by a model in which a decision about an ambiguity is delayed in just that case where there is semantic information in opposition to the structural bias. Again there are limits past which a decision cannot be delayed, although the limits are set by other processing load considerations rather than by simple length.

One last question concerns the contrast between the pattern of results seen in the reduced relative clause and complement clause cases and those reported by Taraban and McClelland (1989). There semantic information appeared to be used to actually make a decision, producing garden paths in both directions, as predicted by the Immediate Semantic Decision model. I would like to suggest that this case is different than those that I have just discussed and an immediate decision is indeed made. The cases differ in that for both of the sets of experiments that supported the Limited Delayed Decision model, additional information is always eventually available to disambiguate between the two alternatives: that is, there is something to delay for. With the prepositional attachment case, no syntactically disambiguating information can be expected and therefore, in terms of the syntactic structure, there is nothing to delay for.

One implication of this argument is that Minimal Attachment, as such, is not the reason for the syntactic preference found in reduced relative clauses and complement sentences. There is no overall preference to build the simplest structure. If there were, Taraban and McClelland ought to have found evidence of a conflict of syntactic and semantic bias and a delayed decision, just as for the reduced relatives and complement clauses. In stead of an across the board preference, a syntactic preference seems to be found only where at least one of the alternatives cannot be completed without the presence of additional material, but the other can. Consider *The courses taught at the university*. If that is the entire sentence, then in fact the ambiguous sequence is the main verb phrase, no matter how anomalous the interpretation. It is only locating an alternative main verb phrase that allows the reduced relative interpretation. If it is not found by the end of the sentence (or within a reasonable period of time), then the main verb phrase analysis is chosen. In other words, it is not the simplicity of one structure as opposed to the other that determines the choice, but instead whether additional structure is necessary to validate one of the structures. I have called the alternative that does not require additional material the **syntactic default**.

## Conclusion

To summarize then, I have discussed evidence that when a sentence contains a syntactic ambiguity:

(1) An **immediate choice** is made **if possible.**

(2) Both syntactic preference and semantic information are taken into account in making the choice.

(3) The basis of syntactic preference is not simplicity of structure, but whether the structure requires additional material to be completed. If it does

not, it can serve as the preferred **syntactic default.**

(4) If the syntactic default and semantic bias are in opposition, the **decision is delayed.**

(5) The decision can only be delayed as long as there are processing resources available to do so: **Limited Delay**.. If resources run out, an immediate decision is made, and the basis of the decision is the syntactic default, not semantic bias.

Clearly additional evidence is necessary to fully support this model. However, it seems to be compatible with a large amount of data for which the Minimal Attachment and Immediate Semantic Decision models fail.

## REFERENCES


Crain, S. and Steedman, M. 1985. On not being led up the garden path: The use of context by the psychological syntax processor. In D. Dowty, L. Karttunen, and A. Zwicky, eds., *Natural language processing: Psychological, computational and theoretical perspectives.* Cambridge: Cambridge University Press.

Cupples, L., and Stowe, L.A. 1991. Thematic structure and the disambiguation of reduced relative clauses. Unpublished manuscript, Ohio State University.

Frazier, L., and Rayner, K. 1982. Making and correcting errors during sentence comprehension: Eye movements in the analysis of structurally ambiguous sentences. *Cognitive Psychology* 14:178-210.

Holmes, V.M., Stowe, L.A., and Cupples, L. 1989. Lexical expectations in parsing complement-verb sentences. *Journal of Memory and Language* 28:668-689.

Rayner, K., Frazier, L., and Carlson, M. 1983. The interaction of syntax and semantics during sentence processing: Eye movements in the analysis of semantically biased sentences. *Journal of Verbal Learning and Verbal Behavior* 22:358-374.

Taraban, R. and McClelland, J. 1989. Constituent attachment and thematic role assignment in sentence processing: Influences of content-based expectations. *Journal of Memory and Language* 27: 597-632.

# Feature Diagnosticity as a Tool for Investigating Positively and Negatively Defined Concepts

**Robert L. Goldstone**

Psychology Department
Indiana University at Bloomington
Bloomington, IN. 47405

## Abstract

Two methods of representing concepts are distinguished and empirically investigated. Negatively defined concepts are defined in terms of other concepts at the same level of abstraction. Positively defined concepts do not make recourse to other concepts at the same level of abstraction for their definition. In two experiments, subjects are biased to represent concepts underlying visual patterns in a positive manner by instructing subjects to form an image of the the learned concepts and by initially training subjects on minimally distorted concept instances. Positively defined concepts are characterized by a large use of nondiagnostic features in concept representations, relative to negatively defined concepts. The distinction between positively and negatively defined concepts can account for the dual nature of natural concepts - as directly accessed during the recognition of items, and as intricately interconnected to other concepts.

## Introduction

The central purpose of this paper is to characterize two different methods of defining concepts. The characterization is supported by experiments that systematically affect the relative use of these two definition methods by human subjects. The distinction is between positively and negatively defined concepts. A concept is negatively defined if it is defined in terms of, or depends upon, other concepts at the same level of abstraction. A concept is defined positively if its intension does not make recourse to other such concepts.

Concepts seem to be directly accessed on the one hand, and intricately connected to each other on the other hand. While certain shapes seem to be instantly recognized as *dog* examples, the concept *dog* also depends on concepts such as *mammal*, *tail*, and *domesticated* for its meaning. The complex and structured nature of concepts such as *domesticated* is more consistent with viewing the *dog* concept as relating to an equally rich *domesticated* concept rather than containing it as a feature. There is often a dual-nature of concepts: concepts as bins used in classifying objects, and concepts as the units that define, elaborate, and explain other concepts. Concepts appear to be isolated from each other, each acting as an independent detector polling the world and yet concepts also seem to be intricately and deeply connected to each other in a virtually seamless network.

## Ways to be a Negatively Defined Concept

The original use of "negatively defined" concepts comes from Ferdinand de Saussure (1915/1959) who argued that all linguistic concepts are solely negatively defined. He argued that "Language is a system of interdependent terms in which the value of each term results solely from the simultaneous presence of the others" (p. 114) and that "concepts are purely differential and defined not by their positive content but negatively by their relations with the other terms in the system." (p. 117). For example, the French word "redouter" (dread) is defined by its opposition to words such as "craindre" (fear) and "avoir peur" (Be afraid). If "redouter" did not exist, then "all its content would go to its competitors." (p. 116). Specific examples of negatively defined concepts might include:

1. Relative properties: "Concept X is P-ier than concept Y." Part of our *toy poodle* concept is that they are smaller than *standard poodles*. Cafe Latte is capuccino, but milkier. Often concepts that belong to the same contrast set (Monday, Tuesday, Wednesday... belong to the same contrast set) are defined in terms of relative properties, where the values are relative to the other members in the contrast set.

2. Added or missing properties: "Concept X has property P where concept Y doesn't." Unicorns, as a first pass, are horses with horns. A person might think of the *absence of laces* when thinking about moccasins, and the *absence of flight* when thinking of

penguins. A property can be missing if it was expected (Kahneman & Miller, 1986), and "added" if it was unexpected. The use of "And everything else" concepts also falls into this category. The concept *Gentile* is used as an "And everything else" category that means anybody that is not Jewish. *Jew* could be positively defined, and *Gentile* would gain its meaning by contrast to this category.

3. Functional or theoretical definitions: "Concept X is part of a system with terms A, B, C." Within the game of baseball, the meaning of *strike* depends on such concepts as *batter, ball, strike zone*, and *swing* and, in turn, is necessary to define terms such as *out* and *steal*. Philosophers have been interested in concepts that are defined by their role in a system. Fodor (1983) has recently discussed "Quinean" (where "the degree of confirmation assigned to any given hypothesis is sensitive to properties of the entire belief system" (p. 107)) and "isotropic" (where "the facts relevant to the confirmation of a ... hypothesis may be drawn from anywhere in the field of previously established ... truths." (p. 105)) systems. Theory-dependent concepts are not defined in terms of their contrast set, but in terms of their theory cohabitants.

4. Niche-defined concepts: "Concept X, not concept Y, extends to region R." Saussure gives the example of the French word for Sheep: "mouton." While English speakers have a separate word for sheep that is eaten ("mutton"), French speakers have no equivalent term. As such, "mouton" extends to cover the "eaten sheep" extension whereas "sheep" does not. The governing metaphor is one of regions and feature spaces. A particular term covers a certain conceptual area, and terms that are close neighbors will compete for the right to cover a particular area. Recent models of categorization have given mathematical rigor to this notion by representing category examples as points in a MDS space (notably, Nosofsky, 1986). While Nosofsky's intensional representation of categories is solely in terms of their examples, Saussure argues that the actual intension of concepts changes due to "border wars."

5. Extrinsic definition: "Concept X involves concept Y." Barr & Caplan (1987) have recently made the distinction between intrinsic and extrinsic concepts. Extrinsic concepts make reference to objects outside of themselves; *hammer* is an extrinsic concept because its meaning involves *used to hit nails* where *nails* is a separate concept. Natural kinds such as *robin* are mostly intrinsically defined, because the concepts they make reference to, such as *wings* and *eyes*, are parts of the robin; they are not extrinsic to the bird.

## Degrees of Positive Categorization

Upon reading the list of negative-definition techniques, one may agree with Saussure - all concepts are completely negatively defined. One logical reason to doubt this is that 1) finding a negative definition for concept X first requires finding what neighbors/associates/theories X is related to, but 2) finding these, in most or all cases, requires that X first have some positive characterization. For example, Saussure claims that *mutton* is influenced by its neighboring concepts of *sheep* and *roast*. But, this analysis already assumes that there is a method for determining where a concept is roughly located in a conceptual space. One can have neighbors only if one first has a location. This location can be considered the concept's initial positive definition.

A good question for deciding whether a concept is positively defined is: "Could this concept still be possessed if some/most/all other concepts were eliminated?" Some of the same features associated with Fodorian (Fodor, 1983) modules (fast processing, automaticity, cognitive impenetrability, informational encapsulation) would be associated with positively-defined categories. Going from clear cases of positive definition to less clear cases:

1. Feature Detectors. Hubel & Wiesel (1968) feature detectors are clear cases of positive identification. In order for a feature detector to fire when a line of a particular orientation appears in a certain area, the feature detector needs to know absolutely nothing about other feature detectors or concepts or theories. To the extent that other recognition decisions can be made by higher-level feature detectors, these categorizations are also positively based.

2. Templates. Related to feature detectors, it has been proposed that categorization proceeds by comparing the item to be categorized to a template representing a typical example of the category. The template is often considered to be a nondecomposable or holistic representation. As such, templates are also clearly positive definitions - not decomposed into or dependent upon other concepts. If faces are recognized by comparing them to learned templates, then the concept of *Sam's face* is not

defined in terms of concepts such as *bushy eyebrows* and *thick lips* - the concept is simply the image itself.

3. Decomposable concepts. A concept can still be quasi-positive even if it is decomposable in terms of other concepts, as long as the component concepts seem to be on a lower level, where "level" roughly refers to abstractness or processing distance from perceptual input. Bruner, Goodnow, and Austin's (1956) concept formation stimuli are decomposable into more elementary features: number of borders, shape, size, and color. However, their concept *red or large square* still seems positively-defined. While it is true that the criterial question yields a negative answer ("The concept could not exist without the *red* concept"), it seems quasi-positively-defined because 1) there is some hope that the component concepts can someday be perceptually given (by feature detectors, for instance), and 2) the concept does not make recourse to other concepts at the same level of abstraction.

4. Hermit concepts. While some concepts are involved in tight contrast sets, others seem not to have close, interchangeable neighbors. While some concepts take part in intense competition for a particular region, others seem to have a relatively "off the beaten path" niche. While *spaghetti* and *linguine* fight over small parcels of dear real estate and would therefore be expected to greatly influence each other's definition, *lasagna* has more breathing room and might be relatively more positively defined.

## Methods for Analyzing Conceptual Structure

It is not clear whether NDCs share any deeper similarities other than being defined by other equally-abstract concepts. The taxonomy of NDCs indicates an eclectic assortment of concepts. In this paper, a tool for identifying a subset of NDCs is proposed. The subset corresponds roughly to those NDCs that are defined by added or missing properties. One way to tell whether a concept is positively or negatively defined is by observing the influence of nondiagnostic features on categorization accuracy. The difference between diagnostic and nondiagnostic features is only relevant for negatively defined concepts (NDCs). For positively defined concepts (PDCs) any feature that is present in the concept representation will be used for categorization purposes. For PDCs, category validity, the probability that concept X has feature F, is the representational basis. If instances of X usually have F then F will be part of X's positive definition, and the probability of categorization items into X will increase if the items have F.

Diagnosticity, or cue validity, of features only becomes an issue when there are a set of candidate categories, and the system is looking for ways to distinguish between the possible choices. Nondiagnostic features, features that do not distinguish between the possibilities, should have no effect on categorization for NDCs. If there are only two equally likely categories possible, and both categories have an 80% chance of having feature F, then the cue validity of the feature for either category will be 50% (the feature is uninformative with respect to deciding between the categories). The experimental question, then, is whether features with 50% cue validity and greater than 50% category validity increase the accuracy of categorizing if they are present in the item to be categorized. For PDCs these nondiagnostic features should increase accuracy. For NDCs nondiagnostic features should not increase accuracy. A full range of intermediate results are possible, depending on factors that would bias concepts to be encoded positively or negatively. An informative operational measure is to take the ratio of the effect of a nondiagnostic feature on categorization accuracy divided by the effect of a diagnostic feature. Pure PDCs predict this value to be 1; pure NDCs predict this value to be 0.

Artificial stimuli are used in the experiments because it is important that subjects do not have previous familiarity with the materials to be learned, and features must be identifiable as diagnostic or nondiagnostic. For one experiment, the category items shown here were used.

Subjects see distortions of the examples of Foos and Bars shown on the previous page, and categorize the distortions into the correct group. The distortions are formed by randomly switching, on a certain proportion of trials, each of the 20 line segments (7 black lines, 13 white lines) that make up these prototypical Foo and Bar pictures.

The features are assumed to be the individual line segments. A line segment is nondiagnostic if Foos are as likely to have the segment as Bars are. A feature is diagnostic if the feature is part of one category's prototype but not the other's. Task manipulations hypothesized to make Foo and Bar positively defined should result in a relatively strong influence of nondiagnostic features. Manipulations that tend to make one concept defined in terms of the other concept should yield NDCs and less influence due to nondiagnostic features.

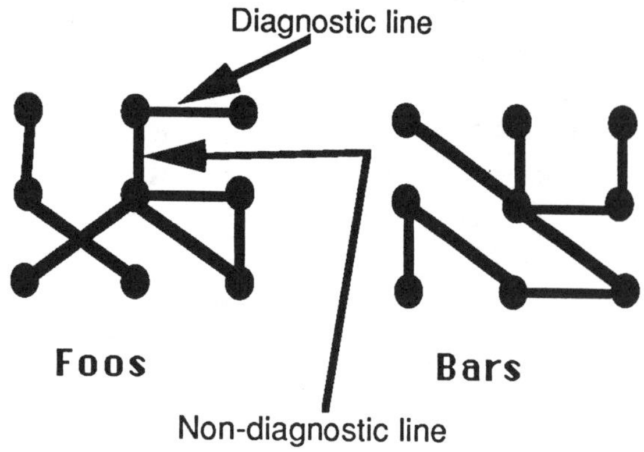


**Manipulation 1: "Form image" vs. "Find distinguishing features" instructions**
Twenty-eight undergraduates from University of Michigan were divided into two groups. One group of subjects (the Image instructions group) were told "While you are learning the two categories, you should try to form an image of what each category looks like." The other groups of subject (the Discriminate instructions group) were told "While you are learning the two categories, you should try to find particular features in the pictures that help you distinguish between the two categories." The first set of instructions was aimed at promoting PDCs; if an image is formed for each group, then there should be little influence of one concept on another's concept's representation. The second set of instructions was aimed at promoting NDCs; a concept's distinguishing features are only diagnostic relative to another concept.

Six hundred pictures were displayed to subjects. On each trial, either a Foo or a Bar prototype was distorted by randomly changing each of its line segments from black to white or from white to black 20% of the time. Eight of the line segments were diagnostic (shared by both Foos and Bars) and twelve were nondiagnostic. Once a picture was displayed, the subject pressed one of two keys, to indicate their category prediction. They received feedback indicating whether their choice was correct and what the correct response was.

Each picture can be analyzed in terms of how many diagnostic and nondiagnostic features are altered from the category's prototype. In general, the more features that are altered, the harder it will be to correctly categorize a picture. The results, shown on the previous page, indicate that this is the case. As we change a greater number of features, categorization performance uniformly decreases (all of the lines slant downwards). The graph also indicates that altering diagnostic features (represented by circles in the graph) is more detrimental to categorization than changing nondiagnostic features (represented by squares). This is indicated by the slope of the lines. Going from 0 to 4 features changed has more of an influence if the features that are changed are diagnostic features than if they are nondiagnostic features. This is an intuitive result because if a concept's diagnostic feature is changed, then the picture becomes more similar to the other concept. If a white line segment in a certain position is diagnostic of Foos, then a black line segment in the same position is diagnostic of Bars. Conversely, altering nondiagnostic segments decreases the resemblance of the distorted picture to both concepts.

A less obvious result is that nondiagnostic lines do increase categorization accuracy if they are present. Even though a nondiagnostic feature, by itself, cannot serve to discriminate between the categories, it does increase the percentage of correct classifications. If line segments are in fact the psychologically relevant features (see below) then the influence of nondiagnostic lines on accuracy suggests that the categories are positively defined to at least some degree. PDCs benefit from the presence of nondiagnostic features because such features increase the similarity of the picture to concept. A categorization rule for PDCs might be "If there are X or more features in common between a picture and a concept's prototype, then the picture belongs to the concept. If both pictures exceed the threshold, or neither do, then categorize the picture randomly." Using this threshold rule, categorization accuracy will often (depending on the stimulus structure and threshold) be greater if nondiagnostic features are present than if they are not. For an alternate mathematical model that predicts an influence of nondiagnostic features on accuracy, see Nosofsky (1991).

In addition to the interaction between segment diagnosticity and number of features changed on accuracy, there is also a three way interaction involving these two factors and the instruction type. When subjects are given the PDC instructions ("form an Image" - represented by black figures in the above graph) then we find that nondiagnostic features are more important than if subjects are given the NDC instructions ("look for distinguishing features" - indicated by white figures). If nondiagnostic features are altered and subjects are given the NDC instructions, there is very little effect on accuracy. These nondiagnostic features have a significantly greater influence when subjects are told to form an image of the concepts. Using categorization accuracies, the ratio

$$D = \frac{(0\ NDA) - (4\ NDA)}{(0\ DFA) - (4\ DFA)}$$

was computed for each subject, where NDA is "nondiagnostic features altered" and DFA is "diagnostic features altered." A large D value indicates a relatively large influence of nondiagnostic features. The ratios were significantly larger for the PDC instruction group than the NDC group (Unpaired T (26) = 2.7, $p<.05$). As such, the instructional manipulation seems to support the functional distinction between positively and negatively defined concepts.

At the end of the experiment, subjects were asked to draw a picture of the "best example of each of the categories" on a 3 X 3 grids. The pictures were analyzed for how many diagnostic and nondiagnostic features were correctly drawn. "Image" instructions yielded higher proportions of correctly drawn features (an average of 14.6 out of 20 correct segments) than "Discriminate" instructions (13.9 correct segments). Importantly, the difference is particularly large for nondiagnostic features (image = 8.2 out of 12 correct segments; discriminate = 6.7 correct). This provides some evidence that the instructional manipulation changes the internal representations of the concepts, and not just the categorization profiles. The PDC group is more likely to represent their concepts in terms of nondiagnostic features than is the NDC group. The following rebuttal is possible: This analysis depends on individual line segments being the correct unit of analysis. What if psychological features do not coincide with the experimenter-determined features? For example, suppose that subjects have "Foo" encoded as "has an 'X' in the lower left hand quadrant." This feature is diagnostic in that Bars do not typically have this feature. But, by taking away so-called "nondiagnostic" lines, we are eliminating this diagnostic feature.

However, the results we obtained validates our choice of line segments as features. Under some circumstances, subjects' analysis of features do coincide with the experimenter-determined analysis. Specifically, when subjects are instructed to look for diagnostic features, the lines we are calling nondiagnostic do not have much influence on categorization accuracy. Simply saying that subjects' features do not agree with the experimenter's features will not explain why sometimes the two do agree, and if we can present a characterization of concepts that explains when and why the two agree, then the PDC/NDC conceptual analysis gains support.

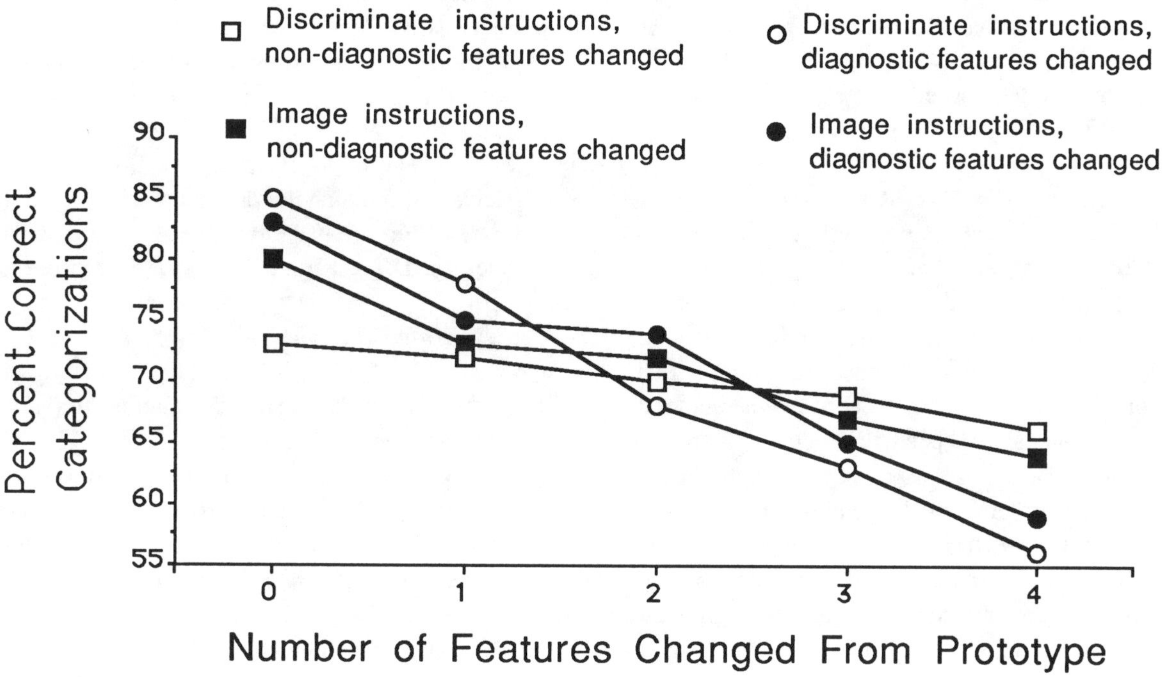

**Manipulation 2: increasing vs constant distortion**

Using stimuli similar to the first manipulation, sixteen subjects saw distortions in which, on average, 1/5 of the line segments were switched on every trial. Sixteen other subjects were shown pictures that became successively more distorted. In the first 100 trials, 1/15 of the line segments were switched; in the second 100 trials, 1/10 of the line segments were switched; in the remaining 400 trials, 1/5 of the line segments were switched. The first group saw pictures at a constant level of distortion. The second group saw pictures highly similar to the prototype early in learning, that became increasingly distorted with practice. The "increasing distortion" group was hypothesized yield more PDCs than the "constant distortion" group because A) it would be easier to form an image of the concepts early in training, and B) the instances from the concepts would be more dissimilar from each other, and consequently would not require the fine distinctions and focused testing that is characteristic of searching for discriminating features.

The hypothesis is supported by the data. Data from trials 200-600 were analyzed. An average D value of 0.42 was found for the increasing distortion condition, compared to a D value of 0.28 for the constant distortion condition (Unpaired $T(24) = 3.51$, $P<.05$). Drawn pictures of the concepts had 9.1 and 7.7 nondiagnostic features correctly depicted for the increasing and constant distortion groups, respectively (Unpaired $T(24) = 3.45$, $p<.05$). Both results indicate that nondiagnostic features are part of concept representations to a greater degree when the concepts

## Conclusions

The above proposal has been a first pass at investigating the seemingly dual nature of concepts: concepts as directly accessed for the purpose of recognition, and concepts as interconnected and defined in terms of one another. Directly accessed concepts, without inter-conceptual relations, cannot account for the elaborate network of conceptual dependencies, competitions, and explanations that humans exhibit. Concepts that are solely defined in terms of their inter-conceptual relations, without any perceptual grounding or conceptual independence, cannot account for the connection between our conceptual structure and our physical world. It is argued that concepts are represented both positively (independent of other concepts) and negatively (in terms of other concepts). The proposed distinction was empirically tested by two task manipulations that, on a priori grounds, were thought to bias subjects toward positively or negatively defined concepts. Task manipulations that bias subjects toward positively defined concepts result in greater relative sensitivity to nondiagnostic stimulus features, and greater incorporation of nondiagnostic features in subject's drawings of concepts. In contrast to previous work on categorization that stresses the learning of diagnostic features, the current work suggests that much of categorization involves comparing objects to representations that incorporate nondiagnostic features. Features that by some analyses cannot distinguish between concepts still can serve to increase categorization accuracy by increasing item-to-concept similarity. Not only do we look to distinguish between choices; we also look for information that is consistent with one possibility, irrespective of other candidates.

## Acknowledgments

This paper has greatly benefitted from suggestions from Dorrit Billman, Frances Kuo, Richard Nisbett, Robert Nosofksy, Edward Smith, and Douglas Medin.

## References

Barr, R. A., & Caplan, L. J. 1987. Category representations and their implications for category structure. *Memory & Cognition* 15:397-418.

Bruner, J. S., Goodnow, J. J., & Austin, G. A. 1956. *A study of thinking*. New York: Wiley.

Fodor, J. A. 1983. *The modularity of mind*. Cambridge, MA: MIT Press/ Bradford Books.

Hubel, D. H., & Wiesel, T. N. 1968. Receptive fields and functional architecture of monkey striate cortex. *Journal of Neurophysiology* 195: 215-243.

Kahneman, D., & Miller, D. T. 1986. Norm theory: comparing reality to its alternatives. *Psychological Review* 93:136-153.

Nosofsky, R. M. 1986. Attention, similarity, and the identification-categorization relationship. *Journal of Experimental Psychology: General* 115:39-57.

Nosofsky, R. M. 1991. Tests of an exemplar model for relating perceptual classification and recognition memory. *Journal of Experimental Psychology: Human Perception and Performance* 17:3-27.

Saussure, F. 1959. *Course in general linguistics*. New York: McGraw-Hill.

# Context-Sensitive, Distributed, Variable-Representation Category Formation*

**Mirsad Hadzikadic and Paul Elia****

Department of Computer Science
University of North Carolina
Charlotte, NC 28223

## Abstract

This paper describes INC2, an incremental category formation system which implements the concepts of family resemblance, contrast-model-based similarity, and context-sensitive, distributed probabilistic representation. The system is evaluated in terms of both the structure of categories/hierarchies it generates and its categorization (prediction) accuracy in both noise-free and noisy domains. Performance is shown to be comparable to both humans and existing learning-from-example systems, even though the system is not provided with any category membership information during the category formation stage.

## Introduction

The issues of category formation and categorization represent an important research topic due to the fact that categories lie at the core of our thought, perception, speech, and action. Researchers from several diverse disciplines (psychology, philosophy, linguistics, anthropology, and computer science) actively work in this area. Computer science, in particular, offers wealth of results under the common term of *concept formation.*

Fisher & Langley (in press) describe concept formation as the incremental unsupervised acquisition of categories. A system which can accomplish this task can be used both as an aid in organizing and summarizing complex data and as a retrieval system which can predict properties of previously unseen objects. Such a system will be useful in domains where knowledge is incomplete or classifications and/or human experts do not exist.

This paper describes INC2, an incremental concept formation system which combines the effectiveness of similarity-based learning methods (computer science) with the plausibility of the modified contrast-model (psychology) and family-resemblance (philosophy) theories. The system uses a context-sensitive, distributed probabilistic representation to store the knowledge about categories, their descriptions, and members. The system has been evaluated in the domains of soybean disease, breast cancer, and primary tumor cases. When compared to humans and existing learning-from-examples systems, INC2 shows comparable performance in terms of prediction accuracy.

## INC2

INC2 shares the object/category representation formalism with its predecessor, the INC system (Hadzikadic & Yun 1989). Also, they both incrementally build a hierarchy (tree) of disjoint categories (although an object may match descriptions of more than one category) in an unsupervised fashion. That, however, is where the similarities end. We will now concentrate on the description of the INC2 system.

### Representation

As already mentioned, INC2 builds a hierarchy of non-disjoint category descriptions. The leaves of the hierarchy are objects (singleton categories). The root of this hierarchy has associated with it a description which is a summary of all descriptions of the objects seen by the system to date. As one traverses the hierarchy, downward pointers lead to nodes with more specific descriptions, while upward pointers lead to nodes with more general descriptions.

A description of each category $C$ is defined as a set of features $f$ (attribute-value pairs). Each feature has a conditional probability $p(f \mid C)$ associated with it. Thus, representing the color feature of red apples would take the form (*color red* 0.25). The 0.25 means that members of this category are red 25% of the time. Consequently, singleton categories will have probabilities equal to 1.0. In addition to nominal attributes, INC2 supports the structural ones as well. Representing a structural fact such as the fact that object $a$ is inside object $b$ would take the form (*contains* ($b$ $a$) 1.0). Finally, INC2 supports structured domains, i.e., (*shape triangle* 1.0) will match (*shape square* 1.0), provided the knowledge that both triangles and squares are specializations of polygons.

* This work was supported by a grant from the UNCC, College of Engineering.

** Currently with the Energy Management Associates, Inc, Atlanta, Georgia.

The notion of *probabilistic concept representations* was introduced by Smith & Medin (1981). However, since members of a given category may reside in distinct portions of the hierarchy, the adopted representation formalism is referred to as *distributed probabilistic concept hierarchies* (Fisher & Langley in press).

## Family Resemblance

In addition to its features and hierarchical pointers, each category description contains an estimate of its cohesiveness given in the form of *family resemblance.* Family resemblance, first advanced by Wittgenstein (1953), is defined here as the average similarity between all possible pairs of objects in a given category. More formally, we define the family resemblance $FR$ of a given category, $C$, to be

$$FR(C) = \frac{\sum_{a,b} [s(a,b) + s(b,a)]}{2 \times \binom{n}{2}}$$

where $a \neq b$, $a$ and $b$ are members of $C$, $s(a,b)$ is any similarity function defined for two objects described with feature sets, $n$ is the number of children of the node associated with the category $C$, and $\binom{n}{2}$ is the number of distinct two-element sets of objects in $C$.

We interpret family resemblance as a measure of the cohesiveness of a category. To save processing time, INC2 compares pairs from the children of $C$ rather than from all the objects stored in the subtree headed by $C$ to approximate the family resemblance for a given category, $C$. A special case arises for categories which do not have any pairs, namely singleton categories. The value we used for the family resemblance of a singleton category is the family resemblance of its least compact sibling. If there are no siblings, the family resemblance of the parent is used instead. In essence, this means that two objects will form a new category if their similarity is greater than either the average similarity associated with the least compact sibling or the average similarity between all pairs of objects within the given context (assuming that no other category represents a better host for the object).

## Similarity Function

The similarity function $s$ used by INC2 represents a variation of the contrast model (Tversky 1977) which defines the similarity between an object and a category as a linear combination of both common and distinctive features. Our modification of the contrast model includes: (a) elimination of a referent since comparisons take place between nodes at the same level of the hierarchy, (b) consequent introduction of the symmetricity property, thus reducing the number of comparisons needed by 50%, and (c) normalization of the function so that the values fall into the (-1.0, 1.0) range, with 1.0 denoting identical objects/categories and -1.0 indicating completely dissimilar ones. The new function is now formally defined as

$$s(A,B) = \frac{\lambda - \alpha - \beta}{\lambda + \alpha + \beta}$$

$$\lambda = \frac{m_A \times \sum_{f_{AB}} p(f_{AB} \mid A) + m_B \times \sum_{f_{AB}} p(f_{AB} \mid B)}{m_A + m_B}$$

$$\alpha = m_A \times \sum_{f_A} p(f_A \mid A)$$

$$\beta = m_B \times \sum_{f_B} p(f_B \mid B)$$

where $A$ and $B$ are (possibly singleton) categories, $\lambda$ represents the contribution of the common features, while $\alpha$ and $\beta$ introduce the influence of the features of $A$ not shared by $B$ and vice versa, respectively, $m_A$ is a number of objects stored under the node associated with the category $A$, $m_B$ is similarly interpreted for the category $B$, $f_{AB}$ is the set of features shared by $A$ and $B$, $f_A$ is the set of features present in the description of $A$ but not $B$, and $f_B$ is the set of features present in the description of $B$ but not $A$.

## Operators

INC2 uses four operators to guide the category formation process: create, extend, merge, and delete. *Create* forms a new category for an object found to be dissimilar to all examined categories, while *extend* adds a new object to the most similar category found.

*Merge* unites two or more categories at the same level in the hierarchy that are found to be similar to a new object to form a new category. The object is then recursively classified with respect to the category which maximizes the increase in its cohesiveness upon incorporating the object.

When the family resemblance of a given category is less than the family resemblance of its parent, that category does not represent a proper specialization of its parent. Such a situation is likely to occur in noisy domains upon application of the merge operator. The *delete* operator rectifies this problem by removing the category and promoting its specializations.

## Algorithm

Figure 1 presents the classification procedure of INC2. It implements a hill-climbing strategy which encourages advancement toward the maximal improvement of the hierarchy as measured by the increase in the family resemblance of every candidate host category.

The algorithm can be paraphrased as follows. To begin, pass to the procedure both an object, *a*, and the root of the hierarchy, *C*. The first action is to update the description of the category *C* based on the description of the object *a*. Next, find the change in the family resemblance measures (ΔFR) that would result from temporarily placing *a* in each of *C's* specializations (subcategories).

If only one subcategory experiences improvement (an increase in its family resemblance measure), then either extend this subcategory (if singleton) or call the classification procedure recursively on *a* and that subcategory, the *besthost*. If two or more subcategories experience improvement, then find the subcategories which are at least as similar to *a* as the family resemblance of *C*, i.e., those subcategories which are more similar to *a* than the degree of compactness in the given context. These categories are merged together to form a new category, with *a* continuing the classification process recursively with respect to the new best host (maximizing the increase in its family resemblance). If no subcategory experiences improvement, then *a* is unique and a new singleton category is created.

Once a home for *a* is found and all category descriptions affected have been updated, the system updates the family resemblance measures of the categories on the path from *a* to the root.

**Classify(a,C)**
Update the description of the category *C* using *a*.
Compute ΔFR for each of *C's* subcategories.
IF a single subcategory (*besthost*) has a positive ΔFR
THEN

IF the best host is singleton

THEN call **Extend(a, best host)**

ELSE call **Classify(a, best host)**

ELSE

IF many subcategories have a positive ΔFR
THEN

(a) find the subcategories which are at least as similar to *a* as the FR of the parent category

(b) call **Merge(a, similar subcategories)**

(c) compute the description of the new category

(d) determine the subcategory *(best host)* that maximizes ΔFR

(e) call **Classify(a, best host)**

ELSE

(no subcategory has a positive ΔFR)
call **Create(a)** (create a new singleton category for *a*)

Update the FRs of the categories on the path from *a* to the root.
IF a merge was done
THEN

search that subtree for any class with a FR less than the FR of its parent, **deleting** if found, promoting its children one level higher, and recomputing the FR of the parent of the deleted category.

Figure 1: The classification algorithm.

Figure 2 presents the retrieval procedure of INC2. We begin by passing the procedure an object *a* and the root of the hierarchy. The next step is to compute the similarity between *a* and each of *C'*s subcategories. Then, the procedure is called recursively with the subcategory that maximizes the similarity function. The process stops after reaching a singleton category. Note that the retrieval procedure utilizes the similarity function rather than the concept of family resemblance since the latter is needed mainly for plausible clustering of objects.

**Retrieve(a,C)**
IF *C* is a singleton category
THEN return *C*
ELSE

(a) compute the similarity between *a* and all the subcategories of *C*

(b) find the subcategory (best candidate) that maximizes the similarity

(c) call **Retrieve(a, best candidate)**

Figure 2: The retrieval algorithm.

## Drop Threshold

Although it is not specified in the algorithms, both the classification and retrieval procedures rely on the *drop threshold.* This threshold allows for category descriptions to be either probabilistic or logical. It can be set between 0.0 and 1.0 and means that any feature which falls below this threshold in conditional probability should be dropped from the description of the given category. A value of 1.0 for this threshold would yield a logical category description.

However, the nature of the classification process calls for a dynamically adjustable threshold rather than a fixed one. For example, at the top level of the hierarchy all features are important no matter how low their probabilities might be, due to the potential noise in object descriptions as well as the diversity of objects in the domain. Therefore, the drop threshold should be set close to 0.0. At the lower levels of the hierarchy, however, certain patterns have been detected, resulting in high conditional probabilities for 'participating features' and, consequently, lower probabilities for the ones not significantly present in those patterns. However, since all categories at the lower levels have few members, all the features found in their descriptions will have relatively high conditional probabilities (1 out of 2 still gains the probability of 0.5). To avoid the interference of unimportant features with the retrieval process, the drop threshold should be set close to 1.0. The intermediate categories will, then, require the drop threshold somewhere between 0.0 and 1.0, depending on the level of the hierarchy (the lower the level, the higher the drop threshold).

In order to alleviate this problem, we rely on family resemblance to provide an estimate of the drop threshold. As we stated earlier, family resemblance can be interpreted as an estimate of the category compactness. It is naturally close to 0.0 at the root (summarizing the whole universe) and to 1.0 at the leaves. Therefore, during both classification and retrieval INC2 sets the drop threshold to the value of the family resemblance of the parent category. It increases with the object traversing the hierarchy downward.

This dynamically adjustable drop threshold and the fact that the expand, merge, and delete operators depend on the family resemblance of the parent category (as well as the candidate subcategory's siblings in the case of merge) represent two important features introduced by INC2. As a result, INC2 performs a context-sensitive classification/retrieval due to its adaptive behavior that changes from level to level of the hierarchy. In that process, consequently, INC2 uses different representations to describe objects/categories at different levels of the hierarchy, possibly moving from the probabilistic representation (drop threshold = 0.0) at the top level to the logical one (drop threshold = 1.0) at the leaves.

## Performance Evaluation

The performance of INC2 was evaluated in the domains of soybean disease, breast cancer, and primary tumor cases. Although a true expert in these domains would have access to much richer data, we show that the knowledge representation and algorithm used by INC2 yield a useful hierarchy of categories. The soybean disease domain was selected as a representative of noise-free domains, while the breast cancer and primary tumor cases included a lot of data with incorrect or missing values (a noisy domain). In each experiment, the information concerning the ideal category for a given object was not given to the system and training sets were randomly both selected and ordered.

The soybean disease domain consisted of forty-seven cases with four ideal categories represented. A high number of features used to describe the cases were common to all forty-seven cases, making the domain very compact. Training sets of sizes five to twenty-five in increments of five were selected randomly from the domain. For assessing prediction accuracy (determining category-membership for previously unseen objects), twenty cases were randomly selected from the set of remaining cases. Table 1 shows the range, mean, and sample standard deviation prediction accuracies for five experiments per training set size.

| Training Size (#) | Range (%) | Mean (%) | Sample Std. Deviation |
|---|---|---|---|
| 5 | 75-95 | 82 | 10.4 |
| 10 | 95-100 | 98 | 2.7 |
| 15 | 95-100 | 99 | 2.2 |
| 20 | 95-100 | 99 | 2.2 |
| 25 | 95-100 | 99 | 2.2 |

Table 1: INC2's prediction accuracies for five experiments per training set size in the soybean disease domain.

With four ideal categories in the soybean disease domain, there is a 25% chance of simply guessing the correct diagnosis. INC2's prediction accuracy was consistently above chance, even at low levels of experience. From the results summarized in the table 1 it is obvious that INC2's performance improves with experience and that it needs a small portion of objects from a domain to make a plausible decision with respect to category membership of previously unseen objects.

The breast cancer domain[1] (prediction of cancer recurrence five years later) consisted of 286 cases with two ideal categories represented, *yes* and *no*. The domain itself can be characterized as very noisy. A total of nine features were randomly missing from the 286 cases, with no more than two features missing from any one case. There were cases from two different categories bearing the same exact description. Five specialists[2] were presented this data and then tested for diagnostic accuracy. They were correct 64% of the time.

[1] Data from this domain were provided by the Institute of Oncology of the University Medical Center in Ljubljana, Yugoslavia.
[2] The specialists were from the Institute of Oncology, Ljubljana.

The other oncology-related example, the primary tumor domain[1] (prediction of tumor locations), consisted of 339 cases with 20 ideal categories represented. A total of 224 features were randomly missing from the 286 cases. Once again, there were cases from two different categories having the same exact description. Four internists and four specialists[2] were tested for diagnostic accuracy. Internists were correct 32% of the time, specialists 42%.

Since the results obtained in the soybean disease domain suggest that 25% of the total number of objects in the domain is sufficient for estimating a maximum achievable performance by the system, we have decided to use 25% cases for training and the remaining 75% cases for prediction. This is in sharp contrast with other learning systems which used 70% of the objects for training and the remaining 30% for prediction.

Table 2 presents the prediction accuracies for two learning-from-examples systems (AQ15 [Michalski et al 1986] and Assistant-86 [Cestnik et al 1987]), human experts, and INC2. Note that in a system which learns from examples, training cases are associated with the correct response, and the goal of the system at that point is to find a set of rules which will cover that data and be useful for classifying previously unseen objects.

| System | Breast Cancer (%) | Primary Tumor (%) |
|---|---|---|
| AQ15 | 66 | 39 |
| Assistant-86 | 78 | 44 |
| INC2 | 69.2 | 30 |
| Human Experts | 64 | 42 |

Table 2: Mean prediction accuracies for the oncology domains. The INC2's accuracy is obtained by averaging values achieved in five runs on randomly chosen sets of objects.

There is a 50% chance of simply guessing the correct diagnosis for the breast cancer domain and 5% for the primary tumor domain. INC2's prediction accuracy was shown to be significantly above chance. Furthermore, INC2 compares to both human experts and learning-from-examples systems given the same task (significantly better in the breast cancer domain than in the primary tumor domain). This is despite the fact that INC2 is never given any help from a teacher.

In order to evaluate the effectiveness of the fixed-value drop threshold compared to its variable, context-sensitive alternative, we have carried out a sequence of experiments in both soybean and breast cancer domains. In the soybean-disease experiments we fixed the size of the training set to ten (25% of the complete set), while varying the drop threshold from 0.0 to 1.0 with the increment of 0.25. The results are summarized in the table 3.

| Drop Threshold | Range (%) | Mean (%) |
|---|---|---|
| 0.00 | 90-100 | 98 |
| 0.25 | 100-100 | 100 |
| 0.50 | 95-100 | 98 |
| 0.75 | 90-100 | 98 |
| 1.00 | 95-100 | 99 |

Table 3: Prediction accuracies for the fixed-value drop threshold in the soybean disease domain.

The results are obviously equal or slightly better than that of the variable drop threshold. That is especially true for the value of 0.25 with its perfect score. However, when we applied INC2 with the drop threshold = 0.25 to the breast cancer domain its prediction accuracy dropped from 69.2% (for the variable drop threshold) down to 63.3%. To make sure that this was no accident we checked the basic case (drop threshold = 0.0), where all the features are taken into consideration at all times, and recorded a similarly reduced performance, 63.6%. It seems that the variable, context-sensitive drop threshold provides a more robust alternative for noisy domains while retaining a good performance in the noise-free ones.

Finally, in order to estimate the effect of ordering of objects on the resulting classification, we evaluated prediction accuracies for five different random orderings of the same set of objects. Table 4 presents the obtained results.

| Ordering (#) | Accuracy (%) |
|---|---|
| 1 | 71.0 |
| 2 | 64.5 |
| 3 | 61.7 |
| 4 | 64.5 |
| 5 | 68.2 |

Table 4: Prediction accuracies for five random orderings of the set of input breast cancer cases.

If we compare the sample standard deviation for the data presented in the table 4 (3.63) and for the data used to derive the value reported for the breast cancer domain for INC2 (4.53; Table 2), a decrease of 20%, then we can conclude that INC2 represents a relatively robust incremental category formation system with respect to the ordering of input objects, which has been a major drawback of all incremental systems. Furthermore, we suggest that the performance of different hierarchies rather than their form should be measured when evaluating the effect of object orderings on the system's performance.

## Previous Work

Most existing category formation systems use hill-climbing methods to find suboptimal clusterings of objects to be characterized and create nondisjoint category descriptions. Five existing systems which share these features are COBWEB (Fisher 1987), CLASSIT (Gennari, Langley, & Fisher 1989), UNIMEM (Lebowitz 1987), CYRUS (Kolodner 1984), and WITT (Hanson & Bauer 1989).

Out of the aforementioned incremental concept formation systems, INC2 is most similar to COBWEB. They utilize similar distributed probabilistic concept representation formalism and operators. However, INC2 possesses several features that significantly distinguish it not only from COBWEB but from all other systems as well:

**I.** Family resemblance provides an estimate of category compactness, which is used by INC2 to: (a) introduce an objective, domain-adaptive, context-sensitive bias for classification; (b) implement a variable, context-sensitive representation of concepts, thus focusing only on important features in the current context; (c) measure an improvement over a single category in addition to the improvement over the whole level of the hierarchy, thus utilizing finer-grained information to guide classification; and (d) achieve a relative robustness with respect to object orderings.

**II.** INC2 seeks tree-structures that optimize not only the top level of the hierarchy, but the hierarchy as a whole.

**III.** It introduces no constraints on input objects (they may be hierarchies of objects on their own).

## Conclusion

INC2 incrementally builds a hierarchy of category descriptions based on a set of objects described with nominal and/or structural attributes. The system is based on our own interpretation of both family resemblance and contrast model theories. It uses a context-sensitive threshold to eliminate all irrelevant features from concept descriptions, thus effectively introducing an adaptive, context-dependent representation of concepts. Performance has been shown to be comparable to both human experts and learning-from-example systems.

Future research will involve three directions: (a) reducing the number of steps required to compute family resemblance for categories with many children, (b) enhancing the representation with continuous (linear) attributes, and (c) performing further experiments in order to objectively evaluate the INC2's results in the light of both fan and typicality effects as well as the structure of categories.

## References


Cestnik, G. et al. 1987. Assistant-86: A Knowledge-Elicitation Tool for Sophisticated Users. *Progress in Machine Learning,* 31-45, Sigma Press.

Fisher, D. (1987). Knowledge Acquisition Via Incremental Conceptual Clustering. *Machine Learning* 2(2):139-172.

Fisher, D. and Langley, P. In Press. The Structure and Formation of Natural Categories. *The Psychology of Learning and Motivation,* G. Bower (Ed.).

Gennari, J. H., Langley, P., and Fisher D. 1989. Models of Incremental Concept Formation. *Artificial Intelligence* 4(1-3):11-61.

Hadzikadic, M. and Yun, D. Y. Y. 1989. Concept Formation by Incremental Conceptual Clustering. In *Proceedings of the Eleventh International Joint Conference on Artificial Intelligence,* August 20-26, Detroit, Michigan.

Hanson, S. J. and Bauer, M. 1989. Conceptual Clustering, Categorization, and Polymorphy. *Machine Learning* 3(4):343-372.

Kolodner, J. L. 1984. *Retrieval and Organizational Strategies in Conceptual Memory: A Computer Model.* Lawrence Erlbaum Associates, Publishers, London.

Langley, P., Gennari, J. H., and Iba, W. 1987. Hill-Climbing Theories of Learning. In *Proceedings of the Fourth International Workshop on Machine Learning,* 312-323, University of California, Irvine.

Lebowitz, M. 1987. Experiments with Incremental Concept Formation: UNIMEM. *Machine Learning* 2(2):103-138.

Michalski, R. S. et al. 1986. The AQ15 Inductive Learning System: An Overview and Experiments. Technical Report, Intelligent Systems Group, University of Illinois at Urbana-Champaign.

Smith, E. E. and Medin, D. L. 1981. *Categories and Concepts.* Harvard University Press, Cambridge, MA.

Tversky, A. 1977. Features of Similarity. *Psychological Review* 84:327-352

Wittgenstein, L. 1953. *Philosophical Investigations.* MacMillan, New York.

# Constraints on Analogical Mapping: The Effects of Similarity & Order

**Mark T. Keane**

Department of Computer Science,
Trinity College, University of Dublin,
Dublin 2, IRELAND
mkeane@cs.tcd.ie

**Tim Ledgeway & Stuart Duff**

School of Psychology,
University of Wales College of Cardiff,
PO Box 901, Cardiff, WALES
duff@cardiff.ac.uk

**Abstract**

One of the central problems in analogical mapping is overcoming the ambiguities which can occur when matching up corresponding concepts in two domains of knowledge; specifically, to ensure that one-to-many and many-to-one matches are resolved to be one-to-one matches. Various analogy theories have attempted to deal with these problems by maintaining that analogical matching is constrained in various ways. For example, that only predicates of the same structural-type are matched, that primacy is given to matches that are similar or identical, and that a match which comes before an alternative match is preferred. Two experiments are reported, involving an attribute-mapping problem, which isolate the effects of similarity and order. The first shows that the semantic similarity of predicates in the two domains has a facilitating effect on analogical mapping when other constraints are held constant. The second experiment shows that analogical mapping is sensitive to the order in which matches are made. The implications of these results for current computational models of analogy are discussed, with a special emphasis on the consequences that order effects have for connectionist models.

## Introduction

When we draw an analogy the hard work is done by the process that performs an analogical mapping between two domains of knowledge (see e.g., Burstein 1986; Carbonell 1986; Gentner 1983, 1989; Gick & Holyoak 1980, 1983; Holyoak 1985; Keane 1985, 1988a). Ranged about this core process are the other processes of analogical thinking; for example, processes concerned with the representation of the domains and the retrieval of analogues from long-term memory. Typically, when an analogical mapping is made two distinct computations occur; first, the corresponding concepts in both domains are *matched* and, second, a portion of the conceptual structure of one domain is *transferred* (or carried over) into the other domain to form the basis of analogical inferences. For example, if you want to understand why electrons revolve around the nucleus in the atom, and are told that the atom is like a miniature solar system (see Gentner 1983), you would *match* the corresponding REVOLVES relations in both domains and *transfer* relations of ATTRACTION from the solar system domain to apply in the atom domain.

In general, people appear to match up the concepts in two domains with an effortlessness that masks some tricky computational problems. Even though two domains may have a one-to-one correspondence between their parts, they can hold a great potential for ambiguous matches; for one-to-many and many-to-one matches between their parts. Psychologically, these problems only surface when we make the mapping task difficult, such as in the following problem from Holyoak & Thagard (1989):

| A | B |
|---|---|
| Bill is smart. | Fido is hungry. |
| Bill is tall. | Blackie is friendly. |
| Tom is timid. | Blackie is frisky. |
| Tom is tall. | Rover is hungry. |
| Steve is smart. | Rover is friendly. |

In this problem, subjects are asked to say which things in list A correspond to which things in list B (ignoring the meaning of the words). Essentially, subjects have to discover a one-to-one mapping between all the individuals and attributes in list A and list B. This is quite a difficult task as a lot of ambiguous matches have to be resolved. For example, *smart* may match *hungry* or *friendly* or *frisky* and the correct match can only be determined by eliminating the inconsistent matches which follow from all but one of these matches. The unique one-to-one mapping which solves the problem involves matching Steve and Fido, Bill and Rover, Tom and Blackie, smart and hungry, tall and friendly and timid and frisky.

Palmer (1989) has pointed out that any adequate theory of analogical mapping, which accounts for problems like the above one, will have to operate at several levels of description (for similar ideas see Marr 1982). At the highest level, one needs to characterise the *informational constraints* implied by the task situation; this level is concerned with describing what an analogy is, what needs to be computed to produce appropriate outputs given certain inputs (like Marr's computational level). Below this level is the level of *behavioural constraints* which have to capture the empirical facts of people's observable analogical behaviour (Marr's algorithmic level). Hence, this level should include constraints that predict when one analogy is harder than another and the sorts of errors that people produce. Finally, there is the level of hardware constraints which aims to capture the hardware primitives of analogical thought (Marr's hardware level).

Current theories of analogy propose many informational constraints but less attention have been given to the behavioural and hardware constraints. In this paper we test one of the well-established, information constraints, and a prediction from behavioural constraints previously proposed by Keane (1990a).

## Informational Constraints on Analogical Mapping

We have already seen, that there are several difficulties associated with achieving optimal analogical mappings. Three informational constraints have been proposed to solve these problems (see e.g., Gentner 1983; Holyoak & Thagard 1989; Keane 1990a).

The most important set of constraints are *structural constraints*. These constraints are used to enforce a one-to-one mapping between the two domains (Falkenhainer, Forbus & Gentner 1986; Holyoak & Thagard 1989). This is done using several techniques:

- *matches are made only between entities of the same type*; for example, only attributes are matched with attributes, objects with objects and two-place predicates with two-place predicates. This reduces the total number of matches that needs to be considered (see Gentner 1983; Holyoak & Thagard 1989).
- *exploit structural consistency*; that is, if the propositions REVOLVES(A B) and REVOLVES(C D) match, then the arguments of both should also be matched appropriately, A with C and B with D. This is especially useful in eliminating many-to-one and one-to-many matches (see Falkenhainer et al. 1986, 1989).
- *favour systematic sets of matches* (Gentner's 1983, systematicity principle); this proposes that if one has two alternative sets of matches then the match-set with the most higher-order connectivity should be chosen. This aids the choice of an optimal match-set from among many match-sets.

These techniques have been shown to be very powerful. In many cases, structural constraints alone can find the optimal mapping between two domains (as in the above attribute-mapping case).

A *similarity constraint* can also be used to reduce the number of matches considered or to disambiguate between alternative matches. When this constraint is applied only identical concepts are matched between the two domains (Gentner 1983) or, more loosely, semantically-similar concepts are matched (Gick & Holyoak 1980). Semantic similarity can be used to disambiguate matches because if one match in a set of one-to-many matches is more similar than the others then it can be preferred.

A final constraint on analogical mapping is the *pragmatic constraint* (e.g., Holyoak 1985; Keane 1985). Again, this may disambiguate a set of matches. For example, if in a certain analogical mapping situation one match is pragmatically more important (or goal-relevant) than other alternatives then it will be preferred over these alternatives.

These informational constraints constitute a very, high-level specification of what makes a particular comparison between two domains an analogical comparison. As such, they can and, indeed, have been implemented by many different algorithms. Falkenhainer et al.'s (1986, 1989) *Structure Mapping Engine* (SME) implements structural and similarity constraints in a serial fashion that generates all the possible match-sets between two domains and then chooses the best match-set according to structural criteria. Forbus & Oblinger (1990) have extended SME to implement the pragmatic constraint. Holyoak & Thagard's (1989) *Analogical Constraint Mapping Engine* (ACME) uses parallel constraint satisfaction to ease a network of possible matches into a single stable set of matches, in accordance with the three constraints. Keane & Brayshaw's (1988; Keane 1991) *Incremental Analogy Machine* (IAM) implements the same constraints, using a different serial algorithm to SME that computes parts of the analogy incrementally.

## Behavioural Constraints on Analogical Mapping

Any model which solely implements the above informational constraints can make certain behavioural predictions. For example, Skorstadt, Falkenhainer & Gentner (1987) have shown that SME can produce outputs that parallel broad trends in subjects' soundness ratings of different comparisons. Holyoak & Thagard (1989) have shown that the number of cycles their network goes through before settling into a correct mapping bears some correspondence to the difficulty subjects find in analogical mapping. However, because these models do not embody behavioural constraints, they fail to capture significant facets of human analogical performance. While they may predict the outputs of analogical mapping they do not predict response-time differences for different analogies (see Experiment 2). They also tend to be silent on the source and nature of errors in analogical performance.

Keane (1988b, 1990a, 1990b) has elaborated two behavioural constraints; first, that analogising should be subject to constraints imposed by *working memory limitations* and, second, that differential *background knowledge about the target domain* affects the ease with which analogical transfers are validated. It should be noted that both of these constraints are predicted to affect response times even though the content output from analogising may be unchanged. Hence, the separation between informational and behavioural constraints. Working memory limitations may also result in errors (see Keane 1990b).

Keane (1990a) has also argued that working memory limitations have wider implications for the nature of analogical processing. In particular, that human analogical processes will reduce the computational overheads in analogising. As such, human analogisers will tend to only map portions of the base domain (see Keane 1985) or will choose one optimal mapping. This is the theoretical rationale for the Incremental Analogy Machine. Rather than generating all possible mappings, IAM tries to build just one mapping resolving ambiguities in a local fashion. It may even use fairly arbitrary means to resolve an ambiguity. For example, if a one-to-many set of matches is encountered and there are no grounds for selecting one match over the others, IAM will simply choose the first match in the set.

Clearly, there will be occasions when IAM will commit itself to one mapping only to find that it is non-optimal. When this happens it will undo this mapping and start constructing an alternative mapping. So, while IAM tries to avoid constructing all possible mappings it may backtrack repeatedly to construct several alternative mappings.

As we shall see in Experiment 2, IAM predicts that the order in which the attributes are presented in versions of the attribute-mapping problem will differentially affect subjects' response times. This prediction is to be contrasted with the outputs of the SME and ACME models. Both of these models do not predict such order effects.

## The Effects of Similarity and Order on Analogical Mapping

Many experiments have shown that varying the similarity of two analogues affects the ease of analogical mapping (e.g., Keane 1985; Gentner & Landers 1985; Gick & Holyoak 1980; Holyoak & Koh 1987). It should also be possible to show this in a very precise fashion in the above attribute-mapping problem.

In Experiment 1, we systematically vary the number of predicates that are similar while controlling the other constraints. For instance, a modified version of the problem with similar attributes should be much easier:

| A | B |
|---|---|
| Bill is intelligent. | Fido is clever. |
| Bill is tall. | Blackie is big. |
| Tom is timid. | Blackie is shy. |
| Tom is tall. | Rover is clever. |
| Steve is intelligent. | Rover is big. |

In Experiment 2, we examine the prediction of IAM that ordering effects should be observed under some conditions in this attribute-mapping problem. Holyoak & Thagard (1989) had randomised the order of presentation of the attributes in both lists thus abolishing any order effects. To appreciate how these effects might arise consider what makes the problem so difficult.

In the abstract version of the attribute-mapping problem, each list has two individuals (e.g., Bill and Tom) with two attributes and a remaining individual (i.e., Steve) who has just one attribute. This is very important because matching up the single individuals in both lists (i.e., Steve and Fido) is the key to achieving the isomorphic mapping. The presence of these single individuals with one attribute (which we will call *singletons*) disambiguates the set of matches between the two lists (this also applies to the single attribute in both lists). We expect, on the basis of simulations using IAM, that order has an effect on matching and that people match items incrementally starting with the first item in list A. Hence, the problem should be easier when the singletons are placed at the beginning of the list where they can be matched first.

## Experiment 1: Similarity Effects

To test the prediction that semantic similarity can facilitate analogical mapping three groups received three different versions of the attribute-mapping problem which had either no similar attributes, one set of similar attributes or three sets of similar attributes. We predicted that subjects would produce the correct one-to-one mapping more rapidly as a function of the increasing similarity between the two domains.

## Method

**Subjects & Design.** Twenty-four undergraduates at the University of Wales College of Cardiff took part voluntarily in the experiment. The experiment used a between-subject design and subjects were assigned randomly to one of the three conditions. Three subjects were dropped from the experiment before data analysis because they misunderstood the experimental instructions. Data analysis was carried out on the remaining 21 subjects, who were equally distributed across the three conditions.

**Materials.** We used three versions of the attribute-mapping problem (see Table 1). Each version had two lists of attributes. In each list, there were three individuals and three attributes; two individuals had two attributes and one individual had a single attribute. The three versions differed in terms of the number of attributes that were similar in both lists. In the None-Similar version none of the attributes were semantically similar; in the One-Similar version one set of the attributes was semantically similar ("intelligent" in list A and "clever" in list B); in the All-Similar version, all the attributes in one list had a semantically-similar parallel attribute in the second list (see Table 1).

| *A* | *B (None-Similar)* |
|---|---|
| Bill is intelligent. | Fido is hungry. |
| Bill is tall. | Blackie is friendly. |
| Tom is timid. | Blackie is frisky. |
| Tom is tall. | Rover is hungry. |
| Steve is intelligent. | Rover is friendly. |
| *B (One-Similar)* | *B (All-Similar)* |
| Fido is clever. | Fido is clever. |
| Blackie is friendly. | Blackie is big. |
| Blackie is frisky. | Blackie is shy. |
| Rover is clever. | Rover is clever. |
| Rover is friendly. | Rover is big. |

**Table 1 The Different Problems Used in the Three Conditions of Experiment 1 (Only B differed in each)**

Subjects were instructed in writing that their "task is to figure out what in the left set corresponds to what in the right set of sentences". A single column below list A listed the names of the individuals and attributes in that list. Next to each was a space for subjects to write the corresponding name or attribute from list B. The order of sentences in each list was randomised with the proviso that sentences with attributes about the same individual were kept together.

**Procedure**. Subjects were first shown the instructions which they were asked to read carefully. They were then shown the problem and asked to solve it. A stop-watch was used to time them from this point to when they solved the problem. If subjects produced an incorrect answer they were told so and asked to continue solving the problem. Only when the correct answer was produced was the clock stopped and the elapsed time recorded.

## Results & Discussion

The results corroborated our expectations. The presence of semantic similarities between the elements of the two domains has an important facilitating effect on the ease of analogical mapping (see Figure 1). The elapsed time taken to solve the problem gradually decreases across the three conditions, with the None-Similar Condition being the slowest ($M$ = 210.9 secs.), the One-Similar Condition being faster ($M$ = 164.9 secs.) and the All-Similar Condition being the fastest ($M$ = 69.7 secs.) [$F(2, 18) = 8.747$, $p < .005$]. This is the first demonstration that domains which systematically-differ in terms of their similarity, give rise to systematically-differing solution times.

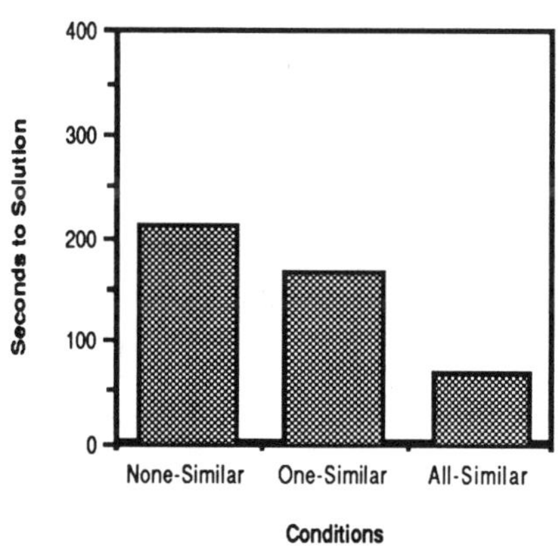


**Figure 1 The Mean Solution Times in the Conditions of Experiment 1**

## Experiment 2: Order Effects

Our second experiment tested whether there were order effects on analogical mapping, as is predicted by the Incremental Analogy Machine. As we said earlier, the key to finding the isomorphism, in the abstract version of the problem, lies in matching the two singletons in both lists. So, if one group receives the singletons first they should match them and solve the problem quicker, than if they form the singleton match later. In two models of analogy, SME and ACME, simple changes of this sort do not significantly change the course or outputs of analogical mapping. However, IAM's processing is radically different for both versions. When IAM matches the singletons first it does not have to undo its matches and so finds the isomorphism faster than when the singletons come later in the lists. Specifically, IAM does not backtrack at all in the Singleton-First case but has to backtrack four times in the Singleton-Last case.

## Method

**Subjects & Design.** Twenty-three undergraduates at the University of Wales College of Cardiff took part voluntarily in the experiment. As before, the experiment had a between-subject design and subjects were assigned at random to one of the two conditions. Again, three subjects had to be excluded from the experiment prior to data analysis because they misunderstood the experimental instructions. Data analysis was carried out on the remaining 20 subjects, who were equally distributed across the two conditions.

**Materials & Procedure.** The materials consisted of two abstract versions of the attribute-matching problem (see Table 2). In the Singleton-First version the singletons were at the top of both lists, while in the Singleton-Last version the singleton in list A was in the last position, while the Singleton in list B was in the first position (see Table 2). The order of the remaining sentences was randomised as before.

The remainder of the materials were as in Experiment 1, with the exception of the additional sentence: "The meaning of the words in the sentences is irrelevant". Holyoak and Thagard used this sentence when they gave subjects the abstract version of the problem.

Subjects were shown the sheet containing the instructions and problem, and were timed, as before.

| *Singleton-First* | |
|---|---|
| A | B |
| Steve is smart.* | Fido is hungry.* |
| Bill is tall. | Blackie is friendly. |
| Bill is smart. | Blackie is frisky. |
| Tom is tall. | Rover is hungry. |
| Tom is timid. | Rover is friendly. |
| *Singleton-Last* | |
| A | B |
| Bill is smart. | Fido is hungry.* |
| Bill is tall. | Blackie is friendly. |
| Tom is timid. | Blackie is frisky. |
| Tom is tall. | Rover is hungry. |
| Steve is smart.* | Rover is friendly. |

**Table 2 The Two Versions of the Problem Used in Experiment 2 (*the singleton)**

## Results & Discussion

The slight change in the ordering of the singletons has a marked effect on the ease of analogical mapping (see Figure 2). Subjects in the Singleton-First condition were almost twice as fast at solving the problem ($M$ = 178.0 secs) compared to the Singleton-Last condition ($M$ = 363.1 secs) [Mann-Whitney $U = 7$, $p < .005$, 1-tailed].

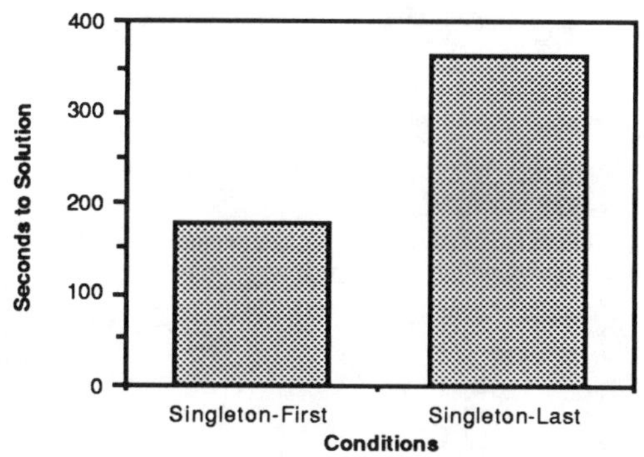


**Figure 2 The Mean Solution Times for the Conditions in Experiment 2**

This result clearly shows that the order in which the attributes are presented affects subjects' response-times. In this experiment, we have looked at a situation in which the 'easy' problem had the two singletons in the *same position*. IAM does not predict that being in the same position *per se* is important. The singletons could be in different positions and still show different response times. For example, if the singleton in list A was first and that in list B was last then IAM would predict a slower response time than, say, a problem where the list B singleton was third. These are predictions that need to be examined in future experiments.

## Conclusions

Both experiments provide strong evidence for the effects of similarity and order on analogical mapping. If the predicates in both domains bear some unambiguous similarity to one another then the mapping is considerably facilitated. Furthermore, these predicates need not be identical to aid matching but may be merely similar (e.g., big and tall).

We have also seen that under certain conditions the order in which the information in a domain is presented can have significant effects on the ease of analogical mapping. As we have seen these results are not predicted by the SME and ACME models. IAM does, however, predict these effects because it has mechanisms that try to reduce the computational overheads on analogical mapping, by heuristically disambiguating one from a set of matches using order, in accordance with the behavioural constraint of working memory limitations.

The present work was deeper implications for connectionist models in cognitive science. ACME was a important program because it showed that an example of high-level cognition, namely analogical thought, could be modelled by connectionist techniques. The present work suggests that ACME is not a wholly adequate model of analogical behaviour. It fails to capture the more serial aspects of analogising manifested in the present study. This, therefore, keeps open the debate in this area on whether the nature of high-level cognition can be captured singularly by connectionist, 'brain-style' AI.

## References


Burstein, M.H. 1986. Analogical learning with multiple models. In T.M. Mitchell, J.G. Carbonell & R.S. Michalski (Eds.), *Machine Learning: A Guide to Current Research*. Lancaster: Kluwer Academic Publishers.

Carbonell, J.G. 1986. Derivational analogy: A theory of reconstructive problem solving and expertise acquisition. In R.S. Michalski, J.G. Carbonell & T.M. Mitchell (Eds.), *Machine Learning II: An Artificial Intelligence Approach* . Los Altos, Calif: Kaufmann.

Falkenhainer, B., Forbus, K.D., and Gentner, D. 1986. Structure-mapping engine. Proceedings of the Annual Conference of the American Association for Artificial Intelligence. Los Altos: Kaufmann.

Falkenhainer, B., Forbus, K.D., & Gentner, D. 1989. Structure-mapping engine. *Artificial Intelligence* 41:1-63.

Forbus, K.D. & Oblinger, D. 1990. Making SME greedy and pragmatic. Twelfth Annual Conference of the Cognitive Science Society. Hillsdale: Erlbaum.

Gentner, D. 1983. Structure-mapping: A theoretical framework for analogy. *Cognitive Science* 7 :155-170.

Gentner, D. 1989. Mechanisms of analogical learning. In S. Vosniadou & A. Ortony (Eds.), *Similarity and Analogical Reasoning*. Cambridge: CUP.

Gentner, D. & Landers, R. 1985. Analogical reminding: A good match is hard to find. Proceedings of the International Conference on Systems, Man and Cybernetics , Tucson, Arizona, November.

Gick, M.L., & Holyoak, K.J. 1980. Analogical problem solving. *Cognitive Psychology* 12:306-355.

Gick, M.L., & Holyoak, K.J. 1983. Schema induction in analogical transfer. *Cognitive Psychology* 15:1-38.

Holyoak, K.J. 1985. The pragmatics of analogical transfer. *The Psychology of Learning and Motivation* 19:59-87.

Holyoak, K.J., & Koh, K. 1987. Surface and structural similarity in analogical transfer. *Memory & Cognition* 15:332-340.

Holyoak, K.J., & Thagard, P. 1989. Analogical mapping by constraint satisfaction. *Cognitive Science* 13:295-355.

Keane, M. 1985. On drawing analogies when solving problems: A theory and test of solution generation in an analogical problem solving task. *British Journal of Psychology* 76:449-458.

Keane, M.T. 1988a. *Analogical Problem Solving.* Chichester: Ellis Horwood (New York: Wiley).

Keane, M.T. 1988b. Analogical Mechanisms. *Artificial Intelligence Review* 2(4):229-251.

Keane, M.T. 1990a. Incremental analogising: Theory and model. In K.Gilhooly et al. (Eds.). *Lines of Thinking*. Chichester: John Wiley.

Keane, M.T. 1990b. Anchoring Analogies in Past Experience. Unpublished ms.

Keane, M.T. 1991. IAM V2: A Re-Implementation in Lisp with Further Additions. CS-TCD Technical Report, May.

Keane, M.T., & Brayshaw, M. 1988. The Incremental Analogical Machine: A computational model of analogy. In D. Sleeman (Ed.), *European Working Session on Machine Learning*. London: Pitman.

Marr, D. 1982. *Vision.* San Francisco: Freeman.

Palmer, S.E. 1989. Levels of description in information processing theories of analogy. In S. Vosniadou & A. Ortony (Eds.), *Similarity and Analogical Reasoning*. Cambridge: CUP.

Skorstadt, J.; Falkenhainer, B.; and Gentner, D. 1987. Analogical processing: A simulation and empirical corroboration. Proceeding of AAAI'87. Las Altos: Morgan Kaufmann.

# Dimensional Attention Learning in Models of Human Categorization

**John K. Kruschke**
Department of Psychology and Cognitive Science Program
Indiana University, Bloomington IN 47405 USA
e-mail: kruschke@ucs.indiana.edu

## Abstract

When humans learn to categorize multidimensional stimuli, they learn which stimulus dimensions are relevant or irrelevant for distinguishing the categories. Results of a category learning experiment are presented, which show that categories defined by a single dimension are much easier to learn than categories defined by the combination of two dimensions. Three models are fit to the data, ALCOVE (Kruschke 1990a,b, in press), standard back propagation (Rumelhart, Hinton & Williams 1986), and the configural-cue model (Gluck & Bower 1988). It is found that ALCOVE, with its dimensional attention learning mechanism, can capture the trends in the data, whereas back propagation and the configural-cue model cannot. Implications for other models of human category learning are discussed.

## Introduction

Imagine learning to classify mushrooms as "poisonous" or "edible." You are shown one mushroom after another and told which category it belongs to. After seeing many examples, your accuracy of classifying new examples improves. One of the key aspects of such learning is the determination of which features of the mushrooms are relevant to the categorization. For example, it might be that all red mushrooms are poisonous, but red mushrooms occur in the same range of sizes as edible mushrooms. In that case it would be wise to weigh information about about color more heavily than information about size. On the other hand, it might be that more than one dimension is relevant to the categorization; e.g., perhaps red or spotted mushrooms are poisonous. In that case, one should pay attention to both dimensions of color and texture.

Posner (1964) called situations in which there was a single relevant dimension *gating* tasks, since the irrelevant dimension(s) could be gated out of consideration, and he called situations in which more than one dimension was relevant *condensation* tasks, since the information from multiple dimensions had to be condensed into a single categorization decision. Posner and others (e.g., Garner 1974) have established that gating tasks are generally easier to learn than condensation tasks. In this article I show that standard back-propagation (Rumelhart, Hinton & Williams 1986) and the configural-cue model (Gluck & Bower 1988) cannot capture that basic result, while another connectionist model called ALCOVE (Kruschke 1990a,b, in press) can. I report results of a category learning experiment and fits of the models to the data. The key difference between the models is that ALCOVE incorporates constraints to reflect the dimensional attention learning abilities of people, whereas the other models do not.

## The ALCOVE Model

ALCOVE is a feed-forward network that learns by gradient descent on error, but it is unlike standard back propagation (Rumelhart *et al.* 1986) in its architecture, its behavior, and its goals. Unlike the standard back-propagation network, which was motivated by generalizing neuron-like perceptrons, the architecture of ALCOVE was motivated by a molar-level psychological theory, Nosofsky's (1986) generalized context model (GCM). The psychologically constrained architecture results in behavior that captures the detailed course of human category learning in many situations where standard back propagation fares less well (Kruschke 1990a,b, in press). And, unlike many applications of standard back propagation, the goal of ALCOVE is not to discover new (hidden-layer) representations after lengthy training, but rather to model the course of learning itself, by determining which dimensions of the given representation are most relevant to the task, and how strongly to associate exemplars with categories.

Like the GCM, ALCOVE assumes that input patterns can be represented as points in a multidimensional psychological space, as determined by multi-dimensional scaling algorithms (e.g., Kruskal 1964; Shepard 1962). Each input node encodes a single psychological dimension, with the activation of the node indicating the value of the stimulus on that dimension. Figure 1 shows the architecture of ALCOVE,

Figure 1: The structure of ALCOVE. The pyramids in the hidden layer indicate the activation profile of hidden nodes, as determined by Equation 1, with $r = q = 1$.

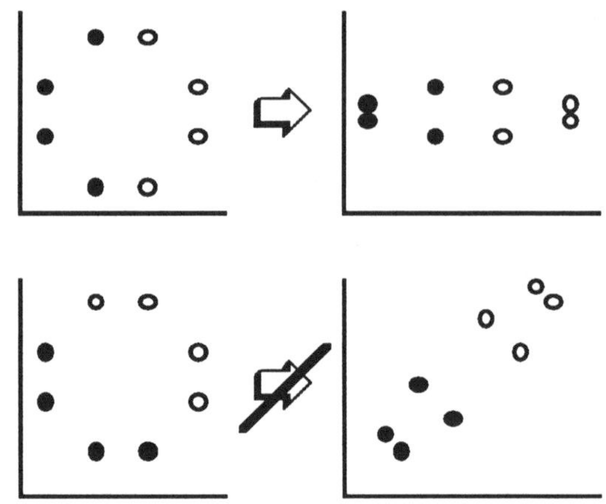

Figure 2: Top panel shows that increasing attention to the horizontal dimension and decreasing attention to the vertical dimension causes exemplars of the two categories (denoted by filled and open circles) to have greater between-category dissimilarity and greater within-category similarity (after Nosofsky 1986, Fig. 2). Lower panel shows that ALCOVE cannot differentially attend to diagonal axes.

illustrating the case of just two input dimensions.

Each input node is gated by a dimensional *attention strength* $\alpha_i$. The attention strength on a dimension reflects the relevance of that dimension for the particular categorization task at hand, and the model learns to allocate more attention to relevant dimensions and less to irrelevant dimensions.

Each hidden node corresponds to a position in the multi-dimensional stimulus space, with one hidden node placed at the position of every training exemplar. Each hidden node is activated according to the psychological similarity of the stimulus to the exemplar represented by the hidden node. The similarity function comes from the GCM and the work of Shepard (1962; 1987): Let the position of the $j^{th}$ hidden node be denoted as $(h_{j1}, h_{j2}, \ldots)$, and let the activation of the $j^{th}$ hidden node be denoted as $a_j^{hid}$. Then

$$a_j^{hid} = \exp\left( -c\left(\sum_i \alpha_i \left|h_{ji} - a_i^{in}\right|^r\right)^{q/r}\right) \tag{1}$$

where $c$ is a positive constant called the *specificity* of the node, where the sum is taken over all input dimensions, and where $r$ and $q$ are constants determining the similarity metric and similarity gradient, respectively. For separable psychological dimensions, the city-block metric ($r = 1$) is used, while integral dimensions might call for a Euclidean metric ($r = 2$; *cf.* Garner 1974, Shepard 1964). A city-block distance metric ($r = 1$) with exponential similarity gradient ($q = 1$) is used here (Shepard 1987).

The dimensional attention strengths adjust themselves so that exemplars from different categories become less similar, and exemplars within categories become more similar. Consider a simple case of eight stimuli that form the corners of an octagon in input space, as in Figure 2. The stimuli are mapped to one of two categories, as indicated by filled or open circles. When only one dimension is relevant, as in the top panel of Figure 2, ALCOVE learns to increase the attention strength on the relevant dimension, and to decrease the attention strength on the irrelevant dimension. By contrast, ALCOVE cannot stretch or shrink diagonally, as suggested in the lower panel of Figure 2. Two points made in this article are (1) to demonstrate that this constraint is an accurate reflection of human performance, in that categories separated by a diagonal boundary take longer to learn than categories separated by a boundary orthogonal to one dimension, and (2) to show that standard back propagation and configural-cue model do not capture this fact.

Each hidden node in ALCOVE is connected to output nodes that correspond to response categories. The connection from the $j^{th}$ hidden node to the $k^{th}$ category node has a connection weight denoted $w_{kj}$, called the *association weight* between the exemplar and the category. The output (category) nodes are activated by the linear rule used in the GCM and the network models of Gluck and Bower (1988):

$$a_k^{out} = \sum_{\substack{hid \\ j}} w_{kj} a_j^{hid}\,. \tag{2}$$

In ALCOVE, unlike the GCM, the association weights

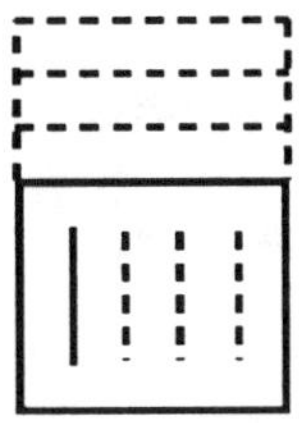

Figure 3: Stimuli used in category learning experiment. The rectangle could have one of four heights (the shortest rectangle is shown with a solid line) and the interior vertical segment could have one of four lateral positions (the leftmost position is shown with a solid line).

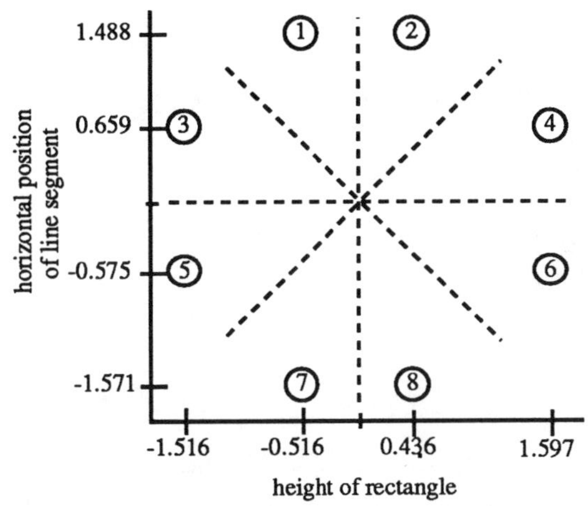


Figure 4: Locations of stimuli in psychological similarity space. Dashed lines suggest the four category boundaries used in the learning experiment.

are learned and can take on any real value, including negative values. Category activations are mapped to response probabilities using the same choice rule (Luce 1963) as was used in the GCM and network models. Thus,

$$\Pr(K) = \exp(\phi\, a_K^{out}) \Big/ \sum_{\substack{out\\k}} \exp(\phi\, a_k^{out}) \qquad (3)$$

where $\phi$ is a real-valued scaling constant. In other words, the probability of classifying the given stimulus into category $K$ is determined by the magnitude of category $K$'s activation relative to the sum of all category activations.

The dimensional attention strengths, $\alpha_i$, and the association weights, $w_{kj}$, are learned by gradient descent on sum-squared error, as used in standard back propagation (Rumelhart *et al.* 1986) and in the network models of Gluck and Bower (1988). Space constraints prohibit further discussion of the model; details can be found in Kruschke (1990a,b, in press).

In fitting ALCOVE to human learning data, there are four free parameters: the fixed specificity $c$ in Equation 1; the probability mapping constant $\phi$ in Equation 3; the association weight learning rate; and, the attention strength learning rate.

## A Human Learning Experiment

To test the implications of attention learning in ALCOVE, human subjects were trained on the category structures shown in Figure 2. The stimuli were geometric forms, as shown in Figure 3. Each stimulus consisted of a rectangle with one of four heights, and an interior vertical segment at one of four lateral positions. Only 8 of the 16 possible combinations of height and position were used, corresponding to the abstract structure in Figure 2.

### Scaling the stimulus space

The first step in modelling this situation is to determine the psychological coordinates of the stimuli. To do this, one obtains similarity ratings of pairs of stimuli, and determines the coordinate values in psychological space that best predict those similarities (Kruskal 1964; Shepard 1962). This process is analogous to generating a spatial map of cities when all you are told is the distances (dis-similarities) between cities.

*Procedure:* The two stimuli of each pair were presented sequentially on a computer screen for 1.75 seconds each, separated by a 0.75 second blank screen. The subject then entered a similarity rating from 1 to 9 on the computer keyboard. Each pair was presented four times in each order, for each of 50 subjects, yielding 400 similarity ratings for each pair.

The best-fitting psychological coordinates of the stimuli accounted for over 98% of the variance in similarity ratings, and are shown in Figure 4. Notice that while the *physical* values of height and horizontal position were equally spaced (see Figure 3), the *psychological* values were not. Note also that the two dimensions are about equally salient, in that the total range is about the same on the two dimensions, but the middle interval on the horizontal-position dimension is bigger than the middle interval of the height dimension. That implies that a category boundary indicated by the horizontal dashed line in Figure 4 should be easier to learn than a category boundary indicated by the vertical dashed line. This will be confirmed in the learning data.

### Category learning results

The strong prediction of ALCOVE is that the four alternative category disctinctions indicated in Figure 4 should not be equally easy; rather, the distinctions for which only a single dimension is relevant (the vertical

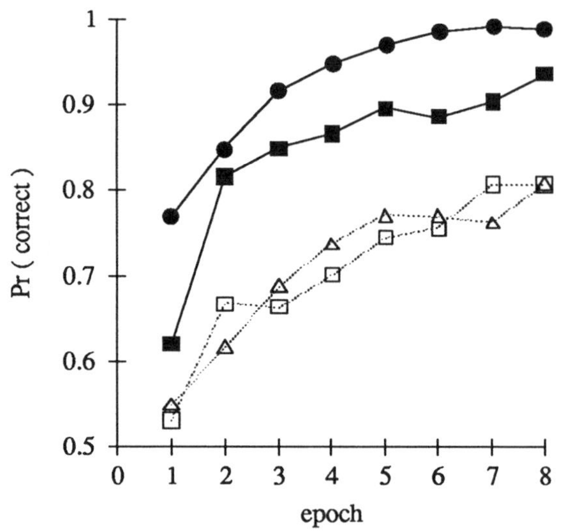


Figure 5: Human learning data. Filled markers correspond to single-dimension categories (filled circle is horizontal position relevant, filled square is height relevant); open markers show diagonal categories.

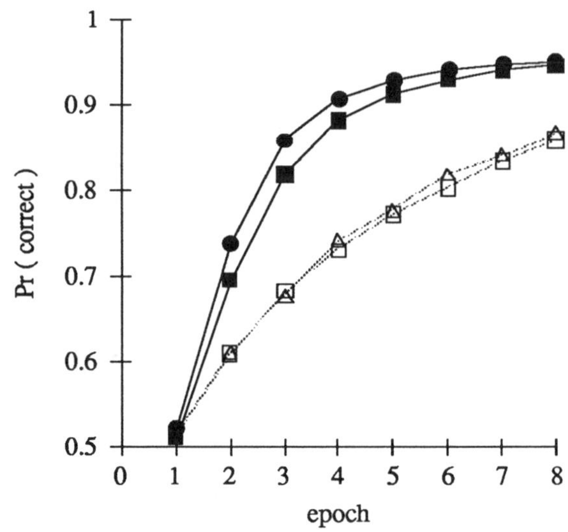


Figure 6: Best fit of ALCOVE to human learning data in Figure 5.

and horizontal dashed lines in Figure 4) should be easier than the diagonal distinctions, because attention learning cannot differentially accentuate diagonal directions. The four category distinctions were given to different groups of subjects. For example, one group learned the position-relevant distinction, for which the stimuli marked 1, 2, 3 and 4 in Figure 4 were given one category label, with the remaining stimuli (numbered 5–8) were given a different category label.

*Procedure:* Subjects were given instructions that included exposure to the eight stimuli without any category feedback. Each training trial consisted of a presentation of a stimulus, which was terminated when the subject pressed a response key. The subject was then given feedback indicating whether the response was correct or incorrect, and the correct response. Each of the four groups saw the same fixed sequence of 64 stimuli. All that varied between groups was the category labels given to the stimuli. A total of 160 subjects were run, 40 in each group. Category labels were counter-balanced within groups.

Results are summarized in Figure 5. Each datum shows the mean percent correct for the preceding 8 trials (one epoch). Two effects are clear: The single-dimension categories are learned much faster than the diagonal categories; and, the horizontal-position dimension is learned faster than the height dimension.

## Modelling the learning

### Fit of ALCOVE

ALCOVE was applied to this situation by using two input nodes, corresponding to the two stimulus dimensions, eight hidden nodes, corresponding to the eight training exemplars, and two output nodes, corresponding to the two categories. The task for the model was to fit the learning curves from the four category types using a single set of parameter values. The discrepancy of ALCOVE from the human data was measured as the squared difference between observed category choice probabilities and predicted choice probabilities, summed across the 64 individual trials in each category type, across both category choices within each type, and across the four types.

The best fit of ALCOVE is shown in Figure 6. The fit produced a root mean squared deviation (RMSD) of 0.116, with parameter values of $\phi = 1.568$, $c = 1.662$, association weight learning rate of 0.08431 and attention learning rate of 0.6593. ALCOVE clearly shows the two main trends seen in the human data: The single-dimension categories are learned much faster than the diagonal categories, and the horizontal-position dimension is learned faster than the height dimension. ALCOVE learns the single-dimension categories faster by virtue of attention learning — the attention strength on the relevant dimension increases rapidly, while the attention strength on the irrelevant dimension decreases rapidly. ALCOVE learns the position dimension faster than the height dimension because the exemplar nodes reflect the greater distinctiveness of the middle interval of the position dimension.

### Fit of back propagation

Standard back propagation (henceforth "backprop") was also applied. Two input nodes and two output nodes were used, as in the application of ALCOVE. In order to equilibrate the backprop architecture with the

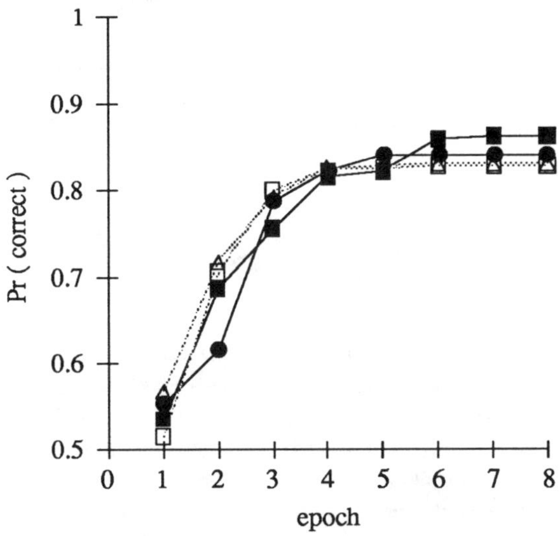


Figure 7: Best fit of standard back propagation to human learning data in Figure 5.

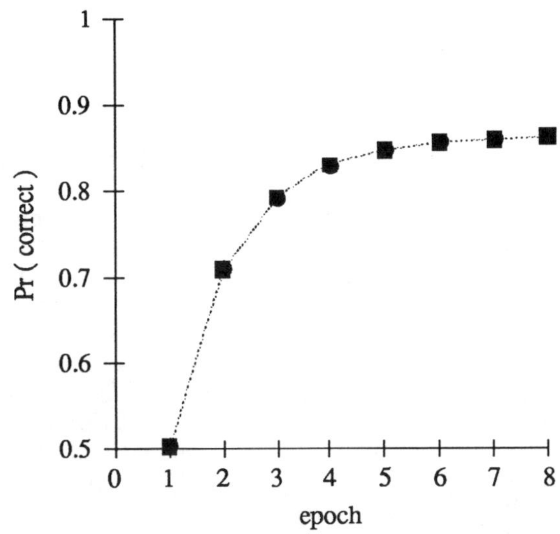


Figure 8: Best fit of the configural-cue model to human learning data of Figure 5. All four curves are exactly superimposed.

ALCOVE architecture as much as possible, the two output nodes were linear, with response probabilities computed using Equation 3, and eight hidden nodes were used (with standard linear-sigmoid activation functions; Rumelhart *et al.* 1986).

The backprop network was given five free parameters: The learning rate on the output weights; the learning rate on the hidden weights; the learning rate on the hidden node thresholds; the response mapping constant $\phi$ in Equation 3; and, the value of a fixed *gain* parameter in the sigmoidal activation function (see Kruschke & Movellan 1991). Initial values of hidden weights and thresholds were drawn from a uniform distribution over the interval $[-1, +1]$.

For any choice of parameter values, the fit was measured in the same way as for ALCOVE, but choice predictions of backprop were computed by first averaging over 200 different random initializations of the hidden weights and thresholds.

The best fit of backprop to the learning data is shown in Figure 7. It yielded an RMSD of 0.152, using $\phi = 0.6636$, output weight learning rate of 0.2049, a hidden weight learning rate of 1.091, a hidden threshold learning rate of 0.04159, and a gain of 2.249. The fit of backprop is much worse than that of ALCOVE, and the qualitative behavior of backprop departs badly from the data. Indeed, backprop learns the single-dimension and diagonal categories at essentially the same pace. The best backprop can do is try to match the mean learning curve across all four category types.

Why does backprop do so poorly? To explain why, first I'll define a hidden node's *weight vector* as the ordered list of connection weights fanning into the node. The weight vector of a hidden node specifies the direction in stimulus space that causes the biggest change in the node's activation value, and thereby indicates the underlying dimension in stimulus space to which the node is responding. The weight vectors can, potentially, point in *any* direction in stimulus space. In particular, they can align with the diagonal axes of the stimulus space just as easily as they can align with the canonical axes. Standard back-propagation learning is *isotropic* in that sense, unlike human learning.

## Fit of the configural-cue model

The configural-cue model (Gluck & Bower 1988) was originally proposed for stimuli with binary-valued dimensions, and therefore is not directly applicable to the present situation. However, it can be reasonably extended as follows: Each value on each dimension is encoded by a separate input node, and each combination of values from the two dimensions is encoded by a separate node, yielding $4 + 4 + 16 = 24$ input nodes. A given stimulus activates a subset of 3 of the 24 input nodes. The input nodes are connected directly to the output nodes, which are governed by Equations 2 and 3. The configural-cue model has two parameters, the learning rate for the connection weights and the mapping constant in Equation 3.

The best fit is shown in Figure 8 ($\phi = 0.9289$, connection weight learning rate was 0.1591, yielding an RMSD of 0.150). Figure 8 shows that all four learning curves are exactly superimposed, quite unlike the human data. The reason is that the (extended) configural-cue model makes no structural distinction between the category types in its input representation; the model has no representation of the fact that some values lie on the same dimension but other values come from different dimensions.

## Discussion

It is possible that back propagation or the configural-cue model could be modified to include some form of dimensional attention learning; this awaits future research. The results reported here pose a challenge for other models of category learning that do *not* incorporate attention learning, such as Hanson and Gluck's (1991) Cauchy-node model. It remains to be seen if such models can fit the data reported here, without extending them to include some form of dimensional attention learning. Other models that *do* include attention learning, such as J. R. Anderson's (in press) rational model, or Hurwitz's (1990) hidden pattern unit model, could, no doubt, capture the difference between single-dimension and diagonal categories.

All of these models, however, would probably fail to match one other prominent aspect of the data: the extreme rapidity with which the single-dimension categories are learned, very early in training (see Figure 5). Some subjects, in fact, made virtually no errors after the first two or three trials. There is little hope that learning algorithms that take small incremental steps on each trial could accomplish that.

What is needed, I believe, is a *rule hypothesizing system*: Early in training the subject might guess the right rule and make no more errors. Perhaps the primary question for such a rule generating system is, Which rules should be hypothesized and tested first? The behavior of ALCOVE suggests that one might generate and test rules using the dimension(s) with the largest attention strength. The notion is that category learners always employ an ALCOVE-like system, and simultaneously try to summarize, generalize, and leap-frog the performance of that system by hypothesizing and testing rules. The underlying ALCOVE-like system steers the rule generating system and acts as a fall-back when adequate rules are not yet found.

## Acknowledgments

I thank Terry Bleizeffer, Steve McKinley, and Rita Randolph for running subjects. This research was supported by Biomedical Research Support Grant RR 7031-25 from the National Institutes of Health.

## References

Anderson, J. R. (in press). The adaptive nature of human categorization. *Psychological Review*.

Garner, W. R. (1974). *The Processing of Information and Structure*. Hillsdale, NJ: Erlbaum.

Gluck, M. A. & Bower, G. H. (1988). Evaluating an adaptive network model of human learning. *J. of Memory and Language*, **27**, 166-195.

Hanson, S. J. & Gluck, M. A. (1991). Spherical units as dynamic consequential regions: Implications for attention, competition and categorization. In: R. P. Lippmann, J. Moody and D. S. Touretzky (eds.), *Advances in Neural Information Processing Systems 3*. San Mateo, CA: Morgan Kaufmann.

Hurwitz, J. B. (1990). A hidden-pattern unit network model of category learning. PhD dissertation, Harvard University.

Kruschke, J. K. (1990a). A connectionist model of category learning. PhD dissertation, University of California at Berkeley. Available from University Microfilms International.

Kruschke, J. K. (1990b). ALCOVE: A connectionist model of category learning. Research Report 19, Cognitive Science Program, Indiana University.

Kruschke, J. K. (in press). ALCOVE: An exemplar-based connectionist model of category learning. *Psychological Review*.

Kruschke, J. K., & Movellan, J. R. (1991). Benefits of gain: Speeded learning and minimal hidden layers in back-propagation networks. *IEEE Trans. Systems, Man and Cybernetics*, **21**, 273-280.

Kruskal, (1964). Nonmetric multidimensional scaling: a numerical method. *Psychometrika*, **29**, 115-129.

Luce, R. D. (1963). A threshold theory for simple detection experiments. *Psychological Review*, **70**, 61-79.

Nosofsky, R. M. (1986). Attention, similarity and the identification-categorization relationship. *J. Exp. Psych.: General*, **115**, 39-57.

Posner, M. I. (1964). Information reduction in the analysis of sequential tasks. *Psychological Review*, **71**, 491-504.

Rumelhart, D. E., Hinton, G. E., & Williams, R. J. (1986). Learning internal representations by back-propagating errors. In: D. E. Rumelhart & J. L. McClelland (eds.), *Parallel Distributed Processing, Vol. 1*, pp. 318-362. Cambridge, MA: MIT Press.

Shepard, R. N. (1962). The analysis of proximities: Multidimensional scaling with an unknown distance function, I & II. *Psychometrika*, **27**, 125-140, 219-246.

Shepard, R. N. (1964). Attention and the metric structure of the stimulus space. *J. of Mathematical Psychology*, **1**, 54-87.

Shepard, R. N. (1987). Toward a universal law of generalization for psychological science. *Science*, **237**, 1317-1323.

# Commonalities, Differences and the Alignment of Conceptual Frames During Similarity Judgments

**Arthur B. Markman**
University of Illinois
Department of Psychology
Beckman Institute
603 E. Daniel St.
Champaign, IL 61820

**Dedre Gentner**
Northwestern University
Department of Psychology
Institute for the Learning Sciences
2029 Sheridan Rd.
Evanston, IL 60208

### Abstract

Tversky demonstrated that the similarity of two objects increases with their commonalities and decreases with their differences. We believe that determining commonalities and differences is a complex task. Using analogical mapping as a guide, we propose the process of *frame-alignment* which can be employed to find the commonalities and differences of structured representations. We then test the predictions of this approach by asking subjects to list the commonalities and differences of word pairs that vary in their degrees of similarity. The results of this study support the predictions of the frame-alignment view.

## Introduction[1]

Similarity is a central aspect of virtually all cognitive processes, whether of categorization (Smith and Medin, 1981), problem solving and transfer (Ross, 1989, Holyoak and Koh, 1987), or even skill acquisition (Logan, 1989). Thus, it is incumbent on cognitive psychologists to detail the algorithm by which similarity is determined.

Tversky (1977) provides evidence that the similarity of two objects increases with their commonalities and decreases with their differences. Given this assumption, the process of calculating the similarity of two objects reduces to the subtasks of finding their commonalities and differences and combining these elements. At first blush, this approach seems straightforward. Consider the pair of circles in Figure 1a. We could describe both circles using lists of features. The features 'circle' and 'medium-sized' are common to both objects, and will increase the similarity of the pair. However, one object has the feature 'shaded', while the other has the feature 'striped'. These distinctive features will decrease similarity.

If we consider the pair in Figure 1b, we see the need for a more complex account. Notice that each of these scenes can be described by exactly the same set of features, though clearly they are not identical. In this case, we must make a decision before counting commonalities and differences: either the two circles are placed in correspondence and 'shading' and 'location' are distinctive features, or the two leftmost objects are placed in correspondence and 'shading' is a common feature, while 'shape' becomes a distinctive feature. The point is, before we can determine what is common and what is distinctive, we must align the representations based on their structure.

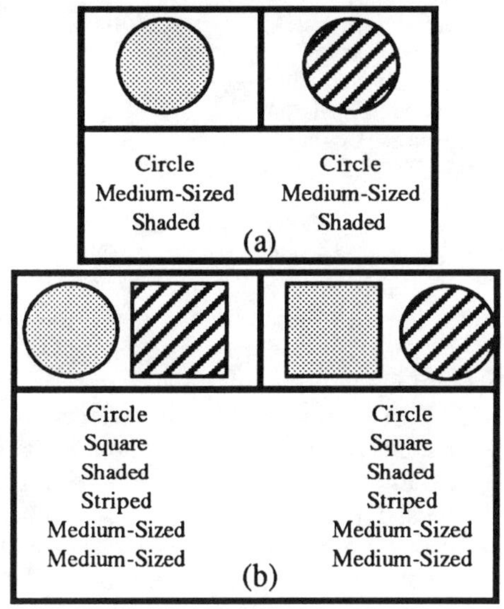


**Figure 1:** Examples of feature alignment.

There is empirical support for the claim that similarity is a process of alignment of structured representations. For example, Goldstone, Gentner and Medin (1989) demonstrated that pairs of objects with consistent relations between their parts are more similar than objects that do not have consistent relations between their parts. Furthermore, Gentner and Rattermann (1987) found that stories with similar plots and different characters (and hence similar structural elements) are often thought to be more similar than stories with different plots and similar characters (thus sharing few structural elements). Taken together, this evidence indicates that any theory of similarity must take relational structure into account.

[1]This work was supported by NSF grant BNS 87-20301 and ONR grant N00014-89-J1272 awarded to the second author as well as a University Fellowship awarded to the first author by the University of Illinois.

One format for structured representation is *conceptual frames* (Minsky, 1981, Schank and Abelson, 1977, Lenat and Guha, 1990, Barsalou, 1990, Norman and Rumelhart, 1972). In a conceptual frame, objects are represented as *slot-value pairs* with relations between them. Slots represent dimensions along which a particular object varies. For example, a car may have a color or a type of engine (see Figure 2). The value associated with a slot represents the way that role is filled for that object. This value may be a single entry (e.g. **powerDevice:** engine) or it may be a list of entries (e.g. **instances:** DatsunB210, ToyotaCorolla, LincolnTownCar...). Furthermore, relations between slots, such as mutual constraints between the values of two slots and information about the function of particular slots, are included. All slots, values and relations are themselves represented by frames.

*Frame: Car*
specializationOf: Vehicle
instances: DatsunB210, ToyotaCorolla, LincolnTownCar. . .
locomotionDevice: Wheels
numberOfWheels: 4
powerDevice: Engine
engineTypes: V-8, V-6, Rotary . . .
structureOfBody: EnclosedBody
colorOfBody: Red, Green, Blue. . .
fuel: Gasoline, Gasahol

**Figure 2:** Sample Frame

Lenat and Guha (1990) detail a long-term project in which all of common sense knowledge is to be encoded into an enormous knowledge-base composed entirely of frames. In this system, slots are organized in hierarchies. Through the creation of a consistent, global ontology, similar objects (which have similar slots) will have a common structure defined by the position of their slots within this organization. In addition, Barsalou (in press) reviews work demonstrating that frames may be used as the basis for diverse cognitive processes including mental-models, categorization, conceptual-combination and planning.

Given structured representations, we require a mechanism for comparison that can be used to determine similarity. We postulate that the commonalities and differences of representations can be found through a process of *frame-alignment*. During the alignment of two frames, we assume that slots are placed in correspondence if they are identical, or if they play the same role within the relational structures of both frames. One candidate for this process would be the local to global computation proposed to mediate analogical reasoning (Gentner, 1983, 1989, Holyoak and Thagard, 1989). Once the slots have been aligned, their values may also be compared.

Figure 3a depicts schematic object representations for two highly similar objects. Every slot in the first representation has a corresponding slot in the second representation (i.e. slot1 in Frame1 matches slot1 in Frame2). Further analysis of the values associated with the corresponding slots indicates that slots 1, 2 and 3 all have identical values in both frames. These matching slot-value pairs represent *commonalities* between the objects. In contrast, slot4 has value4 in Frame1 and value10 in Frame2. These mismatching values represent differences between objects fill corresponding (aligned) slots. Thus, we will refer to this kind of difference as an *alignable difference* (AD).

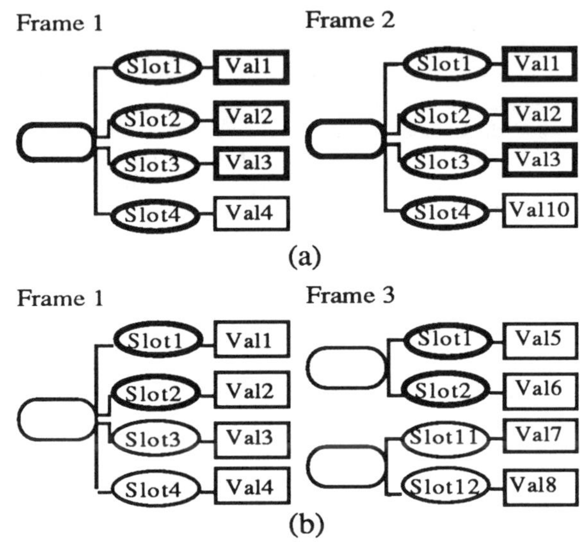


**Figure 3:** Schematic Frames

There is a third possibility, as shown in Figure 3b which depicts the frame representations of two dissimilar objects. Besides the two alignable differences [slot 1:(val1/val5)] and [slot2:(val2/val6)], there are two *non-alignable differences*. These differences between frames arise because the slots do not correspond. For example, slot3 in frame1 has no equivalent slot (and hence no corresponding slot-value pair) in frame2. A similar analysis may be made for slot4 in frame1 as well as slot11 and slot12 in frame2.

The frame-alignment view of similarity gives rise to the ternary distinction of commonalities, alignable differences and non-alignable differences.[2] Each of

[2]Other models of similarity make different partitions of commonalities and differences. Mental-distance models (Kruskal and Wish, 1978) are based on a unary scheme which does not differentiate commonalities from differences. Feature models (Tversky, 1977) are based on a binary distinction of commonalities and differences.

these factors is expected to play a discernably different role in the determination of similarity. Like the model proposed by Tversky (1977), the frame-alignment view assumes that the similarity of two objects increases with the number of their commonalities and decreases with the number of their differences (both ADs and NADs). However, the numbers of commonalities ADs and NADs is expected to differ for objects of high and low similarity.

For example, the comparison of similar objects, like those in Figure 3a, should yield many matching slot-value pairs (commonalities) as well as matching slots that have mismatching values (ADs). Similar pairs should have few slots that have no corresponding slot in the other frame (NADs). For dissimilar pairs the pattern should be quite different. As depicted in Figure 3b, we suggest that dissimilar objects have few common slots. Hence there should be few commonalities and few ADs. Rather, we expect that many slots will have no corresponding slot in the other representation, resulting in many NADs.

This analysis leads to four major predictions of the frame-alignment model of comparison. First, we expect that, as the number of commonalities between two objects increases, the number of ADs should increase as well. Second, because ADs arise from a matching slot, there should be a semantic link between ADs and commonalities. Third, similarity should increase with the number of commonalities and decrease with the number of differences (both ADs and NADs). Finally, if we make the simplifying assumption that the frames being compared are roughly the same size (have the same number of slots), there should be few NADs when there are many ADs and commonalities (and vice versa). Thus, the total number of differences listed need not increase with distance.

To test these predictions, we asked subjects to list commonalities and differences for various pairs of concepts in an effort to tap the output of the comparison process. In order to vary similarity systematically, we used an ontology tree similar to the ones used by Keil (1979) and Lenat (Lenat and Guha, 1990). This tree is shown in Figure 4. Eight basic level object names were listed at each leaf of the tree. These objects were taken from two different categories that fall in that area of the ontology. For example, we used four birds and four reptiles under the animal branch of the ontology. Sixty-four terms were spread evenly among the eight branches of the tree.

Subjects were presented with pairs that differed in their distance within the tree. Distance was calculated by the number of nodes that need to be traversed to get from one term to the other. Thus, two animals (i.e. salamander/robin) are at distance 0, an animal and a plant (i.e. salamander/oak) are at a distance of 1, an animal and an artifact (i.e. salamander/bus) are at a distance of 3 and an animal and any abstract term (i.e. salamander/meeting) are at a distance of 5.[3]

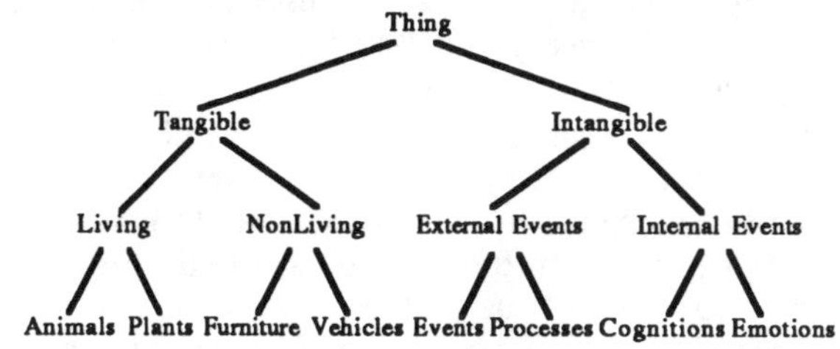


**Figure 4:** Ontology tree used here.

In the classic account, in which objects are represented as lists of features (and we retain the simplifying assumption that the representations are roughly the same size[4]), the prediction is that similar objects should have many commonalities and few differences, while dissimilar objects should have few commonalities and many differences. The frame-alignment model also predicts that similar objects will have many commonalities and few non-alignable differences, while dissimilar objects will have few commonalities and many non-alignable differences. However, an opposite pattern is predicted for alignable differences: similar objects should have many alignable differences, while dissimilar objects will have few. Thus, in this account all differences are not alike.

## Experiment

### Method

Subjects: Subjects were 44 students at the University of Illinois who received course credit for their participation. A second group of 40 subjects produced similarity ratings.

Stimuli: As described above, we use the ontology depicted in Figure 4. Eight basic level terms (4 from each of 2 categories subsumed under that node of the

[3]These a priori distances were highly related to the subjects' similarity ratings, with similarity decreasing as distance increased.For pairs of distance 0, mean similarity was 5.06. For pairs of distance 1, mean similarity was 3.02. For pairs of distance 3, mean similarity was 2.01. Finally, for pairs of distance 5, mean similarity was 1.59.

[4]In the context of this experiment, this assumption requires that size of frame and ontological distance be decorrelated. Although abstract and concrete objects probably differ in the number of slots in their frames, equal numbers of pairs of abstract and concrete items are used at distances 0, 1 and 3. Pairs of distance 5 are made of one abstract item and one concrete item, so this assumption may be less tenable for these items. If abstract or concrete items are analyzed alone the same pattern of results is obtained.

ontology) were listed at each node for a total of 64 terms. Two stimulus sets were created. Each set contained 16 pairs of distance 0, 8 pairs of distance 1, 4 pairs of distance 3 and 4 pairs of distance 5. Each word used in a pair of distance 0 in one set was used in a pair of distance greater than 0 in the other set.

Procedure: Subjects were run individually. Subjects were seated at a table in front of a pile of index cards (face down) and a tape recorder. Each card contained a word pair and instructions to list either commonalities or differences for that pair. Subjects were given one minute to make their listings aloud. If the subject did not make a response for 30 seconds, the trial was terminated and the next word pair was presented. All sessions were taped.

A separate group of subjects was given the pairs with the instructions to rate the similarity of the pairs on a scale from 1 (low) to 9 (high). Twenty subjects rated each stimulus set.

Design: All distance conditions were within subjects. Half of all trials were commonality listings, and half were difference listings. Thus, commonalities and differences were obtained between subjects. Two different stimulus sets were used in this experiment. Stimulus set was a between-subjects factor.

Scoring: All subject tapes were transcribed. The transcripts were scored as follows. For listings of commonalities, one commonality was counted for each distinct item listed (i.e. both a car and motorcycle have wheels).[5] However, grammatical commonalities, commonalities of the actual words and associations between the words were not counted as commonalities, and will not be considered further here.

For difference listings, the items were broken down into alignable differences (AD) and non-alignable differences (NAD). A listing was considered to be an AD if subjects mentioned a contrasting value for both objects in their protocol (i.e. a car has 4 wheels and a motorcycle has 2). Subjects need not have mentioned the exact values for both objects (i.e. cars and motorcycles have different kinds of engines). All other listings (except for differences in grammar, differences in the actual words or associations) were considered to be NADs (i.e. cars have trunks, motorcycles don't). An independent rater reviewed a random 20% subsample of the data. Interrater reliability was 0.89.

**Results**

Eleven subjects listed commonalities and differences for each of the 64 pairs in this study. Table 1 shows sample protocols. As predicted, this subject listed more commonalities for the highly similar pair <*Chair/ Dresser*> than for the dissimilar pair <*Pine/Motorcycle* > (6 vs. 3). Interestingly, the same number of differences was listed for the similar and dissimilar pair: however, the pattern of ADs and NADs differed. The differences for the similar pair (*Car/Motorcycle*) were mostly ADs (i.e. Cars carry more people than motorcycles) while the differences for the pair <*Sofa/Frog*> were mostly NADs (Frogs have bulging eyes, Sofas don't).[6]

This pattern of results was obtained across items as well. We averaged the responses for each item and performed analyses on the average scores for each of the 64 pairs. The first prediction that commonalities would decrease as similarity decreased was borne out. As shown in Figure 5a, the number of commonalities for each pair was greatest for pairs of distance 0 and decreased as distance increased. These differences were significant ($F(3,60)=11.65$ $p<.001$).

Consistent with the predictions of the frame-alignment view, more ADs were listed for similar items than for dissimilar items, while more NADs were listed for dissimilar items than for similar ones (see Figure 5b). A MANOVA on the number of AD and NAD listed for items indicates that these differences are significant ($F(6,118)=5.272$, $p<001$). Strikingly the total number of differences did not change with distance in the ontology ($F(3,60)=0.011$, $p>.5$).

Next, we examined the relationship between the number of commonalities listed and the numbers of ADs and NADs listed. As predicted, there was a positive correlation between the number of commonalities listed for a particular pair and the number of ADs listed for that pair ($r(62)=0.74$, $p<.001$). However, as expected, there is no positive correlation between the number of listed commonalities and the number of listed NADs for each stimulus ($r(62)=-0.12$, $p>.05$).

Finally, we assumed (in common with Tversky's featural account) that commonalities would increase similarity, while ADs and NADs would decrease similarity. We performed a multiple regression of similarity (SIM) on commonalities (COMM), AD and NAD, and found

$$\text{SIM} = 0.676\ (\text{COMM}) - 0.293\ (\text{AD}) - 1.528\ (\text{NAD}).$$

As expected, similarity increases with the number of listed commonalities and decreases with the number of listed ADs and NADs (although the regression coefficient for ADs is not significantly greater than 0). Overall, this equation accounts for 67.6% of the variance in the similarity judgments. Finally, a

[5]Listed commonalities did not have to be true of the items. Half of the subjects considered frogs as amphibians, but half considered them to be reptiles.

[6]The two commonality listings were taken from one subject and the two difference listings were taken from another.

**Commonalities**

| Chair/Dresser (Similarity=5.40) | |
|---|---|
| 1 | Both are pieces of furniture |
| 2 | Both found in the home |
| 3 | Both can be made of wood |
| 4 | Both might be in your room |
| 5 | Both can be bought at the furniture store |
| 6 | Both are common, but chairs are more common |

| Pine/Motorcycle (Similarity=1.35) | |
|---|---|
| 1 | Both are kept outdoors |
| 2 | Both can be dangerous |
| 3 | Both touch the ground |

**Differences**

| Car/Motorcycle (Similarity=6.20) | |
|---|---|
| AD | Car has 4 wheels/Motorcycle has 2 |
| AD | Car has bigger engine than Motorcycle |
| AD | Car can carry more people |
| AD | Motorcycle can go faster than a car |
| NAD | Car can carry luggage/A motorcycle can't |
| NAD | Car protects from rain/Motorcycle doesn't |
| AD | Motorcycle's engine is exposed/Car's engine is under the hood |

| Sofa/Frog (Similarity=1.30) | |
|---|---|
| AD | Frogs are greenish/Sofas come in many colors |
| AD | Frogs found outdoors/Sofas found indoors |
| AD | Sofas are furniture/Frogs are animals |
| NAD | Sofas can be sat in/Frogs can't |
| NAD | Frogs jump/Sofas can't |
| NAD | Frogs have bulging eyes/Sofas don't |
| NAD | Sofas can be jumped on/Frogs can't |
| NAD | Frogs can jump/Sofas can't |

**Table 1:** Sample subject protocols

similar regression performed with only commonalities and the total number of differences (AD + NAD) accounted for only 59.8% of the variance.[7]

## Discussion and Conclusions

From our analysis of the frame-alignment model, we derived four predictions about the nature of commonalities and differences. First, similarity should increase with commonalities and decrease with differences. Second, the number of commonalities and ADs should be positively correlated. Third, commonalities and ADs should be linked conceptually as well as numerically. Finally, the number of NADs should decrease with similarity. The results obtained in this experiment support all four of these predictions.

We found that the number of commonalities and ADs listed for each item is indeed positively correlated. Despite this relationship, commonalities serve to increase similarity, while differences (both ADs and NADs) decrease similarity. However ADs and NADs are distributed differently. Subjects list more ADs for similar items than for dissimilar items, while they list more NADs for dissimilar items than for similar ones. In fact ADs and NADs appear to play a different role in the calculation of similarity. A regression equation that predicts similarity from commonalities and differences accounts for more of the variance when the differences are broken down as ADs and NADs than when they are aggregated together.

In addition to the numerical relationship between commonalities and ADs, there are semantic relationships. Examination of subjects' protocols clearly reveals this link as well. For example, 91% (10/11) of subjects asked to list commonalities of the

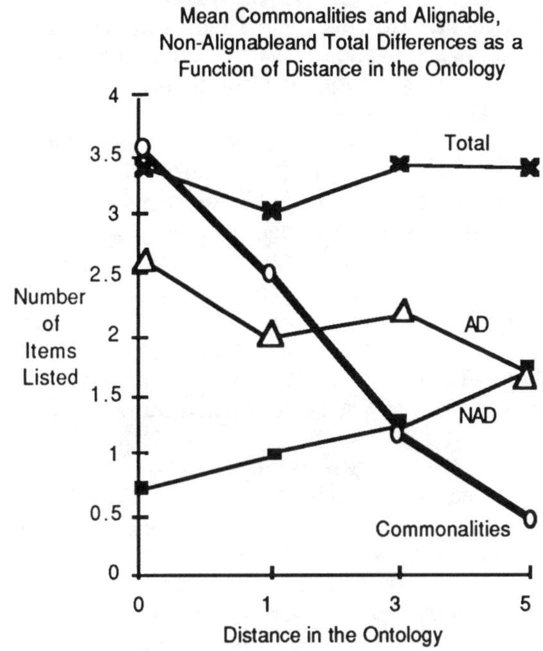


**Figure 5:** Number of Items listed in Experiment

pair <bus/bicycle> stated that both have wheels, while 82% (9/11) of the subjects asked to list differences for this pair noted that a bus has more wheels than a bicycle. In another example, 100% (11/11) of the subjects listing commonalities of the

[7] Since Total Differences=AD + NAD, we can conclude that merely redescribing the difference data as ADs and NADs accounts for more similarity variance than total differences alone. No statistical test is required.

pair <chair/car> noted that you can sit in both, while 73% (8/11) of the subjects listing differences stated that the primary function of a chair is sitting, while the primary function of a car is travelling. This pattern of responses arose often, and indicates that many of the commonalities and differences are strongly linked. This conceptual connection between commonalities and ADs is a direct prediction of the frame-alignment model. On the other hand, feature-based models, which are unambiguous in their prediction that the sets of common and distinctive features are independent, forbid such conceptual associations.

The frame-alignment model draws on earlier work on analogical mapping (Gentner, 1983, 1989, Holyoak and Thagard, 1989) in order to provide an account of how similarity is determined. Our results indicate that the seemingly simple task of partitioning representations into commonalities and differences requires that the alignment process be sensitive to the relations between features. The use of analogical mapping allows features to be placed in correspondence based on their common position within a relational structure.

The debt of the frame-alignment model to featural models of similarity is clear. Tversky's (1977) contrast model (a featural account of similarity) proposed that similarity increases with the commonalities of the pair and decreases with the differences, a result which was obtained in this experiment. However, the frame-alignment model represents an advance over the feature-matching view in three areas. First, the frame-alignment model predicts the relationship (both numerical and conceptual) between commonalities and alignable differences. Second, the ternary distinction of commonalities, ADs and NADs accounts for subjects' similarity ratings better than the binary distinction of commonalities and differences. Finally, and perhaps most importantly, the frame-alignment view provides a mechanism for comparing two conceptual representations that captures the role of common relational structures in determining what counts as a commonality and what counts as a difference.

### Acknowledgments

The authors would like to thank Gary Dell and Brian Ross for helpful comments on earlier versions of this paper. We would also like to thank Evan Heit, Ken Forbus, Rob Goldstone, Doug Medin, Mary Jo Rattermann, Ed Wisniewski and the rest of the Similarity and Analogy Group for many discussions and suggestions during the evolution of this project. A special thanks to Linda May for transcribing the data. Finally, thanks to Bill Parcells for two Super Bowls.

## References

Barsalou, L. (in press) Frames and the Construction of Semantic Fields. to appear in A. Lehrer and E. Kittay (eds.) The Structure of the Lexicon: Semantic Fields and Other Alternatives.

Gentner, D. (1983) Structure-Mapping: A theoretical framework for analogy. Cognitive Science, 7, 155-170.

Gentner, D. (1989) The Mechanisms of Analogical Reasoning. in S. Vosniadou and A. Ortony (eds.) Similarity and Analogical Reasoning. London: Cambridge University Press.

Goldstone, R.L., Gentner, D. and Medin, D.L. (1989) Relations Relating Relations in The Proceedings of the 11th Annual Conference of the Cognitive Science Society. Ann Arbor, MI.

Holyoak, K.J. and Koh, K. (1987) Surface and Structural Similarity in Analogical Transfer. Memory and Cognition. 15(4), 332-240.

Holyoak, K.J. and Thagard, P. (1989) Analogical Mapping by Constraint Satisfaction. Cognitive Science. 13, 295-355.

Keil. F.C. (1979) Semantic and Conceptual Development. Cambridge, MA, Harvard University Press.

Kruskal and Wish, M. (1978) Multidimensional Scaling. Newbury Park: SAGE Publications

Lenat, D. and Guha, R.V. (1990). Building Large Knowledge-Based Systems. San Francisco: Addison, Wesley Publishers, Inc.

Logan, G. (1990). Toward an Instance Theory of Automization. Psychological Review.

Minsky, M. (1981) A Framework for Representing Knowledge. in J. Haugeland (ed.) Mind Design. Cambridge, MA: The MIT Press.

Norman, D. and Rumelhart, D.E. (1972) Explorations in Cognition. San Francisco, Freeman Publishers, Inc.

Rattermann, M.J. and Gentner, D. (1987) Analogy and Similarity: Determinants of Accessibility and Inferential Soundness. in The Proceedings of the Ninth Annual Conference of the Cognitive Science Society. Seattle, WA.

Ross, B. H. (1989) Distinguishing Types of Superficial Similarities: Different Effects on the Access and Use of Earlier Examples. Journal of Experimental Psychology: Learning, Memory and Cognition. 15(3), 456-468.

Schank, R., and Abelson, R. (1977) Scripts, Plans, Goals and Understanding. Hillsdale, NJ: Lawrence Erlbaum Associates.

Smith, E.E. and Medin, D.L. (1981). Concepts and Categories. Cambridge, MA, Harvard University Press.

Tversky, A. (1977) Features of Similarity. Psychological Review. 84(4), 327-352.

# Efficient Visual Search: A Connectionist Solution

**Subutai Ahmad and Stephen Omohundro**

International Computer Science Institute,
1947 Center Street, Suite 600, Berkeley, CA 94704.
ahmad@icsi.berkeley.edu, om@icsi.berkeley.edu

**Abstract**

Searching for objects in scenes is a natural task for people and has been extensively studied by psychologists. In this paper we examine this task from a connectionist perspective. Computational complexity arguments suggest that parallel feed-forward networks cannot perform this task efficiently. One difficulty is that, in order to distinguish the target from distractors, a combination of features must be associated with a single object. Often called the *binding problem*, this requirement presents a serious hurdle for connectionist models of visual processing when multiple objects are present. Psychophysical experiments suggest that people use covert visual attention to get around this problem. In this paper we describe a psychologically plausible system which uses a focus of attention mechanism to locate target objects. A strategy that combines top-down and bottom-up information is used to minimize search time. The behavior of the resulting system matches the reaction time behavior of people in several interesting tasks.

## Introduction

In 1986, Sejnowski wrote: "The binding problem is a touchstone for testing network models that claim to have psychological validity" (Sejnowski 1986). In 1991, the statement is still true. In visual search two aspects of the problem are important: feature integration and localization. Feature integration is concerned with the interference between features of different objects when a parallel representation is used. Consider an image with red, blue, vertical and horizontal objects. By computing a global OR of appropriate feature maps, one can detect in parallel which colors and orientations are present in the image. However, to detect a red and horizontal object in the presence of other objects one would have to pre-compute every possible conjunction of features at every location. Similarly, the interference between objects makes it difficult to recover the locations of individual objects. With a selective attention mechanism that inihibits all but the features of a single object, the interference is removed and the binding problem goes away. Such a model implies that in some situations a serial search is required. This line of reasoning is used to explain experimental results on visual search (reviewed in (Treisman 1988)). The original experiments showed that search for targets defined by a single feature can be computed in parallel but that targets defined by a conjunction of two features required time linear in the total number of objects. (See, however, the section on simulations for exceptions to this rule.)

For a network implementation of visual search to be useful as well as psychologically plausible, a number of constraints must be met. The system should work for high resolution images and must therefore be efficient along several dimensions. The complexity of the network should be low. The time per attention shift should be small. Covert attention shifts in people take about 40-60msecs (Julesz & Bergen 1987). Since neurons can only fire every 5-10 msecs, this leaves time for at most 8-12 sequential steps. The system must be able to deal with objects that vary continuously in size. Finally, the serial component of the search process should be as small as possible, so the number of successive fixations should be minimized.

We have previously described an efficient mechanism for selective attention in the context of a connectionist network for computing spatial relations (Ahmad & Omohundro 1990a). In the following section we describe an extended version for modeling visual search. This network meets the efficiency constraints listed above. A parallel search strategy, SWIFT, is used to minimize search time. In a final section we present simulation results of the system and discuss its relation to recent experimental results on visual search.

## A Model Of Visual Attention

In this section we describe a connectionist model of covert visual attention (see Figure 1). A set of basic features are first computed from the image. The information is then fed to two different systems: a gating network and a priority network. The gating network implements the focus - its function is to restrict higher level processing to a single circular region. The priority network ranks image locations in parallel according to their relevance to the current task. Finally, a set of control networks are responsible for mediating the information flow between these two networks, as well as incorporating top-down knowledge. Each of these parts are described in more detail below.

### The Feature Maps

Feature maps in the network are analogous to the topographic maps early in the visual system. A set of basic features (orientation, color, etc.) are detected at each pixel in

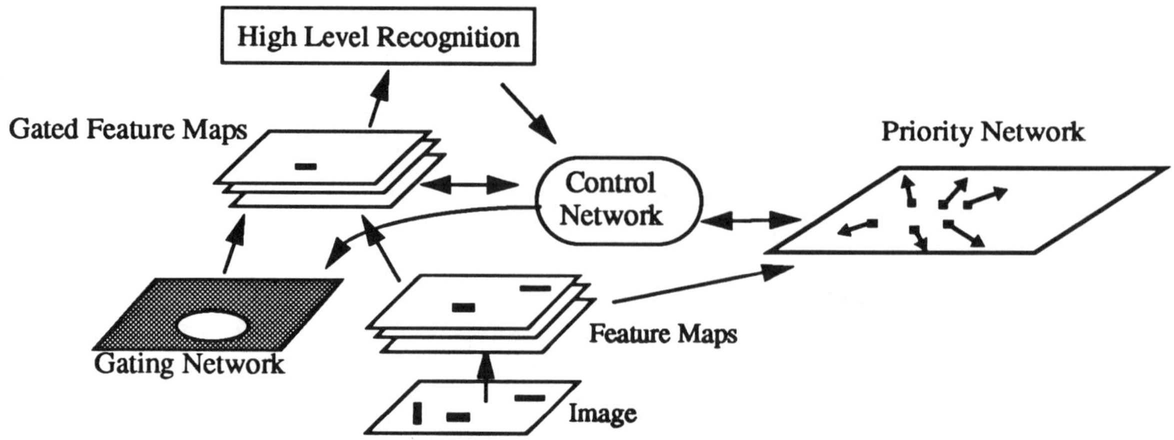


Figure 1 . An overview of the network.

the image in parallel, using one unit at each location for every feature. In addition there is a unit for each feature map which computes the global sum of the activity in the map. Exactly which features should be included is an active area of research. For our purposes, any local feature may be used. Our current implementation uses four feature maps: red, blue, horizontal, and vertical.

## The Gating Network and Gated Feature Maps

To tackle the binding problem the network must be able to inhibit the transmission of features to the recognition stage. This is accomplished by the gating network and the gated feature maps. The gating network contains one unit per pixel. Each gate unit receives as input three parameters ($A_x$,$A_y$, and $A_r$) representing the center and radius of the current circular focus of attention. Only the gate units *outside* the circle turn on (see (Ahmad and Omohundro 1990a) for details). Each unit within the gated feature maps receives activation from the corresponding feature detector and inhibition from a gate unit (Figure 1). Only the portion of the gated feature map that falls within the current focus will respond. The resulting system can filter image properties based on an external control signal. As with regular feature maps, the network computes a global sum for each gated feature map. When attention is focused on an object, the activity of these sum units will reflect the object's features, regardless of distractors. Retrieving the location of an object is simple: with attention centered on the object, the units representing $A_x$, and $A_y$ will reflect the object's location. The complexity of the network is linear in the number of pixels and the time to focus is a small constant.

An alternate architecture would use direct inhibition of the feature maps themselves. This would eliminate the need for a separate set of gated maps. However, in a focused state, such a network would be unable to make global decisions based on the features. With the configuration described above, the network can efficiently access both local and global information simultaneously. As we will see below, this ability is crucial in efficiently carrying out visual search. There is even some direct psychological evidence to support the current architecture. When attention is highly focused, people are able to report primitive features of stimuli appearing outside the focus of attention (Rock et al. 1990) but they are unable to report shape information suggesting that higher level processing is affected.

## The Priority Network

This gating method relies on an external mechanism for determining focus locations and this is provided by the priority network. Its job is to rank image locations in order of importance and to help shift the focus to those locations. The main component is a coarse coded map in which the output of each unit reflects the priority of the region within its receptive field. A simple and efficient way to rank regions is to use the mass of the points within the receptive field (Ahmad & Omohundro 1990a). If this is the only ranking available, however, then search for attributes other than size could be inefficient. Several psychophysical experiments have pointed out other possibilities. (Yantis & Jonides 1990) provide evidence that stimuli which appear abruptly are attended to sooner than persistent stimuli, but that this can be overriden by explicit instructions to the subject. Experiments on visual search suggest that objects with the same features or form as the target object can get higher priority than other objects (Egeth, Virzi, & Garbart 1984). All of this suggests a much more dynamic and flexible priority system than one which simply ranks the locations based on pixel density.

This sort of flexibility can be added to the network without sacrificing efficiency. The priority network contains a unit associated with each feature map, $P_f$, whose value indicates the importance of that map. This value is dynamically adjusted according to the task. The priority of units in the error map is computed as:

$$A_i = G\left(\sum_{x,y \in RF_i}\sum_{f \in F} P_f A_{fxy}\right)$$

$A_{fxy}$ is the activation of the feature unit at location *(x,y)*. $RF_i$ denotes the receptive field of unit *i*, and *G* is a monotonically increasing function of its input (we use a sigmoid). When $P_f$ is *0*, feature *f* has no effect on the priority map. This allows the system to completely shut off the effect of any feature map in parallel.

We also need a way to update the focus of attention to the relevant locations. To accomplish this each location in the priority map contains two additional units whose values encode an "error vector". The error vector is simply the difference between the units' location and the current center of focus. These vectors are constantly updated as the focus moves around. This representation is flexible and efficient. To move the focus to the highest priority location, the control network simply chooses the corresponding error vector and adds its components to $A_x$ and $A_y$. To choose the nearest location, the control network selects the smallest error vector. To choose locations to the right it can select a vector whose first component is positive.[1]

### The Control Network

The control network coordinates the information flow between the gating network and the priority network. It consists of a collection of autonomous sub-networks carrying out independent tasks. There are networks which continually fine tune the scale and location of the focus of attention, networks for storing locations, and a network for updating the focus to the next error vector. These are described in detail in (Ahmad & Omohundro 1990ab). The main addition is the subsystem SWIFT which controls the search process. This is described below.

**The SWIFT search strategy.** The main function of SWIFT is to integrate top-down and bottom-up knowledge to efficiently guide the search process. Top down information about the target features are stored in a set of units. Let *T* be this set of features. Since the desired object must contain *all* the features in T, any of the corresponding feature maps may be searched. Using the ability to weight feature maps differently, the SWIFT network can remove the influence of all but one of the features in *T*. By setting this map's priority to 1, and all others to 0, the system will effectively prune objects which do not contain this feature. (Hence the name SWIFT: Search WIth Features Thrown out.) To minimize search time, it should choose the one corresponding to the least number of objects. Since it is difficult to count the number of objects in parallel, the network chooses the map with the minimal total activity as the one likely to contain the minimal number of objects.

SWIFT was inspired by the experiments in (Egeth, Virzi, & Garbart 1984). These authors present evidence suggesting that serial search can be restricted to objects with a particular feature. For example, if subjects were instructed to attend to red objects, and the number of red objects were kept small and constant, then search time was constant with respect to the number of distractors. In our implementation, the system dynamically computes the best feature.

## Simulations

The simulation proceeds as follows. Initially the network is presented with an image and "shown" the target object by focusing attention on it. The network stores the activity of the gated feature maps in a set of units and these become the target features. For each subsequent image, the total activity of all the feature maps is computed in parallel. Among the target features, the network chooses the one with the least activity and sets its priority to *1* and all others to *0*. Search then proceeds by sequentially visiting locations in order of their saliency. As the focus of attention stabilizes on each location, the control network checks the features of the current object against the stored target representation. This continues until a match is found or there are no more objects.

The output of the simulator is shown in Figure 3(a). The lower left quadrant displays the image (the one shown is 64x64 pixels). The top left quadrant displays the activity of the four gated feature maps. Clockwise from top left the features are: blue, red, vertical and horizontal. In the figure the system is attending to the leftmost red-vertical object, so only the units at that location are active. The bottom right quadrant shows the feature maps that are currently affecting the priority map. Since the target is a blue-vertical object, the system has chosen the vertical map as the minimal feature map. The top right quadrant displays the error vectors in the priority map. Note that only vertical objects have significant priority. The run-time behavior of the network is discussed in the following sections.

### Search Time With SWIFT

Since SWIFT always searches the minimal feature map, the critical variable, *M*, that determines search time is:

$$M = \min_{f \in T}\{O(f)\}$$

where *f* ranges over all the target object's features, and *O(f)* is the number of objects with feature *f*. Search time will always be linear in *M*, but does not necessarily have anything to do with *D*, the number of distractors. For example, in images such as in Figure 3(b), the vertical map will be chosen as the minimal feature map. Search time will not depend on the number of horizontal items. In a sense the search time is dependent on the discriminability of the target object and not on the total number of distractors. Figure 4 plots the actual search time averaged over several trials for various combinations of *M* and *D*. In Figure 4(a), the number of distractors is fixed at *40* as *M* is gradually increased. As expected, mean search time in-

1. The error vector representation was inspired by a similar mechanism for controlling eye saccades in the monkey superior colliculus (Sparks 1986).

creases linearly. Since the search is self-terminating, the ratio of the slopes for the target absent and target present cases is about 2:1. In Figure 4(b), the graphs show that search time can remain relatively flat as $D$ increases, as long as $M$ is held constant. To our knowledge this specific set of experiments has not been performed on people. In the following sections we discuss SWIFT in relation to many of the experiments that have been done.

## Relationship With Psychological Data

**Single and conjunctive feature searches.** We first show that the original search results (Treisman 1988) can be replicated with SWIFT. The experiments showed that targets defined by a single feature (e.g. a red target among blue objects) can be detected in parallel. Targets defined by a conjunction of two features (e.g. a red-horizontal target among red-vertical and blue-horizontal objects) required time linear in the number of objects.

For single feature searches, $T$ contains one feature so SWIFT will always choose it. To detect whether the target is present just requires one step since there can be at most one object with that feature. For conjunction searches $T$ contains two features. If the number of objects with each feature is chosen randomly, on average $M$ will be $1/2D$. Therefore average search time will grow linearly with $D$ (see (Egeth, Virzi, & Garbart 1984) for a similar argument). The ratio of slopes for images with target absent to target present will be *2:1*, consistent with any self-terminating serial scan. More recently it was shown that accurately detecting conjunctions depended on accurate localization of the target (Treisman & Sato 1990). This is also consistent with our architecture (Ahmad & Omohundro, 1990a).

**Triple conjunction search.** Search for an object defined by a conjunction of three features results in different search slopes (Quinlan & Humphreys 1987). There were two situations that were tested: (a) every distractor shares exactly one feature with the target object, or, (b) every distractor shares exactly two features with the target. Both cases resulted in sequential search, but the slope in (b) was always steeper than the slope in case (a). These results are consistent with SWIFT. In case (a), on average the minimal feature will eliminate *2/3* of the distractors. In (b), only *1/3* would be eliminated on average. Thus SWIFT predicts that the slope in (a) should be about half that of (b).

**Search asymmetries.** There is another search paradigm where constant and linear time searches have been reported. Searching for a line oriented 18° among vertical lines can be done in constant time, but searching for a vertical line among these oblique lines takes linear time (Treisman 1988). This asymmetry is explained by assuming that the early representation includes a finite number of orientations that are coarse coded, including vertical and an orientation greater than 18°. Each oblique line is represented as a combination of activity in the vertical map and the map coding a successive orientation. If this is true, then a pattern containing a single oblique line among a field of vertical lines will cause several regions of activity in the vertical map but only a single region of activity in the other map. The presence of the oblique line can therefore be detected in constant time by computing a global OR. However, the image of a vertical line among several oblique lines will generate several active regions in both maps except at one location, where only the vertical map is activated. In this case, the network must bind the presence of activity in one map with the absence of activity at the same location in another map. This requires serial search.

Similar asymmetries are present when detecting curvature, circles vs ellipses, single vs paired lines, etc. In all of these cases, a central question is: how does the brain know what to do? The subject has no knowledge about his/her internal representations. Just knowledge about the target object is insufficient - the map that is searched depends on the particular image. The answer is simple if SWIFT is used: searching the map with the least total activity will always produce the correct results.

**Parallel processing of conjunctions.** Some authors have reported conjunctive feature searches which always result in flat slopes. (McLeod, Driver, & Crisp 1988) report that the detection of a moving X among static X's and moving O's can be done in parallel. (Nakayama & Silverman 1986) tested conjunction searches using the features color, motion, and depth. They found that motion-color conjunctions required serial processing, whereas depth-color and depth-motion conjunctions could be processed in parallel.

Recently (Treisman & Sato 1990) and (Wolfe, Cave & Franzel 1989) have suggested models where conjunctions can be detected in constant time with top-down information. It is possible to implement Treisman and Sato's Feature Inhibition model in our architecture. They suggest that if the features that are *not present* in the target inhibit the priority map (i.e. $P_f$ is negative) then a location containing the conjunction of two features would retain the highest priority. This can be easily modeled in our network, however there is one problem: it cannot explain sequential search! If people can use such a general strategy, why do we get linear search times at all? A related problem is that both models cannot explain why only specific feature combinations give rise to parallel search.

SWIFT can explain these results if one assumes that certain feature combinations are represented in parallel. For example, (McLeod, Driver, & Crisp 1988) mentions that area MT contains cells which are tuned to both direction of motion and orientation. Since a primary feature that distinguishes X's from O's is an oriented line, a moving X should produce a unique pattern of activity in this feature map. If such combinations are present, then SWIFT would select the appropriate feature map and detect the target in constant time.

(a) (b)

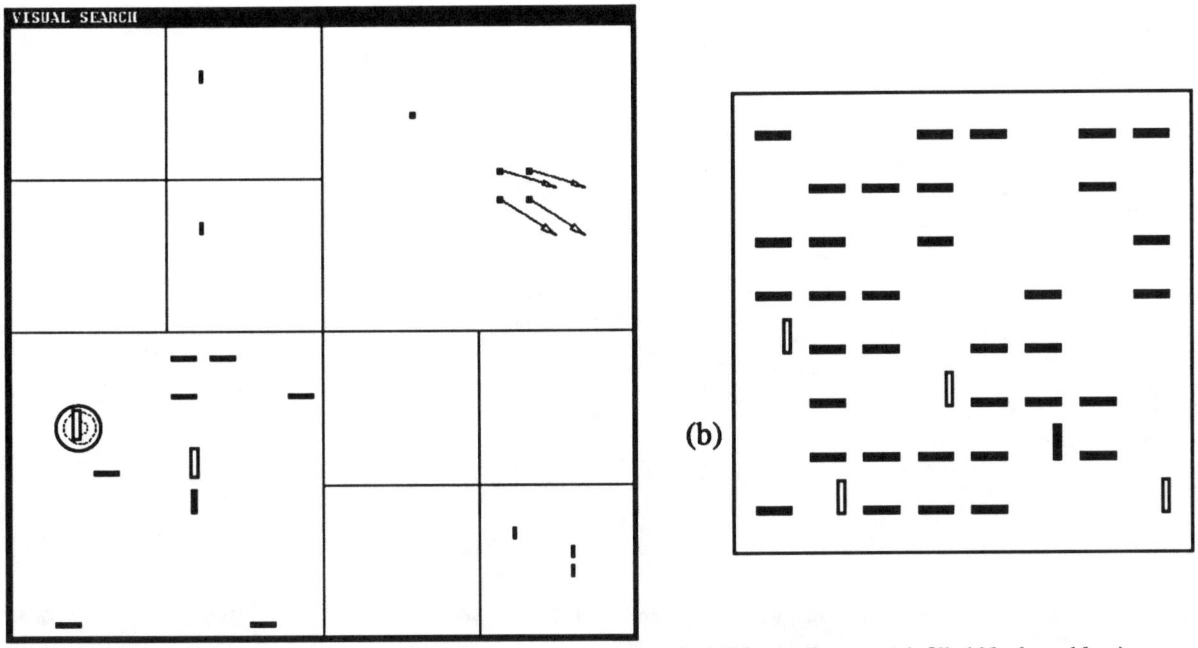


Figure 3. (a) Sample output from our simulator. (b) An image with M=5. (No shading => red, filled black => blue.)

## Concluding Remarks

**Optimal features for visual search.** In light of the above results it is natural to ask what the best set of features should be. If SWIFT is used as a constraint, then we want the set of features that minimize *M* over all possible images and target objects, i.e. that best discriminate objects. It is easy to see that the optimal set of features should be maximally uncorrelated and that the distribution of feature values should be uniform over the space of possible objects. In other words, the optimal features should be the principal components of the distribution of images. It is interesting to note that a single Hebb neuron extracts the largest principal component of the input distribution and with inhibition, sets of Hebbian neurons can extract successively smaller components. Moreover, as some researchers have demonstrated (e.g. Linsker 1989), simple Hebbian learning can lead to features that look very similar to the features in the visual cortex. If the early features in visual cortex are in fact the principal components, then SWIFT is a simple strategy that takes advantage of it.

**Other computational models.** (Chapman 1990) has implemented a pyramid model of attention. It has only been used to replicate the original single feature vs conjunctive feature searches although, in principle, a control strategy like SWIFT could be employed. (Wolfe, Cave, & Franzel 1989) have simulated a model of visual search which they call the Guided Search model. Their model accounts for a wider range of results than Chapman's. Our model is consistent with their philosophy in that a smart parallel strategy is used to rank possible candidates. However their model cannot account for the search asymmetry results or the *2:1* ratios in the slopes. In addition, the model requires complete connectivity of the units in the feature maps. This would require $O(N^2)$ weights (resulting in approximately $10^{12}$ connections for a 1000x1000 image) and is therefore not implementable for high-resolution images. Neither of these models implement a continuous focus of attention: each object is assumed to occupy exactly one pre-determined location.

To our knowledge, the only realistic implementation of visual search that works with pixel-based images is in (Mozer, 1991). The model also implements a continuous focus. There are some differences in our models. In our model every aspect of the control process is made explicit, including the use of top-down information. In their model the differing search slopes are explained by assuming a specific amount of noise in the activations of the feature maps. In our model search time depends on the feature representations and the minimal feature map.

**Conclusions.** We have presented efficient psychologically plausible connectionist mechanisms for visual attention. These mechanisms have been integrated into a complete system for visual search. The resulting network scales well both in terms of the number of connections (linear in the number of pixels) and in the focusing time (constant). The implementation of a single plausible search strategy, SWIFT, was shown to be consistent with the single/conjunctive search, the 2:1 ratio in the target absent/present slopes, and dependence on localization. The strategy extends other sequential integration models in that it is also consistent with search for triple conjunctions, search asymmetries, search within a feature, and possibly the constant time detection of certain feature combinations.

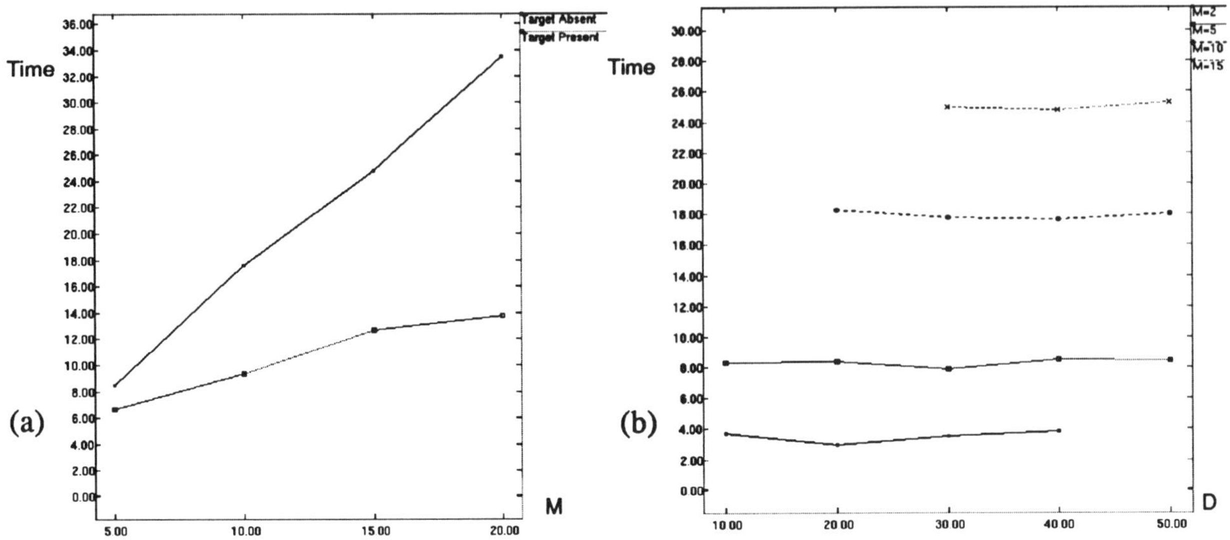


Figure 4. In (a), D is constant at 40 while M is varied. In (b) M is kept fixed and D is varied. (Target absent only. Target present results are similar.)

## Acknowledgments

We thank Peter Blicher, Jerome Feldman, Jitendra Malik, and especially Anne Treisman for their helpful comments and stimulating discussions.

## References

Ahmad, S., and Omohundro, S. 1990a. Equilateral Triangles: A Challenge for Connectionist Vision. In: Proceedings of the 12th Annual Conference of the Cognitive Science Society, MIT, July, 1990.

Ahmad, S., and Omohundro, S. 1990b. A Connectionist System for Extracting the Locations of Point Clusters, Technical Report, TR-90-011, International Computer Science Institute, Berkeley, CA.

Chapman, D. 1990. *Vision, Instruction, and Action.* Ph.D. diss., Dept. of Computer Science, MIT.

Egeth, H.E., Virzi, R.A., and Garbart, H. 1984. Searching for Conjunctively Defined Targets. *Journal of Experimental Psychology: Human Perception and Performance,* **10**(1):32-39.

Julesz, B., and Bergen, J.R. 1987. In: Fischler, M.A. and Firschein, O. (Eds.) *Readings in Computer Vision: Issues, Problems, Principles, and Paradigms.* Morgan Kaufmann.

Linsker, R. 1989. How to Generate Ordered Maps by Maximizing the Mutual Information Between Input and Output Signals. *Neural Computation,* **1**:402-411.

McLeod, P., Driver, J., and Crisp, J. 1988. Visual search for a conjunction of movement and form is parallel. *Nature,* **332**:154-155.

Mozer, M. 1991. *The Perception of Multiple Objects: A Connectionist Approach.* MIT Press, Cambridge, MA.

Nakayama, K., and Silverman, G. 1986. Serial and parallel processing of visual feature conjunctions. *Nature,* **320**:264-265.

Quinlan, P.T. and Humphreys, G.W. 1987. Visual search for targets defined by combinations of color, shape, and size: An examination of the task constraints of feature and conjunction searches. *Perception & Psychophysics,* **41**:455-472.

Rock, I., Linnett, C.M., Grant, P., and Mack, A. 1990. Results of a New Method for Investigating Inattention in Visual Perception. Paper presented at 31'st annual meeting of the Psychonomic Society, New Orleans, LA, November, 1990.

Sejnowski, T.J. 1986. Open Questions About Computation in Cerebral Cortex. In McClelland, J.L., and Rumelhart, D.E. (Eds.) *Parallel Distributed Processing: Explorations in the Microstructure of Cognition.* Cambridge, MA, MIT Press.

Sparks, D. L. 1986. Translation of Sensory Signals into Commands for Control of Saccadic Eye Movements: Role of Primate Superior Colliculus, *Physiological Reviews,* **66**(1).

Treisman, A. 1988. Features and Objects: The Fourteenth Bartlett Memorial Lecture. The Quarterly Journal of Experimental Psychology, **40A** (2).

Treisman, A., and Sato, S. 1990. Conjunction Search Revisited. Journal of Experimental Psychology: Human Perception and Performance, **16**(3):459-478.

Wolfe, J.M., Cave, K.R., and Franzel, S.L. 1989. Guided Search: An alternative to the modified feature integration model for visual search. *Journal of Experimental Psychology: Human Perception and Performance,* **15**:419-433.

Yantis, S., and Jonides, J. 1990. Abrupt Visual Onsets and Selective Attention: Voluntary Versus Automatic Allocation. *Journal of Experimental Psychology: Human Perception and Performance,* **16**(1):121-134.

# Perceptual Simplicity and Modes of Structural Generation

**Jacob Feldman**

Department of Brain and Cognitive Sciences,
Massachusetts Institute of Technology,
Cambridge, MA 02139
jacob@psyche.mit.edu

**Abstract**

This paper describes a formal framework for perceptual categorization that can account for the salient qualitative predicates human observers are willing to ascribe to a closed class of objects, and consequently the simple groupings they can induce from small sets of examples. The framework hinges on the idea of a generative process that produces a given set of objects, expressed as a sequence of group-theoretic operations on a primitive element, thus ascribing algebraic structure to perceptual organization in a manner similar to Leyton (1984). Putatively, perceivers always seek to interpret any stimulus as a formally *generic* result of some sequence of operations; that is, they interpret each object as a typical product of some generative process.

The principle formal structure is a "mode lattice," which a) exhaustively lists the qualitative shape predicates for the class of shapes, and b) defines the inferential preference hierarchy among them. The mechanics are worked out in detail for the class of triangles, for which the predicted qualitative features include such familiar geometric categories as "scalene," "isosceles," and "right," as well as more "perceptual" ones like "tall" and "short." Within the theory it is possible as well to define "legal" vs. "illegal" category contrasts; a number of examples suggest that our perceptual interpretations tend to regularize the latter to the former.

## Introduction

Among all the descriptions of a stimulus configuration that are in principle possible, human observers seem to favor the simplest ones, a generalization that can be traced back as least as far as the Gestaltists (e.g., Koffka, 1935; Köhler, 1947). Consequently, many perceptual theories have attempted to account for preferred descriptions by showing that they minimize some measure of complexity, often description length in some description language (e.g., Buffart, Leeuwenberg, & Restle, 1981; Darrell, Sclaroff, & Pentland, 1990; Hochberg & McAlister, 1953; Simon, 1972; see also Hatfield & Epstein, 1985; Leeuwenberg & Boselie, 1988; Perkins, 1976). This notion, often termed the "minimum principle," parallels the mathematical theory of informational complexity (e.g. Chaitin 1966, Kolmogorov, 1965), in which the complexity of a string is defined as the length of the shortest computer program that can generate it. However, these perceptual theories have often suffered from the fact that the description languages used must be somewhat arbitrary; the atoms of the language—individual features, predicates, and types of spatial relationships that are worth incorporating into descriptions—are difficult to motivate by underlying theory. Rather than finding one "minimum principle" underlying perception, as would be desired, perceptual theorists find many. Moreover, the relationship between descriptive simplicity and descriptive *correctness*—if there is any—has never been made completely clear.

This paper attempts to get around these problems by proposing a framework of stimulus description which explicitly generalizes across different ways of conceiving of the stimulus as having been constructed. A particular conception of the generating process underlying a particular class of stimuli entails a lattice relating the subspaces and half-spaces. This lattice both a) exhaustively enumerates the qualitative structural predicates and features of the class of objects, and b) makes the inferential preference relationships among them explicit. The scheme is worked out here in detail for one class of object, triangles; the resulting legal shape predicates seem, intuitively, both simple and natural. In this way the qualitative shape predicates that human observers are willing to ascribe to triangles, as evidenced in the categories they are able to induce from a handful of examples, are thus characterized formally, and (in theory) exhaustively.

## Mode lattices

Leyton (1984) developed a view of perceptual interpretation in which a stimulus is treated by the observer as the product of a sequence of nested geometric transformations. Perceptual interpretation, then, proceeds by factoring the input, layer by layer, into successively simpler structures to which more complex ones are referenced. The algebraic structure this theory imposes on perceptual organization is intriguing and compelling—explaining, for example, why a parallelogram is perceived as a slanted rectangle, and a rectangle as a stretched square. A line segment, in this framework, might be the visible product of a translation operation on a point in the plane, which we label the origin (giving us translation invariance): $(\mathbf{0})\tau_{r,\theta}$. The space of all line segments constructed this way (the *configuration space*) has two dimensions, $r$ and $\theta$. A generic point in this space—a typical product of the generative process we are assuming to have produced all line segments—is a line with non-zero length that is neither exactly horizontal nor exactly vertical. Non-generic points fall on one of the subspaces (i.e., axes) of the space, and correspond to the class of vertical lines, the class of horizontal lines, and the origin point itself. Line segments that fall into one of these subspace classes are very special, and it is worthwhile making their specialness explicit in the perceiver's representation.

In general, we model each stimulus configuration $u$ as the visible product of the action on the origin of a sequence of one-parameter operations

$$u = (0)\mathcal{T}_1^{\alpha_1} \cdot \mathcal{T}_2^{\alpha_2} \cdot \ldots \cdot \mathcal{T}_d^{\alpha_d},$$

where $d$, the number of one-parameter operations, is the dimensionality of the resulting configuration space, and each operation $\mathcal{T}_i$ is carried out to a magnitude $\alpha_i$. Any operation whose magnitude is zero reduces to the identity operation, and thus drops out of the expression completely; this corresponds to a subspace of the configuration space. Within a given space (or subspace) the generic objects are all produced by generative strings in which none of the operations has dropped out. Recursively, a non-generic object that falls in a given subspace can either be generic in that subspace, or it can actually fall in a subspace of the subspace, in which case it is even more non-generic; that is, it has higher codimension in the overall configuration space. (The codimension is simply the difference between the dimension of an object and the dimension of the space in which it is embedded.) Each proper subspace divides each embedding superspace (of dimension one greater) in half; we call the two half-spaces *modes* of the space. Conversely, each (possibly improper) subspace of dimension $k$ can be divided into two half-space modes by a subspace (of dimension $k$ - 1) in $k$ different ways.

*Generic interpretations are preferred.* Apparently, perceivers always seek to interpret the observed configuration as a *generic* product of some sequence of generating operations. The generating sequence of operations is pruned of identities until a generic sequence can be seen to yield the observed object; that is, if an observed object can be expressed with one of the operations as the identity, the object's interpretation drops down a dimensional level to an embedded subspace. Lower-dimensional (higher codimension) subspaces, since they have more dimensions fixed in a way unlikely to occur by accident, always make stronger inferences of structure. For a closed class of objects in one configuration space, therefore, all the predicates a perceiver is willing to apply to objects in the class should correspond to subspaces or half-spaces of the configuration space.

These spaces can all be connected up schematically in a "mode lattice." The mode lattice is built around the ordinary subspace lattice, to which is added, for each subspace, the two modal half-spaces of the embedding superspace into which the subspace divides it. The resulting construction makes the dimensionality and subspace relationships among interpretations of objects in the class explicit. Then, putatively, all allowable predicates on these objects can be read off the mode lattice in an automatic fashion. The next section describes a mode lattice for triangles.

## Mode lattice for triangles

One way of conceiving a triangle is as two legs joined at a fulcrum, and a third leg joining their ends. We now cast this generating process formally as a sequence of operations on the origin, and construct the corresponding mode lattice. Note that any alternative way of formally conceiving the construction of triangles would lead to a different mode lattice, one which imposes a different geometry on the space of triangles. There are a finite number of such alternatives.

In this way of conceiving triangles, there are two controlling dimensions: the angle $\theta$ between the two sides that meet at the fulcrum, and the difference $\Delta r$ between the lengths of the legs. Assuming scale invariance, the dimension of the configuration space $d$ is thus 2, so that there are 3 ($2^d$ - 1) proper subspaces: $r_1 = r_2$, $\theta = \pi/2$, and the origin, at which both $r_1 = r_2$ and $\theta = \pi/2$. ($\theta = \pi/2$ is an identity, briefly, because only this special angle differs from its complementary angle by the identity transform.) The mode lattice appears in Figure 1; the legend and explanatory schematics are on the left of the figure, and the lattice itself is drawn on the right. A dot has been drawn on all the sample triangles at the reference vertex, both in order to define it and to encourage the reader's perceptual apparatus to treat it as special, thus hopefully creating a bias towards seeing the triangles as products of a generative process of the sort described.

The "scalene" triangle (the term here meaning generic in the overall configuration space) appears at the top of the lattice. An arbitrary triangle, that is, a typical product of this generative process, is scalene. The proper subspaces are also familiar triangle categories: the axes are "isosceles" and "right", two categories traditionally treated specially, and the origin is of course "right isosceles." Necessarily, the origin triangle is shape-invariant, since there are no degrees of shape freedom left to vary. This triangle, via the spatial parameters it has fixed, in a sense implies the entire lattice.

The feature predicates that appear on the mode lattice should all be categories of triangles that human observers can induce easily from even a very small set of examples. This is most acutely seen when categories are contrasted with one another, as in a "Bongard" problem (Bongard, 1970; Feldman, in preparation; Richards, 1988). The next section proposes a definition of "legal" category contrasts, i.e., legal pairings of modes and subspaces from a mode lattice. The criterion for legal contrasts, though motivated on independent formal grounds, turns out to be intimately related to a simple model of naturally-occuring categories.

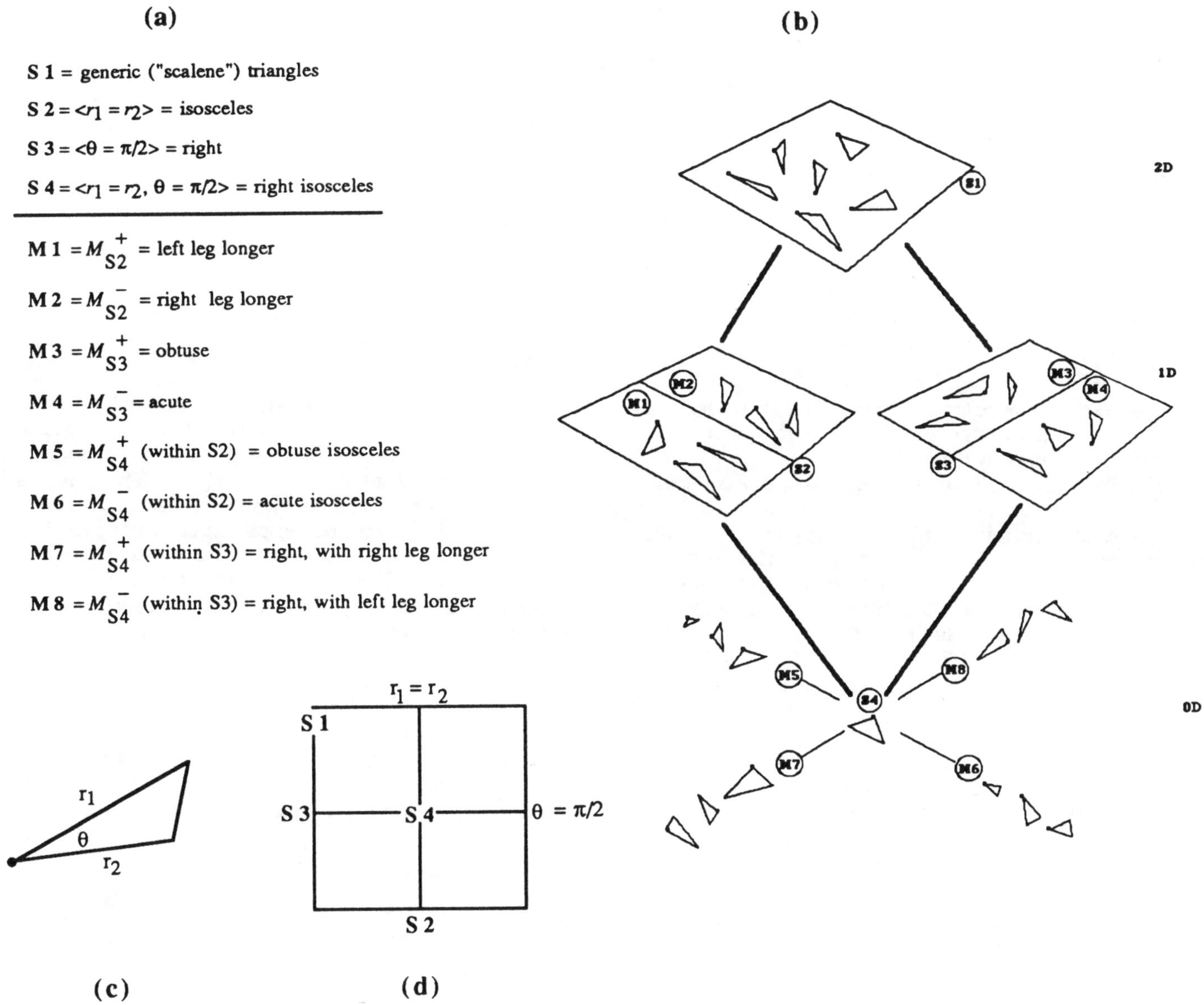


Fig. 1. Mode lattice for triangles: (a) list of subspaces and modes (half-spaces); (b) mode lattice; (c) schematic of a triangle, labeling the angle and sides; and (d) schematic of the configuration space, showing the relation of the four subspaces. The notation $M^{+}_{S2}$ (etc.) indicates the mode that is on the plus side of subspace S2.

## Legal category contrasts

The 2-D triangle space has 4 subspaces (including the entire space) and 8 modes, so there are a total of 12 theoretically reasonable categories; in the absence of constraint any of the 66 pairings of these might make a good contrast. We now cut this number down by proposing constraints.

Legal category contrasts naturally fall into two types: between-space and within-space.

*Between-space contrasts.* Between-space contrasts, such as those between the two axes of the triangle space, or between the entire space and one axis, involve categories that can be written as completely different sequences of generative operations. A subspace of a space is generated by a subsequence of its defining operations, with the identity operation dropping out. Two crossing subspaces may be thought of as the product of two different operations on the same point. In either case, the resulting categories amount to different "species" of object; with different generative identities, a perceiver might justifiably expect them to have qualitatively different structural properties and attributes.

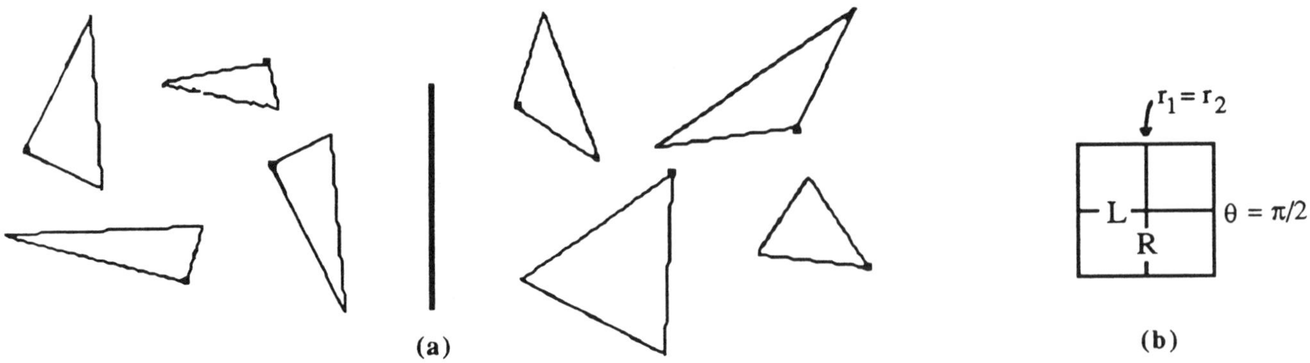


(a) (b)

**Fig. 2.** (a) "Right" (L) vs. "Isosceles" (R) triangles; (b) schematic of the contrast, showing it as two subspaces of triangle configuration space.

An example appears in Figure 2. The reader may confirm that the category inductions are perceptually apparent and intuitively natural.

*Within-space contrasts.* A category within a (sub)space is a collection of quadrants of the space; any two such categories correspond to different classes within the same basic generating process. Though they cannot be expressed as the product of different sequences of operations, the contrasting categories manifest different values along some structural dimensions. That the differences between them are modal—they cross the boundaries delineated by the subspaces—suggests that such categories might, like between-space contrasts, exhibit qualitatively different structural properties or attributes.

To motivate the definition of legal within-space contrasts, consider that some collections of quadrants in a 2-D can be expressed as a contrast along one dimension that is then extruded symmetrically along the other dimension. For example, if the full 2-D space is divided in half, there is contrast between the two half-spaces across one axis (the bisecting one), but symmetry across the perpendicular axis. In such an arrangement, we call the asymmetrical axis the *contrast* component, and the symmetrical axis the *extrusion* component (because the contrast is extruded along it). Legal contrasts are defined to be those that can be expressed as a pure combination of an extrusion component of magnitude $e$ and a contrast component of magnitude $c$, such that $c > 0$ (or else there is no contrast) and $c + e = d$ (all dimensions are accounted for). In the extreme case of pure contrast ($e = 0$), two diagonally opposite quadrants meet at the origin. In this case the contrast between the categories is complete: they differ on every essential structural parameter.

An example of a legal within-space contrast appears in Figure 3. Again, the contrast seems to capture a perceptually intuitive and salient distinction.

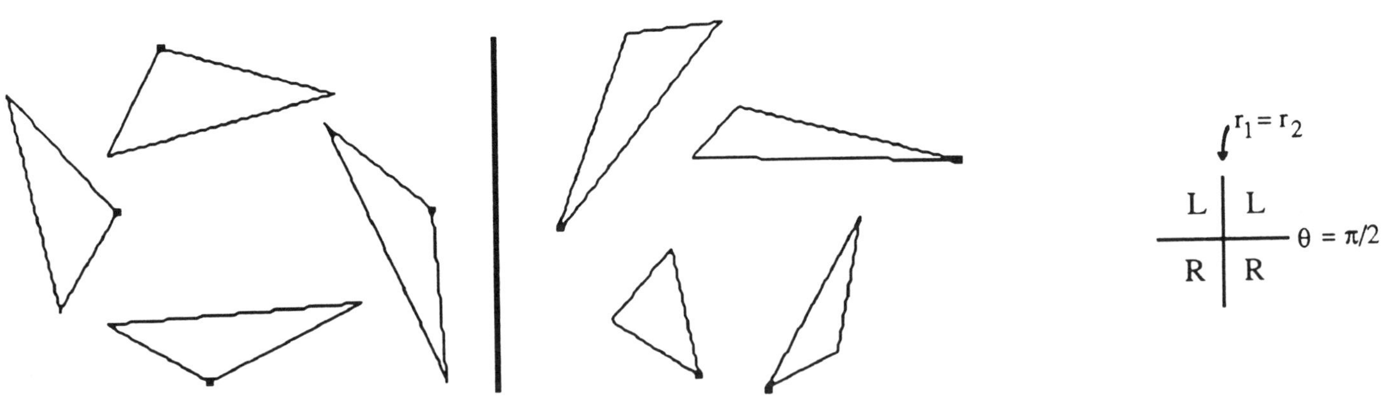


(a) (b)

**Fig. 3.** (a) "Obtuse" (L) vs. "Acute" (R) main angle; (b) schematic of the contrast, showing it as two modal half-spaces of the configuration space. The categories are contrasted along the $\theta$ dimension, and extruded along the $\Delta r$ dimension.

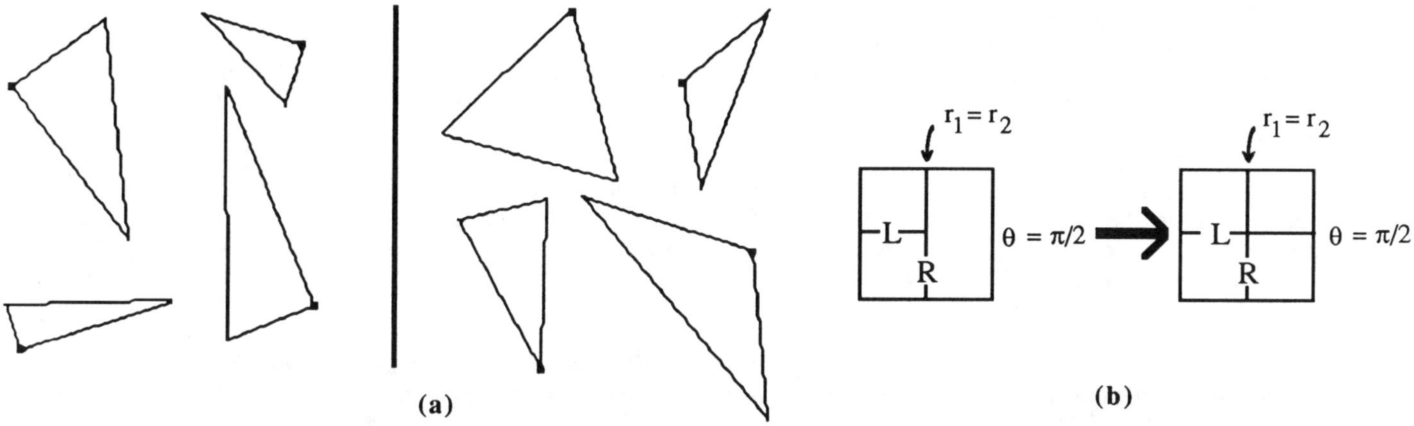


**Fig. 4.** (a) "Right & right leg longer" (L) vs. "Isosceles" (R); (b) schematic showing the regularization of the contrast to "Right" vs. "Isosceles."

These two examples are meant only to give the general flavor of legal contrasts: they capture in a formalism the intuitively natural contrasts that one may exhibit between triangles (at least those that were created by *this* generative process). There are other qualitative types of legal category contrast, for which space does not permit examples. All the basic types can be catalogued exhaustively, so that theory's categorization of hypothetically "natural" perceptual categories is complete.

## Illegal category contrasts: regularization?

*Non-modal contrasts.* Many of the a priori possible category contrasts, both between- and within-space, are illegal as defined. Each of these, apparently, is regularized by our perceptual apparatus to one of the legal types. Figure 4 gives an example. The space of right triangles (L) is incomplete, including only triangles with the right leg longer, but our perceptual apparatus apparently either fails to notice this or considers it insignificant.

*Amodal contrasts.* So far, in considering possible classes of object within a space, we have completely ignored most possible subregions of the configuration space, in favor of the subspaces and modes. Nearly all subregions of the configuration space, of course, are not composed, even illegally, of subspaces and modes. Arbitrary contiguous patches of the space, not to even mention discontiguous parts (such as topologically disconnected or even nowhere continuous point sets), might in the absence of any constraint constitute reasonable stimulus categories. We would not expect our perceptual apparatus, however, to be able to correctly characterize world categories of arbitrary complexity. To reflect this, it should be the case that most topologically possible categories are *not* modal in any parameterization of the object class. Moreover, the total number of modal categories, considering all conceivable generative processes, should be finite. This turns out to be the case; all but a finite family of categories are amodal, that is, are not constructed from legal modal categories in any parameterization.

In pilot work we have looked at a particularly simple case of such amodal categories, which as such is a particularly good candidate to falsify the theory: a pair of simple linear categories that cross at a point (like the two axes of the configuration space), but whose crosspoint is not an origin in any parameterization (unlike the axes'). Such categories cross modal boundaries in a manner inconsistent with any legal category interpretation. The result is a pair of categories that seem not to have a perceptually reasonable distinction between them. When the categories are mixed together, moreover, the perceiver tends to regroup them modally. The "real" category is regularized to one which is incorrect but is at least perceptually reasonable.

## Conclusion: legal categories as world categories

Though the term "qualitatively similar" has some intuitive meaning to human perceivers, a formal definition is elusive; formally, *any* two distinct items in a set share the same number of properties unless some theory constrains the properties (Watanabe, 1985). Formal theories of perception (Bennett, Hoffman, & Prakash, 1989; Bobick & Richards, 1986; Feldman, 1991; Jepson & Richards, 1991; Leyton, 1984, 1986, 1988, 1989; see also Fell, 1976; Witkin & Tenenbaum, 1983, 1986) attempt, in essence, to distinguish the perceptual predicates that perceptual systems such as ours favor from the much wider class of arbitrary predicates, nearly all of which are perceptually useless. The mode lattice theory provides a formal definition, which the examples suggest matches human intuitions.

We may now reasonably ask, why does this definition work for the human perceptual system? In particular, do "legal" categories correspond reliably to extant natural categories in the physical and biological world? It turns out that they do. It can be shown that an extremely simple, formal model of natural categories and processes reduces exactly to the legal categories, as defined above. The model, briefly, assumes that natural categories manifest heavy internal structure within themselves, and correspondingly sharp contrasts (that is, sharper than a priori necessary) between each other. This idea is a particularization of the Principle of Natural Modes, articulated by Bobick (1987) and Richards & Bobick (1988). Space requires that a full discussion, including the requisite formal definition of "structure," be left to a future paper. The thrust is that the apparent psychological preference for modal categories, as evidenced by the various types of regularization demonstrated above, seems to be rooted in sensible inferential logic.

Perceptual interpretations, it has long been assumed in perceptual theories, should (1) be as simple as possible consistent with the image data, and (2) accurately represent the essential structural features of the object. The theory outlined in this paper sheds light on what these two desiderata have to do with each other, and suggests why achieving the second might require achieving the first. If the perceiver is able to identify a formal model of the generation of the stimulus class that more or less correctly mirrors the actual physical process by which the stimulus object was created, then the lowest-dimension description possible within that generation scheme, read directly off the mode lattice—in a very particular sense, the simplest description possible—will accurately represent the structurally important properties of the object.

## Acknowledgements

I wish to thank Don Hoffman, Jan Koenderink, Shimon Ullman and most especially Whitman Richards for much helpful criticism of the ideas expressed in this paper.

## References

Bennett, B., Hoffman, D. & Prakash, C. 1989. *Observer mechanics*. London: Academic Press.

Bobick, A. 1987. Natural object categorization. Technical Report no. 1001, M.I.T. A.I. Lab.

Bobick, A. & Richards, W. 1986. Classifying objects from visual information. A.I. Memo no. 879, M.I.T. A.I. Lab.

Bongard, M. 1970. *Pattern recognition*. New York: Spartan Books.

Buffart, H., Leeuwenberg, E., & Restle, F. 1981. Coding theory of visual pattern recognition. *Journal of Experimental Psychology: Human Perception and Performance*, 7 (2):241-270.

Chaitin, G. 1966. On the length of programs for computing finite binary sequences. *Journal of the Association for Computing Machinery*, 13 (4):547-569.

Darrell, T. Sclaroff, S. & Pentland, A. 1990. Segmentation by minimal description. Proceedings of the Third International Conference on Computer Vision, Osaka, Japan. Los Alamitos, CA: IEEE Computer Society Press.

Feldman, J. 1991. Perceptual decomposition as inference: Continuous curvilinear processes. Occasional Paper no. 42, Center for Cognitive Science, M.I.T.

Feldman, J. in preparation. Perceptual induction in Bongard problems.

Fell, H. 1976. Detectable properties of planar figures. *Information and Control*, 31 (2):107-128.

Hatfield, G. & Epstein, W. 1985. The status of the minimum principle in the theoretical analysis of visual perception. *Psychological Bulletin*, 97(2):155-186.

Hochberg, J. & McAlister, E. 1953. A quantitative approach to figural "goodness." *Journal of Experimental Psychology*, 46:361-364.

Jepson, A. & Richards, W. 1990. What is a percept? Occasional Paper no. 43, Center for Cognitive Science, M.I.T.

Koffka, K. 1935. *Principles of Gestalt psychology*. New York: Harcourt, Brace and Co.

Köhler, W. 1947. *Gestalt psychology*. New York: Liveright Publishing Co.

Kolmogorov, A. 1965. Three approaches to the quantitative definition of information. *Problems of Information Transmission* 1(1):1-7.

Leeuwenberg, E., & Boselie, F. 1988. Against the likelihood principle in visual form perception. *Psychological Review*, 95:485-491.

Leyton, M. 1984. Perceptual organization as nested control. *Biological Cybernetics*, 51:141-153.

Leyton, M. 1986. A theory of information structure. I. General principles. *Journal of Mathematical Psychology*, 30:102-160.

Leyton, M. 1988. A process-grammar for shape. *Artificial Intelligence*, 34:213-247.

Leyton, M. 1989. Inferring causal history from shape. *Cognitive Science*, 13:357-387.

Perkins, D. 1976. How good a bet is good form? *Perception*, 5:393-406.

Richards, W. 1988. *Natural computation*. Cambridge: M.I.T. Press.

Richards, W. & Bobick, A. 1988. Playing twenty questions with nature. In Z. Pylyshyn ed. 1988, *Computational processes in human vision: an interdisciplinary perspective*. Norwood, NJ: Ablex.

Simon, H. 1972. Complexity and the representation of patterned sequences of symbols. *Psychological Review*, 79(5):369-382.

Watanabe, S. 1985. *Pattern recognition: Human and mechanical*. New York: John Wiley & Sons.

Witkin, A. & Tenenbaum, J. 1983. On the role of structure in vision. In H. Rosenfeld ed. 1983, *Human and machine vision*. New York: Academic Press.

Witkin, A. & Tenenbaum, J. 1986. On perceptual organization. In A.P. Pentland ed. 1986, *From pixels to predicates: Recent advances in computational and robotic vision*. Norwood, N.J.: Ablex.

# Featural Priming: Data Consistent With the Interactive Activation Model of Visual Word Recognition

**Robert Kachelski and John Theios**

Department of Psychology
University of Wisconsin
Madison, WI 53706

**Abstract**

The effect of featural priming on word identification was investigated as a test of the interactive activation model of word perception put forth by McClelland and Rumelhart (1981). Observers were presented with a 250 msec featural prime, which was either consistent with or inconsistent with the letters in the target word that immediately followed. Reading latencies were recorded for 96 trials per subject. A neutral prime condition consisting of a random dot pattern was used as a control in order to obtain baseline identification times. The prediction of the interactive activation model that mean reading latency would be significantly longer for words that were primed with inconsistent features than for those that were primed with consistent features was confirmed, adding to the empirical support for the model.

## Introduction

The interactive activation model of visual word perception was proposed by McClelland and Rumelhart in 1981. Essentially, it was a combination of the basic assumptions of the Rumelhart (1977) interactive model with the spreading activation notions of the McClelland (1979) cascade model. The model was designed to represent the processes involved in word recognition. In the discussion of this model that follows, the initial emphasis will be on the assumptions made in its construction, and these assumptions will then be examined in light of the predictions they suggest for the present experiment.

The first main assumption of the interactive activation model is that perceptual processing takes place in a system of several levels of processing. In their original discussion, McClelland and Rumelhart (1981) described three such levels in depth: the feature level, the letter level, and the word level. Each level is made up of a set of parallel nodes, with one node representing each possible element at that level. For instance, at the feature level, there is a node for each possible feature at each letter position in a word. Similarly, there is a letter node for each letter in each position in a word, and there is a word node for each word at the word level. Higher levels of processing, which supply so-called "top-down" information to the word level, were also assumed to exist.

A second major assumption of the model is that visual perception involves parallel processing. This means not only parallelism in terms of processing information about several letters of a word simultaneously, but also in terms of processing information at all of the levels simultaneously.

The third assumption is closely related to the second. It is the idea that since processing in the visual system is massively parallel, there is a great deal of interaction both between and within the levels of processing, and this interaction ultimately determines what is perceived. In the model, there are top-down, conceptually driven aspects of processing, such as knowledge of words and orthographic structure, which occur simultaneously with bottom-up, data-driven aspects, such as featural information. The multiple constraints created by these two types of processing result in perception of a word or letter that is consistent with both (cf. Norman & Bobrow, 1975).

The interaction between the nodes discussed above is achieved through simple excitatory and inhibitory activations of a neural type (Anderson & Hinton, 1981; McClelland, Rumelhart & Hinton, 1986). Each node is connected to a number of other nodes in a bi-directional fashion. These connections may be within a given level and between adjacent levels. A connection between nodes will be excitatory if the nodes are consistent with each other; if not, the connection will be inhibitory. Every word node inhibits all other word nodes, and every letter node inhibits all other letter nodes at a particular position in a word. Connections between the word level and the letter level may be either excitatory or inhibitory, depending on whether the letter is a part of the word in that particular letter position. In the same way, the type of connections between the feature level and the letter level depend on whether or not the features present at a given letter position are consistent with the letter in question.

At any given time, each node has a specific activation level that varies continuously between a minimum and maximum level. When there is no input from neighboring nodes, a node is said to be inactive, and its activation level decays to a resting level. However, when the neighbors of a node are active, the result is either excitation or inhibition of the node, depending on the relationships among the nodes. The net input to the node is a weighted sum of the activation levels of the other nodes in the system and the connection strengths from those nodes. The net input, current activity level, resting level, and decay rate are then used to update the activity level of the node (cf. Golden, 1986; McClelland & Rumelhart, 1981).

In this model, then, when a word is presented, the parallel layer of feature nodes is activated first, presumably by feature detectors in the visual system. Activated feature nodes then send excitatory messages to letter nodes that contain those features and inhibitory messages to those that do not. Activated letter nodes can then activate consistent word nodes, and once a word node becomes activated, it begins to inhibit all other word nodes. Top-down, conceptually driven processing also occurs, since partially activated word nodes excite their component letter nodes and inhibit all other letter nodes. These interactions among the nodes at the different levels continue until a single word node gains superior activation and drives the activation of the other nodes down through massive inhibition. When this occurs, the system has essentially "recognized" the stimulus word as the word represented by the node with the highest activation value at the time of output.

When the interactive activation model was experimentally tested (Rumelhart & McClelland, 1982), the results provided a very close representation of some of the major phenomena in letter and word perception, including the way contextual inputs influence perceptual processing. In particular, Rumelhart and McClelland tested the perceptibility of a single target letter in four-letter displays. They manipulated the onset and offset times of the context letters in order to determine the extent to which they could facilitate perception of the target. Consistent with the predictions of the model, the duration and timing of the presentation of contextual information had significant effects on the perceptibility of a target letter embedded in a word.

Specifically, greater accuracy on a forced-choice task of target-letter identification was associated with longer context durations and greater numbers of letters enhanced within the context. In other words, the longer the context letters (or "primes") were present before the target letter appeared, the easier it was to identify the target letter. Similarly, the more letters that were used as a prime, the greater the percentage of correct target identifications. The finding that these context-enhancement effects were more profound when the extra contextual information came before rather than after the target letter points to the conclusion that the context must affect perception of the target letter itself as it is being processed. Presumably, this occurs because the context spreads activation to the appropriate letter node, and this activation is reinforced when the target letter appears, allowing the node to reach an activation level sufficient for letter identification more quickly than would be possible without a contextual prime.

The spreading activation notions central to the interactive activation model of McClelland and Rumelhart (1981) are by no means new, and they are certainly not exclusive to this model. Ever since Quillian introduced his theory of spreading activation (Quillian, 1962, 1967), studies have been conducted to assess whether this concept is a useful one in building a model of human perceptual processing (e.g. Collins & Quillian, 1969; Meyer & Schvaneveldt, 1971; Ratcliff & McKoon, 1981). Based on the results of such experimentation, the idea of spreading activation has become widely popular, forming the central mechanism in a number of models of memory and information processing (e.g. Anderson, 1976; Collins & Loftus, 1975; Dell & Reich, 1977; Hinton, 1981).

A large portion of the research on spreading activation has centered on semantic priming. Significant semantic priming effects have been reported by a number of researchers (e.g. Collins & Quillian, 1969; Meyer, 1970; Meyer & Schvaneveldt, 1971). This research supports the notion that memory storage is structured so that associated words and concepts can activate each other. Letter recognition tasks have also been used to study spreading activation. In addition to the experiments mentioned earlier that were conducted by Rumelhart & McClelland (1982), other research has shown effects of orthographic structure on letter perception that are consistent with the top-down processing ideas in the interactive activation model (Massaro & Klitzke, 1979; Spoehr & Smith, 1975).

As one can see from the discussion above, research evaluating the notion of spreading activation, and in particular the model put forth by McClelland and Rumelhart (1981), has concentrated almost exclusively on the letter and word levels of processing. To date, little evaluation of the model has been done in terms of how well it can predict and account for the results of experimentation at the level of basic features. Since the processing that takes place at the feature level is vital to the functioning of the model, such evaluation is necessary before an assessment of the model as a whole can be made.

The present experiment is designed to test the interactive activation model by attempting to produce priming effects at the feature level. If the processing of words proceeds according to the interactive activation model, priming an observer with features that are consistent with the letters in a target word should facilitate recognition of that word. In this case, the appropriate letter and word nodes presumably would be activated sooner than if there were no prime. In the words of McClelland and Rumelhart (1981, p.382), "If the input features were close to those for one particular set of letters and those letters were consistent with those forming a particular word, the positive feedback in the system will work to rapidly converge on the appropriate set of letters and the appropriate word."

Similarly, a prime consisting of features that are inconsistent with the letters in a target word should interfere with the recognition of the word. The model would predict that inappropriate letter and word nodes would be activated, and the appropriate letter and word nodes would be initially inhibited. The letters and words would compete with each other and, "perhaps no single set of letters or single word will get enough activation to dominate the others. In this case the various active units might strangle each other through mutual inhibition" (McClelland & Rumelhart, 1981, p. 382).

In terms of reading latencies, then, the model would predict that consistent featural primes would result in shorter reading latencies than neutral primes, and that inconsistent featural primes would result in longer reading latencies than neutral primes.

## Method

### Observers

The observers were 18 graduate and undergraduate students at the University of Wisconsin-Madison. All observers had normal or corrected-to-normal vision.

### Apparatus

The stimuli were presented using a Scientific Prototype model 800F two-channel tachistoscope with a Psionix, Inc. timer (model 1248B) and a LaFayette Instrument voice-activated relay. The sensitivity of the voice-activated relay was set at a level that was effectively activated by human speech, but was sufficiently insensitive to extraneous noise so that there were no inappropriate activations. Observers spoke into a Wollensak 3M microphone that was placed in a stand about 30 cm from their mouths.

### Materials

Ninety-six words were chosen for this experiment. The words used were all four to six letters long. All of the words chosen were relatively common (25 appearances or more in print per million words). The words were chosen such that most of their letters had predominantly vertical, horizontal, slanted, or curved visual features, for example: HELP, THERE, MANY, or DROOP, respectively.

Each of the target words was printed on a card measuring 13.5 cm wide by 24 cm high in order to be presented in the tachistoscope. These stimulus items were generated by a Macintosh Plus computer with an Apple laser printer, using Chicago font. The letters in each word were in upper case and were 1.1 cm wide by 1.2 cm high. When the words were presented in the tachistoscope, each letter subtended .84 degrees of visual angle in width and .96 degrees of visual angle in height.

Stimulus cards to be used as primes in the experiment were constructed for each word. A prime card for a particular word consisted of features consistent with the letters of that word in their appropriate spatial position. For example, the prime for the word "HELP" would consist of the five vertical features of the four letters, each in their correct position. There were four general types of primes: vertical, horizontal, slanted, and curved, corresponding to the words with these predominant features.

A "neutral" prime card was also constructed. It consisted of a random dot pattern of the same dimensions as the featural primes on the other cards.

### Procedure

Observers were first given a vision test in order to ensure that only those with normal vision were included in the experiment. They were then instructed to read each word that was presented to them out loud as quickly as possible without making errors. Observers were given six practice trials in order to eliminate any confusion and to orient them to the task. The 96 experimental trials were then presented, with a five-minute break at the halfway point.

For each trial, the experimenter gave the observer a verbal ready signal and then presented the stimuli. The prime was presented for 250 msec and was immediately followed by the target word, which remained on until the observer read the word, thus activating the voice relay. There were three types of primes: (a) consistent = the prime created for that particular target word using its predominant features ( all features being either vertical, horizontal, slanted, or curved), (b) inconsistent = a prime whose features did not match the predominant features of the target word, and (c) neutral = the random dot pattern. Reading latencies were recorded in milliseconds for each trial.

The 96 trials were presented in a different random sequence to each observer. Both word order and priming condition (consistent prime, inconsistent prime, or neutral prime) were randomized, with the constraint that there was an equal number of trials with each type of prime. Observers saw each target word only once.

## Results

When all of the data were collected, a single-factor, within-subjects ANOVA was conducted on the reading latencies. The result was a significant effect for type of prime [$F(2,34) = 3.18$, $p<.05$]. The mean reading latencies for the words in each priming condition were: consistent prime = 556 msec, neutral prime = 561 msec, and inconsistent prime = 567 msec. A planned comparison between the means of the inconsistent and consistent priming conditions revealed that mean reading latency for words that were primed inconsistently were significantly longer than the mean latency for words that were primed consistently [$F(1,17) = 5.45$, $p<.05$]. The mean latency for the neutral prime condition was not significantly different from either the mean latency for the inconsistent prime condition [$F(1,17) = 1.17$, $p>.05$] or the mean latency for the consistent prime condition [$F(1,17) = 1.12$, $p>.05$].

To examine the data more closely, separate sign tests were conducted between the neutral prime condition and each of the other two priming conditions. It was found that 13 of the 18 observers had shorter mean reading latencies in the consistent priming condition than in the neutral priming condition. This result was significant ($p<.05$). For the sign test comparing the neutral prime condition to the inconsistent prime condition, 11 of the 18 observers had shorter mean reading latencies in the neutral prime condition than in the inconsistent prime condition. This result was not significant ($p=.24$).

## Discussion

The finding that the mean reading latency for consistently primed words was significantly shorter than the mean latency for inconsistently primed words is supportive of the interactive activation model's representation of the role of the feature level of processing in visual word recognition. It appears that having accurate information about the features of the letters in a word before the word is actually presented gives the appropriate letter nodes and the target-word node a "head start" in terms of activation as compared to a situation in which the featural information is inconsistent with the letters in the target word. On the other hand, when the featural information is inconsistent with the target word, the relevant letter nodes and the target-word node itself tend to be initially inhibited, resulting in longer word recognition times. This result fits in nicely with the research findings on priming at the letter and word levels of processing. It lends support to the idea of spreading activation in human information processing, and in particular to the parallel-distributed-processing, connectionist formulation of the interactive activation model of visual word recognition.

It should be stated, however, that the present experiment ws not designed to provide a critical test of the interactive activation model *in comparison to* other models of word recognition. The purpose here was to test a specific prediction about the role of featural input in this model. The present results would also have been predicted by a number of other models of word recognition, such as Morton's (1969) logogen model, Smith's (1971) feature model, and Paap et al.'s (1982) activation-verification model, to name a few. Although these models are very different, they all share the assumption that analysis of the features in words is critical for successful word perception. The featural priming effects reported here should be viewed as supportive of this assumption in general. On the other hand, the present data could not be accounted for explanations of word recogntion that rely on "holistic" or global features of words, such as word "envelope" or shape (e.g. Haber, Haber & Furlin, 1983; Monk & Hulme, 1983) or on the contours of adjacent letters (e.g. Wheeler, 1970). Features of only *some* of the letters were presented as primes in this experiment, so the overall word shape could not have been determined from the prime. In addition, only upper-case letters were used for the stimuli, so these words did not have the distinctive shapes, defined by ascending, descending, and neutral letters, that supposedly underlie word recognition by word shape.

Although the results of the planned comparisons revealed that the mean reading latency for the neutral prime condition was not significantly different from the mean of either the consistent or the inconsistent prime conditions, the sign test showed that the number of observers whose mean for the consistent prime condition was lower than their mean for the neutral prime condition was significantly greater than would be expected if the two types of primes had the same effect. The 5 msec difference between the overall means for the consistent prime condition and the neutral prime condition was quantitatively small, but it appeared fairly consistently across our sample of observers. Therefore, consistent featural priming seems to have a slight facilitative effect on word recognition speed as compared to even a neutral prime.

There are at least three reasons why the reading latencies in the neutral prime condition were not significantly different from those in the inconsistent prime condition. First, observers may recover rather quickly from misleading featural information in a word recognition task. The initial inhibition of appropriate feature, letter, and word nodes due to the inconsistent prime may be easily and quickly overcome once the target word appears. This may be a reflection of more emphasis being given to excitatory connections than inhibitory connections in the lower featural levels of the visual word recognition process (in contrast to the emphasis on inhibition in McClelland and Rumelhart's (1981) simulations). Second, the correct choice of a true "neutral" prime or baseline control in these types of experiments is problematic. The random dot prime itself may cause some degree of inhibition. After all, it is a type of visual noise. Thus, we may have been optimistic to assume that the random dot prime would not have an interfering effect on the perception of the target. Finally, the 250 msec exposure duration we used for the primes may have been too long to achieve maximal excitation and inhibition effects. Given Di Lollo's (1980) work on visual persistence and visual integration, a prime exposure duration of under 100 msec may be a better choice for future research on featural priming. Although further research is necessary to identify the underlying mechanisms are that are responsible for the results of the present study, it is clear that they are generally consistent with the predictions that follow from Rumelhart and McClelland's (1981) parallel-distributed-processing, interactive activation model of visual word recognition.

## References


Anderson, J. A. (1976). *Language, memory, and thought.* Hillsdale, NJ: Erlbaum.

Anderson, J. A., & Hinton, G. (1981). Models of information processing in the brain. In G. Hinton & J.A. Anderson (Eds.), *Parallel models of associative memory* (pp. 9-48). Hillsdale, NJ: Erlbaum.

Collins, A., & Loftus, E. (1975). A spreading activation theory of semantic processing. *Psychological Review, 82*, 407-428.

Collins, A., & Quillian, M. (1969). Retrieval time from semantic memory. *Journal of Verbal Learning and Verbal Behavior, 8*, 240-248.

Dell, G. S., & Reich, P. A. (1977). A model of slips of the tongue. In R. Dipietro & E. Blansitt (Eds.), *The third LACUS forum.* Columbia, SC: Hornbeam Press.

Di Lollo, V. (1980). Temporal integration in visual memory. *Journal of Experimental Psychology: General, 109*, 75-97.

Golden, R. (1986). A developmental neural model of visual word perception. *Cognitive Science, 10*, 241-276.

Haber, L. R., Haber, R. N., & Furlin, K. R. (1983). Word length and word shape as sources of information in reading. *Reading Research Quarterly, 18*, 165-189.

Hinton, G. (1981). Implementing semantic networks in parallel hardware. In G. Hinton & J. A. Anderson (Eds.), *Parallel models of associative memory.* Hillsdale, JN: Erlbaum.

Massaro, D. W., & Klitzke, D. (1979). The role of lateral masking and orthographic structure in letter and word recognition. *Acta Psychologica, 43*, 413-426.

McClelland, J. L. (1979). On the time relations of mental processes: An examination of systems of processes in cascade. *Psychological Review, 86*, 287-330.

McClelland, J. L., & Rumelhart D. E. (1981). An interactive activation model of context effects in letter perception: Part 1. An account of basic findings. *Psychological Review, 88*, 375-407.

McClelland, J. L., Rumelhart, D. E., & Hinton, G. E. (1986). The appeal of parallel distributed processing. In D. E. Rumelhart & J. L. McClelland (Eds.), *Parallel distributed processing: Explorations in the microstructure of cognition. Volume 1: Foundations* (pp. 3-44). Cambridge, MA: MIT Press/Bradford.

Meyer, D. E. (1970). On the representation and retrieval of stored semantic information. *Cognitive Psychology, 1*, 242-300.

Meyer, D. E., & Schvaneveldt, R. W. (1971). Facilitation in recognizing pairs of words: Evidence of a dependence between retrieval operations. *Journal of Experimental Psychology, 90*, 227-234.

Monk, A. F., & Hulme, C. (1983). Errors in proofreading: Evidence for the use of word shape in word recognition. *Memory & Cognition, 11*, 16-23.

Morton, J. (1969). Interaction of information in word recognition. *Psychological Review, 76*, 165-178.

Norman, D. A., & Bobrow, D. G. (1975). On data-limited and resource-limited processes. *Cognitive Psychology, 7*, 44-64.

Paap, K. R., Newsome, S. L., McDonald, J. E., & Schvaneveldt, R. (1982). An activation-verification model for letter and word recognition: The word superiority effect. *Psychological Review, 89*, 573-594.

Quillian, M. R. (1962). A revised design for an understanding machine. *Mechanical Translation, 7*, 17-29.

Quillian, M. R. (1967). Word concepts: A theory and simulation of some basic semantic capabilities. *Behavioral Science, 12*, 410-430.

Ratcliff, R., & McKoon, G. (1981). Does activation really spread? *Psychological Review, 88*, 454-462.

Rumelhart, D. E. (1977). Toward an interactive model of reading. In S. Dornic (Ed.), *Attention and performance IV*. Hillsdale, NJ: Erlbaum.

Rumelhart, D. E., & McClelland, J. L. (1982). An interactive activation model of context effects in letter perception: Part 2. The contextual enhancement effect and some tests and extensions of the model. *Psychological Review, 89*, 60-94.

Smith, F. (1971). *Understanding reading*. New York: Holt, Rinehart and Winston.

Spoehr, K., & Smith, E. (1975). The role of orthographic and phonotactic rules in perceiving letter patterns. *Journal of Experimental Psychology: Human Perception and Performance, 1*, 21-34.

Wheeler, D. D. (1970). Processes in word recognition. *Cognitive Psychology, 1*, 59-85.

# The Analysis of Resource-limited Vision Systems

**Ronald A. Rensink**
Department of Computer Science
University of British Columbia
Vancouver BC, V6T 1Z2, Canada

**Gregory Provan**
Department of Computer and Information Science
University of Pennsylvania
Philadelphia PA, 19104-6389, USA

## Abstract

This paper explores the ways in which resource limitations influence the nature of perceptual and cognitive processes. A framework is developed that allows early visual processing to be analyzed in terms of these limitations. In this approach, there is no one "best" system for any visual process. Rather, a spectrum of systems exists, differing in the particular trade-offs made between performance and resource requirements. [1]

## Introduction

Consider a cat in its natural environment. If it is to catch prey and escape from predators, the cat must not only be able to process visual information, but must also do so in real time. Its visual system is therefore best explained not only in terms of limitations on the information available to the eye, but also in terms of limitations on other resources, such as time and space.

There is an increasing awareness – especially within the more computational sub-disciplines of cognitive science – that these more general resource limitations influence many kinds of perceptual and cognitive processes. For example, Cherniak [1984] argues that classical logics cannot form the basis for cognition because such cognition is computationally intractable; this has led to an examination of heuristics by which fast reasoning could take place [Levesque and Brachman, 1985, Levesque, 1989]. Similarly, Tsotsos [1987, 1990] has argued that the processes of early vision must have at most polynomial-time complexity if they are to be carried out in real time. But although there is an increasing appreciation of the role of resource limitations (e.g. [Bylander *et al.*, 1989, Kasif, 1986, Rosenfeld, 1987]), no general framework for discussing these issues has emerged to date.

This paper discusses some of the issues that must be addressed in developing such a framework. In particular, it focuses on the influence of resource limitations on early visual processing. Marr [1982] has made a beginning in this domain, showing how vision can be analyzed in terms of constraints that allow good use to be made of the information available in the image. We will show how this framework can be expanded to handle other kinds of resource limitations, yielding added insight into the interconnections that exist among task, algorithm and architecture. Since many of these issues are general ones, the framework presented here will contain elements that are also applicable to other areas of perception and cognition.

## Resource Limitations and Explanation

Many of the earlier analyses based on resource limitations (e.g., [Norman and Bobrow, 1975]) focused on limitations in the system architecture, for example, limited memory or channel capacity. These did not yield the insights that had originally been hoped for; indeed, it has been argued [Navon, 1984] that such limitations are inherently incapable of leading to unequivocal insights into the operation of perceptual and cognitive processes.

But architectural limitations are not the only kind that arise – more general limitations also exist, such as limits on the available information, and on the time and space allowed for a computation. These "processor-indifferent" limits are potentially more powerful than those based on architectural limitations, essentially describing the structure of the *task itself*.

Given that these general limitations must be taken into account, how might they be used to analyze the underlying mechanisms? One of the most successful approaches to date has been the computational framework put forward by David Marr [1982], in which visual processing is analyzed in terms of constraints that allow good use to be made of the information available in the image. In what follows, we will show that this framework can be expanded to accommodate not only limits on available information, but other kinds of resource limits as well, and that such a revised framework

[1]The work by R. Rensink was supported by NSERC grant, via R.J. Woodham. Much of the work by G. Provan was done at the University of British Columbia, under NSERC grant A9281. The authors would like to thank Marion Rodrigues and Marc Romanycia for their comments on earlier versions of this paper. Many thanks also to Bob Woodham for his support of R. Rensink

can lead to a new understanding of several aspects of early vision.

## Marr's Framework

According to Marr [1982], a complete analysis of a visual process involves three distinct levels of explanation:[2]

1. *Computational level.* Analysis at this level is entirely concerned with the specification of the task itself. This consists of two parts: (i) describing the constraints that exist between the input of a visual process and its output, and (ii) describing the reasons *why* these constraints have been chosen.
2. *Algorithmic level.* This level views explanation in terms of the representations and algorithms used for the process. More precisely, an "algorithmic" explanation is a constructive demonstration that there exists a formal algorithm sufficient to perform the required task.
3. *Implementational level.* This level is concerned with the physical substrate on which the algorithms are implemented. An "implementational" explanation is a constructive demonstration that there exists a physical system sufficient to carry out the required computations.

One of the great strengths of Marr's approach is its recognition of a "computational" level of explanation, in which emphasis is placed upon determining the *what* and the *why* of the particular operations being carried out. This has helped clarify our understanding of several processes of low-level vision, including edge detection [Marr and Hildreth, 1980], stereopsis [Marr and Poggio, 1979], and motion perception [Hildreth, 1984]. Consider, for example, the computational analysis of stereopsis. Determining the *what* consists of finding the constraints on the acceptable correspondences between features in the left and right images, and constraints on the form of the recovered surface. These constraints must be sufficient to describe a unique mapping between the image and the resulting map of disparity estimates. Determining the *why* essentially consists of a demonstration that these constraints serve to allow a satisfactory recovery of disparity estimates from the image pairs.

Hence, the stereopsis problem can be seen as the specification of a mapping from a given set of image pairs to a set of (reconstructed) surfaces. This mapping can easily be described by its "extension", viz., a list of the pairings made between individual images and surfaces. Such a description, however, does not really provide an *explanation* for the process, any more than a list of planetary positions over some given interval explains their emotions. Explanation must involve a description of the "invariants", or "deep structure" that underlie the particular mapping that is made. The constraints sought for at the computational level provide exactly this kind of explanation. To justify the choice of a particular set of constraints (explaining *why*), requires showing that the constraints lead to an acceptable set of associations between image and scene in the world under consideration.

But although Marr's approach has helped explain several parts of low-level vision, it has not helped in our understanding of many others, e.g., color perception or texture perception [Morgan, 1984]. For example, in texture perception, it is the resources available to the processor (e.g. time and space) which are relatively scarce, rather than the information in the image. Marr's framework cannot handle such matters, since the computational level of analysis (implicitly) assumes that perception relies on processors with unlimited computational resources. [3]

## Resource Limitations and Constraints

To see how these more general kinds of resource limitations influence the operation of a visual process, it is important to note that these limitations fall into three main groups:

1. *Projective limitations.* The available information in the image may be considered a basic resource acquired by the sensors of the system; the limitations on this resource stem from the way in which the scene is projected to the image. The type and amount of available information may strongly influence the kinds of computations that can be performed; if so, the process can be characterized as "data-limited" [Norman and Bobrow, 1975].
2. *Computational limitations.* A processor is also limited by many aspects of the *way* in which it operates, aspects which have no direct connection with its physical composition. Although many of these are specific to the particular computational architecture used (e.g., the particular set of elementary operations available, bandwidths, etc), more general ones also exist. It is this latter set of resources – in particular the time and space required for a computation – that will be considered here. Limitations on these resources will be referred to here as *complexity limitations*.
3. *Physical limitations.* A processor is also governed by limitations stemming from its physical make-up. Again, many of these quantities refer to the particular architecture of the processor. But limitations also arise from more general consideration, such as the matter and energy required for a given task [Bennett, 1982, White, 1988].

[2] It is important to note that Marr considers explanations at each level to be essentially independent of those at the other two [Marr, 1982, Chapter 1]. For example, analysis at the algorithmic level is not concerned with ultimate purpose nor does it depend on any details of implementation.

[3] Marr [1982] does consider efficiency to be important, but only once the task itself has been laid out. As such, it does not enter into the general analysis carried out at the computational level.

In order to overcome these sets of limitations, a processor must impose corresponding sets of constraints on its operation. To completely understand a given process, therefore, is to understand these sets of constraints.[4] Thus, for example, in Marr's framework, projective limitations are the only kind the visual system is considered to grapple with. To make up for such lost information, a corresponding set of *projective constraints*[5] is needed on the mappings between image and scene; essentially, these determine which of the many possible scenes actually corresponds to a given image.

But such "processor-indifferent" explanations need not be restricted to invariants of the *form* of this mapping – there may also exist a set of constraints on the *algorithm* and *representation* used to carry it out. More generally, such "complexity" constraints describe the resources used by a given process. This in turn limits the kinds of mappings that can be made. To completely explain the form of a mapping, then, both projective and complexity constraints will usually be required. Only when computational resources are unlimited (as assumed in Marr's approach) will projective constraints alone be enough to explain a visual process.

Note that in Marr's framework, no general constraints are imposed on an algorithm, so that they often have a large element of the *ad hoc*, being based on current beliefs of psychology and physiology. But complexity constraints can provide such general guidelines, thereby substantially reducing the initial number of candidate algorithms and thereby reducing the need for the *ad hoc* element in any particular model.

## A Revised Framework

A computational explanation of a visual process, then, will include a description and justification of the projective, complexity, and physical constraints imposed to handle the corresponding types of resource limitation. [6] Different levels of explanation still exist, but are now based on the degree of generality of the constraints, rather than on issues of abstract mapping, process, and implementation:

1. *Computational level.* This includes not only the projective constraints, but also those complexity and physical constraints that are "processor-indifferent". Since the mapping from image to scene is completely independent of the structure of the processor, the constraints at this level must be sufficient to uniquely determine its form.
2. *Algorithmic level.* This involves the more specific complexity and physical constraints that are placed on the "internal" structure of the system to give the algorithm and representation a unique determination; since projective constraints have no further bearing on this matter, they are necessarily absent from this level.
3. *Implementational level.* Although not developed here, it is apparent that this level concerns the remaining constraints on the particular system being modelled.

Thus, the three levels of Marr's framework are preserved in large measure – analysis still occurs at each of the computational, algorithmic, and implementational levels. But the constraints required at the computational and algorithmic levels of analysis have been tightened up significantly, due to new sets of constraints. The most important change, however, is that analysis is based on the *generality* of the constraints. Since constraints on algorithms and implementation result from both general and more specific constraints, this provides an interesting linkage between mappings, algorithms, and implementation.

## Analyzing Vision Systems

The explanation of a visual process is essentially the description and justification of the projective, complexity, and physical constraints that govern its operation. Since the use of projective constraints is already an integral part of "conventional" analyses, and physical constraints are not considered here, we will focus on the way in which complexity constraints can be used to analyze the operation of a vision system.

The theory of computational complexity (cf. [Garey and Johnson, 1979]) can be used to formalize many of the concepts pertaining to complexity constraints. Specifically, it can define the time and space requirements of particular tasks, independent of the algorithm or architecture. For the purposes of this paper, it is sufficient to distinguish tasks which can be solved quickly (e.g. the class P of tasks solvable within a time proportional to some polynomial of the input size), from more time-consuming tasks (e.g. the class of NP-complete problems, which – in the worst case – require time that increases exponentially with the size of the input).[7]

[4] In what follows, 'limitations' will be used when referring to resource limits imposed *on* the system, limits over which the process has no control (e.g. total amount of time, space, energy, etc). These must be distinguished from 'constraints', which are imposed *by* the system itself to make good use of its available resources. When talking about a system, the term 'constraint' will only be used in this latter sense.

[5] The term 'projective constraint' is meant to replace 'computational constraint' as used in Marr's framework. The alternate term is used to avoid confusion between constraints placed on the form of the "interpretation mapping" between image and scene, and the constraints on the algorithm used to compute it.

[6] In this paper, much of the focus will be on complexity constraints, since projective constraints are relatively well understood, and physical constraints add little (at least at this stage of development) to what can be learned by discussing complexity constraints.

[7] It is entirely possible that the average resource use is much more representative that the worst case situation, and this may be used as one of the complexity measures. However, worst-case situations must still be dealt with.

## Efficient use of resources

If the computational demands of a task exceed the resources available, it is obvious that the task will need to be reformulated. However, this reformulation may be kept to a minimum by making efficient use of the time and space that is available. As used here, the term "efficient" does not necessarily mean optimal; rather all that is meant is that relatively little time or space is wasted.

The efficient use of resources depends greatly upon the choice of particular algorithms and representations used in a process. However, there exist a few general considerations that are relevant:

1. *Parallelism.* Perhaps the most obvious way of reducing the time required for a task is to carry it out in parallel. However, it must be noted that it is not always possible for processes to take advantage of parallelism – e.g., there can be no reduction in time for Constraint Satisfaction Problems [Kasif, 1986], since they are inherently sequential. Thus, if a task is to take advantage of parallelism, it must be such that most of its computations can be done locally. But each of these local computations must operate within the given time, and – taken as an ensemble – they must also operate within the given space.

2. *Resource trade-offs.* The specification of a desired level of performance does not uniquely determine the exact resources necessary to attain it – trade-offs between various computational resources can still be made. One well-known example of this is the trade-off between time and space: for instance, look-up tables can be used instead of computing values on demand. Thus, Goad [1983] presents an object recognition scheme where the poses of objects are precomputed so that viewpoint determination is speeded up. Alternatively, redundant coding can often be used to decrease processing time (see, e.g., [Arbib, 1987, pp 87–89]). Note that this use of redundant representations contrasts with Marr's approach, in which the goal is to use nonredundant (orthogonal) systems of representation as much as possible.

3. A priori *knowledge.* One final consideration that also enters into the efficient use of resources in visual processing is the possible use of "high-level" *a priori* constraints based on the particular characteristics of the objects in the scene. In many cases, higher-level constraints could significantly reduce the computational complexity of a process; if these constraints could be selectively "loaded into" lower-level processes, this could often achieve a considerable speedup of processing. The relation between early vision and later levels of processing is a complex one, and will not be discussed here, but it is worth pointing out that if such "downloading" of *a priori* knowledge does indeed occur, issues of resource use will prove to be critical for its analysis.

## Performance trade-offs

Even though a process is as efficient as possible, it may still be impossible to carry it out using the available time and space. If so, the process cannot be used; it must be replaced by one that *does* satisfy the resource constraints. The efficiency of such "approximating" processes is obtained by lowering the quality of the mapping between input and output. More generally, there is usually a trade-off between the complexity of the mapping between image and scene, and the resources required to compute the mapping. Thus, depending on available computational resources, the visual process most suitable for a particular task can range from "traditional" processes that use unlimited computational resources, to "quick and dirty" systems that require only a small amount of time and space. Part of a computational explanation of a visual process is therefore to specify what the particular choice of trade-off is, and why it was made.

There appear to be some general aspects to the methods by which performance can be "gracefully" traded off for reduced computational complexity, and it is likely that these strategies will enter into many of the particular processes of early vision. A few of these strategies (together will some possible applications) will now be discussed in regards to the reduction of processing time.

1. *Reducing quantity of input.* In general, more information requires more computation time [Levesque and Brachman, 1985]. Thus, one way to reduce time is to reduce the amount of data in the input that has to be handled. For example, visual search is an NP-complete problem, requiring time that increases exponentially with the size of the input [Tsotsos, 1987]. This time can be reduced (see below) by taking advantage of the coherence and uniformity of the world to represent the original image by a smaller set of coarser-grained patterns that could be comfortably handled with the available resources. As the grain of these patterns increases, the number of distinctions that can be made in the input decreases; however, these distinctions may be quite suitable for many purposes.

2. *Reducing quantity of output.* Given that computational complexity can be reduced by effectively reducing the amount of information in the input, a natural "dual" would be to reduce complexity by reducing the amount of information in the output. Such outputs would contain coarser-grained descriptions of the more important aspects of the scene. Note that this "coarse grain" need not always correspond to a diminished resolution in some property such as spatial location or velocity of motion; instead, the "equivalence classes" of outputs[8]

[8] An *equivalence class* is defined as the set of algorithms and representations which carry out the same mapping while using the same information content and computational resources.

could be based on such things as topological properties. In a very general sense, then, these outputs may be regarded as providing qualitative descriptions of the scene. For example, in Marr's theory it is assumed that the 3-dimensional structures of objects in space are represented as point-by-point mappings of local depth and/or orientation. Such representations are computationally difficult to compute, and it is possible that they are not computed at all. Indeed, it appears that more qualitative descriptions – such as descriptions of affine or ordinal structure – may provide all the information that is required by subsequent processes [Todd and Bressan, 1990].

3. *Reducing quality of the mapping.* Reducing the information in the input and output of a mapping is often sufficient to reduce complexity, but it isn't always necessary. For many processes, the availability of a spatiotopic array of processors is sufficient to allow them to be carried out in constant time. For example, a simple remapping of all intensities in an image can be done immediately on a parallel array, no matter how much information is contained in the input and output. A process can therefore trade off performance against complexity by altering the nature of the mapping itself; essentially, it is the quality of the mapping that is being traded off. This is the strategy adopted in rapid line interpretation (see below) – increased speed is obtained by reducing the validity and global coherence of the recovered scene.

## Examples

To illustrate how resource constraints can help explain various aspects of visual perception, we will briefly sketch how this approach can be applied to two particular processes in early vision: visual search and the preattentive recovery of three-dimensional orientation from line drawings. If described in the "conventional" way, i.e., making optimal use of available information, both problems are NP-complete. But the processes of early vision are generally carried out within several hundred milliseconds, making it unlikely that these problems can be formulated in this way. This suggests a shift in the way these processes should be viewed: instead of making optimal use of information, they appear instead to emphasize "quick and dirty" performance.

### Visual Search

One of the first treatments of complexity in early vision was that of Tsotsos [1987, 1990], who analyzed the process of visual search. Here, the problem is to determine as rapidly as possible the presence or absence of a known target pattern in an image. Tsotsos showed that if optimal decisions are to be made, this problem is NP-complete, requiring an exponential amount of time in the worst case. This is at odds with evidence that many kinds of targets can be reliably detected within several hundred milliseconds, while others require a time directly proportional to the number of items in the image (e.g. [Treisman and Gormican, 1988]).

The first step in the analysis is to determine the extent to which time and space can be reduced while maintaining optimal detection performance. Tsotsos shows that hierarchical coding can help to minimize the resources required, but that the problem still remains NP-complete, since the target must be compared against all possible aspects of all possible subsets of the image. If visual search is to be carried out rapidly, this can only be done for a select group of image subsets (or equivalently, a select group of target patterns).

Even by defining preferred patches as convex patches with uniform properties (arising from the convexity and uniformity of objects in the physical world), the time required is still too high to be compatible with the complexity constraint. Consequently, a more radical step is taken: information is thrown away. This is done both by reducing spatial resolution in the basis set of patterns and by reducing the number of properties that can be considered at any one time. Although the completeness of the system is thereby sacrificed, these constraints do allow an architecture to be specified that is compatible with the time and space limitations generally found in biological systems. Interestingly, this architecture has many of the general characteristics of the human visual system, viz., a small set of physically separated spatiotopic maps, columnar organization of processors, and coarse coding of local properties [Tsotsos, 1987, Tsotsos, 1990].

### Rapid interpretation of line drawings

Another problem in which time limitations play an important part is the rapid interpretation of line drawings. The goal of the line interpretation task is to recover the three-dimensional structure of opaque polyhedral objects from line drawings describing their projection onto the two-dimensional image plane (see, e.g. [Sugihara, 1986]). Interpretations generally take the form of a labeled drawing in which each line element (or region) is assigned a unique interpretation as a three-dimensional structure (e.g., that the line has a particular three-dimensional orientation, or that it forms the boundary of the object being viewed, etc.). This process has been shown to be NP-complete [Kirousis and Papadimitriou, 1985], ruling out the possibility that it is carried out in early vision.

However, it has recently been shown [1990, 1991] that the three-dimensional orientation of some objects can be recovered at early stages of visual processing, within several hundred milliseconds of display onset. As in the case of visual search, then, optimal use of information is not to be expected for such a process; part of the explanation must involve complexity as well as projective constraints.

In the model proposed by Enns and Rensink, interpretation is accomplished via two parallel stages, each of which involves only a small number of steps. The first stage is carried out in parallel on each trilinear junction in the image. These junctions are places in the image where three line segments join up. Any junction may correspond to several three-dimensional structures in the scene; here, however, only the most likely interpretation is assigned. Once these initial interpretations have been established, consistency is tested by comparing the interpretations at each junction against those of their immediate neighbors. This can be done within some constant time simply by propagating the local estimates along the lines that connect the junctions.

If such a "quick and dirty" process is used in the early stages of vision, this should allow some line drawings to "pop out" of a display on the basis of the three-dimensional orientation of the block they describe.[9] Enns and Rensink [1991] show that this is exactly what happens.

These findings suggest that other "quick and dirty" processes may also be used in early vision. For example, it is possible to rapidly determine the concavity/convexity of surfaces, based on the patterns of shading in the image [Ramachandran, 1988]. If these processes are representative of the operations carried out at early levels, this will force a new look at the nature of early visual processing.

## References


[Arbib, 1987] M.A. Arbib. *Brains, Machines, and Mathematics (2nd ed.).* New York: Springer, 1987.

[Bennett, 1982] C.H. Bennett. The Thermodynamics of Computation – a Review. *International Journal of Theoretical Physics*, 12:905–940, 1982.

[Bylander *et al.*, 1989] T. Bylander, D. Allemang, M.C. Tanner, and J. Josephson. Some Results Concerning the Computational Complexity of Abduction. In *Workshop on Principles of Knowledge Representation and Reasoning*, 1989.

[Cherniak, 1984] C. Cherniak. Computational Complexity and the Universal Acceptance of Logic. *Journal of Philosophy*, 81(12):739–758, 1984.

[Enns and Rensink, 1990] J.T. Enns and R.A. Rensink. Sensitivity to Three-Dimensional Orientation in Visual Search. *Psychological Science*, 1:323–326, 1990.

[Enns and Rensink, 1991] J.T. Enns and R.A. Rensink. Preattentive Recovery of Three-Dimensional Orientation from Line Drawings. *Psychological Review*, 98, 1991.

[Garey and Johnson, 1979] M.R. Garey and D.S. Johnson. *Computers and Intractability: A Guide to the Theory of NP-Completeness.* San Francisco: W.H. Freeman, 1979.

[Goad, 1983] C. Goad. Special Purpose Automatic Programming for 3D Model-based Vision. In *Proc. AAAI*, pages 349–353, 1983.


[9] Note that the result of this process will generally not be a valid reconstruction of the scene, since successful recovery can be expected only at the relatively sparse set of locations in the visual field where the projective assumptions describe the actual situation. Such a description, however, would still be valuable, since it might be expected that at least a few of its estimates would be accurate, and so guide fast-acting processes that – on average at least – will be helped more than hindered.


[Hildreth, 1984] E.C. Hildreth. *The Measurement of Visual Motion.* Cambridge, MA: MIT Press, 1984.

[Kasif, 1986] S. Kasif. On the Parallel Complexity of Some Constraint Satisfaction Problems. In *Proc. Conf. of AAAI*, pages 349–353, 1986.

[Kirousis and Papadimitriou, 1985] L. Kirousis and C.H. Papadimitriou. The Complexity of Recognising Polyhedral Scenes. In *26th FOCS*, 1985.

[Levesque and Brachman, 1985] H. Levesque and R.J. Brachman. A Fundamental Tradeoff in Knowledge Representation and Reasoning. In R.J. Brachman and H. Levesque, editors, *Readings in Knowledge Representation*, pages 41–70. Morgan Kaufmann, Los Altos, CA, 1985.

[Levesque, 1989] H. Levesque. Logic and the Complexity of Reasoning. Technical Report KRR-TR-89-2, Dept. of Computer Science, University of Toronto, Toronto, Canada, 1989.

[Marr and Hildreth, 1980] D. Marr and E.C. Hildreth. Theory of Edge Detection. *Proc. Roy. Soc. London B.*, 207:187–217, 1980.

[Marr and Poggio, 1979] D. Marr and T. Poggio. A Computational Theory of Human Stereo Vision. *Proc. Roy. Soc. London B.*, 204:301–328, 1979.

[Marr, 1982] D. Marr. *Vision.* San Francisco: Freeman, 1982.

[Morgan, 1984] M.J. Morgan. Computational Theories of Vision. *Quarterly J. Exp. Psych.*, 36A:157–165, 1984.

[Navon, 1984] D. Navon. Resources: A Theoretical Soup Stone? *Psychological Review*, 91:216–234, 1984.

[Norman and Bobrow, 1975] D.A. Norman and D.G. Bobrow. On Data-limited and Resource-limited Processes. *Cognitive Psychology*, 7:44–64, 1975.

[Ramachandran, 1988] V. Ramachandran. Perceiving Shape from Shading. *Scientific American*, Aug:76–83, 1988.

[Rosenfeld, 1987] A. Rosenfeld. Recognizing Unexpected Objects: A Proposed Approach. In *Proc. of the DARPA Image Understanding Workshop*, pages 620–627, 1987.

[Sugihara, 1986] K. Sugihara. *Machine Interpretation of Line Drawings.* Cambridge, MA: MIT Press; New York: McGraw-Hill, 1986.

[Todd and Bressan, 1990] J.T. Todd and P. Bressan. The Perception of 3-Dimensional Affine Structure from Minimal Apparent Motion Sequences. *Perception & Psychophysics*, 48:419–430, 1990.

[Treisman and Gormican, 1988] A. Treisman and S. Gormican. Feature Analysis in Early Vision: Evidence from Search Asymmetries. *Psychological Review*, 95:15–48, 1988.

[Tsotsos, 1987] J.K. Tsotsos. A "Complexity Level" Analysis of Vision. *International Journal of Computer Vision*, 1:346–355, 1987.

[Tsotsos, 1990] J.K. Tsotsos. Analyzing Vision at the Complexity Level. *Behavioral and Brain Sciences*, 13, 1990.

[White, 1988] I. White. The Limits and Capabilities of Machines – A Review. *IEEE Trans*, SMC-18:917–938, 1988.

# What do feature detectors detect? Features that encode context and the binding problem

**Robert Ward**

Department of Psychology
Carnegie Mellon University
Pittsburgh, PA 15213
ward@psy.cmu.edu

**Abstract**

The representation of visual features is investigated by examining the types of information that are encoded at the feature level which are used for feature binding. Features are often assumed to be bound together by virtue of their common location, but the current study shows that shared context, as well as location, acts to constrain the feature binding process and the formation of illusory conjunctions. Two different sorts of context manipulations are reported. In one manipulation, the context of each item in the display is established by flanking bars, and binding errors are examined as a function of this shared context. Also examined is a more global context manipulation in which the items presented form either a word or nonword. Both sorts of contexts affect feature binding, although in different ways. Finally, some of the computational difficulties in implementing a feature representation that encodes context are considered.

Does a feature detector detect something more than just the presence of a feature? In fact, the answer is probably yes. For example, consider a retinotopic feature map, in which many feature detectors with limited receptive fields are distributed to span a large region of visual space. Since each detector is tied to a particular location within the visual field, an active detector specifies both the presence of a feature and the location of that feature. In such a feature map, it would be correct to say that features are encoded with location information.

The possibility that feature detectors may encode more than the mere presence of a feature is important when considering the problem of *feature binding* -- that is, when determining which features belong to the same object and which features belong to different objects. Figure 1 shows a set of feature detectors that encode only the presence of features. The activation of these detectors by themselves cannot be unambiguously interpreted. If instead we have a set of retinotopic feature maps, and we know the correspondence between locations on the two maps, then the activity of the feature detectors specifies not only the location of features, but also the location of feature conjunctions. Theorists in psychology have differed on how easily the correspondence between two maps might be accessed, or how available this correspondence might be. For example, Treisman and Gelade (1980) assumed that focal attention was required to determine the features present at any single location; more recently, Wolfe, Cave, and Franzel (1989) allowed that in effect, feature maps could be superimposed upon each other, so that the combined activity of relevant feature maps could be determined for all locations simultaneously. But what is important to note is that these different theories share the idea that features are bound together by virtue of their common location.

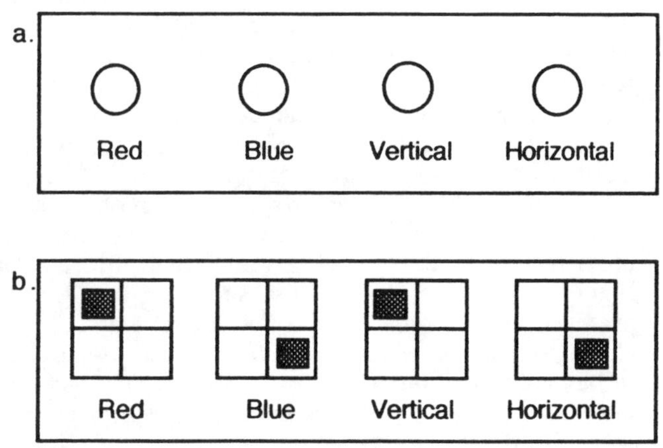


Figure 1. a) This feature representation cannot distinguish between the simultaneous presentation of a red vertical with a blue horizontal line and a red horizontal with a blue vertical line. b) This representation specifies the location of all features as well as the location of the red vertical and blue horizontal lines.

But should location be special? Or might features encode other types of information? Might there be a way of identifying which features belong together besides shared location? Duncan (1989) suggests that the visual system is most likely not limited to location cues for integrating features together. This report will examine the possibility that features encode aspects of their *surrounding context*. In this way, features that share a similar context would be bound together in the same way as features sharing similar locations. For example, consider the letter **A** in the word **CAT**. One constraint on the way the features of the **A** would be conjoined is their shared context: all the features of the A appear in a context consisting a C on one side and a T on the other. This information could be useful in deciding that all the features of the A belong together.

It is already well-established that context plays a central role in our perceptions. The famous example illustrated in Figure 2 shows how the same figure, in this case the ambiguous A/H object, can be interpreted in different ways depending upon the object's surround. However, it is a very large jump from the claim that context can affect our interpretation of objects, to the claim that aspects of context are encoded at the feature level. Furthermore, there is a considerable difference between encoding location information at the feature level and contextual information at the feature level.

Figure 2. An example of the effect of context on the interpretation of an ambiguous object.

It seems intuitively reasonable, or at least consistent with the known anatomy of the visual cortex, to imagine a feature representation consisting of many varieties of retinotopic maps. Such maps can be quite general purpose, in that they can specify all the possible locations in which an object might appear. It seems much more difficult to specify all the different *contexts* in which an object might appear; in fact, it can be hard to visualize the organization of a feature "map" encoding contextual information. Figure 3 shows one possibility. The map specifies the presence of different features within several contexts. Contexts which are similar to each other are close together in this contextual encoding space. Such a map has some interesting properties. The contextual map does not specify *where* in the image a feature may occur, but instead specifies *what* sort of surround the feature occupies. Although this map does not directly encode location, it is still useful for feature binding, since features that share the same contexts will tend to belong together. Also, such a map can explain some sorts of perceptual grouping effects. In the representation used in Figure 3, a set of colinear points will share a more similar representation than a set of haphazardly arranged points. This similarity in encoding could serve as one basis for grouping the points together, so that the perceptual grouping would be explained at the level of feature representation.

The experiment reported here is based upon a principle of encoding similarity. It is assumed that the processes responsible for binding together features use information encoded by the feature detectors to determine likely feature conjunctions. Features that are encoded in similar ways, perhaps due to their similar locations or similar contexts, are therefore likely to be bound together. Likewise, illusory conjunctions, or the false recombinations of features (Treisman & Schmidt, 1982), should be more likely to occur among features given similar encodings. Cohen & Ivry (1989) found higher rates of illusory conjunctions among features that were displayed close together, suggesting that features encode location information. In the same way, if contextual information is encoded at the feature level, then illusory conjunctions should occur more frequently between features sharing similar contexts.

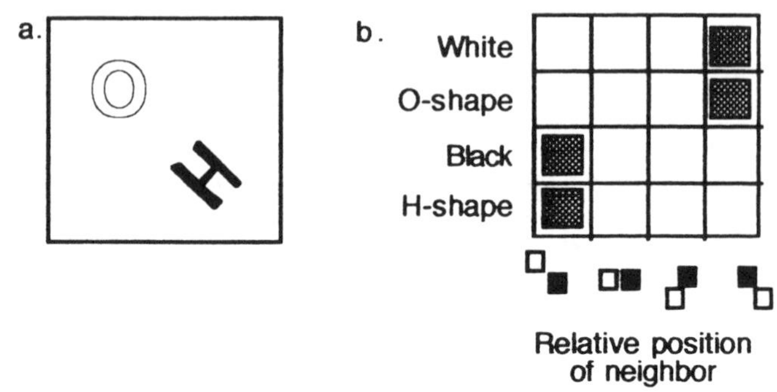


Figure 3. A sample feature map based on surrounding context. a) Sample stimulus. b) Encoding of stimulus using context based feature map. Each detector encodes information about the context of its associated feature, in this case the position of neighboring elements.

Two sorts of contexts are examined in the current experiment. The first sort of context is very simple, consisting of a pair of bars that flank display items either vertically or horizontally. This context manipulation, referred to here as "shared context", is local to the features present in the display; for example, items in the display can be presented in the same or in different contexts, and the effect of shared local context on feature binding can be observed. Based on the principle of encoding similarity, features presented in the same contexts should be more likely to be incorrectly conjoined. This simple context manipulation seems like a good starting point for finding contextually bound representations. The position of neighboring elements has been suggested before as a nonspatial basis for feature binding by Strong and Whitehead (1989). In the Strong and Whitehead model, the relative position of elements outside the focus of attention are used to provide a unique "tag" for the set of features within the focus of attention. The second type of context is a more global property of the stimulus, namely whether the features presented comprise a word or a nonword. An effect of these global properties would imply that the feature representation is determined only after a higher level of processing for which the global properties can be defined. At the same time, the experiment further investigates the encoding of location information by systematically varying the distance between features in these contexts.

## Method

Stimuli. The experimental materials used in this experiment were very similar to those developed by Prinzmetal and Millis-Wright (1984, Experiment 2). Each stimulus display contained a string of three colored letters that appeared unpredictably to the left or right of fixation. There were six possible strings formed by the three letters, all containing the letter P: the words PIE, SPY, and MAP, and the nonwords PLF, BPT, and NVP. Letters were randomly colored in red, blue, green, or yellow, with the constraint that the same color was never used more than once in a letter string.

Each letter appeared in either a vertical or horizontal context. The vertical context consisted of white bars appearing above and below the letter; the horizontal context consisted of bars to the left and right of the letter. The bars forming the context always appeared to "bracket" their associated letter - that is, bars in the vertical context were oriented horizontally and bars in the horizontal context were oriented vertically. Stimuli were constrained so that each display contained letters in both the horizontal and vertical contexts, for a total of six possible arrangements of contexts among the letters. Figure 4 illustrates a sample display.

Figure 4. Sample stimulus. Each letter subtends 0.8 x 0.8 degrees with 2.4 degrees between letters. In this example, the P and F appear in the vertical context, and the L in the horizontal context.

Procedure. The task for subjects was to decide if the target letter P was the same color as a predesignated signal color. Each trial proceeded as follows. A colored square appeared in the center of the display for 1000 msec. The square served both to mark the fixation point and to designate the signal color for the trial. The stimulus display was then presented for 150 msec. Subjects pressed a key to indicate whether the color of the target letter P matched or did not match the signal color. Subjects were instructed to answer as quickly as possible without sacrificing accuracy. However, no feedback was provided during the experiment.

There were a total of 216 trials in the experiment. Pairing each of the six letter strings with each of the six possible context arrangements yielded 36 stimulus items, which appeared both to the left and right of fixation for a total of 72 stimulus displays. Each display was presented three times: once as a positive trial, in which the color of the target letter P matched the signal color; once as a feature trial, in which the signal color was not present in the display; and once as a conjunctive trial, in which the signal color appeared in one of the letters other than the target letter P. The conjunctive trials were the crucial trials of the experiment, since they allowed the possibility of feature conjunction errors. In all, three variables were manipulated in the conjunctive trials of the experiment: (1) whether the display consisted of a word or nonword; (2) the distance in characters between the letter P and the letter appearing in the signal color; and (3) whether the P and the signal color shared the same horizontal or vertical context. Subject to these constraints, the color of stimulus materials was determined randomly.

Before the experiment began, subjects received 32 practice trials in which feedback was provided.

Subjects. Fourteen Carnegie Mellon undergraduates participated for class credit. The data from one subject were discarded due to very high error rates, averaging 48% correct on the conjunctive trials, which are the crucial trials of the experiment, and 57% over all trials.

## Results

The average error rates for positive trials, feature trials, and conjunctive trials were 34%, 6%, and 23% respectively. There were significantly more false alarms in the conjunctive conditions than in the feature conditions, $t(12) = 44.5$, $p < 0.001$. False alarms on the conjunctive trails are assumed to indicate feature binding errors; however, clearly there may be other causes for errors on the conjunctive trials. Therefore, it is not the absolute levels of errors that are interesting, but the effects of word stimuli, distance, and shared context on performance in the conjunctive trials. These effects are illustrated in Figure 5. A rich variety of both spatial and nonspatial influences on feature representation and binding can be examined in the current experiment: the effects of a simple shared context, the effects of distance between the conjoined features, and the effects of a "high level", more global context, created by word versus nonword stimuli.

The effects of shared context will be examined first. Averaging over the word and distance factors, subjects were significantly more likely to make conjunctive errors on the shared context trials than on the different context trials. False alarm rates averaged 27% for shared context trials and 20% for different context trials, $F(1,12) = 5.225$, $p = 0.041$. The probability of subjects conjoining features is therefore affected by the context surrounding the features,

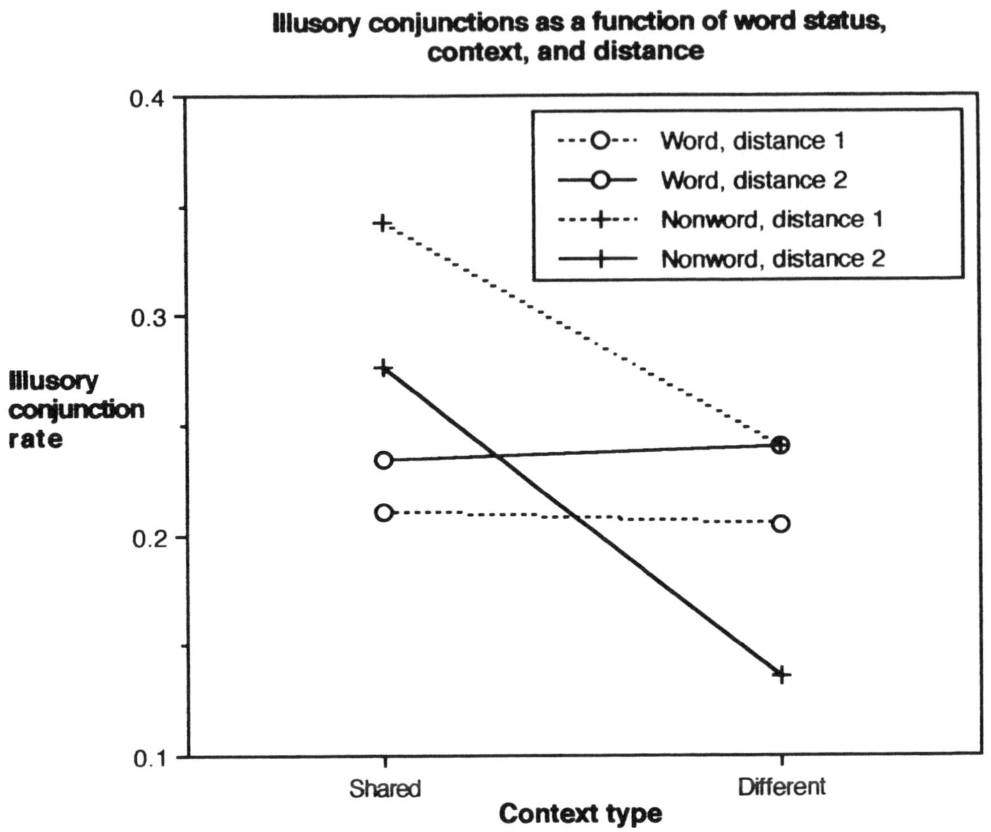


Figure 5. Illusory conjunctions, measured by the false alarm rate for the conjunctive trials.

such that features are more likely to be conjoined if they appear in the same context. This finding suggests that features are encoded in a context sensitive way. But as Figure 5 illustrates, the effect of shared context appears to vary considerably depending upon whether the features appear within a word or nonword. Although the interaction of shared context and word did not quite reach significance, (1,12) = 3.85, p = 0.073, the effect of shared context can be clearly seen for the nonwords, but not at all for the word stimuli.

The effect of distance was also modulated by the word nature of the stimuli. There was a significant interaction between the word and distance factors, F(1,12) = 8.18, p = 0.014, so that the effects of distance were more pronounced in nonwords than in words. The principle of encoding similarity appeared to hold in the nonwords, in which adjacent features were more likely to be conjoined than nonadjacent features. If there is any effect of distance on word items, it is that adjacent features are less likely to be conjoined than nonadjacent ones. However, the magnitude of the distance effect is much smaller for words than nonwords, 3.0% versus 8.5%.

Despite the influence of word displays on other experimental variables, there was no evidence of any main effect of word versus nonword displays, F(1,12) = 0.416. This result may at first seem at odds with the findings of Prinzmetal and Millis-Wright (1984), who reported fewer conjunctive errors in nonword displays than word displays. In the current experiment, most conditions showed fewer errors for word than nonword stimuli; however, when conditions were optimized for error-free performance based on the principle of encoding similarity (different context, distance 2), there were fewer errors on the nonword than the word items. The current finding therefore offer some limits to the generality of the Prinzmetal and Millis-Wright results.

## Discussion

The pattern emerging from these results suggests that the representation of a feature depends both upon local attributes of the feature, like its location and immediate context, as well as more global properties of the feature environment, such as whether the feature appears within a word or nonword. In the current experiment, the representation of features in nonwords included information about feature location and aspects of the immediate context. At least for nonwords, this location and contextual information is used in the feature binding process, as evidenced by the greater number of illusory conjunctions occurring between features sharing similar locations and contexts.

But another sort of feature representation seems to be established when features appear within words. The current results suggest that the final representation of features can be determined at least in part by knowledge about higher order structures, like words. However, the influence of word structure on feature representation is a contextual effect very different from the effect of the flanking horizontal and vertical contexts. It appears that the feature representation established as a result of the interaction with word knowledge is not sensitive to the specifics of feature instances, like location or surrounding context. Instead, it seems that these local attributes of features are overridden by the influence of higher level organization. This raises the interesting question of how feature binding might proceed in cases where top-down influences are relevant. An interesting sidelight is that although the word factor interacted with other experimental factors, a post-experiment survey found only one subject who was aware of any words being presented during the experiment.

The most interesting result of the experiment may be the evidence it provides for context sensitive feature encoding, since it appears that the feature binding processes use both location and contextual information to bind features together. However, implementing a contextually encoded feature representation raises a number of difficult computational problems. One concern is the possible size of the feature encoding space. The horizontal and vertical contexts used here are one simple example of a context that may be encoded at the feature level, but there is a potentially enormous space of possible contextual encodings. Coarse coded representations are one way of reducing the number of detectors required to span a given encoding space, and Hinton (1981) has shown that the savings advantage of coarse coding increases with the number of encoded dimensions. Other savings could be introduced by using dynamically programmed feature detectors. A relatively small set of detectors might be programmed to encode task relevant context information. A purely speculative approach at this point might be an implementation based upon the CID model of McClelland (1985).

A second problem in implementing contextual maps is using information from a nonspatial contextual map in conjunction with feature maps based on other types of spatial and nonspatial metrics. Two spatially organized maps for different features can be meaningfully compared and correlated, since a direct correspondence between locations in the two maps can be established. However, different contextual maps will be laid out according to different metrics, and there may be no direct mapping from positions within one type of context map to those in another. One way around this difficulty may be to link feature maps based upon different metrics indirectly, through a higher, object level representation. In this scheme, each map would act as a source of constraint on the number and description of objects in the environment. The final representation of the environment would be determined by a constraint satisfaction process using both location and contextual information, operating over the entire set of feature maps. In the resulting representation, each object in the environment description would be linked to the relevant values on all the feature maps. In this way, corresponding positions on differently structured maps could be found by specifying a particular object.

## References


Cohen, A. & Ivry, R. (1989). Illusory conjunctions inside and outside the focus of attention. Journal of Experimental Psychology: Human Perception and Performance, 15, 650-663.

Duncan, J. (1989). Parallel processing: Giving up without a fight. Commentary. *Behavioral and Brain Sciences, 12,* 402-403.

Hinton, G.E. (1981) Shape representation in parallel systems. *Proceedings of the Seventh International Joint Conference on Artificial Intelligence*, pp. 1088-1096.

McClelland, J.L. (1985). Putting knowledge in its place: A scheme for programming parallel processing structures on the fly. *Cognitive Science, 9,* 113-146.

Prinzmetal, W. & Millis-Wright, M. (1984). Cognitive and linguistic factors affect visual feature integration. *Cognitive Psychology, 16,* 305-340.

Strong, G.W., & Whitehead, B.A. (1989). A solution to the tag-assignment problem in neural networks. *Behavioral and Brain Sciences, 12,* 381-433.

Treisman, A., & Gelade, G. (1980). A feature integration theory of attention. *Cognitive Psychology, 12,* 97-136.

Treisman, A. & Schmidt, H. (1982). Illusory conjunctions in the perception of objects. *Cognitive Psychology, 14,* 107-141.

Wolfe, J.M., Cave, K.R., & Franzel, S.L. (1989). Guided search: An alternative to the feature integration model for visual search. *Journal of Experimental Psychology: Human Perception and Performance, 15,* 419-443.

# The Role of Input and Target Similarity in Assimilation

**Brian T. Bartell †** **Garrison W. Cottrell †** **Jeffrey L. Elman ‡**

† Department of Computer Science and Engineering
‡ Center for Research in Language
University of California, San Diego
San Diego, CA 92093

## Abstract

We investigate the situation in which some target values in the training set for a neural network are left unspecified. After training, unspecified outputs tend to *assimilate* to certain values as a function of features of the training environment. The roles of the following features in assimilation are analyzed: similarity between input vectors in the training set, similarity between target vectors, linearity versus non-linearity of the mapping, training set size, and error criterion. All are found to have significant effects on the assimilation value of an unspecified output node.

## Introduction

We consider here the case in which the target vectors in the training set for a neural network are not completely specified. That is, for certain output units on some input patterns, the desired (target) values may be indeterminate or irrelevant (*don't-care*). During Back Propagation [Rumelhart *et al.*, 1986] the error at these don't-care outputs is zero, so weight changes do not include terms from these nodes.

We define *assimilation in neural networks* to be the act of a don't-care output unit taking on (assimilating to) a value after training. The focus of this work is on examining the characteristics of the training environment which determine a node's assimilation value.

Assimilation in neural networks is interesting because it has been used to model the articulation of words or phrases occurring in natural language [Jordan, 1986] [Hare, 1990]. Our ability to evaluate these models and apply them to other domains relies heavily on our level of understanding of the underlying assimilation phenomenon in the model.

Our interest in assimilation also derives from its involvement with other more general neural network learning issues. It involves an interesting form of generalization, in which the input patterns for the don't-care targets have been seen by the net during training, rather than having been reserved for a separate test phase. How a don't-care node responds depends on how the representational resources of the hidden units have been employed to perform the task. This, of course, would seem to depend on the similarity structure in the input and target domains and other characteristics of the mapping. The value which an output assimilates to may also suggest values which would be easier for the output to learn if it were required to. Thus, understanding assimilation may offer insights on the general learning process.

## The Assimilation Effect

We briefly review two works which use assimilation to model articulation data from linguistics [Jordan, 1986] [Hare, 1990]. Each work offers an alternative hypothesis to explain the assimilation results which were witnessed. We will present these two explanations, and offer counter-examples demonstrating that neither is accurate in all cases.

Both works use the same Jordan network model[1] trained to generate a sequence of phonemes representing the articulation of a word or phrase. Phonemes are represented as feature vectors, and some of the features for certain phonemes are left unspecified. For example, the nasal feature for the phoneme subsequence */ria/* might be left unspecified in the articulation of *freon* (phoneme sequence */frian/*) [Jordan, 1986]. After training, this

[1]The network can be viewed as a feed forward network except that a portion of the input vector (the "state") is a function of the previous outputs: the state at time $t$ equals the state at $t-1$ times a decay parameter $\mu$, $(0 \leq \mu \leq 1)$ plus the output at time $t-1$. See [Jordan, 1986] for a detailed treatment of the Jordan Network architecture.

| Time | Input/State | | | | | | | Target | | | | | | |
|---|---|---|---|---|---|---|---|---|---|---|---|---|---|---|
| 1 | .5 | .5 | .5 | .5 | .5 | .5 | .5 | 1 | 0 | 0 | 0 | 0 | 1 | 1 |
| 2 | 1.3 | .3 | .3 | .3 | .3 | 1.3 | 1.3 | 0 | 0 | 0 | 1 | 0 | 0 | 1 |
| 3 | .8 | .2 | .2 | 1.2 | .2 | .8 | 1.8 | */0.9 | 0 | 0 | 0 | 0 | 1 | 1 |

Table 1: Jordan's hypothesis based on input similarity predicts that the don't-care output ("*") will assimilate to value 0. This is because the third input pattern is more like the second input than like the first input, and the don't-care output at time $t = 2$ is 0. However, assimilation is to 1.

feature might take on intermediate values (between low /f/ and high /n/), indicating that the model predicts anticipation of the nasal feature for the phonemes before the /n/.

## Input Similarity

Coarticulation in speech is a phenomenon in which the pronunciation of two phonemes overlaps in time. That is, there can be a blurring of articulation features between phonemes which are nearby temporally.

In Jordan's assimilation model of this phenomenon, don't-care outputs also assimilate to nearby articulated features. Jordan's explanation for this observed assimilation effect in the network is that, typically, similar inputs generate similar outputs. We may justify this *input similarity* hypothesis by reasoning that the function computed by a network is a continuous mapping; therefore, two similar inputs will generate two similar outputs, unless trained otherwise. Additionally, Jordan notes that the sequence of states traversed by the network tends to be continuous in time (that is, nearby states in time will tend to be more similar than arbitrary pairs). He calls this the *continuity property* of the next-state function. This, in combination with the hypothesized input similarity effect[2], predicts the observed results: that don't-care outputs will assimilate to specified outputs that are nearby in time.

Although this hypothesis is useful in explaining Jordan's results, a counter-example can be provided. Table 1 depicts such a case (due to Hare[3]). A don't-care target is denoted by a "*" in the table, followed by the mean assimilation value after training. In this case, an input similarity explanation predicts assimilation to 0 since the third input is more like the second than like the first (by euclidean distance). However, assimilation is actually to 1 ($mean = 0.9$, $stddev = 0.05$, 50 samples).

## Target Similarity

In the Hungarian vowel harmony system, suffix vowels will alternate in backness in order to agree with the last vowel of the root. For example, the suffix vowel *a* in *pugo + nak* is a back vowel in order to agree with the back vowel *o*. However, in certain exceptional cases, the last vowel of the root will be *transparent* to the assimilation process. In this case, the suffix vowel will agree for backness with a non-final root vowel, ignoring the final transparent vowel. An example is *taxi + nak*, in which *i* is transparent and the suffix vowel alternates to agree with the back vowel *a*.

In Hare's assimilation model of this phenomenon, the don't-care outputs assimilate to the last vowel of the root in the general case, and also correctly assimilate to the back vowel in the exceptional case. This linguistic data is satisfied without the stipulation of transparent vowels by the model.

Hare's explanation for this observed assimilation effect in the network is that high similarity between target patterns can override the effect of input similarity. Thus, although a suffix vowel will usually assimilate with another vowel that has the most similar input (which, according to Jordan, will be the most recent vowel), the suffix vowel will override input similarity in the case when the suffix and another vowel share many of the same target features. In this case, assimilation is to the similar target, rather than input. A possible justification for this *target similarity* hypothesis is that similar outputs will constrain the hidden unit representation of their inputs to nearby regions in activation space. This is plausible since the hidden unit representation must be linearly separable for the mapping to be learned. Therefore, two hidden unit representations generating very similar outputs may be relatively similar, causing a don't-care output in one pattern to take on the value of the specified output in the other.

As in the case with Jordan's input similarity hypothesis, we can also construct a contra-

[2]Note that in this model, the "state" acts as the input for the mapping performed at each time step. Thus, similar states also generate similar outputs.

[3]Our assimilation results in this case (averaged over 50 samples) differ somewhat from the results reported in [Hare, 1990], although this may be attributable to different initial weights or a larger sample size. Results are achieved with initial state = 0.5, $\mu$ (state decay) = 0.6, $\eta$ (learn rate) = 0.1, and $\alpha$ (momentum) = 0.0.

| Time | Input/State | | | | | | | Target | | | | | | |
|---|---|---|---|---|---|---|---|---|---|---|---|---|---|---|
| 1 | .5 | .5 | .5 | .5 | .5 | .5 | .5 | 0 | 0 | 0 | 0 | 0 | 0 | 0 |
| 2 | .3 | .3 | .3 | .3 | .3 | .3 | .3 | 1 | 1 | 1 | 1 | 1 | 1 | 1 |
| 3 | 1.2 | 1.2 | 1.2 | 1.2 | 1.2 | 1.2 | 1.2 | */0.0 | 1 | 1 | 1 | 1 | 1 | 1 |

Table 2: Hare's hypothesis based on target similarity predicts that the don't-care output will assimilate to value 1. This is because the target pattern at time $t = 3$ is very similar to the target at $t = 2$, and the output at $t = 2$ for the don't-care node is 1. However, assimilation is to 0.

diction to Hare's target similarity explanation. This case is illustrated in Table 2. Although the target at the third time step is exactly like the second target but unlike the first, assimilation of the don't-care output is to the first output ($mean = 0.0$, $stddev = 0.03$, 25 samples).

## Some Factors Affecting Assimilation

We have presented two possible explanations for the assimilation effect in neural networks: assimilation based on similar inputs, and assimilation based on similar targets. Both hypotheses have intuitive justifications and have been used to explain network behavior in the literature. Nevertheless, the counter-examples suggest that other factors are also involved.

We turn now to a set of experiments which examine the effects on assimilation of the following factors:

- input and target similarity,
- linearity versus non-linearity of the mapping,
- the error criterion to halt training, and
- the size of the training set (TS), for fixed hidden layer size.

These factors are considered because they are conspicuous features of the simulations in [Hare, 1990], and therefore may contribute to the observed assimilation effects. Specifically, in Hare's work, input and target similarity are not independently varied while testing the assimilation effect (see section V in [Hare, 1990] – varying the first target in the sequence also varies all subsequent states). Also, the networks tended to be highly trained and have twice as many hidden units as there are patterns to learn. Lastly, all mappings were linear.

We restrict our attention to feed forward binary mappings[4]. In the following experiments, $\eta$ (learn rate) = 0.2 and $\alpha$ (momentum) = 0.9.

### Input and Target Similarity

A 3 x 3 x 3 x 11 experiment was performed in order to test the independent and interactive effects of input similarity, target similarity, training set size, and error criterion, on the assimilation process. For each of the $3 \cdot 3 \cdot 3 \cdot 11 = 297$ cells in the experiment, 50 trials were performed. In each trial, a randomly initialized network was trained on a parameterized randomly initialized training set.

**Design** We use a strategy in which certain patterns (called Key) in a training set serve as potential "magnets" or sources of assimilation for other patterns (called Don't-Care) which contain unspecified units. The training set is parameterized by the input similarity and target similarity factors (each are one of LOW, MEDIUM, or HIGH), and the training set size (one of 4, 8, or 12). Table 3 depicts an example training set for factor values input=HIGH, target=LOW, and TS size=4. The training set is constructed as follows: 1) random binary vectors are generated for the Don't-Care pattern, 2) the Key pattern is generated such that its input and target vectors are as similar to the Don't-Care pattern as is specified by the input and target similarity factors, 3) Neutral patterns are added to make the training set the size of the TS size factor, and 4) the extra target unit for the Don't-Care pattern is made a don't-care ("*"), the target unit for the Key pattern is made a 1, and for all Neutral patterns it is made a 0[5].

The network is a 3-layer feed forward architecture with 10 inputs, 4 hidden units, and 11 outputs. The network is trained until the error criterion factor is reached (one of 11 mean squared error levels between 2.0 and 0.0002 in roughly a logarithmic progression)[6]. Once trained, the Don't-Care input is presented, and the don't-care output unit is examined. A value near 1 indicates high assimilation to the Key pattern; lower values indicate a decreasing

[4]Note that Hare's simulations can be posed as static feed forward tasks, because all inputs are determined by the initial state and target vectors.

[5]There is a bias introduced by making the extra target unit for all of the Neutral patterns a 0. With many Neutral patterns in the training set, the don't-care output is biased towards 0 more than with fewer Neutral patterns.

[6]A trial is aborted if the network takes many cycles without reaching the error criterion. Therefore, there are not 50 samples in each test cell.

| Pattern | Input (10) | | | | | | | | | | Target (10 + 1) | | | | | | | | | | |
|---|---|---|---|---|---|---|---|---|---|---|---|---|---|---|---|---|---|---|---|---|---|
| Don't-Care | 1 | 1 | 1 | 1 | 1 | 1 | 1 | 1 | 1 | 1 | 0 | 0 | 0 | 0 | 0 | 0 | 0 | 0 | 0 | 0 | * |
| Key | 1 | 1 | 0 | 1 | 1 | 1 | 1 | 1 | 1 | 1 | 1 | 0 | 1 | 1 | 1 | 1 | 1 | 1 | 1 | 1 | 1 |
| Neutral-1 | 1 | 1 | 1 | 1 | 0 | 1 | 0 | 0 | 0 | 0 | 0 | 0 | 0 | 0 | 1 | 1 | 0 | 1 | 1 | 1 | 0 |
| Neutral-2 | 1 | 1 | 1 | 0 | 0 | 0 | 1 | 0 | 0 | 1 | 0 | 0 | 0 | 1 | 1 | 0 | 0 | 1 | 1 | 1 | 0 |

Table 3: An example training set. The Key pattern has HIGH input similarity with the Don't-Care pattern, but LOW target similarity. The 2 Neutral patterns all have NEUTRAL input and target similarity with the Don't-Care pattern.

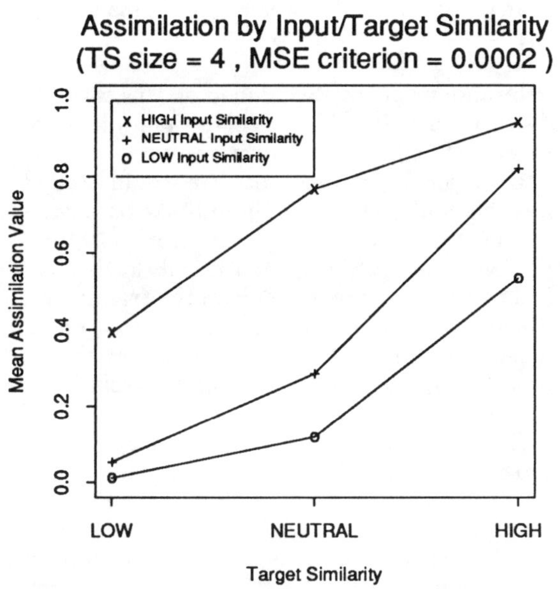


Figure 1: Mean assimilation results by input and target similarity for an underconstrained (TS size = 4) and highly trained (MSE = 0.0002) network.

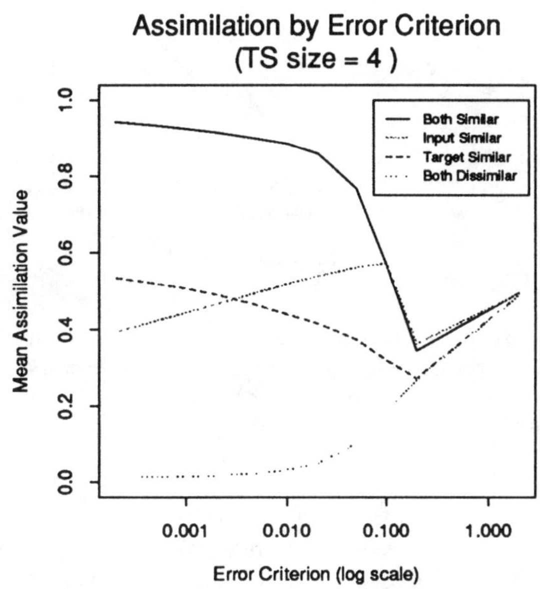


Figure 2: Mean assimilation results by input and target similarity as the MSE criterion varies between 2.0 and 0.0002. High MSE values (corresponding to networks trained for a short time) are on the right; low MSE values (for highly trained networks) are on the left.

assimilatory effect of the Key pattern.

**Results** All four factors examined as well as all possible interactions between factors have significant effects on assimilation (with $p < 0.001$ for all factors, except size x input and size x input x target with $p < 0.01$; the size=12 training set case is excluded from the analysis of variance because at this level, too few of the networks respond in all variations of the other factors). The single largest effect is due to target similarity ($F = 293.9$), followed by interaction between target similarity and error criterion ($F = 197.2$), input similarity ($F = 99.9$), error criterion ($F = 58.0$), and training set size ($F = 52.0$).

Figure 1 displays mean assimilation values by the 3 x 3 combinations of input and target similarity. Note that assimilation is high whenever one of the input or target patterns is highly similar and the other is neutral or similar. The assimilation response is also above the *a priori* rate (0.33) whenever either input or target similarity is high, regardless of the similarity of the other vector. The assimilation response is actually suppressed (below the *a priori* rate) when one of the input or target patterns is dissimilar and the other is neutral or dissimilar.

Figure 2 displays mean assimilation values as the MSE criterion varies between 2.0 and 0.0002. Initially, input similarity alone has a higher mean assimilation value than target similarity alone, and generally mimics the case in which both input and target patterns are similar. However, below MSE= 0.1 input similarity steadily decreases, and is eventually overtaken by the increasing target similarity case just above MSE= 0.001. Thus, a high degree

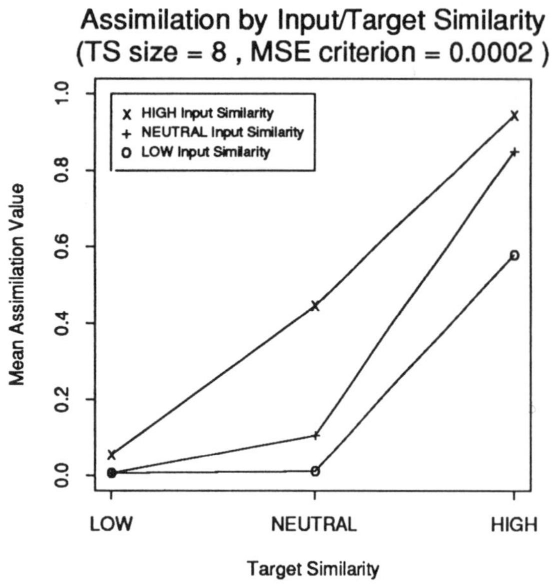


Figure 3: Mean assimilation results by input and target similarity when the training set increases in size (TS size = 8) relative to a fixed number of hidden units (4).

of training seems to accentuate the role of target similarity over input similarity.

Figure 3 displays mean assimilation values by input and target similarity for the case in which the training set has increased in size to 8. In this case, with high target similarity there is some assimilation response regardless of the level of input similarity (and the response gets larger as input similarity increases, as would be expected). However, the same is not true for high input similarity. When input similarity is high but target similarity is low, there is no assimilation response. Thus, the target similarity effect appears to be somewhat robust to an increase in the training set size, whereas input similarity appears less robust.

## Non-linear Task Effect

The relationship between assimilation and the non-linearity of the mapping task is tested using the 2, 3, and 4 bit parity problems with varying numbers of hidden units. Only the 2 hidden unit XOR case is reported; more hidden units and 3 and 4 bit parity yield qualitatively similar results.

Ten output units are used. The 10 targets are identically the XOR of the 2 inputs, except that a single output unit is left unspecified for one of the input patterns (00) or (11). See Table 4 for an example. For 50 trials, a network with random weights is trained on this task. 17 of the 50 samples did not solve the problem at the end of 500 epochs and were discarded.

| Input | | Hidden | | Output (10 units) | | | | |
|---|---|---|---|---|---|---|---|---|
| 0 | 0 | 0 | 0 | 0 | 0 | 0 | ... | 0 |
| 0 | 1 | 0 | 1 | 1 | 1 | 1 | ... | 1 |
| 1 | 0 | 0 | 1 | 1 | 1 | 1 | ... | 1 |
| 1 | 1 | 1 | 1 | */1 | 0 | 0 | ... | 0 |

Table 4: An example of the XOR mapping, including the hidden unit representations learned during one trial. Although target similarity predicts assimilation to 0, the don't-care output never sees the "boxed" hidden units (11) during training and instead assimilates to 1.

Based on target similarity, we would predict that the don't-care value should take on a value 0. This is because the target vector containing the don't care output is all 0's, and this is identical to another target vector in the training set which is also all 0's and specified on all units. In fact, the opposite occurs. In 30 of the 33 trials (91%), the don't-care output assimilates to 1 rather than 0.

These results can be understood by analyzing the internal representations learned to solve the task. The representations learned during one trial, in which assimilation is to 1, are depicted in Table 4. The don't-care output is never trained on the fourth internal state corresponding to input = (11). Thus, it never sees the first hidden unit on. Because of this, the don't-care output simply learns to be on whenever the second hidden unit is on. All other output units learn to be on if the second hidden unit is on *and* the first hidden unit is off.

An interesting fact occurs in the XOR case: assimilation to similar targets can be encouraged by adding another layer for possible recodings. To demonstrate this, an additional hidden layer with 4 units is added between the first hidden layer and the output layer. Of 74 sampled networks which learn the mapping, 42 (57%) assimilate to 0. Recall 0 is the value which would be expected if target similarity was having an effect on assimilation. 22 (30%) assimilate to 1 and 10 take on intermediate values. Thus, "target similarity" based assimilation can occur in the non-linear case, but it requires an extra layer for recoding.

## Non-linear Interaction

It is interesting to examine what happens if similar and dissimilar inputs and targets are placed in a *single* training set. Our experiment to test this has a design quite similar to the 3 x 3 x 3 x 11 design, except that there are no

Neutral patterns, and there are four Key patterns corresponding to the four combinations of similar and dissimilar inputs and targets. The Don't-Care pattern also has four don't-care bits instead of just one, in order to examine the assimilation values in the four cases in a single training set.

Non-linear interactions are observed when similar and dissimilar patterns are grouped in this way. For example, in a small training set (4 Key patterns and 1 Don't-Care pattern), the response when both input and target are similar in the same pattern is larger than the sum of the responses for similar input alone and similar target alone. As the training set size increases (to 12 Key patterns - 3 sets of similar and dissimilar inputs and targets), the effect is magnified tremendously. An assimilation response is only achieved when both input and target are similar in the same pattern, and the response is very strong and consistent. In the other cases, the assimilation response is suppressed to 0.0.

## Discussion

The results presented here suggest the following framework for understanding assimilation in feed forward nets on random boolean tasks:

- both input and target similarity have an effect on assimilation,
- in a non-linear mapping, positive target assimilation responses are minimal unless an extra hidden layer is provided,
- the assimilation response for the target similarity case appears robust to an increase in the number of patterns in the training set; the input similarity response is less so,
- when the error criterion is high, input similarity has a stronger assimilation response; for low error criteria, target similarity is stronger, and
- when similar and dissimilar inputs and targets are in the same training set, non-linear interactions can occur: in our simulations both input similarity and target similarity together are necessary for any assimilation response (for a large training set).

It is possible to reexamine Hare's work in the context of our results. In particular, the work corresponds to the linear, highly trained, small training set case. We can predict that, had the mappings learned in that work been non-linear, similar target assimilation results would have been achieved only using a second layer of hidden units. Our results also suggest that training a single Jordan Network to generate multiple sequences will result in a minimized role for target similarity or input similarity individually on the assimilation value. Rather, assimilation will tend to coincide with patterns in which both the input *and* the target are similar to the assimilation pattern.

Further work is needed before we can provide a complete explanation of the assimilation effect in networks. The current focus on binary mappings and feed forward networks should be widened. Also, although our training set construction is a convenient formulation, it does not cover the range of possible similarity relationships between vectors in a training set. For a truly complete account, we further need an understanding of the effect of individual training patterns on the *learning process*, perhaps through analyzing the induced error surface.

## Conclusion

We have examined two hypothesized explanations from the literature for the effects of similarity on assimilation in networks with don't-care outputs. Both are demonstrated to be incomplete by counter-example. Experimental evidence is provided which suggests factors to be included in a more comprehensive account. These factors are: the similarity of the don't-care pattern to other input and target vectors, the non-linearity of the mapping, the amount of training performed, and the size of the training set for a fixed number of internal units. Results in the literature can be reexamined in light of the current findings. This provides an alternate descriptive framework, and allows predictions of a model's behavior in novel training environments.

**Acknowledgements:** The authors wish to thank Mary Hare and Mark St. John for useful discussions. The first author is supported by a Cubic Fellowship and Peregrine Systems, Inc., Carlsbad CA., and can be reached at *bbartell@cs.ucsd.edu*.

## References

[Hare, 1990] Mary Hare. The role of similarity in hungarian vowel harmony: a connectionist account. *Connection Science*, 2(1), 1990.

[Jordan, 1986] Michael I. Jordan. Serial order: A parallel distributed processing approach. Technical Report ICS Report 8604, Institute for Cognitive Science, May 1986.

[Rumelhart *et al.*, 1986] David E. Rumelhart, Geoffrey E. Hinton, and Ronald Williams. Learning internal representations by error propagation. *Nature*, 323:533–536, 1986.

# Learning the past tense in a recurrent network: Acquiring the mapping from meaning to sounds

**Garrison W. Cottrell***
Department of Computer Science and Engineering
University of California, San Diego

**Kim Plunkett**
Institute of Psychology
University of Aarhus, Denmark

## Abstract

The performance of a recurrent neural network in mapping a set of plan vectors, representing verb semantics, to associated sequences of phonemes, representing the phonological structure of verb morphology, is investigated. Several semantic representations are explored in attempt to evaluate the role of verb synonymy and homophony in deteriming the patterns of error observed in the net's output performance. The model's performance offers several unexplored predictions for developmental profiles of young children acquiring English verb morphology.

## Introduction

Prior attempts to model the acquisition of English verb morphology in connectionist nets (Plunkett & Marchman [1991]; Plunkett, Marchman & Knudsen [1990]; Rumelhart & McClelland [1986]) have focused on the problem of learning the relationships between phonological representations of various forms of the verb. Phonological information is exploited by children and adults when prompted for the past tense form of a novel stem (Bybee & Slobin [1982]; Marchman [1988]). Nevertheless, the phonological form of the verb does not uniquely *determine* its past tense form. Although all verbs which have identical stem and past tense forms possess a dental final consonant (e.g. *hit* → *hit*), not all verbs that end in a dental consonant have identical stem and past tense forms (Pinker & Prince [1988]). Furthermore, connectionist models that learn purely *intra*-level phonological mappings cannot distinguish verb-stem homophones that take different past tense forms. For example, *to brake* and *to break* take past tense forms *braked* and *broke* respectively. Since the inputs to the network in these cases are identical, so will their outputs remain identical.

In this paper, we present a connectionist model of the acquisition of English verb morphology in which a network is trained to map a semantic representation of verbs to phonological representations of their stems and/or past tense forms (cf. Gasser & Lee [1990]). The mapping function to be learnt may be considered analogous to aspects of the production process in which meaning is mapped to sound representations. Phonological regularities between verb classes must be captured by the hidden unit representations generated through training. Homophones in this formulation are unproblematic, as they consitute a many-to-one mapping. However, two new potential problems arise. First, as Pinker & Prince [1988] point out, the semantics of a verb is *not* a good predictor of the type of inflectional mapping that it must undergo. The three verbs *hit, strike* and *slap* are closely related semantically but they have different mapping types relating their stem and past tense forms (*hit* → *hit, strike* → *struck, slap* → *slapped*). The network must learn to *ignore* this similarity in learning the mapping. Second, this same problem arises in general for the network, insofar as there is an arbitrary relationship between the meaning of the verb and its phonological form. Similar inputs do *not* lead to similar outputs. In particular, highly similar inputs, modeling synonyms, provide a potential source of error in these networks.

Our goals in this work are twofold:

1. To examine the performance of the network in solving a mapping problem that is analogous to that of mapping meaning to sound and determine its generalization characteristics.

2. To evaluate the pattern of outputs and errors produced by the network during the course of training and use these errors to predict those produced by children acquiring English verb morphology.

We report on three sets of simulations that differ either in the nature of the semantic representations used to encode verb meanings and/or in the number of meaning/form pairs that the network is required to learn. In each case, we provide an evaluation of the performance of the network on trained verbs and of the ability of the network to generalize to verb forms on which it has not been trained. In one set of simulations, we provide a detailed error analysis.

*We thank Steen Ladegaard Knudsen for his assistance in programming, analysis and running of simulations.

## Methodology

All simulations utilize a simple recurrent network of the type developed by Elman [1990] (see Figure 1). In all simulations the output phoneme consists of a

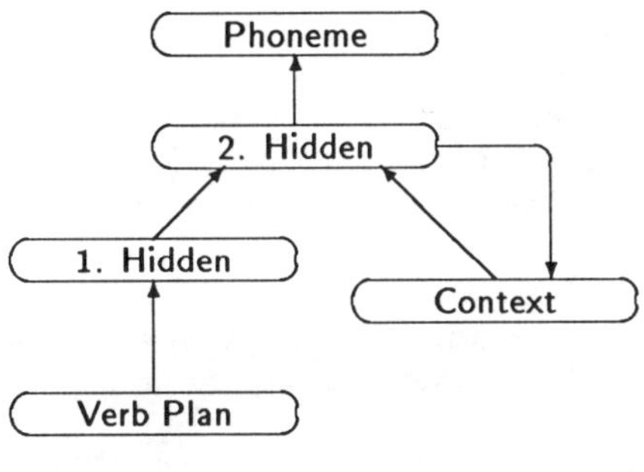


Figure 1: General Network Architecture

15 bit vector that reflects standard phonological contrasts. A noteworthy characteristic of this phonemic representation lies in its attempt capture the sonority relationships between vowels and consonants (see features O1–7 in Table 1).[1] The task of the network

| | | | | LO | MD | HI | GL | SN | FR | ST | PLACE | | | | | |
|---|---|---|---|---|---|---|---|---|---|---|---|---|---|---|---|---|
| | | BK | TN | O7 | O6 | O5 | O4 | O3 | O2 | O1 | LA | CR | VL | NS | SB | VO |
| vat | æ | −1 | −1 | +1 | +1 | +1 | +1 | +1 | +1 | +1 | −1 | −1 | −1 | −1 | −1 | +1 |
| cut | ˆ | +1 | −1 | +1 | +1 | +1 | +1 | +1 | +1 | +1 | −1 | −1 | −1 | −1 | −1 | +1 |
| ate | e | −1 | +1 | −1 | +1 | +1 | +1 | +1 | +1 | +1 | −1 | −1 | −1 | −1 | −1 | +1 |
| bet | ε | −1 | −1 | −1 | +1 | +1 | +1 | +1 | +1 | +1 | −1 | −1 | −1 | −1 | −1 | +1 |
| boat | o | +1 | +1 | −1 | +1 | +1 | +1 | +1 | +1 | +1 | −1 | −1 | −1 | −1 | −1 | +1 |
| bought | O | +1 | −1 | −1 | +1 | +1 | +1 | +1 | +1 | +1 | −1 | −1 | −1 | −1 | −1 | +1 |
| beat | i | −1 | +1 | −1 | −1 | +1 | +1 | +1 | +1 | +1 | −1 | −1 | −1 | −1 | −1 | +1 |
| bit | I | −1 | −1 | −1 | −1 | +1 | +1 | +1 | +1 | +1 | −1 | −1 | −1 | −1 | −1 | +1 |
| boot | u | +1 | +1 | −1 | −1 | +1 | +1 | +1 | +1 | +1 | −1 | −1 | −1 | −1 | −1 | +1 |
| foot | Ʊ | +1 | −1 | −1 | −1 | +1 | +1 | +1 | +1 | +1 | −1 | −1 | −1 | −1 | −1 | +1 |
| | y | −1 | −1 | −1 | −1 | −1 | +1 | +1 | +1 | +1 | −1 | −1 | −1 | −1 | −1 | +1 |
| | w | +1 | −1 | −1 | −1 | −1 | +1 | +1 | +1 | +1 | −1 | −1 | −1 | −1 | −1 | +1 |
| | h | −1 | −1 | −1 | −1 | −1 | +1 | +1 | +1 | +1 | −1 | −1 | −1 | −1 | −1 | −1 |
| | r | −1 | −1 | −1 | −1 | −1 | +1 | +1 | +1 | +1 | −1 | +1 | −1 | −1 | −1 | +1 |
| | l | −1 | −1 | −1 | −1 | −1 | −1 | +1 | +1 | +1 | −1 | +1 | −1 | −1 | −1 | +1 |
| | m | −1 | −1 | −1 | −1 | −1 | −1 | +1 | +1 | +1 | +1 | −1 | −1 | +1 | −1 | +1 |
| | n | −1 | −1 | −1 | −1 | −1 | −1 | +1 | +1 | +1 | −1 | +1 | −1 | +1 | −1 | +1 |
| | η | −1 | −1 | −1 | −1 | −1 | −1 | +1 | +1 | +1 | −1 | −1 | +1 | +1 | −1 | +1 |
| | f | −1 | −1 | −1 | −1 | −1 | −1 | −1 | +1 | +1 | +1 | −1 | −1 | −1 | −1 | −1 |
| | v | −1 | −1 | −1 | −1 | −1 | −1 | −1 | +1 | +1 | +1 | −1 | −1 | −1 | −1 | +1 |
| | s | −1 | −1 | −1 | −1 | −1 | −1 | −1 | +1 | +1 | −1 | +1 | −1 | −1 | +1 | −1 |
| | z | −1 | −1 | −1 | −1 | −1 | −1 | −1 | +1 | +1 | −1 | +1 | −1 | −1 | +1 | +1 |
| theatre | θ | −1 | −1 | −1 | −1 | −1 | −1 | −1 | +1 | +1 | −1 | +1 | −1 | −1 | −1 | −1 |
| mother | δ | −1 | −1 | −1 | −1 | −1 | −1 | −1 | +1 | +1 | −1 | +1 | −1 | −1 | −1 | +1 |
| | p | −1 | −1 | −1 | −1 | −1 | −1 | −1 | −1 | +1 | +1 | −1 | −1 | −1 | −1 | −1 |
| | b | −1 | −1 | −1 | −1 | −1 | −1 | −1 | −1 | +1 | +1 | −1 | −1 | −1 | −1 | +1 |
| | t | −1 | −1 | −1 | −1 | −1 | −1 | −1 | −1 | +1 | −1 | +1 | −1 | −1 | −1 | −1 |
| | d | −1 | −1 | −1 | −1 | −1 | −1 | −1 | −1 | +1 | −1 | +1 | −1 | −1 | −1 | +1 |
| | k | −1 | −1 | −1 | −1 | −1 | −1 | −1 | −1 | +1 | −1 | −1 | +1 | −1 | −1 | −1 |
| | g | −1 | −1 | −1 | −1 | −1 | −1 | −1 | −1 | +1 | −1 | −1 | +1 | −1 | −1 | +1 |
| silence | - | +1 | +1 | +1 | +1 | +1 | +1 | +1 | +1 | +1 | +1 | +1 | +1 | +1 | +1 | +1 |

Table 1: Phonological representation

is to output a sequence of phonemes that correspond to the stem or past tense of the verb whose semantic representation is presented at the input. The distinction between stem and past tense forms is encoded by a 2-bit vector at the input level. The inventory of verbs (both stems and past tense forms) that the network is required to produce at the output is taken directly from previous simulations conducted by Plunkett, Marchman & Knudsen [1990]. In this work 500 stem/past tense pairings are used. Each stem consists of a Consonant-Vowel-Consonant (CVC) string, a CCV string or a VCC string. Each string is phonologically well-formed, even though it may not correspond to an actual English word. Verbs are assigned to one of four classes. Each class corresponds to a different type of transformation analogous to a distinct past tense form in English as in (Plunkett & Marchman [1991]). The first class follows the regular rule in English. The three irregular classes are: (1) ***arbitrary verbs*** have no systematic relationship between the stem and past tense form (*go* → ***went***), (2) ***identity verbs*** are unchanged between forms (***hit*** → ***hit***), and (3) ***vowel change verbs*** undergo a change in vowel in CVC forms (***strike*** → ***struck***). Verbs are assigned randomly to each of the four classes, with the constraint that stems possess the appropriate characteristics of a given class.

Semantic representations of verbs are of two types. In the first set of simulations, each verb is represented in a localist manner in a 500-bit vector. An additional two units encode whether the network is to produce a sequence of phonemes corresponding to the stem of the verb or the past tense of the verb at the output. In the second and third set of simulations a similarity structure is imposed on the semantic representations by using distortions of several prototype vectors (Chauvin [1988]). Distortions may vary in their distance from the prototype. We use 9 or 50 prototype vectors (and thus as many categories) depending on vocabulary size. Two extra inputs specify the stem or the past tense form. For the large simulations, 10 distortions each of the 50 prototypes are generated at 3 levels of distortion.

Training proceeds by randomly selecting a plan vector (the verb's semantic representation) and a tense bit (stem or past tense). This composite vector is then presented at the input units over a number of time steps that correspond to the number of phonemes in the output form. At each time step, the discrepancy between the actual output of the network and the target phoneme is used as the error signal to a back propagation learning algorithm. We use the TLEARN simulator developed by Jeff Elman at UCSD. As part of the teaching signal, the verb plan is trained to produce an end-of-form signal (corresponding to the *silence* phoneme in Table 1). The "context units" are reset between forms.

## Analysis

The performance of the network is analysed at regular intervals in training. In this paper we present two types of analysis. First, we determine the hit rate for stems and past tense form, both on the entire training

[1]The phonological representation presented here was originally designed by Alan Prince and Kim Plunkett.

set and on a class-by-class basis. Hit rate is evaluated by determining which of the phoneme vectors (as defined in Table 1) is closest to the output vector using a least squares measure at each time step. This yields a sequence of output phonemes for each verb plan. In the first two sets of simulations we report on whether the output sequence is a hit or a miss.

We analyse the generalization characteristics of the network by first training the network with 25 verb plans to produce only the stem form of the verb and with another 25 verb plans to produce only the past tense form of the verb. Each verb plan is then tested on the phonological form of the verb to which it has not been trained i.e. 25 stem forms and 25 past tense forms. The output of the network on these novel inputs is used to evaluate the net's generalization properties.

Finally, in the third set of simulations we provide a detailed analysis of the output of the network when trained on just 50 verb plans i.e. 100 phoneme sequences. These analyses relate the role of semantic similarity to the similarity of the phoneme sequences across different verbs, and the syllabic structure that the network extracts over the sequence of output phonemes.

## Experiment One

This experiment reports the results of simulations using a 500 word vocabulary and orthogonal representations of the verb plan. Figure 2 (a) provides a summary of the network performance on all past tense forms and all stem forms while Figure 2 (b) compares the gener-

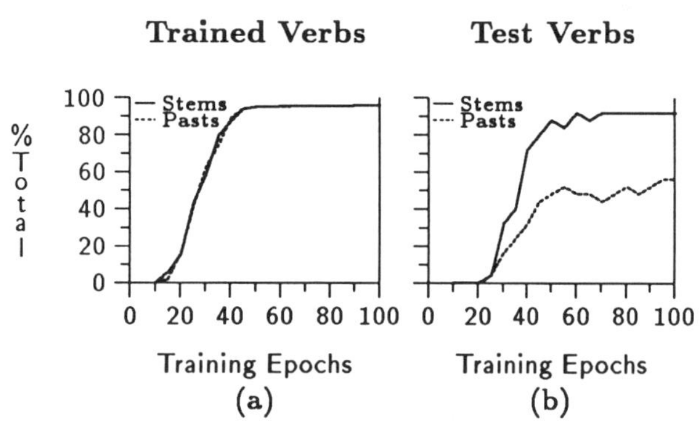


Figure 2: Overall Orthogonal Results

alization characteristics of the network in predicting the past tense forms of the verb when it has only be trained on the stem and *vice versa*. Figure 2 (a) shows that the network is equally fast at learning both stem and past tense forms and that learning undergoes a spurt in growth around the 20 epoch mark. In contrast, the test verbs differ with respect to their performance on stems and past tense forms. Figure 2 (b) shows that when a verb plan is trained to a past tense form, the network is quite accurate in predicting the correct stem ($\geq$ 90% after 70 epochs of training). On the other hand, generalization from stem to past tense never exceeds 55%. It should be noted that we use a very strict criteria for generalization: All past tenses are assumed to be regular. Over several simulations, we find that performance on past to stem generalizations is always good, while stem to past varies. This result is to be expected given that the form which the past tense takes is a better predictor of the stem than the stem is of the past tense form (e.g. if the past tense of a verb is *talked* then its stem form is unambiguous, but if the stem is *hit* then, in principle, its past tense form is underdetermined). Indeed the discrepancy between the generalization curves in Figure 2 (b) can be accounted for in this fashion.

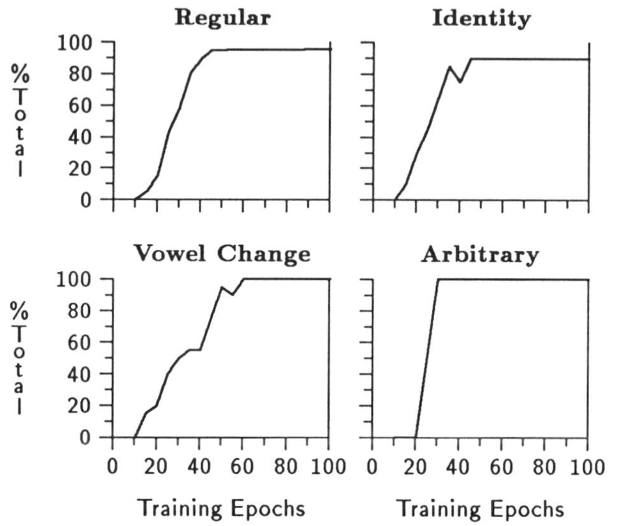


Figure 3: Stem Forms by Class (Orthogonal)

Figure 3 provides a class-by-class breakdown of network performance on stems. These result indicate that the regular, identity mapping and vowel change classes are learned first, while arbitrary mappings are delayed. Figure 4 reveals a similar rank ordering of classes with past tense forms.

## Experiment Two

This experiment reports the results of simulations using a 500 word vocabulary and semantically structured representations of the verb plan. Figure 5 (a) provides a summary of the network performance on all past tense forms and all stem forms while Figure 5 (b) compares the generalization characteristics of the network in predicting the past tense forms of the verb when it has only be trained on the stem and *vice versa*. As with Experiment One, Figure 5 (a) shows that the network is equally fast at learning both stem and past tense forms. Learning undergoes a spurt in growth around the 25 epoch mark. Similarly, there is a contrast between the test stems and the test past tense forms. However, the generalization characteristics for

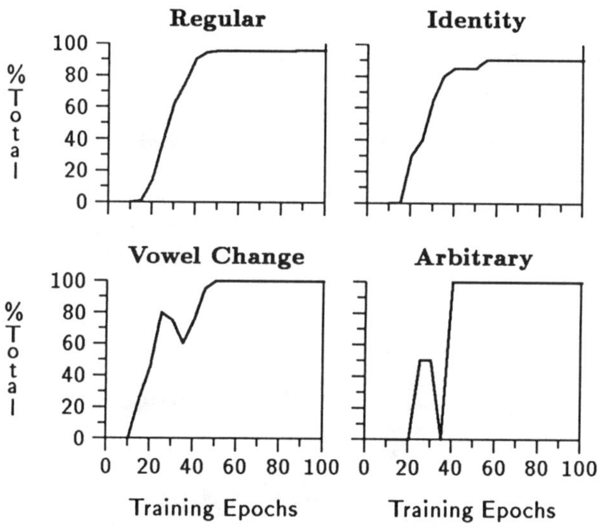


Figure 4: Past Forms by Class (Orthogonal)

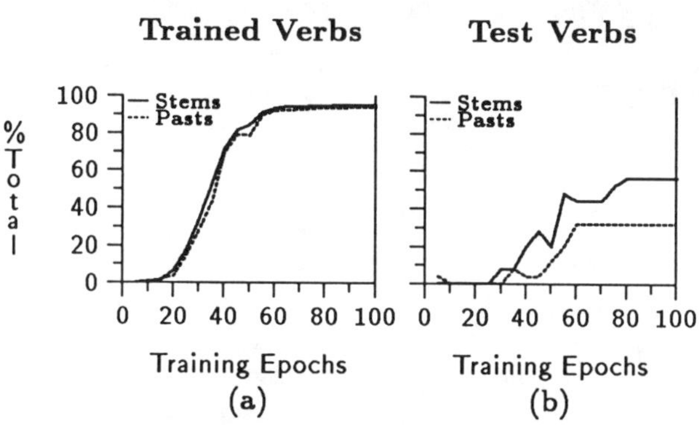


Figure 5: Overall Orthogonal Results

this network are more restricted than in the simulations where an orthogonal verb plan is used. An examination of the output for test stems indicates that the network has difficulty generating the epenthisized form of the *ed* suffix.

A class-by-class analysis of past tense forms and stem forms reveal results similar to the class-by-class analysis in Experiment One and so will not be reported here.

## Experiment Three

This experiment reports the results of analysis of the effects of input and target similarity structure on the forms the network learns. The semantic classes in this experiment are designed to highlight these effects. In this experiment, we use a 54-stem subset (with one new arbitrary) of the larger set, with 31 regular forms, 3 arbitraries, 8 identities and 12 vowel-change verbs. We use 9 prototype vectors, where each exemplar within a class has the same amount of distortion from the prototype. Three of the classes use high distortion, three medium distortion, and three low distortion. Two out of three of these classes has one lexical item more frequently represented than the rest, a regular in one, and an arbitrary in the other.

We perform two kinds of analyses on this network:

1. We measure the changes in the stem output strings during learning from the point of view of the syllabic structure of the target language.
2. We measure the changes in similarity of stem and past output strings during learning with respect to the semantic clusters.

### Syllabic structure changes

In order to assess the vocabulary development of the network, we divide the stem output strings of the network into three classes:

**Words:** Strings that belong to the target vocabulary.

**Pseudo-words:** Strings that are not in the target vocabulary but conform to the syllabic structure of the language — CVC, VCC, or CCV.

**Non-words:** Strings that do not fit the above criteria — CCC, CVV, VCV, VVC and VVV.

A graph of the numbers of unique forms of each kind over learning, along with the total number of unique forms, is shown in Figure 6. Interestingly , long before

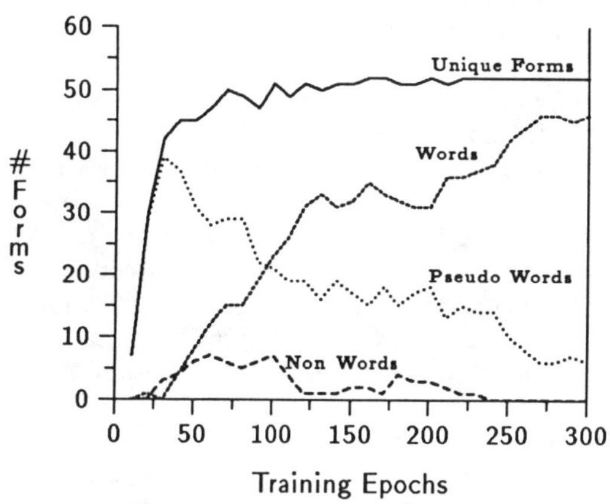


Figure 6: Vocabulary Structure

it has acquired any words in the language, the network has captured the syllabic structure, as evidenced by the high proportion of pseudowords in the set of unique forms. Over learning, there is an inverse relationship between these forms and the correct forms as the pseudo-words migrate into the target vocabulary. This set of curves is reminiscent of a similar set of curves found by Plunkett [1990] in a case study of language acquisition in two Danish children between the ages of 12 and 26 months. The total number of consistently produced and applied non-(adult) Danish phonological forms was inversely related to the number of Danish words over the period studied. That is, the children had their own vocabulary early in development that was eventually replaced by target forms.

Note that in the simulation, the number of non-words actually *increases* during the acquisition of the target vocabulary. Further analysis reveals a simple explanation for this effect. We assign each string of the network's stem outputs to one of 8 classes given by the possible combinations of {C∨V}, {C∨V} and {C∨V}. Three of these classes characterize the target language's syllabic structure which consists of 41 CVC, 7 CCV, and 5 VCC strings (there is one less than 54 due to a homophone). A plot of the number of network outputs that belong to each of these classes over training is shown in Figure 7. The network quickly learns

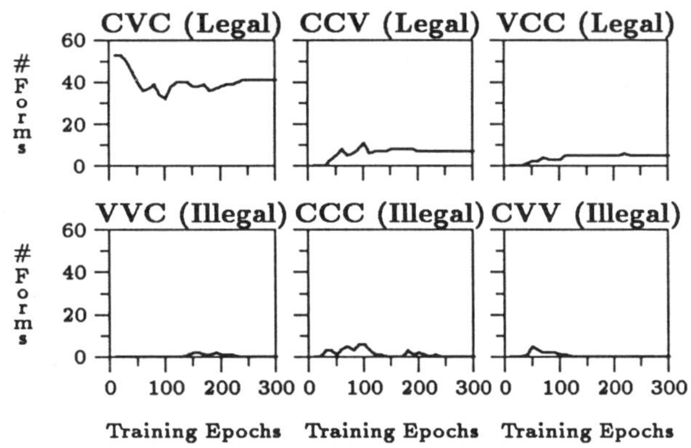


Figure 7: Syllabic Structures

the dominant CVC form of the language, overgenerating strings in this class initially. In order to extend to the target language syllabic structure these strings have to mutate from CVC to CCV and VCC during epochs 30 through 120. Strings changing from CVC to CCV have to change the mid-vowel to a consonant and the *coda* consonant to a vowel. The possible intermediate forms are CCC and CVV. Similarly for CVC to VCC, the possible intermediate forms are CCC and VVC. Indeed, a graph of the number of these forms produced by the network shows that they occur only during the cross-over from pseudo-words to words. The (logically possible) forms VVV and VCV never occur.

## Similarity Effects

We hypothesized that the synonym groups would produce outputs that were more similar to one another than the non-synonym groups and the vocabulary as a whole. In order to test this hypothesis, we use the following measure of within-group similarity:

$$\text{Similarity}(\mathcal{G}) = 1 - \frac{1}{N(N-1)} \sum_{\substack{i,j \in \mathcal{G} \\ i \neq j}} \mathcal{RD}_{i,j}$$

where $i, j$ range over the members of $\mathcal{G}$, and $\mathcal{G}$ is of size $N$. $\mathcal{RD}_{i,j}$ is a relative distance measure given by:

$$\mathcal{RD}_{i,j} = \frac{dist^{O}_{i,j}}{dist^{T}_{i,j} + 1}$$

This is the distance between the output phonemes in a string *relative* to the distance they should be after learning. $\mathcal{RD}_{i,j}$ should tend to 1 as the network learns, so Similarity($\mathcal{G}$) should tend to 0.

We apply this similarity measure to the strings produced by the network for each of the 9 prototype classes. For comparison purposes, we subtract the within-group similarity of the *total* output of the network from each score. Figure 8 (a) shows the average of this measure across the low-distortion (synonym) classes and the average across the high- and medium-distortion classes. The curves show that, in general,

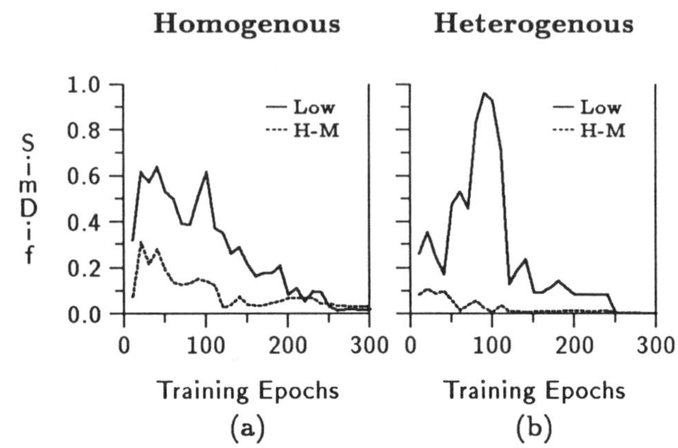


Figure 8: Synonym Analyses

semantic classes produce surface forms that are more cohesive than the forms of the network are as a whole. However, this effect disappears over training. This is the result of the network overcoming the false cue of input similarity. As expected, the synonym classes have higher within-group similarity throughout training than non-synonym classes. It is noteworthy that during the period that the network is actually acquiring the target vocabulary, i.e. from 70–120 epochs (cf. Figure 6), the within-group similarity of the synonyms *increases* compared to the rest of the classes. That is, synonyms are forced to be near-homonyms.

This "squeezing" effect is greatly magnified when there is only one synonym (Low distortion) cluster out of the nine classes (Figure 8 (b)). The explanation in terms of the network's organization is that patterns that are more easily learned (because they have less input similarity) are dominating the error gradient. The patterns may be characterized as *competing for representational resources* at the first hidden layer.

Examination of the network outputs over the training period reveals that the output strings for synonym classes during the steepest rate of target acquisition are within 2 or 3 features of one another. An interesting question here is: What is the string the outputs of a synonym class are pushed towards? Is it a blend of all of the strings of the class, or does one string in the class "capture" the output for that class? We tentatively find that when there is one lexical item that is more frequent than the others, it captures the class. Interestingly, this is not the case if the more frequent item in a synonym class is an arbitrary verb, probably due to the fact that they map similar inputs to very different outputs (see Bartell, Cottrell & Elman, this volume for a thorough discussion of this issue). In the case that all are equally frequent, a blend of all of the outputs for the class is produced.

Taken in combination, these results suggest an unorthodox account of the source of the non-word forms found in Plunkett's subjects. These consistently-used pseudo-words are the result of two constraints or pressures on the child's language production: (1) A pressure to produce forms that are in keeping with the syllabic structure of the language at the output level, and (2) a pressure to produce similar forms based on input similarity. The child is thus producing the best approximation to a word in the language that is a blend of *all* of the words for that semantic class, with a tendency for this blend to be similar to the most frequent element of that class. A second counter-intuitive prediction of this work is that, during acquisition of the correct forms, the child will produce strings that may be inappropriate for the target language because they are *between* a common (over-acquired) form and a less common form.

## Conclusions

We have described a connectionist model of morphology acquisition in which input forms representing the semantics of words are mapped to sequences of outputs representing their phonological forms. The network is successful in producing appropriate forms, even in the case where the input forms have a similarity stucture that is independent of the output similarity structure. Furthermore, the learning curves indicate a spurt-like acquisition profile. There is ample evidence for the spurt-like nature of vocabulary growth (McShane [1979]). It is unclear whether the acquisition of inflectional morphology in children shows a similar non-linear growth to that observed in the network.

The network experiences difficulty in generalizing from the stems to past tense forms. The model predicts that children are better at generalising from past tense forms to stems than *vice versa*. Further analysis is needed to investigate what modifications must be made to the model in order to achieve good generalization in the structured input case.

The analysis of the influence of input and target similarity on the acquisition of phonological form suggests some radical predictions. Children's non-adult forms may be a result of blending words for the same category. Looked at another way, words are distorted by their neighbors in a semantic class. The effects of similarity at the phonological level suggest that children will produce forms that do not belong to the syllabic structure of their language if these forms are between the most common form in the language and other forms. Finally, the model suggests that during the vocabulary burst, synonyms will be forced to be homonyms.

One problem that we have avoided addressing in this work, and suggested by other research (McClelland, personal communication) is that such models have difficulty learning uneven length strings. We plan to investigate ways to overcome this limitation in future research.

## References


Bybee, J. & Slobin, D. I. [1982], "Rules and schemas in the development and use of the English past tense," *Language* 58, 265–289.

Chauvin, Y. [1988], "Symbol Acquisition in Humans and Neural (PDP) Networks," University of California, San Diego, PhD Thesis.

Elman, J. L. [1990], "Finding structure in time," *Cognitive Science* 14, 179–211.

Gasser, M. & Lee, C. [1990], "Networks that Learn about Phonological Feature Persistence," *Connection Science* 2, 265–278.

Marchman, V. [1988], "Rules and regularities in the acquisition of the English past tense," *Center for Research in Language Newsletter* 2.

McShane, J. [1979], "The development of naming," *Linguistics* 17, 879–905.

Pinker, S. & Prince, A. [1988], "On language and connectionism: Analysis of a parallel distributed processing model of language acquistion," *Cognition* 28, 73–193.

Plunkett, K. [1990], "The segmentation problem in early language acquisition," CRL, University of California, Center for Research in Language Newsletter, San Diego.

Plunkett, K. & Marchman, V. [1991], "U-Shaped learning and frequency effects in a multi-layered perceptron: Implications for child language acquisition," *Cognition* 38, 1–60.

Plunkett, K., Marchman, V. & Knudsen, S. L. [1990], "From Rote Learning to System Building: Acquiring Verb Morphology in Children and Connectionist Nets," in *Proceedings of the 1990 Connectionist Models Summer School*, D. S. Touretzky, J. L. Elman, T. J. Sejnowski & G. E. Hinton, eds., Morgan Kaufmann, San Mateo, CA.

Rumelhart, D. E. & McClelland, J. L. [1986], "On learning the past tense of English verbs," in *Parallel distributed processing: Explorations in the Microstructure of Cognition, #2: Psychological and Biological Models*, J. L. McClelland, D. E. Rumelhart & PDP Research Group, eds., MIT Press, Cambridge, MA, 216–271.

# What a Perceptron Reveals about Metrical Phonology

**Prahlad Gupta**
Department of Philosophy
Carnegie Mellon University
Pittsburgh, PA 15213
prahlad@cs.cmu.edu

**David S. Touretzky***
School of Computer Science
Carnegie Mellon University
Pittsburgh, PA 15213
dst@cs.cmu.edu

## Abstract

Metrical phonology is a relatively successful theory that attempts to explain stress systems in language. This paper discusses a perceptron model of stress, pointing out interesting parallels between certain aspects of the model and the constructs and predictions of metrical theory. The distribution of learning times obtained from perceptron experiments corresponds with theoretical predictions of "markedness." In addition, the weight patterns developed by perceptron learning bear a suggestive relationship to features of the linguistic analysis, particularly with regard to iteration and metrical feet. Our results suggest that simple statistical learning techniques have the potential to complement, and provide computational validation for, abstract theoretical investigations of linguistic domains.

## Basics of metrical theory

This section outlines the structure of the syllable and the basics of metrical theory[1]. A syllable is composed of an *onset*, which contains the material before the vowel, and a *rime*. The rime is composed of a *nucleus*, which contains the vocalic material, and a *coda*, which contains any remaining non-vocalic material.

A syllable may be *open* (ends in a vowel) or *closed* (ends in a consonant). Viewing syllable structures as trees, an open syllable has a *non-branching* rime (the rime has a nucleus, but not a coda), while a closed syllable has a *branching* rime (the rime has both a nucleus and a coda). A syllable may also have a *long* vowel, in which case the nucleus is considered to branch.

In many languages, stress tends to be placed on certain kinds of syllables rather than on others; the former are termed *heavy* syllables, and the latter *light* syllables. What counts as heavy or light differs across languages but, most commonly, a heavy syllable is one with a branching rime ([Goldsmith 1990, p. 113]). However, it is possible for other properties to contribute to syllable weight. For example, in some languages, only syllables with a long vowel (i.e., branching nucleus) count as heavy. Closed syllables with short vowels do not count as heavy, as they would in the more commonly-occurring heavy/light distinction ([Goldsmith 1990, p. 179]).

Languages that distinguish between heavy and light syllables are termed *quantity-sensitive*, while languages that do not make this distinction are termed *quantity-insensitive*. There seems to be theoretical agreement that, in quantity-sensitive systems, the placement of stress is sensitive to the structure of the rime, but not the onset. We follow this assumption[2]. Thus rime structure is taken to be the basic level at which accounts of stress systems are formulated. Stress patterns are controlled by metrical structures built on top of rime structures. The version of metrical structure adopted here is *metrical feet*. We assume the *parameters* formulated by Dresher & Kaye ([Dresher & Kaye 1990, p. 142]):

(P1) The word-tree is strong on the [Left/Right]
(P2) Feet are [Binary/Unbounded]
(P3) Feet are built from the [Left/Right]
(P4) Feet are strong on the [Left/Right]
(P5) Feet are Quantity-Sensitive (QS) [Yes/No]
(P6) Feet are QS to the [Rime/Nucleus]
(P7) A strong branch of a foot must itself branch [No/Yes]
(P8) There is an extra-metrical syllable [Yes/No]
(P9) It is extra-metrical on the [Left/Right]
(P10) A weak foot is defooted in clash [No/Yes]
(P11) Feet are non-iterative [No/Yes]

As an example of the application of these parameters, consider the stress pattern of Maranungku, in which primary stress falls on the first syllable of the word and secondary stress on alternate succeeding syllables. Figure 1 shows an abstract representation of a six-syllable word, with each syllable represented as $\sigma$. The assignment of stress is characterized as follows. Binary, quantity-insensitive, left-dominant feet are constructed iteratively from the left edge of the word. Each foot has a "strong" and a "weak" branch (labeled "S" and "W," respectively, in the figure). The strong, or dominant branch assigns stress to the syllable it dominates. Since the feet are left-dominant, odd-numbered syllables are assigned stress. Over the roots of these *metrical feet*, a left-dominant *word-tree* is constructed, which assigns stress to the structure dominated by its leftmost branch. The third and fifth syllables are each dominated by the dominant branch of one metrical structure (a foot), while the first syllable is dominated by the dominant branches of two structures (a foot, and the word-tree). Even-numbered syllables are dominated only by non-dominant branches of feet. The result is that even-numbered syllables receive no stress; the third and fifth syllables receive one degree of stress (secondary stress); and the first syllable receives two degrees of stress (primary stress.) The parameter settings characterizing Maranungku are: [P1 Left], [P2 Binary], [P3 Left], [P4 Left], [P5 No], [P7 No], [P8 No], [P10 No], [P11 No]. Parameters P6 and P9 are irrelevant because of the settings of parameters P5 and P8, respectively.

## Sixteen stress systems

Six quantity-insensitive (QI) languages and ten quantity-sensitive (QS) languages were examined in our experiments. The data, summarized in Table 1, were taken primarily from [Hayes 1980]. Note that the QI stress patterns of Latvian & French, Maranungku & Weri, and Lakota & Polish are mirror images of each other. The QS stress patterns of Malayalam &

*The second author was supported by a grant from Hughes Aircraft Corporation, and by the Office of Naval Research under contract number N00014-86-K-0678.

[1]For an overview of metrical phonology, see [Goldsmith 1990, chapter 4], [Kaye 1989, pp. 139-145], [van der Hulst & Smith 1982] or [Dresher & Kaye 1990, pp. 140-147].

[2]See, for example, ([Dresher & Kaye 1990, p. 141] or [Goldsmith 1990, p. 170]). However, both [Davis 1988] and [Everett & Everett 1984] present evidence that onsets may in fact be relevant to the placement of stress.

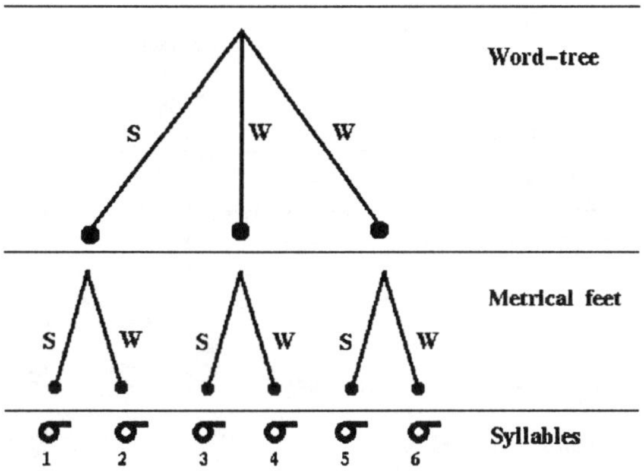


Figure 1: Metrical structures for a six-syllable word in Maranungku.

Yapese, Ossetic & Rotuman, Eastern Permyak Komi & Eastern Cheremis, and Khalka Mongolian & Aguacatec Mayan are also mirror images[3].

## Learning stress with a perceptron

In separate experiments, we taught a perceptron to produce the stress pattern of each of the sixteen languages. The domain was limited to single words, as in the previous learning models of metrical phonology developed by Dresher & Kaye, and Nyberg ([Dresher & Kaye 1990, Nyberg 1989]). Again as in the other models, the effects of morpho-syntactic information (such as lexical category) were ignored, and the simplifying assumption was made that the only relevant information about syllables was their weight.

Words up to 7 syllables long were slid across a 13-element input window as shown in Figure 2. At each time step, the perceptron was trained to predict the stress of the center element. In order to distinguish heavy from light syllables, each input element consisted of two units. Thus the perceptron had a total of 26 inputs, plus a bias connection. The input patterns used were [1 0] for a heavy syllable, [0 1] for a light syllable, and [0 0] for no syllable.[4] The output targets used in training were 1.0 for primary stress, 0.5 for secondary stress, and 0 for no stress.

The input data set for all stress systems consisted of all 255 word-forms of up to seven syllables. Each word was processed one syllable at a time. Connection weights were adjusted at each time step using the back-propagation learning algorithm of [Rumelhart, Hinton & Williams 1986]. One *epoch* consisted of one presentation of the entire training set. The network was trained for as many epochs as necessary to ensure that the stress value produced by the perceptron was within 0.1 of the target value, for each syllable of the word, for all words in the training set. A *learning rate* of 0.05 and *momentum* of 0.90 was used in all simulations. Initial weights were uniformly distributed random values in the range $\pm 0.5$.

[3]We have somewhat simplified the descriptions of Polish and Malayalam compared with those in [Halle & Vergnaud 1987, pp. 57-58] and [Hayes 1980, p. 66, 109]. However, this does not detract from our discussion in any way, as stress systems corresponding to our simplifications are reported to exist: Swahili ([Halle & Clements 1983, p.17]) and Gurkhali ([Hayes 1980, p.66]), corresponding to Polish and Malayalam, respectively.

[4]Note that the distinction between heavy and light syllables is irrelevant in a quantity-insensitive language. However, this representation was used in all simulations to maintain consistency across quantity-sensitive and quantity-insensitive systems.

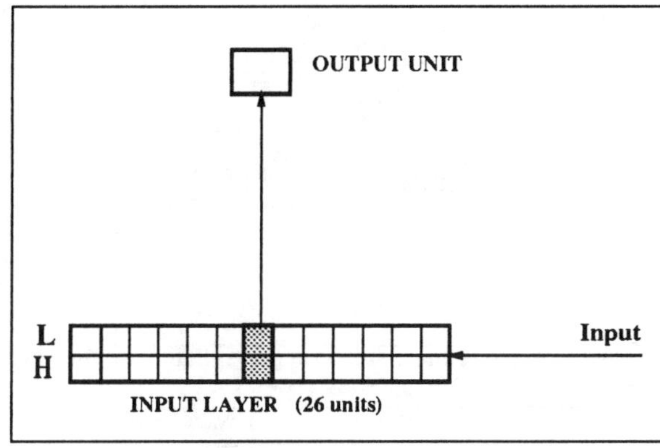


Figure 2: Perceptron model used in simulations.

Each simulation was run at least three times, and the learning times averaged. Results are shown in Table 2.

## Markedness, learnability, and simulations

A universal grammar (UG) of stress should incorporate a theory of markedness, so as to predict which features of stress systems are at the core of the human language faculty and which are at the periphery. The distributional approach to markedness treats as "unmarked" those linguistic forms that occur more frequently in the world's languages. This seems to be the approach taken by, for example, Hayes[5].

Such an approach can be criticized, however, on the grounds that the frequency of occurrence of some linguistic form does not necessarily determine its status as "core" or "peripheral", and the non-occurrence of some form does not show that it is "impossible." The distribution of languages in the world is a function of all kinds of historical, non-linguistic, factors, and does not necessarily have linguistic-theoretic significance[6].

Another approach to markedness is *learnability theory*, which examines the logical process of language acquisition[7]. Thus, for example, Dresher & Kaye take iteration to be the default or unmarked setting for parameter P11, because there is evidence that can cause revision of this default if it turns out to be the incorrect setting: the absence of any secondary stresses serves as a diagnostic that feet are *not* iterative ([Dresher & Kaye 1990, p. 191]). If non-iteration were the default, their learning system might not encounter evidence that would enable it to correct this default setting, if it were in fact incorrect. It should be noted that, while this is a representative application of subset theory, the choice of default parameter values depends on the particular learning algorithm employed.

[5]([Hayes 1980, p. 50]): "In justifying a foot inventory as the unmarked one, a minimal requirement is to show that all the members of the inventory are attested in a fair number of languages ..."

[6]To quote Pullum ([Pullum 1982, p. 343; p. 340]): "... no one has any idea to what extent the history of the human race has skewed the distribution of [linguistic] types by skewing the distribution of people ... to postulate a default assumption that, say, *wh*-movement cannot be rightward, merely because it is commoner (in currently well-studied languages) for it to be leftward, is surely perverse as well as unnecessary. Language acquisition takes place within the infant, not within the context of a statistical survey of currently attested languages ..."

[7]As an example, the Subset Principle ([Berwick 1985], [Wexler & Manzini 1987]) has implications for markedness. Suppose that two possible settings $a$ and $b$ for parameter $P$ result in the learner respectively accepting sets $S_a$ and $S_b$ of linguistic forms. If $S_a$ is a subset of $S_b$, then, once $P$ has been set to value $b$, it will never get re-set to $a$, even if that was the correct setting. Unmarked values for parameters should therefore be the ones yielding the most constrained system.

| LANGUAGE | DESCRIPTION OF STRESS PATTERN | EXAMPLES |
|---|---|---|
| Latvian | Fixed word-initial stress. | $S^1S^0S^0S^0S^0S^0S^0$ |
| French | Fixed word-final stress. | $S^0S^0S^0S^0S^0S^0S^1$ |
| Maranungku | Primary stress on first syllable, secondary stress on alternate succeeding syllables. | $S^1S^0S^2S^0S^2S^0S^2$ |
| Weri | Primary stress on last syllable, secondary stress on alternate preceding syllables. | $S^2S^0S^2S^0S^2S^0S^1$ |
| Lakota | Primary stress on second syllable. | $S^0S^1S^0S^0S^0S^0S^0$ |
| Polish | Primary stress on penultimate syllable. | $S^0S^0S^0S^0S^0S^1S^0$ |
| Koya | Primary stress on first syllable, secondary stress on heavy syllables.<br>(Heavy = closed syllable or syllable with long vowel.) | $L^1L^0L^0H^2L^0L^0L^0$<br>$L^1L^0L^0L^0L^0L^0L^0$ |
| Eskimo | (Primary) stress on final and heavy syllables.<br>(Heavy = closed syllable.) | $L^0L^0L^0H^1L^0L^0L^1$<br>$L^0L^0L^0L^0L^0L^0L^1$ |
| Malayalam | Primary stress on first syllable except when first syllable light and second syllable heavy.<br>(Heavy = long vowel.) | $L^1L^0L^0H^0L^0L^0L^0$<br>$L^0H^1L^0H^0L^0L^0L^0$ |
| Yapese | Primary stress on last syllable except when last is light and penultimate heavy.<br>(Heavy = long vowel.) | $L^0L^0L^0H^0L^0L^0L^1$<br>$L^0H^0L^0H^0L^0H^1L^0$ |
| Ossetic | Primary stress on first syllable if heavy, else on second syllable.<br>(Heavy = long vowel.) | $H^1L^0L^0H^0L^0L^0L^0$<br>$L^0L^1L^0L^0L^0L^0L^0$ |
| Rotuman | Primary stress on last syllable if heavy, else on penultimate syllable.<br>(Heavy = long vowel.) | $L^0L^0L^0H^0L^0L^0H^1$<br>$L^0L^0L^0L^0L^0L^1L^0$ |
| Komi | Primary stress on first heavy syllable, or on last syllable if none heavy.<br>(Heavy = long vowel.) | $L^0L^0H^1L^0L^0H^0L^0$<br>$L^0L^0L^0L^0L^0L^0L^1$ |
| Cheremis | Primary stress on last heavy syllable, or on first syllable if none heavy.<br>(Heavy = long vowel.) | $L^0L^0H^0L^0L^0H^1L^0$<br>$L^1L^0L^0L^0L^0L^0L^0$ |
| Mongolian | Primary stress on first heavy syllable, or on first syllable if none heavy.<br>(Heavy = long vowel.) | $L^0L^0H^1L^0L^0H^0L^0$<br>$L^1L^0L^0L^0L^0L^0L^0$ |
| Mayan | Primary stress on last heavy syllable, or on last syllable if none heavy.<br>(Heavy = long vowel.) | $L^0L^0H^0L^0L^0H^1L^0$<br>$L^0L^0L^0L^0L^0L^0L^1$ |

Table 1: Stress patterns: description and example stress assignment. Examples are of stress assignment in seven-syllable words. Primary stress is denoted by the superscript 1 (e.g., $S^1$), secondary stress by the superscript 2, and no stress by the superscript 0. "S" indicates an arbitrary syllable, and is used for the QI stress patterns. For QS stress patterns, "H" and "L" are used to denote Heavy and Light syllables, respectively.

If learnability arguments should propose the setting $x$ for parameter $P$, then some explanation would be needed if 95 per cent of the world's languages could be analyzed as having the setting $y$ for the same parameter. Although distributional observations may not be an appropriate starting point for theory construction, they do provide a set of additional data points. However, given the previously noted criticisms, it appears they can only provide a weak constraint on metrical theories. Some other source of evidence would be valuable.

It is therefore interesting to note that the simulations described in this paper do provide "learnability" results for a variety of stress patterns. By extension, they make predictions about the learnability of various linguistic forms in metrical phonology. We claim that these results provide a source of data that can complement theoretical investigations.

Table 2 shows the stress systems grouped by their theoretical analyses in terms of the parameter scheme discussed in the first section. The last column of the table shows the average learning time in epochs for each group of stress patterns. As can be seen, there appears to be a fairly systematic differentiation of learning times for groups of stress patterns with different clusters of parameter settings.

First of all, learning times appear to be significantly higher for stress systems in groups 5 through 9, which have non-iterative feet, than for those in groups 1 through 4, which either do not have metrical feet at all, or else have iterative feet. This makes the interesting prediction that non-iterative feet are more difficult to learn, and hence marked. This prediction corresponds with both Halle & Vergnaud's Exhaustivity Condition[8], and with the choice of marked and unmarked settings in Dresher & Kaye's parameter scheme (Parameter P11)[9].

Comparison of learning times for group 1 vis-a-vis groups 2, 3 and 4 suggests that a stress system with only a word-tree (i.e., with no metrical feet) is easier to learn than one with (iterative) metrical feet.

The dramatic difference in learning times between groups 8 and 9 suggests that it is marked for the dominant node to be obligatorily branching[10]. Group 8 differs from group 9 only in not having obligatory branching, and average learning times were 218 epochs vs. 2308 epochs.

This prediction agrees with the distributional view that obligatory branching is relatively marked[11]. However, it runs counter to Dresher & Kaye's choice of default values (parameter P7)[12].

However, comparison of group 6 with group 7 suggests that systems with obligatory branching may be more *easily* learned: group 6, with obligatory branching, has a learning time of 19

[8] "The rules of constituent boundary construction apply exhaustively ..." ([Halle & Vergnaud 1987, p. 15]).

[9] In Dresher & Kaye's model, iteration is the default or unmarked parameter setting, because there is evidence that can cause revision of this default. The absence of any secondary stresses serves as a diagnostic that feet are *not* iterative ([Dresher & Kaye 1990, p. 191]).

[10] This means that the strong branch of a foot must dominate a heavy syllable, and cannot dominate a light one.

[11] [Hayes 1980, p. 113]: "... the maximally unmarked labeling convention is that which makes all dominant nodes strong ... the convention that wins second place is: label dominant nodes as strong if and only if they branch ..."

[12] Obligatory branching is the default because evidence (the presence of any stressed light syllables that do not receive stress from the word-tree) can force its revision ([Dresher & Kaye 1990, p. 193]).

| | LANGUAGE | CHARACTERIZATION | EPOCHS |
|---|---|---|---|
| 1 | Latvian, French | Word-tree, no feet | 1 |
| 2 | Maranungku, Weri | Word-tree, iterative binary QI feet | 2 |
| 3 | Koya | Word-tree, iterative unbounded QS feet | 2 |
| 4 | Eskimo | No word-tree, iterative unbounded QS feet | 3 |
| 5 | Lakota, Polish | Word-tree, non-iterative binary QI feet | 9 |
| 6 | Malayalam, Yapese | Word-tree, non-iterative binary QS feet, dominant node branches | 19 (±1) |
| 7 | Ossetic, Rotuman | Word-tree, non-iterative binary QS feet | 29 (±1) |
| 8 | Komi, Cheremis | Word-tree, non-iterative unbounded QS feet | 218 (±1) |
| 9 | Mongolian, Mayan | Word-tree, non-iterative unbounded QS feet, dominant node branches | 2308 (±4) |

Table 2: Learning times for QI and QS stress patterns grouped by theoretical analysis. Each figure is the average learning time for languages in the group.

epochs, compared with group 7, without obligatory branching, but with a learning time of 29 epochs. This runs counter to the distributional argument, but agrees with the learnability view.

Two points are worth noting. First, it is interesting that where there is a conflict between the distributional and learnability theory predictions of markedness, there is also conflicting evidence from the perceptron simulation. Second, these conflicting perceptron results highlight the fact that it may be infeasible to analyze the effects of different settings for *individual* parameters; it may only be possible to make broader analyses of the effects of *clusters* of parameter settings. Strong interactions between parameters have also been observed in other computational learning models of metrical phonology (Eric Nyberg, personal communication).

However, in view of the greater differential in learning times between Groups 8 and 9 than between Groups 6 and 7, we conclude that the effect of obligatory branching is to *increase* learning time. That is, we view our learning results as supporting the markedness of obligatory branching. This raises the interesting possibility that learning results such as those from the present perceptron simulations can provide a new source of insight into questions of markedness. As previously noted, there is controversy over the relevance of distributional facts to theories of markedness. Moreover, the distributional view of the markedness of obligatory branching ([Hayes 1980, p. 113]) seems to conflict with the learnability view ([Dresher & Kaye 1990, p. 193]). The present simulations seem to agree with the distributional view, and, we suggest, provide a potential means of validation for theoretical analyses.

This contribution to theoretical analysis can be further illustrated for the stress systems of Lakota and Polish, which are mirror images. The analysis so far adopted for Lakota is that it has non-iterative binary right-dominant QI feet constructed from left to right, with a left-dominant word-tree[13]. Let us call this *Analysis A*. As illustrated in Figure 3, this leads to the construction of one binary right-dominant QI foot at the left edge

[13]This is based on Hayes' analysis of *penultimate* stress ([Hayes 1980, p. 55]).

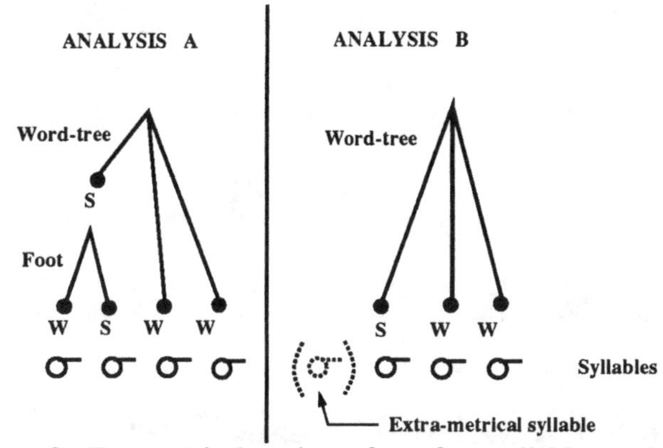


Figure 3: Two metrical analyses for a four-syllable word in Lakota. Strong branches are labeled "S", and weak branches "W".

of the word. This, together with the left-dominant word-tree, results in the assignment of primary stress to the second syllable. As has been shown, under this analysis, the perceptron learning results support the markedness of non-iteration (recall the differing learning times of Groups 1 through 4, vs. Groups 5 through 9).

However, an alternative analysis is that Lakota has a left-dominant word-tree with no metrical feet, and the first syllable is extra-metrical ([Dresher & Kaye 1990, p. 143]). Let us call this *Analysis B*. As illustrated in Figure 3, the leftmost syllable is treated as "invisible" to the stress rules, and the word-tree assigns primary stress to the leftmost of the "visible" syllables. The result is that the second syllable receives primary stress. Under this analysis, Lakota and Polish (Group 5, in Table 2) differ from Latvian and French (Group 1 in Table 2) only in having an extra-metrical syllable. The differing learning times for the two groups (1 epoch vs. 9 epochs) then suggest that extra-metricality is *marked*. However, this runs counter to both the distributional view ([Hayes 1980, p. 82]) and the learnability-theoretic view ([Dresher & Kaye 1990, p. 191]).

To summarize, *Analysis A* views Lakota and Polish as having non-iterative feet, which both the distributional/theoretical and learnability approaches treat as marked. *Analysis B* views these stress patterns as having an extra-metrical syllable, which both approaches treat as unmarked. So far, there is nothing theory-external to help choose between the analyses. We claim that the present simulation results provide such a means: since the learning results are consistent with the theoretical markedness of non-iteration, but not with the unmarkedness of extra-metricality, they provide at least weak support for preferring *Analysis A* over *Analysis B*.

In summary, the present learning results are from a model whose initial state is devoid of any information about the constructs of metrical theory that characterize different stress systems. Nevertheless, the learning results exhibit interesting correspondences with theoretical predictions. These results suggest that computational modeling may have something to contribute to the development of a markedness theory and, more generally, to aspects of linguistic analysis.

## Analysis of weight patterns

In learning a stress pattern, the perceptron has acquired and encoded in its connection weights its "knowledge" of that pattern. Connection weights for all the languages studied are shown in Figure 4. Each display is a representation of the network as

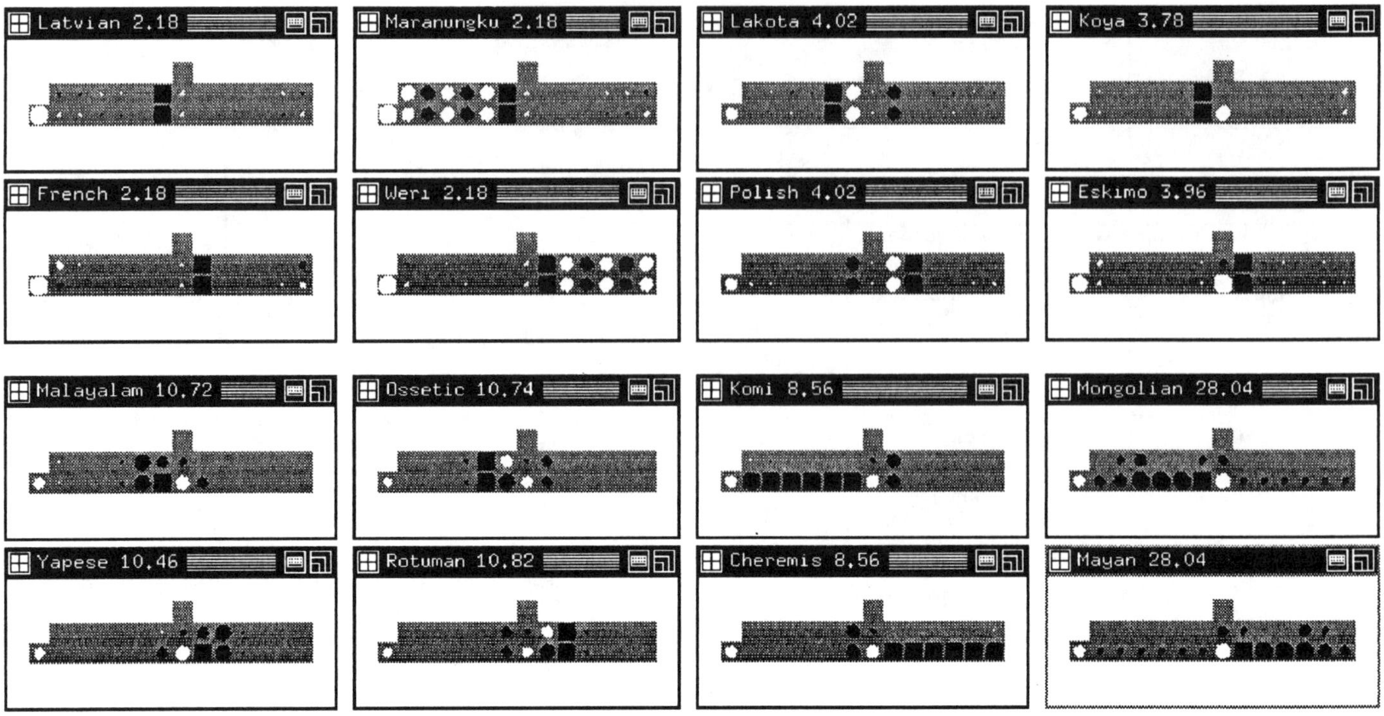


Figure 4: Learned connection weights for the sixteen stress patterns.

a whole. The large grey shaded rectangle represents the input layer of the network, with its two rows of input units. The single square protruding from the left of the input layer is the *bias* connection. The output unit is represented by the protruding square at the top of the input layer.

A blob in a particular position denotes a weight from the unit in that position to the output unit. White blobs denote positive weights, and black blobs negative weights. The size (area) of the blobs is proportional to the absolute magnitude of the weight. Weights are scaled so that the largest absolute magnitude is depicted in each display as a perfect square; other weights in that display appear as blobs of proportionate size. The scale (i.e., absolute magnitude of the largest weight) is shown in the title bar of each display. Thus, for Maranungku, the absolute magnitude of the largest weights is 2.18; these are the large (black) negative weights left of center in the input layer.

Just as with learning times, the fact that two stress patterns are mirror images of each other is reflected in the connection weights. Moreover, there seem to be correspondences between the form of the encoded knowledge and the characterization of the stress pattern in terms of parameters. Maranungku and Weri are the only stress systems with iterative binary feet (Group 2, Table 2). For these systems, but for no others, there is a very clear binary alternating pattern of positive and negative weights (see the alternation of black and white blobs in the weight displays for Maranungku and Weri). If, as is natural, we take a positive weight to correspond to the strong branch, and a negative weight to correspond to the weak branch of a foot, then for Maranungku we see left-dominant binary feet, and for Weri right-dominant binary feet – just as in the theoretical analysis[14]. It does not seem too far-fetched to say that the perceptron has discovered a version of iterative binary feet.

The single set of negative weights for Latvian and French (immediately to the left and right of center, respectively) can perhaps be interpreted as a left-dominant and right-dominant word-tree.

Recall that Lakota has non-iterative binary right dominant QI feet constructed from left to right, and that Polish has non-iterative binary left dominant QI feet constructed from right to left. That is, there will be a single binary tree, constructed at the left edge of the word for Lakota, and at the right edge, for Polish. Under this analysis, the weights to left and right of center for Lakota and Polish can be interpreted respectively as (single) right-dominant and left-dominant binary QI feet.

The weight patterns for Koya and Eskimo are close to mirror images, but not completely symmetric. Koya assigns primary stress to the first syllable and secondary stress to non-initial heavy syllables, while Eskimo assigns only one level of stress to final and heavy syllables. The chief theoretical difference between the two languages is that the former, but not the latter, has a word-tree. This difference is reflected in the fact that there are two magnitudes, or levels, of center connection weights for Koya (the large negative, and the smaller positive weights), whereas for Eskimo, there is only one level of weights (the positive and negative weights at the center are approximately equal.) This can be viewed as analogous to the two levels of metrical structure in Koya (metrical feet and word-tree) vs. the single level of structure in Eskimo (metrical feet only.)

Table 2 shows that Malayalam, Yapese, Ossetic and Rotuman (Groups 6 and 7) are the only languages with non-iterative binary QS feet. These are also the only patterns that have more than two large negative weights grouped together to the left (for Malayalam and Ossetic) or right (for Yapese and Rotuman) of center. We can take these three-or-four negative weight struc-

[14]As discussed previously, Maranungku has binary, left-dominant QI feet constructed iteratively from the left edge of the word. Weri has binary, right-dominant QI feet constructed iteratively from the right edge of the word.

tures to correspond to a non-iterative binary QS foot. There is a clear structural difference as compared with the (analogues of) non-iterative binary QI feet in the weights for Lakota and Polish.

Komi, Cheremis, Mongolian and Mayan are the only languages with non-iterative unbounded QS feet (Groups 8 and 9, Table 2). The connection weights for these systems show a pattern of nearly-identical negative weights spanning a set of several input units, and such a pattern does not occur for any other of the stress systems. Such a set of "spanning" weights seems analogous to an unbounded foot. The pattern of weights for Komi seems to correspond to an unbounded right-dominant QS foot, while weights for Cheremis seem to correspond to an unbounded left-dominant QS foot (note the single positive weight at the *right* and *left*, respectively, of the sets of weights, similar to the dominant branch of the foot). The difference in analysis between Komi & Cheremis and Mongolian & Mayan is that feet in the latter pair have obligatory branching, meaning the strong node of the foot must dominate a heavy syllable. As for Komi and Cheremis, the weights for Mongolian and Mayan show a pattern that can be interpreted as an unbounded QS foot. However, they additionally have a set of weights adjacent to the positive weight (i.e., to the "dominant branch" of the unbounded foot), which are not present for Komi and Cheremis; these additional weights can loosely be interpreted as corresponding to a branching dominant node.

It should not be too surprising that correspondences can be found between learned weight patterns and metrical constructs: both representations are derived from the same data in an attempt to explain or model the same phenomena. Nonetheless, we find it somewhat encouraging that these correspondences exist.

## Conclusions

One of the supposed attractions of connectionist learning is that networks can start out with no theoretical preconceptions, developing their representations *de novo*. In truth, though, any computational model is founded on preconceptions, and ours is no exception. We have preprocessed the data to remove all detail from the words except for the heavy/light syllable distinction. In addition, we assume that all the syllables of the word are simultaneously accessible, i.e., that the word fits in a buffer, and that the stress assignment process can access the entire buffer in parallel. These are not unreasonable assumptions, but neither are they indisputable.

We have tried to show that, within the range of observed human stress systems, our model exhibits interesting behaviors, providing a fresh source of insight about linguistic stress. However, the model does not accurately reflect human capabilities for processing stress systems. For example, it is incapable of learning some existing human stress patterns (e.g., Warao, Southern Paiute), and is capable of learning stress patterns that are unlikely ever to exist amongst human languages. (However, as has been argued here, distributional evidence is in itself only weakly suggestive of what is "impossible"). Thus, our model is not suitable for defining the boundaries of human abilities, but this is scarcely surprising given that we are merely optimizing a linear threshold function.

Nevertheless, this simple model has produced interesting results. Although the simulations did not incorporate the foot and word tree constructs of metrical phonology as computational primitives, there appear to be correspondences between the performance of the model and the analyses of metrical theory. First, the learning results correspond with theoretical predictions of markedness. The theoretical predictions, as they stand, may be based either on distributional facts or on learnability theory. Our simulation results provide predictions of markedness that are based on actual "learning." Second, there seem to be parallels between theoretical characterizations of various stress systems and the knowledge of those stress systems encoded in the perceptron's weights. These parallels are at least suggestive of a mapping between the level of representation at which the perceptron performs its computations and the level of investigation at which metrical phonology is formulated.

In conclusion, these experiments suggest that simple statistical learning techniques such as the perceptron model have the potential to provide computational validation for, and to complement, theoretical investigations of domains of language such as stress systems.

## Acknowledgments

We would like to acknowledge the feedback provided by Deirdre Wheeler throughout the course of this work. The first author would like to thank David Evans for access to exceptional computing facilities at Carnegie Mellon's Laboratory for Computational Linguistics, and Dan Everett, Brian MacWhinney, Jay McClelland, Eric Nyberg, Brad Pritchett and Steve Small for helpful discussion of earlier versions of this paper. Of course, none of them is responsible for any errors.

## References

[Berwick 1985] Berwick, R. 1985. *The Acquisition of Syntactic Knowledge*. Cambridge, Mass.: MIT Press.

[Davis 1988] Davis, S. 1988. *Topics in Syllable Geometry*. New York: Garland.

[Dresher & Kaye 1990] Dresher, B. E. & Kaye, J. D. 1990. A Computational Learning Model for Metrical Phonology. *Cognition* 34: 137-195.

[Everett & Everett 1984] Everett, D. & Everett, K. 1984. On the Relevance of Syllable Onsets to Stress Placement. *Linguistic Inquiry* 15: 705-711.

[Goldsmith 1990] Goldsmith, J. A. 1990. *Autosegmental and Metrical Phonology*. Oxford, England: Basil Blackwell.

[Halle & Clements 1983] Halle, M. & Clements, G. 1983. *Problem Book in Phonology*. Cambridge, Mass.: MIT Press.

[Halle & Vergnaud 1987] Halle, M. & Vergnaud, J.-R. 1987. *An Essay on Stress*. Cambridge, Mass: MIT Press.

[Hayes 1980] Hayes, B. 1980. A Metrical Theory of Stress Rules. Ph.D. diss., Dept. of Linguistics, Massachusetts Institute of Technology.

[Kaye 1989] Kaye, J. D. 1989. *Phonology: A Cognitive View*. Hillsdale, NJ: Lawrence Erlbaum.

[Nyberg 1989] Nyberg, E. 1989. Weight Propagation and Parameter Setting, Ph.D. thesis proposal, Dept. of Philosophy, Carnegie Mellon University.

[Pullum 1982] Pullum, G. 1982. Letter, *Linguistics* 20: 339-344.

[Rumelhart, Hinton & Williams 1986] Rumelhart, D., Hinton, G. & Williams, R. 1986. Learning Internal Representations by Error Propagation. In D. Rumelhart, J. McClelland & the PDP Research Group. *Parallel Distributed Processing, Volume 1: Foundations*. Cambridge, Mass.: MIT Press.

[van der Hulst & Smith 1982] van der Hulst, H. & Smith, N. 1982. An Overview of Autosegmental and Metrical Phonology. In H. van der Hulst & N. Smith (eds.). *The Structure of Phonological Representations, Part I*. Dordrecht, The Netherlands: Foris.

[Wexler & Manzini 1987] Wexler, K., & Manzini, M. 1987. Parameters and Learnability in Binding Theory. In T. Roeper & E. Williams (eds.). *Parameters and Linguistic Theory*, Dordrecht, The Netherlands: Reidel.

# A connectionist model of auditory word perception in continuous speech

**Richard Shillcock**
Centre for Cognitive Science
University of Edinburgh
2 Buccleuch Place
Edinburgh
rcs@cogsci.ed.ac.uk

**Joe Levy**
HCRC
University of Edinburgh
2 Buccleuch Place
Edinburgh
joe@cogsci.ed.ac.uk

**Nick Chater**
Department of Psychology
University of Edinburgh
7 George Square
Edinburgh
nicholas@cogsci.ed.ac.uk

### Abstract

A connectionist model of auditory word perception in continuous speech is described. The aim is to model psycholinguistic data, with particular reference to the establishment of lexical percepts. There are no local representations of individual words: feature-level representations are mapped onto phoneme-level representations, with the training corpus reflecting the distribution of phonemes in conversational speech. Two architectures are compared for their ability to discover structure in temporally presented input. The model is applied to modelling the phoneme restoration effect and phoneme monitoring data.

## Introduction

The recognition of a spoken word involves correctly matching some representation derived from the acoustic input, often of poor quality, with some stored representation. The nature of the recognition problem is partly determined by the frequency distribution of elements in spoken language. This paper presents a model of how these distributional statistics can effect the recognition of spoken words. The aim is to capture as much of the relevant psycholinguistic data as possible in terms of statistical properties of streams of phonetic and phonemic input, without adverting directly to explanation at the word level[1].

## The process of lexical access

Speech sounds arrive over time and must be matched against some kind of stored representation. Psycholinguistic theories have variously proposed that such input is represented in terms of features, phonemes, morphemes and words. The Cohort Model (Marslen-Wilson & Welsh 1978; Marslen-Wilson 1987) has generated a "lexicalist-localist" tradition in which the incoming signal is seen as directly contacting specific lexical representations, which are activated in proportion to their match with the input. These activated representations then compete until one of them is uniquely distinguished by the input and/or is integrated into the ongoing interpretation of the utterance. That is, it is assumed that contacting the lexical entry for the word *automatic*, for instance, makes available all of its associated information (pronunciation, orthography, semantics, and so on) and that once *automatic* is alone in the cohort (at some point during its third syllable), then it exclusively determines processing from that point until the end of the word.

Within this approach, less has been said about the development of the representation which makes initial contact with the lexical entry (/ɔ/, /ɔt/, /ɔtə/... for *automatic*). There are different claims concerning the types of information which may influence the activation of lexical representations, the effect of non word-initial partial matches (/ɔt/ in the input matching *porter* or *short* in the lexicon), the mechanism which mediates competition between activated lexical representations, and the continuing role of representations which cease to match the input. However, one computationally explicit account, the TRACE model (McClelland & Elman 1986), based on the early Cohort Model, has captured many aspects of human spoken word recognition in a principled way and represents a coherent stance on the issues mentioned above. For instance, constraining the activation of lexical representations and segmenting the continuous input are two major issues in spoken word recognition, and TRACE provides a computationally explicit answer to both.

Three aspects of TRACE suggest avenues for further research. First, the implementation of TRACE is limited to 15 different phonemes; it is desirable to be able to model the full scale and richness of human word recognition both to handle real discourse and to be able to assess the model's performance on actual stimulus materials taken from psycholinguistic experiments. Second, TRACE does not learn; TRACE's knowledge of the language is confined to the word frequency values which are built into its lexicon. Third, TRACE's lexical level mediates influence between adjacent phonemic material. An alternative possibility is that processing may be adequately captured just by statistical dependencies at the phonemic level. For example, there is a TRACE/Cohort prediction that monitoring of a word-medial phoneme, like /t/ in *curtail* should be facilitated to the extent that it lies on, or close to, the word's

[1] We would like to thank the following: Geoff Lindsey for advice concerning the phonemic and featural descriptions, Steve Finch for assistance and advice with bigram and trigram statistics, Alex Monaghan for the use of the CSTR text-to-phonemes program, Ellen Bard, Henry Thompson and Richard Rohwer for valuable discussion.

uniqueness point: at this point only one lexical representation remains in the cohort, the listener has all the information necessary to identify the word and its constituent phonemes. In a TRACE simulation, *curtail* would become the most highly activated word node at or just beyond the uniqueness point and would supply top-down information to the phoneme-nodes representing its constituent phonemes. If, however, it could be demonstrated that the perception of the /t/ in *curtail* could be captured by a model which had no local lexical nodes, and had had no previous experience of the word *curtail*, then we would be justified in preferring this simpler explanation.

The model advanced below is motivated by the belief that it is necessary to account for as much of the data as possible on the basis of processing in which there are no explicit exclusive lexical representations. The model makes no distinction between representations of any frequent sequence, whether it be specifically a morpheme, a syllable, a word or an idiom. This model builds on the recent departure from the lexicalist-localist view of lexical access, involving a distributed connectionist model, as described in the next section . This perspective does not rule out the possibility that explicit specific lexical representations might be necessary to account for certain data. Only after investigating exhaustively how much of the data can be accounted for by a model which does not possess such representations can the role of lexical representations in explaining psycholinguistic data be properly assessed.

More recently a second computationally explicit model of spoken word recognition has been presented by Norris (1988), which employs an architecture also investigated by Elman (1988, 1990). In Norris' model feature-level representations of consecutive phonemes are mapped onto local representations of words via one layer of hidden units. Recurrent connections from the hidden units copy their pattern of activity to a set of state nodes which then re-present this pattern to the hidden units at the next time-step. Thus the network has the potential to respond to patterns of phonemes across time. (This approach is described in more detail below.)

Norris's model is architecturally more elegant than TRACE. It learns the frequency of the words it can recognize from its training set, and, as Norris (1988) reports, it captures a range of "cohort behaviours". In simulations with miniature lexica the model generally assigns a spread of activity to all words which are congruent with the input up to and including the current phoneme. At the uniqueness point of a word the model generally opts overwhelmingly for that word and maintains its level of activation until the end of that word in the input. Again, three aspects of the model suggest possible further research. First, the simulations which Norris reports are all with small scale lexica (each word which the model can recognize is given a specific output node). Second, there is a considerable volume of psycholinguistic data which addresses infra-lexical processing (*e.g.* phoneme-monitoring) and which it is not feasible to model using the activation levels assigned to whole words which are the output of Norris' model. Third, the inclusion of a specifically lexical level, the output level of Norris' model, is unduly constraining on any wider model of sentence processing. It prevents the model from learning sub-word regularities, except to the extent that word-initial similarity occasions activation of a cohort, and also super-word regularities, unless strings like *that's*, or *out of* or *good morning* are lexicalized by being given a dedicated output node.

The model we present in Section 4 resembles Norris' model in network architecture and in involving a mapping from feature-level descriptions. Crucially, however, the model does not contain local representations of words; a number of advantages spring from this fact. Before describing the model, however, it is necessary to motivate further the exclusion of local lexical representations.

## Seidenberg and McClelland's model of pronunciation

Considerable coverage of psycholinguistic data has been achieved with Seidenberg and McClelland's (1989) connectionist modelling of word naming and visual word recognition. In modelling naming, orthographic representations are mapped onto phonological representations; there are no local representations of words, only weights between the three layers of nodes. An extension of the model, involving an identity mapping of the orthographic level, captures aspects of visual word recognition. Crucially the training regime reflects the frequency with which the words appear in the language.

How can word recognition data be modelled when there are no local representations of words and therefore no activation levels which might be assigned to specific words? As an example, Seidenberg and McClelland argue that in tasks in which subjects are required to discriminate between orthographically regular words (*e.g. fellow, tanker*) and orthographically irregular nonwords (*e.g. fnrkte, jplerhn*), their performance may be accurately modelled by the accuracy with which an identity mapping may be made between two orthographic levels mediated by a layer of hidden units and trained by back-propagation.

In the model of spoken word recognition described below, an analogous approach is taken within the auditory domain: feature-level representations are mapped onto phoneme-level representations. The training regime is taken from spoken discourse and reflects the frequency with which speech sounds corresponding to phonemes occur and co-occur in spoken language.

## The problem of learning the structure of temporal sequences

Neural network methods have been developed largely to learn to classify static patterns. Since many important aspects of cognition involve processing temporally structured sequences, there has been considerable attention devoted to extending network methods to learn the structure of time-varying sequences. One strategy is simply to represent a "moving window" of past inputs explicitly (used, for example in the NETTALK model of reading (Sejnowski & Rosenberg 1987)). Explicit buffering of past input can be avoided by using recurrent connections which recirculate past input so that it may continue to have an influence on network performance, rather than being "flushed through" the network. There has been much recent work on various ways in which back-propagation can be generalized to recurrent networks (Rumelhart, Hinton & Williams 1986; Almeida 1987; Pineda 1987; Pearlmutter 1990). In some of these regimes, learning occurs when the net has settled into a stable pattern, and in some continuous time signals are used. Rumelhart, Hinton & William's original suggestion, "back-propagation through time", which simply unfolds a recurrent network into many copies arranged in a feed-forward architecture and applies standard back-propagation to the result, is the most appropriate for this kind of task. The more time-steps back the network is unfolded, the better the network will be able to learn to respond to temporally remote information, but at greater computational expense.

## The model

The model, illustrated schematically in Fig. 1, simply consists of a mapping between two levels of representation within the auditory modality – a feature level and a phoneme level.

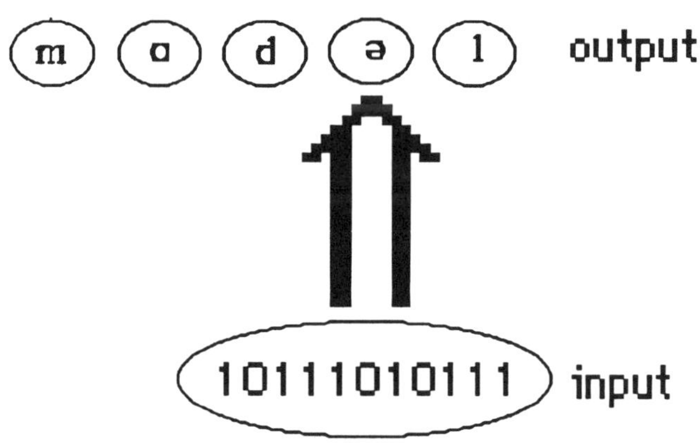


**Figure 1: Feature to phoneme mapping. The input is a binary feature-level description of the current segment. The output is the identity of the current phoneme, the predicted next phoneme and a number of previous phonemes.**

Input is a bundle of features corresponding to the segment at that point in time. This input is replaced by successive bundles of features corresponding to consecutive segments across time. The output is a phoneme-level description of the current segment, together with a prediction of the identity of the next segment and a confirmation of the identity of the last several phonemes. The simplest version of the model possesses a three-segment output window – current, predicted and last.

Two phonemes may share a number of features (/p/ and /t/ are identical on all but one). In the hypothetical case where two phonemes possess the same feature-level instantiation (as a result of noise, for instance), the only way in which they may be distinguished is by their surrounding context: thus the phoneme designated * will be classified as /s/ in /y e */ and as /f/ in /i */, given that its feature-level description is ambiguous between /s/ and /f/ and it has been trained on the words *yes* and *if*. In the model, each phoneme possessed a unique feature-level description. This resulted in "current" identifications approximating to 100% and remaining at that level in the "last" position, in which confirmation was expected. In human speech perception, however, segments are often underspecified since the signal is noisy, and information relevant to the identification of any one segment is often spread over several surrounding segments. In some of the simulations below we rely on the addition of noise to the signal during training to encourage the network to rely on the phonemic context since the information may no longer be encoded in a single segment.

The mapping in Fig. 1 may be achieved by means of any one of a family of networks which are sensitive to structure across time. The minimum sensitivity to such structure involves replicating simple bigram probabilities. The extent to which the networks under study are sensitive to more than bigram probabilities is an empirical question. Comparison with simple bigram and trigram statistics is a powerful means of assessing the performance of the models (although detailed analysis is not reported here). Finally, it is important that the network be tested with the full extent of the feature to phoneme mapping in the language. Below we report results using the Elman/Norris net and a feedforward net incorporating a moving window.

## The networks

The basic mapping between the two levels of representation was achieved using a "cut-down" version of back-propagation through time (Rumelhart, Hinton & Williams 1986), unfolding the network once rather than many times (Servans-Schreiber, Cleeremans & McClelland 1989; Chater 1989) and thus sacrificing the ability reliably to pick up long distance dependencies, in exchange for speed of training. This "copyback" structure (Fig. 2) was introduced by Elman (1988, 1990) and Norris (1988) .

There were 11 input units, 15 hidden units, and hence 15 corresponding "copyback" units (which retained the hidden unit activations from the previous timeslice) and 108 output units (coding the 36 phonemes at the previous, current and next time-step). For a qualitative comparison, a simple "moving-windows" architecture, also with 15 hidden units (with 22 additional input units representing the phonetic features at the previous two time steps), was implemented. The architecture was that of a standard feedforward network, trained with back-propagation, where the input layer represents not just the current phonetic input, but also the phonetic input at previous time steps. In our simulations, the moving window extended over three time-steps.

Three simulations are reported: the recurrent network architecture was trained both on noisy and non-noisy inputs, and the moving window architecture was trained only on non-noisy input data. Using only the bigram statistics of the stream of input data, and assuming that error on the previous and current phoneme is 0, the best total sum of squares error over the training corpus is 8044 steps. Of course, by considering higher order statistics, better performance is possible, though a large corpus is required to obtain reliable higher order statistics.

This figure is comparable with the performance of the networks: 8023 for the recurrent network trained on the noise-free corpus, 9788 when the training data is noisy and 7892 for the moving windows architecture with noise-free input. The slight difference in performance between the recurrent network and the moving windows architecture results from better performance of moving windows at outputting the previous phoneme – in this architecture the phonetic input at the previous time step is presented as part of the input, whereas the recurrent network must learn to buffer this information. That network performance is comparable with the results of a bigram analysis does not of course imply that the network is responding only to bigram structure – in fact, it seems more likely that it is picking up some higher order statistics of the input, while not perfectly accounting for bigram statistics. This is currently being tested by comparing network performance of the net on real data versus *n*th-order approximations to that data.

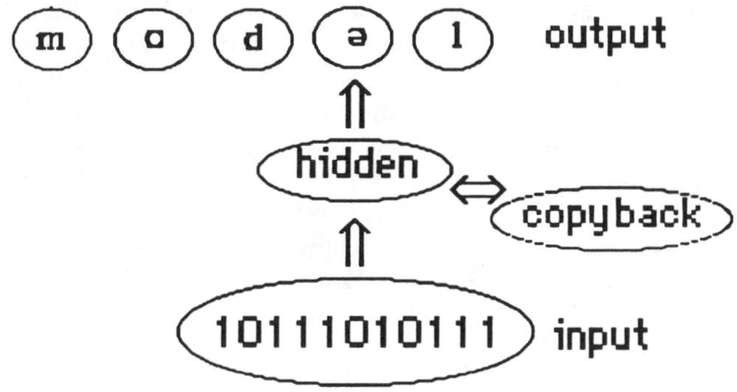


**Figure 2. Structure of the Elman/Norris network. Input is the feature-level representation of the current segment. Output is the classification of the current segment, prediction of the next segment, and confirmation of the last several segments. The copyback units re-present to the hidden units the pattern of activation on those units at t-1.**

In all the simulations reported, a learning rate of 0.1 was used, and momentum was not used. Each network was trained until it began to show signs of overfitting – training that resulted in a decrease of error for the training set but led to increasing error for a separate test set was disregarded. This required between 500 and 600 epochs. In some simulations, the phonetic input was made noisy, by randomly changing 9% of the input values from 0 to 1 or *vice versa*. The noise was generated on-line and was different for every epoch of training. The learning phase of the simulations was quite computationally intensive, using 30-40 CPU hours on a variety of SUN SPARC-based machines, using a customized version of the Rumelhart and McClelland (1988) simulation package.

### The corpus and training regime

The initial, limited training data was derived from some 3490 words of spoken discourse, taken largely from the LUND Corpus (Svartik & Quirk 1980). The discourse was transcribed at the word level and included filled pauses, false starts and corrections. The training set was made up of 9097 phonemes and a test set of 3285 phonemes was used to test generalisation/overfitting. In some later simulations the training data have been generated from a 33,000 word phonemic dictionary, containing frequency information; this allowed better exposure to open-class words while sacrificing some of the character of the distribution of closed-class words.

### The phoneme-level descriptions

The utterances were converted to idealized phonemic representations using the CSTR text-to-phoneme program and employing 36 different phonemes based on those of the CSTR Machine Readable Phonetic Alphabet. The eight dipthongs were each converted to sequences of two phonemes.

### The feature-level descriptions

The phonemic transcription was then converted to an idealized feature-level representation, consisting of the following 11 features based on those of Jakobson, Fant and Halle (1952): *vocalic/non-vocalic, consonantal/non-consonantal, voiced/unvoiced, discontinuous/continuant, strident/mellow, nasal/oral, diffuse/non-diffuse, compact/non-compact, tense/non-tense, grave/acute, flat/plain.* Thus the phonemes /ə/ and /I/ were represented on the respective features as below.

| | | | | | | | | | | | |
|---|---|---|---|---|---|---|---|---|---|---|---|
| ə= | 1 | 0 | 1 | 1 | 0 | 0 | 0 | 0 | 0 | 1 | 0 |
| I = | 1 | 0 | 1 | 1 | 0 | 0 | 1 | 0 | 0 | 0 | 0 |

All 36 phonemes were given a value of 1 or 0 for each of the 11 features. The final form of the training corpus was of a continuous stream of such feature-level descriptions of segments, with no information being given concerning word boundaries.

## Behaviour of the networks

To illustrate the behaviour of the networks with unseen discourse, consider the test string *this is a test of the model*. When this was converted into sequences of feature descriptions and given to the networks, all three networks demonstrated sensitivity to phonotactic constraints, with greater probabilities of predicting the next phoneme in the short sequences of closed-class words and in the unstressed syllable of *model*. The models all predict the next phoneme more accurately when the sequence is from normal discourse than when the same phonemes are presented in random order. The effect of training with noise was to depress the scores given to phonemic hypotheses and to increase the number of hypotheses which received a non-zero score. In the simulation reported here, this increased noise did not result in better performance in terms of the ranking of the correct hypothesis within the total list of hypotheses.

It is easier to see the networks' use of context in the case of the classification of the current phoneme, where all of the information necessary for the unique identification of the phoneme is present on that presentation. For the Elman/Norris net trained with noise, performance was worse in the scrambled phonemes case compared with the normal discourse case (mean phoneme scores were 66.6 and 79.1, respectively; $t = 2.35$, $df = 18$, $p = .031$). We may conclude that in the noise condition this network was relying on previous context to identify the current phoneme; when this context was aberrant, it hindered correct recognition. The Elman/Norris net trained without noise and the moving-window net both generated reduced mean "current" phoneme scores for the scrambled input but neither of the differences was significant.

Human listeners employ context both before and after the phoneme in question. The aim of training with noise was to force the network to rely on both "left" and "right" context. The scores for the phoneme in "past" position were compared on the normal and abnormal discourse. While there was no significant difference between the two mean scores for the Elman/Norris net trained without noise, the version trained with noise was significantly worse on the abnormal discourse (means were 82.7 and 63.2, respectively; $t = 3.88$, $df = 18$, $p = .001$.). The network was sensitive to right context in classifying phonemes, and was misled by abnormal right context.

## Modelling psycholinguistic data

### Phoneme restoration

Listeners' perception of degraded individual speech sounds in words is often restored (Warren 1970). Restoration is strongest when the intended phoneme and the replacing sound (*e.g.* white noise, a click, silence) are similar, and when replacement occurs after the uniqueness point of the word. Otherwise the effect is not compelling.

This was modelled by putting minimally different test words, like *got* and *gop*, in the carrier sentence *...and the next word is x and the next word is y and the....* For frequent words like *got*, *this*, and *yes*, there was no substantial restoration. For example, when the current phoneme was /i/ in *thif*, the prediction for /s/ was 14 and the prediction for /f/ was 4 (*if* is a frequent word); when the current phoneme became /f/, it was scored at 98, compared with 2 for /s/. The frequency of *this* was not enough to overturn the bottom-up information.

Restoration was observed, however, when the value of the critical feature distinguishing /f/ and /s/ was replaced by 0.5. Input, for the *this/thif* case, was then /ð i */ where * was completely ambiguous between /f/ and /s/. In this case, the current scores were 99 for /s/and 5 for /f/, changing to 98 and 0 respectively at confirmation.

The model respects the input. It does not hallucinate phonemes on the basis of word frequency. This captures the effect more accurately than TRACE, in which lexical level reinforcement restores any degraded phoneme and even overturns bottom-up evidence, converting *vocabulaty* to *vocabulary*

Phoneme restoration also occurred based purely on the right context. When the input was /* e s/, in which * was ambiguous between /y/ and /r/, the model restored the /y/.

### Monitoring for word-medial phonemes

Simulations were run to test whether the model predicted the data from an experiment (Shillcock *submitted*) in which subjects monitored for word-medial phonemes like /p/ in *repel* or /p/ in *lapel*. Subjects in the experiment took significantly longer to respond to phonemes in monomorphemic words compared to matched prefixed words.

The stimulus materials from the experiment were embedded in the context *...and the next word is...and the next word is...* and presented to the three trained networks. Activations for the critical phoneme in each word (/p/ in *repel*) were recorded when that phoneme was in "next", "current" and "past" position. Only the moving window network gave significantly different mean activations for the monomorphemic words and the prefixed words, mirroring the human data (97.7 and 98.7, respectively; $df = 14$, $t = 2.137$, $p = .05$). This difference occurred in the "current" position, reflecting the fact that response times were facilitated if the sequence of phonemes up to and including the critical phoneme represented a prefix as opposed to a monomorphemic word beginning. When activations in the simulations for the individual stimulus items were compared with the mean response times from the experiment, there was no significant correlation. Shillcock (*submitted*) reports that the best predictor ($r = -.477$) of the phoneme-monitoring data was the (frequency-weighted) number of times the sequence of phonemes up to and including the critical phoneme (*i.e.* /r I p/ for *repel*) occurred in a large phonemically transcribed dictionary (*i.e.* /r I p/ in *script, report, unrepentant*..., all weighted by word frequency). The networks were therefore not employing information as relevant as the bigram and trigram information available from a large phonemic dictionary. This may reflect most on the size of the training corpus.

## Conclusions

Initial testing of the model gives encouraging results, with the various simulations from the different architectures demonstrating desirable behaviours. Many of its limitations may be due to the modest size of the training corpus: the 3490 words in the corpus represented 905 different words. A very large corpus will be required to ensure adequate coverage of the open-class vocabulary. Training with a corpus of transcribed discourse opens the possibility of studying what special processing of the closed-class vocabulary may emerge; the literature contains numerous claims concerning the special status of the closed-class vocabulary compared with the open-class vocabulary. Yet further work will be required concerning the amount and nature of (idealized) phonological reduction in the corpus.

Regarding network architecture, it may be that the most promising avenue for future research lies in a network which allows back-propagation through time for several rather than just one time-step. We are currently exploring this avenue.

The model described is seen as part of a larger model incorporating semantic representations. The absence of explicit, localist lexical representations is crucial. We envisage a mapping from the phonemic output of the model to semantic representations, mediated only by a layer of hidden units, although this clearly raises a serious binding problem in the absence of word boundary information. This arrangement gives more scope than a lexicalist-localist model for modelling the details of effects in which homophones and partial homophones produce brief erroneous priming.

In conclusion, it may be that many psycholinguistic phenomena which have been taken to involve access to specific representations of spoken words may be explained in terms of the low-level statistical structure of the phonetic/phonemic input, as picked up by a simple neural network account. We are currently applying this model to a range of other experimental phenomena. There is a methodological imperative within psycholinguistic research to allow "higher level" interpretation of empirical data only when low level explanations can be ruled out.

## References


Almeida, L. B. 1987. A learning rule for asynchronous perceptrons with feedback in a combinatorial environment.. In Proceedings of the IEEE First International Conference on Neural Networks, San Diego, California, June, 1987.

Chater, N. 1989. Learning to respond to structures in time, RIPRREP 1000/62/89 Research Initiative in Pattern Recognition, St Andrews Road, Malvern, Worcs., U.K.

Chater, N. Responding to temporal structure. Submitted to *Cognition.*

Chater, N. & Ganis, G. Double dissociation in distributed systems. Submitted to *Psychological Review*

Elman, J. L. 1988. Finding structure in time. Technical Report, CRL TR 8801, Centre for Research in Language, UCSD.

Elman, J. L. 1990. Finding structure in time. *Cognitive Science*, 14:179-211.

Jakobson, R., G. Fant and M. Halle 1952. Preliminaries to speech analysis. Technical Report 13, M.I.T. Acoustics Laboratory, Mit Press.

Jordan, M. 1986. Serial order: a parallel distributed approach. Institute for Cognitive Science Report, 8604, University of California, San Diego.

Juang, B. H. 1984. On the hidden Markov model and dynamic time warping for speech recognition - a unified view. *Bell Systems Technical Journal*, 63, no.7, 1213-1243.

Marslen-Wilson, W. 1987. Functional parallelism in spoken word-recognition. In U. H. Frauenfelder & L. K. Tyler eds *Spoken word recognition*, 71-102, Special Issue of Cognition.

Marslen-Wilson, W. & Welsh, A. 1978. Processing interactions and lexical access during word recognition in continuous speech. *Cognitive Psychology*, 10, 29-63.

McClelland, J. L. & Elman J. L. 1986. Interactive processes in speech perception: the TRACE model. In D. E. Rumelhart & J. L. McClelland eds. *Parallel Distributed Processing,* Vol. 2., 58-121, Cambridge, Mass: MIT Press.

McClelland, J. L. & Rumelhart, D. E. 1988. *Explorations in Parallel Distributed Processing: Models, Programs and Exercises*. Cambridge, Mass: MIT Press.

Minsky, M. & Papert, S. 1969. *Perceptrons*. Cambridge, Mass: MIT Press.

Norris, D. G. 1988. A dynamic-net model of human speech recognition. Talk given at the Cognitive Models of Speech Processing Workshop, Sperlonga 1988. See G. Altmann ed. 1990, *Cognitive Models of Speech Processing*, MIT Press.

Patterson, K. E., Seidenberg, M. S. & McClelland, J. L. 1989. Connections and disconnections: acquired dyslexia in a computational model of reading processes. In R. G. M. Morris ed., *Parallel distributed processing: implications for psychology and neurobiology*. Oxford, Oxford University Press.

Pineda, F. 1987. Generalization of back-propagation to recurrent and higher order neural networks. *Neural Information Processing Systems*, New York.

Seidenberg, M. S. & McClelland, J. L. 1989. A distributed, developmental model of word recognition and naming. *Psychological Review*, 96, 523-568.

Sejnowski, T. J. & Rosenberg, C. R. 1987. Parallel networks which learn to pronounce English text. *Complex Systems*, 1, 145-160.

Servans-Schreiber, D., Cleeremans, A. & McClelland, J. L. 1989. Learning sequential structure in simple recurrent networks in D. Touretsky ed. *Advances in Neural Information Processing Systems*, Vol 1, Morgan Kaufman, Palo Alto, 643-653.

Svartvik, J, & Quirk, R. 1980. *A corpus of English conversation*. Lund: Gleerup.

Shillcock, R. C. 1990. The processing of prefixed words: a connectionist account. Submitted to *Memory & Cognition*

Waibel, A., Hanazawa, T., Hinton, G. E., Shikano, K. & Lang, K . 1987. Phoneme recognition using time-delay neural networks. ATR Technical Report TR-I-0006 Japan: ATR Interpreting Telephony Research Laboratory.

Warren, R. M. 1970. Perceptual restoration of missing speech sounds. *Science*, 167, 392-393.

# A Computational Model of Connotation-Based Revision

Wing-Kwong C. Wong

AI Lab*
Department of Computer Sciences
The University of Texas at Austin
Austin, TX 78712-1188

Center for High Performance Computing
The University of Texas System
Balcones Research Center

## Abstract

Previous studies on text generation and revision seldom consider emotions and social relations as motivations for linguistic variations such as word choice and syntactic structure arrangement. This paper proposes a computational model that uses four attributes *affect, activity, power,* and *emphasis* to revise texts. These attributes express ideological beliefs, the connotations of lexical items, and connotation propagation properties of sentence structures. With these formal tools, an algorithm of backward chaining revises sentences with sensible choices of word and sentence structures. The model provides a basis for future research that can lead to a fully automated text revision system.

**Topic areas:** Text generation and revision, ideology, lexical and sentential connotations.

## Introduction

In the early 1970's, research on linguistic variations during text generation started with the works on paraphrase generation from semantic nets ([Simmons and Slocum, 1972], [Goldman, 1975]). But this approach did not address the problem of how to determine the fitness of paraphrase in a text nor did it consider social context for the text.

Artificial intelligence research on ideology modeling was done in [Abelson, 1973] and [Carbonell, 1978]. This approach focused on the planning of actions with given ideologies but did not consider the ideological effects on linguistic variations when describing the actors and their behavior.

Computational issues on text generation based on pragmatic and social factors are addressed in [Hovy, 1988]. His system PAULINE makes use of conversation setting such as: time availability and formality, the expertise of speaker and hearer, their social relationship, and speaker's goals in affecting hearer's emotions towards speaker. These parameters influence linguistic decisions like topic collection and organization, sentence grammatical arrangement, and word choice. These parameters characterize the relationship between the writer and the reader and say little about the connotations on the topics in the text (with the exception of the parameter of affect). This paper proposes a text revision model that focuses on the connotations imposed on the textual topics. Hence, our model extends the logic of PAULINE to make better writing decisions.

Underlying PAULINE is an important message that many computational linguistic studies have neglected — there is significant relationship between language and society. This is in sharp contrast to the more commonly recognized relationship between language and thought. This paper takes a social perspective of language, as urged by Whorf ([Carroll, 1956]), Fowler, Kress, Hodge and Drew, ([Fowler *et al.*, 1979], [Kress and Hodge, 1979]), etc. From this perspective, language plays a major role in the social construction of reality, manipulates as well as informs, and distorts reality to serve class and individual interests. The sociological linguistic studies cited above provide plenty of empirical texts where people use language to achieve their social objectives and to communicate their "social reality" with systematic choices of words, syntactic and case relations, and information.

Our strategy is to identify a small number of social attributes to express the ideological beliefs of a text, the connotations of individual words and phrases, and the connotations assigned to topics in a sentence structure. The major attributes of interest are **affect, activity, power** and **emphasis.** Based on the author's understanding of lexical connotations from sources such as [Hayakawa, 1968] and the Webster's New World Dictionary, these attributes are defined in the following way. Affect refers to people's general feeling towards events, people, institutions, or ideals, e.g., happiness is positive affect (+aff), tragedy negative (-aff). Activity refers to people's activeness in performing an action, e.g., fighting an enemy is active (+act) and resting is inactive (0act). Power refers to

*This work was partially sponsored by the Army Research Office under contract DAAG29-84-K-0060 for the AI Lab.

any physical, emotional, social or spiritual force, e.g., riot is more forceful than disorder, which is more forceful than trouble. We define riot to have emphasized power (+pow, +emp), disorder non-emphasized power (+pow, 0emp), and trouble no power (0power). Emphasis refers to the degree of connotation, e.g., the adverb 'very' and 'extremely' connote positive emphasis, while 'merely' and 'just' connote de-emphasis(-emp). The subtle distinctions between the above synonyms and hyponyms are undoubtedly relative and dependent on cultural framework and communication situation. So the values of 'positive' or 'plus' (+), 'negative' or 'minus' (-), and 'neutral' (0) assigned to words and concepts are arbitrary to certain degree. Future research must sharpen these distinctions with a refined scale or more attribute types. For the purpose of this paper, we assume a convention on the attribute values of words and concepts exists for our particular communication situation.

## Language and Ideology

Briefly speaking, an ideology is a body of related beliefs about social agents, institutions, and relationships from a specific social perspective. Almost every real-life text is produced with some ideological beliefs and reflects the beliefs. As an illustration, let us examine the news reports from two newspapers, the *Sun* and the *Morning Star* (hereafter referred to as the *Star*) about the disorder at the Notting Hill Carnival in London in 1977 ([Drew, 1979]). Because of ideological differences, two contrasting pictures about the carnival resulted:

1. *Sun* August 30, 1977:

   *INTO BATTLE! RIOT SHIELDS OUT AS THE POLICE STORM CARNIVAL MOB.*
   *Two hundred police carrying riot shields and truncheons last night charged a rioting mob of black youths at London's Notting Hill Carnival. More than 70 policemen were injured, one stabbed, before the Special Patrol Group officers cleared the trouble spot at Acklam Road — flash-point of last year's riot in which 600 were injured.*
   *The 10-minute riot began when youths charged a police cordon. Hurling bottles and bricks they burst through the thin blue line....*

2. *Star* August 30, 1977:

   *FIGHTING MARS END OF CARNIVAL AFTER A DAY OF PEACE.*
   *Police observers hovering in a helicopter above the huge crowds at London's Notting Hill Carnival yesterday estimated between 200,000 and 250,000 people were taking part.*
   *For most of the time and for the majority of the people it was a happy, peaceful occasion.*
   *But by 9p.m. some streets in Notting Hill had become a battlefield, with the police mounting a massive operation to clear them.*
   *The trouble started about 8p.m. at the top of the Portobello Road, near Westway. A scuffle erupted into the police cordon....*

Clearly, the two reports asserted two very different set of connotations about the Carnival, police, and people who fought with the police. Each set reflects a certain network of ideological beliefs about the government, people and other social elements. These beliefs affect many of the linguistic and information choices of the reports.

## Input Text and Writing Decisions

A text generation system usually takes some conceptual structures as inputs, e.g., relational data base ([McKeown, 1985]), conceptual dependency ([Hovy, 1988]), and frames ([Paris, 1988]). These conceptual structures generally contain concrete, denotative meanings such as physical motions and attributes but not connotative meanings such as emotions and social relationships. The text generated from these structures is like a first draft, which is to be revised with considerations on connotative meanings given as additional input. This level of revision is what our model focuses on.

The model accepts input text that is pre-analyzed and represented in terms of syntactic relations and sorted case relations. A major task of the model is to consider word choices to satisfy the given ideological beliefs. If no word choices can satisfy the desired connotations, an alternative sentence structure is tried until success is achieved. If no structures can satisfy the wanted connotations, deletion of the sentence from the text is suggested.

All input sentences from the initial draft are represented with a conjunction of grammatical relations (e.g., subject, object and preposition) and case relations (e.g., agent, affected entity, location, attributeOf). Each object is sorted as some general class (e.g., action, occasion, attribute, entity). For instance, a sentence in the draft *A mob rioted at the end of the Carnival* is represented as:

- **v:act:v1 –subj:agt→ n:entity:a1**
  The general subject:agent relation signifies an action undertaken by an entity, which can be people, organization, non-human objects, etc. The particular relation, for the clause 'mob rioted', specifies a mob a1 who performed the action v1 of rioting.
- **v:act:v1 –p:orient:p1→ n:occ:e2**
  The general preposition:orientation relation signifies a relation between the action and the occasion from a particular perspective such as 'action *at* occasion' or 'action *after* occasion'. The particular relation, for the phrase 'rioted at the end', specifies a rioting action at the end of the Carnival occasion.
- **n:att:e2 –p:attOf:p2→ n:occ:e1**
  The general preposition:attributeOf signifies that

something is an attribute of an occasion. The particular relation, for the phrase 'the end of the Carnival', focuses on the end e2 of the Carnival occasion e1.

During revision, the synonyms and hyponyms of the words in this sentence are considered. For examples, the noun 'mob' has a generalization 'people', the verb 'riot' has a less forceful synonym 'disorder' and an even less forceful synonym 'make trouble.' Also, the phrase 'the end of' is grammatically optional. But pragmatically, the phrase increases the distance between the riot and the Carnival, thus de-emphasizing the negative affect associated with the Carnival due to the riot. Moreover, an alternative structure is 'there-be' construction with nominalization of the action verb with agent deletion. A combination of all the above alternatives generates 18 near-paraphrases, one of which is to be selected during revision. These candidate sentences assign connotations to the agents or occasion mentioned in the sentence in very different ways. A good target sentence must satisfy the given ideological beliefs. We summarize the 18 sentences as below, with the abbreviation notation "{A | B | C}" meaning one of A, B, or C can be selected when revising the sentence:

- **{A mob | People} {rioted | caused disorder | made trouble} at the end of the Carnival.** (6 choices)
- **{A mob | People} {rioted | caused disorder | made trouble} at the Carnival.** (6 choices)
- **There was {a riot | a disorder | trouble} at the end of the Carnival.** (3 choices)
- **There was {a riot | a disorder | trouble} at the Carnival.** (3 choices)

## Representation of Ideological Beliefs

Our revision model expresses ideological beliefs with the attributes of affect (aff), activity (act), power (pow), and emphasis (emp). The two news reports about the Notting Hill Carnival quoted earlier reflect the following ideological beliefs:

1. The *Sun* hated (-aff) the dangerous rioting (-aff, +act, +pow, +emp) mob, who thus carried the attribute values **{-aff, +act, +pow, +emp}**. The *Star* de-emphasized the affect, activity, and power of the people (-aff, +act, +pow) involved in the disorder. They did not have any common characteristics such as "black" or "youth" and they were not (-emp) even identified as causing agents of violent actions (+act, +pow). So for the Star, these people carried **{-aff, +act, +pow, -emp}**.
2. The *Sun* disliked (-aff) the Carnival because it gave the mob a chance to riot (-aff, +act, +pow). So the Carnival carried the social values **{-aff, +act, +pow, 0emp}**. The Star increased the distance between the riot (+act, +pow) and the Carnival (0act, 0pow), and de-emphasized (-emp) the negative affect (-aff) of the Carnival, which thus carried **{-aff, 0act, 0pow, -emp}**.
3. The *Sun* liked (+aff) the police, who were well equipped (+pow) to protect (+act) the people from the violence of the mob. The *Star* disliked (-aff) the armed (+pow) police who attacked (+act) people. So for the Sun, the police carried the attribute values **{+aff, +act, +pow, 0emp}** and for the Star, the police carried **{-aff, +act, +pow, 0emp}**.

From the perspective of text understanding, these ideological beliefs are abstractions of the two newspaper reports. Some time after reading such reports, the reader is likely to forget about the details but still feel these abstracted connotations associated with the Carnival, the police and the people involved in the disorder. It is through numerous exposures to reports with similar ideological beliefs (such as that of a single newspaper or magazine) that the reader is indoctrinated. The ideological manipulation of language is very powerful indeed ([Fowler *et al.*, 1979], [Kress and Hodge, 1979]).

## Word Dictionary of Connotations

Given the ideological beliefs on the topics of a text, the text must be composed in a way to assert these beliefs. The connotations the text carries must finally depend on specific words in the text. Most words have synonyms and hyponyms, which stress different attributes or different values of the attributes. A writer's ability to differentiate these stresses, which are often quite subtle, is a major component of the writer's competence. If a computer is to write, as people do, about people and for people, it must possess knowledge about the social connotations of words.

Our model represents words, especially those carrying heavy connotations and have synonyms, with the social attributes of affect, activity, power, and emphasis. These attributes are undoubtedly not sufficient to distinguish many words. For instance, one other common attribute is formality ([Hovy, 1988]), which has more to do with the writer-reader relationship and communication setting than with the subject matter itself. But the four attributes do exhibit some generality in differentiating many words. In particular, the three attributes of affect, activity, and power follow the ***semantic differential*** model of [Osgood *et al.*, 1957], which found in psychological experiments people differentiated words with different values of evaluation, activity, and potency.

Our revision model uses a dictionary that specifies the social connotations of words. The dictionary is expedient for the purpose of this paper but not universal, as discussed in the introduction section. For the following sample lexical entries, the absence of an attribute implies that the value of the attribute is neutral, and "?" means a variable which can match any sign (+,

-, or 0) during connotation propagation in a sentence (see a later section):

- charge (verb): +act +pow (0aff 0emp)
- disorder (noun): -aff, +pow
- end (noun): ?aff, ?act, ?pow, -emp
- erupt (verb): +act +pow
- fight (verb): +act
- make (verb): +act
- mob (noun): -aff +pow
- people (noun): all neutral
- protect (verb): +aff +act
- riot (verb): -aff +act +pow +emp
- riot (noun): -aff +act +pow +emp
- scuffle (verb): +aff +act
- scuffle (noun): -aff +act
- trouble (noun): -aff
- very (adverb): ?aff, ?act, ?pow, +emp

## Connotation Propagation Properties of Sentence Structure

Consider the eighteen near-paraphrases given in an earlier section. Each candidate sentence assigns different values of social attributes to the Carnival and the people who took part in the disorder. Each particular sentence structure, with its syntactic and sorted case relations, assigns connotations to the sentential topics in its own way. The properties of these sentence structures can be expressed as rules. The precondition of a rule specifies the syntactic relation and the sorted case relations between two constituents of a sentence. The action of a rule specifies how the connotations of one constituent propagates to another. We show four sample rules below.

### V:Action-to-Subj:Agent Rule

If **v:act –subj:agt→ n:entity**,
then **n:entity := v:act ⊕ n:entity:lex**

The above assignment statement (":=") means to assign to the agent (n:entity) the vector addition ⊕ of the connotations of the action (v:act) and the connotations of the actual words used for the subject (n:entity:lex). A vector addition between two n-tuples is defined to be a n-tuple whose sign elements result from the addition of the corresponding sign elements of the original two tuples. A table for addition of two signs is given in Table 1, with the notation '?' referring to a variable that can match any sign during connotation propagation.

This rule applies to clauses like 'mob rioted' and 'police charged.' In the first case, the verb 'rioted' connotes {-aff, +act, +pow, +emp} and the noun 'mob' connotes {-aff, 0act, +pow, 0emp}. According to the rule, the mob is assigned the connotations of {-aff, +act, +pow, +emp}, the result of ({-aff, +act, +pow, +emp} ⊕ {-aff, 0act, +pow, 0emp}).

| Sign1 | Sign2 | Sign1 + Sign2 |
|---|---|---|
| + | + | + |
| + | 0 or ? | + |
| + | - | 0 |
| 0 or ? | + | + |
| 0 | 0 | 0 |
| 0 | ? | 0 |
| ? | 0 | 0 |
| ? | ? | ? |
| 0 or ? | - | - |
| - | + | 0 |
| - | 0 or ? | - |
| - | - | - |

Table 1: Sign Addition

### V:Action-to-Pp:Occasion Rule

If **v:act –p:orient→ n:occ** and
**p= 'at', 'in', 'during', or 'under'**,
then **n:occ := v:act ⊕ n:occ:lex**

This rule applies to phrases like 'made trouble at the end' and 'danced in the Carnival.' In the first case, the verb phrase 'made trouble' connotes {-aff, +act, 0pow, 0emp} and the noun 'end' connotes {?aff, ?act, ?pow, -emp}. The propagation rule states that the 'end' occasion is assigned {-aff, +act, 0pow, -emp}.

### N:Attribute-to-Pp:Occasion Rule

If **n:att –p:attOf→ n:occ**
then **n:occ := n:att ⊕ n:occ:lex**

This rule applies to phrases like 'end of the Carnival' and 'shattering music of the concert.' In the first case, the noun 'end' connotes {?aff, ?act, ?pow, -emp} and the referring expression 'the Carnival' is neutral ({0aff, 0act, 0pow, 0emp}). Thus, the Carnival occasion is assigned the connotation of {0aff, 0act, 0pow, -emp}.

### Adj:Act-to-N:entity Rule

If **adj:act –preModOf→ n:entity**,
then **n:entity := adj:act ⊕ n:entity:lex**

This rule applies to phrase like 'rioting mob.' Since 'rioting' connotes {-aff, +act, +pow, +emp}, and 'mob' connotes {-aff, 0act, +pow, 0emp}, then the mob is assigned the connotations {-aff, +act, +pow, +emp}.

### Multiple Connotation Propagations

Understanding the connotations assigned to topics by a sentence usually involves a series of connotation propagations. For example, consider the sentence 'The mob rioted at the end of the Carnival.' The connotation assignment of the occasion Carnival (e1) involves two steps. First, the condition *v:act –p:orient→ n:occ* matches the verb phrase 'rioted (v1) at end (e2)'. So the V:Action-to-Pp:Occasion Rule applies. The propagation resulted from the rule involves the following derivation, with each step accompanied with a reason at the end of the line:

| | | | |
|---|---|---|---|
| e2 := | | v1:lex ⊕ e2:lex | [Rule] |
| | = | {-aff, 0act, 0pow, +emp} | ['rioted'] |
| | | ⊕ {?aff, ?act, ?pow, -emp} | ['end'] |
| | = | {-aff, 0act, 0pow, 0emp} | [⊕] |

Second, the condition *n:att:e1 -p:attOf→ n:occ:e2* matches the noun phrase 'end (e2) of the Carnival (e1)'. Thus,

| | | |
|---|---|---|
| e1 := | e2 ⊕ e1:lex | [Rule] |
| = | {-aff, 0act, 0pow, 0emp} | ['end'] |
| | {0aff, 0act, 0pow, 0emp} | ['the Carnival'] |
| = | {-aff, 0act, 0pow, 0emp} | [⊕] |

## Selection of Word and Sentence Structure

The above connotation rules can be used for both understanding and generating sentences. Since this paper focuses on revision, it remains to show how words and sentence structures are selected with the given ideological beliefs. From a psychological perspective, such selection can be done actively or subconsciously. A manipulative writer does the selection actively to affect the reader. A casual writer does the selection subconsciously because ideological beliefs limit the way the writer understands and perceives events, people and institutions.

The basic idea of the algorithm for word choices and syntactic structure arrangement is backward chaining with backtracking. This is best illustrated by an example. Consider again the sample draft sentence of **A mob rioted at the end of the Carnival** and the ideological beliefs of the newspapers Sun and Star. The sentence has the following relations:

- **v:act:v1 –subj:agt→ n:entity:a1 ('mob (a1) rioted (v1)')**
- **v:act:v1 –p:orient:p1→ n:occ:e2 ('rioted (v1) at (p1) end (e2)')**
- **n:att:e2 –p:attOf:p2→ n:occ:e1 ('end (e2) of (p2) Carnival (e1)')**

### Star's Ideological Linguistic Choices

When revising this sentence based on the Star's ideological beliefs, two conditions are relevant:

**Condition 1** The affect, activity, and power of the people (a1) involved in the riot have de-emphasis. That is, **a1 must connote {-aff, +act, +pow, -emp}.**

**Condition 2** The Carnival (e1) has de-emphasized negative affect. That is, **e1 must connote {-aff, 0act, 0pow, -emp.}.**

The first condition will eliminate any sentence structure containing the relation *v:act:v1 -subj:agt→ n:entity:a1*, which asserts the existence of a specific group of people actively causing disorder and fails to satisfy the required negative emphasis of the attributes for the people a1 (Condition 1).

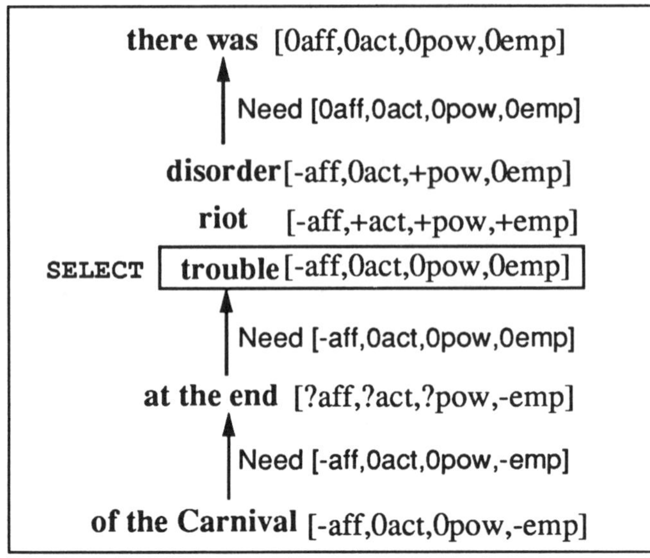


Figure 1: Backward Propagation to select 'trouble'

An alternative structure which satisfies Condition 1 is the "there-be" construction with verb nominalization and agent deletion. One such sentence is **'There was a riot at the Carnival'** where the noun *'riot'*, now normalized to an occasion (occ), can be replaced by synonyms *'disorder'* or *'trouble'*.

The candidate structure contains the noun phrase 'riot at the Carnival', represented as *n:occ:v1 - p:orient:p1→ n:occ:e1* whose propagation rule is **e1 := v1 ⊕ e1:lex.** Below is the derivation of the connotation back-propagation with the rule.

| | | | |
|---|---|---|---|
| **A** e1 | = | {-aff, 0act, 0pow, -emp} | [Cond. 2] |
| **B** e1:lex | = | 'the Carnival' | |
| | = | {0aff, 0act, 0pow, 0emp} | [Neutral] |
| **C** v1 | = | {-aff, 0act, 0pow, -emp} | [A, B] |

Now, the structure 'there was' is neutral, and consider the word choices:

| | |
|---|---|
| v1:lex= 'riot'= | {-aff, +act, +pow, +emp} |
| v1:lex= 'disorder'= | {-aff, 0act, +pow, 0emp} |
| v1:lex= 'trouble'= | {-aff, 0act, 0pow, 0emp} |

No word choice can satisfy the back-propagated Condition C. Hence this sentence structure **fails** to satisfy the top-level ideological beliefs.

Then another structure with the addition of 'the end of' is tried with the same procedure as above — **'There was a riot at the end of the Carnival'**, again with *'trouble'* and *'disorder'* as alternatives to *'riot'* (Figure 1). This sentence involves the phrase 'the end of the Carnival' represented as *n:att:e2 -p:attOf:p2→ n:occ:e1*, whose propagation rule **e1 := e2 ⊕ e1:lex** applies. Thus the back-propagation continues:

| | | | |
|---|---|---|---|
| **D** e1 | = | {-aff, 0act, 0pow, -emp} | [Cond. 2] |
| **E** e1:lex | = | 'the Carnival' | |
| | = | {0aff, 0act, 0pow, 0emp} | [Neutral] |
| **F** e2 | = | {-aff, 0act, 0pow, -emp} | [D, E] |

Now, the sentence also involves the noun phrase 'a riot at the end' represented as *n:occ:v1 –p:orient:p1→ n:occ:e2*, the propagation rule **e2 := v1 ⊕ e2:lex** applies. The back-propagation continues:

| | | | |
|---|---|---|---|
| **G** e2:lex | = | 'end' | |
| | = | {?aff, ?act, ?pow, -emp} | [Lex] |
| **H** v1 | = | {-aff, 0act, 0pow, 0emp} or | |
| | | {-aff, 0act, 0pow, -emp} | [F, G] |

Only the noun *'trouble'*, but not *'riot'* or *'disorder'*, has connotations that satisfy the back-propagated Condition H. So the "there-is" structure with the inclusion of the structure 'the end of' and the word choice of *'trouble'* can satisfy the top-level ideological Conditions (1) and (2). The sentence selected is **'There was trouble at the end of the Carnival.'**

Similar procedure revises **'A mob started a disorder at 8pm'** as **'Trouble started at 8pm'** and **'A mob scuffled and erupted into the police cordon'** as **'A scuffle erupted into the police cordon.'**

### Sun's Ideological Linguistic Choices

For the Sun, the mob carried negative affect, plus activity, plus power, and plus emphasis while the Carnival carried the same social values with no emphasis. The same backward chaining procedure mentioned above does not change the draft sentence **'A mob rioted at the end of the Carnival'** during revision, but revises the sentence **'A mob started a disorder at 8pm'** as **'A mob started to riot at 8pm.'**

## Summary

This paper has demonstrated the use of four attributes affect, activity, power, and emphasis in influencing writing decisions of word choices and syntactic structure arrangement using given ideological beliefs, a lexicon of connotations, and rules of connotation propagation. Since the model is still at the exploratory stage, we expect modifications as the model develops further. Nevertheless, the current model has demonstrated a significant insight that most of the previous works of computational linguistics have ignored — an inherent and often the most interesting part of the meanings of language is its connotations. Our model is a further step in understanding more about the way people communicate linguistic connotations. In addition, the model provides a basis for future research to automate text revision with social intelligence.

## Acknowledgment

I thank the following people for helpful discussions with the model described in this paper: Robert Simmons, Ron Mabbit, James Lester, Wood-Wai Lee, Gregory Kelly, Hwee-Tou Ng, Joycelyn Liu.

## References

[Abelson, 1973] R. P. Abelson. The structure of belief systems. In R. C. Schank and K. M. Colby, editors, *Computer Models of Thought and Language.* Freeman, San Francisco, CA, 1973.

[Carbonell, 1978] J. G. Carbonell. Politics: Automated ideological reasoning. *Cognitive Science*, 2:27–51, 1978.

[Carroll, 1956] J. B. Carroll, editor. *Language, Thought and Reality. Selected Writings of Benjamin Lee Whorf.* Wiley, New York, NY, 1956.

[Drew, 1979] T. Drew. What the papers say: Linguistic variation and ideological difference. In R. Fowler, B. Hodge, G. Kress, and T. Drew, editors, *Language and Control.* Routledge & Kegan Paul, London, Eng., 1979.

[Fowler *et al.*, 1979] R. Fowler, B. Hodge, G. Kress, and T. Drew, editors. *Language and Control.* Routledge & Kegan Paul, London, Eng., 1979.

[Goldman, 1975] N. Goldman. Conceptual generation. In R Schank, editor, *Conceptual Information Processing.* North-Holland, Amsterdam, Neth., 1975.

[Hayakawa, 1968] S. I. Hayakawa. *Choose the Right Word.* Harper & Row, New York, NY, 1968.

[Hovy, 1988] E. Hovy. *Generating Natural Language Under Pragmatic Constraints.* Lawrence Erlbaum Assoc., Hilldale, NJ, 1988.

[Kress and Hodge, 1979] G. Kress and R. Hodge. *Language as Ideology.* Routledge & Kegan Paul Ltd., London, Eng., 1979.

[McKeown, 1985] K. R. McKeown. *Text Generation: Using Discourse Strategies and Focus Constraints to Generate Natural Language Text.* Cambridge Univ. Press, Cambridge, Eng., 1985.

[Osgood *et al.*, 1957] C. E. Osgood, G. J. Suci, and P. H. Tannenbaum. *The Measurement of Meaning.* Univ. of Illinois Press, Urban, IL, 1957.

[Paris, 1988] C. L. Paris. Tailoring object descriptions to a user's level of expertise. *Computational Linguistics*, 14(3), 1988.

[Simmons and Slocum, 1972] R. F. Simmons and J. Slocum. Generating english discourse from semantic networks. *Communications of the Association for Computing Machinery*, 15(10), 1972.

**Richard Catrambone**

School of Psychology
Georgia Institute of Technology
Atlanta, GA 30332
rc7@prism.gatech.edu

### Abstract

A proposal is made for representing the knowledge learners acquire from examples in terms of subgoals and methods. Furthermore, it is suggested that test problems can also be represented in terms of the subgoals and methods needed to solve them. Manipulations of examples can influence the particular subgoals and methods learned. Thus, transfer can be predicted by the overlap in the learned subgoals and methods and those required to solve a novel problem. A subgoal is an unknown entity, numerical or conceptual, that needs to be found in order to achieve a higher-level goal of a problem. A method is a series of steps for achieving a particular subgoal. The experiment presented here suggests that elaborations in example solutions that emphasize subgoals may be an efficient way of helping a learner to recognize and achieve those subgoals in a novel problem, that is, to improve transfer. It is argued that conceptualizing problem-solving knowledge in terms of subgoals and methods is a psychologically plausible approach for predicting transfer and has implications for teaching and design of examples.

## Introduction

This paper makes two basic claims. One, people can learn subgoals for solving problems in a particular domain from the examples they study, and two, a person's success at solving a novel problem is partly a function of whether he or she possesses the necessary subgoals to solve that problem. A subgoal is an unknown entity, numerical or conceptual, that must be found in order to achieve a higher-level goal of a problem. A method is a series of steps for achieving a particular subgoal. No explicit mechanism is proposed here for *how* subgoals are learned. Rather, the emphasis is on identifying subgoals and methods and elucidating the conditions under which they will be learned. This emphasis was chosen since it seems important to gather empirical evidence in a variety of domains concerning the factors that influence whether subgoals and methods are learned and to integrate these findings into the simple scheme proposed here, particularly if this scheme can successfully predict transfer performance and improve the design of examples. Initial experiments have provided some support for the above approach for examining transfer (Catrambone, 1991; Catrambone & Holyoak, 1990). This paper explores one particular way of designing examples to increase the likelihood of conveying subgoals to learners: Briefly labeling or elaborating the subgoals in examples. Work in perceptual category learning has suggested that transfer is improved when subjects are led to focus on useful distinctions (categories) during training (e.g., Medin, Dewey, & Murphy, 1983). Subgoal learning may perhaps be viewed as a form of category learning and thus, may also benefit from manipulations that serve to highlight subgoals.

**Credits**
This research was supported by Office of Naval Research Grant N00014-91-J-1137 and National Institutes of Health Biomedical Research Support Grant 2-SO7-RR07024-24.

It is argued that successful learners acquire useful subgoals and methods for solving problems in a particular domain. The meaning of "useful" is operationally defined as those subgoals and methods that can be used to solve some corpus of target problems. If a learner has learned a particular subgoal, then he or she is more likely to recognize the need to find that subgoal in a novel problem. Even if the method for achieving that subgoal in the novel problem is new or requires a modification of an already-learned method, the learner who recognizes the need to achieve a particular subgoal will at least have some guidance for recruiting useful background knowledge to achieve the modification compared to a learner who does not possess the subgoal. Thus, subgoals provide a control structure to guide problem solving and transfer.

For example, consider a learner who studies a worked-out physics mechanics problem involving a block resting on an inclined plane. Perhaps the problem asks what the angle of incline has to be in order for the block to start sliding down the plane given that other things such as the block's mass and the coefficient of friction were known. A learner could memorize the steps for solving this problem such as picking a coordinate system with the x-axis parallel to the plane (rather than parallel to the problem solver's perspective) and performing the trigonometry for dealing with the angle of incline. However, the learner might not realize that most of these steps collapse into methods for achieving various subgoals. These subgoals might include high-level ones such as finding the opposing forces and equating them as well as lower-level subgoals such as splitting the forces into those acting along the x- and y-axes. Thus, when given a "simpler" problem such as determining how hard one would have to push a block resting on a flat table in order to just get the block to start moving, the learner could be at a complete loss. Many of the steps from the example, such as those involving the trigonometry, are no longer needed, except in a trivial way. If the learner had learned only a series of steps from the example, he or she would have little guidance in determining which ones should be eliminated or adapted for the new problem. Conversely, if the learner did have his or her knowledge organized hierarchically into a series of subgoals, then the learner would be in a better position to separate the steps into those that were necessary for the current subgoals (such as equating the opposing forces) and those that might *not* be necessary (such as choosing a coordinate system that is not horizontal and vertical).

Prior work has demonstrated that when learners study one or two worked-out example problems, such as algebra problems dealing with two workers completing some task together, they can solve isomorphic problems (Reed, Dempster, & Ettinger, 1985). For instance, subjects might study an example in which the workers' rates are given and the goal is to determine how long one of the workers works given that the other worker's time is known. The difficulty emerges when learners are faced with non-isomorphs, that is, problems requiring a modification of the old procedure. For example, in Reed et al.'s study, a non-isomorphic test problem might express one worker's rate in terms of the other worker's rate. Subjects typically had difficulty solving non-isomorphs. Part of the difficulty appeared to be that subjects formed subgoals such as "find the two numbers associated with worker A and multiply them together" rather than subgoals such as "find each worker's rate" and "find each worker's time." In problems which did not have two numbers associated with a worker, perhaps because the rate was a variable, subjects were often unable to solve the problem.

The poor transfer performance that is typically found in studies may partly be due to a failure to identify the necessary knowledge components that should be conveyed to learners to help them solve novel problems. It is claimed here that these components can be viewed as subgoals. The goal of the current experiment was to manipulate the subgoals conveyed to learners via examples and examine transfer effects. Cognitive architectures such as ACT* (Anderson, 1983), PUPS (Anderson & Thompson, 1986), and SOAR (Laird, Newell, & Rosenbloom, 1987) have mechanisms for predicting transfer. For example, ACT* predicts transfer of procedures by measuring overlap of old and new production rules (see also Singley & Anderson, 1989). However, these models are quite complex and many aspects of the models, such as chunking and proceduralization, are not necessary to derive useful predictions about learning from examples. Research utilizing the approach advocated here should allow the development of a relatively simple model to predict learning from examples as well as provide constraints on more complete models of cognition. This approach, if successful in terms of making useful predictions, could be used by researchers and textbook writers fairly easily as a practical guide for designing better examples (to improve learning) and test items (to improve diagnosticity).

In some sense, the greatest value of the approach described here may be its emphasis on the effort required by the teacher or researcher to identify the useful subgoals and methods in a domain. That is, assuming that examples do convey subgoals and methods, it is necessary to first determine what the useful ones are, and then to create examples to convey them. It is not claimed that the useful subgoals and methods can be identified in any algorithmic way. The approach advocated here for determining subgoals and methods is to identify a target set of problems that one wants students to be able to solve. Then solutions to these problems should be written out. These solutions are then analyzed in terms of the steps for achieving certain unknowns, either numerical or conceptual, in a hierarchical fashion. For example, in the physics problems discussed earlier, a high-level subgoal would be to equate the forces along the x-axis. This high-level subgoal will be achieved by first satisfying a lower-level subgoal such as finding the various forces acting along the x-axis. This subgoal can be satisfied by first satisfying the subgoal of identifying all the forces in the problem and breaking them into their x- and y-components. This subgoal decomposition can continue until one reaches the level of mathematical operations. It is a judgment call as to how much decomposition one feels is necessary to represent subgoals and methods at a useful level. In any case, the subgoals and methods identified by the researcher or teacher are at least open to inspection by others and can be debated.

## Experiment

The experiment presented here examined subjects' ability to solve permutation and combination problems in probability. This domain was chosen because prior work indicated that learners were strongly influenced by superficial features of training and test problems. For example, Ross (1989; Experiment 1B) had subjects study example probability problems, such as ones involving permutations, and then solve several test problems. The entities (e.g., computers, cars, scientists) in the examples and test problems were manipulated. For instance, Table 1a presents a permutation problem involving the determination of the probability of scientists choosing particular computers. The equation presented to subjects in Ross (1989) in order to solve this example was $1/[n(n-1)...(n-r+1)]$. Test problems included some subjects finding the probability of students choosing particular cars while other subjects had to find the probability of particular students being assigned to particular cars (i.e., inanimate objects "choosing" humans).

Corresponding roles are held by the humans and inanimate objects in the example in Table 1a and the test problem involving students choosing cars. That is, the mathematical roles of the humans (scientists and students) and the inanimate objects (computers and cars) are identical: the inanimate objects provide the values for $n$ and for $r$ while the humans provide the value only for $r$. The second type of test problem, however, has reversed object correspondences: the humans provide the values for $n$ and $r$ while the inanimate objects provide the value only for $r$. Ross found that subjects were more successful solving the first type of test problem than the second, presumably because their problem solving was guided to some degree by an object mapping approach. That is, if humans provide the value of $n$ in the examples, the subjects were likely to assign them this role in test problems even if it was really an inanimate object that should be providing the value of $n$.

Ross also had subjects study and solve combination problems. An example combination problem might ask for the probability of the seven hooks nearest the classroom door being picked by the seven tallest students in a class (see Table 1b). The equation presented to subjects in order to solve combination problems of this sort was $[h!(j-h)!]/j!$, where $h$ is the number of objects (e.g., students) doing the choosing and $j$ is the number of objects in the pool from which objects are chosen (e.g., hooks). Again, Ross found that if the roles of the objects were switched from the training examples to the test problems, subjects would not solve the test problems correctly (that is, they would switch the values for $j$ and $h$).

While permutation and combination problems can be solved using quite different equations, a further examination of these problem types indicates that they are similar at a fundamental level. Both types of problems can be analyzed by considering the individual event probabilities that contribute to an overall probability. For example, in the permutation problem involving scientists choosing computers (Table 1a), the overall probability can be calculated by explicitly considering each of the individual probabilities. This approach is demonstrated in the solution provided in Table 2c. The combination problem in Table 1b involving students and coathooks can be analyzed in a similar way:

Probability of one of the seven tallest students getting a hook near door = 7/17

Probability of one of the remaining six tallest students getting a hook near door = 6/16

Probability of one of the remaining five tallest students getting a hook near door = 5/15, etc

So, $\frac{7}{17} * \frac{6}{16} * ... * \frac{1}{11} = \frac{7!}{17*16*...*11}$ = overall probability

Table 1
**Permutation and Combination Problems**

a.) The supply department at IBM has to make sure that scientists get computers. Today, they have 11 IBM computers and 8 IBM scientists requesting computers. The scientists randomly choose their computer, but do so in alphabetical order. What is the probability that the first 3 scientists alphabetically will get the lowest, second lowest, and third lowest serial numbers, respectively, on their computers?

b.) The Happy House Nursery School has had 17 hooks put up in the hall for the coats of their 14 students, with each student using one hook. The students are each randomly assigned a hook as they come in one morning. What is the probability that the 7 hooks closest to the classroom door are assigned to the 7 tallest students?

Table 2
**Solution Types Used in Experiment for Problem in Table 1a**

a.) Equation Solution

The equation needed for this problem is $\frac{1}{n * (n - 1) * ... * (n - r + 1)}$. In this problem $n$ = 11 and $r$ = 3. So, $\frac{1}{11 * 10 * 9} = \frac{1}{990}$ = overall probability

b.) Equation+Elaboration Solution:

The equation needed for this problem is $\frac{1}{n * (n - 1) * ... * (n - r + 1)}$. This equation allows one to determine the probability of the above outcome occurring. In this problem $n$ = 11 and $r$ = 3. The 11 represents the number of computers that are available to be chosen while the 3 represents the number of choices that are being focused on in this problem. The equation divides the number of ways the desired outcome could occur by the number of possible outcomes. So, inserting 11 and 3 into the equation, we find that $\frac{1}{11 * 10 * 9} = \frac{1}{990}$ = overall probability

c.) Subgoal Solution:

The equation needed for this problem is $\frac{1}{n * (n - 1) * ... * (n - r + 1)}$. In this problem $n$ = 11 and $r$ = 3. However, another way of approaching the problem is to think of it in the following way:

Probability of the first scientist (who comes first alphabetically) getting the computer with the lowest serial number = 1/11.
Probability of second scientist getting second lowest serial number = 1/10.
Probability of third scientist getting third lowest serial number = 1/9.

So, $\frac{1}{11} * \frac{1}{10} * \frac{1}{9} = \frac{1}{990}$ = overall probability

d.) Numerator/Denominator-Subgoal Solution:

The equation needed for this problem is $\frac{1}{n * (n - 1) * ... * (n - r + 1)}$. In this problem $n$ = 11 and $r$ = 3. However, another way of approaching the problem is to think of it in the following way:

Of the 11 computers that the first scientist has to choose from, only 1 of them would be acceptable, since only 1 has the lowest serial number. So, the probability of the first scientist getting the computer with the lowest serial number = 1/11.

Of the 10 remaining computers that the second scientist has to choose from, only 1 of them would be acceptable, since only 1 has the lowest remaining serial number. So, the probability of the second scientist getting a computer with the lowest remaining serial number = 1/10.

Of the 9 remaining computers that the third scientist has to choose from, only 1 of them would be acceptable, since only 1 of has the lowest remaining serial number. So, the probability of the third scientist getting the computer with the lowest remaining serial number = 1/9.

So, $\frac{1}{11} * \frac{1}{10} * \frac{1}{9} = \frac{1}{990}$ = overall probability

The permutation problem solution presented in Table 2c might help learners form the subgoal of finding each event probability (e.g., the probability of the first scientist getting the computer with the lowest serial number, the probability of the second scientist getting the computer with the second-lowest serial number, etc). In particular, it should help learners form the subgoal of finding the entity being chosen and using the value associated with that entity as the starting value for the denominator. Thus, subjects who study examples using this solution approach are predicted to be more likely to be able to find event probabilities in cases where the roles of the inanimate objects and humans are different from the examples (see Table 3b) compared to subjects who study examples using only the permutation equation (Table 2a). This latter group presumably learns only to place numbers into an equation. Furthermore, learners who study examples emphasizing each event probability might be more likely to represent the event probabilities correctly in combination problems for which the numerators are no longer simply "1" (such as the combination problem in Table 1b). That is, because these subjects are assumed to be more likely to be able to focus on the individual event probabilities, they may have a better chance of representing them correctly compared to learners who do not have this focus. However, this generalization may be too difficult for learners when solving combination problems. Since the solution presented in Table 2c does not emphasize finding the numerator, but rather just involves placing a "1" in each numerator, subjects who study this solution type might not form the subgoal of finding the numerator from the number of "acceptable" outcomes. Subjects may require example solutions that *explicitly* highlight how the numerator is chosen in the permutation examples in order to form a subgoal for dealing with the numerators. A solution approach that might accomplish this is presented in Table 2d.

Another concern in this study was whether subjects might misinterpret the combination problems as permutation problems. If that were to happen, then subjects would be likely to use "1" in the numerators of the probabilities for all problems regardless of the examples they studied. In order to investigate this issue, some subjects were given combination problems that contained a clarification indicating that order did not matter. Subjects who study examples demonstrating the solution approach in Table 2d might outperform subjects who study examples using the approach in Table 2c on combination problems containing a clarification. This would occur because the latter group will presumably not have the subgoal for finding the number of acceptable outcomes to put in the numerator for each probability and thus, they would be more likely to simply use "1" in the numerator.

## Method

**Subjects.** Subjects were 90 students from introductory psychology classes at the Georgia Institute of Technology.

**Materials and Procedure.** All subjects studied three isomorphic example problems dealing with permutations that involved humans choosing objects (including the problem in Table 1a). Two factors were manipulated. The first was the type of solution provided to the examples subjects studied. The examples either provided a solution using the permutation equation (the Equation-Only group; see Table 2a for an example), the permutation equation plus an elaboration of what the numbers in the equation represented (the Equation-Elaboration group; Table 2b), a subgoal-oriented solution that emphasized the rationale for the denominator in each probability (the Denominator-Subgoal group; Table 2c), or a subgoal-oriented solution that contained a rationale for the denominator *and* numerator in each probability (the Numerator/Denominator-Subgoal group; Table 2d).

The second factor manipulated was whether the test combination problems contained a clarification about order. The clarification consisted of the following sentences added on to the end of the first and second combination problems (see Table 3c, d), respectively: "It does not matter which of the particular seven hooks closest to the door these students get, just as long as it is any one of the seven closest" and "It does not matter which of the particular six front seats the pitchers get, just as long as it is any one of the six in the front."

After studying the examples subjects solved the four problems in Table 3. Subjects could not refer back to the examples when working on the problems. The first problem was isomorphic to the examples (see Table 3a). The second problem was a permutation problem isomorphic to the examples but had humans playing the roles of $n$ and $r$ and objects playing only the role of $r$ (Table 3b). The third and fourth problems were combination problems. The solutions to these problems involve both a numerator and denominator that start at some value and then are decremented. The first combination problem involves humans picking objects (Table 3c) while the second combination problem involves objects picking humans (Table 3d).

## Results and Discussion

Each problem was scored for whether a subject used the correct starting number in the denominator and whether the denominator was decremented appropriately. For instance, the solution to the second permutation problem is 1/11 * 1/10. If subjects wrote 1/14 * 1/13, confusing the roles of the chairs and secretaries, they would be scored as having the incorrect starting number of the denominator but the correct number of decrements. For combination problems, the numerator and denominator were both scored for whether the correct starting number was used and whether they were decremented appropriately. The frequencies for the various categories listed above were analyzed using the likelihood ratio chi-square test ($G^2$; Bishop, Fienberg, & Holland, 1975).

There was no significant difference in performance among the groups on the first permutation problem that was isomorphic to the training examples and had humans and objects playing the same roles as in the examples (Equation-Only: 73%, Equation-Elaboration: 83%, Denominator-Subgoal: 91%,and Numerator/Denominator-Subgoal: 91%), $G^2(3)=3.77, p=.29$.

The second permutation problem had objects picking humans, a reversal from the training examples (see Table 3b). The major error subjects made was to use the wrong starting value for the denominator. Many subjects used 14 (the number of chairs) as the starting point rather than 11 (the number of secretaries). It was expected that the two subgoal groups would be more likely to use the correct starting point for the denominator than the equation groups. However, while the subgoal groups tended to perform the best, the most striking result was the poor performance of the Equation-Elaboration group (Equation-Only: 32%, Equation-Elaboration: 13%, Denominator-Subgoal: 48%, and Numerator/Denominator-Subgoal: 41%), $G^2(3)=7.63, p=.054$. It is not clear why the Equation-Elaboration group did so poorly.

**Table 3**
**Test Problems**

a.) As part of a new management policy, the Campbell Company is allowing the 20 company-owned vacation cottages to be used for vacations by their 14 plant managers. If the managers, in order of seniority, randomly choose a cottage from a list, what is the probability that the manager with the most seniority gets the most lavish cottage, and the manager with the second most seniority gets the second most lavish, and the manager with the third most seniority gets the third most lavish, and the manager with the fourth most seniority gets the fourth most lavish cottage?

b.) The secretaries at city hall are supposed to get new chairs this week. Today, city hall received 14 new chairs and there are 11 secretaries requesting them. For inventory purposes, the property manager wants to assign the chairs in the order that they are unpacked. So, starting with the chair that is unpacked first, he randomly chooses a secretary to receive it, and continues until all the secretaries have chairs. What is the probability that the first 2 secretaries alphabetically will get the first and second chairs that are unpacked, respectively?

c.) The Happy House Nursery School has had 17 hooks put up in the hall for the coats of their 14 students, with each student using one hook. The students each choose a hook at random as they come in one morning. What is the probability that the 7 tallest students get the 7 hooks closest to the classroom door?

d.) The Nashville Gnats Baseball team has a bus that has 30 seats. There are 25 players that are going on a road trip to play in a nearby town. To avoid arguments, the manager randomly chooses a player for each seat, starting with the seats in the front. What is the probability that the 6 pitchers get the 6 front seats?

**Table 4**
**Performance (Percent Correct) on Various Aspects of Combination Problems as a Function of Examples Studied and Presence of Clarification**

| | Group | | | | | | | |
|---|---|---|---|---|---|---|---|---|
| | Clarification | | | | No Clarification | | | |
| | Eq Only (n=11) | Eq + Elab (n=12) | Denom Subgoal (n=12) | Num/Denom Subgoal (n=11) | Eq Only (n=11) | Eq + Elab (n=11) | Denom Subgoal (n=11) | Num/Denom Subgoal (n=11) |
| Problem | | | | | | | | |
| Combination Problem #1 (people choose objects) | | | | | | | | |
| Correct Start for Denominator | 82 | 70 | 75 | 80 | 91 | 100 | 91 | 70 |
| Correct Denominator Decrement | 82 | 60 | 92 | 90 | 82 | 91 | 91 | 80 |
| Correct Start for Numerator | 0 | 0 | 25 | 40 | 9 | 0 | 20 | 10 |
| Correct Numerator Decrement | 0 | 0 | 33 | 40 | 0 | 0 | 18 | 10 |
| Combination Problem #2 (objects choose people) | | | | | | | | |
| Correct Start for Denominator | 27 | 30 | 33 | 36 | 27 | 36 | 9 | 27 |
| Correct Denominator Decrement | 82 | 70 | 92 | 64 | 82 | 82 | 82 | 64 |
| Correct Start for Numerator | 0 | 0 | 33 | 46 | 9 | 0 | 9 | 9 |
| Correct Numerator Decrement | 0 | 0 | 33 | 36 | 9 | 0 | 9 | 9 |

The third and fourth problems were combination problems (see Tables 3c and d). It was hypothesized that subjects might interpret these combination problems as permutation problems and, as a result, the subgoal manipulation would not have as strong an effect as it could. In order to investigate this possibility, half of the subjects in each instructional group received modified versions of the two combination problems that included a clarification about order. If subjects needed the clarification in order to be more likely to interpret these problems correctly, then the instructional manipulation might only produce an effect for subjects who receive the clarification.

The first combination problem involved humans picking objects. Thus, as in the examples, humans provide the value for the denominator. The presence of a clarification did not seem to differentially influence performance for the various instructional groups on the likelihood of choosing the correct starting value for the denominator (see Table 4; Clarification: $G^2(3)$=.49, $p$=.92, No-Clarification: $G^2(3)$= 5.29, $p$ =.15). The clarification also did not differentially affect performance on decrementing the denominator (Clarification: $G^2(3)$=4.04, $p$=.26, No-Clarification: $G^2(3)$ =.91, $p$=.82). These results are not surprising since the clarification has little relationship to choosing the denominator and decrementing it appropriately.

Subjects were much less successful choosing the correct numerator on the first combination problem. Unlike the examples, the numerator is not simply "1," rather it is a value starting at 7 and decrementing to 1. Most subjects simply used "1" as the numerator, as in the training examples. For non-clarification subjects, there was no significant difference in the groups' success in choosing the correct starting value for the numerator as a function of the examples they studied, $G^2(3) = 3.21, p = .36$. However, for clarification subjects, the subgoal groups outperformed the equation groups with the Numerator/Denominator group performing the best, $G^2(3) = 11.25, p = .01$. This pattern of results is repeated for decrementing the numerator.

Subjects were less likely to choose the correct starting value for the denominator in the second combination problem, presumably due to the switch in roles of humans and objects; in this problem objects were picking humans. As in the prior problem, though, the presence of a clarification did not seem to differentially influence performance for the various instructional groups on choosing the correct starting point for the denominator (Clarification: $G^2(3) = .24, p = .97$, No-Clarification: $G^2(3) = 2.58, p = .46$). The clarification also did not differentially affect performance on decrementing the denominator (Clarification: $G^2(3) = 3.21, p = .36$, No-Clarification: $G^2(3) = 1.45, p = .69$).

As in the first combination problem, subjects were not very successful choosing the correct numerator. For non-clarification subjects there was no significant difference in the groups' success in choosing the correct starting value for the numerator, $G^2(3) = 1.80, p = .62$. However, for clarification subjects, the subgoal groups outperformed the equation groups with the Numerator/Denominator group performing the best, $G^2(3) = 12.03, p = .007$. This pattern of results is repeated for decrementing the numerator.

## Conclusions

The results from this study are important because they suggest that training subgoal recognition facilitates their flexible application. This conclusion is supported by the fact that the Numerator/Denominator-Subgoal group was the most likely to recognize that the numerators of the individual event probabilities in the combination problems were not simply 1 but rather were a value that represented the number of acceptable outcomes. Even though these subjects had studied examples that only used 1 in the numerators, they were sensitized to the role this value played--presumably via the elaboration during training--and were able to generalize when faced with novel problems. This supports the claim that if useful subgoals can be conveyed to learners then they will be able to solve novel problems, that is, problems that involve modified or new methods for achieving those subgoals.

## General Discussion

The approach for predicting transfer described in this paper views problem-solving knowledge in terms of subgoals and methods. An important task then is to develop rules for predicting when these subgoals and methods will be learned from examples and to develop a more theoretically-guided motivation for the rules. It has been argued that the relatively simple scheme presented here can account for transfer effects without considering the additional complexities of a full model of cognition. Clearly this simplicity is a virtue only if the approach is reasonably accurate and can be applied fairly easily to a variety of domains by different researchers and teachers. While full models of cognition are extremely valuable, a simpler approach to predicting transfer, if successful, can help improve instruction in a direct way. The quality of examples and the diagnosticity of tests can be improved if instructors and textbook writers become sensitive to the subgoals and methods that students need to learn in a particular domain. The exercise of identifying these components is valuable in its own right since it would make researchers and instructors aware of the building blocks learners need to acquire in order to solve novel problems successfully. Examples can then be devised that convey the subgoals and methods. Although converging evidence is certainly needed, the present results are consistent with predictions that learners who acquire a more hierarchical subgoal structure will be more successful on novel problems than learners who learn a linear series of steps.

## References


Anderson, J.R. (1983). *The architecture of cognition.* Cambridge, MA: Harvard University Press.

Anderson, J.R., & Thompson, R. (1986). *Use of analogy in a production system architecture.* Paper presented at the Illinois Workshop on Similarity and Analogy, Champaign-Urbana, IL, June.

Bishop, Y.M.M., Fienberg, S.E., & Holland, P.W. (1975). *Discrete multivariate analysis: Theory and practice.* Cambridge, MA: MIT Press.

Bovair, S., Kieras, D.E., & Polson, P. (1990). The acquisition and performance of text editing skills: A production system analysis. *Human Computer Interaction, 5(1)*, 1-48.

Catrambone, R. (1991). *Using labels to help learners acquire subgoals for solving novel problems.* Paper presented at the Fourth European Conference for Research on Learning and Instruction, Turku, Finland.

Catrambone, R., & Holyoak, K.J. (1990). Learning subgoals and methods for solving probability problems. *Memory & Cognition, 18(6)*, 593-603.

Laird, J.E., Newell, A., & Rosenbloom, P.S. (1987). Soar: An architecture for general intelligence. *Artificial Intelligence, 33(3)*, 1-64,

Mawer, R.F., & Sweller, J. (1982). Effects of subgoal density and location on learning during problem solving. *Journal of Experimental Psychology: Learning, Memory, & Cognition, 8*, 252-259.

Medin, D.L., Dewey, G.I., & Murphy, T.D. (1983). Relationships between item and category learning: Evidence that abstraction is not automatic. *Journal of Experimental Psychology: Learning, Memory, & Cognition, 9(4)*, 607-625.

Newell, A., & Simon, H.A. (1972). *Human problem solving.* Englewood Cliffs, NJ: Prentice-Hall.

Reed, S.K., Dempster, A., & Ettinger, M. (1985). Usefulness of analogous solutions for solving algebra word problems. *Journal of Experimental Psychology: Learning, Memory, and Cognition, 11*, 106-125.

Ross, B. (1989). Distinguishing types of superficial similarities: Different effects on the access and use of earlier problems. *Journal of Experimental Psychology: Learning, Memory, and Cognition, 15*, 456-468.

Singley, M.K., & Anderson, J.R. (1989). *The transfer of cognitive skill.* Cambridge, MA: Harvard University Press.



### Acknowledgement

I thank Alana Anoskey for her help in collecting and coding the data and Sharon Carter and two anonymous reviewers for their helpful comments.

# Strategy Shifts Without Impasses: A Computational Model of the Sum-to-Min Transition

**Randolph M. Jones**
**Kurt VanLehn**

Department of Computer Science
University of Pittsburgh
Pittsburgh, PA 15260
JONES@CS.PITT.EDU
VANLEHN@CS.PITT.EDU

**Abstract**

The SUM-to-MIN transition that children exhibit when learning to add provides an ideal domain for studying naturally occurring discovery processes. We discuss a computational model that accounts for this transition, including the appropriate intermediate strategies. In order to account for all of these shifts, the model must sometimes learn without the benefit of impasses. Our model smoothly integrates impasse-driven and impasse-free learning in a single, simple learning mechanism.

## Introduction

This paper discusses models for the well-known SUM-to-MIN transition (Ashcraft, 1982, 1987; Groen & Parkman, 1972; Groen & Resnick, 1977; Kaye, Post, Hall, & Dineen, 1986; Siegler & Jenkins, 1989; Svenson, 1975). This is an ideal domain for studying naturally occurring discovery processes (Siegler & Jenkins, 1989). When young children first learn to add two small numbers, they use the so-called SUM strategy. They create sets of objects to represent each addend, then count the objects in the union of the two sets. For instance, in order to solve 2 + 3, the child says, "1, 2" while raising two fingers on the left hand; then "1, 2, 3" while raising three fingers on the right hand; then "1, 2, 3, 4, 5," while counting all the raised fingers. This strategy is called the SUM strategy because its execution time is proportional to the sum of the two addends. Older children use a more efficient strategy, the MIN strategy, whose execution time is proportional to the minimum of the two addends. In following this strategy, the child first announces the value of the larger addend, then counts on from it. For instance, in order to solve 2 + 3, the child would say "3" then say "4,5" while raising two fingers on one hand.

Although the SUM strategy is taught in school, the MIN strategy appears to be invented by the children themselves. The best evidence for this comes from a longitudinal study by Siegler and Jenkins (1989). They interviewed 8 children weekly for 11 weeks, each time asking them to solve about 15 orally presented addition problems. After each problem, the child was asked how they got their answer. The child was also told whether their answer was correct, and was given a gold star if it was. Videotapes were analyzed and the child's behavior on each problem was classified according to the strategy used. As far as Siegler and Jenkins could determine, the only instruction that the subjects received during this period was their school's normal instruction on the SUM strategy. Nonetheless, 7 of the 8 children began to use the MIN strategy. Moreover, they appear to have discovered it during the video-taped sessions. The tapes make it clear that they received no help from the experimenter, so the MIN strategies appear to have been invented by the subjects themselves.

The issue addressed by this paper is explaining how children discover the MIN strategy. This particular discovery presents a challenge for current theories of learning. Although a variety of learning methods have been proposed, many of them are triggered when the problem solver reaches an impasse, and yet Siegler and Jenkins found no signs of impasses during the discovery of the MIN strategy. The exact definition of "impasse" depends on the problem-solving architecture, but roughly speaking, an impasse occurs whenever the solver has a goal that cannot be achieved by any operator that is believed to be relevant to the task at hand. The essential idea of impasse-driven learning is to resolve the impasse somehow, then store the resulting experience in such a way that future impasses will be avoided or at least handled more efficiently. Many systems use impasse-driven learning, including SWALE (Schank, 1986), OCCAM (Pazzani, Dyer & Flowers, 1986), LPARSIFAL (Berwick, 1985), SOAR (Newell, 1990), SIERRA (VanLehn, 1990), and CASCADE (VanLehn & Jones, in press).

Siegler and Jenkins looked specifically for signs of impasses and found none. In particular, they designed some of the problems to cause impasses by making one of the addends very large (e.g., 23 + 1). They found that "The specific problems on which the children first used the MIN strategy were 2 + 5, 4 + 1, 3 + 1, 1 + 24, 5 + 2, and 4 + 3. These problems did not deviate from the characteristics of the overall set in any notable way."(p. 67) In fact, some of the children had earlier successfully solved exactly the same problem that they were working on when they discovered the MIN strategy. Although the impasse problems did cause subjects who had already invented the MIN strategy to start using it more frequently, the problems did not cause those who had not invented the strategy to do so.

Siegler and Jenkins sought signs of impasses by examining solution times and errors in the vicinity of the discovery events. Solution times were longer than normal for the problems where the discovery occurred and for the problems immediately preceding the discovery trial. This might suggest some kind of impasse, but error rates indicated that the problems themselves were not particularly difficult. Siegler and Jenkins suggest

This research benefited from discussions with Jeff Schlimmer and Bob Siegler. It was supported in part by contract N00014-88-K-0080 from the Office of Naval Research, Cognitive Sciences Division, and a postdoctoral training grant from the Department of Health and Human Services.

a reconciliation between their findings and impasse-driven learning theories:

> Two types of strategy changes can be distinguished: changes in which the main difference between the strategies is in the answers themselves, and changes in which the main differences are not in the answers that are generated but rather in the efficiency with which answers are generated and/or the aesthetic appeal of the procedures. The first type of strategy change may occur primarily as a result of encountering impasses, but the second may typically occur for other reasons. (p. 104)

One model in the literature effects a strategy change whenever it detects an opportunity for improving the efficiency of a procedure. HPM (Neches, 1987) keeps a complete trace of its processing, which is constantly monitored by heuristics such as: "If a subprocedure produces an output, but no other subprocedure receives that result by the time the overall procedure finishes, then modify the overall procedure to eliminate the superfluous computation." Neches demonstrated that this heuristic and two others sufficed for changing the SUM strategy into the MIN strategy.

Enroute, two transitional strategies are necessarily produced by HPM. Siegler and Jenkins sought evidence for these transitional strategies in their data. One of the strategies occurred 6 times, all in the protocol of the same subject. Moreover, all these instances occurred *after* the MIN strategy was invented. The second transitional strategy predicted by Neches did not appear at all. These unfulfilled predictions cast doubt on the HPM model. The model itself, as Neches noted, "assumes the relative accessibility of extremely detailed information about both on-going process and related past experiences. How can this be reconciled with known limitations on the ability to report this information?" (p. 213) Although HPM is computationally sufficient to produce the SUM-to-MIN transition, it makes dubious empirical and mneumonic assumptions.

The objective of this research has been to generate a computationally sufficient account of the SUM-to-MIN transition that produces only observed transitional strategies and makes plausible demands on memory. The result is a problem solver called GIPS (General Inductive Problem Solver). This paper describes GIPS and its account of the strategy shifts observed by Siegler and Jenkins (1989).

GIPS is a generalized means-ends analysis (MEA) problem solver (Jones, 1989) whose primary learning mechanism is based on Schlimmer's (1987; Schlimmer & Granger, 1986a, 1986b) STAGGER system, which uses a probabilistic induction technique to learn concept descriptions from examples. Other systems have combined inductive concept learners with problem solvers (e.g., Langley, 1985; Mitchell, Utgoff & Banerji, 1983), but they acquire only search-control knowledge: concepts that indicate which operator to select when several operators match the current goal. GIPS modifies the descriptions of the operators themselves as well as the heuristics for selecting operators. Both types of learning play crucial roles in making the SUM-to-MIN transition.

Following a brief description of GIPS (see Jones & VanLehn, 1991, for a complete description) is the main section of this paper, which presents the GIPS account of the SUM-to-MIN transition.

## The General Inductive Problem Solver

GIPS' basic problem-solving algorithm is a generalized version of MEA borrowed from the EUREKA system (Jones, 1989). It is based on trying to achieve a state change. The desired change is represented by a TRANSFORM, which is simply a pair consisting of the current state and some goal conditions. In order to achieve this transformation, GIPS selects an operator and attempts to apply it. If the operator's preconditions are met, it is executed and the current state changes. If some of the preconditions are not met, a new TRANSFORM is created with them as the goals. When this transformation is achieved, GIPS returns to the old TRANSFORM and attempts again to apply the operator. So far, this is just the standard MEA algorithm, but GIPS adds two important differences.

In standard MEA, operators are selected if they reduce the difference between the current and goal states. In GIPS, selection is determined by *selection concepts*. Each operator has a concept that indicates when it should be selected. If the concept depends mostly on the current state of the TRANSFORM, then the operator will act like a forward-chaining inference rule and fire whenever the state is appropriate, regardless of the current goals. If the concept depends mostly on the goals of the TRANSFORM, then it will act like a backward-chaining inference rule. Typically, forward and backward operators intermingle during problem solving, yielding a psychologically plausible blend of goal-directed and opportunistic behavior.

Each operator has a selection concept. The concept is represented as set of literals (predicates that may or may not be negated), and each literal has two values associated with it: its sufficiency and its necessity. In order to evaluate the worth of selecting an operator, GIPS matches the literals against the current TRANSFORM. It determines the subset of literals that match ($M$) and fail to match ($F$), then calculates

$$Odds(C) \prod_{L \in M} sufficiency(L) \prod_{L \in F} necessity(L),$$

where $Odds(C)$ is the prior probability that the concept is worth selecting. This is the same formula used by STAGGER, Schlimmer's (1987) propositional concept formation system, to estimate the probability that a given object is an instance of a particular concept. When all the operators have been matched and their worth has been calculated, GIPS chooses the one with the highest rating.

GIPS adjusts its selection concepts on the basis of its successes and failures while solving problems. When a problem is finally solved, for each operator along the solution path, GIPS adjusts the sufficiency and necessity values so that the operator will be rated even higher the next time a similar TRANSFORM occurs. For each operator that initiated a failure path, GIPS adjusts the values in its selection concept so that it will receive a lower value next time. In order to learn, GIPS must store the solution path and every operator that led off it. However, as soon as the problem is finished and the updating is completed, this information can be forgotten. HPM (Neches, 1987) must store the whole search tree, not just the solution path, for an indefinite period. Holland, Holyoak, Nisbett, and Thagard (1986) describe a technique, called the bucket brigade algorithm, that achieves the same kind of update as GIPS without storing any of the solution path.

The mechanism described so far can learn search-control knowledge, as do many other machine learning systems. In order to make the SUM-to-MIN transition, the system must modify the preconditions of operators as well. In standard MEA, after an operator has been selected, its preconditions are matched to the current state. If all of them match, the operator is executed. If some do not match, they become subgoals. Thus, preconditions determine the goal structure of the problem solving. GIPS takes an indirect approach to adjusting preconditions. In addition to the selection concept, GIPS provides a second concept, called the *execution concept*, which serves as a repository for its experience in attempting to execute the operator. When values in the execution concept cross a threshold, appropriate modifications are made to the preconditions. Thus, GIPS only modifies preconditions when warranted by a great deal of experience. The next two paragraphs describe exactly how execution concepts are used.

When an operator is selected, the system matches the execution concept to the current TRANSFORM in the same way that selection concepts are matched. If the calculation returns a value greater than 1, GIPS attempts to execute the operator. If the value is less than 1, GIPS follows the standard MEA practice of matching the preconditions of the operator to the current state and setting up subgoals for the unmatched ones.

When GIPS attempts to execute an operator, it has no arms and eyes so it cannot tell if its attempt succeeds, thus it asks the user. If an operator succeeds, the values in the execution concept are updated appropriately and the new state becomes current. If an operator fails, the values are updated and the system follows the standard MEA routine of matching preconditions and setting up subgoals.

When the sufficiency value for a literal in the execution concept crosses a threshold, that literal is added to the operator's preconditions. Thus, GIPS only adds a literal when that literal has been found time and time again to be present when the operator executes. However, GIPS is quick to remove a literal from the preconditions if it ever finds that the literal is not actually necessary for the execution of the operator. If the system successfully executes an operator and not all of the preconditions are satisfied, it removes the unmatched literals. There is no sense continuing to subgoal for an unmatched literal if that literal is not in fact necessary for executing the operator.

## Representation of the Addition Domain

GIPS describes the world as a set of relations between objects. In the addition domain, these objects and relations include the numbers that are part of the problem, the state of the problem solver's "hands" while it is adding, and the value of a counter that the problem solver keeps "in its head."

GIPS requires 16 operators to represent the addition domain. There are two particular operators, which we refer to as the END-COUNT operators, that are involved in most of the strategy shifts. For future reference, the series of preconditions that the LEFT-END-COUNT operator acquires appears in Table 1. The 16 operators' selection concepts were initialized so that the system generates the SUM strategy. The literals of each operator's selection concept were the preconditions and the goals that the operator could satisfy. The necessity and sufficiency of these literals were set so that they would be retrieved in either a backward-chaining or forward-chaining fashion, depending on the role of the operator in the domain.

**Table 1.** A series of preconditions for LEFT-END-COUNT.

SUM strategy (a):
```
Raising(Lefthand)
Counting(Lefthand)
Assigned(Lefthand,=Value)
Counter-value(=Value)
```

SUM strategy (b):
```
Raising(Lefthand)
Counting(Lefthand)
Assigned(Lefthand,=Value)
Counter-value(=Value)
Raised-fingers(Lefthand,=Value)
```

SHORTCUT SUM strategy (c):
```
Raising(Lefthand)
Counting(Lefthand)
Assigned(Lefthand,=Value)
Raised-fingers(Lefthand,=Value)
```

FIRST strategy (d):
```
Raising(Lefthand)
Counting(Lefthand)
Assigned(Lefthand,=Value)
```

## Strategy Acquisition in the Addition Domain

This section presents GIPS' behavior through a series of different strategies for adding numbers. These strategy shifts arise from the learning algorithm incorporated into the system, and they correspond well with the shifts observed by Siegler and Jenkins. Siegler and Jenkins classified their subjects' behavior into 8 strategies, of which 4 were based on counting (the others involved various kinds of recognition and guessing). In this section, we describe each of the 4 counting strategies in the order in which they generally appear. However, it is important to note that children always intermingle their strategies, sometimes even on a trial by trial basis. We will discuss the issue of strategy variability in the next section.

### The SUM Strategy

GIPS' initial strategy for addition is the SUM strategy. The first thing the system does is assign an addend to each hand. For example, when adding 2 and 3, the system may assign the number 2 to the left hand and the number 3 to the right hand. However, in this strategy the order of the addends does not make a difference, so it could just as easily have switched them.

Next, the system begins its procedure of counting out a set of fingers on each hand. To accomplish this task the END-COUNT operators initially use a counter to determine when a hand is finished being counted out. For example, the preconditions of LEFT-END-COUNT demand that the system be raising fingers on the left hand, and that the value of the counter be equal to the value of the left-hand addend. These preconditions are set up as subgoals, causing the selection of the START-RAISE and START-COUNT operators, which initialize the forward-chaining procedure of raising and counting fingers one at a time. These operators execute alternately until LEFT-END-COUNT can execute, when the correct number of fingers have been counted on the left hand.

After the left hand has been counted, the CLOBBER-COUNTER operator immediately executes. This operator executes when all the fingers of a hand have been raised along with a running count. Its effects are to zero the value of the counter to prepare it for the next hand, and to mark the current hand as *uncounted*, because the counter's value has been changed. This entire procedure then repeats with the right hand.

After both hands have been counted, DETERMINE-ANSWER checks whether it can execute. It can only execute if both hands are marked as *counted*, but execution of CLOBBER-COUNTER has caused this to be false. Therefore, the system again attempts to count up fingers on each hand, this time marking fingers that are already raised. For this procedure no CLOBBER-COUNTER is necessary, because the number of raised fingers (rather than the value of the counter) is used to terminate the count for each hand. Finally, after each hand has been counted for the second time, GIPS announces the answer.

As the system repeatedly solves addition problems, it continuously updates the execution concepts for the END-COUNT operators. After a while, the concept encodes several regularities that are always true when these operators execute. For example, there are always two addends in the problem description, and the number of "marked" fingers is always zero. Most importantly, however, the concept encodes the number of raised fingers is always equal to the counter value (which in turn is equal to the goal value for counting an addend). Thus, these literals eventually get added into the preconditions for the END-COUNT operators (see Table 1(b)). This action alone does not change the system's outward behavior, but it proves important for later strategies.

## The SHORTCUT SUM Strategy

After new preconditions have been added and a number of addition problems have been solved, the new literals in the system's execution concepts for LEFT-END-COUNT and RIGHT-END-COUNT become strong enough that GIPS attempts to execute the operators earlier than usual. At some point, it thinks that the operators should execute when the number of fingers raised on a hand is equal to the goal value even though the system has not yet incremented its count for the last finger. It turns out that the system can successfully solve the addition problem even if it executes this operator prematurely, so it deletes the condition that the current counter value must be equal to the goal value in the preconditions of the END-COUNT operators (see Table 1(c)).

This change has a direct effect on GIPS' behavior. When attempting to apply LEFT-END-COUNT, the value of the counter no longer appears in the preconditions, so it is not posted as a subgoal. This means that the START-COUNT operator is no longer selected. Thus, a running count is still kept while raising fingers, but the counter is not marked for use as the termination criterion. This means that CLOBBER-COUNTER will not fire, and that leads to two changes in strategy. First, the counter is not reset to zero after counting the left hand, and counting continues from the left hand's final value. Second, the hands are not marked as *uncounted*, so there is no need to count up the raised fingers again after the two hands have initially been counted. This behavior corresponds to the SHORTCUT SUM strategy, which was invented by all 8 of Siegler and Jenkins' subjects.

## The SHORTCUT MIN Strategy

The next shift leads to an intermediate strategy between SHORTCUT SUM and MIN, which we call SHORTCUT MIN. Although Siegler and Jenkins do not classify SHORTCUT MIN as a distinct strategy from SHORTCUT SUM, they do note (p. 119) that some of their subjects begin to switch addends during SHORTCUT SUM so that they start counting with the larger addend on the left hand, rather than just picking whichever addend appears first in the problem. GIPS can also account for this behavior.

An important feature of the SHORTCUT SUM strategy is that the problem solver's counter value is not equal to the number of fingers being raised on the right hand (i.e., the second hand). We hypothesize that this causes interference and subsequent failure. Such interference would not occur with the left hand, because the number of raised fingers in the SHORTCUT SUM strategy is always equal to the value of the counter for that hand. We simulated interference between the value of the counter and the number of fingers raised on the right hand by causing GIPS to fail sometimes during the SHORTCUT SUM strategy when it decided to count the larger addend on its right hand. This caused the system to update the selection concept for the operator that initially assigns an addend to each hand, so that it would prefer to count the smaller addend on the right hand.

## The MIN Strategy

The final strategy shift occurs in a similar manner to the shift from SUM to SHORTCUT SUM. At this point, GIPS has attempted to execute the END-COUNT operators at various times and has been given feedback each time as to whether it would be able to solve the current problem if it executed the operator at that time. Thus, it is slowly learning a "good" concept for when the END-COUNT operators are executable. One of the things that proves to be true every time these operators are executed is that the goal value for counting out a hand is equal to the addend assigned to that hand.

Eventually, the system attempts to fire the LEFT-END-COUNT operator without having raised any fingers at all. When it succeeds by doing this, it deletes the precondition that the number of fingers raised on the hand be equal to the goal value (see Table 1(d)). The system has learned that it can simply start counting from the goal value for the left hand rather than starting from zero. GIPS also attempts to execute the RIGHT-END-COUNT operator early, but this leads to failure. Thus, the system begins to exhibit the MIN strategy, in which the largest number (the left-hand number) is simply announced and used to continue counting the smaller number as in the SHORTCUT MIN strategy.

## The First Strategy

The only other counting strategy found by Siegler and Jenkins is the FIRST strategy. It was used on only 6 trials, all by the same subject. FIRST is similar to the MIN strategy, except that it does not assign the larger addend to the left hand. Rather, it starts with whichever addend is presented first, and continues counting with the second. In GIPS, this strategy follows from the SHORTCUT SUM strategy when the system does not learn about ordering the addends. While using the FIRST strategy, the system can still eventually generate the MIN strategy through the same type of failure-driven learning that leads from SHORTCUT SUM to SHORTCUT MIN.

## Summary and Discussion

Both the SUM strategy and the MIN strategy have three main subgoals: to represent one addend, to represent the the other addend, and to count the union of the representations. The SUM-to-MIN transition involves three independent modifications to the SUM strategy: (1) The subgoal of representing one addend changes from explicitly constructing a set of objects to simply saying the addend. (2) The order of addends is made conditional on their size so that the larger addend is represented by the easier process. (3) The process of representing the other addend is run in parallel with counting up the union. In the SUM strategy, representing the other addend is finished before counting up the union begins.

The GIPS account for each of these transitions is as follows. The first transition is caused by correlational learning of preconditions. GIPS keeps track of which literals in the situation are correlated with the final achievement of the goal of representing the first addend. Eventually, it considers these correlated literals to be just as essential as the originally specified preconditions. It eventually discovers that the originally specified preconditions can be ignored as long as the correlated literals are achieved.

The second transition is caused by normal failure-driven learning. The system uses two apparently equivalent methods, but persistent errors in one of them causes the other one to eventually dominate.

The third transition again involves a correlational type of learning. The COUNT operator is responsible for incrementing the oral counter. Initially, it is selected only when the subgoal of counting-up an addend is present. Eventually, correlated relations that are present in the current state (i.e., that a finger has just been raised) come to dominate the selection concept, and the operator becomes a forward-chaining operator. Basically, the person has developed the habit of counting whenever they raise a finger even if that count doesn't serve any direct purpose. Although GIPS could learn this habit, we actually gave COUNT a forward-chaining selection concept in our experiments in order to save time. Given this habit, it serendipitously achieves the goal of counting the union even when the counter is no longer used to represent the second addend.

Although this summary leaves out some crucial details, it makes it clear that correlational learning is crucial to the GIPS account for the first and third transitions. Ordinary failure driven learning can handle the second.

Our analysis with GIPS helps clarify several important, general issues about strategy change. Siegler and Jenkins observe that, "Not one child adopted a strategy that could be classified as indicating a lack of understanding of the goals of addition." (p. 107) In this respect, the subjects are similar to those of Gelman and Gallistel (1978) who found that very young children would invent correct strategies for counting a set of objects even when unusual constraints were placed on them to thwart their normal strategy.

The initial knowledge given to GIPS does not include any explicit principles of addition or counting. As far as it is concerned, the SUM strategy is just a song that has to be sung the right way. How then does it avoid developing bad strategies? In the Siegler and Jenkins study, students were told after each trial whether they got the problem right. This kind of feedback is crucial to GIPS' learning. GIPS occasionally attempts to execute an operator in situations that would produce a wrong answer. If it were not told that the execution was wrong, it would develop wrong strategies. This demonstrates that an innate understanding of addition is not necessary for a computationally sufficient account of the observed competence.

A common misconception about discovery is that a newly discovered strategy or concept instantly and totally supplants its predecessor. In all protocol-based studies of discovery (e.g., Kuhn, Amsel, & O'Laughlin, 1988; Siegler & Jenkins, 1989; VanLehn, in press), the transition between the old strategy and the new one is gradual. We have not tried to model the gradual transition to the use of the MIN strategy with GIPS because doing it right would require implementing several memory-based strategies. However, it is clear that the probabilistic nature of GIPS' selection and execution concepts would tend to predict a gradual transition.

Starting in the eighth week of the study, Siegler and Jenkins began including "impasse problems," such as 2+23. They had hoped that these would encourage discovery of the MIN strategy, but they did not, for no child first used the MIN strategy on an impasse problem. However, children who had already discovered the MIN strategy began to use it much more frequently on the impasse problems and even on the non-impasse problems that followed the eighth week. GIPS would tend to do the same thing if it were given impasse problems. The larger addend would invite errors during the SHORTCUT SUM and FIRST strategies, which would lower the values of their selection concepts. The inclusion of impasse problems would not effect the error rate of the MIN strategy, so it would gradually become the preferred strategy for all counting trials.

Siegler and Jenkins noticed that some children were consciously aware that they had invented a new strategy in that they could explain it on the first trial where they used it, and some even recognized that it was a "smart answer," in the words of one child. Other children denied using the MIN strategy even when the videotape showed that they had used it. Siegler and Jenkins divided children into those who seemed conscious of the strategy and those who did not, and measured the frequency of their subsequent usage of the MIN strategy. The high awareness group used the MIN strategy on about 60% of the trials where they used any counting strategy. The low awareness group used the MIN strategy on less than 10% of the trials. This suggests that being aware of a newly discovered strategy facilitates subsequent usage of it.

This finding cannot be modeled by GIPS because GIPS has no way to distinguish a strategy that can be explained from one that is inaccessible to consciousness. However, the finding could probably be modeled by combining GIPS with a symbolic example-learning system such as CASCADE (VanLehn & Jones, in press; VanLehn, Jones, & Chi, 1991). In the new system, GIPS would discover a strategy and store a trace of the strategy's actions in memory. This trace would be used as an example to be explained by the second system. If enough of the trace can be recalled for the explanation to succeed, it annotates the steps in the trace and perhaps the operators whose executions produced the steps. These elaborations would make it easier to retrieve the modified operators from memory, and perhaps help in assigning credit and blame, thus speeding the adjustment of the preconditions, selection, and execution concepts. These influences would increase the usage of the new strategy on subsequent problems.

To summarize, GIPS achieves its main research objective, providing a computational account of the several strategy shifts observed during the SUM-to-MIN transition. It uses plausible local processes, rather than global optimization techniques as required by the HPM system. In addition, GIPS uses modest amounts of storage, in contrast to HPM, which stores complete solution traces for indefinite periods. Most importantly, GIPS produces all and only the transitional strategies observed in the Siegler and Jenkins study. The GIPS analysis solves a number of puzzles raised by the Siegler and Jenkins study. These include the ability to make significant strategy shifts without impasse-driven learning, and to avoid inventing bad strategies without assuming innate knowledge of the principles of addition. Thus, GIPS provides a plausible, computationally sufficient account of the discovery of the MIN strategy. However, Siegler and Jenkins produced a second set of findings on the gradual increase in usage of the newly discovered strategy. We have not yet tried to model these findings, but GIPS seems to provide an appropriate framework for doing so.

## References


Ashcraft, M. H. (1982). The development of mental arithmetic: A chronometric approach. *Developmental Review, 2*, 213–236.

Ashcraft, M. H. (1987). Children's knowledge of simple arithmetic: A developmental model and simulation. In C. J. Brainerd, R. Kail, & J. Bisanz (Eds.), *Formal methods in developmental psychology.* New York: Springer-Verlag.

Berwick, R. (1985). *The acquisition of syntactic knowledge.* Cambridge, MA: MIT Press.

Gelman, R., & Gallistel, C. R. (1978). *The child's understanding of number.* Cambridge, MA: Harvard University Press.

Groen, G. J., & Parkman, J. M. (1972). A chronometric analysis of simple addition. *Psychological Review, 79*, 329–343.

Groen, G., & Resnick, L. B. (1977). Can preschool children invent addition algorithms? *Journal of Educational Psychology, 69*, 645–652.

Holland, J. H., Holyoak, K. J., Nisbett, R. E., & Thagard, P. R. (1986). *Induction: Processes of inference, learning, and discovery.* Cambridge, MA: MIT Press.

Jones, R. M. (1989). *A model of retrieval in problem solving.* Doctoral dissertation, University of California, Irvine.

Jones, R. M., & VanLehn, K. (1991). *The Gips model of strategy acquisition.* Manuscript submitted for publication.

Kaye, D. B., Post, T. A., Hall, V. C., & Dineen, J. T. (1986). The emergence of information retrieval strategies in numerical cognition: A development study. *Cognition and Instruction, 3*, 137–166.

Kuhn, D., Amsel, E., & O'Laughlin, M. (1988). *The development of scientific thinking skills.* New York: Academic Press.

Langley, P. (1985). Learning to search: From weak methods to domain-specific heuristics. *Cognitive Science, 9*, 217–260.

Mitchell, T. M., Utgoff, P. E., & Banerji, R. (1983). Learning by experimentation: Acquiring and refining problem-solving heuristics. In R. S. Michalski, J. G. Carbonell, T. M. Mitchell (Eds.), *Machine learning: An artificial intelligence approach.* Los Altos, CA: Morgan Kaufmann.

Neches, R. (1987). Learning through incremental refinement of procedures. In D. Klahr, P. Langley, & R. Neches (Eds.), *Production system models of learning and development.* Cambridge, MA: MIT Press.

Newell, A. (1990). *Unified theories of cognition: The William James lectures.* Cambridge, MA: Harvard University Press.

Pazzani, M., Dyer, M., & Flowers, M. (1986). The role of prior causal theories in generalization. *Proceedings of the Fifth National Conference on Artificial Intelligence* (pp. 545–550). Philadelphia: Morgan Kaufmann.

Schank, R. (1986). *Explanation patterns: Understanding mechanically and creatively.* Hillsdale, NJ: Lawrence Erlbaum.

Schlimmer, J. C. (1987). Incremental adjustment of representations for learning. *Proceedings of the Fourth International Workshop on Machine Learning* (pp. 79–90). Irvine, CA: Morgan Kaufmann.

Schlimmer, J. C., & Granger, R. H., Jr. (1986a). Beyond incremental processing: Tracking concept drift. *Proceedings of the Fifth National Conference on Artificial Intelligence* (pp. 502–507). Philadelphia: Morgan Kaufmann.

Schlimmer, J. C., & Granger, R. H., Jr. (1986b). Incremental learning from noisy data. *Machine Learning, 1*, 317–354.

Siegler, R. S., & Jenkins, E. (1989). *How children discover new strategies.* Hillsdale, NJ: Lawrence Erlbaum.

Svenson, O. (1975). Analysis of time required by children for simple additions. *Acta Psychologica, 39*, 289–302.

VanLehn, K. (1990). *Mind bugs: The origins of procedural misconceptions.* Cambridge, MA: MIT Press.

VanLehn, K. (in press). Rule acquisition events in the discovery of problem solving strategies. *Cognitive Science.*

VanLehn, K., Jones, R. M., & Chi, M. T. H. (1991). *A model of the self-explanation effect.* Manuscript submitted for publication.

VanLehn, K., & Jones, R. M. (in press). Integration of explanation-based learning of correctness and analogical search control. In S. Minton & P. Langley (Eds.), *Proceedings of the symposium on learning, planning and scheduling.*

# Learning, Memory, and Search in Planning

**Pat Langley** and **John A. Allen**

AI Research Branch (MS 244-17)
NASA Ames Research Center
Moffett Field, CA 94035

## Abstract

*This paper describes* DÆDALUS, *a system that uses a variant of means-ends analysis to generate plans and uses an incremental learning algorithm to acquire probabilistic search heuristics from problem solutions. We summarize* DÆDALUS*' approach to search, knowledge, organization, and learning, and examine its behavior on multi-column subtraction. We then evaluate the system in terms of its consistency with known results on human problem solving, comparing it to other psychological models of learning and planning.*

## Introduction

A central aspect of human intelligence is the ability to plan, that is, to generate action sequences that achieve one's goals. As a result, planning and problem solving are important topics within both cognitive psychology and artificial intelligence. However, relatively few planning systems have been designed with an eye toward explaining psychological findings, and even fewer provide accounts of human learning on planning tasks.

In this paper we describe DÆDALUS, a planning system that we designed with human behavior in mind. Unlike most recent work on learning and planning, DÆDALUS employs a combination of forward chaining and means-ends search, represents knowledge in a probabilistic framework, stores both cases and abstractions, and learns through an incremental process of concept formation. In the following section we discuss these aspects of the system, relating them to high-level knowledge of human behavior. After this, we consider its behavior on the domain of multi-column subtraction. Finally, we present a brief evaluation of DÆDALUS' and other models' abilities to explain robust psychological findings.

## An Overview of DÆDALUS

Like most planning systems, DÆDALUS must solve problems that involve transforming an initial state into a desired state through the application of operators. The system describes each state as a set of literals (predicates with arguments), and it describes each problem or subproblem as an initial state conjoined with a set of differences that must be eliminated. DÆDALUS represents each operator in a similar manner, specifying its preconditions as a set of state descriptors and its effects as a set of differences. Later we present examples of states and problems from subtraction, but first let us consider how the system uses this representation to solve problems, organize knowledge, constrain search, and learn from experience.

### Organization of Search

One of the most pervasive phenomena in problem solving is that humans carry out search through some problem space (Newell, 1980), and computational work on problem solving has focused on two basic approaches to organizing this search. In *forward chaining* or state-space search, one applies an operator to an initial state, another operator to its successor, and so forth. At each stage of this process, one considers an operator only if its preconditions exactly match the current state. In *means-ends analysis*, one chooses a difference between the current and desired state, selects an operator which reduces that difference, and attempts to apply the operator. If the operator's preconditions are not satisfied, the method is recursively called with the task of changing the current state into one that satisfies them. Once the operator's preconditions are met, a new state resulting from its application is generated. If the new state satisfies the goals, the method exits successfully; otherwise it recurses with the task of changing the new state into one that does satisfy the goals.

There is considerable evidence that humans use means-ends analysis when information is available about the desired state (Newell & Simon, 1972), letting them focus on operators relevant to goals. However, they can also solve problems for which there is no explicit goal description. Moreover, traditional means-ends systems examine only one difference at a time, and it seems unlikely that humans operate in such a non-gestalt manner. In response, DÆDALUS uses an alternative control scheme – *flexible means-ends analysis* – that prefers operators which reduce more differences and operators whose preconditions more closely match the current state. Thus, the retrieval process incorporates ideas from both approaches, biasing the system toward operators that have more effects and that are more nearly applicable. On tasks with explicit goals, the system takes both differences and state features into account; on less well-defined problems, it retrieves operators based on state descriptors alone. As we will see below, DÆDALUS can also place weights on each difference and state descriptor, letting the system attend to the features appropriate for a given domain.[1]

The system uses these biases to direct a heuristic depth-first search through the space of problem-solving traces. Figure 1 shows a successful trace that transforms the initial into the desired state for a problem involving multi-column subtraction. Each node in this trace corresponds

[1] DAEDALUS borrows the notion of flexible means-ends analysis from Jones' (1989) EUREKA system, which used a very similar idea with a different retrieval method.

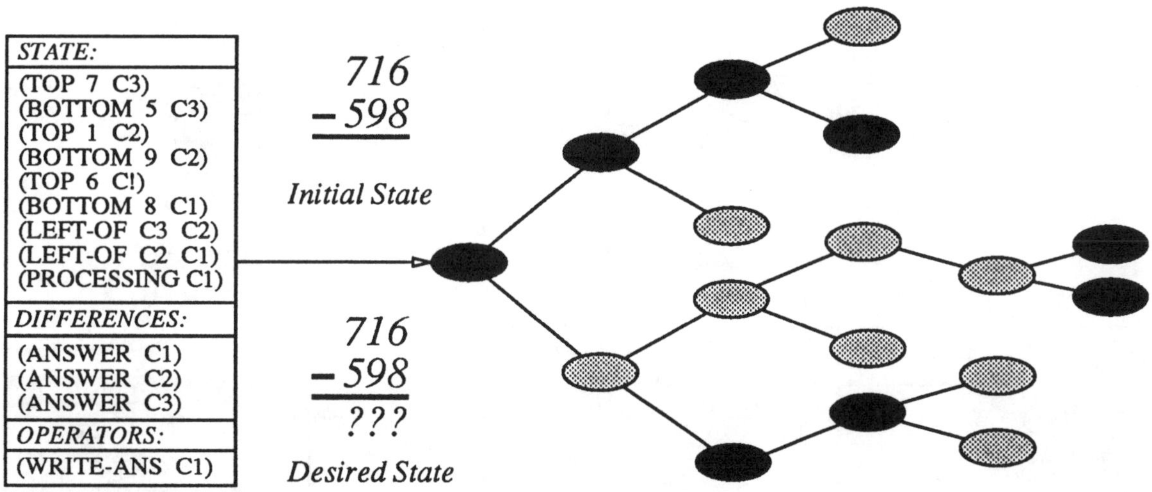


Figure 1: A multi-column subtraction problem, along with a problem-solving trace generated by DÆDALUS. Each node consists of a state description, a set of differences, and the selected operator. Black nodes correspond to problems on which the system initially selected the incorrect operator; gray nodes specify problems on which it made the right selection.

to a problem or subproblem that is described as a set of state predicates, a set of differences, and the operator selected to solve it. DÆDALUS constructs such traces from left to right and from top to bottom. For instance, it first selects an operator for the overall (leftmost) problem, which creates the two subproblems immediately to its right. The system then selects an operator for the first (topmost) subproblem, generates subsubproblems, and recurses. After solving this subproblem, it tentatively applies the original operator and attempts to solve the second subproblem of the original task. In situations where DÆDALUS cannot solve a subproblem, it backtracks and selects another operator. However, user-specified parameters limit the number of operators it considers at each level and the depth of the resulting trace; these bound the overall amount of search.

## Organization of Plan Knowledge

One common approach to encoding plan knowledge involves the use of abstract rules or schemas. For instance, Minton et al.'s (1989) PRODIGY uses abstract search-control rules and Mooney's (1990) EGGS employs general plan schemas. Each rule or schema covers many specific situations, letting these systems use a simple matching or unification algorithm to determine their applicability. Another approach encodes knowledge as specific cases from the domain, including particular problems or subproblems, desirable and undesirable approaches to these problems, and possibly the reasons for their desirability. Researchers in this *case-based* paradigm have proposed a variety of methods for using this information (e.g., Carbonell & Veloso, 1988; Hammond, 1990; Jones, 1989), many of them with direct mapping to techniques that assume abstractions. This approach relies on more sophisticated matching schemes than needed for abstract knowledge, often requiring relational partial matching (i.e., structural analogy).

However, anecdotal evidence suggests that humans store both cases and abstractions in long-term memory. Early knowledge of a domain takes the form of specific problem-solving traces, which must be accessed through some form of analogical retrieval. Later, additional experience forms the basis for abstract schemas, which 'blur together' a number of similar cases. DÆDALUS takes such an integrated view, storing both cases and abstractions in a single probabilistic concept hierarchy. Figure 2 shows the probabilistic hierarchy for the subtraction domain after DÆDALUS has incorporated its experience with the problem from Figure 1. The terminal nodes in gray represent components of the problem-solving trace generated during solution of this task. As we saw earlier, each such trace component corresponds to a problem described as a set of state predicates, a set of differences, and the operator used to solve it.

The figure includes full descriptions for two of these cases (nodes N2 and N3). The additional terminal nodes (in white) represent the original operator schemas that were already present in memory. The hierarchy also contains some abstractions (in black) that DÆDALUS created during the process of storing the trace components, as we describe later. The figure also shows the full description of one abstraction (node N1), which provides a probabilistic summary of the nodes (N2 and N3) below it. Each such description includes an overall probability of occurrence, together with a conditional probability for each difference, state descriptor, and operator. Briefly, the interpretation of each node is that, given the probabilistically specified state descriptors and differences, one should select the operator with the highest probability. This is the form of knowledge that DÆDALUS uses to constrain operator retrieval and thus to direct search during planning.

## Constraining Search with Knowledge

Another basic finding is that experts use their knowledge of a domain to reduce or eliminate search (Chi, Glaser, & Rees, 1982). Early research on planning and problem solving (e.g., Newell et al., 1960; Fikes et al., 1971) had little to say about this topic, and in time a separate tradition emerged that focused on the retrieval of relevant plans or components from memory (e.g., Hammond, 1990). This work has emphasized solution of familiar problems using

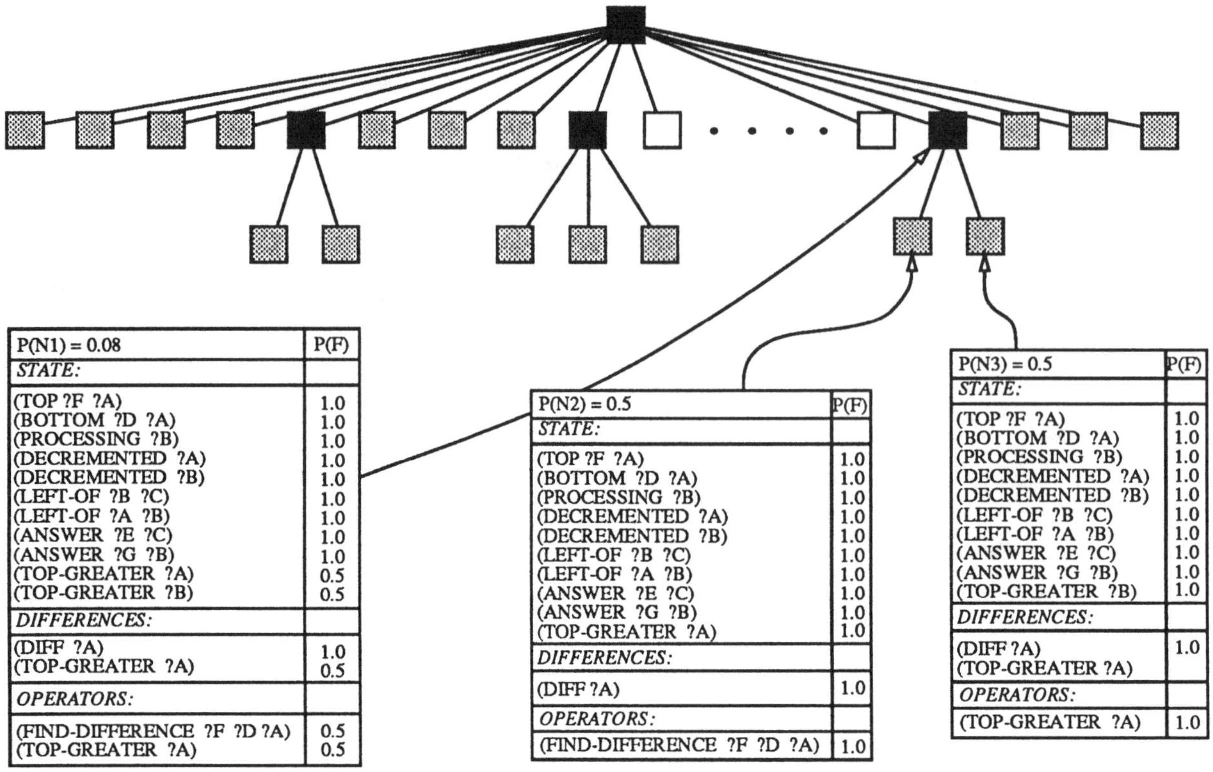


Figure 2: A DÆDALUS concept hierarchy that incorporates cases (gray) and abstractions (black) resulting from storage of components from a problem-solving trace for subtraction, along with the original operator schemas (white) for this domain.

stored knowledge, but downplayed results about problem solving in novel domains, where search plays a major role.

DÆDALUS attempts to model the importance of both search and knowledge in planning. The system operates within a problem-space framework, using domain-specific knowledge stored in its probabilistic concept hierarchy to constrain search when available. However, upon encountering new problems for which it has little knowledge, DÆDALUS gracefully falls back on a more search-intensive approach to plan generation. In both cases, the hierarchy plays the same role for DÆDALUS as does the table of connections for Newell et al.'s GPS (1960), letting it select operators for flexible means-ends analysis.

In order to select an operator, the system invokes COBWEB$_R$, a variant of Fisher's (1987) COBWEB that carries out heuristic classification on relational descriptions. The COBWEB$_R$ module accepts a problem description, which consists of the current state and a set of differences, and sorts it through the concept hierarchy in an attempt to retrieve the most relevant piece of knowledge to bias operator selection. As Allen and Thompson (1991) describe in detail, the system uses the conditional probabilities stored with each node to carry out a heuristic search through the space of partial matches, attempting to find the best match in each case. COBWEB$_R$ invokes a reduced version of *category utility*, the evaluation function used in Fisher's system, to guide search through the space of matches.

After determining a best match between the problem description and each node at a given level of the hierarchy, COBWEB$_R$ selects the node with the highest-scoring match and then recurses to the next level. It continues in this manner until it reaches a terminal node, then selects the operator associated with that node and uses the partial match for the node to determine the operator's bindings. The usefulness of operators retrieved in this manner is related to the amount of knowledge stored in long-term memory. If DÆDALUS has access only to the operators for a domain, it will often select an unprofitable one, leading to backtracking or failure. However, if memory contains detailed knowledge about the situations that occur in a domain, the system is more likely to select an operator on the solution path, reducing search and increasing success rate.

## Acquisition of Plan Knowledge

Human problem solvers learn from experience, reducing their search as they become familiar with a domain. Most recent work on learning in planning domains has taken an analytical approach (e.g., Laird, Rosenbloom, & Newell, 1986; Minton et al., 1989; Mooney, 1990), which involves compiling existing knowledge into new forms. Although elegant, such mechanisms tend to predict faster learning than occurs in humans, who must often work through a problem many times to eliminate search (Anzai & Simon, 1979). Anderson (1983) describes one response, which involves combining an analytical learning mechanism with an empirical technique for strengthening rules. DÆDALUS takes a different approach, employing an inductive learning method to index cases and to construct probabilistic abstractions. Some research has focused on inductive approaches to learning search-control knowledge (e.g., Langley, 1985), but unlike earlier work on this topic, DÆDALUS supports the acquisition of probabilistic plan knowledge. Fisher and Langley (1990) review psychological evidence for probabilistic representations of concepts, and we believe that plan knowledge is stored in the same manner.

Learning in DÆDALUS occurs whenever the system finds a solution to a problem or subproblem, at which point it incorporates the problem description and the successful operator into long-term memory. For instance, each of the nodes from the problem-solving trace in Figure 1 would be sorted through the concept hierarchy and stored in the hierarchy. Like the COBWEB system on which it is based, DÆDALUS makes use of two main learning operations. If a trace component reaches a terminal node in memory, the $\text{COBWEB}_R$ module extends the hierarchy downward, creating a new node $N$ that is a probabilistic summary of the case and the terminal node, and making them both children of $N$. In contrast, if the trace component is summarized poorly by all the children of node $K$, $\text{COBWEB}_R$ creates a new child of $K$ based on the case. During learning, this module uses the full version of category utility to determine when a new branch is justified; this differs from retrieval, in which problems are always sorted to terminal nodes.

In addition, when in learning mode, the system also considers merging and splitting existing concepts, in an attempt to minimize effects due to training order. Again, $\text{COBWEB}_R$ uses the full version of its evaluation function to select between these learning operations and those described above. If the routine decides to incorporate a training case into a given node $N$, it simply updates the probability for $N$ and the conditional probabilities for each of the features (both problem descriptors and operators) associated with it. If a feature occurs in the training instance that is not present in the node, $\text{COBWEB}_R$ adds it to the node's description with a low probability. This averaging process occurs for each node through which the instance passes. One important point is that DÆDALUS incorporates each plan component into long-term memory during problem solving. This scheme is consistent with the incremental nature of human learning.

## DÆDALUS on Multi-column Subtraction

In the previous section we referred to an example involving multi-column subtraction; here we consider DÆDALUS' behavior in this domain in more detail. We focus on subtraction tasks because they have educational relevance, they have been widely studied, and they can be clearly defined yet cause difficulty for many students. The task used in Figure 1, 716 − 598, provides a relatively simple example. Given two rows of digits, one must find their overall difference – in this case 118 – and write this in a third row. Tasks that involve borrowing (i.e., in which the top number in a column is smaller than the lower one) require more operators and are more difficult for humans; this holds especially for tasks containing zeroes in the top row.

For this domain we initialized DÆDALUS' memory with nine operators, each defined in terms of preconditions, add lists, and delete lists, as in STRIPS (Fikes et al., 1971) and in many production-system models (e.g., Anderson, 1983). Three of the operators – FIND-DIFFERENCE, FIND-TOP, and SKIP-ZERO – are responsible for finding the answer associated with a particular column, but are applicable under different conditions and involve slightly different actions. Two other operators – ADD-TEN and DECREMENT – implement borrowing, whereas SHIFT-LEFT and SHIFT-RIGHT change the column on which attention is focused. Finally, the operators WRITE-ANSWER and TOP-GREATER make inferences that support the solution process.

The operators are quite similar to those used by Langley, Wogulis, and Ohlsson (1990) to model subtraction errors, which in turn were based on the rules used in error models proposed by Young and O'Shea (1981). However, this earlier work assumed that students solved subtraction problems in a forward-chaining manner. This runs counter both to intuitions about borrowing and to strategies presented by expert teachers (S. Ohlsson, personal communication), which suggest that means-ends reasoning occurs in multi-column subtraction tasks. DÆDALUS employs its knowledge in this fashion, selecting operators (e.g., FIND-DIFFERENCE) that would make progress toward the desired state, noting that preconditions are not met, and then selecting other operators (e.g., ADD-TEN) to achieve these preconditions. This framework also differs from the problem-reduction scheme used by Brown and Burton (1978) and by VanLehn (1990). Both approaches decompose problems into subproblems, but problem-reduction methods specify such decompositions in advance, whereas DÆDALUS generates them dynamically.

Given the correct operators for multi-column subtraction, DÆDALUS cannot explain the errors commonly observed in students' behavior on this domain (Brown & Burton, 1978; VanLehn, 1990). Because the system only applies operators when their preconditions have been met, its answers are guaranteed to be correct. However, unlike the previous models of subtraction behavior, DÆDALUS may require search to solve a problem even when its operators are correct; this occurs because it may not *select* the best operator on its first attempt. Combined with the system's constrained search algorithm, which considers only a few alternatives at each level, this means DÆDALUS may fail entirely to solve certain problems. This differs from the behavior observed in many students who, upon encountering a difficult problem, produce syntactically correct answers with incorrect digits. In future work, we plan to extend DÆDALUS to generate abstract plans that ignore the violation of certain operator preconditions. The execution of abstract plans would explain some catalogued subtraction errors, such as students' tendency to subtract the top number from the lower one when the latter is larger.

Although DÆDALUS does not model subtraction errors, we believe it provides a viable model of learning in this domain. To demonstrate this, we presented the system with worked solutions to eight problems taken from VanLehn (1990, p. 55), in each case providing DÆDALUS with a problem-solving trace like that in Figure 1. The system stored each component of these traces in long-term memory for use on future tasks. We then presented DÆDALUS with the original tasks and a separate set of twelve test problems taken from the same source, which it had to solve in the absence of further learning. For comparison, we also let the system work on the same problems with no knowledge except the correct operators. Table 1 gives the results: DÆDALUS solved all of the training problems after learning, but poor operator selection let it solve only one of them before learning. Similarly, it solved two of the twelve test problems after learning but none before, showing modest transfer. Moreover, the problems solved after learning were handled with no search. Thus, the system suggests one

Table 1: DÆDALUS' behavior on multi-column subtraction problems taken from VanLehn (1990).

| | PROBLEMS SOLVED | |
|---|---|---|
| | TRAINING | TEST |
| BEFORE LEARNING | 12% | 0% |
| AFTER LEARNING | 100% | 17% |

possible mechanism through which students improve their ability by training on worked sample problems. However, our purpose here is not to claim that DÆDALUS provides a better model of subtraction behavior than other frameworks, but simply to demonstrate that it supports learning in an interesting domain. Elsewhere (Langley & Allen, in press) we report similar results on the blocks world.

## Psychological Adequacy of DÆDALUS

Earlier we noted some high-level aspects of human cognition that are reflected in DÆDALUS, but this was only in passing. Now let us reconsider the system's psychological plausibility in more detail, drawing on VanLehn's (1989) excellent review of the major findings with respect to human problem solving. These phenomena are qualitative in nature, but they still provide constraints on the operation of cognitive simulations. Table 2 lists most of the behaviors that VanLehn reports which are directly related to problem solving, along with others noted by Jones (1989). The table also shows how DÆDALUS fares relative to three other models of problem solving and learning: Anderson's (1983) ACT, Laird, Rosenbloom, and Newell's (1986) SOAR, and Jones' (1989) EUREKA. Langley and Allen (in press) present a more extensive comparison.

The first three phenomena address issues about basic problem-solving strategies rather than learning. Both DÆDALUS and EUREKA incorporate a version of means-ends analysis, which humans appear to use in novel domains; in contrast, SOAR can produce means-ends behavior with preference rules, but the process is not built into the architecture, and ACT provides support for backward chaining but not true means-ends analysis. Of the four frameworks, only EUREKA mimics the nonsystematic strategy of humans, who often explore a search path in depth, then return to the initial state to consider an alternative (Newell & Simon, 1972). All four architectures should account for the relative difficulty of problem isomorphs (Kotovsky, Hayes, & Simon, 1985), but this ability relies on representational assumptions outside the systems themselves.

Some additional behaviors concern changes in performance as humans gain experience in a problem-solving domain. The most basic finding is that learning leads to reduced search on a class of problems. As we report elsewhere (Langley & Allen, in press), DÆDALUS generally carries out less search with experience, as do EUREKA, ACT, and SOAR, although the latter two employ explanation-based learning methods, Jones' system relies entirely on analogical reasoning, and DÆDALUS uses concept formation. The different architectures also differ in their rate of learning. Another phenomenon involves the asymmetry of *transfer* across problems, in which experience on difficult tasks aids the solution of simpler problems more than the reverse situation. Presumably this occurs because the structures needed for the simpler task are subsumed by the more difficult one. Since all four models decompose problems into subproblems, then learn methods for solving these subproblems, all should produce this result. Another well-established transfer effect, *Einstellung* (Luchins, 1942), occurs when one is trained on problems with complex solutions and then given problems with analogous solutions but also with simpler ones. Under such conditions, subjects inevitably solve the new problems in the complex way that has worked in the past. EUREKA and ACT have been explicitly shown to produce this behavior, and we expect SOAR and DÆDALUS to generate similar results.

Table 2: Psychological adequacy of four models of learning in problem-solving domains in terms of whether they account (⊕), fail to account (⊖), or partially account (⊙) for phenomenon from VanLehn (*) and Jones (◇).

| | ACT | SOAR | EUREKA | DÆDALUS |
|---|---|---|---|---|
| Means-ends◇ | ⊙ | ⊕ | ⊕ | ⊕ |
| Nonsystematic◇ | ⊖ | ⊖ | ⊕ | ⊖ |
| Isomorphs* | ⊙ | ⊙ | ⊙ | ⊙ |
| Reduced search* | ⊕ | ⊕ | ⊕ | ⊕ |
| Asymmetries* | ⊕ | ⊕ | ⊕ | ⊕ |
| Einstellung* | ⊕ | ⊕ | ⊕ | ⊕ |
| Verbalization* | ⊕ | ⊕ | ⊖ | ⊖ |
| Reduced time* | ⊕ | ⊕ | ⊖ | ⊖ |
| Rare analogy* | ⊖ | ⊖ | ⊙ | ⊙ |
| Superficiality* | ⊖ | ⊖ | ⊕ | ⊕ |

In addition, the skills of experienced problem solvers are more *automatized* than those of novices, in that they can carry them out with little attention. Experts typically solve problems much more rapidly, even when their solutions involve the same number of steps, and they tend to verbalize much less, suggesting that they have lost access to intermediate subproblems. Both DÆDALUS and EUREKA have difficulty explaining these phenomena, in that they never change the steps taken in generating a solution; learning may eliminate poor choices, but each node in the problem-solving trace must still be constructed one step at a time. In contrast, ACT and SOAR actually eliminate subproblems through learning, which explains the reduction in verbalization and some of the observed speedup.

A final set of empirical results concern problem solving by analogy. Experiments reveal that spontaneous cross-domain analogy is quite rare (e.g., Gick & Holyoak, 1980). People can solve problems by analogy when given an explicit mapping between source and target problems, but they cannot always find such a mapping on their own. Moreover, in cases of spontaneous retrieval, the reminding is usually based on some superficial, surface similarity that may produce a misleading analogy (e.g., Ross, 1984). EUREKA relies on a form of analogical retrieval that operates on surface-level descriptions, and Jones (1989) has shown that it replicates the basic finding that experience with one problem increases the likelihood of solving an analogous problem, but that spontaneous analogy across domains is far from guaranteed. Because DÆDALUS also uses a form of analogical retrieval, it should produce similar results, but (as with EUREKA) only provided it is given states and operators that share some surface features across the domains. ACT and SOAR have more difficulty with these phenomena, since neither has any architectural mechanism for analogy, though both could mimic analogy using explicit rules.

## Conclusions

In the previous pages we described DÆDALUS, a planning system that was designed to be consistent with knowledge of human problem-solving behavior. We found that the system directs problem-space search using a flexible version of means-ends analysis, organizes both cases and abstractions in a probabilistic concept hierarchy, employs domain knowledge to constrain its generation of plans, and acquires plan knowledge through an incremental learning process. In particular, DÆDALUS retrieves relevant operators by sorting new problems through its concept hierarchy, and it stores the components of successful plans in this hierarchy for future retrieval. We demonstrated the system's behavior in the domain of multi-column subtraction.

We also examined DÆDALUS' ability to explain aspects of human cognition at a qualitative level, finding that the system is consistent with a variety of robust phenomena that have been observed in human problem solving. However, three previous models also explain roughly the same behaviors. DÆDALUS differs from Laird et al.'s SOAR and Anderson's ACT in its partial coverage of analogical reasoning, an area it shares with Jones' EUREKA system. However, DÆDALUS fails to explain the reduction of verbalization and the automatization observed in highly-skilled problem solvers, and its search organization does not mimic the nonsystematic behavior found in human problem solving, which only EUREKA has attempted to handle.

In addition, the system fails in the broader sense that humans are physical agents who interleave planning with other processes. A fuller model of human behavior would explicitly link cognition with action and perception. Such issues arise even in constrained domains like subtraction, in which students cannot hold the entire problem in short-term memory, and thus must perceive and alter a physical display. Elsewhere (Langley & Allen, in press) we have described our ideas for extending DÆDALUS along these lines, and developing this augmented model remains an important direction for future research.

## Acknowledgements

We thank Randy Jones, Doug Fisher, Steve Minton, Wayne Iba, Kevin Thompson, John Gennari, Kathleen McKusick, and Deepak Kulkarni for lengthy discussions that led to many of the ideas in this paper.

## References

Allen, J. A., & Thompson, K. (1991). Probabilistic concept formation in relational domains. *Proceedings of the Eighth International Workshop on Machine Learning*. Evanston, IL: Morgan Kaufmann.

Anderson, J. R. (1983). *The architecture of cognition*. Cambridge: Harvard University Press.

Anzai, Y., & Simon, H. A. (1979). The theory of learning by doing. *Psychological Review, 86*, 124–140.

Brown, J. S., & Burton, R. R. (1978). Diagnostic models for procedural bugs in basic mathematical skills. *Cognitive Science, 2*, 155–192.

Carbonell, J. G., & Veloso, M. (1988). Integrating derivational analogy into a general problem solving architecture. *Proceedings of the DARPA Workshop on Case-based Reasoning* (pp. 104–121). Clearwater Beach, FL: Morgan Kaufmann.

Chi, M. T. H., Glaser, R., & Rees, E. (1982). Expertise in problem solving. In R. J. Sternberg (Ed.), *Advances in the psychology of human intelligence*. Hillsdale, NJ: Lawrence Erlbaum.

Fikes, R. E., Hart, P. E., & Nilsson, N. J. (1971). STRIPS: A new approach to the application of theorem proving to problem solving. *Artificial Intelligence, 2*, 189–208.

Fisher, D. H. (1987). Knowledge acquisition via incremental conceptual clustering. *Machine Learning, 2*, 139–172.

Fisher, D. H., & Langley, P. (1990). The structure and formation of natural categories. In G. H. Bower (Ed.), *The psychology of learning and motivation: Advances in Research and Theory* (Vol. 26). Cambridge, MA: Academic Press.

Gick, M. L., & Holyoak, K. J. (1980). Analogical problem solving. *Cognitive Psychology, 12*, 306–355.

Hammond, K. J. (1990). Case-based planning: A framework for planning from experience. *Cognitive Science, 14*, 385–443.

Jones, R. (1989). *A model of retrieval in problem solving*. Doctoral dissertation, Department of Information & Computer Science, University of California, Irvine.

Kotovsky, K., Hayes, J. R., & Simon, H. A. (1985). Why are some problems hard? Evidence from tower of Hanoi. *Cognitive Psychology, 17*, 248–294.

Laird, J. E., Rosenbloom, P. S., & Newell, A. (1986). Chunking in SOAR: The anatomy of a general learning mechanism. *Machine Learning, 1*, 11–46.

Langley, P. (1985). Learning to search: From weak methods to domain-specific heuristics. *Cognitive Science, 9*, 217–260.

Langley, P., Wogulis, J., & Ohlsson, S. (1990). Rules and principles in automated cognitive diagnosis. In N. Fredericksen, R. Glaser, A. Lesgold, & M. G. Shafto (Eds.), *Diagnostic monitoring of skill and knowledge acquisition*. Hillsdale, NJ: Lawrence Erlbaum.

Langley, P., & Allen, J. A. (in press). A unified framework for planning and learning. In S. Minton (Ed.), *Learning and planning*. San Mateo, CA: Morgan Kaufmann.

Luchins, A. S. (1942). Mechanization in problem solving: The effect of Einstellung. *Psychological Monographs, 54* (248).

Minton, S., Carbonell, J. G., Knoblock, C. A., Kuokka, D. R., Etzioni, O., & Gil, Y. (1989). Explanation-based learning: A problem solving perspective. *Artificial Intelligence, 40*, 63–118.

Mooney, R. J. (1990). *A general explanation-based learning mechanism and its application to narrative understanding*. San Mateo, CA: Morgan Kaufmann.

Newell, A. (1980). Reasoning, problem solving, and decision processes: The problem space hypothesis. In R. Nickerson (Ed.), *Attention and performance VIII*. Hillsdale, NJ: Lawrence Erlbaum.

Newell, A., & Simon, H. A. (1972). *Human problem solving*. Englewood Cliffs, NJ: Prentice-Hall.

Ross, B. H. (1984). Remindings and their effects in learning a cognitive skill. *Cognitive Psychology, 16*, 371–416.

VanLehn, K. (1989). Problem solving and cognitive skill acquisition. In M. I. Posner (Ed.), *Foundations of cognitive science*. Cambridge, MA: MIT Press.

VanLehn, K. (1990). *Mind bugs: The origins of procedural misconceptions*. Cambridge, MA: MIT Press.

Young, R. M., & O'Shea, T. (1981). Errors in children's subtraction. *Cognitive Science, 5*, 153–177.

# Memory for Problem Solving Steps

**Stellan Ohlsson**

Learning Research and Development Center
University of Pittsburgh,
Pittsburgh, PA 15260, USA
stellan@vms.cis.pitt.edu

**Abstract**

A widely adopted theory of procedural learning claims that people construct new problem solving rules through induction over past problem solving steps. The underlying assumption that people store information about problem solving steps in memory was tested by measuring subjects' memory of their own problem solving steps in four different ways. The results support the assumption that people store enough information in memory to enable induction of new problem solving rules.

## Memory in Procedural Learning

Theoretical analyses of procedural learning have converged on three general principles for the acquisition of problem solving rules through practice: (a) Learners solve unfamiliar problems through weak problem solving methods such as hill climbing, planning, and search. The function of a weak method is to generate task relevant behavior in the absence of knowledge about the task. (b) Information about the problem solving steps generated by the weak method--context, type of action, outcome--is stored in long-term memory. (c) New problem solving rules are induced from the resulting memory record.

This general theory of procedural learning has been instantiated in several computer simulation models. The models differ with respect to problem solving method, type of information stored in memory, and induction mechanism. For example, the ACT* model (Anderson 1983) solves problems through planning, stores information about the temporal order of problem solving steps, and learns through rule composition (as well as in other ways). The UPL model (Ohlsson 1987) solves problems through search, stores state-operator pairs in memory, and learns through generalization (as well as in other ways). The SAGE model (Langley 1987) solves problems through search, stores information about good and bad outcomes, and learns through discrimination. The Soar model (Newell 1990) also solves problems through search, stores sequences of search steps, and learns through chunking. Although these (and many other) models of procedural learning differ in their specifics, they nevertheless instantiate the same three general principles stated above.

The most common approach to empirical validation of such models is to test their *predictions* against data. For example, validation might involve deriving the power law of learning (Newell & Rosenbloom 1981) or observed error patterns (VanLehn 1990) from a given model. An alternative approach is to provide empirical support for the *assumptions* underlying the entire class of models, preferably support which is independent of the learning data which inspired the creation of the models.

One central assumption of the theory of procedural learning is that people store detailed information about their own problem solving steps in memory while solving unfamiliar problems. Because procedural learning is most rapid at the outset of learning, the hypothesis of internal induction implicitly assumes that the memory record builds up rapidly, i. e., that the memory record is not only detailed but also relatively complete. This assumption is dubious. Both problem solving and memory storage are capacity demanding processes, so it is reasonable to expect them to interfere with each other. If it can be shown that learners do not store information about problem solving steps in memory, the plausibility of the the current theory of procedural learning, and of the entire class of learning models that build on it, is undermined. Conversely, if it can be shown that people indeed store a detailed and complete memory record of their own problem solving steps, the plausibility of the theory of procedural learning is strengthened.

In a study of the Missionaries and Cannibals Puzzle, Reed and Johnsen (1977) found relatively good recognition memory for problem states (63% correct recognition, on the average), but rather poor ability to recall which step was taken in a given problem state. However, as Reed and Johnsen (1977, p. 198) point out, measuring memory for problem solving steps is not straightforward. Unlike the stimuli used in ordinary memory experiments, problem solving steps are generated by the subject, rather than by the experimenter. Hence, the experimenter cannot prepare the stimuli in advance. Problem solving steps can be recorded and used as stimuli later. But accurate assessment of the memory trace has to occur immediately after problem solving, before the trace is obliterated by decay or interference.

In the present study, the power of the computer was used to overcome this methodological difficulty. The experimental problems were presented to the subject via a computer terminal. The computer recorded the subjects' steps. Immediately after a subject indicated that he[1] was finished with a problem, the computer retrieved the subject's solution path, constructed a memory test on the basis of it, and presented the test to the subject.

## General Method

The subjects' memory for their own problem solving steps were measured in different ways in four experiments which followed the same general format.

### Problem Solving Tasks

In the type of problem used in this study, the subject was shown three unmarked jars of different capacities. Initially, the largest jar is full of water and the two smaller jars are empty. The task is to redistribute the water so that the two larger jars have equal amounts (and the smallest jar is empty). The only available action is to pour water from one jar into another jar, until either the source jar is empty or the receiving jar is full, whichever happens first. A solution consists of a sequence of pouring actions that transforms the initial distribution of water into the desired one. Problems like these are known as *water jar problems.* They were introduced into the psychological laboratory by Luchins (1942) and have been studied frequently (Atkinson, Masson, & Polson 1980).

The subjects in all four experiments solved the same six water jar problems in the same order. The simplest problem (with jar capacities 6, 5, and 3) required only two steps and served as a training problem. The second problem was the 8-5-3 problem studied by Atkinson, Masson, and Polson (1980). The remaining four problems had capacities 16-10-6, 24-15-9, 32-20-12, and 40-25-15, respectively. They are isomorphs of the 8-5-3 problem, i. e., all five experimental problems had exactly the same problem space. The graph of this space has 16 states and the shortest path to solution is eight steps long (see Atkinson, Masson, & Polson 1980, p. 183). There are many opportunities to unintentionally return to previously visited states in this space, making the problems non-trivial for novices.

### Presentation System and Procedure

The problems were presented on a computer terminal. The computer represented a problem in an alphanumeric, as opposed to graphical, format: Each jar was labelled with a single letter, and the computer showed a problem state by printing a table with the letters and the current water volumes of the corresponding jars on the screen. The first screen showed both the initial state and the goal state; subsequent screens did not show the goal state but only the current state. Although the initial state could be reviewed at any time (see below), there was no way to review other past problem states without returning to them.

The subject indicated which pouring action he choose by first pressing the keyboard key corresponding to the jar he was pouring from and then the key corresponding to the jar he was pouring into. Illegal moves, e. g., attempting to pour from an empty jar, were rejected with a short explanation (e. g., "You cannot pour from an empty jar").

When the subject reached the goal state, the computer system waited for the subject to press a key that indicated that he knew that he was done with the problem. If the subject instead performed yet another pouring action, thus moving away from the goal state, the system made no objection. The subject could also press a key that indicated that he gave up. The subject also had the options of backing up, i. e., returning to a state preceeding the current state, and of restarting, i. e., of returning to the initial problem state. The initial problem state could also be reviewed at any point along the way. Finishing, giving up, backing up, restarting, and reviewing were each signalled by a single key press, except backing up, which required a second key press indicating the number of steps the subject wished to back up.

When the subject signalled that he was done with a problem, the system constructed a memory test based on its recording of the subject's behavior. The memory items were then presented one at a time. On each item, the subject was asked for a response--different for different memory tests--and a confidence judgment. The confidence judgment was made on a five point scale, from 1 ("just guessing") to 5 ("completely certain"). The subject was then given feedback about the correctness of his response, and the next memory item was presented. After the memory test was completed, the next problem solving task was presented, and so on, through the five trials. Each trial thus consisted of one problem solving attempt followed by a memory test.

The subject was given a brief description of the purpose of the experiment, and he was told that he would be subject to a memory test after each problem. The instructions emphasized that the memory test should be answered by recalling what was done during problem solving, rather than by re-thinking the problem solving task. The experimenter explained the idea behind water jar problems and the operation of the computer system, and remained in the laboratory while the subject solved the training problem and the memory test that followed it. He then waited outside the laboratory while the subject

[1] For the sake of brevity, the pronouns "he", "his", "himself, etc. are used generically to mean "he or she", "his or her", "himself or herself", etc.

solved the five experimental problems and responded to the associated memory tests.

### Subjects

Eight subjects, four psychology students and four computer science students, participated in each of the four experiments, for a total of 32 subjects. The psychology students participated in order to fulfill a course requirement, while the computer science students were payed a small amount for their participation. Half the subjects were men and half were women. The two different subject populations were included in the experiment to ensure a wide range of puzzle solving ability among the subjects.

### Evidence for Procedural Learning

The purpose of the study was to measure memory for problem solving steps *during procedural learning*. Hence, it is important to verify that that the subjects' problem solving performance improved across the five trials. Percent correct, time to solution, and number of steps to solution are commonly used performance measures. Percent correct does not accurately reflect knowledge about water jar puzzles, because the goal state can be found by exhaustive search, or even by a random walk, through the (rather small) problem space. Time to solution is only applicable on correctly solved problems. Worse, because different (correct) solutions have different lengths, time to solution confounds the question of which solution the problem solver knows with the question how fast he is executing it. Neither measure is an accurate indicator of procedural learning. Number of steps to solution have similar disadvantages.

The acquisition of a procedure for a puzzle problem should lead to better solutions. The *solution quality* was measured in the following way. If the subject gave up without having reached the goal state, his solution was assigned the value 0. If the subject reached the goal state, but not through the shortest path, then his solution was assigned the value 1. Shortest path solutions were assigned the value 2. Measured in this way, solution quality improved significantly across trials (Kruskal-Wallis one-way analysis of variance by ranks, corrected for ties; chi-square = 10.65, df = 4, $p < .05$).

A second consequence of procedural learning is that the problem solver has to think less in order to decide what to do in each problem state. Therefore, the *residence time*, i. e., the time the problem solver remains in a problem state before taking an action[2], ought to decrease. The computer system recorded the residence time for each subject and problem state. There was a statistically significant decrease in the average residence time across the five trials (two-way ANOVA, subject population x trials, F = 4.53, df = 4, $p < .01$).

In summary, both the quality of the solutions and the speed of decision making improved across trials, indicating that the subjects learned something about how to solve water jar problems in the course of the study.

## Experiment 1: Path Retracing

The purpose of the first experiment was to measure the subjects' ability to retrace their own solutions. After the subject declared himself finished with a problem, the computer system retrieved the subject's solution path, presented the initial state of the problem and asked "Which action did you take in this situation?". If the subject responded with the action which he did, in fact, take while solving the problem, the system responded "Yes, that is what you did" and presented the next state on the subject's path. If the subject responded with an action different than the one he executed while solving the problem, the computer system answered "No, that is not what you did. Instead, you poured ... into ...", and presented the next problem state *on the subject's original solution path* (rather than the state that would result from applying the operator the subject responded with). In short, the *path retracing* test measured the subject's ability to recreate his solution by stepping through his path and asking him to recall which action he took in each successive problem state. Memory for state-operator pairs is crucial for learning mechanisms such as generalization (Anderson 1983; Ohlsson 1987) and discrimination (Langley 1987).

The results of this memory test are shown in Table 1, row 1. On the average, the subjects retraced 81% of their steps correctly. Although every subject performed better on trial 2 than on trial 1, Table 1 nevertheless suggests that the subjects' memory performance did not improve systematically across trials. A two-way ANOVA (subject population x trials) confirms that the effect of trials in Table 1 is not significant (F = 2.145, $p > .10$).

## Experiment 2: Operator Recall

It is possible to solve the path retracing task used in Experiment 1 by re-solving the problem, rather than by retrieving the previous solution from memory. In order to prevent subjects from answering the memory items through problem solving rather than through memory retrieval, the subjects in Experiment 2 had to recall the operators in random order. After the subject indicated that he was finished with a particular problem, the computer system retrieved the subject's solution path, presented a *randomly* chosen state on

2 The concept of residence time was introduced by Newell and Simon (1972, pp. 811-814).

**Table 1.** Memory performance, in terms of % correct, on each trial for each experiment (memory measure), averaged across subjects.

| Memory measure | Trial 1 | 2 | 3 | 4 | 5 | M |
|---|---|---|---|---|---|---|
| Path retracing | .62 | .81 | .86 | .93 | .82 | .81 |
| Operator recall | .72 | .77 | .74 | .68 | .90 | .76 |
| State recognition | .73 | .84 | .88 | .83 | .87 | .83 |
| Order recognition | .87 | .86 | .86 | .87 | .91 | .87 |

the path, and asked "Which action did you take in this situation?". The system then presented another state selected at random, and so on, until the subject had responded to each state on his solution path. In short, the *operator recall* test measured the subjects' ability to recall which operator he executed in a randomly chosen problem state. This memory test is similar to the memory task used in Experiment II of Reed and Johnsen (1977).

The results of this memory test are also shown in Table 1, row 2. On the average, the subjects recalled 76% of their operators correctly. The memory performance did not improve systematically across trials ($F = 1.251$, $p > .30$).

## Experiment 3: State Recognition

In the third experiment the subjects were tested for memory of problem states rather than operators. After the subject indicated that he had finished with a problem, the computer system first retrieved all problem states on the subject's solution path. It then retrieved from the problem space an equal number of problem states which the subject had *not* visited while solving the problem. The two sets of states were mixed and arranged in random order. Thus, the subject was presented with a sequence of problem states, half of which he had seen while solving the problem and half of which he had not seen before (but which belonged to the same problem space). The subject responded by answering "yes" or "no" to the question whether he had encountered the shown state during problem solving. In short, the state recognition test measured the subjects' ability to discriminate the problem states they had visited during problem solving from those they had not. This memory test is similar to the recognition task used in Experiments I and II of Reed and Johnson (1977). Memory for problem states is crucial for learning mechanisms such as constraint violation (Ohlsson & Rees 1991) and subgoal discovery (Ohlsson 1987).

The results from this memory test are shown in Table 1, row 3. The subjects recognized 83% of the states correctly, on the average. There was no systematic improvement in state recognition across trials ($F = 1.368$, $p > .10$).

## Experiment 4: Order Recognition

In the fourth experiment, the subjects were tested for memory of the temporal order in which they encountered particular problem states. After the subject indicated that he was finished with a problem, the computer system retrieved all the problem states the subject had visited during problem solving and constructed a set of memory items by pairing the problem states. The pairs were presented in random order. One state was presented on the left side of the terminal screen and the other on the right side. The computer system asked the subject "Which of these states did you encounter before the other?". In half the items, the left state was, in fact, encountered earlier in the solution path than the right state; in the other half, the reverse was the case. In short, the *order recognition* test measured the subjects' ability to remember in which order they encountered problem states. Memory for temporal order of problem solving events is crucial for learning mechanisms such as composition (Anderson 1983) and chunking (Newell 1990).

The results of this memory test are shown in Table 1, row 4. The subjects remembered the temporal ordering of 87% of the problem states correctly. As in the previous experiments, there was no significant training effect for the memory performance (F = .153, p > .10).

## Summary and Discussion

The results of the four memory tests were uniformaly high, indicating that the subjects stored approximately 80% of the information probed for while solving the experimental problems. The performance was equally high across the four different ways of accessing memory: retracing a solution, recalling which operator was taken in a given problem state, distinguishing problem states encountered during problem solving from other states in the same problem space, and judging the temporal order in which problem states were encountered. Because the problem states and the operators in the water jar problem are visually impoverished and syntactically similar to each other, the interference effects ought to be strong, biasing the study against high memory performance. Hence, the results presented here might underestimate the amount of memory storage that is typical during problem solving.

The performance of the subjects in the present study is significantly higher than the performance of the subjects in Reed and Johnsen (1977). One possible explanation for this difference is that the states and operators of water jar puzzles are inherently easier to remember than the states and operators of river crossing puzzles. A second possible explanation is that the computer system used in the present study delivered the memory test more promptly, thus minimizing decay of the memory trace. Finally, the difference in performance level might only represent normal sampling variation. These hypotheses cannot be evaluated without further data.

Given that memory for problem solving steps is good but not perfect, it would be useful to know if some types of problem solving steps are remembered better than others. Table 2 shows memory accuracy and confidence judgments from Experiments 1 and 2 for four different types of steps. "Final step" is the action of indicating that the goal state has been reached, "moves on path" are pouring actions on the subject's path to the goal state, "moves off path" are pouring actions that lead away from that path; "corrections" include both backups and restarts. The same pattern is reproduced in both experiments and in both variables: Finishing actions are remembered almost perfectly; moves on path are remembered better than moves off path, which in turn are remembered better than correction steps. This pattern, if replicated in future studies, supports learning mechanisms which induce new problem solving rules from correct steps (e. g., composition, generalization) over mechanisms which derive new rules from incorrect steps and errors (e. g., constraint violation, discrimination).

**Table 2.** Memory accuracy and confidence judgements for steps that finish a problem, moves on path, moves off path, and corrections, for all subjects in Experiments 1 and 2. Confidence was rated on a five-point scale on which 5 indicated highest confidence.

| | Type of problem solving step | | | |
|---|---|---|---|---|
| Experiment | Final step | Moves on path | Moves off path | Correction steps |
| Exp. 1: Path retracing | | | | |
| Accuracy | .96 | .76 | .73 | .43 |
| Confidence | 4.3 | 3.6 | 3.0 | 2.0 |
| Exp. 2: Operator recall | | | | |
| Accuracy | .97 | .79 | .56 | .24 |
| Confidence | 4.7 | 3.3 | 3.0 | 2.0 |

It is possible that the subjects changed their behavior on the problem solving tasks so as to maximize their performance on the memory tests. For example, they might have rehearsed each problem solving step, something they probably would not normally do during problem solving. This possibility cannot be conclusively ruled out on the basis of the present data. But three aspects of the data argue against this possibility. First, the subjects learned to solve water jar problems; their problem solving performance improved across trials. Hence, they did not ignore the problem solving task. Second, if the subjects used various strategies for maximizing their memory performance, then we would expect their memory performance to improve across trials, but this was not the case. Third, although one can imagine that rehearsal of steps would help memory performance on the first three experiments, it is not obvious what kind of memory strategy one could use to improve performance on the order recognition test. But the subjects performed well on this test also.

It is also possible that the subjects ignored the instruction to respond to the memory items by trying to remember what they did during problem solving, and instead answered them by re-solving the problem. This possibility cannot be conclusively ruled out. However, re-solving the problem is obviously easier to do while retracing a solution path than while recalling operators in random order, which implies that the subjects should have performed significantly better in Experiment 1 than in Experiment 2. But this was not the case.

The results of this study support the notion that people store a detailed description of their own problem solving behavior while solving problems. The conclusion implies that it is *possible* that people learn new rules through induction over past problem solving steps. The present study does *not* address the question whether, in fact, there is a causal link between memory storage and procedural learning. If there is such a link, one would expect rate of procedural learning to be predictable from memory performance. But the data presented here cannot be used to investigate the correlation between memory performance and problem solving performance, for two reasons. First, it is not clear how to compute a measure of rate of learning from these data. Second, the uniformly high performance on the memory tests causes a restriction of range on the presumed predictor variable, precluding finding a significant correlation, even if it exists. Trying to predict rate of procedural learning from memory performance is a task for the future. The present study does not prove that our current theory of procedural learning is true, but it provides support for one of its basic assumptions, support which is independent of the learning curves that prompted the formulation of the theory.

## Acknowledgements

Preparation of this manuscript was supported in part by grant no N00014-89-J-1681 from the Cognitive Science Program of the Office of Naval Research. Nancy Bee and Joyce Holl did accurate and reliable work on the data files.

## References

Anderson, J. R. 1983. *The architecture of cognition.* Cambridge, MA: Harvard University Press.

Atwood, M. E., Masson, M. E. J., and Polson, P. G. 1980. Further explorations with a process model for water jug problems. *Memory & Cognition* 8:182-192.

Langley, P. 1987. A general theory of discrimination learning. In D. Klahr, P. Langley, and R. Neches, eds. *Production system models of learning and development* (pp. 99-161). Cambridge, MA: MIT Press.

Luchins, A. S. 1942. Mechanization in problem solving. The effect of Einstellung. *Psychological Monographs,* 54(6), Whole No. 248.

Newell, A. 1990. *Unified theories of cognition.* Cambridge, MA: Harvard University Press.

Newell, A., and Rosenbloom, P. S. 1981. Mechanisms of skill acquisition and the law of practice. In J. R. Anderson, ed. *Cognitive skills and their acquisition* (pp. 1-55). Hillsdale, NJ: Erlbaum.

Newell, A., and Simon, H. A. 1972. *Human problem solving.* Englewood Cliffs, NJ: Prentice-Hall.

Ohlsson, S. 1987. Transfer of training in procedural learning: A matter of conjectures and refutations? In L. Bolc, ed. *Computational models of learning* (pp. 55-88). New York, NY: Springer-Verlag.

Ohlsson, S., & Rees, E. 1991. Adaptive search through constraint violation. *Journal of Experimental and Theoretical Artificial Intelligence* 3:33-42.

Reed, S. K., and Johnsen, J. A. 1977. Memory for problem solutions. In G. H. Bower, ed. *The psychology of learning and motivation. Advances in research and theory. Vol 11* (pp. 161-201). New York, NY: Academic Press.

VanLehn, K. 1990. *Mind bugs. On the origins of procedural misconceptions.* Cambridge, MA: Bradford Books.

# Knowledge Transfer among Programming Languages

**Quanfeng Wu & John R. Anderson**
Department of Psychology, Carnegie Mellon University
Pittsburgh, PA 15213
wu@psy.cmu.edu, anderson@psy.cmu.edu

**Abstract**

Two experiments were conducted to investigate knowledge transfer from learned programming languages to learning new ones. The first experiment concerned transfer from knowing LISP to learning PROLOG; the results showed that subjects who knew LISP had significant advantages over subjects who did not. Moreover, among the subjects who knew LISP those who knew LISP better seemed to learn PROLOG faster. The second experiment studied transfer from knowing either PASCAL or PROLOG to learning LISP; attention was specifically focused on transfer of knowledge of writing recursive and iterative programs in these languages. The results indicated that PROLOG programmers, who were usually more knowledgeable on recursion, were more ready to learn the recursive part of the LISP language. Some general theoretical discussion about knowledge transfer among programming languages is also presented in the paper.

## Introduction

Two kinds of knowledge transfer among programming languages are possible and can be separately studied. One is problem solving transfer; it involves people who already know both languages and transfer knowledge from solving a certain problem in one language to that in the other. Katz (1988) and Wu & Anderson (in preparation) have reported some evidence of transfer of this type among LISP, PROLOG, and PASCAL; these studies demonstrated significant positive transfer among the three languages. The other type is learning transfer; i.e., transfer from knowing one or more programming languages to learning a new one. The present paper concerns this second type.

This article will report two experimental studies of such transfer, independent from the tutoring work. The first experiment investigated learning transfer from LISP to PROLOG. The results showed significant positive transfer. Moreover, the first experiment also revealed that subjects who knew LISP better seemed to learn PROLOG faster. The second experiment studied learning transfer from either PASCAL or PROLOG to LISP. This experiment reinforced the evidence of positive transfer among the three languages. Also, in this experiment attention was specifically focused on transfer of knowledge of writing two types of repetition programs -- namely, recursive and iterative programs -- among these languages. The results indicated that PROLOG programmers, who were usually more knowledgeable on recursion, were more ready to master the recurive part of the LISP language.

It is our conviction that commonalities in knowledge representation form the basis for transfer. That is, it is the common knowledge shared by two domains that makes transfer possible from one domain to the other. PASCAL, LISP, and PROLOG are generally classified into three different categories of programming languages -- namely, imperative (or procedural), functional, and logical languages. However, these three languages actually have a great deal in common. These commonalities include similar basic arithmetic operations and predicates, and similar basic data types and operations on them. Some comparative analyses about commonalities shared by the three languages are given in Wu & Anderson (in preparation). Following the analyses there, it is reasonable to expect that learning transfer would occur among them, just as problem solving transfer occurred among them.

## Experiment 1: Transfer from LISP to PROLOG

As this experiment was to investigate learning transfer from knowing LISP to learning PROLOG, the experimental design was basically a between-subject type. By comparing the learning performance of Know-LISP and Not-Know-LISP subjects we can study whether and to what extent LISP experience would facilitate learning PROLOG.

### Method

***Subjects.*** 24 subjects participated in the experiment; they were recruited from CMU (Carnegie Mellon University), either graduates or undergraduates. Among them, sixteen knew LISP beforehand and the rest did not know LISP. From the information gathered from questionnaire, subject's averaged GRE/SAT math scores were 708. For those who knew LISP, their averaged self-ratings of LISP proficiency were 3.44 (along a scale from 0 to 5) and their self-ratings of PASCAL proficiency were 3.63 (all of them knew PASCAL). Among those who did not know LISP only 6 knew PASCAL and their self-rating of PASCAL proficiency were 2.67. The GRE/SAT math scores for the Know-LISP condition and the Not-Know-LISP condition were comparable.

***Design.*** Table 1 shows the design of the experiment. There were three groups of subjects with eight in each group. The subjects who knew LISP were divided into Group 1 and Group 2; the subjects who did not know LISP constituted Group 3. There were four experimental sessions, roughly about two hours for each.

Group 1 and Group 2 were treated differently only in Session 4: one group was shown LISP solutions only on odd-number problems while the other was shown LISP solutions only on even-number problems. The manipulation of Know-LISP versus Not-Know-LISP was aimed to get a global measure of how LISP knowledge facilitates learning PROLOG; the manipulation of shown versus not-shown LISP solutions was intended to get some specific understanding of how LISP knowledge helps PROLOG programming on some particular problems. These two manipulations can be seen at the two different levels of transfer between LISP and PROLOG -- that is, leaning transfer and problem solving transfer.

**Table 1. The design of Experiment for transfer from knowing LISP to learning PROLOG.**

| | Know-LISP conditions | | Not-Know-LISP condition |
|---|---|---|---|
| | Group 1 | Group 2 | Group 3 |
| Session 1 | Questionnaire; Two LISP programming problems for pretesting LISP knowledge; Read introductory material which is based on LISP knowledge. | | Questionnaire; Read introductory material to PROLOG which is independent from LISP knowledge. |
| Sesssion 2 & Session 3 | Use the PROLOG tutoring system to go through a series of problems covering basic concepts and syntax in PROLOG. | | The same as the knew-LISP conditions. |
| Session 4 | Use real PROLOG environment to do eight problems, mostly dealing with list processing; | | The same as the knew-LISP conditions; |
| | For odd number problems, shown LISP solutions; for even number problems, not shown LISP solutions. | For even number problems, shown LISP solutions; for odd number problems, not shown LISP solutions. | No LISP solutions shown at all. |

***Materials and Procedure.*** In Session 1, two LISP programming problems were used as a pretest of subject's LISP knowledge for Know-LISP subjects. Subjects did LISP programming on the Andrew system, a campus network at CMU; the particular language version used was CommonLisp. Also in Session 1, all the subjects were asked to read an introductory material to PROLOG. For most of the subjects participating in the experiment, that was their first time exposed to PROLOG language; few of them had already learned some concepts about PROLOG, but none had ever actually programmed in PROLOG. Two different kinds of instructional materials were used for Know-LISP subjects and Not-Know-LISP subjects. For Know-LISP subjects, the material was written by the authors based on a comparison of LISP and PROLOG; for Not-Know-LISP subjects, the instructional material was a chapter taken Mayer (1988) which assumes no prerequisite knowledge in LISP.

In Session 2 & 3, subjects used a PROLOG tutoring system developed here in the Psychology Department at CMU to go through a series of problems with the purpose of reinforcing their knowledge about the basic concepts and syntax of PROLOG they had studied in Session 1. In Session 4, subjects again used the Andrew system to solve eight problems in a real PROLOG environment; the particular language version used was CPROLOG. Session 4 was particularly intended as a post-test of subject's PROLOG knowledge learned in the first three sessions. The problems were mostly dealing with list processing as that is a prominent feature of both LISP and PROLOG. These eight problems were: Membership, Append, Attach, Reverse, Powerset, Binary-Code, Flatten, and Skeleton; structurally, the odd-number problems are isomorphic to the following-up even-number problems (e.g., Powerset to Binary-Code).[1]

When subjects were using real LISP or PROLOG programming environments on the Andrew system, they had two windows -- one was EMACS editor window to edit the program and the other was used to debug and test the program. Each time when they made revisions to the program they save the program into different files so that the experimenter could keep track the significant errors made by subjects in the process.

## Results and Discussion

Here two measures were used for the comparison made between the two groups of Know-LISP subjects: the total problem solving time and the algorithmic similarity between the shown LISP solutions and the first drafts of subject's PROLOG programs.

[1] The problems and instructuctional materials used in the experiment can be obtained by writing to the authors.

Figure 1 shows the mean total problem solving times subjects spent on the last six problems in Session 4. Note that there are some dramatic contrasts between these two groups for odd-number problems, but less dramatic contrasts for even-number problems; this is because the even-number problems are isomorphic to the just-previous odd-number problems. Thus in this experiment, we not only got the results of transfer from studying LISP solution to PROLOG programming, we also got the manifestation of transfer between isomorphic problems (However, since this transfer is not the primary goal of the present study, it is not investigated any further here). To test the statistical significance of the data, a three-way 2 X 2 X 3 ANOVA (Shown/Not-shown LISP solutions, Odd/Even problems, 3 pairs of problems) was performed. The results showed significant main effects due to whether or not LISP solutions were shown: $F(1, 7) = 48.4$, $p < 0.001$ . Averaging the data over the two groups and the six problems, there was 42% saving of the total problem solving time for shown LISP solutions than not shown LISP solutions.

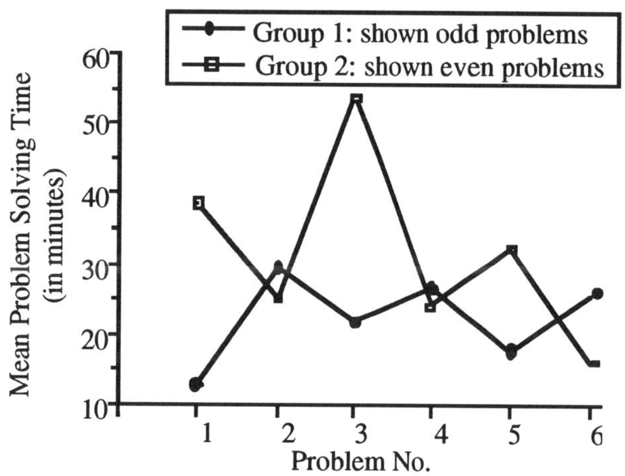


**Figure 1. The mean problem solving time of two groups of know-LISP subjects.**

With respect to the algorithmic similarity between LISP and PROLOG programs, we first broke the shown LISP programs into basic algorithmic components and then scored how many of these basic components appeared in the first drafts of the PROLOG programs written by subjects. By such a measure we can determine how many algorithmic components subjects imported from the shown LISP solutions to PROLOG programming. The data for this measure are presented in Figure 2. The transfer pattern of the results is the same as the patterns appearing in the first measure; this again demonstrated significant transfer from the shown LISP solutions to PROLOG programming. Again, a three-way ANOVA was performed on the data; the test results once more showed significant main effect due to whether or not LISP solutions were shown: $F(1, 7) = 47.53$, $p < 0.001$.

The above two measures largely contribute to our understanding of problem solving transfer from particular

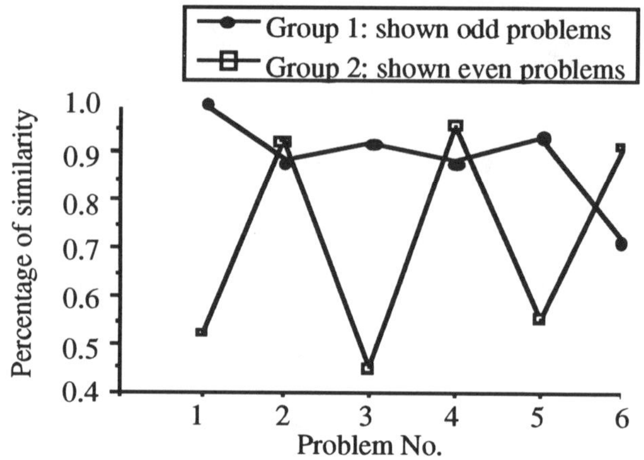


**Figure 2. Algorithmic similarities between shown LISP solutions and first drafts of PROLOG programs by Know-LISP subjects**

LISP programs to PROLOG programming. To see the effect of learning transfer of LISP knowledge upon PROLOG learning, we used the problem solving time for two comparisons: one between Know-LISP subjects versus Not-Know-LISP subjects, and the other between those who had stronger LISP background versus those who had weaker LISP background among Know-LISP subjects. However, the Not-Know-LISP subjects only finished on average 3.5 problems in this session. Thus, here we only used the problem solving time for the first three problems. Table 2 show the data for these two comparisons; as only six Not-Know-LISP subjects finished three or more problems, the data presented for this condition were only averaged over them. For Know-LISP subjects, the two classes of Strong-LISP and Weak-LISP were determined by the mean problem solving time spent by them on the two LISP pretest problems. As seen from the data, Know-LISP subjects spent less time on solving all the three problems; that is, they were learning faster than Not-Know-LISP subjects. The results also indicated that stronger LISP background helped subjects to learn PROLOG faster. Overall, these results tend to indicate that LISP knowledge indeed facilitates PROLOG learning. However, as the questionnaire information indicated, those who were good in LISP were usually also good in PASCAL or other programming languages; therefore, in general, we may say it is the knowledge of previously learned languages which helps learning a new programming language.

Although there were some syntactic interferences from LISP knowledge to PROLOG programming such as confusing list conventions in the two languages, these interferences seemed to be insignificant and could be avoided by paying more attention in both programming and debugging. Stylistically, LISP knowledge also seemed to make subjects come up with PROLOG solutions which were more procedural that they might have otherwise, and led to other confusions such as the notions of a predicate in

PROLOG and a function in LISP. However, these negative effects could be easily overcome by conscious practice. To some extent, since our introduction to PROLOG based on LISP has emphasized the differences between the two languages, subjects were prepared to avoid difficulties.

**Table 2. The comparison between Know-LISP and Not-Know-LISP subjects and the comparison between Strong-LISP and Weak-LISP subjects.**

| | | GRE/SAT Math scores | Lisp Experience | Time for problem 1 | Time for Problem 2 | Time for Problem 3 | Mean |
|---|---|---|---|---|---|---|---|
| Know-LISP | Strong-LISP | 714 | 41.0 | 16.3 | 19.5 | 30.5 | 22.1 |
| | Weak-LISP | 740 | 79.8 | 35.7 | 34.5 | 45.5 | 38.6 |
| Not-Know-LISP | | 710 | | 53.3 | 36.8 | 55.2 | 48.4 |

## Experiment 2: Transfer from PASCAL/PROLOG to LISP

The purpose of this second experiment was to investigate learning transfer from either PASCAL or PROLOG to LISP. Here we focused our attention specifically on transfer of knowledge of writing repetition programs in these languages. In each of these languages there are two ways of implementing repetition, i.e., iteration and recursion. However, in PASCAL iteration is usually stressed while in PROLOG recursion is usually emphasized.[2] We would expect that PASCAL programmers might be more ready to transfer knowledge of iteration to LISP programming while PROLOG programmers might be more ready to transfer knowledge of recursion to LISP programming. The experimental design used here was specifically aimed at testing this hypothesis.

### Methods

***Subjects.*** 9 subjects participated in this experiment. They were all from CMU; among them 2 were undergraduates and 7 were graduates. Eight of them knew PASCAL before taking part in the experiment, while only three knew PROLOG beforehand (i.e., two knew both PASCAL and PROLOG). From the information collected from questionnaire, subject's averaged GRE/SAT math scores were 734. For those who knew PROLOG, their averaged self-ratings of PROLOG skill were 3.29; for those who knew PASCAL, their averaged self-ratings of PASCAL skill were 3.33; besides, all but one subject knew C, their averaged self-ratings of C skill were 3.00. The GRE/SAT math scores of the Know-PASCAL condition and that of the Know-PROLOG condition were comparable.

***Design.*** The experimental design is shown in Table 3. Basically, six subjects who knew PASCAL but did not know PROLOG were used as the Know-PASCAL condition; while the other three, all of whom knew PROLOG but two of them also knew PASCAL, consisted the Know-PROLOG condition (the small number of subjects in this condition was the result of our difficulty of recruiting PROLOG programmers as subjects from the CMU community). There were three experimental sessions, each approximately one and half hours long. The manipulation of Know-PASCAL versus Know-PROLOG, together with the manipulation of presenting recursive vs. iterative problems in Session 2 & 3, were aimed to test the hypothesis mentioned above.

***Materials.*** In Session 1, three problems were used to pretest subject's knowledge of either PASCAL or PROLOG programming; these problems are Summorial, Read-month, and Fibonacci. Also in Session 1, subjects were asked to read an introduction to LISP based on either PASCAL or PROLOG according to their conditions. The materials covered in these introductions included basic concepts and syntax, arithmetic expressions, flow-of-control, and some intermediate data types in LISP. Subjects normally spent about one hour to cover the major portion of the material in Session, and spent about twenty minutes in Session 2 & 3 to cover the rest. There were 12 problems used in Session 2 & 3 for LISP programming. These problems were: Simple-Expression, Circle-Area, Line-Slope, Add-Fraction, Least-Common-Multiplier, (the next four on recursion) Summorial, Integer-Power, Add-Together, Ackerman, (the next three concerning iteration) Summorial, Add-Together, Exponential.[3]

***Procedure.*** As in Experiment 1, subjects were asked to fill out the questionnaire form. For programming in the three languages, almost the same procedure as used in Session 4 of Experiment 1 was followed in all the sessions of this experiment. However, here instead of using the standard EMACS editor we adopted a specifically designed editor based on EMACS; this editor was used to time-stamp each step of interaction subjects made with the computer. Each time when the subject began to edit or modify a program he was requested to turn on the timing editor; and each time when he quit the editing window to

[2] The PROLOG interpreter employs a back-tracking mechanism which inherently implies recursive programming is natural for most of problems which involve repetition.

[3] Again, the problems and instructional materials used in this experiments can be obtained by writing to the authors.

go to the debugging window he was asked to turn off the timing editor. This editor enabled us to collect some fine-grained measures of the time course of programming.

**Table 3. The design of Experiment 2 on transfer from PASCAL/PROLOG to LISP.**

| | Know-PASCAL condition | Know-PROLOG condition |
|---|---|---|
| Session 1 | Questionnaire;<br>Three PASCAL problems for pretesting PASCAL knowledge;<br>Read an introduction to LISP which is based on PASCAL knowledge. | Questionnaire;<br>Three PROLOG problems for pretesting PROLOG knowledge;<br>Read an introduction to LISP which is based on PROLOG knowledge. |
| Session 2 | Five LISP problems for basic syntax and simple arithmetic expressions;<br>Two LISP problem for recursion. | The same as the know-PASCAL conditions. |
| Session 3 | Two LISP problem for recursion;<br>Three LISP problems for iteration. | The same as the know-PASCAL conditions. |

## Results and Discussion

For the time measure, we dissected the problem solving time for each problem into programming (coding) time and debugging time; this could be easily done by counting the time spent either in the editing window or in the debugging window. Taking the method from Singley & Anderson (1990) and Katz (1988), we further decomposed the programming time into thinking time (planning time) and keystroking time (execution time). The keystroking time was defined as follows:

Keystroking time = $\Sigma$ over all keystrokes T

Where T is either the interval between two consecutive keystrokes if that is less than or equal to 2 seconds, or just 2 seconds if the interval is greater than 2 seconds. The thinking time is simply the rest of the programming time; that is,

Thinking time=Programming time- Keystroking time.

We could further decompose the thinking, keystroking and debugging times into first-draft (initial-draft) time and rest-draft (subsequent-draft) time; this could be done by considering the transitions made between the two windows.

In Wu & Anderson (in preparation) we found that time-savings in problem-solving transfer were largely localized in the first-draft coding and debugging. This finding was verified also to be true for our present situation of learning transfer. So in this article we will limit our report of data only to the first-draft time; however, the data pattern also applies to the total problem solving time. [4]

Table 4 presents the data for the first-draft time for a comparison between Know-PASCAL and Know-PROLOG subjects. A two-way ANOVA (recursion/iteration X PASCAL/PROLOG) was performed on the data of PASCAL/PROLOG programming time. The effect due to recursion/iteration was insignificant: $F(1, 8) = 0.51$, $p > 0.1$; the effect due to PASCAL/PROLOG was also insignificant: $F(1, 8) = 0.09$, $p > 0.1$; however, the interaction between them was significant: $F(1, 8) = 8.04$, $p < 0.05$. Another similar two-way ANOVA was performed on the data of LISP programming time for both PASCAL and PROLOG knowers. The effect due to recursion/iteration was significant: $F(1, 32) = 6.72$, $p < 0.05$; the effect due to PASCAL/PROLOG was also significant: $F(1, 32) = 13.08$, $p < 0.05$; however, now the interaction between them was insignificant: $F(1, 32) = 0.21$, $p > 0.1$. Also from the table we see that the time ratio between these two conditions on recursion problems remained almost the same from PASCAL/PROLOG to LISP. Know-PROLOG subjects were better on recursion that Know-PASCAL subjects in the initial test. This advantage was maintained in recursive programming in LISP. That is, PROLOG programmers indeed transferred what they learned about recursion in PROLOG to LISP programming. On the other hand, our Know-PROLOG subjects also did better on iteration in LISP even though they had not in the original test. Concerning iteration, it seems that no general conclusion could be reached about transfer between the three languages from this experiment.

[4] A more detailed report can be obtained by writing to the authors.

## General Discussion and Conclusion

In explaining knowledge transfer, the theory of identical elements had been proposed very early in the century ( Thorndike & Woodworth, 1901). Basically, the theory postulates that it is the common elements shared by two domains of knowledge that enables the knowledge acquired in one domain to be transferred to the other. As there are indeed a great many commonalities between LISP and PROLOG, it is not surprising that in the first experiment we witnessed substantial transfer from LISP knowledge to learning PROLOG. With respect to recursion, as PROLOG and LISP have more in common than PASCAL and LISP have, so in the second experiment we demonstrated that PROLOG programmers had relative advantage over PASCAL programmers in learning the recursive part of LISP.

**Table 4. Data comparison between the Know-PASCAL and the Know-PROLOG conditions in Experiment 2 (time in seconds).**

| | Total time on 3 PASCAL/ PROLOG problems | Time on one PASCAL/ PROLOG iteration problem | Time on one PASCAL / PROLOG recursion problem | Total time on 12 LISP problems | Total time on 3 iteration LISP problems | Total time on 4 recursion LISP problems |
|---|---|---|---|---|---|---|
| Knew-PASCAL | 2173 | 547 | 917 | 8051 | 2780 | 2628 |
| Knew-PROLOG | 2152 | 922 | 352 | 3778 | 1313 | 980 |
| PROLOG/PASCAL | 99% | 168% | 38% | 47% | 47% | 36% |

In Wu & Anderson (in preparation), we reported three experiments demonstrating the existence of problem solving transfer among PASCAL, LISP, and PROLOG. To account for the results, there we proposed that there are basically three levels of transfer across programming in different languages -- namely, the syntactic, algorithmic, and problem levels. There we found that at the syntactical level there were minor interferences (a type of negative transfer) among programming in different languages; at the algorithmic level, substantial positive transfer was manifested as on most occasions subjects used the same algorithm for the same problem in different languages and a great deal of time-saving was exhibited in reprogramming. Besides these two levels, there we also found that sometimes, although different algorithms were used for the same problem in different language, there was still positive transfer manifested (as time-saving); this transfer must have resulted from problem understanding and we postulated it as transfer at the problem level.

In the present investigation we have demonstrated transfer both for problem solving and for learning. Here again we showed that syntactic transfer is largely negative but only plays a minor role in learning and problem solving. We also showed that algorithmic transfer constitutes the major portion of transfer, as manifested by saving the total problem solving time, reducing the number of program drafts needed for revising the program, and importing algorithmic components from LISP solutions to PROLOG programming. However, in both experiments we did not see that different algorithms were used for the same problem in different languages, thus transfer at the problem level was not demonstrated in this study.

On the other hand, as this study was focused on learning transfer, we also witnessed a higher level transfer than the syntactic and algorithmic level transfer; and this transfer is also different from the transfer at problem level. This level may be called learning transfer -- the transfer of the most general programming knowledge which one does not need to learn again and again when learning new programming languages. One example of such transfer would be the notion of recursion; after learning how to implement repetition in a recursive way in one language one can easily transfer that knowledge to learning a new language provided that the new language also easily facilitates recursion.

W would conclude from the present study:

1. There was significant transfer from learned LISP knowledge to learning PROLOG; that is, LISP knowledge helped to learn PROLOG faster. Moreover, the more LISP knowledge the more facilitation to learning PROLOG.

2. Although syntactically and stylistically, there were some interferences from LISP to PROLOG programming, these negative effects were greatly overwhelmed by positive transfer at other higher levels. At the algorithmic level, transfer was manifested as substantial time saving, fewer revisions of programs, and importing algorithmic components from shown LISP solutions to PROLOG programming.

3. As PROLOG programmers usually would be more knowledgeable than PASCAL programmers on recursive programming, PROLOG programmers seemed to have relative advantages over PASCAL programmers in learning the recursive part of LISP programming.

4. Aside from the three levels of transfer we proposed for problem solving transfer among programming languages -- the syntactic, algorithmic, and problem levels, one more level of transfer was proposed there -- the learning transfer. The knowledge responsible for this transfer is the most general programming knowledge across several languages and across a large number of problems, such as the knowledge about recursive programming.

## References


Anderson, J. R. 1983. *The Architecture of Cognition.* Cambridge, Mass: Harvard University Press.

Katz, I. R. 1988. Transfer of Knowledge in Programming. Ph.D. Diss., Dept. of Psychology, Carnegie Mellon Univ.

Mayer, H. G. 1988. *Programming Languages.* New York: MacMillan Publishing Company.

Singley, M.K.; Anderson, J. R. 1990. *The Transfer of Cognitive Skill.* Cambridge, Mass: Harvard University Press.

Thorndike, E. L.; Woodworth, R. S. 1901. The influence of improvement in one mental function upon the efficiency of the other function. *Psychological Review*, Vol. 8.

Wu, Q.; Anderson, J. R. in preparation. Problem solving transfer among programming languages (submitted to *Human-Computer Interaction*).

# A Decision Support System for Generalized Negotiations*

**Michael A. Harris**
Inst. for Advanced Computer Studies
and Dept. of Applied Mathematics
University of Maryland
College Park, MD 20742
e-mail: harris@umiacs.umd.edu

**Sarit Kraus**†
Graduate School of Library Studies and
Dept. of Computer Science
Hebrew University
Jerusalem 91904 Israel
e-mail: sarit@humus.huji.ac.il

**Jonathan Wilkenfeld**
Dept. of Government and Politics
University of Maryland
College Park, MD 20742
e-mail:wilkenfeld@umd2.umd.edu

**Elizabeth Blake**
Dept. of Government and Politics
University of Maryland
College Park, MD 20742
e-mail:blake@umd2.umd.edu

## Abstract

This paper reports on the development of a Decision Support System GENIE that aids participants in crisis negotiation simulations. GENIE is an instance of a general DSS which can be used to support a large class of decision problems. The major function of GENIE is to provide the user with on-line information about a complex decision scenario. To this end, GENIE utilizes a combination of graphic and textual information presentation formats to create an environment in which a user can develop a mental picture of the decision problem facing him/her and then dynamically formulate an effective negotiating strategy. The design and development of GENIE are described along with an explanation of the major features of the system.

Experimental results from user evaluations and system log files are also discussed. These results allowed us to gain insight into the decision processes of the users and rate the effectiveness of our DSS design strategy. Experimental results indicate that simulation participants who had access to the system performed on average better than participants without access to the system.

## Introduction

Decision makers today are frequently overwhelmed by the vast amounts of information which they must consider. Often, they are forced to make partially informed decisions which ignore critical issues because of the complexity of the situation being analyzed. Decision support systems (DSS) can play a crucial role in the decision making process by allowing the decision maker to navigate large amounts of information and to explore interrelationships between factors that may influence his/her decision. A DSS can also facilitate the simultaneous evaluation of multiple negotiating positions. This can play a decisive role in real time negotiations by allowing the supported parties to rapidly formulate dynamic strategies and quickly evaluate opponent proposals. DSS systems, when used in the context of negotiation support, are called negotiation support systems (NSS).

This paper describes the design and implementation of GENIE, a specific NSS/DSS developed jointly by the University of Maryland Institute for Advanced Computer Studies and Project ICONS of the Department of Government and Politics. GENIE is an instance of a general NSS/DSS which can be used to support a large class of decision problems. The problem domain corresponding to the GENIE class of DSS/NSS is formally described in section 2. GENIE is specifically customized to support a subset of this domain, namely n-player time variant negotiations with full information. See [S. Kraus and J. Wilkenfeld, 1990b, S. Kraus and J. Wilkenfeld, 1991b, S. Kraus and J. Wilkenfeld, 1991a] .

Several systems have previously been developed to aid decision makers in negotiation scenarios. Some examples of these systems are NEGO [Kersten, 1985], MEDIATOR [M. Jarke *et al.*, 1987], and NEGOPLAN [S. Matwin *et al.*, 1989] [1] Of the three systems mentioned, GENIE is most similar to NEGOPLAN in that it allows a user to explore various negotiation positions without making "judgments" about the utility of the positions.

When describing GENIE, major emphasis will be placed on GENIE's model visualization capabilities, user interface, and negotiation support tools. We will also discuss the simulation experiments which tested the effectiveness of GENIE.

*The authors wish to acknowledge the support of the University of Maryland Institute for Advanced Computer Studies, the Instructional Computing Program of the Computer Science Center, and Project ICONS.

†Part of this work was done while the author was visiting the Institute for Advanced Computer Studies and the Department of Computer Science, University of Maryland, College Park.

[1] The following work is also relevant: [B. Shneiderman, 1987, Andriole, 1989, Heymann and Bloom, 1988, Tufte, 1983, Bui, 1987, Mittra, 1986].

## GENIE: (A Simulation Environment)

GENIE was developed as part of a simulation environment of a real world situation in which negotiators can be trained and where experiments can be conducted.[2] GENIE is designed to be used as an aid to players in a hostage crisis simulation. This simulation is a hypothetical model of an international crisis [S. Kraus and J. Wilkenfeld, 1990b, S. Kraus and J. Wilkenfeld, 1990a, S. Kraus and J. Wilkenfeld, 1990c]. Each party has a list of objectives with associated payoffs. An objective of one party may or may not conflict with the objectives of either of the other two parties. Each party knows the objectives and associated payoffs of the other players (i.e., the model assumes full information.) During the simulation players have access to information about their situation. This information changes over time. Based on this information, players are able to negotiate with other players or take actions which they believe will result in their own payoff maximization. A successful player must consider information about his/her own objectives as well as those of the other players. There are 35 possible outcomes to the simulation in each of 50 time periods (1,750 total possible outcomes). If a player were to consider each of these outcomes from his own point of view and from those of the other two players he would have to consider 5,250 different point values. Clearly, this is an overwhelming amount of information. We hope to show that our NSS/DSS both helps simulation participants to develop clear mental pictures of the situation facing them as well as helping them to create logical and well-informed plans of action.

By studying the actions of undergraduate political science students at the University of Maryland using the DSS, we were able to gain insight into the decision processes of the students and rate the effectiveness of the DSS design strategy. Preliminary experimental results are from a group of simulation runs of a hostage crisis scenario involving, Israel, Egypt, and hijackers. Testing of the interface is underway using an India, Pakistan, Sikh scenario.

After each simulation, participants are given a questionnaire asking them to rank the relative importance of each of the features of the system. The DSS keeps log files on each player to determine the frequency and duration of use of each of the DSS features.

Although GENIE was customized to work with the hostage crisis model, it can be easily modified to support many other types of models. The following section describes the general class of models which can be handled by a GENIE-type system.

[2]Another aspect of the project is the creation of a prototype automated negotiator that will take roles in such simulations [S. Kraus and J. Wilkenfeld, 1990c].

## Model Description

Formally, we describe the class of situations or problems that can be handled by GENIE as follows: Let $\mathcal{P} = \{P_1, ..., P_N\}$ be the set of agents involved in the situation. Let $W_j =< f_1, f_2, ...., f_m >$ define a state of the world, where $f_i$ is the $i^{th}$ factor that influences this state. We assume that there is a finite number of factors and that each of these factors can range over a finite set of values. Therefore we assume a finite number of world states $W_i$. We denote the set of world states by $\mathcal{W}$. Note, as the complexity of the model increases, $\mathcal{W}$ can become very large. Let $A_i$ be the set of possible actions available to agent $i$. Also if $X$ is a state, then let $Pr(X)$ be the probability that state $X$ will occur. If $a \in A_i$ then let $g(a, W_j) = W_j \circ a$ where $\circ$ is an operator that has the effect of applying action $a$ to the state of the world $W_j$ to produce a lottery $((W_k, Pr(W_k)), (W_l, Pr(W_l)), (W_m, Pr(W_m)), \ldots)$ where $Pr(W_k) + Pr(W_l) + Pr(W_m) + \ldots = 1$. In the simplest case $g(a, W_j) = ((W_k, 1))$, i.e., application of action $a$ to the state $W_j$ produces the state $W_k$.

The class of models that can be handled by GENIE assumes that each agent $i$ places a numerical payoff value on each world state $W_j$. Let $V_i(W_j)$ be the payoff value that agent $i$ places on the world state $W_j$.

These payoffs can be ordered to represent the preferences over world states of player $i$. Then without loss of generality we can assume that for a fixed agent $i$ we have the ordering: $V_i(W_1) \geq V_i(W_2) \ldots \geq V_i(W_P)$. In a model with full information, it is assumed that an arbitrary agent $i$ knows the preference orderings of all of the agents $i = 1 \ldots N$. [3]

Even though we are assuming that an agent has full information about the other agents' preferences over different states, we cannot assume that the agent knows the other agents' preferences over the set of possible actions given an arbitrary state $W_k$. It is often the case that while the world is in some state $W_k$ agent $i$ wants to predict the actions of agent $j$ by studying the set of possible outcomes $g(a, W_k)$ for all $a \in A_j$. For example, suppose that $\{a_1, a_2\} \subseteq A_j$ and that there exist world states $W_l, W_m, W_n$ such that $g(a_1, W_k) = ((W_l, Pr(W_l)), (W_m, Pr(W_m)))$ and that $g(a_2, W_k) = ((W_n, 1))$. Also assume that $V_j(W_l) > V_j(W_n)$ and that $V_j(W_m) < V_j(W_n)$ and $Pr(W_l) + Pr(W_m) = 1$.

Now agent $i$ wants to decide which action agent $j$ will take. The problem is that agent $i$ doesn't know agent $j$'s attitude towards risk. [4] Later in this paper

[3]Note, we cannot assume that if $V_i(W_k) \geq V_j(W_k)$ then world state $W_k$ is more preferable to agent $i$ than it is to agent $j$.

[4]To clarify the idea of risk attitude, consider the following short example. Consider the CEOs of two competing tire manufacturing companies trying to formulate production strategies for the next fiscal year. Assume that $CEO_1$ is trying to predict the production plan of $CEO_2$. Assume further that $CEO_1$ knows that $Y$ units of rubber are worth $Z$ dollars on the open market. However, $CEO_1$ does not

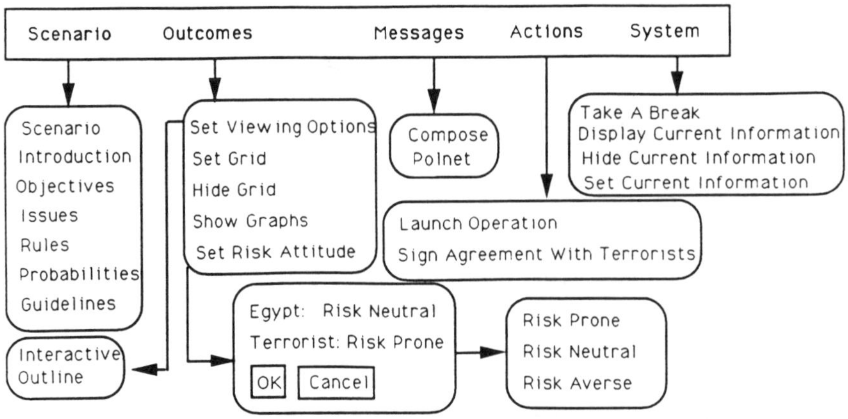


Figure 1: Schematic of the GENIE Menu System

we discuss a method to assist the user in forming a model of his/her opponents' preferences over actions.

## GENIE: (A Functional Description)

GENIE is composed of three main components: the interface, the knowledge base, and support tools. The interface and tools platform provide an environment for the decision maker to access, view, manipulate, and act based on information contained in the knowledge base. The knowledge base consists of information which models the decision problem scenario. Detailed information about the specific model used in GENIE can be found in [S. Kraus and J. Wilkenfeld, 1990b]. Figure 1 is a schematic of GENIE's menu system.

Fumás [Fumas, 1987] states that future DSSs should have the following characteristics: modularity, interfaces, graphics, ease of use, and availability on mini-computers. From the discussion below it can be shown that GENIE meets all of these criteria.

Ghiaseddin [Ghiaseddin, 1987] gives an in depth description of the characteristics of a successful DSS. He states that a successful DSS should have the following functional capabilities: modeling, data management, and support of all decision making activities. In addition he states that a DSS implementation should provide personalized support, security and integrity, transferability, and evolving capabilities for the support of increasingly complex demands. [5] Within this framework we will describe the major features of GENIE.

### Modeling and Data Management

GENIE is designed to provide the user with a clear mental picture of the model contained in the knowledge base. This is a difficult task since often the model of the scenario being analyzed is complex with numerous interdependencies among data objects. The task is further complicated when the model contains large amounts of numeric information. As stated in [Pracht, 1990] research has shown that images and symbolic representations are much more easily assimilated by humans as compared with textual information. This would suggest representing the structure and content of the model in some symbolic form. The problem is then transformed to finding a concise symbolic representation for the model. [6]

GENIE combines its data management and modeling capability in one mouse-supported screen which enables a user to quickly set parameters for information viewing. This screen not only provides quick access to information items in the knowledge base, but also allows the user to form a mental picture of the entire simulation. This screen serves to organize the complex details of the scenario into one easy-to-digest outline. With this outline, the user can than brainstorm and play "what if" to form a personalized strategy for outcome maximization.

One data management feature of GENIE which is critical to any successful DSS is the ability of the user to control the complexity of the queries and responses

know whether $CEO_2$ would rather have the $Y$ units of rubber and risk going into production or play it safe and keep the $Z$ dollars.

[5] Ghiaseddin also includes ability to learn as a non-essential feature. Currently, our system does not have this capability.

[6] A previous approach to this problem is taken by Pracht [Pracht, 1990] who uses a frame based knowledge representation. This scheme partitions knowledge into discrete structures called frames. Each frame has a set of slots for holding clusters of related knowledge. Hierarchical relationships among frames are expressed as arcs between nodes in a tree. Our approach differs from that of Pracht in that we are exclusively concerned with aiding the decision maker in model visualization. Pract's system devotes substantial resources to assisting the model developer.

from the DSS. Novice users who ask the DSS simple questions should not be flooded with screens of complex charts and graphs. At the same time, advanced users should have the facilities to develop sophisticated strategic models. GENIE's interface allows a user to define one or more hypothetical state(s) of the world and then investigate possible future actions based on these states. The user can explore outcomes resulting from his/her own actions as well as those of his/her opponents. Also, a user can switch viewpoints to see things from the point of view of one or more of his/her opponents. This is possible, since the model assumes full information. A simultaneous display of these viewpoints allows the user to formulate a strategy which takes into account possible opponent actions.

The system employs a model specific interactive outline with information categories which a user can select to see graphic information about the scenario. This outline gives the user access to the major model visualization and data management features of the system which include: multiple frames of reference, time variant outcome projections, and multiple world state definitions. The following section gives a detailed description of the interactive outline.

**The Interactive Outline** The interactive outline is organized into three main categories: viewpoints, decisions, and world states. By selecting items and defining parameters in each of these categories, a user is able to view the pieces of the model that are directly relevant to formulation of his/her negotiation strategies. [7] The outline not only allows users to define choices and parameters, but also uses heuristics to determine whether the choices are reasonable in terms of the complexity of the graphic output produced. A powerful feature of the system is that a user can select to view multiple frames of reference, decisions, and world states simultaneously. This allows a user to compare numerous negotiating positions concurrently. [8]

The viewpoints section of the outline gives the user a list of the parties[9] involved in the negotiations. The user can interactively select one or more of these parties to define a set of viewpoints. Once selected, these viewpoint(s) become the frame(s) of reference for all subsequent queries to the knowledge base. When more than one viewpoint is selected, information corresponding to each viewpoint is displayed simultaneously. This facilitates the development of strategies which incorporate the goals of all involved parties. Having fixed the viewpoint(s), the user can then move to the decisions section of the outline.

The decisions section lists all possible types of decisions that a player can make during the course of the negotiations. The user selects the types of decisions to investigate by moving the mouse cursor to the appropriate position on the outline and clicking on the desired choices. Obviously, the system must provide a method for the user to specify parameters that will uniquely identify the exact decision(s) to be investigated. Note that since the model contains information about decisions that can be made by any of the parties, each party that is supported by the system should be able to ask for information about all decisions, not just those that he/she can personally make. Once the frame(s) of reference and decision(s) are fixed, the user can then move to the world states section.

In the world states section a user fills in parameters that define the hypothesized state of the world at the time the decision(s) in the previous section take effect. This organization of the outline reveals the main structure of the model by emphasizing the dependence of the decision(s) on the state of the world at the time they are made. Parameters of the world state may be positions held by other negotiating parties, economic conditions, time, etc.

One of the powerful features of this system is that it allows a user to project the outcomes resulting from one or more decisions over a range of time periods. Using this feature, users are given information that helps them to decide not only what decision to make, but when to make it.

Another feature which is helpful for developing negotiating strategy is that of varying world state parameters. Often decision makers have some influence over the state of the world (i.e., they have influence over some of the parameters in the world state definition). During strategy formulation, a decision maker may want to find out whether it is worth it for him/her to expend energy to change the world state. [10] GENIE lets a user select multiple values for a given parameter and then simultaneously view the selected decision outcomes based on the different world states.

Figure 2 shows the outline developed for the Israel-Egypt-hijacker hostage crisis model. Here the user has fixed the frames of reference to be Israel and Egypt. The decisions to be investigated are set to Israeli operation, terrorists giving up, a deal which involves the exchange of 400 prisoners in Israeli jails for the release by the terrorists of all hostages, and a deal which involves the exchange of 500 prisoners.

Here under the Egyptian behavior section of the outline the user has defined a state of the world that assumes that Egypt has not given the press access to

[7] Interactive outlines also have application to scenarios outside the realm of negotiation support.

[8] The reader can refer to figure 2 for a concrete example of the concepts explained in the following discussion.

[9] Or stakeholders in the language of management decision support.

[10] For example, in the hostage crisis scenario, the hijackers could decide that they will blow up the plane and kill all of the hostages. In this case they should try to influence the state of the world so that at the time of the action, they (or their cause) receive maximum benefit. Influencing the state of the world could mean waiting a certain amount of time or convincing Egypt to allow press coverage.

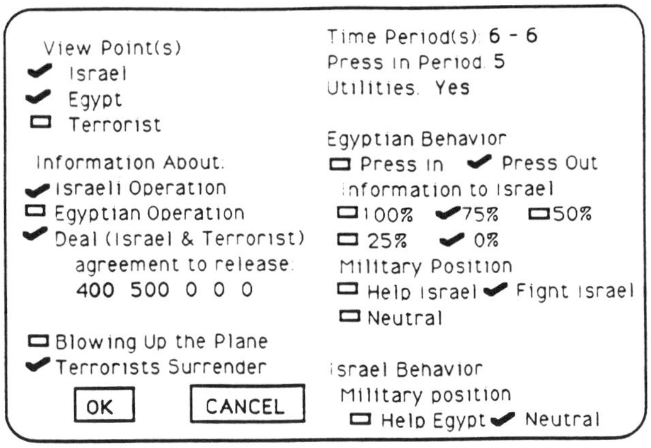


Figure 2: Interactive Outline: Middle East Scenario

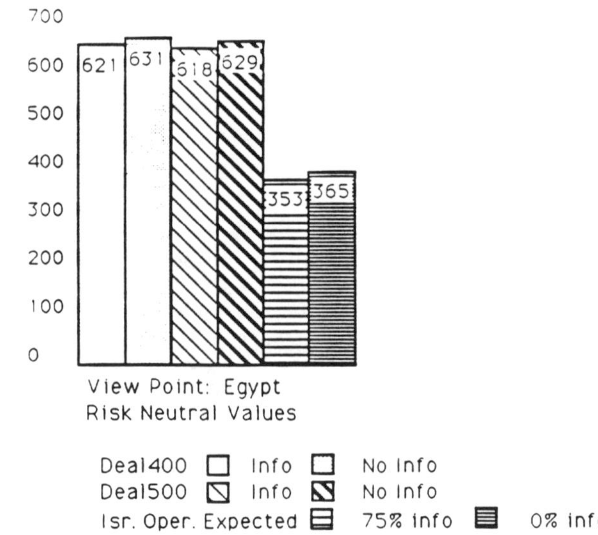


Figure 3: Genie output from the viewpoint of Egypt.

the terrorists' and that Egypt will fight Israel in the event of an Israeli operation. Two options have been selected under the "Information to Israel" category, 0% and 75%. These numbers correspond to percentage of information provided by Egypt to Israel that might aid in an Israeli operation. Two world states are defined by the specified parameters, one that assumes that Egypt provides no information and one that assumes that Egypt provides 75% information. Figure 3 shows the system output from the point of view of Egypt. [11]

**Attitude Towards Risk** Attitude towards risk becomes an important consideration when a player is trying to predict the actions of an opponent. Seeing an opponent's payoff for the outcome of a given decision or action does not give the user any idea about how the opponent views or interprets these point values. GENIE has a feature which helps the user form an approximation of an opponent's view of the world. It allows a user to fix an attitude towards risk for each opponent in the negotiations. The values for risk attitude are: risk neutral, risk averse, and risk prone. A player is **risk averse** if he always prefers payoffs equal to the expected payoff value of a lottery. He is **risk prone** if he always prefers entering a lottery to receiving a payoff equal to the expected payoff value of the lottery. He is **risk neutral** if he is indifferent between entering a lottery and receiving a payoff equivalent to the expected payoff of the lottery.

When an opponent's risk attitude is set to risk prone, we apply a convex function to his payoffs, to generate appropriate preferences. When an opponent risk attitude is set to risk averse, we apply a concave function on his payoffs [French, 1986].

[11] Numbers on the bars correspond to payoff values to Egypt for various outcomes. From the figure it is apparent that for the options considered it is most advantageous for Egypt to try to encourage some type of deal between the Israel and the hijackers.

**Graphic Output** After the user has selected the frame(s) of reference, decision(s), and world state(s), the system sends the request to the evaluation and display routines. These routines determine the format, content, and emphasis of the graphic output of the system. Point values are displayed in graphic form. Histograms are used to display values that are not projected over multiple time periods. When values are projected, piecewise linear graphs are used.

If the result of a decision is a lottery of world states, then the user must be informed both of the payoff of each of these states and the probability that each state will occur.

When the outcomes are not projected over time, these probabilities are simply presented in numeric form next to the associated label in the display key. The user also has the option of seeing expected payoffs. These values are calculated by taking the product of the probability that a given state will occur with the payoff value from this state.

When objectives are projected over time, the probability values can no longer be easily presented to the user. In this case we decided to factor the probabilities into the payoff values to give the user a display of the expected payoffs from an outcome.

Figure 3 is an example of output from a typical query to the knowledge base.

## Experimental Results

The simulation experiments were run with undergraduate students in the political science department. Many of the students had little or no experience using computers.

Our preliminary results are from eight simulation runs, each involving three students. Two of these runs involved players who were participating in the simulations for a second time.

In the six runs involving new students, we gave a thirty to forty five minute oral introduction to the system. The students were then allowed about fifteen minutes to experiment before the simulations began.

The results of the simulation runs indicated that multiple frames of reference, projection into the future, and multiple world state definition are the most important features of our system.

Currently, we are running more extensive and controlled experiments which involve comparing the outcomes from simulations run on participants who have access to GENIE versus simulations where participants do not have such access. [12] Among the controls that we have introduced are a modification in setting from the Middle East to India/Pakistan, and a comparison of participants with access to the interface with students with no access. This second round of experiments involves two sets of runs with approximately 13 simulations per set. Preliminary analysis of total payoffs earned by system users versus total payoffs earned by users without access to the system indicate that the system users had on average higher payoffs. From further analysis of these results we hope to be able to draw conclusions not only about the success of our design strategy, but also gain insight into the decision processes of the simulation participants.

## Future Work

In the future we plan to add a feature that will allow the user to view multiple outcomes in a tree structure. The leaves of the tree will contain numeric values that represent the payoff to the user from a certain outcome. Interior nodes will represent parameters in the world state on which these payoffs depend. In this way, a user will have access to a symbolic representation that summarizes all of the dependencies of a specified outcome or group of outcomes.

## References


[Andriole, 1989] S. J. Andriole. *Handbook of Decision Support Systems.* Tab Books Inc., 1989.

[B. Shneiderman, 1987] B. Shneiderman. *Designing the User Interface: Strategies for Effective Human-Computer Interaction.* Addison-Wesley Publishing Company, 1987.

[Bui, 1987] T. X. Bui. *Co-oP, A Group Decision Support System for Cooperative Multiple Criteria Group Decision Making.* Springer-Verlag, 1987.

[French, 1986] S. French. *Decision Theory: An Introduction to the Mathematics of Rationality.* Ellis Horwood Limited, 1986.

[Fumas, 1987] V.S. Fumas. Strategy planning: Implications for the design of dss. In C.W. Holsapple and A.B. Whiston, editors, *Decision Support Systems: Theory and Application*, pages 429–449, 1987.

[Ghiaseddin, 1987] N. Ghiaseddin. Characteristics of a successful dss user's needs vs. builder's needs. In C.W. Holsapple and A.B. Whiston, editors, *Decision Support Systems: Theory and Application*, pages 159–184, 1987.

[Heymann and Bloom, 1988] H. G. Heymann and Robert Bloom. *Decision Support Systems in Finance and Accounting.* Quorum Books, 1988.

[Kersten, 1985] G. Kersten. Nego: Group decision support system. *Information and Management*, 8:237–246, 1985.

[M. Jarke *et al.*, 1987] M. Jarke, M. Jelassi, and M. Shakun. Mediator: Towards a negotiation support system. *European Journal of Oprerational Research*, 31:314–334, 1987.

[Mittra, 1986] S. S. Mittra. *Decision Support Systems, Tools and Techniques.* John Wiley and Sons, 1986.

[Pracht, 1990] W. Pracht. Model visualization : Graphical support for dss problem structuring and knowledge organization. *Decision Support Systems*, 6:13–27, 1990.

[S. Kraus and J. Wilkenfeld, 1990a] S. Kraus and J. Wilkenfeld. An automated strategic model of negotiation. In *AAAI-90 Workshop on Reasoning in Adversarial Domains*, Boston, 1990.

[S. Kraus and J. Wilkenfeld, 1990b] S. Kraus and J. Wilkenfeld. Modeling a hostage crisis: Formalizing the negotiation process. Technical Report UMIACS TR 90-19 CS TR 2406, Institute for Advanced Computer Studies, University of Maryland, 1990.

[S. Kraus and J. Wilkenfeld, 1990c] S. Kraus and J. Wilkenfeld. A strategic negotiations model with applications to an international crisis. Technical Report UMIACS TR 90-20 CS TR 2407, Institute for Advanced Computer Studies, University of Maryland, 1990.

[S. Kraus and J. Wilkenfeld, 1991a] S. Kraus and J. Wilkenfeld. Negotiations over time in a multi agent environment: Preliminary report. In *Proc. of IJCAI-91*, Australia, 1991. To appear.

[S. Kraus and J. Wilkenfeld, 1991b] S. Kraus and J. Wilkenfeld. The function of time in cooperative negotiations. In *Proc. of AAAI-91*, California, 1991. To appear.

[S. Matwin *et al.*, 1989] S. Matwin, S. Szpakowicz, Z. Koperczak, G. E. Kersten, and W. Michalowski. Negoplan: An expert system shell for negotiation support. *IEEE Expert*, 4(4):50–62, 1989.

[Tufte, 1983] E. Tufte. *The Visual Display of Quantitative Information.* Graphics Press, 1983.


[12] The authors wish to thank J. Hewitt and M. Lebrun for helping to conduct the simulation experiments.

# The Structure of Reasoning in Conversation[1]

**Lauren B. Resnick**
Learning Research and
Development Center
University of Pittsburgh, 15260

**Merrilee H. Salmon**
History & Philosophy of Science
University of Pittsburgh, 15260

**Colleen M. Zeitz**
Learning Research and
Development Center
University of Pittsburgh, 15260

## Abstract

The findings of philosophers, linguists, and psychologists are conjoined here in an effort to develop a descriptive/analytic account of reasoning as it actually occurs in social settings. The primary focus of this paper is the reasoning that occurs in discussions of controversial social issues by groups of peers. Preliminary analysis indicates that rather sophisticated argument structures emerge from these informal settings, and that conversational interaction stimulates the development of arguments. Although participants did not always fully state their arguments, the investigators felt justified in attributing implicit premises because these were referred to later in the conversation *as if they had been stated.* In this respect at least, instead of merely assuming that subjects are good at reasoning, this study provides evidence for the claim.

In this paper, a system for coding elements of reasoning and a method for displaying the interactive structure of reasoning in conversations are developed. A long-term goal is to use the information gained from this study to help understand and correct two problems that arise in the teaching of reasoning: the difficulty many students have with acquiring the principles of reasoning in standard logic courses and the difficulty of transferring whatever reasoning skills are acquired in the classroom to new situations.

## Introduction

George Herbert Mead called thought a "conversation with the generalized other," implying that as we think individually we attempt to respond--internally and vicariously--to the imagined responses of others to our ideas and arguments. His formulation suggests the study of actual conversation in order to analyze the public forms of reasoning that are thought to lie at the roots of private forms of competence. The study of reasoning as a form of social practice has only recently entered the mainstream of psychological research, influenced heavily by the rediscovery of Vygotsky. He, like Mead, offered a vision of human thinking grounded in social life. Both saw private reasoning as an internalization of processes originally carried out publicly and in interaction with others. In examining the actual processes of shared reasoning, an obligation is incurred to understand human reasoning in broader terms than those offered by formal logic and statistics. This work thus departs from studies by psychologists in which a major concern is the explanation of why human reasoning does not always conform to accepted normative standards (Braine & Rumain 1963; Johnson-Laird 1983; Nisbett & Ross 1980; Tversky & Kahnemann 1974). Piaget, whose work traces the acquisition of normative logico-mathematical structures and the effects of their absence on children's reasoning, also falls within this normative tradition.

Within cognitive science, challenges to the dominant role of formal logic as an account of human reasoning are producing growing concern for the role of argument, contradiction, and negotiation processes. Doubts that thought can be fully accounted for in terms of an individual's processing a set of symbols with fixed meanings are leading some cognitive scientists to press for theories of cognition that are more connected to both the physical and the social world. The metaphor of cognitive systems as social systems in connectionist and blackboard models of thinking makes the entire cognitive science community more open than it was a decade ago to the idea of knowledge as distributed across several individuals whose interactions determine decisions, judgments, and problem solutions. Sharing with the Soviet-originated *activity theory* (Leonte'ev 1981) an antifunctionalist point of view in which intentionality and affect are components of cognitive activity, American theorists of situated cognition question whether a cognitive core that is independent of context and intention can be found (Brown, Collins, & Duguid 1989; Greeno 1988; Lave 1988; Suchman 1987). Our project attempts to show how the forms of discourse are intimately related to reasoning, and in doing so forges multidisciplinary links to rhetoric, informal logic (Grize & Pieraut-LeBonniec 1983;

[1]**Credits**
The authors are grateful to the Andrew W. Mellon Foundation for supporting this work.

Toulmin 1958), pragmatics (Grice 1989; Searle 1983) and the structure of conversation (Clark 1978; Schegloff, in press; Sperber & Wilson 1986).

**Situations of shared reasoning.** In studying reasoning as a form of social practice, we assume that the structure of reasoning depends significantly not only on the content of the conversation, but also on the nature of the situation in which that reasoning occurs. For example, features likely to affect the course of reasoning include the relative social status of the participants, the goal that the group adopts for the conversation, the goals of each participant, the kinds of physical displays available, and who can see or manipulate these displays. This project examines several conversational situations in which reasoning occurs: classroom discussions led by a teacher, joint problem-solving sessions related to schoolwork (some involving manipulable computer displays), and discussions of controversial social issues by groups of peers.

An analysis of a conversation among three university students is reported here. Volunteers from a course on philosophy of science and public policy were organized into groups of three with two initially opposed to nuclear power and one in favor of it. They were instructed to discuss nuclear power for about twenty minutes with the goal of reaching a consensus. Their discussions were videotaped and transcribed.

**Preliminary analysis.** After viewing the tape several times, the investigators framed three broad categories of analysis: Translation (an interpretation by the researchers of the actual utterances of the subjects); Logic (an outline of the patterns of reasoning); and Discourse (other linguistic, social, and interpersonal aspects of conversation).

*A. Translation.* Each speaker was identified by a letter. Grammatical units that approximated sentences were given separate line numbers. In "translating" sentences, referents to pronouns, dates, and times were secured. When false starts and qualifiers seemed not directly relevant to the line of reasoning, they were transferred to the Discourse category for later analysis. To check reliability, investigators independently constructed translations. The degree of intersubjective agreement obtained provided reasonable assurance that the translations articulated more fully the intentions of the speaker instead of imposing new meanings that the speaker did not intend.

Subjects often made elliptical claims that could be construed as parts of arguments. Whenever missing elements were supplied, the interpolations were presented alongside the actual words of the participants. The appended transcription is a reprinting of part of the protocol with actual words of the participants in the left column and translations and interpolations in the right column. The label "T" is used to indicate a translation. For example, TC4.2 denotes our translation of speaker C's words in 4.2. Interpolated elements of arguments are marked "L."

*B. Logic.* In accordance with the goal of understanding the structure of reasoning under conditions of conversation, but also recognizing that only a limited number of variables could be considered in the first round of analysis, a coding system (summarized in appended Logic Key) was devised to pick out salient features of reasoning.

*C. Discourse.* Analysis of pragmatic features of conversations less obviously relevant to reasoning than those coded in the logic category is scheduled to begin this summer.

**Analysis of the structure of reasoning.** Two questions motivated this research. One was whether people actually engage in reasoning in conversational settings, or, as many critics complain, people simply state opinions and beliefs without attempting to mount arguments for them. The second, more open-ended, question focused on how the conversational setting and interplay between several participants encouraged or discouraged the use of reasoning skills.

Most studies of reasoning, even the recent literature on informal reasoning and critical thinking, ignore social aspects of the topic while emphasizing patterns of individual reasoning. With few exceptions, such as "brainstorming" sessions in which people try to gain insights to solve specific problems, and the presentation of scholarly papers, where arguments are offered to peers for critical evaluation, reasoning is conceived as a solitary activity engaged in rather infrequently, and then as a response to puzzle-solving pressure. Nevertheless, in response to the first question, it was apparent that the subjects, who might have used various nonargumentative techniques such as social pressure to reach consensus, instead constructed fairly elaborate arguments that were recognized and responded to appropriately by their peers. Although not trained as debaters, subjects employed standard techniques such as offering concessions to preempt criticisms of their own arguments or, alternatively, capitalizing on opponent's concessions to support a counterargument.

Some insight into the problem of unstated premises emerged from the study. Social rules governing conversation (cf. Grice 1989) suppress the explicit statement of claims that are believed obviously acceptable to parties in a conversation. In trying to reconstruct conversational arguments, however, making implicit claims explicit is problematic. Does one always assume that the person presenting the argument has in mind a premise that will make the argument succeed, or does one count failure to make a premise explicit a failure in reasoning? In this study, assignment of some unstated premises and conclusions was amply justified by specific references made at later points in the discourse to these unstated claims. Because no participant protested that these had not been stated or were not intended, one could safely assume that no party disputed their role in the arguments.

Figure 1 represents the first minute of the transcribed conversation. Premises of arguments are indicated by triangles and conclusions by rectangles. Either may be stated explicitly (solid lines) or implied (dotted lines). *Factual* premises, which are specific to a topic under discussion, are distinguished from *theoretical* premises, which are taken to be general beliefs shared by most members of the participants' culture. Premise LC4.3i ("Energy choice is limited to nuclear power or fossil fuels") is an example of a factual implied premise. Premise LC4.3ii ("The less costly alternative should be chosen") is a theoretical premise. Because theoretical premises form part of a widely shared cultural system of beliefs, they are rarely challenged or even acknowledged. As part of this project, additional experiments have been undertaken to validate attributions of implied premises. Challenges and responses to them (diamonds) and concessions (hexagons) are also identified as basic elements of reasoning. This represents a departure from most studies of reasoning in which only premises and conclusions are regarded as components of arguments. Premises are generally understood to be claims that could be either true or false, whereas challenges and concessions can be questions, doubts, or exclamations to which it may not even make sense to supply a truth value. Nevertheless, challenges and the like form parts of coherent structures of reasoning. Far from being extraneous or irrelevant, they form and guide reasoning in clearly recognizable ways. In our structural diagrams of conversational reasoning, the challenges and concessions are the most densely interconnected components. This indicates that they are doing important work in moving arguments forward in ways that are not captured by canonical formal logical categories.

Bolded rectangles and ovals surround whole argument structures, each made up of several of the basic elements just described. Ovals indicate that an argument structure constitutes an attack on a previous argument structure. One argument structure can be embedded within another. For example, premises B8.3 and TB8.4a, leading to conclusion TB8.4b, constitute an argument structure inside an attack. Arrows show how sentences refer to one another or how arguments are related to other parts of the conversation.

Figure 1 displays two argument structures, one presented by participant C, the second by participant B. Participant C builds an argument for her opening statement in favor of using nuclear power (LC4.1). She begins by conceding the negative costs of nuclear power (TC4.2) and then employs three premises. Only the premise that fossil fuels are more costly than nuclear power is explicit (TC4.3). Two (implicit) premises are supplied by the researchers. All four elements--the concession, the stated premise, and the two implicit premises--constitute support for the conclusion that nuclear power should be used.

Because C holds the minority position in this group, this argument comes under considerable attack. Initially B attacks the argument as a whole (B8.1) and supports the position with an embedded argument. B's stated premise in B8.2, that ". . . there are alternative methods we can use," is a negation of C's implicit premise (LC4.3i) that energy choice is limited to nuclear power or fossil fuels. This constitutes the first validation of our assignment of the implicit premise to C. Other instances of this type of validation occur in later parts of the conversation. Premises B8.3 and TB8.4a, leading to the conclusion that the consequences of nuclear power are too harsh to consider its use (TB8.4b), intensify B's initial position against the use of nuclear power (LB8.1), which is in turn a negation of C's conclusion.

## Concluding Discussion

The present methodologically oriented paper is but an early report on a new line of work. With the tools developed here, further analysis can help expose how people reason cooperatively in social settings. This analysis goes beyond those that have become conventional, even in studies of informal logic, in treating as integral parts of arguments not only premises and conclusions but also other elements such as challenges, responses to challenges, and concessions. Some progress toward solving the troublesome problem of identifying and attributing implicit premises has resulted from this work.

In the small sample of conversations analyzed thus far, the coherence of the reasoning displayed is impressive. Participants listen carefully to each other and construct their arguments in relation to what others say. More specifically, they appear to build complex argument and attack structures. People appear to be capable of recognizing these structures and effectively attacking component parts as well as whole arguments. The discussants disarm anticipated (or previous) arguments by first making concessionary

## Group A Nuclear

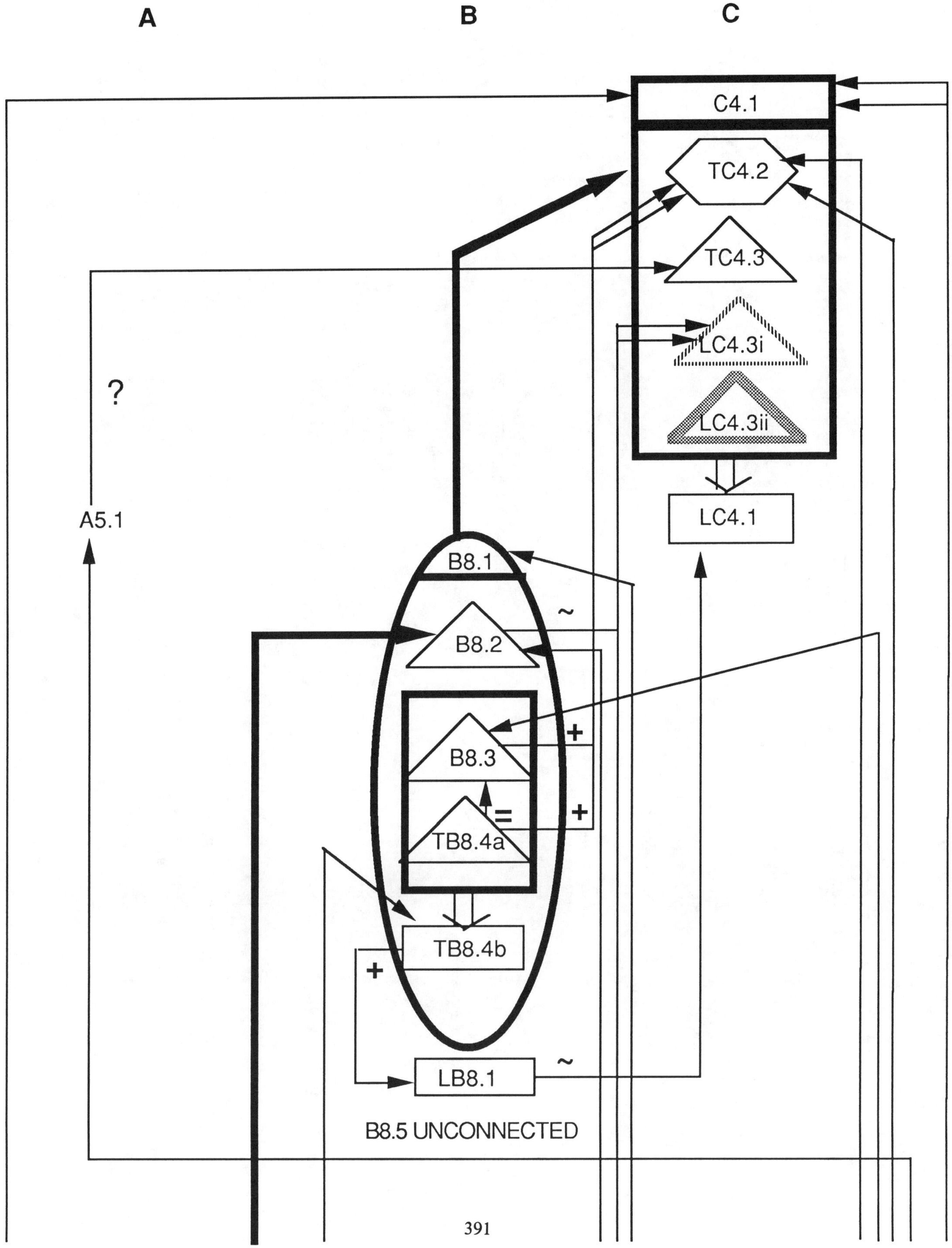

Transcription of Discussion

| Actual Statements | Added translations and premisses |
|---|---|
| A1: What do you think about nuclear power? | |
| C2: Well! | |
| A3: What's your position? | |
| C4.1: My position is that I advocate it | LC4.1: Nuclear power should be used |
| C4.2: 'n I realize that there are definite costs | TC4.2: I realize that there are defintite costs to nuclear power. |
| C4.3: but I also feel that there are costs . um . I think more prevalent when um . you're talking about fossil fuels | TC4.3: There are costs that are more prevalent for fossil fuels than for nuclear power.<br>LC4.3i: Energy choice is limited to nuclear power or fossil fuels.<br>LC4.3ii: The less costly alternative should be chosen. |
| A5: What do you...think about fossil fuels what do they do? | |
| C6: Well why don't you guys give your positions first before [ ]// | |
| A7: Ok | |
| B8.1: Allright, um well I'm against it | LB8.1: Nuclear power should not be used. |
| B8.2: because I believe there are alternative methods we can use.. | |
| B8.3: um the consequences of nuclear power are just too disasterous | |
| B8.4: they're just they're too harsh to even consider it | TB8.4a: The consequences of nuclear power are so harsh<br>TB8.4b: that you cannot even consider using nuclear power. |
| B8.5: and you can't predict the probability of what's gonna happen and what's not gonna happen.. | |



## LOGIC KEY

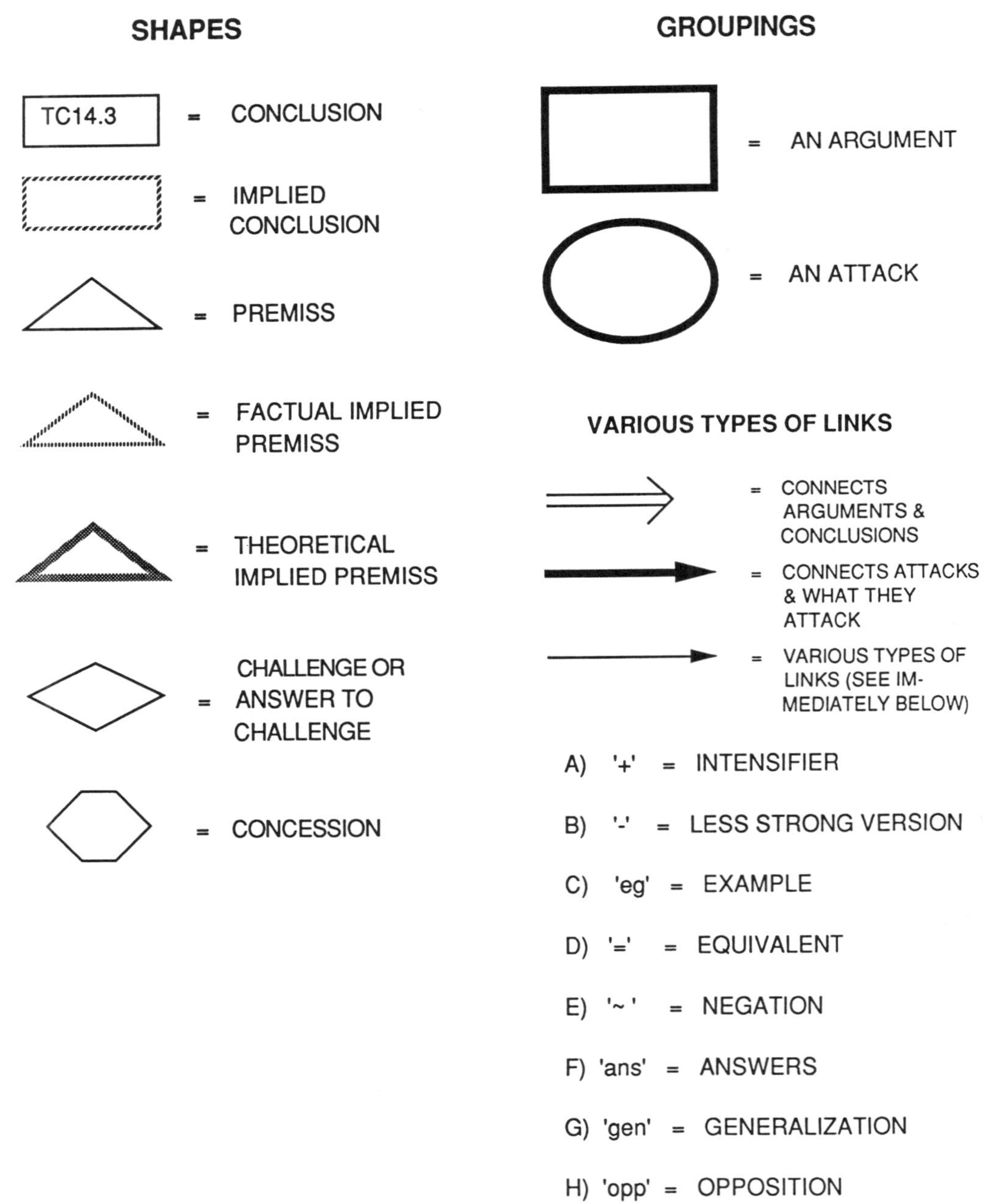

statements and then weakening or attacking these statements themselves.

Although plausible alternatives to the clever argumentative strategies we describe exist, subjects in our studies appeared to avoid them. Very few assertions occur that do not play some role in an argument structure, and rarely are claims asserted without some supporting reasons. This indicates that our subjects were interactive as they built argument structures. The discussants might have dropped arguments midstream or inserted numerous irrelevant utterances into an otherwise coherent argument. Subjects could have attacked positions that none of the other discussants was willing to defend ("straw men"). If our subjects had not apprehended the force of arguments, they might have failed to attack others' arguments or failed to rebut attacks of their own arguments. In fact, the protocols examined to date contain very few instances of these types of error.

These observations point to a number of questions that will be the object of future investigations. To test how widespread the capacity for shared reasoning is, the structure of discussions of social issues by groups of different ages and different levels of education will be compared.

The question of how specific linguistic and conversational structures might support or inhibit reasoning will also be pursued by analyzing the "Discourse" features of the present corpus of material. Other experiments will set up conditions of "discussion" in which the normal face-to-face conditions are partly interrupted. For example, a study is underway in which subjects watch videotapes of several minutes of discussion and then respond as if they were members of the discussion group they watched. Computer-mediated conversations, which are an intriguing newer form of social practice, may also be investigated.

## References


Braine, M. D. S., and Rumain, B. 1963. Logical Reasoning. In P. H. Mussen (Ed.), *Handbook of Child Psychology* (Vol. III, pp. 263-340). N.Y.: Wiley.

Brown, J. S., Collins, A., and Duguid, P. 1989. Situated Cognition and the Culture of Learning, *Educational Researcher* 18(1):32-42.

Clark, H. 1978. Inferring What is Meant. In W. Levelt and G. Flores d'Arcais (Eds.), *Studies in the Perception of Language*. Chichester: Wiley.

Greeno, J. G. 1988. Situations, Mental Models, and Generative Knowledge, Report No. IRL88-0005. Palo Alto, Calif.: Institute for Research on Learning.

Grice, H. P. 1989. *Studies in the Way of Words.* Cambridge, Mass.: Harvard University Press.

Grize, J-B., and Pieraut-LeBonniec, G. 1983. *La Contradiction: Essai sur les Operations de la Pensee.* Paris: Presses Universitares de France.

Johnson-Laird, P. 1983. *Mental Models.* Cambridge: Cambridge University Press.

Lave, J. 1988. *Cognition in Practice.* Cambridge: Cambridge University Press.

Leonte'ev, A. N. 1981. The Problem of Activity in Psychology. In J. V. Wertsch (Ed.), *The Concept of Activity in Soviet Psychology* (pp. 37-71). Armonk, N.Y.: M. E. Sharpe.

Nisbett, R., and Ross, L. 1980. *Human Inference: Strategies and Shortcomings of Social Judgment.* Englewood Cliffs, N.J.: Prentice-Hall.

Schegloff, E. A. in press. Conversation Analysis and Socially Shared Cognition. In L. B. Resnick, J. Levine, and S. D. Behrend (Eds.), *Socially Shared Cognition.* Washington, D.C.: American Psychological Association.

Searle, J. 1983. *Intentionality.* Cambridge: Cambridge University Press.

Sperber, D., and Wilson, D. 1986. *Relevance: Communication and Cognition.* Cambridge, Mass.: Harvard University Press.

Suchman, L. A. 1987. *Plans and Situated Actions: The Problem of Human-Machine Communication.* Cambridge: Cambridge University Press.

Toulmin, S. 1958. *The Uses of Argument.* Cambridge: Cambridge University Press

Tversky, A., and Kahnemann, D. 1974. Judgment Under Uncertainty: Heuristic and Biases. *Science* 185:1124-1131.

# Active Language in the Collaborative Development of Cooking Skill

**Jeff Shrager**
System Sciences Laboratory
Xerox Palo Alto Research Center

**Maureen Callanan**
Psychology Department
University of California at Santa Cruz

### Abstract

It is crucial to approach a cognitive account of development with an accurate picture of the parent-child system in hand, otherwise one will tend to underestimate the richness of support and dynamics of that system and so will tend to overestimate the complexity of the learning processes of the child. In order to understand the developmental functions of collaborative action and its accompanying linguistic activity we examine the verbal and physical activity in parent-child cooking. We present an analysis of the physical collaborative structure of a baking soda measurement task from 36 parent-child dyads in three (child) age groups: 3, 4, and 5-years old, and a qualitative analysis of some phenomena of active language in this setting. Active language is discussed in terms of its function in providing clues to lexical semantics, to the structure of the task, and to contextual cues and non-obvious aspects of the situation.

This work was supported in part by NICHHD grant HD26228 to the second author, by funds from the Institute for Research in Learning, Palo Alto, CA, and by The University of California at Santa Cruz. Julie Nicholson, Arlene Holloway, and Carol Capelli helped collect the data. Julie Morrisett, Anneliese Heyl, Susan Christopher, and Gail Nelson transcribed the data. Arlene Holloway helped carry out the analyses. Arlene Holloway and Marti Hearst commented on drafts of this paper. We had helpful discussion with Carol Capelli, Arlene Holloway, Joyce Moore, and other colleagues. Correspondence should be addressed to Jeff Shrager at Xerox PARC, 3333 Coyote Hill Rd., Palo Alto, CA, 94304, (415)494-4338, Shrager@Xerox.com

## The Resource of Active Language

Cognitive development is largely concerned with how children operate in the complex physical and social world, and how their basic ontologies and cognitive skills, such as planning, analogy, and problem-solving, arise. Many developmental theories consider language, action, thought, and the physical setting as interacting phenomena – each serving to organize the others. In these theories the physically and socially embedded activity of parent-child dyads (and, later, of other social systems) is a crucial locus of learning (cf. Bruner, 1983; Kaye, 1982; Rogoff, 1990; Vygotsky, 1978). However, there is less agreement upon the methods and theories by which psychologists can study and understand parent-child systems, in part because this requires studying a system with complex long-term dynamics that cannot be confined to the laboratory.

We think it crucial to approach a cognitive account of development with an accurate picture of the parent-child system in hand, otherwise one will tend to underestimate the richness of support and dynamics of that system and so will tend to overestimate the complexity of the learning processes of the child. One can view the early strategy choice work of Siegler & Shrager (1984) and VanLehn's (1987) work on procedure induction as examples of both the advantages of utilizing an accurate analysis of the learner's social and physical environment, and as examples of the limitations imposed by our still shallow understanding of that environment. Siegler and Shrager modeled the dynamics of subtraction skill acquisition and the pattern of use of strategies by simulating the strategies observed in children and, more importantly, by presenting problems to the simulation in accord with the frequency with which problems were presented to children by their parents (as observed by Hamman and Siegler). VanLehn's simulation was given ordered sequences of worked examples of long subtraction problems from which it learned a general subtraction procedure. Although both of these models rely upon time-extended activity of the child in executing strategies or in reading the worked examples, the child's *interaction* with the parent or teacher, or with the task context was not considered. Nor was there consideration given to the linguistic context that one would expect to infuse such settings. The present work contributes to efforts at developmental modeling by offering a description of parent-child collaborative work that includes their interaction physically and communicatively.

One can describe the embedded parent-child system in terms of resources that are available to the child. In making use of, or by influence of a complex of resources, a relatively simple learner can develop complex modes of activity. The frequency of problems given the simulation by Siegler and Shrager is a resource of practice guidance, and the sequence of examples seen by VanLehn's simulation is an resource of induction examples. Here we examine another sort of resource: the verbal correlates of collaborative activity. We want to understand the learning function of the utterances that accompany collaborative work. We shall refer to this as the social resource of "active language."

## Collaborative Cooking

The study of active language requires simultaneous analysis of activity and of the language functions that takes place during that activity. Shrager, Callanan, and Holloway (1991) videotaped 38 roughly upper-middle-class parent-child pairs baking muffins in their home kitchens from a recipe in a children's cookbook. The children occupied three age groups; mean ages 2;11, 3;9, and 4;11. One step of this recipe calls for one-and-one-quarter teaspoons of baking soda to be added to the dry materials being collected in a bowl. The recipe book comes with a set of colored plastic measuring spoons, whose colors are coded to measures by the pictures in the recipe book. In the case of the baking soda, a picture of a box of baking soda is shown with a red and a blue spoon next to it; the red spoon is 1/4-teaspoon and the blue spoon is 1-teaspoon.

### *Analysis of Physical Collaboration*

Shrager, et al. studied the baking soda segment from each tape. The time taken for this step varied widely, ranging from 10 to 238 seconds, but there was no age difference between these durations (mean=70.7, stdev=23.8 at 3-years, 81.5, 50.7 at 4-years, and 81.6, 31.9 at 5-years; F(2,33)=0.34, $p > .7$). The baking soda task was divided into six steps: each of the two spoons has three substeps: filling the spoon with baking soda, optionally leveling the spoon (if it is too full, or adding more if there is too little), and dumping the baking soda into a receptacle. Each of these steps was coded according to who *physically* carried out the step: the parent, the child, both, or the step was excluded or uncodable (p, c, b, or e/x), leading to a six-part code such as: PBC/CBC (the parent fills the first spoon, they jointly level it and the child dumps it into the bowl; the child fills the second spoon, they again jointly level it, and the child again dumps it). We would expect to find a number of different sorts of organizations of collaboration: the child might passively watch the parent doing the activity; the parent might support the child's doing parts of the task; the parent might demonstrate with the first spoon and then give the second to the child; the parent might explain what to do and the child do it; etc. When analyzed without the E and X codes there was a significant interaction between age (3, 4, 5) and contributor (p, b, c) only for the first spoon leveling step ($\chi^2(4) = 10.08, p < .05$). These effects reveal that whereas the leveling activities for the first spoon are apparently being taken up by the child, this is not the case for the first spoon's filling, nor for the steps in the second spoon. (Dumping is generally carried out by the child throughout ages and spoons.) The cross-step effects (fill-level-dump by parent-both-child) are significant at 3- (spoon 1: $\chi^2(4) = 21.08, p < .001$; spoon 2: $9.52, p < .05$) and 4-years ($24.37, p < .001$; $12.11, p < .02$), but not at 5. The effects demonstrate that whereas there is a difference between who is carrying out the activity (parent vs. child) in the first two age groups, there is no difference in the oldest group (5-year olds). That is, a "division of labor" is eventually reached wherein the child is a full collaborator, in the sense that we do not find the tasks differentially distributed to either parent or child. (More detail can be found in Shrager, Callanan, & Holloway, 1991.)

*Qualitative Analysis of Active Language*

The second step in our method is to analyze the functions of the utterances that occur in this setting, leading eventually to a joint analysis that should enable us to study the relationships between the language and the development of the collaborative activity. As a first attempt to capture these details we present a qualitative analysis of the active language of three subjects in the baking soda step. In the next section we discuss the possible functions of the active language observed here, and in the conclusion we ask how this analysis can inform cognitive theories of development.

Let us examine one dyad from each age group: Colin (2;9) and his mother, Cindy[1], took about 85 seconds to complete the task, and their collaborative structure was coded as: "PPC/CBC" (the parent filled and leveled the first spoon and the child dumped it into the bowl, the child filled the second spoon, they both leveled it, and the child again dumped it); Sarah (3;7) and Sue took about 60 seconds, and were coded as: "PPC/CCC"; Jennifer (4;10), her mother, Joan, (and Jennifer's sister, Amy) took about 98 seconds, and were coded as: "BBC/CCC".[2] The activity of each of these parent-child pairs, the children about a year apart, is in some ways very similar: one can see from the given collaboration codes that in each the parent helps the child out more with the first spoon than with the second, but also that the older children are able to accomplish the activity by themselves, whereas Colin's mother helps him with the leveling of the second spoon. Our summary data suggests that this distribution of collaboration is typical.

We will focus here upon the way in which the parents introduce the procedure for leveling the material in the spoons, and how they guide the child through this procedure. We will see that in these case the parents of younger children seem to combine both physical and verbal guidance while parents of older children rely on verbal guidance more than upon physical guidance. Eventually, for the oldest children, parents are able to simply explain and demonstrate the activity, leaving the child completely unmonitored. Older children become more articulately and physically skilled in the task, as well as in their engagement with their parents, for instance replying to guidance and inquiry and taking initiative by greater degrees. Thus the structure of the interaction between parent and child differs in significant ways.

Colin's mother, Cindy, guides him physically through the leveling activity on the second spoon:

> At the point in the second spoon where it is filled with baking soda, and as Colin moves to dump the material into the big bowl, Cindy says: "We only want..." and as Colin pulls out the spoon, apparently going to dump it into the bowl, Cindy says: "Woop woop woop woop..." and takes the spoon handle in her right hand just below the point at which Colin is holding it. She says: "We're gonna do one more thing before we put it in there, okay? We're gonna scrape it. Scrape it." While saying this she is guiding the spoon back into the box (with Colin still holding it) and then she changes grip so that she is holding his hand rather than the spoon itself and says: "Scrape the top so it comes real flat....just like that...see?" They draw the spoon out of the box, scraping it along the top of the box so the baking soda level in the spoon becomes (real) flat. Cindy says, more loudly and brightly: "Okay?" and releases Colin's hand near the bowl. He moves the spoon fully over the bowl and dumps the material in as she backs away. She says: "Then we won't get too much [in]."

Sarah's mother, Sue, verbally guides Sarah through the procedure with the second spoon:

> Sue takes back the spoons and picks out the blue one and hands it to Sarah. Sue holds the box, turning it toward Sarah saying: "Okay stick it in there" which Sarah does. Sue goes on: "and fill it up". Sarah digs around. Sue touches Sarah's hand briefly in

[1] These are, of course, not their real names.

[2] A complete transcript of Jennifer and Joan's (and Amy's) baking soda segment is included as an appendix. [Square brackets indicate utterances that were difficult to transcribe, and so are questionable.]

gentle direction: "That's it." Sarah continues to dig around. Sue takes Sarah's hand, gently guiding it, saying "Turn it around." Sue releases Sarah's hand: "and then scrape it off [unintelligible]" which Sarah also does. Sue says: "Oh, you didn't fill it all the way." and moves apparently to take Sarah's hand but switches to pointing to the material in the spoon. Sarah digs back into the box of baking soda and this time, without further conversation or guidance, completes the operation including scraping and dumping the material into the bowl.

Jennifer, the oldest child, is able to complete the leveling activity on the first spoon with verbal guidance and physical guidance and is then able to conduct the entire activity on her own with the second spoon (see Appendix). She also seems to be engaged with her mother's directive verbalizations, answering "yeah" to her mother's question: "See how it makes it flat?", whereas Sarah and Colin do not reply to their parents' directives and Colin seems to require physical guidance. (Whereas Jennifer's mother eventually physically guided her activity as well in the first spoon, Jennifer had simply got the notion of scraping the spoon backwards – scraping it upside-down on the box top instead of rightside up.)

It is worth noting a closely related example of differential language use in the way that the parents indicate which spoons are to be used in the activity. Cindy selects the spoons for Colin and does not indicate the color or label. Sue selects the spoons for Sarah but also indicates the colors as the selection method. Joan selects the first spoon for Jennifer, indicating both the color and label. Jennifer selects the second spoon by herself, the appropriate color having been indicated by Joan (see Appendix).

## Functions of Active Language

We have seen in these examples that the parents' physical guidance of the child's activity seems to decrease whereas the parent-child connection through verbal communication seems to increase. What learning functions are served by active language in enabling this change? The observed changes presumably entail (at least) improvements in the child's domain specific skill as well as changes in the child's ability to engage with the parent in a verbally-mediated interaction. How might active language serve as a *resource for learning* about this task, as well as about how to collaborate? Some potential learning functions include:

*Object and action labels:* What is being manipulated, as well as the action taking place, is spoken about continuously, making explicit reference available to the child in both understanding and in production (e.g., in order to ask questions).

*Sequencing of expectations (procedure organization):* In part of its role in mediating joint attention, the language provides sequencing information for the child. What is to come next is often mentioned before the action takes place. Thus, if one learns a verbal narrative, and a set of action labels (as above), one can reproduce the action sequence. (See Bruner, 1983, for an excellent discussion of closely related functions of joint active language in game playing.)

*Task structuring articulations (and activities):* Various common ways of speaking, intonation, and repeated activities cue the beginnings and ends of action units, and the goal-structure of the task. Utterances such as: "Okay" ... "So"... "What are we gonna do now?"

*Explication of non-obvious aspects and focusing on relevant aspects:* Explanations articulate non-obvious or cross-situational aspects of the setting (e.g., causal relations), and can guide attention to relevant aspects from among the many features in a complex situation. Some such aspects include the conditions that indicate when it is appropriate to carry out an action, and explicit task goals, and the purposes of actions with respect to those goals.

*Interaction facilitating articulations (and activities):* In addition to learning about the task itself, the child learns ways of operating in this domain which facilitate the collaboration. For instance, some of the task structuring articulations, such as saying "Okay" at various points, are picked up and used by the child, and elicit guidance from the parent, as when the child says "Okay" but it is not the end of the activity (to the parent). Ways of asking for instruction refinements (e.g., "Where?" and "What?") are also important facilitators of the collaboration, and come to be used by children with cooking-specific content.

A given utterance and action context may serve a number of functions. Consider, again, Cindy's introduction of the leveling procedure to Colin, quoted above. As she is guiding Colin through the procedure, Cindy says: "We're gonna do one more thing before we put it in there, okay? We're gonna scrape it. Scrape it...Scrape the top so it comes real flat....just like that...see?...Okay?...Then we won't get too much [in]." This series of utterances, combined with the details of the action (see above) potentially provide: a name for the procedure ("scrape"), a focus for the procedure (the spoon they are holding, called "it"), the relevant conditions and local purposes of scraping ("so it becomes real flat") as well as the purpose with respect to the task goals ("then we won't get too much in"). Further, the structure of the active language is so timed that it brackets the procedure itself (The interruptions of "Woop, woop..." through "Okay?").[3]

Interestingly, all of the aspects of cooking available in active language – the names of things, the task goal and structure, causal structure, conditions and purposes, etc – are *interpretations* of the activity. In engaging the child in cooking activity, the parent is simultaneously engaging the child in *active interpretation about the activity.* Thus, the process of learning about cooking involves the child in learning to become a *culturally appropriate interpreter.*

We are presently working to develop coding schemes, like the collaboration coding scheme of Shrager, et al, (described above) through which we can test hypotheses about the these functions in the observed developmental changes.

## Discussion

The collaborative work settings in which the child participates are extremely rich in resources for the local support of action as well as for learning. Although models such as those of Siegler & Shrager and VanLehn take some account of the structure of the child's context, they have barely grasped the nature of the parent-child system – a system characterized by close dynamically managed interaction, and including a richly woven system of active language. One promising approach to understanding collaborative development adopts the metaphor of the child-as-apprentice (e.g., Kaye, 1982; Rogoff, 1990) in contrast to the Piagetian view of the child-as-scientist. Substantiating the apprenticeship metaphor requires theoretical frameworks that focus on the mutual dynamics of the learner and the context of change (e.g., Agre & Shrager, 1990; Lerner and his coworkers, see Lerner, 1986; Kaye, 1982, Rogoff, 1990), as well as a modeling technology that integrates individual mechanisms of cognition (e.g., learning), attentional and perceptual mechanisms, and the collaborative mechanisms of jointly managed attention, negotiated guidance, etc. There is already some movement on this direction with the theories of Bruner (1983), Callanan (1991), and Rosenthal & Zimmerman (1978), and the models of Kaye (1982). With details of collaboration such as those described here in hand, we will be in a better position to build process models in which the developmental work is appropriately apportioned among the three areas of cognition, perception, and collaboration.

[3] Intonations are very relevant in this argument, as in all of the above, but we cannot delve into this topic in the present limited space.

## References

Agre, P. & Shrager, J. (1990) Routine evolution as the microgenetic basis of skill acquisition. In the proceedings of the 12th annual meeting of the Cognitive Science Society (August, 1990). Hillsdale, NJ: Lawrence Erlbaum Associates.

Bruner, J. (1983). Child's talk. New York: W.W. Norton & Co.

Callanan, M. A. (1991). Parent-child collaboration in young children's understanding of category hierarchies. In J. Byrnes & S. Gelman (Eds.), Perspectives on Language and Thought: Interrelations in Development. NY: Cambridge University Press.

Kaye, K. (1982). The mental and social life of babies; How parents create persons. The University of Chicago Press.

Lerner, R.M. (1986). Concepts and theories of human development. Hillsdale, NJ: Lawrence Erlbaum Associates.

Rogoff, B. (1990). Apprenticeship in thinking. Oxford University Press.

Rosenthal, T.L. & Zimmerman, B.J. (1978). Social learning and cognition. New York: Academic Press.

Shrager, J., Callanan, M. & Holloway, A. (1991). The dynamics of collaboration in the development of complex activity. Presented at the Semi-Annual meeting of the Society for Research in Child Development. Seattle, April 1991.

Siegler, R.S. & Shrager, J. (1984). Strategy choice in addition and subtraction: How do children know what to do? In C. Sophian (Ed.), Origins of Cognitive Skills. Hillsdale, NJ: Lawrence Erlbaum Associates.

VanLehn, K. Learning one subprocedure per lesson. Artificial Intelligence, 31(1), 1-40.

Vygotsky, L. S. (1978). Mind in society: The development of higher psychological processes. Harvard University Press.

## Jennifer (4;10), Joan (and Amy)

There is a younger child (Amy) standing between Jennifer and Joan. Joan picks up the spoons and reads the recipe: "Okay, for the baking soda we need to put one and a quarter teaspoons." She sorts through the spoons. Jennifer reaches for the baking soda, which is in a plastic baggie. Joan selects the blue and red spoons and shows them to Jennifer: "So here's the one." Pointing to the blue spoon's label: "See the one?" Jennifer: "Uh huh" Pointing to the red spoon's label Joan says: "And that's a quarter...one fourth" Joan begins to open the baggie: "So you put a blue one and a red one." As she says "blue" and "red" she touches the blue and red spoons respectively. She opens the baggie, saying: "of this is here...let's see if we can...okay it's opened up..." Jennifer and Amy watch Joan's activity. Joan continues: "And the way you do that...." Jennifer reaches for the blue spoon that Joan is beginning to put into the box while holding it with her other hand. Joan: "...remember how you level off the spoon?" She gives the spoon to Jennifer who digs into the box. Joan continues to hold the box with her left hand: "With this one we don't use a knife, you can just kind of rub it against the top of the box." Joan gestures a scraping motion with her free right hand near the box top. Jennifer does the appropriate leveling action, withdrawing the spoon. They inspect the spoon contents. Joan: "Up, let's see, get it all the way full." Jennifer repeats the operation, digging in for baking soda and scraping the spoon along the top of the box, while Joan continues to steady the box. Joan says "That-a-girl" as Jennifer withdraws the spoon. Joan says: "Okay and kind of scrape it..." Jennifer turns the spoon *upside down* on the box top and scrapes it backward, dumping the baking soda back into the box. Joan takes Jennifer's hand with the spoon: "Okay. I'll show you." They dig in together. As they do the activity together, Joan says: "You kind-a...put it in...and you fill it up...and then you kind of rub it against the top here." They dig in again: "Get it nice and full. Rub it against the top. See how it makes it flat?" Jennifer replies: "Yeah." Joan: "That makes it exactly right." Joan releases Jennifer's hand and the child dumps the material into the bowl. Joan says: "Okay." Jennifer begins to refill the blue spoon. Joan says: "Now we need a quarter, which is the red one." and touches the red spoon, which has been dangling in front of Amy, the younger child. Jennifer stops filling the blue spoon and begins to switch to the red one. Amy has meanwhile grabbed the dangling spoons. Jennifer makes a frustrated plea: "Amy!" for her to release them as she (Jennifer) pulls them away. Joan picks up Amy and puts her aside on the floor saying: "Okay. You know what? I'll get her her own spoons." Joan walks out of the scene. Meanwhile, Jennifer has filled and scraped the red spoonful of baking soda and holds it out to show (apparently no one as Joan has walked off camera), smiling and saying: "I got a [tiny]." She laughs and moves to dump the baking soda into the bowl but stops and says over her shoulder to about where Joan has gone: "I got it, mom." Joan calls from the hallway: "Okay. Did you get it..." and looks into the room at Jennifer who is holding the full spoon out toward the mother. Joan says: "...that-a-girl" and Jennifer turns back to the bowl to dump in the baking soda, Joan continues: "why don't you put the...that right in there..." Jennifer dumps the soda into the bowl.

# Forming Shared Mental Models

**Katia P. Sycara**
The Robotics Institute
Carnegie Mellon University
Pittsburgh, PA 15213
katia@cs.cmu.edu

**C. Michael Lewis**
Department of Information Science
University of Pittsburgh
Pittsburgh, PA. 15260
ml@pigeon.lis.pitt.edu

## Abstract

As problems increase in complexity it becomes impossible for any one person to know all the things necessary to make good decisions. A group of specialists, even if they possess the requisite knowledge, remain a collection of individuals until their expertise can be jointly brought to bear. The problem of fusing expertise where individuals have very detailed knowledge in their own areas and much weaker understanding of others is that no one knows what anyone else needs to know. This impasse cannot be broken until shared mental models are developed to provide the common perception needed to focus the activity of the group. This paper presents characteristics of shared mental models and a model of the effect, nature, and process of the formation of shared mental models in cooperative problem solving by a team of specialists.[1]

## Introduction

Decision making in a variety of complex tasks requires the cooperation of many decision makers. Very often, and when the task is especially complex, each decision maker is also a specialist in a particular area. At the same time decision making in a team of specialists is frought with difficulties. A major difficulty is that different specialists lack (1) a shared language for communication and (2) shared perceptions of the task. Evidence supporting this assumption comes from a variety of sources. Case studies of decision making in organizations (e.g., Bond's [Bond 89] Lockheed study of aircraft design) have found that specialists do not understand the details of each other's models and language, but through cooperation and interaction are somehow able to produce designs of very complex artifacts, such as aircraft. Research in group decision making has found that the variability of approaches to decision making across groups appears to be even greater than across individuals [DeSanctis 87]. The only consistently reported difference between successful and unsuccessful problem-solving groups has been that successful groups devote adequate time to *problem formulation* and *planning of meeting strategy*, whereas unsuccessful groups immediately begin to search for alternative solutions [Hirokawa 83, Hackman.Kaplan 74]. In other words, successful groups spend adequate time to build shared vocabulary and mental models of the nature of the decision problem, strategies, significance of information, and participants' roles. Shared mental models have also been hypothesized by [Athans 82] and [Fischhoff 86]. The importance of the formation of shared mental models has also been experimentally observed in time critical high-stress decisions such as the air crew emergencies studied by [Orasanu 90].

Although the evidence exists that the formation of shared mental models is a major factor in successful group decision making by humans, little has been done to characterize what "shared mental models" may be, the processes by which they are formed, or their role in group decision making.

Similarly, Distributed AI research sheds little light on this problem. Most work in DAI and team decision making has focused on homogeneous agents [Durfee 87, Cammarata 83, Rosenschein 85, Sullivan 89]. In such situations, the agents already share a common perception of the task and based on this understanding they have to coordinate their *actions*. If they communicate at all, communication of high level agent plans [Durfee 87], based on which agents form expectations of other agents' behavior, ensure coordination. In the team of specialists situation, the agents are by definition heterogeneous, they do not share a common perception of the problem, or each others' expertise. Yet they have to make highly interdependent decisions with uncertain outcomes. A prerequisite to coordinating their actions is to coordinate their *thoughts* through the formation of shared mental models.

In particular, we present the following characteristics of shared models:

- Shared models act as a basis for inferring opportunities for cooperation

[1]This research has been supported by DARPA and AFOSR under contract number F49620-90-C-0003.

- Shared models form the basis for identifying another's needs resulting in helpful intervention
- Shared models act as a basis for inferring another agent's capabilities (resulting in the agent asking for help) and limitations and taking them into consideration in communicating with other agents
- Shared models serve as a basis for knowing *what* to communicate (i.e. what will be understood)
- Shared models act as basis for conflict resolution
- Shared models act as basis for optimal solution integration

The formation of shared mental models is an incremental process consisting primarily of information communication and successive updating and refinement of a sketchy understanding of the ramifications of the problem that the team is solving. In this paper, we present a model of agent models that are active during group decision making and their interactions to forming shared mental models in problem solving by a team of specialists. The domain of our study is concurrent engineering design [Sycara 89a]. The model is currently being implemented in CADET, a design problem solving system [Sycara 90].

## Cooperative Problem Solving by a Team of Specialists

The present model of decision making by a team of specialists extends earlier work [Sycara 89b, Sycara 90] characterizing group decision making as a negotiation process. A team of specialists is considered to be a group of individuals with common goals, each with highly specialized knowledge in a particular area but with less precise knowledge of other areas. A normative group decision, therefore, is one which integrates the relevant portions of this specialized knowledge within a common model. The key to this definition lies in the determination of relevance. We presume that in a group decision process of this sort, it is neither feasible nor desirable to form a common model incorporating all of the group's expertise. This would require making M.D.'s of physicists and engineers of accountants. Instead, a normative group decision is one generated by a negotiation process which efficiently elicits some small subset of the group's expertise which determines a decision as good or better than other possible negotiations of comparable length.

Group decision making by cooperating specialists can be viewed as a multi-agent planning task where each agent has (a) incomplete knowledge of the environment, (b) limited knowledge of the constraints and intentions of other agents, and (c) limited number and amount of resources that are required to produce a system solution. When specialists cooperate, they bring together multiple viewpoints and diverse knowledge on a single problem. Bringing together diverse knowledge is a source of robustness and balance. The team can solve problems that are beyond the scope of any of the individual members. Furthermore, the solutions are generated from a rich and varied body of knowledge which provides potentially, a bigger set of good solutions to choose from and the potential for creativity. On the other hand, there are difficulties with resolving the conflicts that arise when trying to merge multiple goals, priorities, and evaluation criteria that are the results of individual expertise. Thus, each agent has a limited and egocentric view of the problem. This may result in misunderstandings, goal conflicts and solution suboptimalities. Because of the lack of appropriate expertise in all areas needed to solve the problem, and because of the presence of conflicting constraints, goals and possibly evaluation criteria, it is impossible for each agent/expert/specialist[2] to reach an optimal solution using only local information. Typically the decisions of one agent impact the decisions of another and vice versa. Thus, we cannot simply model each agent. Rather we must augment an agent's problem solving process by including yet another model, the "shared model" that captures his interactions and decision-coordination with the other agents.

Decision making by a team of specialists has the following characteristics:

- The global system goal is to produce a decision that is synthesized from contributions of different expertise, concerns and constraints.
- During the process, conflicts in the form of constraint and goal violations could arise.
- Disparate evaluations of a situation could surface as a result of different criteria used to evaluate the situation from different perspectives. Typically, these criteria cannot be simultaneously and optimally satisfied.
- Another kind of conflict is over approaches used to achieve a goal.
- The system goal is achieved by making the best tradeoffs on conflicting goals and constraints.

As a result of these characteristics, there must be a process that helps incrementally incorporate in the final decision the various types of available expertise, and the best tradeoffs that could be made in a given situation.

The dilemma confronting a team of experts is that the subset of individually held expertise that is relevant to solving the problem at hand cannot be known in advance. Each expert has detailed knowledge of some aspect of the task and beliefs, based in part on inexpert understanding of other aspects, about how this expertise may be related to the groups goals. Because no member knows precisely

[2]We use the words "agent", "expert" and "specialist" interchangeably for the purposes of this paper.

what any other knows, none can determine the relevance of their own expertise, if that relevance depends on things known only by others. The situation is similar to that of a group attempting to assemble a complex puzzle. Each member has some pieces mixed in with pieces from a great many other puzzles. Players can only view their own pieces and there are costs associated with making pieces public by placing them on the table. If players hold pieces without prominent identifiers such as matching background color, the relevance of these pieces can only be established through an iterative process of advancing and retracting pieces until a context is developed allowing the players to "see" which ones fit.

In the design domain for example, a tricycle designer knows (from his expert knowledge) that using cotter pins and caps to hold the rear wheels and pedals of a tricycle together results in much higher performance (leading to higher quality) than using press-on caps. Because of his ignorance of costs, and his approximate knowledge that quality is linear with costs, the designer assigns a higher cost to a design that uses cotter-pins than to one that uses press-ons, thus making him indifferent to the choice. Similarly, the manufacturing engineer knows (from his expert knowledge) that drilling round stock to make the hole for the cotter pins is a more expensive process than fitting the press-on caps, so (having the inaccurate general knowledge that quality linear with cost), he is also indifferent to the choice. The union of their expertise, however, may strongly favor one or another of these alternatives. For instance, if the designer rated the cotter pin design as four times more desirable than the press on caps and the manufacturing engineer rated the cost at less than double, the cotter pin design is favored. This representation of the compartmentalization of expertise captures the features we believe are essential to describing the pooling of expertise. Each expert has specialized knowledge about his part of the problem but misconceptions about other agents' parts. As a consequence, no expert knows which parts of his expertise are really needed by other members of the group.

We claim that expert team decision making solves this problem through iteratively updating these naive models via the communication of the group members' expertise resulting in the development of "shared models".

## Models of Agents

The model we are proposing operationalizes private and shared knowledge in the form of expert domain knowledge, utility functions and constraints and the formation of shared models as a negotiation [Sycara.aor, Sycara.eurjor90] modifying these structures. The model is quite general and independent of decision protocols, describing with equal ease consensus, voting, or advisory regimes. The model provides normative references for group decisions by assessing product and process simultaneously. A group decision can be judged either with respect to the publically available model at the time of decision of the distribution of normative decisions for possible models resulting from negotiations of equivalent or shorter length. The assumption that group decision making is a satisficing process trading off time and effort for decision "quality" is a central tenet of this approach.

In complex domains, such as design, attributes of concern (e.g., design attributes such as material out of which an artifact is made) interact in complex non-linear ways with the objective to be maximized, profits. This interaction happens through mediating variables, such as functionality, durability, aesthetic attractiveness etc. The question that arises is, what is suitable information aggregation for inclusion in the "shared models"? How does a "shared model" interact with individual specialized knowledge? What is a suitable representation of individual expertise and "shared models"? What are the variables used for communication among group members?

Our hypothesis is that each agent:

- Has a model of his individual unique expertise, called the *"expert model"*, characterized by detailed knowledge about some particular aspect of the task.
- Has a naive understanding of aspects of the problem outside of his area of expertise, called the *"naive model"*. The naive model characterizes weak commonly held beliefs such as, "the more expensive a material is, the more durable it will be."
- Develops through interaction with other specialists a more comprehensive model of the problem at hand, called the *"shared model"*, which incorporates elements of others' expertise. The "shared model" between two agents also defines the common vocabulary that the two agents can use to communicate in an intelligible way.

In general, "shared models" might not be the same across all agents. For the group to make a common decision, the potentially different"shared models" that have been formed through interactions of subsets of agents must converge to a common view of the global problem which captures the important variables and decisions that must be coordinated. The parts of different agents' "shared models" that are common across the group comprise the *"common (group) model"*.

During decision making, agent models appropriate for the task are composed by substituting the corresponding parts of an expert model for the naive model. The utility an agent associates with a particular alternative, therefore, reflects both components. As group problem solving progresses, agent models are modified to incorporate expertise imparted by other agents. These modifications lead to the incremental building of shared models among sets of agents.

## Model Characteristics

Models of agents are hierarchically organized in the form of increasingly refined domain concepts, constraints and utility functions. The complex representation at the bottom of this hierarchy corresponds to the full expertise of the agent which may or may not be needed by the group for a particular decision. At the top of the hierarchy there is sketchy knowledge, the naive models shared by the members of the group. The naive models are required to be accurate in aggregating across situations but may be imprecise in capturing distinctions among situations made by the expert model. In other words, using only his specialized knowledge and the naive model, each specialist would be led to erroneous decisions. Naive model imprecisions may arise through naive approximation of the values, relations and constraints among the attributes of the problem variables.

The mental models we are investigating have a variety of characteristics. At the beginning of problem solving, an agent's mental model consists of his expertise and a naive understanding of other specialties which is refined in the course of the consultation developing into a (set of) shared models. The naive model contains aggregate variables that constitute the vocabulary through which enables the different specialists to begin the problem solving interaction.

An agent's expertise consists of facts, constraints, their relations and utilities that lie in his specialty. For example, a designer's expertise contains the constraints among design attributes such as artifact components, component materials, types of allowable structural connections among components, and artifact dimensions. Utilities are associated with particular values of these artifact-descriptive variables that operate in the design space. Utilities are in the range of [0, 1]. For example, the utility of making a tricycle out of light steel may be .8, making it out of heavy steel .6, making it out of wood 0 and so on.

The naive model consists of a set of relations among aggregate variables. In design, these relations relate the quantities profits, design quality, cost, price and saleability. These relations are at an aggregate level and are therefore only approximately accurate. For example, the naive model may say that production quality is proportional to production costs, unit sales inversely proportional to price, and willingness to pay (saleability) (for the decreasing number who could) proportional to quality. As a result, the naive model predicts that saleability will be independent of design decisions balancing cost and quality. (The group's profit margin can be maximized by choosing the design which minimizes the cost while maximizing the quality). In other words the naive model is indifferent among design options. This evaluation made by the naive model is not accurate since it does not take into consideration precise relations between variables of the problem. The naive model constitutes a common and unbiased basis which is transformed and refined through the group interaction.

The naive model and an agent's expertise are connected through a set of relations that map variables in the *expert space* to the variables in the *naive space*. The mappings contain both accurate and approximate knowledge. The accurate knowledge is the set of relationships that map variables in the expert space to *mediating* aggregate variables that are within the area of agent expertise. For example, the designer's expert space variables, such as the artifact-specific attributes, their constraints, utilities and interrelationships are mapped into the inexpert space that contains the variables profits, cost, price and quality through mediating variables, such as durability, structural soundness etc. For the designer, these intermediate variables map to design *quality* through a set of relationships. Design quality is an aggregate variable in the domain of expertise of the designer that is also present in the naive model. The set of intermediate relationships that map the variables within an expert's specialty to the naive model is available only to the particular expert and is not used for intra-group communication. On the other hand, relations between aggregate variables, within and outside the domain of expertise of a specialist are incompletely known to him. For example the relation between quality (an aggregate variable within the domain of a designer's expertise) and cost is only approximately known to the designer, since cost is not one of the aggregate variables in his domain of expertise. This interaction of accurate and inaccurate knowledge may lead the expert to incorrect inferences. Expert team decision making solves this problem through iteratively updating the naive models via the communication of the group members' expertise resulting in the development of a shared model.

## An Example

Consider the decision situation for a team of specialists in a manufacturing enterprise tasked with concurrently engineering the design of a tricycle. The team objective is to optimize tricycle design. A design will be optimal if it maximizes profits, under certain assumptions relating design attributes to cost, price, and ease of selling the tricycle.

The design attributes have associated utilities that express the desirability of particular attribute values with respect to different evaluation criteria. For example, in terms of tricycle durability, high grade plastic has utility .8, whereas low grade plastic .4. In terms of cost, high grade plastic has utility .6 whereas low grade .8 (high grade plastic is more expensive than low grade). Each team member has expertise in terms of optimizing only part of the design (e.g., cost of material, ease of manufacturing, reliability). Because of the inherent tradeoffs of attribute values with respect to different criteria not all parts of the design can be simultaneously optimized. Furthermore, the overall evaluation function that incorporates cost tradeoffs is not precisely known. In a group decision making situation, each expert knows the relation among attributes in his domain of expertise and the payoff associated with each attribute value alternative, but he is unaware of (a) the relations among attributes in someone else's domain of expertise, or (b) how the attributes in his domain of expertise impact the overall cost function. For example, the designer knows that in terms of strength of the tricycle frame, braced and welded frame has the highest payoff, bolted the next highest, and integral the lowest. In terms of weight, high grade plastic has the highest payoff, with heavy steel tubing the lowest.

The group's goal is to maximize profit and may be expressed as:

profit = unit_sales x (unit_price - unit_cost)

A negotiation over this decision might proceed as:

*Manufacturing Agent:* Do we want to use press-on caps or cotters for the wheels and pedals The drill press operations will add another 3% to manufacturing costs. (This is an informational exchange. The manufacturing agent is indifferent to this decision because it does not affect the group's goal as predicted by his agent model.)

*Design Agent:* In that case, I think we should use cotters. The press-on caps are likely to start falling off after 6 months-1 year while the cotter pins will hold the wheels on indefinitely. (The reliability of the artifact which supports the fulfillement of the public attribute of quality is increased by 500% at a manufacturing cost of only 3%, therefore by the naive portions of the designer's model relating quality to profits through unit sales this alternative should more than double profits furthering the group's goal.)

*Sales Agent:* I don't think we should do it. A buyer can't see that there is a cotter under the cap and therefore it has no effect on perceived quality. (Perceived quality is the sale's experts private version of the public attribute quality and is the attribute in the group's expert model which relates quality to unit sales. It therefore takes precedence.)

All three agents now agree because with this modification, their shared model predicts that the increased production cost associated with cotters will be detrimental to the goal of increasing profits because there is no offsetting influence of this form of quality on unit sales.

## Concluding Remarks

The model proposed in this paper is intended as a research tool for the study of group decision making by people and machines. The model provides the first normative treatment of group decisions which incorporates the very features motivating this form of decision making. The importance of shared mental models has been recognized in the cognitive decision making literature [Athans 82, Fischhoff 86] but no investigation as to the process of forming shared mental models has been proposed to date. We have proposed the existence of a variety of mental models (the expert model, the naive model, shared models and a potentially common model) during group decision making and presented their interactions during the problem solving process. In addition, we have presented a concrete approach to describing those models and the process by which they are formed. The shared mental models discussed in this paper define the nature and minimal level of aggregation of the information necessary to be common knowledge to agents so that suitably coordinated decisions can be made. It is our hope that this model can provide a tool for the study of bias in group decision making which may prove as fruitful as normative models have been in the study of individual decisions.

## References


[Athans 82] Athans, M.
The Expert Team of Experts Approach to Command-and-Control Organizations.
*IEEE Control Systems Magazine* September:30-38, 1982.

[Bond 89] Bond, A., and Ricci, R.
Cooperation in Aircraft Design.
In *Proceedings of the MIT-JSME Workshop on Cooperative Product Development.* Cambridge, Mass., 1989.

[Cammarata 83] Cammarata, S. McArthur, D. and Steeb, R.
Strategies of cooperation in distributed problem solving.
In *Proceedings of IJCAI-83*, pages 767-770. IJCAI, Karlsruhe, W. Germany, 1983.

[DeSanctis 87] DeSanctis, G., and Gallupe R. B. .
A Foundation for the Study of Group Decision Support Systems.
*Management Science* 33:589-609, 1987.

[Durfee 87] Durfee, E.H.
*A Unified Approach to Dynamic Coordination: Plannign Actions and Interactions in a Distributed Problem Solving Network.*
PhD thesis, COINS, University of Massachusetts, 1987.

[Fischhoff 86] Fischhoff, B.
Decision making in complex systems.
In Hollnagel (editor), *Intelligent Decision Support in Process Environments.* Springer Verlag, New York, N.Y., 1986.

[Hackman.Kaplan 74] Hackman, J.R. and Kaplan, R.E.
Interventions into Group Process: An Approach to Improving the Effectiveness of Groups.
*Decision Science* 5:459-480, 1974.

[Hirokawa 83] Hirokawa, R.Y., and Pace, R.A.
A DEscriptive Investigation of the possible Communication-based Reasons for Effective and Ineffective Group Decision Making.
*Communication Monographs* 50:363-379, 1983.

[Orasanu 90] Orasanu, J.
Shared Mental Models and Crew Performance.
In *Proceedings of the 34 Annual Meeting of the Human Factors Society.* Orlando, Fla., 1990.

[Rosenschein 85] Rosenschein, J.S.
*Rational Interaction: Cooperation Among Intelligent Agents.*
PhD thesis, Stanford University, 1985.

[Sullivan 89] Sullivan, D., and Kleinman, D.L.
A Normative Descriptive Approach to Team Distributed Scheduling.
In *Proceedings of the 1989 IEEE International Conference on Systems, Man and Cybernetics.* IEEE, Cambridge, Mass., 1989.

[Sycara 89a] Sycara, K. and Navinchandra D.
A Process Model of Experience-Based Design.
In *Proceedings of the Eleventh Annual Conference of the Cognitive Science Society.* Ann Arbor, MI., 1989.

[Sycara 89b] Sycara, K.
Argumentation: Planning Other Agents' Plans.
In *Proceedings of the Eleventh International Joint Conference on Artificial Intelligence (IJCAI-89).* Detroit, Mich, 1989.

[Sycara 90] Sycara, K.
Viewing Design as a Cooperative Task.
In *Proceedings of the Twelfth Annual Conference of the Cognitive Science Society.* Boston, Mass., 1990.

# Avoiding Mis-communication in Concept Explanations

**Ingrid Zukerman**
Department of Computer Science
Monash University
Clayton, VICTORIA 3168, AUSTRALIA

## Abstract

This paper offers a mechanism for the generation of expressions that convey a concept to a hearer. These expressions often include rhetorical devices, such as Descriptions and Instantiations. Our mechanism is based on the representation of the preconditions for the understanding of a concept in terms of failures in communication. We distinguish between two types of communication failures based on their cause: failure due to a hearer's inability to evoke an intended concept, and failure due to the hearer's lack of expertise with respect to the aspects of a concept which are necessary for the comprehension of the discourse. This categorization supports the principled selection of rhetorical devices which are tailored to the prevention of undesirable situations that result in the failure of a communicative goal. These ideas are being implemented in a discourse planner in the domain of high-school algebra.

## Introduction

In knowledge acquisition settings, one agent transfers information which is believed by him/her to another agent, with the intent that the second agent learn this information. In these settings, the speaker's communicative goal is *KNOW(Hearer,item)*, where the meaning of the goal *KNOW* is that the hearer learn prescribed aspects of the item in question. This item may be a single concept, such as a procedure or an object, or a message composed of predicates and concepts. For example, the message [Distributive-Law apply-to Algebraic-Terms] is composed of the predicate apply-to and the concepts Distributive-Law and Algebraic-Terms. In an introductory algebra class, when the teacher demands that the students know this message, s/he usually means that they should know how to apply Distributive Law to Algebraic Terms. In a different setting, the requirements posed by the goal *KNOW* may differ. Educational settings are representative of situations where the item being conveyed is a single concept, e.g., *KNOW(Hearer,Conceptual-Dependency)*, since in these settings, a lesson often revolves around teaching different aspects of a concept, such as its representation and its semantics.

A necessary but not sufficient precondition for the satisfaction of the goal *KNOW(Hearer, message)* is that the subgoals *RETRIEVE(Hearer,concept)* and *KNOW(Hearer,concept)* be satisfied with respect to the concepts in the message. That is, the hearer should be able to recall the concepts intended by the speaker, and s/he should also have a level of expertise with respect to these concepts which is sufficient to understand the message. For instance, if in the above message, the speaker uses the term 'Algebraic Terms' to refer to the concept Algebraic-Terms, this term should enable the hearer to reach the intended concept in his/her memory. In addition, the hearer should know enough about Algebraic Terms to understand how Distributive Law can be applied to them. A necessary and sufficient condition for the fulfillment of the goal *KNOW(Hearer,concept)* stipulates that the subgoal *KNOW (Hearer,message)* be satisfied with respect to a set of messages the speaker considers sufficient to define the concept.

Following [Webber 1983], we distinguish between two different aspects of the goal *RETRIEVE* according to its place in the discourse: *EVOKE* – where a concept is retrieved for the first time, and *ACCESS* – which pertains to subsequent retrievals. However, we depart from Webber's *internal* perspective to discourse, which deals with relations among discourse entities, and adopt Appelt's *external* perspective, which pertains to relations among the actual objects referred to by the discourse entities [Appelt 1986].

In this paper, we specify conditions for the satisfaction of the subgoals *KNOW* and *EVOKE* with respect to concepts, and present a mechanism which generates expressions that satisfy these subgoals. In this context, we address the generation of a class of *Rhetorical Devices (RDs)* which includes Descriptions, Instantiations and Analogies. Our mechanism follows the Impairment Invalidation paradigm [Zukerman 1990a, Zukerman 1990b] which postulates that RDs are generated in order to preclude anticipated impairments to a hearer's comprehension process. Thus, the preconditions for the attainment of the above subgoals are cast in terms of failures of the communication process. This approach generalizes the principle of *least collaborative effort* [Clark & Schaefer 1989] to the generation rhetorical devices. According to this principle, a speaker corrects his/her own mistakes in an effort to minimize the anticipated effort the hearer will make to understand the discourse.

In the following sections, we present the procedures for planning expressions which satisfy the subgoals *EVOKE* and *KNOW*, we examine the impairments to the satisfaction of these subgoals, and discuss the RDs which invalidate these impairments. Next, we present a simple algorithm which combines these procedures. Finally, we compare our approach with the traditional planning approach.

## Planning Expressions for the Subgoal EVOKE

People use different words to refer to the same thing, and they also use the same word to refer to different things. Thus, for a particular hearer, some wordings may be more felicitous than others in achieving successful communication. Further, sometimes it may be best not to refer to an item by its name, rather, indirect means, such as similes or descriptions, may be preferable. The mechanism presented in this section takes these observations into consideration for the generation of evocative expressions.

An evocative expression is composed of two parts: (1) A *Simple Referring Expression (SRE)*, which may be a lexical item or *nil*; and (2) An *Indicative RD* which complements the SRE if it is insufficient to evoke the intended concept. Thus, an evocative expression may comprise an SRE by itself, e.g., "Look at *Mary*," an Indicative RD by itself, e.g., "Look at *the girl with red hair*," or both "Look at *Mary, the girl with red hair*." The following procedure proposes evocative expressions which combine these parts.

1. Consult the speaker's knowledge base and a model of the hearer's beliefs to propose a list of candidate SREs for a given concept (only correct SREs are extracted from the model of the hearer's beliefs).
2. For each candidate SRE, consult a model of the hearer's beliefs to ascertain whether the SRE is likely to cause an impairment to the evocation of the intended concept. If so, propose an Indicative RD that invalidates the impairment. If there is more than one suitable Indicative RD, a set of candidate evocative expressions is built by combining each Indicative RD with the SRE.

For instance, in the above example, the candidate SREs are: "Mary" and *nil*. Now, if the speaker is uncertain that the hearer knows that the name of the referent is Mary, the speaker will propose an Indicative RD, such as a partial Description, which includes sufficient attributes of Mary to enable the hearer to identify her. A null candidate SRE implies that the speaker does not know or does not wish to use the name of a concept, and thus forces the generation of an Indicative RD.

### Impairments to the Evocation of a Concept

We have identified two impairments which preclude the evocation of a concept: *Lack of Connection* and *Misunderstanding*.

**Lack of Connection** occurs when the hearer cannot link the expression used by the speaker with the concept intended by the speaker. We anticipate Lack of Connection when one of the following conditions holds: (1) a null SRE is given by the speaker, (2) the SRE used by the speaker to refer to an intended concept does not exist in the user model; (3) the SRE exists in the user model, but it is not connected to a concept; (4) the SRE is weakly connected to the intended concept (and no other concept); or (5) the SRE is connected to the intended concept (and no other concept), but the SRE is not *accessible* in the user model, i.e., the SRE is not sufficient to activate the corresponding term in the user model[1]. The following dialogue illustrates a situation were the last condition holds:

S1: I saw Mary yesterday.
S2: [*blank stare* ]
S1: You know, the one with red hair and glasses.
S2: Oh! *That* Mary!

**Misunderstanding** occurs when the concept retrieved by the hearer is not the one intended by the speaker. This impairment is predicted if one of the following conditions holds: (1) the SRE mentioned by a speaker is strongly connected (either correctly or incorrectly) to concepts which are different from the intended one, and either (1.1) the intended concept does not exist in the user model, or (1.2) it is not connected to the SRE, or (1.3) it is weakly connected to the SRE; or (2) the SRE is connected to several concepts, including the intended one, and either more than one concept is primed, i.e., the proposed SRE is ambiguous, or the 'wrong' concept is primed. For instance, in the above dialogue, the first condition would hold if the hearer and the speaker did not know the same Mary.

### Invalidating Evocation-related Impairments

The invalidation of an evocation-related impairment requires an Indicative RD which identifies the intended concept uniquely. This RD is composed of a set of concepts which are related to the intended concept. The level of detail of an Indicative RD must be sufficient to satisfy the subgoal *EVOKE* with respect to the concepts mentioned in it, i.e., the fulfillment of the subgoal *KNOW* is not necessary with respect to these concepts.

[1]This condition is stated for the sake of completeness, but has not been implemented.

We distinguish between two types of Indicative RDs: *positive* and *negative complement*. An Indicative RD is positive, if the selected concepts indicate what the intended concept *is*, as opposed to what it *is not*, e.g., "Knowledge Representation, *where we studied Predicate Calculus and Semantic Nets.*" It is negative complement, if it contains the complement of the intended concept, i.e., all the concepts that the intended concept is not, e.g., "Knowledge Representation, *which is not Search or Automated Reasoning.*"

In order to generate a positive Indicative RD for an intended concept, we must present a set of concepts which identify uniquely this concept, i.e., we must find a set of concepts, $S$, such that the intersection of the concepts outside $S$ to which all the concepts in $S$ are connected consists only of the intended concept and no other concept. According to Grice's Maxim of Quantity [Grice 1975], $S$ should be the smallest set which satisfies this constraint. However, this requirement makes our problem NP-complete (provable by reduction to the graph covering problem). In this research, we apply a heuristic similar to the one used in [Appelt 1982, Dale 1990], whereby a uniquely identifying set of concepts is built by iteratively adding concepts to an initial identifying concept, until the intended concept is the only one connected to all the concepts in the resulting set of concepts.

Negative complement RDs may be generated only if the following conditions hold: (1) the hearer is aware of the existence of the intended concept, and (2) the situation at hand is a 'closed world' situation, where the number of candidate concepts to which an SRE could be referring is small. For example, if the only topics covered in an Artificial Intelligence course were Knowledge Representation, Search and Automated Reasoning, a mediocre student may believe that all three topics are possible meanings of the SRE 'Knowledge Representation'. In this case, a lecturer may identify the intended concept by saying "Knowledge Representation, *which is not Search or Automated Reasoning,*" provided the student is unlikely to experience evocation-related impairments with respect to the terms in the Indicative RD.

In Lack of Connection, the hearer has no link between the given SRE and other unintended concepts. Hence, this impairment can be invalidated only by means of a positive Indicative RD. Misunderstanding may be invalidated either by means of a positive Indicative RD or a negative complement Indicative RD. Both types of RDs single out the intended concept. However, negative complement RDs also address directly a hearer's beliefs which are related to the intended concept and the SRE, while positive RDs address these beliefs only indirectly. For instance, in the Artificial Intelligence example, the negation of the erroneous meanings will both single out the intended concept and contradict the erroneous beliefs. On the other hand, a perspective, such as "Knowledge Representation, *where we studied Predicate Calculus and Semantic Nets,*" does not directly contradict the erroneous beliefs. Note that if all the possible meanings of a lexical item are correct, a negative complement RD should be limited to a negation of intent, e.g., "He was holding a crook, *not the criminal kind.*" However, if some of the possible meanings are correct and some are not, the generated text must address these cases explicitly in order to avoid erroneous presuppositions. For example, consider a situation where the hearer correctly believes that 'crook' means both shepherd's staff and criminal, and s/he incorrectly believes that it also means cooking utensil. In this case, if the speaker means 'crook' in the sense of shepherd's staff, s/he should say something like "*I don't mean a criminal, and a crook is not a cooking utensil.*"

At this stage of the generation process, all the Indicative RDs which single out an intended concept are considered possible alternatives.

## Planning Expressions for the Subgoal KNOW

The expertise required from the hearer for the goal *KNOW(concept)* to be satisfied is determined by the speaker's communicative intent, which in turn is influenced by the following factors: (1) the relationship between the concept and the message; (2) the relevance of the message to a main idea the speaker wishes to convey; (3) the social setting, e.g., classroom or casual conversation; and (4) the speaker's perception of the hearer's conversational goals. The speaker's communicative goal affects the expertise requirements along two dimensions: *Content* and *Strength*.

**Content** determines the aspects of a concept which must be known by the hearer to satisfy the goal *KNOW*. For instance, the aspects of a procedure are: sequence of steps, goal, type, objects to which it applies and conditions for its application. The aspects of an object are: components, attributes, function, internal process, type and subclasses. However, depending on the speaker's communicative goal, only some of these aspects may be required to satisfy the subgoal *KNOW*. Consider, for example, the following dialogue:

S1: I cleaned up in 500.
S2: What is 500?
S1: It is a card game.

In this dialogue, when S1 is questioned by S2 about the concept 500, he does not provide an explanation which conveys a full understanding of this concept. Rather, S1 presents only one aspect of the concept, namely its type, since he believes that this information is sufficient to enable S2 to understand the message.

**Strength** determines the degree of expertise the user should have with respect to the aspects selected by Content. For example, a low or medium degree of expertise would usually be required for a concept which is mentioned in passing.

Like for the subgoal *EVOKE*, we consult a model of the hearer's beliefs in order to ascertain whether an expertise-related impairment is anticipated in relation to the concept in question. If so, a *Creative RD* is proposed to prevent this situation. Each Creative RD which can perform this task is maintained as a candidate until the final selection process. For instance, an Analogy, a Description and an Instantiation are all possible candidates to explain an unknown concept. Note that the content of Creative and Indicative RDs may overlap, since a Creative RD which describes a concept is likely to identify it uniquely as well.

## Expertise-related Impairments

We have identified two impairments which preclude the goal of knowing a concept: *Lack of Understanding* and *Insufficient Understanding*.

**Lack of Understanding** is anticipated when there does not exist in the user model a concept which corresponds to an intended concept. It entails an evocation-related impairment, since an SRE cannot point to a concept which is unknown by the hearer.

**Insufficient Understanding** is predicted when there exists a concept in a hearer's memory which corresponds to an intended concept, but the hearer has a low level of expertise with respect to some aspects which are required by the speaker for the comprehension of this concept. For example, the hearer is unaware of a step of a procedure or a part of an object. This impairment may occur in conjunction with a evocation-related impairment or by itself.

## Invalidating Expertise-related Impairments

The invalidation of Lack of Understanding requires the generation of a Creative RD which satisfies the subgoal *KNOW(concept)*. The information included in this RD depends on the type of the concept, e.g., procedure or object, the aspects of the concept considered relevant by the speaker, the user's ability and his/her expertise. For example, if the concept in question is a procedure, such as Distributive Law, the speaker may be interested only in the sequence of its steps and in the objects to which it applies, requiring a Creative RD which conveys these aspects, such as a Description or an Instantiation. However, different users are likely to draw different inferences from the presented information, depending on their ability and expertise. These inferences must be taken into consideration when selecting an RD. For example, if the selected Creative RD is an Instantiation which shows how Distributive Law applies to an Algebraic Term, such as $2(x+y)$, a talented student will probably make all the correct inferences regarding the other objects to which Distributive Law applies, and will also generalize the procedure from the Instantiation. On the other hand, a mediocre student cannot be relied upon to make these inferences. Furthermore, s/he may draw some erroneous inferences which have to be addressed. Thus, for a mediocre student, all the relevant information must be spelled out. Finally, the subgoals *KNOW* and *EVOKE* have to be satisfied with respect to the propositions and concepts presented in a selected RD. This may require additional Creative and/or Indicative RDs. For instance, if the hearer is not proficient with respect to the multiplication of Algebraic Terms, a Creative RD must be generated to satisfy the subgoal *KNOW(*multiplication apply-to Algebraic-Terms*)* at a level of expertise considered adequate to understand the application of Distributive Law to Algebraic Terms.

Like for Lack of Understanding, a Creative RD is required to invalidate Insufficient Understanding. However, in this case, the lack of expertise may be localized to some weak, missing or erroneous constituents. Thus, the RD may be focused on these constituents. However, in order to prevent wrong implicatures, structural constraints must be applied. These constraints stipulate that if the hearer is not proficient with respect to some information items in a particular aspect, and the system decides to mention these items, then all the information items which pertain to this aspect must be mentioned. For instance, if the speaker is interested in the objects to which Distributive Law applies, and the hearer knows only that Distributive Law applies to Algebraic Terms, simply saying "Distributive Law applies to Numbers" may carry the wrong implicature that Numbers is *all* that Distributive Law applies to. This implicature is prevented by repeating what the hearer already knows, e.g., "*In addition to Algebraic Terms,* Distributive Law applies to Numbers."

The process of generating Creative RDs is implemented by presenting the information corresponding to each aspect in a list of aspects which must be known by the hearer in order to satisfy the subgoal *KNOW*. At present, this list of aspects is hand coded for each concept in a message, but at a later stage, it may be inferred from the predicates in the message. For instance, if the intended concept is a chair, and the aspects type, function and components are specified, the following text may be generated: "A chair *is a piece of furniture* (type) *which is used for sitting* (function), *and has a seat, a back and usually four legs* (components)." The main distinction between this list of aspects and schemas [McKeown 1985] or process traces [Paris 1988] is that schemas and process traces list the types of information to be presented in order to convey a concept, whereas our list of aspects constitutes desiderata which the hearer should know in order to understand a concept. These desiderata may be fulfilled in a straightforward manner by means of a Description, as in the above example, or by indirect means, such as Analogies or Instantiations. Partial descriptions are generated by omitting from a proposed Creative RD relations which, according to our model, are known by a hearer, subject to the above structural constraints. In addition, if based on a hearer's profile, a concept is considered too abstract or difficult for the hearer, an Instantiation may be proposed.

At present, the only Creative RDs which have been implemented are Descriptions, where the constituents of a concept are also described if this is required by the subgoal *KNOW*. In order to avoid indefinite recursion, this process is repeated only once, for the constituents of the main concept. The implementation of indirect Creative RDs, such as Analogies, Similes and stand-alone Instantiations, requires rules of inference which stipulate the conditions for inferring the desiderata from these RDs.

## The Main Algorithm

The main algorithm combines the procedures described in the previous sections to generate an expression which accomplishes the goals *KNOW* and *EVOKE* with respect to the concepts in a message. The algorithm receives as input a concept $C$ and performs the following steps.

1. Create a set called *CRDs* which includes Creative RDs that achieve the goal $KNOW(C)$. If no expertise-related impairments are anticipated with respect to $C$, this set is empty.
2. Create a set of pairs called *SRE-IRDs*, where each pair contains an SRE and an Indicative RD that complements the SRE to achieve the goal $EVOKE(C)$. If no evocation-related impairments are anticipated between a particular SRE and $C$, this SRE will appear in *SRE-IRDs* with a nil RD.
3. Create a new set of RDs called *JOINRDs*, in which each element in *CRDs* is combined with each element in *SRE-IRDs*.
4. For each element in *JOINRDs*, check whether other beliefs maintained by the hearer are adversely affected by this element. If so, add to this element an RD which prevents this effect [Zukerman 1990a].
5. Select the most succinct element in *JOINRDs*.

In this procedure, expertise-related impairments are checked first, followed by evocation-related impairments. If Lack of Understanding is anticipated in the first step and Lack of Connection is anticipated in the second step, no Indicative RD is necessary, since the Creative RD proposed to invalidate the Lack of Understanding is sufficient to identify uniquely the intended concept. However, other combinations of evocation- and expertise-related impairments may require both Indicative and Creative RDs.

Let us illustrate the workings of our procedure by means of the following example. Given the message [Distributive-Law has-goal Bracket-Elimination], our procedure is activated with respect to the concepts Distributive-Law and Bracket-Elimination. Now, let us focus on Distributive Law. We assume that the only aspect of Distributive Law which is necessary to understand the message is the sequence of its operators. Hence, if this sequence is not well known by the hearer, Insufficient Understanding is anticipated, and a Creative RD containing this sequence must be presented. This may be an Instantiation, a Description, or a combination of both. Further, if an individual step is not well known, then a Creative RD must be generated for it as well. Next, candidate SREs are proposed, and evocation-related impairments are considered. In this case, the SREs are *nil* and 'Distributive Law'. The null SRE triggers Lack of Connection, calling for a positive Indicative RD. However, since the sequence of steps of Distributive Law is sufficient to identify it, no additional Indicative RD is necessary. A text such as the following may be generated if a null SRE is used: "We apply *the following procedure* to eliminate brackets." The SRE 'Distributive Law' results in the anticipation of Misunderstanding, if the hearer is conjectured to believe that the name 'Distributive Law' points to another concept, say Bracket Simplification. In this case, like for Lack of Connection, no positive Indicative RDs need to be proposed, but a negative complement Indicative RD is still an option, e.g., "*which is not Bracket Simplification.*" After the third step in our algorithm, the RDs in *JOINRD* include: (1) all the proposed Creative RDs, where each RD is accompanied by a null SRE, (2) all the proposed Creative RDs, where each RD is accompanied by the SRE 'Distributive Law', and (3) all the proposed Creative RDs, where each RD is accompanied by the SRE 'Distributive Law' and by a negative complement Indicative RD.

If a candidate SRE is a lexical item and an RD was proposed for it, the inferences from the presentation of this information may cause adverse effects with respect to a hearer's beliefs which are related to the intended concept and the SRE. These effects take place if one of the following conditions holds *and the inferences conflict with this condition*: (1) the intended concept is connected (either correctly or incorrectly) to another lexical item, i.e., it is identified with another name; or (2) the lexical item is connected (either correctly or incorrectly) to another concept, i.e., there is more than one concept associated with the lexical item. The anticipation of such comprehension impairments calls for the generation of a Revision of the information in question if it is correct, and a Contradiction otherwise [Zukerman 1990a]. Note that these conditions do not conflict with the understanding and retrieval of the intended concept, rather, they affect related beliefs. For example, if the speaker says "She was holding a crook, *which is a shepherd's staff*," a hearer who is not proficient in English may experience confusion with respect to his/her previous belief that a crook is a criminal, since a possible implicature from this statement is that a crook is *only* a shepherd's staff. Indeed, when this example was given to a non-native English speaker, he voiced this concern. The generation of a negative complement Indicative RD to invalidate Misunderstanding prevents the second of the above conditions. However, as stated above, these RDs may be generated only if the hearer is aware of the intended meaning of the SRE.

Our algorithm terminates by selecting the most succinct of the proposed expressions. At present, we are considering a measure of succinctness based on the probability of the components of a message.

## Planning versus Failure Prevention

The plan-based approach consists of applying goal-based reasoning to generate utterances that satisfy increasingly refined communicative goals [e.g., Appelt 1982, Hovy 1988, Moore & Swartout 1989, Cawsey 1990]. The failure-prevention approach, on the other hand, proposes remedial actions to prevent or overcome anticipated failures in the communication process. Its explicit representation of communication failures supports the use of a uniform mechanism to perform tasks that require special purpose mechanisms in the plan-based approach. For example, the problem of satisfying the subgoal *KNOW* may be cast both as a planning problem and as a failure-prevention problem. For this subgoal, both approaches are equivalent in the sense that they address the problem of planning discourse by trying to satisfy the preconditions for the attainment of a communicative goal. However, while in the plan-based approach the presentation of background information is addressed by adding domain-dependent preconditions to the plan operators [Cawsey 1990], in the failure-prevention approach, this task is handled through the recognition and subsequent invalidation of expertise-related impairments. Further, the generation of discourse to satisfy the subgoal *EVOKE* is an associative task which is not amenable to handling by planning. Thus, special purpose mechanisms are invoked by a planner to perform this task [Dale 1990]. In the failure-prevention approach, this task is performed by means of the recognition and subsequent invalidation of evocation-related impairments. Finally, the impairment invalidation paradigm also accounts for the generation of additional RDs that address impairments to the comprehension process due to inferences which affect other beliefs maintained by the hearer.

## Conclusion

The discourse planning procedure offered in this paper generates expressions which include rhetorical devices that address comprehension problems with respect to concepts. To this effect, our mechanism relies on a characterization of preconditions to the comprehension of a concept in terms of possible failures in communication. These failures, in turn, are expressed in terms of conditions which pertain to a hearer's beliefs.

The proposed mechanism requires an accurate model of certain aspects of a user's beliefs. Thus, it is particularly suitable for highly interactive situations, and for situations where a large amount of information is available about common user beliefs and misconceptions, such as educational settings.

Our mechanism was used successfully to analyze texts in a variety of domains ranging from educational textbooks to everyday dialogues. At present, steps 1 and 4 of the main procedure have been implemented with respect to a small subset of high-school algebra, and step 2 is currently being implemented. The output of our mechanism is in the form of predicates.

## References


Appelt, D.E. (1982). Planning Natural Language Utterances to Satisfy Multiple Goals, Technical Note 259, SRI International, March 1982.

Appelt, D.E. (1986). Towards a Plan-based Theory of Referring Actions. Kempen, G. (Ed.), *Natural Language Generation: New Results in Artificial Intelligence, Psychology and Linguistics*, Kluwer Academic Publishers, 63-70.

Cawsey, A. (1990). Generating Explanatory Discourse. Dale, R.; Mellish, C.; and Zock, M. (Eds.), *Current Research in Natural Language Generation*, Academic Press, 75-102.

Clark, H.H. and Schaefer, E.F. (1989). Contributing to Discourse. *Cognitive Science* 13(2):259-294.

Dale, R. (1990). Generating Recipes: An Overview of Epicure. Dale, R.; Mellish, C.; and Zock, M. (Eds.), *Current Research in Natural Language Generation*, Academic Press, 229-255.

Grice, H.P. (1975). Logic and Conversation. Cole, P.J. and Morgan, J.L. (Eds.), *Syntax and Semantics, Volume 3: Speech Acts*, Academic Press.

Hovy, E.H. (1988). Planning Coherent Multisentential Text. In Proceedings of the Twenty-Sixth Annual Meeting of the Association for Computational Linguistics, 163-169.

McKeown, K.R. (1985). Discourse Strategies for Generating Natural Language Text. *Artificial Intelligence* 27:1-41.

Moore, J.D. and Swartout, W.R. (1989). A Reactive Approach to Explanation. In Proceedings of the Eleventh International Joint Conference on Artificial Intelligence, 1504-1510.

Paris, C.L. (1988). Tailoring Object Descriptions to a User's Level of Expertise. *Computational Linguistics* 14(3):64-78.

Webber, B.L. (1983). So What Can We Talk About Now? Brady, M. and Berwick, R.C. (Eds.), *Computational Models of Discourse*, MIT Press.

Zukerman, I. (1990a). A Predictive Approach for the Generation of Rhetorical Devices. *Computational Intelligence*, 6(1):25-40.

Zukerman, I. (1990b). Towards a Failure Driven Mechanism for Discourse Planning: a Characterization of Learning Impairments. In Proceedings of the Twelfth Conference of the Cognitive Science Society, 495-502.

# A Connectionist Model of Intermediate Representations for Musical Structure

**Edward W. Large**
Computer and Information Sciences Department
2036 Neil Avenue
large@cis.ohio-state.edu

**Caroline Palmer**
Psychology Department
1885 Neil Avenue
cpalmer@magnus.acs.ohio-state.edu

**Jordan Pollack**
Computer and Information Sciences Department
2036 Neil Avenue
pollack@cis.ohio-state.edu

The Ohio State University
Columbus, OH, 43210

**Abstract**

The communication of musical thoughts and emotions requires that some musical knowledge is shared by composers, performers, and listeners. Computational models of musical knowledge attempt to specify the intermediate representations required to generate adequate predictions of musical behavior. We describe a connectionist model that encodes the rhythmic organization and pitch contents of simple melodies. As the network learns to encode melodies, structurally more important events tend to dominate less important events, as described by reductionist theories of music (Lerdahl & Jackendoff, 1983; Schenker, 1979). We describe an empirical study in which improvisations on a tune by a skilled music performer are compared with the encodings produced by the network. The two are examined in terms of the relative importance of the musical structure they posit at intermediate levels of representation.

## Introduction

A primary goal of music cognition is to specify mental representations for musical knowledge. Computational models of music composition, performance, and perception often posit multiple levels of structural description in mental representations. We refer to these levels of structural description as *intermediate representations*, because they mediate between the perception of a musical surface (the score or acoustic signal) and the resulting musical behavior (the performance or memory of a piece). However, the proposed structural descriptions often fall short of adequately specifying the relative importance of the musical events. We propose a connectionist model of mental representations for music that emphasizes the hierarchical nature of musical structure, and we compare its predictions of relative importance with evidence from skilled music performance.

Theoretical accounts of mental representations for musical structure often emphasize the importance of hierarchical organization. Hierarchical models of rhythmic organization, for example, describe the way in which musical events are combined to form larger structural units in a nested fashion (Cooper & Meyer, 1960). In a particular musical context, certain pitch events are heard as being dominant in the hierarchy and others are heard as elaborations of the dominant events (Lerdahl & Jackendoff, 1983; Schenker, 1979). Some perceptual cues to hierarchical organization are present in performed music, such as expressive variations in timing and dynamics. However, the models described above posit intermediate levels of mental representation that are based on information not necessarily present in the musical input. These intermediate levels of description are thought to reflect statistical regularities, derived from the input with the aid of general knowledge about the roles that events play in a particular musical idiom (Palmer & Krumhansl, 1990; Knopoff & Hutchinson, 1978).

Computational models of music cognition attempt to specify the intermediate representations required to generate adequate predictions of human behavior. Several researchers have adopted connectionist models which provide general-purpose learning algorithms capable of responding to the statistical regularities of the learning environment. However, connectionist models have been notoriously weak at representing hierarchical relationships, such as those found in music or language (Fodor & Pylyshyn 1988). Recursive Auto-Associative Memory (RAAM) is a connectionist architecture which develops distributed representations of hierarchical structures, directly attacking this problem of representational adequacy (Pollack, 1988; Pollack, 1990).

In this paper, we describe a RAAM model that encodes the rhythmic organization and pitch contents of simple melodies. As the network learns to encode melodies, structurally more important events tend to dominate less important events, as described by reductionist theories of music (Lerdahl & Jackendoff, 1983; Schenker, 1979). We then describe an empirical study in which improvisations on a tune by a skilled music performer are compared with the encodings produced by the RAAM network. The two are examined in terms of the relative importance of the musical structure they posit at intermediate levels of representation.

In the next section we propose a connectionist model for encoding representations of hierarchically nested time-spans. By examining the representations, we can predict the relative importance of musical events within each time-span. We then describe evidence from a skilled music performance that tests the predictions of relative importance made by the connectionist model. A pianist's improvisations on a theme are contrasted with the model's predictions of relative importance of different musical events.

## Time-Span Reductions

One theory emphasizing intermediate levels of representation attempts to model an experienced listener's intuitions of Western tonal music (Lerdahl & Jackendoff, 1983). The theory describes many types of hierarchical representations, including metrical structure, grouping structure, and time-span reduction. Metrical structure describes the way in which a series of pulses are mentally combined to create nested hierarchical levels of alternating strong and weak pulses. Grouping structure describes nested groups of events forming motives, phrases, and larger sections of music. The outputs of metrical and grouping structures combine to segment a piece into hierarchically nested rhythmic units called time-spans. At the lowest levels time-spans are determined primarily by metric structure, and at the highest levels by grouping structure.

A *time-span reduction* organizes all musical events in a piece into a single coherent structure that reflects a strict hierarchy of relative importance. Within each time-span a single most important event is identified and all other events are heard as subordinate to it. In Figure 1, we show a time-span reduction for the melody "Hush Little Baby". The top staff shows the melody, and the brackets show how the piece is segmented into time-spans. The staves below show the intermediate levels of the reduction. At each level, less important events are eliminated, leaving a "skeleton" of the melody.

## Recursive Auto-Associative Memory

RAAM is a connectionist architecture that develops distributed representations of variable sized, compositional data structures. It has been used to model the encoding of hierarchical structures such as those found in linguistic syntax and logical expressions (Pollack, 1990). Conceptually, a RAAM consists of two machines, a compressor and a reconstructor. The compressor is trained to recursively encode sets of fixed-width patterns into single patterns of the same size. The reconstructor is trained to recursively decode the patterns produced by the compressor into facsimiles of the original sets of patterns. These mechanisms are co-evolved by linking their training sets together using an auto-associative form of back-propagation. Our current work is based on an implementation of a RAAM as a 3-layer feed forward neural network where the input-to-hidden layer transformation is the compressor and hidden-to-output layer transformation is the reconstructor.

In order to find the intermediate distributed representations for a set of melodies, we segment the melodies into time-spans as shown in Figure 1, and use these hierarchies as the training set to a RAAM. The primitive events in each melody are represented as binary feature vectors. We have chosen the encoding strategy shown in Figure 2. One set of units encodes pitch class, and a second set encodes local implied harmony

Figure 1: A Partial Time-Span Reduction for "Hush Little Baby."
A) Original melody; B) Tactus level reduction; C) Measure level reduction.

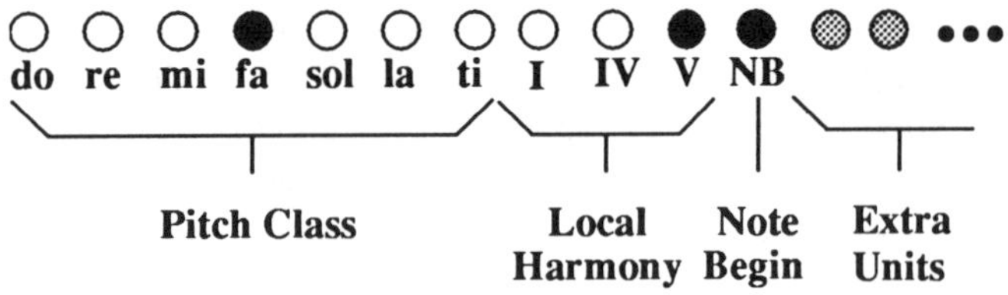


Figure 2: The representation of pitch events.

similar to Lerdahl and Jackendoff's use of local harmony in creating time-span reductions. An additional unit designates the beginning of an event. When this unit is turned on, it indicates that the event has its attack at that particular point. When turned off, it indicates that event is a continuation of a previous event.

After training, the compressor and reconstructor are treated as separate networks for the processes of encoding and decoding intermediate representations. Figure 3 shows a a short melodic excerpt segmented into time-spans, and depicts the processes of encoding and decoding it. First, the compressor encodes a set of the lowest level time-spans, and these encodings are recursively encoded to produce a representation for the entire structure. Next, the reconstructor decodes the compressed representation to retrieve a facsimile of the original structure. In order to capture the distinctions between binary and ternary groups found in music, we use a quaternary RAAM, that is, one with four fields of input units (11 input units per field) activated as shown in Figure 4.

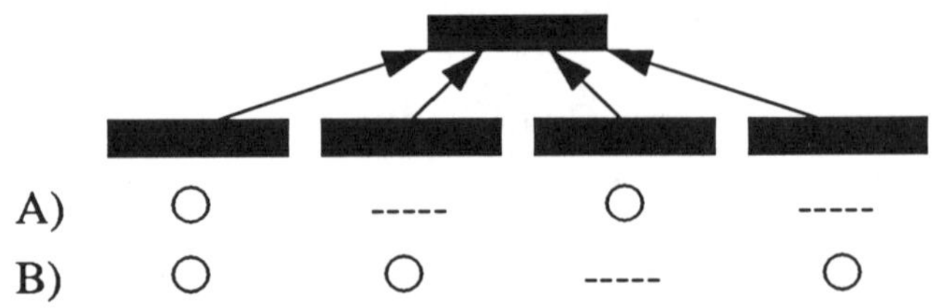


Figure 4: Encoding both binary and ternary groups in a quaternary RAAM

When a binary group is encoded (A), the first and third fields are activated. The second and fourth are set to zero, simulating a binary RAAM. When a ternary group is encoded (B), the first, second and fourth fields are activated. The third is set to zero, simulating a ternary RAAM.

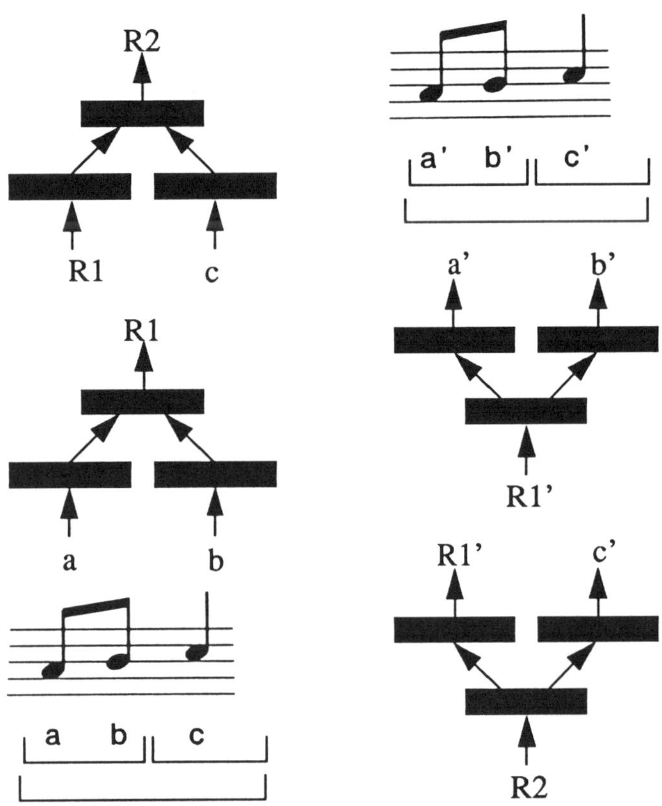


Figure 3: Network encoding and decoding of a time-span

The time-span [[a b] c] is encoded by compressing the primitive event representations of a and b, producing representation R1. R1 is then compressed with the next event, c, producing R2, which is a representation of the higher level structure. Next, R2 is decoded by reconstructing facsimiles R1' and c'. R1' is then reconstructed to produce facsimiles a' and b'. Thus a facsimile of the original structure, [[a' b'] c'], is produced.

The encodings developed by the network reflect the temporal embedding of the time-span hierarchy. The duration of each encoding produced by the network is considered to be the sum of the durations of its component events. This allows us to represent the rhythmic structure of the melody without pre-specifying a smallest possible time-slice (see Todd, 1989). Instead, we follow Lerdahl and Jackendoff's description of the tactus as the most salient metrical level (i.e. the level of foot-tapping tempo). We require that the tactus be continually represented throughout the piece, but time-spans derived from smaller metrical levels are represented only when actually present in the melody.

The compressor and reconstructor networks, taken together, comprise a well-formedness test for novel structures. Given a novel structure, the compressor network is used to create a representation. The reconstructor network is then applied to the representation to retrieve its constituents. If the reconstructed structure matches the input structure, either exactly or within some tolerance, this novel structure can be considered to be well-formed. We will use the difference between the constituents of the input structure and those of its reconstruction to determine the relative weighting of each musical event within the representations developed by the network.

## Measures of Relative Importance

### RAAM Network

Sixteen nursery tunes (such as "Mary Had a Little Lamb") were chosen as a training set because they provide a simple, natural musical case for study. Each tune was a simple melody between 4 and 12 measures in length, with a meter of 2/4, 4/4, 6/8, or 12/8. The tunes comprised ten unique melodies;

four of these ten melodies had variations (tunes with similar pitch or rhythmic properties). Although the event representations required only 11 bits, we used 23 units, allowing 12 extra "degrees of freedom" for the system to use in arranging its intermediate representations. These extra dimensions of representation were set to 0.5 on input, and trained as don't-care's (Jordan 1986) on output. Thus, for the quaternary RAAM, our network had 92 input and output units, and 23 hidden units. The network was trained on the time-spans for the melodies with length less than or equal to the measure level. The network was not trained until it memorized the tunes, but instead for 1500 cycles with a learning rate of 1.0 and momentum of 0.5. The network was therefore not able to reconstruct every melody in the training set note-for-note. The network's output representations are interpreted from the output values of the pitch-class units produced by the decoder for each event in the sequence. If all outputs are less than some threshold value at any given point, the event is interpreted as a rest (a null event). Otherwise the pitch class unit with the greatest activation at any given point is interpreted as the reconstructed event. With this interpretation the reconstructor was able to reproduce fairly accurate facsimiles of input melodies, although some events were "forgotten" and others were "remembered" incorrectly (in the reconstruction but not from the original melody).

To test the network's ability to encode novel tunes, we exposed it after training to a melody dissimilar to the original set of 14 tunes. Figure 5A shows the melody "Hush Little Baby" after its reconstruction by the network, as described above. Because this tune was dissimilar to those in the training set, we used the sensitive threshold activation value of 0.1. The network's actual activation values for each event in the sequence are shown below the reconstruction.

## Improvised Performances

To compare the network's predictions of relative importance with those of skilled musicians, we recorded improvisations on a tune by a skilled pianist. Improvisation in Western tonal music commonly requires a performer to identify a framework of important melodic and harmonic events, and apply procedures to create elaborations and variants on them (see Pressing, 1988 for a review of improvisational models). Thus, improvisation on a musical tune allows the pianist freedom to determine which musical events should be retained (those of primary importance), and which should not (those of less importance).

A professional pianist from the Columbus, Ohio area was asked to perform three nursery tunes on a computer-monitored acoustic upright piano. The pianist was experienced in improvising in a contemporary/popular musical style. The pianist performed three melodies: one included in the network's training set, one not in the training set but similar to it in pitch and timing (a variation on one of the melodies), and a third melody unrelated to any in the training set. The pianist first performed each melody as it was notated, to become familiar with it. He then improvised five simple melodic (single-line) variations on each melody. All pitch, intensity, and timing information in the performances was recorded on computer, and compared with the original melody. Only performances of the third melody (the most stringent test of the network's ability to predict relative importance) will be described here.

According to our application of the time-span reduction hypothesis to improvisation, structurally important events should be less likely than unimportant events to change in variations of a melody. To test this hypothesis, the number of musical events identical to the original melody in pitch were summed across variations. The sums ranged from 5 (the same

Figure 5: "Hush Little Baby."
A) Network reconstruction; B) 0.66 threshold criterion reduction;
C) Lerdahl & Jackendoff style tactus level reduction; D) Lerdahl & Jackendoff style measure level reduction.

Figure 6: "Hush Little Baby."
A) Original melody with performance ratings; B) Rating criterion reduction.

pitch in each of the five performances) to 0 (the same in no performance) for each event location. Figure 6 shows the original melody, with the ratings for each location.

### Comparison of RAAM Network and Improvised Performances

First, we compare the original tune (Figure 6A) and the network's reconstruction (Figure 5A). The reconstructed tune is a reasonable facsimile, considering that it was not a variation of any tune from the training set, but a distinctly novel melody. This indicates the network's ability to generalize beyond its input. We then compare the network's predictions of relative importance with the pianist's improvisations by developing reductions for both the reconstructed melody and the improvisations. An intermediate representation was developed for the improvisations by including only events from the original melody that were retained often across performances (those scoring 4 or higher). Using this method, we included only the highest third of the range of values, and we applied the same criterion to the network output by raising the activation threshold to 0.66. The reductions obtained from this criterion are shown for the network in Figure 5B and for the improvisation data in Figure 6B. At this level the reductions show significant agreement. In general, the network tended to retain more events than the improvisations.

Finally, we can compare more abstract levels of representation to those predicted by Lerdahl and Jackendoff's theory. It is difficult to produce further reductions for the improvisations because of the resolution of our measurements. However, we can produce further reductions of the network's encodings. Instead of reapplying the threshold criterion, we chose the event in each time-span with the highest activation, similar to Lerdahl and Jackendoff's (1983) method of computing reductions. By comparing Figures 5D and C to Figures 1B and C, we see strong agreement at intermediate levels of reduction. The network's ability to generalize well enough to develop a representation for this melody may be related to its weighting of important events at each representational level.

## Conclusions

The similarities seen here between improvisational music performance and a connectionist model of simple melodies may result from similar computational constraints. The skill of improvising on a theme has been described as a largely unconscious process of identifying important structural elements and applying creative procedures to elaborate on those elements (Johnson-Laird, in press; Steedman, 1982). The resulting variations are related to each other by transformational rules that generate the possible improvisations on that theme. One important consequence of this approach is that it reduces the memory demands that can accompany the use of multiple intermediate representations. Instead of retaining each element at each representational level (thereby increasing the necessary storage capacity), only a reduced set of elements is stored, from which other representational levels are generated. The RAAM network produces a compressed representation, in which the structurally more important events dominate, such that they are more likely to be reliably reconstructed than structurally less important events. These similar constraints on processing demands may account for the similarities seen here in the improvised variations in music performance and in the network reconstructions.

Reductionist theories of music cognition have inspired other computational models of intermediate representations. Scarborough et. al. (1989) describe a parallel constraint satisfaction approach for the perception of metric structure, modelled as the response of independent metronome-like agents to individual musical events. Based on inter-agent constraints that enforce Lerdahl & Jackendoff's rules for metric structure, a hierarchical representation of the metric structure of a piece emerges. Rosenthal (1989) has described the perception of grouping structure as the process of constructing "recognizer agents". The processes which construct recognizers operate in accordance with Lerdahl and Jackendoff's rules for producing grouping structure analyses. Once constructed, agents recognize the repetition of rhythmic structures, thus implementing a restricted form of parallel structure recognition. The program's output is a mental representation of the piece stored as symbolic data structures. However, additional mechanisms must be posited to determine the similarity of two elements in the model (Rosenthal, 1989). One of the advantages of using RAAM is that the intermediate representations admit simple similarity measures, such as euclidean distance, capturing the statistics of the input environment.

Other researchers in music cognition have focused on the importance of expectation. Meyer (1956) argues that "...the inhibition of the tendency to respond or, on the conscious level, the frustration of expectation (is) the basis of the affective and intellectual response to music." Bharucha and Todd (1989) have described a computational model of musical expectation using Jordan nets (Jordan, 1986) and Todd (1989) has described how similar networks may be used for music composition. Todd notes, however, that these networks produce melodies "...high in local structure, but lacking in overall global organization". These computational approaches to expectation may fail to capture the intermediate representations required for hierarchically structured events such as music.

Another observation we would like to make regards the adequacy of the RAAM architecture for developing representations of the hierarchical structure of melodies. Although we have only reported the model's results for the encoding of one tune, this result represents the most stringent test of the network's ability: to encode novel sequences in a manner similar to that of skilled musicians. We did not expect that this tune would be correctly encoded by the network because it was not closely related to the melodies in the training set. The fact that the network was able to reconstruct a reasonable facsimile of this melody shows that not only is the network capable of encoding melodic structure, but it is capable of generalizing in a robust manner. The final observation regards the nature of the representations that the network develops. The fact that the representations weighted events in a way similar to both the musician's choices of events to retain in improvisations and predictions of relative importance based on Lerdahl and Jackendoff's theory (1983) indicates that the computational model may capture the relevant hierarchical properties of humans' mental representations for musical melodies.

## Acknowledgments

The research was partially supported by NIMH grant 1R29-45764 to the second author and ONR Grant N00014-89-J1200 to the third author. Thanks to Carolyn Drake and Mari Jones for their comments on an earlier draft of this paper, and to Peter Angeline, Kory Klein, and Carla van de Sande, for their assistance.

## References

Bharucha, J. J. & Todd, P. M. (1989). Modeling the perception of tonal structure with neural nets. *Computer Music Journal*, 13 (3), 44-53.

Cooper, G. & Meyer, L.B. (1960). *The rhythmic structure of music*. Chicago: University of Chicago Press.

Fodor, J. & Pylyshyn, Z. (1988). Connectionism and cognitive architecture: A critical analysis. *Cognition*, 28, 3-71.

Johnson-Laird, P.N. (in press). Jazz improvisation: a theory at the computational level. In P. Howell, R. West, and I. Cross (Eds.), *Representing musical structure*. NY: Academic Press.

Jordan, M. I. (1986). Attractor dynamics and parallelism in a connectionist sequential machine. Proceedings of the Eight Annual Conference of the Cognitive Science Society. Hillsdale, N.J.: Erlbaum Press.

Knopoff, L. & Hutchinson, W. (1978). An index of melodic activity. *Interface*, 7, 205-229.

Lerdahl, E. & Jackendoff, R. (1983). *A generative theory of tonal music*, Cambridge: MIT Press.

Meyer, L.B. (1956). *Emotion and meaning in music*, Chicago: University of Chicago Press.

Palmer, C. & Krumhansl, C.L. (1990) Mental representations of musical meter. *Journal of Experimental Psychology: Human Perception & Performance*, 16, 728-741.

Pollack, J. B. (1988). Recursive auto-associative memory: Devising compositional distributed representations. Proceedings of the Tenth Annual Conference of the Cognitive Science Society, Montreal, 33-39.

Pollack, J. B.(1990). Recursive distributed representations. *Artificial Intelligence*, 46, 1, 77-105.

Pressing, J. (1988). Improvisation: methods and models. In J. Sloboda (Ed.),*Generative processes in music: the psychology of performance, improvisation, and composition*. NY: Oxford University Press.

Rosenthal, D. (1989). A model of the process of listening to simple rhythms. *Music Perception*, 6 (3), 315-328.

Scarborough, D. L., Miller, B. O., & Jones, J. A. (1989). PDP models for meter perception. Proceedings of the Eleventh Annual Conference of the Cognitive Science Society. Hillsdale, N.J.: Erlbaum Press.

Schenker, H. (1979). *Free composition* (E. Oster, Trans.). NY: Longman

Steedman, M. (1982). A generative grammar for jazz chord sequences. *Music Perception*, 2, 52-77.

Todd, P. M. (1989). A connectionist approach to algorithmic Composition. *Computer Music Journal*, 13 (3), 29-43.

# Combining a Connectionist Type Hierarchy with a Connectionist Rule-Based Reasoner*

**D. R. Mani** and **Lokendra Shastri**
Department of Computer and Information Science
University of Pennsylvania
Philadelphia, PA 19104, USA
shastri@cis.upenn.edu

## Abstract

This paper describes an efficient connectionist knowledge representation and reasoning system that combines rule-based reasoning with reasoning about inheritance and classification within an *IS-A* hierarchy. In addition to a type hierarchy, the proposed system can encode generic facts such as 'Cats prey on birds' and rules such as 'if $x$ preys on $y$ then $y$ is scared of $x$' and use them to infer that Tweety (who is a Canary) is scared of Sylvester (who is a cat). The system can also encode qualified rules such as 'if an *animate* agent walks into a *solid object* then the agent gets hurt'. The proposed system can answer queries in time that is only *proportional* to the *length* of the shortest derivation of the query and is *independent* of the *size* of the knowledge base. The system maintains and propagates variable bindings using temporally synchronous – i.e., in-phase – firing of appropriate nodes.

## Introduction

In [Shastri & Ajjanagadde 1990a, Shastri & Ajjanagadde 1990b, Ajjanagadde & Shastri 1991], Ajjanagadde and Shastri have described a solution to the variable binding problem [Lange & Dyer 1989, Smolensky 1987] and shown that the solution leads to the design of a connectionist reasoning system that can represent systematic knowledge involving $n$-ary predicates and *variables*, and perform a broad class of reasoning with extreme efficiency. The time taken by the reasoning system to draw an inference is only proportional to the *length* of the chain of inference and is independent of the number of rules and facts encoded by the system. The reasoning system maintains and propagates variable bindings using temporally synchronous – i.e., in-phase – firing of appropriate nodes. The solution to the variable binding problem allows the system to maintain and propagate a large number of bindings *simultaneously* as long as the number of *distinct* entities participating in the bindings during any given episode of reasoning, remains bounded. Reasoning in the proposed system is the transient but systematic flow of *rhythmic* patterns of activation, where each *phase* in the rhythmic pattern corresponds to a distinct *constant* involved in the reasoning process and where variable bindings are represented as the synchronous firing of appropriate argument and constant nodes. A fact behaves as a temporal pattern matcher that becomes 'active' when it detects that the bindings corresponding to it are present in the system's pattern of activity. Finally, rules are interconnection patterns that propagate and transform rhythmic patterns of activity.[1]

In this paper we describe how the above reasoning system may be combined with an *IS-A* hierarchy. Such an integration allows the occurrence of types (categories) as well as instances in rules, facts, and queries. This has the following interesting consequences. First, the reasoning system can combine rule-based reasoning with inheritance and classification. For example, such a system can infer that 'Tweety is scared of Sylvester', based on the generic fact 'Cats prey on birds', the rule 'If $x$ preys on $y$ then $y$ is scared of $x$ and the *IS-A* relations 'Sylvester is a Cat' and 'Tweety is a Bird'.[2] Second, the integrated system can use category information to *qualify* rules by specifying restrictions on the type of argument fillers. An example of such a rule is: $\forall x{:}animate,\ y{:}solid\text{-}obj\ [\ walk\text{-}into(x,y) \Rightarrow hurt(x)\ ]$, which specifies that the rule is applicable only if the two arguments of 'walk-into' are of the type 'animate' and 'solid-object', respectively.

An overview of the rule-based reasoning system is followed by a description of the *IS-A* hierarchy realization and its interface with the reasoning system. A detailed discussion of the reasoning system and the type hierarchy interface may be found in [Shastri & Ajjanagadde 1990b] and [Mani & Shastri 1991] respectively.

*This work was supported by NSF grant IRI 88-05465 and ARO grant ARO-DAA29-84-9-0027.

[1] It may be worth stating that the system does *not* require a central controller or a global clock.

[2] Observe that this kind of reasoning combines 'relational inheritance' [Fahlman 1979] with rule-based reasoning: While relational inheritance can support the inference 'Sylvester preys on Tweety' by using the *IS-A* relationships on the generic fact 'Cats prey on birds', it cannot support the inference 'Tweety is scared of Sylvester', because doing so also requires the use of the rule 'If $x$ preys on $y$ then $y$ is scared of $x$'.

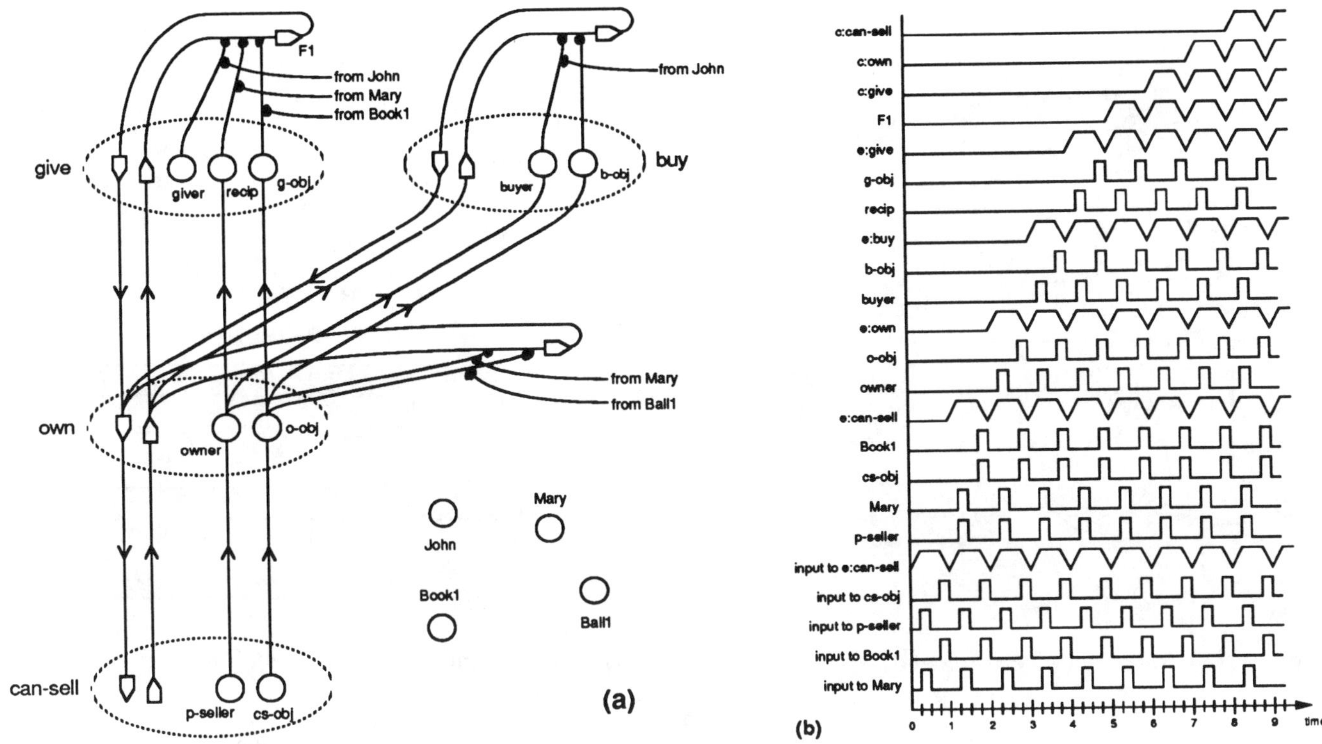


Figure 1: (a) Encoding rules and facts. (b) Activation trace for the query *can-sell(Mary,Book1)?*

## The rule-based reasoning system

Fig. 1a illustrates how long-term knowledge is encoded in the rule-based reasoning system. The network encodes the following *rules* and *facts*: i) $\forall x,y,z$ [ *give(x,y,z)* $\Rightarrow$ *own(y,z)* ], ii) $\forall x,y$ [ *buy(x,y)* $\Rightarrow$ *own(x,y)* ], iii) $\forall x,y$ [ *own(x,y)* $\Rightarrow$ *can-sell(x,y)* ], iv) *give(John,Mary,Book1)*, v) *buy(John,x)*, and vi) *own(Mary,Ball1)*.

The encoding makes use of two types of nodes: $\rho$-btu nodes (depicted as circles) and $\tau$-and nodes (depicted as pentagons). The computational behavior of these nodes is as follows: A $\rho$-btu is a phase-sensitive binary threshold unit. When such a node becomes active, it produces an oscillatory output in the form of a pulse train that has a period $\pi$ and pulse width $\omega$. The timing (or the *phase*) of the pulse train produced by a $\rho$-btu node depends on the phase of the input to the node. A $\tau$-and node acts like a *temporal* AND node. Such a node also oscillates with the same frequency as a $\rho$-btu node except that it becomes active only if it receives *uninterrupted* activation over a whole period of oscillation. Furthermore, the width of the pulses produced by a $\tau$-and node equals $\pi$.[3] The maximum number of distinct entities that may participate in the reasoning process equals $\pi/\omega$ (assume integer divide). The encoding also makes use of *inhibitory modifiers* – links that impinge upon and inhibit other links. A pulse propagating along an inhibitory modifier will block a pulse propagating along the link it impinges upon. In Fig. 1a, inhibitory modifiers are shown as links ending in dark blobs.

Each constant in the domain is encoded by a $\rho$-btu node. An $n$-ary predicate is encoded by a pair of $\tau$-and nodes and $n$ $\rho$-btu nodes, one for each of the $n$ arguments. One of the $\tau$-and nodes is referred to as the *enabler* and the other as the *collector*. As a matter of convention, an *enabler* always points upwards and is named *e:<predicate-name>*. A *collector* always points downwards and is named *c:<predicate-name>*. The *enabler e:P* of a predicate $P$ becomes active whenever the system is being queried about $P$. Such a query may be posed by an external process or by the system itself during an episode of reasoning. On the other hand, the system activates the *collector c:P* of a predicate $P$ whenever the system wants to assert that the current dynamic bindings of the arguments of $P$ are consistent with the knowledge encoded in the system. A rule is encoded by connecting the *collector* of the antecedent predicate to the *collector* of the consequent predicate, the *enabler* of the consequent predicate to the *enabler* of the antecedent predicate, and by connecting the arguments of the consequent predicate to the arguments of the antecedent predicate in accordance with the correspondence between these arguments specified in the rule. A fact is encoded using a $\tau$-and node that receives an input from the enabler of the associated predicate. This input is modified by inhibitory modifiers from the argument nodes of the associated predicate. If an argument is bound to a constant in the fact then the modi-

[3] Later we will introduce a third type of node, namely the $\tau$-or node. A $\tau$-or node becomes active on receiving *any* activation but its output is like that of a $\tau$-and node.

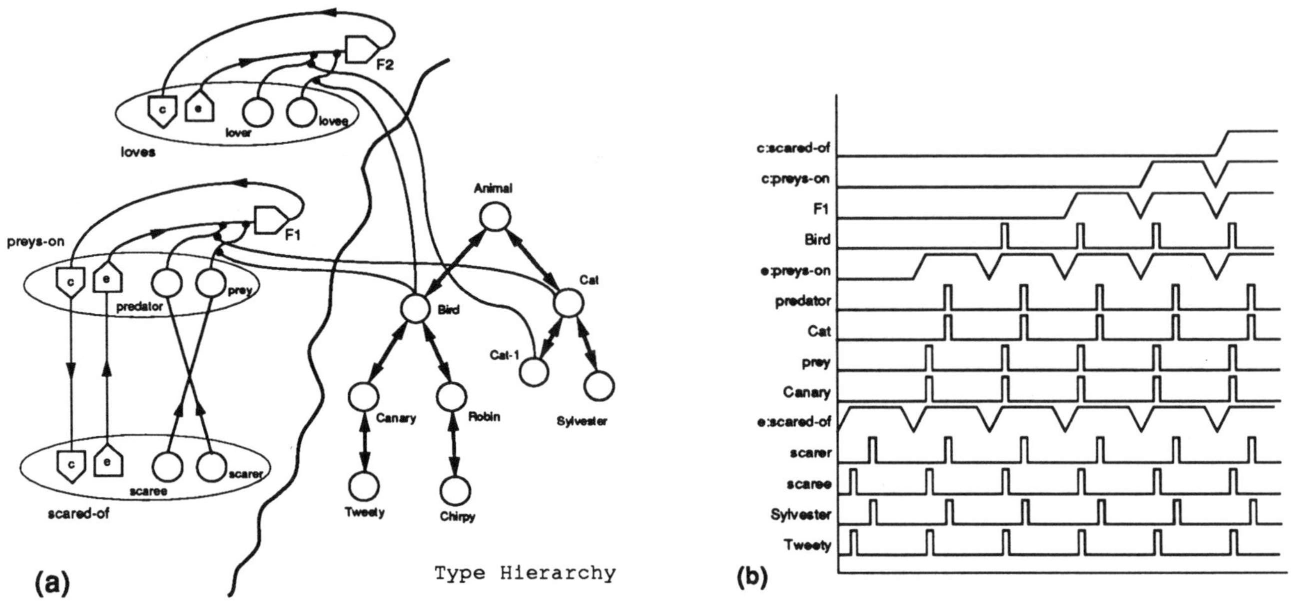


Figure 2: (a) An example network. (b) Trace of spreading activation for the query *scared-of(Tweety,Sylvester)?*

fier from such an argument node is in turn modified by an inhibitory modifier from the appropriate constant node. The output of the $\tau$-and node is connected to the *collector* of the associated predicate (refer to the encoding of the fact *give(John,Mary,Book1)* and *buy(John,x)* in Fig. 1a.)

## Inference Process

Posing a query to the system involves specifying the query predicate and the argument bindings specified in the query. In the proposed system this is done by simply activating the relevant nodes in the manner described below. Choose an arbitrary reference point in time – say, $t_0$ – for initiating the query. We assume that the system is in a quiescent state just prior to $t_0$. The query predicate is specified by activating the *enabler* of the query predicate, with a pulse train of width and periodicity $\pi$ starting at time $t_0$.

The argument bindings specified in the query are communicated to the network as follows: Suppose the argument bindings in the query involve $n$ distinct constants $c_1, ..., c_n$. With each of these $n$ constants, associate a delay $\delta_i$ such that no two delays are within $\omega$ of one another and the longest delay is less than $\pi - \omega$. Each of these delays may be viewed as a distinct *phase* within the period $t_0$ and $t_0 + \pi$. Now the argument bindings of a constant $c_i$ are indicated to the system by providing an oscillatory pulse train of pulse width $\omega$ and periodicity $\pi$ starting at $t_0 + \delta_i$, to $c_i$ and all arguments to which $c_i$ is bound. This is done for each constant $c_i$ ($1 \leq i \leq n$) and amounts to representing argument bindings by the *in-phase or synchronous activation of the appropriate constant and argument nodes.*

We illustrate the reasoning process with the help of an example. Consider the query *can-sell(Mary,Book1)?* (i.e., Can Mary sell Book1?) The query is posed by i) activating the *enabler* node *e:can-sell* ii) activating *Mary* and *p-seller* in the same phase (say, phase-1), and iii) activating *Book1* and *cs-obj* in some other phase (say, phase-2). As a result of these inputs, *Mary* and *p-seller* will fire synchronously in phase-1 of every period of oscillation, while *Book1* and *cs-obj* will fire synchronously in phase-2 of every period of oscillation. The node *e:can-sell* will also oscillate and generate a pulse train of periodicity and pulse width $\pi$ (Fig. 1b). The activations from the arguments *p-seller* and *cs-obj* reach the arguments *owner* and *o-obj* of the predicate *own*, and consequently, starting with the second period of oscillation, *owner* and *o-obj* become active in phase-1 and phase-2, respectively. At the same time, the activation from *e:can-sell* activates *e:own*. The newly created dynamic bindings for the arguments of *own*, in conjunction with the activation of *e:own* can be thought of as encoding the query *own(Mary,Book1)?* The $\tau$-and node associated with the fact *own(Mary, Ball1)* does not match the query and remains inactive. As the activation propagates from the arguments of *own* to the arguments of *buy* and *give* (Fig. 1b), new bindings for the arguments of *give* and *buy* get established which, in effect, encode two new queries: *give(x,Mary,Book1)?* (i.e., Did *someone* give Mary Book1?), and *buy(Mary,Book1)?* The $\tau$-and node F1, associated with the fact *give(John,Mary,Book1)* becomes active as a result of the uninterrupted activation it receives from *e:give*. Observe that the inhibitory modifiers from *recip* and *g-obj* incident on the link from *e:own* to F1 are blocked by the in-phase inputs from *Mary* and *Book1*, respectively. The activation from F1 causes *c:give*, the *collector* of *give*, to become active and the output from *c:give* in turn causes *c:own* to become active and transmit an output to *c:can-sell*. Consequently *c:can-sell*, the *collector* of the query predicate *can-sell*, becomes active indicating an affirmative answer to the query *can-sell(Mary,Book1)?* (Fig. 1b).

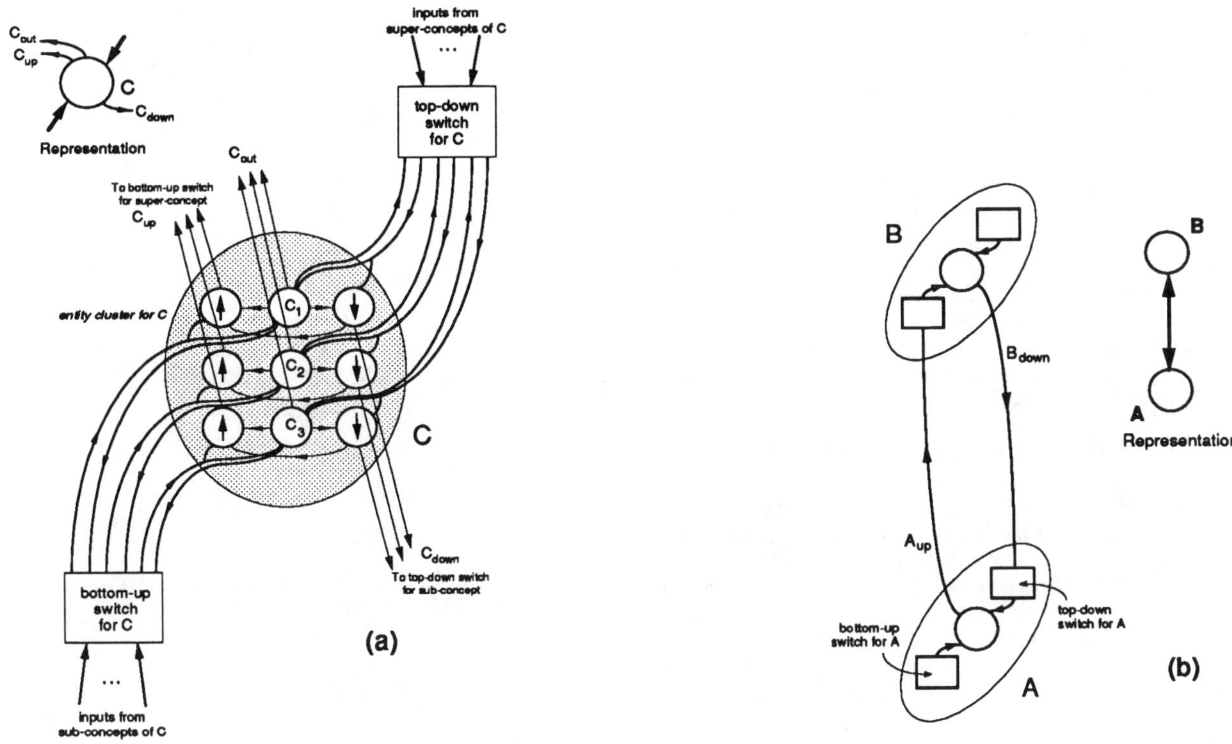


Figure 3: (**a**) Structure of the entity cluster for $C$, and its interaction with the bottom-up and top-down switches. The $\uparrow$ and $\downarrow$ nodes have a threshold $\theta = 2$. The multiple instantiation constant, $k = 3$. (**b**) Encoding of the *is-a* relation *is-a(A,B)*. A bundle of $k$ wires is represented by a single link.

## Combining the rule-based reasoner with an *IS-A* hierarchy

Fig. 2a gives an overview of the combined reasoning system. The rule-based part of the network encodes the rule $\forall x,y$ [ *preys-on(x,y)* $\Rightarrow$ *scared-of(y,x)* ] (i.e., if $x$ preys on $y$, then $y$ is scared of $x$), and the facts $\forall x$:*Cat*, *y*:*Bird preys-on(x,y)* and $\exists x$:*Cat* $\forall y$:*Bird loves(y,x)*. The former is equivalent to *preys-on(Cat,Bird)* and amounts to 'Cats prey on Birds'. The latter amounts to 'there is a cat that loves all birds'. The network on the right encodes the *IS-A* relationships: *is-a(Bird, Animal)*, *is-a(Cat, Animal)*, *is-a(Robin, Bird)*, *is-a(Canary, Bird)*, *is-a(Tweety, Canary)*, *is-a(Chirpy, Robin)*, and *is-a(Sylvester, Cat)*.

Facts involving typed variables are encoded in the following manner: A typed, universally quantified variable is treated as being equivalent to its type. Thus $\forall x$:*Cat*, *y*:*Bird preys-on(x,y)* is encoded as *preys-on(Cat,Bird)*. A typed, existentially quantified variable is encoded using a unique subconcept of the associated type. Thus in Fig. 2a, $\exists x$:*Cat* $\forall y$:*Bird loves(x,y)* is encoded as *loves(Cat-1,Bird)*, where *Cat-1* is a unique instance of *Cat*. For now let us assume that i) each concept (type or instance) is encoded as a $\rho$-btu node ii) each conceptual *IS-A* relationship such as *is-a(A,B)* is encoded using two connectionist links – a bottom-up link from $A$ to $B$ and a top-down link from $B$ to $A$, and iii) the top-down and bottom-up links can be enabled selectively by built-in control mechanisms (missing details are provided below).

The time course of activation for the query *scared-of(Tweety,Sylvester)?* is given in Fig. 2b. The query is posed by turning on *e:scared-of* and activating the nodes *Tweety* and *Sylvester* in synchrony with the first and second arguments of *scared-of*, respectively. The bottom-up links emanating from *Tweety* and *Sylvester* are also enabled. The net result of spreading activation (Fig. 2b), in the conceptual hierarchy and the rule-base is that the query *scared-of(Tweety,Sylvester)?* is transformed into the query *preys-on(Cat,Bird)?* The latter query matches the stored fact *preys-on(Cat,Bird)* and leads to the activation of *c:preys-on*. In turn, *c:scared-of* becomes active and signals an affirmative answer to the query.

### Two technical problems

There are two technical problems that must be solved in order to integrate the conceptual hierarchy and the rule-based component. First, the encoding of the *IS-A* hierarchy should be capable of representing multiple instantiations of a concept. For example, in the query discussed above, we would like the network's state of activation to represent both 'the animal Tweety' and 'the animal Sylvester'. This is problematic because the node *Animal* cannot be in synchrony with both *Tweety* and *Sylvester* at the same time. Second, the encoding must provide *built-in* mechanisms for controlling the direction of propagating activation in the *IS-A* hierarchy so as to correctly deal with queries containing existentially and universally quantified variables. Thus i) Activation originating from an instance or a concept that corresponds to a universally quantified variable in the query should propagate upwards to all its ancestors, and ii) If the *IS-A* hierarchy is a taxonomy, then activation originating from a concept $C$ that corresponds to an existentially quantified variable in the query should propagate to the ancestors as well as descendents of $C$. If however, the *IS-A* hierarchy permits multiple inheritance then the activation must *also* propagate to the ancestors of the descendents of $C$. The following solution to these problems does not require an external controller to monitor and control the nodes in the network, during the reasoning process.

## Implementing the Type Hierarchy

Each entity (i.e., type or instance) $C$, is represented by a group of nodes called the *entity cluster* for $C$. Such a cluster is organized as shown in Fig. 3a. The entity cluster for $C$ has $k$ *banks* of $\rho$-btu nodes, where $k$, the *multiple instantiation constant*, refers to the number of dynamic instantiations a concept can accommodate. Each *bank* $C_{B_i}$ consists of three $\rho$-btu nodes: $C_i$, $C_{i\uparrow}$, $C_{i\downarrow}$. Each $C_i$ represents a distinct (dynamic) instantiation of $C$. If this instantiation is in phase $\rho$, then, $C_i$ fires in phase $\rho$. The *relay nodes* $C_{i\uparrow}$ and $C_{i\downarrow}$ control the *direction of propagation* of the activation represented by $C_i$. The $C_{i\uparrow}$ and $C_{i\downarrow}$ nodes have a threshold $\theta = 2$. As shown in Fig. 3a, $C_i$ is connected to both $C_{i\uparrow}$ and $C_{i\downarrow}$. $C_{i\downarrow}$ is linked to $C_{i\uparrow}$, but not vice versa. Directional control of propagating activation is exercised using a suitable modification of the *relay-node* scheme discussed in [Shastri 1988].

Every entity $C$ is associated with two *switches* – a *top-down* switch and a *bottom-up* switch. The switches, both of which are identical in structure, control the flow of activation in the type hierarchy. Each switch has $k$ outputs. *Output*$_i$ from the bottom-up switch connects to $C_i$ and $C_{i\uparrow}$ while the corresponding output from the top-down switch goes to the $C_i$ and $C_{i\downarrow}$ nodes, $1 \leq i \leq k$. There is also a feedback from the $C_i$ nodes to both the switches (See Fig. 3a and Fig. 4.)

The interaction between the switches and the entity cluster (Fig. 3a) brings about efficient and *automatic dynamic allocation of banks* in an entity cluster, by ensuring that:

- Activation is channeled to the entity cluster banks only if the entity cluster can accommodate more instantiations; the maximum number of instantiations is therefore limited to $k$.
- Each $C_i$ picks up a *unique* phase; thus new instantiations are always in a phase not already represented in the entity cluster.

The architecture of the switch (with $k = 3$) is illustrated in Fig. 4. The $k$ $\rho$-btu nodes, $S_1, \ldots, S_k$, with their associated $\tau$-or nodes form the basic components of the switch. Every input to the switch makes two connections – one excitatory and one inhibitory – to each of $S_2, \ldots, S_k$; as a result of these excitatory-inhibitory connections, all these nodes are disabled to begin with, and cannot respond to incoming activation. Input activation will have an effect only on the $S_1$ node, since the

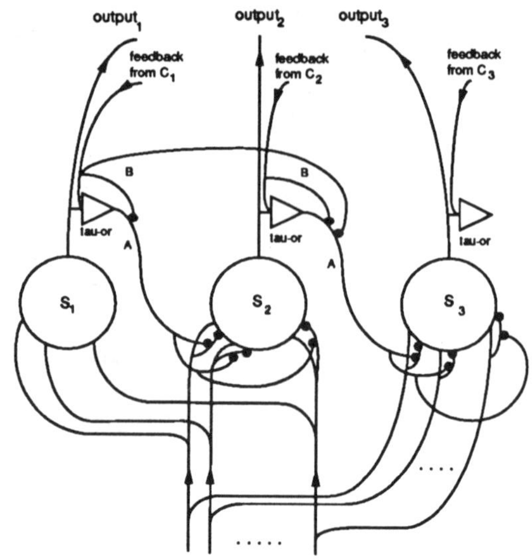


Figure 4: Architecture of the switch. The multiple instantiation constant $k = 3$.

inputs to the switch directly connect to $S_1$ (Fig. 4.) In keeping with the behavior of $\rho$-btu nodes, $S_1$ becomes active in response to the first available input and continues to fire in phase with that input as long it remains active. As $S_1$ goes active, the $\tau$-or node associated with $S_1$ turns on, thereby enabling $S_2$. Inhibitory feedback from $C_1$ ensures that $S_2$ is *not* enabled during the phase $\rho$ in which $C_1$ is firing[4]. Thus $S_2$ selects and starts firing in a phase *other than* $\rho$. Once $S_2$ has made its selection, $S_3$ gets its turn, and so on.

As instantiations are deputed to the entity cluster, the $\rho$-btu nodes in the switch are progressively enabled from left to right. If $C_1, \ldots, C_{i-1}$ are firing in phases $\rho_1, \ldots, \rho_{i-1}$, then $S_i$ always picks a distinct phase $\rho \notin \{\rho_1, \ldots, \rho_{i-1}\}$, since inputs in phases $\rho_1, \ldots, \rho_{i-1}$ are inhibited by the feedback links from $C_1, \ldots, C_{i-1}$. At any stage, if $C_i$, $1 \leq i \leq k$, picks up activation channeled by the *other* switch, feedback from $C_i$ into the $\tau$-or node associated with $S_i$ causes $S_{i+1}$ to be enabled, even though $S_i$ has not picked a phase. This ensures that at most $k$ instantiations are selected *jointly* by the bottom-up and top-down switches; hence, only $k$ instantiations can be channeled to $C$, at worst.

A fact of the form *is-a(A,B)* is represented by (Fig. 3b): (i) connecting the $A_{i\uparrow}, i = 1, \ldots, k$ nodes to the bottom-up switch for $B$; (ii) connecting the $B_{i\downarrow}, i = 1, \ldots, k$ nodes to the top-down switch for $A$.

Consider a concept $C$ in the type hierarchy. Suppose $C_i$ receives activation from the bottom-up switch in phase $\rho$. $C_i$ starts firing in synchrony with this activation. The $C_{i\uparrow}$ node (with threshold $\theta = 2$) is now receiving *two* inputs in this phase (from the bottom-up switch and from $C_i$; see Fig. 3a) and begins to fire in phase $\rho$. This causes activation in phase $\rho$ to eventually

[4]In general, $C_i$ could receive input in *two* phases – one from the bottom-up switch for $C$, and another from its top-down switch. $C_i$ being a $\rho$-btu node, picks one of these phases to fire in.

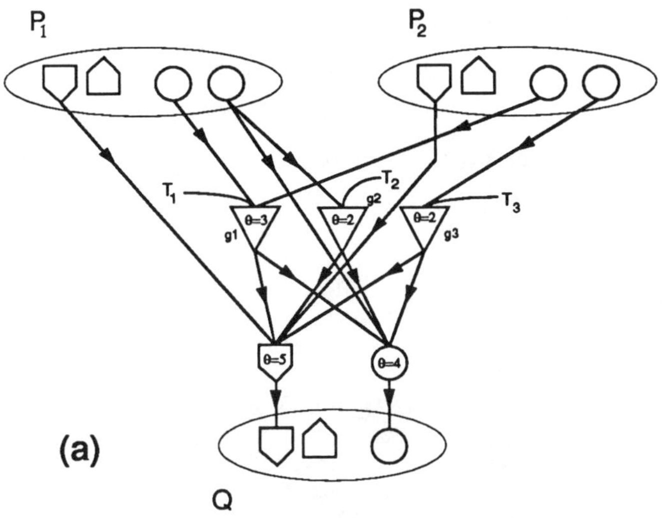


Figure 5: Network encoding the rule $\forall x{:}T_1,\ y{:}T_2\ \exists z{:}T_3$ [ $P_1(x,y)$ & $P_2(x,z) \Rightarrow Q(y)$ ] in a forward reasoning system. Links from $T_1$, $T_2$ and $T_3$ are actually bundles of $k$ wires carrying the $k$ instantiations of these constants.

spread to the super-concept of $C$. Hence, any upward traveling activation continues to travel upward – which is the required behavior when $C$ is associated with a universal typed variable. Similarly, when $C_i$ receives activation from the top-down switch in phase $\rho$, both $C_i$ and $C_{i\downarrow}$ become active in phase $\rho$. $C_{i\uparrow}$ follows suit, because of the link from $C_{i\downarrow}$ to $C_{i\uparrow}$, so that the whole bank $C_{B_i}$ now fires in phase $\rho$. This mechanism allows a concept associated with an existential typed variable to eventually spread its activation to its ancestors, descendents and ancestors of descendents.

## Typed Variables in Rules

The type hierarchy can be used to impose type restrictions on variables occurring in rules, for both forward and backward reasoning systems. To utilize this feature, we need to modify the implementation of rules: In a *forward reasoning system*, the rule is encoded by introducing a $\tau$-or node to perform the type checking for the argument under question. For example, in Fig. 5, which encodes the rule $\forall x{:}T_1,\ y{:}T_2\ \exists z{:}T_3$ [ $P_1(x,y)$ & $P_2(x,z) \Rightarrow Q(y)$ ], $g_3$ would turn on if and only if the second argument of $P_2$ and $T_3$ are in synchrony – which is to say that the argument is bound to an object of type $T_3$. In the forward reasoner, typed variables are allowed only in the antecedent of the rule.

In a *backward reasoner*, the strategy is similar, except that i) type checking for a typed universally quantified variable is enforced by an inhibitory link from the concept representing the type of the concerned argument; ii) For a typed, existentially quantified variable, an inhibitory link derived from a unique subconcept of the associated type performs the type enforcement – in a manner similar to the interpretation of a typed existential variable in a fact [Mani & Shastri 1991]. In the backward reasoner, typed variables are allowed only in the consequent. Both in the forward and backward reasoners, a rule fires only if *all typed arguments are firing in synchrony with their respective types.*

## Conclusions

Adding a type hierarchy allows the connectionist reasoning system to handle rules, facts, and queries with typed variables. The simulator described in [Mani 1990] has been used to test the system. Several extensions to the system proposed here are being investigated. The current system assumes that any fact or query with both existentially and universally quantified variables is such that all the universal quantifiers are within the scope of the existential quantifiers. Work is being done on handling more general forms of facts and queries. We are also working on the design of an expanded system that would allow property-value attachments to concepts [Shastri 1988]. We also wish to combine the *IS-A* hierarchy with a reasoning system that allows multiple instantiation of predicates. Lastly, we hope to combine a forward and backward reasoner that can make use of both long-term and dynamic (temporary) facts during reasoning.

## References


[Ajjanagadde & Shastri 1991] Ajjanagadde, V., and Shastri, L. 1991. Rules and variables in neural nets. To appear in *Neural Computation* 3(1).

[Fahlman 1979] Fahlman, S. 1979. *NETL: A System for Representing Real-World Knowledge*. Cambridge, Mass.: MIT Press.

[Lange & Dyer 1989] Lange, T. E., and Dyer, M. G. 1989. High-level inferencing in a connectionist network. *Connection Science*, 1(2):181-217.

[Mani 1990] Mani, D. R. 1990. *Using the Connectionist Rule-Based Reasoning System Simulator.*

[Mani & Shastri 1991] Mani, D. R., and Shastri, L. Combining a Type Hierarchy with a Connectionist Rule-Based Reasoner. Technical Report MS-CIS-91-33, Department of Computer and Information Science, Univ. of Pennsylvania, May 1991.

[Shastri 1988] Shastri, L. 1988. *Semantic networks: An evidential formulation and its connectionist realization.* Los Altos: Morgan Kaufman.

[Shastri & Ajjanagadde 1990a] Shastri, L., and Ajjanagadde, V. 1990. An optimally efficient limited inference system. In *Proceedings of AAAI-90, the Twelfth National Conference of the American Association of Artificial Intelligence*, 563-570. Cambridge Mass.: American Association for Artificial Intelligence.

[Shastri & Ajjanagadde 1990b] Shastri, L., and Ajjanagadde, V. 1990. From Simple Associations to Systematic Reasoning: A Connectionist Representation of Rules, Variables and Dynamic Bindings. Technical Report MS-CIS-90-05, Department of Computer and Information Science, Univ. of Pennsylvania.

[Smolensky 1987] Smolensky, P. 1987. On variable binding and the representation of symbolic structures in connectionist systems. Technical Report CU-CS-355-87, Department of Computer Science, Univ. of Colorado at Boulder.

# The Connectionist Scientist Game: Rule Extraction and Refinement in a Neural Network

**Clayton McMillan, Michael C. Mozer, & Paul Smolensky**

Department of Computer Science and
Institute of Cognitive Science
University of Colorado
Boulder, CO 80309-0430

### Abstract

Scientific induction involves an iterative process of hypothesis formulation, testing, and refinement. People in ordinary life appear to undertake a similar process in explaining their world. We believe that it is instructive to study rule induction in connectionist systems from a similar perspective. We propose an approach, called the Connectionist Scientist Game, in which symbolic condition-action rules are extracted from the learned connection strengths in a network, thereby forming explicit hypotheses about a domain. The hypotheses are tested by injecting the rules back into the network and continuing the training process. This extraction-injection process continues until the resulting rule base adequately characterizes the domain. By exploiting constraints inherent in the domain of symbolic string-to-string mappings, we show how a connectionist architecture called RuleNet can induce explicit, symbolic condition-action rules from examples. RuleNet's performance is far superior to that of a variety of alternative architectures we've examined. RuleNet is capable of handling domains having both symbolic and subsymbolic components, and thus shows greater potential than purely symbolic learning algorithms. The formal string manipulation task performed by RuleNet can be viewed as an abstraction of several interesting cognitive models in the connectionist literature, including case role assignment and the mapping of orthography to phonology.

## Introduction

A cognitive behavior may be characterized in terms of a transformation from an initial cognitive state to a target cognitive state. There are many ways of specifying this transformation. At one extreme, an enumeration of input/output pairs provides a description of the transformation, but such a list is not particularly satisfying or concise because it does not explicitly capture regularities of a domain. An alternative is a set of symbolic condition-action rules that provides an algorithm for performing the transformation. The notion of explicit rules has played an important role in cognitive modeling because rules are a descriptive, readily comprehensible, and powerful representational language, and because people appear to exhibit rule-governed behavior when performing higher cognitive tasks. However, rule-based characterizations are often brittle and incomplete, and mechanisms for learning condition-action rules from scratch have not been extensively studied.

To model the process by which people acquire rules, it is instructive to examine theory construction in scientific domains. Scientists approach a domain first by observation, and then, armed with initial intuitions, formulate explicit hypotheses concerning regularities in the domain. Such hypotheses can be tested by experimentation, allowing for an iterative refinement of the hypotheses until eventually the hypotheses are consistent with the observations. We call this process of iterative hypothesis formulation and refinement the *Scientist Game.* We conjecture that rule acquisition by people in ordinary life involves a very similar process, one in which people play the role of the scientist and the hypotheses take the form of explicit rules in a given domain.

This paper proposes a modification of the scientist game, called the *Connectionist Scientist Game*, as a technique for hypothesizing and refining a set of rules through induction in a connectionist network. The Connectionist Scientist Game is an iterative process that involves first training a network on a set of input/output examples, which corresponds to the scientist developing intuitions about a domain. After a certain amount of exposure to the domain, symbolic rules are extracted from the connection strengths in the network, thereby forming explicit hypotheses about the domain. The hypotheses are tested by injecting the rules back into the network and continuing the training process. This extraction-injection process continues until the resulting rule base adequately characterizes the domain.

Although the view held by many neural net researchers is that explicit rules are unnecessary (e.g., Rumelhart and McClelland, 1986), the Connectionist Scientist Game suggests that rules can be exploited in connectionist networks in several important ways: 1) as a means for better understanding the inner workings of a

This research was supported by NSF Presidential Young Investigator award IRI-9058450, grant 90-21 from the James S. McDonnell Foundation, and DEC external research grant 1250 to MM; NSF grants IRI-8609599 and ECE-8617947 to PS; by a grant from the Sloan Foundation's computational neuroscience program to PS; and by the Optical Connectionist Machine Program of the NSF Engineering Research Center for Optoelectronic Computing Systems at the University of Colorado at Boulder.

network (McMillan & Smolensky, 1988), 2) as a technique for increasing learning speed, 3) as a means of constraining – and thereby improving – generalization, and 4) as a way of bridging the gap between sub-conceptual and conceptual levels of cognitive processing (Smolensky, 1988).

We describe an architecture called RuleNet, which, based on a characterization of the task domain, plays the Connectionist Scientist Game. Although connectionist networks are often conceptualized as embodying *implicit* rules, the RuleNet architecture embodies rules explicitly. In the following sections we describe the domain towards which this work is currently directed, the task and network architecture, simulations that demonstrate the potential for this approach, and finally, future directions of the research leading towards more general and complex domains.

## Domain

We are interested in intrinsically rule-based domains that map input strings of $n$ symbols to output strings of $n$ symbols. Rule-based means that the mapping function can be completely characterized by explicit, mutually exclusive condition-action rules. A *condition* is a feature or combination of features necessarily present in each input in order for a given rule to apply, and an *action* describes the mapping to be performed on each input if the rule does apply. For example, a typical condition might be that the symbol A must be present at the beginning of the input string. The string ABCD meets this condition, while the string BCDA does not. A typical action might be to switch the first two symbols in the input, replace the third symbol with the symbol A, and copy the remainder of the input exactly to the output string, e.g., ABCD → BAAD.

In these simulations we allow three types of conditions: 1) a *simple* condition, which states that symbol $s$ must be present in slot $x$ of the input string in order for a rule to apply, 2) a *conjunction* of two simple conditions, and 3) a *disjunction* of two simple conditions. The action performed by a rule is to produce an output string in which the value of each slot is either a constant or a function of a particular input slot, with the additional constraint that each input slot maps to at most one output slot. In the present work, this function of the input slots is the identity function, and a constant is any symbol in the output alphabet.

The number of possible rules in a given domain is dependent upon the length of the strings, $n$, and the size $k$ of the input/output alphabets. The number of possible rules is:

$$2\left((k+1)^n - \frac{nk}{2}\right) \sum_{i=0}^{n} k^i \left(\frac{n!}{i!}\right)^2$$

which is an exponential function of $n$.

An example of such a rule for strings of length three over the input/output alphabet {A, B, C, D, E, F, G, H} is:

**if** ($input_1 = A \wedge input_3 = G$) **then**
($output_1 = input_3$, $output_2 = input_2$, $output_3 = B$)

where $input_\alpha$ denotes slot $\alpha$ of the input string, and $output_\beta$ denotes slot $\beta$ of the output string. As a shorthand for this rule, we write [∧ A_G → 32B], where the square brackets indicate this is a rule, the "∧" denotes a conjunctive condition, and the "_" denotes a *wildcard* symbol. A disjunction is denoted by "∨".

This formal string manipulation task can be viewed as an abstraction of several interesting cognitive models in the connectionist literature. For example Miikkulainen and Dyer (1988) have built a case role assignment network that maps syntactic constituents of a sentence into their semantic roles. This amounts to copying the representation of say, a sentence subject, such as *boy*, to a role slot in the output, such as *agent*. NETtalk (Sejnowski and Rosenberg, 1987) and McClelland and Rumelhart's past tense model (1986) are two further examples of problems that might be viewed as string mapping problems, although implementing them in the framework described above requires slight modifications to the original approach.

## Task

The task RuleNet must perform is to induce a compact set of rules that accurately characterize a set of training examples. Note that any corpus of examples has at least one set of rules that correctly characterizes the mapping function, namely the set containing one rule per example. However, this set of rules is trivial in that it is no more interesting than an enumeration of the patterns themselves. We consider the appropriate set of rules to be the minimal set that completely describes the training corpus. In this work, the training set for a given simulation is generated using a set of prespecified rules as templates for creating valid examples. Input strings that meet the conditions of several rules in the template set are excluded. The rules are over strings of length four and an alphabet of eight symbols, {A, B, C, D, E, F, G, H}. For example, the following rules may be used to generate these exemplars:

[∨ _EG_ →30E2]: GEAF→FGEA, BECE→EBEC
[∧ _BF_ →3G12]: EBFF→FGBF, DBFG→GGBF
[ _F__ →30B1]: GFAB→BGBF, CFHD→DCBF

The patterns above are similar to those that form the training corpus used in the simulations described later. In general there is no guarantee that the underlying rule base represents the minimal set that describes the training corpus. However, given that in our experiments the number of rules is very small relative to the number of examples, it is unlikely that a more compact set of rules exists that describes the training set. Therefore, in our simulations the target number of rules to be induced is the same as the number used to generate the training corpus.

There are several traditional, symbolic systems, e.g., COBWEB (Fisher, 1987), that induce rules for classifying inputs based upon training examples. The power of these systems lies in their ability to build complex conditions that recognize relevant features of an input. It seems likely that, given the correct representation, a system such as COBWEB could learn rules that would *classify* patterns in our domain. However, it is unclear how such a system could also learn the action associated with each class. Classifier systems (Booker, Goldberg, & Holland, 1989) learn both conditions and actions, but the actions are limited to passing an input feature through unchanged, or writing a fixed feature to the output. Although classifiers could easily be made to represent the symbolic strings of our domain, there is no obvious way to map a symbol in slot $\alpha$ of the input to slot $\beta$ of the output.

## Architecture

We propose a neural network architecture, called RuleNet, based on the work of Jacobs, Jordan, Nowlan, and Hinton (1991), that can implement string manipulation rules of the type outlined above. As shown in Figure 1, RuleNet has three layers of units: an input layer, an output layer, and an intermediate layer of *condition units*. There are *m* rules represented in the network and one condition unit per rule. The input layer activates the condition units, which, to first approximation, participate in a winner-take-all competition. The winning condition unit enables one set of connections from the input layer to the output layer through a set of multiplicative or gating connections. The input units then pass their activity through the enabled set of weights to activate the output units.

The condition and action parts of a rule are implemented in different subnets. In the *condition subnet*, the net input to each condition unit *i* is computed,

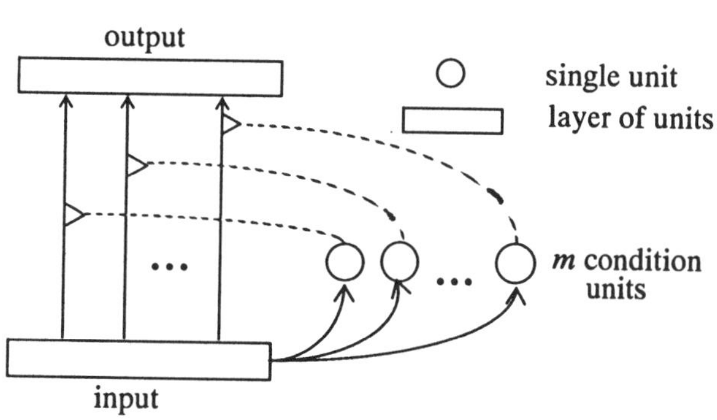


**Figure 1**
The RuleNet architecture.

$$net_i = 1/(1+e^{-\mathbf{c}_i^T\mathbf{x}})$$

where $\mathbf{x}$ is the input vector and $\mathbf{c}_i$ is the incoming weight vector to condition unit *i*. The activity of condition unit $i$, $p_i$, is then determined by a normalization:

$$p_i = \frac{net_i}{\sum_j net_j}$$

The normalization enforces a competition among condition units. The activation of condition unit *i* represents the probability that rule *i* applies to the current input. In the *action subnet* there are *m* weight matrices $\mathbf{A}_i$, one per rule. A set of multiplicative connections between each condition unit *i* and $\mathbf{A}_i$ determines to what extent $\mathbf{A}_i$ will contribute activation to the output vector $\mathbf{y}$, calculated as follows:

$$\mathbf{y} = \sum_i^m p_i \mathbf{A}_i \mathbf{x}$$

Ideally, one condition unit is fully activated by a given input.

Although this architecture is based on the work of Jacobs et al., it is independently motivated in our work by the demands of the task domain. The Jacobs architecture divides up the input space into disjoint regions and allocates a different *local expert* network to each subspace. A *gating network* determines the probability that expert $\alpha$ will produce the correct output for a given input vector $\mathbf{x}$. During learning, weights are adjusted using back propagation to increase the probability that expert $\alpha$ produces the correct output and decrease the probability that other experts produce any output at all for $\mathbf{x}$. In effect, each network competes to be the sole network responsible for input $\mathbf{x}$. RuleNet has essentially the same structure as the Jacobs network, where the action substructure of RuleNet serves as the local experts and the condition substructure serves as the gating network. However, their goal—to minimize crosstalk between logically independent subtasks—is quite different than ours.

Input/output strings, such as AEG, are encoded as binary vectors. We use a local representation of each symbol and concatenate them together. The representation of the *i*th symbol in an alphabet of *k* symbols is a subvector of length *k* in which the *i*th bit is 1 and the remaining bits are 0. With the eight symbol alphabet {A, B, C, D, E, F, G, H}, then, the representation of AEG is composed of three subvectors of length eight concatenated together: (10000000 00001000 00000010).

To precisely implement rules of the type discussed in the previous section, it is necessary to impose some restrictions on the values assigned to the weights in $\mathbf{c}_i$ and $\mathbf{A}_i$. In $\mathbf{c}_i$, the first *k* weights detect the appropriate symbol in the first slot of the input string, the next $k$ weights detect the symbol in the second slot, and so on, up to the length of the string. Since this is a local representation, one bit, at most, in each $k$-bit subvector should be nonzero. For example, using the eight-symbol alphabet, the vector $\mathbf{c}_i$ that detects the simple condition $input_1$ = A is: (10000000 00000000 00000000 0). Condition unit $i$ will be activated only if there is an A in the first slot of the input string. The final weight in $\mathbf{c}_i$ is the condition bias, $\theta_i$, which is required to detect a conjunctive condition. If the condition is a conjunction, as in [∧ A_G → 021], $\theta_i$ must be negative to compensate for the effect of one input; that is, the net input will be positive only if both symbols are present. If the condition is simple or a disjunction, $\theta_i$ should be zero to allow the net input to be positive if any input line is active. To encourage the condition weights to develop such a structure, it is necessary to initialize all weights in $\mathbf{c}_i$ except $\theta_i$ to non-negative values, and then set a lower bound of zero on those weights during training.

Similarly, if we wish the actions carried out by the network to correspond to the string manipulations allowed by our rule domain, it is necessary to impose some restrictions on the values assigned to the weights in $\mathbf{A}_i$. An action matrix $\mathbf{A}_i$ has an $n \times n$ block form, where $n$ is the length of input/output strings. Each block is a $k \times k$ submatrix, and must be either the identity matrix or the zero matrix. The block at block-row $\alpha$, block-column $\beta$ of $\mathbf{A}_i$ copies $input_\alpha$ to $output_\beta$ if it is the identity matrix, or does nothing otherwise. An additional constraint that only one block may be nonzero in block-row $\alpha$ or block-column $\beta$ of $\mathbf{A}_i$ ensures that there is a unique mapping from $input_\alpha$ to $output_\beta$. If $output_\beta$ is to be a constant, then block-column $\beta$ must be all zero except for the action bias weights in block-column $\beta$. Further, because the output is a local representation, at most one bias in block-column $\beta$ should be nonzero.

To ensure that during learning every block approaches the identity or zero matrix, we constrain the off-diagonal terms to be zero and constrain weight changes within a block to be equal for all weights on the diagonal, thus limiting the degrees of freedom to one parameter within each block. By imposing these restrictions on weight changes, the hope is that the resulting $\mathbf{c}_i$-$\mathbf{A}_i$ pairs are close enough to the block structure described above that it will be easy to extract a symbolic description of the mapping.

The constraints described above, however, do not guarantee that learning will produce weights that correspond exactly to symbolic rules. However, using a process we call *projection*, it is possible to transform the $\mathbf{c}_i$ and $\mathbf{A}_i$ weights such that the resulting network can be interpreted as a set of symbolic rules.

Projection of $\mathbf{c}_i$ involves setting non-essential weights to zero, setting essential weights to 1, and setting $\theta_i$ to zero if there is only one essential weight or if a disjunction is indicated, -1 otherwise. Because of the constraint that only one unit per slot in the input can be on, it is possible to subtract a value $\varepsilon_\alpha$ from slot $\alpha$ of $\mathbf{c}_i$ and add that value to $\theta_i$ without affecting the net input to condition unit $i$. This $\varepsilon_\alpha$ can be thought of as an irrelevant component of each weight in slot $\alpha$, a by-product of learning. We could estimate $\varepsilon_\alpha$ with a least squares procedure. However, in our simulations, we use an approximation to least squares which involves taking the average of all weights in slot $\alpha$ other than the maximum, subtracting this estimate from $\alpha$ and adding it to $\theta_i$. The resulting $\mathbf{c}_i$ is compared to prototype models of simple, disjunctive, and conjunctive conditions, and the closest model is taken as the projected condition vector.

Projection on a matrix $\mathbf{A}_i$ requires finding the largest block diagonal in $\mathbf{A}_i$, located say, at block-row $\alpha$ and block-column $\beta$, setting it to 1, and setting all other blocks in block-row $\alpha$ and block-column $\beta$ to zero. The process is repeated until $n$ blocks have been found (one per symbol in the input/output strings). Action bias strengths in a block-column are summed and treated as any other block during this process. The biases themselves are projected by setting the maximum bias in each slot to 1, and setting all other biases to zero. All biases in a block-column $\beta$ with a non-zero identity block are also set to zero.

## Simulations

Having described the projection technique for extracting explicit rules from RuleNet, we turn to simulations of the network. The simulations allow RuleNet to play the Connectionist Scientist Game by iteratively extracting rules and injecting them back into the network. To illustrate how this extraction-injection cycle improves RuleNet's performance, we compare learning performance using this technique with learning using several simpler techniques:

1) Technique J: Start with a fixed set of $r$ rules, random initial weights, and minimize the error function described by Jacobs et al.

2) Technique JC: Learn as in technique J, but constrain weights in $\mathbf{A}_i$ to a single parameter on block diagonals as described above.

3) Technique JCA: Learn as in technique JC, but start with one rule, learn for $m$ epochs, then add a new rule with random initial weights, and repeat until $r$-$1$ rules have been added or perfect performance achieved.

4) Technique JCAP (the Connectionist Scientist Game): Learn as in technique JCA, but before adding a new rule, *project* weights in $\mathbf{c}_i$ and $\mathbf{A}_i$ to conform exactly to the closest valid rule.

In all simulations, the maximum number of rules allowed was 10. In simulations using techniques JCA and JCAP, a new rule was added after every 500 epochs. All simulations ran for 5000 epochs and used a learning rate of .009 on both $\mathbf{A}_i$ weights and $\mathbf{c}_i$ weights. We tested four different rule bases, consisting of eight, three, three, and five rules (see Figure 2 for the rule templates) using strings of length four over the alphabet {A, B, C, D, E,

F, G, H}. For each rule base, a training set was formed by generating fifteen random instances of each rule template. Figure 2 shows the error curve for the four simulations using learning techniques J, JC, JCA, and JCAP. Each curve is an average over five runs with different initial weights. The spikes in the curves for techniques JCA and JCAP reflect the additional error when RuleNet is allowed to make use of a new rule. In all four cases the error for techniques J and JC is monotonically decreasing; however, JC arrives at a consistently small error, while the pure Jacobs network, technique J, does not. There is a similar distinction between techniques JCA and JCAP: JCAP arrives at a reasonable solution, while JCA does not. In terms of learning speed and convergence, techniques JC and JCAP are clearly preferable.

The question is, can a given learning technique be relied upon to discover a valid set of rules? Table 1 shows the average number of valid rules extracted from RuleNet after 5000 epochs and the percentage of the training set mapped correctly to the output string in each case. The critical measure of performance should take both percentage of patterns and number of valid rules into account. Recall that our criterion for an appropriate set of rules was to find at most $q$ rules, where $q$ is the size of the set used to generate the training corpus, such that those rules completely describe every pattern in the training set. Learning technique JCAP is certainly the most effective judging by this criterion. In fact, technique JACP induces exactly the target number of rules in every simulation except one, mapping virtually all the patterns correctly. JC maps patterns correctly, but at the expense of the number of rules. In this domain at least, it appears that RuleNet makes use of approximately as many rules as one allows it. Technique JCA comes closer to inducing the target number of rules, but at the expense of consistent accuracy. Finally, technique J leaves quite a bit to be desired in both the number of valid rules extracted and the percentage of patterns covered.

Technique JCAP is essentially an implementation of the Connectionist Scientist Game within the framework of the RuleNet architecture. The Connectionist Scientist Game involves iteratively forming a hypothesis set of rules, testing these rules on exemplars, and then refining the rules to yield better coverage of the exemplars. The

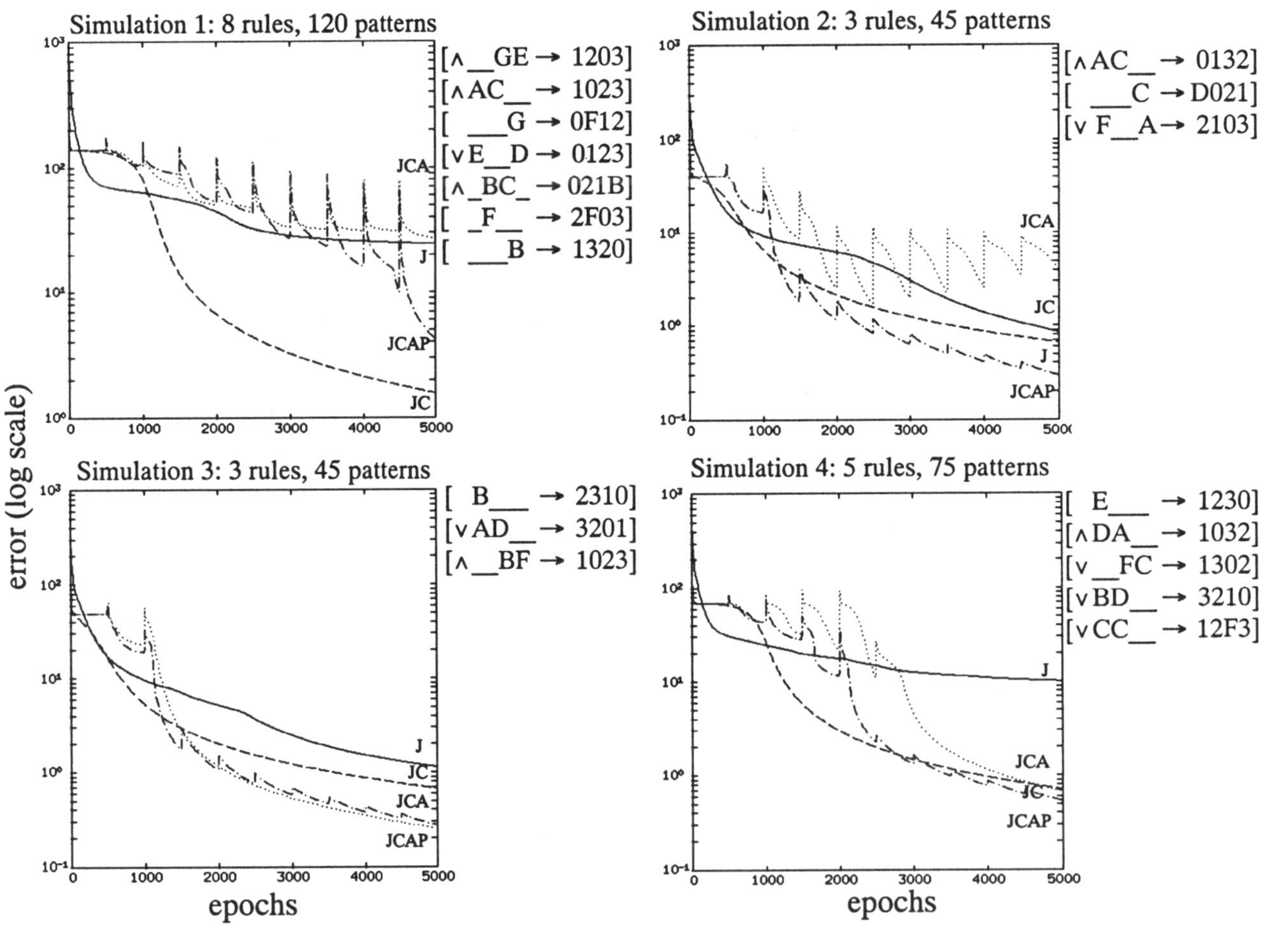


**Figure 2**

Average RuleNet performance on four simulations using learning techniques J, JC, JCA, and JCAP. Key: J = Jacobs algorithm, C = constrain $A_\alpha$ to diagonals, A = add new rules, P = project.

| | | Learning Technique | | | | | | | |
|---|---|---|---|---|---|---|---|---|---|
| | | J | | JC | | JCA | | JCAP | |
| Sim. | Target # rules | rules | % | rules | % | rules | % | rules | % |
| 1 | 8 | 2.4 | 17 | 9.5 | 99 | 6 | 69 | 9 | 96 |
| 2 | 3 | 1.4 | 10 | 9.5 | 100 | 3.8 | 95 | 3 | 100 |
| 3 | 3 | 1.4 | 28 | 9 | 100 | 3 | 100 | 3 | 100 |
| 4 | 5 | 3.6 | 34 | 9 | 100 | 5.2 | 100 | 5 | 100 |

**Table 1**
Average number of valid rules extracted and percentage of patterns covered to within an error of .5.

game is played until a satisfactory number of exemplars is covered by the rules. Figure 3 shows a template of five rules, used to generate a set of 75 exemplars, and the evolution of the hypothesis set of rules learned by RuleNet over 2500 training epochs. As in previous simulations, projection was done every 500 epochs. At the first projection the initial rule is not syntactically valid. The second projection, at 1000 epochs, results in two valid rules which correctly mapped 3% of the patterns. Subsequent iterations result in a more refined hypothesis. It is interesting to note the metamorphosis of both the set of rules and individual rules. For example, the first valid rule learned, [∧ B_H_→1230], is modified between the projection at 1000 and 1500 epochs to [∨BD__→3210]. As new rules are added, the condition and/or action components of old rules may be modified to reflect a different set of exemplars for which they must be accountable.

We have argued that learning rules can greatly enhance generalization. Using technique JCAP, in three of the four simulations reported above, ruleNet not only learned a set of rules that correctly mapped the training set, it also learned the *exact* rule templates used to generate that set. Even in simulation 1, seven of the eight template rules were induced exactly, while one simple condition, [ _F__→2F03], was induced as the disjunction [∨ BF__→2103]. In cases where RuleNet learns the original rule templates, it can be expected to generalize perfectly to any pattern that can be generated by those templates, as was shown in Table 1.

The degree to which generalization can be enhanced is clearly illustrated in a comparison of the performance of a standard three layer back-propagation network with the performance of RuleNet using the JCAP learning technique, as summarized in Table 2. In each of the four simulations, the back-prop network had the same input and output representation as RuleNet, with 10 hidden units per rule. The learning rate used was .05, with a momentum of .9. The training sets were the same as reported in the RuleNet simulations and the numbers for RuleNet are taken from Table 1. The tests sets represent the complete set of patterns that can be generated with the rule templates, excluding patterns that fire more than one template rule. During the back-prop simulations, outputs were processed by setting the maximum unit in each slot to 1 and all others to zero. The *cleaned up* outputs were compared to the targets to determine which were mapped correctly. Values in Table 2 represent the average over five runs with different initial weights. RuleNet's performance on the training set was significantly better than the three layer back-prop net, and as much as 2000% better on the test set.

As the results in Figure 2 and Tables 1 and 2 indicate, playing the Connectionist Scientist Game allows RuleNet to discover a set of symbolic rules that describe a set of examples, and to do so more rapidly, reliably, and with greater power to generalize than the other methods examined.

| | Valid Rules Learned and Percent of Training Set Covered Perfectly | | | |
|---|---|---|---|---|
| Template rules | epoch 1000 | epoch 1500 | epoch 2000 | epoch 2500 |
| [ E___ → 1230] | [∧B_H_ →1230] | [∨BD__ →3210] | [∨BD__ →3210] | [∨BD__ →3210] |
| [∧DA__ → 1032] | [∧_FC →1302] | [∨ _FC →1302] | [∨ _FC →1302] | [∨ _FC →1302] |
| [∨_FC → 1302] | | [∧ CC__ →1230] | [∨ CC__ →12F3] | [∨ CC__ →12F3] |
| [∨BD__ → 3210] | | | [∧ EA__→ 1230] | [ E__ → 1230] |
| [∨CC__ → 12F3] | | | | [∧ DA__→ 1032] |
| coverage of training set | 3% | 40% | 64% | 100% |

**Figure 3**
Evolution of RuleNet's hypothesis set of rules over 2500 training epochs while playing the Connectionist Scientist Game.

| Architecture | Simulation 1 train | Simulation 1 test | Simulation 2 train | Simulation 2 test | Simulation 3 train | Simulation 3 test | Simulation 4 train | Simulation 4 test |
|---|---|---|---|---|---|---|---|---|
| | % of patterns correctly mapped | | | | | | | |
| RuleNet (JCAP) | 96 | 77 | 100 | 100 | 100 | 100 | 100 | 100 |
| 3-layer back-prop | 88 | 22 | 58 | 5 | 84 | 8 | 72 | 27 |
| # of patterns in set | 120 | 1635 | 45 | 1380 | 45 | 1380 | 75 | 1995 |

**Table 2**
Generalization performance of RuleNet compared to a standard 3-layer back-prop network with 10 hidden units per rule (80, 30, 30, and 50 hidden units for simulations 1-4, respectively).

## Conclusion

We have proposed an iterative discovery process for connectionist networks called the Connectionist Scientist Game, and have explored an architecture, RuleNet, that plays the Connectionist Scientist Game. RuleNet is able to learn a set of explicit, symbolic rules *from scratch*. These rules are useful because they accurately and concisely characterize the domain from which training examples are drawn.

Although the end product of RuleNet is a set of rules, RuleNet benefits from intermediate stages of learning in which it is allowed to construct hypotheses that do not correspond exactly to rules. That is, being a neural network, RuleNet can represent a broader range of input/output mappings than those permitted by symbolic rule-based descriptions. We conjecture that RuleNet's exploration in this subsymbolic hypothesis space gives it an additional degree of flexibility that facilitates learning, even if the end product does not require this flexibility. It is for this reason that we feel that connectionist learning techniques offer great promise in domains even where the objective is to discover symbolic representations.

Our future work will concentrate on how to expand this architecture to more challenging and complex domains. One sense in which the symbol mapping domain we have considered is too simplistic is that no type/token distinction is made. The rules we have defined are formulated directly in terms of input symbols. However, much of the power of rules lies in their ability to apply to a general symbol type, rather than just to symbol tokens. For example, rather than inducing rules that simply enumerate a collection of tokens, such as **if** (*subject* = boy) **then** (perform action *x*), **if** (*subject* = man) **then** (perform action *x*), **if** (*subject* = woman) **then** (perform action *x*), it would be more useful to induce a single rule that covers a type of condition, e.g., **if** (*subject* = human) **then** (perform action *x*). This task requires that RuleNet learn not only the conditions and actions that form a rule, but also a language in which to redescribe the input symbols.

We are currently extending RuleNet to handle this additional level of complexity. With this extension, RuleNet will learn to both classify input tokens (e.g., classifying instances such as boy, man, and woman into the category human) and perform transformations of the input that are based on the induced categories. This involves subsymbolic processing—classification—as well as symbolic processing—executing condition-action rules. A unified framework that allows both types of processing should be a significant step toward an explanation of complex cognitive behaviors.

## References


Booker, L.B., Goldberg, D.E., and Holland, J.H. 1989. Classifier Systems and Genetic Algorithms, *Artificial Intelligence* 40:235-282.

Fisher, D.H. 1987. Knowledge Acquisition via Incremental Concept Clustering. *Machine Learning* 2:139-172.

Jacobs, R., Jordan, M., Nowlan, S., Hinton, G. 1991. Adaptive Mixtures of Local Experts. *Neural Computation, 3*:79-87.

McMillan, C. & Smolensky, P. 1988. Analyzing a Connectionist Network as a System of Soft Rules. In Proceedings of the 10th Conference of the Cognitive Science Society, 62-68. Hilsdale, NJ: Laurence Erlbaum and Associates.

Miikkulainen, R. & Dyer, M. 1988. Encoding Input/output Representations in Connectionist Cognitive Systems. In Proceedings of the 1988 Connectionist Summer School.

Rumelhart, D., & McClelland, J., 1986. On Learning the Past Tense of English Verbs. In J.L. McClelland, D.E. Rumelhart, & the PDP Research Group, *Parallel Distributed Processing: Explorations in the microstructure of cognition. Vol. 2: Psychological and biological models*, 216-271. Cambridge, MA: MIT Press/Bradford Books.

Sejnowski, T. J. & Rosenberg, C. R. (1987). Parallel Networks that Learn to Pronounce English Text, *Complex Systems*, 1: 145-168.

Smolensky, P. (1988). On the Proper Treatment of Connectionism. T*he Behavioral and Brain Sciences.* 11 (1).

# Hybrid Encoding: The Addressing Problem

**Jon M. Slack**

Istituto per la Ricerca Scientifica e Tecnologica (I.R.S.T.)
38050 Povo (TN)
ITALY
e-mail: slack@irst.it

**Abstract**

Locality constraints are generally assumed to have their source at the hardware, or neuronal, level. However, the paper shows that the way symbols address constituent structure represented at the connectionist level limits their access to the encoded information. These limitations are expressed as the constructs of local and address domain and provide an explanatory basis for a wide range of cognitive constraints.

## Introduction

Cognitive Scientists have recently turned their attention to the exploration of hybrid symbolic-connectionist architectures (Hendler 1991). Fodor and Pylyshyn (1988) have claimed that the explanation of mental functioning is properly located at the symbolic level and that connectionism provides little more than an implementation paradigm for such models, making issues of hybrid architecture irrelevant. This claim is refuted by showing that the hybrid nature of cognitive architecture is itself the source of constraints which have a powerful influence on shaping mental abilities.

In outline, the structure of the argument to be presented is as follows: A number of researchers have shown that constituent structure can be encoded in terms of connectionist representations and combinatorial operators (see Hinton 1990; Pollack 1990; Touretzky 1990; Smolensky 1990). Fodor and Pylyshyn (1988) view this as a demonstration that Connectionism merely provides an implementation paradigm for the symbolic-level explanation of cognition. But even as an implementation it is necessary to specify how symbols make contact with the underlying connectionist encodings. This is a problem because, as Fodor and McLaughlin (1990) argue, such encodings are not directly accessible for symbolic addressing. Instead, it is proposed that symbolic access is only possible through the use of a particular communication protocol, one based on the general view of structural configurations as comprising an address structure and a set of constituents. This particular protocol is forced on hybrid systems by the nature of information representation at the connectionist level. The next section outlines these properties.

Generally, locality constraints are explained in terms of hardware limitations or neuronal properties. An alternative explanation can be cast in terms of a restrictive communication protocol mediating between the levels of an hybrid symbolic-connectionist architecture. The third section outlines the form of this communication protocol.

The basic thesis being proposed is that certain forms of locality imported into symbolic accounts of cognition derive from the fact that symbolic structures are encoded in a medium that imposes natural local restrictions. In the fourth section, these constraints are derived from the symbolic addressing protocol. Finally, it is suggested that such constraints might underlie a wide range of locality properties originally invoked to account for the observed features of human cognition.

## E-Space: Encoding Space

The question of the most appropriate assignment of cognitive phenomena to symbolic and connectionist levels of explanation is left as open. What is being proposed is that the mapping between the two levels provides an appropriate basis for the explanation of locality constraints, even when only minimal assumptions are made concerning the nature of the relationship between them.

Encoding structure in an hybrid symbolic-connectionist architecture involves showing how the notion of combinatorial structure can be preserved in passing from the symbolic level to an encoding medium comprising points in a vector space. Slack (1990) shows how the operations of association and superposition can be used to define a connectionist representation medium, referred to as *Encoding Space*, E-space, consisting of composed connectionist states. Formally, the structure of E-space can be defined as a *semiring*, as follows:

**Definition** E-space comprises the quintuple $(V,+,**,\mathbf{0},\partial)$
1. $(V,+,\mathbf{0})$ is a commutative monoid defining the superposition operator;
2. $(V,**,\partial)$ is a monoid defining the association operator;
3. $**$ distributes over $+$.

V is the set of points, or corners, of a vector space, and the two identity elements, $\mathbf{0}$ and $\partial$, correspond to identity vectors for the operations of superposition and association, respectively. It is important to note that the set V is closed under both operations.

We can now define a representational mapping, referred to as the *encoding homomorphism*, $f_e$, which preserves combinatorial structure in mapping the set of symbolic languages $L_X$, that can be defined over an alphabet of symbols, X, using the combinatory operations of concatenation, denoted by '.', and union, denoted by '∪', into E-space. The homomorphism is defined as follows:

**Definition** The encoding homomorphism, $f_e$, maps the semiring $(L_X, \cup, ., \emptyset, \{\varepsilon\})$ into the semiring $(V, +, **, \mathbf{0}, \partial)$, such that

$$f_e(x \cup y) = f_e(x) + f_e(y)$$

$$f_e(x.y) = f_e(x) ** f_e(y), \text{ where } x, y \in L_X.$$

Thus, through $f_e$ the set of languages that can be defined on a symbolic alphabet can be encoded in E-space, preserving their combinatorial structure.

This representational framework establishes a formalism for describing the composition of distributed representations which can encode constituent structure. It is not intended as a specific connectionist representation scheme in that it does not specify the details of particular connectionist operators. Rather, it captures the general properties of an encoding medium comprising a vector space and the operations of association and superposition.

## Addressing Encoded Structures

The encoding mapping, $f_e$, can be used to build composite E-space vectors but it isn't clear how the encoded information can be accessed. For example, consider the structured vector built using the operations of association and superposition whose composition is described by the expression given in (1)

$$[\mathbf{a}**[\mathbf{c}**w + \mathbf{d}**x] + \mathbf{b}**[\mathbf{f}**y + \mathbf{g}**z]] \quad . . (1)$$

where the bold letters denote elements of the set **V**. This symbolic expression describes the vector composition tree shown in figure 1.

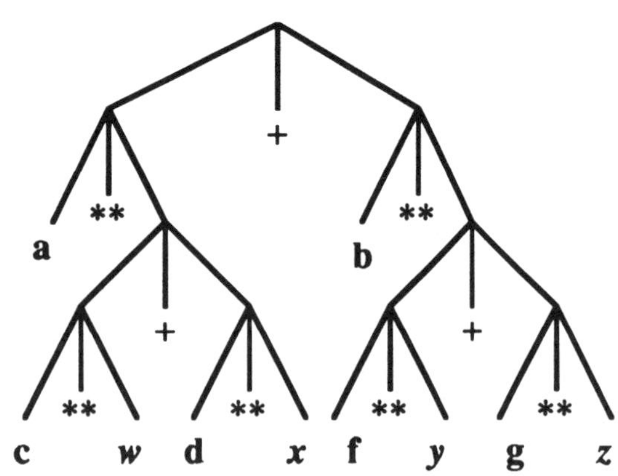


Figure 1. Vector Composition Tree

As an E-space encoding this composite structure is realised as just a point in vector space, its implicit structure is not apparent in its encoding. This problem is at the heart of Fodor and McLaughlin's (1990) argument against connectionist structure. As Fodor and McLaughlin put it,

> ...when a complex Classical symbol is tokened, its constituents are tokened. When a tensor product vector or superposition vector is tokened, its components are not .......the components of tensor product and superposition vectors can have no causal status as such.
> (Fodor & McLaughlin 1990, p. 198)

This is illustrated by the complex symbol [[John] [gave [Mary] [the ball]]] where each component is tokened as part of the composed structure's token. With composed vectors, on the other hand, the constituents are not directly accessible. The problem is to specify how the constituents of a composed E-space encoding can be accessed such that mental processes can be sensitive to their structure.

In an hybrid architecture mental processes can be defined at two levels; symbolic processes and connectionist processes. The latter class of processes might operate on compound vector encodings exhibiting a form of *systematicity* equivalent to that underlying symbolic processes and structures, as suggested by van Gelder,

> A second and more radical approach is to devise models in which structure-sensitive processes operate on the compound representations themselves without first stopping to extract the basic constituents. These processes must capitalize directly on the inherent and systematic structural similarities among nonconcatenative representations. In such models it is not only storage that takes place in the nonconcatenative domain, but the primary processing responsible for systematic cognitive behavior as well.
> (van Gelder 1990, p. 381)

However, the potential for such processes remains to be explored and the focus of this paper is on the problem of how structures encoded in E-space can be made accessible to processes defined at the symboilic level. Some view this as a trivial problem of implementation (Fodor and McLaughlin, 1990; van Gelder, 1990), but it will be shown that important cognitive constraints have their origins in this problem of access. To solve this problem it is necessary to show how symbols make contact with, or address, the nodes of vector composition trees. The problem is viewed as one of defining an address system for composed E-space encodings.

With complex symbols such as [[John] [gave [Mary] [the ball]]] the constituents are made explicit by the parentheses; each pair of parentheses denotes a non-terminal node in the symbolic derivation tree. When Fodor and Pylyshyn (1988) speak of mental processes being *sensitive* to the structure of such symbols, they mean that the processes can access the individual constituents on the basis of their structural configuration. The configuration of the symbol is given by the embedding of the parentheses, which expresses the derivation of the symbol string. The problem with E-space encodings is that the equivalent configuration is not directly accessible to symbolic-level processes. What we need to develop is a general concept of *address structure* which specifies how information encoded in a structure can be accessed, regardless of whether it is encoded symbolically or as an E-space vector.

In general, symbolic structures support two basic forms of addressing, *direct addressing* and *relative addressing*. The former involves direct access to constituents. For example, in the complex symbol the pairs of parentheses distinguish the different components and can be used to access them directly provided they are individuated through some form of indexing or labeling scheme. This is equivalent to uniquely labeling each nonterminal node of the symbol's derivation tree. In addition, the individual words label the terminal nodes of the tree. The parentheses can also be used to define a relative addressing system in terms of their depth of embedding and/or sequential order. For example, the component *Mary* can be located using the address *the first set of paretheses at the second level of embedding*. Such an addressing scheme requires (i) a pre-defined origin to function as a 'zero' address, and (ii) a set of labels for distinguishing the elements of the path connecting the origin to the addressed component. From these examples, it is clear that the address structure comprises a structural basis, the symbol's derivation tree, plus some form of addressing scheme, that is, a set of labels over which a formal language can be defined, for accessing the individual constituents. Whereas direct addressing is insensitive to the structural configuration of the symbol, relative addressing schemes utilise it to encode addresses. Thus, a direct addressing scheme merely comprises a set of labels plus a mapping taking elements from this set onto the set of constituents. A relative addressing scheme, on the other hand, requires both a set of labels and a structural configuration over which to define paths, plus a mapping which assigns the labels to elements of the configuration. Both addressing schemes are necessary for symbols to access constituent structure encoded in E-space.

## Dual Addressing

E-space encodings require a direct symbolic addressing scheme so that the encoded constituents can be processed as 'chunks'. This type of distal access (Newell 1980) which is the capacity to address a structure in some remote, abbreviated manner, is fundamental to any notion of compositionality as without it a structure could only be referenced through its full specification. In a classical complex symbol, the parentheses function to indicate the 'chunking' of the symbol into constituents. However, each chunk needs to be uniquely labeled for it to be accessed directly. This reference capability also instantiates *the principle of explicit naming* (Marr 1982). According to this principle if you want to refer to a 'thing' so that you can describe 'it' or reason about 'it', then begin by giving 'it' a label. Once an entity has a label, it can be used in repeated applications of structure building operations to derive larger entities, which themselves require labeling if they are to be processed as single entities.

Relative addressing schemes are important for expressing the address of one constituent relative to another. This referential capacity is crucial to the notion of constituent structure as it captures the mother-daughter, or dominance, relationship expressed in composition trees. 'Daughter' nodes can be addressed relative to their 'mother' nodes. This capacity is utilised in building composite structures where a constituent can only be identified relative to an existing node in the structure. In such circumstances only a relative addressing scheme is available.

Symbolic access to E-space encodings needs to support both of these referential capacities and thus, an address structure incorporating both direct and relative addressing schemes is required.

## Address Structure DAGs

Symbolic addresses access E-space encodings through the operation of *retrieval*. This operation delimits the communication protocol between the levels of the hybrid architecture. In most connectionist systems retrieval is the inverse operation of association. We can define a general retrieval operator, denoted #, as follows: Given the composite vector, $\mathbf{v_1} = [\mathbf{a}**\mathbf{b} + \mathbf{c}**\mathbf{d}]$, then

$$\mathbf{a} \# \mathbf{v_1} = \mathbf{b} \quad \ldots (2)$$

Because the association operator is noncommutative we can distinguish between its arguments, similar to the role-function distinction maintained in Smolensky's representation scheme (1990). The arguments are labeled *address* and *constituent* according to the interpretation given in (3),

**[address vector] ** [constituent vector]** . (3)

Within this protocol, **a** is interpreted as the address of the constituent vector, **b**, within the composite encoding, $\mathbf{v_1}$. Thus, when the composite vector is accessed using the address vector, **a**, it retrieves the constituent vector, **b**.

This address-constituent protocol is independent of the form of the address and thus supports both direct and relative addressing. Direct addressing is achieved by associating each constituent vector of an E-space encoding with an address vector corresponding to a symbolic label. This label can then retrieve the constituent vector through its assigned address vector. Symbolic labels that function in this way are referred to as *global labels*.

Relative addressing is the natural mode of operation of the retrieval operator in the sense that an address vector specifies the address of a constituent vector relative to a particular composite encoding. For example, **a** is the address of **b** relative to $\mathbf{v_1}$, but would not retrieve **b** given a different composite encoding. In this sense, global labels can also be regarded as defining relative addresses; they specify the direct address of constituents relative to the top-level composite encoding, that is, the vector that encodes the complete structure. Symbolic labels assigned to address vectors that retrieve constituents relative to other constituent vectors are referred to as *local labels*.

Via the retrieval operator, the constituents of an E-space encoding can be accessed using both direct and relative symbolic addresses. The full set of such addresses comprises the *address structure* of the composite encoding. Because symbolic addressing is mediated through the address-constituent protocol these structures are restricted to the class of *regular languages*. This allows address structures to be represented as directed acyclic graphs (DAGs), in which the edges and nodes are assigned local labels and global labels, respectively. Figure 2 shows the address structure DAG (AS-DAG) for the compound vector given in (1) and whose E-space encoding is shown in figure 1. The local labels are denoted by the lower-case letters, a-g, and the global labels by the upper-case letters, A-C. The global labels function as direct addresses for the constituent vectors within the composite E-space encoding. The local labels, on the other hand, specify the address of a constituent relative to its superordinate ('mother') constituent.

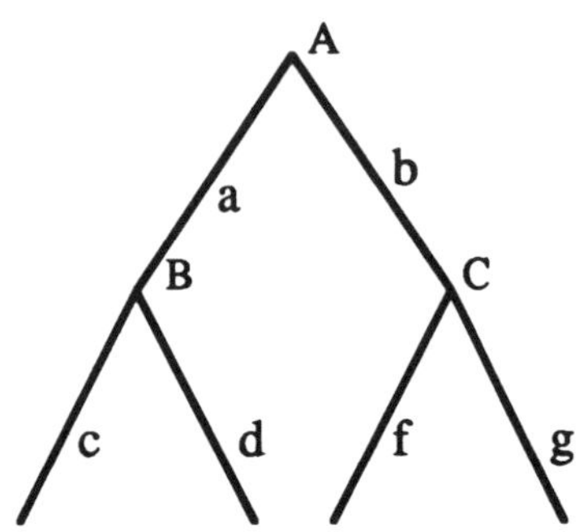


Figure 2. Address Structure Tree (AS-DAG)

In fact, the AS-DAG shown in figure 2 represents the union of two separate address structures both of which take the global label *A* as their root-node. It is important to remember that vector composition trees, such as that shown in figure 1, specify the encoding of composed vectors, but are not necessarily accessible at the symbolic level as connected configurations. An AS-DAG corresponds to a symbolic level 'view' of a composite E-space encoding. However, because they have a restricted form, limited to the class of regular languages, the full address structure for a particular E-space encoding may comprise the union of a set of AS-DAGs, each providing a view of a different part of the composite configuration.

There remains the problem of how the labels of an AS-DAG are assigned vector values under the encoding mapping. Because AS-DAGs incorporate a dual-addressing system the resulting redundancy requires the dynamic assignment of vector values to at least one of the two sets of labels. Local labels capture local structure and the assignment should therefore be both *faithful* and *consistent*, that is, elements of the set of local labels, denoted by $\mathcal{L}$, should map to different elements of $\mathcal{V}$, the set of vectors, and the mapping should be constant, at least local to a given address structure. Global labels are unique relative to a given address structure by definition and so the problem of consistency does not arise. However, the mapping should be *faithful* (elements of the set of global labels, denoted by $\mathcal{G}$, should map to different elements of $\mathcal{V}$)[1].

These constraints give a higher priority to fixing the vector values of local labels under $f_e$. In this case, the values of the global labels are determined by (i) assigning a vector value to the root-node of the AS-DAG, and (ii) propogating the vector values down the tree through association with the fixed local label vector values. Thus, an AS-DAG functions both as the component of $f_e$ that maps global labels onto vector values, and as the symbolic 'view' of the local structural relations existing between the constituents the global labels reference.

In summary, hybrid encoding captures combinatorial structure through the association of a composite E-space encoding with a corresponding address structure, the latter specifying the symbolic-level access to constituent encodings. If the address structure is well-formed, then the E-space encoding has a unique decomposition.

**An Example Address Structure**

If the idea of hybrid encoding is valid then symbols functioning as global and local labels should be implicit in Classical symbolic cognitive representations. In particular, linguistic representations should embody such functional labels as language is the prime example of a symbolic address system. To illustrate the use of these labels in the

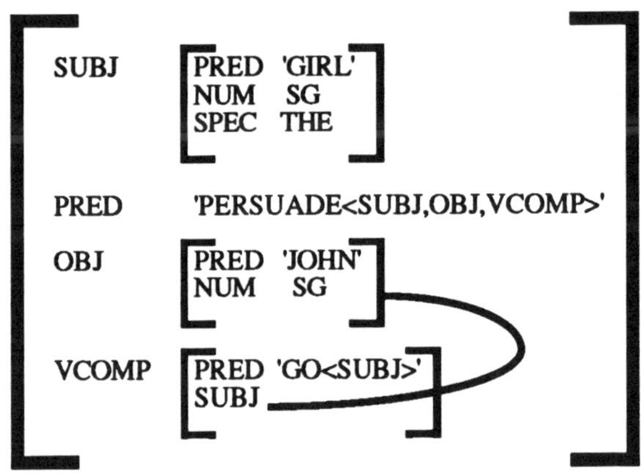


Figure 3. LFG representation of the sentence "The girl persuades John to go"

mental representation of syntactic structure, consider the

1 As defined, the sole function of global labels is to retrieve constituents directly from composite E-space encodings and as such the relationship between their vector values and the constituent vectors they access can be quite arbitrary. There is no necessity for the address labels to comprise partial descriptions of their associated constituents as implemented in some hybrid architectures (Touretzky 1990). This independence allows the vector values of global labels to be determined as a function of the AS-DAG configuration rather than being related to the constituent vectors they access.

LFG[2] representation of the sentence *The girl persuades John to go* shown in figure 3. In this structure, each pair of parentheses signifies a functional structure ($f$-structure), thereby designating a constituent of the overall structure. The grammatical functions, SUBJ, OBJ, etc., label local structure in that they specify the relationships between constituent $f$-structures. Such symbols function as local labels and in the hybrid encoding framework are assigned to AS-DAG edges. This is shown in figure 4 which depicts the AS-DAG for the E-space encoding of the $f$-structure shown in figure 3. Thus, AS-DAG edges designate local symbolic structure, that is, the structural relations holding between adjacent constituents.

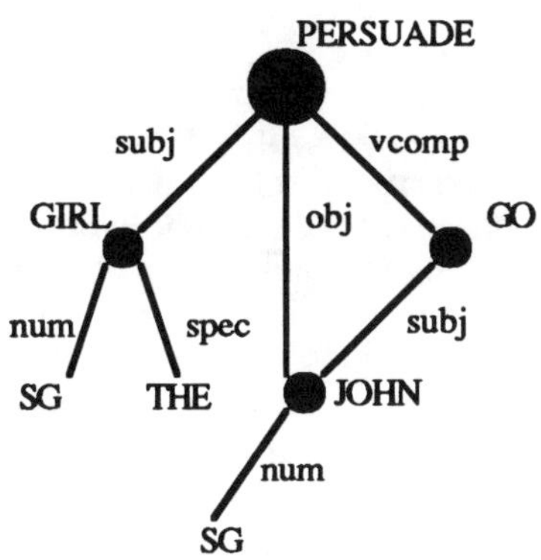


Figure 4. AS-DAG for the LFG representation in figure 3

The AS-DAG edges leading to terminal nodes represent a different form of local structure corresponding to basic featural relations, such as, *number* or *specification*, and are assigned feature labels (num, spec, etc.). The terminal nodes of the AS-DAG correspond to the values of the basic features, for instance, SG (singular) for the feature *number*, and so on. Both feature labels and function as local labels and thus, have fixed E-space encodings.

Global labels correspond to any set of symbols within a representation that function to distinguish the individual structural constituents. In figure 3, the parentheses denote the constituent $f$-structures, but they are themselves undifferentiated as labels. However, associated with each set of parentheses is a predicate label which uniquely identifies each component $f$-structure and these function as global labels, being assigned to the AS-DAG nodes in figure 4.

2 LFG, or Lexical Functional Grammar, is a syntactic formalism which encodes predicate-argument structures in terms of the grammatical functions specified in the sentence. For more details see, for example, Kaplan and Bresnan (1982).

## Locality Constraints on AS-DAGs

A number of important locality constraints emergent from the properties of hybrid encoding can be defined on AS-DAGs. These constraints provide the basis for defining the constructs of *local* and *address domain*.

As AS-DAG representations are drawn from the class of regular languages they can be generated by right-linear grammars where each production has the form,

$$A \rightarrow \omega B \quad \text{or} \quad A \rightarrow \omega$$

and where $A,B \in \mathcal{G}$, $\omega \in L_{\mathcal{F}}$, and $\mathcal{F}$ is the set of labels which have faithful and consistent encodings under $f_e$. This allows each level of an AS-DAG structure to be defined as a domain local to a particular global label.

Complex relative addresses can be defined on AS-DAGs as strings comprising a global label, functioning as the origin node, followed by a string of local labels of arbitrary length. For example, given figure 4, we can define the address PERSUADE-vcomp-subj-num to reach the terminal node with the value SG. However, this structural referential capacity is in fact restricted to *nearest neighbour addressing*. That is, AS-DAG nodes can only be addressed relative to their mother node, and relative addresses defined on AS-DAGs comprise the global label of the mother node plus at most one local label (that is, one AS-DAG edge). This is because structural AS-DAG addresses connect points in E-space and between any two such points there are an infinite number of possible paths linking them, but only the minimal path is distinguishable. Thus, unless the intermediate nodes of an E-space path are differentiated through labeling, that is, direct addressing, the path is not recoverable, only the end point is accessible.

These different constraints can be integrated into a single construct, referred to as a local domain, incorporating the following properties:

(a) A local domain is indexed by a unique **global label**;
(b) associated with each global label is a **finite local structure** comprising local labels (AS-DAG edges) which have vector values fixed relative to the value of the global label; and in addition,
(c) a local domain can dominate upto **one** subordinate local domain.

The full definition is as follows,

**Definition** A local domain is a labeled AS-DAG comprising a global label at its root node, a fixed local structure, that is, a finite set of descending edges, such that each edge is assigned a local label, with a maximum of one such label addressing a non-terminal AS-DAG node.

In fact, we can distinguish between terminal and non-terminal local domains, with only the latter having property (c). At the intuitive level, local domains can be thought of as the permanent building blocks used to compose structured E-space encodings. All the information encoded in a local domain is accessible from its global label, as the vector value of each daughter node is known relative to the root node.

### Address domain.

Address structures are built up through the composition of

local domains. This operation is schematized in figure 5 in which local domains are represented as productions. In this figure, the domain associated with the global label $A_1$ specifies the address of a local domain, X, the identity of which is 'unknown' prior to composition. In composing the next level of structure, this address is attached to the new local domain labeled $B_1$ through the process of *address unification*. In assigning the address of X to $B_1$, the relative addresses encoded in its local domain become instantiated as E-space encodings. This is because the domain contains local labels which specify E-space addresses relative to the vector value of the domain's global label. Once the global label receives a value, the relative addresses are automatically established.

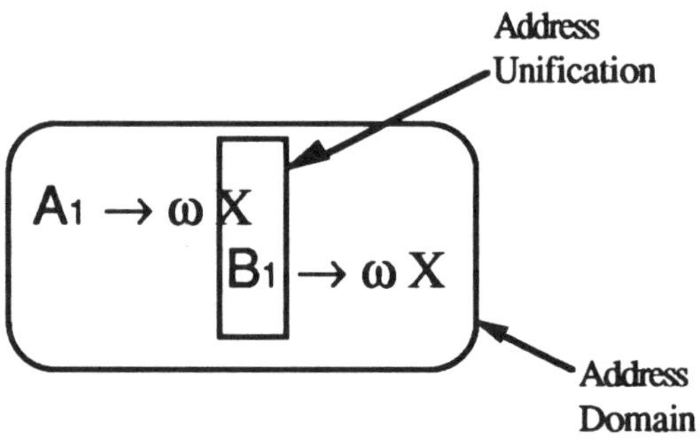


Figure 5. Address domain for A1

The operation of composition can be viewed from the opposite direction and seen as linking the contents of the constituent $B_1$ to the address specified by X. In this sense the composition operator instantiates the address-constituent protocol. Hence the protocol spans adjacent local domains. In figure 5, the address part of the protocol is specified relative to $A_1$ and the constituent part is the local domain $B_1$. Thus, the protocol through which symbolic processes access E-space encodings can be defined on AS-DAGs as a window spanning two adjacent local domains, indexed to the global label of the most dominant of the two domains. This construct of a symbolic addressing window is referred to as an *address domain* and is defined as follows:

**Definition** The address domain indexed to an AS-DAG node, A, comprises the local domain of A plus the subordinate adjacent local domain attached through composition.

To illustrate the importance of this construct, consider the representation shown in figure 3, where the OBJ *f*-structure is co-indexed with the *f*-structure of the (VCOMP SUBJ) function, as indicated by the link. In formalisms such as LFG, these indexations are expressed as control equations of the form

(vcomp subj) = (obj) . . . (C1)

specified as part of the lexical entry for the predicate (in this instance, the predicate PERSUADE). Assuming that LFG predicates have the properties of global labels, then the AS-DAG address referred to by the left-hand term of the equation is within the address domain of the predicate PERSUADE, and hence, accessible for symbolic indexing. However, such an address cannot contain more than two local labels as this would overflow the address domain that limits symbolic level access This explains why no known natural language requires control equations of the form

(vcomp vcomp subj) = (obj) . . . (C2)

where the left-hand term involves control across three levels of nested structure. In other words, the control equations attached to a predicate can only be defined within its address domain. While the LFG framework can capture such constraints it cannot explain them in any principled way. The present framework, however, provides a natural basis for such a constraint.

## Conclusions

Elsewhere the construct of a local domain has been used to explain unbounded dependency phenomena (Slack 1990). In the longer version of this paper (Slack 1991) a wide range of locality constraints inherent in natural language phenomena are accounted for in terms of the functions of local and global labels. However, the full range of applicability of these ideas remains to be explored.

## References


Fodor, J.A., and McLaughlin. B. P. 1990. Connectionism and the problem of systematicity : Why Smolensky's solution doesn't work. *Cognition* **35**: 183-204.

Fodor, J.A., and Pylyshyn. Z.W. 1988. Connectionism and cognitive architecture: A critical analysis. *Cognition* **28**: 3-71.

Hendler, J. 1991. Special issue of *Connectionist Science* on hybrid symbolic-connectionist architectures. Forthcoming.

Hinton, G.E. 1990. Mapping part-whole hierarchies into connectionist networks. *Artificial Intelligence* **46**: 47-77.

Kaplan, R.M. and Bresnan, J. 1982. Lexical-Functional Grammar: A formal system for grammatical representation. In J. Bresnan (ed.), *The Mental Representation of Grammatical Relations*. Cambridge: MIT Press.

Marr, D. 1982. *Vision*. San Fransisco:Freeman.

Pollack, J.B. 1990. Recursive distributed representations. *Artificial Intelligence* **46**: 77-107.

Slack, J.M. 1990. Unbounded Dependency: Tying Strings to Rings. In *Proceedings of COLING-90*, Helsinki, Finland.

Slack, J.M. 1991. Hybrid Encoding as a Source of Cognitive Constraints. Forthcoming.

Smolensky, P. 1990. Tensor product variable binding and the representation of symbolic structures in connectionist systems. *Artificial Intelligence* **46**: 159-217.

Touretzky, D.S. 1990. BoltzCONS: Dynamic symbol structures in a connectionist network. *Artificial Intelligence* **46**: 5-47.

# Connectionist Models of Rule-Based Reasoning

**Ron Sun**
Brandeis University
Computer Science Department
Waltham, MA 02254
rsun@cs.brandeis.edu

## Abstract

We investigate connectionist models of rule-based reasoning, and show that while such models usually carry out reasoning in exactly the same way as symbolic systems, they have more to offer in terms of commonsense reasoning. A connectionist architecture for commonsense reasoning, CONSYDERR, is proposed to account for commonsense reasoning patterns and to remedy the brittleness problem in traditional rule-based systems. A dual representational scheme is devised, utilizing both localist and distributed representations and exploring the synergy resulting from the interaction between the two. CONSYDERR is therefore capable of accounting for many difficult patterns in commonsense reasoning. This work shows that connectionist models of reasoning are not just "implementations" of their symbolic counterparts, but better computational models of commonsense reasoning.

## Introduction

Rule-based reasoning is the most prominent paradigm of symbolic AI. Whether connectionist models can be a viable alternative to symbolic AI depends, to some extent, on their ability to account for rule-following behaviors and rule-based reasoning.

To account for rules in connectionist models, various approaches have been explored, and several different systems have been implemented that can carry out rule-based reasoning rather completely. However, most of them are straight "implementations", without any fundamental difference from symbolic systems in terms of logical capabilities or reasoning capacities.

Can connectionist models do more and better in terms of accounting for robust, flexible, and multi-faceted human commonsense reasoning, especially in the types of reasoning that are most often dealt with by rule-based paradigms? Can those data (such as in Collins & Michalski 1990) that are difficult to account for by symbolic rule-based systems be explained by connectionist rule-based reasoning models? We try to answer these questions by building a connectionist reasoning system to account for common patterns in commonsense reasoning. This work shows that connectionist models of reasoning are not just an "implementation" of their symbolic counterparts; rather, they are better computational models of commonsense reasoning, taking into consideration the approximate, evidential and flexible nature of rule-based reasoning, and similarity-based reasoning (or a limited form of analogy), and also accounting for the spontaneity and parallelism in reasoning processes.

Below we will first look into various existing connectionist models of rule-based reasoning. Then we will identify some problems that are difficult for them to solve. We will move on to develop a new connectionist architecture that can better deal with these problems. In this paper, we will concentrate more on motivational issues than on technical details.

## Background Reviews

Let us look into some previous work in rule-based reasoning, especially connectionist ones. Rule-based systems have a long history in AI and cognitive science. The early successful work such as those described in Buchanan & Shortliffe (1984) demonstrated the promise of this overall approach, which adopts a simple representation with modular units called rules composed of a small set of conditions and conclusions. Many elaborate cognitive theories of learning, problem solving and memory, etc. were built based on this paradigm (cf. Klahr et al 1987) [1].

Because of this success, when connectionism came along, one of the main challenges for connectionism was how to implement rule-based reasoning in a network fashion. Touretzky & Hinton (1985) is the first work that tackles this problem. They basically emulate the structure of a symbolic rule-based (production) system, with separate modules for working memory, rules, and facts. An elaborate pull-out network is designed to match working memory data against rules

[1] Nevertheless the paradigm has long been plagued by the *brittleness* problem for large scale systems, as will be discussed.

and to decide which matching rule is to fire. Competition and winner-take-all are used for the matching purpose. The result is the equivalent of a simple sequential symbolic rule-based system.

Barnden (1988) represents another early attack on this problem. In his system, data reside in grid-like networks (called Configuration Matrices), coded with the help of adjacency relations and highlighting techniques. Hardwired rules are used to detect the presence of data that match particular rules, and an "Action Part" module can be used to add a new data structure representing the conclusion from the matched rule. Although there is some parallelism, it is mostly a sequential rule-based system, carrying out symbolic processing.

From these examples, it is quite clear now that connectionist models are capable of implementing rule-based reasoning in a variety of ways [2]. Now the question is: Can connectionist models account better for *commonsense* reasoning? The evidential, robust, flexible, and multifaceted nature of commonsense reasoning is evident from various studies (such as in Collins & Michalski 1990) yet they are all absent from these above models. What is really needed for a connectionist model of rule-based reasoning to be able to model commonsense reasoning adequately? We confront the above questions, by analyzing real protocol data and then actually building a new kind of connectionist models as a computational mechanism for commonsense reasoning.

## Common Reasoning Patterns

Allan Collins collected a number of protocols of commonsense reasoning. He indicated the inadequacy of traditional logic in explaining those reasoning patterns, and argued for the use of different formalisms or frameworks in the study of *common reasoning patterns* found in various commonsense reasoning tasks. Collins & Michalski (1990) have done an impressive job in terms of analyzing the data and establishing a unifying framework for explaining them. What is needed is a computational mechanism from which various inference patterns contained in the data can emerge into existence. We believe that the mechanism ought to be analytically simple, structurally unified, and mechanistically sound. Connectionist models in general fit these above descriptions very well, so they might provide such a mechanism.

Let's look at some examples from Collins & Michalski (1990). One protocol is as follows:

Q: Do you think they might grow rice in Florida?

R: Yeah. I guess they could, if there were an adequate fresh water supply, certainly a nice, big, warm, flat area.

[2] Also, Dolan & Smolensky (1988), Sun (1989) (see also Sun & Waltz 1991), Ajjanagadde & Shastri (1989), and Lange & Dyer (1989) are works basically belonging to this same category.

In this example, the person answering the question deduced an uncertain conclusion based on partial information, with a piece of crucial information (fresh water) missing. This example also indicates the need for an additive procedure for accumulating evidence.

Another case is as follows:

Q: Is the Chaco the cattle country?

R: It is like western Texas, so in some sense I guess it's cattle country.

Here because there is no known knowledge (or no applicable rules), an uncertain conclusion is drawn based on similarity with known knowledge (rules).

Yet another case is

Q: Are there roses in England?

R: There are a lot of flowers in England. So I guess there are roses.

Here the deduction is based on property inheritance (flower LOCATION England; rose IS-A flower; so rose LOCATION England), and the conclusion is partially certain and can be drawn only when there is no information to the contrary (i.e. no cancellation).

Existing connectionist models, or any computational models for that matter, so far cannot deal very well with these above patterns in a single unified model.

## The Brittleness Problem

Those above examples are actually instances of a general problem that has been plaguing symbolic AI for long, namely the brittleness problem.

The brittleness problem can be defined, for the purpose of this research, as the inability of a system to deal with the following aspects of reasoning in a systematic way within a unified framework: (1) partial information (for example, the first protocol), (2) uncertain or fuzzy information, (3) no matching rules (for example, the second protocol), (4) rule interactions, i.e. lack of consistency and completeness in a fragmented rule base, (5) generalization, (6) bottom-up inheritance, (7) top-down inheritance (for example, the third protocol), and (8) learning new rules.

Detailed analysis of these aspects (with lots of examples) shows that, while they look like a disparate set of problems, they can all be characterized as reasoning with rules supplemented with (feature) similarity-related inferences. We define a measure of **conceptual similarity** (cf. Tversky 1977) as

$$(A \sim B) = \frac{|F_A \cap F_B|}{|F_A|} \in [-1, 1]$$

such that [3]

$$if\ ACT_A = a,\ then\ ACT_B = a * (A \sim B)$$

where $F_i$ is the feature representation of node "i", and $ACT_i$ is the activation of node "i". We define a measure of **knowledge links** (i.e. rules) as

$$(A \longrightarrow B) = r \in [-1, 1]$$

[3] We assume there is nothing else affecting $ACT_B$

such that

$$if\ ACT_A = a,\ then\ ACT_B = a * (A \longrightarrow B)$$

where $ACT_i$ is the activation of node "i", and r is the knowledge link (rule) strength [4] between A and B. Each of these above cases can be analyzed and dealt with utilizing these two concepts, for example, the no-matching-rule situation can be described as:

$$A \sim B$$

$$B \longrightarrow C$$

and A is activated ($ACT_A \neq 0$). So we have

$$ACT_C = (B \longrightarrow C) * ACT_B$$

$$= (B \longrightarrow C) * ACT_A * (A \sim B)$$

Other cases can be described similarly, except learning new rules, which is a separate issue (see Sun & Waltz 1991).

These two mechanisms are embedded in our new architecture: CONSYDERR [5], so each of these aspects of brittleness can be handled by our system.

## A Sketch of the Model

The CONSYDERR architecture consists of two levels: CL and CD. CL is a connectionist network with localist representation, or roughly reasoning at the conceptual level (cf. Smolensky 1988). Rules are represented in CL as links between two nodes representing the condition and the conclusion respectively. The scheme proposed, FEL or Fuzzy Evidential Logic, can handle a superset of Horn clause logic and Shoham's modal logic (or Causal Theories, cf. Shoham 1990), so that it can fully accommodate traditional rule-based reasoning and capture commonsense causal knowledge. Moreover, it is capable of approximate and cumulative evidential reasoning and works with partial and uncertain information. Unlike Horn clause logic, it can deal with negative as well as positive evidence. It can handle variable bindings by utilizing the DN/PDN formalism as in Sun (1989) and Sun (1990). The basic operation of this scheme is simply weighted-sum computations, therefore this scheme can be implemented, with ease, in a connectionist network with weighted-sum node activation functions. Because of the limited space and the need to emphasize the main points in this short presentation, we will not discuss in detail the above points regarding rule representations (see Sun 1991 for details).

CD is a connectionist network with distributed representation, roughly corresponding to reasoning at the subconceptual level. Concepts and rules are diffusely represented by sets of units overlapping each other. The amount of overlapping of two sets of units representing two different concepts is proportional to the degree of similarity between these two concepts. I call this a *similarity-based representation*, in which units can be features, perceptual primitives, internal goals or affect states. Concepts are "defined" in terms of their similarity to other concepts in these primitive representations. I will utilize these primitives only as a substratum for similarity-based representation of higher level concepts.

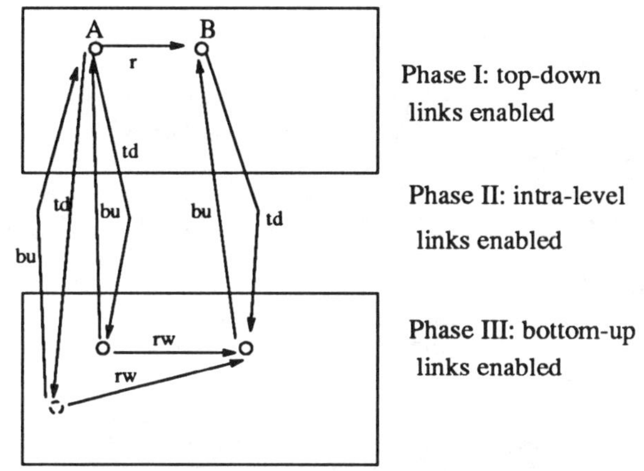


Figure 1: A two-level architecture

Now we can link the localist network (CL) with this distributed network (CD), by linking each node in CL representing one concept to all the nodes in CD representing the same concept, and assign them appropriate weights (see Fig.1). Those cross-component links are moderated by a latch mechanism. The rule links in CL are duplicated (diffusely) in CD. The interactions of the two components are in fixed cycles: first the latch opens to allow the activation of CL nodes to flow into corresponding CD nodes, and then the two parts start settling down on their own simultaneously, and finally the latch opens to allow the activation of nodes in CD to flow back into CL to be combined with the activation of corresponding CL nodes.

From the above description it is clear that the system is a combination of rule-based and similarity-based components, interwoven together. It implements naturally the functions defined above for knowledge links and conceptual similarity. The synergy of the two types of representation and reasoning helps to deal with the brittleness problem listed above and, therefore, to account for the aforementioned common reasoning patterns.

For example, in order to solve the no-matching-rule situation, we can explore the similarity between the current situation and the rule conditions as represented in the CD part of the system. Consider the following case:

Cars are for traveling on ground. Airplanes are for

[4]When there are multiple conditions in a rule, this measure becomes a vector, and the multiplication used here is generalized to inner-products.

[5]It stands for a CONnectionist SYstem with Dual representation for Evidential Robust Reasoning

traveling in air. Are buses for traveling on ground or in the air?

In the context of deciding modes of transportation, based on the similarity of relevant features [6] as represented with the amount of overlapping in corresponding sets of units in CD, buses are closer to cars than to airplanes. So "traveling on ground" is activated more strongly. One concludes that buses are for traveling on ground.

Another example concerns the problem of inconsistent/incoherent rule bases (rule interaction), we can utilize the rule interactions in CD:

If carrying cargo, buy utility vehicles. If carrying passengers, buy passenger vehicles. If carrying both cargo and passengers, what shall one buy?

Different types of vehicles are represented as features in CD. When the above two rules are both activated (in response to the question), all features corresponding to both utility and passenger vehicles will be activated in CD. All this information will go up to CL, and the things corresponding to the intersection of utility and passenger vehicles will be activated strongly (because they have all the features). So something like "van" will win.

Other aspects of the brittleness problem can be solved in a similar fashion, including the common patterns identified by Collins & Michalski (1990) [7]. This solution is quite different from Collins & Michalski. My contention is that this model is conceptually simpler and computationally more efficient (by combining and eliminating many parameters).

Formal mathematical analyses were performed for each of these aspects, and we came up with a set of requirements and constraints for each aspect regarding the parameters of a system that can deal with that particular aspect. After analyzing how these requirements and constraints imposed by each of these aspects interact with one another, a synthesis is achieved, so that a unified system is formed with a unique set of parameter settings satisfying all requirements. Based on that, a large-scale system consisting of about two hundred nodes was built to test, in a realistic setting, how these fragments combine. The system utilizes geographical knowledge extracted from encyclopedias and performs commonsense reasoning based on that knowledge (see Sun 1991).

## A Detailed Example

### The Problem

Look at the "Chaco" example (Collins & Michalski 1990).

Q: Is the Chaco the cattle country?

R: It is like Western Texas, so in some sense I guess it's cattle country.

We can put it another way to straighten out the reasoning:

Western Texas is cattle country.
Chaco is similar to western Texas (in some relevant aspects).
So Chaco is cattle country.

### An analysis

In this example, because there is no known knowledge (or no applicable rules), an uncertain conclusion is drawn based on similarity with known knowledge (rules). Using the formalism we developed, it can be described as:

$$Chaco \sim WesternTexas$$

$$WesternTexas \longrightarrow cattlecountry$$

Given "Chaco" with $ACT_{Chaco} = 1$, "cattlecountry" is concluded with $ACT_{cattlecountry}$ calculated as follows:

$$ACT_{cattlecountry} = (WesternTexas \longrightarrow cattlecountry)$$
$$*ACT_{Chaco} * (Chaco \sim WesternTexas)$$

where the similarity measure is chosen to facilitate later implementations in the CONSYDERR architecture:

$$Chaco \sim WesternTexas = p*\frac{|F_{Chaco} \cap F_{WesternTexas}|}{|F_{Chaco}|}$$

$p \in [0,1]$ is a parameter used for adjusting the system's behavior, from absolute rigidity to free-floating thinking (see Sun 1991).

These equations can be readily translated into the CONSYDERR architecture: Links between nodes in both CL and CD represent rule strength measures (the link weights are defined to be the corresponding rule strengths), and similarity measures are implemented with CD representations (we use a set of nodes to represent all features in CD and the amount of overlapping between representations of two concepts expresses the conceptual similarity of these two concepts). See Fig.2.

### The Working of the System

After starting to receive input data, the CONSYDERR system operates in fixed cycles: (1) Top-down phase, (2) Settling phase, and (3) Bottom-up phase. This cycle can be repeated to continuously track inputs.

In Top-down phase, the computation is as follows:

$$x_i(t+1) = \max_a ACT_a(t)$$

where a is any node in CL that has $x_i \in CD_a$.

In Settling phase, the computation is as follows:

$$\Delta ACT_a = \alpha \sum W_i I_i(t) - \beta ACT_a(t)$$

[6] Such as Having-wings, Having-tails, Wheels-on-both-sides, Aerodynamic-shapes, Landing-gears, etc.

[7] We are certainly not implying that we solved the brittleness problem completely. Rather, we are aiming for a simple and elegant model that can deal with some important and predominant aspects of the problem very effectively and efficiently.

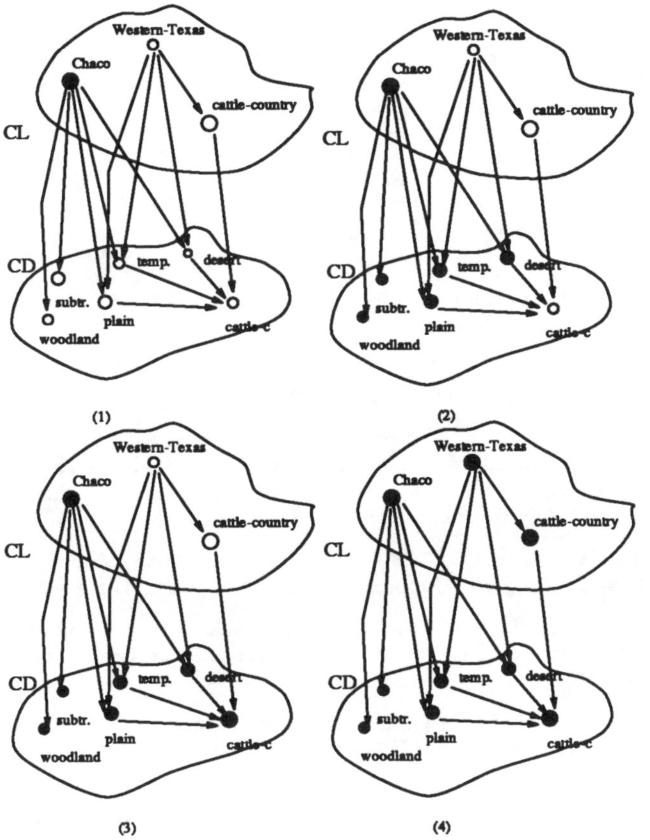


Figure 2: The Reasoning Process for the Chaco Protocol: (1) Receiving inputs, (2) Top-down, (3) Settling (rule application), and (4) Bottom-up.
(To save space, unrelated nodes are not shown here).

and

$$\Delta x_i = \mu \sum w_i i_i(t) - \nu x_i(t)$$

where $W_i, w_i$ are rule strength (weight) measures, and $I_i, i_i$ are the activations of related concepts or features (premises or logical predecessors).

In Bottom-up phase, the computation is as follows:

$$ACT_b(t+1) = \max(ACT_b(t), \sum_{x_i \in CD_b} \frac{x_i(t)}{|CD_b|})$$

where b is any node in CL.

Applying this cycle to the example: Top-down phase will activate the CD representation of "Chaco" and activate partially the CD representation of "WesternTexas" based on their similarity; then in Settling phase, rules (links) take effect and this amounts to applying in CD the rule: *WesternTexas is cattle-country*, so the CD representation of "cattle country" is partially activated; finally in Bottom-up phase, the partially activated CD representation of "cattle country" will percolate up to activate the "cattle country" node in CL. The result can be read off from CL. See Figure 2.

Comparing it with other systems: CONSYDERR utilizes parallelism inherent in the data to the maximum extent, especially when compared with Touretzky & Hinton (1985) or Dolan & Smolensky (1989). While most other connectionist rule-based systems (Lange & Dyer 1989, Ajjanagadde & Shastri 1989, etc.) are functionally comparable to the CL part of CONSYDERR, the CD part is unique in that it provides an efficient way for similarity matching to supplement rule-based reasoning; the CL/CD dual representation scheme constitutes a principled way of accounting for the dichotomy of conceptual level and subconceptual (intuitive) level reasoning (Smolensky 1988, Sun 1991). More recently Barnden & Srinivas (1990) utilize connectionist rule-based systems to explore similarity in reasoning (i.e. connectionist case-based reasoning); while the idea is very similar to ours (see Sun 1991 for details), their system requires a complex retrieval/matching process.

## The Initial Setup

The question of how we can gather data and set up a large system can be divided into two questions: how do we obtain rule weights and how do we obtain similarity measures?

In our large-scale system (Sun 1991), rules are obtained by going through geography sourcebooks, picking out the relevant information and integrating it into the network with the CFRDN procedure (see Sun 1991). The rules being put into the system include *WesternTexas is cattlecountry*, etc. [8]

Similarity measures are obtained by an indirect means: we first obtain all the relevant features needed for representing the concepts involved, and then naturally the amount of feature overlap is the similarity between concepts involved. In order to come up with detailed feature representations for concepts, we preestablish a set of feature nodes, and we then go through sourcebooks, establishing links (cross-level links) between a concept in CL and its features in CD, based on what we read in the sourcebooks. The features include: altitude, rainfall, vegetation, population, temperature, terrain, etc. with various ranges. [9]

One important issue is how the system focuses on relevant features and ignore or discount somehow the irrelevant ones, given the context (or the query, in the above-mentioned cases). This is done by the attention focusing module external to the system, in which a set of "context rules" are used to pick out all relevant features and suppress others when a certain context

[8] In general, weights representing rules can be obtained by reading textbooks, instructions, or by using learning algorithms through interactions with the environment. There is no universally applicable way to do this, or in other words, it is domain-specific.

[9] Another possible way of obtaining similarity measures is to conduct a test, asking a group of subjects to rate the similarity of concepts concerned and then construct CD representations based on the collected test scores with the STSIS procedure (see Sun 1991).

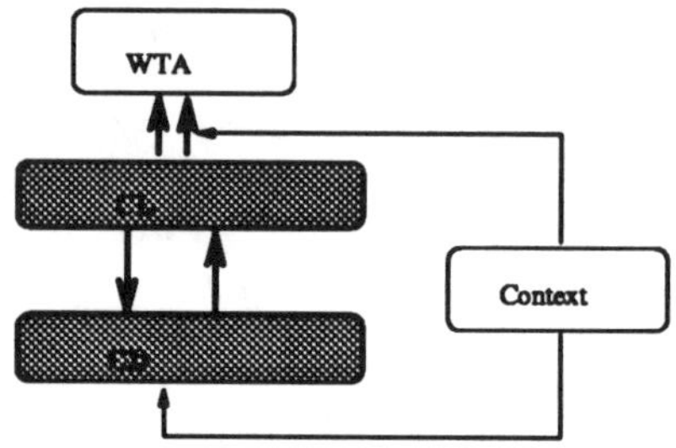


Figure 3: The Overall Architecture with Feature and Result Selections

is given (by activating the node representing the particular context, in ways described by Sun 1989). See Figure 3 for a sketch of these mechanisms. Because of the fact that these mechanisms are domain specific, they are not part of the CONSYDERR architecture, but add-on mechanisms (cf. Sun 1991).

## Summary

We analyzed connectionist models for rule-based commonsense reasoning. A connectionist architecture is proposed to account for some common patterns found in commonsense reasoning and to remedy to a certain extent the brittleness problem found in typical symbolic systems. Different from other existing connectionist systems, a dual representational scheme is devised, which has extensional objects (localist representation) as well as intensional objects (distributed representation with features). By using feature-based distributed representation in addition to the localist representation, we are able to explore the synergy resulting from the interaction between these two types of representations and between rule-based reasoning and similarity-based reasoning. This synergy helps to deal with problems such as partial information, no exact matching, property inheritance, rule interaction. and therefore the CONSYDERR system is capable of accounting for many difficult reasoning patterns in one unified system. This architecture also demonstrates that connectionist models equipped with symbolic capabilities are powerful tools for modeling reasoning capacities as well as for constructing efficient practical systems (by utilizing massive parallelism), and they are not mere implementations of their symbolic counterparts.

**Acknowledgments** I wish to thank Dave Waltz, James Pustejovsky, and Tim Hickey for many helpful discussions, and Don Smith for commenting on the paper.

## References

V. Ajjanagadde and L. Shastri, 1989. Efficient Inference with Multi-place Predicates and Variables in a Connectionist System. In Proc.11th Cognitive Science Society Conference, 396-403. Hillsdale, NJ: Erlbaum.

J. Barnden, 1988. The right of free association: relative-position encoding for connectionist data structures. In Proc.10th Conference of Cognitive Science Society, 503-509. Hillsdale, NJ: Erlbaum.

J. Barnden and K. Srinivas, 1990. Overcoming Rule-Based Rigidity and Connectionist Limitations Through Massively Parallel Case-based Reasoning, Technical Report MCCS-90-187, Computing Research Laboratory, New Mexico State University, Las Cruces, NM.

B. Buchanan and E. Shortliffe, 1984. Rule-Based Expert Systems. Reading, MA: Addison-Weslley.

A. Collins and R. Michalski, 1990. The logic of plausible reasoning, Cognitive Science, 13(1):1-49.

C. Dolan and P. Smolensky, 1988. Implementing a connectionist production system using tensor products. In D. Touretzky et al. eds. Proc.1988 Connectionist Summer School, 265-272. San Mateo, CA: Morgan Kaufman, 1989

J. Klahr et al, eds. 1989. Production System Models of Learning and Development. Cambridge, MA: MIT Press.

T. Lange and M. Dyer, 1989. Frame selection in a connectionist model. In Proc.11th Cognitive Science Conference, 706-713. Hillsdale, NJ: Erlbaum.

L. Shastri, 1988. A Connectionist Approach to Knowledge Representation and Limited Inference, Cognitive Science, 12:331-392.

Y. Shoham, 1990. Non-monotonic Reasoning and Causation, Cognitive Science, 14:213-252.

P. Smolensky, 1988. On the proper treatment of connectionism, Behavioral and Brain Sciences, 11:1-43.

R. Sun, 1989. A Discrete Neural Network Model for Conceptual Representation and Reasoning. In Proc.11th Cognitive Science Society Conference, 916-923. Hillsdale, NJ: Erlbaum.

R. Sun, 1990a. Rules and Connectionism. In Proc.INNC-Paris, p.545. Netherlands: Kluwer.

R. Sun, 1990. The Discrete Neuronal Models. In Proc.INNC-Paris, 902-907. Netherlands: Kluwer.

R. Sun and D. Waltz, 1991. A Neurally Inspired Massively Parallel Model of Rule-Based Reasoning. In B. Soucek ed. Neural and Intelligent System Integration. New York, NY: John Wiley & Sons.

R. Sun, 1991. Integrating Rules and Connectionism for Robust Reasoning: An Architecture with Dual Representation. Ph.D Thesis, Computer Science Department, Brandeis University, Waltham, MA.

D. Touretzky and G. Hinton, 1985. Symbol Among Neurons. In Proc.9th IJCAI, 238-243. San Mateo, CA: Morgan Kaufman.

A. Tversky, 1977. Features of Similarity, Psychological Review, 84(4):327-352.

D. Waltz and J. Pollack, 1985. Massive parallel parsing, Cognitive Science, 9:51-74.

# Incremental learning, or
# The importance of starting small

**Jeffrey L. Elman**

Departments of Cognitive Science and Linguistics
University of California, San Diego
elman@crl.ucsd.edu

## Abstract

Most work in learnability theory assumes that both the environment (the data to be learned) and the learning mechanism are static. In the case of children, however, this is an unrealistic assumption. First-language learning occurs, for example, at precisely that point in time when children undergo significant developmental changes. In this paper I describe the results of simulations in which network models are unable to learn a complex grammar when both the network and the input remain unchanging. However, when *either* the input is presented incrementally, *or*—more realistically—the network begins with limited memory that gradually increases, the network is able to learn the grammar. Seen in this light, the early limitations in a learner may play both a positive and critical role, and make it possible to master a body of knowledge which could not be learned in the mature system.

## INTRODUCTION

One of the things which makes language learning such an interesting phenomenon is what has been called the 'projection problem'. The idea is just that, if the problem of the language learner is to figure out the underlying regularities—that is, the grammar—which are responsible for the language he or she hears, then the data which are available to the learner may not be sufficient to uniquely determine the correct grammar.[1]

This problem of the apparent insufficiency of the data has been discussed in many contexts (e.g., Baker, 1979; Bowerman, 1987; Pinker, 1989; Wexler & Culicover; 1980) but one of the simplest demonstrations comes from Gold's (1967) work. Gold shows that if a language learner is presented with positive-only data (what he calls 'text presentation'), only regular languages can be learned. Regular languages are languages which can be generated by finite state automata. The rub is that, on the one hand, natural languages appear to belong to a more powerful class than this (Chomsky, 1957); and on the other, there is no good evidence that children receive or use negative data during learning (Brown & Hanlon, 1970; Hirsh-Pasek, Treiman, and Schneiderman, 1984; Braine, 1971).

Gold advances several suggestions in order to account for the fact that, despite his findings, children *do* learn language. Although children do not appear to receive explicit negative evidence, they may receive indirect negative evidence. Or possibly, some of what children know is innate; thus they need not infer the grammar solely on the basis of positive data.[2]

Almost certainly both of the possibilities outlined by Gold are true to some extent. That is, the child is not an unconstrained learning mechanism in the sense of being able to learn any and all possible languages. Rather, innate predispositions narrow the range of what can be learned. Of course, it is very much an open (and controversial) question exactly what form that innate knowledge takes. A number of investigators have also proposed that although *direct* negative evidence may not be available, there are subtler forms of negative evidence. For example, the non-occurrence of an expected form constitutes an indirect sort of negative evidence. Just how far this sort of evidence can be used has been challenged (Pinker, 1989). Thus, although innateness and indirect evidence plausibly participate in the solution of the learnability problem, their contribution is not known and remains controversial.

In this paper, I want to pursue what may be a third factor in helping account for the apparent ability of learners to 'go beyond the data'. This factor hinges on the simple fact that first language learners (children) are themselves undergoing significant developmental changes during precisely the time that they learn language. Indeed, language learning after these developmental changes have completed seems to be far less successful. This is often attributed to the passing of a 'critical period' for language learning. But this is no more than a restatement of facts. What I would like to consider here is the question of what it is about the so-called critical period

[1] I say 'may' because much hinges on exactly what one believes the nature of the input to be: bare sentence strings? strings accompanied by semantic interpretations? strings accompanied by information about the environment in which they were uttered?

[2] Gold mentions a third possibility, which is that if the text is ordered, then positive-only presentation is sufficient to learn even the most complex set of languages he considers. The details of this proposal are not well-developed however.

that might facilitate learning language.

Interestingly, with the notable exception of work by Newport (1988, 1990) almost all learnability work ignores this basic fact. It is typically assumed that both the learning device and training input are static. One might wonder what the consequences are of having either the learning device (network or child) or the input data not be constant during learning? Recent results from the connectionist literature (Allen, 1990; Cottrell & Tsung, 1989; Plunkett & Marchman, 1990) suggest that incremental training strategies may play an important role in the successful mastery of a domain. We might also ask what the consequences are when the learning mechanism itself changing.

In this paper, I will report the effect of staged input on learning in a connectionist model. The network fails to learn the task when the entire data set is presented all at once, but succeeds when the data are presented incrementally. I then show how similar effects can be obtained by the more realistic assumption that the the input is held constant, but the learning mechanism itself undergoes developmental changes. Finally, I examine the network to see what the mechanism is which allows this to happen and suggest what conditions are necessary for incremental learning to be useful.

## Simulations

This work was originally motivated by an interest in studying ways in which connectionist networks might use distributed representations to encode complex, hierarchically organized information. By this I mean just the sort of relationships which typically occur in language. For example, in the sentence *The girls who the teacher has picked for the play which will be produced next month practice every afternoon*, there are several events which are described. Some are backgrounded or subordinate to the main event. This has grammatical consequences. Thus, the main verb (*practice*) is in the plural because it agrees with *the girls* (not *the teacher*, nor *the play*). And although *picked* is a transitive verb which often takes a direct object following it, no noun appears after the verb because the direct object (*the girls*) has already been mentioned.

These sorts of facts (specifically, the recursive nature of embedded relative clauses) led many linguists to conclude that natural language could not be modeled by a finite state grammar (Chomsky, 1957), and that statistical inference as a learning mechanism for language was untenable (Miller & Chomsky, 1963). These conclusions about the representational and learnability requirements of natural language seem to pose real problems for connectionist networks, which typically rely heavily (though not necessarily exclusively) on statistical inference, and which are more similar to finite state machines than other sorts of computational devices (e.g., pushdown automata; but see Pollack, 1990).

My first approach was to try construct a semi-realistic artificial language which had some of the crucial properties that were cited by Chomsky and his colleagues as being problematic for FSA's and statistical learning, and to train a neural network to process sentences from this language. The network was a simple recurrent network (Elman, 1990, in press; Jordan, 1986; Servan-Schreiber, Cleeremans, & McClelland, 1988). The salient property of this architecture is that it is a kind of dynamical system which allows inputs to be processed in sequence, and in which the internal states are fed back at every time step to provide an additional input. The network must learn to develop internal states (i.e., the hidden unit activation patterns) which encode temporal information in ways which enable the network to produce the correct outputs. The network architecture that was used is shown in Figure 1.

The input corpus consisted of sentences which were generated by a grammar that had certain critical properties: (a) there was number agreement between subject nouns and their verbs; (b) verbs differes with regard to argument expectations (some required direct objects, others optionally permitted objects; others precluded direct objects); (c) sentences could contain multiple embeddings in the form of relative clauses (in which the head could be either the subject or object of the subordinate cluase). The existence of these relative clauses considerably complicated the set of agreement and verb argument facts. (See Elman, in press, for details of this language.)

The results of the first trials were quite disappointing. The network failed to master the task, even for the training data. Performance was not uniformly bad. Indeed, in some sentences, the network would correctly coordinate the number of the main clause subject, mentioned early in a sentence, with the number of the main clause verb, mentioned after many embedded relative clauses. But it would then fail to get the agreement correct on some of the relative clause subjects and verbs, even when these were close together. (For example, it might produce *The boys who the girl chase see the dog*, getting the number agreement of *boys* and *see* right, but failing on the more proximal—and presumably, easier—*girl chase.*) But even this pattern was idiosyncratic.

This result, of course, is exactly what might have been predicted by Chomsky, Miller, and Gold.

### Incremental input

In an attempt to understand where the breakdown was occurring, and just how complex a language the network might be able to learn, I devised a regimen in which the

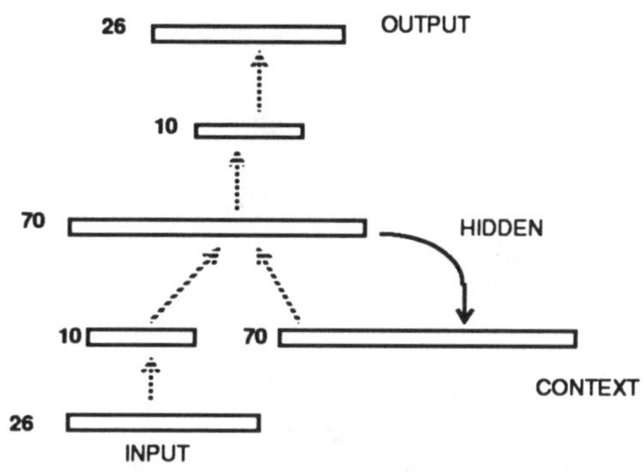


Figure 1

training input was organized into corpora of increasing complexity, and the network was trained first with the simplest input. There were five phases in all. In the first phase, 10,000 sentences consisting solely of simple sentences were presented. The network was trained on five exposures ('epochs') to this database. At the conclusion of this phase, the training data were discarded and the network exposed to a new set of sentences. In this second phase, 7,500 of the sentences were simple, and 2,500 complex sentences were also included. As before, the network was trained for 5 epochs, after which performance was also quite high, even on the complex sentences. In phase three, the mixture was 5,000 simple/5,000 complex sentences, for 5 epochs. In phase four, the mixture was 2,500 simple/7,500 complex. And in phase five, the network was trained on 10,000 complex sentences.

Since the prediction task—given this grammar—is non-deterministic, the best measure of performance is not the extent to which the literal prediction is correct (measured thus, 0 error would require the network to memorize the training data) but rather the degree to which the network's predictions approximate the empirical probability distributions. Performance using this metric was high at the conclusion of all phases of training, including the final phase: final performance had an error of 0.177, with network output measured against the empirically derived likelihood estimates. (Alternatively, one can measure the cosine of the angle between these two vectors. Mean cosine at the end of traing was 0.852; perfect performance would have been 1.00.) Furthermore, the network's high performance generalized to a variety of novel sentences which systematically test the capacity to predict grammatically correct forms across a range of different structures.

This result contrasts strikingly with the earlier failure of the network to learn when the full corpus was presented at the outset.[3] Put simply, the network was unable to learn the complex grammar when trained from the outset with the full 'adult' language. However, when the training data were selected such that simple sentences were presented first, the network succeeded in not only mastering these, but then going on to master the complex sentences as well.

In one sense, this is a pleasing result, because the behavior of the network partially reesembles that of children. Children do not begin by mastering the adult language in all its complexity. Rather, they begin with the simplest of structures, and build incrementally until they achieve the adult language.

There is an important disanalogy, however. In this simulation, the network was placed in an environment which was carefully constructed so that it only encountered the simple sentences at the beginning. As learning and performance progressed, the environment was gradually enriched by the inclusion of more and more complex sentences. This is not a good model for the situation in which children learn language. Although there is evidence that adults modify their language to some extent when interacting with children, it is not clear that these modifications affect the grammatical structure of their speech. Unlike the network, children hear exemplars of all aspects of the adult language from the beginning.

If it is not true that the child's environment changes radically (as in this first simulation), what is true is that the *child* changes during the period he or she is learning language. A more realistic network model would have a constant learning environment, but some aspect of the network itself would undergo change during learning.

### Incremental memory

One developmental change which is plausibly relevant to learning is the gradual increase in memory and attention span which is characteristic of children. In the network, the analog of memory is supplied by the access the network has (via the recurrent connections) to its own prior internal states. The network can be given a more limited memory by depriving it of access, periodically, to this feedback. The network would thus have only a limited temporal window within which patterns could be processed.

3. Both this result and the earlier failure were replicated several times with different starting conditions, a variety of different architectures, and various settings of the learning parameters (learning rate, momentum, bounds on beginning random weight initialization).

A second simulation was therefore carried out with the goal of seeing what the effect would be, not of staging the input, but of beginning with a limited memory and gradually increasing memory span. The rationale was that this scenario more closely resembled the conditions under which children learn language.

In this simulation, the network was trained from the outset with the full adult language (i.e., the target corpus that had previously been shown to be unlearnable when it was presented from the beginning). However, the network itself was modified such that during the first phase, the recurrent feedback was eliminated after every third or fourth word (randomly).[4] In the second phase, the network continued with another set of sentences drawn from the the adult language (the first set was discarded simply so the network would not be able to memorize it); more importantly, the memory window was increased to 4-5 words. In the third phase, the memory window was increased to 5-6 words; in the fourth phase, to 6-7 words; and in the fifth phase, the feedback was not interfered with at all.

Under these conditions, it turned out that the first phase had to be extended to much longer than in the previous simulation in order to achieve a comparable level of performance (12 epochs rather than 5; for purposes of comparison, performance was measured only on the simple sentences even though the network was trained on complex sentences as well). However, once this initially prolonged stage of learning was over, learning proceeded as quickly through the remaining stages (5 epochs per stage). At the end, performance on both the training data, and also on a wide range of novel data, was as good as in the prior simulation. If the learning mechanism itself was allowed to undergo 'maturational changes' (in this case, increasing its memory capacity) during learning, then outcome was just as good as if the environment itself had been staged.

Before discussing some of the implications of this finding, it is important to try to understand exactly what the basic mechanism is which results in the apparently paradoxical finding that learning can be improved under conditions of limited capacity. One would like to know, for example, whether this outcome is always to be expected, or whether this result might be obtained in only special circumstances.

We begin by looking at the way the network eventually solved the problem of representing complex sentences. The network has available to it, in the form of its hidden unit patterns, a high-dimensional space for internal representations. It is well known that in such networks these internal representations can play a key role in the solution to a problem. Among other things, the internal representations permit the network to escape the tyranny of a form-based interpretation of the world. Sometimes the *form* of an input is not a reliable indicator of how it should be treated; put another way, appearances can deceive. In such cases, the network uses its hidden units to construct a *functionally-based* representational scheme. Thus, the similarity structure of the internal representations can be more reliable indicator of 'meaning' than the similarity structure of the bare inputs.

In this simulation, the network utilized the various dimensions of the internal state to represent a number of different factors which were relevant to the task. These include: individual lexical item; grammatical category (noun, verb, relative pronoun, etc.); number (singular vs. plural); grammatical role (subject vs. object); level of embedding (main clause, subordinate, etc.); and verb argument type (transitive, intransitive, optional). Principle component analysis (Gonzalez & Wintz, 1977) can be used to identify the specific dimensions associated with each factor. The internal representations of specific sentences can then be visualized as movements through this state space (one looks at selected dimensions or planes, chosen to illustrate the factor of interest).

One can also visualize the representational space more globally by having the network process a large number of sentences, and recording the positions in state space for each word; and then displaying the overall positions. This is done in Figure 2a. Three dimensions (out of the 70 total) are shown; the $x$ and $y$ coordinates together encode depth of embedding and the $z$ coordinate encodes number. (See Elman, in press, for details.)

At the outset of learning, of course, none of these dimensions have been assigned to these functions. If one passes the same sentences through a network prior to training, the internal representations have no discernible structure. These internal representations are the important outcome of learning; they are also the necessary basis for good performance.

The state-space graph shown in Figure2a was produced under conditions of incremental training, which, we have seen, was crucial for successful learning. What does the state-space look like under conditions of failure, such as when we train a fully-mature network on the adult corpus from the beginning? Figure 2b shows such a plot.

Unlike Figure 2a, Figure 2b reveals a less clearly organized use of the state space. There is far greater variability, and words have noisier internal representations. We do not see the kind of sharp distinctions which are associated with the encoding of number, verb argument type, and embedding as we do when the network has succeeded in mastering the language. Why might this

[4] This was done by setting the context units to values of 0.5.

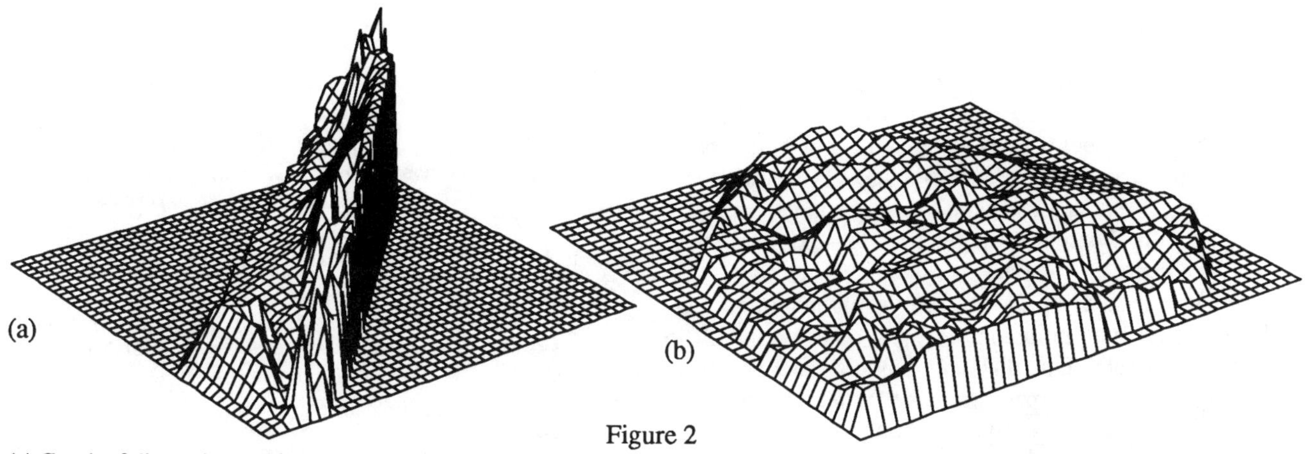


Figure 2
(a) Graph of dimensions which encode embedding (x,y) and number (z) from a network which succeded in learning the grammar. (b) The same graph from a network which failed in the learning task.

be?

When the network is confronted from the beginning with the entire adult corpus the problem is this. There are actually a relatively small number of sources of variance (number, grammatical category, verb-argument type, and level of embedding). However, these sources of variance interact in complex ways. Some of the interactions involve fairly long-distance dependencies. For example, in the sentence *The girls who the dogs that I chased down the block frightened, ran away*, the evidence that the verb *frightened* is transitive is a bit obscure, because the direct object (*the girls*) not only does not occur after the verb (the normal position for a direct object in simple English sentences), but occurs 10 words earlier; and there are several other nouns and verbs in between. The simple recurrent network does not have perfect memory. All things being equal, information decays exponentially. What happens is that the network finds a solution to the task which works enough of the time to yield reasonable performance. However, the solution is imperfect and results in a set of internal representations which do not reflect the true underlying sources of variance. The outcome is, as already pointed out, consistent with the claims of Chomsky (1957) Miller & Chomsky (1963), and Gold (1967).

When learning proceeds in an incremental fashion—either because the environment has been altered or because the network itself is initially handicapped—the result is that the network only sees a subset of the data. When the input is staged, the data are just the simple sentences. When the network is given a limited temporal window, the data are the full adult language but the *effective* data are only those sentences, and portions of sentences, which fall within the window. These are the simple sentences. (Now we see why the initial phase of learning takes a bit longer in this condition; the network also has to wade through a great deal of input which it is essentially noise.)

This subset of data, the simple sentences, contain three of the four sources of variance (grammatical category, number, and verb argument type) and there are no long-distance dependencies. As a result, the network is able to develop internal representations which encode these sources of variance. When learning advances (either because of new input, or because improvements in the network's memory capacity give it a larger temporal window), all additional changes are constrained by this early commitment to the basic grammatical factors.

The effect of early learning, thus, is to constrain the solution space to a much smaller region. The solution space is initially very large, and contains many false solutions (in network parlance, local minima). Whether or not it is really the case that the data truly underdetermine the solution, it does seem to be true that the chances of stumbling on the correct solution are small. However, by selectively focusing on the simpler set of facts, the network appears to learn the basic distinctions—noun/verb/relative pronoun, singular/plural, etc.—which form the necessary basis for learning the more difficult set of facts which arise with complex sentences.

Seen in this light, the early limitations on memory capacity assume a more positive character. It is natural to believe that the more powerful a network, the greater its ability to learn a complex domain. However, this appears not always to be the case. If the domain is of sufficient complexity, and if there are abundant "false solutions", then the opportunities for failure are great. What is required is some way to artificially constrain the solution space to just that region which contains the true solution. The initial memory limitations fill this role; they act as a filter on the input, and focus learning on just that subset of facts which lay the foundation for future success. This conjecture is in fact just what Newport (1988, 1990) has previously suggested, under what she has termed the 'less is more' hypothesis.

Higher primates are distinguished by their extended period of development, and by their relatively diminished capacities at early stages of development. It has been suggested that this is an evolutionary development which represents a compromise between large brain size and the narrow pelvic girdle required for upright posture. This is a plausible hypothesis; however, there may be additional benefits to a prolonged period of maturation. Early developmental limitations may play an essential role in learning complex domains, and may ultimately be what enable the higher primates to achieve their characteristically high level of cognitive function.

Thus, the so-called critical period may be critical not by virtue of special capacities which are present during childhood and magically and lamentably lost at puberty. Rather, the critical period may be special because the later abilities which are found in adults have *not* yet fully developed. The simpler view of the world which this affords makes learning tractable.

Finally, it is worth pointing out that the advantage which accrues to incremental learning does not arise in all circumstances. Consider the extreme case where what is to be learned is a completely random collection of facts. In such circumstances, undersampling the data (which is what incremental learning involves) runs the risk of establishing faulty generalizations. Incrememental learning can only be useful if (a) the environment contains structure; and (b) the material learned early embodies the major generalizations in a simpler form. In practice, the world is not a random place, and the sorts of things children have to learn about typically contain a great deal of structure. In the specific case of language, it may also be that the filtering which occurs as a result of limited memory picks out just that evidence which is required for successful language learning, and which allow the child to indeed go beyond the data.

## ACKNOWLEDGEMENTS

This research was supported by contract DAAB-07-87-C-H027 from Army Avionics, Ft. Monmouth. I m am grateful to Dick Aslin, Liz Bates, Mary Hare, Mike Maratsos, and Virginia Marchman for their comments on an earlier draft of this paper.

## REFERENCES

Allen, R.B. 1990. Connectionist language users. *Connection Science*, 2:279-311.

Baker, C.L. 1979. Syntactic theory and the projection problem. *Linguistic Inquiry*, 10:533-581.

Bowerman, M. 1987. The 'no negative evidence' problem: How do children avoid constructing an overly general grammar? In J.A. Hawkins (Ed.), *Explaining language universals.* Oxford: Basil Blackwell.

Braine, M.D.S. 1971. On two types of models of the internalization of grammars. In D.I. Slobin (Ed.), *The ontogenesis of grammar: A theoretical perspective.* New York: Academic Press.

Brown, R., & Hanlon, C. 1970. Derivational complexity and order of acquisition in child speech. In J.R. Hayes (Ed.), *Cognition and the development of language.* New York: Wiley.

Chomsky, N. 1957. *Syntactic structures.* The Hague: Mouton.

Cottrell, G., & Tsung, F-S. 1989 Learning simple arithmetic procedures. In *Proceedings of the Eleventh Annual Cognitive Science Society Conference,* Ann Arbor, MI, 58-65.

Elman, J.L. 1990. Finding structure in time. *Cognitive Science,* 14:179-211.

Elman, J.L. In press.. Distributed representations, simple recurrent networks, and grammatical structure. *Machine Learning.*

Gold, E.M. 1967. Language identification in the limit. *Information and Control*, 16:447-474.

Gonzalez, R.C., & Wintz, P. 1977. *Digital image processing.* Reading, MA: Addison-Wesley.

Hirsh-Pasek, K., Treiman, R., & Schneiderman, M. 1984.Brown and Hanlon revisited: Mothers' sensitivity to ungrammatical forms. *Journal of Child Language,* 11: 81-88.

Jordan, M. I. 1986. Serial order: A parallel distributed processing approach. Institute for Cognitive Science Report 8604. University of California, San Diego.

Miller, G.A., & Chomsky, N. 1963. Finitary models of language users. In R.D. Luce, R.R. Bush, & E. Galanger (Eds.), *Handbook of Mathematical Psychology, Volume II.* New York: John Wiley.

Newport, E.L. 1988. Constraints on learning and their role in language acquisition: Studies of the acquisition of American Sign Language. *Language Sciences,* 10:147-172.

Newport, E.L. 1990. Maturational constraints on language learning. *Cognitive Science,* 14:11-28.

Pinker, S. 1989. *Learnability and cognition.* Cambridge, MA: MIT Press.

Plunkett, K., & Marchman, V. 1990. From rote learning to system buildling. Center for Research in Language, TR 9020. University of California, San Diego.

Pollack, J.B. 1990. Language acquisition via strange automata. *Proceedings of the Twelfth Annual Conference of the Cognitive Science Society.* Hillsdale, NJ: Erlbaum.

Rumelhart, D.E., Hinton, G.E., & Williams, R.J. 1986. Learning internal representations by error propagation. In D.E. Rumelhart & J.L. McClelland (Eds.), *Parallel distributed processing: Explorations in the microstructure of cognition (Vol. 1).* Cambridge, MA: MIT Press.

Servan-Schreiber, D., Cleeremans, A., & McClelland, J.L. 1986. Encoding sequential structure in simple recurrent networks. CMU Technical Report CMU-CS-88-183. Computer Science Department, Carnegie-Mellon University.

Wexler, K., & Cullicover, P. 1980 *Formal principles of language acquisition.* Cambridge, MA: MIT Press.

# An On-Line Model of Human Sentence Interpretation*

**Daniel Jurafsky**
571 Evans Hall
Computer Science Division
University of California at Berkeley
Berkeley, California 94720
jurafsky@magnolia.Berkeley.EDU

### Abstract

This paper presents a model of the human sentence interpretation process, concentrating on modeling psycholinguistic data through the use of rich semantic and grammatical knowledge and expectations. The interpreter is an on-line model, in that it reads words left-to-right, maintaining a partial interpretation of the sentence at all times. It is strongly interactionist in using both bottom-up evidence and top-down suggestions to access a set of constructions to be used in building candidate interpretations. It uses a coherence-based selection mechanism to choose among these candidate interpretations, and allows temporary limited parallelism to handle local ambiguities. The interpreter is a unified one, with respect to both representation and process. A single kind of knowledge structure, the *grammatical construction*, is used to represent lexical, syntactic and semantic knowledge, and a single processing module is used to access and integrate these structures.

## Criteria for a Theory

Although the sentence interpretation process has received a great amount of attention in the cognitive science community, earlier models address very limited subparts of the problem of interpreting an utterance, and do not extend well to other significant parts of the task. As a guideline for designing and judging any model of sentence interpretation, this section presents three criteria of adequacy. The first criterion of **Functional Adequacy** constrains the nature of the interpretation.

> **Functional Adequacy:** *An interpreter must produce a representation which is rich and complete enough to function as an interpretation of the sentence in a larger model of language understanding.*

The functional adequacy criterion is the most important one for an interpreter which is intended to model human processing. It is the necessity of meeting this criterion which distinguishes an interpreter, which must meet semantic and functional constraints on its representation, from a parser, which need not.

The second criterion for an interpreter is that of **Representational Adequacy:**

> **Representational Adequacy:** *An interpreter must include a declarative and linguistically motivated representation of linguistic knowledge.*

This criterion insures that the representational basis of the processing model meets independent linguistic criteria for linguistic knowledge, particularly the need to capture relevant linguistic generalizations and account for the creativity of the language faculty. This knowledge must include more than just phonological or syntactic information. An interpreter must bring to bear a large and rich collection of semantic, pragmatic, and world knowledge in order to be complete enough to meet the constraint of Functional Adequacy.

The final criterion concerns psychological validity:

> **Cognitive Adequacy:** *An interpreter must meet standards of psycholinguistic and general cognitive validity.*

The principle of Cognitive Adequacy requires that the theory account in a principled manner for psycholinguistic results. Such results include the on-line nature of the language interpretation process, the nature and time course of lexical access, on gap filling, the use of thematic roles, on inference, anaphora, and work on grammatical interpretation phenomena such as attachment preferences and garden path effects.

As mentioned above, most models have focused on limited parts of sentence interpretation process, and do not meet all the adequacy criteria. Many processing models which emphasize Representational Adequacy, particularly those associated with linguistic theories, such as Ford *et al.* (1982) (LFG), Marcus (1980) (EST) or Pritchett (1988) (GB), include no semantic knowledge and do not meet the criterion of Functional Adequacy. Alternatively, some models such as Riesbeck & Schank (1978) have emphasized semantic knowledge but ignored syntactic knowledge. In general, much of the cognitive modeling community has emphasized either syntactic parsing or lexical access. Very few cognitive models of interpretation have been proposed, although these models (such as Hirst (1986) and Kurtzman (1985)) have proved extremely important.

## Architectural Principles

The first architectural principle, the **On-Line Principle**, follows directly from the criterion of cognitive adequacy:

> **On-Line Principle:** *Maintain a continually-updated partial interpretation of the sentence at all times in the processing.*

There is a great amount of psycholinguistic evidence for the on-line nature of natural language processing, including Swinney (1979), Tanenhaus *et al.* (1979), Tyler

*Thanks to Peter Norvig, Nigel Ward, and Robert Wilensky. Sponsored by the Defense Advanced Research Projects Agency (DoD), monitored by Space and Naval Warfare Systems Command under N00039-88-C-0292, and by the Office of Naval Research under contract N00014-89-J-3205.

& Marslen-Wilson (1982), Marslen-Wilson *et al.* (1988) and many others cited below.

The on-line principle has two implications for the interpreter. The first is that it must produce an interpretation incrementally, that is in a strictly left-to-right manner while the sentence is being processed. This rules out the traditional depth-first or backtracking control structure for parsers. Thus, for example, depth-first ATN's do not conform to the principle, even those like RUS (Bobrow & Webber 1980) which eliminate some backtracking through semantic constraints.

The second implication of the on-line principle is that the interpreter cannot maintain all possible interpretations of a sentence during the processing. It is required, fairly frequently, to choose a single interpretation with which to continue processing. Evidence from lexical access, (Swinney (1979) and other cited above), inference generation (Swinney & Osterhout 1990) gap filling (Kurtzman (1989) and Carlson & Tanenhaus (1987)) show that selection among candidate interpretations is done on-line. This rules out the use of parallel parsers which maintain every possible syntactic or semantic structure in parallel, such as the active chart parser of Winograd (1983), the breadth-first ATN parser (Woods 1970), or the table-driven parser of Tomita (1987).

Unfortunately it is not possible to follow the on-line principle by simply choosing an interpretation immediately whenever an ambiguity arises. This is due to the fundamental conflict in human language understanding between the need to produce an interpretation as soon as possible, and the need to produce the correct interpretation. Because evidence for the correct interpretation may be delayed, any on-line interpreter must choose a method for integrating this late evidence.

Our model uses limited local parallelism to represent these local ambiguities while waiting for further evidence. At any point, multiple possible candidate interpretations are entertained, but only for a short time, and the interpreter is forced to choose among them quickly. The use of temporary local parallelism in interpretation can be considered an extension of results which show temporary local parallelism in lexical processing (Swinney (1979), Tanenhaus *et al.* (1979), and Tyler & Marslen-Wilson (1982)). In this sense our model resembles Gibson (1990a) and (1990b) and Kurtzman (1985).

Alternative approaches for modeling local ambiguity fall into two classes. The first is the "wait-and-see" approach proposed by Marcus (1980) (used also by the shift-reduce parser of Shieber 1983). In this approach, the model waits to build structure until it can be certain it is building the correct interpretation, although the delay is strictly limited. The second class of models use global heuristics (such as Minimal Attachment) to resolve local ambiguity immediately. Example of these include Kimball (1973), Frazier & Fodor (1978), Wanner (1980), and Pritchett (1988).

The second principle, the **Expectation Principle**, calls for a knowledge-based approach to sentence processing.

**Expectation Principle:** *Make use of syntactic, semantic, and higher-level expectations to help integrate new information into the interpretation.*

The use of expectations to guide the integration of linguistic knowledge in processing is well-grounded in artificial intelligence. Language interpretation should make use of frames (Hirst 1986), thematic roles (Carlson & Tanenhaus 1987 and Stowe 1988), and other high-level semantic information (Riesbeck & Schank 1978) to build interpretations.

The final principle, the **Uniformity Principle**, makes more specific claims about the algorithm used to produce the interpretation.

**Uniformity Principle:** *A single processing module accounts for access, integration, and selection of structures at all levels of sentence processing.*

The uniformity principle proposes a single, integrated mechanism to replace the traditional informationally encapsulated lexical analyzer, syntactic tree-builder, morphological analyzer, and interpretation mechanisms. This has the advantages of parsimony and elegance.

The functions of access, integration, and selection apply uniformly across the lexical, syntactic, and semantic domains. For example, the access function accounts for the access of lexical items as well as syntactic rules (as we will see below, this is because both are represented as "grammatical constructions"). The integration function builds structures by combining component structures at each level (in building words, syntactic phrases, or semantic interpretations). The selection function resolves both lexical and higher-level ambiguities.

## Introducing the Algorithm

Any model must implement the basic functions of the interpretation process: *access*, *integration*, and *selection*. This section provides a sketch of the algorithm used to implement each function.

The *access* function suggests possible constructions to use in an interpretation, using a number of knowledge sources. These include bottom-up information from the input or from previously accessed constructions, and top-down information and semantic expectations. Constructions can also be annotated with access clues. The *integration* function combines information from various constructions to build complete interpretations. The function is currently implemented using an extension of unification. The *selection* function chooses the best interpretation from the disjunction of candidates being considered at any time. Selection takes place quite soon after a set of parallel candidates is proposed, using a simple coherence metric to choose among interpretations.

In the remainder of the paper, I will further describe these three theories and the representation of linguistic knowledge in the interpreter. A prototype interpreter has been implemented, using a small grammar of about 60 constructions and interpreting a set of questions from an earlier consultant project.

## Representation

The interpreter's linguistic knowledge consists of a collection of structures which represent information from various domains of linguistic knowledge: phonological, morphological, syntactic, semantic, and pragmatic. These uniform collections of constraints are called *grammatical constructions*, following the construction grammar of Fillmore *et al.* (1988). A construction is an abstraction over some complex pairing of meaning and form. Because both lexical items and syntactic rules are constructions, there is no distinction between the lexicon and the syntactic rule base.

The representation language and the grammar have a number of distinguishing features, two of which will be discussed here. First is the ability to define constituents of constructions semantically as well as syntactically. Second is the definition of two kinds of constructions, *strong constructions* and *weak constructions*, characterized by their positions on a grammatical abstraction hierarchy.

Semantic constraints on a constituent are part of the definition of a construction. If an instance of a construction violates either syntactic or semantic constraints on its constituents it is unacceptable. Following Stucky (1987), I call such unacceptable constructions *uninterpretable* rather than *ungrammatical*, using the term *interpretability* for this extension of the idea of grammaticality. This distinguishes constructions from Montague Grammar rule-pairs, whose semantic rules play no role in grammaticality.

As an example of a construction which requires semantic constraints, consider the **How-Scale** construction first defined in Jurafsky (1990) which occurs in examples like the following:

*(1a)* **How wide** *is the ocean?*
**How accurate** *is her prophecy?*
*(1b)* **How much** *does she love him?*
**How often** *does it rain in San Francisco?*
**How quickly** *did she finish her work?*
*(1c)* **How many** *angels can dance?*

The **How-Scale** construction has two constituents. The first constituent is the lexical item "how". The second may be an adjective, such as "wide" or "accurate" in (1a), an adverb such as "quickly" or "often" in (1b), or even a quantifier like "many". Specifying this constituent syntactically requires a very unnatural disjunction of adverbs, quantifiers, and adjectives. Furthermore, such a disjunctive category is insufficient to capture the constraints on this constituent. For example, not every adverb or adjective may serve as the second constituent in the construction. Note the uninterpretability of (2abc), which have respectively an adverb, an adjective, and a quantifier as their second constituent.

*(2a) *How abroad ... ?*
*(2b) *How infinite ...?*
*(2c) *How three ...?*

The commonality among the grammatical uses of the construction can only be expressed semantically: the semantics of the second constituent must be *scalar.* A *scale* is a semantic primitive in the representational system, and is used to define traditional scalar notions like *size* or *amount* or *weight.* Note that in (1) above all the elements which are allowable as second constituents for the **How-Scale** construction have semantic components which are scales. Terms like "wide", "strong", and "accurate" meet the traditional linguistic tests for scalar elements (such as co-occurrence with scalar adverbs like "very", "somewhat", "rather", and "slightly"). The elements in the ungrammatical examples (2) do not have any sort of scalar semantics. The second constituent of the **How-Scale** construction may be an adjective, an adverb, or a quantifier so long as it has the proper semantics.

A theory which could not use semantic information to constrain a constituent would be unable to represent the **How-Scale** construction completely. Even a theory such as HPSG (Pollard & Sag 1987) which allows semantic constraints for lexical constructions could not represent it because HPSG does not allow semantic constraints on *syntactic* constructions. Figure 1 presents a sketch of the representation of the **How-Scale** construction. The semantic representation language used is a frame-like one, where the operator 'a' creates an instance of a concept.

```
(lexicalconstr How-Scale
    [(a Identify $t
        (Unknown $x)
        (Background $s)
     SuchThat
     (a Scale $s
        (Location $z $x)))]
  ->
    ["how"]
    [(a Scale $s
        (On $z))])

  "Given two constituents, one the word "how"
   and one a scale $s, construct a question
   about the location of object $z on scale $s"
```

**Figure 1**

A second feature of the grammar is the novel use of an abstraction hierarchy to structure the grammar. The grammar includes two kinds of constructions: *strong constructions*, which express productive grammatical knowledge, and *weak constructions*, which express abstractions over strong constructions. The use of these weak constructions allows us to capture the traditional notion of *semi-productive rules* or *subregularities.* Psycholinguistic arguments for the presence of these kinds of constructions are summarized in Cutler (1983). Weak constructions provide a structure for the linguistic knowledge base, rather than acting as regular, productive constructions. This use of the abstraction hierarchy is different than other uses of abstraction in linguistic knowledge (Bobrow & Webber (1980), Jacobs (1985), Flickenger *et al.* (1985), Pollard & Sag (1987), Jurafsky (1988)) in making this epistemological distinction between terminal (strong) and non-terminal (weak) nodes in the hierarchy.

## The Access Theory

The history of models of access is an extremely rich one, but has tended to be somewhat balkanized. Psycholinguists have studied lexical access extensively, so much so that a rather broad consensus has arisen on at least the general nature of the lexical access process, while very little psycholinguistic work has been done on syntactic access. Some work on syntactic access has been done by computational linguists who have studied the computational properties of various algorithms for syntactic rule-access in parsing, but with no attempt to model human behavior.

By conflating the lexicon, the syntactic rule-base, and the semantic interpretation rules into a single linguistic knowledge base, and by using a uniform processing module, we are able to propose a single access algorithm which accounts for psycholinguistic data and meets computational criteria.

Our access algorithm generalizes the lexical access algorithm indicated by results such as Swinney (1979) and Tanenhaus *et al.* (1979) which show that when an ambiguous input is read, every sense of the ambiguous word is activated. All of these candidate senses of the word are maintained for a short period of time, after which various contextual factors can help select a single sense. Our access function similarly activates multiple candidate constructions, using a number of knowledge sources for clues. Two sources of access suggestions are the traditional bottom-up and top-down ones. Constructions are suggested if their first constituent matches the left-hand side of a recently applied rule, or if their left-hand side matches the current position of some previously suggested construction.

Following Wilensky & Arens (1980), access can also be delayed until more than one constituent of a construction has been seen. Thus constructions like "The Big Apple", which occur rarely but begin with common constituents like "the", are not accessed whenever "the" appears in the input. Instead such constructions are augmented with information specifying how much of the construction should be seen before it is accessed.

The richest source for access suggestions is semantic expectations. When a verb has a particular semantic subcategorization for one of its arguments, the access mechanism suggests a construction which builds that semantics. For example when parsing the verb "know", which semantically subcategorizes for a proposition, the mechanism suggests the Proposition construction, whose syntax builds a finite clause with a "that" complementizer. This is consistent with the experimental results of Shapiro *et al.* (1987) (also suggested by Kurtzman (1989)) that a verb's semantic arguments rather than syntactic ones are used immediately in processing. In this sense, our model is a *strongly interactionist* one, in allowing top-down and contextual information to directly affect the access of constructions. Although many researchers have not found evidence of these contextual effects, Tabossi (1988) has found such evidence by using particularly strong contexts. In general, construction access in our approach resembles the broader problem of conceptual access in memory (see also Riesbeck (1986).

The access theories of most earlier analyzers are quite straightforward. For example in a traditional bottom-up parser, the access algorithm recursively selects any rule for which a handle is found in the input. Access with top-down parsers works the same way, although indexing a rule by its left-hand side instead of by the first constituent. Some parsers (Tomita 1987) allow the use of both top-down and bottom-up access methods. In all these cases, access is very simple, since the number of syntactic categories is quite small. In the system described here, a construction's constituents may include any set of semantic relations rather than being restricted to a small, finite set of syntactic symbols. Thus these simple access methods used for parsers are insufficient. Most semantic analyzers have also simplified the access problem. For example in ELI (Riesbeck & Schank 1978) and the Word Expert Parser (Small & Rieger 1982) all constructions are lexical, and thus there are no higher-level constructions to consider accessing.

## The Integration Theory

Partial interpretations are built by combining the information from constructions with an operation called *integration*. In its simplest form, integration is an extension of the unification operation to a richer semantic domain. Integration also differs from unification in being asymmetric, because it gives privilege to constructions over proposed constituents. Any proposed constituent must meet the constraints established by a construction, but not vice versa. Integration also allows modifying the structure of the elements being integrated, such as by binding the value of one element to some open variable in another. Where unification merely binds two elements together, integration allows one to fill some semantic gap inside the other.

The remainder of this section will work through the representation and integration of the two constituents of the **Wh-Non-Subject-Question** construction (Jurafsky (1990)). This construction accounts for sentences which begin with certain wh-clauses, where these clauses do not function as the subject of the sentence. Examples include:

*(3a)* **How** *can I create disk space?*
*(3b)* **What** *did she write?*
*(3c)* **Which book** *did he buy?*

The construction has two constituents. The first, indicated in bold type in the examples above, is a wh-element. The second is an instance of the **Subject-Second-Clause** construction, which consists of an auxiliary, a subject, and a verb phrase. The representation for the construction appears in Figure 2 below:

```
(constr Wh-Non-Subject-Question
  [(a Question $q
      (Queried $var)
      (Background (Integrate $/pre $/v)))]
  ->
  [(a Identify $t
      (Unknown $var)
      (Background $pre))]
  [(a Subject-Second-Clause $v)]  )

"Given two constituents, the first a questioned
element $t and the second a subject-second-clause
$v, build a question about $t which includes as
background knowledge the information in $v."
```

**Figure 2**

Note in Figure 2 that the background knowledge for the question is formed by integrating the variables \$pre and \$v. These contain the information from the two constituents. Note also that each of these variables is preceded by a slash. A slash on a variable means that the variable may be the matrix structure in which to search for a gap. Thus to integrate an argument into a verb, the variable representing the verb is slashed, while the argument is not. In Figure 2, both variables are slashed, indicating that the semantic gap could be in the structures bound to either of these variables. The gap could be inside the **Subject-Second-Clause** or inside the **Identify** structure.

For example, in the sentence "*What did she write?*" the gap is located in the second constituent, the **Subject-Second-Clause**, because the verb "write" has an unfilled semantic slot for the object written. The integration algorithm will bind the semantics of "what" to the unfilled "written-object" slot of the verb "write". Note that gap-filling is a semantic, not a syntactic process.

For the sentence "*How can I create disk space?*", the gap is in the first constituent, the word "how". The semantics of this "how" construction in Figure 3 below are concerned with specifying the means or plan by which some goal is accomplished. The gap in this construction is the goal $g.

```
(a Identify $t
     (Unknown $p)
     (Background $x)
  Such-That
  (a Means-For $x
     (Means $p)
     (Goal $g)))

"The means $p for achieving
some goal $g is unknown."
```

**Figure 3**

The second constituent, the **Subject-Second-Clause** "can I create disk space", produces the semantics shown in Figure 4.

```
(a Ability-State $x
   (Actor (a Speech-Speaker))
   (Action
       (a Creation-Action
           (Actor (a Speech-Speaker))
           (Theme (a Disk-Freespace))))

"the ability of the speaker in the
discourse to create disk free space"
```

**Figure 4**

In order to build the correct interpretation of the sentence, the integration algorithm realizes that the goal $g in Figure 3 is a semantic gap which can be filled by the Ability-State $x in Figure 4, and it binds the Ability-State to the variable $g. The final result of this integration looks like Figure 5.

```
(a Question $q
   (Queried $p)
   (Background
     (a Means-For
       (Means $p)
       (Goal
         (a Ability-State  $x
            (-Actor (a Speech-Speaker))
            (-Action
                 (a Creation-Action
                     (Actor (a Speech-Speaker))
                     (Theme (a Disk-Freespace)]
"A question about the means for achieving the
goal of being able to create some disk space."
```

**Figure 5**

This example of integration highlights a number of linguistic features of this model. First, the gap in the sentence "*How can I create disk space*" is in the word "how" rather than in the **Subject-Second-Clause**. Other linguistic analyses require wh-phrases to fill a syntactic gap in the matrix clause, which requires them to include traces or empty categories corresponding to each possible syntactic modifier position in the **Subject-Second-Clause**. By placing the gap inside the semantics of "how", we eliminate these numerous empty categories. Because the gap is semantic rather than syntactic, there is no need for the grammar to contain constructions with syntactic gaps (such as the slash-constructions of GPSG). Using semantic gaps also allows long-distance dependencies (WH-movement, Right-Node Raising, Topicalization, etc) to be treated with the same mechanism that is used to link verbs and other predicates with their arguments.

Using semantic gaps to act as expectations for binding WH- constructions is also consistent with a number of experimental results from psycholinguistics, including Kurtzman (1989), and Carlson & Tanenhaus (1987). Although these results were originally interpreted as support for the on-line location of gaps, they are consistent with a semantic integration of the wh-element into the verb. Our model can be distinguished from these by considering syntactic gaps which are located before the verb: we predict that subject gaps, for example, should not cause processing difficulty because the interpreter integrates the wh-element directly into the following verb without proposing a gap. This is exactly the result found by Stowe (1986).

The integration operation would need more inferential power to make the metaphoric inferences of Martin (1990), or the abductive inferences of Charniak & Goldman (1988) or Hobbs *et al.* (1988). Making these inferences in the integration algorithm allows them to be made on-line, unlike these earlier inferencing mechanisms.

## The Selection Theory

The selection theory is based on assigning each candidate interpretation a confidence measure based on a simple coherence metric with semantic and constructional expectations (Ng & Mooney 1990). As soon as the measure of one of the candidates exceeds a threshold that interpretation is selected. If no candidate is ahead after a certain time threshold, the top-ranked interpretation is chosen, forcing the parallelism to be limited. Exactly when this selection takes place depends on the nature of the candidate interpretations. The current model uses a confidence threshold which is similar to the memory-capacity model of Gibson (1990b) and (1990a). Because an interpretation may be selected before all possible evidence has come in, the analyzer may choose an incorrect interpretation, discarding the correct one, and producing the well-known *garden path* phenomenon.

The selection model is still incomplete. A more complete metric would need to include syntactic distance, as well as take into account the relative frequencies of constructions.

## Conclusions

I have proposed an on-line, incremental, strongly-interactionist model of human sentence interpretation, with a unified linguistic knowledge base and a single algorithm to use this knowledge to build interpretations. The model focuses on the use of semantic knowledge in each part of the interpreter: the use of semantic information to improve construction *access*, to do more intelligent *integration* of constructions, and to *select* constructions based on their coherence with expectations. Similarly, the representation language focuses on the use of semantic constraints on constituents and a semantic account of long-distance dependencies.

## References


Bobrow, Robert J. & Bonnie Webber. 1980. Knowledge representation for syntactic/semantic processing. In *Proceedings of the First National Conference on Artificial Intelligence*, pp. 316–323. Morgan Kaufmann.

Carlson, Greg N. & Michael K. Tanenhaus. 1987. Thematic roles and language comprehension. In Wendy Wilkins, editor, *Thematic relations*. San Diego: Academic Press.

Charniak, E. & R. Goldman. 1988. A logic for semantic interpretation. In *Proceedings of the 26th ACL*.

Cutler, Anne. 1983. Lexical complexity and sentence processing. In G. B. Flores d'Arcais & R. J. Jarvella, editors, *The Process of language*, pp. 43–79. New York: Wiley.

Fillmore, Charles J., Paul Kay, & M. C. O'Connor. 1988. Regularity and idiomaticity in grammatical constructions: The case of let alone. *Language*, 64(3):501–538.

Flickenger, D., C. Pollard, & T. Wasow. 1985. Structure-sharing in lexical representation. In *Proceedings of the 23rd ACL*, pp. 262–267.

Ford, M., J. W. Bresnan, & R. M. Kaplan. 1982. A competence-based theory of syntactic closure. In Joan Bresnan, editor, *The Mental Representation of Grammatical Relations*. Cambridge: MIT Press.

Frazier, L. & J. D. Fodor. 1978. The sausage machine: A new two-stage parsing model. *Cognition*, 6:291–295.

Gibson, Edward. 1990a. Memory capacity and sentence processing. In *Proceedings of the 28th ACL*.

Gibson, Edward. 1990b. Recency preference and garden-path effects. In *Proceedings of the 12th Annual Conference of the Cognitive Science Society*, pp. 372–379.

Hirst, Graeme. 1986. *Semantic Interpretation and the Resolution of Ambiguity*. Cambridge: Cambridge Univ Press.

Hobbs, Jerry R., Mark Stickel, Paul Martin, & Douglas Edwards. 1988. Interpretation as abduction. In *Proceedings of the 26th ACL*, pp. 95–103, Buffalo, NY.

Jacobs, Paul. 1985. A knowledge-based approach to language generation. Technical Report 86/254, University of California at Berkeley Computer Science Division, Berkeley, CA.

Jurafsky, Daniel. 1988. Issues in relating syntax and semantics. In *Proceedings of 12th Coling*, pp. 278–284, Budapest.

Jurafsky, Daniel. 1990. Representing and integrating linguistic knowledge. In *Proceedings of 13th Coling*, Helsinki.

Kimball, J. 1973. Seven principles of surface structure parsing in natural language. *Cognition*, 2(1):15–47.

Kurtzman, Howard S., 1985. *Studies in Syntactic Ambiguity Resolution*. PhD thesis, MIT, Cambridge, MA.

Kurtzman, Howard S., 1989. Locating wh- traces. Draft, to appear.

Marcus, Mitchell P. 1980. *A Theory of Syntactic Recognition for Natural Language*. Cambridge: MIT Press.

Marslen-Wilson, W. D., Colin M. Brown, & L. K. Tyler. 1988. Lexical representations in spoken language comprehension. *Language and Cognitive Processes*, 3(1):1–16.

Martin, James H. 1990. *A Computational Model of Metaphor Interpretation*: Academic Press.

Ng, Hwee Tou & Raymond J. Mooney. 1990. On the role of coherence in abductive explanation. In *Proceedings of the Eighth AAAI*, pp. 337–342. Morgan Kaufmann.

Pollard, Carl & Ivan A. Sag. 1987. *Information-Based Syntax and Semantics: Volume 1: Fundamentals*. Chicago: University of Chicago Press.

Pritchett, Bradley. 1988. Garden path phenomena and the grammatical basis of language processing. *Language*, 64(3):539–576.

Riesbeck, Christopher K. 1986. From conceptual analyzer to direct memory access parsing: An overview. In *Advances in Cognitive Science 1*, pp. 236–258. Chichester: Ellis Horwood.

Riesbeck, Christopher K. & Roger C. Schank. 1978. Comprehension by computer: Expectation-based analysis of sentences in context. In W. J. M. Levelt & G. B. Flores d'Arcais, editors, *Studies in the perception of language*, pp. 247–293. London: Wiley.

Shapiro, L. P., E. Zurif, & J. Grimshaw. 1987. Sentence processing and the mental representation of verbs. *Cognition*, 27:219–246.

Shieber, Stuart M. 1983. Sentence disambiguation by a shift-reduce parsing technique. *ACL Proceedings, 21st Annual Meeting*, pp. 113–118.

Small, S. L. & C. Rieger. 1982. Parsing and comprehending with word experts. In Wendy G. Lehnert & Martin H. Ringle, editors, *Strategies for Natural Language Processing*, pp. 89–147. New Jersey: Lawrence Erlbaum.

Stowe, L. A. 1986. Evidence for on-line gap location. *Language and Cognitive Processes*, 1(3):227–245.

Stowe, L. A. 1988. Thematic structures and sentence comprehension. In G. N. Carlson & M. K. Tanenhaus, editors, *Linguistic structure in language processing*. Dordrecht, Holland: Kluwer Academic.

Stucky, Susan U. 1987. Configurational variation in english. In Geoffrey J. Huck & Almerindo E. Ojeda, editors, *Syntax and Semantics 20: Discontinuous Constituency*, pp. 377–404: Academic Press.

Swinney, D. A. 1979. Lexical access during sentence comprehension: (re)consideration of context effects. *Journal of Verbal Learning and Verbal Behavior*, 18:645–659.

Swinney, David A. & Lee Osterhout. 1990. Inference generation during auditory language comprehension. In Arthur C. Graesser & Gordon H. Bower, editors, *Inferences and Text Comprehension*: Academic Press.

Tabossi, P. 1988. Sentential context and lexical access. In G. W. Cottrell & M. K. Tanenhaus S. L. Small, editor, *Lexical ambiguity resolution*. San Mateo: Morgan Kaufmann.

Tanenhaus, Michael K., James M. Leiman, & Mark S. Seidenberg. 1979. Evidence for multiple stages in the processing of ambiguous words in syntactic contexts. *Journal of Verbal Learning and Verbal Behavior*, 18:427–440.

Tomita, Masaru. 1987. An efficient augmented context-free parsing algorithm. *Computational Linguistics*, 13(1–2):31–46.

Tyler, L. K. & W. D. Marslen-Wilson. 1982. Speech comprehension processes. In E. C. T. Walker & M. Garrett J. Mehler, editor, *Perspectives on mental representation*. Hillsdale, NJ: Erlbaum.

Wanner, Eric. 1980. The ATN and the sausage machine: which one is baloney. *Cognition*, 8:209–226.

Wilensky, Robert & Yigal Arens. 1980. Phran: A knowledge-based natural language understander. In *Proceedings of the 18th ACL*, pp. 194–204, Philadelphia.

Winograd, Terry. 1983. *Language as a Cognitive Process: Volume 1 Syntax*: Addison-Wesley.

Woods, William A. 1970. Transition network grammars for natural language analysis. *Communications of the ACM*, 13(10):591–606.

# Why Do Children Say "Me do it"?

Rick Kazman
Department of Philosophy and
Software Engineering Institute
Carnegie Mellon University
Pittsburgh, PA 15213
rnk@sei.cmu.edu

## Abstract

A common feature of early speech is that children use case marking incorrectly. Several researchers have proposed that the child's mistakes are limited to the misuse of nominative case, and are corrected once the child acquires verbal morphology. In this paper I will show that this characterization of the problem is incorrect: children misuse *all* case forms, not just nominative case. In addition, I will show that the child's use of case is related to the acquisition of *nominal* morphology, not verbal.

Case marking can be better understood as a result of the child learning the *productive agreement processes* of his language. This characterization accounts for the acquisition of case and the "waffling" which children exhibit, and does so within a unified theory of lexical and syntactic acquisition.

## Introduction

Many language researchers have noted a salient characteristic of early speech:[1] children appear to assign case roles incorrectly. They say things like: *Me go home, Her too cold*, and *I knock he over*. Several researchers have attempted to explain why children have difficulty marking case ([Radford, 1990], [Aldridge, 1989]). Their explanations center around the connection between the child's acquisition of verbal morphology and the assignment of nominative case. They claim that the child cannot assign nominative case until he has first gained knowledge of the language's verbal morphology, paralleling the link between these two phenomena proposed in Government Binding theory [Chomsky, 1981]. I will show that this characterization of child language is incorrect—children do not simply lack the ability to assign nominative case, they lack the ability to assign *all* syntactic case—and I will present a model of language acquisition which relates the child's acquisition of case to the acquisition of *nominal* morphology, and does so within a unified framework of the acquisition of the lexicon and syntax.

[1]The period of acquisition which I will be concerned with is roughly ages 18 to 36 months. During this time, most of the major components of language—word order, argument structure, agreement and case marking—are acquired.

The computational model of the acquisition of morphology and its relationship to syntax which I will be adopting has been discussed elsewhere ([Kazman, 1990], [Kazman, 1991]). In this paper, I will show that by extending the notion of *agreement* in this model to include case marking, the acquisition of case—both the time course of acquisition and the mistakes which children commonly make—is explained. In this way the child's acquisition of case is accounted for by positing a minimal change to a language acquisition system which already demonstrably accounts for much of the child's early syntax and inflectional morphology.

In addition, this proposal will account for the waffling stage through which children seem to pass on their way to the acquisition of a particular facet of the adult language. Waffling refers to the tendency in the child's speech, at this time, to use lexical items and grammatical constructions intermittently, and seemingly randomly.

Consider, for example, the following data taken from the speech of Adam, at age 2;10:[2]

Adam: Look my telephone.
You don't like me tow truck?
Mine expresso.
I making noise.
I'm swimming.

In these utterances—all of which were spoken in a period of approximately 20 minutes—Adam sometimes uses the correct possessive *my*, and at other times uses *mine* or *me* as a possessive. Similarly, he sometimes uses the auxiliary verb *be* with progressive forms of verbs, and other times does not (compare *I'm swimming* and *I making noise*).

The acquisition procedure proposed here can account for the waffling in both case marking and auxiliary usage because they are really different manifestations of the same thing—the child's incomplete understanding of the agreement processes in his language.

[2]All of the data used in this paper was taken from the CHILDES database [MacWhinney and Snow, 1985]. The data for Adam was taken from [Brown, 1973], the data for Peter was taken from [Bloom, 1973], the data for Shem came from [Clark, 1978], and the data for Nina was taken from [Suppes, 1973]. Some further examples from a number of children will be quoted from [Radford, 1990]. The age notation that I will be using throughout this paper is years;months.

## The Acquisition Model

### The Induction of the Lexicon

The acquisition model which I am proposing is based upon the assumption that the basic principles of the grammar of a language can be determined by an analysis of the morphology of the target language, and by an examination of the interface between morphology and syntax in this language. The analysis of the lexicon and the development of rules of grammar appear to be significantly different, although related, classes of problems. Accordingly, different learning strategies are applied to them: I am advocating an inductive approach to the acquisition of the lexicon and a deductive approach to the acquisition of syntax.

The lexicon is constructed inductively, through a pseudo-statistical ranking procedure which postulates hypotheses about lexical items based upon their input frequency and other properties such as their phonological salience. The main purpose of the lexical acquisition procedure is to discover the root word forms and the productive affixes of a language. Once the inductive procedure has discovered the form and meaning of the productive affixes, the *structure* of the lexicon can be said to be acquired. Certainly, there is a great deal more to lexical acquisition than this, but once the structure of the lexicon is learned, much of lexical acquisition is reduced to a relatively rote process of adding new entries which conform to previously learned paradigms. That is, once the procedure has learned *bites, walks, throws, wants, gives*, etc., learning *brings* or *takes* involves only the application of the rule that the child already knows, i.e. add *-s* to the verb root.

The lexical acquisition procedure proposes hypotheses for word roots and affixes. These hypotheses are ranked, and their ranks are continually updated based upon how well they account for the model's input. In this way, the most productive morphological rules will be highly reinforced by the input, whereas less productive rules will either be pruned or simply listed in the lexicon as irregular forms. Initially, the ranks for all affixes will be low. This corresponds to the period in which words are typically produced by the child uninflected. As the model receives more input, the ranks of the productive affixes increase relatively quickly. During this stage, waffling and overgeneralization[3] occur. Overgeneralization occurs because the child learns about the productive affixes more quickly than he learns about irregular affixes (since productive affixes are reinforced by many different words within the language) and, in this model, roots and affixes are combined exhaustively and then unproductive combinations are pruned by the ranking procedure.

Waffling occurs because the child possesses roots and affixes which have a rank which has not yet reached certainty. The highly productive inflectional affixes encode agreement information on their host category—for instance, the verbal *-ed* affix in English encodes verbal past tense. This piece of information about *-ed* is not innate however. It represents an inductive generalization by the child based upon exposure to a large number of verbs. The rank of an affix is a quantification of the child's belief that that word undergoes a *productive agreement process*. The model of language generation being assumed here assumes that all semantically valid forms of a word activated in parallel (following [Simpson, 1984] and [Carlson and Tannenhaus, 1988]), and the child chooses stochastically among them, based upon their normalized ranks. Assume, for example, that the child has a root, *swim*, and two affixes: the *-ed* affix, with a rank of 0.25, and the i→a ablaut, with a rank of 0.50. These ranks are a statistical measure of the child's degree of belief in these affixes (or agreement process) and so the child will, at this stage, produce *swimmed* 1/3 of the time and *swam* 2/3. The result of this ranking procedure, to the external observer, will be a waffling in the child's morphological productions, agreement processes and case marking.

[3]Overgeneralization refers to the tendency of children to overapply productive affixes, producing novel forms such as *foots, ated, goed* and so on. This subject is treated more fully elsewhere ([Marcus *et al.*, 1990], [Kazman, 1990])

### The Deduction of Syntax

The deductive portion of the acquisition model is logically dependent upon the results produced by the inductive portion. The knowledge-base for making deductions about the language is the lexicon. The lexicon, when properly structured, contains much detailed information about the agreement properties of the categories of a language.

Intuitively, this dependence upon the inductive lexical acquisition process can be seen as a way of buffering a deductive model of syntax from the enormous amount of variation and inconsistency in language. Once the inductive analysis has provided a clear model of the lexicon, then the deductive portion of the model may safely make inferences based upon this information.

The deductive portion of the model is structured as a rule-based system, like a production system—a system of if-then rules tied to an associated database of facts, where any rule may be activated when its preconditions are met in the database.[4] To give a concrete example, we might postulate the following rules for the deductive model of syntax:

IF a category X exhibits regular agreement
THEN hypothesize an *Agr* node dominating XP

IF a category X has an *Agr* node and there is
a function word which subcategorizes for X
THEN analyze this function word in XP's *Agr* node

These rules provide a means by which the target grammar can be gradually refined, based upon an analysis of grammatical relations revealed in the lexicon. In the above example, these rules determine if and when two categories

[4]A general introduction to production systems can be found in [Davis and King, 1984]. Production systems have been used in the past to model human cognitive processes. Newell, for instance, used it to model human performance in a memory scanning task [Newell, 1973]. Furthermore, rule-based systems form the backbone of most expert systems.

| Morpheme | Order of Acquisition | | |
|---|---|---|---|
| | Brown | *Babel* | De Villiers |
| Present Progressive | 1 | 1 | 2 |
| on | 2 | 2 | 2 |
| in | 2 | 3 | 4 |
| Plural | 4 | 4 | 2 |
| Possessive | 5 | 6 | 6 |
| Articles | 6 | 5 | 5 |
| Past Regular | 7 | 7 | 7 |
| 3rd Person Regular | 8 | 8 | 7 |

Figure 1: Brown vs. *Babel* vs. De Villiers

should be constrained to agree in the syntax. Rules like the above must be included in any serious model of grammar, since cross-linguistically different languages exhibit different agreement facts, and developmentally, children begin their linguistic careers with little or no knowledge of agreement and gradually develop this knowledge over time.

This model has been implemented as a computer program, written in C, called *Babel*. *Babel* has been shown to correctly simulate the acquisition of English and Polish—two historically unrelated languages. To be more precise, this model of lexical and syntactic acquisition makes certain predictions about the rate and order of acquisition of productive affixes and syntactic agreement phenomena. The results of simulating acquisition with this model[5] are shown in figure 1, compared with the order of acquisition of the same morphemes by children, as documented in Brown's seminal work *A First Language: The Early Stages* [Brown, 1973], and in De Villiers & De Villiers *The Acquisition of English* [de Villiers and de Villiers, 1985]. The Pearson product moment coefficient of correlation[6] was calculated for the three sets of results. The order of acquisition produced by *Babel* is correlated at a level of 0.99 with Brown's results, and 0.96 with the De Villiers' results.

In addition, the predictions which the model makes for Polish have been shown to conform to the order of acquisition presented in *The Acquisition of Polish* [Smoczyńska, 1985]. Furthermore, these results also hold for Dutch, French and Hebrew, as I have discussed elsewhere ([Kazman, 1990]), although I have not, as yet run computer simulations of the acquisition of these languages.

## Case marking

Several researchers have suggested that case marking is missing in early child language, and is acquired at the same time as function words like determiners and auxiliaries (see [Radford, 1990], [Kazman, 1988], [Aldridge, 1989], [Guilfoyle and Noonan, 1988]). Different theories have been proposed to account for this, but they largely rely on case being a property of a phrasal category, Infl, which is a projection of a verb's inflectional information. One problem with these theories is that they force acquisition to rely on an innate knowledge of putative linguistic particulars (such as the notion that nominative case is assigned by a tensed Infl), or on the maturation of some linguistic property, such as Case marking. This amounts to little more than stipulation.

For instance, Radford and Aldridge suggest that children acquire nominative case marking as a result of the creation of Infl. Nominative case is assigned, they claim, once the child has created an Infl node in his syntactic representations. In order to justify this claim, they must not only show that nominative case is not operational in the child's grammar before Infl has been acquired, but that accusative case *is* operational (since accusative case is assigned by a verb to its direct object—a construction which the children have already acquired). Radford provides evidence suggesting that this is the case. He first provides examples of children using pronouns marked for accusative case:

Paula: Paula put *them*. Throw *them* in.
Daniel: Mum, Wayne hit *me*. Want kick *him*.
Put *him* bed.
Hayley: Help *me* out. [Radford, 1990, p. 133]

He then provides examples of children using accusative pronouns in positions which would call for nominative ones:

Angharad: *Me* have biscuit.
Neil: *Him* on there.
Hayley: *Her* do that. [Radford, 1990, pp. 133-4]

On closer inspection, however, it does not appear that this analysis can be maintained. If Radford and Aldridge are correct in assuming that the child can use accusative case (because this is assigned by a verb) before being able to use nominative case (which is assigned by Infl), then they would predict that accusative case should be much more productive than nominative case in the child's spontaneous speech. Radford clarifies this point as follows: "Our central claim here is that children at the lexical stage in their categorial development have not *mastered* the conditions under which nominative case marking takes place in adult English ... However, the hypothesis that children have not *mastered* nominative case assignment should not be taken as implying that we never find any examples of (what might appear to be) nominative forms in the speech of children at this stage. ... It is certainly true to say that we find the sporadic occurrence of nominative pronouns in the transcripts of some children who otherwise appear to be at the lexical stage in their categorial development." [Radford, 1990, pp. 134-5] To paraphrase, we might find some examples of nominative case marking in the child's speech, but they should be infrequent and unproductive; simply the result of the child's use of rote-learned phrases such as *Off he goes*.

However, it appears that Radford and Aldridge's assumption can not be substantiated. The frequencies of pronoun usage for Nina and Adam are given in figure 2. These are counts of the number of utterances in which the children use a given pronoun, taken from the samples before Infl has

[5]These results were based upon an input sample of over 44,000 words of transcribed spoken adult English taken from the CHILDES database.

[6]The Pearson product moment coefficient of correlation is a scaleless measure of correlation between two variables. Its value varies from -1 (completely unrelated) to +1 (completely related).

been acquired by the children (as indicated by metrics such as the child's use of auxiliary verbs and productive verbal affixes).

| Pronoun | Adam | Pronoun | Nina |
|---|---|---|---|
| I | 719 | I | 154 |
| me | 494 | you | 92 |
| you | 452 | him | 62 |
| he | 50 | her | 49 |
| him | 46 | me | 41 |
| them | 25 | he | 39 |
| we | 14 | them | 4 |
| her | 10 | they | 4 |
| she | 8 | she | 3 |
| us | 6 | we | 3 |
| they | 3 | us | 0 |

Figure 2: Adam and Nina's Pronoun Usage

If the Radford/Aldridge hypothesis were true, we would expect to see significantly higher usage of accusative pronouns than nominative. However, there does not appear to be any tendency to use accusative pronouns more frequently. In fact, just the opposite effect appears to be the case (although only slightly): the most productive pronoun, *I*, is used significantly more often by both children than its accusative counterpart, *me*.

Radford and Aldridge also claim that children use accusative pronouns in cases where nominative ones are called for, thus indicating that nominative case marking has not been acquired. However, another logical possibility may also be considered: that children simply have *no* notion of case marking at this stage, and so they use their pronouns fairly randomly. If this hypothesis were true, what you would expect to see in the child data is a wide variety of misuses of pronouns, indicating that no case marking of any type has been acquired. So, in addition to the examples above which show children using accusative pronouns in nominative contexts, you would expect to find nominative pronouns in accusative contexts and genitive pronouns in nominative contexts. This is exactly the case, as shown in the following representative examples:

Adam: Let he walk.
See he walk like dat.
Her sing it.
I going knock they down.

Nina: My close it.
My get my car.
Hold him eyes.
Him can't see.

Shem: You heard he climbing in the doggie.
Me no have juice.
Him go meow.

As a consequence of examples such as these, the Radford/Aldridge hypothesis—that children have mastered only accusative case before developing Infl, and master nominative case only after Infl is acquired—cannot be maintained. Another hypothesis can be suggested given the theoretical framework I have described: case is first acquired by the child as a another feature of lexical items. Whereas inflectional affixes are an audible form of agreement indicating some intrinsic feature of a word (such as number or tense) case is an audible form of agreement indicating an *extrinsic* feature of a word: its function in the sentence.

In effect, this is another application of the notion of "bootstrapping", proposed by [Macnamara, 1982] and [Pinker, 1984]. The child starts out with a simplistic but natural notion of case—that it is another agreement feature of a word—and uses this to "lift himself by the bootstraps" into the more complex adult notion of case. The hypothesis that case is first acquired as a type of agreement not only accounts for the acquisition facts of case, but does so by positing no new mechanisms, and by claiming no reliance on either innate knowledge or maturation.

This proposal carries some empirical consequences with it. It suggests that: 1) we should see the same waffling in the acquisition of case that we see in the acquisition of morphology; 2) case marking should not be productive in the stage before inflectional morphology has been acquired; 3) case marking should be acquired at the same rate as inflectional morphology. These predictions appear to be substantiated. First, we do see waffling in both case marking and inflectional morphology, as pointed out earlier. Second, case marking *is* acquired at the same time as inflectional morphology. The table given in figure 3 shows Peter's usage of plurals, pronouns and genitive *'s*. The numbers of plurals, genitive *'s* and pronouns appear to be increasing over time[7] in lock step. In fact, the Pearson coefficient of correlation for plurals and pronouns is 0.87, for pronouns and *'s* is 0.83 and for plural and *'s* is 0.96.

| Age | Plural | Pronouns | 's |
|---|---|---|---|
| 1;9 | 6 | 1 | 1 |
| 1;10 | 20 | 1 | 2 |
| 1;11 | 27 | 17 | 2 |
| 2;0 | 28 | 33 | 5 |
| 2;1 | 16 | 50 | 5 |
| 2;2 | 18 | 80 | 11 |
| 2;3 | 93 | 219 | 32 |
| 2;4 | 21 | 137 | 19 |
| 2;5 | 83 | 224 | 26 |
| 2;6 | 36 | 169 | 14 |

Figure 3: Peter's Acquisition of Nominal Elements

What we can infer from this high level of correlation

[7]The numbers do not monotonically increase, in part, because of the different number of utterances collected for Peter each month. For instance, at age 2;3 3300 utterances were recorded for Peter, compared with less than 1600 for 2;4. The other part of the variation is attributable to the stochastic nature of the child's speech.

is that case marking, as evidenced in pronouns and in the genitive marker *'s*, is acquired along with agreement, as evidenced in plurals. In a theory such as the current one, where case marking and agreement are simply different manifestations of the same process, this correlation is predicted.

Finally, the third prediction—that the *rates* of acquisition of case marking and inflection are related—can also be substantiated. In a language like English, where inflection and case markings are relatively rare and typically phonologically reduced, the acquisition of inflection and case is relatively late. Most children do not achieve anything like adult mastery until about 3 years of age. In Polish, on the other hand, which has abundant, phonologically salient inflection and case marking, the acquisition of inflection and case is precocious. Smoczyńska reports: "as far as the acquisition of specific case endings is concerned, most of them are used correctly from the very moment of emergence of a given category" and "the acquisition of gender is precocious in Polish children. Most of the children have acquired this distinction before the age of 2" [Smoczyńska, 1985, pp. 618, 645].

## Conclusions

This proposal explains why children use case marking inconsistently in the early stages. Since case marking is simply another type of agreement, the child, in the stage before he has identified the productive agreement processes of his language, will have several competing hypotheses for the structure of nouns in his language, each of which is weighted by its rank, as determined by the lexical acquisition procedure. Thus, if a pronoun is associated with two possible forms, one correct and one incorrect, each of which have a rank of 0.5 then we would expect that the child will use the correct form about 50% of the time and the incorrect form about 50% of the time—this will appear, to the listener, to be waffling. As the child learns more about his language, the rank of correct forms will near 1, the child will use case and agreement with adult proficiency, and waffling will disappear. The waffling stage will be the same time that the child will be overgeneralizing morphology and use other types of agreement inconsistently. For instance, during this time the child will produce utterances such as "a shoes" (where the number of the determiner does not agree with the number of the noun), "that's feet" (where the demonstrative and copula don't agree in number with the noun) and "sit on the lap" (where the determiner and noun don't agree in definiteness).

So, why do children say *Me do it*? For the same reason that they say *He want some*: because they have not learned nominal and verbal morphology. In particular, they have not learned that a verb agrees with its subject, and they have not learned that a pronoun "agrees" with its position in the sentence. This is also the reason why children produce utterances like *I making noise*. The auxiliary verb *am* is missing from this sentence because the child is not enforcing verbal agreement in his syntax.

The unified lexical and syntactic acquisition procedures presented here not only learn the productive agreement processes of the target language, but do so in a way which is consistent with the acquisition of the lexicon, as recorded in longitudinal studies: words are first learned unanalyzed, then some words are produced incorrectly and gradually, over time, an adult-like knowledge of the language's syntax and morphology, including case marking, is learned.

## References


[Aldridge, 1989] Michelle Aldridge. *The Acquisition of Infl.* PhD thesis, University College of North Wales, Bangor, 1989.

[Bloom, 1973] Lois Bloom. *One Word at a Time.* Mouton, The Hague, 1973.

[Brown, 1973] Roger Brown. *A First Language: The Early Stages.* Harvard University Press, Cambridge, Mass., 1973.

[Carlson and Tannenhaus, 1988] Greg N. Carlson and Michael K. Tannenhaus. Thematic roles and language comprehension. In Wendy Wilkins, editor, *Syntax and Semantics, Volume 21: Thematic Relations.* Academic Press, San Diego, 1988.

[Chomsky, 1981] Noam Chomsky. *Lectures on Government and Binding.* Foris, Dordrecht, Holland, 1981.

[Clark, 1978] Eve Clark. Discovering what words can do. In D. Farkas, W. M. Jacobsen, and K. W. Todrys, editors, *Papers from the parasession on the lexicon*, pages 34–57. Chicago Linguistic Society, Chicago, 1978.

[Davis and King, 1984] Randall Davis and Jonathan J. King. The origin of rule based systems in AI. In Bruce G. Buchanan and Edward H. Shortliffe, editors, *Rule Based Expert Systems.* Addison Wesley, Reading, Mass., 1984.

[de Villiers and de Villiers, 1985] Jill G. de Villiers and Peter A. de Villiers. The Acquisition of English. In Dan I. Slobin, editor, *The Cross-linguistic Study of Language Acquisition, Volume 1: The Data*, pages 27–139. Lawrence Erlbaum, Hillsdale, New Jersey, 1985.

[Guilfoyle and Noonan, 1988] Eithne Guilfoyle and Maire Noonan. Functional categories and language acquisition. Presented to the Boston University Conference on Language Acquisition, 1988.

[Kazman, 1988] Rick Kazman. Null subjects and the acquisition of Case and Infl. Presented to the 13th Boston University Conference on Language Development, 1988.

[Kazman, 1990] Rick Kazman. The Genesis of Functional Categories. Presented to the 15th Boston University Conference on Language Development, 1990.

[Kazman, 1991] Rick Kazman. On building a model of grammar from information in the lexicon. Working Notes: AAAI Spring Symposium Series, Workshop on Machine Learning of Natural Language and Ontology, 1991.

[Macnamara, 1982] John Macnamara. *Names for Things, A Study of Human Learning.* Bradford Books/MIT Press, Cambridge, Mass., 1982.

[MacWhinney and Snow, 1985] Brian MacWhinney and Catherine Snow. The child language data exchange system. *Journal of Computational Linguistics*, 12:271–296, 1985.

[Marcus *et al.*, 1990] Gary F. Marcus, Michael Ullman, Steven Pinker, Michelle Hollander, T. John Rosen, and Fei Xu. Overregularization. Technical Report Center for Cognitive Science Occasional Paper #41, Massachusetts Institute of Technology, 1990.

[Newell, 1973] Alan Newell. Production systems: Models of control structures. In W. G. Chase, editor, *Visual Information Processing*, pages 463–526. Academic Press, New York, 1973.

[Pinker, 1984] Stephen Pinker. *Language Learnability and Language Development*. Harvard University Press, Cambridge, Mass., 1984.

[Radford, 1990] A. Radford. *Syntactic Theory and the Acquisition of English syntax*. Blackwell, Oxford, UK, 1990.

[Simpson, 1984] Greg B. Simpson. Lexical ambiguity and its role in models of word recognition. *Psychological Bulletin*, 96(2):316–340, 1984.

[Smoczyńska, 1985] Magdalena Smoczyńska. The Acquisition of Polish. In Dan I. Slobin, editor, *The Crosslinguistic Study of Language Acquisition, Volume 1: The Data*, pages 595–686. Lawrence Erlbaum, Hillsdale, New Jersey, 1985.

[Suppes, 1973] Patrick Suppes. The semantics of children's language. *American Psychologist*, pages 103–114, February 1973.

# Integrating Knowledge Sources in Language Comprehension

**Jill Fain Lehman** and **Richard L. Lewis** and **Allen Newell**
School of Computer Science
Carnegie Mellon University
Pittsburgh, PA 15213

## Abstract

Multiple types of knowledge (syntax, semantics, pragmatics, etc.) contribute to establishing the meaning of an utterance. Immediate application of these knowledge sources is necessary to satisfy the real-time constraint of 200 to 300 words per minute for adult comprehension, since delaying the use of a knowledge source introduces computational inefficiencies in the form of overgeneration. On the other hand, ensuring that all relevant knowledge is brought to bear as each word in the sentence is understood is a difficult design problem. As a solution to this problem, we present NL-Soar, a language comprehension system that integrates disparate knowledge sources automatically. Through experience, the nature of the understanding process changes from deliberate, sequential problem solving to recognitional comprehension that applies all the relevant knowledge sources simultaneously to each word. The dynamic character of the system results directly from its implementation within the Soar architecture.

## Introduction

Language comprehension is the process by which knowledge is used to map an utterance to a meaning. Many sources of knowledge contribute to this process: morphology, syntax, semantics, pragmatics, and discourse conventions are those most often considered. In addition to providing knowledge that may be critical in finding the correct mapping, each source may contribute significantly to reducing the effects of local ambiguity on the search for a meaning. Reducing search is a computational necessity; without multiple knowledge sources even common constructions such as prepositional phrase attachment can lead to an exponential number of interpretations. More important, search reduction is necessary to satisfy the real-time constraint of 200 to 300 words per minute required to model adult comprehension. To guarantee the correct mapping we need all knowledge sources. To guarantee a minimal search space we must be able to bring each type of knowledge to bear as soon as it is relevant.

Immediate application of multiple knowledge sources is a key component in many psychological theories of comprehension and parsing (e.g., [Altmann and Steedman, 1988, Marslen-Wilson and Tyler, 1980, Thibadeau *et al.*, 1982]). However, constructing a system to bring its knowledge to bear at the right moment is difficult. There seems to be a trade-off between search efficiency and ease of engineering. At one end of the spectrum is a pipeline design in which knowledge sources are ordered (morphology, syntax, semantics, etc.), and each source generates a set of structures based on the previous phase's output. A system with separated knowledge sources is relatively easy to create and extend, and exhibits few surprising or inconsistent behaviors as the grammar becomes large. Yet, by delaying the application of a type of knowledge, initial stages overgenerate, producing extra interpretations and unnecessary search[1]. At the other end of the spectrum is a fully-integrated design which encodes all the knowledge in the generator. In short, all knowledge sources contribute at once to the meaning structure produced as each word is comprehended. Although overgeneration is prevented, building such a system requires anticipating and keeping track of all

This research was sponsored in part by the Avionics Laboratory, Wright Research and Development Center, Aeronautical Systems Division (AFSC), U.S. Air Force, Wright-Patterson AFB, Ohio 45433-6543 under Contract F33615-90-C-1465, ARPA Order No. 7597 and in part by The Markle Foundation, and a National Science Foundation Graduate Fellowship.

The views and conclusions contained in this document are those of the authors and should not be interpreted as representing the official policies, either expressed or implied, of NSF, The Markle Foundation or the U.S. government.

[1]The work in principle-based parsing [Fong and Berwick, 1990] is an attempt to control the overgeneration in a syntactic pipeline while preserving its ease of engineering. However, since there is no single optimal ordering of modules, computing the optimal ordering for each sentence, while inexpensive, is heuristic. Consequently, overgeneration may still result. Similarly, blackboard models like Hearsay [Erman *et al.*, 1980] and READER [Thibadeau *et al.*, 1982] may reduce inefficiency by allowing for more flexible interaction than the pipeline approach, but cannot guarantee against overgeneration to the same degree full integration can. Furthermore, the same flexibility that can increase efficiency makes creating and extending the knowledge sources difficult.

the points at which each type of knowledge is relevant. With multiple knowledge sources, fully specifying the points of interaction becomes quite complex, and unexpected inconsistencies tend to arise as the grammar becomes large. In fact, no fully-integrated system has been attempted to date. Instead, most attempts at integration have combined only syntax and semantics, generally through the use of semantic grammars [Birnbaum and Selfridge, 1981], domain-specific syntactic grammars [Sager, 1981], or semantic annotations to syntactic rules [Shieber, 1986].

We believe that NL-Soar [Lehman *et al.*, 1991], the model of language comprehension we have implemented in Soar [Laird *et al.*, 1987, Lewis *et al.*, 1990, Newell, 1990], is a significant step toward a fully-integrated system. It comes closer to achieving ideal search behavior than previous approaches by bringing all sources of knowledge to bear as soon as possible. At the same time, the system maintains the modularity of a pipeline design. We achieve this combination of advantages by virtue of Soar's learning mechanism, *chunking*. Initially, comprehension proceeds by deliberate problem solving using independently represented knowledge sources. Through chunking, new knowledge is added to the system that applies all these sources in a single processing step for each word.

## Overview of NL-Soar

Soar distinguishes between knowledge made available through search in problem spaces and knowledge available immediately through recognition. The computational model underlying Soar supports this view by committing to a long-term *recognition memory* (realized as a parallel production system) that yields knowledge in constant time. Problem solving proceeds recognitionally when Soar can select or apply operators in a problem space by simply drawing on this memory. Given the mapping of Soar onto human cognition [Newell, 1990], there is time for only a few (1-3) operators per word within the 200-300 millisecond per word constraint. Therefore, achieving recognitional comprehension is cast as the problem of creating a *comprehension operator* [Newell, 1990] that is directly implemented by productions in the recognition memory. A comprehension operator applied to a word must simultaneously bring to bear all the knowledge necessary to understand the word in the given context. Thus, the comprehension operator is a realization of the *immediacy of interpretation principle* [Thibadeau *et al.*, 1982]. Fully-integrated comprehension for the whole sentence proceeds via word-by-word application of comprehension operators.

Comprehension operators map an utterance to its meaning. The representation for meaning used in NL-Soar is a *model* of the situation that the utterance is about [Johnson-Laird, 1983, Newell, 1990, Polk *et al.*, 1989]. The essential property of models is that the elements in the representation map one-to-one into the situation. The right-hand side of Figure 1 shows the situation model that results from comprehending the sentence *The star above the circle is red.* To aid in building the situation model, NL-Soar also constructs an *utterance model* (Figure 1, left-hand side). This model is

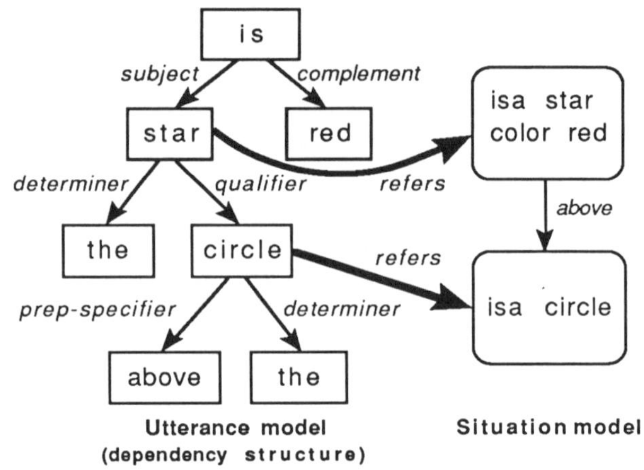


Figure 1: **The models for** *The star above the circle is red.*

a type of dependency structure [Hays, 1964, McCord, 1989, Mel'čuk, 1988], in which nodes correspond to words, and arcs label grammatical relations between words. Thus, the utterance model reflects the syntactic structure of the utterance, while the situation model reflects the structure of the situation being discussed. The relationship between the two models is one of *reference*; parts of the utterance model may refer to objects in the situation. Both models provide context for the comprehension process[2].

If models come from the application of comprehension operators, where do comprehension operators come from? As noted above, hand-coding fully-integrated operators for each word is an extremely difficult and complex task. In NL-Soar, however, hand-coding is unnecessary—comprehension operators arise automatically. To understand how, examine the system's problem spaces in Figure 2. The Comprehension space is where the recognitional comprehension capability is to reside. When Soar first attempts to apply the comprehend operator for a word in a new context, however, an *impasse* arises because the knowledge is not immediately available in recognition memory. Soar responds to this impasse (as it does all impasses) by creating a new subgoal to acquire the needed knowledge through search in another problem space. Knowledge in NL-Soar then chooses the Language space to deliberately implement comprehension by searching the space of partial utterance and situation models. For the operators in the Language space to construct and modify the two models, various syntactic, semantic, and pragmatic constraints on those operators must be satisfied. Ensuring that the constraints are met happens through the sequential application of operators in the Constraint space. If an impasse arises while checking semantic constraints, the Semantics space accesses the current situation model or general world knowledge to deliberately justify a semantic relation.

The result of the deliberate problem solving in the lower three spaces is to resolve the impasse for the comprehend

[2]Additional models may be needed (e.g. a discourse model) as new knowledge sources are added to the system.

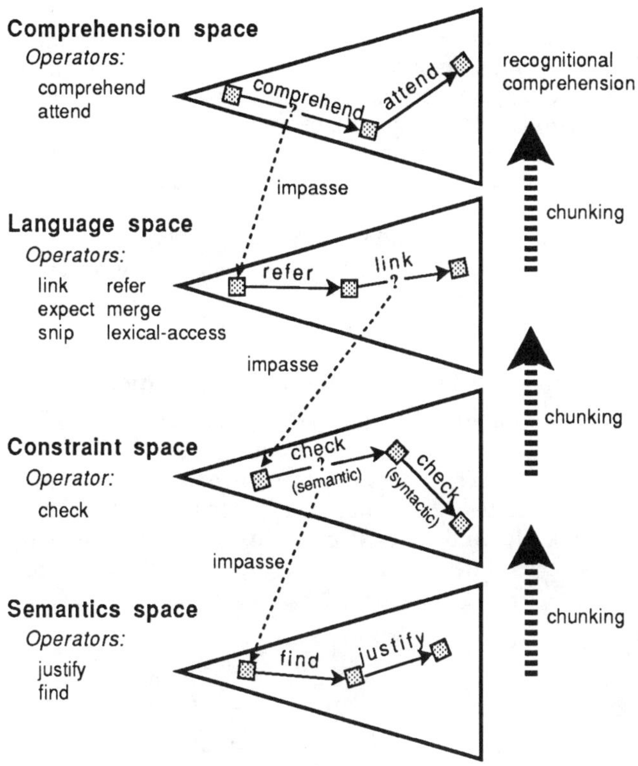


Figure 2: **The NL-Soar tower of problem spaces.**

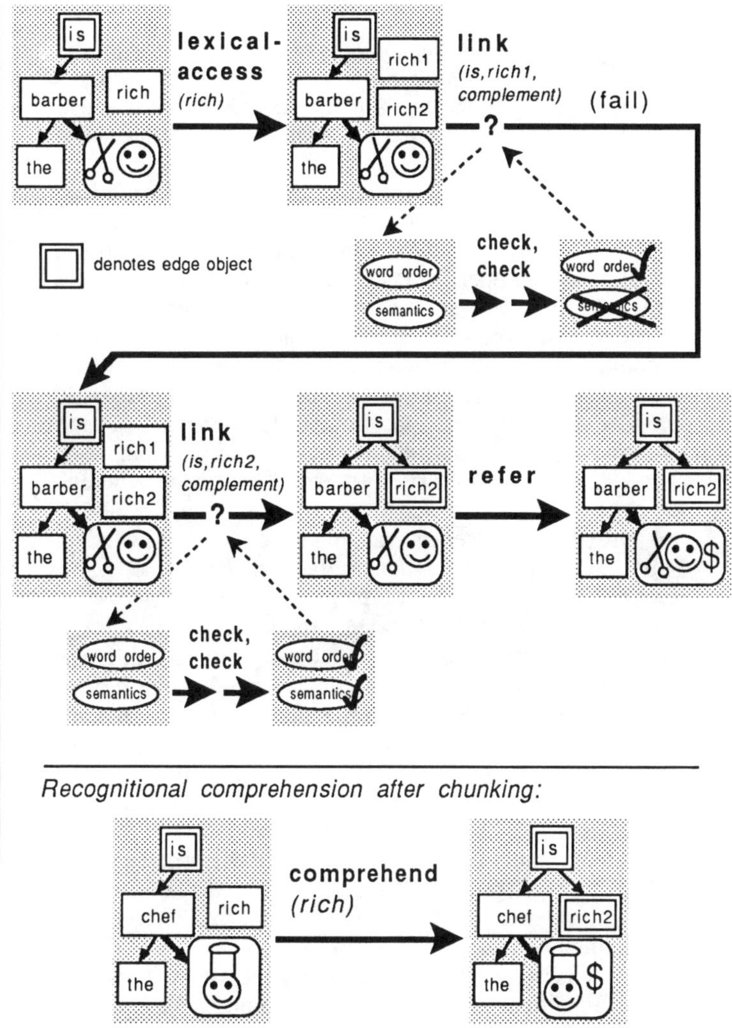


Figure 3: **Deliberate and recognitional comprehension of *rich*.**

operator, that is, to determine the appropriate changes to the two models in response to the word being comprehended. As these changes are made, Soar's chunking mechanism stores away the results of the problem solving and their pre-impasse conditions in the form of new productions (chunks) in recognition memory for the Comprehension space. These chunks form a recognitional comprehension capability, so that in future similar situations, comprehension can proceed without deliberate search. The chunks simultaneously bring to bear all the knowledge sources that were sequentially accessed in the lower problem spaces[3].

## Examples of NL-Soar's behavior

The description of NL-Soar's problem spaces given above says little about the specific knowledge in the system or the particular type of processing that results. In this section, we make these ideas concrete by tracing the system's behavior with three simple examples. The characterization that emerges is that of a single-path comprehension system that combines both *bottom-up* (data-driven) and *top-down* (expectation-driven) knowledge with a limited capability for repairing misinterpretations.

[3]We have concentrated on how chunking creates new productions in the Comprehension space. However, chunking takes place whenever an impasse is resolved. Thus, knowledge is constantly moving up the tower of problem spaces in Figure 2, making processing more efficient in all spaces.

### Constructing the utterance and situation models

Consider comprehension of the word *rich* in *The barber is rich.* When the system cannot recognitionally implement the comprehend operator for *rich*, an impasse arises and problem solving continues in the Language space. Figure 3 illustrates the chain of operator applications that follows the creation of the Language space's initial state. As shown in the first grey box, the state consists of a partial utterance model and a single object in the situation model representing the barber. The first operator to be applied is lexical-access which retrieves two senses of *rich*: an adjective describing food (*rich1*), and an adjective meaning wealthy (*rich2*). Next, link operators are proposed for each sense. A link operator specifies a relationship between the word being comprehended and a word on the active edge of the utterance model[4]. Thus, functional, structural, and lexical ambiguity contributes to the parallel generation of multi-

[4]The active edge is a subset of the nodes in the utterance model corresponding to non-closed constituents. When *rich* is

ple link operators. In the absence of other knowledge, the system arbitrarily chooses which link operator to try.

In this case an attempt is made to link the first sense of rich (food). Associated with each link operator is a set of constraints which must be met before the operator can be applied. When knowledge of the success or failure of these constraints is not immediately available, the constraints are checked independently in the Constraint space. The current link operator evokes only two constraints: one for word order (the verb assigning a predicate complement must come before the complement), and another requiring semantic consistency between the predicate and the subject. Semantic consistency may be justified by using pragmatic knowledge (finding an appropriate referent already in the situation model) or by simple inference in the Semantics space. Since *rich1* describes food and not people, there is neither a barber who is rich (in the food sense) in the model, nor a justifying inference available from world knowledge. Thus the semantic constraint fails, and the link operator terminates with failure as well.

Next, the link for *rich2* is tried. Again, word order passes, and this time the semantic check passes as well (although no instance of a rich barber could be found in the model, knowledge exists in Semantics that allows a referent that isa person to be rich in the sense of wealthy). Since both constraints pass for *rich2*, this link operator succeeds, making the correct addition to the utterance model. This new piece of structure then triggers the refer operator, which updates the situation model by adding a "wealthy" property to the barber object. In general, the refer operator may create new objects, properties, and relations or resolve a reference to an existing object in the situation.

These changes to the utterance and situation models complete the application of the top-level comprehend operator for *rich*. Soar's chunking mechanism now integrates all four types of knowledge—lexical, syntactic, semantic, and pragmatic—used in the lower spaces into new productions in recognition memory. Figure 4 shows an example of one of these productions. These new productions directly im-

| | |
|---|---|
| IF comprehending *rich* | [lexical] |
| and there is a verb on the edge, | [syntactic] |
| and the verb has an unfilled pred-complement role, | [syntactic] |
| and the subject of the verb isa person, | [semantic] |
| and the subject's referent isn't already wealthy | [pragmatic] |
| THEN the sense of *rich* is wealthy, | [lexical] |
| assign *rich* to be the pred-complement of the verb, | [syntactic] |
| assert the wealthy property of the subject's referent | [pragmatic] |

Figure 4: **A new production learned for *rich*'s comprehension operator.**

plement the comprehension operator for *rich* in utterances similar to the example. As shown at the bottom of Figure 3, this part of *rich's* comprehension operator will transfer to all predicate complement utterances in which the subject has a referent in the model with the appropriate semantic features.

comprehended, only *is* is in the edge set available for attachment.

NL-Soar differs from previous attempts at integration (e.g., semantic grammars, domain-specific syntactic grammars, and syntactic grammars augmented with selectional restrictions) in three important ways. One, the single comprehension operator avoids generating any of the intermediate structures of the deliberate search. Two, the integration of referent resolution permits the context (the situation model) to aid in disambiguation. Three, the system is designed to be extendible to other knowledge sources while maintaining both modularity and recognitional comprehension.

## Adding expectations and top-down knowledge

Often it is impossible to link an incoming word directly into the dependency structure. Consider the word *the* in *Mary loves the barber*. It does not properly attach to either *Mary* or *loves*; it must attach to an as yet unseen noun. Rather than buffer such words, NL-Soar uses expectations to maintain a connected utterance model. Figure 5 shows the creation

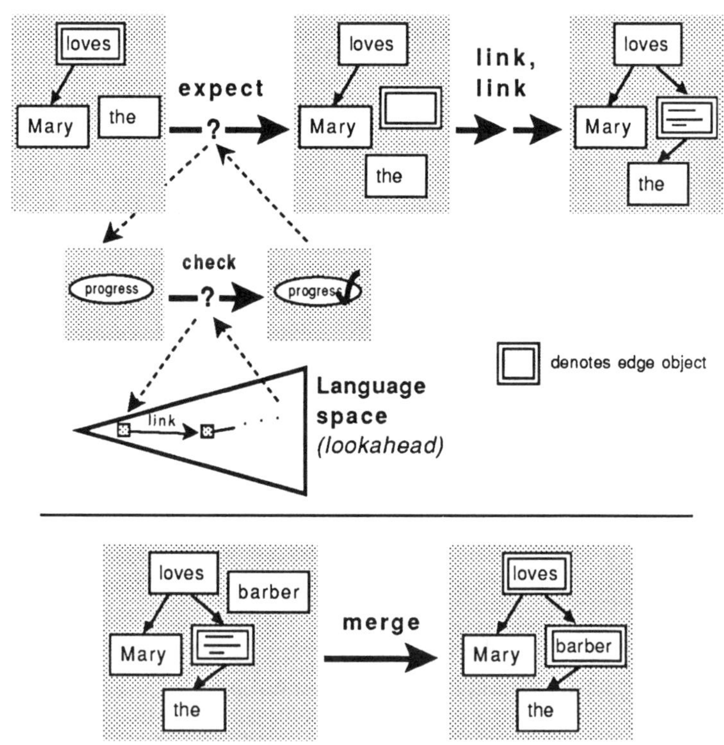


Figure 5: **Creating and merging expectations.**

of an expectation during comprehension of *the* (we omit the situation model in this and subsequent figures). Ignore for the moment the impasse on the expect operator and follow the states along the top row of the figure. First, expect produces an empty node—the expectation—that reserves a place for the future word. Next, two link operators are applied sequentially: the first assigning *the* as the specifier of the expectation, and the second assigning the expectation as the object of *loves*. As the links are made, the expectation accretes the properties necessary to satisfy the constraints of the links. Thus, there is no special knowledge encoded in the expect operator about what to expect; instead, this knowledge is generated dynamically as a side effect of linking the expectation into the existing model. When the expected word finally arrives, the merge operator (Figure 5, bottom) merges it with the expectation, making sure that the new word is consistent with the constraints previously recorded.

How does the system know *when* to create an expectation? There are no specific conditions for proposing the expect operator—it may be attempted at any time. But it is only appropriate when it makes progress toward the successful comprehension of the incoming word as a cohesive part of the utterance to this point. This idea is cast as a *progress* constraint in the Constraint space (the impasse on the expect operator in Figure 5). To determine whether positing an expectation makes progress, NL-Soar searches in a copy of the Language space. This lookahead search reveals that *the* can be linked to the expectation, and the expectation linked to *loves*. Therefore a cohesive structure integrating *the* is possible. Chunks learned during lookahead search encode the conditions that determine when an expectation is needed.

All operators in the Language space use local, bottom-up knowledge. But like expect, all operators in the Language space also have a progress constraint. Thus, the scheme outlined above provides a general mechanism for automatically acquiring top-down knowledge in a comprehension operator via lookahead to pass the progress constraint. If an expectation is created in order to comprehend the current word, then the creation of the expectation and its links is combined with the necessary syntactic, semantic, and pragmatic constraints in the chunks that define the comprehension operator for that word. By integrating over all available knowledge sources, this conception of "top-down" subsumes and moves beyond that found in standard syntactic parsing strategies.

### Adding recognitional repair

Since NL-Soar maintains a single interpretation during comprehension of a sentence [Altmann and Steedman, 1988, Frazier and Rayner, 1982, Pritchett, 1988], some mechanism is needed to recover if the system is led astray by a local ambiguity. Consider the sentence *Mary knows Sharyn loves the barber*. *Sharyn* is initially assigned the object role of *knows*. At the point of comprehending *loves*, *Sharyn* must be reassigned to be the subject of *loves* so that the embedded clause may serve as the object. When a previous interpretation proves to be incorrect, the snip operator is applied to undo part of the utterance model. An interpretation was incorrect if no link or expect operator can make progress. A snip operator is proposed for each link connected to a node in the edge set (lookahead may be evoked to determine which snip will be successful). Figure 6 shows the processing during comprehension of *loves*. First, snip cuts the object link between *knows* and *Sharyn*. Next, link operators apply, attaching *Sharyn* as the subject of *loves* and then *loves* as the clausal direct object of *knows*. The snip operator itself only disconnects a piece of the utterance; the

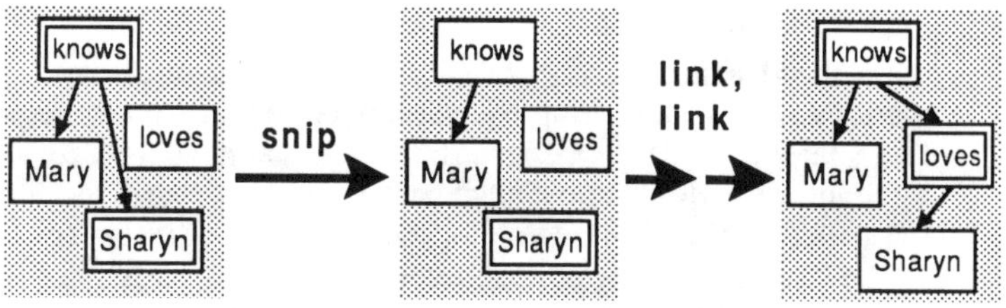


Figure 6: **Repairing a misinterpretation in** *Mary knows Sharyn loves the barber.*

actual repair is carried out by the same link and refer operators we saw above. In contrast to an arbitrary backtracking scheme, the repair mechanism does not require maintaining a complete memory of previous choice points[5].

As in the case of creating expectations, if the comprehension of a word in the current context requires a repair, the effects of all the separate actions that went into making the repair will be captured in new productions for a single operator in the Comprehension space. In this way, limited *recognitional repair* becomes another part of NL-Soar's comprehension capability.

## Summary

We have seen examples of sentences in which NL-Soar's processing proceeded strictly bottom-up (*The barber is rich*), included a top-down component (*Mary loves the barber*), or required repair (*Mary knows Sharyn loves the barber*). In each case, the operators that were applied accessed many different types of knowledge at different times. What is important to note about these examples is that the modularity of both the search control strategies and the knowledge sources is irrelevant in the Comprehension space. The comprehension operators produced by chunking are fully integrated, simply making changes to the models directly (and without search) for each word. Currently in NL-Soar, chunking is a compilation mechanism rather than a language acquisition mechanism. What changes is not the set of utterances that can be assigned meaning structures but the process by which a meaning is assigned. Chunking transforms the mapping process from a search-intensive sequential application of operators representing different but necessary types of knowledge, to the simultaneous application of all knowledge sources by a single operator.

Up to this point, the system's development has been driven by two types of demands. The first type of demand has been the satisfaction of basic psychological criteria: word-by-word, incremental comprehension under the real-time constraint. The second (and related) type of demand has been functional: how to compose meaning structures, how to prevent overgeneration, how to commit to structures before all relevant information is available, how to undo incorrect commitments. As a result of this emphasis, certain knowledge sources have been completely ignored to

[5]The limited nature of the repair means that garden path phenomena may arise. We are currently investigating NL-Soar's predictions in this area.

date (the discourse level, for example). Given the automatic way in which comprehension operators arise within NL-Soar, however, we believe the system can be extended to include additional knowledge sources without sacrificing the benefits of fully-integrated processing. In addition to the lack of some knowledge sources, both the system's syntactic and semantic coverage remain underdeveloped. The system's grammatical coverage is currently limited to about 75% of Allen's basic English grammar [Allen, 1987]. Its world model is a simple class hierarchy in the Semantics space. So far this degree of coverage has been adequate for preliminary work in three areas: a natural language interface to a robot arm [Laird *et al.*, 1990], instruction taking for simple reasoning experiments in psychology [Lewis *et al.*, 1989], and predicting garden path phenomena. To continue to extend the system's coverage in realistic and useful ways, we are exploring integrating NL-Soar with other Soar systems. For a more detailed description of the system, see [Lehman *et al.*, 1991].

**Acknowledgements.** The content and clarity of this paper were improved by the comments of Angela Kennedy Hickman. The integration of NL-Soar with the Robo-Soar project at the University of Michigan was done by Scott Huffman.

## References

[Allen, 1987] J. Allen. *Natural Language Understanding*. Benjamin/Cummings, Menlo Park, CA, 1987.

[Altmann and Steedman, 1988] G. Altmann and M. Steedman. Interaction with context during human sentence processing. *Cognition*, 30:191–238, 1988.

[Birnbaum and Selfridge, 1981] L. Birnbaum and M. Selfridge. Conceptual analysis of natural language. In R. C. Schank and C. K. Riesbeck, editors, *Inside Computer Understanding*. Lawrence Erlbaum Associates, Hillsdale, New Jersey, 1981.

[Erman *et al.*, 1980] L. D. Erman, F. Hayes-Roth, V. R Lesser, and D. R. Reddy. The Hearsay-II speech-understanding system: Integrating knowledge to resolve uncertainty. *Computer Surveys*, 12(2), 1980.

[Fong and Berwick, 1990] S. Fong and R. Berwick. The computational implementation of principle-based parsers. In M. Tomita, editor, *Proceedings of the First International Workshop on Parsing Technologies*, pages 75–84, Pittsburgh, PA, 1990. Carnegie Mellon University.

[Frazier and Rayner, 1982] L. Frazier and K. Rayner. Making and correcting errors during sentence comprehension: eye movements in the analysis of structurally ambiguous sentences. *Cognitive Psychology*, 14:178–210, 1982.

[Hays, 1964] D. G. Hays. Dependency theory: a formalism and some observations. *Language*, 40(4):511–525, 1964.

[Johnson-Laird, 1983] P. N. Johnson-Laird. *Mental Models*. Harvard, Cambridge, MA, 1983.

[Laird *et al.*, 1987] J. E. Laird, A. Newell, and P. S. Rosenbloom. Soar: An architecture for general intelligence. *Artificial Intelligence*, 33(1):1–64, 1987.

[Laird *et al.*, 1990] J. E. Laird, E. S. Yager, M. Hucka, and C. M. Tuck. Robo-Soar: An integration of external interaction, planning, and learning using Soar. To appear in Van de Velde (ed.) *Machine Learning for Autonomous Agents*, 1990.

[Lehman *et al.*, 1991] J. F. Lehman, R. L. Lewis, and A. Newell. Natural language comprehension in Soar. Technical report, School of Computer Science, Carnegie Mellon University, 1991. Forthcoming technical report.

[Lewis *et al.*, 1989] R. L. Lewis, A. Newell, and T. A. Polk. Toward a Soar theory of taking instructions for immediate reasoning tasks. In *Proceedings of the Eleventh Annual Conference of the Cognitive Science Society*, pages 514–521, 1989.

[Lewis *et al.*, 1990] R. L. Lewis, S. B. Huffman, B. E. John, J. E. Laird, J. F. Lehman, A. Newell, P. S. Rosenbloom, T. Simon, and S. G. Tessler. Soar as a unified theory of cognition: Spring 1990. In *Twelfth Annual Conference of the Cognitive Science Society*, pages 1035–1042, 1990.

[Marslen-Wilson and Tyler, 1980] W. Marslen-Wilson and L. K. Tyler. The temporal structure of spoken language understanding. *Cognition*, 8:1–71, 1980.

[McCord, 1989] M. C. McCord. A new version of slot grammar. Technical Report RC 14506, IBM Research Division, Yorktown Heights, NY 10598, 1989.

[Mel'ĉuk, 1988] I. A. Mel'ĉuk. *Dependency Syntax: Theory and Practice*. State University of New York Press, Albany, 1988.

[Newell, 1990] A. Newell. *Unified Theories of Cognition*. Harvard University Press, Cambridge, Massachusetts, 1990.

[Polk *et al.*, 1989] T. A. Polk, A. Newell, and R. L. Lewis. Toward a unified theory of immediate reasoning in Soar. In *Proceedings of the Eleventh Annual Conference of the Cognitive Science Society*, pages 506–513, 1989.

[Pritchett, 1988] B. L. Pritchett. Garden path phenomena and the grammatical basis of language processing. *Language*, 64:539–576, 1988.

[Sager, 1981] N. Sager. *Natural Language Information Processing*. Addison-Wesley, Reading, MA, 1981.

[Shieber, 1986] S. M. Shieber. *An Introduction to Unification-Based Approaches to Grammar*. Center for the Study of Language and Information, Stanford, CA, 1986.

[Thibadeau *et al.*, 1982] R. Thibadeau, M. A. Just, and P. A. Carpenter. A model of the time course and content of reading. *Cognitive Science*, 6:157–203, 1982.

# A Graph Propagation Architecture for Massively-Parallel Constraint Processing of Natural Language

**Hideto Tomabechi**
Carnegie Mellon University
Smith Hall 109, Pittsburgh PA 15213
tomabech@cs.cmu.edu

## Abstract

We describe a model of natural language understanding based on the notion of propagating constraints in a semantic memory. This model contains a massively-parallel memory-network in which constraint graphs that represent syntactic and other constraints that are associated with the nodes that triggered activations are propagated. The propagated constraint graphs of complement nodes that collide with the constraint graphs postulated by the head nodes are unified to perform constraint applications. This mechanism handles linguistic phenomena such as case, agreement, binding and control in a principled manner in effect equivalent to the manner that they are handled in modern linguistic theories.

## Motivation

The so called 'direct memory access' (or memory-based) paradigm of natural language recognition seems to have established itself as an important approach to natural language. Originated by Quillian ([Quillian, 1969]), this model of natural language processing views natural language understanding as the activity of identifying the input utterance with existing memory structures. While these memory structures have taken various forms depending upon underlying theoretical approaches (such as MOPs of [Riesbeck and Martin, 1985] and Cases in [Martin, 1989]), the underlying control structure has remained shared, namely, spreading activation and marker passing. The advantage of this control structure is that since spreading activation takes place directly in the memory network, direct memory-based inferential processing can be employed at any point of recognition. A large number of systems have taken advantage of this control structure in order to perform recognition and inference directly on a semantic memory. One well known weakness of this approach is its inability to handle complex syntactic constraints. Since the memory network represents the hierarchical organization of conceptual relations and packaging of thematic and experiential memory units, adequate representation of syntactic constraints has been very difficult, if not impossible. It is because syntactic constraints are often configurational and are assigned based upon particular structural patterns of the constituent buildup at the time of utterance. For example, a syntactic *case* of a particular noun phrase is dynamically determined at the time of the utterance based upon the particular syntactic configuration and is essentially impossible to capture *a priori* in the semantic network (otherwise we will have to specify infinitely many sentential patterns as well as many redundant subcategories of concepts based upon *case* variety simply to capture the *case* agreement phenomenon). Similarly a recognition of *Sue said that Mary ran* under the memory-based paradigm would require instantiation of two sequences of concepts in memory [*PERSON *MTRANS-word that *ACTION] and [*PERSON *RUN] (i.e., a typical sentential recognition methodology currently employed under this paradigm using patterns of concepts). The first sequence is probably attached to a root node such as *MTRANS-EVENT and the second sequence is attached to *RUN-ACTION which is a subclass of *ACTION. The last element of the first sequence *ACTION will be activated by the recognition of any subclass of *ACTION including *RUN-ACTION, *SLEEP-ACTION, etc.. However, there is no way to ensure that if the *MTRANS-word is *said*, the set of entities that caused the activation of *ACTION contain (be headed by) a finite verb (unless we create a possibly infinite number of concepts for different combinations of syntactic features such as *FINITE-FORM-RUNNING-ACTION-TAKING-NOMINATIVE-SUBJECT). Thus, it is essentially impossible to capture the grammaticality of *John said Mary runs* and the ungrammaticality of *John said Mary run* (or *John said Mary to run*) within the constraints capturable inside the semantic memory network. Another well known syntactic phenomenon that is essentially uncapturable within the semantic network methodology is the so called *obligatory control* phenomenon in which a subject (or object) of the external clause controls the unexpressed subject of the internal clause. For example, in *John persuaded Mary to* *give Sandy sushi*, the subject of *give* is unexpressed but is 'controlled' to be the object of the external clause, i.e., *Mary*. This is called *object control* and in order to handle this, a configurational formulation of the syntactic constraints will be needed, which is non-trivial in a straightforward semantic network.[1]

Other problematic constraints include *long-distance dependency* in which the dependency relation between the two noun phrase positions is postulated for an unbounded length and depth of the syntactic tree configuration. Some binding of anaphors, such as the binding of reflexive pronouns in English, also requires configurational constraint processing. Thus, many syntactic phenomena require constraint processing based upon syntactic constituents and configurations that are dynamically created during the recognition and *a priori* formulation of such constraints within the semantic memory network is very difficult, if not impossible.

## Constraint Propagation in Memory

The central notion of the proposed model is the **propagation of constraints** in the semantic memory network. This is in contrast to past models of semantic memory-based natural language recognition in which activations and pointers to sources of activations (i.e., *markers*) are propagated in the memory network. Instead of propagating markings on memory nodes, we explicitly propagate constraints. As a representational scheme for capturing constraints to be propagated, we find directed graphs to be appropriate. It is because their expressivity has been well supported by linguists ([Bresnan and Kaplan, 1982], [Pollard and Sag, 1987], etc.) and the operations on them (i.e., graph unification) have been extensively researched[2]. In fact the problematic linguistic phenomena for the memory-based paradigm such as *agreement*, *obligatory control*, and *long-distance dependency* have all been captured in the unification-based grammar theories in which graph representation and graph unification are assumed to be the tool for capturing the constraints required for handling these phenomena. Thus, we propose a model of natural language recognition based upon propagation of constraints in a semantic memory network in which constraints are captured using directed graphs. We will call the proposed semantic network with capacity to propagate graphs *Graph-based Constraint Propagation Network or GCPN* in this paper. The graphs[3] propagated in the GCPN may point to any location in the network and may contain complex paths including convergent arcs and cycles. The existential nature of the graphs is different from that of the constraint propagation network itself[4] in the sense that the semantic memory network constitutes (following the tradition since [Quillian, 1966]) the hierarchical organization of concepts that represents a semantic memory and the constraint graphs represent mostly syntactic constraints such as head features and configurational constraints which do not belong to the semantic memory.

The expressivity of the graph-based constraint propagation scheme is significantly greater than that of traditional memory-based schemes and their extensions. As we mentioned above, syntactic expressivity of this scheme is essentially equivalent to the expressivity of unification-based theories (such as HPSG, [Pollard and Sag, 1987] and LFG [Bresnan and Kaplan, 1982]) while the semantic and pragmatic expressivity of the memory-based paradigm is maintained.

Below is a sample lexicon from our experimental system called MONA-LISA ([Tomabechi, ms]) which uses the GCPN as the central representational network. We have adopted HPSG ([Pollard and Sag, 1987]) as the basis of providing the linguistic constraints to the system as graphs. The first sample entry is the specification for the concept *JOHN with the constraints specified here using path equations similar to the PATR-II notation ([Shieber, *et al*, 1983]).

```
(def-frame *JOHN
  (inherits-from *MALE-PERSON)
  (type :lex-comp)
  (spelling   John)
  (synsem
   (def-path
    (<0 loc cat head maj> == n)
    (<0 loc cat marking> == unmarked)
    (<0 loc cont para index> == [[per 3rd]
                                 [num sng]
                                 [gend masc]])
    (<0 loc cont restr reln> == *JOHN)
    (<0 loc context backgr> == [[reln naming]
                                [name JOHN]])
    (<0 loc context backgr bearer>
                        == <0 loc cont para index>)
    (<0 mem> == <0 loc cont para index iden>))))
```

When this lexical definition is read into the system, the path equations are converted into graphs as shown below.

```
(*JOHN
  (INHERITS-FROM *MALE-PERSON)
  (TYPE
```

[1] In fact, the DMA systems that handle these phenomena rely on either prestored functional constraint applications specifically prepared for particular sentential constructions (such as [Tomabechi and Levin, 1989]) or on prestored (hard-wired) relations between conceptual entities for particular sentential constructions (such as [Kitano, 1990]). In either case, formulation of prestored relational constraints are at best *ad hoc* and no generality is guaranteed.

[2] Such as [Pereira, 1985], [Karttunen, 1986], [Kasper, 1987], [Wroblewski, 1987], and [Tomabechi, 1991].

[3] Implementationally, they are pointers to graphs instead of graphs themselves.

[4] Of course, the GCPN itself is a graph; however, here, 'graphs' means the constraint graphs and not the memory-network.

```
    (VALUE
      (COMMON :LEX-COMP)))
  (SPELLING
    (VALUE
      (COMMON JOHN)))
  (SYNSEM
    (VALUE
      (COMMON
 X01[[0 X02[[MEM X03[]]
           [LOC X04[[CONTEXT
                     X05[[BACKGR
                          X06[[BEARER
                               X07[[IDEN  X03]
                                   [GEND  X08 MASC]
                                   [NUM  X09 SNG]
                                   [PER  X10 3RD]]
                                   [NAME  X11 JOHN]
                                   [RELN  X12 NAMING]]]
                     [CONT X13[[RESTR
                                 X14[[RELN X15 *JOHN]]
                               [PARA  X16[[INDEX  X07]]]
                     [CAT  X17[[MARKING  X18 UNMARKED]
                               [HEAD
                                 X19[[MAJ X20 N]]]]]]))))
```

In the current formulation of GCPN, the constraint graphs are stored in synsem values of the nodes and the top level number 0 arc represents constraints to the node itself. If a node has its complement nodes the constraints are specified by numbers higher than 0 (in the order of obliqueness). For example, the lexical specification for the node *GIVE looks as follows:

```
(def-frame *GIVE
 (inherits-from *GIVE-ACTION)
 (type :lex-head)
 (spelling give)
 (synsem
  (def-path
   (<0 loc cat head> == [[maj v]
                         [vform bse]
                         [aux -]
                         [inv -]
                         [prd -]])
   (<0 loc cat marking> == unmarked)
   (<0 loc cat subcat 1> == <1>)
   (<0 loc cat subcat 2> == <2>)
   (<0 loc cat subcat 3> == <3>)
   (<0 loc cont reln> == *give-action)
   (<1 loc cat head maj> == n)
   (<1 loc cat head case> == nom)
   (<0 loc cont agent> == <1 loc cont para index>)
   (<1 loc cont restr reln> == *person)
   (<2 loc cat head maj> == n)
   (<2 loc cat head case> == acc)
   (<0 loc cont goal> == <2 loc cont para index>)
   (<2 loc cont restr reln> == *person)
   (<3 loc cat head maj> == n)
   (<3 loc cat head case> == acc)
   (<0 loc cont theme> == <3 loc cont para index>)
   (<3 loc cont restr reln> == *matter))))
```

The equations are converted into directed graphs when read into the system. This way, instead of simply storing simple case-frame type lexical specifications in the lexical nodes, we provide full graph-based lexical constraints in the lexical level nodes in the constraint propagation network. Let us provide a sample lexical node definition for the object control verb *persuaded*:

```
(def-frame *PERSUADED
 (inherits-from *PERSUADE-ACTION)
 (type :lex-head)
 (spelling persuaded)
 (synsem
  (def-path
   (<0 loc cat head> == [[maj v]
                         [vform inf]
                         [aux +]
                         [inv -]
                         [prd -]])
   (<0 loc cat marking> == unmarked)
   (<0 loc cat subcat 1> == <1>)
   (<0 loc cat subcat 2> == <2>)
   (<0 loc cat subcat 3> == <3>)
   (<1 loc cat head maj> == n)
   (<1 loc cat head case> == nom)
   (<1 loc cont restr reln> == *person)
   (<0 loc cont agent> == <1 loc cont para index>)
   (<0 loc cont persuadee> == <2 loc cont para index>)
   (<0 loc cont persuadee>
              == <0 loc cont circumstance agent>)
   (<2 loc cat head maj> == n)
   (<2 loc cat head case> == acc)
   (<2 loc cont restr reln> == *person)
   (<3 loc cat head maj> == v)
   (<3 loc cat head vform> == inf)
   (<3 loc cat head aux> == +)
   (<3 loc cat subcat 1 loc cat head> == [[maj n]
                                          [case nom]])
   (<3 loc cat subcat 2 loc cat head> == saturated)
   (<3 loc cat subcat 3 loc cat head> == saturated)
   (<0 loc cont circumstance> == <3 loc cont>)
   (<0 loc cont reln> == *PERSUADE-ACTION)
   (<3 loc cont restr reln> == <3 loc cont reln>)
   (<3 loc cont restr reln> == *action))))
```

Thus the two equations:

$(\langle$ 0 loc cont persuadee $\rangle$ == $\langle$ 2 loc cont para index $\rangle)$

$(\langle$ 0 loc cont persuadee $\rangle$ == $\langle$ 0 loc cont circumstance agent $\rangle)$

can easily specify the control constraints lexically in the network.

## The Graph Propagation Recognition

Currently, our graph-based constraint propagation inheritance network has 5 types of nodes: conceptual-class nodes, lexical-head nodes, lexical-complement nodes, memory-instance nodes, and phonological-activity nodes. The conceptual-class nodes are nodes in the high levels of abstraction and are used for discourse and episodic recognition. Lexical-head nodes are nodes that are phonologically invoked with lexical activations and they package the complement nodes.

Lexical-complement nodes are the nodes that are lexically invoked and do not have their own complements. Memory-instance nodes are actual instances of lexical-heads and lexical-complements that are specific to the current utterance. Phonological-activity nodes are the nodes that represent phonemic units and are connected to the phonemic recognition modules. The activity of phonological-activity nodes is not discussed in this paper and the input is assumed to be already hypothesized as words (Please refer to [Tomabechi, ms] for activities of those nodes).

Below is the central part of our sentential recognition algorithm for word level input.

```
function sentential-recognize (input-stream)
  create-process (recognize-lexical (input-stream));
  invoke-global-incidents;
  for all NODE in DecayingLayer do
     print-node(NODE);

function recognize-lexical (input-stream)
 reset activities in Activation Layer
                  and Decaying Layer
 for word-hypothesis in input-stream do
    create-process (activate-lex-node
                              (word-hypothesis));
 invoke-global-incidents;

function activate-lex-node (word-hypothesis)
 create instance of word-hypothesis
 and make a copy of constraint graph
 with addition of an 'mem' arc pointing
 to the created instance;
 if the node type is lexical-complement
    then propagate copied (and modified)
         constraint graph upward;

function invoke-global-incidents ()
 for head-instance in ActivationLayer do
    create-process (grab-subcats (head-instance)) ;

function grab-subcats (head-instance)
 for arcs specified in subcat graph
                (i.e, <0 loc cat subcat>) do
     if conceptual restriction node exists
                (i.e, <loc cont rest reln> has value)
     and if that node has received
            the constraint graph propagation
         then unify the subcat graph
              with the propagated graph;
              if unify succeeds
                 and obliqueness order is met
                 then store result destructively
                      in head-instance;
                      propagate synsem graph upward;
```

Originally, the GCPN is configured hierarchically in terms of conceptual inheritance. Graph propagation occurs only upward in the inheritance hierarchy and never horizontally. Conceptual relations (other than inheritance) between lexical nodes are specified through constraint graphs (as seen in the sample entry in the previous section). All nodes originally belong to StaticLayer and when they are lexically invoked they move into an activated state (the nodes move to ActivationLayer). Only the saturated lexical-head nodes move to the DecayingLayer. So at the end of the recognition, the constraint 0 graph (i.e. the constraint to itself) of the nodes in the DecayingLayer contains information that can be used for further processing (such as generation). Actually, the printout of the constraint 0 graph of the Decaying node should look exactly like that of the output of a unification-based parser[5]. Let us briefly look at the constraint propagation activity with the recognition of *John persuaded Mary to give Sandy sushi.* First when the word *John* activates the lexical-complement node *JOHN, an instance *JOHN001 is created.[6] When the instance is created, the synsem graph of *JOHN is copied while putting the created instance as the destination of the ⟨ 0 mem ⟩ path of the newly created synsem graph. This synsem graph is propagated upward in the abstraction hierarchy (through *MALE-PERSON, *PERSON, ...). When the activation reaches the top of the abstraction hierarchy, then global-incidents is triggered. All currently activated head-instances concurrently check their subcat graph paths and if their destination nodes have currently received a graph propagation, then the head-instance's constraint graph and the propagated graphs are unified. Since at this moment no head-instance is activated nothing happens. When the next word *persuaded* is hypothesized, a head-instance *PERSUADE001 is created and a constraint-graph containing *PERSUADE001 in the ⟨ 0 mem ⟩ graph path is propagated upward in the abstraction hierarchy. When the activation reaches the top, global-incidents is invoked and PERSUADE001 which is a head-instance tries to grab its subcat fillers by checking the content of ⟨ loc cont rest reln ⟩ path for each of the subcategorization elements. Since *PERSON has already received a graph propagation at the recognition of *John* the propagated constraint graph of *JOHN001 is unified against the ⟨ 0 loc cat subcat 1 ⟩ graph of the constraint graph contained in *PERSUADE001. This succeeds and the result graph is stored in *PERSUADE001. When the rest of the sentence is input, *GIVE gets partially saturated with *SANDY and *SUSHI and while the subject is still unfilled is grabbed by the subject control verb head *to* which is in turn grabbed by object control verb head *persuade* (*PERSUADE001). With the graph con-

[5]Except for one significant difference which is that the constraint 0 graph actually contains pointers to the real instances in memory (such as Mary001) instead of a simple string (such as "Mary") found in unification-based parsers.

[6]If an instance of *JOHN already exists then both are considered as candidates with currently existing instance with less cost assigned. Cost (and reverse cost) information is used for phonemic confusions, lexical ambiguity, and multiple instance candidates and recurrent net subsymbolic conceptual predictions in the MONA-LISA system which is not discussed in this paper.

vergence specified in the constraint graphs representing control relations both *TO001 and *PERSUADE001 get their subject and object positions filled respectively with the appropriate instance MARY001. At the end of the recognition only *PERSUADE001 moves to the DecayingLayer (only fully saturated head-instance) and the synsem graph of *PERSUADE001 contains the fully built result of the recognition.

## Discussion:

We have seen in the previous sections that our model is capable of taking advantage of constraints that are postulated in modern linguistic theories by fully supporting arbitrary graphs to be used as constraints in the network. In fact our current implementation uses HPSG which is represented by path equations (which are automatically converted to directed subsumption-ordering graphs[7]) that are no different from the grammar that we are using for the unification-based parsers. This is in contrast to the past attempts in augmenting the memory-based paradigm with syntax (such as [Tomabechi and Levin, 1989]) where due to the architectural limitation of the memory-based system, full configurational constraint processing could not be supported. Given that our model is basically memory-based (i.e., spreading activation on memory) advantages of the memory-based recognition (such as direct memory access inferences on episodic memory discussed extensively in the existing literature) are maintained on top of the capacity to handle syntactic constraints. Moreover, clearly memory-based (DMA) network would significantly overgenerate without strong syntactic constraints as demonstrated in our example of *object control* phenomena. Without the *control* constraint, the subject of an embedded clause could be any instance that meets semantic requirements (for example *JOHN001 and *MARY001). Seemingly simpler phenomena such as agreement of *case* are equally problematic for the memory-based systems since the *case* of a noun phrase (entity) is not determined until the actual time of utterance and often *case* requirements are postulated from external clauses. In other words, without strict syntactic processing, there will be a massive overgeneration of acceptable candidates with semantics and phrasal lexicon alone with a realistic grammar. This means that naive acceptance of the DMA style marker-passing scheme as an extremely fast way of natural language recognition needs to be reexamined. Application of fragments of knowledge could be very fast in a simple marker intersection search, but with a realistic linguistic coverage that in return means massive overgeneration which cannot be solved by increase in number of processing units and parallelism.

Another important advantage of our model is that the singularity of the control structure and the uniformity of the processing are maintained. Past attempts to enhance spreading activation marker passing recognition to handle syntactic constraints either inevitably involved introduction of arbitrary functional applications to handle syntax (such as [Tomabechi and Levin, 1989]) or required a separate control structure (i.e., syntactic parsers) to handle syntax (such as [Norvig, 1989] and [Hendler, 1989]). In the graph-based constraint propagation network, the control structure is the propagation of constraint graphs which is uniform. The constituent build-up for subsumption-ordering graphs and the constraint applications are also uniformly performed by graph-unification requiring no prestored and arbitrary functional constraint checking mechanism. One rather practical advantage due to the support of graph unification as the constraint application mechanism is that the result of configurational buildup of the subsumption-ordering graph (as postulated in unification-based linguistic theories) is incrementally created during the recognition. These results of constituent buildup are stored in the path 0 graph in our implementation. (Path 0 represents the constraint to the node itself, whereas paths 1, 2, and 3 represent the constraints for its complements). Thus, at the end of the sentential recognition, the constraint 0 graph of the head-instance which survived through constraint propagation and moved into the DecayingLayer represents the feature-structure representation of the result of the recognition. This result graph can be extracted out and be passed to other natural language and inferential modules if necessary.

## Conclusion

The paradigm of natural language recognition based upon the direct recognition in a semantic memory has been appealing from different view points. These included the appeal from the cognitive view-points as well as the phenomenological ones (especially the assumed massive-parallelism). The paradigm's practical appeal as a parsing architecture has been strong as well due to its strength in handling contextually sensitive inputs. However, one significant weakness of this paradigm has been its lack of capacity to handle syntactic constraints in a manner that is consistent with the rest of the memory structure. Linguistic and psychological communities have long accepted the significance of syntactic constraints as playing an important role in many types of linguistic phenomena. On the other hand, with few exceptions, syntactic constraints have been either ignored or handled in an *ad hoc* manner. Our model attempts to model interactions among various syntactic constraints (as represented by graphs) as well as between syntax, semantics and pragmatics in a principled manner through a uni-

[7] Subsumption ordering is an assumed condition for the graph-based representation in unification-based grammar formalisms. Simply put, the information content represented by the directed graph node which is closer to the root of the graph subsumes the information content of the nodes further from the root in the same graph.

form representation and a singular control structure of constraint propagation. It should be emphasized that the syntactic processing is fully integrated into semantic processing since both semantic and syntactic constraints are represented through directed graphs. The processing itself is semantic-driven since the graph propagation in the semantic network is the underlying control structure.[8] In our model, phenomena such as *case*, *agreement*, *control*, *binding*, and *long-distance dependency* can be handled in a generalized manner most straightforwardly as formulated by the modern linguistic theories based upon feature structures and unification. The difference between this model and past 'syntactic daemon' models is substantial both in terms of the expressivity and generality of constraint applications. Generalized linguistic phenomena such as those mentioned above are represented and processed in a principled manner compared to an *ad hoc* and specialized manner associated with the prestored *daemon* models[9]. With the expressivity of our model in capturing the syntactic constraints as postulated in the modern linguistic theories and uniformity of processing them that is integrated to a spreading activation memory-based natural language recognition in a principled manner, our model seems viable as one paradigm of natural language recognition.

## Acknowledgments

The author would like to thank Masaru Tomita, Jaime Carbonell, David Evans, Alex Waibel, Hiroaki Kitano, and Margalit Zabludowski for their comments. Special thanks are due to Akira Kurematsu, Hitoshi Iida and the members of the ATR Interpreting Telephony Research Laboratories for their support while I was visiting ATR implementing the central part of the architecture.

## Implementation

We have implemented our demo system on a Sequent Symmetry shared-memory parallel machine with 16 CPUs by simulating massive-parallelism by taking advantage of micro-tasking parallelism using light-weight processes in parallel CommonLisp. The activation in the GCPN is only propagated upward in the inheritance hierarchy and never horizontally. Since the increase in the grammar size takes place horizontally, the complexity increase can be essentially countered by an increase in the number of processing units.

## References

[Bresnan and Kaplan, 1982] Bresnan, J. and R. Kaplan "Lexical-Functional Grammar: A Formal System for Grammatical Representation". In J. Bresnan (ed) *The Mental Representation of Grammatical Relations*, MIT Press, 1982.

[Hendler, 1989] Hendler, J. "Marker-Passing over Microfeatures: Towards a Hybrid Symbolic / Connectionist Model". In *Cognitive Science*, Vol. 13. Num 1, 1989.

[Karttunen, 1986] Karttunen, L. "D-PATR: A Development Environment for Unification-Based Grammars". In *Proceedings of COLING-86*. 1986.

[Kasper, 1987] Kasper, R. "A Unification Method for Disjunctive Feature Descriptions". In *Proceedings of ACL-87*. 1987.

[Kitano, 1990] Kitano, H. *Phi*DMDIALOG *A Speech-to-Speech Dialogue Translation System* Machine Translation 5 (1990), 301-338. Kluwer Academic Pub., 1990.

[Martin, 1989] Martin, C. "Case-based Parsing" in Riesbeck, C. and R. Schank eds., *Inside Case-based Reasoning*, Lawrence Erlbaum Associates.

[Norvig, 1989] Norvig, P. "Marker Passing as a Weak Method for Text Inferencing". In *Cognitive Science*, Vol. 13, Num. 4, 1989.

[Pollard and Sag, 1987] Pollard, C. and I. Sag *Information-based Syntax and Semantics*. Vol 1, CSLI, 1987.

[Pereira, 1985] Pereira, F. "A Structure-Sharing Representation for Unification-Based Grammar Formalisms". In *Proceedings of ACL-85*. 1985.

[Quillian, 1966] Quillian, M.R. *Semantic Memory*, Doctoral Dissertation, Carnegie Institute of Technology (Carnegie Mellon University), 1966.

[Quillian, 1969] Quillian, M.R. *The teachable language comprehender*. BBN Scientific Report 10, 1969.

[Riesbeck and Martin, 1985] Riesbeck, C. and C. Martin *Direct Memory Access Parsing*. Yale University Report 35, 1985.

[Shieber, *et al*, 1983] Shieber, S., H. Uszkoreit, J. Robinson, and M. Tyson *The Formalism and Implementation of PATR-II*. In Research on Interactive Acquisition and Use of Knowledge. Artificial Intelligence Center, SRI International, 1983

[Tomabechi, 1991] Tomabechi, H. "Quasi-Destructive Graph Unification". In *Proceedings of ACL-91*, 1991.

[Tomabechi, ms] Tomabechi, H. *MONA-LISA: Multimodal Ontological Neural Architecture for Linguistic Interactions and Scalable Adaptations*, forthcoming as Technical Report, ATR Interpreting Telephony Research Laboratories, 1991.

[Tomabechi and Levin, 1989] Tomabechi, H. and L. Levin "Head-driven Massively-parallel Constraint Propagation: Head-features and subcategorization as interacting constraints in associative memory", In *Proceedings of The Eleventh Annual Conference of the Cognitive Science Society*, 1989.

[Wroblewski, 1987] Wroblewski, D."Nondestructive Graph Unification" In *Proceedings of AAAI87*. 1987.

[8]In other words, syntax is not simply acting as a filter for marker intersection; instead, both syntactic and semantic constraints are applied through graph propagation and unification constraint satisfaction activities.

[9]which normally works for one or two typical sample sentences alone.

# Cognitive plausibility of a conceptual framework
## *(for modeling qualitative prediction of behaviour)**

**Bert Bredeweg** and **Cis Schut**
Department of Social Science Informatics (S.W.I.)
University of Amsterdam, Roetersstraat 15
1018 WB Amsterdam (The Netherlands)
Telephone: +31-20-5256788
Telefax: +31-20-5256896
E-mail: bert@swi.psy.uva.nl

## Abstract

*In this paper we investigate the cognitive plausibility of an integrated framework for qualitative prediction of behaviour. The framework is based on the KADS expertise modeling approach and integrates different approaches to qualitative reasoning. The framework is implemented in a program called GARP. To test the cognitive plausibility a physics problem involving qualitative prediction of behaviour was constructed. The behaviour prediction of this problem generated by GARP was compared to think-aloud protocols of human subjects performing the same problem solving task.*

## Introduction

Developing conceptual frameworks for modeling problem solving expertise is considered to be an important aspect of research on Knowledge Based Systems (KBS) [15; 8; 12]. It is a generally accepted hypothesis that such frameworks can usefully support the knowledge acquisition process. However, little research has been published that investigates this hypothesis. In particular little attention has been given to test the cognitive validity of these frameworks. After all they are used to interpret verbal data of human problem solving behaviour.

In this paper we investigate the cognitive plausibility of the conceptual framework proposed by the KADS methodology. Within this framework we focus on the reasoning task *qualitative prediction of behaviour* as the specific object of our research. We compare think-aloud protocols of human subjects predicting the behaviour of a complex configuration of balances with an implemented computer model of the same reasoning task.

The contents of this paper are structured as follows. The next two sections describe the KADS conceptual framework and the reasoning task qualitative prediction of behaviour. The paper then moves on to a protocol analysis of this problem solving task carried out by human subjects and describes to what extent the framework for qualitative prediction of behaviour fits the think-aloud protocol data. Finally, the discussion summarises and further discusses the main results.

## Modeling problem solving expertise

KADS proposes a methodology for building KBS's [6; 7] and is based on the premise that knowledge used by people during reasoning processes can be distinguished according to several types, corresponding to the different roles the knowledge plays in the reasoning process (see figure 1). First, the *domain layer*

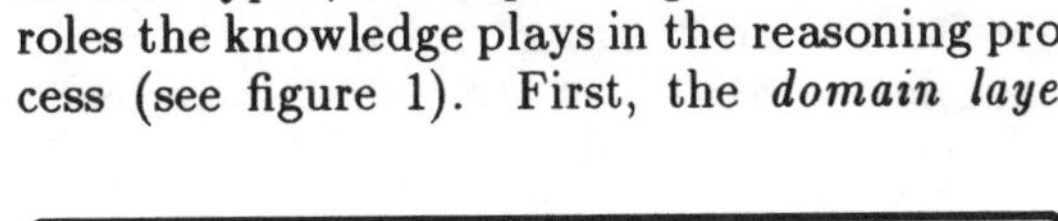


Figure 1: The KADS four layers

describes the domain knowledge in terms of

*The research reported here was partially funded by the Esprit Basic Research Programme of the Commission of the European Communities as project number 3178 (Reflect). The partners in this project are the University of Amsterdam (Amsterdam, The Netherlands), the Netherlands Energy Research Foundation ECN (Petten, The Netherlands), the National German Research Centre for Computer Science GMD (St. Augustin West-Germany) and BSR Consulting (Munich, West-Germany).

domain concepts, relations and complex structures such as models of processes or devices. This knowledge is static in nature, that is, it is not specified what control regime is needed in order to use the knowledge effectively

The second type of knowledge, the *inference layer*, concerns the canonical problem solving actions (*knowledge sources*) that are the basis of reasoning. Knowledge sources are elementary in the sense that other parts of the problem solver can not influence their internal control. They represent the way in which a domain relation can be used to make inferences. For example: a *specification* and an *abstraction* inference might both use a *subsume* relation at the domain layer to make their inference.

The different *roles* the domain knowledge plays in the reasoning process is described by *meta-classes*. For example, a domain concept like *faulty-transistor* may play the role of a *finding*, but may also play the role of a *hypothesis*. In this case finding and hypothesis represent domain independent use of domain concepts for diagnostic reasoning.

The third type of knowledge, the *task layer*, describes what goals are involved in a particular problem solving task and how the available knowledge sources can be ordered and applied to satisfy those goal. For example, the problem solving task *monitoring* can be carried out *data-driven* or *model-driven*, both representing different task structures that can be used to do monitoring.

The fourth type of knowledge concerns strategic knowledge. At the *strategic layer* strategies are described that control the overall reasoning process. It should, for instance, plan a particular task structure, monitor its execution, and, if needed, diagnose, repair or even substitute the current task structure with another task structure, until the desired problem solving goal is reached.

There is a tradeoff between the task and the strategic layer. In case of non-standard problems the problem solving process will use more strategic reasoning. This in contrast to more routine problems in which case the problem solver is more likely to execute an already known task structure.

## Qualitative prediction of behaviour

The qualitative reasoning community has delivered a number of implemented AI programs along with theoretical insights of what constitutes qualitative reasoning [2; 13]. Qualitative prediction of behaviour is an analysis task during which new properties of a system, namely states of qualitatively distinct behaviour (=output), are derived from a structural description of the system (=input). We have developed a description of this prediction task, in terms of the conceptual framework provided by KADS [5; 3], that unifies the theoretical insights mentioned above. To test the validity of our framework we have implemented a program called GARP [4][1] that can simulate qualitative reasoning as described by the original approaches. GARP is therefore a qualitative reasoning shell that integrates the basic approaches to qualitative reasoning.

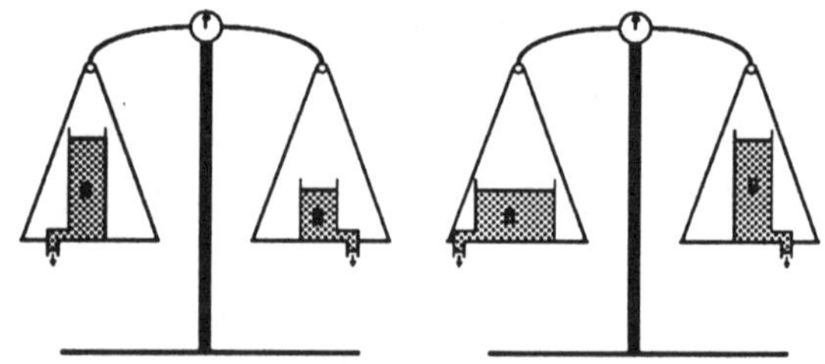

Figure 2: Two of the six balance problems

To investigate the cognitive validity of the theory underlying GARP, the qualitative reasoning task was operationalised with six balance problems (see for examples figure 2). The problem was to predict the behaviour of balances with containers filled with water on each balance-arm. Through outlets near the bottom of the containers the water gradually flowed out of the containers. In each version of the problem the two containers differed in shape and in the amount of water they contained.

In the next sections the 4-layer model for solving these balance problems is described.

### Domain and inference layer

The original approaches to qualitative reasoning provide specific ontologies for structuring the knowledge at the domain layer (cf. [3]). At the inference layer the meta-classes describe how the domain knowledge is used and the knowledge sources describe the inferences that can be made. The framework distinguishes between the following meta-classes:

- *System model description*
  A description of the system during a period of time in which the behaviour of the

[1] GARP -A Generic Architecture for Reasoning about Physics- is implemented in SWI-Prolog [14]. Both GARP and SWI-Prolog are distributed without charge for non-commercial purposes. Please contact the first author for more details.

system does not change (=SMD). The notion of change is rather subtle because the actual (real-world) system may change whereas from a qualitative point of view its behaviour remains constant.

- *System elements*
  Entities (like physical objects) from the real-world explicitly represented in the problem solving process.
- *Parameters*
  The properties of system elements.
- *Parameter values*
  The values a parameter can have.
- *Quantity spaces*
  An ordered set of values a specific parameter can have.
- *Parameter relations*
  Dependencies between parameters.
- *Qualitative calculi*
  Define the semantics of a parameter relation.
- *Mathematical model*
  A set of relations that holds during a particular SMD.
- *Partial models*
  Partial models[2] are used to further specify an SMD. They represent the general knowledge of the domain, like for example the properties of a liquid contained by a container of some sort.
- *Transformation rules*
  Represent knowledge about how to find successive SMD's. Three types of rules have been identified: *termination*, *ordering* and *continuity* (they are explained in more detail below).
- *Behaviour descriptions*
  A set of SMD's ordered in time. It represents the possible behaviour of some real-world system.

There is not enough space in this paper to completely describe the domain knowledge needed to represent the balance problems. Some of the most important entities of the domain knowledge are therefore enumerated below, together with the meta-class that describes their role.

- *System elements*
  balance, container, water, balance-arm, outlet, contain-relation support-relation, connect-relation.
- *Parameters*
  pressure, amount, mass, volume, flow-rate, height, width.
- *Parameter values*
  zero, minus, plus, absent, present, low, high, equilibrated
- *Parameter relations*
  pressure → flow-rate, height → pressure, amount = volume, amount = mass.
- *Partial models*
  water-flow-from-container-to-world, balance-with-weights, contained-liquid

The partial models in the domain layer specify several possible viewpoints that subjects can use in solving the balance problems. In particular we expect the partial model 'water flow from container to world' to differ for subjects with respect to the factors that determine the speed of the water outflow:

A size of the outlet

B height of the water column

C volume of the water column

D width of the water column

Given the list of meta-classes, the canonical inferences (knowledge sources) used in qualitative prediction of behaviour can be described. Figure 3 gives an overview of these inferences.

The purpose of the *compound specify* inference is to develop a 'complete' description of a particular state of behaviour of a system (=augmented SMD), which means assigning a qualitative value to each parameter that is used to describe the system. Two inferences must be carried out in order to arrive at such a description, *assembling* a mathematical model and *computing* the corresponding qualitative values. Assembling a mathematical model (=partially augmented SMD) is done by finding partial models that apply to the input system (=partial SMD). Once the SMD is augmented with a mathematical model, the computation of qualitative values can be done. There are two outputs possible, (1) contradiction, which means there is no set of values consistent with the current set of relations, (2) solution, which means one, or more sets of values are consistent with the current set of relations.

The *compound transformation* inference is concerned with identifying successive states of behaviour. *Termination* rules specify the conditions under which a particular state of behaviour will terminate and are used to *select* the possible terminations. *Precedence* rules specify the order in which changes take place

[2] In previous work we referred to partial models as 'system structures'.

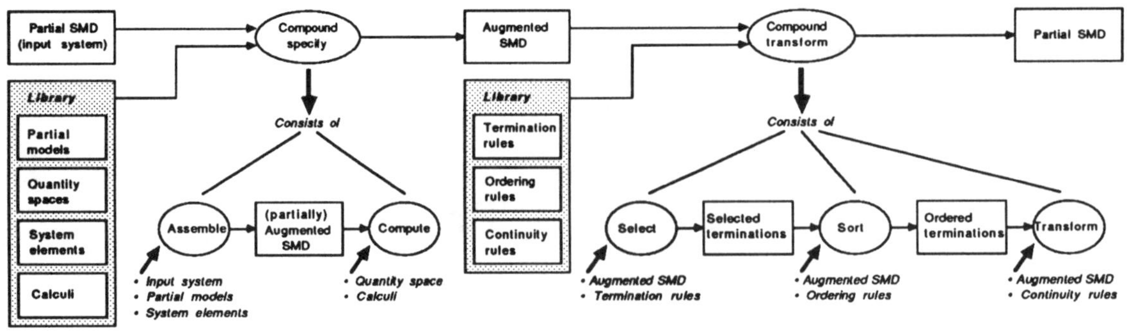


Figure 3: The inference structure for prediction of behaviour

and are used to *sort* the available terminations. Finally, *transition* rules specify the continuity conditions that need to be satisfied by the new state of behaviour in order to be a valid successor. They are, together with the ordered terminations, the input for *transforming* the current SMD into its successor.

### Task and strategic layer

In the prevaling approaches to qualitative reasoning the task layer is only minimally filled. There are in fact only two tasks: 1) finding all states of behaviour and 2) finding one specific trace of behaviour. In case of the latter additional input parameter values are taken into account to limit the number of SMD's that can be found.

Strategic knowledge is not present in the original approaches. It is not possible to monitor the inference process and modify or change the reasoning process automatically.[3] However, it is likely that the reasoning process will encounter difficulties, i.e. that the problem solving goal can not be reached. We expect two types of difficulties to emerge (cf. [1; 10]): first, problems with the available knowledge (what we refer to as 'knowledge conflicts'), and second, shortage of processing capacity. We will not discuss processing capacity further in this paper. For the knowledge conflicts concerning missing, ambiguous, or contradictory facts in the domain knowledge, we expect the following repairs:

- **Missing knowledge**
  *Practical repairs:*
  - read the question again
  - ask for additional information
  *Repairs by reasoning:*
  - continue after making an assumption
  - try extreme values
  - use analogy
  - use different/other domain knowledge
- **Ambiguous knowledge**
  - try one (randomly or based on heuristic)
  - try all
  - reason backwards from a known state
- **Contradictory knowledge**
  - try again
  - check computations
  - use other domain knowledge

Once a repair is selected its plan is tuned to fit on the specific instance of the impasse. This means that strategic reasoning may involve a revision at the task layer.

## Protocol analyses of the conceptual framework

To test the conceptual framework we compared problem solving activities as manifested by human subjects in think-aloud protocols with those predicted by the framework. The problem solving model that was built and implemented in GARP was therefore translated into a *coding template* with which the think-aloud protocols could be analysed. For every inference activity in the reasoning process that was distinguished in the model, a category was created and a code assigned. The subjects were ten psychology sophomores who had taken high school courses in physics. Three protocols were coded entirely, resulting in 673 coded expressions. All ten protocols were screened on the occurrence of following three phenomena:

- *The model is incomplete*
  There are expressions that can not be coded.
- *The model is too detailed*
  There are coding categories that are not

[3] There is some work going on as an extension of the constraint centered approach (cf. [11]) to filter (and thereby reduce) the number of generated states.

used.

- *The model is wrong*
  There are deviations of the expected order in which expressions appear.

Below the results of this analysis are described for each knowledge layer of the model.

## The domain layer

In going through the sequence of problems subjects seemed to have a tendency to abstract from the details of the system. While solving the first problems specific *system elements*, like container and water, were referred to. In the later problems the behaviour of the system was described as if only one object was present, the balance, with a mass-loss on two sides, one losing mass faster than or with equal speed as the other. In a similar way many *parameters* were used while solving the first problems. For example, height (or amount etc.) was seen as determining the pressure, which in turn determines flow-rate which equals mass-loss. In abstracted form the height determined the mass-loss directly. Although each specific model of the balance problems can be represented in GARP, the abstraction process itself can not.

Subjects used different *partial models*. One subject used viewpoint A (size of outlet) but shifted on a later version to D (width of water column). One subject used viewpoint B (height of water column) (although three other subjects also considered it). Six subjects used viewpoint C (volume of the water column), three of whom after showing doubts about other viewpoints. Only one subject used viewpoint D from the outset. The remaining subject used an alternative viewpoint: the weight of the air over the water-surface determines the water-pressure.

## The inference layer

*Compound specify* made up 37 percent of the protocols statements, with a ratio of 2:3 between *assemble* and *compute*. Assembling the mathematical model typically appeared in the protocols as expressions in which was stated that a parameter had a value or that a relation or inequality applied. Computing values made implicit use of a qualitative calculus. It appeared in remarks like: "because the pressure in A is higher the water flows out with more force". Below some examples of these inferences from the protocols are listed:

- *Specify*
  It just matters ...
  In a balance it just matters what the weights are
- *Compute*
  Well, I think first it goes down at the side of container A
  Because at the outset it is heavier off course
  Because here [ A ] is more water
- *Specify/Compute*
  Let me think if it flows here [ A ] faster than here [ B ]
  Here [ A ] the pressure is, let me think
  Yes, here [ A ] it is higher of course at the bottom of the container
  Does it influence the flow-rate, let me think
  Yes, the force of outflow will be bigger, undoubtedly

The sequence of assemble followed by compute was regularly violated and as such differed from what the model predicted. First for one part of the system, applicable partial models were found and computable parameter values were calculated. This was repeated for the next part of the system. It seemed as if founded partial models focussed the search for new ones (cf. [9]).

*Compound transform* made up 25 percent of the protocols statements. *Selecting* terminations was expressed in ways a state of behaviour ends: "then the amount of water in container A becomes equal to that in B". The *sorting* of terminations was only present in the protocols when the reasoning process was disrupted, i.e. if ambiguity in a situation was detected. Below some examples of these inferences from the protocols are listed:

- *Select (terminations)*
  A goes down ...
  Until B is entirely empty ...
- *Sort (terminations)*
  Yes, they either are empty at the same time or ...
  B is already ...
  I can not imagine A being sooner empty, but why can't I ?
  I just, it's more a feeling, let me think
  A has got greater pressure but ...
  No, I wouldn't know if they'd become empty at the same moment

The limited occurrence of the sorting inference can be explained in two ways. First, sorting occurs only if more than one termination is found. If so, the subjects did not find impossible terminations and terminations that were aspects of the same change, they just found possible ones (sometimes comprising different aspects). Second, impossible terminations and different aspects *were* found but the sorting of terminations in unambiguous cases is an automated process and therefore not reportable. The fact that the knowledge used to make this inference is not domain specific, makes an automated process plausible. However, additional research is needed to resolve this issue.

Little evidence was found for the third inference that constitutes the *compound transformation* (note that in figure 3 this inference, which follows sort, is also called transformation). We must however assume that subjects make this inference in order to preserve the continuity between one state and the next: values and inequalities that do not change in the next state are propagated. The lack of expressions confirming this inference can be explained by regarding it as a default mechanism. It provides a good example of how subjects deal with the frame problem, that is, unless something explicitly changes, they consider things to be constant over states.

### The task and strategic layer

A total of 40 percent of the protocol expressions was coded to involve strategic reasoning, of which 7 percent was devoted to the detection of impasses, 33 to the overcoming of these impasses. This includes 12 percent specification and 11 percent transformation tasks that were executed in, for example, reasoning backwards from the derived final state, starting all over again and re-checking the computations. The remaining 10 percent covered other repairs such as making assumptions, comparing competing knowledge, using analogies and reading the question again. Below follow examples of these repairs in the protocols:

- *Extremes*

  I wonder if in case of a narrow, very narrow and very high column ...

  If the pressure is higher than in a very wide container with a very thin layer of water

  Just to imagine if it matters, if it's volume that matters or just the height of the column

- *Assumption*

  I think it's volume that matters

  So they flow out with equal speeds and the balance remains equilibrated

- *Analogy*

  Maybe it's like, a bit strange maybe:

  Taller people have bigger feet, so the pressure is more spread

- *Back from end*

  So, after a while it will ...

  Yes, in the end it will come back to equilibrium, no doubt, but ...

  Let me see

The expressions that could not be coded, were all of one type: they referred to an activity that related different problems of the sequence. Apparently the subjects were able to see similarities in two versions of the problem. This also explained the occurrence of the immediate production of an answer without any apparent specification and transformation in a few places. When the model, built for one version of the problem, was applied to another in the sequence, the complete behaviour description was applicable so the answer could be produced immediately. This type of inferencing could not be accommodated for by the GARP model.

## Discussion

The model presented in this paper extends previous descriptions of qualitative prediction of behaviour by distinguishing between domain, inference, task, and strategic knowledge. The conceptual framework implemented in GARP, which is based on these knowledge types, appears to be very useful for describing and interpreting the reasoning processes involved in this problem solving task. Both the different viewpoints subjects have on the domain knowledge as well as their reasoning process can easily be modeled by the framework. The canonical inferences and the metaclasses defined in the model, provide strong means to interpret the steps of the reasoning process in the protocols. The notion of strategic reasoning explains disruptions and changes of order in which new states are determined.

However, from the presented experiment it also becomes clear that the model of the dynamics of the qualitative reasoning process of human subjects needs to be refined. In particular it is unclear what determines the order of the assemble and the compute inference during the specification of an SMD (=state of behaviour).

The modeling framework also does not account for learning over a number of problem solving sessions. As the subjects moved from one problem to another they abstracted from irrelevant details in the description and the analysis of the problem. The framework does not support this abstraction process. It seems however that this lack of support is part of a larger problem, namely the lack of structuring principles at the domain layer. The modeling framework provides no normative support to choose a level of abstraction at which the knowledge in the domain layer should be modeled.

The notion of a task structure was less useful in analysing the protocols. Experts performing a particular problem solving task may develop, through the repeated execution of the same sequence of inferences, a trace of this sequence and as such learn a task structure. That is, an instance of a strategic reasoning process. Non-experts, like the subjects in the experiment reported on here, just use general strategic reasoning.

In conclusion we believe that the model presented in this paper provides a strong basis for further research on the way humans perform qualitative prediction of behaviour. Such research should in particular focus on the learning aspects and the knowledge structuring principles that people use to develop their domain knowledge.

**Acknowledgement**

Bob Wielinga and Maarten van Someren provided helpful comments on earlier drafts of this paper. Also we would like to thank the referees who reviewed our paper for their detailed and accurate comments.

## References

[1] B. Bartsch-Sporl, M. Reinders, H. Akkermans, B. Bredeweg, T. Christaller, U. Drouven, F. van Harmelen, W. Karbach, G. Schreiber, A. Voß, and B. Wielinga. A tentative framework for knowledge-level reflection. ESPRIT Basic Research Action P3178 REFLECT, Deliverable IR.2 RFL/BSR-ECN/I.3/1, BSR Consulting and Netherlands Energy Research Foundation ECN, August 1990.

[2] D.G. Bobrow, editor. *Qualitative Reasoning about Physical Systems*. Elseviers Science Publishers B.V., Amsterdam, 1984.

[3] B. Bredeweg. Introducing meta-levels to qualitative reasoning. *Applied Artificial Intelligence*, 3-2:85–100, 1989.

[4] B. Bredeweg, M. Reinders, and B.J. Wielinga. Garp: A unifying approach to qualitative reasoning. Technical report, University of Amsterdam, January 1990.

[5] B. Bredeweg and B.J. Wielinga. Integrating qualitative reasoning approaches. In *Proceedings of ECAI-88, Munich*, pages 195–201, 1988.

[6] J. Breuker, B. Wielinga, M. van Someren, R. de Hoog, G. Schreiber, P. de Greef, B. Bredeweg, J. Wielemaker, J-P Billault, M. Davoodi, and S. Hayward. Model Driven Knowledge Acquisition: Interpretation Models. ESPRIT Project P1098 Deliverable D1 (task A1), University of Amsterdam and STL Ltd,, 1987.

[7] J. A. Breuker and B. J. Wielinga. Model Driven Knowledge Acquisition. In P. Guida and G. Tasso, editors, *Topics in the Design of Expert Systems*, pages 265–296, Amsterdam, 1989. North Holland.

[8] B. Chandrasekaran. Towards a functional architecture for intelligence based on generic information processing tasks. In *Proceedings of the 10th IJCAI*, pages 1183–1192, Milano, 1987.

[9] A.A. DiSessa. Phenomenology and the evolution of intuition. In D. Gentner A.L. Stevens, editor, *Mental Models*. Lawrence Erlbaum, Hillsdale, 1983.

[10] W. Jansweijer, J. Elshout, and B. Wielinga. The expertise of novice problem solvers. In *Proceedings ECAI'86, Brigthon*, 1986.

[11] B. Kuipers. Qualitative simulation. *Artificial Intelligence*, 29:289–388, 1986.

[12] L. Steels. Components of expertise. *AI Magazine*, (summer issue), 1990. Also as: AI Memo 88-16, AI Lab, Free University of Brussels.

[13] D.S. Weld and J. De Kleer. *Readings in Qualitative Reasoning about Physical Systems*. Morgan Kaufmann Publishers Inc., 1990.

[14] J. Wielemaker. SWI-Prolog 1.4 Reference Manual. Technical report, SWI, University of Amsterdam, Amsterdam, The Netherlands, January 1990.

[15] B. Wielinga and J. Breuker. Models of expertise. In *Proceedings ECAI'86*, pages 306–318, 1986.

# WHY DO THOUGHT EXPERIMENTS WORK?

Nancy J. Nersessian
Program in History of Science
Princeton University
Princeton, NJ 08540

## Abstract

Thought experiments have played a central role in historical cases of major conceptual change in science. They are important in both constructing new representations of nature and in conveying those representations to others. It is proposed that research into the role of mental modelling in narrative comprehension can illuminate how and why thought experiments work. In constructing and "running" the thought experiment, we make use of inferencing mechanisms, existing representations, and general world knowledge to make realistic transformations from one possible physical state to the next and this process reveals impossibility of applying existing concepts to the world and pinpoints the locus of needed conceptual reform.

## Introduction

Throughout the history of science there are numerous instances where scientists who have created major conceptual innovations employed thought experiments and used them as a means of conveying the new representations to others. Some, such as Einstein (1945), have claimed that a specific thought experiment was essential in their initial construction of the new representation. Using Einstein as an example, Wertheimer (1945) confirmed the centrality of that thought experiment in Einstein's "*Gestalt* switch", but did not analyze how and why it worked to create the conceptual change. Indeed, despite widespread recognition of the use and importance of thought experiments in conceptual change in science, there has been little analysis - by either psychologists or philosophers and historians of science - of how they function and why they are effective in bringing about change.

Within philosophy and history of science, earlier scholarship presented two poles of interpretation of their role in creating conceptual change. Duhem dismissed them as bogus precisely because they are "not only not realized but incapable of being realized" (Duhem 1914, p.202), i.e., they are not "experimental" in the customary sense. Koyré (1939, 1968), on the other hand, argued that their logical force is so compelling that they supplant real experimentation in the construction of new representations for phenomena. The "thought" part of the experiment predominates and shows the synthetic *a priori* nature of scientific knowledge.

Contemporary historians and philosophers of science by and large reject both extremes of empiricism and rationalism. They acknowledge that thought experiment, while not eliminating the need for real experiment, is an important heuristic for creating conceptual change in science. There have been a few recent attempts to analyze the function of thought experiments. While these are sketchy and limited, some do yield useful insights.

Two analyses focus on the empirical force of thought experiments. The notion that an experiment having real empirical consequences can take place in thought seems paradoxical. Yet, Kuhn claims:

"Thought experiment is one of the essential analytical tools which are deployed during crises and which then help to promote basic conceptual reform" (Kuhn, 1964, p.263).

He argues that thought experiments perform this function by showing that there is no consistent way, *in actual practice*, of using accepted existing concepts. That is, the thought experiment reveals that it is not possible to apply the conceptualizations we have of pehnomena consistently to real-world phenomena and that this practical impossibility translates into a logical requirement for conceptual reform. Gooding (1990) in his analysis of Faraday's experimental practices, picks up on Kuhn's analysis, rendering the empirical force of thought experiments in terms of their demonstration of what he calls the "impracticability of doing". Gooding is concerned to show how the real world experimenter's "procedural knowledge", i.e., tacit and explicit knowledge of practical skill, is utilized in constructing and manipulating thought experiments. Both of these analyses contain important insights about the relationship between the conceptual and the experiential dimensions of knowledge.

Two other analyses focus on the logical force of thought experiments. First, Brown (1987) claims that thought experiments are a species of *a priori* reasoning, but they get at something that cannot be derived from logical argumentation. His positive suggestion is that they provide a special window through which the mind grasps universals. It seems that Brown is trying to capture the idealizing function of thought experiments. However, his approach through linguistic analysis alone does not afford the possibility of understanding their experimental nature and why they have empirical consequences. Second, Norton (1986) claims that thought experiments can, in essence, be reconstructed as, and replaced by, arguments. This analysis is the most sympathetic for philosophers who wish to restrict reasoning to logical argumentation - whether deductive or inductive. Certainly thought experiments contain an argument. However, as Norton himself acknowledges, the argument can only be constructed *after the fact*, i.e., the argument is not evident until after the thought experiment has been executed. Additionally, he supports his claim by noting the presentation of a thought experiment contains particulars irrelevant to the generality of the conclusion. While this is correct, his emphasis reveals that he has failed to see the *constructive function of the narrative form* in which thought experiments are customarily presented.

This paper proposes that construing thought experiments as a species of mental modelling offers the best possibility of explaining how it is that an experiment made in thought can have both the logical and empirical force to compel and communicate conceptual change that the historical record seems to indicate. While certainly not uncontroversial, the notion that mental simulation is a central aspect of cognition is widely accepted among cognitive scientists. However the potential role of "mental modelling" in conceptual change, both in science and in general, has not been investigated. My aim is to show that bringing thought experiments under the perview of mental models research will enable us to better understand their function and their relationship to other aspects of scientific thinking. The weakness of the analyses of Brown and Norton lie in their taking the traditional philosophical route of construing a thought experiment as a form of reasoning with propositions. What Kuhn and Gooding are trying to capture may, in fact, not be able to be represented propositionally. My hypothesis is that propositional representations cannot capture the experimental dimension of a thought experiment. Running a *mental simulation is required for a it to be both "thought" and "experimental"*. The analysis presented here conceives of the original thought experiment as the construction of a mental model by the scientist who imagines a sequence of events. She then uses a narrative form to describe the sequence in order to communicate the experiment to others, i.e., to get them to run the corresponding simulation. While thought experiments are used extensively in scientific thinking and teaching, we will focus on a few that have been instrumental in creating major conceptual change in science.

## Case Studies: Galileo and Einstein

Although there are many great thought experimenters in the history of science, the ones who have attracted the most attention are Galileo and Einstein. Bowing to tradition and hoping to capitalize on what may already be familiar to the reader, I will give a brief presentation of a few of the thought experiments devised by Galileo and by Einstein to convey some sense of the variety such experiments display and to elicit some common features for further analysis.

Galileo's importance as a pivotal figure in the transition from the qualitative categories of aristotelian and medieval theories of motion to the quantitative representation of motion provided by Newton's mechanics is widely recognized. As shown in analyses by Koyré (1939) and Clavelin (1968), among others, Galileo drastically transformed the problem of how to go about constructing a mathematical representation of the phenomena of motion. That process, which Koyré called "mathematization", required constructing an idealized representation, quantifying this representation, and mapping the quantified representation back onto the real world. While it is now clear that Galileo must have performed many more real-world experiments than Koyré would have liked (See, e.g., Drake 1973; Naylor 1976; and Settle 1961), no one would deny the importance of his use of thought experiment in the mathematization process. Take, as example, his analysis of falling bodies (Galilei 1683, pp. 62-86).

<u>THOUGHT EXPERIMENT G1</u>: According to the aristotelian theory heavier bodies fall faster than lighter ones. This belief rests on a purely qualitative analysis of the concepts of 'heaviness' and 'lightness'. Galileo argued against this belief and constructed a new, quantifiable representation through a sustained analysis using several thought experiments and limiting case analyses. The outline of his use of these procedures is as follows. He calls on us to imagine we drop a heavy body and a light one, made of the same material, at the same time. We would customarily say that the heavy body falls faster and the light body more slowly. Now suppose we tie the two

bodies together with a very thin - almost immaterial - string. The combined body should both fall faster and more slowly. It should fall faster because a combined body should be heavier than two separate bodies and should fall more slowly because the slower body should retard the motion of the faster one. Clearly something has gone amiss in our understanding of 'heavier' and 'lighter'. Having pinpointed the problem area, Galileo then goes on to show that it is a mistake to extrapolate from what is true at rest to what happens when bodies are in motion. That is, he has us consider that when the two bodies are at rest, the lighter will press on the heavier, and therefore the combined body is heavier. But when we imagine the two bodies are in motion, we can see the lighter does not press on the heavier and thus does not increase its weight. What Galileo has done up to this point is use the thought experiment to reveal the inconsistencies in the medieval belief, the ambiguities in the concepts, and the need to separate the heaviness of a body from its effect on speed in order to analyze free fall. He then goes on, using the methods of thought experiment and limiting case analysis in tandem to show that the apparent difference in the speed of falling bodies is due to the effect of the medium and not to the difference in heaviness between bodies.

As the historian Clavelin has pointed out, it is crucial for quantifying the motion of falling bodies that 'heaviness' not be the cause of the difference in speed because then we could not be sure that motion would be the same for all bodies. Galileo went on to use a further thought experiment to demonstrate that the observed differences in speed should be understood as being caused by the unequal way media lift bodies.

THOUGHT EXPERIMENT G2: Galileo asks us to suppose, for example, that the density of air is 1, that of water 800, of wood 600 and of lead 10,000. In water the wood would be deprived of 800/600th of its weight, while lead would be deprived of 800/10,000th's. Thus, the wood would actually not fall (i.e., would float) and the lead would fall more slowly than it would in a less dense medium, such as air. If we extrapolate to a less dense medium, such as air, we see that the differential lifting effect is much less significant (e.g., 1/600 to 1/10,000 in air). The next move is to consider what would happen in the case of no medium, i.e., in extrapolating to the limiting case. With this move, Galileo says "I came to the opinion that if one were to remove entirely the resistance of the medium, all materials would descend with equal speed." (Galilei 1638, p. 75). Having performed the extrapolation in this way we can quantify this idealized representation of the motion of a falling body and know that it is relevant to actual physical situations; we need only add back in the effects of a medium.

Galileo repeatedly used thought experiments and limiting case analyses in tandem as shown by this example both in constructing a quantifiable representation of bodies in motion and in attempting to convey this new representation to others. Later I will propose the cognitive function of thought experiments and of limiting case analysis are much the same.

Einstein employed several thought experiments in developing the special and general theories of relativity. These thought experiments were central in his reconceptualization of 'space', 'time', and 'simultaneity'. He began his paper, "On the Electrodynamics of Moving Bodies" (1905), with the following thought experiment.

THOUGHT EXPERIMENT E1: Einstein asks us to consider the case of a magnet and a conductor in relative motion. There are two possibilities for explaining how a current is produced in the conductor. In the first case, the magnet is at absolute rest and the conductor moving. According to electromagnetic theory, the motion of the conductor through the magnetic field produces an electromotive force that creates a current in the conductor. In the second case, the conductor is at rest and the magnet moving. In this case, again according to electromagnetic theory, the motion of the magnet creates a changing magnetic field which induces an electric field that in turn induces a current in the conductor. However, with respect to the relative motions, it makes no difference whether it is the magnet or the conductor that is considered to be in motion. But according to the Maxwell-Lorentz electromagnetic theory, the absolute motions do create a difference in how we would explain the production of a current in the conductor. Since the explanatory asymmetry could not, in principle or in practice, be accounted for by the observable phenomena - the measurable current in the conductor - Einstein argued that this supported his conclusion that "the phenomena of electrodynamics as well as of mechanics possess no properties corresponding to the idea of absolute rest" (p.37).

Although we cannot discuss them here, two more thought experiments figure crucially in the complete analysis. The second thought experiment in the paper is the most famous one in which Einstein constructed an operational definition for the concept of simultaneity.

In a similar manner many thought experiments figured in Einstein's constructing and communicating the general theory of relativity, i.e., the field representation of gravitational action. We will just consider one he presented in various formats but claims to have first conceived in 1907.

THOUGHT EXPERIMENT E2: Einstein (1917, pp.66-70) asks us to imagine that a large opaque chest, the interior of which resembles a room, is located in space far removed from all matter. Inside there is an observer with some apparatus. In this state, the observer would not experience the force of gravity and would have to tie himself with strings to keep from floating to the ceiling. Now imagine that a rope is connected to the outer lid of the chest and a "being" pulls upward with a constant force, producing uniform acceleration. The observer and any bodies inside the chest would now experience the very same effects, such as a pull towards the floor, as in a gravitational field. The experiment demonstrates that the behavior of a body in a homogeneous gravitational field and one in a uniformly accelerated frame of reference would be identical. Once we see that there is no way of distinguishing these two cases we can understand the importance of the Newtonian law that the gravitational mass of a body equals its inertial mass: these are just two manifestations of the same property of bodies. That is, we have a different interpretation for something we already knew.

## Common Features

Before beginning to sketch a way of understanding thought experiments and their role in conceptual change in terms of the mental models framework, we need first to glean some common features of thought experiments from the narratives presented above. While there is great variety among thought experiments, in general, the ones presented here do exemplify important salient features.

FEATURE 1: By the time a thought experiment is public it is in the form of a narrative. The narrative has the character of a simulation. It calls upon the reader/listener to imagine a dynamic scene - one that unfolds in time. The invitation is to follow through a sequence of events or processes *as one would in the real world*. That is, even if the situation my seem bizarre or fantastic, such as being in a chest in outer space (E2), there is nothing bizarre in the unfolding: objects float as they would in the real world in the absence of gravity. The assumption is that if the experiment could be performed, the chain of events would unfold according to the way things usually take place in the real world.

FEATURE 2: A thought experiment embodies specific assumptions - either explicit or tacit - of the representation under investigation. It usually exposes inconsistencies or exhibits paradoxes that arise when we try to apply certain parts of that representation to a specific situation, such as 'heavy' and 'light' to falling rocks (G1). The paradox can take the form of a contradiction in the representation, e.g. it requires that an object be both heavy and light, or of something being not physically possible, e.g., observing the asymmetry required by electromagnetic theory (E1).

FEATURE 3: By the time a thought experiment is presented it always works and is often more compelling than most real-world experiments. We rarely, if ever, get a glimpse of failed thought experiments or avenues explored in the construction of the one presented to us. Some experiments, such as G2, could potentially be carried out - at least until the analysis extrapolates to the limit. Others, such as E1, underscore that doing a real-world experiment could not provide the data the theory requires. While others, such as E2, are impossible to carry out in practice, either in principle or because we do not yet have the requisite level of technological achievement. However, once understood, a thought experiment is usually so compelling in itself that even where it would be possible to carry it out, the reader feels no need to do so. The constructed situation is apprehended as pertinent to the real world either by revealing something in our experience that we did not see the import of before - e.g., the measurable current in the stationary and in the moving conductor is the same, so on what basis can we support the difference in theoretical explanation? - or by generating new data - e.g., in the case of no medium, lead and wood would fall at the same speed - or by making us see the empirical consequences of something in our existing representation - e.g., the attributes called 'gravitational mass' and 'inertial mass' are the same property of bodies.

FEATURE 4: The narrative presentation has already made some abstraction from the real-world phenomena. For example, certain features of objects that would be present in a real experiments are eliminated, such as the color of the rocks and the physical characteristics of the observers. That is, there has been a prior selection of the pertinent dimensions on which to focus, which evidently derives from our experience in the world. We know, e.g., that the color of a rock does not effect its rate of fall. This feature strengthens our understanding of the depiction as that of a prototypical situation of which there could be many specific instances. In more colorful narratives there may be more irrelevant features in the exposition, but these most often serve to reinforce crucial aspects of the experiment. For example, in one version of the chest - or "elevator" - experiment

(E2), Einstein depicts the physicist as being drugged and then waking up in a box. This colorful detail served to reinforce the point that the observer could not know beforehand if he were falling in outer space or sitting in a gravitational field. And, in the version discussed above, the opacity of the chest is to prevent the observer from seeing if there are gravitational sources around.

## Thought Experimenting as Mental Modelling

Rendering thought experiments as a species of mental modelling supports the interpretation that when they are employed in conceptual change, they are "essential analytical tools" in the process. While there is an extensive "mental models" literature, only recent work by Qin & Simon (1990) attempts, explicitly, to capture the simulation process of a thought experiment, which they characterize as a "mental image". There are many things we do not yet understand about the processes involved in mental simulation and how these differ from propositional reasoning. Thus, I can only hope the sketch I present will persuade the reader that following this direction does offer good prospects for accounting for both the "thought" and the "experiment" aspects of thought experiments, and for explaining how they can be "essential tools" in the process of conceptual change.

We can only speculate about what goes on in the mind of the historical scientist in the original thought experiment. Scientists have rarely been asked to discuss the details of how they went about setting up and running such experiments. However, reports of thought experiments are always presented in the form of narratives that call upon the reader/listener to simulate a situation in his or her imagination. Thus, drawing on what we think we know both about the processes through which we imagine or "picture" in general, and through which we comprehend any narrative, may help us to answer that most perplexing question about thought experiments: *how can an "experiment" carried out in thought have such powerful empirical force?* The most pertinent aspect of mental models research for this analysis is that which investigates the hypothesis that understanding a narrative involves the construction of a mental model. In the case of thought experiments we need to understand how: (1) a narrative facilitates the construction of an experimental situation in thought and (2) thinking through the experimental situation has real-world consequences. Framed in mental models terms, the "thought" dimension would include constructing a mental model and "running" a mental simulation of the situation depicted by the model, while the "experimental" dimension comprises the latter and the connection between the simulation and the world.

Briefly, the mental models thesis about text comprehension is that understanding the meaning of a narrative involves relating linguistic expressions to models; i.e., the relationship between the words and the world is mediated by the construction of a structural analog to the situations, processes, objects, events, etc. depicted by a text (Franklin and Tversky 1990; Mani and Johnson-Laird 1982; Johnson-Laird 1983; McNamara and Sternberg 1983; Morrow *et al.* 1989; and Tversky 1990). What it means for a mental model to be a "structural analog" is that it embodies a representation of the spatial and temporal relationships between, and the causal structure connecting, the events and entities of the narrative. In constructing and updating a representation, the reader would call upon a combination of conceptual and real-world knowledge and would employ the tacit and recursive inferencing mechanisms of her cognitive apparatus.

That the situation is represented by a mental model rather than by an argument in terms of propositions is thought to facilitate inferencing. We can actually generate conclusions without having to carry out the extensive computations needed to process the same amount of background information propositionally. The conclusions drawn are limited to those that are directly relevant to the situation depicted. The ease with which one can make inferences in such simulative reasoning has suggested to some that mechanisms either used in - or similar to those used in - perception may be involved. If we do employ perception-like mechanisms here many inferences would be immediate.

To date, most empirical investigations of the hypothesis have focused on the representation of spatial information by means of mental models. The main disagreement has been over whether the representation involves a perception-like image or is "spatial", i.e., allows different perspectives and differential access to locations (See, Tversky 1990). It is not clear that this debate has any bearing on the function of thought experiments. Far more important is the question of how knowledge and inferencing mechanisms are employed in running and revising the simulation, and on this subject there is nothing beyond some explorations of how implicit knowledge of causal relations influences updating a model (Morrow *et al.* 1989).

Returning to thought experiments, the proposal offered here is that the cognitive function of the narrative form of presentation of a thought experiment to others is to guide the construction of a structural analog of the prototypical situation depicted in it. Over the course of the narrative, we are led to run a simulation that unfolds the events and processes by constructing, isolating, and manipulating the pertinent dimensions of the analog phenomena. The constructed situation inherits empirical force by being abstracted from both our experiences and activities in, and our knowledge, conceptualizations, and assumptions of, the world. In running the experiment, we make use of inferencing mechanisms, existing representations, and general world knowledge to make realistic transformations from one possible physical state to the next. In this way, the data that derive from a thought experiment, while constructed in the mind, are empirical consequences that at the same time pinpoint the locus of the needed representational change.

Characterizing thought experiments in this way also provides new insight into limiting case analysis, which has figured prominently in conceptual change in science as well. This form of idealization can be construed as a species of thought experiment. In this species the simulation consists of abstracting specific physical dimensions to create an idealized representation, such as of a point particle falling in a vacuum. The isolation of the physical system in thought allows us to manipulate variables beyond what is physically possible. Just what dimensions produce the variation and how to extrapolate from these may be something we determine initially in real-world experimentation, but the last step can only be made in the imagination. In physics, it is the idealized representation that is quantifiable. However, the idealized representation is rooted in and relevant to the real world because it has been created by controlled extrapolation from it. We get from imagination to application to the real world by adding in some of the dimensions we have abstracted, again in a controlled process.

## Conclusion

Thought experiments play a central role in conceptual change in science. In constructing and "running" the experiment, a scientist is able to demonstrate how and why specific concepts are not applicable to the world. This understanding forms the basis of problem-solving efforts to construct an empirically adequate conceptualization. However the initial thought experiment is constructed and run, it is conveyed to others in the form of a narrative. I have proposed that we understand thought experiments as a species of simulative reasoning and that we draw upon research into the construction and running of mental models during narrative comprehension to understand how they can function to create change. In linking the conceptual and the experiential dimensions of human cognitive processing, the mental models analysis offers the possibility of explaining how thought experiments demonstrate the undesirable real-world consequences of a conceptualization, thereby compelling change. To develop and test this hypothesis, it will be necessary to have a better understanding of the processes through which mental simulations are "run".

There are further implications of this proposal for cognitive science. Historical cases provide valuable data on the processes through which individual scientists change their representations of nature (Nersessian 1991a). If - as a number of cognitive psychologists are claiming - the processes of conceptual change in cognitive development and in learning are indeed like those of major scientific revolutions, the potential role of thought experiments in creating these kinds of conceptual change should be explored (See, Nersessian 1991b).

## References


Brown, J.R. 1986. "Thought Experiments Since the Scientific Revolution." *International Studies in the Philosophy of Science* 1: 1-15.

Clavelin, M. 1968. *The Natural Philosophy of Galileo: Essay on the Origins and Formation of Classical Mechanics.* Cambridge, MA: MIT Press, 1974.

Drake, S. 1973. "Galileo's Experimental Confirmation of Horizontal Inertia: Unpublished Manuscripts." *Isis* 64: 291-305.

Einstein, A. 1905. "On the Electrodynamics of Moving Bodies." in *The Theory of Relativity*, New York: Dover, 1952.

Einstein, A. 1917. *Relativity: The Special and the General Theory*. London: Methuen, 1977.

Einstein, A. 1946. "Autobiographical Notes." in *Albert Einstein: Philosopher - Scientist.* (ed.) P.A. Schilpp. Evanston, Il.: Open Court, 1949.

Franklin, N. and Tversky, B. 1990. "Searching Imagined Environments." *Journal of Experimental Psychology* 119: 63-76.

Galilei, G. 1638. *Two New Sciences.* (trans.) S. Drake. Madison: University of Wisconsin Press,1974.

Gooding, D. 1990. *Experiment and the Making of Meaning: Human Agency in Scientific Observation and Experiment.* Dordrecht: Kluwer Academic Publishers.

Johnson-Laird, P.N. 1983. *Mental Models*. Cambridge: Harvard University Press.

Koyré, A. 1939. *Galileo Studies*. Atlantic Highlands, N.J.: Humanities Press, 1979.

--------. 1968. *Metaphysics and Measurement*. Cambridge.MA: Harvard University Press.

Kuhn, T.S. 1964. "A Function for Thought Experiments." in *The Essential Tension: Selected Studies in Scientific Tradition and Change*. Chicago: University of Chicago Press, 1977.

Mani, K. and Johnson-Laird, P.N. 1982. "The Mental Representation of Spatial Descriptions." *Memory and Cognition* 10: 181-187.

McNamara, T.P. and Sternberg, R.J. 1983. "Mental Models of Word Meaning." *Journal of Verbal Learning and Verbal Behavior* 22: 449-474.

Morrow, D.G., Bower, G.H., and Greenspan, S.L. 1989. "Updating Situation Models during Narrative Comprehension." *Journal of Memory and Language* 28: 292-312.

Naylor, R. 1976. "Galileo: Real Experiment and Didactic Demonstration." *Isis* 67: 398-419.

Nersessian, N. J. 1991a "How Do Scientists Think? Capturing the Dynamics of Conceptual Change in Science." in *Cognitive Models of Science*, (ed.) R. Giere. *Minnesota Studies in the Philosophy of Science 15*. Minneapolis: U. of Minnesota Press.

Nersessian, N.J. 1991b. "Constructing and Instructing: The Role of 'Abstraction Techniques' in Developing and Teaching Scientific Theories." in *Philosophy of Science, Cognitive Science, and Educational Theory and Practice*, (eds.) R. Duschl and R. Hamilton. Albany: SUNY Press.

Norton, J. 1986. "Thought Experiments in Einstein's Work." Unpublished manuscript.

Qin, P. and Simon, H.A. 1990. "Laboratory Replication of Scientific Discovery Processes." *Cognitive Science* 14: 281-308.

Settle, T. 1961. "An Experiment in the History of Science." *Science* 133: 19-23.

Tversky, B. 1990. "Induced Pictorial Representations." Unpublished manuscript.

Wertheimer, M. 1945. *Productive Thinking*. New York: Harper.

# Representing Aspects of Language

**Robert F. Port**
Departments of Linguistics and Computer Science
*port@ucs.indiana.edu*

**Timothy van Gelder**
Department of Philosophy
Indiana University, Bloomington, Indiana 47405
*tgelder@ucs.indiana.edu*

## Abstract

We provide a conceptual framework for understanding similarities and differences among various schemes of compositional representation, emphasizing problems that arise in modelling aspects of human language. We propose six abstract dimensions that suggest a space of possible compositional schemes. Temporality turns out to play a key role in defining several of these dimensions. From studying how schemes fall into this space, it is apparent that there is no single crucial difference between AI and connectionist approaches to representation. Large regions of the space of compositional schemes remain unexplored, such as the entire class of active, dynamic models that do composition in time. These models offer the possibility of parsing real-time input into useful segments, and thus potentially into linguistic units like words and phrases.

## Introduction

What is the relationship between the kinds of symbolic representations deployed in "classical" cognitive models and representations in their connectionist counterparts? The study of human linguistic capability is a particularly appropriate domain in which to discuss this issue. But we must not let theoretical linguistics define the problem. Linguistics presumes the separability of linguistic knowledge and language use, the distinction of competence from performance. There is much more to the linguistic cognition than syntax. Still, language implies a system of complex representational structures for various purposes. And one of these is *orthography*, the level of an author or editor producing text in a "received" form. The discrete, static, concatenated nature of printed language does imply a model for the underlying cognitive architecture. The intuitive appeal of this view of language has inhibited consideration of the many problems in accounting for speech perception and production, the lexicon, slips of the tongue, etc, that imply consideration of other kinds of models. Both AI theorists and linguists have regarded natural language as basically consisting of concatenated structures of discrete, arbitrary symbols, and have constructed models of language processing that postulate internal representations of the same general kind. Connectionists working on natural language have been developing alternative distributed representational schemes, but tend to agree with traditionalists regarding certain crucial ingredients of the traditional picture — such as the need for compound representations systematically constructed out of standard parts. So what, if anything, really differentiates traditional symbolic and connectionist forms of compositional representation?

There is a widespread tendency to suppose that there must be some one crucial issue which neatly separates traditional symbolic representations from connectionist representations. We believe, however, that no such magic key exists. Rather, there is a variety of ways in which connectionist schemes of representation differ, to a greater or lesser extent, from traditional schemes of representation. This is true even when we restrict attention to specifically compositional schemes.

We consider a representational scheme to be compositional if it systematically constructs complex representations from basic compoundable units, such that the semantic and causal significance of the compound whole is a function of the significances of its basic parts. We describe individual representations belonging to a compositional scheme as *symboloids* in order to stress that the category of representational schemes we are interested in includes, but is considerably more general than, traditional symbolic representations. There must also be included various kinds of fuzzy and dynamic distributed patterns that serve a representational function for nervous systems.

To understand this situation we find it useful to view traditional and connectionist schemes as varying along a number of abstract dimensions defining a space of possible schemes of compositional representation. Classical systems occupy one relatively restricted region (centered on prototypical symbolic schemes such as the structures in LISP or formal logic) while various connectionist schemes tend to be widely scattered across other regions. Even different instantiations of natural language, such as printed English and spoken utterances, are found at quite distinct points. The kinds of neural representations that underlie natural language capacities most likely occupy several quite different regions in the space depending on which linguistic skills one looks at (e.g., interpreting speech directed at you, editing a text, learning a new word, mimicking someone's dialect, etc.).

Our research is aimed at isolating and clarifying the most important dimensions of this space — to distill out the conceptual "principal components" (or so we hope). We seek a picture of compositional representation in cognitive systems that is more general and inclusive than has previously been available. An important result of this search is that, from the perspective afforded by an understanding of these various dimensions, one can see that standard AI-style schemes of representation have no monopoly on compositionality. Instead, they merely occupy one relatively narrow corner of the space of possible systems that allow composition of tokens. Further, we think that exploring other regions of this space will be more fruitful for cognitive science research than developing variants within this narrow region. The following is a tentative list of the most general dimensions of the space of symboloidal representational schemes. The issue should not be which scheme is the correct one, but rather which scheme is most appropriate for which phenomena?

## Typology of Representations

The major dimensions fall into at least three groups: (1) intrinsic properties of the compoundable tokens, (2) the manner in which composition of tokens takes place and (3) the functional role of symboloids in the operation of the system. We do not claim these are a minimal set of dimensions. (For example, there are probably differing possible assumptions about semantics, that is, about the nature of the world itself.) We have tried to choose dimensions that are independent and which clarify the range of possible solutions to the representation problem. In addition to describing particular computational models, we will make reference both the linguists' models of language and to various aspects of knowledge about language exhibited by skilled users.

### Properties of Basic Tokens

The tokens are the most basic entities with representational significance which can be compounded into derived structures. In natural language processing, individual words are assumed (by both linguists and high-school English teachers) to serve as the basic units. As a first-order model, this is satisfactory. Otherwise, orthographies and typographical conventions would be much more problematic than they are. But Bolinger among others has frequently pointed out how difficult it is to nail down the actual list of the particular items that have just the right properties (Bolinger, 1975). Language is not just a set of morphemes plus rules of syntax. Speakers use and 'know' linguistic fragments that come in many sizes: from submorphemic ideophones to words, idioms, and Bolinger's 'collocations', as well as 'cliches' and even entire sentences and paragraphs of boilerplate (as in genres like wills and academic recommendation letters). The nature of these mysteriously constrained yet mysteriously flexible lexicalized 'units' lies well beyond the grasp of current representational schemes employed in linguistic theory. Although an orthography can afford to ignore these data, a model of human linguistic representation must support all of these kinds of linguistic knowledge.

**Discrete vs. Continuous Token Set.** Basic tokens are physical entities that gain their type-identity (e.g., being an instance of the word *cat*) by exhibiting characteristic variation along certain critical dimensions (e.g., placement of ink on the page). These dimensions can be thought of as delineating a space of possible physical items which might count as tokens (e .g., ink marks on a page or a set of letter strings). This first issue concerns how the basic tokens fall in this space. Are they sparsely or densely packed into it? That is, are there entities which count as basic tokens in the scheme 'between' any two tokens of the scheme? For example: the so-called 'phonemes' or 'phonological segments' used to describe particular languages are assumed to be discrete (that is, located far apart in stimulus space — at least one feature step apart), while the acoustic and articulatory phenomena themselves are clearly densely packed or continuous.

**Static vs. Dynamic Tokens.** Standard symbolic models invariably assume that basic tokens are static. By this we mean that the tokens exist indefinitely and change only when explicitly altered — like the basic vocabulary of predicate calculus. They provide a conceptual model for standard orthographies of, say, English and French. The inventory of units is relatively static and the objects (the letters and words) endure fairly indefinitely once converted to the printed medium.

These should be contrasted with the representation that presumably must underlie spoken words. This must change continuously over time as the pattern is either produced or recognized. Acoustic changes are essential to a word's being the word that it is, and to its being correctly perceived (Liberman, et al, 1967). Sometimes temporal specifications can be very subtle indeed (Port and Crawford, 1989). It is likely that internal representations of spoken words are also dynamic in this sense, as suggested by some (Browman and Goldstein, 1986; Saltzman and Munhall, 1989).

With dynamic tokens, the type-identity of tokens is based on physical change over time. Another example is the differentiation of smells in the olfactory bulb of the rabbit (Skarda and Freeman, 1987). Familiar smells each exhibit a characteristic limit cycle through the activation state-space of a set of cells. There appears to be no single point in time when the instantaneous state 'represents' the identity of the scent.

Currently one can find connectionist models that employ both kinds of representations. Most networks produce static output vectors, that is, the dynamics of the network leads it toward a point attractor (eg, (Sejnowski and Rosenberg, 1987; Elman and Zipser, 1988)). These two properties of basic tokens, discrete/continuous and static/dynamic seem to be the most important ones, [1] but models also differ in the way composition of basic units takes place.

## Manner of Composition

How are representational tokens of whatever form actually combined to make compounds or higher-level representations? This ability is clearly essential to any representational scheme that is to be generally useful (Fodor and Pylyshyn, 1988).

**Concatenation vs. Superimposition.** Connectionist work has revealed the importance of distinguishing several ways of combining tokens into complex wholes (van Gelder, 1990). In traditional logic and AI, there is, of course, no change whatever in tokens themselves when composed, as when LISP atoms are strung together to form a complex statement. At the other extreme is superimposition, or simultaneous combination, as found, for example, in a Recursive Auto-Associative Memory or RAAM (Pollack, 1991; Blank et al., 1992). Such systems can be trained to push and pop the elements of hierarchical trees from a single distributed representation in a fixed-size group of nodes. Input tokens are systematically combined into completely distributed compounds by a sequence of learning processes. The result is a static representation from which the entire tree can be constructed even though no specific physical characters of the constituents is present. It is even possible to demonstrate structure-sensitive transformations on distributed representations using a RAAM (Chalmers, 1990). These demonstrations suggest that symbolic composition is not the only kind possible.

[1]There is at least one other property of tokens as well that may be relevant. How are basic tokens paired with what they represent? Are basic tokens with similar meaning also similar physically? Natural and formal languages use arbitrary symbols. That is, tokens the typically have no intrinsic relation to what they represent. This lack of meaning is a key ingredient in the notorious "grounding problem". Connectionist models often use arbitrary patterns as the basic compoundable units but not always. On the other hand, the stimulation patterns on a sensory surface of an animal (like the finger tips, retina and basilar membrane) illustrate highly non-arbitrary pairing of a stimulus pattern with meaning.

Between these simple physical concatenation and superimposition lies Context Sensitive Concatenation (Smolensky, 1988). Tokens with distinct physical identities are concatenated but they are affected by their combination with particular other tokens. These are illustrated by Elman's recent models (Elman, 1989) and the dynamic memory of Anderson and Port.

**Static vs. Temporal Combination.** A crucial issue that cuts across the previous one is whether the actual act of combining basic units to form compound representations occurs in time or only statically. (The temporality of combination should not to be confused with the temporal vs. static nature of basic tokens themselves that we discussed in the previous section.) Both standard computer languages and linguistic models (Chomsky, 1965) depend on static hierarchical structures.

Of course, if modelling of time as a sequence of symbols proves inadequate, one can always model time as 'just another parameter' and deposit some kind of time measurements within the representation itself (Klatt, 1976; Port, 1981). But eventually, before addressing the real world, the artificiality of this maneuver must be faced by any model of representation in a nervous system (Port, 1990).

Certain 'recurrent' connectionist models (eg, (Anderson and Port, 1990; Elman, 1989)) deploy complex representations that are essentially temporally extended. The basic components, attractor states, are appended in time, so the structure of a sequence is encoded in characteristic trajectories through state space. Systems along this line can be extended to exhibit limit-cycle dynamics. It is likely that a dynamic analog of concatenation in time can be achieved by passing across the saddle points that separate one attractor basin from the next for each sequential component. We suspect that complex representations formed by temporal combination must underlie a wide range of real-time cognitive processes in speech and language.

Thus, there appear to be several possible kinds of tokens. And each may support composition in several ways.

## Functional Role of Representations

Symboloidal schemes of representation exhibit certain kinds of properties that only become apparent when we consider how representations in that scheme are handled within the context of processing. It is to be expected that the radical differences in the nature of representational elements we are proposing will have consequences for their use.

**Passive vs. Active.** Representations are normally seen as inert structures that are distinct from the processes that manipulate them. This powerful assumption underlies the notion of a Universal Turing Machine. The embedded Turing machine is described as a set of rules, which the universal Turing machine can execute at its convenience. The representations of the embedded model sit on the tape waiting to be acted upon by the host machine.

Another possibility is that representations are self-executing processes. That is, a symboloid might propel itself toward new representational stabilities without need of an external executive that reads the states and executes the steps specified in a static rule table. Self-execution can be driven either by external input or by internal dynamic representations from elsewhere in the system. As a primitive example, the Interactive Activation models (McClelland & Rumelhart,1981; Elman and McClelland, 1986) for speech recognition have representations that are (weakly) active in this sense. As input feature vectors become active during word presentation, they excite phoneme-level units and eventually word-level units. Information flows up, down and laterally within the model such that evolving representations at one level drive the development of representations at other levels. No external observer (or universal Turing machine) reads the activation levels and to decide what to do next. The control of the model has been decentralized or distributed. The Anderson-Port model for auditory pattern recognition that will be described below is also active in this sense.

**Digital vs. Analog.** Do the representations enable operations upon them that are 'positive' and 'reliable' in the terms of (Haugeland, 1985)? In classical symbol systems, representations are digital in the sense that the most basic identifying and transforming processes can always be carried out with complete, unambiguous success (e.g., the executive system can tell that the symbol in the buffer is either *foo* or it is not; there is no question of its being "somewhat *foo*"). This digital character is, of course, supported by other properties of the representational scheme such as discrete basic tokens and the strict concatenation property.

In some connectionist schemes, however, we find representations that are not digital in this sense. For example, if one attempts to use the RAAM architecture to represent many sequences by sequential superimposition, the representations of stack states become so closely packed in the activation space of the relevant units that operations on the representations eventually cannot be carried out positively and reliably (Pollack, 1991). Such a system is essentially analog. As trees are reconstructed from the stack, reliability deteriorates. Things that are different are eventually forced into equivalence classes.

## Dynamic Models and Time

We think that these dimensions are useful in revealing both the possible diversity in symboloidal schemes of representation and some of the similarities that exist between the classical symbolic model and connectionist schemes. But the connectionist framework also lends itself to dynamic models (Hopfield, 1984; Grossberg, 1980) and these open up opportunities for a different way of conceptualizing representations (see (van Gelder, 1991) for further discussion). Thus, we now explore that possibility in a little more detail in the context of a model for recognition of auditory patterns. These patterns might be as short and complex as a familiar chair squeek or bird chirp, of intermediate length, like a syllable or a strum on a guitar, or they might be hierarchically structured auditory objects like words, melodies or sentences.

A crucial step in dealing with real-time input to a cognitive model is the step from *milliseconds* to abstract *event categories*. This a very difficult to achieve because abstract events typically occur with variable durations. Thus temporal measurements in milliseconds (even if nervous systems did have a way of measuring them) are simply the wrong kind of description (Rosen, 1978). It appears that dynamic systems provide a way to obtain abstract categories from real-time input.

**Dynamic Memory.** Processing of temporal patterns as dynamic representations has been demonstrated to a limited extent. For example, Simple Recurrent Networks (Elman, 1990; Elman, 1989) respond to learned sequential patterns by following a characteristic trajectory in activation space. We view these trajectories as dynamic representations, distinct for each input sequence. In this case, states of the system near the end of each sequence distinguish each learned pattern. This can be illustrated by looking a little closer at the model described by Anderson and Port.

A network was trained (by a form of supervised gradient-descent learning) to recognize sequential patterns of tones. The network has 6 input frequency bins of which only one was strongly active on each time frame (since the input was effectively sinusoidal over this frequency range). There were 7 nodes in the fully recurrent Dynamic Memory, as shown in Figure 1. The activation of each node on the next time step is the decayed value from the previous time step plus the squashed sum of weighted activations from all other nodes in the Dynamic Memory clique. In addition, each of these nodes receives inputs from a layer that indicates the energy in the six frequency bands. The system learned to recognize a particular sequence of 8 input spectra comprising a melody-like pattern (Anderson and Port, 1990).

The model was trained to recognize one or more melody-like target patterns by using one of the dynamic memory nodes as an identification node for each target. The weight vector after training enabled it to produce stable trajectories for target sequences ending in a particular corner of the space (where an identification node approached 1) for the last two time cycles of presentation of any target. The model achieved better than 90% correct identification on each of several tar-

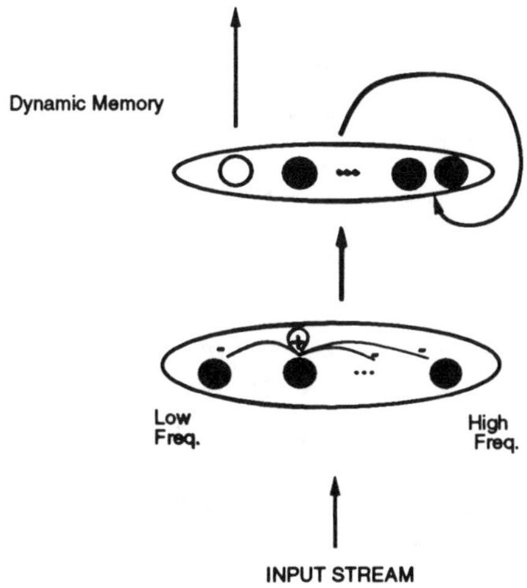


Figure 1: The network that recognizes melody-like 8-step sequential patterns. There are six input nodes, one for each frequency bin, and seven dynamic memory nodes that are fully connected (to itself and all neighbors. One (or more) of these nodes also serves as a category identification node and is trained to reach a target value for just the last 2 time frames of the pattern. The input-feature group learned to sharpen the spectrum by lateral inhibition of its neighbors.

gets. This contrasts with the response of the system to distractor patterns (that is, to non-trained patterns). In this case, trajectories are not distinctive. As shown in the left hand panel of Figure 2, when a target is presented, the memory followed a characteristic path through activation space (represented here by its two most significant principal components).

What if the rate of presentation is altered? When the patterns are presented at half-tempo, as shown in the right hand panel of Figure 2, essentially the same trajectory is produced. This suggests the possibility that chains of abstract objects could be learned by such a dynamic network. If the system waits to allow the input patterns themselves determine when the representational state should change, a major problem in control of cognitive models might be solved — the problem of how to parse time into useful pieces.

The critical property of rate invariance is achieved because the representation consists of a sequence of dynamically linked states. These states are not forced to proceed from one to the next at a rate clocked by the machine itself, but are instead controlled by changes in the inputs (in interaction with the internal state). These results show how static fixed points can be combined into a learned basic sequence when driven by external input (which might be sensory or produced by another group of nodes). It is likely such tokens can be combined into larger structures by a process that nests them into higher-level, more slowly changing, representations. The incoming pattern itself en-

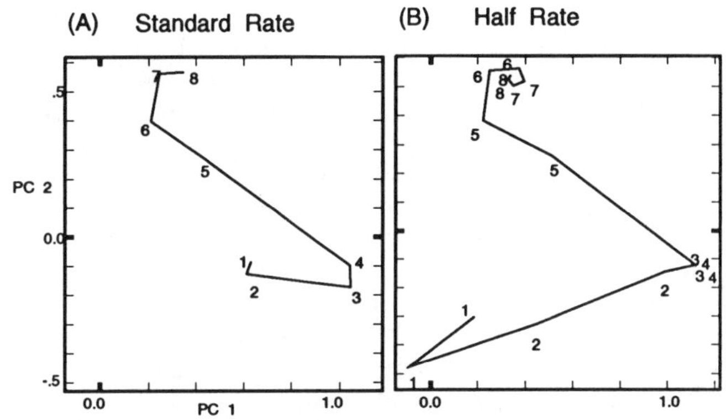


Figure 2: The trajectory of a target pattern plotted on the first two principal components of the 7-D activation space of dynamic memory. The patterns were 8-frame tone sequences (see Anderson and Port, 1990). The left panel shows the response after the presentation at the standard rate (the rate at which training was performed). The right panel shows the same network's response to the target when each spectral frame was presented twice.

trains the dynamics of the trajectory determining the entire composite representation. The representations of the constituents of these melody-like patterns actively lead toward those states characteristic of the learned pattern as a whole. This is the sense in which the tokens are dynamic and play an active role in the system. They determine, at the appropriate point in time, what will happen next.

Interestingly, when attention is addressed to the dynamics of recurrent networks as a way of accounting for interesting behavior, then one notices that connectionism itself is important in this enterprise particularly for its support of learning and distributed representations. Other, non-connectionist models might also support interesting dynamics.

## Conclusions.

One goal of this paper was to suggest that there are many more possible schemes that can serve a representational function than have been well-explored thusfar. Without a clear idea of the degree of independence of the parameters discussed here, it is difficult to estimate the size of this space. We hope that the framework offered here will help fix thoughts on this problem.

A second goal was to emphasize the possibility of dynamic, active representations that are meaningful because of their grounding in sensory surfaces in real-time (Gibson, 1968). Although symbolic models for language assume timeless static symbols *apriori*, systems embedded in the world must obtain appropriate symbolic functions on their own. Dynamic representations in the auditory pattern-recognition model exhibit natural rate invariance by entraining their dynamics to the temporal structure of stimulation. This is an important step toward an understanding of how the ordered, nested, symbol-like representations of language could be learned by a system that functions in real time.

Returning to language, we have suggested that dynamic models have many useful properties – especially for aspects of linguistic skill that are ignored by linguistics. Clearly, further technical developments will be necessary to demonstrate composition beyond the most basic level for dynamic representations. Many aspects of human speech have proven intractible to symbolic models. Perhaps symboloids can do better.

**Acknowledgements.** Thanks to Sven Anderson for important contributions to the development of these ideas and to David Chalmers as well as Lisa Meeden, Doug Blank and other members of the Mind, AI and Computation group at Indiana for helpful comments on various drafts. This research was funded in part by the Office of Naval Research, grant number N00014-91-J1261.

## References

Anderson, S. and Port, R. (1990). A network model of auditory pattern recognition. Technical Report 1, Indiana University, Institute for the Study of Human Capabilities.

Blank, D. S., Meeden, L. A., and Marshall, J. B. (1992). Exploring the symbolic/subsymbolic continuum: a case study of RAAM. In Dinsmore, J., editor, *Closing the Gap: Symbolism vs. Connectionism.* L. Erlbaum Assoc.

Browman, C. and Goldstein, L. (1986). Towards an articulatory phonology. *Phonology Yearbook*, 3:219–252.

Chalmers, D. J. (1990). Syntactic transformations on distributed representations. *Connection Science*, 2:53–62.

Chomsky, N. (1965). *Aspects of the Theory of Syntax.* MIT Press, Cambridge, Mass.

Elman, J. (1989). Representation and structure in connectionist models. Technical Report 8903, Center for Research in Language, UCSD, La Jolla, CA 92093-0108.

Elman, J. (1990). Finding structure in time. *Cognitive Science*, 14:179–211.

Elman, J. and Zipser, D. (1988). Learning the hidden structure of speech. *Journal of the Acoustical Society of America*, 83:1615–26.

Fodor, J. A. and Pylyshyn, Z. W. (1988). Connectionism and cognitive architecture: A critical analysis. *Cognition*, 28:3–71.

Gibson, J. J. (1968). *The Senses Considered as Perceptual Systems.* Harcourt Brace, New York.

Grossberg, S. (1980). How does a brain build a cognitive code? *Psychological Review*, 87(1):1–51.

Haugeland, J. (1985). *Artificial Intelligence: The Very Idea.* MIT Press.

Hopfield, J. (1984). Neurons with graded response have collective computational properties like those of two-state neurons. *Proceedings of the National Academy of Sciences*, 81:3088–3092.

Klatt, D. (1976). Linguistic uses of segmental duration in English: Acoustic and perceptual evidence. *Journal of the Acoustical Society of America*, 59:1208–21.

Pollack, J. (1991). Recursive distributed representations. *Artificial intelligence*, page in press.

Port, R. and Crawford, P. (1989). Pragmatic effects on neutralization rules. *Journal of Phonetics*, 16(4):257–282.

Port, R. F. (1981). Linguistic timing factors in combination. *Journal of the Acoustical Society of America*, 69:262–274.

Port, R. F. (1990). Representation and recognition of temporal patterns. *Connection Science*, 1-2:151–176.

Rosen, R. (1978). *Fundamentals of measurement and representation of natural systems.* North-Holland.

Saltzman, E. and Munhall, K. (1989). A dynamical approach to gestural patterning in speech production. *Ecological Psychology*, 1:333–382.

Sejnowski, T. and Rosenberg, C. (1987). Parallel networks that learn to pronounce English text. *Complex Systems*, 1:145–168.

Skarda, C. and Freeman, W. (1987). How brains make chaos in order to make sense of the world. *Behavioral and Brain Sciences*, 10:161–195.

Smolensky, P. (1988). On the proper treatment of connectionism. *Behavioral and Brain Sciences*, 11:1–74.

van Gelder, T. (1990). Compositionality: A connectionist variation on a classical theme. *Cognitive Science*, 14:355–384.

van Gelder, T. (1991). Connectionism and dynamical explanation. In *Proceedings of the 13th Meeting of Cognitive Science Society*, page in press, Hillsdale, NJ. L. Erlbaum Assoc.

# Intentions, Commitments and Rationality

**Munindar P. Singh***
Center for Cognitive Science
and Dept of Computer Sciences
University of Texas
Austin, TX 78712
USA

## Abstract

Intentions are an important concept in Cognitive Science and Artificial Intelligence (AI). Perhaps the salient property of (future-directed) intentions is that the agents who have them are *committed* to them. If intentions are to be seriously used in Cognitive Science and AI, a rigorous theory of commitment must be developed that relates it to the rationality of limited agents. Unfortunately, the available theory (i.e., the one of Cohen & Levesque) defines commitment in such a manner that the only way in which it can be justified reduces it to vacuity. I present an alternative model in which commitment can be defined so as to have more of the intuitive properties we expect, and be closely connected to agent rationality. This definition is intuitively obvious, does not reduce to vacuity, and has useful consequences, e.g., that a rational agent ought not to be more committed to his means than to his ends.

## 1 Introduction

Intentions, along with beliefs and desires, are an important component of the folk psychological concepts of intelligence and agency, especially as these concepts are used in Cognitive Science and Artificial Intelligence (AI). Recently, there has been some interest in the formalization of the semantics of intentions—i.e., of the conditions under which an agent may or may not be said to have an intention [Cohen and Levesque, 1990, McDermott, 1982, Singh, 1990, Singh and Asher, 1990].

The modern philosophical view is that intentions cannot be reduced to desires and beliefs (e.g., see [Brand, 1984, pp. 121–125], [Bratman, 1987, pp. 18–23] and [Harman, 1986, pp. 78–79]). Intentions are most often seen as being mutually consistent, compatible with beliefs, and direct or immediate causes of action (e.g., [Brand, 1984, p. 46]). This is a useful property for the purposes of this paper, since it helps relate intentions to rationality via actions. Intentions come in at least two shades: present-directed ones and future-directed ones. It is the latter that will interest us here.

Perhaps the salient property of future-directed intentions is that they involve *commitment* on the part of agents. This view has been gaining ground in the philosophical and AI literatures recently (e.g., see [Bratman, 1987, ch. 2], [Harman, 1986, p. 94] and [Cohen and Levesque, 1990, p. 217]). The idea here is that an agent who has an intention is in some way committed to it—not only does he intend to achieve the relevant condition right now, but would also intend to achieve it later, even as the circumstances changed, perhaps for the worse. Thus there is a certain amount of irrationality built into the very idea of commitment.

Yet there are philosophical as well as practical advantages to the view that agents are (or should be) committed to their intentions. While it admits present-directed intentions (e.g., for actions being done intentionally now), it gives primacy to future-directed ones. This is important since it allows an agent's intentional state *now* to influence his actions later. When conceived of as involving commitments, future-directed intentions allow an agent to coordinate his activities, both with his other activities, and with those of other agents. This is also practically important since it simplifies the design and analysis of complex agents, an important issue in AI.

The commitment-based view of intentions suggests that an agent reconsider his intentions only occasionally, rather than at every step. This allows even a computationally and perceptually limited agent to carry on fairly effectively in a world that, relative to his cognitive and physical capacities, is highly complex and changing rapidly. I take this much as granted in this paper.

In §2, I describe the notion of commitment as applied to intentions in philosophy; in §2.1, I describe how it is formalized in the theory of intentions of Cohen & Levesque [1990]; in §2.2, I point out the major problem with their approach, which seems to trivialize the notion of commitment; and in §2.3, I present my own intuitions about how commitment can be reconciled with rationality, and propose a definition of it that does not reduce to vacuity. Next, in §3, I explain the ontological framework of this paper. In §5, I present the formal language, and in §6, the formal model. In §7, I return to commitment and try to place it in the context of the formal model presented in §6. The approach presented here is general and independent of the exact semantics given to intentions—be it possible worlds based, sentential or any other.

## 2 What is Commitment?

An agent's commitment to his intentions differs from his commitment to his beliefs in that only the former can cause the agent to act. Following Bratman and Harman, I consider a mental or internal notion of commitment, rather than a social or external one—an agent is committed (by himself, as it were) to his intentions, not to anyone else. Commitment is thus a purely psychological concept, but has some obvious ramifications on the behavior of agents. As we have seen, it entails that the agent continue to hold on to his intentions over time, even as things get worse. If the circumstances change for the worse, he might try harder, i.e., spend more energy and time on it. E.g., if you are committed to being at the airport at 6:00pm, you would make more than one attempt to hail a taxi; if no taxis are forthcoming you might walk to a better location, rent a car, or request a friend for a ride, and so on.

Commitments help limited agents pursue complex goals that would otherwise be beyond their capacities. Thus, while commitments might prove quite irrational in some cases (e.g., where they lead the agent to do actions that are too expensive, or whose side-effects are too damaging), overall, in at least ordinary circumstances, they are quite rational for agents who cannot think too fast on the fly. E.g., your commitment to be at the airport might make you hijack a bus there (something that you might regret the rest of your life), but such cases of over-commitment are rare (or ought to be rare among rational agents). However, having the commitment saved you from repeatedly planning during the day to be in a neighborhood cafe at 6:00pm.

The moral of this is that (1) if you do not know too much about the present and future state of the world, and have too little time to think, then, on the average, commitments are a good way of being able to get something done; and (2) while you may have commitments, it is not a good idea to over-commit. It is these opposing intuitions about commitment that make it difficult to capture in a reasonably rigorous framework.

### 2.1 Commitments a la Cohen & Levesque

Cohen & Levesque (hereinafter C & L) agree that commitment is one of the most important characteristics of intentions. They capture commitment as a form of persistence over time in the definition of "persistent goal" [1990, p. 236] and define intentions as special kinds of persistent goals [1990, pp. 245, 248]. This takes care of the positive part of commitments. C & L recognize that this could easily lead to over-commitment and define intentions as a persistent goal that the agent persists with precisely till the point where either (1) he comes to believe that it has been satisfied; or (2) he comes to believe that it will never be satisfied. This is obviously too strong: in many cases an agent should not persist with an intention even though neither of (1) and (2) hold—e.g., Joe can intend to go to Mars, but give up that intention when he realizes that he does not want to suffer through the training. Later in their paper, C & L also allow intentions to be dropped in a third way—when the "reason" for adopting them is no longer valid [1990, pp. 254–255]. However, this would not help in Joe's case: he might still persist with his reason for his original intention, which is to be mentioned in the history books as one of the pioneers of interplanetary travel. Thus it is not easy to give up an intention in this theory.

We can try to weaken C & L's requirement for dropping an intention by (1) generalizing believing that an intention has been achieved to believing that the intended condition would hold even if the agent does not perform any (costly) actions to achieve it; and (2) generalizing believing that an intention is impossible to achieve to believing that it is too expensive to try achieving. Without some motivation on grounds of rationality, C & L would have no legitimate basis for their definitions. Indeed, they are quite explicit that their goal is to capture the *normative* criteria for the "rational balance" between (among other things) (a) committing to, and (b) dropping intentions [1990, p. 214].

### 2.2 Critique of C & L's Approach

Reasonable though the idea of treating commitment as temporal persistence may seem, it has some major philosophical and technical shortcomings. Remember that all we intuitively wanted as a property of commitment was that it lead to persistence under ordinary circumstances, *not* that it be *identified* with plain persistence. As I described above, C & L's solution to the problem of intentions never being dropped is a special case of the maxim *"intend something as long as it is useful to do so"*; in other words, as long as the intention (or acting for the intention) has a positive expected utility—the expected utility is negative or zero if the agent believes that the intention has already been achieved or believes that it never will be.

This maxim, which seems to underlie C & L's proposal, is not just true; it is *tautologous!* A rational agent, it says, should have an intention only so long as he believes it to be beneficial to him, all things considered. I.e., at any given time (in any given situation), whether an agent ought to persist with an intention or not depends on whether or not it has positive expected utility for him. But this is really saying that the concept of intentions is redundant in the theory—if the agent is going to look at what is best for him in each situation, then of what use are his intentions to him? And what use is the concept of intentions to us, as theoreticians? In this framework, agents have some memory in that they can continue with their intentions, but must deliberate about these intentions from moment to moment. Thus one of the major philosophical intuitions about the concept of (future-directed commitment-based) intentions is lost (see §1).

In other words, I am arguing that (1) the notion of commitment-based intentions is an important one for Cognitive Science and AI; and (2) C & L's formalization of it (even if weakened and generalized) does not do justice to it. The claim of this paper is that one can, however, understand commitment in a way that can be felicitously formalized, and which avoids the criticism just levied on C & L's theory. In the next subsection, I try to formulate some intuitions about commitment from the point of view of agent rationality.

### 2.3 Commitment and Rationality: Conative Entrenchment

Intentions are attitudes of rational agents. For agents who are limited, but are rational to some extent, having a commitment is a means of making the effort and time spent on deliberation have a longer term effect than on just the current action—if an agent can commit to an intention or a course of action, he does not have to repeatedly rethink some issues from first principles. By thus committing, the agent would certainly miss out some opportunities that he could have noticed by rethinking, but this comes at the advantage of not having been swamped by deliberation. In many cases, careful deliberation once in a while is better than poor reasoning done repeatedly. And in the long run, the limited agent ought to come out ahead in terms of effort expended and benefits accrued. I take this for granted in this paper.

However, the question I shall address is related to it. Given that commitments are a good idea for the kind of agents and environments that we are considering, one can naturally focus on the normative criteria for determining how committed an agent *should* be to an intention of his. Now the commitment of an agent to an intention is really a measure of the effort he is willing to put in to achieve it, or of the risk he is willing to take in trying to achieve it, or of something along those lines.

Ideally, the commitment of an agent to an intention should depend on its *utility* to him, "utility" here being a normative concept. For a real-life agent, the commitment would actually have to be set equal to the utility he subjectively expects from the intention. This approach has the advantage that once an agent has adopted an intention and decided his level of commitment for it, he does not have to repeatedly reconsider his commitment—he would need to reconsider it only when he had put in effort for it well above his initial commitment, or had tried all the sufficiently low-risk and low-cost means he knows of. At that point he could either drop the intention altogether or reinstate it with a new commitment. Thus, the greater (i.e., larger) the agent's commitment to an intention, the less frequently he would need to reconsider it. To coin a phrase analogous to the well-known for beliefs, an agent's commitment to an intention is a measure of its *conative entrenchment.*

In this paper, I consider only the sense of conative entrenchment in which the expected utility of an intention is involved (rather than risk, or some other such potentially useful criterion). For concreteness, I now turn to a formal model involving action and time in which intentions and commitments can be formalized. This model is quite abstract, is derived from models for branching time temporal logic, has previously been developed, and has been applied to the formalization of intentions and know-how [Singh, 1990, Singh, 1991, Singh and Asher, 1990].

## 3 The Model, Intuitively

For concepts such as intention, commitment and expected utility to even be formalized, we need a formal model that includes not just time and action, but also possibility, probability and choice. The model I propose here is based on possible worlds. Each possible *world* has a branching *history* of *times.* Histories are sets of times, partially ordered by temporal precedence, $<$. They branch into the future, and are assumed to never end. The sets of times in the history of each world are disjoint. A world and time are a "situation." A *scenario* at a world and time is any maximal set of times containing the given time, and all times that are in a particular future of it; i.e., a scenario is any single branch of the history of the world that begins at the given time, and contains all times in some linear subrelation of $<$. Different scenarios correspond to different ways in which the world may develop as a result of the actions of agents.

Even though a world may develop in several different ways, only one scenario can be actualized. I take probability as applying to scenarios and denoting their objective chance of being *actualized*. The probability of a scenario is given relative to a world and time and some description of an agent's cognitive state. Even for the same world and time, the description can vary—this allows us to express probability before and after an agent's intentions are considered, and is crucial to the goals of this paper. Only objective probability is considered, but it is seen as being dependent on the agent's internal state, since actions can influence what occurs later, and actions are chosen by agents depending on their internal state. An agent *may* do several basic actions at any world and time; for simplicity, I assume that on each scenario he would do exactly one, intuitively, the one he *chooses* to do.

## 4 Primitive Concepts

I take *Commits* as a primitive notion here and consider intention as derived. $Commits(x, p, c)$ means that agent $x$ is committed to achieving $p$ to a level of $c$. Then $Intends(x,p) \equiv (\exists c > 0 : Commits(x,p,c))$. Note that even though commitments can be of different degrees, these degrees just represent the entrenchment of the corresponding intention—an intention itself is treated as being either ON or OFF, i.e., as binary. This is crucial since the motivational component of intentions, which is what makes agents act, is needed fully, if at all, for an agent to act for it—how much effort an agent expends is a different matter (I do not consider actions done half-heartedly: even our pretheoretic intuitions are unclear about such cases).

Each agent deliberates from time to time. $Deliberates(x)$ is true at precisely the situations where $x$ deliberates. The process of deliberation is not studied here, and the theory presented applies only between successive deliberations (on any scenario). Each action when done at a given time along a given scenario has a certain cost attached to it—this cost can vary between different instances of the same action, and equals the value of $Cost(x, a)$ on a given world, time and scenario.

Objective probability is needed in the model to take care of the notion of objective chance. Many actions, e.g., coin tosses or rolls of dice, have several possible outcomes which have (perhaps, different) objective probabilities associated with them. These outcomes may also have different utilities for an agent. In the model, objective probability is treated as a function, $\pi(\cdot)$, from scenarios to the unit interval, $[0 \ldots 1]$ and, utility is captured by a function $\Omega(\cdot,\cdot)$ applied to agents and scenarios. In the language, utility is expressed by a function $Utility(\cdot,\cdot)$ applying to an agent and a condition, and is meant to take the objective chance of different scenarios on which that condition is true into account.

The key feature of intentions that we need is that they lead to action. Intentions here are future-directed and allow not just immediate actions, but also those in the future. Therefore, another useful primitive is *acting for an intention*: an agent acts for an intention when his action is a part of what he would do in order to satisfy it—the details of this process are not focused on here. Acting for an intention is a cognitive concept—it depends on the agent's internal state rather than the world. An agent acting for an intention may be doing so even if it would be impossible or unlikely for him to ever succeed (by doing that action). The same action could be done for two different intentions; of course, several temporally isolated actions may have to be done for a single intention. I notate this concept as a three place predicate $Acts\text{-}for(\cdot,\cdot,\cdot)$: applied to an agent identifier, basic action and a condition. In order to connect the agent's cognitive state with the world, we need the concept of *performing* an action in the world—this is notated by a predicate $Performs(\cdot,\cdot)$. *Performs* of an agent $x$ and action $a$ is true over a subscenario on which $x$ does $a$. I assume that an agent who acts for a condition intends it, and also immediately performs the action by which he acts for that condition; i.e., $Acts\text{-}for(x,a,p) \rightarrow Performs(x,a)$ is always true.

Commitments (and therefore intentions) and beliefs are given a simple semantics for ease of exposition, and to focus on the matters of interest—a *Commits*, *Believes* or *Acts-for* formula is true over a subscenario or interval if it belongs to the agent's cognitive state during that subscenario. These concepts are treated purely qualitatively; they can straightforwardly by analyzed using subjective probabilities, if one wishes. I consider it a strength of this approach that it does not require the notion of subjective probability, while being compatible with it. Note that agents can have beliefs, and even intentions, that involve objective probability and utility statements.

## 5 The Formal Language

The formal language of this paper, $\mathcal{L}$, is CTL* (a propositional branching time logic [Emerson, 1989]) augmented with quantification over basic actions; functions: *Prob*, *Utility*, *Cost*; and predicates: *Believes*, *Commits*, *Intends*, *Acts-for* and *Performs*; and the arithmetic required. Let $x$ be an agent; $p$, $q$ propositions; $a$ an action; and $v$ a probability.

A formula can be any of the following: an atomic formula ($\psi$), a conjunction of formulae ($p \wedge q$), a negation of a formula ($\neg p$), an until-expression ($p\mathsf{U}q$), an action-expression or any of the four special predicates applied to the appropriate kinds of arguments, or a path-quantifier followed by a formula. A path-quantifier is one of A and E. A denotes "in *all* scenarios at the present time," and E denotes "in *some* scenario at the present time"—i.e., $\mathsf{E}p \equiv \neg\mathsf{A}\neg p$. $\mathsf{F}p$ denotes "$p$ holds sometimes in the future on this scenario" and abbreviates "trueU$p$." G denotes "$p$ always holds in the future on this scenario" and abbreviates "$\neg$F$\neg p$." Implication ($p \rightarrow q$) and disjunctions of formulae ($p \vee q$) are defined as the usual abbreviations. An action-expression is of the form $\langle a \rangle p$ and means that action $a$ is done on the given scenario at the given time by agent $x$, and that $p$ holds as soon as $a$ is done.

## 6 The Formal Model

The semantics of $\mathcal{L}$ is given relative to intensional models (as described informally in §3): it is standard for CTL*. The formal model is. Let $M = \langle F, N \rangle$ be an intensional model, where $F = \langle \mathbf{W}, \mathbf{T}, <, \mathbf{A}, \mathbf{U} \rangle$ is a frame, and $N = \langle [\![]\!], \mathbf{B}, \mathbf{C}, \pi \rangle$ an interpretation. Here $\mathbf{W}$ is a set of possible worlds; $\mathbf{T}$ is a set of possible times ordered by $<$; $\mathbf{A}$ is the class of agents in different possible worlds; $\mathbf{U}$ is the class of basic actions; as described below, $[\![]\!]$ assigns intensions to atomic propositions and actions. Scenarios as described in §5 can be defined easily from $<$ [Singh, 1991]; $\mathbf{S}_{w,t}$ is the class of all scenarios at world $w$ and time $t$: $(w \neq w' \vee t \neq t') \Rightarrow \mathbf{S}_{w,t} \cap \mathbf{S}_{w',t'} = \emptyset$. $\langle S, t, t' \rangle$ is a ***subscenario*** of $S$ from $t$ to $t'$, inclusive. $\mathbf{B}$ assigns basic actions to the agents at different worlds and times. $\mathbf{C}$ assigns a cognitive state to each agent at different subscenarios, as defined below. A coherence requirement is that the cognitive state for a subscenario cannot be different than for a subscenario containing it. $\pi$ assigns probabilities to scenarios in $\mathbf{S}_{w,t}$, for each world $w$ and time $t$.

The intension of an atomic proposition is the set of worlds and times where it is true; that of an action is, for each agent $x$, the set of subscenarios in the model in which an instance of it is done (from start to finish) by $x$; e.g., $\langle S, t, t' \rangle \in [\![a]\!]^x$ means that agent $x$ does action $a$ in the subscenario of $S$ from time $t$ to $t'$. I assume that $[\![]\!]$ respects $\mathbf{B}$; i.e., $a \in \mathbf{B}_{w,t}(x)$. For the models to be coherent, we need to constrain them so that (1) an action begun at a time ends at most once on any scenario there; (2) subscenarios are uniquely identified by the times over which they stretch; (3) there is a always future time; and (4) something must be *done* by each agent along each scenario in the model, even if it is a dummy action. Restrictions on $[\![]\!]$ can also be used to express the limitations of agents; e.g., $x$ cannot pick up three glasses at once.

The semantics of formulae is given relative to a model as defined above and a world and time in it. $M \models_{w,t} p$ expresses "$M$ satisfies $p$ at $w, t$." $M \models_{S,t} p$ expresses "$M$ satisfies $p$ at time $t$ on scenario $S$," and is needed for some formulae as defined in §5. $p$ is *satisfiable* iff for some $M$, $w$ and $t$, $M \models_{w,t} p$. $p$ is *valid* in $M$ iff it is satisfiable at all worlds and times in $M$. The satisfaction conditions for the temporal operators too are adapted from those in [Emerson, 1989]. Formally, we have the following definitions:

1. $M \models_{w,t} \psi$ iff $\langle w, t \rangle \in [\![\psi]\!]$
2. $M \models_{w,t} p \wedge q$ iff $M \models_{w,t} p \wedge M \models_{w,t} q$
3. $M \models_{w,t} \neg p$ iff $M \not\models_{w,t} p$
4. $M \models_{w,t}$ E$p$ iff $(\exists S : S \in \mathbf{S}_{w,t} \wedge M \models_{S,t} p)$
5. $M \models_{w,t}$ A$p$ iff $(\forall S : S \in \mathbf{S}_{w,t} \rightarrow M \models_{S,t} p)$
6. $M \models_{S,t} \langle a \rangle p$ iff $(\exists t' : \langle S, t, t' \rangle \in [\![a]\!]^x \wedge M \models_{S,t'} p)$
7. $M \models_{S,t} p$U$q$ iff $(\exists t' : M \models_{S,t'} q \wedge (\forall t'' : t \leq t'' \leq t' \rightarrow M \models_{S,t''} p))$
   $p$U$q$ is satisfied at time $t$ on scenario $S$ iff there is a time such that $q$ holds at it, and for all times between now and then, $p$ holds at them.
8. $M \models_{S,t} p$ iff $M \models_{w,t} p$, if $p$ is not of the form $q$U$r$ or $\langle a \rangle q$, and $w$ is the (unique) world such that $S \in \mathbf{S}_{w,t}$
9. $M \models_{S,t}$ *Believes*$(x, p)$ iff $(\exists t'$ : *Believes*$(x, p) \in \mathbf{C}_x(\langle S, t, t' \rangle))$
10. $M \models_{S,t}$ *Commits*$(x, p, c)$ iff $(\exists t'$ : *Commits*$(x, p, c) \in \mathbf{C}_x(\langle S, t, t' \rangle))$
11. $M \models_{S,t}$ *Acts-for*$(x, a, p)$ iff $(\exists t'$ : *Acts-for*$(x, a, p) \in \mathbf{C}_x(\langle S, t, t' \rangle))$
12. $M \models_{S,t}$ *Performs*$(x, a)$ iff $(\exists t' : \langle S, t, t' \rangle \in [\![a]\!]^x)$
13. $M \models_{w,t}$ *Utility*$(x, p) = u$ iff
    $$[\sum_{S \in \mathbf{S}_{w,t} \wedge M \models_{S,t} p} \pi_{w,t}(S) \times \Omega_{w,t}(x, S)] = u$$
    The utility of $p$ to $x$ is the weighted sum of the utilities of the scenarios on which $p$ holds.
14. $M \models_{w,t}$ *Prob*$(p) = v$ iff
    $$[\sum_{S \in \mathbf{S}_{w,t} \wedge M \models_{S,t} p} \pi_{w,t}(S)] = v$$
    The probability of $p$ holding is the sum of the probabilities of the scenarios that satisfy it.

## 7 Commitment Formalized

Now I return to commitments. An important property of intentions that connects them to action is captured by the following constraint on our models: an agent who has a positive commitment to achieving a condition must eventually act on (unless he deliberates again in the meantime). A stronger formulation would require that an agent did not hold an intention infinitely without acting for it, but there is no space to include such "fairness" conditions here [Emerson, 1989].

1. A[*Intends*$(x, p) \rightarrow$ F*Deliberates*$(x)$ $\vee$ F$(\exists a$ : *Acts-for*$(x, a, p))$]

Another constraint that we need is the following which essentially "uses up" a part of the commitment to an intention. Here the metaphor of commitment as a measure of the resources committed to an intention is especially attractive. As the agent does actions for his intention, he uses up resources for it, and his intention becomes progressively less entrenched. Finally when his commitment becomes too little, constraint 1 will no longer apply; the agent will no longer be required to act for that condition (and ordinarily, he would not). He might adopt an intention for the same condition again, in which case he will again be able to do some actions for it.

2. $\mathsf{A}[(Commits(x,p,c) \quad \wedge \quad Acts\text{-}for(x,a,p) \quad \wedge \quad Cost(x,a)=u) \rightarrow \langle a\rangle Commits(x,p,c-u)]$

For contrast, C & L's persistence condition translated to the framework of this paper would be

3. $\mathsf{A}[Intends(x,p) \rightarrow (Intends(x,p)\mathsf{U}(Believes(x,p) \vee Believes(x,\mathsf{AG}\neg p)))]$

This requires that an agent not deliberate about an intention once he adopts it. The proposed framework does not prevent or require such deliberation; e.g., it does *not* require constraint 4, which says that if an agent comes to believe that an intention of his has been satisfied, he has to deliberate immediately.

4. $\mathsf{A}[Intends(x,p) \wedge Believes(x,p) \rightarrow Deliberates(x)]$

Instead, we can have the much weaker constraint 5, which says that when an intention is believed to have succeeded, the agent would eventually deliberate. Essentially the same improvement can be made for the constraint calling for intentions to be dropped when it is believed that the intended is impossible in the future.

5. $\mathsf{A}[Intends(x,p) \wedge Believes(x,p) \rightarrow \mathsf{F}Deliberates(x)]$

Deliberation is also called for when the commitment falls below a certain threshold required to do something.

6. $\mathsf{A}[Commits(x,p,c) \wedge (\forall a : (\mathsf{E}Acts\text{-}for(x,a,p)) \rightarrow Cost(x,a) > c) \rightarrow \mathsf{F}Deliberates(x)]$

This leaves one important requirement still to define. This requirement concerns the definition of commitment as the expected utility of the corresponding intention to the given agent. The following constraint says that if an agent deliberates and adopts an intention, his commitment to that intention equals what he believes is his objectively expected utility of achieving that condition sometimes in the future.

7. $\mathsf{A}[(Deliberates(x) \wedge Commits(x,p,c) \wedge c > 0) \rightarrow Believes(x, Utility(x,\mathsf{F}p)=u)]$

One important aspect of intentions and rationality is that intentions serve as both the *ends* of agents and as their *means* for their other intentions. Joe, in §2.2, has an end of being famous and his means for that end is to go to Mars. Both the end and the means are intentions; however, a rational agent, such as Joe, ought to be more committed to his (ultimate) ends than to the specific means he has chosen—there are usually more than one means to an end, after all. I do not focus here on how an agent might perform the appropriate kind of means-ends reasoning. All that is needed is a formal description of when condition $q$ is seen as a means for $p$. I propose that $q$ is one of $x$'s means for $p$ iff $x$ intends both and on all scenarios, until he redeliberates all his actions for $q$ are also for $p$.

8. $Means(x,q,p) \equiv [Intends(x,p) \wedge Intends(x,q) \wedge \mathsf{A}[(Acts\text{-}for(x,a,q) \rightarrow Acts\text{-}for(x,a,p))\mathsf{U}Deliberates(x)]]$

Combining this definition with constraint 2 above, we can easily see that, as desired, the commitment for a means to an end must be less than or equal to the commitment to the end itself.

## 8 Conclusions and Future Work

I have taken an important property of intentions, commitment, and shown how it can be felicitously reconciled with the idea of rationality for limited agents. It should be clear that limited agents, for whom deliberation is expensive, benefit from being committed to their intentions. Future work planned includes formally deriving that condition, and expressing rationality postulates that relate planning and intentions. Another interesting idea is to define habits as sequences of actions whose cost is lower than the sum of the costs of the individual actions that compose it.

## References


[Brand, 1984] Myles Brand. *Intending and Acting.* MIT Press, Cambridge, MA, 1984.

[Bratman, 1987] Michael E. Bratman. *Intention, Plans and Practical Reason.* Harvard University Press, Cambridge, MA, 1987.

[Cohen and Levesque, 1990] Philip R. Cohen and Hector J. Levesque. Intention is choice with commitment. *Artificial Intelligence*, 42:213–261, 1990.

[Emerson, 1989] E. A. Emerson. Temporal and modal logic. In J. van Leeuwen, editor, *Handbook of Theoretical Computer Science.* North-Holland Publishing Company, Amsterdam, The Netherlands, 1989.

[Harman, 1986] Gilbert Harman. *Change in View.* MIT Press, Cambridge, MA, 1986.

[McDermott, 1982] Drew V. McDermott. A temporal logic for reasoning about processes and plans. *Cognitive Science*, 6(2):101–155, 1982.

[Singh and Asher, 1990] Munindar P. Singh and Nicholas M. Asher. Towards a formal theory of intentions. In *European Workshop on Logics in Artificial Intelligence*, September 1990.

[Singh, 1990] Munindar P. Singh. Group intentions. In *10th Workshop on Distributed Artificial Intelligence*, October 1990.

[Singh, 1991] Munindar P. Singh. A logic of situated know-how. In *AAAI*, July 1991.

# Connectionism and Dynamical Explanation

Tim van Gelder
Department of Philosophy
Indiana University
Bloomington IN 47405
*tgelder@ucs.indiana.edu*

## Abstract

A distinctive feature of connectionism as a research paradigm in psychology is use of a form of scientific explanation here termed dynamical explanation. In dynamical explanation, the behavior of a system is explained by reference to points and trajectories in an abstract state space. This paper contrasts dynamical explanation with some other major forms of scientific explanation, and discusses how dynamical explanation of the behavior of artificial neural networks can constitute genuine psychological explanation.

What is distinctive about connectionism as a research paradigm in psychology? This question has been approached from many different directions. Many have pointed to novel connectionist methods of representation; others, meanwhile, have focused on such issues as the connectionist emphasis on learning, the level at which it operates, whether or in what sense it employs rules, and so forth. Here I will be suggesting that, whatever else may be distinctive about it, connectionists are developing - or at least, importing into psychology - a novel form of scientific explanation. Since this kind of explanation proceeds by constructing models conceived of as dynamical systems, the most appropriate term for this approach is dynamical explanation. In what follows I will give an intuitive introduction to the concept of dynamical explanation, contrasting it with other generic styles of explanation widely used in science, and will then discuss how dynamical explanation is deployed in connectionist modeling of psychological phenomena.

There is, of course, nothing new in the idea that connectionist networks can be thought of as dynamical systems, or in the idea that doing so is a key element in connectionism's distinctive perspective on cognition. (On both these points, see, e.g., (Smolensky, 1988).) I am not arguing for these points, but rather attempting to make explicit one form of scientific explanation one will be using if one describes networks as dynamical systems and uses such descriptions in accounting for psychological data. Making explicit the explanatory strategies in use is but one step in a general project of clarifying the conceptual foundations of connectionism.

## Styles of Explanation.

In a well-known article outlining "cognitivism" (now more widely known as "classical" cognitive science), John Haugeland (Haugeland, 1981) pointed out that scientists commonly utilize at least three broad kinds of explanation - the deductive-nomological, the morphological, and the systematic. Deductive-nomological explanation is the classic; in this approach, scientists develop general, abstract, mathematically formulated laws, and then explain a given phenomenon by showing that it is merely an instance of the operation of such a law - i.e., by subsuming the phenomenon under the law. In morphological explanation, by contrast, an ability or disposition "is explained through appeal to a specified structure and to specified abilities of whatever is so structured" (p.247). For example, the uncanny ability of an eggshell to resist breaking when force is applied at the ends can be explained by pointing to the shape of the shell and the way such a shape distributes pressure. Here there is, at least in the first instance, no subsuming under laws; there are of course laws in the background, but knowledge of such laws is certainly not prerequisite for finding the explanation illuminating. In systematic explanation, the ability or disposition of some whole, in this case known as a "system," is also explained by pointing to a particular kind of internal structure, but this time one made up of distinct parts which participate in a "complexly organized pattern of interdependent interactions". For an example illustrating the systematic approach, consider an explanation of how a car engine works. Long before any laws become relevant, one has to describe the various parts of the engine (carburetor, cylinders, radiator etc.), how they individually function, and how each of their functions is integrated to produce a smoothly and powerfully rotating driveshaft.

These different explanatory strategies can be regarded as alternative ways of conceptualizing or "organizing" the world such that we can make sense of it. More than this, however: they tend to reflect real differences in the way the world itself is constituted and events unfold. The difference between morphological and systematic explanation, for example, reflects the structural fact that systems, but not the kinds of mechanisms we describe morphologically, are made up of distinct parts which interact with each other in complex ways. In general, whether a particular explanatory strategy is the most appropriate in a given case depends on both the specific character of the situation under consideration and our explanatory interests and resources.

From this perspective, an obvious question arises: how many significantly different kinds of scientific explananda, and corresponding different explanatory strategies, are there? Or, less ambitiously: can we think of further kinds of situations, to which some further explanatory strategy is most appropriate? In the current context, of course, the ultimate goal of this line of questioning is cognition. What kind of mechanisms are cognitive mechanisms? Which explanatory strategy, or strategies, are the most appropriate in describing how cognition arises?

In fact, I think that there is at least one more major, generic form of explanation, one that is suggested by study of how explanations often proceed in distributed connectionist work, though it is most certainly not limited to that context. The kind of mechanism to which this species of explanation applies is the dynamical system. A dynamical system is any closed system whose state at a given time can be adequately captured by specifying the values of each of a set of parameters. As the dynamical system changes over time, the values of these parameters evolve in interdependent ways. In studying and explaining the behavior of dynamical systems one aims at formulating equations which describe the evolution of the system, and which can consequently be used to explain why the system is in the state it is in, or to predict what states it will come to be in. A classic example of a dynamical system is a pendulum. Parameters pick out the displacement and velocity of the bob, and relatively simple general equations govern how the values of these parameters change over time, capturing the periodic swinging motion of the pendulum.

The state space of a dynamical system is all the possible states which the system can be in. A state space can be represented using a vector space containing a point for every possible combination of values of all the relevant parameters; thus, the state space has as many dimensions as there are parameters. If we know the current state of the system - i.e., the point in state space it currently occupies - then we can use the equations governing the behavior of the system to determine what point it will occupy next. A succession of such points is a trajectory in state space, and amounts to a picture of how the dynamical system changes as time goes on. Every point that the system might occupy lies on some trajectory or other, and shapes of the all the trajectories are fixed by the equations. Crucially, to understand how the system works is to have a sense for how the system changes over time, or, equivalently, to understand the general dynamical "topography" of the system. What do the trajectories look like? If the system is in a given state, how will it evolve?

In the case of the pendulum, the equations which tell us how the state of the system changes over time are given by general laws of classical mechanics. For this reason it might seem that dynamical explanations are just special instances of deductive-nomological explanation. In general, however, this is not true. For one thing, we can use dynamical explanations in situations where the equations governing the evolution of the system are not themselves general physical laws. In principle, of course, one supposes that the equations are ultimately derivable from general physical laws, and it may even be possible to provide such a derivation. However, the crucial point is that as long as one has the equations for the system, such a derivation is not in practice part of the actual explanation of the behavior of the system, and none of the explanatory force is lost if no such derivation is forthcoming.

Second, dynamical explanations may proceed without making explicit use of the equations governing the system. If the system is complex enough, one may not actually have the full equations in hand; alternatively, one might have them, but nevertheless find more perspicuous ways of explaining how the system works that proceed independently of the equations. This seems to be the situation on occasions when connectionists explain the performance of their networks by sketching a picture of the trajectories through which the overall patterns of activity evolve. By presenting a series of carefully selected two-dimensional snapshots of trajectories, they provide a sense of the dynamical structure of the system - i.e., an understanding of how the system behaves. Subsequent explanations of particular features of the behavior of the network can proceed by referring to that topography without needing to advert to the full equations which, formally, govern the behavior.

Dynamical explanations are not morphological or systematic either. In its pure form, dynamical explanation makes no reference to the actual structure of the mechanism whose behavior it is explaining. It tells us how the values of the parameters of the system evolve over time, not what it is about the way the system itself is constituted that causes those parameters to evolve in the specified fashion. It is concerned to explore the topographical structure of the dynamics of the system, but this is a wholly different structure than that of the system itself. This point is crucial to understanding dynamical explanations and how they differ from systematic explanations in particular. Dynamical explanations turn on the way the values of the parameters of the system interdependently evolve over time, whereas systematic explanations turn on the complex interactions among the parts of the system itself. Consider the car engine again. Supposing a dynamical systems-style explanation were possible, it would begin by picking out the crucial parameters: engine temperature, r.p.m., gas level, timing advance, cylinder pressure - indeed, many of the quantities measured on by instruments on the dash, and no doubt a host of others besides - and then proceed to show how variations in one parameter affects the others, or how the engine typically proceeds through various states corresponding to regions of the state space (e.g., from cold to warm to out of gas), possibly according to rough equations describing the behavior of the particular engine. Explanations formed along these lines are quite different from systematic explanations which advert to the various parts of the engine and how they interact. In particular, the parameters utilized in dynamical explanations do not in general pick out parts of the system under study. Temperature and r.p.m. are not parts of the engine which interact with other parts. You can remove the carburetor leaving the rest of the engine behind, but you cannot remove the r.p.m..

The general point here is that dynamical explanation, which proceeds in terms of parameters, equations and state spaces, takes place at one level of remove from the actual mechanisms which produce the behavior quantified and explained in the dynamical account. To be sure, if one wanted an explanation of why one equation rather than some other governs a system, of why the state of the system travels through some trajectories and not others, one may be able to go on to offer a further, presumably systematic, explanation; this does not however make the initial dynamical explanations themselves either morphological or systematic, and the usefulness or validity of the dynamical explanation does not depend on one's being able to provide such further explanation.

I claim, then, that dynamical explanation constitutes a genuine alternative to other common forms of scientific explanation. The dynamical approach to explanation seems to have been neglected in the philosophy of science, but not in scientific practice itself, for a moment's reflection suggests that explanations fitting this general mold are widely used. What remains to be shown here is how dynamical explanation figures in connectionist work. This involves two steps: first, illustrating how the behavior of connectionist networks can be explained in dynamical terms; and second, showing how dynamical explanation of connectionist networks combines with the technique of modeling to construct genuinely psychological explanations.

## Dynamical Explanation of Network Behavior.

It is probably obvious enough how connectionist networks are conceived as dynamical systems. In the simplest and most familiar case, the parameters specify activity levels for each of the units, and equations based on the processing characteristics of the units and the way they are interconnected describe how activity levels will evolve over time. Patterns of activation over the network as a whole can be regarded as points in an activation state space, and processing can be thought of as traveling along a trajectory in activation space. From this perspective, to explain the behavior of a network is to provide a sense for the dynamical structure of the network as a whole, such that any particular state it might be in can be related to other states that it was or will be in; or such that one trajectory can be contrasted with others. This "sense" can be obtained either formally, by providing equations capturing the dynamical structure, or intuitively, by (for example) sketching trajectories in a selected few of the dimensions. (Note that explanation of the behavior of the network in terms of its dynamical structure must be carefully distinguished from another common explanatory task in connectionism, namely, explaining how a network comes to have the dynamical structure that it does.)

Hopfield nets (Hopfield, 1982) and Boltzmann machines (Hinton and Sejnowski, 1986) are excellent examples of connectionist networks whose behavior is naturally understood in dynamical terms. In general it is helpful to see processing, over many time steps, as tracing out a somewhat erratic trajectory in the activation state space of the network. Indeed, if we add to the space a dimension for another parameter such as global energy, the settling process characteristic of such networks can be visualized as a downward slide to a resting place. Explanations of particular aspects of the networks behavior typically refer to the general structure of these dynamics. Why did the network settle at a particular point? Because it began its settling process at another point which happened to be in the basin of attraction for the settling point.

Many other connectionist explanations proceed in basically dynamical terms, though the dynamical character is often concealed by the fact that they are often highly simplified varieties of the strategy. Explanations in this latter category tend to consider only a few time steps (i.e., highly truncated "trajectories") and, from step to step, shift attention from one sub-space of the overall activation space to another. Consider for example the explanation of generalization phenomena in the well-known past tense learning model of Rumelhart & McClelland (Rumelhart and McClelland, 1986). The ability of the network to produce the correct past tense form of an unseen, irregular verb such as *bid* is explained in terms of the fact that the trained network is "sensitive to the subregularities as well". That is, there are regularities even among irregular verbs; irregular verbs with similar present tense forms often also have similar past tenses. To say that the network is "sensitive to the subregularities" is to say that it treats members of these subgroups in a similar fashion. Thus, explaining the network's correct performance on the verb *bid* is a matter of (a) pointing out that its input representation is similar (i.e., close in the space of input patterns) to other irregular verbs, and (b) adverting (in this case, in a highly informal way) to the overall dynamical structure of the network. This explanation considers a "trajectory" of only one time step and its description of the topological structure of the dynamics of the network is framed only in terms of how points in the input-unit activation sub-space are mapped to the output-unit subspace.

A particularly provocative example of dynamical explanation at work is found in Jeff Elman's descriptions of his SRN models of sentence processing (see e.g., (Elman, 1989)). Elman made effective use of an increasingly common technique exploiting principal components analysis to select an appropriate two-dimensional "window" onto the activation state space - or rather, in this case, the sub-space corresponding to hidden unit activity. In this window one plots a series of points, corresponding to states that the hidden units go through upon presentation of a series of inputs; this series of points amounts to a picture of the activation trajectory for those inputs. Collectively, a series of such trajectory pictures yields at least a glimmer of understanding of the complex dynamical structure of the network. Explanations of the successful performance of the network at its word prediction task make crucial reference to these trajectories; thus, the network is held to be able to correctly predict the next word in a complex sequence precisely because the structure of the sequence up to that point had been encoded in the activation trajectory (see (Port and van Gelder, 1991)

Why are connectionists increasingly using dynamical explanation in preference to, for example, any of the other forms of explanation described by Haugeland? The quick and easy answer is that neural networks are themselves the kinds of systems for which the most natural explanations are dynamical. Deductive-nomological explanation requires general covering laws, yet the systems of differential equations governing the behavior of complex networks are not general laws, and if they can be formulated at all for such networks, are typically unwieldy and unilluminating. Morphological explanation is inappropriate since it does not make room for the change and interdependence of the parameters; and systematic explanation is of little help since the "parts" of the structure are so many and so similar, and key parameters (e.g., "energy") do not refer to parts of the system at all. In dynamical explanation one abstracts away from any consideration of how the system under study is actually put together, and focuses only on how various parameters change in interdependent ways. It seems that this distancing from the implementation is an essential simplifying step in attaining a deep understanding of how these particular kinds of highly complex systems work.

## Dynamical Explanation in Psychology.

These examples of dynamical explanation in connectionism only involve explanation of the behavior of networks themselves; psychological data has not yet entered the story. How then can dynamical explanation constitute psychological explanation?

To answer this question, the concept of dynamical explanation of neural networks has to be combined with an understanding of how those networks are supposed to function as models of psychological phenomena. In modeling we explore one relatively familiar, often artificial structure or mechanism as a means of exploring another less familiar one. Thus, in psychology, and speaking in the broadest possible terms, one thing we wish to do is understand that mechanism which produces our behavior - at least, our behavior as codified in the data of experimental psychology. That mechanism, of course, is the brain. (Some prefer to distance themselves more from the messy neurobiological details, and so call the mechanism the mind.) Yet that mechanism is so awesomely complex, and the gap between neuroscience and psychology still so large, that we cannot yet explain how the brain (or mind) produces our behavior directly. As an intermediate step, we construct a model - a substitute mechanism which is supposed to be relevantly similar to the original, at least so far as the latter is described at some suitably abstract level. By exploring and understanding the properties of the model, we hope to reach some understanding of the original mechanism.

An obvious test of the adequacy of a model is that it account for the data - that is, it should produce the same overall behavior as the mechanism that is the ultimate explanatory target. Most psychologists producing models of human performance direct virtually all their attention at the constraints provided by this requirement. In principle, however, we can judge the adequacy of the model by reference to any knowledge we happen to have of the original mechanism. For example, a model of internal processes in the sun can be judged not only according to whether it generates the right behavior (corresponding to light and heat output, flares, sunspots etc.) but also according to whether it relies on processes that accord with other quite general knowledge from chemistry and physics. In the psychological case the "top-down" constraints of matching performance can, at least in principle, be supplemented with "bottom-up" constraints provided by our increasing knowledge of the brain, its structure and modes of functioning.

No model will be identical, in all its features, to the mechanism that is to be explained; only some aspects of the model are relevant to whether it meets the constraints provided by the performance data, and only some aspects are relevant to whether it accords with neuroscientific evidence of the basic constitution of the neural mechanism. Thus, whenever a model is proposed, it ought to be accompanied by what Wilfred

Sellars called a commentary, specifying which features of the model are to be taken as relevant to whether it satisfies the various constraints. For example, implicit in the discussion of the past tense learning model was the claim that which set of Wickelfeatures is output for a given input is taken to be relevant to whether the model accounts for the psychological data, while how fast it produces that set is not relevant. Other things being equal, a better model is one in which more of its features are judged relevant to its acceptability as a model.

Now, connectionist explanations of the behavior of their networks in dynamical terms constitutes dynamical explanation in psychology if (a) the network is being used as a model for the mechanism producing the psychological data, and (b) it is explicitly claimed or at least implicitly assumed that the abstract dynamical structure of the network corresponds to the abstract dynamical structure of the original mechanism. If these conditions are satisfied, then a connectionist's explanation of the performance of her network in terms of locations and trajectories in unit activation space is, at the same time, explanation in dynamical terms of how humans exhibit the performance they do. Note that for connectionist explanations to count as genuine dynamical explanations of psychological data, it is not required that the processing units of the model be interpreted as corresponding to real neurons, the connections as real synapses, etc.; the commentary may simply leave the interpretation of the units themselves undetermined. What matters is the overall structure of change, not what it is that is changing.

In short, connectionists are, increasingly, producing dynamical explanations of psychological data by combining dynamical explanation of the behavior of their networks with a certain way of interpreting those networks as models of the actual mechanisms underlying human performance. It may eventually be possible to produce dynamical explanations in psychology without first producing and explaining artificial neural network models; until the actual mechanisms are much better understood, however, modelling of some kind is an essential intermediate stage.

## Conclusion.

According to Haugeland, understanding how mainstream symbolic cognitive science uses systematic explanation is essential to understanding its distinctive approach to cognition and how it can constitute respectable scientific investigation in psychology. It would probably be premature to assert, at this stage, that the concept of dynamical explanation will play a similar role for connectionism. It is safe to say that dynamical explanation is increasingly common in connectionist practice, arising in response to at least two pressures - the increasing complexity of connectionist models, and the developing interest in the temporal structure of cognitive processes. I am not claiming that the use of dynamical explanation is any kind of criterion that can be used to distinguish connectionist approaches from others; in general, it is a methodological blunder to suppose that there is any fail-safe distinguishing marker. Further, I am not claiming that dynamical explanation is the only kind of explanation connectionists use. It is, nevertheless, a distinctive and novel feature of the connectionist perspective on cognition.

## References


Elman, J. (1989). Representation and structure in connectionist models. Technical Report 8903, Center for Research in Language, UCSD, La Jolla, CA 92093-0108.

Haugeland, J. (1981). The nature and plausibility of cognitivism. In Haugeland, J., editor, *Mind Design*, pages 243–281. Bradford/MIT Press, Cambridge, Mass.

Hinton, G. and Sejnowski, T. (1986). Learning and unlearning in boltzmann machines. In Rumelhart, D. and McClelland, J., editors, *Parallel Distributed Processing, Vol. 1*, pages 282–317. The MIT Press, Cambridge, MA.

Hopfield, J. J. (1982). Neural networks and physical systems with emergent collective computational abilities. In *Proceedings of the National Academy of Sciences*, volume 79, pages 2554–2558. National Academy of Sciences.

Port, R. and van Gelder, T. (1991). Representing aspects of language. In *Proceedings of the 13th Meeting of Cognitive Science Society*, Hillsdale, NJ. L. Erlbaum Assoc.

Rumelhart, D. and McClelland, J. (1986). On learning the past tense of English verbs. In McClelland, J. and Rumelhart, D., editors, *Parallel Distributed Processing*, volume 2, pages 216–271. MIT Press, Cambridge, MA.

Smolensky, P. (1988). On the proper treatment of connectionism. *Behavioral and Brain Sciences*, 11:1–74.

# MAC/FAC: A Model of Similarity-based Retrieval

**Dedre Gentner** and **Kenneth D. Forbus**
The Institute for the Learning Sciences, Northwestern University
1890 Maple Avenue, Evanston, IL, 60201

## Abstract

We present a model of similarity-based retrieval which attempts to capture three psychological phenomena: (1) people are extremely good at judging similarity and analogy when given items to compare. (2) Superficial remindings are much more frequent than structural remindings. (3) People sometimes experience and use purely structural analogical remindings. Our model, called MAC/FAC (for "many are called but few are chosen") consists of two stages. The first stage (MAC) uses a computationally cheap, non-structural matcher to filter candidates from a pool of memory items. That is, we redundantly encode structured representations as *content vectors*, whose dot product yields an estimate of how well the corresponding structural representations will match. The second stage (FAC) uses SME to compute a true structural match between the probe and output from the first stage. MAC/FAC has been fully implemented, and we show that it is capable of modeling patterns of access found in psychological data.

## Introduction

Similarity-based remindings range from the sublime to the stupid. On one extreme is being reminded by octaves in music of the periodic table in chemistry. On the other extreme are times when a bicycle reminds you of a pair of eyeglasses. Most often, remindings are somewhere in between, such as when a bicycle reminds you of another bicycle. Our theoretical attention is inevitably drawn to spontaneous analogy, i.e., structural similarity unsupported by surface similarity, partly because it offers perhaps our best entree to studying the creative process. However, a good model must also capture the frequency of different outcomes, and research on the psychology of memory retrieval points inescapably to a preponderance of the latter two types of similarity – (mundane) literal similarity, based on both structural and superficial commonalities – and (dumb) superficial similarity, based on surface commonalities. Rare events are hard to model. A major challenge for research on similarity-based reminding is to devise a model that will produce chiefly literal-similarity and superficial remindings, but still produce occasional analogical remindings.

This paper presents MAC/FAC, a model of similarity-based reminding which attempts to capture these phenomena. We first review psychological evidence on retrieval and mapping of similarity comparisons and describe the design of MAC/FAC. We then describe computational experiements which simulate the patterns of access found in a psychological experiment, and close by describing further avenues to explore.

## Framework

Similarity-based transfer can be decomposed into subprocesses. Given that a person has some current *target* situation in working memory, transfer from prior knowledge requires at least (1) *accessing* a similar (*base*) situation in long-term memory, (2) *creating a mapping* from the base to the target, and (3) *evaluating* the mapping. In the structure-mapping framework (Gentner, 1983, 1988), mapping is the process by which two representations present in working memory are aligned and further inferences imported. The process of computing a mapping from one situation to another is governed by the constraints of *structural consistency* and *one-to-one mapping.*

This account differs from most psychological treatments by defining similarity in terms of correspondences between structured representations. Matches can be distinguished according to the kinds of commonalities present. An *analogy* is a match based on a common system of relations, especially involving higher-order relations.[1] A *literal similarity* match includes both common relational structure and common object descriptions. *Surface matches* are based primarily on common object descriptions along with some shared first-order relations.

There is considerable evidence that people are good at mapping. People can readily align two situations, preserving structurally important commonalties, making the appropriate lower-order substitutions, and mapping additional predicates into the target as *candidate inferences.* For example, Clement & Gentner (in press) showed people analogies and asked which of two lower-order relations,

[1] We define the order of an item in a representation as follows: Objects and constants are order 0. The order of a statement is one plus the maximum of the order of its arguments.

both shared by base and target, was most important to the match. Subjects chose relations that were governed by shared higher-order relations. In a second study, subjects showed the same sensitivity to connectivity and systematicity in choosing which predicates to map as candidate inferences from base to target. Further, people rate metaphors as more apt when they are based on relational commonalities than when they are based on common object-descriptions (Gentner & Clement, 1988) and they rate pairs of stories as more sound when they share higher-order relational structure than when they share object-descriptions (Gentner & Landers, 1985; Rattermann & Gentner, 1987). We also find effects of relational structure on judgments of similarity (Goldstone, Medin & Gentner, in press; Rattermann & Gentner, 1987) and on the way in which people align perceptually similar pictures (Markman & Gentner, 1990).

An adequate model of human similarity and analogy must capture this sensitivity to structural commonality, by involving structural representations and processes that align them. This would seem to require abandoning some highly influential models of similarity: e.g., modeling similarity as the intersection of independent feature sets or as the dot product of feature vectors. However, we show below that a variant of these nonstructural models can be useful in describing some aspects of access.

*Similarity-based Access from Long-term Memory:* There is psychological evidence that access to long-term memory relies more on surface commonalities and less on structural commonalities than does mapping. For example, people often fail to access potentially useful analogs (Gick and Holyoak, 1980). Ross (1984, 1987) further showed that, although people in a problem-solving task are often reminded of prior problems, these remindings are often based on surface similarity rather than on structural similarities between the solution principles.

In our research we used the "Karla the hawk" stories to investigate the determinants of similarity-based access. We put people in the position of trying to access analogy and similarity matches from long-term memory and asked which kinds of comparisons were easiest to retrieve (Gentner & Landers, 1985; Rattermann & Gentner, 1987). Subjects first read a large set of stories. Two weeks later, they were given new stories which matched the original ones in various ways. Some were true analogs of the first stories; others were surface matches, sharing lower-order events and object descriptors but not higher-order relational structure. Subjects were asked to write out any prior stories recalled while reading the new stories. Afterwards, they rated all the pairs for soundness: i.e., how well inferences could be carried from one story to the other.

The results showed that, although subjects rated the analogies as much more sound than the surface matches, they were more likely to retrieve surface matches. Surface similarity was the best predictor of memory access, while similarity in relational structure was the best predictor of subjective soundness and also of subjective similarity. This dissociation held not only between subjects, but also within subjects. That is, subjects given the soundness task immediately after the cued retrieval task judged that the very matches that had come to their minds most easily (the mere-appearance matches) were highly unsound (i.e., unlikely to be useful in inference). This suggests that analogical access may be based on qualitatively distinct processes from analogical inferencing[2].

*Comparison to Current Approaches.* Some models of similarity assume smart processes operating over richly articulated representations. Most case-based reasoning models have this character (Schank, 1982; Kolodner, 1988). These models are rich enough to capture processes like case alignment and adaptation. But their models of memory access involve intelligent indexing of structured representations, which can predict superhuman access behavior; that is, that people should typically access the best structural match, even if it lacks surface similarity with the current situation. Further, models that assume that elaborate structural mapping processes are used to compare the current situation with stored situations have the disadvantage of being hard to scale up to large data bases. The reverse set of advantages and disadvantages holds for approaches that model similarity as the result of a dot product (or some other operation) over feature vectors, as is commonly done in mathematical models of human memory (e.g., Medin & Schaffer, 1978) and in connectionist models of learning (Smolensky, 1988). These models, with their nonstructured representations and relatively simple processes, do not allow for the structural precision of people's similarity judgments and inferences. However, they provide an appealing model of access since: (1) these computations are simple enough to make it feasible to compute many such matches and choose the best (the scaling criterion); and (2) being simple, these models will not always produce the best match (the fallibility criterion). While this might be a disadvantage in a normative model, it could be an advantage in modeling human similarity-based access, provided that the best match is sometimes produced. Next we propose an approach that we think may offer the best of both kinds of models.

## The MAC/FAC model

The complexity of the phenomena in similarity-based access suggests a two-stage model. Consider the computational constraints on access. The large number of cases in memory and the speed of human access suggests a computationally cheap process. But the requirement of judging soundness, essential to establishing whether a match can yield useful results, suggests an expensive match process. A common solution is to use a two-stage process, where a computationally cheap filter is used to pick out a subset of likely candidates for more expensive processing (c.f. Bareiss & King, 1989). MAC/FAC uses this strategy. The

[2]The finding is not that higher-order relations do not contribute to retrieval. Adding higher-order relations led to non-significantly more retrieval in two studies and to a small but significant benefit in the third. The point is simply that higher-order commonalities have a much bigger effect on mapping once the two analogs are present than they do on similarity-based retrieval.

puzzling phenomena noted previously, we claim, can be understood in terms of the interactions of its two stages.

Figure 1 illustrates the components of the MAC/FAC model. The inputs are a pool of memory items and a *probe*, i.e., a description for which a match is to be found. The output is a memory description and a comparison of this description with the probe.

There is little consensus about the global structure of long-term memory. Consequently, we assume only that at some stage in access there is a pool of descriptions from which we must select one (or a few) which is most similar to a probe. We are uncommitted as to the size of this pool. It could be the whole of long-term memory, or a subset of it if one postulates mechanisms for restricting the scope of search, such as spreading activation or indexing[3].

Both stages consist of a *matcher*, which is applied to every input description, and a *selector*, which uses the evaluation of the matcher to select which comparisons are produced as the output of that stage. Conceptually, matchers are applied in parallel within each stage. Since the role of the MAC stage is to produce plausible candidates for the FAC stage, we discuss FAC first.

## The FAC stage

The FAC matcher is simply the literal similarity computation defined by structure-mapping. Its output is a set of correspondences between the structural descriptions, a numerical *structural evaluation* of the overall quality of the match, and a set of candidate inferences representing the surmises about the probe sanctioned by the comparison. In subsequent processing, the structural evaluation provides one source of information about how seriously to take the match, and the candidate inferences provide potential new knowledge about the probe which must be tested and evaluated by other means. We implement this computation using SME, the Structure-Mapping Engine (Falkenhainer, Forbus & Gentner, 1989).

We use literal similarity rather than analogy in order to get the high observed frequency of surface remindings, which would mostly be rejected if FAC were strictly an analogy matcher. We believe this choice is ecologically sound because mundane matches are often the best guides to action. Riding a new bicycle, for instance, is often just like riding other bicycles (Gentner, 1989; Medin & Ortony, 1989). Associating actions with particular complex descriptions makes good computational sense because such associations can often be made before one can delinate exactly which aspects of a situation are relevant.

Currently FAC selects as output the best match, based on its structural evalution, and any others within 10% of it. In pilot studies we have experimented with various criteria, such as broadening the percentage, selecting a fixed number, and so forth. We settled on the 10% criteria because it generally returns a single result, only producing multiple results when there are two extremely close candidates. Depending on the assumptions one makes about subsequent processing, a modification which places a strict upper bound on the number produced (say, two) may also be appropriate.

Sometimes a probe reminds us of nothing. There are several ways this can arise in the MAC/FAC model. First, the FAC stage may not receive any candidates from the MAC stage (see below). Second, FAC might reject all candidates provided. This shows up by no match hypotheses being created; this has occurred, albeit rarely. Third, there could be a threshold on structural evaluations, so that matches below a certain quality simply were not considered. We view this as psychologically plausible, but do not include such thresholds currently because we have not yet found good constraints on them.

## The MAC stage

Even though the FAC stage is reasonably efficient[4], it is too expensive to consider running it exhaustively on realistic-sized memories as the "inner loop" in an analogical processing system. The MAC stage uses an extremely cheap matcher to estimate how well FAC would rate comparisons, to filter candidates down to a manageable number.

One estimate is the number of match hypotheses that FAC would generate in comparing a probe to a memory item, the *numerosity* of the comparison. If very few local matches are hypothesized, then clearly the best global interpretation cannot be large. On the other hand, numerosity is not a perfect estimator, since having a large number of local matches does not guarentee a large global interpretation. This is true because (1) match hypotheses can end up being ungrounded because some of their arguments cannot be placed into correspondence (and are thus ignored), and (2) the mutual incompatibilities introduced by the 1:1 constraint may prevent a single large interpretation from forming, yielding instead several small ones.

The most straightforward way to compute numerosity is to actually generate and count the match hypotheses. This is what our original version of MAC/FAC did (Gentner, 1989). It also partly what ARCS (Thagard et al 1990) does. ARCS builds much of the network which ACME would build between target and base but between the probe and every item in memory. We view these solutions as psychologically and computationally implausible. Even with parallel and/or neural hardware, it is hard to see how the expense of generating match hypothesis networks between a probe and everything in a large pool of memory can provide realistic response times. Instead, we turn to a novel means of estimating numerosity.

Let P be the set of functors (i.e., predicates, functions, and connectives) used in the descriptions that constitute

[3]In current AI systems indexing often yields a unique description; we view this property as unlikely to scale. For example, there could be dozens or even hundreds of experiences which are similar enough to be put in the same index entry, yet different enough to make it worthwhile to save them as distinct memories.

[4]$O(n^2)$ for match hypothesis generation, where $n$ is the number of items in base or target, and roughly $O(log(n^2))$ to generate a global interpretation, using the greedy merge algorithm of Forbus & Oblinger (1990).

Figure 1: The MAC/FAC model

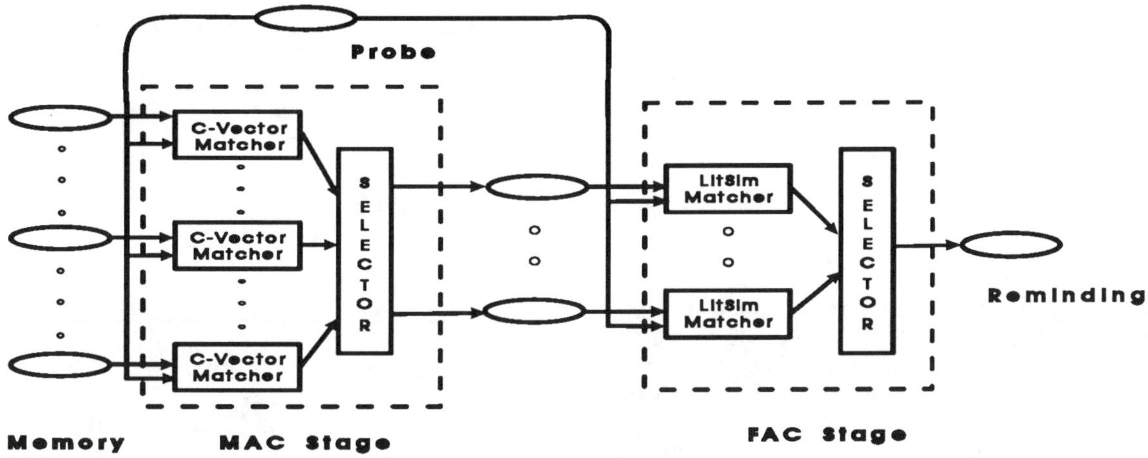


memory items and probes. We define the *content vector* of a structured description as follows. A content vector is an n-tuple of numbers, each component corresponding to a particular element of P. Given a description D, the value of each component of its content vector indicates how many times the corresponding element of P occurs in D. Components corresponding to elements of P which do not appear in statements of D have the value zero. One simple algorithm for computing content vectors is to simply to count the number of occurrences of each functor in the description. Thus if there were four occurrences of IMPLIES in a story, the value for the IMPLIES component of its content vector would be four[5]. Thus content vectors are easy to compute from a structured representation and can be stored economically.

The MAC matcher works as follows: Each memory item has a content vector stored with it. When a probe enters, its content vector is computed. A score is computed for each item in the memory pool by taking the dot product of its content vector with the probe's content vector. These scores are fed to the MAC selector, which produces as output the best match and everything within 10% of it, as in the FAC stage. (We plan to add a threshold so that if every match is too low MAC returns nothing.)

Clearly, measuring similarity using content vectors has critical limitations, since the actual relational structure is not taken into account. But the dot product can be used to estimate relative similarity, since it is a good approximation to numerosity. (Essentially, the product of each corresponding component is an overestimate of the number of match hypotheses that would be created between functors of that type.) Content vectors are insufficient because they do not provide the correspondences and candidate inferences which provide the power of analogy. But by feeding MAC's results to the structural matcher of the FAC stage, we obtain the required inferential power.

This MAC matcher has the properties we desire. It is cheap, and could be implemented using a variety of massively parallel computation schemes, including connectionist. Next, we demonstrate that MAC/FAC provides a good approximation of psychological data.

## Computational Experiments

We have successfully tested MAC/FAC on a variety of descriptions, including simple metaphors and physics scenarios. Here we compare the performance of MAC/FAC with that of human subjects, using the "Karla the Hawk" stories. For these studies, we wrote sets of stories consisting of base stories plus four variants, created by systematically varying the kind of commonalities. All stories share first-order relations, but vary as follows:

| | Common h.o. relations | Common object attributes |
|---|---|---|
| LS: | Yes | Yes |
| SF: | No | Yes |
| AN: | Yes | No |
| FOR: | No | No |

As discussed above, subjects rated analogy (AN) and literal similarity (LS) as more sound than surface (SF) and FOR matches (matches based only on common first-order relations, primarily events). Previously, we tested SME running in analogy mode on SF and AN matches and found that it correctly reflected these human soundness rankings (Forbus & Gentner, 1989; Skorstad et al, 1987). Here we seek to capture human retrieval patterns: Does MAC/FAC duplicate the human propensity for retrieving SF and LS matches rather than AN and FOR matches. The idea is to give MAC/FAC a memory set of stories, then probe with various new stories. To count as a retrieval, a story must make it through both MAC and FAC.

[5]We have also experimented with normalized content vectors, to minimize the effects of size discrepancies. So far we have seen no significant empirical difference between these algorithms, but we suspect that normalization will be necessary when adding retrieval thresholds.

Table 1: Proportion of correct retrievals given different kinds of probes

1. Memory contains 9 base stories and 9 FOR matches; probes were the 9 LS, 9 SF, and 9 AN stories.
2. The rows show proportion of times the correct base story was retrieved for different probe types.

| Probes | MAC | FAC |
|---|---|---|
| LS | 1.0 | 1.0 |
| SF | 0.89 | 0.89 |
| AN | 0.67 | 0.56 |

Table 2: Mean numbers of different match types retrieved when base stories used as probes

1. Memory contains 36 stories (LS, SF, AN, and FOR for 9 story sets); the 9 base stories used as probes
2. Other = any retrieval from a story set different from the one to which the base belongs.

| Retrievals | MAC | FAC |
|---|---|---|
| LS | 0.78 | 0.78 |
| SF | 0.67 | 0.44 |
| AN | 0.33 | 0.11 |
| FOR | 0.22 | 0.0 |
| Other | 1.33 | 0.22 |

Table 3: Mean numbers of different match types retrieved with base stories as probes

1. Memory contains 27 stories (9 SF, 9 AN, 9 FOR); 9 base stories used as probes.

| Retrievals | MAC | FAC |
|---|---|---|
| SF | 0.89 | 0.78 |
| AN | 0.56 | 0.45 |
| FOR | 0.22 | 0.11 |
| Other | 1.11 | 0.11 |

In the psychological experiment, the human subjects had a memory set consisting of 32 stories, of which 20 were base stories and 12 were distractors. They were later presented with 20 probe stories which matched the base stories as follows: 5 LS matches, 5 AN matches, 5 SF matches and 5 FOR matches and told to write down any prior stories of which they were reminded. The proportions of remindings for different match types were .56 for LS, .53 for SF, .12 for AN and .09 for FOR. Across three variations of this study, this retrievability order has been stable: LS $\geq$ SF > AN $\geq$ FOR.

For the computational experiments, we encoded predicate calculus representations for 9 of the 20 story sets (45 stories). These stories are used in all three experiments described below.

*Simulation Experiment 1.* In our first study, we put the 9 base stories in memory, along with the 9 FOR stories which served as distractors. We then used each of the variants – LS, SF, and AN – as probes. This roughly resembles the original task, but MAC/FAC's job is easier because (1) it has only 18 stories in memory, while subjects had 32, in addition to their vast background knowledge; (2) subjects were tested after a week's delay, which may have caused some memory deterioration.

Table 1 shows the proportion of times the base story made it through MAC and through FAC. MAC/FAC's performance is much better than that of the human subjects, perhaps partly because of the differences noted above. However, its results show the same ordering as those of human subjects: LS > SF > AN.

*Simulation Experiment 2.* To give MAC/FAC a stronger challenge, we put the four variants of each base story into memory. This made a larger memory set (36 stories) and also one with many competing similar choices. Each base story in turn was used as a probe. This is almost the reverse of the task subjects faced, and is more difficult.

Table 2 shows the mean number of matches of different similarity types that succeed in getting through MAC and through FAC. There are several interesting points here. First, the retrieval results (i.e., the number that make it through both stages) ordinally match the results for human subjects: LS > SF > AN > FOR. This degree of fit is encouraging, given the difference in task. Second, as expected, MAC produces some matches that are rejected by FAC. This number depends partly on the criteria for the two stages. Here, with MAC and FAC both set at 10%, the mean number of memory items produced by MAC is 3.3, and the mean number accepted by FAC is 1.5. Third, as expected, FAC succeeds in acting as a structural filter on the MAC matches. It accepts all of the LS matches MAC proposes and some of the partial matches (i.e., SF and AN), and while rejecting most of the inappropriate matches (i.e., FOR and matches with stories from other sets).

*Simulation Experiment 3.* In the prior simulation, LS matches were the resounding winner. While this is reassuring, it is also interesting to know which matches are retrieved when there are no perfect overall matches. Therefore we removed the LS variants from memory and repeated the second simulation experiment, again probing with the base stories. As Table 3 shows, SF matches are now the clear winners in both the MAC and FAC stages. Again, the ordinal results match well with those of subjects: SF > AN > FOR.

*Summary of Simulation Experiments.* The results are encouraging. First, MAC/FAC's ordinal results match those of human subjects. In contrast, the closest competing model, Thagard et al's (1991) ARCS model of similarity-based retrieval, when given the Karla the hawk story in memory (along with 100 fables as distractors) and the four similarity variants as probes, produced two violations in its order of asymptotic activation. Its asymptotic activations were LS (.67), FOR (-.11), SF (-.17), AN (-.27). Thus MAC/FAC explains the data better than ARCS. This is especially interesting because Thagard et al argue that a complex localist connectionist network which integrates semantic, structural, and pragmatic constraints is required to model similarity-based reminding. While such models are intriguing, MAC/FAC shows that a simpler model can provide a better account of the data.

Finally, and most importantly, MAC/FAC's overall pattern of behavior captures the motivating phenomena: (1) it produces a large number of LS matches, thus satisfying the *primacy of the mundane* criterion; (2) it produces a fairly large number of SF matches, thus satisfying the fallibility criterion; (3) it produces a small number of analogical matches, thus satisfying the *existence of rare events* criterion; and finally, (4) its algorithms are simple enough to apply over large-scale memories, thus satisfying the *scalability* criterion.

## Discussion

We have presented MAC/FAC, a two-stage similarity-based model of access. The MAC stage uses content vectors, a novel summary of structured representations, to provide an inexpensive "wide net" search of memory, whose results are pruned by the more expensive literal similarity matcher of the FAC stage to arrive at useful, structurally sound matches. We demonstrated that MAC/FAC can simulate the patterns of access exhibited by human subjects. We believe that the psychological issues MAC/FAC raises are worth further study. MAC/FAC is reasonably efficient, even on serial machines, so we believe it could be a useful component in performance-oriented AI systems also.

In addition to the psychological issues raised earlier, there are several computational studies in preparation using MAC/FAC. These include:

*Experiments with larger knowledge bases:* A crucial question for any access model is how well it scales to substantially larger memories. Two avenues we are exploring are: (1) using the CYC knowledge base as a source of descriptions and (2) using MAC/FAC as a tool on the ILS Story Archive Project to aid in spotting potentially relevant links between stories.

*Larger-scale process models:* Several psychological questions about access cannot be studied without embedding MAC/FAC in a more comprehensive model of analogical processing. For example, there is ample evidence that subjects can "tune" their similarity judgements when the items being compared are both already in working memory. While it seems clear that MAC is impenetrable, it is hard to tell whether or not FAC is tunable or whether a separate similarity engine is required. Order effects in analogical problem solving (Keane, in press) suggest the latter. How can the access system be used to incrementally construct abstractions and indexing information to help structure long-term memory (c.f. Skorstad, Gentner, and Medin 1988)?

## Acknowledgements

This research was supported by the Office of Naval Research (Contract No. N00014-89-J-1272) and by the National Science Foundation. We thank Ray Bareiss, Mark Burstein, Gregg Collins, Brian Falkenhainer, Rob Goldstone, Art Markman, Doug Medin, and Mary Jo Rattermann for discussions of these issues.

## References

Bareiss, R. and King, J. 1989. Similarity Assessment in Case-Based Reasoning, Proceedings of the DARPA Case-Based Reasoning Workshop, Pensacola Beach, FL, pages 67-71.

Clement, C. A., & Gentner, D. (in press). Systematicity as a selection constraint in analogical mapping. *Cognitive Science.*

Falkenhainer, B., Forbus, K., Gentner, D. 1989. The Structure-Mapping Engine: Algorithm and examples, *Artificial Intelligence*, 41:1-63.

Forbus, K. and Gentner, D. 1989. Structural evaluation of analogies: What Counts?, Proceedings of the Cognitive Science Society.

Forbus, K. and Oblinger, D. 1990. Making SME greedy and pragmatic. Proceedings of the Cognitive Science Society.

Gentner, D. 1983. Structure-mapping: A theoretical framework for analogy. *Cognitive Science* **7**(2).

Gentner, D. 1988. Analogical inference and analogical access. In A. Prieditis (Ed.), *Analogica* (pp. 63-88). Los Altos, CA: Morgan Kaufmann.

Gentner, D. 1989. Finding the needle: Accessing and reasoning from prior cases. Proceedings: Case-based reasoning workshop, DARPA ISTO, Pensacola, Florida.

Gentner, D., & Clement, C. 1988. Evidence for relational selectivity in the interpretation of analogy and metaphor. In G. H. Bower (Ed.), *The psychology of learning and motivation* (Vol. 22, pp. 307-358). New York: Academic Press.

Gentner, D., & Landers, R. 1985. Analogical reminding: A good match is hard to find. *Proceedings of the International Conference on Cybernetics and Society* (pp. 607-613), Tucson, AZ. New York: IEEE.

Gick, M. L., & Holyoak, K. J. 1980. Analogical problem solving. *Cognitive Psychology*, **12**:306-355.

Goldstone, R. L., Medin, D. L., & Gentner, D. (in press). Relational similarity and the non-independence of features in similarity judgments. *Cognitive Psychology.*

Keane, Mark T. (in press) Similarity and Ordering Constraints on Analogical Mapping.

Kolodner, J.L. (Ed.) 1988. *Proceedings of the First Case-Based Reasoning Workshop*, Morgan Kaufmann, Los Altos, CA.

Markman, A. B., & Gentner, D. 1990. Analogical mapping during similarity judgments. Proceedings of the Twelfth Annual Conference of the Cognitive Science Society, 38-44, Cambridge, MA. Hillsdale, NJ: Erlbaum.

Medin, D. L., and Ortony, A. 1989 Psychological Essentialism. In S. Vosniadou and A. Ortony (Eds.), *Similarity and Analogical Reasoning*,179-195. New York: Cambridge University Press.

Medin, D. L., & Schaffer, M. M. 1978. Context theory of classification learning. *Psychological Review*, **85**:207-238.

Rattermann, M. J., & Gentner, D. 1987. Analogy and similarity: Determinants of accessibility and inferential soundness. In Proceedings of the Ninth Annual Meeting of the Cognitive Science Society, 23-34, Seattle, WA.

Ross, B. H. 1984. Remindings and their effects in learning a cognitive skill. *Cognitive Psychology*, **16**:371-416.

Ross, B. H. 1987. This is like that: The use of earlier problems and the separation of similarity effects. *Journal of Experimental Psychology: Learning, Memory, and Cognition*, **13**:629-639.

Schank, R. 1982. *Dynamic Memory.* Cambridge University Press, Cambridge, England.

Skorstad, J., Falkenhainer, B., and Gentner, D. 1987. Analogical processing: A simulation and empirical corroboration. Proceedings of AAAI-87.

Skorstad, J. Gentner, D. and Medin, D. 1988. Abstraction processes during concept learning: A structural view. Proceedings of the Cognitive Science Society.

Smolensky, P. 1988. On the proper treatment of connectionism. *Behavior and Brain Sciences.* **11**:1-74.

Thagard, P., Holyoak, K.J.,Nelson, G., and Gochfeld, D. 1990. Analog retrieval by constraint satisfaction. *Artificial Intelligence*, **46**:3.

# A Functional Perspective on Reminding*

**Tom McDougal**
and
**Kristian Hammond**
**Department of Computer Science**
**University of Chicago**
**1100 E. 58th Street**
**Chicago, IL 60637**

**Colleen Seifert**
**Department of Psychology**
**University of Michigan**
**330 Packard Road**
**Ann Arbor, MI 48104**

### Abstract

This paper explores the relationship between human activity and remindings. We argue that the type of activity in which a person is engaged influences the kinds of features that trigger remindings, and, conversely, that remindings can change situated behavior over time. We argue further that the types of indices used in memory are not uniform, but depend upon the nature of the task that a person is engaged in as well as his history of interaction with the world.

## Introduction

A rift currently divides the field of analogical reasoning, with researchers in psychology on one side and those in Artificial Intelligence on the other. This rift comes from a basic difference in research philosophies within the two camps. Researchers in AI generally try to solve problems; they determine the functionality they need, develop theories of behavior to provide that functionality, and then correlate those theories with human behavior. Researchers in psychology, work in the other direction. They try to progress from observed human behavior to a functional theory.

Specifically, a debate has long brewed with regard to people's ability to be reminded on the basis of abstract features. The AI functionalists have constantly promoted the need to find a "best match" between a new problem description and an existing example in memory (Kolodner & Simpson 1989, Hammond 1989, Bareiss 1989). They have concentrated on examples that push the limits of their ability to *represent* relevant features, let alone organize memory around them (Birnbaum and Collins 1989, Schank 1982).

Conversely, the empiricists in psychology claim that the experimental data shows that retrieval occurs only on the basis of surface features (Gentner & Landers 1985, Ross 1987, Gick & Holyoak 1983, Holyoak & Thagard 1985).

This either/or debate hides the diversity of remindings. Memory serves many masters, and thus we should *expect* it to be sensitive to the activity in which the agent is engaged. Sometimes remindings occur on the basis of surface features, sometimes on the basis of deep structure. Evidence exists for each, and each has its unique functionality. Likewise, learning from remindings is not always the one-shot deal often postulated by those in case-based reasoning, but can arise from a dynamic interaction between repeated remindings and activity in the world.

In this paper, we will present several examples of remindings in different contexts, examining each in detail to see what functional role the remindings play, and what they tell us about the structure and dynamic of memory. Our argument will be that, while memory is driven by a single process, the different functions it serves cause it to exhibit different behaviors.

* This work was supported in part by the Defense Advanced Research Projects Agency, monitored by the Air Force Office of Scientific Research under contract F49620-88-C-0058, and the Office of Naval Research under contract N0014-85-K-010.

## Two Examples of Remindings

We begin with two examples which address the abstract feature/surface feature dispute, with implications for planning. Both are of naturally occurring remindings of the sort studied by Schank in *Dynamic Memory* (Schank 1982).

**Flight of the Phoenix**

X was working late on a project that was due in a matter of days. As he saw the deadline approach, he considered the following two plans: either continue to work straight through the night (and the next day) or get a good night's sleep and come back to the office refreshed. The first plan allows the use of all of the time for work on the project. The second provides less time, but the time it does provide is of a better quality.

While thinking about his problem, X was reminded of a scene from the movie "Flight of the Phoenix" in which a character played by Jimmy Stewart had to start a damaged plane in the desert. The plane's ignition used explosive cartridges and Stewart had only seven left. The plane's exhaust tubes were filled with sand—which could be blasted out using the same explosive cartridges.

Jimmy Stewart's character was faced with a choice: either try to start the plane using the cartridges directly, or use some of them to clear the exhaust tubes thus enhancing the overall utility of the other cartridges.

This is a classic problem-solving situation involving resource conflicts. The reminding is of a problem analogically similar to the problem facing X, and it carries with it information about how to make the decision. In the next example, the reminding is much more concretely related to the task at hand.

**Missed Exit**

Y was driving along an expressway in the left lane, because traffic was moving faster there. He spotted the exit where he wanted to get off. Unfortunately, by the time he worked his way into the right-hand lane, he had missed the exit.

The next time Y was driving on the same expressway in the left lane, intending to get off at the same exit, he remembered the exit earlier than before, but still too late to make the exit.

The third time, Y made the exit.

This example, due to Phil Agre (1990, personal communication), demonstrates the role that remindings can play in repeated situated activity. By remembering earlier failures, Y is able to modify his behavior incrementally until he reaches an optimal solution for balancing the conflicting subgoals of driving fast and catching the exit.

In addition to their relevance to the abstract versus surface feature dispute, these two examples represent opposite ends of the planning spectrum in AI. Together, they challenge us to devise a coherent model of memory that supports both behaviors.

## Analysis

Although we use the words "remember" and "reminding" to describe the actions of memory in each of the above examples, the dynamics which give rise to the remindings are very different, and the remindings themselves play different functional roles. These differences have important implications for the features used for memory indexing and the mechanisms by which memory functions.

In the Flight of the Phoenix example, X faces a choice between two plans to satisfy a single goal. The reminding from the movie comes about as a by-product of contemplating the two choices, and uncovering the nature of the sleep-then-work plan. The situation faced by Jimmy Stewart shares no surface features with the write-or-sleep decision. On the level of abstract problem-solving, however, the two situations match exactly. Each protagonist has a goal and a limited resource. The protagonists must decide between two plans for using the resource; one plan uses the resource directly in service of the goal, while the other splits the resource between an optimization step and a direct-use step.

The reminding plays a valuable functional role in that it gives X a different perspective on the problem at hand. In the movie, Jimmy Stewart went with the clear-and-start plan over the objections of his companions, but from X's point of view the important information comes from the debate over whether the charges that remained after clearing the exhaust would be sufficient to start the plane. This debate focuses X's attention on the question of whether sleeping would leave enough time to complete the project.

In the Missed Exit example, remindings play a different functional role, contributing to a change in behavior within a repeated activity. Structurally, the problem is to optimize over two goals: drive fast (hence drive in the left lane), and get off at the right place. However, Y can act only on the basis of immediate environmental cues and memory.

The classical planning approach to this problem, the kind of approach used successfully in the Flight of the Phoenix example, would require an uncovering of the deep structure of the problem. However, the structure is minimal in this case; a structure-level "solution" would be look like "Move into the right-hand lane only when you need to," which is clearly no solution at all.

Failing a solution on the abstract level, the classical planning approach would use projection. For projection, a planner would require knowledge of the speed of traffic in each lane. It would require knowledge of the traffic density, since that affects how quickly one can change lanes. It would require exact knowledge of distances between landmarks.

Yet distance between landmarks is not commonly known with any precision, and traffic changes chaotically, making precise measurements and predictions impossible. A classical planner cannot hope to decide *a priori* where to begin changing lanes, because the necessary information just isn't available.

No useful solution to this problem can be defined in terms of the structure of the situation. The solution must necessarily be defined in terms of the surface features that define the activity. Those features are detectable only at the time of execution, and imperfectly even then, rendering projection impossible.

Runtime trial-and-error based on immediately perceptible environmental cues is the only practical way to solve this problem. We need to remember where we began our move last time, what happened, and how traffic today compares to traffic last time. That kind of information is exactly what the remindings in this example provide.

## Implications for Memory

The Flight of the Phoenix and the Missed Exit examples tell us that memory responds to the functional needs of the situation. The functional needs of the two examples place different vocabulary and structural demands on memory. This section defines what those demands are.

As he remembers "Flight of the Phoenix," X is able to recall many surface-level details. These include the leading actor, the setting, the plane, the means for starting the plane, the two plans for starting the plane. From this we draw the uncontroversial conclusion that the vocabulary of memory includes the common semantic vocabulary of people, planes, and deserts.

Additionally, X remembered more complex aspects of the particular scene. He remembered the rationale for the clear-and-start plan and the concern that, after clearing the exhausts, the remaining cartridges would be insufficient. Thus memory includes a vocabulary associated with planning, including notions of resource limitations and of optimization.

We should point out that deep knowledge of the situation can reduce the overhead in storing some of the details. For example, the relationship between the setting (the desert) and the problem (sand in the exhaust) makes it easy to reconstruct either detail if one can remember the other.

By what features is the scene from "Flight of the Phoenix" indexed? X could probably recall that movie in the context of movies starring Jimmy Stewart, or set in the desert, or involving damaged planes. However, what makes the Flight of the Phoenix example interesting is that X's current dilemma (whether to write or sleep) contains none of these surface details, but matches the movie perfectly on the abstract planning level. We can only conclude that the movie scene is also indexed by and retrieved on the basis of those abstract planning characteristics.

Unfortunately, empirical psychologists have been unsuccessful at obtaining deep-structure remindings in a controlled setting (Gentner and Landers 1985, Ratterman and Gentner 1987, Ross 1987, Gick and Holyoak 1980, 1983, Holyoak and Thagard 1985, Holyoak and Koh 1987). Their experience is that memory retrieval occurs primarily on the basis of surface features; retrieval on the basis of abstract structure is rare and hard to reproduce.

One possible explanation for their results is that the structure and functioning of memory itself limits our ability to have such remindings. However, the Flight of the Phoenix example contradicts this explanation. We believe that memory is equally well-equipped for deep remindings and surface-feature remindings; any limitation arises from the task at hand.

We find that much of the psychologists' data comes from shallow story understanding and problem-solving tasks. In (Gentner and Landers 1985), the task was to read stories generally two paragraphs in length, and to write down any remindings the subjects had of an earlier group of stories. The stories themselves were not difficult to read, and from the subjects' point of view, *any* reminding was sufficient.

The predominance of surface-feature remindings in this context does not surprise us. Seifert et al (1986) also observed the absence of priming effects (implicit activation) during story processing on the basis of thematic structure similarity. However, when the subjects were instructed to attend to thematic structure, priming effects were discovered.

In (Gick and Holyoak 1983), subjects were given a story, referred to as "The General," in the context of a recall task. Later they were presented with a problem to solve, known as the "radiation problem."

In the story, a general wishes to overthrow a dictator. The dictator occupies a fortress in the country.

Several roads lead into the fortress; however, all of them have been mined such that while small groups can pass safely, any large force will detonate the mines. The general's solution is to partition his army among the roads, so that each group is small enough to pass safely, yet when they converge on the fortress they will be strong enough to capture it.

The "radiation problem" posits an inoperable tumor which must be killed with radiation. Any radiation strong enough to damage the tumor will also damage the surrounding tissue. The problem is to destroy the tumor with radiation while preserving the surrounding tissue; the solution is to radiated the tumor from several directions with rays weak enough that they do not damage the healthy tissue. Converging on the tumor, however, they sum together to kill the tumor.

The "radiation problem" bears little surface similarity to the story of the general, yet it was intended to be structurally analogous. Subjects found it difficult enough that few of them were able to solve it without a hint directing them to use the solution in the story.

Our problem with this study is that although the solution to the radiation problem and the solution used by the general are beautifully analogous, the *problems* are not. The goal in the story is to upset the status quo. The goal in the problem is to remove an invasive entity to *restore* the status quo. Seifert and Gray (1990) changed the General story so that the fortress is occupied by terrorists instead of by the dictator, spontaneous analogical recall was doubled.

In their conclusion, Seifert and Gray noted that the ability to be reminded based on abstract features requires that both episodes be encoded the same way. The abstract features must be available at both storage and retrieval times. What makes the Flight of the Phoenix reminding possible is that the structure of the situation is explicitly stated in the movie, allowing that structure to be used as an storage index, and the structure is uncovered by X as he ponders his dilemma, allowing it to be used as a retrieval index. Simple processing tasks will not evoke an encoding sufficiently rich for deep-structure analogical retrieval.

The remindings in the Missed Exit example key off of surface features, and we have seen that there are good functional reasons for that. What is peculiar about this example, however, is the change in behavior over time. How does the reminding occur farther back along the expressway each time?

As explained above, the circumstances are too complicated to determine an exact point at which to begin changing lanes. This is true *a posteriori* as well as *a priori*, and for the same reasons. We hypothesize, therefore, that when the failure occurs it gets indexed by associated landmarks. "Associated landmarks" would include landmarks near the exit, and probably landmarks near where Y began to change lanes. A conscious version of this indexing would be something like "The bridge wasn't soon enough. Next time I need to start moving earlier."

How does this help next time? We note that a good driver does not focus his attention directly in front of her car, but scans the road far ahead. In situations where his view of a familiar road is blocked, as around a corner or in fog, the good driver does a small amount of projection, anticipating what will come next. Thus, before he gets there, Y will either see the landmark with which the failure is associated, or will imagine it. Seeing it or imagining it is enough to trigger the reminding. If the reminding occurs far enough back along the expressway, Y will catch the exit. Otherwise, he has another episode to remember next time, indexed by a landmark farther back.

## More Examples, More Functionalities

The people involved in the Flight of the Phoenix and the Missed Exit examples had different functional requirements: X had to decide between two alternative courses of action; Y had to remember to begin moving toward the right-hand line earlier. Those different functional requirements caused caused memory to behave in different ways. Next we consider two examples which seem to place the same functional requirements on memory, but yield different behavior. These behaviors reveal additional subtleties in the organization of memory.

**Milk**

J ran out of milk during breakfast. He couldn't buy any then because he had to go to work.

While running an errand later that day, J passed a grocery store. He remembered that he needed milk. Deciding that the milk would keep until he went home, he stepped into the store.

J has a goal to get milk. He remembers the goal at the best possible moment for satisfying it, when the conditions in the world are ripe for buying milk.

The next example shows that we can't always recognize even the most obvious of opportunities.

**Steve's Loan**

Steve loaned his good friend Tom $20 one Friday night. The next morning, Tom found the $20 bill, unspent, and told himself to give it to Steve when he saw him next. Later, Tom went to Steve's apartment to watch a ball game, but didn't remember the money until Steve happened to take some dollar bills out of his pocket.

Tom knew in advance the conditions of a good opportunity to give Steve the money: seeing Steve. For some reason the effort to remember at the right time failed, however, and he remembered thanks only to an unexpected event.

## Analysis

The remindings in these two stories share the common theme of remembering to do something. The interesting difference between them is that, in the first case, the reminding came at a "good" time; in the second, the reminding was triggered by a causally unrelated and unpredictable event.

Running out of milk gave J the goal to get milk, but he couldn't couldn't pursue the goal immediately. What happened to that goal? We could imagine him forgetting about it, not remembering that he was out of milk until the next time he needed some. Or, we could imagine him planning to buy milk at a specific time of day, such as during lunch, or perhaps anticipating the errand. The former is an unfortunate lapse which happens to all of us sometimes; the latter requires projection, which has been proved intractable in the general case (Chapman 1985).

In this case, however, J made no effort to plan how a trip to the grocery store might fit into the tumultuous events of the day, nor was the goal to get milk forgotten. Instead, it remained in memory indexed by features which would indicate a good opportunity to buy milk, i.e. the image of a grocery store. This is an example of *opportunistic memory,* an idea examined in (Converse, Hammond, and Martin, 1990).

According to the opportunistic memory theory, goals and plans are governed by *policies* (McDermott 1978). A policy describes a state of the world which, while desirable to an agent, does not require constant attention. Having milk is a typical policy. We don't spend our days asking if we have milk, but if we run out, we notice it and make some effort to get more.

Well-worn goals like obtaining milk have well-worn plans indexed under them. In opportunistic memory, when the agent decides that he cannot immediately pursue the goal, the goal is *suspended,* or saved in memory. The preconditions of the plans for satisfying it are used as indices, so that if the world presents a situation in which many of the preconditions are satisfied, the agent can take advantage of the opportunity.

Getting milk is a common task which J has accomplished many times, and almost always at a grocery store. The precondition of being at a store are well-known and easily recognized. Thus when the goal of getting milk is suspended, J knows under which the features to index it. When the opportunity arises to buy milk, J can recognize it and take advantage of it.

The example of Steve's Loan provides an interesting counterpoint to the first because it seems to betray the opportunistic memory theory. Tom had a goal to return the money. He made an explicit, conscious attempt to index that goal under the conditions that would make it easy to satisfy: seeing Steve. Yet, the goal was not triggered when those conditions were met, but only on the basis of surface features which could not have been anticipated.

The trouble arises out of a need to avoid spurious remindings. Some features are too common to be used as reliable indices, unless there is a close two-way correlation between the feature and what is to be remembered. For Tom, the sight of Steve is associated with too many things to be a good feature on which to index anything. They commute to work, cook dinner, play pool, and watch TV together. Grocery stores, on the other hand, have a unique association with purchasing food. Although the sight of a grocery store is common, the role grocery stores fill in our lives is narrowly enough defined that they make a good feature for remindings about purchasing food.

## Simple Remindings

In contrast to the complex, structure-rich reminding in the Flight of the Phoenix example, many, probably most of our remindings are extremely simple.

**Hawks**

X says to Y, "I saw a hawk yesterday." Y says, "Oh, I was just reading about hawks. Do you know what kind it was?"

Here is a surface-level reminding with no apparent functionality. Consider the task, however: Y has only to understand what X has said to him and react to it in an appropriate way. The subject of the utterance is a hawk-sighting; the context of the utterance is informal conversation about recent personal events in one's life. Hence the reminding is doubly appropriate. In the process of relating the utterance to his relevant knowledge of hawks, Y is reminded of a recent activity suitable for conversation. Simple remindings of this kind are a part of basic human understanding.

## Summary

This paper presented five reminding examples. First, a problem-solving situation generated an reminding which, while sharing no important surface features, provided insight into the nature of the problem. In the second example, remindings occurred in the midst of repeated, situated activity on the basis of perceptually evident features. The remindings helped the agent avoid failure and optimize over conflicting subgoals. Third, a goal that could not be pursued immediately was suspended and indexed in memory by the features that would indicate a good opportunity for its satisfaction. The fourth example revealed a limitation of this functionality: goals cannot be indexed by features which are extremely common and associated with many other goals. Finally, informal conversation provoked a simple reminding nonetheless well-suited to the task of understanding an utterance and generating an appropriate reply.

These five examples reveal the diversity of situations in which remindings occur, and the diverse roles they play. Remindings help us solve problems, they help us optimize between conflicting goals, they help us notice opportunities, they help us do what we intend to do, and they make it possible for us to relate new information to what we already know.

In response to the debate over whether remindings occur on the basis of surface features or deep structure, we have argued that the context and type of activity is the important factor, rather than any inherent limitation of memory. Tasks requiring abstract analysis yield abstract remindings; tasks that depend on features in the world index on the basis of readily-identifiable surface features. Finally, we have argued that this influence of activity on memory is functionally significant: abstract remindings are more likely to carry the kind of information we need to solve complex problems; surface-level remindings are more likely to help us react quickly to our environment.

## References


Bareiss, R. (1989) *Exemplar-Based Knowledge Acquisition.* Academic Press.

Birnbaum, L, & Collins, G. (1989). Remindings and Engineering Design Themes: A Case Study in Indexing Vocabulary. In Proceedings of the Second DARPA Workshop on Case-based Reasoning in AI. Pensacola Beach, FL. DARPA.

Chapman, D. (1985) Nonlinear Planning: A Rigorous Reconstruction. In Proceedings of the Ninth International Joint Conference on Artificial Intelligence, pp.1022-24. IJCAI, 1985.

Converse, T., Hammond, K., & Martin, C. (1990). Integrating Planning and Acting in a Case-based Framework. In Proceedings of the Eighth Nation Conference on Artificial Intelligence, pp.292-297. AAAI, 1990.

Gentner, D. & Landers, R. (1985). Analogical Reminding: A Good Match is Hard to Find. In Proceedings of the International Conference on Systems, Man and Cybernetics, Tucson, AZ.

Gick, M. & Holyoak, K. (1983). Schema Induction and Analogical Transfer. *Cognitive Psychology,* 15:1-38.

Hammond, K. (1989). *Case-based Planning: Viewing Planning as a Memory Task.* Academic Press.

Holyoak, K. & Thagaard, P. (1985). Discovering the Wave Theory of Sound: Inductive Inference in the Context of Problem Solving. In Proceedings of the Ninth International Joint Conference on Artificial Intelligence, pp 610-612, Los Angeles, Ca. AAAI.

Kolodner,J. & Simpson, R. (1989). The Mediator: Analysis of an Early Case-based Problem Solver. *Cognitive Science Journal.*

McDermott, D. (1978). Planning and Acting. *Cognitive Science,* **2**, 1978.

Rattermann, M. & Gentner, D. (1987). Analogy and similarity: Determinants of accessibility and inferential soundness. In Proceedings of the 9th Annual Conference of the Cognitive Science Society. Hillsdale, NJ: Lawrence Erlbaum Associates.

Ross, B. H. (1987). This is Like That: The Use of Earlier Problems and the Separation of Similarity Effects. *Journal of Experimental Psychology: Learning, Memory and Cognition,* 13(4):629-639.

Schank, R. (1982). *Dynamic Memory: A theory of reminding and learning in computers and people.* Cambridge University Press.

Seifert, C. & Gray, K. (1990). Representational Issues in Analogical Transfer. In Proceedings of the Twelfth Annual Conference of the Cognitive Science Society.

Seifert, C., McKoon, G., Abelson, R.P., & Ratcliff, R. (1986). Memory connections between thematically similar episodes. *Journal of Experimental Psychology: Learning, Memory, and Cognition,* 12, 220-231.

# Improving Case Retrieval Through Observing Expert Problem Solving

**Michael Redmond***
College of Computing
Georgia Institute of Technology
Atlanta, Georgia 30332-0280
E-mail: redmond@cc.gatech.edu

## Abstract

As case-based reasoners gain experience in a domain, they need to improve their case retrieval so that more useful cases are retrieved. One problem in doing this is that the reasoner who most needs to learn is least able to explain successes or failures. A second problem is that uncontrolled pursuit of an explanation could be very expensive. There are three keys to the approach presented. First, the student observes expert problem solving and sets up expectations for what the expert will do next. When expectations fail, the reasoner has its failure isolated to a single step, and the correct action for the situation has been provided. Second, if the student can retrieve part of a case that would have suggested a correct prediction, then that case snippet can be used to limit the explanation process, making the process more efficient. Third, when no explanation can be found, the reasoner resorts to empirical adjustment of feature importance.

Case-based reasoning (CBR) is based on the observation that experience, retained in the form of cases, can be used to efficiently and effectively solve future, similar problems. A case-based reasoner can improve in a number of different ways. It can acquire new cases. Or it can improve its case retrieval, so that more useful cases are retrieved. In Redmond (1989b) we discussed our approach to acquiring new cases through apprenticeship. Part of apprenticeship involves observing and understanding expert problem solving. This same kind of experience can be used to improve retrieval of cases.

Improving case retrieval is one of the key issues in case-based reasoning. Novice reasoners are frequently most influenced by surface features in retrieving previous experiences (Ratterman and Gentner 1987; Ross 1987). In becoming more expert, a reasoner must learn to retrieve *usefully* similar cases.

Much of the work on improving case retrieval has focused on learning indices though explaining a reasoner's own successful or unsuccessful problem solving. Learning indices based on explanations requires at least three things:

1. Realizing the need to learn.
2. Determining what the correct result should be.
3. Assigning credit for successes or blame for failures.

The problem is that these can be hard. A novice reasoner is the *most* in need of improvement and the *least* prepared to learn. A novice may not be able to generate the correct result. A novice may also have trouble assigning credit or blame. How can a novice get around these problems?

Apprenticeship can provide assistance with this problem. A novice can observe an expert solving a problem. A good student, who actively follows along with the example, sets up expectations at each point in the expert's problem solving. When the expectations are incorrect then he has a failure. The student thus has immediate feedback. The student realizes the need to learn. The student has been given the correct result. Most important, the failure has been isolated to a single step. This tighter feedback loop enables learning when it otherwise might not be possible, and makes learning more efficient.

An example will illustrate some of the issues involved in improving case retrieval through observing an expert. A student is observing an instructor solve an automobile diagnosis problem. This is part of an ongoing mentor relationship. The instructor has checked whether the car stalls when cold. The instructor has adjusted the idle mixture screw and determined that the car still stalls. The instructor has tightened any loose spark plugs, and determined that the car still stalls. He has checked if the throttle dashpot is out of place (it wasn't), and if the fuel level in the carburetor was too high (it was). Figure 1 shows the effect of all this on the problem solving context and also shows some of the more general features of the problem that were elicited by the instructor. At this point the reader need only note that there are a

*This research has been supported by the Army Research Institute for the Behavioral and Social Sciences under Contract No. MDA-903-86-C-173, and Contract No. MDA-903-90-K-0112 and by DARPA contract F49620-88-C-0058 monitored by AFOSR. The author wishes to thank Janet Kolodner for her advice and guidance, and Ashwin Ram for helpful comments on earlier versions of the paper.

**Current Context**

| | |
|---|---|
| Complaint: | Car Stalls Out |
| Other Symptoms: | None Reported |
| Frequency of Problem: | intermittent |
| Temperature When Fail: | Cold |
| Type of Car: | 1980 Chrysler LeBaron |
| Car Owner: | Julie Crider |
| Mechanics Involved: | Tom Davis |

**Ruled In:**
- Spark Plug Connections Ok
- Idle Mixture Ok
- Carburetor Fuel Level High ***

**Ruled Out:**
- Low Idle Speed
- Lean Idle Mixture
- Incorrect Position Throttle Dashpot

**Tests Done:**
- Car Stalls When Cold?
- Spark Plugs Loose?
- Car Still Stalls?
- Too Small a Distance Between Throttle Dashpot Stem and Throttle Lever?
- Too High Level of Fuel in Carburetor Float Bowl?

**Test Results:**
- Car Stalls When Warm
- All Loose Spark Plug Connections Tightened
- Car Still Stalls
- Distance Between Throttle Dashpot Stem and Throttle Lever = 2cm
- Distance Between Carburetor Float and Choke Chamber Surface = 3cm

**Fixes Done:**
- Increase Position Idle Mixture Screw
- All Loose Spark Plug Connections Tightened

**Figure 1: Part of Current Context.**

number of contextual features, including complex values for things that had been ruled in and ruled out, tests done etc.

At this point, the student has set up an expectation using a part of a case from his episodic memory. The student had previously been a participant or an observer of that step. Figure 2 shows most of the context at the time of the predicted action's previous occurrence. For short, we call this the *'incorrect' context.* The student's experience suggests the hypothesis that the choke linkage is sticking. We call this the *'incorrect' prediction* to indicate that it is not an appropriate prediction for the current situation. The instructor makes a different hypothesis, that the carburetor needle valve leaks. The student could have made this prediction, which we call the *'correct' prediction.* In the past he had observed the instructor taking that action. The context from that previous time is shown in Figure 3. We will call that the *'correct' context.* Why is the instructor's action a better choice for the current problem solving? The incorrect context matches a good number of the features of the current problem. The key difference favors the instructor's action, however. The information in the current context that the carburetor fuel level is high is the key difference. This information favors the correct hypothesis that the carburetor needle valve leaks. How can the student improve his case retrieval so that he will make the correct prediction in the future?

**Incorrect Context**

| | |
|---|---|
| Complaint: | Car Stalls Out |
| Other Symptoms: | None Reported |
| Frequency of Problem: | intermittent |
| Temperature When Fail: | Cold |
| Type of Car: | 1979 Chrysler Cordova |
| Car Owner: | Bill Moss |
| Mechanics Involved: | Tom Davis, Kevin Cousins |

**Ruled In:**
- Spark Plug Connections Ok
- Idle Mixture Ok
- Idle Speed Ok

**Ruled Out:**
- Lean Idle Mixture
- Low Idle Speed
- Incorrect Position Throttle Dashpot
- Idle System Leak Air

**Tests Done:**
- Car Stalls When Cold?
- Spark Plugs Loose?
- Car Still Stalls?
- Too Small a Distance Between Throttle Dashpot Stem and Throttle Lever?
- Idle System Leak Air?

**Test Results:**
- Car Stalls When Cold
- All Loose Spark Plug Connections Tightened
- Car Still Stalls
- Distance Between Throttle Dashpot Stem and Throttle Lever = 2cm
- No Apparent Air Leaks

**Fixes Done:**
- Increase Position Idle Mixture Screw
- Increase Position Idle Speed Screw
- All Loose Spark Plug Connections Tightened

**Figure 2: Portions of incorrect context.**

**Correct Context**

| | |
|---|---|
| Complaint: | Car Stalls Out |
| Other Symptoms: | None Reported |
| Frequency of Problem: | daily |
| Temperature When Fail: | Cold |
| Type of Car: | 1981 Chrysler Cordova |
| Car Owner: | Paul Crider |
| Mechanics Involved: | Kevin Cousins |

**Ruled In:**
- Spark Plug Connections Ok
- Idle Mixture Ok
- Carburetor Fuel Level High ***
- Idle Speed Ok

**Ruled Out:**
- Lean Idle Mixture
- Idle System Leak Air

**Tests Done:**
- Car Stalls When Cold?
- Spark Plugs Loose?
- Car Still Stalls?
- Idle System Leak Air?

**Test Results:**
- Car Stalls When Cold
- All Loose Spark Plug Connections Tightened
- Car Still Stalls
- No Apparent Air Leaks

**Fixes Done:**
- Increase Position Idle Speed Screw
- All Loose Spark Plug Connections Tightened

**Figure 3: Portions of correct context.**

When the student's expectations are not met, then the student realizes the need to learn. Apprenticeship also takes care of the need for the student to know the correct action. The action taken by the instructor is assumed to be correct. The student still must determine what features of the current situation make the correct prediction appropriate, and/or make the incorrect prediction inappropriate. The student's exposure to that action in a previous example helps. He knows the context in which the action was previously taken. There are a number of similarities and differences among the contexts. A purely empirical approach will place some of the credit or blame on some spurious features that don't make a difference. On the other hand, if the student tries to analyze all of the feature values in the contexts and explain why one prediction is more relevant than the other, that will be expensive. Also, we certainly can't assume that a student will always be able to explain why one prediction is more appropriate. How can these opposing forces be reconciled?

Our solution makes use of a combination of analytical and empirical methods. Similarity-based methods are used to focus explanation. When the student is able to explain the appropriateness or inappropriateness of a prediction, the associated case part is marked with that indication. The indication can be positive, that the prediction made by that part of the case is appropriate in a situation, in which case we call it an index. The indication can also be negative, that the prediction is not appropriate in a situation. We call the latter a *censor*. In addition, the student empirically adjusts the importance of matching different features. We first present the analytical approach to learning indices and censors. Then we discuss the empirical adjustment.

## Analytical Approach - Learning Indices and Censors

The purpose of learning indices and censors is to improve prediction and diagnosis when they are carried out through CBR. Redmond (1990a) discussed our case representation. Briefly, cases are divided into *snippets*, each of which contain the information relevant to the pursuit of one primitive goal. Predictions are generated by retrieving a case snippet from memory. Each snippet contains the context in which it occurred. This enables similarity assessment, as well as the comparisons between correct and incorrect contexts mentioned above. An index is a particularly salient set of features of the context in which the snippet occurred. During retrieval, if a situation matches some of the indices, then that significantly increases the possibility that that snippet will be retrieved to provide guidance. The indices can be *parts* of contextual features. For example, *one* of the test results found in the problem solving leading up to the snippet might be marked as an index. A censor is a state (part of a feature) that suggests that a case snippet is not appropriate in that situation. If in a future situation that state exists, then the snippet can be rejected during retrieval.

We call our method *analytical feature comparison*. It has five steps which involve distinguishing responsible features through first comparing feature values, then trying to explain the relevance of differences. The initial feature comparisons are similarity-based, the analysis comes into play in latter steps. The opportunity to apply the process occurs when the student, observing the instructor, uses a case snippet to incorrectly predict the instructor's action. The process starts out by retrieving a snippet that would have predicted the instructor's action. For short, we will call this snippet the *'correct' snippet*. The action itself is an effective additional cue that enables the correct snippet to be retrieved now. We will call the snippet that the student used to make the incorrect prediction the *'incorrect' snippet*. After the correct snippet has been retrieved, the process is ready to begin.

---

Current Context Ruled In:
Spark Plug Connections Ok    Idle Mixture Ok
Carburetor Fuel Level High

**(a): Part of the current problem solving context.**

---

Correct Snippet (case-generate-hypoth-305) Context
Ruled In:
Spark Plug Connections Ok +++    Idle Mixture Ok +++
Carburetor Fuel Level High +++    Idle Speed Ok

**(b): Part of the correct snippet's problem solving context.**

---

Incorrect Snippet (case-generate-hypoth-312) Context
Ruled In:
Spark Plug Connections Ok +++    Idle Mixture Ok +++
Idle Speed Ok

**(c): Part of the incorrect snippet's problem solving context.**

---

**Figure 4: Portions of snippets involved in learning an index.**

**1. Eliminate from consideration features which are the same in both snippets' contexts.** Features compared include aspects of the diagnosis, such as the tests done and their results, things ruled out, etc., as well as domain dependent features, such as car type. Given the contexts in Figures 2 and 3, the complaint, the other symptoms, and the ambient temperature when the failure occurs are the same in both snippets. Therefore, they don't provide a way of distinguishing why the correct snippet is appropriate in the current context. These are therefore no longer considered candidates for explanation.

**2. Compare the remaining features with the current problem solving context.** Features in which the current context better matches the correct context than the incorrect context are selected as candidates. It is more likely that something in those features would indicate that the correct snippet is appropriate. For example, in Figure 1 the current context's 'ruled in' feature had the value shown in Figure 4(a). The correct context and the incorrect context had the 'ruled in' values shown in Figure 4(b) and 4(c). Matches to the current context are shown by '+++'s. The incorrect context's 'ruled in' was not as close a match as the correct snippet's 'ruled in'. Thus 'ruled in' will continue to be included in the candidates. At the same time, the incorrect snippet's 'fixes done' is a better match to the current context than the correct snippet's 'fixes done'. Therefore, 'fixes done' will be eliminated from consideration for explanation. In the example, this step eliminates the frequency of the problem, the items 'ruled out', and the mechanics involved, since they favor retrieval of the incorrect piece.

```
Correct Snippet (case-generate-hypothesis-305) Context
  Type of Car:          1981 Chrysler Cordova
  Car Owner:            Paul Crider ***
  Ruled In:
      Spark Plug Connections Ok
      Idle Mixture Ok
      Idle Speed Ok
      Carburetor Fuel Level High ***
  Tests Done:
      Car Stalls When Cold?
      Spark Plugs Loose?
      Car Still Stalls?
      Idle System Leak Air?
      Too High Level of Fuel in Carburetor Float Bowl? ***
  Test Results:
      Car Stalls When Cold
      All Loose Spark Plug Connections Tightened
      Car Still Stalls
      No Apparent Air Leaks
      Distance Between Carburetor Float and
      Choke Chamber Surface = 2cm ***
```

**Figure 5: Remaining portions of correct snippet after second step of analytical feature comparison.**

**3. Isolate the parts of the correct snippet's feature values that cause the remaining features to better match the current context.** This is to narrow the responsibility to *parts* of features which favor the retrieval of the correct snippet in the current context. These parts of features are the best candidates for distinguishing why the correct snippet is appropriate in the current context. In the current example, for items 'ruled in', the isolated part is that the carburetor fuel level is high. This is the clause that makes the correct snippet's 'ruled in' better match the current 'ruled in'. Figure 5 shows the features remaining after the second step of the process. The parts of features isolated by this third step are marked with '***'s. This is where the explanation process will be focussed. This completes the initial feature comparison, which limits the analytical reasoning done.

**4. Try to explain the significance of each of the remaining parts of features.** The learner can try to relate each to the current action taken by the instructor. The student in the example tries to explain the relationship between the 'Ruled In' that the carburetor fuel level is high and the hypothesis that the carburetor needle valve leaks. An explanation that the student could (depending of his level of knowledge) construct is that a leaking carburetor needle valve could enable fuel to keep flowing into the carburetor float bowl, thus causing the fuel level to become high.

For learning censors a greater variety of relationships are useful. The learner can try to relate each remaining feature part to the current action taken by the instructor, or to the action suggested by the incorrect snippet. The relationships themselves can be different. A part of the current context relating to a 'contradiction' of the incorrect action is a good indication of a need for a censor. For example, if in a second problem *Carburetor Fuel Level Normal* has been 'ruled in', that contradicts an incorrect snippet's hypothesis that the carburetor needle valve leaks.

**5. If such a relationship can be found, then the partial feature value is marked as an index or censor.** In the main example, *Ruled In: Carburetor Fuel Level High* can be saved as an index to the correct snippet. In the same or similar situation in the future the snippet will then be more likely to be retrieved as a source of predictions or diagnostic actions.

In the second example, the incorrect snippet can be annotated with the censor *Ruled In: Carburetor Fuel Level Normal.* In the same or similar situation in the future the snippet will not be retrieved as a source of predictions or diagnostic actions.

We should note some limitations of our approach to learning indices and censors. First, the matching of parts of features has to be exact. Some form of knowledge-based matching like that used by Bareiss (1989) and Koton (1988) is needed. Fortunately, substituting knowledge-based pattern matching for simple matching doesn't affect the method. It just substitutes better reasoning capability in steps 1 through 3. Second, there is no attempt to learn indices that involve a conjunction of feature parts. There are certainly situations in which such an index would be necessary. With better explanatory capability, the student could learn more sophisticated indices. Some of the limitations on explanation may need to be relaxed though, costing the process some efficiency. Both of these limitations will be addressed in future work.

## Empirical Feature Comparison

Sometimes a learner may not be able to explain why a prediction is wrong, or why another one might be better. The learner is not an expert. When he cannot, he needs to resort to less powerful, less knowledge-intensive methods. He can attempt an empirical approach to improving case retrieval. As with learning indices and censors, apprenticeship helps identify the need to learn. It also isolates the problem to a single step, and provides the correct action for the situation. As with learning indices and censors, the learner can attempt to retrieve part of another case that would have suggested a correct prediction. If such a case snippet can be retrieved, then the contexts can be compared. The differences can be used to empirically adjust the weights on features in the matching function. The adjustment method places greater importance on features that match and lead to correct predictions. It places less importance on features that match and lead to incorrect predictions. We should note that there are results that suggest that people can learn to distribute their attention among features, giving different weights to each (Nosofsky 1987). We originally discussed this approach in (Redmond 1989a). We have now integrated it with the analytical approach discussed in the previous section.

| Feature | 'Incorrect' | 'Correct' | Change in Importance |
|---|---|---|---|
| CAR-TYPE | Partial | No Match | less important |
| CAR-OWNER | Match | No Match | less important |
| COMPLAINT | Match | Match | no change |
| HOW-LONG | Match | Partial | less important |
| RULED-OUT | Partial | Match | more important |
| TESTS-DONE | Partial | Match | more important |
| FIXES-DONE | Partial | Match | more important |
| CURRENT-HYP | Match | Match | no change |
| WHEN | Partial | Slight | less important |

**Figure 6: Example empirical saliency adjustment.**

Figure 6 shows how the empirical change is done on an example incorrect prediction. Those features of the current context that match the context of the correct snippet more closely than the context of the incorrect snippet are made more important. Those features of the current context that match the context of the incorrect snippet more closely than the context of the correct piece are made less important.

The student can also make empirical adjustments when he is successful. When the student correctly predicts the instructor's action, the features of the current problem solving context that matched the features in the previous case are made slightly more important. In future problem solving these empirical adjustments lead to the features considered important by the instructor having more influence on the case pieces retrieved.

## Related Work

We have addressed the issue of making case retrieval better. We use a form of apprenticeship to isolate a failure, and to obtain the correct solution step. Retrieval of other previous case snippets and comparison of features is used to focus explanation. Through explanation, indices and censors are learned.

Other approaches have been suggested for learning indices and censors. Hammond (1986), and Simpson (1985)'s approaches explained instances of their own successful problem solving. This requires a significantly more expert reasoner because the correct steps must be generated. Hammond and Simpson also emphasized creating indices to avoid failures. However, in trying to explain failures their approaches had to consider all features. Our approach limits the consideration of features to a smaller set. In addition, our approach avoids having to match the situation to an abstraction, as in Hammond's approach, an expensive proposition. Therefore, our approach saves explanatory effort.

In Barletta and Mark's (1988) Explanation-Based Indexing approach, the reasoner attempts to explain the relationships of features of the case to the actions taken. Those features for which an explanation can be formed are made indices of the case. Our approach takes more information into account in order to limit explanatory effort. Thus our approach is more efficient.

Bareiss (1989)'s PROTOS created both censors that specified when a case was not appropriate, and 'difference links' which specified the condition in which one solution should be avoided and another used. But the instructor did all the distinguishing of which features should be considered in explaining the failure. PROTOS doesn't include the initial consideration of an extra case that our approach does. So our approach offers the benefit of increased efficiency of learning through initial similarity-based comparison to other cases.

At a high level, our approach shares similarity with Lebowitz (1986)'s suggestion to use similarity based measures to focus explanation. However, we have applied an instantiation of that general idea to learning indices and censors.

Empirical adjustment of feature importance was also suggested by Aha (1989) and Salzberg (1988) in the context of purely empirical instance-based learning. The contribution here is its use in conjunction with analytical approaches, as a fallback strategy when the reasoner doesn't possess the knowledge necessary to distinguish why an action is appropriate or inappropriate.

## Evaluation

Through the process of observing an expert, a reasoner can significantly improve its ability to predict the expert's actions, and thus its ability to diagnose. The improvement comes through acquisition of new cases, learning indices and censors, and through adjusting feature salience. Our system, CELIA, which is based on the model, has been evaluated through presentation of a sequence of 24 examples of expert problem solving. Twenty random orders of the examples were presented. The performance measure was the average accuracy of the system's predictions of the expert's actions. The improvement over the course of exposure to examples is seen in the data presented in Figure 7.

An ablation study showed that the by the end of the example sequences, the effect of the learning methods is starting to be seen. We lesioned the functions which learn indices and censors, and which adjust feature saliency. By the last 6 examples in the sequences, the average performance advantage for including these approaches was two percent. This difference is not that large, but for some problems the advantage was as high as 10 percent.

We expect that the effects will be larger with greater experience. To this experience level, not that many indices are learned. [1] Many of these are learned late in the sequences, leaving little time for them to be useful. The system doesn't have a strong domain model, so in many instances it cannot explain the importance of different feature values. With greater experience, this problem will gradually become less of a factor. In addition,

[1] An average of 1.2 are learned per observed example. An average example has 37 steps, which become 37 snippets. The range is from 18 steps to 79 steps.

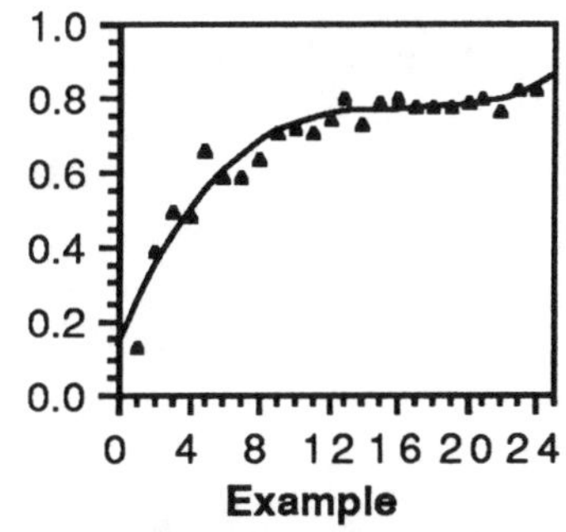


**Figure 7: CELIA: Average results of exposure to sequence of examples.** Average accuracy for reasoner predicting observed expert problem solving actions.

the student will have opportunities to use the indices and censors learned late in the test period.[2]

The approach has several advantages over previous approaches. The use of instruction through worked out examples, in conjunction with the student making predictions, shortens the feedback cycle, making learning more likely and more efficient. The retrieval and comparison of an additional case helps limit efforts at explanation. Third, the use of empirical adjustment of feature importance allows the student to become better at retrieving appropriate case pieces even when he is unable to explain failures.

## Conclusions

In order for a case-based reasoner to improve its performance, it must acquire more cases and improve its case retrieval so that the right cases are retrieved at the the right times. When the reasoner uses part of a previous case to suggest an incorrect action, that case snippet needs to be marked as to what makes it not relevant in the current situation. If a part of a case would have suggested the correct action, that case snippet should be marked with the features of the current situation that make it relevant. The student must try to form an explanation based on the incomplete knowledge that he has.

Our approach to learning indices and censors offers the benefit of increased efficiency of learning through initial similarity based comparison to other cases and through immediate direct feedback available through observation of expert problem solving. Our case representation facilitates the learning process. Each snippet contains the pursuit of one primitive goal and the context in which it occurred. This enables the necessary comparisons between correct and incorrect contexts. The general approach allows the student to obtain the correct action for the current situation from the instructor, and attempts to distinguish when the correct and predicted actions are appropriate or not.

Learning indices, learning censors, and adjusting feature salience result in improving case retrieval. All three of these types of learning fit neatly within our general approach of learning by observing expert problem solving. Much of the power of CBR comes from retrieving useful cases, so this learning improves the case-based reasoner. Indices help the reasoner retrieve case snippets when they would be particularly relevant. Censors help the reasoner avoid being misguided by a snippet which may seem to be relevant but isn't. Improved knowledge of feature salience will lead to less consideration of superficial and spurious features during similarity measurement prior to use of indices and censors.

[2]It should be noted that some orders of presentation of examples will facilitate index learning. This ablation study did not take that into account. If example order can be controlled then learning can have a greater effect.

## References

D. W. Aha. Incremental, instance-based learning of independent and graded concept descriptions. In *Proceedings of the Sixth Annual International Workshop on Machine Learning*, San Mateo, CA, 1989. Morgan Kaufmann.

R. Bareiss. Exemplar-based knowledge acquisition: a unified approach to concept representation, classification, and learning. Academic Press, New York, NY, 1989.

R. Barletta and W. Mark. Explanation-based indexing of cases. In *Proceedings of a Workshop on Case-Based Reasoning*, San Mateo, CA, 1988. Morgan Kaufmann.

K. J. Hammond. Learning to anticipate and avoid planning problems through the explanation of failures. In *Proceedings of the National Conference on Artificial Intelligence (AAAI-86)*, San Mateo, CA, 1986. Morgan Kaufmann.

P. Koton. *Using experience in learning and problem solving*. PhD thesis, Massachussetts Institute of Technology, Cambridge, MA, 1988.

M. Lebowitz. Concept learning in a rich input domain: Generalization-based memory. In R. Michalski, J. Carbonell, and T. Mitchell, editors, *Machine Learning: An Artificial Intelligence Approach, Volume II*. Morgan Kaufmann, Los Altos, CA, 1986.

R. M. Nosofsky. Attention and learning processes in the identification and categorization of integral stimuli. *Journal of Experimental Psychology: Learning, Memory, and Cognition*, 13:87–108, 1987.

M. J. Ratterman and D. Gentner. Analogy and similarity: Determinants of accessibility and inferential soundness. In *Proceedings of the Ninth Annual Conference of the Cognitive Science Society*, Hillsdale, NJ, 1987. Lawrence Erlbaum Associates.

M. Redmond. Combining explanation types for learning by understanding instructional examples. In *Proceedings of the Eleventh Annual Conference of the Cognitive Science Society*, Hillsdale, NJ, 1989. Lawrence Erlbaum Associates.

M. Redmond. Learning from others' experience: creating cases from examples. In *Proceedings of the Second Workshop on Case-Based Reasoning*, San Mateo, CA, 1989. Morgan Kaufmann.

M. Redmond. Distributed cases for case-based reasoning; facilitating use of multiple cases. In *Proceedings of the National Conference on Artificial Intelligence (AAAI-90)*, Boston, MA, 1990. Morgan Kaufmann.

B. H. Ross. This is like that: The use of earlier problems and the separation of similarity effects. *Journal of Experimental Psychology: Learning, Memory, and Cognition*, 13:371–416, 1987.

S. Salzberg. Exemplar-based learning: Theory and implementation. Technical Report TR-10-88, Harvard University, Center for Research in Computing Technology, 1988.

R. L. Jr. Simpson. *A Computer Model of Case-Based Reasoning in Problem Solving*. PhD thesis, Georgia Institute of Technology, Atlanta, GA, 1985.

# Explanation-Based Retrieval in a Case-Based Learning Model

**Gerhard Weber**

Department of Psychology
University of Trier
D-5500 Trier, Germany
weber@uni-trier.dbp.de

**Abstract**

Retrieving previous similar cases from a memory of cases is central to case-based reasoning systems. In most systems, this retrieval is done by a detailed indexing mechanism. Thagard and Holyoak argue that indexing is the wrong way to retrieve analogues. They propose a retrieval model (ARCS) based on a competing constraint-satisfaction approach. In this paper, an explanation-based retrieval method (EBR) for retrieving analogues from a case-base with cases stored with respect to an interpretation of these cases as analyzed by a cognitive diagnostic component is described. The system is designed to the domain of problem solving in LISP. In a simulation study, it can be shown that the EBR-method performs equally well or even better than the ARCS-method.

## Analogical Retrieval and Indexing

Both case-based reasoning (CBR) and finding analogies are concerned with retrieving previous cases and problem solutions. In CBR-systems, cases are usually retrieved by elaborated indexing mechanisms. Thagard & Holyoak (1989) argue in their contribution to the 1989 Case-Based Reasoning Workshop that indexing is the wrong way to retrieve analogues. They make their points on two different levels of argumentation. On a more general level, they claim

- that CBR when viewed as a cognitive model is based only on anecdotes,
- that CBR generally is concerned with analogies within a single domain only, and
- that their own model for analogical reasoning is a component of a more general cognitive architecture.

On a more detailed level they claim about indexing,

- that it is excessively serial instead of simultaneously probing into memory,
- that the retrieval process is not competitive,
- that indexing overemphasizes pragmatic features such as goals and prediction failures, whereas semantic features and structural similarities are underemphasized, and
- that indexing requires to much pre-processing.

In this paper, it will be shown how a CBR-system can be built in a way that most of these arguments can be rejected. Such a system retrieves analogues as good as the ARCS-program (Thagard et al. 1990) does, or even better.

As Thagard et al. (1990) point out, five aspects central to analogue retrieval in human memory are implemented in their ARCS-program. These are the constraints of (1) semantic similarity, (2) structural consistency, and (3) pragmatic centrality. Additionally, (4) a parallel algorithm has to be used for determining stored analogues fitting best these three constraints and (5) there must be a competition between potential candidates for source analogues.

In analogical retrieval (compared to mapping analogues) finding semantic similarities between elements and relations in the target analogue and in the source analogues in memory is essential (Holyoak & Koh 1987; Holyoak & Thagard 1990). But, structural and pragmatic constraints also contribute to the retrieval process. For most analogue retrieval models, it is a problem to consider all constraints simultaneously like it is done in ARCS.

In case-based reasoning systems, the retrieval of similar cases depends on indexes used to probe the case base. These indexes consist of relevant features of the input problem determined in a first analysis phase (Riesbeck & Schank 1989). For CBR-systems, it is critical how indexes can be found, how specific they are to identify the most similar cases, and how common they are, so that they apply to a wide range of possible cases.

To address these problems of indexing in CBR-systems we have built an 'Episodic Learner Model' (ELM), a case-based learning model in the domain of learning LISP. In this model, previous cases are stored such that information from stored cases can be used to analyze new cases. These interpreted input problems can be used directly to access similar previous cases. The ELM-model is used as a student model in an intelligent tutoring system for the programming language LISP (Weber 1988). One of the intelligent features to support tutoring is an analogical component. This component searches for similar cases to a given situation to give the tutorial component the opportunity to explain to the student his or her bugs and misconceptions related to previous errors and to solutions for these errors. One of the central tasks of this component is to retrieve analogous cases. As the ELM-LISP-Tutor is an on-line system, the retrieval process has to be fast enough to respond to the student within acceptable time. So, retrieving a candidate for an analogue with an acceptable effort in memory space and time is one further constraint for the retrieval process.

In the next section, the ELM-model is described to show how cases (LISP-programs students have completed to solve programming problems) are interpreted and stored. In the following section, the algorithm of the explanation-based retrieval method (EBR) is described. Finally, this method is compared to the ARCS-model by a simulation study.

## ELM: A Case-Based Learning Model

The ELM student model is a type of user model containing knowledge about the user (student) in terms of a collection of episodes. In the sense of 'case-based learning', such episodes can be viewed as cases. To construct the student model, the student's code is analyzed related to the domain knowledge on the one side and to a task description on the other side. This cognitive diagnosis results in a derivation tree of concepts and rules the student might have used to solve the problem. These concepts and rules are instantiations of units from the knowledge base. The episodic student model is made up by these instantiations and later generalization based thereupon. To understand this form of episodic student model, a short description of the knowledge representation and the diagnostic process will be given.

Students have to program function definitions in a structured LISP-editor (Köhne & Weber 1987). So, the function code is at least syntactically correct. A series of a student's attempts to solve a programming problem from the exercises is shown in Tab. 1. The "cognitive analysis" of the program code works like an explanation-based generalization (EBG) method (Mitchell, Keller, & Kedar-Cabelli 1986; DeJong & Mooney 1986). It starts with a task description related to higher concepts, plans, and schemata in the knowledge base. For every concept, a set of rules is stated describing different ways to solve the goal given by the plan of the concept. These rules are comparable to the notation of implementation methods for goals in PROUST (Johnson 1986). There are "good", "suboptimal", and "buggy" rules explaining correct, correct but suboptimal, and incorrect solutions of the current goal or subgoal. The set of buggy rules is comparable to an error library in other ITSs (e.g. in Anderson's LISP-tutor (Anderson & Reiser 1985)). Applying a rule results in comparing the current plan description to the corresponding part of the student's code. In the plan description, further concepts may be addressed. The cognitive diagnosis is called recursively until a function name, a parameter, or a constant is matched.

The cognitive diagnosis results in a derivation tree (Weber 1989) built from all concepts and rules identified to explain the student's solution (Tab. 2). This derivation tree is an explanation structure in the sense of EBG and is the basis to build up the episodic learner model. Concepts and rules addressed in the derivation tree are the basis to create episodic frames. These frames are integrated into the knowledge base as instances of their concepts and rules. Slots in these instances refer to the context where they were used (especially the current task), to the type of transformations of concepts, to the observed rules, and to the argument bindings. The set of these instances (and further generalizations) constitutes the episodic student model.

In the next step, episodic frames are generalized. These generalizations represent the class of types of plans and corresponding data which will be used in further cognitive analyses to interpret the student's code. Generalizations refer to structural and semantic similarities of observed data for related concepts and rules. So, this is a form of similarity-based generalization within the framework of an EBG-method. If an episodic frame is the first instance under a concept of the knowledge base, this single case is generalized from structural and semantic aspects in the data so that this generalization will explain further matching data. In Fig. 1, a hierarchy of episodic instances and generalizations under the concept "THIRD-ELEM" is shown. The first generalization frame THIRD-ELEM.GEN-1 was obtained by a single-case generalization after diagnosing Solution 1 from Table 1. The frame THIRD-ELEM.GEN-2 was generalized after the student's second attempt to solve the problem with a different plan. The third generalization frame THIRD-ELEM.GEN-3 resembles the communalities of the second and third attempt to solve the problem. In Tab. 3 the generalized structures for DATUM-slot and for the PLAN-RULE-SEQUENCE-slot are shown.

With increasing knowledge about a particular student, hierarchies of generalizations and instances are built under the concepts and rules of the knowledge base. They constitute the episodic student model. This generalization mechanism is comparable to the single-case generalization in the "explanation-based learning" approach (Mitchell et al. 1986). One single event (or case) is interpreted considering the knowledge base and the student model and the result of this interpretation is integrated into the student model.

**Problem:**
"Define a function that returns the third element of a list."

**Solution 1:**
```
(DEFUN THIRD-EL (EXPR)
   (REST (REST (FIRST EXPR))))
```
**Solution 2:**
```
(DEFUN THIRD-EL (EXPR)
   (SECOND (FIRST (REST EXPR))))
```
**Solution 3:**
```
(DEFUN THIRD-EL (EXPR)
   (SECOND (REST EXPR)))
```

**Table 1:** A series of subsequent attempts to solve a programming problem from the exercises used in our LISP-course.

```
(THIRD-ELEM
 Third-Elem-With-Second-Rule
  (SECOND-ELEM-CALL
   Unary-Func-Rule
    (SECOND-ELEM-OPERATOR
     Correct-Coding-Rule)
    (REDUCE-LIST
     Unary-Func-Rule
      (REDUCE-LIST-OPERATOR
       Correct-Coding-Rule)
      (PARAMETER
       Correct-Param-Rule))))
```

**Table 2:** Explanation structure for the body of the function definition of solution 3 in Table 1.

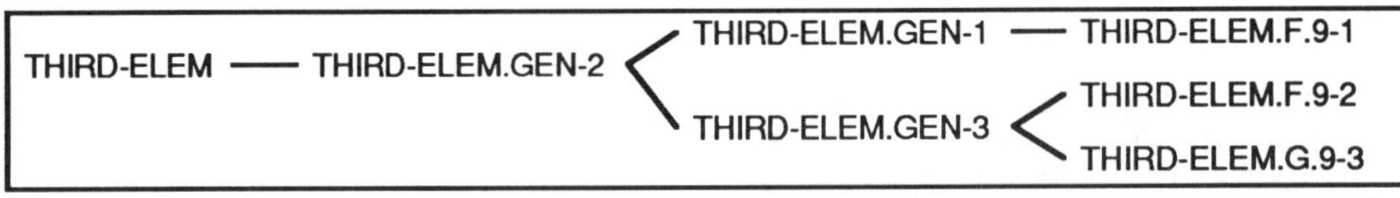


**Figure 1:** Hierarchy of episodic instances and generalizations under the concept THIRD-ELEM for the three solutions in Table 1.

```
?DATUM
  (SECOND (<SELECTOR-OPERATOR> ?ARG-273))
?PLAN-RULE-SEQUENCE
  (THIRD-ELEM Third-Elem-With-Second-Rule !SEQ-273)
```

**Table 3:** Part of the slots of the generalization frame THIRD-ELEM.GEN-3 from Fig. 1.

This EBL-method implies that not only pragmatic aspects are stored in the case memory (as in most other case-based reasoning systems). The hierarchy of stored solutions for different subproblems resembles the structure of the solution. Stored and generalized data structures emphasize semantic similarities and may also be considered to observe superficial aspects of similarity.

In the context of the ELM-model in the ELM-LISP-tutor, two typical situations emerge where analogous cases are retrieved. On the one side, analogues are needed by the tutorial component to offer the student remindings and analogies to examples in the materials when an error or a suboptimal solution is detected by the diagnostic component in the student's program code. Such remindings can be used to explain to the student how the error is similar to a previous error and how the problem can be solved, or how the incorrect part of the code resulted from superficial similarities to previous solutions which were incorrectly mapped to the current problem. On the other side, if the student asks for help while developing the code for the current problem the tutorial system should be able to offer the student analogies to similar previous problems or examples. Here, the analogical component has to look for analogies for the solution of the next subproblem to be solved in the context of subgoals solved up to that point.

In both cases, existing program code is the basis to retrieve analogous cases from the case-base. In the next section, it is explained how an explanation-based retrieval method (EBR) is suited to find analogues in a case-memory of a case-based learning system.

## The Explanation-Based Retrieval Method

On the basis of the explanation structure stored into distributed frames for the steps of the solution path, it is easy to retrieve analogical problem solutions for the current problem. There are two different situations where an analogue may be retrieved.

*Similar solution situation.* In this situation, the retrieval component is given a solution of a known problem and has to look in the case memory for the best analogues to the given solution. These analogues may help to interpret how the person solved the problem in the given way or why an observed error may have happened.

*Similar problem situation.* In this situation, the retrieval component is given a problem and possibly, a partial solution. By its knowledge about the problem, a solution can be generated considering the case memory and the partial solution. On the basis of the concepts and rules used to generate this solution, the case memory can be probed for cases most similar to this solution and analogues can be found for missing parts of the partial solution. As different subproblems have to be solved and gathered for the final solution there may be several different analogues which may match best the subproblems. This is a by-product of the distributed storage of problems in terms of solutions for the different subproblems. So, analogies to the sub-problems may be detected in different previous cases which, therefore, may be retrieved as analogues.

Both situations differ with respect to the question whether the problem is already solved or not. In the first case, the already existing solution is used for analogy, in the latter case, a solution generated automatically is used for an interpretation and for retrieving analogues afterwards.

The algorithm of the EBR-method works in five steps as follows:

*Step 1*: Diagnose the solution of a problem which is either submitted by the problem solver or generated automatically by the system. This will result in an explanation structure of all concepts and rules which may be used to solve the problem.

*Step 2*: Store all elements of the explanation structure into episodic frames of the case memory and generalize if possible. If the solution was generated automatically by the system, store episodic frames and generalizations only temporarily.

*Step 3*: Scan all episodic frames of the new (target) episode and look for similar episodic frames in previous episodes. All those episodic frames subsumed under the same direct generalization frame are given the highest similarity weight. For other similar episodic frames subsumed under higher levels of abstraction or indexed by related rules assign lower similarity weights, accordingly.

*Step 4*: For each episode considered in Step 3, sum the similarity weights of the similar frames assigned in Step 3. For each episodic frame of the source episode only one similar frame of a previous episode may be considered.

*Step 5*: Sort the considered target episodes by their summed weights of episodic frames similar to the source episode computed in Step 4.

In the first place, this algorithm pays attention to semantic similarities of concepts, plans, and rules identified by the analysis of the program code. But, as structural similarities play an important role in the generalization of episodic frames, structural consistencies are considered, too. Pragmatic aspects are not considered directly in the current version of the EBR-algorithm. But, it would be easy to give special pragmatic weights to observed buggy rules and to poorly solved concepts so that corresponding episodic frames from previous cases can dominate the retrieval of these episodes. So, tutorial goals could give pragmatic constraints to the retrieval process. An algorithm to retrieve analogues for special subgoals is still under development.

Using an explanation structure in the EBR-method to retrieve analogues cases is comparable to the construction of a hypothesis tree in the explanation-based indexing (EBI) approach (Barletta & Mark 1988). But, the EBR-method differs from the EBI-method in the distributed representation of episodic instances of concepts and rules.

## A Simulation Study Comparing the EBR-Method with the ARCS-Model

The ARCS-model (Thagard et al. 1990) is supposed to be a general model for analogue retrieval. As the algorithm of this retrieval method is fully described in Thagard et al. (1990) and easy to compute, it can directly be compared to other methods in a simulation study. One of the basic assumptions of the ARCS-model is to represent the target and the potential source candidates in terms of the predicate calculus. Therefore, it is critical for this method how the description of a problem is decomposed into predicates.

For retrieving analogues of LISP-programs from a case-library of stored LISP-programs, the decomposition into predicates can be computed by an explicit algorithm. As LISP-programs are composed of hierarchically nested function calls, they can be decomposed into predicates as shown in Holyoak & Thagard (1989). Function calls are represented as predicates with n+1 arguments where n arguments stand for the arguments of the function and the additional argument for the result of the function call. Only special forms (e.g. schemata for defining functions with "defun" (special decomposition for variable list and function body) or schemata for case decisions with "cond") are handled specifically to reflect their semantics. An example for a decomposition of a simple function definition into a set of predicates for the ARCS-method is shown in Tab. 4A.

Identifying semantic similarities is the second critical point for the ARCS-method. As mentioned in several studies (e.g. Holyoak & Koh 1987; Ratterman & Gentner 1987; Ross 1987) semantic similarities are most important for retrieving analogues. In their ARCS-model, Thagard et al. (1990) use WordNet (Miller, Fellbaum, Kegl, & Miller 1988) to identify different kinds of semantic relations (e.g. synonyms, superordinates, coordinates, subordinates, holonyms, antonyms, etc.) and assign to them different weights determining their influence on the retrieval process.

Since analogs of LISP-programs are retrieved, the case of within-domain analogies is given. Therefore, most semantic similarities stem from the identity of function names. Further degrees of semantic relations can be obtained from hierarchically clustering functions by their meaning.

**A) Decomposition of code into propositions:**

```
(DEFUN (THIRD-EL Varlist-931 form-933) P-937)
(VARLIST (EXPR Varlist-931) P-932)
(SECOND (arg-934 form-933) P-936)
(REST (EXPR arg-934) P-935)
```

**B) Decomposition of diagnosed code into propositions:**

```
(DEFINE-PROCEDURE
    (Proc-Def-Rule arg-938) P-948)
(BODY-FORMS
    (Rule-For-Body arg-939 arg-938) P-947)
(THIRD-ELEM
    (Third-Elem-With-Second-Rule arg-940 arg-939) P-946)
(SECOND-ELEM-CALL
    (Unary-Func-Rule arg-941 arg-940) P-945)
(REDUCE-LIST
    (Unary-Func-Rule arg-942 arg-941) P-944)
(PARAMETER
    (Correct-Param-Rule arg-942) P-943)
```

**Table 4:** An example for a decomposition into propositions for A) LISP-code and B) the result of the explanation-based diagnosis for Solution 3 from Tab. 1.

For example, all predicates or all arithmetic functions are grouped into clusters of lower similarity, type predicates, arithmetic functions for performing addition, types of case-decisions, and so on, within these clusters are building smaller clusters with medium similarity, and synonyms (e.g. car and first) are linked on a high level of similarity.

One can argue that there is much preprocessing in the EBR-method diagnosing the LISP-code by the ELM-model to get the explanation structure. Also, the explanation structure may contain more information so that the EBR-method is able to retrieve better analogues than the ARCS-method based only on the pure LISP-code. As the explanation structure consists of a hierarchy of nested concepts and rules, this nested list can be decomposed into propositions, too. Different weights of similarity for concepts and rules were obtained from the hierarchy of concepts and rules in the knowledge base. A decomposition of diagnosed code into propositions is shown in Tab. 4B.

Our implementation of the ARCS-method was run with the same values for parameters as described in Holyoak & Thagard (1989) and Thagard et al. (1990) with decomposing LISP-programs and assigning weights for semantic similarities as described above.

Data for the simulation study were gathered from interaction protocols of 13 students computing function definitions for exercises of an introductory LISP-course. All function definitions were programmed in a structured LISP-editor (Köhne & Weber 1987), so they were at least syntactically correct. The range of programming problems comprised 5 lessons beginning with the first exercises defining LISP-functions to programming recursive functions. In these five lessons, 35 different problems could be solved by the students. All student's attempts to evaluate a complete function definition were considered as candidates for targets of analogies.

For each student, the sequence of solutions for different tasks within each lesson was matched step by step against all previous solutions within the same lesson and against the examples from the materials for these lessons. Retrieving analogues was restricted to analogies within one lesson for two reasons. First, examples and solutions within a lesson are thematically more similar than those between lessons. In simulations considering data from all lessons, only very few analogies were found between lessons. Second, the computational effort to retrieve analogues by the ARCS-method from a basis of more than 10 possible candidates increased dramatically, consuming too much time and space on a serial computer. All in all 406 cases were observed where analogue cases could be retrieved.

The same procedure to retrieve analogues for observed problem solutions step by step within each lesson was performed by the EBR-method. Time to retrieve an analogue by this method ranged from 0.05 seconds to 2 seconds for the range of problems described above. This is more than 200 times faster than the time needed to build up the network of hypotheses only. If we consider a minimum of 7 cycles to retrieve the best analogue, the ARCS-method takes more than 2000 times longer on a serial computer.

The results comparing both methods by the described procedure are shown in Tab. 5. In about 3 out of 4 cases, both methods (EBR and ARCS) retrieved the same analogue as their best result. This was computed by an algorithm comparing rank orders of retrieved episodes. As expected, the correspondence between both methods is slightly higher if the ARCS-method is run on a propositional decomposition of the explanation structure (77.3%) compared to a decomposition of the LISP-code (72.7%). When there was a discrepancy between the results of both methods two experienced LISP-programmers rated which of both methods retrieved the best analogue or whether both methods failed.

In less than 5% of all cases, the ARCS-method retrieved the better optimal analogue. With 4.7%, the ARCS-method worked slightly better with a propositional decomposition of the LISP-code (compared to 2.7% with interpreted code). In 1.7% of all cases, this resulted from errors or combinations of errors in the student's LISP-code which could not sufficiently be interpreted by the diagnostic component of the ELM-model. So, a retrieval process must be preferred which is based particularly on semantic similarities of predicates without knowledge about intentions and meaning of the program code and interpretations of errors. In most other cases, the ELM-method failed because the current generalization mechanism in the ELM-model is slightly too sensitive to structural differences in the student's LISP-code.

In about 6% of all cases, different analogues were retrieved as best results, but both analogues could be accepted equally well. In these cases, there was a very similar (and possibly buggy or incomplete) previous solution and additionally, a similar good example. In most cases, the correct similar example was preferred by the EBR-method whereas the similar, but possibly incorrect, previous solution was preferred by the ARCS-method.

| | | diagnosis | code |
|---|---|---|---|
| same | | 77.3 | 72.7 |
| different | both methods equally well | 6.4 | 5.9 |
| | EBR better | 11.6 | 15.0 |
| | ARCS better | 2.7 | 4.7 |
| | both methods failed | 2.0 | 1.7 |

**Table 5:** Percentage of cases showing the same or different best retrieved analogues by the EBR- and the ARCS-method for a total of 406 cases. In the "diagnosis" column, the ARCS-method was run with a propositional decomposition of the explanation structure from the cognitive diagnosis in the ELM-model, in the "code" column, the ARCS-method was run with a propositional decomposition of the LISP-code.

In more than 10% of all cases, the best analogue was retrieved by the EBR-method. In many of these cases, this could be attributed to two problems inherent to the constraint-satisfaction algorithm of the ARCS-method. In at least 8 cases (2%), the so-called "gang-effect" (McClelland & Rumelhart 1981) was observed. In these cases, very similar or identical solutions for a similar problem existed in the case base because the student solved the same problem twice with the same solution or recoded an example from the materials. These identical solutions grouped to a gang and supported each other while another better analogue was suppressed. In at least 15 cases (3.7%), a problem occured which we call the "additional-proposition-effect". This means that if two solutions differ in that one has an additional proposition, this solution was preferred though the other one was more similar. This must result from more supporting relations in the net of hypotheses in the case of the additional proposition.

In 2% of all cases both methods failed. That is, the retrieved cases could not be accepted as good analogues. In most cases, this resulted from the absence of appropriate examples in the case-base or from very buggy and uncommon solutions.

## Conclusion

The ARCS-model (Thagard et al. 1990) is a very general and well suited program to model human retrieval. But, there are some questions about the efficiency of its algorithm. Thagard & Holyoak (1989) argue that indexing in case-based reasoning systems needs too much preprocessing and therefore, may not be appropriate. But, the ARCS-model needs very much time to build up the network of hypotheses. Even if the following process of settling the network is performed on a parallel computer and therefore, will be very fast, the network is temporarily built up only once for one retrieval process. For the next retrieval of another analogue a new network has to be constructed and for a little bit more complex problems than usually shown in examples, this is very much space and time consuming. For example, in the data of our students it is very typical that in the final lesson more than 40 previous solutions and examples have to be considered as possible candidates for source analogues. Probing all candidates in parallel with the ARCS-model resulted in more than 3500 nodes and more than 200,000 symmetric links to be built up. Even on a fast 12 MIPS computer this requires more than 30 MByte of memory and more than 10 minutes to build up the network. If we extrapolate to situations in typical human retrieval situations where many more previous situations may be candidates for analogue retrieval because of overlap in their semantics, it cannot be imagined how such a problem could be handled by the ARCS-algorithm. An incremental model (Keane 1990) may be better suited to retrieve analogues from a large set of cases.

In this study, it could be demonstrated that the ARCS-methods has some problems if analogues have to be retrieved from a case base where possibly very similar previous cases exist. As shown for the "gang-effect" and the "additional-proposition-effect", the ARCS-method may fail if some special constellations of very similar cases exist. Maybe, this could not be observed in other studies with the ARCS-method because all cases in these examples differed in many aspects. But, in realistic, non-experimental situations, for example while learning to program, it is typical that many very similar cases are observed. In such a situation, the EBR-method seems to yield generally better results.

The EBR-method can be seen as an approximation to a competitive, parallel method of analogue retrieval as realized in the ARCS-model. As powerful parallel computers cannot be used in most cases, the EBR-method may be preferred in on-line situations (e.g. in tutorial systems). Additionally, information can be gathered about the mechanism how concepts and episodic structures are stored and retrieved from human memory.

## Acknowledgments

This research was supported by the Deutsche Forschungsgemeinschaft under Grant We 498/12.

## References

Anderson, J. R., and Reiser, B. J. (1985). The LISP tutor. *Byte* 10(4):159-175.

Barletta, R., and Mark, W. (1988). Explanation-Based Indexing of Cases. In J. L. Kolodner (Eds.), Proceedings of the DARPA-workshop on Case-Based Reasoning, 50-60. Los Altos, CA: Morgan Kaufmann Publishers.

DeJong, G., and Mooney, R. (1986). Explanation-based learning: an alternative view. *Machine Learning* 1:145-176.

Holyoak, K. J., and Koh, K. (1987). Surface and structural similarity in analogical transfer. *Memory & Cognition* 15:332-340.

Holyoak, K. J., and Thagard, P. (1989). Analogical mapping by constraint satisfaction. *Cognitive Science* 13:295-356.

Holyoak, K. J., and Thagard, P. (1990). A constraint-satisfaction approach to analogue retrieval and mapping. In K. J. Gilhooly, M. T. G. Keane, R. H. Logie, & G. Erdos (Eds.), *Lines of thinking* (Vol. 1, pp. 205-220). Chichester, England: Wiley.

Johnson, L. W. (1986). *Intention-based diagnosis of novice programming errors.* London: Pitman.

Keane, M. T. G. (1990). Incremental analogizing: theory and model. In K. J. Gilhooly, M. T. G. Keane, R. H. Logie, & G. Erdos (Eds.), *Lines of thinking* (Vol. 1, pp. 221-236). Chichester, England: Wiley.

Köhne, A., and Weber, G. (1987). STRUEDI: a LISP-structure editor for novice programmers. In H. J. Bullinger, & B. Schackel (Eds.), *Human-Computer Interaction INTERACT '87* (pp. 125-129). Amsterdam: North-Holland.

McClelland, J. L., and Rumelhart, D. E. (1981). An interactive activation model of context effects in letter perception: Part 1. An account of basic findings. *Psychological Review* 88:375-407.

Miller, G. A., Fellbaum, C., Kegl, J., and Miller, K. (1988). WORDNET: An electronic lexical reference system based on theories of lexical memory. *Revue Québécoise Linguistique* 17:181-213.

Mitchell, T. M., Keller, R. M., and Kedar-Cabelli, S. T. (1986). Explanation-based generalization: a unifying view. *Machine Learning* 1:47-80.

Ratterman, M., and Gentner, D. (1987). Analogy and similarity: determinants of accessibility and inferential soundness. In Proceedings of the Ninth Annual Conference of the Cognitive Science Society, 23-35. Hillsdale, NJ: Erlbaum.

Riesbeck, C. K., and Schank, R. C. (1989). *Inside case-based reasoning.* Hillsdale, NJ: Lawrence Erlbaum Associates.

Ross, B. H. (1987). This is like that: the use of earlier problems and the separation of similarity effects. *Journal of Experimental Psychology: Learning, Memory, and Instruction* 13:629-639.

Thagard, P., and Holyoak, K. J. (1989). Why indexing is the wrong way to think about analog retrieval. In K. J. Hammond (Ed.), *Proceedings of the Second Workshop on Case-Based Reasoning* (pp. 36-40).

Thagard, P., Holyoak, K. J., Nelson, S., and Gochfeld, D. (1990). *Analog retrieval by constraint satisfaction* (CSL-Report No. 41). Princeton University.

Weber, G. (1988). Cognitive diagnosis and episodic modelling in an intelligent LISP-tutor. In Proceedings of Intelligent Tutoring Systems ITS-88, 207-214. Montreal, June 1-3.

Weber, G. (1989). Automatische kognitive Diagnose in einem Programmier-Tutor. In D. Metzing (Ed.), *Künstliche Intelligenz GWAI-89* (pp. 331-336). Berlin: Springer.

# Retrieval Competition in Memory for Analogies

**Charles M. Wharton[1], Keith J. Holyoak[1], Paul E. Downing[1], Trent E. Lange[2], and Thomas D. Wickens[1]**

[1]Department of Psychology and [2]Department of Computer Science
University of California, Los Angeles, CA 90024

**Abstract***

An important question for cognitive models of human memory is the question of how analogical similarity affects memory retrieval. While the importance of surface lexical and semantic similarities between reminding cues and memory targets has been well-documented, clear empirical evidence that human memory retrieval is influenced by analogy has proven difficult to demonstrate. We report two experiments in which subjects used a series of single sentences as reminding cues for previously-seen mini-texts. Some cue sentences contained nouns and verbs that were hyponyms (i.e., words subordinate to the same category) of those in corresponding target sentences presented in one or two earlier passages. The role of analogical similarity in reminding was examined by varying the correspondence of noun case-role assignments of cue/target homonyms. Results indicate that retrieval competition and analogical similarity influence reminding. Recall of semantically-related passages was significantly greater for structurally consistent (i.e., analogical) cues. Retrieval access was impaired when two semantically related passages were present in memory. Access to the passage with analogical resemblance to the cue was decreased by retrieval competition to an extent consistent with a ratio rule.

## Retrieval Competition in Memory for Analogies

One of the most intriguing qualities of human memory is its capacity to allow novel experiences to activate relevant prior knowledge, even though the objects and events in the new situation have never been directly associated with those involved in the remembered ones. A person who sees the movie *West Side Story* for the first time is likely to be reminded of the play *Romeo and Juliet*, notwithstanding the displacement of the characters over centuries and continents. Although such remindings are surely related to general principles of memory retrieval, they differ significantly from standard retrieval tasks studied by psychologists. Unlike standard retrieval tasks, such as free or cued recall, there is no clear "right answer" as to what one should be reminded of in response to a given cue. Further, reminding may be unintentional (although as Schank, 1982, observed, one may also seek to be reminded).

The above example illustrates *analogical* reminding, in which the relations embodied in the cue and in the stored situation correspond in systematic ways, despite notable differences between their constituent elements. In other words, two episodes are structurally consistent (analogous) when the overall conceptual relationships between their actors, actions, plans, goals, and themes are similar. In spite of the centrality of structural consistency in proposed definitions of the "soundness" or "goodness" of analogies (Falkenhainer, Forbus, & Gentner, 1989; Holyoak & Thagard, 1989), it has proved surprisingly difficult to provide empirical evidence that human memory retrieval is influenced by analogy. The importance of surface lexical and semantic similarities has been well-documented, but clear evidence demonstrating the effect of structural consistency on reminding (such as the overall plot similarity between *Romeo and Juliet* and *West Side Story*) has been difficult to achieve. The answer to the question of whether structural consistency influences reminding would have important implications for psychological and artificial-intelligence models of analogical, episodic, and case-based retrieval. Do the individual elements of a story simply serve as independent retrieval cues, or do the relations among those elements also have an impact on what is recalled? Is the effect of structural consistency influenced by whether or not the target situation is semantically related to more than one episode in long-term memory?

### Previous Studies

The evidence concerning the role of structural consistency in analogical reminding is mixed. Indeed, the most robust finding in the analogy literature is that people often fail to retrieve relevant, but superficially-dissimilar, source analogs (Gick & Holyoak, 1980; Gentner & Landers, 1985; Ratterman & Gentner, 1987; Seifert, McKoon, Abelson, & Ratcliff, 1986). Such negative findings suggest that the process by which analogs are retrieved may not be sensitive to their configural properties, even though such properties are crucial in ensuring that the retrieved information is actually useful.

However, there is evidence that if the source and target share similar features, then degree of isomorphism has an impact on retrieval (e.g., Holyoak & Koh, 1987; Ross, 1989). Subjects in Ratterman and Gentner (1987) were given a set of stories and were asked to write down any stories that they were reminded of from a previous session,

### * Acknowledgements

Preparation of this article was supported by Contract MDA 903-89-K-0179 from the Army Research Institute, as well as by a grant from the Keck Foundation to the UCLA Cognitive Science Research Program. We thank Michael Anderson, Robert Bjork, Eric Melz, Paul Thagard, and Bobbie Spellman for their valuable insights.

one week earlier. Two types of cue/target similarity were manipulated: direct similarity of objects and actions (i.e., similar vs. dissimilar concrete nouns and verbs), and configural similarity of causal structure (i.e., similar vs. dissimilar causal connections among the objects and actions). Direct similarity of objects and actions was the dominant determinant of reminding; however, consistent, though nonsignificant, trends suggested that cue/target analogical similarity may also influence remindings (see also Gentner & Landers, 1985; Johnson & Seifert, 1990; Read & Cesa, 1991).

## Analogical Reminding Within a Constraint-Based Retrieval System

The present experiments were designed to test a recent artificial-intelligence model of analogical reminding, ARCS (Thagard, Holyoak, Nelson, & Gochfeld, 1990). Thagard et al. assert that semantic similarity and isomorphism (i.e., structural consistency and one-to-one mapping) influence reminding. Semantic similarity concerns taxonomic relations between individual concepts, such as superordination (e.g., dog - animal) and hyponymy (e.g., dog - cat). The similarity constraint favors mappings between concepts with close taxonomic relationships. The isomorphism constraint depends crucially on case-role relations (i.e., mapped elements should fill corresponding case roles across propositions in which they appear). Analogs are represented in ARCS as symbolic structures within a semantic network; during the retrieval process, a "mapping network" is formed to enforce the constraints, and a connectionist settling process then guides retrieval of the stored structure(s) that best satisfy the constraints.

Retrieval in ARCS is fundamentally competitive in that evidence favoring access to one stored structure serves to reduce (via inhibitory links) the likelihood of retrieving other stored structures. A central prediction, then, is that reminding will be competitive, so that if a cue has strong semantic links to multiple stored structures, reminding of each will be impaired relative to the case in which it is the sole semantically-related candidate. In addition, ARCS predicts that the effect of structural consistency will interact with competition. A disanalogous passage (relative to a given cue) might have a high probability of retrieval if it were the only semantically-related structure in memory. However, it would have a greatly decreased retrieval probability if a semantically-related analogous structure was also in memory.

The basic qualitative predictions of ARCS are mimicked by a simple mathematical model, the Bradley-Terry-Luce choice model (Bradley & Terry, 1953; Luce, 1959; Wickens, 1989a). The choice model states that the probability of making a given choice or response is a function of the "strength" (e.g., preference) of that choice divided by the strength of all choices or responses. As applied to memory retrieval, the choice model, like ARCS, predicts that the probability of retrieving a given item is determined by how strongly that item is associated to a given cue relative to the strength of all other items associated to that cue (see Raajimakers & Shiffrin, 1981). According to the choice model, the probability that a subject makes response type *i* to cue type *j* is:

$$p_{ij} = S_i \Big/ \sum_{k=1}^{n} S_k$$

$S_i$ is the strength of response type i, and the summation is over the strengths of all the response types available to the subject from cue j. The choice model captures two aspects of the competitive responding inherent in ARCS. First, choices are made in comparison to available alternatives so that relative weighting should apply. Second, the dependence of these weights on the chosen alternatives implies that a consistent set of estimates should be obtained over all conditions. Accordingly, we used the choice model to derive quantitative predictions for reminding performance in our experiments.

## Experiments 1 and 2

We performed two experiments to investigate the influence of structural consistency and competition on analogical reminding. The general design was to have subjects encode a series of very simple (2-3 sentence) passages, in the context of various incidental tasks designed to ensure semantic processing. After a brief delay, subjects were given a series of single-sentence cues in the forms *The subject verb the object* (SVO) or *The object was verb by the subject*, and asked to recall as much as they could from any and all passages of which they were reminded by each cue sentence. Some of the cue sentences contained nouns and verbs that were hyponyms of those in corresponding target sentences in one or two of the earlier passages; other cue sentences were unrelated to any of the prior passages.

Table 1 presents an example of a set of two passages and a related cue sentence. The target sentence in each passage is identified by italics. We will refer to the non-target portion of each passage as the "surround". One passage contains an analogous target, *The pastor calmed the businessman,* in which the nouns fill case roles parallel to those of their respective hyponyms in the target-cue sentence, *The rabbi reassured the chairman.* The other passage contains a disanalogous target, *The executive soothed the priest*, in which the case roles of the noun hyponyms are reversed between the target sentence and cue. Of the cues which shared noun and verb homonyms with target sentences, half were matched to a single previously-studied analogous or disanalogous target sentence (singleton condition) and half were matched to two previously-studied analogous and disanalogous target sentences (competitor condition).

We predicted (a) passages in the competitor condition would be recalled less often than in the singleton condition, (b) passages cued with analogous sentences would be retrieved more frequently than passages cued with disanalogous ones, and (c) there would be an interaction of analogy with retrieval competition such that the effect of cueing with analogous sentences would be greater in the

**Table 1.** Example of Paired Passages with Consistent Versus Inconsistent Targets in Relation to Cue Sentence

**Consistent Target**

Having just been fired from a high level job, he decided to go to his church for counseling. *The pastor calmed the businessman.*

**Inconsistent Target**

The church was having trouble approaching local corporations for contributions to the shelter. *The executive soothed the priest.*

**Target-Cue Sentence**

*The rabbi reassured the chairman.*

competitor than in the singleton condition.

Several points about the general nature of the materials deserve emphasis. First, structural consistency was simply manipulated by a single cross-mapping between a pair of SVO sentences. Second, semantic overlap was equated as structural consistency was varied. Third, the cue and target sentences had no direct lexical overlap, but were nonetheless semantically related by virtue of shared superordinates linking corresponding hyponyms. Fourth, because the cue sentences were directly related only to fragments of the passages (i.e., to the target sentences but not the surrounds), any reminding would have to be based on only partial overlap between cues and passages (cf. Johnson & Seifert, 1990).

## Method

### Materials

There were 24 sets of materials similar to the example in Table 1, each consisting of two target passages and one cue sentence. All sets of materials appeared in all conditions. Each of the two passages within a set contained a target sentence that was related to the set's cue sentence. The matched target and cue sentences all shared two sets of associated nouns (e.g., *pastor, priest, rabbi*; *businessman, executive, chairman*) and a single set of associated verbs (e.g., *calm, soothe, reassure*). The nouns and verbs within a set were chosen so that the nouns would jointly make sense in either the object or subject position. Within each passage pair, verbs were randomly assigned to pairs of target nouns, after which each target sentence was randomly assigned to one of the two passages. In order to avoid confounding cue/target consistency with surface order of the noun hyponyms, an equal number of active and passive cue sentences and target sentences were constructed in each condition. Random assignment was used to decide whether cue and target sentences would be active or passive and which target passage would be analogously cued.

*Materials ratings:* In order to determine whether people are sensitive to our structural consistency manipulation when retrieval is *not* required, 96 undergraduates attending the University of California, Los Angeles (UCLA) were asked to assess the perceived similarity of the cue sentences and the passages. Subjects were told to rate, "How similar are the scenes being described in the story and the sentence?" for consistent, inconsistent, and unrelated (i.e., cues for a different passage) cues and targets. Responses were recorded on a 6-point Likert scale (1 = completely dissimilar, 6 = completely identical). There was a significant difference between subject ratings for analogous ($M = 3.98$) and disanalogous ($M = 2.75$) cue/passage pairs, $min\ F'(1, 100) = 28.04$, as well as disanalogous and unrelated cue/passage pairs ($M = 1.59$), $min\ F'(1, 75) = 36.10$, both $p < .0001$. (In order to be able to generalize our findings beyond the specific materials we created, the conservative $min\ F'$ formulation of analysis of variance was calculated for all tests of mean differences; Clark, 1973.)

### Subjects and Design

All subjects were UCLA undergraduates, with 36 subjects in Experiment 1 and 72 in Experiment 2. Two within-subject factors were manipulated: (1) whether cues were analogous or disanalogous with respect to a corresponding target sentence in a passage, and (2) whether one (singleton condition) or two (competitor condition) passages related to a given cue had been studied. For example, if both the passages in Table 1 were presented to a subject, the two passages would represent the competitor condition; whereas if only one of the paired passages had been presented, it would represent the singleton condition. Cue sentences thus might be semantically related to (a) both an analogous passage and a disanalogous passage, (b) a single analogous passage, or (c) a single disanalogous passage. This manipulation enabled us to test the prediction that competition would decrease reminding, especially for disanalogous passages. In Experiment 1, subjects first read a set of target materials that contained 4 unrelated and 8 related target passages. After completion of a distractor task, subjects were given booklets containing 4 unrelated and 6 related cue sentences (2 competitor, 2 analogous, and 2 disanalogous).

In Experiment 2, subjects initially read 4 unrelated, and 4 related target passages along with 3 other target passages for an additional factor that will not be discussed in the present paper. After completion of a distractor task, subjects were given 4 unrelated and 3 related cue sentences (1 competitor, 1 analogous, and 1 disanalogous) and 2 cues for the additional factor.

### Procedure

Subjects were told that they would be asked to read and rate a series of passages for plausibility, meaningfulness, and imagibility. Subjects rated one attribute on each of three passes through a booklet. For the first pass, subjects were given 30 s to read and rate each passage; for the second and third passes they were given 20 s for each. After completion of the third pass through the booklets, subjects were given a 4 min distractor task.

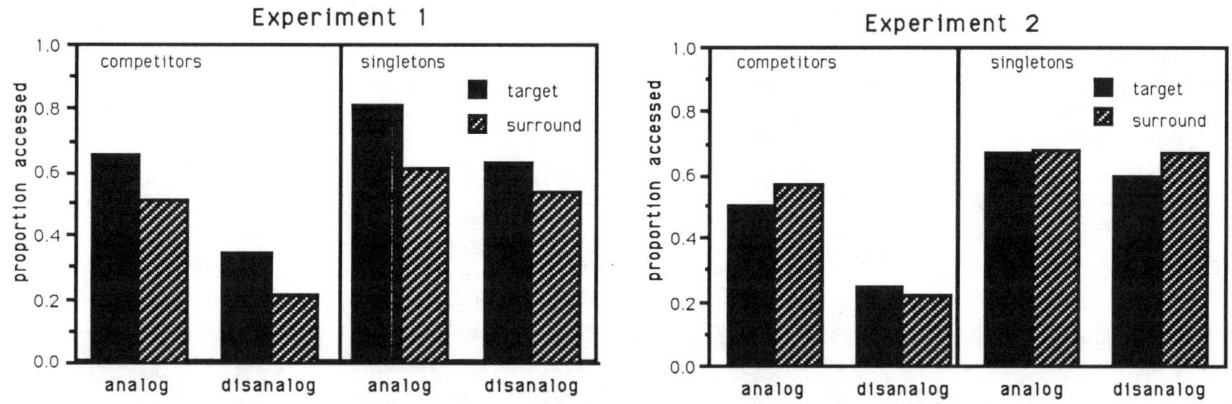


**Figure 1.** Proportion of Passages Accessed as a Function of Competition and Analogy.

**Table 2.** Min F' Analysis of Sentence-Access Main Effects and Interactions

| | Competition | | Structural Consistency | | Interaction | |
|---|---|---|---|---|---|---|
| | *min F'* | *df* | *min F'* | *df* | *min F'* | *df* |
| **Experiment 1** | | | | | | |
| target sentence | 10.95** | *34* | 9.27** | *42* | < 1 | |
| *surround sentences* | 10.22** | *42* | 6.56* | *38* | 2.54 | *39* |
| **Experiment 2** | | | | | | |
| *target sentence* | 20.15*** | *68* | 3.56[a] | *54* | 1.19 | *39* |
| *surround sentences* | 19.99*** | *53* | 4.33* | *48* | 4.87* | *47* |

[a] $p < .05$ for both subject and item ANOVAs but not for *min F'*, * $p < .05$, ** $p < .01$, *** $p < .001$

After completion of the distractor task, subjects were told that they would now read a group of sentences, some of which were similar to one or more passages that they had read in the first part of the experiment. They were instructed to rate each sentence for plausibility, meaningfulness, and imagibility and write down the passage or passages they were reminded of while they made these ratings. No time limit was imposed, but subjects were told that after they turned the page of the booklet, they should not turn back to that page even if they were later reminded of something. The entire experimental session lasted approximately 30 min.

## Results

The written reminding protocols were scored with respect to which of the studied passages was accessed in response to a given sentence cue. Access to the target sentence and the surround of each passage was scored separately. For each separate attempt at recalling a passage (i.e., each attempt to report what the subject considered a single passage), credit for access was given to whichever studied passage had content words recalled; if content words from two passages were included, credit was given to the passage that contributed the greater number of content words. If subjects wrote down content words from two passages in *separate* retrieval attempts in response to a single cue, access credit was given for each. Because synonym substitutions could be confused with interchanges of hyponyms across paired passages, a criterion of literal recall was used in scoring content words. An important feature of this access measure is that it is not itself sensitive to the structure of the accessed passage, thus providing a strong test of the hypothesis that structural consistency actually facilitates access to passages, regardless of whether the relational structure of the accessed passage can be reported correctly.

Summaries of the main effects and interactions in the retrieval probabilities are as follows:

*Competition:* As can be seen in Figure 1 and Table 2, for a given cue sentence the probability of retrieving a target passage greatly decreased when there was more than one lexically-related target passage in memory.

*Structural Consistency:* Analogously-cued target passages were recalled more often than disanalogously-cued target passages. In Experiment 2, there was a reliable competition by analogy interaction for surround sentences. In the competition condition, analogously-cued surround sentences were recalled significantly more often than disanalogously-cued passages, *min F'* $(1, 43) = 6.60, p < .05$.

### Choice Model

Each panel of Table 3 presents a 4 x 4 matrix of access proportions for each response type (e.g, consistent, inconsistent, unrelated, and no recall) that can occur in each experimental condition (e.g., competition, consistent, inconsistent, and unrelated). Only 12 of the combinations are possible. For example, an inconsistent passage cannot be recalled in response to a consistent cue in the singleton condition, because no such inconsistent passage was presented to the subject. To fit the choice model, it is assumed

**Table 3.** Observed and Predicted Proportions of Recalled Sentences by Experimental Condition

| Response type | Competition | | Consistent | | Inconsistent | | Unrelated | | Strength |
|---|---|---|---|---|---|---|---|---|---|
| **Experiment 1 target sentences** | | | | | | | | | |
| Consistent | .55 | _.56_ | .78 | _.78_ | – | – | – | – | 1.00 |
| Inconsistent | .30 | _.28_ | – | – | .60 | _.63_ | – | – | .49 |
| Unrelated | .01 | _.03_ | .03 | _.04_ | .07 | _.06_ | .17 | _.16_ | .05 |
| No recall | .13 | _.14_ | .19 | _.19_ | .33 | _.31_ | .83 | _.84_ | .24 |
| **Experiment 1 surround sentences** | | | | | | | | | |
| Consistent | .48 | _.46_ | .59 | _.61_ | – | — | – | – | 1.00 |
| Inconsistent | .18 | _.26_* | – | – | .55 | _.47_ | – | – | .56 |
| Unrelated | .01 | _.04_* | .04 | _.05_ | .07 | _.07_ | .16 | _.14_ | .09 |
| No recall | .32 | _.25_ | .37 | _.33_ | .38 | _.46_ | .84 | _.86_ | .54 |
| **Experiment 2 target sentences** | | | | | | | | | |
| Consistent | .47 | _.48_ | .67 | _.66_ | – | – | – | – | 1.00 |
| Inconsistent | .22 | _.28_ | – | – | .60 | _.54_ | – | – | .59 |
| Unrelated | .00 | _.05_* | .01 | _.06_* | .07 | _.08_ | .21 | _.18_ | .09 |
| No recall | .30 | _.20_* | .32 | _.28_ | .33 | _.38_ | .79 | _.82_ | .41 |
| **Experiment 2 surround sentences** | | | | | | | | | |
| Consistent | .54 | _.50_ | .66 | _.71_ | – | – | – | – | 1.00 |
| Inconsistent | .21 | _.29_* | – | – | .66 | _.58_ | – | – | .58 |
| Unrelated | .00 | _.04_* | .01 | _.06_* | .07 | _.08_ | .23 | _.20_ | .08 |
| No recall | .25 | _.17_* | .32 | _.23_* | .26 | _.33_* | .77 | _.80_ | .33 |

*Note.* Italicized values are expected proportions, non-italicized values are observed proportions.
* Difference between observed and expected cell frequency is ≥ 1 standardized cell deviation

that each of the 4 possible response types has a certain "strength" and that all 12 response outcomes can be predicted by these strengths as a function of the choice rule.

The italicized values in Table 3 are the predicted access proportions derived by fitting the choice model, and the rightmost column in the table reports the strength parameters for the four response types. (The statistics reported here were obtained from TWOWAY: Wickens, 1989b; see pp. 109-116, Wickens, 1989a). As can be seen , the choice model accurately describes our data; 92-99% of variance is accounted for relative to a model in which all response types have equal strength. Corroborating ARCS, the estimated strength parameters for inconsistently-cued passages are almost half of those for consistently-cued passages.

## General Discussion

We have shown that reminding can be described as a competitive selection among alternatives as represented by the choice model. Further, the present findings clarify a number of basic constraints on the mechanisms of analogical reminding and support several predictions from ARCS (Thagard et al., 1990). We found that varying the structural consistency of single sentence cues influenced access to semantically-related passages stored in memory. This effect on access was not only observed for the target sentence that was semantically linked to the cue, but it also extended to the indirectly-related surround portion of the passage. These results were obtained under conditions in which similarity of individual concepts was equated (because the same words were used in both the consistent and inconsistent cues), and surface correspondences at the level of word order were controlled (by random assignment of active versus passive voice to cues and target sentences). Our measure of access to passages did not depend on successful completion of post-access processes such as problem solving or explanation, thus localizing the impact of consistency at the initial access stage itself. Finally, the observed effect of configural overlap manifested itself in an access measure based on recovery of content words from the passage, rather than recovery of relational structure (although subjects did accurately recall the case-role structure). Our results therefore support the conclusion that structural relations between a cue and stored memory representations constitute one of the basic constraints on initial access.

The present study also provided clear evidence that memory access is fundamentally competitive, as the ARCS model postulates. We found that the presence in memory of two representations containing a target sentence semantically-related to the cue impaired access to either of the competitors. Access suffered even though instructions to subjects stressed that they should report all passages of which they were reminded.

As predicted by the ARCS model, the effect of structural consistency on memory access is greater when competing representations are activated by the cue. The fact that previous studies of analogical reminding only examined the equivalent of our singleton conditions, for which the influence of structural consistency tended to be weaker than in our competitor conditions (see Table 3), may have contributed to some of the failures to find robust effects of configural overlap on reminding (e.g., Seifert et al., 1986; Ratterman & Gentner, 1987). As analogical reminding in everyday life presumably takes place in the context of competition among multiple partially-activated memory representations, it would seem that laboratory studies of reminding have probably tended to understate the influence of analogical similarity on memory retrieval. Also, the insignificant difference ($M = .02$) between recall of analogous and disanalogous singleton surround sentences has important implications for artificial intelligence models of reminding— simple recall schemes that rank-order indexed targets by their degree of similarity and select the "best" one will likely not make accurate models of human reminding.

Future work needs to address the role of comprehension and inference processes in reminding. ARCS, which uses only taxonomic relations between individual concepts as retrieval paths, lacks any capacity to infer, for example, that the overall theme of a passage is "retaliation". As a result, the model is unable to use such implicit abstractions to guide retrieval by indexing episodes with similar abstract themes. We have attempted to address this problem with SAARCS (Lange, Melz, Wharton, & Holyoak, 1990), a hybrid connectionist model that integrates text inferencing and analogical reminding.

Studies on the role of comprehension and inference processes in reminding should also help determine exactly *why* structural consistency is a factor in recall. It is possible that the main benefit of structural consistency between a cue and a target is that similar cue/target structures guide the inferences made and thereby result in more matching of abstract concepts. For example, the inferences needed to fully comprehend the sentences *The priest calmed the chairman* and *The chairman calmed the priest* could be very different (e.g., religious vs. financial counseling). While the present study has demonstrated that the structural differences between cues such as the priest/chairman do indeed affect reminding, it has not determined whether that effect is due to the structural differences *per se*, the different inferences the structural differences engender, or a combination of the two.

## References


Bradley, R. A., & Terry, M. E. (1953). Rank analysis of incomplete block designs. I. The method of paired comparisons. *Biometrika, 39*, 324-345.

Clark, H. H. (1973). The language-as-fixed-effect fallacy: A critique of language statistics in psychological research. *Journal of Verbal Learning and Verbal Behavior, 12*, 335-359.

Falkenhainer, B., Forbus, K. D., & Gentner, D. (1989). The structure-mapping engine: Algorithm and examples. *Artificial Intelligence, 41*, 1-63.

Gentner, D., & Landers, R. (1985). Analogical reminding: A good match is hard to find. *Proceedings of the International Conference on Systems, Man and Cybernetics* (pp. 607-613). Tucson, AZ.

Gick, M. L., & Holyoak, K. J. (1980). Analogical problem solving. *Cognitive Psychology, 12*, 306-355.

Gick, M. L., & Holyoak, K. J. (1983). Schema induction and analogical transfer. *Cognitive Psychology, 15*, 1-38.

Holyoak, K. J., & Koh, K. (1987). Surface and structural similarity in analogical transfer. *Memory & Cognition, 15*, 332-340.

Holyoak, K. J., & Thagard, P. (1989). Analogical mapping by constraint satisfaction. *Cognitive Science, 13*, 295-355.

Johnson, H. M., & Seifert, C. M. (1990). Predictive utility in case-based memory retrieval. In *Proceedings of the 12th Annual Meeting of the Cognitive Science Society* pp. 757-763). Hillsdale, NJ: Erlbaum.

Lange, T., Melz, E., Wharton, C., & Holyoak, K. J. (1990). Analogical retrieval within a hybrid spreading-activation network. In D. S. Touretzky, J. L. Elman, T. J. Sejnowski, & G. E. Hinton (Eds.), *Proceedings of the 1990 Connectionist Models Summer School*. San Mateo, CA: Morgan Kaufmann.

Luce, R. D. (1959). *Individual choice behavior: A theoretical analysis*. New York: Wiley.

Raajimakers, J. G. W., & Shiffrin, R. M. (1981). Search of associative memory. *Psychological Review, 88*, 93-134.

Ratterman, M. J., & Gentner, D. (1987). Analogy and similarity: Determinants of accessibility and inferential soundness. *Proceedings of the Ninth Annual Meeting of the Cognitive Science Society* (pp. 22 -34). Hillsdale, NJ: Erlbaum.

Read, S. J., & Cesa, I. L. (1991). This reminds me of the time when...: Expectation failures in reminding and explanation. *Journal of Experimental Social Psychology, 27*, 1-25.

Ross, B. H. (1989). Distinguishing types of superficial similarities: Different effects on the access and use of earlier problems. *Journal of Experimental Psychology: Learning, Memory, and Cognition, 15*, 456-468.

Seifert, C. M., McKoon, G., Abelson, R. P., & Ratcliff, R. (1986). Memory connections between thematically similar episodes. *Journal of Experimental Psychology: Human Learning and Memory, 12*, 220-231.

Schank, R. C. (1982). *Dynamic memory*. New York: Cambridge University Press.

Thagard, P., Holyoak, K. J., Nelson, G., & Gochfeld, D. (1990). Analog retrieval by constraint satisfaction. *Artificial Intelligence, 46*, 259-310.

Wickens, T. D. (1989a). *Multiway contingency tables analysis for the social sciences*. Hillsdale, NJ: Lawrence Erlbaum.

Wickens, T. D. (1989b). *Twoway* [computer program]. University of California, Los Angeles.

# Attention, Automaticity, and Priority Learning*

**Prahlad Gupta**
Department of Philosophy
Carnegie Mellon University
Pittsburgh, PA 15213
prahlad@cs.cmu.edu

**Walter Schneider**
Learning Research & Development Center
University of Pittsburgh
Pittsburgh, PA 15260
schneider@vms.cis.pitt.edu

## Abstract

It is widely held that there is a distinction between *attentive* and *automatic* cognitive processing. In research on attention using *visual search tasks*, the detection performance of human subjects in *consistent mapping* paradigms is generally regarded as indicating a shift, with practice, from serial, *attentional*, controlled processing to parallel, *automatic* processing, while detection performance in *varied mapping* paradigms is taken to indicate that processing remains under attentional control. This paper proposes a *priority learning* mechanism to model the effects of practice and the development of automaticity, in visual search tasks. A connectionist simulation model implements this learning algorithm. Five prominent features of visual search practice effects are simulated. These are: 1) in consistent mapping tasks, practice reduces processing time, particularly the slope of reaction times as a function of the number of comparisons; 2) in varied mapping tasks, there is no change in the slope of the reaction time function; 3) both the consistent and varied effects can occur concurrently; 4) reversing the target and distractor sets produces strong interference effects; and 5) the benefits of practice are a function of the degree of consistency.

## Introduction

Human performance changes dramatically as *practice* develops, leading to improved performance and a decrease in demands made on attentive resources. Many frameworks have discussed this transition (see [Shiffrin & Schneider 1977]). However, there is no generally accepted model of how performance changes with practice. Moreover, although theoretical frameworks have been proposed, they have generally not provided a detailed computational account of hypothesized processing.

In this paper, we present a computational model of practice effects in *visual search tasks* (see discussion below), which have played an important role in research on attentional processes. The term *attention* is generally used to indicate aspects of human cognitive processing that the subject can control, and that involve capacity or resource limitations. Attentional processing is taken to be a slow, serial activity, with the focus of attention limited to being one thing at a time. However, with practice in consistent tasks, *automatic processes* develop allowing parallel processing that is faster and not as limited by attentional resources[1].

The distinction between controlled and automatic processes has been the focus of much research in the field, particularly in visual search tasks. In such tasks, stimuli are presented visually to the subject, who is required to detect the presence of members of a set of target stimuli (the *memory set*). Non-target stimuli are termed *distractors*. In *consistent mapping* paradigms, a target stimulus will never appear as a distractor on any trial – for example, the subject has to search the display for the memory set digits (e.g., the digits {8 4 5 2}) on all trials, with the remaining digits appearing as distractors. In *varied mapping* paradigms, targets on one trial may be distractors on another trial (e.g., the targets that the subject has to search for on a particular trial have been selected in the experiment by randomly sampling four of the ten digits, with the non-selected items being potential distractors; since a new sample is taken every trial, individual digits change roles as targets/distractors over trials).

A reliable finding from varied mapping studies is that reaction time increases roughly linearly with memory set size ([Kristofferson 1972a]). Consistent mapping studies have found that, with practice, search becomes much faster and memory set size has less impact on reaction times than in the varied mapping condition ([Neisser 1963, Neisser, Novick & Lazar 1963, Kristofferson 1972b, Schneider & Shiffrin 1977]).

These findings have led to the hypothesis that, in varied mapping, each item in the display is serially compared with items in the memory set, with each comparison taking on the order of 50 msec. In contrast, practice in the consistent mapping condition leads to a switch from serial, attentional, to parallel, automatic processing ([Shiffrin 1988]).

Our focus in this paper is on *priority learning*, which we propose as a mechanism that models the effects of practice, and the development of automaticity, in visual search tasks. We review specific experimental results that have been obtained in studies involving visual search tasks, and present simulations indicating how priority learning can provide an account of the observed practice and automaticity effects.

## An architecture for attentional effects in visual search tasks

This section describes briefly the larger architecture for attentional processing within which the priority learning model is embedded. The modular architecture (Figure 1) combines standard *connectionist components* such as connectionist units in a multi-layer organization (see [Rumelhart et al. 1986], for example) with *control elements* that modulate the flow of information between modules. The control involves a *gating unit* (Unit 1 in the figure) that provides a scalar multiplication of a module's output vector. The gating unit receives input from *priority units* within the module, and from an external *attentional control* that coordinates activity between modules. This architecture parallels certain aspects of neurophysiology, and is detailed elsewhere (see [Schneider & Detweiler 1987, Shedden & Schneider 1991, Schneider & Oliver, forthcoming]).

Figure 1 illustrates the overall model of a visual search task. "Visual" modules V1 and V2 represent areas of cortical visual processing, each corresponding to the small area of the visual field in which one stimulus appears. Each module consists of an *input layer* (labeled *I* in the figure) and an *output layer* (labeled *O*). The input layer projects via weighted connections to the output layer, which projects to a module at the next level of processing (Level 2). Each module also has a layer

*This work was supported in part by ONR contracts N00014-87K-0397 and N00014-86-K-0678 and ARI contract MDA903-89-K-0174 to the second author.

[1]See [Shiffrin 1988] for an excellent review of research on attention.

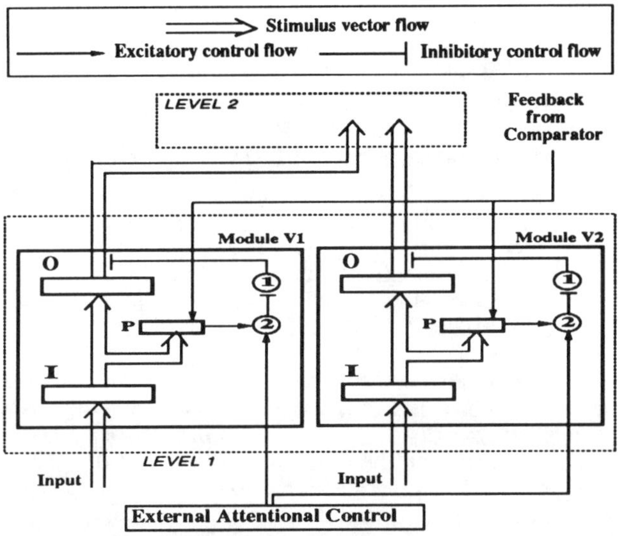


Figure 1: **Architecture of the model.** Stimuli from earlier stages of visual processing enter the input layers *I* of visual modules V1 and V2 and propagate forward to the output layers *O* (Level 1 of processing). The output layer vectors are sequentially transmitted (under external attentional control) to a comparator module (not shown), where they are serially compared with memory set items. Stimuli in the input layers of V1 and V2 also evoke a *priority level* activation over the layer of priority units *P*, which also receive feedback from the comparator module. The priority levels provide the basis for a transition from serial transmission and comparison of stimuli (*attentive* processing) to parallel, *automatic* processing (see text).

of priority units (labeled *P*), which receives input from the module's input layer.

Unit 1 within each module "gates" the forward propagation of activation from the output layer. When this gating is in effect, no vector *transmission* from the output layer of the module takes place, although the *activation* of the output layer is unchanged (e.g., as in the inhibition of the axon initial segment, but not the soma, of a neuron – see [Douglas & Martin 1990]). However, the external attentional control can (by exciting Unit 2, which inhibits Unit 1) inhibit Unit 1, which will release Unit 1's gating effect and lead to onward transmission of the vector in the output layer of that module. The attentional control thus provides for the sequential transmission of vectors from the visual modules, which models the shifting of attention from one area of the visual field to another, i.e., from one displayed stimulus to another, in the visual search task (see [Detweiler & Schneider, forthcoming]).

When the output layer vector of each module propagates forward in turn, it is compared serially with memory set items in a "comparator" module (not shown in Figure 1)[2]. If a visual stimulus matches a memory set item, then the output vector of that visual module also propagates forward to a "motor" module at Level 2 , initiating the motor sequence necessary to produce the detection response.

This paper describes the *priority learning* aspects of the model, whereby learning alters the activity of the priority units and allows automatic transmission of the output in the absence of attentional control.

## Priority learning

A sensory system that includes parallel input at early stages with serial processing at later stages would benefit from a filtering system that prioritizes inputs for limited serial processing. Early visual input processing is parallel ([Ericksen & Spencer 1969]). With a priority filtering scheme, a serial processing system can processes stimuli in their order of importance following the rating of stimuli based on a parallel low level interpretation of stimuli ([Norman 1969]). This allows the most important stimulus to be processed first. Humans appear to have such a priority filtering scheme ([Yantis & Johnson 1990]).

Our computational implementation of priority filtering is in the form of a connectionist network that associatively maps each input vector to a scalar priority. For example, in visual search, each letter would evoke a unique input vector of length 50 in a module (see Figure 1). Each input vector would associatively evoke a scalar value specifying the priority (on a scale of 0 through 10, represented by activation of the priority units) for that input vector. Input vectors evoking a high priority are transmitted to higher levels whereas low priority vectors are gated off, blocking the transmission. Many modules in the visual system can receive input simultaneously, with the respective input vectors each evoking a priority level, in parallel. If only one module contains a high priority target, it will "pop out" of the display ([Triesman 1988]). This occurs by having the priority units inhibit the gating units, thus allowing the output vector to be transmitted *automatically* to the next level of processing[3].

The priority units must be trained to associate input vectors to the appropriate priority levels. This is accomplished in the current model via feedback from the comparator stage. We assume that processing at the comparison stage results in higher output for more important stimuli ([Schneider & Detweiler 1987]). A scalar transform of this output signal is fed back to all modules at the earlier stage. This becomes the target level for the priority units of the transmitting module. Shortly after a vector is transmitted out of a module under External Attentional Control, the priority units are re-trained to approximate the feedback signal via associative learning.

The net result of priority learning is that if the stimulus in a visual module turns out to be a target, then the priority learning network is re-trained to evoke a *higher* priority than was currently evoked; if the stimulus turns out to be a distractor, then the priority learning network is re-trained to evoke a *lower* priority than was currently evoked.

The predicted effects of the learning of priorities are as follows. In a *consistent mapping* task over time, stimuli that enter modules V1 and V2 and correspond to targets will evoke increasingly high priorities in those modules, while stimuli

[2]This comparison can be performed by summing the transmitted vector with each memory set item into a layer of connectionist units. The sum of the squared activation of each unit in the receiving layer then provides a measure of the correlation, or *similarity* between the two vectors. If activation is above a criterion level, there is a *match* of the stimulus with a memory set item, otherwise no match. See [Shedden & Schneider 1991] for details.

[3]Figure 1 shows the excitatory connections from the priority units to Unit 2, which inhibits Unit 1. If the priority level evoked by a stimulus is above some threshold, Unit 2 can inhibit Unit 1 sufficiently to release its gating effect on the output layer, resulting in an "automatic" transmission from the visual module to the motor module. In this case, the motor response is triggered without any sequential scanning of modules under attentional control, or comparison of the stimulus in a module with memory set items. For simplicity, we have omitted discussion of control signals from the priority units to the external attentional control. These are described in [Schneider & Detweiler 1987], and provide a mechanism for external attentional control to entirely ignore a very low priority stimulus, i.e., to *fail* to initiate transmission from that module to the comparator.

corresponding to distractors will have increasingly low priorities. Once the priorities evoked by a target stimulus in modules V1 or V2 have crossed threshold level $\tau$, the presentation of that stimulus in either of those modules will lead to *automatic transmission* from the output layer of that module to the motor module, without sequential scanning of the modules, or serial comparison of their "contents" with memory set items. For distractor stimuli, low evoked priorities result in their being ignored by the attentional control. Thus, serial, attentional comparison is supplanted by parallel, automatic processing, resulting in reduced reaction times for the detection of targets, and in the independence of reaction time from memory set size.

In a *varied mapping* task, however, since a particular stimulus is sometimes a target and sometimes a distractor, there is no gradual increase/decrease of priorities for different sets of stimuli. Instead, the priority levels of stimuli will tend to fluctuate around the default priority level. That is, the evoked priority levels of presented stimuli do not cross threshold, and so there is no automatic over-ride of attentional processing, and no "ignoring" of stimuli. The production of the detection response has to continue to be through serial attentional comparison. Accordingly, there will be no decrease in reaction times, or in their linear relationship with memory set size[4].

## Implementation of priority learning

We ran simulations of the priority learning component of the architecture outlined above, using a single two-layer connectionist network incorporating the input layer and priority layer of one visual module shown in Figure 1. Our aim was to examine whether a priority learning scheme would provide a basis for the practice effects observed in consistent mapping and varied mapping paradigms.

Stimuli entering the network's input layer (*Input*, corresponding to layer *I* in Figure 1) are vectors of length 50, with each element being in the range $\pm1.0$; that is, the input layer consists of 50 units. The priority layer (*Priority*, corresponding to layer *P* in Figure 1) consists of 10 units[5]. A stimulus in *Input* is transmitted in feed-forward fashion to *Priority*. Net input to each priority unit is computed as the weighted sum of inputs from all the input units. The *priority level* evoked by a stimulus is computed by a simple count of the number of priority units in *Priority* that have positive net input. Thus if all ten priority units have negative net input, the priority level is computed as 0; if *n* priority units have positive net input, the priority level is *n*, with the highest possible priority being 10. Bias to the priority units is distributed linearly over the range $\pm0.40$, so that five priority units have a positive bias, and five a negative bias. In the absence of input to the priority units from *Input*, therefore, the evoked priority level has the default value of 5.

When a stimulus is presented to the network, a priority vector is evoked at the priority layer. The network is then re-trained to evoke, for the current stimulus, either a higher priority level than currently evoked (if the current stimulus is designated a target), or a lower priority level than currently evoked (if the current stimulus is designated a distractor)[6]. This will be referred to as the *incrementing* and *decrementing* of priority, respectively. The simulations described here have employed the Widrow-Hoff learning algorithm ([Widrow & Hoff 1960]). One *epoch* consisted of presentation of the entire set of stimuli, with appropriate incrementing/decrementing of priority after presentation of each stimulus.

## Simulations of priority learning

**Simulation 1: Consistent mapping.** In consistent mapping studies with human subjects, the rate of search has been shown to become much faster with practice ([Neisser 1963]). Studies with well-practiced human subjects have exhibited reaction times that varied relatively little with memory set size ([Schneider & Shiffrin 1977, Experiment 2], [Neisser, Novick & Lazar 1963]).

In our simulations of consistent mapping, the training set for a module consisted of 16 random vectors, which were partitioned into two disjoint sets of 8 stimuli each, one set being designated *targets*, and the other *distractors*. During priority learning, the evoked priorities were consistently incremented or decremented for the two sets of stimuli respectively.

The results of the consistent mapping simulations are shown in Figure 2a. The slopes show the evoked mean priority level of targets and distractors, as a function of training (epochs). With increased epochs of training, target stimuli come to evoke increasingly high priorities, while distractor stimuli have increasingly low priorities. As discussed in the section on priority learning, a visual module in which a stimulus evokes a priority level of greater than some threshold $\tau$ will automatically transmit the vector in its output layer.

Figure 2b shows reaction times for targets in consistent mapping experiments with human subjects, who were trained over a period of 36 days ([Kristofferson 1972b]). Memory set sizes of one, two and four were used. For each set size, there was a decrease in reaction time over the training period, which is interpreted as being largely due to speeding up of non-attentional components such as the motor response component of the task, which is also the case in varied mapping tasks ([Shiffrin 1988, page 748]). Figure 2b also illustrates the non-linear set size functions characteristic of consistent mapping.

The priority levels of targets in our simulations of consistent mapping give rise to similar reaction times. On each epoch, search time *S* for a target was simulated as follows. If the priority level evoked by the stimulus was greater than a threshold level $\tau$, then the stimulus would evoke automatic detection, taking "automatic response time" *A*. If the priority level was below threshold, then the target would have to be compared serially with memory set items, with each such comparison taking a constant "comparison time" *C*. We assumed non-terminating search, so that the number of comparisons required is equal to memory set size *m*. Total reaction time for a target was therefore calculated as (i) a base time *B*, representing non-attentional factors, *plus* (ii) search time *S*, which was either (a) automatic detection time *A* (if the priority level was greater than $\tau$), or (b) controlled response time equal to comparison time *C* times memory set size *m* (if the priority level was less than $\tau$).

The time per comparison *C* was taken to be 50 msec, and the time for an automatic response, 40 msec. We used priority threshold $\tau = 8.0$. We used a base reaction time *B* of 290 msec, which decayed to about 230 msec over 200 epochs of training, simulating the speed-up of non-attentional components of reaction times[7]. Figure 2c shows the simulated reaction times,

---

[4]Atkinson & Juola ([Atkinson & Juola 1973]) have presented a similar model of recency learning, in which stimuli of intermediate familiarity require attentional scanning while novel stimuli do not.

[5]Although our simulations have used a *layer* of 10 priority units, it should be possible to achieve the same effects with a single priority unit.

[6]The higher or lower priority levels correspond to modulation of priority layer activations by feedback from the comparator module.

[7]These figures were derived from the human subject data by subtracting the time for one attentional comparison (50 msec) from total reaction time for memory set size one (a) at Days 1-6 (approximately 340 msec), and (b) at

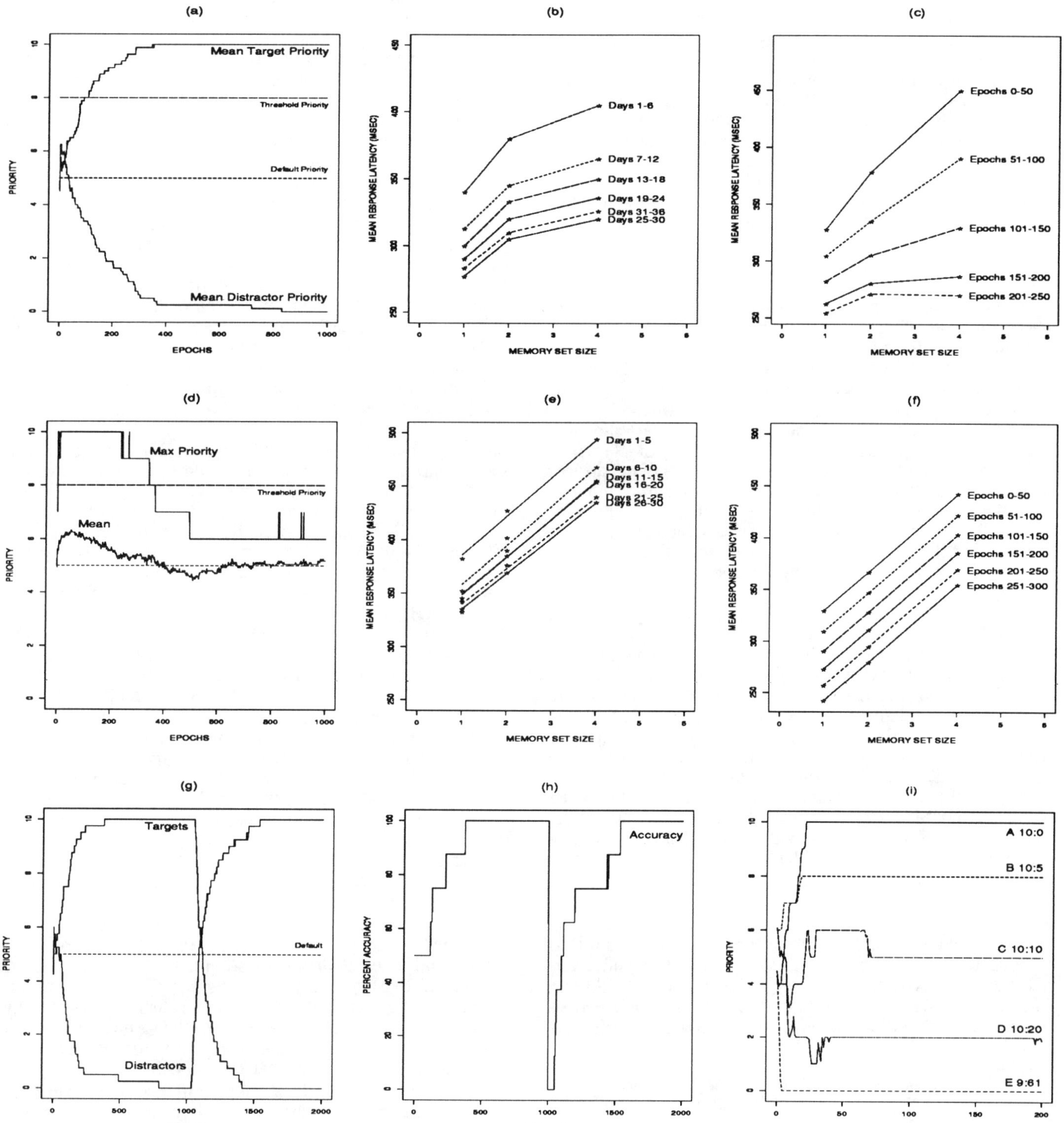


Figure 2: **Priority learning simulation results.** (a) Priority levels in consistent mapping simulation (Simulation 1). (b) Human subject reaction times for target stimuli in a consistent mapping task, after M. Kristofferson, "When Item Recognition and Visual Search Functions are Similar", *Perception and Psychophysics, 12*, p.381. (c) Simulated reaction times for targets, averaged over five consistent mapping simulation runs (Simulation 1). (d) Priority levels in varied mapping simulation (Simulation 2). (e) Human subject reaction times for all stimuli in a varied mapping task, after M. Kristofferson, "Effects of Practice on Character Classification", *Canadian Journal of Psychology 26*, p.57. (f) Simulated reaction times for all stimuli, in varied mapping simulation (Simulation 2). (g) Priority levels with reversal of targets/distractors in consistent mapping (Simulation 4). (h) Detection accuracy with reversal of targets/distractors in consistent mapping (Simulation 4). (i) Priority levels of stimuli with varying degrees of consistency (Simulation 5).

for memory set sizes of one, two and four. Levels of practice are shown in terms of average reaction time over blocks of 50 epochs, corresponding to the average reaction times over blocks of days in Figure 2b.

As in the human data (Figure 2b), the simulation results (Figure 2c) show a downward shift (with practice) of reaction times for each memory set size; this is the combined result of decreasing base response times as well as of developing automatic responses. Moreover, the reaction times for different set sizes at a particular level of practice exhibit the non-linearity expected in a consistent mapping task. Overall, the simulation results appear to model the human data quite well[8].

**Simulation 2: Varied mapping.** Varied mapping studies with human subjects have found that search time continues to be a linear function of memory set size, even after practice ([Kristofferson 1972a]). In our simulations of varied mapping, the training set for a module consisted of 16 random vectors. The priority level for each stimulus was incremented or decremented at random, modeling the effect of stimuli not being treated consistently either as targets or distractors.

Results from the varied mapping simulations are shown in Figure 2d, which plots the mean evoked priority of all stimuli. The mean priority levels fluctuate around the "default" priority level of 5, even with extended practice, as a result of the non-consistent treatment of stimuli. Note that in the simulation results shown, there is an initial random increase in mean priority; in other simulations, there was an initial decrease in priorities. But in all cases, the mean priority settled around the value of 5.

Human reaction times from a varied mapping task ([Kristofferson 1972a]) are shown in Figure 2e (all stimuli), which shows that detection times remain linear with memory set size, even with extended practice. The priority levels of stimuli in our varied mapping simulation give rise to similar reaction times, calculated in the same way as for consistent mapping (Figure 2f). As in the human subject data, practice does not change the linearity of set size functions. The reason for this is that, since few stimuli achieve threshold priority levels, serial comparison remains the only basis for target detection. Automatic detection does not develop, and therefore search functions remain linear with set size. Thus our simulation provides a good match with the observed varied mapping results[9].

**Simulation 3: Consistent & Varied mapping combined.** Schneider and Fisk conducted an experiment in which subjects were asked to carry out a consistent mapping task and a varied mapping task simultaneously ([Schneider & Fisk 1982a]). The findings were that both tasks could be accomplished simultaneously about as well as when each task was carried out alone.

We ran simulations in which 32 stimuli were partitioned into two sets of 16 stimuli each. One set of 16 stimuli represented a consistent mapping task: 8 of the stimuli were treated as targets, and the other 8 as distractors. The other set of 16 stimuli represented a varied mapping task, and the stimuli were treated randomly as targets and distractors. One epoch involved presentation of all 32 stimuli, and represented the simultaneous performance of both the consistent and varied mapping tasks on different sets of stimuli.

The priority levels of consistently mapped stimuli (which were consistently either targets or distractors) separated as in the consistent mapping task alone. The average evoked priority of the varied mapping stimuli remained close to the default priority level of 5, as in the varied mapping task alone. Thus the priority levels of both the consistent and varied mapping stimuli were similar to those in each of the tasks performed independently, indicating that practice effects in the combined consistent and varied mapping task can be modeled in terms of priority learning under the same assumptions as for the independent tasks, in conformity with the findings of the experiment described above.

**Simulation 4: Target/distractor reversal in consistent mapping.** Implicit in the priority learning scheme we have presented is the hypothesis that, in consistent mapping, with practice, target stimuli come to "attract attention" to themselves. Shiffrin and Schneider empirically verified the prediction that if the set relationships are reversed after the priorities have been established, subjects' performance deteriorates below that of novice subjects. For example, in initial practice, targets might be {A B C}, with {X Y Z} as distractors. After reversal, {X Y Z} would be the targets, and {A B C} the distractors. This reversal is expected to produce deterioration because the stimuli that were previously targets attract attention away from the previous distractors that become the new target stimuli.

In the experiments, subjects practiced a consistent mapping task ([Shiffrin & Schneider 1977, Experiment 1]). Once performance had stabilized (2100 trials), targets and distractors were interchanged. On reversal, detection accuracy dropped to below the level it had been at the beginning of the original consistent mapping task (approximately 50 percent), and 900 trials were needed for accuracy to return to the novice level with targets and distractors reversed. Subsequently, gradual re-learning occurred, and accuracy reached the previously established level.

We simulated this reversal in a consistent mapping task. A consistent mapping simulation was run with a set of target stimuli and a set of distractor stimuli for 1000 epochs. At the end of 1000 epochs, the targets and distractors were reversed, i.e., the priorities of the former distractors were now consistently incremented, and those of the former targets were now consistently decremented. The simulation was run in this reversed condition for a further 1000 epochs.

Figure 2g shows that the mean priority levels of the original targets/distractors increase/decrease over the first 1000 epochs of consistent mapping training. When targets and distractors are reversed at epoch 1001, the priority levels of former targets start dropping sharply, and those of former distractors start increasing sharply.

We computed an accuracy measure as follows: a given stimulus presentation was considered to have evoked an accurate response either if (i) the stimulus was a target, and evoked a priority of above threshold, or (ii) it was a distractor, and evoked a priority of below threshold. The threshold priority level used was 9.50. Figure 2h shows the variation of this

---

Days 31-36 (approximately 280 msec). This decrease in base time, from 290 msec to 230 msec, was fitted by an exponential function with constant 0.9985, as a first approximation.

[8]Our model predicts that reaction time functions will become perfectly flat at some point. This does occur in our simulations, by approx. 400 epochs of training.

[9]Note that the behavior of the *slopes* of simulated reaction times fit the human data quite well, which was the major focus of interest in this paper. By contrast, the changes in intercept (i.e., base reaction time) in simulations do not fit the human data well. This is because simulation of the base time practice effect did not include an asymptote effect and hence overestimated the decrease in base time. With humans, decreases in base time seem to asymptote at approximately 200 msec. Adding an additional parameter to the model would allow a good fit of both the constant slope and declining intercept effect in varied mapping.

accuracy measure. At the start of initial consistent mapping training, accuracy was approximately 50 percent, and this improved steadily, asymptoting by about 500 epochs of training. When targets and distractors were reversed at epoch 1001, accuracy dropped sharply, to below the level at the start of the simulation, and it required approximately 100 epochs of training to reach that level (50 percent). Beyond this point, re-learning occurred gradually, and by epoch 1500, accuracy had returned to the previous maximum. Thus the simulation models the human data fairly well.

**Simulation 5: Degree of consistency.** Schneider and Fisk found that consistency has a graded rather than an all-or-none effect. In one experiment, subjects searched for one letter (i.e., memory set size 1) in frames containing four letters ([Schneider & Fisk 1982b, Experiment 1]). Subjects were required to indicate the frame position in which the target had appeared, at the end of each trial. The degree of consistency of appearance of letters as targets or distractors varied. Thus one letter had the consistency ratio 10:0, meaning that it appeared as a target in 10 trials per block of trials, and never as a distractor. Three other letters had the consistency ratios 10:5, 10:10 and 10:20, respectively. Five other letters each had degree of consistency 9:61.

Throughout the experiment, detection accuracy was highest for the 10:0 letter, followed closely by the 10:5 condition. Detection accuracy was lowest for the 10:20 and 9:61 conditions, which did not differ substantially from each other. The 10:10 letter showed intermediate detection accuracy.

In our simulation, our training set consisted of five stimuli, which we designated A, B, C, D and E. One epoch consisted of the appearance of each of the letters A, B, C and D 10 times as a target, and 0, 5, 10 and 20 times, respectively, as a distractor; thus these letters corresponded to the consistency ratios 10:0, 10:5, 10:10 and 10:20, respectively. The E stimulus appeared 9 times as a target and 61 times as a distractor, corresponding to the consistency condition 9:61. The simulation involved presentation of this training set to a network representing the priority learning component of a single module. By approximately 70 epochs of training, the priority levels of the A, B, C, D and E stimuli had settled to 10, 8, 5, 2 and 0, respectively (Figure 2i).

Although we have not constructed simulations corresponding to the overall architecture that would be required for a simulation of the experiment described above, the priority learning results from the simulation with a single module suggest accuracy would be a graded function of the degree of consistency. The overall model would incorporate four modules V1, V2, V3 and V4, corresponding to the visual fields in which each of the four letters in a frame appear. The process of training would lead to each of the modules having the priority levels described above, which developed for a single module. A winner-take-all network between modules would lead to processing of stimuli in their order of priority, so that attention would be preferentially attracted to a higher-priority item even when it is not the target. Consequently, accuracy would be greater for higher-priority items.

## Conclusions

This paper has examined the viability of proposed "priority learning" mechanisms as the basis of a computational model of a number of the phenomena observed in visual search tasks performed by human subjects. The simulation results that have been presented are consistent with a fairly broad set of experimental findings. We therefore conclude that priority learning provides a fairly good account of practice and automaticity effects in visual search tasks.

## References


[Atkinson & Juola 1973] Atkinson, R. & Juola, J. 1973. Factors influencing speed and accuracy of word recognition. In S. Kornblum (ed.), *Attention and performance IV*. New York: Academic Press.

[Detweiler & Schneider, forthcoming] Detweiler, M. & Schneider, W. Modeling the acquisition of dual task skill in a connectionist/control architecture. In D. Damos (ed.), *Multiple-task performance: Selected Topics*. London:Taylor & Francis. Forthcoming.

[Douglas & Martin 1990] Douglas, R. & Martin, K. 1990. Control of neural output by inhibition at the axon initial segment. *Neural Computation* 2: 283-292.

[Ericksen & Spencer 1969] Ericksen, C. & Spencer, T. 1969. Rate of information processing in visual perception: Some results and methodological considerations. *Journal of Experimental Psychology Monograph 79* (2, Part 2).

[Kristofferson 1972a] Kristofferson, M. 1972a. Effects on practice on character classification performance. *Canadian Journal of Psychology* 26: 54-60.

[Kristofferson 1972b] Kristofferson, M. 1972b. When item recognition and visual search functions are similar. *Perception & Psychophysics* 12: 379-384.

[Neisser 1963] Neisser, U. 1963. Decision time without reaction time: Experiments in visual scanning. *American Journal of Psychology*. New York: Appleton-Century-Crofts.

[Neisser, Novick & Lazar 1963] Neisser, U., Novick, R., & Lazar, R. 1963. Searching for ten targets simultaneously. *Perceptual and Motor Skills* 17: 955-961.

[Norman 1969] Norman, D. 1969. *Memory and attention: An introduction to human information processing*. New York: John Wiley.

[Rumelhart et al. 1986] Rumelhart, D., McClelland, J. & the PDP Research Group 1986. *Parallel Distributed Processing, Volume 1: Foundations*. Cambridge, Mass.: MIT Press.

[Schneider & Shiffrin 1977] Schneider, W. & Shiffrin, R. 1977. Controlled and automatic human information processing: I. Detection, search, and attention. *Psychological Review* 84: 1-66.

[Schneider & Fisk 1982a] Schneider, W. & Fisk, A. 1982a. Concurrent automatic and controlled visual search: Can processing occur without resource cost? *Journal of Experimental Psychology: Learning, Memory and Cognition* 8: 261-278.

[Schneider & Fisk 1982b] Schneider, W. & Fisk, A. 1982. Degree of consistent training: Improvements in search performance and automatic process development. *Perception & Psychophysics* 31: 160-168.

[Schneider & Detweiler 1987] Schneider, W. & Detweiler, M. 1987. A connectionist/control architecture for working memory. In G. Bower (ed.), *The Psychology of learning and motivation*, Vol. 21. Orlando, FL: Academic Press.

[Schneider & Oliver, forthcoming] Schneider, W. & Oliver, W. An instructable connectionist/control architecture: Using rule-based instructions to accomplish connectionist learning in a human time scale. In K. Van Lehn (ed.), *Architecture for intelligence*. Hillsdale, NJ: Lawrence Erlbaum. Forthcoming.

[Shedden & Schneider 1990] Shedden, J. & Schneider, W. 1990. A connectionist model of attentional enhancement and signal buffering. In Proceedings of the Twelfth Annual Conference of the Cognitive Science Society. Hillsdale, NJ: Lawrence Erlbaum.

[Shedden & Schneider 1991] Shedden, J. & Schneider, W. 1991. A connectionist simulation of attention and vector comparison: The need for serial processing in parallel hardware. In Proceedings of the Thirteenth Annual Conference of the Cognitive Science Society. Hillsdale, NJ: Lawrence Erlbaum.

[Shiffrin & Schneider 1977] Shiffrin, R. & Schneider, W. 1977. Controlled and automatic human information processing: II. Perceptual learning, automatic attending, and a general theory. *Psychological Review* 84: 127-190.

[Shiffrin 1988] Shiffrin, R. 1988. Attention. In R. Atkinson, R. Herrnstein, G. Lindzey & R. Luce (eds.), *Steven's handbook of experimental psychology*. New York: John Wiley.

[Triesman 1988] Triesman, A. 1988. Features and Objects: The 14th Bartlett Memorial Lecture. *Quarterly Journal of Experimental Psychology* 40A: 201-237.

[Widrow & Hoff 1960] Widrow, G. & Hoff, M. 1960. Adaptive switching circuits. In Institute of Radio Engineers, Western Electric Show and Convention, Convention Record, Part 4, 96-104.

[Yantis & Johnson 1990] Yantis, S. & Johnson, D. 1990. Mechanisms of attentional priority. *Journal of Experimental Psychology: Human Perception and Performance* 16: 812-825.

# Constraints on the Interaction Between Context and Stimulus Information

**Michael E. J. Masson**

Department of Psychology
University of Victoria
Victoria, British Columbia
Canada V8W 3P5
bitnet: mmasson@uvunix.ca

### Abstract

A central issue in the development of models of context effects, such as the interactive activation model, concerns the relationship between contextual and stimulus information. Empirical evidence regarding perception of spoken and printed words indicates that context and stimulus information make independent contributions to perceptual identification. Recent research on visual object recognition, however, suggests that context may have a direct influence on the rate or accuracy of visual analysis. These results imply that contextual influences in language comprehension and object recognition may operate in fundamentally different ways. A series of experiments is described that lead to a reinterpretation of the object recognition results. It is concluded that contextual information contributes to the interpretation of stimulus input without altering its form or the rate of its acquisition.

## Introduction

The influence of context on the identification of stimuli is very powerful and has been demonstrated in a number of domains including auditory and visual word identification, letter identification, and picture identification (e.g., Biederman, Mezzanotte, & Rabinowitz 1982; McClelland & Rumelhart 1981; Stanovich & West 1983; Connine 1987). Two major theoretical views of how context influences stimulus identification have emerged. The *interactionist* view emphasizes the interactive nature of cognitive and perceptual processes and suggests that contextual knowledge directly enhances the speed or accuracy with which visual perception operates, thereby increasing the discriminability of alternatives (McClelland & Elman 1986; Rumelhart 1977). The *classic* view adheres to a modular approach in which it is assumed that processes devoted to visual and contextual analysis operate autonomously and that evidence independently obtained from each process is integrated to complete stimulus identification (Fodor 1983; Massaro 1989; Oden & Massaro 1978; Rueckl & Oden 1986). According to the classic view, context does not directly influence featural analysis of a stimulus or discriminability of alternatives, but instead affects the criterion applied when selecting among stimulus alternatives.

Massaro (1989) challenged the class of interactive activation models, and in particular the TRACE model of speech perception (McClelland & Elman 1986), on the grounds that such models predict a contextual influence on acquisition of stimulus information, whereas data from a speech perception paradigm clearly indicated that asymptotic performance reflected independent contributions of contextual and stimulus information. Contrary to the suggestion that these results rendered interactive models invalid as descriptions of perceptual identification, a revision to the interactive activation model proposed by McClelland (1991) reproduced the independence effect obtained by Massaro. A critical alteration to the original interactive activation model was an architectural constraint that prevented direct interactions between contextual units and stimulus input units. Interactive processing was retained in the sense that units representing alternatives among which a choice must be made (e.g., letters in a letter identification task) could influence activation among lower level feature units. Contextual information (e.g., the word in which a target letter is embedded) could indirectly affect stimulus information, through its effect on units representing response alternatives.

The question addressed here is whether, in the early stages of perceptual identification, context can influence the rate or accuracy of stimulus analysis, even indirectly as suggested by McClelland's (1991) revised approach. An example of such an effect offered by McClelland is the finding that forced-choice identification of a letter is enhanced when the target letter is presented in the context of a word or pronounceable pseudoword (Reicher 1969). This discriminability effect, however, is not at the level of letter features, but at the level of letter identity which under even the revised interactive model has bidirectional links to the word level. Furthermore, tests of a possible interaction between sentence context and word identification have shown that sentence context serves to influence response criterion, not discriminability among alternatives (Masson 1988). In a related study Kroll (1990) demonstrated that sentence context has similar effects on identification latencies for both printed words and drawings of objects. The cross-modality nature of this result suggests that sentence

This project was supported by grants from the Natural Sciences and Engineering Research Council of Canada and from the University of Victoria.

context interacted with stimulus processing at the level of meaning rather than featural analysis. Moreover, the influence of sentence context on auditory word perception appears to operate post-perceptually, rather than on the encoding of the auditory stimulus (Connine, 1987).

More convincing evidence for an interaction between context and stimulus processing has been obtained with visual object identification tasks. In general, observers are more accurate at identifying objects when presented as part of or following a relevant context as opposed to an irrelevant context (e.g., Biederman 1972; Biederman et al. 1982; Boyce, Pollatsek, & Rayner 1989; Reinitz, Wright, & Loftus 1989). Boyce et al. and Reinitz et al. used methods to correct for decision bias effects and still obtained an advantage for objects presented with relevant contextual information.

In this paper I report a series of experiments designed to explore the source of the context effects obtained in the Reinitz et al. (1989) and Boyce et al. (1989) studies. The general issue was whether these context effects arise from the influence of context on the rate or accuracy of stimulus processing. The alternative hypothesis underlying these new experiments was that enhanced object identification is a result of using context to interpret visual information without actually modifying the amount or accuracy of that information. In the case of the Reinitz et al. study, the method of assessing decision bias effects was not deemed appropriate, and in the Boyce et al. experiments it was not clear whether contextually enhanced identification was accompanied by improved acquisition of visual details. These two issues are addressed in the experiments reported here.

## Experiment 1

The purpose of the first experiment was to replicate the Reinitz et al. (1989) results using a set of materials that could be altered for use in a paradigm amenable to signal detection analysis. A set of line drawings of objects served as the stimuli rather than photographs of real objects as in the Reinitz et al. experiments. This format allowed objects to be redrawn with alterations for use in Experiment 2, which examined the possible influence of context on detection of visual detail.

### Method

The subjects were 24 university students. A set of 72 objects were selected from the Snodgrass and Vanderwart (1980) norms and an artist reproduced a line drawing of each one. Eight of the objects were used as practice items and the remaining 64 were treated as critical items. In addition, each object's name served as its appropriate context word. Half of the critical and practice items were presented with their appropriate context word and half with an inappropriate context word. Inappropriate context words were obtained by randomly reassigning words to objects with the constraint that the resulting pairs were unrelated. Assignment of critical items to context conditions was counterbalanced across subjects.

Stimulus materials were shown using slide projectors and tachistoscopic shutters controlled by an Apple II microcomputer. On each trial the subject was shown a context word for 1 s followed by the target object which was in view for 75 ms before being replaced by a pattern mask. The subject's task was to name the briefly presented object. Subjects were informed that sometimes the word would name the object and sometimes it would not. Eight practice trials were presented first, followed by a randomly ordered presentation of 64 critical trials.

### Results and Discussion

The mean proportions of correctly identified critical objects were .64 in the appropriate context condition and .36 in the inappropriate context condition. These means were significantly different, $t(23) = 7.45$, $p < .001$. The probability of subjects offering the context word as their response in the inappropriate condition was .08. When this value was used to correct for guessing in the appropriate condition (as did Reinitz et al. 1989) the corrected probability was .59, which was significantly different from the inappropriate condition, $t(23) = 5.57$, $p < .001$. The results of Experiment 1 successfully replicated the Reinitz et al. (1989) finding of an advantage of appropriate context even when a correction for guessing was applied.

## Experiment 2

To test the hypothesis that the context advantage observed in Experiment 1 stemmed from more efficient visual analysis in the appropriate context condition, a second experiment was conducted. In this experiment the displays were very similar to those of Experiment 1, except that the task was to examine a clearly visible comparison drawing to determine whether it was identical to the briefly displayed object or whether it was an altered version of the object. If context serves to enhance the accuracy or rate of visual analysis, as claimed by Reinitz et al. (1989), subjects should be more accurate in this task when objects are presented following an appropriate context word.

### Method

The subjects were 64 university students. The materials and procedure were the same as in Experiment 1 with the following exceptions. For each of the objects a second version was drawn with a visual detail altered (see Figure 1 for an example). On each trial the pattern mask was followed by a comparison item that represented the same object as the one that had appeared before the mask. On half of the trials the comparison item was physically identical to the target object and on

*Figure 1.* An example of an object in its original (left) and altered (right) form.

half of the trials the altered version of the target object was shown. The task was to decide whether the comparison item was identical to the target. For half of the trials of each type the context word was appropriate for the target object and for half of the trials it was inappropriate. In order to bring performance on this discrimination task above chance, it was necessary to raise the exposure duration. Two different durations were used, 180 ms and 230 ms, with half of the subjects tested at each duration.

### Results and Discussion

For the purposes of signal detection analysis a hit was defined as correctly classifying an identical comparison item as being the same as the target object, and a false alarm was defined as classifying an altered comparison item as the same as the target object. Performance did not vary significantly as a function of exposure duration, so the data presented here were collapsed across that factor. The mean hit and false alarm rates for the appropriate and inappropriate context conditions are shown in Table 1. A nonparametric, two-high threshold signal detection analysis was also applied to the data. This analysis was chosen because it maintains independence between measures of sensitivity and bias and does not rely on assumptions regarding normal distributions with equal variance (Snodgrass & Corwin 1988). The sensitivity measure, $P_r$, is computed as hit rate minus false alarm rate, with a value of 1.0 representing perfect discrimination and 0.0 representing chance. The bias measure, $B_r$, represents the probability of a "same" response when the subject is in a state of uncertainty. The mean $P_r$ and $B_r$ values are shown in Table 1.

Analyses of the sensitivity and bias scores indicated that there was no significant effect of context for either

Table 1
*Mean Values of Signal Detection Measures as a Function of Context in Experiment 1*

| Context | Hit rate | FA rate | $P_r$ | $B_r$ |
|---|---|---|---|---|
| Appropriate | .74 | .40 | .34 | .61 |
| Inappropriate | .72 | .42 | .30 | .61 |

measure, $t(62) = 1.34$, $p > .15$, and $t < 1$, for sensitivity and bias, respectively. Given the failure to find a reliable effect of context on sensitivity, an analysis of the power of the study to find such an effect was carried out. The size of the context effect in Experiment 1 was approximately .8, a large effect size (Cohen 1977). It was assumed that the effect of context on detection of a visual alteration likely would be smaller, so the power analysis was carried out assuming a medium effect size of .5. The power of the study to detect an effect of this size was approximately .86.

Despite adequate power in Experiment 2 to detect a medium effect of context on a visual discrimination task, no reliable evidence for such an effect was obtained. It is conceivable that a smaller effect, not detectable by this experiment, exists. Even if such an effect exists, however, it would appear to be so small as to be incapable of accounting for the large effect of context observed in Experiment 1. By this reasoning, it seems plausible that a different process is responsible for generating much if not all of the context effect found in Experiment 1. In particular, it is possible that the context effect arises not from increased efficiency in visual processes, but from improvement in the accuracy of the interpretation of the results of visual analysis.

## Experiment 3

Experiment 3 was designed to test the hypothesis that the effect of a context word on object identification is due to the availability of an appropriate interpretation or schema. Further, if contextual and stimulus information are integrated in some way *after* both are constructed, it should not matter whether the contextual information is supplied before or after the object is presented. The procedure for this experiment was similar to that of Experiment 1, except that the test format was a two-alternative forced choice. It was predicted that if the main function of the context word was to supply an interpretive schema, rather than a means of enhancing visual processing, then the availability of an appropriate schema in the form of a test alternative might serve a similar function. Consequently, the forced choice format was expected to eliminate the context effect observed in Experiment 1.

### Method

A group of 24 university students was tested. The same materials and procedure as those in Experiment 1 were used for Experiment 3, with one exception. On each trial after the mask was removed, the subject was shown two alternative object names and was asked to select the one that corresponded to the briefly presented object.

### Results and Discussion

The mean proportion of correct responses in the appropriate and inappropriate context conditions were

.80 and .79, respectively. These means were not significantly different, $t < 1$. Given that the design of Experiment 3 was identical to that of Experiment 1 it would be expected that similar context effects should have been found had the the context word served to increase the rate of visual information acquisition. The power of Experiment 3 to detect a context effect of the size observed in Experiment 1 was estimated at .98.

The outcome of Experiment 3 confirmed the hypothesis that the advantage provided by an appropriate context word that precedes the target object can be virtually eliminated by using a forced choice testing format. The presence of a correct alternative following the object on every trial created a level playing field for both context conditions, apparently by always providing a valid interpretive schema for the object. The results of Experiments 1-3 generally support the conclusion that the effect of a preceding context word on object identification is not to increase the rate of visual processing. Rather the influence of context appears to be on the interpretation of the results of visual analysis, as suggested by information integration models (e.g., Massaro 1989).

## Experiment 4

An alternative method of providing context for object identification is to embed an object in an appropriate scene. The Boyce et al. (1989) experiments clearly demonstrated an improvement in object discriminability (sensitivity in signal detection analyses) under these conditions. The critical issue examined in the remaining experiments described here was the source of this enhanced identification accuracy. Experiment 4 was designed to produce a replication of the effect of context on identification accuracy and Experiment 5 tested a hypothesis regarding the source of that enhancement.

### Method

The subjects were 24 university students. The 64 critical objects used in the earlier experiments were arranged as pairs and a background scene relevant to each object was created with the constraint that an object could be replaced in its scene by its partner to create a clearly inappropriate combination of object and context. An example of an object in an appropriate and an inappropriate scene is shown in Figure 2. For each scene a relevant and an irrelevant object name representing objects not in the scene were chosen for use as foils on the identification test. The 32 pairs of objects were arranged as four sets of eight pairs. Two sets were assigned to a condition in which the scene contained an appropriate target object, and two were assigned to a condition in which the scene contained an inappropriate target object. In each of these conditions, one set of scenes was probed with the name of the target object it contained and one set of scenes was probed with with the name of a foil. In the case of foils, the foil appropriate to the scene was used for scenes that contained an appropriate target and the inappropriate foil was used for scenes that contained an inappropriate target (see Figure 2). Assignment of the sets of scenes to context conditions was counterbalanced across subjects. Scenes and probes were also created for the eight practice items.

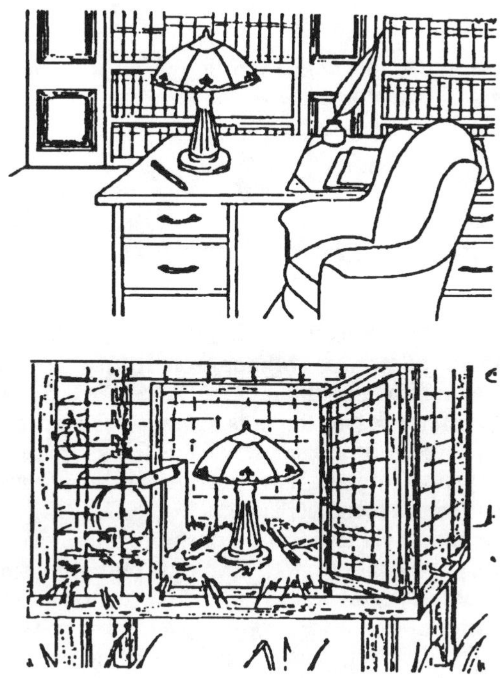

*Figure 2.* An example of a target object (lamp) embedded in an appropriate (upper) and an inappropriate (lower) scene used in Experiment 4. An appropriate scene was probed with the name of the target or the name of a foil object appropriate to the scene. An inappropriate scene was probed with the name of the target or the name of a foil object inappropriate to the scene.

The same apparatus was used as in earlier experiments. On each trial the subject was shown a scene for 130 ms followed by a pattern mask. The mask was followed by a probe consisting of the name of an object. The subject's task was decide whether the object named by the word was present in the scene. A subject was first shown a series of eight practice trials then 64 trials involving critical items presented in a random order.

### Results and Discussion

A hit was defined as correctly claiming that a probed object was present in the scene and a false alarm was defined as making such a claim when a foil probe was presented. The mean hit and false alarm rates for the appropriate and inappropriate context conditions are shown in Table 2, along with the results of a two-high threshold signal detection analysis. Significance tests

Table 2
*Mean Values of Signal Detection Measures as a Function of Context in Experiment 4*

| Context | Hit rate | FA rate | $P_r$ | $B_r$ |
|---|---|---|---|---|
| Appropriate | .80 | .43 | .36 | .66 |
| Inappropriate | .52 | .31 | .21 | .38 |

were applied to the sensitivity and bias scores and it was found that there was higher sensitivity in the appropriate context condition, $t(23) = 2.52, p < .02$. In addition, differences in the bias scores indicated that there was a reliably greater tendency for subjects to claim a probe object was present when it was appropriate to the scene, $t(23) = 5.05, p < .001$.

The results of Experiment 4 constitute a replication of the Boyce et al. (1989) findings regarding the contribution of a relevant contextual scene to the identification of constituent objects, and also provide evidence for a strong bias on the part of subjects to claim an object was present if it were relevant to the scene. It is reasonable to conclude that these two effects of context, sensitivity and bias, occurred independently because the signal detection analysis applied to the data allows the separation of these two influences. One might argue, then, that the bias effect reflects the influence of context on interpretation processes and the sensitivity effect is evidence for the role of context in enhancing the acquisition of visual information. Before accepting this conclusion, however, an alternative explanation for the source of the sensitivity effect must be considered. It is conceivable that context serves to enable the identification of an object with *less* visual analysis than might be required for an object placed in an inappropriate scene. In this view, visual analysis may proceed at the same rate when viewing objects regardless of the scene in which they are placed, but less visual information may be needed in order to identify an object appropriate to the scene. When an inappropriate object is encountered more visual analysis may be needed in order to identify it. If viewing time is restricted, an observer would be less likely to identify an object that is not appropriate to the scene.

A counterintuitive prediction follows from this alternative view. If subjects are required to report on visual details of objects presented in appropriate versus inappropriate scenes, they may have a tendency to engage in less visual analysis of objects in the former case because they fit so well into the interpretive schema instantiated by the scene. Consequently, subjects might be expected to show *lower* sensitivity to objects appearing in appropriate scenes when the task requires discrimination of visual details of objects appearing in appropriate as compared to inappropriate scenes.

# Experiment 5

## Method

A group of 48 university students served as subjects in the experiment. The materials consisted of the scenes from Experiment 4 and the drawings of the target objects in their original and altered forms that were used in Experiment 2. The design and procedure was the same as in Experiment 4, with target objects appearing in appropriate or inappropriate scenes, except that the probe item consisted of a drawing of the target object in its original or altered form. The subject's task was to decide whether the drawing was visually identical to the version of the object that had appeared in the preceding scene. In order to bring performance above chance on this task it was necessary to increase the exposure duration of the scenes to 280 ms and to dispense with presentation of the pattern mask following the scene.

## Results and Discussion

A hit occurred when the probe drawing was identical to the target that appeared in the scene and the subject claimed that it was identical. A false alarm occurred when the same claim was made about a probe that was an altered version of the target. The mean hit and false alarm rates, and mean sensitivity and bias scores are shown in Table 3. An analysis of the sensitivity scores revealed a marginally reliable difference in favor of the inappropriate context condition, $t(47) = 1.69, p < .10$, and no reliable difference on the bias measure, $t < 1$.

The results of Experiment 5 provided weak support for the prediction that subjects would tend to apply more visual analysis to objects that do not meaningfully fit in a scene, and clearly contradicted the notion that context enabled faster or more efficient acquisition of visual information. Given this outcome, the favored view of the influence of contextual scenes on object identification appears to be one in which context is assumed to enable earlier identification of objects that fit the schema invoked by the context. Rather than directly affecting the rate of visual information acquisition, then, context appears to reduce the amount of such information that is necessary for an accurate identification response to be produced.

Table 3
*Mean Values of Signal Detection Measures as a Function of Context in Experiment 5*

| Context | Hit rate | FA rate | $P_r$ | $B_r$ |
|---|---|---|---|---|
| Appropriate | .70 | .53 | .17 | .65 |
| Inappropriate | .73 | .48 | .24 | .64 |

## Conclusion

The experiments reported here examined two very different kinds of context, both of which were shown to have large effects on the ability of observers to identify target objects. One type of context consisted of an object name, presented asynchronously with a target object. The other type of context comprised a scene presented contemporaneously with the target object. There was no evidence, however, to support the claim that presentation of appropriate contextual information enhanced the rate or accuracy of visual information acquisition. Moreover, it was shown that (a) when an object name served as context, its contribution was equally powerful when presented before or after visual processing had taken place, and (b) the contribution of appropriate contextual scenes may consist of enabling observers to identify objects with less visual information rather than increasing the rate or accuracy of visual processing. All of these effects were obtained under very brief exposure durations, enabling snapshots of processing as it occurs at very early stages of perceptual identification.

The present results converge with recent work involving context effects on visual and spoken word identification in localizing the influence of context at a level removed from direct perceptual analysis of a stimulus. In general, this research supports the view that context serves to influence the interpretation of stimulus information rather than to affect the rate at which stimulus processing goes forward or the accuracy with which it is conducted.

The empirical evidence regarding early stages of perceptual identification is consistent with a view of a connectionist architecture that includes bidirectional movement of activation between "intermediate level" processing units that represent words, letters, or objects and the units representing contexts in which they occur, but that constrains the connections between units representing featural details and units at the intermediate level to a bottom-up movement of activation. The similarity of the results across both linguistic units and objects suggests that this constraint may be a fundamental aspect of the cognitive system.

## References


Biederman, I. 1972. Perceiving real-world scenes. *Science* 177: 77-80.

Biederman, I., Mezzanotte, R. J., and Rabinowitz, J. C. 1982. Scene perception: Detecting and judging objects undergoing relational violations. *Cognitive Psychology* 14: 143-177.

Boyce, S. J., Pollatsek, A., and Rayner, K. 1989. Effect of background information on object identification. *Journal of Experimental Psychology: Human Perception and Performance* 15: 556-566.

Cohen, J. 1977. *Statistical power analysis for the behavioral sciences*. New York: Academic Press.

Connine, C. M. 1987. Constraints on interactive processes in auditory word recognition: The role of sentence context. *Journal of Memory and Language* 26: 527-538.

Fodor, J. A. 1983. *The modularity of mind*. Cambridge, MA: MIT Press.

Kroll, J. F. 1990. Recognizing words and pictures in sentence contexts: A test of lexical modularity. *Journal of Experimental Psychology: Learning, Memory, and Cognition* 16: 747-759.

Massaro, D. W. 1989. Testing between the TRACE model and the fuzzy logical model of speech perception. *Cognitive Psychology* 21: 398-421.

Masson, M. E. J. 1988. The interaction of sentence context and perceptual analysis in word identification. *Memory and Cognition* 16: 489-496.

McClelland, J. L. 1991. Stochastic interactive processes and the effect of context on perception. *Cognitive Psychology* 23: 1-44.

McClelland, J. L., and Elman, J. L. 1986. The TRACE model of speech perception. *Cognitive Psychology* 18: 1-86.

McClelland, J. L., and Rumelhart, D. E. 1981. An interactive activation model of context effects in letter perception. Part I: An account of basic findings. *Psychological Review* 88: 375-407.

Oden, G. C., and Massaro, D. W. 1978. Integration of featural information in speech perception. *Psychological Review* 85: 172-191.

Reicher, G. M. 1969. Perceptual recognition as a function of meaningfulness of stimulus material. *Journal of Experimental Psychology* 81: 274-280.

Reinitz, M. T., Wright, E, and Loftus, G. R. 1989. Effects of semantic priming on visual encoding of pictures. *Journal of Experimental Psychology: General* 118: 280-297.

Rueckl, J. G., and Oden, G. C. 1986. The integration of contextual and featural information during word identification. *Journal of Memory and Language* 25: 445-460.

Rumelhart, D. E. 1977. Toward an interactive model of reading. In S. Dornic (ed.), *Attention and performance VI* (pp. 573-603). Hillsdale, NJ: Erlbaum.

Snodgrass, J. G., and Corwin, J. 1988. Pragmatics of measuring recognition memory: Applications to dementia and amnesia. *Journal of Experimental Psychology: General* 117: 34-50.

Snodgrass, J. G., and Vanderwart, M. 1980. A standardized set of 260 pictures: Norms for name agreement, image agreement, familiarity, and visual complexity. *Journal of Experimental Psychology: Human Learning and Memory* 6: 174-215.

Stanovich, K. E., and West, R. F. 1983. On priming by a sentence context. *Journal of Experimental Psychology: General* 112: 1-36.

# A Connectionist Simulation of Attention and Vector Comparison: The Need For Serial Processing in Parallel Hardware[1]

**Judith M. Shedden & Walter Schneider**
Learning Research and Development Center[2]
University of Pittsburgh, Pittsburgh, PA 15260

**Abstract**

Given the massively parallel nature of the brain an obvious question is why are so many information processing functions serial? In particular, this paper addresses the issue of the comparison process. Behavioral data show that in perceptual matching tasks (such as memory scanning and visual search) performance is systematically affected by stimulus load, in that required processing time increases with each additional comparison item. It is arguable whether this indicates a processing system that performs serial comparisons, or a system for which comparisons are done in parallel but reaction time is affected by load because of other system limitations. In this simulation we show that in a modular connectionist system vector transmission is possible in parallel, but the comparison process within a module must be done serially unless accuracy is sacrificed.

This paper examines the question of the serial or parallel nature of the comparison process, and describes the implementation of a connectionist model designed to test the efficiency of parallel multiple comparisons. Despite the fact that neocortex is massively parallel in its architecture there are many tasks for which behavioral data illustrate serial processing. It is important to determine what processing limitations induce serial processing even in the presence of parallel hardware. In some cases this can be explained by limitations in the number of responses that can be made at one time (Deutsch & Deutsch, 1963), by crosstalk or vector transmission interference (Schneider & Detweiler, 1987), or by competition for limited processing resources or operators (Kahneman, 1973). Connectionist based modeling has emphasized the parallel nature of processing. However, even in connectionist based vector processing systems there is a need to serialize operations. With respect to the comparison process the limitation may be due, not to limitations of the system to *make* multiple comparisons, but to the increased error that results from multiple comparisons in the same comparator.

## Theories of Comparison Processes: Background

Psychologists have been studying the perceptual comparison process for years using a multitude of tasks, for the most part analyzing measures of accuracy and reaction time to respond to a target in a field of distractors. In a memory scanning and visual search task subjects compare one or more items in memory to one or more items on visual display, looking for a match between items (Sternberg, 1969; Schneider & Shiffrin, 1977; see Shiffrin, 1988 for review). In a same/different response task subjects compare items in two lists looking for a mismatch between items (Proctor, Healy & Van Zandt, 1991; Ratcliff & Hacker, 1981). In a conjunction search task subjects search for a target consisting of a conjunction of particular features in a field of distractor items which consist of the same features but not the conjunction (Treisman & Gelade, 1980).

A robust finding of memory or display scanning is that reaction time increases in a nearly linear fashion as a function of the number of comparisons that must be performed (e.g., Sternberg, 1969, Schneider & Shiffrin, 1977). This linear increase typically occurs when there is a varied mapping between stimuli and responses (e.g., the subject's responses to the same stimuli change from trial to trial, see Schneider & Shiffrin, 1977) or under conditions of high accuracy and low discriminability. These data have been interpreted by some as indicating a serial repetition of the perceptual comparison process (Sternberg, 1969; Schneider & Shiffrin, 1977; Treisman & Gelade, 1980). Another interpretation of the linear increasing reaction time function is that all items are compared in parallel, and the effect of load on reaction time is due to other limitations of the parallel processing system (Pashler & Badgio, 1987; Proctor, Healy, & Van Zandt, 1991; Ratcliff, 1988). If there is a well practiced consistent mapping between stimuli and responses then comparisons can be performed in parallel (see Schneider & Shiffrin, 1977) indicating the hardware can support parallel comparisons. A reaction time function that does not increase with load indicates preattentive, or automatic parallel processing in which the target "pops out," and is thought to be independent of the comparison process (Schneider, 1985; Treisman, 1985). The modeling of consistent search is detailed elsewhere (Gupta & Schneider, 1991). This paper focuses on the varied mapping search in which serial processing occurs.

There have been a variety of modeling techniques used to explain the linear increase in reaction time (see Townsend & Ashby, 1983; Luce, 1986). The models take the form of either assuming the comparisons are performed sequentially, or they are performed in parallel but at a reduced rate due to the need to share the resources that enable parallel processing. In general, examining the mean data can not distinguish between the serial or parallel resource limited processing models. However, if the data (e.g., Schneider & Shiffrin, 1977, experiment 2) show a self-terminating scan in which the slope of the positive mean function increases at half the rate of the negative function, and the slope of the variance function increases faster for the positive than the negative responses,

1: This research was supported in part by ONR contract N00014-87K-0397 and N00014-86K-0678 and ARI contract MDA903-89K-0174, and in part by a Natural Sciences and Engineering Research Council of Canada Postgraduate Scholarship to the first author.

2: Email: shedden@vms.cis.pitt.edu schneider@vms.cis.pitt.edu then there is strong support for a serial comparison process (see Townsend & Ashby, 1983; Luce, 1986). In this paper we examine why the processing must sometimes be serial even with parallel hardware.

Existing models generally do not provide an interpretation of why processing should be serial or resource limited. From the physiological perspective there is little justification for a requirement that visual processing be serial. The retina and early levels of the visual system certainly operate in parallel with different retinal locations activating topographically distinct sections of tissue in multiple visual maps (Desimone & Ungerleider, 1989). It may be that these parallel channels must converge to a single comparator which becomes inaccurate when receiving multiple inputs. The present model explicitly models such a comparator in a connectionist simulation, and maps out accuracy as a function of the number of concurrent inputs. The simulations described in this paper provide evidence for serial comparisons within a single comparator module. Parallel comparisons are still conceivable if more than one comparator is available. However, the behavioral data supports the view that human visual and memory processing is serial, and is likely to represent processing by a single comparator in varied mapping search tasks.

## Modular Organization of Cortical Anatomy

The present model utilizes a common modular architecture and parallel processing incorporating salient features of cortical processing. Cells throughout cortex (post the initial sensory areas such as visual area V1) show similar patterns of layering, types of cells, and local connections. The structure of cortex is modular, with processing occurring in identical columns, or hypercolumns, which are highly connected within and sparsely connected between (Mountcastle, 1979). Studies of V2 cortex show a structured layering system of cells and connections (Lund, Hendrickson, Ogren, & Tobi, 1981). Information is transmitted through a column in a feed forward direction through two layers of pyramidal cells. An excitatory signal is input to Layer 4 pyramidal cells, which project to layer 2-3 pyramidal cells, which in turn project out of the column. In addition, there are recurrent connections within a column, in which excitatory pyramidal cells feedback to themselves. Inhibitory interneurons are primarily local connections within a column, and it is thought they perform gating and modulatory functions. A special class of axon-axon inhibitory cell is the chandelier cell, which connects to the axon initial segments of sets of pyramidal output cells (Peters, 1984). Chandelier cells have fast inhibitory effects, and possibly function as attentional gating devices (see Shedden & Schneider, 1990, Douglas & Martin, 1990).

## A Modular Connectionist Model

The simulation under discussion was implemented in the CAP2 computer simulation environment. The general model incorporates modules, units, layers, and control elements which can be combined in various architectural configurations. A module consists of an input layer of units, a recurrently connected auto-associative matrix, a feed-forward associative matrix, and an output layer of units (see Figure 1). Modules can be added to the system in breadth, so that several input modules (Figure 1: 1a and 1b) connect to one higher level module (Figure 1: 2a). In addition, modules can be added to the system in depth, creating several hierarchical layers. Associated with each module are control elements (gain and feedback) which manipulate signal strength within the system, and report elements (activity and priority) which manipulate attentional effects.

Each module effectively has three layers making up a connectionist network, including the module **input layer** (traditionally called the hidden layer), the module **output layer**, and the **data input** (which may be the output from the previous level of modules). The current simulation consists of two modules connected hierarchically, so that the output from one module feeds forward through an associative matrix to become the data input to the other module.

The model incorporates the recurrent nature of cortical within-column connections as seen on Layer 4 pyramidal cells. The modules are implemented differently from the standard three-layer connectionist network, in that the input layer of each module is recurrently connected through an auto-associative matrix to itself. In this way information input to the module on each iteration is a function of the external input plus internal feedback from previous transmissions. In a hierarchically organized architecture the external input is received from an input module (or modules) on the processing layer below, and the strength of the external signal is controlled by a scalar gain control element associated with the input module. The strength of the internal feedback signal in a module is controlled by a scalar feedback control element associated with that module. Thus, the gain and feedback control elements function to modulate the output of a population of units in an analogous fashion to the hypothesized function of the inhibitory chandelier cell discussed earlier. In general, the net input to a module is:
net input = (feedback * internal input) + (gain * external input),
where feedback and gain are scalar values and the internal and external inputs are vectors.

**Simulations.** The vector space for one simulation consists of ten pairs of input and target vectors, each having a length of 50 units, with correlations of 0.15 or below within the members of each input and target vector set. This is done by generating random vectors and discarding those with correlations above 0.15. Activation levels for each vector unit range from -1.0 to 1.0, with a resting activation of 0.0. Input vectors and target vectors have unit activations set randomly to -1.0 or 1.0, and the target vector units are then clipped to -0.9 or 0.9 respectively.

During training an input vector is presented to the system, activation is allowed to spread through the network, and the error is calculated between the output vector and the target vector. Activation of the hidden layer units are allowed to range freely between -1.0 and 1.0; activation of the output layer units are subjected to the nonlinear logistic function

$$activity = -1 + 2/(1 + e^{-activity * 0.9}).$$

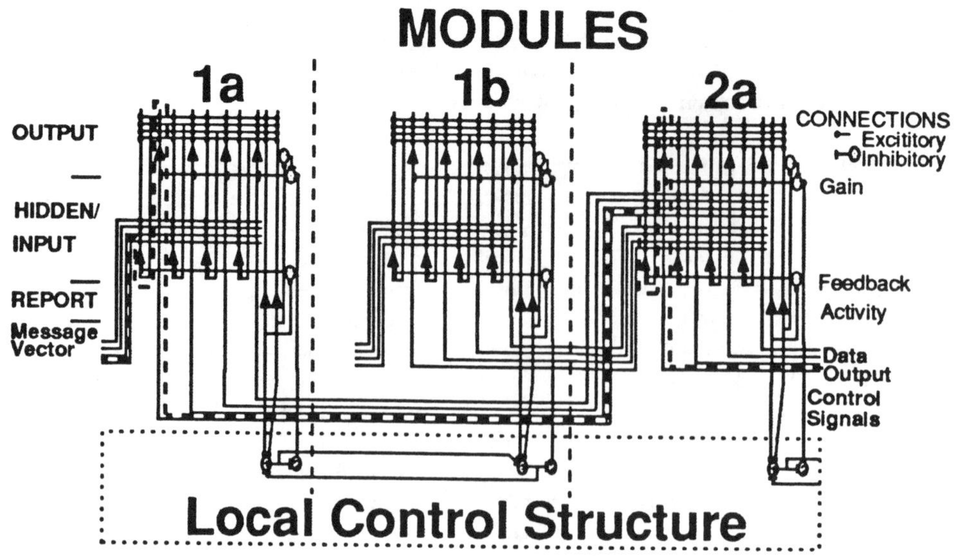


**Figure 1:** A connectionist model of cortical processing. This model consists of a two layer structure that parallels the cortical input layer 4 pyramidal cells, and the cortical output layer 2,3 pyramidal cells. Inhibitory modulation units control the strength of the internal feedback and output signals (feedback and gain). Activity and priority control information is carried by report units (layer 5,6 pyramidal cells) which transmit to central control structures as well as directly to the inhibitory modulation units. The thick dotted line illustrates the information flow from left to right through two modules. The CAP2 environment is described in more detail elsewhere (Schneider & Detweiler, 1987; Detweiler & Schneider, in press; Shedden & Schneider, 1990; Gupta & Schneider, 1991).

Connection weights are changed by the back propagation of the error after each presentation (Rumelhart, Hinton, & Williams, 1986). At the same time the recurrent connection weights of the auto-association matrix at the hidden layer are changed using the delta learning rule (see McClelland & Rumelhart, 1988). The network is trained for 20 epochs, each of which consist of one presentation of all input/target vector pairs in random order. At the completion of training the system has reached the criterion of 100% accuracy and correlations of 0.94 to 0.99 between the output and target vectors for each input pattern.

**The Vector Comparison Task.** The task set for the system is a simple matching task, in which vectors to be matched are presented along with 1, 2, 3, or 4 comparison vectors, thus requiring either 1, 2, 3, or 4 parallel comparisons. During testing every input vector is used in turn as the vector to be matched (sample vector), and the comparison vectors are chosen randomly without replacement from the remaining vectors in the set. On positive trials, one of the comparison vectors is identical to the sample vector. An equal number of positive and negative trials are presented. On each trial the comparison set of input vectors are added, and scaled as a function of the number of inputs (e.g., the addition of from 2 to 5 vectors are scaled from 0.5 to 0.2 of the single vector input). Activity is allowed to build in the system for 5 iterations of external input plus internal recurrent feedback. The data presented below are from 5 different sets of 10 input/target vector pairs, processed at 4 levels of feedback (discussed below), for a total of 20 simulations and 50 datum points per condition.

The measure of evaluation the network uses to determine a match is *vector activity*. The vector activity is the strength of the evoked vector following the summation of two or more input vectors. For example, in Figure 1, modules 1a and 1b output and activate vectors in module 2a. An input vector evokes a specific pattern of activity over the input and output layers of the module. When multiple vectors are presented their effect on the input layer is additive, thus when the vectors are similar the overall level of activity will be higher than when they are uncorrelated (see Schneider & Oliver, 1991).

One measure of vector activity is the average sum of the squared activity of each unit, which can be thought of in geometric terms as the length of the vector. When two vectors are added together, the length of the resultant vector is a measure of the similarity of the two vectors. The resulting activity is equal to:

$$activityreport = \sum_{i=1}^{n} x_i^2 + \sum_{i=1}^{n} y_i^2 + 2|x|\,|y|\cos\theta$$

where n is the number of vector units, theta is the angle between the vectors, and |x| and |y| are the Euclidian lengths of the vectors x and y.

In a matching task a system attempts to detect a match if the measure of activity (or vector length) is above a set criterion, and reject a match otherwise. This analysis examines the function of accuracy at this task with an increasing number of parallel comparisons. A measure of comparison accuracy is provided by the *d'* metric. The d' (from Signal Detection Theory) is a measure of the signal detection sensitivity of a system, and takes into account possible response biases (Tanner & Swets, 1954). There are two distributions of possible vector activity, one in which no match occurs (noise), and one in which a match does occur (signal plus noise). To achieve high accuracy the system must not only detect the signal, but must make a correct rejection of noise in which no signal occurs. Thus a match criterion must be chosen so that both the probability of missing a signal and the probability of making a false alarm to noise are low. This is only possible if sensitivity to the signal is high enough, that is, if there is enough distance between the two distributions of noise, and signal plus noise. The d' is a measure of the distance between the means of the two distributions in normal standard deviations, and is therefore a measure of sensitivity that is not affected by the possible positive or negative response biases for which humans are prone. From d', if one assumes a non-biased criterion it is possible to determine the probability for error, which is simply the area under the overlapping tails of the two distributions. A d' of 4, 3, 2, 1, or 0.5 normal standard deviations corresponds to an error probability of 0.02, 0.07, 0.16, 0.31, or 0.4 respectively, assuming the subject makes an equal number of misses and false alarms. In scanning experiments humans are generally expected to maintain accuracy above 95% and hence a d' of above 3 is expected.

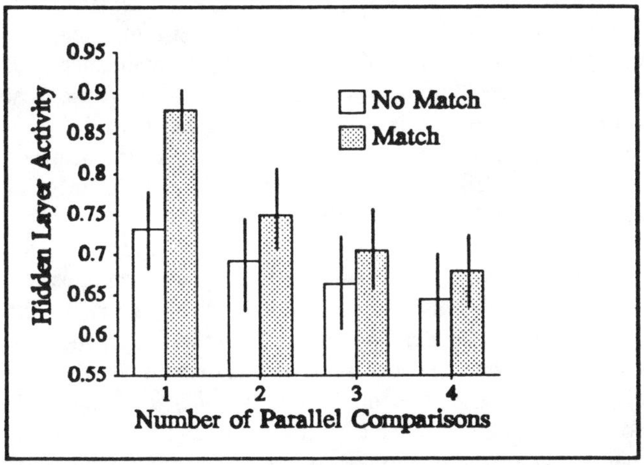


**Figure 2:** Activity on the hidden layer is presented as a function of the number of parallel comparisons. When only one comparison is necessary the difference between the Match and No Match distributions is large, but decreases dramatically if multiple comparisons must be made. The measure of activity is the average sum of the squared activity of each vector unit. Error bars indicate one standard deviation above and below the mean.

There is much more information contained in a vector than its length, and it would be possible to train a network specifically to distinguish between distributions of noise and signal plus noise. However, we are interested the human capacity to successfully perform comparisons on the first trial.

**Simulation Results.** Figure 2 shows the activity level of the hidden layer of the network as either 1, 2, 3, or 4 parallel comparisons are made (50 trials per condition). The measure of activity is the average sum of the squared activity of the vector units. When only one comparison is required there is clear separation between the Match and the No Match distributions (means of 0.88 and 0.73 respectively). With one comparison an activity criterion can be set which results in a d' of 3.48 and a 4% error rate. However when two or more comparisons are performed the difference between the distributions becomes much smaller and the region of possible error becomes much larger. For two comparisons the d' drops to 0.98 with an error rate of 31%, which would not be acceptable for most search tasks.

These severe decrements in accuracy with parallel comparisons are robust for different metrics of activity. Figure 3 shows d' values determined for distributions based on the average absolute value of vector activity as well as the average sum of squares, for comparisons based on the hidden and output layer.

An important issue relating to cortical architecture is whether comparisons can be performed on the hidden or output layer. There are three reasons to suggest that a module would monitor the hidden rather than the output layer. First, the output pyramidal cells often do not make synaptic connections within the layer. Second, if the output layer is gated to control the output to the next level, the comparison could not be performed until the vector is transmitted to the next level of modules. Third, when the output layer is transmitting, the transmission to the higher level of modules will interfere with any other potential signals. Performing the comparison within the module allows other modules to transmit to the higher level modules. This is analogous to the problem faced with data bus arbitration in computer architectures. Each device on the bus limits its transmission in order to allow other devices to transmit on the bus. Typically in computers, each device makes a priority assessment of its internal state without transmitting on the bus. Data is transmitted only after the device activates a bus request and is granted permission from the bus arbitration logic to transmit the data. For all of the above reasons it is important to determine if the match could be performed on the hidden versus the output layer.

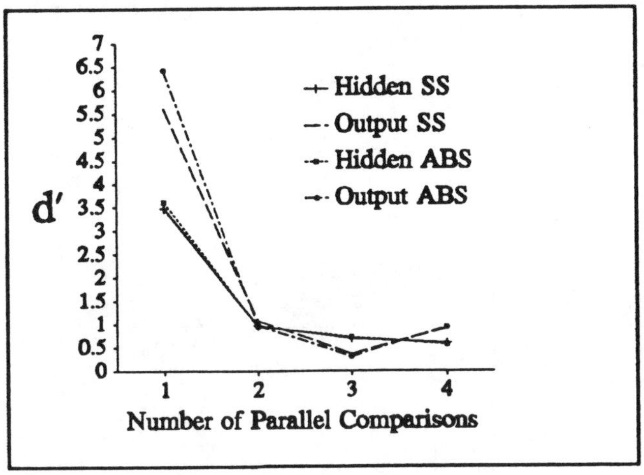


**Figure 3:** The d' for the hidden and output layers is shown as a function of the number of parallel comparisons. Two metrics of activity are graphed: The average sum of the squared activity of the vector units (SS), and the average sum of the absolute value of the vector units (ABS). In all cases, d' decreases dramatically if more than one comparison is made.

Figure 3 graphs the d' values for both the hidden layer and the output layer of the system. For both layers there is a robust deficit for multiple comparisons. The output layer does show better match sensitivity for single comparisons (for example, the d' of the average absolute value metric is 6.5 versus 3.62). It is unclear how relevant the increased detection sensitivity is because human sensitivity in search tasks is usually below 4. In future investigations we will determine how the hidden/output layer d' differences vary as a function of the nature of the squashing function (we used a logistic on the output layer and a step function on the hidden layer), vector size, and correlations among distractors and targets. In the current data, d' was higher on the output layer, but comparisons based on the hidden layer are in the range typical of human performance. Perhaps in cases where crosstalk can be managed and accuracy is extremely important, the more effortful comparison on the transmitted output is beneficial. In any case, as Figure 3 illustrates, both the hidden and the output layer show the harmful effects of multiple comparisons on the probability for error.

**Levels of Feedback.** The recurrent connections that are ubiquitous throughout cortex are represented by the auto-association on the hidden layer in the model. As described above, the feedback control element modulates the strength of the recurrent signal within a module. Four different levels of feedback were tested for each number of required parallel comparisons, and results for the different conditions are shown in Figure 4. When only one comparison is made, the d' is reasonably high for both the hidden and output layers at a feedback level of 0.1, but falls off at feedback levels above and below 0.1. The d' is below 1.6 for any case where multiple comparisons are made, and feedback has very little effect.

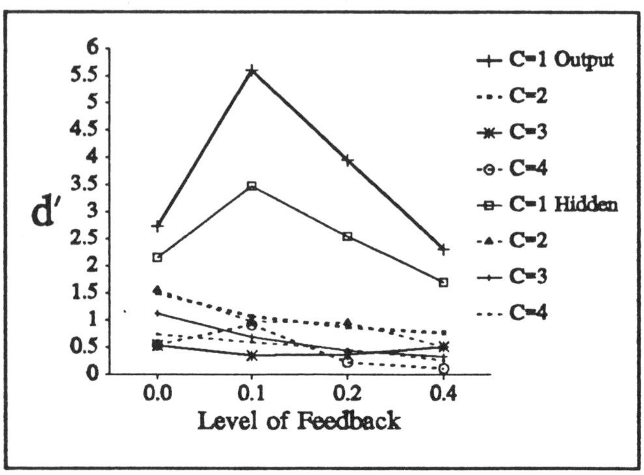


**Figure 4:** The d' for the hidden and output layers for each number of parallel comparisons is shown as a function of level of feedback. The feedback element controls the strength of the internal recurrent signal. For the optimum feedback level of 0.1 the d' measure is greatest, but it falls off with higher and lower levels of feedback.

There are several reasons to include feedback in the module making comparisons. First, input or hidden layer feedback is common in cortex. Second, feedback is critical for latching and holding signals, for the categorization of incoming information, and for signal buffering during concurrent transmissions (see Shedden & Schneider, 1990). Third, feedback may enhance d' because the associative feedback will strengthen previously learned vectors. Matching vectors have a close resemblance to previously learned vectors. In contrast, mismatching pairs represent a blending of features that have not been learned in the auto-associative matrix.

Figure 4 shows the effect of the strength of auto-associative feedback on detection sensitivity. Without feedback sensitivity levels are low (d' of 2.74 and 2.15 for output and hidden layers) and the error rates are higher (about 10%) than those typically observed in scanning experiments. Increasing feedback to 0.1 improves comparison sensitivity to d' measures of 5.60 and 3.48 respectively for the output and hidden layer for a single comparison. When feedback levels are too high distortion of signals begins to occur, and correlations between the actual and desired output vectors drop. The noise in the distributions increases and it becomes more difficult to detect a match. There appears to be an optimum level of internal feedback which is high enough to maintain signal strength and low enough to maintain signal accuracy. The 0.1 feedback range that provided the best comparison sensitivity in these simulations has been shown in previous simulations to be best for signal maintenance properties as well (see Shedden & Schneider, 1990 for other simulations dealing with feedback).

**Serial and parallel processing.** In the present architecture, to respond accurately, the system must serialize the comparisons. For example, if four display items must be compared to one memory item, the system inputs the first display item and the one memory item into a single comparator module. Then the second display item and the one memory item are input to the comparator, continuing until the fourth display item has been compared. In this way accuracy can be maintained although processing time increases linearly (a similar argument was made by Luce (1986, p.444) for serializing comparisons in a limited short term memory). The need for the serial processing of comparisons predicts the varied mapping search data.

How can the transition to parallel processing in consistent search (see Shiffrin, 1988) be explained in this architecture? We assume that each module can associate a priority tag for each learned vector at the hidden layer level. If there is a consistent relationship in which certain stimuli are always targets, they come to evoke a high priority relative to the distractors. Each module makes the priority assessment internally in parallel. If only one module has a high priority it transmits first. The reaction time for the first transmission does not increase with the number of stimuli. This model is detailed elsewhere (Gupta & Schneider, 1991) and provides a good fit to practice effects in consistent search tasks.

## Conclusions and Summary

Although the architecture of cortex is very parallel there are operations that must be performed serially. One of these operations is the comparison process. This paper described simulations in a modular connectionist architecture incorporating central features of cortical structure. The results illustrate why a single comparator module cannot make multiple comparisons and still maintain high accuracy. Accurate performance requires high sensitivity (d') to the presence of a signal, and multiple comparisons generate too much noise for this to occur. In the present simulations matches and mismatches could only be discriminated at human performance levels if the comparisons were performed serially with modest auto-associative feedback (0.1) on the hidden layer. For single comparisons d' is high when measured on either the hidden layer or the output layer of the network. The d' measure is robust for different metrics of activity, and is sensitive to different levels of auto-associative feedback. None of the simulations involving parallel comparisons produced accuracy levels compatible with the human data. In the present architecture, even if all the modules could transmit in parallel to higher levels, the system would have to serialize the comparisons to maintain acceptable accuracy levels. The serialization necessitated in the current architecture matches the apparent serial processing of humans in varied mapping memory and visual search. This set of simulations shows the need for serial processing in parallel hardware.

## References


Desimone, R. & Ungerleider, L.G. (1989). Neural mechanisms of visual processing in monkeys. In F. Boller and J. Grafman (Eds.), *Handbook of Neuropsychology*, Vol. 2:267-299. Netherlands: Elsevier.

Detweiler, M. & Schneider, W. (1991). Modeling the acquisition of dual task skills in a connectionist/control architecture. In D.Damos (Ed.) *Multiple-task performance: Selected topics.* London: Taylor & Francis.

Deutsch, J. & Deutsch, D. (1963). Attention: Some theoretical considerations. *Psychological Review*, 70:80-90.

Douglas, R.J. & Martin, K.A.C. (1990). Control of neuronal output by inhibition at the axon initial segment. *Neural Computation*, 2:283-292.

Gupta, P. & Schneider, W. (1991). Attention, automaticity, and priority learning. *Proceedings of the Thirteenth Annual Conference of the Cognitive Science Society.*

Kahneman, D. (1973). *Attention and effort.* Englewood Cliffs, NJ: Prentice-Hall.

Luce, R.D. (1986). *Response times.* New York, NY: Oxford University Press.

Lund, J. S., Hendrickson, A. E., Ogren, M. P., & Tobin, A. E. (1981). Anatomical organization of primate visual cortex area VII. *Journal of Comparative Neurology*, 202:19-45.

McClelland, J. L., & Rumelhart, D. E. (1988). (Eds.), *Explorations in parallel distributed processing: A handbook of models, programs, and exercises.* Cambridge, MA: MIT Press.

Moran, J., & Desimone, R. (1985). Selective attention gates visual processing in the extrastriate cortex. *Science*, 229:782-784.

Mountcastle, V.B. (1979). An organizing principle for cerebral cortical function: The unit module and the distributed system. In F.O. Schmitt & F.G. Worden (Eds.), *The neurosciences: Fourth Study Program* (pp.21-42). Cambridge, MA: MIT Press.

Pashler, H. & Badgio, P.C. (1987). Attentional issues in the identification of alphanumeric characters. In M. Coltheart (Ed.), *Attention and Performance XII* (pp. 63-81). Hillsdale, NJ: Erlbaum.

Peters, A. (1984). Chandelier cells. In A. Peters & E. G. Jones (Eds.), *Cerebral cortex: Vol. 1: Cellular components of the cerebral cortex* (pp. 361-381). New York: Plenum.

Proctor, R.W. & Healy, A.F. (1985). Order-relevant and order-irrelevant decision rules in multiletter matching. *Journal of Experimental Psychology: Learning, Memory, & Cognition*, 11:519-537.

Proctor, R.W. & Healy, A.F. (1987). Task-specific serial position effects in comparisons of multiletter strings. *Perception & Psychophysics*, 42:180-194.

Proctor, R.W., Healy, A.F. & Van Zandt, T. (1991). Same-different judgments of multiletter strings: Insensitivity to positional bias and spacing. *Perception & Psychophysics*, 49:62-72.

Ratcliff, R. (1988). Order information and distributed memory models. In *Proceedings of the Ninth Annual Conference of the Cognitive Science Society* (pp. 474-486). Hillsdale, NJ: Erlbaum.

Ratcliff, R. & Hacker, M.J. (1981). Speed and accuracy of same and different responses in perceptual matching. *Perception & Psychophysics*, 30:303-307.

Rumelhart, D.E., Hinton, G.E., & Williams, R.J. (1986). Learning internal representations by backpropagating errors. *Nature*, 323:533-536.

Schneider, W. (1985). Toward a model of attention and the development of automaticity. In M. Posner & O. S. Marin (Eds.), *Attention and performance XI* (pp. 475-492). Hillsdale, NJ: Erlbaum.

Schneider, W. & Detweiler, M. (1987). A connectionist/control architecture for working memory. In G. H. Bower (Ed.), *The psychology of learning and motivation, Volume 21* (pp. 54-119). New York: Academic Press.

Schneider, W. & Oliver, W.L. (1991) An instructable connectionist/control architecture: Using rule-based instructions to accomplish connectionist learning in a human time scale. In K. VanLehn (Ed.) *Architecture for intelligence.* Hillsdale, NJ: Erlbaum. Forthcoming.

Schneider, W. & Shiffrin, R.M. (1977). Controlled and automatic human information processing: I. Detection, Search, and Attention. *Psychological Review*, 84:1-66.

Shedden, J.M. & Schneider, W. (1990). A connectionist model of attentional enhancement and signal buffering. In *Proceedings of the Twelfth Annual Conference of the Cognitive Science Society* (pp. 566-573). Hillsdale, NJ: Erlbaum.

Shiffrin, R. M. (1988). Attention. In R. C. Atkinson, R. J. Herrnstein, G. Lindzey & R. D. Luce (Eds.), *Steven's handbook of experimental psychology, Volume 2: Learning and cognition* (739-809). New York: John Wiley & Sons.

Shiffrin, R. M., & Schneider, W. (1977). Controlled and automatic human information processing. II: Perceptual learning, automatic attending, and a general theory. *Psychological Review*, 84:127-190.

Sternberg, S. (1969). Memory-scanning: Mental processes revealed by reaction-time experiments. *American Scientist*, 57:421-457.

Tanner, W.P. & Swets, J.A. (1954). A decision-making theory of visual detection. *Psychological Review*, 61:401-409.

Townsend, J.T. & Ashby, F.G. (1983). *Stochastic modeling of elementary psychological processes.* New York, NY: Cambridge University Press.

Treisman, A.M. (1985). Preattentive processing in vision. *Computer Vision, Graphics, & Image Processing*, 31:156-177.

Treisman, A.M. & Gelade, G. (1980). A feature-integration theory of attention. *Cognitive Psychology*, 12:97-136.

Treisman, A. M., & Riley, J. G. A. (1969). Is selective attention selective perception or selective response? A further test. *Journal of Experimental Psychology*, 79:27-34.

# An Operator-Based Attentional Model of Rapid Visual Counting*

**Mark Wiesmeyer**
Artificial Intelligence Laboratory
The University of Michigan
1101 Beal Ave.
Ann Arbor, MI 48109-2110
wiesmeye@caen.engin.umich.edu

## Abstract

In this paper we report on the use of our operator-based model of human covert visual attention [Wiesmeyer and Laird, 1990] to account for reaction times in counting tasks in which a stimulus is presented and left undisturbed until a response is made. Previous explanations have not employed an attentionally-driven model. Our model, which is based on the Model Human Processor [Card *et al.*, 1983], is an early selection model in which an attentional "zoom lens" [Eriksen and Yeh, 1985] operates under the control of cognition in order to both locate features in visual space and improve the quality of featural information delivered to short-term memory by perception. We have implemented our model and the control structures to simulate rapid counting tasks in the Soar cognitive architecture [Laird *et al.*, 1987], which has been suggested as the basis for a unified theory of cognition [Newell, 1990]. Reaction times in the counting task are explained using operator traces that correspond to sequences of deliberate acts having durations in the 50 msec range.

## Background

Rapid visual counting has been a recurring focus of interest in psychology for many years, and still seems fertile ground for research—new phenomena continue to be forthcoming and no theory adequate to explain all phenomena has yet been found. Visual counting is generally agreed to be of three varieties: immediate apprehension, item-by-item counting, and estimation. Immediate apprehension, most often labeled ***subitizing*** [Kaufman *et al.*, 1949], is rapid, confident, error-free counting of small numbers of items where "small" is defined to be from 1-3 items [Klahr and Wallace, 1976] to 1-6 items [Kaufman *et al.*, 1949]. Item-by-item counting, which is slower and less accurate, must be employed for displays exceeding the subitizing limit, while estimation, which is fast yet quite inaccurate, is employed in time-limited situations.

Of the three modes of counting, subitizing has generated the most interest and, hence, the most controversy. Besides the question of the maximum number of items that can be subitized, opinions differ as to whether it is a serial [Klahr and Wallace, 1976, Folk *et al.*, 1988] or a parallel [Mandler and Shebo, 1982, Sagi and Julesz, 1984] process. More evidence is needed before a clear determination can be made, however with the notable exception of the Mandler and Shebo data, data supporting the parallel position were derived using blocked, limited choice, or forced choice paradigms, or were dependent on subjects' intuition of stimulus countability. These conditions may allow strategy to play an increased role and, thus, lessen the role of a counting component, which makes the subitizing process appear more parallel than serial.

We have combined a serial subitizing component and an attentionally controlled, item-by-item counting component to account for reaction times in counting tasks in which a stimulus is presented and left undisturbed until a response is made. We use data gathered by Chi [Chi and Klahr, 1975] that has been previously modeled [Klahr and Wallace, 1976]. Klahr and Wallace's model is similar in many respects to ours: it provides reaction time estimates; has a similar set of memories; has a notion of operators; and is implemented in a production system. Their model differs from ours most strongly in that it employs a different set of primitive actions; the level of description and analysis is the individual production; and it is not driven by the constraints of attention as we know them today. This last difference is a significant advantage of our model given the central importance that attention has been shown to have in visual tasks. Further, since our analysis is at the level of operators, we avoid giving too much credence to implementational details, such as the number of productions that are used to represent a process. Operators correspond to deliberate acts which are posited to be independent of implementation.

*This research was sponsored by grant NCC2-517 from NASA Ames.

## Methodology

We would like our model of covert visual attention to be capable of predicting *average reaction time behavior* in a wide range of visual tasks. Currently, we limit application of our model to tasks requiring only covert visual attention in two dimensions (hereafter simply "visual attention"); however, in future work we may expand our model to include eye and head movements. Our model is task-independent and implemented on a computer. Task independence, as we think of it, means that—with the exception of task-specific control—operators, memories and the information flow assumed by our model are unchanging across the range of tasks it seeks to cover. Implementation is a check of sufficiency, allows more complex models, and may facilitate integration with other models of behavior, since demonstrably sufficient models can better serve as foundations for more complicated cognitive models.

The level of detail (grain size) at which cognitive models are described determines the sort of explanations or predictions that they can make. The grain size of information in our model is the individual visual feature with accompanying descriptive information, and our model offers explanations of how featural information arrives from perception and is transformed as it travels through short term memories[1]. The grain size of actions in our model is the deliberate act. We define a deliberate act to be an action whose selection may differ according to the current task, given the same stimuli. Thus, deliberate acts are active as opposed to passive, top-down as opposed to bottom-up, and can be expected to be sensitive to knowledge and experience. We model deliberate acts as the selection and application of operators. Figure 4 shows a sequence of operators (actually, the major result of this paper), which we call an "operator trace." Each operator application has an associated duration and the total amount of time that an operator requires to accomplish an action may be increased by interactions with perceptual or motor subsystems. Total reaction time (as in the Model Human Processor) is determined by summing the durations of constituent operator applications and their perceptual and motor dependencies.

We verify the utility of the operators that we use in our model through *coverage*. Coverage in terms of our work means testing our model on a large number of tasks, while only allowing small variations for task-specific control to occur. To date we have applied our model to precuing tasks [Colegate *et al.*, 1973], search tasks [Treisman and Gelade, 1980], tasks that produce illusory conjunctions [Treisman and Schmidt, 1982], as well as visual decay [Sperling, 1960] and crowding experiments [LaBerge and Brown, 1989]. The application of our model to these experiments is described in [Wiesmeyer, 1991].

A critical part of our modeling, once a sensible operator trace for a task has been found, is to determine operator application times that make the model fit observed behavior as well as possible. We do this by systematically adjusting operator application times to minimize the average error of the model's prediction with respect to the experimental data. Times for operator creation, shifting attention, perception, and motor processes are kept constant because determining the best fit for operator traces is an underconstrained problem—unless some times are held constant, there will be many possible solutions. Holding these particular times constant allows us to see if all operators and application times in the model are "in the ball park." Reasonable times are defined in terms of the nominal times that are specified in the Model Human Processor. This methodology may seem ad hoc, but in fact it is consonant with the goal of coverage, since we seek operators and application times that are suitable across the full range of tasks that the model has been applied to. If an operator or application time is not generally suitable, then it must be rejected.

## The Model Human Processor and Soar

The Model Human Processor (MHP) is a cognitive model that has been used successfully in Human Computer Interaction (HCI) research for estimating reaction time performance [Card *et al.*, 1983]. Figure 1 shows a schematic of the MHP that omits (for economy of space) all sensory modalities except vision and includes some of our extensions for visual attention. The MHP splits the human system up into three subsystems: Perception, Cognition, and Motor. Each subsystem operates semi-autonomously and has its own domain of specialization and set of performance characteristics. Perception is composed of the lower-level processes of each of the sensory modalities. Each modality within Perception delivers information to Working Memory (essentially the same as "short-term memory") independently and at a particular rate (i.e., cycle time). Thus, the perceptual cycle time that we are interested in for our model is the time required for a stimulus to travel from the retina to Working Memory. Cognition is composed of Working Memory, Long-term Memory, and the Cognitive Processor. Cognition functions at the level of operators, and its cycle time is the time required for an operator to be applied. Motor executes commands that Cognition deposits in Working Memory, and motor cycle time is the time required for a Working Memory change to affect an overt motor response. Cycle times are specified in terms of average values and ranges: Perception, 100 msec (50-200 msec); Cognition, 70 msec (25-170 msec); and Motor, 70 msec (30-100 msec). Tasks modeled using the MHP are cast as sequences of operators applications. We have found the MHP to be a good conceptual and the-

[1] Features are iconic representations of both shape and color stimuli in our model; however, since the rapid counting task does not make use of color information, details about color are omitted in this paper.

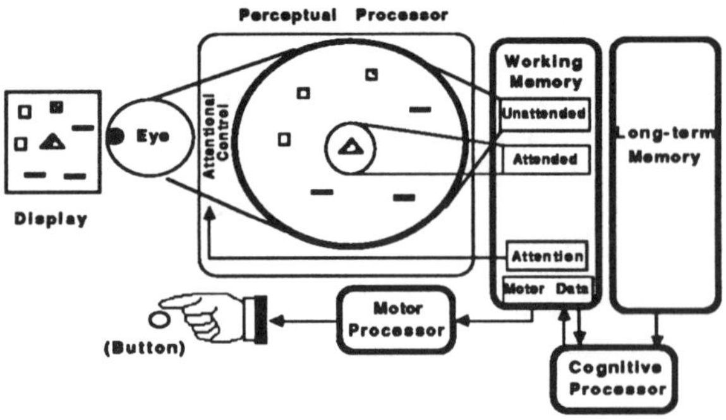


Figure 1: The Extended Model Human Processor

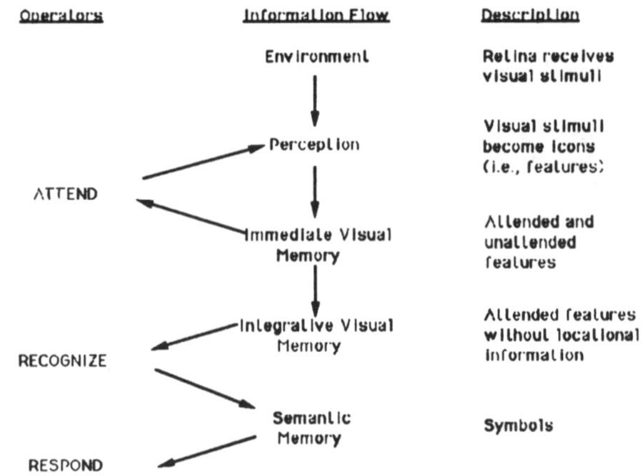


Figure 2: Information flow and operators in the Extended Model Human Processor

oretical framework to start from, since it is operator-based, already has a well-established and proven set of performance parameters, and is structurally similar to Soar (which makes implementation relatively easy).

Soar also has Perception, Cognition, and Motor subsystems[2]. Cognition is composed of a Working Memory and a Long-term Memory, which is a parallel production system. Instead of deliberation through conflict resolution at the level of productions, deliberately selected operators provide the basis of action. Thus, operators in our model map directly onto operators in Soar. Both the selection and application of operators is controlled by productions in Long-term Memory matching against Working Memory and suggesting changes to it. In Soar, specific operators (that is, instantiations of operator types) are created as soon as the data needed to instantiate them are available. In general, this means that several new operators will be created and ready for selection, during the application of the currently selected operator; however, only one operator is applied at a time. Perception is implemented as Lisp functions that transduce environmental stimuli and send input to Working Memory, while Motor is implemented as Lisp functions that receive output from Working Memory and then act on the environment. Low-level Perception occurs in parallel with the firing of productions, as does Motor.

## Attentional Model

The MHP is a much more complete theory than has been presented in the previous section, however our model is only dependent upon those aspects already discussed. We capitalize on the general organization and timing of the MHP and seek to further define aspects such as timing, operators, memories, and information flow that support covert visual attention.

We use the average cycle times of both Perception (100 msec) and Motor (70 msec) in our modeling, and *usually* employ a somewhat faster value of about 50 msec for Cognition. Neither operator creation nor operator selection play any part in the MHP, its emphasis being wholly on operator application. The concepts of creation and selection in our model are derived from Soar which has been proposed as a unified theory of cognition [Newell, 1990]. Extending both the MHP and Soar, we propose that operator creation is independent of operator application and requires time. However, since next operator creation most often takes place during current operator application, these functions often overlap in a pipelined manner. Thus, creation times are often eliminated from the operator trace timing calculations. Operator selection time is assumed to be minimal and not to affect timing.

Visual attention is controlled by Cognition, as shown in Figure 1. A single ovoid region of attentional focus separates the visual field into attended and unattended stimuli. Features deposited by Perception in Working Memory may have both identity and locational information [Mishkin and Appenzeller, 1987]. Features derived from attended stimuli are guaranteed to have good identity and locational information, while those from unattended stimuli are likely to have poor identity and locational information. Differences in the quality of attended and unattended featural information create the need for attention in both recognition tasks (as in Treisman's "Feature Integration Theory" [Treisman and Gelade, 1980]) and in the counting task described later in this paper.

We posit that application of an attention shift operator, which we call **ATTEND**, causes activity at a site in the visual cortex called V4, which has been shown to be a sort of attentional gate that splits the visual field into attended and unattended regions [Moran and Desimone, 1985]. Exact timing for activity related to V4 has not been determined experimentally, but an estimate of 50 msec for deliberately changing the gate (applying the operator) and another 50 msec for receiving new visual features in Working Memory has

[2]In addition, Soar has capabilities for planning and learning that are not needed for the tasks described in this paper.

seemed to work well in our simulations. An example of using these times appears in the first five steps of Figure 4, which shows the sequence of actions that occurs when Cognition shifts visual attention to a new stimulus. It must get features from Perception that signal that a new stimulus is present (100 msec); react to the new stimulus by creating an operator (50 msec); shift attention to that new stimulus by selecting and applying the operator (50 msec); and receive new features from Perception in Working Memory that reflect a change in visual attention (50 msec), requiring a total of 250 msec. An example of shifting attention to features already in Working Memory is shown in the same figure using times at 500 and 550 msec as markers. The total time required to shift is 100 msec, since the new information appears in Working Memory at 600 msec. Another such example occurs between 700 and 800 msec. Note that the **ATTEND** operator that is applied in these latter episodes is created at some point earlier in the trace and is not shown in order to keep the trace simple.

Figure 2 shows our model from the perspective of information flow and operators. The column marked "Description" is intended to elaborate either the process that is occurring or the type of information that is available in each stage of the "Information Flow" column. To start out the information flow, light from the environment stimulates the retina and causes Perception to deposit features in *immediate visual memory.* Both immediate visual memory and *integrative visual memory,* the next memory stage, are part of iconic (i.e., pre-symbolic) Working Memory. Immediate visual memory is composed of all currently attended and unattended features. Each unattended feature in immediate visual memory causes a **ATTEND** operator to be created, which represents the possibility of deliberately shifting attention to that feature. Attended features from immediate visual memory are automatically copied into integrative visual memory. Thus, integrative visual memory is composed of features that are currently or have previously been attended. Integrative visual memory allows iconic visual memories to linger after a shift of attention or change of stimulus without deliberate recognition as discussed below [Intraub, 1985]. Although out of the scope of this paper, the fact that locational information is missing in integrative visual memory is posited to be the cause of "illusory conjunctions" [Treisman and Schmidt, 1982].

Arrows pointing to operators indicate the memories that they are dependent upon, while arrows pointing from operators indicate where their effects are felt. Operators fall into two classes: *visual operators*, such as **ATTEND**, **RECOGNIZE**, and **COUNT**, and *semantic operators*, such as **RESPOND**. It is assumed that all operators (except **RESPOND**) function identically in the tasks in which they apply, except for minimal timing variations to account for individual differences. Visual operators key off of iconic information to either affect future iconic input to Working Memory, though the **ATTEND** operator, or transform iconic memories into symbolic memories through, for instance, the **RECOGNIZE** operator. Once in semantic memory, symbolic stimuli may influence future operator selection, thus affecting behavior, or be reported through operators like **RESPOND**.

## Chi's Experiment and Simulation

Twelve adults observed random dot patterns of from one to ten dots displayed on a standard video monitor that was controlled by a computer. They responded by saying the number of dots that they saw. A voice-activated relay allowed the computer to measure latencies to the nearest msec. Latencies were measured as the amount of time from when the dot pattern first appeared on the display until the relay was activated.

Subjects were told to determine how may dots were present in each trial as quickly and accurately as possible. At the start of each trial the word "**READY**" was displayed in the center of the monitor. Subjects fixated on the central "A" and pressed a button when they felt ready for the test stimulus to appear. After 1.5 sec, the stimulus appeared and the subject responded vocally and response time was recorded by the computer. Immediately afterwards (this part of the trial was not timed), the word "**ENTER #**" appeared on the monitor and the subject typed the number of dots seen. Stimuli were centered and at all times less than 1.8 degrees of visual angle, so they always fell entirely within fovea. Thus, eye movements during counting, although not tested for, were not likely and explanation of reaction times using a model of covert visual attention is appropriate.

Figure 3 shows a plot of the best straight line fits for Chi's results: there is a shallow slope of 46 msec with an intercept of 495 msec for the first three items, and a steeper slope of 307 msec with an intercept of -442 msec for subsequent items. Since there are two major slopes, there must be two major types of processes employed. We assume that the first process, as with most theories of rapid counting, is subitizing. In our model of subitizing, a counting operator iterates over attended items (at about 50 msec per iteration), thus generating a shallow slope. We further assume that counting a large number of items is accomplished by an initial subitizing stage followed by an item-by-item counting stage. In our model of item-by-item counting, attention shifts to each individual, uncounted item which is then counted using a counting operator. Although there is no definitive evidence from the literature, we assume that at least part of the need for individual shifts of attention is due to the inaccuracy of the shifting process. Inaccurate shifts of attention might result in items being counted more than once or not at all.

We present two attempts at fitting Chi's data that have been implemented in Soar. The initial attempt employs a single family of operator traces with default operator application times and shows that these traces

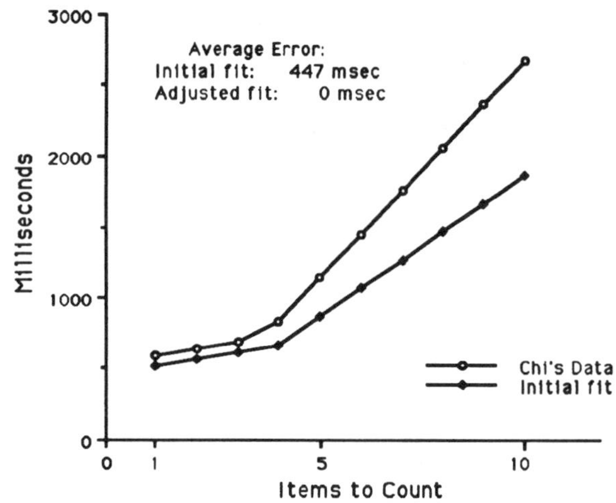


Figure 3: Comparison of Chi's data to modeling attempts

| Initial fit | Operator Event | Event/Comment |
|---|---|---|
| 0 msec | None | Stimulus at retina |
| 0+ msec | None | Stimulus in P |
| 100 msec | ATTEND(s) created | Stimulus in WM |
| 150 msec | ATTEND applies | Shift to stimulus group |
| 200 msec | ATTEND completed | Attention message at V4 |
| 250 msec | COUNT (4 created) | Stimulus attended in WM |
| 300 msec | COUNT applies | Current count is one |
| 350 msec | COUNT applies | Current count is two |
| 400 msec | COUNT applies | Current count is three |
| 450 msec | COUNT applies | Current count is four |
| 500 msec | ATTEND applies | Shift to singleton |
| 550 msec | ATTEND completed | Attention message at V4 |
| 600 msec | COUNT* created | Stimulus attended in WM |
| 650 msec | COUNT* applies | Count is five |
| 700 msec | ATTEND applies | Shift to singleton |
| 750 msec | ATTEND completed | Attention message at V4 |
| 800 msec | COUNT* created | Stimulus attended in WM |
| 850 msec | COUNT* applies | Count is six |
| 900 msec | RESPOND applies | |
| 950 msec | RESPOND completed | Motor sequence begins |
| 1020 msec | | Announce "six" |

Figure 4: Operator trace for rapid counting of six items

can generate a reaction time profile that is similar to the experimental data. (A family of operator traces is required, because counting different numbers of items requires operator traces with different numbers of operator applications.) The second attempt combines the weighted predictions of two *different* families of operator traces, which are similar to those used in the initial attempt, but employ systematically adjusted operator application times.

In the initial attempt to fit to the data, we assume that counting operators for both subitizing and item-by-item counting are identical. Further, we place the subitizing limit at four, rather than the three items implied by the best straight line fits for Chi's data. This is to limit our error, since the first really big skip in Chi's times that needs to be accounted for by a shift of attention (part of the item-by-item stage) occurs at five items. Figure 4 shows an operator trace for counting six items that incorporates subitizing between 250 and 500 msec and item-by-item counting between 500 and 900 msec. (Please read **COUNT*** as **COUNT** for this initial attempt.) Due to space limitations, it was impractical to present the complete family of operator traces for counting different numbers of items. However, subitizing traces can be produced from Figure 4 by removing the item-by-item stage and traces for larger numbers numbers of items can be created by adding shifts of attention with counting operations. Figure 3 shows a plot of reaction times predicted by such traces for this initial attempt.

The initial attempt does not fit Chi's data well for two reasons. First, if deliberate acts employed vary between identical task instances, a single family of operator traces is usually insufficient to predict reaction times. The initial fit has a single discontinuity at four items, while Chi's best fit data has discontinuities at both three and four items. Thus, Chi's data is likely the result of sometimes subitizing three items and sometimes four when there are four or more items to be counted. To account for the intermediate slope between three and four items, the reaction time predictions of a family of operator traces employing three **COUNT** operators in the subitizing stage must be averaged, weighted by frequency of occurance, with the predictions of a family of operator traces employing four **COUNT** operators in the subitizing stage. Second, a new operator, **COUNT***, is required because the item-by-item slope is about 100 msec too shallow. Since the 50 msec operator application time of **COUNT** creates just about the right slope for subitizing we would like to keep it. In order to tailor the model to the experimental data, we derived equations for reaction times of both subitizing and item-by-item counting based on new families of operator traces that use the new **COUNT*** operator. (The traces again are not shown, but can easily be derived from Figure 4.)

$$T_{Subitizing} = T_{InitialShift} + T_{CreateCOUNT} + i * T_{ApplyCOUNT} + T_{ApplyRESPOND} + T_{Motor}$$

$$T_{Item-by-item} = T_{Subitize} + (i - NumberSubitized) * (T_{NewShift} + T_{CreateCOUNT*} + T_{ApplyCOUNT*})$$

where $i$ is the number of items to be counted, $T_{InitialShift}$ is 250 msec, $T_{CreateCOUNT}$ and $T_{CreateCOUNT*}$ are both 50 msec, $T_{NewShift}$ is 100 msec, $T_{Motor}$ is 70 msec, and $T_{Subitize}$ is the total amount of time required to subitize either three or four items and respond. In order to get the best fit to Chi's data, we systematically altered $T_{ApplyCOUNT}$, $T_{ApplyCOUNT*}$, and $T_{ApplyRESPOND}$ in the equations and the weights by which three item and four item subitizing versions of the equations were averaged until the minimum average error was found. Times for operator creation, shifts of attention, Perception, and Motor were kept constant.

Operator application times found for the best fit: $T_{ApplyCOUNT}$, 46 msec; $T_{ApplyCOUNT*}$, 157 msec; and

$T_{ApplyRESPOND}$, 125 msec. All of these times are within the MHP nominal ranges and significantly decrease the average error of the model (from 446 msec to 0 msec per item). This fit requires that four item subitizing occurs six out of ten times (59%) and three item subitizing occurs on the rest of the trials, when there are four or more items to be counted. The adjusted fit is not shown in Figure 3 because it is identical to the experimental data.

Since the best application time for **COUNT*** (157msec) is about three times the usual operator application time used in our modeling (50 msec), it is likely that **COUNT*** is a complex operator composed of simple operators. It would be too speculative to guess exactly what those operators might be, but it is likely that at least one of those operators is a semantic operator. Some evidence that **COUNT*** has a semantic component derives from the fact that 157 msec is very close to the silent counting rate, which has been found to be 167 msec [Landauer, 1962]. On the other hand, **COUNT** is clearly a simple visual operator since its application time (46 msec) is too short for anything other than a single iconic to semantic transformation to take place.

Klahr and Wallace (1976) also used Landauer's results in the analysis of their model and had similar conclusions about the memories used in subitizing and item-by-item counting. However, their timing explanations were not as accurate as ours and did not predict a ratio of three item to four item subitizing.

## Acknowledgements

I would like to thank Tony Simon for discussions that piqued my interest in modeling visual counting, and John Laird and Allen Newell for much of the guidance and inspiration that has motivated my model of visual attention.

## References

[Card *et al.*, 1983] S.K. Card, T.P. Moran, and A. Newell. *The Psychology of Human-Computer Interaction.* Erlbaum, Hillsdale, NJ, 1983.

[Chi and Klahr, 1975] M.T.H. Chi and D. Klahr. Span and rate of apprehension in children and adults. *Journal of Experimental Child Psychology*, 19:434–439, 1975.

[Colegate *et al.*, 1973] R. Colegate, J.E. Hoffman, and C.W. Eriksen. Selective encoding from multielement visual displays. *Perception and Psychophysics*, 14:217–224, 1973.

[Eriksen and Yeh, 1985] C.W. Eriksen and Y.Y. Yeh. Allocation of attention in the visual field. *Journal of Experimental Psychology*, 11:583–597, 1985.

[Folk *et al.*, 1988] C.L. Folk, H. Egeth, and H. Kwak. Subitizing: Direct apprehension or serial processing? *Perception and Psychophysics*, 44:313–320, 1988.

[Intraub, 1985] H. Intraub. Visual dissociation: An illusory conjunction of pictures and forms. *Journal of Experimental Psychology: Human Perception and Performance*, 11:431–442, 1985.

[Kaufman *et al.*, 1949] E.L. Kaufman, M.W. Lord, T.W. Reese, and J. Volkmann. The discrimination of visual number. *American Journal of Psychology*, 62:498–525, 1949.

[Klahr and Wallace, 1976] D. Klahr and J.G. Wallace. *Cognitive Development: An Information-processing View.* Erlbaum, Hillsdale, NJ, 1976.

[LaBerge and Brown, 1989] D. LaBerge and V. Brown. Theory of attentional operations in shape identification. *Psychological Review*, 96:101–124, 1989.

[Laird *et al.*, 1987] J. E. Laird, A. Newell, and P. S. Rosenbloom. Soar: An architecture for general intelligence. *Artificial Intelligence*, 33, 1987.

[Landauer, 1962] T.K. Landauer. Rate of implicit speech. *Perception and Psychophysics*, 15:646–650, 1962.

[Mandler and Shebo, 1982] G. Mandler and B.J. Shebo. Subitizing: An analysis of its component processes. *Journal of Experimental Psychology: General*, 111:1–22, 1982.

[Mishkin and Appenzeller, 1987] M. Mishkin and T. Appenzeller. The anatomy of memory. *Scientific American*, June:80–89, 1987.

[Moran and Desimone, 1985] J. Moran and R. Desimone. Selective attention gates visual processing in the extrastriate cortex. *Science*, 229:782–784, 1985.

[Newell, 1990] A. Newell. *Unified Theories of Cognition.* Harvard University Press, Cambridge, MA, 1990.

[Sagi and Julesz, 1984] D. Sagi and B. Julesz. Detection versus descrimination of visual orientation. *Perception*, 13:619–628, 1984.

[Sperling, 1960] G. Sperling. The information available in brief visual presentations. *Psychological Monographs*, 74:1–29, 1960.

[Treisman and Gelade, 1980] A. Treisman and G. Gelade. A feature integration theory of attention. *Cognitive Psychology*, 12:97–136, 1980.

[Treisman and Schmidt, 1982] A. Treisman and H. Schmidt. Illusory conjunctions in the perception of objects. *Cognitive Psychology*, 14:107–141, 1982.

[Wiesmeyer and Laird, 1990] M.D. Wiesmeyer and J.E. Laird. A computer model of visual attention. In *Twelfth Annual Conference of the Cognitive Science Society, Boston*, 1990.

[Wiesmeyer, 1991] M.D. Wiesmeyer. *An Operator-based Model of Covert Visual Attention.* PhD thesis, The University of Michigan, Ann Arbor, 1991.

# Educational Tools for What You Wanted to do Anyway

**Beth Adelson**
Dept. of Psychology
Rutgers University
adelson@cogsci-1.rutgers.edu

## Abstract

**Approach, Technologies and Goals**
This paper describes a set of educational tools designed to support central cognitive skills such as argument analysis and construction, cooperative negotation, collaborative writing and scientific inquiry. In building these tools we drew upon multi-media technologies, interactive video and object-oriented programming techniques. Our approach however, was not motivated by the technologies but rather by a desire to embed our educational objectives in situations which were intrinsically interesting to the students. This approach, alluded to in the title of the paper, came out of our intention to have the student feel good about her personal interests, and further to have her feel good about her intellectual competence as a vehicle for furthering her interests.

## Outline

Three tools will be described[1]. The first supports learning to analyze and construct arguments which are effective at both a psychological and sociological level. The second is a tool for collaborative negotiation. The third tool, supports: scientific inquiry; the building of theories/models; and the use of explanatory theories to help the student effect her environment.

[1]This work was supported by grants from NSF's Engineering and CISE directorates; MIT's Center for Coordination Science and Tufts' FRAC and Dean's awards.

Thanks to: MIT's VCG; Prof. Tom Malone & Gilberte Furstenberg; RAs: Kris Bartol, Mike DiBlasio, Troy Jordan and Mark Zudeck; and Boston Latin School

## Psychological and Social Analysis and Argument Construction

This application [2] was conceived of as a tool which would take the task of learning a foreign language (French) beyond the task of only learning vocabulary and syntax and into the task of learning to be effective in the culture in which the language was spoken.

In using this application the student interacts with a program in which she encounters and analyzes some powerful 'cultural icons' and then constructs arguments for the purpose of effecting their actions.

### Technology and Scenario

Using the language *Muse*, (Hodges, Sasnett, & Ackerman, 89; Hodges, Sasnett, & Harward, 90) we created an interactive, multi-media tool which runs on both MicroVax and RT workstations. Students using the tool were presented via video disk[3], with a story whose protagonist was named Philippe. Although quite likeable, Philippe had, through his flakiness just lost his job, his girl and his apartment.

As a result, Philippe was faced with a set of challenges: He had to convince an appropriately sceptical realtor, M. Jacot, of the "Paradise Realty Co.", that he would be a responsible tenant. As a fallback position, he felt that he also needed to convince his aunt, a staunch pillar of the bourgeoisie, that he is the perfect choice for house-sitter during her up-coming three month African safari. Additionally, he had to deal with a plumber over the leak in the kitchen sink in the apartment he had shared with his lost love Elisabeth.

[2]Thanks to Evelyn Schlusselberg and Gilberte Furstenberg

[3]The video disk was produced for an earlier application in which Gilberte Furstenberg was director of educational design; Ayshe Farman-Farmaian was producer; and Janet H. Murray was executive producer.

Philippe as a character is too discombobulated to deal effectively with these situations. This creates the opportunity for the student to step in and argue on Philippe's behalf[4]. The intrinsic interest here is that these are challenges which the college-age students using the program are just beginning to face in their own lives.

## Student Exercises

For reasons explained below we have created two versions of the tool, each calling for analysis and explanation:

1. Direct Version:
   Here the student directly argues with the realtor, the aunt and the plumber on behalf of Philippe. As each situation arises the program presents the student with a list of topic sentences each of which could form the core of an argument. The student chooses the topic which she feels would be most effective given the station and nature of the character Philippe is currently confronting. She additionally goes on to construct an argument for Philippe. The full text of her argument is sent off to an instructor for later grading; however, the topic she has chosen has an immediate effect because multiple outcomes are stored on the disk. For example, Philippe's inability to get the plumber to come and fix the kitchen sink over the last six months has contributed to the break up of his relationship. Elisabeth has finally made the appointment herself and has asked Philippe simply to let the plumber in to the apartment and to pay him. However, Philippe has not paid attention to where Elisabeth has left the check. If the student can convince the plumber to, on this one occasion, extend credit to Philippe Elisabeth becomes convinced that Philippe has turned over a new leaf; the story therefore ends with Elisabeth giving the relationship another try. If the student is unsuccessful with the plumber but convinces either the aunt and/or the realtor of Philippe's trustworthiness Philippe writes Elisabeth a rather cool 'Dear Genevieve'. If none of the characters are convinced, Philippe is left at the conclusion of the story to face another difficult day.
2. Agent Version:
   This version differs from the Direct Version in that at the outset of the story the student chooses an agent to argue on Philippe's behalf (Laurel, 90; Oren, Salomon, Kreitman & Don, 90). The agent is either General Charles De Gaulle or the torch singer Edith Piaf. The student chooses the agent after seeing a description both of the agent's role in French society and of the agent's (strong) character. Additionally the student sees a video segment of Piaf singing her theme song "La Vie en Rose" and historical photos of DeGaulle's imposing presence.

   As the situations with the realtor, the aunt and the plumber present themselves the agent offers to make one of three arguments. The topics of the arguments are the same as the ones in the Direct Version; here however, the student sees the full argument, rather than just a topic sentence. The arguments, written by a French historical novelist, are very much in the distinctive 'voice' of the agent. The student chooses the argument which she feels would be most effective if presented by the agent. The student then *explains* her choice and sends it off for later feedback. Again the choice of argument effects the outcome of the story.

**Examples of the Arguments: Cultural and Psychological Analyses** To give the reader a sense of the nature of the student's task we present first some of the topics from the Direct Version and second some of the arguments from the Agent Version among which the student has to either choose or elaborate upon.

1. Direct Version: Topic Sentences from which to construct arguments for the realtor

(a) "Philippe has always paid his rent on time."

(b) "Philippe comes from an old and respected family."

(c) "Philippe is a well connected journalist."

The argument that will convince the Parisian realtor is different from the one which would be most likely to convince his American counterpart[5]. Journalists enjoy a privileged position in French society and their presence is welcome even when their credit history is shaky. The student needs to call upon this kind of knowledge of cultural differences when choosing and constructing an argument. (This is not an impossible task since cultural mores are discussed in the

[4]The story, the tasks presented to the student, and the responses required from the student are entirely in French.

[5]Notice there is room for the student to successfully disagree with our sense of what constitutes a winning argument during the later interaction with the instructor.

French course in which the users of this tool are enrolled).

2. Agent Version: DeGaulle's arguments for the plumber

(a) "Monsier Morel, looking around, you can surely see that Monsieur [Philippe] Vitaz is perfectly capable of paying you. He clearly has the means to make good on his debt. A man born as Philippe is will not fail to discharge his debt! It is a question of honor!"

(b) "Monsieur Morel, for the French, each must trust the other. What would happen to our country? Our country without the solidarity which has always been its strength? I tell you, to trust Philippe, it is your obligation."

(c) "Monsieur Morel, your reputation is known. It is said that no plumber is better than you. Monsieur Vitaz, by all evidence, is in need of your good services. May I allow myself to ask you, in this one case, to have the goodness to make this repair?"

Again the winning argument is not obvious. Having seen the video of the plumber the student sees the degree of Plumber Morel's pride. However, from the way in which he takes command of the situation, savors the difficulties of a leaky sink and wields the tools of his trade it can be inferred that his pride in his craft outweighs his pride as a Frenchman. This kind of careful arguing from the evidence is what will allow the student to make a successful choice.

We claim that both versions of the tool ask the student to develop a type of understanding which is central in language learning. Further we believe that the activities of analysis, and careful arguing from evidence have general utility. Our hope is that the student will also gain confidence as she develops the skills associated with using the tool.

## Motivation Behind the Two Versions

Both versions have the strength that they provide rich analysis and explanation tasks. However, each has a weakness. In the Direct Version the program knows which topic the student has chosen to argue on and takes a path based on that choice of topic without regard to the *goodness* of the argument. We were uncomfortable with the possibility of the students noticing this limitation. That is, if a student was unable (or unwilling) to create an explanation which she felt was adequate and the outcome of the situtation was still positive the tool would suffer in the student's eyes. For this reason we constructed the Agent Version where, complete and well formed arguments were provided within the program. In order give the students using the Agent Version an equally rich task we asked them to *explain* their choice rather than construct the argument[6].

The possible limitation of the Agent Version is that since the student effects the outcome of the story in a less direct way she may feel less engaged and therefore she may learn less. Currently we are comparing the two versions of the tool to assess the relative strengths and weaknesses of each.

## Collaborative Negotiation Tool

Here we describe a tool which allows parties to negotiate in a non-adversarial way. It is implemented in Object Lens (Lai, Malone, Yu, 88; Malone, Grant, Lai, Rao, Rosenblitt, 89), an object-oriented system running on both Apple and Xerox machines.

The tool guides collaborators through a cooperative (as opposed to an adversarial) negotiation. At the same time, it provides a sort of spread sheet for keeping track of the state of the negotiation. The tool purposely does not make any decisions for the parties since the goal is to have the parties develop those skills for themselves.

The process is cooperative in that the two (or more) parties *jointly* build a solution by considering the needs as well as the resources which they bring to the table. They can assume cooperative positions because they are allowed to back off from any initial unilaterally proposed solutions as they are prompted through the following stages (Susskind & Cruikshank, 87; Brockner & Rubin, 85; Raiffa, 82; Fisher & Uri, 81; Fisher & Brown, 88):

1. Problem Statement: Both parties state the problem as they see it.
2. Initial Solution: Both parties propose an initial solution which is satisfying to their side.
3. Underlying Needs and Resources: Based on their 'Problem Statement', *rather* than on their 'Initial Solution', the parties jointly construct a list of what they need from the situation and what resources are available.
4. Collaborative Solution: The parties jointly construct a new solution by considering how the 'Resources' match the 'Needs'.

[6]Of course the problem still remains that the students could construct half-hearted explanations but we felt that this was less serious. That is, it would still make sense to the student that the story was being positively affected by the agent's cogent argument, even if the student's explanation of the argument was less than it should have been.

5. Iterating: Both parties can now examine the fit between their needs and the new collaborative solution. They may accept the solution, further clarify their needs or rework the solution.

At each stage in the process the system offers the parties a workspace which is designed to make it easy to enter and consider the needed information. The work spaces are object-oriented windows.

Because the solution process involves making the resources of the situation explicit, the result is often a solution which more fully meets both parties' needs. However, a different kind of result may also obtain. On reflection, it too is desirable; because the parties have made their needs explicit, they can quickly see when they are in a situation which does not allow a joint resolution. This allows them to retreat from the particular effort, thereby preserving the collaborative relationship for future ventures (Fisher & Uri, 81; Fisher & Brown, 88).

**Example of How the System Guides and Supports Negotiation: 'Hiring Henry Staller'** One negotiation example on which our system has been run proceeds as follows[7]:

- **Scenario:**
  Two managers, Carl Creative and Edgar Efficient have received funding for a propsal which has put them in charge of the development of a new system. The system is potentially hot enough to put their company on solid footing. They are at the point of making some staffing decisions and have reached a deadlock over a prospective candidate. The system prompts them to state the following:

1. Problem Statement:
   - Carl: "We have funding for the big new system. Our group has a chance to save the company if we can come up with a breakthrough. Staller would be perfect for lead programmer. He's full of ideas. And he only wants half of what we budgeted for, so the financial pressure is off".
   - Edgar: "If we can get the new system out the door by the dead line we can save the company. Staller is brilliant but he doesn't follow through."
2. Initial Solution:
   - Carl: "Hire Henry."
   - Edgar: "Over my dead body."
3. Underlying Needs and Resources:
   - Needs
   (a) Carl: Get a creative breakthrough
   (b) Edgar: Get the system built on time
   - Resources
   (a) Money saved on Henry
   (b) Henry's creative ability
4. Collaborative Solution:
   - Hire Henry, and with the money saved hire a second less creative, less expensive, more reliable programmer to make sure the work gets done.

Clearly the collaborative solution reached here is better than either of the initial unilateral solutions. But to what extent did the negotiation tool help? Coming up with the collaborative solution is easy if (and only if) the parties can perceive that 'Needs' (a) and (b) can be met by 'Resource' (a).

Matches between needs and resources which may initially be non-obvious become clear because the system prompts the parties to make both explicit, and then asks the parties to consider how the resources can provide for the needs.

Additionally, the system facilitates the discovery of successful solutions because it asks the parties to retreat from their adversarial 'Initial Positions' and then find alternative ways to satisfy their 'Underlying Needs'. That is, the system allows solutions and needs to be listed separately and therefore to be decoupled.

**Implications of the Negotiation Tool** The negotiation tool is intended to build *ongoing* collaborative relationships among the parties. It does so by allowing the parties to back off from adversarial positions while saving face by turning their attention to their needs and resources. Additionally the tool helps in the building of continuing relationships because it allows for early and graceful exits from situations in which collaboration has a low probability of success.

In terms of anecdotal data, within our research group we have reached successful agreements on, for example, division of labor and work schedules, using this negotiation process.

## Teaching Scientific Inquiry

The goal here was to devise a course of study[8] in which seventh grade students would: gain some scientific inquiry skills; increase their understanding of theories and models and the the relation of these abstractions to the physical

[7]The example is taken from a domain outside of collaborative authoring and shows the generality of the negotiation tool.

[8]Thanks to the kids, teachers and adminstrators of Boston Latin School

world; learn about the tension between affecting and being affected by their environment; and see that science and technology can help them in this resolving this tension.

## Constructing the Course of Study

Again we took the approach of attempting to embed the educational tasks within a framework which was intrinsically interesting to the students. Therefore, to obtain the goal of teaching scientific inquiry we devised a curriculum which called upon students to conduct inquiries in the course of learning geography. What is notable about the curriculum is that it takes the form of exercises which are carried out for the purpose of working out issues, such as increasing their basketball skill, making new friends, etc; issues which *they* had defined as being important.

Below we sketch out the curriculum and then present two student-inspired exercises for working through the curriculum.

**Geography Curriculum** We would want students who had learned about the geography of a given place to have learned what that place was like both physically and culturally. We would also want them to have gained an understanding of the relationship between the physical and the cultural elements. In some sense, here we want the students to have formed a *descriptive theory* of the place.

Additionally, we would want the students to look at how this description was changing. Further we wanted them to be able to ask why the change was occuring; whether they liked the change; and, if not, what could they do about it. To address these questions the students would need an *explanatory theory.*

## Developing the Exercise/Inquiry

In order to develop an exercise for learning scientific inquiry and geography skills we visited Boston Latin School and interviewed a number of seventh grade students on the subject of the interests and challenges which they faced outside of school. Many of the stories had a common element; the students were trying to deal with issues of loneliness and popularity. We chose one story as both striking and representative and built two exercises around it. Clearly the exercises generalize.

> Mike's story:
> Mike told us that he was particularly lonely since not only was it his first year at Boston Latin, but he had also just moved to a new neighborhood and was sorely lacking friends. He also told us that he loved to play basketball and was pretty good at it.

Descriptive Theory: Wed. 5 PM Norfolk St. Game

1. Forwarding is weak
2. Many shots missed
3. Potholes on North boundary of court
4. Half court game

Elements of the Explanatory Theory: Wed. Game

1. Forwarding is weak because
   the Essex St. team can't get around the strong guarding by Sam & Bill of the Broadway team.
2. Many shots missed because
   both teams are shooting as if they had the nets which they have in the school gym.

Elements of the Practice Plan

1. Getting thru Sam & Bill :
   They don't expect your left hand approach.
2. Not missing shots:
   Practice netless shots.

Figure 1: Descriptive & Explanatory Theories (top & center). Practice Plan (bottom).

**Exercise 1: Creating a Descriptive Theory** Mike's story prompted us to create an exercise in which he could make friends by excelling in the 'pick-up' games around his neighborhood. The exercise was designed so that he would excel because of his scientific inquiry and geography skills. That is, we wanted to have Mike ask what the neighborhood basketball courts were like through the use of survey and observation techniques. This would result in his forming a descriptive theory of the neighborhood games (Figure 1, top).

We claim that simply reading about theories does not allow students to understand what they are. We believe that they have to be built and used before they can be understood. We hope therefore, to give these students an early insight into what a theory is.

Once the students are back in the classroom they will be provided with map-making and data base tools that will allow them to record their measurements and observations.

By sending students out into the field and then having them record their measurements back in the classroom we hope to decrease the kind of behavioral problems which the teachers of these bright,[9] energetic students are noticing. That is, we plan not only to burn up the students' healthy energy but also to reinvest it in the technology we are providing in the classroom. Our hope is that this investment of energy will make these technological skills personal and desirable to the students.

**Exercise 2: Creating an Explanatory Theory and a Plan of Practice** Recall that our second goal in teaching scientific inquiry was to allow students to have an effect on their environment. In this context that means building a plan of practice for the skills which are needed in each of the pick up games. In order to derive a plan of practice Mike will need to derive an explanatory theory from his previously built descriptive theory. This can be done by systematically asking what underlies or causes each element in the descriptive theory (Figure 1, top).

After generating the explanatory theory (Figure 1, center) and entering it into the appropriate data base, the student can generate a practice plan (Figure 1, bottom). This is done by systematically asking how to affect each element in the explanatory theory. Our intention here is to have students gain an understanding of the relationship between a model and the world. That is, by using the explanatory theory to create a plan which will have an effect on their lives the students will come to see that these abstractions can bear on reality.

### Future work on Scientific Inquiry

Some parts of the scientific inquiry process are well understood. For example, explanatory theories lacking in detail need to have possible mechanisms proposed and tested. It is not too difficult to envision the exercises that would teach these scientific inquiry skills. However, much of the scientific inquiry process is not understood. Here we plan to look at the great empirical scientists and ask what phenomena grabbed their attention and how did they notice them? We will then look for the analogous phenomena for our students and attempt to engage their curiosity.

## Summary

We have described three projects involving educational technology. Our goal in each of these projects was to create educational situations which grew out of the students' interests so that the students' would feel good both about their lives and their intellectual abilities.

[9] Students are admitted to Boston Latin School through a competitive city-wide test.

## References


Brockner, J. and Rubin, J. Entrapment in Escalating Conflicts NY: Springer-Verlag. 1985.

Fisher, R. and Uri, W. Getting to Yes. NY: Penguin. 1981.

Fisher, R. and Brown, S. Getting Together. NY: Penguin. 1988.

Hodges, M. Sasnett, R. and Ackerman, M. A construction set for multi-media applications. *IEEE: Transactions in Software Engineering.* January, 1989.

Hodges, M. Sasnett, R. and Harward, V. J. *Unix Review.* Vol. 8, 2.

Laurel, B. Interface Agents. In The Art of HCI Design. (B. Laurel, ed.) NY: Addison-Wesley. 1990.

Oren, Salomon, Kreitman and Don, In The Art of HCI Design. (B. Laurel, ed.) NY: Addison-Wesley. 1990.

Raiffa, H. The Art and Science of Negotiation. Cambridge: Harvard Press. 1982.

Susskind, L. and Cruikshank, J. Breaking the Impasse. Basic Books: NY. 1987.

Lai, K.Y., Malone, T., Yu, K. Object lens: A spread-sheet for cooperative work. *ACM Trans. on Office Information Systems* October, 1988.

Malone, T., Grant, K. Lai, K. Rao, R. and Rosenblitt, D. Information lens: An intelligent system for information sharing and coordination. In Tech. Support for Work Group collaboration. M.H. Olson, ed. Hillsdale NJ:Erlbaum 1989.

# The VCR Tutor: Evaluating Instructional Effectiveness

**Mary A. Mark** **Jim E. Greer**

ARIES Laboratory
Department of Computational Science
University of Saskatchewan
Saskatoon, Canada S7N 0W0
greer@cs.usask.ca

## Abstract

People use a wide variety of devices. Operation of a device can usually be described in terms of knowledge of specific procedural sequences. However, execution of procedures may also depend upon knowledge of the device, its behaviour, and the relationships between device features and device actions. A video cassette recorder (VCR) is one commonly used device. Programming a VCR to automatically record a chosen television program is an example of a device manipulation task. In designing a device tutor, it is relevant to ask how instruction about device operation should be designed, and to ask whether knowledge engineering for a device tutor should focus on procedural knowledge or involve factual and referential knowledge as well. Four versions of a tutoring system for the VCR device and programming task have been implemented, incorporating different tutorial approaches using different types of knowledge. The effectiveness of these versions has been examined experimentally. Subjects who used the knowledgeable tutoring version learned to program a VCR simulation using fewer steps and with fewer errors and error types than subjects who used a prompting version of the tutor.

## Introduction

Operation of a device involves explicit procedural knowledge, knowledge of sequences of actions that can be consciously described and physically executed. Execution of such procedures may also depend upon referential and factual knowledge about the appearance, behaviour, and operation of the device. Different perspectives about the knowledge involved in operation of a device lead to significantly different instructional strategies for a device tutor. If a pure procedural view of device use is correct, then prompting or correctness feedback systems may be a sufficient to teach. If a knowledgeable perspective is correct, then purely procedural approaches to instruction are likely to be less effective than knowledgeable instruction.

A device and task have been chosen as suitable for research in this area. The device is a video cassette recorder (VCR). VCRs are commonly used, but sufficiently complex that many people have difficulty operating them. Setting a VCR to automatically record a selected program is one of many procedural tasks involving VCRs. This task has been chosen as representative of tasks in many domains which involve manipulation of devices.

Alternate versions of a tutor can be used as a basis for empirical studies to evaluate instructional effectiveness (Mark, 1990). Four versions of a VCR device tutor have been implemented. Each version utilizes a different instructional approach. An experimental study has been carried out, comparing the effectiveness of the four prototypes. Results are discussed in terms of their relevance to theories of knowledge and instruction.

## Theories of Knowledge

Knowledge can be categorized into types, based upon the degree of conscious awareness of knowledge and the purposes for which knowledge appears to be used. Knowledge of symbols and meanings can be called *referential* knowledge, verbal information (Gagné, 1974), or semantic knowledge (Mayer, 1987; McGraw & Harbison-Briggs, 1989). *Factual* knowledge is specifiable knowledge about objects and relationships between objects within the world. Mayer (1987) includes factual knowledge within his category of semantic knowledge, while McGraw and Harbison-Briggs (1989) classify it as part of a larger category of declarative knowledge, knowledge which can be readily described. *Procedural* knowledge of how to do things can be categorized as explicit or implicit. *Explicit* procedural knowledge can be consciously described in terms of an algorithm, rules, or procedures, and used to guide performance (Mayer, 1987). This has also been referred to as intellectual skill (Gagné, 1974) and declarative knowledge (McGraw & Harbison-Briggs, 1989). *Implicit* procedural knowledge is knowledge of how to do something that cannot be easily described verbally. Mayer (1987) refers to this as skills. McGraw & Harbison-Briggs (1989) reserve the term procedural knowledge to indicate implicit procedural knowledge. *Metacognitive* knowledge (Borkowski & Cavanaugh, 1979), also referred to as metaknowledge (McGraw & Harbison-Briggs, 1989), strategic knowledge (Mayer, 1987), or learning strategies (O'Neil, 1978), is increasingly believed to be important in the conscious monitoring of human information processing behaviour (Baron & Sternberg, 1987).

The knowledge type most obviously relevant to operation of devices is procedural knowledge. One viewpoint is that to achieve a goal such as the automatic recording of a program by a VCR, a person only needs to know a correct sequence of steps which can be executed to achieve that goal. Such sequences can be described explicitly. Once learned, they can be repeated in exactly the same way whenever the same goal is desired. One of several possible procedures for programming a VCR to record a specific television program is outlined in Figure 1. This sequence has been adapted from the documentation of a particular brand and model of VCR. Each step can be described in terms of simple manipulations of device features, such as pressing buttons and changing switches. The only conditions necessary for execution of this sequence are that the VCR and television are connected and that the VCR clock is set to the current time.

1 Power switch -> on
2 Open subpanel door
3 Clock-timer switch -> start
4 Select program number (press Prog Nbr button)
5 Select the channel you want to record (press Channel)
6 Select the day of the week (press Day button)
7 Set the recording start time (press Hour, 10Minute, Minute)
8 Clock-timer switch -> end
9 Set the recording end time (press Hour, 10Minute, Minute)
10 Clock-timer switch -> set
11 PROG REC switch -> on
12 Close subpanel door
13 Insert video tape into VCR

**Figure 1**: Procedural steps in programming a VCR

According to another view, more complex knowledge is necessary to operate devices. Norman (1988) discusses the operation of devices such as VCRs in terms of conceptual models which humans may use to relate visible device features to underlying device actions that may or may not be readily visible. The referential and factual knowledge contained in such models enables a user to identify significant features of a device and to choose appropriate actions to manipulate that device. Cognitive models of the world and of objects may be extremely important when the mapping between visible features and operations is hidden or poorly represented, when the outcomes of manipulations are not readily visible, or when there is sufficient variation in the world to make the ability to identify relevant features essential.

According to Norman's view, the procedural description of Figure 1 assumes referential and factual knowledge on the part of the VCR user. Referential knowledge is involved in the labelling of features. The VCR tutor displays similarly shaped and labelled buttons, for example Program, Rec and Prog Rec. Referential knowledge can be important in identifying and distinguishing between these buttons.

Factual knowledge about features can be described in terms of affordances, constraints, and mappings. Affordances of manipulatable features describe how features can be used. Buttons can be pressed, movable switches can be moved between settings, and doors can be opened and closed. Constraints describe how features cannot be used. Buttons cannot be twisted. Mappings describe the relationship of manipulations to system changes. Such knowledge is important in enabling the user to manipulate the device and interpret the results of such manipulations, and in assisting the user to identify similar and differing features of machines. For example, two types of VCR buttons may appear similar and have different effects when they are manipulated. A binary button will have alternating effects on alternating button presses. The Power button works in this way, switching the power off and on. Another type of button can be repeatedly pressed, producing the same effect each time. The Prog Nbr button increments a value each time it is pressed.

Other factual knowledge relates to the meaning and purpose of the different steps, and their relationships to one another. Such knowledge can help the learner to organize and monitor the sequence of steps which is being executed, to remember what is being done at any given point in the sequence, to remember what to do next, to correct errors when correctable errors are made, and to restart the sequence when errors cannot be corrected. It can also enable the user to identify and execute variations of a procedural sequence which will achieve a desired goal. Relating Clock-Timer settings to manipulations of the Hour, 10Minute, Minute, Day and Channel buttons can be important in organizing and carrying out the programming task taught by the VCR tutor.

Metacognitive skills related to learning and memory may affect cognitive tasks such as memorizing and monitoring execution of the procedural sequence. Metacognition may also influence users' strategies for using the tutor and responding to the behaviour of the device and the tutoring system. Without being specific to the device or task, metacognitive knowledge may indirectly influence learning about device operation.

## Implementation

Four versions of the VCR tutoring system have been implemented on a Macintosh computer in object-oriented Allegro Common Lisp. Versions of the tutor differ in instructional approach. The level of understanding considered sufficient for knowledge communication is one criterion for categorizing instructional approaches. *Rote* learning is oriented towards memorization. It presents information that is to be repeated or mimicked without any other understanding on the part of the learner. In contrast, *knowledgeable* learning approaches emphasize the importance of understanding underlying qualities and relationships between information. Knowledgeable learning is similar in meaning to "instruction that supports cognitive mediation" (Winne, 1985). It indicates the explicit presentation of factual information to the learner. Another way to categorize instructional approaches is to look at the techniques that are used to teach and the processes in which students are expected to engage while learning (Gagné, 1985). *Prompting* explicitly tells or shows the learner what to do, but gives no further information. *Practice and feedback* is an instructional technique in which students are presented with tasks to execute and given feedback on their attempts to execute them. Such feedback may or may not be knowledgeable.

The four versions of the VCR tutoring system are identical except in their instructional approach. The versions, named Mark-I through Mark-IV, use the instructional approaches of *prompting with a fixed programming procedure* (Mark-I), *practice and feedback on correctness with a fixed procedure* (Mark-II), *practice and feedback on correctness with a flexible programming procedure* (Mark-III), and *practice with knowledgeable feedback* (Mark-IV). The fixed procedure for programming the VCR is shown in Figure 1. In reality there are many ways to achieve the desired final setting for recording a TV program. Mark-III and Mark-IV accept any valid procedural sequence which results in the desired settings. Mark-II and-III provide feedback only on correctness or incorrectness of each step, while Mark-IV provides knowledgeable advice and attempts to diagnose and correct deeper misconceptions that the learner might have about the device.

The four versions of the VCR tutor are identical except in instructional approach. The tutor architecture conforms to the intelligent tutoring system (ITS) model proposed by McCalla and Greer (1990). The core of the VCR tutor is a device simulation consisting of a manipulatable interface which resembles a VCR (Figure 2), and a device automaton whose settings and states describe the underlying structure and behaviour of a VCR. The VCR interface forms a major part of the communications component of the VCR tutor. It is the focus of the learner's interaction with the system. Through manipulation of interface features, the learner affects the states and settings of the underlying device automaton. The control component of the VCR tutor system collects events from the interface and notifies other components of those incoming events. As the learner interacts with the VCR simulator, events are reported to the device automaton of the simulation, and to the instructional components. The instructional components embody knowledge about the task to be taught and the instructional approach to be used. Events reported from the VCR interface are the observable learner behaviour upon which instructional behaviour of the VCR tutor is based. The instructional components control the selection of varying types of feedback, to guide the learner's performance in accordance with the tutoring approach. The communication component manipulates the VCR interface and presents textual information in accordance with the selected feedback. In addition to this dynamic feedback, subjects have voluntary menu access to a summary of procedural steps for the task (similar to Figure 1).

The tutoring approach underlying Mark-I, the first version of the system, is *prompting*. Gagné (1985) suggests that the repeated presentation of prompts for steps in a procedural sequence can support the memorization and subsequent use of that sequence. This view of learning assumes that any prerequisite knowledge for carrying out the steps of the action sequence is already available to the learner. Since only information about the correct operations to carry out is presented to the learner, this approach is unlikely to lead to generalization or discrimination.

Mark-I restricts a learners' possible actions so that only one predetermined correct step can be carried out at any time. This is done by manipulating the VCR tutor interface, disabling all displayed images except for the one the user is to manipulate next. The correct next step is visually apparent, remaining in normal video while disabled images are greyed. Only the desired event can occur. The teaching component of this prompting version of the tutor requires knowledge of a single correct sequence of manipulations which lead to a particular set of device conditions. Teaching knowledge is represented as a task automaton, whose states represent steps in the procedural sequence of Figure 1. Each state contains information about its expected event and related interface restrictions, and about device conditions relating to state transitions. A state may have several possible transitions with varying destinations, depending on conditions of the device automaton following execution of the correct step. A knowledge model of a single correct procedural task is sufficient to direct the tutorial component of the VCR tutor when prompting. Knowledge of the learner's most recent

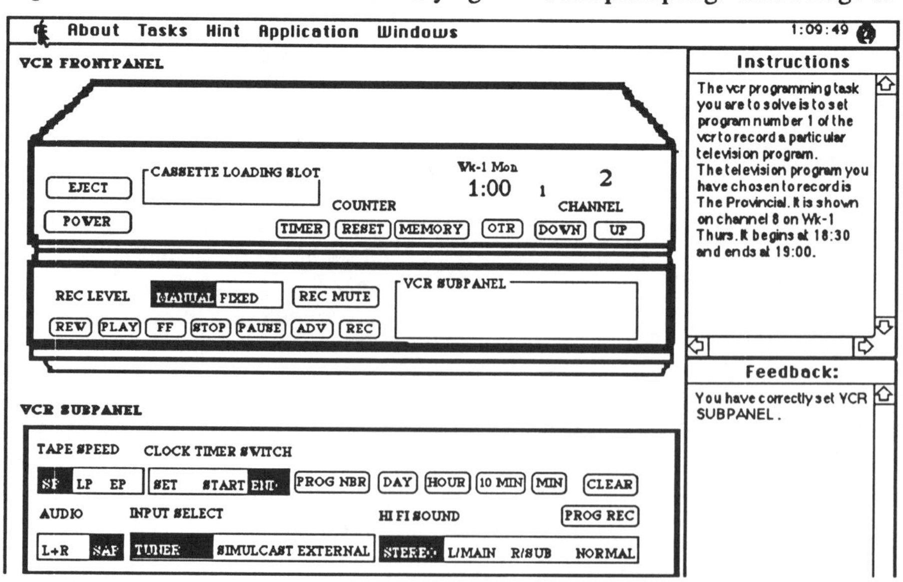


**Figure 2:** Sample VCR tutor display

action enables the task automaton to respond appropriately at any point in the task, so student modelling and diagnosis components are not required by Mark-I.

One approach to practice and feedback presents immediate feedback about the correctness or incorrectness of the last step of the procedural sequence executed by the student. Instead of being prompted for each correct step, the learner is allowed to try a variety of appropriate or inappropriate operations while manipulating the device. In response, the learner is informed of the correctness or incorrectness of those actions. This is similar to the rigid model tracing approach used in early versions of Anderson's LISP Tutor (Anderson & Reiser, 1985). Like prompting, this is a learning approach in which knowledge of a sequence of actions is believed to become proceduralized through practice. Correctness feedback is considered sufficient because it is assumed that necessary prerequisite knowledge is either already available to the learner, or discoverable through trial and error. Given that the learner can carry out a variety of actions and determine their correctness, this tutoring approach may foster discrimination and generalization. However, presentation of cases may not be varied or purposeful enough to indicate important similarities and differences characteristic of the domain.

Mark-II uses this type of rote learning approach while restricting acceptable user behaviour to a *single correct procedural sequence*, comparable to that allowed by the prompting approach (Mark-I). Mark-II presents only those visual cues which would be available from the actual device. All displays can be manipulated by the learner, who is given immediate textual feedback about the correctness or incorrectness of these actions relative to the fixed procedural sequence. Feedback indicates only that an action is or is not acceptable, without indicating why. If the learner makes a slip, such as going past the correct value when changing a setting, feedback indicates the mistake and instructs the learner to correct the slip. If the learner makes an error, such as pressing a button outside the correct sequence, the simulation retracts (undoes) this incorrect action before allowing the learner to continue. In this way, attention is directed to the required procedural sequence at the same time that realistic visual cues are presented.

Knowledge requirements for this single procedure correctness feedback version of the tutor are similar to those for prompting. An explicit knowledge model of the procedural task is the basis of the teaching component. The task automaton is extended by adding transitions to states to deal with incorrect events. Since the learner is not allowed to alter underlying device conditions through incorrect actions (results of such events are retracted), the set of states is not extended. The instructional component is different in that states send textual messages about the correctness of events to the communications component of the tutor when transitions occur, rather than triggering modifications of the simulation displays. Local knowledge remains sufficient to determine whether a subject's current action is correct or incorrect, and to identify the correct next step at a given time. An individual student model is not necessary.

Mark-III is similar to Mark-II in that it provides *practice with correctness feedback*. However, Mark-III expands the range of acceptable learner behaviour, allowing the learner to carry out *any potentially correct procedural sequence*. This approach resembles Anderson's later, more flexible, model-tracing tutors. In Mark-III a task automaton is used to model knowledge of the procedural task and to monitor learner behaviour, but it is extended to accept a variety of procedures for accomplishing the task. This requires alteration of both the set of states and the state transitions in the task automaton. Less importance is placed upon the sequence in which events occur, and more importance upon the examination of conditions affecting the device automaton. This local knowledge is used to determine whether the learner's current action is correct or incorrect. Feedback is only concerned with whether each individual step is consistent with reaching the desired final state, so an individual student model is not necessary.

Mark-IV is also built around *practice and feedback*. It attempts to promote *knowledgeable* rather than rote learning. Knowledgeable learning makes few assumptions about the learner's understanding of the domain. It attempts to identify missing knowledge and explicitly remediate misconceptions. Learner behaviour must not be restricted in ways that invalidate the device model to be taught. A student should be able to carry out any procedural sequence compatible with the VCR device and the ultimate solution of the task. A knowledgeable tutor must therefore deal with a wide range of learner behaviour. This includes slips, errors related to faults in learning, problems in reasoning and misconceptions resulting from incorrect models of the device (Norman, 1988), as well as correct behaviour. A knowledgeable feedback system should identify such behaviour and present relevant feedback. Instruction and feedback are concerned with conceptual models of the device as well as learner behaviour when manipulating the device. Generalization of knowledge and discrimination of relevant similarities and differences can be fostered explicitly.

Mark-IV is the most knowledge-intensive of the four versions of the VCR tutor. Extensive conceptual models of the task to be performed and of the learners' behaviour in manipulating the device are required. An extended task automaton accepts a variety of possible correct procedures. Local knowledge of the learner's last action is insufficient. Student modelling and individualized error diagnosis are introduced. A student modelling component tracks student behaviours and identifies characteristic behaviour patterns indicating problems. A catalog of known conceptual and procedural bugs is used to diagnose slips and errors related to procedural sequence, errors in reasoning, and underlying misconceptions reflecting incorrect models of the device. Slips, errors and misconceptions for the bug catalog were identified through detailed observation of ten individuals who used the Mark-III VCR tutor prototype. Mark-IV's knowledgeable feedback is targeted at correcting a learner's conceptual model of the device, as well as correcting the procedural sequence that the learner is attempting to execute.

A number of hypotheses about the effectiveness of different methods of tutoring follow from the theories of knowledge and instruction previously outlined. If a purely procedural view of the VCR programming domain is correct, then the only knowledge that is relevant in the VCR programming domain is procedural knowledge. Therefore, instructional methods such as prompting and practice with correctness feedback should be sufficient for instruction. It should be possible to learn and apply the desired procedures without difficulty. Knowledgeable feedback, while it may also lead to learning of the procedure, should not promote more effective application of procedural knowledge. In contrast, if a knowledgeable view of the VCR programming domain is correct, prompting may have limited value, since needed generalization and discrimination will be unlikely to occur. For practice with correctness feedback, the degree to which generalization and discrimination will occur is uncertain. Those who have used the prompting and correctness feedback versions of the tutor may learn how to carry out the procedural sequence, but they may also have difficulty in effectively applying that knowledge. If a knowledgeable view of the VCR programming domain is correct, knowledgeable instruction should teach students the VCR programming task, and enable learners to apply their procedural knowledge more effectively and avoid mistakes.

In the absence of a priori evidence about differences between the tutoring versions, the following null hypotheses were formed. It was hypothesized that no differences in VCR programming achievement would be observed when comparing groups of subjects who used the four versions of the VCR tutor. It was also hypothesized that no differences in time to complete the training regimen would be observed when comparing groups of subjects who used the four versions of the VCR tutor.

## Experimental Methods

The subjects for this experiment were 80 undergraduate students at the University of Saskatchewan, who volunteered for the study. Most were registered in undergraduate computing classes. Subjects did not have previous experience in programming a VCR. They were told that the study was intended to examine the effectiveness of a VCR tutor, but were not given information about the instructional approaches involved, or the existence of differing versions of the tutor. Each subject was randomly assigned to one of the four instructional versions of the VCR tutor for training. Following testing, the test results of 4 subjects were found to be more than two standard deviations away from their group means. Those subjects were removed from the analysis, reducing the sample to 76.

Initial preparation for all subjects in the experiment was identical. Each subject was asked to read a written description of the domain, the programming procedure, and the use of the tutor. Details of the method of instruction were not given. The programming procedure described (similar to Figure 1) was procedural rather than conceptual. The programming procedure was also available to the student during training from a HINT menu. Training consisted of three trials. In each trial the subject chose a target TV program from an automated TV Guide, and then programmed the VCR simulation to record that TV program, receiving guidance from the tutor in accordance with the instructional strategy for that version.

Following training, each subject used a testing version of the tutor. The testing version of the tutor presented a VCR simulation similar to the Mark-III training version, accepting all possible correct variants of the procedure for accomplishing the task. The testing version analysed subject behaviour by tracking progress and identifying bugs (as in Mark-IV) and thus automatically collected data during the test. However, no feedback was given to the subjects, and the HINT menu describing the procedure was not available. During the testing period, subjects were asked to complete a single trial, involving the programming of the VCR simulation to record an assigned TV program.

Three measures of achievement were selected as indicators of learning in the VCR programming domain. The first achievement measure was the number of steps required to correctly program the VCR, without assistance, during the test. This measure captures notions of both accuracy and efficiency in carrying out the procedure. The second achievement measure was the total number of bugs diagnosed during the test. The third measure was the number of categories in the bug catalog in which errors were identified. These measures summarize the conceptual and procedural errors made by the student. The training time was the number of seconds that students spent in carrying out three practice runs at programming their version of the VCR tutor to record particular TV programs.

## Results

Table 1 shows means and standard deviations for each group on each dependent variable. It also displays the F-ratio and significance levels from a one-way analysis of variance for each experimental measure. Significant differences among groups were found on all experimental measures. Post-hoc analysis using the Scheffe Method showed that subjects trained with the knowledgeable version of the tutor (Mark-IV) used fewer steps ($\alpha <=.01$).and had fewer total errors ($\alpha <=.05$).and types of errors ($\alpha <=.01$) than students trained with the prompting version (Mark-I). Students trained with a tutor providing practice with correctness feedback on all correct paths (Mark-III) also had significantly fewer types of errors than students who used the prompting version (Mark-I) ($\alpha <=.05$).. Post-hoc analysis of training time showed that training time with Mark-III was significantly greater than with Mark-I ($\alpha <=.05$).. There was no significant difference in training time between Mark-IV and the other versions.

Certain trends in the data, although not significant with such a small sample, merit some discussion. It is interesting to note that the two model tracing versions, Mark-II and Mark-III, led to very similar mean achievement but surprisingly different standard deviations. It is also interesting that Mark-IV with its knowledgeable feedback seems to promote marginally better achievement than any other version of the tutor.

| | Mark-I (n=19) | Mark-II (n=18) | Mark-III (n=20) | Mark-IV (n=19) | F | P |
|---|---|---|---|---|---|---|
| Steps: | | | | | | |
| Mean | 107.7 | 89.6 | 69.4 | 54.6 | 5.26 | 0.002 |
| St.D. | 55.0 | 61.5 | 29.8 | 16.4 | | |
| Total Errors: | | | | | | |
| Mean | 28.5 | 21.0 | 10.0 | 6.1 | 4.06 | 0.01 |
| St.D. | 33.5 | 27.4 | 9.7 | 6.9 | | |
| Distinct Error Types: | | | | | | |
| Mean | 6.1 | 4.6 | 3.0 | 2.2 | 6.63 | 0.001 |
| St.D. | 3.7 | 3.9 | 1.8 | 1.6 | | |
| Training Time: | | | | | | |
| Mean | 1053 | 1360 | 1716 | 1330 | 3.37 | 0.02 |
| St.D. | 357 | 728 | 890 | 501 | | |

Table 1: ANOVA Results for Experimental Measures

## Discussion

Given that all students, regardless of tutoring approach, were able to correctly program the VCR simulation during testing, it appears that any version of the VCR tutor is sufficient to train potential VCR users. However, knowledgeable feedback promoted more effective acquisition of procedural knowledge than did prompting, at no additional cost in training time. These results support the idea that referential and factual knowledge are important in this seemingly procedural domain. The results lend credence to the claim that knowledgeable tutoring provides better advice to the learner and thus leads to improved achievement on procedural tasks. This is consistent with Norman's theories of knowledge and device use (Norman, 1988).

Only the Mark-IV version of the VCR tutor can be considered an intelligent tutoring system since it incorporates both individualized student modelling and knowledgeable feedback. This research suggests that knowledgeable tutoring is preferable in this procedural domain and shows that it can be provided by an intelligent tutoring system utilizing individualized student modelling. There is, however, a cost-benefit issue to be considered. Constructing the knowledgeable version of the tutor required cognitive modelling to develop a reasonably comprehensive and accurate catalog of bugs. Another important concern was to ensure that the knowledgeable version of the tutor would continue to run in real time, so that the computationally expensive diagnosis and student modelling would not render the simulation unusable. Additional modelling, programming, and performance enhancement costs are the norm in developing intelligent tutoring systems: "the cost of the I in the ITS". This research suggests that the investment can be worthwhile.

This research makes a number of contributions relevant to research about instructional systems.

- It surveys principles of knowledge and instruction that relate to device operation.
- It applies these principles to the design and implementation of various versions of a particular device tutor for a particular task.
- The resulting tutor versions are evaluated to investigate relationships between knowledge and instructional strategy in this domain.
- By comparing different versions of a single tutor, a tightly controlled experiment was assured.
- This evaluation methodology is applicable to the assessment of tutoring systems and the examination of theoretical assumptions underlying their design.
- The results support the claim that knowledgeable tutoring can lead to greater achievement in procedural domains, and so may be a preferable method of training.
- The research demonstrates that intelligent tutoring systems can provide effective instruction.

## Acknowledgements

We would like to thank the Natural Sciences and Engineering Research Council of Canada. We would also like to thank the many students who acted as subjects.

## References

Anderson, J. R., Reiser, B.J. The LISP Tutor. *BYTE*, 1985, 10(4), 159-175.

Baron, J. B., Sternberg, R. J. *Teaching Thinking Skills: Theory and Practice*. New York: W.H. Freeman & Co., 1987.

Borkowski, J.G., Cavanaugh, J.C. Metacognition and intelligence theory. In M.P. Friedman, J.P. Das, and N. O'Connor (Eds.) *Intelligence and Learning*. New York: Plenum Press, 1979.

Gagné, E.D. *The Cognitive Psychology of School Learning*. Toronto: Little, Brown, and Company, 1985.

Gagné, R. M. *Essentials of Learning for Instruction*. Hinsdale, Illinois, U.S.A: Dryden Press, 1974.

Mark, M.A. *Evaluation of intelligent tutoring systems*. (ARIES Laboratory Research Report 90-2). Saskatoon, Saskatchewan: University of Saskatchewan, Department of Computational Science, 1990.

Mayer, R.E. *Educational Psychology: A Cognitive Approach*. Toronto: Little, Brown and Company, 1987.

McCalla, G.I., Greer, J.E. The practical use of artificial intelligence in automated tutoring. In C.K. Leung, B.S. Randhawa (Eds.) *Understanding Literacy and Cognition: Theory, Research and Application*. New York: Plenum Publishing, 1990.

McGraw, K.L., Harbison-Briggs, K. *Knowledge Acquisition: Principles and Guidelines*. Englewood Cliffs, New Jersey: Prentice-Hall, 1989.

Norman, D.A. *The Psychology of Everyday Things*. New York: Doubleday, 1988.

O'Neil, H.F. Jr. (Ed.) *Learning Strategies*. New York: Academic Press, 1978.

Winne, P.H. Steps toward promoting cognitive achievements. *The Elementary School Journal*, 1985, 85(5), 673-693.

# ASK TOM: An Experimental Interface for Video Case Libraries

**Roger C. Schank, William Ferguson, Lawrence Birnbaum, Jorn Barger, and Mathew Greising**

Northwestern University
The Institute for the Learning Sciences and
Department of Electrical Engineering and Computer Science
Evanston, Illinois

## Abstract

ASK TOM represents a new approach to structuring access to a newly emergent kind of knowledge base, the video case library. It is based on two premises: First, that cases and stories in the form of video clips can provide much of the viscerality and memorability that is lacking from textual forms of presentation; and second, that AI theories, specifically the approaches to memory organization derived from work in case-based reasoning, can provide the structure that is essential to achieving the shared context that makes communication possible. The aspect of AI research that is crucial in providing this structure does not concern algorithms. Rather, it is the *content* of domains and tasks that is paramount.

## 1 Introduction

Wouldn't it be nice if, whenever you had a question about something, an expert were available for you to talk with? An expert can answer your tough questions. Perhaps more importantly, an expert can help you to *ask* the right questions, and can fill in exactly the background you need to understand the answers. An expert can help you organize the information you are getting in a way that will be useful in solving your specific problem. And an expert can deliver all of this information in a way that is natural, memorable, and if you are lucky, perhaps even entertaining: An expert can tell you *stories.*

Unfortunately, expertise is expensive, and so access to experts is severely limited.[1] It therefore makes sense to explore the possibility of constructing computer-based systems that would enable users to gain access to expertise in a way that is as natural, memorable, and efficient as a real conversation with an expert. In the ideal case, such a system would be the ultimate artificial intelligence program: It would be able to understand spoken natural language, it would be a true expert—armed with powerful reasoning capabilities, a diverse stock of useful models of the domain of expertise, and a large memory of specific cases indexed and organized intelligently—and it would be able to communicate its answers to you in both verbal and visual form. In the meantime, though, can anything be done with existing technology to produce something that is close enough to the cognitive equivalent of a conversation with an expert to be useful?

We think the answer is yes. We have developed an experimental hypermedia system, called ASK TOM, for accessing a videodisc database of short (one- to four-minute) film clips culled from interviews with an expert consultant in the trust banking industry. Our goal was to construct a system that would enable naive users to gain access to clips containing information pertinent to their questions, and to structure their interaction with the system in a way that realized the most important benefits of a conversation with a human expert. In particular, we wanted the system to provide users with a coherent model of the domain, one that would enable them to assimilate the answers to their questions in such a way that they could then use those answers to solve problems.

In order to achieve this coherence, the video clips in the system's data base are indexed via nodes in an associative network that is organized to reflect both the objects and relationships in the domain and the task of solving problems in that domain. The network includes both hierarchical links for quickly pinpointing potentially relevant neighborhoods within the data base—what we call *zooming*—and a dense set of links between cases which facilitate *browsing* by providing context and managing follow-up questions. A user interacts with ASK TOM by navigating through a hypertext-like network whose links and nodes are labelled with useful questions and

[1] This is, of course, the rationale behind the construction of expert problem-solving systems.

important features, and viewing video clips that seem likely to provide answers to his or her questions.[2]

## 2 Using content-based AI to structure human-computer interaction

The basic interface principle underlying the construction of ASK TOM is that even a rudimentary AI theory of a task or a domain provides constraints that can be used to structure human-computer interactions in a way that may be more meaningful to the human user.[3] To the extent that an interaction so constrained corresponds to the user's understanding of the task and the domain, the context provided by the system at any given point will agree with the context that the user naturally expects at that point. This shared context provides the fundamental basis for meaningful and useful communication. Thus, the information provided by the system will correspond better to the information needed by the user; the choices made available to him or her will be more relevant, more useful, and easier to resolve; and the information conveyed in the interaction will be associated with the appropriate context in the user's memory.

It is important to make clear that ASK TOM contains only a trivial algorithm: The key to its functionality and the substance of our results is the *content* of the network contained in the system. The theoretical basis underlying the system's organization is provided by *case-based reasoning* (CBR), a paradigm for understanding and problem-solving which takes memory to be the foundation of both human and artificial cognition. CBR is based on the idea that solving a problem or understanding a situation involves retrieving similar prior experiences, and adapting them to provide guidance in the current situation (see, e.g., Schank, 1982; Riesbeck and Schank, 1989). It posits that a significant portion of human expertise derives from specific experiences, and is best represented in that form in memory to provide guidance in future problem-solving situations.

Two major teaching principles follow from such an account of human cognition: First, if expertise is best represented in terms of specific experiences, then a natural way to facilitate the assimilation of new information by humans is to present that information in the form of cases or *stories*. The great majority of video clips in ASK TOM's database therefore consist of an expert telling stories. Second, because the usefulness of information depends crucially on the ability to retrieve it when appropriate, such cases are best presented in a context that enables the listener to determine where they should be indexed in his or her memory, and how they should be connected to other relevant cases. In ASK TOM, this context is provided by the network that enables access to particular video clips, and by the clips that the user has viewed along the way.

The two basic processes that underlie case-based reasoning are *indexing*, or finding appropriate cases in memory, and *adaptation*, or altering the specific information in those cases to fit the current circumstances. In ASK TOM, we rely heavily on the user to carry out these processes. In particular, the human user guides the traversal of the indexing network in directions that seem likely to yield information relevant to his or her needs. The system itself knows virtually nothing about the contents of the clips other than the links that connect them to each other and to the network as a whole. However, the user's choices are constrained by a model of link types that relate cases to each other in a way that is intended to reflect their potential uses in planning and problem-solving, such as background, results, warnings, and so on.

A key issue in the approach we are pursuing here is whether it will be possible to anticipate all, or at least most, of the questions that novices would ask of a human expert in a domain such as trust bank consulting. This becomes especially critical in the case of follow-up questions, where the absence of the desired information would be extremely irritating to users, and where the design of the system interacts with physical constraints of screen layout to limit the number of choices that can be offered. Two observations provide some basis for optimism on this issue. First, experts, when consulted, repeatedly offer their favorite "war stories" as carriers of advice in response to a wide variety of different questions (Schank, 1990). Second, the paths people follow in conversations are, in many respects, predictable given the appropriate level of analysis (Schank, 1978). Not surprisingly, in conversations about planning and problem-solving tasks, the level at which we can make such predictions is precisely the level at which it is possible to represent abstract planning knowledge, as exemplified by Sussman's (1975) *planning critics* or Schank's (1982) *thematic*

---

[2] A good overview of hypertext systems and concepts may be found in Conklin (1987).

[3] This view is gaining increasing acceptance in AI and cognitive science; see, for example, Winograd and Flores (1986), or the announcement for the 1990 AAAI Spring Symposium on "Knowledge-Based Human-Computer Communication" organized by Gerhard Fischer, Clayton Lewis, Jim Miller, and Elaine Rich.

*organization points.*[4] It is precisely at such a level of analysis, therefore, that we strove to label and link stories in ASK TOM's indexing network.

The design of ASK TOM was also greatly influenced by the *lack* of a usual constraint, in that we did not need rigor, specificity, or rigid consistency in our representations. They were not, after all, going to be manipulated by any natural language generator or parser, nor by any automated reasoning system. The mind of the user could supply the knowledge of language and of the world needed to weed out minor inconsistencies and, more importantly, to specify vague denotations appropriately. In fact, in this system *vagueness* is a resource we used liberally to enable our representation system to function in many contexts and across many categories in what is, after all, a large and rich domain.

## 3 Indexing system description

ASK TOM consists of a case-base of video clips about consulting in the trust banking industry, together with an indexing system as described in general terms above. The scenario for using ASK TOM is that the user, having a question in mind, will navigate through the network to an appropriate clip and then play it. He or she may then go back to the beginning and find the answer to further, unrelated questions, or else move on through the network to locate and play clips which answer followup questions.

To describe the structure of the indexing scheme which lies at the heart of ASK TOM, three indexing levels need to be distinguished. They are (1) the *big-picture models*, (2) the *themes*, and (3) the *low-level links*. We will describe these three levels and their function by walking through a prototypical interaction with the system.

### 3.1 Big-picture models

When ASK TOM is first started, the user is presented with a diagram showing the major agents in the world of trust bank consulting—e.g., the trust department of a bank, a consulting firm, the individual consultant, the trust industry as a whole, and so on—with links between them showing how they interact. This is the top-level big-picture model (see figure 1). Clicking on the nodes or links in this diagram causes the screen to display either a more specific big-picture diagram, describing, e.g., the temporal layout of a typical consulting engagement at a trust bank, or else a screen showing a set of themes and the clips that they organize (discussed below). The set of all big-picture models in the system forms a conceptual map of the domain, organized around its most important agents, relationships, and processes. These big-picture models mediate the first (and often the second) choices made by the user during the initial process of zooming in on a region of specific interest. When the user makes choices on these diagrams, he or she is implicitly categorizing the information desired in terms of the model denoted by the diagram. Thus, the user is implicitly preparing to incorporate that information into his or her own memory in a way that reflects its proper relationship to the rest of the model.

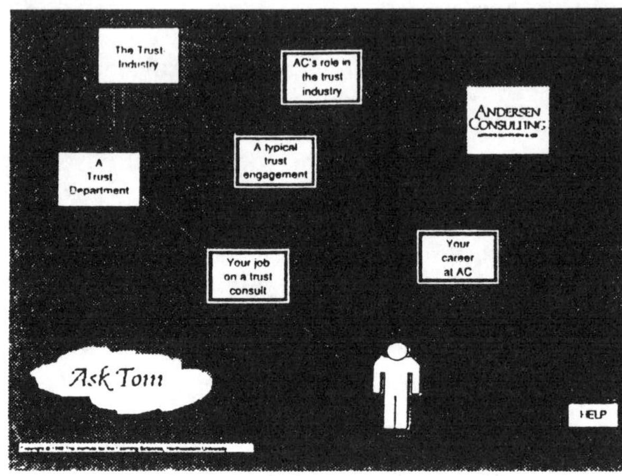


Figure 1: The opening screen of the ASK TOM system.

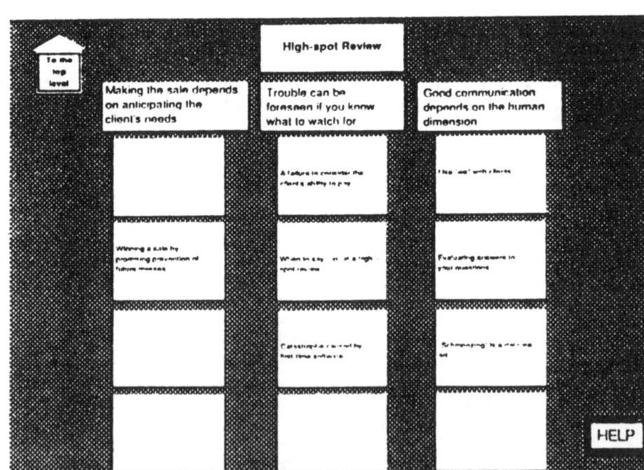


Figure 2: The theme screen under the High-spot review button.

### 3.2 Themes

After one or two choices, the user will arrive at a theme screen (Figure 2). This screen displays the clips relevant to the previously selected elements of the big-picture models, partitioned by the themes which they embody. A theme, as the term is used in

[4] See also Sacerdoti (1977), Wilensky (1983), and Hammond (1989), among others.

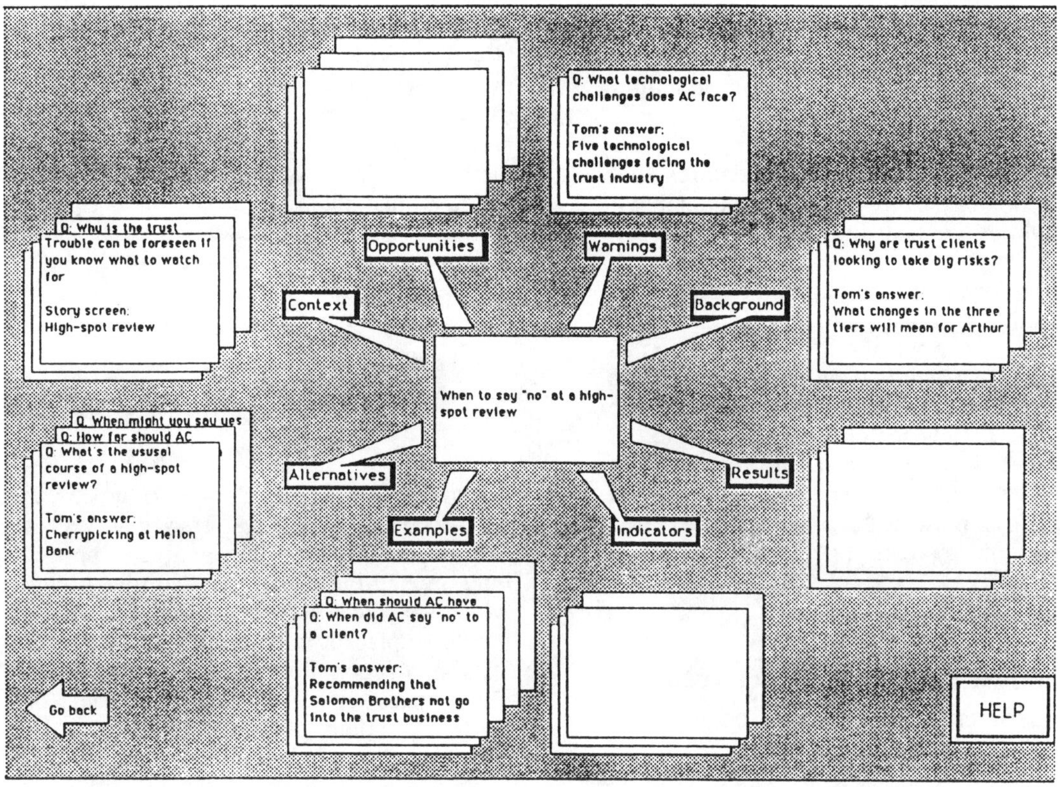


Figure 4: The clip screen reachable from the theme screen above.

ASK TOM, is a lesson or point relevant to the big-picture node under which it is found, e.g., "Correct trouble at its source." Themes thus serve as a bridge between domain models and case knowledge. They are not, however, specializations of, nor are they components of, the big-picture nodes under which they are indexed. Rather, they serve to categorize the contents of the system's clips according to how the information they convey is useful in problem-solving. The selection of a clip from a theme screen concludes the zooming process. (Themes are discussed in more detail below.)

### 3.3 The low-level links

When a clip is selected from a theme screen the system displays a clip screen (see Figure 4). In the center is the selected clip, which can now be clicked on and run, at which point it will appear on the screen. From the central clip node, the low-level links radiate toward eight "stacks" of clips which are the neighbors of the center clip. Each clip node displayed on the screen is tagged with a question that relates it to the information presented by the central clip, along with a one sentence summary of its contents. The eight link types are the same throughout the network of clips, and always appear in each clip screen. As discussed earlier, they are intended to relate the system's cases to each other in ways that are relevant to the tasks of planning and problem-solving, and thus to suggest, and offer answers to, useful follow-up questions. The eight link types are:

**Background**: This links a clip to antecedent information about that clip, e.g., the history behind the story it contains, or the causes of the situation it describes.

**Results**: This provides a link between a clip and the consequences of, or events subsequent to, the situation it describes.

**Examples**: This link relates a clip to more detailed examples of the situation it describes.

**Context**: This links a clip to a context which enables its point to be appreciated.

**Warnings**: This links a clip to negative prospects that might arise within the context of the situation it describes.

**Opportunities**: This link relates a clip to positive prospects that might arise within the context of the situation it describes.

**Alternatives:** This provides a link between the situation described in a clip and alternative or contrasting situations that might arise in a similar context.

**Indicators:** This links a clip to features that facilitate detection or diagnosis of the situation that it describes within the contexts in which it might arise.

After the user has watched the clip in the center—or, perhaps, instead of watching that clip—he or she may move to one of the peripheral clips by clicking on its clip node. When this happens, the center clip absorbs its other neighbors and moves to a point on the periphery; then the chosen clip node moves to the center, and a new clip screen unfolds from it, displaying its neighbors. This process can be repeated as the user browses through the clip network, travelling over links representing questions arising in each context.

## 4 Themes as a bridge between domain and case knowledge

The ASK TOM project, because of its scope, prodded us to look at some large-scale representation problems that have often been lost in the details in existing knowledge-intensive systems. In designing the system's indexing network, we worked simultaneously from two directions: We looked at the domain information in the clips from the top down, attempting to pick out the major agents, techniques, situations, and relationships, and to construct out of these elements some high-level models that would span the entire domain. Pragmatically, at least, we succeeded at this endeavor. Simultaneously, we worked bottom-up, starting from the individual clips and the relationships that existed among them. Again we had some success.

At this point, however, we had a problem, insofar as the nodes in our high level models could not be linked, in any meaningful way, to the clips in the low level network. A fairly crude partitioning was possible—we could, for instance, pick out twenty or so clips and say that these belonged under the node denoting the relationship between a trust department and an individual at a consulting firm. But we still needed a well-motivated approach to structuring this set itself.

Our approach to this problem, as described above, was to categorize these moderately-sized sets of clips into *themes* or *lessons*. We were steered toward this answer, in part, by our investigations of story-based models of human memory. As we examined the clip database, we realized that certain lessons cropped up repeatedly, lessons like "Know enough not to ask stupid questions," "The trust banking industry is changing in ways that we must be prepared for," "Correct trouble at its source," and so on. Often, these lessons were explicitly and consciously drawn as morals by the expert telling the story. As described earlier, these themes provided a division of the clips which was independent of, and often very different than, that provided by the big-picture nodes. By placing these themes as an additional indexing level between the big-picture models and the clips themselves, we were able to introduce enough distinctions to easily distinguish and locate most clips.

## 5 The need for an indexing methodology

By far the biggest impediment to the construction of the first version of ASK TOM has been the sheer magnitude of the data-analysis phase. After interviewing the expert, we had nine hours of video which we edited into about 200 clips. Each clip then had to be placed properly into the indexing network. This was accomplished in two stages: First, each clip was associated with all appropriate big-picture nodes[5]; then the clip's neighbors had to be determined, and the appropriate low level links established. We looked for neighbors in four ways: First, clips indexed under the same big-picture node were likely candidates. Second, clips that were near each other in the original interview were also liable to be linked by whatever thread led from one to the other in the interview. Third, we sorted all the clips into piles based on "subject" (a term which we never tried to define), and generated possible neighbors that way. Finally, we worked backwards: For each clip, we considered each of the eight low-level links in turn, and probed our own memories of the corpus, i.e., we tried to get reminded of other clips that would attach to the current one via the given link. This last method proved especially fruitful, providing us with some confidence that we had indeed hit upon a semantically significant set of low level links.

This whole process was arduous, especially in that it required several members of the project team to internalize the contents of the entire corpus. The construction of larger video data bases along the lines of ASK TOM will necessitate the development of alternative approaches, because it will ultimately be impossible for any one person to be familiar with, let alone internalize, all of the cases in such a system.

---

[5]Each clip was stored in an average of two places, but there was considerable variance, with some clips appearing four or five times. Each location corresponds roughly to a distinct high level question that the clip was perceived to answer.

In addition, if the construction of these indexing networks is to become commercially viable on a large scale, the process must be codified and realized in a set of software tools so that it can be carried out by people who do not have extensive training in developing representations for artificial intelligence systems.[6] The successful codification and implementation of such a process in software tools would provide evidence for the transportability and generality of the indexing vocabulary used in ASK TOM.

## 6 Progress on the implementation

We have completed a prototype of the ASK TOM system containing a twenty-six clip database, which is a limit imposed by using a single video disc. We have also substantially completed the analysis of our entire corpus of approximately 200 clips. This process has gone far enough that we feel confident that the big-picture models we have developed will be adequate, and that the same eight low-level links can be used across the entire corpus without alteration. The hardware consists of a Pioneer LC-V330 video jukebox connected to a Macintosh II workstation. The software is written in Supercard, using Voyager VideoStack video drivers.

## 7 Conclusion

ASK TOM represents a new approach to structuring access to a newly emergent kind of knowledge base, the video case library. It is based on two premises: First, that cases and stories in the form of video clips can provide much of the viscerality and memorability that is lacking from textual forms of presentation; and second, that AI theories, specifically the approaches to memory organization derived from work in case-based reasoning, can provide the structure that is essential to achieving the shared context that makes communication possible. The aspect of AI research that is crucial in providing this structure does not concern algorithms. Rather, it is the *content* of domains and tasks that is paramount.

Our eventual goal is to situate systems like ASK TOM in teaching environments which also include a simulator and a tutorial planner. The scenario we envision is one in which the planner sets up the simulator to produce an interesting situation that the student must address. In response to the developing situation and the student's actions, we want to be able to provide, when appropriate, expert advice in the form of video clips of stories which bear on the problem at hand. We view this as a crucial step towards the development of a new generation of computer-based educational software systems, which takes as its fundamental principle the idea that learners must be actively engaged in realistic problem-solving tasks, embedded in realistic situations, in order to properly assimilate lessons in a way that will enable their application in appropriate situations in the future.

**Acknowledgments:**

We thank Ryan Lanham, Chris Riesbeck, Diane Schwartz (who did the graphics design for the system), and especially our anonymous experts at Andersen Consulting for their help. The Institute for the Learning Sciences was established in 1989 with the support of Andersen Consulting, part of The Arthur Andersen Worldwide Organization.

## References

Conklin, E. 1987. Hypertext: An introduction and survey. *IEEE Computer*, vol. 2, pp. 17-41.

Hammond, K. 1989. *Case-Based Planning: Viewing Planning as a Memory Task.* Academic Press, San Diego.

Riesbeck, C., and Schank, R. (eds.) 1989. *Inside Case-Based Reasoning.* Erlbaum, Hillsdale, NJ.

Schank, R. 1977. Rules and topics in conversation. *Cognitive Science*, vol. 1, pp. 421-441.

Schank, R. 1982. *Dynamic Memory: A Theory of Reminding and Learning in Computers and People.* Cambridge University Press, Cambridge, England.

Schank, R. 1990. *Tell Me a Story: A New Look at Real and Artificial Memory.* Charles Scribner's Sons, New York.

Sacerdoti, E. 1977. *A Structure for Plans and Behavior.* American Elsevier, New York.

Sussman, G. 1975. *A Computer Model of Skill Acquisition.* American Elsevier, New York.

Wilensky, R. 1982. *Planning and Understanding.* Addison-Wesley, Reading, MA.

Winograd, T., and Flores, F. 1986. *Understanding Computers and Cognition: A New Foundation for Design.* Ablex, Norwood, NJ.

[6]In some ways, our problem here is similar to the knowledge-acquisition problems faced in the construction of expert problem-solving systems.

# Language Differences in Face-to-Face and Keyboard-to-Keyboard Tutoring Sessions*

**Jai Seu** and **Ru-Charn Chang** and **Jun Li** and **Martha Evens**
Computer Science Department
Illinois Institute of Technology
Chicago, IL 60616
csevens@harpo.iit.edu

**Joel Michael** and **Allen Rovick**
Department of Physiology
Rush Medical College
Chicago, IL 60612

## Abstract

Face-to-face and keyboard-to-keyboard tutoring sessions were recorded and analyzed as a first step in building a machine tutor that can understand and generate natural language dialogue. There were striking differences between these modes of interaction. The number of turns in an hour-long session dropped and so did the length of the sentences, although the number of sentences per turn stayed roughly the same. Students contributed 37% of the words in the face-to-face sessions; their share dropped to 25% in the keyboard sessions. Sentence structure is simpler in the keyboard sessions. The tutors ask more questions in the keyboard sessions; they explain this as a deliberate strategy to keep the dialogue going. The tutors also use a much wider range of expressions of acknowledgement in the keyboard sessions in a deliberate attempt to communicate verbally the kind of encouragement that is often expressed nonverbally in a face-to-face situation.

## Introduction

This paper compares aspects of the language used in face-to-face and keyboard-to-keyboard tutoring sessions. These sessions were recorded as part of a project to design and develop a machine tutor that can generate tutorial dialogues. The tutor is intended to help first-year medical students understand the complex negative feedback system that regulates blood pressure (Kim et al. 1989).

The first step in our research was the recording of human tutoring sessions and the analysis of the transcripts of those sessions with regard to both language and content. The language analysis involves the study of vocabulary, syntax, and discourse for use in the understanding of ill-formed input (Lee et al. 1990; Seu et al. 1991) and the generation of responses (Zhang et al. 1990). The work described here formed part of the first phase in the language analysis.

Studies of keyboard-to-keyboard sessions in the literature had led us to expect that the language in keyboard sessions would be terser and more elliptical (Thompson 1980; Grosz 1978). It was. Many of our ideas about what phenomena to look for come from Grosz or from Grishman & Kittredge (1986). From Grosz (1978) we expected frequent use of "OK" with multiple meanings. In fact, it was our most frequent single word sentence and it seems to mean "I understand," "You are on the right track," and "Get on with it" in about equal proportions. The use of "you" also parallels that described by Grosz, but "I" is used more often as an abbreviation for "increase" than as a personal pronoun. When it appears as a personal pronoun, it is often used in pseudo-questions by students, such as "I don't understand about contractility."

Our tutors responded to Fox' (1988b) description of repair situations in tutoring by trying very hard to avoid ambiguity, even though they do not seem to be as verbose as the tutors described in (Fox 1990).

## Capturing the Tutoring Sessions

The data used in this study comes from two face-to-face sessions and eight keyboard-to-keyboard sessions. Each session lasted approximately one hour. The tutor in every case was either Joel Michael or Allen Rovick, faculty members in the Department of Physiology at Rush Medical College. The students were first-year medical students at Rush, who were learning this material as part of their first-year curriculum. The students volunteered to take part in our research and were aware that we were recording these sessions.

Each tutoring session begins with a description of a procedure that perturbs the system; then the student is asked to predict how certain important parameters will change. The procedures used in these examples were

*This work was supported by the Cognitive Science Program, Office of Naval Research under Grant No. N00014-89-J-1952, Grant Authority Identification Number NR4422554, to Illinois Institute of Technology. The content does not reflect the position or policy of the government and no official endorsement should be inferred.

haemorrhage (loss of 0.5 or more liters of blood) and a malfunctioning pacemaker. The student is asked to predict what will happen to these parameters at three different stages: at the direct response stage immediately after the perturbation takes effect, at the reflex response stage when the baroreceptor reflex provides negative feedback, and in the steady state.

The data on face-to-face tutoring sessions comes from two hour-long sessions both conducted by Joel Michael with two different first-year medical students at the time when they were studying this material in class.

The keyboard-to-keyboard sessions were captured by a program we wrote for the purpose. To run this program (called STS) requires two PC's or compatibles and a Hayes smart modem or an equivalent and a telephone line from one site to another. With this program we are able to capture 60-90 turns an hour; the limitation seems to be the typing abilities of the student and the tutor and not the equipment. We would be happy to give the STS program and the data to anyone who would like to use it.

In the first version of the program we had a formal turn-release mechanism, but it was possible for both tutor and student to type at the same time. This resulted in several unintelligible passages of intermingled characters. In the sessions (K1–K8) where this occurred we edited these passages, where we could reconstruct the words, into sequential turns. The interruption is marked TI or SI, depending on whether the tutor or the student performed the interruption.

In the second version of the program we made it impossible for the tutor and the student to type at the same time. Barbara Fox' work suggests that these interruptions play an important part in face-to-face sessions, so in the third version of the program we buffer what the interrupter types and display it on the recipient's screen. It appears in the transcript after the turn that it interrupted. We have now captured 24 keyboard-to-keyboard tutoring sessions, each one hour in length.

An important first step in our research was identifying each item in the data. Jai Seu wrote a program that numbers each sentence or fragment in a way that identifies the session, who is typing, the turn, and the sentence within the turn. For example, the number K24-ST-057-01 indicates that this sentence comes from Keyboard session 24, that the student is typing, that this is turn 57, and that it is the first sentence in that turn. The second component may be TU (tutor), TI (tutor interrupting the student), ST (student), or SI (student interrupting the tutor). The data described here comes from the first eight keyboard sessions.

## Differences in Session Length and Sentence Length

Both the face-to-face and keyboard-to-keyboard sessions were roughly an hour long, but many more words were exchanged in an hour of face-to-face tutoring than in a keyboard session. This is consistent with results reported by Gamoran (1991) using somewhat different experimental conditions. She found three times as much tutoring in face-to-face sessions as in remote sessions.

The two face-to-face sessions contained a total of 6,410 words divided into 489 turns. The eight keyboard-to-keyboard sessions had a total of 7,303 words divided into 699 turns. Thus, the number of words per turn dropped from an average of 13.1 in the face-to-face sessions to 10.1 in the keyboard-to-keyboard sessions.

The students were affected by this drop much more strongly than the tutors, with the result that the tutors controlled the floor much more in the keyboard sessions than in the face-to-face sessions. In the face-to-face sessions the tutor spoke 4,094 (62.8%) of the 6,410 words and the student spoke 2,386 (37.2%). In the keyboard sessions the tutor typed 5,452 (74.7%) of the 7,303 words while the student typed only 1,851 (25.3%).

The sentence length dropped for both the tutor and the student. The average sentence length in the face-to- face sessions was six and a half words; this average dropped to five words in the keyboard sessions. The distribution in sentence lengths is shown in more detail in Tables 1, 2, and 3. The number of sentences per turn stayed roughly constant, however. There were 1.86 sentences per turn in the face-to-face sessions and 1.74 sentences per turn in the keyboard sessions.

**Table 1. Average Sentence Length**

| | Face-to-Face | Keyboard-to-Keyboard |
|---|---|---|
| Tutor | 8 | 6 |
| Student | 5 | 4 |
| Combined | 6.5 | 5 |

**Table 2. Distribution of Sentence Lengths**

| | Face-to-Face | | Keyboard-to-Keyboard | |
|---|---|---|---|---|
| | N | Perc | N | Perc |
| **Tutor** | | | | |
| 1 to 3 words | 187 | 37.6% | 266 | 32.7% |
| 4 to 10 words | 155 | 31.1% | 355 | 43.6% |
| above 10 words | 155 | 31.1% | 193 | 23.7% |
| total | 497 | | 814 | |
| **Student** | | | | |
| 1 to 3 words | 186 | 45.1% | 273 | 67.7% |
| 4 to 10 words | 155 | 37.6% | 69 | 17.1% |
| above 10 words | 71 | 17.2% | 61 | 15.1% |
| total | 412 | | 403 | |
| **Combined** | | | | |
| 1 to 3 words | 373 | 41.0% | 539 | 44.2% |
| 4 to 10 words | 310 | 34.1% | 424 | 34.8% |
| above 10 words | 226 | 24.9% | 254 | 20.9% |
| total | 909 | | 1217 | |

**Table 3. Average Number of Sentences in a Given Length Category per Session**

| | Face-to-Face | Keyboard-to-Keyboard |
|---|---|---|
| **Tutor** | | |
| 1 to 3 words | 93.5 | 33.2 |
| 4 to 10 words | 77.5 | 44.3 |
| above 10 words | 77.5 | 24.1 |
| total | 244.5 | 102.0 |
| **Student** | | |
| 1 to 3 words | 93.0 | 34.1 |
| 4 to 10 words | 77.5 | 8.6 |
| above 10 words | 35.5 | 7.6 |
| total | 206.0 | 50.0 |
| **Combined** | | |
| 1 to 3 words | 186.5 | 67.3 |
| 4 to 10 words | 155.0 | 53.0 |
| above 10 words | 113.0 | 31.8 |
| total | 455.0 | 602.0 |

## Differences in Sentence Complexity

With the difference in sentence length came a marked drop in sentence complexity. We counted the numbers of subordinate clauses. The frequency of sentential complements and relative clauses was cut in half. The frequency of adverbial clauses (mostly "if" and "when" clauses) also decreased but not as markedly. We assume that this happened because adverbial clauses are used to express the underlying causal relationships that are the focus of the tutoring effort. The numbers of subordinate clauses and the percentages obtained by dividing these numbers by the number of sentences are shown in Table 4.

**Table 4. Subordinate Clauses**

| | Face-to-Face | | Keyboard-to-Keyboard | |
|---|---|---|---|---|
| Clause Type | #(Cl) | Perc | #(Cl) | Perc |
| Sent.Comp. | 96 | 10.6% | 57 | 4.7% |
| Adverbial | 81 | 8.9% | 86 | 7.1% |
| Relative | 48 | 5.3% | 32 | 2.6% |
| #(Sent) | 909 | | 1217 | |

Although the tutor use of subordinate clauses stayed roughly the same over the different keyboard sessions, the students varied a great deal, so we have given the data for individual students in Table 5.

**Table 5. Student Use of Subordinate Clauses in the Keyboard Sessions**

(S1-S8 are the eight different students involved. Clause Types (CT) are Sentential Complement (SC) Adverbial (Ad), and Relative (Rl).)

| CT | S1 | S2 | S3 | S4 | S5 | S6 | S7 | S8 | Stu Tot | Tut Tot |
|---|---|---|---|---|---|---|---|---|---|---|
| SC | 1 | 4 | 0 | 3 | 3 | 0 | 0 | 0 | 11 | 46 |
| Ad | 0 | 3 | 1 | 12 | 8 | 0 | 0 | 0 | 24 | 72 |
| Rl | 0 | 0 | 0 | 2 | 0 | 0 | 0 | 0 | 2 | 30 |

## Differences in Vocabulary

We expected to find a wider range of vocabulary items in the face-to-face sessions. We were wrong. There are 740 types (different individual words) in the face-to-face sessions and 915 types in the keyboard sessions. There were 6410 tokens (total words) in the face-to-face sessions, so the type-to-token ratio is 740/6410 = 0.115. There were 7303 tokens in the keyboard sessions; the keyboard type-to-token ratio is 915/7303 = 0.125. Details of the type distribution are shown in Table 6.

**Table 6. Distribution of Word Types**

| | Face-to-Face | | Keyboard-to-Keyboard | |
|---|---|---|---|---|
| Total Used | 740 | | 915 | |
| Used Once | 432 | 58.4% | 573 | 62.6% |
| More Often | 308 | 41.6% | 342 | 37.4% |
| Tutor | 611 | 62.4% | 770 | 66.6% |
| Student | 368 | 37.6% | 386 | 33.4% |
| Both | 239 | | 241 | |

Our reading and conversations with Bozena Thompson had led the computer scientists among us to expect phatics (particularly expressions of rage and frustration). Our colleagues at Rush predicted that no phatics would appear and they were largely right. The only phatics that we found were "Well" and "Oh." They predict that Rush students will not express rage and frustration when communicating with our machine tutor either. We are eager to find out. The distributiion of "Well" and "Oh" is shown in Table 7. It does not seem surprising that these expressions appear more often in spoken language than in typed language. We cannot explain why the tutors did not stop using them as completely as the students did.

**Table 7. Distribution of Phatics**

| | Face-to-Face | | Keyboard-to-Keyboard | |
|---|---|---|---|---|
| | Well | Oh | Well | Oh |
| Tutor | 18 | 2 | 14 | 2 |
| Student | 49 | 22 | 3 | 0 |

## Differences in Discourse Strategy

We have only begun to explore this dimension but there are certain fairly obvious differences between the two media here. The tutors asked more questions in the keyboard sessions (217) than in the face-to-face sessions (129). They explain this as a conscious attempt to get students to respond.

The tutors actively sought verbal means to express approval and enthusiasm in the keyboard-to-keyboard sessions, information that they express nonverbally in face-to-face seessions. Thus, the words "absolutely," "super," and "excellent" appear four or more times in the keyboard sessions and not at all in the face-to-face sessions. They also softened disconfirmations (cf. Fox 1988a) with "not exactly," or "not quite."

Our tutors responded to the description of repair situations in (Fox 1988b) with a determination to avoid such situations. Examination of repair situations arising in three talk-aloud sessions not included in this study suggested that the most common source of repair problems was "How" questions. They have made a deliberate effort to replace these questions with more specific questions beginning "By what mechanism ... " or "In which direction ... " For whatever reason we have too few repair situations in our current sessions to make any interesting generalizations about them.

The tutors expressed a feeling of loss of control after the keyboard sessions. We suspect that this feeling may explain why the tutors uttered only ten imperative sentences out of the 909 sentences in the face-to-face sessions (3 in one and 7 in the other), but typed in 90 imperative sentences (out of 1217) in the keyboard-to-keyboard sessions. The students produced no imperative sentences in either medium.

## Future Research

Clearly we need to collect and analyze more data. We have already recorded six more face-to-face tutoring sessions, but they have not yet been transcribed. We are making plans to record eight more. We actually have twenty-four keyboard-to-keyboard sessions in three groups of eight; we used the first eight in these analyses because the protocols involved are closer to those used in the face-to-face sessions.

The numbers of subordinate clauses of different types were recorded by hand. These numbers are probably not quite accurate. We are constantly expanding our parser to handle ill-formed input and we hope in a few months to be able to parse all the sentences produced by both students and tutors. At that point we will be able to undertake a more sophisticated analysis of syntactic differences and carry it out more accurately.

We have already written software to count differences in vocabulary between the student and the tutor. We want to examine other differences in language usage. We are particularly interested in studying how student language changes over a series of sessions and in finding out whether it becomes more like that used by the tutor. We need to record a series of sessions with the same student carrying out a number of different procedures. This project should also contribute to the design of the student modeler.

Our tutors reacted to (Fox 1988b and 1990) with a deliberate and largely successful attempt to avoid repair situations in the keyboard-to-keyboard sessions. We are still trying to figure out how they accomplished this, because we do not know how to make the machine tutor recognize the need for repair or perform it.

## Summary

We have recorded human tutoring sessions carried out both face-to-face and keyboard-to-keyboard in the process of gathering information about language, tutoring strategies and modeling techniques for a machine tutor. In carrying out the language analysis we discovered some major differences in language use.

Predictably, the number of turns and the sentence length dropped, but the students were affected much more than the tutors, so that the percentage of words contributed by the student dropped from 37% to 25%. As sentences grew shorter there were fewer subordinate clauses of all kinds, but there were still a fairly large number of adverbial clauses that express the basic causal relationships in the underlying domain.

The language used by the tutors shows evidence of several strategies for overcoming the lack of nonverbal communication. The tutors ask more questions and use more imperatives in an apparent effort to prompt student response. They also provide more verbal encouragement.

## References


Fox, B. 1988a. Cognitive and Interactional Aspects of Correction in Tutoring. Technical Report #88–2, University of Colorado.

Fox, B. 1988b. Repair as a Factor in Interface Design. Technical Report #88–4, University of Colorado.

Fox, B. 1990. Final Report to the Office of Naval Research. University of Colorado.

Gamoran, M. 1991. The Tutor Behind the Curtain: Face-to-Face and Remote Tutoring. Paper presented at the Annual Meeting of the American Educational Research Association.

Grishman, R., and Kittredge, R. 1986. *Analyzing Language in Restricted Domains: Sublanguage Description and Processing.* Hillsdale, New Jersey: Lawrence Erlbaum Associates.

Grosz, B. 1978. Discourse Analysis. In Donald Walker ed., *Understanding Spoken Language*, New York, New York: Elsevier-North Holland. 138–176.

Kim, N., Evens, M., Michael, J.A., Rovick, A.A. 1989. CIRCSIM–TUTOR: An Intelligent Tutoring System for Circulatory Physiology. In Mauer, H. ed., *Computer Assisted Learning, Proceedings of the International Conference on Computer Assisted Learning.* 254–266. Berlin: Springer–Verlag.

Lee, Y.H., Evens, M., Michael, J.A., Rovick, A.A, 1990. IFIHS: Ill-Formed Input Handling System. In Proceedings of Second Midwest Artificial Intelligence and Cognitive Science Society Conference. 93–97. Carbondale, Illinois: Computer Science Dept., Southern Illinois Univ.

Seu, J.H., Evens, M., Michael, J.A., Rovick, A.A. 1991. Understanding Ill-Formed Input to an Intelligent Tutoring System in an LFG Framework. In Proceedings of the Third Midwest Artificial Intelligence and Cognitive Science Society Conference. 36–40. Carbondale, Illinois: Computer Science Dept., Southern Illinois Univ.

Thompson, B.H. 1980. Linguistic Analysis of Natural Language Communication with Computers. In Proceedings of COLING 80, Tokyo, Japan. np.

Zhang, Y., Evens, M., Michael, J.A., and Rovick, A.A. 1990. Extending a Knowledge Base to Support Explanations. In Proceedings of the Third IEEE Conference on Computer-Based Medical Systems. 259–266. :IEEE.

# Decision Making Connectionist Networks

**Yves Chauvin**
Psychology Department
Stanford University
Stanford, CA 94305
chauvin@psych.stanford.edu

## Abstract

A connectionist architecture is proposed and provides representations for probabilities and utilities, the basic elements of formal decision making theories. The outputs of standard feedforward feature-extraction networks then become inputs to this decision making network. A formalism shows how the gradient of expected utility can be back-propagated through the decision making network "down" to the feature extraction network. The formalism can be adapted to algorithms which optimize total or minimum expected utility. Utilities can be either given or estimated during learning. When utility estimation and decision making behavior adapt simultaneously, learning dynamics show properties contrasted to "puzzling" observations made in experimental situations with human subjects. The results illustrate the interest in computational properties emerging out of the integration of elements of decision making formalisms and connectionist learning models.

## Introduction

By tradition, formal theories of decision making build concepts of utility and probability out of a set of normative axioms characterizing "rational" behavior (e.g., Slovic, Lichtenstein & Fischhoff, 1988). They usually do not emphasize adaptive behavior, notions of representation, or information processing properties. Inspired by connectionist "principles", this paper suggests a general connectionist architecture for decision making which can be integrated with pre-existing feature extraction feedforward networks. The architecture provides representations for probabilities and utilities. Learning integrates elements of decision making formalisms in a computational framework and then operates through the resulting representations.

Using this architecture, a variant of the back-propagation learning algorithm can be derived which propagates the gradient of a response expected utility through the decision network "down" to the feature extraction network. This principle generates two algorithms, one maximizing total expected utility, the other maximizing the minimum expected utility computed over the set of possible responses. The goal of this paper is then to examine learning dynamics and convergence properties of these algorithms.

These properties are described through simulations of two-response environments. In particular, when utilities are estimated during behavior adaptation, simulations show interesting results which are then compared to experimental data observed with human subjects. The results illustrate the relevance of computational models of decision making.

## Probability and Utility Representations

### Probability Representation

Consider a multinomial distribution where a given input pattern has to be classified in 1 of $n$ categories. We would like the $n$-unit output layer of a network to represent the set of classification probabilities associated with this multinomial distribution. If we call $p_i$ the activations of these units, we need to have $p_i \geq 0$ and $\sum_i^n p_i = 1$. One way to satisfy these constraints is to make sure that unit activations of a previous layer are positive and to normalize these activations at the output layer. If we write $s_i$ the "net input" to the first layer, positivity can be obtained with exponential *e-units*: $e_i = e^{\beta s_i}$. Normalization is then obtained with *p-units*:

$$p_i = \frac{e_i}{\sum_j^n e_j} = \frac{e^{\beta s_i}}{\sum_j e^{\beta s_j}} \tag{1}$$

This corresponds to the familiar ratio rule used in numerous decision tasks (e.g., Busemeyer and Myung, 1989; Luce, 1959; Estes, 1987). The parameter $\beta$ can be seen as a sensitivity parameter. For $\beta = 0, p_i = 1/n$: the response distribution is uniform across categories. When $\beta \rightarrow \infty, p_M \rightarrow 1$ where $M$ corresponds to $s_M = max_i\{s_i\}$: the network response is deterministic and corresponds to the largest net input. A simple computation yields the derivative of the *p-units* with respect to the activations of the *e-units*:

$$\frac{\partial p_i}{\partial e_j} = -\frac{p_i . p_j}{e_j} \quad j \neq i \tag{2}$$

$$\frac{\partial p_i}{\partial e_i} = \frac{(1-p_i).p_i}{e_i}$$

We can then write the derivative of the *p-units* with respect to the net inputs of the *e-units*:

$$\frac{\partial p_i}{\partial s_j} = \beta(\delta_{ij} - p_i).p_j \qquad 1 \le i, j \le n \qquad (3)$$

where $\delta_{ij}$ is the Kronecker symbol ($\delta_{ij} = 1$ for $i = j$, 0 otherwise).

## Subjective Expected Utility

We now consider the activations $p_i$ of the *p-units* correspond to network "propensities" to classify an input pattern in category $i$. We assume these response propensities (or "beliefs" or "subjective probabilities") are computed so as to maximize a given performance index. We consider payoffs are obtained when the network classifies an input pattern in category $j$ and when the environment responds with category $i$. We write $u_{ij}$ (utilities) the perceived values of the terms of the $n.n$ payoff matrix (e.g., utility might typically be a bounded, monotonically increasing function of payoff; utility functions are not examined in this paper). We assume for now the $u_{ij}$ are known. We will see below how they can also be estimated during learning.

The network expected utility for category $i$ can be written as $E[U_i] = \sum_j^n u_{ij}.p_j$. This equation can be implemented in a network by adding a layer of $n$ linear *u-units* with a connectivity matrix between *p-units* and *u-units* corresponding to the utility matrix. Outputs of a feature extraction network can then serve as inputs to the decision making network computing probabilities and expected utilities. The resulting network architecture is presented in Figure 1.

Learning consists in maximizing the expected utility ($EU_t = E[U_t]$) associated with a targetted category $t$. The network learns by adjusting its set of "beliefs" $p_j$ through parameter (weight) adaptation. We can use gradient ascent on $EU_t$ to compute weight changes after each pattern presentation:

$$\Delta w_q = \eta \frac{\partial U_t}{\partial w_q} \qquad (4)$$

Using the chain rule, we get:

$$\frac{\partial E[U_t]}{\partial w_q} = \sum_j^n \frac{\partial E[U_t]}{\partial s_j} . \frac{\partial s_j}{\partial w_q} \qquad (5)$$

The terms $\partial s_j / \partial w_q$ might be computed with the standard back-propagation algorithm (Rumelhart, Hinton & Williams, 1986), considering a given feature-extraction architecture "below" the *e-units*. We then use the chain rule to obtain the back-propagated "error" term $\delta e_j$ associated with the *jth e-unit*:

$$\begin{aligned} \delta e_j &= \frac{\partial E[U_t]}{\partial s_j} = \sum_k^n \frac{\partial E[U_t]}{\partial p_k} . \frac{\partial p_k}{\partial s_j} \\ &= \sum_k^n u_{tk} \beta(\delta_{kj} - p_k).p_j = \beta p_j.(u_{tj} - U_t) \quad (6) \end{aligned}$$

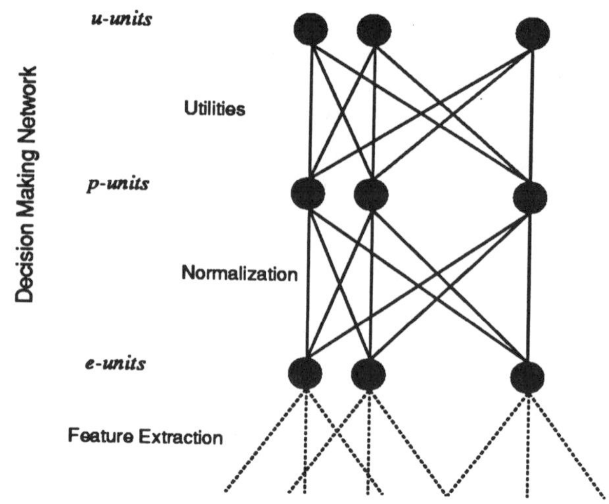


Figure 1: Decision making network architecture: The architecture provides representations for probabilities (*p-units*) and utilities (weights from *p-units* to *u-units*). The decision making network might be connected to a feature extraction network, for example a standard multi-layer back-propagation network.

Note since $\sum_j p_j = 1$, $\sum_j \delta e_j = 0$. Moreover, if $u_{tj} > E[U_t]$, the network "receives more than expected": $p_j.(u_{tj} - U_t) > 0$ and $p_j$ will increase, as it should.

Given a set of $u_{ij}$, the previous equation computes the gradient of expected utility $EU_t$ with respect to the input of the decision making network. This gradient can then be back-propapaged "further down" in hidden layers to extract features relevant to the decision strategy. The parameters (weights) of the feature-extraction network can then be changed by gradient ascent. The goal of this paper is not to investigate how the decision making network influences the properties of the feature extraction network. Rather, we look at properties of the decision network itself, at various decision strategy implementations and at utility estimations.

## Maximizing Total Expected Utility: Fixed Utilities

Considering a set of training patterns, the total expected utility is $EU = \sum_t EU_t$: maximizing the total expected utility consists in maximizing the sum of response expected utilities. If the training sample is known, batch learning can be implemented to compute $\partial EU / \partial w_q$. Alternatively, the well-known Widrow-Hoff procedure can be used to adjust the weights after each pattern presentation.

### Sample Learning Dynamics

We consider the $u_{ij}$ are known *a priori*. Suppose the network is now being trained with a finite data base in which a given input pattern is presented $N$ times and targetted $N_i$ times in each category $i$ ($\sum_i N_i = N, f_i = N_i/N, \sum_i f_i = 1, f_i \neq f_j$ for $i \neq j, f_M = max_i\{f_i\}$). We can then compute the total sample gradient of $EU$ (associated with that given input pattern) with respect to the net input $s_j$:

$$\delta e_j = \frac{\partial EU}{\partial s_j} = \sum_i N_i \frac{\partial E[U_i]}{\partial s_j} \tag{7}$$

Assuming a utility matrix equal to the identity matrix ($u_{ij} = \delta_{ij}$), we get:

$$\begin{aligned}\frac{\partial U_i}{\partial s_j} &= \beta(\delta_{ij} - p_i)p_j \\ \delta e_j &= N\beta p_j(f_j - \sum_i f_i p_i) \\ &= N\beta p_j \sum_{i \neq j} p_i(f_j - f_i) \qquad (8) \\ \delta e_M &= N\beta p_M \sum_{i \neq M} p_i(f_M - f_i)\end{aligned}$$

Equilibria are obtained for $\delta e_j = 0$ for all $j$. But $\delta e_M = 0$ implies $p_M = 0$ or $p_i = 0$ for $i \neq M \Leftrightarrow p_M = 1$. The first equilibrium is unstable; the second one is stable. Therefore the response probability associated with the biggest response frequency of the sample will increase to 1 during training. Using the same arguments, the response probability associated with the smallest response frequency will decrease to 0. All other response probabilities may either decrease or first increase then decrease depending on the initial values of $p_j$. These results may be generalized for any value of $u_{ij}$.

The learning principle involved in this framework is gradient ascent on the total expected utility when utilities are known in advance. It is well-known that resulting so-called Bayesian optimality can be obtained when response probabilities correspond to "pure" decisions ($p_j = 0$ or 1), even in stochastic environments (e.g., DeGroot, 1970).

### Illustration: Two-response Environment

The algorithm was implemented for a simple two-response environment without feature extraction. In this case, the two weights to be learned in the network are simply the biases of the *e-units*. The environmental binomial distribution can be characterized by a single parameter $p^* = p_1^*$, the probability of occurrence of the first response/category. Rewards from the environment are received as a function of the network responses. The previous $EU$-back-propagation algorithm should find the optimal values of the probabilities as a function of the utility matrix.

As expected, simulations show the network always learns how to make pure optimal decisions, independently of the utility matrix and environmental probabilities (except in obvious trivial cases). Convergence speed is essentially dependent on the environmental probabilities, on the range of the $u_{ij}$, on the learning rate $\eta$ and on the sensitivity parameter $\beta$.

Standard Bayesian techniques might first estimate posterior probabilities and then compute expected costs from given a *priori* cost matrices. Input patterns are then classified in categories associated with low expected costs. The network presented above directly estimates the model parameters so as to directly maximize total expected utility. The $EU$-back-propagation might "bypass" the estimation of posterior probabilities to compute response behavior. From a performance point of view, it is quite possible to imagine that the model might actually make better use of its resources (weights) by not "wasting" them in estimating posterior probabilities. Performance characteristics are not investigating further in this paper.

## Maximizing Total Expected Utility: Estimated Utilities

We now assume the $u_{ij}$ have to be estimated during learning. This task can be considered more typical of human learning situations in decision making environments. In this framework, the $EU$-back-propagation algorithm can be used with the $EU$ gradient being back-propagated through currently estimated utilities. The task of the decision network is still to maximize $EU$ through adaptation of the response probabilities $p_j$ given current $u_{ij}$.

### Trial Estimation

Let $R$ (with elements $r_j$) be the current $n$-dimensional response vector generated by the network. The response $R$ is a random variable with a multinomial distribution characterised by the $p_j$. At each trial, we assume the network response $R$ is chosen in function of this distribution and that the environment responds with a "target" vector $R^*$ (with elements $r_i^*$) drawn from the environmental multinomial distribution characterised by a set of probabilities $p_i^*$. We write $u_{ij}$ the current estimates of the utilities and $u_{ij}^*$, the perceived values of the true rewards obtained after each network response $R$ and environment response $R^*$.

Trial learning can be implemented in a network having the set $r_j$ as input units and the set $r_i^*$ as output units. The received utility predicted by the network after each trial is then $U_i = \sum_j u_{ij} r_j$; the truly received utility is $U_i^* = \sum_j u_{ij}^* r_j$. Estimation of the received utilities is performed by gradient descent on $L$, the sum of squares of the differences between predicted and truly received utilities. Considering a single trial, we obtain:

$$\begin{aligned}L &= \sum_i r_i^*(U_i - U_i^*)^2 \\ \Delta u_{ij} &= -\alpha_1 \frac{\partial L}{\partial u_{ij}} = -\alpha(U_i - U_i^*)r_i^*.r_j \qquad (9) \\ E[\Delta u_{ij}] &= -\alpha(u_{ij} - u_{ij}^*)p_i^* p_j\end{aligned}$$

Equilibrium is obtained for $E[\Delta u_{ij}] = 0 \Leftrightarrow u_{ij} = u^*_{ij}$. Speed of convergence will not only depend on the learning rate $\alpha$ but also on the current set of beliefs $p_j$ and on the the environmental probabilities $p^*_i$. After each trial, utilities are evaluated, and the $EU$-back-propagation algorithm is used with the currently estimated set of utilities. Eventually, the estimated utilities will converge to the environmental utilities (for non-zero $p_j$ and $p^*_i$). At this point, response behavior adaptation will become identical to the case where utilities are known *a priori*. Therefore, the network will end up making optimal pure decisions. However, in general, response probabilities being learned simultaneously with utility estimation, the complete learning dynamics might become quite complex. These characteristics are examined in more detail below.

## Average Estimation

With trial learning, $u_{ij}$ is reevaluated after each trial, in function of the network response $R$ and the environment response $R^*$. Because a specific element of the utility matrix has to be estimated after each trial, trial learning makes the implicit assumption that the learner already understands the deterministic property of trial obtained rewards. Trial learning also makes the computational assumption the learner stores and estimates each elements of the utility matrix independently. We now turn to a maybe more natural learning hypothesis where expected utilities are estimated in the average. Utilities are evaluated in function of current response probabilities because the learner does not assume (or trust) the determinism of the utility matrix or does not have the information processing capacity to estimate each element of the utility matrix.

After each trial, the expected utility (in the mathematical sense of expected) is $E[U_i] = \sum_j u_{ij} p_j$. For given $p_j$, it is possible to estimate the new $u_{ij}$ by gradient descent on the squares of the differences between expected utility $E[U_i]$ and the truly received utility $U^*_i$. Considering a single trial, we obtain:

$$
\begin{aligned}
L &= \sum_i r^*_i (E[U_i] - U^*_i)^2 \\
\Delta u_{ij} &= -\alpha_1 \frac{\partial L}{\partial u_{ij}} \\
&= -\alpha (E[U_i] - U^*_i) r^*_i . p_j \\
E[\Delta u_{ij}] &= -\alpha (E[U_i] - E[U^*_i]) p^*_i p_j \qquad (10)
\end{aligned}
$$

Equilibrium is reached when $E[U_i] = E[U^*_i] \Leftrightarrow \sum_j u_{ij} p_j = \sum_j u^*_{ij} p_j$. An obvious solution to this equation is found for $u_{ij} = u^*_{ij}$. Again, in this case, the network will then converge to pure decision optimality. However, the equilibrium condition might lead to other estimations of $u_{ij}$ which might correspond to various response behaviors. For each run and for a given limited learning time, obtained behaviors will depend on the network initial conditions and on the random order of category presentations during training.

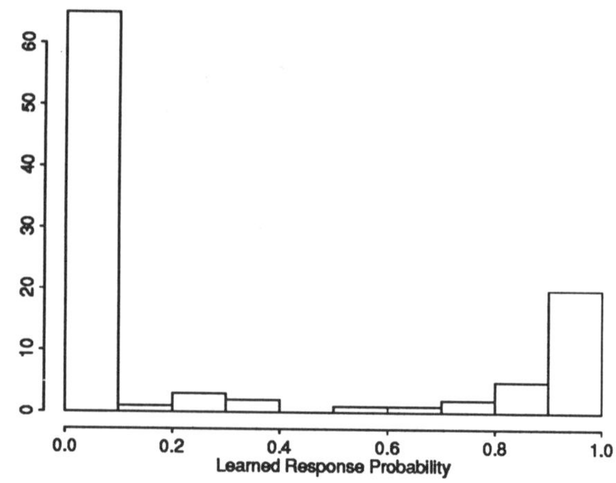


Figure 2: Histogram of learned response probability $p$ after 200 trials for 100 runs. The decision network is trained with the $EU$-back-propagation algorithm using average estimated utilities.

Simulations of the algorithm were then performed in the previous two-response environment with $u^*_{ij} = \delta_{ij}$. The initial response probability parameter $p$ was randomly chosen at each run. The following general results were obtained, that we will contrast to observations made with humans in two-response experiments.

A great majority of runs are stable after a fixed number of trials: for a sufficient number of runs, average performance seems to have reached an asymptotic level. Settlement occurs in two modes: pure decision optimality where $p$ converges to 1 if $p^* > .5$ and to 0 otherwise; a less common "reversed" optimal mode where $p$ converges to 0 if $p^* > .5$ and to 0 otherwise. The mean of the resulting bimodal distribution depends of the values of the parameters, in particular of $u^*_{ij}$, of the initial values for $u_{ij}$ and of the parameters $\alpha$ and $\beta$. Larger learning rates for the estimation of utilities or more accurate initial estimations lead to improved estimations of $u_{ij}$ during learning and to higher proportion of cases converging to optimal behavior. Figure 2 represents the histogram of learned response probability $p$ after 200 trials for 100 runs (varying the initial values of $p$; $\alpha = 1$, $\beta = 4$, initial $u_{ij} = .25\delta_{ij}$, $p^* = .2$). The mean value of the distribution is .27 in this case, roughly matching the environmental $p^*$.

A model predicting "reversed" optimality seems a *priori* to deserve rejection as a potential model of human behavior. However, Luce and Suppes (1963), reviewing a large number of two-response studies, have the following warning:

> The data reported in the literature are always averages for groups of subjects, [...]. Considerable evidence (not much of it published) indicates that these group results can be quite misleading. The

distribution of estimated $p_\infty$ over subjects often seems not to be binomial, and sometimes there is little doubt but that it is bimodal, with the valley of the distribution coinciding roughly with the group mean probability. (Luce & Suppes, 1963, p. 391).

Until now, these experimental individual data still remain a puzzle for theoretical psychologies of learning and decision making. The present approach suggests an explanation for these data in the simultaneous average estimation of rewards and adaptation of response behavior. The mean result (probability matching) is then observed along with the puzzling bimodal distribution of individual responses. Precise simulations of experimental results (e.g., 25 studies examined in Luce et al., 1963) are beyond the scope of this paper.

## Minimax Learning

So far, we have assumed the goal of the decision making network was to maximize the total expected utility over the environmental distribution. We now turn to the other well known framework for decision making: we would like the network to maximize the minimum expected utility over the set of possible decisions (e.g., von Neumann & Morgenstern, 1944).

The architecture of the decision network presented in Section 2 is still valid since it simply provides a representation for probabilities and utilities (usually written as costs in game setups) and a formalism for computing the gradient of a given response expected utility. The difference is now in the definition of optimality (*minimax* for costs, equivalently *maximin* for utilities) and in the resulting learning algorithm.

### Back-propagation for Maximum Minimum Utility

We would now like the *p-units* to compute probabilities maximizing the minimum of expected utilities over a set of possible responses. For any given input pattern, the expected utility associated with each response is $E[U_i] = \sum_j u_{ij}p_j$. Learning then consists in gradient ascent on the minimum expected utility computed over the set of responses:

$$\Delta w_q = \eta \frac{\partial min_i\{E[U_i]\}}{\partial w_q} \tag{11}$$

The expression $min_i\{E[U_i]\}$ is not differentiable (but still continuous) with respect to the weights of the network. The proposed implementation consists in computing $E[U_t]$ for each training pattern and to check if $E[U_t] = min_i\{E[U_i]\}$. If $E[U_t] = min_i\{E[U_i]\}$, the gradient of $E[U_t]$ is computed (Section 2) and weights are changed accordingly. If $E[U_t] \neq min_i\{E[U_i]\}$, weights remain unchanged.

Theoretical minimax probabilities usually correspond to mixed decisions ($p_j \neq 0, 1$) function of the cost (utility) matrix. These probabilities do not depend on environmental distributions. However, if network resources are limited (limited number of weights in the feature-extraction network), the computations of the network minimax probabilities might depend on the training distribution, justifying the relevance of a training sample. Again, the consequences of decision strategies on feature extraction are not examined in this paper.

|  | Net. Resp. 1 | Net. Resp. 2 |
|---|---|---|
| Env. Resp. 1 | 1 | -2 |
| Env. Resp. 2 | -3 | 3 |

Table 1: Cost/utility matrix for the two-response and game situations as a function of environmental and network responses.

### Example Implementation

The minimax-back-propagation procedure was implemented in a two-response situation using the cost/utility matrix shown in Table 1. The theoretical minimax equilibrium is found for $p_1 = p = 5/9$ and $p_2 = 1 - p = 4/9$. Simulations show the algorithm always converges to the optimal minimax probabilities, independently of initial $p_j$. The non-differentiability of the algorithm creates an angular point when $p$ reaches the desired equilibrium but the network remains stable around equilibrium point. This behavior is described in more detail in the following situation.

### Game Learning

The minimax-back-propagation algorithm was implemented in a two-player zero-sum game playing situation. The cost/utility matrix of Table 1 was used for player 1 and its opposite for player 2. The well-known *minimax theorem* (von Neumann & Morgenstern, 1944) shows there always exists a minimax equilibrium for such games. With the given cost/utility matrix, the minimax equilibria are found for the mixed strategies $p^1 = p_1^1 = 5/9$ for player 1 and $p^2 = p_1^2 = 2/3$ for player 2.

The game was implemented using two networks playing each other, simultaneously trained with the minimax-back-propagation algorithm. The "environmental" probability of one network/player now corresponds to the response probability of the other network/player: $p_i^{*1} = p_j^2, p_i^{*2} = p_j^1$. Figure 3 shows $p^1$ and $p^2$ as a function of the number of plys played by the networks for various initial conditions.

Simulations show the networks always converge to optimal minimax response behavior. When equilibria are reached, the response dynamics present an angular point but the equilibria remain stable in all cases. Furthermore, the network/player learning dynamics show interdependence since the response probabilities of one network/player correspond to the environmental prob-

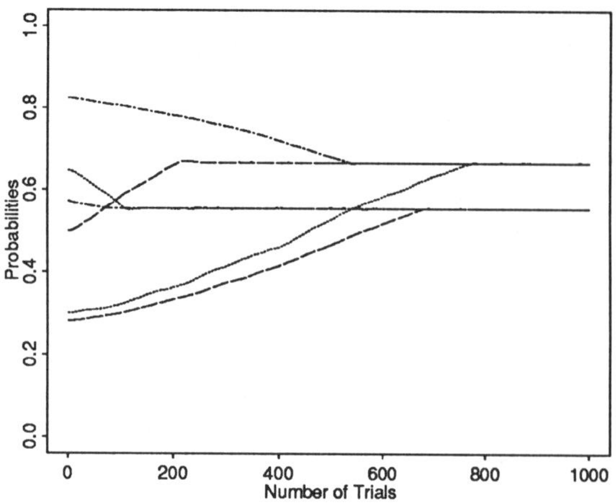


Figure 3: Examples of learning dynamics for the minimax-back-propagation algorithm during a game playing situation. Both network/players reach and remain at optimal minimax probabilities (5/9 and 2/3) independently of initial conditions (3 runs are shown, one per line type).

abilities of the other network/player. This result provides interesting hypotheses and predictions on learning dynamics where players learn from each other's playing strategy. The implications on observed experimental data in game learning situations are beyond the scope of this paper.

## Conclusion

A general decision making network architecture is suggested along with representations for probabilities and utilities. The involved principles integrate formal properties of optimal decision making and connectionist computational characteristics. Variants of the back-propagation learning algorithm are then obtained to maximize total and minimum expected utility. The dynamic properties of the resulting algorithms are examined through simulations of a two-response environment.

Obviously, the complex properties of human decision making are barely mentioned in this paper. Nonetheless, in the case of simultaneous utility estimation and behavior adaptation, the approach provided interesting insights on observed experimental data. In general, the proposed architecture seems able to provide a rich computational framework which might lead to a number of interesting hypotheses and predictions about human learning and decision making behavior.

**Acknowledgements.** Thanks to David Rumelhart for helpful discussions and to Julie Holmes for her help throughout this project.

**References**

Busemeyer, J. R. & Myung, I. J. (1989). An adaptive theory of human decision making. In D. Vickers & P. L. Smith (Eds.), *Human Information Processing: Measures, Mechanisms and Models.* North-Holland: Elsevier.

DeGroot, M. H. (1970). *Optimal Statistical Decisions.* N. Y.: McGraw-Hill.

Estes, W. K. (1987). Application of a cognitive-distance model to learning in a simulated travel task. *JEP: Learning, Memory and Cognition, 13,* 380–386.

Luce, R. D. (1959). *Individual choice behavior.* N. Y.: Wiley.

Luce, R. D. & Suppes, P. (1963). Preference, utility, and subjective probability. In R. D. Luce, R. R. Bush, & E. Galanter (Eds.), *Handbook of mathematical psychology (Vol. 3).* New York: Wiley.

Rumelhart, D. E., Hinton, G. E., & Williams, R. J. (1986). Learning internal representations by error propagation. In D. E. Rumelhart & J. L. McClelland (Eds.), *Parallel Distributed Processing: Explorations in the Microstructures of Cognition (Vol. 1).* Cambridge, MA: MIT Press.

Slovic, P., Lichtenstein, S., & Fischhoff, B. (1988). Decision making. In R. C. Atkinson, R. J. Hernstein, G. Lindsey, & R. D. Luce (Eds.), *Steven's handbook of experimental psychology.* New York: Wiley. 2nd. Edition

von Neumann, J. & Morgenstern, O. (1944). *Theory of games and economic behavior.* Princeton, N.J.: Princeton University Press.

# Emergency Decision-Making by Nurses in the Context of Telephone Interactions

**Judith Leprohon and Vimla L. Patel**

Cognitive Studies in Medicine: Centre for Medical Education
McGill University
3655 Drummond Street, Room 529
Montreal, Quebec, Canada H3G 1Y6
INM8@MusicB.McGill.CA

**Abstract**

In Montreal, nurses respond to 9-1-1 emergency calls for medical help, backed up by physicians when needed. In this context, they have to make rapid decisions based on limited and sometimes unreliable information. The purpose of this study was to describe the decision-making processes used by nurses in telephone triage and to examine the relations among these processes in relation to nurses' characteristics and performance. The study was conducted in real emergency conditions. The sample included 34 nurses and 50 calls. Each call was transcribed and subjected to performance evaluation and content analysis. This paper focuses on the cognitive analyses of two protocols associated with different outcomes. The results show that nurses' decision-making in triage situations are often based on surface features (patterns of symptoms) rather than the underlying pathophysiology, particularly in high urgency cases. High performance was related to decisions based on the evaluation of the whole emergency situation. The contribution of training and the effects of experience on triage performance are discussed.

## Introduction

Urgences santé, the Montreal system of Emergency Medical Services (EMS), introduced nurses into the pre-hospital chain of intervention in order to deal simultaneously with the rising cost of health care and the need to insure quality of care. Nurses assume the first-line response to emergency calls for medical help from the population through the 9-1-1 emergency line, assisted by an attending physician when needed. Established nine years ago, Urgences santé receives an average of a thousand calls a day, of which less than ten percent are serious emergencies and 35% result in the dispatching of emergency vehicles[1].

This task, referred to as triage, consists in assessing the level of urgency of a situation presented by a patient or someone calling for him[2], and in determining the appropriate intervention and delay[3]. It involves making decisions in the presence of several constraints. First, the telephone reduces the communication bandwidth. Second, triage represents an ill-structured domain and involves a high level of uncertainty. Third, time limitations constitute a "vital" constraint in this setting since the urgency and severity of any unattended call remain unknown and a delay can potentially cost a life.

Baumann (1987) defines decision-making in critical care situations in reference to "situations in which a choice is made among a number of possible alternatives, often involving trade-offs among the values given to diffferent outcomes" (p.1). She illustrates this distinction using the domain of medicine where problem-solving can lead to one solution, such as a diagnosis, while decision-making better describes the process of choosing among various therapeutic alternatives. Triage also involves both problem-solving and decision-making. Indeed, every triage situation represents a real-life problem of varying complexity whose solution must be reached in minimal time because of the impending urgency, but it also involves deciding among six alternatives[4].

Significant research has been done on clinical reasoning and decision-making processes in the medical domain where the diagnostic model is based on pathophysiological considerations (Barrows & Feltovich 1987; Elstein et al. 1978; Patel & Groen 1986). The clinical reasoning process used in medical diagnosis has been related to a

The authors would like to thank Urgences Santé for their support throughout this research.

[1] Emergency vehicles include ambulances with two technicians and Md-vehicles with a physician and a technician.

[2] In order to simplify the text, the feminine gender will be used to refer to a nurse an the masculine gender for a patient or a physician.

[3] The appropriate intervention may be "no" intervention and the implementation of this intervention may require some negotiation with the patient.

[4] These include: 1) sending a physician and an ambulance immediately; 2) sending an ambulance immediately; 3) sending an ambulance within 20 minutes; 4) sending an ambulance within 45 minutes; 5) referral and 6) advice. These decisions represent an hierarchy of the alternate solutions according to the associated levels of urgency.

hypothetico-deductive process (Barrows & Feltovich 1987; Feltovich et al. 1984). When solving ill-structured problems within their domain of expertise, medical experts formulate an initial diagnostic hypothesis, usually accurate (Patel 1988), which can be obtained from their highly organized knowledge base using pattern recognition and forward reasoning (Patel & Groen 1988). Medical experts also distinguish relevant from irrelevant information and attend only to the former (Patel & Groen 1988). As a result, their search is limited and their problem-solving efficient. However, experts may have insufficient domain-specific knowledge to solve an unfamiliar or novel problem, which brings them to rely more on backward reasoning (Joseph & Patel, 1986).

Triage represents a grey area between medicine and nursing where medical diagnoses may not be essential in order to make a decision about the appropriate interventions, and where the determination of the level of urgency becomes the primary focus of the reasoning process. The purpose of the present paper, drawn from a larger research on nurses' triage decision-making (Leprohon, 1991) is two-fold: 1) to identify the processes used by nurses in triage decision-making using a combination of cognitive methods; 2) to determine to which extent the relations among these processes, nurses' characteristics and triage performance confirm existing theories of problem-solving, decision-making and expertise.

## Methodology

Based on ethnomethodological aspects of Suchman's work (1987), our methodological framework takes into account both the actor's "pre-understanding" (i.e. past experience and training) and the situation as it evolves in its context. Triage decision-making is dynamic and situated. Therefore, we chose to study it in its natural setting, in order to capture its authenticity and maintain its representativeness. We also combined several methods to get a more comprehensive view on decision-making processes and to enhance internal validity. These methods included discourse analysis of the calls, based on theories and methods developed within cognitive science for the analysis of reasoning processes during doctor-student (Evans and Gadd, 1989) and doctor-patient (Patel, Evans & Kaufman, 1989) dialogues. Another level of analysis represented information in terms of semantic networks based on discourse analysis and propositional representation of written discourse. Although the methodology used to evaluate the outcome of triage decision-making was derived from decision analysis theory, this paper focuses on the cognitive analyses that were used to describe telephone triage decision-making by nurses. The analysis of two calls associated with accurate and inaccurate decisions will serve to illustrate how the various methods have been assembled and lwill ead us into the discussion of some aspects of nursing expertise in the light of our present knowledge of medical decision-making. Data obtained in the larger study (Leprohon, 1991) will be brought in to corroborate some of these findings.

The final sample included 34 nurses actively working at Urgences santé who had a minimum of two years of clinical experience and who agreed to participate. Since the purpose of this study was to describe the decision-making processes involved, the calls were chosen by the nurses working in the triage setting based on three pre-determined exclusion criteria: 1) interactions where communication constraints interfered with decision-making; 2) interactions where the decision was not made exclusively by the nurse; and 3) interactions involving psychiatric cases. Each nurse was asked to select two calls within a specific period of time, approximately half a shift (i.e. three to four hours). After each selected call, she disconnected herself from the system in order to write down a summary and explanation of the case without being interrupted by another call coming in. The final sample included 50 calls.

Since all interactions were routinely audiotaped for legal purposes, the selected calls were transcribed verbatim and analyzed. Calls which resulted in the transportation of the patient by ambulance to the hospital or in a medical visit by an Urgences santé physician were followed up by consulting the hospital or Urgences santé's medical records, whereas for the other calls, the follow-up consisted of a telephone call within ten days from the initial call. Nursing records were also used to identify the decisions that were made. At the end of each call, nurses were asked to summarize the case and explain their decisions in terms of the patient's health status or prognosis. They were also asked to fill in a questionnaire including such information as their age, their experience and their working status.

## Data Analysis

Although data analysis in the present paper focuses on the cognitive methods used for the fine-grain analysis of individual cases, the three levels of data analysis used in the larger study will be enunciated.

First, we used a methodology developed by Champagne et al. (1988) to determine the accuracy or inaccuracy of triage decisions. Each decision was thus compared to the optimal decision which experts determined by consensus based on the patient's medical diagnosis and/or his health status one week after the initial call. False negatives, generated by a lack of sensitivity, resulted in subobtimal interventions, and possibly in costs in terms of patients' health; whereas false positives, related to a lack of specificity, represented a "waste" of resources which could have been needed for other patients.

Second, at the process level, verbal protocols of real telephone interactions were first analyzed using discourse analysis in order to derive information about the nurse's reasoning from the flow of information going back and forth between the nurse and the patient. A medical expert was then asked to elaborate a canonical network linking the final diagnosis and/or the patient's condition on follow-up to the relevant information obtained in the course of the real call. Finally, summaries and written explanations obtained retrospectively from nurses (immediately after completion of the call in order to minimize distortion), were analyzed using propositional analysis and represented in a semantic network.

These cognitive methods were combined as follows. Based on a summary of the discourse analysis, relevant information elicited by the nurse or volunteered by the caller and/or patient was mapped onto the causal network corresponding to the expert's representation in order to identify underlying reasoning stategies and to determine which information was available at the time the decision was made. The propositional representation of the nurse's explanation was used to elaborate another causal network which corresponded to the nurse's rationale for making a particular decision, or changing her decision, and for performing particular interventions. The final step of the analysis consisted of comparing these two causal networks and to examine any discrepancies since they may point out particular aspects of nurses' reasoning.

In the larger study, explanations were also subjected to content analysis which allowed for the macroscopic determination of decision-making processes, and decision times were measured on the basis of the number of exchanges comprising each call. Results from the latter analyses will only be reported punctually when they support the current findings.

Third, results from the survey on nurses' characteristics were analyzed and relations between these characteristics, process and outcome variables were derived. These will also only be reported when they support the findings resulting from the cognitive analyses.

## Results and Discussion

Comparison of cases A and B illustrates the use of this multimethod approach. Case A, with a medical diagnosis of "subendocardial infarct", had a high urgency level and was associated with an accurate decision while Case B, with a medical diagnosis of "parietal chest pain", had a low urgency level and was associated with an inaccurate decision characterized by a lack of specificity.

Figure 1 presents excerpts from the summary of the discourse analysis of Case A. This call included two interactions: the first one between the nurse and the caller (i.e. the patient's wife) and the second one between the nurse and the patient. Figure 1 shows the opening statement of the wife and a few exchanges which preceded the nurse's decision. The black arrows point to observations elicited by the specific foci of the nurse's questions while the grey arrows indicate which of the nurse's questions refer to observations volunteered by the patient. The numbering of each piece of information according to the exchange containing it allows us to determine the timing of the decision with respect to the information available. In this case, the decision to send a physician and an ambulance immediately was reached at the 20th exchange (D20).

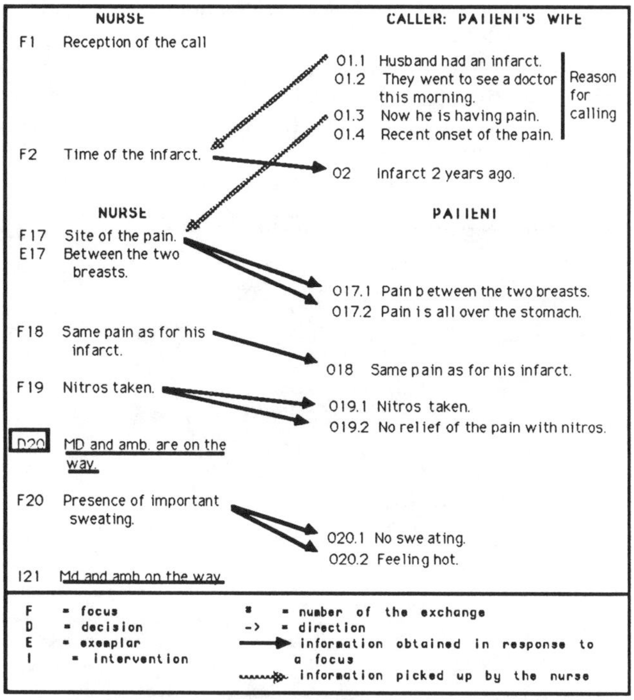


Fig. 1: Excerpts from the summary of Case A's discourse analysis

According to the ECN for Case A (Figure 2), the symptoms caused by the infarct and the supporting evidence had been obtained by the nurse prior to the decision. Although the information concerning the patient's heart failure condition only came up after the decision had been made, it was not directly related to the present diagnosis, hence not essential for decision-making.

According to the ECN for Case B (Figure 3), all the pieces of information from the discourse analysis corresponding to elements of the ECN had been obtained prior to the decision (D36) and were thus available to be used in the decision-making process.

At this stage of the analysis we have more information about the information available at the time of the decision and we can estimate whether the essential information pertaining to the diagnosis, as established by the medical expert (ECN), had been gathered when the nurse made her decision. However, we don't know exactly which of these pieces of information have actually been used by the nurse in making the decision nor do we know why she made that decision. In order to find out, we must examine the rationale underlying each decision.

The NCN for Case A (Figure 4) shows that the nurse considered cardiac arrest a direct potential consequence of the symptoms of cardiac distress that the patient was experiencing. This suggests that she made her decision based on symptoms that she associated with a potentially fatal prognosis; there is no trace of any diagnostic hypothesis in her explanation. Also, retrospectively, the nurse included in her rationale information which had been gathered after her decision had been made. This suggests that she considered the heart failure condition of the patient confirming evidence for her decision and that she used this information in building a coherent representation of the situation. It also means that her rationale does not reflect her decision-making as it actually took place.

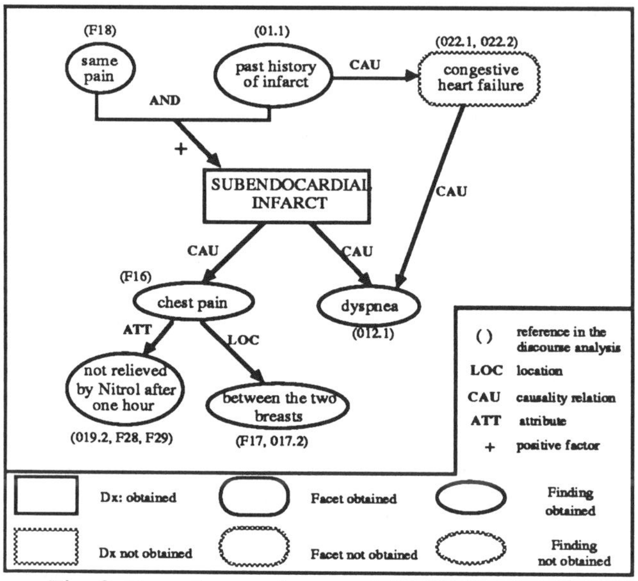

Fig. 2: Expert's causal network (ECN) for Case A

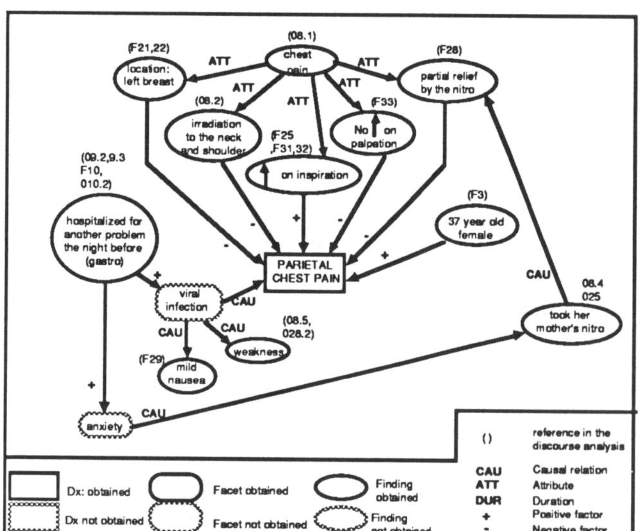

Fig. 3: Expert's causal network (ECN) for Case B

The NCN for Case B (Figure 5) shows that the propositions referring to the absence of modification on deep inspiration [10.1 and 10.3] contradict the information collected during the interaction and appearing in the discourse analysis (see ECN, Fig. 3: F31, 031; F32, 032). In the course of the interaction with the patient, the nurse seemed to collect information as if she had been guided by a prototype of cardiac chest pain, following a checklist of the important symptoms. Most of them supported the hypothesis of cardiac chest pain and she may have missed this divergent symptom. This error suggests that the nurse adopted the hypothesis corresponding to the schema or prototype of cardiac chest pain too soon in her interaction with the patient, orienting the remaining of the interview to gather confirming evidence for her hypothesis and ignoring information which did not fit the prototype, such as the fact that the pain increased on deep inspiration. This reflects a premature closure on the hypothesis of cardiac chest pain.

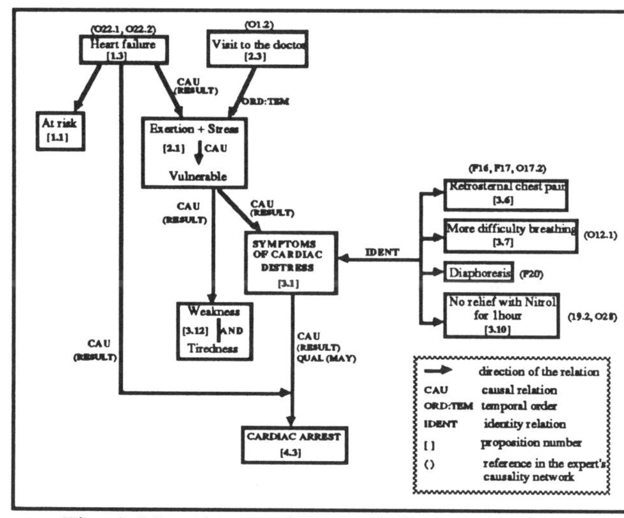

Fig. 4: Nurse's causal network (NCN) for Case A

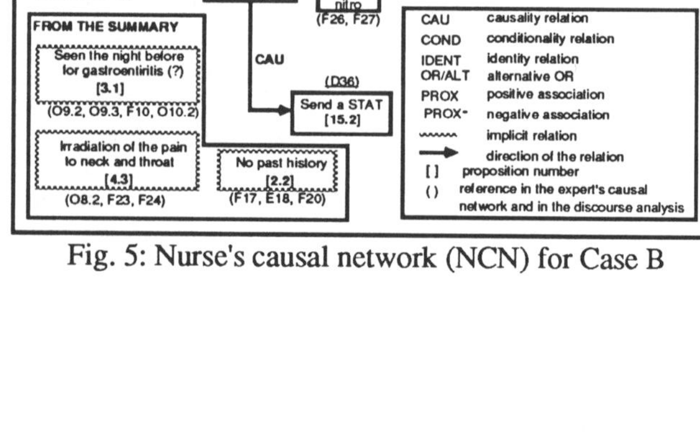

Fig. 5: Nurse's causal network (NCN) for Case B

When making her decision, the nurse could not establish beyond any doubt that the pain was not cardiac based on the extent to which the criteria for cardiac chest pain were met. As a result, she insured the sensitivity of her decision at the risk of reducing its specificity, conformingly to the organization's philosophy, and sent resources according to the worst situation, cardiac chest pain.

Since the ECN represents the pathophysiological explanation of the final diagnosis and/or the patient's condition while the NCN represents the pathophysiological explanation of the nurse's decision and interventions, the mapping of these two networks provides more information about the decision-making processes of the nurse. Hence, the two causal networks should overlap to the extent that the nurse bases her decision and interventions on a hypothesis corresponding to the patient's real condition. In Case A, we find that the symptoms caused by the infarct (O12.1, F16, F17, 017.2, F19, 019.2) according to the ECN (Figure 2) can be mapped onto the NCN (Figure 4), since they correspond to the symptoms of cardiac distress [3.6, 3.7, 3.10] identified by the nurse as potentially leading to cardiac arrest [4.3] and on which she based her decision to send a doctor and an ambulance immediately. On the other hand, a comparison of the ECN (Figure 3) and the NCN (Figure 5) for Case B shows that the nurse did not have a holistic perspective on the situation. She did not pick up the information pertaining to the patient's last hospitalization, although she did include it in her summary of the relevant information [3.1]. Consequently, she never considered the possibility that the pain may be related in some way to her recent gastroenteritis. Also, she did not detect the patient's anxiety, or at least did not attend to it, even though she mentioned reassurance as one of her interventions [8.1]. The addition of a false but important one to the list of criteria being met by this patient for the presence of cardiac chest pain has probably contributed to the inaccuracy of the nurse's decision. Indeed, she sent a doctor and an ambulance immediately while, according to the gold standard, the patient should have been referred to her doctor. In fact, this error represents one of the manifestations of the premature closure that the nurse demonstrated in her interaction with this patient. Her limited perspective on the situation constitutes another one.

Comparing the characteristics of the nurses who treated these two calls provides evidence supporting the role of experience in expert performance. The nurse in Case A had 20 years of experience in nursing, some emergency room and intensive care experience and she had been working in this organization for six years and a half. Her participation in the study involved two calls, one of high urgency and one of low urgency, and her decisions were accurate in both cases. The nurse in Case B, on the other hand, had seven years of experience in nursing, two years in an emergency department, and she had been working in this organization for four years. She also contributed two cases, one moderately urgent which was associated with an accurate decision, and this one, of low urgency, where her decision lacked specificity.

In fact, the nurse's explanation for Case B resembles to some extent the explanation provided by the other nurse for Case A (which really involved a cardiac problem), except for the type of reasoning involved. Case A involved forward reasoning, whereby a decision was made directly from the symptoms of cardiac distress. Case B, on the other hand, involved backward reasoning since the hypothesis of cardiac chest pain was directing the collection of supporting evidence. Another problem with Case B is the fact that the premature closure limited the scope of the triage assessment by adopting too soon an erroneous schema in a low urgency situation where several factors should have been considered in order to grasp the whole situation and make an accurate decision.

These results are in agreement with those obtained in the context of the content analysis and the interrelations observed among nurses' pre-understanding, calls' characteristics, process and outcome indicators. Indeed, general nursing experience of ten years or more has been associated with an accuracy rate of 64.9% as compared to 26.1% when experience is below this threshold. Also, decisions in the two calls (100%) whose domains were directly related to specific experience of the nurse and three out of the four calls (75%) where the relation was indirect were accurate as compared to only two calls out of four (50%) when it was unrelated.

These results suggest that nursing decision-making, at least in triage situations, is of a different nature than medical decision-making. 34% of the decisions were based on symptoms with an accuracy rate similar to the total sample's, i.e. 52.9%. Of these decisions, 58.8% were associated with calls in the high urgency category. This supports the finding that the decision can be made directly from symptoms in very urgent situations. These decisions must rely on prior instances in order to allow the nurse to determine, based on minimal information, the urgency of the situation and to make her decision.

Decisions based on a hypothesis, although there were only five, had an accuracy rate of 0%; while decisions based on the situation, with the same number, had an accuracy rate of 100%. Training may explain this difference. As opposed to physicians, nurses have not been trained to do medical diagnoses which involve the generation of hypotheses, but they have extensive training in assessing the patient's situation, inventorying the resources available and considering alternate interventions.

These results suggest that the nurse uses various approaches to decision-making, depending on the urgency of the situation. While ensuring that the intervention is implemented represents the priority in high urgency situations, representing exactly the situation, with the inventory of all available resources and reviewing alternate solutions is more important in low urgency situations.

Domain-specificity could also be found in relation to specific processes, but its effect seemed to some extent subsumed under the urgency-specificity. Clarifying the relationship between these two factors will require further investigation.

## Conclusion

The predominant findings concern the decision-making processes used by nurses to make their decisions, which are based on symptoms, especially in situations involving a high level of urgency. Three main factors have been related to performance: the nurse's experience, the urgency and the domain of the call.

These results have important implications for practice and for training. First, hiring selection criteria, particularly with respect to experience, should be revised. Second, feedback on performance should be provided to ensure the acquisition of experience which, being specific to triage situations, may contribute significantly to increase triage performance. Training should as much as possible be situated and experiential. This could be achieved in the context of preceptorship which would also contribute to the evaluation of triage processes and performance.

These results thus enhance our knowledge and understanding of nursing clinical decision-making and this methodological approach should be applied to other domains.

## References

Barrows, H.S., and Feltovich, P.J. 1987. The clinical reasoning process. *Medical Education* 19: 86-91.

Baumann, A.O. 1987. Rapid Decision Making in Critical Care. Doctoral thesis, University of Toronto.

Champagne, F.,; Contandriopoulos, A.P.; Eeckhoudt, L.; Kalina, M.; Farand, L.; Leprohon, J., and Preker, A. 1988. L'Evaluation du tri des appels telephoniques de la population aux services de santé. Projet de recherche soumis au FRSQ, Mars 1988.

Elstein, A.S.; Shulman, L.S., and Sprafka, S.A. (1978). *Medical Problem Solving: An Analysis of Clinical Reasoning*. Cambridge MA: Harvard University Press.

Evans, D.A., and Gadd, C.S. (1989). Managing coherence and context in medical problem-solving discourse. In D.A. Evans and V.L. Patel eds. *Cognitive Science in Medicine: Biomedical Modeling*. Cambridge, MA: MIT Press, pp. 211-256.

Feltovich, P.J.; Johnson, P.E.; Moller, J.H., and Swanson, D.B. 1984. LCS: The role and develoopment of medical knowledge in diagnostic expertise. In W.J. Clancey & E.H. Shortliffe eds., *Readings in Medical Artificial Intelligence*. Reading, MA: Addison-Wesley.

Joseph, G.M., and Patel, V.L. 1986. Specificity of expertise in clinical reasoning. In Proceedings of the Eight Annual Conference of the Cognitive Science Society. Hillsdale, N.J: Lawrence Erlbaum Associates.

Leprohon, J. 1991. Nursing Clinical Decision-Making in the Context of Clinical Interactions. Doctoral dissertation, Dept. of Educational Psychology, McGill University.

Patel, V.L. 1988. The Nature of Understanding and Problem Solving in Medical Expertise. Technical Report # CME8-CS4, McGill University, Montreal.

Patel, V.L.; Evans, D.A., and Kaufman, D.R. 1989. Cognitive framework for doctor-patient communication. In D.A. Evans and V.L. Patel eds., *Cognitive Science in Medicine: Biomedical Modeling*. Cambridge, MA: MIT Press, pp. 257-312.

Patel, V.L., and Groen, G.J.(1986. Knowledge Based Solution Strategies in Medical Reasoning. *Cognitive Science* 10:91-116.

Patel, V.L. & Groen, G.J. 1988 Toward a unified theory of problem solving: Views from the content domains. A view from medicine. Paper presented at the Annual Conference of American Educational Research Association, April 4-9,1988, New Orleans, Louisiana.

Suchman, L.A. 1987. *Plans and Situated Actions: The Problem of Human-Machine Communication.* New York: Cambridge University Press.

# Goal-based Decision Strategies

**Stephen Slade***
Yale Computer Science Department
P.O. Box 2158
New Haven, CT 06520
slade@cs.yale.edu

**Abstract**

We present a process model of decision making mediated by goals and relationships. The model is implemented in the VOTE computer program which simulates Congressional roll call voting. In this paper, we focus on VOTE's decision strategies, which are based on the need not only to arrive at a vote, but also to produce an explanation for each decision. We describe several typical strategies, as well as an indirect strategy, DEEPER ANALYSIS, that is invoked when the normal strategies fail to arrive at a decision.

## 1 The Process of Decision Making

Decision making is a complex cognitive process. A realistic model of decision making must account for a multitude of goals and limited resources. Many problems will not admit to perfect solutions. Practical decisions often involve trade-offs.

We have developed a goal-based model of decision making in which the relative priorities of goals drive the decision process. This model serves as the basis for the VOTE program which simulates Congressional roll call voting decisions on thousands of votes, based on the relative priorities of goals and constituent relationships, and the need to provide an explanation for the decision.

Alternative models of decision making have used prescriptive decision analysis (Raiffa 1968), which uses estimates of payoffs and uncertainty of outcomes to arrive at an optimal result. The assumptions of decision analysis are inappropriate in many domains for which our goal-based, descriptive model may apply. Our model is an extension of the theory of goals developed by Schank and Abelson (1977), Wilensky (1978) and Carbonell (1979), and is compatible with the use of goals in case-based reasoning systems (Simpson 1985, Hammond 1986).

We define the priority or importance of a goal to be proportional to the resources that an agent may be willing to allocate in service to that goal. Resources are viewed quite broadly to include not only time or money, but also cognitive resources, such as attention and memory. For example, an agent should think more about important matters than about trivial ones.

*Author's address as of fall 1991: Stern School of Business, New York University, New York, NY 10006

| Predictions | Individual | | Deeper Analysis | |
|---|---|---|---|---|
| Correct | 1738 | 77% | 124 | 69% |
| Incorrect | 532 | 23% | 56 | 31% |
| Predicted Votes | 2153 | 100% | 181 | 100% |
| All Votes | 2814 | | 238 | |

Table 1: VOTE Summary Statistics

We extend the model of goals to include interpersonal relationships (Slade 1990). An agent adopts the goal agenda of another agent with a priority proportional to the importance of the relationship. For example, an agent is more likely to help a friend than a stranger. This extension permits the processing of adopted goals in a manner uniform with individual goals.

The VOTE program does not rely on a single method for ranking alternatives, but rather implements a collection of strategies that relate decisions to explanations. In this paper, we discuss a number of those decision strategies.

## 2 The VOTE Program

VOTE, written in T (Slade 1987), comprises over 14,000 lines of code, and over 9,000 lines of data. The VOTE program relies on a set of interrelated databases, including issues (over 200 currently in the database), constituency groups (150), bills (43), members (65), and decision strategies (16). We note that many decision strategies are required since the explanation of the decision depends on the strategy. It is not enough to use one simple strategy of summing the weights of the conflicting issues and relationships.

Table 1 provides a summary of VOTE's performance. We note that we do not have actual votes for all of the possible decisions. This is the case for hypothetical bills, for bills that precede or follow a member's term, or for bills on which the member chose not to vote.

VOTE correctly predicts individual members' votes 77% of the time. Approximately 8% of those votes involve DEEPER ANALYSIS (discussed below), which arrives at the correct result 69% of the time. The net accuracy for bills not requiring DEEPER ANALYSIS is 82%.

In VOTE, we use the names and records of real members of Congress, and attribute relationships and issue stances to them. The coding decisions, though subjective, reflect the author's political experience as a campaign consultant and White House staff member. By way of disclaimer, we should state that these data are merely illustrative, and are not meant to represent the beliefs of actual members of Congress. The accuracy of the program is a reflection of the coding decisions. The primary purpose of VOTE is not to predict individual voting decisions, but rather to demonstrate the feasibility of a particular model of goals, relationships, and decision making.

## 3 Stances: Preferences and Consequences

In VOTE, goals are issues, such as gun control or affirmative action. A member's preferences are expressed as **stances**, which comprise an **issue**, a **side** (pro or con), and a level of **importance** (A, B, or C, where A is high and C is low). For example, a member who strongly supports affirmative action while opposing gun control might have preference stances such as the following.

```
(PRO A MEMBER:MEMBER.2319 AFFIRMATIVE-ACTION)
(CON B MEMBER:MEMBER.2319 GUN-CONTROL)
```

While VOTE prints stances as lists, the program actually uses a more complex object-oriented representation. Constituency groups have similar sets of preference stances to represent their issue agendas. Furthermore, VOTE associates normative stances with issues to reflect typical public opinion. For example, the flag burning norm is one of mild opposition:

```
    (CON C ISSUE:ISSUE.1077 FLAG-BURNING)
```

Stances are also used to represent justification consequences of legislation. For example, HR-2978, the 1989 bill which banned flag desecration, is represented with the following set of stances respectively favoring and opposing passage of the bill.

```
Stance-FOR:
  (((CON B BILL:HR-2978 FLAG-BURNING)
    (PRO B BILL:HR-2978 PATRIOTISM)))
Stance-AGN:
  (((PRO B BILL:HR-2978 CONSTITUTION)
    (PRO B BILL:HR-2978 FREE-SPEECH)))
```

VOTE has a natural language generation facility which can express the above stances in English.

> *Support of the flag desecration bill stands firmly against the right of burning the American flag in protest. It upholds patriotism and devotion to this country. Opposition to HR-2978 upholds the belief in the United States Constitution and the Bill of Rights, in addition to the right of freedom of speech.*

The process of decision making requires matching preference stances with consequence stances. In VOTE, members ignore consequence stances that do not match any of their preference stances. That is, members only pay attention to those consequences about which they or their constituents care.

## 4 Choices and Explanations

Stated generally, VOTE tries to choose the outcome that most closely matches a member's preferences. However, given the conflicts that often arise, VOTE must be able to resolve, or at least address, the conflicts. The VOTE model of decision making assumes that there are many ways of arriving at a decision which may differ significantly in their underlying structure and rationale. Our decision making model focuses on two questions.

- *Choice:* What does the agent want to do?
- *Explanation:* How can the agent justify his decision?

These questions depend on an agent's preferences and an action's consequences. The basic VOTE decision making algorithm is as follows.

1. Extract preference stances from voting record and constituent relationships, and combine with member's personal ideology or "credo" stances.
2. Match Bill consequence stances for/agn with the member's composite preference stances.
3. Analyze relative importance of for/agn stances with respect to norms, voting record, groups, and personal beliefs.
4. Apply decision strategies until one fits.
5. If no strategy fits, then use DEEPER ANALYSIS to expand the implications of the for/agn stances to include the important consequences of the current decision stances. Go to step 2.
6. Update decision database, adding new vote, and print the result.
7. Produce an English summary of the decision.

In this paper, we focus on steps 4 and 5: the application of decision strategies. We note the following general points.

- A specific strategy does not always guarantee a specific result. The same strategy can be used on the same bill by different members of Congress to arrive at different conclusions.
- Decision strategies are driven by the need to explain a decision. Each strategy is tied to a specific type of explanation.
- Given a member of Congress and a bill, different strategies may result in different conclusions. That is, in some cases, VOTE can generate plausible reasons for both sides of a bill for the same member of Congress.
- One decision strategy, DEEPER ANALYSIS, is an indirect strategy used in step 5, above. It provides a mechanism for VOTE to consider further consequences of a decision. The DEEPER ANALYSIS strategy takes advantage of the cognitive assumptions behind the importance accorded issue stances.
- We do not claim that the strategies presented here are either necessary or sufficient. We present them as being illustrative to demonstrate the feasibility of our decision making model.

A key feature of these strategies is the role of explanation. Not all decisions require explanations. However, we can identify two broad categories of decisions that generally do require justification.

- Decisions that violate expectations.
- Decisions that have adverse consequences.

If an agent makes a decision that differs from what others have expected, then it is reasonable for the agent to offer an explanation. For example, if we observe someone walking around the office barefooted, we might reasonably ask the person for an explanation. We would not require someone to explain why he is wearing shoes at the office. Wearing shoes is the expected behavior, while going barefoot is unusual.

If an agent chooses a course of action which has adverse consequences for another agent (and even for himself), he might be expected to justify his actions. For example, if a man decides to refuse a promotion and raise, his spouse might expect him to provide an explanation, e.g., he would have to travel too much. If he accepts the promotion and raise, no justification would be required – unless of course his wife had *expected* him to refuse the promotion.

In the Congressional roll call voting domain, explanation plays a central role. In our model, the adverse consequences of a vote are represented as downside stances from constituency groups. Unexpected votes are those in which the downside stances include positions from the member's voting record or credo stances. That is, we expect a Congressman to vote in a manner consistent with his voting record and his personal beliefs. Thus, decisions involving downside stances suggest the need for an explanation. The political science literature (Kingdon 1973, Fenno 1978) indicates that members of Congress are cognizant of the role of explanation in decision making.

We note that this type of explanation is different from the usual sense of explanation found in the case-based reasoning literature (Schank 1986, Ram 1989, Kass 1990, Leake 1990). Previous researchers have focused on explanation of anomalous observed events as part of the process of learning. Our present use of explanation is complementary to that process: decision makers offer explanations for the benefit of observers who may find the decision to be anomalous.

Strategy application proceeds with the following tests for choice.

1. *Unanimous.* The first step is always to check for a unanimous decision – one in which all the stances fall on one side of the bill. In such a case, the explanation is the self-evident one that the choice was a popular decision.
2. *Consensus.* We define a consensus to obtain when the most important preference stances all fall on the same side of the bill. That is, the most important group stance, member credo stance, and member voting record stance are all either FOR or AGN the bill. At this stage of the process, we do not have unanimity, so we try to develop a suitable explanation for the consensus choice. There are five possible strategies, the last of which is simply to state that there is a consensus.
3. *Majority.* In the absence of a consensus, there may still be a majority of stances on one side or the other. As with the consensus, we again try to develop an appropriate explanation to justify the choice. Again, there are five possible strategies, the last of which states that there is a majority of opinion for this bill.
4. *Other.* There are three other strategies that use special conditions for choice. We shall discuss one of those below.
5. *Deeper Analysis.* If none of the previous strategies have triggered, then VOTE tries an indirect strategy, DEEPER ANALYSIS, which infers additional consequences of the bill. VOTE then begins the strategy application process over again.
6. *No Decision.* The DEEPER ANALYSIS process will always terminate. If it has not succeeded in arriving at a decision, then VOTE will return the failure condition of NO DECISION.

In this paper, we present examples of several representative individual decision strategies, beginning with POPULAR DECISION.

## 5 Popular Decision

The first strategy tried for any decision is POPULAR DECISION.

**Test:** All stances are on one side of the bill. There is either no opposition or no support for the given measure. The resulting vote is consistent with all constituencies. There are no conflicts.

This strategy is used for the clear cases, not the close calls. Members of Congress report that 80% to 90% of their votes fall in this category (Kingdon 1973). It is important to note that a given bill can result in an easy choice both for or against, depending on the given member.

Below we present two votes on the same bill by different members of Congress. Both votes use POPULAR DECISION, however they arrive at opposite conclusions.

```
> (vote 'gephardt 'clean-water)

* Member:        Richard A. Gephardt
* Bill:          Water Quality Renewal Act
* Amendment to freeze funding for the Clean
* Air Act through 1990.
— OMITTING INTERMEDIATE OUTPUT —
Decision strategy: Popular decision
All stances are AGN this bill:
(((PRO B GROUP:CONSERVATION-VOTERS CLEAN-AIR)
  (PRO B GROUP:DEMOCRATS AIR-POLLUTION)
  (PRO B GROUP:CONSERVATION-VOTERS CLEAN-WATER)
  (PRO B GROUP:DEMOCRATS CLEAN-WATER)))
There are no reasons to vote FOR this bill.
Adding current vote to DECISION database
— OMITTING INTERMEDIATE OUTPUT —
```

> *Richard A. Gephardt is opposed to bill HR-8, the Water Quality Renewal Act. He believes this bill not to be in the best interests of the people. He feels strongly in favor of the policy of increased funding for air pollution control. Gephardt is committed to the program of increased funding for water pollution control.*

VOTE's natural language generation facility produces the English rationale.

Congressman Gephardt finds it easy to oppose HR-8. We see below that Representative Gingrich has no problems supporting the bill. We note that the reasons for opposing the bill are not merely the opposite of the stances in support of the bill.

```
> (vote 'gingrich 'clean-water)

* Member:           Newt Gingrich
* Bill:             Water Quality Renewal Act
* Amendment to freeze funding for the Clean
* Air Act through 1990.
— OMITTING INTERMEDIATE OUTPUT —
Decision strategy: Popular decision
All stances are FOR this bill:
(((PRO B GROUP:CHAMBER-COMMERCE DEREGULATION)
  (PRO B BILL:S-557 DEREGULATION)
  (PRO B MEMBER:MEMBER.3055 DEREGULATION)
  (PRO B MEMBER:MEMBER.3055 FREE-ENTERPRISE)
  (PRO B GROUP:REPUBLICANS DEREGULATION)
  (PRO C BILL:HR-5210 DEREGULATION)))
There are no reasons to vote AGN this bill.
Adding current vote to DECISION database
— OMITTING INTERMEDIATE OUTPUT —
```

> *Newt Gingrich votes for bill HR-8, the Water Quality Renewal Act. He believes this bill to be in the best interests of the people. He feels strongly in favor of limiting federal regulation of industry and society. Gingrich is committed to free enterprise and capitalism.*

The explanation associated with POPULAR DECISION is not very involved given that the member's constituencies should agree with the vote in the first place. That is, a member has to pay more attention to an explanation when he knows that people will disagree with his vote. Other decision strategies must deal with the disaffected constituencies and the resulting downsides of decisions.

## 6 Not Constitutional

One of the few strategies that is clearly specific to the Congressional voting domain is the NOT CONSTITUTIONAL strategy. Decisions dictated by this strategy have a clear explanation. However, members may often find it convenient to vote in favor of a bill they deem unconstitutional, knowing that they can get symbolic milage from a measure that will be thrown out in court.

**Test:** There is a consensus opposing the measure. One issue among the opposing stances is the Constitution. The member votes against a measure that would be struck down by the Supreme Court.

The bill prohibiting desecration of the American flag provides a good example of this strategy.

```
> (vote 'schroeder 'hr-2978)

* Member:           Patricia Scott Schroeder
* Bill:             Flag Desecration
* Bans the desecration of the American flag.
— OMITTING INTERMEDIATE OUTPUT —
Decision strategy: Not constitutional
Found a consensus AGN this bill.
Most important stances are all AGN this bill:
Group: (AGN (PRO A GROUP:ACLU CONSTITUTION))
Record:(AGN (PRO B BILL:HR-5345 CONSTITUTION))
Norm:  (AGN (PRO B ISSUE:SELF CONSTITUTION))
Constitutional grounds for opposing this bill:
      (((PRO A GROUP:ACLU CONSTITUTION)
        (PRO B BILL:HR-5345 CONSTITUTION)
        (PRO B GROUP:COUNTRY CONSTITUTION)))
Adding current vote to DECISION database
— OMITTING INTERMEDIATE OUTPUT —
```

> *Patricia Scott Schroeder votes against bill HR-2978, the flag desecration bill. She believes that provisions of this bill are not constitutional. She is unwavering in her support of the United States Constitution and the Bill of Rights. However, Schroeder sees that members of the Democratic party object to the right of burning the American flag in protest.*

Clearly, many members were willing to vote in favor of this bill, as it passed by a 10 to 1 margin. However, the resulting law was subsequently appealed, and declared unconstitutional by the Supreme Court.

## 7 Balance the Books

BALANCE THE BOOKS relies on the fact that over time, a member's record can be used to support most any position. This strategy suggests that the downside of a vote can be mediated through reference to previous votes in support of the relevant issues.

**Test:** There is a majority opinion on this bill. The member's record has positions that fall on both sides of the bill. The member may offset the current vote with past or future votes.

```
> (vote 'boggs 'hr-3)

* Member:           Corinne Claiborne Lindy Boggs
* Bill:             Child Care
* Provide greater funding for child care.
— OMITTING INTERMEDIATE OUTPUT —
Decision strategy: Balance the books
Found a simple majority FOR this bill.
There are 7 FOR stances:
     (((PRO B GROUP:AFLCIO-COPE CHILD-CARE)
       (PRO B GROUP:AFLCIO-COPE LABOR)
       (PRO B GROUP:ADA LABOR)
       (PRO B GROUP:DEMOCRATS LABOR)
       (PRO B GROUP:DEMOCRATS CHILD-CARE)
       (PRO C BILL:HR-1234 LABOR)
       (PRO C GROUP:COUNTRY CHILD-CARE)))
```

```
There is only 1 AGN stance:
    (PRO B BILL:HR-3299 FISCAL-RESPONSIBILITY)
The record supports positions on both sides:
FOR:(PRO C BILL:HR-1234 LABOR)
AGN:(PRO B BILL:HR-3299 FISCAL-RESPONSIBILITY)
Adding current vote to DECISION database
— OMITTING INTERMEDIATE OUTPUT —
```

> *Corinne Claiborne Lindy Boggs is in favor of bill HR-3, the child care bill. She believes that the people who disagree with this vote may find reassurance in the record. She is eager to support the proposal of increased funding for child care. Boggs strongly supports the legitimate concerns of decent working people. However, Boggs believes that her voting record shows support for fiscal responsibility.*

In this case, Congresswoman Boggs is already on the record as supporting fiscal responsibility. Therefore, she can state that her vote on this bill does not necessarily imply that she opposes fiscal responsibility.

## 8 Shifting Alliances

SHIFTING ALLIANCES is a complex strategy. The member is faced with a dilemma. There are constituencies on both sides of the bill with equally compelling stances. In order to break the tie, the member looks beyond the present case to find another issue that divides these two groups, to see if he has a position on that other issue that would tend to make him politically or philosophically closer to one of the groups.

**Test:** The member has no credo stance on this bill. There are constituencies on opposite sides of the issues. One of the constituencies is in conflict with the member on another issue, or one of the constituencies is in agreement with the member on another issue. Then, the member sides with the constituency that is ideologically closer. The member may lower the importance of the other relationship.

```
> (vote 'morrison 'hjr-350)

* Member:          Bruce A. Morrison
* Bill:            Flag Burning Amendment
* Constitutional amendment to ban flag burning
— OMITTING INTERMEDIATE OUTPUT —
Decision strategy: Shifting alliances
Conflict with FOR groups:
 (CON C MEMBER:MEMBER.872 DEATH-PENALTY)
There is no credo stance involved in this vote.
There are groups on either side of this bill:
FOR:(CON C GROUP:DEMOCRATS FLAG-BURNING)
AGN:(PRO A GROUP:ACLU CONSTITUTION)
The member has belief conflicts with the
FOR group (noted above),
so the decision is with the AGN group.

Adding current vote to DECISION database
— OMITTING INTERMEDIATE OUTPUT —
```

> *Bruce A. Morrison votes against bill HJR-350, the Flag Burning Constitutional Amendment. He believes that sincere people have honest differences of opinion on this bill. He is deeply committed to the principle of the United States Constitution and the Bill of Rights. Even so, Morrison knows that members of the Democratic party are against the right of burning the American flag in protest.*

In this case, the groups are the Democrats on one side and the ACLU on the other. Morrison decides that since he disagrees with the Democrats on the issue of the death penalty, he will side with the ACLU on this matter. Another pragmatic strategy would be to vote with the Democrats on this bill, assuming that the member would be able to assuage the ACLU on a later death penalty vote.

## 9 Deeper Analysis

DEEPER ANALYSIS is a "think harder" strategy. None of the other strategies have worked, for whatever reason. The agent now tries to come up with more stances on both sides of the bill to resolve the question.

**Test:** This is an indirect strategy. First expand the bill stances through inference, then run through the preceding decision strategies again. If that fails, then try a deeper level of inference. Inferencing is triggered by the level of importance of the stances. Process stops at the D level.

The DEEPER ANALYSIS strategy depends on inferences made from the core bill stances. Typically, DEEPER ANALYSIS is invoked if none of the bill consequence stances match any of the member's preference stances. That is, the member has no position on any of the relevant issues. Associated with each issue is a set of pro and con consequence stances. These are stances that might be used to justify a pro or con stance on that issue. For example, the issue of trade tariffs and restrictions has the following pro and con stances.

```
PRO Stances:
((PRO B ISSUE:TRADE-TARIFFS JOB-PROTECTION))
CON Stances:
((PRO B ISSUE:TRADE-TARIFFS TRADE)
 (PRO B ISSUE:TRADE-TARIFFS FREE-MARKET)
 (PRO B ISSUE:TRADE-TARIFFS FOREIGN)
 (PRO A ISSUE:TRADE-TARIFFS FREE-TRADE))
```

VOTE expresses these positions in English as follows.

> *Support of trade tariffs and restrictions is important for efforts to protect domestic jobs. Opposition to protectionism is always part of free trade among countries. It upholds stimulating foreign trade and investment, free enterprise and capitalism, and this nation's foreign policy.*

The method of generating new stances is based on the importance of the initial stances, and is analogous to spreading activation models of memory. The DEEPER ANALYSIS algorithm proceeds as follows.

1. Expand all A consequence stances both FOR and AGN the bill, and then reprocess the decision.

2. If no decision results from step 1, expand all A or B stances (including those derived from step 1) and reprocess the decision.
3. If no decision results from step 2, expand all A, B, or C stances (including those derived from steps 1 and 2), and reprocess the decision. At this stage, all remaining consequence stances will be expanded, regardless of level.

The point of this strategy is to consider the more important consequences first, and to give them greater influence on the decision making process. Our view of cognitive resources suggests that importance provides us with a principled means for organizing and searching memory. DEEPER ANALYSIS is one method of implementing an importance-based search.

In the following example of DEEPER ANALYSIS, an A level analysis results in a SIMPLE MAJORITY decision against the bill.

```
> (vote 'foley 'hr-1154)

* Member: Thomas S. Foley
* Bill:   Textile and Apparel Trade Act
* Passage, over Reagan's September 28 veto,
* of the bill to limit imports of textiles,
* apparel, and footwear.
— OMITTING INTERMEDIATE OUTPUT —
Decision strategy: Deeper analysis...
results in new bill stances at level A...
New AGN stances:
 (PRO A ISSUE:TRADE-TARIFFS FREE-TRADE)
* Re-Analyzing alternative positions
— OMITTING INTERMEDIATE OUTPUT —
Decision strategy: Simple Majority
Found a simple majority AGN this bill.
There are 2 AGN stances:
 ((PRO B BILL:HR-1234 FREE-TRADE)
  (CON C BILL:HR-1234 TRADE-TARIFFS))
There is only 1 FOR stance:
 ((PRO B GROUP:COPE TRADE-TARIFFS))
Adding current vote to DECISION database
— OMITTING INTERMEDIATE OUTPUT —
```

> *Thomas S. Foley is opposed to bill HR-1154, the Textile and Apparel Trade Act / Veto Override. Upon reflection, he believes that the majority of people oppose this bill. He is eager to support free trade among countries. Foley cares about free trade. Still, Foley knows that the Committee on Political Education of the AFL-CIO is eager to support the policy of trade tariffs and restrictions.*

Foley required a one-step, A level analysis. He ended up voting against the bill using a SIMPLE MAJORITY strategy.

## 10 Conclusion

We have discussed the role of decision strategies in the VOTE program. VOTE is a robust demonstratation of the computational feasibility of our model of decision making. In contrast to prescriptive decision analysis, VOTE provides a descriptive framework for examining decision making as a process. The VOTE model does not assume that there is one right or best result, but rather that the final decision depends on a range of goal-based cognitive processes. VOTE makes explicit many of the underlying issues of reasoning about goals, consequences, and explanations in decision making.

## References


Carbonell, J. 1979. Subjective Understanding: Computer Models of Belief Systems. PhD thesis, Yale University. Technical Report 150.

Fenno, R. 1978. *Home Style: House Members in their Districts.* Little, Brown, Boston.

Hammond, K. 1986. Case-based Planning: An Integrated Theory of Planning, Learning and Memory. PhD thesis, Yale University. Technical Report 488.

Kass, A. 1990. Developing Creative Hypotheses by Adapting Explanations. PhD thesis, Yale University.

Kingdon, J. 1973. *Congressmen's Voting Decisions.* Harper and Row, New York.

Leake, D. 1990. Evaluating Explanations. PhD thesis, Yale University. YALEU/CSD/RR 769.

Raiffa, H. 1968. *Decision Analysis: Introductory Lectures on Choices under Uncertainty.* Addison-Wesley.

Ram, A. 1989. Question-driven understanding: An integrated theory of story understanding, memory, and learning. PhD thesis, Yale University. YALEU/CSD/RR 710.

Schank, R. 1986. *Explanation Patterns: Understanding Mechanically and Creatively.* Lawrence Erlbaum Associates, Hillsdale, NJ.

Schank, R. and Abelson, R. 1977. *Scripts, Plans, Goals and Understanding.* Lawrence Erlbaum Associates, Hillsdale, New Jersey.

Simpson, R. 1985. A Computer Model of Case-based Reasoning in Problem-solving: An Investigation in the Domain of Dispute Mediation. PhD thesis, School of Information and Computer Science, Georgia Institute of Technology.

Slade, S. 1987. *The T Programming Language: A Dialect of LISP.* Prentice-Hall, Englewood Cliffs, NJ.

Slade, S. 1990. A goal-based model of interpersonal relationships. In Proceedings of the Twelfth Annual Conference of the Cognitive Science Society, pages 836–843, Boston, MA. Cognitive Science Society.

Wilensky, R. 1978. Understanding Goal-Based Stories. PhD thesis, Yale University. Technical Report 140.

# Preschoolers' Understanding of Gravity

Eric Amsel
Psych. & Cog. Sci.
Box 292, Vassar College
Poughkeepsie, NY 12601
Eramsel@vaxsar.vassar.edu.

Dallas Savoie
Dept. of Psychology
U. of Saskatchewan
Saskatoon, Sask
S7N-0W0 Canada

Gedeon Deak
Inst. of Child Develop.
U. of Minnesota
Minneapolis, MI
55455 USA

Megan Clark
Dept. of Psychology
U. of Saskatchewan
Saskatoon, Sask.
S7N-0W0 Canada

**Abstract**

Previous research suggests that preschoolers have logical or cognitive deficits that limit their understanding of gravity as an explanatory concept. Four experiments were designed to test whether, in contrast to the results from previous research, preschoolers have a coherent, consistent and theoretical understanding of gravity. In each study, preschoolers made judgments regarding objects' behavior in at least one gravity-related event (e.g., speed of falling objects, trajectory of thrown objects, the behavior of balance scales). Predictions were made about children's performance based on the hypothesis that preschoolers understand gravity to be a property of objects. Predicted age-related changes in causal judgments were found on each task, as were positive correlations in performances across the tasks. The results support the claim that preschoolers understand gravity as a property of objects, an understanding that undergoes conceptual change.

## Introduction

Research from a variety of theoretical perspectives within cognitive-developmental psychology suggests that children's knowledge about the nature of the gravity develops over age. However, there is disagreement about the nature and process of the developmental change. Selman, Krupa, Stone, and Jacquette (1982) found a contingency between 5- to 20-year-olds' responses to an interview measure of gravity knowledge and measures of logico-mathematical reasoning. It was claimed that children in Piaget's preoperational stage, who lack logically reversible assimilatory structures, are only able to understand gravity as a cause-and-effect relation between an act (e.g., releasing an object) and an outcome (e.g., an object falling). These authors claimed that a complete understanding of gravity as an explanation of the behavior of objects is acquired only by adolescents who have previously acquired formal operations. Research from an information-processing perspective has focused on developmental changes in performance on gravity-related tasks. For example, Kaiser, Proffitt, and McCloskey (1985, Exp. 1) found that preschoolers incorrectly predicted that an object launched from a table would fall in a straight-down or inverted-L rather than a parabolic trajectory, regardless of whether the object was launched passively from a moving object or actively from an inclined plane. They also found (Exp. 2) that preschoolers predicted a straight-down trajectory for objects launched from a table at different speeds. Kaiser et al. claimed that young children lacked the necessary cognitive capacity and relevant experience to recognize that an object moves in a parabolic trajectory when launched from a table. Young children, they claimed, are only able to make predictions on the basis of the rule that objects fall straight-down when unsupported. Nussbaum and Novak (1976) also found that young children hold a straight-down direction rule when predicting the trajectory of objects falling in space.

These representative studies suggest that young children's understanding of gravity as an explanation of the behavior of objects is limited or constrained by cognitive or logical deficits. In this paper we present evidence from 4 studies testing the results from previous research. The major goal of this research is to demonstrate that young children have a theoretical, general, and consistent, albeit incorrect understanding of gravity as a property of objects.

## Experiment 1: Preschoolers' Causal Knowledge About Gravity and Light

Children's ability to form theories has been questioned by studies showing that preschoolers lack the cognitive tools to explain many physical phenomena. For example, it is claimed that preschoolers do not understand that gravity (Selman et al., 1982) and light (DeVries, 1979; Piaget & Garcia, 1974) involve the action of a causal force. These studies were based on children's verbal explanations of weightlessness and free-fall (gravity-related events), and projection and shadows (light-related events). The first study tests this claim by assessing preschoolers' causal judgments (not verbal explanations) about the status of four variables with regard to gravity and light. Subjects made judgments regarding each variable's effect on the speed of falling objects and the length of objects' shadows. If children lack causal knowledge about light or gravity, then they should show an inconsistent pattern of judgments regarding the causal status of the variables.

Nine 4- and 5-year-olds (4 girls and 5 boys) were the subjects in the experiment. The children judged whether variation in height, distance, orientation, and weight influences the speeds of objects in free-fall and the length of objects' shadows. The free-fall task involved dropping two objects that landed on the table or an elevated box (distance), from one of two positions on a stand (height). The shadow task involved illuminating two objects that were placed on the table top or elevated on a box (distance), with flashlights placed in one of two positions on a stand (height). Each task was presented twice to subjects, using small rectangular wooden blocks and a six-piece lego structure. Both height and distance are causal variables (they influence the outcomes), whereas orientation and weight are not causal.

These latter variables were manipulated by turning the block or lego structure around (orientation), and by adding either another block or lego piece (weight). Subjects were randomly presented with either the shadow or gravity task first, then completed the second task. Use of the blocks or lego first was also randomly assigned for each subject. Each task was completed with the blocks and legos before the second task was presented.

Each problem was posed in the following way; for the free-fall task: "When I let them go, will the blocks (legos) hit the ground at the same time or at different times?" and for the shadows problem: "When I turn on the flashlight, will the blocks (legos) have shadows of the same length or different lengths?" If subjects judged that the objects' lengths or speeds would be different, they were then asked to specify the object that would hit first or have the longer shadow. Subjects' responses were coded as correct or incorrect. Subjects' mean percentage of correct judgments on the free-fall task (M= .65) was significantly higher than their mean on the shadow task (M= .38). Statistically, subjects' mean percentage of correct responses on the shadow task did not differ from the percentage expected from random responding (M= .33), whereas it did differ on the free-fall task. Moreover, the means for all variables on the free-fall task, except distance, were significantly different from random responding (*p*'s<.01), whereas none of the variables on the shadow task was significantly different (see Figure 1). That children predicted that weight was a noncausal factor was particularly interesting since many adults use a "heavier-faster" prediction strategy regarding the behavior of objects in free-fall (Gunstone & White, 1981).

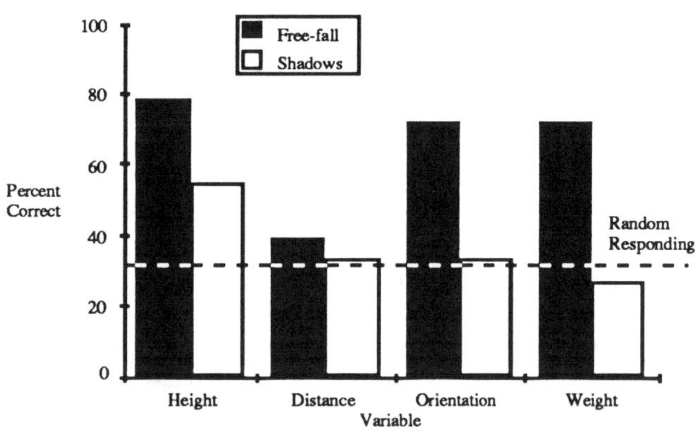


Figure 1: Percentage of correct predictions, Experiment 1

The results show that preschoolers make consistent judgments about the causal status of the variables regarding free-fall but not shadows. Two explanations can explain the task differences in preschoolers' performance. Preschoolers may have had more experience with gravity- (e.g., free-fall) than light-related events (e.g., shadows), which can account for the differences in task performance. Experience alone is likely an insufficient basis for making consistent judgments about the non-causal status of variables. However experience alone could be a necessary, but not a sufficient basis for making non-causal judgments because there are an infinite number of variables that are non-causally-related to an outcome. The second explanation of the results is that preschoolers have different concepts available to understand free-fall and shadows. It is proposed that childrens' performances are based on a materialistic understanding of gravity and a non-materialistic understanding of light. In a materialistic understanding it is assumed that light and gravity have a physical, material basis (Reiner, Chi, & Resnick, 1988). Without a materialized concept of light, children may know that light causes or produces shadows but have no basis for understanding how it does so. As a consequence, they would have no basis for inferring how variation in height and distance influences the length of shadows, or why weight and orientation are non-causal variables. On the other hand, children's material concept of gravity involves understanding that there is a force or mechanism that makes things fall. Shultz (1982) has demonstrated that an understanding of mechanisms is the fundamental basis of preschoolers' causal inferences. With respect to gravity, a material understanding supports inferences about the conditions under which the force or mechanism is the same for two objects (e.g., the non-causal variables of orientation and weight) or different for two objects (e.g., the causal variable of height).

## Experiment 2: The Nature of Preschoolers' Understanding of Gravity

Despite having a material understanding of gravity, young children may still understand gravity not as an attraction between masses, but as a property of objects (Reiner, Chi, & Resnick, 1988). Experiment 2 tests the claim that young children have a material, object-oriented understanding of gravity. We hypothesized that with an object-oriented understanding of gravity, children would judge that the gravitational force within an object changes when its appearance or structural integrity is altered. In particular, it was predicted that children with an object-oriented understanding of gravity would make different predictions regarding the gravity-related behavior of objects that differ in appearance or integrity.

To test the hypothesis, 24 preschool, 24 first grade, 12 third/fourth grade, 12 sixth/seventh grade and 12 college students were given a free-fall task and a weight conservation task. The tasks were presented in random order. In the weight conservation task, subjects saw three transformations (rolled into sausage, cut in half, and cut into many pieces) of one of two identical play-doh balls, and, after each transformation, were asked: "Which one weighs more, or do they weigh the same?" The free-fall task involved presenting two play-doh "bricks" of equal size. The subject was then asked, "If the identical bricks are dropped at the same time and from the same height, will they land at the same time or at different times?" Upon agreeing that the bricks would land at the same time, the experimenter cut one brick into two, stuck the two pieces back together, and again asked "If the bricks are dropped at the same time and from the same height, will they land at the same time or at different times?" The subject could answer that the two would land at the same time (correct), that the intact object would land faster (intact faster), or that the altered object would land faster (altered faster). The task was repeated 10 times, with subjects being told that the brick would be cut into hundreds of pieces on the last repetition. It was predicted that if children conceive of gravity as a property of objects, then by destroying the integrity of one brick, the two bricks would be predicted to hit the table at different times.

Subjects were very consistent in their judgments, with over 95% of the subjects giving the same response on 6 out of 10 items. There was a statistically significant age-related increase in the percentage of subjects who consistently made correct predictions, and an age-related decrease in the percentage of subjects who consistently made intact faster predictions (see Figure 2). Perhaps predictions on the gravity task simply reflect a misunderstanding that the weight of the play-doh brick changes when it is cut up. That is, subjects may judge that the transformed object weighs a different amount than the intact object, and because of the difference in weight, predict that the intact and transformed objects behave differently. However, there was no contingency between performance on the conservation of weight task and the free-fall task within any age group.

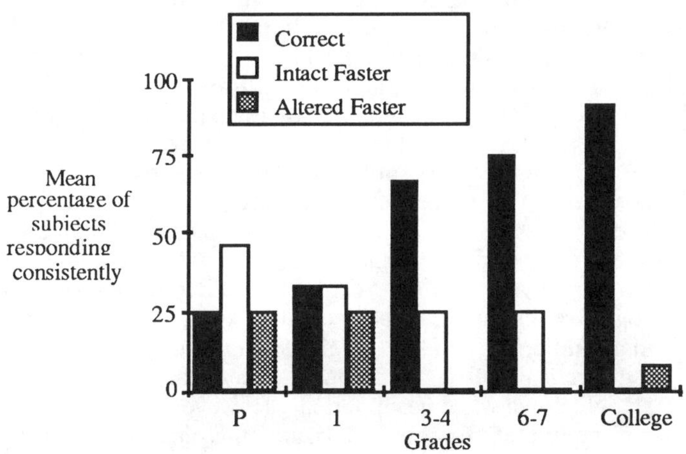


Figure 2: Subjects' responses on the free-fall task, Exp. 2

The results suggest that altering the integrity of objects influences preschoolers' predictions regarding the behavior of the objects in free-fall. But, by the elementary school years, children are no longer substantially influenced by the manipulation of the structural integrity of objects. When children make erroneous free-fall predictions, they are not based on a simple weight strategy (i.e., object perceived as heavier falls faster), since those subjects within each age group who conserved weight were no more likely to make correct free-fall judgments than those subjects who failed to conserve weight. Most significantly, the results suggest that preschoolers make predictions that could not be derived from experience alone. We suggest that the predictions are based on an understanding of gravity that is derived from theoretical assumptions as well as perceptual experiences. The theoretical assumption that children make is that gravity is a property of objects. No doubt preschoolers are capable of discovering properties of objects that affect their behavior; for example, that rubbery objects bounce, brittle objects break, and that spongy objects are pliable. In this case, children infer (rather than observe) that certain properties of an object are related to its behavior in free-fall.

## Experiment 3: Predictions on the Free-fall and Trajectory Tasks

Experiments 1 and 2 provide evidence that children have a material and theoretical understanding of gravity as a property of an object. This result challenges Selman et al.'s claim that children are capable only of an empirical understanding of gravity due to general logical deficits (e.g., a lack of reversible operations). Experiment 3 focuses on Kaiser et al.'s claim that preschoolers employ a straight-down rule when making predictions regarding the trajectory of falling objects launched from a table. They claim that the straight-down rule results from preschoolers' cognitive limitations in integrating the horizontal and vertical vectors of motion and their lack of perceptual experience. We examine whether erroneous predictions regarding the trajectory of falling objects can be explained as due to children's understanding of gravity as a property of an object. If preschoolers' use of the straight-down rule reflects a material and theoretical understanding of gravity as a force internal to objects, then (a) use of the straight-down rule is theoretically deduced and not the result of cognitive limitations or perceptual inexperience, and (b) preschoolers will believe that there is a force internal to rather than operating on objects that causes them to fall in the straight-down trajectory. To test whether children believe that the straight-down rule reflects a theory-based belief regarding a force internal to objects, we tested whether preschoolers predict that objects launched from a horizontal plane can follow trajectories other than straight-down. If preschoolers predict that objects can follow a variety of different trajectories (including the parabolic trajectory), then use of the rule does not merely reflect a cognitive or perceptual deficit (they can integrate horizontal and vertical vectors) or a belief in an external force that operates on objects. In Experiment 3, preschool and college students made predictions regarding whether an object launched from a horizontal plane could follow a variety of different trajectories. Another purpose of Experiment 3 was to examine the generality of preschoolers' understanding of gravity. If their understanding of gravity is general, then there ought to be empirical relations between performances on the free-fall and trajectory tasks. This prediction is consistent with Carey's (1985) claim that children's knowledge of the world is structured in terms of theories with general explanatory concepts and causal beliefs used to understand a domain.

Nine preschoolers and 8 college students were given a revised version of Kaiser et al.'s trajectory task and a revised version of the free-fall task. The revised trajectory task involved presenting subjects with a block sitting on a cardboard box; near the box were arrangements of two hoops. Subjects were required to judge whether or not they could push the block off the box so that it would go though the arrangements of two hoops. The two hoops were arranged in five different ways to reflect three different trajectories. The two Violation trajectories involved arrangements of hoops that required the block to change its trajectory in midflight. The two Straight trajectories (straight-down and inverted "L") involved arrangements of hoops that required the block to follow a straight-down trajectory. Finally, the Parabolic trajectory involved the block following a parabolic trajectory. (See Figure 3.) The only feedback subjects were given was during a practice session in which they pushed the block through one hoop. In the revised free-fall task, subjects made 20 predictions regarding whether two objects (legos, blocks, and play-doh bricks) would hit the table at the same time or different times. The predictions were made in three conditions, reflecting whether (a) the objects were released from the same heights and there were no other differences between them (Identical condition), (b) the objects were released from different heights and there were no other differences between them (Relevant condition), and (c) the objects were released from the same height but one object was transformed in a way that altered its structural integrity, appearance, status as an object, or connection between components but not its mass (Irrelevant condition). Each subject made four predictions in the identical condition and eight each in the relevant and irrelevant condition.

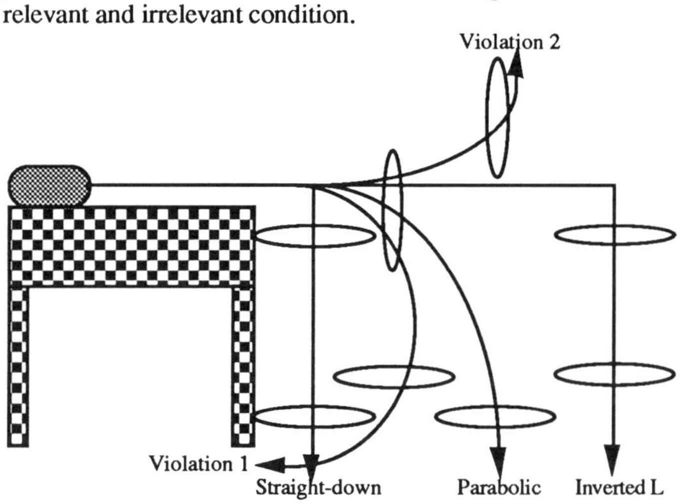

Figure 3: Arrangements of hoops used in Experiment 3

For the free-fall task, preschoolers' (M= .55) mean percentage of correct responses was significantly lower than adults' (M= .96). However, the mean percentage of correct responses by preschoolers was significantly above random responding in the identical (M= .66) and relevant (M= .61) conditions, but not in the irrelevant (M= .45) condition. Analyses of individual subjects suggested that 4 children consistently made the intact-faster judgment in the irrelevant condition, 1 made the transformed-faster, and 4 made correct judgments. A higher percentage of subjects consistently made correct predictions in this experiment (44%) compared to Experiment 2 (25%), although a chi-square analysis revealed no difference between the percentage of subjects in each experiment who made correct free-fall judgments.

The frequency of correct responses by subjects for the Violation, Straight and Parabolic trajectories was analyzed. Only the mean frequency of correct responses for the two Violation trajectories showed a significant age effect; the mean frequency of correct responses of college students (M= 1.62) for Violation trajectories was significantly higher than the mean for preschoolers (M= 0.44). The college students' means for the two Straight (M= 0.38) and the one Parabolic (M= 1.00) trajectories were not significantly different from the preschoolers' means (M= 0.33 and M= 0.78 respectively). The results show that the children believed that the block could be pushed through the hoops no matter what their arrangement, but they were only correct in their predictions on the Parabolic trajectory. In contrast, the college students correctly recognized the impossibility of pushing the block through the hoops in the Violation trajectory. Finally, there was a significant positive correlation (r= +.46) between subjects' total percentage of correct responses on the free-fall and the trajectory tasks.

The finding that children and adults were no different in predicting that objects can fall in straight or parabolic trajectories casts some doubt on Kaiser et al.'s claim that children's use of the straight-down trajectory rule reflects cognitive limitations and perceptual inexperience. Contrary to Kaiser et al.'s conclusion, the present results show that (a) preschoolers are not completely insensitive to the possibility that objects fall with integrated horizontal and vertical vectors, and (b) college students do not completely reject the possibility that objects fall without integrated horizontal and vertical vectors. Rather than reflecting cognitive limitations or perceptual inexperience, we propose that children's use of the straight-down trajectory rule reflects a theoretical belief: Under most circumstances objects fall straight down when unsupported. However, this belief does not imply that preschoolers assume that there is an object-independent gravitational force at work -- a force operating on objects rather than one internal to objects. Children also may believe that the internal force can be compromised by the way in which an object is set in motion.

The positive correlation between subjects' performance on the free-fall and trajectory tasks is a basis for claiming that an understanding of gravity underlies performance on the tasks. Such generality in children's understanding of gravity seems to be what Carey (1985) has in mind when she describes children as having theories constituted by explanatory concepts, causal beliefs, and a domain of application. Preschoolers' understanding of gravity as a property of objects may well be an explanatory concept, supporting causal beliefs regarding the role of object integrity in predictions of free-fall rates and the role of how objects are set in motion in predictions of object trajectory. However, the positive correlation can only be treated as preliminary evidence in support of such a explanation.

## Experiment 4: Predictions on the Free-fall and Balance Scale Task

Experiment 3 provides preliminary support for Carey's claim that preschoolers have general explanatory concepts (e.g., gravity) that are used in explanations of phenomena (free-fall rates, object trajectories) within a domain (behavior of objects). The purpose of Experiment 4 was to extend the analysis of phenomena that are influenced by preschoolers' understanding of gravity as a property of objects. In particular, we examine whether children's understanding of gravity is related to their knowledge of the operation of balance scales. Siegler (1976) examined developmental changes in reasoning about balance scales and found that between the ages of 5 and 10 years, children change in making predictions about the behavior of balance scales that are based exclusively on the weights of blocks placed on the arms (Rule I), to making predictions based on the blocks' weights and distances from the fulcrum (Rule II), although not based on integrated information regarding weight and distance. We propose that there are direct connections between changes in reasoning on the balance scale and free-fall tasks, each reflecting a change in children's understanding of gravity as a property of an object. We hypothesized three ways in which children's performance on the free-fall task would be related to performance on the balance scale task. First, children's use of Rule I or Rule II on the balance scale task would be related to their percentage of correct judgments on the free-fall task. Second, just as preschoolers judge that alterations in the appearance but not the mass of objects makes them fall at different rates in the free-fall task, so will they predict that alterations in the appearance of objects on the arms of a balance scale will cause the scale not to balance. Third, there will be a relation between predictions regarding the behavior of balance scales and objects in free-fall.

Twelve preschool, 12 first-grade, 12 third-fourth grade, and 12 sixth-seventh grade students who were used in Experiment 2 were also given a revised version of Siegler's (1976) balance scale task (Amsel & Goodman, submitted). In the revised balance scale task, subjects make judgments regarding the causal status of the causally-relevant (weight and distance) and causally-irrelevant (color and orientation) variables. Weight was manipulated by placing a different number of equally weighted rectangular blocks on the arms of the scale. Distance was manipulated by having an equal number of blocks on the arms of the scale placed at different distances from the fulcrum. Color was manipulated by having an opposite number of black and white blocks (i.e., the total number on each arm was the same) placed equidistant from the fulcrum. Orientation was manipulated having an opposite number of rectangular blocks "lying down" and "standing up" placed equidistant from the fulcrum. Subjects made 8 different weight, distance, orientation, and color judgments. The balance scales were drawn on cards, but subjects were shown a working balance scale during a practice session. The free-fall task was the one described in Experiment 2. Approximately half the subjects received the balance scale task first and the other half received the free-fall task first. Subjects' percentage of correct, intact-faster, and altered-faster judgments on the free-fall task was computed, as was their frequency of correct predictions regarding the Weight, Distance, Orientation, and Color variables on the balance scale task. Subjects were judged as using Siegler's Rule I if at least 6 out of 8 of their Weight predictions were correct, and if less than 6 out of 8 of their Distance predictions were correct. Twenty-nine subjects (with a mean age of 7 years, 3 months) fit the description of using Rule I. Subjects were judged as using Siegler's Rule II if at least 6 out of 8 of their Weight and Distance predictions were correct. Thirteen subjects (with a mean age of 10 years, 7 months) fit the description of using Rule II. The 6 subjects who fit neither Rule I or Rule II tended to be from the preschool and first grade groups. These findings replicate Siegler's results.

To test Hypothesis 1, Rule I and Rule II subjects' mean frequency of correct (out of 10) responses on the free-fall task was compared. The mean frequency of correct free-fall judgments by subjects using Rule I on the free-fall task (M= 3.90) was significantly lower than the mean of subjects using Rule II (M= 7.54). However, this statistically significant difference between the groups drops to a difference that only approaches statistical significance ($p$=.09) when frequency of conservation of weight scores is used as a covariate in an ANCOVA. The ANCOVA procedure was used so that the relation between the tasks do not reflect shared variance due to other general factors associated with cognitive development. To test Hypothesis 2, developmental changes in performance on the Color and Orientation variables on the balance scale task were examined. Preschoolers' mean frequency (out of 8) of correct color judgments (M= 6.67) was significantly lower than the means for the 5th/6th-graders (M= 8.00), but not different than the 1st- (M= 7.73) and 2nd/3rd-graders (M= 7.75). Similarly, preschool (M= 3.92) and 1st-grade (M= 4.18) students' mean frequency of correct Orientation predictions were significantly different then the 6th/7th-grade students' mean (M= 7.50), but not different than the 2nd/3rd-grade students' mean (M= 6.00). To test Hypothesis 3, children's frequency of correct predictions on the free-fall task (maximum score of 10) was correlated with their frequency of correct predictions regarding the causal status of the orientation and color variables (maximum score of 8 for each variable). A partial correlation procedure was used which removed children's performance on a conservation of weight task. Subjects' frequency of correct free-fall judgments was significantly ($p$<.05) and positively correlated with their frequency of correct judgments regarding the noncausal status of Color (r= .25) and Orientation (r= .31).

The results suggest that age-related changes in predictions regarding the behavior of balance scale variables are directly related to age-related changes in predictions regarding the rate of objects in free-fall. Why should this be? We propose that it is best explained as a restructuring in children's understanding of gravity. Children who believe that gravity is a property of objects should make incorrect free-fall predictions because the change in the appearance of the blocks marks different amounts of gravity in the blocks. Moreover, such children should predict that variation in blocks' orientation and color (an appearance-altering transformation) makes a difference in the behavior of the balance scale, whereas variation in distance (a transformation that does not alter appearance) makes no difference. This pattern of results seem to capture the performance of 5-year-olds, who made correct free-fall, distance, orientation and color judgments less often than older subjects. In contrast, 11-year-olds made correct free-fall, distance, orientation and color judgments more often than younger subjects. By 11 years of age it seems that children recognize that gravity involves some relation between an object and the earth, rather than exclusively being a property of objects.

While many other phenomena no doubt develop during the years between 5 and 11 that may underlie development on both the free-fall and balance scale tasks, there is some reason to believe that the relation is direct and not due to a third variable associated with cognitive development. There were significant correlations between correct free-fall judgments and correct orientation and color judgments, despite removing the influence of subjects' conservation of weight scores. However, the difference between Rule I and Rule II subjects in their frequency of correct free-fall judgments only approaches significance when conservation of weight score is used as a covariate.

## Conclusion

The major goal of this research was to demonstrate that young children have a theoretical, consistent, and general, albeit incorrect, understanding of gravity as a property of objects. Evidence for the theoretical nature of preschoolers' understanding of gravity as a property of objects comes from Experiment 1, in which preschoolers were shown to have a material understanding of gravity and Experiments 2 and 3, in which preschoolers were shown to believe that free-fall is influenced by an object's structural integrity, a variable which in reality does not make a difference. Evidence for the consistency of preschoolers' understanding of gravity as a property of objects comes from Experiment 2, in which (a) almost all subjects made consistent predictions regarding the rate of free-fall of two objects that were identical except that one object's structural integrity was damaged, and (b) a majority of the preschoolers predicted that the intact object would fall faster. This latter result was replicated in Experiment 3, in which 4 of 9 preschoolers predicted that an object's free-fall behavior is affected when its structural integrity (and other transformations in an object's appearance) is altered. Evidence for the general nature of preschoolers' understanding of gravity as a property of objects comes from Experiments 3 and 4, in which performance on the free-fall task was correlated with performance on the trajectory and balance scale tasks. The claim that preschoolers have a theoretical, consistent, and general, albeit incorrect, understanding of gravity as a property of objects is inconsistent with hypotheses based on previous research that preschoolers' understanding of gravity is the result of a deficit. Hypotheses following from the notion that children's understanding of gravity was due to logical limitations, or both a cognitive deficit and perceptual inexperience were tested and rejected in Experiments 1 and 3. In general we prefer a conceptual-change explanation of the data (Carey, 1985; Vosniadou & Brewer, 1987). According to this explanation, preschoolers' understanding of gravity as a property of objects serves as an explanatory concept within a theory regarding the behavior of objects and also serves as a basis for causal and non-causal beliefs across a variety of tasks.

## References


Amsel, E., & Goodman, G. (Submitted). Developmental changes in children's understanding of the behavior of balance scales.

Carey, S. (1985). *Conceptual change in childhood.* Cambridge MA: MIT.

DeVries, R. (1979). Children's concept of shadow phenomena. *Genetic Psychology Monographs, 112*, 479-530.

Gunstone, R., & White, R. (1981). Understanding gravity. *Science Education, 65*, 291-299.

Kaiser, M., Proffitt, D., & McCloskey, M. (1985). The development of beliefs about falling objects. *Perception and Psychophysics, 38*, 533-539.

Nussbaum, J., & Novak, J.D. (1976). An assessment of children's concepts of the earth utilizing structured interviews. *Science Education, 60*, 535-550.

Piaget J., & Garcia, R. (1974). *Understanding causality.* New York: W.W. Norton.

Reiner, M., Chi, M., & Resnick, L. (1988). Naive materialistic belief: An underlying epistemological commitment. *Conference of the Cognitive Science Society, 10*, 544-551.

Selman, R., Krupa, M., Stone, C., & Jacquette, D. (1982). Concrete operational thought and the emergence of the concept of unseen force in children's theories of electromagnetism and gravity. *Science Education, 66*, 181-194.

Shultz, T.R. (1982). Rules of causal attribution. *Monographs of the Society for Research in Child Development, 47*, (1, Serial No. 194).

Siegler, R. (1976). Three aspects of cognitive development. *Cognitive Psychology, 8*, 481-520.

Vosniadou, S., & Brewer, W. (1987). Theories of knowledge restructuring in development. *Review of Educational Research, 57*, 51-67.

# The Invention of the Airplane

**Gary Bradshaw**
**Marsha Lienert**

Institute of Cognitive Science
Campus Box 345
University of Colorado
Boulder, Colorado 80309
(303) 492-1496
gbradshaw@clipr.colorado.edu

**Abstract**

The invention of the airplane spans a period of 110 years from 1799 when Cayley first described the design of fixed-wing aircraft to 1909 when practical craft were flown at the Reims Air Show. At least 100 different designs were built and tested during this period, often at great expense, and occasionally at the cost of the pilot's life. With the exception of the Wright Brothers, progress was slow and sporadic. The Wrights needed only four years to develop their first airplane. Their efficiency is unique: No other inventor was able to duplicate the steady rapid progress of the Wright Brothers or duplicate their results until details of the Wright craft became available in 1906.

Numerous reasons have been advanced to explain this efficiency. Among these are the Wright Brothers' sense of materials, their focus on the control problem, the availability of gasoline engines, the synergy of the brothers in their work, and constraints imposed by limited funding. Although these factors are important, a crucial difference can be found in the manner in which the Wright Brothers and their contemporaries approached the task of invention. Most inventors spent their time constructing and testing complete craft; the Wright Brothers worked by isolating a problem, finding a system to test potential solutions, and integrating their solution back into an airplane design.

The problem-solving model of scientific discovery in turn explains why the Wright Brothers method was so efficient. The Wright Brothers were exploring a *function space* while their contemporaries were exploring a *design space* of aircraft. From the properties of the two spaces, it can be shown that search in the function space is far more efficient than search in the design space. Many of the same properties hold for the invention of other complex devices such as intelligent programs.

## Introduction

Sir George Cayley began the process of the invention of the airplane in 1799 with the publication of his paper *On Aerial Navigation*. This paper distinguished between the lifting and propulsion functions of bird wings and showed how these functions could be separated in a fixed-wing craft with propellers. One hundred and ten years later several practical craft were demonstrated at the Reims Air Show in France, completing the period of invention.

Figure 1 depicts the flight distances of 85 gliders and powered craft developed during this period, representing the vast majority of well-documented attempts to conquer the air. These attempts have been grouped into three categories: the Wright Brothers efforts, the efforts of other inventors from 1809 through 1905, and the efforts of other inventors from 1906 to 1909. Regression lines depicting progress over time are shown for each group: the long dashed line running across the figure represents early efforts by non-Wright inventors, the solid line rising sharply from 1900 represents the Wright Brothers, while bold line furthest right represents the efforts of other inventors from 1906 to 1909. In computing these regression lines unsuccessful craft that did not fly or were not tested have been excluded, as were 7 craft for which performance data was uncertain. Each point represents the maximum distance achieved by a particular design.

Trends in Figure 1 are obvious: for the most part, progress was slow and sporadic between 1809 and 1905. The Wright Brothers are a dramatic exception to this tendency. Their progress was rapid and steady. By 1905 they were making flights of 24 miles while their contemporaries were lucky to fly 300 feet. In 1906 the Wrights received a patent covering their control system, and details of the construction of their craft were widely published. Using this information, other inventors were able to copy the important features of the Wright craft and make rapid progress. By 1909 several were able to duplicate the Wright Brothers' success.

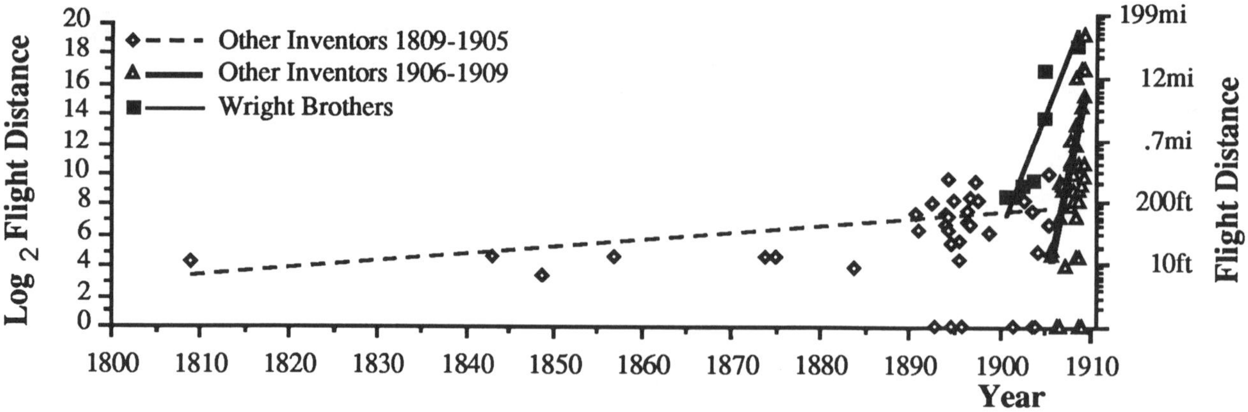


**Figure 1**
**Flight Attempts 1809 - 1909**[1]

## On the efficiency of the invention process

The issue we wish to address is: What made the Wright Brothers so efficient in their invention? This question has been discussed by historians and biographers, and many others as well. As with many other questions, the answer is not simple: Doubtless many factors contributed to the efficiency of the Wright Brothers. Yet many factors that contributed to their efficiency were held in common with other inventors. Our purpose is not to develop a comprehensive list of contributing factors, but to identify what set the Wright Brothers apart from their contemporaries.

Many historians, biographers, and others have suggested reasons for the Wright Brothers' efficiency. Two reasons are frequently suggested by people unfamiliar with the details of the invention of the airplane: the synergy provided by working together as a team and the constraints imposed by limited funding. Doubtless both factors contributed to the Wright Brothers' success, but these are far from unique factors. Many inventors worked in teams: Samuel Langley and Charles Manley, Gabriel and Charles Voisin, Otto and Gustav Lilienthal. Blériot assembled a large and talented team headed by Peyret. Indeed, working in teams was almost a necessity given the task of constructing a person-carrying craft. Along the same lines, we find most of the inventors struggling with limited resources: the Voisin brothers, Herring, even Blériot. None of the serious accounts of the invention of the airplane focuses on these as pivotal factors for the Wright Brothers' success.

Charles Gibbs-Smith, one of the foremost authorities on the history of the invention of the airplane, advances a different reason for the Wright Brothers' efficiency. He divided inventors into two camps: *chauffeurs of the air* and *airmen* (Gibbs-Smith, 1966). A characteristic of the chauffeur approach was the belief that an airplane would be similar to a car. Given sufficient power, a plane could be 'driven' into the air much as a car drives along a freeway. In the chauffeur mentality, muscle substituted for aerodynamics and control. This group focussed their efforts toward the development of large power plants. Sir Hiram Maxim, the inventor of the machine gun, illustrates the chauffeur mentality. Squandering £20,000 of his own funds, Sir Hiram constructed a four-ton steam-powered behemoth in 1894. The wingspread was over a hundred feet. Eighteen-foot propellers drew the unwieldy craft down an 1,800 foot track. On its first test, the plane struggled two feet into the air, broke a guard rail on the track, and was shut down, never to be tested again.

Airmen focussed their attention on the problems of flight with less regard to the development of powered craft. Airmen often built and tested gliders in an attempt to master the art of flight. The Wright Brothers clearly fall into the airmen category. They developed three gliders from 1900 to 1902 before attempting to construct an airplane. Gibbs-Smith's distinction is a useful one: we now recognize the importance of aerodynamics and control in airplane flight. Airmen have an obvious advantage over chauffeurs of the air in developing airworthy craft. Unfortunately, once again we find that this distinction does not differentiate the Wright Brothers from all other inventors. Lilienthal, Pilcher, Chanute, and Herring were airmen, sharing this advantage with the Wright Brothers.

[1] This figure was compiled using the following sources: Crouch (1982, 1989a, 1989b), Dollfus & Bouché (1980), Howard (1987), Jarrett (1987), Magoun & Hodgins (1931), Gibbs-Smith (1966), McFarland (1953), and Schwipps (1979). Models developed during the period are not shown due to the difficulty in equating their performance with larger-scale craft. The figure is not comprehensive: as many as 25 additional attempts are mentioned in Gibbs-Smith (1966) without sufficient detail for inclusion in the table. Few of these craft would have flown as far as 50 feet.

Gibbs-Smith also draws attention to the Wright Brothers' concern with active control instead of passive inherent-stability. Here we find one factor that distinguishes the Wright Brothers from their contemporaries. Even when details of the Wright system of control became available, few appreciated the importance of this contribution to the development of practical craft. Significantly, the Wright Brother's close friend Octave Chanute, who witnessed the success of the Wright gliders and flyers, was unable to appreciate the importance of an active system of control.

Yet the Wright Brothers' concern with active control does not explain all of the efficiency of their invention process. Indeed the control problem was largely solved by the Wrights in 1899 when they constructed a 5' biplane kite to demonstrate the efficacy of wing-warping as a method of lateral control. This kite also had an elevator for horizontal control. During the next two years the Wrights spent little time on this problem; the controls for wing-warping were tied off in most of the trials of the 1900 and 1901 gliders. The 1901 glider had a nearly-complete control system, yet Wilbur was so dismayed by the performance of the craft that he told Orville on the trip back from Kitty Hawk that man would not fly within their lifetime (Crouch, 1989a). In addition to working out the control problem, the Wrights revised lift tables of the day, determined efficient and stable shapes for wings, developed the modern theory of propeller function, and tested wing spars for drag. These developments were not motivated by a concern for active control, yet all were necessary for the first powered flight.

Biographers often mention the Wright Brothers' skill in the construction of lightweight craft. They were good with their hands and knew how to develop strong lightweight structures. In contrast, Samuel Langley would create engineering blueprints of craft, then turn the designs over to workmen for construction. The resulting products were often weak and unstable. Once again this factor is not unique to the Wright Brothers. Lilienthal, Chanute, and Herring were all accomplished at the construction of lightweight, strong craft. Lilienthal and Chanute both had engineering backgrounds. Chanute introduced to the field the biplane design strengthened by the Pratt truss used in bridge construction. This design was used by the Wright Brothers in all their early craft.

Without a doubt all of these factors contributed to the Wright Brothers' efficiency and ultimate success. Yet with the exception of their focus on active control, none of the factors is unique to the Wright Brothers, and so cannot explain their advantage over contemporary competitors. Part of their success derives from their focus on active control, yet this factor is too specific to explain all of the Wrights' success. Of course, success may derive from a subtle synergy of several factors working together, rather than being the product of a single factor. Rather than pursue this line, I will discuss one final factor mentioned by historians: the method used by the Wright Brothers. Although many give the Wrights credit for their method of invention, few comprehensive accounts have been provided that describe the nature of their method, and document why it was so different from their contemporaries.

## Methods of invention

Consider first the style of invention prevalent at the time. A would-be inventor would design and construct a new glider or plane. The craft usually embodied some novel feature that the inventor hoped would enable the craft to succeed where others had failed. The craft would be taken out to the field and tested. In many cases modifications or adjustments would be made to the craft, while at other times the craft would be quickly destroyed in an accident. Eventually the inventor would tire of the design; sometimes he would return to the drawing board to try again, sometimes he would give up in discouragement. Central to this approach is a concern with the performance of a working craft. Inventors measured their success in how high and far the craft flew, how long it remained in the air, and so on. Usually the results were not impressive: only 4 non-Wright craft out of 39 tested up to 1906 broke the 300' barrier, flying further than the length of a football field. Those who went back to the drawing board attempted to utilize their performance data to improve the design of a newer model, although many simply abandoned one ill-considered design in favor of another equally-poor choice.

This approach is termed *design-space* search for the prominence given to the design, construction, and testing of complete crafts. The inventor has available a number of different features that may be included in a design, including the number of wings, their shape and placement, the position of the rudder and elevator, propulsion systems, and so on. An inventor might explore a monoplane, biplane, or triplane design. Multiple wings could be arranged in a vertical, staggered, or tandem configuration. The rudder and elevator could be integrated in one unit or divided into two. Wings and tails could be mounted on springs, and so on. To construct a craft, the inventor has to make a decision about each of these factors, which are largely independent of one another. Table 1 shows some of the many features that were manipulated to create different designs for aircraft. Even though the list is far from comprehensive, at least 12,960,000 designs can be realized through different decisions about craft design. The task of the inventor is to determine a configuration of airplane components that leads to a working craft.

Octave Chanute developed an interesting craft in 1896 that illustrates the design-space approach. His glider had six pairs of wings that could be independently placed on the fuselage. The first configuration tested had all six pairs mounted in the front. This configuration was top-heavy. The number of wing pairs was reduced from six to four. Then it was tested with two pairs in front and four in the rear. The wings were shifted about mercilessly; finally the arrangement was fixed with five pairs in front and one pair at the rear. In this configuration it flew a distance of 82 feet. Later an umberella-like appendage, called an aerocurve, was added above the wings. The modified craft made a flight of 188 feet. In designing a glider with

## Table 1: Design space of gliders

| Characteristics of craft | Variants explored by early researchers |
|---|---|
| Number of wings | 1 - 80 |
| Wing configuration | monoplane, biplane, triplane |
| Wing placement | stacked, tandem, staggered |
| Wing angle | anhedral, flat wing, dihedral |
| Camber of wings | 1/20, 1/15, 1/12, 1/8, 1/6 |
| Wingspan | 6' to 104' |
| Chord | 3' to 10' |
| Shape of wings | bird-like, rectangular, bat-like, insect-like |
| Tail placement | forward (canard), rear, mid |

80*3*3*3*5*20*5*4*3 = 12,960,000 different designs

movable wings, Chanute was attempting to expedite the search through the space of glider designs: with a single craft he was able to explore a number of different designs.Even still, the different configurations that were tested represent only a fraction of the possible configurations shown in Table 1.

For the most part, the Wright Brothers did not work this way. Rather than building whole craft, they explored problems related to flight in isolation. This style is apparent from the very beginning of their efforts. Recognizing the need for a method to control the lateral orientation of an airplane to prevent rolls, Wilbur realized that by twisting the wings of an airplane it could be turned into an 'animated windmill' to right itself. The brothers built a 5' biplane kite to test this method. Twisting the wings of the kite proved effective at controlling the lateral movement of the plane. This solution was embodied in every craft the Wright Brothers built through 1909, without significant change.

Having solved the problem of lateral control, the Wrights built and tested their first glider in 1900. Given the innovative nature of the new system for controlling lateral roll via warping the wings, it might be expected that much of the testing that year would center around the controls. Nothing could be further from the truth: the new glider for the most part was flown as an unmanned kite with the wing-warping controls tied off. Wilbur and Orville were confident in their solution to the control problem, and were far more concerned with the amount of lift generated by their glider. They obtained crude estimates of lift using wind-speed data provided by the nearby Kitty Hawk weather station and measuring the pull of the ballasted craft with a spring balance. The obtained lift was far below their expectations, although some of the problem arose because they were forced to substitute short spars in their original design, and the curvature of the wing did not match that used by Lilienthal in determining his tables of lift.

Unhappy with the compromised construction of their first craft, the Wrights built a second glider in 1901. This craft had adequate spars and wing curvature matching the semicircular camber favored by Lilienthal. Before returning to Kitty Hawk in 1901, the brothers realized they needed their own anemometer to obtain reliable measures of wind speed at the test site. Once again the craft was routinely tested as an unmanned kite. Using a spring balance and measurements of wind speed, the brothers quickly realized that the lift developed by their craft was below the expected value. Recognizing the futility of attempts to construct further gliders in an effort to solve the problem of lift, a discouraged Wilbur told his brother on the trip back from Kitty Hawk that man would not fly for fifty years.

Recognizing the limited success of the 1901 glider, an inventor adhering to design space search might have continued to construct and test new 'improved' versions of the craft. The Wrights never appear to have given a moment's thought to this idea. After returning to Dayton, the brothers soon realized they needed another approach to explore the problem of lift. They wanted a way to quickly explore a large set of wing designs, something impractical with full-scale craft. Between September and November of 1901 the brothers built three different instruments to measure lift. One was a simple balance wheel mounted on a bicycle, two others were wind tunnels. Using the third wind tunnel, they were able to measure the lift and drag of fifty different wing designs in three weeks, and found a wing design far more efficient than that used in their 1900 and 1901 gliders.

Armed with this newfound knowledge, the brothers built a third glider in 1902, returning to Kitty Hawk to test it late that year. Although Lilienthal had flown hang gliders that achieved equal distances, the 1902 glider was the first successful aircraft with three-axis control. It represented the solution to most of the fundamental problems of powered flight. From this success, the Wright Brothers turned to the task of developing a powered aircraft. Here they were faced with another problem: how to build effective propellers. In late 1902 and early 1903 the brothers developed the modern theory of the propeller as a moveable wing with forward lift, and returned to their wind tunnel to explore effective propeller shapes. This led to the development of their 1903 flyer, the first successful airplane.

The Wright Brother's method can be understood as a search through the *function space*. For a plane to fly, the wings must have sufficient lift, the power plant must generate sufficient thrust, and the operator must be able to control the orientation of the craft. Rather than exploring the design space for a configuration that met these functions, the Wrights tackled each problem independently, and found effective solutions for each problem. Each feature could be integrated back into the design of an aircraft. This led to a pattern of steady progress: with the exception of the 1901 glider, each craft flew further than its predecessor[2].

## The efficiency of different search methods

The efficiency of the Wright Brothers was a natural outcome of their method of invention. For many inventions, search in the function space is easier than search in the design space. This inherent efficiency stems from two factors: feedback in the function space is more direct than feedback in the design space, and the function space is smaller than the design space.

To illustrate the role of feedback, we must recognize that inventors often use a hill-climbing strategy in their development process: they attempt to use the performance of an existing design to improve upon that design. Consider the problem of applying a hill-climbing strategy to make improvements to a glider or plane. When the current design is tested in the field, the performance of the craft is a function of a number of factors. Many factors are related to the design of the plane, such as the lift of the wings and the drag of the fuselage. Yet external factors play a role in overall performance as well. Is the air still or turbulent? Are updrafts or downdrafts present? Even the density of air and air temperature can have a measurable influence. The plane must be tested extensively to eliminate the influence of external factors. Even still, there is no simple way to attribute characteristics of flight performance to features of the design. Did the craft fly 80 feet because it was a biplane configuration or because it had a rear-mounted elevator? The global performance metrics give few clues to the individual contributions of the design features. Thus, even having an accurate picture of the performance of a craft in the field, it is difficult to develop an improved design. This can be seen in the many setbacks suffered by most inventors: it was not uncommon for a moderately-successful craft to be replaced by one that could not be flown.

Feedback in the function space is far more direct. When the Wright Brothers measured lift in their wind tunnel, they were obtaining direct evidence relevant to wing shape. From this information, they were quickly able to identify an efficient wing. Compared to the design space, feedback in function space is far more diagnostic, so that improvements are easier to discover.

The function space is a smaller space to search than the design space. This is evident when considering the small number of functions a plane must possess to fly: develop sufficient lift to overcome gravitational attraction, generate sufficient thrust to overcome drag, and control the plane in 3 axes. The vast array of contemporary aircraft illustrates that many different designs can execute the same basic functions, evidence that the design space is larger than the function space.

A curious aspect of function space search is that it often involves design-space search in a smaller design space. Consider for a moment the manner in which the Wright Brothers solved the problem of lift. In testing wing shapes, the Wright Brothers engaged in a search of a smaller design space. They had little theory of aerodynamics to guide their construction and testing of fifty wing shapes. The necessary mathematics and physics were not developed until after the airplane was a practical reality. Yet in investigating the design space of wing shapes, the Wrights did not have to concern themselves with other factors of glider design, such as fuselage shape, wing placement, and so on. The design space of wing shapes is a part of the larger design space of gliders and planes, because all wing shapes can be used in gliders or planes, yet gliders and planes involve other design decisions as well. By searching through the more restricted design space, the Wrights were quickly able to identify an efficient wing design, and preserve the efficiency of the function space approach in developing a working airplane.

Given the combination of direct feedback and a reduced search space, the method of investigating the functions of aircraft is far more efficient than constructing and testing full designs. Out of all factors contributing to the Wright Brothers' efficiency of invention, this factor is far more important than others that have been discussed. This method enabled the Wright Brothers to make sustained progress that eluded their contemporaries. How and why the Wrights happened upon this approach is not known, but the difference between Wilbur and Orville and their contemporaries is striking.

## Conclusion

This analysis is based upon the problem solving model of scientific discovery put forward by Herbert Simon (1973). Although invention of a complex device is not synonymous with the discovery of scientific laws, it shares an important component of creativity. The scientist or inventor is understood to be engaged in heuristic search. Drawing upon the general model of problem solving (Newell & Simon, 1972), we see that the Wright Brothers were searching through a different space than their contemporaries. Given the attributes of this search space, their search was inherently more efficient than other erstwhile inventors. As a result of this efficiency of

[2] The 1900 craft flew about 400' with assistance from the ground crew, who ran alongside and stabilized the craft in flight. The 1901 glider flew a distance of 389' in free flight.

method, the Wrights were quickly able to invent the airplane.

Not every invention has the properties of the airplane, and can be subjected to the same type of analysis. However, it is likely that the airplane is not an isolated case. Consider the development of complex intelligent systems, such as speech recognition programs. Inventors can focus their efforts on the construction and testing of complete systems and receive global performance feedback that is not very diagnostic, or they might focus their efforts on functions that must be carried out by the system, then develop and test modules that carry out each function. An advantage the Wright Brothers enjoyed was that the functional decomposition had already been provided by Cayley one hundred years before they began their research. Yet the efficiencies of the function-space method are so great that inventors might consider a program of determining the functional decomposition and working in the function space. Given the advantages of function-space search, this method is likely to be more effective than blind attempts to construct complex systems.

## Acknowledgements

The authors thank Lewis Harvey and Evelyn Ferstl for their assistance in translating German sources, and to the Undergraduate Research Opportunities Program for financial support. Additional funds were provided by the Department of Psychology and the Institute of Cognitive Science.

## References

Crouch, T.D. 1982. *Blériot XI The Story of a Classic Aircraft.* Washington, D.C.: Smithsonian Institution Press.

Crouch, T.D. 1989a. *The Bishop's Boys A Life of Wilbur and Orville Wright.* New York: W.W. Norton & Co.

Crouch, T.D. 1989b. *A Dream of Wings Americans and the Airplane 1875-1905.* Washington, D.C.: Smithsonian Institution Press

Dollfus, C., and Bouché, H. 1980. *Histoire de L'Aéronautique.* Reprint of 1942 edition published by L'Illustration. New York: Arno Press.

Howard, F. 1987. *Wilbur and Orville A Biography of the Wright Brothers.* New York: Alfred A. Knopf.

Gibbs-Smith, C. 1966. *The Invention of the Aeroplane, 1799-1909.* London: Faber & Faber.

Jarrett, P. 1987. *Another Icarus: Percy Pilcher and the Quest for Flight.* Washington, D.C.: Smithsonian Institution Press.

Magoun, F.A., and Hodgins, E. 1931. *A History of Aircraft.* New York: McGraw-Hill.

McFarland, M.W. ed. 1953. *The Papers of Wilbur and Orville Wright: Including the Chanute-Wright Letters and Other Papers of Octave Chanute.* New York: McGraw-Hill.

Newell, A., and Simon, H.A. 1972. *Human Problem Solving.* Englewood Cliffs, N.J.: Prentice-Hall.

Schwipps, W. 1979. *Lilienthal.* Berlin: Arami-Verlag GmbH.

Simon, H.A. 1973. Does scientific discovery have a logic? *Philosophy of Science 40,* 471-480.

# Language Evolution and Human-Computer Interaction

**Jonathan Grudin**

Department of Computer Science
Aarhus University
8000 Aarhus C Denmark

**Donald A. Norman**[1]

Department of Cognitive Science
University of California, San Diego
La Jolla, CA 92093

### Abstract

Many of the issues that confront designers of interactive computer systems also appear in natural language evolution. Natural languages and human-computer interfaces share as their primary mission the support of extended "dialogues" between responsive entities. Because in each case one participant is a human being, some of the pressures operating on natural languages, causing them to evolve in order to better support such dialogue, also operate on human-computer "languages" or interfaces. This does not necessarily push interfaces in the direction of natural language—since one entity in this dialogue is not a human, this is not to be expected. Nonetheless, by discerning where the pressures that guide natural language evolution also appear in human-computer interaction, we can contribute to the design of computer systems and obtain a new perspective on natural languages.

## Introduction

A "dialogue" does not require natural language, or even words. Animals engage in sustained interactions that can be characterized as dialogues. A mime is engaged in dialogues with real or imaginary objects and with the audience. Two individuals who do not share a common language can work out a means of communication, perhaps as a step to developing a shared "pidgin language."

In this paper, we address extended human-computer interactions that are "dialogues" in this general sense. We include *all* forms of human-computer interaction, not just "conversational" interfaces to computers. Consider, for example, this sequence of events in discarding a document on a Macintosh. As you move the mouse, the arrow or pointer moves across the display. When the pointer is above the icon that represents the document, you press and hold down the button on the mouse. The icon switches to "reverse video" (interchanging black and white), signalling that you have succeeded in selecting it. You move the mouse and an outline image of the icon moves, indicating your position and telling you that you remain in control of the document. When the pointer and the outline image reach the vicinity of an icon in the shape of a trash can, that image spontaneously switches to reverse video, signalling that when you release the mouse button, the document icon will disappear and effectively be discarded. You do so, and the sides of the trash can bulge slightly, indicating that the document is inside. The bulging sides signal that there are now documents in the trash can that can be retrieved if desired. This entire sequence can be considered to be a dialogue, although no words are used. The system and you monitor one another; you communicate by mouse movements and button presses, the system communicates by moving objects, switching them to reverse video, making them appear and disappear, and changing their shape.

An interface designer is really a designer of interaction languages. Computer systems are unique among artificial devices in allowing for a substantive, intelligent interaction between person and artifact. The development of interaction techniques is still in its infancy. Certain design guidelines are widely endorsed with little critical examination, such as "build consistent interfaces." The inconsistencies in natural languages—the naturally occurring, continually evolving communication media used for everyday interaction among people—*have* been analyzed, revealing tradeoffs among competing pressures on dialogue. By contrasting the two domains, by finding where analogues exist and where they do not, we may obtain insight into the nature of computer system design and a richer perspective on the constraints on natural languages.

Below, we examine the "design guidelines" proposed by Slobin (1977) for this rich and complex natural system—human language. Obviously, languages have not been designed; they have evolved over thousands of years subject to numerous competing pressures, including political, cultural, and religious factors. Nonetheless, a natural system such as language has much in common with artificially designed computer systems. Many of the requirements are similar. Each must act as a communication medium to transmit intentions, actions, and results among the participants, each must be learnable by beginners, yet efficient for skilled performers. The ability of naturally evolving systems such as language to deal with these conflicting pressures can be revealing for the design of computer systems.

In this paper we examine the changes in both natural and computer languages, the latter including high-level interface languages, operating systems, and even programming languages. We restrict ourselves to word choice and form and syntactic structure. Thus, we do not deal with speech acts or other subtleties of language.

1 Norman's research was supported by grant NCC 2-591 to Donald Norman and Edwin Hutchins from the NASA Ames Research Center in the Aviation Safety/Automation Program. Everett Palmer served as technical monitor. Additional support was provided by funds from the Apple Computer Company and the Digital Equipment Corporation to the Affiliates of Cognitive Science at UCSD.

## Four Design Characteristics for Language

Slobin (1977) has analyzed what we might call the "design characteristics" of language, aspects of the usability and functionality of language that lead toward language development and change and that affect the ease of acquisition by children. He identifies four constraints on language:

1. Language should be clear;
2. Language should be quick and easy;
3. Language should be expressive;
4. Language should be processible.

We examine the application of each of these rules both to natural language and to human-computer interaction.

### 1. Language Should Be Clear

**Natural language.** Slobin defines clarity to be a consistent "one-to-one mapping between underlying semantic structures and surface forms." Thus, Slobin's concept of clarity corresponds to consistency as it is generally applied in human-computer interaction. Consistency in a language facilitates learning, both in children and adults. Children not only learn more quickly where it is found, but they enforce consistency by ignoring alternative constructions (using "I will" or "I will not" where adults would say "I'll" or "I won't") or by using a consistent form even where it is considered to be ungrammatical (using "hitted" rather than "hit" for the past tense).

All natural languages have inconsistencies, the irregularity of verbs being a well-known example. These irregularities cause the language learner great difficulty, because violations of consistency mean that a single rule no longer applies to a wide class of instances, and instead, many cases have to be learned individually. Although people have created more consistent, artificial languages (e.g., Esperanto), it is significant that none of the thousands of known naturally-forming languages is completely consistent. If consistency were as primary a design rule as some have argued, one might have expected to find a greater appearance of consistency in natural languages.

**Human-computer interaction.** Computer systems can accrue the same benefits as natural language systems from a clear, consistent mapping between underlying semantic structures (or actions) and surface forms (or commands and system output). Here, too, consistency has been shown to facilitate learning (e.g., Polson, 1988).

However, despite heavy rhetoric advocating consistent design and its prominent place in the standard guidebooks, consistency is often violated. This is not solely due to oversight—in the best of systems, this violation can improve performance (Grudin, 1989). A major point of this paper is to show that some of the same pressures that militate against consistency and an emphasis on clarity in computer systems are found in natural language as well, where they are clearly seen to serve important purposes.

### 2. Language Should Be Quick and Easy

**Natural language.** A language principle that often conflicts with consistency and clarity is the desire to be quick and easy. This tendency shows up in numerous ways. Most common words are short and monosyllabic, even in languages that relish long words, such as German. Language is further simplified through abbreviation or other shortening, obtaining efficiency at the expense of learnability, regularity or even clarity. Irregular verbs and plural nouns are generally shorter than their regular counterpart would be—inconsistency is introduced in the service of efficiency.[2]

Often, as a word increases in frequency of use, it is given an abbreviated form: "automobile" becomes "auto," "television" becomes "TV," "picture element" becomes "pixel." Pronouns shorten utterances, but at the cost of introducing ambiguity, reducing clarity. Entire phrases may be eliminated in the cause of efficiency. Although such utterances can technically be ambiguous, usually, when interpreted in context, they are not.

Note that irregular constructions that simplify and shorten will work only if everyone is familiar with them. Therefore, irregularity is most often found with frequently occurring constructions—it is the most frequently occurring verbs that tend to be irregular.

Irregularities cause difficulty during learning, but once learned, they simplify the language process, making the constructions quick and easy to use. As long as the irregularities are frequently encountered, they stay learned. Thus, the mature native speaker seldom has difficulties with irregularities: It is only the learner or the novice user who has trouble.

An interesting demonstration of the relationship between irregular language forms and frequency of usage occurs as language evolves and words change in their frequency of usage. When the frequency of usage of an irregular verb drops, the verb also drops its irregularities and reverts to a regular form (Bybee, 1988). Thus, speakers are not burdened with the task of keeping track of language exceptions that rarely occur.

**Human-computer interaction.** Do we find the same push toward non-standard, abbreviated structures in computer interactions? Yes, a frequent user's desire for quick and easy means to carry out operations results in simplification, abbreviation—and, therefore, inconsistency. Many computer systems allow their users to create short keystroke sequences as substitutes for longer command sequences: Some systems even provide these "shortcuts" as standard features: shell commands, aliases, scripts, macros, dedicated function keys, option-key equivalences, or "power-keys." Much as the shorter

[2] For example, from a list of 173 irregular English verbs, the irregular written form is shorter in 166 cases, the same length in six, and only longer in one ("bought" is longer than "buyed" would be, although equally "quick and easy" to pronounce).

constructions in natural language tend to be those that are used with higher frequency, shortcuts in computer systems are used primarily for high-frequency operations.

In the Macintosh computer, users wanted a quick way to eject a diskette from the drive and to free the memory that the system had reserved for it. Initially, two operations were required: an "eject" command and the action of moving its remaining, "greyed-out" icon into the trash can. In a triumph of usage over consistency, an imaginative programmer combined these into one operation, carried out by moving the diskette icon to the trash can icon. The operation violated many people's notions of consistency and confused first-time users, but due to its overwhelming efficiency it became widely accepted.[3]

Computer users who create their own shortcuts often produce namesets that are efficient, but so inconsistent that they themselves subsequently forget the names that they devised (Grudin & Barnard, 1985). They may misjudge the frequency with which these terms will be accessed. Other users, of course, are likely to find these personal shortcuts to be incomprehensible. Natural language handles the corresponding problem through several mechanisms.

With a computer system, if a user invents a new command name or other shortcut, this innovation is kept relatively private: Only the user and the computer system need know. Similarly, if a computer designer creates a poor name or shortcut, a user may be able to fix it with an alias, but again this remains a private adjustment.[4] With natural language, however, a neologism is only effective if it is used with others. This shared social use provides for a natural evolutionary process. Successful innovations are those that are kept alive through usage within a language community—we see examples in the way that some slang terms maintain their existence through frequent usage, whereas others die natural deaths. In language evolution, one natural tendency is towards consistency, and only frequently used constructions maintain an inconsistent form. Today's computer systems provide neither the extensive shared social use of innovations nor an equivalent evolutionary process that will rescue users from poorly devised names or procedures.[5]

**The "Law of Least Effort" in human performance.** The pressure to increase efficiency is observed in many domains of human skill. Zipf (1949, 1965) postulated that a general "law of least effort" applied to much of human behavior. Zipf showed that a power law applies between the length or size of an instance and its relative rank of frequency occurrence.[6] Ellis and Hitchcock (1986) have found that experienced computer users create command abbreviations ("aliases") that follow Zipf's Law, with shorter terms used for higher-frequency commands. As expertise develops, people modify the task, system, language or method of operation in order to produce smooth, effortless, and efficient performance (Agre & Shrager, 1990; Newell & Rosenbloom, 1981).

## 3. Language Should Be Expressive

**Natural language.** Natural languages must have powerful expressive capability "... to communicate effectively, engagingly, appropriately, and so forth. The speaker must be able to direct the listener's attention, to take account of his knowledge or expectations" (Slobin, 1977, p. 187). The central point here is that language must function in a wide range of contexts, requiring a versatility that often comes into conflict with the other constraints. In order to be both expressive and efficient, language must be compressed—thereby sacrificing a clear, consistent mapping between form and function. Slobin writes, "it is the charge to be expressive which introduces much of the complexity into language."

Miller (1951) observed that "the social pressure for a common vocabulary and the convenience of monosyllabic words tend to restrict the variety of our responses, whereas the attempt to differentiate between similar statements expands the vocabulary and leads to the occasional use of polysyllabic words" (p. 94). This captures the opposing pressures of Slobin's maxims "be expressive" and "be quick and easy." The various tensions push the solutions in opposing ways.

**Human-computer interaction.** The range of expression is narrower in computer interaction than in language, but as applications mature the demands for a wider range of expressiveness grow, and we find analogous sources of inconsistency. Information retrieval systems provide a wide range of search capabilities, whereas a simple string search is sufficient for a word processor; a professional typographer requires a degree of layout precision not needed for most document preparation. The result is often inconsistent interaction languages of varying complexity. In general, large applications may have hundreds of commands to satisfy the requirements of thousands of different users, who oftentimes require very different system performance. We expect this issue to be of increasing importance as computer systems become richer and more powerful.

3 The inconsistency has always bothered the design team, however, who plan to phase out this "slang" shortcut if a more consistent but equally efficient solution is found.

4 A major use of customization features the "undoing" of designer innovations in new releases (Mackay, 1990).

5 The best analogy between linguistic and computational neologisms may be private abbreviations used in personal diaries or notebooks. Here there are no social interactions, and here the analogy holds well: Much as in the computer world, a diarist's clever abbreviations and phrases may prove undecipherable to the writer turned reader.

6 This abbreviated description fails to do justice to Zipf's numerous observations of the relationship between ranking, size, and relative frequency. His observations, largely forgotten, may be worth reexamination as indicative of some general principles of human action.

### 4. Language Should Be Processible

**Natural language.** The rate at which the speaker and listener can accurately encode and decode language utterances must be comparable. If they proceeded at rates that were too discrepant or led to too much error, communication would suffer[7]. This is a particular challenge in spoken language, because of the non-persistence of sound—the listener has a limited ability to review what has been spoken, and thus must process it in "real time."

**Human-computer interaction.** Computer communication is persistent: The computer can preserve a record of input and can provide persistent output by means of a static visual display or by allowing ready repetition of an otherwise transient auditory or visual signal. Even so, a general constraint to be humanly processible operates in the visual medium as well as the acoustic. For example in the design of visual icons, the relative size difference of a trash can and a document in the real world is not mapped onto the interface (it would make one icon too large or the other too small); similarly, one may enhance the users' ability to distinguish among objects by exaggerating differences (Hollan Hutchins, McCandless, Rosenstein, and Weitzman, 1987). Thus, if it is crucial for the users of a system that controls an industrial process to distinguish between 200-gallon and 220-gallon boilers, a designer might use icons that vary in size by 50% rather than a precisely-mapped 10%. This violates a clear mapping of semantic information onto surface form, but provides greater human processibility.

## Contextual Factors in Language Change

In addition to the design rules, Slobin (1977) discussed four different means by which natural pressures can change natural languages:

1. Gradual evolutionary processes;
2. Contact with other languages;
3. Creolization;
4. Individual development.

In this section we briefly examine these four avenues of language change and the way similar factors influence computer interaction design.

[7] The issue is not simply that the listener be able to keep up with the speaker, but that the processing rates be comparable. If the "speaker" were forced to handwrite instead of speak, the mismatch of rates would disrupt communication: Normal reading rates are about twenty times as fast as normal writing rates. We handle this discrepancy by having most writing take place "off-line," asynchronously with reading. In normal spoken conversation, the mismatch between speaking and listening rates is not sufficiently great to affect communication, although especially slow speakers can tax especially quick listeners.

### 1. Gradual Evolutionary Processes

Broad shifts in a language occur that are independent of specific external pressure on it. Slobin presents evidence that these primarily improve how well language can be processed: "At each point in its history the language has apparently been strongly constrained by the charge to conform to perceptual strategies." He also discusses constraints that facilitate production (speech).

These language changes correspond in a sense to broad changes in computer interactions that also have moved toward conformance with perceptual-motor abilities. One step in this direction is the shift from simple "glass teleteype" interactions—single line statements displayed on terminals, and typewriter keyboard input— to full-screen graphical interfaces with input through pointing and gesture. Future computer systems promise to enhance the perceptual mapping through increasing use of graphical displays, including large screens, color, and three-dimensions, and the use of motion and sound. Change in production is manifested in the proliferation of input mechanisms including pointing devices, gesture recognition, and even voice recognition and eye-tracking. Interestingly, Slobin notes that language shifts are accompanied by an initial focus on increasing consistency, a pattern also found in computer system design.

However, there is generally little evolutionary force upon specific computer systems apart from slow pressures of the market and innovation that lead to new releases. These artificial systems are relatively immutable: Once designed, one is unchanged until a new system takes its place. A major exception is in the evolution of inherently extensible computer language systems such as Lisp and Unix, in which new constructions or commands that are added by any user become relatively indistinguishable from the original language primitives. But as noted earlier, computer systems lack a feedback or "natural selection" mechanism. The result of evolution for both Lisp and Unix has been an amazing proliferation of commands and structures, so that a new user faces daunting sight of manuals and documentation whose size is measured in meters. Instead of simplifying a user's task, this form of evolution has increased the learning burden.

But signs of evolution are indeed there. Some of the original constructs of Unix and Lisp are no longer taught to newcomers and are replaced instead with more efficient and useful evolutionary appendages. But we suspect that computer systems suffer from the lack of social interaction and communication. Children learn a language by existing and interacting within a community, and what these new learners acquire then determines what they will pass on to their children. The related process in the acquisition of computer languages and systems has a much different character.

### 2. Contact with other Languages

When two societies that speak different languages come into contact, the languages change, in part to make communication between the language groups more efficient. Over time, each language may import elements of the other language: Individual words are the first to cross over, but eventually whole syntactic structures can be incorporated to allow quicker and easier speech (Slobin, 1977). Part of the price of this merger of the two languages and the overall improvement in communication is the introduction of inconsistencies.

A clear analog is found in the computer domain. Operating systems, applications, and application domains can be thought of as independent language families. Contact among these language groups takes place as users move among them or when a single computer comes to support several systems (e.g., as the stand-alone word processor, personal computer, transaction processing, and other worlds come together). Different names are suddenly being used for the same thing or the same name has different meanings in different contexts. This seems a particularly promising topic for further exploration.

In computer programming languages, as with other human-computer interfaces, the clash of different cultures has meant changes to all languages. Thus, elements of structured programming have come to even the least structured languages of all: Basic and Fortran; and algebraic languages have made their impact upon such deviant structures as Lisp and Prolog, which in turn, have led to changes in the algebraic languages.

### 3. Creolization

The term "Creolization" refers to the creation of a new language by the expansion of a "pidgin" or barter language. Pidgins are communicative systems developed to make it possible for groups that use widely different language systems to interact. These are used primarily for bartering and they tend to be simple and not very expressive. When children acquire the pidgin as a first language, this starts its evolution into a full-fledged language—a Creole. Children first make the language more regular, then expand it to apply it to all situations, adding vocabulary, verb tense, and so forth.

Erickson (1990) notes parallels between pidgin languages and many of today's simple computer interaction languages. As functionality is added, a point is reached where the language form cannot support the desired functions: It is time for the pidgin to become a full-fledged language. Erickson notes that the lack of tense—our restricted ability to refer to past and future events—is shared by pidgins and computer languages. Such limitations are often most apparent to new users of a system who may feel that the existing structures are needlessly complex yet insufficiently expressive for their needs. New users provide the pressure to develop a full-fledged language—Creolization.

### 4. Individual Development

Slobin notes that the language learner is first most concerned that language be consistent and processible. Later, the language learner is willing to sacrifice consistency for expressiveness and efficiency. Speakers of natural languages share their knowledge of the language by propagating their innovations to other speakers.

In the computer world, one finds similar processes. Consistency is of most importance for learners, whereas advanced computer users may welcome or develop shortcuts, even at the expense of consistency. Advanced users do tend to share their special knowledge with others, trading macros, scripts, hints, and shortcuts (Mackay, 1990). Computer magazines usually have columns devoted to hints for the use of specialized systems. And informal tutoring networks develop.

Even so, there is far less sharing in computer usage than in language, because most dialogues involve only one person, and the computer does not learn from the experience. Innovations in speech are immediately passed on to the people with whom we speak, but innovations in computer use only affects one computer system's interaction with the innovator. We have to make a special effort and use a special forum to communicate this innovation to others. To complete the analogy with language, it is the computer that needs to change: As we develop shortcuts, the computer system must make them available to other users of similar computer systems.

## Language Evolution and the Design of Computer Systems

The analyses of natural languages and the design of interactive computer systems reveal many of the same pressures. In both communication media, these pressures lead to innovations in the structure of the medium, inconsistencies, and a continual tension between expressiveness, ease of use, ease of understanding, and ease of learning.

Computer systems lack the human ability to interpret context and are thus unable to take full advantage of mechanisms for promoting efficiency. Computer systems and designers could make better use of contextual effects to interpret people's actions, allowing simplification of the actions required of the user. A good example of the use of context is in the specification of Unix files. The full name of a file includes its compete "path" (the entire directory hierarchy), but Unix allows for considerable abbreviation by using the current location of a user in the file hierarchy as the default context. There may be many files in the computer system named "notes," but a user who types just the name "notes" is assumed to be referring only to files in the current directory. Unfortunately, this nice use of context is more the exception than the rule in current system design.

Spoken human communication inevitably contains errors. Listeners often do not even notice these errors because the context makes the utterance interpretable even when ambiguous or erroneous. When listeners do have troubles, the speaker can often detect this through the listener's nonverbal and verbal reactions. Language is an example of a system that seems designed for error—it tolerates a good deal of imprecision and it provides error-correcting mechanisms that are so effective that, after the fact, sometimes neither listener nor speaker is aware of the error. Computer systems' general lack of sensitivity to context means that developers must take the initiative by building in safeguards and confirmation steps to prevent catastrophic errors (Lewis and Norman, 1986)—which can, of course, add complexity or inconsistency to the dialogue.

Today, the proper analogy with computers is perhaps not full-fledged natural languages, but rather pidgins. Like pidgins, human-computer interaction deals with exchanges between users and system that are restricted in domain. Pidgins are restricted in expressive power. But the "pidgin" used for human-computer interaction must develop toward a Creole as greater range is sought, thus bringing into play all the issues discussed in this paper.

Computer systems are still small and limited. Unlike natural language systems, they do not last for multiple generations of users, and they do not provide mechanisms for the sharing of developments among the user community. Unlike human listeners, they do not evaluate innovations and propagate good ones. More important, perhaps, is that there is none of the richness of natural language that allows for heavy use of context, a relative insensitivity to error, and efficient error correcting mechanisms.

An understanding of how naturally evolving, intensely social systems such as languages cope with conflicting pressures can help the designers of artificial systems. But if we are to adapt some of the lessons, we must move beyond today's systems which have relatively limited capabilities and limited lifetimes and that are static and unresponsive. Instead, we must learn to develop systems that have long lifetimes of gradual evolution, and that are adaptive, flexible, and robust.

## Acknowledgment

We thank Karen Courtenay, Rod Owen, Larry Parsons, and Steve Pinker for helpful discussions and literature references on the role of language change and irregularity. Tom Erickson, Don Gentner, John Paulin Hansen, Jim Hollan, Phyllis Reisner, Hank Strub and John Sullivan also provided helpful comments.

## References

Agre, P. and Shrager, J. 1990. Routine evolution as the microgenetic basis of skill acquisition. *Proceedings of the Conference of the Cognitive Science Society.* Hillsdale, NJ: Lawrence Erlbaum.

Bybee, J. 1988. Morphology as lexical organization. In M. Hammond and M. Noonan (Eds.), *Theoretical morphology.* New York: Academic.

Ellis, S.R. and Hitchcock, R.J. 1986. The emergence of Zipf's Law: Spontaneous encoding optimization by users of a command language. *Transactions on systems, man, and cybernetics, 16,* 3, 423-427.

Erickson, T. 1990. Interface and evolution of pidgins: Creative design for the analytically inclined. In B. Laurel (Ed.), *The art of human-computer interface design.* Reading, MA: Addison-Wesley.

Grudin, J. 1989. The case against user interface consistency. *Communications ACM, 32,* 1164-1173.

Grudin, J. and Barnard, P. 1985. When does an abbreviation become a word? and related questions. In *Proceedings of CHI '85.* New York: ACM.

Hollan, J.D., Hutchins, E.L., McCandless, T.P., Rosenstein, M., and Weitzman, L. 1987. Graphic interfaces for simulation. In W.B. Rouse (Ed.), *Advances in man-machine systems research, Vol. 3,* Greenwich, CT: JAI, 129-163.

Lewis, C. and Norman, D. A. 1986. Designing for error. In D. A. Norman and S. W. Draper (Eds.), *User centered system design.* Hillsdale, NJ: Lawrence Erlbaum.

Mackay, W.E. 1990. Patterns of sharing customizable software. *Proceedings of CSCW '90.* NY: ACM.

Miller, G. A. 1951. *Language and communication.* New York: McGraw-Hill.

Newell, A. and Rosenbloom, P. S. 1981. Mechanisms of skill acquisition and the law of practice. In J. R. Anderson (Ed.), *Cognitive skills and their acquisition.* Hillsdale, NJ: Lawrence Erlbaum.

Polson, P. (1988). The consequences of consistent and inconsistent user interfaces. In R. Guindon (Ed.), *Cognitive science and its applications for human-computer interaction.* Hillsdale, NJ: Lawrence Erlbaum.

Slobin, D. I. 1977. Language change in childhood and history. In J. Macnamara (Ed.), *Language learning and thought.* NY: Academic, 185-214.

Zipf, G. K. 1949. *Human behavior and the principle of least effort; an introduction to human ecology.* Cambridge, MA: Addison-Wesley.

Zipf, G. K. 1965. *The psycho-biology of language; an introduction to dynamic philology.* (Introduction by George A. Miller.) Cambridge, MA: MIT.

# Human discovery of laws and concepts; An experiment *

**Jan M. Żytkow**
Computer Science Department
Wichita State University
Wichita, KS 67208, USA
zytkow@wsuiar.wsu.ukans.edu

**Anna N. Żytkow**
Institute of Astronomy, Madingley Road
Cambridge CB3 0HA, England
anz@ast-star.cam.ac.uk

**Keywords:** law discovery, concept generation, experimentation heuristics

## Abstract

In order to understand the relationship between human and machine discovery, it is necessary to collect data about human discoveries which can be compared with machine discovery performance. Historical data are difficult to obtain, are very sparse, and arguably do not reflect a typical human performance. We contend that cognitive experiments can provide meaningful data, because the discovery tasks can be carefully defined and tailored to the comparison task, and the subject selection can be controlled in different ways.

We describe an experimental study of the human processes of concept formation and discovery of regularities. In our experiments human subjects were allowed to interact with three world models on the computer. Our results demonstrate that humans use heuristics similar to those used in computer discovery systems such as BACON or FAHRENHEIT on comparable tasks which include finding one dimensional regularities, generalizing them to more dimensions, finding the scope of a regularity, and introduction of intrinsic concepts. Virtually all our subjects made some relatively simple discoveries, while some of them were able to develop a complete theory of simple world models. The progress made by human subjects on comparably simple tasks was impeded significantly when the domain became richer, as measured by the number of regularities, their dimensionality, and the number of intrinsic concepts involved. This can be called a contextual complexity phenomenon. Our subjects demonstrated a definite pattern of chaotic experimentation and lack of theoretical progress when the level of complexity was too high.

*The work described in this paper was supported by the Office of Naval Research under grants #N00014-88-K-0226 and #N00014-90-J-1603.

## The task

Little systematic study is available on human attempts at discovery tasks (Mynatt, Doherty, and Tweney, 1977; Shrager, 1985, 1987; Qin and Simon, 1990). Studies that concentrate on concept creation, such as Bruner, Goodnow, and Austin, (1956) do not consider concepts in the context of law discovery. Discovery has traditionally been perceived as an exceptional achievement that happens rarely and to special people. Recent experience with computer discovery systems demonstrates that the capability for making discoveries can be decomposed into a number of different "small" capabilities, each devoid of mystery and applicable to a broad range of cases. Relatively simple discovery systems like BACON (Langley, Simon, Bradshaw, and Zytkow, 1987) or FAHRENHEIT (Koehn and Zytkow, 1986) are successful in solving simple discovery tasks and in combining them into more complex discoveries of concept and law. FAHRENHEIT has been applied in controlling chemistry experiments and collecting data in a real laboratory, and to infer theories based on those data (Żytkow, Zhu, & Hussam, 1990). Studies on human subjects done by Shrager (1985, 1987), Qin and Simon (1990), and casual experiments reported in Langley et al. (1987, p.53) demonstrate similar discovery behavior in humans. It is reasonable to conjecture that many humans are endowed with the skills needed to become successful discoverers. World models which can be explored both by humans and discovery systems provide a controlled environment for studying human discoveries.

In order to construct and justify a framework for describing human discoveries, it is necessary to collect a lot of data in exploratory mode. In our experiments we wanted to see how humans combine different discovery skills, and how they select methods appropriate to different tasks, in particular the tasks which involve the accumulation of several discoveries. We were able to keep our subjects motivated up to several hours. Thus, simple multidiscovery tasks that could be concluded in a matter of hours (covering perhaps a few days in separate sessions) were chosen. In the future we foresee longer experiments.

Our subjects were confronted with computer simulations of simple world models. These world models were interactively encoded, allowing subjects to conduct experiments of their choice. Yet, how does one create a world of appropriate difficulty? We did it empirically, by trial and error. In the first series of experiments (section on the ZOO), we underestimated the difficulty of discovery problems. In effect, our subjects discovered very little, and the protocols were not useful for comparisons because different people explored different small subspaces of all possibilities. We have learned several interesting things, however, which we briefly report in the section on the ZOO experiments. The next two experiments (GOBLINS and GREMLINS) were more successful and we report on them in this paper.

## The method

Our subjects were computer science graduate students and undergraduate seniors who had enrolled in machine learning courses. They were told that their participation in the experiment would help them understand the problems of machine learning and discovery. Although there were no time limits set on their discovery task, they were asked to continue their research as long as needed to develop a complete theory of each world, a theory that allows one to predict the result of any experiment. They were allowed to conduct as many sessions as they wished, and to resume them when they wished. Only after all subjects finished their work, the issues of machine discovery have been addressed in class. To explore the experimental worlds, the students were placed in a loop in which they designed an experiment, were provided with a result, designed the next experiment, and so forth. Each experiment took a few keystrokes, and the result was immediately available. Students developed their theories on paper.

We did not want to create abstract problems and, at the same time, we wanted to minimize the interfering background knowledge. The world models were introduced in brief stories. The stories introducing GOBLINS and GREMLINS are quoted in the next two sections. After completing their experiments, many subjects acknowledged that they quickly forgot about the stories, and started to think in terms of abstract problems. The worlds of GOBLINS and GREMLINS required at least several hours of study before a subject could develop a more or less complete theory. The level of difficulty was a little too high, but the ability to keep people interested was just about right for the selected group of subjects.

Two types of protocols were kept. First, each experiment was automatically recorded on a computer: time of experiment, input data and the result were stored. Secondly, we asked students to record as many ideas that came to their mind as possible, including all hypotheses that they entertained, and to mark the time at which they occurred. The information about time was helpful in cross-referencing both types of protocols. As we expected, the notes of students differed considerably. Still, we did not want to specify in detail what they were supposed to record. We felt that any instruction that would result in a more uniform collection of data would provide hints that might influence their discoveries.

After the students' protocols were turned in, only some clarifying questions were asked. In this way we sought explanation of unfinished phrases and obscure formalisms that they invented. No additional comments were accepted that could reflect knowledge gained at a later stage. As a result, the protocols were quite clear and adequately complemented the listings of experiments collected on the computer. We believe that the open format we selected best fits our exploratory studies. However, other methods could also be used (Ericsson and Simon, 1984).

## The experiment with GOBLINS

The following story introduced the world of GOBLINS:

**There are many goblins in the fairyland. Each goblin is characterized by intelligence and appetite. Both can be ANY number, and are limited to numbers. Each goblin has its energy, too, that is measured in numbers. Does energy depend on intelligence and appetite? If so, in what way? This is the problem for you. Find a complete theory, applicable to all goblins. You can develop your theory as a result of experimentation with goblins. You can also make hypotheses if you wish.**

The subjects could study the world of GOBLINS by making "experiments". Each experiment consisted of selecting numerical values for intelligence $i$ and appetite $a$ for which the value of energy $e(i,a)$ was returned. The world model of GOBLINS is described by a numerical function $e(i,a)$ of two independent numerical variables $i$ and $a$:

$$e(i,a) = \begin{cases} (a-1)(0.5i+1) & \text{for } a > 2i \\ i * a^2 & \text{for } a > 10 - i^2 \\ a^2 + 2 & \text{in all other cases} \end{cases}$$

Possible discoveries include (1) three regularities for energy $e(i,a)$ as a function of intelligence $i$ and appetite $a$, that hold in three areas of the 2-dimensional plane $(i,a)$, and (2) two boundaries that separate those areas. Each regularity can be discovered by BACON and FAHRENHEIT. FAHRENHEIT can also discover the boundaries $a = 2i$ and $a = 10 - i^2$ that separate the areas of different regularities.

Eleven students worked on this problem. The number of experiments varied between 71 and 1227. The

Table 1: Summary of experiments and theories generated within the GOBLINS experiment

| Number of experiments | 1227 | 1067 | 300 | 289 | 95 | 224 | 137 | 71 | 180 | 80 | 96 |
|---|---|---|---|---|---|---|---|---|---|---|---|
| Formulae discovered | 3 | 3 | 3 | 3 | 3 | 3 | 2 | 2 | 1 | 1 | 0* |
| Boundaries found | 2 | 2 | 2 | 1 | 1 | 0 | 1 | 0 | 0 | 0 | 0 |

* Two regions of different behavior were discovered

time spent on experiments varied from 16 minutes, corresponding to 137 experiments, to 924 minutes, corresponding to 1067 experiments, plus additional time on building theories. Everybody had some success in making discoveries (cf. Table 1). Three students developed a complete and correct theory, including each of the three regularities and their boundaries. Columns 3 and 5 of Table 1 stand out in comparison to other entries because those subjects needed fewer experiments to achieve successful results. In both cases, early experiments were centered on a particular idea. It is not really possible to generalize the typical route to success, because the protocols reveal a variety of approaches. For instance, for the entry in column 6 with 224 experiments, the records show that the first conclusions were reached after one experiment, and all subsequent experiments were very systematic with a good range of positive and negative numbers as input data. The first correct formula was thought up after about 15 experiments, the second after 93 experiments and the third after 144 experiments. A long time was then spent in trying to understand the boundaries for each of the three formulae; this issue, however, was never fully resolved. Much value was attached to the predicting power of new ideas.

Protocols of students' data collection reveal a frequent use of two experimentation patterns which can be described as (1) sequential and (2) boundary search, usually bisectional. The first pattern includes all sequences of experiments in which one variable is varied whereas other variables are kept constant. A sequence of experiments can be further characterized by the initial value of the varied variable, the increment, and the number of experiments in sequence. Typically, a student kept one of the variables, say appetite, constant and varied intelligence in a sequence of 4-6 experiments, using a small integer as a starting value, and a small increment, usually 1. Then he varied appetite by a small increment and made another sequence of experiments by varying intelligence. This can be called "recursive sequential experimentation" and it occurs when the same pattern is used for varying other variables. Both BACON and FAHRENHEIT use this strategy for data collection.

Sequential experimentation was used in the search for regularities, whereas the second experimental strategy was used almost exclusively in the search for a boundary after a regularity had been detected. The bisectional search narrows the distance between two values of an independent variable (other variables held constant): the last known value for which a particular regularity holds, and the first for which that regularity fails. FAHRENHEIT uses this strategy, systematically dividing the difference in half, and terminating when the difference is less than the experimental error. Some of our subjects terminated that search when the difference was one, but some others carried the divisions to the limit of the floating point operations.

## Experiments with GREMLINS

The world of GREMLINs was described in the following way:

**There are many gremlins in the fairyland. There is a distinct species of gremlins for each character from a to z. For instance 'c' gremlins, or 'v' gremlins. Each gremlin belongs to one species, while there may be many gremlins in one species. Gremlins like to eat, and each gremlin has its (or his, or her) taste denoted by a positive number. Gremlins like to fight each other, too. Find the law that for ANY TWO gremlins, tells which one wins if they fight one against the other. You can develop your theory in result of experimentation with gremlins. You can also make hypotheses if you wish.**

The subjects specified the type and taste for both gremlins: a letter $l_i$, $(i = 1, 2)$, and a positive number $t_i$, $(i = 1, 2)$ for each, and in return received a message about win, tie or loss by the first gremlin. Each pair $(l_i, t_i)$ defines a particular gremlin. Invisibly to the experimenter, each letter is mapped to a number:

```
a b c d e f g h  i  j  k  l m  n  o  p
2 3 4 5 6 8 9 10 11 12 1 12 12 12 12 4.6666

q  r  s  t  u      v  w  x  y  z
12 12 12 12 8.3333 12 12 12 12 12
```

Then, the products $n_i t_i$ are computed and compared. If $n_1 t_1 > n_2 t_2$, then the first gremlin wins; if $n_1 t_1 = n_2 t_2$ , then both tie; if $n_1 t_1 < n_2 t_2$ , then the first gremlin loses.

Two major discoveries include (1) the introduction of an intrinsic property for each gremlin, defined by a mapping from letters to numbers, and (2) finding that there is a boundary between the wins and loses, and that $n_1 t_1 = n_2 t_2$ determines the tie, while two inequalities determine the win and loss.

Fourteen students worked on this problem. The number of experiments performed varied between 55 and 589, with the amount of time spent at a terminal between 62 minutes (corresponding to 60 experiments) and 460 minutes (corresponding to 482 experiments). Three persons developed a theory that accurately predicts all experimental results, each theory based on a different formalism, in 179, 326, and 589 experiments respectively. Three other students developed almost the full theory, but were plagued by errors in assigning numbers to letters. Seven students assigned numerical values to letters or to pairs of letters. Seven never tried to map letters to numbers.

Contrary to the world of GOBLINS, where everybody detected a meaningful regularity, GREMLINS presented insurmountable problems for those who did not introduce an intrinsic numerical property. Those subjects detected only "weak" regularities, such as "the higher letter wins" (at the conclusion of 55 experiments), and "species in category 1 are stronger than species in category 2; in the same category it requires higher taste to beat someone farther away alphabetically" (at the conclusion of 125 experiments).

An intrinsic property can be introduced in various ways. Our subjects used three ways of assigning numbers to letters: (1) a number to each letter, (2) a fraction to a pair of letters, (3) express each letter by a multiplier for a selected single letter, say k: a is 2k, b is 3k, and so forth. Within (1) two mappings were used. One was the *ordinal* mapping of letters to consecutive numbers: a-1, b-2,c-3,...,z-26. The other was identical or close to the actual mapping used in GREMLINS, and described above. Below we give a sample of various experimental strategies, formalisms, and theories related to the introduction of intrinsic concepts by individual subjects.

S-1 introduced the ordinal mapping after 31 experiments, then switched to the mapping derived from experiments, but still some remnants of the original mapping remained in the final solution reached after 164 experiments.

S-2 introduced an almost correct mapping at the end of his study, after 258 experiments, as a result of theoretical analysis.

S-3 performed about 100 experiments and then introduced a "shifted" ordinal mapping, by assignments: a-2, b-3, c-4,..., z-27. Later he realized that k is special, and letters past k are equivalent, but he never gave up the initial assignment, producing as a result a very complex theory of 21 rules. The cost of maintaining a wrong paradigm is obvious, but the result was empirically adequate, with some exceptions.

S-4 performed 150 experiments and then started to use fractions of letters, and equated them with fractions of numbers, for instance $a/b = 3/2$. Very soon S-4 realized that $\frac{a}{c} = \frac{a}{b}\frac{b}{c} = \frac{3}{2}\frac{4}{3} = \frac{2}{1}$, and quickly generalized this computation to all letter fractions. Afterwards, S-4 used only fractions of consecutive letters. Finally, after 326 experiments, S-4 derived a theory that perfectly predicts results of all experiments, but never attempted a mapping to single letters.

S-5 wrote down the ordinal mapping entirely apriori, before starting any experiments. But in 178 experiments he never used that mapping and he did not produce any regularities, ending up "frustrated at the fact that the system contained so many inconsistencies."

S-6 assigned fractions after 28 experiments, then after 54 experiments switched to a "multiplier" for each number, at the same time producing a theory that perfectly predicts all results. From then, until he completed 179 experiments, he concentrated on assigning numbers to letters.

S-7 came up with another perfectly predicting theory, making 589 experiments, and using still another formalism. All letters were grouped and all groups were ordered according to different multipliers '¡of k. For instance b belongs to 2k, c to 3k. k plays a "privileged" role and does not belong to any group. No attempt to assign numbers to letters was made.

Recovering from the erroneous mapping of letters into numbers (the ordinal mapping) provided us with interesting accounts of scientific crisis and theory revision. S-3 responded to the empirical inconsistencies with complicating his theory, while S-1 changed his theory gradually. The first law had the form $t_2 = 2t_1/(N_1 + N_2)$, and was very nicely confirmed within the first 39 experiments. Then, in response to the growing counter-evidence (around experiment 93), S-2 introduced another law, not changing the ordinal mapping. $(N_1 + 1)t_1 = (N_2 + 1)t_2$. This law works nicely for letters a through e. After that, rather than creating new "small" laws as did S-3, S-2 used the new law for assigning numbers to letters. He eventually changed his law to the final form: $N_1 t_1 = N_2 t_2$. His protocols offer a captivating example of gradual recovery from a false theory.

Two very useful experimental strategies were introduced. One consisted in using the same taste value for both gremlins, while the letters were varied. This allowed for efficient ordering of all letters according to their "strength". S-2 used it in 84 experiments, before switching to another strategy which he used for 86 final experiments. The second strategy, used by all six successful subjects can be described as a search for the boundary between win and loss, in a sequence of experiments where two letters and one number are fixed. The second number, usually the taste of the first gremlin was varied until a draw was produced, or until the boundary was determined with the integer accuracy. The assumption that the tie is a state between win and lose was responsible for the bisectional boundary search at the beginning, while the method was later strengthened by its successful results. Several unsuccessful subjects also used this strategy, but they tried it unsystematically, abandoning it for more chaotic experimentation. Only the successful subjects systematically varied three other variables in the four-dimensional space of all experiments. Two discoveries made by all successful subjects reduced their experimentation. First, varying the other number was quickly abandoned when the law of proportionality for the tastes was detected. Then, after the symmetry between the first and the second gremlin was detected, or a mapping between letters and numbers, only selected combinations of letters were used.

## The ZOO experiment

This experiment was chronologically first. The task was too difficult, however, allowing the subjects to make only sparse discoveries. The ZOO world encodes Black's law and is isomorphic to the processes in which the temperature of equilibrium is reached for a mixture of two samples (each can be water or mercury), characterized by their initial temperatures and masses. Intrinsic concepts include specific heat, latent heat, and the phase. Each experiment requires a choice of value for six variables: one nominal and two symbolic for each sample. The dimensionality of regularities and their boundaries is much higher than in the case of GOBLINS and GREMLINS. The cover story talked about animals rather than samples. Curiosity played the role of mass, while intelligence doubled for temperature. Animals stayed in pairs, and in each pair the intelligence of both animals became equal. The task was to find the intelligence of animals in a given pair.

Eight students, different from those who worked on GOBLINS and GREMLINS, worked on the ZOO problem. Nobody developed the full theory governing the ZOO (i.e. discovered the full form of Black's law) or even came close to discovering the latent heat law, the specific heat concept, etc. Some lesser things were discovered though. For instance, several students came across the numbers 0 (freezing point of water), 100 (boiling point of water), 357 (boiling point of mercury), and noted their significance, describing them as *magic numbers* (2 persons), telling that "Strange things happen for 357, 100", or that "some ranges of numbers always seem to give you the intelligence of 357". Only three persons attempted to fit some formulae to their experimental results. The rest gave qualitative descriptions only: "beta animals seem to be more intelligent", and the like.

Two phenomena differentiate the work on the ZOO from the simpler domains of GOBLINS and GREMLINS. First, it seems that the overall difficulty encountered in the ZOO impeded even partial results. It was much easier to find regularities and concepts if they occurred in the simpler worlds of GOBLINS and GREMLINS, even if they were equally difficulty as many partial results in the ZOO. The second phenomenon refers to the experimentation in the ZOO world. Everyone started with a very systematic, sequential experimentation pattern (see Section on GREMLINS), which for all but two subjects quickly degenerated into chaos. Four subjects showed a pattern of random choice of input parameters in at least the last 15% of their experiments. Chaotic experimentation patterns seem to be due to the complexity of the problem and should be further investigated as a response to non-progressing research, which does not provide new regularities and generalizations of the existing laws. Chaos emerged towards the end of most ZOO experimentation, whereas only two students working with GREMLINS started to show a bit of chaotic pattern towards the end of their efforts, complaining about complexity of the problem: "this is definitely getting complex".

## Dimensions of discovery

The results of our study can be analysed along many dimensions. In the current section we will briefly summarize selected results.

**Experimentation strategies:** We have already reported on these in the sections on GOBLINS, GREMLINS, and ZOO. The exploration of all worlds benefited from the use of orderly experimental methods, even though some people needed hundreds of experiments to achieve success. In particular, the strategy of two persons who best described the world of GOBLINS can be characterized as that of a calculator programmed to check every detail step by step as opposed to reaching any conclusions in a more daring way.

**Experimental Limitations:** Most subjects used only a very limited range of input values in experimentation. Small integers ($n < 13$) were used almost exclusively. Six persons never used fractions. Most of the others used them sporadically if they could not complete the bisectional search otherwise. Five people never used negative numbers in their experiments with GOBLINS. Small integers were almost exclusively used as increments for generating experimental sequences.

**Testing:** Very little time was spent on testing the discovered theories. We noted an intriguing phenomenon. After making a major discovery, a student commonly switched to another task rather than going on to test the validity of the discovery. The same phenomenon can be noticed in protocols from experiments conducted by Shrager (1985). Positive thinking prevailed over critical thinking, and there is little evidence that falsified theories were rejected. In general, subjects did not change their theories in response to contradicting experiments, even if performed immediately afterwards. Positive evidence, however, was remembered for a long time.

**Complexity:** First conclusions were drawn very quickly, on average after 10 experiments. This is compatible with discovery systems, which also draw conclusions that involve one independent variable after examining only a few results. Another phenomenon was quite clear: If two laws of similar complexity hold in two world-models, $wm_1$ and $wm_2$, $wm_2$ being overall more complex than $wm_1$, then it is far more difficult for human discoverers to find such a regularity in $wm_2$ than in $wm_1$. Discovery systems can succeed equally well in both models, if they are put on the right track.

**Meta-principles:** The modest support sufficient to draw meta-level conclusions was remarkable. Subjects who made metalevel statements used just one or few experiments to claim symmetry ("order of pairing is unimportant"), time independence ("time is not a factor"), or determinism ("no randomness in the domain"). It is actually true that both domains were deterministic, and the world of GREMLINS is symmetric.

**Beyond empirical equivalence:** Additional criteria can be applied to further rank the final theories. Three subjects reached the experimentally complete theory of GREMLINS, but one result ranks higher because each gremlin was represented by a number, whereas in the remaining two a number is associated with each pair of gremlins. The first theory is more comprehensive. It implies both theories of the second type, but not the other way.

## Conclusion

Our research confirms that many students have been able to combine useful experimentation strategies with elementary discoveries, and to combine elementary discoveries into more complex theories. This confirms conclusions of Shrager (1985), and Qin and Simon (1990). Similarly to Qin and Simon, we found many analogies between human subjects and discovery systems in the use of operators and heuristics However, it is difficult to go beyond superficial comparisons to related work because the space of possible experiments on discovery is enormous, and each existing study explores an "orthogonal" direction. Qin and Simon (1990), for instance, used Kepler's third law to study in depth the process of one-dimensional equation finding. Although discoveries in our worlds involved several one-dimensional equations, none of them has been complex enough to allow comparison to the results of Qin and Simon. Moreover, while Qin and Simon provide students with data, we require them to make experiments. Experimentation would simplify the problem, because it would be easier to find Kepler's law in data obtained for small integers, which the subjects would undoubtedly use.

Our work can be extended in many directions. We would like to better understand the mechanisms by which humans introduce intrinsic variables and use symmetry heuristics. BACON's mechanisms for both (Langley et al., 1987) were not used, but BACON's symmetry operators require a particular experimentation strategy (Langley et al., 1987, p.171-4). If that strategy is not followed, the comparison is difficult. Experiments on new versions of GREMLINS may help in solving these problems.

Another interesting problem is the proportion of effort spent on experimentation and theories. In a yet unreported experiment, we found that even a few seconds' delay in computer response to the requests for experimental results stimulates subjects' interest in theory formation, limiting the number of experiments they perform before they start building theories.

**Acknowledgements.** Many thanks to Mary Edgington, Lance Petrie, and two anonymous reviewers for their helpful suggestions.

## References

Bruner, J.S., Goodnow, J.J., & Austin, G.A. (1956). *A Study of Thinking.* New York, NY: Wiley.

Ericsson, K.A., & Simon, H.A. (1984). *Protocol Analysis; Verbal Reports as Data.* Cambridge, MA: MIT Press.

Koehn, B.W., & Zytkow, J.M. (1986). Experimenting and theorizing in theory formation. In: Ras Z., Zemankova M. eds.*Proceedings of The ACM Sigart International Symposium on Methodologies for Intelligent Systems* New York, NY: ACM SIGART, 296-307.

Langley, P., Simon, H. A., Bradshaw, G. L. & Zytkow, J. M. (1987). *Scientific discovery: Computational explorations of the creative processes.* Cambridge, MA: MIT Press.

Mynatt, C.R., Doherty, M.E., & Tweney, R.D. (1977). Confirmation bias in a simulated research environment: an experimental study of scientific inference. *Quarterly Journal of Experimental Psychology* 29: 85-95.

Qin, Y., & Simon, H.A. (1990). Laboratory replication of Scientific Discovery Processes. *Cognitive Science* 14: 281-312.

Shrager, J. (1985). Instructionless learning: Discovery of the mental model for a complex device. Ph.D. diss., Dept. of Psychology, Carnegie-Mellon Univ. Pittsburgh, PA.

Shrager, J. (1987). Theory change via view application in instructionless learning. *Machine Learning* 2:247-276.

Żytkow, J.M., Zhu, J., & Hussam, A. (1990). Automated Discovery in a Chemistry Laboratory, *Proceedings of the AAAI-90.* AAAI Press, 281-287.

# Hypothesis Generation and the Coordination of Theory and Evidence in Medical Diagnostic Reasoning

**José F. Arocha and Vimla L. Patel**

Cognitive Studies in Medicine
Centre for Medical Education
3655 Drummond Street, Room 529
McGill University
Montreal, Quebec, Canada
H3G 1Y6
AX53@MUSICA.McGill.CA

**Abstract**

This paper investigates the process of hypothesis generation and the coordination of hypothesis and evidence in medical diagnostic tasks. Two issues are addressed: the generation of hypothesis and the directionality of reasoning. Two problems whose initial presentation suggested an initial hypothesis were presented to subjects with different degrees of expertise in clinical medicine. When faced with contradictory evidence against the initial hypothesis, 1) early novices either modified the initial hypothesis, or ignored, or reinterpreted the cues in the problem to fit the hypothesis; 2) intermediate novices generated concurrent hypotheses to account for different sets of data; and 3) advanced novices generated several initial hypotheses and subsequently narrowed the hypothesis space by generating a single coherent diagnostic hypothesis. All subjects, used a mixture of forward reasoning and backward reasoning. A more forward-directed reasoning was related to diagnostic accuracy. These results on diagnostic reasoning are discussed in relation to findings on scientific reasoning.

The study of diagnostic reasoning has had a long history in various fields, especially in troubleshooting and medicine. In medicine, this study has had a number of different forms. One is the normative study of diagnostic reasoning based either on the rules of deductive logic or on the experiential knowledge and intuition of physicians (Feinstein, 1967). Another is the psychometric study of the factors affecting physicians' diagnostic skills (Rimoldi, 1961). Still, another form is the comparison with the processes assumed to be characteristic of scientific reasoning. The cognitive tradition in the study of medical reasoning, which started with the publication of the book by Elstein, Shulman, and Sprafka (1978), gave support for the view that diagnosticians reasoned in way similar to the way that scientists are assumed to reason; that is, by applying hypothetico-deductive methods. This view was later challenged (Groen & Patel, 1985) given arguments from philosophy of science and results from studies in cognitive psychology. This led to the conceptualization of reasoning as analogical, rather than deductive. Philosophical argument as well as cognitive science research supports this position.

This work was supported by a grant from the Medical Research Council of Canada (MH004) to Vimla Patel and by a fellowship award from the Royal Victoria Hospital Research Institute (McGill University) to José Arocha.

Criticism of the work of Elstein, Shulman, and Sprafka (1978) was made because they seemed to have proposed hypothetico-deductive reasoning as psychologically real, rather than as an intellectual tool. In actuality, people do not think logically; rather, they think by means of "retrieving" analogous cases from prior experience. Furthermore, as it is now widely known, people make logical mistakes in their reasoning; and obviously, scientists or physicians are no exception. However, this does not mean that they do not *use* hypothetico-deductive methods. They may use it in circumstances where it is required.

## Hypothesis Generation in Scientific and Diagnostic Reasoning

Several attempts have been made by researchers working within different theoretical approaches to investigate the reasoning processes in scientific discovery tasks: Langley, Simon, Bradshaw, and Zytkow (1987) in studies and simulation models of how scientists solve scientific problems; Dunbar (1989), Dunbar and Schunn (1990), Klahr and Dunbar (1989), Kuhn (1989), who have worked on simulated scientific problems; and Tweeney and Hoffner (1987) in their reconstruction of the scientific reasoning from Faraday's diaries.

Kuhn (1989) addressed the issue of the conceptual understanding and its relationship to processes and strategies in scientific problem solving. She argued for a developmental shift in the processes used in coordinating theory and evidence from childhood and adulthood. Children, more than adults, fail to differentiate evidence and theory; these become integrated, undifferentiated, in a single whole. Adults, contrarily, are able to differentiate between the two and can therefore accommodate or revise their theories when faced with inconsistent evidence, something that children are unable to do. Kuhn (1989) has shown that subjects tend to ignore evidence that contradicts the theories they believe to be true. She explains this by hypothesizing that subjects fail to separate their beliefs about the world from the evidence that support them. This separation is, in part, determined by the type of evidence the subjects have to coordinate. The weaker the evidence the more difficult it is to achieve such a separation. The difficulty of problem solving increases with the the presence of weak evidence. The process of solution with weak evidence requires combining the pieces of evidence such that they add up to the equivalent value of a strong evidence. Strong evidence more directly suggests a solution.

A theoretical distinction has been made by Klahr and Dunbar (1988) in their work on scientific discovery: working on a science domain involves a search on, at least, two spaces, the space of the *hypotheses* and the space of the *experiments*. They have found that subjects' doing scientific discoveries do a search on these two spaces to solve problems. In a recent study, Dunbar and Schunn (1990) presented two problems with the same underlying structure to subjects in a scientific discovery task. The authors concluded that in scientific discovery, 1) subjects search for a hypothesis that can account for a previous instance which is similar to the present problem; 2) the hypothesis searched is then tested; 3) if this test fails, then the subject searches for an alternative hypothesis. This search is for an underlying mechanism and if a mechanism is primed then it is selected ; 4) if no hypothesis is found, then the subject *adjusts the present hypothesis to conform to the data*.

In a medical context, the process of hypothesis generation in the diagnostic process has been studied by Joseph and Patel (1990). They compared low-knowledge and high-knowledge subjects in the process of evaluating a clinical case in endocrinology. Low-knowledge subjects were physicians who had general knowledge of the domain but lacked specialized knowledge; high-knowledge subjects were specialists in endocrinology. The results showed that it is not the process of generating hypotheses what distinguishes low and high knowledge subjects. It is the process of hypothesis evaluation. High-knowledge subjects evaluated their hypotheses by searching for confirmation and refining their initial hypothesis. Low-knowledge subjects generated alternative hypotheses after producing the correct hypothesis, even though the data could not discriminate among them. Joseph and Patel (1990) concluded that diagnostic reasoning can be described as a two stage process: a hypothesis generation phase, which is dependent on general domain knowledge; and a process of evaluating the hypotheses thus generated. This stage is dependent on the specific domain knowledge possessed by the subjects.

## Directionality of Diagnostic Reasoning

At a more detailed level of analysis, the processes involved in diagnostic reasoning have been shown to make use of two types of inferences. These have been termed *forward* and *backward*. The term forward reasoning refers to an inference pattern that acts on evidence and generates a hypothesis from it. Research has shown that forward reasoning is characteristic of successful expert performance in some domains, and that it is used when subjects possess the relevant knowledge to solve the problem. It is highly efficient and fast. Backward reasoning is a pattern of inference in which a hypothesis serves to generate the evidence. Its direction, then, is from hypothesis to data. An alternative, but related definition is that it goes from the goal to the problem data. In such a instance, it may describe the directionality of a procedure. In either case, there is first the generation of a hypothesis and then the checking of the evidence against the hypothesis. Backward inference is used mainly when there is lack of sufficient knowledge to reach the solution from the data alone. It is slow and makes heavy demands on working memory. It is most characteristic of the novice and it is frequently tied to unsuccessful performance. However, its use is a function of both the problem solver and the problem to be solved. Nonetheless, it has been found to characterize the problem solving of experts in various domains (Larkin, McDermott, Simon, & Simon, 1980; see Anderson (1987) for a curious situation in the domain of programming).

Although forward reasoning is frequently associated with expert performance, there is evidence that it is also used by novices. Recent research (Patel, Groen, & Norman, in press) has provided evidence that the directional pattern of reasoning may be linked to curricular factors in novices. Patel, Groen, and Norman carried out two studies which they compared the diagnostic reasoning process of students who were trained in two different curricula. One group (the PBL group) was trained in a problem solving environment, which emphasizes the utilization of a hypothetico-deductive approach to medical diagnosis, and where the acquisition of basic science is concurrent with clinical practice. The other (the CC group) was trained in a traditional medical curriculum, where most of basic science is taught independently of, and prior to, clinical training. The results showed that whereas the PBL group relied more on backward reasoning and produced more elaborations, the CC group relied more on the use of forward reasoning.

This paper presents some results from a larger research project (Arocha, 1991) on the process of problem solving by novice diagnosticians. The results reported here regard the process by which medical students, trained in a traditional curriculum, generate and evaluate diagnostic hypothesis.

## Method

**Subjects:** The subjects were second, third, and fourth year medical students at McGill University (4 subjects per group) and one cardiology resident. Second year students had just started an introduction to clinical medicine; they had no experience in the hospital and, therefore, their knowledge of disease comes principally from lectures and textbooks. Third year students were in their first year of clinical training in the hospitals. Fourth year students had just completed their medical. degree requirements with two years of biomedical sciences and two years of clinical training, and were preparing for their provincial and federal certification exams.

**Materials:** Two real-life clinical problems in cardiology selected were modified in such a way that the relevant information was presented later in the case. Both cases were similar in their initial presentation, but differ in subsequent information. The cases were divided into three segments:

(1) presenting complaint; (2) past history; and (3) results of the physical examination.

**Procedure:** The subjects were all tested individually in a single 90 minute session. Their task was to explain each case as completely as possible. The cases were presented on three consecutive cards, each containing one segment of information. The subjects were instructed to think out loud while they read the segments. They were asked to make a tentative diagnosis after each segment. Then, they followed to the next card and repeated the procedure for the second and third cards.

**Analysis:** The analysis consisted of representing the cases and protocols in terms of propositions (Frederiksen, 1975; van Dijk & Kintsch, 1983). From these propositions, a reference frame for each case was constructed as a semantic network, which has been successfully used in cognitive research in medicine by Patel and Groen (1986). The reference frames so developed served as a comparison for the assessment of individual frames constructed from the subjects' protocols. The network representations contained mostly conditional and causal relationships. Other relationships were attributive, locative, and temporal. Coordination of theory of evidence was assessed by confirmation (i.e., a positive cue serves to generate a hypothesis) and disconfirmation (a positive or negative cue serves to rule out a hypothesis) strategies. Directionality of reasoning was assessed by the use of backward (i.e., hypothesis is matched to cue) and forward (cue suggests hypothesis) inferences.

## Results

A summary of the results for all groups of subjects is presented first. The results include accuracy of the diagnosis, the hypothesis generation and the evaluation strategies the subjects used, and the directionality of their reasoning. A detailed analysis of two subjects' schematic representations follows. This allows us to see the process by which these subjects dealt with the evidence for or against the diagnostic hypotheses in the cases and describe the processes they used to solve the problems, specially the strategies they followed and the directionality of their reasoning.

**Diagnostic Accuracy:** Five subjects reached the correct diagnostic hypothesis: Two second year students, two fourth year students and the resident. Only the resident was successful in both problems. No third year student generated the correct solution. Although there was little success in reaching accurate diagnosis in most cases, the subjects were able to select all the important cues in the case for interpretation.

**Hypothesis Generation:** The number of hypotheses generated varied across groups; specially, those generated at the first segment, with the second year students generating the fewer diagnostic hypotheses ($\bar{x}$= 1.5) and third and fourth year students generating a larger number ($\bar{x}$= 3 and $\bar{x}$=3.4, respectively). The resident generated the same number of diagnostic hypotheses for both cases (n=5). In terms of the quality of the diagnoses, all subjects generated the initially suggested hypothesis at the first segment.

**Evaluation Strategies:** The most widely used evaluation strategy was confirmation. A confirmation strategy was one which linked a positive cue to a hypothesis as support for the hypothesis. Seventy one percent of the cues by second year students were of this kind; 88% were confirming relations for both third year and fourth year students. The resident generated 92% confirming relations. All second year students kept their initial hypothesis even when contradictory data were presented. They ignored contradictory evidence or modified slightly the initial hypothesis to fit the data. Third and fourth year students used concurrent hypotheses to account for different cues in the case. These were of two kinds: cardiac hypotheses to account for the initial cues and pleuro-pulmonary to account for cues that did not fit the initial diagnosis. The expert subject generated more diagnostic hypothesis, but was able to produce a single hypothesis in the third segment by selecting the one that accounted for more of the data.

**Directionality of Reasoning:** Most reasoning inferences were forward inferences. In the cases in which the correct diagnoses were reached, 4 subjects used forward inferences to conclude the final diagnosis. All these corresponded to case 2 (AD45). The resident also used backward inference in reaching the solution to case 1 (VP45), when provided with contradictory evidence.

**Detailed Analysis of Two Protocols:** This section presents a comparison of the solution processes given by a second year student and the resident to case 1 (VP45) . Figure 1 presents the schematic representation of the problem solving process generated from the protocol of subject 2.1. The subject starts by generating the hypothesis of myocardial infarction, given the cues of *severe central crushing chest pain* and *chest pain of four hours*. The subject adds support for this hypothesis by noting that the age of the patient is in keeping with a related affection: coronary artery disease. A second hypothesis, which the student disconfirms, is rheumatic fever (streptoccocal infection), based on the finding of *chest pain of 4 hours*. The negative findings in the case are used to disconfirm pulmonary edema secondary to left ventricular failure, which may result from myocardial infarction. Lastly, the subject concludes that the *cough* is irrelevant to the patient's main condition. The subject bases her conclusion on the negative findings of *no hemoptysis* and *no sputum*, which are associated with cough in myocardial infarction.

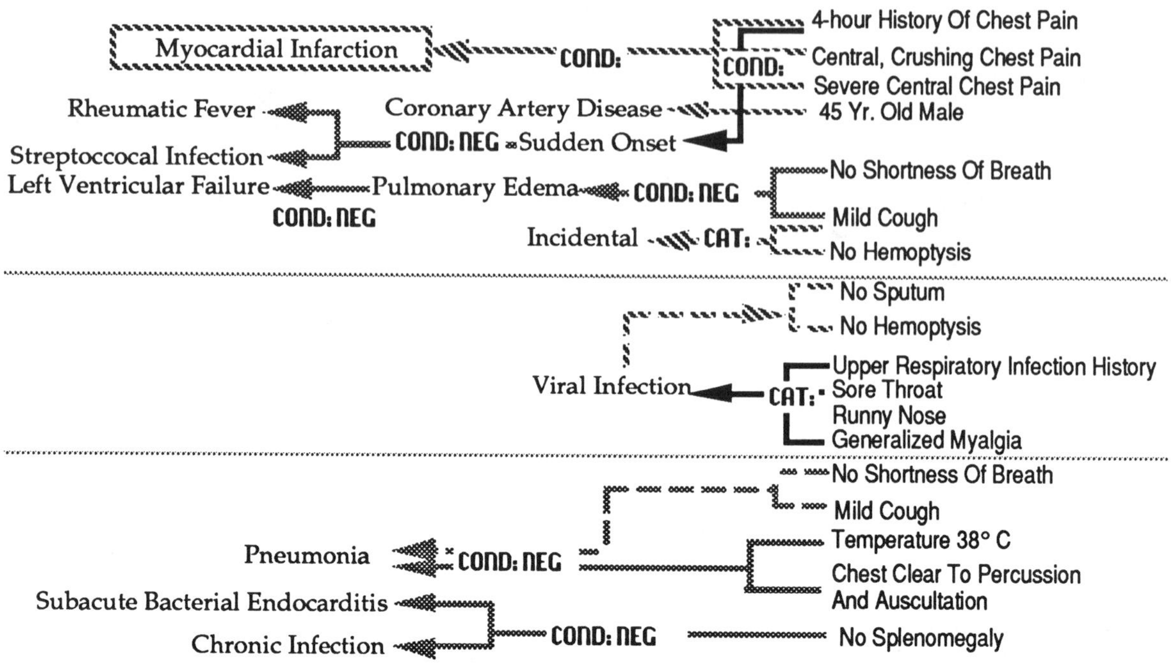


**Figure 1.** Schematic representation of the problem solving process of case 1 (VP45) by the a novice. A — represents a confirming relation towards the correct diagnosis; a ----- represents a confirming relation towards and alternative diagnosis; a ----- represents a disconfirming relation. A dashed line represents a relation linking a cue in the current segment to either a hypothesis or another cue in previous segments. **COND:** represents a conditional relation; **CAU:** represents a causal relation; **NEG** represents a negation; **CAT:** represents a category relation. A box indicates the final diagnosi reached by the subject.

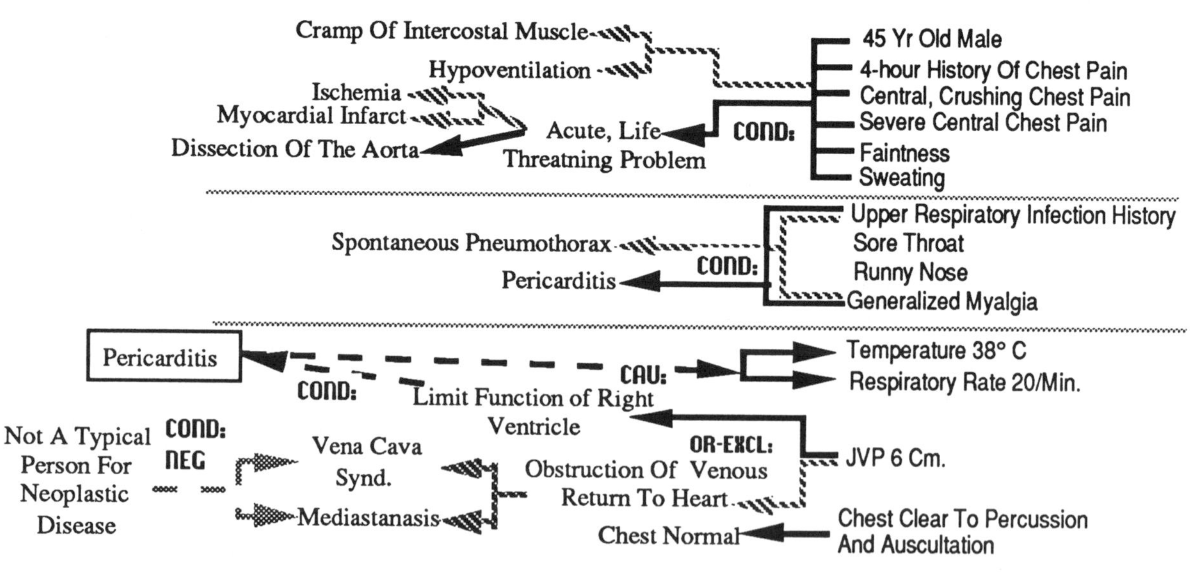


**Figure 2.** Schematic representation of the problem solving process of case 1 (VP45) by the resident. A — represents a confirming relation towards the correct diagnosis; a ----- represents a confirming relation towards and alternative diagnosis; a ----- represents a disconfirming relation. A dashed line represents a relation linking a cue in the current segment to either a hypothesis or another cue in previous segments. **COND:** represents a conditional relation; **CAU:** represents a causal relation; **NEG** represents a negation; **OR-EXCL:** represents an exclusive disjunctive relation. A box indicates the final diagnosis reached by the subject.

Thus, up to this point, this second year student has generated a single diagnosis, that of myocardial infarction, with one modification: no left-ventricular failure. The rest of the hypotheses she generates are disconfirmed: they are ruled out by the evidence. The second segment introduces the findings of *history of upper respiratory infection*, *sore throat*, *runny nose* and *generalized myalgia*, which the subject correctly interprets as a viral infection. However, she uses these findings to conclude that is not a streptoccocal infection, which is related to the hypothesis, disconfirmed in the first segment, of rheumatic fever (streptoccocal infection precedes the appearance of rheumatic fever). This subject then concludes that the infection is unrelated to the chest pain. Thus, the subject at this segment only rules out the relevance of the infection while maintaining the diagnosis of myocardial infarction. The third segment introduces 7 new findings of which the subject selects 3 for comment. These are *temperature 38°C*, *chest clear to percussion and auscultation*, and *no splenomegaly*. The same strategy she used in the previous segment is used here: she uses the findings to disconfirm alternative diagnoses. This time, pneumonia and subacute bacterial endocarditis are ruled out: Pneumonia involves congested chest and subacute bacterial endocarditis involves enlargėd spleen. In summary, this subject generated the initial diagnosis of myocardial infarction and maintained it in the presence of contradictory evidence. Further information, which go against such initial diagnosis served to modify the initial hypothesis without rejecting it (i.e., right ventricular rather than left ventricular failure). Other information is either acknowledged, but interpreted as unrelated to the episode of chest pain, or ignored.

Consider now the solution given by the resident to the same case. First, the resident considers a larger number of diagnoses: from a cramp of the intercostal muscles to hypoventilation to myocardial infarction or dissection of the aorta, to pericarditis. Thus, instead of assuming that the central chest pain corresponds to myocardial infarction, the resident acknowledges that this presentation may be produced by other diseases. The subject employs a breadth-first search by considering the possibility of different alternative diagnoses of varied origin, quality, and seriousness. The second segment suggest to the resident the possibility of an infection, but the main focus is on ruling out any serious disease. The infection does not strongly suggest any particular diagnosis. However, he pays more attention to diagnoses that may relate an infectious process and chest pain: pneumothorax or pericarditis At the third segment, the subject focuses on the findings of *fever*, *jugular venous distension* and *chest clear to percussion and auscultation*. To account for the case, then the resident looks for hypothesis that would compatible with the pattern of chest pain, the jugular venous distension, and the clearness of the chest. He generates two intermediate, non-diagnostic, hypotheses that can produce jugular venous distension and chest pain: one is diseases that involve an obstruction of the venous return to the heart (*vena cava syndrome*, *mediastinitis*) the other is diseases that limit of the diastolic function of the right ventricle (*pericarditis*). For the former, the subject finds no evidence in the case presentation. He apparently matches the features that identify a typical person with neoplastic disease with the features of the present case, from which he discards the hypothesis. The other type of underlying process that may cause such a pattern is one that limits the diastolic function of the right ventricle, such as the case of pericardits. Therefore, pericarditis must be the correct diagnosis. The resident does not consider the possibility of myocardial infarction because the *chest is clear*, the patient does not have pulmonary edema, which he would have if he had an myocardial infarction and because the finding *respiratory rate of 20* is not consistent with heart failure.

In summary, the strategy employed by this subject differs substantially from the one used by the first subject. First, he starts by generating several initial diagnoses, even though the case strongly suggests only one. The subject uses subsequent information to narrow the possible hypotheses: in the second segment, by focusing on diseases of infectious character; and in the third segment, by trying to fit all the main case cues into a single hypothesis that covers them all. He uses a form of backward inference which serves to generate the correct diagnostic hypothesis. By postulating two possible causal mechanisms, either an obstruction of venous blood flow to the heart or a limiting of the diastolic function of the right ventricle, and matching the pattern of cues to what would be produced by the two such processes, the subject concludes the correct diagnosis. Such use of bacward inference requires a vast background of biomedical knowledge to be successful (cf. Patel, Groen, Arocha, 1990).

## Discussion

First, the results show that degree of training does not have an important effect on what findings are chosen as being relevant; that is, all subjects focused their attention on the most important findings. This is consistent with findings in previous research (Braccio, 1988). Also, it has been found that novices are frequently able to recall the important findings in a case but are, nonetheless, unable to use those findings correctly in the solution of the case (Patel & Frederiksen, 1984). Second, second year students seemed to have been strongly influenced by the initial hypothesis and were unable to change their diagnosis in light of contradictory evidence. More experienced students seem to have recognized the importance of contradictory evidence but were also unable to override their initial hypothesis; instead, they generated concurrent hypotheses to explain sets of cues. The resident was the only subject who changed the initial diagnosis as contradictory evidence was presented and who attempted to provide a single coherent hypothesis to account for all the case cues. These results are compatible with those obtained in research on the scientific discovery process (Dunbar, 1989; Dunbar & Schunn, 1990; Klahr & Dunbar, 1988). However, there may be specific differences between levels of expertise, as the present results suggest. As expertise with a domain increases, subjects seem to be more aware of the role of providing coherence to the problem interpretation. The three general processes reported here, namely, 1) generating a major hypothesis and either reinterpreting or ignoring contradictory data; 2) generating concurrent hypotheses to account for different sets of cues; and 3) generating various initial hypotheses and then narrowing the hypothesis space and interpreting the data in terms of a single hypothesis by imposing coherence, may be three general strategies linked to the development of expertise. Further investigation is required to validate this conclusion.

The two schematic representations presented on the problem solving process of two subjects serve to highlight some points. Previous research has found the use of forward inferences linked to successful performance of expert subjects. In this paper, as it has been reported elsewhere (Patel, Groen, & Norman, in press), the use of forward reasoning was the most consistently used type of inference. However, it consisted of a "shallow" form of forward inference as it was not supported by knowledge of the underlying mechanism. Backward reasoning was used in a few occasions mainly by third year students, who generated the longest protocols and gave the largest number of explanations for the cases, and by the resident.

The case of the resident shows an instance where the use of a backward inference is successfully used to generate the correct diagnosis. This case is also interesting because it presents the use of biomedical knowledge to reach the correct diagnosis. It has been argued that biomedical knowledge forms a separate "world" from the clinical knowledge. This interpretation is, to some extent, supported in the present study. Recall that second year students had no clinical training; their only knowledge of clinical medicine was through lectures, but they had 2 years of basic science teaching. Even these subjects, who one may expect to use their knowledge of basic science to solve the clinical problem, did not do so (cf. Patel, Groen & Norman 1990). On the contrary, they relied almost exclusively on the association between clinical cues and diagnostic categories. The resident, however, was the subject with longest clinical training. Nonetheless, he used knowledge of basic science to reach the diagnosis.

## References


Anderson, J.R. (1987). Skill acquisition: Compilation of weak–method problem solutions. *Psychological Review, 94*, 192–210.

Arocha, J.F. 1991. Clinical Case Similarity and Diagnostic Reasoning in Medicine. Unpublished Ph.D. diss. Department of Educational Psychology and Centre for Medical Education., McGill Univ.

Braccio, A. 1988. On-line Analysis of Novice Problem Solving on Medicine. M.A. thesis. Dept. of Educational Psychology, McGill Univ.

van Dijk, T. and Kintsch, W. 1983. *Strategies of Discourse Comprehension*. New York: Academic Press.

Dunbar, K. 1989. Scientific Reasoning Strategies in a Simulated Molecular Genetics Environment. In Proceedings of The 11th Annual Conference of the Cognitive Science Society. Hillsdale, NJ: Lawrence Erlbaum.

Dunbar, K. and Schunn, C.D. 1990. The Temporal Nature of Scientific Discovery: The Roles of Priming and Analogy. Proceedings of The 12th Annual Conference of the Cognitive Science Society. Hillsdale, NJ: Lawrence Erlbaum

Elstein, A.S., Shulman, L.S., and Sprafka, S.A. 1978. *Medical Problem Solving: An Analysis of Clinical Reasoning*. Cambridge, MA: Harvard University Press.

Evans, D.A. and Gadd, C.S 1989. Managing Coherence and Context in Medical Problem-solving Discourse. In D.A. Evans and V.L. Patel , eds. *Cognitive Science in Medicine*. Cambridge, MA: MIT Press.

Feinstein, A.R. 1967. *Clinical Judgement*. Baltimore: The Wilkins and Williams Company.

Frederiksen, C. H. 1975. Representing Logical and Semantic Structure of Knowledge Acquired From Discourse. *Cognitive Psychology, 7*, 371-458.

Groen, G.J., and Patel, V.L. 1985. Medical Problem Solving and Cognitive Psychology: Some Questionable Assumptions. *Medical Education, 19* , 95-100.

Joseph, G.-M., and Patel, V. L. 1990. Domain Knowledge and Hypothesis Generation in Diagnostic Reasoning. *Medical Decision Making, 10*, 31-46.

Klahr, D. and Dunbar, K. 1988. Dual Space Search During Scientific Reasoning. *Cognitive Science, 12*, 1-55.

Kuhn, D. 1989. Children and Adults as Intuitive Scientists. *Psychological Review, 96*, 674-689.

Langley, P., Simon, H., Bradshaw, G.L., and Zytkow, J.M. 1987. *Scientific Discovery: Computational Explorations of the Creative Process*. Cambridge, MA: MIT Press.

Larkin, J.H., McDermott, J., Simon, D.P., and Simon, H.A. 1980. Expert and Novice Performance in Solving Physics Problems. *Science, 208*, 1335-1342.

Patel, V.L., and Groen, G.J. 1986. Knowledge-based Solution Strategies in Medical Reasoning. *Cognitive Science, 10*, 91-116.

Patel, V.L. and Frederiksen, C.H. 1984. Cognitive Processes in Comprehension and Knowledge Acquisition by Medical Students and Physicians. In. H.G. Schmidt, and M.C. DeVolde, eds. *Tutorials in Problem-Based Learning*. Assen, Holland: Van Gorcum.

Patel, V.L., Groen, G.J., and Arocha, J.F. 1990. Medical Expertise as a Function of Task Difficulty. *Memory & Cognition, 18* (4), 394-406.

Patel, V.L., Groen, G.J., and Norman, G.R. 1991. Two Modes of Thought: A Comparison of Effects of Conventional and Problem-based Medical Curricula. *Academic Medicine*, July.

Rimoldi, H.J.A. 1961. The Test of Diagnostic Skills. *Journal of Medical Education, 36*, 73-79.

Tweeney, R.D. and Hoffner, C.E. 1987. Understanding the Microstructure of Science: An Example. In Proceedings of The Ninth Annual Conference of The Cognitive Science Society. Hillsdale, NJ: Lawrence Erlbaum.

# The Heuristics of Spatial Cognition

**Stephen C. Hirtle**

Department of Information Science
University of Pittsburgh
Pittsburgh, PA 15260
sch@icarus.lis.pitt.edu

**Michael F. Mascolo**

Department of Psychology
Merrimack College,
North Andover, MA 01845.
mascolo@merrimack.edu

### Abstract

Distance estimation has been used extensively in the investigation of cognitive maps, yet it is not well understood as a cognitive process in its own right and, as a result, has been viewed as a simple read-out from a spatial representation. In contrast, this paper considers distance estimation to be a complex mental process in which heuristics guide the choice of strategies. Specifically, verbal protocols were collected on a distance estimation task for 20 undergraduates using a variety of city pairs in U.S. and Canada. On the basis of these data, distance estimation is shown to be a constructive process, using a relatively limited number of heuristics, such as addition, hedges and ratios. The choice of heuristics and the time to make a judgment are shown to be related to variables such as the familiarity of locations and the distance to be judged. The advantage of viewing distance estimation as a constructive process rather than a passive readout off an internal map is argued.

## Introduction

Spatial reasoning and spatial cognition provides an important domain for the study of reasoning processes (e.g., Furnas 1990, Byrne & Johnson-Laird 1989), narrative comprehension (e.g., Bower & Morrow 1990), expertise (e.g., Chase & Chi 1981), artificial intelligence (e.g., Kuipers 1978; Kuipers & Levitt 1988) and the representation of knowledge (e.g., McNamara, Hardy, & Hirtle 1989). Within this domain, the methodological advances during the past decade have been notable. There has been a positive shift from hand-drawn sketch map methodologies to the use of indirect measurements (Evans 1980; Siegel 1981). However, just as early work on spatial memory assumed that sketch maps directly tapped one's internal representation (e.g., Lynch 1960), the later work makes equally restrictive assumptions of cognitive processes. Specifically, there has been a recent reliance on the use of distance estimation as a primary measure, examined either alone (e.g., Allen 1981), in conjunction with MDS (e.g., Baird 1979; Magaña, Evans, & Romney 1981), or through paired comparisons (e.g., Baum & Jonides 1979; Evans & Pezdek 1980; Hirtle & Jonides 1985).

An implicit assumption in using distance estimation, particularly in conjunction with MDS, is that one may characterize the internal representation of space in a two dimensional, continuous, holistic, picture-like form. Distance estimation as a process acting on that mental representation can then be thought of as a mental scanning of the internal map, yielding behaviors characteristic of an analog code (Kosslyn, Ball, & Reiser 1978; Thorndyke & Hayes-Roth 1981). However, while a map may be an appropriate first approximation, there is a simplicity in considering the representation to be continuous and distance estimation to be a direct task. Surely, given our understanding of other cognitive processing domains, the schemata of space should play an equally important role. Pipkin (1982) offers a robust argument along just these lines, suggesting that geographers recognize the variety of ways in which surface schemata are consciously processed in the course of spatial problem solving.

Direct experimental evidence leads to a similar conclusion. Spatial processing is influenced by hierarchical structure (McNamara, Hardy, & Hirtle,1989; Stevens & Coupe 1978), by reference points (Holyoak & Mah 1982) and by heuristics of perceptual organization (Tversky 1981). Thus, distance estimation could be viewed as a type of problem-solving in which cognitive structures are as important as spatial distance.

As an analogous situation consider the task of mental arithmetic. Mentally adding 2 or 3 digit numbers for most individuals includes both automatic "table look up" (e.g., 9 + 7 is 16) and effortful manipulation (e.g., carry the one). The speed (and accuracy) of performing an arithmetic problem is a function of the processes involved. Likewise in distance estimation, we can immediately generate certain distances while others we need to calculate.

This research was sponsored by NSF grant BNS-8617732.

The analogy would suggest that techniques common to problem-solving, such as protocol analysis (Ericsson & Simon 1984), could prove beneficial in examining distance estimation. Thus, we directed our subjects to produce on-going verbalizations of distance estimation. Our purpose is to delineate specific heuristics used in distance computation, such as time estimation, triangulation, segmentation and the use of reference points, in order to understand the process of distance estimation. Furthermore, we examined the time to make distance estimates as a function of the referentiality and the familiarity of the locations. It is our belief that the time to make distance estimates is longer for nonreference points due to a constructive process rather than merely a difference in familiarity.

We began by collecting norms of national cognitive maps for the subject pool. 200 cities were rated by 60 undergraduates on familiarity and referentiality. These data were then used in construction of the protocol stimulus space for this experiment.

## Method

### Subjects

Fifteen male and five female undergraduates attending the State University of New York at Albany participated in order to fulfill a course requirement. The session lasted for one hour.

### Materials

Sixty city pairs were constructed on the basis of the previous data obtained on referentiality and familiarity of cities. The sample of cities was chosen to represent a broad range of judgments on both the referentiality and familiarity dimensions. These cities were paired in order to form a heterogeneous sample in terms of distance, angle of orientation, and geographical location. The actual distance of the pairs varied from 37 miles to 2553 miles. Each of the sixty city pairs was printed in bold typeface in the center of an index card.

Three arithmetic problems and one physics problem served as practice items. Each practice problem was printed on a separate sheet of paper. Responses were recorded through use of a cassette tape recorder and lapel microphone.

### Procedure

Subjects were seated in a sound proof room in front of the papers containing the practice problems. The experimenter read the practice instructions to the subject, which asked the subject to solve problems out loud, rather than work silently. The experimenter read the practice problems as they were presented to the subject.

After completing the four practice problems, the experimenter attached the microphone and read the instructions for the distance estimation task. Subjects were asked to calculate mentally the distance between pairs of cities, giving all answers in miles, while being as accurate as possible. Subjects were told that the cross-country distance is roughly 3000 miles, but were not told any other mileages. The pairs of cities were presented on 3 in x 5 in cards. Subjects were asked to think out loud, describing any steps used in arriving at the distance estimation. The experimenter read the city pairs to the subject on each trial. Index cards were shuffled before presentation to each subject.

## Results

The verbal protocols were coded independently by two coders using a common coding scheme. The coding scheme was designed to include a variety of judgment strategies. On the occasion that a judgment did not fit any of the codes, an additional code was established. As a result, twenty different strategies were identified from the protocols. These are listed in Table 1. These strategies were divided into two main classes, level one and level two. Each of the level one strategies could result in a direct numerical estimate, while the level two strategies required additional estimates in order to generate a numerical answer. For example, the strategy Add was classified as a level two strategy, as two or more distances were added together. The subject using Add would need to generate these component distances by using either a level one strategy or another level two strategy. Eventually, each component distance would need to be generated by a level one strategy. Differences in the level of strategy will be examined later.

Our first concern was to make sure that the subject's estimates were reasonable and that in fact they were carrying out the original task of distance estimation. Overall, there was a high correlation between the actual distance and mean distance estimate ($r = .965$, $p < .001$), with subjects showing a slight overestimation. The best-fitting regression line was given by the equation:

Distance Estimate = 1.08 * Actual Distance + 117.36

The correlations for individual subjects were slightly lower, which was to be expected. They ranged from .687 to .953, with a mean of .887 and a standard deviation of .0627.

We then summarized the data by city to see if there were specific strategies with specific pairs of cities. We did this first in a qualitative manner, noting which strategies were most common and noting under which conditions they were used. The use of strategies was not evenly distributed. The predominant strategies were Add, Analogy, Hedge, Ratio, Subtract and Time Retrieval. Choice of strategy was in part based on the distance to be estimated. Time Retrieval was most commonly used with distances up to 400 miles, while Ratios were used most commonly with cross-country distance of 600 miles or more. The other strategies of Add, Hedge, Analogy, and Subtract were evenly distributed across all distances.

Table 1
*Heuristics for Distance Estimation*

| | | | |
|---|---|---|---|
| **Level 1 strategies** | | | |
| 1. | DR | Direct Retrieval | "I know Boston to NYC is 180 miles" |
| 2. | DRU | Dir. Retr.-Unspecified | "Boston to NYC is 180 miles" |
| 3. | TR | Time Retrieval | "Boston to NYC is 3 hours, or 180 miles" |
| 4. | LD | Limit Distance | "It can't be more than 200 miles" |
| 5. | LT | Limit Time | "It can't be more than 3 hours" |
| 6. | LDG | Limit Distance General | "They're pretty close, say 180 miles" |
| 7. | PR | Prior Reference | "I remember that I said..." |
| 8. | CDU | Compare Dist. Unclear | "Comparing it to the distance across, I'd say..." |
| **Level 2 strategies** | | | |
| 9. | R | Ratio | "That's about halfway across the US" |
| 10. | MULT | Multiplication | "It's about 3 times..." |
| 11. | ADD | Addition | "50 plus 60 is 110" |
| 12. | SUB | Subtraction | "100 minus 30, I'd call it 70" |
| 13. | AD | Analogy-Distinct | "A to B is the same as C to D" |
| 14. | AC | Analogy-Common End | "A to B is the same as A to C" |
| 15. | HA | Hedge-Add | "It's a little more than the distance..." |
| 16. | HS | Hedge-Subtract | "It's a little less than the distance..." |
| 17. | T | Triangulate | Uses two sides of a triangle to calculate the hypotenuse |
| **Miscellaneous Strategies** | | | |
| 18. | /M/ | Modify | Subject changes strategy |
| 19. | TR-G | Time Retrieval-Guess | "I guess that Las Vegas is about 5 hours..." |
| 20. | I | Imagery | Subject is explicit about visualizing |
| 21. | U | Unclear | Unable to classify |

Add was typical with city pairs in which there is a intermediate reference point. For example, Springfield, Mass. and Atlantic City, NJ span the intermediate location of New York City. Subtract was typical with city pairs in which there is a reference point outside the pair. For example, Las Vegas and New York City were often judged in relationship to Los Angeles or the West Coast. Ratio was typical with long distances in which the cities are well-known, such as Dallas and San Francisco. In this case, the ratio was based on a cross-country distance. Finally, Time Retrieval was typical with short, well-travelled routes, such as Albany to Boston.

The qualitative analysis suggested certain relationships, which were further analyzed using several sources of quantitative data. For each city pair, we measured the response time (RT) between the point in time that the experimenter completed the verbal presentation of the city pairs and the point in time that the subject provided the final distance estimate; the number of intermediate distance estimates used in the computation (INT.EST), the number of extra locations used in the computation (EXT.LOC), and the number of hierarchic levels used in completion of the distance estimation was calculated (LEVELS). Thus, there were 4 different indices of the amount of processing involved in each calculation, RT, INT.EST, EXT.LOC, and LEVELS. Note that INT.EST and EXT.LOC are related, but not identical. A calculation from city A to city B, based on the distance from city A to a new city C, plus the distance from city C to city B, would include one extra location and two intermediate estimates, whereas a calculation from city A to city B, based on the distance from a new city C to new city D, would include two extra locations and one intermediate estimate.

Additional data pertaining to various aspects of the city pairs, not provided by subjects in this experiment, were available for inclusion in the data analysis. Previous data, obtained from different subjects in the same subject pool, had established norms for the referentiality and familiarity of each city. Referentiality referred to the extent to which a city would serve as a useful reference point for locating other cities, while familiarity referred to the extent to which a city was known. The mean referentiality score (MREF) and the mean familiarity score (MFAM) for each city pair were calculated. Finally, the actual distance (AD) was calculated by measuring the straight line distance on a U.S. map.

Table 2
*Correlation Matrix Between Quantitative Variable*

| | AD | DE | RT | INT_EST | EXT_LOC | LEVELS | MREF |
|---|---|---|---|---|---|---|---|
| DE | .965*** | | | | | | |
| RT | .042 | .075 | | | | | |
| INT.EST | .404*** | .436*** | .321** | | | | |
| EXT.LOC | .808*** | .858*** | .309** | .652*** | | | |
| LEVELS | .621*** | .655*** | .363** | .758*** | .758*** | | |
| MREF | -.070 | -.047 | -.235* | .206 | -.135 | .022 | |
| MFAM | -.430*** | -.430*** | -.207 | .036 | .433*** | -.252* | .607*** |

*$p < .05$
** $p < .01$
*** $p < .001$.

The correlations between these eight values are displayed in Table 2. The results indicate that the response time (RT) did not covary as a function of DE ($r = .075$, n.s.). However, RT did increase as the mean referentiality (MREF) of the city pairs decreased, (r's = -.235, $p < .05$). Not surprisingly, RT increased with the complexity of the calculation, as indicated by INT.EST, EXT.LOC and LEVELS ($r$'s = .321, .309, .363, respectively, $p < .01$). It is interesting to note that the use of extra locations (EXT.LOC) was related to familiarity ($r = -.433$, $p < .001$), but not to referentiality ($r = -.135$, n.s.). That is to say, if a location is unfamiliar, then the subjects refer to extra locations to create an estimate. In contrast, the number of intermediate estimates (INT.EST) was related to neither the referentiality ($r = .206$, n.s.), nor the familiarity ($r = .036$, n.s.). However, the distance estimates from unfamiliar cities tend to be more complex in terms of the number of levels (LEVELS) in the hierarchy of calculations ($r = -.251$, $p < .05$).

Finally, we looked at the relationship between level one and level two strategies. Table 3 presents the pairwise frequency data for level one and level two strategies. The cells of the table indicate the number of times that the lower level strategy appeared in conjunction with the higher level strategy. Only frequently chosen strategies were included. These were Direct Retrieval-Unspecified (DRU), Direct Retrieval (DR), and Time Retrieval (TR), which accounted for 516 out of 585 total level one strategies, and Add, Sub, Analogy, Hedge and Ratio, which were the only level two strategies, with 50 or more occurrences.

The 5 x 3 contingency matrix reported in Table 2 was tested for independence of rows and columns using the chi square statistic. The test indicated that the level one and level two strategies are not independent, ($\chi^2(8) = 235.12$, < .001). In order to determine the source of this result, we partitioned data into a series of seven non-independent sub-tables. In order to maintain the experiment-wise error rate at .05, we restricted the error rate to each analysis to .05/7, or approximately .01. The analyses showed that Analogy and Hedge showed identical distribution across the three sub-strategies. Furthermore, Direct Retrieval was most common with Ratio, Hedge and Analogy, while Time Retrieval was most common with Add and Sub (each significant at $p < .001$).

Table 3
*Relationship between Level 1 and Level 2 Strategies*

| | DRU | DR | TR | Total |
|---|---|---|---|---|
| ADD | 87 | 28 | 120 | 235 |
| SUB | 26 | 26 | 44 | 96 |
| ANALOGY | 6 | 47 | 11 | 64 |
| HEDGE | 14 | 60 | 19 | 93 |
| RATIO | 10 | 92 | 2 | 104 |
| Total | 143 | 253 | 196 | 592 |

## Discussion

In examining distance estimation as a constructive task, we found a limited number of heuristics were used by subjects in generating distances. The choice of heuristics was shown to be related to variables such as the familiarity of locations and the distance to be judged. The time to make a judgment was related, not only to the complexity of the calculation in terms of intermediate estimates and extra locations, but also to the mean referentiality. Furthermore, it was not related to actual distance or the distance estimate. Finally, times were often added and subtracted, yet rarely was ratio of times taken. Distances on the other hand were used for ratios, as well as addition, subtraction and modification. Our goal in conducting this research was to examine closely the distance estimation process. It has been shown that distance estimation is a complex task, dependent on the nature of the comparison and prior knowledge of the subject, rather than a simple dependent measure based on readout from a cognitive map.

We see two implications for this research. First, for researchers who use distance estimation or reaction time of distance estimation, we voice a concern that the output may be more reflective of the complexity of the task rather than an internal cognitive map. Assurances that variables like prominence and familiarity are controlled adequately should be sought. Second, and more important, the analysis of distance estimation has implications for theories of the mental representation of space. There is evidence that familiar locations can be accessed without accessing related reference points, as seen by the fact that extra locations were only chosen for city pairs with low familarity scores, regardless of the referentiality score.

These results suggest that we may need to modify the hierarchical models of cognitive maps proposed over the past few years (e.g., Stevens & Coupe 1978), in favor of a partially hierarchical model (e.g., McNamara 1986). That is to say that any familiar point in a space may have equal access to any other point through the hierarchical nets, which do not necessarily pass through the primary, reference nodes. Therefore, the cognitive map is more like a system of highways in which highways cluster around populous areas, but, in addition, there are bypasses which connect areas to one another. This would be opposed to a model of represented space analogous to an airline route map, in which all cross-country flights pass through a few hub cities. In the later case, it would be impossible to travel from one small city to another small city without going through the hub city. The hierarchies exist in both models, but the equal access model has greater flexibility in choosing relationships. Through examination of protocols, this study provides evidence of an "equal access" hierarchical map. Of course, the evidence presented is correlational in nature and additional experimental work is needed to confirm such a model.

To summarize, we have shown the importance of considering distance estimation to be a constructive task, based on series of heuristics. The choice of heuristics and the time to make a judgment was shown to be related to variables such as the familiarity of locations and the distance to be judged. The advantage of viewing distance estimation as a constructive process rather than a passive readout off an internal map is argued. This point of view leads to a computational approach that suggests a distinct research program in the area of spatial cognition from the experimental agenda put forth thus far, and suggests that knowledge gained in the fields of problem solving, expert decision making, and mental models can be applied in a fruitful manner to the study of cognitive maps.

## References


Allen, G.L. 1981. A developmental Perspective on the Effects of "Subdividing" Macrospatial Experience. *Journal of Experimental Psychology: Human Learning and Memory* 7:120-132.

Baird. J. C. 1979. Studies of the Cognitive Representation of Spatial Relations. *Journal of Experimental Psychology: General* 108:90-106.

Baum, D. R., & Jonides, J. 1979. Cognitive Maps: Analysis of Comparative Judgments of Distance. *Memory & Cognition* 7:462-468.

Byrne, R. M. J., & Johnson-Laird, P. N. 1989. Spatial Reasoning. *Journal of Memory and Language* 28:564-575.

Chase, W. G., & Chi, M. T. H. 1981. Cognitive Skill: Implications for Spatial Skill in Large-Scale Environments. In J. H. Harvey (Ed.) *Cognition, Social Behavior, and the Environment,* 111-136. Hillsdale, NJ: Erlbaum.

Ericsson, K.A., & Simon, H. A. 1984. *Protocol Analysis: Verbal Reports as Data*: Cambridge: Mass.: MIT Press.

Evans, G. W. 1980. Environmental Cognition. *Psychological Bulletin* 88:259-287.

Evans, G.W., & Pezdek, K. 1980. Cognitive Mapping: Knowledge of Real-World Distance and Location Information. *Journal of Experimental Psychology: Human Learning and Memory* 6:13-24.

Furnas, G. W. 1990. Formal Models of Imaginal Deduction. Proceedings of the Twelfth Annual Conference of the Cognitive Science Society, 662-669. Hillsdale, NJ: Erlbaum.

Hirtle, S. C. & Jonides, J. 1985. Evidence of Hierarchies in Cognitive Maps. *Memory and Cognition* 13:208-217.

Holyoak, K. J., & Mah, W. A. 1982. Cognitive Reference Points in Judgments of Symbolic Magnitude. *Cognitive Psychology* 14:328-352.

Kosslyn, S. M., Ball, T. M., & Reiser, B. J. 1978. Visual Images Preserve Metric Spatial Information: Evidence from Studies of Image Scanning. *Journal of Experimental Psychology: Human Perception and Performance* 4:47-60.

Kuipers, B. 1978. Modelling Spatial Knowledge. *Cognitive Science* 2:129-153.

Kuipers, B.J. & Levitt, T.S. 1988. Navigation and Mapping in Large-Scale Space. *AI Magazine* 9(2):25-43.

Lynch, K. 1960. *The Image of the City*. Cambridge, Mass: MIT Press.

Magaña, J. R., Evans, G. W., & Romney, A. K. 1981. Scaling Techniques in the Analysis of Environmental Cognition Data. *Professional Geographer* 33:294-301.

McNamara, T.P. 1986. Mental representations of spatial relations. *Cognitive Psychology* 18:87-121.

McNamara, T. P., Hardy, J. K., & Hirtle, S. C. 1989. Subjective Hierarchies in Spatial Memory. *Journal of Experimental Psychology: Learning, Memory, and Cognition* 15:211-227.

Pipkin, J.S. 1982. Some Remarks on Multidimensional Scaling in Geography . In R.G. Golledge and J. N. Rayner (Eds.) *Proximity and Preference*. Minneapolis: University of Minnesota Press.

Siegel, A. W. 1981. The Externalization of Cognitive Maps by Children and Adults: In Search of Ways to Ask Better Questions. In L. Liben, A. Patterson, & N. Newcombe (Eds.), *Spatial Representation and Behavior across the Life Span*. New York: Academic Press.

Stevens, A., & Coupe, P. 1978. Distortions in Judged Spatial Relations. *Cognitive Psychology* 10:422-437.

Thorndyke, P. W. & Hayes-Roth, B. 1982. Differences in Spatial Knowledge Acquired from Maps and Navigation. *Cognitive Psychology* 14:560-589.

Tversky, B. 1981. Distortions in Memory for Maps. *Cognitive Psychology* 13:407-433.

# A Cascade-Correlation Model of Balance Scale Phenomena

**Thomas R. Shultz and William C. Schmidt**

Department of Psychology
McGill University
1205 Penfield Avenue
Montréal, Québec, Canada H3A 1B1
shultz@hebb.psych.mcgill.ca

### Abstract

The Cascade-Correlation connectionist architecture was used to model human cognitive development on balance scale problems. The simulations were characterized by gradual expansion of the training patterns, training bias in favor of equal distance problems, and test problems balanced for torque distance. Both orderly rule stages and torque difference effects were obtained. Analyses of the development of network structure revealed progressive sensitivity to distance information. It was noted that information salience effects, such as that for torque difference, are particularly difficult to capture in symbolic level models.

## Introduction

An emerging benchmark for detailed computational modeling in cognitive development is the balance scale. The clarity and replicability of balance scale phenomena with humans, coupled with the classical developmental appeal of its stage-like character, have led to both connectionist (McClelland 1988) and rule based (Newell 1990) models.

Psychological assessments present the child with a rigid balance beam in which differing numbers of weights are placed on pegs at various distances to the left or right of a fulcrum. The child's task is to determine which side of the scale will go down when supporting blocks are removed. A 5 position, 5 weight version of the balance scale is presented in Figure 1. Ordinarily, all of the weights on one side are placed on a single peg. *Balance* problems have equal numbers of weights placed at equal distances. In *weight* problems, the side with more weights goes down since the distances are equal. In *distance* problems, the side with greater distance goes down since the sides have equal weights. The *conflict* problems have greater weight on one side and greater distance on the other side. The side that goes down is the one with greater weight for *conflict-weight* problems, and the one with greater distance for *conflict-distance* problems. The scale balances in *conflict-balance* problems.

Siegler (1976, 1981) has indicated that children's performance on the balance scale progresses through 4 distinct rule based stages: (1) use weight alone to determine if the scale will balance, (2) emphasize weight, but consider distance (correctly) in the event that the weights to the left and right of the fulcrum are equal, (3) consider both weight and distance but get confused when one side has greater weight and the other has greater distance, (4) multiply distance by weight for each side and compare the products. Siegler has noted that each rule makes specific predictions about the kinds of problems that will be solved as illustrated by the predicted percentages correct in Figure 1.

McClelland (1988) reported a pioneering simulation of balance scale stages using a connectionist network with a back-propagation learning rule. This model required a number of limiting assumptions: a strong bias in the training patterns favoring equal distance problems, a local binary representation of weight and distance information, and a forced segregation of weight vs. distance information in connections to the hidden units. Even with these assumptions satisfied, McClelland reported that there was a great deal of shifting back and forth between rules 3 and 4, with stage 4 never being clearly established.

Interestingly, the leading rule learning program, Soar, has also been applied to balance scale phenomena (Newell 1990). Soar acquired rules 1-3 but, like the back-propagation model, did not manage rule 4. It is unclear how dependent the Soar model was on getting balance scale problems in a certain order. It may well be that different problem orders yield different orders of rules.

Neither the back-propagation model nor the Soar model attempted to capture the other major balance scale phenomenon, the torque difference effect (Ferretti & Butterfield 1986). The torque on each side of the fulcrum is the product of weight x distance. Torque difference is the absolute difference between the torques on the two sides. The larger the torque difference, the easier the problem is for children to solve. This could be regarded as an effect of information salience. It is not explainable by Siegler's rules since any such rule should apply regardless of the torque difference. Nor is the torque difference effect explainable by the additive or multiplicative rules of information integration theory (Wilkening & Anderson 1982).

In this paper, we report our attempt to cover rule stages and the torque difference effect in the balance scale using a relatively new connectionist architecture called Cascade-Correlation (Fahlman & Lebiere 1990). Cascade-Correlation builds its own network topology by recruiting new hidden units as it learns to solve problems. Thus, it affords a more principled approach to network construction than is typical in connectionist research. It starts with a minimal topology containing only the input and output units defined by the programmer. In what is called the output phase, connections to the output units are trained until error can no longer be reduced. Then, in the input phase, a pool of candidate hidden units receives trainable input from the input units and any existing hidden units. Outputs from the candidate hidden units are not yet connected to the output units in this phase. The purpose of the input phase is to identify the candidate unit whose

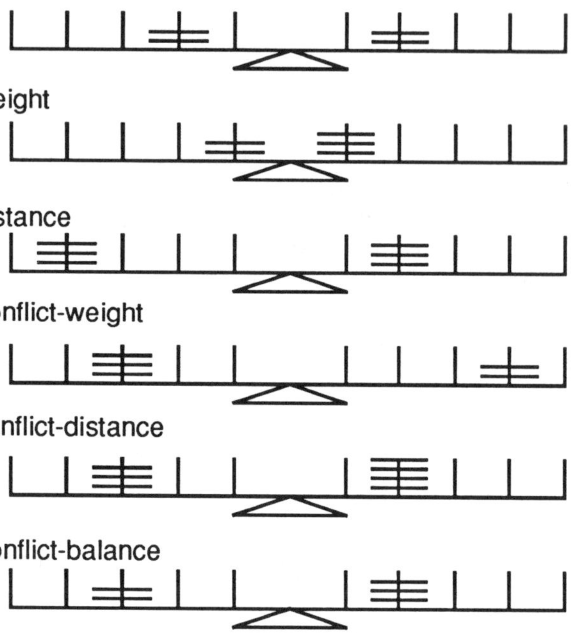

| | Rules 1 | 2 | 3 | 4 |
|---|---|---|---|---|
| Balance | 100 | 100 | 100 | 100 |
| Weight | 100 | 100 | 100 | 100 |
| Distance | 0 | 100 | 100 | 100 |
| Conflict-weight | 100 | 100 | 33 | 100 |
| Conflict-distance | 0 | 0 | 33 | 100 |
| Conflict-balance | 0 | 0 | 33 | 100 |

Figure 1. Balance scale problems and predicted success.

activations correlate best with the output errors. This best candidate unit is then installed in the network, receiving input from all input units and any hidden units already in place and sending output to all of the output units. Once installed, the input side weights to the new hidden unit are frozen, and its output side weights are allowed to change with learning (output phase). Because Cascade-Correlation uses second order error minimization in computing weight changes and learns only one level at a time, it is typically 10-50 times faster at learning than back-propagation.

Although Cascade-Correlation has not yet been applied to cognitive developmental phenomena, it appears to afford a novel and natural interpretation of both qualitative and quantitative developmental changes. Qualitative changes occur through the recruitment of new hidden units; quantitative changes through the adjustment of network weights.

## The Network and the Training

We report two experiments, one on rule diagnosis and the other on the torque difference effect. Both experiments used a 5 peg, 5 weight version of the balance scale, as did McClelland (1988). Our initial network had 4 input units, the obligatory bias unit (which is always on), and 2 output units. Of the 4 input units, one encoded left-side distance, a second encoded left-side weight, a third encoded right-side distance, and the fourth encoded right-side weight. The input coding of distance and weight information was done using integers from 1 to 5. There were 2 sigmoid output units which represented actual balance scale results in a distributed fashion. Left-side down was conveyed by excitation of the first output and inhibition of the second output; right-side down was conveyed by the reverse pattern; and balance was conveyed by neutral values on both outputs. Any recruited hidden units also used a sigmoid activation function.

The initial training patterns were 100 randomly selected without replacement from the 625 possible 5 peg, 5 weight problems, subject to a 0.9 bias in favor of equal distance problems (balance and weight problems, as illustrated in Figure 1). This bias ensured that the probability of drawing an equal distance problem during construction of the training patterns was 0.9. On each epoch in the output phase, another training pattern was randomly selected with replacement, also subject to the 0.9 equal distance bias, and added to the training patterns. We call this Expansion training of the 1+ type. The training set gradually expanded, with 1 new pattern added each output phase epoch. Expansion training conforms to our assumptions that the child's environment changes gradually and that these changes are marked by exposure to more aspects of the environment. The constant bias for equal distance problems reflects the assumption that children have lots of experience lifting differing numbers of objects but relatively little experience placing objects at different distances from a fulcrum.

Pilot experiments had established that rule stages 1 and 2 could not be obtained without a strong bias for equal distance problems; the network went directly to stages 3 and 4. Other pilot experiments indicated that learning was extremely difficult when 100 training patterns were randomly selected each epoch.

We used default parameter values for Cascade-Correlation (Fahlman & Lebiere 1990), with two exceptions. We lowered the input and output Epsilons (learning rates) by 1/2 in order to reduce the bounce in errors from epoch to epoch. We have noticed such error bounce often in using integer-coded input. Also, we used a score-threshold of 0.25. Normally, Cascade-Correlation continues training until all the activations of output units are within score-threshold of their targets in the training patterns. The default score-threshold of 0.4 is appropriate for sigmoid units on a threshold and margin criterion. But because our output units were also coding neutral (balance) patterns, we lowered score-threshold to 0.25 in order to achieve non-overlapping scoring ranges. An output activation had to be equal to or greater than absolute 0.25 in order to count as anything but balance.

Each experiment involved 16 runs. Each run was terminated at 300 epochs because pilot testing had established that most runs were well within stage 4 by that time. With Expansion training, complete mastery of the training patterns is quite difficult to achieve until most of the training patterns have been seen.

## Experiment 1: Rule Diagnosis

Each of the 16 runs used distinct, randomly selected training and test patterns. The 24 test patterns in this experiment were balanced for both problem type and torque difference, such that there were 4 patterns from each of the 6 problem types in Figure 1, 1 pattern representing each of 4 levels of torque difference: 1, 2-5, 6-9, and 10-20. On each epoch during the output phase, the network was tested with the 24 test patterns. A test problem whose output activations were both within score-threshold of their correct targets was coded as correct; any other test problems were coded as incorrect.

The patterns of correct and incorrect problems were used to diagnose rule use. A diagnosis of rule 4 required 20 or more test problems correct; rule 2 required 13 or more correct on balance, weight, distance, and conflict-weight problems and less than 3 correct on conflict-distance and conflict-balance problems; rule 3 required 10 or more correct balance, weight, and distance problems and less than 10 correct on conflict problems; rule 1 required 10 or more correct balance, weight, and conflict-weight problems and less than 3 correct distance, conflict-distance, and conflict-balance problems. Scoring priority for these rules, in decreasing order, was 4, 2, 3, and 1. Rule 2 was given a higher priority than rule 3, because rule 2 produces fewer errors on conflict-weight problems, as shown in Figure 1.

Figure 2 shows a plot of the rule diagnosed at each output epoch for a representative computer subject. It shows a typically orderly progression through Siegler's rules. The *H* on the bottom of the plot signifies where a hidden unit was added to the network. Tabulation of rule diagnosis results revealed that 11 of the 16 subjects showed the predicted 1 2 3 4 ordering. Two other subjects showed rules 1 2 3; 1 showed rules 1 2 4; 1 showed rules 1 2 4 with regression to 3 and 2; and 1 showed rules 1 2. It is quite likely that all subjects would have reached rule 4 with continued training. The overlap between the diagnoses of adjacent rules near transition points reflected the tentative nature of each transition.

Of the 16 computer subjects, 9 recruited 1 hidden unit, 6 recruited 2 hidden units, and 1 recruited 3 hidden units. Of these 24 hidden units, 13 were associated with a very quick progression from one rule to the next: 5 moved up to rule 4, 7 to rule 3, and 1 to rule 2.

To better understand developing network structure, we drew Hinton diagrams in the middle of each rule stage. Each such diagram shows the size and sign of incoming weights at a particular epoch. Size of weight is indicated by the size of the square; sign of the weight is indicated by the color of the square, with white indicating positive and black negative. Hinton diagrams for a representative subject are presented in Figure 3. The first epoch number in each diagram excludes input phases; the second epoch number (in parentheses) includes input phases.

During rule 1, which uses only weight information, the output units were highly sensitive to weight information. The right-side down output received a positive signal from the right-side weight input, whereas the left-side down output received a positive signal from the left-side weight input. During rule 2, which continues to use weight but begins to use distance when the weights on each side are equal, the network's outputs became more sensitive to distance information. The differential sensitivity to sides was retained, and the new hidden unit was particularly sensitive to weight information. During rule 3, which is characterized by the use of both weight and distance information but confusion when these are in conflict, the outputs became about as sensitive to distance as to weight. And in rule 4, which signifies nearly correct performance, a new hidden unit emerged that was particularly sensitive to distance information. The two hidden units, one representing mainly weight and the other mainly distance, sent opposite signals to the outputs.

More generally, we found that, of the 21 hidden units with Hinton relevance, 8 were especially sensitive to side information, 11 were mainly sensitive to side x distance or side x weight, and 2 were mainly sensitive to the bias unit or to older hidden units.

## Experiment 2: Torque Difference Effect

This experiment employed exactly the same techniques as Experiment 1, except that the principal interest was in recording errors for 4 torque difference levels: 1, 2-5, 6-9, and 10-20. For each run, 4 sets of test patterns were randomly selected with 4 problems of each of the 6 problem types in Figure 1. Each set of test patterns contained only problems representing 1 of the 4 torque difference levels.

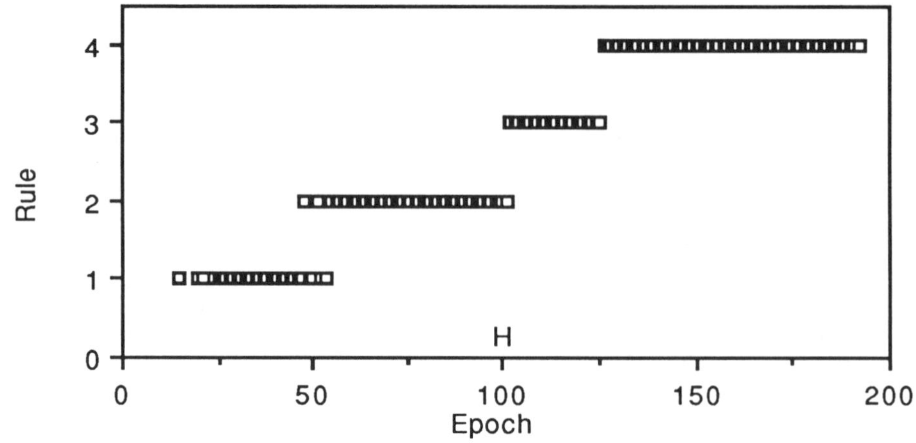


Figure 2. Rule diagnosis for 1 subject.

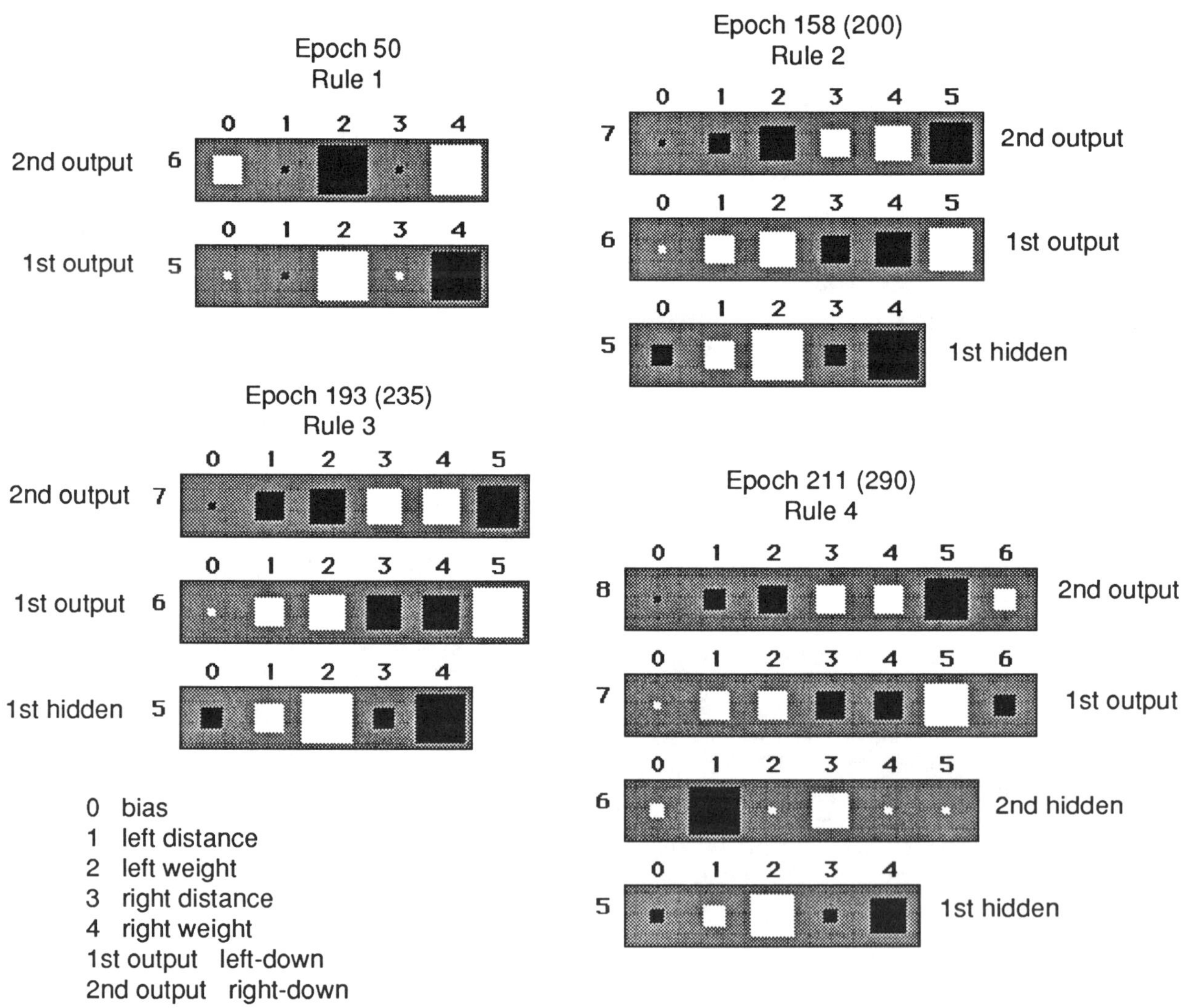


Figure 3. Hinton diagrams of incoming weights for 1 subject.

Errors are plotted over epochs for a representative subject in Figure 4. As expected, this subject showed faster and deeper error reduction with increasing torque difference.

An ANOVA of these error signals midway (epoch 75) and late (last epoch) in learning was performed for all subjects, yielding only a main effect for torque difference level, $F(3, 42) = 48.57$, $p < .001$, with a strong negative linear trend, $F(1, 42) = 140.45$, $p < .001$. The mean errors at these two epochs for the 4 torque difference levels are presented in Figure 5. The larger the torque difference, the smaller the error.

## Discussion

In these simulations, Cascade-Correlation networks learned to perform on balance scale problems as if they were following rules, including clear performance at the level of stage 4. Further, these rules emerged in the psychologically correct order, almost without exception. Some developmental regressions and stage-skipping were observed, just as in human subjects. Unlike previous models, these nets also captured the torque difference effect. The present models are, of course, highly simplified compared to the environment and computational resources of children.

The Cascade-Correlation networks covered these psychological phenomena without at least some of the restrictive assumptions of McClelland's (1988) back-propagation networks. We didn't need to encode weight and distance inputs in local binary form, or implant segregated hidden units for weight vs. distance information, or indeed implant any hidden units at all. We did, however, follow McClelland's lead in strongly biasing the training patterns in favor of equal distance problems. Such input bias may not be the only way to obtain human-like stages in connectionist models of the balance scale, but it's effectiveness in producing stages may encourage researchers to examine biases in the child's environment.

Like other early connectionist attempts to model phenomena in cognitive development (Chauvin 1989; McClelland 1988; Plunkett & Marchman 1989), the present simulations suggest that the connectionist approach deserves serious consideration as a means of studying transition mechanisms for higher level reasoning. Connectionist networks appear capable of reproducing classic developmental phenomena such as rules and stages, as well as more subtle effects such as information salience that explicit symbolic rule systems have particular difficulty with.

An explicit symbolic rule-based model trying to capture the torque difference effect would presumably find itself in the paradoxical position of having to compute torque differences well before stage 4. It might require rules of the form *if torque difference is greater than x then apply rule i*, where $x$ is some integer between 1 and 20 that decreases with age, and $i$ is the current stage. Such a model would apparently have to compute and use torque differences to mimic the torque difference effect well before it could compute and use torques to solve balance scale problems. This would possibly fit the psychological data, but would be extraordinarily awkward.

## Acknowledgements

This research was supported by a grant from the Natural Sciences and Engineering Research Council of Canada. We are grateful to Scott Fahlman for providing the code for the Cascade-Correlation algorithm and for suggestions on running the simulations, and to Chris Schunn for providing the code for automatic drawing of Hinton diagrams.

## References

Chauvin, Y. 1989. Toward a connectionist model of symbolic emergence. In Proceedings of the Eleventh Annual Conference of the Cognitive Science Society, 580-587. Hillsdale, NJ: Lawrence Erlbaum.

Fahlman, S. E., & Lebiere, C. 1990. The Cascade-Correlation learning architecture. Technical Report, CMU-CS-90-100, School of Computer Science, Carnegie-Mellon University.

Ferretti, R. P., & Butterfield, E. C. 1986. Are children's rule assessment classifications invariant across instances of problem types? *Child Development* 57:1419-1428.

McClelland, J. L. 1988. Parallel distributed processing: Implications for cognition and development. Technical Report, AIP-47, Department of Psychology, Carnegie-Mellon University.

Newell, A. 1990. *Unified theories of cognition.* Cambridge, MA: Harvard University Press.

Plunkett, K., & Marchman, V. 1989. Pattern association in a back propagation network: Implications for child language acquisition. Technical Report, #8902, Center for Research in Language, University of California, San Diego.

Siegler, R. S. 1976. Three aspects of cognitive development. *Cognitive Psychology* 8:481-520.

Siegler, R. S. 1981. Developmental sequences between and within concepts. *Monographs of the Society for Research in Child Development* 46:Whole No. 189.

Wilkening, F., & Anderson, N. H. 1982. Comparison of two rule-assessment methodologies for studying cognitive development and knowledge structure. *Psychological Bulletin* 92:215-237.

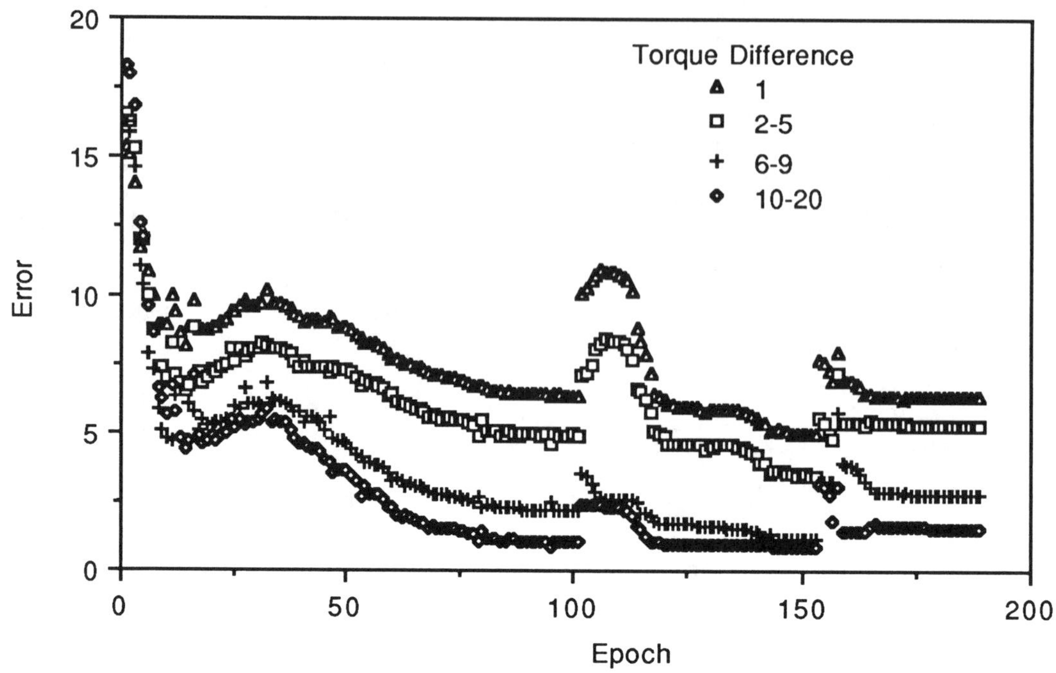


Figure 4. Errors on test problems at 4 torque difference levels for 1 subject.

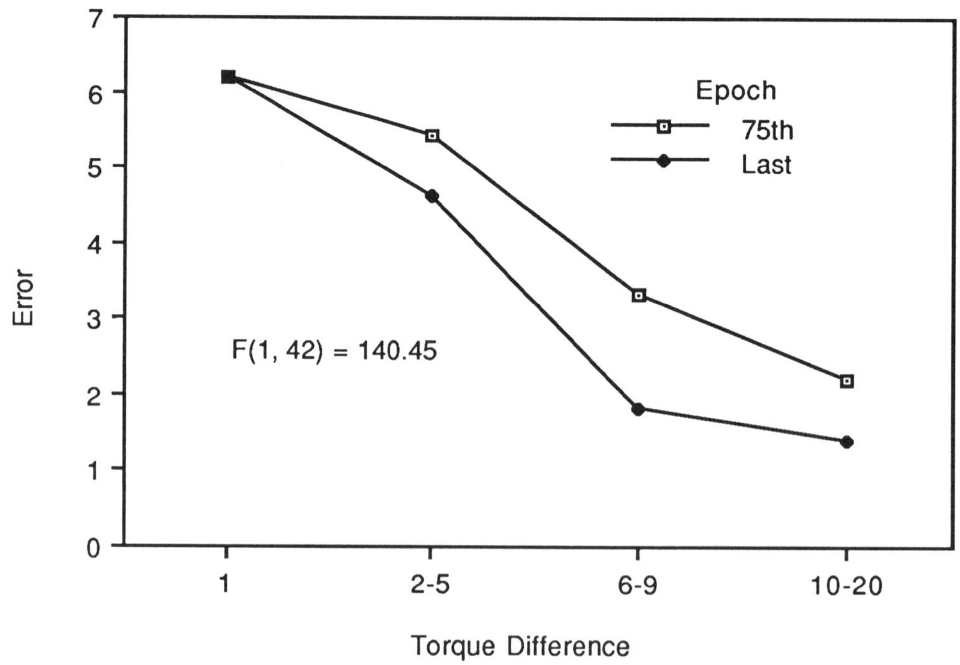


Figure 5. Mean errors on test problems at 4 torque difference levels.

# Posters

# Dynamic Fact Communication Mechanism: A Connectionist Interface

**Jeffrey S. Aaronson**
Department of Computer and Information Science
University of Pennsylvania
Philadelphia, PA 19104
aaronson@linc.cis.upenn.edu

## Abstract

Shastri and Ajjanagadde have proposed a biologically plausible connectionist rule-based reasoning system (hereafter referred to as a knowledge base, or KB), that represents a dynamic binding as the simultaneous, or in-phase, activity of the appropriate units [Shastri & Ajjanagadde 1990]. The work presented in this paper continues this effort at providing a computational account of rapid, common-sense reasoning. The Dynamic Fact Communication Mechanism (DFCM) is a biologically plausible connectionist interface mechanism that extracts a temporally-encoded fact (i.e. a collection of dynamically-encoded bindings) from a source KB and incorporates the fact into a destination KB in a manner consistent with the knowledge already represented in the latter. By continually interpreting source KB activity in terms of target KB activity, DFCM is able to transfer facts between distinct KBs on the same time scale needed to perform a single rule application within a single KB. Thus, DFCM allows the benefits of decomposing a phase-based reasoning system into multiple KBs, each with its own distinct phase structure, while rendering the inter-module communications costs negligible. A simple modification to DFCM allows the unit of transfer to be *groups* of facts. Finally, the number of units that compose DFCM is linear in the size of the KB.

## Introduction

Shastri and Ajjanagadde have suggested a computational account of how a human agent, seemingly without conscious intervention, can perform a broad class of inference with remarkable efficiency (within a few hundred milliseconds) [Shastri & Ajjanagadde 1990]. This class of reasoning is described as *reflexive* reasoning, since the inferences seem to be drawn automatically and effortlessly [Shastri 1990]. Reflexive reasoning proficiency is independent of the size of the body of knowledge upon which the reflexive inferences are drawn.[1] As argued in [Shastri & Ajjanagadde 1990], this strongly suggests that reflexive reasoning is a manifestation of a massively parallel reasoning system in which response time is governed by the length of the chain of inference, and that multiple potentially relevant inference paths are explored simultaneously. In light of results from complexity theory, this implies that human agents are not general purpose reasoners, as all valid inferences cannot be derived proportional to the length of the proof. The connectionist knowledge base described by Shastri and Ajjanagadde not only offers a biologically plausible, computational account of reflexive reasoning, but suggests a limited class of inference by which humans do reason.

A brief description of their system follows. Each constant and predicate argument in the KB is represented by a unit. Long-term, or static, facts are physically encoded as an ensemble of units with structured interconnections that bind together the appropriate filler/role pairs. Rules are represented as a mapping between arguments of the antecedent predicate(s) and arguments of the consequent predicate. As such, there is a physical link reflecting argument mappings for each rule. Inferences are drawn as a result of activation spreading through the system in parallel, by way of the links between predicate arguments. During each episode of reasoning, many transient facts are created, propagated, and eventually discarded. Any attempt at modeling reflexive reasoning must have a method of creating and propagating these dynamic bindings (see [Shastri & Ajjanagadde 1990] for an in-depth discussion) that allows the inference process to perform in real-time, without succumbing to a complex form of cross-talk called the *variable binding* problem. A hard-wired solution to this problem is adopted in order to represent *static, long*-term facts. However, this method is not a feasible means of representing *dynamic* bindings in *short*-term memory. Consider that an episode of reasoning takes on the order of a few hundred milliseconds. This suggests that the creation of the new bindings introduced by each rule application can take only on the order of tens of milliseconds. It has yet to be shown that synaptic modification occurs anywhere near this quickly. Hence, reflexive reasoning does not permit us the luxury of allowing synaptic modification to recruit new units and build up the necessary connectivity "on the fly".

The solution chosen by Shastri and Ajjanagadde represents a dynamic binding as the simultaneous, or in-phase, activity of the appropriate units [Shastri & Ajjanagadde 1990]. Note that a system in which nodes characteristically fire in short slices at a given frequency *does not require a controlling global clock*. Rather, the only requirement is that we have units that fire with a fixed period of oscillation. Thus the inference mechanism consists of a rhythmic spread of activation through the network, each set of bindings "lighting up" during its distinct phase within a fixed-length period, dynamically creating a set of bindings with more member units as the inference process continues. An analogous spatial solution would be realized by creating $\rho$ copies of each unit in the KB, one for each of the $\rho$ phases. If a unit were active in the former system in phase $i$, then the $i^{th}$ copy of the unit would be active in the latter system. While this method would increase the size of the KB by a factor of $\rho$, the speed of the system would presumably increase by a factor of $\rho$ as well. Were we not concerned with a biologically plausible system, this might

[1] If this were not the case, then it would take you longer to understand this sentence today than it did 5 years ago!

well be an attractive tradeoff. However, the firing characteristics of neurons are such that they fire in short bursts followed by longer periods of latency. Thus, by adopting the temporal synchrony approach we reduce the size of the knowledge base without sacrificing speed. Furthermore, there is mounting biological evidence of synchronous neural activity in the brain, and that we encode the unity of objects via the proposed temporal method [Freeman 1981, Eckhorn *et al.* 1988, Gray & Singer 1989, Eckhorn *et al.* 1989]. Note that this paper proposes an idealized biological system. We do not claim that facts, rules, and dynamic bindings are encoded by the brain via the methods described herein. However, we do think it is significant that there are mechanisms in the brain that support synchronous activity, and that this temporal synchrony may be an encoding utilized by the animal brain to process information.

The bottleneck in such a phased system is $\rho$, the ratio between period duration and phase duration. This ratio determines the number of distinct individuals (or entities) that can be attuned to during an episode of reasoning. Furthermore, this ratio is KB dependent; different KB, with different characteristic ratios, will be able to support a different number of distinct individuals at one time during an episode of reasoning. Hereafter, a binding will refer to a single entity filling a single role. A Binding Set (BSet) will refer to a set of bindings involving the same entity. Thus the characteristic ratio determines the number of temporal BSets that may be supported during an episode of reasoning. It is important to note that the number of roles to which an entity can be dynamically bound is unlimited. Hence, each BSet may include an arbitrary number of bindings.

## Dynamic Fact Communication Mechanism

The animal brain is a highly structured, modular architecture in which specific areas are delegated toward processing one type, or one form, of information. Despite this clear separation, intelligent agents have the ability to draw reflexive inferences that can only be reached as a result of an integration of data from distinct sensory processing "stations". An interface mechanism that is able to communicate between 2 distinct repositories of information, for e.g., the "visual" KB and the "auditory" KB must be a component in not only the reasoning process performed by an intelligent agent, but in the *reflexive* reasoning process performed by an intelligent agent. A typist glancing down at his keyboard is reflexively aware that the repetitive clacks he hears are the result of the depression of keys on his keyboard. This conclusion could not be reached without a mechanism that allows the visual and auditory stations to confer with the requisite efficiency that enables such an inference to be drawn reflexively.

The Dynamic Binding Communication Mechanism (DBCM) [Aaronson 1991] makes the first attempt at designing such an interface mechanism. DBCM is a biologically plausible, connectionist communication system that operates between 2 distinct phase-based KBs. The unit of transfer is a BSet (i.e. a single entity and all of the roles to which it is bound). The interface mechanism presented in this paper continues the effort begun by DBCM. The Dynamic Fact Communication Mechanism (DFCM) is a biologically plausible connectionist interface mechanism that extracts a temporally-encoded fact (i.e. a collection of dynamically-encoded bindings) from a source KB and incorporates the fact into a destination KB in a manner consistent with the knowledge already represented in the latter. DFCM continually monitors the entity-to-phase assignments within both KBs in order to interpret source KB activity in terms of target KB activity. Effectively, DFCM processes each potential request *prior to the arrival of the request itself.* The dynamic bindings that compose each interpreted fact flow along the DFCM links, poised at the threshold of the target KB. The arrival of a predicate transfer request acts much like the opening of a door to that target KB predicate, enabling the transfer to take place instantly. DFCM is thus able to transfer facts between distinct KBs *on the same time scale needed to perform a single rule application within a single KB.* This allows us to enjoy the benefit of decomposing a phase-based reasoning system into multiple KBs, each with its own distinct phase structure, while rendering the inter-module communications costs negligible. Moreover, the number of units that compose DFCM is linear in the size of the KB.

Later, we discuss a simple modification to DFCM that allows the transfer unit to be *all* active facts associated with a predetermined group of predicates. Despite this increased bandwidth, the time needed to effect a transfer remains unchanged. Such a system could naturally operate between KBs that allow multiple instantiations of predicates [Shastri & Ajjanagadde 1990]. A more drastic modification allowing dynamic grouping of predicates is also discussed.

### Setting the Scene

All links in DFCM are unidirectional. Arrows and filled circles designate excitatory and inhibitory links, respectively. Links between sets of units indicate pairwise connections. DFCM makes use of 2 simple processing elements:

**$\rho$-btu** $\rho$-btu nodes take the weighted sum of their inputs. If this total meets or exceeds the threshold of the node, the node begins to fire synchronously with the driving input. Node continues to fire in this fashion as long as driving input persists. $\rho$-btu nodes are represented diagrammatically by a circle.

**$\tau$-or** $\tau$-or nodes take the weighted sum of their inputs. If this total meets or exceeds the threshold of the node, the node fires with a duration equal to the period of oscillation. $\tau$-or nodes are represented diagrammatically by a triangle.

As stated in the introduction, the representation and propagation of dynamic bindings can be realized without the use of a global clock. Units firing synchronously belong to the *same* BSet; units firing asynchronously belong to *different* BSets. A global clock is not needed to support this encoding; rather, all we require are units which fire with a fixed period of oscillation, according to the $\rho$-btu activation function described above. Hereafter, we will make reference to a synchronous set of units occupying a distinct phase within a period of oscillation. This is done merely for explanatory convenience, and in no way implies the presence of a global clock regulating unit activity.

Let $\rho_s$ and $\rho_t$ be the number of active phases in the source and target KBs, respectively. For simplicity, we assume that both KBs have the same period of oscillation and thus can encode the same number of phases. This characteristic ratio will be denoted by $\rho$. If an entity is active in more than one KB, we say that the phases which encode the entity in each KB are *entity equivalent.* An active source KB phase for which there is no entity equivalent target KB phase will be called a *foreign phase*. Bindings represented in a foreign phase are called *foreign bindings*.

DFCM delivers target KB input into a special-purpose buffer predicate associated with each target KB predicate. A single rule application will instantiate a target KB predicate with the contents of its buffer. A rule application is performed every period of oscillation [Shastri & Ajjanagadde 1990]. The $i^{th}$ role of the buffer predicate e.g., GIVE, will be denoted $\mathcal{B}_i^{GIVE}$.

The responsibility of generating transfer requests is application dependent; both source and target KBs may play this role. Signal duration should equal period of oscillation. Multiple requests should not be signaled simultaneously.

## Architecture

There are 2 categories of bindings that must be processed. Bindings in entity equivalent phases require supplementing the number of bindings within a target KB BSet. Foreign bindings require allocating a new phase within the target KB, thereby creating a new BSet. Interpreting these latter bindings requires care. Note only must an *unused* phase be allocated for each binding, but we must satisfy a higher-level encoding as well. A given dynamic fact may require several foreign phases to represent its collection of bindings. Properly interpreting such a fact includes allocating a distinct, new phase *for each such foreign phase*. Failure to do so would result in cross-talk being introduced into the target KB. The interpretation mechanism of DFCM reflects these issues. DFCM is functionally divided into $p+1$ planes, where $p$ is the arity of the predicate of maximum arity in the source KB. Let $\mathcal{P}_i$ denote the $i^{th}$ plane. $\mathcal{P}_0$ has 2 responsibilities: $i$) interpreting all bindings in entity equivalent phases, $ii$) distributing the foreign phases amongst the remaining planes such that the distinct foreign phases which compose each dynamic fact are assigned to distinct planes. $\mathcal{P}_1, \ldots, \mathcal{P}_p$ are responsible for interpreting foreign bindings. Each of these planes is assigned a currently available phase in which to re-encode its assigned foreign bindings. In order to ensure that cross-talk is not introduced into the target KB, a phase may not be assigned to more than one plane. Figure 1 illustrates how the responsibility of interpreting the

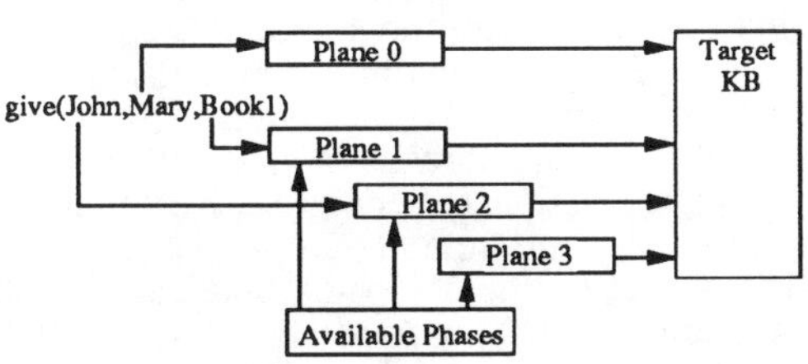


Figure 1: Here, $p = 3$. *give(John,Mary,Book1)* is a dynamic source KB fact. The entity *Mary* is active in the target KB; the entities *John* and *Book1* are inactive in the target KB. $\mathcal{P}_0$ interprets the *recipient* role in the entity-equivalent phase given by the activity of its filler *Mary* within the target KB. $\mathcal{P}_1$ interprets the filler/role pair *Book1/give-object*, $\mathcal{P}_2$ interprets the pair *John/giver*. Each of the latter planes are assigned a unique, available target KB phase in which to encode their assigned bindings.

filler/role pairs is distributed amongst the different planes for the dynamic fact *give(John,Mary,Book1)*.

DFCM relies on several subnetworks that monitor the content of both the source and target KBs in order to provide each plane with the phase information it needs to act as desired. The Free Phase Map (FPM) communicates unique free phases to $\mathcal{P}_1, \ldots, \mathcal{P}_p$. The Entity Map (EM) provides $\mathcal{P}_0$ with a mapping between entity equivalent phases across the source and target KBs.

### Free Phase Map

The function of the Free Phase Map is to monitor the target KB and assign unique, unused phases to distinct planes. There are 2 sets of units that compose the FPM. See figure 2.

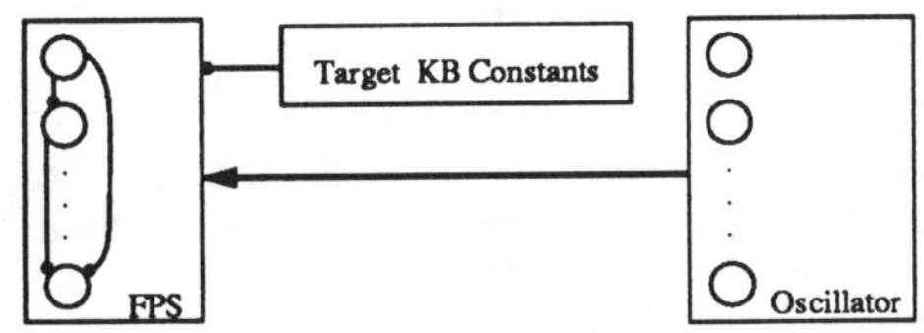


Figure 2: The Free Phase Map

First, there is a set of $\rho$ $\rho$-btu units that oscillate within each period, according to the phase structure of the target KB. The $i^{th}$ unit of this Oscillator, denoted $Osc_i$, fires during the $i^{th}$ phase of each period. The behavior of these units never varies. They are present in order to communicate the phase structure of the target KB to other parts of DFCM. One of these parts is the second set of units composing the FPM. This group consists of $p$ "free phase selector" (FPS) units, one for each plane $\mathcal{P}_1, \ldots, \mathcal{P}_p$. $FPS_i$ firing in synchrony with its (lone) driving input $Osc_j$ informs $\mathcal{P}_i$ to interpret bindings using phase $j$. Each of the $\rho$ oscillating units connect to each FPS unit. In the absence of any inhibition, activation along each link is sufficient to incite the destination unit to fire. There are 2 constraints that must be placed on the firing of the FPS units to prevent cross-talk from being introduced into the target KB. First, we cannot allow an FPS unit to signal its associated plane with a phase that is already in use within the target KB. Second, we cannot allow multiple FPS units to fire in concert, thereby signaling multiple planes with the same phase. The former constraint is achieved by establishing a strong inhibitory connection from each target KB constant to each FPS unit. The latter constraint is enforced via a downstream inhibition technique among the FPS units, whereby there is a strong inhibitory connection from $FPS_i$ to $FPS_j$, for all $j > i$.

### Entity Map

DFCM monitors the entity-to-phase mapping within both source and target KBs, maintaining a record of entity equivalent phases. $\mathcal{P}_0$ relies on this mapping to provide the correct phase in which interpret its assigned roles. The Entity Map has $\rho$ units, $EM_0, \ldots, EM_{\rho-1}$. $EM_i$ firing in phase $j$ indicates that phase $i$ in the source KB and phase $j$ in the target KB are entity equivalent. That is, the entity belonging to the BSet represented in phase $i$ in the source KB is the same entity belonging to the BSet represented in phase $j$ in the target KB. Since each BSet includes a constant, it is sufficient for the Entity Map to monitor the constants in the source and target KBs. See figure 3 to see how the EM monitors the source and target KB representations of a sample constant, labeled $S$ and $T$, respectively. There is a link from $T$ to each EM unit. There are $\rho$ $\tau$-or units for each constant. The $i^{th}$ $\tau$-or unit fires in response

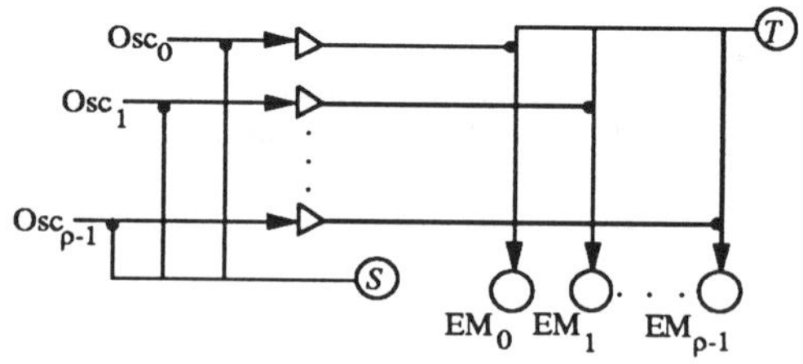


Figure 3: Connections for only a single constant are shown.

to $Osc_i$, provided that the $Osc_i$ signal is not blocked by an inhibitory connection from $S$. Similarly, $EM_i$ fires in response to $T$, provided that activation along the incoming link is not blocked via activation traveling along an inhibitory connection from the $i^{th}$ $\tau$-or unit. The net result of this architecture is that $EM_i$ fires in phase $j$ if and only if some constant is active in phase $i$ of the source KB and phase $j$ of the target KB.

## Initial Plane

$\mathcal{P}_0$ is responsible for interpreting roles that are active in entity equivalent phases. The activity of the EM units precisely identifies these roles. If $EM_i$ is active then $\mathcal{P}_0$ must interpret all source KB roles active in phase $i$. Moreover, the entity equivalent phase in which we want to introduce these roles is given by the phase in which $EM_i$ fires. Figure

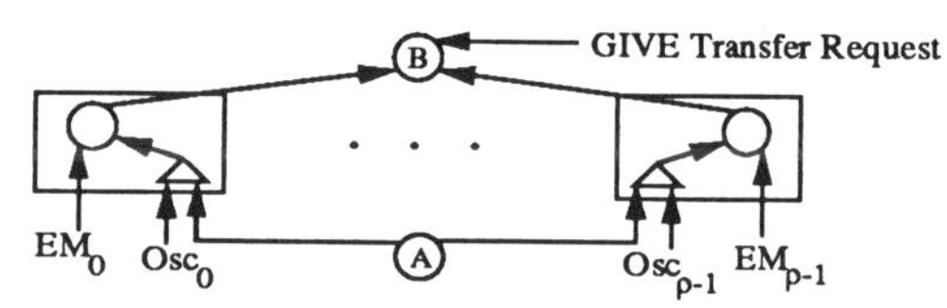


Figure 4: Unit A is GIVE$_k$ in the source KB. Unit B is $\mathcal{B}_k^{GIVE}$.

4 illustrates how $\mathcal{P}_0$ processes the $k^{th}$ role of the source KB predicate GIVE. There are $\rho$ subnetworks for each role, each acting as a temporal pattern matcher. Thresholds and connection weights are such that each unit must receive input from 2 sources in order to become active. Thus, if a role is active in phase $i$, then only the $\tau$-or unit in the $i^{th}$ subnetwork will fire. If $EM_i$ is firing in phase $j$, then the associated $\rho$-btu unit will fire in phase $j$, thereby communicating its activation to $\mathcal{B}_k^{GIVE}$. The onset of the GIVE transfer signal enables $\mathcal{B}_k^{GIVE}$ to respond in this entity equivalent phase. Later, we see how $\mathcal{B}_k^{GIVE}$ may be incited by activity from other planes.

The second task facing $\mathcal{P}_0$ is to distribute the foreign phases to the remaining planes. Since the DFCM unit of transfer is a fact, it is sufficient if this distribution is internally consistent at the fact-level. Accordingly, there is a dedicated director network, composed of $n$ $\rho$-btu units, for each $n$-ary predicate.

Figure 5 details the director $\mathcal{D}^{GIVE}$. The idea is analogous to that used in the Free Phase Map. Whereas there we wanted the FPS units to latch onto unique available phases, here we want the director units to latch onto unique foreign phases. Each predicate's role units are connected pairwise to the corresponding director units. Each director

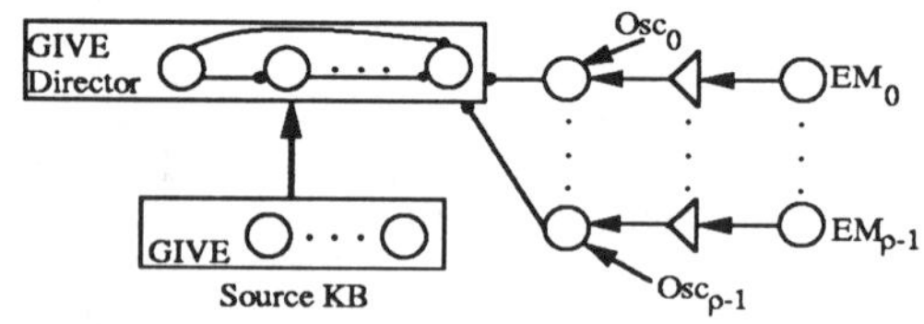


Figure 5: The director network for predicate GIVE

is endowed with downstream inhibition so that no 2 planes will be assigned the same foreign phase. We use the Entity Map to inhibit director units during the the entity-equivalent phases processed by $\mathcal{P}_0$. If $\mathcal{D}_i^{GIVE}$ fires in phase $j$, then $\mathcal{P}_i$ is responsible for interpreting *all* the phase $j$ foreign bindings that contribute to the representation of the dynamic fact GIVE in the source KB. These will be encoded in the phase specified by $FPS_i$.

## Remaining Planes

$\mathcal{P}_0$ processes bindings in entity equivalent phases, thereby only needing to incorporate *roles* into the target KB. However, the remaining planes process foreign bindings, thereby needing to incorporate *fillers* and *roles* into the target KB.

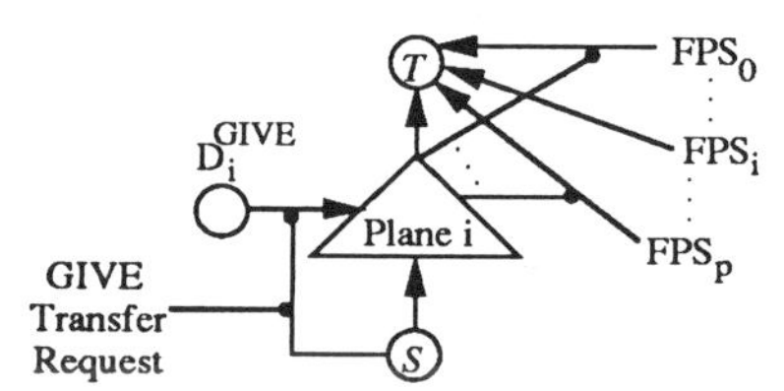


Figure 6: Predicates other than GIVE are suppressed.

Figure 6 shows how $\mathcal{P}_i, i > 0$ processes a sample constant. There is a $\tau$-or unit in each plane for each constant in the source KB. Each such $\mathcal{P}_i$ unit receives activation from its corresponding source KB constant as well as from the $\mathcal{D}_i$ unit corresponding to each source KB predicate. Activation along each of these latter links is effectively blocked via inhibition from the source KB constant *should the phases coincide*. This inhibition may itself be blocked for a sample predicate should that predicate transfer signal be active. Weights and thresholds are such that the $\tau$-or unit in $\mathcal{P}_i$ corresponding to a constant $S$ fires if and only if $i)$ $S$ is active in the source KB in phase $j$, $ii)$ some $\mathcal{D}_i$ is firing in phase $j$, and $iii)$ the predicate transfer signal corresponding to that $\mathcal{D}_i$ is active. A simple inhibitory technique ensures that the target KB constant is incited in the free phase specified by the appropriate FPS unit. Each FPS unit is connected to each target KB constant. Each $\mathcal{P}_i$ $\tau$-or unit is connected upwards to the corresponding target KB constant, and inhibits all but the $FPS_i$ unit from sending activation to that constant. Weights and thresholds are such that a target KB constant cannot fire in the absence of input from a $\tau$-or plane unit.

Figure 7 illustrates how $\mathcal{P}_i$ processes the $k^{th}$ role of the source KB predicate GIVE. There is a $\tau$-or unit in each plane for each role in the source KB. This unit receives activation from the corresponding source KB role as well as from the director unit indexed by that predicate and plane. Weights and thresholds are such that the $\tau$-or unit fires in

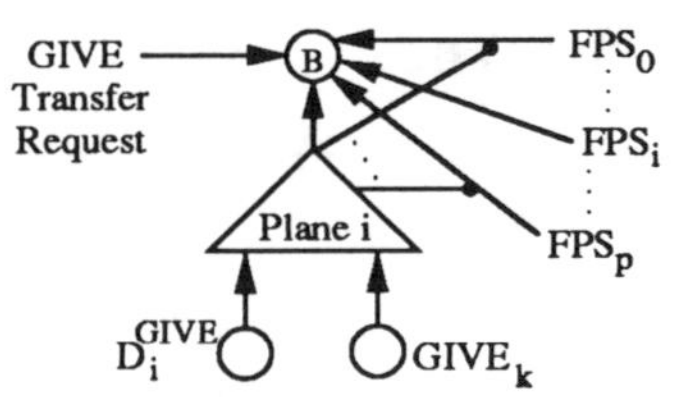


Figure 7: Unit B is $\mathcal{B}_k^{GIVE}$. $\mathcal{P}_0$ connections are suppressed.

response to the simultaneous activation of both sources. Activity of this unit indicates $\mathcal{P}_i$ is responsible for interpreting the source KB role $GIVE_k$. Each $\tau$-or units connects to the appropriate $\mathcal{B}$ role. We use the analogous inhibitory technique as above to ensure that each $\mathcal{B}$ role is incited in the free phase specified by the appropriate FPS unit.

## Performance Evaluation

A buffer predicate is fully instantiated one period following the arrival of the predicate request signal. New target KB entities are activated (if required) 2 periods following the arrival of a predicate request signal.

The director corresponding to a source KB fact whose dynamic bindings require representation in $f > 0$ foreign phases needs $2f - 1$ time periods *following onset of the fact* to reach desired state. New phase allocation within the target KB requires the resetting of each director that has a unit firing in the entity equivalent phase within the source KB. In the worst case, resetting requires $2f - 1$ time periods. Since dynamic facts are typically active for tens of time periods during an episode of reasoning [Shastri & Ajjanagadde 1990], and $f$ will characteristically be quite small, this delay should not detract from the effectiveness of the system. The FPS units require $2 * (\rho - \rho_t) - 1$ time periods to be *fully* functional following new phase allocation within the target KB (although multiple phases may be allocated seamlessly within one request). Since this allocation rate will typically be quite slow, this should not detract from the effectiveness of the system. We hope to develop quicker methods of resetting the FPS and director networks, although these limitations are likely to become inconsequential in future versions of DFCM, in which the unit of transfer is increased from a *single* fact to a *group* of facts.

## Extension

The foreign phase to plane assignments made by each predicate director ensures that DFCM interprets each fact in an internally consistent fashion. However, arbitrary sets of facts, in general, will not be interpreted consistently, as there is no cooperation between directors. This limits the bandwidth of transfer to a single fact. The unit of transfer, however, can be increased to be *all* active facts associated with a predetermined group of predicates by allowing each group to *share the same director*. Moreover, increasing the transfer bandwidth does not increase the transfer time, and the number of units needed is still linear in the size of the KB. Such a system could naturally operate between KBs that allow multiple instantiations of predicates [Shastri & Ajjanagadde 1990]. The increased bandwidth limits the urgency of being able to process successive requests quickly. A more drastic modification of DFCM would permit dynamic grouping of predicates, allowing for arbitrarily defined groups at run-time. Such a change would alter the philosophy of DFCM, rendering it more of a reactive mechanism, as the entity-to-phase assignments would have to wait until the predicate composition of the requested group is known. Despite this, DFCM would still be able to satisfy a transfer request in small constant time (proportional, but bounded by $\rho$, to the number of distinct foreign phases per group).

## Conclusion

The Dynamic Fact Communication Mechanism (DFCM) is a biologically plausible connectionist interface mechanism that operates between 2 KBs that dynamically encode variable bindings by means of temporal synchrony. DFCM continually monitors the entity-to-phase assignments within both KBs in order to interpret source KB activity in terms of target KB activity. Effectively, DFCM processes each potential request *prior to the arrival of the request itself*. This active approach enables a fact to be transferred between distinct KBs *on the same time scale needed to perform a single rule application within a single KB*. The interface thus allows the benefits of decomposing a phase-based reasoning system into multiple KBs, each with its own distinct phase structure, while rendering the inter-module communications costs negligible. Moreover, the number of units composing DFCM is only linear in the size of the KB. We believe this system to be an important component in a general model of reflexive, i.e., rapid common sense reasoning.

## Acknowledgements

I'd like to thank Lokendra Shastri for the many discussions that provided the motivation for this work, and for his guidance along the way.

## References

Aaronson, J.S. 1991. Dynamic Binding Communication Mechanism, Technical Report MS-CIS-91-16, CIS Dept., University of Pennsylvania.

Eckhorn, R., Bauer, R., Jordan, W., Brosch, M, Kruse, W. Munk, M., and Reitboeck, H.J. 1988. Coherent Oscillations: A mechanism of feature linking in the visual cortex. *Biological Cybernetics* 60:121-130.

Eckhorn, R., Reitboeck, H.J., Arndt, M., and Dicke, P. 1989. A Neural Network for feature linking via synchronous activity: Results from cat visual cortex and from simulations. In Model of Brain Function, Cotterill RMJ (ed), Cambridge Univ. Press.

Freeman, W.J. 1981. A physiological basis of perception. In Perspectives in Biology and Medicine. Univ. of Chicago Press.

Gray, C.M. and Singer, W. 1989. Stimulus-specific neuronal oscillations in orientation specific columns of cat visual cortex. *PNAS*, 86:1698-1702.

Shastri, L. 1990. Connectionism and the Computational Effectiveness of Reasoning. *Theoretical Linguistics* 16(1):65-87.

Shastri L, and Ajjanagadde, V. 1990. From Simple Associations to Systematic Reasoning: A Connectionist Representation of Rules, Variables and Dynamic Bindings, Technical Report MS-CIS-90-05, CIS Dept., Univ. of Pennsylvania.

# Mechanisms of Temporal Integration and Knowledge Representation with Simple Recurrent Networks

Robert B. Allen
Bellcore, Morristown NJ
rba@bellcore.com

Knowledge representation was studied for simple recurrent networks which were trained to respond to the semantics of entire statements. In the first set of studies, networks were trained to recall category membership for objects. The representations showed clear differentiation by category. Moreover, simple inheritance of features was demonstrated. In other studies, networks were trained to respond to kinship relationships each of which was presented sequentially. The networks were shown to recognize indirect descriptions of people. An analysis of the hidden units showed high correlations for related questions. Moreover, analysis of individual bits showed that some seem to detect features.

## Knowledge Representation with Neural Networks

While Simple Recurrent Networks (SRNs) have been shown to learn several interesting temporal integration tasks, including tasks that are relevant to natural language processing, little is known about how they perform those tasks. Although some previous work has explored the representations of SRNs, it is unclear what part, if any, the semantics of the entire statement played in these effects. Specifically, Elman (1990) has shown that networks predicted successive terms in pseudo-sentences, and terms which were likely to occur together were clustered together. In addition, Harris and Elman (1989) studied trajectories in principle components space for this task. Furthermore the predictions for these tasks are of a low level, and it seems likely that the computational complexity of the predictions would swamp any practical use of the higher-level effects. On the other hand, it seems possible that the human language system incorporates a variety of sources of feedback, including successive words, and it is possible that some combination of the different types of feedback work together in human language performance.

In this research, a variety of simple tasks were presented to modified SRNs and the activations of the hidden units were examined. In all of the studies, error correction was made for the entire statement. The activations were examined directly, without possibly obfuscating high-level statistics.

## Network Architecture

Figure 1 illustrates the network employed in most of the following research. This network is different from the basic SRNs in a number of ways. First, a memory term has been added to the state units. Second, the entire question and one additional cycle with null inputs are presented (see also Sect. 9 of Allen 1990) prior to any error correction. Furthermore, as shown in Figure 1, the network is connected with fixed one-to-one weights between the inputs and hidden units. These weights were fixed because error correction occurred only after all of the input patterns had been presented and the inputs were "off". Parametric pilot studies showed that fixed weights of 6 were most effective for "on" inputs and -1 for "off" inputs. The network parameters were: momentum, $\alpha$=0.9; learning rate, $\eta$=0.01; hidden-state copy, $\beta$=1.0; and state memory, $\mu$=0.5.

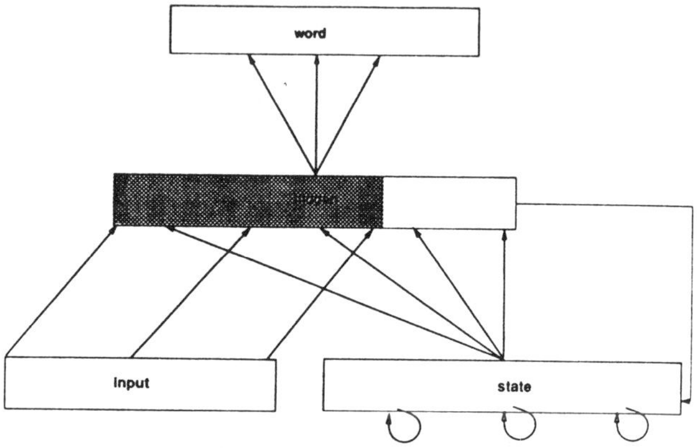


Figure 1. Basic Network Architecture

## Simple Categorization and Inheritance

One area in which a hierarchical structure might be expected to develop in the hidden unit activations is with categorization of objects. Thus, a corpus was developed of true/false questions with 4 categories identified and 3 "objects" assigned to each. Moreover, each category was associated with a single "feature". Altogether there were 270 sentences in the corpus. The four of these describing the relationship between **object 1** and **category 0** were withheld and used for the transfer test. Networks with 25 input units, 40 hidden and state units, and 2 output units, 25-40*-2, learned this task without error after 325K pattern presentations.

The networks correctly answered the transfer questions and thus appeared to show limited 'inheritance'. The reason for this effect, as shown in Table 1, is that similar patterns of weights were generated by the object and category questions. While this result seems simplistic it has some appeal, in that the effects of category names, prototypical objects, both types of questions produce similar transformations of the state vectors. On the other hand, other simple operations are not supported by this model. For instance, networks were not able to transfer when training sentences linking the categories to features were withheld. The networks showed 'inheritance' but not abstraction. The addition of a small amount of jitter (e.g., 0.0000001 * *net input*) to the hidden unit activations reduced the number of cycles to reach low error (about 0.05) by about 10%. As a test of the importance of the memory term, $\mu$, on the state units, a network was run without a memory term (Elman, 1990) and no learning was observed.

Table 1. Roughly Orthogonal State-Unit Weights

| | category | | object | |
|---|---|---|---|---|
| | bit 28 | bit 29 | bit 28 | bit 29 |
| 0 | -36 | 47 | -27 | 37 |
| 1 | -16 | -20 | 6 | -14 |
| 2 | 16 | -41 | 11 | -27 |
| 3 | 11 | 33 | 19 | 38 |

## Scaling Features in Categories

SRNs have been applied only to toy problems. The issue of whether the technique will scale to large stimulus sets is crucial for practical applications. As a problem gets larger we would hope find that each component of the problem gets easier. Thus, the function of the number of presentations to convergence for large problems should be sublinear. The categorization task above may be scaled by increasing the number of objects per category and giving the network a surplus of hidden units. Figure 2 shows the scaling is S-shaped for 10 averaged runs with from 2 to 6 objects in each of 5 categories.

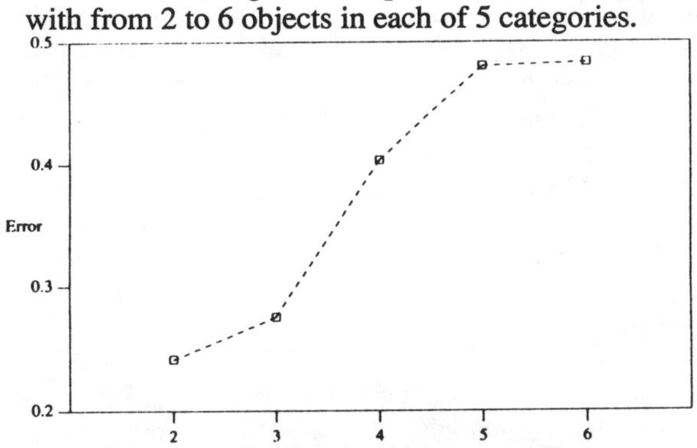


Figure 2. Scaling by Objects per Category

## Kinship

The representation of kinship relationships can be a highly structured knowledge-representation task. Back-propagation training has been applied to the learning of kinship relationships (Hinton 1986) but only with simple three-term relationships which had no temporal processing of the inputs. Nonetheless, Hinton demonstrated that features relevant to the kinship relationships were observed in the hidden layer, at least for the highly constrained architecture he employed. In comparison to Hinton's approach, SRNs have the potential for arbitrary length descriptions. Moreover, the SRNs may be trained to answer questions about kinship or to respond to the truth of the assertions.

*Procedure*

Questions were developed based on the kinship relationships shown in Figure 3 which describe 12 people in one family. In the corpus, questions involving a second family, with an identical structure, were also included. The total lexicon was 40 input terms and 27 output terms. Some of the terms were compressed phrases; for instance, the phrase "the father of" was compressed to "fatherof". The kinship relations were "motherof", "fatherof", "sonof", "daughterof", "sisterof", "brotherof", "wifeof", and "husbandof".

Table 2. Examples of Training Statements

| type | number | example question | answer |
|---|---|---|---|
| whois | 80 | whois motherof mA2 | fA0 |
| kinrel | 40 | kinrel fB2 fB3 | motherof |
| agerel | 46 | agerel fA3 fA2 | youngerthan |
| gender | 24 | gender mA5 | male |
| true | 205 | daughterof fA2 fA3 | true |
| false | 205 | wifeof mA5 fA5 | false |

The corpus included both questions and true or false statements. As shown in Table 2, four types of base questions were employed. This resulted in 190 base questions for each of the two families. Moreover, individuals could be specified indirectly. For instance, "mA2" (i.e., male 2 in family A) could also be described as "fatherof fA3". Thus, the questions "kinrel motherof fA3 mA2" and "kinrel fA2 fatherof fA3" are equivalent to asking for the kinship relationship between fA2 and mA2 , and the answer is that they are married. The expanded corpus included all possible indirect descriptions of the individuals and had a total of 4391 questions. Presumably this is because the feedback from the true/false responses was inadequate to develop the relatively complex discriminations required.

A 40-80*-38 network (40 locally encoded input units, 80 hidden units, 80 state units, and 38 locally encoded output units) was trained for 650K patterns on the basic corpus (without variations). Indirect specifications for the people were introduced and the network was trained for an additional 1250K patterns.

*Results*

The values of hidden units were inspected during the responses to individual questions. Because the procedure in which hidden-unit values were copied in a one-to-one correspondence with individual state units, a large positive weight between the state unit and the corresponding hidden unit effectively latched those units. The network had 32.0% such weights (37

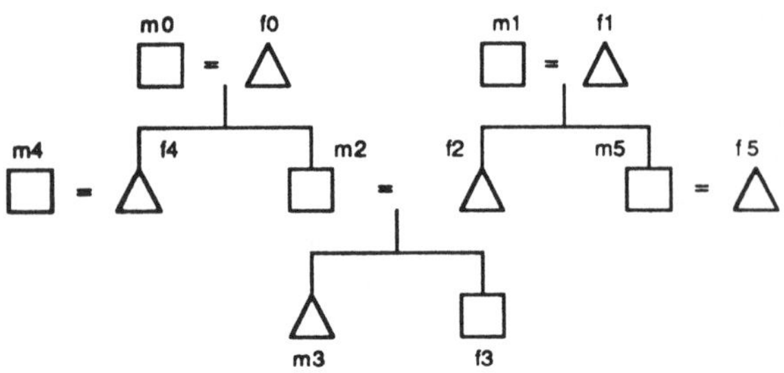


Figure 3. Template for Kinship Relationships

out of a possible 116) which were above an arbitrary threshold (2.5) while only 1.3% of the other weights (79 of 6320) exceeded that threshold.

It is also of interest to compare the representations across related questions. The Pearson *r* correlation statistic was used as a measure of the similarity of sets of hidden units. Correlations were computed between hidden unit vectors at the point when the response was made. Correlations for a typical question are shown in Table 3. It can be seen that many, but not all of the most highly correlated sentences are semantically related to the target.

Table 3. Examples of Question Correlations with Targets

| correlation | question | answer |
|---|---|---|
| target | mA2 husbandof fA2 | true |
| 0.9752* | sonof mA0 husbandof fA2 | true |
| 0.9652 | mA4 husbandof fA4 | true |
| 0.9508 | husbandof fA2 mA2 | true |
| 0.9432 | sonof husbandof fB1 mB5 | true |
| 0.9408 | sonof mB1 husbandof fB5 | true |
| 0.9365 | daughterof fB1 sisterof mB5 | true |
| 0.9306 | mB4 husbandof daughterof mB0 | true |
| 0.9264 | daughterof fB1 daughterof fB1 | true |
| 0.9198 | daughterof fB1 daughterof mB1 | true |
| 0.9184* | mA2 husbandof daughterof fA1 | true |

The starred correlations are for questions which are identical in meaning to the target question. While the starred statements suggest that there is some similarity between the target and its variations, there are many variations and it is worth compiling summary statistics. The mean rank across all variations of the target statement is shown in the center of Table 4. In addition, the corpus contains other statements that are essentially identical in meaning with the target sentence, and the means of the variations of those sentences are also shown in the center column of Table 4. Because the total number of true statements is 1576 it can be seen that the statement forms are above the mean rank. Because the SRN successively compounds the input activations, it might be speculated that the similarity results were not due to training, but only to compounding of the input patterns. Thus, the weights of the network were scrambled within layers and the mean rank of sentence variations are shown on the right side of Table 4. The ranks are generally lower; hence there is more to the similarity results than simply compounding of initial inputs. However, the effect is not overwhelming, especially for the last two statements, which begin with the female rather than the male.

Table 4. Ranks of Variations of Question Set

| question template | trained rank | scrambled rank |
|---|---|---|
| mA2 husband fA2 | 191.7 | 460.3 |
| husband fA2 mA2 | 160.8 | 539.3 |
| fA2 wifeof mA2 | 731.6 | 1065.9 |
| wifeof mA2 fA2 | 592.0 | 357.7 |

Rather than looking at an entire statement, it is possible to correlate activations for partial statements. The activations for all person descriptions were obtained. The hidden unit vector correlations for mA0 were highest with other members of the same generation. and smallest with males of lower generations and females in the other family.

Figure 4 shows the hidden unit activations for a simplified stimulus. In the upper section of the figure are activations averaged across descriptions of males and females in families A and B. The lower portion of the figure shows activations for individual relationship terms (e.g., fatherof). Examination of the figure suggests that bit 26 (marked with a tick) codes for gender. Moreover, the relationship operators are differentiated in the same positions. Thus, the statement 'fatherof mA2' would set the generation bit in 'mA2' to produce a representation similar to 'mA1'. Beyond individual terms, it is possible to consider how the truth of an entire statement is determined.

A simplified model of what the network is doing to process these statements is summation of activations. If the same person is being described as both the left and right sides of the statements and the activations sum, then the veracity of the statement can be determined by checking the threshold of the activations. However, that is not an adequate model since the network is able to take temporal order into account and process statements with the same words in different ways. For instance, "mA2 husbandof fA2" is responded to as "true" while "fA2 husbandof mA2" is correctly responded to as "false".

*Convergence of Kinship Representations*

A direct test was devised of whether associations of the hidden unit activations reflect similarity in semantics of the input. Specifically, two input units were given exactly the same meaning and the representations generated by those units were compared. The procedure was similar to the previous experiment except that a simpler family network was employed (with only 8 members) and only the base questions were used. Hence, the network required 33 input units (32 units for the family relationships and 1 extra for the redundant input). Duplicate "kinrel" questions were generated. The networks were trained for 600K pattern presentations. Across five runs, the average correlation of the hidden units for the two versions of kinrel was 0.56. As a control, the correlations of the activations in five untrained networks were obtained; the average correlations had a mean of 0.15. Clearly, the training of the network has caused the activations for the semantically related terms to become more similar.

## Implications and Challenges

SRNs were shown to be able to generate responses for highly structured knowledge representation tasks. The development of bits which code for specific features was striking. Moreover, there is considerable other evidence that the representations reflected the semantics of the questions and statements. In addition, the results show clearly the state-like nature of recurrent network processing. The results also have implications for models of the semantic space. High correlations for semantically related statements may imply that surface features are lost during processing.

*Bit Overlays*
Similar mechanisms of simple bit operations were found in both the categorization and kinship studies. While the representations identified in this work were not always clear, the most common mechanism appeared to be operations on bit vectors. The process is unlike the usual slots of frame-based approaches to representation; however it is consistent with many psychological models. For instance, the vectors shown in Table 1 are reminiscent of psychological spaces derived from multivariate statistical methods.

*Analogy*
The view of cognitive processing as feature bit activation and masking is consistent with models of analogy (see Rumelhart and Abramson, 1973). Thus an analogy may be seen as applying similar operators to similar, though meaningfully different, vectors.

*Information Retrieval*
Information retrieval (IR) has received only cursory attention with neural networks (Mozer, 1984). However, it might be expected that networks would be effective for IR because many of those procedures are statistically based, and there are similarities between several of these IR models and neural models. Moreover, there is a variety of ways in which neural networks may be applied to IR. For instance, judgments about the relevance of a document may be used as a source of feedback for error correction of a neural network. However, it is difficult to obtain enough relevance judgments to make this practical. Few of the existing statistical IR techniques take advantage of the temporal integration of terms and SRNs have that potential. Indeed, since SRNs have been shown to be able to handle simple grammars (Allen 1990; Servan-Schreiber, et al., 1989), it is possible that they could process some aspects of

Figure 4. Hidden Unit Activations

Hidden Unit Vectors By Gender with Operators.

```
MA  011302000022603000010010121051001000000002100300111030010010001000000002710710026
MB  011211000022612000000001011041002000000003200200010040111010001001100001620721016
FA  010401000016603000020001101020001000001010002000100401100000010020000001810800036
FB  020201000022602000010002101040003000000002100200000050121000001004000001720801037
                              |                                                 |
fa  000001000022502000000000261040004100000000400400100040000000010002000000630810001
mo  000301000010702000040002611051032000003013004025310208100000010080000003920801005
br  010003000014813000000001006100002210000200400100000020305000011000000000830823036
si  121201000020702000000100121000001000000000510200010050060000000000100000920801037
hu  011903000023702000000000170011013000001007002001101402000300010000000004810821026
wi  010201000013603000010001100040002000000001000300000040010000001000000001820801026
so  000100000026703000000002001000002100001025003000010300020000010091000003530812005
da  011100000038702000000111004002001000000003730200010030040000001000000003020700017
```

Hidden Unit Vectors By Age with Operators.

```
oA  000400000013702000010000300300000000000103004000000402000000000001000101420700002
oB  000300000011802000100000200300000000000212003000000401000000000001000200420700012
mA  000300000015702000000000010000000000000001200300000060111000000000000000310700003
mB  000200000013701000000000100100010000000130020000006012100000000020001002207000003
yA  000200001016802000000000000000000000000130020000006002100000000010001003107000005
yB  000000000012701000100000011000100020000024001000100701240000000002100000120800005
                                 |               |                                  |
fa  000001000022502000000000261040004100000000400400100040000000010002000000630810001
mo  000301000010702000040002611051032000003013004025310208100000010080000003920801005
br  010003000014813000000001006100002210000200400100000020305000011000000000830823036
si  121201000020702000000100121000001000000000510200010050060000000000100000920801037
hu  011903000023702000000000170011013000001007002001101402000300010000000004810821026
wi  010201000013603000010001100040002000000001000300000040010000001000000001820801026
so  000100000026703000000002001000002100001025003000010300020000010091000003530812005
da  011100000038702000000111004002001000000003730200010030040000001000000003020700017
```

natural language syntax. In addition, SRNs also seem well suited for IR because of their ability to model users (Allen 1990).

A compact and efficient knowledge representation could be useful for IR. Thus, it is possible to imagine that a representation for a query could be developed and this could be matched with representations for statements from a document. If SRNs are trained to make responses based on the meaning, they may have that potential (Allen 1990). For the generation of responses to questions, (as is the case with the studies reported in this paper) the activations would tend to converge. Declarative statements would simply generate activations without resolution; thus at the end of a statement or document, the residual feature activations might reflect the overall meaning of the document.

Unfortunately, the results in these studies do not provide much encouragement. For instance in the kinship studies, not all of the semantically related statements showed high correlations with the targets. Of course, it would be desirable to find ways to improve the quality of the performance. Clearly, for a practical applications, such as IR, a much larger lexicon would be required and in those, facts will be sparser than in the training sets used here. On one hand that would require greater training time; however, it might also make retrieval much easier since there would be fewer easily confused cases. Of course, there is still the problem of getting a corpus of statements which had an appropriate level of feedback (e.g., true/false) for the type of training employed here.

*Massive Knowledge Bases*

If clear-cut representations are not able to be identified, it may be more useful to think of developing massive question-answering systems (Allen, 1990). Of course, there are also substantial difficulties in that direction. For instance, better techniques are needed for integrating sequential output with sequential inputs. Another problem is finding a corpus with sufficient feedback for training.

*Architectural and Procedural Issues*

The studies here demonstrate the advantage of the modifications to the standard SRNs. Indeed SRNs without state memories were not able to complete the categorization task. Furthermore, additional variations of the basic network proved advantageous in some cases. In addition, small amounts of noise apparently helped the network in the categorization task avoid temporal local minima.

*Extensions*

Previous work and the present results suggest that there is a great deal that is right with this model for language processing. For instance, representations were found which were similar to those proposed for other models of psychological space; scaling, at least in one case, was found to be sublinear; limited inheritance was demonstrated; and feature-detecting bits were identified in the representations. However there are also substantial limitations, the representations often seemed contorted and the processing mechanisms seemed too simple for complex language. There are two directions to consider for moving to practical models of language processing with neural networks. On one hand, we could change expectations as to what is required for a language processing system and attempt to get networks to learn with fairly simple modifications to the existing mechanisms. On the other hand the network mechanisms may be embellished. For instance ways might be sought to integrate them with perceptual processes (see Allen 1990), or to train them to develop more complex structures such as stacks and variable bindings. The present results suggest that the latter strategy will probably be more successful that the former.

**References**

Allen, R.B. 1990. Connectionist language users. *Connection Science,* 2, 279-311.

Elman, J.L. 1990. Finding structure in time. *Cognitive Science, 14,* 179-211.

Harris, C. and Elman, J. 1989. Representing variable information with simple recurrent networks. *Proceedings of the Cognitive Science Society,* Ann Arbor, 635-642.

Hinton, G. 1986. Learning distributed representations of concepts. *Proceedings of the Cognitive Science Society,* Amherst, MA. 1-12.

Mozer, M.C. 1984. *Inductive information retrieval using parallel distributed computation.* ICS Technical Report 8406, La Jolla, UCSD.

Rumelhart, D.E. and Abramson, A.A. 1973, A model for analogical reasoning. *Cognitive Psychology, 5,* 1-28.

Servan-Schreiber D., Cleeremans, A., and McClelland, J. 1989. Learning sequential structure in simple recurrent networks. In: D. Touretzky (ed.) *Advances in Neural Information Processing Systems, 1,* Morgan-Kaufmann, San Mateo, CA, 643-652.

# Reading Instructions *

**Richard Alterman** and **Tamitha Carpenter**
Computer Science Department and Center for Complex Systems
Brandeis University, Waltham, Massachusetts 02254
tamitha@chaos.cs.brandeis.edu (617) 736-2718

## Abstract

This paper describes a model for reading instructions. The basic framework is that an agent engages in an activity and resorts to using instructions only when "all else fails". That is, by reading the instructions *during* the period of engagement, the meaning of the instructions can be clarified by feedback from the world. This model has been implemented in a computer program, IIMP. IIMP is the instruction reading component of FLOABN (Alterman et al., 1991), an integrated architecture whose domain is reasoning about the usage of mechanical and electronic devices.

## Introduction

In general, instructions can convey a variety of information to an agent, ranging from short fixes to a given plan (imagine encountering a detour sign on the road), to the specification of an entirely new plan (e.g., instructions for assembling a piece of furniture). Instructions are difficult to understand in the abstract because they are schematic. Most instructions (written or verbal) omit an enormous number of shared details, with the assumption that an agent sharing general background and an understanding of the situation can fill them in (e.g., LeFevre & Dixon, 1986). Outside of the context of use, the agent can grasp only the general sense of what the instructions mean and the operations they depict. Examples of previous computational approaches to instruction usage include Mannes & Kintsch, 1991; Kieras & Bovair, 1986; and Chapman, 1990.

This paper develops a model of instruction usage where instructions are used after the agent engages in the activity. The basic idea is that instructions are used only when "all else fails". This model has been implemented in a computer program called IIMP. IIMP relies on the fact that, by reading the instructions *during* the period of engagement, the meaning of the instructions can be clarified by feedback from the world. IIMP is the instruction reading component of FLOABN[1], an integrated architecture for an intelligent agent. The core of FLOABN is an adaptive planner which interacts in a situation largely by a process of constructing an interpretation of that situation (Alterman, 1988). When IIMP is invoked, it has access to the interpretation of the current state of the situation of engagement that was constructed by the adaptive planner.

IIMP combines marker passing (e.g., Norvig 1989, Charniak 1986, and Hendler 1986) with summarization and importance techniques (e.g., Lehnert, 1981; Kintsch & van Dijk, 1978; Graesser, 1981; Trabasso & van den Broek, 1985) to make the connection between text and situation. This yields two results. First, it builds a representation of the instructions which the planner can proceduralize and then act upon. Second, IIMP learns the 'purpose' of the instructions and adds this to semantic memory to aid future interpretations in similar situations.

## Domain and an Example

This paper explores a model of text understanding in the domain of learning from instructions. This model is developed in the context of FLOABN, an integrated system for reasoning about electronic and mechanical devices. The core of the system is an adaptive planner and plan-learning system, SCAVENGER (Zito-Wolf & Alterman, 1990). When all of SCAVENGER's plan-adapting techniques fail, the second component of FLOABN, IIMP, interprets the instructions provided on or near the device. The capability to use this sort of human intervention greatly reduces the possibility for failure and the amount of problem-solving the system must do as it deals with real-world situations. This is augmented by a third component, SPATR (Goodman, Waterman, & Alterman 1991), which uses a case-based approach to reason about the spatial content of the instructions.

IIMP uses marker passing techniques to build a coherence analysis of the relationship between the ongoing activity and the instructions. IIMP then uses a summary of the instructions to learn the main idea they are trying to express. If this idea is not already

*This work was supported in part by the Defense Advanced Research Projects Agency, administered by the U.S. Air Force Office of Scientific Research under contract #F49620-88-C-0058.

[1] For Lack Of A Better Name. See [Alterman *et al.*, 1991] for further details

| | |
|---|---|
| 1. Please hang up, | 3. deposit the required coin, |
| 2. listen for dial tone, | 4. dial your call again. |

Figure 1: The instructions heard on the pay phone.

in semantic memory, IIMP adds it. By doing so, IIMP will be able to use the current instructions to ease the understanding process for future examples.

The centerpiece example in this paper will be a FLOABN simulation of an agent interacting with a pay telephone for the first time. When he approaches the pay phone, the agent has a semi-complete set of procedures for using a home phone and the assumption that the procedures for using the pay phone will be similar. The sequence of steps that occurs is as follows: the agent lifts the receiver, dials a number, and hears two rings followed by a beep. Immediately following the beep, a recording gives the instructions shown in Figure 1. During this sequence of events, SCAVENGER is building a representation of the ongoing activity in semantic memory, which is available to IIMP when SCAVENGER asks for help. Using this representation, the recorded instructions, and the instruction 'For local cash calls deposit ten cents before dialing' from the front of the pay phone, IIMP is able to accomplish two things. First, it builds a representation of the content of the instructions, which SCAVENGER is then able to *proceduralize*[2] and incorporate into its plan. Second, it learns that the main point of the instructions is: "payment on a pay phone involves inserting a coin", which IIMP adds to semantic memory. In subsequent encounters with the pay phone, this concept is used and refined (e.g., refinements include different types of payment on a pay phone).

## Three Stages of IIMP: An Overview

IIMP works in three stages. The first stage of IIMP is:

- Use semantic memory to build a coherence representation that characterizes the relationship between the instructions and the ongoing activity.

IIMP constructs a 'copy' of a coherent, internally consistent piece of instantiated semantic memory that represents the coherence of the instructions in relation to the ongoing activity. Each node in the coherence graph (or 'interpretation graph') is either explicitly mentioned in the text of the instruction, a concept encountered during the building of the coherence representation, or a concept developed in the understanding of the ongoing activity as developed by the adaptive planner, SCAVENGER.

The second stage of IIMP is:

- Reason about this graph using summarization technology.

The set of procedures that accomplish this are based on the summarization techniques of the program, SSS (Alterman & Bookman, 1991). IIMP uses these techniques to find the conceptual roots of the interpretation graph built in the first stage, generate a graded list of the relative importance of each node in the graph, and, in stage 3, to produce a base-line summary of the interpretation graph.

The third stage of IIMP is:

- Build a new concept in semantic memory which represents what is learned from the instructions.

This stage uses a modification of Explanation-based Learning (EBL) (e.g., DeJong & Mooney, 1986; Mitchell et al., 1986). Since no *goal concept* is provided with instructions, IIMP uses the information found in stage 2 to find the goal concept of the instructions.

## Stage 1: Building a Coherence Graph

Artificial intelligence treats text understanding as a problem of representation; a machine 'understands' a piece of text to the extent to which it can use its representation for such tasks as summarization and question answering. IIMP builds this representation by using the 'marker passing' scheme described in Norvig, 1989. (Other examples of marker passing methods for inferencing on text are described in Charniak 1986 and Hendler 1986.)

IIMP's marker passing technique works from a semantic memory structure, recognizing general classes of inference as follows:

1. For each node **N** in semantic memory representing the ongoing activity and each word in the text of the instructions, do:
   (a) Find the nodes related to **N** via any of a relatively small number of predefined patterns of connectivity (*path shapes*).
   (b) Identify nodes where these paths meet (*collisions*).

Path shapes are defined by the types and order of links between the nodes. Only a few collision types are defined or needed; all undefined collisions are ignored.

Again, consider the pay phone example. This stage begins with a semantic memory containing relevant background knowledge, including knowledge about the current situation of engagement, but with no representation of a pay telephone (since this is the first time IIMP has encountered a pay phone). IIMP passes markers from each significant word in the instructions (i.e., words other than 'please', 'the', and 'and'), as well as from the current plan being adapted: 'telephone plan'. 'Hang up' is found to indicate that the call should be restarted. 'Listen for dial tone' and 'dial' are found to be steps from the original plan.

The instruction 'deposit the required coin' causes a collision at the 'payment' node, as illustrated in Figure 2. The path between 'insert-coin' and 'payment' is labelled as a 'goal of action' path shape. The path between 'telephone plan' and 'payment' is labelled as a 'goal of plan' path shape. The collision defined by these two path shapes is called a 'new step' collision. This indicates that 'insert coin' is valid as a 'new step'

[2]To convert the "understanding" into plan modifications.

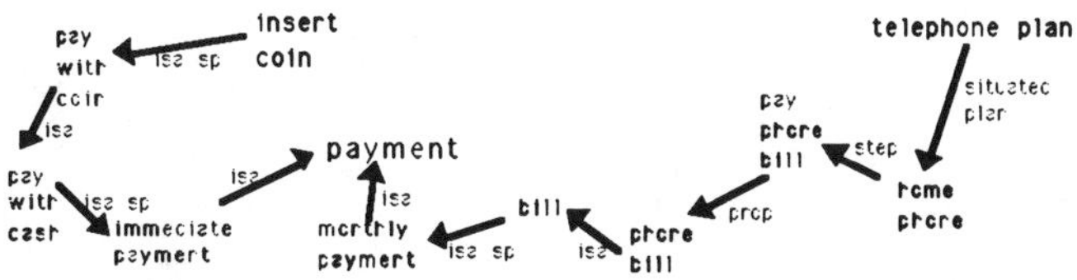


Figure 2: The collision between telephone-plan and insert-coin

for the 'telephone plan', because it satisfies the goal of 'payment', which was satisfied in the 'home phone' situation by paying a monthly bill.

When the marker passing is completed, all the paths that were involved in defined collisions, such as those in Figure 2, define the *episode*: a relevant subset of semantic memory which represents the coherence representation of the text. The episode will be used in the next two stages. This representation has three features that are important for the functionality of the following stages: unconnected features will not be included, since they cause no collisions, thus eliminating irrelevant information; it is a 'copy' of a piece of semantic memory (i.e., the representation is built from the 'vocabulary' and 'structure' of semantic memory); and it is in the form of a directed acyclic graph (DAG).

## Stage 2: Analysis of the Coherence Graph

During this stage of processing, IIMP analyzes the coherence representation of the instructions that was produced in stage 2. This is done using summarization techniques described in Alterman & Bookman (forthcoming). These techniques take advantage of the DAG structure imposed on the input as a result of the semantic memory encoding during the coherence building stage.

The first step is to find the **conceptual roots** of the coherence graph; roughly, these correspond to the basic notions of the understanding. The conceptual roots are the minimal set that *covers* the interpretation graph, where coverage is defined in terms of *reachability* in graph theory. Since the interpretation graph is a DAG, the conceptual roots are simply all the nodes unreachable from any other node. The conceptual roots for the pay phone example are PAYMENT, INSERT, RESTART-CALL, and TELEPHONE-PLAN. Using the conceptual roots, IIMP is able to succinctly explain the connection between each instruction and the ongoing activity (see Figure 3). The adaptive planner is then able to proceduralize this connection and incorporate it into a new plan.

Also derived from the coherence graph is a measure of importance that quantifies the conceptual emphasis of the understanding. The importance of any given node is roughly the number of nodes that node covers. For example, the most important node from the pay phone example is 'payment' with an importance of

```
((hang-up :type instruction :isa restart-call
        :modifier ((momentarily)))
 (listen-for-dial-tone :type instruction :step-of telephone-plan)
 (insert-coin :type instruction :isa payment :with
        ((:value * ten-cents)))
 (dial :type instruction :modifier ((again)) :step-of telephone-plan))
```

Figure 3: The results produced for SCAVENGER.

12, the next is 'insert with an importance of 9, and so on. Alterman & Bookman (*ibid*) prove that each of the most important nodes is either a conceptual root or is covered by one of the conceptual roots. In the next stage, the importance measure will be used in conjunction with the conceptual roots to determine the goal concept of the example.

## Stage 3: Explanation-based Learning

The instructions provided with a device provide sufficient information, when read in the context of the situation of engagement, to perform the task required. However, the instructions provide more information than simply mindless commands. That is, not only do the instructions say *how* to accomplish a goal, but they say, implicitly, *why* the steps given will accomplish that goal. While understanding the reasons for taking a step is not necessary to perform the step, this understanding makes it possible to learn from this implicit information in a way that can aid in the use of instructions in future situations.

Traditional EBL is unable to cope with the possibility of learning from instructions. This is because EBL 1) begins with a functional description of a goal concept, then 2) builds an explanation of why a training example demonstrates this goal concept, and 3) generalizes this explanation. The problem with directly applying the EBL approach is that, frequently, instructions do not explicitly provided a goal concept. IIMP deviates from EBL by first performing text inferencing to construct an explanation, in the form of a coherence graph. IIMP then determines the goal concept, using the most important conceptual root from this graph. Moreover, the summary of the coherence graph serves as the generalization.

### Generalization as summarization

Many techniques have been used for forming the generalization in the EBL literature (*partial evaluation*: van Harmelen & Bundy; *goal regression*: Mitchell et al.; *explanation patterns*: Schank & Leake). The key to generalization is to determine the relevant features of the training example while building the explanation — which is exactly the role of summarization. Based on the results from stage 2, IIMP constructs a summarized version of the basic event content of the instructions and treats it as the generalization. This summary consists of those conceptual roots with above average importance (and, implicitly, the subset of the DAG which these conceptual roots cover). Alterman & Bookman

(1991) empirically show that this summary has four interesting properties:

- **coherence**: The summary holds together and make sense.
- **coverage**: The summary covers, at least implicitly, many of the events of the original text.
- **importance**: The summary includes the important parts of the text.
- **workload**: It takes less work to construct an interpretation of the summary than of the original understanding.

Each of these properties has significance for the generalization stage of EBL. Ensuring that the generalization constructed by EBL provides coherence and coverage means that the generalization will include the important points of the training example and will have internal consistency.

Creating a generalization that decreases the workload for the system is important for ensuring the operationality of the result. Operationality requires two properties: usability and utility. That is, the resulting concept definition must be usable by the system, and the result must improve the system's performance. It is important to explicitly show operationality in traditional EBL systems, so the final result will be strictly distinguishable from the original statement of the goal concept. From the perspective of summarization, the property of utility is seen as equivalent to the property of reduced workload. Thus, the utility of the explanation is assured by the 'simplification' property of the summary.

Below, we will show that the question of importance has additional significance for EBL: the most important concept of the generalization provides valuable information for determining the goal concept when one is not explicitly provided.

## Finding the goal concept

The information that IIMP needs to derive the goal concept comes from the analysis of the coherence graph and the summarization. What remains of the task of finding the goal concept is to give it a name and decide where it belongs in semantic memory.

**The most important concept is *related* to the goal concept**. The goal concept from any text is going to be related to the most important concept. This is assured by the importance measure itself, which was developed to choose the main thrust of a piece of text. For the pay phone example, the relationship between the goal concept and the most important conceptual root is that 'payment on a pay telephone' is an adjustment for the ongoing activity on how 'payment' is achieved.

**The most important concept is *not* the goal concept**. When IIMP began reasoning about the pay phone instructions, the notion of payment and its connections to the telephone plan and coin insertion were already in semantic memory. Analyzing the coherence graph identified 'payment' as the most important conceptual root, but IIMP does not need to learn about payment. What IIMP does need to learn about is payment *on a pay telephone*, and how it is related to using a phone at home. If this notion were already in semantic memory, the graph built in stage 1 would have a different character, and the EBL mechanism would act differently. Alternately, if the background knowledge were too specific, where some EBL systems might build an entirely new concept for the new example, this system would relate the new example to the too-specific piece of memory.

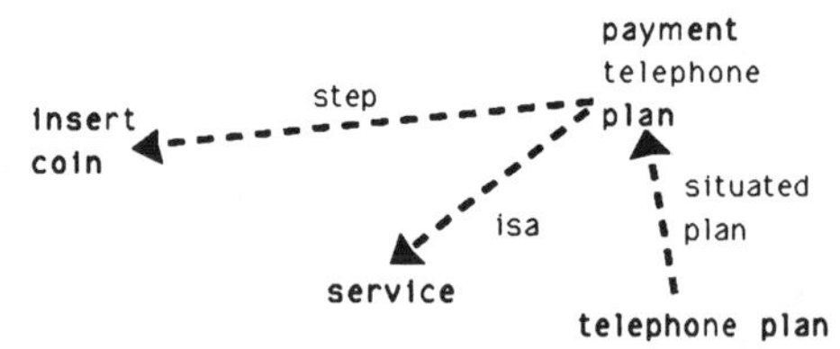


Figure 4: The results of operationalizing and classifying the goal concept

It is true in general that the most important concept is not the goal concept. Simply put, IIMP already knows about the concept expressed by the conceptual root, since it was copied directly from IIMP's original memory. The concept IIMP is trying to learn should be a new concept that somehow encapsulates the understanding of the example. In other words, semantic memory has a gap that IIMP needs to fill.

**Technique**.

IIMP's EBL technique uses the results from the previous stages as follows:

1. Let **N** be the most important concept as determined in stage 2 (e.g., 'payment').
2. Let **C** be the collision occurring at N (this includes the piece of semantic network defined by the collision as well as the collision type, e.g., 'new step')
3. if **C** is a 'new step' collision:
   (a) Let **N1** be the beginning of the 'goal of plan' path shape (e.g., 'telephone plan').
   (b) Let **N2** be the beginning of the 'goal of action' path shape (e.g., 'insert coin').
   (c) Create a new node, **X**, using **N1** and **N** (e.g., 'payment telephone plan'). **X** is a 'situated plan' of **N1**, and **N2** is a 'step' for **X**.
4. if **C** is a 'similar action' collision:

⋮

The new piece of semantic built by this technique is consistent with the rest of semantic memory, and will maintain the integrity of the marker passing in later scenarios.

## Classifying the goal concept

The previous step operationalized the goal concept, 'payment on a pay telephone', adding all the relevant information from the instructions to semantic memory. This involved giving it a name and a place in semantic memory. However, the structure of semantic memory provides some implicit information which can be made explicit now. This information could be found when doing text inferencing (stage 1), but making semantic memory more explicit now will make future inferences more efficient.

In particular, IIMP finds the relationship between the goal concept and the nodes covered by the most important concept. By passing markers from the new node in semantic memory, IIMP is able to learn more from how it is related to other concepts in semantic memory. In the case of the pay phone example, the new concept, 'payment telephone plan', has a relationship with the original telephone plan at the 'service' node, so an 'isa' link is added from the new concept to 'service'.

The final result of operationalizing and classifying the goal concept for the pay phone is shown in Figure 4.

## Summary and Remarks

The problem of reading an instruction in the context of the ongoing activity is a problem of text inferencing. IIMP combines the advantages of marker passing and summarization to produce a coherent interpretation that is built in relation to plan notions and actions and is given in a form that a planner can use to construct a new plan. This approach is able to provide information that is unavailable when instruction usage is separated from action, and, thereby, increases the domains in which a planner is able to act.

Using an augmented version of explanation-based learning, IIMP is able to build new concepts in semantic memory, thus making the content of the instructions available for use when interpreting future instructions for the device. For example, a subsequent encounter with the pay phone involves payment for a long distance call. Rather than constructing the interpretation of payment from scratch, IIMP simply augments the interpretation built during the first encounter with the pay phone.

## References


[Alterman and Bookman, forthcoming] Richard Alterman and Lawrence A. Bookman. Reasoning about a semantic memory encoding of the connectivity of events. *Cognitive Science Journal*, forthcoming.

[Alterman *et al.*, 1991] Richard Alterman, Roland Zito-Wolf, and Tamitha Carpenter. Interaction and instruction usage. *To appear in* The Journal of the Learning Sciences, 1991.

[Alterman, 1988] Richard Alterman. Adaptive planning. *Cognitive Science*, 12:393–421, 1988.

[Chapman, 1990] David Chapman. Vision, instruction, and action. Technical Report AI-TR 1024, MIT Artificial Intelligence Laboratory, 1990.

[Charniak, 1986] Eugene Charniak. A neat theory of marker passing. In *Proceedings of the Fifth National Conference on Artificial Intelligence*, pages 584–588. Morgan Kaufmann, 1986.

[DeJong and Mooney, 1986] Gerald DeJong and Raymond Mooney. Explanation-based learning: An alternative view. *Machine Learning*, 1(2):145–176, 1986.

[Goodman *et al.*, 1991] Marc Goodman, Scott Waterman, and Richard Alterman. Interactive reasoning about spatial concepts. To appear in *Thirteenth Annual Meeting of the Cognitive Science Society*, 1991.

[Graesser, 1981] Arthur Graesser. *Prose comprehension beyond the word.* Springer-Verlag, 1981.

[Hendler, 1986] James Hendler. *Integrating marker-passing and problem-solving: A spreading activation approach to improved choice in planning.* PhD thesis, Brown University, 1986.

[Kieras and Bovair, 1986] David E. Kieras and Susan Bovair. The acquisition of procedures from text: A production-system analysis of transfer of training. *Journal of Memory and Language*, 25:507–524, 1986.

[Kintsch and van Dijk, 1978] Walter Kintsch and Teun van Dijk. Toward a model of text comprehension and production. *Psychological Review*, 85:363–394, 1978.

[LeFevre and Dixon, 1986] Jo-Anne LeFevre and Peter Dixon. Do written instructions need examples? *Cognition and Instruction*, 3(1):1–30, 1986.

[Lehnert, 1981] Wendy Lehnert. Plot units and narrative summarization. *Cognitive Science*, 5:293–331, 1981.

[Mannes and Kintsch, 1991] Suzanne M. Mannes and Walter Kintsch. Routine computing tasks: Planning as understanding. Unpublished manuscript, 1991.

[Mitchell *et al.*, 1986] T. Mitchell, R. Keller, and S. Kedar-Cabelli. Explanation-based generalization: A unifying view. *Machine Learning*, 1:47–80, 1986.

[Norvig, 1989] Peter Norvig. Marker passing as a weak method for text inferencing. *Cognitive Science*, 13(4):569–620, 1989.

[Schank and Leake, 1989] Roger C. Schank and David B. Leake. Creativity and learning in a case-based explainer. *Artificial Intelligence*, 40:353–385, 1989.

[Trabasso and van den Broek, 1985] Tom Trabasso and Paul van den Broek. Causal thinking and representation of narrative events. *Journal of Memory and Language*, 24:612–630, 1985.

[van Harmelan and Bundy, 1988] Frank van Harmelan and Alan Bundy. Explanation-Based Generalisation=Partial Evaluation. *Artificial Intelligence*, 36:401–412, 1988.

[Zito-Wolf and Alterman, 1990] Roland Zito-Wolf and Richard Alterman. Ad-hoc fail-safe plan learning. In *Proceedings of the Twelfth Annual Conference of the Cognitive Science Society*, pages 908–914. Lawrence Erlbaum Associates, 1990.

# The Time course of Metaphor Processing: Effects of Subjective Familiarity and Aptness

Dawn G. Blasko
Cynthia M. Connine

State University of New York at Binghamton
Binghamton, New York 13902-6000

## Abstract

A cross-modal priming paradigm was used to investigate the time course of figurative activation for metaphors which varied in familiarity. In Experiment 1 the target was presented immediately at the offset of the vehicle. For high familiar metaphors, both literal and figurative interpretations showed evidence of immediate availability. For low familiar metaphors, the literal interpretation was available but the figurative target showed inhibition. Experiment 2 delayed presentation of the target 300 ms. and similar results were found, although inhibition of the figurative target decreased. Together, Experiments 1 and 2 showed the figurative meaning is more readily available in highly familiar metaphors. The results of Experiment 3 suggest metaphor aptness is especially important for low familiar metaphors. The implications of these findings for models of non-literal language are discussed.

## Introduction

A comprehensive theory of language processing must be capable of explaining how individuals comprehend the meaning of utterances that differ in their literal and non-literal interpretations. Non-literal language is an extremely pervasive component of natural language, therefore, the understanding of metaphors and other non-literal language is a critical theoretical problem.

The purpose of this paper is to address two questions concerning the processing of one form of non-literal language, namely metaphors. First, what is the time course of metaphorical activation? Is the figurative interpretation of a metaphor processed directly, or is it necessary for a literal interpretation to be computed prior to the metaphorical interpretation? Secondly, what are the factors that influence this time course? Two specific factors, subjective familiarity and aptness were investigated.

One approach to the comprehension of metaphors that makes clear predictions about the time course of metaphorical activation is Searle's (1979) three stage model. According to the three-stage model all metaphors are comprehended in a series of three stages. In stage 1 an attempt is made to interpret the metaphor literally. In stage 2 the utterance is found to be in some way defective when taken literally. This, in turn, triggers stage 3, which is a search for a non-literal interpretation (Searle, 1979). One prediction of the three-stage model is that a statement should take longer to comprehend when intended metaphorically than when it is intended literally. This is because the process of metaphorical interpretation can only begin after the literal interpretation has been computed and rejected. However, a number of researchers (Ortony, Schallert, Reynolds, and Antos, 1978; Inhoff, Lima and Carroll, 1984) found that this was not the case if a metaphor is presented in a context sufficient to specify both its component parts (eg. the appropriate topic) and its metaphoricity (the fact that it was not meant literally). Other experiments, (Glucksberg, Gildea, and Bookin, 1982) have demonstrated that under some conditions metaphor processing is obligatory. Subjects were asked to decide if a sentence was literally true or false. For example, the sentence SOME JOBS ARE JAILS is literally false, but metaphorically true. The automatic comprehension of the non-literal truth of a metaphor interfered with judgments of literal falseness, slowing responses.

These studies suggest that metaphor comprehension may not require an initial literal interpretation. However, Janus and Bever (1985) have suggested that this conclusion may be premature. They argued that the total sentence reading times used by Ortony et al. (1978) were not sensitive enough to detect differences in metaphorical and literal processing. When Janus and Bever (1985) replicated Ortony et al. (1978)

using phrase by phrase reading times, they found slowing of the critical metaphorical phrase as predicted by the three-stage model. Although the controversy is by no means settled, this study does point out the need for on-line measures of metaphor processing.

One factor which may well influence the efficiency of metaphor comprehension is subjective familiarity. The role of subjective familiarity has not been previously examined, however, it is likely that experience with a particular metaphor should facilitate subsequent comprehension. A second factor, metaphor goodness or aptness, has been investigated (eg., Trick and Katz, 1986). It appears that a highly apt metaphor is one in which the domains of the topic and vehicle are relatively distant but the within domain features are relatively close in semantic space (Trick and Katz, 1986). The between domain distance must be fairly large for the metaphor to be effective because close distances provide little interaction or surprise. For example, A SQUIRREL IS A CHIPMUNK carries little or no metaphorical meaning because SQUIRREL and CHIPMUNK are both from the same domain (small furry mammals) and therefore share most features. This suggests that a certain degree of aptness is likely to be critical to metaphor comprehension.

There are two major questions to be addressed in the present paper. First, what are the temporal parameters of a metaphorical interpretation? Is the figurative meaning of the metaphor available immediately along with the literal interpretation of the vehicle or is the literal interpretation available earlier as suggested by the three-stage model. Second, does the subjective familiarity and aptness of a metaphor influence the availability of the metaphorical interpretation?

## Method

### Materials and Procedures

In three cross-modal priming experiments a metaphorical phrase embedded in a neutral sentence context was presented auditorily and the subject's task was to make a lexical decision (word or non-word) about a letter string (Target) presented visually for 250 ms. To assure that the sentences were being attended to, a comprehension test was built into the experiment. On each trial, an individual subject had a 50% chance of receiving a word vs. a non-word target, and 50% chance of hearing a metaphorical vs. a literal sentence. Three visual target conditions included a word related to the metaphorical meaning of the vehicle (FIGURATIVE TARGET), a word related to the vehicles literal meaning (LITERAL TARGET) and an unrelated word (CONTROL TARGET). Targets were selected in 3 individual norming studies. Target words were matched to unrelated control targets that were approximately equated for word frequency, word complexity and length. Metaphors were selected from a corpus of 260 non-literary metaphors compiled by Katz, Paivio, Marschark and Clark (1988) and had a consistent X is Y structure. The topic and vehicle were explicitly stated and no literal reading of the statement was plausible. For example:

Jerry first knew that LONELINESS WAS A DESERT when he was still very young.

Figurative Target: ISOLATE
Literal Target: SAND
Control Target: MUSTACHE

Faster lexical decisions in comparison to the control targets are assumed to indicate on-line activation. In Experiments 1 and 2, 12 high familiar metaphors (mean familiarity 4.16 - 5.28) and 12 low familiar metaphors (mean ratings 2.29 - 3.39) were chosen according to independent ratings completed by 44 subjects.

### Subjects

The subjects for all three experiments were undergraduate students enrolled in psychology classes at the State University of New York at Binghamton who received class credit or monetary payment.

## Experiment 1

In Experiment 1, the visual target was presented immediately at the offset of the metaphorical vehicle. The three stage model predicts faster reaction times to literal targets than to controls, whereas, the figurative targets should not differ from the controls. This would indicate the literal interpretation was available prior to the figurative interpretation. If the metaphorical interpretation is available along with the literal interpretation, both literal and figurative targets should be faster than the controls. Also if subjective familiarity influences the availability of a metaphorical interpretation then facilitation for figurative targets should be primarily for highly familiar metaphors.

Results. The overall correct lexical decision rates for experiment 1 was 98% correct, experiment 2 was 97% correct and experiment 3 was 99% correct. Reaction times less than 250 ms and greater than 1750 ms were excluded from all analysis.

The results shown in Table 1 differ depending on familiarity. In the high familiar condition, facilitation is seen for both literal and figurative targets in comparison to the controls. This indicates that both the literal and metaphorical interpretation of the vehicle was available. In contrast, the low familiar condition showed facilitation for the literal target, but the figurative target is slower than it's control.

Table 1
Experiment 1: Means and SDs of correct responses

| | Low Familiar | | | High Familiar | | |
|---|---|---|---|---|---|---|
| | MEAN | SD | DIF* | MEAN | SD | DIF* |
| FIGUR. | 983 | 208 | -54 | 887 | 173 | 157 |
| LIT. | 881 | 194 | 48 | 880 | 181 | 164 |
| CONT. | 929 | 189 | | 1044 | 180 | |

* difference from control

The analysis of variance was conducted for both subjects (F1) and items (F2). A two way ANOVA (Familiarity: high vs. low X Target: literal, figurative, control) showed a significant main effect of target ($F1(2,160) = 24.82$, $p < .0001$; $F2(2,22) = 3.44$, $p < .05$,). The effects of familiarity were seen in the interaction between familiarity and target ($F1(2,160) = 28.42$, $p < .0001$; $F2(2,22) = 6.13$, $p < .01$.). In the high familiar condition both the literal and figurative targets were significantly faster than their control indicating activation of the metaphorical interpretation simultaneous with literal activation of the vehicle. Planned comparisons using Bonferroni t-tests to control for familywise error ($p < .05$) support this conclusion (figurative vs. control $t(80) = 7.58$, $p < .0001$; literal vs. control ($t(80) = 8.33$), $p < .0001$).

In the low familiar case the findings are quite different. The literal target is faster than its control, whereas, the figurative target appears to show a pattern of inhibition (literal target vs. control, $t(80) = 2.67$, $p < .01$); figurative target vs. control.($t(80) = 2.78$, $p < .01$).

## Experiment 2

In Experiment 2 the target was presented 300 ms downstream of the vehicle. Results for the high familiar condition should replicate Experiment 1 with both literal and figurative targets showing patterns of facilitation. If additional time is needed for the comprehension of low familiar metaphors, the figurative targets should show facilitation.

Results. The results of Experiment 2 were quite similar to Experiment 1 (see Table 2). It is clear that the pattern of reaction times differs based on familiarity. The high familiar condition continues to show faster lexical decisions to the literal and figurative targets in comparison to their unrelated controls. In contrast, in the low familiar condition only the literal target is facilitated whereas the figurative target is once again slower than it's control. A two way ANOVA (Familiarity: high vs. low X Target: literal, figurative, control) showed a main effect of target in the subject analysis that narrowly misses significance in the items analysis ($F1(2,70) = 17.59$, $p < .0001$; $F2(2,22) = 3.15$, $p < .06$). Once again, the effects of familiarity can be seen in the interaction between familiarity and target ($F1(2,70) = 18.88$, $p < .0001$; $F2(2,22) = 4.43$, $p < .05$).

The results of Experiment 2 appear to be similar to those found in Experiment 1, however, closer comparison of the means (see Table 2) does show one important difference. With the additional time the figurative targets in the low familiar condition show less inhibition. Unlike Experiment 1, there is no longer a significant difference between the figurative and control targets in the low familiar condition ($t(35) = .72$). However, there is significant facilitation of the literal target ($t(35) = 3.43$, $p < .002$) that is almost twice the size of the facilitation found for the literal target in the low familiar case in Experiment 1. It is possible that the greater facilitation with additional time may reflect the continued search process for features of the vehicle's domain that are necessary for comprehension of the figurative interpretation.

Table 2
Experiment 2: Means and SDs of correct responses

| | Low Familiar | | | High Familiar | | |
|---|---|---|---|---|---|---|
| | MEAN | SD | DIF* | MEAN | SD | DIF* |
| FIGUR. | 926 | 210 | -24 | 795 | 129 | 172 |
| LIT. | 811 | 141 | 91 | 820 | 141 | 147 |
| CONT. | 902 | 221 | | 967 | 136 | |

*difference from control

In the high familiar condition, facilitation is strong for both the figurative ($t(35) = 9.45$, $p < .0001$) and literal targets ($t(35) = 8.01$, $p < .0001$). There is now a 25 ms advantage for the figurative over the literal target. Although the difference is small, it may indicate that the literal meaning is beginning to fade as comprehension of the metaphorical phrase is completed.

The continued lack of availability of the low familiar figurative interpretation prompted a closer look at the item means in terms of their aptness. In the high familiar case the pattern is consistent. All high familiar metaphors showed activation regardless of their aptness. On the other hand, for the low familiar metaphors the high apt group showed facilitation, whereas the moderate apt group showed inhibition. This suggests that aptness may play an especially important role in low familiar metaphors.

## Experiment 3

Experiment 3 was designed to investigate whether the processing of low familiar metaphors may depend on aptness. Twelve additional metaphors were selected to create high and moderate apt groups of low familiar metaphors (12 in each of the aptness categories). As in Experiment 1, the visual target was presented immediately at the offset of the metaphor vehicle. If the post hoc observations of Experiments 1 and 2 are viable, then the moderately apt metaphors should exhibit inhibition of the figurative targets and the high apt metaphors should show evidence of facilitation.

Results. Overall the results of Experiment 3 were consistent with predictions (Table 3). A two way ANOVA (Aptness: high vs. moderate X Target: literal, figurative, control) shows a marginally significant main effect of aptness attributable to faster times overall for the high apt targets ($F1(1,38) = 3.13$, $p < .08$; $F2(1,11) = 4.22$, $p < .06$. There is also a main effect of target ($F1(2,76) = 18.96$, $p < .0001$; $F2(2,22) = 9.78$, $p < .001$). The predicted interaction between aptness and target is significant in the analysis by subjects ($F1(2,76) = 3.64$, $p < .05$ and marginally significant in the items analysis ($F2(2,22) = 2.66$, $p < .09$).

As seen in Table 3, in the moderate apt low familiar condition the figurative target is slower than the control target although the difference does not reach significance in Bonferroni t-tests ($t(38) = 1.45$, $p < .16$. In contrast, the literal target ($M = 780$) is clearly faster than the control ($M = 855$), $t(38) = 4.07$) demonstrating activation of the vehicles' literal meaning.

Table 3
Experiment 3: Means and SDs of correct responses

| | Moderate Apt | | | High Apt | | |
|---|---|---|---|---|---|---|
| | MEAN | SD | DIF* | MEAN | SD | DIF* |
| FIGUR. | 889 | 145 | -34 | 816 | 117 | 50 |
| LIT. | 780 | 121 | 75 | 779 | 123 | 87 |
| CONT. | 855 | 130 | | 866 | 144 | |

*difference from control

The overall pattern of the high apt low familiar condition is also consistent with predictions. Both the literal target and the figurative target are faster than the control but the figurative difference is only marginally significance by Bonferroni criteria (literal vs. control $t(38) = 3.46$, $p < .001$; figurative vs. control $t(38) = 2.14$, $p < .039$).

The results of Experiment 3 suggest immediate availability of both literal and figurative interpretations of a low familiar metaphor if the metaphor is highly apt. When a person encounters a metaphor with which they are not familiar, the aptness of the metaphor may be especially critical to comprehension.

## General Discussion

The three experiments presented here suggest that it is possible for the metaphorical interpretation of a phrase to be processed as rapidly as the literal meaning of one of its elements. There are two important qualifications to this statement. First, rapid processing seems to depend on previous experience with the metaphor, shown by high ratings of subjective familiarity. Highly familiar metaphors seem to be more easily and rapidly comprehended than relatively novel, low familiar metaphors. Second, even if the metaphor is less familiar, it's figurative interpretation may still be available relatively early if it has been judged to be highly apt.

If all metaphors are comprehended using a stage-like process in which an attempt at a literal meaning must first be made in order for a non-literal meaning to be computed (Searle, 1979), then we would expect only the literal interpretation to be available in the 0-delay condition. This is clearly not the case. The results of Experiment 1 and 2 show that when metaphors are high in subjective familiarity they can be processed as rapidly as the literal meaning of one of their components. These findings can not be readily explained by the three-stage model.

One alternative to the three-stage model is the Categorization model of metaphor processing (Glucksberg & Keysar, 1990). In this approach the literal and metaphorical interpretations of an X is a Y assertion are both comprehended as simple class-inclusion statements. For example, the metaphor MY HUSBAND IS A BABY is understood by assigning the metaphor's topic, HUSBAND, to the class exemplified by the metaphor's vehicle BABY, as well as infants and other dependent creatures. In contrast, if the contextual environment favored a literal interpretation of the word BABY then the class would include a different membership, restricted to those of a very young age. Consistent with the data presented here, the categorization approach suggests direct processing of the metaphorical interpretation without the need to first compute and reject a literal interpretation.

It is also possible to view the roles of aptness and familiarity within the categorization approach. Aptness has been shown to be reflected, in part, within the structure of the metaphor itself. An apt metaphor is one in which the domains of the topic and vehicle are relatively distant while the within domain features are relatively close in semantic space (Trick & Katz, 1986). Subjective familiarity, on the other hand, is a measure of how familiar the subject is with the metaphor. A highly familiar metaphor may be one in which considerable experience with the individual domains of the topic and vehicle has made the appropriate features of the vehicle more salient and therefore, more quickly and easily activated. As experience with the metaphorical usage of the vehicle increases, the relevant features may become more and more central to the domain, therefore requiring less time and effort for comprehension. In direct contrast, less familiar metaphors, which are also less apt, may have vehicles in which the critical features needed to define the category, are not highly salient. This may account for the pattern of inhibition found for the low familiar moderate apt metaphors. When embedded in discourse, contextual information probably highlights the features critical for comprehension.

In conclusion, in order to explain the comprehension of metaphors it may not be necessary to postulate either separate metaphorical stages or separate mechanisms. However, this does not imply that metaphors are unimportant or uninteresting. Quite the contrary, the study of non-literal language may be crucial to the understanding of a constantly evolving natural language system. Throughout the ordinary evolution of any natural language, novel usages that are capable of expressing a great deal of information powerfully and succinctly will be invented, and if used frequently, may eventually become part of the literal usage.

## Acknowledgements

This research was supported by the Center for Cognitive and Psycholinguistic Sciences and NINCDS Grant R29 26587 and NIH BRSG Grant SO7RR07149-17 to the second author.

## References

Glucksberg, S., and Keysar, B. 1990. Understanding Metaphorical Comparisons: Beyond Similarity. *Psychological Review* 97:3-18.

Glucksberg, S., Gildea, P., and Bookin, H. 1982. On Understanding Nonliteral Speech: Can People Ignore Metaphors? *Journal of Verbal Learning and Verbal Behavior* 13:512-521.

Inhoff, A.W., Lima, S.D., and Carroll, P.J. 1984. Contextual Effects On Metaphor Comprehension in Reading. *Memory & Cognition* 12:558-567

Janus, R.A. and Bever, T.G. 1985. Processing of Metaphoric Language: An Investigation of the Three-Stage Model of Metaphor Comprehension. *Journal of Psycholinguistic Research* 14:473-487.

Katz A.N., Paivio A., Marschark, M. and Clark J. 1988. Norms for 204 Literary and 260 Nonliterary Metaphors on Psychological Dimensions. *Metaphor and Symbolic Activity* 3:191-214.

Ortony, A., Schallert, D., Reynolds, R.B., and Antos, S.J. 1978. Interpreting Metaphors and Idioms: Some Effects of Context on Comprehension. *Journal of Verbal Learning and Verbal Behavior* 17:465-477.

Trick, L., and Katz, A.N. 1986. The Domain Interaction Approach to Metaphor Processing: Relating Individual Differences and Metaphor Characteristics. *Metaphor and Symbolic Activity* 1: 185-213.

Searle, J. 1979. Metaphor. In A. Ortony (Ed.), *Metaphor and Thought*. Cambridge, England: Cambridge University Press.

# The Effects of Feature Necessity and Extrinsicity on Category Contrast in Natural Language Categories

**Leslie J. Caplan**
**National Institute of Mental Health**
National Institutes of Health
Federal Building, Room B1A-14
Bethesda, MD 20892

**Robin A. Barr**
**National Institute on Aging**
National Institutes of Health
Building 31, Room 5C-32
Bethesda, MD 20892

**Abstract**

This experiment tested two hypotheses: 1) that categories represented by features that many people believe to be necessary will demonstrate stronger category contrast than those represented by features that few people believe to be necessary, and 2) that categories that people believe are represented primarily by intrinsic features (i.e., features true of an entity in isolation) will have stronger category contrast than those that people believe are represented primarily by extrinsic features (i.e., features that represent relations between an entity and other entities). The findings support only the second hypothesis.

## Introduction

In this paper, we examine contrast among hierarchically organized categories. Consider a classical nested hierarchy containing one superordinate node (A), and two subordinate nodes (B and C). The hierarchy should exhibit the property of class inclusion (a B is an A, and a C is an A), as well as mutual exclusivity among other categories (a B cannot be a C, and a C cannot be a B). In the experiment reported here, we examine the second kind of relation: category contrast. To use Kay's (1971) terms, we investigate categories that should demonstrate either direct contrast (i.e., two categories that share the same immediate superordinate), or indirect contrast (i.e., two categories that contrast because their superordinates are themselves in direct contrast). In the experiment reported here, we explore two possible mechanisms for inter-category differences in the degree to which such contrast holds: the extrinsicity of features in category representation (Barr & Caplan, 1985, 1987; Caplan & Barr, in press), and probabilistic features (e.g., Rosch, 1973, 1975).

### Feature extrinsicity and category contrast

In previous work (Barr & Caplan, 1985, 1987; Caplan & Barr, in press), we have presented a model of category representation that accounts for inter-category differences in gradients of membership. According to that model, the features that comprise a category's representation may be considered to be either *intrinsic* or *extrinsic*. Intrinsic features are features that are true of a potential category member in isolation (e.g., "has leaves", "has hair"). Extrinsic features are features that represent relations between potential category members and other entities (e.g., "is used for transportation", "has a child"). We have demonstrated that class-inclusion relations are stronger for categories represented by intrinsic features than for those represented by extrinsic features (Caplan & Barr, in press).

Our model of category representation also predicts that inter-category contrast will be more likely to characterize relations among intrinsically represented than among extrinsically represented categories. Why should this be the case? Consider a person responding to questions of the form, "Can a B be a C?" (e.g., "Can a weapon be a tool?", "Can a plant be an animal?"), as subjects in the experiment reported below were asked to do. In order for a member of Category B to become a member of Category C, it must somehow acquire the features included in Category C's representation, and/or lose those of Category B's representation. Such feature change or acquisition should be more likely to occur for extrinsic than for intrinsic features. Because intrinsic features are represented as true of an object in isolation, whether or not an item is represented as possessing an intrinsic feature will remain constant across situations. Although it is possible that an entity might lose the feature, that possibility is not expressed in the feature's representation. Therefore, intrinsic category representations do not include the possibility for feature change, and strong inter-category contrast should result. On the other hand, membership in extrinsically represented categories depends not on the attributes of the possible category member itself, but on the relations it holds with other entities or objects. Because such relations may change over different contexts (e.g., a possible weapon may sometimes be used to harm people, and sometimes not), entities may be conceived to slip in and out of membership. In other words, extrinsic representations include the possibility for change in category membership, since whether an object is a category member depends on the relations it holds with other objects, and those relations may change. Accordingly, extrinsically represented categories should show much weaker contrast.

### Probabilistic features and category contrast

Over the last 20 years, "family resemblance" or probabilistic approaches to semantic category representation have challenged traditional approaches to category representation (see Smith & Medin, 1981, for a review). These theories (e.g., Rosch, 1973, 1975) posit that category exemplars typically exhibit some, but not all, of the pool of features associated with a category. According to most versions of probabilistic theory, category membership is determined by the sum of the weighted features possessed by the exemplar (see Smith & Medin, 1981). One of the major characteristics of probabilistically represented categories is the fact that categories may, through possessing some subset of the relevant features, be more or less good members of the category. Accordingly, categories may include many "borderline" members – members which may also be considered members of other, related categories. Presumably, borderline members of one category can also be members of another category from the same hierarchy if they possess the appropriate subsets of features for both categories; therefore, one would *not* expect to find strong evidence of inter-category contrast. For example, a large serrated knife may also be a saw. It possesses enough features to qualify for membership in both categories (knife and saw), without being a very good or very typical member of either.

Recently, Markman (1989) has suggested that categories differ in whether they are probabilistically or classically represented, with some categories represented by sets of features necessary for category membership, and others represented probabilistically. [1] In fact, we have demonstrated that such differences exist, and that hierarchies of categories represented by features which many subjects believe to be necessary are characterized by stronger class inclusion than are those represented by features that few subjects believe to be necessary for category membership (Caplan & Barr, in press). Similarly, whether a category is represented "classically" or probabilistically should determine whether or not strong inter-category contrast obtains among members of a hierarchy. Probabilistically represented categories should reveal weak contrast, and categories represented by sets of necessary features should be characterized by strong contrast.

## Method

### Subjects

Subjects were 96 undergraduates from a midwestern state university.

### Stimuli

Twenty-eight categories, representing four hierarchies, each of which consisted of three levels of abstractness, were used. The four hierarchies were: (a) Man-made Objects (Artifacts), which included Tools (Hammers, Saws) and Weapons (Guns, Bombs), (b) Living Things, which included Animals (Cats, Dogs) and Plants (Trees, Cacti), (c) Occupations, which included Lawyers (Real-estate Lawyers, Corporate Lawyers) and Physicians (Dermatologists, Surgeons), and (d) Relatives, which included Parents (Mothers, Fathers) and Grandparents (Grandmothers, Grandfathers). Within each hierarchy, every subordinate (i.e., every term with the exception of Man-made Objects, Living Things, Occupations, and Relatives) was paired with every other subordinate in a question of the form "Can an x be a y?". Because this method of question construction yielded three pairs in the Relatives hierarchy that were actually questions of class inclusion (mother-grandmother, father-grandfather, and parent-grandparent), these pairs were excluded from analysis. In order to keep the length of the task manageable, reversals (i.e, "Can a y be an x?") were not included among the stimulus items for any hierarchy.

For each of these categories, we had collected necessity and extrinsicity information from previous groups of subjects. The extrinsicity ratings were provided by one group of subjects (N = 15) who rated each category in terms of how much the category's rules for membership relied on information about the relationships that category members hold with other things, on a nine-point scale. Each category's extrinsicity value was the mean rating (across subjects) for that category. To collect necessity information, we first had a second group of subjects (N = 48) list the features they believed defined the category. The categories were divided into three sets; each set was "defined" by a subgroup of 16 subjects. We then

[1] An extrinsic feature may be necessary for category membership, but because the possession of the feature is situation-dependent, may not characterize even very good category members under all circumstances (see Caplan & Barr, in press, for more detail). For example, most of the subjects in that study asserted that "covers the body" is a necessary characteristic of Clothing. However, a given shirt may have other uses, depending on the situation (e.g., one may use it as a dustcloth). Thus, a feature may be necessary for category membership (e.g., in order for something to be clothing, it must cover the body), but not necessarily true of a particular exemplar in a given situation (e.g., a given shirt need not always cover the body).

combined the features listed by at least five of the 16 subjects in each subgroup with features collected from definitions in two dictionaries (*Webster's New World Dictionary of the American Language*, 1980; *Thorndike-Barnhart Desk Dictionary*, 1952). These feature lists were our feature sets. A third group of subjects (N=19) then judged each feature from the feature sets as necessary, typical, or neither for category membership. Each category's "necessity" measure was the mean (across all features from the category's feature set which received at least one "necessary" judgment) proportion of subjects who gave "necessary" judgments.[2]

### Procedure

The stimuli were presented in random order in a stimulus booklet. Subjects answered "yes" or "no" to each question by filling in a circle on a computer scorable answer sheet. Every subject received every question. The subjects were tested in groups of about 20.

## Results

Table 1 shows the number of subjects who assented to each question. The more subjects who assented, then the less the members of the two categories contrast. For example, 89 out of the 96 subjects agreed that a weapon could be a tool. Therefore, the members of these two categories do not contrast well. On the other hand, just three out of 96 subjects agreed that a dog could be a cat. The members of these two categories, then, do contrast well.

We had predicted that the more extrinsic the representations of the two categories, or the more probabilistic, then the more subjects would agree to questions of the form "Can an x be a y?" However, the number of features by which the two categories differ must also be taken into account, because only distinctive features can contribute to contrast. In order to make this estimate, we counted the total number of features that had received at least one "necessary" judgment from the feature sets for the two categories, and subtracted all features that the two categories shared. We then conducted a multiple regression analysis in which the number of subjects assenting to each question was the criterion variable, and the predictor variables were: (1) the number of features by which the two categories differed, (2) the degree to which the representations of the two categories were extrinsic (expressed as the mean of the two categories' mean extrinsicity ratings), and (3) the mean of the two categories' necessity values. The multiple $R$ resulting from this analysis was .70, $F\ (3,33) = 10.74$, $p < .0001$. Of the three predictor variables, only extrinsicity accounted for a significant portion of the variance, $ß = .78$, $t\ (33) = 3.48$, $p < .01$ in the final regression equation; the correlation between extrinsicity and number of subjects assenting was .70.

## Discussion

All pairs of categories used in this experiment do contrast, in the sense that they are drawn from different branches of the same hierarchy. Nevertheless, as the results of the analysis showed, the extent to which the *members* of the two categories contrast is determined by the degree to which the representations of the two categories are extrinsic. Extrinsically represented categories showed much weaker contrast than did intrinsically represented categories.

We were somewhat surprised to find that people's judgments of feature necessity did not predict category contrast, particularly in light of our earlier findings that categories represented by features judged by many people to be necessary showed stronger class inclusion than categories represented by features judged to be necessary by fewer subjects (Caplan & Barr, in press). If anything, our present findings regarding the necessity measure were the reverse of predictions: category pairs with higher necessity values tended to have high numbers of subjects assenting to the questions (i.e., less inter-category contrast), $r = .35$. In combination with our earlier findings, these results suggest that feature necessity is more critical for class inclusion than for direct and indirect contrast relations among categories of a hierarchy.

In addition, these findings suggest that the inter-category relations implied by traditional hierarchies may not always describe psychological inter-category relations. In fact, it may be inappropriate to assume that relations among categories are characterized by unqualified "is a" links, or by unqualified contrast relations. Instead, these relations depend at least on the degree to which the categories' representations rely on intrinsic or extrinsic information.

Many authors have identified difficulties in the assumption that truly necessary features exist as criteria for category membership (e.g., Medin, 1989; Murphy & Medin, 1985). We do not mean to suggest that such features have been identified, or that they are the only source of information included in people's representations of natural language categories (see Caplan & Barr, in press, for further detail). Nevertheless, the findings reported here suggest that feature-level mechanisms, such as extrinsicity, may play a useful role in explaining psychological aspects of categorization phenomena.

---

[2] In previous research (see Caplan & Barr, in press, for more detail), we collected judgments of both feature extrinsicity and of feature necessity. Some examples follow. For the category Clothing, the features "covers the body" and "keeps the body warm" were both judged to be extrinsic by more than half the subjects; however, the former was judged to be a necessary feature by more than half the subjects, and the latter was judged necessary by fewer than half of the subjects. For the category Birds, the features "has wings" and "sings or chirps" were both judged to be extrinsic by less than half the subjects; however, the former was judged necessary by more than half of the subjects and the latter was judged necessary by less than half the subjects.

| Category Terms in Question X | Y | Necessity | Extrinsicity | Number of Subjects Assenting |
|---|---|---|---|---|
| Weapon | Tool | .48 | 7.00 | 89 |
| Weapon | Hammer | .40 | 6.23 | 92 |
| Weapon | Saw | .44 | 5.86 | 94 |
| Tool | Bomb | .56 | 6.14 | 58 |
| Tool | Gun | .56 | 5.94 | 62 |
| Bomb | Gun | .56 | 5.00 | 31 |
| Bomb | Hammer | .49 | 5.36 | 21 |
| Bomb | Saw | .53 | 5.00 | 15 |
| Gun | Hammer | .49 | 5.16 | 15 |
| Gun | Saw | .53 | 4.80 | 14 |
| Hammer | Saw | .46 | 5.16 | 13 |
| Mother | Father | .60 | 6.20 | 15 |
| Mother | Grandfather | .82 | 6.30 | 15 |
| Father | Grandmother | .62 | 6.36 | 14 |
| Grandfather | Grandmother | .84 | 6.46 | 14 |
| Plant | Animal | .52 | 3.14 | 14 |
| Plant | Dog | .56 | 3.07 | 5 |
| Plant | Cat | .54 | 3.14 | 9 |
| Animal | Cactus | .48 | 2.80 | 10 |
| Animal | Tree | .56 | 2.94 | 9 |
| Tree | Cactus | .50 | 3.14 | 43 |
| Tree | Cat | .57 | 2.94 | 10 |
| Tree | Dog | .60 | 2.87 | 3 |
| Dog | Cactus | .51 | 2.74 | 5 |
| Dog | Cat | .58 | 2.54 | 3 |
| Cat | Cactus | .48 | 2.80 | 7 |
| Physician | Lawyer | .72 | 6.07 | 80 |
| Physician | Real-estate lawyer | .72 | 6.00 | 86 |
| Physician | Corporate lawyer | .67 | 5.97 | 83 |
| Lawyer | Surgeon | .76 | 6.38 | 80 |
| Lawyer | Dermatologist | .78 | 6.07 | 80 |
| Dermatologist | Corporate lawyer | .72 | 5.97 | 82 |
| Dermatologist | Real-estate lawyer | .78 | 6.18 | 80 |
| Dermatologist | Surgeon | .84 | 6.18 | 87 |
| Surgeon | Real-estate lawyer | .74 | 6.32 | 80 |
| Surgeon | Corporate lawyer | .69 | 6.28 | 74 |
| Corporate lawyer | Real-estate lawyer | .64 | 6.10 | 85 |

**Table 1.** Number of Subjects (N = 96) Assenting to Each of the 37 Questions of the Form "Can an X be a Y?", As a Function of Mean Necessity Value and Mean Extrinsicity Rating Across the Two Categories

## References


Barr, R. A., & Caplan, L. J. 1985. Two Kinds of Feature? A Test of Two Theories of Typicality Effects in Natural Language Categories. In Proceedings of the Seventh Annual Cognitive Science Society. Hillsdale, NJ: Erlbaum.

Barr, R. A., & Caplan, L. J. 1987. Category Representations and Their Implications For Category Structure. *Memory & Cognition* 15:397-418.

Caplan, L. J., & Barr, R. A. in press. The Effects of Feature Necessity and Extrinsicity on Gradedness of Category Membership and Class Inclusion Relations. *British Journal of Psychology.*

Kay, P. 1971. Taxonomy and Semantic Contrast. *Language* 47: 866-887.

Markman, E. M. 1989. *Categorization and Naming in Children.* Cambridge, MA: MIT Press.

Medin, D. L. 1989. Concepts and Conceptual Structure. *American Psychologist* 44:1469-1481.

Murphy, G. L., & Medin, D. L. 1985. The role of theories in conceptual coherence. *Psychological Review* 92:289-316.

Rosch, E. 1973. On the internal structure of perceptual and semantic categories. In T. E. Moore (Ed.), *Cognitive Development and the Acquisition of Language.* New York: Academic Press.

Rosch, E. 1975. Cognitive Representations of Semantic Categories. *Journal of Experimental Psychology: General* 104:192-233.

Smith, E. E., & Medin, D. L. 1981. *Categories and Concepts.* Cambridge, MA: Harvard University Press.

*Thorndike Barnhart Comprehensive Desk Dictionary.* 1952. New York: Doubleday.

*Webster's New World Dictionary of the American Language.* (Second College Edition) 1980. New York: Simon & Schuster.

# Double dissociation and isolable cognitive processes

**Nick Chater**
Department of Psychology
University of Edinburgh
7, George Square,
Edinburgh, EH8 9JZ, U.K

**Giorgio Ganis**
Department of Cognitive Science
University of California at San Diego
9500 Gilman Dr.
92093-0515, La Jolla, CA.

## Abstract

Data from Neuropsychology have been widely used in order both to test pre-existing cognitive theories and to develop new accounts. Indeed, several theorists have used dissociations, and in particular double dissociations, both in theory testing and in developing new theoretical accounts

Double dissociations are indeed believed to be a key tool in revealing the gross structure or "modularity" of cognitive processes. In this paper, in the light of a case study in which a simple electrical system is systematically lesioned, we argue that double dissociation in an arbitrary modular system need not, and typically will not, reveal that modularity. These results suggest that the observation of a double dissociation implies little about the structure of the underlying system. We finish arguing that the weakness of the methods described involves that neurobiological data have to be seriously taken into account in order to uncover the real structure of the cognitive system.

## Introduction

Evidence from brain damaged patients has been widely used in order to inform and constrain theories of normal functioning. Such evidence has served both to test existing psychological theories and to suggest how new theories can be developed.

The major aim of cognitive neuropsychology is to construct theories of normal function by studying impaired function (Caramazza 1986; Shallice 1988). The main assumption here is that patterns of impaired performance are not random but are constrained by the actual structure of the cognitive system itself. This is a typical inverse problem which, by definition, allows more than a single solution. The strategy followed by cognitive neuropsychologists in order to solve this problem has been to look for particular patterns of impaired performance., like associations and dissociations (Caramazza 1986; Shallice 1988). Double dissociations, in particular, are considered to be extremely valuable, since they may reveal the gross structure, or 'modularity', of the cognitive system

In a previous paper (Ganis & Chater 1991), we argued that inference from double dissociation to a particular modular structure of the underlying system is uncertain because double dissociations may be observed in a fully distributed system -that is, a system which does not decompose into isolable subsystems.

In this paper, we show that even in a modular system, double dissociation data, as conventionally interpreted, may be entirely misleading about that modularity.

## Single and double dissociations

The range of characterizations of the method of double dissociation (Shallice 1988) makes exposition of the method difficult. We shall assume what we take to be a typical modern "functional" formulation.

A patient with a lesion exhibits a single dissociation between tasks I and II when performance on task I is very poor, whereas performance on task II is either close to or at a normal level, or at least very much better than performance in task I (Shallice 1988). It was once thought that such dissociations allowed one to infer that the set of isolable processes underlying the two tasks must be different. However, it has been argued that this inference is not licensed, since task I may make greater demands on a single damaged subsystem(s) than does task II. A subsystem working at, say, 50% capacity might be adequate for task II, but not sufficient for task I. This is often referred to as the problem of resource artefacts (Shallice 1988). An analogy with locomotion may be helpful. There is a frequently observed dissociation between running and walking -many people who can walk normally can run hardly at all. This is not because running and walking use different bodily structures, but simply that running places greater resource demands upon those structures. Hence people with limited air intake, stamina and so on are able to walk much better than they can run. In response to such difficulties, it has been proposed that double rather than single dissociations are required to infer that two tasks draw on different subsystems. Tasks I and II doubly dissociate if there are patients A and B, such that A is more impaired than B in task I and, conversely, B is more impaired than A in task II. For example, a double dissociation between running and throwing is found between patients with broken limbs. Those with broken legs have impaired running, but throwing is preserved, and those with broken (dominant) arm have selectively impaired throwing. The double dissociation inference is that there must be separate systems underlying throwing and running. *Prima facie*, these correspond to the sites of the damage, the arms and legs.

If double dissociation is to constitute an advance on single dissociation as a method for uncovering isolable subsystems, then it should not be obtained between running and walking, which are subserved by the same processing systems. Indeed, while there are patients whoa can walk but not run, there are presumably no patients who can run but not walk. Thus, unlike single dissociation, double dissociation cannot be generated with a resource artifact explanation. If task I makes greater demands on a single processing subsystem than task II, then task II may be selectively preserved (generating a single dissociation), but the reverse cannot occur. For if the subsystem is sufficiently impaired to damage task II, then task I, which relies on it even more heavily, will be even more severely impaired.

The method of double dissociation has pleasingly distinguished between the two examples, one in which the two tasks are subserved by different processes, subsystems or modules and one in which they are subserved by the same system. However, we shall argue that this apparently clear-cut outcome may be the exception rather that the rule.

## Can double dissociations suggest the wrong modularity ?

In a previous paper (Ganis & Chater 1991) we have observed that the double dissociation inference may be misleading, since at least some double dissociations are consistent with a non-modular underlying architecture. We now argue that the inference can be misleading for modular systems too - the modularity postulated by the inference from double dissociation may have no relation to the actual modularity of the damaged system. In particular, we show that even under the weaker definition of modularity e.g. functional modularity (Shallice, 1988) the double dissociation method can be seriously misleading.

Let us consider a non-psychological example. Consider a dissociation between javelin throwing and rope climbing, due to blisters on the hands. Certain blisters (perhaps caused by overzealous javelin throwing on the previous day) will make throwing impossible, while leaving rope climbing more or less unimpaired. On the other hand, blisters in slightly different locations (perhaps caused by excessive recent rope climbing) will make climbing impossible will leaving throwing relatively intact. The double dissociation inference is that there must be separate processes or modules underlying rope climbing and javelin throwing. However, both involve the same hand-arm system (and indeed both involve the cooperative action of the whole of the body). It may be relatively easy to avoid being mislead, if the source of damage is visible and its effects relatively easy to discern, as in the case of blisters and broken limbs. It may be very much harder when the damage, and the functional impact that the damage has, is only indirectly detectable. Of course, this is no more than a suggestive analogy.

To better assess whether the analogy is instructive or misleading, let us consider the specific example of a simple mechanism which performs two tasks. Task I is to repeat on-line each member of a sequence of "a"s and "b"s which are given as input. Task II is to rehearse whichever is presented at a specific time, and to ignore subsequent input. In its normal state, the system under study is able to switch between tasks I and II at will by varying the value of a parameter. After subjecting the system to a full range of possible lesions, the following pattern of performance was obtained (Table I):

| | | Task I (Repetition) | |
|---|---|---|---|
| | | normal | impaired |
| Task II (Rehearsal) | normal | normal behaviour | 5 |
| | impaired | 9 | 16 |

Table I

The table shows a clear double dissociation between tasks I and II. 5 lesions resulted in the selective loss of the ability to perform Task I, and 9 lesions resulted in the selective loss of the ability to perform Task II. 16 lesions result in the loss of performance on both tasks. As a first attempt to explain this data, the inference from double dissociation suggests that we can posit two distinct modules, one responsible for repetition and one for rehearsal. We might, for example, follow the logic which lead Marshall & Newcombe (1973) to postulate a "two-route" model of the reading process, giving a diagram such as (Figure 1).

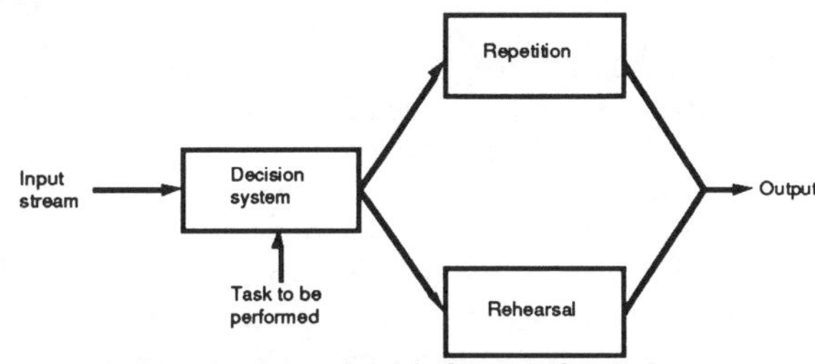


Figure 1 A boxes model inferred from the patterns of impaired performance in Table 1.

Selective damage to either the repetition or rehearsal modules would lead to the double dissociation between performance in tasks I and II. Damage to the decision system, which selects one of the other two modules as appropriate, might be the most likely explanation for the loss of both capabilities. This model may be assessed more closely by looking in more detail at the kind of pathologies found.

[1] Task I lost; Task II intact (5) No repetition (5): Task II intact; Task I locks to b after the first b in the input stream.

[2] Task II lost; Task I intact (9) No rehearsal (4): Task I intact; continuous stream of "a"s (4) or "b"s (5) in Task II.

[3] Both Task I and Task II lost (16)

[3a] Fixation (11): system may produce either only "a"s (10) or only "b"s (1), for both tasks.

[3b] Fixation and oscillation (1). Continuous stream of "a"s in Task I. Continuous stream of "...abababababa..." in Task II.

[3c] Mirroring and oscillation (1). In task 1, the output is a "mirror image" of the input. For example, "aabab" becomes "bbaba". Continuous stream of "...abababababa..." in Task II.

Let us consider each of [1]-[3]. in turn, with respect to our hypothetical model.

The basic pattern of [1] is explained by selective damage to the repetition module. That the damaged module produces only a continuous stream of output in Task I (b's) may also suggest various hypotheses about the possible internal organization of the module. If "a"s and "b"s are produced by different components of the module, then selective impairment of either, or a loss of the ability to switch between them, or some problem concerning the levels of their mutual inhibition might be responsible for this pathology. The fact that streams of "a" are not found may perhaps be a consequence of some asymmetry in this mechanism.

[2] is taken to be a consequence of selective damage to the rehearsal module. Again, that the damaged module produces only a continuous stream of output suggests that there may be separate components or submodules which produce "a"s and "b"s. As before, selective damage to either component, a loss of the ability to switch between them, or some problem concerning the levels of their mutual inhibition might be responsible for fixation on a single output. This suggests that, in key respects, the rehearsal and repetition modules may have a similar structure.

[3] Loss of the ability to perform both tasks may be due either to damage to both the repetition module and the rehearsal modules, or to the decision box. [3a] A continuous fixation on the same letter for both tasks, a very common pathology observed with nine distinct lesions, might result from damage to the decision box disconnecting the modules from current input, and any instructions to change between tasks. The current output cannot be disable, and may hence be repeated indefinitely. There are just two lesions with puzzling behavior for this account. [3b] and [3c] produce continuous oscillations between "a" and "b", on task II. [3c] also "mirrors" rather than repeats its input on task I. It may be hoped that a fuller understanding of the fine details of each module and its component parts might reveal the origins of such pathologies, but that the basic structure of the hypothetical model is not impugned.

So, in the light of this reasoning we have derived a model with the following gross structure (fig 2):

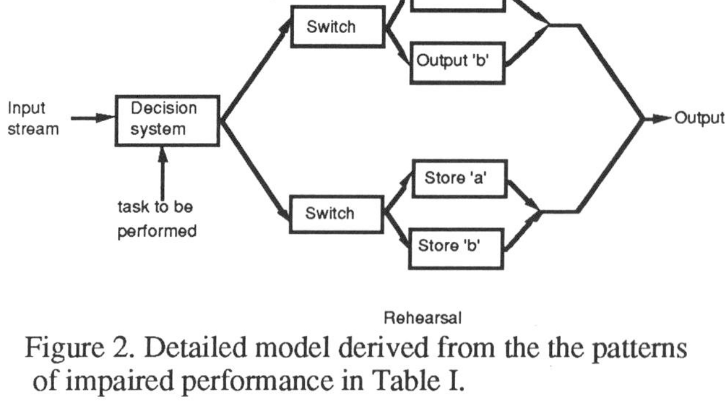


Figure 2. Detailed model derived from the the patterns of impaired performance in Table I.

A possible alternative is shown in figure 3 and there are, of course, many others.

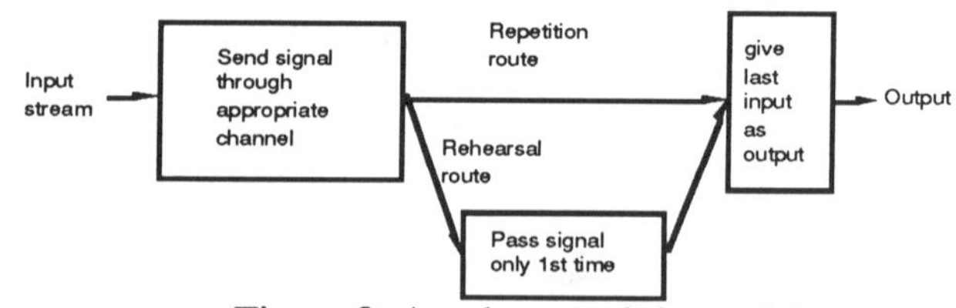


Figure 3. Another possible model.

If the task is repetition, the input stream is directed either directly to the output module. To perform rehearsal, input is directed to the the "block" module, which passes only its first input and blocks all future inputs. The output module produces an "a" if its most recent input was an "a", and a "b" if its most recent was a "b". So, if the system is in repetition mode, the output module receives a continuous stream of input and simply reproduces that stream. When the system is in rehearsal mode, only a single input is received, from the block module, and the output module repeats that input endlessly. The particular models that we have suggested are of illustrative significance only. In each, the basic finding of double dissociation and certain other aspects of the data may be accommodated, with suitable auxiliary assumptions. Certain aspects of the data are left unexplained by all (for example, oscillations in output, and the production of the "mirror" of the input stream). What is significant about the models that we have discussed, and variants like them, also derived to explain the double dissociation is that they bear no relation to the modularity of the mechanism which generated the data: a 'pseudo D' flip-flop. Figure 4 shows the circuit diagram of such a 'pseudo D' flip-flop, the basic component of memory registers in digital computers, which is able to hold a single bit of information.

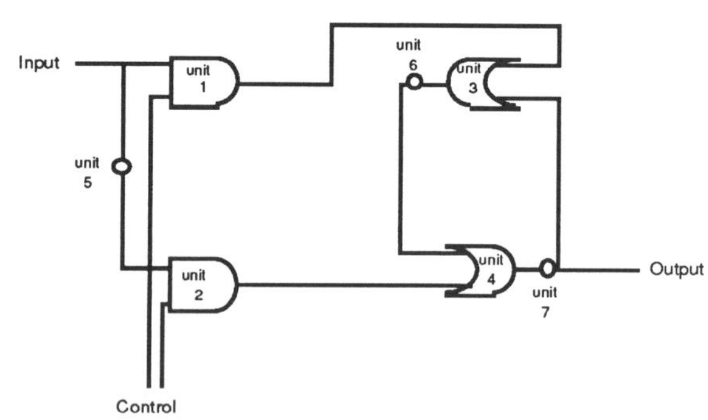


Figure 4. Diagram of a 'pseudo D' flip-flop.

The input and output streams of "a"s and "b"s is, in this context, usually thought of as streams of "0"s and "1"s. When the control signal is on (the device is in "repetition" mode) the input is passed unchanged to the output. When the control signal is off (the device is in "rehearsal" mode), the input is cut off, and the output is frozen to its previous value. The circuit can be interpreted either in a functional or in an anatomical way (i.e., each unit might not correspond to any particular piece of hardware; alternatively, it could correspond to a silicon or neural gate).

This system is highly modular and has 7 component units or gates, connected by a number of links. Values for links and units may be idealized as having only two discrete values, 0 and 1. The two AND gates on the left hand side of the diagram, output a 1 just when both their inputs are 1 and a 0 otherwise. The two OR gates on the right output a 1 is at least one of their inputs is a 1, and 0 otherwise. There are also three NOT gates (denoted by small unfilled circles) which take a single input, mapping 0 onto 1 and 1 onto 0.

*Prima facie*, the number, function and interconnection of the components of the actual system bear little resemblance to the models proposed above, in an attempt to account for the data on pathological function. Crucially, there appear to be no separate subsystems which subserve the two tasks of rehearsal and repetition, and the double dissociation inference - the system is composed of four components, rather than the two or three postulated above, and each is a simple logic gate rather than having more elaborate functions such as gating the input or printing out the last item received.

It might be thought that the circuit diagram representation is simply too detailed to reveal a clear modularity. Perhaps there is some subsystem of units and connections particular to each task.

Consideration of how the flip-flop operates does indeed suggest, for example, that the right hand side of the diagram - the OR units 3 and 4 - is concerned with rehearsal (since it embodies a loop in which activation values can be stored), and the left hand side - the AND units 1 and 2 - is concerned with passing on, or gating out, the incoming input, depending on whether the task is repetition or rehearsal (input is only passed on if the control signal is 1). This gives us Figure 5, which, hearteningly, is not so very different from Figure 3.

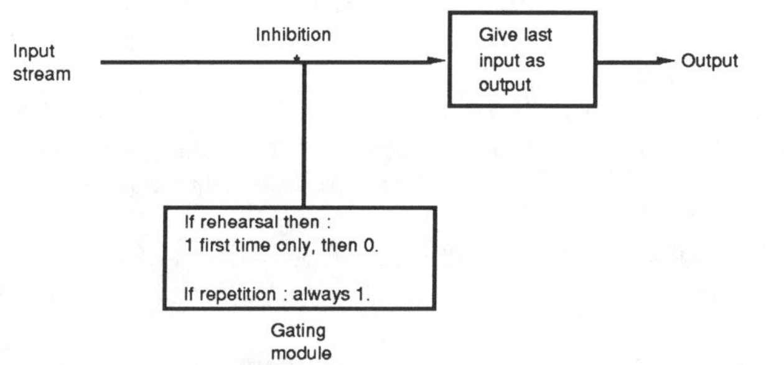


Figure 5 A possible 'higher' level description of the flip-flop in Fig 3.

However, Figure 5 is precisely the kind of diagram that appears to be incompatible with double dissociation evidence (still less can it explain the detailed patterns of pathologies observed - oscillations, mirroring the input stream, and so on) . Damage to module A, cutting off or distorting the input stream, seems likely to impair both repetition and rehearsal (since the single input which must be stored may not be passed successfully). Damage to module B, which is also implicated in both tasks, should impairing the ability to printing of the most recent input, whether in repetition or rehearsal. In short, to the extent that there is a higher level modular structure, each part of that structure is involved in both tasks, pace the double dissociation evidence. So, not only can the double dissociation inference lead to the postulation of high level descriptions unrelated to the real components of the system, but it may specifically preclude the high level descriptions which are closest to the truth.

## Conclusion

We have shown in this paper that double dissociations can suggest the wrong modularity even if the underlying system is modular.

It is worth discussing why the double dissociation method does not work with our simplified examples. The double dissociation method is based on the assumption that the cognitive system is composed of a set of modules the effectiveness of operation of which can be measured by a single real variable -the resource. The 'neurological disease depletes the amount of each particular resource by an unknown value, which varies with the patient)' (Shallice 1988). The performance on a given task, say delayed repetition, is then assumed to be a function of these real- valued numbers, one for each module. Even with a simple example like the flip-flop one, this approach has severe problems. Take, for example, an AND gate. There are different ways in which it can be damaged (Chater & Ganis 1991). Each kind of damage can produce a particular 'deficit' in the AND. Theoretically, the AND can become one of the 16 possible two-arguments binary gates (the extreme cases being a gate which gives as output always 1 or always 0). The impact of these different damages on the performance of the flip-flop in the two tasks is dramatically different even when the severity of the damage (in terms of percentage of spared input-outputs mappings) is kept constant. This is why a resource/ performance analysis cannot account for the behavior of the flip-flop. More realistic systems seem to be even more sensitive to the details of the damage (Getting, 1989). We think that the examples so far discussed are just instances of the difficulties which top-down functional investigations of the cognitive system have run into. In fact, the claim that a pure functional approach has to work in parallel with neurobiological investigation is often made (see for example Caramazza(1990)). Unfortunately, these claims are not followed by concrete efforts in trying to base neuropsychological theories on neurobiological data. A possible reason is that, as in our examples, the wrong modularity has been inferred, rendering simply impossible any link with neurobiological data (see also Seidenberg (1988) and Sereno (1991)).

## Appendix

The sites of the lesions are shown in figure 6.

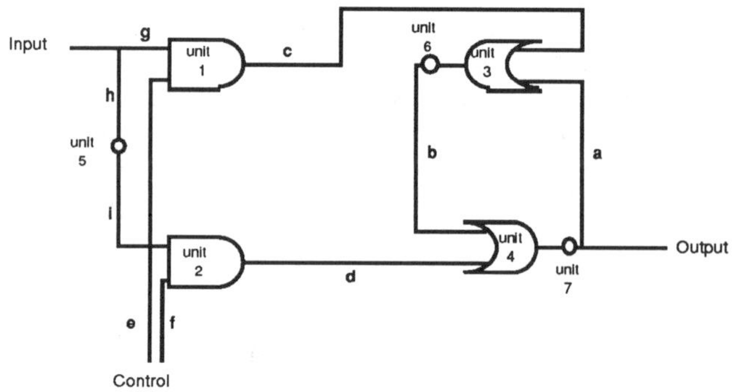

Figure 6. Sites of the 'lesions'.

A summary of the pathologies obtained with the various lesions is shown in table II

| Pattern of performance | Details | Lesion responsible |
|---|---|---|
| Selective loss of task I (no repetition) | In task I, output locks to I after first T' in input | 2.0 4.i d f i |
| Selective loss of task II (no rehearsal) | In task II, output has a fixed value | Fixation on '0' : a 3a 3.i 2.i<br>Fixation on '1' : b 1.1 3.1 4.ii 1.i |
| Loss of tasks I and II | The output has a fixed value, either '0' or '1' in both tasks. | Fixation on '0' : c e g h 1.0 2.1 2.ii 3.0 3.ii 4.1<br>Fixation on '1' : 4.0 |
| | Fixation and oscillation : Fixation on a single value in one task; instability and oscillation in the other | 1.ii 2.ii n2 n3 |
| | Mirror and oscillation: in task I the output is the negation of the input. Oscillation in task ii. | n4 |

Table II

We have attempted to pay enough regard to the details of electrical circuitry. The lesions correspond, at least roughly, to the kinds of damage that can occur in a real circuit. However, since illustrative value rather than electronic realism is of our goal, certain simplifications have been introduced. The behavior of the circuit is considered with each of the wires cut in turn. We make the simplifying assumption that cutting a wire corresponds to a zero input into the component into which it feeds Damage to both AND and OR gates gives rise to the same four distinct pathologies (although for rather different reasons). Output may be fixed to 0 or 1 or may simply pass one of the inputs i or ii. Such damage, for unit X, is denoted X.0, X.1, X.i and X.ii respectively For simplicity, only one form of damage to NOT gates is considered - a short-circuit from input to output wire. The damaged NOT gate thus passes the input unchanged. Ablation of the unit, so that its output is fixed at 0 produces no new patterns of behavior of the circuit and is so not considered. Table II summarizes the input/output functions of the damaged gates.

| i | ii | X0 | X1 | Xi | Xii |
|---|---|---|---|---|---|
| 0 | 0 | 0 | 1 | 0 | 0 |
| 0 | 1 | 0 | 1 | 0 | 1 |
| 1 | 0 | 0 | 1 | 1 | 0 |
| 1 | 1 | 0 | 1 | 1 | 1 |

Table II.

The details of the various kinds of damage are described elsewhere (Chater & Ganis 1991).

## Acknowledgements

This work was carried out while the authors were at the Department of Psychology, University College London. We should like to thank Peter Passmore for discussion and help with maths packages, and thanks also to Margaret and Marty Sereno for comments on the manuscript.

## References

Caramazza, A. 1986 . On drawing inferences about the structure of normal cognitive systems from the analysis of patterns of impaired performance: The case for single patient studies. *Brain & Cognition* 5:41-66.

Chater, N; Ganis, G. 1991 . What double dissociation tell us about the modularity of the cognitive system. Submitted to *Psychological Review.*

Ganis, G.; Chater, N. 1991 . Can double dissociation uncover the modularity of the cognitive system?. In Proceedings of the 13th Annual Meeting of the Cognitive Science Society. Forthcoming.

Getting, P. A. 1989 Emerging principles governing the operation of neural networks. *Annual Review of Neuroscience* 12:185-204.

Kinsbourne, M. 1971. Cognitive deficit: Experimental analysis. In J. L. McGaugh (ed.). *Psychobiology* New York, Academic Press.

Marin, O. S. M., Saffran, E. M. & Schwarz, D. F. 1976. ' Dissociation of language in aphasia: Implications for normal functions. *Annals of the New York Academy of Sciences*, 280:868-884.

Marshall, J.C.; Newcombe, F. 1973. Patterns of paralexia: a psycholinguistics approach. *Journal of Psycholinguistics Research*, 2:175-199.

Seidenberg, M.S 1988. Neuropsychology of language. Paper presented at the Venice Conference on Connectionism and Neuropsychology.

Sereno, M.I. 1990. Language and the primate brain .In Proceedings of the 13th Annual Meeting of the Cognitive Science Society. Forthcoming.

Shallice, T. 1984 . More functionally isolable systems but fewer "modules" . *Cognition* 17:243-252.

Shallice, T. 1988. *From neuropsychology to mental structure* . Cambridge:Cambridge University Press.

# Neuro-Soar: A Neural-Network Architecture for Goal-Oriented Behavior*

**Bonghan Cho** and **Paul S. Rosenbloom**
Computer Science Department
University of Southern California
Los Angeles, CA 90089
Email: bcho,rosenbloom@ISI.EDU

**Charles P. Dolan**
AI Center, Hughes Research Labs
3011 Malibu Canyon Rd.
Malibu, CA 90265
Email: cpd@aic.hrl.hac.com

## Abstract

The ability to set and achieve a wide range of goals is one of the principal hallmarks of intelligence. The issue of goals, and of how they can be achieved, has been one of the major foci of Artificial Intelligence (AI), and the understanding of how to construct systems that can accomplish a wide range of goals has been one of the major breakthroughs provided by the study of symbolic processing systems in AI. Neural networks, however, have not shared this focus on the issue of goals to any significant extent. This article provides a progress report on an effort to incorporate such an ability into neural networks. The approach we have taken here is to implement a symbolic problem solver within a neural network; specifically we are creating Neuro-Soar, a neural-network reimplementation of the Soar architecture. Soar is particularly appropriate for this purpose because of its well-established goal-oriented abilities, and its mapping onto levels of human cognition — in particular, the ways in which it already either shares, or is compatible with, a number of key characteristics of neural networks.

## Introduction

An ability to set and achieve a wide range of goals is one of the principal hallmarks of intelligence. The issue of goals, and of how they can be achieved, has been one of the major foci of Artificial Intelligence (AI), and the understanding of how to construct systems that can accomplish a wide range of goals has been one of the major breakthroughs provided by the study of symbolic processing system in AI. Neural networks have not shared this focus on the issue of goals to any significant extent. However, if neural networks are to be serious contenders as the basis for a complete intelligent system, they must eventually incorporate the ability to set and achieve the full range of goals faced by such systems.

This article provides a progress report on an effort to incorporate such an ability into neural networks. The approach we have taken here is to implement a symbolic problem solver within a neural network; specifically we are creating Neuro-Soar, a neural-network reimplementation of Soar — a symbolic architecture for intelligence that integrates basic mechanisms for problem solving, use of knowledge, learning, and perceptual-motor behavior [Laird *et al.*, 1987]. Soar[1] is particularly appropriate for this purpose because of its well-established goal-oriented abilities, and its mapping onto levels of human cognition [Newell, 1990] — in particular, the ways in which it already either shares, or is compatible with, a number of key characteristics of neural networks [Rosenbloom, 1989]. Work to date on Neuro-Soar covers the lowest three levels of Soar — memory accessing, decision making, and (problem space) operations. Together these capabilities are sufficient to enable goal-oriented behavior in a single problem space, but are not yet sufficient to enable higher-level search and learning behaviors. One of the key developments in accomplishing this is the implementation of a novel neural-network-based production system that allows multiple variables per production, and parallel matching and firing of multiple productions.

## From Soar to Neuro-Soar

In Soar, all symbolic goal-oriented behavior is formulated as steps in problem spaces. A problem space is a uniform task representation that allows Soar to be applied to a wide range of goals. The problem space determines the set of states and operators that can be used during the processing to attain a goal. Goals, problem spaces, states, and operators exist as data structures in Soar's working memory — a transient declarative memory. Each goal defines a problem solving context which contains roles for a problem space, a state, and an operator.

Problem solving is driven by the acts of selecting problem spaces, states, and operators for the appropriate roles in the context; where each selection is accomplished via a two-phase decision cycle. First, during the *elaboration phase*, the description of the current situation (that is, the content of working memory) is elaborated with relevant information from long-term memory — an associative memory, constructed as a parallel production system. The rules contain variabilized conditions that are matched against the contents of working memory, and actions that add elements to working memory when all of the conditions are met. One important type of knowledge that may be added during the elaboration phase is *preferences*. There is a fixed

*This research was partially supported by a gift from the Artificial Intelligence Center of the Hughes Aircraft Company.

[1] In this article, we focus on Soar version 4.

language of preferences which is used to describe the acceptability and desirability of the alternatives being considered for the selection. There are two types of preferences, unary and binary. The former states an absolute preference to all others (e.g., "best"), while the latter states a preference between two objects (e.g., "better"). In the second phase, the *decision phase*, the preferences in working memory are interpreted by a fixed decision procedure. If the preferences uniquely specify an object to be selected for a role in a context, then a decision can be made, and the specified object becomes the current value of the role. However, if the preferences for a decision are either incomplete or inconsistent, then an impasse occurs. Soar automatically generates a subgoal for the task of resolving the impasse and creates a new problem solving context within which it can be resolved. Soar learns by acquiring new productions which summarize the processing that leads to results of subgoals, a process called chunking.

Soar has been used to implement goal-oriented systems in a wide variety of domains: immediate reasoning tasks, simple puzzles and games, classical expert system domains, robotic control tasks, and natural language understanding. It has also demonstrated the ability to utilize a wide variety of problem solving methods. However, despite all of this high-level symbolic capability, much of the architectural functionality described above can be mapped to corresponding neural network concepts (at least at a gross level). In particular, Soar's production memory is just a fine-grained parallel associative memory — though it does work with graph-structured objects and does allow multiple variables — and its decision procedure is just a method for locally integrating together multiple constraints on selection, as is generally done via activation combination in neural networks. The need to allow parallel matching and firing of productions over graph-structured objects with multiple variables adds some complexity to this mapping; however, in contrast to traditional production systems, a novel formulation — that restricts attributes to having at most one value at a time (referred to as the ***unique-attribute*** formulation) — has been developed for Soar that greatly simplifies implementation in neural networks by guaranteeing both linear-time match complexity and a maximum of one binding per variable [Tambe *et al.*, 1990].

In the remainder of this article we focus on how Soar's production system and decision procedure can be implemented in a neural network, and how these two capabilities can be combined to enable the selection and application of sequences of problem-space operators that solve simple blocks world problems. Impasse detection, subgoal generation (with the concomitant abilities of using multiple problem spaces and searching), and learning are left for another day.

## Representing Problem Solving Knowledge

In problem solving behavior in a simple blocks world problem (Sussman's anomaly) three blocks are arranged on a table, and there are operators that allow blocks to be moved onto either other blocks or the table. In the first situation, three operators are suggested, and sufficient preferences are retrieved from long-term memory to allow the correct operator to be selected. Once it is selected, further memory retrieval effectively applies the operator

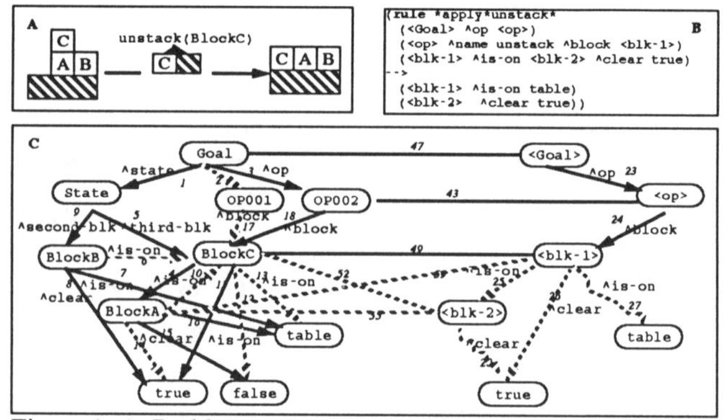


Figure 1. A. Problem space representation for operator application. B. Corresponding production rule. C. The graph view. One unit will be assigned to each number. Some edges for the working memory and bindings are not shown for clarity.

by modifying the problem solving state (moving block C to the table) Problem solving then continues by repeating the entire process.

Figure 1A focuses on a fragment of this overall behavior — the application of the selected operator to modify the state. In the remainder of this section we focus on how the knowledge is represented in Soar and the transformations that lead from this representation to the one used in Neuro-Soar. Figure 1B provides a *text view* of how this is represented in Soar (in a slightly simplified form) as a production. The conditions are prior to the arrow and the actions after. Symbols preceded by " ^ " and bounded by " <> " are attributes and variables, respectively.

To get from this text view of the information to the *neural view*, it is useful to go through an intermediary *graph view*, as shown in Figure 1C. In the graph view, each individual triple of an object, attribute, and value is represented by a labeled directed edge from the object to the value. In the working memory, a solid edge indicates a working memory element and a dashed edge indicates an element that potentially could be in working memory, but currently isn't. This distinction is not needed in the text view, but is crucial for the neural view. In production memory, there are two types of edges, one for the conditions (denoted with straight edges), and the other for actions (denoted with jagged edges). Solid and dashed edges denote matched and unmatched triples, respectively. The bindings are represented by undirected edges between the vertices of the working memory graph and the production memory graph. A solid edge indicates a binding of a production variable to an object in working memory, and a dashed edge indicates a potential binding.

A match between the working memory and a production occurs when there is an exact match between a subgraph of working memory and the production graph through a set of bindings[2]. For a successful match, there will be exactly one solid edge from each vertex in the production to some vertex in the working memory.

Production systems in the neural view are composed of relatively simple, neuron-like processing units ("units" for

[2]In contrast to Soar 4, preferences are not in working memory, instead they exist in preference memory, which cannot be matched by productions.

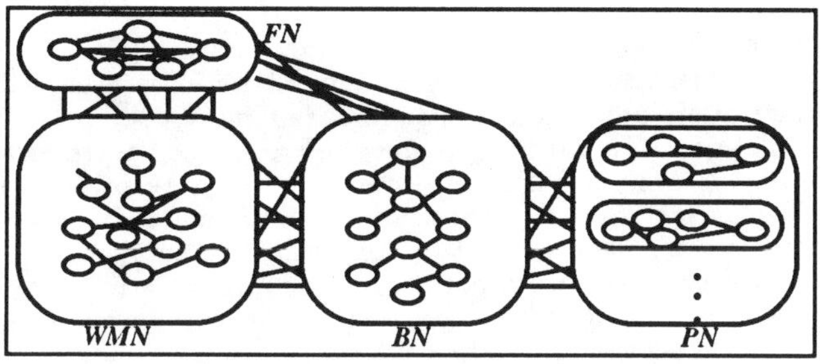


Figure 2. Network Architecture of Neuro-Soar.

short) and weighted connections between them. In the neural view, there is one unit for each edge in the graph view. The details are provided in the following section.

## Representation in Neuro-Soar

Neuro-Soar is composed of four sub-networks and connections between them (see Figure 2). Three of the sub-networks correspond to the three components described in the previous section: the working memory network, the production network, and the binding network. The fourth sub-network is the preference network, which did not appear in the previous section because it is used in the selection of operators, but not in their application.

The role of the working memory network (WMN) is to represent the contents working memory as a combination of unit activations. In the neural view, each WMN unit corresponds to an edge in the graph view. WMN units are activated by production actions, and deactivated when there is no longer an activated path between them and the goal node. This deactivation process corresponds to a garbage collection process in the symbolic implementation of Soar.

The production network (PN) is actually a set of networks, one for each production. Each network can be activated individually, in parallel, by the active segments of WMN via the binding network. To create a PN for a particular production, condition and action units are created for the condition and action edges in the graph view (one unit per edge). Each production also needs a rule unit to detect the joint activations of all the condition units. Each condition unit is connected to the rule unit, and the threshold of the rule unit is simply the number of condition units. The rule unit activates the action units, which in turn activate working memory units.

The binding network (BN) is the vehicle by which conditions are matched to working memory, and by which actions have their effects on working memory. In so doing, it computes variable bindings between the PN and WMN, and ensures their consistency — each instance of the same variable in a production must bind to the same object, each object must bind to no more than one variable[3], and no more than one object can be bound to any variable at the same time. The basic binding units are obtained from edges in the graph view of binding. Each binding unit represents a particular instantiation of a variable of a production to a (potential) object in working memory.

Figure 3 shows how consistent bindings are computed.

[3]Soar actually enforces this restriction in learned rules, but not in programmed rules. In Neuro-Soar it is enforced for all rules.

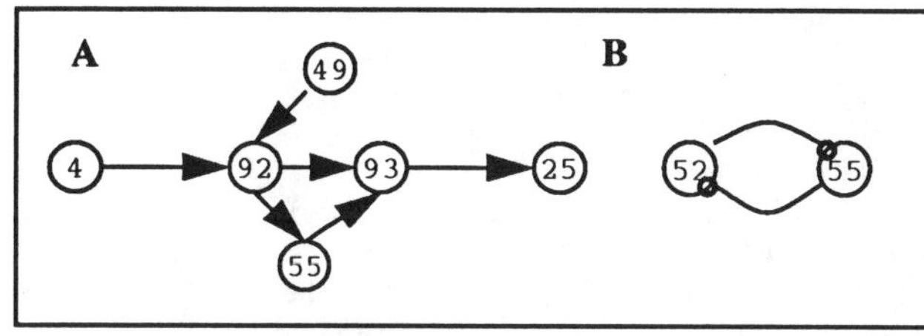


Figure 3. **A.** Matching mechanism in Neuro-Soar. Units 4 and 25 are WMN and PN units, respectively. Units 49 and 55 are binding units. Units 92 and 93 are instantiation units. **B.** Mutual inhibition among units whose bindings are contradictory to each other, through which consistent variable bindings are computed.

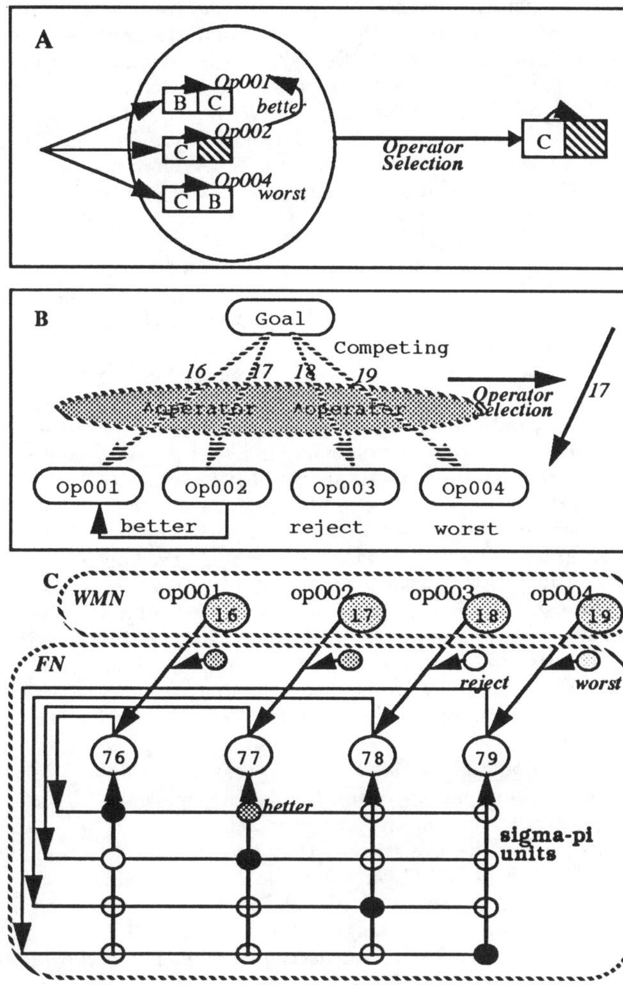


Figure 4. Neuro-Soar decision procedure. *A*. Problem space representation for operator selection. **B.** Corresponding graph view. **C.** Neural network implementation of corresponding preference information. A new layer of units (76,77,78,79) is introduced where mutual inhibitory connections exist by default. Preferences are stored in sigma-pi units (small units). A selection of unit 77 by the neural decision procedure will feed back to unit 17 in the WMN.

The binding of at most one object to a variable is enforced by mutually inhibitory connections among the binding units for the variable (Figure 3B). Likewise, the binding of an object to at most one variable is enforced by mutual inhibition among the binding units for the object. Consistent bindings across multiple occurences of the same variable are enforced as shown in Figure 3A. This fragment of the binding network exists to match the working memory triple (BlockA ^ on BlockB) to the condition (<blk-1> ^ on <blk-2>). As before the units in the figure are identified by their numbers: 4 is the working memory unit, 25 is the condition unit (in PN), 49 and 55 are binding units, and 92 and 93 are (partial) instantiation units. In the current situation, both unit 4 (representing that BlockA is on BlockB) and unit 49 (representing the binding of BlockA to <blk-1>) are active. This combination then activates unit 92, which represents a partial instantiation of condition unit 25; that is, that the appropriate element is in working memory, and its first symbol is bound to the first variable in the condition. To pursue the match further, unit 55 (representing the binding of BlockB to <blk-2>) needs to be activated. This unit receives activation from unit 92, so if there is no competition from other binding units for <blk-2> — or if unit 55 simply wins the competition — unit 55 will become activated. Unit 93 is analogous to unit 92, but it represents a complete instantiation of condition unit 25 — there would be one unit like 93 for each possible instantiation of condition unit 25. This then activates condition unit 25, thus successfully completing this segment of the match.

The overall match starts with an initial binding between the goal object in working memory and the first condition in each production. From there, the match ripples down the networks, with each new binding that is established enabling further matches to proceed (or be inhibited). When all of the conditions of a production are matched, the production unit becomes active, and in turn activates the action units. The action units then modify working memory activations through an inverted binding network.

In addition to the straightforward effect of action units on working memory, action units can also modify working memory indirectly by asserting preferences among units in WMN — this is how problem solving operators are selected. The role of the preference network (FN) is to reflect these current preferences (during the elaboration phase) and to select a unit out of the competing candidate units, based on their preferences (during the decision phase).

Figure 4A shows the operator selection episode, elaborated with the preferences that lead to the selection of Op002 (put Block C on the table). This operator is selected because it is *better* than Op001, and the other candidate (Op004) is *worst.* Figure 4C shows how this behavior is implemented as a neural decision procedure in the preference network. The FN is built on top of the WMN as an additional layer that stores and processes preferences (see units 76, 77, 78, 79 in Figure 4C)[4]. The FN is basically formed as winner-take-all network with a guide of preferences. Unary preferences are encoded as wieghts between the WMN units and the FN units, while binary preferences are encoded as weights between the competing units in FN. The use of Sigma-Pi units[Rumelhart *et al.*, 1986] is required for this (see small units in Figure 4C).

At first, each competing unit in FN receives a level of activation from the correponding unit in WMN that is a function of its unary preferences. Prodded by these activations, competition then occurs in the binary preference layer, which then feeds back to the WMN units. If a selection can be made, processing continues with another elaboration phase. Otherwise, the resulting impasse leads to termination (though ultimately Neuro-Soar must be extended perform subgoal generation instead).

The Neuro-Soar approach of encoding preferences as constraints on activation combination is not guaranteed to yield the same outcome as would be achieved through Soar's decision procedure, but it does capture the spirit of their meaning. The extent to which the Neuro-Soar scheme is actually better or worse than the symbolic approach is still to be determined.

## Experiments and Analysis

So far, Neuro-Soar has been applied in two domains: the blocks world and the tower of Hanoi. In this section, we focus on results from the blocks world. The implementation consists of 8 productions. Two test problems have been run — Sussman's anomaly and a simpler problem of unstacking a tower of three blocks. In both cases Neuro-Soar solved the problem by replicating the sequence of operator selections and applications that were produced by Soar when it was given the same knowledge.

Since Neuro-Soar employs a local representation scheme, a sufficient number of units in working memory must be preallocated. This number is dependent on the number of objects in working memory and their attributes. The number of units in PN is dependent on the number of conditions and actions in the productions. Likewise, the number of units in the binding network is related to the number of objects in working memory and the productions. Only a small number of units are required in FN since they are built on top of a small part of working memory. The blocks world implementation requires 135 WMN units, 119 PN units (of which there are 8 production units, 48 condition units, and 63 action units), 2148 BN units (of which there are 185 binding units and 1963 instantiation units), and 10 FN units. More generally, the worst space complexity of a task implemented in Neuro-Soar is as follows. Let $nUNIT_{netwrok}$ be the number of units in network $network$.

$$nUNIT_{PN} = O(nTrp)$$
$$nUNIT_{WMN} = O(\sum_{attr} nObjHas_{attr} * nObjOf_{attr})$$
$$nUNIT_{BN} = O(nOBJ_{WMN} * nOBJ_{PN} + \sum_{attr} nWMN_{attr} * nPN_{attr})$$
$$nUNIT_{FN} = O(nOBJ_{WMN})$$

where $nTrp$ is the number of triples in productions; $nObjHas_{attr}$ and $nObjOf$ are the number of objects of working memory that can have attribute $attr$ and can be the value of $attr$, respectively; $nOBJ_{network}$ is the number of objects in $network$; and $nWMN_{attr}$ and $nPN_{attr}$ are the number of units which have attribute $attr$.

The time complexity for matching a production in Neuro-Soar is on the order of the maximum depth of the graph representation of the production. The depth of the graph representation is approximately $log(nTrp)$. The matching process is relatively fast and efficient in terms of neurosimulation cycles, not only because the rules are fired in parallel, but also because the valid partial matches

[4] In reality, there is also one additional layer that handles *indifferent* preference.

are kept (analogous to state saving in the Rete algorithm [Forgy, 1982]).

## Related Work

The two classes of related work to which it is most important to compare Neuro-Soar (other than Soar itself) are: other attempts at constructing neural production systems, and other attempts at constructing neural problem solvers. On the former, two previous systems have implemented production systems using neural networks, the Distributed Connectionist Production System by Touretzky and Hinton[Touretzky and Hinton, 1988] and the Tensor Product Production System by Dolan and Smolensky[Dolan and Smolensky, 1988], both of which used distributed representations. Neuro-Soar, though simpler in that it uses a local representation, goes significantly beyond these two systems in allowing parallel matching and execution of productions, (unique-attribute) graph-structured representations, and multiple variables per production.

Ajjanagadde and Shastri[Ajjanagadde and Shastri, 1990] have developed a connectionist system that represents knowledge as rules and facts with multi-place predicates. A multi-phased clock is proposed as a solution to the variable binding problem. The system produces answers, when given 'yes/no' questions, based on the knowledge. However, the system lacks a capability to create and alter the content of its temporary knowledge so that situation dependent firing of rules (or planning) is not possible.

This is by no means the first effort to construct a neural-network-based problem solver (see, for example, [Sutton and Barto, 1981, Anderson, 1989]). However, the current effort differs from most of the earlier approaches in adding representational and structural complexity — specifically, graph-structured objects and the decomposition of the overall memory into four specialized subnetworks — in service of domain generality. The hope is that this will allow Neuro-Soar to be extended up to Soar's full level of capability, without ultimately sacrificing learnability or neural plausibility.

## Conclusions and Future Work

Work to date on Neuro-Soar has demonstrated that it is possible to reproduce key aspects of Soar's general goal-oriented ability even when the implementation technology is shifted from "symbolic" to "neural network". In particular, we have demonstrated a domain-independent approach to knowledge-driven operator selection and application. To accomplish this, we have developed a preference-based neural-network decision procedure, and a novel neural-network production system that allows parallel matching and firing of rules, graph structured objects, and multiple variables per rule.

However, much remains to be done in three general areas. The first area is the completion of Neuro-Soar with respect to Soar's functionality. This primarily involves the incorporation of impasse detection and subgoaling, use of multiple problem spaces, and learning. The second area is the loosening up of the design so that it less slavishly follows how things are currently done in Soar, and more takes advantage of the strengths of neural networks. We have begun with the conservative approach of trying to model the existing Soar architecture as closely as possible; however, we eventually want to get to where we can ask whether the specification of Soar ought to be altered because of the advantages accruable from using neural networks. For example, what would happen if we were to shift to a distributed representation for memory, based perhaps on a tensor-product formulation [Smolensky, 1990], or if we were to shift from a neural-network analogue of chunking to a more typically neural learning algorithm. Would such changes result in a more robust system that can learn more effectively? The third area is to see whether Neuro-Soar provides leverage in interfacing Soar to external environments. Would encoding the required signal-to-symbol transducers in neural-networks that are directly connected to Neuro-Soar provide a more seamless interface than is now provided by the use of Lisp code for the transducers?

## References


[Ajjanagadde and Shastri, 1990] V. Ajjanagadde and L. Shastri. An optimally efficient limited inference system. In *Proceedings of AAAI 90*, pages 563–570, 1990.

[Anderson, 1989] C. W. Anderson. Tower of hanoi with connectionist networks: Learning new features. In *Proceedings of the Sixth International Workshop on Machine Learning*, pages 345–349, 1989.

[Dolan and Smolensky, 1988] C. P. Dolan and P. Smolensky. Implementing a connectionist production system using tensor products. Technical Report UCLA-AI-88-15, University of California, Los Angeles, 1988.

[Forgy, 1982] C. L. Forgy. Rete: A fast algorithm for the many pattern/many object pattern match problem. *Artificial Intelligence*, 19:17–37, 1982.

[Laird *et al.*, 1987] J. E. Laird, A. Newell, and P. S. Rosenbloom. Soar: An architecture for general intelligence. *Artificial Intelligence*, 33:1–64, 1987.

[Newell, 1990] A. Newell. *Unified Theories of Cognition.* Harvard University Press, Cambridge, MA, 1990.

[Rosenbloom, 1989] P. S. Rosenbloom. A symbolic goal-oriented perspective on connectionism and soar. In R. Pfeifer, editor, *Connectionism in Perspective.* North-Holland, 1989.

[Rumelhart *et al.*, 1986] D. E. Rumelhart, G. E. Hinton, and J. L. McClelland. A general framework for parallel distirbuted processing. In D. E. Rumelhart, J. L. McClelland, and the PDP Research Group, editors, *Parallel Distributed Processing, Volume 1: Foundations*, pages 45–76. Bradford Books/MIT Press, 1986.

[Smolensky, 1990] P. Smolensky. Tensor product variable binding and the representation of symbolic strucutre in connectionist systems. *Artificial Intelligence*, 46:159–216, 1990.

[Sutton and Barto, 1981] R. S. Sutton and A. G. Barto. An adaptive network that constructs and uses an internal model of its world. *Cognition and Brain Theory*, 4:217–246, 1981.

[Tambe *et al.*, 1990] M. Tambe, A. Newell, and P. S. Rosenbloom. The problem of expensive chunks and its solution by restricting expressiveness. *Machine Learning*, 5:299–348, 1990.

[Touretzky and Hinton, 1988] D. Touretzky and G. E. Hinton. A distributed connectionist production system. *Cognitive Science*, 12:423–466, 1988.

# Interpretation of Definite Reference with a Time-Constrained Memory

**Jean-Pierre Corriveau**
Department of Computer Science
University of Toronto,
Toronto (Ontario) CANADA M5S 1A4
(819) 682-1279
jeanpier@bnr.ca

## Abstract

In this paper, I demonstrate how complex cases of definite reference resolution can be processed within the independently-motivated framework of *time-constrained memory*, and, most importantly, without having to resort to the complex mechanisms assumed by Haddock (1987). The key idea of the solution is to initiate a referent search for the complex definite NP formed by the attachment of a prepositional phrase to a definite noun phrase.

## Haddock's Problem

### Introduction

The goal of this paper is to solve a problem in the interpretation of definite reference that was first put forth by Haddock (1987). The linguistic framework provided by *time-constrained memory* (Corriveau, 1991) offers an approach that avoids the complexity and drawbacks of the mechanisms that Haddock assumed to be necessary for solving this problem.

The problem is this: Suppose we have a context in which there are three rabbits, R1, R2, and R3, two hats, H1 and H2, and one box, B1. One of the rabbits, R2, is in one of the hats and another one is in the box, as illustrated in figure 1. Rabbit R2 can be referred to by means of the following complex NP:

**0.1** *the rabbit in the hat*

What is interesting about this expression is that even though as a whole it is perfectly natural, the NP *the hat* uses the definite determiner *the*, despite the fact that there are not one but two h ats in context. Compositional accounts of NP semantics would judge (1.1) to be infelicitous. Haddock uses this fact to argue for an incremental approach

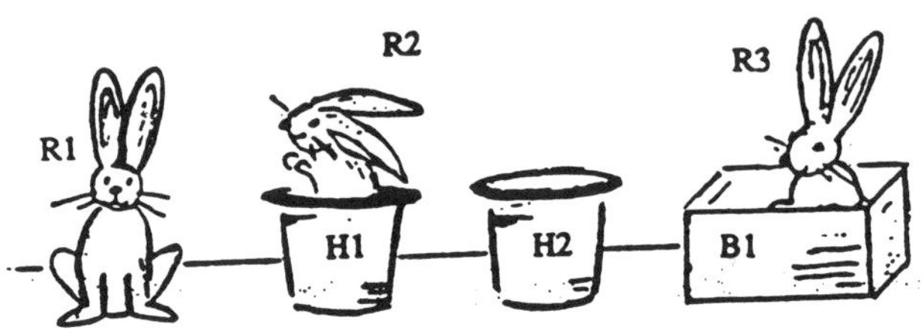


Figure 1: The context for Haddock's example

that "evaluates a semantic representation—after each word, say" and will never co nsider the empty hat as a "viable candidate for the inner NP" (1987, p.661).

> When the word *rabbit* is reached, a hearer can collect together in his mind the set of rabbits in the context. After the preposition, this set is refined to contain only rabbits which are *in* something and, most importantly, the hearer can start thinking about another set of objects, those which have rabbits in them. T here is only one hat in this new set and so by the time the inner NP is processed a definite determiner sounds natural. (*ibid.*)

In more explicit computational terms, Haddock suggests that, given the predicates and variables corresponding to the input discourse, the alg orithm first determines, for each variable, all entities in the discourse that satisfy the predicates, and second, collects these into candidate sets for each variable. From the fragment *the rabbit in*, we get the predicates $rabbit(e1)$ and $in(e1, e2)$.

Given the context of (1.1), the algorithm will label the variable $e1$ with the set $\{R2, R3\}$ and $e2$ with $\{H1, B1\}$. Haddock emphasizes that the advantage of this scheme

> resides in that satisfying the constraints of the two predicates determine a can didate set for $e2$ as well as

*e*1. Because the phrase is evaluated strictly word-by-w ord, in left-to-right order, this scheme is also supported by psycholinguistic eviden ce (Steedman, 1987).

Haddock assumes that it is syntax that controls the assembling of semantic repre sentations, and thus, in order to provide incremental semantics, adopts the framework of Combinatory Categorial Grammar (Ades and Steedman, 1982). My intent in this paper is to present an alternative approach to the interpretation of (1.1), one that does not require all the compl exities, such as *type-raised* categories, of Combinatory Categorial Grammar.

## Difficulties with Haddock's Solution

Unfortunately, Haddock's approach exhibits drawbacks. First, because his approach is syntax-driven, it depends heavily on the unambigu ous attachment of *in* to *the rabbit.* But it is commonly accepted that structural disambiguation requires more than syntax and often involves semantics and inference (Hirst, 1987). Thus, because later in ferences, for example, may influence the attachment of the PP, it seems incorrect to have the semantic processing of the preposition *immediately* restrict the set of rabbits to those that are *in* somethi ng. Furthermore, it may not even be the case that the preposition heads a locative p roposition. In other words, further input text might be required for correct attachment.

Second, a solution based on the attrition of arbitrary large sets is not cogniti vely appealing. If, for example, there were a multitude of rabbits in various objects, the real- time processing constraint of human comprehension would not allow the user to mentally assemble the set of all rabbits nor the set of all containing objects. Instead, the attentional m echanisms of the comprehender's memory would in practice restrict the set of possible refe rents.

Third, the comprehender does not necessarily start thinking about objects that h ave rabbits in them after processing *in.* Not only could this approach lead to a large, unmanageable set of rabbit containers, but it also seems to depend on the *inference strategy* of the comprehender (Granger and Holbr ook, 1983). Haddock's approach corresponds to the particular case in which a comprehender, o n processing the preposition *in*, sets up an expectation about its attachment to the NP *the rabbit* and, consequently, comes to expect that what follows the preposition must refer to one of the objects that contain a rabbit. This is just one of several possible reference resolution strategies. Another tactic may be to wait for further input and for correct attachment, and only then look for a referent for the resulting complex NP. In other words,

another valid resolution strategy is to wait for the complex NP *the rabbit in the hat* to be constructed before looking for a referent. The precise reference resolution s cheme can vary across comprehenders, and even, for a single comprehender, can change d epending on his/her state of mind.

Finally, though I will not elaborate on this topic, I have serious concerns abou t any linguistic approach that assumes a more or less direct correspondence between wo rds and simple predicates (see Corriveau, 1991, chapter 2).

## The Proposal

The approach I suggest circumvents the identified drawbacks of Haddock's. In essence, I assume that each definite NP is given a short amount of *time* to find a referent. Finding a referent is taken to consist in searching the cogniti ve structures that have been built from the processing of previous input for an ent ity that 'matches' the definite NP. The set of possible referents is dynamically dependent on the amount of time allocated to the search, as well as the traversal order of co gnitive structures in memory. If, at the end of the time-constrained search, a single referent is found, the definite NP is resolved. If no referent is found, then the definite NP is left unresolved. And, most importa ntly, if ever a second possible referent is found during the search, the search is immediately abandoned and the definite NP is taken to be ambiguous.

So, assuming all possible referents are in context, the processing for (1.1) wou ld be the following:

- The NP *the rabbit* is assembled and the search for a referent is init iated. This search finds a first possible referent, then a second one, at which point it is abandoned.
- The preposition *in* is input, and initiates a process that will event ually attach the NP it governs, namely *the hat,* to the NP *the rabbit.* Such an attachm ent would be prevented in sentences such as *Put the rabbit in the hat.*
- The words *the* and *hat* are input and assembled into an NP. The search for a referent for this NP is initiated. This search fails because there are two hats in context, that is, two possible referents.
- Eventually the PP *in the hat* is attached to the NP *the rabbit.* This forms a new definite NP for which a new search for a referent is initiated. This searc h reaches R2, which is considered a possible referent because it is a rabbit in a hat. Because no other possible referent is found before the end of the search, it succeeds and the complex NP *the rabbit in the hat* is resolved.

## Overview of Time-Constrained Memory

Let me now express this solution in more explicit computational terms.

The solution I propose for the interpretation of (1.1) is rooted in the notion of a strictly quantitative *time-constrained memory.* In Corriveau (1991), I motivate and develop a linguistic framework specified in terms of a computational model of memory that, although following a simple, knowledge-independent algorithm, is able to account for many problems of understanding written language. Here, I can provide only a brief review.

Of the multitude of possible factors that constrain linguistic comprehension, I focus primarily in this work on the role and importance of quantitative *time*, *i.e.,* time as it pertains to memory management and memory processes such as retrievals (see Corriveau, 1987). More specifically, the fundamental and most pervasive hypothesis of this work is that, generally, linguistic comprehension is a *time-constrained process*—a *race.*

I abandon the notion of a correct interpretation of a text and, rather, adopt a *reader-based* approach to text understanding, one in which 'meaning ' is constituted by the interaction between text and reader (Holub, 1984). The reader is not seen as an idealized competent entity, but rather as an individual (with all the idiosyncrasies implied by this term) that comes to a personal interpretation of a text with respect to a private idiolect. Therefore, the concept of linguistic *competence* (which is inaccessible by definition) is abandoned. I focus only on actual *performance,* which is taken to be idiosyncratic.

Thus, this approach shifts the focus from rules and algorithms to the underlying cognitive architecture which is taken to be strictly quatitative: "With the abandonment of algorithmic control, the brain can be hypothesized to follow a strictly quantitative 'trivial algorithm' that fixes only the general form of operations; all qualitative information (*e.g.,* rules, concepts, etc.) is treated as data."

I choose to root the cognitive architecture that I propose in the basic metaphor of human memory. The system is partitioned into static and dynamic components of memory. Static memory consists of a large network of simple computing units that operate in parallel and exchange simple signals. These units are called *knowledge units* (KUs) and have the ability to modify the current context, which is simply taken to be the set of *clusters* in dynamic memory. More specifically, a KU has its *expansion procedure* execute upon it becoming activated. Expansion procedures consist of ordered sequences of primitive cluster operations, which are built-in instructions of the model and are independent of any linguistic consideration. These operations provide the basic functionality required for generic data structures: access, addition, deletion, comparison, traversal, etc. Among these operations, one that is particularly relevant to a solution to the problem of definite reference: *findInclusiveReference.* This operation is intended for general matching. It takes as an argument a cluster, and searches dynamic memory for another cluster whose set of features includes all those of the argument. The operation is given a limited amount of time for its search, and it fails either if multiple referents can be found during its time-span or if no referent at all can be found within the time-span. A straightforward approach to reference resolution can be developed in a single KU using this instruction.

A cluster is defined as a non-empty set of knowledge units, each

governing a (possibly empty) set of clusters. In other words, a cluster is a hierarchical structure whose 'leaves' are knowledge units. Clusters are general enough to allow the construction of almost any kind of representation (*e.g.,* parse trees, scripts, etc.) that might be hypothesized for comprehension.

Among the six kinds of signals that KUs can exchange, the *expectation signal* allows the specification of exactly the sort of expectation that Haddock assumes of the hearer who starts thinking about objects that contain rabbits.

Within the framework of my basic metaphor:

- All memory processes are taken to be strictly quantitative, *i.e.,* mechanical and deprived of any linguistic and semantic knowledge.
- All 'knowledge', that is, all qualitative information, is assumed to specified separately and act as data for the underlying system.
- The processes of linguistic comprehension must be defined only in terms of memory processes.

A prototype of time-constrained memory has been implemented and a knowledge base of more than 300 knowledge units has been specified to address the problems of syntax, reference resolution, word-sense and structural disambiguation, and bridging inferences required for written text comprehension.

## Examples

I now show a number of working examples of definite reference resolution in this system.

- **0.2** *R1 is a rabbit. R2 is a rabbit. The rabbit eats.*

  Multiple referents are found, and thus there is no resolution.
- **0.3** *R1 is a rabbit in a hat. The rabbit in the hat eats.*

  In the second sentence, the PP is attached to the NP *the rabbit* and then a unique referent is found.
- **0.4** *R1 is a rabbit. R2 is in a box. R1 is in a hat. R2 is a rabbit. The rabbit in the box eats.*

  In the last sentence, the disambiguation of *the rabbit* fails because of mu ltiple referents. Then a unique referent is found for *the rabbit in the box.*
- **0.5** *The young rabbit eats. The rabbit sleeps.*

  A unique referent is found, although the second NP is not identical to the first .
- **0.6** *Put the rabbit in the hat.*

  Simple VP-PP attachment occurs, which blocks, in this case, the NP-PP attachemen t.

- **0.7** *R1 is a rabbit in a hat. Put the rabbit in the hat.*

  In the second sentence, a reference is found for *the rabbit* and the PP is attached to the verb, not to *the rabbit.*

- **0.8** *R1 is a rabbit. R2 is a rabbit. Put the rabbit in the box.*

  In the third sentence, no referent is found for *the rabbit*, and there is o nly simple VP-PP attachment.

- **0.9** *R1 is a rabbit in a hat. R2 is a rabbit. Put the rabbit in the hat in the box.*

  The third sentence is interpreted as: *Put in the box the rabbit that is in the hat.*

- **0.10** *Put the rabbit.*

  A syntactic conflict is detected, which prevents any attempt at reference resolu tion.

## Conclusion

In this paper, I have demonstrated that complex cases of definite reference resolution can be processed in a straightforward way within the independently motivated framework of *time-constrained memory*, and, most importantly, without having to resort to the complex mechanisms assumed by Haddock (1987). Moreover, the proposed strategy hinges on a strictly quantitative memory operation, which is to say it is not a special-purpose linguistic rule. The key idea of the solution is to initiate a referent search for the complex definite NP formed by the attachment of a PP to a definite NP. This solution is very efficient because it immediately abandons a search and leaves a definite NP unresolved if more than one possible referent can be found.

## References


Ades, A. and Steedman, Mark J. (1982). "On the Order of Words". *Linguistics and Philosophy*, 4:517–518.

Corriveau, Jean-Pierre (1987). "On the Role of Time in Reader-Based Comprehension", *Proceedings of the Ninth Annual Conference of the Cognitive Science Society*, Seattle, WA, pp. 794-801.

Corriveau, Jean-Pierre (1991). *Time-Constrained Memory for Reader-Based Text Comprehension*, Doctoral Dissertation, Department of Computer Science, University of Toronto, Toronto, Ca nada.

Granger, Richard and Holbrook, Jennifer (1983). "Perseverers, Recencies and Deferrers: New Experimental Evidence for Multiple Inference Strategies in Understanding", *Proceedings of the Fifth Annual Conference of the Cognitive Science Society*, Rochester, NY.

Haddock, Nicholas (1987). "Incremental Interpretation and Combinatory Categorial Grammar", *Proceedings of the Tenth Joint Conference on Artificial Intelligence*, Milan, Italy, pp. 661-663.

Hirst, Graeme (1987). *Semantic Interpretation and the Resolution of Ambiguity*, Cambridge University Press, Cambridge, England.

Holub, Robert (1984). *Reception Theory—A Critical Introduction*, Methuen, New York.

Malsburg, Charles von der (1985). "Algorithms, Brain, and Organization", In: Demongeot, J., Golès, E. and Tchuente, M. (eds.) *Dynamical Systems and Cellular Automata*, Academic Press, London.

Steedman, Mark J. (1987). "Combinatory Grammars and Human Language Processi ng", In: Garfield, J. (ed.) *Modularity in Knowledge Representation and Natural Language Processing*, MI T Press, Cambridge, MA.

# Action Planning: The Role of Prompts in UNIX Command Production

**Stephanie M. Doane, Danielle S. McNamara, Walter Kintsch, Peter G. Polson, R. Gary Dungca, Deborah M. Clawson**

Institute of Cognitive Science and Department of Psychology
Campus Box 345
University of Colorado, Boulder, CO 80309-0345
sdoane@boulder.colorado.edu

## Abstract

Our goal is to provide empirical support for assumptions of the Doane, Kintsch, & Polson (1989;1990) construction-integration model for generating complex commands in UNIX. In so doing we designed a methodology that may be used to examine the assumptions of other cognitive models. The planning task studied was the generation of complex sequences of UNIX commands. The sequences were novel, and as such could not be recalled from memory. We asked users whose UNIX experience varied to produce complex UNIX commands, and then provided help prompts when the commands they produced were erroneous. The help prompts were designed to assist the subjects with both knowledge and processes which our UNIX modeling efforts have suggested were lacking in less expert users. There are two major findings. First, it appears that experts respond to different prompts than do novices. Expert performance is helped by the presentation of abstract information, while novice and intermediate performance is modified by presentation of concrete information. Second, while presentation of specific prompts aids the less expert, it does not appear to be sufficient information to obtain optimal performance. To do this, the less expert subjects require information about the ordering of the items in a command. Our analyses suggest that information about the ordering of prompts helps the less expert with memory load problems in a manner consistent with skill acquisition theories.

## Theoretical Background

The goal of this study is to provide empirical support for assumptions about user knowledge of UNIX and user memory processes which are proposed in the Doane, Kintsch, & Polson (1989;1990) UNICOM construction-integration model for generating complex commands in UNIX. In so doing, we present a general methodology that may used to examine the complex knowledge and process assumptions made by other cognitive models.

Previous empirical work gave UNIX users at varied levels of expertise textual descriptions to produce legal UNIX commands that required the redirection of standard input and output (i.e., "composite" commands) (Doane, Pellegrino, & Klatzky, 1990). Their data suggest that novices and intermediates have knowledge of the elements of the system; that is, they can successfully produce the single and multiple commands that make up a composite. They could not, however, put these elements together using pipes and/or other redirection symbols to produce the composite commands. The symbols that enable input/output redirection are fundamental design features of UNIX, and these features are taught in elementary computer science courses. Doane et al (1990) demonstrate that these features can only be used reliably after extensive experience (e.g., experts had, on the average, 5 years of experience with UNIX).

UNICOM (Doane, Kintsch, & Polson, 1989;1990), was developed to determine why the Doane et al. 1990 users had such difficulties generating composite commands. To facilitate discussion of the UNICOM knowledge base, we will use the command "nroff -ms ATT2>ATT1" as an example. This command formats the contents of the file ATT2 using the utility nroff and the -ms macro package, and then stores the formatted contents of ATT2 in the file ATT1. To correctly produce composite action plans like this, the model must first identify the items that must be used to produce the command, and then it must sequence these items using the properly placed redirection symbol to construct an action plan. Thus, the Doane et al (1989;1990) model assumed that expert users use a two-stage process to complete action plans and that this process requires knowledge of both items and order. To complete the first phase, the model requires four different types of item knowledge to produce composites. UNICOM must know two syntax specific types of knowledge (a) command syntax (e.g., the nroff command and the -ms flag) and (b) I/O redirection syntax (e.g., the ">" redirection symbol). The model must also have two general types of knowledge that are syntax independent. The model must know (c) general facts about the redirection of input and output (e.g., that redirection of input an output can occur between commands). This is separate from the syntax specific knowledge of I/O redirection symbols. That is, some users appear to know that redirection can occur, and not know the specific syntax. UNICOM must also have (d) general facts about command redirection (e.g., that the output of nroff can be redirected to a file). The latter form of knowledge is necessary, because UNIX commands are not consistent in their I/O structure (e.g., an "ls" cannot take input from another command, but "sort" can).

In the second phase of action planning, the model has to sequence items properly and then keep track of the intermediate stages of its action plan in order to produce a successful command. For example, to produce the command "sort file|head||lpr", the model had to first select the items to be used (sort,|,|,lpr,head,file) and then determine the order of the items and keep track of which command was first, second, and third, and what the output of the first command was, where its output was redirected, and so on. Doane et al (1989;1990) assumed that the mechanisms involved in integrating and ordering the items to produce a composite would produce serious memory load problems. In addition, it was assumed that experts have knowledge structures which decrease this memory load, whereas novices do not.

It is important to our research that explanation of the item knowledge and ordering processes fits in a more global context than simply UNIX command production. There are reasons to believe that this is the case. First, UNIX composites have characteristics that are important for understanding problem solving and skill acquisition in general. The literature on expertise and skill acquisition (e.g., Chi, Feltovich, & Glaser, 1981; Larkin, 1983) suggests that less expert individuals may know the separate facts in a domain, but not be able to use them productively in the sense described by Wertheimer (1982/1945). Thus, this issue is not unique to the UNIX composite problem. Understanding how individuals utilize textual instructions to develop their factual knowledge into effective problem solving procedures is an important skill acquisition issue in general. Second, we are analyzing the knowledge and processes required to chain elements together in the context of a theory of comprehension that has been used to explain algebra story problem comprehension (Kintsch, 1988), and solution of simple computing tasks (Mannes & Kintsch, in press). Thus, rather than developing an idiosyncratic knowledge analysis, we are performing our research in the context of a general architecture of cognition. As such, we contribute to the goal suggested by Newell (1987) to develop a "unified theory of cognition".

## Experiment

The goals of this study, then, are to determine more precisely what users at different levels of expertise know about UNIX, what information is lacking when users produce erroneous commands, and what information (i.e., prompt contents) helps. The experiment uses a prompting paradigm to assess the knowledge and processes of users at various levels of expertise. We assume that users have different amounts of the required four types of knowledge and that displaying the prompts that help with each respective type of knowledge will impact subsequent user performance. We hold similar assumptions about the process knowledge (i.e., keeping track of intermediate results).

### Method

**Subjects.** Twenty-two computer science and electrical engineering majors with between six months and 6 years experience with UNIX were paid $20 for their participation. Novices had less than 1.25 years of experience with UNIX and no operating systems courses; intermediates had between 1.25 and 3.0 years experience with UNIX and some had operating systems courses; and experts had greater than three years experience with UNIX and all had taken an operating systems course.

Task Description: Format the text in ATT2 using the -ms macro package and store the formatted version in ATT1

Prompts:

Prompt 1. You will need to use the following command
One that will format the contents of a file using the -ms macro package

Prompt 2 You will need to use this command
nroff -ms will format the contents of a file using the -ms macro package

Prompt 3 You will need to use a special symbol that redirects command output to a file

Prompt 4 You will need to use the arrow symbol ">" that redirects output from a command to a file

Prompt 5 You will need to use the arrow symbol ">" and the command nroff -ms

Prompt 6 You'll need to use an nroff -ms on ATT2 (which will output the formatted contents of ATT2), and you'll need to redirect this output as input to ATT1

Prompt 7 You will need to use exactly the following command elements (though not necessarily in this order):
>, nroff -ms

Prompt 8 You'll need to use the command nroff -ms followed by the arrow symbol ">"

Prompt 9 The correct production is
nroff -ms ATT2>ATT1
Please enter this production now

Table 1. Example of task description and prompts for problem nroff -ms ATT2> ATT1.

**Apparatus and Materials.** All tasks were performed on a Macintosh with large-screen monitor. The stimuli were task statements, a fixed directory of file names and a series of help prompts, all displayed on the screen, as well as three "error cards" presented by the experimenter. Subjects responded by typing at the keyboard a command or series of commands which would accomplish a given task and then using the mouse to "click" on a displayed button.

The task instructions described actions that could best be accomplished by combining two or three commands through the use of redirection (i.e., composite problems). An example is the task shown in Table 1 for which the command sequence would be "nroff -ms ATT2>ATT1." A fixed directory listing of file names was displayed on the screen at all times.

For incorrect responses, there was a series of help prompts designed to address specific deficits in the subjects' UNIX knowledge. These prompts were displayed on the screen one at a time in a fixed order regardless of the type of error that the subject had made. The help prompts were developed to address the most probable causes of error, as suggested by the UNICOM model of UNIX command production. There are seven different areas in which the help prompts could assist the subjects, and these are best described by referring to the example in Table 1.

Finally, we used error cards, describing the type of error made by the subject on an incorrect attempt. The three types were "Illegal," "Different: Legal but does something other than what the problem asks for," and "Keystrokes; legal and does what the problem asks for but uses more than the minimum possible keystrokes." (This last error category was intended to force subjects to use composite commands rather than a series of commands with temporary files in order to be correct.)

**Procedure.** Subjects were run individually for a single two-hour session. The experiment was made up of three sections: a questionnaire, an orientation, and completion of the actual UNIX problems.

**Composite Command Problems.** The subjects were given a series of 21 composite command problems. Together, the problems used 10 different utilities (e.g., lpr, cmp, cat) and 4 different input/output (I/O) redirection symbols (e.g., <, |, >>, >). In this activity, subjects were presented with a task statement on the computer screen, such as that shown in Table 1 and asked to type a command or series of commands that would accomplish the task. For each composite problem, the task statement appeared first along with the directory listing of all file names used in the experiment. The subjects would then type their first attempt at a command or command series to accomplish the task. After the subject clicked the button image labelled "DONE", the program would evaluate the subject's attempt and give feedback of "Correct" or "Try again". If the subject's answer was not correct, the experimenter evaluated the subject's attempt and handed the subject the appropriate error card. After reading the error card, the subject clicked another button image; the program then erased the subject's attempt and revealed the first help prompt.

Once the help prompt was revealed, the subject tried again to give a command sequence that would accomplish the task. If incorrect, the subject received another error card and was shown another help prompt. Throughout a given task, the task statement, directory listing, and all previous prompts remained visible; only the subject's previous attempts and error cards were removed. If at any time the subjects felt completely stymied, they were able to type "no help" rather than a solution attempt in order to get the next help prompt. The procedure of subject attempt, error card, help prompt continued until the subject typed the correct response. When the subject typed the required command sequence, the program displayed a "CORRECT!" screen and then revealed the next composite task instructions.

## Results

**Scoring Correct Command Productions.** Productions were scored as correct by the Mac computer if they matched the syntax of the idealized command (spaces were not parsed). Thus, a subject had to produce the command that required the least number of keystrokes (i.e., subjects could not substitute "sort>file1; head file1>file2" for the command "sort file1|head>file2"). Problems were broken into two groups, where problems 1-4 constituted the first group where most of the learning took place, and problems 5-21 made up the second, more stable group. In this paper we report only the data from the first group of problems.

**Correct Productions as a Function of Prompt.** Figure 1(a) shows the cumulative percentage of correct composite productions at each prompt level for the three expertise groups. Experts have the highest percentage overall, followed by the less expert groups. We expect experts to exceed novice performance. Prompts have a differential influence on correcting performance for the three expertise groups. For example, the change in percent correct performance from Prompt 3 to Prompt 4 is very small for experts, leading to the inference that Prompt 4 provided little or no new information for them. Conversely, the same change between prompts 3 and 4 for the less expert groups is relatively large, suggesting that this prompt does provide them with some new information. To analyze this effect, we study the four types of knowledge at each prompt level in detail.

**Scoring of Knowledge.** Each of the 21 problems given to subjects required certain amount of the four types of knowledge. For example, the problem described in Table 1 requires two command syntax facts: nroff -ms as a command name, and that nroff takes a file argument. It requires knowledge of one I/O redirection syntax fact, that ">" redirects output from a command to a file. Conceptual I/O knowledge required is that redirection of input and output can occur, and that I/O redirection can occur from commands to files. The required command redirection knowledge includes the fact that nroff output can be redirected to a file. Answers for each task were scored for the percentage of each type of knowledge displayed by

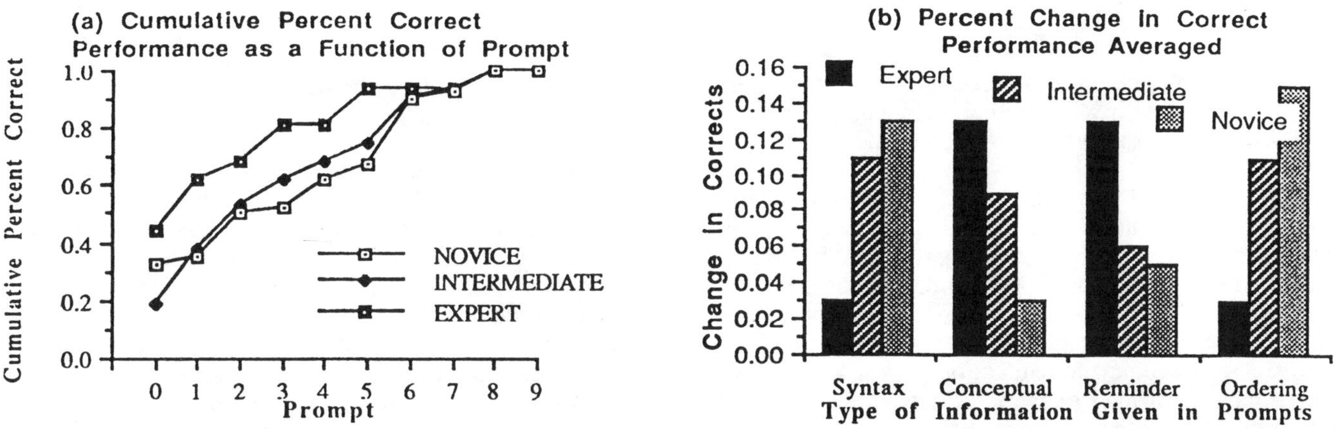


Figure 1. (a) Cumulative percent correct composite answers as a function of prompt; (b) Percent change in correct composite answers averaged across certain prompts.

each subject at each prompt level. For example, if a command requires two command syntax facts, and the attempt at prompt 0 (before any prompts) shows evidence of only one of these facts, then the subjects is credited for having 50% of the required command syntax knowledge for that task. When a production was correct, all knowledge types were at 100% levels.

**Knowledge Analyses.** Figures 2 (a) - 2 (d) show the mean knowledge scores for the three expertise groups for prompts 0-9. The arrow markers signal the reader what prompt first provided subjects with information relevant to the knowledge type displayed in the graph. For example, in Figure 2(a), Prompt 2 is the first prompt that describes all of the command syntax knowledge required to complete the task (see Table 1 for an example of all prompt types described in this section). To determine whether amount of relevant knowledge was influenced by expertise and prompt, these data were subjected to a 3 x 4 x 10 analysis of variance (ANOVA) with group (expert, intermediate, novice) as a between-subjects variable, and knowledge type and prompt as within-subjects variables. [In all cases, Greenhouse-Geisser degrees of freedom are used for within-subject variables.] This analysis resulted in a marginal effect of level, $F(2,19)= 3.24$, $p < .06$. Experts possess significantly higher knowledge scores than do the other two groups. There was also a main effect of knowledge type, $F (3,57) = 24.52$, $p<.01$, with the command redirection conceptual knowledge showing the lowest scores overall and the command syntax knowledge showing the highest scores overall. We did not obtain a knowledge type by level interaction, $F < 2$, seemingly due to the expert group ceiling effect. A main effect of prompt exists, $F (9,171) = 29.04$, $p < .01$, and a prompt by level interaction, $F (18,171) = 3.74$, $p < .01$. Performance improves as users receive more prompts, though the amount of improvement is different for experts than for novices. Experts appear to reach ceiling performance more quickly than do novices. Prompts influence the four types of knowledge differently, as evidenced by knowledge by prompt interaction, $F (27,513) = 8.12$, $p < .01$.

It is interesting to look at the three expertise groups separately to determine how the prompts influence their knowledge scores. Looking at Figures 2 (a-d), it appears that less expert group's knowledge scores differ from the expert's. To determine this, each of the three groups' knowledge scores were subjected to separate 4 x 10 ANOVAs with the knowledge types and the prompts as within-subject variables. The novices and intermediates show a main effect of knowledge type ( Novice $F ( 3,27) = 21.99$, $p < .01$; Intermediate, $F (3,21) = 13.02$, $p < .01$) and a main effect of prompt (Novice, $F (9,81) = 26.82$, $p < .01$; Intermediate, $F (9,63)= 18.65$, $p < .01$). Both groups show different levels of the four knowledge types, and an improvement in performance with the prompts. They also each show a knowledge type by prompt interaction (Novice, $F (27,243) = 6.44$, $p < .01$; Intermediate, $F (27,189) = 4.84$, $p < .01$). This suggests that the influence of the prompts on the four knowledge types is not equivalent for these two groups. The experts also show a main effect of knowledge, $F (3,9) = 5.94$, $p < .05$, and prompt, $F (9,27) = 6.90$, $p < .02$. Their data do not show a knowledge type by prompt interaction ($F < 3$). The results may reflect expert ceiling effects. Their knowledge level is so high to begin with that there is minimal variance in their data.

The arrow markers on Figures 2(a-d) show some of the most interesting interactions, especially when compared with the percent correct performance shown in Figure 1(a). Figure 2(a) suggests that for novices and intermediates, presentation of Prompt 2, which helps with command syntax (see Table 1 for an example), improves command syntax knowledge and percent correct performance (see Figure 1(a)). This is less the case for the experts. Figure 2(b) shows groups with improved I/O conceptual knowledge scores for their attempts following presentation of Prompt 3, which gives this information. However, Figure 1(a) suggests that while the experts can use conceptual I/O information to increase their chances for a correct production, the novices cannot. This leads to the inference that the novices can use conceptual I/O information to make a subsequent attempt that demonstrates greater I/O conceptual knowledge, but they are still lacking other information which would allow them to produce a correct command. Figure 2(c) shows that Prompt 4 appears to assist the novice group I/O syntax knowledge, and that the same trend holds to a lesser degree for the intermediates. Figure 1(a) suggests that this prompt increases the percent correct performance for the less expert groups. The experts, however, find this information useless. Figure 2(d) and Figure 1(a) show similar trends in the the change in command redirection knowledge for less expert groups following Prompt 6. This prompt gives both command redirection information and ordering information. Note that experts find this information useless.

A summary of the trends found in correct performance as a function of prompt is found in Figure 1(b). This figure shows the percent change in correct performance averaged together for the first prompts that provide syntax knowledge (both I/O and command syntax, Prompts 2 & 4), for the conceptual I/O prompt (Prompt 3), Prompt 5, which is the first reminder of the items required for the production, and the command redirection prompts, which also give ordering information (Prompts 6 & 8). Overall, there appears to be a strong interaction, with novices and experts showing very different changes in performance as a function of exposure to different types of prompts, and intermediates falling somewhere between the other two groups.

It appears that novices and intermediates need help with both command and I/O syntax, but that this is not a problem for the experts. Experts get help from an abstract statement of I/O conceptual knowledge (see Table 1 for an example),while this statement is practically worthless to the less expert groups. This finding is in line with the expertise literature (e.g., Chi et al., 1981) which suggests that experts can process more abstract information than can novices. Certainly Prompt 3 is more abstract that Prompts 2 and 4. Experts also successfully utilize a reminder prompt that lists all of the command and I/O elements together in one prompt. Novices and intermediates don't seem to utilize this information. Anecdotally, it is our view that the experts that have not solved a problem correctly by this point have often represented the problem with one command reversed or dropped, and this prompt reminds them that they have forgotten an element, and they can use this information to make a correct production. The less expert subjects seem to have more problems at this point than just knowing the elements, they can't put them together with just a list of the items. Their response to the ordering prompts (6 & 8) make this clear. These are the first prompts that tell users

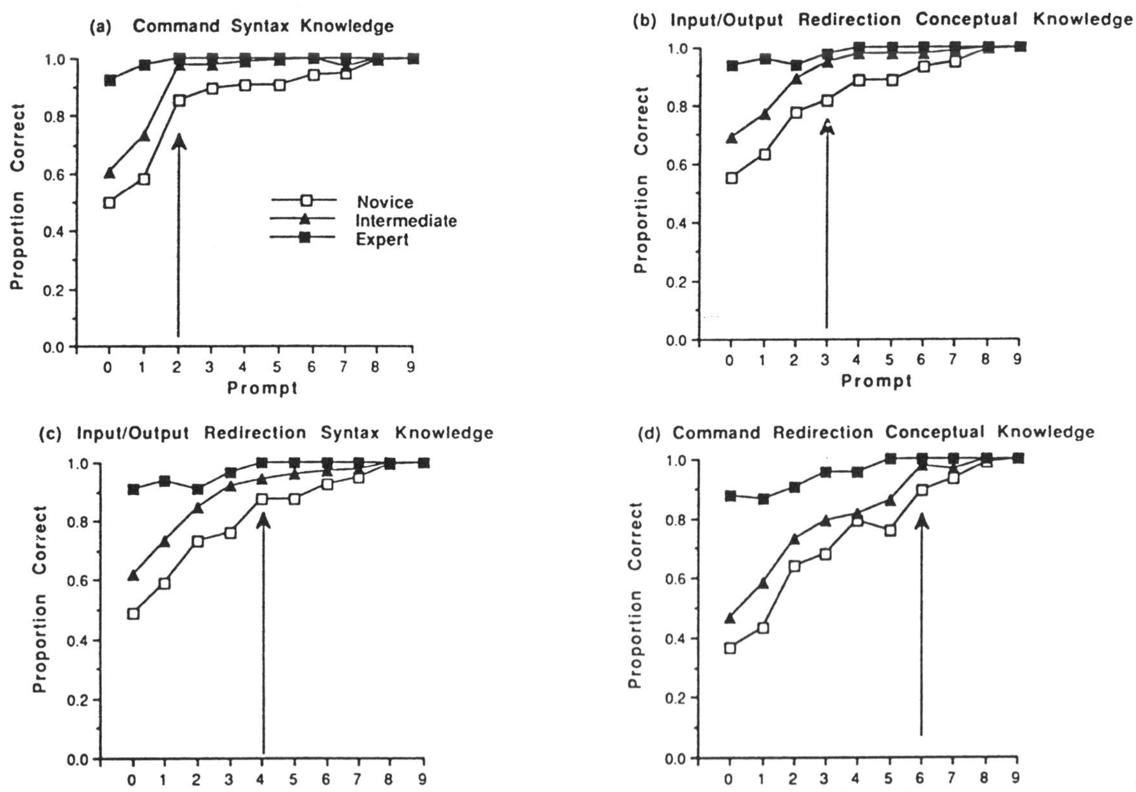


Figure 2. (a) Command syntax knowledge; (b) I/O conceptual knowledge; (c) I/O redirection syntax knowledge; and (d) Command redirection conceptual knowledge as a function of prompt.

what order they must use for the commands. For example, that first nroff -ms must be performed on ATT2 and then that this output must be redirected to the file ATT1, and so on. This ordering information helps the less expert groups quite a bit; it does almost nothing for the experts.

## Discussion

Our findings indicate that users vary markedly in their response to very different classes of prompt as a function of expertise. The results suggest that novices and intermediates lack both specific syntax knowledge and more general conceptual redirection knowledge. If prompted with specific syntax knowledge, novices and intermediates can increase their chances of producing a correct composite. However, novices and intermediates cannot utilize our more abstract prompt (Prompt 3) which provides conceptual redirection information knowledge that they need (see Figure 2(b)) to increase their correct performance (see Figure 1(a)). Our experts, in contrast, can use this information effectively to increase their performance. This finding is consistent with the literature on expertise which suggests that experts can utilize abstract information, while novices cannot (e.g., Chi et. al, 1981).

Novices and intermediates seem to need help retrieving the elements that go into making a composite; but for many of them, this is not enough. They also need help ordering the elements -- or constructing an action plan. When they receive a prompt that helps them order the items, their performance improves. There is evidence that ordering the elements taxes their working memory. They delete and substitute many more symbols than do experts, and these memory errors decrease markedly once ordering information has been given (Doane et al. 1991). Thus, the hypothesis suggested by the UNICOM construction-integration model was supported. Retrieving the elements is only part of the problem in composing a composite. To chain these elements together is another aspect of the problem, and this requires relating the pieces of knowledge in an ordered fashion and tracking the intermediate results.

While these data provide support for the assumptions of the Doane et. al (1989;1990) model, more direct support is provided by extending our modeling work to account for these prompt data. We are currently building models of individuals in this study, and running simulations by giving UNICOM prompts, and evaluating prompt activation within the individual's network. We hope to determine whether the overlap between the knowledge in a prompt and the individual network of knowledge dictates the usefulness of the prompt. This work will have clear implications computer aided instruction and for system design.

## Acknowledgements

This research was supported by grant IRI-8722792 from the National Science Foundation.

## References

Anderson, J. R., & Jeffries, R. 1985. Novice LISP errors: Undetected losses of information from working memory. *Human-Computer Interaction, 1*, 133-161.

Chi, M. T. H., Feltovich, P. J., & Glaser, R. 1981. Categorization and representation of physics problems by experts and novices. *Cognitive Science, 5*, 121-152.

Doane, S. M., McNamara, D. S., Kintsch, W. Polson, P. G., Clawson, D. M., Dungca, R. G. 1991. *Prompt Comprehension in UNIX Command Production.* Technical Report 91-4. Institute of Cognitive Science, University of Colorado, Boulder.

Doane, S. M., Kintsch,, W. & Polson, P. 1989. Action Planning: Producing UNIX commands. *Proceedings of the Eleventh Annual Conference of the Cognitive Science Society.* (pp. 458-465). Ann Arbor, Michigan: Lawrence Erlbaum Associates.

Doane, S. M., & Kintsch, W., & Polson, P. G. 1990. *Understanding UNIX Commands: What Experts Must Know.* Technical. Report. 90-1. Institute of Cognitive Science, University of Colorado, Boulder.

Doane, S. M., Pellegrino, J. W., & Klatzky, R. L. 1990. Expertise in a computer operating system: Conceptualization and performance. *Human-Computer Interaction, 5*, 267-304.

Kintsch, W. 1988. The use of knowledge in discourse processing: A construction-integration model. *Psychological Review, 95*, 163-182.

Larkin, J. H. 1983. The role of problem representation in physics. In D. Gentner & A. Stevens (Eds.), *Mental models* (pp. 75-98). Hillsdale, NJ: Lawrence Erlbaum Associates, Inc.

Mannes, S. M. & Kintsch, W. in press. Routine computing tasks; Planning as understanding. *Cognitive Science.*

Murdock, B. B. 1974. Human memory: Theory and data. Lawrence Erlbaum Associates.

Newell, A. 1987. Unified theories of cognition. *The 1987 William James Lectures.*

van Dijk, T. A. & Kintsch, W. 1983. *Strategies of discourse comprehension.* New York: Academic Press.

Wertheimer, M. 1982/1945. *Productive thinking.* Chicago, IL: University of Chicago Press.

# Recovering Structure from Expression in Music Performance

**Carolyn Drake and Caroline Palmer**

**Ohio State University**
**Psychology Department**
**1885 Neil Ave**
**Columbus OH 43210**
**cdrake or cpalmer @ magnus.acs.ohio-state.edu**

**Abstract**

Mental representations of structural content in music can be communicated to listeners by expressive variations in performance. We attempt to recover structural content from patterns of expression in skilled music performance, and we contrast possible mappings between structure and expression that allow communication of musical ideas. Three types of musical structure are investigated: metric, rhythmic grouping, and melodic accent structures. Skilled pianists performed musical sequences which were examined for expressive variations that coincide with each accent structure. The mapping of structure to expression is compared for music in which the accent structures are presented singly, are combined to coincide or conflict, or naturally co-occur. The findings suggest that associated sets of expressive variations in performance provide an unambiguous and flexible system for communicating musical structure.

A basic problem in domains such as speech, vision, and music, is how a perceiver is able to recover structure from a continuous input. One possible solution for music perception is that a performer emphasizes structurally important events, drawing the listener's attention to them and thus aiding the perception of structure. Successful communication between performer and listener requires that the listener distinguish between musical structure (information usually indicated in a musical score) and musical expression (particular methods of emphasizing structure) in music performance. This task is complicated by the fact that both are coded in the same acoustic dimensions such as frequency and intensity. However, systematic expressive marking of structural features in music performance may aid the perceiver's recovery of structure.

This paper focuses on the communication of musical structure through expressive variations in performance. We address how the mapping of structure to expression may be mediated by performers' mental representations. Events interpreted as structurally important are often performed in a way that draws attention, by playing them louder, longer, or more legato than surrounding events (Clarke,1988; Gabrielsson,1974,1982;Palmer,1989). Hypothetically, it should be possible for the listener to identify events played in this manner and infer their structural importance. Unfortunately, this direct recovery of structure from expression is problematic because music is typically composed of several superimposed structures, some of which contradict each other (Lerdahl & Jackendoff, 1983). Thus, a particular combination of expressive variations may arise from multiple structural descriptions, making the recovery of structure ambiguous.

The question arises as to which structural features of music receive emphasis through expression. Accents, events that stand out and capture a listener's attention, are possible candidates. Based on related findings in music perception and performance, we have investigated three accent structures (Drake, Dowling & Palmer, 1991; Drake & Palmer, in prep.). The first, metric accent structure, partitions the musical sequence into a periodic pattern of alternating strong and weak beats, with important events at the beginning of each measure. A second accent structure, rhythmic grouping, segments the sequence into groups of temporally proximal events: an event of long duration separates one rhythmic group from the next and is considered to be structurally important. A third accent structure, melodic jumps, segments the sequence on the basis of pitch height, and an event after a large change in pitch is structurally important. These three accent structures are explicitly marked in the musical score and examples are shown in Figure 1, with X's indicating accents. We predict that these events will receive emphasis in performance. Pianists can emphasize events in at least three ways: by changes in intensity, timing, and articulation. Intensity is influenced by the keypress force correlated with piano hammer velocities. Timing, in particular interonset timing, reflects event durations relative to those indicated by the musical score. Articulation is affected by the time between successive keypresses and releases, resulting in legato (smooth) or staccato (disjointed) events.

How do expressive variations reflect a performer's interpretation of musical structure? Theoretically, a performer could use all methods of expression (intensity, interonset timing, and articulation) for each of the

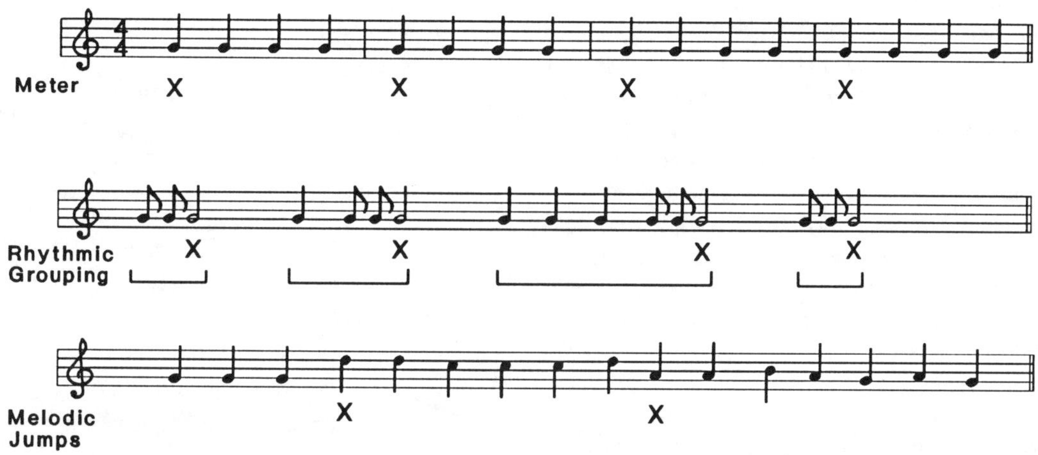


Figure 1. Examples of accent structures. X denotes an accent.

accent structures (meter, rhythmic grouping, and melodic jumps) so that each type of accent would be associated with expressive variations in all three performance measures. This is indicated in Figure 2A where lines join all possible accents with all possible expressive variations. In this type of mapping, the recovery of structure from expression can be ambiguous: an event may be expressively marked as structurally important, but it may be difficult to distinguish which accent structure is implicated.

A second class of mappings associates each accent structure with variations in only one expressive dimension. For instance, Figure 2B shows metric structure associated with variations in intensity, rhythmic grouping with temporal distortions, and melodic jumps with articulation. This would be a completely unambiguous communication system: for instance, a loud event could be interpreted unambiguously as being metrically important.

In a third class, each accent structure would be related to an associated set of expressive variations unique to each accent structure. For instance, Figure 2C shows metrically important events played both louder and longer in duration, whereas rhythmic grouping accents are played longer and more legato. To distinguish between musical structures giving rise to the expressive output in this mapping, listeners would have to be sensitive to associated combinations of variations.

We present evidence from skilled music performance that allows us to distinguish among these mappings of structure to expression. We will contrast findings from empirical studies presented in detail elsewhere (Drake & Palmer, in prep.). First, we examine performances of musical sequences in which only one accent structure is presented at a time, to determine expressive variations used in the absence of

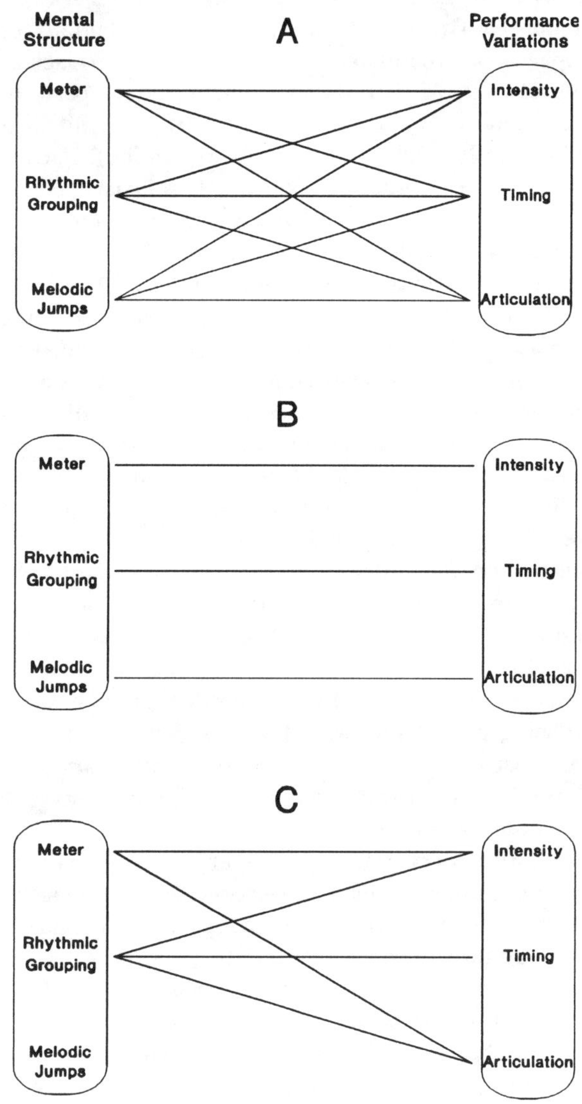


Figure 2. Three theoretical mappings of structure to expression.

other accents. Second, we examine the performance of simple musical sequences in which the three accent structures are combined in different ways so that they conflict or coincide, to determine how the presence of each structure influences the expression of other structures. Third, we analyze a performance of complex music in which the three accent structures naturally co-occur. By using progressively more complex stimuli, we hope to distinguish among possible mappings of structure to expression in skilled music performance, and to test whether or not this mapping is dependent on the musical context in which the accent structures are presented.

## A. Isolated accent structures

Ten skilled pianists performed simple musical sequences in which each accent structure was presented in isolation. Examples of the sequences are shown in Figure 1. To evaluate the relative contributions of emphasis of structurally important events versus lack of control (noise), pianists were instructed to perform each sequence in two ways: first musically and then mechanically (without adding any expression). Performances were recorded on a Yamaha Disklavier acoustic upright piano monitored by a personal computer. Only results reaching statistical significance ($\underline{p} < .05$) from analyses of variance are described here. For each accent structure, the expressive variations coinciding with accented events in Figure 1 were compared with those on surrounding events: for instance, events on the first beat in the measure were compared with all other events, the first and last events in a rhythmic group were compared with the other events, and events on melodic jumps and turns were compared with events preceding and following them.

**Musical performances.** The expressive variations observed in the musical performances are shown in Figure 3. Events on the first beat in the measure were played louder and more legato than other events in the measure. The end of a rhythmic group was emphasized by being played louder, delayed, and preceded by a short pause. Thus, consistent expressive variations were related to metrical and rhythmic grouping accents but not to melodic jumps.

**Mechanical performances.** The expressive variations observed in the mechanical performances of the same musical sequences are shown in Figure 4. No expressive variations were recorded for either meter or melodic jumps, but the rhythmic groups were emphasized with the same timing and overlap variations as seen in the musical performances. With these two exceptions, the mechanical performances demonstrate that the mapping of structure to expression is voluntary and related to the performers' emphasis of musical structure.

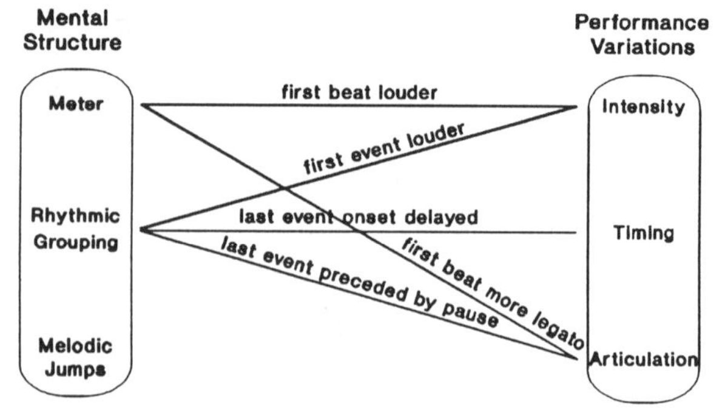


Figure 3. Observed mappings for musical performances of isolated accent structures.

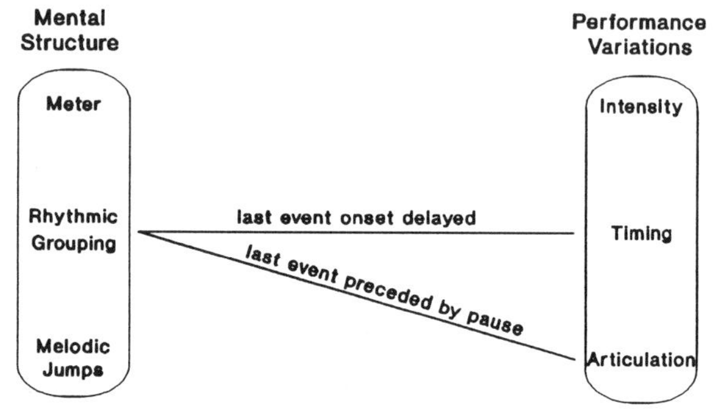


Figure 4. Observed mappings for mechanical performances of isolated accent structures.

## B. Combined accent structures

The same pianists then performed musical sequences (in a musical fashion only) in which all three accent structures either coincided or conflicted. Examples of tunes in which accents coincide and conflict are shown in Figure 5, and the expressive performance variations are summarized in Figure 6. The same variations were observed for rhythmic grouping as seen in the musical performances of the isolated accent structures (last event played louder, preceded by a pause, and delayed). No systematic variations were observed for meter, but melodic jumps were preceded by a lengthened duration delaying their onset. These findings were consistent across performances of stimuli containing coinciding and conflicting accent structures. Note that three of the six expressive variations observed in the musical performances of the isolated accent structures (Figure 3) were also observed here, suggesting that they are unaffected by the presence or absence of the other accents.

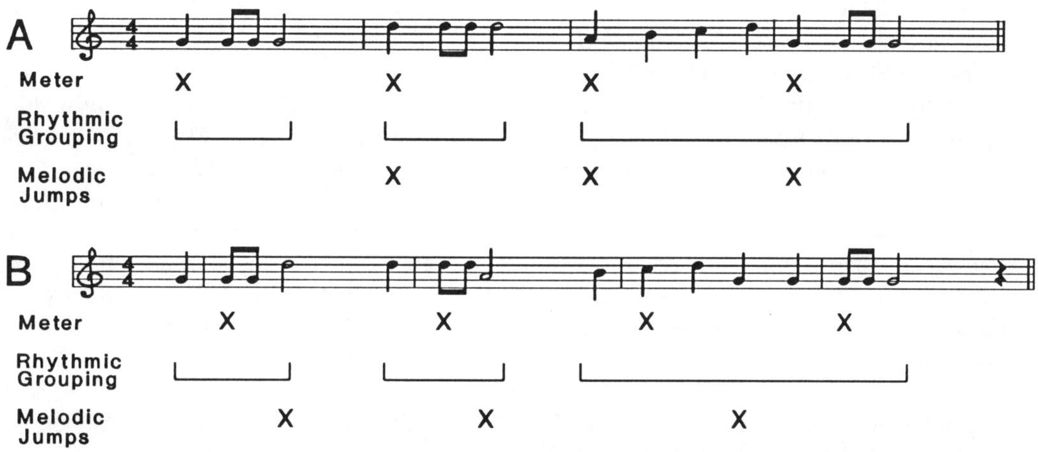


Figure 5. Examples of musical sequences in which accents (A) coincide or (B) conflict.

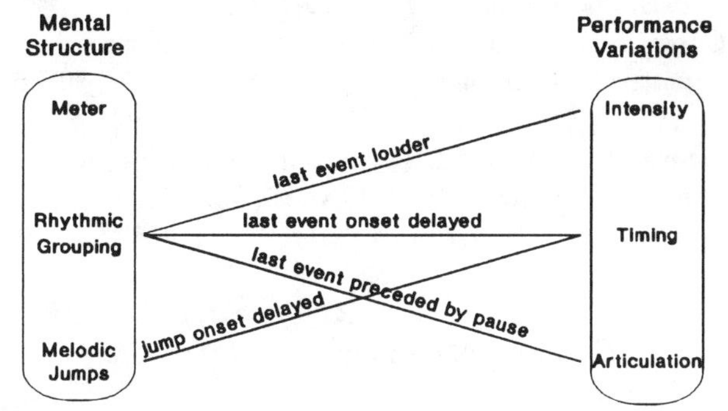


Figure 6. Observed mappings for performances of combined accent structures.

## C. Complex music with co-occurring accent structures

An experienced concert pianist performed Beethoven's Piano Sonata Opus 111, with which he was familiar, on a computer-monitored grand piano. This piece was chosen because it presents a more complex case in which the three accent structures naturally co-occur: a significant correlation was found between the rhythmic grouping and melodic jump accents in the composition (r = .38, p <.01). The performance of the melody in the first movement was chosen for analysis because it contains many metric, rhythmic grouping, and melodic accents.

The expressive variations in this performance are presented in Figure 7. The metric accent structure was marked by variations in intensity, with structurally important events played louder. Rhythmic grouping was emphasized again with the last event played louder, delayed, and preceded by a pause. In addition, the first events in a rhythmic group were preceded by a pause. There was also a more complex variation for intensity within the rhythmic group: the first event was the softest and the last event the loudest, with a slow increase throughout the rhythmic group. Melodic jumps were preceded and followed by a pause and a longer interval. Thus, similar expressive variations were observed in the performance of this complex music as in the performances of the isolated accent structures (Figure 3): five of the six variations were identical, as well as one additional variation related to rhythmic grouping and two related to melodic jumps. Therefore, the mapping of structure to expression is highly consistent in the presence or absence of other co-occurring accent structures.

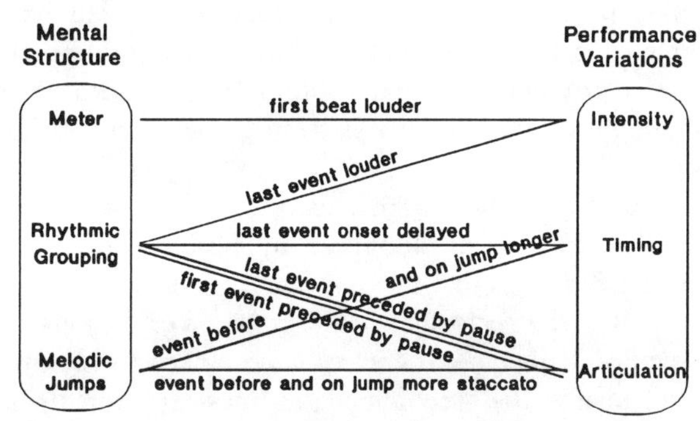


Figure 7. Observed mappings for performances of co-occurring accent structures.

Returning to the three possible classes of mapping presented earlier, these analyses provide support for the third class of mapping; associated sets of expressive variations reflect structural content. The other two classes of mappings are not supported because each accent structure is not related to variations in only one or all of the expressive dimensions.

## Discussion

The recovery of structure from a rich continuum of input is a critical problem for many perceptual domains, including music. We have demonstrated that, in the case of music, performers use a wide range of expressive variations to disambiguate musical structure. This mapping of musical structure to expression in music performance must be sufficiently informative and robust across musical contexts to account for communication from performer to listener. The mapping described here, an associated set of expressive variations, is encouraging for theories that view music perception as the recovery of structure from expression because music performances provide the primary source of input for perception of music. To complete the chain of communication, listeners must be able to detect and interpret these expressive variations. Those described here fall within the range of perceivable changes (Drake, 1990), and there is evidence to suggest that listeners do detect and use these variations to interpret musical structure (Clarke, 1989; Nakamura, 1987; Sloboda, 1983).

In summary, the consistent findings across musical contexts suggest three important properties of the mapping of musical structure to expression. First, structurally important events are indeed emphasized through a limited set of expressive variations. In piano performance, important structural features may be performed louder, longer, and more smoothly. One or more of these variations may be applied at the same time; a single expressive variation may be used in simple musical contexts and multiple variations in more complex musical contexts, as evidenced by the increased complexity in the mapping from performances of simple musical sequences to more complex musical forms.

Second, the finding that many expressive variations greatly decrease or disappear in mechanical performances indicates that they are under voluntary control and reflect the performers' interpretation of the musical structure. This type of control is necessary for flexible performance that accommodates musical contexts in which structure covaries, yet maintains a determinate mapping of structure to expression.

Third, these findings suggest that an associated set of expressive variations consistently signals at least some structural content across musical contexts. The complex interplay between expression and structure reflects the flexibility attributed to skilled performance, in which an expressive dimension can be adjusted to accommodate different structural features in different performances of the same music. This view requires that both explicit expressive variations and the structural features implied by them contribute to the mental representations of musical knowledge shared by performer and listener.

## References


Clarke, E.F. 1988. Generative principles in music performance. In J.A. Sloboda (Ed.) *Generative processes in music: The psychology of performance, improvisation and composition* (pp. 1-25). Oxford: Oxford University Press.

Clarke, E.F. 1989. The perception of expressive timing in music. *Psychological Research* 51:2-9.

Drake, C. 1990. Processus cognitifs impliques dans l'organisation du rythme musical. Ph.D. diss., Psychology Department, University of Paris V.

Drake, C., Dowling, W.J, & Palmer, C. 1991. Accent structures in the reproduction of simple tunes by children and adult pianists. *Music Perception* 8:313-332.

Drake, C. & Palmer, C. 1991. Accent structures in music performance. In preparation.

Gabrielsson, A. 1974. Performance of rhythmic patterns. *Scandinavian Journal of Psychology* 15:63-72.

Gabrielsson, A. 1982. Perception and performance of musical rhythms. In M. Clynes (Ed.), *Music, mind, and brain: the neuropsychology of music* 159-169. New York: Plenum Press.

Lerdahl, F. & Jackendoff, R. 1983. An overview of hierarchical structure in music. *Music Perception* 1:229-252.

Nakamura, T. 1987. The communication of dynamics between musicians and listeners through musical performance. *Perception & Psychophysics* 41:525-533.

Palmer,C. 1989. Mapping musical thought to musical performance. *Journal of Experimental Psychology: Human Perception and Performance* 15: 331-346.

Sloboda, J. A. 1983. The communication of musical metre in piano performance. *Quarterly Journal of Experimental Psychology* 35: 377-396.


## Acknowledgments


This research was supported by NIMH grant 1R2945764 to the second author. We are grateful to Mari Jones and Carla van de Sande for comments on an earlier version of this paper.

# Diagnostic Reasoning of High- and Low-Domain Knowledge Clinicians: A Re-analysis

**Arthur S. Elstein, Benjamin Kleinmuntz, Mitchell Rabinowitz, Robert McAuley, James Murakami, Paul S. Heckerling, James M. Dod,**
Department of Medical Education (M/C 591)
University of Illinois at Chicago
PO Box 6998, Chicago, IL 60680

**Abstract**
Thinking aloud protocols previously obtained by Joseph and Patel were re-analyzed to determine the extent to which their conclusions could be replicated by independently developed coding schemes. The data set consisted of protocols from 4 cardiologists (low domain knowledge = LDK) and 4 endocrinologists (high domain knowledge = HDK), individually working on a diagnostic problem in endocrinology. Both analyses found that HDK physicians related data to potential diagnoses more than the LDK group, and that there were trends for HDK physicians to be more focused on the correct diagnostic components and to employ more single-cue inference and less multiple-cue inference. However, the re-analysis found no meaningful differences between groups in diagnostic accuracy, speed of diagnosis, or in the breadth of the search space used to seek a solution. The generalizability of results of protocol-analysis studies can be assessed by using several complementary coding schemes.

Analyzing the verbalization ("thinking aloud") of physicians as they respond to a sequentially presented clinical problem has been a popular strategy for studying the study of medical cognitive processes. These studies typically present a diagnostic or treatment planning problem in a series of discrete steps, thereby assuring that each subject is exposed to the same information in the same order, and ask subjects to think aloud as the clinical situations unfolds.

Comparing, contrasting and then generalizing across studies has been difficult. Investigators have used different problems, different approaches to or frameworks for protocol analysis, and different groups of subjects (medical students, residents, academic physicians, practitioners in various practice settings). Since there is ample evidence that medical problem solving is partly contingent on the content of the problem, the phenomenon variously called domain- or content- or case-specificity (Elstein et al., 1978, 1990; Grupen, Wolf, Van Voorhees, 1988; Norman, 1988; Swanson, 1990), generalizing about cognitive processes from a limited sample of cases is clearly a problem. Substantial within-group variability has also been observed in many studies (e.g., Grant & Marsden, 1987; Johnson, et al., 1982). Yet protocol analysis is so labor-intensive that these studies invariably employ small samples of both cases and subjects.

In addition to variation across studies due to cases and subjects, investigators have employed protocol analysis schemes that were not subsequently used by other research teams. These analytic methods usually contain many qualitative or subjective elements, and it is often unclear that the conclusions asserted would have been reached by a different team of investigators, even had they employed the same approach to protocol analysis. Consequently, it has been difficult to compare and contrast conclusions of different studies, and to determine to what extent the conclusions of any investigation are independent of the method of protocol analysis, or what aspects of the data are highlighted by each method of analysis.

One approach to these difficulties is to re-analyze protocols obtained by one set of investigators using another framework. This paper reports one such analysis, presenting a re-analysis of data originally collected and analyzed by Joseph and Patel (1990). Its aim is to determine the reproducibility of their conclusions by re-analyzing the raw data with independently constructed schemes.

## Method

### Subjects

Joseph and Patel's study employed a sample of 9 senior physicians, all associated with the Faculty of Medicine at McGill University, 4 endocrinologists and 5 cardiologists. The clinical problem concerned the

This research was supported in part by a grant from the National Library of Medicine, RO1-LM-4583. We are grateful to Vimla Patel for providing materials from her files for this re-analysis.

diagnosis of an endocrine disorder, so the endocrinologists were considered as High Domain Knowledge (HDK) subjects and the cardiologists as Low Domain Knowledge (LDK) subjects. All of the subjects were board-certified, practicing physicians with five to ten years of experience in their fields. Our re-analysis was performed on protocols of 8 subjects, all that were made available.

## Clinical Problem

The complete text is presented in Joseph & Patel (1990). The vignette was organized into three parts, corresponding to the traditional partition of a case presentation into history, findings from the physical examination, and x-ray and laboratory test results. These parts were further subdivided into 25 segments presented sequentially. The first part consisted of 4 segments of medical history, the next 14 presented physical exam information, and the final 7 segments included x-ray and laboratory test information. According to Joseph and Patel, the accurate diagnosis of this case is Hashimoto's hypothyroidism with myxedema pre-coma. The diagnosis can be further divided into three components of increasing specificity: hypothyroidism, myxedema pre-coma, and an autoimmune condition called Hashimoto's thyroiditis. The cognitive task is diagnosis, not treatment or management. The time elapsed during the events in this case is not specified in the vignette, but is probably not more than 3 hours.

## Original Procedure

All subjects were individually tested. The stimulus material was presented on a computer monitor one segment at a time. Subjects were asked to think aloud after each segment, verbalizing about the role and importance of the information in each segment in reaching the correct diagnosis. They controlled the rate of presentation of each segment and there was no time limit for their responses. Transcripts of tape recorded verbalizations provided the raw data.

## Protocol analysis procedures

In prior studies, we had developed two coding schemes as part of an investigation of expert-novice differences in clinical reasoning about cases in a medical intensive care unit. One focused upon problem solving processes (Kleinmuntz & Elstein, 1990), the other upon categories of knowledge used in problem solving (Rabinowitz, Cottrell & Elstein, 1990).

The analysis of problem solving processes utilized 5 constructs: 1) formulates new diagnostic hypothesis; 2) confirms or strengthens diagnostic hypothesis using new facts or findings; 3) rejects or weakens diagnostic hypothesis using new facts or findings; 4) bases inference on multiple cues; and 5) bases inference on single cue. Each process was coded every time it occurred in the 25 segments, except that diagnostic formulations (construct #1) were noted only the first time they were mentioned, as only new formulations were coded. To facilitate coding, specific inclusionary and exclusionary criteria were developed and representative examples were used to define further the characteristics of each construct. Scores were summed for each subject within and across the three sections. For the present study, the coding scheme was revised by adding a distinction between diagnoses (more specific and precise formulations) and problems (more vague and global). The diagnostic hypotheses were reviewed by one of the authors (PSH) and classified into diagnoses and problems. The coding was done by a single rater (JM).

The knowledge utilization analysis focused on the categories of knowledge used in diagnostic reasoning and on the question of whether differences in category utilization can distinguish groups of clinicians. The protocols were divided into units consisting of either a single sentence or a sentential phrase with subject and object. No unit was longer than a sentence; some were even briefer. These units were identified and then coded by a single rater (RM) utilizing a scheme designed to identify the categories of knowledge and information in each unit. Four primary categories were employed: context (personal and social history, risk factors, circumstances of acute onset), findings (relevant medical history and physical findings), tests, and diagnosis. Additional "relational" categories were used to identify units in which two primary categories were related to each other.

## Analysis of Data

While quantitative analytic methods were employed, statistical tests of significance are not very useful. We have chosen to describe the quantitative findings and to suggest trends in the data.

Our analysis focussed upon 4 central questions. For each, we present initially the findings of Joseph and Patel, and then our own conclusions, noting agreements and divergences, as appropriate. The questions are:

1. Diagnostic accuracy: Did HDK and LDK physicians differ in diagnostic accuracy?

2. Timing and course of diagnostic hypotheses: When in the sequence of data presentation were the correct diagnostic components generated? Were HDK physicians correct earlier?

3. Diagnostic process: What are the differences between groups in diagnostic process, particularly in generating and testing alternative hypotheses? Do they differ in the number of alternative hypotheses considered? Do HDKs and LDKs differ in the breadth of the search space used in seeking a solution?

4. Knowledge utilization: Are there differences between groups in the categories of knowledge used in problem solving, particularly in relating data to diagnostic hypotheses?

## Results

### Diagnostic Accuracy

Joseph and Patel did not provide any data regarding the number of clinicians in each group who generated each component of the correct diagnosis, but they reported that "those LDK subjects who produced accurate diagnostic components produced them later than did the HDK subjects." Our analysis shows that the first diagnostic component, hypothyroidism, and the second diagnostic component, myxedema, were generated by all clinicians. The third diagnostic component, Hashimoto's thyroiditis was generated by two HDK and two LDK clinicians. Diagnostic accuracy is not a function of the number of hypotheses proposed. We conclude that, with respect to diagnostic accuracy, there is no meaningful difference between groups.

### Timing and Course of Hypotheses

The common wisdom about diagnostic skill is that experts reach conclusions more rapidly and efficiently than those less experienced in a particular domain. Joseph and Patel's findings accord with this general view. They reported that HDK clinicians generated each diagnostic component earlier than LDK clinicians, and illustrated this conclusion by providing data from one subject from each group rather than by providing aggregate data from all.

Our analysis identified the point in the case at which each subject generated each diagnostic component. The conclusions of Joseph and Patel are affirmed in broad outline with respect to the first and third diagnostic component, hypothyroidism and Hashimoto's thyroiditis, although there are minor differences in our analyses. We found that HDKs generated these components somewhat earlier than LDKs (although not so much earlier as to be clinically significant). However, our interpretation of the data regarding the second diagnostic component, myxedema, is quite different. Two HDK clinicians developed this component very early (by segment 4), while two did not mention it (even with very liberal criteria) until segment 8 and 17, respectively. The LDK clinicians all generated this aspect of the diagnosis in segment 5 or 6. Thus, HDK physicians were more variable in the timing of this diagnostic component and are not necessarily earlier, although the lag is clinically unimportant. Both groups tend to move from general diagnoses to more specific diagnoses.

### Diagnostic Process

The cumulative sum of the mean number of new diagnoses produced was the dependent variable Joseph and Patel used to examine diagnostic process. They reported that after HDK physicians had produced the accurate diagnostic components, they generated few new diagnostic hypotheses and spent the rest of their time confirming hypotheses. They sporadically ruled out some of the hypotheses generated earlier, and most often used the new findings from the physical examination to confirm the diagnosis and determine secondary problems. In contrast, LDK subjects continued to generate new diagnostic hypotheses throughout the case. They focused primarily on associating clinical findings and test results with new possibilities. They generated very few secondary problems, and did not rule out the diagnostic hypotheses that they had generated earlier, even when the new hypotheses were contradictory to some of the earlier ones. In general, the HDK subjects narrowed uncertainty while LDK subjects increased it. Their results imply that the HDK subjects "coned down" on a diagnosis, working in a progressively narrower problem space.

Our reanalysis hypothesis and problem generation found that the number of new problems identified levels off for both groups after segment 15. The mean number of new diagnostic hypotheses continued to climb slowly for both groups. Both HDK and LDK clinicians considered similar numbers of new diagnostic hypotheses across all segments. In contrast to the view of Joseph and Patel, this analysis argues that the two groups do not differ quantitatively in the breadth of the search space used to seek a solution.

The diagnostic reasoning process may be further examined by referring to measures of hypothesis confirmation, disconfirmation, and types of inference. The mean number of attempts to confirm and to rule out diagnoses was the same for both groups. We tabulated the numbers and percentages of efforts at hypothesis confirmation that involved the correct diagnostic components vs. all other formulations. HDKs produced slightly more instances of confirmation of the correct components.

## Single-Cue and Multiple-Cue Inferences.

A single-cue inference relates a single item of information to a diagnostic hypothesis, in an effort either to confirm or rule out; a multiple-cue inference relates two or more cues to a diagnostic hypothesis in a semantic unit such as a sentence. The data show that LDK physicians produced slightly more multiple-cue inferences than did HDKs, and a lower percentage of this output was concerned with the correct components. This pattern is reversed for single-cue inferences: HDKs produced more such processes and a higher percentage was focused on the correct diagnostic components. Thus, there is a consistent trend for HDK clinicians but this trend is not statistically significant. The mean ratio of single-cue to multiple-cue inferences was 2.65:1 for HDK physicians and 1.86:1 for LDKs. These ratios show a trend for HDK clinicians to rely more on single-cue inference and less on multiple-cue inference. These results might suggest that as the case moves along, HDK subjects rely more on direct associations from data to cognitive schemata and less on step-by-step aggregation of findings to build a clinical picture than do the LDKS.

## Knowledge Utilization

Joseph and Patel found that HDK subjects used more relations to connect important information and ignored irrelevant information. The evidence presented in support of this claim was largely anecdotal.

The quantitative analysis of knowledge utilization focused on the ratios between the primary categories of "Findings" (History in part 1 of the case, physical findings in part 2) and "tests" (Laboratory test data and x-rays, in part 3) and the relational categories Findings-to-Diagnosis,Diagnosis-to-Findings,Test-to-Diagnosis, and Diagnosis-to-Test.

Both groups handle Part 1 in a fairly similar fashion, with statements about findings alone equalling the number of relational statements. In parts 2 and 3, however, the HDKs move toward a much more relational mode of discourse, consistently relating findings or test results to diagnoses much more than do the LDKs. The LDKs' statements typically stayed closer to the data, and related them less often to the diagnoses under consideration. These measures provide a concise quantitative picture of information processing by HDK and LDK clinicians and in general support the results of Joseph and Patel. At the level of a sentence-by-sentence analysis of the discourse of these physicians, HDK clinicians do relate the data to diagnoses more than do LDKs. Our analysis supports Joseph & Patel's conclusions that HDK physicians on the average used more relations to connect important information, but we also found substantial within-group variation.

Joseph & Patel reported that HDKs concentrated more on the correct diagnostic components while LDKs were more concerned with incorrect hypotheses. We analyzed the content of the relational diagnostic statements made by the members of each group to determine if they referred to one of the three correct diagnostic components of the case. 82% of the HDKs' diagnostic relational statements and 76% of the LDKs' concerned one or more of the three diagnostic components. Our interpretation is that there is probably no meaningful difference between groups.

## Discussion

In this paper, data previously reported by another group of investigators have been independently reanalyzed by a set of methods for protocol analysis that are quite different than the techniques they employed. Joseph and Patel had studied two small groups of experienced clinicians on one endocrinology case. One group was quite expert in the domain (High Domain Knowledge, HDK) and the other was not (Low Domain Knowledge). They identified several differences in the reasoning of the two groups. According to their account, HDK clinicians produced the components of an accurate diagnosis earlier than the LDK group and were also more efficient and focused. HDK clinicians generated fewer new diagnostic alternatives, spent more time confirming prior hunches, used more relations to connect important information, and ignored irrelevant information.

We found some trends in quantitative measures of diagnostic reasoning that are partial support for their overall conclusions. Both analyses agree that HDK physicians related data to potential diagnoses more than the LDK group, and that there was a trend for HDK physicians to be more focused on the correct diagnostic components. It can be argued that HDKs know the links between findings and diagnoses, while LDKs recognize clinical abnormalities both in the physical examination and laboratory data, but do not have the linkages between the data and diagnoses needed for understanding and solving difficult diagnostic problems.

On the other hand, some of their conclusions are not supported by our reanalysis. Specifically, we found no important or clinically meaningful differences between the groups in overall accuracy of diagnosis, speed of accurate diagnosis (as measured by the segment number at which each component is first mentioned), or the quantitative breadth of the search space (a measure of diagnostic focus). We found that both groups considered comparable numbers of new diagnostic hypotheses and more general problem formulations, continued to generate a small but steadily increasing number of diagnostic alternatives as the problem unfolded, and were very similar in mean numbers of efforts at hypothesis confirmation or disconfirmation.

## References


Barrows, H. S., & Feltovich P. 1987. The clinical reasoning process. *Medical Education*, 21, 86-91.

Elstein, A. S. 1988. Cognitive processes in clinical inference and decision making. In D. C. Turk and P. Salovey (Eds.), *Reasoning, inference and judgment in clinical psychology* (pp. 17-50). New York: Free Press/Macmillan.

Elstein, A. S., Shulman, L. S., & Sprafka, S.A. (1978). *Medical problem solving: An analysis of clinical reasoning*. Cambridge, MA: Harvard University Press.

Elstein, A. S., Shulman, L. S., & Sprafka, S. A. (1990). Medical problem solving: A ten-year retrospective. *Evaluation and the Health Professions*, 13, 3-36.

Grant, J., & Marsden, P. (1987). The structure of memorized knowledge in students and clinicians: An explanation for diagnostic expertise. *Medical Education*, 21, 92-98.

Gruppen LD, Wolf FM, Van Voorhees C, Stross JK. The influence of general and case-related experience on primary care treatment decision making. Arch Intern Med 1988;148:2657-2663.

Johnson, P.E., Duran, A.S., Hassebrock, F., Mollar, J., Prietula, M., Feltovich, P., & Swanson, D.B. (1981). Expertise and error in diagnostic reasoning. *Cognitive Science*, 5, 235-285.

Joseph, G. M., & Patel, V. L. (1990). Domain knowledge and hypothesis generation in diagnostic reasoning. *Medical Decision Making*, 10, 31-46.

Kassirer, J. P., (1989). Diagnostic reasoning. *Annals of Internal Medicine*, 110, 893-900.

Kassirer, J. P., & Kopelman, R. I. (1989). Cognitive errors in diagnosis: Instantiation, classification and consequences. *American Journal of Medicine*, 86, 433-441.

Kassirer, J. P., Kuipers, B. J., & Gorry, G. A. (1982). Toward a theory of clinical expertise. *American Journal of Medicine*, 73, 251-259.

Kleinmuntz, B., & Elstein, A.S. (1990). Toward medical decision making expertise. Paper presented at the Annual Meeting of The Judgement and Decision Making Society, New Orleans, November 18, 1990.

Norman, G. R. (1988). Problem-solving skills, solving problems and problem-based learning. *Medical Education*, 22, 279-286.

Swanson, D. B. (1990). Issues in assessment of practical skills in medicine. *Professions Education Researcher Quarterly*, 12, 3-6.

# Symbolic Action, Behavioral Control and Active Vision

**R. James Firby and Michael J. Swain** *
Computer Science Department
University of Chicago
1100 East 58th Street, Chicago, IL 60637
*firby@cs.uchicago.edu, swain@cs.uchicago.edu*

## Abstract

This paper is about the interface between continuous and discrete robot control. We advocate encapsulating continuous actions and their related sensing strategies into structures called ***situation specific activities***, which can be manipulated by a symbolic reactive planner. The approach addresses the problem of turning symbolic actions into continuous activities, and the problem of mapping continuous input into discrete symbols for use in planning and modeling.

## Introduction

Planning is generally regarded as a symbolic task. An abstract model of the world is used as the starting point from which to construct a sequence of actions that will achieve some set of goals. Goals, actions, states, and the arguments they apply to, are all treated as discrete entities that can be given names for identification, representation, and reasoning.

In contrast, real robotic systems act over time in a complex, dynamic world. Sensors must be pointed and their data processed quickly and continuously for use as feedback to control motors and other actuators. Continuous control is the only way to actually get robots to do things.

This paper discusses the problems involved in connecting symbolic planning with continuous control. In particular, the issues involved in realizing discrete actions using fast concurrent behaviors are addressed. Task-specific, real-time perception is a fundamental part of these behaviors. While researchers have successfully used primitive touch and sonar sensors in such situations, it is more problematic to achieve reasonable performance with complex signals such as those from a video camera. Active vision routines are suggested as a means of incorporating visual data into real time control and as one mechanism for designating aspects of the world in an indexical-functional manner. Fast concurrent behaviors coupled with real-time indexical-functional designation bridge the gap between symbolic models and continuous action.

*This work has been supported in part by the Office of Naval Research under contract N0014-85-K-010.

## Discrete Reactive Execution

Symbols give a system the power to build models and keep track of facts about the world that cannot be directly sensed. For example, a model might contain assertions that `item-23` is a hamster, that it is colored white, and that it is in `box-3`. Should a planner ever need a hamster or a box, it can consult these assertions and construct an appropriate plan. Given the task "get a hamster", a robot could decide to `(goto box-3)` and `(pickup item-23)`.

Actually executing such a plan requires filling in a great deal of detail at execution time. There may be obstacles near the box, there may be things on top of the box, and there may even be more than one hamster in the box. In each of these cases, the agent must refine its plan to include steps to deal with the difficulties as they arise. This refinement is often called reactive planning, or reactive execution, and a large amount of work has been done in the area recently [Firby, 1989a, Schoppers, 1987]. While these approaches differ in philosophy and implementation, they all agree on one point: planning and reactive execution bottom out in a set of discrete, symbolic actions.

## Defining Symbolic Actions

Symbolic actions are intended to be the interface between discrete and continuous control. Unfortunately, the realization of discrete actions as real continuous processes is quite complex. In our example above, `(pickup item-23)` must be left as a single action because it must be realized in continuous motions and not a sequence of finer discrete tasks. Grabbing a moving hamster requires a sophisticated connection between tracking the hamster and moving the arm.

In fact, implementing symbolic actions involves two major problems in general: (1) Defining control algorithms that will implement the action on real hardware. (2) Mapping the symbolic arguments of the actions into sensory features that can be used easily by the control algorithms. A reactive planner, such as the RAP system [Firby, 1989a], which uses symbolic actions must address each of these problems explicitly. Rather than simply generate a symbolic primitive action with symbolic arguments, such a system must select a control algorithm and appropriate sensor features instead.

This paper discusses the issues involved in constructing control algorithms and sensing routines so that they can be combined into continuous activities that appear discrete to a symbolic planner.

## Situation Specific Control

Control is typically *situation specific*, by which we mean that input is not interpreted in complex ways. The relevance of any input to the system's actions is direct and unambiguous, being computed by a fixed function that maps inputs directly into outputs. The behavior of the control system does not depend on the context in which input is received but only on the input itself. Thus, control systems are usually designed to work in a single, well-defined situation.

Controllers are structured this way because they must be fast. To ensure rapid response they are specifically constructed to avoid relying on the recognition of abstract states that require any significant sensor data processing. Information must flow directly from input to output without dependence on a world model or past history.

The most successful examples of implemented situation specific control systems for real robots are organized around the idea of behaviors. Rather than generate a single function that maps all inputs into all outputs, the control system is broken down into relatively independent processes that monitor a set of inputs and produce a specific result. These results are then combined to drive the outputs in a coherent fashion.

Behavior-based control systems have been very successful on quite sophisticated robots. Brooks has made particular progress using his subsumption architecture [Brooks, 1986]. Using subsumption, each behavior is described by a finite automaton and the overall behavior of the system arises from the interaction of the automata with the environment and each other. Typical behaviors in a mobile robot might be "move away from obstacles", "move forward", and "turn towards the light". The more complex task of "move toward a light without hitting anything" is achieved by running these simple behaviors together.

This architecture has been used by Horswill to build a robot which follows arbitrary moving objects [Horswill and Brooks, 1988] and by Mataric to build a robot that does navigation and learning in hallways [Mataric, 1990]. Gat has used a similar notion of interacting simple behaviors to construct a robot that gathers up small toys [Gat, 1990] and Soldo uses interacting simple behaviors to navigate robustly in office hallways [Soldo, 1989].

The advantages of behavior-based situation specific control systems are:

- Bounded response time from change of input to change of output.
- Modular representation and construction.
- Robustness in the face of small uncertainties and perturbations in the world.

These advantages are gained by using specialized sensing strategies and a single control program. As a result, the system as a whole exhibits behavior that might be ascribed to a single plan.

Our goal is to construct a control system that retains the advantages of situation specific control but can be used to implement multiple, varied and changing plans. We wish to construct activities with the moment to moment characteristics of continuous control but which have emergent discrete properties over longer timescales.

### Situation Specific Activities

Using situation specific control systems for multiple plans in various contexts requires a shift in viewpoint. Rather than see the individual behaviors in the system as parts of an immutable whole, one can look at them as the building blocks for many different control systems that implement many different tasks. A symbolic action then becomes a particular set of primitive behaviors that can be activated together to achieve a specific goal in the outside world. Different actions activate different sets of behaviors.

Therefore, to define a symbolic action means to construct a set of behaviors that will predictably achieve a particular task. Sometimes these definitions will be easy because the desired goal will require only a single behavior. Usually, however, the symbolic actions useful in plans will be more complex and will require a whole set of behaviors for their definition. For example, an action like "go down the hallway" might require a behavior to move the robot forward, another to direct the robot away from obstacles, and yet another to keep the robot a constant distance from the right-hand wall. Executing "go down the hallway" as a single action means invoking all of these behaviors together. Once activated, their interaction with each other and the world will have the *emergent* property of satisfying the desired goal. Moreover, during the time it takes to move down the hallway, the control system doesn't change — the same behaviors are running doing the same thing all of the time. A short term stable situation has been generated. From the robot's point of view, a single control state persists over an interval of time and a discrete action has been carved out of the continuous world.

We define a situation specific activity as a collection of situation specific behaviors that work together over an interval of time to achieve a clear symbolic goal.

### Guarded Situation Specific Activities

Situation specific activities are not enough to define symbolic actions by themselves. A critical feature of symbolic actions is that they are discrete and finite. Each action must be implemented as a stable control state with a well-defined beginning and end. When a situation specific activity is used to define an action, it begins when it is invoked but the constituent behaviors will run happily forever.

To be useful, activities must be finite; they must signal when they have successfully completed or when they discern that they can no longer achieve their intended goal. When the activities involve hardware actions with well defined completions this requirement is fulfilled simply by having the hardware signal when the operations are complete. The end of a sonar reading or a camera motion is unambiguous and clear.

However, when the purpose of an activity is emergent, there is no clear sign that it has been achieved. Moving forward, avoiding obstacles, and staying a constant distance from the wall has the effect of moving an agent down the hallway, but when is that action complete? Completion may mean many things: having traveled a specific distance, having reached the end of the hall, or having passed three doorways. Regardless of which of these goals is desired, noticing that it has been achieved requires monitoring states in the world beyond those needed just to carry it out. These additional monitoring behaviors must be included in any situation specific activity that implements a symbolic action.

The activities that define actions must not just signal completion; they must also signal reliably when they fail. For example, suppose a robot is executing a set of behaviors designed to move it down a hall 10 ft. What happens if the hallway is only 5 ft. long? Presumably it will eventually come to a halt under the direction of its various obstacle avoidance behaviors but its completion monitor will not yet be satisfied. At that point something should happen or the robot will just sit there forever, secure in the fact that its situation specific behaviors will move it reliably down the hall, not realizing that they have been stymied by the context the robot unwittingly finds itself in.

The solution to this problem is to augment the behaviors implementing the action to include monitor behaviors for failure states as well as for success states. This procedure is mandatory for every activity that is to define a reliable symbolic action. In the hallway, this may involve something simple like signaling that the robot has been stationary for 10 seconds, or something more complicated like signaling that the robot moving back and forth to go around the obstacle making up the end of the hall but is making no measurable progress. The important point is that the activity must *always* signal that it is wedged.[1] Only with such assurances can the planning system trust that its symbolic actions will interact with the world in a sensible fashion.

We define a guarded situation specific activity to be a set of situation specific behaviors that together achieve a particular goal and always signal when the goal has been achieved, or when it has become impossible. As the symbolic system knows which behaviors it has activated to signal success, and which it has activated to signal failure, it can tell whether an activity has completed normally or been interrupted simply by knowing which behavior is signaling. There is no need for the control system to understand the implication or purpose of its constituent behaviors in order for the symbol system to understand what has occurred.

## Situation Specific Sensing

Developing guarded situation specific activities is only half the story in defining realizable symbolic actions. Authors of control systems for real robots routinely make the claim that their systems work because sensing is restricted to precisely those features of the world that are relevant to the task at hand [Horswill and Brooks, 1988, Soldo, 1989]. Underlying this claim is the realization that processing all available sensor data is both too slow to support control and unnecessary in most cases. Sensing strategies can and should be specialized to the particular needs of the situation specific behaviors that use them. Building a world model, or matching sensor readings against an existing world model, is typically not a bounded time procedure and would endanger behavior response time.

However, the interpretation of many forms of sensor data depends heavily on context. In particular, computer vision algorithms have typically required a significant application of previous knowledge, expectations about what is being viewed, and complex processing directed by this knowledge and these expectations. This would suggest that vision cannot be used as a reliable sensor in behavior-based control systems. However, recent research in *active vision* has suggested that, in fact, it can.

### Active Vision

Active vision algorithms are designed specifically to be part of situation specific activities [Aloimonos *et al.*, 1988, Ballard, 1989, Swain and Ballard, 1991]. They can take advantage of the constraints provided by the motion of the observer, and are designed to extract precisely the knowledge that is required for the activity to function successfully, within the time constraints required by the agent and its environment.

For example, Ballard and Ozcandarli [Ballard, 1988] showed how an observer that maintains its gaze on an object in the world while moving perpendicular to the line of site can easily tell what was in front and behind the plane of fixation. The algorithm requires only local image computations and therefore can be executed in real-time. Similarly, Nelson [Nelson, 1989] showed that divergence (expansion) of image flow vectors can be used to detect objects that the agent is in danger of colliding with. These computations do not involve the

[1] This requirement may seem more stringent than it actually is; taking longer than expected to complete an activity can always be used as a default signal that things are going wrong.

intermediate step of calculating the three-dimensional position or relative velocity of the objects.

Active visual routines share many similarities with Ullman's idea of visual routines [Ullman, 1984] (see also [Chapman, 1990]), which have been proposed as computational models of human intermediate visual processing. Ullman's visual routines are a set of basic task-specific operations functioning on low-level representations generated from the image array which can, in theory, be composed to extract an unbounded variety of shape properties and spatial relations.

The data generated by active vision routines can be used directly by behaviors without complex, context dependent interpretation. For example, the center of an object being tracked can be used as a position for the robot to travel towards. The resulting emergent behavior has the robot follow the object even though following is not explicitly represented and neither recognition nor interpretation of the object is required.

### Extracting Functional Attributes

Attributes of an object like its center, direction, or speed are important because they are useful in many different situations for accomplishing many different tasks. For example, the direction to an object can be used by behaviors designed to approach the object, to avoid the object, and to point in the object's direction. As long as the direction is known, these activities can proceed. It is the direction attribute that has a functional use in the behaviors, not the object itself or the way that the attribute is extracted.

Functional attributes are the features of the world that need to be computed in real time for use in feedback loops. Each attribute that a robot can extract, corresponds to a particular interpretation of its sensor data that is relevant to some activity that it might pursue. When behavioral control researchers suggest adding a new sensor when a new behavior is to be encoded, what they really mean is to add a new real-time functional attribute extractor. Active vision routines are a particularly useful idea in this respect because they are designed to extract different attributes from the world using the same sensor.

In fact, there will often be many different active vision routines for extracting the same functional attribute from the world. This allows an agent substantial leeway to instantiate its activities in different ways under different circumstances using different active vision routines. For example, tracking the location of an obstacle might be done using color, shape, or motion, depending on what is most reliable in the specific context.

### Indexical-Functional Designation

Guarded situation specific activities define emergent symbolic actions that can be carried out in the world. Active vision routines compute the functional attributes of items in the world that are used within activities. Thus, if we think of an activity as describing a symbolic action, we can think of active vision routines as describing the arguments to that action.

Consider our robot's desire to fetch a hamster. The robot moves itself into position so that it can reach into the box containing the hamster, and it refines its plan to the action `(pickup item-23)` where `item-23` is the hamster. To instantiate this action, the robot looks up its guarded activity description for `pickup` and finds behaviors to move its hand to a location $x$, to warn if location $x$ is out of reach, to signal when motion is complete, *etc.* These behaviors all refer to a location but do not define it.

The robot now examines its world model and determines that `item-23`, the hamster, is white and the box holding it is brown. Therefore, from the many possible active vision routines for tracking the hamster, the robot chooses a routine to track the position of a colored object (in this case white). The output of this tracking routine designates the position of the hamster. Putting these all together, the robot sets $x$ to be the output of the a sensing behavior that generates constant updates of the hamster's location and activates action behaviors that move its hand toward that location. With luck, the hand will eventually end up at the hamster and the pickup will succeed.

[Agre and Chapman, 1987] describe attributes of the world like the hamster's position as functional-indexical references. The robot's choice of an active vision routine to extract an attribute from the world is precisely the generation of a functional-indexical designation for the symbolic description of that attribute in the robot's world model. The active vision routine operationalizes exactly what is needed to reliably refer to and use an abstract symbolic property of the world. The hamster becomes "the-position-I-am-tracking" and the pickup activity takes "the-position-I-am-tracking" as its argument.

## Previous Work in Robotic Control

Several researchers have designed architectures for robot control, but few bridge the gap between symbolic and continuous control. If they do, they do not consider the problems of interacting with the world with high-bandwidth sensors that require context dependent interpretation.

Brooks's subsumption architecture [Brooks, 1986], Maes's competence modules [Maes, 1989], and Kaelbling's Gapps architecture [Kaelbling, 1988] contain no symbolic plan or model, and therefore are restricted in the tasks they can carry out and the flexibility of their responses. On the other hand, Robo-Soar [Laird, 1990] contains no continuous control, and often has lethargic, coarse-grained responses to events in the world. Theo [Mitchell, 1990] is an example of one system that does attempt to bridge the gap, using a set of control rules to define the continuous system augmented by a planner, which is invoked when there is no appropriate rule. However, the simplicity of Theo's sensors (a pair of sonar depth sensors) and actions (rotate a sonar, move the robot) make the discrete/continuous interface relatively straightforward.

Dean's recent survey of control theory and AI planning techniques for robot control shows a clear dichotomy between techniques for continuous and symbolic control [Dean, 1991]. We believe that the dichotomy is not a historical accident, but that it is a result of differing requirements of speed and flexibility at different levels of representation. Only by building an architecture that spans the gap will intelligent yet nimble robots be built.

## Summary and Future Work

This paper proposes a model of symbolic action that includes a realistic translation between symbolic and continuous control regimes. Symbolic actions map into guarded situation specific activities which taken together have the emergent property of achieving the desired state in the world. The behaviors making up the activity refer to the world, not directly through sensors, but rather through functional-indexical attributes that can be computed quickly and easily.

The arguments to actions map into active perception strategies, such as active vision routines [2], designed to compute functional-indexical attributes in real time. The choice of both the activity for the action and the sensing strategies for designating its arguments are up to the robot's symbolic or reactive execution system. As long as the execution system has a reasonable idea of the context surrounding the agent, it will be able to assemble situation specific strategies that will achieve appropriate results.

Adopting the approach to planning and control advocated in this paper emphasizes research areas at the boundary between symbolic and continuous action that have traditionally been ignored in both the planning and vision communities. We must define situation specific activities and functional-indexical strategies that can be used together to achieve semantically meaningful discrete goals in the world. In other words, we must define a reasonable set of primitive actions for real life — a language for describing the steps we take to do ordinary things.

We believe that the ideas we have outlined in this paper provide a framework for developing such a language. Our current work is aimed at defining the activities needed to move a simple robot around our laboratory and search for specific items using visual feedback. Defining such a language will bring us significantly closer to understanding adaptable, reliable behavior, and building useful robots.

[2] Or any other fast method for extracting specific interpretations of data. As we mentioned, touch sensors and sonars compute some attributes of the world directly.

## References


[Agre and Chapman, 1987] Philip E. Agre and David Chapman. Pengi: An implementation of a theory of activity. In *AAAI '87*.

[Aloimonos *et al.*, 1988] Yiannis Aloimonos, Isaac Weiss, and Amit Bandyopadhay. Active vision. *IJCV*, 1: 333–356, 1988.

[Ballard, 1989] Dana H. Ballard. Animate Vision. *Artificial Intelligence*, 48: 57–86, 1991.

[Ballard, 1988] Dana H. Ballard and A. Ozcandarli. Eye Fixation and early vision: Kinetic Depth. In *ICCV '88*.

[Brooks, 1986] Rodney A. Brooks. A robust layered control system for a mobile robot. *IEEE Journal of Robotics and Automation*, RA-2(1), 1986.

[Chapman, 1990] David Chapman. Intermediate vision: Architecture, implementation, and use. Technical Report TR-90-06, Teleos Research, 1990.

[Dean, 1991] Tom Dean. Control theory and artificial intelligence. Manuscript in preparation, January 1991.

[Firby, 1989a] R. James Firby. Adaptive execution in complex dynamic worlds. Technical Report YALEU/CSD/RR #672, Computer Science Department, Yale University, January 1989.

[Firby, 1989b] R. James Firby. Task-directed sensing. In *Sensor Fusion II: Human and Machine Strategies*, Philadelphia, PA, November 1989. SPIE.

[Gat, 1990] Erann Gat. BDL: A language for programming reactive robotic control systems. Unpubl. doc., 1990.

[Horswill and Brooks, 1988] Ian Douglas Horswill and Rodney Allen Brooks. Situated vision in a dynamic world: Chasing objects. In *AAAI '88*, 796–800.

[Kaelbling, 1988] Leslie Pack Kaelbling. Goals as parallel program specifications. In *AAAI '88*, 60–65.

[Laird, 1990] John E. Laird. Integrating planning and execution in Soar. In *AAAI Spring Symposium Series: Planning in Uncertain, Unpredictable, or Changing Environments*, 1990.

[Lawton *et al.*, 1990] Daryl Lawton, Ron Arkin, and Jonathan Cameron. Qualitative spacial understanding and reactive control for autonomous robots. In *IROS 90*, Tsuchiwra, Japan, June 1990.

[Maes, 1989] Pattie Maes. The dynamics of action selection. In *IJCAI '89*, 991–997.

[Mataric, 1990] Maja J. Mataric. *A Model for Distributed Mobile Robot Environment Learning and Navigation.* PhD thesis, Department of EE & CS, MIT, 1990.

[Mitchell, 1990] Tom M. Mitchell. Becoming increasingly reactive. In *Proceedings of AAAI '90*, 1051–1058.

[Nelson, 1989] R. C. Nelson and J. Aloimonos. Obstacle avoidance using flow field divergence, *IEEE PAMI*, 11: 1102–1106, 1989.

[Schoppers, 1987] M. Schoppers. Universal plans for reactive robots in unpredictable environments. In IJCAI '87.

[Soldo, 1989] Monnett Hanvey Soldo. Fusion without representation. In *SPIE Vol. 1198 Sensor Fusion II: Human and Machine Strategies*, 1989.

[Swain and Ballard, 1991] Michael J. Swain and Dana H. Ballard. Color indexing. *International Journal of Computer Vision*, to appear 1991.

[Ullman, 1984] Shimon Ullman. Visual routines. *Cognition*, 18:97–159, 1984.

# Modeling human memory retrieval and computer information retrieval: What can the two fields learn from each other?

**Peter W. Foltz**[1]

Department of Psychology
University of Colorado
Boulder, CO, 80309
pfoltz@boulder.colorado.edu

**Abstract**

Models of human memory and computer information retrieval have many similarities in the methods they use for representing and accessing information. This article examines the methods and representations used in both human memory modeling and computer information retrieval and discusses similarities and differences From these similarities and differences, the features that lead to successful retrieval in both human memory and computer information retrieval domains can be determined. An analysis of these features can then help in the future design of both human and computer retrieval models.

## Introduction

A recent estimate puts the amount of information accumulated by an adult over a lifetime to be on the order of $10^9$ bits (Landauer, 1986). Modern computers are also able to store about this same amount of information. While this great capacity permits humans and computers to store much information, one of the keys to the flexibility of both the human brain and the computer is the ability to retrieve the relevant information quickly and accurately. Without fast access to information, both humans and computers could not make the almost instantaneous actions for which they are both known. Thus, a primary issue in both computer and human information retrieval is how to access the correct information in an efficient manner.

Human and computer memories differ in many ways. While human memory is neurally based and has evolved over millions of years, computer memory is silicon based and has existed for only half a century. The approaches to the study of the two also differ. In human memory retrieval, the primary approach has been to predict memory retrieval phenomena through development of psychological models of memory and to test these models via laboratory experiments. However, in computer retrieval, rather than concerns of trying to explain how the retrieval works, the primary issues involve designing the most efficient manner to store and retrieve the information.

While both fields have different goals and work on different types of storage, they have similarities in the methods used for representing and accessing information. This article examines methods and representations used in both human memory modeling and computer information retrieval From their similarities and differences, the features that lead to successful retrieval in both domains can be determined. An analysis of these features can aid in the design of both human and computer retrieval models. This article is a summary of findings from Foltz (1991).

## Stages of Retrieval

The retrieval of information from both human memory and computers occurs in three stages: generating the retrieval cues, using the cues to retrieve information, and verifying that the retrieved information is what is desired. These stages differ in the type of cognitive activity used. To retrieve information, cues first need to be generated. This process is strategic, involving controlled processing. A person develops cues that describe the information desired based on the current context. The generated cues are then used for the retrieval. In the case of information retrieval, the cues are transmitted to the computer, while in human retrieval, the cues are used automatically. In both cases, the actual retrieval process is automatic, i.e. not under a person's strategic control. Once information has been returned from the retrieval process, it is again under the control of a strategic process which evaluates the information to determine if it is what was desired. The retrieval of information may not be just a single cycle of these three stages, but may involve several iterations. Information retrieved in a previous iteration may be added as additional cues for the next retrieval. In many human and computer retrieval models, the automatic and the strategic components are often treated separately. In this article, the models are organized based on whether an automatic or strategic component of retrieval is being examined.

## Psychological approaches to retrieval

Human memory models are developed with the primary goal of trying to explain how our memory works. The approach tries to answer questions about how we store, retrieve and represent information, and why we may fail at retrieval. While there have been many models of memory retrieval, this section focuses on three different classes of models of automatic retrieval from long term memory.

[1]This research was supported by the Army Research Institute, project MDA 903-86-CO143. The author thanks Walter Kintsch, Anders Ericsson, Reid Hastie, Gerhard Fischer, Peter Polson, Adrienne Lee, Susan Dumais, and Thomas Landauer for helpful discussions and comments on versions of this paper.

### Automatic memory models

**Compound cue.** The compound cue model of memory (Raaijmakers & Shiffrin, 1981; Ratcliff & McKoon, 1988), assumes associative connections between retrieval cues and items in memory. The models serve as a general theory of retrieval from long-term memory and has been used to predict such phenomena as the recall of word lists and the range and decay of priming. In the model, each item in memory has a probability of being sampled that is dependent on the associative strength of that item to a set of probe cues in comparison to the strengths of all other items to the same set of probe cues. While cues are connected to memory items, there are no connections between the items in memory. Thus, connections between memory items are only mediated through connections from two memory items to the same cue. Retrieval of information is treated as an iterative process, adding retrieved memory items as new retrieval cues until a certain number of failures to retrieve any new items occurs. Thus, retrieval in the compound cue model is entirely dependent on the cues used.

**Spreading Activation.** The concept of spreading activation has been widely used as a search mechanism of semantic networks (e.g. Anderson, 1983). Spreading activation models of retrieval use a semantic network of interconnected memory items, with connections representing semantic relationships. As in the compound cue model, initial items are activated by the cues given. The activation of these items then spreads along the connections to activate other associated items. The amount of spread from one item to another is typically based on the amount of activation received by the first item, the number of connections out of the first item and the connection strength between the items.

In terms of retrieval, compound cue and spreading activation models differ primarily on the range of activated nodes. In compound cue retrieval, the only activated nodes would be those directly associated with the cues. In spreading activation, retrieval is not only cue-dependent, but also dependent on the connections between memory items.

**Distributed models.** Distributed models of memory (Murdock, 1983; Hintzman, 1986), differ from the previous models in terms of their representation of information in memory. In these models, information is represented as a vector of features rather than discrete nodes in which a concept is stored. In retrieval, a vector representing the cues and context is compared to the stored vectors. The result represents the best match between the cues and the stored information while also capturing some of the similarities between the information that has been previously stored. In this manner, given a set of features, the retrieval process returns items that best match the set of features, although the items are slightly abstracted from when they were encoded due to their associations to other memory items.

One primary difference from the previous models is that the distributed models match based on features of the item and on the associations made with other encoded items. While the compound cue and spreading activation models have associations between similar items, they do not directly match based on the encoded features. However, the models all encode information items using some form of associations between the items and other information encoded with them and therefore all exhibit context dependence. They also all display cue-dependence in that the retrieval of information is primarily dependent on the cues given, although it can be mediated by the patterns of associations of other stored information.

### Strategic retrieval

While the models described above address retrieval issues once the cues have been provided, they do not address many of the issues of human strategic control of retrieval. When we need to retrieve information, we must decide what are the best cues to use for the retrieval and evaluate what is returned to determine if it is what we want. In a protocol study of people recalling the names of their high school classmates, Williams (1978) identified some of the strategies used in retrieval. He found that people initially identified a context in which to search, then searched their memory, and then verified what was retrieved. The process was applied recursively, with people using the whole cycle to retrieve contexts for another retrieval. The Williams study highlights the need for an appropriate context for retrieval. Context effects have been widely researched in connection with Tulving's encoding specificity principle (Tulving & Thompson, 1973). Thus, information can only be retrieved given the appropriate context with which it was stored.

In addition to context, encoding strategies also play a major role in retrieval. Chase and Ericsson's (1981) studies on expert memory show that retrieval is not only dependent on effective retrieval strategies, but also on effective encoding strategies. Thus, specific strategies are developed for both encoding and retrieving information, with these strategies being highly person and domain specific.

The strategic aspects of retrieval described in this section point to the fact that retrieval is not just the process of searching a series of memory items given certain cues to see which items best match the cues. Instead, a context for the retrieval must be established. Individual cues, such as the few words used in memory models for cued recall, are seldom the only items used for retrieval. A person not only uses the word cue that was provided, but also the additional context of the situation when the original word was encoded for the retrieval. As information is retrieved, it too serves as a context for retrieving additional information. Specific strategies are similarly used when initially encoding the information, thereby facilitating the use of context to help retrieve it.

## Computer information retrieval

The goal of Information Retrieval (IR) is to retrieve what is wanted while minimizing the retrieval of undesired information. There are again, the two interacting sub-processes of retrieval. There is a controlled process in which humans must decide what cues to provide to the IR system and evaluate what is returned. Once they have decided on the cues to use, they provide the cues to the automatic process of the system to retrieve the information.

### Automatic retrieval models

As in human memory retrieval, retrieval of an item is dependent on the similarity between that item and the cues given. The variety of information retrieval models represent different methods of calculating and representing these similarities in order to maximize the effectiveness of the retrieval given the tasks and environment. In the retrieval of textual information, each document is treated as a set of features, in which each feature corresponds to a term used in the document. A standard method of retrieval is to represent each term as a vector with each vector element representing whether a particular document contains the term. Given a query consisting of terms, the best matching documents can be retrieved through Boolean operations.

In the vector space model (Salton & McGill, 1983), each document is represented as a vector of terms in a high dimensional space based on the indexing terms. Any query can also be represented as a vector in the space. The similarity between a query vector and document vectors can then be measured by the distance between the two vectors. Like the distributed human memory models, this similarity is calculated using the sum of the products of the vectors.

Another retrieval method used is probabilistic retrieval (Bookstein & Swanson, 1975). Documents are ranked based on two probabilistic relations between terms and documents, the probability that a given term occurs in the relevant set of documents and the probability that a term occurs in the non-relevant set of documents. By estimating probabilities from a collection of documents, the probability of the relevance of each document given the terms can be calculated.

The fact that many IR methods require using the exact words from the document to retrieve it highlights one of the deficiencies in current techniques. People seldom know which words will describe a document and there is a great variability in the choice of words between people. Thus, keyword matching can fail due to *polysemy* (multiple meanings for a word) and *synonymy* (multiple ways of refering to one concept).

While the models described above are based on relations between individual documents and the cues given to retrieve them, some retrieval models are more *network* based, in that they emphasize the connections between the documents. In this way, the documents form a semantic network of information in which documents on similar topics tend to cluster. A variety of methods have been used for creating network representations using the terms to determine feature similarity such as clustering and factor analytic methods. Retrieval of information can then be done by returning information occuring in regions of the semantic space. The problems of exact word usage are avoided since semantically similar documents can be retrieved together.

### Strategic retrieval

In retrieval it is often not clear to the user how or what can be retrieved. Users may not know which terms to use due to problems of synonymy and because they are not familiar with what type of information they can retrieve from the database. There are also problems with the actual interaction with the system; users may not know how to form a query or use the query language. Thus, a user interacting with a retrieval system may need to use some conscious strategies. To ease these strategic problems, IR uses methods for interpreting what a user wants and ways of letting the user browse through the information.

In relevance feedback, (Salton & Buckley, 1990) the IR system returns a list of documents after an initial query. A user can then rate the relevance of the retrieved documents. This information is then used to perform additional retrievals. Since the new query contains more refined information about what the user considers relevant, it tends to return more relevant items. Users thus, are no longer dependent on providing accurate terms for their queries, but can just indicate documents that appear relevant to their interests. This reduces the effort of having to know what is in the database and allows users to have a lot more control of the iterative search process.

Relevance feedback is also similar to information browsers and hyptertexts. Information browsers use a set of rich connections between documents to allow a user to navigate through the space of information. Through navigation, a user can see the relationships between items in the semantic network and move to areas where the most relevant items are located. Thus, as in human retrieval, strategies can be used for finding the desired information.

## Hybrid models

With the many similarities between human retrieval and computer information retrieval, there are some models and systems that have incorporated features of both. These models could be considered *hybrids* in that they do information retrieval, but apply psychological models of retrieval. These models cross the boundaries of being just an information retrieval or a human memory model, but use features from both fields in order to be an effective retrieval system. They therefore illustrate the complementary nature of human and computer retrieval models.

One of the primary ideas from the human memory literature that has been used in information retrieval is the concept of differential associative connections between items of information. For this reason, there have been a variety of retrieval models using spreading activation or connectionist models (Jones, 1986; Rose & Belew, 1989). In these systems, a user can activate certain terms for their query. Activation then flows from these terms to the documents and other terms until the network settles. The highest activated documents would then be retrieved.

Some information retrieval systems incorporate psychological models of users' retrieval strategies. One such system was Williams' (1984) RABBIT system, based on his research on memory retrieval. strategies. With RABBIT, people used an iterative process of giving a partial description of what they wanted, retrieved a general context and then used the information to narrow the cues to get the information. Thus, the system allowed users to do computer retrieval using familiar memory retrieval strategies.

A very different hybrid approach is Anderson's (1990) rational model of human retrieval, which combines features of both information retrieval with the constraints of human memory. It serves as a model of human retrieval, although it is built using features of a probabilistic model of library borrowing. Incorporating history of usage and context factors, his model uses some of the environmental features that have typically not been considered in other psychological models. This differs from the other hybrid models, which applied psychological principles to design an information retrieval model because it applies some of the principles from information retrieval to help constrain the design of a psychological model of human retrieval.

## Conclusions

### Models and representation of information

The retrieval models described in this article use a limited number of methods for creating the connections between cues and memory items, matching the information and representing the memory items. These features are outlined for some of the models in Table 1.

For memory items to be retrievable, there must be connections from the items to the all the cues that could be used to retrieve it. As seen in Table 1, the origin of these connections for all of the models derive from either similarity of the features in the cues and items or based on the co-occurrence of cues and items during encoding. In this manner, the connections in all models are based on temporal or featural co-occurence. Once these connections are created, the models use matching methods to retrieve items given cues. In some models, the matching methods use the connections directly as the retrieval method (e.g. compound cue model). However in most models, the retrieval is also based on what other information is encoded in memory. In spreading activation, retrieval is based both on the cue connections and the inter-item connections. Similarly in distributed models with a composite associative trace, retrieval is based not only on feature similarity between a cue and information items, but also on relationships between items added to the composite associative trace.

Based on the models discussed, there are two ways of representing the memory information, as a network, or as a set of features. In a network representation, the information is represented as discrete items. These items are connected to each other through representing activations between the items. In the feature representation, information is represented as a set of features rather than discrete elements. Rather than representing item A as strongly connected to B and weakly connected to C, it could be represented as sharing many features of B and few features of C. In this manner, connections between information in network representations are typically made through examining feature overlap. Therefore a network representation can be derived from a feature representation, as is typically done for spreading activation models.

### Strategic retrieval

Humans are experts at using strategies to store and retrieve information from their own memory. They are familiar with the structure of the information stored and the retrieval cues that can be used since they did the initial encoding. This is not the case in computer retrieval. Users are seldom familiar with what information is available and how it is organized. This unfamiliarity hinders their ability to develop good retrieval cues to give to the system. Users are also not familiar with the ways of specifying the cues. Since most retrieval systems are term based, the exact terms must be specified to get the desired information.

Information retrieval systems are currently adding features to aid strategic retrieval. Iterative retrieval allows users to refine their query interactively as they become more familiar with the structure and semantics of the stored information. Relevance feedback similarly lets users narrow the context to where they want to search by providing additional context cues for the search. Browsing also permits users to navigate through a set of information, typically staying within the same context. Thus, a user is in control of the search process, making the decisions on which direction to go to find a relevant piece of information.

### Implications for the two fields

The similarities between human and computer retrieval indicate the complementary nature of the two systems. Information retrieval systems function as extensions of our own memory. For this reason, there should be some similarities in how information retrieval systems are used and designed. One such feature is to allow users to employ familiar retrieval strategies when using the system. As described above, some systems have implemented these strategic aids. Nevertheless, computer retrieval systems do not make the large number of associations between information items as do humans. This is due to the fact that systems have minimal encoding skills and must rely on term based encoding as opposed to the human's semantic encoding. The systems also do not use much context in retrieval. Typically, retrieval is performed using just a set of words provided by a user, whereas human retrieval uses many additional contextual cues such as temporal and locational information. Thus, improvement in information retrieval models can be made through tailoring the systems to incorporate semantic relationships in encoding and to use contextual information for retrieval such as user profiles. Psychological models of memory and of retrieval strategies highlight the current abilities of the human retrieval system and can provide directions for information retrieval systems to augment the human's ability to find information.

Conversely, human memory retrieval can learn from the insights into computer information retrieval. Information retrieval researchers have developed efficent systems, taking into account such features as the frequency of occurrence of information, associative and feature based representations, and human retrieval strategies. These models have become fairly similar to those used in psychological memory modeling. This suggests that both fields are searching for analogous solutions. Developments of efficient information retrieval models can provide guidance for psychologists as to types of representations and algorithms to be used to model efficient memory retrieval. The improvements in retrieval when using relevance feedback suggests that the automatic process of retrieval need not be entirely efficient for human memory. With strategic intervention of relevance feedback, a query can be quckly narrowed down to the relevant information. This may be the case too in human memory. The actual automatic retrieval process need not be entirely efficient, since we have strategic skills for developing the context and revising the cues to narrow down on the desired information. There are few memory models that incorporate both the automatic retrieval component and the wealth of strategies used to develop and revise the cues.

In information retrieval, one of the key methods to developing a retrieval system is based on knowledge of task and environmental factors. As suggested by Anderson's (1990) work and from the ecological approaches to human memory, human memory models in the future may examine more of what types of tasks human memory is designed to do, and what are some of the environmental constraints in which the memory must operate.

The insights from the two fields can provide guidelines to aid development of better retrieval models. These models will more likely fall under the category of hybrids incorporating the best features and constraints of the two types of retrieval. In the long run, these insights can provide both a better understanding of how our own retrieval system works and how to develop better external retrieval systems.

## References


Anderson, J. R. 1983. *The architecture of cognition.* Cambridge, MA: Harvard University Press.

Anderson, J. R. 1990. *The adaptive character of thought.* Hillsdale, NJ: Erlbaum.

Bookstein, A., & Swanson, D. R. 1975. Probabilistic models for automatic indexing. *Journal of the American Society for Information Science, 26,* 1 45-50.

Chase, W. G., & Ericsson, K. A. 1981. Skilled memory. In J. R. Anderson (Ed.), *Cognitive skills and their acquisition.* Hillsdale, NJ: Erlbaum.

Foltz, P. W. 1991. Models of human memory and computer information retrieval: Similar approaches to similar problems. Tech Report 91-03, Institute of Cognitive Science, Univ. Of Colorado, Boulder. (Submitted for publication)

Hintzman, D. 1986. "Schema abstraction" in a multiple-trace memory model. *Psychological Review, 93,* 411-428.

Jones, W. P. 1986. On the applied use of computer models of memory: The Memory Extender personal filing system. *Intl. Journal of Man-Machine Studies,* 25, 191-228.

Landauer, T. K. 1986. How much do people remember? Some estimates of the quantity of learned information in long-term memory. *Cognitive Science, 10,* 477-493.

Murdock, B. B. 1983. A distributed memory model for serial-order information. *Psychological Review, 90,* 316-338.

Raaijmakers, J. G. W., & Shiffrin, R. M. 1981. Search of associative memory. *Psychological Review, 88,* 93-134.

Ratcliff, R., & McKoon, G. 1988. A retrieval theory of priming memory. *Psychological Review. 95,* 385-408.

Rose, D. E., & Belew, R. K. 1989. Legal information retrieval: A hybrid approach. *Proceedings of the International Conference on AI & Law,* Vancouver, BC.

Salton, G., & Buckley, C. 1990. Improving retrieval performance by relevance feedback. *Journal of the American Society for Information Science. 41,* 4, 288-297.

Salton, G., & McGill, M. J. 1983. *Introduction to Modern Information Retrieval.* McGraw-Hill: NY.

Tulving, E., & Thompson, D. M. 1973. Encoding specificity and retrieval processes in episodic memory, *Psychological Review, 80,* 352-373.

Williams, M. D. 1978. The process of retrieval from very long term memory. Tech report 7801. Center for Human Information Processing, Univ. California, San Diego.

Williams, M. D. 1984. What makes RABBIT run? *Intl. Journal of Man-Machine Studies, 21,* 333-352.


| Retrieval Model | Example | Domain | Matching Method | Origin of connections | Representation |
|---|---|---|---|---|---|
| Compound Cue | SAM Gillund & Shiffrin (1984) | Human | Activation of items associated with cues | Cues to item strength based on STM co-occurrence | Network |
| Distributed | TODAM Murdock (1982) | Human | Correlation of item feature vectors to cue vector | Feature similarity | Feature |
| Spreading Activation | ACT* Anderson (1983) | Human | Activation of items spread from cues and other items | Based on degree of association between items | Network |
| Rational Memory model | Anderson (1990) | Human | Based on probabilities of history and context | Based on probabilities of history and context | Network |
| Inverted index | Salton & McGill (1983) | IR | Boolean operations on feature vectors | Feature similarity (term overlap) | Feature |
| Vector Space | SMART Salton & McGill (1983) | IR | Distance between item and cue vectors | Feature similarity (term overlap) | Feature |
| Probabilistic | Bookstein & Swanson (1975) | IR | Prob. a cue is in a relevant item vs. prob. in a non relevant item | Feature similarity (term overlap) | Feature |
| Spreading Activation | Memory Extender Jones (1986) | IR | Activation of cues spreads to cues and other items | Semantic network based on feature (term) similarity | Network/ Feature |
| Connectionist | Rose & Belew (1989) | IR | Activation of cues spreads to other cues and items | Feature (term) similarity and user feedback | Network/ Feature |

Table 1. Representations and methods used for human and computer retrieval models.

# Tabletop: An Emergent, Stochastic Model of Analogy-Making

Robert M. French and Douglas R. Hofstadter
Center for Research on Concepts and Cognition
Indiana University
510 North Fess
Bloomington, Indiana 47408
e-mail: french@cogsci.indiana.edu, dughof@cogsci.indiana.edu

## Abstract

This paper describes Tabletop, a computer program that models human analogy-making in a micro-world consisting of a small table covered with ordinary table objects. We argue for the necessity, even in this simple domain, of an architecture that builds its own representations by means of a continual interaction between an associative network of fixed concepts (the *Slipnet*) and simple low-level perceptual agents (*codelets*), that relies on local processing and (simulated) parallelism, and that is fundamentally stochastic. Several problems solved by the Tabletop program are used to illustrate these principles.

## Introduction

Tabletop, the program described in this paper, is a computer model of analogy-making. Unlike many analogy programs, it does not attempt to discover analogies between political situations, concepts in science, or plots in literature. Rather, it operates in a microdomain, the Tableworld, consisting of ordinary table objects on an ordinary table. Imagine there are two people, Henry and Eliza, seated at a table facing each other. They play the following game: Henry touches some object on the table and says to Eliza, "Do this!" Eliza must respond by doing "the same thing". In other words, she must find the object that, from her vantage point, seems to correspond most closely to the object that Henry touched. In a sense, this is analogy-making at its most prosaic. And yet, the psychological mechanisms that underlie human analogy-making in this microdomain are, we believe, of the same kind and complexity as in any real-world situation.

## Competing pressures

A human Eliza intuitively takes into account a number of factors when determining her choice. These factors include: the positions of the objects on the table, the category to which they belong, their size, and their functional association with other objects (for example, cups and saucers are related in a way that cups and plates are not). Each of these factors exerts some pressure on her decision. Some pressures, such as the ones just mentioned (except for position), are relatively context-independent. Other pressures are evoked by the situation itself and cannot reasonably be anticipated either by a human Eliza or by a program simulating Eliza. Rather, they emerge gradually as the person or program develops a representation of the configuration of objects on the table.

Consider the three table configurations in Figure 1. In the simplest situation (Figure 1*a*), the two competing pressures are category membership and position on the table. In this situation, the chances are good that Eliza would touch the cup in the diagonally opposite corner of the table. But when both the touched cup and the glass directly opposite it are surrounded by several objects, this changes the pressures (Figure 1*b*). The similar groups of objects around the touched cup and the glass generate contextual pressures that make it considerably more likely that Eliza will touch the glass instead of the cup. This probability is strongly enhanced by the fact that a glass is conceptually similar to a cup. Now what if we modified these pressures further, by replacing the glass by a saucer? Would this shift the balance of pressures to favor the isolated cup? Probably, but not necessarily; this is because even though cups and saucers, strictly speaking, do not belong to the same category, they are conceptually related as "objects that are used together", in much the same way that knives and forks are normally thought of as being used together. Finally, what if the saucer were replaced by a salt shaker (Figure 1*c*)? This time the pressures seem sufficiently in favor of the isolated cup that Eliza would probably touch it.

Tabletop is an attempt to model the way humans respond to these (and other) pressures in the Tableworld. We have argued in detail elsewhere [Chalmers, French, & Hofstadter 1991] that a satisfactory computer model of analogy-making must be able to build its own representation of the situation that it is faced with. We have also argued against the possibility of a separate "representation module" that would act as a preprocessor to an "analogy-making module". In accordance with these precepts, the processes of representation-building

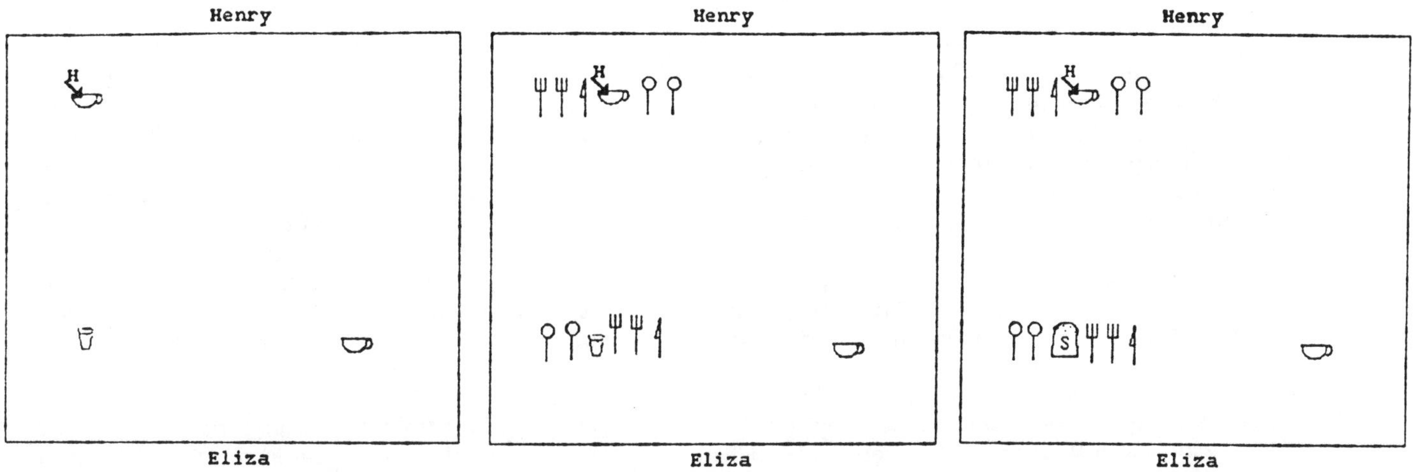


Figure 1*a* Figure 1*b* Figure 1*c*
Various competing pressures in three table configurations

and correspondence-finding are completely integrated in Tabletop. This represents a significant departure from current analogy-making programs [e.g., Burstein & Adelson 1987; Kedar-Cabelli 1988; Falkenhainer et al. 1989; Thagard et al. 1991] with the exception of Copycat [Mitchell 1990; Hofstadter & Mitchell 1991].

We will begin our description of Tabletop with a brief discussion of its architecture. We then explain why we believe this type of architecture to be essential to models of analogy-making. We conclude with several runs of the program to illustrate how it develops representations of situations and uses them to make analogies. We argue that preprogramming explicit mechanisms to handle the kinds of difficulties posed by these situations is psychologically implausible. Rather, implicit pressures should emerge that accomplish the same things as explicit mechanisms. Our architecture allows this to happen.

## High-level perception and the architecture of Tabletop

We define high-level perception as that stage of perceptual processing where concepts begin to play a significant role. High-level perception ranges from our ability to recognize objects to our capacity to grasp grasp relationships among objects and, ultimately, includes our ability to understand entire complex situations, such as a love affair or a war [Hofstadter 1985; Chalmers, French, & Hofstadter 1991]. Tabletop simulates this range of high-level perception in a number of ways, including how it scans the table looking for objects and groups of objects, how it discovers relationships among objects and, finally, how it gradually builds up a coherent view of a complete situation.

Three central features of the Tabletop architecture provide the basis for its high-level perceptual abilities. Before illustrating several runs of the program we will briefly focus our discussion on these features, which are:

- continual interaction between the associative concept network (the Slipnet)and many low-level perceptual agents (codelets) whose job it is to examine and structure the Tableworld;
- local processing and (simulated) parallelism;
- stochastic mechanisms.

There are other features — most significantly, computational temperature mechanisms [Mitchell & Hofstadter 1989] — that distinguish Tabletop from other models of analogy-making. However, in this paper we will restrict ourselves to a discussion of the three features above.

### *Interaction between the Slipnet and codelets*

Tabletop, like Copycat [Hofstadter 1984; Mitchell 1990; Hofstadter & Mitchell 1991] before it, has an associative concept network, called the *Slipnet*, and a *Workspace* filled with simple perceptual agents, called codelets, responsible for observing the Tableworld and building the structures needed to understand it. The division of the program into these two aspects reflects the traditional philosophical distinction between types and tokens: the Slipnet contains Platonic concepts, while the Workspace contains instances of those concepts, which codelets observe and manipulate. Examples of codelets include agents that look for groups of objects on the table; agents that look for neighbors of a particular object; agents that, given a particular group of objects, look for the same type of group elsewhere on the table; and so on. These perceptual agents are low-level observers and builders in the sense that they do not have a global view of the table at any time.

The Slipnet and the Workspace continually interact with one another. The Slipnet is reminiscent of a typical spreading-activation concept network [Collins & Loftus 1975] with a number of differences, possibly the most significant being that the distances between concepts in the Slipnet vary according to pressures evoked by the situation (and conveyed to the Slipnet by codelets). In a reciprocal manner, the numbers and types of active perceptual agents are governed by the activations of concepts in the Slipnet.

### *Local processing and (simulated) parallelism*

At no time does any codelet have a global view of the table. At the start of a run, the program knows only which object was touched by Henry and where on the table that object is located. The initial codelets that run may be thought of as scouts examining the table for information (e.g., finding an object in a particular location, discovering the neighbor of a particular object, etc.) and looking for various structures (e.g., groups of objects) that might be useful to its subsequent understanding of the Tableworld. Depending on their relation to the touched object, certain areas of the table are given preference and will generally be examined first. This can be thought of as a visual scan of the table, a scan with top-down control, one in which the salience of objects preferentially determines which objects will be attended to first, and, significantly, one in which some objects, or even groups of objects, might sometimes be altogether ignored.

Consider, for example, *group-finding* scout codelets. Suppose that the object touched is in the left-hand corner of the far side of the table: then these codelets will preferentially (but not deterministically) look in the corner of the table diagonally opposite the touched object to determine if there are any groups of objects there. This bias is built into the program deliberately in order to imitate typical human behavior.

It is important to note that group-finding codelets, like all codelets, neither have an overview of the table that would allow them to immediately spot all of the groups of objects, nor do they conduct a systematic search for groups, say from the upper left to the lower right-hand corner of the table. Instead, they search the table probabilistically. Codelets are biased to look at different parts of the table with different probabilities. Factors influencing the probability of an area being looked at include the presence of objects known to be conceptually similar to the touched object and being diagonally or directly opposite the touched object.

One might object that such a probabilistic search technique might allow the program to overlook some structure that could be useful later. This is true, but there is a trade-off — namely, it allows the program to focus its resources on avenues that seem most promising. The ultimate reason for this strategy is to make the program psychologically realistic; in particular, its mechanisms should remain valid in scaled-up situations. True real-world situations are generally far more complex than these simple caricatural situations, and brute-force search techniques are entirely inappropriate. There have to be strong pressures that serve to focus resources on promising avenues. Any probabilistic search technique is a risk-taking strategy and as such, sacrifices guaranteed success for greater efficiency.

Active concepts in the Slipnet dispatch agents to look for instances of themselves. For example, if the "group" concept is active, codelets will be dispatched that search for groups in the Tableworld. If one of these codelets succeeds in finding a group, this causes more activation to be sent to the concept "group" in the Slipnet. This amounts to a signal to the system to hunt for even more groups; thus more codelets are sent out to look for groups. However, as the rate of group-finding falls (i.e., the rate of change of activation of "group" decreases), proportionately fewer codelets are sent out to look for groups. Eventually, the program concludes that there are probably no more groups to be found on the table.

This process is rather like a person who wants to pick all of the apples from a tree. Initially, it is easy to see the apples. She will look first in the most salient places on the tree (e.g., on the outer, leafy branches). The sight of apples remaining in the tree will encourage her to look for more. As the number of apples in the tree gets smaller, she will start to look in less likely places (e.g., on the lower branches, closer to the main trunk). Gradually, as fewer and fewer apples remain, it becomes harder to determine if there are any left at all. Ultimately, there will come a point when she will peer into the tree several times and decide that she has gotten all the apples. But did she *really* get them all? Probably, but not necessarily. Ultimately, though, the energy (to say nothing of the eye-strain) involved in attempting to find any additional apples becomes prohibitive, and she stops looking.

The group-finding codelets find groups on the table in a similar way. It is true that some groups may be missed, but, on the other hand, the program does not expend needless energy covering every last square centimeter of the table to make sure that every single group is found, even those that have very little chance of playing any role in a potential analogy. In such a local approach, as more group-finding codelets fail to find groups, fewer of these codelets run, until finally none run at all. The program has then "decided" — without any given codelet seeing the whole table, any more than the person could take in the entire apple tree at one time — that it has found all of the groups on the table, or at least all groups worth looking for.

This might seem somewhat strange at first blush. Why not allow the program to search the table *systematically* to make sure that it has gotten *all* instances of groups of objects? The answer is that such a brute-force technique would lead to unacceptable and psychologically unrealistic performance in scaled-up situations. Tabletop's strategy of having a multitude of small processes running in parallel provides an efficient focusing mechanism that allows the program's attention to be probabilistically shifted to promising areas of exploration.

A fundamental assumption behind Tabletop is that high-level cognitive processes are the emergent result of myriad low-level, quasi-independent processes. It is our belief that it is the interaction of these low-level processes that give cognition its flexibility. It is therefore important to attempt, insofar as possible, to model analogy-making by relying on low-level processes to gradually build up the high-level structures necessary to solve an analogy problem.

In Figure 2, we give some examples of Tabletop configurations where a particular structure — in this case, a diagonal correspondence — unexpectedly blocks a very reasonable answer. There are, however, no special "blockage-checking mechanisms" built into Tabletop to deal with such situations. It would be unreasonable to have a mechanism that invariably looked for blockages of this type. This illustrates a general principle about Tabletop — namely, the lack of special-purpose mechanisms. Building in a raft of diverse special-purpose mechanisms all of which invariably get "wheeled out" for every situation not only is psychologically implausible but also poses severe problems of scaling-up. Rather than routinely invoking special-purpose mechanisms, Tabletop evokes context-dependent pressures that accomplish the same results but without brute-force techniques.

### *Justification for stochastic mechanisms*

One of the most surprising features of the Tabletop architecture is that, at all levels, almost all of the processes are stochastic. Hofstadter and Mitchell have argued for the necessity of stochastic mechanisms [Hofstadter & Mitchell 1991] in emergent architectures modeling cognitive processes. Two significant advantages of stochastic mechanisms for exploration and structure-building are the following:

- they allow, on average, promising avenues to be explored before less promising ones;
- they also, on occasion, allow improbable paths to be explored, whereas in normal heuristic search, these paths might never be chosen at all.

In the language of traditional artificial intelligence, the stochastic mechanisms in Tabletop are reminiscent of probabilistic heuristic search. In deterministic heuristic search, if the program must chose between action A, which will give an estimated result of 60, and action B, which will give 40, the program will invariably select action A. In probabilistic heuristic search, by contrast, A would be selected 60 percent of the time, B 40 percent.

Why is this sensible? Shouldn't the program *always* explore the most promising paths first? The answer is: no, sometimes it should not. Tabletop problems in particular, and analogy problems in general, have the property that often several solutions of different levels of quality exist. Pathways that don't look promising at the outset may conceal very high-quality answers, and conversely, pathways that look very promising at the outset may lead to mediocre answers. A strategy that always pursued the most promising avenues until it found a solution would therefore miss any hidden high-quality solutions (if they existed). On the other hand, it would not be wise to totally ignore the estimated promise of pathways. Standard AI strategies would deterministically explore a vast number of possible pathways, leading to many, if not all, answers, and then would select the best answer. This type of strategy is appealing; however, once again, it poses severe scaling-up problems in real-world situations. Tabletop's strategy is poised between these two deterministic types of strategy — it will *sometimes* choose less promising routes, but the less promising the route, the less often it will be chosen. This probabilistic technique leaves all routes at least theoretically open and so, on some runs of the program, the hidden gems (if they exist) will be uncovered, although on most runs the most obvious solutions will be chosen. This non-deterministic strategyavoids any combinatorial explosion while at the same time avoiding the trap of always following the most obvious pathways. Note that in the microdomain we could have opted for a brute-force solution; however, since our research aims for psychological realism, we eschewed this technique.

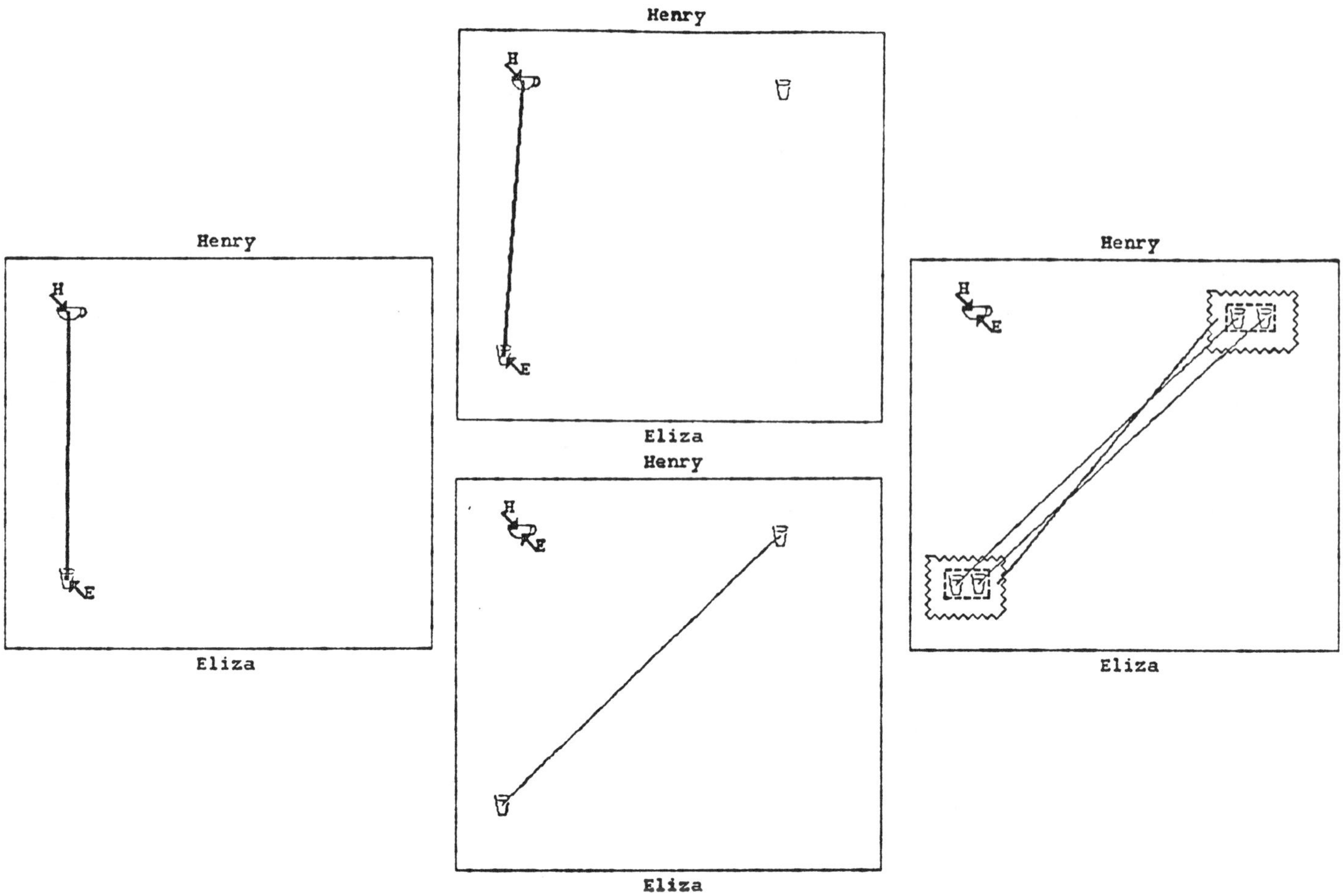


Figure 2*a* Figure 2*b*, parts *i* and *ii* Figure 2*c*
Three Tabletop problems (note the two solutions to the second problem)

## Three analogy problems illustrating these mechanisms

In the second section of this paper we discussed some of the competing pressures in the Tabletop world that are involved in deciding which table objects and relationships among objects are examined and which structures are built in order to arrive at an answer to the "Do this!" challenge. These pressures include object category; whether or not an object is in a group; what the neighbors of an object are; whether or not the object is on the edge of a group; whether there are other similar or identical objects elsewhere on the table, especially if they are in salient positions with respect to the touched object (in particular, diagonally or directly opposite it); whether an object or group of objects is close to a particularly salientobject or group of objects; and so on. These groupings, descriptions, and relations are discovered only after the program has begun scanning the table. Descriptions (neighbors, in a group or not, etc.) are gradually attached to the various objects, groups of objects, and correspondences.

In order for the program to give an answer, it needs to have built up one or more mappings between objects or groups of objects across the table (called *correspondences*). However, the erection of such structures is not separated in time from the build-up of other representations (descriptions, groupings, etc.); rather, it is part and parcel of a single high-level perceptual process, in which correspondences can become elements of descriptions of objects and, conversely, descriptions become the underpinning of new correspondences. This fusion of representation-building with correspondence-building marks a fundamental difference with many other current analogy-making programs.

In Figure 2 we illustrate three separate Tabletop problems. In all three problems, Tabletop is capable of producing more than one distinct answer, but in the first and third problems, there is one answer that swamps the others in frequency, so it is almost as if there is just one answer. However, in the second problem, there are two distinct answers both of which are produced with fairly high frequency. These three problems illustrate how an unanticipated pressure can emerge that will significantly bias the program's responses.

In the first problem, the program, like people, will usually touch the glass opposite the cup that Henry touched. This is quite straightforward.

In the second problem, a single glass is added in the upper right-hand corner of the table. Even though one would not *a priori* expect it to affect the description of any of the objects (e.g., it is not the neighbor of any object; it is not part of a group of objects; it is not in a particularly salient location on the table with respect to the touched object, etc.), its presence nonetheless significantly biases the choice of the second object to be touched. As the program runs, it sometimes builds a correspondence between the two glasses (Figure 2*b*, part *ii*). When it does so, the new pressure that this creates will often be enough to cause Eliza to touch the very cup that Henry touched. Of course, had the second glass not been on the table, such a choice would seem rather strange.

In the final example, there are two glasses in the lower left-hand corner and two in the upper right-hand corner of the table. The correspondence between these two *groups* of glasses is very salient and very strong, usually strong enough to cause the program to touch the cup that Henry touched.

Thirty-one human subjects were given these three problems and, like the program, in the second configuration, more subjects touched the cup in the upper left-hand corner than in the first configuration (40% vs. 33%); in the final configuration, twice as many touched the touched cup (66% vs. 33%).

## Conclusion

We have described a microdomain for computer analogy-making consisting of ordinary objects on an ordinary table. This world has a decidedly more "real-world" feel to it than the only other program of its kind, Copycat [Hofstadter & Mitchell 1991; Mitchell 1990]. We argue that it is necessary, even in this simple domain, to use a model that builds its own representations by means of a continual interaction between an associative network of fixed concepts and simple low-level perceptual agents, that relies on local processing and (simulated) parallelism, and that is fundamentally stochastic.

## Acknowledgments

Thanks to David Chalmers for his helpful comments on this paper.

## Bibliography

Anderson, J. R. (1983), *The architecture of cognition.* Cambridge, MA: Harvard University Press.

Burstein, M. & Adelson, B. (1987). Mapping and integrating partial mental models. In *Proceedings of the ninth annual conference of the Cognitive Science Society.* Hillsdale, NJ: Lawrence Erlbaum Associates.

Chalmers, D. J., French, R. M., & Hofstadter, D. R. (1991). *High-level perception, representation, and analogy: A critique of artificial intelligence methodology.* CRCC Technical Report 49-1990, Center for Research on Concepts and Cognition, Bloomington, Indiana. Also submitted to *Journal of Experimental and Theoretical Artificial Intelligence.*

Collins, A. M. & Loftus, E. F. (1975). A spreading activation theory of semantic memory. *Psychological Review,* **82** 407–428.

Falkenhainer. B., Forbus, K. D., & Gentner, D. (1989). The structure-mapping engine. *Artificial Intelligence,* **41** (1), 1–63.

Kedar-Cabelli, S. (1988). Towards a computational model of purpose-directed analogy. In A. Prieditis (ed.), *Analogica.* Los Altos, CA: Morgan Kaufmann.

Hofstadter, D. R. (1984). *The Copycat project: An experiment in nondeterminism and creative analogies.* AI Memo No. 755, Massachusetts Institute of Technology, Cambridge, MA.

Hofstadter, D. R. (1985). Analogies and roles in human and machine thinking. In *Metamagical themas,* 547–603. New York: Basic Books.

Hofstadter, D. R. & Mitchell, M. (1991). *An overview of the Copycat project.* CRCC Technical Report 51-1991, Center for Research on Concepts and Cognition, Bloomington, Indiana.

Mitchell, M. (1990). *Copycat: A computer model of high-level perception and conceptual slippage in analogy-making.* Ph.D. dissertation, University of Michigan, Ann Arbor, Michigan.

Thagard, P., Holyoak, K. J., Nelson, G., & Gochfeld, D. (1991) (in press). Analog retrieval by constraint satisfaction. *Artificial Intelligence.*

# Can double dissociation uncover the modularity of cognitive processes?

**Giorgio Ganis**
Department of Cognitive Science
University of California at San Diego
9500 Gilman Dr.
92093-0515, La Jolla, CA.

**Nick Chater**
Department of Psychology
University of Edinburgh
7, George Square,
Edinburgh, EH8 9JZ, U.K

## Abstract

Neuropsychological evidence has proved influential both in testing pre-existing cognitive theories and in developing new accounts. It has been argued that dissociations, and, in particular, double dissociation are particularly valuable in developing new theoretical accounts, since they may reveal the gross structure or "modularity" of cognitive processes. In this paper, we show that even fully distributed systems -i.e. systems with no modularity - can give rise to double dissociations. We give the example of a recurrent neural network which draws loops and spirals which shows a double dissociation between the two tasks when lesioned. This result suggests that the observation of a double dissociation implies little about the modularity of the underlying system. In the final section we argue that a dual task technique can give additional hints about the structure of the underlying system because the class of distributed systems we describe are not able, in general, to perform two tasks at the same time. Finally, we argue that neurobiology has to be taken into account in order to interpret purely behavioral data.

## Introduction

For several centuries neurological patients has been used to inform and constrain psychological accounts of normal function. Such evidence has served both to test existing psychological theories and to suggest how new theories can be developed. The value of neuropsychological evidence in theory testing is at least relatively uncontroversial and has been carefully analyzed (Caramazza, 1986). Much of the current upsurge of interest in neuropsychology within cognitive psychology and cognitive science has, however, stemmed from the hope that studying impaired function can play a role in the construction of theories of normal function. For example, a central theme of Shallice's recent and important book (Shallice, 1988) is that cognitive neuropsychology can have an important and proactive input into building theories of the processes involved across the range of cognitive domains, from language, memory and thinking to perception and action. We share the hope that evidence from impaired function may be an important and much needed source of constraint on cognitive theory, but suspect that inferences from neuropsychological data should be used to guide the development of new theories only with considerable trepidation Other authors (Henderson 1981; Crowder 1982), express of similar sentiments, though for different reasons). In the context of theory testing, Caramazza (Caramazza 1986) has argued that dissociations and associations are equally important. In the context of theory construction, Shallice (Shallice 1988) has argued that dissociations, and, in particular, double dissociation are particularly valuable, since they may reveal the gross structure or "modularity" of cognitive processes. In this paper, we argue that inference from double dissociation to a particular modular structure of the underlying cognitive system is problematic because double dissociations can be observed in a fully distributed system - that is, a system which does not decompose into isolable subsystems.

Even if double dissociations *per se* are not, as we argue, a sure guide to the existence of separable underlying subsystems, it may be that for example, a dissociation between "phonological" and "lexical" reading strategies do indicate the existence of distinct phonological and lexical routes. The plausibility of the dual route versus a single route model can only be decided in the light of the relative merits of specific models attempting to account for the range of data from impaired and normal function (for example, Patterson, Seidenberg & McClelland1989; Coltheart 1990). - one piece of data for which reading models must account will be the dissociation of phonological and lexical reading. Our contention is not that double dissociations do not amount to interesting data for theory development and evaluation - rather we argue that they have no special status as a means of directly uncovering the modularity of the cognitive system.

## Single and double dissociations

The range of characterizations of the method of double dissociation (Teuber 1955; Kinsbourne 1971; Shallice 1988) makes exposition of the method difficult. We shall assume what we take to be a typical modern "functional" formulation.

A patient with a lesion exhibits a single dissociation between tasks I and II when performance on task I is very poor, whereas performance on task II is either close to or at a normal level, or at least very much better than performance in task I (Marin, Saffran & Schwarz 1976; Beauvois & Derouesne 1979; Shallice 1988). It was once thought that such dissociations allowed one to infer that the set of isolable processes underlying the two tasks must be different. However, it has been argued that this inference is not licensed, since task I may make greater demands on a single damaged subsystem(s) than does task II. A subsystem working at, say, 50% capacity might be adequate for task II, but not sufficient for task I. This is often referred to as the problem of resource artefacts (Shallice 1988; Dunn & Kirsner 1988). In response to such difficulties, it has been proposed that double rather than single dissociations are required to infer that two tasks draw on different processes, subsystems or modules. Tasks I and II doubly dissociate if there are patients A and B, such that A is more impaired than B in task I and, conversely, B is more impaired than A in task II. The point is that, unlike single dissociation, double dissociation cannot be generated with a resource artefact explanation. If task I makes greater demands on a single processing subsystem than task II, then task II may be selectively preserved (generating a single dissociation), but the reverse cannot occur. For if the subsystem is sufficiently impaired to damage task II, then task I, which relies on it even more heavily, will be even more severely impaired.

Original formulations of the inference from double dissociation (Teuber 1955; Kinsbourne 1971) assumed distinct and consistent lesion sites for patients with each kind of selective impairment. This anatomical assumption has been dropped in more recent "functional" formulations (Marin Saffran & Schwarz 1976; Shallice 1979; Shallice 1988). We argue that this less stringent criterion, although widely used (Caramazza 1990), may suggest an entirely misleading picture of the modularity of the underlying system.

## Double dissociations in distributed systems

Shallice (Shallice 1988) observes that "to make the inference [from observed double dissociation to separate underlying subsystems] valid, one would need to add the assumption that [double] dissociations do not arise from damage to non-modular systems" (Shallice, 1988:248). *Prima facie*, this claim runs counter to evidence for double dissociation in lesioned distributed neural networks (Wood 1978; Wood 1980; Wood 1982; Gordon 1982; these studies are based on the "brain state in a box" model of Anderson, Silverstein, Ritz & Jones 1977).

Wood specifies two patterns for the network to learn, which differ in activation at just two of the units. Selective ablation of each unit produces selective damage to the memory for each pattern. Thus the memory for the two patterns doubly dissociates, even though the memory for each pattern is distributed through entire set of network weights. Such examples rely on a close relationship between the structure of the task and particular units. Shallice argues that this may reduce their relevance to discussion of effects of damage on real neural networks which "...will be composed of millions of neurons... [of which] no individual neuron is likely to have much importance in determining what output occurs" (Shallice 1988:255). He thus adds two conditions that a lesion in a distributed system would have to meet to be threatening to the double dissociation inference: "First, before the lesion is made, the influence of any particular neuron on what output is produced should be small. Second, the neurons affected by the lesion should not be selected by some complex algorithm that is determined by the dissociation to be explained and that is not typical of those that arise naturally. It seems most unlikely that if these conditions are satisfied, a ...double dissociation could be demonstrated in a properly distributed memory system." (Shallice 1988:256).

While this claim is couched in terms of memory, the domain in which Wood's network showed a double dissociation, it is clearly intended to apply more generally. We now describe a fully distributed system which performs two tasks - drawing loops or spirals, and which is intended to satisfy Shallice's two conditions. We shall introduce the network in three stages. Firstly, we describe the non-linear equation which the network implements. Secondly, we give a simple "local" network implementation of the equation. Finally, a totally distributed version of this network is described, and the effect of damage to this network discussed.

### The equation

The equation that we chose is a simple iterative difference equation, a variant of the logistic difference equation, x(n+1) = lambda.x(n).(1-x(n)), used in classical population genetics (e.g., Maynard-Smith 1968) which was unexpectedly found to produce chaotic behavior (May 1976). Our equation is the delayed logistic map: x(n+1) = lambda.x(n).(1-x(n-1)) , which, when x(n+1) is plotted against x(n) for each n produces either spirals if lambda is less than 2 and loops if lambda is greater than 2 (fig 1).

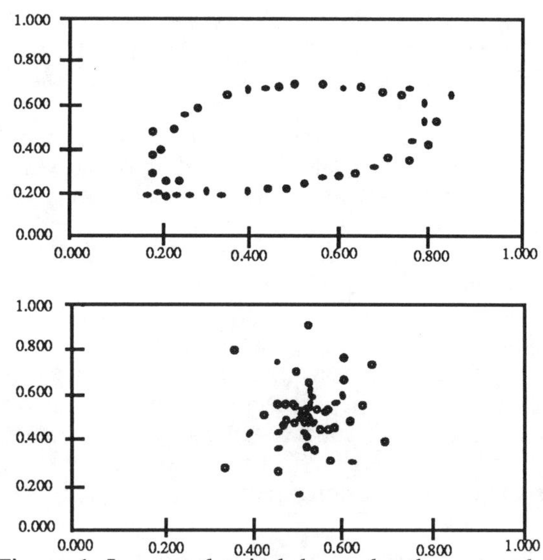


Figure 1. Loop and spiral drawn by the network (lambda equal to 1.925 and 2.075 respectively).

This critical value, at which the topology of the output is, in dynamical systems terminology, "structurally unstable", is know as the Niemark bifurcation. This equation is a standard example in non-linear dynamics (for example, Thompson & Stewart 1986). The parameter lambda, which determines whether a loop or a spiral is drawn, and the particular form of each, is what we shall term the "global" feedback parameter of our distributed network, and perturbation of this parameter due to damage can lead to the selective loss of either spiral or loop drawing.

### A local network implementation

A natural implementation of this equation in a simple neural network is shown in Figure 2.

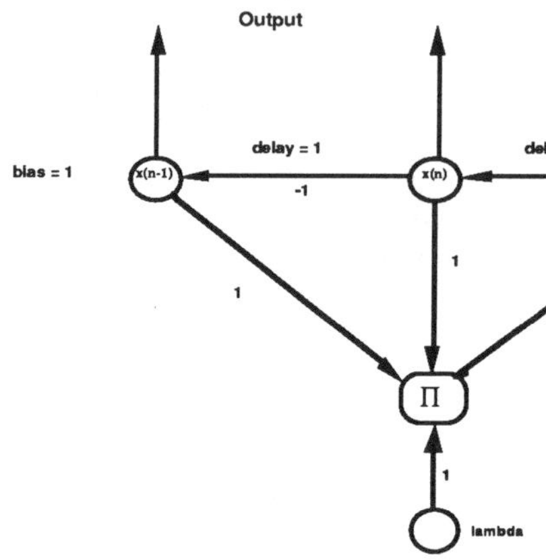


Figure 2. Local implementation of the delayed logistic function.

To draw a spiral a lambda of less than 2, say 1.95, is necessary. A lambda of 2.05 might be used to draw a loop.

The structure of the network mirrors the fact that the equation embodies both non-linearity and feedback. All units are linear, except for the single multiplicative unit (Hinton 1981), which takes the product of the value of the input, and the two previous values x(n-1) and x(n). The delay-lines feed the activation of the x(n) unit to the x(n-1) unit at the next time step, and the activation of the x(n+1) back to the x(n) unit. All other lines propagate immediately (with delay 0). The delay lines serve to "feed back" the output of the units at a given time-step as inputs at the next time-step. The combination of non-linearity and feedback is required for a system to exhibit a range of dynamically interesting behaviors, including chaotic behavior. It is not important here that we are using a discrete iterative map, rather than a system which continuously evolves in time according to a set of differential equations. The advantage of iterative maps, exploited both in the study of non-linear dynamics and in our illustrative example, is simplicity.

### A distributed network implementation

In the local implementation, the loss of any unit or link would lead to catastrophic failure. There are, of course, many ways in which this function could be distributed across a larger number of units - we choose one of the simplest. For each local element, a corresponding distributed network will have the following properties (Chater & Ganis 1991) :

i) Each single unit of the local network is replaced by a group of units in the distributed network. The mean activation value of all the units in a group corresponding to the activation of the single unit in the local implementation.

ii) Each single link between a pair of units is replaced family of links, such that there is a connection between each unit in the "source" group to each unit in the "target" group. Thus, each unit in the target group is not fed a substantial input from a single unit in the input group, but rather receives a small amount of input from every unit in the source group (fig3).

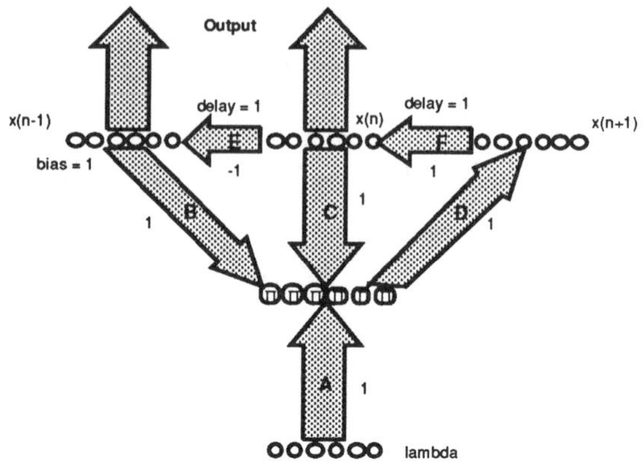


Figure 3. Distributed implementation of the delayed logistic function.

An average (Chater & Ganis 1991) of the values of these links corresponds to the value on the corresponding single link of the local implementation (there is the additional constraint that the values of the links from each unit are the same, to keep analysis tractable).

Thus, the distributed implementation (figure 2) simply reduplicates hardware so that the role of each unit is replaced by a set of units. The values of x(n-1), x(n) and x(n+1) are represented by the average of the activation levels of each group of unit, and the large arrows denote totally connected sets of links between groups of units with the ensemble averages (1 or -1) as shown.

By design, the distributed network behaves in the same way as the corresponding local network - the ensemble average values of groups of units and links correspond to the values of single units and links in the local implementation. If we consider the various ways in which the network may be impaired by ablation of a subset of a single group of links or set of units, then the damaged network will again be equivalent to a local network. Suppose that some proportion of a set of links with ensemble average 1 are ablated. If the links damaged are mostly positive, then the the ensemble average of the set will be reduced - the equivalent local network will have a single link with a value of, say, 0.8. On the other hand, if it happens that the links ablated are mostly negative then the ensemble average of the set (and hence the value on the corresponding link in the local implementation) may be increased - say to 1.2. The way in which positive (excitatory) and negative (inhibitory) connections are anatomically organised (whether, for example, they are separated or together), and depending on the typical size of the link values (e.g., whether the inputs to a typical unit are, say, 0.2, -0.2, 0.4, -0.1, 0.4, -0.2, 0.3,

0.2 (= 1) or 1, -5, 6, 2, 3, -3, 4, -7 (=1)) will have a dramatic impact on the distribution of changes in ensemble average that a lesion is likely to cause. We shall consider a full range of lesions at each of the links A, B, C, D, E & F and ablations of units in each set of the units. Notice that the ablation of a unit is equivalent to the removal of the connections which feed out of that unit. Thus unit ablation is simply a special case of the ablation of connections A-F, except for the ablation of the x(n) group, which feeds into both x(n-1) and the multiplicative units - thus the ablation of these units is equivalent to the ablation of subsets of two sets of links - C & E. Suppose that the value of a link is reduced/increased by a factor of mu (mu1 and mu2 in the C & E case). Then the equations governing the impaired system is no longer x(n+1) = lambda.x(n).(1-x(n-1)) but:

A, B, C, D, F: x(n+1) = [lambda*x(n)*(1-x(n-1)]*mu

E: x(n+1) = lambda * x(n) * (1-mu) * x(n-1)

C & E: x(n+1) = [lambda*x(n)*(1-mu2*x(n-1)]*mu

Thus the general form of the equations after damage is:

x(n+1) = [alpha * lambda] * x(n) * (1-beta * x(n-1)

The beta term, which differs from 1 only in two cases, serves only to slightly distort the spirals or loops drawn, rather than changing their underlying structure. In particular, it does not disturb the value of the Niemark bifurcatiom. So, if the feedback parameter lambda is less than 2, a spiral will be drawn; if lambda is greater than 2, a loop will be produced. Hence, damage to the distributed network has the same effect as setting lambda to lambda * mu, in the corresponding local network. Now it is clear how a double dissociation may arise.

Suppose spirals are drawn with lambda 1.95, and loops with lambda 2.05. Damage to the network which ablates more excitatory than inhibitory connections will reduce the amount of feedback, and mean that alpha is less than 1 - say 0.95. In this case, a lambda of 2.05 would correspond to lambda*0.95, that is 1.95 : the network will no longer be able to draw loops.

Notice that our example conforms with Shallice's strictures. Firstly, the influence of each neuron on task performance is small. Whereas in Wood's examples, particular neurons were especially significant for remembering certain patterns, in this example each neuron has the same influence every other neuron, in both loop and spiral drawing. Damage produces a dissociation by changing global system parameters rather selectively impairing particularly important individual units. Secondly, the kinds of damage that we have suggested do not involve any complex procedure for selecting which parts of the net should be ablated. Any kind of damage which alters the amount of feedback in the system, whether as a result of a chemical changes, loss of some external non-specific input or some other pathology, is liable to generate a dissociation of loop and spiral drawing.

## Conclusion

We have shown in this paper that fully distributed systems can generate double dissociations. The fact that one distributed system can produce a double dissociation does not necessarily mean that a large and neurally plausible class of distributed systems can do so, and only in the latter case will the inference from double dissociation to modularity be impugned in practice. Certainly neural systems do fall into the class of non-linear dynamical systems with feedback, and will exhibit far more interesting and elaborate dynamics than that generated by the logistic function. In particular, they may have a far more elaborate range of distinct structural configurations rather than just two, and the operative configuration is likely to be determined by a complex set of global parameters rather than the value of a single parameter, as here. Further, which of these structural configurations is operative may well importantly affect the task performed, and the relevant parameter values may be altered by a range of neurologically plausible forms of damage - chemical imbalances, localized lesions, loss of non-specific input from other centers, and so on. There is strong evidence that real neural networks can be multifunctional; this means that a single anatomically defined neural network can generate more than one behavior , depending on the value of one or more global parameters. 'Modulation of the network, synaptic, and cellular building blocks can serve to adapt the output pattern to ongoing needs or may dramatically reorganize a network into an entirely new mode mediating a different behavior' (Getting 1989). Therefore, there seems to be plenty of scope for double dissociations in real distributed neural systems. It may be countered that, whereas it is easy to see how loop drawing and spiral drawing may be products of the same system, with different global parameter values, it is less easy to conceive of how different global parameter values might transform a system from, say, phonological reading, to lexical reading. On the other hand, it is only easy to see how different parameter values of the same underlying system can produce loops and spirals in retrospect, and that our understanding of the properties of complex non-linear dynamical systems is too slight to put much weight on intuition. It may be that confidence in the inference from observed dissociation to underlying modular organization is based as much on our current lack of understanding of distributed systems, as on their underlying properties.

The question of how we can reduce the uncertainty of inferences from double dissociations obviously arises. We suggest two answers, one tied to our particular example, the other more general.

With regard to our specific example, we suggest that data from dual task techniques might be useful in interpreting double dissociations. Indeed, the class of multifunctional networks we have described is not able to perform two tasks at the same time (namely, drawing loops and spirals). Therefore, if there is a double dissociation between two tasks, and the subjects are able to perform both tasks at the same time, it seems unlikely that the underlying system be a multifunctional network like the one we put forward. In practice, results from dual task experiments may be difficult to interpret; for example, it is obvious that it is impossible to draw spirals and loops with the same hand at the same time; this, however, does not imply anything about the existence of one or more cognitive isolable subsystems for drawing spirals and loops.

With regard to the general problem of interpreting double dissociations, we think that the only possible way to reduce uncertainty is to take into account evolutionary constraints and advantages (Weiskrantz 1990) and data from neuroscience (Weiskrantz 1990; Sereno 1990). Indeed, we think that the purely 'functional approach', still dominant in cognitive neuropsychology, has led to '...a kind of candy floss neuropsychology, brightly labelled, complexly reticulated, full of growth but shifting in substance' (Weiskrantz 1990), the reason for this being the systematic overlooking of neurobiological data'.

## Acknowledgements

This work was carried out while the authors were at the Department of Psychology, University College London. We should like to thank Peter Passmore for discussion and help with maths packages, and thanks also to Margaret and Marty Sereno for comments on the manuscript.

## References

Anderson, J. A., Silverstein, J. W., Ritz, S.A. & Jones, R. S. 1977 . 'Distinctive features, categorical perception, and probability learning: Some applications of a neural model. '.*Psychological Review* 84:413-451.

Beauvois, M. F. & Derouesne, J. 1979 'Phonological alexia: Three dissociations', *Journal of Neurology, Neurosurgery and Psychiatry* 42:1115-1124.

Chater, N.; Ganis, G. 1991. Can double dissociations uncover the modulairy of the cognitive system? Forthcoming.

Caramazza, A. 1986 . On drawing inferences about the structure of normal cognitive systems from the analysis of patterns of impaired performance: The case for single patient studies. *Brain & Cognition* 5:41-66.

Coltheart, M. 1990. Computational models of the human reading system . Paper presented at the British Psychology Society: Cognitive Psychology Section Annual Conference, University of Leicester, September, 1990.

Crowder, R. G. 1982. General forgetting theory and the locus of amnesia . In L. S. Cermak (ed.), *Human memory and amnesia* Hillsdale, N.J.:Earlbaum.

Dunn, C.J. & Kirsner, K. 1988. Discovering functionally independent mental processes: the principle of reversed association. *Psychological Review*. 95:91-101.

Getting, P. A. 1989 Emerging principles governing the operation of neural networks. *Annual Review of Neuroscience* 12:185-204.

Gordon, B. 1982. Confrontion naming: Computational model and disconnection simulation In M. A. Arbib, D. Caplan and J. C. Marshall (eds.).*Neural models of language processes* New York : Academic Press.

Henderson, L. 1981. Information processing approaches to acquired dyslexia. *Quarterly Journal of Experimental Psychology* 35A:507- 522.

Hinton, G. E. 1981. A parallel computation that assigns canonical object-based frames of reference. In Proceedings of the 7th JointConference on Artificial Intelligence, 683-685.

Kinsbourne, M. 1971. Cognitive deficit: Experimental analysis. In J. L. McGaugh (ed.). *Psychobiology* New York, Academic Press.

Marin, O. S. M., Saffran, E. M. & Schwarz, D. F. 1976. ' Dissociation of language in aphasia: Implications for normal functions. *Annals of the New York Academy of Sciences*, 280:868-884.

May, R. 1976 . Simple mathematical models with very complicated dynamics . *Nature* 261:459-467.

Maynard Smith, J. 1986. *Mathematical ideas in biology* . Cambridge: CambridgeUniversity Press.

McClelland, J. L. 1986. A programmable blackboard model of reading. In J. L. McClelland and D. E. Rumelhart (eds.) *Parallel distributedprocessing: explorations in the microstructures of cognition. Volume 2: Psychological and biological models.* MIT Press, Cambridge, Mass..

Patterson, K. E., Seidenberg, M. S. & McClelland, J. L. 1989 . Connections and disconnections: acquired dyslexia in a computationalmodel of reading processes . In R. G. M. Morris (ed.) *Paralleldistributed processing: Implications for psychology and neurobiology.* Oxford:Oxford University Press.

Sereno, M.I. 1990. Language and the primate brain .In Proceedings of the 13th Annual Meeting of the Cognitive Science Society. Forthcoming.

Shallice, T. 1984 . More functionally isolable systems but fewer "modules" . *Cognition* 17:243-252.

Shallice, T. 1988. *From neuropsychology to mental structure* . Cambridge:Cambridge University Press.

Teuber, H.L. 1955. Physiological psychology. *Annual Review of Psychology.* 9:267-296.

Thompson, J. M. & Stewart, H. B. 1986. *Nonlinear dynamics and chaos* . Chichester, Wiley.

Weiskrantz, L. 1990. Multiple memory systems... *Proceedings of the Royal Society of London, Series B: Biological Sciences.* 1253:99-107.

Wood, C. C. 1978. Variations on a theme of Lashley: Lesion experiments on the neural network of Anderson, Silverstein, Ritz & Jones. *Psychological Review* 85:582-591.

Wood, C. C. 1980. Interpretation of real and simulated lesion experiments . *Psychological Review* 87:474-476.

Wood, C. C. 1982. Implications of simulated lesion experiments for the interpretation of lesions in real nervous systems . In M. A. Arbib, D. Caplan and J. C. Marshall (eds.), *Neural models of language processes.* New York:Academic Press.

# The Development of the Notion of Sameness: A Connectionist Model

**Michael Gasser and Linda B. Smith**
Indiana University

## Abstract

Comparison is of two types, the implicit sort that is behind all categorization and the explicit sort by which two object representations are compared in short-term memory. Children learn early on both to categorize and to compare explicitly, but they only learn to use dimensions in these processes considerably later. In this paper we present a connectionist model which brings together categorization and comparison, focusing on the development of the use of dimensions. The model posits (1) a general comparison mechanism which is blind to the nature of its inputs and (2) the sharing of internal object and dimension representations by categorization and comparison processes. Trained on the two processes, the system learns to use dimension inputs as filters on its representations for objects; it is these filtered representations which are matched in comparison. The model provides an account of the tendency for early comparison along one dimension to be disrupted by similarities along other dimensions and of the process by which the child might overcome this deficiency.

## Background

### Comparison and Cognition

Generalization from past to present experience involves a measure of the similarity of present perceptual input to what has been perceived before. The likelihood that we call some object a *dog* is a function of how similar that object is to other objects known to be *dogs*. But humans do more than categorize objects; we also compare objects along a wide array of perceptual dimensions. For example, we judge a dog to be the same color as our cat or to be large for dogs in general. Indeed, what we consider higher mental functioning—metaphor, poetry, science itself—involves pointing to and discovering novel kinds of similarity.

The problem of how a child develops a system of multiple kinds of perceptual similarity together with devices for linguistically communicating about similarity is clearly of great importance to cognitive science. This is an area in which there is rich and detailed data about human development but no current theory that adequately explains it.

In this paper we describe a connectionist model of some of the basic facts of comparison along perceptual dimensions. The workings of the model are based on the idea that categorization (*what color?*) and comparison (*same color?*) make use of the same dimension representations and the same internal representations for objects. We propose that these representations develop in response to the demands of the two tasks.

### Categorization and Comparison

**Categorization** involves comparing a stimulus in short-term memory to representations of previously encountered stimuli in long-term memory. A simple pattern associator performs this implicit form of comparison through the connection weights that make up its long-term memory. Categorization can be in terms of either complex categories such as DOG and CHAIR or dimensional attributes such as RED and BIG.[1]

But the implicit comparison between an item in short-term memory and long-term representations may be quite different from the comparison of two items in short-term memory. We shall call the latter **explicit comparison.** If we take the evidence from language seriously, this process goes on often in human cognition. A sentence such as *my ball is the same color as yours* requires speaker and hearer to maintain representations of both objects in short-term memory, where they can be compared. In this paper we will only be concerned with that subtype of explicit comparison which is signalled in English by the word *same* along with a perceptual dimension noun such as *size.*

In order for an abstract comparison device which looks for symmetry in its two input patterns to make judgements of "same thing," "same color," and "same size," it must have access to representations in which only the relevant dimension manifests itself. Irrelevant dimensions need somehow to be "filtered" out.

[1] We are concerned here with categorization in the sense of naming an object or an attribute and not in the sense of the underlying meaning of concepts.

## Developmental Facts

There is a well-documented trend in the development of object and dimensional comparisons and object and dimensional language. The following specific facts are those that we are interested in accommodating.

1. Early object categorizations are principally across all dimensions at once [Smith 1989a, 1989b]. The sensory features are somehow compressed into a single representation in which all constituents are weighted more or less equally.
2. The comparison of objects by overall similarity—the judgement that two identical cups are alike in the same way as two identical dogs—appears very early [Smith, 1984]. By 24 months, children comment on the similarity of objects through iterative naming, counting, and use of the plural [Sugarman, 1983]. The productive use of a form such as the plural at this age suggests that there is a comparison component that operates early and that is independent of specific perceptual properties.
3. The ability to make judgements of sameness along a single dimension—to know that two green objects are alike in the same way as two blue objects—develops later, after the acquisition of the words by which we talk about the perceptual properties of objects [Smith, 1984, 1989b].
4. Early judgements of sameness along a dimension appear to be contaminated by overall similarity. That is, 3- and 4-year-old children will call a big red square and a big orange square *the same size* but will refuse to call a big red square and a big blue square *the same size* [Kemler, 1982].

## The Model

We model these phenomena using a connectionist network which takes as inputs "pre-perceived" images and dimension words such as *size* and yields lexical outputs such as *big* and *same*. The architecture of the model is shown in Figure 1. Boxes denote banks of units and solid arrows complete connectivity between banks. The main features of the model are the following:

1. Global comparison and comparison along various dimensions are handled by the same subnetwork.
2. The dimension and the internal object representations are shared by the comparison and categorization components of the system.
3. Following training on categorization and comparison, input from dimension words "filters" out irrelevant dimensions in the representations of objects.

### Categorization

The CATEGORIZATION component, shown on the right side of Figure 1, is composed of a simple pattern associator. Input to this component comes in in the form of a "pre-perceived object" (hereafter "PPO"),

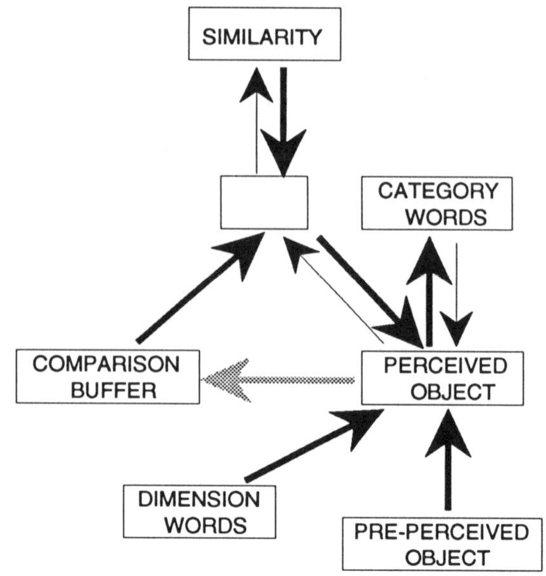


Figure 1: Architecture of the Model

corresponding to the output level in a theory such as Treisman and Gelade's [1980] (see also Smith [1989b]). The PPO has been segregated from background objects and contains information about its perceptual features. There are separate sets of units for each of several "pre-dimensions" at this level. However, the system does not recognize these dimensions as such; they cannot be used in making categorizations or comparisons along particular dimensions. The PERCEIVED OBJECT ("PO") layer corresponds to what is perceived or experienced. A pattern on this layer is a compression of the PPO pattern and may be influenced by input from other layers, in particular from the DIMENSION WORDS ("DW") layer. This influence takes the form of the focusing of attention on one or more dimensions. For example, the question *what color is it?* should cause color information to dominate the representation.

On the CATEGORY WORDS level, attribute words like *red* and *big* and complex category nouns such as *dog* and *chair* are each assigned a single unit. In categorization this layer is the output of the network; the system "sees" an object and names it or assigns it an attribute. While the network is designed to learn both complex categories such as DOG and attributes such as BIG, we will be concerned only with the latter. The CATEGORY WORDS layer may also function as an input layer (indicated by the thin arrow in the figure), e.g., in modeling the system's response to an utterance like *the marble is green.* In this case, it influences the pattern of activation on the PO layer.

During training and testing on categorization, the network is presented with a PPO on its input layer, consisting of a pattern of features on the various pre-dimensions. A single unit is also turned on on the DW layer, corresponding to a question about one dimension, e.g., *what color is the object?*. The network is trained using backpropagation [Rumelhart *et al.*, 1986]. In an effort to make the degree of supervision realistic, output targets are provided only for those units which are above a response threshold, unless no unit goes above the threshold, in which case a target (1.0) is provided for a single appropriate unit.

### Comparison

Alongside the CATEGORIZATION subnetwork, we propose a "dumb" COMPARISON component which does not know what objects are being compared or the dimensions on which they are to be compared. It simply compares patterns of activation. Selective attention—changes in the dimensions along which the comparison is to made—is accomplished by the same mechanisms that are involved in categorization by dimension. The overlap between these mechanisms is suggested by the fact that languages refer to dimension categorization (*what color?*) using the same nouns as are used for sameness comparison (*same color*).[2]

The COMPARISON component is shown on the left side of Figure 1. This is another pattern associator, similar to the symmetry network described by Rumelhart et al. [1986], with a hidden layer to handle inputs which are not linearly separable. The compared patterns appear on two input groups. One is just the PO layer, which also participates in categorization. The other is a short-term memory buffer which contains a copy of a recent pattern from the PO layer.

This component implements two sorts of processes. Run in one direction, it compares two input objects. Input from the PPO and DW layers produces a pattern on the PO layer. If a DW unit is on, the PO pattern is a "filtered" version of the object. This pattern is copied to the COMPARISON BUFFER layer, and a second PPO is fed to the network together with the same DW pattern. Finally the two (possibly filtered) object representations are compared at the SIMILARITY layer, which consists of a single unit.

Run in the other direction (the thick arrows in the figure), the network models responses to assertions about the sameness of objects (*my doll is the same as yours*). The input is a pattern on the COMPARISON BUFFER units representing one object, a pattern on the SIMILARITY unit representing sameness, and a pattern representing (possibly incomplete) knowledge about the second object on the PO layer. The output is an updated representation on the PO layer.

### Dimension as a Filter

The CATEGORIZATION and COMPARISON components share the PO representations, which are subject to the filtering effects of dimensional input. In order for the CATEGORIZATION subnetwork to succeed on a dimensional categorization task (*what color is it?*), the input from the DW layer should highlight the relevant dimension and attenuate the other dimensions to the extent that only the appropriate output unit (e.g., RED) reaches the response threshold. It is thus possible for the network to produce an appropriate response even with some contamination from irrelevant dimensions. That is, training on categorization may not result in DW-to-PO connection weights which completely eliminate irrelevant dimensions from a representation.

The comparison task is more demanding. Consider the case of two objects which are the same on the dimension in question but significantly different on all others. Any contamination from the irrelevant dimensions at all would adversely affect the output on the SIMILARITY unit.

How might the system's performance on the comparison task vary with time? We assume the COMPARISON component is first trained simply to detect similarity between pairs of input patterns. At this point, the system would be unable to make use of dimension information. Next, training on the categorization task would result in some filtering out of dimensions other than the one that is input from the DW layer. Now the network should also begin to be able to detect similarity between two objects along a given dimension. But, as in children, similarity judgements at this stage should still depend on the overall similarity between the objects. Training on the comparison task itself would then refine the behavior of the dimension filter. Given two objects and the assertion that they are the same on a given dimension, their filtered representations should be identical. Thus the filtered representation of the first could be used as a target for the filtered representation of the second. Together with continued training on the categorization task, this should result in an adequate representation of dimension.

## Experiment

We ran an experiment in which the same network was trained on categorization and comparison tasks. The procedures described below were repeated six times with different initial random connection weights.

### Categorization Task

A categorization network was first set up with random initial weights. The PPO layer consisted of 28 units, 7 for each of 4 simple linear "pre-dimensions". The PO layer consisted of 20 units; that is, there was some compression of the patterns from the input layer. The DW layer contained 3 units, one for each of the output dimensions, that is, those for which there were target categories. There were 9 CATEGORY WORDS units, one for each of the target categories.

The network was trained to perform dimension categorization on 2500 randomly generated "pre-

[2]It remains to be established whether children know the use of words such as *color* in comparison once they have acquired their use in categorization.

Table 1: "Same" and "Different" Pattern Distances

| | Same | Different |
|---|---|---|
| Before training | 1.464 | 0.966 |
| After categorization training | 1.172 | 1.211 |
| After comparison training | 0.314 | 0.519 |

perceived" input objects. Input objects were constrained in ways designed to model in a gross fashion the structure that is present in the world; the details need not concern us here. Also given to the network was input from a single unit in the DW layer. Thus the network's task corresponded to a question such as *what color is this object?*. Output targets were provided using the procedure described above. That is, targets depended on the system's own output, in ways that seem to correspond to what goes on in actual language acquisition contexts.

The performance of the network on categorization improved overall, as would be expected, though with the output-generated targets, improvement was not as smooth as it would have been with completely supervised learning. For most of the runs, the network succeeded in correctly categorizing at least 25 consecutive input objects by the end of the training.

The critical question, however, is how well the network learns to selectively attend to single dimensions. To determine this, we created a set of 45 test pattern pairs. These were of two types, those in which the objects were the same on the input dimension and different on the other three dimensions (hereafter referred to as the "same" pairs) and those in which the objects were different on the input dimension and the same on the other three dimensions (hereafter referred to as the "different" pairs). Testing the network consisted in running it with the objects in the test pairs as inputs and determining the Euclidian distance between the PO responses to the inputs for each pair. Of interest is the relative distance between the pairs. To the extent that the dimension input is behaving like a filter, as described above, the distance between the hidden-layer patterns for the "same" pairs should be smaller than that between the "different" pairs.

We made these comparisons before the network was trained, after the categorization training on 2500 inputs, and again following the second, comparison phase of training (described below) on 2500 additional inputs. Table 1 shows the results of the comparisons, averaged over the 6 runs.

The "same" pairs start out considerably further apart than the "different" ones because they differ on three out of four, vs. one out of four, input dimensions. The effect of training on the categorization task is to significantly ($p < .01$) diminish this difference, though the "same" pairs are only slightly closer than the "different" pairs. Although the dimension filter is not doing a very good job of eliminating irrelevant dimensions from the input, the network has learned to categorize quite well. Good categorization along dimensions and an inability to ignore overall similarity in comparing objects on one dimension are precisely the behaviors exhibited by 4- and 5-year-old children.

## Comparison Task

During the second phase of training, the network was trained on more categorization on half of the trials and on explicit comparison on the other half. The comparison task was designed to fit the real task in which there are two objects in front of the child and the adult says *X is the same color as Y*. This task conforms to running the network in the direction indicated by the thick arrows in the COMPARISON part of Figure 1. Given an input object which is red, big, round, and smooth, and another which is red, small, square, and rough,[3] the system was expected to use the information that the objects were the same color to help it later make sameness judgements of its own.

This task was implemented in the following manner. The input pattern for one object was presented to the PPO layer together with one lexical dimension on the DW layer, just as for the categorization task. The pattern this yielded on the PO layer was then saved. Next the second input object, identical to the first on the input dimension, was presented in the same way. Now the stored pattern was treated as a target for the PO layer. Note that the idea is not that the response to the first object is somehow superior to the response to the second, only that training in this manner should bring the patterns closer together. The important point is that an effective filtering mechanism is learned via the explicit comparison of objects along single dimensions.

We did not actually use the COMPARISON component for the implementation of this task. We assumed that the COMPARISON network, given a filtered representation of one object in the COMPARISON BUFFER and an indication of sameness to another object on the SIMILARITY unit, could generate its own internal target for the filtered representation of the second object by simply copying the pattern for the first object.

Results of comparisons between the "same" and "different" pairs following this phase are shown in Table 1. As predicted, the "same" distances have decreased significantly ($p < .01$) relative to the "different" distances. There is also a significant overall decrease in distances for both pairs. Because we trained the network only on comparison of objects that were meant to be the same, this is not surprising. Though it learned to treat some as more similar than others (those that are the same on the input dimension), in general it moved object representations closer together.

[3]Labels for the various dimensions are used for convenience only.

## Discussion

In our model dimension words such as *color* have the same internal representations whether they apply to categorization or to comparison of objects. The two tasks place similar demands on the dimension representations: within the distributed PO representations, features of the input dimension must be played up and features of other dimensions played down. As we have seen, however, the comparison task is more demanding in this regard. This aspect of our model, which fits with one current mathematical model of children's similarity judgements [Smith, 1989b], may help us understand why young children are able to categorize objects even seemingly by a single attribute long before they can make explicit comparisons along single dimensions.

Training on categorization is insufficient for the formation of an effective dimensional filter. Following categorization training only, the distance between pairs of dimension-filtered representations is about the same when the two objects are the same only on the input dimension as it is when they differ only on the input dimension. It is training on explicit comparison in our model that gives rise to effective dimension filters. The developmental implications of this finding are clear. Training children on the language of dimensional comparison may be a causal force in the emergence of the ability of children in the late preschool period to selectively attend to single dimensions.

One possible criticism of our experiment is the sequencing that we imposed on the learning. Comparison training began only after categorization was learned. This order fits the developmental facts [Macnamara, 1982]. Nonetheless, determining whether (and how) our results depend on the sequencing of training will be an important aspect of future research.

This research makes three contributions. First, it provides a model of one of the major trends in human development, from wholistic object comparison to dimensional comparisons. Second, the model distinguishes categorization and comparison in ways which clarify the theoretical issues and suggest new experiments. For example, it suggests that children's early use of plural and iterative naming may depend on global similarity, in addition to category identity, a hypothesis that could be tested empirically. Third, our model may bring insights to connectionist modeling of cognitive development. For example, the idea of using one internal representation as a target for another may be applicable generally when there is reason to posit representations that are shared by processes which constrain them in different ways.

## Conclusions

The central problem in understanding development is understanding how new behaviors emerge. The inherent difficulty of this problem has led much of the best work in cognitive development to be essentially adevelopmental. The dominant empirical strategy consists of describing behavior at different developmental points. We know for example that 5-month-olds can discriminate colors, that 2-year-olds have difficulty learning color words relative to other words, that 5-year-olds have a rudimentary mapping of color words to the color space, and that adults exhibit sophisticated and highly structured color concepts. But we do not know how the abilities of babies translate into the difficulties of toddlers, the minimal competence of children, and the sophistication of adults.

While the model described in this paper is still primitive, it already demonstrates how a system can get from a stage at which it judges two objects to be the same color only if they are similar overall to a stage at which it can make the judgement without paying attention to irrelevant features. It does this by adjusting its connection weights in such a way that dimensional input has the effect of filtering out the irrelevant features in its internal representations for objects. Thus, for the crucial area of comparison and categorization, this connectionist model provides a starting point for understanding developmental change.

## References


[Kemler, 1982] D. G. Kemler. The ability for dimensional analysis in preschool and retarded children: Evidence from comparison, conservation, and prediction tasks. *Journal of Experimental Psychology*, 34:469–489, 1982.

[Macnamara, 1982] J. Macnamara. *Names for Things: A Study of Human Learning.* MIT Press, Cambridge, MA, 1982.

[Rumelhart *et al.*, 1986] D. E. Rumelhart, G. Hinton, and R. Williams. Learning internal representations by error propagation. In D. E. Rumelhart and J. L. McClelland, editors, *Parallel Distributed Processing*, volume 1, pages 318–364. MIT Press, Cambridge, MA, 1986.

[Smith, 1984] L. B. Smith. Young children's understanding of attributes and dimensions: A comparison of conceptual and linguistic measures. *Child Development*, 55:363–380, 1984.

[Smith, 1989a] L. B. Smith. From global similarities to kinds of similarities: The construction of dimensions in development. In S. Vosniadou and A. Ortony, editors, *Similarity and Analogy*, pages 146–178. Cambridge University Press, Cambridge, 1989.

[Smith, 1989b] L. B. Smith. A model of perceptual classification in children and adults. *Psychological Review*, 96:125–144, 1989.

[Sugarman, 1983] S. Sugarman. *Children's Early Thought.* Cambridge University Press, Cambridge, 1983.

[Treisman and Gelade, 1980] A. Treisman and G. Gelade. A feature-integration theory of atttention. *Cognitive Psychology*, 12:93–136, 1980.

# Concept Formation and Attention

**John H. Gennari**
Department of Computer Science, Keio University
3-14-1 Hiyoshi, Kohoku-ku; Yokohama, JAPAN 223
⟨gennari@aa.cs.keio.ac.jp⟩

## Abstract

In this paper, I combine the ideas of *attention* from cognitive psychology with *concept formation* in machine learning. My claim is that the use of attention can lead to a more efficient learning system, without sacrificing accuracy. Attention leads to a savings in efficiency because it focuses only on the relevant attributes, retrieves less information from the environment, and is therefore less costly than a system that uses every piece of information available. I present a working algorithm for attention, built onto the CLASSIT concept formation system, and describe results from three domains.[1]

## 1. Motivation

In cognitive psychology, *selective attention* refers to focusing one's cognitive effort (or processing power) on only a fraction of the perceptual input. Intuitively, this is the ability to preferentially 'concentrate' on a single task. This phenomena can be observed with visual input – we can easily attend to part of a scene, allowing unimportant areas and details to be ignored. This is a well-studied phenomena, and it appears in a wide variety of tasks and applications (Treisman, 1969).

Unsupervised *concept formation*, like selective attention, is a basic aspect of human intelligence. Concept formation can be viewed as a general problem that appears in a variety of practical, engineering tasks. In statistics, this problem is known as *cluster analysis* (Anderberg, 1973), and can be defined: given a set of instances or objects, find or impose some classification scheme on those objects. Note that the learning system is not given a set of classified training objects; thus it carrys out *unsupervised* learning. In machine learning, concept formation systems include Lebowitz's (1987) UNIMEM and Fisher's (1987) COBWEB systems; these contrast with Quinlan's (1986) ID3 and Aha's (1989) IBL systems. The former do "learning from internal feedback" (Billman & Heit, 1988), while the latter "learning from example" systems use classified training instances to learn concept definitions.

In this paper, I describe a system that applies selective attention to concept formation. Machine learning is an appropriate place for such a wedding: a discipline with roots in both engineering and psychology. Most previous concept formation research (Lebowitz, 1987; Fisher, 1987) has used all available information to make decisions. In contrast, I will assume that some of the available input should be ignored; the task for the attention mechanism is to find the relevant attributes and allow the system to work with these alone.

My goal here is simply to demonstrate the practical application of attention to concept formation: I expect that attention can provide a tangible reward in terms of *efficiency*. In most applications, there is a cost associated with retrieving information about an instance. This may be negligible in some applications, but in others, such as diagnosis, information retrieval can be very expensive. If an attention mechanism focuses the system on a small percentage of the available input, this can lead to a more efficient, cost-effective system, without sacrificing performance accuracy.

The attention mechanism I introduce here is built on the CLASSIT system (Gennari, Langley & Fisher, 1989; Gennari, 1990). Although this system is not a model of human learning, it is closely related to Billman and Heit's (1988) psychological model of concept formation with attention. Their task is similar to mine: to acquire knowledge about which features are most helpful for classification. However, their model uses a rather different representation for knowledge, and is limited to learning rules about pairs of features. In contrast, CLASSIT uses probabilistic concepts for representation (as in Smith and Medin, 1983), and builds a concept hierarchy to organize its acquired knowledge.

## 2. An overview of the CLASSIT system

For the CLASSIT system (or for COBWEB) an important characteristics of the learning task is that it

[1]This research grew out of work with the ICARUS research group at UCI: Pat Langley, Kevin Thompson, Wayne Iba and John Allen. Mike Pazzani also contributed some key ideas about attention. This research was supported by Contract MDA 903-85-C-0324 from the Army Research Institute, IBM Japan, and by Keio University.

occurs in an *incremental* manner. This means that the set of input instances is treated as a sequence over time, and that the system must learn from each new instance without reprocessing the previously seen instances. This restriction seems intuitive from a psychological perspective: humans do not remember every instance, and are able to learn despite a virtually infinite sequence of instances. This requirement is also reasonable for robotic or real-time applications, where a response may be needed at any point during learning. In this way, CLASSIT differs from statistical methods in cluster analysis and from non-incremental learning systems such as CLUSTER/2 (Michalski & Stepp, 1983).

## Representation

Instances for CLASSIT are described by a simple list of attribute-value pairs. Attributes may be symbolic, with a finite (and usually small) range of values, or continuous, with an infinite set of possible values. Additionally, CLASSIT allows for missing attributes: some instances may not have values for every attribute. This form of representation is reasonably general, but it cannot handle relational or structured information (see Iba & Gennari, in press).

Like COBWEB, CLASSIT uses probabilistic concept descriptions to represent acquired knowledge. Thus, rather than all-or-none conjunctive concept definitions, this approach uses probabilities to build approximate concepts. For symbolic attributes, these probabilities can be computed by counting the number of times each attribute-value appears and the number of member instances. For continuous attributes, where the probability of any single value is zero, the system uses the mean and the standard deviation over member instances.

These concepts are organized in memory in a general-to-specific hierarchy. Toward the top of the tree are general concepts, summarizing many instances. Lower in the tree are more specific concepts, and the leaves may be single instances. A leaf may also summarize a number of (very similar) instances. In this case, the system has *forgotten* those individual instances, and can only retrieve the summarizing leaf concept.

## Algorithm

CLASSIT begins with an empty hierarchy, and adds to its concept hierarchy as it classifies each new instance. The algorithm presented in Table 1 is an overview of how learning (modifying the hierarchy) and performance (classification) occur. The system classifies instances by sorting them through the concept hierarchy from the root node down to the leaves. At each level, CLASSIT can place the node in an existing concept or decide that the instance is sufficiently different to warrant the creation of a new class. The other two choices (operators c and d in Table 1) reorganize the concept hierarchy; they give the system some ability to recover from a misleading sequence of instances.

```
1. Incorporate x into the root class.
2. Choose the best of four operators:
      a) incorporate x into a child class.
      b) create a new disjunct based on x.
      c) merge two child classes.
      d) split a class into its children.
3. If the new class has no children, or
   If the match is close enough, end.
   Else, recurse on the chosen class.
```

Table 1: The incremental algorithm used by CLASSIT.

In order to avoid storing all instances, and to avoid overfitting in noisy domains (see Gennari, 1990), CLASSIT uses a parameter called the *recognition criterion* as it classifies instances. This tells the system that the new instance matches the current concept well enough to consider that instance as recognized and to halt the classification process (Step 3 in Table 1).

## Evaluation function

In order to choose among the four operators, CLASSIT uses an *evaluation function*. This is an expression that evaluates the quality of a set of concepts and returns a numeric score, allowing the system to choose the operator that leads to the highest score. CLASSIT uses a version of *category utility* for its evaluation function. This function is designed to maximize the predictive ability of classes and was originally developed by Gluck and Corter (1985). For CLASSIT, category utility is:

$$\frac{\sum_i^I \sum_j^J P(C_j) Info(C_{ij}) \; - \; Info(C_{ip})}{I \cdot J} \quad , \qquad (1)$$

for $I$ attributes and $J$ classes, where $P(C)$ is the probability of class $C$ and $C_{ip}$ refers to attribute $i$ in the parent class. $Info(C)$ is a function that measures the value or quality of class $C$. For a symbolic attribute $i$ (with $V$ values)

$$Info(C_i) = \sum_v^V P(x_{iv}|C)^2 \quad ,$$

and for a continuous attribute $i$

$$Info(C_i) = 1/\sigma_{iC} \quad ,$$

where $\sigma_{iC}$ is the standard deviation of an attribute in class $C$.[2] To summarize, this function sums over every

[2]With singleton classes, this standard deviation is zero, leading to an infinite $1/\sigma$. To solve this problem, CLASSIT uses an *acuity* parameter that specifies a minimum (non-zero) standard deviation. This limit corresponds to the notion of a 'just noticeable difference' in psychophysics – the lower limit on our ability to make perceptual discriminations.

```
1. Select an unseen attribute with probability
   based on its salience.
2. Update the salience of the selected attribute.
3. Compute the category utility score for the best
   classification, X, based only on observed
   attributes.
4. Consider all remaining unseen attributes and
   compute scores for 'worst-case' scenarios:
   where these attributes might match either
        a) An alternative concept.
        b) A new disjunct.
5. If either of these scores is better than X,
   then go to step 1.
   Else, ignore remaining attributes.
```

Table 2: An algorithm for attention

child concept, $C_j$, and subtracts the information at the parent, $C_p$. Thus, it measures the gain in $Info(C)$ from parent to child levels of the hierarchy.

## 3. Attention applied to CLASSIT

In order to choose only the 'important' attributes, the system must learn the relative *salience* of attributes. Note that this is not given *a priori*, so attention adds a second learning task to the system: CLASSIT must learn both concept descriptions and saliences of attributes. Salience is defined as the per-attribute contribution to category utility (see Equation 1). Hence, for a given attribute $i$,

$$Salience_i = \frac{\sum_j^J P(C_j)Info(C_{ij}) - Info(C_{ip})}{J} .$$

These scores produce a dynamic ordering of the attributes from most salient (attributes that should be inspected first) to least salient (attributes that probably need not be inspected). However, attributes are not always inspected in exact order of salience; instead, the system chooses attributes probabilistically as a function of their salience. This allows the system to recover from 'incorrect' scores: even a low-scoring attribute may be occasionally inspected. If such an attribute is 'noticed' in this way, and if that attribute actually is salient, then its score will improve, and it will be more likely to be inspected in the future.

In addition to an ordering, the system must decide how many attributes to inspect before making a clustering decision. CLASSIT resolves this 'stopping condition' problem by imagining a worst-case scenario where the unobserved attributes match some other concept perfectly, and then considering whether this information would change the current clustering decision. If so, then it must continue inspecting attributes; if not, it has inspected sufficient attributes to make a decision.

Table 2 presents the attention algorithm used in CLASSIT. This mechanism is embedded within the basic concept formation algorithm as described in Table 1. Attention is used whenever making a clustering decision: choosing one of the four operators at a level in the hierarchy. If attention chooses to ignore some attributes, these attributes are treated as missing: the system simply makes classification decisions based on partial information. When CLASSIT descends to the next level, attributes that were observed earlier are 'remembered' and added to the list of known attributes. Thus, the system must inspect more attributes in order to make more specific classifications.

With little or no previous information, all attributes are equally salient, and the system must inspect most or all attributes before choosing a clustering operator. However, as more instances are observed, concepts should emerge in which some attributes contribute heavily to the total category utility score, while others contribute less. This means that the salience scores for attributes become more disparate, letting the system inspect only those attributes that have high scores. The attribute learning process is synchronous with the concept learning process: as the system defines concepts, it learns which attributes are more salient.

As stated earlier, the purpose of an attention mechanism is to improve efficiency by looking at fewer attributes. Yet there is no improvement in computational efficiency with this algorithm. In particular, by applying the halting condition after observing each of $n$ attributes (Step 4 in the table), I have added an $O(n^2)$ cost to the algorithm. However, I assume that the cost of observing an attribute is far greater than the time required for internal computation. This is reasonable if one imagines an application to diagnosis or robotics, where considerable work and real time may be needed to observe features (Tan & Schlimmer, 1990).

## 4. Experimentation with attention

The basic claim to be verified by experimentation is that attention can increase efficiency without a loss of accuracy. More specifically, it should decrease the number of attributes the system must inspect without decreasing its predictive accuracy. In addition, I should verify that the attention mechanism behaves as expected: the number of attributes inspected should decrease over time, and the system should focus only on the "relevant" attributes.

In the following experiments, I measure efficiency by counting the total number of attributes observed during classification of each instance. For accuracy, I measure the predictive ability of the concept hierarchy on unseen test instances. Both measures are taken with the system in a 'testing' mode. Learning (the modification of the concept hierarchy) is turned off during testing, and the recognition criterion is set at a low level. The recognition criterion represents a trade-off between careful learning and rapid recognition. Low-

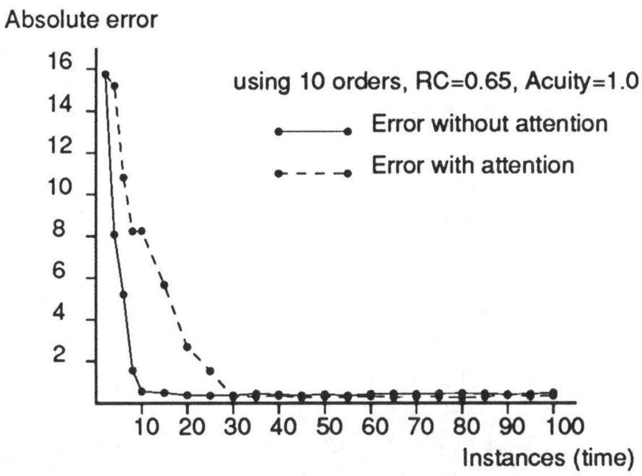


Figure 1: Predictive accuracy – artificial domain

ering the parameter allows the system to quickly recognize an instance at a high, general level of the tree, allowing the attention mechanism to use only a small number of attributes.[3]

## Results with artificial data

As an initial experiment, it is useful to investigate CLASSIT's performance with an artificial database. With such a domain, one can guarantee that some attributes are truly 'irrelevant' – their values do not depend on class membership. In this experiment, I created a database with 20 continuous attributes, where 12 of these had generating distributions that did not depend on class. Any of the remaining eight attributes can be used to distinguish among the classes, but four are *noisy*: the differences between classes are small, and the standard deviations within a class are high.

Figure 1 shows a learning curve for accuracy with and without the attention mechanism. The predicted attribute is attribute 13, one of the clean, relevant attributes, and accuracy is measured as a simple average of the absolute error of predictions. As expected, both systems reach the same asymptote, although the use of attention seems to result in a slower rate of learning. This may be due to the additional learning task for attention.

Figure 2 characterizes the behavior of the attention algorithm. In this figure, the frequency of attribute inspection is shown over time, for all attributes. Note that the system almost always inspects the three clean, relevant attributes (nos. 14, 15 and 16).[4] CLASSIT next prefers the noisy attributes (nos. 17 through 20) and only rarely does the system inspect the irrelevant attributes (nos. 1 through 12). This figure also shows that the system begins by inspecting many attributes (about 12 to 15), and only after about 30 instances does it 'settle' on the best attributes (inspecting about 7 attributes per instance). It is interesting that this point in time matches the place in Figure 1 where the two systems reach the same level of accuracy.

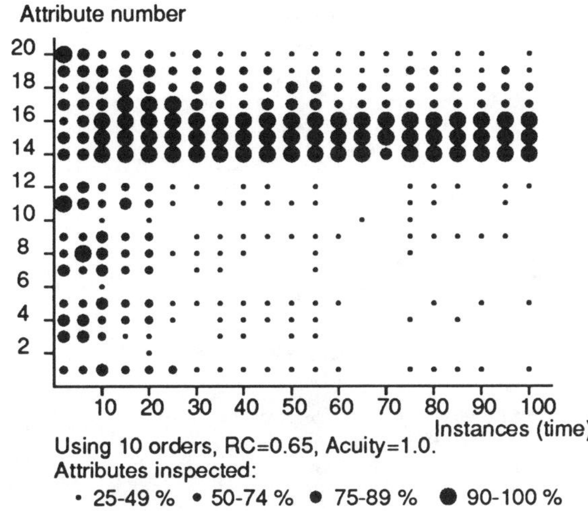


Figure 2: The frequency of attribute inspection in the artificial domain

## Results with real databases

In order demonstrate the practical use of a learning system, one must demonstrate performance of the system in real-world domains. Here, I present results for two databases from the UCI machine learning database repository: [5] the voting database and the heart-disease database. The voting database was collected from 1984 congressional records by Jeff Schlimmer, and consists of 17 symbolic attributes per instance: 16 votes and party affiliation. The heart disease database encodes patient information from the Cleveland Clinic Foundation and was collected by Robert Detrano (see Detrano et al., 1990). This data includes eight numeric attributes and six symbolic ones, including a binary sick/healthy attribute.

Figure 3 shows the accuracy of CLASSIT on these databases with and without the attention mechanism. In both domains, the predicted attribute is binary and symbolic: for the voting database, party affiliation; for the heart-disease database, the sick/healthy attribute. Thus, performance is measured by percentage error – how often the prediction is incorrect. Although the use of attention results in some loss of accuracy in the heart-disease domain, this appears to be a minor effect.

[3] Without the use of the recognition criterion, attention tends to observe all (or almost all) attributes. This is because different attributes are relevant at each level of the tree. By the time classification reaches the leaves of the tree (singleton classes), all or most attributes are 'known'.

[4] Attribute 13 is not available because it is used for the prediction task.

[5] To obtain these databases or information about them, send e-mail to ml-repository@ics.uci.edu, or contact Patrick Murphy, ICS Department, UC Irvine 92717.

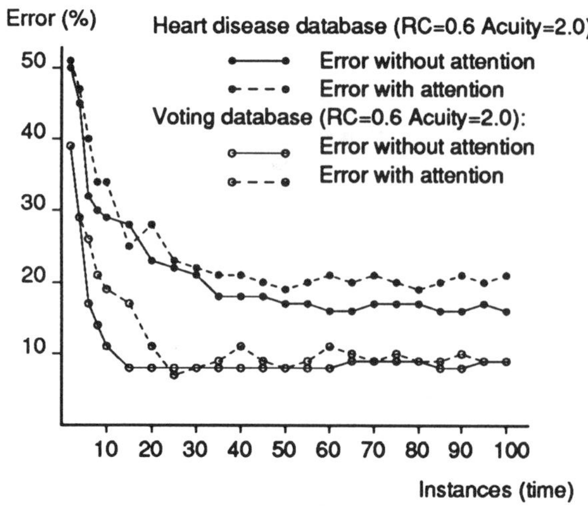


Figure 3: Predictive accuracy – real domains

The attribute frequency graph for the voting database (not shown) provides additional evidence for the attention mechanism. Although not as dramatic as Figure 2, it shows that CLASSIT focuses on partisan votes such as 'aid to El Salvador' and 'aid to the contras', at the expense of non-partisan attributes such as a vote about immigration. In this domain, CLASSIT focuses almost immediately onto about 9 attributes. A simple explanation for this rapid learning is that it is very easy (requires few instances) to distinguish between the two parties in this database.

The heart disease database has an attribute frequency graph that indicates a more even distribution among attributes. This suggests that all or most of the attributes in this domain are relevant for prediction. This is not surprising since the attributes were selected by an expert in the domain. However, even in this domain, CLASSIT learns to use only about 7 attributes out of a total of 13.

## 5. Discussion

One important domain for attention is visual processing. This is an interesting application since real-world images have thousands of attributes and a large amount of irrelevant information. However, before applying attention to vision, I must improve the 'stopping condition' currently used by CLASSIT. Experiments to date have confirmed that the the current method is too costly and too conservative about not risking accuracy.

The results presented here demonstrate that the attention mechanism does work as it should: it increases the efficiency of the system without reducing accuracy. However, experimental results do not "prove" that this is the best approach, nor support this research as a model of human attention and learning. This is simply an effort to use a known phenomena in human psychology to achieve improvement in a learning system. In turn, I hope that as these computer systems evolve and improve, they can lead to advances in developing models of human cognition.

## References

Aha, D. (1989). Incremental, instance-based learning of independant and graded concept descriptions. *Proceedings of the Sixth International Workshop on Machine Learning* (pp. 387–391). Ithaca, NY: Morgan Kaufmann.

Anderberg, M. (1973). *Cluster analysis for applications.* New York: Academic Press.

Billman, D., & Heit, E. (1988). Observational learning from internal feedback: A simulation of an adaptive learning method. *Cognitive Science, 12*, 587–625.

Detrano, R., et al. (1989). International application of a new probability algorithm for the diagnosis of coronary artery disease. *American Journal of Cardiology, 64*, 304–310.

Fisher, D. (1987). Knowledge acquisition via incremental conceptual clustering. *Machine Learning, 2*, 139–172.

Gennari, J. H. (1990). *An experimental study of concept formation* (Doctoral dissertation, also Technical Report 90-26). Irvine: University of California, Department of Information and Computer Science.

Gennari, J. H., Langley, P., & Fisher, D. (1989). Models of incremental concept formation. *Artificial Intelligence, 40*, 11–61.

Gluck, M., & Corter, J. (1985). Information, uncertainty and the utility of categories. *Proceedings of the Seventh Annual Conference of the Cognitive Science Society* (pp. 283–287). Irvine, CA: Lawrence Erlbaum.

Iba, W., & Gennari, J. H. (in press). Learning to recognize movements. In D. Fisher & M. Pazzani (Eds.), *Computational approaches to concept formation.* San Mateo, CA: Morgan Kaufmann.

Lebowitz, Michael. (1987). Experiments with incremental concept formation: UNIMEM. *Machine Learning, 2*, 103–138.

Michalski, R. S., & Stepp, R. (1983). Learning from observation: Conceptual clustering. In R. S. Michalski, J. G. Carbonell, & T. M. Mitchell (Eds.), *Machine learning: An artificial intelligence approach.* San Mateo, CA: Morgan Kaufmann.

Quinlan, J. R. (1986). Induction of decision trees. *Machine Learning, 1*, 81–106.

Smith, E. E., & Medin, D. L. (1981). *Categories and concepts.* Cambridge, MA: Harvard University Press.

Tan, M., & Schlimmer, J. C. (1990). Two case studies in cost-sensitive concept acquisition. *Proceedings of the Eighth National Conference on Artificial Intelligence* (pp. 854–860) Boston, MA: MIT Press.

Treisman, A. M. (1969). Strategies and models of selective attention. *Psychological Review, 76*, 282–299.

# On the Problems with Lexical Rule Accounts of Argument Structure

Adele E. Goldberg
University of California, Berkeley
eva@cogsci.berkeley.edu

## 0. Abstract

It has recently been suggested that the different valence possibilities of a single verb stem can be accounted for by postulating lexical rules that operate on the semantic structure of verbs, producing different verb senses. Syntactic expression is then taken to be predicted by general linking rules that map semantic structure onto syntactic form (Alsina and Mchombo 1990, Bresnan and Moshi 1989, Levin 1985, Pinker 1989, Rappaport, Laughren, and Levin 1987). In this paper, general problems with such approaches are discussed, including the following: a) such theories require a large number of both distinct verb senses and lexical rules, b) *ad hoc* and often implausible verb senses are required, c) an unwarranted asymmetry between different argument structures is posited, and d) many generalizations are obscured.

An alternative is suggested that involves considering the various valences as templates or constructions that are paired with semantics independently of the verbs that may occur with them. For example, abstract semantics such as "X causes Y to receive Z," "X causes Y to become Z" etc. are associated directly with the skeletal syntactic ditransitive and resultative constructions, respectively, allowing the verbal predicates to be associated with richer frame-semantic representations.

## 1. Introduction

There is a widespread assumption in current linguistic theories that syntax is a projection of lexical requirements. This claim is explicit in The Projection Principle (GB), the Completeness and Coherence Principles(LFG), and in all current "linking theories" which attempt to predict overt syntax from semantic role or theta role arrays. In all of these cases, it is the *verb* which is taken to be of central importance. That is, it is assumed that the verb determines how many and which kind of complements will cooccur with it. In this way, the verb is analogized to the predicate of formal logic which has an inherent number of distinct arguments. The verb is taken to be an n-place relation waiting for the exactly correct type and number of arguments.

At the same time, it is also recognized that the same verb stem often occurs with more than one valence. These two factors that 1)syntax is taken to be a projection of lexical, and in particular, verbal requirements, and that 2)the syntactic valence associated with particular verbs is not uniquely determined, have recently led many theorists to postulate lexical rules which are designed to operate on the semantic structures of lexical items. The overt complement structure is then to be predicted by general linking rules that map semantic structure onto syntactic form (cf. Alsina and Mchombo 1990, Bresnan and Moshi 1989, Levin 1985, Pinker 1989, Rappaport, Laughren, and Levin 1987).

By postulating rules that operate on semantic structure as opposed to rules or transformations that are purely syntactic, these theories manage to incorporate important insights. First, different valence expressions are typically accompanied by slightly different semantic interpretations and secondly, alternations are commonly constrained in semantic ways (cf. Pinker 1989 for review and discussion).

In this paper, however, I argue that semantic changing lexical rules are subject to many general criticisms, despite the fact that their semantic basis is well-motivated. An alternative is suggested which attempts to incorporate the semantic insights while avoiding many of the problems associated with postulating rules that change the inherent semantics of the main verb. The alternative involves considering the various valences as templates or constructions that are paired with semantics independently of the verbs that may occur with them. For example, abstract semantic representations such as "X causes Y to receive Z," "X causes Y to become Z" can be associated directly with the skeletal syntactic ditransitive and resultative constructions, respectively, allowing the verbal predicates to be associated with richer frame-semantic representations. This approach is currently being developed as part of the theory of Construction Grammar (Fillmore, Kay and O'Connor 1988, Fillmore 1987, Kay 1990, Lakoff 1987). Aspects of this type of approach have been recently suggested by Jackendoff 1990 and Wierzbicka 1988 as well.

## 2. Problems with (Semantic) Lexical Rule Approaches

There are many problems with lexical rule accounts. First, although such accounts only appeal to pair-wise alternations, it should be recognized that the number of distinct argument structures that can be used with a verb stem is typically greater than two. For example, an ordinary verb such as *kick* can appear with at least eight distinct argument structures:

1. Joe kicked the wall.
2. The horse kicks.
3. Joe kicked the football into the stadium.
4. Joe kicked at the football.
5. Joe kicked Bob black and blue.
6. Joe kicked his foot against the chair.
7. Joe kicked Bob the football.
8. Joe kicked his way out of the operating room.

To relate these valences, seven lexical rules would be required, to allow for eight distinct senses of *kick*. It is clear that this approach leads to a proliferation of both verb senses and lexical rules.

Moreover, despite claims that the linking rules mapping arguments to grammatical functions or to syntactic positions are general and language independent, the proposed rules are commonly quite construction-specific; i.e. they are often not general across construction types even within a single language. For example, in English the *theme* argument (*the ball(s)* in each of the following) is mapped to subject in intransitive unaccusative expressions, e.g. *The ball dropped*, to direct object in simple transitive expressions, e.g. *Joe dropped the ball*, to the second object position in ditransitives: *Joe threw Mary the ball*, and to an oblique complement in *She loaded the carton with balls*.

### 2.1. The Problem of Circularity

A fundamental problem that arises from assuming that syntax is a projection of lexical requirements, is that such an assumption often results in circular analyses. For example, the theory is forced into the position of claiming that *kick* is a binary relation, and therefore must have two and only two complements, except in *The horse kicks* in which it is unary, and therefore has one complement, and except in *Joe kicked Bob the football* in which it is ternary and therefore has three complements.

A constructional approach to argument structure allows us to avoid the circularity of arguing that a verb is an *n*-ary predicate and "therefore" has *n* complements when and only when it has *n* complements. Instead, the ternary relation for example, can be directly associated with the skeletal ditransitive construction, and we can understand that relation to be *imposed* on *kick* when *kick* occurs with that particular construction. This idea is discussed more fully below.

### 2.2. Sentential Effects

Another difficulty with lexical rule approaches is that it is often not possible to determine whether a given argument structure can be used on the basis of the verb alone. For example, holding the verb constant, the following a) sentences are better than the corresponding b) sentences:

9a. ?TV is watched.

9b. TV is watched an average of six hours per day.

10a. *She sent the countryside a package.

10b. She sent the Countess a package.

These effects are well known, but there is no natural way to capture these types of constraints in the lexical semantics of the main verb. On a constructional account, however, it is possible to associate constraints on the subject, direct object, or on the overall interpretation of the expression directly to the construction. Intuitively it makes sense to ascribe constraints that occur only with particular argument structures directly to those argument structures, instead of postulating new verb senses for each of the verbs and then needing to stipulate that these particular senses only occur in the particular argument structure. For example, the constraint that the first object must be animate when verbs are used with the ditransitive construction (cf. exs. 11a-b) can be understood as a constraint on the construction; it should not be taken to imply that there is a special verb sense *send2* which constraints its goal to be animate. If a new verb sense were postulated, we would also have to also posit an *ad hoc* stipulation that this sense could only occur in the ditransitive valence. Notice we cannot readily claim that the ditransitive syntax is determined by general linking rules, because verbs which do *lexically* constrain their goals to be animate--e.g., verbs such as *give* or *hand*-- retain the constraint when they are used with other valences.

### 2.3. Implausible Verb Senses

Another serious problem with the type of entirely bottom-up analysis of the lexical rule approach is that it encounters problems when faced with examples such as:

12. He sneezed the napkin off the table.
13. She baked him a cake.
14. Dan talked himself blue in the face.

In none of these cases does the verb intuitively require the direct object complement. To account for 12, for example, we would have to say that *sneeze*, a parade example of an intransitive verb, actually has a three-argument sense, "to laugh in such a way as to cause motion." To account for 13, we would need to claim that there exists a special sense of *bake* that has three arguments, an agent, a theme, and an intended recipient. This in effect argues that *bake* has a sense which involves something like "intend to cause to have." To account for 14, we would need to postulate a special sense of *talk*, "to talk in such a way as to effect a change in the talker."

On a constructional approach, we can understand aspects of the final interpretation involving caused motion, intended transfer, or caused result to be contributed by the respective constructions. That is, we can understand skeletal valence constructions to be capable of contributing arguments. For example, we can define the ditransitive construction to be associated directly with the semantic roles of agent, patient and recipient, and in addition associate the class of verbs of creation with the ditransitive construction. We do not need to stipulate a specific sense of *bake* unique to this construction. In general we can understand the direct objects found in the above examples to be licensed, not directly as an argument of the verb, but by the particular constructions.

### 2.4. Lack of Explanation

In addition to these problems, there is reason to doubt the explanatory nature of lexical rule accounts. It is well-known that there are cases which meet the structural description of a particular alternation, but which do not alternate. For example, in the case of the ditransitive, we find *donate, explain,* and *choose* which do not appear ditransitively; in the case of the locative alternation, we find *pour, drip,* and *coil* which do not occur in the *She loaded the wagon with hay* frame. Because of these cases, many theorists have sought to add constraints to the relevant structural descriptions. Attempts in this direction can be found throughout the literature. However, to date, no general constraints have been discovered which distinguish the verbs which allow certain alternations from those that do not. (see Pinker 1989 for a full discussion of the failure to find adequate ways to constrain, in a general way, the application of particular alternations). Most theories solve this problem by positing some type of designation on either the verbs which allow the alternation or on the verbs which do not.

Complementary classes of cases are also recognized to exist. These are cases which have the "output" structure, But not the "input" structure. For example for the ditransitive, *envy, forgive, cost* appear ditransitively, but not with prepositional paraphrases. In the case of the locative, *fill, cover, inundate, adorn, etc.*, cannot occur with directional phrases, but do occur in the *with* variant.

These two classes of cases, that the rules do not apply to all verbs with the correct structural description, and that not all expressions with the "output" structure have the same "input" structure, cast doubt on the explanatory value of the alternation account. The alternation cannot be stated as an implication in either direction. Instead, the statement of the alternation can only say: some verbs with argument structure X also have argument structure Y. This is true, but it tells us nothing about either of the argument structures or about the particular verbs; it simply describes the fact that there is some overlap between the set of verbs associated with two distinct valences.

### 2.5. Generalizations are Obscured

Further undermining the explanatory value of alternation accounts is the fact that they, by their nature, obscure certain generalizations. Because the rules are written with reference to one particular basic argument structure, other rules with the same output, but a different input structure are postulated as independent rules. E.g. In the case of the ditransitive, alternation accounts of the construction typically assume the existence of two distinct constructions: the "*to*-dative" and the "*for*- dative", because of the fact that two different "input" argument structures are involved. In this way the alternation accounts typically ignore the fact that *to*-datives and *for*-datives are intimately related both syntactically and semantically (for discussion, see e.g., Goldberg 1989). A related problem is that the class of expressions that share the same output argument structure, but that do not have any other relevant argument structure are often not accounted for in any systematic way. Instead, cases such as

15. He gave her an idea.
16. The music lent the party a festive air.

are said to be idiomatic, and are often ignored.

In the case of the locative alternation, similar generalizations have been overlooked. Accounts of "locative" verbs have been systematically delimited to those cases which alternate, or in some cases (e.g. Rappaport and Levin ms., Pinker 1989) to those cases which are of the same general kind as those which alternate. However, motion verbs such as: *push, move, run, throw,* and metaphorical cases are all excluded from analysis. (Talmy 1980 and Jackendoff 1985 are notable exceptions to this generalization in examining all types of verbs of motion as a class. However, perhaps tellingly, their discussions are not made in the framework of describing alternations).

A broader perspective of the construction as an entity in its own right leads to interesting generalizations. For example, the expressions:

17. The wind blew the ship off course.
18. She forced him into the room.

can be assimilated into an account of a construction involving caused-motion, although they are not at all relevant to the "locative" *with* variant. Similarly, we can account for the metaphorical extensions:

19. He drove her into a state of madness.
20. He forced the idea out of his mind.

which are even more removed from the *with* variant, as metaphorical extensions of the construction.

### 2.6. Assumptions of Asymmetry

Alternation accounts generally assume an asymmetry between the two argument structures (Pinker 1989 is an exception in allowing lexical rules to operate in both directions). In a few cases this is desired, as it is may be meant to capture a difference in markedness between the two argument structures. For example, having a lexical rule which transforms an active argument structure into a passive argument structure captures the idea that the active argument structure is in some sense more basic or less marked. However, active/passive pairs represent perhaps the clearest case of an asymmetrical relation. In many other cases, the asymmetry is not warranted and often is not defended.

In the case of the ditransitive, *He gave the book to her* is usually supposed to be more basic than *He gave her the book.* (however, decisions as to which is more basic are not unanimous, see Dryer 1986 for the opposite analysis). A typical reason that is given is that the verbs which allow ditransitives are a proper subset of those that allow prepositional paraphrases. However, this is not actually so since *refuse*, and *deny* do not have paraphrases with *to* or *for*, and neither do many metaphorical expressions, for example:

21a. She gave me a headache/a kiss/an idea.

21b. *She gave a headache/a kiss/an idea to me.

Oehrle(1976), moreover, has argued that there is no principled way to distinguish those cases which have prepositional paraphrases from those that do not.

Developmental data (Gropen et al. 1989) shows that the ditransitive and prepositional paraphrases occur at roughly the same time in children's speech, neither construction reliably preceding the other, so that evidence for an asymmetry cannot be grounded in evidence from children's acquisition of the forms.

The asymmetry implicit in alternation accounts is even more difficult to defend in the case of the locative alternation. In general, the valence associated with *He loaded hay onto the wagon* is supposed to be more basic than *He loaded the wagon with hay.* However, when different verbs are examined, this asymmetry is clearly unwarranted. So, although *stack* and *plaster* all allow both argument structures, there is no intuition that the *into* variant is more basic than the *with* variant. That is, the following appear to have equal status in terms of being basic or unmarked:

22a. He stacked the shelves with boxes.

22b. He stacked boxes onto the shelves.

23a. He plastered the wall with posters.

23b. He plastered posters onto the wall.

Moreover, *adorn, blanket, block, cover, dam, enrich, fill, dirty, litter, smother, soil, trim, endow, garnish, imbue, pave, riddle, saturate* to name a few, only occur with the *with* variant. In fact in a detailed study of locative verbs, Rappaport and Levin(ms) found that out of 142 verbs studied, only 34 alternate.

On a constructional account, we need not assume an asymmetrical relationship between two constructions that are found to be related. Instead, we can describe instances of partial-overlap of syntax, semantics, or pragmatics as such, without necessarily assuming that one or the other is basic, the other derived. So, for example we can state that the semantics associated with the ditransitive construction is related to the semantics of the paraphrase with *to*; we do not need to assume the primacy of one over the other. Likewise, we can describe the relationship between paraphrases with *to* and the caused motion construction.

**3. The Interaction of Verb and Constructional Semantics**

A constructional approach to argument structure allows us to avoid many of the problems discussed above. At the same time, we should not assume that all of the analysis is top-down, that constructions simply *impose* their meaning on unsuspecting verbs. In point of fact, there are reasons to think that the analysis must be *both* top-down and bottom-up. That is, the meanings of constructions and verbs can be seen to interact in non-trivial ways, and some cross reference between verbs and argument structures will be necessary.

Minimal entries for verbs will need to make reference to:

1)idiosyncratic facts(although typically motivated by 2) about its syntax, semantics, and use, and

2)frame semantic knowledge of the concept associated with the verb.

*Kick*'s syntactic properties are not idiosyncratic, so no syntactic properties will have to be listed in this case; any idiosyncratic semantic or pragmatic properties that *kick* may have are not immediately relevant, so I will for now avoid discussion of them. The frame semantic knowledge of *kick* must include something to the effect:

> *kicking* involves an abrupt forceful motion of the leg such that the leg is brought away from the body and then returned.

Argument structure constructions will need to refer to particular verb classes. Thus, for example, the ditransitive construction will need to make reference to verbs of ballistic motion, verbs of creation, verbs of future having, verbs of refusal, etc. (see Green 1974, Oehrle 1976, Wierzbicka 1988, Pinker 1989, and Goldberg to appear, for discussions of such classes). The caused-motion construction will make reference to verbs of motion, verbs of directed force, verbs of manner, verbs of instrument, etc. (see Talmy 1976, Jackendoff 1985, Shibatani 1976, Pinker 1989 for discussion of such classes).

Consider again *kick*, to take a particular case. The ditransitive construction will make reference to the fact that it can occur with verbs which can cause ballistic motion. The fact that *kick* is such a verb will be inferred from its frame semantics, i.e. from the fact that it kicking can be used to cause an object to move by an abrupt motion of the leg. We do not need to claim that the semantics of *kick* undergoes any change when used ditransitively.

The fact that *kick* can be used with the caused-motion construction as in:

24. She kicked the ball into the end zone.

can be accounted for by the fact that the caused-motion construction will make reference to verbs of forceful directed action, and it will be inferred from the frame semantics of *kick* that it is such a verb.

The fact that *kick* can appear in the transitive construction will follow from the fact that the transitive construction will make reference to verbs of contact, and again, the fact that *kick* can be interpreted as a verb of contact will be inferred from its frame semantics.

In this way, we do not need to postulate different senses of *kick* because of its different valence possibilities. Instead we can understand the sentential interpretation to arise from a combination of verb and constructional semantics. Valence constructions will be associated with one or more abstract semantic representations, and will make reference to classes of verbs. Whether *kick* falls into the appropriate classes will be determined by general inferences made on the basis of the frame-semantic representation of *kick*. That is, the semantic representation of the verb may underdetermine argument structure expression.

A benefit to assigning the skeletal semantics sometimes described as the "syntactically relevant" aspect of verb meaning directly to argument structure constructions, is that Pinker's (1989) observation that the syntactically relevant aspects of verb meaning are like that of closed class elements is predicted since argument structure constructions *are* closed class elements.

**4. Issues**

The suggestion being made here brings several tricky questions to the fore. If argument structure expressions are viewed as constructions, and we wish to claim that essentially the same verb is involved in more than one argument structure expression, we need to deal with several issues. Exactly how is the verb represented if not by some direct representation of an argument structure? Another question that arises is, when and why does the same verb occur with more than one argument structure construction? And finally, if the verb does not explicitly specify the correct number and type of arguments, how then do we prevent the generation of:

25. *Joe touched Bob a cupcake.

26. *Joe murdered.

I.e. how does this approach accomplish what was supposed to be accomplished by principles such as Projection Principle of GB or the Completeness and Coherence Principles of LFG?

However it should be made clear that these problems are not created by a constructional approach. In fact, although these issues are somewhat obscured by the mechanisms of traditional theories, they are by no means solved by those theories. For example, the question of how to capture the meaning of a verb independently of the particular argument structures in which it appears has its direct analog in more traditional accounts: the question that requires explanation is, in what way are the lexical items created by lexical rule related to each other? One non-answer to this question would be to stipulate that the meanings are not related semantically, that each is an independent lexical entry. Another uninteresting answer would be that the meaning of the verb stem is a disjunction of all of its argument structures; this would not answer the question of how the disjuncts are related to eachother.

It may be suggested that the lexical rule itself explains the relationship between two argument structures. For example, in the case of the causative lexical rule, one might argue that that the relationship is captured because the semantic decomposition of an inchoative predicate is contained in the semantic decomposition of a causative predicate. That is, the predicates may be said to be related by a containment relation. However, it is clear that simply stating that one representation properly contains another, does not explain the relationship between the two. Moreover, not all containment relationships form the basis of potential lexical rules, and conversely, the relationship that is intended to be captured by lexical rule is often not one of containment. Typically, the lexical rule merely stipulates that there *is* a relationship, it does not shed light on what that relationship is.

The second question as to when and why a particular verb stem can occur in more than one argument structure, is also not explained by alternation accounts. As was discussed above, alternation rules simply say: some verbs with argument structure X also have argument structure Y. This does not at all account for why a verb stem occurs in both, it simply stipulates that it does.

Finally, concerning the question of disallowing the following in a principled way:

27. *Joe touched Bob the cupcake.

28. *Bob murdered.

As has been noted, appealing to the fact that a verb has *n* number of arguments and therefore must have exactly *n* complements begs the question, since the way *n* is determined is by considering how many complements the verb actually does occur with.

I take it to be a valuable outcome of the constructional approach that these issues are made clear. Further research will be necessary to fully address these issues, however. For example, we will need inter and intra linguistic studies to determine which verb classes tend to be associated with which constructions, how constructions are related to eachother, and the exact way in which verbal, and more generally, lexical meaning is to be represented. What we stand to gain from such studies is a deeper understanding of the nature of argument structure than has heretofore been proposed.

## References


Alsina, Alex and Sam Mchombo. 1990. The Syntax of Applicatives in Chichewa: Problems for a Theta Theoretic Asymmetry. to appear in *Natural Language and Linguistic Theory*.

Bresnan, Joan and Lioba Moshi. 1989. Object Asymmetries in Comparative Bantu Syntax. *LI*. 20.4

Dryer, Matthew. 1986. Primary objects, secondary objects, and antidative. *Language Vol. 62 No. 4*. p. 808-845.

Fillmore, Charles, Paul Kay and Catherine O'Connor. 1988. Regularity and Idiomaticity in Grammatical Constructions: The Case of *Let Alone*. *Lg 64*. *501-538*.

Fillmore, Charles. 1987. Stanford Summer Linguistics Institute Lectures.

Goldberg, Adele. 1989. A Unified Account of the Semantics of the English Ditransitive in *BLS 15*.

__________. to appear. The Inherent Semantics of Argument Structure. *Cognitive Linguistics*.

Green, Georgia. 1974. *Semantics and Syntactic Regularity*. Bloomington: Indiana University Press.

Gropen, Jess Steven Pinker, Michelle Hollander, Richard Goldberg, and Ronald Wilson. 1989. The Learnability and Acquisition of the Dative Alternation in English. *Language*.

Kay, Paul. 1990. Even. *Linguistics and Philosophy 13*:59-111.

Jackendoff, Ray. 1990. *Semantic Structures*. Cambridge: MIT Press.

__________ 1985. *Semantics and Cognition*. Cambridge, Mass: MIT Press.

Lakoff, George. 1987. *Women, Fire and Dangerous Things*. Chicago: University of Chicago Press.

Levin, Beth. 1985. Lexical Semantics in Review. MIT's Lexicon Project Working Papers #1. pg 1-62.

Oehrle, Richard. 1976. The Grammatical Status of the English Dative Alternation. MIT diss.

Pinker, Steven. 1989. *Learnability and Cognition*. MIT press.

Rappaport, Malka and Beth Levin. ms. A Case Study in Lexical Analysis: The Locative Alternation.

Rappaport, Malka, Mary Laughren and Beth Levin. 1987. Levels of Lexical Representation. MIT's Lexicon Project Working Papers #20.

Shibatani, Masayoshi ed. 1976. *Syntax and Semantics The Grammar of Causative Constructions*. Vol 6. New York: Academic Press.

Talmy, Leonard. 1980. Lexicalization patterns in language. in Shopen et al. eds. *Language typology and syntactic fieldwork*

__________. 1976. Semantic Causative Types. in *Syntax and Semantics Vol 6*. ed. Masayoshi Shibatani.

Wierzbicka, Anna. 1988. *The Semantics of Grammar*. Benjamins: Philadelphia.

# Interactive Reasoning about Spatial Concepts *

**Marc Goodman, Scott Waterman, and Richard Alterman**

Computer Science Department, Center for Complex Systems
Brandeis University, Waltham, MA 02254
Email: goodman@chaos.cs.brandeis.edu

## Abstract

Spatial relations and spatial language form an important part of everyday reasoning. This paper describes SPATR, a system which addresses the labelling of components of objects and the interpretation of spatial relations between objects within the framework of adaptive planning. SPATR implements a model of *spatial reasoning*, which mediates among language, memory, and perception. Using a case-based approach for reasoning from past experience, SPATR makes use of spatial relationships corresponding to closed class terms, as well as a 3D, hierarchical representation of objects for retrieving relevant past experience.

## Introduction

An agent using instructions to interact with a device must resolve several types of Natural Language references to spatial concepts. For example, when using an Airphone (a special type of payphone found on airplanes), an agent may be instructed to 'insert credit card, face up, with card name to the right' into the Airphone. Fully understanding the spatial aspects of this instruction requires the system to decide what is meant by "insert" (i.e., is it the kind of insert one uses when one wishes to insert a key into a lock, a pin into a voodoo doll, a bank card into an ATM, etc.), what is meant by the "face" of a credit card, what "up" means with respect to this face, where the "card name" is and whether right refers to the agent's right (a deictic reference) or some intrinsic right of the card or the Airphone.

We view many of these problems within the framework of Adaptive Planning [Alterman, 1986; Alterman, 1988]. Adaptive Planning suggests that *engagement*, or interacting with an environment, is a memory intensive task wherein an agent confronted with a 'typical' situation will behave in a manner consistent with previous experiences of that situation, and that an agent confronted with a novel situation will attempt to adapt strategies used in a previous similar experience. An implication of this approach is that experience serves to guide the direction of attention within the environment (i.e., you look for things you expect to find from past experience) and that noticing novel features of the environment guides the retrieval and adaptation of experience (e.g., noticing an airport shuttle sign when you need to get from one airport gate to another to make a connection suggests taking a shuttle rather than walking).

Within the context of the Airphone example, we view the process as follows:

- The agent approaches the Airphone for the first time. At this point in the interaction, the agent has not conducted an exhaustive visual inspection of the device, and therefore has only a skeletal representation for the Airphone.
- The agent is instructed to insert the credit card, face up with card name to the right. The 'figure' of the insert (see [Talmy, 1983] for explanations of the terms 'figure' and 'ground') is the credit card, and an appropriate geometric representation of the card is retrieved from memory. The ground of the insert is the Airphone, and the skeletal representation for the Airphone is used. The semantics of 'insert' suggest that an 'into' relationship is desired and this causes the retrieval of a likely previous experience, based on known features of the figure, and known and assumed features of the ground. For example, the agent may be reminded of inserting a bank card into an ATM.
- Inserting a bank card into an ATM consists, in part, of finding a slot on the front of the ATM, and this information is used to constrain visual search for such a slot on the front of the Airphone.
- This visual search fails and a new retrieval is performed based on different assumptions about the ground (and perhaps additional known features gained through the visual search). The agent may then be reminded of inserting a library book return

*This work was supported in part by the Defense Advanced Research Projects Agency, administered by the U.S. Air Force Office of Scientific Research under contract #F49620-88-C-0058.

card into the envelope on the front of a library book. This suggests a vertically oriented slot with an opening along its top, into which the card is inserted with a downward motion.

- Visual search succeeds for this feature and this experience of insert is chosen as the case to be adapted.
- 'Face' of a card metaphorically refers to the front of the card. We do not provide a model for linguistic metaphor, however we feel that [Martin, 1990] is more than adequate for these purposes. If the agent does not already know which side of a credit card is its front, then this decision could be handled by retrieving previous cases of labelling, and adapting these cases. For example, the agent may be reminded of a playing card where the front of the card is the side with the picture and writing on it, and label the credit card accordingly.
- Face 'up' actually means face towards the agent, given that insert requires a downward, sliding motion with the card oriented in a vertical position. This could be handled as a metaphoric meaning of up, an adaptation on the insert, or as a problem similar to the selection of the proper meaning of insert.
- We are assuming that the geometric representation of the credit card has adequate ties to semantic memory to find the 'card name' of the card.
- 'To the right' requires the agent to resolve whether 'right' refers to deictic or intrinsic right. For a full discussion of this process the reader is referred to [Alterman *et al.*, 1991].

## SPATR

This paper will describe SPATR (pronounced "spatter"), a SPATial Reasoner which models the above processes. SPATR is implemented as a distributed case-based reasoning system built on top of Cognitive Systems' CBR Shell [Goodman, 1989; Goodman, 1990]. There are three components to SPATR: the I/D Labeller, CLTER, and SPACL. These components communicate through a blackboard. Each component is geared to handle a specific type of spatial problem.

### Geometric Representations of Objects

In [Marr, 1982] a model for the geometric representation of objects based on a hierarchically organized 3D model is presented. For example, the overall shape of a person is cylindrical with a generating axis running from top to bottom, and this overall shape can be decomposed into a cylinder for the torso, cylinders for the arms and legs, etc. Arms can be further decomposed into forearms and biceps, forearms can be decomposed into hands, which can be decomposed into fingers, and so on. [Biederman, 1987] advocates a similar hierarchical decomposition into 'geons' which are a fundamental set of conics. Both approaches allow the recognition and categorization of perceptual information to occur at both coarse-grained and fine-grained levels.

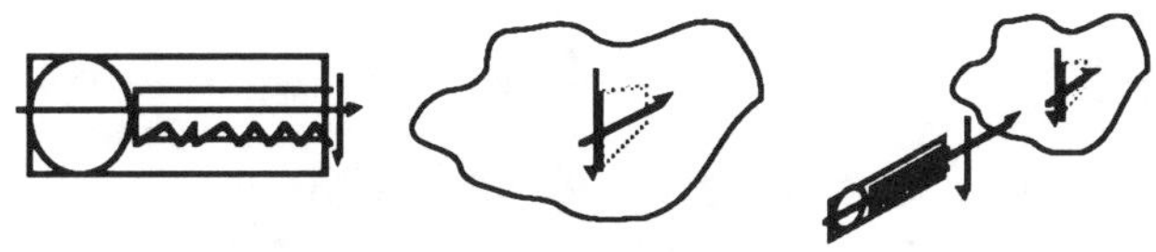

Figure 1: Pictorial Representations of Key, Lock, and Key into Lock.

In [Jackendoff and Landau, 1990] a set of extensions to Biederman's approach to account for certain linguistic terms is given. These extensions allow:

- The addition of secondary directing axes in addition to a main generating axis to account for characteristic orientation of objects.
- Two-dimensional as well as three-dimensional subcomponents of objects (which offers a qualitative distinction between things which are fundamentally 'sheet-like' as opposed to 'very thin').
- Hollow objects (as opposed to shapes with slightly smaller negative shapes inset).
- Texture or character for objects (such as smooth, rough, jagged).
- Surface markings and/or color on objects (for example, writing on one side of a credit card).

SPATR's representational scheme for objects is adopted from this approach. For example, SPATR's representation for an (asymmetric) key would be a frame that could be represented pictorially as shown in Figure 1. The representation specifies that the overall shape of the key is strip-like, with a directed main generating axis. For convenience, we arbitrarily label this axis as a "Front→Back" axis (we will designate directed axes from X to Y as "X→Y" and undirected axes as "X↔Y"). The shape is further decomposed into a disk for the "back" part (actually, one end of the main axis) and a strip for the "front" part. This "front" strip itself contains a secondary directed axis which we arbitrarily label the "Top→Bottom" axis. The "top" of the strip is an edge whose character is straight and the "bottom" of the strip is an edge that is jagged.

Similarly, a lock could be represented as shown in Figure 1. This representation consists of a generic (or unspecified) 3D shape which contains a negative strip (i.e. the hole where the key fits). This strip has a directed main generating axis (which we label "Front→Back" for convenience). The "front" of this strip is a negative edge which has an undirected "Top↔Bottom" axis. Note that since our representation must account for world knowledge which is not necessarily perceptual in nature (for example, that there is, in fact, a negative strip inside the lock rather than a spherical shape or a slab, when only the leading edge of that strip is visible) SPATR must either deduce or explicitly represent the distinction between what is known through direct observation (e.g. that a lock has a negative edge) and what is known through other methods (e.g. that a lock has a negative strip).

## Motion and Change in Orientation

Taking action in the world requires a representation of motion and change in orientation over time. Inserting the card, lowering the door handle, removing and inserting the handset, and removing the card all require specifying paths between objects (as in [Jackendoff and Landau, 1990] or [Talmy, 1983]). Representing these paths may require specifying starting points (as in removing the handset and credit card), one or more midpoints (as in inserting the credit card into its holder and the handset into its cradle), and/or end points of the path (as in lowering the door handle over the card and the location to which the handset is returned). An example of changing the orientation of an object is specifying the rotation of a credit card so that it is "face up with name to the right."

Actions required to place a figure into a ground are decomposed into specifications of axial alignment, rotational alignment, and movement along paths in a manner akin to [Herskovits, 1985]. For example, inserting a key into a lock is represented as a frame which specifies that the "Front→Back" axis of the key must be aligned with the "Front→Back" axis of the negative strip of the lock, the key should be rotated until its "Top→Bottom" axis is parallel to the "Top↔Bottom" axis of the negative edge of the lock, and the key should be moved along the path specified by its "Front→Back" axis. This process can be represented pictorially as shown in Figure 1.

An issue of primary importance is whether spatial relationships should be represented as abstract generalizations over related meanings (as is practiced by [Herskovits, 1985], c.f. [Talmy, 1983], and [Chen, 1987]) or as discrete, though related instances of differing relationships (*Extended Category Structure*: [Lakoff, 1987]). [Badler *et al.*, 1990] use an abstract notion of the meaning of spatial prepositions, such as 'on,' to resolve specific instructions, such as "put the block on the table."

This approach, however, requires significant work when given an instruction like "put the picture on the wall," or "put the teakettle on the stove." When placing a picture on a wall, the supporting surface is not horizontal, and when placing a kettle on a stove, a specific sub-area of the stove (i.e. the burner) is usually meant. SPATR's approach, following Lakoff, is to treat putting a block on a table, a picture on a wall, and a teakettle on a stove as distinct senses of 'on,' and to adapt the most similar sense given a new instruction.

CLTER, (pronounced as "clutter") the CLosed-class TERm library, represents each spatial interaction between a figure and ground as a distinct case, containing geometric representations of the figure and ground, and decompositions of actions required to affect the spatial relation. A working hypothesis of CLTER is that these cases are organized primarily into less than 100 spatial categories of interaction corresponding to the closed-class terms of a language (e.g. prepositions like 'in,' 'over,' 'at,' 'into,' etc. [Talmy, 1983]). Within these spatial categories, concepts form an extended category structure [Lakoff, 1987] with the leaf nodes for a category like 'into' being cases of, for example, 'inserting key into lock' or 'inserting pen into pocket.' The position of a node in the extended category structure is defined by its spatial characteristics (i.e. features of the figure and ground which account for differences in the spatial relationships).

When given a complete spatial representation of a figure and a partial spatial representation of a ground, CLTER retrieves previously stored spatial relationships, whose grounds are used to constrain visual search for necessary components of the ground. If this visual search fails, CLTER retrieves additional cases within the given spatial category and the process repeats until the search succeeds or the possibilities are exhausted.

## Orientation and Labelling of Components

Many objects have salient features which allow us to distinguish between various orientations and sides of those objects. For example, the instruction which requests that the credit card be oriented "face up with name to the right" requires us to identify the card's face and to orient it accordingly. A further example is the instruction which requests that the handset be replaced "heel first." Such instructions refer to the fronts, backs, tops, bottoms, and left and right sides of objects. SPACL, the SPAtial Component Labeller (pronounced as "spackle"), maintains a case library of object representations with associated labellings based on the encoding of 'directing axes' in those representations (e.g. the top of a bottle is generally the side with an opening, and the front of a credit card is generally the side with the picture and bank name). These cases are used to resolve the labelling of components of objects based on previous, similar labellings (e.g. The front of the Airphone is the side with the phone hanging on it, based on our knowledge that the front of a payphone is the side with the phone hanging on it). More information on SPACL may be found in [Alterman *et al.*, 1991].

The I/D (Intrinsic/Deictic) labeller distinguishes between deictic and intrinsic reference systems given a referenced object. The I/D Labeller uses a case-based implementation of an algorithm suggested by Miller and Johnson-Laird [Miller and Johnson-Laird, 1976] to determine whether spatial descriptions imply intrinsic or deictic reference systems. Semantic knowledge of the referenced object is used to decide if the object has such characteristics as inherent perceptual apparatus, a normal orientation with respect to people, a characteristic direction of motion, etc. When these criteria indicate that the referenced object has an intrinsic ref-

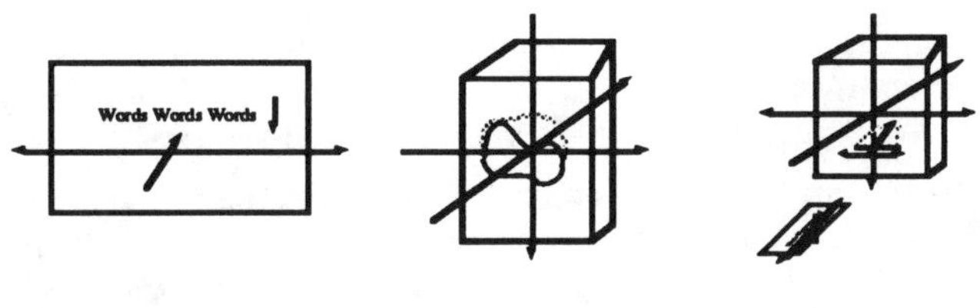


Figure 2: Pictorial Representation of Card, Airphone, and Card into ATM.

erent for the relation specified, the intrinsic system is used. A deictic system of reference is used when no intrinsic referent exists, or when insufficient semantic knowledge is present. Additional information on the I/D Labeller is also in [Alterman *et al.*, 1991].

## FLOABN

SPATR has been applied to spatial problems within the context of the FLOABN (For Lack Of A Better Name) project (*Instruction Usage*: [Alterman and Zito-Wolf, 1990], *Adaptive Planning and Learning*: [Zito-Wolf and Alterman, 1990; Alterman *et al.*, 1991]). The FLOABN project attempts to adapt previously devised plans for the use of devices to new types of devices. In doing so, it may need to make use of instructions as well as generic semantic knowledge of device use.

SPATR will be invoked either through a failure in the adaptive planner (e.g. the planner tries to insert the credit card in a certain way, fails, and posts its failure to the blackboard causing SPATR to suggest different types of insertion), through text inferencing on instructions (e.g. the inferencer posts a partial interpretation of "Face up with card name to the right" and SPATR activates to resolve references to the card's face, and the orientation of the card [Carpenter and Alterman, 1991]), or through external reference resolution (e.g. the text inferencer interprets "door handle" to refer to a specific door handle, causing the reference resolver to look for a door handle, causing a request to provide a schematic description of a door handle, which activates SPATR).

## SPATR Example

Given a new device, such as the Airphone, and an instruction to "insert credit card face up, with card name to right," into it, CLTER starts with a full representation for the credit card and a partial representation for the Airphone as shown in Figure 2. This representation states that the Airphone is overall slab-shaped, with directed axis from top to bottom, front to back and left to right (the directed axis follows from the device being mounted against a wall, in a particular orientation). The representation also states the expectation (from semantic memory) that the Airphone will have a negative space of undefined shape somewhere on its front where the card may be inserted.

Indices are traversed and the most similar previous

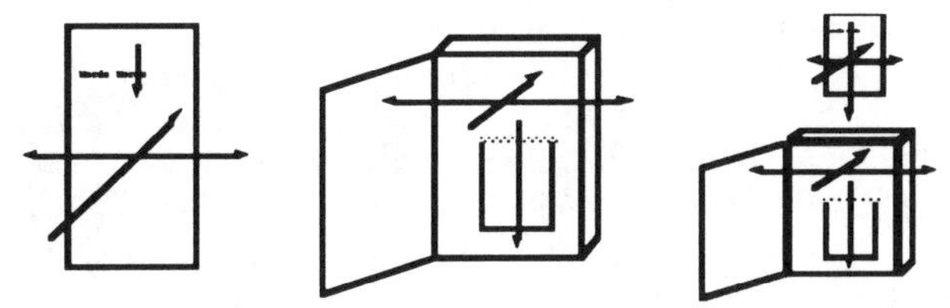

Figure 3: Pictorial Representation of Return Card, Library Book, and Insertion.

case of insert is retrieved. The indices used for retrieval can be interpreted as:

- The Airphone has a main or secondary axis from top to bottom. This is true since the Airphone is mounted in a particular orientation.
- The Airphone is three dimensional. This is true from a cursory examination of the Airphone.
- There is a subcomponent of the Airphone on its front or back. This is true from our assumption that there will be a place on the front of the Airphone in which to insert the credit card.
- There is not a subcomponent of the Airphone with a top or bottom that has further subcomponents. This index serves to discriminate between cases where there is a slot or hole on the top of the object (e.g. a piggy bank or trash can) and other cases. Since we can't know whether the Airphone has or lacks this feature, we assume the Airphone doesn't have it. Note that our strategy will be to jump to conclusions to get a reminding, use the specific information off of the reminding to test its validity, and reverse assumptions if we fail.

The case retrieved is that of inserting a bank card into an ATM. The actions represented by this insertion are shown pictorially in Figure 2. SPATR posts a request to the blackboard for the (unimplemented) perceptual system to find a slot similar to the ATM's, which fails. Based on this failure we reverse our assumption about slots or holes on top, and re-retrieve. The result of this new retrieval is inserting a return card into a library book. A pictorial representation for this insertion is shown in Figure 3. Based on this retrieval, SPATR asks the perceptual system to find a slot oriented vertically on the front of the Airphone, with an opening along its top. This request succeeds and this meaning of 'into' is chosen as the appropriate precedent for adaptation.

Next, SPATR must identify the front of the credit card. When given a schematic representation of a credit card, as in Figure 2, SPACL retrieves a previous, labelled representation for a similar object (in this case a library book return card, shown in Figure 3). The return card is selected since both objects are sheets (i.e. flat rectangles) and both objects have words on a side referenced by a directed axis (i.e. there is a differentiation between a side which has words and the opposite side). The return card specifies that the front of the return card is the side with the words (return dates) on it. SPACL uses this case to identify that the front of the credit card is the side with the words on it.

Finally, SPATR must decide whether "name to the right" refers to deictic or intrinsic right. The I/D labeller [Alterman *et al.*, 1991] decides that right is deictic (i.e. the viewer's) right, by analogy with mirrors, pictures, and playing cards, all of which are sheet-like objects with images and/or words on one side.

The final interpretation for "insert card face up, with name to the right" is, therefore, orient the credit card with the side which contains words towards the agent, with the edge of the card that contains the name toward the agent's right, and insert the card into a vertically positioned slot of the airphone with an opening towards the top of the slot with a downwards, sliding motion.

## Conclusions

We have identified several problems of spatial reference and provided a framework for their solution which is consistent with Adaptive Planning. We have implemented systems to label components of objects, resolve references to intrinsic or deictic coordinate systems, and choose appropriate instances of actions and spatial relationships to constrain visual search and drive action.

## References


[Alterman and Zito-Wolf, 1990] Richard Alterman and Roland Zito-Wolf. Planning and Understanding: Revisited. In *Proc. AAAI 1990 Spring Symposium*, Computer Science Department, Brandeis University, 1990.

[Alterman *et al.*, 1991] Richard Alterman, Roland Zito-Wolf, Tamitha Carpenter, Marc Goodman, and Scott Waterman. Readings from the FLOABN Project, Vol. 1. Draft CS-91-157, Brandeis University, Computer Science Department, Brandeis University, Waltham Massachusetts 02254-9110, 1991.

[Alterman, 1986] Richard Alterman. An adaptive planner. In *Proceedings of the Fifth National Conference on Artificial Intelligence*, pages 65–69, Philadelphia, PA, August 1986.

[Alterman, 1988] Richard Alterman. Adaptive planning. *Cognitive Science*, 12:393–421, 1988.

[Badler *et al.*, 1990] Norman Badler, Bonnie Webber, Jugal Kalita, and Jeffry Esakov. Animation from instructions. In N. Badler, B. Barsky, and D. Zeltzer, editors, *Making Them Move: Mechanics, Control and Animation of Articulated Figures*, chapter 3, pages 51–93. Morgan Kaufmann Publishers, Los Altos, CA, 1990.

[Biederman, 1987] I. Biederman. Recognition-by-components: A theory of human image understanding. *Psychological Review*, 94:115–147, 1987.

[Carpenter and Alterman, 1991] Tamitha Carpenter and Richard Alterman. Explanation-based learning and summarization. To Appear in *Proceedings of the Thirteenth Cognitive Science Conference*, 1991.

[Chen, 1987] Su-Shing Chen. A geometric approach to multisensor fusion and spatial reasoning. In Avi Kak and Su shing Chen, editors, *Proceedings of the 1987 Workshop on Spatial Reasoning and Multi-Sensor Fusion*, pages 201–210. Morgan Kaufmann Publishers, Inc., 1987.

[Goodman, 1989] Marc Goodman. CBR In Battle Planning. In *Second Proceedings of a Workshop on Case Based Reasoning*, pages 312–326, 1989.

[Goodman, 1990] Marc Goodman. Prism: A case-based telex classifier. In Alain Rappaport and Reid Smith, editor, *Proceedings of the Second Conference on Innovative Applications of Artificial Intelligence*, pages 86–90, 1990.

[Herskovits, 1985] Annette Herskovits. Semantics and pragmatics of locative expressions. *Cognitive Science*, 9:341–378, 1985.

[Jackendoff and Landau, 1990] Ray Jackendoff and Barbara Landau. Spatial Language and Spatial Cognition: A Swarthmore FestSchrift for Lila Gleitman. In J. Kegl D. J. Napoli, editor, *Bridges Between Psychology and Linguistics*. Lawrence Erlbaum Associates, Inc., 1990.

[Lakoff, 1987] George Lakoff. *Women, Fire, and Dangerous Things*. The University of Chicago Press, 1987.

[Marr, 1982] D. Marr. *Vision*. W. H. Freeman, San Francisco, California, 1982.

[Martin, 1990] James H. Martin. Computer understanding of conventional metaphoric language. Technical Report CU-CS-473-90, University of Colorado, Boulder, Computer Science Department and Institute of Cognitive Science, 1990.

[Miller and Johnson-Laird, 1976] George A. Miller and Phillip N. Johnson-Laird. *Language and Perception*. The Belknap Press of Harvard University Press, Campbridge, Massachusetts, 1976.

[Talmy, 1983] Leonard Talmy. How language structures space. In Herbert Pick and Linda Acredolo, editors, *Spatial Orientation: Theory, Research, and Application*, pages 225–282. Plenum Press, 1983.

[Zito-Wolf and Alterman, 1990] Roland Zito-Wolf and Richard Alterman. Ad-Hoc, Fail-Safe Plan Learning. In *Proceedings of the Twelfth Cognitive Science Conference*, Computer Science Department, Brandeis University, 1990. Lawrence Erlbaum Associates.

# The Microgenetic Analysis of an Origami Task

**Daniel M. Gruen, Giorgio Ganis, & Randy Gobbel**

Department of Cognitive Science 0515
University of California at San Diego
La Jolla, CA 92093
Phone: (619) 534-2088
email: gruen@cogsci.ucsd.edu

## Abstract

An experiment was conducted in which subjects repeatedly constructed Origami boxes, following example models displayed by the experimenters. A microgenetic analysis was performed on videotapes of the experiment. Results show an increase in speed generally following the power law of practice, and the rearrangement and combination of operations into larger units. They give evidence for the importance of external information in the tasks people perform, and contain a possible example an occasion of insight prompted by earlier breakdowns. Most importantly, the experiment shows how an apparently straightforward improvement in performance can be dissected to uncover the myriad factors and effects that underlie it.

## Introduction

Many researchers have investigated the ways people improve when performing repeated tasks. Newell and Rosenbloom (1980) showed that the power curve of practice can be applied broadly to many aspects of performance, and applied the chunking theory of psychology as a partial explanation for this phenomenon. Gentner (1982), studying the evolution of typewriting skill, and Agre and Shrager (1990), who performed a microgenetic analysis of a copying machine task, demonstrated the ways in which tasks undergo qualitative changes as they are performed over time. Norman (1988) and Zhang (1990) discussed the ways in which external information is used in problem solving tasks, not only as a memory aid, but also as a mechanism by which the nature of the task itself is changed. There has also been considerable investigation into the ways in which people's mental models of the tasks they perform develop and evolve over time (Johnson-Laird (1988), Simon (1985), Kintsch and Greeno (1985), and others).

We set out to see how many of the phenomena identified in the substantial literature on skill acquisition and development would appear in a relatively simple, short term study based on observing subjects as they repeatedly constructed origami boxes. This task was chosen for several reasons. Firstly, the task was complex enough so that the task could evolve over time, yet simple enough so that it could be learned quickly by novices. Secondly, the task involves easily observable manual manipulations which would, it was hoped, provide clues to the subjects' thinking as they performed the task. Finally, the task contains three distinct yet related stages, and it was hoped that subjects would come to understand the differences and relationships among the stages. The central, and most complex, stage was comprised of a set of repetitive steps which had to be repeated in each of four places, but allowed the subjects some flexibility in determining the specific order in which they were performed. Finally, the complete task could be performed in several minutes, so many trials could be run and analyzed in the time available for this research. It was felt that many repeated trials were desired to increase the likelihood that patterns in performance would emerge and be identified with some assurety.

Even in this relatively simple, short-term experiment, we found evidence for many of the features of skill acquisition and development identified by previous researchers. The time in which each trial was performed decreased, roughly following the power curve law of practice. The task evolved qualitatively over time. There was evidence of chunking as operations were combined into identifiable, frequently repeated patterns. We also observed a clear occurrence of insight, and the context in which it took place and the effects it had on overall performance gave credence to established theories on the role of insight in problem solving and skill development.

## Experiment

### Procedure

Subjects were presented with a set of ten models showing each stage in the construction of an origami box. The models were arrayed in order from left to right on a table at which the subjects sat; each model was also numbered to indicate the order in which the steps could be performed to complete the origami box. This method of instruction was chosen after attempts to devise a set of clearly understandable, written instructions proved quite difficult. Also, experimentation with published origami instructions showed them to be confusing and hard to follow. It was also hoped that it would be easy to monitor the use of the models by the subjects, by observing when they looked up at the models or took them in hand. The subjects were given a stack of 8.5 inch square sheets of used copier paper with printing of some sort on one side. An experimenter sat next to the subjects at all times, and was available to answer questions and provide assistance as requested.

The subjects were asked to repeatedly make origami boxes, using the models in front of them as guides. The subjects were told that their times were being recorded, and were given the stated goal of completing a box in under 75 seconds as the criterion for terminating the experiment. The subjects were encouraged to ask questions as needed if they encountered difficulties in understanding the instructional models. They were also told to feel free to deviate from the exact sequence of steps represented by the models if they felt that doing so would be faster or easier for them; it was emphasized that the time in which they completed the boxes was important and not the specific steps they followed to make them.

The entire experiment was videotaped with the permission and knowledge of the subjects. (The videotape camera was clearly visible in the room, and red lights visible to the subject indicated when the machine was recording.)

After the main trials, the subjects were asked to describe the procedure for making an origami box from memory, with neither paper nor the instructional models in front of them. The subjects were then given a sheet of paper and asked to make an origami box while describing aloud the steps they were taking.

## Subjects

Two subjects participated in the experiment. Both were female college graduates in their mid-twenties, and both were left-handed. Neither subject had extensive prior experience with origami, and neither subject knew before the experiment how to construct an origami box.

## The Task

The process of making a box was divided into nine steps, grouped into the stages of *preparation*, in which scoring folds are made which facilitate later structural folds, *folding*, in which the major structural folds of the box are made, and *inflation*. The usual way of turning the flat structure produced by folding process into a three-dimensional box is to blow sharply into one end. These steps are diagrammed in Appendix A. The folding stage (steps 5-8) involved four folds that must be made in each of four places on the paper shape. The nine steps used to make the origami box are:

*preparation:*
1. horizontal fold in half
2. diagonal fold
3. diagonal fold
4. "collapse" to a triangular shape

*folding:*

5. fold bottom corners up (in four places)
6. fold side corners to middle (in four places)
7. fold top corners down (in four places)
8. fold corners into "pockets" formed in step 6 (in each of four places)

*inflation:*

9. inflate into box

We devised a notational scheme to represent the sequence of steps in the order in which they occurred, as it is in the changes to this order that some of the more interesting results appeared. Diagrams of selected trials using this notation accompany the text. Figure 1. illustrates this notation.

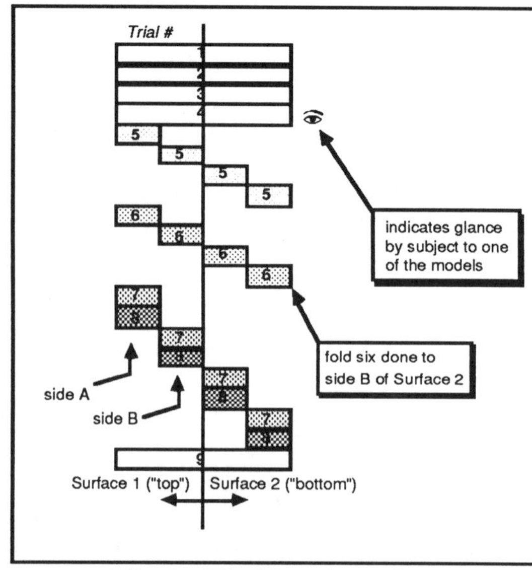


**Fig. 1 The Notational Scheme**

*In this example, the subject performed steps 1 through 4, then step 5 in all four places, then step 6 in all four places, then steps 7 and 8 together in each of the four places, and finally step 9 to complete the box. The "eye" symbol (👁) indicates that the subject glanced at one of the models at some point while performing step 4.*

## Results

The first subject was analyzed in the most detail, and it is her performance that will be discussed in the most depth here. Subject one showed a clear increase in overall speed, roughly following the power curve of practice. Speeds for individual stages within the trials also tended to increase, though at different rates and following different trends than did the overall speedup. These results are plotted on a log-log scale in figure 2.

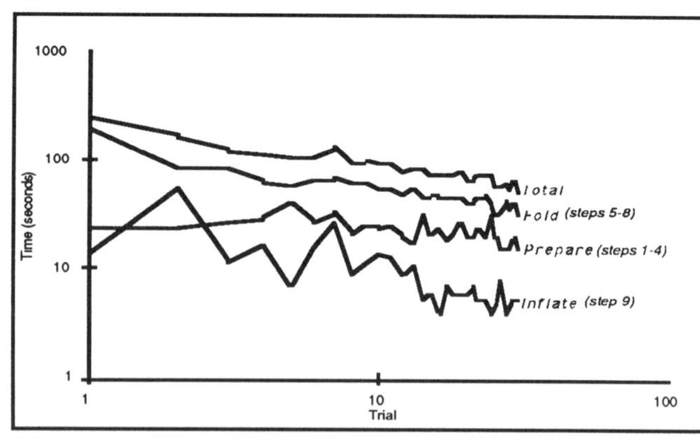


**Figure 2. Times for subject one on a Log-Log Scale**

At first, the subject glanced frequently at the models, but these glances dropped off rapidly to only three by the fifth trial, and to only one in trials seven, eight and ten. After this they were eliminated completely except for during trial twenty-four, which will be discussed further below. The glances were mostly made during the first six steps; in fact, after the third trial the subject did not glance at the models at all during the last three steps of making the boxes (steps 7-9). Many of the glances were "look-ahead" glances, in which the subject looked up at a model as she was finishing the fold she was making, in preparation for beginning the next fold (Figure 3).

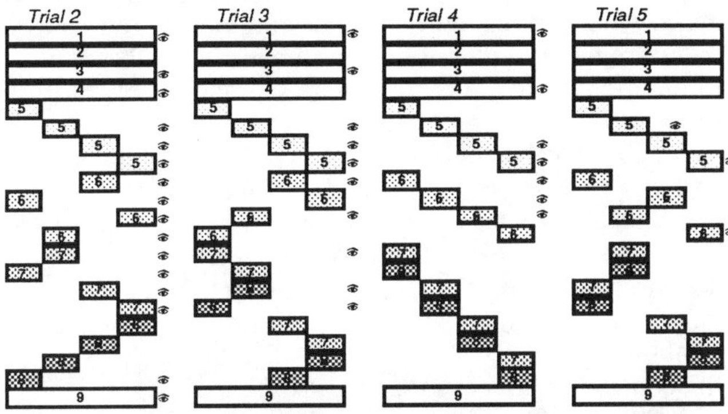


**Fig. 3 Glances during early trials (shown by 👁's).**

The subject started to make errors in only two of the trials. In the first trial, when the subject was first learning the task, and later in the ninth trial, the subject started to make a fold on a corner out of sequence (eg. fold 7 before fold 6 had been made), but she almost immediately realized her mistake and made the proper fold.

There was clear evidence of qualitative changes in the task over time. The incidence of "chunking", by which we mean places in which operations were combined and performed in groups rather than singularly for each of the corners, increased, as can be seen in the detailed summary of trials in the appendix. At first, these chunks consisted of two operations being performed on a given corner before moving on to another corner. In the seventh trial, a larger chunk emerged, which was used for both the top and bottom surfaces of the construction (Figure 4).

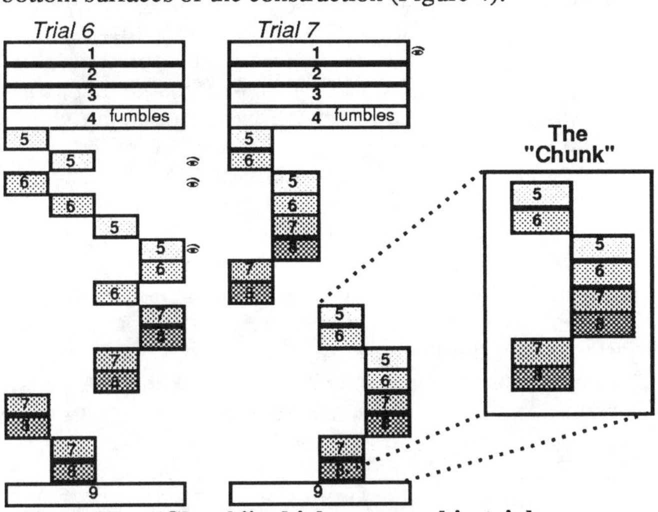


**Fig. 4 The "Chunk" which appeared in trial seven.**

This chunk was then used consistently for *each and every one* of the subsequent trials on the bottom, or second, surface. It was not, however, followed again for the top surface until trial thirteen, which was preceded by a twenty-five minute break during which the subject walked around, talked with the experimenters about topics unrelated to the experiment, and drank a cup of decaffeinated coffee (Figure 5).

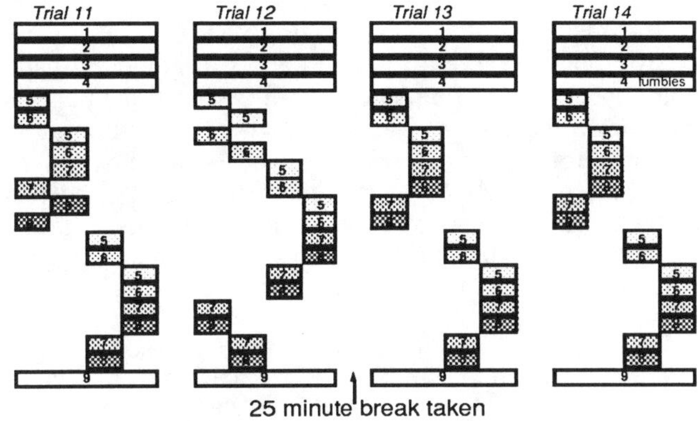


**Fig. 5 Increased use of the "chunk" after a break.**

From trial thirteen on, this chunk was used most of the time for the top surface up to trial twenty-four. In each of the three cases during these trials in which this chunk was not used for the top surface, the subject had encountered some difficulties in an earlier step. This included fumbling during step four (trials 19 and 24) and a corner jamming during step five (trial 17).

The fourth step, which consisted of collapsing the paper to a triangular shape, caused the subject difficulties in many of the trials. This step was often accompanied by noticeable fumbling. The subject noticed this problem early on, as seen by the following exchange which occurred after trial six:

S: Did you guys also have trouble with that middle folding part, in step 4?

E: Uh... Different people have trouble with different parts.

S: That's interesting...

There was some evidence that the subject understood the purpose of the first three steps as creating creases to make this easier and neater. In trial five, the subject fumbled during step 4 and *then went back to rescore the diagonal folds she had made in steps 2 and 3*. In fact, the subject had, from the beginning, been making the first fold in the wrong direction. Instead of helping the paper neatly fall into place, the fold was actually working against the subject, at best not helping and occasionally hindering her efforts to collapse the paper into the triangular shape she needed to proceed. This was something the subject did not realize until the twenty-fourth trial (Figure 6).

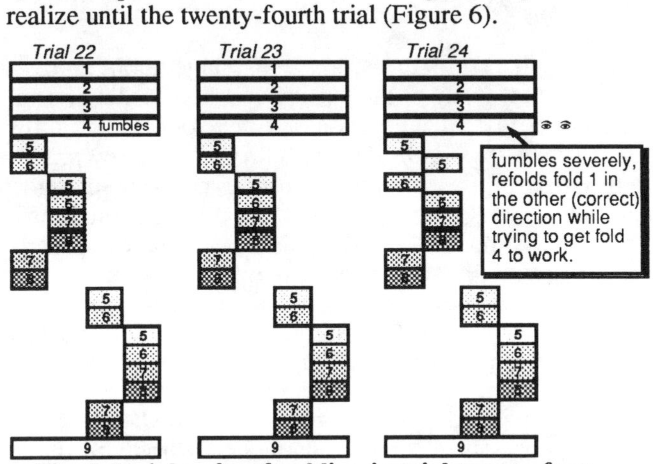


**Fig. 6 Insight after fumbling in trial twenty-four.**

In this trial, the subject fumbled severely while performing step 4 and in her attempts to guide the paper into place, opened it up and forcefully folded the paper in half down the center. The subject then glanced up at the models (for the first time since trial ten), collapsed the paper more easily into the triangular shape, and glanced up at the models again before proceeding with steps 5 through 9. After completing this trial, the subject had an exchange with the experimenter in which she said:

S: That cost me time, but I just realized something. Either I've been misunderstanding how this fold was supposed to be made, or it's just simply easier if you do it the other way. It may eliminate my fumbling.

The subject immediately incorporated the corrected step 1 fold direction into her procedure. Interestingly, this new addition seemed to perturb the chunking the subject had been using, causing the subject to deviate from using the chunk described above. As before, this deviation only affected the first surface; the subject continued faithfully to use the chunk on the second surface. Discovering the proper direction for the fold in step one helped the subject in two ways. Firstly, it reduced the time required, on average, for step 4. Secondly, it made the times for step 4 much more consistent than they had been. This is an effect that the subject herself realized. After trial twenty-seven, the subject commented about the newly discovered fold direction:

S: I was continuing to do this .../demonstrates the wrong direction fold/... which meant this fold, this part, was inside-out, and I had to play with it, sometimes it worked easily and sometimes I fumbled quite a bit.

After performing the trials, the subject was asked to describe, from memory, the steps taken to make an origami box, with neither paper nor any of the models in front of her. The subject expressed the feeling that this would be difficult, and that she was not sure if she could do it accurately. While describing the steps she took, the subject used her hands extensively, following many of the motions she would actually use to construct an origami box.

## Discussion

The results from the trials with subject one showed evidence for many of the phenomena discussed in the vast literature on the acquisition and development of skills. Performance overall increased, roughly following the power law of practice, yet this overall increase represented the cumulative impact of several different qualitative effects. As mentioned above, there was clear evidence for the combination of individual operations into larger procedural units or chunks. This chunk was first used consistently for the first surface after a break had been taken, lending some support to the idea that the chunk was stored as a single unit, and retrieved as a unit by the subject after she returned from her break. The hypothesis that this chunk represented a rather automated procedural unit is strengthened by the fact that subsequent deviations from use of this chunk were traceable to fumbling of some sort in an earlier stage. It appeared as if these instances of fumbling caused the subject to think consciously again about what had become a rather automatic process. She therefore neglected to use the automatic chunk but returned to the more deliberate operations she had followed earlier in the task.

It was after the chunks had been used repeatedly on both surfaces that the subject remarked that the task was "just now becoming mindless." The fact that this remark happened to be made after a trial in which she did, in fact, deviate from using the chunk on the first surface need not minimize its significance. It is even possible that the fumbling which occurred in that trial, by requiring the need for conscious intervention, led to the realization that she had been performing the task automatically. It is also possible that she felt that she had been performing the task *too* automatically, causing her performance to become more prone to mistakes and fumbling. She seems to express the belief that the task becoming more automatic was a factor in its becoming sloppier when she later states that she was "starting to, like, just do it now, without thinking" but that "it's getting sloppier also though." Interestingly, the fastest time was recorded for the subject in the last trial, in which she used the chunk for both surfaces and which she performed, in part, while talking, strengthening the impression that the task had truly become an automatic one.

The instance of insight observed in step twenty-four, in which the subject discovered that she had been incorrectly folding folds 2 and 3 in the same direction as the first fold, contained many of the characteristics identified early on by Ruger (1910), Woodworth (1939), and others both in its causes and its effects on subsequent performance. This insight was led to by the subject's manipulations while trying to overcome severe problems in completing step 4. It is significant that the subject had encountered similar difficulties with this step several times before, but never severely enough to cause the additional folding which led to the subject's insight. The time required for the "prepare" stage (steps 1-4) was therefore higher in trial twenty-four than it had been in the immediately preceding trials. This insight improved performance in subsequent trials in two ways – the prepare stage was made not only faster but also more consistent.

Several insights into the mental models the subject was using to complete the task can be drawn from observations of her behavior, and through protocol analyses of the things she said during and after the task. As mentioned above, the subject showed an understanding of the relationships between the tasks, realizing, for example, that the folds made in the first steps would assist with collapsing the paper in step 4. The subject seemed to realize early on that steps 5-8 could be combined in any way she wanted, as long as they were performed in the correct sequence within each corner. The subject moved her hands through the motions of constructing an origami box when she attempted to describe the procedure from memory, implying that a somewhat kinesthetic procedure was involved in the subject's internal representation.

There was also substantial evidence for the role external knowledge played in the subject's performance of the task, in many of the same ways that were identified by Norman (1988) and Zhang (1990). The paper being folded itself provided an indication of where within the procedure the subject was. It also provided certain affordances, helping to insure that the proper sequence was followed since it made some steps impossible before others had been performed. It also provided constant feedback, in that a wrong fold would lead to an unusual shape the subject could recognize as one not normally encountered when the procedure was followed correctly. This was seen in the way the subject quickly recovered after starting to make an error in trials one and seven when she realized that either she could not yet make the fold she was intending to make, or that it would not look right if she did. The subject alluded to the use of these affordances when recalling the procedure from memory. She mentions folding "the four pieces you can" and talks about folds meeting and lining up with other folds that had been made previously. A careful reading of the transcript of this portion of the protocol analysis shows obvious mistakes made by the subject in describing the shapes encountered at stages of the procedure that caused her no difficulties during the actual trials. It seems likely, therefore, that the subject was making use of information beyond that contained in her internal representation. It appears that the subject herself acknowledged the importance of this external information, in that she was not sure if she could accurately describe the task totally from memory, and that after she attempted to do so she asked, "Did I get it?".

## Summary

The illustrative experiment discussed in this paper demonstrated many of the phenomena previously described in the literature on skill acquisition. It showed how operations can be rearranged and combined into larger chunks, and how these chunks can then be stored and utilized almost automatically. It revealed a clear example of the effect that failure can have in prompting moments of insight that lead to qualitative changes in the tasks that are performed. It gave evidence for the importance of external information in the tasks people perform. Most importantly, the experiment showed how an apparently straightforward improvement in performance can be dissected through microgenetic analysis to uncover the myriad factors and effects that underlie it.

## References

Agre, P. & Shrager, J. (1990). Routine evolution as the microgenetic basis of skill acquisition. *Submitted for the Proceedings of the Cognitive Science Society*. Cambridge, MA. Hillsdale, NJ: Erlbaum.

Gentner, D.R. (1982). *The Development of Typewritten Skill*, Center for Human Information Processing, University of California, San Diego CA

Johnson-Laird, P. N. (1988). Mental models. In M. Posner (Ed.), *Foundations of Cognitive Science.* Cambridge MA: MIT Press.

Kintsch, W. and Greeno, J.G. (1985). Understanding and Solving Word Arithmetic Problems. *Psychological Review, 92, 109-129*

Newell, A. & Rosenbloom, P. S. (1981). Mechanisms of skill acquisition and the law of practice. In J. R. Anderson (Ed.), *Cognitive skills and their acquisition.* Hillsdale, NJ: Erlbaum.

Norman, D.A. (1982). *Learning and Memory,* San Francisco CA: W.H.Freeman and Co.

Norman, D.A. (1988). *The Psychology of Everyday Things,* New York NY: Basic Books

Ruger, H.A. (1910) The Psychology of Efficiency. *Archives of Psychology*. 1910, 19, 15. Quoted in Norman (1982) and Woodworth (1939)

Simon, H. (1985). *The Science of the Artificial.* 2nd ed. Cambridge MA: MIT Press

Woodworth, R.S.(1939). *Experimental Psychology*. New York NY: Henry Holt and Co.

Zhang, Z. (1990). *The Interaction of Internal and External Information in a Problem Solving Task,* Department of Cognitive Science, University of California, San Diego CA

## Appendix A: Steps to Make an Origami Box.

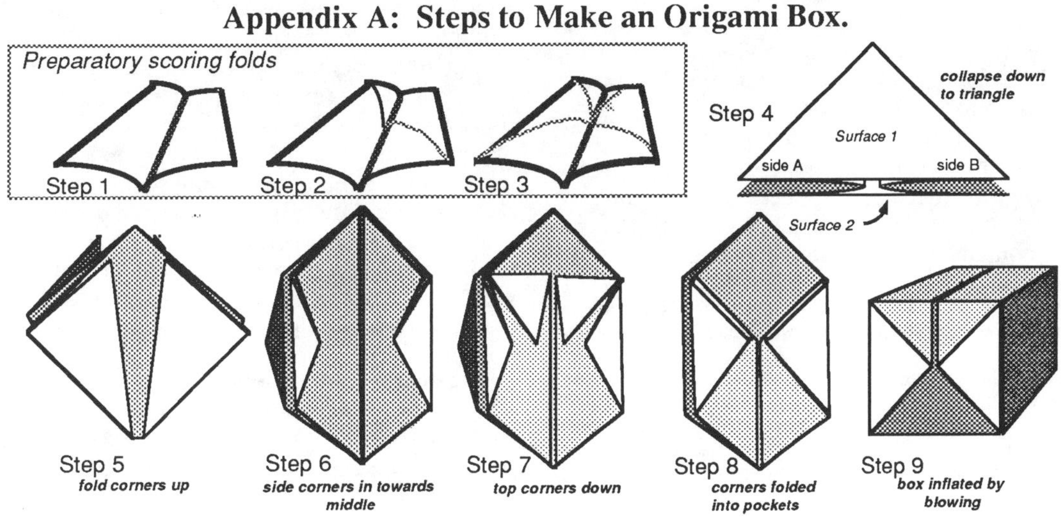

# Partial Match and Search Control via Internal Analogy*

**Angela Kennedy Hickman**
School of Computer Science
Carnegie Mellon University
Pittsburgh, PA 15213
ach@cs.cmu.edu

**Marsha C. Lovett**
Department of Psychology
Carnegie Mellon University
Pittsburgh, PA 15213
lovett@psy.cmu.edu

## Abstract

In a previous study (Hickman & Larkin, 1990), we introduced a within-trial analogy mechanism called *internal analogy* that transfers both success and failure experiences between corresponding parts of the search tree for a single problem. In this paper, we describe powerful extensions to the learning procedure and their consequences on problem solving behavior. First, we explain how our similarity metric can be naturally augmented to provide a more flexible partial match. To overcome the need for a static measure, however, we propose a mechanism that *learns* the appropriate level of partial match through feedback from previous analogical reasoning. Second, we show how this partial match mechanism controls the problem solver's search. Protocol data from a subject working in a geometry theorem-proving domain provide support for the psychological fidelity of the extended internal analogy model.

## Introduction

Analogical reasoning is a psychologically-motivated computational mechanism for efficient problem solving. In essence, it reduces search by modifying and transferring components of previous solution instantiations to guide the current problem solving. Unlike macro-operators (Anderson, 1983), chunking (Laird & Newell, 1986), and EBL (Mitchell, Keller, & Kedar-Cabelli, 1986), analogy employs partial match and adaptation of past knowledge instead of exact match and direct replay. This allows a previous experience to provide some search guidance even if it cannot provide an exact solution. Although the process is not completely understood, evidence suggests that people reason analogically when learning a new domain (Faries & Reiser, 1988) or to reduce the amount of effort required to solve a problem in a known domain.

In a previous study (Hickman & Larkin, 1990), we introduced a psychologically motivated technique called *internal analogy* that transfers both success and failure experiences from a previously solved subgoal to a current subgoal that is contained in the same problem. Most previous studies of analogy in problem solving have investigated transfer between problems from similar domains (within-domain analogy) (Carbonell & Veloso, 1988). Psychological models of analogy have concentrated on the transfer of knowledge between similar problems in different domains (cross-domain analogy) (Gentner & Toupin, 1986). In general, both of these types of analogy require an acceptable partial match between entire problem instances and then transfer entire problem solutions. On the other hand, EUREKA (Jones, 1989) uses analogical reasoning to choose only the next problem solving step. Unlike all of these systems, internal analogy takes advantage of whatever regularity exists between the two problems in as efficient a manner as possible. Since the process may be applied to multiple subgoals while solving a single problem but does not require the entire problem definition to acceptably match a source problem, it should be better able to exploit symmetries and other regularities that exist in the search space. Being able to accomplish transfer at varying levels of granularity is an important mechanism for optimally reducing search. As a result, internal analogy appears more widely applicable, providing greater search reduction.

The current implementation of internal analogy provides only for exact match and direct replay. In this paper, however, we first describe how the *information content metric*, our similarity metric, can be modified to implement a more flexible partial match. In order to overcome the need for a static measure, though, we describe a mechanism that *learns* the appropriate level of match through feedback from previous analogical reasoning. Using this new match criterion, internal analogy transfers the solution in an efficient compiled form and also automatically identifies what information is missing and requires adaptation. The adaptation process is carried out by recursively applying internal analogy or other general problem solving methods to find the missing segments of the solution. Secondly, we show how this new partial match mechanism will provide search control for the problem solver by using it to model protocol data of a subject solving geometry problems. This analysis provides even more support for the psychological fidelity of the internal analogy model.

The next section presents the model of the extended internal analogy process, and the following section provides evidence from the protocol that supports the psychological fidelity of the extended model. The final section summarizes the results and discusses plans for future work.

*This research was supported in part by DARPA contract number F33615-87-C-1499 Amendment 20, and AIP-ONR contract number N00014-86-K-0678-N123. Angela Kennedy Hickman was partially supported by a Zonta International Amelia Earhart fellowship. Marsha Lovett was supported by a NSF Graduate Research Fellowship. We thank our collaborators, John Anderson, Jaime Carbonell, Jill Larkin, and Peter Shell for their insights, and Peter for his invaluable help in the implementation of the model.

## Internal Analogy Model

Before describing the extensions to the model, we briefly review the internal analogy process. When a new subgoal is opened, the system examines previously solved goals of the same type, the candidate source goals. It then calculates something similar to the weakest preconditions for a candidate source's solution from the instantiated productions that compose that solution.[1] If these preconditions are all satisfied in the current target problem (information-content(candidate) $\leq$ information-content(target)), then the candidate is chosen as the actual analogical source. Otherwise, it is rejected. If the candidate is chosen as the actual source, its solution is appropriately mapped by instantiating the productions that solved it in the current working memory and replaying them to solve the current subgoal. This process operationally defines the relevant structure that is important to be mapped. Since this newly solved subgoal is contained in the current problem, it automatically becomes available as a future candidate source. The system learns from its failures using the same mechanisms. The interested reader can find more details on the algorithm in (Hickman, Shell, & Carbonell, 1991).

Empirical studies and a theoretical analysis demonstrate that this version of internal analogy produces significant savings in search during problem solving (Hickman, Shell, & Carbonell, 1991). However, it is extremely conservative; it errs by undergeneralizing, not overgeneralizing. This conservative strategy may inhibit the learner from using source solutions that actually could provide some, if not complete, search guidance. To reach its full effectiveness, the internal analogy process must be extended to perform a more flexible partial match and adaptation. In the following discussion, we show how natural extensions to the mechanisms already in place can produce this behavior.

The two similarity metrics the system currently uses are the comparisons of 1) the quantity types and 2) the information contents of the source and target goals. One way to relax the current exact match is by allowing goal types that are "close" in the type hierarchy to match. Additionally, we can expand our definition of compatible information contents. For instance, if in the initial state of the candidate source, a variable had a known value which is not known in the current state, the candidate source may still be determined to be compatible with the target goal. In that case, internal analogy can apply the source solution until the value of the unknown variable is needed. At that point, it can recursively use the problem solver's operators (which include internal analogy) to find the value of the unknown variable. This affords internal analogy a way of transferring compiled solutions at the correct level of granularity to achieve maximal efficiency. The ability to adapt the solution where needed and to fail it completely if necessary are capabilities which chunking, EBL, and macro-operators do not possess. All of these methods require complete justification before applying the compiled knowledge. Thus, by using partial match and adaptation during the reuse of experience, internal analogy will provide search control in many cases that the other methods cannot.

Next, we refine the notion of partial match even further. First, we associate a deviance measure with each pair of types in the type hierarchy. Second, we cache two deviance weights for each operator precondition. The purpose of these weights is to reflect how closely the weakest preconditions contained in the source information content are required to match the target. The first one of these weights denotes how deviant it is for an information content precondition to be totally absent from the current working memory. The second weight denotes how deviant it is for this precondition to be filled by the same type working memory element instead of by the exact same working memory element. To compute the deviance between the candidate source and target subgoals, we combine the quantity type deviance measure with the information content deviation weights, using the least deviant instantiation of the target information content.

With the partial match mechanisms in place, we can extend the functionality of the system to *learn* the correct circumstances under which solutions can transfer. Alternately, one could view this as learning the correct notion of partial match for the domain. This learning is accomplished by the system as it tunes the deviance measures described above according to feedback from the previous analogical reasoning. The feedback contains information about whether the analogy failed or succeeded and how much effort was required to adapt the retrieved solution. In summary, not only do these deviance measures provide the system with a method for choosing the best source, they provide a way of learning the definition of "best" for any given domain.

The consequence of this new flexible partial match mechanism is that it provides even greater search control for the problem solver than the mechanism did previously. Comparing the deviance measures for all the candidate sources, internal analogy can perform a least-deviant-first search over its memory to find the best candidate source to choose as the actual source. The lower the total deviance of the source, the less adaptation that will be required. As such, there is a tradeoff between the amount of search the system does to find the best source and the amount of adaptation that source requires. In addition, the acceptability of the best candidate source is determined by a threshold that rejects any source whose total deviance is too high. This threshold is based on how much search the problem solver has already undertaken. If the best source is determined to be acceptable, the problem solver reuses the solution stored with that source. As the deviance measures become "tuned" through learning, the search for the best source gets even smarter. Thus, internal analogy finds increasingly better analogical sources by a smarter search through memory instead of by a constant reorganization of it.

In addition to providing search control for backward

[1]This can be viewed as using a form of guided lazy EBL for retrieval. However, the presence of the target goal provides guidance as to 1) what to learn and 2)what amount of generalization to perform. Since our original study in 1990, Veloso has implemented a closely related mechanism for PRODIGY, a general purpose planner (Veloso & Carbonell, 1991).

chaining as described above, internal analogy utilizes the new partial match metric in a different way to provide search control for forward chaining. In forward chaining, there is no target goal driving the search. However, there is a decision point at each operator application. Therefore, instead of fixing the target goal and finding the source goal via least-deviant-first search, the forward chaining problem solver picks a source goal, most likely via recency, and uses the partial match mechanism to find the least deviant target goal. In this way, the problem solver uses a known strategy to try to easily elucidate more knowledge about a problem. As can be seen in the next section, this new partial match mechanism provides complex but useful search control guidance to the problem solver, which is not easily explained using any existing theories.

## Evidence from the Protocol

The original computational model was compared with the protocols of four subjects solving DC-circuit and fluid statics problems, and traces of its behavior matched the subjects' use of internal analogy. In the current paper, we explain the behavior of one subject, WB, solving a problem in the domain of geometry theorem-proving using the extended internal analogy model. This domain and the resulting problem space is more complex than the DC-circuit domain, and thereby provides an even better test of the extended model's psychological fidelity and its general applicability.

To begin the experiment, the subject, WB, studied a geometry review sheet, which remained available to him throughout the study. Next, four problems were presented to WB that had the potential to elicit both internal and inter-problem analogy. (In the analysis below, we describe his protocol for the third problem.) After the subject finished each problem, the experimenter removed the review sheet from view and presented the next problem.

WB's search for a solution to the problem in Figure 1 can be segmented into three parts: the opening, the intermediate segment, and the final path. These segments are distinguished by the types of goals WB verbalizes, the strategies he employs, and the degree of success he attains.

During the opening, WB restates the problem and weighs the potential of various solution strategies. He then embarks on a forward chaining sequence that includes all the conservative applications of internal analogy. Even though WB's opening strategy does not seem to be directed at fulfilling any specific goal, it does provide a relatively easy way of elucidating a lot of new knowledge about the problem, some of which is needed for his solution.

The intermediate segment of the protocol is generally characterized by unsuccessful backward chaining and exhaustive search. WB attempts to backward chain on the current goal three times during this segment, yet he never gets beyond the first level of goal decomposition. Instead, WB spends the main portion of this segment doing exhaustive search through the entire set of operators. He accomplishes this twice by reading through the list of theorems on the review sheet and sequentially checking the applicability of each one. During this segment, however,

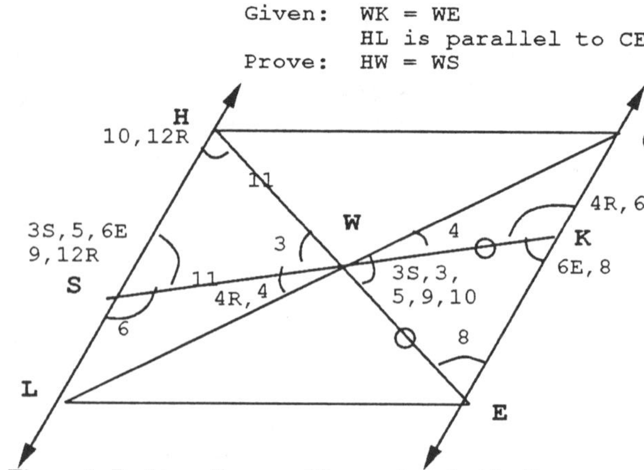


Figure 1: Problem diagram. The numbers in this diagram represent the result of WB's problem solving at each step; numbered arcs denote congruent angles, and numbered lines denote congruent line segments.

WB only gains two new pieces of knowledge.

WB's final path to solving this problem includes backward chaining with multiple applications of internal analogy. In this segment, WB solves the problem directly with no extraneous search.

In the following two subsections, we present a more detailed analysis of WB's protocol. It provides evidence for the psychological fidelity of the internal analogy model's extensions - the new partial match metric and the associated least-deviant-first search.

### Partial Match

During the forward chaining opening segment of the protocol, WB first uses internal analogy as his fourth search step (See Figure 2). His third step, then, is important because it serves as a source for the subsequent internal analogy. (Indeed, it is the only available source at this point in WB's problem solving.) In the third step of the protocol, WB tries to use the Alternate Interior Angles Corollary (AIAC, Figure 3). However, there are two bugs in his application of AIAC. First he states that <HSW and <KWE are alternate interior angles with respect to the parallel lines HL and CE and the transversal SK (Line 3S, Figure 2; and the corresponding step, Figure 1). This is an incorrect definition of AIAC, which WB reuses throughout the protocol; we call it AIAC-av. Second, WB does not correctly write down his verbalization of this step on the solution sheet. Instead, he transcribes that <SWH = <KWE, thus, introducing the second bug[2] (Line 3, Figure 2). We call this incorrect definition of alternate interior angles AIAC-vv. (According to the AIAC rule in Figure 3, these bugs correspond to incorrect definitions of the precondition *alt-int(<a>, <b>, <line1>, <line2>, <t>)*.)

Next, WB immediately applies internal analogy saying, "through using the same theorem [AIAC], I can say that CWK, angle CWK, is congruent to, not sure if this is going

[2]This second bug fortuitously yields a correct piece of knowledge since <SWH = <KWE by the vertical angles theorem.

| Step | Operator | Instantiation | Added knowledge |
|---|---|---|---|
| 0. | Read problem | Current problem | WK=WE, HL//CE |
| 1. | CTCP | tri-SHW, tri-WKE, ?, ? | Suspend |
| 2. | AIAT | HL//CE | Already given |
| 3S. | AIAC-av | HL//CE, SK, <HSW, <KWE | <HSW = <KWE |
| 3. | AIAC-vv | HL//CE, SK, <SWE, <KWE | <SWH = <KWE |
| **4.** | **IA** | HL//CE, SK, <CWK, <SWL | <CWK = <SWL |
| 4R. | AIAC-av | HL//CE, SK, <CKW, <SWL | <CKW = <SWL |
| **5.** | **IA** | HL//CE, SK, <HSW, <KWE | <HSW =<KWE |
| **6.** | **IA-aa** | HL//CE, SK, <CKW, <WSL | <CKW =<WSL |
| 6E. | Told | Replace knowledge from 5. | <HSW = <WKE |

Figure 2: The Opening Segment of the Protocol. CTCP = Congruent Triangles Congruent Parts, AIAT = Alternate Interior Angles Thm., AIAC = Alternate Interior Angles Cor., IA = Internal Analogy.

```
IF      angle(<a>)
        angle(<b>)
        goal(congruent(<a>,<b>))
        parallel(<line1>,<line2>)
        transversal(<t>)
        alt-int(<a>,<b>,<line1>,<line2>,<t>)

THEN    congruent(<a>,<b>)
```

Figure 3: The AIAC Rule.

to help me but it never hurts, to angle SWL."* (Line 4, Figure 2) The target in Line 4 is a partial match to the source in Line 3, and its information content deviance with respect to the source is very low. To understand why, consider the AIAC rule shown in Figure 3. Since this internal analogy application transfers only a simple operator, the information content of the source includes instantiations for all the variables in the left-hand side of the rule. WB's target instantiation contains all the same working memory elements, except for the pair of angles. This produces a deviation value whose only non-zero terms correspond to these new angle instantiations. Due to the physical constraints of the geometry domain, changing any other variable in this rule (e.g. the parallel lines or the transversal) would necessarily imply a change of angles as well, and hence, a greater deviation value. Therefore, the optimal choice for the target instantiation is one that only changes the instantiations of the angles. This is exactly WB's choice in his internal analogy of Step 4.

As shown above, the partial match metric models WB's choice of a target for this internal analogy because it uses a deviance metric that is very sensitive to the similarity of the source and target instantiations. This sensitivity eases the problem solver's task not only by providing search control (as discussed in the next section), but by easing the burden of instantiation. By using a least-deviant-first search strategy, the problem solver is guaranteed to choose an analogy that requires making the fewest changes to the target instantiation. WB's protocol supports this efficiency argument, for when he first applies the AIAC rule, he verbalizes the entire instantiation: "Okay, and the what I want to say in here is use line... Which do I want to use as the transversal? ... Okay, first of all, I'm going to say that HL and CE are bis- are cut by a transversal line SK." Upon applying the internal analogy in step 4, however, he merely states the changes to the source instantiation, the new pair of angles. (See *'d protocol excerpt above.) This same behavior was observed in the forward chainer from the previous study.

## Search Control

Now we describe how the new partial match metric explains WB's complex but useful search strategy. WB begins his problem solving by stating the givens and the goals of the problem and then considers whether to use a generalized schema that consists of proving Congruent Triangles and then proving Congruent Parts (CTCP; Line 1, Figure 2). In weighing the merits of this strategy, he says, "Ok. There's a long way of doing and proving congruent triangles and then showing there are corresponding parts of congruent triangles which might be easier because I've just done that [referring to an earlier problem in the study]. But then there's also this alternate interior angles theorem which actually probably gets me the same way just a little bit quicker." With the statements, "long way," and "quicker," WB weighs the complexity, which relates to the size of the information content, and the optimality of this solution (Carbonell & Veloso, 1988). In opposition to this, he says, "might be easier because I've just done that." With this statement, the subject is estimating the efficiency of finding a solution by referring to the last problem he completed, during which he successfully used the CTCP schema. Thus, WB is considering using inter-problem analogy, which can also be explained using our notion of partial match. At this point in the problem solving, none of the preconditions of this schema are known; the total deviance measure exceeds WB's acceptability criterion; and the analogy is abandoned.

After reaching a dead end in Line 2, WB produces the buggy internal analogy of Lines 3 and 4 described in the previous subsection. At this point, the experimenter points out that <CWK and <SWL are not alternate interior angles with respect to lines HL, CE, and SK. The subject agrees and produces the behavior in line 4R. Notice that he now has returned to his original buggy definition of alternate interior angles first exhibited in Line 3S. WB's second application of internal analogy now occurs as he reapplies in Line 5 the immediately preceding AIAC-av operator to obtain the congruence of <HSW and <KWE. For the same reasons explained in the preceding section, this is the least deviant target with respect to the source in Line 4R; WB's behavior supports the predictions of the model once more. WB's third application of internal analogy occurs in Line 6 when he says, "Through the same reasoning <CKW is congruent to <WSL." In this least-deviant application, WB introduces yet another bug into his original buggy definition of alternate interior angles (AIAC-av). This bug yields the correct definition, and the subject proves <CKW = <WSL. Finally, the experimenter corrects the knowledge WB derived in Line 5 and establishes the correct knowledge <HSW = <WKE in Line 6E. At this point, there are no more analogous applications of AIAC left that involve the transversal SK. Since it is still early in the problem solving, changing the

| Step | Operator | Instantiation | Added knowledge |
|---|---|---|---|
| **9.** | IA | HL//CE, SK, <HSW, <EWK | <HSW = <EWK |
| **10.** | IA | HL//CE, HE, <SHW, <EWK | <SHW = <EWK |
| 11. | ITC | <HSW, <SHW, HW, WS | HW=WS |
| 12R. | TP | <HSW, <EWK, <SHW | <HSW=<SHW |

Figure 4: The Final Path Segment of the Protocol. IA = Internal Analogy, ITC = Isosceles Triangle Cor., TP = Transitive Property.

transversal raises the deviation level too much, and it is not an acceptable target. As a result, the subject has exhausted his available applications of internal analogy and must pursue another strategy.

During this opening segment, WB performs all possible applications of internal analogy, with the constraint that he allows only small deviations between the source and target. This strategy allows him to easily elucidate knowledge about the current problem, some of which is relevant to his solution. The forward chainer in the previous DC-circuit experiment employed this same strategy. However, the search space in that experiment was smaller due to the smaller number of operators (yielding smaller breadth) and the shorter solution (yielding smaller depth). As a result, this strategy was sufficient for that subject to solve the problem. WB, on the other hand, is still missing some crucial knowledge at this point in the protocol. Also, when WB uses a buggy source in this segment, he produces a buggy target solution (Lines 4 and 5, Figure 2). This too replicates a result from our previous study where Subject One used internal analogy to transfer a buggy equation to a new situation.

At the beginning of the intermediate segment, WB reconsiders using the CTCP schema from Line 1 in Figure 2. Like the forward chainer in the previous study, WB also verbalizes the information content as he says, "I do have two angles in each of these triangles." At this point, he realizes that the target's current information content is less deviant to that of its source than it was when he considered using CTCP in the opening segment. As a result, he spends most of the intermediate segment trying to use this strategy and eventually fails. The bulk of the segment is characterized by large amounts of exhaustive search through the entire set of operators.

At the beginning of the final path segment (Figure 4), WB posts the goal <SHW = <HSW and backward chains using the transitive property. As a result, he produces the analogical behavior in Line 9 of Figure 4, reestablishing the erroneous knowledge that was originally created in Line 5 of Figure 2 and later deleted by the experimenter. In this step, he chooses AIAC-av as a source from the opening segment and instantiates it with HL//CE, SK, <HWS, <EWK. This instantiation is chosen because it involves the same transversal as the source, SK, and, thus, has a low deviance level. Now, however, its deviance is even lower because one of the goal angles driving the backward chaining, <HSW, is an exact match.

Next, internal analogy explains WB's behavior in Line 10 of Figure 4 for several reasons. A different goal angle, <SHW, is an exact match this time, adding zero to the deviance measure for this instantiation. Unlike the internal analogies of the opening segment, the parallel lines and the transversal of this internal analogy are now only a type match. Originally, this mismatch caused the total deviation to be too high, making this instantiation unacceptable. (i.e. WB never took this step in the opening segment, even though he employed internal analogy many times.) However, after the large amount of exhaustive search in the intermediate segment, WB's threshold allows for more deviation and accepts the instantiation in Line 10. Line 11 completes the subject's buggy solution, which resulted from his buggy AIAC rule. And finally, in Line 12 the experimenter asks WB to fill in the transitivity step he left implicit in his solution.

## Conclusion and Future Work

In this paper, we have first shown how a powerful adaptive partial match can be integrated into the internal analogy model by extending the basic mechanisms already provided. This extended model also differentiates to a finer granularity the general notion of exact match as previously defined by macro-operators and chunking. Second, the extensions have proven useful for explaining the search control of a complex protocol in a complicated domain. They were sufficiently general to model both the exploratory forward chaining of the subject's opening segment and the efficient goal-directed backward chaining of his final path segment. In the future, we plan to implement and demonstrate the computational efficiency afforded by the new extensions.

## References


Anderson, J. R. (1983). Acquisition of proof skills in geometry. In *Machine Learning, An Artificial Intelligence Approach*, pages 191–220. Morgan Kaufmann, San Mateo, CA.

Carbonell, J. G. & Veloso, M. M. (1988). Integrating derivational analogy into a general problem solving architecture. In *Proceedings of the First Workshop on Case-Based Reasoning*, pages 104–124, Clearwater Beach, FL. Morgan Kaufmann.

Faries, J. M. & Reiser, B. J. (1988). Access and use of previous solutions in a problem solving situation. In *Proceedings of the Tenth Annual Conference of the Cognitive Science Society*, Montreal, Canada. Lawrence Erlbaum Associates.

Gentner, D. & Toupin, C. (1986). Systematicity and surface similarity in the development of analogy. *Cognitive Science*, 10:277–300.

Hickman, A. K. & Larkin, J. H. (1990). Internal analogy: A model of transfer within problems. In *The 12th Annual Conference of The Cognitive Science Society*, Cambridge, MA. Lawrence Erlbaum.

Hickman, A. K., Shell, P., & Carbonell, J. G. (1991). Internal analogy: Reducing search during problem solving. In Copetas, C., editor, *The Computer Science Research Review 1991*. The School of Computer Science, Carnegie Mellon University.

Jones, R. (1989). *A Model of Retrieval in Problem Solving*. PhD thesis, Department of Information and Computer Science, University of California, Irvine.

Laird, J. E., R. P. S. & Newell, A. (1986). Chunking in soar: The anatomy of a general learning mechanism. *Machine Learning*, 1(1):11–46.

Mitchell, T. M., Keller, R. M., & Kedar-Cabelli, S. T. (1986). Explanation-based learning: A unifying view. *Machine Learning*, 1(1):47–80.

Veloso, M. M. & Carbonell, J. G. (1991). Variable-precision case retrieval in analogical problem solving. In *Proceedings of the Third Workshop on Case-Based Reasoning*, Washington, D. C. Morgan Kaufmann.

# Some Principles for Route Descriptions Derived from Human Advisers

**Kristina Höök and Jussi Karlgren**
SICS, Knowledge Based System Laboratory, Box 1263, 164 28 Kista, Sweden
+46 8 752 15 00 (phone) +46 8 751 72 30 (fax)
kia@sics.se and jussi@dsv.su.se

## Abstract

Through a study with experienced driver-navigators, we have deduced some principles as to how route descriptions are constructed and expressed by humans. Some of these principles are implementable, and a rough outline of a program is presented. Given a plan of how to go from A to B in a city, the program produces a non-linguistic object that represents all the route information needed to present the route to a specific driver. A verbal description of that object is then produced[1]. The goal is to incorporate verbal descriptions in route guidance systems, primarily aimed at driver-navigators with some knowledge of the city.

Furthermore, we speculate into what kind of cognitive processes are involved when humans choose and describe routes.

## 1. Introduction

The purpose of the work described here, is to investigate route descriptions given by human advisers, in order to find principles as to how route descriptions are built and expressed. We wanted to find principles that could be implemented so that, given a plan, we would be able to generate a description of that plan automatically.

In the area of Interactive Route Guidance (IRG), systems aimed foremost at drivers with no or very limited experience of a city exist. These systems are able to guide the driver from one point to another in a city, for example see [von Tomkewitsch 1990], through giving detailed information in each intersection of what to do next. We wanted to deduce principles as to how we could describe routes to drivers with some or a lot of experience of a city (we shall hereafter refer to this group of people as *residents*). These drivers would not be satisfied with a route description that keeps giving detailed instructions in every intersection, and does not tell them where they are heading. Instead an overview of what route has been chosen, and a few helpful instructions while driving, give enough guidance for these drivers. The overview must, for safety reasons, be given before the trip starts, as we do not want to interfere with the already complicated driving task. This puts high demands on the presentation *form*, *length* and *content*. One question raised on content was: How can we abstract away from all the tedious details of a route description and still keep the description unambiguous?

The starting point for the study reported here was a pilot study [Höök 1991] where we tried to investigate the influence of form (map or verbal) on content of human route descriptions. Apart from demonstrating some differences due to the mode of presentation, this prestudy also showed that people are able to "chunk" a number of roads into one unit. For example, driving from the starting point to the nearest big road, in our experiment the E4, involved four roads. These four roads were "chunked" into the expression: "get down to E4".

We found this chunking mechanism interesting, since it could provide us with a means of reducing the length of a route description while keeping it unambiguous. We know from other studies on route guidance that humans choose routes depending upon at least two important factors: firstly, the route should, if possible, follow a pattern of trying to find a big road nearby, going on it for a while and then getting off (even if there exists a route on smaller roads that would be shorter) [Streeter 1985, Streeter and Vitello 1986]; secondly, the hierarchical level of a road (local, secondary and major roads) influences whether it is chosen or not [Pailhouse 1970, Chase 1986, Streeter 1985]. We were interested in seeing whether the principles for route choice would be reflected in the route description, and whether the reasons for route choice are the same that underlie the choice of route chunks that humans describe to one another.

In the study described here, we asked persons who have long experience of Stockholm and who from time to time are asked to give route descriptions, to give us two kinds of descriptions, one for a tourist, and another aimed at a Stockholm resident. Contrasting these two descriptions would give us a key to which principles had been used for doing the latter description.

Apart from finding out about the underlying principles, we were also hoping to explore the kind of concepts people use when they describe routes in order to get ideas how to generate, automatically, verbal descriptions of routes. We expected to find a restricted language of route descriptions; a limited set of expressions with quite definite rules of how these expressions could be combined.

It must be emphasized that the study should foremost be looked upon as a data collecting study rather than an investigation into human behaviour. We are interested in finding results that can be implemented, and used as an interface to a route guidance system. In section 4, we roughly outline of how we imagine that the results from the study can be used to make such an implementation. Even so, in section 3 we have allowed ourselves to speculate upon what kind of mental processing is taking place when the subjects chose and describe routes.

## 2. Experiment set-up

The experiment was conducted with routes in Stockholm, and the set-up consisted of 10 persons that were paid. The

This work was sponsored by IT4 research programme, within Prometheus, PRO-ART European Cooperation.

[1]The description does not necessarily have to be verbal, it could also, for instance, be a combination of maps and words.

subjects were taxi drivers, traffic policemen and residents who all were experienced drivers used to giving directions. The interviews took approximately 1 - 2 hours per subject. The subjects were given six different starting points and a destinations and were asked to do the following for each:

1. First find a route between the two points, either by head or using a map provided by us. Subjects were encouraged to think aloud during this phase.
2. Give a verbal route description of the chosen route to another experienced navigator (one of the authors). The map was taken away so that no "pointing" at the map would clutter up the results.
3. Give a verbal route description of the chosen route to a "tourist" driver (the other author).

(Throughout the rest of the document, we shall refer to the experimental leader who played the role of experienced driver as the *resident*, and the other experimental leader will be called the *tourist*.) The whole session was videotaped. The starting and end points were chosen so that routes between would vary in terms of:

- hierarchical level of the roads involved in the route, where hierarchical is taken to mean whether the road was an alley, big road or expressway;
- "patterns", where we expected different patterns to emerge, a pattern being, for example, drive from starting point to some big road, from big road to destination;
- distance, where two of the pairs had similar or identical starting points, and differed mainly in the distance to the destination.

## 3. Results

The influence of hierarchical level proved to be an important factor. We confirmed that the route descriptions follow a pattern usually consisting of driving from the starting point to some known big road, and then driving off the big road to the goal point. The subjects also chose different roads for the tourist and the resident, sometimes explicitly stating that the reason was that the route described to the tourist was more easily described. We found some interesting properties of the descriptions aimed for the resident, some of which explain when and why something is turned into a chunk. We were also able to construct a grammar that together with some "heuristic" rules defines the route description language.

### 3.1 Three kinds of descriptions

When analysing the experiment we found three different kinds of descriptions. Firstly, there was the tourist description, which we call the *procedural* description. The character of the procedural description (see also section 3.2) is primarily that it is a description of a *procedure*, namely, how to drive from A to B. Every road in the route is mentioned, and the subjects try to find properties of the road that the tourist can use for recognition. Then there were two kinds of descriptions aimed at residents, one which we could obtain just from taking the description aimed for the tourist and cross out some of the information, hereafter called the *mixed* description, and another based on an entirely different way of thinking about the routes, which we call the *declarative* description. In figure 1 we can see one example of the difference between these three kinds of descriptions. (The example descriptions have been paraphrased according to the grammar in figure 2, for a full transcript turn to [Höök and Karlgren 1991].)

The mixed description seemed to come up whenever the subject felt that the resident interviewer did not know the route the subject was about to describe. The route could contain some new twitch not commonly used, or it could be that a part of the route was at a lower hierarchical level than the rest of the route. Usually, this kind of description was only used for a minor part of the whole route, but sometimes the entire route would be explained at this level. This kind of description is foremost characterized by the fact that it used 'spatial markers' like 'left', 'right' etc., and that we could obtain this description simply from crossing out certain extra information, like landmarks, lane information, etc., from the procedural description. Usually, no *objects* (roads, intersections or landmarks) were skipped from this description.

The declarative description aimed at residents, is most easily characterized as an attempt to only mention enough road names to exclude any other possible route, or rather any other candidate alternative route. It is a kind of reflection of the search space that the subjects traverses in order to decide upon a route. All the intersection and some roads have disappeared from this description, and there are no 'spatial markers' left.

### 3.2 What Does A Novice Description Look Like?

A procedural description can be seen as a sequence of identify-act instructions. Firstly, a point along the route is identified and then the action that should take place at that point is mentioned. Of course, the main place where actions happen while driving is in intersections. Thus intersections are central to procedural route descriptions.

| **Procedural** | **Mixed** | **Declarative** |
|---|---|---|
| Drive Valhallavägen past two roundabouts up to a small divider that "stands out into the street a little" on Valhallavägen, where you turn left onto Artillerigatan. Drive it all the way down to Karlavägen where you turn right. Drive Karlavägen which is a boulevard past a park on your left-hand side up to the second left after the park where you turn left on Rådmansgatan. Drive it down past two red-lights, all the way until Sveavägen where you turn left. Drive immediately right onto Tégnergatan. Drive up a hill to Tégnerlunden. | Find your way up to Valhallavägen. Drive Valhallavägen up to Artillerigatan. Drive Artillerigatan down to Karlavägen where you turn left. Drive Karlavägen up to Rådmansgatan where you turn left. Drive Rådmansgatan down to Sveavägen where you turn left. Drive up to Tégnergatan where you turn right. Drive up the hill and there is Tégnerlunden. | Drive Karlavägen to Rådmansgatan. Drive Rådmansgatan to Sveavägen. |

**Figure 1.** From Gustav Adolfs kyrka to Tegnérlunden; examples of procedural, mixed and declarative route descriptions .

In procedural route descriptions we also found two other kinds of actions, namely to continue to drive along a road, and to pin-point a position. We call those the *maintain* and *placement actions*.

**Identifying intersections.** There are several methods of identifying an intersection. The most common way of identifying an intersection is by identifying the *intersecting road*. Another methods is to identify it by *name*: "Brommaplan", but very few intersections have names in themselves, and if they do, the names only very seldom are poster. The *type* of an intersection can be used for identification: "a roundabout". If the intersection is of an unusual type it is more likely to be used as an identifying feature: roundabouts, T-crossings, forks, and merges are all used in the interviews.

Intersections can also be referred to by using *landmarks* situated near or in the intersection. The *size* of the intersection is sometimes used, which is derived from the hierarchical level of the roads intersecting. An intersection is referred to as large when the intersecting road is at least as large as the street being traversed. One method (which more often than not proved faulty), was to identify the intersection by the *number* of intersections before it. Sometimes subjects would use several of these methods to identify the intersection. These methods can be (partially) ordered: name of the intersection is most important, then comes type, landmark and size and thereafter intersecting road and number of intersections.

**Identifying roads.** Roads need to be identified as well. Roads are identified by *name* by *size* or by *type*. The priority between the three is somewhat different from the identification of intersections. The name is very important. Even if the name of a road is new to a tourist it will be used. *Lane* information is very seldom used at all, but when it is, it seems to take on two different functionalities. It is either a description of a road to help determine the size or it is used to describe a turn in more detail.

**Maintenance and placement actions.** In the maintenance action, where the driver is instructed to continue following a road, there are several different ways for indicating distance: number of blocks passed, distance in kilometres, or general expressions like "for a bit", "all the way", and so forth. Maintenance actions appear whenever the road is going to be driven for an unexpectedly short or long distance.

The placement action is not really an action but rather a general description of a location: a description of a place, an intersection, or a view. Placement instruction can be compositions of several landmark references. Placement instructions tend to occur in the beginning of procedural descriptions, and then, later in the interview, when some characteristic object or view shows up, most often an object that is so visible that it absolutely must be mentioned. In the latter situation the placement instruction serve as a pause in the description. After the pause, the last part of the trip is often repeated.

**Beginning and end of the route.** In the interviews, tourists tended to be guided out from the starting point, and also received placement instructions for the starting point: the subject assumed that the tourist did not know what the starting point looked like.
Sometimes the end point would only be described by "then you follow the signs", which is only possible when the end point is big enough to be posted. In other cases, the whole end part of the trip was described and a placement instruction was given so that the tourists would know when they had reached the end.

**A descriptive grammar.** Bringing together what has been said above we find that we have a small grammar for route descriptions, see figure 2. It turns out that most regularities of procedural route descriptions can be captured even with this relatively crude formalism.

### 3.3 Differences between novice and expert

Given that we know what a procedural description looks like, a mixed description can be obtained by deleting irrelevant information that the resident navigator already knows. The basic structure of this mixed description is not identify-act but rather identify as simply as possible and then act. By identify as simply as possible we understand that no hierarchical information, descriptions of landmarks, traffic, traffic regulations etc, are included in the description. What will be left is simply the backbone of a procedural route description; "drive Road1 until Intersection1, turn left and drive Road2". There are a few exceptions from this pattern. The road name can be substituted by a known landmark, and subjects sometimes feel inclined to describe a road on a very low hierarchical level as "small". Otherwise, mixed descriptions can be obtained by removing the "IntersectionId" and the "RoadId" from the grammar rule "Action", and take away the placement and maintain instructions from the grammar in figure 2.

**Route_descr** → **Start Instruction* Goal**
**Instruction** → **Action** | **Maintain** | **Placement**
**Start** → **Placement**
**Goal** → **Placement** | **Sign**
**Action** → drive **Road** (**RoadId**)
    (**Until Intersection** (**IntersectionId** ))
    (where you **Do**)
    (**Direction**)
    (into **Road** (**RoadId**))
**Maintain** → drive **Road** (**RoadId**)
    {past **Landmark** | through **Area** | **Sign** |
    towards **NSWE** | until namechange to **Name** | **Distance**}

**Placement** → You can see {**Landmark** | **Road** | ... }
**RoadId** → which has **Feature**
**Until** → until | all the way until | right to the end | ...
**Intersection** → **Name** | **Type** | **Size** | an intersection
**IntersectionId** → which has **Feature** | which is by **Landmark**
**Do** → turn left | turn right | follow the traffic | go right through
**Direction** → towards **Area** | towards **NSWE** | "down" | "up" | **Sign**
**Distance** → **N** {meter | intersections | red_lights | blocks | ... }
    | for a while | for a short while | directly
**Sign** → following sign towards {**Area** | **Road** ...}
**N** → 1 | 2 | ...

**Figure 2.** A descriptive grammar for tourist route descriptions.

Declarative descriptions on the other hand, are fundamentally different from procedural descriptions. We cannot simply take a procedural description and cross out irrelevant information and obtain a declarative description. Actions for residents do not happen in intersections, instead the description consist of a number of simple instructions that in most cases simply is a road name, see figure 3.

| | |
|---|---|
| **Route_descr →** | **Start Instruction* Goal** |
| **Instruction →** | **Resident** |
| **Start →** | [] |
| **Goal** | → [] \| **Sign** \| **Mixed** |
| **Resident →** | drive **Road** \| drive past **Landmark** |
| **Sign →** | following sign towards {**Area** \| **Road** ...} |
| **Mixed →** | see above. |

**Figure 3.** A descriptive grammar for resident route descriptions.

**Gaps in the declarative description.** In the procedural descriptions, every road, every intersection where actions take place, and some of the landmarks were mentioned. In the resident description, all the intersections, and almost all the landmarks disappear. It is even the case that some of the roads are left out. Which ones?

Firstly, subjects seem to take away the first part of the trip. The subjects apparently assume that the reason for asking about a route between the starting point and the destination, is that you know where you are, but you do not know exactly where the destination is situated. The principle seems to be that all roads before a road with quite a high hierarchical level, can be taken away, but not more than 2 - 3 roads.

The end part of the trip is not described with the same pattern, instead the mixed description is often used by the subjects. Exceptions arise when the goal is on a very high hierarchical level or close to a road which is. Only then are roads left out.

There are also gaps in the middle of the description where roads have been left out, and the principle behind those gaps is much more complicated to depict. It seems that the roads that are mentioned serve the purpose of excluding other possible routes. By other possible routes, we do not mean *any* possible route, but rather any route that would have been a likely candidate to get to the destination. The subjects also add roads that make the route description complete, usually roads with a high hierarchical level. A road which is only going to be followed for a minor part of its length, or that is on a lower hierarchical level than the rest of the roads, is only mentioned if it gives the route an unexpected twitch or if it leads to the goal. The roads that are followed to their end are usually mentioned.

**The use of landmarks.** Subjects use landmarks in quite different ways for residents as opposed to tourists. The obvious difference is that some landmarks are only visible/usable to a driver if she knows the city. Tourists in general were provided with a lot more landmarks than were residents.

With tourists we found that landmarks are used for three different purposes. First, they make it possible for the tourist to recognize an intersection where an action is supposed to take place. Second, they can be used to keep the tourist on a certain road for a while, in maintenance instructions. Third, placement instructions usually consists of one or more landmarks, used in a more checkpoint-like manner, to indicate that a new series of instructions will follow after that checkpoint.

With resident navigators on the other hand, landmarks are used as choice points rather than as recognizable items. Instead of using a road name, a landmark that uniquely determines that road can be used. The landmark is therefore not described in any detail, but is only referred to by name.

For the tourist, landmarks are used whenever there is a noticeable landmark in an intersection. An intersection without landmark is described by some other means, like name, intersecting roads, hierarchical level, or traffic intensity. Landmarks along the road only appear when you have to travel for some longer time on that road, as in maintenance instructions.

In the resident descriptions, landmarks quite frequently turn up when the subject cannot remember the name of the road, or when the landmark is such a known item that its name would supersede the road name.

## 3.4 Route Choice

One of the pairs of starting and end points did not allow different route choices, but the other ones did. We found that in quite a few cases, the subjects would choose different routes for the tourist and the resident, and they would mention the fact that they did so, and sometimes why. Usually the reason was that the route chosen for the tourist was more easily described.

Now, what makes a route more easily described than another one? [Davis and Trobaugh 1987] have defined some properties of easily described routes; they should not contain too many turns, and it should, when possible, go via important landmarks. One of our subjects explicitly exemplified the second property. He consistently tried to get the tourist past big landmarks that were visually attractive. The first hypothesis could also be verified, the tourist routes had in general less turns.

We would like to add another property to do with hierarchical level of roads. Our subjects frequently tried to choose roads on the highest hierarchical level. The concern seems both to be that the route be the short in terms of time, but also that drivers, especially tourists, should not get lost. A bigger road is easier to recognize, it has better signs, and you can usually stay on it longer than on a road on a lower hierarchical level. When analysing the material further we found that the number of roads appearing in the resident route is slightly larger, but the interesting difference is that the hierarchical level is in general lower for the resident.

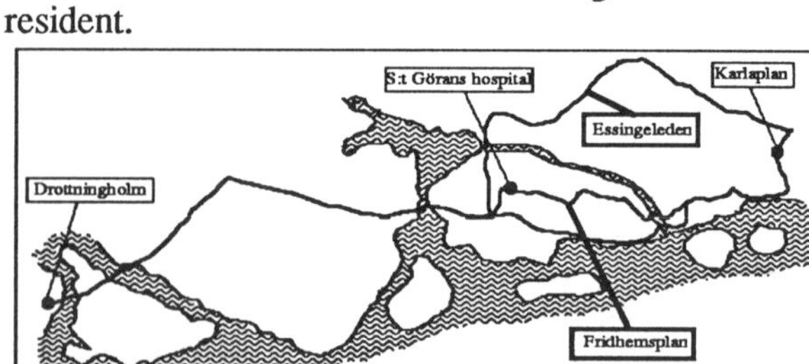


**Figure 4.** Going from Karlaplan to S:t Görans hospital or Drottningholm.

Different routes were also quite frequently chosen for goals along the same line, but with one goal further away than the other. An interesting pattern becomes apparent. When a route is chosen, the subjects seemed first to try backtracking from the goal in the direction of the starting point, to a well-known spot or road to which they knew they could find their way from the starting point.

Now, in figure 4, we see that one of the destinations that we choose for our study, lies very close to a big road, marked as Essingeleden. This big road was very frequently chosen for the second goal, Drottningholm, but not as often for the first goal, S.t Görans hospital, that lies so close to it. We found that since chosing this road would mean going in the wrong direction a couple of blocks (in the direction from the starting point), subjects were unwilling to make this connection, both for tourists and for residents. Instead they connected the point Fridhemsplan with the goal, and tried to find a route from the starting point to Fridhemsplan. The route through Fridhemsplan is much more complicated both to describe and to travel. Essingeleden is a route at the highest level of hierarchy, and the speed-limit is here 70 km/h instead of 50 km/h .

To some extent the way that subjects construct routes in their minds, will manifest itself in their description. In the example above, Fridhemsplan was mentioned as an important point in the resident description if that route was chosen.

## 4. Automatic generation of messages

We aim to do a 'sensible' implementation of the principles extracted from the study. Since 'sensible' is not necessarily equal to how humans describe routes [Riesbeck 1980], we also need to look at how humans understand and misunderstand route descriptions. Luckily, there have been quite a number of such studies [Streeter and Vitello 1986, Labiale 1989, etc.]. We know that humans are bad at reading and understanding maps, bad at estimating distances and number of blocks, that humans tend to forget instructions with more than 7±2 items, etc.

Let us explore a system for plan presentation. Given a starting point, a destination, knowledge of the user's experiences in the area, and a database of the route to be traversed, the strategy for description will be based on a three step process, see figure 5.[1] First the planner constructs a route. It consist of nodes and segments, where the nodes roughly corresponds to intersections, and the segments to roads.

Secondly, the route chunking mechanisms, using the plan, the user model, and the map database, process the plan into suitable chunks according to the results obtained in our study. It will in the tourist case change the plan into a "Road1 Intersection1 Road2, Road2 Intersection2 Road3" pattern with the intermediate road repeated. It will then adorn this pattern with names of roads and intersections, landmarks, lane information, etc., where needed. Placement and maintain instructions will be added where needed.

In the resident case, the route chunking mechanism will change the plan by deleting all intersections and as many roads as possible. It will also serialise the objects to be described into chunks of reasonable size, and decide which roads to substitute with landmarks or nicknames. Here we will use the same principles as did our subjects, for instance: which roads to delete will be deduced from their hierarchical level and place in the pattern; all intersections will be deleted; if an extremely low level road has been chosen, we will describe it if comes in the middle of the trip; the first and last parts of the trip will be taken away, etc. We shall look out for any strange twitch in the plan where we might need to use the mixed description rather than the declarative description. The roads are given names taken from the map database.

Lastly, the output from the generator is fed into a NL generator. In view of the fact that the descriptions constitute such a limited subset of natural language, as can be seen from our descriptive grammar, even a relatively small and unsophisticated generator can be expected to produce satisfactory results.

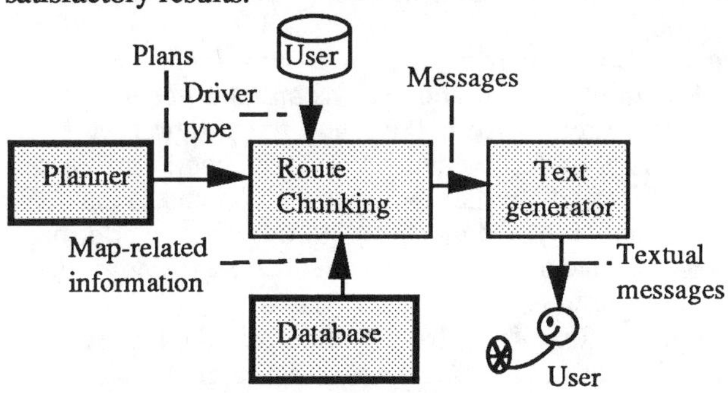


**Figure 5.** Route chunking mechanism.

The output from the route chunking mechanism, is roughly equivalent to a *message* as defined by [Mellish and Evans 1989]. It is a nonlinguistic object, at a more abstract level than the linguistic level. A message is built from a limited message language specifically devised to express a certain domain, in Mellish and Evans' case, the planning domain.

Our *message language* is much more limited than Mellish and Evans' since it only will deal with sequential plans instead of nonlinear, there is only one actor able to perform the instructions instead of several, and the utterances we would like to express are fewer. Still we would like to make our message language rich enough to later on be able to express explanations, and as a basic datastructure for dynamic replanning[2] reasoning.

A *message* is firstly an utterance, but also various ways of combining messages. An *utterance* in turn is either an instruction to perform an action, an explanation of an action, or an expansion of a complex action into subactions, etc. An *action* in our case is what the driver can do while following the route, i.e. turn, maintain, placement or drive actions. The purpose of the route chunking mechanisms

[1] In figure 5 we have left out a number of components necessary for a route guidance system, for instance, a route monitor, some integrated dialogue management, dynamic replanning etc. For a complete picture of the system architecture turn to [Brown et al. 1990].

[2] Dynamic replanning comes into play when the system can replan due to traffic intensity or other variables. A new plan then might have to be presented to the driver. Some reasoning as to how much of this new plan needs to be presented, has to take place.

will therefore be to take a plan and decide what is to be said, in what order and to whom, and then produce a message with that information.

In figure 5 above, the user model only influences the route chunking. It is important to point out that in this case, and many others, a user model should really be allowed to influence earlier stages as well. Here, that would mean influencing the planning process. If the planner can be adopted for the tourist, so that it produces plans with fewer turns, as in [Elliott and Lesk 1982], and includes roads at higher hierarchical levels, it would indeed be easier to describe the route to the tourist. The resident, would benefit from the shortest route, even if it is hard to describe and drive. In [Lindevall and Höök 1991] we investigate this issue further and also look at other user groups, as for instance the commuters.

## 5. Conclusions

We have investigated the underlying principles of resident and tourist route description done by humans, both in terms of how they are described and, to some extent, why. We have also seen how the actual route choice differs for the two groups.

Furthermore we have indicated how these principles could be implemented and used as an interface to a route guidance system. We believe that the solution we have outlined is good for residents, but not as good for tourists, though this needs further inquiries. We intend to make an implementation of the ideas, and to use a real world map. This would enable us to test the obtained route descriptions with humans.

In figure 2 page 4, we have summarized the tourist route description language. There are some heuristic rules as to when the grammar rules come into play, like, for instance, when describing an intersection, the name of the intersection is most important, then type, landmark, and size and thereafter intersecting road and number of intersections.

The mixed description (as described above) can be based on the grammar in figure 2, by simply removing some of the grammar rules.

For the resident descriptions, the grammar rule for the surface language is very simple, see figure 3 page 5. What is interesting about resident descriptions is that they are very short: the instructions are not adorned with as much extra information, and are *declarative* rather than *procedural*. The other reason to being short is that roads are left out in the middle of the trip, the start of the trip, and the end of the trip. Those gaps can be expressed with some heuristic rules:

- The first 1 - 3 roads before a road on a higher hierarchical level, can be removed from the resident description.
- If the goal is on a high hierarchical level, the last 1 - 3 roads can be skipped.
- In the middle of the trip, only roads that help exclude other alternative routes are mentioned, plus some roads that makes the route description 'complete'.

There are also some principles for when a switch to mixed descriptions is made. It occurs when a 'strange' route has been chosen. Something seems to be a strange route whenever roads on a low hierarchical level are chosen, or when the chosen route is much longer than the "as the crow flies" distance.

In terms of route choices, the underlying principles seem to be that:

- Tourist routes are chosen by subjects for ease of explanation. A route is easy to explain when it is on a high hierarchical level, contains few turns, and goes by important landmarks.
- Routes are constructed by tracing the route backwards from the goal to a point that is easy to connect with the starting point.

## 6. Acknowledgements

We would like to thank Carl Brown, Annika Wærn, Yvonne Wærn, Manny Rayner, Per Lindevall and Björn Gambäck for many valuable comments on this paper and the study in general. Thanks also to Lynn Streeter and Michael Lesk for taking time to discuss their research results in the field. Last but not least we thank the 10 subjects who took part in the study.

## 7. References

Brown, C.; Höök, K.; Lindevall, P. and Wærn, A. 1990. "A System Architecture for an Interactive Route Guidance System", SICS-KBS working document PS-9007.

Chase, W.G. 1983. "Spatial Representation of taxi drivers", In D.R. Rogers and J.A. Sloboda (eds) *Acquisition of Symbolic Skills*, 391-405, New York: Plenum.

Davis, J.R. and Trobaugh, T.F. 1987. "Direction Assistant", Speech Research Group Technical Memo 1, The Media Laboratory, MIT.

Elliott, R.J. and Lesk, M. E. 1982., "Route Finding in Street Maps by Computers and People", AAAI-82 Conf. pp. 258-261, Los Altos, CA:American Association for Artificial Intelligence.

Höök, K. 1991. "Maps versus verbal route descriptions", Forthcoming.

Höök, K. and Karlgren, J. 1991. "Some Principles Derived from Human Advisors", SICS Research Report R91:06.

Labiale, G. 1989. "Influence of In Car Navigation Map Displays on Drivers' Performance", SAE Tech. Paper Series, 891683.

Lindevall , P. and Höök, K. 1991. "Humans Routes - Routes for Humans", SICS-KBS working document PS-9003.

Mellish, C. and Evans, R. 1989. "Natural Language Generation from Plans", *Computational Linguistics*, vol. 15, no 4.

Pailhouse, J. 1970. "La Représentation de l'Espace Urbain", Paris: Presses Universitaires de France.

Riesbeck, C. J. 1980. "'You Can't Miss It!': Judging the Clarity of Directions", *Cognitive Science* 4, pp. 285-303.

Streeter, L.A. 1985. "Automatic Street Route Generation: Modeling Expert Direction Givers", TM-ARH-001369, Bell Communications Research.

Streeter, L.A. and Vitello, D. 1986. "A Profile of Drivers' Map-Reading Abilities", *Humans Factors*, 28(2).

von Tomkewitsch, R. 1990. "ALI-SCOUT (LISB) - a Beacon based System for RTI and Dynamic Traffic Management", proc. Nässlingen -90, available from SICS.

# Adult Age Differences in Visual Mental Imagery: Evidence for Differentially Age-Sensitive Components

by

**Scott H. Johnson**
Department of Psychology
Uris Hall, Cornell University
Ithaca, NY 14853-7601
Bitnet: huc@cornella

**Abstract**

A set of tasks developed in accordance with Kosslyn, Van Kleeck, & Kirby's (1990b) neurologically plausible model of visual mental imagery was used to explore effects of aging on specific component processes involved in image generation and maintenance. Contrary to the widely held belief that age differences in cognition are attributable to a single mechanism of global effect (e.g., a reduction in critical processing resources, Salthouse, 1988), results indicate that processes involved in visual mental imagery are differentially age-sensitive. More precisely, components required to actively maintain images are particularly sensitive to effects of aging, while those that access visual information from memory are not especially affected. Advantages of a componential approach to understanding age differences in cognitive processing are disscussed, as well as the potential for such age-related changes to be a readily exploitable source of information regarding the functional architecture of cognition.

## Age Differences in Visual Mental Imagery

Research into effects of aging on visual mental imagery has focused primarily on the ability to "mentally rotate" disoriented objects (for a comprehensive review of such tasks see Shepard & Cooper, 1982). In general, results from these studies are consistent with age differences found on virtually all speeded cognitive tasks: That is, older adults perform more slowly than do their younger counterparts (Salthouse, 1985). In addition to this generalized, or task-independent, slowing, there is reason to believe that specific imagery processes differ in their sensitivity to effects of aging. An unpublished task analysis of the six most recent chronometric studies of aging and mental rotation (Berg, Hertzog & Hunt, 1982; Cerella, Poon & Fozard, 1981; Gaylord & Marsh, 1975; Jacewicz & Hartley, 1979; Puglisi & Morrell, 1986; Sharps & Gollin, 1987) suggested that observed age-related slowing may be primarily attributable to specific deficits in shifting visual attention. Furthermore, in the one mental rotation task reviewed where attentional demands were minimized (by presenting to-be-compared stimuli sequentially rather than simultaneously), older adults performed as quickly and accurately as the young adult group (Jacewicz & Hartley, 1979). Similarly, Dirkx and Craik (1989) found that older adults were impaired on certain tasks of mental image rotation (East-West Test) and image generation (Clock Test), yet demonstrated normal levels of performance on another version of the mental rotation task (Brook's Letter Test). Finally, in their neuropsychological study of various imagery abilities, Farah and Hammond (1988) reported age differences in the ability of healthy, older control subjects to perform judgements based on self-generated images in some, but not all tasks.

In sum, previous research suggests that some aspects of imagery are particularly sensitive to effects of aging, while others are not uniquely affected. The equivocal nature of these findings is not easily reconciled with the prevailing wisdom that age differences in performance are attributable to a single mechanism of global effect (e.g., a decline in the availability of critical processing resources (reviewed in Salthouse, 1988), slowed rate of neural conduction (Birren, 1974), or undifferentiated information-loss arising from increased 'neural-noise' (Myerson, Hale, Wagstaff, Poon, & Smith, 1990). If this is so, performance of all imagery tasks should be similarly affected by aging. Variability in effects of aging across imagery tasks is to be expected, however, if: (a) performance of visual imagery tasks demands contributions from numerous, relatively independent processing components (e.g., Kosslyn, 1987; Kosslyn, et al., 1984; Kosslyn, et al., 1990b); and, (b) such components are differentially sensitive to effects of aging (e.g., Johnson, 1990, Johnson & Kosslyn,1991; Johnson & Rybash, 1991). The present studies undertook to investigate these possibilities within the context of two basic imagery abilities -- image generation and image maintenence.

## A Componential Approach

Kosslyn and his colleagues (Kosslyn,1988; Kosslyn, et al., 1988; Kosslyn et al., 1990a) have developed a set of tasks based on Podgorny and Shepard's (1978) paradigm that provide a means of examining the efficiency of particular component processes involved in generation and maintenance of visual mental images.

This research was funded in part by a College Grant from Cornell Universitiy's College of Human Ecology. This manuscript was prepared while the author was a trainee on NIMH grant # T32-MH19389-02.

Because these tasks allow one to study particular components in isolation, they are ideal for evaluating hypotheses regarding mechanisms responsible for age differences in imagery behavior. According to Kosslyn et al. (1990b), image generation and image maintenence do not depend entirely on the same set of processing components. Therefore, if components involved in imagery differ in their sensitivity to effects of aging, then image generation and image maintenance processes need not be similarly affected. If age differences in imagery arise from a generalized deficit that does not respect boundaries between component processes, however, image generation and image maintenance processes should be similarly affected.

## Isolating Components

Eighteen older adults (ages 63 - 81) and 18 younger adults (ages 18 - 30), of comparable educational and intellectual status, were tested on three tasks -- Image Generation, Perceptual Control, and Image Maintenance. Tasks each consisted of 48 experimental trials preceded by 12 practice trials and were administered in an order counterbalanced within each age group.

**Image Generation Task.** Subjects studied a black, two-dimensional block pattern within a 4 x 5 cell grid displayed against a white field on a standard Macintosh SE20 CRT (Figures depicting similar displays can be found in Kosslyn, 1988; Podgorny & Shepard, 1978). When the figure was committed to memory, subjects depressed the space bar. One-hundred ms. later, a homogenous black mask appeared for 500 ms.; followed immediately by a blank 4 x 5 cell grid with an "X" probe in one of the cells. Subjects were required to decide, as quickly and accurately as possible, whether the probe would have fallen on or off the block pattern, were it still present in the grid. Reaction times (RT) were measured from onset of the probe until subjects responded by depressing either the "yes" or "no" key.

According to Kosslyn et al's. (1990b) neurologically-plausible model of visual mental imagery, image generation requires two subsets of processes: one to access stored visual information from memory, and a second to activate a pattern based on this retrieved information within a temporally-limited visual buffer.

In an elegant series of experiments employing tasks similar to those presently discussed, Kosslyn, et al. (1988) demonstrated that images are generated sequentially - - one segment at a time (see also Kosslyn, 1988). Concomitantly, stimulus complexity can be manipulated to affect varying degrees of processing load precisely on those component processes involved in generating images, and effects of these demands can be observed to determine the efficiency of these processes. To this end, an equal number of trials consisting of either one, two or three segment stimuli were presented.

In addition to imagery, response latencies on this Image Generation task reflect non-imagery processes used to encode the probe and execute a motor response. Because demands on these non-imagery processes do not vary with complexity of the stimulus, however, comparison of performance between trials with one, two, and three segment stimuli should reveal only rates of imagery processes (Kosslyn, et al., 1990a).

**Perceptual Control Task.** Differences in RT for stimuli of different complexity, however, reflect not only time to generate the image, but also time to search the image for probe location (Kosslyn, 1988; Kosslyn et al., 1988; Kosslyn et al., 1990a), and it is possible that rates of image search may covary with stimulus complexity. A Perceptual Control task was therefore included to control for rate of image search. This task was identical to the Image Generation task, except that both the probe and the stimulus pattern were visible simultaneously so as to obviate the need to generate an image: The subject merely had to determine whether the probe was on or off the pattern as quickly and accurately as possible while both were fully visible.

Given the preceding logic, effects of stimulus complexity on Image Generation and Perceptual Control tasks can then be compared, and any differences should reflect contributions of processes specific to image generation, apart from contributions of those involved in either non-imagery or image search processes. Likewise, age differences in efficiency of image generation components can be examined by comparing differences in effects of stimulus complexity on Image Generation and Perceptual Control tasks between younger and older adults.

**Image Maintenance Task.** Finally, an Image Maintenance task was used to evaluate age differences in the ability to actively maintain visual mental images for varying amounts of time. Procedurally speaking, the Image Maintenence task was identical to the Image Generation task except that: (a) no mask was used to disrupt subjects' maintenance of stimulus patterns over the delay, or interstimulus interval (ISI) interval; and, (b) probes on half on the trials were presented 500 ms. after the stimulus frame ( the same ISI used in the Image Generation task), while the ISI on remaining trials was lengthened to 3000 ms.

According to Kosslyn et al. (1990b), image maintenence requires a subset of processes used to generate visual images from memory -- those that activate (or re-activate) patterns within a temporally-limited visual buffer. Since no mask was used during the ISI, it was reasoned that an image of the stimulus pattern would be actively maintained and, consequently, components used to access stored visual information would not be required. Manipulation of stimulus complexity and ISI should, therefore, place additional processing demands precisely on image maintenance components (Kosslyn et al., 1990a).

## Results and Discussion

**Image Generation.** As intended, there was a significant two-way interaction between task (Image Generation vs. Perceptual Control) and stimulus complexity(one, two, or three segments) for both RT [F(2,64) = 21, p = .0001] and accuracy [F(2,64) = 4.8, p = .01] performance. This finding reflects the fact that as stimulus complexity increased performance on the Image Generation task decreased, while performance on the Perceptual Control task was unaffected (see Figure One). Manipulation of stimulus complexity was,therefore, successful in affecting additional processing demands precisely on those components involved in image generation apart from either perceptual-motor or image search processes.

There was also evidence for substantial age differences in performance: Older adults were slower than younger adults on both the Perceptual Control and Image Generation tasks, F(1,32) = 40.81, p<.0001). Older adults also committed significantly more errors than younger adults [(F(1, 32) = 9.52, p = .004], indicating that the main effect of age on RT was not attributable to speed/accuracy tradeoff (see Figure One).

Consonant with the observation that age differences tend to be more pronounced on tasks of greater difficulty (e.g., Cerella, Poon, & Williams, 1980), age differences were of greater magnitude on the Image Generation than the Perceptual Control task [F(1,32) = 7.46, p = .01]. Likewise, older adults were less accurate on the Image Generation than the Perceptual Control task [(F(1,32) = 6.31, p = .017)], again providing assurance that age differences in RT were not attributable to speed/accuracy tradeoff (see Figure One).

Contrary to what is expected if components involved in image generation are particularly sensitive to effects of aging, variations in stimulus complexity failed to exacerbate age differences in RT performance, F(2,64) = 2.0, p = .14. Age differences in accuracy did, however, show a tendency to increase with stimulus complexity, F(2,64) = 3.18, p = .05 (see Figure One). This raises the possibility that older adults may have sacrificed accuracy for speed. Since error rates were quite low (< 6% in all cells), it seems unlikely that speed-accuracy tradeoff alone was responsible for this finding.

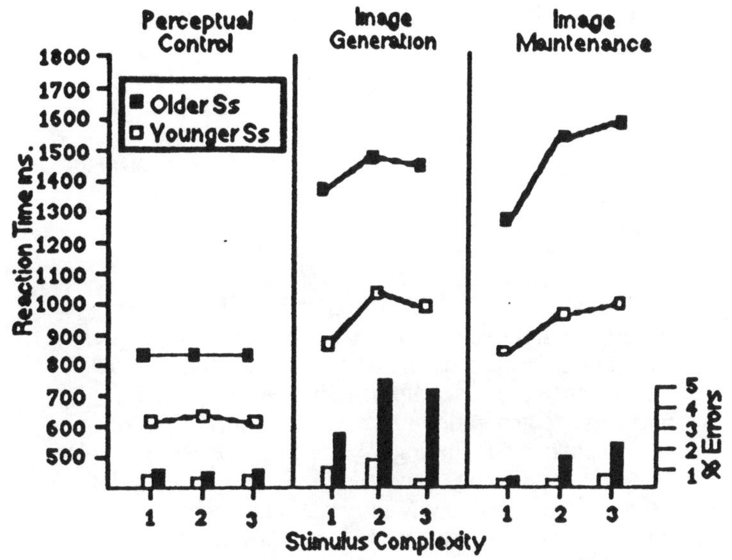


**Figure One depicts effects of stimulus complexity on RT and accuracy performance of younger and older adults on the Perceptual Control, Image Generation, and Image Maintenance Tasks.**

**Image Maintenance.** As expected, older subjects were also significantly slower [F(1,32) = 29.02, p = .0001)] and less accurate [F(1,32) = 8.69, p = .006)] than younger adults on the Image Maintenance task. In contrast to image generation, however, there was evidence that components involved in image maintenance are particularly age-sensitive. More precisely, age differences in RT [F(2,64) = 3.2, p = .047] and accuracy [F(2,64) = 5.48, p = .006] performance increased as a function of stimulus complexity (see Figure One). Age differences in RT or accuracy were not, however, affected by a six-fold increase in the interval over which images were maintained for RT or accuracy performance [F < 1 in both cases].

## General Summary

Support was garnered for the hypothesis that aging has differential effects on specific components involved in visual mental imagery: More precisely, processes required to activate stored visual information during image generation are not particularly sensitive to effects of aging, while those needed to actively maintain patterns of visual information are uniquely affected.

The finding that imagery components are differentially age-sensitive is difficult to reconcile with the contemporary wisdom that attributes age differences to a single mechanism of global effect (e.g., Birren, 1974; Myerson et al., 1990; Salthouse, 1985).

These results instead suggest that age differences in information-processing may, in part, be attributable to highly selective mechanisms. Failure of previous attempts to identify such mechanisms may be due to employment of methodologies limited to examining cognitive performance at the relatively gross level of processing stages, rather than at the level of specific processing components of which such stages consist (for a comprehensive review of this issue see Johnson & Rybash, 1991).

The fact that age differences obtain for image maintenance but not image generation also provides additional support for the hypothesis that image generation and image maintenance do not recruit entirely the same set of processing components. That image maintenance processes are particularly age-sensitive, while image generation processes are not is, however, a somewhat curious finding. According to Kosslyn et al's. (1990b) model image generation requires one subset of processes to access visual information from memory and another to activate patterns based of this information in a temporally-limited visual buffer. In contrast, image maintenance requires only the latter subset. On one hand, processes that activate visual memories during image generation may somehow offset age differences in processes that activate patterns in the visual buffer, thereby minimizing age differences in performance on the Image Generation task. On the other hand, Image Generation and Image Maintenance may require entirely orthogonal sets of processing components, with only the latter set being particularly sensitive to effects of aging. Further experimentation will be required to discriminate between these alternatives.

To conclude, the present findings suggest that a componential approach may provide a better understanding of mechanisms responsible for age differences in behavior, and a potentially fertile avenue for exploring the functional architecture of cognition.

## Acknowledgements

The author wishes to gratefully acknowledge the invaluable contributions of Stephen M. Kosslyn to this research project, as well as the helpful support of Steven W. Cornelius, John M. Rybash, Stephen Ceci, and James E. Cutting.

## References

Berg, C., Hertzog, C. & Hunt, E. (1982). Age differences in the speed of mental rotation. *Developmental Psychology, 18,* 95-107.

Birren, J. E. (1974). Psychophysiology and the speed of response. *American Psychologist, 29,* 808-815.

Cerella, J., Poon, L.W. & Fozard, J. L. (1981). Mental rotation and age reconsidered. *Journal of Gerontology, 36,* 620-624.

Cerella, J., Poon, L.W. & Williams, D.M. (1980). Age and the complexity hypothesis. In L.W. Poon (Ed.), *Aging in the 1980's: Psychological issues* (pp. 332-340). Washington, D.C.: American Psychological Association.

Dirkx, E. & Craik, F. (1989). *Age-related changes in memory and imagery processes.* Forthcoming.

Farah, M. & Hammond, K.M. (1988). Mental rotation and orientation invariant object recognition: Dissociable processes. *Cognition, 29,* 29-46.

Gaylord, S. A. & Marsh, G. R. (1975). Age differences in the speed of a spatial cognitive process. *Journal of Gerontology, 30,* 674-678.

Jacewicz, M. M. & Hartley, A. A. (1979). Rotation of mental images by young and old college students: The effects of familiarity. *Journal of Gerontology, 34,* 396-403.

Johnson, S. H. (1990). Reaction-time slowing in older adulthood: Evidence for qualitative changes in cognitive efficiency. *Brain & Language, 39,* 600-601.

Johnson, S. H., & Kosslyn, S. M. (1991). *Age-sensitive components in generation and maintenance of visual mental images.* Cornell University Manuscript.

Johnson, S.H., & Rybash, J. M. (1991). A Cognitive Neuroscience approach to adult age differences in cognitive efficiency. In J. Cerella, W. J. Hoyer, J. M. Rybash, & M. L. Commons (Eds.), *Adult information processing: Limits on loss.* New York: Academic. Forthcoming.

Kosslyn, S. M. (1987). Seeing and imagining in the cerebral hemispheres: A computational approach. *Psychological Review, 94,* 148-175.

Kosslyn, S. M. (1988). Aspects of a cognitive neuroscience of mental imagery. *Science, 240,* 1621-1626.

Kosslyn, S. M., Brunn, J., Cave, K. R. & Wallach, R. W. (1984). Individual differences in mental imagery ability: A computational analysis. *Cognition, 18,* 195-243.

Kosslyn, S. M., Cave, C. B., Provost, D., & Von Gierke, S. (1988). Sequential processes in image generation. *Cognitive Psychology, 20,* 319-343.

Kosslyn, S. M., Margolis, J. A., Barrett, A. M., Goldknopf, E. J., & Daly, P. F. (1990a). Age differences in imagery abilities. *Child Development, 61,* 995-1010.

Kosslyn, S. M., Van Kleeck, M. C., & Kirby, K. N. (1990b). A neurologically plausible theory of individual differences in visual mental imagery. In P. J. Hampson, D. F. Marks, & J. T. E. Richardson (Eds.), *Imagery: Current Developments.* London: Routledge.

Myerson, J., Hale, S., Wagstaff, D., Poon, L. W., & Smith, G. A. (1990). The information-loss model: A mathematical theory of age-related cognitive slowing. *Psychological Review, 97,* 475-487.

Podgorny, P., & Shepard, R. N. (1978). Functional representations common to visual perception and imagination. *Journal of Experimental Psychology: Human Perception and Performance, 4,* 21-35.

Puglisi, T. J. & Morrell, R. W. (1986). Age-related slowing in mental rotation of three-dimensional objects. *Experimental Aging Research, 12,* 4, 217-220.

Salthouse, T. A. (1985). Speed of behavior and its implications for cognition. In J. E. Birren and K. W. Schaie (Eds.), *Handbook of the Psychology of Aging* (2nd ed. pp. 400-426). New York: Van Norstrand Reinhold.

Salthouse, T. A. (1988). Resource-reduction interpretations of cognitive aging. *Developmental Review, 8,* 238-272.

Sharps, M. J., & Gollin, E. S. (1987). Speed and accuracy of mental image rotation in young and elderly adults. *Journal of Gerontology, 42,* 3, 342-343.

Shepard, R.N. & Cooper, L.A. (1982). *Mental Images and Their Transformations.* Cambridge, MA: MIT Press.

# A Framework for Opportunistic Abductive Strategies*

**Todd R. Johnson**

Laboratory for Artificial Intelligence Research
Dept. of Computer and Information Science
The Ohio State University
2036 Neil Avenue Mall
Columbus, Ohio 43210
Email: tj@cis.ohio-state.edu

**Jack W. Smith**

Division of Medical Informatics
The Ohio State University
571 Health Sciences Library
376 W. 10th Ave.
Columbus, Ohio 43210
Email: smith.30@magnus.ohio-state.edu

**Abstract**

Any single algorithm for abduction requires specific kinds of knowledge and ignores other kinds of knowledge. A knowledge-based system that uses a single abductive method, is restricted to using the knowledge required by that method. This makes the system brittle, because the single fixed method can only respond appropriately in a limited range of situations and can only make use of a subset of the potentially relevant knowledge. In this paper, we describe a framework from which abductive strategies can be opportunistically constructed to reflect the problem being solved and the knowledge available to solve the problem. We also describe ABD-Soar, a Soar-based implementation of the framework, and demonstrate its behavior.

## Introduction

Abduction, finding a best explanation for a set of data, is an important part of many knowledge-based (KB) systems, particularly those concerned with diagnosis. In recent years, several different algorithms for doing abduction have been devised. Any single algorithm for abduction, however, requires specific kinds of knowledge and ignores other kinds of knowledge. Hence, a KB system that uses a single abductive method, is restricted to using the knowledge required by that method. This makes the system brittle, because the single fixed method can only respond appropriately in a limited range of situations and can only make use of a subset of the potentially relevant knowledge. To remedy this we have endeavored to develop a framework from which abductive strategies can be opportunistically constructed at run-time to reflect the problem being solved and the knowledge available to solve the problem. In this paper, we present this framework and describe ABD-Soar, an implementation of the framework. We show how ABD-Soar can be made to behave like the abductive strategy used in Red (Josephson, et al., 1987), a system for red cell antibody identification. Finally, we discuss the differences between ABD-Soar and 2 tools for abduction: Peirce (Punch III, et al., 1990), and Mole(Eshelman, 1988).

This work contributes both to our understanding of KB systems and abduction. First, it illustrates how the problem-solving capabilities of KB systems can be increased by using mechanisms that permit the use of all relevant knowledge. ABD-Soar requires little domain knowledge to begin solving a problem, but can easily make use of additional knowledge to solve the problem better or faster. Second, the framework can be used to provide a flexible abductive problem-solving capability for KB systems. Third, ABD-Soar gives Soar an abductive capability such that many systems written in Soar can begin to solve abductive problems. Fourth, the framework provides a very simple and general mechanism for abduction that is capable of generating the behavior of various fixed methods. Fifth, ABD-Soar can be used to experiment with different abductive strategies, including variations of existing strategies and combinations of different kinds of strategies. Finally, the framework provides guidelines for building abductive systems because it provides a theory of abduction and specifies the kinds of knowledge needed to do abduction.

## The Abductive Task

Most Artificial Intelligence research on abduction is concerned with strategies for a type of abduction called *hypothesis assembly*. In hypothesis assembly, a complete explanation of the data must be formed by composing a number of smaller explanations. For example, in medical diagnosis the data to be explained are the signs and symptoms of the patient. If the patient has multiple diseases, the best explanation for the data will consist of all of the diseases present. Each disease might only explain a subset of the data, but taken together they must explain all of the data.

Most models of abduction can be described using four criteria to define the best explanation: *coverage* (the number of explanata[1] explained or covered by the explanation), *belief* (some measure of plausibility), *parsimony*, and *consistency* (Josephson and Goel, 1988). The models vary in their interpretation of these criteria. For instance, parsimony can be defined in terms of *minimal cardinality*, i.e., the fewest number of component hypotheses, or in terms of *irredundant covers*, i.e., an explanation in which no proper subset of the component hypotheses can explain the data. Belief is also subject to different interpretations. Some models view the most plausible explanation as that with the highest overall plausibility. Other models view the most plausible explanation to be that in which each component hypothesis is the most plausible explanation for at least one datum (this has been called *best-for-some* (Bylander, et al., 1988)). In general, there is probably no single model of abduction appropriate for all abductive tasks. Ultimately, the goals of the specific abductive task and characteristics of the task domain define what counts as a best explanation.

An analysis of the computational complexity of abduction has shown that the problem is, in general, computationally intractable (Allemang, et al., 1987). There are, however, some conditions under which the problem is computationally tractable. Bylander, et. al. writes:

> Our primary conclusion is that very restrictive conditions must be satisfied for abduction to be tractable: determining the plausibility and explanatory cover-

* This research is supported by National Heart Lung and Blood Institute grant HL-38776, and National Library of Medicine grant LM-04298.

1. An *explanatum* is a datum to be explained. *Explanata* are all of the data to be explained.

age of hypotheses must be tractable, there cannot be substantial incompatibility and cancellation interactions between hypotheses, and plausibility comparison between composite hypotheses must be logically weak. Specific domains may escape some of these requirements if composite hypotheses are guaranteed to be small or if strong domain knowledge can rule out most individual hypotheses. (Bylander et al., 1988)

Thus, it is very unlikely that any single strategy for abduction is appropriate to all tasks. The efficient solution of most real-world abduction problems requires the careful application of domain knowledge to transform the problem into either one that is tractable or one that is small enough so that tractability concerns are unimportant. Hence, it is very important to build systems that can make use of all of the available knowledge when solving abductive problems.

## Specifying Problem-Solving Methods

Before we describe the framework for abduction, we must first explain how we are going to specify it. Here we address this in the context of the general problem of specifying problem-solving methods.

One view of a problem-solving method is that it consists of a set of operators (operations on data) with preconditions and knowledge that indicates the order in which to apply those operators. Since we desire an opportunistic system, we need to be able to specify a set of operators without necessarily specifying a complete ordering of those operators. An opportunistic system works by enumerating possible operators to apply to the immediate situation and then selecting one of those operators based on the current goal and situation. Furthermore, a flexible system must be capable of generating or using additional control knowledge. Thus, if a system must decide between several operators, it must be possible for the system to engage in complex problem-solving to determine which operator is best.

To achieve these results we have been using the problem-space computational model (PSCM) (Newell, 1990) to specify methods. In the PSCM, all problem-solving is viewed as search for a goal state in a problem-space. Knowledge about when operators are applicable to a state can be specified independent of knowledge about which operator to select. Operator selection knowledge, called *search control knowledge*, is expressed in terms of preferences for or against applicable operators. If at any time during the problem-solving the search control knowledge is insufficient to indicate which operator to select, a subgoal is set up to generate additional knowledge so that a single operator can be selected. This subgoal is achieved by searching another problem-space. Operators can either be implemented by directly available knowledge or by using an operator-specific problem-space. Implementation in a problem-space is similar to using a subfunction to implement an operator in lisp.

## The Framework

The general abductive framework can be described by a single problem-space its goal, the knowledge content of its states, the initial state, operators, and search control knowledge. Following the description of the problem-space, we describe the minimal knowledge required to use the framework.

*State Description:* The state contains knowledge of explanata, explanations and an indication of the explanata they explain, knowledge about inconsistent or redundant objects, information about whether an explanation is the only possible way to explain an explanatum, and knowledge about whether the implications of a newly added object have been processed. In addition, any other information necessary for solving the problem can be kept in the state.

*Initial State Schema:* The initial state need contain only the explanata. Additional information can be provided.

*Desired State Schema:* The desired state must meet 6 conditions: 1) The explanation must be *complete*, i.e., all the explanata must be explained. 2) The hypotheses in the explanation must be at the desired level of detail for the problem being solved. For example, in diagnosis a disease hypothesis must be at a level of detail such that a therapy can be recommended. 3) No part of the explanation can be redundant. 4) No part of the explanation can be inconsistent. 5) All parts of the explanation must be certain. 6) All parts of the explanation must be processed. That is, the implications of adding the object and its effect on the rest of the explanation must have been considered.

There are 7 operators in the abductive space. The last 3 are used to determine the logical implications of adding an object to the explanation. 1) **Cover *explanatum*** is proposed for each explanatum that is not yet explained by a hypothesis (the explanation is not yet complete). Its goal is to add to the composite explanation one or more hypotheses that explain the explanatum. 2) **Resolve-redundancy** is proposed whenever there are redundant objects in the explanation. Its goal is to make the explanation irredundant. 3) **Resolve-inconsistency** is proposed whenever there are inconsistent objects in the explanation. Its goal is to make the explanation consistent. 4) **Determine-certainty *object*** is proposed for each uncertain object in the explanation. The operator is successfully applied when the object is deemed to be certain, or else the object is deemed to be not present in the explanation. 5) **Determine-accounts-for *hypothesis*** is proposed for each new hypothesis. Its goal is to determine what explanata the hypothesis can account for. 6) **Mark-redundancies *object*** is proposed for each new object. Its goal is to indicate which objects in the explanation are redundant with the newly added object. Hypotheses that offer to explain an identical explanatum (or explanata) are considered redundant unless other knowledge indicates that they are not. 7) **Mark-inconsistencies *object*** is proposed for each new object. Its goal is to indicate which objects are logically inconsistent with (or contradict) the newly added object.

**Search control knowledge** is specified as 1) *Determine-accounts-for* is better than all other operators. 2) *Resolve-redundancy, resolve-inconsistency,* and *determine-certainty* are indifferent to each other. [Note that an order can still be imposed on these operators by preferring one over another for a particular domain.] 3) *Mark-redundancies* and *mark-inconsistencies* are equal to one another and better than all other operators except for *determine-accounts-for*.

This is the minimum search control needed to ensure correct operation of the abductive mechanism. However, this search control only specifies a partial ordering of the operators. Any control decisions that can be based on domain-dependent knowledge have been left unspecified. This allows the designer of the system to add appropriate search control for the task being done. For example, there is no knowledge about what to do when multiple cover operators tie since this decision can be based on domain-dependent knowledge.

To use the framework, a minimum of 6 kinds of knowledge is required: (1) knowledge mapping each possible explanatum to potential explanations (to implement *cover*); (2) knowledge mapping an explanation to the explanata it can explain (to implement *determine-accounts-for*); (3) knowledge to determine the certainty of the explanations (to implement *determine-certainty*); (4) knowledge about the consistency of the model (to implement *mark-inconsistencies*); (5) knowledge about redundant objects (to implement *mark-redundancies*); and (6) additional search control knowledge to sequence the operators.

As a result of the operators and search control, the basic method is to pick an explanatum to cover, add one or more explanations for that explanatum, determine what each new explanation explains, and then pick another (unexplained) explanatum to explain. This continues until all explanata are explained. If at any time the model becomes inconsistent, redundant, or uncertain, an operator is proposed to resolve the problem. At that time, a decision must be made about whether to fix the problem or continue covering explanata.

## Implementing the Framework: ABD-Soar

ABD-Soar is a Soar-based implementation of the abductive framework. Soar is used because it directly supports the PSCM; however, the framework does not absolutely require Soar—other architectures that support PSCM-like functionality can also used. ABD-Soar supplies all of knowledge specified in the framework: knowledge to propose the abductive operators and detect their successful application, and the minimal search control specified in the framework. This knowledge is encoded as a set of Soar productions that can apply to any problem-space. This means that the complete body of abductive knowledge can be brought to bear during any problem-solving activity. ABD-Soar also provides default implementation knowledge for *cover*, *resolve-redundancy*, and *resolve-inconsistency* and a default method for generating additional search control knowledge.

The default knowledge for operator implementation is encoded in three problem-spaces with names identical to the operators they implement: *cover*, *resolve-redundancy*, and *resolve-inconsistency*. *Cover* generates possible explanations and then applies knowledge to select one of the candidates. The candidates are generated in response to an *explain* operator that must be implemented using domain specific knowledge. *Resolve-redundancy* and *resolve-inconsistency* remove each redundant/inconsistent object until the explanation is irredundant/consistent. If multiple irredundant/consistent explanations are possible, then all are found and the best one is used. This makes use of lookahead and the evaluation function described below. Any explanation that is an *absolute essential* (i.e., the only possible explanation for an explanatum) will not be removed. Also, it is possible to resolve a redundancy by explicitly indicating that particular redundant objects are not a problem.

The default method for generating additional search control knowledge is to use lookahead and an evaluation function to determine which operator to take when multiple operators are applicable. This is implemented in two problem-spaces: *find-best* and *evaluate-state*. *Find-best* evaluates each operator and selects the operator with the best evaluation. To evaluate an operator *find-best* applies the operator to a copy of the original state and then continues to do problem-solving from that state until a state is found that can be evaluated. The default evaluation function for *evaluate-state* is a summation of the number of explanata left to explain, the number of explanations, the number of inconsistent objects, the number of explanata explained by inconsistent explanations, the number of redundant objects, and the number of uncertain objects. The model with the lowest evaluation is chosen as the best alternative.

ABD-Soar uses irredundant covers as the parsimony criterion; however, when the default evaluation function is used the system also uses minimal cardinality to choose between competing composite explanations. The belief criterion (by default) is best-for-some since the method attempts to pick the best explanation for a finding being covered. In the absence of plausibility ratings, the evaluation function is used to rate competing explanations for a finding and the one with the best evaluation is selected.

| Red Cells | | | |
|---|---|---|---|
| 1 | 2 | 3 | 4 |
| (r1)1+ | (r2)3+ | (r4)1+ | (r5)1+ |
| 0 | (r3)1+ | 0 | 0 |

Anti-K explains r2 and r3, ie, the 3+ and 1+ on red cell 2
Anti-$Fy^a$ explains r1, r4, and r5, ie, the 1+ on red cells 1, 3, and 4
Anti-C explains r1 and r4, ie, the 1+ on red cells 1 and 3
Anti-N explains r4, ie, the 1+ on red cell 3
0 reactions do not need to be explained

Figure 1: Red cell antibody identification case

To use ABD-Soar the designer of a system must provide knowledge to implement *explain* and *determine-accounts-for*. When necessary, knowledge must also be provided for *mark-inconsistencies*, *mark-redundancies*, and *determine-certainty*. Given just this knowledge and the built-in defaults, abductive problems can be solved. An example of the behavior produced under these conditions is given below in the section on demonstrating ABD-Soar.

Optionally, the designer can choose to add additional search control knowledge and/or knowledge to override any of the default knowledge. Adding knowledge beyond the minimum requirements can greatly increase the efficiency of the abductive system. ABD-Soar is completely open with respect to the addition of new knowledge. Additional operators can be added to any space. New ways of implementing existing operators can be added. Search control knowledge can be added so that lookahead can be avoided. Furthermore, the additions can be made general so that they work for all tasks, or specific so that they only work for a single task or problem. Every addition of knowledge will alter problem-solving behavior. In this framework the available knowledge shapes the strategy, unlike the traditional approach where strategies must be designed to use pre-specified kinds of knowledge.

## Demonstration of ABD-Soar

The problem-solving behavior of ABD-Soar changes according to the task and the knowledge available to solve the task. Here we describe the default behavior of the system and show what knowledge must be added so the system will behave similar to Red's strategy for abduction. The first example illustrates the system's default behavior. The second example illustrates behavior similar to that of Red.

Both of the examples use the red cell antibody identification case shown in Figure 1. The details of the domain are unimportant—only the knowledge of what each antibody explains is necessary to understand the examples. The correct answer for this case is anti-K and anti-$Fy^a$.

### Example 1: Default Behavior

To solve the case in Figure 1 using default behavior requires a domain space with an initial state containing the reactions (the explanata) and knowledge to implement *explain* for each reaction and *determine-accounts-for* for each antibody.

*Explain* is implemented using a single production for each reaction and explanation pair. These are of the form: *if trying to explain 1+ on cell 1 then consider Anti-Fy$^a$*. *Determine-accounts-for* requires two productions for each antibody: one to enumerate the reactions explained by the antibody and one to detect that the operator has been applied. Overall, the case consists of 31 productions.

A fragment of the system's behavior is shown in Figure 2. The numbers indicate the problem-solving sequence and are used in the following description to indicate what part of the figure is being discussed. Five cover operators (one for each reaction) are applicable to the initial state. This leads to an *operator tie* impasse (1) and the selection of *find-best* to break the tie (2). Find-best evaluates each operator by applying it to a copy of the initial state of the domain space. The figure shows a fragment of the lookahead process for *cover r2* (3). The domain space is selected (4) and *cover r2* is applied to the state (5). The cover operator is implemented in the *cover* space (6) by applying *explain* (7) which generates a single explanation and then *add* (8) to add that explanation to the model. Once anti-K is added, *cover r2* has been successfully implemented (9). Lookahead problem solving then continues in the domain space by processing the newly added explanation (10) using four operators: *determine-accounts-for*, *mark-redundancies*, *mark-inconsistencies*, and *new-to-processed*. Their location is indicated with an ellipse in the figure. Next, three cover operators can apply to the current state so an *operator tie* impasse arises (11). This impasse is resolved using lookahead (not shown) resulting in the selection of *cover r4*. This reaction is explained using anti-Fy$^a$ which, when processed (12), results in a model that satisfies the abductive criteria. At this point the state is evaluated (13) and the result of the evaluation is returned to the *find-best* space (14). The system then repeats this process with the remaining *cover* operators (15-18). In this example, all of the *cover* operators evaluate to 2. This means that no matter which finding is picked to *cover* first, the resulting explanation will be equally good. Based on these evaluations *find-best* generates *indifferent* preferences for each of the *cover* operators (19). This allows the system to choose a *cover* operator at random. In this run, the system decides to cover the 1+ on red cell 1 (20). This reaction can be explained by anti-Fy$^a$ or anti-C, so the system uses lookahead to rate each explanation (not shown). Anti-Fy$^a$ results in a better evaluation since it explains more than anti-C, so anti-Fy$^a$ is added to the model and processed (21). Next the system randomly chooses to cover the 3+ on red cell 2 (22). Anti-K is the only antibody that will explain this reaction, so it is added to the model and processed (23). This results in a best explanation that satisfies all of the abductive criteria so the system halts. A total of 1038 decision cycles were required to solve the problem. A decision cycle corresponds to the selection of a goal, problem-space, state, or operator.

This example illustrates how lookahead can be used to generate knowledge about what to do when the designer of the system does not or cannot supply that knowledge. However, lookahead is quite expensive so it is desirable to add additional knowledge whenever possible. The next example illustrates how the addition of some simple search control knowledge can greatly decrease the number of decision cycles needed to solve the problem.

### Example 2: Red-like Abduction

Red uses two heuristics to help prune the search space: (1) The system prefers to cover stronger reactions before weaker reactions. If reactions are equal, then one can be selected

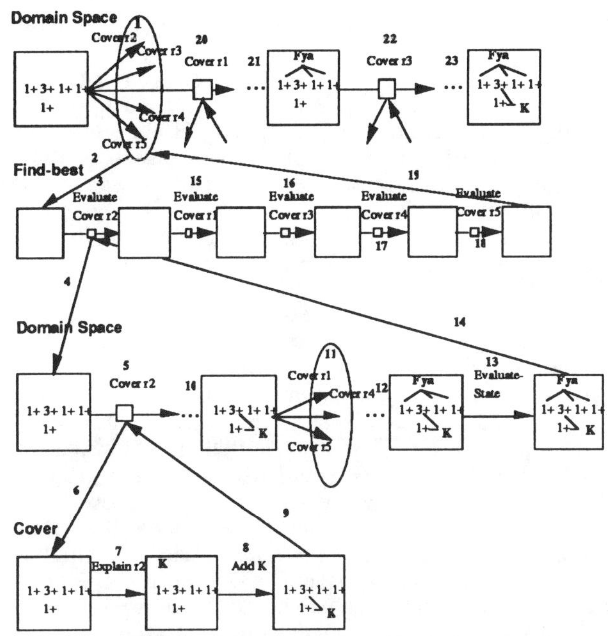


Figure 2: Default behavior of ABD-Soar

at random to cover; and (2) Whenever multiple antibodies can explain the same reaction, the antibodies are ordered according to plausibility and the one with the highest plausibility is used. If multiple antibodies have the same plausibility, then one is selected at random. This knowledge can be added to ABD-Soar by rating antibodies with a plausibility and by adding three search control rules: ID1) If $r_1 > r_2$ then *cover* $r_1$ is better than *cover* $r_2$. ID2) If $r_1$ is equal to $r_2$ then *cover* $r_1$ is indifferent to *cover* $r_2$. ID3) If $a_1$ has a higher plausibility than $a_2$ for explaining a reaction then $a_1$ is a better explanation than $a_2$. Since it has not been specified that equally plausible antibodies are indifferent, the system will do lookahead to differentiate between them.

As a result of this knowledge, the system solves the case in 23 decision cycles (versus 1038 for example 1). First the system decides, because of ID1, to cover the 3+ on red cell 2. It does this using anti-K, the only antibody that explains the 3+. Next, because of ID2 the system randomly selects the 1+ reaction on red cell 1 to explain. This can be explained using either anti-Fy$^a$ or anti-C, but anti-Fy$^a$ is more plausible so, because of ID3, anti-Fy$^a$ is selected.

This example illustrates how small changes to search control knowledge can radically alter the behavior of the system. The system in this example exhibits behavior very much like Red. Furthermore, whenever the search control is inappropriate, the system can fall back on the default knowledge to make progress.

## Comparison to Other Tools

### Peirce

Peirce is a tool built to do hypothesis assembly (Punch III, et al., 1990). It provides a control mechanism, called *sponsors and selectors* (Brown and Chandrasekaran, 1989), to encode search control knowledge for sequencing and achieving its abductive goals in a somewhat flexible manner. Flexible subgoal sequencing is accomplished by using knowledge about the applicability of subgoals to the current situation (*sponsors*) and knowledge about how to choose a subgoal to pursue based on the applicability of all the subgoals (*selectors*).

Because of the generality of its goal/subgoal hierarchy and its control mechanism, Peirce can be used to encode many different strategies for abduction. However, it ultimately restricts flexibility by limiting the knowledge that a Peirce-based system can use. First, there is no way to generate additional search control knowledge in Peirce at run-time. There is also no way to add new goals or methods at run-time. This is not a problem in ABD-Soar because any subgoal or impasse can be resolved using the complete processing power of the PSCM. Thus, search control knowledge, evaluation knowledge, or operators for a problem-space can be generated just like any other kind of knowledge in ABD-Soar. The second limitation is that control knowledge is limited in Peirce because methods are encoded as opaque lisp functions. Thus, once a method is invoked it is no longer controlled by Peirce. This makes monitoring of problem-solving progress difficult if not impossible for Peirce. It can also be difficult to look at the local state of the methods since the local variables of a lisp function cannot be seen from outside the function. This is not a problem for ABD-Soar because it is meant to be used within Soar, where all "methods" are implemented within the architecture as problem-spaces whose states are accessible by knowledge associated with any problem-space. Finally, the goal/subgoal structure in ABD-Soar is much finer grained than the one used in Peirce. This means that the abductive strategy can be controlled at a finer level of detail in ABD-Soar.

### Mole and Cover-and-Differentiate

*Cover-and-differentiate* (McDermott, 1988) is a method for a form of abduction. It is implemented in Mole (Eshelman, 1988), a tool for building cover-and-differentiate systems. McDermott defines its method as 1) Determine the events that potentially explain the symptoms. 2) If there is more than one candidate explanation for any event, then identify information that will differentiate the candidates by ruling out an explanatory connection, ruling out an explanatory event, confirming an explanatory event, or preferring one explanatory connection over another. 3) Get this information and apply it (in any order). 4) If step 3 uncovers new symptoms, go to step 1(McDermott, 1988).

This method and Mole, in particular, have several limitations from a flexibility perspective. First, the method is extremely rigid—it specifies a specific fixed sequence of actions for solving the problem. Hence the search control knowledge is pre-specified and cannot be altered or extended. Second, the "grain-size" of the actions are quite large. For example, all possible explanations for every finding must be enumerated in a single step. This eliminates all methods that decompose the problem into a number of smaller problems. This, in turn, means that knowledge about what to cover first (something that can increase the efficiency of problem-solving) cannot be used by the method. Third, all possible candidate explanations must be statically pre-enumerated—Mole does not allow the candidates to be generated or constructed at run-time. Finally, the method is most applicable when there is only a single fault—the method is not designed for hypothesis assembly.

## Limitations

ABD-Soar has two limitations. First, to use ABD-Soar, a system builder must be well acquainted with Soar. Second, the current version of ABD-Soar does not use Soar's built-in learning mechanism. This is because the low level representation of annotated models we used causes Soar to learn overgeneral productions. This can be solved by using a different representation for annotated models. Such a version of ABD-Soar is being planned.

## Conclusion

ABD-Soar implements an extremely general framework for building abductive systems. It can use a wide range of knowledge and alter its behavior based on that knowledge. Instead of programming a method for abduction, a system builder can give the system the knowledge available to solve the problem and the system will behave appropriately. In the absence of specific knowledge the system can fall back to default knowledge to make progress, however slowly. Along with its flexibility and generality, ABD-Soar provides guidance for building systems and acquiring knowledge because it provides a theory that specifies specific kinds of domain and search control knowledge.

## Acknowledgments

We thank B. Chandrasekaran, John Josephson, and Kathy Johnson and the AIM research group for their assistance and comments on this paper and the work it reports. We also thank the members of the Soar community for their intellectual and technical support.

## References

Allemang, D., Tanner, M., Bylander, T., &Josephson, J. R. 1987. On the Computational Complexity of Hypothesis Assembly. In Procs. of the Tenth International Joint Conference on Artificial Intelligence, :1112-1117. Milan, Italy.

Brown, D. C., &Chandrasekaran, B. 1989. *Design Problem Solving: Knowledge Structures and Control Strategies*. San Mateo, CA: Morgan Kaufmann Publishers.

Bylander, T., Allemang, D., Tanner, M. C., &Josephson, J. R. 1988. Some results concerning the computational complexity of abduction (Technical Report #88-TB-COMPLEXITY). Laboratory for Artificial Intelligence Research, Dept. of Cptr. & Information Science, The Ohio State Univ.

Eshelman, L. 1988. MOLE: A knowledge-acquisition tool for cover-and-differentiate systems. In S. Marcus (Eds.), *Automating Knowledge Acquisition for Expert Systems* (37-80). Kluwer Academic Publishers.

Josephson, J., Chandrasekaran, B., Smith, J., &Tanner, M. 1987. A mechanism for forming composite explanatory hypotheses. *IEEE Transactions on Systems, Man, and Cybernetics*, 17(3):445-454.

Josephson, J. R., &Goel, A. 1988. Tractable Abduction (Technical Report) The Ohio State University, Laboratory for Artificial Intelligence Research.

McDermott, J. 1988. Preliminary steps toward a taxonomy of problem-solving methods. In S. Marcus (Eds.), *Automating Knowledge Acquisition for Expert Systems* (225-256). Kluwer Academic Publishers.

Newell, A. 1990. *Unified Theories of Cognition*. Cambridge: Harvard University Press.

Punch III, W. F., Tanner, M. C., Josephson, J. R., &Smith, J. W. 1990. Peirce: A tool for experimenting with abduction. *IEEE Expert*, 5(5):34-44.

# Button Theory: A Taxonomy of Student-Teacher Communication for Interface Design in Computer-Based Learning Environments*

**Menachem Jona, Benjamin Bell, and Lawrence Birnbaum**

Northwestern University
The Institute for the Learning Sciences
Evanston, IL 60201

**Abstract**

This paper introduces Button Theory whose two principle goals are, first, to provide a taxonomy of the ways that students might usefully interact with and control a computer-based teacher, and second, to provide a natural mechanism by which they may exercise that control. We have developed a small but comprehensive set of messages that students would find it useful to convey to a teacher during a tutorial interaction. and have associated each message with a button presented iconically on the computer screen We describe our experience with the use of Button Theory in a prototype computer-based teaching system, and demonstrate how, even with rather simple mechanisms, this framework enables surprisingly rich interactions.

One of the more serious problems facing education today is that students lack control over the instruction they receive (see, *e.g.*, Schank & Jona, 1991). When students feel that their learning needs are not being attended to, the result is boredom, frustration, and a dislike for learning in general. This is especially true in computer-based learning environments, where technological limitations, aggravated in some cases by poor design decisions, have historically put the student in an extremely passive role (see, *e.g.*, Papert, 1980). This paper introduces *Button Theory,* developed in order to allow students using computer-based learning environments to have as much control as possible over what they see, hear, and learn.

The two principle goals of Button Theory are, first, to provide a taxonomy of the ways that students might usefully interact with and control a computer-based teacher, and second, to provide a natural mechanism by which they may exercise that control. We have developed a small but comprehensive set of messages that students would find it useful to convey to a teacher during a tutorial interaction. These messages are organized into three groups, **feelings**, **questions**, and **control**, reflecting the three natural domains of discourse in a computer-based learning environment: The student may wish to talk about himself (**feelings**), about the domain or task of instruction (**questions**), or about the teaching itself (**control**). The **feelings** group contains messages about how the student is feeling about the lesson, allowing expressions of boredom, excitement, or disbelief. The **questions** group, as its name implies, contains various questions a student is likely to ask (*e.g.* Why?, What's the point?, etc.). The third group, **control**, allows the student to change the progression of the lesson to suit his or her needs, speeding up when the material is too easy, backing up when it is too difficult, or getting more detail about something interesting.

As the name of the theory indicates, rather than using natural language, the mechanism we provide to enable students to express these messages is a set of buttons presented iconically on the computer screen, each button corresponding to one message. The user can "press" any of these buttons using a mouse-pointer. This "Button Pad" can be thought of as analogous to the remote-control clicker of a TV or VCR, except that instead of controlling the volume or the channel, it allows the student to manage the instruction he or she is receiving.[1]

As described above, the types of communications that might take place between a student and a computer teacher can be divided into the three categories of **feelings**, **questions**, and **control**. These three categories are based on the three naturally-occurring domains of discourse that exist in any interaction between a student and a computer-based teacher, namely (a) the student, (b) the teacher/program, and (c) the domain and/or task being taught (see Figure 1).

The kinds of communications that fall into each of the three categories can be characterized as follows:

- **Feelings** – This category concerns the student's internal state. By "internal state" we mean the states of the student's emotional and belief systems. These are influenced by, among other things, various attributes of the domain and/or task, the teaching program, and the idiosyncrasies of the individual student. The student may feel bored or excited by the domain, may find his or her beliefs challenged, or may be frustrated by his or her inability to understand the material. Each student will also have different reactions to different teaching styles and subject matter.
- **Questions** –This category concerns the domain and/or task being taught. Communicative acts in this group involve efforts by the student to comprehend the subject

* This research was supported in part by the Defense Advanced Research Projects Agency, monitored by the Air Force Office of Scientific Research under contract F49620-88-C-0058 and the Office of Naval Research under contract N00014-90-J-4117, by the Office of Naval Research under contract N00014-J-1987, and by the Air Force Office of Scientific Research under contract AFOSR-89-0493. The Institute for the Learning Sciences was established in 1989 with the support of Andersen Consulting, part of The Arthur Andersen Worldwide Organization. The Institute receives additional support from Ameritech, an Institute Partner, and from IBM.

[1] See Schank & Jona (1991) for a fuller discussion of this analogy and its implications.

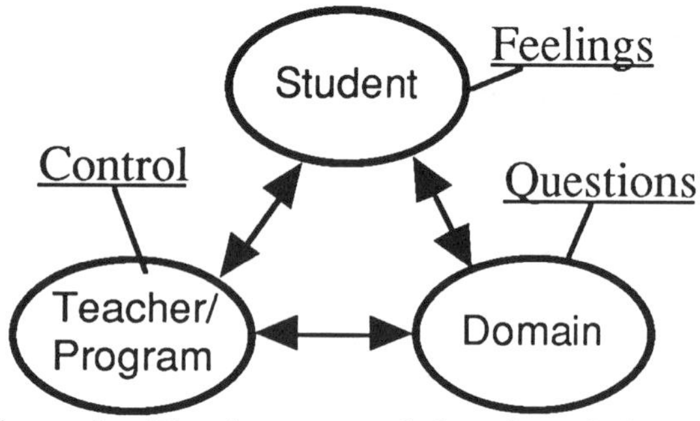


Figure 1. The three natural domains of discourse in a student-teacher interaction and their corresponding categories of communicative acts (underlined).

or task which he or she is engaged in learning. The student may be seeking to understand the reasons for doing or knowing something, or may want to know what to do next.

- **Control** – This third category contains expressions about the progress of the lesson. The student may wish to speed up or slow down the rate of instruction, or may wish to have material presented in greater detail. The structure of a lesson is often tightly linked with the domain being taught, and hence the two are often difficult to tease apart. Nevertheless, this category is intended to deal with the mechanics and style of teaching rather than with the domain.

Figure 2. A condensed version of the Button Pad.

Within each of the above categories, we have selected what we believe to be the five most important, most frequently used communicative acts, and assigned each to an iconic button on the Button Pad (see Figure 2). The five messages we have chosen for each category are as follows:

- **Feelings** — "Awesome!", "Boring", "No way", "Huh?", "Too hard"
- **Questions** — "Why?", "How do I do that?", "Now what?", "What's the point?", "History"
- **Control** — "Change task", Back up", Big picture", "More detail", "Skip this"

It should be clear that the first step in developing a theory of communicative interaction in tutorial situations must be to develop a taxonomy of kinds of communicative acts that a student might be expected to perform in such situations, as we have described above. This is true regardless of whether the ultimate application of the theory will be in the development of the sort of iconic interface we are pursuing, or in natural language processing. That is to say, the first step in developing any theory of this communicative process must be to characterize the class of expected inputs. Moreover, this characterization must be functional in nature, that is, it must facilitate a determination of the intended meaning of a communicative act, and of an appropriate response. In other words, a representation theory must be developed that enables us to categorize student inputs in terms of both content and intent. The taxonomy described above embodies one such theory.

It is certainly arguable that our choices within each category are, to a certain extent, arbitrary, or that we should have chosen more or fewer items in each category. Our response to such arguments is similar to that offered in defense of any knowledge representation theory.[2] We may not have the right number of buttons, nor do we claim that we have identified the only possible "correct set." Rather, we claim that this is one reasonable solution, and that any equally reasonable set of choices would be roughly equivalent in a functional sense. To put this another way, the choice of buttons in each category is not a question of right or wrong: The real question is, how well, and how easily, does this comparatively narrow channel of communication allow students to express themselves? Our experience using the Button Pad in several computer-based learning environments indicates that Button Theory provides a good balance of ease of use and expressivity.

## The Phone Tutor: A simple case study of Button Theory in action

To get a better feel for the kinds of interactions supported by Button Theory, let's look at a concrete example of the theory in action. The example comes from a prototype system called the Phone Tutor which is designed to help phone company customers learn how to use special phone

[2] For example, see Schank (1975) for a discussion of this issue as it arises in the context of the Conceptual Dependency meaning representation for natural language.

services such as call waiting or call forwarding.[3] The idea behind the system is to provide users with an enhanced telephone keypad containing a few additional buttons that enable them to interact with an on-line tutor. The three additional buttons on the phone keypad are a small subset of the fifteen buttons discussed above, one from the **feelings** category ("Huh?"), and two from the **questions** category ("Why?" and "Now what?").

The system responds to button presses with recorded audio messages. A transcript of a sample interaction is shown below. The student's actions are shown in boldface.

**Practice**

Thanks for calling the Phone Tutor. The Phone Tutor can help you learn to use Custom Calling Services. Choose the service about which you are confused. Press 1 for call forwarding, press 2 for call waiting, press 3 for speed dialing, or press 4 for three-way calling.

**1**

With call forwarding, you can have your calls -- local or long distance -- transferred to another number. That means your calls can follow you when you leave home or work. Is this the first time you have practiced call forwarding?

**Yes**

OK, we will start by practicing the procedure for setting up call forwarding. The first step is to dial 72. Do that now.

**Why?**

Pressing 72 is the special code that tells our computers that you are about to set up call forwarding. Other special services use different codes. Now type 72.

**72**

Good. Now press the # key.

**Huh?**

The # key is just to the right of the 0 key. It is also labelled "No". Now press the # key.

**Why?**

Pressing the # key is needed to indicate that you want to access a special service instead of dialing a regular number. All special phone services, including call forwarding, require you to dial # after the special code. Now press the # key.

**#**

That's right. Now type the number you want to forward to.

**Huh?**

After dialing 72#, indicating that you want to set up call forwarding, the next step is to enter the number to which you wish to have your calls forwarded. Now type the number you want to forward to.

**Now what?**

Try it.

**Now what?**

Type the number you want to forward to.

**5553500**

So far you've practiced typing 72# and the number separately. Now try to do it all at once. Press 72, then the # key, then the number.

**725553500**

Let's try that again. Press 72, then #, then the number you want to forward to.

**72#5553500**

OK, for practice, pretend you would like to forward a call to 555-8989.

[dialog continues...]

The program that produces these dialogs is rather simple in structure: It is essentially an augmented finite state machine, in which button presses and student responses are used to determine what sort of action the program should take, within the context provided by a network representing the hierarchical goal/sub-goal structure of the task, the student's current position in that task, and a few other features, *e.g.*, the last button they pressed. The relative richness and coherence of the dialog presented above is not really due to the algorithm employed. Rather, it reflects the utility of the three queries permitted by the available buttons in conjunction with the content of the task network. The message here is that with a good representation of what needs to be taught, and a sensible taxonomy of student queries, even rather simple mechanisms will suffice to produce relatively natural and flexible interactions with a teaching system.

## Buttons as conceptual interface

An important goal of Button Theory is to shift the burden of mapping from the learner's needs to the system's responses from the learner to the system itself. For example, a great deal of research in human-computer interaction is concerned with how easily a user can associate a screen icon with a system action or component (see, *e.g.*, Pejtersen & Goodstein, 1988; Norman & Draper, 1986). In contrast, Button Theory is concerned with making it easy for the student to associate screen icons (*i.e.*, buttons) with conceptual categories that directly express his or her feelings, questions, or desires. Determining how to interpret and respond appropriately to a button press is up to the system itself; the user need not—indeed, often cannot—specify the particular action to be performed by the system.

As Norman (1986) points out, the chief task of any interface is "to bridge the gap between [user] goals and [the] system." He continues, "there are only two ways to do this: move the system closer to the user; move the user closer to the system. Moving from the system to the user means providing an interface that matches the user's needs, in a form that can be readily interpreted and manipulated" (p. 43). Herein lies the strength of the Button Pad interface. By incorporating the basic messages provided by our taxonomy

[3] The Phone Tutor system was built by the first author.

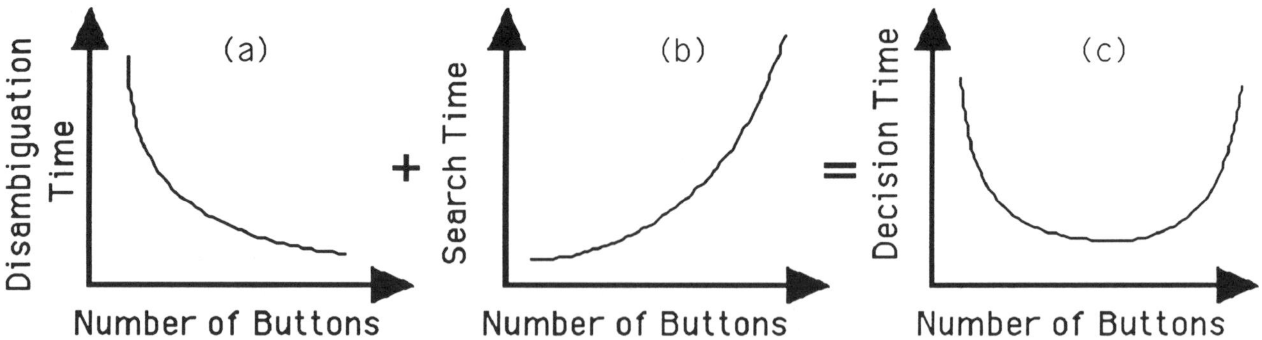


Figure 3. Schematic depiction of how disambiguation time and search time combine to shape the student's ease of use as a function of the number of buttons.

of student-teacher communication, the Button Pad interface enables the learner to express his or her needs directly, in a way that is nearly as simple and natural as natural language itself. This close match provides the Button Pad interface with what has been called *semantic directness*, the ability to easily express things of interest using the interface (Hutchins, Hollan, & Norman, 1986). Interfaces that provide users with these important characteristics we term *conceptual interfaces*.

The Button Pad (Figure 2) always displays the complete range of expressions possible. The visual presence of the Button Pad on the screen is a subtle but important aspect of the efficacy of this type of interface. The sequence of mappings that the user must construct from the thought he or she wants to express, to a screen icon, to a program command, is made simpler and more direct. The availability of the buttons offers a palette of general utterances from which the student need only select the one closest to what he or she wants to say.

A related question, given the restricted vocabulary provided by Button Theory, is the effect that the interface has on how the learner thinks about the domain, "because the interaction language demanded by the physical system comes to color the thoughts of the person" (Norman, 1986, p. 39). The manner in which the way we think about something is shaped by the the vocabulary we have for talking about it is known as *linguistic determinism* (Hutchins et al., 1986).[4] This is problematic in that the restrictions imposed by the interface may come to restrict the user's ability to think flexibly about the domain (Hutchins et al., 1986). Our expectation is that, because the Button Pad provides the learner with an interface that embodies natural categories of expression, it is unlikely to force an overly limited perspective on the learner. In fact, because it always displays the entire range of queries possible, the Button Pad interface turns linguistic determinism to its advantage, by assisting learners in the often difficult process of question formation. Indeed, it may well lead to better, more creative thinking, because it encourages students to ask more and better questions than they might ordinarily ask (Schank, 1986, 1988).

## Tradeoffs in designing a set of buttons

There exist a number of underlying tradeoffs in designing a set of buttons. From the perspective of building a program that can respond intelligently to button presses, clearly the more buttons the better. More buttons, in general, means that there will be less ambiguity to resolve for each individual button press. The range of possible meanings associated with each button press will decrease as the number of buttons increases, since more buttons are being used to cover the same range of possible meanings. Unfortunately, an indefinitely large set of buttons is not a plausible option when building a user interface.

Moreover, from the user's perspective the situation is not so clear-cut. With a small number of buttons, the student is faced with the task of fitting his intended message to the button that he thinks best subsumes that message. Whether or not this is a severe bottleneck will depend upon the details of the decision procedure the student employs in this task. If the student first formulates an intended message and then attempts to determine the button that best expresses it, as seems plausible (especially in the case of novice users), then his task will be the inverse of that facing the interpreter of the button press, and he will face a similar problem of ambiguity (Figure 3a). With a large number of buttons, the student is faced with another difficult task, that of finding the button that best expresses his or her question (Figure 3b). Presumably there is a middle ground (analogous to the basic level for categories described by Rosch et al., 1976) which closely matches the kinds of things students typically wish to express, minimizing the cognitive effort the student must devote to expressing his question by reducing the amount of searching and fitting that is required (as reflected by decision time; see Figure 3c). We have selected a set of buttons which we believe corresponds closely to the needs of the student while at the same time providing sufficient clarity to facilitate implementing the appropriate responses in a computer-based learning environment.

## Interpreting button presses

Another prerequisite for the success of Button Theory is that computer-based teachers employing the buttons be capable of interpreting what the learner means by a given button

[4] This is, of course, closely related to the Whorf-Sapir hypothesis.

press in the context of a tutorial dialog. The tractability of this interpretive task is dependent upon the availability of contextual information that can significantly constrain the meaning of button presses. We believe that this contextual information can be made available in tutorial dialogs.

In computer-human interactions, context acts to reduce ambiguity in two distinct ways. First, it can make the meaning of the buttons in a specific situation clearer to the learner (Jones, 1987; Pejtersen & Goodstein, 1988), thus reducing the likelihood that a user pushes the "wrong" button. Second, the context naturally constrains the range of possible meanings the learner may be expected to have in mind when posing a question to the program. For example, Moore (1989) has investigated appropriate responses to "huh" questions, Moore & Swartout (1990) present an analysis of "why" and "how" questions, and Kahneman & Miller (1986) discuss "why" queries. This work, in conjunction with our own experience to date, supports the claim that the disambiguation of button presses by a computer-based teacher is tractable.

## Future work and conclusions

To investigate the usefulness of Button Theory we have incorporated the Button Pad interface in several prototype computer-based learning environments. In addition, we are currently investigating a number of empirical questions raised by Button Theory, including whether all of the buttons we have devised are indeed useful, and what kind of response a teacher should make to each one. We have begun to design a suite of experiments to investigate these questions, which are intended to shed light on such issues as the optimal number of buttons, which messages are most necessary to include, what kind of response students expect when they press a given button, what kind of response human teachers actually give, and how well these latter two correlate. We would also like to investigate the extent to which the buttons affect both the quantity and quality of questions asked by students. The hypothesis to be investigated is whether, rather than constraining the kinds of questions asked, the Button Pad will actually encourage not only more question asking, but also better question asking (Schank, 1986, 1988). We hope to report the findings of these experiments in the near future.

It should be apparent that the problem of correctly interpreting and responding to a given button press by a given student in a given context is, in its full extent, as difficult a problem as natural language understanding. In fact, it is the same problem. Our hope is that the framework provided by Button Theory will permit us to tackle this problem incrementally, starting with simple responses for a subset of the buttons, and gradually increasing the sophistication of the interpretive mechanisms underlying each button as necessary. The key point is that even with rather simple mechanisms, this framework enables surprisingly rich interactions. These interactions are only as successful as the taxonomy upon which they are based; but even the rather straightforward analysis of student communicative needs offered by Button Theory provides a great deal of leverage.

## References


Hutchins, E.L., Hollan, J.D., & Norman, D.A. (1986). Direct manipulation interfaces. In D.A. Norman & S.W. Draper (Eds.), *User Centered System Design: New Perspectives on Human-Computer Interaction.* Hillsdale, NJ: Lawrence Erlbaum.

Jones, W.P. (1987). How do we distinguish the hyper from the hype in non-linear text? In H.J. Bullinger & B. Schackel (Eds.), *Human-Computer Interaction: INTERACT '87.* Amsterdam: North-Holland.

Kahneman, D., & Miller, D.T. (1986). Norm theory: Comparing reality to its alternatives. *Psychological Review, 93,* 136-153 .

Moore, J.D. (1989). Responding to "huh?": Answering vaguely articulated follow-up questions. In *Proc. of the Conference on Human Factors in Computing Systems* , Austin, TX, April 30-May 4 1989.

Moore, J.D., & Swartout, W.R. (1990). Pointing: A way toward explanation dialogue. *Proc. of Eighth National Conference on Artificial Intelligence* (pp. 457-464). Cambridge: MIT Press.

Norman, D.A. (1986). Cognitive engineering. In D.A. Norman & S.W. Draper (Eds.), *User Centered System Design: New Perspectives on Human-Computer Interaction.* Hillsdale, NJ: Lawrence Erlbaum.

Norman, D.A. & Draper, S.W. (Eds.) (1986). *User Centered System Design: New Perspectives on Human-Computer Interaction.* Hillsdale, NJ: Lawrence Erlbaum.

Papert, S. (1980). *Mindstorms: Children, computers, and powerful ideas.* New York: Basic Books.

Pejtersen, A.M., & Goodstein, L.P. (1988). Beyond the desktop metaphor: Information retrieval with an icon-based interface. In P. Gorny & M.J. Tauber (Eds.), *Visualization in Human-Computer Interaction.* Berlin:Springer-Verlag.

Rosch, E., Mervis, C. B., Gray, W. D., Johnson, D. M., & Boyes-Braem, P. (1976). Basic objects in natural categories. *Cognitive Psychology, 8,* 382-439.

Schank, R.C., & Jona, M.Y. (1991). Empowering the student: New perspectives on the design of teaching systems. *The Journal of the Learning Sciences, 1,* 7-35.

Schank, R.C. (1988). *The Creative Attitude: Learning to ask and answer the right questions.* New York: Macmillan.

Schank, R.C. (1986). *Explanation Patterns: Understanding Mechanically and Creatively.* Hillsdale, NJ: Lawrence Erlbaum.

Schank, R.C. (1975). *Conceptual Information Processing.* New York: American Elsevier.

# Human Performance in Visually Directed Reaching Results in Systematic, Idiosyncratic Error

**Michael L. Kalish**
Department of Cognitive Science
UCSD, La Jolla, CA 92093
(619) 534-4348
mlkalish@ucsd.edu

## Abstract*

We study the performance of human subjects in a task which requires multi-jointed reaches to be made to targets spaced over a wide area. In accordance with established research, we find that subjects' reaches are not accurate when they cannot see either their hands or the targets. The errors subjects make are different at different targets, suggesting that they are due to an error in the planning of movements. However, contrary to existing models of this error, we find that it is highly idiosyncratic. This leads to the rejection of the most straightforward model of how reaching is learned, and poses problems which a future model must address.

## Introduction

When humans reach to visually located targets without visual feedback, they make errors. The nature of these errors is a reflection of the way the motor system is controlled, and is the focus of this paper. We investigate visually directed reaches made under a variety of conditions, and find that subjects produce idiosyncratic errors which vary with the target's location. If the responses of a group of subjects is averaged together, the individual idiosyncrasies cancel each other, and the average error at any target is very small, approaching zero. The behavior of the group is thus not representative of any individual, suggesting that a model of motor control should attempt to explain the variations observed to exist between individuals.

Reaching, in the experiments described here, is the process of controlling the arm in order to position the finger in the environment. What variables are controlled, and how, is the topic of a variety of models. These models all assume that the final position of the finger is known, and that the motor system works to achieve this goal state. If the variables under control are joint angles or muscle lengths, then clearly a transformation of the visual target into an internal, postural one must precede movement planning. Even if the position of the hand is directly controlled, the joints and muscles must still be told what to do, and this requires that the desired hand position or trajectory be turned into desired postures.

Reaches made without visual feedback require a mapping from visual information about target location to an arm configuration. This mapping is ill posed since there are multiple arm configurations which can produce any single finger position. One way to learn an approximation of the mapping, called flailing or direct inverse modelling (Jordan and Rumelhart 1990), is to pick large numbers of arm configurations at random and then combine all random movements terminating in the same spatial region by a mechanism that approximates the average or *expected* values for each independent degree of freedom (Kuperstein 1988). These expected values can then be used to associate a unique arm posture with each visual location. These postures point to locations which approximate the associated visual locations but do not necessarily correspond to them exactly, resulting in an error field of four dimensions for targets in a plane that can be compared to experimental data. Prediction of minimum possible error from the flailing model does not depend on the details of the implementation of the model, as the properties sufficient to produce error predictions are embodied in the geometry of the arm and body and the description of the task.

The flailing model accounts to some degree for the gross quantitative properties of the error observed here, but it does not account for the idiosyncratic variations by subjects. The overall average magnitude of error predicted by the expected values is close to that seen in the experiments, differing by at most a factor of two from the observed error magnitude. Both observed and predicted error differ with target position, and not in any simple way. An individual subject will make different errors at different targets, but the direction of those errors is not consistently related to any outside variable. While this is true also for the predicted errors, the flailing model predicts that there is one unique actual response for each target, when in fact the responses are very different for each subject.

The specific expected values of a flailing model depend on the structure of the model arm and its task. Where Kuperstein's model was formulated using only a two-jointed artificial arm, we have attempted to construct a more realistic model. Using data from a kinesiology text (Luttgens and Wells, 1982) to establish the ranges of

* Research supported by NIMH grant NH45271 to David Zipser, and an NDSEG fellowship to the author.

motion for the joints of the arm, we have modelled a reach as composed of thirteen independent variables. Expected values for errors in finger location after reaching to targets throughout the range of the arm are generated by simulated flailing. Since the arm almost never reaches to exactly the same place twice, the reaching space is divided into cells and all reaches terminating within a cell are grouped together for purposes of averaging. For each joint the angles of the postures which reach to a cell are averaged, computing a single angle. The posture resulting from all the average joint angles represents the expected values that would result from associational learning. The location of the finger tip resulting from the average posture for a cell is compared to the location of the center of the cell, resulting in a difference or error arising from the model. The mean magnitude of the errors made by the model is 1.75cm over the entire reachable surface, equivalent to approximately 2% of average arm length. The overall error magnitude is within the range of Kuperstein's original formulation of the flailing model and the available data from adult visually directed reaches.

## Experiments

The flailing model predicts both large and small scale features of the error distribution, predicting that movements made to different targets should differ in their error, regardless of variable starting position and speed, and independent of the individual reacher. Our first two experiments examine the structure of the error distribution across a surface, the effect of starting position and individual difference. Our third experiment focuses on individual differences, studying three subjects' errors with respect to starting position and movement speed.

### Method

**Subjects:** Fourteen undergraduates from UCSD participated in the first experiment for partial fulfillment of course requirements, or for payment. Nine such undergraduates performed in the second experiment, although one was discarded when equipment failure was detected after the experiment. The third experiment employed three volunteer research associates who were naive to the hypothesis being tested.

**Apparatus:** Subjects were seated at a 160cm x 160cm x 77.5cm high table so that their sternums were abutted against a center mark. At the four corners of the table top were set microphones comprising a collection array for a three-dimensional audio digitizer (Graph Pen GP-6-3D-P). The subjects wore two small, Plexiglass-encased sound-emitters on their right index finger.

On the table, fifteen two cm target circles were selected to cover the reaching surface. These targets were selected to be within range of all the subjects. Mounted to the right of the subjects' chair was a large four sided box, open to the subject and the table, set at table level. In front of the subjects at an elevation of 15cm was an 8cm wide, 100cm long shelf. The experimenter sat behind the subject, in order to operate the digitizer.

**Procedure:** The subject's task was to move their right forefinger as fast as possible to land on a given target. The arm movement involved in this action was made ballistically, with no visual guidance or feedback. Subjects began each trial with their arm either inside the box to their right or beneath the shelf in front of them.

To initiate each trial, the experimenter identified one of the fifteen targets by reading aloud the number which labelled it. The subject was then given an indefinite period in which to orient to the target and prepare the movement. When ready, the subject shut their eyes and reached in one fast motion to place their finger on the target. When the movement ceased, the digitizer was activated and the position of the finger recorded.

After the experimenter told the subject that the recording had taken place, the subject was free to return their hand to the starting position, and only then to re-open their eyes. Since the box and shelf both obscured the hand from view, this assured that the subject could not see their hand for the entire experimental session.

Subjects in experiment one moved to each target four times from each starting position with a different random order within counterbalanced starting position blocks for each subject. This yielded 120 trials for each of the fourteen subjects. After the experimental session, the location of the targets were digitized by having the subjects tour the targets with their eyes open.

Experiment two was identical to experiment one, except that each of the eight subjects toured the targets both before and after the session.

In experiment three, an additional speed condition was added. Subjects were instructed to either move as quickly as possible (as above) or were told to take as much time as was needed to move as accurately as possible. Each subject moved to each target four times from each starting position in each speed condition with a different random order for each subject. This yielded 240 trials for each of the three subjects.

## Results

These experiments showed that individuals produce systematic error when making visually directed reaches, and that the errors made at particular targets varied from individual to individual. Similarly, the errors made by any one individual varied from target to target. The observed idiosyncrasy in the error makes it difficult to conclude much about the group performance, but is in itself an important finding.

In order to determine if subjects made systematic error, a test of each mean response was made. For the first two experiments, a mean response was computed for each subject at each target for both initial positions. The third experiment included a speed condition, doubling the number of mean responses for each subject. Each subject made multiple responses in each condition, and so a mean response was calculated.

$$\text{mean response} = \left(\frac{\Sigma x}{n}, \frac{\Sigma y}{n}\right) = (\bar{x}, \bar{y}) \quad (1)$$

Considering each response as a vector directed from the target to the finger's location, the mean response is the vector average of all responses in a single condition. To establish a confidence level in the distance of the mean response from the target, the standard error of the mean was computed in each condition. To compute the standard error, we first computed the distance of the mean response from the target.

$$D = \sqrt{\bar{x}^2 + \bar{y}^2} \quad (2)$$

The standard error is a measure of the dispersment of the individual data points around the mean. To measure this, we computed the distance of each individual response from the mean for that condition.

$$d = \sqrt{(x - \bar{x})^2 + (y - \bar{y})^2} \quad (3)$$

The actual standard error was then computed, using the difference between the individual distances and the mean distance of error.

$$\sigma_d = \sqrt{\frac{\Sigma(d-D)^2}{n-1}} \quad (4)$$

$$SEM = \frac{\sigma_d}{\sqrt{n}} \quad (5)$$

Dividing the distance of the mean response by the standard error gives a Z score, which can be used to measure the certainty with which the distance is non-zero. If the mean response is far from the target, and the standard error is small, then there will be a high level of confidence in the mean response not being equal to zero, indicating that moving from one initial position to one target, that subject makes a systematic error.

The perceptual-motor system is noisy, and this noise, even if unbiased, can be expected to produce error in reaching. In order to compare the observed results to the result of noise, we ran a Monte Carlo experiment. Unbiased error follows from two assumptions, first that the error has mean zero and second that it is distributed with equal variance in all directions around the target. Data was generated from these two assumptions. For each of 850 separate simulated targets, eight simulated reaches were made. Error in the X and Y dimensions was generated separately, with a single value being selected from each of two normal distributions of equal variance and each with mean zero. For each target a simulated mean response and standard error was computed in the same manner as for the observed data, leading to a judgement of confidence in the mean response error being non-zero. Overall, more than one third of all mean experimental responses (276 out of 835) were more than 95% likely to not be zero for the observed data. This compares with only 0.35% more than 95% likely for the Monte Carlo experiment.

While there were a large number of non-zero errors, the average across subjects of all responses at each target was not discernible from zero for all three experiments. This suggests that few of the non-zero responses contained the same systematic error, as otherwise the mean across subjects would show a bias. To investigate the possibility that the biases were idiosyncratic, a pairwise comparison of all non-zero mean responses at each target was made. The likelihood that any two means were drawn from the same distribution was computed to determine how different each mean response was from the others. If the likelihood that two means come from the same distribution is less than five percent, then the difference between the means was taken to be significant.

Overall, there were 907 pairwise comparisons possible, of which 768, or 85%, proved significant. This percentage was nearly constant across the three experiments, as 84% (495/587) of the comparisons in experiment one, 86% (197/230) in experiment two and 84% (76/90) in experiment three were significant. The source of these differences was predominantly the difference between one subject and another, rather than one subject reaching from two different initial positions.

## Discussion

The results of these experiments demonstrate that there is a mechanism governing visually directed reaching which produces different systematic error for each individual. Reaching is a complex operation, and many factors could contribute to the systematic error. Factors such as the veridicality of visual and proprioceptive perception, the appropriate use of feedback signals and an accurate representation of the target for arm control are all candidates, but only the last is a possible source of error in these experiments. The flailing model predicts systematic error due to an inaccurate representation of the target, but it predicts the same error for each individual. Without extensive modification the flailing model cannot account for the observed reaching behavior.

Given that many factors participate in the control of movements, what is the source of the error observed in the experiments? We postulate a control structure with four basic components, any of which could introduce error into a reach. The first step in making a movement is acquiring the spatial target, which involves visual perception and memory. The perception of the visual target must then be transformed into a desired posture. A feedforward attempt to reach the desired posture is then begun, and, conditions permitting, feedback can be used to correct the reach. Visual information about hand position can be used directly, or indirect feedback can be produced from the proprioceptively perceived arm posture by a feedback controller using a forward model of the arm's kinematics.

Perceptual errors will result in a reach directed to the wrong spatial location, as will memory errors in the time between acquisition of the target and completion of the movement. If the inverse kinematics employed by the feedforward controller are incomplete, the commanded posture will not lead to the desired location. Once the motor system tries to achieve the desired posture, the interaction of the arm with the environment can introduce more errors in the execution of the command. Inaccurate direct feedback about posture can result in the arm taking on a posture different from the desired one. If the forward model used for indirect feedback is incomplete, then this is one last source of error which could cause reachers to believe their hands are not where they really are.

These four types of error: perception of the goal, noise in the plant, inaccurate feedback control and incorrect transformation of the perceived goal into an internal one, each are candidates for explaining the observed errors. Before considering each of these in turn, we can distinguish between two general classes of error which can be introduced into the movement system. Systematic error in reaching must be the result of a bias by one or more of the elements of movement control. Random error, or noise, introduced into the system will effect motor performance on any given reach, but will disappear when multiple reaches are averaged together. This is true whether the noise comes from vision, proprioception, motor execution or a control command.

The perception of the goal can introduce error through either perception or memory, as mentioned above. Any systematic error in visual perception will be transparent to the motor control system, since motor behavior operates in the visual world. This means that if an individual always perceives targets to be to the left of where they are, say, then they will also perceive their hand to be to the left. The result of seeing the hand and the target in the same space is to negate any biases in the visual system, as is made clear through experiments in which the visual world is displaced, but where motor control quickly adapts (Harris, 1965). Foley and Held (1972) showed that large systematic errors result from reduced visual information about target location. Since the experiments presented here do not change the relation between the visual and motor systems, perceptual mistakes cannot account for the observed biases in responses. Systematic error can also be observed if movement is delayed for an extended time (Schoecting and Flanders 1989, vonHofsten and Rosblad 1988). This is most likely due to the forgetting of target location, a process which can introduce biases into the goal command. However, in the experimental paradigm presented here, subjects are not required to remember the target location for longer than it takes them to voluntarily initiate movement. They are allowed to keep their eyes open until they begin moving, and so have little chance to forget the target location before making the movement.

Errors in motor function are similar to errors in perception, in that systematic failures would be overcome by learning. If one's elbow always extended farther than desired, the extension command would be reduced to produce the desired movement. While we have not presented the mechanism for these corrections explicitly, they lie implicit in our discussion of learning through association. In addition, the ability to achieve desired postures is well documented in the context of posture duplication. If an individual moves a finger to a certain angle, and the finger is then displaced, the finger can be returned to the initial angle with a high degree of accuracy (Kelso and Holt 1980, Bizzi, et al 1982). This indicates that motor system can achieve its desired postures well.

Both visual perception and motor execution are essentially accurate, so we must look to the control mechanisms which govern reaching for the source of its errors. There are two potential proprioceptive feedback systems which can operate during movement. The first is direct feedback, which compares afferent information about the current posture with the desired posture to help achieve the desired posture. Success in duplicating postures points to the veridicality of this system. The second use of postural information is for indirect feedback effecting the desired hand position itself. Using a model of the forward kinematics of the arm, an indirect feedback controller maps posture into hand position. Actual hand positions can be compared to desired ones, and new desired position commands sent. This replicates the process by which visual feedback produces accurate reaches. If the forward map is inexact, then the computed hand position will not be the actual one and 'corrective' signals will result in additional error. There is evidence, reviewed by Olson and Hanson (1990), to suggest that the forward mapping is in fact inexact. Researchers have found varying degrees of constant error and directional biases in experiments involving perception of hand position from postural information. A failure to locate the position of one's hand after a passive movement indicates that proprioception is only poorly transformed by the forward map used by a feedback controller. However, the high speed required for the movements in our experiments and the inability to correct movements suggest that a closed loop system does not underlie fast visually directed movements. Thus, the possible errors due to feedback cannot appear in at least our first two experiments.

In order to produce fast movements, a feedforward controller is a necessity. The transformation from desired hand position to desired posture is potentially a source of error due to the properties of the inverse kinematics of a redundant manipulator like the arm. The excess degrees of freedom of the arm means that there is no unique mapping from hand position to arm postures. The flailing model describes how a mapping between desired spatial locations (targets) and postures can be formed, and provides expected values for the errors made during visually directed reaching. In the experiments we collected data which can be directly compared to the model's predictions.

As a model of average performance, the expected value model makes a little progress toward explaining the observed data. It predicts successfully the magnitude of error and the dependence of error on differences in target location. The model does not predict the individual biases, which were a major result of the experiments. Our experimental finding of an average overall error magnitude of 4% of arm length for visually directed reaching is in line with other findings (Olson and Hanson 1990). The flailing model predicts an overall expected value error of approximately 2% of arm length, which Kuperstein's implementation approached with a 4% error. When the flailing model is tested at only the experimental targets the predicted error increases to nearly 8% of arm length, suggesting that the inverse transformation is about as accurate as individual subjects.

The failing of the flailing model lies in its determinacy. Since predictions come from geometric properties of the arm and from principles of association taken to be constant between individuals, all individual difference must be explained only be differing bodily geometries. However, the observed individual performance differences are much greater than the model can accommodate, especially inasmuch as individual differences extend to sensitivity to initial position and movement speed. In order to account for individual difference, there must be parameters of the model which are free to vary during learning, and which do not deterministically relate to the history of the learning period. This requirement stems from our assumption that many learning trials are required for the acquisition of reaching skill, and that these trials are essentially identical across individuals.

What then do these comparisons say about the flailing model as a theory of how people make visually directed reaches? The gross characteristics of the data, which are matched by the model, suggest that a feedforward inverse transformation is applied by people when they reach. The model makes predictions detailed enough to be rejected, but by the same token it demonstrates that theoretical model at this level should and can be held closely to complex aspects of the data for comparison. Attempts have been made to identify a variable responsible for the idiosyncratic differences (eg., Foley and Held 1972, Jeannerod 1988), but none have been satisfactory. The prevalence of idiosyncratic error in our experiments leads us to conclude that it is real, and that it must be explained.

## References


Bizzi, E., Chapple, W., and Hogan, N. (1982). Mechanical properties of muscles: Implications for motor control. *Trends in Neuroscience,* **5**, 395-398.

Foley, J.M. and Held, R. (1972). Visually directed pointing as a function of target distance, direction and available cues. *Perception and Psychophysics,* **12**, 263-268.

Harris, C.S. (1965). Perceptual adaptation to inverted, reversed, and displaced vision. *Psychological Review,* **72**, 419-444.

Jeannerod, M. (1988). *The Neural and Behavioral Organization of Goal-Directed Movements*, Clarendon Press, Oxford.

Jordan, M. and Rumelhart, D (1990) Forward models: Supervised learning with a distal teacher. MIT Center for Cognitive Science Occasional Paper #40.

Kelso, J.A.S. and Holt, K.G. (1980). Exploring a vibratory systems analysis of human movement production. *Journal of Neurophysiology,* **43**, 1183-96.

Kuperstein, M. (1988). A Neural model of adaptive hand-eye coordination for single postures. *Science*, **239**, 1308-10.

Luttgens, K. and Wells, K (1982). *Kinesiology: Scientific Basis of Human Motion.* Saunders College Pub., Philadelphia.

Olson, C.R. and Hanson, S.J. (1990). Spatial representation of the body. In S. J. Hanson and C. R. Olson (Eds.) *Connectionist Modeling and Brain Function*, MIT Press, Cambridge.

Soechting, J.F. and Flanders, M. (1989). Errors in pointing are due to approximations in sensorimotor transformations. *Journal of Neurophysiology*, **62**, 595-608.

von Hofsten, C. and Rosblad, B. (1988). The integration of sensory information in the development of precise manual pointing. *Neuropsychologia*, **26**, 805-821.

# Assessing Transfer of a Complex Skill

**Irvin R. Katz**

Division of Cognitive and Instructional Science
Educational Testing Service
Princeton, NJ 08541

katz@clarity.princeton.edu

**Abstract**

While recent studies have demonstrated various ways that transfer might be achieved in a domain, the measures used to assess transfer rarely stray from time and error data. This paper examines transfer in the complex skill of computer programming in order to explore more flexible and sensitive methods of assessing transfer. In the experiment, subjects wrote both a PASCAL and a LISP version of two programming problems. Although a simple accuracy measure provides evidence for knowledge transfer between the two programming languages, measures based on analyses of the task domain (i.e., partial-credit accuracy, strategy use) provide much stronger evidence. Curiously, these measures target different subjects as exhibiting transfer, suggesting that more than one type of knowledge may be available for transfer.

## Introduction[1]

Transfer is the phenomenon in which knowledge or skills learned in one context affect the learning or performance of another task. Not too long ago, an article on transfer would start with an argument for why transfer might be found between two domains and end with inconclusive results (Gray & Orasanu, 1987). However, the recent literature has seen a wave of studies demonstrating the various ways that transfer can be achieved through careful analysis of the task domains (e.g., Singley & Anderson, 1990).

While the methods used to achieve successful results have progressed, there has been little change in the ways that transfer is measured. In a typical experiment, subjects receive training in one domain and then are asked to perform tasks in another domain. Evidence for transfer is reported when subjects perform the transfer task either faster or more accurately than subjects without the benefit of training. One reason for this use of time and error data might be because transfer has been investigated in fairly simple, structured domains (e.g., Tower of Hanoi: Smith, 1986; text editing: Polson, Muncher, & Engelbeck, 1986). Those studies that use other measures of transfer, such as subjects' use of a particular problem-solving strategy, tend to involve more complex domains like physics and algebra (e.g., Bassok & Holyoak, 1989).

Unlike most studies of transfer, the motivating question for this research is not "Does transfer occur?" but rather "How should transfer be assessed?". This paper describes an experiment investigating transfer of computer programming knowledge. By using computer programming as the task domain, we can explore a greater variety of transfer measures than that possible in more structured domains. Computer programming has the added advantage that although it is a complex, semantically-rich domain, it nevertheless has been intensively analyzed by many researchers (Pennington, 1985). These prior analyses form a basis for understanding how transfer might occur in the computer programming domain.

Because the motivation for this work is an interest in the nature of any knowledge transfer that might occur, the ideal experimental task would be one in which the subjects are likely to try and use previously generated knowledge. As discussed by Gick & Holyoak (1987), when the surface characteristics of two tasks are very similar, it is likely that subjects will attempt to use information learned from one task while solving the second task. Thus, in the experiment, subjects were asked to write two versions of the same programming problems, the versions differing only in the required implementation language, PASCAL or LISP.

Most models of programming skill would predict that transfer should occur in this situation. When writing a program, typically a programmer devises some initial mental plan of how the program should work, then begins implementing that plan. Through the process of coding the program, difficulties with the plan may be uncovered,

[1]This research was supported in part by a National Science Foundation Graduate Fellowship and by Contract MDA903-85-K-0343 from the Army Research Institute and the Air Force Human Resources Laboratory. Preparation of this manuscript was supported by a grant from the Japan Society for the Promotion of Science. I thank John Anderson, Jerry Feigenbaum, and Claudius Kessler for their comments on various portions of this work.

requiring the plan to be refined or changed (Green, 1987). The final result of this processing is not only a program, but also the refined plan for writing the program, the latter of which forms the programmer's mental representation of the just-written program This representation is often characterized as containing abstract knowledge independent of the particular computer language used to implement the program (Rist, 1989).

The programmer's representation serves as the starting point for the implementation of the second version of the program in the new language. The coding of the second version proceeds as if this transferred information were the initial plan developed specifically for the second version. As a result, in coding the second version, transfer should occur because the program's plan need not be generated from scratch. This notion of transfer of planning knowledge is supported by verbal protocol data (Katz, 1988). Subjects make less verbalizations referring to the program plan while writing the second version of a program, but produce an equal number of verbal statements referring to the actual program code for both versions of a program.

## Method

**Subjects.** Subjects were 21 students who had just completed a four week LISP mini-course. All subjects had already known PASCAL before entering the LISP course.

**Materials.** Subjects were given two problems, called Printnums and Addfract, and wrote a LISP and PASCAL version of each problem. The Printnums problem was to write a program that translates a given number between 1 and 999 into its corresponding words. Thus, if given "115," the program should output "ONE HUNDRED FIFTEEN." Subjects were provided with functions that returned the appropriate units-, teens, or tens-word given a digit (e.g., the tens-word function returned "FIFTY" when given "5"). The Addfract problem was to write a program that accepts two fractions in numerator-denominator form and returns the reduced sum of the fractions, also in numerator-denominator form. Thus, given "5 / 6" and "7 / 15" the program should return "13 / 10." Preliminary analyses revealed that these two problems are of approximately equal difficulty.

**Design.** As stated above, all subjects wrote both LISP and PASCAL versions of the two problems — four programs total. The ordering of the actual problems received was counter-balanced. Thus, when discussing the results, the particular serial order of programs will not be considered.[2]

[2]An additional manipulation — whether or not subjects' previous programs were available for reference — was included in this experiment, but did not affect any of the reported measures.

**Procedure.** After writing two simple warm-up programs (one each in PASCAL and LISP), subjects were given the first problem description. The programs were written using a standard screen editor, and subjects were able to test their programs whenever they wished. If a subject did not write a working program within 30 minutes, the subject was stopped and then given the instructions for the next problem. Thus, at the most, the experiment lasted approximately 2.5 hours, including the warm-up exercises.

## Results and Discussion

While subjects overall had trouble writing perfectly working programs (approximately 65% of the programs written did not execute perfectly), by ignoring syntax errors and mis-understandings of the problem descriptions, there is evidence for knowledge transfer between programming languages (Table 1). For the first version of the programs, only 48% of the programs were correct. In contrast, 71% of the second versions were correctly written, and this interaction is significant by a chi-squared test ($\chi^2(1) = 4.94, p<.05$). Clearly, subjects gained knowledge while writing the first version of their programs that they could apply in writing the second version.

Considering the closeness of the two tasks, however, it is strange that so little transfer occurred. Only about one-quarter of the subjects showed the expected accuracy improvement. Overall accuracy might not be a particular sensitive measure of transfer. We need more flexible assessments of subject performance, ones that are based on an analysis of the problem-solving tasks.

| | Correct | Incorrect |
|---|---|---|
| First Version | 20 | 22 |
| Second Version | 30 | 12 |

**Table 1: Program accuracy**

**Partial Credit.** Using a partial credit scheme for scoring accuracy provides a more sensitive measure of transfer. Thirty minutes turned out not to be enough time for subjects to write a completely working program. Thus, some of the programs written by subjects were incomplete (i.e., missing some essential component(s) of a working algorithm). However, if subjects reused their knowledge in writing the second version of a program, then they should have been able to progress further in writing the second version than they did in writing the first. In other words, they should have been able to "pick up where they left off" writing the first version.

In order to judge the completeness of the programs, a general algorithm was enumerated for each of the two problems, Printnums and Addfract. As an example, the Printnums algorithm is shown in Table 2. While there are alternative algorithms for the problems, the other algorithms include approximately the same pieces as the ones enumerated, except in a different order. The parts of the algorithm that the subject tried to achieve were recorded for each subject's program. It was not necessary for a particular algorithm piece to be completely correct — the important point is that the subject tried to achieve that piece. This scoring of each program pair (first and second versions) was done by a judge blind as to which of the two programs in a pair was actually written first.

1. Get number
2. Is there a hundreds digit?
3. Calculate hundreds digit
4. Call units function with above value
5. Print/include result of (4)
6. Print/include "hundred"
7. Remove hundreds digit
8. Is there a separate tens digit?
9. Calculate tens digit
10. Call tens function with above value
11. Print/include result of (10)
12. Remove tens digit
13. Handle separator ("-")
14. Is the number a teen?
15. Call teens function on number
16. Print/include result of (15)
17. Is there a units digit?
18. Call units function on number
19. Print/include result of (18)

**Table 2: Printnums algorithm**

If subjects reused information, then those subjects who wrote an incomplete program for the first version should have written a second version including all the pieces of the first version, plus some more. In contrast, a lack of transfer would show as the two versions being incomplete in different ways. For example, the two versions might consist of different patterns of missing and implemented algorithm pieces, or the second version may be more incomplete than the first. Finally, if the two versions are both complete, it is ambiguous as to whether or not transfer has occurred.

As shown in Table 3, in 77% of the unambiguous cases (i.e., excluding the "Both Fully Complete" column), subjects' second version was more complete than the first version, even though the two versions were written in different computer languages. This result is a strong indication of knowledge transfer; subjects used information learned in the writing of the first version when writing their second version of the program in a new language. By re-using this knowledge, subjects built on work done previously to produce a more complete program.

| Pgm. Name | More Complete | Less Complete | Both Fully Complete |
|---|---|---|---|
| Printnums | 11 | 3 | 7 |
| Addfract | 12 | 4 | 5 |

**Table 3: Partial credit analysis**

**Strategy Use.** Another method for assessing transfer is to observe subjects' use of problem-solving strategies. In the simplest case, possible strategies would include one correct method for solving the problem and several incorrect methods (e.g., "balance" vs. "switch" strategies in the Missionaries-Cannibals problem and its isomorphs; McDaniel & Schlager, 1990). In a complex domain such as programming, there may be a number of strategies that all lead to a correct solution. Each of these strategies would result in programs with different overall organizations. For investigating transfer of strategy use, the problem Printnums was analyzed because subjects solved this problem in two very different ways.

The first algorithm used to solved this problem is to categorize the given number into one of a set of mutually exclusive categories. For example, some subjects would categorize the number as either a one, two, or three digit number and then do the appropriate actions. The important feature of this solution is that, at the top-level, control is given to only one set of actions. Subjects usually implemented this algorithm using a single, nested conditional statement, so this algorithm will be referred to as the "nested" condition structure.

The second solution involves breaking the number into its component digits and performing appropriate actions for each of the hundreds, tens (or teens), and units digits. Control-flow might proceed where first a decision is made about the hundreds digit, and different actions are performed depending on whether or not that digit is zero (i.e., if a hundreds digit exists). In either case, however, control comes back together to handle the remaining two digits of the number. The distinguishing characteristic of this algorithm is the recombination of control. Subjects usually implemented this algorithm using multiple, separate conditional statements, so this algorithm will be referred to as the "separate" condition structure.

For the analysis, each of the 21 pairs of programs written by subjects was categorized as reflecting either a nested or a separate algorithm, or as ambiguous. As described above, the categorization of a program as nested or separate is objective: if the program has a single, top-level conditional statement, it is categorized as nested. If at the top-level of the program there are more than one conditional statements, the program is categorized as separate. Ambiguous programs were so incomplete that a distinction between the algorithms could not be made.

Two subjects wrote one ambiguous program apiece, and the pairs of programs from these subjects were eliminated from all further analyses, resulting in 19 pairs of programs.

The argument for transfer in this case goes as follows. First one must demonstrate subjects' preferences for the two algorithms under normal circumstances. That is, without having written another program previously, which algorithm are subjects more likely to use? When writing the second version, transfer would be evident if, as a result of writing the first version, subjects show a different pattern of biases toward the two algorithms.

For the first versions of each program pair, there was an effect of language on algorithm choice. As shown in Table 4, subjects programming in PASCAL showed a strong tendency to use a separate condition structure while the LISP subjects showed no such preference. Although the PASCAL subjects' bias is significantly different from chance ($\chi^2(1)=6.4$, $p\leq.025$), the interaction is only marginally significant ($\chi^2(1)=2.9$, $p<.10$).

| | Nested | Separate |
|---|---|---|
| PASCAL | 1 | 9 |
| LISP | 4 | 5 |

**Table 4: Condition structure choices in each language (First Version)**

As for performance on the second program, subjects clearly preferred to use the same structure that they had used on their previous program, even though that program was written in a different language (Table 5). If no transfer had occurred, the previously mentioned language-induced biases would have been evident on the second program as well. Instead, 14 subjects kept the same general algorithm on both programs, while only 5 subjects switched to the other algorithm type ($\chi^2(1)=4.26$, $p<.05$).

| | Second Version | |
|---|---|---|
| First Version | Nested | Separate |
| Nested | 4 | 1 |
| Separate | 4 | 10 |

**Table 5: Condition structure choices for each version**

So far, we have seen evidence for significant levels of transfer along two fronts. Subjects' programs were more complete on the second versions and the subjects tended to use the same methods for implementing each version of their programs. The measures based on analyses of the programming problems, the partial credit and strategy use transfer measures, provide much stronger and unambiguous evidence of transfer as compared with a simple accuracy measure.

Clearly subjects are generating knowledge while writing their first versions that they are able to use in writing the second versions of their programs. Thus, there should be a perfect correlation between the two measures of transfer. Subjects who re-use information should show transfer on both measures while other subjects should show a lack of transfer no matter how transfer is assessed.

As shown in Table 6, this simple model of transfer is simply wrong; there is no evidence for inter-dependence between the two measures of transfer. Approximately half of the subjects who demonstrated accuracy transfer in the partial credit scheme did not demonstrate strategy transfer — even though the subjects wrote more complete programs for their second versions, those programs did not always reflect the same algorithm as was used on the first version. The same is true for subjects who did not demonstrate accuracy transfer. Finally, subjects who wrote fully complete programs for both versions, and thus did not provide evidence either for or against transfer, used the same algorithm each time.

| | Algorithm used on First vs Second Version | |
|---|---|---|
| | Same | Different |
| More Complete | 6 | 4 |
| Less Complete | 1 | 1 |
| Both Fully Complete | 7 | 0 |

**Table 6: Partial credit and strategy transfer**

The two transfer measures do not correlate probably because they are measuring different kinds of transfer. When writing a program in one language, subjects use knowledge previously generated while writing the same program in another language, but that knowledge may consist of more than one type of information. This is an important result because most transfer studies attempt to assess only one type of transferred information for a given problem situation. What are the two sorts of knowledge that are indicated here?

During programming, a programmer breaks the main goal of the program into a set of less complex, more manageable subgoals. Through this activity, a programmer can focus on implementing just certain subgoals, which is an easier task than trying to keep in mind all of the requirements of the programming problem (Fisher, 1986). Furthermore, these subgoals must be combined so as to represent how the program works. For example, Kant & Newell (1984) argue that the subgoals are organized in a way representing the flow of data through the program. Thus, the programmer's representation should consist of at least two kinds of information: programming subgoals and the way that those subgoals are organized. McDaniel & Schlager (1990) note a similar distinction in their study of a Missionaries-Cannibals isomorph. Solving that puzzle involves knowing both the correct general strategy and the particular moves needed to implement the steps (subgoals) of the strategy.

These two types of program knowledge are separately assessed by the partial credit and strategy use transfer measures. Through the partial-credit measure, we assess whether or not subjects transfer information about the specific subgoals used in the first version of the problem. The strategy measure shows whether or not subjects transfer information about how subgoals are organized. Thus, the assessment methods based on analyses of the task domain demonstrated that more than one type of transfer may occur in a single problem-solving situation.

## References


Bassok, M., & Holyoak, K. J. 1989. Interdomain transfer between isomorphic topics in algebra and physics. *Journal of Experimental Psychology: Learning, Memory, and Cognition* 15(1):153-166.

Ericsson, K. A., & Simon, H. A. 1984. *Protocol Analysis: Verbal Reports as Data.* Cambridge, MA: MIT Press.

Fisher, C. 1986. How do programmers program: Coping with complexity. Unpublished manuscript, Department of Psychology, Carnegie Mellon University.

Gick, M. L., & Holyoak, K. 1987. The cognitive basis of knowledge transfer. In S. M. Cormier and J. D. Hagman eds., *Transfer of Learning: Contemporary Research and Applications.* New York, NY: Academic Press.

Gray, W. D., & Orasanu, J. M. 1987. Transfer of cognitive skill. In S. M. Cormier and J. D. Hagman eds., *Transfer of Learning: Contemporary Research and Applications.* New York, NY: Academic Press.

Green, T. R. G. 1987. Parsing and gnisrap: A model of device use. In G. M. Olson, S. Sheppard, and E. Soloway eds., *Empirical Studies of Programmers: Second Workshop.* Norwood, NJ: Ablex.

Kant, E., & Newell, A. 1984. Problem solving techniques for the design of algorithms. *Information Processing & Management* 20(1-2):97-118.

Katz, I. R. 1988. Transfer of knowledge in programming. Ph.D. dissertation, Department of Psychology, Carnegie Mellon University.

McDaniel, M. A., & Schlager, M. S. 1990. Discover learning and transfer of problem-solving skills. *Cognition and Instruction* 7(2):129-159.

Pennington, N. 1985. Cognitive components of expertise in computer programming: A review of the literature. *Psychological Documents* 15(2702).

Polson, P.G., Muncher, E., & Engelbeck, G. 1986. A test of the common elements theory of transfer. In Proceedings of the CHI '86 Conference on Human Factors in Computing Systems, 78-83. New York, NY: ACM.

Rist, R. 1989. Schema creation in programming. *Cognitive Science* 13:389-414.

Singley, M. K., & Anderson, J. R. 1990. *The Transfer of Cognitive Skill.* Cambridge, MA: Harvard University Press.

Smith, S. B. 1986. Transfer of learning between Tower of Hanoi isomorphs. Ph.D. dissertation, Department of Psychology, Carnegie Mellon University.

# Interaction of Deductive and Inductive Reasoning Strategies in Geometry Novices

**Kenneth R. Koedinger**
**John R. Anderson**

Psychology Department
Carnegie Mellon University
Pittsburgh, PA 15213
koedinger@psy.cmu.edu

**Abstract**

This paper is part of an effort to extend research on mathematical problem solving beyond the traditional focus on formal procedures (both in the classroom and in problem solving research). We are beginning to investigate students' inductive discovery-oriented strategies and the interaction between these and formal deductive strategies. In contrast to typical classroom problems in math and science which demand the application of a learned formal procedure (e.g., prove X), we gave students more open-ended problems (e.g., is X true?) for which the formal deductive procedure is useful, but other, possibly informal or inductive, strategies are also potentially useful. The normative approach for solving these problems, in fact, requires the use of both a deductive strategy, which is definitive only when X is true, and an inductive search for examples, which is definitive only when X is not universally true. When presented with these problems we found that geometry students have some limited facility to perform the deductive strategy (though, less so in this context than when they are directly asked to write a proof) and use a degenerate version of the inductive strategy. Instead of considering multiple examples and looking for a counter-example, students tend to read off the conclusion from the single example (or model) we provided.

## Introduction

This paper presents a preliminary investigation into the nature of novice reasoning in geometry. This area is interesting theoretically as it relates to both research on human reasoning biases and research on novice problem solving, particularly in physics. Both areas have identified a number of situations in which humans behave in apparent contrast to normative expectations. Nickerson, Perkins, and Smith (1985) provide a good review of the empirically established human reasoning biases. The research on naive physics has shown that students' prior conceptions of the physical world are often at odds with the more detailed conceptions of the science of physics. Both to add to these bodies of research and for pedagogical purposes, we have begun to explore novice reasoning behavior in mathematics. While the focus in naive physics research has been on students' conceptions, the focus here is on reasoning strategies, both deductive and inductive. In contrast to typical classroom problems in math and science which demand the application of a learned formal procedure (e.g., prove X), we gave students more open-ended problems (e.g., is X true?) for which the formal deductive procedure is useful, but other, possibly informal or inductive, strategies are also potentially useful. We were interested in seeing to what extent students would bring the deductive strategy to bear, to what extent they would use informal strategies, and, in particular, if we could identify biases in these strategies that lead to systematic errors.

In previous research (Koedinger & Anderson, 1990), we found that skilled geometers initially plan proofs using informal knowledge (perceptually based) and deal with the formal details of writing down the proof as a secondary issue. To better evaluate students' proof abilities in terms of this distinction, we wanted to design a task that tapped only the informal planning abilities and did not require the detailed proof execution skills. The geometry truth judgment task poses questions of the form "If <givens>, must <goal>?". Figure 1 shows an example item.

One approach to answering such questions is to try to find a proof of the <goal> from the <givens>. Subjects don't need to write down this proof, but if they can come up with an accurate one, however informal, they can reliably answer YES.

Besides our initial intention of using this truth judgment task to measure informal planning skills, we have found this task interesting in its own right. From a pedagogical standpoint, it gives students an idea of the role of proof in mathematics, that is, to help determine what is true. From a scientific standpoint, it is analogous to other reasoning tasks and thus, may elicit similar strategies to those observed in everyday reasoning or in other laboratory reasoning tasks. And lastly, it allows us to further explore the complementary roles of informal and formal reasoning strategies.

11. If ∠**TQU** ≅ ∠**RQU** and ∠**QTU** ≅ ∠**QRU**, must ∠**STU** ≅ ∠**SRU**?

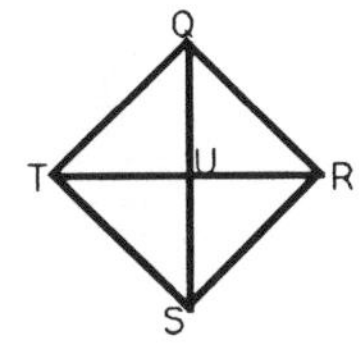


**YES**

**NO**

**Figure 1.** An example "truth judgment" problem. Subjects circled YES or NO.

### A Study of Novice Reasoning in Geometry

Geometry truth judgment problems are much like the logical syllogism problems which have received much attention in the human reasoning literature. A logical syllogism has two premises and a conclusion – these are analogous to our <givens> and <goal>. A typical task is to have subjects judge whether a given syllogism is VALID (must be true) or INVALID (not necessarily true) – analogous to YES and NO, respectively, in the truth judgment task. A leading theory on how humans solve logical syllogisms is Johnson-Laird's (1983) mental model theory. (Polk and Newell, 1989, have refined this theory and implemented it in the Soar architecture.) Essentially, the idea is that subjects convert what is nominally a deductive problem into an inductive problem by thinking in terms of possible models or instances of the premises and testing whether the conclusion holds true for these instances.

A deductive strategy for solving syllogism problems would involve applying logical rules to the premise statements in an effort to derive the conclusion. In contrast, the mental model theory proposes that people pursue a strategy of imagining one or more specific instances which are consistent with the premises. Then they check the problem conclusion to see if it is consistent with the instances they have imagined. In other words, they check whether the conclusion is a proper induction from the instances. If it is, they answer VALID, otherwise they answer INVALID.

What is unique about our geometry problems is that while in typical syllogism experiments subjects imagine their own instance(s), here subjects are given a candidate instance[1]. The diagram that goes along with a typical geometry problem is an instance of the problem premises and, like any model, it has specific features that are not explicitly stated in the premises. Some of these features may actually follow from the premises, for example, ∠**STU** ≅ ∠**SRU** in the problem Figure 1, while others may be fortuitous, for example, ∠**STU** ≅ ∠**QTU** in the same diagram.

By giving subjects a model, the truth judgment task provides a different kind of test of the mental models theory. In the typical experiment, the match between the theory and the subjects' actual reasoning process is performed indirectly by comparing the error patterns predicted by the theory with the subjects' error patterns. In contrast, we designed the truth judgment test, in part, to see if we might directly influence the reasoning process (and the resulting answer) by changing the characteristics of the model (diagram) we provide. Figure 2 shows the item design and an example group of 4 matched items varying on whether the correct answer is YES or NO and whether the problem givens and goal "look true" in the diagram or the givens and goal "look false". Consider item b. In this case, the diagram is more specific (has more features) than the given requires and the problem goal $\overline{AB} \perp \overline{CD}$ appears true when, in fact, it does not follow from the given $\overline{AC} \cong \overline{CB}$. In this case, a subject reasoning purely from the provided model would incorrectly conclude the goal was true. In contrast, a subject reasoning purely deductively would not be influenced by this instance, would fail to find a proof and thus, would be led to answer NO (though not with complete certainty since failure to find a proof does not guarantee one does not exist). Of course, the ideal subject would search for a counter-example at this point and upon finding one would be certain that the correct answer is NO.

Item c provides a second kind of misleading or inconsistent diagram. In this case the correct answer is YES yet the goal looks-false in the model (and so does the given necessarily). To the extent that subjects are working from this model, they should be biased to incorrectly respond NO since the goal $\overline{LQ} \perp \overline{RM}$ does not appear true in this diagram. In item types a & d, the model is consistent with the correct answer, so model-based reasoners who rely on this model only will perform better on these item types than on the matched items c & b. Stated another way, subjects who tend to use the provided diagram as a model will answer YES on the looks-true items more often than on the looks-false items.

Our hypothesis was not that subjects would rely purely on the diagram provided. In fact, we were hoping to find evidence of proof planning done in service of solving these problems. While finding evidence that subjects perform above chance on this task would be consistent with the hypothesis that they were bringing proof to bear, it would not rule out other reliable strategies like considering multiple instances of the given statements. However, we can get more discriminating evidence on the use of proof by picking problems for which the underlying proof varies in difficulty. If students are doing proofs, they should be more successful on problems which have "easy" proofs than on problems that require "hard" proofs. We defined difficulty with respect to the conceptual steps in the psychological model of informal proof planning identified by our prior research (Koedinger & Anderson, 1990).

[1] Subjects were certainly free to construct other models of the premises. However, subjects did this rarely, if at all.

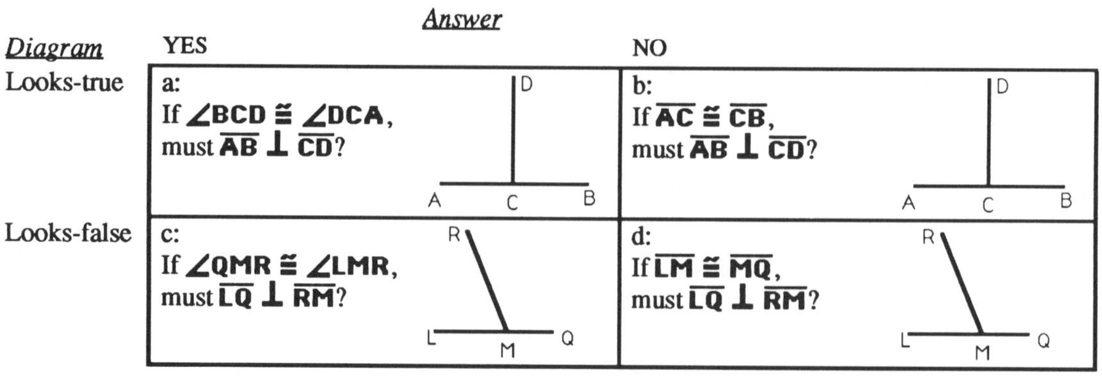


**Figure 2.** Truth judgment item types. Item types a & c have the same diagram, respectively, as item types b & d, while apart from point labels, a & b have the same problem statement, respectively, as c & d.

The shortest proof that verifies a YES answer on "easy" problems requires one conceptual step, while the shortest proof that verifies a YES answer on "hard" problems requires two or more conceptual steps. This categorization is only directly relevant to the YES items (a & c), but in reporting the data the matching items (b & d) will be included in this categorization.

We designed eight groups (1-8) of four items each (a-d) for a total of 32 items. Item groups 1-4 are "easy" problems, while item groups 5-8 are "hard" problems. The item in Figure 1 is 8a: it requires a "hard" proof, the givens and goal look-true in the diagram, and the correct answer is YES.

## Method

*Subjects.* 30 students participated for pay in the summer *after* completing a high school geometry course.

*Procedure.* This data was collected as part of the pre and post testing in a study comparing two intelligent tutoring systems (ITS) for geometry proof problem solving (see Koedinger, 1990). The complete set of 32 items was split into two test versions: 16 items for version A and 16 matching items for version B. Half the subjects received version A as a pre-test and B as a post-test; the other half did version B first and A as the post-test. Between the pre and post tests, subjects received 8 hours of proof instruction using one of the ITSs.

The test instructions (1) explained that any points which appear colinear (on the same line) in a diagram can be assumed to be colinear, (2) warned that "*the diagram may be misleading*", and (3) provided two example problems (types c & d) that both contained misleading diagrams and that, in one case, illustrated an informal proof as a good reason to answer YES and, in the other, illustrated a counter-example as a good reason to answer NO.

## Results

The first thing to note is that on a separate test of proof skill, we observed a significant improvement in students' ability to write two-column proofs. The mean score went from 36% on a pre-test before the ITS instruction to 68% on the post-test ($p < .001$). Given that the proof problems on the proof test were equally difficult as the proofs underlying the "hard-YES" problems on the truth judgment test, we might expect to see similar improvements on the truth judgment task. However, truth judgment performance showed only marginal improvement, going from 67% on the pre-test to 72% on the post-test ($p > .09$)[1]. While there may have been some transfer, it appears that students were not fully appreciating the relevance of proof to solving truth judgment problems. This is not to say that students were not trying proofs in the service of making truth judgments, in fact, as demonstrated below it is clear they were. Rather, it appears there was some slippage in the transfer from the structured problems in the training where a proof is explicitly required to the open-ended truth judgment problems where students must determine for themselves whether and when proof is relevant. These results are not changed by focussing the data analysis on the items for which improved proof skills might be most relevant (e.g., the hard-YES problems): The improvement trends of about 5% are uniform across all problem types.

Given that there are no qualitative differences between students' performance on the truth judgment pre- and post-tests, the remainder of the analysis uses subjects' scores on the pre- and post-tests combined. The goal is to get a general sense of the reasoning strategies students used to make truth judgments. In particular, to what extent did they bring proof to bear? To what extent are they influenced by the provided diagram? Is there evidence for other strategies? The results are displayed in the left hand graph in Figure 3. This graph shows the mean probability of answering YES for the 8 cells in the 2x2x2 item design. There are three significant main effects on each of three dimensions (correct answer, difficulty, and diagram) and one significant interaction between correct answer and difficulty. They are summarized as follows by indicating which item types lead to a higher probability for answering YES:

[1]This improvement is not as quite as small as it seems since getting 50% on these YES or NO problems is not difficult.

1. *Main effect*: YES items > NO items
2. *Main effect*: looks-true items > looks-false items
3. *Main effect*: easy items > hard items
4. *Interaction*: easy-YES > hard-YES,
but easy-NO ≈ hard-YES items

All of these effects are significant across both subjects and items ($p < .01$). We will consider them in more detail.

Looking at the answer dimension (represented by the two lines in the graph), students respond YES significantly more often when the correct answer is YES (see the dark line) than when the correct answer is NO (the light line). This indicates that they were not behaving randomly and had some sense for correct answers: they got 66% of the YES-problems correct and 71% of the NO-problems. That proof skill was an important factor in this success is indicated by the fact that performance significantly diminishes going from easy problems (76% correct) to hard problems (61%) and, more importantly, that this difference occurs for the YES-problems but not for the NO-problems. Averaging over diagram appearance, subjects are 83% correct on easy-YES items but 49% correct for hard-YES items while they are 69% correct for easy-NO items and 73% for hard-NO items. In other words, the evidence that students are doing proofs is that students' performance varies in a predictable way on the problems for which proof is required, but not on the problems for which proof is not required.

Though smaller in magnitude, the diagram had a clear and consistent effect on student performance. Independent of the correct answer, students were more likely to say YES on the looks-true items than on the looks-false items. In Figure 3, note how the points drop from the first column (looks-true) to the second (looks-false) and from the third to the fourth. This meant that they made significantly more errors on problem types b & c where the correct answer and the looks of diagram were in conflict than on problem types a & d where the answer and diagram were in accord. Stating this result in a different way, students showed a significant tendency to reason from the provided diagram (or model) making the inductive leap that if the goal looks-true (looks-false) in the model, then YES (NO) it must (not) be true in general.

### Modeling Reasoning Strategies and Biases

In this section, we first present a task analysis of what strategies might be brought to bear on this task and the circumstances in which they are effective. With that background, we next discuss what strategies these students appear to have used and how they interact. We propose a mathematical model as a concise summary of our claims about what students are doing. The model is supported both by its close fit to the quantitative data as well as from some preliminary protocol data.

## Task Analysis

Effectively solving these truth judgment problems, requires two complementary strategies. A *deductive rule-based* strategy that involves looking for a proof (i.e., a sequence of rules) which derives the problem goal from the givens. If one finds a correct proof, one can conclude YES with certainty. However, if one fails to find a proof, it is not necessarily that case that there isn't one. In this case, one can answer NO heuristically with some better than chance probability of being right. An *inductive example-based* strategy is complementary in that when properly applied one can answer NO with certainty, but can answer YES only heuristically. This strategy, like Johnson-Laird's mental model approach, involves considering examples of the given statements and checking whether the goal is true in these examples. If it is in all cases, one can answer YES but only heuristically since there is no guarantee that the next example won't contradict the problem statement. If the goal is ever found to be false, this example is counter-example to the problem statement and one can answer NO with certainty.

The normative overall strategy is to alternatively pursue one and then the other of these strategies until one of them yields a certain result, that is, until either a proof or a counter-example is found.

## A Mathematical Model

By looking at the statistical results from the study, it appeared that students' behavior could be characterized by the interaction of a proof strategy and a degenerate version of the inductive example-based strategy as follows:

1. Subjects initially attempt to find a proof and answer YES if they find one.
2. If they don't find a proof, they "guess" at the answer based on an induction from the provided diagram.

In mathematical terms, this is:

$$prob(YES) = p + (1-p)g$$

where $p$ = prob of finding a proof
$g$ = prob of guessing YES

This general form is instantiated in different ways depending on the problem characteristics. If the correct answer to the problem is NO, the probability of finding a proof is 0. If the correct answer is YES, the model claims that students find a proof with probability $e$ for an easy problem or probability $h$ for a hard problem. If they don't find a proof, they will guess YES with probability $t$ if the problem goal looks-true in the diagram or $f$ if the goal looks-false. In summary, $p=0$ for NO problems, $p=e$ for easy-YES problems, and $p=h$ for hard-YES problems while $g=t$ for looks-true problems and $g=f$ for looks-false problems.

Note that the model implies that there should be no difference between the easy and hard problems when the correct answer is NO. This is how the model captures the interaction between correct answer and difficulty. The graph on the right in Figure 3 shows the model predictions for the following set of parameter values, $e=.77$, $h=.29$, $t=.37$, $f=.19$. The model provides a close fit to the data. A chi-square test indicates that it does not deviate significantly from the data ($X^2(4, N=30) = 2.19$, $p>.5$)

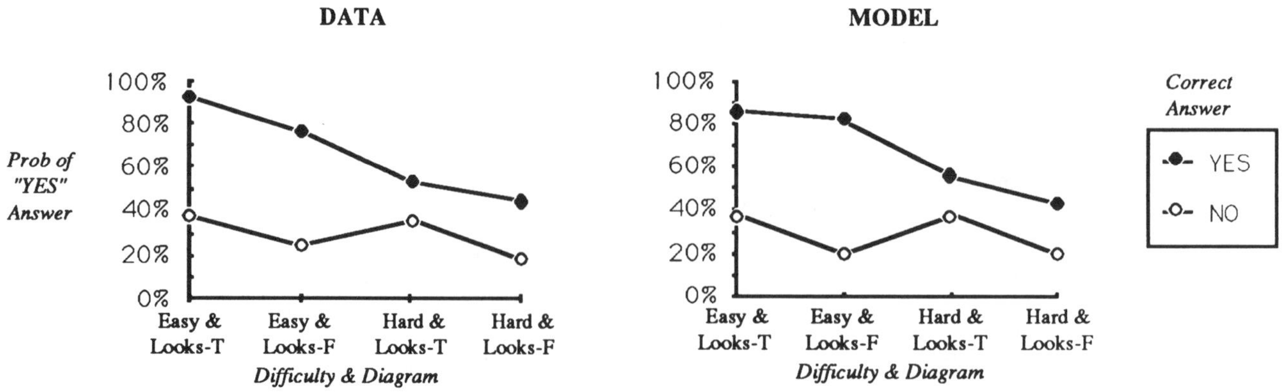


**Figure 3.** Eight data points and model predictions for the 2x2x2, answer by difficulty by diagram, item design. In the data, all three main effects are significant across both subjects and items and so is the interaction between answer and difficulty. The model does not deviate significantly from the data ($X^2(4, N=30) = 2.19, p>.5$).

This model of student performance suggests three deviations from the normative one. First, the deductive strategy component is quantitatively different in that ideally students should be able to find all the proofs for the YES problems, not just 77% (the value of e) for the easy-YES problems and 29% of the hard-YES problems. Second, the model's inductive example-based strategy is different in that students are not considering multiple examples, but are only considering the provided example and also in that they are not realizing that the looks-false problems are not good examples. Third, there is no alternation in the model between the deductive and inductive strategy.

## Other Evidence for the Model

One source of additional evidence for the model comes from the notes students wrote on the test handout provided. In support of the claim in the model that students are not pursuing a search for counter-examples and are not recognizing that the looks-false problems are not good examples, we looked to see how often subjects drew another diagram in addition to the one provided. There were only three cases out of 480 possible opportunities in which a subject redrew a looks-false diagram. While this was a NO problem, none of these diagrams were a counter-example, that is, the goal still looked-true in the diagrams the students drew. Two other instances of redrawing occurred on a different looks-true problem apparently to match up corresponding points of potentially congruent triangles.

We recently gave a version of the test in which we asked students to give reasons for their answers. A large number of the reasons students provide are idiosyncratic, along the lines of "because I think so". But there are also a number of more sensible responses. The reasons students provide for problems they answer YES are often informal proofs, citing intermediate steps and/or theorems from their geometry class. On problems they answer NO, we have yet to find a single instance where the student provided a counter-example. Instead, the most typical, non-idiosyncratic reason for saying NO is that the student thinks there is no way to prove the goal.

## Conclusion

While these results provide some partial support for Johnson-Laird's mental model theory, there are two key differences from that theory. First, student reasoning was not exclusively model-based (inductive), in fact, more prominent in their reasoning is a deductive strategy to find a proof. Second, while subjects solving syllogism problems appear to consider multiple models, students in this study showed little or no evidence of considering any alternative models beyond the given one. This is particularly surprising given that half of the provided models (the looks-false items) did not accurately represent the problem statement. From a pedagogical standpoint, this study suggests that students need extra instruction both 1) in recognizing the applicability of the proof strategy outside the standard context and 2) in understanding and executing the inductive model-based strategy.

## References

Johnson-Laird, P. N. (1983). Mental Models. Cambridge, MA: Harvard University Press.

Koedinger, K. R., & Anderson, J. R. (1990). Abstract planning and perceptual chunks: Elements of expertise in geometry. *Cognitive Science, 14*, 511-550.

Koedinger, K. R. (1990). Tutoring concepts, percepts and rules in geometry problem solving. Unpublished doctoral dissertation. Carnegie Mellon University.

Nickerson, R. S., Perkins, D. N., and Smith, E. E. (1985). *The Teaching of Thinking*. Hillsdale, NJ: Lawrence Erlbaum Associates.

Polk, T. A., & Newell, A. (1988). Modeling human syllogistic reasoning in Soar. *Program of the Tenth Annual Conference of the Cognitive Science Society*. Hillsdale, NJ: Lawrence Erlbaum Associates.

# Multiassociative Memory[1]

**John F. Kolen**
**Jordan B. Pollack**

The Laboratory for AI Research
Department of Computer and Information Science
The Ohio State University
Columbus, OH, 43210
Telephone:(614)292-7402
kolen-j@cis.ohio-state.edu
pollack@cis.ohio-state.edu

## Abstract

This paper discusses the problem of how to implement many-to-many, or multi-associative, mappings within connectionist models. Traditional symbolic approaches wield explicit representation of all alternatives via stored links, or implicitly through enumerative algorithms. Classical pattern association models ignore the issue of generating multiple outputs for a single input pattern, and while recent research on recurrent networks is promising, the field has not clearly focused upon multi-associativity as a goal. In this paper, we define multiassociative memory MM, and several possible variants, and discuss its utility in general cognitive modeling. We extend sequential cascaded networks (Pollack 1987, 1990a) to fit the task, and perform several initial experiments which demonstrate the feasibility of the concept.

## Introduction

In associative memory models, a function is applied to a pattern which transforms it to another pattern. This transformation, or association from a domain set of noisy or partial patterns to a range set of correct patterns is often offered as a model of memory retrieval or recall. Several connectionist models of associative memory exist, such as linear associators (Kohonen, 1972;Anderson, 1972), Hopfield networks (Hopfield, 1982), and feed-forward networks (Rumelhart, Hinton & Williams, 1986), each assuming a many-to-one mapping from domain to range Figure 1a. For noisy or partial memory retrieval, or for perceptual categorization, this assumption might prove valid, however, many other tasks exist, each indescribable within associative frameworks. Consider associating a word with its possible lexical categories or meanings, a chess position to possible next moves, or a category to prototypical members; each input can have a multitude of possible outputs. Even categorization itself is known to vary over time in individual subjects (McCloskey and Glucksberg, 1978).

In an attempt to address the issue of multiple output responses, this paper discusses the nature of *multiassociative* memories capable of responding with one output from a set of possible outputs to an input. The theoretical basis of this class of models is the many to many mapping. Standard associative models use many to one mappings, where several input patterns can elicit the same output pattern. Any system with an internal state, such as a recurrent network, can be viewed as a many to many mapping, and this is our initial choice of model (Figure 1b).

This paper begins by presenting formal definitions of mappings, establishing the relationship between many to many mappings and multiassociative memories, and identifying several problems in cognitive science which exhibit many to many mapping properties. The discussion will then turn to the relative computational and psychological merits of two opposing implementation strategies, namely explicit storage and enumeration. After establishing the practical superiority of enumeration, this strategy will guide the implementation of a multiassociative memory system using a sequential cascaded network, a high-order recurrent connectionist architecture. Reflection on the implementation process and its results suggests possible difficulties with the recurrent network, indicating directions for further work.

## Many to Many Mappings

Associative memories take input patterns, process them, and return output patterns (Kohonen, 1984). One way of describing this operation is as a mapping from input patterns of the domain set to the generated output patterns the range set. Mappings come in four varieties based on the relations between domain and range: one to one, many to one, one to many, and many to many. A simple example of a many to many mapping is (A-> {B, C}, B-> {C, E}, C->{E, F}, D->{E, F}) with domain set {A, B, C, D} and range set {B, C, E, F}.

To capture such an association, a multiassociative memory system must be correct and complete. Thus the multiassociative memory acquisition problem is:

> Given a many to many mapping from a set of input patterns to output patterns, find a system that can associate each domain element to at least one element from its range set during any memory access *(correctness)* and also eventually generate all correct range elements within a finite number of memory accesses (*completeness)*.

It is quite trivial to implement this concept exercising conventional symbolic means, employing either explicit storage or enumeration of the associations. One just explic-

1. This work supported by office of Naval Research grant number N00014-89-J1200.

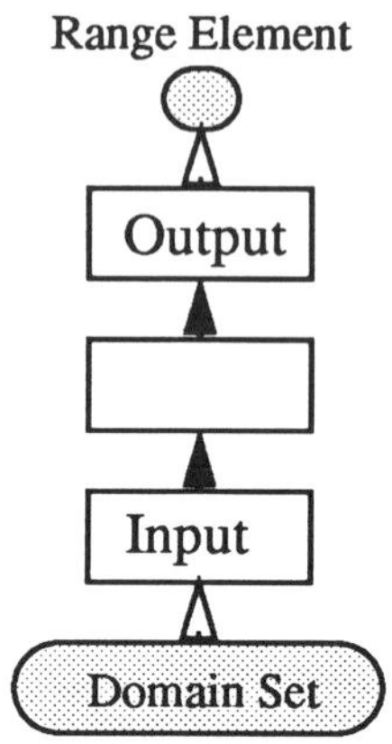


Figure 1a:
Associative Memory
Many to One

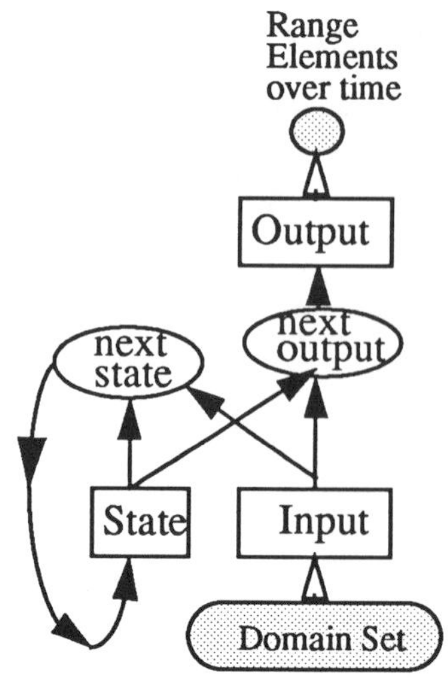


Figure 1b:
Multiassociative Memory
State Machine Conceptualization

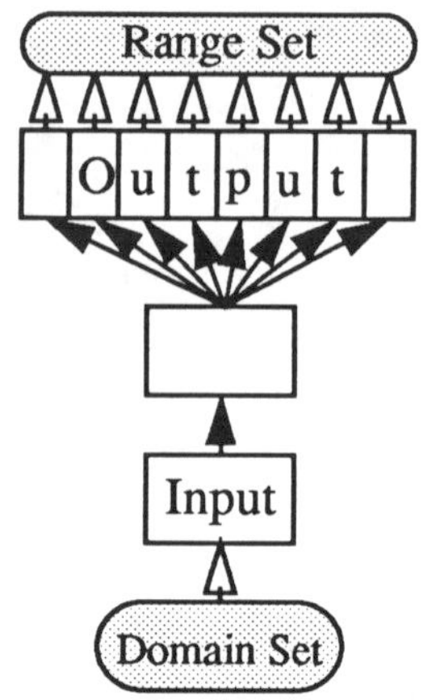


Figure 1c:
Multiassociative Memory
Power Set Approach

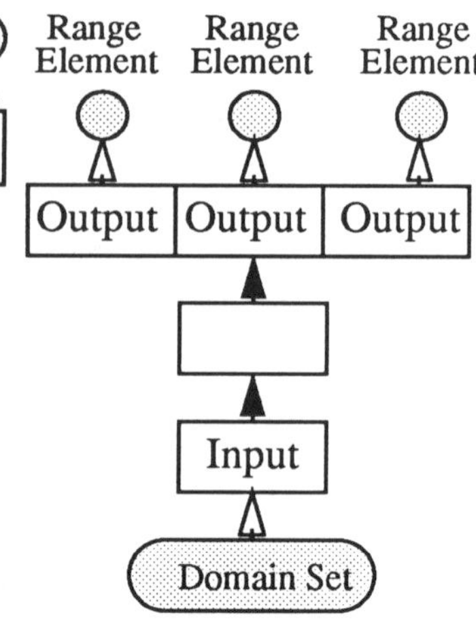


Figure 1d:
Multiassociative Memory
Field Approach

itly stores associations as a linked-list in random-access memory, and retrieve either all the outputs, or keep a counter and get one at a time. This "explicit storage" method does not generalize to very large systems, like chess moves, where it is impractical to store all associations in a memory. Secondly, it is too powerful to be a good cognitive model in that it has no distinction between "recognition" and "recall". Consider the daunting task of recalling the names of all fifty states, as opposed to the trivial problem of recognizing "New Mexico" as one of them.

Connectionist models also have explicit storage solutions. Through the manipulation of output patterns, the *power set* approach involves allocation of one output node for each element in the range (Figure 1c). Thus, any possible subset of the range can be captured by this representation. This localist approach has been used to encode possible moves in a tic-tac-toe game (McClelland and Rumelhart, 1986) and the spreading of activation in a semantic network (Collins and Loftus, 1975). Unconstrained representational power, however, is not without its costs. Since individual elements of a range of $n$ units can be adequately represented with a binary pattern of length $lg(n)$, using $n$ units is exponentially more expensive than a sequential retrieval model.

A possible circumvention to this exponential explosion is to limit the size of the subsets of the range by allocating a fixed number of output *fields* (slots, registers, or buffers) to hold the representations of the elements. Figure 1d illustrates a system capable of outputting at most three range elements for a given domain element. But determining the needed resources *a priori* may be difficult, or in principle impossible: What's the maximum number of possible moves from any given chess board configuration, or the maximum number of words in a sentence? Overestimation wastes resources and underestimation might fail at a critical point.

Enumerative approaches bypass this storage problem by not explicitly operating on the possible responses themselves, rather they calculate the desired responses sequentially. Since only one range element has to be stored at a time, memory usage will be lower. Such a system must have an "internal state" (or counter) that can be used to keep track of the enumeration algorithm, and suitably powerful computational means to compute the output set.

## Connectionist Implementations

A connectionist implementation of Multiassociative Memory (MM) using enumeration would have an underlying state machine, which can be implemented as a recurrent network. Many such networks exist (*e. g.* Jordan, 1987; Elman, 1988; Pollack, 1987) and already exhibit many to many mapping ability. When the input remains stable long enough to establish an periodic attractor (a fixed point or limit cycle) in the output, each element of the attractor corresponds to one of the correct associations to the input pattern. Such a "temporal MM" accesses the range set in a particular order for each input, whereas an "atemporal MM" just guarantees access to the proper range subsets through an aperiodic attractor (aka strange attractor).

To illustrate temporal/atemporal distinction, consider an implementation of a many to many mapping using a recursive auto-associative memory (RAAM) (Pollack, 1990b). A RAAM consists of two functions, an encoder and a decoder. The encoder takes two input patterns and combines them into a single output pattern. The decoder takes one input pattern and generates the two patterns the encoder used to build it. Since all patterns have the same size, the strategic application of the encoder and decoder can construct representations of recursive data structures such as trees and stacks. A MM can be built around a RAAM by assigning non-terminal symbols to the domain elements which eventually decode into elements in the image (see Figure 2). A counter, or random number generator, could be used as the internal state mechanism specifying a unique decoding paths, *i. e.* whether the left or right decoding would be re-

A -> {A, B, C, D}

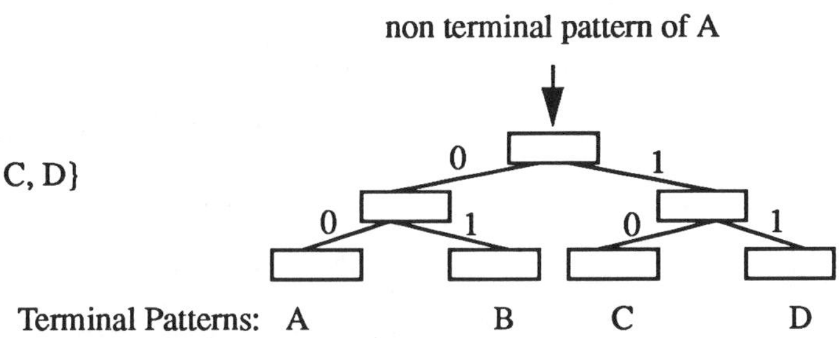


| State | Output |
|---|---|
| 00 | A |
| 01 | B |
| 10 | C |
| 11 | D |

Figure 2: Using a RAAM for Multiassociative Memory

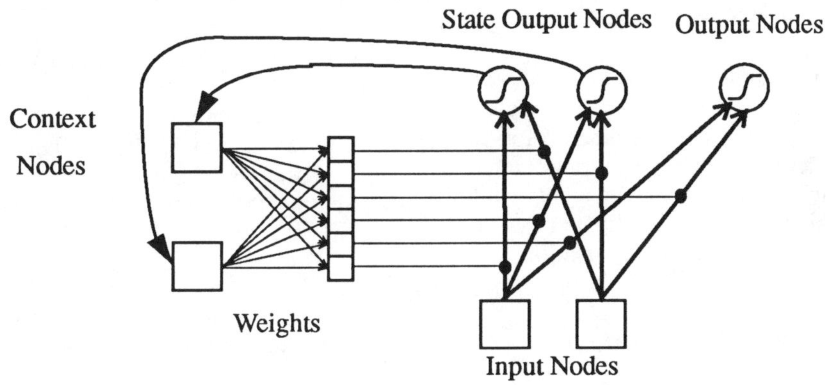


Figure 3: A Sequential Cascaded Network SCN

tained for further expansion. The counter implementation introduces temporal relationships between outputs since it would periodically visit every decoding path, while the random number generator ensures little or no correlation between two successive outputs by randomly varying the paths.

A drawback of this implementation is that tree structure and state organization must be specified *a priori.* A more interesting approach involves the evolution of state organization through learning. Learning, for a MM problem, has two constraints. First, no "reset" information can be provided to the network between changing inputs. Second, only a small number of examples will be available to the learning procedure. The system must be able to generalize its internal state machine to have acceptable behavior over finite but unbounded sequences of inputs. Sequential cascaded networks have exhibited such inductive capabilities, as demonstrated in (Pollack, in press), and therefore been selected for this initial study.

The sequential cascaded network (SCN) is a higher order (sigma-pi) recurrent network. The weights are stored in a three-d array (Context Network) that is multiplied by the current state yielding a weight matrix. This matrix, called the Function Network, is multiplied by the input vector resulting in a net input vector. A sigmoid function is applied to elements of this vector, yielding the next state and output vectors. SCN's were originally used in a formal language decision task, in which the network observed a sequence of symbols from a two symbol alphabet (0 and 1) and calculated a single output value plus a recurrent state vector. The desired output value was 1 if the input string seen so far was contained in the language, and 0 otherwise. Figure 3 contains a schematic representation of a SCN.

This paper extends the sequential usage of the cascaded network by utilizing an output vector longer than a single element. Increasing the output width from a single value necessitates a slight modification to the learning rules described in (Pollack, 1990). More output nodes imply each will contribute error to the context node error. The new error function derivative with respect to the weights feeding into the state output nodes is

$$\frac{\partial E}{\partial w_{ijk}} = \Big(\sum_{a \in A} (z_a(n) - d_a)\, g(z_a(n))\, \Big(\sum_l y_l(n) w_{alk}\Big)\Big)$$
$$g(z_i(n-1))\, y_j(n-1)\, z_k(n-2)$$

and the error derivative for the weights leading to the output units is

$$\frac{\partial E}{\partial w_{ajk}} = (z_a(n) - d_a)\, g(z_a(n))\, y_j(n) z_k(n-1).$$

In both equations, $z_i(n)$ is the output of unit $i$ at time $n$, $y_i(n)$ is the $i^{th}$ input at time $n$, $g(x)$ is the derivative of the nonlinear squashing function in terms of the activation of the unit, $d_a$ is the desired output for output unit $a$ from the set $A$ of output nodes, and $E$ is the standard sum of squared error function. The simulations below use baseline back

Input Sequences: AAAA BBBB CCCC DDDD

Training Sequences: BCBC DEDE EFEF EFEF

Figure 4a: Sample MM training set

Input

Sequence: AA**AABB**BBCC**CCDD**DDAA**AA

Output: BCBCBCDEDEDEEFEFEFEFEFEFBCBCBC

** stands for 16 more input cycles

Figure 4b: Output From SCN as a MM

propagation (gradient descent with momentum) to modify the weights in the network.

The training paradigm used in this paper attempts to exploit the inductive ability of this system by limiting the number of exemplar sequences. One way to develop an SCN training set for a particular mapping is to train the network sequence through the associated range elements while the input to the SCN remains constant with the input vector for the domain element. For instance, the association A -> {B, C} give rise to the sequences

Input: AAAAAAAAAAAA...

Output: BCBCBCBCBCBC...

Since SCN's have shown induction of similarly unbounded sequences from short initial sequences, only two such cycles for each domain element are presented to the network during training. Figure 4a shows the training sequences used to capture a small many to many mapping. This training set, along with domain and range element pattern vectors, was presented to the SCN learning program and after 10 passes through the training set (weight update occurred after presentation of all sequences) the network demonstrated that it could reproduce the training set within tolerance. Tolerance was one half the difference of the maximum and minimum values. Table 1 describes network parameters used for this experiment and the next. In this experiments, the input and output symbols were converted to fixed width random binary patterns. At this point the network captures only the explicit sequences in the training set. To generalize, the learning program must **overtrain** the network. Overtraining forces the network to go through phase changes, drastic alterations in its behavior, as it tries to separate the outputs more and more. Pollack (In press) demonstrates the characteristic shift in behavior with a network which learns successively larger finite subsets of the parity language (even number of 1's in the input string) until it suddenly acquires the parity language for any length input string. In this experiment, the learning procedure stopped only when 10 consecutive overtraining passes occurred.[2] Figure 4b illustrates the response of the SCN to the string $A^{20}B^{20}C^{20}D^{20}A^{20}$.

2. During an overtraining pass weights are modified, even though the all of the outputs are with tolerance of their desired values.

| | Letters | Figures |
|---|---|---|
| Learning rate | 0.1 | 0.1 |
| Momentum rate | 0.3 | 0.4 |
| Overtraining | 10 | 20 |
| High Symbol Value | 1.0 | 1.0 & see fig. 4 |
| Low Symbol Value | -1.0 | -1.0 & see fig. 4 |
| Squashing Function | tanh | tanh |
| Error Tolerance | 1.0 | 0.125 |
| State Width | 3 | 4 |
| Pattern Width | 3 | 2 |
| Initial State Value | 0.9 | 0.25 |

Table 1: Training Parameters

The next experiment demonstrates the possibility of storing visual images as many to many mappings. The domain of this mapping is the set {TRIANGLE, SQUARE}, represented as binary patterns, {(0,1), (1,0)}. The range set of each of these symbols is the vertices of that object as shown in Figures 5a and 5b. A similar training method was employed and Figures 5c, 5d, and 5e illustrate the results of presenting the input string $\text{TRIANGLE}^{20}\text{SQUARE}^{20}\text{TRIANGLE}^{20}$.

## Discussion and Conclusion

In both studies, the network created for itself a periodic output attractor for each input pattern utilizing hidden state information (which will be explored in a longer paper). We observed different organizations of internal state in different training runs due to back propagation's sensitivity to the initial weights (Kolen & Pollack, 1990). In most cases, we also observed that quickly switching between symbols can lead to several cycle delays of the correctness condition as the output dances around the target response. We believe this parallels the "tip of the tongue" phenomena and suggests that enumerative multiassociative memories could provide a useful foundation for modeling this psychological phenomena and others. Nature's solution to the scaling issue, which affects all connectionist models, may turn out to be tied to the distinction between recall and recognition. As has been pointed out in the past, inverting a categorization model or an associative memory is a difficult problem, because it requires a many-to-many mapping, not supplied by the standard architectures (Williams, 1986). The inversion of a multiassociative memory would remain a multiassociative memory.

The main difficulty with our initial MM model, and with most recurrent back propagation systems is the reliance on global state information and synchronous update of that information (Grossberg, 1987). As nature teaches us, it is much cheaper to build asynchronous and local parallel processing, which unfortunately can be quite unstable. The multiassociative task is well-suited for an asynchronous and local treatment. Following Hanson 1990, we have been experimenting with a model in which each weight in a system is a dynamical system, loosely coupled to nearby connections, asynchronously changing over time. Learning does not constrain the network to taking on particular values, but of meeting the necessary functionality of the network in terms of an atemporally multiassociative memory.

By focusing on the many-to-many mapping as a cogni-

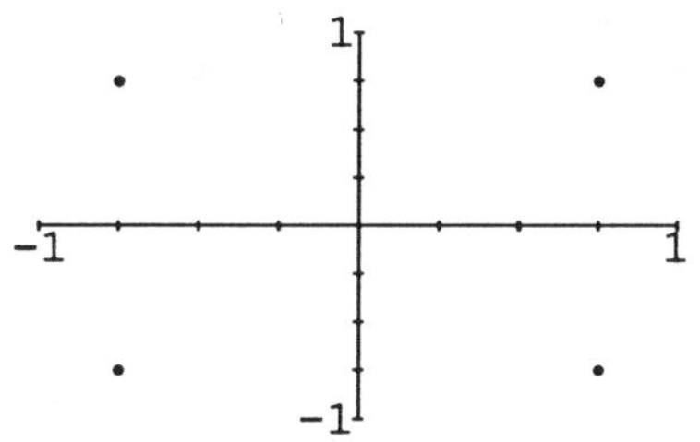


Figure 5a: SQUARE training data

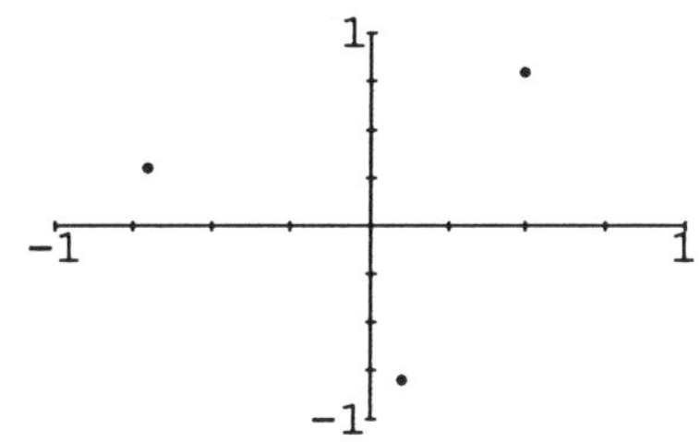


Figure 5b: TRIANGLE training data

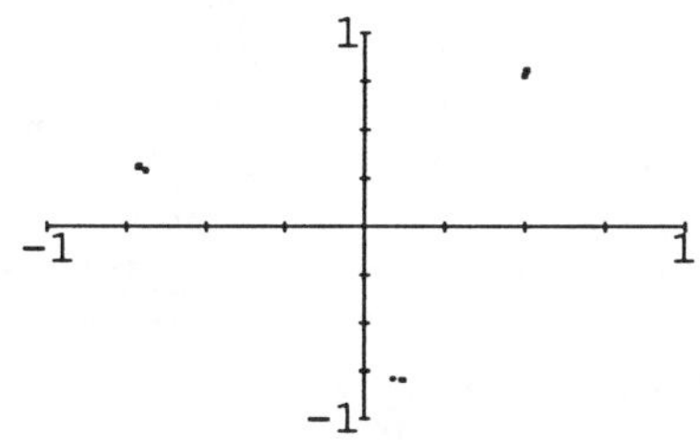


Figure 5c: cycles 1-20

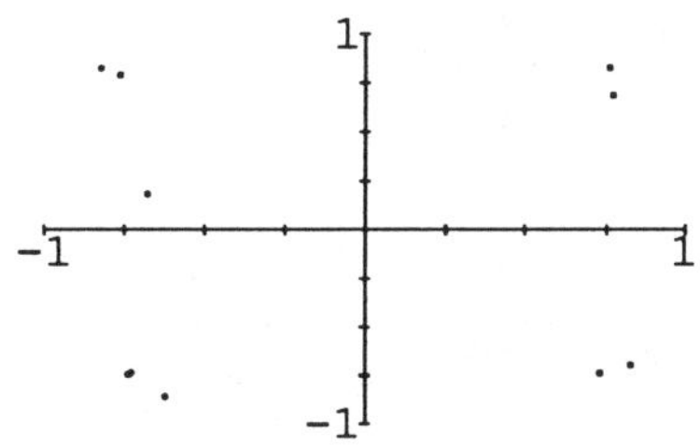


Figure 5d: cycles 21-40

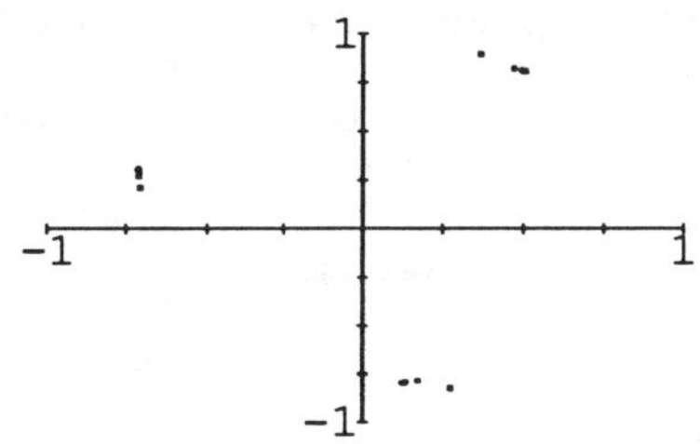


Figure 5e: cycles 41-60

Figure 5: Training Sets and Output displays for the {TRIANGLE, SQUARE} problem.

tive task, implemented in an enumerative rather than explicit storage memory, we see a new general-purpose connectionist model beyond the categorization and associative memory of feed-forward and relaxation models, and the prediction and sequence generation of recurrent networks.

## Acknowledgments

Discussions with members of the LAIR connectionist group have been invaluable. Comments from Mary Jo Carnot helped round out the paper and make it more accessible.

## References

Anderson, J. A. 1972. A simple neural network generating an interactive memory. *Mathematical Biosciences*, 14:197–220.

Collins, A. M. and Loftus, E. F. 1975. A spreading activation theory of semantic processing. *Psychological Review*, 82:240–247.

Elman, J. L. 1988. Finding structure in time. Technical Report CRL 8801, University of California, San Diego.

Grossberg, S. 1987. Competitive learning: From interactive activation to adaptive resonance. *Cognitive Science*, 11:23–63.

Hopfield, J. J. 1982. Neural networks and physical systems with emergent collective computational abilities. *Proceedings of the National Academy of Sciences USA*, 79:2554–2558.

Jordan, M. I. 1987. Supervised learning and systems with excess degrees of freedom. Technical Report COINS Technical Report 88-27, Massachusetts Institute of Technology, Boston.

Kohonen, T. 1972. Correlation matrix memories. *IEEE Transactions on Computers*, C-21:353–359.

Kohonen, T. 1984. *Self-Organization and Associative Memory*. Springer-Verlag.

Kolen, J. and Pollack, J. 1990 Back-Propagation is sensitive to Initial Conditions. *Complex Systems* 4, 269-280.

McClelland J., & Rumelhart, D. 1986 *Parallel Distributed Processing: Explorations in the Microstructure of Cognition*. Cambridge, MIT Press, II, 48-53.

McCloskey, M. and Glucksberg, S. 1978. Natural categories: Well defined or fuzzy sets? *Memory and Cognition*, 6:462–472.

Pollack, J. B.1987. Cascaded back propagation on dynamic connectionist networks. In *Proceedings of the Fourth Annual Cognitive Science Conference*. Seattle, WA, 391-404.

Pollack, J. B. 1990. Language acquisition via strange automata. In *Proceedings of the Twelfth Annual Conference of the Cognitive Science Society*. Cambridge, MA, 678-685.

Pollack, J. B. 1990. Recursive autoassociative memories. *Artificial Intelligence*, 46, 1, 77-105.

Pollack, J. B. In press The acquisition of dynamical recognizers *Machine Learning*, 1991.

Rumelhart D. Hinton, G., and Williams R. 1985. Learning representations by back-propagating errors. *Nature*, 323, 533-536.

Williams, R. 1986 Inverting a connectionist network mapping by back-propagation. *Proc. Eight Annual Conf. of the Cognitive Science Society*, Amherst. 859-865.

# Processing Constraints and Problem Difficulty: A Model

**Kenneth Kotovsky**
kotovsky@psy.cmu.edu

**Nicholas Kushmerick**
kushmerick@psy.cmu.edu

Psychology Department
Carnegie Mellon University
Pittsburgh, PA 15213

### Abstract

In this paper we examine the role played by working memory demands in determining problem difficulty during the solution of Tower of Hanoi Problem isomorphs. We do so by describing a production system model that accounts for subjects' performance on these problems via a dynamic analysis of the memory load imposed by the problem and of changes in that load during the problem solving episode. We also present the results of detailed testing of the model against human subject data. The model uses a highly constrained working memory to account for a number of features of the problem solving behavior, including the dichotomous (exploratory and final path) nature of the problem solving, the relative difficulty of the problems, the particular moves made in each state of the problem space, and the temporal patterning of the final path moves.

## Introduction: Human Problem Solving Performance

One of the large issues in problem solving research is the question of what makes a problem hard or easy to solve. This issue is of interest both to educators interested in designing instruction so as to more effectively understand and promote the growth of problem solving skills, and to cognitive scientists interested in the information processing involved in problem solving. It is from the latter perspective that we attempt here to examine how the processing demands imposed by different problem features interact with limited working memory to determine problem difficulty. We do so by describing a model of problem solving that accounts for subject performance on a set of Tower of Hanoi Problem isomorphs. The model incorporates a dynamic analysis of problem processing demands in the form of memory load, the effects of changes in that load over the course of problem solving, and the limitations inherent in short term memory. In attempting to account for problem difficulty by means of the processing demands or "load" imposed by various features of the problem, the model includes design parameters based on fairly precise data obtained from a number of experiments with human subjects. The model is tested via detailed comparisons of its behavior against that of human subjects.

While the work presents a detailed analysis of the model's application to a particular domain, a set of isomorphs of the Tower of Hanoi Problem, the implications of the work are not limited to that problem domain, but rather address a set of issues applicable to a wide variety of problem solving situations.

In 1974, Hayes and Simon published the first of a series of articles on Tower of Hanoi problem isomorphs (Hayes & Simon 1974, 1977, Simon & Hayes 1976). The set of isomorphs they investigated consisted of problems that had the same problem state space as the three disk Tower of Hanoi Problem. Two problem types were labelled either "Monster Move" problems because they involved monsters passing globes back and forth, or "Monster Change" problems because they had monsters changing the sizes of globes they were holding. A major finding was that isomorphic problems could differ significantly in difficulty. In particular, they obtained difficulty (solution time) ratios averaging about 2:1 between Monster Change and Monster Move problems. Kotovsky, Hayes, and Simon (1985), extended the investigation to a broader array of problems, obtaining difficulty ratios of up to 16:1 for their hardest/easiest pair of isomorphs; the easiest being the original Tower of Hanoi disk/peg problem.

The different problems are defined by the subjects' internal problem representation which included the move operators that define move legality. This representation is engendered by either an external depiction of important features of the problem such as the physical pegs and disks of the Tower of Hanoi problem, or a "cover story" that describes the monsters, their globes and the moves or changes the monsters can make with their globes. Table 1 presents the rules for two problems that were used to produce much of the data discussed here, a Monster Move Problem, and a Monster Change Problem. The cover story that defined the problem had three different sized monsters trying to move or change globes to convert an initial arrangement of monster—globe pairings into a final arrangement.

<u>Monster Move Problem</u>

1. Only one globe may be transferred at a time.
2. If a monster is holding two globes, only the larger of the two may be transferred.
3. A globe may not be transferred to a monster who is holding a larger globe.

<u>Monster Change Problem</u>

1. Only one globe may be changed at a time.
2. If two globes have the same size, only the globe held by the larger monster may be changed.
3. A globe may not be changed to the same size as the globe of a larger monster.

**Table 1: Problem Rules**

The fact that the problems are isomorphic in the above work removes problem search space structural features (branchiness of the problem space, length of the minimum solution path, etc.) as possible sources of the large difficulty differences that were discovered. In an attempt to discover what were the sources of the difficulty differences, two findings emerged from the Kotovsky, Hayes and Simon (1985) work that are the starting points for the research reported here. Those findings are: (a) the crucial role played by the move

This research was supported in part by the Personnel and Training Research Programs, Psychological Sciences Division, Office of Naval Research, under contract N00014-85-K-0696.

operator and its interaction with human information processing limitations in determining problem difficulty, and (b) the discovery of a dichotomous pattern of moves that occurred as people moved through the problem space to reach a solution to the problem.

The first finding was that more difficult problems employed move operators that imposed more of a processing load via the number of entities (monsters and globes) and the number of separate loci that had to be imaged in testing the legality of a move. A ranking of individual moves or problems by the number of separate entities and loci that had to be simultaneously held in working memory was predictive of both the difficulty of making individual moves and also of overall problem difficulty. Thus, in one experiment, when subjects were asked to judge the legality of single moves that were presented tachistoscopically, their response latencies were correlated with the number of entities that had to be imaged in order to make the judgement. For example, in the Move Problem, subjects' judgements were relatively fast when they compared the sizes of two globes held by the same monster (Rule 2) and relatively slow when they compared the sizes of two globes held by different monsters (Rule 3). The explanation was that comparing two globes when they were at two separate loci entailed carrying one in memory and thus imposed an added unit of memory load. An even harder comparison occurs in the Change Problem where subjects had to imagine changing the size of a globe, and then test the imaged size against the size of another globe that was held by either the same monster (Rule 2) or another monster (Rule 3). In that case, the imposed load was higher because of the need to imagine the size change before doing the comparison. There were positive correlations between the processing load imposed by the move operators, operator application time, and problem difficulty. In the current work, this load is termed the "image" load.

The second finding was that subjects' move making was dichotomous, consisting of an initial, "exploratory phase" (whose length differed across problems) in which moves were slow, particularly in the harder isomorphs, and resulted in no net progress toward the goal, followed by a subsequent "final path" phase (virtually identical across problems) where moves were uniformly fast, relatively error free, and quickly led to the solution. The difference was due to subjects in the final path phase being able to plan and execute subgoals (as evidenced by final path patterns of move latencies), whereas earlier in the problem, they were not able to.[1] The empirical evidence suggested that subjects attained this ability to subgoal by reducing the memory demands involved in learning and applying the problem move legality tests during the course of solving the problem (Kotovsky et al 1985).

This "rule learning" effect on the memory load imposed by learning and remembering the problem rules or move legality tests occurred as people made moves. They started out making errors (illegal moves) because of starting to move before they knew the rules, but became progressively better at using the problem rules as they received practice. The difficulty of learning the move rules presented in isolation was determined, and on that basis and primarily via a linguistic analysis of the statement of the problem rules, the relative processing demands of the various problems' rules were determined. This "rule load" decreased as the problem solving proceeded and the rules were learned.

The result of this rule learning was that subjects' moves in the state space regularly exhibited the surprisingly dichotomous "exploratory" and "final path" pattern described above. The exploratory moves were made slowly, they occupied the major phase of the problem solving time, and they were more difficult (took much longer) in the harder isomorphs. Furthermore, subjects were as far from the goal after making these moves as they were at the beginning of the problem. In contrast, the final path moves (after subjects had compiled the problem rules) were relatively error free, were made very rapidly, were executed at a similar speed across all problem isomorphs, and led almost immediately to a problem solution. This dichotomous pattern of slow or difficult move making that made no net progress, and whose length reflected the relative difficulty of the problems, followed by a rapid dash to a solution in the last minute or so of the solution process, regardless of isomorph, is characteristic of a number of other problems as well (Kotovsky & Simon 1989).

This discovery of distinctive exploratory and final path phases provided a plausible link between move operator difficulty and problem difficulty. The issue was that although the processing load imposed by the move operators predicted the ordering of isomorph difficulty, the differences in move time were not great enough to account for the very large differences in problem solution time. The linkage was that during the exploratory phase, the difficulty of remembering and applying the rules (legality tests) prevented subjects from planning move sequences; even those of length two, and prevented their making progress toward the goal (moving down the final path).[2] An information processing analysis of the load imposed in making goal-subgoal pairs of moves showed that the harder, Monster Change problems imposed much higher memory loads than the Monster Move problems. A single move in the Monster Change problem always requires one more entity to be imaged than an equivalent move in the Monster Move problem. The direction of this difference is what would be expected from

---

[1]There is evidence that subjects, just prior to the final path, were not able to successfully execute a subgoal-goal pair of moves when they were in a situation that called for such a move pair. The evidence was obtained from an analysis of moves made when subjects were in a subgoal situation just prior to the final path. This was defined as a locus in the problem state space that required a subgoal move before a progress-making move could be made. Even when the analysis was restricted to the end of the exploratory phase, the move latencies indicate that the subjects did not execute the moves as a move pair at a frequency greater than chance. If they encountered the same situation during the final path phase, it was correctly executed as a move-pair with a frequency well above chance (Kotovsky & Fallside, 1989).

[2][2]An alternative explanation for the exploratory-final path dichotomy has been provided by Anderson (1990) as part of his rational analysis of cognition. His very interesting analysis is predicated on the relative costs of hill-climbing versus means-ends analysis, rather than the near-impossibility of means-ends analysis within the memory limitations we believe are operating. A detailed comparison of the two models is beyond the scope of this paper.

the pattern of move operator difficulties found for individual moves in the various isomorphs. When subjects tried to plan pairs of moves, the load differences were magnified.

To test the hypothesis that subjects were planning move-pairs during the final path phase, the move latencies of the final path were analyzed for evidence of subgoal-goal pairs of moves.[3] The analysis indicated that the subjects solved these five-move minimum path problems in two rapid sequences of moves, corresponding to two goal-subgoal pairs of moves, followed by a final fast move of the last globe to its final goal position. The pattern of move latencies was long-short followed by long-short-short. This is what we would expect if the subject attained the ability to plan and execute a subgoal-goal move pair, as contrasted with the pattern if they made individual moves, or planned and executed all five final path moves as a compiled whole. The long-short pattern of move latencies is presumably due to a planning-plus-move step, followed by a move step. Not only did the subjects for whom final path move latencies were recorded exhibit the long-short and long-short-short temporal patterns, but as noted earlier, subjects in the exploratory phase did not.

## The Model

The model we report on attempts to account for the above findings with a small number of mechanisms and assumptions. The model consists of a production system[4] containing 21 productions that incorporates the following features: a limited working memory, a state memory that records recent moves and acts to prevent looping, a representation of the relative difficulty of the problem rules, a chunking or learning mechanism that operates on the problem rules and a derived measure of move latencies. These major features of the model are implemented as follows:

*Working Memory* is the common "work-space" of the model. It is very limited, able to hold, in the "standard" version of the model, about five entities or chunks. Its contents are the model's current subgoals (the overall goal of getting each monster its own sized globe is assumed to be in a longer term store), the problem rules, and the entities (globes, monsters) that have to be held in mind or imaged in order to be compared as part of the rule-legality checks during move planning and execution. The previously described image load acts here. The load imposed by various items in its contents at any point in time are calculated on the basis of empirical data acquired from the experimental work discussed above. An additional feature is that if this memory is overloaded, every item it contains is a candidate for being displaced, with equal likelihood. There is a working memory *breakpoint*, an inflection point near which each entity currently in working memory becomes separately eligible for displacement, with a probability obtained as follows:

$$p(\text{displacement}) = \frac{1}{1 + e^{\mathit{breakpoint} - \mathit{load}}}$$

where

*load* = "image load" (the number of imaged entities involved in the move or planned move)
+ "spatial load" (the number of loci that had to imaged in the move or planned move)
+ "subgoal load" (the number of subgoals)
+ "rule load" (remembering the current move rule)

The breakpoint is thus the point where the p(displacement)=$^1/_2$. The function rapidly goes to zero when the load is less than the breakpoint and to one when the load moves above the breakpoint. This relatively steep function is meant to approximate the behavior obtained in tests of memory span where the break is fairly sharp as the limit of a person's span is approached. For the data presented here, the breakpoint was chosen to be five. (It is possible to vary this parameter to match the performance of different subjects.) This stringently limited ability to manage information in making or planning moves is a major determinant of the model's behavior, as it is the behavior of the subjects. This is particularly true during the exploratory phase of the problem solving.

*State Memory* is posited to be a somewhat longer term recognition memory for recently made moves in the problem space. It functions to prevent looping (repeatedly traversing the same portion of the problem state space) or perseverating on trying the same move over and over again. It is implemented as a separate store to which the current move (and the state from which it is being attempted) is added (with probability = p) each time a subgoal or goal move is planned or made, and from which the oldest element is displaced (with probability = q) on each production firing. The probabilities, p and q, were set equal to 0.5 and 0.1 in the model data reported here.

The *rule load* designates the immediate memory load (number of chunks) imposed by remembering the problem move-legality rules. This was calculated primarily on the basis of a clausal analysis of the statement of the rules, bolstered by some data on the learning time required by subjects to learn to paraphrase or repeat the rules. The rule "loads" imposed by the problem rules in the order of source rule and destination rule, were for the Move Problem, 3 chunks and 4 chunks, and for the Change Problem, 4 chunks and 5 chunks respectively. These were the initial loads imposed by the rules. As subjects (human or computer) started making moves, they often made errors in which case they were forced to take back the illegal move and were re-exposed to the statement of the violated rule. This learning experience, by which subjects eventually learned to remember and apply the rule could eventually reduce the rule load to one. It was implemented via the following mechanism.

*Rule chunking* is a learning mechanism that is invoked each time a move rule was violated. Upon each violation, the relevant rule was "chunked" (ie. its imposed load was reduced to 1 from its initial value of 3, 4 or 5) with some probability. For the results reported here, the probability was set equal to 0.6 and 0.3 for the Move and Change problem respectively. This chunking in the model corresponded to the automatic restatement of any violated rule to human subjects each time it was violated.

*Try Something Else* is a mechanism that gets invoked

---

[3]An example of a subgoal-goal move pair is that encountered in the Move Problem when trying to move the medium globe to the medium monster when the medium monster is already holding the large globe. The completion of the move requires the subgoal of "clearing" the medium monster by moving the large globe elsewhere, followed by the goal move of moving the medium globe to the medium monster.

[4]The production system is written in a modified version of Grapes (Sauers & Farrell, 1982).

whenever the model's planning function resulted in its looping (perseverating) when trying to plan a particular subgoal-goal move pair or it's trying a move that had recently been tried (ie. was in state memory). The mechanism operated by randomly choosing a move from among the set of all "top goal" moves (ie. moves that would move a globe to its final or "home" position), + all trivial moves (ie. any moves that do not require legality checks at either the source or destination), + one randomly chosen move that is not currently in state memory.

*Move latency* is determined by assuming that the amount of time it takes a production to fire is a function of the complexity of a production (the number of elements on its condition side) and the working memory load extant at the time of the productions firing. The calculation assumes that the larger the current contents of working memory (as a percentage of the breakpoint capacity) the less resources available for the demands of the particular production and as a result, the slower the firing. This calculation of a memory-conditioned production firing time allows us to derive a measure of move latency by simply adding the times of all the productions that fire between a pair of moves. A move latency measure is needed if we are to compare the temporal patterns of the model's moves with the characteristic temporal patterns obtained from human subjects' data.

## Results

The operation of the model consists of trying to reach the specified goal state, which has each monster ending up holding his/her corresponding sized globe, by trying to make moves that will get the correct globes to their appropriate monsters. When moves are blocked by the problem's rules, the model tries, as subjects do, to plan and execute subgoal moves that will clear the way for desired goal moves. In the beginning of solving the problem, the model often makes illegal moves (by forgetting one or more of the move legality checks). Further along in the problem solving, the model, as it remembers to check the legality of contemplated moves, tries to "unblock" desired moves by executing subgoal moves, but is unable to accomplish this because of the memory load involved in planning and making the pair of moves. This results in repeated attempts to make moves that are repeatedly frustrated, more error prone move-making (when a legality check is one of the items displaced from the overloaded working memory), or a retreat to an easier but less useful move. Toward the end of the problem-solving episode, the model frequently exhibits the human subjects' "final path" type of behavior of efficaciously and quickly making moves that achieve the goal state. This is accomplished when it has learned the rules well enough so that they do not overload working memory, thus allowing some planning to occur. As this verbal summary indicates, the model's behavior is roughly similar to that reported for human subjects. We turn now to a more detailed comparison of the behavior of the model with that of people solving the same problems.

*Move data* is presented in Table 2 which contains the numbers of legal and illegal moves in each of four problems. The problems are change and move problems, each with two different starting positions. The data show that the model solves the various problems in about the same number of moves, making about the same number of errors, as the human subjects, with the fit somewhat better for the Change problems than for the Move problems.

| Problem: Start State<br>*Move Type* | Model | Human Subjects |
|---|---|---|
| Change: LSM | | |
| *Legal* | 12 | 13 |
| *Illegal* | 10 | 10 |
| Change: MLS | | |
| *Legal* | 17 | 18 |
| *Illegal* | 11 | 14 |
| Move: LSM | | |
| *Legal* | 12 | 16 |
| *Illegal* | 6 | 9 |
| Move: MLS | | |
| *Legal* | 17 | 15 |
| *Illegal* | 11 | 8 |

**Table 2: Solution Path Length (Number of Moves)**

*Move latencies:* Another way in which the behavior of the model can be compared with the behavior of human subjects was by means of the temporal patterning of moves and the rate of progress toward a solution during various phases of the problem solving. The issue is whether the model exhibited the exploratory and final path dichotomy found in most subjects' behavior; that is, whether the model exhibited a pattern of non-progress making moves followed by a rapid dash to a solution. A further question is whether the model also exhibited the evidence of subgoaling that was so crucial an element in the human subjects becoming able to move down the final path to the goal. The derived time measure is at best only an indicator of relative times, and its results should be interpreted with caution. Observation of the model's performance did reveal the characteristic dichotomy of exploratory and final path phases of moving. That is, there was a period, often quite extended, during which the model made non-optimal moves that contained a high frequency of illegal moves and resulted in no net progress toward the goal. This was followed by a rapid dash to a solution, during which the model made very few errors, and the moves were optimally directed toward the goal. Further, the temporal patterning of moves at the end of the problem solving episode for an absolute majority of solutions in which the model exhibited final path behavior did exhibit the L-S-L-S-S pattern of move latencies that indicated the subgoal-goal move pattern so evident on the subjects' final path. This evidence of final path behavior captures a major feature of the subjects' behavior; the learning of the problem rules and resultant acquisition of a bit of expertise at using them to plan moves. This learning occurs during the exploratory phase of the problem solving episode and allows the planning of subgoal-goal move pairs, thus permitting the subject, human or machine, to plan moves within the sharp constraints imposed by a very limited working memory.

*Move pathways and state visits:* One question that can be asked of the model's performance, in addition to how many moves it took to solve the problem, is whether it moves through the same portions of the problem state space as are traversed by the human subjects. The answer to this question is complicated by the fact that there is great diversity in the pathways chosen by different subjects and by different runs of the model. We constructed a display system that mapped out sets of subjects and sets of model runs and displayed the moves

**Figure 1: Number of Visits to Each State of the Problem Space, Model and Subjects**

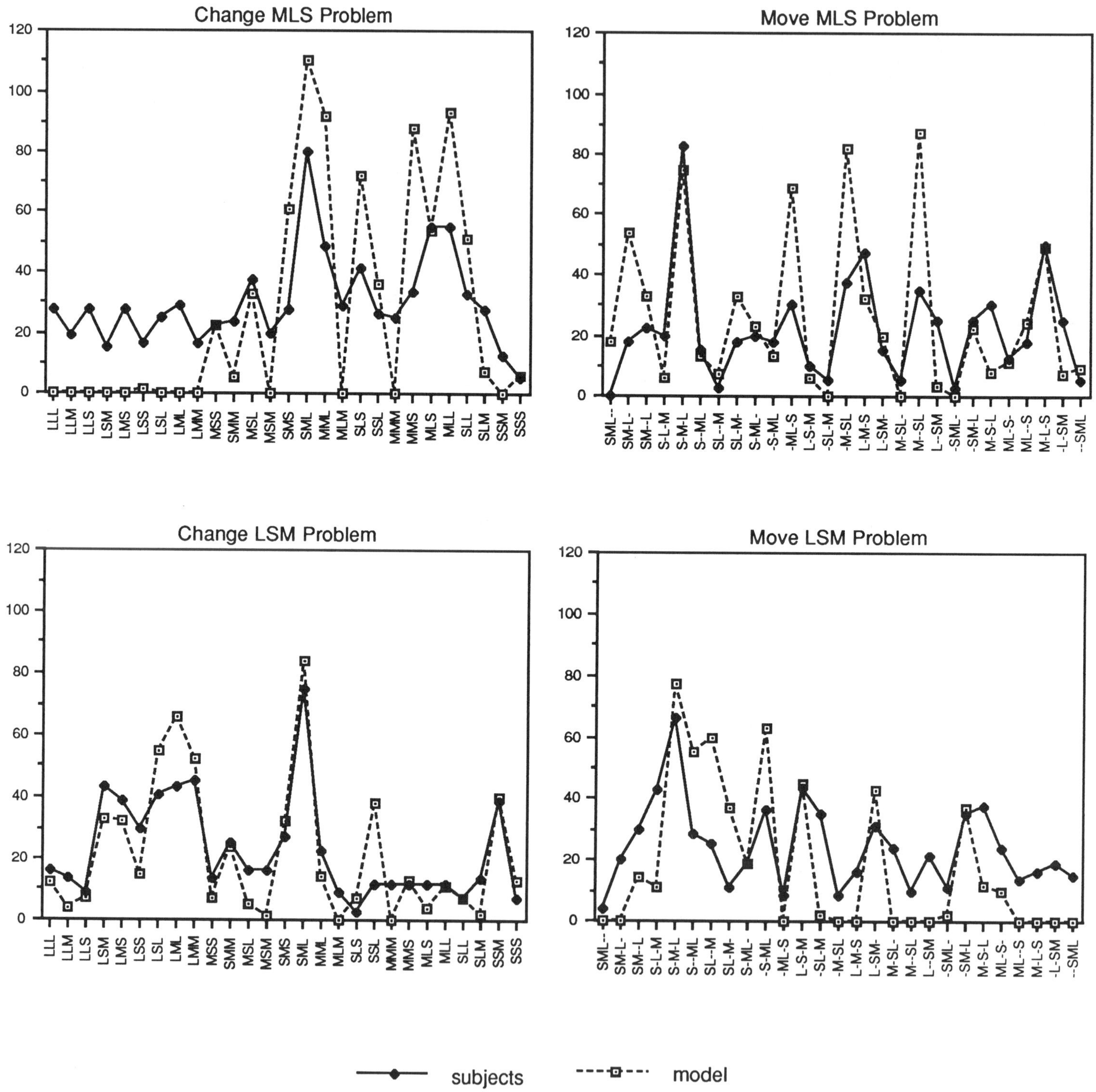

(preserving their latencies) through the problem space from start to goal. In general, the agreement seemed to be high; the diversity in the subjects' moves was matched by the diversity found among different runs of the model. A more controlled comparison was performed by comparing the number of visits to each of the 27 states (monster-globe arrangements) in the problem space made by our human subjects with those made by a set of model runs for four different Monster Problems. The four problems were two Move Problems and two Change problems, where the pairs of each type were defined by different problem start states.[5] The data is contained in Figure 1, where it can be seen that the state visits of the model closely approximate those of the subjects.[6]

*Move comparisons:* An even more detailed comparison of model and subject behavior can be made by comparing the particular moves made from each of the 27 problem space locations. This provides 105 comparisons (the number of each of the three legal moves together with the number of illegal moves from each state in the problem space).[7] The average correlation between moves chosen for the model and humans across the four isomorph types is 0.73, with the four problems' correlations ranging from 0.60 to 0.87. An even finer grain analysis can be made by considering each type of illegal move separately. That is, from each state in the problem space, there are three possible illegal moves that can be made by moving one globe. (If we allow the simultaneous moving of more than one globe, then a move can be made between any state in the search space and any other state, producing a total of 27×26 possible moves). The experiment constrained subjects so they could not move two globes at once (something that rarely happens even in unconstrained experiments), so the real limit on the number of possible moves is 27×6 (3 legal + 3 illegal moves) from each state, except for the three states where only two legal moves possible. Comparing the model with the subjects on the choice of moves for all 159 possible moves, the correlations range from 0.57 to 0.86, with a mean of 0.70. In evaluating this quite positive result, one possibility is that the structure of the state space itself constrains both model and subjects to certain states and moves because of the particular transitions that are allowed. In order to test this possibility, we constructed a random move generating model that randomly chose one of the six possible one-globe moves, and correlated its performance with that of the subjects. The result was that there is a correlation between the random model and the human performance data, but it is small compared to that found with the actual model. Thus the correlations between the random model and the human data ranged from 0.37 to 0.50 for the different problems, with a mean of 0.44. In addition, the random model accounted for none of the temporally patterned final path behavior that the model reproduced. Thus while the structure of the state space does play some role in constraining the behavior of those moving through it, it is not nearly constraining enough to serve as the explanation of the results we have obtained.

Our model, which incorporates the findings that have been made about the difficulties imposed by various features of the problem together with a sharply limited working memory provides a good account of a number of aspects of people's behavior on these problems. This is true whether the comparison is of the portion of the problem search space visited on a subject by subject basis, the frequency of particular state visits, the aggregate frequency of particular moves, the relative frequency of legal and illegal moves, the frequency with which all possible moves are made or the temporal patterns exhibited by the model in the final stages of its solution of the problem. These and other detailed comparisons of the model with human data allow a rigorous testing of the hypotheses that underlie the model's design, and provided convergent evidence for the conclusions reached in previous work on these problems. The work with the model is still in its early stages. We have yet to investigate transfer from one problem to another, nor have we used the power of the model to systematically explore factors that account for individual differences in performance or for performance on a wider array of problem isomorphs. Work on these issues is currently in progress.

[5]The subject data consists of sets of between 10 and 20 subjects per problem condition. The number of runs of the model was set at 25 to insure an adequate sampling, and the subject data were normalized to 25 in order to make the data sets comparable.

[6]State visits included visits to a state in the problem space resulting from either a legal or an illegal move. If only visits resulting from legal moves are considered, the fit is comparable.

[7]Three of the problem space loci allow only 2 legal moves, thus accounting for the 105 instead of 108 possibilities.

## Acknowledgements

We wish to express our gratitude to Herbert A. Simon and J.R. Hayes for their contributions to the conceptual development of the work reported here, and to Aaron Benjamin, for his help in running the simulations and analyzing the data.

## References

Anderson, J.R. (1990). *The Adaptive Character of Thought.* Hillsdale, NJ: Erlbaum.

Hayes, J.R., & Simon, H.A. (1974). Understanding Written Problem Instructions. In Gregg, L.W. (Ed.), *Knowledge and Cognition.* Hillsdale, NJ: Erlbaum.

Hayes, J.R., & Simon, H.A. (1977). Psychological Differences Among Problem Isomorphs. In Castellan, N.J., Pisoni, D.B., & Potts, G.R. (Eds.), *Cognitive Theory.* Hillsdale, NJ: Erlbaum.

Kotovsky, K., & Fallside, D.C. 1989. Representation and Transfer in Problem Solving. In Klahr, D., & Kotovsky, K. (Eds.), *Complex Information Processing: The Impact of Herbert A. Simon.* Hillsdale, NJ: Erlbaum.

Kotovsky, K., Hayes, J.R., & Simon, H.A. (1985). Why Are Some Problems Hard? Evidence From the Tower of Hanoi. *Cognitive Psychology* 17:284-294.

Sauers, R., & Farrel, R. November, 1982. Grapes User's Manual, Technical Report, Department of Psychology, Carnegie Mellon University.

Simon, H.A., & Hayes, J.R. 1976. The Understanding Process: Problem Isomorphs. *Cognitive Psychology* 8:165-190.

# A Revision-Based Model of Instructional Multi-Paragraph Discourse Production*

**James C. Lester** (lester@cs.utexas.edu)

**Bruce W. Porter** (porter@cs.utexas.edu)

Artificial Intelligence Laboratory
Department of Computer Sciences
University of Texas at Austin
Austin, Texas 78712

## Abstract

To communicate effectively, intelligent tutoring systems should be able to generate clear explanations of phenomena in their domain. To explain complex phenomena in scientific domains, they must be able to produce extensive multi-paragraph discourse. Traditionally, discourse planners have taken a *monotonic* approach to generation: once they make a decision, that decision is never revoked. Because these approaches make no provision for evaluating and modifying a plan after it has been constructed, their flexibility is limited. This inflexibility is particularly acute when attempting to generate multi-paragraph discourse.

We propose a *revision-based* model of discourse planning that constructs instructional multi-paragraph discourse plans, evaluates them, and restructures them. This is accomplished by delaying organizational commitments as long as possible and interleaving the planner's content determination and organization activities. This model accords well with research on writing. It has been implemented in an experimental system, KNIGHT, a discourse generator for intelligent tutoring systems. KNIGHT generates multi-paragraph explanations in the domain of biology. A domain expert has analyzed KNIGHT's explanations and found them to be clear and accurate.

## Introduction

The principle mission of an intelligent tutoring system (ITS) is to communicate knowledge to a student [Wenger, 1987]. The explanation generator of an ITS should generate explanations that are appropriate for the student's current mastery of domain concepts. Furthermore, an ITS for scientific domains must be able to explain very complex phenomena, so its explanation generator must be able to produce extensive (multi-paragraph) explanations.

For the past two decades, researchers in natural language generation have been exploring the problem of constructing text from a formal representation of knowledge. Generators that produce multi-sentential or multi-paragraph text have traditionally taken a *monotonic* approach to discourse planning. [1] We say that a discourse planner is monotonic if, once it makes a decision, that decision is never revoked. For example, schema-based planners generate discourse by instantiating the rhetorical predicates of schemata [McKeown, 1985, Paris, 1988, McCoy, 1989]; once the schemata are selected and their predicates are instantiated, the content and organization of the discourse are fixed. More recent, hierarchical discourse planners [Hovy, 1988, Moore and Swartout, 1991] are also monotonic. Given a rhetorical goal, these planners subgoal to construct a tree whose leaves are specifications for producing small pieces of text; they never revoke their selection of operators or change their binding of operator variables.

Because these approaches to discourse planning make no provision for evaluating and modifying a plan once it has been constructed, their flexibility is limited. Therefore, they rely on designers to develop schemata and planning operators that always "get it right" the first time. As the length of the text to be generated expands, it becomes increasingly difficult to anticipate all planning contingencies. This is especially true for a discourse planner that is supposed to be *context-sensitive*, e.g., the discourse planner for an ITS that customizes the content and organization of its explanations for a student's current mastery of domain concepts.

What's needed is a more fluid method of discourse production, such as the one employed by human writers. Researchers who study the writing process argue that revision plays a crucial role in composition [Hayes

*This research is supported by the National Science Foundation under grant IRI-8620052; the Air Force Human Resources Laboratory under grant F33615-90-C-0014; and donations from the research laboratories of Hughes and GTE.

[1] We follow the typical decomposition of natural language generation into *planning*, which is concerned with content determination and high-level organization issues, and *realization*, which is concerned with sentential organization, pronominalization, ellipsis, and lexical choice.

and Flower, 1986]. Although a good case has been made that natural language generators can benefit substantially from revision [Yazdani, 1987], only a few projects have explored revision computationally [Mann and Moore, 1981, Vaughan and McDonald, 1986, Wong and Simmons, 1988, Gabriel, 1988]. These projects contribute to our understanding of the revision process, but they focus on local sentential transformations, such as forming complex sentences from simple ones and combining two sentences with the same verb phrases, rather than on methods for globally restructuring discourse.

We have designed a *revision-based* model of discourse planning that constructs multi-paragraph discourse plans, evaluates them, and globally restructures them. The planner can revise its plans because it delays organizational commitments as long as possible and interleaves the content determination and organization activities. The model has been implemented in an experimental system, KNIGHT[2], a context-sensitive discourse generator for ITSs. KNIGHT generates multi-paragraph explanations in the domain of biology.

## Revision-Based Production

The *revision-based* planning algorithm controls the assembly of discourse plans. To provide organizational flexibility, it delays organizational commitments as long as possible. By delaying organizational commitments, it can more easily revise plans to improve their organization and to include additional content. Initially, the *discourse-elements*—specifications for sentences to be generated—of an evolving plan are placed in the *planning space* in a loosely organized fashion (Figure 1). For organizational purposes, the planning space is partitioned into *clusters*, each of which will eventually result in a paragraph of text. As the plan develops, the planner adds new elements to the plan, transfers elements from one cluster to another, and creates new clusters as needed. Thus it interleaves the construction and organization of elements. It gradually imposes increasing order on the elements until they are sequential, at which time it passes the elements to the *realizer*, which translates them into text.

The planning algorithm is shown in Figure 2. Planning begins when a query is posed. The planner first constructs the initial group of discourse-elements by identifying the relevant knowledge in the Botany Knowledge Base, a large-scale, frame-based representation that we and our colleagues have been constructing for the past four years [Porter *et al.*, 1988]. This knowledge base contains over two thousand frames about botanical objects (i.e., plant anatomy) and the processes that change them (i.e., plant physiology and development). Next, the planner evaluates the initial plan to determine if some elements can be rearranged to improve the organization. If so, it considers forming new clusters and transferring elements among clusters. If the plan was restructured, the planner determines if the reorganized plan should include additional content. If so, the planner applies *elaborators* that construct additional elements and inserts them into the plan. It then re-evaluates the plan to determine if any coherence has been lost. If repairs are called for, it installs the necessary patches, which take the form of outline statements, transition statements, and summaries. Finally, it schedules the elements in each cluster and passes them to the realizer.

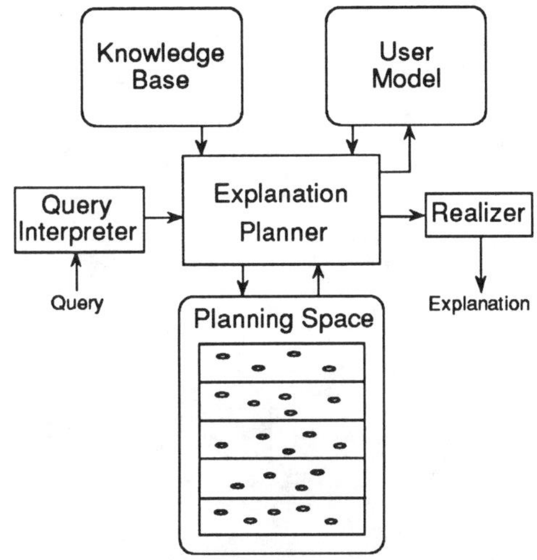


Figure 1: The architecture

```
Plan ← construct-elements(Query);
repeat
        for each restructuring operator R do
            if applicable(R, Plan) then
                Plan ← restructure-plan(Plan, R);
                for each elaborator E relevant to R do
                    if applicable(E, Plan) then
                        Plan ← elaborate-plan(Plan, E);
        if changed Plan then
            for each patching operator P do
                if applicable(P, Plan) then
                    Plan ← patch-plan(Plan, P)
until no more changes
for each Cluster in Plan do
        Cluster ← schedule-elements(Cluster);
        realize(Cluster)
```

Figure 2: The explanation planning algorithm

## Planning a Discourse

To demonstrate KNIGHT's operation, it was posed the query, "How does an embryo sac form?" The relevant portion of the knowledge base is shown in Fig-

[2]KNowledge-Integration-based Generator of History-sensitive Text

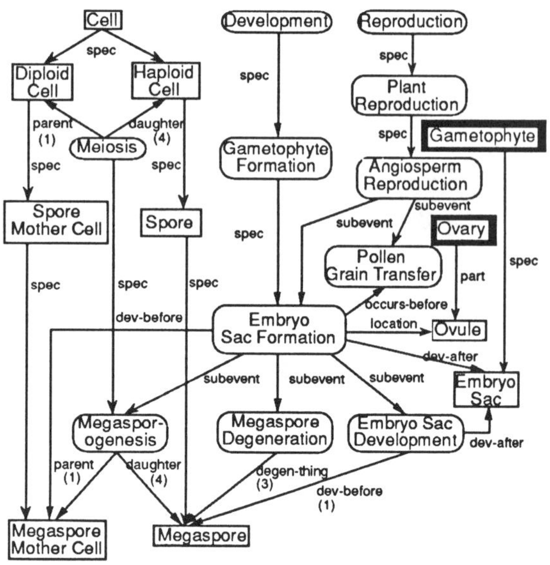


Figure 3: Portion of knowledge base and student model on embryo sac formation

ure 3. (The concepts that were familiar to the student are indicated by the darkened outlines.) The explanation KNIGHT constructed is shown in Figure 4.

## Initial Plan Construction

The planner constructs the initial plan by selecting subgraphs of the knowledge base, constructing discourse-elements from these propositions, and inserting the elements into clusters. To determine the initial content of the discourse, KNIGHT uses *graph schemata* to select the core of the explanation. It instantiates the graph schema associated with the type of the query to extract subgraphs of the knowledge base. For example, to explain a process $P$, KNIGHT applies a *process-description* graph schema to define a graph consisting of $P$ itself, the actors in $P$, the location of $P$, the subprocesses of $P$, and the subprocesses' actors.[3]

As it selects the relevant subgraphs of the knowledge base, KNIGHT constructs discourse-elements from the subgraphs' constituent propositions, and it assigns them to clusters in the planning space. The types of clusters, as well as their order, depend on the type of query issued. For example, the cluster types associated with the `Describe process` query type are the *Motivation* cluster, which conveys the overall significance of the process; the *Introduction* cluster, which provides an overview of the process; and the *Mechanism* cluster, which discusses the mechanism by which the process

[3]Because KNIGHT is a context-sensitive planner, it uses specialized selection strategies to determine additional subgraphs that are appropriate for the student. We discuss these methods for context-sensitive discourse planning elsewhere [Lester and Porter, 1991].

**STUDENT:** How does an embryo sac form?
**KNIGHT:**
Reproduction is a central function of biological entities. One kind of reproduction is angiosperm sexual reproduction. Embryo sac formation is a step of angiosperm sexual reproduction. Before explaining embryo sac formation, we will first discuss some prerequisite information about cells.

Two kinds of cells are a diploid cell and a haploid cell. A diploid cell has 2 sets of chromosomes in its nucleus, but a haploid cell has 1 set of chromosomes in its nucleus. During meiosis, a diploid cell reproduces, resulting in 4 haploid cells. One kind of diploid cell is a megaspore mother cell, which is involved in embryo sac formation. One kind of haploid cell is a megaspore, which is involved in embryo sac formation.

Now we can explain embryo sac formation. Embryo sac formation is a kind of development. It takes place in an ovule. An ovule is part of an ovary. During embryo sac formation, a megaspore mother cell develops into an embryo sac, which is a gametophyte.

Embryo sac formation consists of megasporogenesis, megaspore degeneration, and embryo sac development. First, during megasporogenesis, a megaspore mother cell reproduces, resulting in 4 megaspores. Later, during megaspore degeneration, 3 megaspores degenerate. During embryo sac development, 1 megaspore develops into an embryo sac.

Let's review the key points. Reproduction is a central function of biological entities. One kind of reproduction is angiosperm sexual reproduction. Embryo sac formation is a step of angiosperm sexual reproduction. Embryo sac formation is a kind of development. During embryo sac formation, a megaspore mother cell develops into an embryo sac.

Figure 4: Explanation of embryo sac formation

occurs. These cluster types were derived from an analysis of discourse patterns found in science textbooks. Figure 5 shows a snapshot of the planning space after the initial construction phase. Monotonic planners terminate after this phase, but KNIGHT evaluates and revises plans to improve them.

## Restructuring Plans

Once the planner has completed the initial construction phase, it evaluates the plan to determine if it can make improvements. It decides if a revision should be made by determining the conceptual similarity of discourse-elements in different clusters. If elements from different clusters satisfy the similarity criteria, it forms a new cluster by grouping the similar elements and positioning the new cluster in an appropriate location in the plan.

KNIGHT's criteria for conceptual similarity are based on our theory of *view types* [Souther *et al.*, 1989], which gauge the coherence of groups of concepts. Each view type aggregates concepts according to a particu-

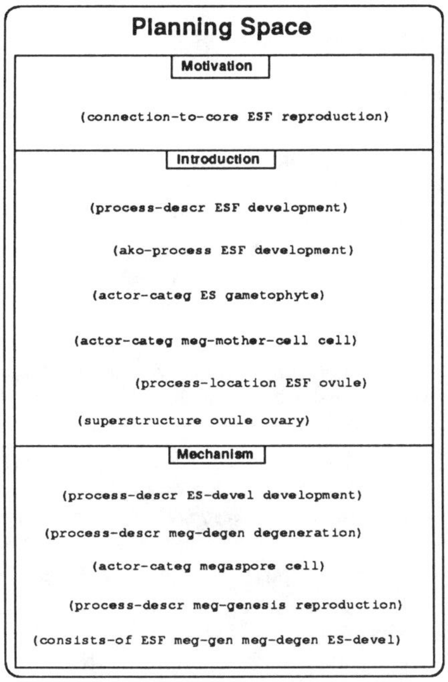


Figure 5: Plan after initial construction phase

lar perspective. For example, the *categorical* view type aggregates a concept and the concepts in its taxonomic heritage. Hence, KNIGHT uses a restructuring operator to group "categorically" similar information. Categorical restructuring is accomplished by finding two discourse-elements, each of which discusses two concepts as being specializations of a common concept, and grouping them together. In attempting to apply the categorical restructuring operator, KNIGHT's search through the planning space reveals that a categorical element in the *Introduction* cluster explains that a *megaspore mother cell* is a kind of *cell*, and another categorical element in the *Mechanism* cluster explains that a *megaspore* is a kind of *cell*. Rather than distributing the discussion across two possibly distant paragraphs, grouping the elements together provides a coherent forum for presenting a top-down taxonomic discussion of cells.

Of course, if the designers of the construction methods could have anticipated this situation, the categorical cluster could have been included in the initial plan. However, the need for such a cluster must be determined at run-time; only then can it be known that there are elements that fortuitously explain two concepts in terms of the same generalization.

## Elaborating Plans

In addition to improving the organization of the discourse, restructuring also provides opportunities for including additional content. This is particularly significant for discourse planners used by ITSs because it allows them to interject important supplementary information that is important in the domain. The planner uses *elaborators* for this purpose.

For example, after a categorical restructuring has been performed, as described in the previous section, the plan will include a cluster that discusses the subclasses of a particular concept. An appropriate elaboration might add discourse-elements that compare the subclasses. KNIGHT uses such a *comparison* elaborator. Given two categorical discourse-elements that explain two concepts in terms of the same generalization, KNIGHT searches for the crucial differences between the two concepts. When it applies the comparison elaborator to the discourse-elements of the categorical cluster, it determines that *megaspore mother cell* and *megaspore* are both kinds of a *cell*, and that these concepts begin to differ in the taxonomy at the concepts *diploid cell* and *haploid cell*. It constructs the discourse-elements that discuss the two kinds of cells and their differences and adds these elements to the newly created *Categorical* cluster.

## Finalizing Plans

The ability to dynamically restructure discourse plans and to generate elaborations on the fly provides high flexibility. However, flexibility is sometimes achieved at the expense of coherence. In these cases, the planner applies *patching* operators to restore coherence. For example, it constructs an outline element and adds it to the *Motivation* cluster, it constructs a transition element and adds it to the *Introduction* cluster, and it constructs a summary cluster and attaches it to the end of the plan. The final plan is shown in Figure 6.

The final step is to linearize the elements within a cluster. The planner invokes its *scheduling* operators to provide this service. Each scheduling operator is based on a commonly occurring pattern of text. For example, textbooks typically discuss the subprocesses of a process in the order in which they occur, so KNIGHT's *Mechanism* scheduling operator imposes a temporal order on the elements that discuss the subprocesses of a process.

# Conclusion

ITSs in scientific domains require explanation generators that can produce coherent multi-paragraph explanations. The revision-based model of discourse planning satisfies this requirement. KNIGHT's output was favorably evaluated by a domain expert, who analyzed three explanations it generated, including the one discussed here. He found the explanations to be both accurate—their statements were scientifically correct—and clear—their presentations were easily understandable.

The planner achieves these results by dynamically restructuring discourse plans. Restructuring is possible because it delays organizational commitments. In

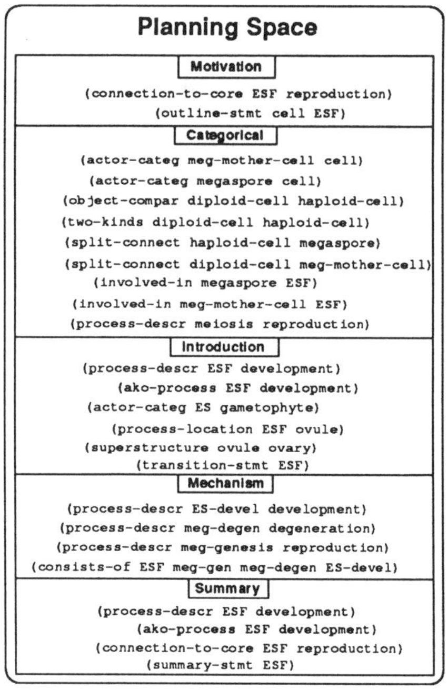


Figure 6: The final plan

contrast to monotonic discourse planners, the revision-based model allows decisions about organization to be evaluated and revised at run-time. This accords well with research on writing. In short, revision-based planning offers cognitive plausibility and constitutes a promising approach to multi-paragraph discourse generation.

## Acknowledgements

We are grateful to those responsible for the construction of the Botany Knowledge Base, including Art Souther, Karen Pittman, Ken Murray, Liane Acker, Brad Blumenthal, Dave Severinson, and Tom Jones. KNIGHT was implemented using Theo, a knowledge base access tool provided by Tom Mitchell. Thanks to Erik Eilerts for extending Theo's capabilities, and to Brad Blumenthal for his Lisp wizardry. Thanks also to Rich Mallory, Kathy Mitchell, Ken Murray, Hwee Tou Ng, Jeff Rickel, Wing Wong, and especially to Liane Acker for helpful discussions and comments on earlier drafts of this paper.

## References

[Gabriel, 1988] Richard P. Gabriel. Deliberate writing. In David D. McDonald and Leonard Bolc, editors, *Natural Language Generation Systems*, pages 1–46. Springer-Verlag, New York, 1988.

[Hayes and Flower, 1986] John R. Hayes and Linda S. Flower. Writing research and the writer. *American Psychologist*, 41:1106–1113, 1986.

[Hovy, 1988] Eduard H. Hovy. Planning coherent multisentential text. In *Proceedings of the 26th Annual Meeting*, pages 163–169, Buffalo, 1988. Association for Computational Linguistics.

[Lester and Porter, 1991] James C. Lester and Bruce W. Porter. A student-sensitive discourse generator for intelligent tutoring systems. In *Proceedings of the International Conference on the Learning Sciences*, 1991.

[Mann and Moore, 1981] William C. Mann and James A. Moore. Computer generation of multiparagraph english text. *American Journal of Computational Linguistics*, 7(1):17–29, 1981.

[McCoy, 1989] Kathleen F. McCoy. Generating context-sensitive responses to object-related misconceptions. *Artificial Intelligence*, 41:157–195, 1989.

[McKeown, 1985] Kathleen R. McKeown. *Text Generation: Using Discourse Strategies and Focus Constraints to Generate Natural Language Text.* Cambridge University Press, 1985.

[Moore and Swartout, 1991] Johanna D. Moore and William R. Swartout. A reactive approach to explanation: Taking the user's feedback into account. In Cécile L. Paris, William R. Swartout, and William C. Mann, editors, *Natural Language Generation in Artificial Intelligence and Computational Linguistics*, pages 3–48. Kluwer Academic Publishers, Boston, 1991.

[Paris, 1988] Cécile L. Paris. Tailoring object descriptions to a user's level of expertise. *Computational Linguistics*, 14(3):64–78, September 1988.

[Porter *et al.*, 1988] Bruce Porter, James Lester, Kenneth Murray, Karen Pittman, Art Souther, Liane Acker, and Tom Jones. AI research in the context of a multifunctional knowledge base: The botany knowledge base project. Technical Report AI Laboratory AI88-88, University of Texas at Austin, Austin, Texas, 1988.

[Souther *et al.*, 1989] Art Souther, Liane Acker, James Lester, and Bruce Porter. Using view types to generate explanations in intelligent tutoring systems. In *Proceedings of the Eleventh Cognitive Science Society Conference*, pages 123–130, 1989.

[Vaughan and McDonald, 1986] Marie M. Vaughan and David D. McDonald. A model of revision in natural language generation. In *Proceedings of the 24th Annual Meeting*, pages 90–96, Columbia University, 1986. Association for Computational Linguistics.

[Wenger, 1987] Etienne Wenger. *Artificial Intelligence and Tutoring Systems: Computational and Cognitive Approaches to the Communication of Knowledge.* Morgan Kaufmann, Los Altos, California, 1987.

[Wong and Simmons, 1988] Wing-Kwong C. Wong and Robert F. Simmons. A blackboard model of text production with revision. In *Proceedings of the AAAI Workshop on Text Planning and Realization*, St. Paul, Minnesota, August 1988.

[Yazdani, 1987] Masoud Yazdani. Reviewing as a component of the text generation process. In Gerard Kempen, editor, *Natural Language Generation*, pages 183–190. Martinus Nijhoff, Dordrecht, The Netherlands, 1987.

# The Role of Conventionality in the Real-time Processing of Metaphor

**Zhihua Long**

Institute of Cognitive Science
Department of Linguistics
University of Colorado
Boulder, CO 80309-0295
zlong@clipr.colorado.edu

**James H. Martin**

Institute of Cognitive Science
Department of Computer Science
University of Colorado
Boulder, CO 80309-0430
martin@cs.colorado.edu

**Abstract**

This project is intended to ascertain the role of conventionality in the use of metaphors in natural language processing. It examines the relationship between the degree of conventionality of a metaphor and the degree of difficulty in processing metaphorical meanings. The overall purpose is to obtain evidence regarding the metaphoric knowledge approach (Martin 1990) which asserts that the interpretation of novel metaphors can be accomplished through systematic extension, elaboration, and combination of knowledge about already well-understood metaphors. Subjects were tested on parsing sentences with different degrees of metaphorical novelty. Reaction times along with their responses were analyzed. The results suggest that a) degrees of conventionality in metaphorical use have a significant effect on the processing of the metaphor, b) degrees of novelty are proportionally related to the degrees of difficulty in processing, and c) conventional metaphors are as privileged in sentence processing as the "literal meaning" uses.

Recent work on metaphor in linguistics (Lakoff and Johnson 1980, Lakoff 1987, Lakoff and Turner 1989) and in natural language processing (Martin 1990) has emphasized the role of conventional metaphor. Conventional metaphors are taken to mean metaphors that are already commonly accepted as a part of the language and novel metaphors are those that are newly "created". The main thrust of Martin (1990) is that normal processing of metaphoric language proceeds through the direct application of specific knowledge about the metaphors in the language, i.e. when processing metaphoric language, people will make *direct* use of their knowledge of the conventional metaphoric meaning and apply it in the process of interpreting a sentence rather than having to give priority to the so-called "literal meaning" and resorting to metaphoric use only when the "literal meaning" is not compatible. Correspondingly, the interpretation of novel metaphors can be accomplished through the systematic extension, elaboration, and combination of knowledge about already well-understood metaphors.

There are, of course, approaches in which metaphors are not considered to be a conventional part of the language, e.g. the "literal meaning first" approach. (cf. Wilks 1975, 1978, Fass 1983, 1988, Russell 1976). The principle characteristic of these approaches is that metaphor is treated as a novel departure from conventional language. In these approaches the only conventional meaning is the literal meaning and a metaphorical interpretation is attempted only if the input is clearly not compatible with a literal interpretation. This incompatibility is typically manifested by a violation of one or more of the semantic constraints posed by the literal interpretation, usually in the form of selectional restriction violations. Because metaphor is not considered to be a conventional part of the language, explicit knowledge about the metaphors in the language is not available to these approaches.

The "literal meaning first" approach goes hand in hand with the literal meaning hypothesis, which assumes that sentences have well-defined literal meanings and that computation of the literal meaning is a necessary first step on the path to understanding speakers' utterances (Gibbs 1984). One version of the literal meaning hypothesis is found in Searle (1975, 1979), where he developed a set of principles by which a hearer is able to infer what a speaker means when using metaphors, ironies, idioms, or indirect speech acts. His principles are 1) the hearer first computes the literal meaning of the sentence; 2) the hearer decides if the literal meaning is defective, given the context; and 3) if the literal meaning is inappropriate, the hearer is led to seek an alternative meaning, which, depending on the principle of convention and his/her knowledge of speech acts, should lead him/her to the speaker's conveyed meaning.

A number of experimental studies have been conducted which suggest that literal meaning does not enjoy a privileged status in sentence processing. Gibbs' studies (1979, 1984, 1986) show that indirect requests take *no* longer to read than either literal sentences or direct requests. Ortony et al. (1978) and Inhoff, Lima and Carroll (1984) have shown that the computation of metaphorical sentences in contexts does not take longer than the computation of comparable literal sentences.

Another implication of the assumption is that a non-literal interpretation is optional because it depends on special triggering factors such as the violation of cooperative maxims. Glucksberg, Gildea and Bookin (1982) suggest that metaphorical meanings are not optional. Instead, metaphorical meanings seem to be computed involuntarily, even when the task requires only literal interpretations.

Keysar (in press) further supports Glucksberg, Gildea and Bookin's position by concluding that metaphorical and literal interpretations are functionally equivalent in comprehension. Firstly, metaphorical interpretations are constructed non-optionally in reading, just as are literal interpretations. They are produced when they make sense in context. Secondly, the metaphorical interpretations do not take on an implied simile form. Like literal interpretations, they do not need a transformation into a simile form. And therefore models that give priority to literal interpretation in sentence computation may be rejected, especially stage models which assume that metaphorical interpretation occurs only as a secondary or optional stage.

Note that all of these studies are designed to address the literal/non-literal distinction. As such, they do not shed any direct light on the issue of the effect of degrees of conventionality. It is to achieve evidence with regard to this and related issues that brings us to conduct the present study. By examining the role assumed by conventionality in a metaphoric use, we can find out about the mechanism that relates and/or distances the metaphors within a set.

Assuming metaphor is a basic, conventional part of the language, the conventionality of a metaphor should be privileged in processing a sentence containing the metaphor and that the conventionality of the metaphor in question is proportional to the ease of its processing, i.e. the more conventional a metaphor is, the easier it should be to process it. So the present study boils down to testing the following hypotheses:

1) Conventionality is privileged in metaphor processing;

2) Degree of conventionality is proportional to the ease of metaphor processing;

3) Literal meaning may not be privileged in real-time processing of a metaphoric use. Instead, what is privileged is conventional metaphor.

## Method

### Subjects

Seventy freshman psychology students from the University of Colorado at Boulder served as subjects. All were native speakers of English and none had participated in any similar studies.

### Materials and Procedures

The test sentences were designed to exhibit metaphors with different degrees of novelty. There were five sentence types: least novel use of metaphor, less novel use of metaphor, most novel use of metaphor, literal meaning use, and a control sentence. The five types of sentences were designed with the closest approximation possible in terms of syntactic structure and complexity. An example of the five sentence types is given below:

---

- *Paraphrase:* *John caught his cold from Mike.*
- Least Novel: Mike gave John his cold.
- Less Novel: Mike passed his cold to John.
- Most Novel: Mike threw his cold to John.
- Literal: Mike infected John with his cold.
- Control: Mike died of John's cold.

---

As is clear from the example, the control sentence is interpretable as inconsistent with paraphrase sentences and is apparently incompatible with the meaning that the other four sentences share. There is a decreasing conventionality in the metaphoric sentences, from the most conventional to the least conventional. The degrees of conventionality are designed with serious consideration of word use frequency. Also considered are the levels from which a metaphorical meaning is extended. Three linguistic professors were asked to judge the so designed degrees of conventionality of metaphoric use and they all agree on the designs.

There were five of these sets, each designed this way. For each set, we designed a *paraphrase sentence* that is not the same in form to any of the five sentences in the set. Take the above set for example, we designed the paraphrase sentence "*John caught his cold from Mike*". The paraphrase sentence is then paired to each sentence in the set. So, there were 25 pairings of the stimuli sentence and test sentence.

The pairings were presented *randomly.* The subject first sees on the CRT a paraphrase sentence, which stays on the screen for 2 seconds. Then comes the test sentence. When the subject has made his/her judgments, he/she presses the space bar and proceeds to the next pair. Every subject went through the same procedure, the only difference being the order in which the sentence pairs were presented.

Presumably, for sentences other than the control sentences, we would ideally hope for a "yes" response, and for the control sentence, a "no" response. The control sentences were designed to determine the sincerity of the subjects in performing the task and to assure them that the answers were not all "yes"-- they do have to do some thinking!

Reaction times as well as correct responses in making the judgments were recorded and analyzed for interpretation.

### Apparatus

Five IBM personal computers in the research lab of the Department of Psychology, University of Colorado, Boulder were used to conduct this experiment.

## Results

The result obtained confirms that conventionality of metaphor does have significant effect on processing the metaphoric use. This is reflected in the mean percentages of correct responses for sentences of the five different types, which are shown in Figure 1.

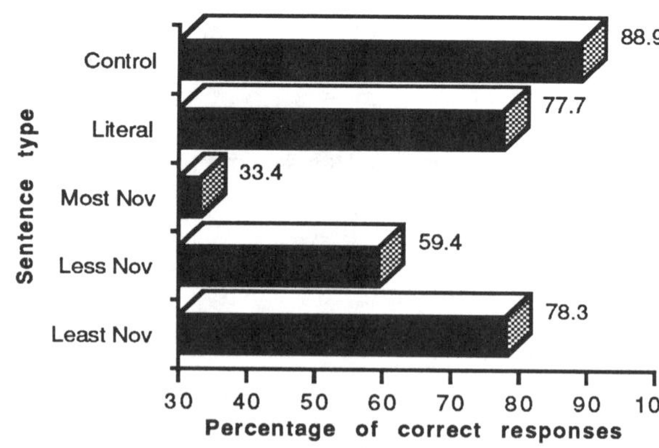


*Figure 1.* Average percentage of correct responses in all sentence types

As illustrated, there was the highest percentage of correct responses for the least novel (78.3%). As the metaphor gets more novel, the percentage gets lower (59.4%) and for the most novel metaphor, it reaches the lowest point (33.4%).

When it comes to literal meaning sentences, the percentage goes up again (77.7%) and for the control sentences, the percentage jumps to the highest point (88.9%). A one-factor analysis of variance yields significant effect of sentence types on correct responses, F (2.37) = 71.34, $p <$ .05. This meets our prediction that the more conventional a metaphor is the more agreements on its conveyed meaning one would expect and, moreover, it supports the claim that the more conventional a metaphor is the easier it is to process it and that the conventionality of a metaphorical meaning does affect the processing of a sentence containing the metaphor.

A 3-factor repeated measures analysis of variance of sentence type, set number and reaction time/response yields *no* significant interaction between sentence type and set number, F (1.64) = .646, $p > .05$.

The results of this analysis also show that no significant difference exists among set numbers, F (2.37) = 2.318, $p <$ .05. However, significant effect is found to exist between different sentence types, F (2.37) = 4.965, $p < .05$.

The results at this stage of analysis are certainly not totally reliable for any meaningful conclusion because what has been analyzed includes reaction times for both correct and incorrect responses. Nevertheless, looking at the percentage of correct responses in each of the five sets, one could see that they all fall close to the average line, as can be well illustrated in Figure 2.

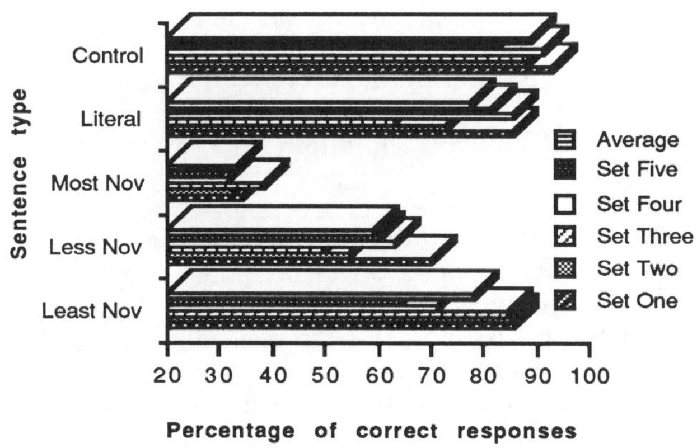


*Figure 2.* Percentage of correct responses for each of the five sets.

And, a two-way analysis of variance in terms of the percentage of correct responses yields *no* significant interaction between set numbers and sentence types, F = (1.64) 1.53, $p > .05$. The difference between sentence sets are not significant either, F (2.37) = 2.325, $p > .05$.

A similar phenomenon is found when looking at the means of reaction times for each of the five sets. They also tend to cluster together along the average, as is shown in Figure 3.

As with percentage of correct responses, we performed a two-way analysis of variance in terms of reaction times, since the percentage of correct responses of the five sets all fell along the average. Incidentally, we find no significant interaction existing between sentence type and set number,

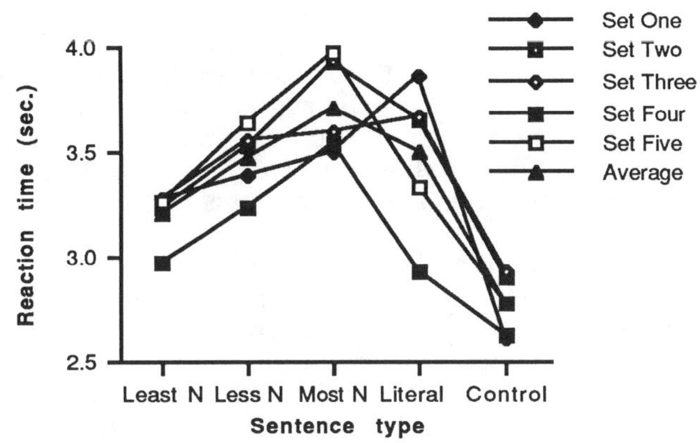


*Figure 3.* Mean reaction times for each of the five sets.

F (2.37) = .693, $p > .05$, though slightly significant effect is found among set numbers, F (2.37) = 2.737, $p < .05$. Significant differences, however, are found between different sentence types, F (2.37) = 13.988, $p < .05$.

Given that no significant interaction exists between sentence type and set number we can disregard the factor of set difference, since we are not interested in the differences between sentence sets anyway. We now look at only the correct responses for their reaction times. A one-factor analysis of variance of the correct responses yields a significant effect existing between different sentence types, F (2.37) =3.374, $p < .05$ . The means of subjects' reaction times to the five sentence types are illustrated in Figure 4.

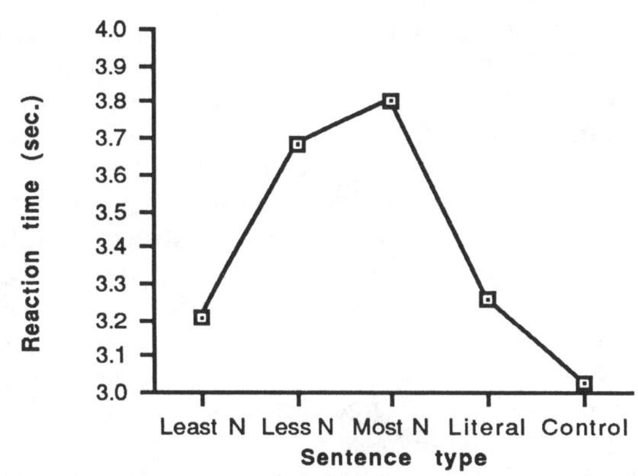


*Figure 4.* Mean reaction times of correct responses in all sentence types.

Interestingly, the plot of mean reaction times for the correct responses to the five sentence types shapes approximately like a mirror image of the plot of average percentages of correct response: for the most conventional use the mean reaction time is the lowest of the three metaphoric uses (3.208 sec.); the less conventional ones take a bit longer (3.686 sec.) and for the least conventional ones, reaction time climbs up to highest point (3.801 sec.). When it comes to literal meaning use, reaction time drops down again (3.256 sec.), but not as low as the most conventional. For the control sentence, the reaction time sinks to the lowest point (3.025 sec.).

Also, the reaction time tends to show a proportional increase as the metaphor gets more novel. So a positive correlation is found to exist between the novelty of a metaphoric use and the difficulty in processing it. As is shown by Figure 4, difficulty in processing as measured by reaction times increases as the novelty of a metaphoric use increases.

Further analysis of the least significant difference (LSD) indicates that the difference between each sentence type is significant (LSD = .113). Not considering the literal meaning sentence type, the rest of the four are all significantly different, the smallest difference (.115) being larger than the LSD (.113). The fact that reaction time of literal meaning sentences comes close to the most conventional metaphoric sentences fits right into our prediction: they should enjoy similar ease of processing. What is interesting is that the mean reaction time for the most conventional metaphoric sentences is somewhat less than that of the literal meaning sentences (3.208 sec. vs. 3.256 sec.). This may indicate which of the two enjoys more ease in real-time processing.

Also note that the percentage of correct responses, as shown in Figure 1 previously, is another dimension which demonstrates the fact that degrees of difficulty in processing increases as the conventionality of the metaphoric use decreases.

## Discussion and Conclusions

The purpose of this research was to examine the role and the nature of conventionality in the use of metaphors in natural language processing. Our particular focus was 1) the relationship between different degrees of conventionality and degrees of difficulty in processing sentences of metaphoric use; and 2) the assessment of processing status of conventional metaphoric use and "literal meaning" use. To summarize the results, conventionality appeared to have a significant effect on subjects' performance when interpreting the sentence. Given the same syntactic structure and the same syntactic complexity, the more conventional a metaphor a sentence contains, the shorter it takes to process the sentence and the fewer errors the subjects make. There is a positive correlation between the degree of novelty of a metaphoric use and the degree of difficulty in processing it. This suggests that there exists a larger conceptual distance for interpreting the more novel metaphoric use, i.e. a novel metaphor "travels" more conceptual distance before it is understood. These results are supportive of the assertion that the interpretation of novel metaphors can be accomplished through systematic extension, elaboration, and combination of knowledge about already well-understood metaphors, which could well be demonstrated by the longer reaction times that are taken to process those sentences. The larger the conceptual distance there is, the longer the reaction time it takes to "travel through".

The differences that lie in the ease of processing sentences with different degrees of novelty in metaphoric use can be accounted for in terms of spreading activation (as defined in Anderson 1976) in the metaphoric associations between the source domain and target domain. The mapping of the metaphoric associations is a highly interconnected structure. The basic mechanism at work is this: some of the dimensions of the source domain get extended and sprout and connect to a target domain, just in the same way a pointer connects to a different node. That target domain may later acquire the status of being a source domain when one of its dimensions gets extended to a newer node, which then becomes a target domain. Hence the process of interpreting a novel use of a metaphor involves the making of a massive metaphoric association net. Take the following examples from Martin (1990).

a) Mary *has* a cold.
b) John *got* his cold from Mary.
c) Mary *gave* John a cold.

The conventional metaphor Infection-As-Possession as in *Mary has a cold* serves as a source domain. It gets extended to the target domain and acquires new metaphorical uses, i.e. Becoming-Infected-As-Getting as in *John got his cold from Mary*, and Causing-Infection-As-Transferring as in *Mary gave John a cold*. Both extended meanings are built on the basis of the source domain. The primary metaphorical meaning is Infection-As-Possession. Becoming-Infected-As-Getting and Causing-Infection-As-Transferring are secondary to the primary metaphorical meaning. The target domain exists on the basis of the source domain.

That the conventional metaphor can usually serve as the source domain from which a newer metaphoric use gets extended means that novel metaphors are *extended* from an existing metaphor rather than *created* out of nowhere. The connection exists during the process of extension, even though the reader/hearer may not be able to make it, in which case they tend to fail to capture the targeted metaphorical meaning. When processing novel metaphors, people have to make use of their existing knowledge about the already well-understood metaphors, the ones that serve as source domains. For the novel metaphors, people need to make the connections. Thus the more nodes there are, the more difficult it is to process the metaphoric use and the longer time it takes. Just as a larger spatial distance would take more time to cover, a larger conceptual distance also requires more time to complete the "travel".

The results showed no apparent advantage in processing "literal meaning" sentences as compared to conventional metaphors. In fact, the most conventional metaphoric use took less time to process than the literal meaning sentences, even though the difference may not be significant. (3.208 sec. vs. 3.256 sec.). The results of the percentage of correct responses also demonstrated this fact (78.3% vs. 77.7%). These results refute the claim that metaphorical meaning is secondary to literal meaning in processing. They also delimit the boundary between the so-called "literal meaning" and the most conventional metaphoric meaning.

The present results on the role and nature of conventionality of metaphoric use lend support to the metaphoric knowledge approach in natural language processing. In particular, they reveal the connections of metaphoric uses that are otherwise thought to be unconnected, and more importantly, they suggest that the connections are systematic and meaningful. These results would certainly help build knowledge representations to account for those otherwise seemingly distinct and unrelated word senses.

**Acknowledgements.** We would like to thank Lyle Bourne, Kathie Carpenter, Barbara Fox and Ernie Mross for their insightful criticisms and suggestions. Any remaining mistakes are of course our responsibility.

**Appendices:** Stimuli and Test Sentences

*Set 1*

| | |
|---|---|
| *Paraphrase* | *John caught his cold from Mike.* |
| Least Novel | Mike gave John his cold. |
| Less Novel | Mike passed his cold to John. |
| Most Novel | Mike threw his cold to John. |
| Literal | Mike infected John with his cold. |
| Control | Mike died of John's cold. |

*Set 2*

| | |
|---|---|
| *Paraphrase* | *Beth created the idea.* |
| Least Novel | Beth produced the idea. |
| Less Novel | Beth manufactured the idea. |
| Most Novel | Beth assembled the idea. |
| Literal | Beth formulated the idea. |
| Control | Beth disliked the idea. |

*Set 3*

| | |
|---|---|
| *Paraphrase* | *That politician was accused of promoting dissent.* |
| Least Novel | That politician was accused of spreading dissent. |
| Less Novel | That politician was accused of sowing dissent. |
| Most Novel | That politician was accused of planting dissent. |
| Literal | That politician was accused of motivating dissent. |
| Control | That politician was accused of opposing dissent. |

*Set 4*

| | |
|---|---|
| *Paraphrase* | *Can you tell me how to terminate the process?* |
| Least Novel | Can you tell me how to kill the process? |
| Less Novel | Can you tell me how to wipe out the process? |
| Most Novel | Can you tell me how to slay the process? |
| Literal | Can you tell me how to stop the process? |
| Control | Can you tell me how to start the process? |

*Set 5*

| | |
|---|---|
| *Paraphrase* | *Teachers help their students to learn.* |
| Least Novel | Teachers cultivate their students. |
| Less Novel | Teachers garden their students. |
| Most Novel | Teachers fertilize their students. |
| Literal | Teachers teach their students. |
| Control | Teachers spoil their students. |

## References

Anderson, J. R. 1976. *Language, Memory, and Thought.* Hillsdale, NJ: Erlbaum.

Fass, D. and Wilks, Y. 1983. Preference Semantics, Ill-Formedness and Metaphor. *American Journal of Computational Linguistics 9,* 178-187.

Fass, D. 1988 Collative Semantics: A Semantics for Natural Language Processing. Ph.D. diss., Computing Research Laboratory, New Mexico State University, NM.

Gibbs, R. 1979. Contextual Effects in Understanding Indirect Requests. *Discourse Processes, 2,* 1-10.

Gibbs, R. 1984. Literal Meaning and Psychological Theory. *Cognitive Science 8,* 275-304.

Gibbs, R. 1986. Skating on Thin Ice: Literal Meaning and Understanding Idioms in Conversation. *Discourse Processes. 9,* 17-30.

Glucksberg, S., Gildea, P., & Bookin, H. 1982. On Understanding Nonliteral Speech: Can People Ignore Metaphors? *Journal of Learning and Verbal Behavior, 21,* 85-98.

Inhoff, A.W. , Lima, S.D & Carroll, P.J 1984. Contextual Effects on Metaphor Comprehension in Reading. *Memory & Cognition, 12,* 558-567.

Keysar, B. (in press). On the Functional Equivalence of Literal and Metaphorical Interpretations in Discourse.

Lakoff, G. 1987. *Women, Fire, and Dangerous Things: What Categories Reveal about the Mind.* Chicago: University of Chicago Press.

Lakoff, G. & Johnson, M. 1980. *Metaphors We Live By.* Chicago: University of Chicago Press.

Lakoff, G. & Turner, M. 1989. *More Than Cool Reason: A Field Guide to Poetic Metaphor.* Chicago: University of Chicago Press.

Martin, J. 1990. *A Computational Theory of Metaphor.* San Diego: Academic Press.

Ortony, A., Schallert, I. Reynolds, R., & Antos, S.1978. Interpreting Metaphors and Idioms: Some Effects of Context on Comprehension. *Journal of Learning and Verbal Behavior, 17,* 465-477.

Russell, S. W. 1976. Computer Understanding of Metaphorically Used Verbs. *American Journal of Computational Linguistics.* Microfiche 44.

Searle, J. 1975. Indirect Speech Acts. In P. Cole & J. Morgan. (Eds.), *Syntax and Semantics 3: Speech Acts.* New York: Academic Press.

Searle, J. 1979. Metaphor. In A. Ortony (Ed.), *Metaphor and Thought.* Cambridge: Cambridge University Press.

Wilks, Y., 1975. Preference Semantics. In Keenan (Ed.), *The Formal Semantics of Natural Language.* Cambridge.

Wilks, Y., 1978. Making Preferences More Active. *Artificial Intelligence 10.*

# Perception–mediated Learning and Reasoning in the CHILDLIKE System*

**Ganesh Mani** and **Leonard Uhr**
Computer Sciences Department
1210 W. Dayton St., Madison, WI 53706
{ganesh,uhr}@cs.wisc.edu

## Abstract

Intelligent agents interacting with their environments combine information from several sense modalities and indulge in tasks that have components of perception, reasoning, learning and planning. Traditional AI systems focus on a single component. This paper highlights the importance of the integrated perceive-reason-act-learn loop, and describes a system designed to capture this loop. As a first step, it learns about simple objects, their qualities, and the words that name and describe them. The visual-linguistic associations formed serve as a bias in acquiring further knowledge about actions, which in turn aids the system in satisfying its internal needs (e.g., hunger, thirst, sleep, curiosity). Learning mechanisms that extract, aggregate, generate, de-generate and generalize build a hierarchical network (that serves as internal models of the environment) with which the system perceives and reasons.

## Introduction

Intelligent agents (embedded in an environment) routinely sense information through one or more sensory channels, engage in various kinds of perceptual reasoning in the process of recognizing a stimulus or a pattern of stimuli, and respond, acting appropriately. Agents also exhibit the capability to learn from repeated interactions with external stimuli. This continuing perceive-reason-act-learn-perceive-... loop is almost certainly central to intelligence, as evidenced by the behavior of animals, including humans. Systems that fully capture this loop will exhibit the essence of mundane everyday reasoning.

The acquisition of knowledge about the environment starting with visual–linguistic learning has begun to gain attention among researchers (e.g., see Nenov & Dyer, 1988; Weber & Stolcke, 1990). Feldman et al. (1990) have dubbed this a touchstone task for cognitive science.

This paper discusses a system that we have designed and are implementing for capturing the perceive-reason-act-learn loop. First, the system learns to recognize and name simple objects, and their parts and qualities. It can then learn, from strings of these perceptually grounded words, about classes, relations, and actions.

Section 2 describes the system's architecture and goals (also see Mani & Uhr, 1991). Section 3 examines the learning mechanisms employed and Section 4 gives details of the acquisition of internal models of the environment.

## The Architecture of CHILDLIKE

The CHILDLIKE[1] system is a computational information-processing model (implemented in Common Lisp) designed to learn about objects, their qualities, and the words that name and describe them; and, further, to use this knowledge to satisfy its internal needs (e.g., hunger, thirst, sleep, curiosity).

The system is input sequences of simple "experiences" from which it attempts to learn. An experience contains several different components— for example, a visual pictorial scene, a short language utterance, an abstracted action.

The visual input is a snapshot at a single moment of time. It consists of a 4-by-4 or 8-by-8 image — typically one that a low- or intermediate-level computer vision system might output — that encodes information such as edges, or colors and textures, or simple shapes. These are pre-processed as needed, using a network of convolution-like mask-matching-plus-thresholding operations, to obtain primitive features such as long vertical or horizontal lines, the texture in a large segment of the image, or the color of these significant regions.

The language input consists of short (typically 2-5 words in the current implementation) English language strings (although any language could be used without changing any aspect of the system). An example sequence of inputs to the system is shown in Figure 1.

| Visual Input: | Language Input: |
|---|---|
| a) [Picture of an apple] | red apple |
| b) [Picture of a banana] | banana |
| c) [Picture of an apple and a banana side by side] | apple and banana |
| d) [Picture of a table] | brown table |
| e) [Picture of an apple on a table] | apple on table |

**Figure 1: A simple input sequence**

*This research was supported in part by a Grant-in-Aid of research from Sigma Xi.

[1] which stands for Conceptual Hierarchies In Language Development and Learning In a Kiddie Environment.

Each visual input to the system is a snapshot of the environment in time. Using such simple ordered sequences of inputs, the system is taught how to associate names of objects with their shape, size and other features, and further to associate relational words with visual features that imply them. Once the system is able to ground language symbols in terms of information perceived through the visual channel, it can be further trained by verbal input alone. Thus the system bootstraps itself, by first grounding words in perceptual information, to the point where it can learn from these grounded words only.

CHILDLIKE's parallel-hierarchical structure was chosen because it is general; successive compounding can produce any possible set of functions, starting from a universal set of primitives. It also has the potential of great speed and efficiency. And it is a widely used structure for perceptual recognition of objects in 2-dimensional images (Hanson & Riseman, 1978; Li & Uhr, 1987; Tanimoto & Klinger, 1980; Uhr, 1987), parsing trees for 1-dimensional language strings (Chomsky, 1986; Osherson & Lasnik, 1990), and the hierarchical building up of logical functions that accept the combinations of 0-dimensional terms handled by concept formation and similarity-based learning systems (Hunt, 1962; Michalski, 1983; Mitchell, 1982; Quinlan, 1986). It appears that the multilayer, converging "recognition cone" structure of micro-modular processes being developed for perceptual recognition tasks (Uhr, 1978; 1987) can also be used for building linguistic structures and visual-linguistic associations. Building on these visual-linguistic associations, the system can further acquire memory structures that encode the effects of actions and reason about need-fulfillment.

## Learning Mechanisms in CHILDLIKE

CHILDLIKE's learning mechanisms extract and incorporate information learned via interactions with the environment. These learning mechanisms can be grouped into the following major types: *extraction*, *aggregation*, *generation*, *de-generation* and *generalization*. Extraction is involved with the process of carving out potentially useful pieces of information from the visual and verbal input fields. Aggregation puts together pieces. Generation creates new links that encode, contain, and apply this new information.

Extraction takes place by imposing windows on the input array(s) containing perceptual information. This embodies a local-receptive field heuristic that favors extracting connected, compact features from which nodes that detect these features are generated. Evidence that brains' neurons predominantly interact with near neighbors, and empirical evidence that such local receptive fields are superior to random receptive fields for visual tasks in the recognition cone framework (Honavar and Uhr, 1989) support the choice of this heuristic. This locality heuristic is used to aggregate compound features at subsequent levels also.

The network of nodes created as a result of extraction, aggregation and generation serves as an internal representation of the environment. Each node encodes some feature or microfeature (e.g., one leg of a chair, the word "leg"), or some compound feature or class (e.g., four legs, the phrase "four-legged chair," furniture).

De-generation and generalization mechanisms simplify and speed up processing, and combat potential combinatorial explosions.

De-generation involves the discarding of nodes and links that appear to encode useless or wrong information. The value of a link is assessed (e.g., by its associated weight, or processes that estimate how useful it has been), and the network pruned accordingly.

Generalization may involve removing certain links (this corresponds to the *dropping condition rule* of Michalski, 1983, in symbolic similarity-based learning), replacing links to a number of nodes representing explicit entities with a link to a single node which may stand for any of these entities (*turning constants into variables rule*), replacing links to nodes $n_k$ at level $i$ with a link to a node $n$ at level $i+1$ such that a majority of $n_k$ are linked to node $n$ (*climbing generalization tree rule*), or replacing a sub-network with a node that stands for the sub-network (*constructive generalization rule*).

The order of the training sequences also plays a significant role in knowledge acquisition. At each step, experiences are judiciously chosen to slowly build on prior learning. For example, training sequences aimed at teaching relations such as "on," "above" and "and" should contain visual information portraying these relations between already-learned objects, along with the associated language string. An example of such graded training can be found in Figure 1. It should be noted that a number of training sequences of the sort shown in Figure 1 may be required before the relations are effectively learned. Training sequences may also force the learner to focus on a particular aspect of the input by varying all other aspects of the input (see below).

## Acquiring Internal Models of the Environment

Reasoning about the world is greatly facilitated by having an internal model of that world. The CHILDLIKE system's internal model consists of subnetworks encoding associations among visual features, words, actions, needs and compound features derived from them.

### Learning Words About Objects

CHILDLIKE learns about simple objects by "seeing" them through the visual channel and simultaneously "hearing" a linguistic description of the object through the language channel. Often, the object is not named in isolation but is named along with words describing other properties (e.g.,"green apple") of the object or the scene. Through repeated extractions from the sensory channels and generation of associations, CHILDLIKE learns the word that corresponds to a particular object in the visual field. It conjectures that other words (which do not appear to correspond to whole objects) may refer to parts

BEHAVING PHASE :

1. Input one of the arrays (visual or linguistic).
2. Extract primitive features, aggregate them into compound features, find the best-match for the entities in this channel and find corresponding entities in the other channel.
   {More formally, the extractions and aggregations can be expressed as
   ($P$ stands for a primitive feature and $C$ for a compound feature):
   Level 0 : $C_j = A_j(P_i, S)$
   Level 1 : $C_k = A_k(C_i, S)$ and so on for subsequent levels.
   Here, $i$ ranges over a local, connected window, S is an associated strength or weight vector (which gets modified based on errors made), and $A_j$, $A_k$, etc., are aggregation functions. Matching is performed by aggregating evidence hierarchically and collecting entities implied at each level.}

LEARNING PHASE:

1. Input visual and linguistic arrays ($V$ and $L$ respectively).

2a. Extract primitive features and aggregate them into compound features hierarchically.

b. Match known entities across the visual and linguistic channels (call the matched parts $v$ and $l$ respectively).

c. Generate all possible links, subject to a resource constraint, between novel features (corresponding to $(V - v)$ and $(L - l)$) across the visual and linguistic channels, plus $q$ (a changeable parameter) links involving already known entities (corresponding to $v$ and $l$).

3. If normal-training
   Change link weights using a Hebbian-like learning mechanism.
   {More precisely, the learning mechanism can be expressed as $W_{new}(N1, N2) = W_{old} + \eta f(N1, N2)$
   where $N1$ is a (primitive or compound) visual feature and $N2$ is a linguistic feature. $f(N1, N2)$ is non-zero only if the features $N1$ and $N2$ are both present in the current input instance; currently a normalized frequency count is used.}
   else (attention-focusing training)
   Modify weights explicitly using a high learning rate.
4. Adjust link weights, delete nodes and links, and form compact structures or subnets using de-generation and generalization.
   {De-generation heuristics and generalization mechanisms are described in the text.}

**Figure 2: The algorithm for learning about objects, object qualities and words that refer to them**

of objects, their qualities, or even relationships among objects.

## Learning Words About Parts and Qualities of Objects

An explicit feature in the visual field may correspond to certain qualities of the object (such as color or texture) or may constitute a sub-part of the object. Words about these features often occur in the language field of the input. Associations between words and their corresponding features are generated by the initial experience, and their weights strengthened (or weakened) by subsequent ones. Qualities and parts of objects will often also tend to be highly correlated with (hence will be linked to) the names of the object; for example, the color "yellow" from the visual field — in addition to being linked to the word "yellow" with a strong weight — may also be linked to the word "banana" and to the words for other yellow objects with moderate weights. But to the extent that the system is input experiences where an object's name is always present, and the names of qualities and parts are present for other objects as well, it will learn the information needed to assign names correctly.

The association between something like the color "yellow" and the word "yellow" can be further strengthened by a process called *attention-focusing* training. This involves making the system concentrate on a particular sub-concept via a training sequence or set of examples that focuses on the particular sub-concept. For example, to force the learner to form a reliable association between the color "yellow" and the word "yellow," one of the following strategies can be used: A) a yellow object is input through the visual channel and explicit feedback (which is specially marked, or input through a specific explicit-feedback channel) refers to the word "yellow" on the language channel. Such specially marked training instances are processed in a special way, resulting in the required link's weight being strengthened rapidly (using a higher learning rate). Note that although ad hoc, this is simply a fast alternative to repeating a large number of training examples such as those described next. B) a number of yellow objects, with other attributes (such as shape and texture) entirely unrelated to those of the objects experienced so far, are input through the visual channel along with the word "yellow" on the language channel. This helps in strengthening the required association without adversely affecting other useful, previously learned associations. Any new associations that are formed (embedding the irrelevant features) get de-generated eventually because they are not reinforced, and hence fall below a dynamically-adjusted threshold.[2]

[2]Threshold adjustments are done when it is evident that the system has made an error (e.g., the visual node for a ba-

## Learning Spatial Relations

Relations among object parts or between objects are much harder to learn if the notion of a relation (and its arity) is not built into the system. In CHILDLIKE (which is designed to learn as much as possible, rather than use built-in knowledge), priority is given to associating words from the language channel with explicit features or objects in the visual channel. Where a word and a sub-object have already been learned and linked together, they are considered to be accounted for. Each of the additional words is tentatively linked to each of the additional sub-objects. The arity of the relation is tentatively assumed to be the number of structures in the visual image already linked to words that are present.

When an example such as

Visual input:
Picture of yellow banana and brown table, with the center of the banana having a y-coordinate 2 units (a unit is a certain number of rows depending on the resolution of the input image) higher than that of the table (the x-coordinates of the 2 objects are the same or differ by less than 0.1 unit, a parameter which could be a function of things such as object-size).

Language input:
banana on table

is presented for learning the relation "on," the system finds the best match (the highest level structure implied across the visual and language input) for each component, and hypothesizes "on" to be the quality of one object and also to be a relation between the two constituent structures in the visual image that had matching words. Subsequent training using examples such as

Visual input:
Picture of red apple and white table, with the center of the apple having a y-coordinate 1 unit higher than that of the table (same x-coordinates).

Language input:
apple on table

reinforces the latter hypothesis.

Finally, rules like the following one are acquired as a result of the training regime explained above (the rule is a hierarchical subnetwork within the system; we have re-expressed the rule here in a more symbolic format for clarity).

on(L,M) $\longleftrightarrow$
L(y-pos = Y + (0..2), x-pos = X $\pm$ delta, color = *)
M(y-pos = Y, x-pos = X, color = *)

When necessary, spatial relations are also taught the system in a more explicit way, using the attention-focusing training mechanism described above. Several experiences are input that contain the same objects, but

nana being active along with the node for the word "apple").

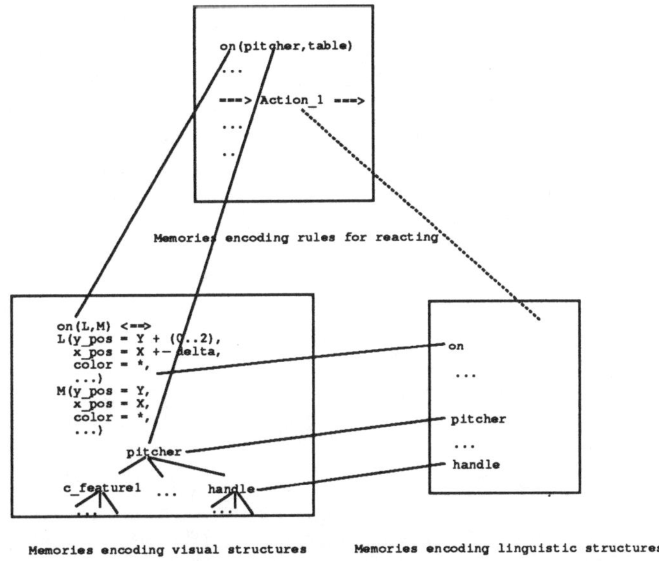


**Figure 3: An overview of the memory structures acquired by CHILDLIKE**

at different locations. Since other properties (such as color and texture) do not change when an object moves, the system learns to associate the (relative) positional information with relational words.

This will handle scenes whose descriptions are, for example, "glass on three-legged stool," as well as scenes involving non-spatial relations, for example ones described by "apple and yellow banana."

## Learning About Actions and Need-fulfillment

Reasoning about actions and their effects is an important aspect of intelligent behavior. In CHILDLIKE, sequences of visual frames are used to learn about the effect of actions. From two visual frames, and an intervening action, the system learns to associate an action (say $A1$) with the description of the two frames (based on a *temporal proximity* heuristic). Needs are sensed internally (in the same way that visual and linguistic features are sensed externally) and action sequences that change need levels favorably (indicative of need-fulfillment) are learned and stored for future use.

## Discussion and Conclusions

We have successfully experimented with the current version of CHILDLIKE using 5–10 object classes (such as fruit, food and furniture) with 5–10 distinct objects in each class (a large number of instances of each object are possible since objects can vary in position and also in their features and feature values). We are in the process of testing the learning abilities of the system on larger sets of objects. Our preliminary explorations appear to indicate that the underlying approach is sound; the system should not exhibit brittleness as the number of objects and their properties grows.

A snapshot of the memories acquired by CHILDLIKE

is shown in Figure 3. Distinct memories are used to encode the actions-related rules and their components; these are linked to memories containing encodings of related visual structures and words. Thus a pre-condition like "on(pitcher,table)" can occur in the action memories as part of a rule, and is connected to the corresponding visual structures and through them to words. (Only the highly weighted linkages are shown in Figure 3; links with small weights exist, for example, between the visual structure for "pitcher" and the word "table" since they co-occur in the same training instance. Links whose weights fall below a dynamically adjustable threshold are periodically removed from the memory structures.) The dotted line in Figure 3 represents an example of the kind of links that would be formed after words about actions are also learned.

An important feature of the system is its ability to learn and reason using language input alone (such as "apple is fruit," "banana is fruit," and "fruit is sweet"). Once a few words and what they refer to have been learned, new concepts can be learned using these words alone. This is salient in human learning, and can greatly speed up learning.

It should be noted that the performance of the system is not tied to specific object classes, specific words about them, or specific needs. The same program can be used with different objects, object-classes, needs, or/and words from a different language.

In CHILDLIKE, the effects of actions are learned from experience. The representation in CHILDLIKE's action memories is akin to that used by planning systems (starting with STRIPS, Fikes & Nilsson, 1971). This acquired knowledge, which is already in a convenient representation, can be used by a powerful planning module to enhance CHILDLIKE's reasoning capabilities.

We have described a system that attempts to capture the perceive-reason-act-learn loop which is central to intelligence, as evidenced by animal and human behavior. A perception–mediated approach enables efficient acquisition of concepts that are descriptive as opposed to those that simply classify and discriminate. The multi-level description of objects, including parts and qualities and words about them, facilitates different kinds of reasoning. Words and small sentences about the visual scene are grounded in terms of visual entities, and vice versa. The system can also learn to reason about simple action sequences that can potentially satisfy needs. All of this learning and reasoning is performed using micro-modular structures and general-purpose mechanisms such as extraction, aggregation, generation, de-generation and generalization.

We are extending CHILDLIKE in several ways. These include using a more sophisticated inheritance paradigm, refining the way actions are handled, and the inclusion of a deliberation module to arrive at approximate plans for fulfilling needs. Lastly, we note that the current functionality of the system is very good considering how little information is built in to the system a priori; all the rules employed in recognizing objects and their parts, in grounding words in objects and relations, and in reasoning are learned from experiences, starting with the most simple ones.

## References


1. Chomsky, N. 1986. *Knowledge of language: Its nature, origin and use.* New York: Praeger.
2. Feldman, J.A., Lakoff, G., Stolcke, A. and Weber, S.H. 1990. Miniature Language Acquisition: A touchstone for cognitive science. *Proc. of the 12th Annual Conference of the Cognitive Science Society.*
3. Fikes, R. E. and Nilsson, N. J. 1971. STRIPS: A new approach to the application of theorem proving to problem solving. *Artificial Intelligence 2 (3/4).*
4. Hanson, A.R. and Riseman, E.M. 1978. Visions: A Computer System for Interpreting Scenes. In: *Computer Vision Systems* (A. R. Hanson and E. M. Riseman, Eds.). New York: Academic Press.
5. Hunt, E.B. 1962. *Concept Formation: An Information Processing Problem.* New York: Wiley.
6. Honavar, V. and Uhr, L. 1989. Generation, Local Receptive Fields and Global Convergence Improve Perceptual Learning in Connectionist Networks. *Proc. of the Eleventh IJCAI,* San Mateo, CA: Morgan Kaufmann.
7. Li, Z.N. and Uhr, L. 1987. Pyramid vision using key features to integrate image-driven bottom-up and model-driven top-down processes. *IEEE Trans. on Systems, Man, and Cybernetics 17.*
8. Mani, G. and Uhr, L. 1991. Integrating Perception, Language Handling, Learning and Planning in the CHILDLIKE System. Working notes of the *AAAI Spring Symposium on Integrated Intelligent Architectures.* Also to appear in *SIGART.*
9. Mitchell, T.M. (1982). Generalization as Search. *Artificial Intelligence 18(2).*
10. Nenov, V.I. and Dyer, M. G. 1988. DETE: Connectionist/Symbolic Model of Visual and Verbal Association. *Proc. of the IEEE International Conference on Neural Networks. Vol. 2.*
11. Osherson, D.N. & Lasnik, H. 1990. *Language: An Invitation to Cognitive Science, Vol. 1.* Cambridge, MA: MIT Press/Bradford Books.
12. Quinlan, J.R. 1986. Induction of Decision Trees. *Machine Learning 1.*
13. Tanimoto, S. and Klinger, A. (Eds). 1980. *Structured Computer Vision.* New York: Academic Press.
14. Uhr, L. 1978. 'Recognition Cones' and some test results. In: *Computer Vision Systems* (A. R. Hanson and E. M. Riseman, Eds.). New York: Academic Press.
15. Uhr, L. (Ed.) 1987. *Parallel Computer Vision.* Boston: Academic Press.
16. Weber, S. H. and Stolcke, A. 1990. $L_0$: A Testbed for Miniature Language Acquisition. ICSI TR-90-010. Berkeley, CA.

# Generating Natural Language Expectations from a Reactive Execution System

**Charles E. Martin and R. James Firby**
Department of Computer Science
University of Chicago
1100 East 58th Street
Chicago, IL 60637
martin@cs.uchicago.edu

## Objectives

We are developing an integrated system for planning, language understanding, and learning through *advice-taking* (Martin and Firby, 1991). Our work brings together two reasonably mature pieces of research: the Reactive Action Package (RAP) execution system of Firby (1989), and the Direct Memory Access Parser (DMAP) of Martin (1991). Specifically, we are reformulating the RAP plan representation and execution algorithm in terms of the semantic network architecture used by the DMAP system. The result is a system with a uniform knowledge representation and processing algorithm that can both act in the world and process natural language. Furthermore, acting and understanding use a single process so the system can change its actions based on language inputs. That forms a basis for learning new plans and actions by being told.

Although "learning by being told in natural language" might strike one as a natural area of research in machine learning, a look through the proceedings of the last three Machine Learning Workshops (ML 1988, 1989, and 1990) and *Readings in Machine Learning* (Shavlik and Dietterich, 1990) will demonstrate that there is in fact *no* current machine learning research in this area. Most research on "advice-taking" has concentrated on the operationalization of *already-interpreted* input; an excellent example is Mostow's (1983) research. Broadly construed, this notion could be expanded to include *any* system that attempts to improve its decision-making data structures using an oracle, guided examples, etc. This is not our goal.

Instead, we have two complementary hypotheses:

1. The natural language understanding process can be greatly simplified by embedding the understander in another task-oriented system, where "embedding" refers to the use of *identical* data structures and a *single* algorithm for both the task and the natural language understanding process.
2. The performance of a task-oriented system can be extended by learning specific local plan modifications, derived by the overall system on the basis of high-level "hints" supplied in natural language.

In our research, the reactive execution system is the task-oriented system in which we embed the natural language understander, and the human expert provides high-level "hints" that are the basis for extending the reactive execution system's performance.

## Background

The DMAP system is designed to function within a large body of existing knowledge. Rather than determine the meaning of a text, DMAP uses the text to recognize relevant existing knowledge structures and then modify them to create specific new instances in memory that represent what is unique about the given text. New instances remain in memory to aid in the interpretation of future texts.

The RAP system is designed to carry out vague goals by waiting until execution time and then expanding the goal into primitive actions based on the details of the situation actually encountered. Possible plan expansions are stored in a hierarchical library of methods indexed by the goal they satisfy. The RAP interpreter selects a method from the library based on existing context and either instantiates the method as a new set of goals to achieve or executes the method directly if it consists of a primitive action. When a method completes (or fails) the interpreter checks to see whether the goal was actually satisfied in the world and, if not, selects and instantiates another method.

In bringing these two systems together we are directed by two "implementation" constraints:

1. There must be a uniform representation for all knowledge in the system, regardless of whether it is used for natural language understanding or planning and execution.
2. There must be a uniform algorithm to operate over the system's knowledge, regardless of whether the system is engaged in natural language understanding or planning and execution.

The first constraint is hardly new to devotees of KRL and its many descendants, though the constraint is honored more often in the breach than in the observance. The second constraint is more radical; in effect, it states that language understanding and planning and execution must be the *same* process. This is our hypothesis, with the caveat that differences between acting and understanding do exist, and they will manifest themselves in the system's *representation*.

The RAP model of execution provides a single, coherent representation for planning and execution knowledge with a simple interpretation semantics that does not require the maintainence of complex dependency structures. The DMAP model of language understanding relies on expectations about language that arise as a result of moving around in a semantic net. To combine the two, we have reformulated the RAP knowledge representation in terms of a DMAP-style semantic network that preserves the execution semantics when traversed in accord with the DMAP algorithm. The result is a single representation and processing algorithm for both acting on goals and processing natural language.

We have implemented our initial system in the robot-truck simulator domain described in Firby and Hanks (1987) and used by the RAP system as described in Firby (1989). We are currently building initial representations for the same RAPs used in that work and we anticipate working in this domain for some time to create a large DMAP system and refine our ideas. Ultimately, we intend to demonstrate these ideas on the robotic platform being built at the University of Chicago.

## Overview of the system

In keeping with the DMAP approach, the combined system expresses interpretation, planning, and execution as *recognition tasks*. Intuitively, the idea is that the system must have something similar to what it supposed to be doing already in its knowledge base. For natural language understanding, the system must already know a concept similar to that being communicated for the text to be understood correctly. For task execution, the system must already know methods that will work for the current goal in common circumstances. Naturally, new elements must be added to account for the differences between what exists and what is newly expressed or executed. For interpretation, these elements represent the content of the communication; for execution, these elements represent the intention and record of action.

### Representation

Representation is in the form of a semantic network, organized by abstraction and labelled packaging relationships. This representation was chosen primarily for the efficiency of the algorithm. Extensions to this basic framework include the specification of constraints on the packaging relationships of sub-structures to ensure that variable bindings make sense. There is also a system of *indices* used by the processing algorithm as a means of maintaining ordering relationships between concepts.

For example, the following memory units represent a decomposition of the task of picking an object up with a robot arm. There two subtasks: moving the arm to the location of the object and grasping the object with the arm. Constraints (the non **isa:** and **index:** forms) assure that variable binding is handled correctly. The index form indicates that in order to *recognize* (or satisfy)[1] this unit, the two sub-units, **task1** and **task2** must be recognized in the given sequence, followed by a recognition of the success condition for the task. Some of this information (such as the success condition) is inherited from its abstraction.

```
(define-unit task
  (success state))

(define-unit arm-pickup
  (isa: task)
  (arm ?arm)
  (object ?object)
  (success holding (arm ?arm) (object ?obj)))

(define-unit arm-pickup-simple
  (isa: arm-pickup)
  (task1 arm-move  (arm ?arm) (object ?obj))
  (task2 arm-grasp (arm ?arm) (object ?obj))
  (index: (task1) (task2) (success)))
```

The following is a more complex task to pick up an object, which involves first picking up the appropriate tool for the object and then the object itself. Note the use of the **instr** constraint to assure that the **?tool** is appropriate for the ?obj. Notice also that this unit and the **arm-pickup-simple** unit both inherit the same success condition from **arm-pickup**.

```
(define-unit arm-pickup-w/tool
  (isa: arm-pickup)
  (instr instrument (tool ?tool) (object ?obj))
  (task1 arm-pickup-simple (arm ?arm) (object ?tool))
  (task2 arm-pickup-simple (arm ?arm) (object ?obj))
  (index: (task1) (task2) (success)))

(define-unit instrument
  (tool tool)
  (object object))
```

### Algorithm

The basic DMAP algorithm is quite simple:

RECOGNIZE (*concept*):

[1] Processing in the DMAP memory is driven by memory unit indices. In language understanding, indices represent expectections for particular words, phrases, and abstract concepts. Processing thus consists of recognizing words, phrases, and concepts as they arrive from a text. In extending DMAP to include execution, indices represent intentions as well as expectations. Processing thus consists of both recognizing expectations and satisfying intentions. The algorithm remains the same, but the intuitive nomonclature shifts a little. We will use the words recognize and satisfy interchangably while discussing the processing algorithm.

If *concept* is a primitive operation
then apply it and process pending sensor inputs;
else gather *indices* to recognize *concept*,
for each *index* in *indices*,
for each *element* of the *index*,
call RECOGNIZE (*element*).

Primitive operations can be either sensor expectations (such as measurements or words) the robot must wait for or effector actions that the robot can execute. Effector actions include movements and active perception. When a sensor expectation is part of a memory unit index, recognition of that unit is blocked until data matching the expectation is generated by a sensor. When an effector action is part of an index, it is executed immediately and that execution may generate sensor data.

Successful execution of an effector operation or the satisfaction of a sensor expectation both result in the "recognition" of the corresponding *<element>* of the *<index>*; if all such *<elements>* have been recognized, the corresponding *<concept>* is recognized. Recognition of a unit causes all other units with the recognized unit as an index element to *specialized* the next element in their index. Specializing a unit involves moving down the `isa:` hierarchy to find the most specific memory unit that satisfies existing binding constraints. Any newly discovered specializations are *instantiated* by creating new versions of them in memory with variables bound according to the constraints. Units without indices are always specialized when instantiated and that is what moves activation of `arm-pickup` to either `arm-pickup-simple` or `arm-pickup-w/tool` depending on the binding of `?obj`.

When "gathering indices," all indices associated with the *<concept>* and any of its abstractions are collected. The system attempts to recognize all such indices through "recursive" calls to RECOGNIZE. If a complete unit index is recognized at any level of abstraction, that unit and all of its subunits are recognized together. That way, goals and concepts can be recognized when they occur even if no specific method or expectation set produces the required result.

For more details of the algorithm, see Martin (1990). For more background on this research, see Martin and Firby (1991).

## Execution Problems and Advice

Consider the following situation: The robot has task (1) "pick up a rock." It decomposes this task into two subtasks: (1A) "release anything currently held in the arm," and (1B) "pick up the rock with the arm." The latter task is in turn decomposed into the two subtasks: (1B1) "move the arm to the rock," and (1B2) "grasp the rock." Unfortunately, the grasping action (1B2) fails when executed because the arm is not well constructed for picking up rocks.

A friendly human wanders by and sees the robot's plight. The human instantly sees the problem, and knows that the grasping action would have succeeded if the robot had previously attached the shovel tool to its arm. There is precedent for this in the robot's memory; had the original task been to pick up a fuel-drum, the robot would have decomposed task (2) "pick up a fuel drum" into (2A) "release anything currently held in the arm," and (2B) "pick up the fuel-drum with the arm." The decomposition of (2B) would have been into three tasks: (2B1) "attach refueling tool," (2B2) "move the arm to the fuel-drum," and (2B3) "grasp the fuel-drum."

In other words, the general plan for (1B) and (2B), "pick up *<object>* with the arm" is able to discriminate between task decompositions on the basis of the *<object>*. This discrimination, of course, is represented in the vocabulary of plan descriptions. From the point of view of the robot, what is required to fix the rock problem is for someone to sit down and reprogram the plan library so that the general plan for (1B) and (2B) knows about rocks (and their associated gripper) as well as fuel-drums.

Unfortunately, the human who wanders by is not a robot programmer! (Perhaps the human builds robot gripper-tools.) From the point of view of the human, what is required is for the robot to somehow comprehend that this handy shovel tool should be used to pick up rocks in the same way that the refueling tool is used to pick up fuel-drums *without* having to have it hand-coded into the plan library. What the human would like to do, of course, is to simply say, "use the shovel."

Our goal is a system that lets the human do just that.

## The Context of Advice-Taking

There are a number of extra-sentential elements that are important to understanding "use the shovel" as useful advice. Chief among these is the fact that the robot is actively engaged in a specific task which it is *failing* to accomplish. This is a crucial piece of information; imagine the interpretation of "use the shovel" when the robot is not having a problem: "for what?" might well be the most appropriate response. However, in this case, the clear implication is that the shovel will be of use for the current task.

This is essentially a statement of the advice-taking problem our system is designed to address. The remainder of the paper describes the way we have approached the problem of establishing the situational context for the interpretation of an utterance, involving as it does tasks of both interpretation and execution. The context we will discuss is the internal state of the robot and not the external state of the world, except for knowledge that a current action is not succeeding.

### Execution Failures

In our example, an execution failure occurs when the robot attempts to grasp the rock without the appropriate tool.

```
(define-unit primitive
  (isa: task)
  (primop $primitive)
  (index: (primop) :ok (success)))

(define-unit arm-grasp
  (isa: task)
  (arm ?arm)
  (object ?object)
  (success holding    (arm ?arm) (object ?object))
  (primop $arm-grasp (arm ?arm) (object ?object)))
```

The goal of picking up the rock results in an attempt to recognize **arm-grasp** which results in an attempt to recognize, in sequence, the primitive **$arm-grasp**, the **:ok** message from the effectors stating that the primitive has been executed correctly, and the *success* condition of the task. If the robot hasn't picked up the appropriate gripper, the sequence of activity will approximate the following:

1. The attempt to recognize **$arm-grasp** succeeds, since an *attempt* at a primitive action always succeeds.
2. The attempt to recognize **:ok** fails, since the *result* from the primitive action is a notification that the grasp failed . In this case, the failure occurs because robot did not have the proper tools for success.

All recognition attempts which fail—whether from an attempt to recognize an execution sequence or a natural language utterance—result in the automatic recognition of a **failure** structure. This will result in the specialization of concepts that have the appropriate failure as an index and the eventual instantiation of a repair memory unit.

The **failure** hierarchy parallels every hierarchy which has indices; that is to say, virtually every hierarchy in the system. Every **failure** has a *source*, which is the structure whose recognition failed. Moreover, because each recognition task is the result of a "recursive call" to RECOGNIZE, the failure of one such task implies the failure of its parent and so on, up the chain of calls. Thus, the failure of the arm-grasp operation results in the recognition—and hence construction—of multiple failure structures.
Assume that the specific instances of the active pickup units involved in our failure are the following:

```
(instance arm-pickup-simple-1
  (isa: arm-pickup-simple)
  (arm arm-2)
  (object rock-3)
  (task1 arm-move-4)
  (task2 arm-grasp-5))

(instance arm-grasp-5
  (isa: arm-grasp)
  (arm arm-2)
  (object rock-3))
```

In this case, the failures that are generated and recognized are:

```
(instance failure-6
  (isa: failure)
  (source arm-grasp-5))

(instance failure-7
  (isa: failure)
  (source arm-pickup-simple-1))
```

### Generating Expectations from Failures

The recognition of these failure structures may, as with any successful recognition, cause the refinement and activation of further structures. The only structures that make reference to **failures** in their indices are those in the **repair** hierarchy. When a **failure** structure is recognized, the DMAP system attempts to refine the corresponding **repair** structures. One such candidate for refinement is **repair-task/advice**, given below:

```
(define-unit repair
  (failure failure))

(define-unit repair-task
  (isa: repair)
  (failure failure (source task)))

(define-unit repair-task/advice
  (isa: repair-task)
  (failure failure (source ?old-task (success ?success))
  (advice  mtrans (info ?new-task (success ?success)))
  (attempt ?new-task)
  (index: (failure) (advice) (attempt)))
```

The index to this structure implies that three recognition tasks take place. The first, recognition of the failure which has happened already, results in creating a context that includes a constraint defining a *communication event*, signified here by **mtrans**, in which the content of the communication is a **task** that shares the success condition of the original task—that is, a piece of advice.

This second element of the index is not an action that can be undertaken by the robot, but instead represents a prediction that natural language input might be forthcoming and that it will satisfy the advice constraint. In other words, upon the robot's failure to execute a task, it will always expect that it may receive a communication of a task sharing the success condition of the task that failed. This communication will be a piece of advice on how to achieve the success condition another way.

### Interpretating Advice

In our example, the advice "use a shovel" is meant to convey to the robot that the shovel is a tool which the robot should pick up before trying to pick up rocks. Note that the "tool-nature" of the shovel is not explicit in the utterance, nor is the fact that it is to be used for rocks. The expectations set up by refining failures to repairs sets up the context necessary for the robot to understand that the phrase, "use a shovel" implies this advice. Consider the following memory structures:

```
(define-unit mtrans
  (info mobject))

(define-unit mtrans-task-w/tool
  (isa: mtrans)
  (info task-w/tool (instr instrument (tool ?tool)))
  (index: use the (info instr tool)))

(define-unit task-w/tool
  (isa: task)
  (instr instrument))
```

The task to recognize the **(advice)** of the earlier repair structure results in instantiation of (among other things) **mtrans-task-w/tool** and an attempt to recognize its index. The nature of this index and memory unit is that the word "shovel" is interpreted as a specification of the **?tool** in the task structure.

Having interpreted "shovel" as a tool, recognition of the **mtrans-task-w/tool** is complete which completes recognition of the **advice** element in the repair structure index. The next element is **attempt** which must be instantiated as a specialized memory unit with the following constraints:

1. It must be a **task**.
2. Its *success* condition must be that of the original task.
3. It must have an *instr* role whose *tool* is a **shovel**.

These conditions are satisfied by the **arm-pickup-w/tool** structure, so the recognition algorithm creates a specific instance of this structure.

## Using Advice

Now that a task has been instantiated for the **attempt** element of the index associated with **repair-task/advice**, the next step is to recognize it. This task is an instance of **arm-pickup-w/tool**, and will require picking up the tool (the **shovel**) first, followed by picking up the object. Presumeably, this attempt will succeed.

The general recognition algorithm described above operates by creating specific instances of recognized memory structures. Many of these are too specific to be of later use to the system; for example, those that deal with picking up **rock-38**. As a consequence of instantiating these specific structures, however, more general concepts such as the fact that **shovels** are useful **instruments** for picking up **rocks** are also created.

At this point, not only has the helpful human's advice caused the robot to change to a more successful subtask, the process of interpreting "use a shovel" has instantiated an **instrument** relation in memory which can be used *anytime* in the future. Specifically, the system has learned that shovels are used for picking up rocks.

(Note that too-specific memory structures might as well be garbage-collected, except that determining which concepts are too specific is not a trivial task. A consequence of the recognition algorithm is that it will always use the most specific concept that is relevant to its inputs, but it is difficult to determine *in advance* what that level will be. Since performance does not degrade if these extra structures remain in memory, no attempt is made to remove them.)

## References


Firby, J. 1989. *Adaptive Execution in Complex Dynamic Worlds.* Ph.D. dissertation, Yale University.

Firby, J. and Hanks, S. 1987. *The simulator manual.* Technical report YALEU/CSD/RR #563, Computer Science Department, Yale University.

Laird, J., Rosenbloom, P., and Newell, A. 1986. Chunking in Soar: The Anatomy of a General Learning Mechanism. *Machine Learning,* **1**:11–46. Also collected in Shavlik and Dietterich (1990).

Martin, C. 1990. *Direct Memory Access Parsing.* Ph.D. dissertation, Yale University.

ML 1988. *Proceedings of the Fifth International Conference on Machine Learning,* Ann Arbor, MI.

ML 1989. *Proceedings of the Sixth International Conference on Machine Learning,* Ithaca, NY.

ML 1990. *Proceedings of the Fifth International Conference on Machine Learning,* Austin, TX.

Mostow, J. 1983. A problem-solver for making advice operational. *Proceedings of AAAI-83,* Washington, D.C., pages 279–283.

Shavlik, J. and Dietterich, T. 1990. *Readings in Machine Learning,* Morgan Kaufmann, San Mateo, CA.

# ACKnowledge: An integrated workbench supporting Knowledge Acquisition.

**Jean-Charles Marty - Fano Ramparany**
**Marie-Stephane Doize - Cleo Jullien**

Cap Gemini Innovation - ZIRST
7, chemin du vieux chene, 38240 Meylan FRANCE

Tel. +33 76 76 47 36 - e-mail marty@capsogeti.fr

## Abstract

*Knowledge Acquisition is a critical and time-consuming phase in the development of Knowledge Based Systems. The ACKnowledge project aims to improve the efficiency of the knowledge acquisition process. The approach is to analyze and evaluate a variety of existing knowledge acquisition techniques, including machine learning methods. Taking into account their complementarities, we integrate these techniques into a Knowledge Engineering Workbench that supports the Knowledge Engineer in his various tasks. This approach is tested on real life applications, simple ones (e.g. analysis in metal fractures) and more complex ones (e.g. failures in the Spanish data communications network).*

**Key words:** Knowledge Acquisition (KA), Integration of KA techniques, Knowledge Engineering Workbench, Integration of Knowledge.

## 1 Introduction

Currently, much work in the AI community is related to the development of **Knowledge Based Systems (KBS)**. We already know that the real bottlenecks in the development of such systems are the acquisition and modelization of the expert's knowledge. Knowledge acquisition refers to the initial phase of KBS development where the **Knowledge Engineer** gathers, analyzes and models the expert's knowledge. The efficiency of knowledge acquisition and the quality of the resulting model of expertise are crucial for the subsequent development phases.

Much work aims at supporting the knowledge engineer in this knowledge acquisition process. Some interview techniques have been developed, some of them together with a software support. But most of them only address particular problems and are not general enough to cover the entire process.

During the European Knowledge Acquisition Workshop 89 (EKAW), one of the conclusions drawn was that the scientific community should concentrate more on putting these tools together and studying their synergy rather than developing new and more sophisticated tools. It was also mentioned that the two fields of "Knowledge Acquisition" and "Machine Learning" should be more closely related.

This is perfectly in line with the work that was going on in the ACKnowledge project[1] [1]. We believe that the efficiency of the **Knowledge Acquisition (KA)** process will be improved if we take advantage of the complementarity of existing techniques. We are thus developing a framework that allows the fruitful combination of a large range of existing KA techniques, including machine learning techniques.

The mere inclusion of these techniques into a single system constitutes the first step for the construction of a **Knowledge Engineering Workbench (KEW)**. But this is not enough. If we want our workbench to be useful, we also need some knowledge about the usability of these techniques. With this new knowledge about the nature and applicability of the techniques, the KEW will actively assist knowledge engineers in their tasks, carrying out some parts of the work by itself, and providing the user (knowledge engineer) with some advice and guidance.

In this paper, we first describe in detail our vision of the KEW, the functions it must provide and its general architecture. We then explain the knowledge integration process, central theme in our approach. Relation to other work in the domain is described. Finally, we give some information on the real-life applications used for the validation of KEW.

## 2 The Knowledge Engineering Workbench

In this section, we describe several aspects of the workbench that we are developing in the project.

### 2.1 The KEW Vision

As we said above, the goal of ACKnowledge is to construct a workbench, that **incorporates different knowledge acquisition tools** within a single integrated system. No new knowledge acquisition tools will be developed within ACKnowledge. However, new implementations will be needed to provide a coherent integration of techniques. Moreover, the workbench will have an open architecture which will leave room for the addition of new tools to the system.

The construction of such a workbench meant that the project had to develop a conceptual framework adequate for characterizing, in a systematic and principled way, the function of

[1] The ACKnowledge (ACquisition of Knowledge) Project is a 3 year project partially funded under the auspices of the ESPRIT Programme of the Commission of the European Communities as project N°2576. The project Consortium consists of the following institutions; Cap Gemini Innovation(F), Marconi Command and Control Systems(UK), GEC-Marconi Research Centre(UK), Telefonica(E), Universidad Politecnica de Madrid(E), Computas Expert Systems(N), Veritas Research(N), the University of Amsterdam(NL), Sintef(N), and the University of Nottingham(UK).

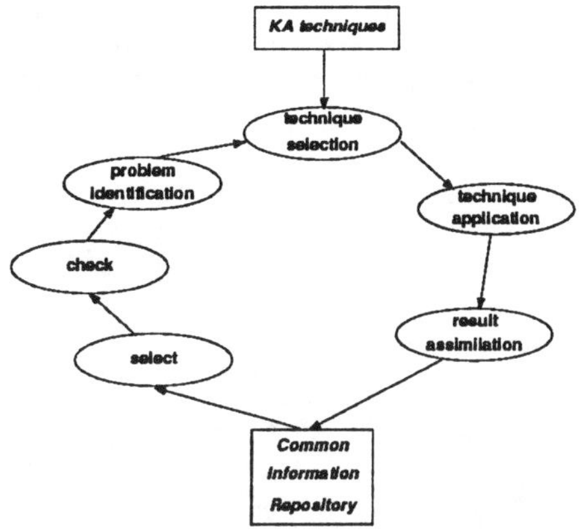


**Figure 1** Basic knowledge acquisition cycle

various techniques and tools. The workbench must be able to deal with a wide range of tools, input and output types, and user interactions.

Ideally, the user obtains a certain type of information with one tool and other kinds of information with other tools. Furthermore, we wanted the information acquired by one tool to be usable by other tools. This required research into the maintenance of knowledge base coherence and methods to effect **transformations between various knowledge representations**. Part of this research is presented in section 3.

The knowledge engineer must consider the knowledge acquisition process as an **incremental process** [2]. At each iteration of the process, information acquired so far is evaluated. Based on the result of this evaluation, an appropriate technique is selected and applied to obtain new information from sources of domain knowledge such as human experts and documents. This information is intended to refine and enrich the current version of the integrated knowledge base.

Further decomposition of one iteration of the KA process reveals the following basic operations:

- **select:** identify which part of the acquired knowledge to evaluate.
- **check:** check the selected piece of knowledge in order to find defects in quality[2].
- **problem identification:** identify and select which of the defects to pursue at the current stage of the KA process.
- **technique selection:** find the adequate KA technique for fixing the problem.
- **technique application:** apply the KA technique to the piece of knowledge selected in the first step.
- **result assimilation:** find out how results of the technique application can be assimilated into the entire existing knowledge.

Obviously from the above decomposition, the control of the KA process is mostly **cyclic** as described in figure1.

### 2.2 Levels of KEW support in the KA process

We envisage workbench support to the KA process at four different levels:

- The first level provides the knowledge engineer with a support for his **clerical activities.** Editing examples for KA tools, reporting on the current status of the KA process, and all this kind of activities are gathered in this level.
- The second level concerns the **execution of the various KA tools** included in KEW. Data inputs and presentation of outputs of these techniques are also considered at this level.
- The third level gathers the support functions that allow the user to assimilate the results obtained with one particular tool to other knowledge obtained previously. Section 3 explains this assimilation of knowledge in more details.
- The fourth level is the control part of KEW that will provide the user with **advice and guidance**. For instance, it will give information about what technique or sequence of techniques is best adapted to the current phase of the KA process.

Each of these levels enriches the preceding ones: If one considers the environment including the functions of the first level only, the knowledge engineer has basically a tool for his project management. With the second level, tools become available either for his interviews with the expert or for his analysis of these interviews. By adding the third level, the synergy among all the tools appears and a real integration of the knowledge obtained from these various tools is possible. Finally, the fourth level gives intelligence to the workbench. The knowledge engineer is not in front of a range of tools without really knowing which one to apply: instead, he receives advice on what the possibilities are at this stage.

### 2.3 The KEW Architecture

The above analysis of the KA-process has been used as a basis for the KEW system architecture, which is shown in figure 2.

Central to the architecture is the **Common Information Repository (CIR)** which stores all information needed during the Knowledge Acquisition process: partial or intermediate piece of knowledge, such as protocol transcript or structured cases; project management documents; and the **Integrated Knowledge Base (IKB)** which is the final and complete knowledge base that the KA process aims to produce. Different **Knowledge Acquisition tools (KA-tools)**[3] support the KEW user during his basic KA-tasks. One such task yields a **Tool Specific Knowledge Base (TSKB)** whose content is to be assimilated later on into the IKB using appropriate **Knowledge Assimilation (K-assimil)** mechanisms. "Tool Specific" means that the knowledge is represented using a language that is specific to the KA-tool. Knowledge assimilation mechanisms will be detailed further in section 3. The **reasoning** component embodies an interpreter of the IKB knowledge representation and allows the user to perform reasoning on the knowledge currently contained in the IKB for testing or simulation purposes.

The **User Interface (UI)** allows the user to browse through the CIR and access its constituents (via the **CIR server**), to

[2] By qualities we mean properties like completeness, correctness, comprehensibility, modularity, etc...

[3] The list of the tools included in KEW is given in figure 4.

use the KA-tools (possibly under the control of the Advice and Guidance module), and to reason on the knowledge acquired so far.

An **Advice & Guidance (A&G)** module provides assistance in decomposing the overall KA-process into KA-activities and further on into basic KA tasks, in selecting the appropriate KA-tool for the current KA task. This module includes a planner component which instantiates the generic KA-cycle presented in section 2.1 and which feeds an **agenda** containing remaining operations of the current KA-cycle plan, and activatable operations ordered with respect to their relevance. The agenda content is derived by using knowledge about the KA process (KA activity breakdown, constraints due to domain/application characteristics), knowledge of generic models (models of problem solving methods, domain structures and content,...), knowledge of the KA-tools themselves (functionalities, input constraints...) and knowledge on the results of KA activity (evaluative knowledge: consistency, coverage, generalization). This knowledge is stored in the **Knowledge Engineering Knowledge Base (KEKB)**. More details about this component can be found in [3].

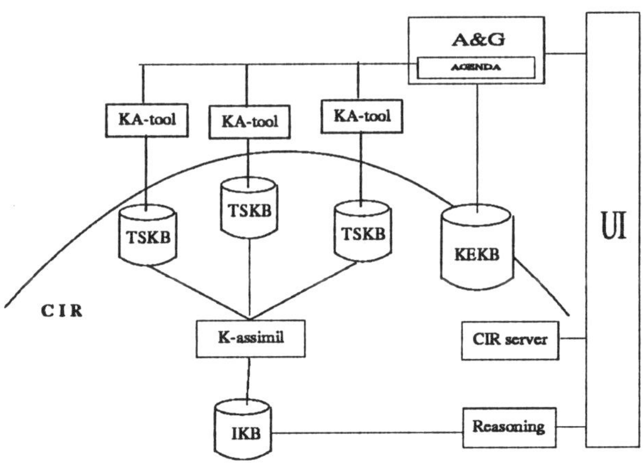


**Figure 2** KEW architecture

## 3 Knowledge Assimilation

Knowledge assimilation is a key issue addressed in the project [4], [5].

As mentioned in section 2.3, the KA process generates a collection of Tool Specific Knowledge Bases, resulting from the different KA technique sessions. Due to the multiplicity of KA techniques those TSKBs will have heterogeneous knowledge representation. In order to assimilate the contents of those knowledge bases, we need functions which transform expressions written in the technique specific knowledge representation formalism into a common knowledge representation formalism. This first transformation step produces **Temporary Knowledge Bases** (Temp KB) that are images of TSKB represented in the common formalism. Functions are also required which integrate Temp KB into the central **Integrated Knowledge Base**.

The contribution of both **knowledge transformation** (K-transf) and **knowledge integration** (K-integr) functions with regard to the knowledge acquisition process is shown in figure 3:

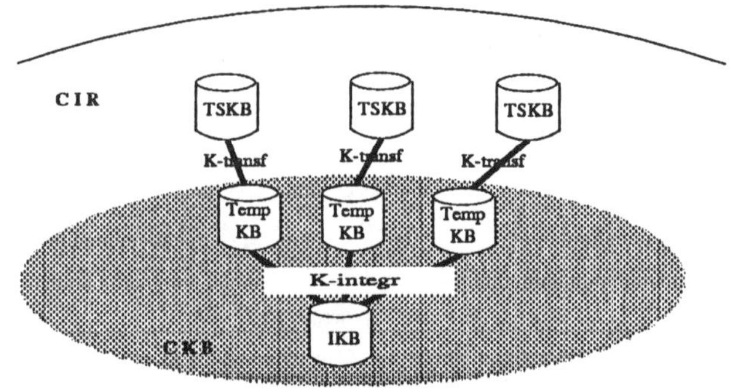


**Figure 3** Role of knowledge assimilation mechanisms

The common information repository has been partitioned in two areas:

- The non shaded area contains heterogeneous knowledge bases that contain data used and produced by knowledge acquisition tools.
- The shaded area is called **Core Knowledge Base (CKB)** and contains knowledge bases expressed using the common knowledge representation language.

Transformation functions produce temporary knowledge bases as part of the Core Knowledge Base, by transforming the contents of tool specific knowledge bases into common knowledge representation structures. Integration functions then merge the different temporary knowledge bases into the integrated knowledge base.

In the following sections we analyze the main issues involved in knowledge transformation and knowledge integration.

### 3.1 Knowledge transformation

We have based our analysis of knowledge transformation on the type of data used and produced by the KA tools that are to be integrated within KEW. Figure 4 enumerates those techniques and their corresponding data types.

The diversity of knowledge representations is quite striking. As one may guess, some knowledge transformation are straightforward, due to the similarity (formally and functionally) of the expressions. For example, entailment rules produced by the repertory grid are easily transformed into CKB production rules. Transformation mechanisms mainly involve syntactic manipulation.

On the other hand, when knowledge representations are very different from one another (mapping between entities defined within the two representation formalisms is not trivial), additional knowledge is required to bridge the semantic gap. Usually, this implies user interaction.

More generally, we distinguish two aspects of knowledge transformation: The syntactic aspect and the semantic aspect.

The **syntactic aspect** of knowledge transformation refers to how terms and constructs from one language are transformed

| *KA Techniques* | *Knowledge representations* |
|---|---|
| protocol editor | fragments of text<br>conceptual model |
| concept editor | hiearchies of concepts |
| laddered grid | hierarchies of concepts |
| repertory grid | cases<br>grids<br>ratings<br>entailment rules |
| similarity based learning tool | structured cases<br>frame hierarchies<br>classification rules |
| case based reasoning tool | cases<br>attributes<br>relations<br>causal net |
| case sort tool | hierarchies of concepts<br>identifiers |
| core knowledge base editor | frames<br>production rules<br>relations |

**Figure 4** Various KA techniques and their specific knowledge representation

in the other language. At this level one is concerned mainly with symbol manipulation. For languages based on the same paradigms, knowledge transformation consists mainly of syntactic transformation.

The **semantic aspect** of knowledge transformation refers to what primitives from one representation means with regard to the other representation. In a more practical view, semantical aspect of the transformation concerns the way basic primitives of the original representation scheme will be interpreted by the target representation scheme. For example, will a knowledge source be transformed into a set of rules or into a procedure within the common knowledge representation language?

Purely syntactic transformation is automatic and does not require the interaction of the user, whereas semantical considerations often require user interaction to guide the transformation.

Once a piece of knowledge resulting from a KA technique session has been expressed in the common knowledge representation language (via the appropriate knowledge transformation application), it is ready for integration into the integrated knowledge base.

### 3.2 Knowledge integration

We define knowledge integration as the operation of merging two different pieces of knowledge to form a unique and coherent knowledge base.

The main requirements for knowledge integration are that the semantics of pieces of knowledge to be integrated be preserved and that some quality criteria be fulfilled. This requirement means that if two KA tools are to use the same piece of knowledge (refer to an unique object in the real world), it should have the same meaning within each tool framework.

Within KEW, we have tackled this requirement by designing an algorithm for integration [4]. The algorithm involves four steps:

- **Ordering:** Ordering the knowledge pieces into an integration order;
- **Recognition:** Identification of the knowledge in the initial base relevant to the knowledge to be integrated;
- **Elaboration:** Modification of the new knowledge according to expectations provided by relevant knowledge in the KB;
- **Adaptation:** Modification of the knowledge base to accommodate the elaborated information.

## 4 Relation to other work

The KBS development process can be looked at from three viewpoints: **activity-oriented, result-oriented** and **internal view.** These views enlighten particular aspects of the process that must be considered for building a support tool [2].

The ***activity-oriented view*** is similar to the KBS life-cycle view. The KBS development process appears as a set of activities ordered in time. Examples of such activities are: *task analysis, static domain analysis, expertise modeling or target knowledge base construction.* Activities give a macroscopic view of the knowledge engineering process.

**Activity support tools** are dedicated to one or more activities. Expert system building tools such as *KEE, ART* [6] *or Knowledge-Craft* [7] belong to this class of tools. Obviously, all the tools supporting a KBS development methodology are also activity support tools (e.g. KPT [8] and more recently Shelley [9] both supporting the KADS methodology [10]).

KEW covers the **activity-oriented view** because it gives global guidelines for structuring the KBS development process. It provides the knowledge engineer with advice on which activities are to be performed and in which order. It also indicates which techniques are available for achieving a particular activity.

The ***result-oriented view*** is concerned with the results to be achieved during the development process. *Lexicon, conceptual model* or *knowledge base* are examples of such results. In this view, we are mainly concern with the contents and representation of these results. Indeed, the same result can be represented in different ways. For instance, a knowledge base may be represented with rules, classification trees or logical expressions.

**Model/result driven tools** guide the user with models of the results to be obtained in the KBS development. *Teiresias* [11], *Roget* [12], *Opal* [13] or *Mole* [14] belong to this class of tools. For instance, the model of rules to be obtained at the implementation level is known in Teiresias.

In KEW, the **result-oriented view** is used as the basis for representations that KEW has to maintain during the KA process. In particular, KEW is equipped with models of results that should be achieved during certain stages of the KBS development.

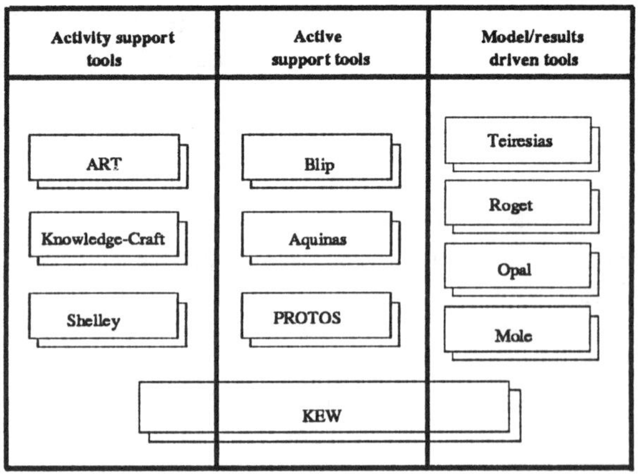


**Figure 5** Classification of knowledge engineering support tools

The ***internal view*** is a microscopic view of the process in the sense that it shows its internal structure. It highlights the elementary tasks that compose the global process and how they interact in order to achieve activities and results. Examples of such tasks are *vocabulary construction, concept editing inference modeling.*

**Intelligent support tools** use some knowledge about the knowledge engineering internal process itself and are able to perform parts of that process autonomously. *Blip* [15], *Aquinas* [16] or *PROTOS* [17] belong to this category. For instance, Blip provides support by using knowledge about generalization and integration processes in KB construction.

From this internal view, the KA process can be split into elementary tasks that allow the identification of the appropriate KEW components to be activated.

Figure 5 shows a summary of the mentioned tools classification.The originality of KEW is that it is based on a synthesis of the KA process views. This is a good ground for obtaining a more complete and generic workbench.

# 5 Applications

In our project, we are testing and validating our KEW on several applications. We have chosen real life applications in order to make a valuable evaluation of our KEW.

## 5.1 Simple applications

We first investigated the area with several "simple" applications, in order to provide data for testing our first prototype. This first prototype was a collection of KA tools running in the same environment but without any integration (level 2 of section 2.2).

Different partners provided the consortium with such applications. We have worked on various domains such as analysis of metal fracture, object recognition and the "Tomato Plant Pathology domain" [18].

The **analysis of fracture damages** in steel constructions is used for instance in shipping where there is a wide range of equipment prone to such failures.

Due to its large set of cases, the application on **recognition of objects** (viz cars/trucks etc.) in 2D images was especially suitable for the machine learning component of the KEW.

The diversity of these applications gave us a clear idea of the kind of data usually available to the Knowledge Engineer.

## 5.2 The telecommunication network failure application

The application that is being used throughout the project is more complex. It deals with the Spanish Data Communications network [19]. The final KB system must **assist the operators in initiating corrective actions** when network anomalies occur. For making such decisions, the operators usually use network diagnostic aids and tools.

The operators are in charge of ensuring good network efficiency at all times. This problem is difficult because operational procedures are complicated and cases often arise which are not completely described in the manuals. Furthermore, modern telecommunications networks are increasing in complexity. Effective maintenance thus requires assimilation of vast amounts of operation and maintenance knowledge.

This application often guided the specifications of our final KEW. For example, the list of tools to be incorporated in the KEW was drawn up from this application requirements.

## 5.3 The Ship Bridge application

A new application is going to be implemented during the last year of the project. The area we have selected is new within classification of ship and deals with **bridge design.** The rules for this classification are aimed at verifying the design of a ship bridge in order to facilitate the shipmaster's work and thereby make the ship movements as safe as possible.

There exists only a handful of experts in the area of bridge design, and they spend much of their time traveling around the world, teaching local surveyors how to use the new rules. This is both a tedious job for the experts and quite expensive for the company. If a computer could do at least parts of this job, it would be a major achievement.

The rules for bridge design are implemented as an electronic rulebook. These rules can be used as a skeleton for adding new knowledge, and/or be the object of improvement itself. This application will thus be a very good testbed for the behavior of KEW in the **refinement phase.**

# 6 Status of work and conclusion

Viewing KA as an incremental construction process based on various elementary tasks and intermediate results allows us to design a knowledge engineering workbench that actively supports knowledge engineers.

KEW is much more than a simple tool box that allows the execution of a collection of KA techniques. KEW provides support for various integration mechanisms that allow an effective combination of the tools: storage and representation of acquired knowledge, translation and integration facilities for partial results obtained with different tools, advice and guidance for the selection of techniques taking into account both applicability and purpose of KA techniques, as well as advice for scheduling KA tasks.

From what we have already achieved, we believe that our approach to the problem is viable and that the main functionalities of the KEW can be implemented shortly.

Currently the project is at mid-course. Until now, we have devoted most of our effort to the conceptual framework and to the development of two prototypes written in Sun Common Lisp [20]. The first gathers some KA techniques without integration of knowledge. The second has allowed us to test some of the KA transformations that we intend to perform in the final KEW. Simultaneously we have analyzed a large range of existing KA techniques [21]. The difficult part of our project remains the "Advice and Guidance module". Most of the knowledge needed on the applicability or efficacy of tools is still to be acquired. Evaluation work is therefore going on for this purpose. We have recently delivered the functional specifications of the final KEW to the CEC. This final KEW will be demonstrated in October 91 at ESPRIT Conference, in Brussels.

## Acknowledgements

We are grateful to members of the ACKnowledge team for the help in developing the ideas in this paper. In particular, we would like to thank Jorunn Brække, Sergio Calabretta, Jorun Eggen, Francisco Garijo, Inge Nordbø, Nigel Shadbolt and Bob Wielinga. Particular thanks to Didier Bronisz, Thomas Grossi, Bertrand Rousseau, Maurice Schlumberger and Philippe Vauquois for their valuable comments on this paper.

### References

[1] J.-C. MARTY, N. SHADBOLT, and B. WIELINGA, "The acknowledge project improving the knowledge acquisition process.," in *Results and Progress of ESPRIT Projects in 1989, XIII/416/89*, November 1989.

[2] B. WIELINGA and al., "Conceptualisation of a knowledge engineering workbench," Report ACK-UvA-T1.4-DL-010-A, ACKnowledge project, ESPRIT, september 1989.

[3] N. SHADBOLT and B. WIELINGA, "Knowledge-based knowledge acquisition: The next generation of support tools.," in *Current Trends in Knowledge Acquisition*, (Amsterdam), June 1990.

[4] J. EGGEN, A. LUNDTEIGEN, and M. MEHUS, "Integration of knowledge from different knowledge acquisition tools," in *Current Trends in Knowledge Acquisition*, (Amsterdam), June 1990.

[5] I. NORDBO and J. EGGEN, "Information pool for knowledge engineering support - the common information repository.," in *ESPRIT Conference 90*, November 1990.

[6] B. CLAYTON, "Art, automated reasoning tool, programming tutorial," 1985.

[7] C. Group, "Knowledge craft 3.0. reference manual," 1985.

[8] A. ANJEWIERDEN, "Knowledge acquisition tools," *AI Communications*, vol. 0, pp. 29–38, August 1988.

[9] A. ANJEWIERDEN, J. WIELEMAKER, and C. TOUSSAINT, "Shelley - computer aided knowledge engineering," in *Current Trends in Knowledge Acquisition*, (Amsterdam), June 1990.

[10] G. SCHREIBER, B. GREDEWEG, P. D. GREEF, P. TERPSTRA, B. WIELINGA, E. BRUNET, N. SIMONIN, and A. WALLYN, "A KADS approach to KBS design," Tech. Rep. Delivrable B6 Esprit Project P1098, University of Amsterdam, September 1989.

[11] R. DAVIS, "Interactive transfer of expertise," *Artificial Intelligence*, vol. 2, no. 12, pp. 121–157, 1979.

[12] J. BENNETT, "ROGET : a knowledge-based system for acquiring the conceptual structure of a diagnostic expert system," *Journal of Automated Reasoning*, vol. 1, pp. 49–74, 1985.

[13] M. LIQUIERE and J. QUINQUETON, "OPAL : object-oriented programming and learning," Rapport de Recherche 32, Centre de Recherche en Informatique de Montpellier, Montpellier, France, Janvier 1987.

[14] L. ESHELMAN and J. McDERMOTT, "MOLE : a knowledge acquisition tool that uses its head," in *Proceedings of the 5th National Conference on Artificial Intelligence*, (Philadelphia, PA), pp. 950–955, August 11-15 1986.

[15] K. MORIK, "Sloppy modelling," in *Knowledge Representation and Organisation in Machine Learning* (K. Morik, ed.), Springer Verlag, 1989.

[16] J. H. BOOSE and J. M. BRADSHAW, "Expertise transfer and complex problems: using aquinas as a knowledge acquisition workbench for knowledge-based systems," in *Knowledge acqusition for knowledge based systems (vol. 2)* (J. H. Boose and B. Gaines, eds.), pp. 39–64, Academic Press, 1988.

[17] E. BAREISS, B. PORTER, and C. WIER, "PROTOS : an examplar-based learning apprentice," in *Proceedings of 2nd AAAI Workshop on Knowledge Acquisition for Knowledge-Based systems*, (Banff, Canada), pp. 1–16, October 1987.

[18] G. COLL and al., "Simple application contexts: Xfrac, danae, vehicle image and tomato plant pathology domains," Report ACK-CX-T5.1-DL-002-A, ACKnowledge project, ESPRIT, June 1989.

[19] L. SOLORZANO and al., "Complex application context for kew-1," Report ACK-TID-T5.2-DL-001-A, ACKnowledge project, ESPRIT, February 1990.

[20] F. RAMPARANY, "KEW: a Knowledge Engineering Workbench that integrates cooperative Knowledge Acquisition Techniques," in *Proc. Third Intern. Symposium on AI*, (Monterrey, N.L. Mexico), October 22-26 1990.

[21] M. V. SOMEREN and al., "A classification of techniques for knowledge acquisition," Report ACK-UvA-T1.2-DL-004-C, ACKnowledge project, ESPRIT, august 1989.

# Analogical Transfer by Constraint Satisfaction

**Eric R. Melz** **Keith J. Holyoak**
Department of Psychology
University of California, Los Angeles, CA 90024

### Abstract*

The robustness of analogical transfer based on the ACME modeling of mapping by constraint satisfaction (Holyoak & Thagard, 1989) was investigated in a series of computational experiments using Hinton's (1986) "family tree" problem. Propositions were deleted randomly from the full representations of either both analogs (descriptions of an English and an Italian family) or just the target, and after mapping a "copy with substitutions" procedure was used to generate transfer propositions intended to restore the full representational structures. If as many as 50% of the propositions in the target analog were deleted, the system was able to recreate all of the missing information without error; significant recovery was obtained even if as many as 80% of the target propositions were deleted. Robustness was only slightly reduced when the two analogs lacked any similar predicates, so that mapping depended solely on structural constraints. Transfer was much more impaired when deletions were made from both analogs, rather than just the target. The results indicate that for isomorphic representations, analogical transfer by constraint satisfaction can exceed the regenerative capacity of general learning algorithms, such as back-propagation.

## Introduction

Theoretical accounts of analogical transfer distinguish various component processes, of which *mapping* is considered central. Mapping entails establishing systematic correspondences between the objects and relations of a source analog and a target analog. But in addition to mapping, analogical transfer depends on post-mapping processes in which new inferences are drawn using the correspondences established by the mapping, in conjunction with the structure of the analogs. Such procedures can effectively fill in gaps of missing information when knowledge from a well-understood domain is transferred to an analogous but less well-understood domain. By analogy to perceptual mechanisms for "filling in" missing information, we will refer to the generation of inferences based on analogical mapping as *analogical pattern completion*. The simplest basic mechanism for analogical pattern completion is *copying with substitution* (CWS), which in some form has been included in all computational models of post-mapping analogical transfer (e.g., Carbonell, 1983; Falkenhainer, Forbus & Gentner, 1989; Winston, 1980). The basic idea of CWS is simple: for a known proposition about the source, construct an inference about the target by substituting the corresponding predicate and argument(s) in the target for those in the source proposition.

Although several models of analogical transfer that include the CWS principle have been proposed, no systematic tests of robustness have been reported for models of either mapping or post-mapping transfer processes. Mapping models have typically been applied to representations that are nearly isomorphic; it is unclear how well the systems could map less orderly representations. Naturalistic use of analogy typically involves situations in which at least one of the analogs -- the novel target -- is imperfectly understood. A model of human analogical transfer must be sufficently robust as to be able to identify systematic correspondences between analogs despite gaps in the initial representations, and then proceed to generate plausible inferences to fill those gaps. In this paper we examine the inferential power of a CWS mechanism as an extension of ACME, an model of analogical mapping by constraint satisfaction (Holyoak & Thagard, 1989).

## Learning Family Trees by Back-Propagation

Hinton (1986) describes a back-propagation network which he trained on propositional representations of family trees. The primary purpose of Hinton's study was to determine if the hidden units of the network could develop intuitively meaningful representations of abstract features of a corpus of propositions. The basic family trees that the network learned, also used in the present study, are depicted in Figure 1. As is visually apparent, these English and Italian families have an isomorphic structure (e.g., Christopher enters into the same pattern of kinship relations as does Roberto). Using 12 common relational terms (father, mother, husband, wife, son, daughter, brother, sister, uncle, aunt, nephew and niece), each family can be described by a set of 56 propositions about relationships among the 12 individuals.

In Hinton's project, propositions of the form (person1 relation person2) (e.g., (Emilio has_father Roberto)) were translated into a localist connectionist representation and presented to a multi-layer network. Using back-propagation, the hidden units were able to abstract useful features, such as nationality and generation, from the training set. When tested on its ability to complete the 4% of the propositions that had not been used in training, the network was correct on 100% of these in one run and 75% in a second run.

* This research was supported by Contract MDA 903-89-K-0179 from the Army Research Institute, NSF Biological Facilities Award BBS 87-1420, and a grant from the Keck Foundation to the UCLA Cognitive Science Research Program.

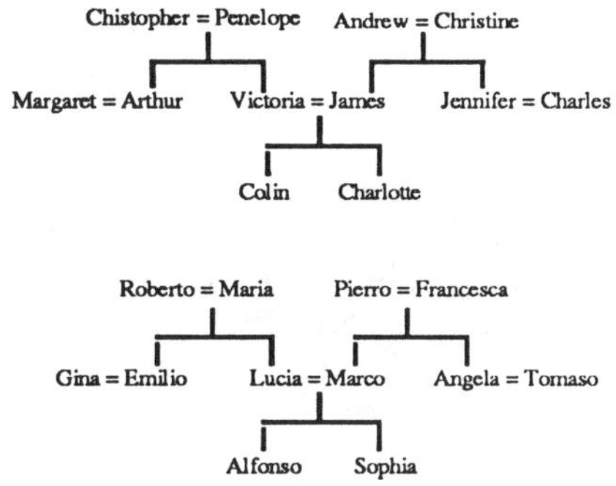


Figure 1: Two isomorphic family trees. The symbol "=" means "married to". (From Hinton, 1986)

## Analogical Mapping and Pattern Completion in ACME

### Mapping

Holyoak and Thagard (1989) frame the problem of analogical mapping in terms of parallel satisfaction of multiple constraints that jointly determine the optimal correspondences between elements of the source and target analogs. Their ACME (Analogical Constraint Mapping Engine) model receives as input a source analog and a target analog represented in predicate-calculus notation. Each analog consists of a set of propositions, where each proposition consists of an $n$-place predicate, a list of constituent arguments of the predicate, and a proposition label. For example, the fact that Emilio's father is Roberto would be represented by the proposition

*(has_father (Emilio Roberto) I1).*

The full representations of each family would consist of 56 propositions, each involving a 2-place relation.

ACME constructs a network of units representing potential elemental correspondences between the two analogs. The constraint that good analogies should tend to be isomorphic is enforced by installing excitatory links between consistent mapping hypotheses and inhibitory links between inconsistent hypotheses. Units representing mappings of semantically similar predicates are linked to a "semantic unit", the activation of which is clamped to a maximum activation value of 1. In the experiments reported in this paper, only mappings of identical predicates have connections to the semantic unit. Units representing correspondences presumed on the basis of prior knowledge are linked to a special "pragmatic unit". The network is then allowed to settle using a relaxation algorithm (see Holyoak & Thagard, 1989). The updating algorithm is implemented in *LISP on a CM2 Connection Machine.

### Pattern Completion

Once a set of mappings for objects, predicates, and propositions have been obtained by relaxing the system, we invoke a simple CWS procedure to generate candidate inferences based on the mappings and the structure of the analogs. The intuition behind this procedure is that in analogical transfer, there is a pressure for propositions in the source to have corresponding propositions in the target, and vice versa. If some proposition exists in one analog, but has no corresponding proposition in the the other, yet all of the constituent elements of the existing proposition map to elements in the other analog, we may reasonably conjecture a new proposition. Formally, the procedure we use to generate propositions in the aftermath of mapping is as follows:

- *If a proposition P consisting of relation r and objects* $o_1, o_2, \ldots, o_n$ *(notated* $P{:}r(o_1, o_2, \ldots, o_n)$*) exists but does not map to any corresponding proposition, and*
- *if P's relation and objects have the mappings*
  $r \to r'$
  $o_i \to o_i'$ *(for* $i=1 \ldots n$*)*
- *then create the new proposition* $P'{:}r'(o_1', o_2', \ldots, o_n')$.

The criteria for generateinga proposition are that (a) the best mapping of P must have an activation below some threshold value, and (b) the relation and objects of P must have activations above the threshold. For all the experiments reported in this paper, this threshold was chosen to be .2. Note that the above procedure for inference generation is inherently symmetrical: the new proposition can be added to either the source or the target analog.

## Transfer Tests

The family-tree problem has several virtues as the basis for computational tests of an analogical transfer model. First, the two complete family structures are in fact isomorphic, so analogical mapping should be possible. Second, the full representations are a well-specified set of propositions, so we can quantify the degree to which analogs have been corrupted by eliminating propositions from the input representations. By deleting propositions from the inputs, we can systematically reduce the degree to which the inputs are actually isomorphic, and hence examine the robustness of the mapping and pattern completion mechanisms.

### Mapping the Intact Family Trees

The first requirement was to demonstrate that ACME could in fact map the two analogs if the complete representations (i.e., 56 propositions for each family) were provided as inputs. This is a non-trivial computational problem simply because the mapping network formed is very large (3424 mapping units interconnected by 381224 symmetrical links), due to the fact that all propositions involve 2-place relations (so that any proposition in one analog could potentially map to any proposition in the other). The major parameter values used were .005 for decay, excitation, and similarity of identical predicates, and -.16 for inhibition. The network settled into a stable asymptotic state after 196 cycles of activation updating, producing a complete and correct set of correspondences between elements of the two structures. All the correct mappings had asymptotic activations close to the maximum possible value of 1.

### Reconstruction of Damaged Analogs With Identical Predicates

Having established that ACME can map the intact analogs, we next performed a series of computational experiments in which we randomly deleted propositions from the family-tree representations provided as inputs to ACME, mapped the damaged representations, and then applied the CWS pattern-completion mechanism to attempt to reconstruct the complete analogs. The results of a series of experiments on analog reconstruction are presented in Figures 2 and 4.

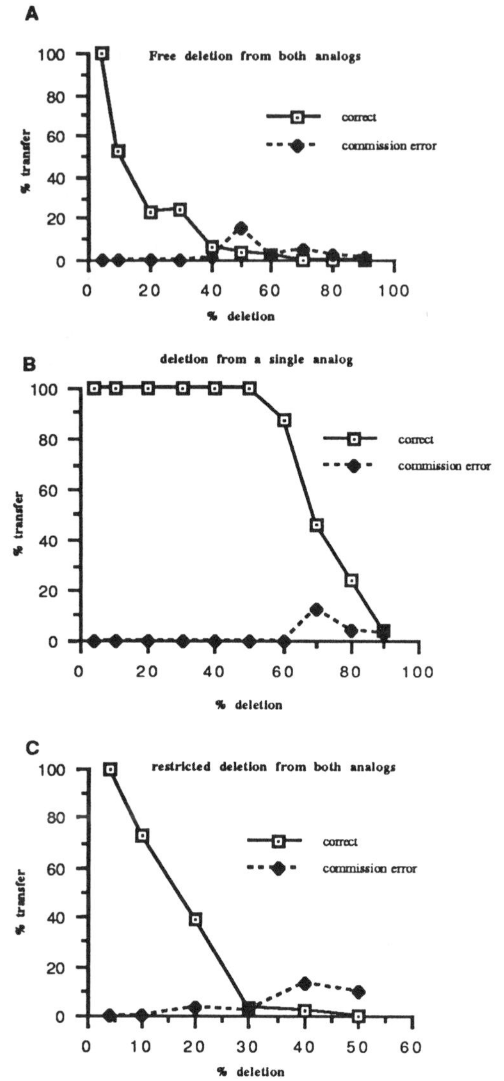


Figure 2: Robustness of transfer (identical predicates)

In our first experiment, we randomly deleted propositions from both analogs, and observed the proportion of deleted propositions that were correctly or incorrectly created by the CWS mechanism. The results of this experiment are shown in Figure 2A. The abscissa represents the proportion of propositions that were deleted from the entire set of propositions in the two analogs. The ordinate represents the proportion of the deleted propositions that were either created correctly (labeled "correct"), as well as the frequency with which incorrect propositions (labeled "commission error") were generated, also expressed as a percentage on the deletions in the inputs. Each data point on this and subsequent graphs represents the average of the results of two runs. ACME was able to reconstruct 100% of the missing propositions when 4% had been deleted, and 53% of those missing when 10% had been deleted, without making any commission errors. Correct restorations diminished to 23% when the deletion rate was increased to 20-30%, and at higher levels of deletion correct inferences were essentially eliminated. Commission errors were very infrequent even at the highest levels of damage to the analogs.

The above experiment involved symmetrical damage to the two analogs, with bidirectional transfer between the two analogs. In contrast, naturalistic analogical transfer typically involves asymmetric transfer from a well-understood source to a poorly-understood target. To more closely approximate the naturalistic asymmetry of analogical transfer, we ran a second experiment in which we restricted proposition deletions to only a single analog (in this case, it happened to be the Italian family). As the results in Figure 2B clearly indicate, transfer was far better than in the previous experiment. Full recovery of deleted propositions was possible at deletion rates of up to 50%, and even at a deletion rate of 80% ACME was able to recover 24% of the missing propositions.

The difference in robustness between the two deletion procedures is extreme indeed. For example, deleting 60% of the propositions in one analog produces the same quantity of missing information as does deleting 30% of the propositions across both analogs. Yet the former procedure yields almost perfect recovery (Figure 2B), whereas the latter procedure allows recovery of less than 1/4 of the missing information (Figure 2A). We next explored potential structural explanations for this difference in robustness as a function of whether deletions were made from one or two analogs.

One possible explanation is that when random deletions occur across both analogs, it is possible that corresponding propositions (e.g., E1 and I1) may both be deleted, in which case the CWS mechanism is guaranteed to fail (because there will be no proposition from which to generate an analogous inference). (The proportion of deleted propositions for which we can expect such an event to occur is $1 - (1 - p)^{d}$, where $p$ is the proportion of the total propositions that are deleted, and $d$ is the number of propositions that are deleted.) In contrast, if one analog is left intact, it is guaranteed that one member of each proposition pair is available (namely, the proposition in the source). To test the effect of this structural advantage for the latter procedure, we introduced a third deletion scheme that allowed propositions to be deleted from both analogs, with the restriction that at most one proposition from each pair of corresponding propositions could be deleted. The results, shown in Figure 2C, indicate that although this procedure produces somewhat more robust transfer than does free deletion from both analogs (Figure 2A), it remains much worse than when deletions are performed from only a single analog (Figure 2B). These results indicate that some other structural factor must account for the greater robustness of transfer when deletions are restricted to a single analog.

Another structural factor that varies when deletions are made from one versus two analogs involved differences in the potential for generating incorrect proposition mappings, as illustrated in Figure 3. In the case where deletion is restricted to a single structure (Figure 3a), for any proposition such as I5 in the target structure that is deleted, the strongest mapping of its corresponding source proposition, such as *E5=I10*, will be inhibited by the correct mapping unit, here *E10=I10*. Since *E10=I10* represents the mapping of corresponding propositions, it will have a high activation, and drive the mapping *E5=I10* well below the threshold required for generating transfer candidates. Hence, proposition E5 in the source will be left unmapped, making it a candidate for generation of an inference about the target by CWS.

In contrast, when proposition deletions occur in both analogs (Figure 3b), incorrect proposition mappings do not necessarily experience the devastating inhibition described above. For example, consider the case in which proposition E10 is deleted from the source and I5 is deleted from the target. If E10 and I5 share a common relation or object, they may produce a reasonably strong mapping. Since neither *E10=I10* nor *E5=I5* exists, the mapping *E5=I10* faces no serious competition, and thus is able to produce an activation level above the transfer threshold, preventing subsequent transfer from E5. The greater potential for erroneous proposition mappings to emerge when deletions are made from two analogs rather than one appears to be the main reason for the reduced robustness of transfer for the former case.

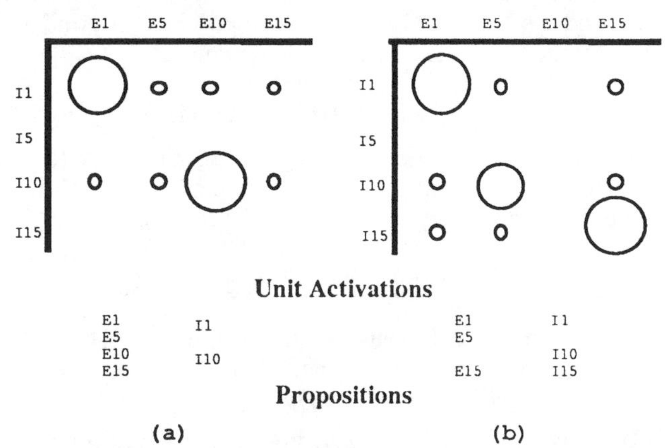


Figure 3: The units created for (a) single-analog deletion and (b) two-analog deletion are represented by cirlcles in the Unit Activation tables. Activation levels of the proposition mapping units *column=row* are indicated by the size of the circles. Mapping units on the diagonal, which represent correct mappings, strongly inhibit the units in their row and column. Hence in (a), the unit E5=I10 is strongly inhibited by E10=I10, so that E5 is left unmapped, and hence a candidate for generating an inference. In contrast, in (b) the unit E5=I10 does not encounter any such inhibition, and hence its activation is above the threshold required for inference candidacy.

## Reconstruction of Damaged Analogs Without Identical Predicates

The previous three experiments have shown that missing information can be reconstructed under a variety of deletion conditions. Such transfer is critically dependent on the quality of mappings ACME produces. As previously discussed, the mapping process is influenced by both structural and semantic pressures. Since in this problem we are mapping two isomorphic structures, the semantic pressure generated by the identical predicates in the two analogs (e.g., *has_father=has_father*) might be regarded as superfluous, or as unduly "setting up" the mapping.

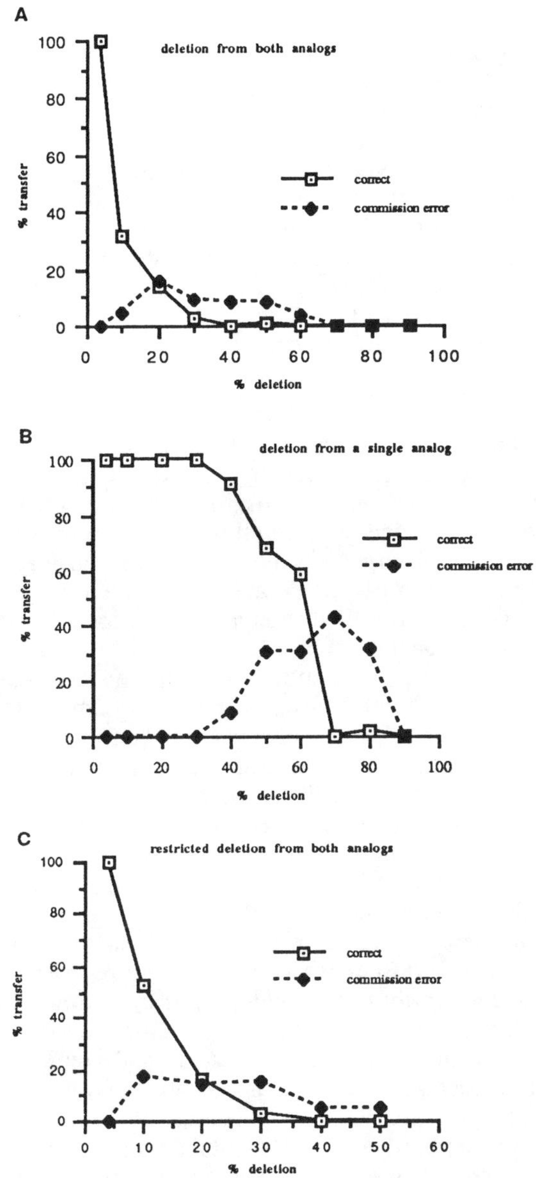


Figure 4: Robustness of transfer (no identical predicates).

In order to eliminate the semantic pressure, we translated all the relation names in the Italian family tree into their Italian equivalents. For example, the relation "has_father" became "ha_padre". Since ACME only creates a strong link between the semantic unit and a predicate mapping units if the mapped predicates have the same name, this modification prevents any useful semantic links from being created. In addition, since the English family tree can be consistently mapped to either the Italian family tree shown in Figure 1 or a vertical mirror image of the Italian family tree[1] (e.g., Christopher maps to Francesca, etc.), we weakly "presumed" one of the appropriate mappings (by giving it a weight of .005 on its link from the pragmatic unit) in order to provide a minimal impetus towards the "correct" set of mappings.

[1] We thank Ed Stabler for making this observation.

We next reran our three basic deletion experiments, eliminating semantic pressure. The results, displayed in Figure 4A-C, represent a qualitative replication of the previous patterns of transfer, with some degradation in the amount of transfer. This degradation is expected for two reasons. First, the overall activations of the proposition, relation and object mapping units are weaker when semantic pressures are removed. Second, clusters of mappings which are locally consistent but globally inconsistent can emerge under this scheme, especially when a significant amount of relational information has been removed from the representations.

## Conclusion

We found that transfer was far more robust when deletions were restricted to a single analog than if deletions were made in both. This result leads to the prediction that transferring knowledge from a well-understood source to a poorly-understood target will be easier than transferring knowledge between two moderately-understood analogs. As far as we know, this prediction has not yet been directly tested for human analogical transfer.

It would be interesting to test the robustness of other models of analogical transfer using the same basic materials used here. It does not appear likely, for example, that comparable results could be obtained using Falkenhainer et al.'s (1989) SME program with their structure-mapping rules. ACME is able to perform well even when the analogs lack any similar or identical predicates, whereas SME with structure-mapping rules requires that mapped relations be identical.

It is of interest to compare the performance of ACME with Hinton's (1986) learning by back-propagation as applied to the family-tree problem. Our system appears better able to recover implicit missing information for the family tree problem; however, obvious and significant differences exist between the two systems. On one hand, ACME requires that the propositions about the two families be explicitly separated into target and source analogs, whereas Hinton's system received all propositions intermixed, and in fact *learned* that the distinction between English and Italian people was an important regularity. On the other hand, Hinton's generalization task involved giving the system the first argument and relation and asking it to generate the second argument, whereas ACME was asked to generate entire new propositions without any explicit partial cues. In addition to these differences in the transfer tasks performed by the two systems, the general aims of each system are quite different. Hinton's system is primarily intended to abstract general features from a body of propositions. ACME, on the other hand, has no such generalization capability, but rather conjectures the existence of unstated information based solely on structural correspondences between two sets of propositions.

The present version of ACME has a number of significant limitations as a model of the post-mapping processes involved in analogical transfer. The success of the CWS procedure depends upon the model's tacit assumption that the source and target situations, despite any apparent "gaps" in their representations, are in fact isomorphic. If the source includes propositions that actually lack parallels in the target situation, commission errors (i.e., erroneous inferences about the target) are likely to result. Successful transfer between non-isomorphic analogs will depend upon additional pragmatic information that either prevents erroneous inferences from being generated, or weeds them out after they are generated. In addition, post-mapping transfer processes must also deal with cases in which some useful source propositions are based on predicates or objects that are initially left unmapped (in contrast to CWS, which can apply to a proposition only if all its elements have been mapped).

Finally, a full model of analogical transfer between non-isomorphic situations will require processes of *adaptation* (Novick & Holyoak, 1991) that modify the inferences generated by analogical pattern completion in order to take account of unique requirements of the target. For example, an analogical solution to a target problem may have to be adapted to incorporate extra constraints required in the target but not the source, or to accommodate variations in which problem elements are specified as unknown by the goal description (Carbonell, 1983). We view the present extension of ACME as only a modest initial step toward accounting for the full range of transfer processes that depend upon, but potentially go far beyond, the information provided by analogical mapping.

## Acknowledgements

We thank Ed Stabler for suggesting the idea for this project. Reprint requests may be directed to K. J. Holyoak, Dept. of Psychology, UCLA, Los Angeles 90024-1563.

## References

Falkenhainer, B., Forbus, K. D., & Gentner, D. (1989). The structure-mapping engine: Algorithm and examples. *Artificial Intelligence, 41*, 1-63.

Carbonell, J. G. (1983). Learning by analogy: Formulating and generalizing plans from past experience. In R. Michalski, J. G. Carbonell & T. M. Mitchell (Eds.), *Machine learning: An artificial intelligence approach.* Palo Alto, CA: Tioga Press.

Hinton, G. E. (1986). Learning distributed representations of concepts. In *Proceedings of the Eighth Annual Conference of the Cognitive Science Society*. Hillsdale, NJ: Erlbaum.

Holyoak, K. J., & Thagard, P. (1989). Analogical mapping by constraint satisfaction. *Cognitive Science, 13*, 295-355.

Novick, L. R., & Holyoak, K. J. (1991). Mathematical problem solving by analogy. *Journal of Experimental Psychology: Learning, Memory, and Cognition, 17*, 398-415.

Winston, P. H. (1980). Learning and reasoning by analogy. *Communications of the ACM, 23*, 689-703.

# A Constraint-Motivated Model of Concept Formation*

**Craig S. Miller** and **John E. Laird**
Artificial Intelligence Laboratory
The University of Michigan
1101 Beal Ave.
Ann Arbor, Michigan 48109-2110

## Abstract

A cognitive model for learning associations between words and objects is presented. We first list basic constraints to which the model must adhere. The constraints arise from two sources. First they stem from observed psychological phenomena including typicality effects, extension errors observed from children and belief-dependent behavior. Secondly they arise from our choice to integrate the model in a unified theory of cognition. In presenting the constraints to the model's construction, we motivate our design decisions while describing our algorithm that takes a symbolic, production-based approach. The model's adherence to the constraints is further supported by some empirical results.

## Introduction

In this paper, we present a model for concept acquisition similar to the learning tasks in systems such as ID3 [Quinlan, 1986] and COBWEB (when used to predict features) [Fisher, 1987]. However, our model has been subjected to further constraints in its role of a psychological model. In particular, we account for phenomena observed in children acquiring new words as well as the more general robust phenomena observed in conceptual acquisition. Furthermore, in order to ground our model in a broader, existing theory, we propose to implement the model in a unified theory of cognition.

In this paper, the task requires our model to predict object names of unlabeled objects based upon training examples that include the object name. For our task, we choose to represent object instances as lists of attribute-value pairs. For example, a particular instance of a ball may be represented as follows: [shape:spherical color:blue texture:smooth size:large]. While limiting instance descriptions to lists of attribute-value pairs admittedly restricts expressibility, the representation still permits specifications for an adequate assortment of objects and easily suffices for illustrating some of the model's interesting behaviors.

*A similar version of this paper appeared as [Miller and Laird, 1991].

## Design constraints

In this section, we present the major constraints that led us to the model's current design. The psychological motivations are discussed for each constraint. In the next section we will explain our model and how these constraints influenced its design.

**Constraint 1** *The word-learning model must be consistent with a unified theory of cognition.*

A unified theory of cognition is a proposed system of mechanisms capable of producing the full range of human cognition [Newell, 1990]. Ideally a model should not operate in isolation but rather play an integral and cooperative role with other cognitive activities. By conforming to a unified theory, the relationship of the various cognitive faculties becomes more apparent.

**Constraint 2** *Learning must be incremental.*

A child learning words does not have the opportunity to learn her entire vocabulary before applying it. Instead, learning and performance are interleaved. The learning of words is an ongoing, *incremental* process. For our purposes, we will consider a learning model to be incremental if the computational complexity of integrating a new instance into the system's knowledge is roughly constant relative to the number of training trials.

**Constraint 3** *The model must exhibit typicality effects.*

Typicality effects include: 1) typical instances are generally processed faster than less typical ones, 2) typical instances lead to fewer errors in category prediction and 3) typical instances are frequently given as an example to a category [Rosch, 1978]. In this paper, we limit ourselves to showing how typical instances are processed faster than less typical ones.

**Constraint 4** *The model must exhibit underextension errors in the early stages of learning.*

Underextensions result when the child's application of a label is too narrow in contrast to the full adult category. In regard to our model, an underextension occurs when no object name prediction is delivered for a particular object instance even though objects of the same category have been previously encountered. For example, a child may be able to identify that a softball is a 'ball' but fail to identify a football as being one. These underextension errors often occur during the early stages of lexical development [Dromi, 1987].

**Constraint 5** *The model must be sensitive to knowledge as determined by the system's beliefs, goals and theories.*

It is well documented that beliefs, goals and theories play an important role in human categorization [Murphy and Medin, 1985, Schank *et al.*, 1986]. Our model addresses this constraint by operating at the deliberate level, i.e. it can potentially bring to bear all pertinent knowledge in guiding its decisions.

## Constraining the design

In applying our first constraint, we choose to construct our model within the confines of Soar, a proposed unified theory of cognition [Newell, 1990]. Soar presents its theory in the form of an architecture—a system of mechanisms that applies knowledge, represented as productions, in creating intelligent behavior. We will further motivate our choice of Soar, but first let us focus on the new constraints that result from this choice. They include: 1) all learning occurs through an experience-based learning mechanism, 2) the system does not have direct access to its long-term memory and 3) processing is fully symbolic with no native support for frequency counts or probabilities. A basic understanding of the system and its assumptions follow within this section. In particular, a short review explains some of the basic ideas of how induction should proceed in Soar. We will then explain our model and how it coincides with the rest of our constraints.

### Learning in Soar

Problem solving in Soar [Laird *et al.*, 1987] consists of the sequential selection and application of operators to a representational state within a problem space. Both the selection and application of operators is determined by the knowledge represented in a long-term recognition memory encoded as productions. Learning involves the construction of new productions, called chunks, which summarize the results of a subgoal. A chunk in Soar is similar to an operationalized result in explanation based learning [Mitchell *et al.*, 1986, Dejong and Mooney, 1986]. Subgoals are born out of impasses caused by conflicting knowledge or a lack of knowledge. Subgoaling relies on additional knowledge to consider the alternatives, often requiring further search. During subgoaling, the system may try out the alternatives or it may recast the problem. In any case, the subgoal resolves the problem that caused the impasse. Furthermore, the architecture traces back through the conditions that led to the impasse's resolution. Using these conditions, a new production is created that summarizes the subgoal's result so that, in analogous applications, the summarizing information is retrieved. This avoids the impasse and thus the costly search encountered when the chunk was first created. In this scenario, the knowledge in the problem space has been partially operationalized in that an increase in efficiency has occurred. Although counter-intuitive, previous work with Soar has demonstrated that this technique can also increase the total knowledge of the system [Rosenbloom *et al.*, 1988]. This approach, called data-chunking, is a precurser to the work we present here.

### Describing the model

For our approach, we rely exclusively on chunking to monotonically add productions that predict a concept name given a partial description of an object. The definition of a concept is not localized to a single production, or even a small set of productions. Instead, a concept definition is distributed across the set of productions that predict the concept name, plus a process that creates abstractions of object descriptions, attempting to find an object description that can be matched by the productions. The search plays an integral role in defining the concept; for it is here that Soar brings to bear its knowledge in determining which abstractions should be considered and in what order. In the model presentation that follows, we will see how this fully symbolic learning process adheres to the constraints described in the previous section.

When attempting to predict the object's category, the object's description serves as the initial state in a problem space. For example, if we want to classify a blue, smooth, spherical object as a ball, the initial state would contain the following object description: [shape:spherical color:blue texture:smooth].

The operators of the problem space recognize descriptions and predict categories. Initially, the system will have very few operators and be able to predict only a few categories, but with experience, more operators will be learned. For our example, let's assume that the system already has acquired the following productions for creating word-prediction operators:

```
[spherical, red, smooth] -->
  create-operator(predict word:ball)
[spherical, red] -->
  create-operator(predict word:ball)
[spherical] -->
  create-operator(predict word:ball)
```

If one of these productions exactly matched the object description, it would fire, create an operator which would then be selected and predict a word and we would be done. However, for a production to apply, we insist that it matches every feature in the object description, so that a general production will only match an abstracted description.[1] This means that even the third production is not applicable for our initial object description because it has no feature match for color and texture.

If an object can not be classified by the existing knowledge, no operators are created and an impasse results. Within the ensuing subgoal, a search is performed to find

[1]Although Soar's production matching generally does not have this property, by representing object descriptions appropriately (i.e., as linked lists), the object recognition productions will have this property.

an abstraction that modifies the object description in such a way as to allow it to be categorized. For our example, the abstraction operators render an object description more general by removing one of the attribute-value pairs from the description. Let us assume the chosen operator removes the texture feature from the description, producing the description (spherical, blue). Following an abstraction, there is the potential that the new description can be categorized. However in this example, no match occurs, and another impasse and subgoal are created. Only after the color has been abstracted out, will the object description be recognized by the third production and a prediction for the word will be made. For performance trials, the correctness of the prediction will depend on the object recognition productions store in long-term memory, as well as the abstraction performed during the search for a match.

In this formulation, there is no explicit abstraction hierarchy; instead the abstractions are generated dynamically. This allows knowledge to be used to determine which abstractions should be made first. The ramifications of this approach are that novel objects will be classified into existing categories if it differs only in features that are abstracted early on.

For training trials, the purpose is not to predict a word, but instead to learn the association between an object description and a word. Consistent with earlier work in data chunking, our approach is to treat a training trial like a performance trial, so that the system will attempt to predict which word it already would associate with the given object. The prediction will be verified by comparing it against the given word before accepting it. If necessary, abstraction will continue until a very abstract production for predicting the word is found. When the correct prediction is finally made, the training trial is finished.

How has the training resulted in any learning? Recall that in Soar, learning happens when a result is produced in a subgoal. In this case, the result is the correct prediction created in the final subgoal where the object is correctly categorized. For a performance trial, when no word accompanies the description, there is no attempt at verification and no chunk is created. For training, learning occurs for the subgoal lying just above the subgoal where recognition was achieved and verified. There the object description [spherical, blue] eventually led, following abstraction, to the results created in the lower subgoal (predict word:ball). Chunking summarizes the abstraction and recognition processing in the subgoal, and builds a new production:

```
[spherical, blue] -->
  create-operator(predict word:ball)
```

In future training trials, this production will be available and allow a prediction to be made even earlier, and thus allow an even more specific production to be learned. As practice continues, more and more specific productions are acquired. So that, although the search for a category proceeds from a specific example through varying levels of abstraction, the learning proceeds from most abstract to more specific. Thus, this learns the same way as discrimination learning and previous data-chunking work in Soar [Rosenbloom *et al.*, 1988] (abstract to specific), but this scheme attempts to perform classification by moving from specific to general (through abstraction operators) while in a discrimination test it is from general to specific (through successive tests for the values of more and more features).

Before continuing with our model's description, let us address how the approach already fulfills some of our initial constraints. First of all, since training and performance runs can be interleaved, learning is incremental, thus fulfilling our second constraint. In a sense, the Soar architecture already enforces this by disallowing direct access to long-term memory, i.e. productions cannot directly access, delete or modify productions. This simply prohibits any action of massively reorganizing the long-term memory every time a new learning instance is encountered. The model itself has a surprising consequence in that learning time per instance can potentially decrease as new instances are acquired. This is a result of having each learning trial build upon chunks acquired on previous learning trials. The more instances the system has encountered, the less likely a new object description will have to be fully abstracted in order to be integrated with the current set of productions.

The typicality effects (Constraint 3) are also a result of how the model searches for the best and most specific match first and then slowly generalizes until a match is found. Here queries using frequent and typical objects will find a match that is not only a better match, but they will also find it quicker. This assumption stems from the notion that typical instances will already have several stored exemplars that nearly (if not completely) match them. Therefore little abstraction is required in order to find the match.

So far we have assumed that there was always some way of classifying an object correctly, even if it had not been seen before. Within our model, that translates into productions that do not test any features and predict the words. Thus, the system must be "bootstrapped" with productions of this form:

```
[] --> word:ball
[] --> word:book
[] --> word:apple
```

Where do these productions come from? They in turn must be learned from lower-level components such as phonemes. This recursion bottoms out in some primitive set of components that the system can inherently generate.

The consequences of starting with these very general productions could cause one to think that many overgeneralized predictions during performance would initially result, thereby contradicting our fourth constraint, that underextensions must initially result. Some overextensions will be produced for this reason, but surprisingly, the choice of these very generalized productions actually lead to many initial underextensions. During performance, these very general productions will often suggest conflicting predictions. Since the system has no way to choose which prediction is correct, *no prediction is made when-*

*ever there is a conflict in predictions.*[2] Because the productions are so general, initially there will be many abstractions in trying to categorize a word, and thus there will be many conflicts. Therefore, there will initially be many underextensions.

We have already discussed how our choice of Soar has led to an incremental learning model. Our choice has also contributed to fulfilling some of our other constraints. First of all, we argue that placing the task on Soar's deliberate processing level is the appropriate for modeling typicality effects. In several psychological experiments, it was shown that differences in response times due to differences of similarity varied on the order by several hundreds of milliseconds [Rosch *et al.*, 1976]. Results from other research and analysis of Soar has led us to conclude that the application of a primitive operator in Soar corresponds to a human time of approximately 50 milliseconds [Newell, 1990, Wiesmeyer and Laird, 1990]. Since difference in response times in our model vary by several operator applications, these results suggest that we are in the ball park and that our explanation of typicality effects on an *algorithmic* level is reasonable. This contrasts with other previous work that seeks a quantitative approach (i.e. typicality is represented in the form of frequency counts or probabilities) for explaining typicality effects. In taking these approaches, strong assumptions are required concerning the underlying implementation in order to explain similarity-dependent response times. For example, in COBWEB, category prediction involves traveling to a node that explicitly represents the instance's category. Typicality timing effects come only with the strong implementational assumption that "the time required to reach a node is inversely proportional to the total predictiveness towards that node." [Fisher, 1988].

Finally Soar provides us with the support of fulfilling our last constraint—goal and belief dependent behavior in categorization. In Soar, the selection of an operator is a deliberate act, i.e. all applicable knowledge can potentially influence the choice of an operator. This knowledge, expressed through productions preferring some operators over others, can also be learned through the result of subgoaling. Although this paper does not emphasize the learning of this knowledge, the issues presented here remain a central concern of our research work. Briefly we can say that we are concentrating on three kinds of knowledge: 1) default preferences that prefer abstracting out some features before others, 2) abstraction preferences learned from prediction failures and 3) abstraction preferences favoring features that lead to the system's goals.

## Empirical results

In this section, we present two experiments that confirm our predictions on extensional errors and typicality effects.

| Training Trials | Average time in decision cycles: Similarity groups $S <= 2.0$ | $2.0 < S <= 3.0$ | $3.0 < S$ |
|---|---|---|---|
| 1 | 23.0 | 18.2 | 17.0 |
| 2 | 23.0 | 12.2 | 9.8 |

Table 1: Response time according to similarity group

For both of these, we chose an assortment of objects that are simply described by attribute-value pairs, similar to the examples given so far. For all our data, we chose objects that roughly correspond with physical objects children are likely to encounter. The abstraction operators generalize the instance descriptions by removing a feature. Furthermore, we have included additional heuristics that guide the selection of these abstraction operators.

Our first experiment demonstrates the effect that similarity has on response time. A training set was presented twice to the system. It is a collection of object descriptions, all of them representing different kinds of balls. With each description, the appropriate French word for ball was given (either *ballon* or *balle*). In French, *ballon* is used to refer to balls that are inflated (basketballs, volleyballs, footballs, etc.), while *balle* refers to solid balls (baseballs, golf balls, tennis balls). Thus, the system must learn to distinguish between objects in the two categories even though the critical feature (inflatable or not) is not included in the input. Between the training trials, a set of object descriptions were run on the system. For each of the performance object descriptions, the similarity was mathematically calculated based upon the number of shared features the object description has with the training object descriptions of the same category. We have divided the performance descriptions into three groups of low, medium and high similarity and average their response times in terms of decision cycles. Table 1 shows that the instances with a lower similarity measure require a longer response time. This table also shows how learning times can improve as new knowledge is acquired. The performance is faster after the second training trial because more specific chunks have been acquired. This means less abstraction need occur in order to make a prediction.

In the second experiment, we trace extensional errors by testing performance on one category of objects. Here training instances drawn from several categories are given one at a time. Between each training instance, a set of testing instances, all from the same category, are run on the system. Figure 1 shows the results of the trace. Here we see an initial onslaught of underextensions and overextensions before correct predictions dominate. We consider the category to be *underextended* whenever the system fails to make a prediction. The category is *overshadowed* by another category whenever a wrong response is given for the category we are testing. The latter is a case of *overextension*, and we have included in our figure a trace of these errors for the sake of completeness. The par-

[2] Conflicting predictions do not arise during training because operators that suggest incorrect predictions are rejected. This is possible because the correct answer is provided.

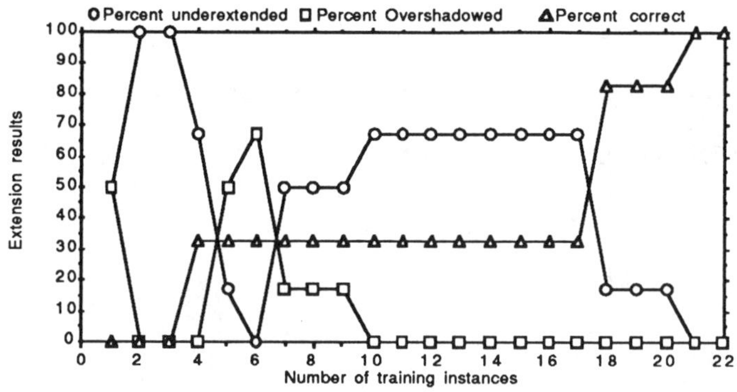


Figure 1: Trace of extensional errors

ticular conclusion that can be drawn from this figure is that several initial underextensions result from our model even though the system starts by learning more general productions first. As we have said before, this is the result of several general conflicting rules matching at the same time.

## Future work

The model we have presented here is a first pass at a system which conforms to our selected set of constraints. For future work, we intend to improve the model by expanding its capabilities while also having it adhere to more constraints. The expansion of capabilities include the learning of abstraction heuristics, extending learning to predict missing features, and increasing instance description expressibility. For adding more constraints, we are looking at covering additional typicality effects and basic-level effects.

**Acknowledgements** This work was supported by a grant from the W.K. Kellogg Foundation through The Presidential Initiative Fund of The University of Michigan. We would like to thank Melanie Mitchell, Steve Lytinen and Mark Wiesmeyer for their helpful comments on earlier drafts of this paper.

## References

[Dejong and Mooney, 1986] G. Dejong and R. Mooney. Explanation-based learning: An alternative view. *Machine Learning*, 1:145–176, 1986.

[Dromi, 1987] E. Dromi. *Early Lexical Development*. Cambridge University Press, New York, 1987.

[Fisher, 1987] D. H. Fisher. Knowledge acquisition via incremental conceptual clustering. *Machine Learning*, 2:139–172, 1987.

[Fisher, 1988] D. H. Fisher. A computational account of basic level and typicality effects. In *Proceedings of the Seventh National Conference on Artificial Intelligence*, pages 233–238, 1988.

[Laird *et al.*, 1987] J. E. Laird, A. Newell, and P. S. Rosenbloom. Soar: An architecture for general intelligence. *Artificial Intelligence*, 33:1–64, 1987.

[Miller and Laird, 1991] C. S. Miller and J. E. Laird. A constraint-motivated model of lexical acquisition. In L. Birnbaum and G. Collins, editors, *Machine Learning: Proceedings of the Eighth International Workshop*, San Mateo, CA, 1991. Morgan Kaufman.

[Mitchell *et al.*, 1986] T. M. Mitchell, R. M. Keller, and S. T. Kedar-Cabelli. Explanation-based generalization: A unifing view. *Machine Learning*, 1:47–80, 1986.

[Murphy and Medin, 1985] G. L. Murphy and D. L. Medin. The role of theories in conceptual coherence. *Psychological Review*, 92:289–316, 1985.

[Newell, 1990] A. Newell. *Unified Theories of Cognition*. Harvard University Press, Cambridge, MA, 1990.

[Quinlan, 1986] J. R. Quinlan. Induction of decision trees. *Machine Learning*, 1:81–106, 1986.

[Rosch *et al.*, 1976] E. Rosch, C. Simpson, and R. S. Miller. Structural bases of typicality effects. *Journal of Experimental Psychology: Human Perception and Performance*, 2:491–502, 1976.

[Rosch, 1978] E. Rosch. Principles of categorization. In E. Rosch and B. B. Lloyd, editors, *Cognition and Categorization*, pages 27–48. Erlbaum, Hillsdale, NJ, 1978.

[Rosenbloom *et al.*, 1988] P. S. Rosenbloom, J. E. Laird, and A. Newell. The chunking of skill and knowledge. In B. A. G. Elsendoorn and H. Bouma, editors, *Working Models of Human Perception*. Academic Press, London, 1988.

[Schank *et al.*, 1986] R. C. Schank, G. C. Collins, and L. E. Hunter. Transcending inductive category formation in learning. *Behavioral and Brain Sciences*, 9:639–686, 1986.

[Wiesmeyer and Laird, 1990] M. Wiesmeyer and J. Laird. A computer model of 2d visual attention. In *The Twelfth Annual Conference of the Cognitive Science Society*, pages 582–589, 1990.

# A Computational Basis for Brown's Results on Morpheme Order Acquisition

**Sheldon Nicholl** and **David C. Wilkins**
Dept. of Computer Science
University of Illinois
1304 W. Springfield Ave.
Urbana, IL 61801

## Abstract

This paper presents the result that a computer program can mimic the acquisition by children of a selected set of grammatical morphemes. Roger Brown [Brown, 1973] studied the acquisition of 14 morphemes, and showed how a set of partial order relations describes this aspect of child language learning. We show that these relations can be given a computational basis. They follow directly from a class of Boolean learning algorithms which have three simple constraints in the manner in which they consider hypotheses. I will call these three constraints the CAM constraints. CAM constraint 1 is to increase the length of the conjuncts one term at a time. The second CAM constraint is to consider all hypotheses of the same length simultaneously. Finally, CAM constraint 3 is to collect all single-term hypotheses involving noun features into a single conjunction prior to Boolean learning.

## Introduction

The problem of how language is acquired is still one of the core problems in understanding cognition. Contributions to an understanding of language acquisition have come not only from Psychology and Linguistics [Brown, 1973], [Pinker, 1984] but also from Computer Science [Gold, 1967] and artificial intelligence [Berwick, 1985]. Therefore we would expect further progress in the study of language acquisition to draw more and more from multiple fields in the cognitive sciences, rather than to draw from purely one area.

Brown's results [Brown, 1973] provide a fine example of this interdisciplinary transfer. In studying the language development of three children, Brown focussed his attention on the acquisition of 14 particular grammatical morphemes. Brown was able to write down the acquisition order of a significant subset of these grammatical morphemes in terms of partial orders. This still stands as a major result in the study of language acquisition.

The goal of this paper is to exhibit a computer model of some of Brown's acquisition order results (see Table 3). This computer model is based on a set of constraints for whose plausibility we hope to argue. I will call these constraints the CAM constraints. The CAM constraints are part of a larger model of language acquisition called CAM (Categories, Agreement, and Morphology) which has been implemented in Common Lisp.

| Morpheme | Meaning |
|---|---|
| Present Progressive | Temporary duration; |
| Plural | Number |
| Past Irregular | Earlierness |
| Third Person regular | Number; Earlierness |
| Uncontractible Auxiliary | Temporary duration; number; earlierness; (process-state) |

Table 1: Excerpt from Brown's Table 60, p. 369 (Brown, 1973)

## Brown's Results

Brown used the idea of a complexity ordering to lend an underlying paradigm to his order-of-acquisition data. Different grammatical constructions are ranked on a scale of increasing complexity; the more complex the construction, the longer it takes to learn. Brown considered two complexity orderings, one based on syntax, the other based on semantics. The syntax-based ordering is stated in terms of a theory which is seriously out of date (Chomsky's Standard Theory [Chomsky, 1965]) and also depends on transformations, which CAM does not represent; this ordering will therefore not be further considered here.

What is of interest here is Brown's semantic complexity ordering. It is appropriate to begin with a description of the actual semantics that form the basis of the ordering.

Table 1 shows the meanings of the plural and past morphemes.[1] Note that although Table 1 refers to both

[1] As Brown points out, the subjunctive use of the past,

regular and irregular forms, this distinction is not of consequence here, since both forms encode the same meaning. This is true of the past forms as well as the third person forms. Now the third person forms refer to both Number and Earlierness, since this morpheme can only be used in singular, present tense contexts.[2] Finally, the auxiliary refers to three features: temporary duration,[3] Number, and Earlierness. The temporary duration feature arises from the correlation between the auxiliaries with the progressive, at least in the data Brown considered:

| I | am | walk | -ing |
|---|---|---|---|
| 1S | AUX | V | progressive |

The other features reflect the particular choice of the auxiliary: *is*, *are*, *was*, and *were* each make different commitments to Number and Tense; hence the need for the features.

Now that the meanings of the morphemes have been explained, Brown's partial orders based on those meanings can be described. Here is Brown's actual listing of the partial orders, redrawn slightly to fit in this column:

$$\left.\begin{array}{l}\text{plural}(x)\\ \text{past irregular}(y)\end{array}\right\} < \begin{array}{l}\text{Uncont. copula}(x+y)\\ \text{3rd-person reg.}(x+y)\end{array}$$

$$\left.\begin{array}{l}\text{Uncont. copula}(x+y)\\ \text{3rd-person reg.}(x+y)\end{array}\right\} < \text{Un. Aux}(x+y+z)$$

$$\text{Progressive}(z) < \text{Uncontractible Aux}(x+y+z)$$

Table 2: Brown's Table 61: "A partial ordering in terms of cumulative semantic complexity."

In a formal sense, the less-than symbol '<' means "less complex than". So for example, plural($x$) < third-person regular($x+y$), a relation implied by Table 2, means that the plural, which depends on only one variable, $x$ (number), is strictly less complex than the third person regular, which depends on two variables, $x$ and $y$, where $x$ is number and $y$ is earlierness. The plus sign '+' just means "in some combination with"; it has no relation to arithmetic addition.

Fortunately, the purely formal, "less complex than" sense of '<' corresponds to an important real-world sense of '<': "acquired before". This means, for instance, that noun plural forms like *boys* are acquired before third-person regular forms like *John walks.* Now "acquired" in Brown's usage refers to the first speech sample that is "the first speech sample of three, such that in all three the inflection is supplied in at least 90 percent of the contexts in which it is clearly required." (Brown quoting Cazden, p. 258).[4] For most of the children a speech sample consisted of two hours of taped speech every second week (Brown, p. 52). The predictions of Table 5 are largely confirmed (Brown, p. 371) according to the 90% criterion. The goal of this paper is to provide a computational model of these results. The specific results addressed by this paper are the subject of the next section.

which is used to refer to unactualized events, need not be considered here, since the children learn this meaning of the past long after the morphemes of interest here have already been learned.

[2]Brown has apparently suppressed Person from his Tables.

[3]I am suppressing "process-state" from my exposition because there are too many exceptions to the notion that the progressive somehow encodes the process-state distinction.

## CAM

CAM is a semantics-based program in that it gets not only a linguistic string as input, but a representation of the string's meaning as well. In the interest of greatest generality, the representational power of the meaning formalism has been restricted as much as seems possible. Only words referring to clearly perceptible physical objects, events involving physical objects, or states involving physical objects are labelled as such in the meaning representation; all other words in the string regardless of their syntactic category are given a null lexical meaning. CAM's learning is interesting because it can still form a grammar despite the impoverished nature of its input.

As mentioned before, CAM is a language acquisition program concentrating primarily on Categories, Agreement, and Morphology. CAM has separate modules for Category learning, e.g., see [Nicholl and Wilkins, 1990], CFG rule learning, affix order learning, and agreement rule learning, among others. It is the agreement rule learner that is of primary interest in this paper, since the agreement rule learner is implemented with a simple bottom-up nominal-feature Boolean learning algorithm that obeys the three CAM constraints.

### The Set of Morphemes Studied Here

For a variety of reasons I will explain shortly, CAM cannot at present acquire all 14 of Brown's morphemes. Fortunately, the CAM constraints are not affected by any of these considerations, so it is entirely possible that with appropriate refinements in other parts of CAM, all 14 morphemes may someday be covered.

A quick examination of Table 1 shows that Brown makes a distinction between the contractible and uncontractible forms of a morpheme. For example, *you're walking* is a contracted version of *you are walking*, but *are you walking?* allows no contraction, and is hence uncontractible. Although syntactic factors do have a bearing on contractibility, it is also clear that phonological factors are very important. CAM makes the important but certainly not unprecedented distinction between syntax and phonology and hence has no

[4]This is the "90% criterion" which I will refer to later.

phonological component as yet. The fact that CAM can still duplicate some (but not all) of Brown's major results, i.e., Table 3 (but not Table 2), *is* unprecedented, as far as I know, and is the major result of this paper. The clear implication is that syntax and semantics are the principal (but not total) determinants of acquisition order in this domain.

Similarly, Table 1 makes a distinction between regular and irregular forms. This distinction is largely lexical. Since CAM focusses on the learning of general rules, irregular forms are neglected. This is borne out in the data to some extent, since the learning of irregular forms is different from that of regular forms. Brown also distinguishes between copulas and auxiliaries. CAM has no provision for copulas as such, due to the semantic indistinguishability of stative verbs and adjectives: both are states; both are predicates. But since CAM can still learn the full agreement rules for the auxiliary versions of all the copulas, this may not be much of a problem, especially if the acquisition of copulas is regarded as "parasitic" on the acquisition of auxiliaries. Possessives are excluded from the present implementation of CAM due again to the difficulty of learning states. For an enlightening discussion of the ambiguous representation of states in natural language, see [Lyons, 1977]. Finally, the prepositions *in* and *on*, like all other prepositions, fail the Relevance Property [Nicholl and Wilkins, 1990] for frequently grammaticized features, and are therefore deliberately ignored by CAM.

$$\left.\begin{array}{l}\text{plural}(x)\\ \text{past regular}(y)\end{array}\right\} < \text{ Third-person regular}(x+y)$$

$$\text{3rd-person reg.}(x+y) < \text{Uncont. Aux}(x+y+z)$$

$$\text{Progressive}(z) < \text{Uncontractible Aux}(x+y+z)$$

Table 3: Order of child language acquisition experimentally observed by Brown, a subset derived from his table 61, p. 370 (also Table 2, this article). The '<' sign is Brown's notation for "acquired before." The relations here in Table 3 have been computationally duplicated by using the CAM constraints. For examples, see Table 5.

Of Brown's original 14 morphemes, this leaves the following morphemes learnable by CAM: the progressive, plural, past, third-person regular, and auxiliary verb morphemes. Fortunately, these are almost exactly the same morphemes covered by Brown's acquisition order results (Table 2) so little is lost.

The forms of the morphemes learnable by CAM are as follows. For the progressive, *-ing* is the form learned. For the plural, *-s* is the allowable form because irregular plurals are excluded. The same is true for the past: *-ed* is learned, while the irregular past forms are neglected here. As one might expect, *-s* is learned for the third-person regular. The case of auxiliary verbs is more interesting. CAM learns *is*, *are*, *was*, *were*, and *am*, but not *be*.[5] Although *been* and certain forms of *have* are also learnable by CAM, they won't be addressed further here since they are irrelevant to the subject of this paper.

Table 3 therefore summarizes the target orderings CAM should be able to learn.

## Input

The relevant excerpt of a typical input to CAM is the following:

```
((the) (man)          (is) (walk ing))
(()    (3rd-person ()   (present
         singular)        progressive))
```

The first line shows the linguistic string, which is segmented into words. If a word contains inflectional morphemes, these are segmented from their stems.

The second line of the input shows the features of the perceptually salient words in the sentence. These features are assumed to be perceptually recoverable: a vision system capable of the object recognition required for this task could presumably recover these additional features with little difficulty.

The input file that generated Table 5 consists of a sequence of 24 inputs as just described. While this input format may seem a bit simplistic, it is sufficient for the acquisition of most of the subject-verb agreement rules in English. Also, CAM's inputs needn't be watered down in order to avoid the occurrence of a construction that might interfere with its ability to construct a grammar. CAM is capable of filtering out inputs too complex to handle. The ability to filter out irrelevant features, such as those encoding animacy or shape in English for example, has not yet been implemented in CAM, but there is no theoretical problem with using the Relevance Property [Nicholl and Wilkins, 1990] to do so, since no morpheme that falls under the scope of the Relevance Property is relevant for animacy or shape in English. The speed of the Relevance Algorithm used to implement the Relevance Property is such that many such irrelevant features can also be filtered out.

## Procedure

CAM is actually a system of several learning procedures each devoted to learning a different aspect of language. Before agreement learning proceeds, categories are formed and CFG rules are built: this is why the whole CAM system is necessary; a Boolean learner cannot handle syntax independently. For example, the formation of one closed category (AUX) is discussed in

[5] This is simply because modals have not yet been given to CAM in experimental trials; CAM has no theoretical learnability problems with modals.

1. Increase the length of the conjuncts one term at a time.
2. Consider all hypotheses of the same length simultaneously.
3. Collect all single-term hypotheses involving noun features into a single conjunction prior to Boolean learning.

Table 4: Boolean learner constraints. These constraints are sufficient to enable a Boolean learner to duplicate Brown's complexity orderings. See text.

[Nicholl and Wilkins, 1990]. Agreement rules are constructed after an input like that above in section 3.2 is fully parsed by previously learned CFG rules.

Once an input can be unambiguously parsed by previously acquired CFG rules, the features in line 2 of the input (see section 3.2) are percolated (or projected) to the maximal category of the word they correspond to. CAM uses a traditional grammar formalism, so the features `(3rd-person singular)`, which correspond to `man`, get propagated all the way to NP. A different grammatical theory might propagate to $N^{max}$ instead.[6]

Once the features are propagated, they are read off the maximal categories in the sentences along with their syntactic positions and collected together into a set. Then, the grammatical morphemes already learned and recognized by CAM are collected into another set. So in the case of the example input in the previous section, the feature set would look like the following, although with more precision as to the exact syntactic position of the elements in question:

```
(NP 3rd-person)
(NP singular)
(V present)
(V progressive)
```

According to CAM constraint 3, the noun-related terms are collected together into one conjunction, shown below. This not only leads to faster acquisition, it also leads to more accurate results than otherwise, since if this is not done, a "race condition" can develop between noun-related terms and verb-related terms, occasionally leading to acquisition orders that violate Brown's orderings.

```
(and (NP 3rd-person) (NP singular))
(V present)
(V progressive)
```

[6]'NP' is of course a more traditional representation for 'Noun Phrase', while '$N^{max}$' is from X-bar theory. My intention here is to keep the exposition as general as possible, and avoid making a commitment to any specific syntactic theory so workers in different theories can still make use of what is presented here.

The set of grammatical morphemes would look like this:

```
verb-ending = -ing
AUX = is
```

The term verb-ending is just for exposition; in actuality an internal syntactic representation is used instead. Same for AUX. I shall avoid more precision in the interest of greatest generality.

## Results

```
----------------------
(= (S 4 V 3 R 1) ING)
(= (S 2 V 2 R 1) ED)
(= (NP 3 N 2 R 1) S)
(= (S 4 AUX 2) (AM))
(= (S 2 V 2 R 1) S)
(= (S 4 AUX 2) (IS))
(= (S 4 AUX 2) (WAS))
(= (S 4 AUX 2) (ARE))
(= (S 4 AUX 2) (WERE))
----------------------

----------------------
(= (S 2 V 2 R 1) ED)
(= (S 4 V 3 R 1) ING)
(= (NP 3 N 2 R 1) S)
(= (S 4 AUX 2) (IS))
(= (S 4 AUX 2) (AM))
(= (S 2 V 2 R 1) S)
(= (S 4 AUX 2) (WAS))
(= (S 4 AUX 2) (ARE))
(= (S 4 AUX 2) (WERE))
----------------------
```

Table 5: Edited output of CAM showing the order in which it acquires morphemes.

One can see clearly that this algorithm will duplicate Brown's acquisition order formalism, since his results show that children learn shorter conjunctions before longer ones. For example,

$$\text{3rd-person reg.}(x+y) < \text{Uncont. Aux}(x+y+z)$$

is not only (1) one of Brown's acquisition order results obtained from observations of children (Table 3), but (2) the order resulting from the CAM constraints above (see Table 5). All the other results in Table 3 follow from the CAM constraints in the same way.

Now the actual runs of CAM shown in Table 5 manifest the CAM constraints as well as Brown's acquisition order results (Table 3). This is what we wanted to demonstrate.

## Discussion and Future Work

This paper has shown how the three CAM constraints are sufficient to duplicate Brown's complexity orderings when applied to a Boolean learner embedded in a larger language acquisition program. We suspect that prior work in formal language acquisition, e.g., [Anderson, 1983], [Pinker, 1984], cannot duplicate Brown's results because these systems do not obey CAM constraints 2 and 3. Both systems examine one feature at a time in sequence; if this sequence happens to be incorrect, their acquisition orders will deviate from Brown's results. CAM constraint 2 prevents this from happening.

```
verb-ending: ING
     (V PROGRESSIVE)

verb-ending: ED
     (V PAST)

verb-ending: S
     (((AND (NP 3RD-PERSON)
            (NP SINGULAR)
            (V PRESENT))

AUX: AM
     (AND (NP 1ST-PERSON)
          (NP SINGULAR)
          (V PRESENT))

AUX: WERE
     (((OR
         (AND (NP 3RD-PERSON)
              (NP PLURAL)
              (V PAST))
         (AND (NP 2ND-PERSON)
              (NP PLURAL)
              (V PAST))
         (AND (NP 2ND-PERSON)
              (NP SINGULAR)
              (V PAST))
         (AND (NP 1ST-PERSON)
              (NP PLURAL)
              (V PAST)))
```

Table 6: Some of the morpheme rules learned by CAM. Lightly edited.

The CAM constraints are not *logically necessary* constraints because there are potentially many other constraints (or algorithms) that could still duplicate Brown's results. One possibility would be to simply to replace CAM constraints 1 and 3 with: "add exactly one verb feature at a time." This might work, since in all of Brown's orderings (see Table 3), the only differences are due to the presence or absence of (1) the past feature and (2) the progressivity feature.

The most obvious direction for future work would be to extend these results to all 14 of Brown's morphemes. The irregulars are a clear target, in that it might be possible to extend the input to the Boolean learner to include lexical forms. Possessives might come next, especially if CAM's semantic input is extended: presently CAM has no way of representing possession or any other adjectival property. Finally, the copulas and prepositions might be mastered, but only after cross-linguistic research has identified the various syntactic realizations of states, and whether there might be tests for these.

## References


[Anderson, 1983] John R. Anderson. *The Architecture of Cognition.* Cambridge, MA: Harvard, 1983.

[Berwick, 1985] Robert C. Berwick. *The Acquisition of Syntactic Knowledge.* Cambridge, MA: MIT Press, 1985.

[Brown, 1973] Roger Brown. *A First Language: The Early Stages.* Cambridge, MA: Harvard University Press, 1973.

[Chomsky, 1965] Noam Chomsky. *Aspects of the Theory of Syntax.* Cambridge, MA: MIT Press, 1965.

[Gold, 1967] E. Gold. Language identification in the limit. *Information and Control,* 16:447–474, 1967.

[Lyons, 1977] John Lyons. *Semantics,* volume 2. Cambridge: Cambridge University Press, 1977.

[Nicholl and Wilkins, 1990] S. Nicholl and D. Wilkins. Efficient learning of language categories: The closed-category relevance principle and auxiliary verbs. In *The Twelfth Annual Conference of the Cognitive Science Society,* 1990.

[Pinker, 1984] Steven Pinker. *Language Learnability and Language Development.* Cambridge, MA: Harvard, 1984.

# An Alternative To Deduction

Daniel Oblinger
Beckman Institute, University of Illinois, 405 N. Mathews, Urbana, IL 61801
Email: oblinger@cs.uiuc.edu Phone: (217) 244–1503

Gerald DeJong
Beckman Institute, University of Illinois, 405 N. Mathews, Urbana, IL 61801
Email: dejong@cs.uiuc.edu Phone: (217) 333–0491

## Abstract

Deductive representations are well defined, easily inspected, and precise. However, they are also brittle, inflexible and difficult to debug. We propose a *plausible* representation whose inference mechanism is weaker than its deductive counterpart. This will allow it to reason with knowledge which is less precise, and replaces the notion of global consistency, with the weaker constraint of local consistency in its explanations[1].

## 1. PI–EBL approach

There are many advantages to a reasoning model based on a declarative representation of world knowledge. Such representations may be leveraged by many knowledge intensive mechanisms: it can be communicated to other agents, multiple models can be combined, parts of a model can be isolated and independently verified, etc. Because deduction is well understood it is natural to ground these models in a deductive framework. Requiring all knowledge to be expressed in a deductive form is restrictive, and causes systems to be very brittle. Connectionist approaches, on the other hand, are not based on such a restricted language, which is a source of their flexibility [Rumelhart86]. The symbolic/connectionist distinction may appear unimportant since the sources of knowledge potentially available are the same. However, research in connectionism is impeded by its representation—its very difficult to understand what abstract principles are revealed by a connectionist solution to a problem. The advantages of both approaches can be achieved by a reasoning paradigm based on an explicit world model (like symbolic approaches) using a non–deductive representation (like connectionist approaches). Research is facilitated by such a paradigm, high–level knowledge intensive mechanisms can be more easily constructed, and the principles underlying their success are much more available for inspection. In this paper we introduce a logic of plausible reasoning with these desired properties. As in distributed connectionist systems, our approach crucially involves learning through observations of the world..

The Explanation Based Learning paradigm is extended to support learning over our logic of plausible reasoning. EBL is a learning mechanism which combines a model of the world with examples to construct explanations which can be later used to reason about the world [DeJong86, Mitchell86]. Like other knowledge intensive mechanisms, it gains its strength from its explicit world model. Much of EBL's potential as a methodology has not been tapped, however, because the knowledge representation it uses is too restrictive. For example, EBL's restriction to symbol level learning is due to its base representation, *not* an inherent restriction on the EBL paradigm. (Symbol level learning adds no new knowledge but simply makes some knowledge easier to derive [Dietterich86].) EBL's limited use of its training examples is also traceable to its base representation. The flexibility of any architecture depends, in large part, on the representation language it uses, no matter what high level task is being performed. We present a plausible representation and inferencing mechanism as an alternative to classical deduction. The flexibility of this representation is demonstrated by using the Plausible Inferencer as the explanative component for an Explanation Based Learner (PI–EBL). Through its flexible representation PI–EBL overcomes EBL's deficiencies mentioned above, and acquires some advantages associated with both connectionism and traditional induction by greater use of its training examples.

## 2. Motivation Of Plausible Theories

Deductive frameworks are not sufficient as a knowledge representation mechanism. Their deficiency stems from their precision. Knowledge placed in a deductive framework must be valid—true for all possible examples. This is a problem since much of the knowledge a reasoner must deal with is true for most but not *all* examples; such knowledge cannot be used in a deductive framework. The impossibility of representing real world knowledge in a deductive theory is noted by McCarthy as the qualification problem [Genesereth87, McCarthy69]. The qualification problem states that any universally quantified implication will need a large (infinite) number of preconditions to exclude *all* possible exceptions. The reader may be convinced by considering the implication: $Bird(x) \rightarrow Fly(x)$ . Correcting this rule to handle exceptions would require adding $Alive(x)$ , $\neg Penguin(x)$ , $Sane(x)$ , etc. as conjuncts. Of

1. This research was supported by the National Science Foundation under grant NSF–IRI–87–19766

course such a deductive theory is not practical. The traditional approach to this problem is to translate the original domain into a micro–world which is consistent with the original domain to a fixed level of detail. Because the micro–world does not reflect all details of the real world, it is finitely describable using a deductive theory, thus a deductive theory may be employed. The disadvantage to this approach is that the model is really a model of the micro–world *not* a model of the original domain. The reasoner using such a theory has no recourse if some important portion of the domain is not modeled in the micro–world. Thus, the micro–world must be complex enough to handle the most detailed example encountered by the reasoner. This forces the entire system to represent and reason using this detailed and complex theory for *all* examples, even those which can be handled using a much simpler theory. Adapting the theory to handle one additional special–case example can easily result in many previously tractable problems to becoming inaccessible. Furthermore, simple theories which make different assumptions (based on different micro–worlds) cannot be directly combined because the resulting theory would likely be internally inconsistent. This is a stifling restriction since different micro–worlds (consistent micro–theories) will be useful in reasoning about different problems. Combining inconsistent micro–theories into a *single* consistent micro–world will result in a theory which is too precise, it is too precise because the closure of these micro–theories contain many facts inconsistent with other micro–theories, and the combined theory must be consistent with only *one* of these choices. Because the combined theory is deductive it must specify *exactly* how its component theories interact, even when the interactions are uninteresting parts of the deductive closure. Using a micro–world forces the theory to make predictions (by way of its deductive closure) about hypothetical situations, most of which are not relevant to the goals of the reasoner. No deductive chain of inference no matter how complex may be inconsistent with any other over the entire theory. Requiring the theory to be internally consistent without gaining consistency with the world gains us little power at the great expense of scalablity. A reasoner using deductive knowledge cannot accept *any* knowledge about the world until all the interactions with the *closure* of its current knowledge are precisely specified. This restriction is unacceptable. Avoiding this restriction and its fixed level of detail forces us to abandon the micro–world approach.

Non–monotonic theories provide a mechanism for reasoning with incomplete knowledge [McCarthy86, Reiter80]. McCarthy's *Abnormal*($x$) predicates allows one set of rules which derives a conclusion to override another rule. Like deductive theories, however, the sentences entailed by a nonmotonic theory are still precisely defined without any examples needed to disambiguate inconsistencies. The whole notion of circumscription makes precise the closure of a non–monotonic theory. In order for a representation to be an effective alternative to the micro–world approach, it must incorporate a means of trading the complexity/precision of a theory against its faithfulness to the world. This means the reasoner cannot close knowledge base (it cannot assume it has complete knowledge of the domain). Instead the world must be left open. Inferred knowledge should be used as *suggestions* to be empirically verified, rather than theorems with are necessarily true.

## 3. Representation Of Plausible Theories

Plausible domain theories are syntactically similar to a set of implications. Semantically they are quite different, however, since the theories are not deductive. The theories must *suggest* possible relationships between concepts without *entailing* them. Semantically the plausible domain theory may be interpreted as deductive implications with missing preconditions. The missing preconditions are collectively called the *implicit context* of the implication. The implicit context is the set of additional constraints sufficient to guarantee that the plausible implication deductively holds. Thus we refer to plausible implications as *influents* to emphasize the idea that the consequent is influenced, but not entailed, by the preconditions. Asserting an influent specifies that its preconditions are relevant to determining its consequent in some contexts. An influent also specifies the direction of influence between the preconditions and consequent, and they may be positively or negatively related. These influents provide the basis for representing PI–EBL domain theories.

Influent: $P \succ Q$

Meaning: $I \wedge P \rightarrow Q$

($I$ ↖ Implicit Context)

Figure 1: Semantics of an Influent.

## 4. Plausible Inferencing

To build explanations from a plausible domain theory (a set of influents) we need an inference mechanism. Influents may be combined with other influents in the same way deductive implications are chained together. Unifying a precondition of one rule with the consequent of another provides the first influent inference mechanism. Chaining influents in this fashion may result in explanations with important preconditions not explicitly represented since each influent has an implicit context which also must be satisfied. Other chainings which derive this conclusion or its negation will also have missing preconditions based on the implicit contexts of its influents. The missing preconditions of the other explanations are likely to be very different, thus together these chainings could provide an explanation which is predictive over a greater range of examples. This is the motivation for the second inference mechanism, the join, which allows combining several influents with the same consequent (or its negation) into an *influent set*. The influent set contains no information regarding the applicability of its component influent chains but it does guarantee that there is some combination of these chains which correctly predicts the conclusion of the influent set over the relevant problem distribution. Unlike the chaining inference mechanism, the join mechanism only partially specifies the function mapping preconditions to conclusions. *Plausible explanations* are generated by repeated application of these two inference mechanisms. Because of the influent sets within a plausible explanation it cannot be used to directly make predictions about its conclusion. The meaning of a plausible explanation is analogous to the semantics of Stuart Russell's determinations [Grosof89], both serve as a specification of the set of determining factors for some value in the domain. A plausible explanation specifies that its leaf preconditions determine its consequent, without completely specifying the mapping. Unlike Russell's determinations, however, there is no guarantee that a plausible explanation actually *does* determine its consequent because nothing is known about its implicit contexts. So a plausible explanation must be empirically verified by checking that it is consistent with examples from the domain before it can be accepted as a plausible theorem. Many different explanations can be conjectured for the same observed behavior. Ideally, the most plausible explanations are generated first. If these are empirically rejected (ie. the observed data contradicts them) the next most plausible explanation is hypothesized and so on until some hypothesis is empirically confirmed. Generation of plausible *theorems* is more complex than its deductive counterpart since is combines both analytical and empirical constraints.

In order to inductively learn a plausible theorem our reasoner must accept and reject potential explanations based on examples. This is challenging since the influents that makeup the explanations can be neither confirmed nor rejected based on empirical evidence since we can never know if their implicit context is satisfied. As yet we have no restriction that a plausible explanation accurately predicts the world, it is this restriction which makes plausible explanations testable. Thus we define a plausible theorem to be a plausible explanation which contains influent sets which satisfies the three properties shown in figure 2 relating it to the set of observations. As shown in the figure an

Let $O$ be the set of observations.
Let $A$ and $D$ be the preconditions of the affirming and denying influents in the influent set I.
Let c be the conclusion of the influent set I.
Let $P \equiv A \cup D$.
Sat(predicate, observation) is true iff the predicate was satisfied in the particular observation.

**Def:** I is *deterministic* on O iff

$$\forall o_i, o_j \in \mathbf{O} \ \left(\forall p \in P \ \ Sat(p, o_i) = Sat(p, o_j)\right) \rightarrow \left(Sat(c, o_i) = Sat(c, o_j)\right)$$

**Def:** I is *predictive* on O iff

$$\neg \exists o_i, o_j \in \mathbf{O}; A' \subseteq A$$
$$\left(\forall p \in P \backslash A' \ \ Sat(p, o_i) = Sat(p, o_j)\right) \wedge$$
$$\left(\forall a \in A' \ \ \neg Sat(a, o_i) \wedge Sat(a, o_j)\right) \wedge Sat(c, o_i) \wedge \neg Sat(c, o_j)$$

$$\neg \exists o_i, o_j \in \mathbf{O}; D' \subseteq D$$
$$\left(\forall p \in P \backslash D' \ \ Sat(p, o_i) = Sat(p, o_j)\right) \wedge$$
$$\left(\forall a \in D' \ \ \neg Sat(a, o_i) \wedge Sat(a, o_j)\right) \wedge \neg Sat(c, o_i) \wedge Sat(c, o_j)$$

**Def:** I is *relevant* on O iff

$$\forall p \in P \ \ \exists o_i, o_j \in \mathbf{O}$$
$$\left(\forall r \in P \backslash p \ \ Sat(r, o_i) = Sat(r, o_j)\right) \wedge$$
$$\left(Sat(p, o_i) \neq Sat(p, o_j)\right) \wedge \left(Sat(c, o_i) \neq Sat(c, o_j)\right)$$

Figure 2. Consistency properties of an influent set.

influent set is deterministic if its consequent is completely determined by its preconditions with respect to the set of observations. There cannot be two observations which agree on the truth value of all of the preconditions but do not agree on the truth value of the consequent. This is the same constraint which Stuart Russell's determinations have. The second consistency property is predictiveness. When an influent is needed to distinguish between two observations, the influence it has on the consequent must be in the predicted direction. If two observations differ only in the truth values of the preconditions of several positive influents then the consequent must be true when those preconditions are true. This constraint ensures that positive values assigned to the preconditions can only have a positive effect on the consequent. The analogous negative relationship must also hold for negative influents. The final consistency property is relevance. An influent set is relevant over a set of observations if the preconditions of each influent are necessary for uniquely determining the value of the consequent. In effect, no influent superfluous to the set of observations is permitted; each influent must be the determining factor in some circumstance otherwise it could be eliminated from the explanation with no effect on the learned function. Relevance is required since it disallows plausible theorems which are correct in the sense that their preconditions determine their consequent, but are useless since they have so many irrelevant influents. Any explanation that satisfies these three properties determines a function which can be used to model the domain. This is PI–EBL's ultimate goal. Thus the PI–EBL approach is a search for plausible theorems—explanations which are empirically shown to be deterministic, predictive, and relevant.

## 5. Reasoning With PI–EBL

A fault found with traditional EBL is its limited use of training examples. Examples are much more central to PI–EBL learning. The PI–EBL approach has aspects of traditional induction, but it is not a hybrid system in the sense that it can be easily decomposed into these separate components. Because of the homogeneity of the approach, aspects of both approaches are intertwined in its learning. The plausible inferencer is used to explain the value of the predicate of interest in the system's observations. Observations are also used to verify or reject hypotheses generated by the inferencer. This is done in a generate and test fashion. If an observation presented to the system causes failure of the current hypothesis to be deterministic, predictive, or relevant over the set of observed examples, then the hypothesis is rejected and the inferencer is used to generate another plausible explanation. As plausible explanations are being generated and monitored for correctness the observed examples are used for yet another purpose. For every influent set in the plausible explanation there is a function mapping the the preconditions of the influent set to its consequent. Observations are used to constrain these functions. This use of the observations is more similar to traditional induction since they are used to learn a function with fixed attributes (preconditions of the explanation). Even here, however, the explanative structure is relevant since it constrains the function being induced; it must be consistent with the plausible explanation.

The plausible inferencer builds explanations of the observation in much the same way that traditional EBL explains the observations it is given. Unlike EBL, the PI–EBL approach admits multiple incompatible explanations of the same goal concept. Because of their implicit context, the predictiveness of these explanations will vary greatly, which places great importance on the *order* that the plausible inferencer uses in generating its explanations. Since these explanations bias PI–EBL's learning, constraints on their order provide an ordering bias for the entire PI–EBL system. The a priori chance that an explanation is a theorem can be approximated in several ways. The size of the explanation is useful, larger explanations generally must satisfy a larger implicit context so they are more likely to fail. Another promising estimator of predictiveness can be derived by combining the predetermined predictiveness of each of the influents in the explanation.

## 6. Learning Example

We present a simple example of the PI–EBL learning algorithm to supplement its description and as a concrete example of its advantages over traditional EBL.

$$Clawed(x) \succ Bird(x)$$
$$Feathered(x) \succ Bird(x)$$
$$Immobile(x) \underset{-}{\succ} Fly(x)$$
$$Dead(x) \succ Immobile(x)$$
$$Bird(x) \succ Fly(x)$$

Figure 3: Plausible Domain Theory

Figure 3 shows a plausible domain theory which is used to predict which objects fly. The theory is plausible since it draws contradictory conclusions about immobile birds. Each of the influents are plausible in the sense that they have many exceptions. We will demonstrate how an agent can use the remainder of these influents to rectify one of the missing preconditions in the $Bird(x) \succ Fly(x)$ influent. For reasons external to the PI–EBL algorithm assume the reasoner needs to learn which objects are likely to fly. The first example seen by the agent is a clawed, feathered bird which can fly. This is explained using: $Clawed(x) \succ Bird(x) \succ Fly(x)$. This explanation is rejected when the agent sees a clawed, hairy bear which cannot fly. The next most a priori plausible explanation chosen is consistent with the examples seen: $Feathered(x) \succ Bird(x) \succ Fly(x)$. Even this explanation is insufficient in explaining a hunters trophy: a dead bird with feathers and claws which cannot fly. As before, the explanation is rejected and the next most a priori plausible explanation is generated. Figure 4 shows the next explanation, it has an influent set com-

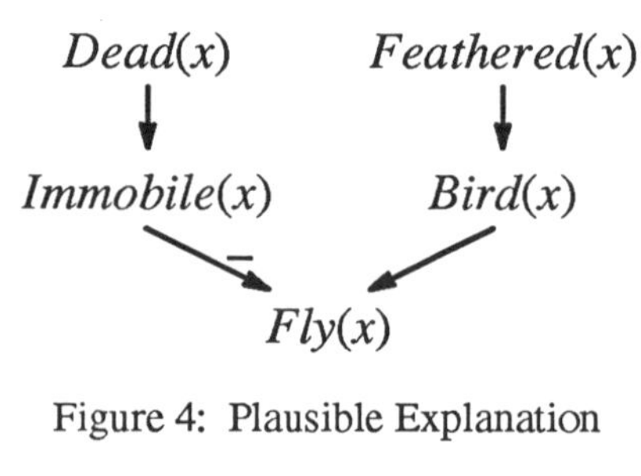


Figure 4: Plausible Explanation

posed of two influences at its top level. There are two functions mapping deadness and featheredness to the ability to fly which are consistent with this plausible explanation. One predicts dead, feathered objects fly, and the second function predicts that they do not. Both are consistent with the theory but only the first is consistent with the hunter's trophy so it is accepted. There are actually sixteen functions which accept two boolean variables, but only two of the sixteen functions would cause the explanation to be deterministic, predictive, and relevant over examples consistent with the function. If the explanation is insufficient to cover some example (it is not deterministic over the examples) then a new explanation is generated, and so on, until enough examples are seen that are consistent with this explanation. At this point it is accepted as a plausible theorem along with the classification function mapping the leaves of the explanation to its consequent.

$Bird(x) \succ Fly(x)$ is missing the precondition "not dead". This is rectified by combining it in an explanation dealing with immobility. The missing precondition was "added" by building the composite explanation. If $Bird(x) \succ Fly(x)$ is later used in another explanation its simpler form may be used. This is important since it is quite likely that the dead bird exception will be irrelevant (as will the hundreds of other exceptions). As we discussed earlier the deductive version: $\ldots \wedge \neg Dead(x) \wedge Bird(x) \succ Fly(x)$ would force the system to reason about *all* those exceptions at *all* times. Plausible explanations allow the reasoner to correctly handle these cases without forcing all reasoning to be done at this level. Notice the imprecision in the plausible theory, the theory does not specify which of the two competing sub–explanations would prevail if both had their preconditions satisfied. Of course if that level of detail were known it could easily be specified. The important point is its possible to specify that featheredness relates to flying and deadness relates to flying without precisely specifying *all* possible interactions between the knowledge. The importance of this simplification increases as the theory increases in complexity because the number of possible interactions increases dramatically with complexity, and many of these interactions need not be dealt with by the reasoner since they will never occur in practice. Even if all necessary distinctions were known, expanding a plausible theory into a deductive theory would cause duplication. As it stands now $Dead(x) \succ Immobile(x)$ can be used as an exception to flying, walking, swimming, dancing, etc. If it were written as a deductive theory the same exception would need to be placed explicitly in the preconditions of many rules.

## 7. Conclusions

The advantages of knowledge intensive approaches are well known. A reasoner using a plausible model instead of a deductive model will retain these advantages, but avoids the unnatural constraints of global consistency and precision of traditional deductive systems. This is very important because the agent may accept and reason with new knowledge without completely understanding all possible interactions with existing knowledge. The PI–EBL approach, because of its representation, allows very simple and imprecise theories to be made precise by learning how the plausible knowledge interacts from actual world observations (examples). In this way the agent can boot–strap itself into specialized domains by combining relevant portions of very general theories, thus building a specialized theory which is adapted empirically.

References


[DeJong86] G. F. DeJong and R. J. Mooney, "Explanation–Based Learning: An Alternative View," *Machine Learning 1*, 2 (April 1986), pp. 145–176.

[Dietterich86] T. G. Dietterich, "Learning at the Knowledge Level," *Machine Learning 1*, 3 (1986), pp. 287–316.

[Genesereth87] M. Genesereth and N. Nilsson, *Logical Foundations of Artificial Intelligence*, Morgan Kaufmann, Palo Alto, CA, 1987.

[Grosof89] B. N. Grosof and S. J. Russell, "Declarative Bias for Structural Domains," *Proceedings of the Sixth International Workshop on Machine Learning*, Ithaca, NY, June 1989, pp. 480–482.

[McCarthy69] J. McCarthy and P. J. Hayes, "Some Philosophical Problems from the Standpoint of Artificial Intelligence,"in *Machine Intelligence 4*, B. Meltzer and D. Michie (ed.), Edinburgh University Press, Edinburgh, Scotland, 1969.

[McCarthy86] J. McCarthy, "Applications of Circumscription to Formalizing Common–Sense Knowledge," *Artificial Intelligence 28*, (1986), pp. 89–116.

[Mitchell86] T. M. Mitchell, R. Keller and S. Kedar–Cabelli, "Explanation–Based Generalization: A Unifying View," *Machine Learning 1*, 1 (January 1986), pp. 47–80.

[Reiter80] R. Reiter, "A Logic for Default Reasoning," *Artificial Intelligence 13*, 1–2 (April 1980), pp. 81–113.

[Rumelhart86] D. E. Rumelhart, G. E. Hinton and J. L. McClelland, "A General Framework for Parallel Distributed Processing,"in *Parallel Distributed Processing: Explorations in the Micro–Structure of Cognition*, D. E. Rumelhart and J. L. McClelland (ed.), MIT Press, Cambridge, MA, 1986, pp. 46–73.

# Searching an Hypothesis Space When Reasoning About Buoyant Forces: The effect of feedback

**Takeshi Okada & David Klahr**
Department of Psychology
Carnegie Mellon University
Pittsburgh, PA 15213

### Abstract

This study addressed the following three questions: (1) To what extent can people's naive, complex, and idiosyncratic knowledge about a real physical domain be captured in a formal representation of an hypothesis space? (2) How does exposure to increasingly complex instances affect subjects' search through the hypothesis space? (3) What is the effect of feedback on hypothesis revision? Six adult subjects solved a series of physics problems involving buoyant forces and liquid displacement. An analysis of subjects' verbal protocols suggests: (1) Naive, complex and idiosyncratic knowledge can characterized by an hypothesis space and changes in that knowledge can then be described as a search through the hypothesis space; (2) People who receive feedback from experimental outcomes change their hypotheses and reach a higher level in the hypothesis space. Mere problem exposure, without feedback, did not lead to hypothesis revision.

## Background and Purpose

Questions about the cognitive psychology of scientific discovery have been approached from two relatively distinct perspectives. One approach focuses on the formal, domain-general, processes supporting the cycle of hypothesis induction, evidence generation and hypothesis evaluation. These processes have been characterized as a problem-solving search in a space of hypotheses and a space of experiments (e.g. Dunbar, 1989; Klahr, Dunbar & Fay, 1989; Klahr & Dunbar, 1988; Kuhn, Amsel & O'Loughlin, 1988; Kuhn & Phelps, 1982; Schauble, 1990). All of these process-oriented approaches have used either arbitrary domains or very simple physical domains. The other approach focuses on domain-specific content knowledge about aspects of the physical world ( e.g., mass, density, heat, temperature, kinematics, etc.), especially in the area of science education (e.g. Brewer & Samarapungavan, 1990; Chi, Glaser & Rees, 1982; Reif, 1987; Vosniadou & Brewer, 1990). These studies focus on people's content knowledge of the physical world, but they do not articulate the process of hypothesis formation and revision.

This study attempts to integrate the two perspectives by studying people's naive, complex, and idiosyncratic knowledge about a real physical domain, and by representing this knowledge in terms of search through an hypotheses space. Our methodological contribution is a new representation for capturing different levels of knowledge about the physics of buoyant forces. The psychological contribution is a preliminary comparison of the effects of problem exposure plus feedback versus problem exposure alone on people's tendency to revise their hypotheses. If there were a "test effect" from repeated exposure to problems that varied along different physical dimensions, then one would expect to see improvement even in the absence of feedback. In contrast, we found that simply exposing people to a variety of problems, without providing feedback about their predictions was insufficient to cause substantial changes in their hypotheses. Only when feedback was provided did people tend to move toward increasingly correct and consistent hypotheses about the domain under investigation.

## Method

**Tasks.** The basic physical domain involved reasoning about buoyant forces and liquid displacement. Subjects' hypotheses were assessed by a series of pairwise comparisons adapted from work by Itakura et al (1988). (See Figure 1). There were two cubes and two beakers with liquid on a balance scale in all tasks. Subjects were required to predict which side of the scale would go down, or if it would balance, if it were free to move. They were also required to state the reason for their prediction. In Task 1, there is a clay cube in the bottom of the water in the left beaker and a styrofoam cube floating on the water in the right beaker. For Tasks 2 to 10, cubes of the both sides are supported by stiff wire connected with iron stands. Each task contrasts a different set of properties of cubes or liquid (e.g. different volumes, or sizes, or densities, or shapes of cubes, different depths of submersion in liquid, different shape, different densities of liquid).

**Design.** There were three conditions. In all conditions, subjects were presented with a series of physics problems about which they had to make a prediction based on their understanding of the underlying forces. In the Hypothesis - Feedback (HF) condition, subjects were required to state an hypothesis before receiving feedback on each task. In the Feedback-Hypothesis (FH) condition, subjects were given feedback before being asked to state their current hypothesis. In the Hypothesis - No Feedback (HN) condition, subjects were not given any feedback. In order to analyze the hypothesis space search process in detail, this investigation was designed as a series of related case studies, with two subjects in each condition.

**Subjects.** Six native Japanese non science graduates of Japanese universities were used. Subjects had not studied Physics since graduating high school. One male and one female were in each condition. The experimenter was a native-speaking Japanese. All experiments were conducted in Japanese.

**Procedure.** Subjects were run individually. They were instructed in how to give verbal protocols, and all of their task responses were recorded on a VCR and an audio tape recorder. Problems were presented both in a diagram and on a video monitor. Following each prediction, the actual outcome was presented on the monitor (in the HF and FH conditions).

Subjects were told that the ten tasks were related to each other to some extent. The first task had two parts. For the Buoyancy pre-test subjects were asked: "Why do you think the clay sinks, while the styrofoam floats? Please talk aloud about what you are thinking." Following the subject's explanation, the experimenter asked: "On what did you base your explanation?" After the subject's explanation of the basis, the experimenter asked the subject for a confidence rating on a 7 point scale.

The next question was a balance scale question (Task 1: Balance scale pre-test). The experimenter displayed a card with a picture and an information list of the features of this task. Then the experimenter asked the subject to predict which side would go down or if they would balance. The experimenter also asked the subject why he/she thought so. After the subject's prediction and explanation, the experimenter again asked for the subject's basis and confidence level. Cards were displayed so that subjects could refer to them at any time.

Then, task 2 was presented. In the H-F condition and the H-N condition, the subject was asked the same question as in the balance scale pre-test in task 1. Then, each subject in the H-N condition went to the next task. Each subject in the H-F condition was given feedback about the experimental outcome through the display of the videotape and was asked to state (a) a reason why he/she thought this result occurred, (b) the basis of his/her explanation, and the confidence in that explanation. In the F-H condition, subjects were not asked to predict the outcome. Instead, they were first shown feedback on the experimental outcome and then asked to give a reason for the outcome, the basis of the explanation, and the confidence level.

`The same procedure continued through task 10, after which subjects were asked if they could state a consistent theory that could account for all the tasks. Subjects were also asked to state the basis of their explanation and their confidence in it. Subjects were asked to sort these ten tasks based on the similarity of the underlying physical principle determining their outcomes. Then they were asked to subdivide each category, and finally to combine the first category to a higher group if possible.

Then subjects were asked the original buoyancy question and the balance scale question in task 1 again (Buoyancy post-test and Balance scale post-test). Finally, each subject was asked about his/her background knowledge about physics and science.

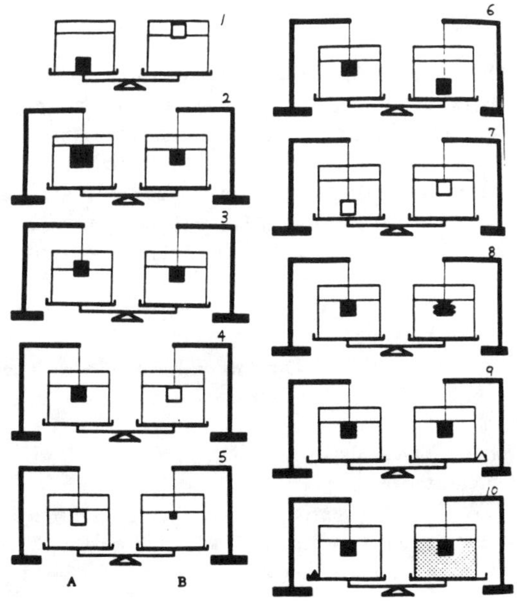


**Figure 1. Tasks in this experiment.**

## Results

**Coding of hypotheses.** Subjects' protocols were transcribed from tapes and coded at a high enough level to describe the subjects' hypotheses. The coding categories are shown in Table 1 and the hypothesis space is described as a hierarchical tree structure (See Figures 3, 5, 7 and 9, which depict individual subjects' movement through the hypothesis space). Each node in the hypothesis space represents a rule about how to determine the net force on each side of the scale. The top node shows three types of "Expert Physicist" theories.

1) <u>Tension theory:</u> F=Wc+Wl+Wb-T. In this case, external force is the force on the container and the water and the block.
2) <u>Buoyancy theory:</u> F=Wc+Wl+Dl*Vb.l*g. Dl*Vb.l*g means buoyancy force, which is determined by the density of the liquid, the volume of displaced liquid (which is equal to the volume of the block in the water) and the gravitational constant. In this case, external force is the force on the container and the block.
3) <u>Hydrostatic Pressure theory:</u> F=Dl*Hl*A*g. In this case, external force is the force on the container.

The hypothesis space is divided into three frames derived from the expert theories: the Tension frame in the left (which includes the Tension Hypotheses group), the Buoyancy frame in the middle (which includes the Buoyancy Hypotheses group), and the Hydrostatic Pressure frame in the right (which includes the Hydrostatic Pressure Hypotheses group). Because there is insufficient information about the tension force, the Tension theory can not predict the right answers. The SAME BLOCK hypothesis does not belong to these three frames, because it is more basic than them.

There are sub-categories in each frame. Levels are loosely connected to the vertical dimension. Hypotheses depicted near the top of Figure 3 can be regarded as being closer to the Expert solutions. The nested boxes within a larger box can be regarded as in the same level. Solid lines between nodes mean that there is a direct relationship and dashed lines mean that there is an indirect relationship.

<u>**Table 1 Coding Categories of Protocol**</u>

T ^ Wb (F x; Wb) The tension from the wire supports the weight of the block and the scale force is unrelated to the weight of the block.
F : Wb-T The scale force is determined by the weight of the block minus the tension force.
F=Wc+Wl+Wb-T (Tension theory). The scale force equals the sum of the container weight plus the liquid weight plus the block weight minus the tension from the wire. This is a physicist theory.
F : Sb The scale force is determined by the size of the block.
F : Sb.l The scale force is determined by the size of the block in the liquid.
F : Wb The scale force is determined by the weight of the block.
F : Wb.l The scale force is determined by the weight of the block in the liquid.
F : SAb The scale force is determined by the surface area of the block.
F : SAb.l The scale force is determined by the surface area of the block in the liquid.
F : VSAb The scale force is determined by the volume (which is related to the surface area) of the block.
F : VSAb.l The scale force is determined by the volume (which is related to the surface area) of the block in the liquid.
F : Vb The scale force is determined by the volume of the block.
F : Vb.l The scale force is determined by the volume of the block in the liquid.
F : Dl,,Vb.l (= F : SGl,,Vb.l) The scale force is determined by the density of the liquid and the volume of the block in the liquid. In this case, subject may not know the clear relationship between the density of the liquid and the volume of the block in the liquid (that is, the density of liquid times the volume of the block in the liquid determines the scale force).
F : Dl*Vb.l (= F : SGl*Vb.l) The scale force is determined by the density of the liquid times the volume of the block in the liquid.
F=Wc+Wl+Dl*Vb.l*g (Buoyancy theory). The scale force equals to the weight of the container plus the weight of the liquid plus the density of the liquid times the volume of the block in the liquid times the gravitation force. Dl*Vb.l means the buoyancy force.
F : Hl The scale force is determined by the height of the liquid.
F : Dl,,Hl (= F : SGl,,Hl) The scale force is determined by the density of the liquid and the height of the liquid. In this case, subject may not know the clear relationship between the density of the liquid and the height of the liquid (that is, the density of the liquid times the height of the liquid determines the scale force).
F : Dl*Hl (= F : SGl*Hl) The scale force is determined by the density of the liquid times the height of the liquid.
F=Wc+Dl*Hl*A*g (Hydrostatic pressure theory). The scale force equals the weight of the container plus the density of the liquid times the height of the liquid times the bottom area times gravitation force.
SAME BLOCK If the same sort of blocks are put on the both sides of a balance scale, the scale balances. This is a basic idea about balance scale.
<u>NO IDEA</u> The subject did not have any clear idea about the task

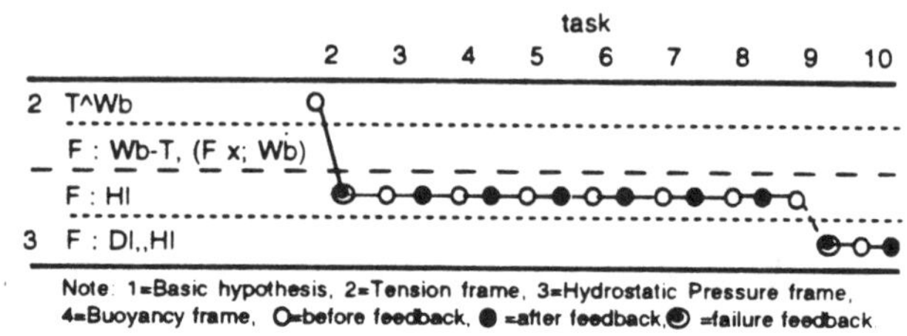


**Figure 2. Hypotheses change following each task for S1.**

## Individual subjects

We will describe in detail only four subjects. Space constraints preclude a description of the other two. Each subject's search through the hypothesis space is depicted in two ways. The tabular form (Figures 2, 4 and 6) shows the temporal dynamics of each problem and the effect of feedback (if any). In these Figures, a plain line between dots means that the subject changed hypotheses without keeping previous hypothesis for previous tasks. Dashed lines between dots mean that subjects searched for new hypotheses while maintaining a previous hypothesis for previous tasks. In addition to the temporal representation, we use the hierarchical representation of the hypotheses space shown in Figures 3, 5 and 7. In these figures, plain lines with arrows mean that the subject changed hypotheses without keeping the previous hypothesis. Dashed lines with an arrow mean that the subject searched for a new hypothesis while keeping a previous hypothesis.

***H-F condition, S1 (male).*** S1's pre-test results showed that he had basic knowledge about buoyancy and balance scales. S1 changed his hypothesis twice; after failure feedback in task 2 and task 9. (See Figure 2) Failure feedback was a key factor in his theory revision. His confidence rating was also changed from 7 to 3 after the first failure feedback and 5 to 3 after the second failure feedback. Figure 3 shows the change of hypotheses through each task. S1 moved from the Tension frame to the Hydrostatic Pressure frame in the first change and moved from lower level to higher level in the Hydrostatic Pressure frame in the second change. For his final "integrative" theory, S1 actually maintained two distinct theories - one to account for Tasks 2-8, and another for 9 and 10.

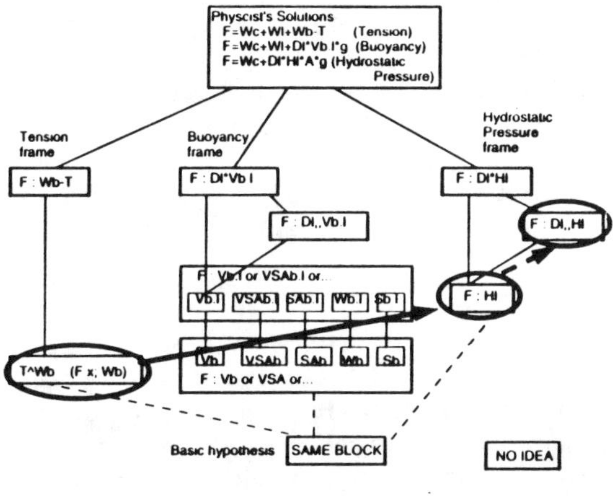


**Figure 3. The Hypothesis Space Trajectory for S1.**

*H-F condition, S2 (female).* S2 incorrectly answered the buoyancy pre-tests using only the concept of weight. She did have basic knowledge about balance scales. S2 changed her hypotheses seven times; (See Figures 4 and 5) after failure feedback in task 3, before receiving feedback and after failure feedback in task 4, before receiving feedback and after failure feedback in task 8, and before receiving failure feedback and after failure feedback in task 9. She changed from a lower level of hypothesis to a higher level of hypothesis in task 3. She changed from a less correct hypothesis to a more correct hypothesis in task 4 and task 8. However, in task 9, S2 could not find any hypothesis at all because she lacked a clear idea about liquid density. Although she also changed her hypotheses when she faced new tasks before receiving failure feedback, the changes were small or in the direction from a higher level to a lower level. That is, she changed from the concept of size in which the concept of weight and volume are not distinguished in task 3 to the concept of weight which became distinguished from the concept of volume.

The change from V to VSA in task 8 can also be regarded as a small change. In this case, S2 did not attend to the relationship between volume and surface area before being faced with task 8. On task 8, her knowledge about that relationship was activated from her memory as she stated in her basis. The change from V in task 8 to SAME BLOCK in task 9 can be regarded as a change of application of her existing hypothesis to the new task rather than a discovery of a new hypothesis, because SAME BLOCK hypothesis is regarded as a basic hypothesis that even children have. It is clear that S2 moved from a lower level to a higher level and moved from a less correct hypotheses to a more correct hypotheses in the Buoyancy frame. When she was faced with task 9, she applied a basic hypothesis. However, she could not interpret last two tasks. S2 had three types of final hypotheses which applied to different tasks. Although the first two hypotheses were in the same category, she was unable to integrate them into a unified hypothesis at the higher level.

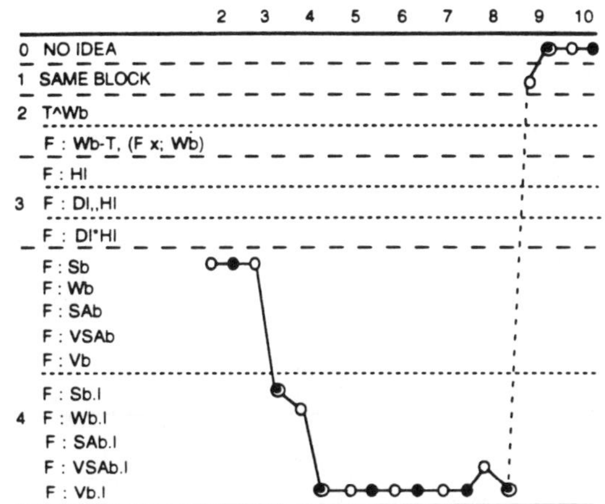


**Figure 4. Hypotheses change following each task for S2.**

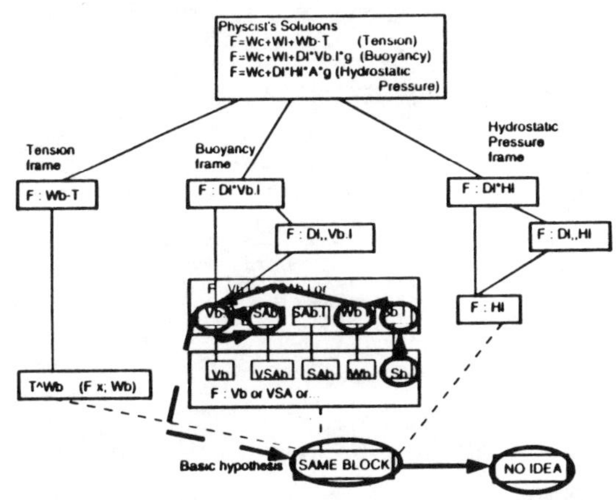


**Figure 5. Hypothesis Space Trajectory for S2.**

*F-H condition, S3 (male).* S3 had basic knowledge about buoyancy and balance scales. He changed his hypothesis twice; in task 2 and in task 9. (Figures 6 and 7) Although subjects in the F-H condition were not required to state their hypotheses before receiving feedback, we infer their implicit prediction from their hypothesis on the previous task. Thus, we infer that S3 received failure feedback in task 2 and task 9. In task 2, he asked, "While the wire supports blocks, the left side went down?" Our inference is that he had T^Wb hypothesis and had predicted that the scale should have balanced, but it didn't. In task 9, because his hypothesis in task 8 predicts the wrong outcome, we also infer that he received failure feedback. In that sense, failure feedback influenced him to change his hypotheses. Figures 6 and 7 show that he moved from the Tension frame to the Buoyancy frame in the first change and moved from a lower level to a higher level in this frame in the second change. He had a single hypothesis to explain tasks 2-10.

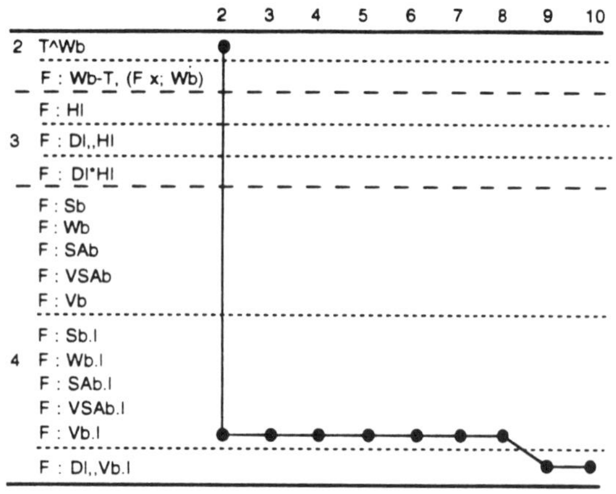


**Figure 6. Hypotheses change following each task for S3.**

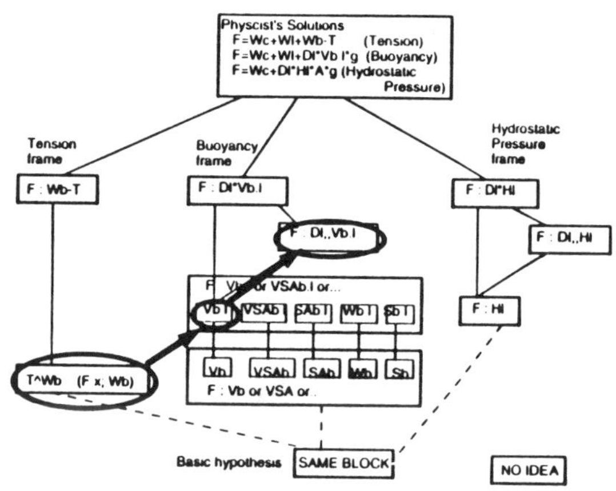


**Figure 7. Hypothesis Space Trajectory for S3.**

***H-N condition, S4 (female).*** S4 had basic knowledge about buoyancy and balance scales. S4 did not change her hypotheses at all, although she stated a different hypotheses vaguely at the same time in task 6 and 7. She remained in the Tension frame throughout. She had one hypothesis to account for all tasks.

## Summary of Findings

1) People's knowledge about physical domains can be characterized as an hypothesis space, and changes in that knowledge can be described as a search through the hypothesis space.

2) There was a dramatic difference between the H-F condition and the H-N condition. Subjects who received feedback changed their hypotheses from a frame to another frame or from a lower level to a higher level, especially after receiving failure feedback. On the other hand, subjects who did not receive feedback did not change their hypotheses. That is, simply being exposed to a variety of problems was insufficient to make subjects search the hypothesis space. However, in the H-F condition (S1, S2), subjects did make small changes in their hypotheses even when they faced new tasks before receiving feedback.

3) There is no clear difference between the H-F condition and the F-H condition.

4) Subjects in the feedback conditions used feedback information from earlier tasks to solve subsequent tasks. However, subject's explanations did not necessarily generalize to all tasks. Instead, they used several different explanations for different groups of tasks.

## References


Brewer, W. F., & Samarapungavan, A. (1990). Children's theories versus scientific theories : Differences in reasoning or differences in knowledge? In R. R. Hoffman & D.S. Palermo (Eds.), *Cognition and the symbolic process: Vol. 3. Applied and ecological perspectives.* Hillsdale, Nj: Lawrence Erlbaum Associates.

Chi, M. T. H., Glaser, R., & Rees, E. (1982). In R. J. Sternberg (Ed.) *Advances in the psychology of human intelligence.* Hillsdale, Nj: Lawrence Erlbaum Associates.

Dunbar, K. (1989). Scientific reasoning strategies in a simulated molecular genetics environment. In *Proceedings of the Eleventh Annual Meeting of the Cognitive Science Society.* Hillsdale, NJ: Lawrence Erlbaum Associates.

Itakura, K., and the hypothesis experiment method research group (1988). *Buoyancy and density, weight and force, friction force.* Tokyo: Kokudosha. (in Japanese)

Klahr, D., Dunbar, K., & Fay, A. (1989). Designing good experiments to test bad hypotheses. In J. Shrager & P. Langley (Eds.), *Computational models of discovery and theory formation.* Hillsdale, Nj: Lawrence Erlbaum Associates.

Klahr, D., & Dunbar, K. (1988). Dual space search during scientific reasoning. *Cognitive Science*, 12, 1-55.

Kuhn, D., Amsel, E., & O'Loughlin, M. (1988). *The development of scientific thinking skills.* NY: Academic Press, Inc.

Kuhn, D., & Phelps, E. (1982). The development of problem-solving strategies. In H. Reese (Ed.), *Advances in child development and behavior* (Vol.17, pp.1-44). NY: Academic Press, Inc.

Reif, F. (1987). Interpretation of scientific or mathematical concepts: Cognitive issues and instructional implications. *Cognitive Science*, 11, 395-416.

Schauble, L. (1990). Belief revision in children: The role of prior knowledge and strategies for generating evidence. *Journal of Experimental Child Psychology*, 49, 31-57.

Vosniadou, S., & Brewer, W. F. (1989). The concept of the earth's shape: A study of conceptual change in childhood. Manuscript submitted for publication.

# Classifying faces by race and sex using an autoassociative memory trained for recognition

**Alice J. O'Toole*, Hervé Abdi*, Ken A. Deffenbacher**, James C. Bartlett***

* School of Human Development
The University of Texas at Dallas
Richardson, TX 75083-0688

** Department of Psychology
The University of Nebraska at Omaha
Omaha, NE 68182-0274

**Abstract**[1]

We examine the ability of an autoassociative memory trained with faces to classify faces by race and by sex. The model learns a low-level visual coding of Japanese and Caucasian male and female faces. Since recall of a face from the autoassociative memory is equivalent to computing a weighted sum of the eigenvectors of the memory matrix, faces can be represented by these weights and the set of corresponding eigenvectors. We show that reasonably accurate classification of the faces by race and sex can be achieved using only these weights. Hence, race and sex information can be extracted in the model without explicitly learning the classification itself.

## Introduction

Adaptive interaction with people in daily life requires proficiency in processing faces in a variety of ways. In addition to our ability to recognize faces, we are also proficient at categorizing faces along a number of dimensions including sex, age, and race. Despite the ease with which we perform these tasks, face processing presents a number of difficult computational problems. As spatial patterns, faces are highly similar in that they all contain the same "features", (e.g., nose, mouth, eyes) arranged in roughly the same configuration. Thus, the information available for distinguishing previously seen faces from new faces, and for classifying faces along other important dimensions, must be found in subtle variations in this prescribed feature set and configuration.

Quantification and representation of the information in faces in a way that makes the tasks of recognition and classification possible is a paradoxical problem. The purpose of recognition is to classify a face as familiar or unfamiliar. Thus, a successful computational model of face recognition should be capable of storing a face in a way that makes it distinguishable from faces which were not learned. This needs to be done regardless of the similarity relationships among familiar and unfamiliar faces. It has been known since Kohonen (1977) that an autoassociative memory, composed of faces represented simply by pixel values, acts as a content addressable memory. O'Toole, Millward, & Anderson (1988) used this type of model in a two-alternative forced choice face recognition task and showed that the model distinguished learned from new faces with a good degree of accuracy.

The problem of classification of faces, on the other hand, requires that faces be categorized according to a set of defining characteristics. For simplicity, we will confine the discussion to categorization by race. But, the present approach holds analogously for the problem of face categorization by sex and age as well. In Bruce & Young's (1986) model of face processing, the information needed to do this kind of classification is called "visually-derived semantic". The term makes the point that these categorizations do not require familiarity with the face, but rather are based on perceptual information and so can be done for unfamiliar as well as familiar faces.

To say that certain categorizations of faces are based on visual information, however, does not mean that learning is not important. While it may not be evident in categorizing a face as Caucasian, Black, or Oriental, it is clear that more subtle race categorizations, such as categorizing an oriental face as Chinese, Japanese, or Vietnamese, does require some learning. We believe that perceptual learning is responsible for fine-tuning our ability to discriminate faces both within and between categories (O'Toole et al., in press).

There is abundant support for an other-race effect in which people are better able to recognize faces of their own race than faces of another race (e.g., the meta-analysis of Shapiro & Penod, 1986). O'Toole et al. (in press) modeled the "other-race effect" in face recognition as a problem in perceptual learning. With greater exposure to faces of a "majority" race, the model was able to make finer discriminations between faces within the majority race and was better able to recognize faces (i.e., discriminate learned from new faces) in the majority race than in the minority race.

Returning to the problem of face classification by race, one approach would be to train a connectionist model to do the classification (cf. Cottrell & Fleming, 1990; Golomb, Lawrence, & Sejnowski, 1991), reinforcing the model for correct categorization. The difficulty with this strategy, in its simplest form, is that learning the classification amplifies the differences between face groups at the expense of de-emphasizing the differences between faces

1. Thanks are due to June Chance and Al Goldstein for providing the Caucasian and Japanese faces used in the simulations, and to Peter Assmann and Barbara Edwards for helpful comments on an earlier version of this manuscript.

within a category. This makes the representations of faces within the model less suitable for recognition. Using back-propagation, Cottrell & Fleming (1990) and Golomb et al. (1991) avoid this problem by preprocessing faces with an image compression network before teaching the categorization. Both models perform well on the task of sex discrimination. In the present work, we show that ability to classify faces into one of two race categories is present implicitly in a model that has not been explicitly taught to do this classification. We use an autoassociative memory that is trained for face recognition and show that information about the race and sex of the face can be extracted with a reasonable degree of accuracy.

**The Autoassociative Memory.** The model is defined first and then its application to the categorization problem is presented. An autoassociative memory was created as follows. A digitized image of each face was coded as a vector of pixel elements concatenated from the rows of the face image. Thus, the $i$th face was represented by a $J \times 1$ vector (where $J$ equals the number of pixels in the image) denoted by $\mathbf{f}_i$. The vectors were normalized. The faces were stored in an autoassociative memory composed of $J$ completely inter-connected units. The connection strengths were stored in a $J \times J$ matrix $\mathbf{A}$ constructed as follows:

$$\mathbf{A} = \Sigma\, \mathbf{f}_i \mathbf{f}_i^T \quad (1)$$

Recall of individual faces from the matrix was done as:

$$\hat{\mathbf{f}}_i = \mathbf{A}\,\mathbf{f}_i \quad (2)$$

where $\hat{\mathbf{f}}_i$ is the system estimate of $\mathbf{f}_i$. The quality of this estimate is measured by comparing the reconstructed image with the original image. This is done by taking the cosine of the angle between $\hat{\mathbf{f}}_i$ and $\mathbf{f}_i$. Interestingly, the comparison may be done visually by displaying the system output, since it will also be a vector of pixels. System performance was improved by using the Widrow-Hoff error-correction rule that iteratively changes the weights in $\mathbf{A}$ to optimize the quality of the recall across the stimulus set:

$$\mathbf{A}_{(t+1)} = \mathbf{A}_{(t)} - \gamma(\mathbf{f}_i - \mathbf{A}_{(t)}\mathbf{f}_i)\mathbf{f}_i^T \quad (3)$$

where $i$ is randomly chosen, and where $\gamma$ decreases as an inverse monotonic function of the iteration number $t$.

While the formulation above shows intuitively what the model is doing, clearly the eigen-decomposition of this matrix is equivalent to principal component analysis (Anderson, et al., 1977; Abdi, 1988). Further, since $\mathbf{A}$ can be expressed as a weighted sum of the outer products of its eigenvectors, recall of a given face from the matrix (i.e, Equation 3) is equivalent to producing a weighted sum of eigenvectors:

$$\hat{\mathbf{f}}_i = \mathbf{A}\mathbf{f}_i = \mathbf{P}\mathbf{\Lambda}\mathbf{P}^T\mathbf{f}_i = \sum_k^N \lambda_k \mathbf{p}_k\mathbf{p}_k^T\mathbf{f}_i = \sum_k^N \lambda_k \cos(\mathbf{f}_i, \mathbf{p}_k)\mathbf{p}_k \quad (4)$$

where $\mathbf{P}$ is the matrix of the $N$ eigenvectors of $\mathbf{A}$ with eigenvalues greater than zero, and $\Lambda$ is the diagonal matrix of the eigenvalues in decreasing order, and $\mathbf{p}_k$ is the $k$th eigenvector. The Widrow-Hoff rule can be expressed via the eigen-decomposition of $\mathbf{A}$ (Abdi, in press):

$$\mathbf{A}_{(t)} = \mathbf{P}[\mathbf{I} - (\mathbf{I} - \gamma\mathbf{\Lambda})^t]\mathbf{P}^T \quad (5)$$

When converged, Equation 4 reduces to (Kohonen, 1977):

$$\hat{\mathbf{f}}_i = \tilde{\mathbf{A}}\mathbf{f}_i = \mathbf{P}\mathbf{P}^T\mathbf{f}_i = \sum_k^N \mathbf{p}_k\mathbf{p}_k^T\mathbf{f}_i = \sum_k^N \cos(\mathbf{f}_i, \mathbf{p}_k)\mathbf{p}_k \quad (6)$$

(i.e., the eigenvalues are then equal to one). As the dimensionality of our images is extremely large by comparison to the number of stimuli, the capacity limit of the system was not a problem for these simulations.

Using this latter formulation of the problem, O'Toole & Abdi (1989) have pointed out the similarity of this approach to traditional multidimensional scaling approaches to representing human similarity data. The axes of the multidimensional scaling solution are analogous to the eigenvectors of the autoassociative matrix. The goal of multidimensional scaling is to represent the similarity relations among stimuli using the smallest number of dimensions possible. Frequently, human similarity data can be represented with a good degree of accuracy using only a few dimensions. Analogously, with an autoassociative matrix composed of a pixel-based representation of faces, Sirovich & Kirby (1987) have shown that faces can be reconstructed to a quite recognizable form (to within 3% error) using only 40 parameters and the corresponding 40 eigenvectors. O'Toole et al. (in press) showed that for the task of simply discriminating learned and unlearned faces, even less parameters are needed. Here reconstructions based on as few as ten eigenvectors were sufficient to produce excellent discrimination of learned and unlearned faces. Recently, Turk & Pentland (1991) have elaborated on this approach in several ways including the addition of an algorithm to locate faces in an image.

The purpose of the present work is to examine the usefulness of these coefficients (i.e., the set of weights used for reconstructing a given face) for predicting class membership. Specifically, we will look at classification by race and sex.

## Method

**Stimuli.** One-hundred and sixty-seven faces were digitized from slides to 16 gray levels using a Fotovix digitizer attached to a 286-based computer with a 16-bit TARGA board (True Vision). Faces were of young adults, with roughly half Caucasian faces and half Japanese faces. Within each race set, faces were roughly half male faces and half female faces. None of the slides pictured people

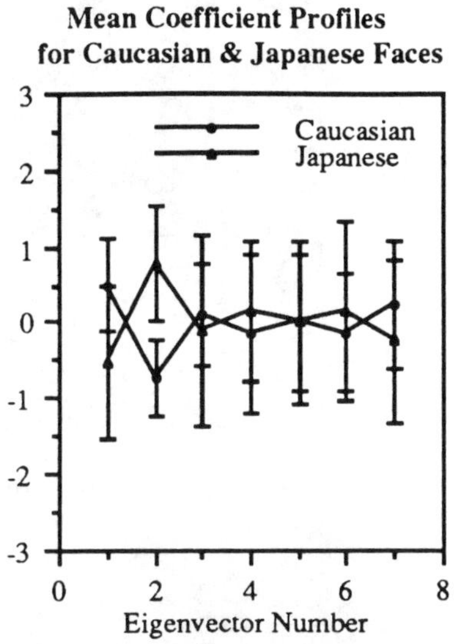


*Figure 1.* Mean coefficient *z*-score profiles of Caucasian and Japanese faces. Error bars show the standard deviations of the *z*-scores. The best race separation is achieved with the second eigenvector.

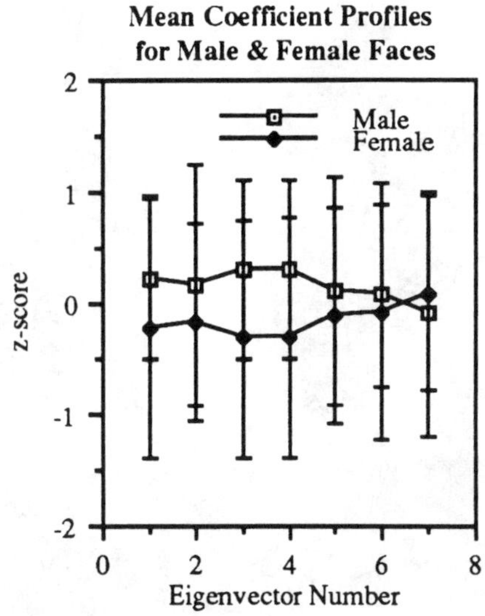


*Figure 2.* Mean coefficient *z*-score profiles of male and female faces. The error bars indicate the standard deviations of the male and female *z*-scores for each race of faces.

with facial hair or glasses. The images were aligned, so that the eyes of all faces were at about the same height. The images were cropped to eliminate clothing. Each face was 151 pixels wide and 225 pixels long (i.e., a 33,975-pixel vector consisting of the concatenation of the pixels row). We used a spatial differentiation encoding that enhanced lines prior to the extraction of the pixel vector, (cf. O'Toole et al., 1988).

**Apparatus.** Simulations were performed on a Sun SparcStation and a Convex C-1 Vector computer.

**Procedure.** An autoassociative memory was created using 40 Caucasian and 40 Japanese faces. The eigenvectors were extracted from the matrix and reconstructions of all 167 faces (i.e., 80 old faces and 87 new faces) were made using Equation 4, with the first seven[2] eigenvectors.

**Analysis and Discussion - Race Discrimination.** A coefficient profile was constructed for each face using the seven coefficients used to reconstruct the face. Since the absolute magnitudes of these coefficients were different for different eigenvectors, the face profiles were converted into *z*-scores. Thus, each coefficient in a face profile was the *z*-score of the coefficient with respect to that coefficient for all faces. An average coefficient profile was then calculated for Caucasian faces and for Japanese faces. The mean coefficients for Japanese and Caucasian faces were significantly different for the first [$t$ (165) = 7.87, $p < .001$], second [$t(165) = 15.25$, $p < .001$], and seventh [$t$ (165)= 3.32, $p < .01$] eigenvectors. This indicates that information relevant to race discrimination is present in each of these eigenvectors. Figure 1 displays these average profiles along with error bars indicating one standard deviation.

We looked at the power of the coefficients for eigenvectors one and two, independently, for making race predictions about the faces. The simplest scheme for doing this is to take the mean of the mean coefficients for the Japanese and Caucasian faces, and to create a decision rule using the mean of these two means as a criterion. Race membership predictions are made by assigning faces with coefficients exceeding this criterion to one race (i.e., the race with the larger of the two coefficient means) and faces with coefficients less than this criterion to the other race (i.e., the race with the smaller of the two coefficient means). Using only the coefficients for eigenvector one yielded correct race predictions for 74.8% of the faces. Basing decisions on the coefficients for the second eigenvector fared better, yielding correct race predictions for 88.6% of the faces.

**Analysis and Discussion - Sex Discrimination.** Average face profiles were created for male and female faces using the procedure described above for race profiles. Figure 2 shows this analysis. These data show that the information for sex discrimination in the model is not localized in one or two eigenvectors. The mean coefficients for male and female faces were significantly different for the first [$t$ (165) = 3.01, $p < .01$], second [$t$ (165) = 2.29, $p < .05$], third [$t(165) = 4.21$, $p < .001$], and fourth eigenvectors [$t$ (165) = 4.07, $p < .001$]. Figure 2 indicates that with the large overlap of the mean distributions, no eigenvector

2. The choice of seven was made somewhat arbitrarily to yield moderate recognition performance. The model discriminated learned from new faces with a *d'* of 1.08 using seven eigenvectors.

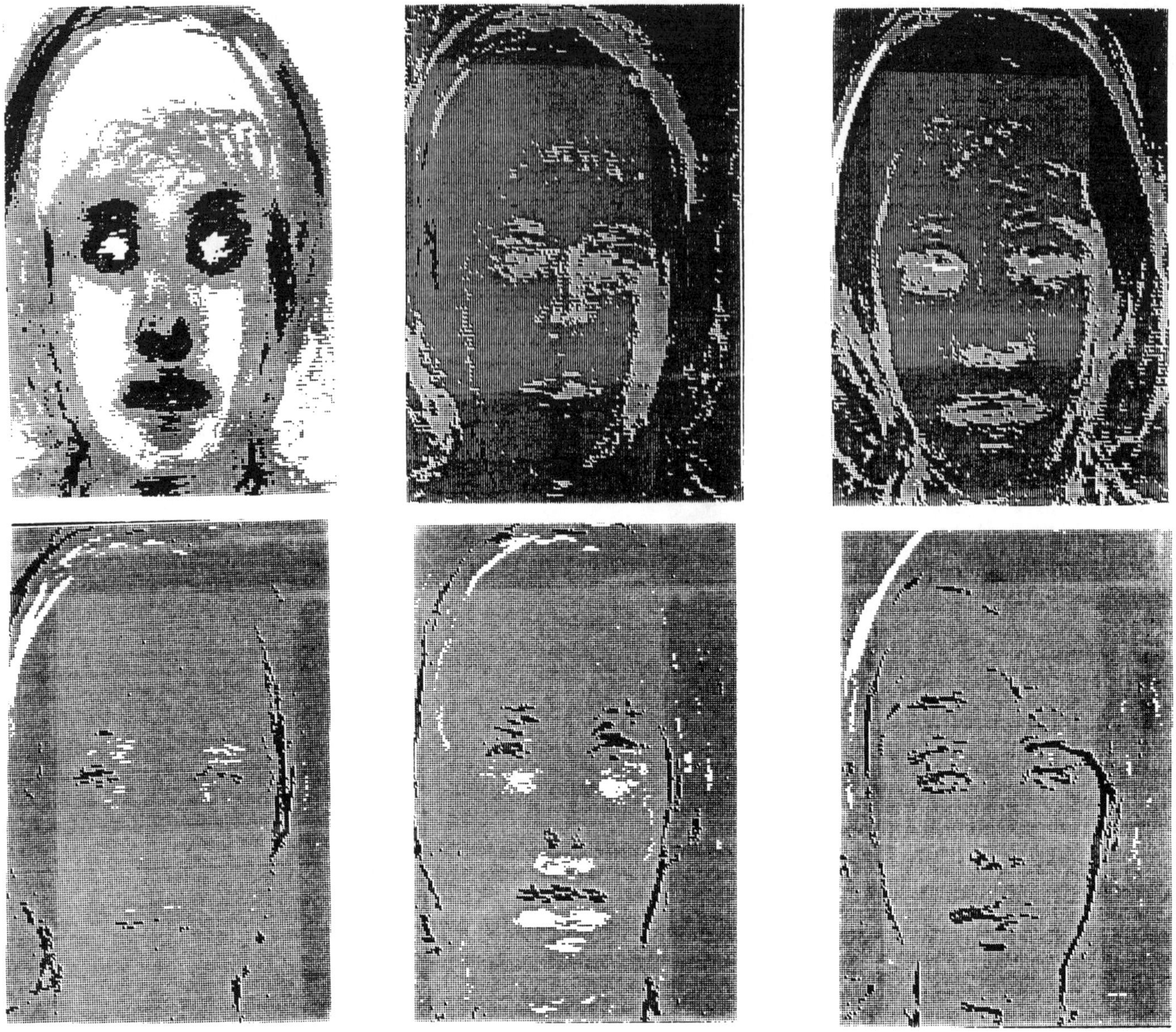

***Figure 3.*** The first six eigenvectors used to reconstruct the faces.

independently will be sufficient to discriminate male and female faces. Since for all of these eigenvectors the mean coefficient for male faces was greater than for female faces, the simplest method for combining information across eigenvectors was to add the coefficients for the four eigenvectors for each face and to use this to predict sex classification. Again, we took the mean of the means of this predictor for male and for female faces and used it as a criterion sex value, assigning male to faces exceeding the criterion value and female to faces less than the criterion value. Using the sum of these coefficients yielded 74.3% correct sex predictions.

**Eigenvectors as Features.** Since the face codings are pixels vectors, the eigenvectors of the memory matrix are also pixel vectors and some examples appear in Figure 3. As noted previously (Abdi, 1988; O'Toole & Abdi, 1989), these vectors are face-like. Further, since any face can be expressed as a weighted combination of eigenvectors, the eigenvectors qualify in some ways as "features" of the face with the coefficients indicating the degree of presence of the feature. The idea of using eigenvectors as features has been around for a relatively long time (Anderson, et al., 1977). These are not localized features but rather more global features. Since the eigenvectors reflect the statistical structure of the stimulus set, a more race-homogeneous matrix will yield eigenvectors that are different from those we have extracted from a matrix made of equal numbers of stimuli from two races. For example, Figure 4 shows the first eigenvector from an autoassociative matrix created from 95% Caucasian and 5% Japanese faces and from 95%

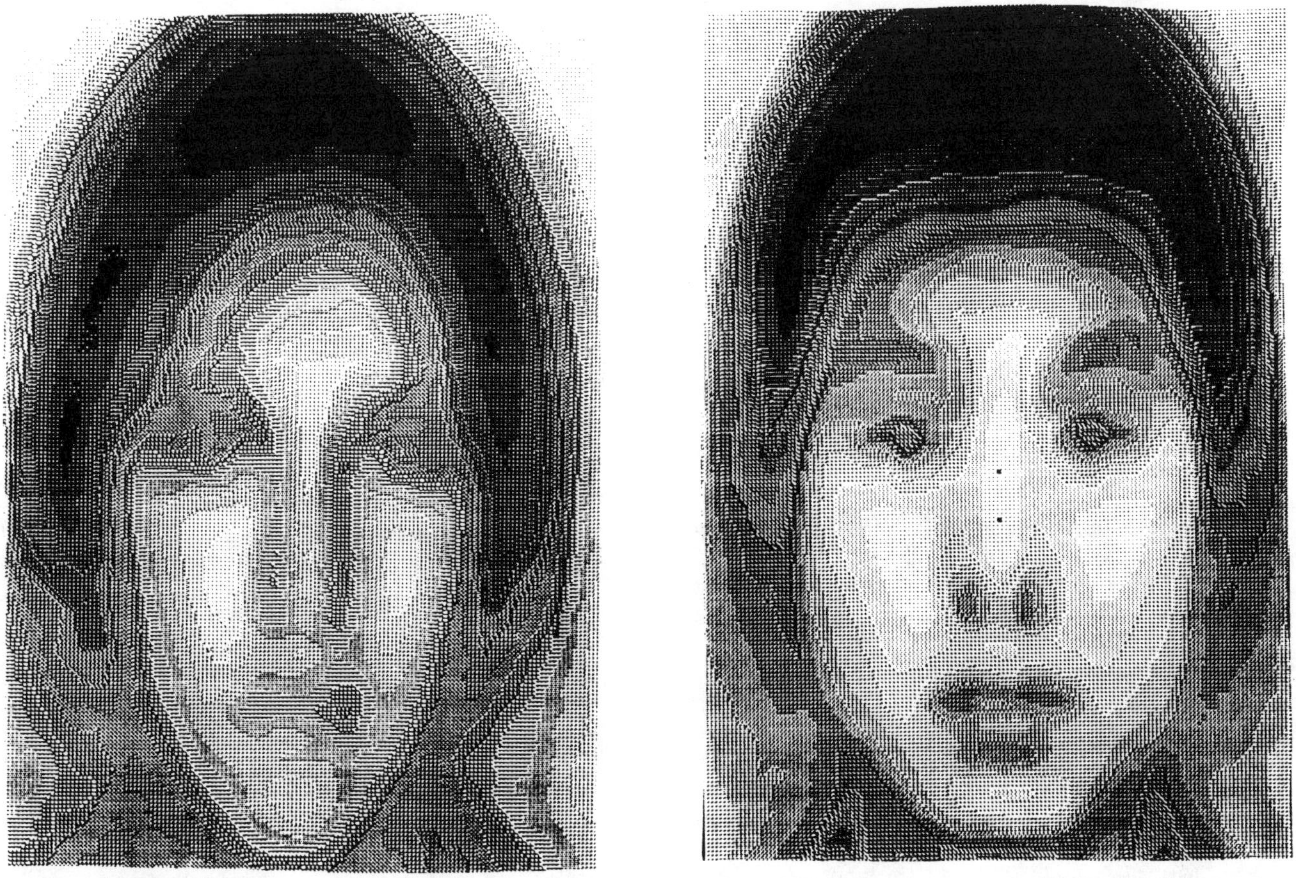

***Figure 4.*** The first eigenvector from an autoassociative matrix created from 95% Caucasian faces and 5% Japanese faces and from 95% Japanese faces and 5% Caucasian faces , respectively.

Japanese and 5% Caucasian faces, respectively. Here the first eigenvectors are typical of the "majority" race in the matrix.

**Summary.** This work shows that the information needed to make race and sex classifications is present in a model trained to perform face recognition. While the discrimination ability of the model is not perfect, it is well above chance. We think that the competing concerns of recognition and classification should be addressed with a unified representation of the stimuli. Future work needs to address the question of how a single model can represent stimuli in a way that answers competing task needs.

**References**


Abdi, H. 1988. A generalized approach for connectionist auto-associative memories: interpretation, implications and illustration for face processing. In J. Demongeot (Ed.), *Artificial Intelligence and Cognitive Sciences.* Manchester: Manchester University Press.

Abdi, H. in press. *Les Réseaux de Neurones.* Presses Universitaires de Grenoble .

Anderson, J. A., Silverstein, J. W., Ritz, S. A., Jones, R. S. 1977. Distinctive features, categorical perception, and probability learning: some applications of a neural model. *Psychological Review, 84,* 413-451.

Bruce,V., Young, A. W. 1986. Understanding face recognition. *British Journal of Psychology, 77,* 305-327.

Cottrell, G. W., Fleming, M. 1990. Face recognition using unsupervised feature extraction. *IJCNN-90, 2,* 65-70

Golomb, B. A., Lawrence, D. T., Sejnowski, T. J. 1991. SEXnet: A neural network identifies sex from human faces. In D. S. Touretsky, R. Lippmann (Eds.) *Advances in Neural Information Processing Systems, 3,* CA:San Mateo: Morgan Kauffmann.

Kohonen, T. 1977. *Associative memory: A System Theoretic Approach.* Berlin: Springer-Verlag.

O'Toole, A. J., Abdi, H. 1989. Connectionist approaches to visually based feature extraction. In G. Tiberghien (Ed.) *Advances in Cognitive Psychology,* (Vol 2). London: John Wiley.

O'Toole, A. J., Deffenbacher, K. A., Abdi, H., Bartlett, J. A. in press. Simulating the "other-race effect" as a problem in perceptual learning. *Connection Science.*

O'Toole, A. J., Millward, R. B., Anderson, J. A. 1988. A physical system approach to recognition memory for spatially transformed faces. *Neural Networks, 1,* 179-199.

Shapiro, P.N., Penod, S.D. 1986. Meta-analysis of face identification studies. *Psychological Bulletin, 100,* 139-56.

Sirovitch, L., Kirby, M. 1987. Low-dimensional procedure for the characterization of human faces. *Journal of the Optical Society of America, 3,* 519-524.

Turk, M., Pentland, A. 1991. Eigenfaces for recognition. *Journal of Cognitive Neuroscience, 3,* 71-86.

# Effect of format on information and problem solving*

**Mukesh J Patel, Benedict du Boulay** and **Chris Taylor**
School of Cognitive and Computing Sciences,
The University of Sussex,
Falmer, Brighton BN1 9QH, UK.
Email: mukesh@cogs.sussex.ac.uk

## Abstract

This study reports the effect of differences in format of Prolog tracers on Prolog problem solving tasks. Three different tracers (Spy, TPM, and EPTB) in different formats were tested to check for their relative effectiveness in solving five different Prolog problems. 43 subjects attempted to solve each problem with each trace (15 problems in total). Preliminary analysis of solution times and response data indicate that EPTB performed best across all problems. An account for this finding is presented, as is one for a number of interesting interactions between the effects of problem type and trace format, which supports the general conclusion that while format is a significant determiner of access to information, it can also constrain the sorts of problems that could be solved readily with that information.

## Introduction

The basic aim of this study was to examine differences in user performance due to differences in Prolog tracer format. The results of the effect of format, particularly on the text-graphics dimension, on the utility of information in solving specific problems encountered by novice programmers are reported. A tracer is a basic tool employed in a wide variety of programming tasks as well as being used by beginners to help them learn how to program. In particular, a tracer can explicate the relation between a program as a piece of text and the actions that it describes. So a trace of specific program is a sequence of changes in the data structures and control flow decisions that have caused changes during program execution. In Prolog the tutorial function of the tracer is more marked than in other programming languages because of the complex nature of the underlying mechanisms of this language. Important aspects of any tracer are (i) what information about internal mechanisms and transient states it is designed to show, (ii) the notation or format in which the information is shown, (iii) the method of controlling the tracer i.e. its interface, and (iv) the intended users and expected tasks for which it will be employed (Taylor, du Boulay & Patel, 1991).

In order to be fully conversant with Prolog it is necessary to understand its complex, internal and largely hidden mechanisms. These operations can be described and/or explained in more than one way (Pain & Bundy, 1987) with different perspective emphasising different aspects, such as variable binding, flow of control, recursion, search space, etc., though most tracers usually provide the minimum information necessary for reconstructing the whole 'story' of a program's execution. Tracers are typically evaluated in terms of variance in *explicit* information about these aspects, and not whether such information is present or absent. The explicitness of information is partly affected by format, and that interaction was taken into account in designing this study.

## Differences between Prolog Tracers

This study concentrates on investigating two dimensions on which Prolog tracers vary; the overall amount of explicit information, and, the format of presentation of the information. Three tracers were evaluated. The first tracer, Spy (or Byrd Box) is rather basic with limited explicit information presented in a textual format which is not always easy to comprehend (Byrd 1980). The second one, Enhanced Prolog Tracer for Beginners (EPTB), also textual, was developed by Dichev and du Boulay (1989) to overcome some of the obvious shortcomings of Spy. The third tracer, Transparent Prolog Machine (TPM), is an *idealised* version of a commercial product based on work by Eisenstadt and Brayshaw (1988) which displays certain information such as flow of control and backtracking in graphical representations of AND/OR trees.

How do tracers vary in terms of explicit information? In this context, the term *information* refers to information about when, how and which clauses are matched,

*This work was supported by a grant from the UK Joint Research Council Initiative in Cognitive Science/HCI. The experimental work was conducted using the POPLOG programming environment.

how variables are bound to (and unbound from) particular values at certain points in the running of the program, and the overall flow of control, including backtracking, together with the success or failure of goals. Tracers provide such information more or less explicitly in different ways. For example, Spy does not explicitly indicate which clause of a predicate is being used at any point, whereas EPTB and TPM do. And, Spy refers to program variables by their internal names such as '_405' while, TPM systematically labels variables with letters, and unlike both, EPTB uses the names chosen by the programmer appended with a numerical subscript to distinguish between copies. Though all three methods serve the same function, they are not equally useful.

So tracers can provide useful information in different ways depending on emphasis on particular perspective of Prolog, which affects their relative informativeness. Because information is partly implicit (e.g., down the screen ordering) and partly explicit (numerical goal numbers) it is hard to determine the unavailability of information. In order to avoid confusion due to this it was assumed that, as far as the experimental task was concerned, all three tracers provide the same sort of information, and that any main differences in their usefulness is due to format. More realistically, it is obvious that in most cases there would be some interaction between format and information content, and so any explanation of helpfulness of tracers would have to give an account of such an interaction. But our strong assumption of information equivalence is justified because the stimuli problems were designed and tested to ensure a fair evaluation of similar features of each tracer. Further, the same rigour in designing the task material enabled us to determine more clearly the source of such interaction.

So assuming that the information content of tracers is fixed for all relevant aspects of Prolog, how can they vary in terms of format? Tracers can present information in a *mainly* graphic or text format. Aside from trivial cases, the distinction between graphic and textual format is rarely clear cut; even in a fully textual format the serial ordering of information about states at each step in a program run reflects an implicit notion of temporal ordering; information at any step $n$ is partly dependent on information at step $n-1$ and will itself determine the information content of step $n+1$. Serial ordering information is such a case is not textual but *graphical*. Conversely, a graphical tracer would be of limited value without textual information such as the value of bound variables at particular steps in the run. However, for our purposes a crude distinction between the two categories of tracers is sufficient because our aim is to concentrate on differences in usefulness due to very broad parameters; for example, the difference due to the representation of flow control as graphical AND/OR trees or as linear ordering of textual information.

The following list of general advantages of graphic format also serves to highlight the disadvantages of text format. It is included to anchor some of the basic differences between graphic and text formats. Some of the advantages such as items 7 and 8 are dependent on the extra degrees of freedom provided by computers and bit-mapped screens.

1. Impact.
2. Proportional spatial display can represent degree of relatedness.
3. Clarity in display of non-linearly ordered information such as loops and backtracking.
4. A degree of abstraction can help overcome ambiguity in information.
5. Can display more than one perspective, if information points are related between dimensions. Also possible in textual format.
6. Enables zooming in and out to access essential relationship between information points.
7. Greater possibility of displaying dynamic process (animation) with real time updates. Though possible in text formats they would lack clarity.
8. More *direct* access to information though this depends on the versatility of the display method.

Note that some advantages are not exclusive to graphic formats, and others are of limited value unless augmented with textual information. Aesthetic aspects of graphic representations can improve communication efficiency (Shu 1988), but such improvement often depends on the nature of information being displayed. In the case of Prolog programs there is a limited number of ways in which information about flow of control and backtracking can be graphically displayed. For example, flow of control can be represented as AND/OR trees, OR-trees (Hook, Taylor and du Boulay, 1990) or in terms of flow of satisfaction arrows (Clocksin & Mellish 1981). In such cases the spatial arrangement of certain aspects of information plays a crucial role in capturing essential and important relationships between different parts of a program. This would be expected to be more helpful in solving problems related to flow of control and backtracking. However, major advantages of graphic formats combined with graphic terminals, may be adversely affected by modality mismatch between textual source code and graphic trace, which is a possible source of ambiguity that could affect ease of access to information. Our results indicate that modality mismatch has some adverse effect on subjects' problem solving performance. Thus advantages of graphic format should not be regarded as uniformly beneficial.

## Details of tracers

While it seems that a graphic format has certain advantages over more conventional textual format, other factors can affect their usefulness for particular tasks. Tracers used in this evaluation study are briefly described to illustrate major differences between them. Most of the emphasis is on highlighting the effect of format on information about certain aspects of Prolog, because, though the degree of differences between tracers is partly influenced by how explicit an aspect of Prolog is presented in a trace output, solutions to problems used in this study depend to a very large extent on the effect of format on ease of access to information. Note that the tracers were not used truly as tracers in a dynamic way. In each case an appropriate screen dump of the relevant trace output was shown in its entirety, so users could not 'grow' the trace nor add or delete information from it. Thus, variation between method of control between tracers was eliminated. Our intention was not to examine the tracers 'in the round' but to focus on the influence of format on clarity and accessibility of information.

### Spy

Spy is a very basic textual (linear) tool and is included in this study because subjects were familiar with it. The version used in this study did not show system goals. This tracer provides most of the basic information necessary for programming or debugging in Prolog, but much of it is implicit. In particular, the relationship between the source code and the trace output is not as clearly displayed as it is in TPM and EPTB, which are both designed to overcome some of the obvious shortcomings of Spy. Given the basic lack of clarity in information presentation, Spy was not expected to perform better than TPM and EPTB.

### Transparent Prolog Machine

TPM tracer makes use of a modified and extended AND/OR tree representation, and is an interesting illustration of how graphical techniques can be utilised. The trace outputs used in this study were modified to include all the relevant details which are normally optionally selected by the user. The spatial layout of TPM provides a great deal of information at a glance, particularly on the flow of control and search space. The trace also *seems* a lot less cluttered than Spy or EPTB; it gains in clarity by exploiting some of the advantages of graphic format outlined above. However, the use of a graphical representation of AND/OR trees restricts the screen space available for information about predicates with a large number of arguments and/or long lists, or variables with long names. This constraint is a consequence of the tracer format. This problem can be overcome by including a scrolling facility but the display of essentially textual information is still poor compared to more conventional textual tracers such as EPTB and Spy.

### Enhanced Prolog Tracer for Beginners

EPTB is a textual (linear) tracer (Dichev and du Boulay, 1989) with a strong emphasis on giving information about data structures and variable binding in the trace output. It also has some other features not present in Spy, such as information on reasons for failure of goals. The tracer generally makes more use of labels and symbols to describe different sections of a trace output, and a wide range of options for adjusting the degree of detail shown. It is expected to be more helpful (than Spy or TPM) for solving problems involving data structures and variable binding.

## Task and Motivation

The task was designed to evaluate differences in performance due to tracer type. Problems used in this study can be roughly divided into two distinct groups. Their grouping was crudely determined by how the solution depended on certain aspects of Prolog. Solution to three problems depended on information related to backtracking, clauses tried and undefined predicates. And solution to two more problems depended on aspects of information about recursion, system goals, goals with variables, and list manipulation. The reader need not be fully conversant with these aspects of Prolog. It will suffice to comprehend that these categories refer to distinct aspects of Prolog and that ease of access to information about them is helpful in programming and debugging tasks in Prolog. The experiment examined differences in performance due to tracer (or format) type. Since information about certain aspects of Prolog varied across tracers, as far as possible problems were designed to address information about aspects of Prolog which were common to all three tracers. So, for example, no problems directly involving cuts were included because the big difference between the information provided by Spy (which is implicit) and TPM (which is very clear). Similarly solutions to problems that depended on explicit information about system goals were excluded because Spy and EPTB, unlike TPM, do not provide explicit information about them. Problems related to such obvious differences in the extent and quality of information were avoided. Reference to quality is important because often what is not explicitly stated in a tracer can often be derived from available information on other aspects of the trace. This makes it more difficult to distinguish clearly between explicit and implicit information in trace outputs, which is also affected by format.

The null hypothesis is that there will be no difference in time or response due to trace output. In other words, format is expected to have no effect on either the speed or accuracy of performance. If, however, a significant difference in subjects' performance is evident then it was assumed that this would be due to format and its effect on accessibility to information. This follows from the assumption outlined above that none of the problems were biased in favour of a partic-

| Problem Tracer | Group 1 1 | 2 | 3 | Group 2 4 | 5 | Mean |
|---|---|---|---|---|---|---|
| TPM | 55.2 | 87.1 | 60.2 | 75.3 | 131.9 | 81.9 |
| Spy | 63.8 | 102.7 | 77.3 | 107.4 | 163.9 | 103.1 |
| EPTB | 70.4 | 75.8 | 81.7 | 69.7 | 112.4 | 82.0 |
| Mean | 63.1 | 88.6 | 73.1 | 84.2 | 136.1 | - |

Table 1: ST's (secs.) of Problem by Tracer (n=43)

| Problem Tracer | Group 1 1 | 2 | 3 | Group 2 4 | 5 | Mean |
|---|---|---|---|---|---|---|
| TPM | 90.7 | 88.4 | 72.1 | 34.9 | 41.9 | 66.6 |
| Spy | 93.0 | 58.1 | 76.7 | 11.6 | 27.9 | 53.5 |
| EPTB | 86.1 | 76.7 | 79.1 | 81.4 | 72.1 | 79.1 |
| Mean | 89.9 | 74.4 | 76.0 | 42.6 | 47.3 | - |

Table 2: % Correct Response of Problem (n=43)

ular type of tracer. The preliminary results presented here suggest an interesting interaction between format and information.

## Method

All subjects had been exposed to the Spy tracer and were given a computer-based tutorial designed to explain TPM and EPTB. On reaching a required level of competence in using these tracers subjects had to solve 15 Prolog problems with the aid of trace outputs. The same problem, suitably disguised, was presented three times; each time with a different trace output. Problems were presented in a pseudo-random order. All problems had multiple choice responses. Subjects were asked to solve the problems by referring to the accompanying trace output and to do so as accurately and as quickly as possible. Forty-three subjects completed this task, which together with the initial tutorial took about one hour. Data on time spent on solving each problem and correct response were collected.

## Results

Solution times (ST) ANOVA was carried out with subjects as the random factor and Tracer (3 levels) and Problem (5 levels) as fixed factors. All the data was included in the analyses. There was a significant main effect of tracers, $\mathbf{F}(2,84) = 11.32$, $p \leq 0.001$, indicating that ST is dependent on tracer type. Overall, subjects spent significantly more time on attempts to solve problems with Spy, than TPM or EPTB. There was a significant main effect of problems, $\mathbf{F}(4,168) = 42.14$, $p \leq 0.001$. Variance in problems difficulty would be expected to have this effect on ST. Broadly speaking, Group 1 problems (1, 2 and 3) were easier to solve than Group 2 problems (4 and 5), as is evident from ST's in Table 1. However, far more interesting is the significant interaction between tracers and problems, $\mathbf{F}(8,336) = 3.53$, $p \leq 0.001$. ST's vary according to tracers for each problem, and these differences are not consistent across tracers. For example, TPM aided solutions to problems 1 and 3 require less time EPTB . The reverse is the case for the remaining 3 problems. Subjects are particularly slow at solving problems 2, 4 and 5 with Spy. These sorts of differences in mean ST's reflect differences in format, and therefore, ease in access to information necessary to solve problems. For instance, subjects' faster solutions to problems 1 and 3 with TPM is probably due to their superior display of information which can be taken in 'at a glance' which is particularly useful for solving those problems.

Differences in ST's are more interesting when jointly considered with response data. An ANOVA similar to that of ST data was carried out on responses. There was a significant main effect of problems, $\mathbf{F}(4,8) = 32.89$, $p \leq 0.001$, which reflects the varying level of difficulty of problems. On average subjects found Group 1 problems easier to solve. There was also a significant main effect of tracers, $\mathbf{F}(2,8) = 23.06$, $p \leq 0.001$. Overall subjects performed best with EPTB and worst with Spy, as illustrated in Table 2. There was a significant interaction between tracer and problem, $\mathbf{F}(8,336) = 8.92$, $p \leq 0.001$. Subjects performed less well in solving Group 2 problems with with TPM and Spy than they did with EPTB. Taking into account the ST's for EPTB, the correct response data indicates that the extra time was well spent. For Spy, the results show an inverse relationship; higher ST's for problems 2, 4 and 5 are reflected in far fewer correct responses. Even when trying hard subjects were encountering severe difficulties in solving problems (particularly, problem 4) with Spy. Similarly, TPM was less helpful than EPTB for solving problems 4 and 5, though there is no consistent correlation (inverse or otherwise) with ST's.

Apart from Spy accompanied with problem 2, all tracers seem to be equally useful for solving Group 1 problems. But they take different ST's, which indicates that though all three traces have the information necessary to solve these problems, access to it is significantly affected by format. Given that Group 1 problems are less difficult to solve than Group 2 problems, this indicates EPTB is particularly helpful when solving more difficult or intricate problems in Prolog.

## Discussion

The results show a major effect of tracer format on subjects' ability to solve Prolog problems. Overall, and as expected Spy had the most shortcomings. It does not provide information on system goals, and the simple, linear text format is not good at conveying information about backtracking, more complicated retry clauses, and information about variables involving list manipulations. TPM is good at giving information about system goals and provides list manipulations' information that *looks* less cluttered than Spy or EPTB. However, EPTB is better at displaying information about retry clauses which explains the high frequencies of correct solutions to Group 2 problems. EPTB's good performance across all problems reflects the benefits of a format that eases access to a range of useful information. This results in a modest increase in ST's as users have to work through a linear presentation of information, not allowing them to be as selective about focusing on specific chunks as in TPM. EPTB cannot exploit the advantages of spatial representations for those aspects of Prolog which can be better presented in graphic formats such as TPM. However, this shortcoming is offset by other advantages such as better display of information on list manipulations and retry clauses, both of which do not suffer from the constraints of graphic format - this is particularly true for *long* list manipulations.

The linear nature of textual formats imposes fewer constraints which partly explains EPTB's better performance, and the remarkable difference in TPM's performance between Group 1 and Group 2. Preliminary analyses of ST's and responses highlight two major effects of format on information accessibility. First, graphic formats benefit from greater impact and clarity, and perform best for solving problems associated with the relationship with various clauses in a Prolog program. Second, this advantage only holds for simple programs with short list manipulations and relatively few retry clauses; programs with long list manipulations or many retry clauses severely test the limitations of graphic tracers such as TPM version evaluated in this study. Graphic formats are far more constrained by a lack of space than linear textual formats. In the latter more space is created by simply adding another line; things are rarely that easy to fix in a graphic format tracers, particularly those, like TPM, which represent certain basic information in an AND/OR trees. Augmenting AND/OR trees with textual information can help overcome some of these constraints to a limited extent. However it is not clear that this difference would scale up to large programs written by experts where it has been argued (e.g., Eisenstadt & Brayshaw 1988) that information about the overall *shape* of the Prolog execution tree is advantageous.

Given the findings reported here it seems that for the presentation of information about Prolog programs executions, a textual format tracer would be adequate for novice programmers. This does not imply that graphic formats are inherently unsuitable, but the results indicate that advantages of graphic format vary between different aspects of Prolog. For complicated and inter-related information such as that of a Prolog program execution, gains in clarity of graphic format result in less than adequate treatment of related aspects (though a graphic based approach is probably far more effective for teaching Prolog). Linear textual formats are not similarly constrained; more than one aspect of a complex piece of information can be simultaneously presented.[1] We are at present working on a prototype (mainly textual) tracer tool which incorporates the relative advantages of each format type (Taylor, du Boulay & Patel 1991). It seems that though format has a significant effect on ease of access to information for problem solving, much of this effect is modified by other variables such as the amount and quality of information and its relevance to problem solutions, and until we have a much better understanding of this complex interaction, linear textual tracers such as EPTB are more effective in helping Prolog programmers solve programming problems.


## Acknowledgement

We acknowledge comments from R. Noble and M. Eisenstadt.


## References


Byrd, L. 1980. Understanding the control flow of Prolog programs in Tarnlund S. ed. Proceedings of the Logic Programming Workshop, 127-138.

Clocksin, W.F. and Mellish, C. 1981. *Programming in Prolog,* Springer Verlag.

Dichev, C., and du Boulay, J.B.H. 1989. An Enhanced Trace Tool for Prolog. In Proceedings of the Third International Conference, Children in the Information Age, 149-163. Sofia, Bulgaria.

Eisenstadt, M. and Brayshaw, M. 1988. The transparent Prolog machine (TPM): An execution model and graphical debugger for logic programming. *Journal of Logic Programming* 5(4):277-342.

Hook, K., Taylor, J. and J.B.H. du Boulay. 1990. Redo "Try Once And Pass": the influence of complexity and graphical notation on novices' understanding of Prolog. *Instructional Science* 19 (4-5):337-360.

Pain, H. and Bundy, A. 1987. What stories should we tell novice Prolog programmers? In Hawley, R. ed. Artificial Intelligence Programming Environments. Ellis Horwood.

Taylor, C., du Boulay, J.B.H., and Patel, M.J. 1991. Outline Proposal for a Prolog 'Textual Tree Tracer' (TTT), Cognitive Sciences Research Paper-177, School of Cognitive Sciences, The University of Sussex.

Shu, N. C. 1988. *Visual Programming,* New York: Van Nostrand Reinhold.


[1] This approach could increase the time required to assimilate all the information but that is a separate issue.

# A Modular Natural Language Processing Architecture to Aid Novel Interpretation

**Justin Peterson**
College of Computing
justin@pravda.gatech.edu

**Dorrit Billman**
School of Psychology
billman@pravda.gatech.edu
Georgia Institute of Technology
Atlanta, GA 30332

## Abstract

Successful and robust natural language processing must efficiently integrate multiple types of information to produce an interpretation of input. Previous approaches often rely heavily on either syntax or semantics, verb-specific or highly general representations. A careful task analysis identifies principled subsets of information from across these spectra are needed. This presents challenges to efficient and accurate processing. We present a modular architecture whose components reflect the distinct types of information used in processing. Its control mechanism specifies the principled manner in which components share information. We believe this architecture provides benefits for processing sentences with novel verbs, ambiguous sentences, and sentences with constituents placed outside their canonical position.

## Introduction

The task of natural language processing is to produce an interpretation of the linguistic input. This task is extremely difficult because the mapping from input to interpretation is complex: a single form can have multiple meanings and a single meaning can be expressed in several different ways. Further, language comprehension occurs over a very wide range of circumstances. We comprehend novel utterances, with novel verbs, *Fax me a better copy* as well as close variations on familiar phrases *Hope you're feeling better.* As a consequence, it is hard to identify the information and processes sufficient for the task across varied circumstances. Is the critical information syntactic or conceptual? Is it highly specific, such as a verb's meaning and argument structure, or quite general, such as specifying that direct objects must immediately follow the verb? Are sentence interpretations retrieved from memory or constructed anew?

Our research attempts systematically to identify what distinctions in the input and in memory provide the information needed. By taking a task-analysis approach we attempt to specify just what distinctions of each type are important. We propose that very selective yet heterogeneous clues are used in interpretation. A discussion of this proposal as well the assumptions we have adopted from the work of others can be found in (Peterson & Billman 1990). The focus of this paper, however, is to outline an architecture for sentence interpretation which efficiently operates on these varied sources of information to produce a sentence interpretation.

## Prior Work

Natural language processing systems vary in their assumptions about the importance of syntactic versus conceptual structure. They differ in emphasizing general principles versus specific content knowledge. And they contrast in treating language processing as retrieval of an interpretation from memory versus treating it as generation of a novel structure from much smaller, known, components. Two quite opposite sets of choices have dominated prior work. Some natural language processing systems emphasize detailed, highly specific representations that contain primarily conceptual (or semantic) information and treat interpretation as retrieving the familiar events or situation being described (Riesbeck & Martin 1986). Semantic expectations are central to this approach. We shall call this extreme the conceptual recognition view. Other systems emphasize highly schematic principles of broad scope that are primarily syntactic and that treat sentence processing as generation of the novel structure specified in a unique utterance (Woods 1978). We shall call this the syntactic construction view.

The conceptual recognition view fares well when the sentence does indeed refer to a familiar event using a familiar form. For example, a familiar verb can be accessed from the lexicon, specify slots for its arguments, and thus allow arguments to be bound to a completely prespecified structure in a simple and stereotyped way (Birnbaum & Selfridge 1981). Where expectations and background knowledge are both complete and correct, this provides a fast and effective processing strategy. Difficulties arise if the detailed expectations mismatch the utterance. This might be because 1) the event is unfamiliar, metaphorical, or in some way discrepant from expectation, 2) the sentence form is unfamiliar or novel or 3) a novel verb is used for which the system has no lexical entry. For example, utterances such as *I stared the man out of the room*, a new sentence form, or *Lucia zorched the book into the fire*, a new verb, would be uninterpretable.

The syntactic generation view has a complementary set of faults and virtues. Here the system can process

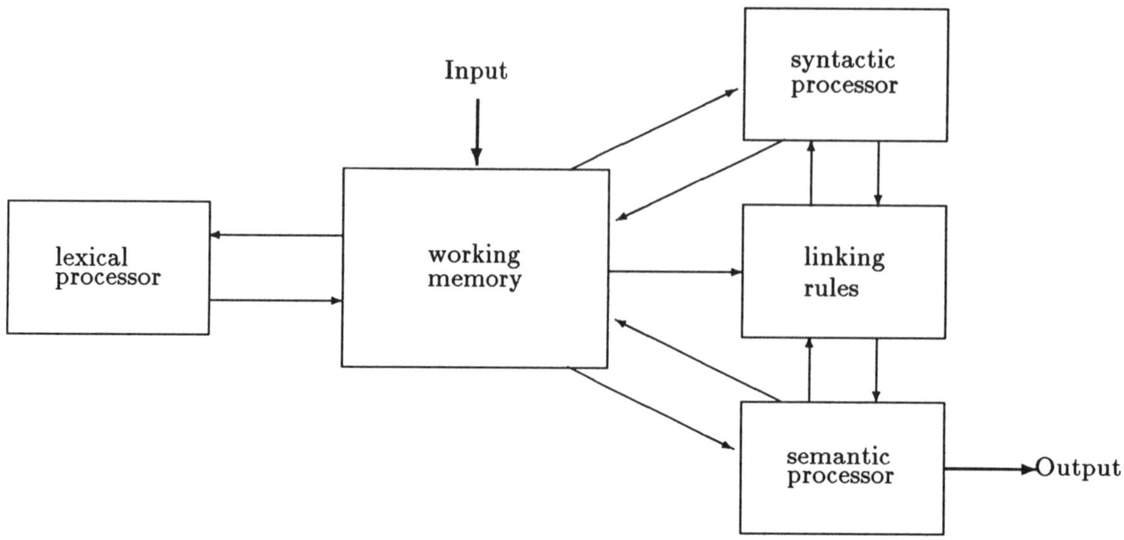


Figure 1: System Architecture

novel utterances as well as stereotyped ones. However, the generality is purchased at the cost of efficiency: a large number of structures might be consistent with the initial segments of input and a forest of possibilities generated. Since many forms can encode similar meanings a multitude of parsing rules are needed and a multitude of partial structures generated. Consider what would be needed to interpret *Which man did Pat punch in the mouth?*: 1) all possible functional roles for *which man* could be generated, 2) the identification of what element is being questioned must be delayed to the end of the sentence, and 3) many sets of rules are needed since the rules needed to process this sentence are not those used for related sentences such as *Pat punched the man in the mouth.*

People do not seem afflicted with either set of problems. New verbs and new uses of old verbs are constantly being introduced into the language; introspection and anecdote show people are quite capable of interpretation even under these circumstances. Psycholinguistic evidence shows that people do begin interpretation of a clause as information accrues and are not hamstrung prior to the final word.

Of course, the field is not as simply or as strongly polarized between the two possibilities as sketched here. Head-driven Phrase Structure Grammar (Pollard & Sag 1987) relies on syntactic information to drive interpretation, yet it specifies this in a very specific, word by word manner, not as general rules. (Lytinen 1986) proposed an approach in which semantic constraints are used to perform syntactic analysis. We believe this type of integration must be taken much further. Successful NLP systems will require a careful combinations of the available techniques and a careful analysis of what information needs to be supplied. Both syntax and semantics, both general and specific patterns, and both construction and memory will play comparably important roles in language interpretation. However, the following questions remain open: What syntactic distinctions map reliably onto subtle differences in meaning? When and in what form are highly specific forms of information used? What form need to be constructed and what can be retrieved from memory? Our research program divides in two parts: 1) a task analysis for identifying critical distinctions that constrain meaning and 2) a processing mechanism that can combine the multiple cue and constraints to provide an interpretation. Clearly both tasks are substantial and in both we are able to capitalize on many parts of analyses done by others. We focus here on describing an architecture that can derive lexical, semantic, and syntactic information and use these distinct but principled sets of cues to effectively drive sentence interpretation.

## Architecture

An implementation of our approach needs to satisfy a number of requirements. First, it must provide a means of mutual communication between syntactic and semantic processing. Second, it must access and use word-specific information when it is available. Third, it must have a general information to resort to when word-specific information is absent. Fourth, both syntactic and semantic information must be used to guide structure composition.

We propose the architecture pictured in figure 1. The architecture has been modularized to reflect the structure of the task. Each module contains a distinct sort of information and distinct function. Within this model, data flows along two paths. Word-specific information flows from the lexical processor to the working memory where it is made available to the semantic, syntactic, and linking rules processor. Structural information flows through the linking rules processor. The syntactic processor and the semantic processor communicate their structural decisions to one another by sending their products to the linking rules processor. The final interpretation is provided by the semantic processor. Below, we describe each of these modules as well as the system control.

## Lexical Processor

The lexical processor provides the word-specific information to the system. It accepts a character string as input, and produces all the lexical entries associated with the word. Each lexical entry is denoted by a unique identifier. This allows individual components to ascertain which entry is being referred to in the working memory. Lexical entries contain syntactic, semantic, and correspondence information.

The syntactic information consists of a lexical category and an argument structure. For verb entries, the syntactic information includes a predicate argument structure which represents the argument structure of semantic functions (Rappaport & Levin 1988). Predicate argument structures perform two services. First, they denote the number of arguments a syntactic structure takes. For instance, the verb *put* has the following arguments

$$put : \begin{bmatrix} external & x \\ direct & y \\ indirect & z \end{bmatrix}$$

This representation indicates that the verb has a subject, a direct object, and a prepositional complement as arguments. Second, they denote the syntactic relations the verb licenses.

$$put : VP \rightarrow NP\ PP[loc]$$

This indicates the verb-phrase can be the mother node for both a noun phrase and a locative prepositional phrase.

The semantic information is either a semantic function or term. Verbs denote semantic functions. Consider the semantic entry for *put*

$$put : \begin{bmatrix}_{event} & ACT(x\ y) \\ & effect\,[_{event}GO(y\ [_{path}TO\,[_{place}z]])] \end{bmatrix}$$

It can be roughly characterized as 'an actor acts upon an object causing it to move along some path'. Variables such as **y** are used to maintain identity relations among the semantic arguments. So for example in the lexical entry for **put**, the thing being acted upon **y** is also the thing being moved. These variables are also used to specify semantic constraints. Above the semantic term assigned to the variable **z** must be a **place**.

The correspondence information maintains the linkings between the syntactic and semantic variables specified in the lexical entry. *Put* specifies the following correspondences

$$put : \begin{array}{lcl} external & = & x\ in\ ACT(x\ y) \\ direct & = & y\ in\ ACT(x\ y) \\ indirect & = & z\ in\ GO(y\ [_{path}TO\,[_{place}z]]) \end{array}$$

## Syntactic Processor

The syntactic processor contains general rules of syntax and constructs syntactic structures. It accepts as input the lexical entries from the lexical processor and any argument bindings already assigned by the linking rules processor (see below). It produces as output a parse tree and predicate argument bindings.

The syntactic processor is an adaptation of Abney's Licensing-Structure Parser (Abney 1989). It has two basic modes of operation: attachment and argument binding. The attachment operation constructs parse trees. The argument binding operation fills the predicate argument structure. Each attachment operation evokes a binding operation updating the predicate argument structure of the mother node.

As an example, consider how parser processes the follow phrase segment

put Mike ...

The *put* is the head of the verb phrase, and *Mike* is a noun phrase. An attachment is licensed by **put**

$$put : VP \rightarrow V\ NP$$

So the noun phrase *Mike* is attached to the verb phrase and assigned to **y**, and the processing of this fragment is complete. In cases where the predicate argument structure for *put* is not available, the syntactic processor invokes a general set of rules that license argument attachment (i.e., lexical redundancy rules).

The syntactic processor must check for internal inconsistency. Parsing inconsistencies are simply misparses. Misparses are identified with the aid of a parsing state. This state is used to indicate how much of the phrase has been parsed and what the possible continuations are. When the input is not a possible continuation, a misparse has occurred.

## Semantic Processor

The semantic processor contains the general semantic information and constructs semantic representations. It accepts as input lexical entries from the lexical processor and semantic function bindings from the linking rules processor. It produces both complete semantic interpretation for system output and partial products used by the linking rules processor.

Like the syntactic processor, the semantic processor combines partial information. The semantic processor has two modes of operation: semantic function argument binding and function composition. Function argument binding assigns values to semantic functions. As an example consider the processing of the following fragment

put Mike

$$\left\{ \begin{array}{l} put : \begin{bmatrix}_{event} & ACT(x\ y) \\ & effect\,[_{event}GO(y\ [_{path}TO\,[_{place}z]])] \end{bmatrix} \\ Mike : [HUMAN\ ANIMATE\ THING] \end{array} \right\}$$

Since **Mike** has no arguments, the possible assignments are to an argument of **put**. There are two possible assignments **y = Mike and z = Mike**. This assignment may be initiated by either the information provided by the linking rules processor or the semantic processor itself. Semantic function composition constructs complex function structures from two functions and will be discussed in an example following this section.

The semantic processor must check for internal inconsistency. There are two types of inconsistencies, violations of composition constraints and of argument binding constraints. The assignment **z = Mike** violates the *place* constraint on **z**; without a locative preposition, **Mike** does not designate a place. When a violation occurs, the semantic processor attempts to propose an alternative.

## Linking Rules Processor

The linking rules processor executes the bi-directional mapping between syntax and semantics. Given a predicate argument binding (produced by the syntactic processor)as input, it produces a semantic function binding as output. Operating in the opposite mode, given function binding (produced by the semantic processor) as input, it produces an predicate argument binding as output. Linking rules allow the system to capitalize on regular correspondences between syntactic arguments and semantic roles, as in lexical redundancy rules. It has a general set of linking rules that can be extended by the correspondences specified in individual lexical entries. When these are provided, it prefers the execution of these lexically specific mappings over the general rules.

As an example consider the following linking rule

$$direct = y \ in \ ACT(x \ y)$$

If the direct object is bound by the syntactic processor, the linking processor can derive a semantic binding from the direct object assignment. The semantic processor executes this binding when it is provided. If it results in an semantic inconsistency, the semantic processor can propose an alternative (which is then communicated to the linking rules processor.) In the case of a conflict, the semantic processor takes precedence.

## Working memory

Working memory executes two roles. It provides a means of communication between the lexical processor and the other processors, and it maintains the current lexical entries, a set of unique lexical items. The lexical processor provides all the lexical entries associated with each particular word. Processing is initiated with most-preferred lexical entries. If either processor finds a lexical entry unacceptable it may select another (i.e., the syntactic processor communicates its lexical category decisions, and the semantic processor specifies its selection of lexical entries.) Working memory must notify the processors of changes in the current lexical entries.

## Communication

Information flows along two paths in the system. Word-specific information flows from the lexical processor to the working memory where it is made available to the semantic, syntactic, and linking rules processor, and structural information flows through the linking rules processor to the semantic and syntactic processors. Consider flow of information during the processing of the text fragment we have been pursuing in this section.

1. *lexical processor* $\rightarrow$
   $[Put.1][Mike.1]$
2. *syntactic processor* $\rightarrow$
   $direct = [Mike.1]$
3. *linking rules processor* $\rightarrow$
   $y/ACT/EVENT = [Mike.1]$
4. *semantic processor* $\rightarrow$
   $$\left[_{event} \begin{array}{l} ACT(...Mike.1) \\ effect\,[_{event}GO(Mike.1...)] \end{array}\right]$$

In step 1, the lexical processor produces the lexical entries for *put* and *Mike*. In step 2, the syntactic processor assigns [Mike.1] to the direct object position. This is communicated to the linking rules processor. This processor produces the semantic argument specification in step 3. In step 4, the semantic processor makes the assignment and executes the inference that Mike is both the thing moving and thing being acted upon.

Global consistency is maintained by keeping the semantic and syntactic processors internally consistent and requiring that the semantic and syntactic processors be consistent with the linking rule products. In cases where the products of the linking rule processor conflicts with the state of either the semantic or syntactic processors, the linking rule products take precedence; linking rule products can only be overridden when they cause syntactic or semantic inconsistencies or violate semantic preferences.

## Novelty Example

As an example of how such information could be used to process novel verbs, consider

Lucia zorched the book into the fire.

Since the predicate argument structure for *zorched* is unknown, the syntactic processor must construct the parse tree with the lexical redundancy rules, producing

$$\left[\begin{array}{ll} external & [Lucia.1] \\ direct & [book.1] \\ indirect & [into\,[fire.1]] \end{array}\right]$$

The linking rules processor executes the mappings which are ambiguous. The subject-direct object pair could correspond to a relation of possession or causality.

$$x/ACT/EVENT = [Lucia.1]$$
$$x/HAVE/STATE = [Lucia.1]$$

$$y/ACT/EVENT = [book.1]$$
$$y/HAVE/STATE = [book.1]$$

$$z/GO/EVENT = TO(IN([fire.1]))$$

Given the mappings, the semantic processor attempts to compose the predicates. The mapping ambiguity is resolved by the constraints on function composition, when the partial results are combined. The information associated with *into* is an effect of **GO**ing and this must co-occur with actions; hence, the **HAVE** mapping (a nonaction) can be ruled out. Only effects such as **GO** co-occur with actions, so the **HAVE** mapping can be ruled out.

$$\left[_{event} \begin{array}{l} ACT([Lucia.1]\ [book.1]) \\ effect\,[_{event}GO([book.1]\ TO(IN([fire.1])))] \end{array}\right]$$

The resulting interpretation refers to an event in which Lucia's acting on the book causes it to go into the fire.

## Summary

Our approach to processing provides a means of handling novelty. It also helps with some other difficult problems in natural language comprehension. Consider the case of prepositional attachment ambiguity. Giving the semantic processor the ability to propose structure allows it not only to identify semantic anomalies in the attachments proposed by the syntactic processors but also propose alternative bindings. It also provides a means of making early commitments in wh-questions. Rather than waiting for a gap in the parse tree, the syntactic processor can tentatively assign the wh-phrase to a syntactic argument. This assignment is sent to the semantic processor, allowing it to evaluate the assignment and to propose a new one if the assignment does not satisfy the semantic constraints. Finally, it aids in resolving lexical ambiguity. Changes to the lexical semantic representation of verbs can be partial. We need not discard the assignments that have already been made, if they still hold for the altered verb representation.

## References


Abney, S. 1989. A Computational Model of Human Parsing. *Journal of Psycholinguistic Research* 18(1):129-144.

Birnbaum, L. & Selfridge M. 1981. Conceptual Analysis of Natural Language In R. Schank & C. Riesbeck (Eds.), *Inside Computer Understanding.* Hillsdale, NJ: Lawrence Erlbaum Associates.

Rappaport, M. & Levin, B. 1988. What to do with Theta-roles. In W. Wilkins (Ed.), *Syntax and Semantics, Volume 21: Thematic Relations.* New York: Academic Press.

Lytinen, S. 1986. Dynamically combining syntax and semantics in natural language processing. *Proceedings of the Fifth National Conference on Artificial Intelligence*, Los Altos, Ca.: Morgan-Kaufman. 574-578.

Peterson, J. & Billman D. 1990. I'm With her and the butler did it With the Knife. Unpublished Manuscript.

Pollard, C. & Sag, I. 1987. *An Information-based Syntax and Semantics, Volume 1: Fundamentals.* Stanford, CA: Center for the Study of Language and Information.

Riesbeck, C. & Martin, C. 1986. *Towards Completely Integrated Parsing and Inferencing.* Proceedings of the Eighth Annual Conference of the Cognitive Science Society. Hillsdale, NJ: Lawrence Erlbaum Associates.

Woods, W. 1978. *Semantics and Quantification in Natural Language Question Answering.* In M. Yovits (Ed.), *Advances in Computers, Volume 17*, 2-64. New York: Academic Press.

# Information Gathering as a Planning Task

**Louise Pryor and Gregg Collins***

The Institute for the Learning Sciences
Northwestern University
1890 Maple Avenue, Evanston IL 60201
pryor@ils.nwu.edu, collins@ils.nwu.edu

**Abstract**

Existing planners fall into two broad categories: reactive planners that can react quickly to changes in the world, but do not project the expected results of a proposed sequence of actions, and classical planners that perform detailed projections, but make assumptions that are unrealistic when operating in a complex and dynamic world. Ideally, a planning agent in such a world should be able to do both. In order to do this, the agent has to be able to differentiate between those situations in which detailed information would aid it in making its decisions, and and those in which such information would not materially improve its performance. We propose an approach to this problem, using well-characterized heuristics to decide what information would be useful, whether to gather it and if so, how.

## Introduction

There have been two main approaches to the problem of planning in complex worlds: ignoring the complexity of the world, and ignoring the complexity of the reasoning required to operate in it. So-called "classical" planners are designed to operate in very simple worlds of which they have certain knowledge. (Sussman 1975, Sacerdoti 1977, Chapman 1987). Planning is assumed to occur in advance of execution, and to be subject to no limitations on the time allowed to produce a plan. When there are several possible courses of action, such a planner can thus predict the results of each and choose the one with the best pay-off. These planners can produce plans that are optimal, given the assumptions under which they are working; however, those assumptions are highly unrealistic. So-called "reactive" planners are designed to operate in more complex worlds (Brooks 1986, Agre & Chapman 1987, Firby 1989). Rather than planning in advance, such systems decide *at execution time* on the actions to be undertaken, based solely on the current circumstances. Of course, such systems are not immune from the problematic interactions that projection is designed to discover; but it is assumed that the system's action rules have been written in a way that largely avoids such interactions. The issue of *how* these rules are written has largely gone unaddressed.

* This research was supported in part by DARPA, monitored by AFOSR under contract F49620-88-C-0058, and by AFOSR under contract AFOSR-89-0493. The Institute for the Learning Sciences was established in 1989 with the support of Andersen Consulting, part of The Arthur Andersen Worldwide Organization. The Institute receives additional support from Ameritech, an Institute Partner, and from IBM.

Ideally we would like a planning agent to be able to move quickly enough to take advantage of opportunities as they arise in the world, while at the same time deciding what to do based on projection of the expected consequences of proposed action sequences. If we view the two approaches to planning described above—complete projection on the one hand, pure reactivity on the other—as the ends of a spectrum of possible approaches, we would thus like a planning agent to fall somewhere in the middle, and to be able to adjust its position in the spectrum according to the demands of the moment. Such an agent would thus be required to balance the benefits of projection, in the form of making better decisions, against the costs, in the form of the time and information required. A trade-off must be made between increased accuracy in projection, and the increased cost of planning. In this paper we discuss an approach to this process, concentrating especially on the problem deciding under what circumstances the agent should try to gather more information.

## Rock delivery

To see what issues are involved in performing plan projection on the fly, let's consider the robot delivery truck domain used by Firby (1989) and Hanks (1990). Suppose that the truck is parked next to a red rock and a green rock, and has a goal to deliver a red rock to the red rock store, and another goal to deliver a green rock to the green rock store. Should it pick up both rocks now, or should it pick up only one? There are opportunities to further both delivery goals, and it must be decided whether to take advantage of either or both of them. If, for example, the truck had another goal, involving, say, blue rocks, that could not be achieved if the time were spent on picking up red or green rocks, it might be a good decision to pick up neither rock. If the truck were running out of fuel, picking up both rocks (which uses little fuel) then going off to refill its tanks would make sense.

When a planning agent has recognized an opportunity to achieve a goal, it must consider whether taking advantage of the opportunity would be worthwhile. If taking advantage of the opportunity would be relatively costly compared to other possible methods of pursuing the goal whose achievement is facilitated by the opportunity, it may be best to ignore it. On the other hand, if there is an opportunity to do something very cheaply that would provide a large pay-off, this opportunity may be worth pursuing even at the cost of suspending other ongoing plans.

### Utility

Traditionally, a goal is a logical expression describing two regions of the outcome space: that in which the goal is satisfied (the set of states with high utility) and that in which it is not (the set of states with low utility). Haddawy & Hanks (1990) extend this notion by viewing a goal as describing a partition of the outcome space such that within each region of the partition, all utility values fall within given bounds. Using this view, we can express the idea that many outcomes, while failing to achieve the stated goal completely, nonetheless have a high utility with respect to that goal. The more complex characterization allows a planning agent to consider the effects of different ways of failing to achieve a goal, as well as providing a simple way of expressing goals that may not have absolute criteria for achievement (such as a goal to deliver as many rocks as possible), and weighing them against other goals in a rational manner.

Although the decision procedure can be summarized by saying "choose the course of action with the highest expected utility" (Feldman & Sproull 1977), the calculation of the expected utilities is often no simple matter, as the following examples demonstrate.

### Deadlines

If our delivery planner has to decide whether to pick up the red rock, the green rock, or both, it should compare the utility values arising from the expected delivery times for both rocks in all three cases, and choose the action with the highest associated value. The notion of a deadline can be expressed by saying that all states in the outcome space which involve delivering the rock before a certain time have a high utility associated with them.

Given the expected utility values for the various courses of action, choosing the one that results in the highest expected utility value is a simple optimization problem. Unfortunately, the specific utility functions for, say, the delivery of the red and green rocks will not usually be explicitly available. In order to calculate the expected utility of a particular course of action, we must calculate the probability of each possible outcome associated with that course of action, and the associated utility. Such calculations, which are based on Bayesian theories (Hanks 1990), depend in turn on many values of both prior and conditional probabilities.

In order to decide what to do, a planner must formulate a set of possible courses of action, then project the consequences of each. Each such projection requires information about the possible results of all the component actions. The real problem in choosing a plan is therefore one of information. If sufficient information is available, the decision-making process becomes simple.

### Resources

Many of the conditional probabilities involved in evaluating a proposed action sequence are based on reasoning about the resources that will be consumed in performing the sequence. For example, if the red and green rocks are to be delivered to different locations, the relative positions of those locations can influence the decision as to which rocks to pick up now. If one destination is directly on the way to the other, or requires only a slight detour, picking up both rocks at once is likely to provide significant savings in time, fuel consumption, and wear and tear on the truck over delivering one rock before returning to pick up the other. If, on the other hand, the two locations are far apart and a good route between them passes through the place at which the decision is being made, it may be better to deliver one and then return for the other.

In such cases there are many factors to be balanced, all involving a detailed domain theory. The knowledge that is needed includes facts such as:

- Higher speeds use more fuel, but may help meet a deadline.
- A bad hilly road may be shorter than a good flat road.
- Speeds are lower on hilly roads than on flat roads.
- There is more wear and tear on the truck on bad roads, and fragile contents may be damaged.

Again, we can see that the more information, in the form of a detailed domain theory, that the agent has, the better its decisions can potentially be.

### Alternative achievement

When deciding whether to take advantage of an opportunity to achieve a goal, the agent must also consider other available opportunities to achieve that goal, and whether they would have higher overall utility. For example, if the planning agent knows that there is a red rock at the green rock store, an effective course of action might be to deliver the green rock, then pick up a red rock at the time of delivery, rather than pick up a red rock now and carry it around.

The factors that should be taken into consideration in cases like these include:

- Likelihoods of other opportunities existing as expected, and unexpected opportunities arising.
- Costs of taking advantage of the various opportunities.
- Creation of opportunities through substitution: if the top level goal is to fill a container with red objects, for example, the extra benefit obtained by delivering a (large) red rock instead of a (small) red apple might not be worth the extra costs.

In such cases the type of information that helps in the decision-making process is information about the state of the world, including information about the probabilities of various facts being true.

## Information and decisions

Although the specification of factors relevant in decision making in the examples given above is not nearly complete, in most cases we would not need to consider the decision in even this much detail. Typically we would believe, for example, that the extra load on the truck from a single rock would affect its performance very little, that the time taken to pick up one rock is unlikely to affect our ability to deliver the other rock on time, and that there are few green rocks in the world so we should take advantage of the opportunity to acquire this one. If these beliefs are true, performing more detailed reasoning will not improve the decision. In many cases, then, it appears that a decision can be made without considering many of the details of the situation.

However, in cases where decisions are more finely balanced, performing more detailed reasoning could make all the difference. Detailed estimates of timings, routes to be taken and so on, are needed when the deadlines are tight, for example. In order to decide on a course of action and be *sure* that it is the best one available, all possible courses of action must be considered and their possible outcomes projected. In an uncertain world, this projection cannot be performed deterministically, but must use probabilistic reasoning. The projection process needs information about probabilities for a wide range of possible events, about the actual state of the world as observed by the planning agent, and about the utilities of all the actions and outcomes to be considered. The projection process also requires time.

An agent operating in a complex, dynamic world frequently has to make decisions, and often crucial information is unavailable and time is severely limited. There are therefore many occasions when the agent simply cannot perform the detailed projections needed in order to make the best decision. There may well be occasions on which the benefit gained from making the right decision is less than the cost of making the decision. In such cases, there is no net benefit derived from the effort that is put in to making the decision, and overall a better result would be achieved by simply picking a course of action at random.

## Gathering information

A planning agent can choose actions in the service of the goal to decide which action to perform next as well as those in direct service of its explicit goals. Among these actions are methods of gaining more useful information about the situation. For instance, suppose the agent has decided to deliver the red rock and is now trying to decide whether to pick up the green one. If there were a telephone available at the destination of the red rock, it might be worth calling ahead to find out whether there is a green rock there, too. Using the telephone presents an alternative to cogitating about the possible presence or absence of green rocks based on the agent's prior knowledge. Of course, the planning agent probably does not want to call all possible places in the world to see if there are green rocks available, but should confine its calls to those places which fit in with its other plans. It needs to have a characterization of the type of information that will be useful in making the decision, and the likelihood of gaining it through performing various actions, so that it can decide which, if any, to perform.

## Planners and information gathering

Purely reactive planners choose actions based on features of their environment, with no explicit projection. In effect, they have a set of rules that guess what the result of the projection would be if it were done, in that the rules are intended to make the agent react appropriately. Firby's RAPs (reactive action package) system (1989), for example, has no explicit knowledge of any probabilities that might affect its behavior. Instead of knowing that an action is especially unreliable, the RAPs system simply doesn't try one that has failed often in the recent past. Agre & Chapman's Pengi (1987) also performs no explicit projection. The system has no preconceived notion of the results of performing an action, but has rules that arbitrate between possible actions. The result of these rules is intended to be that the most appropriate action is chosen, but no representation or search of future world states is performed when applying them. Instead the system's designer has built the results of such reasoning into the rules. Classical planners, on the other hand, explicitly project the results of all the actions that they consider. However, it is assumed that all the required information is available to the system at no cost. Gathering information is thus not an issue for either of these types of systems.

## Deciding to gather information

We need some way of looking at the utility obtained from gathering information in terms of the role played by information in improving decisions. We also need to be able to estimate the costs involved in gathering it. Horvitz *et al* (1989), have looked at the possibility of controlling decision-theoretic inference by considering the expected value of such inference in relation to the decision to be made. Their method uses the properties of *belief networks* (Howard & Matheson 1981, Pearl 1986), and in particular the properties of the algorithm they use to evaluate the networks.

Unfortunately, reliance on a particular algorithm for evaluating belief networks is unlikely to be very useful in more dynamic domains, where the agent can find itself in totally new situations. As we discussed earlier, there is a large amount of information required in order to make good projections of the expected utility of the possible outcomes. If belief networks are used to perform the projection, a new network would have to be constructed for each decision in order to take the relevant influences into account. Constructing the necessary dependencies would be prohibitively expensive, especially as the decisions must be made at execution time.

An alternative approach is suggested by viewing information gathering as just one of the many possible tasks that can be performed at any particular time, and looking at how existing planners decide between tasks. Many systems use *heuristics* to make decisions when they cannot perform a complete analysis. Work has also been done on approximate analysis (Hanks 1990, Wellman 1990). The success of these systems suggests that heuristics and approximations can provide useful information that can help in making a decision. If we view information gathering as a planning task, it is reasonable to ask whether the sorts of decision heuristics employed by planners can also be applied to the problem of deciding when to gather more information. We believe that the answer is yes.

However, in general the heuristics used by existing planners are at best *ad hoc*, and are often implicit in the control structure. We intend to characterize useful heuristics and approximations in a more principled way, so that we can understand when and how they should be used. This characterization will have to specify the results produced by the heuristics in terms of the quality of the decisions that use them, and the costs of using them in terms of time and other resources. We will investigate their use in the context of planning in a complex and dynamic domain so that their applicability can be demonstrated.

## Effective independence

When a planner is considering whether to perform an action, it should consider the benefits and costs associated with the action. The benefits of the action can be measured through its (positive) effect on the goal that it directly furthers and on other goals. The costs can be measured through its (negative) effects on its own and other goals. The resources that are consumed by performing the action affect goals inasmuch as they cannot then be used by other actions. The calculation of these benefits and costs requires information, as we have discussed above, and we intend to investigate methods of judging what information would be useful, how that information can be gathered, and whether it is worth gathering. We have not yet considered he question of what effect an action has on the goal that it directly furthers, and the information required in order to judge that effect, but have started by looking at an action's effects on other goals.

Our first step is to come up with heuristics that indicate when the action's effects on goals that it does not directly further can be ignored. In this case the planner does not need information enabling it to calculate those components of the utility of the action. On the other hand, if the effects on other goals are significant, the information relating to them is required in order to make a good decision.

Clearly, if an action is independent of all goals that it does not directly further, the planner can indeed ignore its effects on other goals. Most actions are not, unfortunately, independent of all other goals. In our delivery example, for instance, the goals to deliver the red and green rocks interact in important ways. These interactions are responsible for most of the complications we discuss. However, in many cases such interactions are insignificant. When we say that the exact time taken to pick up the green rock is unimportant because it is very short compared to the time available for the delivery of the red rock, and that the weight of the green rock can be ignored as being negligible compared to the overall weight of the truck, we are saying that we can treat the action of picking the green rock up as though it were independent of the goal to deliver the red rock. In this case we say that the action of picking the green rock up is *effectively independent* of the goal to deliver the red rock. When actions are not effectivley independent, we need to reason more about the interactions. Heuristics to decide when actions are effectively independent would therefore be very useful in determining whether we need more information about the situation.

### Violations of effective independence

At present we are considering two principal ways in which independence may be violated: through preconditions and through post-conditions.

Two of the more common types of preconditions are that the agent has sufficient resources to carry out a plan that would achieve the goal, and that certain states of the world are true before a task is performed. In many cases, the resources needed in order to achieve a goal are plentiful. However, conflicts can arise when they are limited:

- If the truck were nearly full, picking up the green rock would leave no room for anything else. This has no effect on the goal to deliver the red rock, which is already in the truck, but would mean that other delivery tasks could not be undertaken.
- If picking up a rock uses a large amount of fuel, and the truck's tank is nearly empty, there may not be enough to pick up the green rock and deliver the red rock without refilling the tank. Thus the goal to deliver the red rock could be affected in a couple of ways: the truck could run out of fuel, so preventing the delivery, or could refuel along the way, thus delaying the delivery.

Another type of independence violation through preconditions arises from the fact that performing an action can sometimes change the expected state of the world in a way that interferes with the achievement of another goal. Elements of the world state that can be affected include the weight of the truck, time, and location:

- Suppose that the truck's being under a certain weight is a precondition of being able to cross a bridge which is on the route to deliver the red rock. If the green rock is very heavy, picking up the green rock would violate that precondition by changing the state of the truck so that the bridge would no longer bear it.
- If there is a deadline for the delivery of the red rock, a precondition for the delivery is that the transporting task should not start too late. Picking up the green rock would delay the time at which the truck could start on the task of taking the red rock to its destination, possibly violating the precondition if the deadline is tight.
- Even if there is no green rock at the same place as the red rock, there is always an opportunity to move elsewhere to collect a green rock before continuing with the task of delivering the red rock. The change of location would normally significantly affect the task of transporting the red rock, which would have to start from a place other than that which was expected.

An action can affect another goal by changing the state of the world by making it impossible to achieve the goal:

- Suppose that there is only one green rock, but that there are also some green apples. The truck has goals to deliver a green rock and a green object to separate locations. If it tries to achieve the second goal by delivering the green rock, it has made the first goal unachievable.

Of course, in the last example, it is still possible to achieve the goal of delivering a green rock. In order to do this, however, the goal of delivering a green object would become unsatisfied: a post-condition of the goal is that the green object should be left at the location to which it is delivered, and not moved away from that place. In this situation the agent would need to acquire information about the ways in which both goals can be satisfied, by using a green apple to achieve the goal to deliver a green object, for example.

### Detecting threats to independence

There are three main ways in which the threats to independence that we have discussed can be detected:

- An action may be known to be liable to have a certain effect on goals. For example, moving to another location will always cause a change of state, and should prompt the agent to look for possible conflicts.
- A goal may be known to be sensitive to certain types of interference. If the red rock is particularly fragile, for instance, the agent should beware of actions involving objects that might damage it. The principle involved is very similar to the notion of protections used in classical planning (Sussman 1975, Sacerdoti 1977). Instead of a violation being fatal, however, in this case it simply means that there may be a problem and that the matter should be investigated further.
- A resource or state may be prone to conflicts. For example, if there is little fuel, or any other resource is in short supply, the agent should be on the look out for actions that affect that resource.

A planning agent can use these strategies in a two-stage process to determine if there is effective independence. The first stage uses heuristics to decide whether to perform the second, more detailed, stage.

In order to judge whether an action and a goal are effectively independent, the planning agent should consider aspects of the action, goal, and world state. The agent should perform more detailed analysis in the following cases:

- Action-driven
  - There is a resource which the action uses heavily.
  - There is a world state which the action changes significantly.
- Goal-driven
  - The achievement of the goal relies on a resource used by the action.
  - There are any world-states that are fatal to the achievement of the goal.
- World-driven
  - There are any resources that are in short supply.

## Conclusion

An intelligent agent must be able to differentiate between important decisions and those that have few or unimportant consequences. The complexities involved in such reasoning in a dynamic world, in which exactly the same situation will probably never arise twice, are enormous. A decision-maker therefore must be able to reason about the type of decision required in a particular set of circumstances and the information that is needed in order to be able to make that decision effectively.

This type of decision behavior depends on the ability to estimate how good we expect a decision to be, and the expected benefits of making it better. This means that we need ways of characterizing the quality of decisions, how the information used in them affects that quality, and the availability of relevant information. All these estimates must in turn be related to the expected cost of gathering the information. An agent reasoning in this way about decision making must be able to so efficiently, otherwise the costs of the reasoning will outweigh the benefits.

As we discussed above, the reasoning involved in making decisions about planning can be very complex and time-consuming, and rely on large amounts of information which may not be readily available. One approach to carrying out this reasoning in a timely manner is to develop heuristics and approximations that can be used when detailed reasoning is inappropriate. Such heuristics must also be used when considering whether it is worth gathering more information, either by performing more reasoning or by other methods or whether the less exact methods will be more appropriate in the circumstances.

We intend to continue our investigation of effective independence as one approach to the problem of characterizing useful information in the context of planning decisions. Effective independence concerns the relationship between an action and a goal it does not directly further, and we will look at other aspects of that relationship as well as the way in which an action affects the goal that it serves.

**Acknowledgements:** Thanks to Larry Birnbaum, Matt Brand, Mike Freed and Bruce Krulwich for many useful discussions.

### References

Agre, P. E., & Chapman, D. (1987). Pengi: An implementation of a theory of activity. Proceedings of AAAI-87.

Brooks, R. (1986). A robust layered control system for a mobile robot. *IEEE Journal of Robotics and Automation*, vol 2, no 1.

Chapman, D. (1987). Planning for Conjunctive Goals. *Artificial Intelligence*, 32, 333-337.

Feldman, J. A., & Sproull, R. F. (1977). Decision Theory and Artificial Intelligence II: The Hungry Monkey. *Cognitive Science* 1, 158-192.

Firby, R. J. (1989). Adaptive execution in complex dynamic worlds. Technical Report YALEU/CSD/RR #672, Dept. of Computer Science, Yale University, New Haven CT.

Haddawy, P., & Hanks, S. (1990). Issues in Decision-Theoretic Planning: Symbolic Goals and Numeric Utilities. Innovative Approaches to Planning, Scheduling and Control: Proceedings of a Workshop. DARPA, San Diego.

Hanks, S. J. (1990). Projecting plans for uncertain worlds. Technical Report YALEU/CSD/RR #756, Dept. of Computer Science, Yale University, New Haven CT.

Horvitz, E. J., Cooper, G. F., & Heckerman, D. E. (1989). Reflection and Action Under Scarce Resources: Theoretical Principles and Empirical Study. Proceedings of the Eleventh International Joint Conference on Artificial Intelligence, Detroit, MI.

Howard, R. A., & Matheson, J. E. (1981). Influence Diagrams. In R. A. Howard & J. E. Matheson, editors, *Readings on the Principles and Applications of Decision Analysis*, vol II, pp721-762. Strategic Decisions Group, Menlo Park, CA.

Pearl, J. (1986). Fusion, Propagation, and Structuring in Belief Networks. *Artificial Intelligence* 29, 241-288.

Sacerdoti, E. (1977). *A Structure for Plans and Behavior*. American Elsevier, New York.

Sussman, G. J. (1975). *A Computer Model of Skill Acquisition*. American Elsevier, New York.

Wellman, M. P. (1990). *Formulation of Tradeoffs in Planning under Uncertainty*. Pitman, London and Morgan Kaufman, San Mateo, CA.

# Evaluation of explanatory hypotheses[1]

**Ashwin Ram**
College of Computing
Georgia Institute of Technology
Atlanta, Georgia 30332-0280
E-mail: ashwin@cc.gatech.edu

**David Leake**
Computer Science Department
Indiana University
Bloomington, Indiana 47405-4101
E-mail: leake@cs.indiana.edu

## Abstract

Abduction is often viewed as inference to the "best" explanation. However, the evaluation of the goodness of candidate hypotheses remains an open problem. Most artificial intelligence research addressing this problem has concentrated on syntactic criteria, applied uniformly regardless of the explainer's intended use for the explanation. We demonstrate that syntactic approaches are insufficient to capture important differences in explanations, and propose instead that choice of the "best" explanation should be based on explanations' utility for the explainer's purpose. We describe two classes of goals motivating explanation: knowledge goals reflecting internal desires for information, and goals to accomplish tasks in the external world. We describe how these goals impose requirements on explanations, and discuss how we apply those requirements to evaluate hypotheses in two computer story understanding systems.

In order to learn from experience, a reasoner must be able to *explain* what it does not understand. When a novel or poorly understood situation is processed, it must be interpreted in terms of knowledge structures already in memory. As long as these structures provide expectations that allow the reasoner to function effectively in the new situation, there is no reason to revise them. However, the expectations may fail to apply. In that case, the reasoner is faced with an *anomaly* — a conflict between expectations and new information — and learning is needed to prevent future failures. In order to revise its knowledge, the reasoner needs to know *why* it made the mistaken predictions, and to explain *why* the failure occurred. In other words, it must identify the knowledge structures that gave rise to the faulty expectations, and understand why its domain model was violated in this situation. Once revised, the knowledge can then be stored in memory for future use. *Abduction*, the construction of explanations, is a central component of this learning process.

Abduction is often viewed as inference to the "best" explanation. However, there is often no one "right" explanation; the definition of "best" is dependent on the *goals* of the reasoner in forming the explanation. In these situations, the "best" explanation must be more than a causal chain that correctly describes the domain; it must also address the reason that an explanation was required in the first place. The extent to which it does so determines how effectively the reasoner can learn from the explanation.

This paper addresses the problem of evaluating the "goodness" of hypotheses in the context of a reasoning task. We focus on *explanatory hypotheses*, which are causal chains that attempt to justify a given anomalous fact in terms of causal relationships with reasoner's prior beliefs. We begin with discussing the nature of explanatory hypotheses. We then discuss two classes of evaluation criteria for such hypotheses. *Structural criteria* rely on structural or syntactic properties of the causal chain. For example, a reasoner might always choose the shortest causal chain as the best explanation. *Utility-based criteria* select hypotheses according to requirements arising from the system's intended use for an explanation, such as forming predictions about future events.

Although structural criteria have received the most attention in artificial intelligence programs, we will argue in favor of utility-based criteria on functional grounds. Since these criteria judge candidate hypotheses with respect to the particular task and reasoning needs that the reasoner is currently faced with, they are more likely to enable the reasoner to pick explanations that are "right" for the occasion.

We divide utility-based evaluation criteria into two subclasses: those based on *knowledge goals*, and those based on *tasks*. Criteria based on knowledge goals attempt to evaluate a hypothesis based on the reasoner's internal needs for knowledge. For example, if a system is trying to form a distinction between two categories in memory, an explanation that provides the information required make such a distinction is more useful even though it may not be the shortest explanation. Task-based criteria evaluate hypotheses from the point of view of the real world tasks that the reasoner is trying to perform. A detective, for example, might need to build a very different explanation for a stain of blood on a carpet than a cleaner trying to figure out how to remove that stain.

Our theory is based on a functional analysis of the purposes for which explanations will be used, and motivated by empirical psychological evidence. People quite clearly have what psychologists often call "goal orientations," which have a significant effect on the inferences that people draw from their experiences. There is a large body of psychological research on goal direction in focus of attention, particularly from social psychology. Zukier's (1986) review concludes: "Experimental studies have clearly demonstrated that a person will structure and process information quite differently, depending on the future use he or she intends to make of it" (p. 495). AI researchers have also proposed theories of "subjective interpretation" (e.g., Carbonell (1979), Ram (1989)) in which a system's prior beliefs and goals influence the interpretations drawn in a given situation. In addition, AI research on operationality of explanations has considered the question of

[1]Ashwin Ram's research was supported in part by the National Science Foundation under contract IRI-9009710. Both authors' research was also supported in part by the Defense Advanced Research Projects Agency and the Office of Naval Research under contract N00014-85-K-0108, and by the Air Force Office of Scientific Research under contracts F49620-88-C-0058 and AFOSR-85-0343.

generating suitable explanations for system use, although the range of uses considered has been limited (see Keller (1988) for a discussion of operationality considerations).

From the functional point of view, these ideas are related to the "goal satisfaction principle" of Hayes-Roth and Lesser (1976), which states that more processing should be given to knowledge sources whose responses are most likely to satisfy processing goals, and to the "relevance principle" of Sperber and Wilson (1986), which states that humans pay attention only to information that seems relevant to them. These principles make sense because cognitive processes are geared to achieving a large cognitive effect for a small effort.

Utility-based evaluation of explanatory hypotheses attempts to maximize the utility of explanations by explicitly considering the needs of the reasoner in forming the explanation. Further examples of the task-sensitive nature of human explanation can be found in psychological research on excuse theory, which demonstrates that people manipulate explanations to displace blame for poor performance (e.g., Mehlman & Snyder (1983)).

## The explanation cycle

The process model for the task of explanation consists of the following steps:

1. **Anomaly detection:** Identification of an unusual fact that needs explanation. (See Leake (1989a; 1989b) and Ram (1989) for our approaches to anomaly detection.)
2. **Explanatory hypothesis construction:** Construction of one or more explanatory hypotheses that would resolve the anomaly and explain the situation. This is typically done by chaining together causal inference rules through a search process (e.g., Rieger (1975), Morris & O'Rorke (1990)), through a weighted or cost-based search (e.g., Hobbs *et al.* (1990), Stickel (1990)), or through a case-based reasoning process in which previous explanations for similar situations are retrieved and adapted for the current situation (e.g., Schank (1986), Kass *et al.* (1986), Leake (1989b), Ram (1990b)).
3. **Hypothesis verification:** Confirmation or refutation of possible explanations or, if there is more than one hypothesis, discrimination between the alternatives. A hypothesis is a causal graph that connects the premises of the explanation to the conclusions via a set of intermediate assertions. At the end of this step, the reasoner is left with one or more alternative hypotheses. Partially confirmed hypotheses are typically maintained in a data dependency network (e.g., AQUA's hypothesis tree (Ram, 1989)).

## Evaluation criteria

Regardless of how explanatory hypotheses are constructed, the evaluation of these hypotheses (step 3 in the explanation cycle above) is a central and difficult problem. We categorize evaluation criteria into structural (or syntax-based) and utility-based (or goal-based) criteria.

### Structural criteria

Structural criteria use the structural or syntactic properties of the causal chain to evaluate hypotheses. A goodness measure for each hypothesis is computed based on the length of the causal chain, the number of abductive assumptions, or other such structural properties.

Most structural criteria appeal to Occam's razor by requiring *minimality* of hypotheses. Simply stated, a hypothesis that is "minimal" with respect to some criterion is preferred over one that is not (e.g., Charniak (1986), Kautz & Allen (1986)). For example, Konolige (1990) argues that "closure + minimization implies abduction." To take another example, the TACITUS system for natural language interpretation merges redundancies as a way of getting a minimal interpretation, which is assumed to be a best interpretation (Hobbs *et al.*, 1990). Minimality criteria include:

- **Length:** Causal chains with the shortest overall length are preferred.
- **Abductive assumptions:** Explanations requiring the fewest abductive assumptions are preferred.
- **Subsumption:** If two candidate hypotheses are found and one subsumes the other, the more general hypothesis is preferred.

Another approach focuses on the structural relationship of propositions in an explanation rather than minimality:

- **Explanatory coherence:** The cohesion of an explanation is measured, based on the form of connections between an explanation's propositions, and the "best connected" explanation is favored (e.g., Thagard (1989), Ng & Mooney (1990).)

While structural criteria provide an easy way to evaluate the goodness of a hypothesis, they are not very useful in real situations. Explanations are not constructed in a vacuum. Typically, there is a real world task that the reasoner is performing that requires the reasoner to seek an explanation. The reasoner may also need an explanation to help it with a piece of reasoning that it is trying of perform. Both these types of motivations for explanation influence evaluation criteria.

### Utility-based criteria

An explainer's motivation for explaining will often place additional requirements on candidate explanations, beyond their form. For example, explanations prompted by anomalies must provide particular information, in order to resolve the anomaly. For example, suppose that we expected team X to win over team Y because of the talent of X's star player, but we are told that team X actually lost. If someone explained the loss by 'Y scored more points than X," the explanation would be inadequate. Although it is a correct explanation, it gives no information about *why* our expectation went wrong. The explanation "X's star was injured and couldn't play" does account for what was neglected in prior reasoning, and consequently is a better explanation. However, this explanation would not be preferred on structural grounds alone. The causal chain underlying that explanation is more complex, so it would not be favored by minimality criteria. Likewise, the explainer of the game has access to only one observation, the fact that team X lost, so coherence metrics that measure how an explanation relates *pairs* of observations, such as those described by Ng and Mooney (1990), give no grounds for preferring the second explanation.

To state our relevance criterion another way, *an explanation must address the failure of the reasoner to model the situation correctly.* In addition to resolving the incorrect predictions, it must also point to the erroneous aspect of the chain of reasoning that led to those predictions. An explanation is *useful* if it allows the reasoner to learn, or to accomplish current tasks. The claim here is that *an explanation must be both causal and relevant in order to be useful.*

An explanation of an anomaly, therefore, must answer two types of questions:

1. Why did things occur as they did in the world? This question focuses on understanding, and learning about, the causal structure of the domain.
2. Why did I fail to predict this correctly? This question focuses on understanding, and improving, the organization of the reasoner's own model of the domain.

The answer to the first question is called a *domain explanation* since it is a statement about the causality of the domain. The answer to the second question is called an *introspective* or *meta-explanation* since it is a statement about the reasoning processes of the system.

Each of the above questions relates to a need to collect or organize the missing information that caused the anomaly, and that utility-based evaluation criteria must address. The first question gives rise to *knowledge acquisition goals*, which are goals to collect information or knowledge about the domain that the anomaly has signaled as being missing. The second question gives rise to *knowledge organization goals*, which are goals to improve the organization of knowledge in memory. Let us consider the second question first.

**Introspective explanations: Addressing knowledge organization goals.** One of the questions an explanation must address is why the reasoner failed to make the correct prediction in a particular situation. This could happen in three ways:

1. **Novel situation:** The reasoner did not have the knowledge structures to deal with the situation.
2. **Incorrect world model:** The knowledge structures that the reasoner applied to the situation were incomplete or incorrect.
3. **Mis-indexed domain knowledge:** The reasoner did have the knowledge structures to deal with the situation, but it was unable to retrieve them since they were not indexed under the cues that the situation provided.

When an explanation is built, the reasoner needs to be able to identify the kind of processing error that occurred and invoke the appropriate learning strategy to prevent recurrence of the error. For example, if an incomplete knowledge structure is applied to a situation, the knowledge activated by the resulting processing error must represent both the knowledge that is missing, and the fact that this piece of knowledge, when it comes in, should be used to fill in the gap in the original knowledge structure. Similarly, if an error arose due to a mis-indexed knowledge structure, the explanation, when available, should be used to re-index the knowledge structure appropriately.

Knowledge organization goals can be categorized by the type of gap that gave rise to them, or by the type of learning that results from their satisfaction:

- **Missing knowledge** – learn new knowledge to fill gap in domain model
- **Unconnected knowledge** – learn new connection or new index
- **Implicit assumption** – learn heuristics for when to check assumption explicitly
- **Calculated simplification** – learn heuristics for when to check assumption in detail
- **Explicit assumption** – learn new knowledge to correct the assumption
- **Conjunctive assumptions** – learn new interactions

A hypothesis is evaluated from the point of view of knowledge organization goals by checking to see if it provides the information necessary for the type of learning that the reasoner is trying to perform. For example, suppose the reasoner reads a newspaper story about a Lebanese teenager who, it turns out, is blackmailed into going on a suicide bombing mission. Even if the reasoner already knows about terrorism, religious fanatics and blackmail, the story may nevertheless be anomalous if the reasoner has never seen this particular scenario before. The difficulty arises from the fact that blackmail is not ordinarily something that comes to mind when one reads about suicide bombing. Here, the reasoner can learn a new connection between the knowledge structures describing suicide bombing and blackmail, respectively. In order to do this, the explanation must provide the information required to identify the conditions under which a suicide bombing is likely to be caused through blackmail.

This type of analysis is essential in determining whether an explanation is sufficient for the purposes of the reasoning task at hand. In this example, the reasoning task is to satisfy a knowledge organization goal, which is a goal to learn by reorganizing existing knowledge in memory.

**Domain explanations: Addressing knowledge acquisition goals.** Knowledge acquisition goals seek new causal knowledge about the domain. A domain explanation is a causal chain that demonstrates why the anomalous proposition might hold by introducing a set of premises that causally lead up to that proposition. If the reasoner believes or can verify the premises of an explanation, the conclusion is said to be explained. Explanations are often verbalized using their premises or abductive assumptions. However, the real explanation includes the premises, the causal chain, and any intermediate assertions that are part of the causal chain.

In order to be useful, a hypothesis must provide the information that is being sought by the knowledge acquisition goals of the reasoner. For example, if the reasoner has a goal to acquire knowledge about the biochemical properties of a particular virus, a description of a sick patient must provide the biochemical information in order to qualify as an explanation from the point of view of that goal. An alternative hypothesis that provides causal information suggesting how some drug might destroy the virus, while useful from the point of view of curing the patient, may not provide the required information.

**Task-triggered knowledge acquisition goals.** Knowledge acquisition goals often arise from an explainer's tasks. A doctor wishing to cure a patient will seek an explanation of the patient's symptoms that suggests the plan of action for a cure; an epidemiologist may seek an explanation in terms of environmental factors, to make similar disease outbreaks less likely. In order to reflect such factors in evaluation, we must identify the basic tasks that can be prompted by an anomaly, in turn prompting explanations that focus on particular aspects of a situation. For example, each of the tasks below is associated with particular knowledge acquisition goals:

- **Choosing a response to an unexpected event** – learn causes that allow discrimination between possible plans, by predicting events or identifying current circumstances
- **Repairing an undesirable state** – learn repairable causes of that state
- **Causing recurrence** – learn achievable causes
- **Preventing recurrence** – learn blockable causes
- **Assigning credit or blame** – learn particular actors' influence on an outcome
- **Replicating another actor's success** – learn motivations of the observed actor's unusual planning decisions

Since each of these tasks requires different information, explanations to serve them must be evaluated in the context of those tasks. Evaluation criteria based on knowledge organization goals, knowledge acquisition goals, and tasks are collectively called *utility-based criteria* in this paper.

## Two case studies

We have developed two computer programs to test our theory. Although space limitations preclude detailed descriptions of these programs, we describe them briefly to illustrate the kind of abductive reasoning that our theory of explanation evaluation supports. It should be noted that both programs deal with real world examples in domains in which provably correct domain theories are unavailable. Both programs construct explanations in service of other tasks (e.g., learning, natural language understanding), and hence must evaluate explanations in the context of these tasks.

### The AQUA program

Our first example program is AQUA, a story understanding program that learns from what it reads (Ram, 1987; Ram, 1989). AQUA reads newspaper stories about terrorism, such as the "blackmailed into suicide bombing" story mentioned above (New York Times, Nov 27, 1985). In order to understand text, AQUA must integrate the text, which is often ambiguous, elliptic and vague, with its world knowledge, which is often incomplete and possibly incorrect. In order to learn from what it reads, it must detect perceived anomalies in the text which may identify flaws or gaps in its model of the domain, formulate explanations to resolve those anomalies, confirm or refute potential explanations, and possibly learn new explanations or modify incorrect ones.

The process of natural language understanding generates reasoning goals or *questions*, representing what the understander needs to know in order to perform an understanding task, be it explanation, learning, or some other cognitive task. These questions constitute the specific knowledge goals of the understander generated during a parsing experience, and are used to focus the reasoning processes on aspects of the input that are actually relevant. These goals are also used to focus the learning process so that the understander learns what it needs to know in order to better carry out its tasks.

AQUA uses the following criteria to evaluate hypotheses. Although AQUA uses a case-based approach using explanation patterns to construct explanations (Ram, 1990b), the criteria listed here are applicable to other kinds of explanation construction methods which rely on domain knowledge in the form of inference rules, cases, schemas, or other types of knowledge structures.

1. **Believability:** Do I believe the domain knowledge from which the hypothesis was derived? This is an issue for any learning program in a realistic domain for which a correct domain theory is not yet known.
2. **Applicability:** How well does the domain knowledge (the particular rules, cases or schemas) apply to this situation? Did it fit the situation without any modifications?
3. **Relevance:** Does the hypothesis address the underlying anomaly? Does it address the knowledge goals of the reasoner? The hypothesis is evaluated in the context of both knowledge acquisition and organization goals.
4. **Verification:** How definitely was the hypothesis confirmed or refuted in the current situation? Does the hypothesis spawn new knowledge goals (requiring further information to help verify the hypothesis)?
5. **Specificity:** Is the hypothesis abstract and very general, or is it detailed and specific? This is a structural criterion in the sense that it is based on the structure, and not the content, of the hypothesis. However, the structure of the hypothesis is evaluated in the context of the organization of causal memory.

Intuitively, a "good" explanation is not necessarily one that can be proven to be "true" (criterion 4), but also one that seems plausible (1 and 2), fits the situation well (2 and 5), and is relevant to the goals of the reasoner (criterion 3).

AQUA is a dynamic story understanding program that is driven by its questions or goals to acquire knowledge. Rather than being "canned," the program is always changing as its questions change; it reads similar stories differently and forms different interpretations as its questions and interests evolve. AQUA judges the interestingness of the input with respect to its knowledge goals (Ram, 1990c), and learns about the domain by answering its questions (Ram, 1990a). Both these processes are goal-based. Here, we are proposing that the evaluation of explanations be goal-based as well. It is important for the evaluation criteria to be sensitive to the current goals of the reasoner.

### The ACCEPTER program

Our second example is ACCEPTER (Leake, 1989b), which was designed as the central understanding component for a case-based explanation system (Kass *et al.*, 1986) to both detect problems in explanation and guide adaptation to resolve them. ACCEPTER processes stories of episodes of death (such as the death of the star racehorse Swale), damage (such as car problems), and destruction (such as the explosion of the Space Shuttle Challenger).

Like AQUA, ACCEPTER is a story understanding program that detects anomalous events in the stories it processes, and evaluates the goodness of candidate explanations for those anomalies. However, ACCEPTER's evaluation of explanations includes task-based criteria, according to user-selected explainer goals. The system's evaluation criteria involve criteria to determine:

1. **Relevance** of an explanation to an anomaly (i.e., whether the explanation accounts for why reasoning failed)
2. **Plausibility** of an explanation
3. **Usefulness** of an explanation for the current task

All three criteria must often be satisfied for an explanation to be satisfactory, but in some cases not all are important (e.g., plausibility might actually be undesirable in a humorous explanation.) Given a task, ACCEPTER examines the causes outlined in a candidate explanation, to determine whether those causes are sufficient to accomplish the task. For example, suppose the explainer's task is to repair a malfunctioning device. An explanation must account for the symptoms of the malfunction, and show how those symptoms result from abnormalities in the device. For example, if a television set is smoking, a good explanation for repair must say which parts are causing the smoke, rather than simply saying the smoke is the result of combustion within the device.

ACCEPTER evaluates explanations for four purposes: predicting an event, controlling its future occurrence, repairing an undesirable state, and assigning responsibility. Depending on which purpose is in effect, the system requires different information to be included in an explanation. For example, one episode for which the system evaluates explanations involves a car recall. It is offered two possible explanations for the recall: that there was bad quality control when the car was manufactured, and that the car has a defective transmission. Both explanations are plausible when taken out of context, but when ACCEPTER evaluates them *for the repair task*, it rejects the first explanation and accepts the second, since only the second provides the information needed for the repair.

ACCEPTER's evaluation of usefulness is dynamic, depending on the current task. It may require that an explanation include causes that are observable, predictive, and distinctive (for the task of predicting and avoiding future problems), repairable (for repairing a bad state), controllable (for preventing or causing an outcome), or desirable/undesirable (for assigning praise or blame). All these criteria reflect the content of the explanation, rather than structure alone. A more detailed description can be found in Leake (1989b; 1990).

## Conclusion

Abduction, or inference to the best explanation, is a central component of the reasoning process. The "best" explanation is not one that is the most "correct," if correctness is even measurable in the domain of interest, but one that is most useful to the process that is seeking the explanation. Consequently, criteria for evaluating the goodness of explanations must depend on a theory of the types of uses to which explanation may be applied. We have argued that two main classes of purposes must be considered: those based on knowledge goals reflecting internal desires for information, and those based on goals to accomplish tasks in the external world.

The role of knowledge goals in explanation evaluation has been explored in the AQUA and ACCEPTOR programs. AQUA is a computer model of the theory of question-driven understanding. AQUA builds explanations in order to find answers to questions raised by gaps in its domain knowledge, and learns by incrementally improving its understanding of the domain. The relationship between tasks and requirements for explanations has been investigated in ACCEPTER, a program to evaluate the usefulness of explanations of anomalies for major classes of overarching goals. We are currently investigating the relationships between tasks and knowledge goals, and exploring how a system might generate different knowledge goals depending on its current tasks. Both AQUA and ACCEPTER take a strongly context-dependent view of evaluation of hypotheses: the final determinant of an explanation's goodness is whether it provides the desired information.

## References


J.G. Carbonell. 1979. *Subjective Understanding: Computer Models of Belief Systems.* PhD thesis, Yale University, New Haven, CT.

E. Charniak. 1986. A Neat Theory of Marker Passing. In *Proc. of the Natl. Conf. on Artificial Intelligence*, Philadelphia, PA.

F. Hayes-Roth & V. Lesser. 1976. Focus of attention in a distributed logic speech understanding system. In *Proc. of the IEEE Intl. Conf. on ASSP*, Philadephia, PA.

A. Kass, D. Leake & C. Owens. 1986. *SWALE: A Program That Explains.* In Schank (1986), 232–254.

H.A. Kautz & J.F. Allen. 1986. Generalized plan recognition. In *Proc. of the Natl. Conf. on Artificial Intelligence*, Philadelphia, PA.

R. Keller. 1988. Defining operationality for explanation-based learning. *Artificial Intelligence*, 35.

K. Konolige. 1990. A General Theory of Abduction. In *Proc. of the AAAI Spring Symp. on Automated Abduction*, Stanford, CA.

J.R. Hobbs, M. Stickel, D. Appelt & P. Martin. 1990. *Interpretation as Abduction.* Tech. Note 499, SRI International.

D. Leake. 1989. Anomaly detection strategies for schema-based story understanding. In *Proc. of the 11th Annual Conf. of the Cognitive Science Soc.*, 490–497, Ann Arbor, MI.

D. Leake. 1989. *Evaluating Explanations.* PhD thesis, Yale University, New Haven, CT.

D. Leake. 1990. Task-based criteria for judging explanations. In *Proc. of the 12th Annual Conf. of the Cognitive Science Soc.*, 325–332, Cambridge, MA.

R. Mehlman & C. Snyder. 1983. Excuse theory: A test of the self-protective role of attributions. *Journal of Personality and Social Psychology*, 49(4):994–1001.

S. Morris & P. O'Rorke. 1990. An Approach to Theory Revision using Abduction. In *Proc. of the AAAI Spring Symp. on Automated Abduction*, Stanford, CA.

H. Ng & R. Mooney. 1990. On the role of coherence in abductive explanation. In *Proc. of the 8th Natl. Conf. on Artificial Intelligence*, 337–342, Boston, MA.

A. Ram. 1987. AQUA: Asking Questions and Understanding Answers. In *Proc. of the 6th Natl. Conf. on Artificial Intelligence*, 312–316, Seattle, WA.

A. Ram. 1989. *Question-driven understanding: An integrated theory of story understanding, memory and learning.* PhD thesis, Yale University, New Haven, CT. Research Report #710.

A. Ram. 1990. Incremental Learning of Explanation Patterns and their Indices. In *Proc. of the 7th Intl. Conf. on Machine Learning*, Austin, TX.

A. Ram. 1990. Explanation Patterns: A Theory of Volitional Explanation. In *Proc. of the 12th Annual Conf. of the Cognitive Science Soc.*, Cambridge, MA.

A. Ram. 1990. Knowledge Goals: A Theory of Interestingness. In *Proc. of the 12th Annual Conf. of the Cognitive Science Soc.*, Cambridge, MA.

C. Rieger. 1975. Conceptual Memory and Inference. In R.C. Schank (ed.), *Conceptual Information Processing*, North-Holland.

R.C. Schank. 1986. *Explanation Patterns: Understanding Mechanically and Creatively.* Lawrence Erlbaum, Hillsdale, NJ.

D. Sperber & D. Wilson. 1986. *Relevance: Communication and Cognition.* Language and Thought Series. Harvard University Press.

M.E. Stickel. 1990. A Method for Abductive Reasoning in Natural-Language Interpretation. In *Proc. of the AAAI Spring Symp. on Automated Abduction*, Stanford, CA.

P. Thagard. 1989. Explanatory coherence. *Behavioral and Brain Sciences*, 12(3):435–502.

H. Zukier. 1986. The Paradigmatic and Narrative Modes in Goal-Guided Inference. In R. Sorrentino & E. Higgins (ed.), *Handbook of Motivation and Cognition: Foundations of Social Behavior*, 465–502, Guilford Press.

# Goal Inference in Information-seeking Environments

**Bhavani Raskutti & Ingrid Zukerman**
Computer Science Department
Monash University
Clayton, Victoria 3168
Australia

## Abstract

In cooperative information-seeking environments, we have observed that the dialogues have the following characteristics: (1) they contain sufficient relevant information, (2) they are coherent, and (3) they are well-structured. In this paper, we describe a mechanism for plan inference which takes advantage of these observed features to reduce the number of alternate interpretations of a user's statements. This reduction is achieved as follows: initially, we take advantage of the relevant information trait by using guiding principles and meta predicates to constrain the number of possible interpretations of a single statement. *Discourse coherence* considerations are then applied to integrate subsequent statements and drop incoherent interpretations. The retained interpretations are evaluated using a measure of *information content*, which is used to prefer the interpretations that have more relevant information. The entire mechanism is based on an approach that takes advantage of the well-structured nature of information-seeking dialogues to arrive at the intended interpretation as efficiently as possible.

## Introduction

In this paper, we present a mechanism which infers the plans and goals of a user in cooperative information-seeking settings, e.g., travel agency. The inference is done by generating the possible interpretations of a user's statements, evaluating them and then selecting the more likely interpretations. An interpretation of a user's statements consists of a sequence of plans that the user proposes to carry out, and a plan consists of an action defined by a number of parameters. For instance, in the travel domain, the proposal to fly from Melbourne to Sydney on December 1st, 1990, is a plan, where *flying* is the action, and the parameters *origin*, *destination* and *departure date* are instantiated.

Our inference mechanism operates in the framework provided by a Natural Language Interface (NLI) and a planner. However, in this paper, we focus on the plan inference mechanism, and the only references to the NLI and planner concern our assumptions about them.

In cooperative information-seeking settings, we have observed that the dialogues have the following features: (1) they contain enough relevant information so that the information seeker's plans may be understood easily by the information provider, (2) they are coherent, and (3) they are well-structured, i.e., the dialogue consists of an initial discourse, followed by a clarification interaction. For instance, the following dialogue[1] at a Melbourne travel agency displays these features:

Traveler: "I need a ticket to Sydney tomorrow.
I am going to Hawaii on the 11:00 am flight.
By the way, I'll be leaving from Adelaide."
Agent: "What about your ticket to Adelaide?"
Traveler: "That is all arranged. Thank you."

Our plan inference mechanism takes into consideration these traits of the dialogue while interpreting a user's statements. It differs from earlier mechanisms for plan inference (Grosz 1977, Schmidt, Sridharan & Goodson 1978, Allen & Perrault 1980, Sidner & Israel 1981, Carberry 1983, Litman & Allen 1987, Pollack 1990) in that it can handle multiple interpretations. The problem of multiple interpretations has been handled by Carberry (1990) by using plausibility factors of alternate hypotheses. However, in domains such as travel, where the plans in an interpretation are defined by a number of parameters, this approach alone does not cope with the problem of multiple interpretations. The mechanism presented in this paper addresses this problem by using the observed features of cooperative dialogues, namely, *relevance*, *coherence* and *structure*.

Initially, we take advantage of the relevance trait by using guiding principles and meta predicates to constrain the number of interpretations of a single statement. In the final stages, the amount of relevant information in an interpretation is measured by its information content, and is used to prefer interpretations with more relevant information.

The coherence of an interpretation is determined by

[1]This dialogue is similar to those found in the transcripts of telephone conversations at travel agencies, which were provided by Prof. J. Roach and D. Sanford from the Virginia Polytechnic Institute and State University.

the relation between the interpretation of a statement and the interpretation of the previous discourse. Litman and Allen (1987) have applied ideas related to discourse coherence in order to select a preferred plan. However, since they maintain only one preferred option, they may choose an interpretation that is locally coherent, but does not fit in with the entire dialogue. In this research, this situation is prevented by maintaining other interpretations with less coherent relations while pursuing the more likely interpretations.

All dialogues, whether well-structured or otherwise, can be handled by a single approach, which we call the *cognitive* approach. In this approach the system draws both direct inferences based on the user's statement, and indirect inferences based on domain and world knowledge from each statement as it is uttered. For instance, upon receiving the first statement in our sample text, the system, after making all the direct inferences, will go on to make the indirect inference that the speaker is probably leaving from Melbourne, where the conversation is being conducted. However, some of the system's indirect inferences may have to be retracted due to information provided later on, as when the traveler says "By the way, I'll be leaving from Adelaide," in our sample text. Evidence provided by written and spoken text, where speakers/writers address inferences presumably performed by their hearers/readers, leads us to believe that this approach models human cognition. For instance, meta comments such as "however," "indeed" and "in fact" indicate violation of expectations, realization of expectations and violation of implicatures, respectively.

The cognitive approach is computationally expensive for the initial discourse, due to the retractions that may have to be performed. These retractions may be avoided by using another approach, namely the *implementational* approach. In this approach, the system first draws all the direct inferences from all the statements in the initial discourse, and only then draws the indirect inferences. In this case, there are less chances of early unwarranted inferences, and hence, from the viewpoint of implementation, this approach is attractive. However, this approach alone cannot cope with a clarification interaction after the initial discourse.

In this paper, we argue for a hybrid approach that takes advantage of the benefits of both approaches by taking into account the observed structure of information-seeking dialogues. In the hybrid approach, the implementational approach is applied to the initial discourse, and the cognitive approach is applied to subsequent statements uttered either after a pause or in response to a query posed by the system. The rest of this paper describes our mechanism for plan inference based on the hybrid approach, with particular reference to the strategies that take into account the characteristics of cooperative information-seeking dialogues in order to reduce the number of alternate interpretations of a user's statements.

## The Main Mechanism

The inference mechanism operates on input provided by the NLI, consisting of predicates, such as FLY and GO, and meta predicates that indicate the modality of the statements, such as WANT. Based on this input, it generates the intended plans of the user. The mechanism is based on the hybrid approach and it consists of two parts: (1) the implementational part for handling the initial discourse $S_i$, and (2) the cognitive part for processing the clarification statements $S_c$. Both parts apply three main processes: (1) Direct inference, (2) Indirect inference, and (3) Evaluation of interpretations. However, they are applied at different times in each part. In the implementational part, direct inference of all statements followed by evaluation precedes indirect inferences followed by another evaluation. In the cognitive part, direct and indirect inferences are interspersed, with the evaluation performed last. These three processes are described below with reference to the two procedures *Implementational-Infer* and *Cognitive-Infer*, which embody the two parts of the mechanism. The set of interpretations, $\{IS\}$, is initially set to *nil*.

Procedure *Implementational-Infer* ($S_i$)

1 For each statement $s$ in the discourse $S_i$ do:<br>
2     Generate a set of interpretations $I(s)$ using guiding principles and meta predicates<br>
3     Infer the set $R$ of possible discourse relations between $I(s)$ and $\{IS\}$ and order them according to the coherence measure<br>
4     Combine $I(s)$ and $\{IS\}$ using $R$ to create a new $\{IS\}$, and determine the likelihoods of the interpretations in the new set<br>
5     Use these likelihoods to prune unlikely interpretations from $\{IS\}$<br>
6 For each interpretation $I'$ in $\{IS\}$<br>
7     Use the information content of $I'$ to revise its likelihood<br>
8 Prune unlikely interpretations from $\{IS\}$<br>
9 For each interpretation $I'$ that is left in $\{IS\}$<br>
10     Repeat until [ $I'$ is fully defined ] or [ the information content of $I'$ cannot be increased ]<br>
11         Infer values for all undefined necessary parameters in all the plans in $I'$<br>
12 Prune $\{IS\}$ using information content (lines 7-10)

Procedure *Cognitive-Infer* ($S_c$)

1 For each statement $s$ in the discourse $S_c$ do:<br>
2     Perform direct inferences as in lines 2-5 in procedure *Implementational-Infer*<br>
3     Perform indirect inferences as in lines 9-11 in procedure *Implementational-Infer* to infer the values of all necessary parameters, and update them so that earlier inferences are not refuted by later weaker inferences<br>
4 Prune $\{IS\}$ using information content as in lines 7-10 in procedure *Implementational-Infer*

### Direct Inference

The direct inference stage (lines 2-5 in procedure *Implementational-Infer* and line 2 in procedure *Cognitive-Infer*) consists of: (1) Inferring a set of interpretations from each of the user's statements (line 2 in procedure *Implementational-Infer*); (2) Inferring a set of relations between each interpretation of a user's statement and the previous discourse (line 3); and (3) Generating a set of new interpretations (line 4). In this process, the likelihood of each interpretation is determined, and less likely interpretations are not pursued.

The inference of interpretations from one statement issued by a user consists of inferring a set of possible actions which match this statement, and computing their likelihoods. This inference is done by using a STRIPS-like operator library (Fikes & Nilsson 1971) and plan inference rules (Allen & Perrault 1980). Since the number of interpretations based on just one statement and the operator library can be quite large in a realistic domain, we control the activation of the plan inference rules by means of guiding principles, thus reducing the number of possibilities. In addition, meta predicates are used during this stage to modify the strength of the plan inference rules.

The inference of the discourse relations between the interpretations of one statement and the interpretations of the previous discourse is done so that the coherence of the discourse is increased.

The inference of a new interpretation, based on an interpretation of a new statement, an interpretation of the previous discourse and one of the inferred discourse relations, is done by updating the interpretation of the previous discourse using the relation and the information in the interpretation of the new statement. The likelihoods of the new interpretations are determined using Bayesian theory of probability, and interpretations whose likelihoods are lower than an acceptable threshold are dropped (Raskutti & Zukerman 1991).

### Indirect Inference

The indirect inference stage (lines 9-11 in procedure *Implementational-Infer*, and line 3 in procedure *Cognitive-Infer*) is used to infer information that has not been explicitly stated by the user. In the implementational part, it is performed after direct inference of a chunk of statements. In the cognitive part, it is performed after direct inference of each statement. The strength of indirect inferences is assessed on the basis of their source of information. For instance, the desired mode of transport between Sydney and Hawaii may be inferred by taking into consideration typical assumptions about the domain. This type of inference is stronger than an inference based on general 'world knowledge', but weaker than a direct inference.

Indirect inferences are applied to infer each of the parameters that are necessary for plan definition, and inferences are drawn until a complete interpretation is inferred or no more inferences can be drawn on the basis of the existing information. In the implementational approach, only the parameters that remain undefined after direct inference are inferred. In the cognitive approach, however, all the necessary parameters are inferred, and previous inferences may be refuted or corroborated in light of new inferences.

### Evaluation of Interpretations

After a chunk of statements is processed, the generated interpretations are evaluated using a measure of information content, and the interpretations that are better defined according to this measure are retained. This evaluation is performed twice in the implementational approach, first after all the direct inferences have been drawn, and then again after the completion of the indirect inferences (lines 6-8 and 12 in procedure *Implementational-Infer*). In the cognitive approach, on the other hand, the evaluation is deferred until all the statements in a chunk are fully processed (line 4 in procedure *Cognitive-Infer*).

## The Strategies in Detail

In this section, we describe the strategies that are used to reduce the number of possible interpretations of a user's statements. These strategies take into account two aspects of observed dialogues in information-seeking settings, namely, coherence and relevance.

### Guiding Principles

Guiding principles restrict the number of interpretations that match a single statement of a user by controlling the search through the operator library that defines the basic actions in the domain. Each operator is defined in terms of its preconditions, effects and composition (also called the body of the operator).

During the inference of interpretations of a statement that has been parsed into an input predicate, all those operators that have the predicate in the definition of their precondition, effect or body, are chosen as possible matches. This can lead to the selection of a large number of operators as possible matches. Hence, we use the following guiding principles, which are based on the relevance trait, to restrict the inference process, so that only justifiable inferences are drawn.

1. *Least Complication* – requires that the most primitive of the matched operators must be selected. For instance, when a user says "I want to go to Sydney," then while both one-way and two-way travel are possibilities, one-way travel is chosen.
2. *Least Commitment* – requires that the operator that needs the least number of additional assumptions be chosen (Sacerdoti 1977). In practice, this means that more general operators are chosen, unless there is an explicit reference to a specific operator. For instance, unless there is an explicit reference to a means of transport, those operators that are uncommitted about the means of transport are chosen.

## Meta Predicates

Meta predicates, such as CAN and WANT, represent the modality or propositional attitude of the statements. They are returned by the NLI in order to retain information that may be lost when parsing from natural language to predicates. The presence of a meta predicate enables us to resolve ambiguity when the input predicate appears in both the precondition and the effect of an operator. For instance, if a user's request about BEing at a place is expressed as "I want to be ... " or "I can be ... ," but the NLI returns it as BE(...), this predicate can refer either to the precondition or the effect of an operator with equal likelihood. By taking into account the presence of a meta predicate, we increase the bias towards the appropriate inference rule, e.g., if the user had said "I can be ... " then the likelihood of the precondition rule is increased.

## Discourse Coherence

Discourse coherence considerations are used during direct inference to bias the inference process towards interpretations that make the discourse more coherent. To this effect, the inferred relations are assigned likelihoods that prefer normal patterns of discourse, e.g., the elaboration of the latest plan is preferred to the elaboration of older plans; the likelihood of an introduction is reduced if elaboration is possible; and if there are cue words to indicate a relation, then these are used during the inference of relations (Raskutti & Zukerman 1991).

A new statement of the user can either *elaborate* on an earlier topic, *introduce* a new topic, *digress* from the last topic, or *correct* some information specified earlier. The different possible relations coupled with the different topics that may be corrected or elaborated give rise to a number of different interpretations. The likelihoods of these interpretations depend on the likelihoods of the inferred relations, which in turn depend on whether the NLI has recognized a discourse-statement relationship and/or the referred topic. Thus, there are four possibilities:

1. If there is an explicit reference to an earlier topic, and a discourse relation is specified, then the system has no uncertainty about the topic or the relation.
2. If only a topic is specified, then we assume an elaboration relation, e.g., "*About the Hawaii trip*, I'll be flying QANTAS."
3. If only a discourse-statement relation is specified, e.g., "*On second thought*, make it 10:00 am" (correction), the system needs to determine which of the earlier topics is being referred to. Notice, however, that the only possible discourse relations pertaining to previous topics are Elaboration and Correction. This is due to the fact that in Introduction, the introduced topic is unrelated to previous topics, and Digression is considered as a special case of Elaboration, where the likelihood of elaborating on the latest topic is substantially reduced. The determination of the topics which are elaborated or corrected is performed by following common patterns of discourse, where the elaboration or correction of topics mentioned later in the discourse is preferred to that of topics mentioned earlier.
4. If neither a topic nor a discourse relation is specified, then the inference of discourse relations is performed by first determining which relations are possible and then calculating the likelihood of each postulated relation. We do not postulate that the new statement is a correction or a digression unless there is an explicit indication from the NLI. Hence, the only possible relations are Elaboration of each topic in the interpretation of the earlier discourse, and Introduction of a new topic. Like above, elaboration of topics mentioned later in the discourse is preferred to that of topics mentioned earlier. Introduction of a new topic is always considered a possibility. However, its likelihood is reduced if Elaboration is possible.

## Strength and Categorization of Inferences

The strength of an inference is directly proportional to the reliability of the information source on which the inference is based. Hence, we categorize different information sources that are used to draw inferences and list them below in decreasing order of reliability.

1. *User's Statements* – Direct inferences from what is explicitly stated. While these inferences can be presumed correct, there is still a degree of uncertainty in relating a new statement to the earlier ones due to the different discourse relations possible.
2. *Domain Knowledge* – Indirect inferences that are derived by using the system's beliefs about the user's domain knowledge. A typical example is the inference of the arrival time at the destination once the departure time is known.
3. *Domain Assumptions* – Indirect inferences that are derived by assuming what is normal in the domain. For example, when the mode of travel is not specified, it is possible to derive this information from the usual mode of transport between two places.
4. *User Model* – Indirect inferences that are made on the basis of the system's model of the user. The user model may be a default model describing a typical user, or it may be more specific. In our system, we have adopted a default model based on the assumption that, typically, the information provider cannot form an extensive user model.
5. *Common-Sense* – Indirect inferences that are derived using common notions outside the domain of interest. Typically, such notions are used when we postulate return journeys based on the assumption that people usually do not move from their residence.

The inference types are assigned a strength in the (0,1] range, and this strength is used during the computation of the information content of a parameter. The inferences derived from the user's statements have a strength of 1 and all other inference types have a progressively decreasing strength, according to the reliability of their source of information. The undefined parameters are assigned a minimum strength. This assignment enables us to distinguish between parameters that are defined inexactly by the user and parameters that are left undefined, and assign less information content to undefined parameters.

During the process of inferring the value of a parameter, we emulate one aspect of human behavior whereby once a conclusion is accepted with a particular degree of confidence, people consider it to be certain when drawing subsequent conclusions (Gettys, Kelly & Peterson 1982). To this effect, each parameter in the plans in each interpretation is tagged with the type of inference that gave rise to the value of the parameter, without taking into account the inference types of other parameters that were used for computing the parameter in question. Thus, like Carberry (1990), we do not compound the uncertainty in chains of inferences.

The strength of the inference type in a parameter's tag is used in the cognitive approach to determine whether a particular parameter should be revised by a new inference. If the strength of the new inference is the same or higher than the strength of the inference type in the tag, the new inference replaces the old one. Otherwise, the old inference is retained. In this manner, a weaker inference is prevented from refuting the results obtained from a stronger inference.

### Information Content

We define the information content of an interpretation as the sum of the information content of all the plans in the interpretation, and the information content of a plan as the sum of the information content of all the parameters that are necessary for the definition of the plan. The information content of a parameter, in turn, depends on two factors: (1) its specificity, which is defined as the reciprocal of the number of possible values assigned to this parameter; and (2) its strength, which depends on the source of information from which this parameter was obtained. Thus, both a parameter with multiple values assigned to it and a parameter derived from an unreliable source of information are deemed to have a low information content (Raskutti & Zukerman 1991).

The information content measure is used to update the likelihood of each interpretation in the set of interpretations. This set is then pruned by dropping those interpretations whose likelihoods fall below a relative rejection threshold. In this manner, the interpretations with more relevant information, i.e., those with a higher information content, are chosen.

## Conclusions

We have offered a mechanism which directs inferences towards more likely interpretations of a user's statements by using strategies which take into account the observed traits of information-seeking dialogues, namely, relevance, coherence and structure. Currently, the algorithm for the implementational part is fully operational, with the indirect inference restricted to a few rules from each category. The cognitive part is in the last stages of its implementation. Finally, our experiments with a few discourse samples have indicated that our algorithm chooses the same interpretation that people choose. For instance, in the example presented in the Introduction, our mechanism considers four possible itineraries, and then chooses the intended one, i.e., *Melbourne* → *Adelaide* → *Sydney* → *Hawaii* → *Melbourne*, as the best interpretation.

## References


Allen, J.F. and Perrault, C.R. 1980. Analyzing Intention in Utterances. *Artificial Intelligence* 15:143-178.

Carberry, S. 1983. Tracking User Goals in an Information-seeking Environment. In Proceedings of the National Conference on Artificial Intelligence, 59-63.

Carberry, S. 1990. Incorporating Default Inferences into Plan Recognition. In Proceedings of the National Conference on Artificial Intelligence, 471-478.

Fikes, R.E. and Nilsson, N.J. 1971. STRIPS: A New Approach to the Application of Theorem Proving to Problem Solving. *Artificial Intelligence* 2:189-208.

Gettys, C.F.; Kelly III, C. and Peterson, C.R. 1982. The Best-guess Hypothesis in Multistage Inference. In Kahneman, D., Slovic, P. and Tversky, A. eds., *Judgment under Uncertainty: Heuristics and Biases*, 370-377. Cambridge University Press.

Grosz, B.J. 1977. The Representation and Use of Focus in Dialogue Understanding. Technical Note 151, SRI International, Menlo Park, CA.

Litman, D. and Allen, J.F. 1987. A Plan Recognition Model for Subdialogues in Conversation. *Cognitive Science* 11:163-200.

Pollack, M. 1990. Plans as Complex Mental Attitudes. In Cohen, P., Morgan, J. and Pollack, M. eds., *Intentions in Communications*, MIT Press.

Raskutti, B. and Zukerman, I. 1991. Generation and Selection of Likely Interpretations during Plan Recognition in Task-Oriented Consultation Systems. To appear in *User Modeling and User Adapted Interaction*, an International Journal.

Sacerdoti, E. D. 1977. *A Structure for Plans and Behavior.* American Elsevier, New York.

Schmidt, C.F.; Sridharan, N.S. and Goodson, J.L. 1978. The Plan Recognition Problem: An Intersection of Artificial Intelligence and Psychology. *Artificial Intelligence* 10:45-83.

Sidner, C.L. and Israel, D.J. 1981. Recognizing Intended Meaning and Speakers' Plans. In Proceedings of the Seventh International Joint Conference on Artificial Intelligence, 203-208.

# Towards Fair Comparisons of Connectionist Algorithms through Automatically Optimized Parameter Sets

**Frank E. Ritter**

Department of Psychology
Carnegie Mellon University
Pittsburgh, PA 15213
Ritter@cs.cmu.edu

**Abstract**

The learning rate and convergence of connectionist learning algorithms are often dependent on their parameters. Most algorithms, if their parameters have been optimized at all, have been optimized by hand. This leads to absolute and relative performance problems. In absolute terms, researchers may not be getting optimal performance from their networks. In relative terms, comparisons of unoptimized or hand optimized algorithms may not be fair. (Sometimes even one is optimized and the other not.) This paper reports data suggesting that comparisons done in this manner are suspect. An example algorithm is presented that finds better parameter sets more quickly and fairly. Use of this algorithm (or similar techniques) would improve performance in absolute terms, provide fair comparisons between algorithms, and encourage the inclusion of parameter set behavior in algorithmic comparisons.

Keywords: Connectionist learning algorithms' parameters, Genetic algorithms, Comparative cognitive modeling, Learning algorithm optimization.

Connectionist networks have been used as a cognitive modeling and machine learning algorithm in numerous ways (for a plethora of examples see Rumelhart, McClelland and the PDP Research Group (1986) or any of the Proceedings of Cognitive Science, AAAI or IJCNN). Work with connectionist learning algorithms such as back-prop (Rumelhart, Hinton, & Williams 1986), Boltzmann learning (Ackley, Hinton, & Sejnowski 1985), the mean field theorem (Peterson & Hartman 1989), and even genetic algorithms (Montana & Davis 1987) have restricted themselves to optimizing connection weights (or additionally, the number of hidden nodes) within a given network, leaving certain crucial parameters of these algorithms, such as learning rate, to be set by the user. These parameters, unlike the weights, are not optimized automatically, yet optimal performance depends on them (Kolen & Pollack 1990, McClelland & Rumelhart 1988). Finding fast yet safe values for these parameters represents an important but often neglected research area. The parameters are also theoretically important because conclusions about the algorithms and the models using them are often based on how quickly the models learn and if they converge (e.g., Hinton & Shallice 1989, Small 1990).

Researchers currently attempt to understand and improve their networks by tuning these parameters by hand. Sometimes this work is reported in the literature (e.g., Fahlman 1988), and attempts are made to teach it (McClelland & Rumelhart 1988). While a certain understanding comes from observing networks as they work, using researchers to do this exclusively will pay an inherently large cost and suffer from the known pitfalls of human thinking, such as functional fixedness, limited memory, and slips.

Comparisons of different algorithms may optimize the parameter set by hand for each algorithm (Fahlman 1988) or may optimize one by hand and use the default values for the other (Montana & Davis 1989). At best, these comparisons are comparing how well each reporting researcher can optimize each algorithm. This is useful information about the algorithms' ease of use, but does not necessarily indicate which algorithm learns faster or better. There is little reason to believe that most algorithm researchers or users are finding or using the best parameters; the brief evidence reported below suggests that this is not the case in practice. (For an excellent counter-example, see Belew, McInerney & Schraudolph 1990. While they don't argue the same theoretical position, their results are consonant with the work reported here.)

Alternatively, researchers may just report the results for a variety of parameter settings and let the reader judge for themselves parameter sensitivity and optimality. Kruschke & Movellan (1991) do this, and the results are compelling because the algorithms are variants of each other and technically speaking, one dominates the other. In general however, this approach breaks down from the combinatorial explosion of possible parameter sets when additional parameters and values are considered.

It should not be initially surprising that connectionist networks are still optimized by hand. The algorithms that connectionist networks exploit and that their designers are

After randomly creating an initial generation of 30 members, Mendel-DP computed their fitness as the number of epochs needed to learn the XOR logic function or the artificial grammar to a stopping criterion (ecrit) of 0.1 TSS. The maximum number of epochs allowed was 400. Network runs that hadn't learned by 400 epochs were assigned 400 as a fitness measure. In this algorithm, lower numbers represented better fitness. After the best genotype was reported, the next generation was created from the members of the current generation. This cycle of creation and evaluation was carried out for 15 generations. The proposed best solutions from each of the subjects and from each run of Mendel-DP were then averaged over 150 trials for comparison.

**The genetic operators used.** The best 20% of the genotypes were directly copied (survived) into the next generation. The remaining 80% were created by breeding with 4 operators. Of the four operators used to create new genotypes, two of these, cross-over and mutation, can be considered traditional (Holland 1975), and two, creep-value and average, are relatively novel (Davis & Ritter 1987). *Cross-over* selected two parents (as described below) and for each parameter in the child, a value was randomly chosen from the two parents. *Mutation* randomly changed a parameter value in a genotype (choosing linearly in its range) and put the modified version into the next generation. *Creep-value* modified an interval parameter in a copy of the parent, creeping up or down 5%. *Average* averaged the interval parameter values from two parents to create a child, and copied the categorical values from one of them.

The probability of applying each operator was adjusted by a set amount each generation. The initial and final values (and thus also the change each generation) for each operator were copied with slight modification, from a similar genetic algorithm (Davis & Ritter 1987). The total probability of an operator applying was normalized to one by appropriately setting the probability of a *pass-through* operator that merely copied a selected parent into the next generation. Cross-over started at .3 and was adjusted to .1; mutate started as .1 and was adjusted down to .01; creep-value started at .01 and was increased to .1; average started at .3, and was adjusted down to .1; pass-through, which normalized the probability, increased from .29 to a final value of .69.

**Parent selection.** After an uniform random selection from the 30 member genotype pool, each potential parent was probabilistically selected as a parent based on their fitness compared to the best in their generation according to the following formula:

$$P(genotype_x)=min\text{-}fitness_{Gen_i}/Fitness(genotype_x)$$

For example, if selected, the best genotype in a generation would be automatically used if selected (P(best) = min/min = 1); poor genotypes with fitness of 400 might have a 80/400 or 1/5 chance of becoming a parent once selected.

**Extensions to genetic algorithms to handle noise.** Each time BP runs, it initializes the network with a different set of random weights. This causes the learning rate to vary wildly. Mendel-DP used two approaches to decrease the effect of this noise on its fitness evaluation: 1) Fitness was computed as the average across ten network runs. 2) Fitness was saved and averaged across generations for genotypes that survived, further smoothing the effect of an individual evaluation.

## Comparison of the Results

Table 1 shows the mean fitness values for the various parameter sets. The top half of the table shows the parameter sets for XOR and the bottom half, the grammar learning network. The results for the default values comes first, then the subjects' results, and then the results from Mendel-DP, our example automatic parameter set optimizer. Interestingly, the results particularly surprised the grammar learning researcher in several ways: he believed that his parameters were optimal and that they gave him learning in approximately 300 epochs.

The XOR network did not converge with the default values. For this problem they are inappropriate, and this is even well known (McClelland & Rumelhart 1988). On the grammar learning network they did better than the researcher (but not reliably better, $t(149)=1.69$, $p> .05$).

The parameter sets found by Mendel-DP lead to faster learning than the default set or those produced by hand. For both problems, this improvement was reliable (in all cases $t(149) >= 3.3$, $p<.005$), and these differences were also large enough to be important, between 30% and an order of magnitude faster learning. The lower standard deviation for the grammar learning task also indicates that the learning was vastly more reliable.

Not only did Mendel-DP produce a better parameter set, it took less time. Even on the grammar learning network, Mendel-DP running overnight took less total time than the researcher did. The incremental cost of modifying Mendel-DP to optimize another parameter or to optimize another network is quite small: less than an hour was needed to extend it from XOR to the grammar task.

**Robustness of results.** Although Mendel-DP is itself a stochastic algorithms, the results reported here should be robust. Genetic algorithms are not sensitive to their own parameters (De Jong 1985) in the same way connectionist algorithms are. The selection of parameters for Mendel-DP should have little influence on its practical (finding good parameter sets) or theoretical results (finding fair parameter sets). On the other hand, a more complete analysis of this most familiar with do not lend themselves to self optimization. These algorithms depend on the fitness function being local and inexpensive to compute. Evaluating parameter sets are neither. When the optimized function becomes expensive to compute, these algorithms do not optimize efficiently, and the user community rightfully drops them for other methods such as human implemented heuristic search. Such is the current place of the Boltzmann machine: it requires too much time to compute to be currently considered a viable architecture.

## Automatic Computation of Parameter Sets

Let us consider a possible algorithm to optimize parameter sets, without comment on whether it is the best possible, just that it will serve as an example (for semi-automatic methods see Nowlan (1991)). Optimizing functions that are expensive to compute, such as the evaluation of network parameters, are the natural domain of genetic algorithms (Goldberg 1989, Holland 1975), a family of algorithms loosely based on Darwinian evolution. They optimize functions without assuming that the search space will be linear. They start with a population of templates for possible solutions (analogous to sets of chromosomes), and evaluate them to determine how well they perform (fitness). After the fitness values are computed, a new population is created. A variety of methods have been used to create the next generation, but in each case the underlying principle has been to including copies of the chromosomes proportional to their fitness, and at each generation to create new combinations by combining two parents' chromosomes. And then repeating the cycle of evaluation and creation.

In the genetic algorithm presented here, Mendel-DP, the chromosomes are sets of parameter values that initialize a network, setting learning rate, learning grain size, and so on. The fitness measure optimized will be how quickly the network learns. In general, this could be any measure of performance or fitness, including how safely the network performs (e.g., misclassification measures favored by Nowlan (1991)) or how many hidden nodes were used. These need not be combined into a single measure, but could be optimized as a vector (Schaffer 1985).

In order to find out how well people can tune parameter sets, several subjects optimized the parameter sets for two networks. Their parameter sets, along with the default set supplied with the algorithm and the parameter sets produced by Mendel-DP, were evaluated by averaging over multiple runs of the networks with each of them. In addition to the ubiquitous XOR, to balance its simplicity we also compared how well an artificial grammar learning network learned with its default parameters, those found and used by an active researcher, and those found by Mendel-DP.

### The Networks and Subjects

**The problems.** There were two networks used. The first was the classic two hidden node XOR network provided as a training problem (McClelland & Rumelhart 1988), the second was an artificial grammar learning network with 9 input nodes, 6 hidden nodes, and 3 output nodes. Both networks learned through the back-propagation (BP) algorithm (Rumelhart, Hinton & Williams 1986) provided as part of McClelland & Rumelhart's (1988) book of examples.

**The subjects and the parameters they modified.** The subjects optimizing the XOR network (students in a graduate level PDP modeling course at CMU) were allowed to vary any parameter to any value. While they did not modify all parameters, they did modify both interval (e.g., learning rate) and categorical (e.g., training regime) parameters. The active researcher quite naturally manipulated a superset of these parameters. The complete lists are displayed as the column headings in Table 1.

In order to create a stronger comparison, Mendel-DP only manipulated the parameters that the corresponding subject(s) did.

**The parameters varied.** Intentionally broad limits were imposed on the automatic search performed by Mendel-DP for the parameters used by the subjects. *Learning rate*, which determines how much each training session influences the node weights, was allowed to vary between 0.05 and 12. *Learning grain size* represents how many patterns are presented in each BP training cycle of presentation of patterns, error accumulation, and then weight adjustment. This parameter specifies that either weight updates occur after every single *pattern* presentation or after every *epoch* (presentation of all input patterns). The *training regime* parameter specified the order of pattern presentation. Patterns could be presented in the same sequential order each epoch *(strain)* or in a permuted order *(ptrain)*. As indicated in Table 1, a slightly different set of parameters were varied for the grammar learning network. The remainder of the parameters were set to the defaults provided with the BP program (McClelland & Rumelhart 1988).

### The optimizing algorithm.

The genetic algorithm, Mendel-DP, optimized networks through selective breeding of parameter sets. The detailed description provided below is for application and explanation purposes; genetic algorithms are not particularly sensitive to their own parameters (De Jong 1985), so this level of description should not be necessary for the replication of these results or to support any of the theoretical arguments.

**The basic algorithm.** Each genotype, or evaluated object, was a set of values for the parameters described above.

Table 1 Average epochs to learn (fitness) for generated parameter sets (N=150)
All other parameters left at default values.

A) XOR Task

| Source | Epochs to learn | SD | Learning Rate | Learning Grain | Training Regime |
|---|---|---|---|---|---|
| BP Defaults | 400 | 0 | 0.05 | Pattern | Strain |
| Subject A | 222 | 175 | 8.0 | Epoch | Strain |
| Subject B | 179 | 125 | 1.0 | Pattern | Ptrain |
| Mendel-DP | 130 | 131 | 4.6 | Epoch | Strain |

B) Grammar Learning Task

| Source | Epochs to learn | SD | Learning Rate | Learning Grain | Training Regime | Wrange | Momentum |
|---|---|---|---|---|---|---|---|
| BP Defaults | 147 | 19 | 0.05 | Pattern | Strain | 1.0 | .90 |
| Researcher | 170 | 165 | 0.15 | Pattern | Strain | 2.0 | .65 |
| Mendel-DP | 12 | 4 | 6.33 | Pattern | Strain | 1.0 | .67 |

new technique on a practical level would require the ability to characterize how long Mendel-DP must be run to optimize a particular network. These differences may be small because finding network parameters may be less difficult than solving the network itself. This may remain an empirical rather than theoretical question because we are examining expected rather than worst case behavior. These results will also depend on the choice of how to penalize networks that get stuck in local minimums.

**Non-normal distribution of learning times.** There is one other regularity in the data worth noting. It is the large standard deviations in the learning rates, which are as large as their mean values and significantly skewed. For example, the distribution of learning times for the first entry in the table had a skewness measure of 1.75 (value/SE=8.75, $p<.001$), indicating a highly non-normal distribution. Highly skewed distributions decrease the sensitivity of significance tests (Mosteller & Tukey 1977) and presumably of genetic algorithms. Future work comparing parameters or network performance should use transformed data, which will improve the ability to discern reliable differences. The evaluation of Mendel-DP's results should be interpreted with this fact in mind. There may exist significant differences between parameter sets that standard statistical tests on untransformed data will not find, and the seemingly large sample sizes may not be large enough to find optimum values for minor parameters.

## Conclusions

Having a computer do our work faster and better, isn't this what we all have dreamed of? We saw above that parameter sets derived by hand were sometimes better than the defaults provided by the algorithm's authors and sometimes worse. But in all cases they were inferior to those derived with the new method presented here. Researchers could use systems like Mendel-DP to find better parameter sets than they could on their own, in a fraction of the time. If Mendel-DP only encourages further explorations in automated algorithm evaluation, it should be considered a success. However, it goes beyond this.

In addition to promising to provide better parameter sets than those found by hand, Mendel-DP and the data it provides make several suggestions for connectionist research. The problem of evaluating learning algorithms because of different comparison criteria has been noted before (Fahlman 1988, Kruschke & Movellan 1991), but the problem of comparing best discovered performances should also be emphasized. Current comparisons often depend on the optimization abilities of the algorithms' designers instead of just the algorithms themselves. The use of Mendel-DP or similar techniques would put algorithms on a level playing field, and provide fair comparisons between algorithms (as one reviewer noted, there may even be biases inherent in this method of comparison, but none are known or theoretically expected, and this approach would still retain the advantage of making the comparison explicit). The large standard deviations and non-normal distributions reported here suggest that hand optimization is particularly difficult and that additional caution has to be taken in making such comparisons. Standard statistics are not valid, and sample sizes need to be large -- two problems that particularly affect human reasoning for the worse.

This approach allows, indeed encourages, new concepts in the evaluation of networks. With automatic processing available, work can go into evaluating the network parameters not only on learning rate, but also for safety and reliability. As noted above, because of the large variance in the data and our limited memory span, these concepts are not necessarily even computable without it. The ability to routinely find optimal or near optimal parameters should encourage these parameters to become part of theories rather than hearsay, and encourage comparisons of the optimum parameters between networks and what influences them. Important regularities of connectionist learning algorithms may not be apparent until finding sets of optimum parameters is routine.

## Acknowledgements

This research was partially funded by a training grant from the Air Force Office of Scientific Research, Bolling AFB, DC. Partial computing support was provided by a grant (N00014-86-K-0678) from ONR. I thank Javier Movellan and two anonymous reviewers for their constructive comments, and an anonymous member of my department for his network and comments.

## References

Ackley, D., Hinton, G. E., & Sejnowski, T. J. 1985. A Learning Algorithm for Boltzmann Machines. *Cognitive Science* 9(1): 147-169.

Davis, L. W., & Ritter, F. E. 1987. Schedule Optimization with Probabilistic Search. In *Proceedings of the Third Conference on Artificial Intelligence Applications*, 231-236. Hackensack, NJ: IEEE Computer Society.

De Jong, K. 1985. Genetic Algorithms: A 10 Year Perspective. In *Proceedings of an International Conference on Genetic Algorithms and their Applications,* 169-177. Grefenstette, J.J. ed. Pittsburgh: Texas Instruments & U.S. Navy Center for Applied Research.

Fahlman, S. E. 1988. An Empirical Study of Learning Speed in Back-propagation Networks (Tech. Rep. CMU-CS-88-162), Computer Science Department, Carnegie Mellon University.

Goldberg, D. E. 1989. *Genetic Algorithms in Search, Optimization, and Machine Learning.* Reading, MA: Addison-Wesley.

Hinton, G. E., & Shallice, T. 1989. Lesioning a Connectionist Network: Investigations of Acquired Dyslexia (Tech.Rep. CRG-TR-89-3), Department of Computer Science, University of Toronto.

Holland, J. H. 1975. *Adaptation in Natural and Artificial Systems.* Ann Arbor, Michigan: University of Michigan Press.

Kolen, J. F., & Pollack, J. B. 1990. Scenes from Exclusive-Or: Back Propagation is Sensitive to Initial Conditions. In *Proceedings of the Annual Conference of the Cognitive Science Society*, 868-875. Hillsdale, NJ: Lawrence Earlbaum Associates.

Kruschke, J. K., & Movellan, J. R. 1991. Benefits of Gain: Speeded Learning and Minimal Hidden Layers in Back-propagation Networks. *IEEE Transactions on Systems, Man and Cybernetics.* 21(1): 273-280.

McClelland, J. L., & Rumelhart, D. E. 1988. *Explorations in Parallel Distributed Processing: A Handbook of Models, Programs, and Exercises.* Cambridge, Massachusetts: Massachusetts Institute of Technology.

Montana, D. J., & Davis, L. 1989. Training Feedforward Neural Networks Using Genetic Algorithms. In *Proceedings of the Eleventh International Joint Conference on Artificial Intelligence*, 762-767. San Mateo, California: IJCAI.

Mosteller, F., & Tukey, J. W. 1977 *Data Analysis and Regression.* Reading, MA: Addison-Wesley.

Nowlan, S. J. 1991. Soft Competitive Adaptation: Neural Network Learning Algorithms Based on Fitting Statistical Mixtures. Ph.D. diss., School of Computer Science, Carnegie Mellon University.

Peterson, C., & Hartman, E. 1989. Explorations of the Mean Field Theory Learning Algorithm. *Neural Networks* 2:475-494.

Rumelhart, D. E., Hinton, G. E., & Williams, R. J. 1986. Learning Internal Representations by Error Propagation. In Rumelhart, McClelland, and the PDP Research Group.

Rumelhart, D. E., McClelland, J. L. & the PDP Research Group. 1986. *Parallel Distributed Processing: Explorations in the Microstructure of Cognition Volume 1: Foundations.* McClelland, J. L., & Rumelhart, D. E. eds. Cambridge, Massachusetts: The MIT Press.

Schaffer, J. D. 1985. Learning Multiclass Pattern Discrimination, 74-79. In *Proceedings of an International Conference on Genetic Algorithms and their Applications,* 169-177. Grefenstette, J.J. ed. Pittsburgh: Texas Instruments & U.S. Navy Center for Applied Research.

Small, S. 1990. Learning Lexical Knowledge in Context: Experiments with Recurrent Feed Forward Networks, 479-486. In *Proceedings of the Annual Conference of the Cognitive Science Society.* Hillsdale, NJ: Lawrence Earlbaum Associates.

# Indexing Cases for Planning and Acting in Dynamic Environments: Exploiting Hierarchical Goal Structures[1]

**Stephen Robinson**
**Janet Kolodner**
College of Computing
Georgia Institute of Technology
Atlanta, Georgia 30332-0280
E-mail: robinson@cc.gatech.edu, jlk@cc.gatech.edu

### Abstract

We examine how acting in dynamic, complex, not entirely predictable environments affects the indexing, storage and retrieval of cases in a memory-based system. We discuss how a hierarchical goal structure can be exploited to provide indices for searching and storage when planning and acting in everyday environments under time pressure. The tradeoffs between the costs and utility associated with attempting to prevent repeating a failure or missing an opportunity are briefly examined. Considering these tradeoffs leads to distinguishing between when failures can be allowed to recur and when they should be anticipated and avoided. The amount of effort expended when handling failures differs for the two situations, but in both cases a hierarchical goal structure can be used to choose effective indices efficiently. This paper describes the approach taken in our EXPEDITER[2] system and briefly compares it to other approaches.

## Acting in Complex, Dynamic Environments

Consider the following example of acting in a complex, dynamic, not entirely predictable environment and the difficulties for indexing and retrieving information that it highlights.

> Bob gets himself and his two children ready for work and school in the mornings the same way every day. The routine includes fixing everyone cereal for breakfast. One day, however, after putting out the bowls and spoons, Bob finds no milk in the refrigerator.

At this point Bob is faced with a number of tasks:

- he still has to fix something to eat;
- he may have to undo some work he has already done (e.g., putting away unused utensils);
- he may want to consider how to avoid running out of milk again;
- he may want to consider how he might reuse the alternative actions he chooses if they succeed.

Bob cannot just stop what he is doing and take time to deeply analyze the situation and generate many options to proceed. He has to continue to act while he is dealing with the failure. The amount of effort he can put into analyzing the future consequences of what he chooses to do depends upon how much time he has available and how much effort it takes to repair the current failure, to analyze costs and payoffs, etc.

Continuing with the example, we see there are many factors to be considered when deciding not only what to do immediately, but how current decisions may affect future actions.

> Bob sees no suitable substitute for milk and cannot think of one. He decides that he does not have time to buy more right now. He needs something else to eat that is as easy and nutritious as cereal. Seeing bread in the refrigerator, he decides to serve toast with jelly and orange juice.

If the toast satisfies everyone and Bob decides the extra work of putting away the unused cereal utensils is not significant, he will probably not be concerned with running out of milk again. All he needs to be concerned with is remembering the "toast substitution" so he can use it if the same failure happens again. He is willing to face the failure of running out of milk in the future because it is not worth expending the effort needed to consciously alter his routine to avoid the problems that arise.

However, suppose Bob's children are upset about having to eat "boring, dry toast" in place of cereal and the ensuing commotion makes Bob late. Not being a strict disciplinarian, Bob decides that he would rather make sure that he can serve cereal than deal with upset children on another morning. Now Bob's motivation to avoid the failure is stronger so he may take some time to think of how he can alter his routines in the future to avoid this problem. He also may think of ways to remind himself to make sure he has milk available at a time when he can get more if he does not.

## Indexing Cases While Acting

At the heart of memory-based systems that reuse plans and learn from experience is the storage and retrieval of cases. Storing information about experiences for later retrieval requires that the episodes be labeled in such a way that they will likely be retrieved when they are most useful in the future. The problem of labeling memory and coming up with

[1]This research is supported in part by DARPA contract F49620-88-C-0058 monitored by AFOSR and in part by a President's Fellowship from Georgia Tech.

[2]EXploring Planning and Execution in Dynamic Environments employing Routines, an exploratory system currently being developed.

labels during problem solving to retrieve useful past experiences is often referred to as the "indexing problem." While considerable research has been done on the indexing problem, little has been in the context of acting in dynamic environments. Planning and acting in complex, dynamic environments adds additional constraints and considerations to indexing processes.

- Time is a limiting factor since the environment may change while an agent is engaged in reasoning.
- Repair of the ongoing situation is necessary; there is no going back.
- An agent is concerned not only with accomplishing tasks and satisfying goals in the current context, but also with continued success in the future.
- The richness of complex environments may require an agent to filter through available information to get usable indices.

While systems like CHEF (Hammond, 1989a), RUNNER (Hammond *et al.*, 1990), PLEXUS (Alterman, 1988; Alterman, 1986) and FLOABN/SCAVENGER (Zito-Wolf & Alterman, 1990) are concerned with repairing failures and with success in the future (through adaptation, anticipation of failure and opportunity, etc.), they do not balance the cost of indexing memory to enable such anticipation against the utility of doing so. At first glance, it seems that an agent would always want to avoid repeating failures and to take every opportunity to optimize its activities. However, in environments where changes can occur and time to act is limited, expending effort to carefully augment plans for future use may cause additional failures or missed opportunities in the current time frame. There are a number of tradeoffs to consider.

- The time spent reorganizing memory takes time away from the current task, but
- reorganizing memory may result in improved actions in the future by contributing to
  - optimizing activity and
  - avoiding failures.
- Knowledge may be gained by adding cases to memory but the knowledge may be erroneous.
- Reorganizing memory may result in higher matching costs when searching for repairs, etc., in the future.

These tradeoffs should be considered when decisions about indexing and search are made. In particular, easy ways of doing the required indexing must be found for those situations in which the harder work of blame assignment needed to anticipate failure situations is deemed inappropriate. In this paper, we look at the ways indexing and search processes can take advantage of the hierarchical goal structures employed in planning to generate indices and annotate memory when expectation failures occur. An expectation failure can be either a failure or a previously unnoticed opportunity to improve. We show how indexing can be accomplished quickly when the cost of repeating a failure is not deemed to be high and time is scarce in the current situation. We concentrate on everyday environments[3] where activity is repetitive, allowing an agent to develop habitual ways of accomplishing goals.

One of our goals is to show how "routines" can be acquired in a memory-based planning and acting system via repairing failures and noticing opportunities for optimization. A routine is a set of actions ("primitive" steps or operators) specific to a particular situation and environment which an agent executes to achieve a set of goals. The execution order of operators in a routine emerges with experience as failing conditions and interactions are dealt with. Employing a routine helps reduce the computational load required to achieve repeated conjuncts of goals by allowing the agent to ignore preconditions, interactions, operator ordering, etc., because those details have been iteratively ironed out during the evolution of that routine. While we do not address the full set of issues associated with learning routines in this paper, it should be noted that the processes described here are part of a larger system that aims toward that capability.

It is also important to note that "routinization" does not simplify an agent's interactions with a dynamic environment. Rather, it simplifies how an agent reasons about its environment. An agent can view a familiar dynamic environment as being more stable than it really is because the agent has learned to reliably predict how the aspects it is concerned with will change and interact. An agent can assume that its expectations about an environment and its own actions will hold without checking their validity. When expectations fail, the experiences are stored to be later retrieved and employed to avoid or cope with similar failures. This is why routinization is iterative. This paper presents a method for choosing indices and organizing case memory under the time pressures of a dynamic environment that uses a hierarchical goal structure to provide readily accessible context.

EXPEDITER operates in a small simulated work area occupied by various objects such as an oven, clothes, washing machine, dryer, bread, etc. Objects, goals and operators are represented in a simple frame system patterned after the organization used in PLEXUS (Alterman, 1988). Each object, including the agent, has associated with it message handlers which govern how the object interacts with the rest of the environment at each "time" tick. The agent is limited in how much it can process at each tick, an important consideration we are working to define more carefully.

## Choosing Indices Under Time Pressure

The best indices are general enough to be applicable in a sufficiently large number of situations yet concrete enough to be recognized. Choosing good indices is hard because it is difficult to determine features that will be predictive in future situations. Part of the difficulty is in not knowing the circumstances in which the case could be useful in the future. The indexing process needs predictive information at a useful level of generality from which to derive indices. An agent acting in everyday environments has some advantages. The routines, failures, repairs and opportunities it will index

[3]See Agre's discussion of everyday environments in (Agre, 1988).

are all likely to be important in the future under the same circumstances due to the cyclic nature of such environments.

The question we must address is what "the same circumstances" means. Consider Bob in the example above. He may discover he has no milk when he goes to the refrigerator, or he may remember there is no milk as he walks into the kitchen to make breakfast or any time up to the time he looks in the refrigerator. He may also discover that he is missing utensils or cereal. In a sense, all of these are the same circumstances (it is breakfast time and cereal cannot be served). Certainly, all can be repaired with the same fix (serve toast and jelly). In other senses, however, the circumstances are different. In the first situation, Bob is looking at the place where the milk should be. In the other situations, he is not. Furthermore, he has completed varying amounts of the routine leading up to getting the milk. Clearly, the first instance of not having milk (the one where he figured out that toast and jelly was a good substitute for cereal) must be indexed so that it can be recalled in all of these circumstances. And clearly, since time is of the essence, the method of choosing indices and storing cases must be fast. Search through memory needs to be efficient for the same reasons. When cases are sought to provide repairs for failures, a quick response is needed to allow action to continue.

## Exploiting Hierarchical Goal Structure

The organization of a hierarchical goal structure can be exploited to meet these needs. In this section, we will illustrate how the goal structure can be used to organize the dynamic memory (Schank, 1982) of which it is a part. Consider Bob again. The circumstances under which he needs to remember substituting toast and jelly for cereal are when he has the goal of having cereal for breakfast and cannot. While he may not have been consciously considering that goal as he was going through the routine of getting spoons and bowls, getting cereal, getting milk, etc., it does organize that set of actions. Using that goal as an index to the "toast and jelly" case makes it accessible no matter when during the action sequence the agent discovers that cereal cannot be served.

Such an index can be chosen quickly if memory is organized in goal/subgoal hierarchies that also organize actions according to the subgoals they achieve. Figure 1 shows part of one of these hierarchies. When an action fails, the subgoals it is instrumental to can be found by traversing up the hierarchy. The subgoals found this way become indices for the new situation. Thus, when Bob looks in the refrigerator and finds no milk, he can quickly choose the indices "trying to `grasp-carton` and there is no milk," "trying to `get-milk` and there is no milk," "trying to `prepare-cereal` and there is no milk," and "trying to `have-cereal` and there is no milk." The "toast and jelly" case will be indexed in all of these ways. The figure also shows this configuration.

The hierarchy can also be exploited during search. If a routine action fails and there is no case indexed under it, traversing up the goal/subgoal hierarchy will quickly find goals under which a case might be found. Thus, if Bob remembers he has no milk while he is getting cereal from the pantry, he can find the "toast and jelly" case by traversing up the hierarchy from `grasp-cereal-box` to `get-cereal` to `prepare-cereal`, which serves as an index along with "no milk" to the "toast and jelly" case.

A goal hierarchy preserves relationships between general and specific goals which can be used to select indices pertinent to a problem solving situation at varying levels of generality. When an agent is executing steps of a routine, it is not concerned with the details of all the goals it is achieving. It is only immediately privy to information about the step it is executing. However, it can follow subgoal links up the hierarchy and generalize its knowledge of the current situation by inspecting the goals it finds at the higher levels. When it needs to search for or to index a way to handle an exception, it has simple, directed access to indices which are both generally applicable yet pertinent to the current context.

In the following sections, we discuss our implementation of this scheme in EXPEDITER and the control issues that must be addressed to make it work.

**Implementing the Hierarchical Memory.** EXPEDITER's memory has nodes corresponding to goals and subgoals that are connected to each other according to goal/subgoal decomposition relationships. The agent knows, for example, that `get-ready-for-work` can be decomposed into `eat-breakfast` and `get-dressed`. Goal/subgoal decomposition links are bidirectional. They can be used both to find subgoal decompositions and to find goals a subgoal is instrumental to. These connections are illustrated in Figure 1 in a small portion of EXPEDITER's memory.

The lowest level goals have action sequences associated with them that, when carried out, achieve the goal. `Get-cereal`, for example, is carried out by the sequence `goto-pantry` followed by `grasp-cereal-box`. Action nodes are connected to each other by sequencing links. `Goto-pantry` is followed by `grasp-cereal-box` which is followed by `goto-cabinet`, etc. Subgoals are connected to each other in context and so are actions. Thus, because `get-cereal` and `get-bowls` are both subgoals of `prepare-cereal` and in that context are carried out in a specific order, the sequencing of the actions of each one is specified in that context. We assume that these sequences come from cases in which the same sequence of actions is carried out over and over again.

Exceptions to routines are indexed from these memory structures. In essence, each node in memory can act as a "branch point," providing access to both normal and specialized ways of carrying out the action or achieving the subgoal. Figure 1 shows how the "toast and jelly" case is indexed from these memory structures. A pointer to it is placed at each appropriate goal node in the memory structure, and the pointer is labeled with the circumstances under which that branch should be taken. For example, when attempting to `grasp-carton` and there is no milk, rather than going to the table (the normal routine), the `have-toast` branch is followed (its sequence of events is not specified in the illustration, but can be easily imagined). This same branch is attached to the `get-milk`, `prepare-cereal`, and `have-cereal` nodes.

**Traversal of the Hierarchy under Normal Circumstances.** During normal activity, goals arise through an agent's inter-

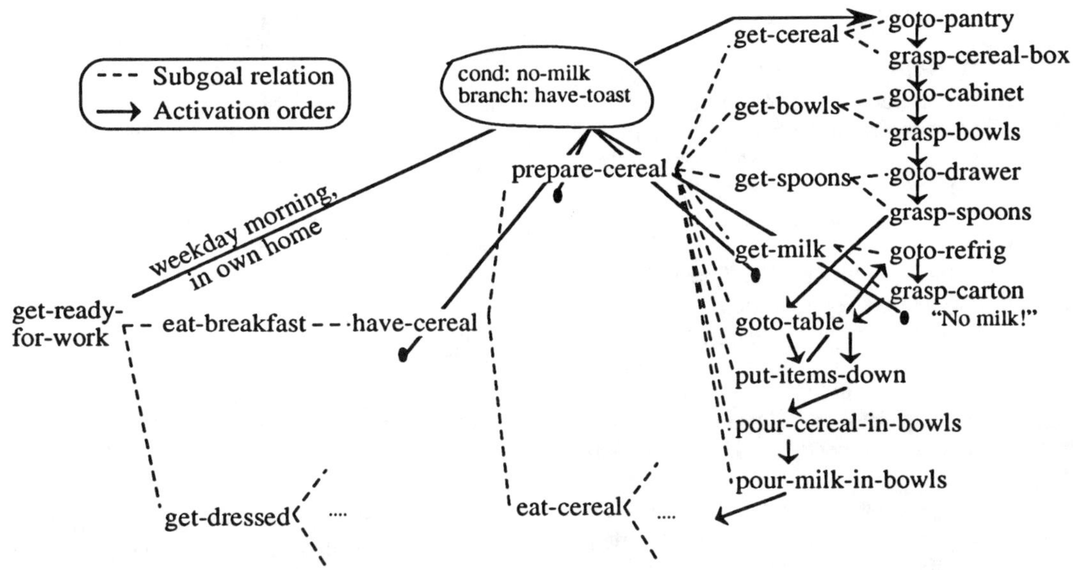


Figure 1: Simplified goal structure and routine with annotations for substituting toast.

actions with the environment. Routine methods for achieving those goals can be found by using features of the goals and the current context as indices to memory. In the normal case of getting up in the morning, for example, the actor's goal of `get-ready-for-work` and the fact that he is in his own home get him to the normal morning routine (beginning with `goto-pantry`). He follows the routine he finds there. When he gets to the `grasp-carton` step, if he has no milk, he follows the `have-toast` branch, otherwise he follows the normal sequencing and continues with `goto-table`.

**Indexing a "New" Case.** New cases must be placed in memory when a failure is encountered or when an alternate way of doing something is found. Indices to the new case must be general enough so that case can be found in a variety of appropriate circumstances. As discussed above, this is accomplished in EXPEDITER by placing indices on appropriate goals in the hierarchy designating that an alternate branch should be taken. Indices either describe circumstances under which the failure can be expected, describe the failure itself (e.g., no milk), or describe an enabling condition that allows the agent to take advantage of an opportunity to improve (e.g., "if a goal to do wash is active, then branch to ..."). They point to the case that provides the set of steps that has previously repaired the failure (or grabbed the opportunity).

Of course, the process must stop at some point in the hierarchy. One way to choose a stopping point is to only go as far up as the goal that was substituted when repairing a failure. In Figure 1 that point is at `have-cereal`. Going farther up is not warranted because there is no indication that goals above `have-cereal` are affected since milk is neccessary for cereal, not for breakfast in general. We are considering other methods for deciding where to stop, especially for cases where a failure should be anticipated and prevented.

**Searching for a Previous Repair.** When EXPEDITER encounters a failure, it attempts to find ways in which it has previously repaired similar failures. It begins by looking for indices associated with the failed action that might point to repairs. If no appropriate one is found, it looks next at the parent goal's indices, first for a direct match to the failure, and then for other branches. If `grasp-carton` fails again, there will be a direct match for the failure annotated at `grasp-carton` so no further search is needed. If EXPEDITER "recalls" that it has no milk at `grasp-cereal-box`, it finds a direct match at `prepare-cereal`. If `grasp-cereal-box` fails, EXPEDITER can find the "have toast" branch at `prepare-cereal` because, although "no milk" is not a direct match, `prepare-cereal` is a failing goal shared with a previous failure.

**Searching for a "New" Repair.** If no particular repair is found, the goal hierarchy can still provide clues that can help in finding a repair somewhere else in memory. Active goals, along with their constraints, reasons, and needed resources can be used to create a more abstract description of the situation needing repair. This more abstract description can be used to search more broadly in memory. For example, if `grasp-carton` fails and EXPEDITER can find no repair, descriptive features of `get-ready-for-work`, `eat-breakfast`, `have-cereal`, etc., can be used to describe the situation. A description including the features "limited time," "easy to fix," and "breakfast food," would be created. In the example scenario, "have toast" was chosen because it was suggested by the presence of bread in the refrigerator, although others may be equally usable. How these features can be selected is an issue we are studying.

In all of this, the goal hierarchy is providing a way to generalize a situation. It allows the choice of indices of varying degrees of generality, starting with very concrete, which are likely to be effective in both the current and future contexts. The indices are available at little cost since the only work is following subgoal links and gathering features from the goals.

### Allowing and Avoiding Repeated Failure

When cost/utility tradeoffs are considered, not all exceptions merit equal treatment. For example, in the first scenario in the first section, the failure is deemed to be a minor problem that could be allowed to happen again. However, in the second scenario where Bob's children get upset, the failure is too "costly" to allow it to recur.

Our discussions so far have addressed the first scenario. The agent decides that there is no need to anticipate a failure (it is willing to let the failure recur, or at least not to take action to try to prevent it at the current time), and effort is put only into recording the ways in which it was repaired. The agent will want to recall how the current failure was handled if it is encountered again, but only needs to be reminded if the failure actually recurs.

However, when a failure is deemed too costly to allow it to happen again, it is not enough to index only on the failed goal and its immediate parents since those annotations will be encountered only if the failure happens again. Indices must include goals that will be active and other context features that will be present when failures and opportunities need to be anticipated. While this is a harder problem, we believe that the hierarchical goal structures can again be used to advantage. The extra step of blame assignment is required when a more costly failure occurs or an important opportunity arises. Blame or credit must be assigned to steps taken earlier that set up the failure or to steps which could have been taken to prevent it. We are not investigating the difficult credit assignment problem at this time; however, once an assignment is made, a process similar to that described in the section on exploiting goal hierarchies can be followed. Limited space does not allow us to add details.

## How EXPEDITER Relates to Other Approaches

Hammond's study of *agency* (Hammond *et al.*, 1990; Hammond, 1989b) is also based on a dynamic memory representation and the acquisition of "well-tuned, default plans" for particular conjuncts of goals. Like EXPEDITER, the RUNNER system is designed to act in a changing environment, to learn from failure and to recognize opportunities. Our approach differs in at least two important aspects: it provides a method that allows it to choose effective indices rapidly and it recognizes the variable importance of analyzing and processing exceptions. RUNNER's spreading activation is good for recognition and recall but it is not clear how its indexing and search would be affected by considering cost/utility tradeoffs. Also, in Hammond's systems, action is always taken to attempt to prevent failure from happening again and every blocked goal results in marking memory to set up a possible opportunity to satisfy the goal. It is not clear that it is always desirable to do this since the effect on current activity may be deleterious. Our approach has a way to choose useful indices from knowledge of its goals quickly and can adjust the amount of effort it expends on indexing repairs to reflect the usefulness of doing so.

Alterman's work on adaptive planning and "ad-hoc" learning (Alterman, 1988; Zito-Wolf & Alterman, 1990) is similar to our work as well. PLEXUS adapts situated plans it already knows to similar situations in order to avoid a more complex planning problem. The FLOABN/SCAVENGER system also adapts situated plans but in addition learns discrimination points and ad-hoc categories which allow it to apply the adaptations it develops when it encounters the same situations again. SCAVENGER addresses choosing indices from features salient to the system's current knowledge and expending effort on reorganizing the plans used the most. SCAVENGER's method of generating indices seems effective but EXPEDITER goes farther by providing an efficient way to choose effective indices along a range of generality. In addition, EXPEDITER addresses the placement of discrimination points to support anticipation of exceptions and reusing repairs for related failures.

## Conclusions

The computational load required for indexing and retrieving cases placed on a memory-based system in a complex domain can be high. Reducing this load becomes more important when the additional constraints placed on a system by dynamic environments are taken into account. However, when everyday, cyclic activity is considered, the hierarchical goal structures used in planning and routine activity may be exploited to lessen the high load.

Our approach addresses difficult problems for indexing cases in dynamic, complex environments. It allows choosing from a broad spectrum of concrete to more general indices, directed by goal knowledge pertinent to current context. It allows good choices to be made when time is limited and better choices to be made as more time is allowed. It is limited to known goal knowledge and may be too closely tied to goal hierarchies but we are working to better understand these issues.

## References


Agre, P. E. 1988. The Dynamic Structure of Everyday Life. Technical Report TR 1085, AI Laborabory, MIT, Cambridge, Massachusetts.

Alterman, R. 1986. An Adaptive Planner. In *Proceedings of the Fifth National Conference on Artificial Intelligence*, pages 65–69, Los Altos, California. Morgan Kaufmann Publishers, Inc.

Alterman, R. 1988. Adaptive Planning. *Cognitive Science*, 12(3):393–421.

Hammond, K.; Converse, T.; and Martin, C. 1990. Integrating Planning and Acting in a Case-Based Framework. In *Proceedings of the Eighth National Conference on Artificial Intelligence*, pages 292–297, Boston, MA.

Hammond, K. J. 1989a. *Case-Based Planning: Viewing Planning as a Memory Task.* Academic Press, Inc. (Harcourt Brace Jovanovich, Publishers), New York.

Hammond, K. J. 1989b. Opportunistic Memory. In *Proceedings of the Eleventh International Joint Conference on Artificial Intelligence*, pages 504–510, Detroit, Michigan.

Schank, R. 1982. *Dynamic Memory.* Cambridge University Press, New York.

Zito-Wolf, R. and Alterman, R. 1990. Ad-Hoc, Fail-Safe Plan Learning. In *Proceedings of the Twelfth Annual Conference of the Cognitive Science Society*, pages 908–914, Cambridge, MA.

# Belief Relativity

**Donald Rose**

Department of Information & Computer Science
University of California
Irvine CA 92717
drose@ics.uci.edu

**Abstract**

This paper describes a model of belief systems called *belief relativity* (BR), which addresses the relationships and structure of knowledge held by multiple interacting agents. This paradigm uses *belief reference frames* (b-frames) as the main unit of belief spaces, within which an agent's beliefs are stored. BR is concerned with how beliefs are created and revised, how they influence each other within or between b-frames, and how one searches for b-frames that are useful (e.g., that remove contradictions). BR also deals with degrees of belief, propagated along the influences that relate beliefs and b-frames to each other. BR attempts to combine the best features of these ideas into a unified, synergistic framework.

## Introduction

This paper presents a model of belief systems called *belief relativity* (BR). The model consists of two main interacting elements. First is the notion of *belief reference frames* (b-frames), within which an agent's beliefs reside along with their *influences* – i.e., supporting beliefs, and any other more indirect evidence. This area of BR is concerned with how beliefs are created and revised, how they influence each other within or between b-frames, and how one searches for useful b-frames (e.g., that remove contradictions). This area has been studied in research on TMS-based systems (Doyle 1979) as well as systems related to ATMS (de Kleer 1984; Rose & Langley 1986). The second notion is *degrees of belief*, propagated along the aforementioned influences between beliefs. This area of BR is concerned with how one can utilize various types of rules to infer the degree to which a belief fits into a given property class. This has been addressed by work in fuzzy systems (Zadeh 1965), as well as other methods for mathematical computation of evidence (Shafer 1976). BR attempts to combine some of the best features of these existing belief models – as well as new ideas – into a unified, synergistic framework, one which can hopefully apply to a wider class of problems, or offer insights into how they may eventually be addressed.

## Beliefs and Reference Frames

One of the basic premises of BR is that the reasoning of intelligent agents should be modelled as reasoning with several types of beliefs, all of which can be grouped into reference frames (b-frames). In BR, beliefs are generally represented as follows.

**<property P>( <b-frame>, <assertion A>) = <degree that P holds for A>.**

The BR model assumes that an assertion is any unit of knowledge that can be attached to a property; for instance, agents can attach the property "relevance" to an assertion. Beliefs can in turn act as assertions to be acted upon by other beliefs, as discussed later. In such nested beliefs, all b-frames are that of the outermost belief, unless stated otherwise. BR allows agents to adopt many b-frames over time, depending on the state of one's reasoning (e.g., when exploring hypotheses within a temporary b-frame that can be deactivated at a later time).

The most basic frame concerns belief *influence* (i.e., the frame that indicates causes, or inferential support, or evidence). A link from a belief X to another belief Y indicates that X influences Y in some fashion; belief X may not directly lead to the inference of Y, yet it may still influence Y is some other indirect manner. The link from X to Y can carry a *degree* of influence, depending on the rules that are used by an agent to perform inferencing. Sets of beliefs, and links that represent the degree of influence, form networks, which have been proposed and analyzed elsewhere (*influence diagrams* (Howard & Matheson 1984) and *belief networks* (Pearl 1988)). For example, the b-frame of a nonsmoker NS might compute that smoking highly affects cancer rates – i.e., **influence( NS, (smoking,cancer) ) = 0.8**, whereas the b-frame of a smoker S might feel the influence is less – i.e., **influence( S, (smoking,cancer) ) = 0.4**.

Note that these examples of *causal influence beliefs* do not indicate inferential support, only causal support. The special case where influences have only 0 and 1 as possible strength values (i.e., degrees of influence are either all or none), and the influences indicate direct inferential support (i.e., can be viewed as inference histories) is analogous to a number of systems for automated reasoning – e.g., TMS, ATMS or REVOLVER (Rose 1989). Examples of *support influence beliefs* might be: **influence( NS, (B1,B2) ) = 0.7**, and **influence( S, (B1,B2) ) = 0.4**, where B1 and B2 are other beliefs. B1 might be **influence( surgeon-general, (smoking,cancer) ) = 0.9** and B2 might be **influence( NS, (smoking,cancer) ) = 0.9**. In other words, NS's believed influence of smoking on cancer *is itself influenced by* (e.g., inferred from) the surgeon-general's belief in this smoking-cancer connection, and NS believes this support with strength 0.7. S does not hold this B1-B2 support as strongly (only 0.4); perhaps S feels the surgeon-general's belief only partly influences NS's stance on the smoking-cancer connection. Later evidence may prove S right (e.g., influences that neither S nor NS immediately recognized, such as old experiential influences that over time have lost nearly all their relevance to more current beliefs). Also note that B1 and B2 need not be influence beliefs, but can also be other types, such as relevance; these new beliefs are discussed below. In general, the nesting of beliefs within beliefs, and the interplay between beliefs of multiple agents, are two of the strongest characteristics and advantages of the BR framework.

Although influence networks are useful tools, the framework proposed in this paper does not limit beliefs to influences. A second type of belief would concern the degree of *certainty* in the beliefs of the influence b-frame. For example, the smoker S above might feel *highly* certain about his belief in the *low* influence of smoking on cancer. This could be represented as:

**certainty( S, influence( S, (smoking,cancer) ) = 0.4 ) = 0.99.**

However, S might *also* feel highly certain about *how NS feels about the influence of smoking on cancer*; i.e.,

**certainty( S, influence( NS, (smoking,cancer) ) = 0.8 ) = 0.9.**

The third main belief type in our framework deals with *relevance.* As an example, NS, being a nonsmoker, might hardly ever think about smoking issues, and hence not find S's beliefs very relevant; e.g.,

**relevance( NS, (influence( S, (smoking,cancer) ) = 0.4 ) = 0.1.** In fact, NS's own beliefs on the subject may also have little relevance to NS:

**relevance( NS, (influence( NS, (smoking,cancer) ) = 0.8 ) = 0.1.**

However, like the other belief types, the degree of relevance might change as new beliefs become part of one's current b-frame. A good example is the inclusion of *goal* beliefs. Goals can be used to increase the degree of relevance of other beliefs. For example, the need to debate a smoker might increase NS's relevance for his beliefs pertaining to smoking issues.

One of the interesting aspects of including relevance and certainty in the representation of agent's belief systems is that they are likely to be intricately *intertwined.* In general, the degree of belief in certain assertions may lead to inferences regarding their relevance, and the degree of relevance can help drive which degrees of belief get propagated and when. Let us look at additional examples which hopefully illustrate these ideas. First, note that relevance may rise if one's degree of certainty rises. For example, the more I believe I have a cold, the more relevant "cold" assertions may get (e.g., remedies; rules for getting well). Alternatively, relevance may decline if degree of belief rises. For example, the more a Mom believes her son is safe, the less she might worry about her son's safety. A third case could arise where, for instance, an agent Y finds agent X's low degree of belief (say in Y's abilities) a very relevant issue.

Note that relevance is a *relative* quality, depending on one's b-frame. As another example, an agent X's assertions (and degrees of certainty in their causal connections) may make a frame F very relevant to X, but agent Y's assertions and beliefs may make it irrelevant to Y. More specifically, suppose an influence frame consists of all beliefs and causal influence links regarding the Kennedy murder conspiracy theory T, and suppose an author A creates, updates, and strongly believes T. T's assertions are thus highly relevant to A, and are almost always active. A kid K might know of T but not believe it much; hence T's assertions have low relevance to K, and are almost always inactive. Lastly, a hippie H might believe T's assertions as highly as A, but have no vested interest in T. Hence T's assertions have low relevance to H, and are almost always inactive.

Having different belief types may also allow more accurate modelling of cognitive phenomena. For example, non-increasing relevance in one part of the influence b-frame's assertions may gradually lead to the virtual deletion/forgetting of these assertions. That is, it may become last on the list of things to think about. This behavior might prove quite valuable. For instance, if one is dealing in limited-memory scenarios (e.g., an autonomous long-term learning system without hardware expandability, such as in space domains), one could implement a Least-Recently-Used-type strategy; this method would explicitly delete those assertions that are either least relevant or least believed, taking into account in how long such a status held.

In summary, utilizing these and other types of interacting belief types should provide a more powerful framework for modelling complex belief reasoning phenomena. By modelling not only how highly one believes an influence, but also the relevance of that influence to one or more agents, the BR model imposes a useful structure on an agent's belief space, which would hopefully enable improved inferencing and revision performance.

## Propagating Degrees of Belief

In a general implementation of the BR model, any belief could be used as input to the degree-of-belief rule of any property (e.g., attractiveness; tallness; relevance). For example, one can have a rule relating the weight of someone (input) to the certainty that that weight is a member of the property class *heavy* (output). (Note, however, that relevance is currently the only property that can apply to any belief.) These "degree-of-property" beliefs can then be described by other properties, to any level of desired nesting. An advantage of BR is that degree-of-property inference rules are independent of the influences that determine which beliefs apply to each other, and can be of any type (e.g., fuzzy/probabilistic membership functions; qualitative ranges; binary decision rules).

For example, a qualitative rule might map beliefs to qualitative values (e.g., if height(X) >= 6'4", then tallness = very-tall; else if height(X) >= 5'8", then tall = tall; else tallness = short). Fuzzy membership functions might map beliefs to a point on along a continuous curve (e.g., tallness( height(X) = 6'4" ) = 0.9). Note that the belief being input *can itself represent a degree of belief.* That is, every application of a belief strength rule by a b-frame results in a new belief, which can potentially become a new input to any other b-frame (e.g., a new domain value in some other membership function). For instance, an agent X might not find the tallness of anybody that relevant; e.g.,

**relevance( X, tallness( X, height(Z) = 6'4" ) = 0.9 ) = 0.2 ).**

In general, the same belief can be processed by *different b-frames* (i.e., different agents, or different hypothetical views by the same user), and these different b-frames may employ different rules for determining degree-of-belief for different properties. Thus, if X worked for the Guinness Book of World Records, and Y told X about an extremely tall person, X might feel this quite relevant; e.g.:

**relevance( X, tallness( Y, height(Z) = ? ) = 0.99999 ) = 0.9 ).**

Note that it is Y's tallness belief that excites X, *not Z's actual height*; X's relevance belief represents the view from X's current b-frame, where exact knowledge of Z's numerical height is not known yet (only exists in Y's b-frame). Once X sees Z directly (height(Z) = 6'6"), X still thinks Z is tall (e.g., tallness( X, height(Z) = 6'6") = 0.9), but this is a lesser degree of tallness than Y believes. Now using her own tallness judgement – the tallness belief about Z from her own b-frame – X revises her excitement downward; e.g.:

**relevance( X, tallness( X, height(Z) = 6'6" ) = 0.9 ) = 0.3 ).**

Let us now look at a more complex example, where agents Jack and Kelly each use a b-frame (call these J and K). K has two fuzzy membership functions: one for determining her certainty in the *heaviness* of someone, another for determining the relevance of heaviness to her. We will assume J has only one membership function, a fuzzy view of K's heaviness-relevance belief about him. That is, J's relevance function determines the relevance of how strongly K finds J's weight relevant to her. A chain of influences has been constructed – from J's weight (belief W), to K's belief in W's degree of heaviness (belief X), to K's belief that X is relevant to her (belief Y), to J's belief that Y is relevant to him (belief Z).

The resultant computations of degrees of belief for W, X, Y and Z can be summarized as follows. If J is 190 pounds, let us suppose K's function finds him quite heavy (level 0.9). This level then becomes the basis to compute the relevance of this heaviness level to her; let us suppose her relevance function indicates her attraction to men over 170 pounds decreases fast; at 190 her attraction level is 0.3 (hence her degree of attraction to J is 0.3). Finally, this degree of attraction that K feels for J is input for J's function for determining his attraction for K. Since K feels attraction for J of only 0.3, J feels attraction for K (assuming a linear function) at the same low level (0.3).

If we were modelling height instead of weight, the influences would remain static over time, since height is not changeable. However, modelling weight allows changes to be made and propagated through the membership functions. If J's weight increases, K increases her certainty in J's heaviness. This belief then causes a decrease in her attraction to J. Finally, this lower degree of attraction that K feels for J causes J to decrease his attraction for K.

Note that J's choice of using an attraction-for-K rule that uses K's-attraction-for-J as its sole influence sets J up for the effects of *cognitive dissonance* (Festinger 1957). That is, J essentially likes K if she likes him back, but once her interest falls, his does too. The richness of allowing multiple interacting b-frames, with interacting degree-of-belief rules, at user-controlled levels of nesting, will hopefully allow agents to model (and, if desired, perform) many qualitative classes of cognitive behavior. An even harder problem is how an agent might decide a belief propagation rule is yielding nonoptimal results, then begin a search for different influences (i.e., change b-frames, and hence change belief propagation rules). The ideas in this paper should provide a starting point from which to build such capabilities into future systems.

## Revision in B-Frames

Several methods currently exist for doing belief revision on influence networks, such as in representations where premise beliefs support other inferred beliefs (e.g., TMS, ATMS, REVOLVER). However, these revision methods tackle only one part of the b-frame proposed above: reasoning within one b-frame, and doing so only on influence beliefs and links. A more general method of belief revision would enable operations that involve multiple b-frames (e.g., how to change from one to another, and when), as well as certainty and relevance beliefs and their associated links. Possible revision operators might be: add or delete a belief (essentially the operators of REVOLVER); modify a belief property (e.g., increase or decrease certainty or relevance); change b-frames (e.g., adopt a larger or smaller set of beliefs and links).

One issue involves the triggering of a *change of frames*. For example, an agent might deem a particular degree of belief to be too neutral (i.e., at or near 0.5). The agent may then decide to form an opinion about this link, one way or the other. Similarly, an agent may want to form an opinion based on beliefs (or degrees of belief) held by other observers; that is, one might analyze beliefs from other agents' b-frames to influence the beliefs or links of one's own b-frame.

To accomplish these revision goals, one might adopt a larger (wider scope) b-frame, in the hope of obtaining more evidence about new beliefs and using this knowledge to revise the beliefs (or belief properties) of one's old b-frame. One method for revising one's b-frame to a larger belief set might be to increase the relevance of other beliefs – in this case, those observed to be present in other agent's b-frames. One could begin by increasing the relevance of those external beliefs whose properties have the highest values in the other agents' b-frames. This embodies the intuitive heuristic: "pay more attention to the strongest outside opinions" (all other factors being equal).

In general, one might view belief revision as a *search through the space of b-frames*, using the operators proposed earlier. In most domains, there should also exist *constraints* one can use to help control this search. For example, one might utilize constraints on what evidence is needed to resolve a current conflict. The need to resolve a conflict between beliefs X and Y should lead an agent to limit revision search to frames involving both X and Y; the goal b-frame is one where the X-Y conflict is resolved. A more complex scenario would be to perform the above search, but do it *so that another agent's conflicts* are resolved; i.e., put oneself into another's head to help resolve a conflict. There are several states that changing one's b-frame can result in. In general, new *types* of beliefs, new *degrees* of belief, or both, can be created or inferred due to a change in b-frame.

Note that allowing degrees of belief on links means a reworking of the concept of contradiction. In the special case where all degrees of belief are either zero or one, a belief B is either held or not; holding both B and not(B) would be a clear contradiction. However, contradictions are more "fuzzy" if beliefs are held in degrees. For example, one might hold B with certainty 0.3 and not(B) with certainty 0.4; should a pair of beliefs with these certainty levels be deemed a contradiction?

To address this, an interesting revision strategy might be to use the degree of difference between two beliefs' certainty levels to influence the degree of relevance of this difference.

That is, as the difference in certainties between two conflicting assertions becomes closer to the standard definition of contradiction (holding B and not(B) with certainty 1.0), this change can be used to increase the relevance of this difference; this acts like an attention mechanism, gradually increasing the importance of this part of an agent's b-frame. Once this relevance passes a threshold, which can be indicated by another belief (perhaps unique to this b-frame), belief revision would begin. In short, we have a striking example of how one can utilize the integration of influence, certainty and relevance; a difference in *certainties* can used to *influence* the *relevance* of this difference, which in turn can be used to trigger revision of any one of these three properties.

A brief example can illustrate the above strategy. A b-frame containing the beliefs certainty(agent,B) = 0.5 and certainty(agent,not(B)) = 0.5 would have a relevance too low to trigger belief revision. Intuitively, the above situation indicates a lack of an opinion about either belief. A different b-frame might have certainties of 0.8 and 0.2, respectively; these new numbers, like the previous ones, sum to 1.0 and hence might be considered non-contradictory, even though there isn't total support for either one. Note that this second b-frame might result after new evidence arrived and led to revision of the two beliefs' certainties. A third b-frame might have certainties of 0.8 and 0.8; this is getting too close to the standard contradiction scenario (i.e., 1.0 and 1.0), and hence might trigger belief revision.

These three b-frames hint that one informal rule of thumb might be to begin increasing the relevance of a belief pair X and not(X) if their total certainties deviate from 1. However, this rule might be fine for one agent (i.e., one b-frame) but replaced by a related yet different rule for another agent (i.e., a different b-frame). This is one illustration of the nature of belief relativity: that b-frames are best viewed as relative states, useful for adopting points of view and exploring the consequences of such views.

## A General View of Belief Relativity

A wider view of BR leads to several interesting issues; specifically, one can produce certain claims based on an analogy between BR and the relativity theory (Einstein 1905) of the physical world (call it PR).

First, just as there is no absolute physical frame of reference from which measurements can be made with complete certainty, in BR *there is no absolute b-frame* from which to perform belief computations with complete certainty. For instance, just as velocity is a relative concept in PR, dependent on the frame one is observing from, the BR view is that *degree of belief is relative*, depending on what b-frame one is believing from. In PR, there exists the relativity of motion; driver A may claim "I'm travelling forward at 5 mph; car B is still", whereas driver B may claim "I'm travelling backward at 5 mph; car A is still". Both are correct, because constructing an answer is relative, and depends on the frame of reference from which one is viewing the situation.

An analogous example in BR is the case where one assertion can be cause different degrees of belief to be computed in different b-frames – e.g., belonging to two different agents. For example, suppose an optimist O and a pessimist P view the same belief X representing the outcome of a mutual friend M's test: X = (grade (M, test) = 79). O may happy about this: happiness( O, X ) = 0.8, perhaps because this shows progress by M. P may not be as happy about the outcome: happiness( P, X ) = 0.5, perhaps because P thinks M's parents will be upset that M didn't ace the test. Who is right? Both, because the answer is relative; multiple b-frames, even if they are constructing beliefs about the same property, can process the same belief with equal validity. That is, different b-frames can have different views of the same situation without contradicting each other. In this case, both happiness levels are "correct", depending on what b-frame one is in.

Note that the qualitative reasons cited above for O's and P's happiness levels might be based on other beliefs in their respective b-frames (i.e., previous experience concerning M's life). In general, BR provides a model for how one processes a belief a certain way (e.g., like an optimist or pessimist) based on the influence of other beliefs in one's b-frame. The same agent can even act like either an optimist or a pessimist, depending on what beliefs are used to compute degrees of certainty for a new belief (i.e., what b-frame one chooses to view a situation from). Finally, whereas this example dealt with how different b-frames view the same property (happiness), earlier examples showed how different b-frames applied to a belief can also utilize different property rules (e.g., tallness and relevance).

The second major claim is that *adopting a higher-level frame can resolve conflicts and increase certainty, but it can also have the opposite effect.* In PR, deciding which of two motions are relative can only be done by comparing them to objects in a larger frame. In the previous PR example, an observer using a higher (more omniscient) frame may decide that car A really was moving forward relative to car B, and that the reverse does not hold, because A is moving forward relative to other objects in the higher frame while B is standing still relative to them. Similarly, the previous BR example might change by adopting a higher frame which includes M's statement that he is as happy about his grade as O claimed; this resolves the relative views of O and P in favor of O. However, adopting still wider frames can again lead to belief revision. A higher frame in the car example might find that the object a believer used as a reference to compare to A and B (e.g., a train) is *itself* moving, and doing so with the same direction and speed as B; in the view of this higher b-frame, where the new reference object is the ground, A is now still while B and the train are moving, thus contradicting the earlier b-frame belief. Similarly, a wider b-frame in the BR example may find that M is lying about his happiness to hide his parents' unhappiness; this now resolves the O-P conflict in favor of P.

In summary, BR theory proposes that one can achieve *local certainty*, but this is the best one can do. Absolute certainty is unattainable, because there is always the potential of finding another frame that invalidates one's "omniscient" point of view. In short, one cannot attain permanent global absolute certainty, since there is no b-frame provable to be absolute over all beliefs for all time. However, finding temporary local certainty among one's beliefs *is* a useful goal, such as to resolve apparent contradictions by going to higher b-frame. In general, an optimal agent must be able to find b-frames that provide relative certainty, while also being smart enough to know when to search for more useful b-frames (either to supplant existing ones, or add new b-frames to one's repertoire). BR can hopefully provide a framework to model and improve such behaviors.

## Related Work

In BR, generating new assertions is performed inside a b-frame, and evidence for these beliefs is also maintained there. In this sense, BR is directly related to systems such as TMS, ATMS, and REVOLVER. If one uses BR only to infer new assertions from old via rules, and all degrees of certainty are either zero or one, and relevance is not represented, then BR is essentially reduced to these frameworks. Like ATMS-based systems, BR systems settle for local coherence or certainty over global, and can opt to keep non-currently-active belief groups in memory. If one augments ATMS-based systems with a method for propagating fractional certainty levels, this type of system resembles influence diagrams, or belief networks. Further, if one uses hillclimbing search for consistent belief sets, this type of system would now resemble systems such as REVOLVER.

However, BR is designed to improve upon the above methods when reasoning about how different agents process beliefs, and how one agent can adopt different points of view. That is, BR provides a better mechanism for reasoning about multiple b-frames. For example, BR provides a stronger framework for dealing with nested beliefs – beliefs about the beliefs of others. Even other b-frames can be the focus of such nested belief (e.g., the degree of certainty agent X has in agent Y's beliefs about some topic). BR also can utilize a more complete set of revision operators than systems such as REVOLVER; the latter could search for and generate new premise beliefs, but BR also deals with creating new b-frames, as well as new influences among them.

Finally, BR utilizes an explicit notion of belief relevance that the systems mentioned above do not. This added concept enables relevance and certainty levels to influence each other to produce interesting behaviors (e.g., by nesting these concepts within one belief). However, degree of relevance can also be used to compute a degree of activeness; that is, relevance can be used to focus system *attention* on the most interesting assertions. An agent can use such knowledge to control the propagation of certainty degrees (e.g., perform such propagation only on those most-interesting assertions). In short, just as other areas of AI utilize knowledge to control search, so too can one use relevance information to control the search for useful beliefs (i.e., those that are relevant to current goals), the beliefs they influence, and the certainty levels of all such *belief families*. The above point is even more important due to the oft-cited problems TMS and ATMS systems have regarding excessive computation required to process even small sets of premises, due to storage of all inferred beliefs in memory. TMS and ATMS systems give beliefs a status of either active or inactive; TMS and ATMS can thus be viewed as as a special case of BR where all relevance levels are either zero or one. By substituting degree of relevance for active/deactive status, BR beliefs can be prioritized for processing, which should lead to more efficient systems. As is the case for certainty levels, relevance levels can be calculated using a method (such as that of Thagard and Kunda (1987)) that will be independent of the methods used to maintain the influence networks.

The second useful comparison is to fuzzy systems (Zadeh 1965). We have seen how agents in BR can propagate degrees of certainty and relevance by using fuzzy membership functions. In addition, like fuzzy logic, BR can deal with the notion that an entity can "have" property P as well as not(P). For example, one agent X can believe that another agent M is both happy and not happy in BR (where X is viewing M from a b-frame that encompasses those of two other agents observing M). However, unlike fuzzy logic, BR can deal with reasoning on the more abstract b-frame level, such as ATMS-type operations (e.g., maintaining chains of inference support), and modelling different points of view as well as what might trigger such changes.

## Conclusion

I have presented a model called belief relativity, a model for reasoning about the structure of belief systems, and how degrees of certainty and relevance are inferred. BR provides a unified framework for ideas from previous research, such as influence/belief networks, TMS and ATMS systems, and hillclimbing systems for belief creation and revision.

An important advantage of BR is that influences, as well as the rules for propagating certainty and relevance values among them, can be represented and processed *independent* of each other. That is, BR belief systems can perform b-frame level tasks (keeping track of influences, trying out new sets of beliefs, creating new influences) yet also perform tasks complementary to those concerning b-frames (e.g., using membership functions to determine degrees of belief, and propagating them along the influences set up by b-frame reasoning). In short, BR should allow the best features from ATMS-style systems and fuzzy/probabilistic systems to be utilized in a synergistic manner.

## References


de Kleer, J. 1984. Choices without backtracking. *Proceedings of the Fourth National Conference on Artificial Intelligence* (pp. 79–85). Austin, TX: Morgan Kaufmann.

Doyle, J. 1979. A truth maintenance system. *Artificial Intelligence, 12*, 231–272.

Einstein, A. 1905. On the electrodynamics of moving bodies. *Annalen der Physik*, 17, 891.

Festinger, L. 1957. *A theory of cognitive dissonance.* Stanford, CA: Stanford University Press.

Howard, R. A. & Matheson, J. E. 1984. Influence Diagrams. In R. A. Howard and J. E. Matheson eds., *Readings on the Principles and Applications of Decision Analysis*, Vol. II, Strategic Decisions Group.

Pearl, J. 1988. *Probabilistic Reasoning in Intelligent Systems.* Morgan Kaufmann.

Rose, D. 1989. Belief Revision and Machine Discovery. Ph.D. diss., Dept. of Information and Computer Science, University of California, Irvine.

Rose, D., & Langley, P. 1986. Chemical discovery as belief revision. *Machine Learning, 1*, 423–451.

Shafer, G. 1976. *A mathematical theory of evidence.* Princeton: Princeton University Press.

Thagard, P. & Kunda, Z. 1987. Hot cognition: Mechanisms for motivated inference. *Proceedings of the Ninth Annual Conference of the Cognitive Science Society* (pp. 753–763). Seattle, WA: Lawrence Erlbaum Associates.

Zadeh, L. A. 1965. Fuzzy sets. *Information and Control, 8*, 338-353.

# Modeling an Experimental Study of Explanatory Coherence

**Patricia Schank and Michael Ranney**

Education in Mathematics, Science and Technology
Graduate School of Education
University of California, Berkeley, CA 94720
schank@garnet.berkeley.edu, ranney@cogsci.berkeley.edu

**Abstract**

The problem of evaluating explanatory hypotheses is to choose the hypothesis or theory that best accounts for, or explains, the given evidence. Thagard (e.g., 1989) and Ranney (in press; Ranney & Thagard, 1988) describe a theory of *explanatory coherence* intended to account for a variety of explanatory evaluations; this theory has been implemented in a connectionist computer model, ECHO. In this study, we examine three questions regarding the relationship between human explanatory reasoning and ECHO's explanatory evaluations: Does ECHO predict subjects' evaluations of interrelated propositions? Are local temporal order differences (not explicitly modeled by ECHO) important to the subjects? Does ECHO predict subjects' inflectional reasoning? We found that subjects often entertain competing hypotheses as nonexclusive and presume an implied backing for certain (superordinate) hypotheses. These tendencies were modeled in ECHO by assigning a fraction of *data priority* (usually reserved for evidence) to the superordinate hypotheses. In sum, the ECHO model helps to interpret subjects' reasoning patterns, and shows continued potential for simulating explanatory coherence processes.

## Introduction

The assessment of explanatory hypotheses involves choosing the hypothesis or theory that best accounts for given evidence. The theory of *explanatory coherence* (e.g., Thagard 1989) accounts for a variety of explanatory evaluations. In the theory, relations of local explanatory coherence and incoherence between *propositions*, interpreted as hypotheses or evidence, are established via several principles. In general, propositions cohere with propositions they explain, propositions they are explained by, or propositions they are analogous with; they incohere with competing or contradictory propositions. The *acceptability* of a proposition depends on the acceptability of the propositions to which it is related, as its acceptability generally increases with the coherence of the sub-system of propositions in which it is embedded. Propositions interpreted as evidence generally accumulate acceptability more rapidly than do hypotheses, though the acceptability of a datum may decrease if it is mostly coherent with propositions that are unacceptable.

The theory of explanatory coherence has been implemented in a connectionist computer model called ECHO. In ECHO, propositions are represented by nodes in a network that send and receive *activation* (the measure of acceptability) to and from related propositions. Hypothesis evaluation is treated as the parallel satisfaction of multiple constraints derived from explanatory and contradictory relations. Once an ECHO network has settled, propositions with high activations are considered accepted (Thagard 1989, in press2). ECHO differs, however, from networks in which node-node link weights are adjusted on each cycle (see Rumelhart & McClelland 1986).

What is the relationship between human explanatory reasoning and explanatory evaluation as implemented in ECHO? ECHO has been used in the past to model juror reasoning (Thagard 1989) in which explanatory coherence plays a crucial role (Pennington & Hastie 1988), to understand mental models of social interactions and relationships (Miller & Read in press), and to model scientific reasoning (Thagard 1989, in press1). Ranney and Thagard (1988), in the first application of ECHO to modeling experimental data, simulated (*ex post facto*) changes in subjects' beliefs and their conceptions of physical motion (Ranney 1987/1988; cf. Nersessian 1989). Ranney (in press) suggests that if ECHO continues to successfully model human reasoning, it might be useful as a tool for teaching coherent argumentation; however, speculation about the relationship between ECHO's descriptive power and prescriptive potential requires that we address several empirical questions about the psychological fidelity of ECHO. We describe here an empirical study that examined ECHO's ability to model explanatory evaluations of textually embedded propositions (Schank 1991).

### The ECHO Model

Coherence involves a relation between two propositions that "hold together," while incoherent propositions resist holding together. Relations of coherence (and incoherence) among propositions are established via the following principles (Thagard in press1; Ranney & Thagard 1988): (1) *Symmetry*: Coherence and incoherence are symmetric relations. (2) *Explanation*: Hypotheses that together explain a proposition cohere with each other, as well as with the explained proposition. (3) *Simplicity*: The coherence of an explanation is inversely related to the number of co-hypotheses used in the explanation. (4) *Analogy*: Similar hypotheses that explain similar evidence cohere. (None of our networks involved analogies.) (5) *Data Priority*: Results of observations, such as evidence and acknowledged facts, have a degree of acceptability on their own. (6) *Contradiction*: Contradictory hypotheses incohere. (7) *Competition*: Competing hypotheses (which explain the same evidence or hypotheses but are not themselves explanatorily related; see Harman 1989) incohere. (This principle did not affect the ECHO networks used in this study since such competitions were explicitly made contradictory; the competition principle is used to *generate* incoherence links.) (8) *Acceptability*: The acceptability (or believability) of a proposition depends on its coherence within the system of propositions in which it is embedded. A proposition's acceptability increases as its coherence with other acceptable propositions increases. (A proposition's acceptability is measured by its activation value, ranging from -1, complete rejection, to 1, complete acceptance.)

Hypothesis evaluation in ECHO is treated as the satisfaction of multiple constraints derived from the explanatory relations, and several parameters provide degrees of freedom (Thagard 1989, in press2). The number of parameters used in computer models often leads to questions: How arbitrary are the simulation findings? How stable is the model under a given set of parameters (i.e., does one set work over a wide range of tasks)? Should some parameters be regarded as variant or invariant over tasks? Thousands of simulations have been done to determine the impact of parameter settings on ECHO's performance, and the values of four parameters—the *excitation*, *inhibition*, *decay*, and *data excitation* rates—appear most critical (Thagard 1989; Ranney & Thagard 1988). However, the criteria used to decide the default values of these parameters have previously been practical rather than empirical. Ranney and Thagard (1988) as well as this study provide the first empirical backing for assigned parameter values.

The excitation value determines the weights on links between cohering propositions. The inhibition value determines the weights on links between incohering propositions. The data excitation value specifies the weight on links between data (usually evidence) and the "special evidence unit" (SEU, a unit with activation clamped at 1.0) to implement principle 4. The decay value specifies the percentage of the (absolute) activation that a proposition loses at each cycle. Thagard (1989) describes interesting psychological interpretations of these parameters. The *tolerance* of the system is the absolute value of the ratio of excitation to inhibition; highly tolerant systems may "believe" several competing hypotheses, while systems with low tolerance tend to accept a single theory. Decay is the *skepticism* of the system; as decay increases, asymptotic activation values of propositions will be compressed toward zero (although the simulations will generally not be qualitatively different). Skepticism applies directly to all propositions, while tolerance is most directly relevant to contradictory or competing propositions.

Given declared input propositions and relations between them, ECHO creates units to represent the propositions, and sets up the following links (with additive link weights): (1) excitatory links between propositions (including co-hypotheses) that are explanatorily or analogously related; (2) inhibitory links between contradictory and (nonevidential) competing propositions; and (3) excitatory links between evidence and the SEU (with the data excitation weight). (The data priority of particular evidence may also be specified separately.) If there are unexplained data, the decay rate is increased appropriately (i.e., unexplained evidence raises skepticism). Unit activations are updated in cycles until the network settles and the change in all units is asymptotic. At each cycle, the activation of a particular unit $u_j$ is determined by the decay rate and the net input to the unit (the weighted sum of the activation of its neighboring units, where a weight is the common link value, $w_{ij}$), and is updated using the following equation (cf. Rumelhart & McClelland 1986):

$$u_j(t+1) = u_j(t)\,(1\text{-}decay) + \begin{cases} net_j\,(max\text{-}u_j(t)) & \text{if } net_j > 0 \\ net_j\,(u_j(t)\text{-}min) & otherwise \end{cases}$$

where $net_j = \sum_i w_{ij}\, u_j(t)$

## Evaluating ECHO

We ask three questions about the relationship between human reasoning and ECHO: How well does ECHO predict differences in subjects' evaluations of different texts? Are local temporal order differences (not explicitly accounted for by ECHO) important to the subjects? How well does ECHO predict inflectional reasoning, in which new information presented in texts (cf. Kintsch 1988) yields significant changes in subjects' beliefs (Ranney in press)?

To investigate these questions we conducted three contrasts between ECHO simulations and subjects' believabilities. Each contrast involved three stages: (1) ECHO simulations were run on given systems of propositions, (2) subjects read textual embodiments of the same propositional system and rated how strongly they believed each proposition, and (3) ECHO's ability to model the subjects' believability ratings was assessed.

**Contrast One: Differential Predictions.** Given textual propositions, how well do subjects' beliefs in the propositions correlate with ECHO's activations for these propositions? For this contrast, two ECHO topologies were used, the second of which includes an extra, critical piece of evidence that qualitatively changes the simulation results. If two groups of subjects are given texts reflecting these topologies (with one group given the extra piece of information), does ECHO predict the differences among the groups' beliefs in the underlying propositions? The major empirical question is: Are subjects' believability ratings reasonably consistent with ECHO's activations?

**Contrast Two: Temporal Order Effects.** Does the order of presentation of theories in a text affect the subjects' believability ratings? Given that all propositions are generally encoded during the same network cycle, ECHO activations are unaffected by the temporal order of propositional encoding. If two groups of subjects are given texts reflecting a single network topology, with the second group given a text that provides the propositions in a different order, will their beliefs in the propositions differ? If so, the finding could be used to suggest modifications to ECHO.

**Contrast Three: Dynamic Modeling.** With its use of data priority, ECHO is somewhat empirically biased. Are people as empirically driven as ECHO? Once subjects are familiar with a text, does adding an additional piece of information—one that produces a "Gestalt switch" in ECHO—induce a Gestalt switch in the subjects? To assess this contrast, an ECHO topology was allowed to run and settle, and then an additional piece of evidence—one that yielded a theory shift reflected by inflected activation trajectories—was introduced (see Ranney & Thagard 1988). Do subjects also shift from accepting one theory to accepting another upon the introduction of the new evidence? Will adding an additional hypothesis, one that strengthens ECHO's acceptance of the newly acceptable theory, further strengthen subjects' beliefs in the theory? Perhaps the new hypothesis will shift subjects' beliefs toward the new theory if the additional evidence did not already do so. The absence of a change when evidence is introduced may suggest that people are more hypothesis-driven than is ECHO. The absence of any shift at all (or a shift only when both the datum and the hypothesis are introduced) may suggest that people are more predictively intransigent (Ranney 1987/1988) than ECHO, and serve as a basis for re-evaluating ECHO's data priority aspect.

## Method and Design

The texts used in this study reflect portions of the topology shown in Figure 1, which represents two conflicting theories, and lends itself to the questions in the above contrasts. In this topology, hypothesis H1 explains evidence E1 and E2, and is in turn explained by hypothesis H0. Hypothesis H2 explains evidence E2, E3, and E4, and is explained by hypothesis H3. Subgraphs of this topology, shown in Figures 2 and 3, are also used for the contrasts of interest.

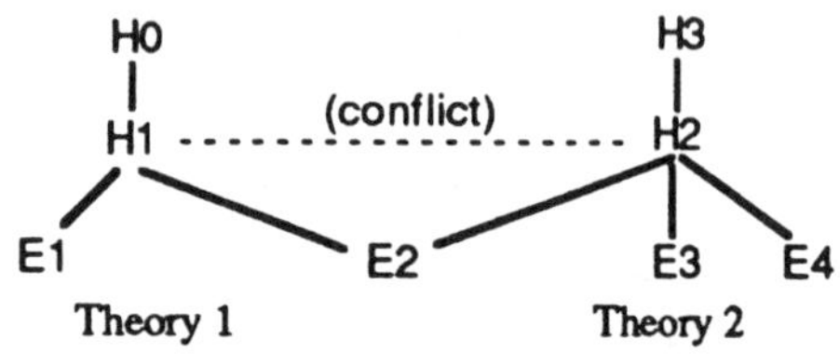


**Figure 1.** Topology of two conflicting explanatory theories; dashes indicate theoretical conflict.

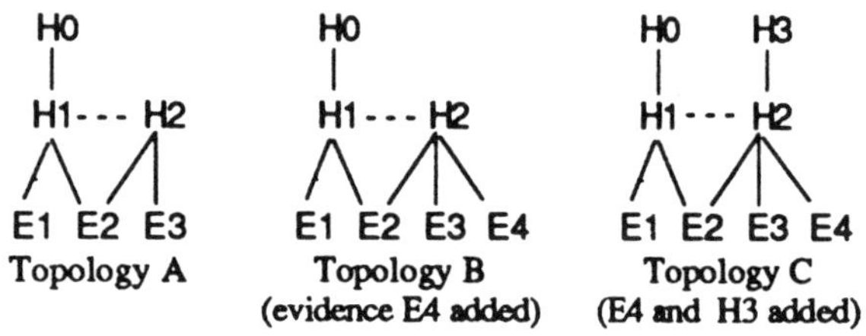


**Figure 2.** Topologies A, B, and C. (Topologies A and B were used for the differential predictions contrast. Topologies A, B, and C were used for the dynamic modeling contrast.)

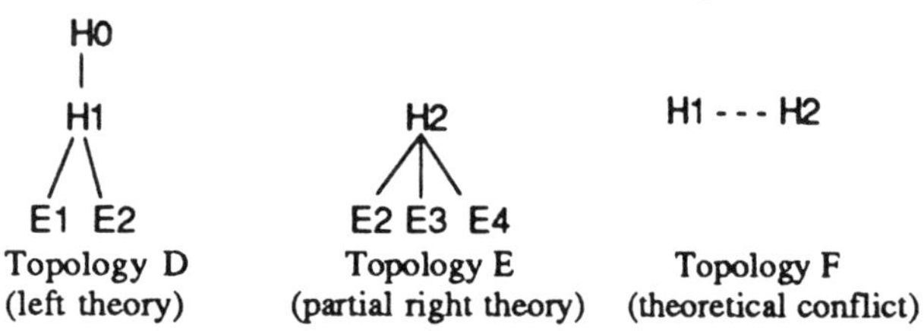


**Figure 3.** Topologies D, E, and F (used to assess temporal ordering effects).

**Table 1.** Four sequences x four texts.

| Topology Sequences | | | |
|---|---|---|---|
| S1 | S2 | S3 | S4 |
| A | D | E | B |
| <rate> | D+F | E+F | <rate> |
| B | D+F+E | E+F+D | |
| <rate> | <rate> | <rate> | |
| C | | | |
| <rate> | | | |
| **Decontextualized Texts** | | | |
| T1 | T2 | T3 | T4 |
| linguistics | physics | wine naming | medical |

Texts involving fictional theoretical controversies were generated to reduce the chance of subjects bringing considerable extraneous information into their analyses. Since dimensions of context are difficult to assess and control, we devised four texts in different domains. Subjects read the texts (given on separate sheets of paper) under no time limit, and then rated (on a separate questionnaire sheet; see Table 1) how strongly they believed the text's statements.

### Materials

The four texts were designed to be fairly context-free, while alternating between science and humanities domains. Four different topological sequences, S1 through S4, were used to constrain the structures of these texts (see Table 1). Sixteen text-sequence combinations resulted (4 sequences X 4 texts, e.g., T1-S1, T1-S2...). The physics text below represents topology C; its ECHO encoding is shown in Table 2.

"Sue and Amy are trying to identify an atomic particle:

Sue and Amy find that it has a charge magnitude of two electrons, high energy, a positive spin, and a baryon number equal to zero. Sue says that physicists think that ziptons are rather common, which explains her belief that the particle is a zipton. Sue also notes that if one assumes that the particle is a zipton, that explains why the particle has a charge magnitude of two electrons and why it has high energy.

Amy disagrees with Sue. Amy believes that the particle is a blinkon, which contradicts Sue's belief that the particle is a zipton. Amy says that physicists think that blinkons are easy to detect, which explains her belief that the particle is a blinkon. Amy also notes that if one assumes that the particle is a blinkon, that explains why the particle has high energy, why it has a positive spin, and why it has a baryon number equal to zero.

What do you think?"

**Table 2.** Encoding of the physics text, topology C.

**Evidence:**
proposition E1 "the particle has the charge of two electrons"
proposition E2 "the particle has high energy"
proposition E3 "the particle has a baryon number of zero"
proposition E4 "the particle has positive spin"
**Hypotheses:**
proposition H0 "physicists think ziptons are rather common"
proposition H1 "the particle is a zipton"
proposition H2 "the particle is a blinkon"
proposition H3 "physicists think blinkons are easy to detect"
**Explanations:**
H0 explains H1
H1 explains E1
H1 explains E2
H3 explains H2
H2 explains E2
H2 explains E3
H2 explains E4
**Contradictions:**
H1 contradicts H2
**Data priority for evidence:**
data E1 E2 E3 E4

### Subjects and Procedure

To control for presentation orders, all 24 possible topological sequence combinations were used, and both forward (T1, T2, T3, T4) and backward (T4, T3, T2, T1) text orders were used. In total, 48 (4! x 2) subjects participated. Subjects had various academic and occupational backgrounds, and ranged in age from 18 to 60 years.

Each experimental session consisted of four parts. In each part, the subject was given a text reflecting one of the text-sequence combinations, never receiving the same text or the same topology sequence twice. For each text-sequence combination, a subject was asked to read the text and then rate the believability of its propositions. Ratings were on a scale from 1, "completely unbelievable," to 7, "completely believable" (4 was unlabeled). When the subjects were finished rating all four text-sequence combinations, they were asked (via audio-taped retrospective protocols), to explain how they came up with their believability assignments—in order to determine if they brought extraneous information into their evaluations.

The ECHO simulations were initially run with parameter settings used by Ranney and Thagard (1988), the first application of ECHO to modeling experimental data. Simulations were subsequently run with slightly adjusted parameter settings, as discussed below.

**Contrast One, Topology A vs. Topology B: Differential Predictions.** Does ECHO differentially predict subjects' responses to different texts? ECHO activations for topologies A and B were compared in a pairwise manner with subjects' ratings on propositions from Sequences 1 (topology A) and 4 (topology B, including its extra piece of evidence, E4). (Subjects were given additional text and questionnaires beyond topology A in Sequence 1; see Contrast Three below and Table 1.) The added datum significantly changes ECHO's simulation: Without the extra evidence E4 and the H2-E4 explanatory link, hypothesis H1 receives higher activation than H2 in ECHO; with the extra evidence and link, the reverse is true.

**Contrast Two, Sequence 2 vs. Sequence 3: Temporal Order Effects.** How sensitive are subjects to the temporal order of propositions? Running ECHO on topology B, hypothesis H2 receives a higher activation than H1, given that all propositions are encoded during the same network cycle. For this contrast, subjects were asked to read three portions of text (presented on separate sheets) reflecting topology B, in which the order of sets of propositions in the texts were varied. Topology B can be decomposed into three subgraphs, representing a "left" theory, a partial "right" theory, and a theoretical conflict (topologies D, E, and F in Figure 3). For Sequence 2, subjects were given three successive texts: the first reflecting the left theory, the next reflecting the left theory and theoretical conflict, and the last reflecting the left and the partial right theories as well as the theoretical conflict (topology order: D, D+F, D+F+E). For Sequence 3, subjects were also given three successive texts, with the order of the left and partial right theories switched (E, E+F, E+F+D).

**Contrast Three, Sequence 1: Dynamic Modeling.** How well does ECHO dynamically model inflectional reasoning? In ECHO, hypothesis H1 receives a higher activation than H2 in topology A (see Figures 2 and 5). After the network settles, adding evidence E4 and the explanatory link H2-E4 causes the activation of proposition H1 to decrease and H2 to increase, surpassing H1, thus simulating a mild Gestalt shift. Adding the hypothesis H3 and the explanatory link H3–H2 causes H2's activation to further increase and H1's to further decrease. For this contrast, all subjects were given a text structurally reflecting topology A, followed by a believability rating questionnaire. They were then given a text reflecting topology B, followed by a new questionnaire. Finally, the subjects were given a text reflecting topology C, followed by a final questionnaire. The subjects' ratings were then compared to activations from ECHO simulations during which propositions E4 and H3 (and their associated links) were added dynamically (i.e., in later network cycles, after successive settlings). If subjects are as empirically driven as is ECHO, they should "switch" from favoring H1 to favoring H2 when E4 is added, and certainly with the addition of H3.

## Results and Discussion

Interviews with the subjects revealed that they adopted some kind of implicit *presumed backing* (i.e., other evidence or beliefs) for the superordinate hypotheses H0 and H3. The ECHO simulations were adjusted to take this into account. Thus, before describing the fit between the subjects' ratings and ECHO's activations, we first discuss the subjects' believability ratings and comments to show how presumed backing affected their ratings.

### Subjects' Believability Ratings

The means of subjects' believability ratings (for propositions in topologies A, B, and C are shown in Table 3. Subjects rated the superordinate H0 higher than its explained H1, and the superordinate H3 higher than its explained H2. Subjects also tended to view H1 and H2 as nonexclusive hypotheses (in spite of the explicitly stated conflicts), particularly on the medical text.

**Table 3.** Mean ratings for all appropriate propositions.

| | Topology (within Sequence) | | | | | |
|---|---|---|---|---|---|---|
| Prop | S1a | S1b | S1c | S2b | S3b | S4b |
| E1 | 6.50 | 6.56 | 6.60 | 6.52 | 6.58 | 6.65 |
| E2 | 6.50 | 6.58 | 6.58 | 6.31 | 6.67 | 6.65 |
| E3 | 6.52 | 6.58 | 6.56 | 6.19 | 6.50 | 6.50 |
| E4 | • | 6.42 | 6.58 | 6.19 | 6.46 | 6.63 |
| H0 | 4.81 | 4.73 | 4.69 | 4.65 | 4.60 | 4.88 |
| H1 | 3.58 | 3.27 | 3.35 | 3.46 | 3.29 | 3.41 |
| H2 | 3.48 | 4.12 | 4.29 | 3.85 | 3.98 | 4.08 |
| H3 | • | • | 4.73 | • | • | • |

**Presumed Backing.** Some subjects justified an implicit backing of the superordinate hypotheses by indicating that the word "think" (in the texts) is used for conversational politeness; i.e., people use the word "think" when they really *know* something. Despite the attempt to generate decontextualized texts, subjects (not surprisingly; Ritter 1991) also considered other knowledge when reasoning about the texts, and this other knowledge likely contributed to their presumed backing of H0 and H3. For instance, one subject hinged his ratings of the wine text on his belief that there are few types of wine, unlike atomic particles, of which there "are many." In contrast, another subject held that there are *not* many types of atomic particles! Although subjects generally gave high ratings to evidence, these ratings still did not boost their beliefs in hypotheses H1 and H2 above their superordinates.

**Text Type and Text Order Effects.** As expected, an analysis of variance on subjects' ratings (on the Topology C texts) revealed that effects of text type and text order were not significant ($p>.05$). There was, however, a (nonsignificant) trend for the superordinate H0 to resist being overcome by H3 with increasing experience with the successive texts. (Nb. Many of this paper's inferential tests are based on standard statistical distributional assumptions that remain to be confirmed for this particular paradigm.)

### ECHO Simulations

In contrast to the subjects, *a priori* ECHO activations of the superordinates H0 and H3 were lower than the hypotheses they explained (H1 and H2) across a wide range of parameter settings. Also, ECHO's activations of the conflicting hypotheses H1 and H2 were slightly more divergent than the subjects' ratings of these hypotheses. This suggested that the apparent differences between subjects' ratings and ECHO's activations were due to a presumed backing of hypotheses H0 and H3 and the tendency of subjects to view H1 and H2 as nonexclusive.

To model subjects' presumed backing, H0 and H3 were given partial data priority by linking them to the SEU with a weight half of that given to evidence. This was encoded by adding these semi–hypotheses to the data statement:

data E1 E2 E3 E4 (H0 .5) (H3 .5)

The parameter values used for four simulations (see below) are shown in Table 4 ("R&T" refers to values used in Ranney & Thagard 1988).

**Table 4.** ECHO parameter settings on simulations.

| Parameters | —no backing— | | —backing— | |
|---|---|---|---|---|
| | Run 1 (R&T) | Run 2 (adjusted) | Run 3 (R&T) | Run 4 (adjusted) |
| excitation | .03 | .005 | .03 | .03 |
| inhibition | -.06 | -.06 | -.06 | -.06 |
| decay | .04 | .15 | .04 | .15 |
| data excitation | .055 | .10 | .055 | .10 |

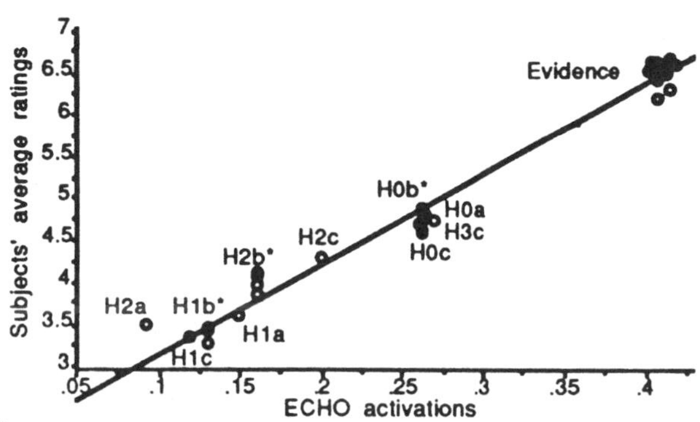


* multiple occurances, from ratings over Sequences S1 to S4

**Figure 4.** Relation between *mean* believability ratings and ECHO's activations with presumed backing.

**Table 5.** Point-by-point ranked correlations between ECHO activations and subjects' belief ratings.

| | —no backing— | | —backing— | |
|---|---|---|---|---|
| | Run 1 | Run 2 | Run 3 | Run 4 |
| SEQUENCE | (R&T) | (adjusted) | (R&T) | (adjusted) |
| S1a | .71 | .67 | .71 | .78 |
| S1b | .71 | .76 | .78 | .78 |
| S1c | .56 | .68 | .73 | .76 |
| S2b | .63 | .70 | .73 | .73 |
| S3b | .73 | .79 | .80 | .80 |
| S4b | .71 | .78 | .80 | .80 |
| TEXT | | | | |
| linguistics | .66 | .72 | .76 | .77 |
| physics | .61 | .67 | .73 | .76 |
| wine | .68 | .76 | .78 | .77 |
| medical | .72 | .79 | .80 | .81 |
| **OVERALL** | **.67** | **.74** | **.77** | **.78** |

## Comparing Subjects and ECHO

Figure 4 shows the relation between the subjects' *mean* ratings and ECHO's predictions when presumed backing is taken into account. Point-by-point correlations between individual subjects' ranked ratings and ECHO's ranked activations were also calculated. Table 5 lists these correlations for different sequences, different texts, and overall.

As shown in Figure 4, subjects' ratings are modeled best when presumed backing is represented in ECHO, and when the decay and data excitation values are both slightly raised, muting the differences between H1 and H2 (see Tables 4 and 5, Run 4). As indicated in Table 5, the overall correlation between the subjects' responses and ECHO is high ($r_{ranked}$=.78).

**Contrast One: Differential Predictions**. An analysis of variance revealed that across topologies A and B, subjects' ratings of hypothesis H1 did not significantly change ($p>.1$; see Table 3). However, as predicted by ECHO, subjects' ratings of hypothesis H2 rose significantly across topologies A and B ($p<.05$) with the introduction of the evidence E4 (which H2 explains). On topology A (S1a), subjects' ratings of hypothesis H1 were not significantly higher than those of H2 ($p>.1$). However, on topology B (S4), subjects' ratings of hypothesis H2 were significantly higher than those of H1 ($p<.05$), as ECHO predicts. When presumed backing for hypothesis H0 is represented in ECHO, the relation between the subjects' ratings and activations on topologies A and B is particularly strong because the subjects' higher ratings of H0, compared to H1, are captured. With presumed backing, the point-by-point correlation between the subjects' averages and ECHO is high on both topology A ($r_{ranked}$=.71-78), and topology B ($r_{ranked}$=.80). (See Table 5, Runs 3 and 4, S1a and S4b.)

**Contrast Two: Temporal Order Effects.** No differences between subjects' ratings of the propositions in Sequences S2b and S3b (and even S1b and S4b) approached significance (see Table 3, S2b and S3b). With presumed backing, the correlation between the subjects' averages and ECHO is high for both S2 and S3 (Table 5, Runs 3 and 4). Nonsignificant hints of subjects' intransigence were shown by ECHO's slightly lower correlation with subjects' ratings for S2 (in which the "losing" theory is given first), and by a smaller variance in subjects' ratings for S3 (in which the "winning" theory is given first). Still, as shown in Table 3, the subjects' mean ratings are almost identical on topology B for S1 and S4 (S1b and S4b), even with different subjects and an intervening rating in S1, indicating few "prediction-freezing" effects. Although one subject remarked after reading the S1 (dynamic) version of the medical text that "the more information that comes up, the less I believe them 'cause they didn't know it before," the local temporal order differences—which ECHO does not automatically take into account—were not generally found to be important.

**Contrast Three: Dynamic Modeling.** Across topologies A (S1a), B (S1b), and C (S1c), subjects' ratings of hypothesis H1 did not significantly change ($p>.1$; see Table 3). However, subjects' ratings of hypothesis H2 rose significantly across topologies ($p<.05$). As predicted by ECHO, the introduction of the new evidence E4 (in S1b) accounted for most of subjects' increased believability in H2 ($p<.05$), while the introduction of the superordinate H3 (in S1c) had little extra effect on subjects' believability of H2 ($p>.1$). On topology A, the subjects' ratings of hypothesis H1 were not quite significantly higher than H2; however, on topologies B and C, subjects' ratings of hypothesis H2 were significantly higher than H1 (both tests p<.05). Thus, both the subjects and ECHO show inflectional reasoning when the new evidence of topology B is introduced. The relation between the subjects' ratings and the ECHO activations is once again particularly strong when presumed backing of the superordinate hypotheses is represented in ECHO. Figure 5 illustrates the inflectional reasoning captured by ECHO (e.g., for H1 and H2). Table 5 shows the high correlations between the subjects' ratings and ECHO's activations for topologies A, B, and C (rows S1a, S1b, and S1c).

Like ECHO, the subjects "switched" and rated H2 higher than H1 when the new evidence was introduced (see Table 6). With presumed backing, the correlation between subjects and ECHO is also high when H3 (of topology C) is added (see Table 5, row S1c).

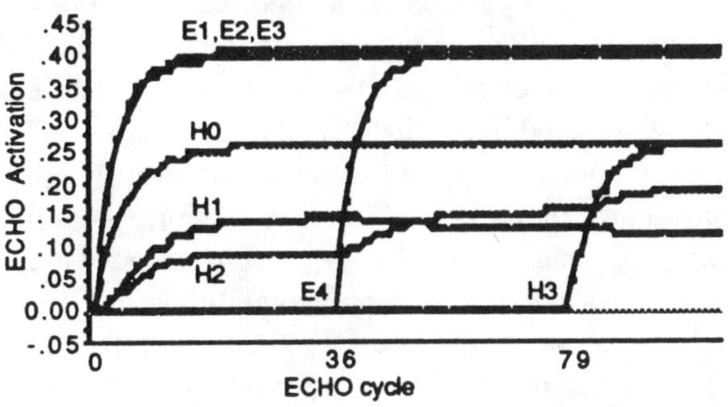


**Figure 5.** ECHO activation trajectories, Topology C, S1 (evidence E4 and hypothesis H3 added dynamically), Run 4.

**Table 6.** Subjects' average believability ratings on S1, suggesting inflectional reasoning.

| | Time 1: S1a | Time 2: S1b (E4 added) | Time 3: S1c (E4 & H3 added) |
|---|---|---|---|
| H1 | 3.58 | 3.27 | 3.35 |
| H2 | 3.48 | 4.12[ab] | 4.29[ab] |

[a]significantly higher than H1 (p<.05).
[b]significantly higher than H2, Time 1 (p<.05).

## Summary and Conclusion

In all contrasts, subjects' believability ratings were modeled well by ECHO. Subjects' tendencies to presume a kind of unstated backing behind the superordinate hypotheses were modeled in ECHO by giving a measure of data priority to these hypotheses and by slightly raising decay (skepticism) and data excitation. The differences in subjects' responses to texts reflecting topologies A and B were captured by ECHO, particularly when the presumed backing was represented. Also, local temporal order differences did not significantly affect subjects' beliefs, and since ECHO does not automatically take such order differences into account, the results do not suggest necessary changes to the model. Finally, ECHO captured subjects' inflectional reasoning—again, particularly well when the presumed backing was invoked. The theory of explanatory coherence and its ECHO model were useful in that they allowed us to interpret patterns in a subject's reasoning process fairly concisely. ECHO's ability to model the subjects' propositional evaluations well, even without the adjustments described above, suggests that ECHO shows continued promise for predicting human explanatory reasoning.

The subjects indicated that they considered some extraneous knowledge when reasoning about the texts, and this implicit knowledge was modeled in ECHO by giving certain hypotheses a fraction of data priority. However, this suggests that the subjects' representations of the situations were not completely captured by the representation encoded into ECHO. Thus, we are now investigating ECHO's ability to model individual subjects' beliefs about physical motion when they make their "implied backings" explicit.

## Acknowledgements

We thank Paul Thagard, Boaz Keysar, Pete Pirolli, Marcia Linn, Mark Wilson, Jeff Schank, Chris Ritter, Peggy Carlock, Ed Lagache, Chris Borton, Mimi Recker, Steve Adams, and Michelle Million for their helpful comments.

## References


Harman, G. 1989. Competition for evidential support. *Proceedings of the Eleventh Annual Conference of the Cognitive Science Society*. Hillsdale, NJ: Erlbaum.

Kintsch, W. 1988. The role of knowledge in discourse comprehension: A construction-integration model. *Psychological Review, 95*, 163-182.

Miller, L. C., & Read, S. J. in press. On the coherence of mental models of persons and relationships: A knowledge structure approach. To appear in G.J.O. Fletcher and F. Fincham Eds. *Cognition in close relationships*. Hillsdale, NJ: Erlbaum.

Nersessian, N. J. 1989. Conceptual change in science and in science education. *Synthese, 80*, 163-183.

Pennington, N. & Hastie, R. 1988. Explanation-based decision making: Effects of memory structure on judgment. *Journal of Experimental Psychology: Learning, Memory, and Cognition, 14*, 521-533.

Ranney, M. 1988. Changing naive conceptions of motion. Ph.D. diss., University of Pittsburgh, Learning Research and Development Center, 1987. *Dissertation Abstracts International, 49*, 1975B.

Ranney, M. in press. Explorations in explanatory coherence. To appear in E. Bar-On, B. Eylon, and Z. Schertz (Eds.) *Designing intelligent learning environments: From cognitive analysis to computer implementation*. Norwood, NJ: Ablex

Ranney, M. & Thagard, P. 1988. Explanatory coherence and belief revision in naive physics. *Proceedings of the Tenth Annual Conference of the Cognitive Science Society*, 426-432. Hillsdale, NJ: Erlbaum.

Ritter, C. 1991. *Thinking about ECHO*. M.A. thesis, University of California, Berkeley, School of Education.

Rumelhart, D. E., McClelland, J. R. & the PDP Research Group. 1986. *Parallel distributed processing: Explorations in the microstructure of cognition*, vols. 1 and 2. Cambridge, MA: MIT Press.

Schank, P. 1991. *An Empirical Investigation of the Psychological Fidelity of ECHO*. M.A. thesis, University of California, Berkeley, School of Education.

Thagard, P. 1989. Explanatory coherence. *Behavioral and Brain Sciences, 12*, 435-502.

Thagard, P. in press1. The dinosaur debate: Explanatory coherence and the problem of competing Hypotheses. In J. Pollock and R. Cummins (Eds.), *Philosophy and AI: Essays at the interface*. Cambridge, MA: MIT Press/Bradford Books.

Thagard, P. in press2. *Conceptual revolutions*. Draft, June, 1990. Princeton, NJ: Princeton University Press.

# Empirical and Analytical Performance of Iterative Operators

**Peter Shell** *(pshell@cs.cmu.edu)*
**Jaime Carbonell** *(jgc@cs.cmu.edu)*

School of Computer Science
Carnegie Mellon University
Pittsburgh, PA 15213
412-268-7651

**Abstract**

Macro-operators and chunks have long been used to model the acquisition and refinement of procedural knowledge. However, it is clear that human learners use more sophisticated techniques to encode more powerful operators than simple linear macro-operators: specifically, linear macro-operators cannot represent arbitrary repetitions of operators. This paper presents a process-model for the acquisition of *iterative* macro-operators, which are an efficient representation of repeating operators. We show that inducing iterative macro-operators from empirical problem-solving traces provides dramatically better efficiency results than simple linear macro-operators. This domain-independent learning mechanism is integrated into the FERMI problem-solver, giving more evidence that humans have a similar learning capability.

## 1. Introducing Iterative Macro-operators

One of the main goals of knowledge compilation is to reformulate or recycle problem solving knowledge in order to reduce the amount of search necessary to solve problems. A number of different methods for reducing search have evolved. For example, if the derivational analogy algorithm (Carbonell, 1986, Hickmanetal90, 1990) can retrieve the solution to a problem similar to the current problem, then it will adapt the previous solution to the new problem. Explanation-based learning (DeJong & Mooney, 1986), on the other hand, reformulates axioms which were used to solve a problem, into operators which can more efficiently solve the problem the next time it is encountered. Another reformulation method is chunking in Soar (Laird *et al*, 1986) and macro-operators (Anderson, 1983, Fikes, 1971). By summarizing the behavior of a sequence of forward-chaining operators, macro-operators also reduce the amount of operator-space search required by the problem-solver.

There are a number of problems with both linear macro-operators and chunks, which iterative macro-operators address. Simple linear macro-operators can never generalize to an arbitrary number of applications of atomic operators. For example, different macro-ops must be learned to solve the 2-disk tower of Hanoi, the 3-disk one, etc. It's clear that people have a more flexible learning method than linear macro-operators, and that they can somehow generalize to an arbitrary number of applications of an operator. Furthermore, both linear macro-ops and Soar chunks which encode multiple applications of a rule turn polynomial-time matches into exponential ones (Tambe, 1988). Iterative macro-operators, as depicted in figure 1-1, would solve the "expensive chunk" problem in SOAR if adapted to that paradigm. They also address the increased match-time search problem which occurs when large numbers of macro-operators are learned (Minton *et al*, 1988).

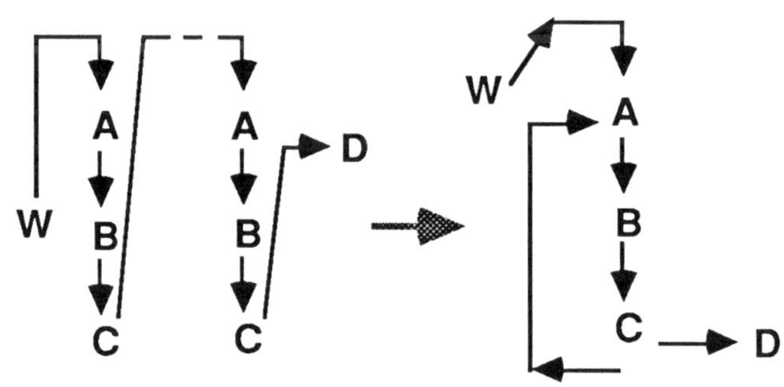


**Figure 1-1:** Formation of iterative macro-operators

In principle, there are two different ways in which macro-operators (iterative or otherwise) can reduce search. First, they can reduce the amount of search necessary in the operator state-space. Secondly, they can compile away temporary computation in the matching process itself. Thus match-level optimizations can reduce "fine-grain search" while operator-level aggregations can reduce "coarse-grain search".

Although previous approaches to iterative operators such as (Cheng & Carbonell, 1986, Riddle, 1988, Shavlik, 1990) address the coarse-granularity search reduction problem, they do not address the equally important issue of optimizing the macro-op itself. If one knows that a set of operators will be applied in a fixed iterative sequence, then the operators themselves can be transformed into more efficient iterative macro-operators.

The transformation algorithm has been generalized since (Shell & Carbonell, 1989) to cover more domains, and is summarized in section 2. Section 4 presents empirical results in these new domains, showing speedups as high as 14-fold. This is significant given that the operators are implemented in a RETE-based system called FRulekit (Shell & Carbonell, 1986), which is already fairly efficient. Furthermore, new theoretical results in section 3 show that iterative operators significantly improve over linear operators in both the coarse-granularity operator search space and in the operator match space. These theoretical and empirical results of the process model give more credence to the idea that people have a similar technique for learning general iterative macro-operators.

## 2. Iterative-Operator Composition

To explain how iterative operators are faster than linear operators, this section will summarize the iterative-operator composition algorithm, and will show a sample iterative operator. A more detailed description of the algorithm is presented in (Shell and Carbonell, 1991).

Generating iterative-operators involves:

1. *Detecting the Repeated Operator.* When a linear operator or macro-operator repeats, it is eligible for transformation into an iterative macro-operator.
2. *Encapsulation of the Invariant Structure.* Variables matched in the preconditions that bind identically each cycle need not be rematched each iteration.
3. *Solution of Recurrence Relations.* Local data updates between each cycle (such as decrementing an array index) are extracted into a single step external to the loop. More complex or open-form updates remain inside the iteration loop.
4. *Composition of the Total Iterative Macro-operator.* The preconditions, iterative body and exit conditions are deduced from the source atomic operator sequence and compiled into production rules.

Iterative operators are efficient because conditions and actions in the source rule that are unnecessary are eliminated. Unnecessary conditions and actions are defined as:

- Conditions that match objects that never change while the iterative operator runs,
- Conditions that match objects that are modified by the operator,
- Actions that modify objects that are matched by the operator,
- And repeated actions which can be induced to a single step as a function of N.

All other conditions and actions are retained.

The iterative operator is always equivalent to the original. The outline of an iterative macro-operator is shown in figure 2-1. This operator is used in the FERMI system for solving systems of simultaneous algebraic equations.

```
RULE Pre-Iterative-solve
 CONDITIONS:
  If all conditions of solve-unknown exist,
  And we are not currently iterating
 ACTIONS:
  Add the goal to iterate solve-unknown

RULE Iterative-solve
 CONDITIONS:
  The goal is to iterate solve-unknown
  And all of the retained tests of
    solve-unknown are still satisfied
 ACTIONS:
  Perform the actions which were retained
    from the original operator.

RULE Post-Iterative-solve
 CONDITIONS:
  The goal is to iterate solve-unknown
  And the retained tests of solve-unknown
    are no longer satisfied
 ACTIONS:
  Remove the goal to iterate solve-unknown
  And update working memory with the actions
```

**Figure 2-1:** Iterative operator to find the next equation to eliminate, generated by FERMI.

## 3. Analytical Results

This section mathematically describes how much iterative operators can reduce search over linear operators. We derive complexity analyses of both the time for searching through a set of operators, and the time to match and execute individual operators. This analysis must be approximate, as there is not sufficient room to look in detail at each test and action of every operator.

### 3.1. Search Speed-up

This section presents an average-case complexity analysis of how much iterative operators reduce search in the operator space. Assume a breadth-first search discipline with **b** operators and a solution at depth **n**. The exponential complexity of search is: $\sum_{i=0}^{n} \mathbf{b}^i$, which is dominated by the last term: $\mathbf{b}^n$ for values of b > 2. If we introduce **c** linear macro-operators we can estimate the dominant term as:

$$(\mathbf{b} + \mathbf{c})^{n/(1+p_m(m-1))}$$

In the formula, **m** is the average length of a macro-operator (i.e., the number of base operators) and $\mathbf{p_m}$ is the ratio of macro-operators to the total (base and macro) operators in the solution sequence. Adding many useless macro-operators increases the base **b + c** considerably but does not reduce the exponent, whereas adding a few very useful operators reduces the exponent without significantly increasing the base.

An iterative macro-operator consists of repeated applications of a base operator or a linear macro-operator. An iterative macro-operator is operationally equivalent to **K** linear macro-operators, where M is the maximal number of iterations reasonable in the given domain. The search reduction, if all macro-operators could be reduced to iterative ones would be calculated by dividing the macro-operator branching factor by the average **K**:

$$(\mathbf{b} + \mathbf{c}/\mathbf{K})^{n/(1+p_m(m-1))}$$

The idea is to reduce search depth while minimally increasing search breadth. This translates into generating the most powerful possible macro-operators.

## 3.2. Match Speed-up

In addition to speeding up the operator search time, iterative operators match and execute faster than their linear counterparts. This section will mathematically analyze the amount of match speed-up realized by individual iterative macro-operators.

### 3.2.1. Sources of Speed-up

There are three different ways that iterative operators optimize the matching of linear operators:

1. They eliminate *condition elements*;

2. They eliminate *actions*;

3. They move tests to earlier parts of the rule.

The first two are the most common and will be the focus of this analysis. Eliminating unnecessary condition elements makes the rule faster by decreasing the number of tests which the rule must perform. Eliminating actions also speeds up the process since the actions cause the system to re-match.

### 3.2.2. The Cost of Performing an Action

The cost of performing an operator is the sum of the costs of the actions in that rule. The costs of those actions in turn depend on how much matching the rules perform. The cost $\mathit{Act\text{-}cost}_a$ of performing any action $a$ in a RETE-based operator is derived in the report (Shell and Carbonell, 1991). It is:

$$\mathit{Act\text{-}cost}_a = \beta_{n-2}C + AC\beta_{n-2}\sum_{i=1}^{L-n}(AD)^{i-1}$$

Where:

$\underline{A}$ is the average size of the input;

$\underline{\beta_n}$ is the number of tuples of objects which have successfully combined at condition element $n$;

$\underline{C}$ is the average cost of testing a condition element;

$\underline{D}$ tells us how discriminating the conditions are. If $D$ is small then few objects pass the tests in the conditions. If $D$ is big then there are more partial matches. I.e., $D$ times $[\beta_n]$ times A equals $[\beta_{n+1}]$;

and $\underline{L}$ is the number of conditions in the rule.

### 3.2.3. Speed-up of the Iterative-Operator

As can be seen from the last equation, the speed of a rule is determined by the ($AD$) product - the rate at which the size of the partial match grows. In expensive rules, the number of partial matches grows as the RETE net is traversed. Thus, ($AD$) is greater than 1 and the cost increases exponentially.

In this case, the second half of the above equation dominates. Suppose that in the iterative operator, one condition element has been removed. Then the cost of an action in this operator would be approximately:

$$AC\beta_{n-2}\sum_{i=1}^{L-n-1}(AD)^{i-1}$$

Thus the speed-up of the iterative operator over the linear operator, is the quotient:

$$\frac{AC\beta_{n-2}\sum_{i=1}^{L-n}(AD)^{i-1}}{AC\beta_{n-2}\sum_{i=1}^{L-n-1}(AD)^{i-1}} = \frac{\sum_{i=1}^{L-n}(AD)^{i-1}}{\sum_{i=1}^{L-n-1}(AD)^{i-1}}$$

which, as L increases, approaches $\underline{AD}$.

When $t$ conditions are removed from the iterative operator, the speed-up is:

$$\frac{\sum_{i=1}^{L-n}(AD)^{i-1}}{\sum_{i=1}^{L-n-t}(AD)^{i-1}}$$

which, as L increases, approaches $\underline{(AD)^t}$.

In efficient rules such as the ones which were profiled for this paper, the product $AD$ is 1 for some conditions and greater than 1 for others. I.e., the partial match does not always grow. Thus, the speed-up will be polynomial or linear instead of exponential, depending on how many expensive tests there are.

# 4. Empirical Results

Timings of the iterative-operator learning method on three different domains is shown here. We first present a summary of the efficiency gains in each domain, and then show graphs of the performance improvement of an iterative-operator in two of the domains.

### 4.1. Speed-up of Iterative Operators

The domains for which iterative operators were tested range from a nine-rule circuits problem solver to a more realistic ninety-rule expert system. The second column shows the total run-time speed-up (time required to run the linear operator divided by the time required to run the iterative operator). The next column shows the speed-up in match time, which is the amount of time spent matching in the RETE net.

| Speed-up of Iterative Operators | | | |
|---|---|---|---|
| Domain | Run-time Speed-up | Match-time Speed-up | # Rules |
| Circuits | 1.69 x | 1.88 x | 9 |
| Algebra | 4.66 x | 4.72 x | 23 |
| Machining | 11.3 x | 14.3 x | 90 |

As the table shows, the larger the domain, the better the speed-up.

### 4.2. Detailed Timings

For two of the above domains, the match times will be graphed for the iterative macro-operator, the linear macro-operator, and the set of base operators without any macro-operators. The input-size is plotted on the X-axis and the match time is on the Y-axis. As we shall see, the iterative operator was always faster than the original ones.

**The Algebra Domain.** The algebra system is a set of productions which solve simultaneous systems of N equations in N unknowns. It is a module of the FERMI general scientific problem-solver (Larkin, Reif & Carbonell, 1986). The X-axis plots the number of equations and unknowns given to the system (figure 4-1).

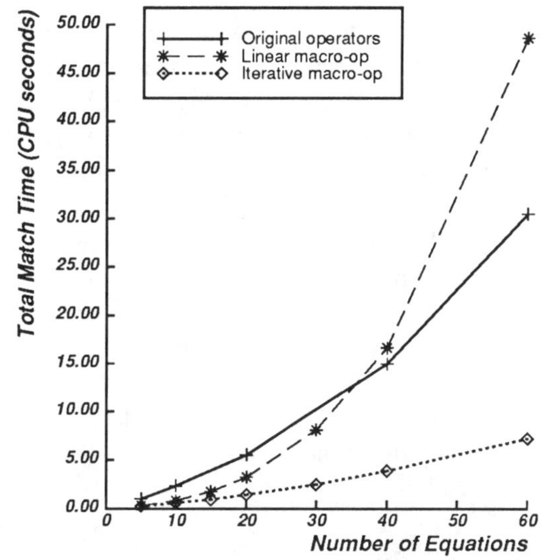


**Figure 4-1:** Time to Solve Systems of Equations

As the graphs show, the "expensive macro-operator" (Minton *et al*, 1988) problem rears its head again. The linear macro-operator is initially faster than the original rules, but as the size of the input increases, it becomes more expensive. However, the iterative macro-op eliminates the expensive part of the linear macro-op and is much faster than both the linear macro-op and the original rules.

**The Machining Domain.** The machining domain is an expert system for process planning. The task is to produce machine parts that have specified characteristics, such as a rectangular block of given dimensions made with a certain material and containing a centered hole. This domain is the largest production set that the iterative-operator algorithm has been tested on yet, and it shows the best improvement over linear operators. The X-axis displays the number of holes it drills (figure 4-2).

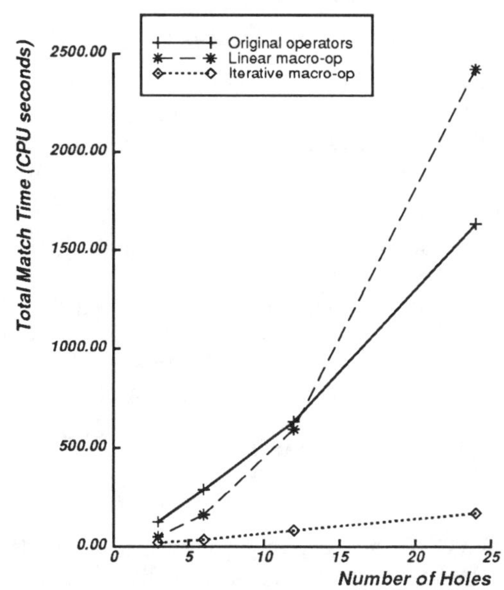


**Figure 4-2:** Time to Drill N Holes

As in the algebra domain, the linear macro-op is more expensive than the original operators, but the iterative operator is faster than both.

## 5. Conclusions

Iterative operators generalize to N by efficiently composing recursive subsequences. They have been shown to improve on linear operators by as much as 14-fold. As we have seen through complexity analysis and detailed empirical results, iterative operators effectively address some fundamental shortcomings of the standard macro-operator and chunking knowledge compilation methods. They do this by requiring a smaller number of total macro-operators, avoiding the "expensive-chunk" and "expensive-macro-operator" problems, and by making the macro-operators more efficient.

Although this paper makes no claims about how people form iterative macro-operators, this production-system model for iterative macro-operators may be a starting point for a cognitive model. Key questions to ask would be:

- Under what conditions do people form iterative rather than linear macro-operators?
- How do people form iterative macro-operators?

## 6. Acknowledgments

We would like to thank Patty Cheng for opening the door to a much more powerful class of macro-operators; and Angela Hickman, Dan Kahn and Ben MacLaren for helpful comments on earlier drafts of this paper.

### References

Anderson, J. A. (1983). Acquisition of Proof Skills in Geometry. In R. S. Michalski, J. G. Carbonell and T. M. Mitchell (Eds.), *Machine Learning, An Artificial Intelligence Approach.* Palo Alto, CA: Tioga Press.

Carbonell, J. G. (1986). Derivational Analogy: A Theory of Reconstructive Problem Solving and Expertise Acquisition. In Michalski, R. S., Carbonell, J. G. and Mitchell, T. M. (Eds.), *Machine Learning, An Artificial Intelligence Approach, Volume II.* Morgan Kaufmann.

Cheng, P. W. and Carbonell, J. G. (1986). Inducing Iterative Rules from Experience: The FERMI Experiment. *Proceedings of AAAI-86.* .

DeJong, G. F. and Mooney, R. (1986). Explanation-Based Learning: An Alternative View. *Machine Learning Journal*, Vol. *1*(2).

Fikes, R. E. and Nilsson, N. J. (1971). STRIPS: A New Approach to the Application of Theorem Proving to Problem Solving. *Artificial Intelligence*, *2*, 189-208.

Hickman, A. Kennedy, Shell P. and Carbonell, J. G. (1990). Internal Analogy: Reducing Search during Problem Solving. Catherine Copetas (Ed.), *The Computer Science Research Review 1990.* , The School of Computer Science, Carnegie Mellon University.

Laird, J. E., Rosenbloom, P. S. and Newell, A. (1986). Chunking in SOAR: The Anatomy of a General Learning Mechanism. *Machine Learning*, *1*(1), 11-46.

Larkin, J., Reif, F. and Carbonell, J. G. (1986). FERMI: A Flexible Expert Reasoner with Multi-Domain Inference. *Cognitive Science*, Vol. *9*.

Minton, S. N., Carbonell, J. G., Knoblock, C. A. and Kuokka, D. R. (1988). Explanation-Based Learning: Improving Problem Solving Performance through Experience. In Carbonell, J. (Ed.), *Paradigms for Machine Learning.* (to appear).

Riddle, Patricia J. (1988). An Approach for Learning Problem Reduction Schemas and Iterative Macro-operators. *Proceedings of the First International Workshop in Change of Representation and Inductive Bias.* .

Shavlik, Jude W. (1990). Acquiring Recursive and Iterative Concepts with Explanation-Based Learning. *Machine Learning*, *5*(1), 39-70.

Shell, P. and Carbonell, J. G. (1986). The FRuleKit Reference Manual. CMU Computer Science Department internal paper.

Shell, P. and Carbonell, J. G. (1989). Towards a General Framework for Composing Disjunctive and Iterative Macro-operators. *Proceedings of IJCAI-89.* .

Shell, P. and Carbonell, J. (1991). *A General Framework for Composing Iterative Macro-Operators.* (Tech. Rep.). Carnegie Mellon University School of Computer Science. Forthcoming.

Tambe, M. and Newell, A. (1988). Some Chunks are Expensive. *Proceedings of the Fourth International Workshop on Machine Learning.* .

# Where am I? Similarity Judgment and Expert Localization

**Kip Smith, Marian Heinrichs, & Herbert Pick, Jr.**

Center for Research in Learning, Perception, and Cognition
University of Minnesota
205 Elliott Hall
75 East River Road
Minneapolis, Minnesota 55455
kip@umnsom.bitnet

**Abstract**

How do skilled map-readers use topographic maps to figure out where in the world they are? Our research addresses this question by studying the problem solving of experienced map-readers as they solve localization - Where am I? - problems. Localization relies upon judgments of similarity and difference between the contour information of the map and the topographic information in the terrain. In this paper we discuss experiments that focus on how map-readers use attributes and structural relations to support judgments of similarity and difference. In our field and laboratory experiments, experienced map-readers implicitly define attributes to be detailed descriptors of individual topographic features. They use structural relations that link two or more topographic features as predicates. The time-course of their problem solving suggests that attributes and relations are psychologically distinct. Attributes like slope, e.g., "steep (hill)", support only initial judgments of difference. Relations like "(this hill) falls steeply down into (a valley)" are more powerful, supporting both judgments of difference and judgments of similarity. Judgments based on relations are used to test hypotheses about location. Experienced map readers exploit the distinction between attributes and relations as they solve localization problems efficiently.

## Localization - the 'Where am I?' problem in navigation[1]

Localization is the familiar task of finding the point on a map that represents your viewpoint in the world. Anyone who has ever been lost knows that localization can pose a difficult problem. It is a fundamental component of all navigation in large-scale space. Diverse professions (e.g., geology and airborne infantry) require individuals to become skilled at localization. The work reported here elucidates the roles of topographic features, attributes of features, and relations among features in the judgments that establish correspondence between map and terrain.

Maps are representations that preserve with fidelity a selected subset of the information available in a section of the world. The information contained in a map provides a context for the map-reader: localization judgments based upon a map can only be made with reference to the type of information it makes available. We restrict our study to topographic maps because they provide a clear, familiar, and pragmatically useful context for constructing a theory of localization that will assist the design of intelligent systems to control vision-based robot navigation.

When using a topographic map to solve a localization problem, the map-reader must find the location among the contours that matches the *viewpoint* in the terrain. The viewpoint is the location in the world where one happens to be standing. It determines what can be seen and what is occluded. The viewpoint dependence of terrain information tightly constrains problem solving. It determines the topographic features and relations among those features that can be used to generate and test hypotheses about location. Map-readers necessarily relate terrain information to their viewpoint.

The constraint of viewpoint dependance and the context provided by topographic maps transform localization into the task of finding the contours on the map that characterize a layout similar to that seen from the viewpoint. Determining the correspondence between map and terrain relies on judgments of similarity and difference between the contour information of the map and the topographic information in the terrain.

There are frequently many locations on a topographic map that appear similar to the viewpoint. Each may be entertained as a hypothesis. Selecting the hypothesis that provides the best match to the terrain relies on judgments that discriminate among competing hypotheses. Thus, there are two basic steps to localization problem solving: (1) generating hypotheses that relate map and terrain information and (2) testing these hypotheses by identifying the best match. Similarity judgment is essential to both.

## Localization and similarity judgment

Gentner (1983) and Medin, Goldstone, and Gentner (1990) emphasize the role of relations among objects in judgments of perceptual similarity. Their notion of structure-mapping holds that the relations among objects constrain judgments of similarity and are, in fact, more central to the process of judgment than are the individual objects themselves. This emphasis on the structure that relations impose on their constituent objects is intuitively consistent with the correspondences that map-readers must make to compare a map to the terrain they see.

Tversky (1977) introduces the notion that judgments of similarity depend on the context of the task in which they are embedded. He specifies a rule for calculating similarity. He implies that application of the rule is dependant on the task context but does not indicate specifically how. Medin et al. (1990) seize on this insight and suggest that the

[1]This work was supported by NSF Grant IRI-8901888 (funded in part by DARPA) and AFOSR Grant 88-0187.

relational structure among objects provides the context missing in Tversky's (1977) model. This too matches the demands of the localization task. Topography provides not only a context but also an intrinsic structure within which a map-reader views both the terrain and the map.

A key component of the structure mapping hypothesis is a fundamental distinction among objects, relations, and attributes. We embrace this distinction. In this paper, we call the topographic objects that capture a map-reader's attention *features*, e.g., "I see a valley". An *attribute* is a property, like gradient, that embellishes the description of a feature, e.g., "I see a steep valley". A *relation* is a connective property that cannot be hung on any one feature; relations span two or more features, e.g., "When I look southeast, I see the ground falls abruptly into a valley". In this example, the relation is a predicate that links the map-reader's viewpoint to a distant feature.

Since localization is a veridical task, individuals who have developed this skill are readily identifiable. They include professionals who make their living finding their way around the world using topographic maps (e.g., geologists and wilderness guides) and serious recreationists (e.g., orienteers and outfitters). By investigating the problem solving of experienced map-readers as they solve localization problems, we gain insight into the methods used in efficient localization problem solving.

These considerations lead us to believe that studies of localization problem solving can shed light on three current issues in similarity judgment: (1) the claim that the structure of relations among features is more vital to these judgments than features taken independently, (2) the utility of making a distinction between relations and attributes, and (3) the processes by which these judgments are made.

## Experiment 1: Field studies

The goal of Experiment 1 was to address these issues using as data the thinking-aloud reports (protocols) of experienced map-readers solving a localization problem (Thompson, Pick, Bennett, Heinrichs, Savitt, & Smith 1990).

### Method

**Subjects.** A total of 29 experienced map-readers including professional geologists, champion orienteers, and wilderness guides participated in Experiment 1.

**Procedure.** Individual subjects were blindfolded and driven approximately 30 miles to a road access about one-quarter mile from the station point: the point in the terrain to be found on the map. They were led across a level field and up a hill to the station point. Once there, the blindfold was removed and they were given a topographic map attached to a clipboard. The map is a cropped U.S.G.S. topographic map from which all non-topographic information (culture) has been deleted. The map contains only contour information about elevation.

The station point is a roughly circular hill that is the westward extension of a larger highland. A distinctive attribute is its steep slope to the southwest. A second round hill with a similar orientation is selected as an alternative hypothesis by all subjects. This hill forms a garden path hypothesis (Johnson, Moen, & Thompson 1988): its similarity to the correct solution and its position in the center of the map lead many subjects to consider it early in their problem solving. Both have a pond to the north. Other alternatives are also considered by most subjects. Some subjects consider as many as eight different alternatives. Selection of the correct solution does not appear to depend on the number of alternatives considered.

The subjects' task was to find their viewpoint on the map. During the drive to the site, they had been briefed on the procedure and instructed to think aloud and to point to what they were talking about as they addressed the task. Subjects spent an average of 45 minutes on the task.

Subjects' verbal reports were recorded as they thought aloud. Simultaneously, their behavior was videotaped to provide information about where they were looking (and pointing) while thinking aloud. The verbal reports were transcribed and the composite audiovisual protocols coordinated and scored. Scoring focused on two aspects of problem solving: on the type of information attended to and how that information was used.

The scoring procedure identifies the source of information, the map or the terrain, and three categories of information - features, relations, and attributes. We define *features* as individual topographic objects that our subjects identify with a familiar count noun, e.g., hill, valley, pond. Each subject's lexicon is small and consistent. The composite lexicon across subjects provides a taxonomy of useful topographic terms. *Attributes* are properties that modify individual features. Subjects tended to use bipolar, qualitative attributes to differentiate among similar features, e.g., narrow or wide, steep or shallow. *Relations* are connectives that conjoin two or more features into a single structural unit we call a *configuration* of features. Some relations are purely topologic connectives, e.g., behind, below. Most configurations are expressed by qualitative predicates, e.g., "and then it (feature 1) gets steep down into (feature 2)". Use of quantitative relations, e.g., higher than, a mile apart, is less common.

Configurations constrain problem solving more effectively than do individual features. For example, there are fewer matches to "a high spot going down steeply to some lakes" than to an individual hill or pond. Distinguishing among features, attributes, and relations is consistent with the arguments made by Gentner (1983) and Medin et al. (1990).

Analysis of the protocols also identifies components of the problem solving process. Three of these processes involve judgments of similarity and/or difference. Localization problem solving is initiated by an extended period of reconnaissance. *Reconnaissance* identifies features, attributes, and relations for subsequent processing. Subjects return to reconnaissance to gather additional information. *Matching* is a form of argument that marshalls evidence that the features or configurations seen in the terrain correspond to those seen in the map, or vice versa. *Hypothesis generation* is an explicit statement about a particular location on the map that may represent the viewpoint. Localization concludes with wholesale acceptance of a hypothesis.

**Condition 1.** The 17 subjects in the first condition were instructed to remain at the station point as they attempted to solve the problem. They were permitted to move a few feet in turning around. As this task proved extremely difficult, a second condition was introduced.

**Condition 2.** In the second condition, the 12 subjects were free to walk about and to explore the terrain.

## Results

**Solution.** Of the 17 stationary subjects, only one arrived at the correct solution. Six of the 12 exploring subjects arrived at the correct solution. This difference in performance is significant, $\chi^2 = 5.60$, $p < 0.05$, $df = 1$.

**Judgments of similarity and difference.** All subjects begin by identifying salient features from the terrain and the map. They may begin with the terrain and move to the map, or begin with the map and move to the terrain. Subjects may identify a large number of features or key on a few salient features. The subject highlighted in Table 1 begins by describing his own position in the terrain as a relatively high area and identifying similarly high areas in the map (lines 4-5). Based on the few features he extracts in the first 25 seconds of reconnaissance, he generates a pair of hypotheses (lines 10-11). One of these hypotheses is the correct location on the map. The second is the garden path hypothesis.

Reconnaissance followed by hypothesis generation is typical of highly proficient subjects. Identification of features appears to be sufficient to generate informed hypotheses. Many subjects spend considerable time identifying features and assembling configurations of features before generating hypotheses. Subjects then proceed to focus on relations and judgments of similarity and difference to evaluate those hypotheses.

Single attributes often provide sufficient information to judge that a map feature cannot stand for a terrain feature. That is, difference judgments are often based on single features. An example of a judgment of difference based upon an attribute is shown in Table 1 (lines 87-90).

As shown in Table 1, the subject follows his generation of hypotheses with the assembly of several configurations, one of which is contained in lines 15 to 19. He conjoins his description of his viewpoint, "a knob", to the "stream valley below" with the predicate "gets pretty steep down into". The steep descent from his knob to the stream valley becomes a structural constraint on similarity judgment.

After assembling several other configurations both in the terrain and the map, he proceeds to attempt to match them. This matching necessarily entails judgments of similarity. One such match is shown in Table 1 (lines 38-43). He begins by reiterating a configuration extracted from the terrain (line 38). He turns his attention to the map to match features constrained by the same relation (lines 41-43). He then judges the two configurations to be sufficiently similar to support the hypothesis.

TABLE 1 SIMILARITY JUDGMENT IN LOCALIZATION

key: 4 - line number; M - map information; T - terrain information

| Stage | Line | Source | Protocol |
|---|---|---|---|
| 1 Identify features including viewpoint, relations, and attributes. Match features to guide assembly of configurations. | 4 | T | All right, well I noticed I'm at one of the higher points within this area, so that's important. |
| | 5 | M | So I'm first looking on the map, for some higher points on the map. |
| 2 Assemble configurations - descriptions of the topographic layout of relations among features including the viewpoint | 15 | T | but what I was actually looking at is how steeply the hill drops off. |
| | 16 | T | And it's kind of a knob right here we're standing on, |
| | 17 | T | and then generally not very steep |
| | 18 | T | and then it looks like it gets pretty steep down into a valley to the east. |
| | 19 | M | Ok, so I'm looking for the same types of things on the map. |
| 3 Generate viewpoint hypotheses | 10 | M | Um, for example say, somewhere here (HYPO 1 - CORRECT), |
| | 11 | M | or on a hill here (HYPO 2 - GARDEN PATH). |
| 4 Eliminating alternatives using attributes to make judgments of difference | 87 | M | And, see I'm kind of looking up here (HYPO 3) |
| | 88 | M | 'cause this also has a hill |
| | 89 | T | but um, ... Now straight to the north it should be quite steep |
| | 90 | M | So, that doesn't seem likely. (REJECT HYPO 3) |
| 5 Matching configurations using relations among features to make judgments of similarity | 38 | T | I'm at a high point. Directly north is a fairly flat area and north of that it get's steep and then there's the lake, ok. |
| | 39 | T | and I'm trying to match those features with what I see here on the map. |
| | 40 | M | Ok, ah, for instance, again. Let's go back to this place (HYPO 1) , ok, so here's a higher area. |
| | 41 | M | Here's a generally flat area. |
| | 42 | M | Then it goes down steeper here and it looks like there's valley coming through here. |
| | 43 | M | And then possibly some lakes or ponds here. |
| 6 Comparing hypotheses using relations to make judgments of difference | 47 | M | Ok, I still kind of like this area (HYPO 1), |
| | 48 | M | But then I was looking up on the map. I also have a high here, (HYPO 2) |
| | 49 | M | and with (a pond?), .. that's not very far at all. |
| | 50 | M | It just doesn't seem to work well |
| | 51 | M | because there's a fairly steep and long gradient here before you get to a flat part |
| | 52 | T | and I don't see that where we're standing. |

This is a consistent pattern in our protocols. Judgments of similarity are used to support hypotheses. They are also used to compare hypotheses. One such comparison is shown in lines 47-52. In this passage, a map configuration relating a high area and a pond is compared with the terrain configuration stated in line 38. This relation is a suitably strong constraint to reject this alternative.

### Summary

The field protocols reveal the critical role of similarity and difference judgments in localization problem solving. Of the six components itemized in Table 1, the final three involve similarity and difference judgments. Topographic relations that link features (including the viewpoint) support both similarity and difference judgments. Attributes of features support difference judgments.

This difference in power between relations and attributes may explain the difference in performance between the stationary and exploring conditions. Subjects who were allowed to explore the terrain had better access to information in general and better information about their viewpoint in particular. They used this information to assemble richer configurations that included the viewpoint. The experimental manipulation cannot distinguish whether the information from the viewpoint or about the terrain at large is the more valuable for successful localization.

## Experiment 2: Laboratory studies

Experiment 2 is a laboratory simulation of the localization task in which the amount of map information available to the subjects was manipulated. In an earlier study, maps were masked so as to obscure a portion of the map (Heinrichs, Montello, Nusslé, & Smith 1989). There were three masking conditions in that study: the mask covered either the inner one-third, the outer two-thirds, or the outer one-third of the map. The control condition used an unmasked map. Subjects were presented with one of the four conditions. Five pairs of masked and control conditions were selected for the present experiment.

The goal of experiment 2 was to determine whether map information around the viewpoint is generally more informative than other regions of the map. A second aim was to elucidate better the different roles of relations and attributes. The third aim was to examine a variety of locations in order to generalize beyond the single location used in the field experiment.

### Method

**Subjects.** Ten subjects from the same pool of experienced map-readers participated in Experiment 2.

**Apparatus.** Subjects were presented with topographic maps of five locations, two in Minnesota, two in New Mexico, and one in Arizona. As in Experiment 1, the maps are enlarged and cropped copies of U.S.G.S. topographic maps with all culture removed. The maps were marked with a single point at the center of the map that identified the location from which color slides were taken. The slides were taken with a camera mounted on a tripod at eye level. The line of sight was horizontal. A complete set of twelve pictures covered the whole $360^{o}$ panoramic view. For the experiment, either one view or three non-overlapping views were selected for presentation to the subject. Slides were presented on a rear projection screen in a darkened laboratory room. The subject sat 120 cm away from the screen with a reading lamp illuminating the map from behind. A remote control allowed the subject to advance or reverse the slides at will.

**Procedure.** Subjects solved two types of localization problems. In the first, one arrow was drawn on the map leading from the center point. Subjects used a remote control to view the three slides. Their task was to select the slide that corresponded to the terrain that would be seen looking in the direction of the arrow on the map. In the second type of problem, three arrows separated by $120^{o}$ were drawn on the map leading from the center point. Subjects were shown only one slide. Their task was to select which of the arrows on the map corresponded to the view of the terrain in the slide. These procedures presented options that subjects could entertain as hypotheses.

As in the field study, subjects were asked to think aloud while solving the problems. Collection of concurrent verbal reports, videotaping, and scoring of the resulting protocols followed the procedures of Experiment 1.

**Condition 1.** To investigate whether information about the viewpoint is favored over information from more distant areas, map information was selectively masked in the first condition. In four of the five trials (one trial for each set of maps and slides) a black circle masked the inner 1/3 of the map. In the fifth (Arizona) a black annulus masked the outer 2/3 of the map.

**Condition 2.** As a control condition, subjects also solved the same set of five tasks in a 'full map' condition in which the masks were removed from the maps. Each subject solved the problems twice, first in the masked condition and in the full map condition. This manipulation allowed within-subjects and within-location comparisons.

### Results

**Solution.** Accuracy was significantly better in the full map condition (66%) than in the masked condition(44%), $t(9) = 3.16$, $p < 0.05$. In the full map condition, performance is significantly different from chance, $t(9) = 6.27$, $p < 0.001$, but not in the masked condition.

**Judgments of similarity and difference.** In this section we compare judgments within subjects and across conditions for three of the five tasks.

In the first task, subjects viewed one slide of rolling terrain typical of southeastern Minnesota. The major discriminating feature in the slide is a prominent hill. Many subjects find the hill so salient that they base their judgments on a match to this feature. They were given a map with three arrows to choose among. In the masked condition, the mask covers the inner 1/3 of the map and totally obscures the prominent hill. One of the three arrows crosses a hill in the unmasked region of the map. Subjects who spend a disproportionate amount of time on the slide select this (incorrect) arrow. The salient hill leads them down a garden path. They ignore information about the relation of distance between the viewpoint and the feature and are led to an incorrect solution.

In the full map condition all subjects select the correct answer. The availability of information near the viewpoint pulls their attention to the distances between the viewpoint and the various hills along the arrows. As only one of these distances is similar to what is seen in the slide, the correct arrow is selected.

In the second task, subjects viewed one slide of mountainous Sonoran Desert terrain and were given three arrows on the map to choose among. In the masked condition, the mask covers the outer two-thirds of the map. This task is unique in that most subjects correctly answer both the masked and full-map conditions.

Subjects focus their attention on the orientation of a series of small ridges and valleys. It is clear that the relation of parallelism among these features (not including the viewpoint) is sufficiently diagnostic to raise only one of the offered choices to the level of a hypothesis.

The full map condition produces a second finding. It reveals a high hill in the distance to the left of one of the arrows. Subjects find that the hills in the slide are not as high and immediately eliminate that arrow from further consideration. This result supports the inference that attributes are sufficient to support judgments of difference.

In the third task, the inner one-third of the map is occluded and one arrow is shown on the map. Subjects viewed three slides of an area adjacent to a large river valley in eastern Minnesota. Their task is to select the slide that contains the terrain they would see looking in the direction indicated by the arrow on the map. The mask covering the viewpoint makes it appear as though the viewpoint is within a valley and the viewing direction is up at a hill. The viewpoint is actually on the crest of a small ridge that is completely obscured by the mask.

One of the slides contains a long gentle slope up to a distant hill. In the masked condition, many subjects make a reasonable assumption and incorrectly select this slide. By occluding the viewpoint, the masked condition eliminates vital information about the distribution and relations among features in the terrain and prevents correct solution. Correct solution of this localization problem clearly requires matching on the basis of relations of features that include the viewpoint.

### Summary

The first task shows the superiority of a judgment of similarity based on a configuration over a judgment based solely on a salient feature. The second task also shows the superiority of a judgment of similarity based on a configuration over a judgment based solely on a feature. In addition, it reveals reliance on the attributes of a feature to justify a judgment of difference. The third suggests that successful localization often requires full knowledge of the relations that tie the viewpoint to nearby features.

## Discussion

Experienced map-readers adopt a basic generate and test strategy to solve localization problems. They move in either direction, from map to terrain and from terrain to map, as they attempt to figure out where in the world they are. Judgments of similarity and difference inform both the generation and testing of hypotheses.

The two experiments reveal that structural relations of features play a key role in both the generation and testing of localization hypotheses. They also show a fundamental difference in the roles played by relations and attributes. The protocols reveal that experienced map-readers make this distinction. Attributes are used to make preliminary judgments about potential hypotheses (Table 1, Section 4) whereas relations are used to scrutinize hypotheses (Table 1, Sections 5 & 6). Features and relations guide the assembly of configurations. Attempts to match map and terrain configurations inform hypothesis testing. These tests rely on judgments of the similarity of relations. Relations are also used in judgments of difference to discriminate among competing hypotheses. In contrast, attributes are used only for judgments that either eliminate an alternative or raise it to the status of hypothesis.

Three questions remain: In judgments of topographic similarity and difference, are some *relations* more important than others? In the judgments of topographic difference, are some *attributes* more important than others? How do these vary with the nature of the terrain?

The roles played by relations and attributes in judgments of similarity and difference are part of a larger theory of localization problem solving. This theory is to be embodied in a system designed to control the navigation of vision-based mobile robots in dynamic environments.

## References


Gentner, D. 1983. Structure-mapping: A theoretical framework for analogy. *Cognitive Science* 7:155-170.

Heinrichs, M.R, Montello, D.R., Nusslé, C.N., and Smith, K. 1989. Localization with topographic maps. In Proceedings of the AAAI Symposium on Robot Navigation, 29-32. Stanford, California.

Johnson, P.J., Moen, J. B., and Thompson, W.B. 1988. Garden path errors in diagnostic reasoning. In Bolc, L., and Coombs, M.J. eds. *Expert Systems Applications*. New York: Springer-Verlag.

Medin, D.L., Goldstone, R.L., and Gentner, D. 1990. Similarity involving attributes and relations: Judgments of similarity and difference are not inverses. *Psychological Science*. 1(1):64-69.

Thompson, W.B., Pick, H.L., Jr., Bennett, B.H., Heinrichs, M.R., Savitt, S.L., and Smith, K. 1990. Map-based localization: The "drop-off" problem. In Proceedings of the DARPA 1990 Image Understanding Workshop, 706-719. Washington D.C.

Tversky, A. 1977. Features of similarity. *Psychological Review* 84:327-352.

# Syntactic Category Formation with Vector Space Grammars

**Andreas Stolcke**

Computer Science Division
University of California
Berkeley, CA 94720

International Computer Science Institute
1947 Center St., Berkeley, CA 94704
stolcke@icsi.berkeley.edu

**Abstract**

A method for deriving phrase structure categories from structured samples of a context-free language is presented. The learning algorithm is based on adaptation and competition, as well as error backpropagation in a continuous vector space. These connectionist-style techniques become applicable to grammars as the traditional grammar formalism is generalized to use vectors instead of symbols as category labels.

More generally, it is argued that the conversion of symbolic formalisms to continuous representations is a promising way of combining the connectionist learning techniques with the structures and theoretical insights embodied in classical models.

## Introduction

Connectionism, and especially Parallel Distributed Processing (PDP) has developed an array of models of learning systems (backpropagation, Boltzmann machines, competitive learning (Rumelhart, McClelland, & The PDP Research Group 1986)). These models typically operate on representations at a rather low and unstructured level (unit activations, bit vectors, microfeatures) relative to the structures used in traditional linguistic descriptions (trees and graphs, case frames, grammar rules, stacks). This is a necessary feature, the algorithms used are powerful and general precisely because they operate on simple and homogeneous representations. Simplicity and uniformity of the representations also allows for mathematical analysis, leading to theoretically well-founded methods such as gradient decent or simulated annealing.

A second prerequisite for these connectionist learning algorithms is that representations be *continuous*. Continuity of the representation space (ofter paired with requirement that some performance measure be *differentiable* with respect to the representations), ensures that *adaptive learning* can take place, i.e., gradual adjustment towards a specified goal. Again, continuity and differentiability are typically not found in traditional linguistic constructs, which tend to be inherently discrete (an exception are Fuzzy Languages (Zadeh 1972)).

It seems desirable, then, to investigate ways to combine connectionist (usually vector-based) representations with structures and concepts developed in traditional theories, especially in cases where those theories have a strong empirical or intuitive appeal. The goal would be to use one or several of the learning techniques mentioned above to learn or develop representations that are meaningful in the framework of the theory at hand.

The remainder of the paper gives an example of this general approach, applied to the theory of context-free grammars and the problem of learning category labels for the phrase types in a language. First we introduce the the hybrid symbolic/connectionist formalism on which the algorithm is based, and then discuss learning by way of a detailed example.

## Vector Space Grammars

The formalism presented here has been dubbed *Vector Space Grammar* (VSG) because it represents a generatization of traditional context-free grammars (CFGs) in which the nonterminals are represented by points in a continuous vector space rather than by symbols. Like in CFG, nonterminal rules (in Chomsky Normal From, CNF) are of the form

$$\mathbf{x} \rightarrow \mathbf{y}\ \mathbf{z} \tag{1}$$

and lexical (terminal) rules of the form

$$\mathbf{x} \rightarrow \mathrm{a} \tag{2}$$

But whereas in CFG categories ($\mathbf{x}$, $\mathbf{y}$, $\mathbf{z}$) are symbols in a space with a binary metric (equality/nonequality), VSG uses *vectors* as nonterminals. This gives a continuous metric on the category space, thus fulfilling one of the prerequisites for an adaptive learning mechanism. Terminals (words) in VSG are still unanalyzed atomic entities, and strings of terminals form the domain over which a language is defined.

A standard non-terminal rule maps two specific symbolic categories into a third symbolic category (the left-hand side of the rule). Similarly, a VSG rule maps two vectors onto a third. From a bottom-up parsing point of view, a traditional CFG rule is applicable if and only if its two right-hand side categories match exactly two other categories (roots of partial parses). In VSG, rule applicability becomes a graded notion, and every rule

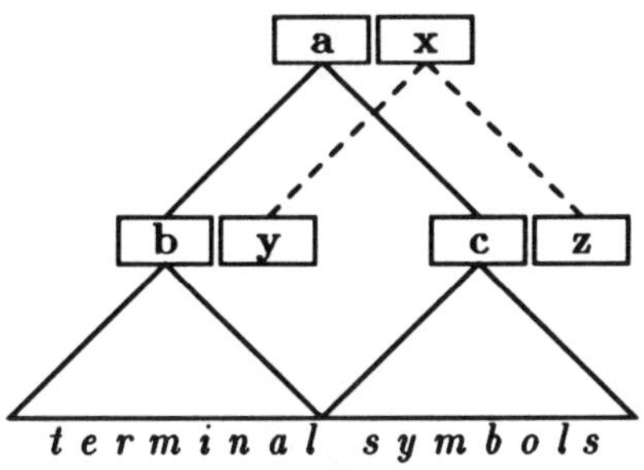


Figure 1: Vectors involved in VSG rule application. The new root vector **a** is a function of the subtree root vectors **b** and **c** and the vectors in the rule **x** → **y z**, e.g., **a** = (**b** · **y**)(**c** · **z**)**x**.

will be applicable to every two categories to some extent. However, the formalism is designed such that well-matching rules give a 'high' output, and poorly matching rules result in a vector close to the zero vector. This is accomplished by the following 'activation function' for VSG rules. Let $\mathbf{x} \rightarrow \mathbf{y}\ \mathbf{z}$ be the rule applied to two categories $\mathbf{b}$ and $\mathbf{c}$ (we use bold letters to denote vector quantities). Then the category resulting from the rule application is defined as

$$\mathbf{a} = (\mathbf{b} \cdot \mathbf{y})(\mathbf{c} \cdot \mathbf{z})\mathbf{x} \tag{3}$$

where · denotes the inner product of the vector space. The two inner products on the right express the match between the categories specified by the rule and the categories assigned to the substrings. Match values range between $-1$ and $+1$ if the category vectors are normalized. Since the right-hand side terms in a context-free rule work conjunctively (all have to match), the match values are multiplied. Choosing the inner product as the measure of matching partly determines the structure of the category space: categories will behave differently to the extent that they are orthogonal. The elements involved in rule application are depicted schematically in Fig. 1.

Traditional CFGs can be mapped into the vector representation such that equation 3 works precisely as traditional grammar rules do (using an all-or-nothing match between categories). This isomorphism maps each nonterminal in the CFG to a dimension in the VSG category space, and demonstrates that VSGs are indeed a formal generalization of symbolic phrase-structure grammars.

*Acceptance of strings* by a grammar can be defined analogously to traditional grammars, although acceptance becomes a non-discrete function (as in fuzzy languages). Since these definitions are not directly relevant to the learning procedure we will omit them here and turn immediately to the learning algorithm (see (Stolcke 1991) for details).

## Learning with Vector Space Grammars

The problem of learning to parse strings of a language can be broken down into two subproblems: finding the structure of the parse tree (i.e., the phrase bracketing), and assigning category labels to the nodes in the tree (the phrases). Current work with VSGs addresses mainly the second of these problems, for several reasons.

There are indications that the two problems might in fact be handled separately by natural language speakers, and that there are cues independent from category assignment that allow speakers to derive the phrase bracketing information. It has been shown that there are very effective statistical methods to find phrase boundaries (without phrase type classification) in text (Magerman & Marcus 1990). Secondly, psycholinguistic data indicates that humans can learn language structures successfully only when they can draw from a rich set of universal intra- and extra-sentential cues to induce phrase structure independently (Morgan, Meier, & Newport 1987; Morgan, Meier, & Newport 1989). Morgan (1986) has also argue for the prior availability of phrase structure information on learnability grounds.

Another source of independent phrase structure information comes from strong correlations between syntactic and semantic structure, i.e., the fact that syntax usually exhibits a structure parallel to one of the conceptual dependencies it expresses. This fundamental 'iconic relationship' between syntax and concepts is understood by some linguists as the very essence of language (Langacker 1985). A learner could capitalize on this principle if one assumes that certain general cognitive capacities are available prior to syntax learning.

For the purpose of this paper, then, we will assume that a learning system has access to phrase-bracketing information from independent sources. We will discuss how the category system and the rules for a language can be learned within the formal framework provided by VSG, given positive (and possibly negative) instances of the language along with their phrase structure boundaries. The kinds of structures available to the learning algorithm are familiar from Levy and Joshi's (1978) *skeletal structural descriptions*, and have been shown to be sufficient for syntax learning (Fass 1983). These learnability results, however, use automata induction techniques with very complex data structures (equivalence classes of trees structures), and are therefore not directly comparable to the methods employed here. VSG learning uses only very simple vector data structures and works on-line, i.e., no history of past samples has to be stored. All the primitive computations performed could be implemented using standard connectionist hardware, except for the global control and structure allocation mechanism.

Two global parameters of the system are the dimension of the category space and the number of rules to be used. These parameters should be set 'large enough' for a given language, and have an effect similar to the number of hidden units in a backpropagation network.

With too little resources, the system will not converge on a solution, and with too many degrees of freedom the solution might be redundant and not express certain generalizations about the input.

At the outset of learning, then, a fixed number of non-terminal rule 'templates' of the form (1) (with a given vector space dimension) are allocated. Additionally, for each terminal symbol, a rule of the form (2) is created. All category vectors, in all rules, are set to random unit-length vectors.

Given a sample string from the language and a parse tree skeleton, we construct a labeled parse tree from the current set of rules. To assign a category vector to a node, the rule whose right-hand side represents the best match for the child node categories is selected and equation (3) is used to compute the output category for that node. 'Best match' is defined according to the same inner product metric as used in equation (3), i.e., using the value $(\mathbf{b} \cdot \mathbf{y})(\mathbf{c} \cdot \mathbf{z})$. Only the rules selected at some node will later participate in the learning process. Since only the currently best rules get selected the whole process strongly resembles the method of *competitive learning* (Rumelhart & Zipser 1985).

By working from the terminal nodes to the root we arrive at a category label for the entire string. If the training sample is a positive instance of the language we know what the target category for the parse should be: the sentence category 'S' that every grammar has to provide. Without loss of generality we can fix S throughout training to be a particular vector, e.g., the unit vector $(1, 0, \ldots, 0)$.

The second idea adapted from connectionist learning methods is that of *error backpropagation* (Rumelhart, Hinton, & Williams 1986). At the root node we can immediately compute an error term for the discrepancy between the desired output and the actual output. For positive examples this is just the difference between S and the root category, for negative examples we compute an error term which tends to make the output category and S orthogonal. A recursive procedure (based on the chain rule) can then compute the derivative of that error with respect to every category vector occurring in some rule (left of right-hand side) applied somewhere in the tree. The computation of error derivatives is straightforward because of the simple linear operations used in equation 3 but omitted here for lack of space (see (Stolcke 1991)).

Derivatives for each category vector are then added up and multiplied by some constant (the 'learning rate') to give the adjustment to be applied to that category. All rules are updated accordingly, all categories are rescaled to unit-length, and the next training example is processed. The algorithm cycles through the training set until the error becomes negligible or no further improvement is observed over a long period of time.

It is important to realize that the backpropation step can not assign errors that are due to choosing the 'wrong' rule at some point, because rule selection is a discrete step that allows no differentiation. Unfortunate rule selection has to be overcome by changing the rules themselves, and competitive rule selection is a heuristic to minimize rule selection errors.

```
+ ((a circle) (touches (a square)))
+ ((a square) (touches (a circle)))
+ ((a circle) (is (below (a square))))
+ ((a square) (is (below (a circle))))
+ ((a circle) (is (above (a square))))
+ ((a square) (is (above (a circle))))
- (a square)
- (a circle)
- (above (a circle))
- (below (a square))
- (touches (a circle))
- (touches (a square))
- (is (above (a square)))
- (is (above (a square)))
- (is (below (a circle)))
- ((a circle) (below (a square)))
- ((a square) (above (a circle)))
- ((a circle) (is (touches (a square))))
- ((is circle) (touches (a square)))
- ((a circle) (a (a square)))
- ((a square) (is (below (is circle))))
- ((a square) (touches (below (a circle))))
- ((a circle) (is (a square)))
- ((a square) (a (above (a circle))))
```

Figure 2: Training set used for the VSG learning experiment. The data is drawn from a fragment of English generated by the grammar given in the text. Positive training instances are labeled with +, negatives ones with -.

## A Sample Grammar

We have run simulations of the algorithm described above to verify that is can indeed converge onto working VSG grammars for a variety of small artificial and 'natural' languages. As indicated in the introduction, one of our main goals was to not only attest learning success (as defined by the error function) but to try to understand how the category vectors formed collaborate to produce a meaningful system of rules.

As an example consider a fragment of English consisting of transitive sentences ('A circle touches a square') and copula sentences ('A circle is below a square') involving the nouns *circle, square*, the verbs *is, touches*, the prepositions *above, below* and the determiner *a* (this fragment is borrowed from the $L_0$ project domain (Feldman et al. 1990), a sample grammar for it is given below).

The algorithm was run over a set of 6 positive and 18 negative samples, listed in Fig. 2. the number of rules was set to 5 and the category dimension to 15. At a constant learning rate of 0.5 the error was typically negligible after 50 passes over the training set.

As a method for analyzing the resulting VSG we used *cluster analysis*, which groups vectors according to a distance metric in a hierarchical fashion. Fig. 3 shows the result of clustering all vectors occurring in rules as well as the fixed S vector.

The graph shows that the vectors fall into nine major clusters of left-hand side and right-hand side rule vectors. We can reconstruct a symbolic rule system from

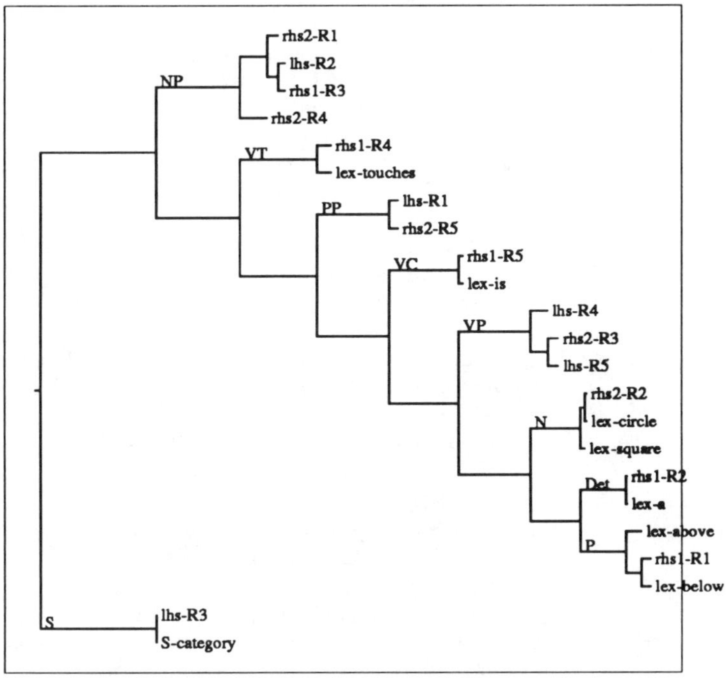


Figure 3: Clusters of category vectors derived from sample language. Left-hand side and right-hand side vectors in non-terminal rules are labeled by one of 'lhs', 'rhs1' and 'rhs2', and the rule number (R1–5). Left-hand side vectors of terminal rules are labeled as 'lex' and the terminal they generate. 'cat-S' is the fixed S category vector. Branches leading to clusters that can be identified with non-terminals from the context-free grammar given in the text are labeled with the corresponding symbol.

the diagram by identifying these clusters with (initially arbitrary) symbols and then filling in the rule templates used. Once this is done, we can interpret the rules on the basis of their interaction with other rules and rename the symbols according to our traditional names for syntactic categories if an unambiguous interpretation is possible.

Analysis of cluster diagram for this example shows not only that a rule system has been formed that accounts precisely for the input sample, but also that these rules and categories can be put into a one-to-one correspondence with a natural standard CFG for the language at hand, such as:

| S | → | NP VP | N | → | square\|circle |
|---|---|---|---|---|---|
| NP | → | Det N | VT | → | touches |
| VP | → | VT NP | VC | → | is |
| VP | → | VC PP | P | → | above\|below |
| PP | → | P NP | *Det* | → | a |

(Fig. 3 explains how CFG symbols map to vector clusters.)

## Discussion

The details of the resulting rule and category structure are highly dependent on the training environment. For the example in the previous section, extreme conditions were intentionally chosen to generate the perfect correspondence between the structure learned and the traditional CFG. Specifically, constraining the number of rules to five forced a parsimonious use of categories. With more rules to work with either redundancies would have developed (several rules serving the same function) or some rules stay useless (never winning a competition and not converging onto meaningful categories). Also, the relatively large number of negative examples ensured that the categories formed were sufficiently discriminatory. With less or no negative examples a grammar develops that accounts for all the positive examples but fails to exclude all the negative ones, due to overly general rules.

The need for negative examples is the most bothersome problem if one is looking for a plausible mechanism for natural language acquisition (and widely acknowledged as a major challenge for many theories of acquisition, see, e.g., (Pinker 1989)). Although our current system is certainly too impoverished to claim to be a model of natural language acquisition (it handles only syntax, for one thing), it would be nice to obviate the need for negative examples.

Experiments show that just dropping the negative examples from the training set produces grammars with too few rules. They account for the training data by clustering a large number vectors together, resulting in categories that are too general and have too little discriminatory power to rule out false negative examples.

This situation arises partly due to a well-known problem with competitive learning schemes. A small number of rules win most of the competitions early, thereby pulling all category vectors into a few large pools. Many rules never get applied and never learn to be useful as a result. To counteract this tendency we have recently modified the learning algorithm to incorporate an idea from learning in *genetic systems* (Holland 1975). In the modified learning schedule, rules that never are used are periodically eliminated from the rule set and replaced by copies ('clones') of rules that are heavily used. These are the ones that tend to be overly general, and duplicating them allows the two copies to specialize into different roles in the grammar. A modified algorithm based on this heuristic does much better in positive-only training, and is able to derive a category structure similar to that in Fig. 3 with only one overly general cluster (merged VP and PP vectors, an error that is in fact motivated by the similar syntactic functions of these two categories).

Before concluding, we would like to contrast the general line followed here with some of the pure PDP approaches to language. Many of these take the view that learning networks have to 'discover' whatever structure is implicit in language, and are reluctant to provide the network with clues to this effect. In this approach, interpreting the results and representation obtained through successful learning (as well as their theoretical implications) becomes a major problem. Frequently researchers resort to post hoc analyses hoping to find familiar structures in their data using techniques similar to the ones used here (Elman 1988; Pollack 1990).

Part of the motivation underlying VSGs is that familiar structures (e.g., context-free rules) can be built into connectionist representations as a bias, allowing a more straightforward interpretation of the results. In the example, traditional categories emerged as a discrete approximation to the cluster structures developed in learning, thereby guiding their interpretation. (The case for symbolic representations as approximations to sub-symbolic entities has been made by Smolensky (1987).)

## Conclusions

We have argued for generalizations of traditional symbolic representations and models to benefit from some of the learning power found in connectionist systems without completely discarding the structural properties and intuitions embodied in traditional theories. As an example, we have introduced a generalization of phrase structure grammars, Vector Space Grammars, that is based on vectors instead of symbols to represent grammatical categories. An algorithm using VSGs based on the principles of adaptation of categories and competition between rules can be used to derive a syntactic category system from exposure to phrase-bracketed sample sentences. The results of learning can be interpreted in traditional notions by interpreting vector clusters as category symbols.

## Acknowledgements

I thank Jerry Feldman, Terry Regier, and Subutai Ahmad for valuable discussions of the ideas presented here. This work is supported by an IBM Graduate Fellowship.

## References

Elman, J. L. 1988. Finding Structure in Time. CRL Technical Report 8801, Center for Research in Language, University of California at San Diego, La Jolla, Calif.

Fass, L. F. 1983. Learning Context-Free Languages from their Structured Sentences. *ACM SIGACT News* 15(3).

Feldman, J. A., Lakoff, G., Stolcke, A., and Weber, S. H. 1990. Miniature Language Acquisition: A touchstone for cognitive science. In *Proceedings of the 12th Annual Conference of the Cognitive Science Society*, 686–693, MIT, Cambridge, Mass.

Holland, J. 1975. *Adaption in natural and artificial systems.* Ann Arbor, Mich.: University of Michigan Press.

Langacker, R. 1985. *Foundations of Cognitive Grammar. Vol. 1: Theoretical Prerequisites.* Stanford: Stanford University Press.

Magerman, D. M., and Marcus, M. P. 1990. Parsing a Natural Language Using Mutual Information Statistics. In *Proceedings of the 8th National Conference on Artificial Intelligence*, Boston, Mass.

Morgan, J. L. 1986. *From Simple Input to Complex Grammar.* Cambridge, Mass.: MIT Press.

Morgan, J. L., Meier, R. P., and Newport, E. L. 1987. Structural Packaging in the Input to Language Learning: Contributions of Prosodic and Morphological Marking of Phrases to the Acquisition of Language. *Cognitive Psychology* 19:498–550.

Morgan, J. L., Meier, R. P., and Newport, E. L. 1989. Facilitating the Acquisition of Syntax with Cross-Sentential Cues to Phrase Structure. *Journal of Memory and Language* 28:360–374.

Pinker, S. 1989. Language Acquisition. In Posner, M. I., ed., *Foundations of Cognitive Science.* Cambridge, Mass.: MIT Press.

Pollack, J. B. 1990. Recursive Distributed Representations. *Artificial Intelligence* 46:77–105.

Rumelhart, D. E., Hinton, G. E., and Williams, R. J. 1986. Learning Internal Representations by Error Propagation. In (Rumelhart, McClelland, & The PDP Research Group 1986), 318–362.

Rumelhart, D. E., McClelland, J. L., and The PDP Research Group 1986. *Parallel Distributed Processing: Explorations in the Microstructure of Cognition*, vol. 1: *Foundations.* Cambridge, Mass.: Bradford Books (MIT Press).

Rumelhart, D. E., and Zipser, D. 1985. Feature Discovery by Competitive Learning. *Cognitive Science* 9:75–112. Reprinted in (Rumelhart, McClelland, & The PDP Research Group 1986), pp. 151–193.

Smolensky, P. 1987. On variable binding and the representation of symbolic structures in connectionist systems. Technical Report CU-CS-355-87, University of Colorado, Boulder, Colo.

Stolcke, A. 1991. Vector Space Grammars and Grammatical Category Acquisition. Technical report, International Computer Science Institute, Berkeley, Calif. in preparation.

Zadeh, L. A. 1972. Fuzzy Languages and their Relation to Human and Machine Intelligence. In *Proceedings of the Conference on Man and Computer, Bordeaux, France, June 1970*, 130–165. Basel: S. Karger.

# Dynamic Inferencing in Parallel Distributed Semantic Networks*

**Ronald A. Sumida**
Artificial Intelligence Laboratory
Computer Science Department
University of California
Los Angeles, CA, 90024

## Abstract

The traditional approach to dynamic inferencing is to represent knowledge in a symbolic hierarchy, find the most specific information in the hierarchy that relates to the input, and apply the attached inferences. This approach provides for inheritance and parallel retrieval but at the expense of very complex learning and access mechanisms. Parallel Distributed Processing (PDP) systems have recently emerged as an alternative. PDP systems use a very simple processing mechanism, but can only access high-level knowledge sequentially and require an enormous amount of training time. This paper presents Parallel Distributed Semantic (PDS) Networks, an approach that integrates the best features of symbolic and PDP systems by storing the content of symbolic hierarchies in ensembles of PDP networks, connecting the networks in the manner of a semantic network, and using Propagation Filters to determine how information is passed between networks. Simulation results are presented which indicate that PDS Networks and Propagation Filters are able to perform pattern completion from partial input, generate dynamic inferences, and propagate role bindings.

## Introduction

In order to dynamically generate inferences, a natural language understanding system must be able to access knowledge structures at varying levels of generality and to use the associated information in making the proper inferences. For example, consider the following texts which involve both general and specific information about various hit actions.

Text 1
John hit Mary because she dumped him for Bill.

Understanding why John hit Mary involves accessing general information about why humans hit one another, whereas understanding:

Text 2
Douglas hit Tyson. He wanted to win the title.

requires utilizing the specific knowledge that one reason why boxers hit one another is to win a competitive activity.

## Previous Work

*Symbolic/Localist systems* represent general and specific knowledge by forming an explicit hierarchy with each concept represented by a symbol or node, and with related inferences attached at the appropriate level. Figure 1 is an example of a semantic network with part of a simplified hierarchy for understanding the texts above. At the top of the hierarchy is the general hit act (HIT), followed by levels for one person hitting a person (HUMAN-HIT-HUMAN), and a boxer hitting his opponent (BOXER-HIT-BOXER). Each of these levels is connected to knowledge about the motivation for the hit act.

The language understander generates inferences by: (1) searching the hierarchy for the most specific node that applies to the input and binding the appropriate roles, (2) applying the attached inferences, and (3) propagating the role bindings. A number of methods for searching the hierarchy have been proposed, such as localist connectionist spreading activation (e.g. [Waltz and Pollack, 1985],[Sumida *et al.*, 1988]).

The advantage of the hierarchical approach is that it provides an economical method for representing structured knowledge (through inheritance) and it allows large amounts of knowledge to be searched in parallel (by applying the search procedure simultaneously to different parts of the network). Unfortunately, this approach suffers from three major shortcomings. First, it is not clear how specific the information represented in the hierarchy should be. Should there be an explicit level for a male hitting a female? Second, a combinatorial explosion problem occurs when new information is added to memory, since there are an enormous number of ways that shared features can be combined to index the new information. Third, the process of searching the hierarchy introduces serious complexities. With spreading activation systems, for example, the search

*This research is supported by a grant from the Keck Foundation.

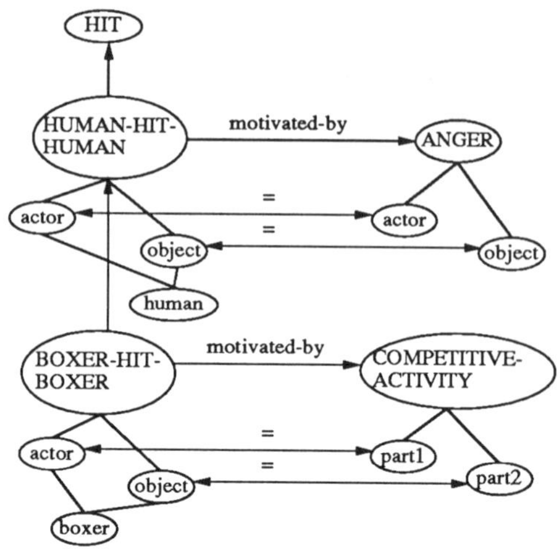


Figure 1: Semantic network with a hierarchy of HIT acts. Single arrows indicate parent/child (is-a) relationships. Straight lines indicate role/filler relationships. Double arrows labeled "=" signify equivalence relationships.

procedure itself is simple but the complexity is hidden in the procedure for determining the link weights.

*Parallel Distributed Processing (PDP) Systems* represent concepts as patterns of activation in a highly interconnected network of units [Rumelhart and McClelland, 1986]. PDP systems provide inheritance between concepts by having the patterns for a general concept and its descendants share parts, so that any effects caused by the pattern for the general concept will be transferred over to its descendants [Hinton, 1981]. Exceptions are encoded by the parts of the pattern that are not shared, so that the unique parts of the pattern can override information from the ancestors. Propositional information is stored in PDP systems by using triples of the form (Role1 Relation Role2). A sequence of propositions can be encoded in a PDP network (e.g. [Pollack, 1988]) by generating a reduced description (over the hidden units) for a triple, and presenting it as input along with the next triple in the sequence.

PDP systems offer a solution to the problems plaguing symbolic/localist systems for storing hierarchical information and generating inferences because they: (1) do not need to make a discrete decision about what knowledge structures to include in the network. The extent to which a concept exists in the hierarchy is determined by statistical correlations between training patterns, (2) avoid combinatoric problems since the network automatically determines (through the learning procedure) which of the shared features are salient, and (3) use an extremely simple processing mechanism.

Unfortunately, PDP systems suffer from their own unique problems because they have a very limited representation of structured knowledge and they store too much information in a single network. The first problem is that it takes an enormous amount of time to train the network because of the incredibly large number of concepts that it must hold. The second problem is the *Knowledge-Level Parallelism Problem* [Sumida and Dyer, 1989]. Since the network can only store or retrieve one triple at a time, it is not possible to search large amounts of knowledge in parallel. This is particularly problematic in inferencing, where a large number of alternatives must be pursued in parallel for efficiency reasons.

## Parallel Distributed Semantic (PDS) Networks

The PDS network approach is to combine the best features of symbolic/localist semantic networks and PDP approaches by: (1) Storing the content of a symbolic hierarchy in an ensemble of PDP units. For example, the HIT hierarchy of Figure 1 is stored in a PDP network, as shown in Figure 2. Similarly, information about humans is stored in a second PDP network. Each level of the hierarchy is represented as a pattern of activity over the appropriate ensemble. For instance, HUMAN-HIT-HUMAN and BOXER-HIT-BOXER are stored as alternative patterns of activation over the HIT ensemble. (2) Connecting the PDP networks according to the structure of a semantic network. In Figure 2, the HIT and HUMAN PDP networks are connected in the same general manner as in Figure 1.

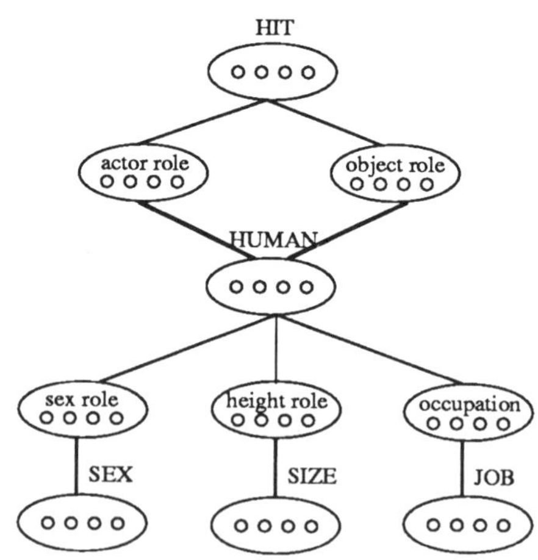


Figure 2: Two PDP networks for storing the hierarchy of Figure 1. The circled dots in the figure indicate an ensemble of units. The conceptual ensembles (i.e., HIT and HUMAN) are labelled with capital letters and the role ensembles are indicated by lower-case labels placed inside the oval of units.

In a previous paper [Sumida and Dyer, 1989], we

showed how PDS networks store *static* role bindings, in which a previously encountered concept is bound to a role of a *known* proposition. This paper demonstrates how PDS networks have been expanded to create generalizations and to generate *dynamic* inferences. In order to generate dynamic inferences, PDS networks: (1) store each hierarchy in the appropriate network by presenting training instances and having the network automatically generate the proper generalizations, (2) classify new input according to the proper level of the hierarchy from the generalizations made during training, and (3) propagate patterns for the role bindings to the proper, related networks. For example, to understand Text 2 using PDS networks: (1) the HIT hierarchy is stored by training the HIT network on a number of instances of humans hitting one another and boxers hitting boxers. The network then generalizes the training data to learn patterns for HUMAN-HIT-HUMAN and BOXER-HIT-BOXER, (2) a new pattern, Douglas hit Tyson, is presented and classified as the BOXER-HIT-BOXER level of the hierarchy that was generalized from the training instances, and (3) the Douglas and Tyson patterns are propagated to the COMPETITIVE-ACTIVITY network.

## Storing Information in PDS Networks

Hierarchical information is stored in a PDS network by generating a unique pattern (for each training instance) that represents a *reduced description* of the input. The network automatically generalizes from these training patterns to generate a pattern that represents a level of the hierarchy. For example, to represent the BOXER-HIT-BOXER level from Figure 1, a number of instances of boxers hitting one another are presented to the network, a unique pattern is generated for each, and the network generalizes from these patterns to generate the pattern for BOXER-HIT-BOXER. A reduced description of the input is generated by using a slightly modified version of the standard encoder network. In a standard encoder network, the input and output layers have the same number of units and are presented with exactly same pattern. The weights are modified so that the input patterns are trained to recreate themselves as output. The resulting pattern over the hidden units is taken to represent the reduced description. In the modified network that we use, the same set of units is used for both the input and output layers. The network can be viewed as an encoder net with the output layer folded back onto the input layer and with two sets of connections: from the single input/output layer to the hidden layer and from the hidden layer back to the input/output layer.

As an example, consider the simplified figure of the HUMAN ensemble shown in Figure 3. The network relates the representation of a human with the features of that person. The sex, height, and occupation role groups cumulatively represent the input/output layer and the HUMAN group represents the hidden units. The black arrows indicate connections from the input/output layer to the hidden layer, and the grey arrows indicate connections from the hidden layer back to the input/output layer. The thick lines in the figure connect a role group to an ensemble which can fill it and indicate links which propagate a pattern without changing it. The jagged lines in the figure are suggestive of the patterns of activation that are presented to the network. Suppose that we want to train the network of Figure 3 on a tall, male boxer. We need to generate a unique pattern in the HUMAN ensemble that is a reduced description of the input and that represents the boxer. The pattern is generated by placing the features for the boxer into the appropriate ensembles (male into the SEX ensemble, tall into the SIZE ensemble, and boxer into the JOB ensemble), propagating the patterns into the input/output role ensembles (male from SEX to sex role, tall from SIZE to height role, and boxer from JOB to occupation role, as indicated by the single dotted arrows in the figure), and training the network by altering the connections between the role groups and the HUMAN units that form the hidden layer (the double dotted arrows indicate the training process). The resulting pattern over the HUMAN units (Boxer1) represents the boxer.

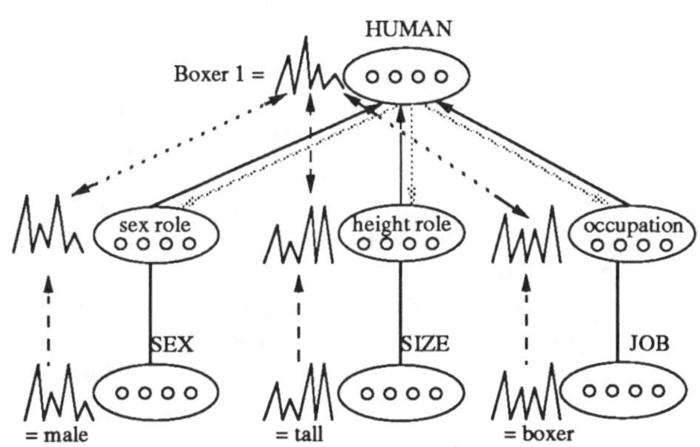


Figure 3: Representing a boxer in the HUMAN network.

Suppose that we now want to represent a proposition for one boxer hitting another, specifically, (Boxer1-HIT-Boxer2). The idea is to use exactly the same method described above for representing a human, but this time applied to the HIT network of Figure 4. Thus, the Boxer1 pattern is propagated from HUMAN to the actor role of HIT, the Boxer2 pattern is generated in the same manner described above and then propagated from HUMAN to the object role of HIT, and the network is trained so that the input/output pattern (the conjunction of Boxer1 and Boxer2) will recreate itself as output. The resulting pattern over the hidden HIT units is the representation and reduced description for the proposition (BOXER1-HIT-BOXER2).

The other training instances are created in exactly

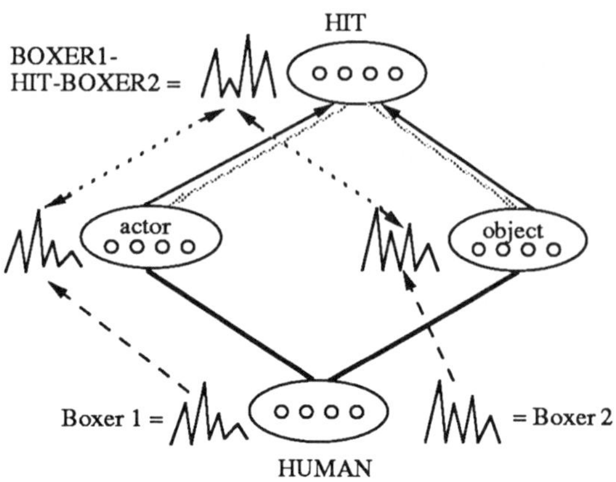


Figure 4: Representing one boxer hitting another in the HIT network.

the same manner. Various humans are represented in the HUMAN network by presenting their features and generating a reduced description. Various hit acts are represented in the HIT network by presenting the actor and object and again generating a reduced description.

## Classifying the Input and Propagating Role Bindings

As a result of the training process, the hidden units learn to classify the input by responding to common features of the training patterns. Particular hidden units develop patterns that generalize from the features shown during training and thus represent a level of the hierarchy. For example, in the simulations described in the next section, two of the HIT units respond with a pattern of "01" when two boxer patterns are presented. The hidden units allow the network to classify a new concept based on its similarity to ones seen during training. When the pattern for the new concept is presented, the hidden units that are sensitive to its features become activated. The pattern over these units serves to classify the input. The hidden units also help the network perform pattern completion from partial or noisy input, so that when an unfamiliar pattern is presented that is similar to a known one, the hidden units will respond to the similar features and recreate the known pattern.

The subset of hidden units that classifies the input acts as a *Propagation Filter* that directs where role bindings are propagated. Propagation filters are based on the idea of skeleton filters [Sejnowski, 1981, Hinton, 1981] which use a pattern over one group of units to enable a restricted subset of a group of filter units. Propagation filters consist of: (1) groups of filter units, each of which gate the connection from a source to a destination and (2) a set of selector units that choose which filter group to enable. The pattern over the selector units opens up the proper filter group by driving its units above threshold. For example, in Figure 5, the "01" pattern over units 2 and 3 of the selector opens up group1 of the filter, which allows the pattern to be passed from source1 to destination1. The pattern in source2 is not propagated because the units in group2 remain well below threshold.

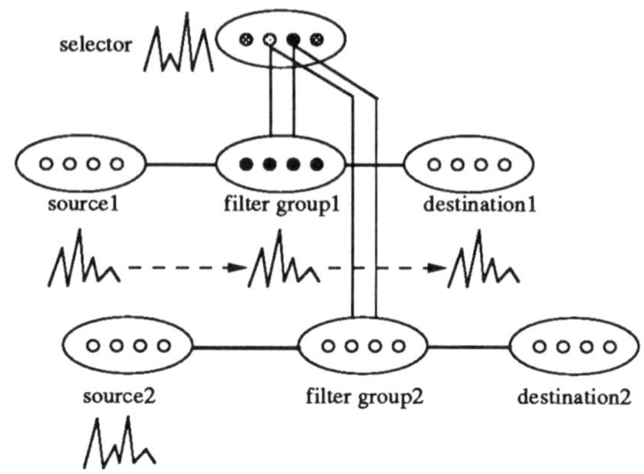


Figure 5: A Propagation Filter with group1 enabled and group2 disabled.

To illustrate how propagation filters are applied to propagating role bindings, suppose that we train the network on numerous examples of boxers hitting one another and of people hitting other people. As we mentioned earlier, one result of the training is that two of the HIT units learn to respond with "01" when two boxer patterns are presented and with "10" or "00" when two non-boxers are presented. These two units constitute a selector that enables the filters connecting the roles of HIT with the roles of COMPETITIVE-ACTIVITY, provided that the pattern over the two units is "01". Similarly, the two units enable the filters connecting the roles of HIT with the roles of ANGER if the pattern is "10" or "00". For example, suppose that we test the network on [Douglas HIT Tyson] by generating the pattern for Douglas in the HUMAN network (by the procedure shown in Figure 3), propagating the Douglas pattern to the actor role of HIT, and generating the Tyson pattern in HUMAN and propagating it to the object role of HIT. As illustrated in Figure 6, the network classifies it as BOXER-HIT-BOXER by generating the "01" pattern over the two HIT units. The "01" pattern enables the filter group that connects the actor of HIT with the first participant role of COMPETITIVE-ACTIVITY and the object of HIT with the second participant role of COMPETITIVE-ACTIVITY. Thus, the Douglas pattern is propagated to the participant1 role and Tyson to participant2. The filters that connect the actor and object roles of HIT with the roles of ANGER (not shown in the figure) are not enabled, since they require that the two selector units have a pattern of "10" or "00". As a result, the Douglas and Tyson patterns are not propagated to ANGER. Note that if the input had been [John HIT Mary] as in Text 1, then the two HIT units would respond with "10", which would instead open the filters from HIT to ANGER and result in John being propagated to the actor role of ANGER and Mary to the object role.

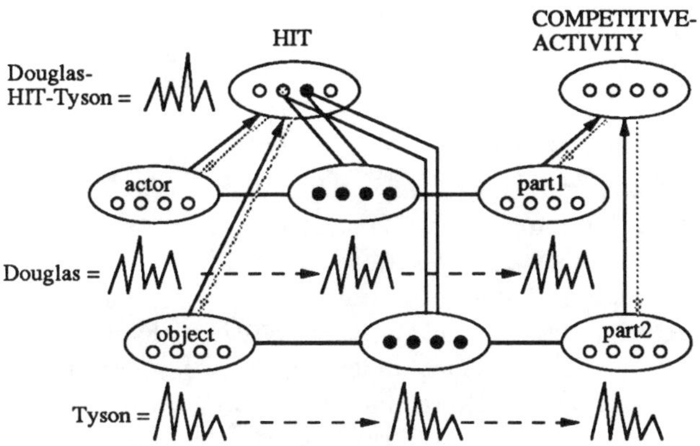


Figure 6: Testing the HIT network on [Douglas HIT Tyson]. The "01" pattern over the two units of HIT opens the filters connecting HIT with COMPETITIVE-ACTIVITY, and allows the Douglas and Tyson patterns to be propagated.

## Implementation Details and Simulation Results

PDS Networks are implemented in a natural language understanding system called DCAIN which is written in the C programming language and is fully implemented on the examples shown above. In our simulations, the HUMAN network was divided into the following 8 role groups: occupation (boxer, businessman, career woman or teacher), strength (strong, average, or weak), height (tall, medium, or short), weight (heavy, average, or light), sex (male or female), hair color (blond, brown, red, or black) and length (short, medium, or long), eye-color (blue or brown), and skin-color (dark or light).

The following is a list of the tasks on which PDS networks were tested. In every case, the network performed successfully: (1) Retrieving the correct (i.e., consistent with the training data) features given a single characteristic. For example, given the boxer occupation, the network correctly filled in the strong, tall and heavy features. (2) Retrieving the correct human given enough distinguishing characteristics. (3) Categorizing familiar HIT acts (i.e., those seen during training) and generating the proper inferences. (4) Categorizing *new* HIT acts involving familiar humans and generating the proper inferences. New hit acts were created by selecting a pair of familiar humans whom the network had not seen hit one another. (5) Categorizing new HIT acts involving new humans and generating the proper inferences. New humans were created by varying irrelevant features (such as hair and eye-color) of the humans that the network had seen during training.

## Future Work

DCAIN needs to be expanded to: (1) dynamically form the overall PDS network structure. DCAIN can add new information and form generalizations within a *single* network while avoiding the problems of localist systems. However, DCAIN does not currently learn the overall structure of the system, and must be expanded to do so. (2) include structural language information and incorporate more conceptual knowledge, (3) address the issue of ambiguity, (4) address timing and sequencing issues, and (5) incorporate elements of explanation-based generalization.

## Conclusions

Parallel Distributed Semantic (PDS) Networks store the content of symbolic hierarchies in ensembles of PDP networks, connect the networks in the manner of a semantic network, and use propagation filters to perform dynamic inferencing. PDS Networks are implemented in DCAIN, a natural language understanding system that: (1) uses a simple procedure for determining link weight values, (2) exploits automatic generalization, (3) represents structure, (4) provides knowledge-level parallelism, and (5) drastically reduces training time. Thus, DCAIN has a number of advantages over previous connectionist systems, and integrates the best features of symbolic and PDP approaches while avoiding their associated problems.

## References


[Hinton, 1981] G. E. Hinton. Implementing Semantic Networks in Parallel Hardware. In *Parallel Models of Associative Memory*, Lawrence Erlbaum, Hillsdale, NJ, 1981.

[Pollack, 1988] J. Pollack. Recursive Auto-Associative Memory. In *Proceedings of the Tenth Annual Conference of the Cognitive Science Society*, Montreal, 1988.

[Rumelhart and McClelland, 1986] D. E. Rumelhart and J. L. McClelland. *Parallel Distributed Processing*, Volume 1. MIT Press, Cambridge, Massachusetts, 1986.

[Sejnowski, 1981] T. J. Sejnowski. Skeleton Filters in the Brain. In *Parallel Models of Associative Memory*, Lawrence Erlbaum, Hillsdale, NJ, 1981.

[Sumida *et al.*, 1988] R. A. Sumida, M. G. Dyer, and M. Flowers. Integrating Marker Passing and Connectionism for Handling Conceptual and Structural Ambiguities. In *Proceedings of the Tenth Annual Conference of the Cognitive Science Society*, Montreal, 1988.

[Sumida and Dyer, 1989] R. A. Sumida and M. G. Dyer. Storing and Generalizing Multiple Instances while Maintaining Knowledge-Level Parallelism. In *Proceedings of the Eleventh International Joint Conference on Artificial Intelligence*, Detroit, 1989.

[Waltz and Pollack, 1985] D. Waltz and J. Pollack. Massively Parallel Parsing. *Cognitive Science*, 9:51–74, 1985.

# Accessing Meaning vs. Form at Different Levels of Comprehension Skill

**David J. Townsend**
Department of Psychology
Montclair State College
Upper Montclair, NJ 07043
townsend@apollo.montclair.edu

**Thomas G. Bever**
Department of Psychology
University of Rochester
Rochester, NY 14627
bever@prodigal.psych.rochester.edu

**Caroline Carrithers**
Department of Psychology
Rutgers University
Newark, NJ 07102

**Abstract**

We examined adult and 10-13 year old skilled and average comprehenders' representation of spoken sentences. In immediate probe tasks, skilled adults were better than average adults at accessing word order information, but they were *poorer* at accessing sentence meaning. After hearing a text, skilled adults were more accurate than average adults in recognizing meaning, but they were *less* accurate in recognizing the wording of test sentences. Speeded speech increased the differences between skill groups *more* for memory for wording than for memory for meaning. The results suggest that comprehenders compute representations of surface form and meaning independently and simultaneously. These representations compete for attention.

## Introduction

Formal models of language comprehension must specify the nature of the hypotheses that comprehenders form about utterances and the mechanisms they use to form them. One of the most striking facts about comprehension is the speed at which comprehenders form these hypotheses. For example, subjects can shadow speech at delays as short as 250 ms and still be sensitive to semantic context (Marslen-Wilson 1975).

The nearly immediate effect of semantic information seems to rule out models in which comprehenders compute successively more abstract representations serially. Highly skilled comprehenders, however, may be able to perform these computations rapidly enough for some aspects of meaning to have immediate effects. If *serial models* are correct, reducing the time that is available for language processing should decrease comprehenders' sensitivity to semantic properties (French 1981; Stine, Wingfield, & Poon 1986), particularly for less skilled comprehenders who may not perform lower level computations as efficiently as skilled comprehenders (cf., Daneman & Carpenter 1980; Lesgold & Perfetti 1978).

A second interpretation of the speed of comprehension is that comprehenders map acoustic information directly onto a relevant discourse representation without any independent representation of sentence information. One *direct model* asserts that comprehenders use expectations based on prior knowledge to by-pass the computation of sentence structure (Bower, Black, & Turner 1979; Schank & Birnbaum 1984). Thus, understanding a sentence is more a matter of confirming expectations than forming syntactic hypotheses. On this view, less skilled comprehenders understand text less well because they fail to recognize the schemata that underlie texts (Schank 1982). If "schema perception theory" is correct, sentence structure should have a <u>smaller</u> role as comprehension skill increases.

Another direct model acknowledges the use of syntactic information during comprehension, but does not distinguish its role from that of discourse-level information. For example, a decrease in either syntactic or semantic constraints may cause comprehenders to build a new structure for subsequent words. Such a "processing shift" might cause comprehenders to forget the surface form and meaning of previous material. According to this model, less skilled comprehenders use sentence- and discourse- constraints less effectively than more skilled comprehenders (Gernsbacher, Varner, & Faust 1990). This model predicts that surface properties of sentences will have a <u>larger</u> role as comprehension skill increases.

A third interpretation of the immediate effects of semantic information emphasizes the on-going interactions between representations at different levels. According to the *representational model*, comprehenders compute form and meaning simultaneously and independently. Access to semantic vs. syntactic constraints depends on how comprehenders shift attentional resources between representations at different levels. Using a dual discourse-judgment and speaker-monitoring task, Townsend & Bever (1991) found that increased sentence-level constraints reduced the time to detect a change of speaker, but increased discourse-level constraints <u>increased</u> it. Thus, making a discourse-level task easier decreases accessibility to acoustic information. Other studies have shown a functional distinction between representations at different levels (Townsend 1983; Townsend, Ottaviano, & Bever 1979; Townsend & Ravelo 1980; Townsend & Saltz 1972). For example, Townsend & Bever (1978) found that the position of *up* in fragments like *Pete called up his aunt each...* versus *Pete called his aunt up each...* influences probe recognition for UP more in some clauses than in others, depending on the discourse-level role of the clause. In those cases in which the position of the target word had a large effect, subjects were slower to judge that USING THE TELEPHONE is synonymous with the clause. These results suggest that representations of word order and meaning compete for attention. Some aspects of comprehension skill may depend on how comprehenders focus attentional resources on already computed representations.

**Hypotheses**

We tested serial, direct, and representational models by comparing skilled and average comprehenders at two age levels. We assessed their access to the meaning and form of sentences in immediate and delayed tests.

Two tasks assessed comprehenders' immediate representation of isolated sentences. In a "meaning probe" task subjects heard a sentence fragment that ended just before the end of a clause, and then a short verb-object phrase (Townsend & Bever 1978). The subjects' task was to say as quickly as possible whether the phrase was similar in meaning to the sentence fragment. In a "word probe" task subjects also heard a sentence fragment, but then a single probe word. Now the task was to say whether the probe word had occurred in the sentence fragment. By varying the location of the target word, we assessed subjects' sensitivity to word order.

If average comprehenders' inefficient lower-level computations prevent them from computing meaning, their deficits will be greater on the meaning probe task. Since there is no linguistic context to activate a schema, a schema perception model predicts that skilled and average comprehenders will perform similarly. If average comprehenders engage in more processing shifts than skilled comprehenders, they should access meaning and word order less effectively than skilled comprehenders. If representations of form and meaning compete for attention, comprehenders who show increased access to word order will show decreased access to meaning, and vice versa.

Two tasks assessed memory for the form and meaning of sentences in text. Subjects heard texts at two different speeds. After each text, subjects took two tests that assessed their recognition of the wording and meaning of sentences.

If average comprehenders inefficient lower-level processes normally disrupt their computation of meaning, a faster rate of speech will increase these deficits. If the presence of a linguistic context allows skilled comprehenders to by-pass syntactic processes, skilled comprehenders will make more errors than average comprehenders on recognizing details about word order. If average comprehenders are less sensitive to constraints of any type, they will make more errors than skilled comprehenders on recognizing the form and meaning of sentences from text. If comprehenders can compute meaning independently of form, a faster rate of speech will not impair their recognition of meaning.

## Method

**Subjects**

The "skilled adults" were twelve college students with Verbal Scholastic Aptitude Test scores of at least 610. The "average adults" were twelve college students with VSAT scores of 400-560. The "skilled children" were twelve 10-13 year old school children who were reading at least one grade level above their current grade, based on the Verbal Standardized Achievement Test. The "average children" were twelve 10-13 year old children who were reading within one grade of their current grade. Skilled and average subjects within age groups were matched on quantitative achievement test scores. All subjects spoke English as their first language, and all were right-handed. Half the subjects in each group were male and half had no left-handed relative.

**Procedure**

The subjects were tested individually. All subjects received the text memory task before the immediate probe tasks. The order of presentation of the immediate probe tasks was counter-balanced. Matched subjects received identical experimental materials and presentation orders.

**Immediate Representation.** For the meaning probe task subjects said as quickly as possible whether a phrase probe was consistent or inconsistent with the meaning of the sentence fragment. For the word probe task subjects said as quickly as possible whether a word probe had appeared in the sentence fragment. Two sets

Table 1
Mean Response Times (ms) on Meaning Probe Task

| | Skilled | Average | Difference |
|---|---|---|---|
| Adults | 2230 | 2124 | -106 |
| Children | 2267 | 2808 | 541 |

Table 2
Mean Response Times (ms) on Word Probe Task

| | Adults | | | Children | | |
|---|---|---|---|---|---|---|
| | Skilled | Average | Mean | Skilled | Average | Mean |
| Early Target | 1561 | 1446 | 1504 | 1551 | 1837 | 1694 |
| Late Target | 1800 | 1411 | 1606 | 1451 | 1951 | 1701 |
| Early Target Advantage | 239 | -35 | 102 | -100 | 114 | 7 |
| Group Mean | 1681 | 1429 | | 1501 | 1894 | |

of six lists of materials were prepared. Subjects received a list from one set for the meaning probe task and a list from the other set for the word probe task. Each list appeared as both meaning probe and word probe across subjects. Each list had six critical positive trials in which the fragment ended before the last word of an initial clause. There were 18 trials that either interrupted the final clause or required a "no" response.

The critical fragments were 9-11 words long. These fragments contained a word that could occupy two different positions without changing the fragment's meaning. At least two words separated the "early" and "late" positions of the target word and the average separation was 2.9 words. An independent group of college students (N=31) rated how similar the meanings of the phrase probes and the fragments were. On a five point scale, the average ratings were 3.65 for positive trials and 1.23 for negative trials.

A male speaker recorded the fragments, and a female speaker recorded the probes. Each fragment ended with a 50 ms, 500 Hz tone. The tone signalled the end of the fragment and triggered a Hunter ms timer that stopped with the subject's vocal response. The probe began 333 ms after the tone. Half the subjects heard the fragments and probes in the right ear, and half heard them in the left ear.

**Memory Representation.** Each subject heard one narrative and one expository text. The texts for adults were selected from a Scholastic Aptitude Study Guide, and those for children from the McCall-Crabbs Standard Test Lessons in Reading. The texts were 550 words long for adults and 200 words long for children. Half the subjects in each group heard the texts in the right ear, and half in the left. Each subject heard one text at a "normal" rate and then one at a "fast" rate. The presentation rates were 3.1 versus 6.2 words per sec for adults and 2.2 versus 4.4 words per sec for children. The tapes for the speeded conditions were prepared by a speech compression system that does not change pitch. Following each text, subjects received two tests. The first test consisted of two-choice meaning recognition questions such as *The writer says that the fog was (a) rising and falling, or (b) like a wall.* The second test consisted of sentences to be classified as having appeared or not appeared in the text. Half of these involved changes in the wording of a text sentence without changing its meaning (Sachs 1967), as in *The carrier's deck...* vs. *The deck of the carrier...* Distractor items were semantically plausible for the text. Each test had 12 items for adults, and 6 items for children. Test items were presented at the "normal" rate of speech.

## Results

**Immediate Representation.** Tables 1 and 2 present the average response times for positive initial clause trials on the immediate probe tasks. Errors occurred on 1.7% of the trials for both tasks. Response times for these trials were replaced by the cell means for correct responses. Tables 1 and 2 show that average adults respond faster than skilled adults on both tasks, but that these skill differences are reversed for children. The skill by age interaction was significant, $F\ (1, 44) = 10.4$, $p < .01$. These results disprove the claim that linguistic contexts cause skill differences. They also disprove the claim that average adult comprehenders access words, meaning, or both, more slowly than skilled adults.

The size of the response time advantage for early targets over late targets assesses sensitivity to the left-to-right order of words in the immediate representation of sentences (Townsend & Bever 1978). The Early Target Advantage in the third row of Table 2 suggests that adults overall are more sensitive than children to the left-to-right order of words, that skilled adults are more sensitive than average adults, and that average children are more sensitive than skilled children. The interaction between age, skill, and target position was significant, $F\ (1, 44) = 4.83$, $p < .05$. Assuming that children as a whole are less skilled comprehenders than adults, it is clear that there is no direct relationship

Figure 1A
Adults' Errors on Memory for Text

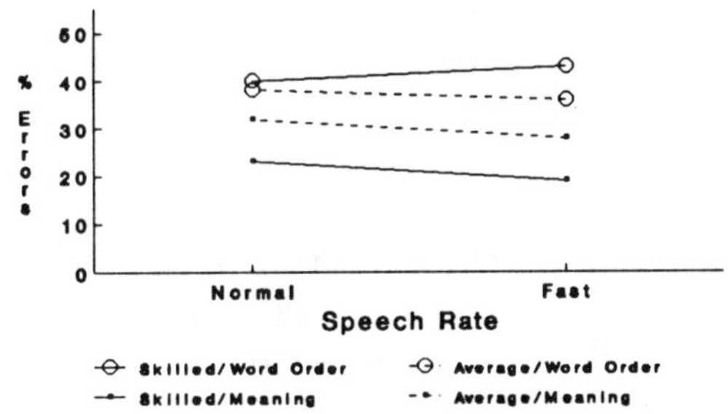


Figure 1B
Children's Errors on Memory for Text

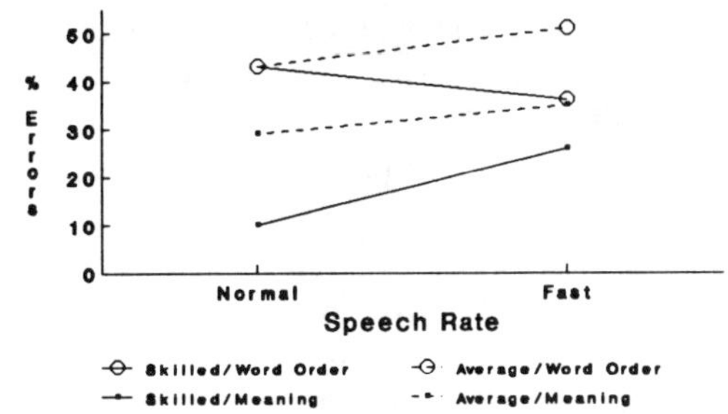


between comprehension skill and immediate access to surface form. However, access to meaning (Table 1) and the Early Target Advantage (Table 2) are related inversely: When a group quickly accesses meaning, it has poorer access to word order, and vice versa.

**Memory Representation.** Figures 1A and 1B show the percentage of errors on memory for word order, shown with circles, and memory for meaning, shown with dots. Errors were more frequent for word order recognition (44%) than for meaning recognition (27%), $F$ (1, 44) = 55.7, $p$ < .001. Skilled comprehenders, shown with solid lines, made fewer errors overall than average comprehenders, shown with dashed lines, (30% vs. 37%), $F$ (1, 44) = 5.86, $p$ < .05. However, Figure 1A shows that skilled adults actually made more errors than average adults in recognizing word order changes.

Figure 1A shows that the fast rate decreases meaning recognition errors equally for skilled and average adults. Figures 1A and 1B show that the fast rate impairs performance more for meaning recognition than for word order recognition only for skilled children. The interaction between age, skill, task, and rate was significant, $F$ (1, 44) = 4.57, $p$ < .05. The results for skilled adults suggest that reducing the time that is available for comprehension can improve access to more abstract levels of representation.

## Discussion

As expected, skilled comprehenders' memory for meaning was better than that of average comprehenders. However, there were several cases in which average comprehenders did better than skilled comprehenders. These results present difficulties for popular views about the organization of language comprehension processes.

Serial models claim that comprehension involves the use of shared processing resources to construct representations at successively higher levels of structure. Less skilled comprehenders understand less well because they have fewer of these resources. Serial models predicted that the fast speech rate would increase the size of skilled comprehenders' advantage over average comprehenders more for recognition of meaning than for recognition of word order. This prediction followed from the serial assumption that computations of meaning depend on first computing a syntactic representation that is sensitive to word order. The serial prediction was not confirmed: the only case in which the fast rate increased skilled comprehenders' advantage was in children's recognition of word order. The fast rate actually reduced the size of skill differences for children's recognition of meaning. Thus, skill differences do not depend on limitations in the resources that are available for serially-ordered computations.

A schema perception model (Schank 1982) claims that as comprehenders learn more about the world, they map linguistic information directly onto memory schemata without determining sentence structure. According to this model, skilled comprehenders should have done more poorly than average comprehenders on remembering the syntactic details of sentences in text. Our text memory results confirmed this prediction only for adults, not for children. In addition, skilled and average comprehenders should have performed similarly on the immediate probe tasks, which provided no linguistic context to activate a schema. We found several differences between skilled and average comprehenders in immediate probe tasks.

The view that skilled comprehenders use sentence and discourse constraints more effectively than average comprehenders predicted that skilled comprehenders would access word order and meaning faster and more accurately (Gernsbacher, Varner, & Faust 1990). We found that skilled children recognized the wording of sentences from text better than average children did, and they responded faster overall on immediate tests of meaning. However, skilled children were less sensitive than average children to the order of words on the immediate word probe task. Since all of these results were reversed for skilled versus average adults, access to word order and meaning is not directly related to comprehension skill.

Our results suggest that comprehenders compute representations of word order and meaning independently and in parallel. However, those representations compete for attention. Comprehenders at different stages of proficiency may have different "habits" for focusing attention on word order versus meaning. The most salient results that support this model are:

*Independent Computations.* The fast rate reduced skilled adults' errors on recognition of meaning and increased their errors on recognition of word order. This result suggests that their computations of meaning do not rely on computations of word order.

*Competition.* On immediate tests skilled adults showed greater sensitivity to word order than did average adults, but they were slower than average adults in accessing meaning. This result suggests that focusing on word order draws attention away from sentence meaning, and vice versa.

*Proficiency.* Children also showed an inverse relation between access to word order versus meaning, but the direction of these skill differences was opposite that for adults. Skill in childhood involves an immediate focus on sentence meaning, but skill in adulthood involves an immediate focus on word order.

This model differs from previous models by distinguishing different kinds of representations and different kinds of processing resources. It also differs from the modular view that distinct physiological processes prevent syntactic and semantic processes from sharing both information and processing resources (cf., Fodor 1983).

## Acknowledgement

This research was supported by NIE G-79-0040, NSF BNS-8120463, and the Separately Budgeted Research Program, the Sabbatical Leave Committee, and the Institute for Critical Thinking at Montclair State College. We are grateful to Marcia Johnson and Princeton University for providing office space during the preparation of this report.

References

Bower, G., Black, J., & Turner, T. 1979. Scripts in memory for text. *Cognitive Psychology* 11:177-220.

Daneman, M., & Carpenter, P. A. 1980. Individual differences in working memory and reading. *Journal of Verbal Learning and Verbal Behavior* 19:450-466.

Fodor, J. A. 1983. *The modularity of mind.* Cambridge, Mass.: MIT Press.

French, P. 1981. Semantic and syntactic factors in the perception of rapidly presented sentences. *Journal of Psycholinguistic Research* 10:581-591.

Gernsbacher, M. A., Varner, K. R., & Faust, M. 1990. Investigating differences in general comprehension skill. *Journal of Experimental Psychology: Learning, Memory, and Cognition* 16:430-445.

Lesgold, A. & Perfetti, C. 1978. Interactive processes in reading comprehension. *Discourse Processes* 1:323-336.

Marslen-Wilson, W. D. 1975. Sentence perception as an interactive parallel process. *Science* 189:226-228.

Sachs, J. S. 1967. Recognition memory for syntactic and semantic aspects of connected discourse. *Perception and Psychophysics* 2:437-442.

Schank, R. C. 1982. *Reading and understanding: Teaching from the perspective of artificial intelligence.* Hillsdale, N.J.: Erlbaum.

Schank, R., & Birnbaum, L. 1984. Memory, meaning, and syntax. In T. Bever, J. Carroll, & L. Miller (Eds.) *Cognitive, philosophical, and computational foundations of language* 209-252. Hillsdale, N.J.: Erlbaum.

Stine, E. L., Wingfield, A., & Poon, L. W. 1986. How much and how fast: Rapid processing of spoken language in later adulthood. *Psychology and Aging* 1:303-311.

Townsend, D. J. 1983. Thematic processing in sentences and texts. *Cognition* 13:223-261.

Townsend, D. J. & Bever, T. G. 1978. Inter-clause relations and clausal processing. *Journal of Verbal Learning and Verbal Behavior* 17:509-521.

Townsend, D. J. & Bever, T. G. 1991. The use of higher-level constraints in monitoring for a change in speaker demonstrates functionally-distinct levels of representation during discourse comprehension. *Language and Cognitive Processes* 6:49-77.

Townsend, D. J., Ottaviano, D., & Bever, T. G. 1979. Immediate memory for words from main and subordinate clauses at different age levels. *Journal of Psycholinguistic Research* 8:83-101.

Townsend, D. J., & Ravelo, N. 1980. The development of complex sentence processing strategies. *Journal of Experimental Child Psychology* 29:60-73.

Townsend, D. J. & Saltz, E. 1972. Phrases vs. meaning in the immediate recall of sentences. *Psychonomic Science* 29:381-384.

# Providing Natural Representations To Facilitate Novices' Understanding in a New Domain: Forward and Backward Reasoning in Programming *

**J. Gregory Trafton** and **Brian J. Reiser**
trafton@clarity.princeton.edu
reiser@princeton.edu

Cognitive Science Laboratory
Princeton University

## Abstract

In many domains, novices exhibit a bias in the direction in which they reason about problems. Earlier studies of LISP programmers using a graphical representation suggested that novice LISP programmers tend to reason forward, working from initial input data toward the goal. We examined novice programmers learning LISP using the GIL programming tutor and manipulated the direction subjects were allowed to reason (forward, backward, or free). Subjects who were required to work backwards (from goal toward givens) exhibited more difficulty solving the problems than subjects working forward or subjects left free to chose their direction. Backward subjects required more time to solve problems, made more errors, and required more time to plan each solution. We suggest that these effects and preferences occur because forward reasoning is more congruent with the way novices reason about computer programs, resulting in an increased working memory load for subjects required to work backward.

## Introduction

Novices reason in a different fashion from those who have acquired more expertise in a domain. Experts are better able to recognize the important structural features of a problem and initiate an appropriate solution procedure, while novices require more search to construct a solution (e.g., Chi, Feltovich, & Glaser, 1981). In some domains, this results in a noticeable contrast in the *direction* of the reasoning steps taken. Larkin and her colleagues have demonstrated that physics experts reason *forward* from given information, applying appropriate equations to infer new quantities until the goal is reached (Larkin, McDermott, Simon, & Simon, 1980). In contrast, novices use means ends analysis, starting with a desired unknown and searching for equations that include that variable, then generating new subgoals to fill in terms unknown in that equation, and so on, thus reasoning *backward* from the goal toward the given information.

In evaluating the source of this intriguing difference, it is important to consider how novices reason in different domains. Rist (1989) found that novice Pascal programmers tended to write their solutions in essentially forward fashion, writing each section of the program from the beginning of the block of code to the end, rather than working backward from the goal of the block to the initial statements. Anderson, Conrad, and Corbett (1989) found that novice LISP programmers tend to construct their solutions "top-down, left-to-right." This corresponds essentially to backward reasoning, constructing the final actions taken on the input data before the initial actions.

What determines the direction in which novices reason in a domain? It may be merely a preference due to students' naive judgments about how to construct solutions. Alternatively, novices may reason in a particular direction because it simplifies the problem solving. Thus, novices may reason backward in physics because this is the simplest way they can select among possible inferences, since they do not have the problem schemas that enable forward inference (Chi et al., 1981).

Research on novice programmers provides an interesting case of forward and backward reasoning. Reiser, Kimberg, Lovett, and Ranney (1991) argued that the graphical representation employed in GIL (Graphical Instruction in LISP), an intelligent tutoring system for LISP programming, provides a more natural external representation for novices to record their reasoning than the traditional text form of programs. Rather than simply retrieving sequences of functions to achieve particular programming goals, novices tend to reason about the behavior of the relevant functions on an example to construct a chain of operations that achieves the desired result. Thus, GIL is designed so that students build a program by constructing a graph

*This research was supported in part by contract MDA903-87-K-0652 from the Army Research Institute, by research grants from the James S. McDonnell Foundation, and the Xerox Corporation University Grant Program. The views and conclusions contained in this document are those of the authors and should not be interpreted as necessarily representing the official policies, either expressed or implied, of these institutions.

of program constructs and the data each manipulates. Students specify the data input to each function in their program and the data that results from it. A complete program consists of a graph specifying how a chain of functions transforms the input data to achieve a particular type of output (see Figure 1).

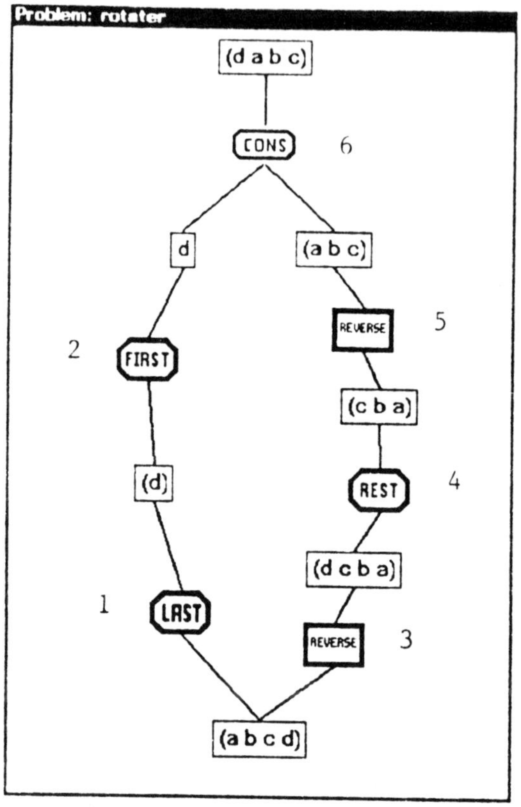


```
(defun rotater (lis)
  (cons (first (last lis))
    (reverse (rest (reverse lis)))))
```

Figure 1: A completed GIL graph and the corresponding text form of the solution. The original GIL input is the list *(a b c d)* at the bottom of the screen; the goal data is the list *(d a b c)* at the top. The modal sequence of solution steps is indicated by the number shown next to each function.

Students are free in GIL to reason at any step either from the goal data toward the original inputs (a backward step) or from the inputs toward the goal data (a forward step). Reiser, Ranney, Lovett, and Kimberg (1989) found that novices exhibited a strong bias for forward reasoning (95% forward steps, 5% backward steps). Reiser et al. argued that novices can reason more easily about the behavior of programs by working forward from the input data rather than backward from the goal. Interestingly, the order of reasoning exhibited by GIL subjects is opposite that of the surface form of the solution. For example, the first function encountered in the body of the text form (the "append") is in fact the last component of the solution assembled by GIL subjects (see figure 1). If students' reliance on forward steps in GIL accurately represents their reasoning, then novices do not appear to plan their solution in the order in which the functions appear in the text form of the solution. If students are led to enter their code in an outside-in left-to-right sequence, as students working with the text representation may be, they may be led to construct a chain of functions for each subgoal before entering it into their solution. Reiser et al. argue that GIL's scaffolding of forward reasoning is one reason why students solve problems more easily with GIL than standard text LISP.

The preference of the GIL students for working forward in LISP appears to conflict with Anderson et al.'s finding that their subjects tended to construct programs using backward reasoning. One possibility is that subjects were not comfortable enough with the editing freedom Anderson et al.'s tutor provided to construct their code from the inside out, whereas GIL's graphical representation is more neutral in biasing between forward and backward steps. Indeed, it seems reasonable to expect that novices will feel compelled to enter the components of their solutions in the "surface order," the order in which these components appear in the final form of the solution. Such a concern with a mismatch between natural ways of reasoning about a problem and the format of a completed solution have been the motivation for the graphical formats used in a number of interactive learning environments (Anderson, Boyle, & Reiser, 1985; Collins & Brown, 1988).

If novices indeed reason more naturally in one direction than another, then it should be possible to facilitate their reasoning by providing a learning situation that supports that direction of reasoning. The initial evidence in programming relies only on subjects' preferences. The present study attempts to explicitly evaluate whether the direction of reasoning affects the difficulty of learning to solve programming problems. We used the GIL tutoring system because it makes it possible to restrict the direction in which subjects' reason, and it allows fine-grained analyses of the time course and errors in subjects' reasoning.

Table 1 summarizes the three versions of our GIL tutor used in this experiment.

Table 1: Versions of GIL

| | |
|---|---|
| *Forward* | Only forward steps permitted. From input data toward goal data. |
| *Backward* | Only forward steps permitted. From goal data toward input data. |
| *Free* | Forward or backward steps allowed throughout. |

The Free condition provides an opportunity to observe how novices choose to reason. The Forward and

Backward conditions enable us to examine whether there is an additional load placed on problem solving by requiring students to reason in the backward direction. We expect the Free and Forward conditions to be similar, and problem solving to suffer in the Backward condition.

## Method

Subjects worked through two chapters of an introductory LISP programming textbook and solved problems assisted by GIL.

*Subjects:* The subjects were 30 undergraduate paid volunteers from Princeton University and other nearby colleges. All had taken no more than one semester of computer programming and had no knowledge of LISP. Ten subjects were assigned to each of the Free, Forward, and Backward conditions, balancing programming ability across conditions, using Math SAT as the predictor of programming ability (Mayer et al., 1986). Math SAT scores ranged from 420 to 760, with a median score of 625 and means of 624, 627, and 627 for Free, Forward, and Backward respectively.

*Instructional Materials:* The textbook explaining data structures and LISP functions was used for all conditions, with several slight wording changes to describe the process of chaining functions together. The Free condition text was neutral, explaining how one might chose to construct a chain either from the givens toward the goal or the reverse, and the Backward and Forward texts described the method for sequencing functions appropriate to that condition. All subjects solved the same fourteen problems, including one problem worked in conjunction with the Experimenter used to demonstrate how to construct programs in GIL.

*Learning Session:* Each subject was given the first of three sections of the text. After reading the first section, subjects were led through a brief demonstration to familiarize them with the learning environment. The Forward condition subjects saw only forward steps; the Backward condition subjects saw only backward steps, and subjects in the Free condition saw steps in both directions. After the demonstration, subjects worked through the remaining text and problems at their own pace.

*Posttest:* Subjects were given a posttest following the learning session, consisting of three near transfer coding problems. Subjects solved the problems on the computer using the same graphical interface with some additional editing features, but received no feedback. The computer did not constrain the subjects' direction of work on the posttest.

## Results

We expected subjects in the Free condition to work primarily forward, so we used a planned comparison to contrast performance in the Forward and Free conditions with the Backward condition. In addition, we performed a second set of analyses to examine whether the performance in the learning conditions differed depending on ability. Subjects below the median SAT score (420-620) were classified as the "Low SAT" group and those above the median (630-760) were classified as the "High SAT" group. We performed an analysis of variance with SAT level and learning condition as independent variables.

We examined the number of forward and backward steps in the Free subjects. As expected, these subjects strongly preferred to work forward ($t(9) = 14.7, p < .001$). Subjects worked 95% of their steps forward and 5% of their steps backward. This finding supports the use of a planned comparison to contrast Forward and Free with Backward subjects.

To examine whether the direction of reasoning affected the difficulty of solving the assigned problems, we first considered solution time. This was defined as the total time required to solve the thirteen assigned problems, not including the demonstration problem or time spent reading the instructional text. Backward subjects took longer to solve the assigned problems than subjects working Forward or Free (47.6 min vs 31.5 and 33.4 min; $F(1,26) = 9.08, p < .01$). Figure 2 shows solution time partitioned into the time spent on correct steps, error steps, initial planning, and cancelled steps. Not surprisingly, Low SAT subjects took longer than High SAT subjects (41.7 and 33.2 min, respectively; $F(1,24) = 3.61, p < .10$). There is a suggestion that the effect of direction is stronger for the lower ability subjects, although this is not statistically reliable ($F(2,24) = 2.23, p = .13$ for the interaction).

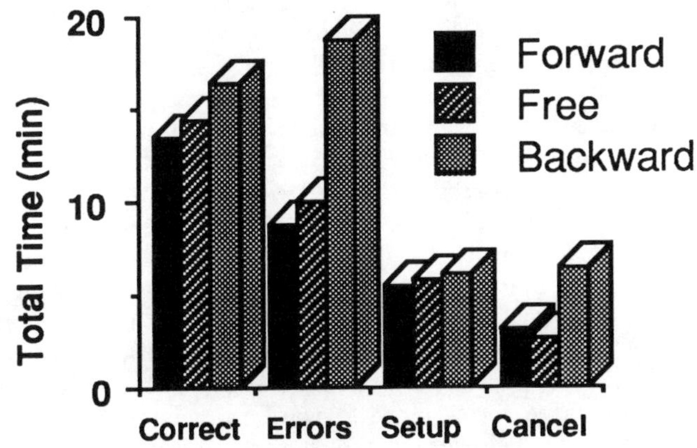


Figure 2: Distribution of learning session time. Time shown is the total amount of time spent on each event type across all 13 problems. Categories are: correct steps, recovery from errors, setting up a plan, and cancelled steps.

There are several possible reasons why subjects in the Backward condition may have taken longer than the other subjects. First, we examined their errors, which can be costly to repair. When an error is made,

GIL prints an explanatory error message, and subjects must read the message and try to fix the error. Indeed, as Table 2 suggests, it appears that the Low SAT subjects in the Backward condition made more errors than either the Forward or Free subjects, while errors produced by the High SAT subjects were not affected by the learning condition ($F(2,24) = 2.89, p < .10$ for the interaction). Differences in solution time may also result from the time subjects spent repairing errors. Subjects who have more difficultly constructing and implementing a plan may spend more time on errors that they do make. Indeed, Backward subjects took more time to repair each error than did Forward and Free subjects (115 vs. 68 and 75 sec; $F(1,26) = 4.36, p < .05$). Furthermore, Backward subjects also spent more time aborting steps ($F(1,21) = 4.8, p < .05$), supporting the hypothesis that these subjects found reasoning more difficult.

Table 2: Mean number of errors in the learning session

| | Low SAT | High SAT |
|---|---|---|
| *Forward* | 8.2 | 8.2 |
| *Free* | 8.0 | 9.2 |
| *Backward* | 17.6 | 6.0 |

Interestingly, subjects in the Backward condition spent more time planning or setting up each problem before they started working on it ($F(1,26) = 7.01, p < .05$). Although the differences are small (mean planning time 37 sec vs. 26 and 28 sec), this suggests Backward subjects found constructing an overall plan more difficult than other subjects.

Finally, we examined subjects' performance on the posttest. Backward subjects spent more time on the posttests than the Forward or Free subjects (19.7 vs. 14.9 and 16.3 min; $F(1,25) = 4.78, p < .05$). However, their scores on the posttest problems did not differ ($F(2,22) = 1.06, n.s.$). The posttest allowed all subjects to work freely, so it was possible to examine the direction subjects chose to work. Not surprisingly, the Forward and Free conditions worked almost exclusively forward as they did during the learning session. However, the Backward subjects performed almost three times more forward than backward steps. Although they were unaccustomed to working forward, they chose to do so, and this may have resulted in their longer solution times on the posttest.

## Discussion

In the domain of LISP programming, subjects given the choice of which direction to work prefer to work forward. This preference is apparently motivated by the difficulty in reasoning backward in this domain. Subjects in the Backward condition took longer to solve the problems than did subjects who worked forward. This difference is primarily due to the number of errors these subjects made and the difficulty they had in repairing them. The effect appears to arise predominantly in the lower ability subjects; the higher ability subjects are less hampered by requiring them to reason backward. The Backward subjects also had more difficulty in constructing and implementing a plan, as evidenced by their increased initial planning times and greater tendency to abort steps.

The preference for forward reasoning appears to be a strong one. Even after working all the learning session problems using backward steps, the Backward condition subjects still chose to perform a majority of their steps on the posttest using forward reasoning. If this strong preference holds in the more advanced parts of the curriculum in this domain, then these subjects' problem solving during the learning session (in which they were forced to use backward reasoning) may not have prepared them for future problem solving as well as the instruction the Forward and Free groups received. Indeed, the Backward condition subjects took longer to solve the posttest problems, presumably because these subjects were relying on less practiced techniques of forward reasoning.

Taken together, these results support the conclusion that, unlike physics, in which novices tend to use backward reasoning, in LISP programming, backward reasoning is more difficult than forward reasoning. Subjects' reliance on forward reasoning is not merely a preference — subjects allowed to use this type of reasoning perform better than subjects who are required to use backward reasoning, despite the fact that backward reasoning better corresponds to the order of solution components in the traditional form of LISP programs.

We consider three types of explanations for these results.

*Naive Causal Models:* Novices tend to reason about how programs behave as they try to construct a program to achieve a particular goal. Functions have an underlying directionality — they take in input data and return output data. Novices can understand how functions embed by viewing an embedded function call as a causal chain, in which the output of one function becomes input to another. This type of causal reasoning naturally starts with the first inputs, the initial "cause," and progresses toward the final result, rather than reasoning backward from the desired output to select a function and necessary input that yields that result. The temporal directionality of operators may have some effect on the direction people choose to reason in other domains. For example, chess operators are typically applied in a forward direction (e.g., castling will develop the rook) rather than backward (e.g., to develop the rook, castle), although both of these processes may play some part in the reasoning process.

*Constraint of the Problem Space:* Because functions are many-to-one mappings, the search of the problem space is more constrained when reasoning forward than when reasoning backward. That is, a function operating on an input has a unique output, but a function could take many possible inputs that would produce a desired output. Furthermore, there are a greater number of applicable functions when reasoning backward than forward. Thus, students can select between fewer potential choices when reasoning forward. This factor may also explain why novices rely upon backward reasoning in physics. In physics, means-ends analysis is profitable because there is only one goal quantity (e.g., velocity), and there are fewer equations that contain that quantity than the equations that could be used to chain forward from all the given information. For other domains, the direction of work with the fewer applicable operators should be preferred.

*Complexity of Planning:* There are several reasons to expect the planning to be simpler when reasoning forward. First, the subgoals are more independent when reasoning forward. Many problems involve combining separately obtained objects. When reasoning backward, it is necessary to know what form the inputs will be to know how to combine the subgoals. This often requires further look-ahead, because the form of the inputs to the step will determine which function to use to combine them. In contrast, when reasoning forward, it is necessary only to determine that a particular subgoal is needed, and then one can reason to obtain that data independently of whatever else will be needed to combine with that subgoal. Students can then make greater use of opportunistic reasoning. The student may realize constraints on a subgoal of the program when building a chain to achieve its sibling subgoal. In contrast, when backward reasoning begins with combining subgoals together, it is necessary to commit to a particular combination of subgoals. Many domains have operators that combine objects in a similar fashion (e.g., SAS or ASA for geometry). This hypothesis predicts that reasoning forward should be easier for those domains that have these types of operators.

A related point concerns the ease of determining whether a particular inference is profitable. GIL students reason using a concrete example. Candidate forward inferences can be evaluated by comparing the results and determining whether each would be useful for obtaining the goal data, which typically entails determining whether the candidate result is a component of the goal. Backward reasoning requires determining whether a particular input to a step can easily be obtained from the original inputs. It is typically easier to see whether a particular object is a component of the goal than to determine whether an object is of the form that can easily be obtained from the inputs.

We suggest that these factors together explain why forward reasoning is simpler in this domain. The greater look-ahead required, the larger number of alternatives to be considered, and the difficulty in mentally simulating the execution of a program when reasoning backward all contribute to an increased working memory load, which results in greater difficulty in constructing and carrying out a plan. It is also possible that the Backward subjects are indeed reasoning forward, but just entering their solutions in the backward sequence as required by the tutor. This too would result in an increased memory load, because subjects would have to construct an entire chain of reasoning before entering the outermost (first backward) step.

In future work, we plan to examine how features of the domain can be used to explain the direction that novices reason, and how problem solvers change their direction of reasoning as they gain expertise in a domain.

*Acknowledgements:* We are grateful to Michael Ranney for many contributions to this research, and to Adnan Hamid for programming assistance.

## References

Anderson, J. R., Boyle, C. F., & Reiser, B. J. (1985). Intelligent tutoring systems. *Science*, 228, 456–462.

Anderson, J. R., Conrad, F. G., & Corbett, A. T. (1989). Skill acquisition and the LISP tutor. *Cognitive Science*, 13, 467–505.

Chi, M. T. H., Feltovich, P., & Glaser, R. (1981). Categorization and representation of physics problems by experts and novices. *Cognitive Science*, 5, 121–152.

Collins, A. & Brown, J. S. (1988). The computer as a tool for learning through reflection. In H. Mandl & A. Lesgold (Eds.), *Learning issues for intelligent tutoring systems.* New York: Springer-Verlag.

Larkin, J. H., McDermott, J., Simon, D., & Simon, H. A. (1980). Models of competence in solving physics problems. *Cognitive Science*, 4, 317–345.

Mayer, R. E., Dyck, J. L., & Vilberg, W. (1986). Learning to program and learning to think: What's the connection? *Communications of the ACM*, 29, 605–610.

Reiser, B. J., Kimberg, D. Y., Lovett, M. C., & Ranney, M. (1991). Knowledge representation and explanation in GIL, an intelligent tutor for programming. In J. H. Larkin & R. W. Chabay (Eds.), *Computer assisted instruction and intelligent tutoring systems: Shared issues and complementary approaches.* Hillsdale, NJ: Erlbaum.

Reiser, B. J., Ranney, M., Lovett, M. C., & Kimberg, D. Y. (1989). Facilitating students' reasoning with causal explanations and visual representations. In D. Bierman, J. Breuker, & J. Sandberg (Eds.), *Proceedings of the Fourth International Conference on Artificial Intelligence and Education* (pp. 228–235). Springfield, VA: IOS.

Rist, R. S. (1989). Schema creation in programming. *Cognitive Science*, 13, 389–414.

# A Schema-based Approach to Cooperative Behavior[1]

**Elise H. Turner**
Department of Computer Science
University of New Hampshire
Durham, NH 03824
E-mail: eht@cs.unh.edu

**Roy M. Turner**
Department of Computer Science &
Marine Systems Engineering Laboratory
University of New Hampshire
Durham, NH 03824
E-mail: rmt@cs.unh.edu

**Abstract**

Agents can rely on the patterns in the world to make their problem solving more efficient. When working with others, agents can also rely on patterns – patterns for communication and group behavior. We discuss how these patterns may be captured in schemas. We present two types of schemas: *procedural schemas* which suggest a course of action for a specific situation, and *contextual schemas* which contain knowledge about specific kinds of problem solving. Both of these types of schemas affect an agent's ability to solve problems and communicate. Both types of schemas also guide the coordination of the groups working together to solve problems. In this paper, we focus particularly on the ways in which a schema-based approach can help agents to work together by integrating their individual problem solving with the constraints of coordinated behavior.

Cooperative distributed problem solving (CDPS) (Durfee *et al.*, 1989) allows artificial intelligence systems to reap the benefits of teamwork. CDPS systems can tackle problems not readily solvable by single agents, such as simultaneous monitoring of environmental conditions in a large area or control of multiple autonomous agents moving through an area and carrying out tasks. From the standpoint of cognitive science, CDPS research sheds light on communication and cooperation between intelligent agents.

For an agent to work well as a collaborator, it must be a competent individual problem solver as well as a helpful colleague. As an individual, the agent must be able to achieve goals (some its own, some related to its shared task) efficiently in a complex, changing world. This means the agent must be able to plan quickly and flexibly, tailoring its behavior to fit its situation. As a collaborator, it must communicate effectively with others and understand its role in the organization of the CDPS system. Its interactions with others must be guided by convention so that all agents' actions can be easily understood by others.

In this paper, we discuss the role of schemas in a design for integrating problem solving, communication, and cooperation. Only parts of this design have been implemented at this time. We adopt a *schema-based* approach to cooperative distributed problem solving, based on our prior work (Turner, 1990; Turner & Cullingford, 1989b; Turner, 1989a; Turner, 1989c), which addresses both of

[1]This work was supported in part by NSF grant number BCS-8905888 (under agreement number 900622CR001-IT1001A with the University of Hawaii PICHTR), ONR grant N00014-89-J-3074, and a grant from the Hubbard Marine Program Endowment. The authors are listed in alphabetical order.

Figure 1: AUV with goal to pick up an object.

these needs. *Procedural schemas*, much like compiled plans, allow reuse of previous planning effort, either the agent's own or another agent's (e.g., a human's). *Contextual schemas* provide knowledge about kinds of problem-solving situations, which allows an agent to automatically behave appropriately in its current context. Procedural and contextual schemas guide an individual's autonomous behavior as well as its behavior within a group.

## An Example of Schema-based Cooperative Behavior

We have chosen the domain of autonomous underwater vehicles (AUVs) for this research. Ultimately, AUVs will replace humans for dangerous or expensive tasks. Many of these tasks either benefit from or require more than one AUV, for example: underwater photography, in which a long baseline is needed between camera and light (Jaffe, 1988); and large-scale environmental monitoring, which benefits from many agents each monitoring a small area. Eventually, our ideas will be tested aboard the University of New Hampshire Marine Systems Engineering Laboratory's EAVE AUVs (Blidberg & Chappell, 1986). The examples in this section are from the AUV domain.

Suppose that an AUV is at position A in Figure 1 and receives a new goal to move the rock at position R. The AUV could simply go to position R and pick up the rock. Here the AUV would use a procedural schema that tells it how to move to the new location. Part of this procedural schema would be to call its path-planner which would guide the AUV from A to R.

Unfortunately, going from place to place underwater is not always "smooth sailing". Figure 2 shows a kelp bed between the AUV and the rock. The kelp bed is an example of a *natural boundary* in the AUV domain. Natural boundaries, such as vegetation, high rocks, outcrops and strong currents, are especially difficult or dangerous for AUVs to cross. The environment can be partitioned into spaces that are surrounded by natural boundaries (Fisher, 1991). Procedural schemas that minimize the number of natural boundary crossings exist for travel among often-visited sets of areas. These schemas' predictions help to decide when and where new goals may best be achieved.

Figure 2: AUV with goal to pick up an object when there is a natural boundary present.

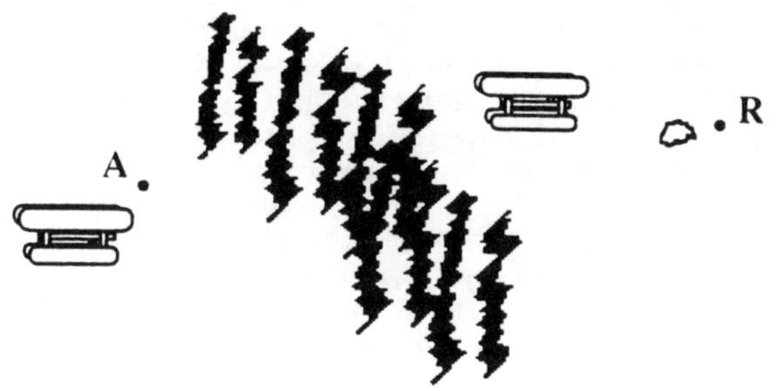


Figure 3: Two AUVs, one with goal to pick up an object, when there is a natural boundary present.

If the new goal can be satisfied in an area that will be visited as part of the schema, whether because the area contains another active location or because it is on the best path between two areas, the goal will be satisfied at that location and can be delayed until the location is reached.

If the AUV is not working alone, as shown in Figure 3, the context of working in a group gives the AUV another option. In this case, the AUV is being guided by a contextual schema representing "working with others", which allows agents to ask for help. Information about the organizational context has priority over information about the agent's individual context, so the agent commits to achieving the goal by asking another agent and the goal's priority is left the same. A procedural schema for communication tells the agent how to send a message to request the other's help.

Communication can be further affected by the organization of the agents. This is mediated by contextual schemas for different kinds of organizations. One kind of organization is a hierarchy. Suppose in this context that the AUV near A sends a message to the AUV near R requesting that the receiver adopt the goal of moving the rock. If the sender has authority over the receiver, then the receiver will raise the priority of this goal over its own goals' priorities. If the receiver has authority over the sender, then it will decide what the goal's priority should be.

## The Relationship between Problem Solving, Communication and Cooperation

Cooperation, communication, and problem solving are all related. Goals related to communicating with another agent can be achieved via problem solving (e.g., (Searle, 1969)) and goals related to problem solving can often be achieved via communication (e.g., (Appelt, 1985)). Coordination between agents' problem solving is also achieved via communication, for example, by communicating partial plans (Durfee *et al.*, 1987). Cooperation, communication, and problem solving are also more deeply related by the way in which they rely on patterns in the world.

Cooperation, communication, and problem solving all benefit from the ability to make predictions about future actions. When an agent commits to a plan of action, its plan can be seen as predictions that allow it to select an order in which to achieve its goals. This allows goals to be grouped together appropriately instead of simply relying on the priority of the goals to provide an ordering. For example, in communication, goals related to the same "topic" should appear together in the conversation (Turner & Cullingford, 1989a); in problem solving, goals that can be achieved near each other should be grouped together (Fisher, 1991).

Commitments to future actions also allow agents to coordinate their behavior with others' and with the external world in a way that cannot be done by simply reacting to immediately-perceivable external conditions (cf. (Agre & Chapman, 1987; Brooks, 1986)). For example, the agent can commit to being at some location at a particular time (e.g., for a rendezvous) or to communicating with another agent at a future time.

Cooperation is facilitated by an agent's ability to predict other agents' behavior. These predictions not only make another's behavior easy to understand, but also allow an agent to coordinate its behavior with another agent's. The most obvious example of this kind of coordination is the use of shared convention in communication. Language users share knowledge of syntax and word meaning as well as knowledge of how conversations are conducted. When language users hear a question, they not only understand what they are being asked, but also that they should answer. We believe that sharing patterns or conventions for problem solving will facilitate cooperation by making it easier for agents to recognize and participate in each others' plans.

## Schemas for Planning and Conversation

In our approach, an agent organizes as much of its knowledge as possible in schemas which reflect the underlying patterns present in the agent's environment, its behavior, and its relationships to other agents. This has the effect of reducing the size of the space in which the agent must search for applicable knowledge by clustering knowledge into schemas. It also reduces the size of the search space involved in planning, since each "operator" (planning schema) spans more of that search space than a single primitive operator would (e.g., (Fikes *et al.*, 1972)).

Two types of schemas are used to represent an agent's knowledge: procedural schemas and contextual schemas. Procedural schemas, or p-schemas, control the actions an agent takes to achieve goals. They are based on procedural schemas (Turner, 1989b) developed in our earlier work on reactive planning and on conversation MOPs (Turner & Cullingford, 1989c) from our work on conversation. The procedural schemas presented here are very similar in structure and function to conversation MOPs (C-MOPs). Contextual schemas, or c-schemas, represent the agent's problem-solving context and provide a background against which the agent's behavior takes place. They are based on the contextual schemas in our earlier work on reactive planning (Turner, 1989c).

## Procedural Schemas

Procedural schemas can be compared to hierarchical plans. Like such plans, they describe how a goal can be accomplished by a (partially-ordered) set of steps, called *scenes.*[2] A scene in a p-schema can be either a goal, another p-schema (sub-schema), or a primitive action that the agent can directly carry out. Unlike hierarchical plans, the order in which scenes are worked on during schema application depends only partially on the order specified by the p-schema. The specified order represents the *conventional* way that the p-schema is applied. However, the execution of p-schemas is flexible enough to allow these conventions to be modified by the agent's current *intentions*, reflected in its goal priorities (Turner & Cullingford, 1989c). The scenes and their ordering provide the predictions that allow an agent to organize its goals effectively (cf. (Georgeff & Lansky, 1987; Firby, 1987)).

P-schemas are organized in a dynamic memory (Schank, 1982) where they are indexed by features of the environment. The most important feature for retrieving p-schemas is the goal that the p-schema can achieve. Once retrieved, a p-schema is instantiated to form a plan (as described in detail in (Turner, 1990)). Briefly, the schema's scenes are specialized based on the current situation using the memory, and attention is focused on a particular scene to execute based on an activation metaphor which takes into account both commitment to intention as encoded in the schema ("MOP-based activation") and the agent's current intentions as encoded in the priorities of its goals ("goal-based activation"). As attention is focused, scenes may be re-specialized to reflect the evolving problem-solving situation.

## Contextual Schemas

Contextual schemas hold knowledge about kinds of problem-solving situations; they can be thought of as generalized *cases* of problem solving (e.g., (Kolodner *et al.*, 1985)). The information in c-schemas is used to control the overall character of the agent's behavior so that it is appropriate for the situation the agent is in.

C-schemas impact several areas of an agent's behavior (Turner, 1989c). They provide suggestions of goals the agent should satisfy in a particular situation and suggestions of how behavioral parameters should be set. For example, a c-schema representing the context of being in a harbor would set bounds on both an AUV's depth (e.g., don't go too near the surface nor near the bottom) and its speed (e.g., don't go too fast, since there are likely to be other vessels and obstacles present). C-schemas also provide information about how unanticipated events should be handled in a context-specific way, generally by creating a new goal in response. They also provide information about the relative importance of kinds of goals in the situation, which aids the agent's attention-focusing process. C-schemas provide information about the appropriate ways to achieve goals in a given context by suggesting which p-schemas should be used to achieve the goals. Finally, c-schemas impact the manner in which p-schemas are applied by changing parameters affecting schema application, such as the relative worth of goal-based versus MOP-based activation.

Contextual schemas are also organized in the agent's dynamic memory, indexed by features describing the kinds of situations they represent. When found, a c-schema causes the agent to automatically adjust its behavior to the context. Two or more c-schemas may be retrieved from memory for the same situation; these in essence correspond to different ways to view the situation. When this happens, the information from the c-schemas is merged to give a representation of the overall context.

# Organizational Schemas for Cooperation

Just as problem solving and communication at the level of the individual agent follows patterns, group behavior also follows patterns. When an agent participates in a group, its behavior will be modified by its role in that group as well as by the current phase of the group problem-solving effort. Group behavior is captured by both p-schemas and c-schemas.

Patterns in group behavior can affect the behavior of an individual in several ways. One way is by impacting the selection of p-schemas used by the agent. Coordinated behavior has patterns that can be captured in p-schemas. The characteristics of the overall CDPS system can provide indices into memory that help to specialize p-schemas chosen by individuals for achieving their goals. This tailors the agent's plan to the constraints of group behavior. Another way group behavior impacts an individual is by affecting the character of the problem solving, consequently affecting which contextual schemas are used by the agent. Some contextual schemas will represent patterns in group behavior that can be thought of as the "global context"; this is in contrast to the "local context" provided by the contextual schemas an agent would use when acting alone. At any given time, due to features of the group problem-solving situation, c-schemas representing both global and local context may be in use by an agent. In our approach, information about the global group effort has precedence over information about the local activity of an individual; consequently, c-schemas related to group activity take precedence over c-schemas related to local activity. Although powerful individuals may achieve their local goals before the goals given to them by the group, their ability to do this is derived from their power in the group—not from simply ignoring the group.

In the remainder of this section, we discuss two types of observed patterns in coordinated behavior and how they can be fit into our schema-based approach.

**Example 1: Phases of coordinated problem solving.** Group problem solving progresses in phases as the agents move from getting to know each other and understanding the task to dividing the task and actually performing it (Bales & Strodtbeck, 1951; Mabry, 1975). These phases can be used to guide the behavior of the agent and predict the behavior of others. Consequently, the phases are represented in a p-schema with the scenes: *information-gathering*, *task-division*, and *goal-attainment*.

However, such a p-schema does more than simply organize related actions into scenes. Each of the p-schema's scenes also influences how problem solving and communication is conducted during that scene. Consequently, the current phase of problem solving, represented by the scene in the organization p-schema that the group is currently executing, provides a kind of context for problem solving and communication, and would figure into the indices the

[2] Since p-schemas grew out of the work on C-MOPs, the terminology of MOPs (Schank, 1982) is retained.

agent uses to select c-schemas to represent its current situation. This would lead to a c-schema being found representing, for example, "being in the information-gathering phase of group problem solving". This c-schema would recommend or facilitate the selection of actions relating to providing information that has not yet been requested by another agent. It would affect problem solving by increasing the priority of goals associated with creating the overall plan as opposed to satisfying some particular subgoal. When problem solving moves to the goal-assignment phase, a c-schema representing that phase would be found and used. Under the direction of this c-schema, priorities would be lowered for goals to communicate unrequested information. This would prevent other agents from being distracted from their assigned task to process incoming messages. The c-schema might also cause agents to lower the priorities of goals not directly related to their individual tasks.

The use of p-schemas to provide indices for c-schema selection is new. We believe that the need for this mostly arises in cooperative behavior, since this behavior is governed by organization p-schemas which span large periods of time and many actions by many different kinds of agents. This means that the scenes of such a p-schema are very general in nature, and that such scenes themselves are important indicators of the agent's context, unlike scenes of p-schemas related to individual problem solving.[3]

**Example 2: Chain of command of the agents.** A great deal of attention in CDPS has been devoted to the way that agents can relate to each other. This is often called the *organization* of the agents and refers to the power structure between the agents (like the organization of a company) instead of to the physical connections between agents. Many organizations have been suggested ranging from a *collective*, or flat structure where all agents are equal, to *master–slave* relationships where one agent has power over another, to *hierarchical* structures where agents can have one of several levels of power (e.g., (Malone, 1987)). In hierarchical organizations, agents may be equal to some agents, in a position of power over others and subordinate to still others. Another way to organize agents is to allow them to choose the organizational structure which best fits their current situation (Gasser *et al.*, 1989) so that their relationship with others is dynamic.

The relationship of one agent in a hierarchy to others affects the nature, method, and timing of problem-solving actions and communication. The relationship can best be represented by contextual schemas, which we can think of as organization c-schemas. Although type of organization and an agent's position in it can be used as indices to the appropriate contextual schema, it is unnecessary to have a separate contextual schema for each possible combination of positions in each organization. Instead, all of the implications of the organizational structure for the agent's behavior can be captured by using a combination of c-schemas: one representing the kind of organization and others, drawn from c-schemas representing generic relationships (*equality*, *authority*, and *subordination*), capturing the relationships between the agent and others in the organization.[4] Each of these contexts affect behavior differently, but the summed contributions from the c-schemas produce behavior appropriate for the agent's role in the organization.

The most obvious example of how these contexts can affect behavior is what happens when one agent needs help in achieving a goal. When an agent makes a request of another agent, it is asking that agent to cooperatively assume some goal. For example, if an agent asks another to move a rock (like the too–distant rock in the example which opens this paper), the agent wants the other to assume the goal of moving the rock. Once we view this request as a new goal for the other agent, it is easy to see how the context will affect the problem solving of the agent that has assumed the goal. If the receiving agent's relationship to the other is captured by the c-schema representing equality, the agent may not change the priority of the goal at all. If the agent is subordinate to the other, the priority of the assumed goal will be increased, to reflect the fact that the authority's goals are more important to the overall problem solving than the subordinate's goals. If the agent has authority over the agent making the request, the assumed goal may have its priority lowered to reflect that the goals of the authority have a higher relative priority to the goals of the subordinate. By changing the priorities based on the context of relative power, the agents can automatically determine how they should respond to requests from other agents.

The relationship between the requesting agent and the agent which will provide aid can also affect the procedural schema chosen by the requesting agent to effect the request. If the requesting agent is in control of the other, it selects a p-schema that simply makes the request; in this case, the agent assumes that the request will be carried out unless it is informed otherwise. On the other hand, if the agent is in a position of subordination, it is in a situation similar to that of reasoning under uncertainty. In this case, the p-schema selected should make the request, then check its environment to see if the goal has actually been achieved (i.e., the request honored) by the other agent.

The context determined by the organization and an agent's position in it also affects the p-schemas the agent chooses to satisfy its goals. So, in our example of needing to request help from others, we can expect that each agent will have several specializations for the general p-schema for getting help from others. An agent needing help might, for example, use a p-schema which simply requests help without giving any rationale or a p-schema which makes the request, but in addition gives reasons why the help is requested or arguments for why the help should be given. The organization context of the agent, represented by a c-schema or set of c-schemas, should provide suggestions about which specialization is appropriate. If the agent is in a position of authority, it can simply send the message that requests help. If it is a subordinate or an equal, it will choose the p-schema which supplies reasons why the other should assume its goal and provide the help.

The use of c-schemas to represent organizational information has another important advantage for cooperative problem solving. It allows the agents' behavior, both individually and collectively, to adapt to situations in which the organization of the agents needs to be dynamic. The simplest way that our approach supports dynamic organization is that agents can change their view of the organi-

[3]However, in the AUV domain, missions of long duration may also be represented by p-schemas with some of the same character, possibly leading to their scenes contributing to the selection of c-schemas (e.g., for the mission phases of going to the work site, working, and returning home).

[4]These contexts could affect the behavior in different degrees, if necessary. For example, extreme authority (like the president of a company talking to the mail clerk) may have a more extreme effect on the context-based decisions.

zational context as easily as they can respond to changes in their environment (Turner, 1989c); in both cases, features of the changed situation provide indices which allow the retrieval of c-schemas representing the new situation. In addition, changes may cause the agent to notice, based on information contained in its contextual schemas, that an organizational change is needed. In this situation, the schema-based approach can allow dynamically adjusting the organization of the problem solvers as well as their individual problem-solving behavior by changing which schemas are in use.

## Conclusion

In this paper, we have sketched a schema-based approach to cooperative distributed problem solving. Schemas are used to represent an agent's knowledge of how to solve problems, communicate, and interact with other agents in the CDPS system. Procedural schemas contain knowledge about sequences of actions. They are used to achieve problem-solving goals, to communicate, and to coordinate an agent's actions with those of others by providing an explicit structure for global problem solving behavior. Contextual schemas contain knowledge about problem-solving and organizational contexts. They are used to adjust the character of an agent's behavior to fit the situation it is in. Contextual schemas impact the use of procedural schemas, both by helping to select them as well as by modifying, in a context-specific manner, how they are applied. Procedural schemas, especially those at the organizational level, impact the use of contextual schemas by providing indices that help the agent select c-schemas to represent the context it is in.

One advantage of our approach is that knowledge is organized so as to capitalize on patterns in the world; this facilitates efficiently bringing the right knowledge to bear at the right time. Another advantage is that schemas allow predictions to be made about how oneself and others will behave in the future. These predictions enable agents to more easily coordinate their actions with those of others. In addition, a schema-based approach to representing and using organizational knowledge potentially allows a CDPS system's organizational structure to change as the situation demands.

This work is still in the early stages of development. As our work progresses, we will refine and evaluate our ideas in the domain of multi-AUV systems, both in simulation and during in-water tests aboard AUVs.

## References


Agre, P. E. and Chapman, D. 1987. Pengi: An implementation of a theory of activity. In *Proceedings of the Sixth National Conference on Artificial Intelligence*, pages 268–272.

Appelt, D. E. 1985. Planning English Referring Expressions. *Artificial Intelligence*, 26:1–33.

Bales, R. F. and Strodtbeck, F. L. 1951. Phases in group problem solving. *Journal of Abnormal and Social Psychology*, 46:485–495.

Blidberg, D. R. and Chappell, S. G. 1986. Guidance and control architecture for the EAVE vehicle. *IEEE Journal of Oceanic Engineering*, OE–11(4):449–461.

Brooks, R. A. 1986. A robust layered control system for a mobile robot. *IEEE Journal of Robotics and Automation*, RA–2(1):14–23.

Durfee, E. H.; Lesser, V. R.; and Corkill, D. D. 1987. Coherent Cooperation between Communicating Problem Solvers. *IEEE Transactions on Computers*, C(36):1275–1291.

Durfee, E. H.; Lesser, V. R.; and Corkill, D. D. 1989. Cooperative Distributed Problem Solving. In Barr, A.; Cohen, P. R.; and Feigenbaum, E. A., eds, *The Handbook of Artificial Intelligence*, volume IV, chapter XVII, pages 83–147. Addison-Wesley Publishing Company, Inc., Reading, MA.

Fikes, R. E.; Hart, P. E.; and Nilsson, N. J. 1972. Learning and executing generalized robot plans. *Artificial Intelligence*, 3:251–288.

Firby, R. J. 1987. An Investigation into Reactive Planning in Complex Domains. In *Proceedings of the Sixth National Conference on Artificial Intelligence*, pages 202–206, Los Altos, California. Morgan Kaufmann.

Fisher, T. L. 1991. Exploiting Natural Boundaries in the Autonomous Underwater Vehicle Domain to Limit Resource Utilization.

Gasser, L.; Rouquette, N. F.; Hill, R. W.; and Lieb, J. 1989. Representing and using organizational knowledge in distributed AI systems. In Gasser, L. and Huhns, M. N., eds, *Distributed Artificial Intelligence, Volume II*, chapter 3, pages 55–78. Morgan Kaufmann Publishers, Inc., San Mateo, CA.

Georgeff, M. P. and Lansky, A. L. 1987. Reactive reasoning and planning: An experiment with a mobile robot. In *Proceedings of the Sixth National Conference on Artificial Intelligence*, pages 677–682, Seattle, Washington.

Jaffe, J. 1988. Underwater Imaging System Characterization. Technical Report H–88–33, Woods Hole Oceanographic Institute.

Kolodner, J. L.; Simpson, R. L.; and Sycara-Cyranski, K. 1985. A process model of case-based reasoning in problem-solving. In *Proceedings of the International Joint Conference on Artificial Intelligence*, Los Angeles, California.

Mabry, E. A. 1975. Exploratory analysis of a developmental model for task-oriented small groups. *Human Communication Research*, 2:66–74.

Malone, T. W. 1987. Modeling coordination in organizations and markets. *Management Science*, 33(10):1317–1332.

Schank, R. C. 1982. *Dynamic Memory*. Cambridge University Press, New York.

Searle, J. 1969. *Speech Acts: An Essay in the Philosophy of Language*. Cambridge University Press, Cambridge, UK.

Turner, E. H. 1990. Integrating Intention and Convention to Organize Problem Solving Dialogues. Technical Report GIT–ICS–90/02, School of Information and Computer Science, Georgia Institute of Technology. Ph.D. thesis.

Turner, E. H. and Cullingford, R. E. 1989a. Making conversation flexible. In *Proceedings of the Eleventh Annual Conference of the Cognitive Science Society*, pages 932–939, Detroit, MI.

Turner, E. H. and Cullingford, R. E. 1989b. Using Conversation MOPs in Natural Language Interfaces. *Discourse Processes*, 12(1):63–90.

Turner, E. H. and Cullingford, R. E. 1989c. Using Conversation MOPs in Natural Language Interfaces. *Discourse Processes*, 12(1):63–91.

Turner, R. M. 1989a. *A Schema-based Model of Adaptive Problem Solving*. PhD thesis, School of Information and Computer Science, Georgia Institute of Technology. Technical report GIT–ICS–89/42.

Turner, R. M. 1989b. Using schemas for diagnosis. *Computer Methods and Programs in Biomedicine*, 30(2/3):199–208.

Turner, R. M. 1989c. When reactive planning is not enough: Using contextual schemas to react appropriately to environmental change. In *Proceedings of the Eleventh Annual Conference of the Cognitive Science Society*, pages 940–947, Detroit, MI.

# A Case-Based Model of Creativity

**Scott R. Turner**

Artificial Intelligence Laboratory
University of California Los Angeles
Los Angeles, CA 90024
srt@cs.ucla.edu

## Abstract

Creating new solutions to problems is an integral part of the problem-solving process. This paper presents a cognitive model of creativity in which a case-based problem-solver is augmented with a set of creativity heuristics. New solutions are discovered by solving a slightly different problem and adapting that solution to the original problem. This model has been implemented in a computer program called MINSTREL.

## 1 Introduction

Creativity is the pinnacle of human achievement. Intellectual heroes such as Albert Einstein, William Shakespeare, and Thomas Edison are revered for the originality and beauty of their ideas. But genius is not the sole province of creativity. We all create on a daily basis: we fix cars using spare change and bailing wire, invent jokes based on the latest domestic crisis, and make up bed-time stories for our children. The ability to invent original, useful solutions to problems is a fundamental process of human thought.

This paper presents a model of creative reasoning as an extension to problem-solving. When problem-solving fails, the creative reasoner invents new solutions by combining old knowledge in new ways. This model has been implemented in a computer program called MINSTREL, which creates stories about King Arthur and his Knights of the Round Table.

## 2 Case-Based Problem-Solving

In case-based reasoning, problems are solved by recalling similar past problem situations and applying the solutions from those situations to the current problem (Kolodner 1987; Hammond 1988; Slade 1991). The problem-solver *recalls* a past problem similar to the current problem, *adapts* the solution from the old problem to the current problem, and then *assesses* the results.

The advantage of case-based problem-solving is that the task of adapting and assessing the recalled solution is simple, because the recalled problem is similar to the current problem. The disadvantage is that case-based problem-solvers cannot solve a problem which is *not* similar to a previous problem. When a new problem is encountered the problem-solver has no way to invent or discover a new solution, and problem-solving fails.

## 3 Creativity

What needs to be added to case-based problem-solving to make it capable of creativity? Consider this example of creativity concerning the seven-year old niece of the author:

> One day, while visiting her grandparents, Janelle was seated alone at the dining room table, having milk and cookies. Reaching for the cookies, she accidently spilled her milk on the table. She decided to clean up the mess herself.
>
> Janelle went into the kitchen, but there were no towels or paper towels available. She stood for a moment in the center of the kitchen thinking, and then she went out the back door.
>
> Janelle returned a few minutes later carrying a kitten. The neighbor's cat had given birth to a litter about a month ago, and she had been over to play with the kittens the previous day. Janelle brought the kitten into the dining room, where he happily lapped up the spilled milk.

This story illustrates three important features of the creative process:

(1) *Creative solutions are both original and useful* (Koestler 1964; Shouksmith 1970). Janelle's solution is original because it has significant differences from all other solutions Janelle knows, and useful because it solves her problem.

(2) *Creativity is driven by the failure of problem-solving* (Weisberg 1986). Janelle could not use the plan she knew for cleaning up a spill because no towel was available. In this case, problem-solving failed because a plan precondition could not be achieved.

(3) *Old knowledge is used to create new solutions* (Koestler 1964; Weisberg 1986). Janelle knew that kittens like milk and that she could find a kitten next door. The creative problem-solver must be able to discover appropriate old knowledge and apply it in new ways to the current problem.

Janelle's example illustrates that a creative problem-solver must be failure-driven, able to use old knowledge in new ways, and able to produce novel and useful solutions.

## 4 Discovering New Solutions

Case-based problem-solving finds solutions by recalling past problems similar to the current problem. To discover new solutions, the problem-solver must (1) recall a past problem *different* from the current problem and (2) adapt the solutions from the recalled problem to the current problem.

MINSTREL implements this process as three part Transform-Recall-Adapt Methods (TRAMs). "Transform" takes a problem and changes it into a slightly different problem. "Recall" takes the new problem description and tries to recall similar past problems from episodic memory. "Adapt" takes the recalled problem solutions and adapts them to the original problem.

TRAM-Cross-Domain-Solution is an example of a Transform-Recall-Adapt Method. TRAM-Cross-Domain-Solution suggests that a new solution to a problem can be found by translating the problem into a new domain, solving the problem in that domain, and then translating the solution back into the original domain.

MINSTREL uses TRAM-Cross-Domain-Solution to create the following story scene:

> One day while out riding, Lancelot's horse went into the woods. Lancelot could not control the horse. The horse took him deeper into the woods. The horse stopped. Lancelot saw Andrea, a Lady of the Court, who was picking berries.

The original specification for this problem is "create a scene in which a knight accidently meets a princess." TRAM-Cross-Domain-Solution maps this specification into the modern domain and recalls this story:

> Walking The Dog
>
> John was sitting at home one evening when his dog began scratching at the door and whining for a walk. John decided to take the dog for a walk. While they were out, John ran across his old friend Pete, whom he hadn't seen in many years. John realized that he would never have run into Pete if his dog hadn't wanted a walk.

This story is adapted for the original problem by mapping the story back into the King Arthur domain, creating a scene in which Lancelot's horse leads him to Andrea. The resulting scene is creative because TRAM-Cross-Domain enabled the problem-solver to discover a solution different from the known solutions.

There are two major benefits to Transform-Recall-Adapt Methods.

First, TRAMs discover new solutions by looking at *slightly* different problems. A slightly different problem shares many of the same constraints as the original problem, and its solution is likely to have some applicability to the original problem. Looking at slightly different solutions also simplifies the adaptation process, since fewer changes will be needed to adapt the solution to the original problem.

Second, TRAMs further simplify the adaptation task because the transformation itself can be used to guide adaptation. Knowing the transformation that led to a solution, the problem-solver can apply a specific adapation to reverse the effects of the problem transformation. In the case of TRAM-Cross-Domain-Solution, the problem-solver knows to adapt the solution by translating back to the original problem domain. In MINSTREL, each TRAM contains both a problem transformation and a specific adaptation to apply to recalled solutions.

TRAMs are added to the model of case-based problem-solving by augmenting the Recall and Adapt steps. During problem-solving, a TRAM is selected from this pool and applied to the current problem. If the TRAM succeeds in discovering a solution to the problem, problem-solving succeeds. If problem-solving fails, the current TRAM is discarded, another selected from the pool of available TRAMs, and the cycle repeated. The first TRAM selected from the pool is always TRAM-Problem-Solving, a special TRAM that implements normal problem-solving by passing the original problem unchanged to the recall step.

## 5 Imaginative Memory

The central step of the Transform-Recall-Adapt Method is recalling a solution from episodic memory. But the process of recalling something from episodic memory can *itself* be viewed as a problem. What happens if creative problem-solving is used to implement recall?

To do this, the Recall step of each TRAM is modified to recursively call the creative problem-solving process. To prevent this from leading to endless recursion, TRAM-Problem-Solving is left unchanged, and continues to recall directly from episodic memory.

To recall something, the problem-solver uses creative problem-solving with the problem specification "Find something in episodic memory that matches these features." TRAM-Problem-Solving is the first TRAM used, and passes the recall features unchanged to episodic memory. If an episode that matches the recall features is found, problem-solving succeeds, and memory behaves as expected.

Something more interesting happens when recall fails. If TRAM-Problem-Solving cannot find an episode in memory that matches the recall features, problem-solving fails and a new TRAM is selected. This TRAM modifies the recall features and recursively calls the problem-solving process with the new recall features.

The first TRAM used on this recursive call is again TRAM-Problem-Solving. If the new features recall an episode, the episode is returned to the previous recursion of problem-solving, where it is adapted to the original problem by the previous TRAM, and recall succeeds. *But because the recalled episode was changed by the Adapt portion of the previous TRAM, it is no longer the episode that was found in memory.* Recall has succeeded in a strange way: by recalling an episode that does not exist in episodic memory. Episodic memory has become imaginative.

Treating recall as problem-solving also enables the problem-solver to apply multiple transformations to a problem when necessary. Each time recall fails another TRAM will be applied to the recall features. In this way, a number of TRAMs can be successively applied to a problem. Each TRAM changes the problem in only a small way, but the cumulative effect may be large, enabling the creative problem-solver to discover new solutions

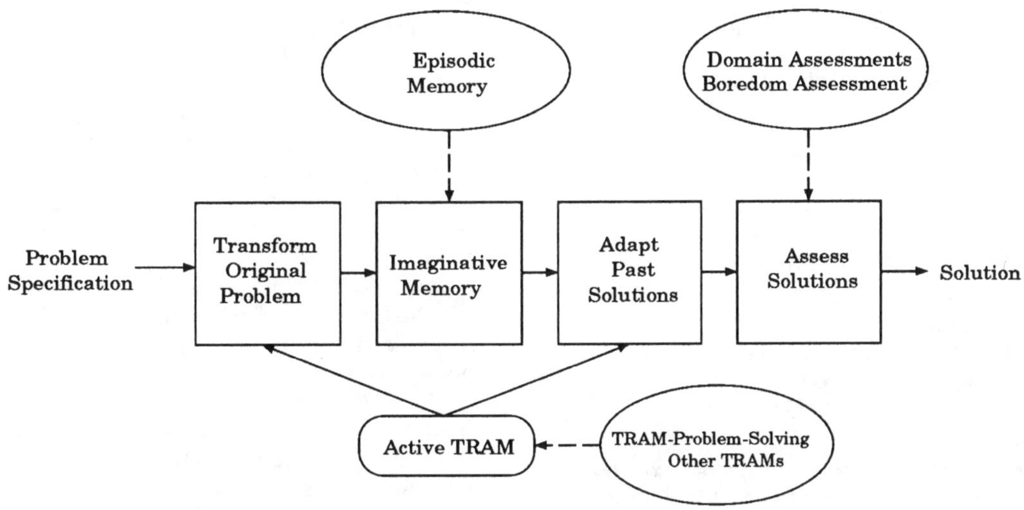


Figure 6.1 MINSTREL's Process Model of Creativity

significantly different from known solutions.

There are two advantages to embedding creativity in the recall process. First, it implements the creative process at a low cognitive level, making creativity transparently available to any cognitive process that makes use of episodic memory for reasoning. Second, it enables the recursive use of creativity heuristics, greatly increasing the power of even simple TRAMs.

## 6 A Model of Creativity

Figure 6.1 show how Transform-Recall-Adapt Methods and imaginative memory are integrated into case-based problem-solving. An active TRAM controls the Recall and Adapt steps. The recall step uses imaginative memory instead of episodic memory.

The problem-solving cycle begins when a problem description enters the recall step. Initially TRAM-Problem-Solving is in control of the recall and adapt steps. The original problem description is used to recall previous similar problem-solving situations from episodic memory. If recall succeeds, the recalled situations are passed to the adapt step. Under TRAM-Problem-Solving, no adaptation is needed because the recalled solutions are very similar to the original problem, so the recalled solutions are passed along to the assess step. In the assess step, all active assessments are applied to the recalled solutions, and if a solution passes all the assessments, it is output as a solution to the original problem. This is simply the normal problem-solving cycle.

If TRAM-Problem-Solving fails, either because no solutions were recalled or because the recalled solutions failed assessment, TRAM-Problem-Solving is discarded and a new TRAM selected as the active TRAM. The selection of a TRAM is based on the type of problem being solved and the TRAMs previously used.

The selected TRAM transforms the original problem specification, creating a new problem specification. The new problem specification is passed to imaginative memory. If recall succeeds, the recalled solutions are passed to the adapt step, where the active TRAM applies a specific adaptation which reverses the problem transformation. If recall fails, then imaginative memory recursively applies a second TRAM, and the problem-solving repeats at the inferior level.

Adapted solutions are passed to the assessment step. If a solution passes all assessments, problem-solving succeeds. If all solutions fail, the active TRAM is discarded, a new TRAM selected, and the Recall-Adapt-Assess cycle repeats.

## 7 Creativity in MINSTREL

The primary creative task in storytelling is scene creation – inventing story events to achieve a particular storytelling goal. This section illustrates how MINSTREL invents three scenes in which "a knight kills himself."

Initially, MINSTREL's episodic memory contains two story scenes:

Knight Fight

> A knight fights a troll with his sword, killing the troll and accidently injuring himself.

The Princess and the Potion

> A lady of the court drank a potion to make herself ill.

Using these scenes and three TRAMs, MINSTREL invents three ways for a knight to commit suicide. Figure 7.1 shows an abbreviated trace of MINSTREL creating these three scenes and generating English descriptions. Except for the deletions marked with "[...]", this trace appears exactly as produced by MINSTREL.

```
==============================================
           MINSTREL Invention
==============================================
Initial specification is #{ACT.105}:
 (A KNIGHT NAMED JOHN DID SOMETHING *PERIOD*
  JOHN DIED *PERIOD*)
RAS Cycle: #{ACT.105}.
 Executing TRAM:GENERALIZE-CONSTRAINT.  [1]
 Generalizing :OBJECT on #{STATE.112}.
 Recalling #{ACT.113}: #{KNIGHT-FIGHT}.
Minstrel invented this story:
 (A KNIGHT NAMED JOHN FOUGHT HIMSELF BY MOVING
  HIS SWORD TO HIMSELF IN ORDER TO KILL HIMSELF
  *PERIOD* JOHN DIED *PERIOD*)
[...]
 [TRAM Recursion: #{ACT.111}.]
    Executing TRAM:SIMILAR-OUTCOMES.     [2]
    Recalling #{ACT.136}:        NIL.
    [TRAM Recursion: #{ACT.136}.]
       Executing TRAM:GENERALIZE-CONSTRAINT.
       Recalling #{ACT.138}: #{PRINCESS-POTION}.
Minstrel invented this story:
 (A KNIGHT NAMED JOHN FOUGHT A DRAGON BY MOVING
  HIS SWORD TO IT IN ORDER TO KILL HIMSELF
  *PERIOD* JOHN DIED *PERIOD*)
[...]
 [TRAM Recursion: #{ACT.112}.]
    Executing TRAM:INTENTION-SWITCH.     [3]
    Recalling #{ACT.126}:        NIL.
    [TRAM Recursion: #{ACT.126}.]
       Executing TRAM:SIMILAR-OUTCOMES.
       Recalling #{ACT.128}: #{KNIGHT-FIGHT}.
Minstrel invented this story:
 (A KNIGHT NAMED JOHN DRANK THE POTION IN ORDER
  TO KILL HIMSELF *PERIOD* JOHN DIED *PERIOD*)
[...]
```

Figure 7.1 Suicide Trace

### 7.1 TRAM-Generalize-Constraint

TRAM-Generalize-Constraint suggests that a new solution to a problem can be found by removing a solution constraint, solving the new problem, and then adding the constraint back to the new solution. For scene creation, the problem constraints are the features of the specification.

In this example, there are four constraints: (1) the actor is a knight, (2) the object of the state is the actor, (3) the type of the state is health, and (4) the value of the state is dead. TRAM-Generalize-Constraint suggests recalling scenes in which one of these constraints has been generalized.

At **[1]** in Figure 7.1, this TRAM generalizes constraint (2), creating the specification "a knight killed something." This recalls the "Knight Fight" episode. "Knight Fight" is adapted to the current problem by replacing the troll with the knight, creating a scene in which a knight kills himself with his sword. TRAM-Generalize-Constraint has used previous knowledge about how knights kill monsters to create a scene in which a knight kills himself.

Two issues which must be addressed to implement TRAM-Generalize-Constraint are (1) how a feature is generalized, and (2) how a feature is adapted back into the created scene.

MINSTREL uses two methods to generalize a feature. First, the feature can be completely removed from the problem specification. This is the broadest possible generalization and is used to select the feature to be generalized. However, it is so broad that it often leads to the recall of scenes which are difficult to adapt to the original problem. A better generalization would ensure some similarity between the original feature and the instantiations of the generalization. One such generalization is based on *class hierarchies.*

Classes group concepts that have many similarities. The "human" class groups a variety of characters – princesses, knights, kings, and hermits – that have many similar features. By generalizing problem features within classes, MINSTREL is more likely to find a useful reminding.

TRAM-Generalize-Constraint uses class hierarchies to create generalizations. In the suicide example, the "knight" feature is generalized to "monster", because both monsters and knights are in the "Violent Characters" class. This generalization recalls the "Knight Fight" episode, in which a knight fights a troll in order to kill the troll.

The second issue in TRAM-Generalize-Constraint is adapting the recalled episode to the original specification. This is achieved by replacing the generalized feature value with the original feature value throughout the recalled episode. In the suicide example, the troll in "Knight Fight" is replaced with the knight, resulting in a scene in which a knight kills himself by fighting himself with his sword.

### 7.2 TRAM-Similar-Outcomes

TRAM-Similar-Outcomes suggests that if an action results in a particular outcome, it might also result in similar outcomes. If riding a horse can carry a knight to the castle, then riding a horse might also carry a knight to the woods.

At **[2]** in Figure 8.1, TRAM-Similar-Outcomes recognizes that being injured is similar to being killed, and transforms the scene description to "a knight purposely injures himself." If MINSTREL can recall a scene which fits this description, it can be adapted to the current problem by guessing that an action which results in injury might also result in death.

However, the description "a knight purposely injures himself" does not recall either of the episodes in MINSTREL's episodic memory. Since recall fails, imaginative memory recurses and applies a new TRAM at the inferior level.

At this new level, MINSTREL applies TRAM-Generalize-Constraint to the description "a knight purposely injures himself" and generalizes the "knight" feature. This description recalls "The Princess and the Potion" in which a lady of the court drinks a potion to make herself ill. This scene is adapted by TRAM-Generalize-Constraint by replacing the "lady of the court" feature with "a knight," resulting in a scene in which a knight makes himself ill by drinking a potion.

This adapted scene is returned to the previous level and is adapted by TRAM-Similar-Outcomes by replacing the illness with death. This results in a scene in which a knight kills himself by drinking a potion, filling the original description "a knight kills himself." Note that in the course of inventing this scene, MINSTREL has also invented the idea of poison – a potion that kills.

The main issue in TRAM-Similar-Outcomes is determining when two outcomes are interchangable. MINSTREL has two methods for deciding this question.

First, MINSTREL assumes that if an action can result in a partial relative change of a state (i.e., drinking a potion results in a partial negative change in health) then the action can also result in a complete relative change of the state (i.e., drinking a potion can make one's health completely negative).

Second, MINSTREL assumes that two outcomes are interchangable if it can recall other scenes in which they are interchangable. MINSTREL assumes that if two outcomes are interchangable in any scene then they are interchangable in every scene.

### 7.3 TRAM-Intention-Switch

TRAM-Intention-Switch suggests that if the effect of an action in a scene was intentional it might just as well have been unintentional, and vice versa.

At [3] in Figure 8.1, TRAM-Intention-Switch creates the new specification "a knight accidently kills himself." Recall on this new specification fails, because MINSTREL's episodic memory does not contain any episodes in which a knight accidently kills himself.

Problem-solving is used recursively, and TRAM-Similar-Outcomes creates the new scene description "a knight accidently injures himself." This recalls "Knight Fight," in which a knight is injured while killing a troll. Both TRAM-Similar-Outcomes and TRAM-Intention-Switch adapt this recalled scene, resulting in a scene in which a knight commits suicide by intentionally losing a fight with a troll.

Since intended and unintended outcomes are always interchangable, TRAM-Intention-Switch is simple and useful in a wide variety of situations.

## 8 Status

MINSTREL is written in Common Lisp and contains about 11,000 lines of code and representation. MINSTREL implements twenty-nine TRAMs, twenty-eight author-level storytelling plans, and has an initial memory of approximately ten story scenes. MINSTREL creates a small number of one-half to one page stories based on three story themes, invents novel story scenes as illustrated in the suicide example, and can do simple mechanical invention.

## 9 Conclusions

Case-based problem-solving cannot invent new solutions to problems. Transform-Recall-Adapt Methods are heuristics that creates solutions by transforming problems into new problems. When integrated into the problem-solving cycle, TRAMs give the problem-solver the ability to invent new solutions and makes creativity transparently available to any cognitive process that uses episodic memory. The power and efficacy of TRAMs has been demonstrated in MINSTREL, a computer program that can tell stories and invent novel story scenes.

## References


Hammond, Kristian 1988. Case-Based Planning. In Proceedings of the Case-Based Reasoning Workshop, May 1988.

Koestler, A. 1964. *The act of creation.* MacMillan, New York.

Kolodner, Janet L. 1987. Extending Problem Solver Capabilities Through Case-Based Inference, In Proceedings of the 4th Annual International Machine Learning Workshop

Shouksmith, George 1970. *Intelligence, Creativity and Cognitive Style.* B.T. Batsford, Ltd., London.

Turner, Scott and Reeves, John 1987. The RHAPSODY Manual, Technical Note UCLA-AI-N-85-87, Artificial Intelligence Laboratory, Computer Science Department, University of California Los Angeles.

Slade, Stephen, 1991. Case-Based Reasoning. *Artifical Intelligence Magazine* 12(1):42-55.

Weisberg, Robert W. 1986. *Creativity: Genius and Other Myths.* W.H. Freeman and Company, New York.

# Efficient Nonlinear Problem Solving using Casual Commitment and Analogical Replay *

**Manuela M. Veloso**
School of Computer Science
Carnegie Mellon University
Pittsburgh, PA 15213
mmv@cs.cmu.edu

## Abstract

Complex interactions among conjunctive goals motivate the need for nonlinear planners. Whereas the literature addresses least-commitment approaches to the nonlinear planning problem, we advocate a casual-commitment approach that finds viable plans incrementally. In essence, all decision points are open to introspection, reconsideration, and learning. In the presence of background control knowledge – heuristic or definitive – only the most promising parts of the search space are explored to produce a solution plan efficiently. An analogical replay mechanism is presented that uses past problem solving episodes as background control guidance. Search efforts are hence amortized by automatically compiling and reusing past experience by derivational analogy. This paper reports on the full implementation of the casual-commitment nonlinear problem solver of the PRODIGY architecture. The principles of nonlinear planning are discussed, the algorithms in the implementation are described in some detail, and empirical results are presented that illustrate the search reduction when the nonlinear planner combines casual commitment and analogical replay.

## Introduction - Why Casual Commitment

Nonlinear planning was developed to deal with problems like Sussman's anomaly, which could not be solved by rudimentary linear planners such as STRIPS [Fikes and Nilsson, 1971, Sussman, 1973]. Least-commitment planners handle this anomaly by deferring decisions while building the plan [Sacerdoti, 1975]. These planners typically output a partially ordered plan as opposed to a totally ordered one, and consequently the term *nonlinear plan* is used. However, the essence of the *nonlinearity* is not in the fact that the plan is partially ordered, but in the fact that a plan need not be a linear concatenation of complete subplans, each for a goal presumed independent of all others [Veloso, 1989].

We advocate a casual commitment approach [Minton *et al.*, 1989], as opposed to a least commitment approach, to the nonlinear planning problem. This paper reports on the full implementation of a casual-commitment nonlinear problem solver, which we refer to as NoLimit, standing for **No**nlinear problem solver using casual com**mit**ment. This work takes context within the PRODIGY integrated intelligent architecture [Carbonell *et al.*, 1990].

NoLimit reasons about *totally* ordered plans that are *nonlinear*, i.e., the plans cannot be decomposed into a sequence of complete subplans for the conjunctive goal set. At choice points, NoLimit commits to a particular alternative, generating the planning steps and testing their consequences, while searching for a solution. All decision points (operator selections, goal orderings, backtracking points, etc.) are open to introspection and reconsideration. We claim that nonlinear planning refers to searching to attain a *set* of goals, allowing interleaving of goals and subgoals at different depths of search. Hence, reasoning about totally ordered plans is not, per se, a characteristic of a linear planner, as reasoning about partially ordered plans is not either, per se, a characteristic of a nonlinear planner [Rosenbloom *et al.*, 1991]. In fact, NoLimit generates a partially ordered plan from a totally ordered solution found, by simply analyzing the dependencies among the steps, and relaxing unnecessary constraints [Veloso *et al.*, 1990].

In a least-commitment planning strategy [Sacerdoti, 1975, Wilkins, 1989], decisions are deferred until forced by constraints. Typically conjunctive goals are assumed to be independent and worked separately, producing unordered sets of actions to achieve the goals. From time to time, the planner fires plan critics that

*This research was supported in part by ONR grants N00014-79-C-0661 and N0014-82-C-50767, DARPA contract number F33615-84-K-1520, and NASA contract NCC 2-463. The views and conclusions contained in this document are those of the authors alone and should not be interpreted as representing the official policies, expressed or implied, of the U.S. Government, the Defense Advanced Research Projects Agency or NASA.

check for interactions among the individual subplans. If conflicting interactions are found, the planner commits to a specific partial ordering that avoids these conflicts. In cases where actions stay unordered during the entire planning process, a final partially ordered plan is produced. In this strategy, it is NP-hard [Chapman, 1987] to determine if a given literal is true at a particular instant of time while planning, when actions are dependent on the state of the world, as all paths through the partial order must be verified.

Using a casual-commitment strategy, in the worst case, also involves an exponential search over the space of solutions. NoLimit uses control knowledge to reduce this exponential search. Provably incorrect alternatives are eliminated and heuristically preferred ones are explored first.

The PRODIGY architecture is a testbed for exploring machine learning approaches to automatically acquiring control knowledge. Casual commitment provides a framework in which it is natural to reason and learn about the control decisions of the problem solver, as successful and failed commitments are explored and can be analyzed. The learned control knowledge transforms a simple casual-commitment search strategy into an efficient one.

In this paper we present how we automatically learn control knowledge by combining the basic nonlinear casual-commitment problem solver with an analogical replay mechanism. The derivational analogy learning mechanism presented consists of organizing and reusing derivational traces of search-intensive problem solving episodes. These search traces are annotated with explicit justifications of successful and failed conditions explored by the casual-commitment problem solver. Subsequent reasoning in similar new problems is driven by the derivational analogy replay machinery.

## NoLimit - The Problem Solving Algorithm

In order to solve problems in a particular domain, PRODIGY must first be given a domain theory, including a set of operators. Each operator has a precondition expression and a list of effects that describe how the application of the operator changes the world. Precondition expressions are well-formed formulas in a form of predicate logic encompassing negation, conjunction, disjunction, and typed-existential and universal quantification. Regular effects are atomic formulas that describe the literals that are added or deleted from the current state when the operator is applied. Conditional effects represent changes to the world that are dependent on the state in which the operator is applied [Minton *et al.*, 1989]. NoLimit follows a means-ends analysis backward chaining search algorithm. NoLimit's nonlinear character stems from working with a **set** of goals in this cycle, as opposed to the top goal in a linearized goal stack.

1. Check if the goal statement is true in the current state, or there is a reason to suspend the current search path.
   If yes, then either, show the formulated plan, backtrack, or take appropriate action.
2. Compute the *set* of *pending goals* $\mathcal{G}$, and the set of possible *applicable operators* $\mathcal{A}$.
3. Choose a goal $G$ from $\mathcal{G}$ or select an operator $A$ from $\mathcal{A}$ that is directly applicable.
4. If $G$ has been chosen, then
   - *expand goal* $G$, i.e., get the set $\mathcal{O}$ of *relevant instantiated operators* for the goal $G$,
   - choose an operator $O$ from $\mathcal{O}$,
   - go to step 1.
5. If an operator $A$ has been selected as directly applicable, then
   - *apply* $A$,
   - go to step 1.

*Figure 1: A Skeleton of* NoLimit*'s Search Algorithm.*

The algorithm in Figure 1 describes the basic skeleton of NoLimit's search algorithm. Dynamic goal selection enables NoLimit to fully interleave plans, exploiting common subgoals and addressing issues of resource contention.

The different commitments along the search algorithm as presented in Figure 1 may lead eventually into dead-end situations. For example, a failure occurs if we reach a subgoal that is unachievable for lack of any relevant operators. NoLimit has several heuristics that propose suspending a search path under various conditions such as when a path becomes unpromising (goal and state loop detections) or when a path becomes too long or costly according to some threshold. Upon failure, NoLimit backtracks to a previous choice point. It has the ability to call backtracking control knowledge that select (or reject) particular backtracking points, thus performing intelligent allocation of resources and permitting dependency-directed backtracking or other interesting disciplines [Drummond and Currie, 1989, Anderson and Farley, 1990].

### Control knowledge

The search algorithm described in Figure 1 involves several choice points, to wit: the *operator* to choose to achieve a particular goal; the *bindings* to choose in order to instantiate the chosen operator; the *goal* to select from the set of pending goals and subgoals; *apply* an applicable operator or continue *subgoaling*; *suspend* the search path being explored; upon failure, the *past choice point* to backtrack to, or the *suspended path* to reconsider for further search.

The casual-commitment problem solver produces a complete search tree, encapsulating all decisions explored - right and wrong ones - as well as the final solution. The analogical reasoner uses this information to automatically generate and store annotated problem solving episodes (cases) into a library of solved plans [Carbonell and Veloso, 1988,

Veloso and Carbonell, 1991b]. This case-based approach combined with the nonlinear planner allows past experience to guide the decision points in similar new planning situations [Veloso and Carbonell, 1990].

## An Example in a Simple Transportation Domain

Consider a generic transportation domain with three simple operators that load, unload, or move a ROCKET as shown in Figure 2. Of course NoLimit solves much more complex and general versions of this domain. The present minimal form suffices to illustrate the casual-commitment strategy in nonlinear planning allowing full interleaving of goals and subgoals. In [Veloso *et al.*, 1990] we show several examples in a complex logistics transportation domain.

```
(LOAD-ROCKET              (UNLOAD-ROCKET            (MOVE-ROCKET
(params                   (params                   (params nil)
 ((obj Object)             ((obj Object)            (preconds
  (loc Location)))          (loc Location)))         (at Rocket locA))
(preconds                 (preconds                 (effects
 (and                      (and                      ((add
  (at obj loc)              (inside obj Rocket)        (at Rocket locB))
  (at Rocket loc)))         (at Rocket loc)))         (del
(effects                  (effects                     (at Rocket locA)))))
 ((add                     ((add
  (inside obj Rocket))      (at obj loc))
  (del                      (del
  (at obj loc)))))          (inside obj Rocket)))))
```

*Figure 2: The ONE-WAY-ROCKET Domain.*

The operator MOVE-ROCKET shows that the ROCKET can move only from a specific location *locA* to a specific location *locB*. This transforms this current general domain into a *ONE-WAY-ROCKET* domain. An object can be loaded into the ROCKET at any location by applying the operator LOAD-ROCKET. Similarly, an object can be unloaded from the ROCKET at any location by using the operator UNLOAD-ROCKET.

Suppose we want to solve a simple two-object problem. In the initial state *obj*1, *obj*2, and the ROCKET are at location *locA*. The problem consists in moving the two objects *obj*1 and *obj*2 to the location *locB*. So the goal statement is the conjunction *(and (at obj1 locB) (at obj2 locB))*. Without any analogical guidance (or other form of control knowledge) the problem solver searches for the goal ordering that enables the problem to be solved. Accomplishing either goal individually, as a linear planner would do, inhibits the accomplishment of the other goal. A precondition of the operator LOAD-ROCKET cannot be achieved when pursuing the second goal (after completing the first goal), because the ROCKET cannot be moved back to the second object's initial position (i.e. *locA*). So interleaving of goals and subgoals at different levels of the search is needed to find a solution.

NoLimit solves this problem, where linear planners fail (but where of course other least-commitment planners also succeed), because it switches attention to the

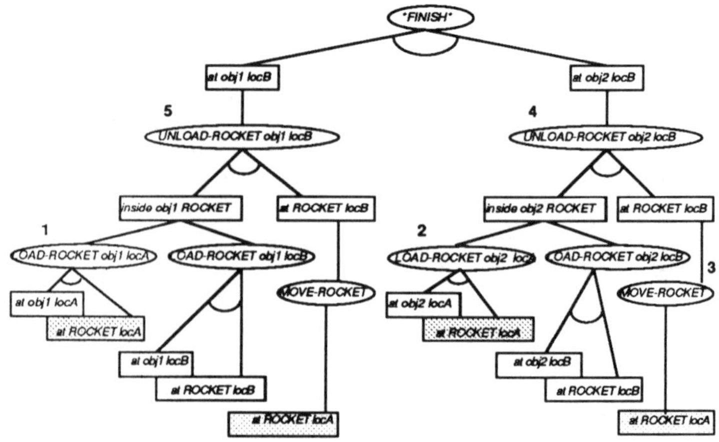


*Figure 3: The Complete Conceptual Tree for a Successful Solution Path. The* **numbers** *at the nodes show the execution order of the* **plan** *steps.*

conjunctive goal *(at obj2 locB)* before completing the first conjunct *(at obj1 locB)*. This is shown in Figure 3 by noting that, after the plan step 1 where the operator (LOAD-ROCKET obj1 locA) is applied as relevant to a subgoal of the top-level goal *(at obj1 locB)*, NoLimit suspends processing and changes its focus of attention to the other top-level goal and applies, at plan step 2, the operator (LOAD-ROCKET obj2 locA) which is relevant to a subgoal of the goal *(at obj2 locB)*. In fact NoLimit explores the space of possible attention foci and only after backtracking does it find the correct goal interleaving. The idea is to learn next time from its earlier exploration and reduce search dramatically.

A solution to this problem is the plan shown in Figure 3: (LOAD-ROCKET obj1 locA), (LOAD-ROCKET obj2 locA) (MOVE-ROCKET), (UNLOAD-ROCKET obj2 locB), (UNLOAD-ROCKET obj1 locB). In [Veloso *et al.*, 1990] we show the algorithm that NoLimit uses to further return the partial plan embedded in this encountered totally ordered plan by simply analyzing the dependencies among the plan steps.

## Replay by Derivational Analogy

Derivational analogy is a reconstructive method by which lines of reasoning are transferred and adapted from similar earlier problem-solving episodes to the new problem to be solved [Carbonell, 1986]. The ability to replay previous solutions requires that the problem solver be able to introspect into its internal decision cycle, recording the justifications for each decision during its extensive search process. These justifications augment the solution trace and are used to guide the future reconstruction of the solution for subsequent problem solving situations where equivalent justifications hold true.

In a casual-commitment search approach these justifications arise in a natural way, covering the set of successful decisions, and pointing out other failed alternatives also explored [Veloso and Carbonell, 1990]. Decision choices according to the algorithm in Figure 1 involve creating goal or operator decision nodes. NoLimit may either apply an operator whose preconditions are satisfied (if any), i.e. its left hand side is true in the current state, or continue subgoaling in an unmatched precondition of a different chosen operator. Figure 4 shows the skeleton of the different decision nodes. The different justification slots capture the context in which the decision is taken and the reasons that support the choice.

| Goal Node | Applied Op Node | Chosen Op Node |
|---|---|---|
| :step | :step | :step |
| :sibling-goals | :sibling-goals | :sibling-relevant-ops |
| :sibling-applicable-ops | :sibling-applicable-ops | :why-this-operator |
| :why-subgoal | :why-apply | :relevant-to |
| :why-this-goal | :why-this-operator | |
| :precond-of | | |
| (a) | (b) | (c) |

*Figure 4: Justification Record Structure: (a) At a Goal Decision Node; (b) At an Applied Operator Decision Node; (c) At a Chosen Operator Decision Node.*

The *step* slots show the selection done. The *sibling-* slots enumerate the alternatives to the choice made. NoLimit annotates the reason why these alternatives were not pursued further according to its search experience (either not tried, or abandon due to a described failure reason). The *why-* slots present the reasons (if any) the particular decision was taken. These reasons can range from arbitrary choices to a specific control rule or guiding case that dictated the selection. These reasons are tested at replay time and are interpretable by NoLimit. Finally the subgoaling structure is captured by the slots *precond-of* at a goal node and the slot *relevant-to* at a chosen operator node.

The problem and the generated annotated solution become a *case* in memory. The case corresponds to the search tree compacted into the successful path as a sequence of annotated decision nodes as presented in Figure 4. As we describe below, a case is not used as a simple "macro-operator" [Fikes and Nilsson, 1971, Minton, 1985] as it *guides* and *does not dictate* the reconstruction process. Intermediate decisions corresponding to steps internal to each case can be bypassed or adapted, if their justifications do not longer hold.

The general replay mechanism involves a complete interpretation of the justification structures in the new context, and development of adequate actions to be taken when transformed justifications are no longer valid. We follow a satisficing paradigm where planning effort is minimized by recycling as much of the old solution as possible. The syntactic applicability of an operator is always checked by simply testing whether its left hand side matches the current state. Semantic applicability is checked by determining whether the justifications hold (i.e. whether there is still a reason to apply this operator). In case the choice remains valid in the current problem state, it is merely copied, and in case it is not valid the system has three alternatives:

1. Replan at the particular failed choice by establishing the current subgoal by other means substituting the new choice for the old one in the solution sequence.
2. Re-establish the failed condition by adding it as a prioritized goal in the planning and, if achieved, simply insert the extra steps into the solution sequence.
3. Attempt to perform the partially unjustified action anyway; if it is successful, the system interacts with the experimentation module to refine its knowledge according to the experiment.

The replay mechanism in the context of casual commitment as opposed to least commitment [Kambhampati, 1989] allows naturally to combine guidance from several past problem solving episodes. Replicated adapted decisions can be interleaved and backtracked upon within the totally ordered reasoning plan. We now provide examples that illustrate the derivational analogy replay mechanism in terms of its effect in problem solving search reduction.

## Pursuing the One-Way-Rocket Example

Let us return to the *ONE-WAY-ROCKET* problem to illustrate the derivational replay process. While solving the two-object problem, NoLimit automatically annotates the decisions taken with justifications that reflect its experience while searching for the solution. As an example, suppose that the correct decision of choosing to work on the goal *(inside obj1 ROCKET)* was taken after having failed when working first on *(at ROCKET locB)*. The decision node stored for the goal *(inside obj1 ROCKET)* is annotated with sibling goal failure as illustrated in Figure 5. *(at ROCKET locB)* was a sibling goal that was abandoned because NoLimit encountered an unachievable predicate, i.e. *(at ROCKET locA)*, as there is no operator that moves the ROCKET back to *locA*.

```
Frame of class goal-decision-node
 :step  (inside obj1 ROCKET)
 :sibling-goals
  (((inside obj2 ROCKET) not-tried)
   ((at ROCKET locB) (:no-relevant-ops (at ROCKET locA))))
 :sibling-applicable-ops NIL
 :why-subgoal NIL
 :why-this-goal NIL
 :precond-of  (UNLOAD-ROCKET obj1 locB)
 step of next-decision-node  (LOAD-ROCKET obj1 locA)
```

*Figure 5: Saving a Goal Decision Node with its Justifications.*

Let NoLimit use the two-object problem to guide similar problems, namely moving three and four objects. We show the empirical results in Table 1. The solution is replayed whenever the same step is a possible step and the justifications hold. For example, in using the two-object case as guidance to the three- (or four-) object problem, the failure justification for moving the rocket - "no-relevant-ops (at ROCKET locA)" - is tested and this step is not replayed until all the objects are loaded into the rocket.

| New Prob | Base Search | Replayed cases Case 2objs | Case 3objs | Case 4objs |
|---|---|---|---|---|
| 2objs | 4.5s | 2s | 2s | 2s |
| 3objs | 14.75s | 4.75s | 3.25s | 3.25s |
| 4objs | 117.5s | 7.75s | 7.75s | 5.75s |

*Table 1: Replaying a Justified Past Solution.*

The improvements obtained are high as the new cases are extensions of the previous cases used for guidance. Maximal improvement is achieved when the case and the new problem differ substantially (two-objects and four-objects respectively). We further show experiments from other two substantially more complicated domains.

## Process-Job Planning and Extended-STRIPS Examples

We ran NoLimit without analogy over a set of problems in the process-job planning and in the extended-STRIPS domains [1]. We accumulated a library of cases, i.e. annotated derivational solution traces. We then ran again a new set of problems using the case library organized as a linear sequence of past problem solving episodes. We used a direct rudimentary similarity metric [Veloso and Carbonell, 1991a] that matched the goal predicates, allowed substitutions for elements of the same type, and did not consider any relevant correlations. Figures 6(a) and (b) show the results for these two domains. We plotted the average cumulative number of nodes searched.

We note from the results that analogy showed an improvement over basic search both for the process-job planning and scheduling domain, and for the extended-STRIPS domain. The test problems in these domains are considerably more complex than in the simple transportation problems shown above. However even the simple similarity metric used can lead to search improvements in the problem solver. In [Veloso and Carbonell, 1991a] we show results of further search reduction upon using a more sophisticated similarity metric. These results illustrate the point that learning from analyzing successful and failed choice points reduces the search effort of the casual-commitment problem solver.

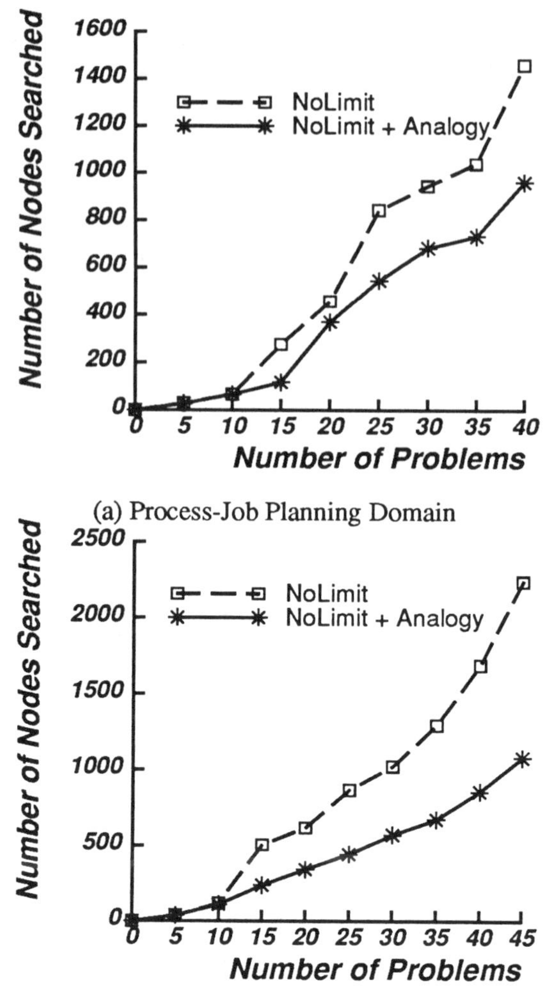


(a) Process-Job Planning Domain

(b) Extended-STRIPS Domain

*Figure 6: Results in the Process-Job Planning and Extended-STRIPS Domains.*

## Conclusions and Future Work

In this paper we reported on NoLimit as a completely implemented nonlinear problem solver that uses an informed casual-commitment strategy to guide its search process. NoLimit has the ability to call user-given or automatically learned control knowledge in all its choice points. NoLimit efficiently solves problems in several other different domains, e.g. multi-agent stripsworld, blocksworld, matrix-algebra, transportation, and process-job planning worlds. The system has additional features not reported here, such as a sophisticated TMS that enables deduction and control of beliefs, and a type hierarchy to organize the objects of the world.

We showed how the casual-commitment planner becomes more efficient by learning by derivational analogy. We only covered in this paper very briefly the derivational analogy full mechanism. We focused on showing the results obtained in terms of search reduction. Our current work in analogical problem solving has new contributions beyond the original derivational analogy framework as presented in [Carbonell, 1986]. Besides the memory model under development [Veloso and Carbonell, 1991a] we refined the initial framework

[1] This set is a sampled subset of the original set used by [Minton, 1988].

in the context of a nonlinear planner. We deal therefore with a considerably larger space of decisions and with more complex planning problems.

Previous work in the linear planner of PRODIGY used explanation-based learning (EBL) techniques [Minton, 1988] to extract from a problem solving trace the explanation chain responsible for a success or failure and compile search control rules therefrom. The axiomatized domain knowledge was also used to learn abstraction layers [Knoblock, 1991], and statically generate control rules [Etzioni, 1990]. We are in the process of extending the nonlinear planner into a hierarchical one by using Knoblock's abstraction hierarchies. We are also analyzing the extension of the EBL and STATIC modules to the nonlinear framework. The use of casual commitment as in the linear planner makes the extension look promisingly successful.

## Acknowledgments

We would like to thank Daniel Borrajo in particular for a major part of NoLimit's implementation, Jaime Carbonell, Oren Etzioni, Yolanda Gil, Craig Knoblock, and Alicia Pérez for many useful comments on this paper, and the whole PRODIGY research group for helpful discussions.

## References

[Anderson and Farley, 1990] J. S. Anderson and A. M. Farley. Partial commitment in plan composition. Technical Report TR-90-11, Computer Science Department, University of Oregon, 1990.

[Carbonell and Veloso, 1988] J. G. Carbonell and M. M. Veloso. Integrating derivational analogy into a general problem solving architecture. In *Proceedings of the First Workshop on Case-Based Reasoning*. Morgan Kaufmann, May 1988.

[Carbonell *et al.*, 1990] J. G. Carbonell, C. A. Knoblock, and S. Minton. Prodigy: An integrated architecture for planning and learning. In Kurt VanLehn, editor, *Architectures for Intelligence*. Erlbaum, Hillsdale, NJ, 1990.

[Carbonell, 1986] J. G. Carbonell. Derivational analogy: A theory of reconstructive problem solving and expertise acquisition. In R. S. Michalski, J. G. Carbonell, and T. M. Mitchell, editors, *Machine Learning, An Artificial Intelligence Approach, Volume II*. Morgan Kaufman, 1986.

[Chapman, 1987] D. Chapman. Planning for conjunctive goals. *Artificial Intelligence*, 32:333–378, 1987.

[Drummond and Currie, 1989] M. Drummond and K. Currie. Goal ordering in partially ordered plans. In *Proceedings of IJCAI-89*, pages 960–965, 1989.

[Etzioni, 1990] Oren Etzioni. *A Structural Theory of Explanation-Based Learning*. PhD thesis, School of Computer Science, Carnegie Mellon University, 1990. Available as technical report CMU-CS-90-185.

[Fikes and Nilsson, 1971] R. E. Fikes and N. J. Nilsson. Strips: A new approach to the application of theorem proving to problem solving. *Artificial Intelligence*, 2:189–208, 1971.

[Kambhampati, 1989] Subbarao Kambhampati. *Flexible Reuse and Modification in Hierarchical Planning: A Validation Structure Based Approach*. PhD thesis, Computer Vision Laboratory, Center for Automation Research, University of Maryland, 1989.

[Knoblock, 1991] Craig A. Knoblock. *Automatically Generating Abstractions for Problem Solving*. PhD thesis, School of Computer Science, Carnegie Mellon University, Pittsburgh, PA, 1991. forthcoming.

[Minton *et al.*, 1989] S. Minton, C. A. Knoblock, D. R. Kuokka, Y. Gil, R. L. Joseph, and J. G. Carbonell. PRODIGY 2.0: The manual and tutorial. Technical Report CMU-CS-89-146, School of Computer Science, Carnegie Mellon University, 1989.

[Minton, 1985] S. Minton. Selectively generalizing plans for problem solving. In *Proceedings of AAAI-85*, pages 596–599, 1985.

[Minton, 1988] Steven Minton. *Learning Effective Search Control Knowledge: An Explanation-Based Approach*. PhD thesis, Computer Science Department, Carnegie Mellon University, 1988.

[Rosenbloom *et al.*, 1991] P. S. Rosenbloom, S. Lee, and A. Unruh. Responding to impasses in memory-driven behavior: A framework for planning. In *Proceedings of the European Working Session on Learning*. Springer-Verlag, March 1991.

[Sacerdoti, 1975] E. D. Sacerdoti. The nonlinear nature of plans. In *Proceedings of IJCAI-75*, pages 206–213, 1975.

[Sussman, 1973] G. J. Sussman. A computational model of skill acquisition. Technical Report AI-TR-297, Artificial Intelligence Laboratory, MIT, 1973.

[Veloso and Carbonell, 1990] M. M. Veloso and J. G. Carbonell. Integrating analogy into a general problem-solving architecture. In Maria Zemankova and Zbigniew Ras, editors, *Intelligent Systems*. Elis Horwood, 1990.

[Veloso and Carbonell, 1991a] M. M. Veloso and J. G. Carbonell. The effect of knowledge-based case retrieval and similarity measure in analogical replay. In *Proceedings of the 1991 DARPA Workshop on Case-Based Reasoning*, May 1991.

[Veloso and Carbonell, 1991b] M. M. Veloso and J. G. Carbonell. Learning by analogical replay in PRODIGY: First results. In *Proceedings of the European Working Session on Learning*. Springer-Verlag, March 1991.

[Veloso *et al.*, 1990] M. M. Veloso, M. A. Pérez, and J. G. Carbonell. Nonlinear planning with parallel resource allocation. In *Proceedings of the DARPA Workshop on Innovative Approaches to Planning, Scheduling, and Control*. Morgan Kaufmann, November 1990.

[Veloso, 1989] M. M. Veloso. Nonlinear problem solving using intelligent casual-commitment. Technical Report CMU-CS-89-210, School of Computer Science, Carnegie Mellon University, 1989.

[Wilkins, 1989] D. E. Wilkins. Can AI planners solve practical problems? Technical Note 468R, SRI International, 1989.

# A Neural Model of Temporal Sequence Generation with Interval Maintenance[1]

**DeLiang Wang and Michael A. Arbib**

Center for Neural Engineering, University of Southern California
Los Angeles, CA 90089-2520, USA

**Abstract**

Based on an interference theory of forgetting in short-term memory (STM), we model STM by a network of neural units with mutual inhibition. Sequences are acquired by combining a Hebbian learning rule and a normalization rule with sequential system activation. As long as sequences are acquired, they can be recognized without being affected by speeds in presentation. The model of sequence reproduction consists of two reciprocally connected networks, one of which behaves as sequence recognizers. Reproduction of complex sequences is shown to be able to maintain interval lengths of sequence components. A mechanism of degree self-tuning based on a global inhibitor is proposed for the model to optimally learn required context lengths in order to disambiguate associations in complex sequence reproduction.

## Introduction

A temporal sequence $S$ is denoted as: $p_1$–$p_2$–...–$p_N$, and the length of a sequence is the number of components in the sequence. Any $p_i$-$p_{i+1}$-...-$p_j$, where $1 \le i \le j \le N$, is called a subsequence of $S$. If $S$ contains repetitions of the same subsequence, like *A-B* in *C-A-B-D-A-B-E*, it is called a complex sequence, otherwise a simple sequence. In complex sequences, the correct successor can be determined only by knowing a subsequence prior to it. We refer to the prior subsequence required to cue unambiguously the current symbol $p_i$ in $S$ as the context of $p_i$, and the length of the context as the degree of $p_i$. The degree of a sequence is the maximum degree of its components.

Neural networks to reproduce a temporal sequence of input stimuli have been previously studied by a number of investigators (among others see Grossberg 1969; Dehaene, Changeux, & Nadal 1987; Kühn, van Hemmen, & Riedel 1989). In most of these models, reproduction of complex sequences poses great difficulty. Recently, we have proposed a new mechanism for learning temporal sequences (Wang & Arbib 1990) in which we model STM by units comprising recurrent excitatory connections between two local neuron populations. Each neuron population is represented by a single quantity corresponding to local field potential. The activity induced by an input signal to a unit oscillates with damping. By applying a Hebbian learning rule at each synapse and a normalization rule among all synapses to a unit, we have demonstrated that the neural networks with this model of STM are able to learn and reproduce complex temporal sequences. What distinguishes our model from others are two basic hypotheses embodied in the model: (1) We assume that there is a common mechanism to process both complex sequences and simple sequences; (2) Reproduction of a component in a sequence is based on recognition of the context of the component.

Since STM is modeled by decay, it has a fixed temporal course, which makes the previous model unable to handle the *time-warp* problem. For a solution to the time-warp problem, we wish that a network can recognize a time-warped sequence for sequence recognition, whereas for reproduction we wish that a network can reproduce a sequence with the same temporal course as the learned sequence. This is the central theme of the present paper.

## A Computational Model of STM

A model of STM must provide the following four basic functions:

(1) Maintaining a symbol for a short time period. What causes forgetting? An *interference* theory proposes that other materials or tasks interfere with memory and thus cause forgetting. A *decay* theory proposes that forgetting occurs even if the subject had to do nothing over the retention interval, so long as the subject did not rehearse the material. (2) Maintaining a number of symbols. Miller (1956) tells us that the number is about seven. (3) Coding the order of input symbols. (4) Coding the length of the presentation of each symbol. The function of STM provides first level information for solving the time-warp problem. When learning a sequence, one can recognize it even though each component of the sequence is presented at considerably different intervals. This function is called *interval invariance*. Yet, a professional musician can recall a multiple-page score, reproducing almost exactly the memorized length of each note. This function is called *interval maintenance*.

Our previous model cannot code the length of each symbol presentation, and therefore cannot solve the time-warp problem. Furthermore, the model conforms with the decay theory of forgetting, whereas the current majority view seems to be that, although some decay may occur, the amount of forgetting caused by decay is substantially less than the amount caused by interference (Murdock 1987). Our following model is based on the interference theory.

Let us assume that there are $n$ memory units, numbered 1, 2, ..., $n$, with each unit inhibited by all the other units,

[1] The research described in this paper was suported in part by grant no. 1RO1 NS 24926 from the NIH (M.A.A, PI).

as shown in Fig.1. Each unit receives an external input $E_i$, which is 1 so long as the external input is on and 0 otherwise. The *internal* state of unit $i$, $s_i$, is defined as

$$s_i(t) = \begin{cases} 1 & \text{if } E_i(t)=1,\ E_i(t-1)=0 \\ 0 & \text{otherwise} \end{cases} \tag{1}$$

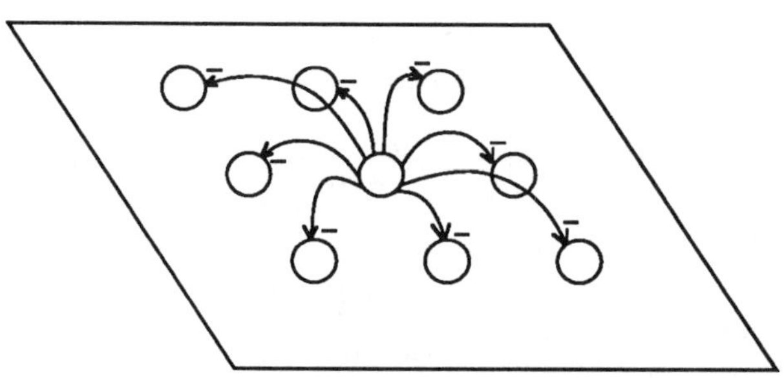

**Figure 1.** Diagram of the STM model. Each unit projects and inhibits all the other units in the model. Shown in the figure is only outgoing projections from one unit. Minus sign indicates inhibition.

From the definition we can see that the internal state is activated only by the beginning of an external input. The *excitation level* of each unit has value range {0, 1, ..., T}, and is defined as

$$x_i(t) = \begin{cases} T & \text{if } s_i(t)=1 \\ x_i(t-1) - 1 & \text{if } x_i(t-1)>0,\ y_i(t)=1 \\ x_i(t-1) & \text{otherwise} \end{cases} \tag{2}$$

where $y_i$ represents overall inhibition that unit $i$ receives from the other units, formulated as

$$y_i(t) = f(\sum_{j \neq i} s_j(t-1) - 1) \tag{3}$$ [2]

$$\text{with } f(x) = \begin{cases} 1 & \text{if } x \geq 0 \\ 0 & \text{otherwise} \end{cases} \tag{4}$$

From the above definitions we see that whenever $s_i(t) = 1$, $x_i(t)$ is brought to its highest value $T$ and unit $i$ is activated. If any of the units is activated, the inhibition that it exerts on the rest of the network will drive all other active units, i.e. those whose excitation levels are larger than 0, down to the next lower level.

This model satisfies the above four requirements for an adequate STM model. It preserves a symbol on a unit whose excitation level codes the item. Let us assume that external inputs arrive at STM serially (it is easy to serialize simultaneous inputs by a competitive network). Any new item input to STM decrements the excitation levels of all active units in STM. Therefore STM can at most code $T$ items so that $T$ is the capacity of the STM model. A symbol gets lost from the STM model because there are other more recent symbols input into the model, conforming with the interference theory. The order of input symbols is coded since the larger the excitation level of a unit, the more recent is the symbol represented by the unit. Finally, the length of a symbol's presentation is reflected by the time period while the corresponding external input is on, and its coding will be given later.

[2] Since the weights of inhibitory connections are the same, the mutual inhibitory connections can be replaced by an global inhibitor. An global inhibitor can reduce the number of connections by one order of magnitute, but results in a less reliable system due to information centralization in the inhibitor.

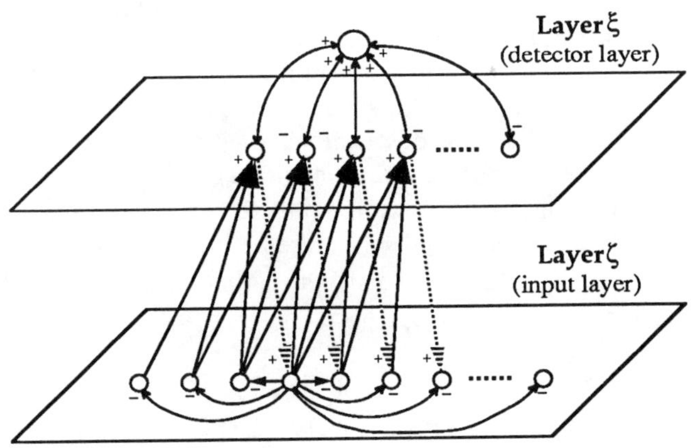


**Figure 2.** Architecture for complex sequence reproduction. Within layer $\zeta$ (the input layer), every unit inhibits every other one to form the STM model shown in Fig.1. Within layer $\xi$ (the detector layer), all units project to a global inhibitor which further projects back to them. Plus sign indicates excitation, and minus sign indicates inhibition.

## Network Architecture

The structure of the model for sequence reproduction has two layers, as shown in Figure 2. Layer $\zeta$ is called the input layer, which basically serves as a STM model shown in Fig.1. Multiple occurences of a particular symbol in a sequence is represented by one single unit in this layer, so different units represent different spatial patterns in layer $\zeta$. Units in layer $\xi$ function as sequence detectors, and there is a global inhibitor within this layer (see footnote 2). These units recognize the contexts of individual components in a sequence, and anticipate the occurence of these components. Layer $\xi$ connects with layer $\zeta$ bidirectionally, and before training connections between them are complete. The projections shown in Fig.2 depict what results from training, such that unit $i$ in layer $\xi$ receives projections only from those units in $\zeta$ that represents symbols in the context detected by unit $i$, and unit $j$ in layer $\zeta$ only receives units in $\xi$ that anticipate the occurence of the symbol represented by unit $j$. This resulted connection pattern is formed through learning. During the training process, a sequence with various component intervals is presented to layer $\zeta$. At the end of each component presentation, a unit in layer $\xi$ is randomly selected (but fixed in successive trainings) to fire. The recurrent connections from layer $\xi$ to layer $\zeta$ are formed according to a Hebbian rule as following. If unit $i$ in layer $\zeta$ (recorded as $<i, \zeta>$) and unit $j$ in layer $\xi$ (recorded as $<j, \xi>$) are firing simultaneously then a connection link from $<j, \xi>$ to $<i, \zeta>$ is established, and its weight will be defined later. All connection weights from units in $\xi$ to ones in $\zeta$ are initially zero.

We proposed in the previous paper (Wang & Arbib 1990) that a unit was represented by an expanded network, such that it has multiple terminals to hold different occurences of a symbol. Each terminal directly connects to other units, and thus a unit has multiple channels to connect to another unit. The following description combines this idea for solving the overwriting problem with the new STM model.

Suppose unit $<i, \zeta>$ has $m$ terminals, and the excitation level of its $r$th terminal is represented by $x_{ir}$. The STM model (Eq.1 through Eq.4) and the definitions of $E_i$, $s_i$, and $y_i$ remain the same except

$$x_{ir}(t) = \begin{cases} T & \text{if } s_i(t)=1, r=1 \\ x_{i,r-1}(t-1) - 1 & \text{if } s_i(t)=1, r>1, x_{i,r-1}(t-1)>0 \\ x_{ir}(t-1) - 1 & \text{if } s_i(t)=0, x_{ir}(t-1)>0, y_i(t)=1 \\ x_{ir}(t-1) & \text{otherwise} \end{cases} \tag{5}$$

The global inhibitor in layer $\xi$ receives input from all units in the layer and projects back to them. A degree parameter $d_i$ is introduced for $<i, \xi>$, and it affects the dynamics of the internal state of $<i, \xi>$ in the following way

$$s_i^{\xi}(t) = f(\sum_{j=1}^{n} \sum_{r=1}^{m} W_{ij}^{r}\, h(x_{jr}(t-1), d_i) + I_i^{\xi}(t-1) - \Gamma_i^{\xi}) \tag{6}$$

$$h(x, y) = \begin{cases} x & \text{if } x > T - y \\ 0 & \text{otherwise} \end{cases} \tag{7}$$

where label $\xi$ in (6) indicates layer $\xi$, $x_{jr}$ is the excitation level of the $r$th terminal of unit $<j, \zeta>$, and $W_{ij}^{r}$ represents the connection weight from the $r$th terminal of unit $<j, \zeta>$ to $<i, \xi>$. Symbols $n$ and $m$ stand for the number of units and the number of terminals for each unit in layer $\zeta$ respectively. The domain of $d_i$ is $\{1, 2, ..., T\}$. Through function $h(x, y)$ the role of $d_i$ is to gate in certain excitation levels of units in layer $\zeta$. Obviously, the larger is $d_i$, the more items can $<i, \xi>$ sense from layer $\zeta$. Learning, or modification of connection weights $W_{ij}^{r}$, follows a Hebbian rule and a later normalization as follows

$$\begin{cases} \widehat{W}_{ij}^{r}(t) = W_{ij}^{r}(t-1) + C_i\, s_i^{\xi}(t)\, h(x_{jr}(t), d_i) \\ W_{ij}^{r}(t) = \widehat{W}_{ij}^{r}(t) \,/ \sum_{j'=1}^{n} \sum_{r'=1}^{m} \widehat{W}_{ij'}^{r'}(t) \end{cases} \tag{8}$$

where $C_i$ is a gain factor of learning. The effect of learning on the detector is to change the distribution of all weights to that unit, so it is reasonable to assume that initially all weights are set equal.

Since without a further activation the excitation level of a unit in layer $\zeta$ is monotonically decreasing, the formal analysis in our previous paper (Wang & Arbib 1990) applies. In particular, if the threshold of unit $<i, \xi>$ in (6) is set as

$$\Gamma_i^{\xi} = \frac{2}{d_i(2T-d_i+1)} \sum_{i=1}^{d_i} (T-d_i+i)^2 \tag{9}$$

then the result of training is to build up activity so as to fire the detector by the presentation of a specific subsequence. Furthermore, after the detector has learned the sequence (simple or complex), only presentation of that sequence induces the maximum activity on the detector unit, regardless of presentation speed of the sequence. The idea behind this interval invariance is that during presentation of a sequence component, only the beginning portion of presentation is captured by the recognition model (cf. Eq.1), and therefore it does not matter how long that presentation lasts. Separation of effective input from external input is an intrinsic property of the STM model, which exhibits differences in computational power resulting from different models of basic brain processes like STM.

## Degree Self-tuning

Let the activity of the global inhibitor of layer $\xi$ be represented by $z$, and $q$ represent the number of units in layer $\xi$. Variable $z$ is defined as

$$z(t) = f[\sum_{i=1}^{q} s_i^{\xi}(t-1) - 2] \tag{10}$$

and therefore the inhibitor will be activated if there is more than one unit firing simultaneously in layer $\xi$. According to (6), the internal state $s_i^{\xi}(t)$ can be triggered either by system input (called *attention*) through $I_i^{\xi}(t-1)$ or by input signals from layer $\zeta$. The latter is called *anticipation*. What the inhibitor actually does is to detect conflicts among those detectors in layer $\xi$. Since system attention is always sequential, the inhibitor can only be activated by conflicting attention and anticipation or just by conflicting anticipation of the detector layer.

Degree $d_i$ ($i = 1, ..., q$) is initially set to 1. Self tuning of $d_i$ is done according to

$$d_i(t) = d_i(t\text{-}1) + 1 \qquad \text{if } s_i^{\xi}(t\text{-}1){=}1,\ z(t){=}1,\ d_i(t\text{-}1){<}T \quad (11)$$

that is, the degree of $<i, \xi>$ increments if this unit together with other units causes activation of the global inhibitor. If the degree of $<i, \xi>$ increments, there will be one more item from the input layer that can be sensed by $<i, \xi>$. Thus the previously learned weight distribution to the unit (see Eq. 8) will have to change its direction of distribution. In the situation, the model re-initiates the weight distribution to $<i, \xi>$ and threshold $\Gamma_i^{\xi}$ is also modified according to (9) based on the new value of $d_i$. From (6), (7) and (8), it is clear that if $d_i(t)$ grew larger than $T$, the STM capacity of layer $\zeta$, it would be equivalent to that $d_i(t) = T$ in the dynamics of the internal state and weight distribution of $<i,\xi>$. That is why $d_i(t)$ has an upper limit of $T$.

A computer simulation of the model was conducted for reproducing a complex sequence $S_C$: *J-B-A-C-D-A-B-A-E-F-A-B-A-G-H-A-B-A-H-I*. Learning a complex sequence is slower than learning a simple sequence, because the complex sequence needs dynamically increasing the degrees of certain detectors, and each time such self organization is done earlier training of those detectors has to be discarded. Roughly speaking, time required for training increases linearly with the degree of a sequence. It took 18 training trials before the model learned to reproduce $S_C$, whereas 6 trials suffice to reproduce a simple sequence. The degree vector acquired by the degree self-tuning mechanism is {1, 2, 3, 1, 1, 2, 3, 4, 1, 1, 2, 3, 4, 1, 2, 2, 3, 4, 2} for those detectors. The ninth component *E*, for example, requires to memorize the prior subsequence of 4 components *D-A-B-A* in order to be generated; and the second component *B*, however, only requires to memorize the previous component *J* in order to be generated.

The above neural algorithm optimally identifies amount of context required to reproduce any complex temporal sequence unambiguously. The same problem of finding minimum amount of context has been studied by Kohonen (1987) for producing unambiguous inference rules in sequence generation. The proposed solution relies on explicit rules for resolving inference conflicts. A basic difference of our proposal from his is that we do not resort to any external rules. Units representing symbols and detectors in our model are connected in a neuron-like manner, and communication among units is typically neural.

## Interval Maintenance

In our model, the interval length of a component presentation is the time period during which the external input of the unit corresponding to that component equals 1. This is equivalent to the period when the excitation level of the unit equals $T$. In sequence reproduction, a unit in layer $\xi$ detects the onset of the context of a component in order to trigger that component in the reproduction process. In $S_C$ above, for example, there is a detector in layer $\xi$ that is trained to detect the context *D-A-B-A* and to anticipate the onset of symbol *E*. According to the model, after training this detector is activated just one time step after the second *A* starts to occur (see Eq.6). But *E* should not be triggered until the whole interval of the *A* occurrence has elapsed. The idea for interval maintenance is to code intervals by connection weights from the detector layer to the input layer. Since the backward projections from layer $\xi$ to $\zeta$ are many-to-one correspondence, the interval of a symbol presentation can be simply coded as the reciprocal of the corresponding connection weight, so that temporal integration of the entire interval is required to trigger the next component.

In general, one interval series of presentation may be different from another one. In order to cope with this situation, instead of storing one interval directly in a weight, two parameters are stored in the connection terminal, one is an average $\mu$ of different training intervals and another is a deviation $\sigma^2$. During reproduction of a sequence, a Gaussian number is generated based on $\mu$ and $\sigma^2$, to control a specific interval. Each generated interval will also modify $\mu$ and $\sigma^2$ like a presentation interval. Therefore, learning is nothing but formation of $\mu$ and $\sigma^2$. Let $e_i$ represent the interval of the $i$th presentation of a symbol. Two factors are taken into consideration for forming $\mu$ and $\sigma^2$. First, each interval should contribute a certain amount. This is called an *averaging factor*. Second, a recent interval should have more impact than a remote one. This is called a *recency factor*. These two factors are embodied in the following learning rules.

$$\begin{cases} \mu_1 = e_1 \\ \mu_{k+1} = (1-\beta)\,\mu_k + \beta\, e_{k+1} \end{cases} \tag{12}$$

where $\beta$ is the recency parameter ranging between 0 and 1, which describes that except the first interval the most recent interval has a constant amount of contribution, regardless of the presentation history.

The following recurrence learning rule for the deviation can be derived from (12)

$$\begin{cases} \sigma_1^2 = 0 \\ \sigma_k^2 = \dfrac{k(1-\beta)}{k-1}\left[\dfrac{k-2}{k-1}\,\sigma_{k-1}^2 + \beta(e_k - \mu_{k-1})^2\right] \end{cases} \tag{13}$$

, and it is easy to see that $\sigma_k^2 = 0$, if $e_1 = ... = e_k$.

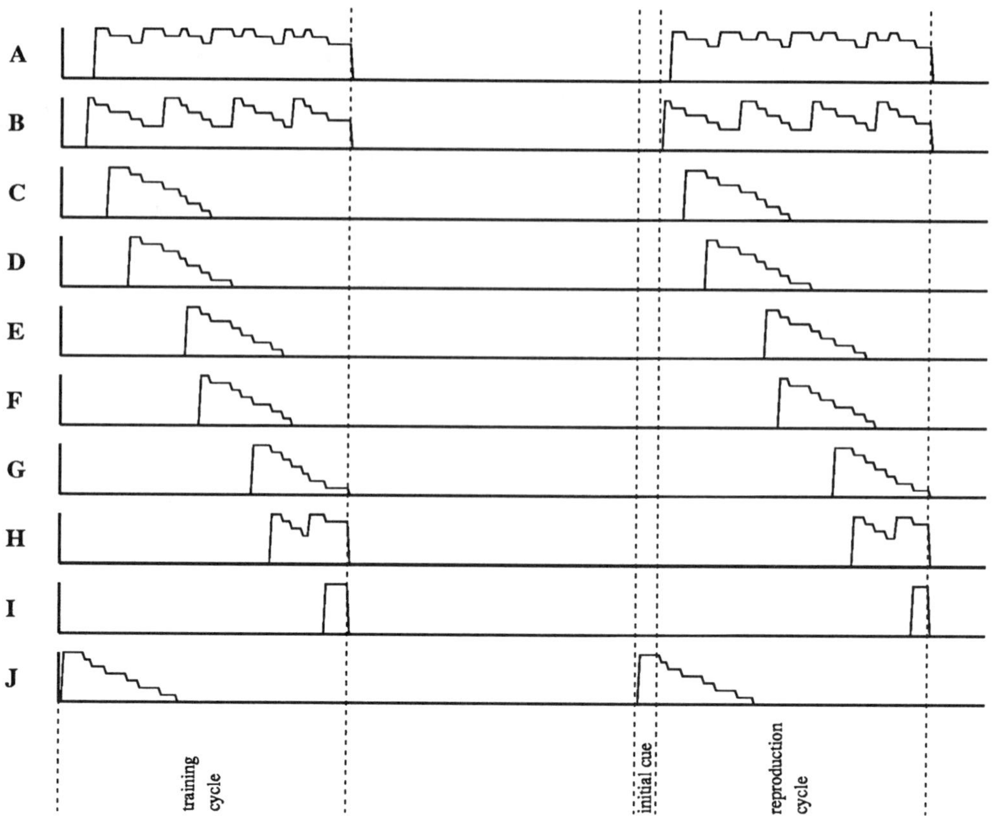


**Figure 3**. Reproduction of the complex sequence $S_C$: *J-B-A-C-D-A-B-A-E-F-A-B-A-G-H-A-B-A-H-I*. The interval series {9,3,6,9,5,9,7,3,6,4,9,4,5,8,5,4,5,3,7,8} was first randomly generated, and fixed in subsequent training trials. All units in layer $\zeta$ have 3 terminals, and $C_i = 0.3$. The other parameters are $\beta = 0.3$, and $T = 7$.

With the learning rule of (12) and (13), interval maintenance defined above is thus achieved. A computer simulation of the model was conducted to reproduce the complex sequence $S_C$. As previously stated, the model took 18 training trials to learn the sequence. The number of trials is basically decided by requirement of degree self-tuning. After learning, the entire sequence with various interval lengths was able to be reproduced by the initial context of the sequence, subsequence *J* in this case. Fig.3 presents the simulation result, which contains a temporal course of the last training trial together with the reproduction process. Since in this simulation the speed of presentation is the same from one trial to another, the acquired deviation for every link interval is zero. Therefore the time course of the sequence is faithfully preserved in reproduction.

In summary, this article presents a neural model of temporal sequence reproduction, which is based on an interference model of short-term memory. The model of neural circuit proposed reproduces any complex temporal sequence which may be distorted in time (time-warped). The new abilities demonstrated in this paper, particularly interval maintenance, demonstrate that a dramatic difference in computational power could be lead to by results from basic studies of cognitive science.

## References

Dehaene, T.; Changeux, J.P.; and Nadal, J.P. 1987. Neural Networks That Learn Temporal Sequences by Selection. *Proceedings of National Academy of Sciences USA* 84: 2727-2731.

Grossberg, S. 1969. Some Networks That Can Learn, Remember, and Reproduce Any Number of Complicated Space-time Patterns, I. *Journal of Mathematics and Mechanics* 19: 53-91.

Kohonen, T. 1987. Dynamically Expanding Context, with Application to the Correction of Symbol Strings in the Recognition of Continuous Speech. In Proceedings of the International Conference on Neural Networks II, 3-9. San Diego, CA.

Kühn, R., van Hemmen, J.L., and Riedel, U. 1989. Complex Temporal Association in Neural Networks. *Journal of Physics A* 22: 3123-3135.

Miller, G.A. 1956. The Magical Number Seven, Plus or Minus Two: Some Limits on Our Capacity for Processing Information. *Psychological Review* 63: 81-97.

Murdock, B.B., Jr. 1987. Serial-order Effects in a Distributed-memory Model. In Gorfein, D.S. and Hoffman, R.R. eds. *Memory and Learning: The Ebbinghaus Centennial Conference*, 227-310. Hillsdale, NJ: Erlbaum.

Wang, D.L. and Arbib, M.A. 1990. Complex Temporal Sequence Learning Based on Short-term Memory. *Proceedings of the IEEE* 78: 1536-1543.

# A Continuum of Induction Methods for Learning Probability Distributions with Generalization

Dekai Wu*
Computer Science Division
University of California at Berkeley
Berkeley, CA 94720 U.S.A.
*dekai@cs.berkeley.edu*

**Abstract**

Probabilistic models of pattern completion have several advantages, namely, ability to handle arbitrary conceptual representations including compositional structures, and explicitness of distributional assumptions. However, a gap in the theory of induction of priors has hindered probabilistic modeling of cognitive generalization biases. We propose a family of methods parameterized along a value $\gamma$ that controls the degree to which the probability distribution being induced generalizes from the training set. The extremes of the $\gamma$-continuum correspond to relative frequency methods and extreme maximum entropy methods. The methods apply to a wide range of pattern representations including simple feature vectors as well as frame-like feature DAGs.

## Introduction

The motivations for this work arise from the shortcomings of existing theoretical frameworks in two fields:

- Many neural network pattern completion models have the desirable characteristic of being inherently biased to generalize from training data. Two drawbacks, however, are: (a) frequently there are no clearly specified desiderata on the nature of statistical distributions to be learned by a neural net, and (b) we are currently unable to efficiently represent compositional structures as feature vectors.[1]

- Probabilistic and statistical inductive models, being symbolic, can easily handle compositional structures. However, there is a lack of models that can be biased to generalize from training data; specifically, the most common methods for inducing prior probability distributions—relative frequency priors and maximum entropy priors—are inadequate.

To address this gap we propose a family of inductive methods, called the $\gamma$-continuum, that can be thought of in several ways:

- From the associationist point of view, a functional specification for a class of pattern completion models.
- From the learning point of view, a non-Bayesian inductive learning method for a probabilistic inference engine. An inductive bias is determined by a set of *abstractive relations.*
- From the probability theory point of view, a method for generating priors from a training set. The method incorporates an a priori *abstractive bias* that causes the model to make generalizations.

Our driving application is probabilistic pattern completion to support integrated natural language parsing and semantic interpretation, where the patterns combine lexical, syntactic, and semantic structures. In a probabilistic pattern completion model, the input is an abstract or partial pattern, and the task is to select the most probable complete pattern.

The proposed methods are more flexible than neural networks with respect to representation constraints; concepts need not be represented as feature vectors, but only need to satisfy a weaker semi-lattice constraint, explained below. Limited forms of compositional conceptual structure are permitted. At the same time, the nature of the probability distributions that can be learned is clearly formulated, and these distributions are better for modeling generalization from a training set than either relative frequency or maximum entropy priors.

Our proposal fills a gap in the existing theory of probability distribution induction. However, it is not intended that the distributions generated by our methods necessarily be evaluable by computationally tractable means. The methods are information-theoretic functional specifications, for which different approximation heuristics may be appropriate depending upon the domain.

## Pattern Structure and the Abstraction Space

We shall only consider examples where the patterns (instances) are encoded using feature-vector and feature-DAG (frame-like) representations, though the internal structure of patterns is of no consequence to the inductive methods and many other representations could be used as well. The set

*This paper has benefitted greatly from helpful discussions with Terry Regier and Steve Omohundro, and I am grateful to Marti Hearst for implementing code to generate the induced distributions. Thanks to Nigel Ward for proofreading, and also to Robert Wilensky, Jerome Feldman, and the members of the BAIR and $L_0$ seminars. This research was sponsored in part by the Defense Advanced Research Projects Agency (DoD), monitored by the Space and Naval Warfare Systems Command under N00039-88-C-0292, the Office of Naval Research under contract N00014-89-J-3205, and the Sloan Foundation under grant 86-10-3.

[1] A number of recent proposals employ recurrent nets to achieve "dynamic compositionality" that can sequentially "expand out" compositional structures (e.g., Pollack 1989, 1990), but there is little consensus as to the limits of such approaches.

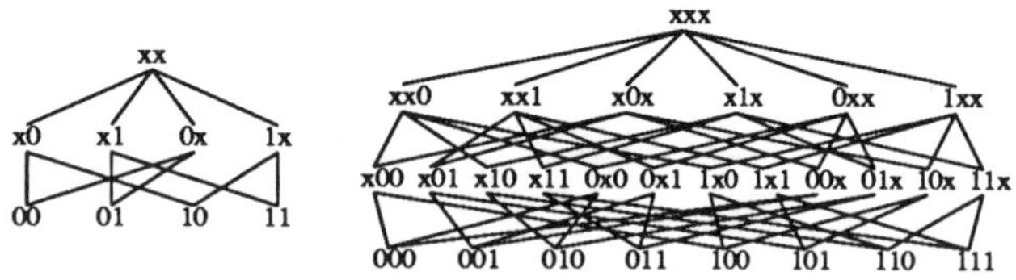


Figure 1: Feature abstraction semi-lattices.

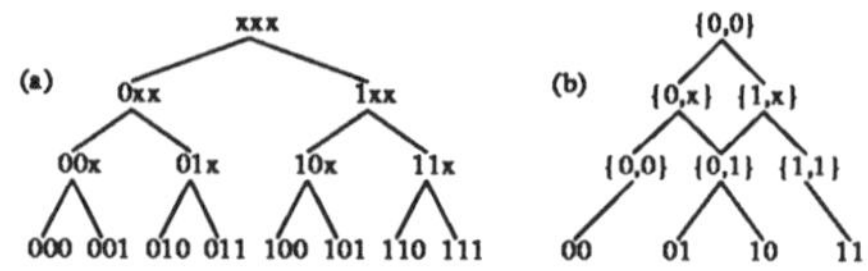


Figure 2: Semi-lattices for (a) incomplete feature abstraction, (b) position-insensitive feature abstraction.

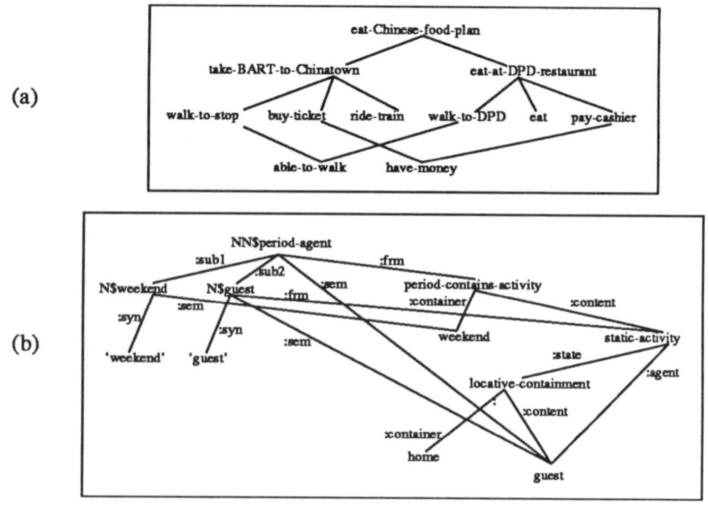


Figure 3: Feature DAGs (see text).

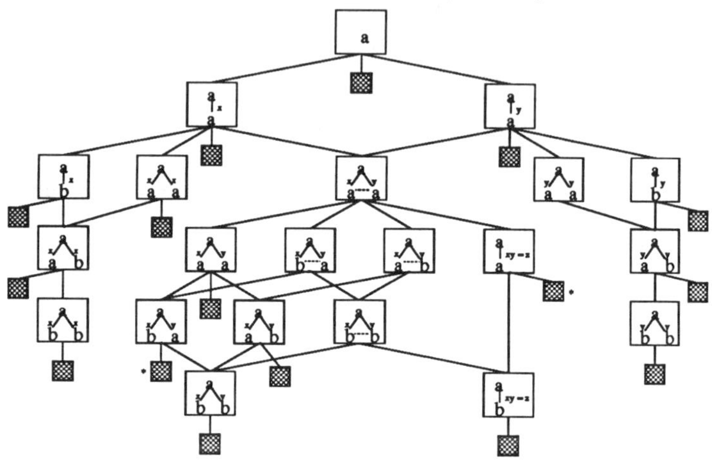

Figure 4: Semi-lattice for a feature-DAG pattern space. The leaf nodes (shaded) are the simple events. Here each feature-DAG is restricted to depth 2 and branch factor 2; there are only two concepts $a$ and $b$, and $a$ is superordinate to $b$; and there are two primitive roles $x$ and $y$ that combine to form a third composite role $z$.

of possible patterns must be finite (though arbitrarily large) and forms the space of simple events. To perform pattern completion using probabilistic inference, we need to know the probability distribution over these events.

Compound events constitute abstractions over groups of patterns. Note that all patterns are defined to be simple events; compound events or abstractions are not proper patterns, but partial or incomplete patterns. This usage should not be confused with the conventional AI use of the notion of abstraction as an epistemological relationship between two concepts.

The shape of the abstraction hierarchy is determined by a set of abstractive relations. Figure 1 shows, for 2- and 3-bit vector patterns, the hierarchy determined by a feature-abstractive relation that substitutes "x" or "don't care" bits for feature values. The leaf nodes are complete patterns (simple events); the internal nodes are incomplete patterns (compound events).

In a pattern completion task, the input is an internal node representing an abstract or partial pattern. The task of completing the pattern corresponds to selecting the most probable leaf node (complete pattern) under the internal node. (The term "simple event" is somewhat counterintuitive when speaking of complete patterns, which are more fleshed out than incomplete ones.)

Minimal constraints are imposed on the shape of the abstraction space. In fact, the only constraint is that the abstractive relations must determine a *semi-lattice* hierarchy, meaning that for any two concepts there must be a unique least upper bound (most specific common ancestor). Figure 2 shows other useful examples of abstractive relations.

The sorts of patterns that motivated development of the $\gamma$-continuum are more complex than feature-vectors. These patterns, which derive from semantic network and predicate logic languages, can be represented as *feature-DAGs*, and allow compositional structures and variable unification. Figure 3 shows two (complete) patterns that demonstrate how feature-DAG representations can be used (the details are unimportant here). An example of a feature-DAG representing plan decomposition is shown in (a). For our parsing and interpretation application, a sample feature-DAG for the nominal compound *weekend guest* is shown in (b). Details on how we map unification-grammar structures into feature-DAGs (for a simpler probability model) are given in an earlier paper (Wu 1990).

Given these sorts of feature-DAG patterns, Figure 4 shows the semi-lattice determined by the abstractive relations we use for our parsing and semantic interpretation model, making some simplifying restrictions. The abstractive relations determine how to generate all the ancestors of any pattern. The four relations used here are: superordinate concept substitution (an ancestor can be generated by replacing any concept with a superordinate concept), sub-DAG partition (an ancestor can be generated by extracting any partition that is itself a DAG), concept unification option (an ancestor can be generated by adding an option to unify any two compatible concepts), and role de-unification option (an ancestor can be generated by adding an option not to unify the primitive roles comprising a composite role). These abstractive relations make it possible to represent such things as the conditional probability of two roles or fillers being unified, something feature-vector models have difficulty with.

## Inductive Methods for Probabilistic Models

The problem of inducing probability distributions from a finite sample, or training set, has a long history. In the past, much debate about the validity of various proposals has arisen from different interpretations of probability theory (see Weatherford 1982; Hacking 1975; Mortimer 1988; Cheeseman 1985). Probability theory, as a mathematical framework, can legitimately be appropriated for different purposes, so long as the interpretation is made clear. Three of the main schools are subjective probabilities, which denote degrees of belief; relative frequency probabilities, which denote real-world physical properties; and logical probabilities, which are purely logical relations. The acquisition of priors is a problem that plagues all probabilistic inference mechanisms, including the widely used Bayesian networks (Pearl 1988; natural language interpretation is done by Goldman & Charniak 1990), but little if any work has attempted to interpret the priors as a model of the *a priori* cognitive biases that give rise to generalization tendencies.

Assume we have a probabilistic pattern completion engine. What should its priors be, i.e., what is a legitimate source of initial probabilities? Any set of priors incorporates biases; there is no such thing as absolutely uninformative priors. (Equiprobability among all events is deceptive: splitting any one event into two causes all other probabilities to be revised for no logically justifiable reason.)

However, the fact that priors incorporate a bias is a plus rather than a negative, given that our purpose is to model human inductive generalization. The important thing is just to match the model's bias as closely to a human's as possible.[2] This puts our use of probability in the subjectivist school, related to logical probabilities but outside the relative frequentist school. We now examine why two of the most commonly used methods for establishing priors are not suitable for our purpose.

There are two main problems with setting the prior distribution equal to the relative frequency distribution in the training set. The first problem holds for both subjective and logical probability models: any event not in the training set is assigned zero probability. For example, in our nominal compound interpretation domain, many nominal compounds like *weekend guest* are novel constructions one would not necessarily expect in a training set. Nonetheless, a nonzero probability should be assigned to the best interpretation. The major contribution of Carnap's (1952, 1962) classic work on logical probability is a solution to the zero-probability problem, called the $\lambda$-continuum of inductive methods. This is a family of methods for inducing a prior distribution from a sample (training set), parameterized by $\lambda$. If $\lambda = 0$ the priors are exactly the relative frequencies, but if $\lambda > 0$ there are no nonzero prior probabilities. At $\lambda = \infty$ equiprobable priors are assigned to all simple events, and there is no sensitivity to the sample.

The second problem holds for subjective probability models: using the relative frequency distribution from the training set permits no generalization. Yet human learners generalize. Neural network research has demonstrated, for a number of different neural models, plausible ways in which generalization biases can be inherent, e.g., restricting hidden layer sizes. This is actually a stronger version of the first problem; the reason we do not want zero probabilities assigned to novel events is that some generalization ought to occur and thus give nonzero probabilities to novel events. None of Carnap's methods perform generalization: an event in the training set never raises the probability for other similar events. The probability for the best interpretation of *weekend guest* should not only be nonzero, but in fact should be greater than that of any other interpretation, because of its similarity to other events that *are* in the training set such as, say, *holiday visitor*.

Of the methods for inducing priors that allow generalization, maximum entropy has been the most popular method (Cheeseman 1987; Jaynes 1979). Given some set of probabilities for compound events (joint probabilities), the probabilities for simple events are computed by choosing the distribution that maximizes an entropy measure

$$H = -\sum_{i=1}^{C} P_i log P_i$$

while still satisfying the given joint probability constraints. In other words, what maximum entropy does is fill in probabilities to complete the joint distribution, given constraints on the values for some of the probabilities. There are information-theoretic arguments that this method minimizes the amount of information assumed. Maximum entropy methods do not specify whether joint probabilities are relative frequencies, but this is usually assumed.

The problem with using maximum entropy methods for generalization is that they do not specify how to choose which compound events to assign probabilities to. Training sets contain simple events, not compound events. The relative frequency distribution for the simple events fully determines the joint distribution for compound events—there is no room for making generalizations. In order to get generalizations, some of the simple events' probabilities must be discarded (as well as some of the compound events' probabilities, for even more generalization) and then recomputed by maximum entropy. (This is generalization because, for example, if for some compound event, maximum entropy replaces all its simple events' probabilities with equal probabilities, and the simple events originally had different training set frequencies, effectively a single generalization about all the simple events comprising the compound event is made.) Maximum entropy does not specify which probabilities to discard, and depending on this, different generalizations will be made. In the extreme, if all relative frequencies are discarded, maximum entropy makes all simple events equiprobable; this is extreme over-generalization because it generalizes every training event to all other simple events.

What we propose is a continuum of methods that vary according to a parameter $\gamma$ that controls how much generalization occurs. The extreme ends of the continuum turn out

[2] It is beyond our present scope to offer methodological guidelines for matching biases.

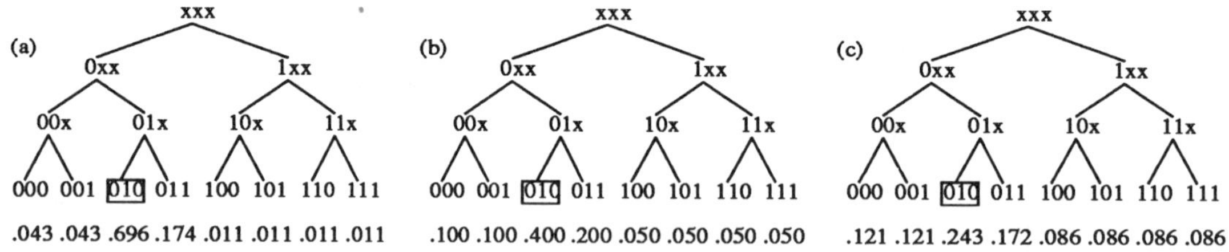


Figure 5: Effect of a training instance 010 for (a) $\gamma = 0.5$, (b) $\gamma = 1$, and (c) $\gamma = 2$.

to be the same as Carnap's $\lambda$-continuum. However, where $\lambda$ dictates the degree of sensitivity to the training set, $\gamma$ dictates the degree of generalization from the training set. At $\gamma = 0$ no generalization is done and the priors are the relative frequencies, and at $\gamma = \infty$ we get the maximum entropy over-generalization extreme.

## The $\gamma$-continuum of Methods

Denote the set of concepts or simple events by $Q = \{q_1, q_2, \ldots, q_C\}$, and let $X$ be a random variable with values ranging over $Q$. Given a training vector $\mathbf{T} = (t_1, t_2, \ldots, t_N)$ where $t_i \in Q$,

$$P_i \stackrel{\text{def}}{=} Pr(X = q_i) \stackrel{\text{def}}{=} \frac{1}{N} \sum_{n=1}^{N} norm[2^{-d(q_i,t_n)/\gamma}]$$

where *norm* means normalization to 1 as follows:

$$P_i \stackrel{\text{def}}{=} \frac{1}{N} \sum_{n=1}^{N} \frac{2^{-d(q_i,t_n)/\gamma}}{[\sum_{k=1}^{C} 2^{-d(q_k,t_n)/\gamma}]}$$

and $d(q_a, q_b)$ is the *logical distance* between two simple events. The logical distance derives from the bias given by the abstractive relations. It is defined in terms of the *logical class cardinality*

$$lcc(q_a, q_b) \stackrel{\text{def}}{=} |leaves(lub(q_a, q_b))|.$$

The logical distance is then

$$d(q_a, q_b) \stackrel{\text{def}}{=} log_2(lcc(q_a, q_b)) = log_2|leaves(lub(q_a, q_b))|.$$

For example, consider the simple binary-tree space from Figure 2(a). The least upper bound $lub(000, 001)$ is 00x, which has only the two leaves 000 and 001. Thus the logical class cardinality $lcc(000, 001) = 2$, and the logical distance $d(000, 001) = 1$. Similarly, the logical distance $d(000, 100) = log_2 8 = 3$. (The logical distance between a node and itself is always $log_2 1 = 0$.) In the general case, logical distances for semi-lattices are usually non-integers; in Figure 4 the logical distance between the two leaves marked with asterisks is $log_2 6$.

Intuitively, logical distances encode an *a priori* semantic distance metric from the built-in inductive bias set up by the abstractive relations. In relative frequency methods, each time a simple event occurs in the training set, its frequency is incremented by 1. We can view this as adding one "unit of count" to the simple event. The $\gamma$-continuum methods instead distribute the "unit" among all simple events, in a proportion that depends on logical distance. For each training instance $t_n$ that is a simple event $q_j$, if the proportion of the "unit" given to the simple event is $u_j$, then the proportion $u_i$ given to any other simple event $q_i$ satisfies the constraint

$$\frac{u_i}{u_j} = 2^{-d(q_i,q_j)/\gamma}.$$

Let us first examine the extreme-case behavior. At $\gamma = 0$, $u_i/u_j = 0$ and so $u_i = 0$ for all $i \neq j$ and $u_j = 1$, thus degenerating into the relative frequency method. At $\gamma = \infty$, $u_i/u_j = 1$ and so $u_i = u_j$ for all $i$, thus incrementing every simple event equally, regardless of what the training instance is.

Now consider again the simple binary-tree space, and examine the effect of a single training instance 010 assuming $\gamma = 1$ as in Figure 5(b). The greatest proportion is $u_{010} = 0.4$, followed by a lesser proportion for the closest simple event $u_{011} = 0.2$, a still lesser proportion for $u_{000} = u_{001} = 0.1$, and finally $u_{100} = u_{101} = u_{110} = u_{111} = 0.05$. Figures 5(a) and (c) show how the value of $\gamma$ controls the degree to which the "unit" is "smeared" toward progressively dissimilar families of events; the more "smear", the more generalization.

As a slightly more complex example, consider again the pattern space of Figure 4. A training set containing 100 instances was used. Figure 6 compares generalization behavior for three different values of $\gamma$. In (a), $\gamma = 0$ and so the distribution is exactly the relative frequency of the training instances. The non-uniform smoothing of the distribution in (b) and (c) shows the effect of the abstractive bias.

The joint probability for any compound event is just the sum of all its simple events' probability. If a compound event is comprised of a set of simple events $\{s_1, s_2, \ldots, s_r\}$ where $s_i \in Q$, then

$$Pr(X \in \{s_1, s_2, \ldots, s_r\}) = \sum_{i=1}^{r} \left[ \frac{1}{N} \sum_{n=1}^{N} \frac{2^{-d(s_i,t_n)/\gamma}}{[\sum_{k=1}^{C} 2^{-d(s_k,t_n)/\gamma}]} \right]$$
$$= \frac{1}{N} \sum_{n=1}^{N} \frac{[\sum_{i=1}^{r} 2^{-d(q_i,t_n)/\gamma}]}{[\sum_{k=1}^{C} 2^{-d(q_k,t_n)/\gamma}]}.$$

## Conclusion

The lack of theoretical tools has hampered the study of how *a priori* biases—especially abstractive biases—in a pattern completor's conceptual representation system affect the tendency to generalize. Generalization is necessary when the size of the training sample is small compared to the size of the domain, a condition that almost always obtains in the real

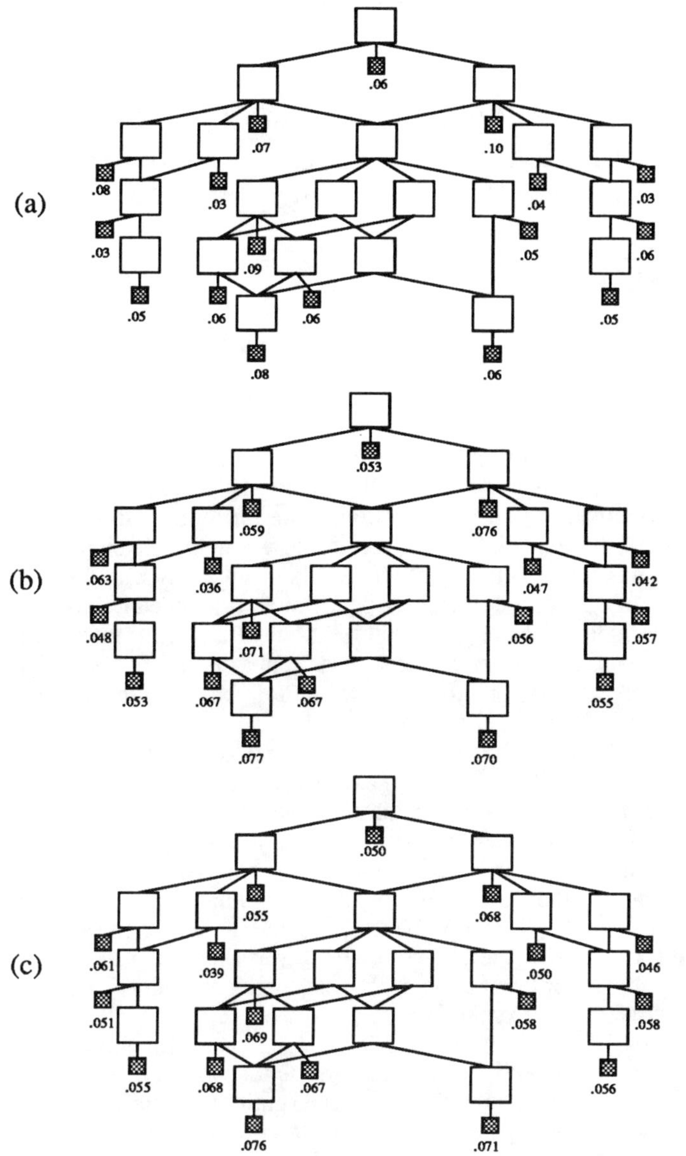


Figure 6: Generalization behavior for (a) $\gamma = 0$, (b) $\gamma = 0.8$, and (c) $\gamma = 1$.

world, in particular for human language acquisition and interpretation. Although neural networks do perform generalization because of their representational biases, the network compression loses so much information that it is difficult to tell what distribution is being learned; and moreover such networks do not (yet) handle compositional structures effectively. The family of methods we have proposed provides a declarative means to model different abstractive biases, and makes all induced distributions explicit.

The investigator must still determine empirically what concept space and abstractive relations are best for modeling cognitive biases in given domains. Moreover, for particular domains and abstractive relations, only empirical tests will tell what values of $\gamma$ are useful. We are currently testing these methods for parsing and interpreting a corpus of nominal compounds, using an abstractive bias deriving from semantic network taxonomies.

Another future direction is the logical distance metric. Weighting schemes can be added to the logical structure to provide more a flexible modeling tool. Also, there are other possible logical distance metrics that possess the same essential characteristics.

Because of the size of the event space for practical domains, heuristic approximation methods are needed to evaluate these distributions. Whether heuristics can be used depends on the types of abstractive relations. In the case of the abstractive relations we use for parsing and interpretation, we are investigating various greedy algorithms including parallel intersection search techniques like marker passing (Wu 1989). Also, we are studying whether existing neural networks or other statistically-based models of generalization can function as heuristic approximation methods for certain types of abstractive relations.

## References


Carnap, R. (1952). *The Continuum of Inductive Methods.* University of Chicago Press, Chicago.

Carnap, R. (1962). *The Logical Foundations of Probability.* University of Chicago Press, Chicago.

Cheeseman, P. (1985). In defense of probability. In *Proceedings of the Ninth International Joint Conference on Artificial Intelligence*, pp. 1002–1009.

Cheeseman, P. (1987). A method of computing maximum entropy probability values for expert systems. In R. C. Smith & G. J. Erickson, editors, *Maximum-Entropy and Bayesian Spectral Analysis and Estimation Problems*, pp. 229–240. D. Reidel, Dordrecht, Holland. Revised proceedings of the Third Maximum Entropy Workshop, Laramie, WY, 1983.

Goldman, R. P. & E. Charniak (1990). A probabilistic approach to text understanding. Technical Report CS-90-13, Brown Univ., Providence, RI.

Hacking, I. (1975). *The Emergence of Probability.* Cambridge University Press, London.

Jaynes, E. T. (1979). Where do we stand on maximum entropy. In R. D. Levine & M. Tribus, editors, *The Maximum Entropy Formalism.* MIT Press, Cambridge, MA.

Mortimer, H. (1988). *The Logic of Induction.* Ellis Horwood, Chichester, England.

Pearl, J. (1988). *Probabilistic Reasoning in Intelligent Systems: Networks of Plausible Inference.* Morgan Kaufmann, San Mateo, CA.

Pollack, J. B. (1989). Implications of recursive auto associative memories. In D. Touretzky, editor, *Advances in Neural Information Processing Systems*, pp. 527–536. Morgan Kaufmann, San Mateo.

Pollack, J. B. (1990). Recursive distributed representations. *Artificial Intelligence*, 46:77–105.

Weatherford, R. (1982). *Philosophical Foundations of Probability Theory.* Routledge & Kegan Paul, London.

Wu, D. (1989). A probabilistic approach to marker propagation. In *Proceedings of the Eleventh International Joint Conference on Artificial Intelligence*, pp. 574–580, Detroit, MI. Morgan Kaufmann.

Wu, D. (1990). Probabilistic unification-based integration of syntactic and semantic preferences for nominal compounds. In *Proceedings of the Thirteenth International Conference on Computational Linguistics*, Helsinki.

# The Interaction of Internal and External Representations in a Problem Solving Task

**Jiajie Zhang**
Department of Cognitive Science
University of California, San Diego
La Jolla, CA 92093-0515

**Abstract**

In these studies I examine the role of distributed cognition in problem solving. The major hypothesis explored is that intelligent behavior results from the interaction of internal cognition, external objects, and other people, where a cognitive task can be distributed among a set of representations, some internal and some external. The Tower of Hanoi problem is used as a concrete example for these studies. In Experiment 1 I examine the effects of the distribution of internal and external representations on problem solving behavior. Experiments 2 and 3 focus on the effects of the structural change of a problem on problem solving behavior and how these effects depend on the nature of the representations. The results of all studies show that distributed cognitive activities are produced by the interaction among the internal and external representations. External representations are not simply peripheral aids. They are an indispensable part of cognition. Two of the factors determining the performance of a distributed cognitive system are the structure of the abstract problem space and the distribution of representations across an internal mind and the external world.

## Introduction

The traditional approach to cognition in general and problem solving in particular focuses on an individual's internal mental states. In the traditional view, cognition is exclusively the activity of an internal mind. External objects, if they have anything to do with cognition at all, are at most peripheral aids. There is no doubt that internal factors are important in cognition. They are not, however, the whole story. From the distributed cognition perspective, cognitive activity is distributed across internal human minds, external cognitive artifacts, groups of people, and across space and time (Hutchins, 1990, in preparation; Hutchins & Norman, 1988; Norman, 1988, 1989, 1990).

In this paper, I develop a framework of distributed problem representations to analyze a set of distributed cognitive tasks. My focus is on the nature of external representations and the interactions among internal and external representations. I show that external objects are not simply peripheral aids--they provide a different form of representation. External representations are interwoven with internal representations to produce distributed cognitive activities.

This research was supported by a grant to Donald Norman and Edwin Hutchins from the Ames Research Center of the National Aeronautics & Space Agency, Grant NCC 2-591 in the Aviation Safety/Automation Program, technical monitor, Everett Palmer. Additional support was provided by funds from the Apple Computer Company and the Digital Equipment Corporation to the Affiliates of Cognitive Science at UCSD.

### Distributed Problem Representations

The basic principle to be explored is that the representational system for a problem can be considered as a set, with some members internal and some external. Internal representations are in the mind, as propositions, mental images, or whatever (e.g., multiplication tables, arithmetic rules, etc.). External representations are in the world, as physical symbols (e.g., written symbols, beads of an abacus, etc.) or as external rules, constraints, or relations embedded in physical configurations (e.g., spatial relations of the digits on a piece of paper, physical constraints in an abacus, etc.). Generally, there are one or more internal and external representations involved in any problem.

Figure 1 shows a representational system for a problem with two internal and two external representations. Each internal representation resides in a person's mind and each external representation resides in an external medium. The internal and external representations involved in a given problem together form a distributed representation space mapped to a single abstract problem space that represents the abstract properties of the problem. Each representation in the distributed representation space sets some constraints on the abstract problem space.

The distributed cognition perspective demands the decomposition of the abstract problem space into its internal and external components. In the traditional studies of problem solving, however, many abstract problem spaces having internal and external components were mistakenly treated as solely internal problem spaces. Generally speaking, the abstract problem space of a problem is not equivalent to its internal problem space.

### The Tower of Hanoi

The Tower of Hanoi (TOH) problem (Figure 2) was chosen as a concrete example to study distributed cognitive activities in problem solving. It is a well-studied problem (Hayes & Simon, 1977; Kotovsky & Fallside, 1989; Kotovsky, Hayes & Simon, 1985; Simon & Hayes, 1976). Much of the research has focused on isomorphs of the TOH and its problem representations. The basic finding is that different problem representations can have dramatic impact on problem difficulty even if the formal structures are the same. External memory aid is one major factor of problem difficulty. Thus, Kotovsky *et al.* (1985) reported that the *Dish-move* isomorph of the TOH, in which all rules had to be remembered, was harder to solve than the *Peg-move* isomorph, in which one of the rules was embedded in physical configurations. Modifications of these two isomorphs were used in two of the three conditions in Experiment 1 of the present study (*I123* and *I12-E3*).

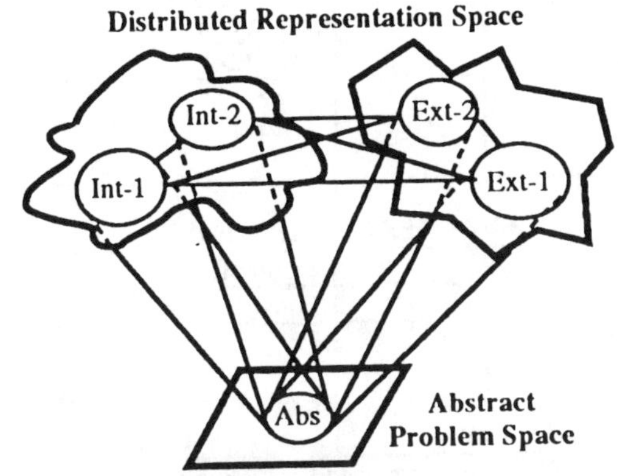


**Figure 1.** The distributed representation space and the abstract problem space of a problem with two internal and two external representations. The abstract problem space is formed by the conjunction of the internal and external problem spaces.

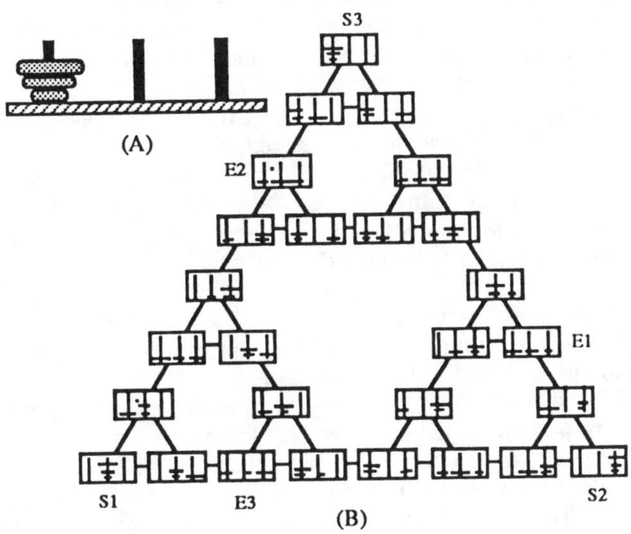


**Figure 2** (A) The TOH. The task is to move the disks from one configuration to another, following two rules: only one disk can be transferred at a time (Rule 1) and a disk can only be transferred to a pole on which it will be the largest (Rule 2). (B) The problem space of the TOH. Each rectangle shows one of the 27 possible configurations of the three disks on the three poles. The lines between the rectangles show the transformations from one state to another when the rules are followed. S1, S2, and S3 are three starting states, and E1, E2, and E3 are three ending states. They will be used later.

**Internal and External Rules.** The TOH problem actually has three rules, not just the two stated earlier. Rule 3 is that only the largest disk on a pole can be transferred to another pole. In the representation shown in Figure 2A, Rule 3 need not be stated explicitly because the physical structure of the disks and poles coupled with Rules 1 and 2 guarantee that it will be followed. But if the disks were not stacked on poles, explicit statement of Rule 3 would be necessary.

In my studies I used four rules:

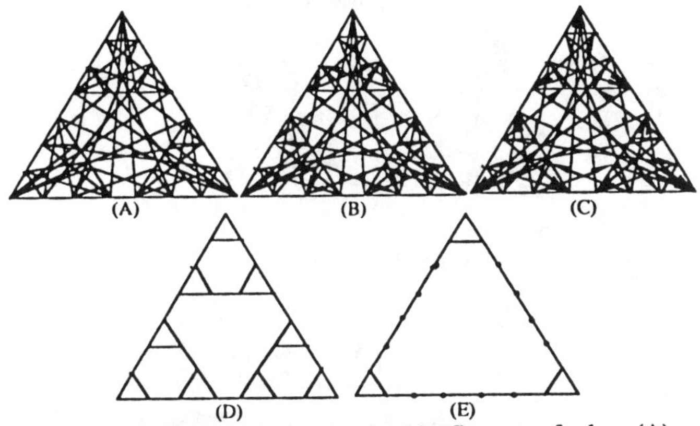


**Figure 3** Problem spaces constrained by five sets of rules. (A) Rule 1. (B) Rules 1+2. (C) Rules 1+3. (D) Rules 1+2+3. (E) Rules 1+2+3+4. Lines with arrows are uni-directional. Lines without arrows are bi-directional. The rectangles (problem states) are not shown in this figure for the reason of clarity.

Rule 1: only one disk can be transferred at a time.

Rule 2: a disk can only be transferred to a pole on which it will be the largest.

Rule 3: only the largest disk on a pole can be transferred to another pole.

Rule 4: the smallest disk and the largest disk can not be placed on a single pole unless the medium sized disk is also on that pole.

Any of these four rules can be either internal, memorized, or external, externalized into physical constraints. In the experiments that follow, I varied the numbers of external rules. In one condition, *I123* (Figure 5A), no rule is external. In a second condition, *I12-E3* (Figure 5B), Rule 3 is external. In the *I1-E23* condition (Figure 5C), both Rules 2 and 3 are external. In the *I1-E234* condition (Figure 6D), Rules 2, 3, and 4 are all external.

**Internal and External Problem Spaces.** A problem space is composed of all possible states and all moves constrained by the rules. Figures 3A-E show the problem spaces constrained by Rules 1, 1+2, 1+3, 1+2+3, and 1+2+3+4, respectively. These five spaces can represent internal problem spaces, external problem spaces, or mixed problem spaces, depending upon how the rules constructing them are distributed. A problem space constructed by external rules is an external problem space, one constructed by internal rules is an internal problem space, one constructed by a mixture of internal and external rules is a mixed problem space. Figure 4 shows the internal, external, and abstract problem spaces of the standard TOH.

## Experiment 1

The standard TOH has three rules which can be distributed among internal and external representations. Different distributions may have different effects on problem solving behavior, even if the formal structures are the same. Experiment 1 investigates these effects. My hypothesis is that the more rules are distributed externally, the easier the problem. There are three conditions, isomorphs of the TOH,

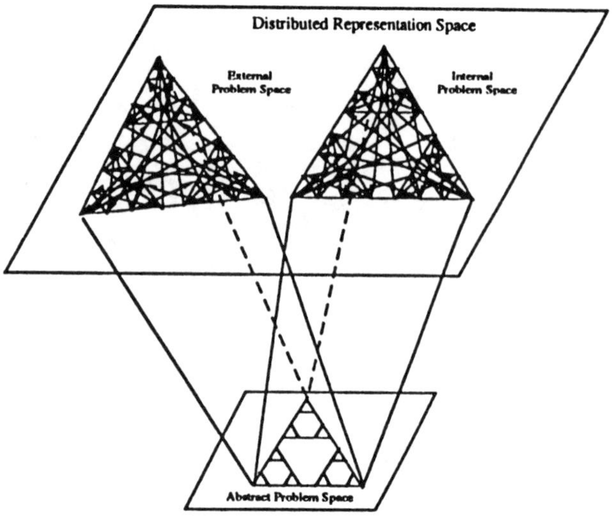


**Figure 4.** The distributed representation space and the abstract problem space for the standard TOH. The distributed representation space is composed of the internal and the external problem spaces, which are constrained by Rules 1+2 and Rules 1+3. The abstract problem space is the conjunction of the internal and the external problem spaces.

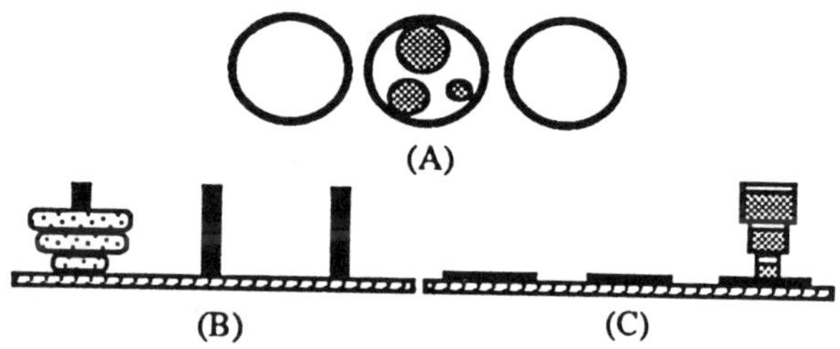


**Figure 5.** (A) *I123*. No physical constraints. (B) *I12-E3*. The physical constraints (coupled with Rules 1 and 2) guarantee that Rule 3 is followed. (C) *I1-E23*. The cups were filled with coffee. A smaller cup could not be placed on the top of a larger cup (Rule 2), as this would cause the coffee to spill. A cup could not be moved if there was another cup on top of it (Rule 3).

which correspond to three different distributions of the three rules.

In Condition *I123* (*I* = Internal rules and *E* = External rules) condition, Rules 1, 2, and 3 were all internal (Figure 5A). In Condition *I12-E3* (Figure 5B), Rules 1 and 2 were internal, and Rule 3 was external. In Condition *I1-E23* (Figure 5C), Rule 1 was internal, and Rules 2 and 3 were external.

### Method

**Subjects.** The subjects were 18 undergraduate students enrolled in introductory psychology courses at the University of California, San Diego who volunteered for the experiment in order to earn course credit.

**Materials.** In the *I123* condition, three plastic balls and three porcelain plates were used. In the *I12-E3* condition, three plastic rings and three plastic poles were used. In the *I1-E23* condition, three plastic cups and three paper plates were used. All three cups were filled with coffee.

**Design.** Each subject played all three games, one for each of the three conditions, once in a randomized order. There were six possible permutations for the three games. Each permutation was assigned to a subject randomly. For each subject, the first, the second, and the third games always started at positions S1, S2, and S3 and ended at positions E1, E2, and E3, respectively (see Figure 2B).

**Procedure.** Each subject read the instructions[3] aloud slowly. Then the subject was asked to repeat all the rules. If a subject could recite all the rules twice without error, he was instructed to start the games. Otherwise he reread the instructions until he reached the criterion. A subject's performance was recorded on a video camera.

TABLE 1. THE RESULTS OF EXPERIMENT 1

| Measurements | Conditions | | |
|---|---|---|---|
| | I123 | I12-E3 | I1-E23 |
| Times (sec) | 131.0 | 83.0 | 53.9 |
| Steps | 19.7 | 14.0 | 11.4 |
| Errors | 1.4 | 0.61 | 0.22 |

TABLE 2. THE *p* VALUES OF EXPERIMENT 1

| Comparisons | Measurements | | |
|---|---|---|---|
| | Times | Steps | Errors |
| Main Effect | < .05 | = .05 | < .005 |
| I123 vs. I12-E3 | < .1 | < .1 | < .03 |
| I123 vs. I1-E23 | < .01 | < .02 | = .001 |
| I12-E3 vs. I1-E23 | > .3 | > .4 | > .2 |

NOTE. Fisher PLSD test was used for the multiple comparisons.

### Results and Discussion

The average solution times, solution steps, and errors are shown in Table 1. The statistics is shown in Table 2. Problem difficulty measured in solution times, solution steps, and errors for the three problems was consistent. The more rules externalized, the easier the task. The order of difficulty was, from hardest to easiest: *I123* > *I12-E3* ≥ *I1-E23*. The difference between *I12-E3* and *I1-E23* was not statistically significant. All errors made were for internal rules: none were for external rules. Rules, once externalized, seem to be error-proof.

## Experiments 2 and 3

Different numbers of rules give rise to different problem spaces. Figure 3 shows that the problem space structure changes with the number of rules. There are at least two rival factors involved. On the one hand, the fewer rules, the more paths there are from an initial state to a final state. Hence, fewer rules might make the problem easier. On the other hand, the more rules, the fewer the choices. The prob-

3 Due to the maximum length limitation of the paper, the instructions are not shown here. The instructions were three restaurant stories: *Waitress and Oranges*, *Waitress and Donuts*, and *Waitress and Coffee*. For example, in the *Waitress and Coffee* condition, Rule 1 was stated as "only one cup can be moved at a time", and Rules 2 and 3 were not stated because they were external.

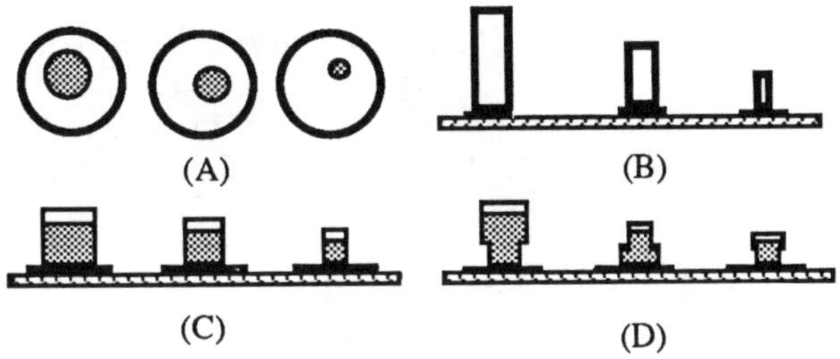


**Figure 6.** The materials used in Experiment 3. (A) *I1*. (B) *I1-E3*. The sizes of the three straws were such that a smaller straw inside a larger one could not be moved out without the larger straw being moved away first (Rule 3). (C) *I1-E23*. The same as the *I1-E23* in Figure 5C. (D) *I1-E234*. The cups were filled with tea. The sizes of the three cups were such that a smaller cup could not be placed on the top of a larger one (Rule 2), only the topmost cup could be moved (Rule 3), and the largest cup and the smallest cup could not be placed on the top of each other (Rule 4).

lem solver can simply follow where the highly constrained structure forces one to go. So, more rules might make the problem easier. This analysis implies that the problem difficulty might not increase monotonically with the number of rules. In addition, the relationship between the problem difficulty and the number of rules might depend on the nature of the rules (whether internal or external). Experiments 2 and 3 investigate these effects, with Experiment 2 focusing on a change of internal rules and Experiment 3 on a change of external rules.

Both Experiments 2 and 3 have four conditions. In Experiment 2, all rules were internal. Condition *I1* has Rule 1, Condition *I13* Rules 1 and 3, Condition *I123* Rules 1, 2, and 3, and Condition *I1234* Rules 1, 2, 3, and 4. Experiment 3 was exactly the same as Experiment 2, except that Rules 2, 3, and 4 were external rather than internal. Condition *I1* (Figure 6A) had only Rule 1 (internal). Condition *I1-E3* (Figure 6B) had Rule 1 (internal) and Rule 3 (external). Condition *I1-E23* (Figure 6C) had Rule 1 (internal) and Rules 2 and 3 (both external). Condition *I1-E234* (Figure 6D) had Rule 1 (internal) and Rules 2, 3, and 4 (all external).

## Method

**Subjects.** The subjects were 48 (24 for each experiment) undergraduate students enrolled in introductory psychology courses at the University of California, San Diego, who volunteered for the experiment to earn course credit.

**Materials.** In Experiment 2, exactly the same materials used in the *I123* condition in Experiment 1 were used in all current four conditions. In Experiment 3, materials for Condition *I1* were the same as for Experiment 2. In Condition *I1-E3*, the straws and tiny plates were made from paperboard. Materials for Condition *I1-E23* were the same as for the *I1-E23* condition in Experiment 1. In Condition *I1-E234*, the four cups were made from metal cans and were all filled with tea.

TABLE 3. THE RESULTS OF EXERIMENTS 2 AND 3

| | Experiment 2 | | | |
|---|---|---|---|---|
| | I1 | I13 | I123 | I1234 |
| Times/min. step | 2.3 | 8.8 | 21.5 | 18.5 |
| Steps/min. step | 1.0 | 1.9 | 2.7 | 1.8 |
| Errors/min. step | 0 | 0.06 | 0.28 | 0.26 |
| | Experiment 3 | | | |
| | I1 | I1-E3 | I1-E23 | I1-E234 |
| Times/min. step | 2.5 | 6.9 | 9.0 | 12.5 |
| Steps/min. step | 1.0 | 1.3 | 1.8 | 1.9 |
| Errors/min. step | 0 | 0 | 0 | 0 |

TABLE 4. THE *p* VALUES OF EXERIMENTS 2 AND 3

| Experiment 2 | | | |
|---|---|---|---|
| Comparisons | Times | Steps | Errors |
| Main Effect | <.0001 | <.0001 | =.0001 |
| I1 vs. I13 | <.05 | <.005 | <.06 |
| I1 vs. I123 | <.00001 | <.00001 | <.001 |
| I1 vs. I1234 | <.00001 | <.01 | <.001 |
| I13 vs. I123 | <.02 | <.01 | <.005 |
| I13 vs. I1234 | <.005 | >.6 | <.01 |
| I123 vs. I1234 | >.36 | <.003 | >.75 |
| Experiment 3 | | | |
| Comparisons | Times | Steps | Errors |
| Main Effect | <.0001 | <.0001 | -- |
| I1 vs. I1-E3 | <.01 | <.1 | -- |
| I1 vs. I1-E23 | <.0001 | <.0001 | -- |
| I1 vs. I1-E234 | <.00001 | <.00001 | -- |
| I1-E3 vs. I1-E23 | >.18 | <.01 | -- |
| I1-E3 vs. I1-E234 | <.0001 | <.0001 | -- |
| I1-E23 vs. I1-E234 | <.03 | >.4 | -- |

NOTE. Fisher PLSD test was used for the multiple comparisons.

**Design.** The design for Experiments 2 and 3 were the same. Each subject played all four games, once each. There were twenty-four possible permutations for the four games. The twenty-four subjects were assigned to these permutations randomly. Due to a limitation in the number of subjects available, the first, second, third, and fourth games always started at positions S1, S2, S3, and S1 and ended at positions E1, E2, E3, and E1, respectively (see Figure 2B).

**Procedure.** The procedures for both Experiments 2 and 3 were the same as in Experiment 1.

## Results and Discussion

The results are shown in Table 3. The minimum number of steps from the starting state to the final state is 2, 4, 7, and 8 for Conditions *I1* and *I1*, *I13* and *I1-E3*, *I123* and *I1-E23*, and *I1234* and *I1-E234*, respectively. In order to make meaningful comparisons, solution times, solution steps, and errors for each condition were normalized by being divided by the number of minimum steps from the starting state to the final state. The statistics is shown in Table 4.

Experiment 2 shows that when all rules were internal the hardest problem was neither the one with the fewest rules (*I1*), nor the one with the most rules (*I1234*), but the one with an intermediate number of rules (*I123*). When solution times or errors are used as the difficulty measurement, the difficulty order was, from easiest to hardest: $I1 < I13 < I1234 \leq I123$. The difference between *I1234* and *I123* was not statistically significant. When solution steps are used as the difficulty measurement, the difficulty order remained the same (*I1* < *I13* ≤ *I1234* < *I123*), but in this case the difference between *I13* and *I1234* was not statistically significant.

Experiment 3 shows that when all but one rule were external problem difficulty increased monotonically with the number of rules. When solution times are used as the difficulty measurement, the difficulty order was, from easiest to hardest: *I1* < *I1-E3* ≤ *I1-E23* < *I1-E234*. The difference between *I1-E3* and *I1-E23* is not statistically significant. If solution steps are used as the difficulty measurement, the difficulty order remained the same, but the difference between *I1-E23* and *I1-E234* is not statistically significant. Subjects didn't make any errors in this experiment.

All four rules in Experiment 2 were internal. Rules 2, 3, and 4 in Experiment 3 were external. Comparing the results in these two experiments, we found that the conditions in Experiment 3 with external rules were easier than their counterparts in Experiment 2. This further supports the claim that the more rules externalized, the easier the problem.

## General Discussion

A problem can be represented among a set of internal and external representations. Given the same set of rules, the more rules were distributed externally, the easier the problem. Given the same initial and final states, the problem difficulty increased monotonically with the number of rules if most rules were external. When all rules were internal, however, the hardest problem was the one with an intermediate number of rules. In addition to memory aids, external representations play other important roles. They provide a different representation. External representations have the following properties.

*External representations provide external memory aids.* For example, for all of the games in the present study, the goal problem states didn't need to be remembered, because they were represented by the diagrams placed in front of the subjects.

*External representations can provide information which can be directly perceived and used without being interpreted and formulated explicitly.* For example, in the *I1-E23* condition, Rules 2 and 3 were not told to the subjects: they were built into the physical constraints and perceived and followed directly. When the subjects were asked to formulate the rules after the games, few could do it.

*External representations anchor and structure cognitive behavior.* The physical structures in the external world constrain the range of possible cognitive behaviors in the sense that some behaviors are allowed and others prohibited. For example, in the *I1-E23* condition, external Rules 2 and 3 could not be violated. They construct the external problem space and hence structure the cognitive behavior.

*External representations change the nature of a task.* Norman (1990) proposed that external representations change the nature of a task from the task performer's point of view and enhance the system's ability from the system's (task performer + external representation) point of view. In the *I123* condition, a problem solver had to process three internal rules, while in the *I1-E23* condition the problem solver only had to process one internal rule. The cognitive processes of the problem solver were different in these two conditions. Nevertheless, the performance of the system *I1-E23* was much better than the system *I123*.

## Conclusion

The framework of distributed problem representations is useful for analyzing distributed cognitive activities. Under this framework, a problem is represented in a set of internal and external representations, and distributed cognitive activities are produced by the interaction of internal and external representations. External representations are not simply peripheral aids, they are an indispensable part of cognition.

## Acknowledgements

I am very grateful to Don Norman for his guidance and extensive conceptual and editorial help in every phase of this project, and to Ed Hutchins for many inspiring comments. I also thank Mark St. John, David Kirsh, Tove Klausen, and Hank Strub for helpful discussions, and Richard Warren for assistance in Experiments 2 and 3.

## References

Hayes, J. R., & Simon, H. A. (1977). Psychological differences among problem isomorphs. In N. J. Castellan, D. B. Pisoni, & G. R. Potts (Eds.), *Cognitive theory.* Hillsdale, NJ: Lawrence Erlbaum.

Hutchins, E. (1990). The technology of team navigation. In J. Galegher, R.E. Kraut, & C. Egido (Ed.), *Intellectual teamwork: Social and technical bases of collaborative work.* Hillsdale, NJ: Lawrence Erlbaum.

Hutchins, E. (in preparation). *Distributed Cognition: a cognitive ethnography of ship navigation.*

Hutchins, E., & Norman, D. A. (1988). *Distributed cognition in aviation: a concept paper for NASA* (Contract No. NCC 2-591). San Diego: University of California, Department of Cognitive Science.

Kotovsky, K., Hayes, J. R., & Simon, H. A. (1985). Why are some problems hard? Evidence from Tower of Hanoi. *Cognitive Psychology, 17,* 248-294.

Kotovsky, K., & Fallside, D. (1989). Representation and transfer in problem solving. In Klahr, D., & Kotovsky, K. (Eds.), *Complex information processing: the impact of Herbert A. Simon.* Hillsdale, NJ: Lawrence Erlbaum.

Norman, D. A. (1988). *The psychology of everyday things.* New York: Basic Books.

Norman, D. A. (1989). Four (more) issues for cognitive science. *Paper presented at the eleventh annual conference of the Cognitive Science Society.* Ann Arbor: Michigan.

Norman, D. A. (in press). Cognitive artifacts. In J. M. Carroll (Ed.), *Designing interaction: Psychology at the human-computer interface.* New York: Cambridge University Press.

Simon, H. A., & Hayes, J. R. (1976). The understanding process: problem isomorphs. *Cognitive Psychology, 8,* 165-190.

# Index